Marschner's Mineral Nutrition

Marschner's Mineral Nutrition of Plants

Fourth Edition

Edited by

Zed Rengel
UWA School of Agriculture and Environment,
University of Western Australia, Perth, WA, Australia

Ismail Cakmak
Faculty of Engineering and Natural Sciences, Sabanci University, Istanbul, Turkey

Philip J. White
Ecological Sciences, James Hutton Institute, Dundee, Scotland, United Kingdom

Academic Press is an imprint of Elsevier
125 London Wall, London EC2Y 5AS, United Kingdom
525 B Street, Suite 1650, San Diego, CA 92101, United States
50 Hampshire Street, 5th Floor, Cambridge, MA 02139, United States
The Boulevard, Langford Lane, Kidlington, Oxford OX5 1GB, United Kingdom

Copyright © 2023 Elsevier Ltd. All rights reserved.

No part of this publication may be reproduced or transmitted in any form or by any means, electronic or mechanical, including photocopying, recording, or any information storage and retrieval system, without permission in writing from the publisher. Details on how to seek permission, further information about the Publisher's permissions policies and our arrangements with organizations such as the Copyright Clearance Center and the Copyright Licensing Agency, can be found at our website: www.elsevier.com/permissions.

This book and the individual contributions contained in it are protected under copyright by the Publisher (other than as may be noted herein).

Notices
Knowledge and best practice in this field are constantly changing. As new research and experience broaden our understanding, changes in research methods, professional practices, or medical treatment may become necessary.

Practitioners and researchers must always rely on their own experience and knowledge in evaluating and using any information, methods, compounds, or experiments described herein. In using such information or methods they should be mindful of their own safety and the safety of others, including parties for whom they have a professional responsibility.

To the fullest extent of the law, neither the Publisher nor the authors, contributors, or editors, assume any liability for any injury and/or damage to persons or property as a matter of products liability, negligence or otherwise, or from any use or operation of any methods, products, instructions, or ideas contained in the material herein.

ISBN: 978-0-12-819773-8

For Information on all Academic Press publications
visit our website at https://www.elsevier.com/books-and-journals

Publisher: Nikki P. Levy
Acquisitions Editor: Nancy J. Maragioglio
Editorial Project Manager: Emerald Li
Production Project Manager: Selvaraj Raviraj
Cover Designer: Matthew Limbert

Typeset by MPS Limited, Chennai, India

Contents

List of contributors xiii
About the editors xv
Foreword xvii

Part I
Nutritional physiology 1

1 Introduction, definition, and classification of nutrients 3
Ernest A. Kirkby

Summary 3
1.1 General 3
1.2 Essential elements for plant growth 4
1.3 Beneficial elements for plant growth 5
1.4 A new definition of a mineral plant nutrient 6
1.5 Biochemical properties and physiological functions of nutrient elements in plants 6
1.6 Variation in the angiosperm ionome 7
References 8
Further reading 9

2 Ion-uptake mechanisms of individual cells and roots: short-distance transport 11
Devrim Coskun and Philip J. White

Summary 11
2.1 General 11
2.2 Pathway of solutes from the external solution into root cells 12
  2.2.1 Influx to the apoplasm 12
  2.2.2 Passage into the cytoplasm 14
2.3 Composition of biological membranes 15
2.4 Solute transport across membranes 18
  2.4.1 Thermodynamics of solute transport 18
  2.4.2 Energy demand for solute transport 25
  2.4.3 The kinetics of solute transport in plant roots 25
2.5 Factors influencing ion uptake by roots 29
  2.5.1 Influx to the apoplasm 29
  2.5.2 Effects of pH 30
  2.5.3 Metabolic activity 32
  2.5.4 Interactions among ions in the rhizosphere 33
  2.5.5 External concentration 40
  2.5.6 Plant nutritional status 41
  2.5.7 Studying nutrition at constant tissue concentration 45
2.6 Uptake of ions and water along the root axis 47
2.7 Radial transport of ions and water across the root 49
2.8 Release of ions into the xylem 53
2.9 Factors governing ion release into the xylem and exudation rate 53
References 56

3 Long-distance transport in the xylem and phloem 73
Philip J. White and Guangda Ding

Summary 73
3.1 General 73
3.2 Xylem transport 74
  3.2.1 Composition of the xylem sap 74
  3.2.2 Xylem loading 76
  3.2.3 Effect of transpiration rate on solute transport in the xylem 82
  3.2.4 Effect of transpiration rate on distribution of elements within the shoot 83
3.3 Phloem transport 84
  3.3.1 Principles of phloem transport and phloem anatomy 84
  3.3.2 Phloem loading and the composition of phloem sap 86
  3.3.3 Mobility in the phloem 88
  3.3.4 Transfer between the xylem and phloem 88
  3.3.5 Phloem unloading 89

3.4 Relative importance of phloem and xylem for long-distance transport of nutrients 89
    3.4.1 General 89
    3.4.2 Nutrients with high phloem mobility 89
    3.4.3 Nutrients with low phloem mobility 90
    3.4.4 Re-translocation and cycling of nutrients 91
3.5 Remobilization of nutrients 93
    3.5.1 General 93
    3.5.2 Seed germination 93
    3.5.3 Vegetative stage 93
    3.5.4 Reproductive stage 94
    3.5.5 Perennials 97
References 97

4 Uptake and release of elements by leaves and other aerial plant parts 105
Thomas Eichert and Victoria Fernández

Summary 105
4.1 General 105
4.2 Uptake and release of gases and other volatile compounds through stomata 106
    4.2.1 Volatile nitrogen compounds 106
    4.2.2 Volatile sulfur compounds 107
4.3 Uptake of solutes 107
    4.3.1 General 107
    4.3.2 Structure of the cuticle 108
    4.3.3 Nutrient uptake through the cuticle 109
    4.3.4 Uptake through stomata 110
    4.3.5 Role of external factors 111
4.4 Foliar application of nutrients 114
    4.4.1 General 114
    4.4.2 Practical importance of foliar application of nutrients 115
    4.4.3 Foliar fertilizers for pest and disease control 118
    4.4.4 Foliar uptake and irrigation methods 118
4.5 Leaching of elements from leaves 119
4.6 Ecological importance of foliar uptake and leaching 119
    4.6.1 Foliar leaching 119
    4.6.2 Foliar water absorption 120
References 121

5 Mineral nutrition, yield, and source–sink relationships 131
Ernest Kirkby, Miroslav Nikolic, Philip J. White, and Guohua Xu

Summary 131
5.1 General 131
5.2 Relationships between nutrient supply and yield 132
5.3 Photosynthetic activity and related processes 133
    5.3.1 Photosynthetic energy flow and photophosphorylation 133
    5.3.2 Photoinhibition and photooxidation 135
    5.3.3 Carbon dioxide assimilation and photorespiration 137
    5.3.4 C_4 pathway of photosynthesis and Crassulacean acid metabolism 140
    5.3.5 Effect of leaf maturation on its sink–source transition 143
    5.3.6 Leaf senescence 144
    5.3.7 Feedback regulation of photosynthesis by sink demand for carbohydrates 147
    5.3.8 Nutrition and photosynthesis 148
5.4 Photosynthetic area 151
    5.4.1 Individual leaf area 151
    5.4.2 Leaf area per plant 152
    5.4.3 Canopy leaf area (leaf area index and leaf area duration) 153
5.5 Respiration and oxidative phosphorylation 154
5.6 Transport of assimilates in phloem and its regulation 156
    5.6.1 Phloem loading of assimilates 156
    5.6.2 Mechanism of phloem transport of assimilates 159
    5.6.3 Phloem unloading 160
5.7 Sink formation 161
    5.7.1 Shoot architecture for grain/seed yield formation 162
    5.7.2 Flower initiation and development 163
    5.7.3 Pollination and seed development 165
    5.7.4 Formation of vegetative sink organs 167
5.8 Sink activity 168
5.9 Role of phytohormones in the regulation of the sink–source relationships 170
    5.9.1 Structure, sites of biosynthesis, and main effects of phytohormones 171
    5.9.2 Phytohormones, signal perception, and signal transduction 175
    5.9.3 Effects of nutrition on the endogenous concentrations of phytohormones 177

	5.9.4 Phytohormones and sink action	181
5.10	Source and sink limitations on yield	183
References		185

6 Functions of macronutrients 201

Malcolm J. Hawkesford, Ismail Cakmak, Devrim Coskun, Luit J. De Kok, Hans Lambers, Jan K. Schjoerring, and Philip J. White

Summary		201
6.1	Nitrogen	201
	6.1.1 Nitrate transport in plants	202
	6.1.2 Ammonium transport into and within plants	206
	6.1.3 Organic N uptake	208
	6.1.4 Nitrogen assimilation	208
	6.1.5 Nitrogen supply, plant growth, and composition	215
	6.1.6 Nitrogen-use efficiency	218
6.2	Sulfur	219
	6.2.1 General	219
	6.2.2 Sulfate uptake, reduction, and assimilation	220
	6.2.3 Metabolic functions of S	221
	6.2.4 Sulfur supply, plant growth, and plant composition	224
6.3	Phosphorus	226
	6.3.1 General	226
	6.3.2 Phosphorus as a structural element	227
	6.3.3 Role in energy transfer	228
	6.3.4 Compartmentation and regulatory role of inorganic phosphate	229
	6.3.5 Phosphorus fractions and the role of phytate	232
	6.3.6 Phosphorus supply, plant growth, and plant composition	234
6.4	Magnesium	235
	6.4.1 General	235
	6.4.2 Binding form, compartmentation, and homeostasis	235
	6.4.3 Chlorophyll and protein synthesis	235
	6.4.4 Enzyme activation, phosphorylation, and photosynthesis	236
	6.4.5 Carbohydrate partitioning	238
	6.4.6 Magnesium supply, plant growth, and composition	240
6.5	Calcium	241
	6.5.1 General	241
	6.5.2 Binding form and compartmentation	241
	6.5.3 Cell wall stabilization	244
	6.5.4 Cell extension and secretory processes	244
	6.5.5 Membrane stabilization	245
	6.5.6 Cation–anion balance and osmoregulation	245
	6.5.7 Calcium as an intracellular second messenger	246
	6.5.8 Calcium as a systemic signal	247
	6.5.9 Calcium supply, plant growth, and plant composition	248
6.6	Potassium	249
	6.6.1 General	249
	6.6.2 Compartmentation and cellular concentrations	250
	6.6.3 Enzyme activation	250
	6.6.4 Protein synthesis	252
	6.6.5 Photosynthesis	252
	6.6.6 Osmoregulation	254
	6.6.7 Phloem transport	257
	6.6.8 Energy transfer	258
	6.6.9 Cation–anion balance	258
	6.6.10 Stress resistance	258
	6.6.11 Potassium supply, plant growth, and plant composition	260
References		260

7 Micronutrients 283

Ismail Cakmak, Patrick Brown, José M. Colmenero-Flores, Søren Husted, Bahar Y. Kutman, Miroslav Nikolić, Zed Rengel, Sidsel B. Schmidt, and Fang-Jie Zhao

Summary		283
7.1	Iron	283
	7.1.1 General	283
	7.1.2 Iron-containing constituents of redox systems	284
	7.1.3 Other Fe-requiring enzymes	287
	7.1.4 Chloroplast development and photosynthesis	287
	7.1.5 Localization and binding state of Fe	290
	7.1.6 Root responses to Fe deficiency	291
	7.1.7 Iron deficiency and toxicity	293
7.2	Manganese	294
	7.2.1 General	294
	7.2.2 Mn-containing enzymes	294
	7.2.3 The functional role of Mn in photosynthesis	294
	7.2.4 Manganese in superoxide dismutase	296

	7.2.5	Manganese in oxalate oxidase	297
	7.2.6	Other Mn-dependent enzymes	297
	7.2.7	Proteins, carbohydrates, and lipids	297
	7.2.8	Cell division and extension	299
	7.2.9	Manganese deficiency	299
	7.2.10	Manganese toxicity	301
7.3	Copper		302
	7.3.1	General	302
	7.3.2	Copper uptake and transport	302
	7.3.3	Copper proteins	303
	7.3.4	Carbohydrate, lipid, and N metabolism	305
	7.3.5	Lignification	306
	7.3.6	Pollen formation and fertilization	307
	7.3.7	Copper deficiency and toxicity	308
7.4	Zinc		310
	7.4.1	General	310
	7.4.2	Zn-containing enzymes	310
	7.4.3	Zn-activated enzymes	313
	7.4.4	Protein synthesis	314
	7.4.5	Carbohydrate metabolism	315
	7.4.6	Tryptophan and indole acetic acid synthesis	316
	7.4.7	Membrane integrity and lipid peroxidation	317
	7.4.8	Phosphorus-zinc interactions	318
	7.4.9	Zinc forms and bioavailability in grains	320
	7.4.10	Zinc deficiency and toxicity	321
7.5	Nickel		323
	7.5.1	General	323
	7.5.2	Ni-containing enzymes	324
	7.5.3	Role of Ni in N metabolism	325
	7.5.4	Nickel concentration in plants	327
	7.5.5	Nickel deficiency and toxicity	327
	7.5.6	Tolerance to Ni toxicity	328
7.6	Molybdenum		328
	7.6.1	General	328
	7.6.2	Molybdenum uptake and transport	329
	7.6.3	Nitrogenase	329
	7.6.4	Nitrate reductase	330
	7.6.5	Other Mo-containing enzymes	331
	7.6.6	Gross metabolic changes	333
	7.6.7	Molybdenum deficiency and toxicity	334
7.7	Boron		336
	7.7.1	General	336
	7.7.2	Boron complexes with organic structures	338
	7.7.3	Function of B	339
	7.7.4	Boron deficiency and toxicity	347
7.8	Chlorine		350
	7.8.1	General	350
	7.8.2	Uptake, transport, and homeostasis	350
	7.8.3	Charge balance	352
	7.8.4	Photosynthesis and chloroplast performance	352
	7.8.5	Cell osmoregulation and turgor	353
	7.8.6	Plant water balance and water relations	354
	7.8.7	Interaction with nitrate	355
	7.8.8	Chloride supply, deficiency, plant growth, and crop yield	356
	7.8.9	Chlorine toxicity	357
	7.8.10	Chlorine as micro- and macronutrient - concluding remarks	358
References			359

8 Beneficial elements — 387

Jian Feng Ma, Fang-Jie Zhao, Zed Rengel, and Ismail Cakmak

Summary			387
8.1	Definition		387
8.2	Sodium		387
	8.2.1	General	387
	8.2.2	Essentiality: Na as nutrient	388
	8.2.3	Role in C_4 species	389
	8.2.4	Substitution of K by Na	391
	8.2.5	Growth stimulation by Na	393
	8.2.6	Application of Na fertilizers	396
8.3	Silicon		397
	8.3.1	General	397
	8.3.2	Uptake, concentration, and distribution	397
	8.3.3	Beneficial effects	399
8.4	Cobalt		402
	8.4.1	Role of Co in plants	402
	8.4.2	Cobalt deficiency and toxicity	403
8.5	Selenium		405
	8.5.1	General	405
	8.5.2	Uptake and translocation	405
	8.5.3	Assimilation and metabolism	407
	8.5.4	Beneficial effects on plant growth	409
	8.5.5	Biofortification	409
8.6	Aluminum		410
8.7	Other elements		411
References			411

9 Mineral nutrition and crop quality — 419

Umit Baris Kutman

Summary		419
9.1	Introduction	419

9.2	Technical quality	420
	9.2.1 Bread and pasta	420
	9.2.2 Sugar and oil crops	421
	9.2.3 Fiber crops	422
	9.2.4 Processing tomatoes	422
	9.2.5 Beer and wine	423
9.3	Sensory quality	423
	9.3.1 Effects of mineral nutrition on visual quality	423
	9.3.2 Effects of mineral nutrition on flavor	425
9.4	Nutritional quality	426
	9.4.1 Mineral nutrients, hidden hunger, and biofortification	427
	9.4.2 Protein concentration and amino acid composition	429
	9.4.3 Vitamins and bioactive phytochemicals	430
9.5	Shelf life of fresh fruits and vegetables	431
9.6	Food safety	432
	9.6.1 Toxic elements	432
	9.6.2 Harmful N compounds	434
9.7	The yield-quality dilemma	435
References		436

10 Relationship between mineral nutrition, plant diseases, and pests 445

Markus Weinmann, Klára Bradáčová, and Miroslav Nikolic

Summary		445
10.1	General	445
10.2	Relationship between susceptibility and nutritional status of plants	447
10.3	Fungal diseases	449
	10.3.1 Principles of infection	449
	10.3.2 Role of Si	450
	10.3.3 Role of N and K	452
	10.3.4 Role of Ca and Mg	453
	10.3.5 Role of phosphate and phosphite	455
	10.3.6 Role of S	455
	10.3.7 Role of Mn	455
	10.3.8 Role of other micronutrients	456
10.4	Bacterial and viral diseases	457
	10.4.1 Bacterial diseases	457
	10.4.2 Viral diseases	457
10.5	Soil-borne fungal and bacterial diseases	458
10.6	Pests	461
10.7	Direct and indirect effects of fertilizer application on plants and their pathogens and pests	464
References		467

11 Diagnosis and prediction of deficiency and toxicity of nutrients 477

Richard Bell

Summary		477
11.1	General	477
11.2	Tools for diagnosis of nutrient disorders	478
	11.2.1 Field responses to nutrient supply	478
	11.2.2 Diagnosis of nutritional disorders by visible symptoms	478
	11.2.3 Plant Analysis	480
11.3	Plant analysis for prognosis of nutrient deficiency	489
11.4	Plant analysis versus soil analysis	490
References		491

Part II
Plant—soil relationships 497

12 Nutrient availability in soils 499

Petra Marschner and Zed Rengel

Summary		499
12.1	General	499
12.2	Chemical soil analysis	499
12.3	Movement of nutrients to the root surface	501
	12.3.1 Principles of calculations	501
	12.3.2 Concentration of nutrients in the soil solution	502
	12.3.3 Role of mass flow	503
	12.3.4 Role of diffusion	504
12.4	Role of root density	509
12.5	Nutrient availability and distribution of water in soils	511
12.6	Role of soil structure	513
12.7	Intensity/quantity ratio, plant factors, and consequences for soil testing	514
	12.7.1 Modeling of nutrient availability and crop nutrient uptake	515
References		516

13 Genetic and environmental regulation of root growth and development 523

Peng Yu and Frank Hochholdinger

Summary		523

	13.1	General	523
	13.2	Genetic control of root growth and development	523
		13.2.1 Root system architecture	523
		13.2.2 Root anatomy and structure: from arabidopsis to crops	525
		13.2.3 Embryonic and postembryonic root branching	526
		13.2.4 Phytohormonal control of root growth and development	527
	13.3	Regulation of root growth and development by environmental cues	530
		13.3.1 Nutritional control of root development	530
		13.3.2 Soil physical and chemical factors	532
		13.3.3 Root–soil biotic interactions	535
		References	537

14 Rhizosphere chemistry influencing plant nutrition 545

Günter Neumann and Uwe Ludewig

	Summary		545
	14.1	General	545
		14.1.1 Rhizosphere sampling	546
	14.2	Spatial extent of the rhizosphere	547
		14.2.1 Radial gradients	548
		14.2.2 Longitudinal gradients	549
		14.2.3 Temporal variability	549
	14.3	Inorganic elements in the rhizosphere	550
	14.4	Rhizosphere pH	552
		14.4.1 Source of nitrogen supply and rhizosphere pH	553
		14.4.2 Nutritional status of plants and rhizosphere pH	557
	14.5	Redox potential and reducing processes	559
		14.5.1 Effect of waterlogging	559
		14.5.2 Manganese mobilization	560
		14.5.3 Iron mobilization	561
	14.6	Rhizodeposition and root exudates	561
		14.6.1 Sloughed-off cells and tissues	563
		14.6.2 High-molecular-weight compounds in root exudates	564
		14.6.3 Low-molecular-weight root exudates	569
		References	574

15 Rhizosphere biology 587

Petra Marschner

	Summary		587
	15.1	General	587
	15.2	The rhizosphere as dynamic system	587
	15.3	Rhizosphere microorganisms	588
		15.3.1 Root colonization	588
		15.3.2 Role in nutrition of plants	590
		15.3.3 Root exudates as signals and phytohormone precursors	591
	15.4	Endophytes	592
	15.5	Methods to study rhizosphere microorganisms	592
	15.6	Mycorrhiza	593
		15.6.1 General	593
		15.6.2 Mycorrhizal groups, morphology, and structure	593
		15.6.3 Root colonization, photosynthate demand, and host plant growth	595
		15.6.4 Mycorrhizal responsiveness	598
		15.6.5 Role of AM in nutrition of their host plant	600
		15.6.6 Role of AM in agriculture	602
		15.6.7 Role of ectomycorrhiza in nutrition of plants	602
		15.6.8 Role of mycorrhiza in tolerance to high metal concentrations	603
		15.6.9 Other mycorrhizal effects	604
		References	606

16 Nitrogen fixation 615

Mariangela Hungria and Marco Antonio Nogueira

	Summary		615
	16.1	General	615
	16.2	Biological nitrogen-fixing systems	616
	16.3	Biochemistry of nitrogen fixation	618
	16.4	Symbiotic systems: how do they work?	621
		16.4.1 General	621
		16.4.2 Range of legume–rhizobia symbioses	621
		16.4.3 Legume root infection by rhizobia	623
		16.4.4 Nodule formation and functioning in legumes	626
	16.5	Effects of nutrients on the biological nitrogen fixation	629
		16.5.1 Nutrients other than nitrogen	629
		16.5.2 Effect of mineral nitrogen	636
	16.6	Soil and environmental limitations	637
		16.6.1 Salinity	637
		16.6.2 Soil water content	637
		16.6.3 Temperature	638

16.7	Methods to quantify the contribution of BNF, amounts of N fixed by legumes, and N transfer to other plants in intercropping and crop rotations	638
16.8	Significance of free-living and associative nitrogen fixation	639
16.9	Microbial inoculation to promote BNF and improve plant nutrition	640
16.10	Final remarks	641
References		642

17 Nutrient-use efficiency 651

Hans Lambers

Summary		651
17.1	General	651
17.2	Calcium and boron requirements of monocots and dicots	652
17.3	Phosphorus and nitrogen requirements of plant species that evolved in severely phosphorus-impoverished landscapes	652
17.4	Micronutrient requirements of plant species that evolved in severely phosphorus-impoverished landscapes	655
17.5	Nitrogen requirements of C_3 and C_4 plants	656
17.6	Calcicole species	656
17.7	Variation in leaf sulfur requirement among plant species	656
17.8	Fluoride in leaves of plants occurring on soils containing little fluoride	658
17.9	Selenium in leaves of some plants	658
17.10	Silicon as a beneficial element in leaves of some plants	658
17.11	Leaf longevity and nutrient remobilization	660
References		660

18 Plant responses to soil-borne ion toxicities 665

Zed Rengel

Summary		665
18.1	Introduction	665
18.2	Acid mineral soils	666
	18.2.1 Major constraints	666
	18.2.2 Proton toxicity	667
	18.2.3 Aluminum toxicity	668
	18.2.4 Manganese toxicity	672
18.3	Mechanisms of adaptation to acid mineral soils	673
	18.3.1 General	673
	18.3.2 Aluminum resistance by avoidance	674
	18.3.3 Aluminum tolerance	677
	18.3.4 Screening for aluminum resistance	677
	18.3.5 Manganese tolerance	678
18.4	Waterlogged and flooded (hypoxic) soils	679
	18.4.1 Soil chemical factors	679
	18.4.2 Hypoxia stress	681
	18.4.3 Phytotoxic metabolites under hypoxia	681
	18.4.4 Phytohormones and root-to-shoot signals	682
	18.4.5 Element toxicity as a component of hypoxia stress	682
	18.4.6 Mechanisms underpinning tolerance to, and avoidance of, hypoxic stress	684
18.5	Saline soils	687
	18.5.1 General	687
	18.5.2 Soil characteristics and classification	687
	18.5.3 Salinity and plant growth	688
	18.5.4 Mechanisms of adaptation to saline substrates	695
	18.5.5 Exploiting salt-affected soils	703
	18.5.6 Genotypic differences in growth response to salinity	703
References		704

19 Nutrition of plants in a changing climate 723

Sylvie M. Brouder and Jeffrey J. Volenec

Summary		723
19.1	General	723
19.2	The changing climate	724
	19.2.1 Historical climate trends	724
	19.2.2 Soil temperature	724
	19.2.3 Precipitation and soil moisture	725
19.3	Plant responses to global climate change	727
	19.3.1 C_3 and C_4 plants	727
	19.3.2 Adaptation of C_3 and C_4 plants to future climates	728
19.4	Nutrient accumulation	728
	19.4.1 C_3 versus C_4 plants	728
	19.4.2 Plant response to fertilization	728
	19.4.3 Nitrogen assimilation in future climates	729
	19.4.4 Leguminous plants and N_2 fixation	730

19.5	Nutrient-use efficiency	731	
	19.5.1 General nutrient-use efficiency concepts	731	
	19.5.2 Nutrient-use efficiency of cereals	731	
	19.5.3 Nutrient-use efficiency of forage and pasture species	731	
	19.5.4 Nutrient-use efficiency of forest species	732	
19.6	Global climate change and root zone nutrient availability	732	
	19.6.1 Impact on coupled carbon-nutrient cycling	733	
	19.6.2 Mycorrhizae and nutrient uptake	735	
	19.6.3 Diffusivity and mass flow	735	
	19.6.4 System-level nutrient inputs and losses	738	
19.7	Mineral composition of food/feed	743	
	19.7.1 Mineral composition of grains and fruits	743	
	19.7.2 Forage and pasture composition and mineral nutrition	744	
	19.7.3 Composition of trees and timber	744	
References		744	

20 Nutrient and carbon fluxes in terrestrial agroecosystems 751

Andreas Buerkert, Rainer Georg Joergensen, and Eva Schlecht

Summary 751

20.1	Microbiological factors determining carbon and nitrogen emissions	751
	20.1.1 CO_2 emission	752
	20.1.2 Fungal and bacterial contributions to CO_2 emissions	752
	20.1.3 CH_4 emissions	753
	20.1.4 N_2 and N_2O emissions	753
20.2	Effects of organic soil amendments on gaseous fluxes	754
20.3	Effects of pH, soil water content, and temperature on organic matter turnover	755
20.4	Global warming effects	756
20.5	Plant–animal interactions affecting nutrient fluxes at different scales	757
	20.5.1 Species-specific relationship between feed intake and excreta quality	757
	20.5.2 Nutrient and carbon losses from livestock excreta	759
	20.5.3 Spatial aspects of livestock-mediated nutrient fluxes and modeling	759
20.6	Scale issues in modeling	763
20.7	Nutrient fluxes in rural–urban systems	763
References		764

Index 775

List of contributors

Richard Bell Centre for Sustainable Farming Systems, Food Futures Institute, Murdoch University, Murdoch, WA, Australia

Klára Bradáčová Institute of Crop Science, University of Hohenheim, Stuttgart, Germany

Sylvie M. Brouder Department of Agronomy, Purdue University, West Lafayette, IN, United States

Patrick Brown University of California–Davis, Department of Plant Sciences, Davis, California, United States

Andreas Buerkert Organic Plant Production in the Tropics and Subtropics (OPATS), University of Kassel, Witzenhausen, Germany

Ismail Cakmak Faculty of Engineering and Natural Sciences, Sabanci University, Istanbul, Turkey

José M. Colmenero-Flores Spanish National Research Council (CSIC), IRNAS, Seville, Spain

Devrim Coskun Department of Phytology, Laval University, Quebec City, Quebec, Canada

Luit J. De Kok Laboratory of Plant Physiology, University of Groningen, Groningen, The Netherlands

Guangda Ding College of Resources and Environment, Huazhong Agricultural University, Wuhan, P.R. China

Thomas Eichert Department of Horticulture, University of Applied Sciences, Erfurt, Germany

Victoria Fernández Systems and Natural Resources Department, Technical University of Madrid, Madrid, Spain

Malcolm J. Hawkesford Rothamsted Research, Harpenden, Hertfordshire, United Kingdom

Frank Hochholdinger Crop Functional Genomics, Faculty of Agriculture, Institute for Crop Science and Resource Conservation (INRES), University of Bonn, Bonn, Germany

Mariangela Hungria Embrapa Soja, Soil Biotechnology Laboratory, Londrina, Paraná, Brazil; INCT-Plant-Growth Promoting Microorganisms for Agricultural Sustainability and Environmental Responsibility, Brazil

Søren Husted University of Copenhagen, Department of Plant and Environmental Sciences, Copenhagen, Denmark

Rainer Georg Joergensen Soil Biology and Plant Nutrition (SBPN), University of Kassel, Witzenhausen, Germany

Ernest A. Kirkby Faculty of Biological Sciences, University of Leeds, Leeds, United Kingdom

Bahar Y. Kutman Gebze Technical University, Institute of Biotechnology, Kocaeli, Turkey

Umit Baris Kutman Institute of Biotechnology, Gebze Technical University, Kocaeli, Turkey

Hans Lambers School of Biological Sciences and Institute of Agriculture, University of Western Australia, Perth, WA, Australia

Uwe Ludewig Department of Nutritional Crop Physiology, Institute of Crop Science, University of Hohenheim, Stuttgart, Germany

Jian Feng Ma Institute of Plant Science and Resources, Okayama University, Kurashiki, Japan

Petra Marschner School of Agriculture, Food & Wine, The University of Adelaide, Adelaide, SA, Australia

Günter Neumann Department of Nutritional Crop Physiology, Institute of Crop Science, University of Hohenheim, Stuttgart, Germany

Miroslav Nikolic Institute for Multidisciplinary Research, University of Belgrade, Belgrade, Serbia

Marco Antonio Nogueira Embrapa Soja, Soil Biotechnology Laboratory, Londrina, Paraná, Brazil; INCT-Plant-Growth Promoting Microorganisms for Agricultural Sustainability and Environmental Responsibility, Brazil

Zed Rengel UWA School of Agriculture and Environment, University of Western Australia, Perth, WA, Australia

Jan K. Schjoerring Department of Plant and Environmental Sciences, University of Copenhagen, Frederiksberg, Denmark

Eva Schlecht Animal Husbandry in Tropics and Subtropics (AHTS), University of Kassel and University of Göttingen, Witzenhausen, Germany

Sidsel B. Schmidt University of Copenhagen, Department of Plant and Environmental Sciences, Copenhagen, Denmark

Jeffrey J. Volenec Department of Agronomy, Purdue University, West Lafayette, IN, United States

Markus Weinmann Institute of Crop Science, University of Hohenheim, Stuttgart, Germany

Philip J. White Ecological Sciences, James Hutton Institute, Dundee, Scotland, United Kingdom

Guohua Xu College of Resources and Environmental Sciences, Nanjing Agricultural University, Nanjing, P.R. China

Peng Yu Emmy Noether Root Functional Biology Group, Faculty of Agriculture, Institute for Crop Science and Resource Conservation (INRES), University of Bonn, Bonn, Germany

Fang-Jie Zhao College of Resources and Environmental Sciences, Nanjing Agricultural University, Nanjing, P.R. China

About the editors

Zed Rengel received BSc (Biology) and MSc (Cell biology) from the University of Zagreb, PhD (agronomy, botany) from Louisiana State University, and Doctor Honoris Causa from Zagreb University. He is a professor of soil science and plant nutrition at the University of Western Australia. He is the editor-in-chief of *Crop & Pasture Science*, *Plants* section of *Biology*, and *Advances in Environmental and Engineering Research*. He is the author/coauthor of more than 400 publications in peer-reviewed international journals. He gave 51 invited keynote addresses at international conferences and presented 328 invited seminars at 79 universities in 28 countries. He was awarded 9 prizes, 38 fellowships and awards (e.g., Fulbright Senior Scholar, Humboldt Lifetime Research Award, Humboldt Fellowship, Campbell Oration-USA, Martin Massengale Lectureship-USA), 4 honorary professorships and 5 visiting professorships. He is Foreign Fellow of Croatian Academy of Sciences and Arts, and is listed as a highly cited researcher (cross-field) by Clarivate Analytics.

Ismail Cakmak received his PhD at the University of Hohenheim in Stuttgart, Germany, under the guidance of the late Prof. Dr. Horst Marschner. Dr. Cakmak had a research collaboration with Prof. Marschner for 12 years, and they published together extensively. He is currently a professor at the Sabanci University-Istanbul, Turkey. He has published 225 peer-reviewed articles and was named among the "highly cited researchers" in agricultural sciences by Clarivate Analytics. He has received the IFA "International Crop Nutrition Award" in 2005, Australian Crawford Fund "Derek Tribe Medal" in 2007, Alexander von Humboldt Foundation-Georg Forster Research Award in 2014, International Plant Nutrition Institute-IPNI Science Award in 2016, and the World Academy of Sciences-Agricultural Science Prize in 2016. He is an elected member of "The Academy of Europe" and "The Science Academy" of Turkey.

Philip John White, FLS, FRSE, graduated from the Oxford University with a BA in biochemistry in 1983 and was awarded a PhD in natural sciences (Botany) from the University of Manchester in 1987, followed by a DSc in 2016. He is currently a research specialist in plant ecophysiology at The James Hutton Institute (UK), a full professor at Huazhong Agricultural University (China), where he lectures on "Mineral Nutrition of Higher Plants—A Phylogenetic Approach", and a visiting professor at the Comenius University (Slovakia). He has published more than 270 refereed papers and over 200 other scientific articles. He was a commissioned contributor to *Trends in Plant Science* and *BioMedNet* (2001–03) and has served on the editorial boards of many periodicals including *Annals of Botany*, *Plant and Soil*, and *Journal of Experimental Botany*. He is listed as a highly cited researcher in plant & animal science by Clarivate Analytics.

Foreword

It is now 61 years ago, in 1962, that I first met Horst Marschner although it seems like only yesterday. At that time I was a postgraduate student in agricultural chemistry (plant nutrition) at Kings College in Newcastle upon Tyne, Durham University, wishing to spend a few weeks working in an appropriate academic institute in Germany. German student friends in Stuttgart whom I had come to know in Newcastle suggested, after taking soundings from local agricultural students, that I write to the professor of plant nutrition, Gerhard Michael, in the nearby Agricultural College of Hohenheim. In response, I was delighted to receive a very friendly informal letter telling me that he had only recently taken up his post moving from Jena in East Germany and inviting me to spend the summer vacation working in the institute. He was also kind enough to arrange for me to have a room living in the institute itself (Fig. 1) and to take breakfast with his colleague and assistant Mr. Marschner together with his wife and their very young son Bernd. The Marschners, like the Michaels, had also recently fled from East Germany to the west via Berlin bringing with them only what they could carry, using an escape route still possible until the building of the Berlin Wall in August 1961.

The Institute of Plant Nutrition at Hohenheim is important in the history of plant nutrition because it was directed from 1922 to 1932 by the first female full professor of plant nutrition and indeed the first female full professor in Germany, the Baroness Margarete von Wrangell (1877–1932). Her brilliant career prior to Hohenheim included working with Nobel prize winners William Ramsay (1852–1916) in London and Marie Curie (1867–1934) in Paris. In 1904 she enrolled to study botany and chemistry at the University of Tübingen, and 5 years later, she was awarded a PhD in organic chemistry with the highest honor (summa cum laude). From 1912 to 1918 she worked in her homeland of Estonia managing the experimental station of the Estonian Agricultural Society in Reval (now Tallinn). Following the 1917 Russian October Revolution, she was imprisoned for some weeks but fortunately liberated by incoming German troops after which she left the Baltics never to return. She was appointed at Hohenheim in 1918—and took over the management of the recently built Institute of Plant Nutrition in 1920—the same year in which she completed her habilitation thesis on "Phosphoric acid uptake and soil reaction." Extending and developing this work was the main

FIGURE 1 The Institute of Plant Nutrition, Hohenheim. Source: *Hochschulführer Hohenheim 1958.*

objective of her research using pot and field experiments—an area of study still relevant today—as discussed in this fourth edition. A plaque standing at the front of the institute describing her fascination with plants translates as "I lived with the plants, I laid my ear to the ground and it seemed to me that the plants were glad to be able to tell me something about their secrets of growth."

My stay during the summer of 1962 as well as being highly enjoyable proved to be life changing in terms of academic development. This was not only by meeting Horst Marschner but also by being fortunate enough to be introduced to Dr. Konrad Mengel with whom I was later to write the book *Principles of Plant Nutrition* (K. Mengel and E.A. Kirkby) which was to run to five editions, be translated into several languages, and serve as a text for students and academics worldwide. As a native English speaker with a reasonable knowledge of German I was also able to contribute to the institute during my stay and later by editing the institute publications before their submission to American and other scientific journals written in English.

Many visits to Hohenheim followed. Over the years, Horst became a very good friend as well as a research colleague and I retained contact with the institute for more than 50 years (Fig. 2). In the latter period following Horst's untimely death in 1996, my visits were to meet up with Volker Römheld, Horst's long-serving assistant and coworker who continued and extended areas of research in plant nutrition. In particular, Volker made a substantial contribution to the third edition in 2012 edited by Horst Marschner's daughter Petra. Sadly, my visits to Hohenheim came to an end after Volker's passing in late 2013.

The basic structure of this fourth edition follows the same lines as all three previous editions, being divided into two parts: (1) Nutritional Physiology and (ii) Plant—Soil Relationships. Importantly, in the fourth edition a new chapter on "Crop Nutrition and Climate Change" is introduced and the longest chapter in the book, Chapter 17 of the third edition "Efficiency of Nutrient Acquisition and Use" which is of considerable significance to practical crop production, is extended to two chapters on "Nutrient-use Efficiency" and "Plant Responses to Soil-borne Ion Toxicities." One of the main changes in this new edition is the inclusion in all the chapters of current knowledge of molecular aspects of plant nutrition, including insights from various -omics approaches, transgenics, and gene editing to enhance our understanding of nutritional physiology and plant—soil interactions. Numerous new figures have been added in full color where appropriate. Notably, a much larger number of contributors have coauthored the revised chapters of this fourth edition bringing a huge wealth of expertise across scientific disciplines. Even though an emphasis in choosing the contributors was on well-established scientists, a number of early career researchers were also included, which undoubtedly would have delighted Horst Marschner who was always supportive in encouraging and educating new generations of students and scientists.

FIGURE 2 Horst Marschner (bottom row, center) and the staff and students of the Institute of Plant Nutrition in the late 1970s. *Source: Berghammer.*

Zed Rengel, Ismail Cakmak, and Philip White are to be thanked and congratulated for their careful selection of contributors as well as for their own contributions and for completing the extremely onerous task of editing the many versions of the chapters of the text. This new edition is to be greatly welcomed in providing a valuable up-to-date account of all aspects of Plant Nutritional Physiology and Plant−Soil Interactions of benefit to everyone from undergraduate students to senior research scientists.

Ernest A. Kirkby
Faculty of Biological Sciences, Life Fellow, University of Leeds,
Leeds, United Kingdom

Part I

Nutritional physiology

Chapter 1

Introduction, definition, and classification of nutrients

Ernest A. Kirkby

Faculty of Biological Sciences, University of Leeds, Leeds, United Kingdom

Summary

This chapter provides a brief overview of the history of plant nutrition. It is currently accepted that plants require 14 essential mineral elements (nutrients), without which they cannot complete their life cycle. The macronutrients nitrogen (N), phosphorus (P), potassium (K), sulfur (S), calcium (Ca), and magnesium (Mg) are required in much larger amounts than the micronutrients iron (Fe), manganese (Mn), boron (B), zinc (Zn), copper (Cu), molybdenum (Mo), nickel (Ni), and chlorine (Cl). Plants also take up beneficial elements, such as sodium (Na), silicon (Si), cobalt (Co), iodine (I), selenium (Se), and aluminum (Al), which can improve resistance to pests, diseases, and abiotic stresses as well as plant growth and crop quality. A scheme to classify plant nutrients according to their biochemical properties and physiological function is presented. A proposal that the current definition of essential nutrients is too narrow to encompass a vision for the future of research in plant nutrition aimed at improving crop production and quality, agronomic practice, and fertilizer use and regulation is discussed. Plant nutrient requirements, and mineral composition, can vary greatly as a consequence of genetic and environmental influences. Extreme variation from the average leaf mineral composition (standard leaf ionome) occurs among some orders of plants: Poales (grasses) are low in Ca, Caryophyllales are high in Mg, and Brassicales are high in S even when sampled from diverse environments. The evolution of such traits has been traced using phylogenetic relationships among angiosperm orders, families, and genera. Variation in the leaf ionome has profound consequences for ecology, mineral cycling in the environment, sustainable agriculture, and livestock and human nutrition.

1.1 General

The beneficial effect of adding mineral elements (e.g., plant ash or lime) to soils to improve plant growth has been known in agriculture for more than 2000 years. Nevertheless, even 200 years ago, it was still a matter of scientific controversy as to whether mineral elements functioned as nutrients in plant growth. The first convincing evidence of the important roles they played came from the work of Carl Sprengel (1787–1859) who was both a chemist and an agronomist teaching at the University of Göttingen, Germany. Following extensive chemical analysis of soils and crops, he concluded that 20 elements [including nitrogen (N), phosphorus (P), potassium (K), sulfur (S), magnesium (Mg), and calcium (Ca)] were required for plant growth. It was he, also, who originally postulated the Law of the Minimum, that growth should be restricted by whichever nutrient resource was in most limiting supply. He published this work in 1831 in the first of his two books on agricultural chemistry, but his valuable contribution to plant nutrition remained largely forgotten until relatively recent times. It was the publication of the book of his fellow countryman, the organic chemist, Justus von Liebig (1803–73), *Chemistry and Its Application to Agriculture and Physiology*, in 1840—a best seller, written in both English and German and reprinted and re-edited many times—that highlighted the importance of the mineral nutrition of plants. Unfortunately, the work of Sprengel, although included in von Liebig's book, was not given due acknowledgment despite Sprengel's protestations. Now, almost 200 years later, the contributions of both pioneering scientists are well recognized. Carl Sprengel is considered the true founder of the doctrine of mineral nutrition of plants and Justus von Liebig the indefatigable advocate for its acceptance. For a detailed account of the origin and theory of the mineral nutrition of plants see van der Ploeg et al. (1999).

The publication of von Liebig's timely book stimulated an enormous interest in the application of mineral nutrients to soils to increase crop yields. In England, beginning in 1843, John Bennet Lawes (1814–1900) and Joseph Henry Gilbert (1817–1901) set up experimental plots treated with varying amounts of what were then described as mineral manures at Rothamsted Manor in Harpenden (Dyke, 1993). One of these manures was monocalcium phosphate [Ca(H$_2$PO$_4$)$_2$], a soluble phosphate fertilizer produced by treating bones or mineral phosphates with sulfuric acid. This process was patented by Lawes in 1842, who named the product "superphosphate" and sold it from his fertilizer factory founded in 1843. Other manures included muriate of potash (KCl), sulfate of soda (Na$_2$SO$_4$), and sulfate of magnesia (MgSO$_4$). Nitrogen was supplied as sulfate of ammonia (NH$_4$)$_2$SO$_4$ or nitrate of soda (NaNO$_3$). Farmyard manure was also applied to the plots. Lawes was very much aware of the importance of N and P for plant growth from his pot experiments testing ammonium phosphate prior to 1840; the benefit of superphosphate in increasing the yields of turnips (*Brassica rapa* subsp. *rapa*) was reported later from the results of the Barnfield plots sown in 1843.

The long-continuing and well-known controversy between Lawes and Gilbert with von Liebig as to how plants obtain N is reviewed in detail by Dyke (1993). In brief, von Liebig, who had little first-hand experience of agronomy, proposed, in the second edition of his book, that plants obtained their N from the atmosphere, whereas Lawes and Gilbert maintained that the soil, and the N fertilizer added to it, constituted the predominant source of N for crops. Results of the Broadbalk winter wheat (*Triticum aestivum*) experiment, sown in 1843, and later findings from the same plots, clearly demonstrated the enormous benefit of applying N fertilizer both in improving the appearance of crops during growth and increasing their yields. Sadly, even when presented with these results, von Liebig flatly refused to admit that he was wrong. To quote Dyke (1993), "If only Liebig had had the grace and humility to admit that, in view of the facts presented by Lawes and Gilbert, he had changed his mind, his reputation with contemporary and later generations of scientists and farmers would have been greatly enhanced."

1.2 Essential elements for plant growth

As evident from the work on field experiments at Rothamsted, the "mineral element theory" as propagated in von Liebig's book went largely untested and the conclusion that mineral elements were required for plant growth was reached mainly by observation and speculation rather than by precise experimentation. This was one of the main reasons for the large number of carefully controlled, experimental studies using solution culture undertaken from the 1840s onwards, and particularly by Wilhelm Knop (1817–1891) and Julius von Sachs (1832–1897) in the 1850s and 1860s. In these experiments plants were deprived of specific elements, and the consequences for growth and development were investigated. Such studies made possible a more precise characterization of nutrient elements and helped determine their role in plant physiology and metabolism. In his book of 1865 on experimental plant physiology, von Sachs identified "essential" and "nonessential" elements. The essential elements included those required for the structure and existence of cells [carbon (C), oxygen (O), hydrogen (H), N, and S] as well as those needed physiologically to maintain growth and the vegetative cycle [K, Ca, Mg, P, and iron (Fe)]. However, it was not until the twentieth century that the requirement of plants for additional elements, present in much smaller amounts in plant tissues, was discovered. Of the 14 mineral nutrients now known to be required for plant growth, two distinct groups are recognized: the macronutrients, N, K, P, S, Ca, and Mg, which are required in large amounts, and the micronutrients Fe, manganese (Mn), boron (B), zinc (Zn), copper (Cu), molybdenum (Mo), nickel (Ni), and chlorine (Cl), which are of equal importance, but required in very much smaller amounts.

Arnon and Stout (1939a) proposed that an element can only be considered essential if the following three criteria are met:

1. a deficiency of it makes it impossible for the plant to complete the vegetative or reproductive stage of its life cycle;
2. such deficiency is specific to the element in question, and can be prevented or corrected only by supplying this element; and
3. the element is directly involved in the nutrition of the plant apart from its possible effects in correcting some unfavorable microbiological or chemical condition of the soil or other culture medium.

These three criteria defining an essential plant nutrient have been more or less universally accepted and appear in numerous publications. However, there are flaws in this proposal, as eloquently summarized by Epstein (1999) and discussed by Brown et al. (2022). Contrary to criterion (1), many plant species may be severely deficient in a nutrient element yet complete their life cycle. Criterion (2) is redundant. Criterion (3) presumes that the designation of an element as essential must entail knowledge of its direct involvement in the nutrition of the plant. This was certainly not the case for B, which has been known to be essential since Warington (1923) reported that broad bean (*Vicia faba*) and other plant species failed to grow and develop without it, but it was not until 1996 that a definitive role for B inplants was established (Brown et al., 2002).

The essentiality of 14 mineral elements for plants is now well established, although the known requirements for the micronutrients Cl and Ni are, as yet, restricted to a few plant species. Progress in establishing the essentiality of elements has been closely related to developments in analytical chemistry, particularly in the purification of chemicals and the development of new analytical techniques. This relationship is reflected in the time-course of the discovery of the essentiality of the micronutrients (Table 1.1).

1.3 Beneficial elements for plant growth

Plants also take up another group of elements that are described as beneficial because they can improve plant growth and crop quality, as well as resistance to pests, diseases, and abiotic stresses. These include sodium (Na), silicon (Si), cobalt (Co), iodine (I), selenium (Se), and aluminum (Al) (for a review of these and other beneficial elements see Barker and Pilbeam, 2015). For several of these elements, whether they should be classified as an essential nutrient or a beneficial element is not clear-cut. For example, Na is required in similar amounts to micronutrients for the C4 photosynthetic pathway (Cheeseman, 2015) and also promotes the growth of euhalophytes, such as beet (*Beta vulgaris*) and saltbush (*Atriplex* spp.), when present in tissues in similar quantities to macronutrients (Debez et al., 2017).

Tissues of many field-grown crops contain Si in similar concentrations (1−100 mg g^{-1} dry weight) to those of the macronutrients (Hodson et al., 2005) and in grass-like monocotyledonous plants, such as rice (*Oryza sativa*) and sugar cane (*Saccharum officinarum*), Si may be the predominant mineral constituent. Silicon, in the form of solid amorphous silica or phytoliths, provides structure and rigidity to aboveground plant parts (Epstein, 1999). Silicon also alleviates deleterious effects caused by abiotic stress (e.g., drought and salinity) and protects plants from biotic stress (e.g., pests and diseases), as reviewed by Luyckx et al. (2017). Recently, the supply of Si has been shown to influence the uptake and translocation of all the macronutrients and some of the micronutrients (Pavlovic et al., 2021). Application of Si fertilizers has been shown to increase yield and quality of both monocotyledonous and dicotyledonous crops grown in the field (Liang et al., 2015). Even in tomato, a plant that contains much less Si than rice or sugar cane, the application of Si in field trials increased yield by 15%−30% by increasing both fruit number and fruit size (Liang et al., 2015).

Nutrient solutions rarely include Si because it is not considered essential for plant growth. Epstein (1994) contends that the omission of Si from nutrient solutions amounts to the "imposition of an atypical environmental stress." This statement is supported by the findings of Gottardi et al. (2012) who observed that the addition of only 30 μM Si to the solution of hydroponically grown corn salad (*Valerianella locusta*) increased edible yield, raised crop quality, and extended shelf life. Similarly, Pozo et al. (2015) demonstrated that the addition of 0.65 mM Si to the nutrient solution of several hydroponically cultivated horticultural crops [lettuce (*Lactuca sativa*), tomato, sweet pepper (*Capsicum annuum*), melon (*Cucumis melo*), cucumber (*Cucumis sativus*)] not only increased vegetative growth but also increased the thickness of plant cuticles, which protected the crops from fungal infection by *Botrytis cinerea*.

Cobalt (Co), in common with the micronutrients Fe, Mn, Cu, and Zn, is classified as a transition metal in the periodic table. It is required for symbiotic N$_2$ fixation in nodulated legumes, where Co is an integral component of cobalamin

TABLE 1.1 Discovery of the essentiality of micronutrients for plants.

Element (chemical symbol)	Year	Discovered by
Fe	1843	E. Gris
Mn	1922	J.S. McHargue
B	1923	K. Warington
Zn	1926	A.L. Sommer and C.B. Lipman
Cu	1931	C.B. Lipman and G. MacKinney
Mo	1939a,b	D.I. Arnon and P.R. Stout
Cl	1954	T.C. Broyer et al.
Ni	1987	P.H. Brown et al.

(vitamin B_{12}) which is required by several enzymes involved in N_2 fixation by rhizobial bacteria. Cobalt is also essential for N_2 fixation by endotrophic and associated bacteria in plant roots, which can contribute substantially to N acquisition by various crops (Reed et al., 2011). Symptoms of Co deficiency (similar in appearance to those of N deficiency) have been reported in both legume and nonlegume plant species, and it seems likely that coenzymes or proteins containing Co participate in plant Co metabolism. The possibility that Co is an essential nutrient for plants was reviewed recently by Hu et al. (2021).

A case for including I among the essential nutrients for plants was presented by Kiferle et al. (2021), who reported impaired growth of arabidopsis (*Arabidopsis thaliana*) plants lacking I and the presence of iodinated proteins in shoots and roots that might function in photosynthetic processes and antioxidant activities, respectively.

1.4 A new definition of a mineral plant nutrient

It has been argued that the current distinction between an *essential nutrient* and a *beneficial element* for plant growth does not support a future vision for plant nutrition that encompasses plant science research, practical agronomy, and fertilizer legislation. Brown et al. (2022) propose that a mineral plant nutrient might be defined more broadly as "an element which is essential or beneficial for plant growth and development or for the quality attributes of the harvested product of a given plant species grown in its natural or cultivated environment". The authors suggest that this new, broad definition, founded in science and relevant in practice, has the potential to revitalize innovation and discovery in plant nutrition. Furthermore, this new definition would align better with nutrients deemed essential for animal and human nutrition (White and Broadley, 2009). The new definition represents a fundamental change in concept, but, nevertheless, it is in keeping with the practical, historic objectives of plant nutrition to improve fertilizer usage and crop production. Brown et al. (2022) see the new concept of a plant nutrient as a starting point for open debate and discussion. They also envisage that an independent global scientific body, such as, for example, the International Plant Nutrition Council, would review the list of essential and beneficial elements periodically in the light of new scientific findings. The inclusion of both essential and beneficial elements in the proposed definition of a plant nutrient is important because in many current legislative jurisdictions "beneficial element" is interpreted as "not a plant nutrient". This does not serve agriculture well and is in urgent need of revision. It is of particular relevance to Si in view of the overwhelming evidence of the importance of Si for crop yield and quality (Liang et al., 2015).

1.5 Biochemical properties and physiological functions of nutrient elements in plants

In addition to their relative concentrations within the plant, elements may also be classified broadly according to their biochemical properties and physiological function. In a scheme proposed by Mengel and Kirkby (2001), all plant nutrients including carbon (C), hydrogen (H), and oxygen (O) as well as two beneficial elements (Na and Si) were classified into four groups (Table 1.2).

The first group incorporates the major constituents of organic plant material: C, H, O, N, and S. These elements are constituents of amino acids, proteins, enzymes, and nucleic acids, the building blocks of life. The assimilation of all these nutrients by plants is closely linked with oxidation−reduction reactions.

Phosphorus, B, and Si constitute a second group of elements. All three occur in the form of inorganic anions or acids in plants or are bound by hydroxyl groups of sugars to form phosphate, borate, and silicate esters. Phosphorus is a key component of the nucleic acids DNA and RNA, as well as membrane phospholipids. Boron is essential for cell wall structure and membrane integrity. Silicon, taken up as silicic acid, is polymerized to solid amorphous hydrated silica, which maintains the structure and rigidity of aboveground parts in many plant species. Silicon also protects plants from a variety of biotic and abiotic stresses (Epstein, 1999).

The third group of elements comprises K, Na, Ca, Mg, Mn, and Cl, all of which are taken up from the soil solution in their ionic forms. Within the plant these ions have non-specific functions, for example, in establishing electrochemical gradients across plant membranes, in maintaining cation−anion balance within cell compartments, and in long-distance nutrient transport. In addition, K activates numerous enzymes, is required for protein synthesis, and is closely involved in photosynthesis at various levels. Potassium fluxes determine stomatal aperture and, thereby, transpiration and C assimilation. Calcium binds strongly to carboxyl groups of pectins in cell walls, plays a fundamental role in membrane integrity, and acts as a second messenger in the form of cytosolic free Ca^{2+} in many developmental and physiological processes (White, 2015). Magnesium, like K, activates numerous enzymes. The biochemistry of adenosine triphosphate, which is central to energy metabolism in plant cells, has an absolute requirement for Mg. Magnesium also

TABLE 1.2 Classification of plant nutrients.

Nutrients	Uptake	Biochemical functions
Group 1		
C, H, O, N, S	as CO_2, HCO_3^-, H_2O, O_2, NO_3^-, NH_4^+, N_2, SO_4^{2-}, SO_2 ions from the soil solution, gases from the atmosphere	• Major constituents of organic material. • Essential elements of cofactors and enzymes. • Assimilation by oxidation–reduction reactions.
Group 2		
P, B, Si	as phosphates, boric acid, and silicic acid from the soil solution	• Esterification with alcohol groups • Phosphate esters involved in energy transfer reactions and phospholipids.
Group 3		
K, Na, Ca, Mg, Mn, Cl	as ions from the soil solution	• Non-specific functions in establishing osmotic potential. • More specific functions for optimal conformation of enzymes (enzyme activation). • Bridging of reaction partners. • Charge balance. • Controlling membrane permeability and electrochemical potential.
Group 4		
Fe, Cu, Zn, Mo, Ni	as ions or chelates from the soil solution	• In chelated form in prosthetic groups of enzymes. • Enable electron transport by valency change.

Source: From Mengel and Kirkby (2001).

functions in protein synthesis. Magnesium can be bound strongly by coordinate and covalent bonds, by so-called chelation, as occurs in the chlorophyll molecule, which allows plants to absorb energy from light in photosynthesis.

The capacity of Mg, Ca, and Mn to form chelates means that these elements resemble those of the fourth group of elements (Fe, Cu, Zn, Ni, and Mo). The latter micronutrients also have specific functions: iron in the synthesis of chlorophyll, Cu in proteins in photosynthesis, respiration and lignification, and Cu, Zn, and Mn in superoxide dismutase isoenzymes that eliminate toxic O_2^- radicals. Additionally, Zn plays a role in membrane integrity, protein synthesis, and the synthesis of the phytohormone indole-3-acetic acid. Molybdenum is a component of nitrate reductase and the nitrogenase required for N_2 fixation (Kopsell et al., 2015). Nickel is a component of urease and hydrogenase. Treatment of soybean seeds with Ni improves biological N_2 fixation and urease activity (Lavres et al., 2016). Cobalt is required for rhizobial N_2 fixation in nodulated legumes.

Critical leaf concentrations of macro- and micronutrients, as well as the functions of these nutrients and their mobility within plants have been reviewed by Grusak et al. (2016). Leaf macronutrient concentrations considered adequate for maximal plant growth are in the range of (mg g^{-1} dry weight): N (15–40), K (5–40), P (2–5), S (1–5), Ca (0.5–10), and Mg (1.5–3.5). Leaf micronutrient concentrations considered adequate for plant growth are in the range of (μg g^{-1} dry weight): Cl (100–6000), B (5–100), Fe (50–150), Mn (10–20), Cu (1–5), Zn (15–30), Ni (≈ 0.1), and Mo (0.1–1.0). The average concentrations of mineral elements are given in Table 1.3. Detailed accounts of the biochemistry and physiology of macronutrients, micronutrients, and beneficial elements are presented in Chapters 6, 7, and 8, respectively.

1.6 Variation in the angiosperm ionome

The ionome is defined as the mineral elemental composition of a subcellular structure, cell, tissue, organ, or organism; it is strongly influenced by interactions between genetics and the environment (Neugebauer et al., 2018). White et al. (2012) confirmed the robustness of shoot ionomes from seven families with 21 plant species growing on six plots of the Rothamsted Park Grass Continuous Hay Experiment, supplied with different fertilizer treatments since 1856 (Dyke, 1993). Regardless of these continuous long-term treatments, plant families could be distinguished by their shoot

TABLE 1.3 Average concentrations of mineral elements in shoot dry matter (d. wt) sufficient for adequate growth.

Element	Chemical symbol	mmol kg^{-1} d. wt	mg kg^{-1} d. wt
Molybdenum	Mo	0.001	0.1
Nickel[a]	Ni	0.001	0.1
Copper	Cu	0.1	6
Zinc	Zn	0.3	20
Manganese	Mn	1.0	50
Iron	Fe	2.0	100
Boron	B	2.0	20
Chlorine	Cl	3.0	100
Sulfur	S	30	1000
Phosphorus	P	60	2000
Magnesium	Mg	80	2000
Calcium	Ca	125	5000
Potassium	K	250	10,000
Nitrogen	N	1000	15,000

[a]Based on Epstein and Bloom (2005) and Brown et al. (1987).

ionomes. The most informative elements for discriminant analysis followed the sequence Ca > Mg > Ni > S > Na > Zn > K > Cu > Fe > Mn > P.

The "standard functional ionome" comprises the concentrations of the 14 mineral nutrients in tissues of the "average" angiosperm when growth is not limited by mineral nutrition. Variations from the standard ionome can be used to compare the effects of genetics and environment on the ionome. The environment exerts a significant influence on the leaf ionome by affecting growth and development as well as the phytoavailability of mineral nutrients in the soil. Particular orders of plant species show specific leaf ionomic characteristics even when sampled from diverse environments. Leaves of Poales (grasses) are relatively low in Ca, whereas those of Caryophyllales [e.g., quinoa (*Chenopodium quinoa*), sugar beet (*Beta vulgaris*)] are high in Mg, and Brassicales [e.g., cabbage (*Brassica oleracea*), oilseed rape (*Brassica napus*)] are high in S. The evolution of ionomic traits have been traced using phylogenetic relationships among angiosperm orders, families, and genera. Consequences of variation in the leaf ionome for ecology, mineral cycling in the environment, sustainable agriculture, and livestock and human nutrition have been discussed by Neugebauer et al. (2018).

References

Arnon, D.I., Stout, P.R., 1939a. The essentiality of certain elements in minute quantities with special reference to coppe. Plant Physiol. 14, 371–375.
Arnon, D.I., Stout, P.R., 1939b. Molybdenum as an essential element for higher plants. Plant Physiol. 14, 599–602.
Barker, A.V., Pilbeam, D.J., 2015. Handbook of Plant Nutrition, second ed. CRC Press, Boca Raton, FL.
Brown, P.H., Welch, R.M., Cary, E.E., 1987. Nickel: a micronutrient essential for higher plants. Plant Physiol. 85, 801–803.
Brown, P.H., Bellaloui, N., Wimmer, M.A., Bassil, E.S., Ruiz, H., Hu, H., et al., 2002. Boron in plant biology. Plant Biol. 4, 205–223.
Brown, P.H., Zhao, F.-J., Dobermann, A., 2022. What is a plant nutrient? Changing definitions to advance science and innovation in plant nutrition. Plant Soil 476, 11–23.
Broyer, T.C., Carlton, A.B., Johnson, C.M., Stout, P.R., 1954. Chlorine – a micronutrient element for higher plants. Plant Physiol. 29, 526–532.
Cheeseman, J.M., 2015. The evolution of halophytes, glycophytes and crops, and its implications for food security under saline conditions. New Phytol. 206, 557–570.
Debez, A., Belghith, I., Friesen, J., Montzka, C., Elleuche, S., 2017. Facing the challenge of sustainable bioenergy production: could halophytes be part of the solution? J. Biol. Eng. 11, 27. Available from: https://doi.org/10.1186/s13036-017-0069-0.
Dyke, G.V., 1993. John Lawes of Rothamsted – Pioneer of Science, Farming and Industry. Hoos Press, Harpenden.

Epstein, E., 1994. The anomaly of silicon in plant biology. Proc. Natl. Acad. Sci. U.S.A. 91, 11–17.
Epstein, E., 1999. Silicon. Annu. Rev. Plant Physiol. Plant Mol. Biol. 50, 641–664.
Epstein, E., Bloom, A., 2005. Mineral Nutrition of Plants: Principles and Perspectives, second ed. Sinauer Associates, Sunderland, MA.
Gottardi, S., Iacuzazo, F., Tomasi, N., Cortella, G., Manzocco, L., Pinton, R., 2012. Beneficial effects of silicon on hydroponically grown corn salad (*Valerianella locusta* (L.) Laterrade) plants. Plant Physiol. Biochem. 56, 14–23.
Gris, E., 1843. Report concerning the action of soluble ferruginous compounds on plants. Compt. Rend. Acad. Sci. (Paris) 17, 679 [in French].
Grusak, M.A., Broadley, M.R., White, P.J., 2016. Plant macro- and micronutrient minerals. eLS. John Wiley & Sons, Chichester. Available from: http://doi.org/10.1002/9780470015902.a0001306.pub2.
Hodson, M.J., White, P.J., Mead, A., Broadley, M.R., 2005. Phylogenetic variation in the silicon composition of plants. Ann. Bot. 96, 1027–1046.
Hu, X., Wei, X., Ling, J., Chen, J., 2021. Cobalt: an essential micronutrient for plant growth? Front. Plant Sci. 12, 768523. Available from: https://doi.org/10.3389/fpls,2021.768523.
Kiferle, C., Martinelli, M., Salzano, A.M., Gonzali, S., Beltrami, S., Salvadori, P.A., et al., 2021. Evidences for a nutritional role for iodine in plants. Front. Plant Sci. 12, 616868. Available from: https://doi.org/10.3389/fpls.2021.616868.
Kopsell, D.A., Kopsell, D.E., Hamlin, R.L., 2015. Molybdenum. In: Barker, A.V., Pilbeam, D.J. (Eds.), Handbook of Plant Nutrition, second ed. CRC Press, Boca Raton, FL, pp. 511–536.
Lavres, J., Castro Franco, G., de Sousa Câmara, G.M., 2016. Soybean seed treatment with nickel improves biological nitrogen fixation and urease activity. Front. Environ. Sci. 4, 37. Available from: https://doi.org/10.3389/fenvs.2016.00037.
Liang, Y., Nikolic, M., Belanger, R., Gong, H., Song, A., 2015. Effects of silicon on crop growth, yield and quality. In: Liang, Y., Nikolic, M., Belanger, R., Gong, H., Song, A. (Eds.), Silicon in Agriculture: From Theory to Practice. Springer, Dordrecht, pp. 209–223.
Lipman, L.B., MacKinney, G., 1931. Proof of the essential nature of copper for higher green plants. Plant Physiol. 6, 593–599.
Luyckx, M., Hausman, J.-F., Lutts, S., Guerriero, G., 2017. Silicon and plants: current knowledge and technological perspectives. Front. Plant Sci. 8, 411. Available from: https://doi.org/10.3389/fpls.2017.00411.
McHargue, J.S., 1922. The role of manganese in plants. J. Am. Chem. Soc. 44, 1592–1598.
Mengel, K., Kirkby, E.A., 2001. Principles of Plant Nutrition, fifth ed. Kluwer Academic Publishers, Dordrecht.
Neugebauer, K., Broadley, M.R., El-Serehy, H.A., George, T.S., McNichol, J.W., Moraes, M.F., et al., 2018. Variation in angiosperm ionome. Physiol. Plant 161, 306–322.
Pavlovic, J., Kostic, L., Bosnic, P., Kirkby, E.A., Nikolic, M., 2021. Interactions of silicon with essential and beneficial elements in plants. Front. Plant Sci. 12, 697592. Available from: https://doi.org/10.3389/fpls.2021.697592.
Pozo, J., Urrestararazu, M., Morales, I., Sanches, J., Santos, M., Dianez, F., et al., 2015. Effects of silicon in the nutrient solution for three horticultural plant families on the vegetative growth, cuticle and protection against *Botrytis cinerea*. HortScience 50, 1447–1452.
Reed, S.C., Cleveland, C.C., Townsend, A.R., 2011. Functional ecology of free living nitrogen fixation: a contemporary perspective. Annu. Rev. Ecol. Evol. Syst. 42, 489–512.
Sommer, A.L., Lipman, C.B., 1926. Evidence for the indispensible nature of zinc and boron for higher green plants. Plant Physiol. 1, 231–249.
van der Ploeg, R.R., Böhm, W., Kirkham, M.B., 1999. On the origin of the theory of the mineral nutrition of plants and the law of the minimum. Soil Sci. Soc. Am. J. 63, 1055–1062.
Warington, K., 1923. The effects of boric acid and borate on the broad bean and certain other plants. Ann. Bot. 37, 629–672.
White, P.J., 2015. Calcium. In: Barker, A.V., Pilbeam, D.J. (Eds.), Handbook of Plant Nutrition, second ed. CRC Press, Boca Raton, FL, pp. 165–198.
White, P.J., Broadley, M.R., 2009. Biofortification of crops with seven mineral elements often lacking in human diets – iron, zinc, copper, calcium, magnesium, selenium and iodine. New Phytol. 182, 49–84.
White, P.J., Broadley, M.R., Thompson, J.A., McNichol, J.W., Crawley, M.J., Poulton, P.R., et al., 2012. Testing the distinctness of shoot ionomes of angiosperm families using the Rothamsted Park Grass Continuous Hay Experiment. New Phytol. 196, 101–109.

Further reading

von Liebig, J., 1840. Chemistry and Its Application to Agriculture and Physiology. Verlag Vieweg, Braunschweig [in English and German].
von Sachs, J., 1865. Handbook of Experimental Plant Physiology: Investigatons of Environmental Effects on Plants and the Functions of their Organs. W. Engelmann, Leipzig [in German].

Chapter 2

Ion-uptake mechanisms of individual cells and roots: short-distance transport

Devrim Coskun[1] and Philip J. White[2]
[1]Department of Phytology, Laval University, Quebec City, Quebec, Canada, [2]Ecological Science, James Hutton Institute, Dundee, Scotland, United Kingdom

Summary

The uptake of nutrients by plants is characterized by selectivity of transport and accumulation in specific tissues, cells, or subcellular compartments. These characteristics are genetically determined, but influenced by the environment, and can differ both between and within plant species. This chapter reviews the environmental, physiological, and developmental factors that affect the entry of nutrients into the extracellular space (apoplasm) of roots, their transport across the plasma membrane and tonoplast of root cells, and the pathways of their movement to the xylem. It describes the structure and composition of cellular membranes, the electrochemical gradients that determine the energetics of solute transport across membranes, and the mechanisms involved and the genetic identity of the proteins that facilitate the transport of nutrients across the plasma membrane and tonoplast of plant cells. The overriding influence of plant nutritional status on the expression of mechanisms by which roots acquire nutrients is emphasized.

2.1 General

As a rule, there is a large discrepancy between the concentrations of mineral nutrients in the soil and the nutrient requirements of plants. Furthermore, soil and, in some cases, nutrient solutions can contain high concentrations of mineral elements not needed for plant growth, or that are potentially harmful to plants. The mechanisms by which plants accumulate nutrients must therefore be selective. This selectivity can be demonstrated particularly well in algal cells (Table 2.1), where the external and vacuolar (cell sap) solutions are separated by only two membranes: the plasma membrane and the tonoplast.

TABLE 2.1 Relationship between ion concentrations in the substrate and in the cell sap of *Nitella* and *Valonia*.

Ion	*Nitella* concentration (mM)			*Valonia* concentration (mM)		
	A	B	Ratio	A	B	Ratio
	Pond water	Cell sap	B/A	Seawater	Cell sap	B/A
Potassium	0.05	54	1080	12	500	42
Sodium	0.22	10	45	498	90	0.18
Calcium	0.78	10	13	12	2	0.17
Chloride	0.93	91	98	580	597	1

Source: Modified from Hoagland (1948).

TABLE 2.2 Changes in the ion concentration of the external (nutrient) solution and in the root sap of maize (*Zea mays*) and bean (*Phaseolus vulgaris*).

Ion	External concentration (mM)			Concentration in root sap (mM)	
	Initial	After 4 days[a]		Maize	Bean
		Maize	Bean		
Potassium	2.00	0.14	0.67	160	84
Calcium	1.00	0.94	0.59	3	10
Sodium	0.32	0.51	0.58	0.6	6
Phosphate	0.25	0.06	0.09	6	12
Nitrate	2.00	0.13	0.07	38	35
Sulfate	0.67	0.61	0.81	14	6

[a]No replacement of water lost through transpiration.

In *Nitella*, the concentrations of K^+, Na^+, Ca^{2+}, and Cl^- ions are higher in the cell sap than in the pond water, but the concentration ratio differs considerably among the ions. By contrast, in *Valonia* growing in seawater, only K^+ is more concentrated in the cell sap, whereas the Na^+ and Ca^{2+} concentrations are lower in the cell sap than in the seawater.

Selective ion uptake is a typical feature of plants. When plants are grown in a nutrient solution of limited volume, the external concentrations of ions change with time (Table 2.2). The concentrations of K^+, $H_2PO_4^-$ (inorganic phosphate, or P_i), and NO_3^- decline markedly, whereas those of Na^+ and SO_4^{2-} can even increase, indicating that water is taken up faster than either of these two ions. Uptake rates, especially for K^+ and Ca^{2+}, differ between plant species [e.g., maize (*Zea mays*) and bean (*Phaseolus vulgaris*), Table 2.2]. The concentrations of ions in the root sap are generally higher than those in the nutrient solution; this is most evident for K^+, NO_3^-, and $H_2PO_4^-$.

These results demonstrate that ion uptake is characterized by:

1. *Selectivity*. Certain mineral elements are taken up preferentially, while others are discriminated against or almost excluded.
2. *Accumulation*. The concentration of elements can be much higher in cell sap than in the external solution.
3. *Genotype*. There are distinct differences among plant species in their ion-uptake characteristics.

These observations lead to many questions. In particular, how do individual cells and plants regulate the uptake of ions both to satisfy plant demand and to avoid ion toxicities? To understand the regulation of ion uptake it is necessary to follow the pathway of solutes from the external solution through the cell wall and the plasma membrane into the cytoplasm and vacuoles of plant cells (Fig. 2.1).

2.2 Pathway of solutes from the external solution into root cells

2.2.1 Influx to the apoplasm

Movement of low-molecular-weight solutes (e.g., ions, organic acids, amino acids, and sugars) from the external solution through the walls of individual root cells is driven by diffusion and mass flow. While the former is dependent on a concentration gradient, the latter is driven by transpiration that can be regulated in response to environmental cues (e.g., nutrient and water availability; Cramer et al., 2009; Maurel et al., 2021; Tardieu et al., 2015). Nevertheless, cell walls can interact with solutes and, thereby, facilitate or restrict passage across the root and uptake across the plasma membrane of individual cells.

The primary cell wall consists of a network of cellulose (accounting for about 15%–30% of its dry weight), cross-linking glycans (generally xyloglucans in Type-I walls, but in the Type-II walls of commelinoid monocotyledons, mostly gluconoarabinoxylans) and glycoproteins, all embedded in a pectin matrix (Burton et al., 2010; Cosgrove, 2018). Type-I cell walls contain more pectin than Type-II cell walls. Both Ca and B are integral components of cell

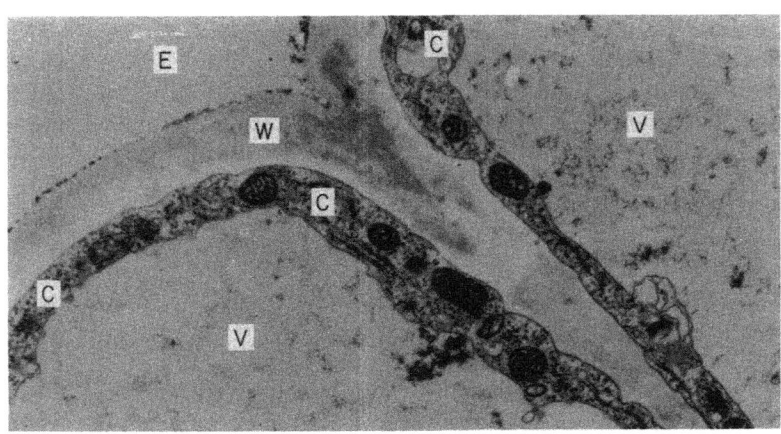

FIGURE 2.1 Cross section of two rhizodermal cells of a maize root. *C*, Cytoplasm; *E*, external solution; *V*, vacuole; *W*, cell wall. *Courtesy C. Hecht-Buchholz.*

TABLE 2.3 Cell-wall thickness and diameter of cell-wall pores, sucrose, and hydrated cations.

Parameter	
	Thickness (nm)
Rhizodermal cell wall (maize; Fig. 2.1)	500–3000
Cortical cell wall (maize)	100–200
	Diameter (nm)
Pores in cell wall	<5.0
Sucrose	1.0
Hydrated ions	
K^+	0.66
Ca^{2+}	0.82

walls, which can additionally be impregnated with Si (Epstein, 1999; White and Broadley, 2003; Wimmer et al., 2020). The cell-wall network contains pores (i.e., interfibrillar and intermicellar spaces) that differ in size. Maximum diameter of the pores in plant cell walls is generally about 5.0 nm (Table 2.3). The diameters of hydrated ions, such as K^+ and Ca^{2+}, are small in comparison. Therefore, the diameter of pores does not restrict the movement of nutrient ions through the cell wall.

In contrast to nutrients and low-molecular-weight organic solutes, the movement of high-molecular-weight solutes (e.g., metal chelates, fulvic acids, and toxins) or viruses and other pathogens through cell walls is severely restricted by the diameter of the pores.

Cell walls have a variable proportion of pectins, mainly polygalacturonic acid in the middle lamella. Their carboxylic groups (COO^-) act as cation exchangers in the cell-wall continuum of plant tissues, the so-called *apoplasm*.

Hope and Stevens (1952) introduced the term *apparent free space* (AFS). This comprises the *water-free space* (WFS) that is freely accessible to solutes, and the *Donnan-free space* (DFS), where cation exchange and anion repulsion take place (Fig. 2.2). Ion distributions within the DFS are characterized by typical Donnan distributions. Trivalent cations (e.g., Al^{3+}) bind more strongly than divalent cations (e.g., Ca^{2+}) that bind more strongly than monovalent cations (e.g., K^+). Plant species differ considerably in their cation exchange capacity (CEC), that is, in the number of cation exchange sites (fixed anions; $R-COO^-$) located in cell walls (Table 2.4).

Generally, the CEC of dicotyledonous species is greater than that of monocotyledonous species (White and Broadley, 2003; White et al., 2018). As the external pH decreases, the effective CEC is reduced, particularly in monocotyledonous species (Allan and Jarrell, 1989) as protons occupy an increasing proportion of the cation-binding sites. Because of apoplasmic barriers within the root, such as the Casparian bands of the endodermis and exodermis, only

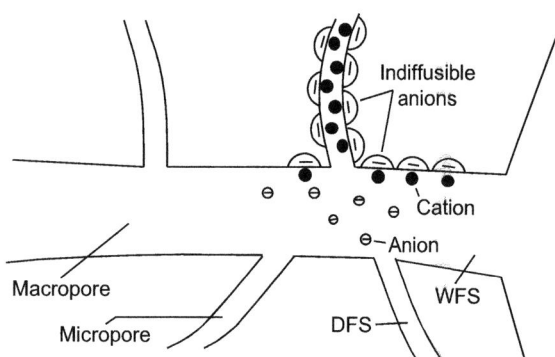

FIGURE 2.2 Schematic diagram of the pore system of the apparent free space. *DFS*, Donnan-free space; *WFS*, water-free space.

TABLE 2.4 Cation exchange capacity of root dry matter of different plant species.

Plant species	Cation exchange capacity [cmol(+) kg^{-1} dry wt]
Wheat	23
Maize	29
Bean	54
Tomato	62

Source: Based on Keller and Deuel (1957).

part of the AFS is directly accessible to cations from the external solution. Exchange adsorption of cations in the apoplasmic AFS is not a prerequisite for ion uptake across the plasma membrane or for the movement of ions within the apoplasm. However, fixed negative charges in the AFS can influence both the absolute and relative concentrations of cations in the apoplasm, especially when roots grow in dilute solutions (White and Broadley, 2003). Thus, root CEC can influence indirectly the rate and selectivity of ion influx into root cells and apoplasmic ion movements.

The apoplasmic AFS can also serve as a transient storage pool for essential mineral elements such as Fe and Zn that can be mobilized, for example, by specific root exudates such as phytosiderophores, and subsequently translocated to the shoots (Cesco et al., 2002; Zhang et al., 1991a, 1991b). For Fe, the size of this storage pool possibly contributes to genotypic differences in sensitivity to iron deficiency in soybean (Longnecker and Welch, 1990).

2.2.2 Passage into the cytoplasm

Despite some selectivity for cation binding in the cell wall, the main site of selectivity in the uptake of cations and anions, as well as solutes in general, is the plasma membrane of individual root cells. The lipid bilayer of the plasma membrane prevents the indiscriminate movement of ions and large polar molecules from the apoplasm into the cytoplasm (influx) and from the cytoplasm into the apoplasm (efflux). Integral membrane proteins facilitate the selective transport of solutes across the plasma membrane.

It can be demonstrated readily that the plasma membrane is a selective barrier to the uptake of ions. For example, when barley plants are placed in a nutrient solution containing Ca^{2+} (^{45}Ca) and K$^+$ (^{42}K), most of the ^{45}Ca that accumulated in the roots in the first 30 min is still readily exchangeable and is almost certainly located in the apoplasmic AFS (Fig. 2.3). By contrast, only a minor fraction of the ^{42}K is readily exchangeable after this period, with most of the ^{42}K having been transported across the plasma membranes of root cells. Furthermore, in the majority of mature plant cells, the vacuole comprises more than 80%−90% of the cell volume, acting as a central storage compartment for ions and solutes (e.g., sugars and secondary metabolites), and it is likely that some of the ^{42}K is sequestered within this cellular compartment after 30 min. Such classical radiotracer analyses can reveal many important kinetic parameters for a given substrate within the "compartments" of the AFS, cytosol, and vacuole, including rates of unidirectional fluxes

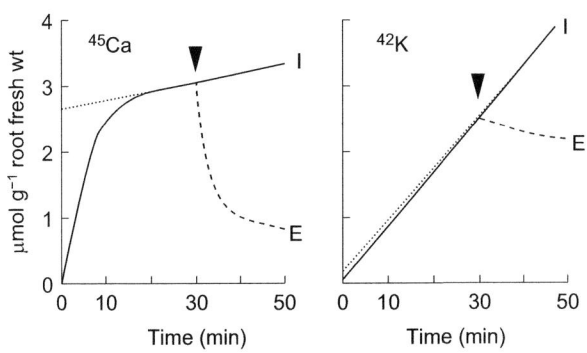

FIGURE 2.3 Time-course of influx (I) and efflux (E) of ^{45}Ca and ^{42}K to isolated barley roots. After 30 min (arrows) some of the roots were transferred to solutions with no labeled Ca^{2+} and K^+. The proportion of the exchangeable fraction (dotted line) in the AFS is calculated by extrapolation to zero time.

(influx and efflux), half-lives of exchange (the time required for half the amount in the compartment to be exchanged), and pool size (quantity/concentration in the compartment) (Britto and Kronzucker, 2001; Coskun et al., 2014; Section 2.4.3).

Within the plant cell, membranes with contrasting lipid and protein composition separate the various cellular compartments. These membranes include the tonoplast (vacuolar membrane), the endoplasmic reticulum (ER), the Golgi apparatus, the nuclear membrane, and the membranes surrounding vesicles, mitochondria, and plastids (Staehelin, 2015). For these membranes, the lipid bilayer provides the barrier to solute movement, and embedded proteins facilitate the selective transport of solutes to provide the unique transport properties required for the function of each compartment. Before the mechanisms of solute transport across membranes are discussed in greater detail (Sections 2.4 and 2.5), some fundamental aspects of the composition and structure of biological membranes will be described.

2.3 Composition of biological membranes

The capacity of plant cell membranes to regulate solute uptake has fascinated botanists since the 19th century. By the early years of the 20th century, some basic facts of solute permeation across biological membranes had been established. High-molecular-weight organic solutes such as polyethylene glycol are not taken up by cells and can be used at high external concentrations as osmotica to induce water deficiency (drought stress) in plants. However, some hydrophobic molecules penetrate membranes much faster than would be predicted on the basis of their size, which is presumably related to their capacity to partition into the lipid bilayer.

Plant membranes are composed of a highly complex matrix of various types of lipids and proteins forming a bilayer (Fig. 2.4; Table 2.5). Membrane lipids generally fall into four main categories: (1) glycerophospholipids, consisting of a glycerol backbone with fatty acids and a phosphodiester polar head group; (2) glycolipids, which have sugars as their hydrophobic headgroup, (3) sphingolipids, which are based around the 18-carbon amino-alcohol moiety, sphingosine; and (4) sterols, which are based around a four-ringed structure (Cassim et al., 2019; Fig. 2.5).

Glycerolipids comprise glycolipids and phospholipids, the latter of which is the most abundant fraction in the plasma membrane (Cacas et al., 2016). The major plant phospholipids of the plasma membrane are phosphatidylcholine and phosphatidylethanolamine, but these also contain phosphatidylglycerol (PG), phosphatidylinositol (PI), phosphatidic acid, and phosphatidylserine. Plasma membranes and mitochondria are enriched in PI and PG, respectively. The fatty-acid moiety in phospholipids varies in both chain length and number of double bonds but is often palmitic (length:double bonds, 16:0), stearic (18:0), oleic (18:1), linoleic (18:2), or linolenic (18:3) acid (Table 2.5). Most glycolipids are found in the chloroplast (Hölzl and Dörmann, 2007), where the thylakoid membrane is predominantly composed of monogalactosyldiacyglycerol, together with digalactosyldiacylglycerol (DGDG) and the sulfolipid, sulfoquinovosyldiacylglycerol (SQDG).

Sphingolipids are abundant and essential components of eukaryotic cells and can represent 10%−20% of the total membrane lipids in plants (Huby et al., 2020; Lynch and Dunn, 2004; Pata et al., 2010). Not only do they perform important structural roles in membranes, sphingolipids are also critical to many cellular and regulatory processes, including vesicle trafficking, plant development, and defense. Sphingolipids in plants comprise four major categories: ceramides, glucosylceramides (gluCer), glycosyl inositol phosphoryl ceramides, and free long-chain bases, representing c. 2%, 34%, 64%, and 0.5% of the total sphingolipids in *Arabidopsis thaliana*, respectively (Markham et al., 2006).

Major plant sterols include campesterol, sitosterol, and stigmasterol. The sterol content of the ER is low, but sterols can make up more than 30% of the total lipids in the plasma membrane and tonoplast (Table 2.5).

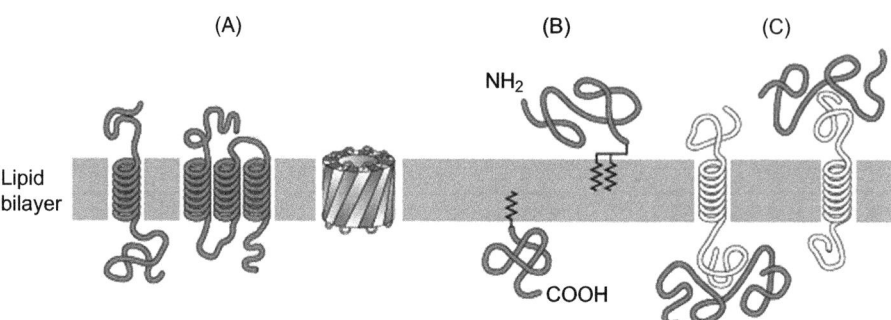

FIGURE 2.4 Protein associations with biological membranes. Integral transmembrane proteins extend through the lipid bilayer in α-helical or β-sheet structures. Peripheral proteins are attached to the membrane either by covalently attached lipid groups or through interactions with integral membrane proteins. (A) Transmembrane, (B) Lipid linked, (C) Protein attached. *Based on Alberts et al. (1998)*.

TABLE 2.5 Lipid and fatty-acid composition of the plasma membrane and tonoplast of mung bean (*Vigna radiata*).

Lipids	Plasma membrane (μmol mg^{-1} protein)	Tonoplast (μmol mg^{-1} protein)
Phospholipids	1.29	1.93
Sterols	1.15	1.05
Glycolipids	0.20	0.80

Fatty-acid composition of the phospholipids

Fatty acid	Chain length	Melting point (°C)	Plasma membrane (% of total)	Tonoplast (% of total)
Palmitic acid	C16	+62.8	35	39
Stearic acid	C18	+70.1	6	6
Oleic acid	C18:1[a]	+13.0	9	9
Linoleic acid	C18:2[a]	−5.5	21	22
Linolenic acid	C18:3[a]	−11.1	19	20
Others	−	−	10	4

[a]*Numeral to the right of the colon indicates the number of double bounds.*
Source: Based on Yoshida and Uemura (1986).

It is now well established that membranes are laterally heterogeneous, giving rise to the concept of "membrane domains" that differ in biophysical properties (e.g., membrane order, surface charge; Gronnier et al., 2018). Cells can also display polar localization of membrane proteins to facilitate the coordinated and efficient transport of solutes (Barberon et al., 2014; Miwa et al., 2009; Raggi et al., 2020; Yoshinari and Takano, 2017). Membrane complexity extends further in terms of bilayer asymmetry. Within the plasma membrane, sphingolipids are primarily located in the outer layer and phospholipids are predominately located in the inner cytosolic layer (Gronnier et al., 2018).

Although the lipid bilayer provides the basic structure of the membrane and forms a permeability barrier, most biological functions of membranes are performed by proteins. The membrane surrounding each cellular compartment has different types of proteins reflecting the particular function of that membrane. Membrane proteins function (1) to anchor the membrane to the cytoskeleton and/or cell wall; (2) as receptors/transducers of compartmentalized signals; (3) as enzymes for specific reactions, such as energy transduction in mitochondria and chloroplasts; and (4) to transport specific solutes across membranes.

There are several ways by which proteins can be associated with the lipid bilayer (Staehelin, 2015). Many membrane proteins extend through the bilayer (Fig. 2.4). These integral transmembrane proteins have both hydrophobic and

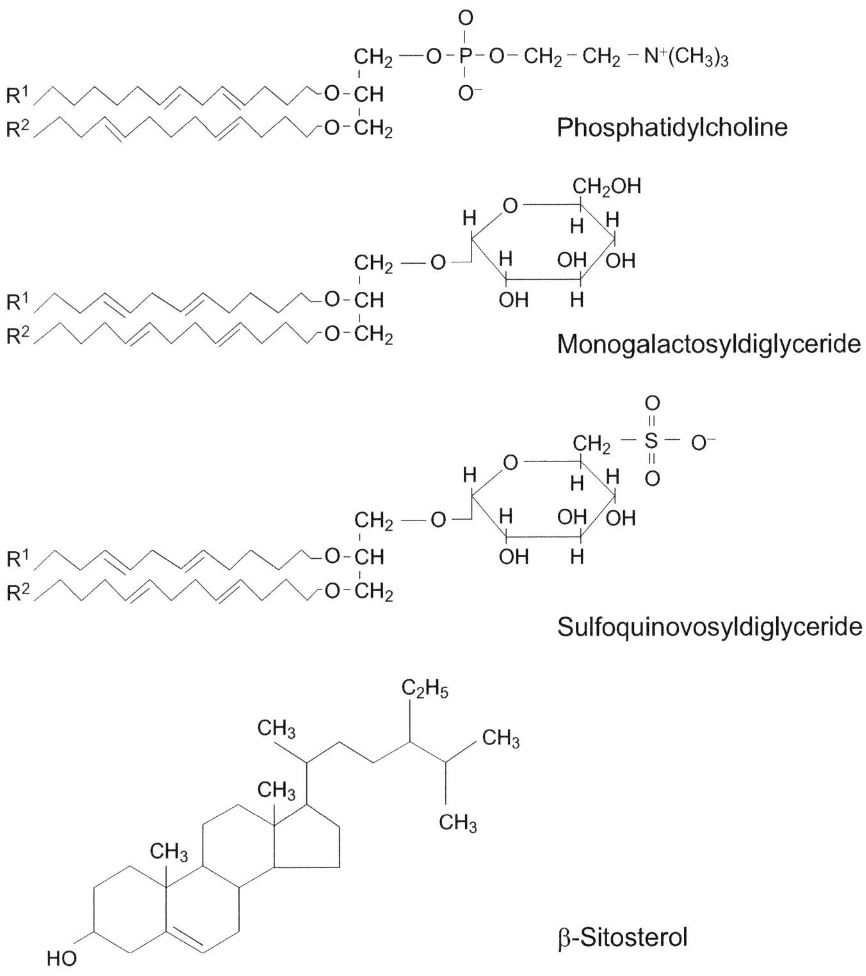

FIGURE 2.5 Chemical structures of selected membrane lipids. Phospholipids are represented by phosphatidylcholine, glycolipids by monogalactosyldiglyceride and sulfoquinovosyldiglyceride, and sterols by β-sitosterol.

hydrophilic portions. Their hydrophobic portions lie within the bilayer, alongside the hydrophobic tails of the lipid molecules, while their hydrophilic portions extend into the aqueous environment on either side of the membrane. Other membrane proteins are located entirely outside the bilayer. These peripheral proteins are bound to the membrane through lipid groups attached covalently through prenylation (attachment of the isoprenoids farnesyl diphosphate or geranylgeranyl diphosphate), S-acetylation (attachment of palmitate or stearate), or N-myristoylation (Sorek et al., 2009) or are associated with other membrane proteins through ionic interactions. It is thought that lipid modification of membrane proteins also facilitates their subcellular targeting and clustering into specific domains. Proteins can also be localized asymmetrically within the plasma membrane bilayer, where glycosylphosphatidylinositol-anchored proteins are present in the outer layer, and acylated and isoprenylated proteins are present in the inner layer (Gronnier et al., 2018).

Lipid composition not only differs between cellular membranes (Table 2.5), plant tissues (Kehelpannala et al., 2021), and plant species (Staehelin, 2015) but is influenced also by environmental factors. In leaves, for example, membrane lipid composition changes during exposure to low temperatures (e.g., Penfield, 2008; Welti et al., 2002), and DGDG and SQDG can replace phospholipids in membranes of P-deficient plants (Hölzl and Dörmann, 2007; Veneklaas et al., 2012). Similarly, the composition of root membranes is influenced by temperature, salinity, and the ionic composition of the external solution (e.g., Lindberg et al., 2005; López-Pérez et al., 2009; White et al., 1990a; Wu et al., 1998). The changes in lipid composition often reflect the acclimation of a plant to its environment. For example, membranes of plants growing at low temperatures have more phospholipids with charged headgroups and shorter fatty-acid chains with lower degree of saturation, and greater sterol content than plants growing at higher temperatures (Penfield, 2008; Staehelin, 2015; Wallis and Browse, 2002). Such changes shift the freezing point (i.e., the transition temperature) of membranes to a lower temperature and may therefore be important for the maintenance of membrane functions at low temperatures.

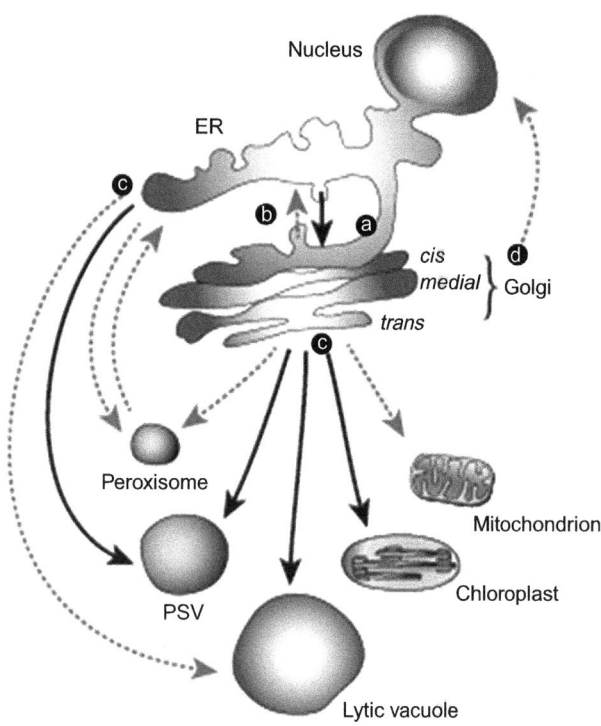

FIGURE 2.6 Pathways of membrane trafficking. The initial pathways are divided into (a) ER and Golgi integration and (b) transport of vesicles between the ER and Golgi. The subsequent pathways (c) involve transport of vesicles between the ER and peroxisomes and lytic vacuoles, and between the Golgi and peroxisomes, lytic and PSV vacuoles, chloroplasts, mitochondria and plasma membrane. Feedback signals from the Golgi to the nucleus (d) are thought to regulate aspects of membrane trafficking. *ER*, Endoplasmic reticulum; *PSV*, protein storage vacuole. *Based on Matheson et al. (2006).*

Cellular membranes are dynamic structures that are remodeled continuously to allow the plant to respond to developmental signals, biotic challenges, and environmental conditions. This remodeling occurs over minutes to months, and is supported by complex trafficking pathways that deliver lipids and proteins to and from cellular membranes (Fig. 2.6). These pathways are functionally linked through the Golgi apparatus to the ER, plasma membrane, peroxisomes, vacuoles, mitochondria, and chloroplasts (Matheson et al., 2006; Robinson et al., 2007). The delivery of secretory vesicles to the plasma membrane can be targeted to specific locations, such as the apex of tip-growing cells (e.g., elongating root hairs or pollen tubes; Cheung and Wu, 2008; Cole and Fowler, 2006; Ishida et al., 2008; Sorek et al., 2009), or to plasmodesmata (Lucas et al., 2009; Maule, 2008; Oparka, 2004). Thus, membranes are not homogeneous but possess domains in which specific lipids and proteins can be clustered, stably or transiently, to improve the efficiency of biochemical and physiological processes (Opekarová et al., 2010).

2.4 Solute transport across membranes

2.4.1 Thermodynamics of solute transport

In the experiment described in Table 2.2, the K^+ concentration in maize root sap (which is approximately equal to the K^+ concentration of the vacuoles) was 80 times higher than in the external solution. By contrast, the Na^+ concentration in the root sap remained lower than that in the external solution. Such phenomena require both a source of energy and selective transport across the plasma membrane of root cells.

Transport across plant membranes is facilitated by transmembrane proteins (Fig. 2.7). These can be classified into three groups: (1) primary active transporters (pumps), in which solute transport against an electrochemical gradient is coupled directly to the hydrolysis of an energy substrate such as adenosine triphosphate (ATP) or pyrophosphate (PP_i); (2) secondary active transporters or "coupled transporters" that harness the electrochemical gradient of (generally) H^+ to the movement of a solute against its electrochemical gradient in either the same (symport) or opposite (antiport) direction; and (3) passive transporters that catalyze the movement of solutes down their electrochemical gradient. The latter group includes a variety of carriers (uniporters) and channels. Channels can be distinguished from carriers by their high catalytic rate that can exceed 10 million ions s^{-1}, which is several orders of magnitude greater than that of carriers (White, 2017).

Under most circumstances, the driving force for the facilitated diffusion of an uncharged solute across a membrane is its concentration gradient, whereas for an ion it is its electrochemical gradient (White, 2017). The Nernst equation

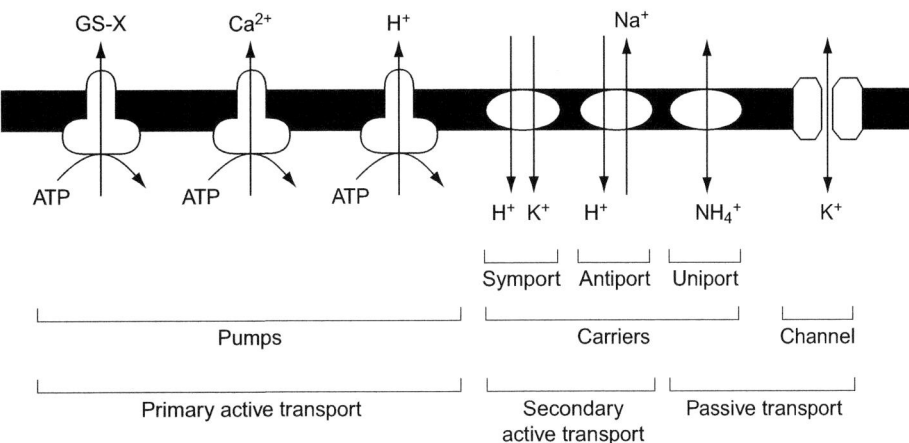

FIGURE 2.7 Nomenclature of transport proteins. Schematic representation of primary active transport mechanisms, such as ABC transporters (e.g., glutathione conjugate pump), metal transporters (e.g., Ca^{2+}-ATPase), and H^+-ATPases, secondary active transport mechanisms, such as the K^+/H^+ symporter or the Na^+/H^+ antiporter, and passive transport mechanisms, such as the NH_4^+ carrier and the K^+ channel. *ABC*, ATP-binding cassette. *Based on White (2017).*

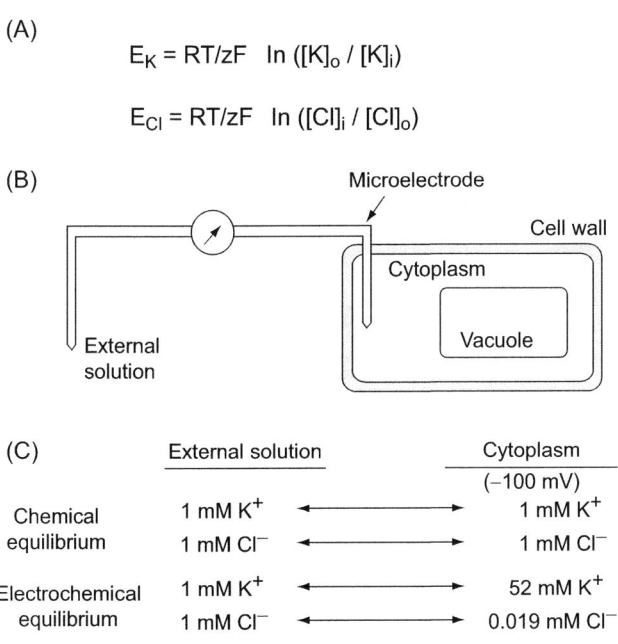

FIGURE 2.8 (A) The Nernst equation. The equilibrium potentials for potassium (E_K) and chloride (E_{Cl}) are given as a function of R [the gas constant (8.314 J K^{-1} mol^{-1})], T (the absolute temperature), z (the valency of the ion), F [Faraday constant, the number of coulombs (C) per mole of electrons], and their activities outside (subscript o) and inside (subscript i) the membrane. (B) Schematic representation of the system used for measuring the membrane potential of plant cells. (C) Example of the calculation of ion distributions at electrochemical equilibrium assuming a membrane potential of −100 mV at 20°C.

(Fig. 2.8) allows the direction of the net diffusive flux of an ion at a given membrane potential and temperature to be determined. When the cell-membrane potential is more negative than the Nernst potential, cations can move into the cell, and anions out of the cell, by facilitated diffusion via specific transport proteins. When the membrane potential is more positive than the Nernst potential, the opposite fluxes are favored. According to the Nernst equation, at 20°C with a membrane potential of −100 mV, K^+ or Cl^- would be in electrochemical equilibrium across the plasma membrane if their concentration in the cytosol were 52 times higher (K^+) or 52 times lower (Cl^-) than in the external solution (Fig. 2.8). At the same temperature and membrane potential, the concentrations of a divalent cation or anion would differ more than 2700-fold between the cytosol and external medium if they were in electrochemical equilibrium.

The resting membrane potential of root cells is often more negative than −100 mV (Coskun et al., 2013a; Maathuis and Sanders, 1993; Walker et al., 1996). It is generated primarily by the activity of plasma membrane H^+-ATPases encoded by members of the *AHA* gene family (Falhof et al., 2016; Gaxiola et al., 2007). These H^+ pumps are clustered in discrete (micro)domains of the plasma membrane and their activity is regulated by phosphorylation-dependent interactions with cytosolic 14-3-3 proteins in response to diverse environmental signals, including salinity stress, nutrient deficiencies, and low temperatures (Falhof et al., 2016; Siao et al., 2020). Under physiological conditions, many cations are in electrochemical equilibrium across the plasma membrane of root cells (White, 2017). However, there is always a large electrochemical gradient driving Ca^{2+} influx to cells, and, in saline environments, there is also a large electrochemical gradient driving Na^+ influx. On the other hand, anion influx to root cells is often catalyzed by symporters coupled to the proton electrochemical gradient (Takahashi, 2019; Wang et al., 2012, 2018a, 2018b).

Numerous passive transporters facilitating the influx of mineral nutrients across the plasma membrane of root cells have been reported (Fig. 2.9; Table 2.6). These include K^+ channels (e.g., AKT1, KC1; Britto et al., 2021; Ragel et al., 2019), Ca^{2+} channels (e.g., cyclic nucleotide-gated channels, GLRs, Mid1-complementing activity channels, annexins, TPC1; Tang and Luan, 2017; White, 2015), NH_3/NH_4^+ transporters (e.g., AMT/MEP/Rh; McDonald and Ward, 2016), NH_3-permeable aquaporins (e.g., AtTIP2;1; Kirscht et al., 2016), Mg^{2+} transporters (e.g., AtMGT1; Tang and Luan, 2017), divalent cation (Fe^{2+}, Zn^{2+}, Cu^{2+} and Mn^{2+}) transporters (e.g., ZRT-IRT-like proteins (ZIPs); Milner et al., 2013), $B(OH)_3$-permeable aquaporins (e.g., AtNIP5;1; Yoshinari and Takano, 2017), ion-conducting aquaporins (e.g., AtPIP2;1; Tyerman et al., 2021), and Cl^- channels (under conditions of salinity stress; e.g., CLCs; Li et al., 2017a).

Similar transport proteins are present in the plasma membranes of other plant cells, where they serve both general and specific functions. In the stele, carriers and channels facilitate the efflux of K^+, NO_3^-, SO_4^{2-}, $H_2PO_4^-$, Cl^-, and organic acid anions from xylem parenchyma cells into xylem vessels in the direction of their electrochemical gradients (Chapter 3). Channels in the plasma membrane of root cells facilitating the efflux of malate or citrate into the rhizosphere, such as members of the Al-activated malate transporter (ALMT) family and the multidrug and toxin extrusion (MATE) protein family, respectively, have been implicated in Al tolerance and improving P availability in acid soils

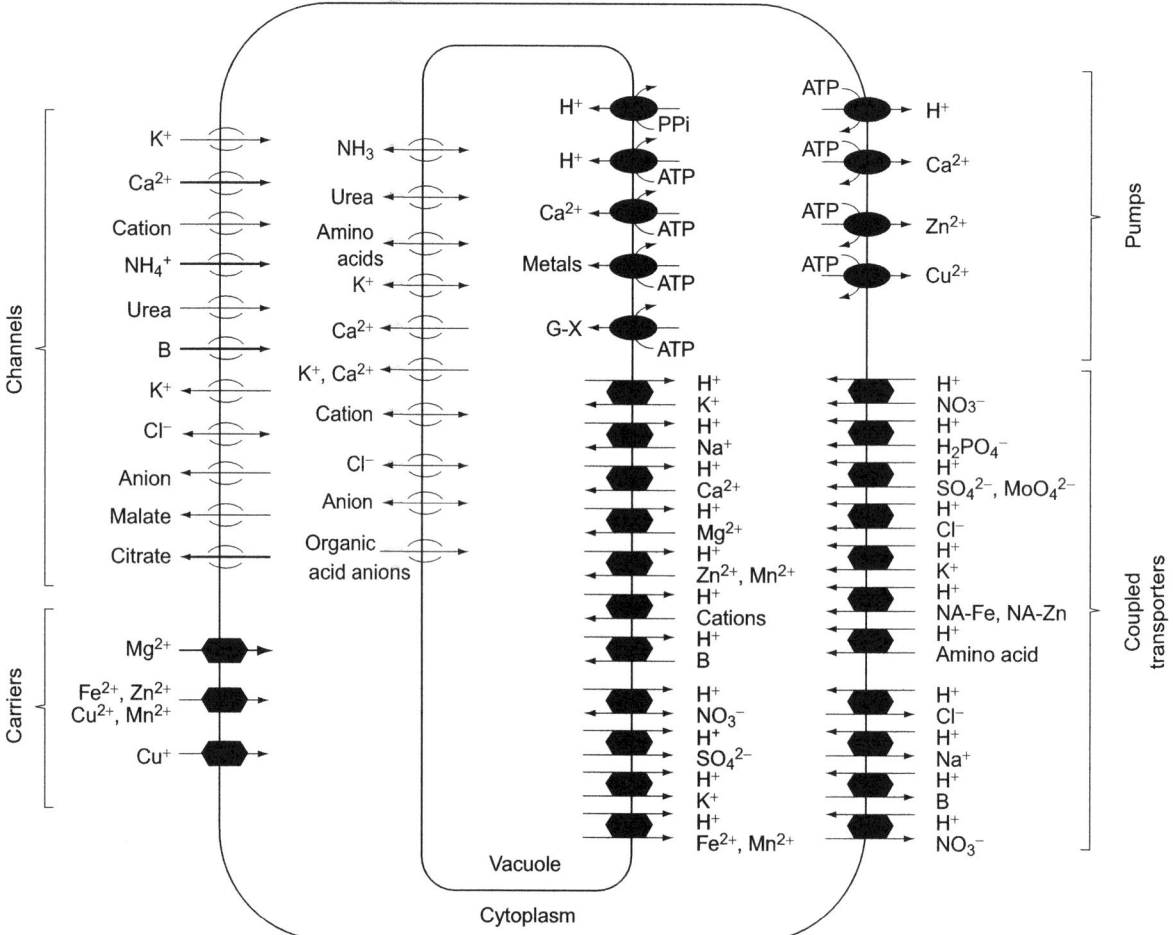

FIGURE 2.9 Transport proteins of the tonoplast and plasma membrane of plant cells. See Section 2.4.1 for details.

TABLE 2.6 Families of selected plasma membrane and tonoplast ion transporters in *Arabidopsis thaliana*, characterized by proposed mechanism, function, gene symbol, and an estimate of the number of functional genes.

Mechanism	Proposed function	Gene symbol	Number of genes	References
1. Primary active transporters				
H^+-ATPase (P_{3A}-type)	Energization of the plasma membrane	AHA	11	Haruta et al. (2015), Falhof et al. (2016)
H^+-ATPase (V-type)	Energization of tonoplast (and other endomembranes)	VHA	14	Forgac (2007), Marshansky and Futai (2008)
Ca^{2+}-ATPase (P_{2A}-type)	Ca^{2+}, Mn^{2+}, and Zn^{2+} efflux across endomembranes	ECA	4	Bossi et al. (2020)
Ca^{2+}-ATPase (P_{2B}-type)	Ca^{2+} efflux from cytoplasm, vacuole (and other organelles)	ACA	10	Bossi et al. (2020)
CPx-ATPase (P_{1B}-type)	Zn^{2+}, Cu^{2+}, Co^{2+}, Cd^{2+}, and Pb^{2+} efflux from cytoplasm	HMA	8	Axelsen and Palmgren (2001), Williams and Mills (2005)
H^+-PPase	Energization of tonoplast and endomembranes	AVP	2	Gaxiola et al. (2007), Segami et al. (2018)
ABC transporter	Detoxification, phytohormone transport, metabolite redistribution	ABCA, ABCB, ABCC, ABCD, ABCE, ABCF, ABCG	130	Verrier et al. (2008), Kang et al. (2011), Lane et al. (2016)
2. Secondary active transporters				
K^+/H^+ symport	K^+ uptake by cells	HAK/KUP/KT	13	Santa-María et al. (2018), Britto et al. (2021)
Monovalent cation/H^+ antiport	K^+ and Na^+ accumulation in vacuole, Na^+ efflux from cytoplasm, pH and K^+ homeostasis	CPA1 (NHX) and CPA2 (CHX/KEA)	42	Sze and Chanroj (2018), Isayenkov et al. (2020), Tsujii et al. (2020)
CCC	K^+ (Na^+)/Cl^- symport; root-shoot translocation; osmoregulation	CCC	1	Colmenero-Flores et al. (2007), Henderson et al. (2018)
H^+/divalent cation symport	Zn^{2+}, Fe^{2+}, Mn^{2+}, Cu^{2+} and Ni^{2+} uptake and redistribution, Cd^{2+} detoxification	NRAMP	6	Mäser et al. (2001), Socha and Guerinot (2014)
H^+/divalent cation antiport	Divalent cation influx to vacuole	CAX	6	Emery et al. (2012), Socha and Guerinot (2014)
H^+/divalent cation antiport	Zn^{2+}, Fe^{2+}, Mn^{2+}, Cu^{2+}, Co^{2+}, Ni^{2+}, and Cd^{2+} uptake and redistribution	MTP (CDF)	12	Gustin et al. (2011), Socha and Guerinot (2014)
H^+/Mg^{2+} antiport	Mg^{2+} influx to vacuole	MHX	1	Shaul et al. (1999), Emery et al. (2012)
H^+/NO_3^- symport	NO_3^- uptake and redistribution	NRT2	7	Krapp et al. (2014), Vidal et al. (2020)
NO_3^- (peptide, K^+)/H^+ transporter	NO_3^- and peptide uptake and redistribution; K^+ efflux to xylem	NPF (NRT1/PTR)	53	Niño-González et al. (2019), Vidal et al. (2020)
Auxin transport, K^+ (Zn^{2+})/H^+ antiport	Zn^{2+} and K^+ uptake into vacuole, K^+ efflux to apoplast	DHA1/ZIF	3	Remy et al. (2013, 2015), Niño-González et al. (2019)

(Continued)

TABLE 2.6 (Continued)

Mechanism	Proposed function	Gene symbol	Number of genes	References
H^+/anion antiport	Cl^- and NO_3^- accumulation in vacuole	CLC	7	De Angeli et al. (2006), von der Fecht-Bartenbach et al. (2010), Fan et al. (2017)
H^+/P_i symport	P_i uptake and redistribution	PHT1	9	Nussaume et al. (2011)
H^+/SO_4^{2-} symport	SO_4^{2-} uptake and redistribution	SULTR	12	Gigolashvili and Kopriva (2014)
H^+/MoO_4^{2-} symport	MoO_4^{2-} uptake and redistribution	MOT1, MOT2	2	Andresen et al. (2018a)
H^+/oligopeptide	Transport of oligopeptides, glutathione, metal chelates of glutathione, phytochelatins, and nicotianamine	OPT/YSL	17	Lubkowitz (2011)
3. Passive transporters				
MIPs/AQPs	Transport of water, small uncharged solutes (e.g., glycerol, urea, NH_3, $B(OH)_3$, $Si(OH)_4$, etc.) and ions (Na^+, K^+, NO_3^-, Cl^-)	PIP, TIP, NIP, SIP	35	Maurel et al. (2015), Tyerman et al. (2021)
NH_4^+ transporter	NH_4^+ uptake by cells	AMT	6	Yuan et al. (2007)
K^+ channel (inward rectifying)	K^+ influx to cells	AKT, KAT, KC	7	Sharma et al. (2013), Britto et al. (2021)
K^+ channel (outward rectifying)	K^+ efflux from cells	GORK, SKOR	2	Sharma et al. (2013), Britto et al. (2021)
K^+ channel	Cellular K^+ homeostasis	TPK, KCO	6	Sharma et al. (2013), Hamilton et al. (2015), Britto et al. (2021)
Na^+ uniport	Na^+ retrieval from xylem	HKT	1	Xue et al. (2011), Hamamoto et al. (2015)
Monovalent cation (and Ca^{2+}) channel	Monovalent cation and Ca^{2+} efflux from vacuole	TPC (SV)	1	Guo et al. (2016), Kintzer and Stroud (2016)
Cation channel (annexin)	Ca^{2+} influx to cell, Ca^{2+} efflux from vacuole	Ann	7	Clark et al. (2001), Davies (2014)
CNGC	Ca^{2+} influx to cell	CNGC	20	Kaplan et al. (2007), Jarratt-Barnham et al. (2021)
GLR	Ca^{2+} influx to cell	GLR	20	Forde (2014)
MSL	Ca^{2+} signaling	MSL	10	Hamilton et al. (2015), Demidchik et al. (2018)
MCA	Ca^{2+} signaling	MCA	2	Hamilton et al. (2015), Demidchik et al. (2018)
MGT	Mg^{2+} uptake and redistribution	MGT/Mrs2	10	Li et al. (2001), Tang and Luan (2017)

(Continued)

TABLE 2.6 (Continued)

Mechanism	Proposed function	Gene symbol	Number of genes	References
Cu^{2+} channel	Cu^{2+} uptake and redistribution	COPT	6	Printz et al. (2016)
Divalent cation transporter	Fe^{2+}, Zn^{2+}, Mn^{2+}, Cu^{2+}, Co^{2+}, and Ni^{2+} uptake and redistribution, Cd^{2+} detoxification	ZIP (ZRT/IRT)	16	Milner et al. (2013), Alejandro et al. (2020)
Anion channel (S-type)	Cl^- and NO_3^- efflux from cell	SLAC/SLAH	5	Vidal et al. (2020)

ABC, ATP-binding cassette; *AQPs*, aquaporins; *CCC*, cation chloride cotransport; *CNGC*, cyclic nucleotide-gated channel; *GLR*, glutamate receptor-like channel; *MCA*, Mid1-complementing activity channel; *MGT*, Mg^{2+} transporters; *MIPs*, major intrinsic proteins; *MSL*, mechanosensitive-like channel.
Source: Adapted from White (2017).

(Delhaize et al., 2007; Ryan et al., 2011). Channels facilitating the efflux of Cl^-, such as the depolarization-activated R-type and S-type anion channels (Roberts, 2006; Teakle and Tyerman, 2010; White and Broadley, 2001) in the plasma membrane of root cells, may be required for charge compensation of other ion fluxes. Homologs of the arabidopsis AtSLAC1 protein have been proposed as candidates for the S-type anion channels of root cells (Hedrich and Geiger, 2017; Teakle and Tyerman, 2010).

In addition to carriers and channels, solute transport across membranes can be catalyzed by primary or secondary active transporters that move solutes against their electrochemical gradient. Several ATPases are present in the plasma membrane of plant cells. These catalyze the efflux of H^+, Ca^{2+}, and metals from the cytoplasm (Bossi et al., 2020; Falhof et al., 2016; White and Pongrac, 2017). The plasma membrane H^+-ATPases catalyze H^+ efflux, which is then coupled directly (through the proton electrochemical gradient) or indirectly (via the cell-membrane potential) to the movement of other solutes. The plasma membrane Ca^{2+}-ATPases remove Ca^{2+} from the cytosol to maintain the low cytosolic Ca^{2+} concentrations required for cell signaling (Bossi et al., 2020; Chapter 6). In the stele, Ca^{2+}-ATPases and CPx-ATPases catalyze the efflux of Ca^{2+} and other divalent cations from the symplasm to the xylem (Bossi et al., 2020; Williams and Mills, 2005; Chapter 3).

A multitude of secondary active transporters are present in the plasma membranes of root cells, coupling H^+ influx to the movement of solutes against their electrochemical gradients (Fig. 2.9; Table 2.6). This includes anion transporters for NO_3^- [NPF (NRT1/PTR) and NRT2 transporters; Wang et al., 2012], $H_2PO_4^-$ (PHT1 transporters; Wang et al., 2018a), SO_4^{2-} (SULTR1 transporters; Takahashi, 2019), Cl^- (CLC transporters; Li et al., 2017a), and MoO_4^{2-} (MOT transporters; Andresen et al., 2018a). In addition, H^+/K^+ symporters, such as those encoded by the *HAK/KUP/KT* gene family, facilitate K^+ uptake by root cells at low external concentrations (Britto et al., 2021; White and Karley, 2010), and homologs of the maize yellow stripe 1 protein (ZmYS1) allow proton-coupled symport of Fe and Zn conjugates into root cells (Amini et al., 2022; Curie et al., 2009; White and Broadley, 2009). Proton-coupled transporters also alleviate element toxicities by removing Cl^- and Na^+ from root cells (Munns and Tester, 2008; White and Broadley, 2001). In the stele, proton-coupled transporters load nitrate into the xylem (Chapter 3). The transport of amino acids, peptides, and sugars across the plasma membrane is also catalyzed by proton-coupled transporters (De Michele et al., 2012; Lalonde et al., 2004; Chapters 3 and 5).

The tonoplast of the vacuole contains a variety of primary and secondary active transporters, carriers, and channels (Martinoia, 2018; Fig. 2.9; Table 2.6). In plant cells, the electrical potential difference between the vacuole and the cytosol is about -20 to -60 mV (Martinoia, 2018). Based on estimates of solute concentrations in the cytosol and vacuole, it is thought that sequestration of K^+, Na^+, Ca^{2+}, Mg^{2+}, Zn^{2+}, Mn^{2+}, and NO_3^- requires active transport into the vacuole, whereas the movement of other anions is likely to be passive (Martinoia, 2018; Martinoia et al., 2007; Teakle and Tyerman, 2010; White and Broadley, 2001, 2003; White and Karley, 2010).

The tonoplast contains two distinct types of proton pumps, the H^+-ATPases (V-ATPases) and the H^+-PPases (V-PPases), that generate the negative electrical potential across the tonoplast and lower the pH of the vacuole (Gaxiola et al., 2007; Martinoia, 2018; Wang et al., 2021). The tonoplast H^+-ATPases of plants are complex oligomeric proteins comprising two subcomplexes: the peripheral V_1 complex that consists of eight subunits (A–H) and is responsible for ATP hydrolysis, and the transmembrane V_0 complex that consists of up to five subunits (a, c, c″, d, and e) and is responsible for proton translocation (Gaxiola et al., 2007). These subunits are encoded by the *VHA* genes. Plants possess two distinct H^+-PPases, both single-subunit enzymes. The Type-I H^+-PPases require K^+ for their activity and are relatively insensitive to inhibition by Ca^{2+}, whereas Type-II H^+-PPases do not require K^+ for their activity and are extremely sensitive to inhibition by Ca^{2+} (Gaxiola et al., 2007; Martinoia, 2018; Martinoia et al., 2007).

Magnesium is essential for both H^+-ATPases and H^+-PPases because their substrates are Mg-ATP and Mg-PP$_i$ (Gaxiola et al., 2007; White et al., 1990b). In addition, the H^+-PPases require Mg^{2+} for their activity (Gaxiola et al., 2007; White et al., 1990b). Inorganic pyrophosphate (PP$_i$) is generated in several biosynthetic pathways, such as starch synthesis or activation of sulfate. Cytosolic PP$_i$ concentrations generally lie in the range of 50–400 μM, which is adequate to drive this proton pump (White et al., 1990b). Under most circumstances, H^+-PPases contribute far less than H^+-ATPases to proton transport into the vacuole. Therefore, it has been suggested that H^+-PPases act as ancillary enzymes to maintain the proton electrochemical gradient across the tonoplast when the activity of the H^+-ATPases is restricted by substrate availability, for example, during anoxia (White et al., 1990b), or at high temperatures that promote protein degradation (Martinoia et al., 2007).

The proton electrochemical gradient generated by the tonoplast H^+-ATPases and H^+-PPases supports the activities of many proton-coupled transporters. These catalyze the efflux of K^+ (e.g., NHX and KEA transporters), Na^+ (e.g., NHX transporters), Ca^{2+} (e.g., CAX transporters), NO_3^- (e.g., AtCLC-a), sucrose (e.g., AtSUT4), and various divalent cations, including Mg^{2+}, Zn^{2+}, and Mn^{2+} (e.g., CAX, MGT, and MTP transporters) from the cytosol to the vacuole, and the influx of NO_3^- (e.g., AtCLC-a), SO_4^{2-} (e.g., AtSULTR4-1, AtSULTR4-2), and iron (e.g., AtNRAMP3) from the vacuole to the cytosol in times of high demand for growth (Amini et al., 2022; Gojon et al., 2009; Martinoia, 2018; Martinoia et al., 2007; Miller et al., 2009; Schumacher, 2014; Shigaki and Hirschi, 2006; White and Broadley, 2009; White and Karley, 2010; Zifarelli and Pusch, 2010). The sequestration of K^+, Cl^-, and NO_3^- in vacuoles is important for turgor regulation, and the sequestration of Na^+, Ca^{2+}, Mg^{2+}, and metals is important to avoid cytoplasmic poisoning. In addition, the sequestration of essential elements and metabolites in the vacuole provides storage for times of need (Martinoia, 2018). A vast Ca^{2+} signaling network made up of calcineurin B-like protein (CBL) Ca^{2+} sensors, their associated CBL-interacting protein kinases, and ion transporter targets has been shown to regulate many of these transport processes (Tang et al., 2020a).

The tonoplast also contains Ca^{2+}-ATPases (e.g., AtACA4) that pump Ca^{2+} into the vacuole (Bossi et al., 2020; White and Broadley, 2003) and a variety of ATP-binding cassette (ABC) transporters that protect the cytoplasm by removing metals, oxidation products conjugated to glutathione, and xenobiotics from the cytosol into the vacuole (Lane et al., 2016; Martinoia, 2018; Martinoia et al., 2007). These transporters are also involved in the sequestration of chlorophyll catabolites and natural pigments in the vacuole (Martinoia et al., 2007).

Several ion channels are present in the tonoplast. These facilitate the movement of K^+, Cl^-, NO_3^-, NH_3, amino acids, urea, Ca^{2+}, SO_4^{2-}, sugars, and organic acid anions in the direction of their electrochemical gradients (Martinoia, 2018; Martinoia et al., 2007; Teakle and Tyerman, 2010; White and Broadley, 2003; White and Karley, 2010). The rapid efflux of K^+ and Cl^- from the vacuole, through fast vacuolar, slow vacuolar (SV) or vacuolar potassium channels, and Cl^- channels, respectively, is required for stomatal closure and other osmotically driven plant movements (Sharma et al., 2013; Tang et al., 2020b; White and Broadley, 2001). The sequestration and release of NO_3^-, NH_3, amino acids, and urea are central to the N economy of plants (Coskun et al., 2017; Tegeder and Masclaux-Daubresse, 2018; Wang et al., 2018b). Aquaporins have been shown to facilitate the transport of NH_3 (e.g., AtTIP2;1, AtTIP2;3) and urea (e.g., AtTIP1;1, AtTIP1;2; AtTIP2;1, AtTIP4;1) across the tonoplast (Martinoia et al., 2007; Miller et al., 2009). The rapid efflux of Ca^{2+} from the vacuole through SV, voltage-regulated, cADPR-regulated, or IP_3-regulated channels is important for cell signaling (Chapter 6). The influx of malate through anion channels is a prerequisite for crassulacean acid metabolism (CAM), in which CO_2 fixation is separated from photoassimilation, allowing plants to restrict water loss in arid environments by closing stomata during the day (Chapter 5). In CAM plants at night, malate enters the vacuoles as malate^{2-} through an anion channel and is accumulated in monovalent and uncharged forms due to vacuolar acidity (Martinoia et al., 2007). Malate is subsequently released to the cytoplasm during the day for photoassimilation to occur.

From the preceding discussion, it is apparent that proton pumps are responsible for energizing the transport of solutes across cell membranes. However, it is important to note that these pumps not only generate the proton electrochemical gradient across the tonoplast and plasma membrane, and the acidic conditions of the apoplasm (pH ~5.5) and the vacuole (pH 4.5–5.9), but also maintain cytosolic pH at its optimal value (pH 7.3–7.6; Cosse and Seidel, 2021; Felle, 2001; Wang et al., 2021).

TABLE 2.7 Respiratory energy costs for ion uptake in roots of *Carex diandra*.

Proportion of total ATP demand required for	Plant age (days)		
	40	60	80
Ion uptake	36	17	10
Growth	39	43	38
Maintenance of biomass	25	40	52

Source: Based on Van der Werf et al. (1988).

2.4.2 Energy demand for solute transport

The energy demand for ion uptake by roots can be considerable, especially during rapid vegetative growth (Table 2.7). Early calculations suggested that, in seedlings, up to 36% of the total respiratory energy cost, expressed as ATP consumption, is required for ion uptake; this proportion declines in older plants in favor of ATP demand for growth and maintenance of biomass (Table 2.7). Kurimoto et al. (2004) subsequently calculated that up to 76% of total respiratory energy cost was required to support low-affinity nitrate influx across the plasma membrane of cereal roots. Although several of the assumptions of these calculations have been criticized, subsequent studies addressing the weaknesses of the original calculations also found very high energy costs of ion uptake (Britto and Kronzucker, 2006, 2009; Teakle and Tyerman, 2010).

High rates of apparently "futile" cycling of K^+, NH_4^+, NO_3^-, and SO_4^{2-} across the plasma membrane of root cells occur when these ions are present at high concentrations in the rhizosphere solution (Britto and Kronzucker, 2006). As the rhizosphere concentration increases, and the rate of unidirectional influx increases, the quotient of the unidirectional rates of ion influx and ion efflux across the plasma membrane approaches unity. This apparently "futile" cycling requires at least two different types of transporters for the same nutrient, which are energized differently, and it has been suggested that this arrangement is required for cytosolic homeostasis of a nutrient (Dreyer, 2021). The energy costs of this "futile" cycling were thought to represent a substantial proportion of the total respiratory energy cost of the root. Similarly, under saline conditions, a considerable proportion (almost all) of the total energy budget of the root was thought to be required for the removal of Na^+ (Britto and Kronzucker, 2006, 2009) and Cl^- (Teakle and Tyerman, 2010) from the symplasm. However, this view has been challenged, at least for K^+ and Na^+ (Britto and Kronzucker, 2015; Coskun et al., 2016; Kronzucker and Britto, 2011; Munns et al., 2020). It has been argued that the apparent K^+ and Na^+ fluxes (both influx and efflux) estimated using radioisotopes at high external cation concentrations may have been misinterpreted as fluxes across the plasma membrane, when, in fact, they were apoplasmic in origin. Although the nature of these apoplasmic fluxes (with different exchange kinetics from the WFS and DFS) remains unknown, if substantiated, this would necessitate a major reevaluation of the energy budgets associated with these fluxes. Whether such a phenomenon also underlies ionic fluxes more generally at high external concentrations has yet to be determined. In addition, it has been observed recently that rapid unidirectional NH_3 fluxes (rather than NH_4^+ fluxes), possibly mediated by aquaporin channels, predominate across the plasma membrane and tonoplast in ammonium-stressed barley roots (Coskun et al., 2013b), and, thus, energy demands in this context would also need to be reevaluated.

2.4.3 The kinetics of solute transport in plant roots

Generally, the rate of solute uptake by plant cells, excised plant tissues, and roots of intact plants saturates with increasing external solute concentration. Emanuel Epstein and colleagues in the early 1950s suggested that this relationship was similar to that between an enzyme and its substrate (Fig. 2.10). In their analogy, the transport protein was an enzyme that catalyzed the movement of its substrate from one side of a membrane to the other. Using the terms from enzymology, they defined the relationship between the rate of transport of a solute and its concentration by the Michaelis−Menten equation:

$$V = (V_{max} \times S)/(K_m + S)$$

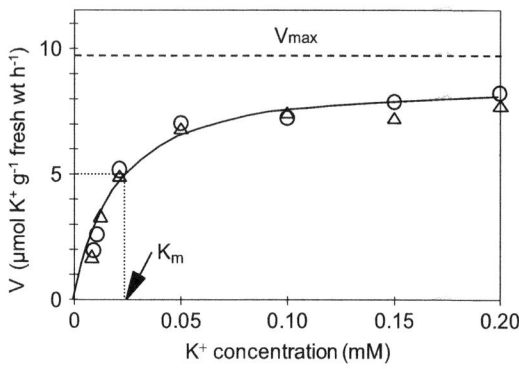

FIGURE 2.10 Rate of K$^+$ uptake (V) by barley roots as a function of the external concentration of KCl (○) or K$_2$SO$_4$ (△); K_m = 0.023 mM. *Adapted from Epstein (1972).*

where *V* is the rate of solute transport at a solute concentration of *S*, V_{max} is the maximal rate of solute transport, and K_m is the Michaelis–Menten constant, which is the solute concentration at which half the maximal transport rate is reached. The K_m value reflects the affinity of the transporter for the solute (in enzymatic reactions, it indicates the affinity of the enzyme for its substrate).

When assayed at low concentrations in the external medium, solute uptake is often described well by this equation, with the relationship between K$^+$ uptake and K$^+$ concentration in the external medium being the same regardless of whether the source of K$^+$ is KCl or K$_2$SO$_4$ (Fig. 2.10). However, as we shall see later, when substrate concentrations are larger, the accompanying anion can have a significant effect on the uptake rate of a cation and vice versa. As a first approximation, in the low concentration range, Michaelis–Menten kinetics can also be applied to describe uptake rates of many other solutes, including the anions NO$_3^-$, H$_2$PO$_4^-$, SO$_4^{2-}$, and Cl$^-$ (e.g., Deane-Drummond, 1987; Epstein, 1972; Li et al., 2007; Siddiqi et al., 1990; White and Broadley, 2001), the cations NH$_4^+$, Ca^{2+}, Mg^{2+}, Mn^{2+}, Zn^{2+}, and Cd^{2+} (e.g., Broadley et al., 2007; Kronzucker et al., 1998; Lux et al., 2011; Rawat et al., 1999; Sadana et al., 2005; Wang et al., 1993) and chelates such as Fe-phytosiderophores (von Wirén et al., 1995). A low-capacity saturable uptake system can sometimes be discerned in B-deficient plants (e.g., Dannel et al., 2000), but in B-replete cells the relationship between the uptake of B and its concentration in the external solution is often reported to be linear (Seresinhe and Oertli, 1991). However, the relationship between solute concentration and its uptake by roots cannot always be fitted to a simple Michaelis–Menten equation. This is a consequence of both the theoretical limitations of Michaelis–Menten kinetics and the presence of multiple mechanisms for the transport of a particular solute (Britto et al., 2021; Dreyer, 2017; Dreyer and Michard, 2020; White, 2017).

The original concept of a single protein-mediated mechanism of ion transport (one carrier system for each ion) does not describe the kinetics of uptake adequately across a wide range of external concentrations. For example, the kinetics of K$^+$ uptake differ considerably at concentrations above 1 mM from those at lower concentrations (Britto et al., 2021; Coskun et al., 2010, 2016; Epstein et al., 1963). The apparent selectivity of transport is lower (Na$^+$ competes with K$^+$) and the accompanying anion has a marked effect on the uptake rate. These observations led to the hypothesis of *dual systems* for K transport, with system I having a higher selectivity than system II.

In view of the usually very low concentrations, particularly of P$_i$ and K$^+$, in soil solutions and the results of ion-uptake studies in the low concentration range (<10 μM), the term C_{min} was introduced to define the concentration at which net uptake of ions ceases before the ions are completely depleted (Fig. 2.11). The C_{min} concentration is an important factor in ion uptake from soils because it is the lowest concentration at which roots can extract an ion from the soil solution. C_{min} concentrations differ considerably among plant species. For P$_i$, for example, a value of 0.12 μM has been found in tomato (Itoh and Barber, 1983), 0.04 μM in soybean (Silberbush and Barber, 1984), and 0.01 μM in ryegrass (Breeze et al., 1984). For K$^+$, the corresponding values were 2 μM in maize (Barber, 1979) and 1 μM in barley (Drew et al., 1984). C_{min} concentrations for NO$_3^-$ can vary from more than 50 μM to less than 1 μM depending not only on the plant species but also on the environmental conditions (Deane-Drummond and Chaffey, 1985; Marschner et al., 1991). For NH$_4^+$, C_{min} concentrations decreased from 30 to 1.5 μM as the root-zone temperature increased (Marschner et al., 1991).

For the kinetics of ion uptake by plants, the Michaelis–Menten equation has been modified to include the parameter C_{min}, and the term I, designating unidirectional influx, has replaced the term V (Fig. 2.11). Very often, though, only the *net uptake* of ions is determined experimentally, which is the net result of influx and efflux across the plasma membrane (Fig. 2.12). Efflux can become similar in magnitude to influx, particularly at extremely low or high external concentrations and therefore can be an important component in determining net uptake (Fig. 2.13). It is also noteworthy

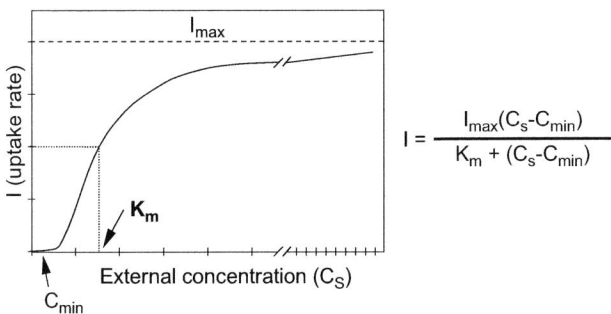

FIGURE 2.11 Schematic presentation of the relationships between uptake rates (net influx = I) of ions and their external concentrations; C_{min} = net uptake zero (influx = efflux).

$$I = \frac{I_{max}(C_s - C_{min})}{K_m + (C_s - C_{min})}$$

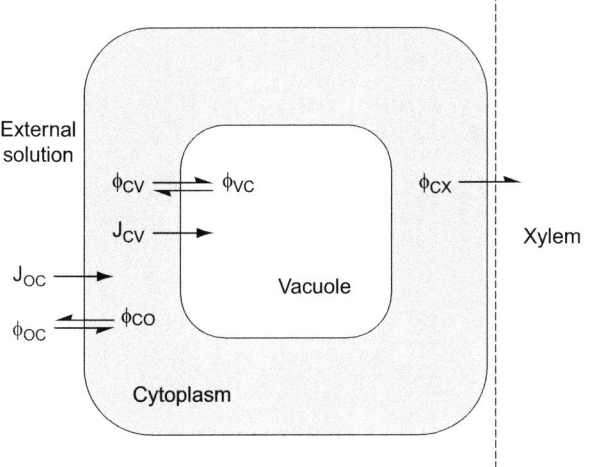

FIGURE 2.12 Nomenclature of unidirectional (ø) and net (J) solute fluxes across the plasma membrane between cytoplasm (c) and the external solution (o) or xylem (x), and across the tonoplast between the cytoplasm and the vacuole (v) of a root cell.

that, at a given external concentration, the efflux of a particular mineral nutrient from roots of plants sufficiently supplied can be many times higher than from roots of deficient plants (e.g., Lee et al., 1990).

The efflux of ions and other solutes is affected by several factors: (1) the integrity of the plasma membrane, (2) the presence of transport proteins allowing efflux, (3) the electrochemical driving force for transport, and (4) the concentration of the solute in the cytoplasm. In pea, for example, the initial high rate of net uptake of SO_4^{2-} by S-deficient roots placed in a solution containing SO_4^{2-} decreases to about 30% within 1 h due to a marked increase in SO_4^{2-} efflux, despite a slight increase in influx (Bell et al., 1995; Deane-Drummond, 1987). Similarly, for NO_3^- and NH_4^+, the efflux component can account for a high proportion of the influx (almost 40%–50%) most probably due to the high concentrations of NO_3^- and NH_4^+ in the cytoplasm (Britto and Kronzucker, 2003, 2006). The rapid exchange of ions between the external solution and the cytoplasm is reflected in short half-times for exchange ($t_{1/2}$), which are 7–14 min for NH_4^+ (Kronzucker et al., 1998), 10–50 min for K^+ (Coskun et al., 2010, 2013a, 2016; Szczerba et al., 2006; White et al., 1991), 7–75 min for Ca^{2+} (White et al., 1992), 10–20 min for SO_4^{2-} (Bell et al., 1995; Deane-Drummond, 1987), 10–20 min for Cl^- (Britto et al., 2004), 4–107 min for NO_3^- (Britto and Kronzucker, 2003; Lee and Clarkson, 1986; Macklon et al., 1990), and 23–115 min for $H_2PO_4^-$ (Lee et al., 1990; Macklon et al., 1996). These rates of exchange with the cytoplasmic pool are usually orders of magnitude faster than the rates of exchange with the vacuolar pool (e.g., Bell et al., 1995; Macklon et al., 1990; White et al., 1991, 1992).

The parameters of ion-uptake kinetics are also strongly influenced by the nutritional status of plants (Section 2.5.6). Roots of plants deficient in a particular nutrient generally exhibit a greater I_{max} and a lower C_{min} for that nutrient than those of plants sufficient in the nutrient. Occasionally, deficient plants also exhibit a lower K_m. An example is given in Table 2.8 for P_i. In plants with greater tissue P_i concentrations, I_{max} for P_i uptake is substantially lower, and K_m is also slightly lower. The I_{max} values were based on net uptake in this experiment, therefore the contribution of increased efflux at higher root P_i concentrations cannot be evaluated.

In the high concentration range (>1 mM), a linear relationship is often found between external concentrations and the rate of ion uptake by roots. This has been observed for the anions NO_3^- (Siddiqi et al., 1990), $H_2PO_4^-$ (Loneragan and Asher, 1967), SO_4^{2-} (Clarkson and Saker, 1989), and Cl^- (Siddiqi et al., 1990; White and Broadley, 2001; Li et al., 2007),

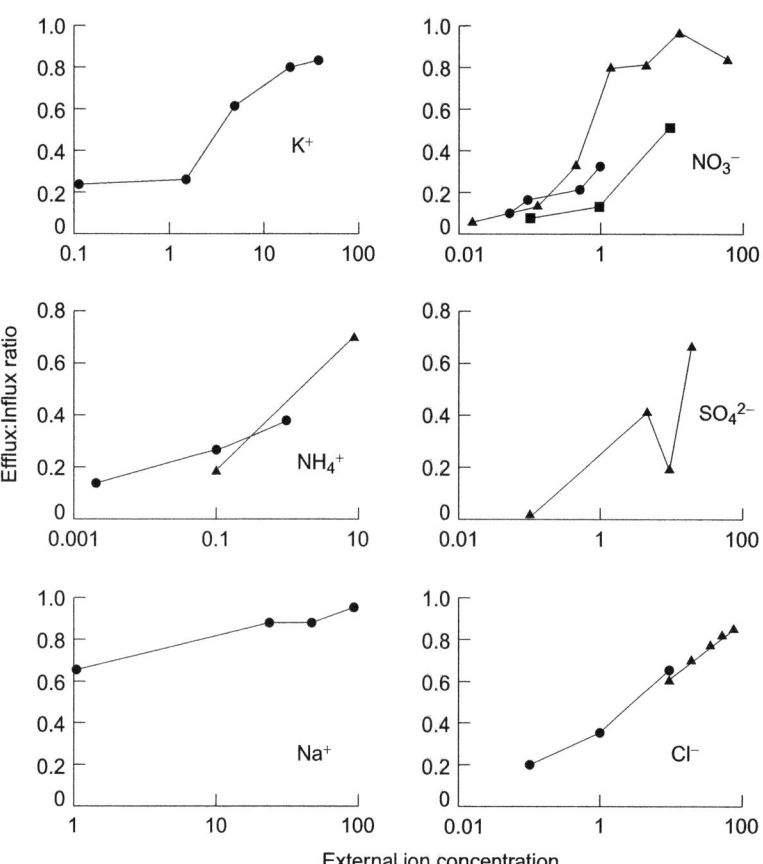

FIGURE 2.13 Ratios of efflux to influx across the plasma membrane for K^+, NO_3^-, NH_4^+, SO_4^{2-}, Na^+, and Cl^-. Data from various studies (represented by different symbols). *Adapted from Britto and Kronzucker (2006).*

TABLE 2.8 Short-term P uptake parameters of soybean plants with different P nutritional status.

Plants grown at P concentration (μM)	P concentration (mg kg^{-1} dry wt) Shoot	P concentration (mg kg^{-1} dry wt) Root	I_{max} (mol cm^{-1} s^{-1}) $\times 10^{-14}$	K_m (μM)
0.03	2.2	2.3	18	1.6
0.3	3.4	3.0	17	1.7
3.0	5.9	5.6	6.5	1.2
30	6.6	9.0	3.7	1.0

Source: Based on Jungk et al. (1990).

the cations NH_4^+ (Rawat et al., 1999), K^+ (Coskun et al., 2016), Na^+ (Hamam et al., 2019), Ca^{2+} (White, 2001), Mg^{2+} (Tanoi et al., 2011), and Zn^{2+} (Broadley et al., 2007), for B (Dannel et al., 2000), and for Fe-phytosiderophores (von Wirén et al., 1995). Several explanations for the linear relationship, formerly defined as system II by Epstein (1972), have been proposed. The first explanation is that it reflects influx through nonsaturating transport proteins, perhaps ion channels, in the plasma membrane of root cells. The second explanation is that it is the consequence of rapid chelation or metabolism of a solute in the cytoplasm, or its removal by sequestration in the vacuole or transfer to the xylem, which maintains the electrochemical gradient, and reduces efflux, across the plasma membrane. The third explanation is that it represents a nonsaturating, apoplasmic flux to the xylem ("apoplasmic bypass flow"), whereby ions bypass apoplasmic barriers, notably the endodermal Casparian band (e.g., Faiyue et al., 2012; Flam-Shepherd et al., 2018; Gong et al., 2006; Lux et al. 2011; Munns et al., 2020; White, 2001; Yeo et al., 1987). However, given the usually low ion concentrations in soil solutions, the ecological significance of the low-capacity, nonsaturating mechanisms for the nutrition of plants grown in natural soils has been questioned. There are, however, at least two exceptions: roots of plants growing in saline soils, and the uptake of mineral nutrients from the apoplasm following their long-distance transport in the xylem and phloem (Chapter 3).

2.5 Factors influencing ion uptake by roots

2.5.1 Influx to the apoplasm

Before reaching the plasma membrane of root cells, ions must pass through the cell wall. In general, neither diffusion nor mass flow of ions or other low-molecular-weight solutes is restricted at the external surface of the roots. The cell walls and water-filled intercellular spaces of the root cortex are, to a certain extent, accessible to these solutes from the external solution.

The main barrier to solute (and water) flux through the apoplasm of young roots is the endodermis, the innermost layer of cells of the cortex (Fig. 2.14). The endodermis forms Casparian bands (ring-like lignin polymers deposited in the middle of anticlinal cell walls between endodermal cells) and suberin lamellae (glycerolipid polymers covering endodermal cells), creating an effective barrier against solute movement into the stele (Doblas et al., 2017; Geldner, 2013; White, 2001). In arabidopsis, suberization of the endodermis starts a few cell layers further away from the root tip than the Casparian bands (Geldner, 2013). Suberin lamellae are particularly important as apoplasmic diffusion barriers to the stele at sites of lateral root emergence where Casparian bands are disrupted (Li et al., 2017c).

Suberization is influenced by numerous environmental factors (Andersen et al., 2015, 2021; Lux et al., 2011; Tylová et al., 2017). In arabidopsis, K or S deficiency and excess NaCl induce endodermal suberization through abscisic acid (ABA)-mediated signaling, whereas deficiency of Fe, Zn, or Mn inhibits suberization via ethylene signaling (Barberon et al., 2016). In rice, suberization of both the endodermis and exodermis is beneficial in saline environments (Krishnamurthy et al., 2009). In barley, the severity of Mn deficiency governs suberization of the endodermis. Mild Mn deficiency decreases suberization of the main root axis, whereas severe Mn deficiency increases suberization, which also prevents the inward radial transport of Ca^{2+} and Na^+ but increases that of K^+ (Chen et al., 2019; Section 2.7).

In most angiosperm species, suberization of the radial and transverse cell walls is also found in the hypodermis or exodermis (Enstone and Peterson, 1992; Ma and Peterson, 2003). Suberization of the exodermis generally occurs after the formation of the endodermal Casparian band, particularly in fast-growing roots (Ma and Peterson, 2003). There are different views on the effectiveness of the exodermis as a barrier to solute movement through the root apoplasm (Clarkson et al., 1987; Enstone and Peterson, 1992; Section 2.7). However, given that the development of the exodermis generally occurs after that of the endodermis, its function is thought to be largely structural. In plants adapted to submerged conditions, the exodermis serves another function, namely, as an effective barrier against oxygen diffusion (leakage) from the root aerenchyma into the rhizosphere (e.g., Kotula et al., 2009; Soukup et al., 2007; Tylová et al., 2017; Voesenek and Bailey-Serres, 2015).

The volume of root tissue accessible for apoplasmic solute movement (the free space) represents only a small fraction of the total root volume. For example, the free space is estimated to occupy 5% of a maize root (Shone and Flood, 1985). The presence of this free space enables individual cortex cells to contribute to solute uptake from the external

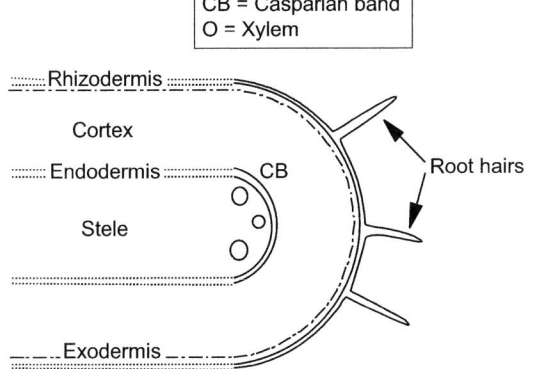

FIGURE 2.14 Schematic representation of cross section of a differentiated root zone of maize.

solution. Solute concentrations in the free space depend on various factors such as the capacity for solute uptake by epidermal cells, the presence of root hairs, the solute concentration in the rhizosphere solution, and the rate of transpiration. As shown more than 50 years ago by Vakhmistrov (1967), at low external concentrations, root-hair formation is usually extensive and the uptake of mineral nutrients is limited mainly to the rhizodermal cell layer (i.e., the outer-most cells of the cortex). This is particularly relevant for roots growing in soil, where the importance of root hairs for the acquisition of nutrients present at low concentrations in the soil solution or with restricted soil mobility, such as P, has been demonstrated clearly (e.g., Brown et al., 2013; Gahoonia and Nielsen, 2004; Zhu et al., 2010).

2.5.2 Effects of pH

The pH of the external solution can have profound effects on the uptake of nutrients by plant roots. These can be divided into three broad categories: (1) effects of solution pH on the chemical species present in solution, (2) effects of apoplasmic pH on the concentrations of ions present in the apoplasm, and (3) effects of rhizosphere pH on the proton electrochemical gradient and the driving force for proton-coupled solute transport. In addition, solution pH can affect ion transport by protonation/deprotonation of amino acids in transport proteins.

The pH of the soil solution influences the availability of cations and anions for root uptake (Läuchli and Grattan, 2012; White and Broadley, 2009). In alkaline soils, the availability of P, Zn, Fe, Mn, Cu, and B is very low, whereas in acid soils, plant growth is mainly limited by toxic concentrations of Al^{3+} or Mn^{2+} in the rhizosphere. In addition, the pH of the external solution also determines the chemical species present in the rhizosphere. This is particularly relevant to the uptake of solutes that can be protonated and are transported across the plasma membrane as specific chemical species, such as boron, phosphate, and ammonium. The rate of B uptake decreases strongly with the increased pH of the external solution (Fig. 2.15). This pattern is closely related to the decrease in the ratio of boric acid [$B(OH)_3$], which is taken up by root cells, to the borate anion [$B(OH)_4^-$] (Miwa and Fujiwara, 2010; Shao et al., 2018; White et al., 2021b). Similarly, the rate of phosphate uptake decreases as the pH of the external solution increases (Fig. 2.16). This can be explained by a decrease in the concentration of $H_2PO_4^-$, the substrate of the proton-coupled phosphate symporter in the plasma membrane, in the pH range of 5.6 to 8.5. By contrast, there is a smaller effect on the uptake rate of

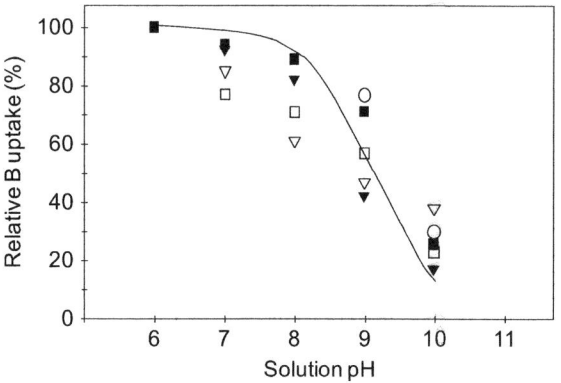

FIGURE 2.15 Relative uptake of B by barley roots as a function of the external solution pH. Uptake at pH 6 = 100% at each supply concentration. Solid line: percentage of undissociated H_3BO_3. Key for B concentrations (mg L^{-1}): 1.0 (*open triangle*), 2.5 (*open square*), 5.0 (*open circle*), 7.5 (*filled triangle*), and 10.0 (*filled square*). Based on Oertli and Grgurevic (1975).

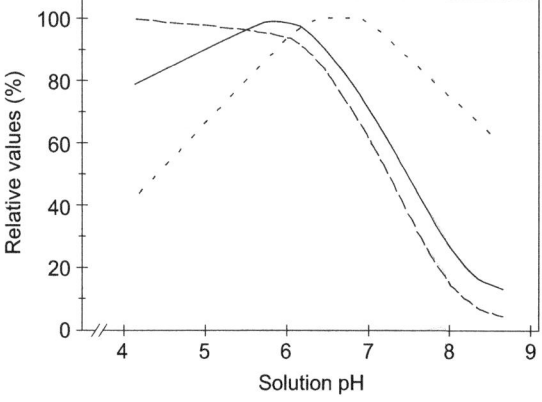

FIGURE 2.16 Relationships between solution pH and the proportion of $H_2PO_4^-$ (*dashed line*) in solution, and the uptake of phosphate (*solid line*) and sulfate (*dotted line*) by bean plants. Data are expressed as relative values. *Adapted from Hendrix (1967).*

sulfate because in this pH range only the divalent anion SO_4^{2-} occurs. The effects of the pH of the external solution on ammonium uptake by plant roots are more complex. At high external pH, ammonium uptake increases sharply in barley (ammonium-sensitive species) due to an increase in the proportion of the uncharged species NH_3 that is probably taken up by root cells through NH_3-permeable aquaporins (Coskun et al., 2013b; Fig. 2.17). However, the opposite effect is observed in ammonium-tolerant rice, where fluxes decline with rising pH (Wang et al., 1993). Whether such different responses of NH_3/NH_4^+ uptake to external pH reflect the underlying mechanisms behind the contrasting sensitivities of these species remains to be determined.

In relation to the effects of apoplasmic pH on the uptake of solutes by root cells, it has been noted previously that both cell walls and biological membranes contain charged groups and that ions interact with these groups (Section 2.2). Generally, the strength of these interactions increases with valency. Fixed negative charges can influence both the absolute and relative concentrations of cations in the apoplasm and, thereby, ion movements in the apoplasm and the rate and selectivity of ion influx to root cells. As the external pH is lowered, the effective CEC of the apoplasm decreases, particularly in monocotyledonous species due to binding of H^+ to the cation exchange sites. Thus apoplasmic pH can affect ion uptake indirectly via cell-wall properties.

The pH of the rhizosphere solution can also affect ion uptake by altering both substrate (H^+) concentration and the electrochemical driving force for proton-coupled solute transport. A decrease in pH can increase the activity of proton-coupled solute transporters and enhance anion uptake. In short-term experiments with maize roots, decreasing the external pH from 8 to 4 increased NO_3^- influx by a factor of about 10 (McClure et al., 1990) and $H_2PO_4^-$ uptake by a factor of about 3 (Sentenac and Grignon, 1985), even when the concentration of $H_2PO_4^-$ was kept constant. By contrast, the efficiency of H^+ efflux decreases as the external solution becomes more acidic and, consequently, the membrane potential of root cells decreases from about -150 mV at pH 6 to -100 mV at pH 4 (Dunlop and Bowling, 1978).

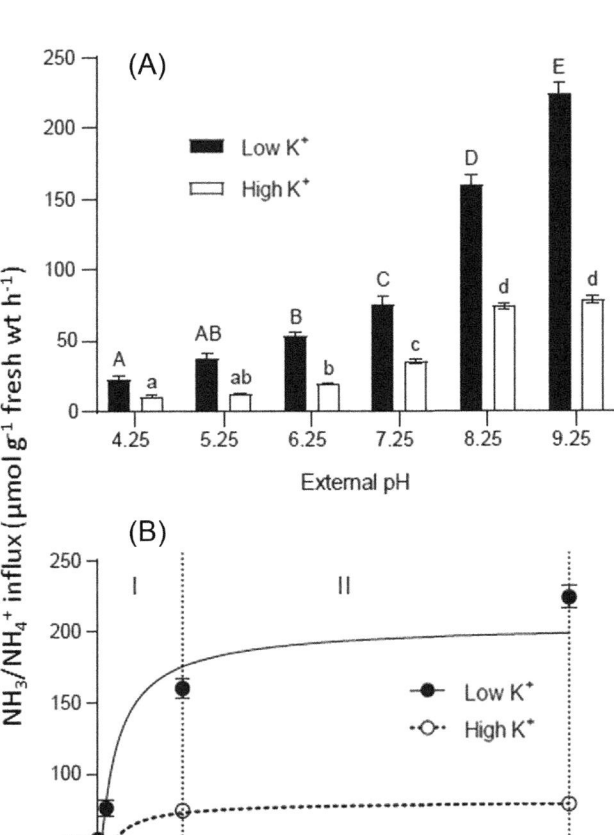

FIGURE 2.17 NH_3/NH_4^+ influx to roots of intact barley seedlings in response to external solution pH. Graph (B) contains data from graph (A) replotted to show dependence of influx on NH_3 concentrations, predicted from solution pH according to the Henderson–Hasselbalch equation (pKa of $NH_3/NH_4^+ = 9.25$). In graph (B), area I represents pH 4.25 to 8.25, and area II represents pH 8.25–9.25. Plants were grown hydroponically with 5 mM $(NH_4)_2SO_4$ and either low (0.01 mM) or high (2.5 mM) K_2SO_4. Error bars represent ± standard error of the mean ($n \geq 3$). Different uppercase and lowercase letters in graph A denote significantly different means ($P < 0.05$) within low- and high-K^+ plants, respectively. *Modified from Coskun et al. (2013b).*

Accordingly, the driving force for cation uptake is decreased. In general, the uptake of cations, such as K$^+$, is inhibited by low pH of the external solution, although Ca^{2+} often has an ameliorating effect.

2.5.3 Metabolic activity

The accumulation of ions and other solutes against a concentration gradient requires an expenditure of energy, either directly or indirectly. The main source of energy in nonphotosynthesizing cells and tissues (including roots) is respiration. Thus, all factors that affect respiration can influence ion accumulation.

2.5.3.1 Oxygen

As oxygen tension is lowered, root respiration is inhibited, and the uptake of ions, such as K$^+$ and H$_2$PO$_4^-$, is decreased. Consequently, oxygen deficiency is one of the factors that can restrict plant growth in poorly aerated substrates (e.g., waterlogged soils; Colmer and Greenway, 2011).

2.5.3.2 Carbohydrates

The main energy substrates for respiration are carbohydrates. Therefore, in roots and other nonphotosynthesizing tissues, under conditions of limited carbohydrate supply from a source (e.g., leaves), a close correlation can often be found between carbohydrate concentration and the uptake of ions. These relationships are of importance ecologically, for example, when leaves are removed (grazing, cutting) or in dense plant stands when light supply to the basal leaves (the main source of carbohydrates for the roots) is limited.

Distinct diurnal patterns in solute uptake (maxima during the day, minima during the night) have been observed for NO$_3^-$ (Clement et al., 1978), NH$_4^+$ (Macduff et al., 1997), K$^+$ (Le Bot and Kirkby, 1992), Fe^{2+} (Cesco et al., 2002), and Zn^{2+} (Zhang et al., 1991b). Root carbohydrate concentration may act as a coarse control for ion uptake and is one of the factors responsible for the diurnal fluctuations in ion uptake. However, in maize roots, for example, diurnal fluctuations in nitrate uptake were only loosely related to root carbohydrate content (Fig. 2.18). It has also been suggested that the delivery of sucrose via the phloem is a systemic signal of plant nutritional status that regulates the expression of genes encoding proteins catalyzing the uptake of NO$_3^-$, H$_2$PO$_4^-$, SO$_4^{2-}$, NH$_4^+$, K$^+$, and Fe^{2+} across the plasma membrane of root cells (Hammond and White, 2008; Hermans et al., 2006; Lejay et al., 2003; Liu et al., 2009; Vance, 2010; Vert et al., 2003; Section 2.5.6).

2.5.3.3 Temperature

Physical processes such as exchange adsorption of cations in the AFS are only slightly affected by temperature ($Q_{10} = 1.1–1.2$, with Q_{10} referring to the change in a reaction or process imposed by a change in temperature by 10°C). However, chemical and biochemical reactions show much greater temperature dependence, with $Q_{10} = 2$ and $Q_{10} > 2$, respectively. The Q_{10} value for the uptake of ions from solutions of low concentration often exceeds 2, at least within

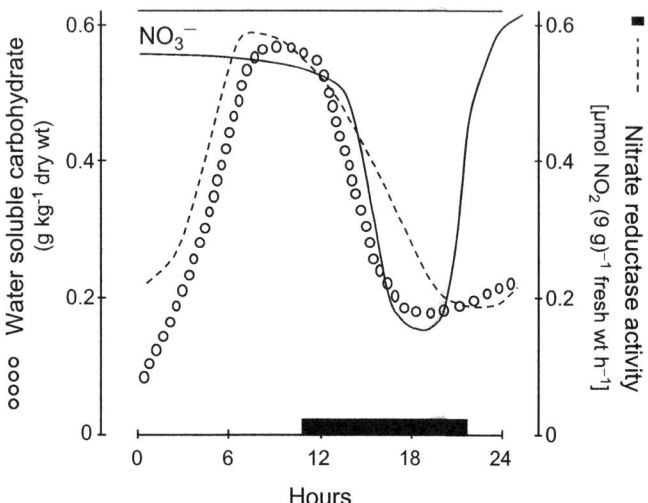

FIGURE 2.18 Diurnal fluctuations in nitrate uptake (*solid line*), nitrate reductase activity, and concentration of water-soluble carbohydrates in maize roots. Nitrate uptake: relative values, uptake at end of the light period = 100. *Adapted from Keltjens and Nijenstein (1987).*

TABLE 2.9 Shoot and root growth and uptake of nitrate and K by maize plants grown at different root-zone temperatures (RZT) and the temperatures at the stem base (shoot growing zone temperature, SGT) for 8 days.

	Temperature treatment (SGT °C/RZT °C)		
	24/24	12/12	24/12
Shoot growth (g fresh wt day^{-1})	1.9	0.32	1.3
Root growth (g fresh wt day^{-1})	0.85	0.20	0.26
Nitrate uptake (pmol g^{-1} fresh wt day^{-1})	6.4	4.2	7.6
K uptake (pmol g^{-1} fresh wt day^{-1})	2.5	1.2	3.1

Source: After Engels and Marschner (1992).

the physiological temperature range (Clarkson et al., 1988; Wang et al., 1993). Ion uptake is more temperature dependent than respiration, especially at temperatures below 10°C. Furthermore, at temperatures above 30°C root respiration further increases whereas ion uptake may decline (Bravo and Uribe, 1981), indicating that membrane transport and respiration are not coupled directly.

In studies of temperature effects on ion uptake, two phenomena are often studied: (1) the immediate effects of an abrupt change in root temperature, which occur within seconds and reflect the direct effects of temperature on the uptake system, and (2) the long-term effects of growing plants at a particular root temperature for several days or weeks. The latter effects include acclimatory responses (e.g., changes in root membrane properties) and are of greater ecological significance. Such studies often compare plants exposed to different root temperatures, but the same shoot temperature. It is noteworthy that the shoot meristem of, for example, graminoid plants, is close to the stem base. Thus, different temperatures in the rooting medium influence cell division and cell elongation in the shoot. In the long term, such experimental systems can affect the growth rates of roots and shoot quite differently (Clarkson et al., 1988). Accordingly, long-term effects of root temperature on ion uptake can include feedback regulation via plant demand. For example, in maize plants the uptake of both K^+ and NO_3^- is determined by the temperature (and growth) of the shoot rather than the temperature of the roots (Table 2.9).

Low root temperatures can affect the uptake of nutrients differently, with P_i uptake usually being decreased more than the uptake of other nutrients (e.g., Engels, 1993; Engels and Marschner, 1992). The uptake of NO_3^- seems to decrease more than the uptake of NH_4^+ when root temperature declines, which might be a consequence of the relatively greater energy requirement for the uptake and assimilation of NO_3^- than NH_4^+ (Britto and Kronzucker, 2013). The uptake of K^+ is often more affected by root-zone temperature than the uptake of Ca^{2+} and Mg^{2+}. In winter wheat, the increase in $K^+/(Ca^{2+} + Mg^{2+})$ ratio in the shoots with increasing root-zone temperature may cause tetany in grazing beef cattle (Miyasaka and Grunes, 1990).

In contrast to plants grown in solution culture, the roots of plants grown in soil must forage for many immobile nutrients (Lynch, 2007, 2011, 2019; Wang et al., 2019; White et al., 2013a, 2013b). Thus, in soil-grown plants, soil temperature (fluctuating daily, seasonally, and with depth) can affect the uptake of nutrients additionally through effects on root growth rate and root system architecture (Koevoets et al., 2016; Rich and Watt, 2013).

2.5.4 Interactions among ions in the rhizosphere

In the preceding sections, for the sake of simplicity, the transport of a particular ion was treated as a singular process. In reality, however, the transporters catalyzing ion uptake are rarely specific, and ions can compete for transport. This competition is influenced by the properties of the transporter itself and by the concentrations of different ions in solution. Solutes that are not transported can also interact with transport proteins altering their activity. In addition, there may be indirect interactions among ions as a result of their transport across the plasma membrane, for example, via effects on membrane potential through the movement of charge, or via effects on the proton electrochemical gradient through the coupling of solute transport to proton movements.

2.5.4.1 Competition

Transport proteins catalyze the movement of nutrients from the rhizosphere solution to the cytoplasm across the plasma membrane of root cells (Fig. 2.9, Table 2.6). Competition between ions of the same valency for entry to a channel protein or for binding to a carrier protein is common. Such competition occurs particularly among ions with similar physicochemical properties (valency and ion diameter), for example, the alkali cations potassium (K^+), rubidium (Rb^+), cesium (Cs^+), and sodium (Na^+), or the Group II divalent cations calcium (Ca^{2+}), strontium (Sr^{2+}), and barium (Ba^{2+}). It is important to note, however, that the inhibition of transport of a particular ion by another ion does not necessarily imply that the inhibitory ion is itself transported.

Radioactive rubidium (^{86}Rb) has often been used as a tracer to study K^+ transport in plants, although this can give misleading results under certain circumstances (Britto and Kronzucker, 2002). In general, transport proteins catalyzing K^+ transport across the plasma membrane of root cells, such as K^+ channels, nonselective cation channels, and proton-coupled K^+ symporters, do not differentiate between K^+ and Rb^+ for transport (Pyo et al., 2010; Vallejo et al., 2005; White, 1997a; White and Karley, 2010).

The major $K+$ channel in roots of arabidopsis (AtAKT1-AtKC1) is relatively impermeable to Cs^+ that inhibits K^+ influx through this channel (Coskun et al., 2013a; White and Broadley, 2000). By contrast, proton-coupled K^+ symporters, such as AtHAK5, transport both K^+ and Cs^+ into root cells when plants are K^+ deficient (Nieves-Cordones et al., 2020; Qi et al., 2008; Rai and Kawabata, 2020; White and Broadley, 2000).

The competition between potassium (K^+) and ammonium (NH_4^+) is difficult to explain simply by competition for a single transport process at the plasma membrane. Even though the presence of NH_4^+ generally inhibits K^+ uptake (Coskun et al., 2013a; Nieves-Cordones et al., 2014; Santa-María et al., 2000), the reverse (inhibition of NH_4^+ uptake by K^+) is observed rarely (e.g., Balkos et al., 2010; Coskun et al., 2013b; 2017; Szczerba et al., 2008) (Table 2.10). In arabidopsis, NH_4^+ competes with K^+ for transport through both AtAKT1 and AtHAK5 (ten Hoopen et al., 2010) and also reduces the expression of *AtHAK5* (Qi et al., 2008), whereas K^+ does not appear to affect the expression or activity of the major NH_4^+ transporter AtAMT1. In addition, at least at high (mM) external ammonium concentrations, a substantial proportion of ammonium may not be taken up as NH_4^+ through transporters such as AtAMT1, but as NH_3 via NH_3-permeable aquaporins (Coskun et al., 2013b; Kirscht et al., 2016; Loqué et al., 2005). Therefore, it is hypothesized that K^+ suppression of NH_3 fluxes (and ultimately K^+-induced alleviation of ammonium toxicity) occurs through the regulation of aquaporin activity (Fig. 2.17; Coskun et al., 2013b, 2017). Thus, the uptake of ammonium by root cells is determined not only by competition of NH_4^+ for cation transporters but also by rhizosphere pH, cellular tolerance of NH_4^+ and NH_3 uptake.

Applications of K and Ca fertilizers often induce Mg deficiency in crop plants (Gransee and Führs, 2013). This is partly a consequence of cations, such as K^+ and Ca^{2+}, inhibiting Mg^{2+} uptake by plant roots (Table 2.11). The presence of Mn^{2+} in the rhizosphere also inhibits Mg^{2+} uptake by roots but has little effect on the uptake of K^+ (Heenan and Campbell, 1981). This presumably reflects the contrasting specificity of the transporters responsible for the uptake of each cation.

Competition also occurs among anions for uptake by root cells. Some well-known examples are competition between sulfate and molybdate, sulfate and selenate, selenite and phosphate, and phosphate and arsenate.

TABLE 2.10 Interactions between the uptake of NH_4^+ and K^+ by maize roots.

(NH_4)$_2SO_4$ (mM)	Concentration in roots (μmol g^{-1} fresh wt)			
	Ammonium		Potassium	
	$-K^+$	$+K^+$	$-K^+$	$+K^+$
0	6.9	6.7	8.2	54
0.15	7.3	7.1	6.7	48
0.5	17	14	8.9	41
5.0	29	32	9.3	27

Duration of the experiment: 8 h; +K indicates addition of 0.15 mM K^+; calcium concentration constant at 0.15 mM.
Source: Based on Rufty et al. (1982).

TABLE 2.11 Uptake of labeled Mg^{2+} (^{28}Mg) by barley seedlings without or with supply of K^+ and Ca^{2+} (each of three ions at 0.25 mM).

	Magnesium uptake [μmol Mg (10 g)$^{-1}$ fresh wt (8 h)$^{-1}$]		
	$MgCl_2$	$MgCl_2 + CaSO_4$	$MgCl_2 + CaSO_4 + KCl$
Roots	165	115	15
Shoots	88	25	6.5

Source: Based on Schimansky (1981).

Sulfate and molybdate are thought to enter root cells through the same proton-coupled symporters (Duan et al., 2017; Fitzpatrick et al., 2008; Shinmachi et al., 2010). Increasing the sulfate concentration in the rooting medium reduces molybdate uptake strongly. Hence, S fertilization may be an effective tool to reduce excessive Mo uptake, thereby improving plant growth and animal nutrition in soils containing high concentrations of Mo (Maillard et al., 2016). However, the competition may become a critical factor in soils containing little Mo.

The interactions between selenate and sulfate are also quite distinct and of considerable practical importance in view of the human and animal nutritional requirement for selenium and the detrimental effects of excessive selenium in the diet (White and Broadley, 2009). Sulfate and selenate are taken up into root cells through the same proton-coupled SULTR symporters (Schiavon and Pilon-Smits, 2017; White, 2016, 2018). Increasing sulfate concentration in the substrate strongly decreases selenate uptake by roots, suggesting direct competition between selenate and sulfate for transport (El Mehdawi et al., 2017; White et al., 2004). On the other hand, increasing selenate concentration in the substrate often increases sulfate uptake by plants (Stroud et al., 2010; White et al., 2004) and upregulates *SULTR* expression by mimicking the sulfate-deficiency response (Schiavon and Pilon-Smits, 2017; White, 2018).

Antagonistic interactions between selenite and phosphate, and also between phosphate and arsenate, are thought to occur because these anions are transported by the same proton-coupled symporters into root cells (Zhang et al., 2014; Zhao et al., 2010). In Yorkshire fog (*Holcus lanatus*), arsenate-tolerant and nontolerant genotypes exist, and arsenate uptake is much lower in the tolerant genotypes (Meharg and Macnair, 1992). The low arsenate uptake rate of tolerant plants is achieved by suppression of a P deficiency-induced high-affinity uptake system. Similar mechanisms of arsenic tolerance have been observed in other plant species. For example, mutants of arabidopsis with defective phosphate transport are more tolerant to arsenate (Shin et al., 2004). Arsenite and undissociated methylated arsenic species are taken up by roots through the silicon transport pathway via members of the nodulin 26-like intrinsic protein (NIP) family (Maurel et al., 2015). Members of this family also transport a range of small neutral molecules including ammonia, urea, boric acid, silicic acid, and selenous acid (Coskun et al., 2019; Maurel et al., 2015; Miwa et al., 2009; Pommerrenig et al., 2020; White, 2018).

The inability of transport proteins to differentiate effectively between potassium and rubidium, calcium and barium, sulfate and selenate, and phosphate and arsenate illustrates that the selectivity of transport proteins in the plasma membrane of root cells does not indicate any essential role for an element in the plant but merely reflects the physicochemical similarities between ions. Plant roots may be unable to exclude many nonessential or toxic ions from the root symplasm. This has important practical implications, for example, for the entry of metals into the food chain via their uptake and accumulation by plants (e.g., White et al., 2012).

Interactions also occur between Cl^- and NO_3^- in their uptake and accumulation. Chloride concentrations in plant tissues, particularly in roots, can be decreased strongly by increasing NO_3^- availability (Glass and Siddiqi, 1985). Similarly, NO_3^- uptake is reduced when roots contain high Cl^- concentrations (Cram, 1973). Several anion channels and proton-coupled symporters in the plasma membranes of root cells appear to facilitate the transport of both Cl^- and NO_3^- (Roberts, 2006; Wege et al., 2017; White and Broadley, 2001), suggesting interactions in the pathways of their uptake and translocation within the plant. In addition, several anion channels facilitate the movement of both Cl^- and NO_3^- across the tonoplast, and the active efflux of Cl^- and NO_3^- from the cytoplasm into the vacuole is catalyzed, in part, by the same proton-coupled transporters, encoded by members of the *CLC* gene family (Martinoia et al., 2007; Teakle and Tyerman, 2010; Wege et al., 2017; White and Broadley, 2001; Zifarelli and Pusch, 2010). Thus, it is possible that the two anions compete for transport across the tonoplast, which affects their accumulation in vacuoles.

An interesting case of the indirect regulation of transport by nutrients is the inhibition of NO_3^- uptake, and stimulation of Cl^- uptake, by NH_4^+ supply (Lee and Drew, 1989; Miller and Cramer, 2004; Xu et al., 2000). By contrast, increasing NO_3^- supply generally has little or no effect on NH_4^+ uptake (Breteler and Siegerist, 1984; cf. Kronzucker et al., 1999). Thus, when nitrogen is supplied as NH_4NO_3, ammonium is often taken up in preference to nitrate. In barley, external NH_4^+ inhibited net influx of NO_3^- within 3 min, and upon removing NH_4^+ from the external solution, net influx of NO_3^- resumed within 3 min (Lee and Drew, 1989). Such quick effects suggest that they arise from the effect of NH_4^+ on the electrochemical gradients supporting NO_3^- uptake across the plasma membrane. A similar phenomenon has been observed for K^+ influx in response to sudden withdrawal of NH_4^+ (Coskun et al., 2013a).

2.5.4.2 Effects of extracellular calcium

An example of synergism, first discovered by Viets (1944), is the stimulation of cation and anion uptake by extracellular Ca^{2+} at low rhizosphere pH (Table 2.12). It is thought that this phenomenon is the result of Ca^{2+} counteracting the negative effects of high H^+ concentrations on plasma membrane integrity or the activity of the plasma membrane H^+-ATPase. Calcium, as a divalent cation, stabilizes membranes through interactions with the negatively charged headgroups of phospholipids and, thereby, influences membrane function. It also contributes to the resealing of the plasma membrane following damage (Schapire et al., 2009). Calcium can be removed readily from its binding sites at the outer surface of the plasma membrane, for example, by high concentrations of H^+ or cations such as Na^+ (Britto et al., 2010; Cramer et al., 1985; Hepler, 2005), which will increase solute efflux. Consequently, the alleviation by Ca^{2+} of NaCl-induced elevations in K^+ efflux in salt-stressed tissues has been proposed to be a result of plasma membrane stabilization (Britto et al., 2010; Coskun et al., 2013c; Cramer et al., 1985).

Rhizosphere Ca^{2+} concentration also influences the selectivity of ion uptake. For example, in the absence of Ca^{2+} there are clear differences in the K^+/Na^+ uptake ratio between the "natrophobic" maize and the "natrophilic" sugar beet (*Beta vulgaris*). However, the presence of Ca^{2+} in the rhizosphere solution shifts the uptake ratio in favor of K^+ in both species (Table 2.13). These shifts in K^+/Na^+ uptake ratio are likely to be due to extracellular Ca^{2+} inhibiting Na^+ influx through voltage-insensitive, non-selective cation channels (Maathuis and Amtmann, 1999; Munns and Tester, 2008; White, 1999) but have little effect on K^+ influx through inward-rectifying K channels (Maathuis and Amtmann, 1999; White, 1997a). High Ca^{2+} concentrations in the soil solution are particularly beneficial for the maintenance of

TABLE 2.12 Rates of net K^+ and Cl^- uptake (means ± SE) by barley roots with or without Ca^{2+} supply (external pH 5.0).

External solution (mM)	Uptake rate [μmol g^{-1} dry wt (2 h)$^{-1}$]	
	K^+	Cl^-
0.1 KCl	117 ± 6	34 ± 4
0.1 KCl + 1.0 CaSO$_4$	140 ± 7	52 ± 4

TABLE 2.13 K^+/Na^+ selectivity of roots from an external solution containing 10 mM each of NaCl and KCl with or without Ca^{2+} supply.

External solution	Uptake rate [μmol g^{-1} fresh wt (4 h)$^{-1}$]					
	Maize			Sugar beet		
	Na^+	K^+	$Na^+ + K^+$	Na^+	K^+	$Na^+ + K^+$
−Ca	9.0	11	20	19	8.3	27
+1 mM CaCl$_2$	5.9	15	21	15	11	26

high K⁺/Na⁺ uptake ratios in saline environments as they increase plant salt tolerance (Kronzucker and Britto, 2011; Manishankar et al., 2018).

2.5.4.3 Cation–anion relationships

The uptake of cations and anions occurs through different transport proteins (Fig. 2.9, Table 2.6); therefore, direct interactions between cations and anions for uptake are rare. However, the uptake of one nutrient can influence the uptake of another indirectly through effects on the membrane potential, the proton electrochemical gradient, or via feedback regulation through plant growth or metabolism. The stimulation of cation uptake by anions, and of anion uptake by cations, is observed frequently and is generally a consequence of the necessity to maintain charge balance. However, synergism in ion uptake can also be the result of a general increase in root metabolic activity when nutrients are supplied after a period of deprivation.

When present at low concentrations in the rhizosphere, the rate of uptake of a cation is not affected by the accompanying anion and vice versa, as shown in Table 2.14 for K⁺ and Cl⁻. At high external concentrations, however, an accompanying ion that is taken up relatively slowly can reduce the uptake of an oppositely charged ion that is transported at a faster rate: for example, SO_4^{2-} depresses K⁺ uptake and Ca^{2+} depresses Cl⁻ uptake from single-salt solutions (Lüttge and Laties, 1966). Moreover, at high (mM) external K⁺ conditions, K⁺ influx has been shown to be stimulated by Cl⁻ and NO_3^-, but not by SO_4^{2-} and $H_2PO_4^-$ (Coskun and Kronzucker, 2013; Coskun et al., 2013a; Epstein et al., 1963; Kochian et al., 1985). Different uptake rates of cations and anions require (1) compensation of electrical charges and (2) regulation of cellular pH. At high external concentrations these requirements become a limiting factor for the uptake of K⁺ when accompanied by SO_4^{2-} and for uptake of Cl⁻ when accompanied by Ca^{2+} (Table 2.14).

Different rates of cation and anion uptake by roots can cause perturbations of intracellular pH. The stabilization of cytosolic pH in the range of 7.3–7.6 is achieved by a cellular pH stat that consists of two components: the biophysical pH stat, characterized by proton transport across the plasma membrane or tonoplast (Fig. 2.19), and the biochemical pH stat that involves production and consumption of protons through metabolism and is achieved by the formation and removal of carboxylic groups (Britto and Kronzucker, 2005; Miller and Cramer, 2004; Peuke, 2010). When K_2SO_4 is supplied, the excess cation uptake is compensated for by an equivalent synthesis of organic acid anions, and when $CaCl_2$ is supplied the excess anion uptake is compensated for by an equivalent decrease in the synthesis of organic acid anions. These changes in organic acid concentrations are also reflected in the rates of CO_2 fixation in the roots (dark fixation).

The main reactions involved in the traditional concept of the biochemical pH stat in relation to different cation–anion uptake ratios are shown schematically in Fig. 2.19. Excessive cation uptake (Fig. 2.19A) results in an increase in cytosolic pH, which increases the synthesis of organic acids. This produces anions (R·COO⁻) for pH stabilization and charge compensation and enables the subsequent transport of cations and anions either into the vacuole or the shoot. By contrast, excessive anion uptake (Fig. 2.19B) is correlated with a decrease in cytosolic pH, which stimulates the decarboxylation of organic acids from the storage pool (i.e., the vacuoles). This causes an increase in pH as decarboxylation consumes protons. In addition to increases or decreases in root concentrations of organic acid anions,

TABLE 2.14 Accompanying anions influence rates of K⁺ and Cl⁻ uptake by maize plants.

Concentration (mM)	Uptake rate (μmol g⁻¹ fresh wt h⁻¹)			
	K⁺ from		Cl⁻ from	
	KCl	K_2SO_4	KCl	$CaCl_2$
0.2	1.6	1.6	0.8	0.7
2.0	2.7	1.9	2.0	1.0
20	5.7	2.2	4.3	2.1

Source: Recalculated from Lüttge and Laties (1966).

38 PART | I Nutritional physiology

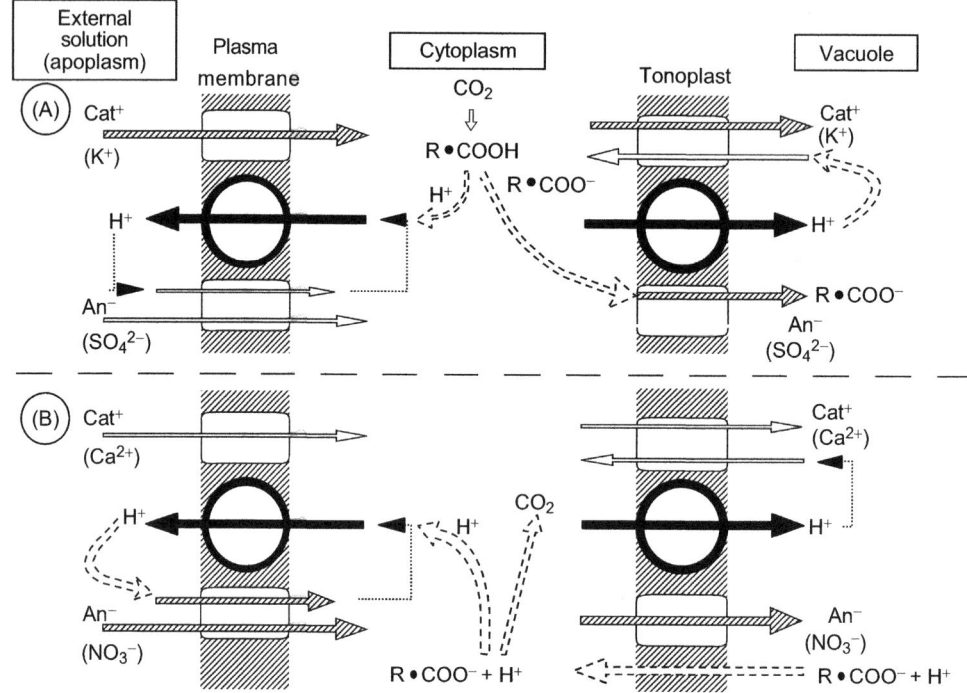

FIGURE 2.19 Model for internal pH stabilization and for charge compensation at different ratios of cation:anion uptake from the external solution. (A) Excessive uptake of cations (Cat$^+$), for example, with K$_2$SO$_4$ supply. (B) Excessive uptake of anions (An$^-$), for example, with Ca(NO$_3$)$_2$ supply.

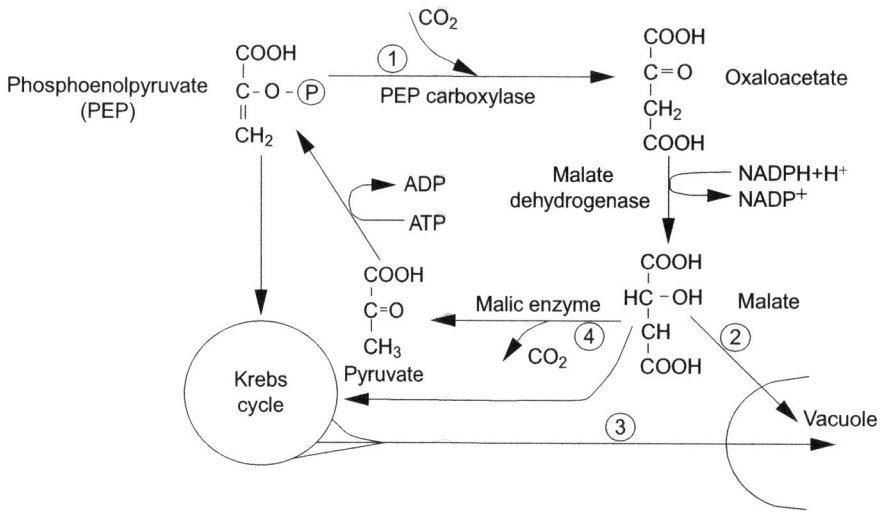

FIGURE 2.20 Model of the pathways of CO$_2$ fixation ("dark fixation") and decarboxylation. Reactions 1–4 are explained in the text.

the biochemical pH stat also influences the pH in the root apoplasm and external solution, with excess cation uptake increasing proton efflux, thus decreasing the external pH, whereas excess anion uptake increases the external pH.

In the cytoplasm, the equilibrium between carboxylation (CO$_2$ fixation) and decarboxylation is thought to be regulated by the pH sensitivity of two enzymes: phosphoenolpyruvate (PEP) carboxylase and malic enzyme (Fig. 2.20). An increase in pH activates PEP carboxylase (reaction 1), and both the rate of CO$_2$ fixation and the synthesis of oxaloacetate are increased. After oxaloacetate is reduced to malate by the enzyme malate dehydrogenase, the malate can be directly transported into the vacuoles (reaction 2), where it acts as a counterion for excess cations (Fig. 2.19A). Alternatively (reaction 3), malate can be incorporated into the cytoplasmic pool of the organic acids of the Krebs cycle, and another organic acid from this pool (e.g., citric acid) can be transported into the vacuole. An oxalate-based biochemical pH stat may play an important role in plant species that accumulate large amounts of oxalate, such as members of the Chenopodiaceae (Davies, 1986). When anions are taken up in excess (Fig. 2.19B), the pH of the cytoplasm decreases and malic enzyme (reaction 4; Fig. 2.20) is activated, leading to the decarboxylation of malate and the production of CO_2. As a result of these reactions, the cytoplasmic pH is stabilized and the cation–anion ratio in the cells maintained. This biochemical pH stat responds rapidly to supply of K_2SO_4, with PEP carboxylase activity being increased by 70% within 20 min (Chang and Roberts, 1992).

Nitrogen nutrition (NH_4^+, NO_3^-, N_2 fixation) has a strong effect on cation–anion relationships in plants because about 70% of the ions taken up by plants are either NH_4^+ or NO_3^- (van Beusichem et al., 1988). Nitrogen nutrition influences both organic acid metabolism and the element composition of plant tissues (Table 2.15). Plants supplied NH_4^+ are generally characterized by a high cation/anion uptake ratio and plants supplied NO_3^- by a low cation/anion uptake ratio. However, the effects of NH_4^+ and NO_3^- on organic acid metabolism differ from those anticipated from Fig. 2.19 because N assimilation in roots is correlated with the production or consumption of protons. The shoots have a limited capacity to dispose of protons; thus, NH_4^+ assimilation takes place largely in roots (Coskun et al., 2017; Engels and Marschner, 1993). The assimilation of NH_4^+ produces protons; thus, despite high total cation uptake by NH_4^+-fed plants, the pH in the cytoplasm decreases during NH_4^+ assimilation and must be stabilized both by enhanced proton excretion and the decarboxylation of organic acids (Fig. 2.19B). By contrast, the uptake of NO_3^- is correlated with an approximately equimolar consumption of H^+ (Raven, 1986). Depending on whether NO_3^- reduction and assimilation take place in the root or the shoot, carboxylates are either produced in the roots or transported in the phloem from the shoots to the roots to maintain charge balance (Peuke, 2010). Legumes dependent on biological N_2 fixation are characterized by a cation/anion uptake ratio >1 and have higher tissue concentrations of organic acid anions and greater proton efflux than plants supplied NH_4^+ (Allen et al., 1988; Li et al., 2013).

The pH of the external solution is strongly influenced by the form of plant nitrogen nutrition due to differences in cation/anion uptake ratio, nitrogen assimilation, and cellular pH stabilization (Fig. 2.21). When plants with preferential NO_3^- reduction in the roots, such as sorghum, are supplied NO_3^-, the external pH usually increases with time. When they are supplied NH_4NO_3, a transient decrease in the pH of the external solution during preferential uptake of NH_4^+ uptake is followed by an increase in pH, as observed for NO_3^--fed plants. However, under conditions where NO_3^-

TABLE 2.15 Ionic balance in shoots of castor oil plants grown with different forms of N supply.

Form of N supply	Cations				Anions					
	K^+	Ca^{2+}	Mg^{2+}	Total	NO_3^-	$H_2PO_4^-$	SO_4^{2-}	Cl^-	Organic acid anions[a]	Total
NO_3^-	99	85	28	212	44	18	11	2	137	212
NH_4^+	55	43	22	120	0	23	33	5	59	120

[a]Calculated from the difference (cations − anions).
Source: From van Beusichem et al. (1988).

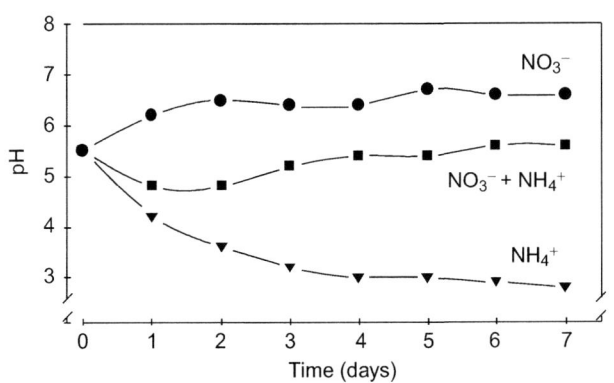

FIGURE 2.21 Time-course of changes in the pH of the external solution containing 300 mg L^{-1} total N as only NO_3^-, only NH_4^+, or both (ratio NO_3^-:NH_4^+ 8:1) as the N source for sorghum plants. *Redrawn from Clark (1982).*

uptake and assimilation are impaired and cation/anion uptake ratio is high, a strong decrease in the pH of the external solution has been observed. This phenomenon occurs in many plant species under, for example, P_i deficiency (Schjørring, 1986), Zn deficiency (Cakmak and Marschner, 1990), and Fe deficiency in dicotyledonous plants.

In plants supplied with NH_4^+, net proton efflux and maintenance of the cellular pH stat becomes increasingly difficult in roots when the pH of the external solution is low; hence, pH decreases in both the cytoplasm and vacuoles of root cells (Gerendás et al., 1990). Poor growth of NH_4^+-fed plants at low external pH is related to the difficulty in maintaining cytosolic pH homeostasis in the face of high NH_4^+ fluxes across the plasma membrane of root cells, cation–anion imbalance, and the potentially high energy costs incurred by the futile cycling of ammonium across the plasma membrane (Britto and Kronzucker, 2006; Coskun et al., 2013b; Miller and Cramer, 2004).

Maintenance of cytoplasmic pH homeostasis involves costs in terms of energy, photosynthate, and water (Britto and Kronzucker, 2005; Cosse and Seidel, 2021; Raven, 1986). This is particularly true in relation to N nutrition. When both NH_4^+ and NO_3^- are supplied, cytoplasmic pH homeostasis may be achieved by similar rates of H^+ production (NH_4^+ assimilation) and H^+ consumption (NO_3^- assimilation) and thus have a low energy requirement (Allen et al., 1988; Britto and Kronzucker, 2005). This may explain, in part, why optimal growth for many plant species is usually obtained with a mixed supply of NH_4^+ and NO_3^- (Kronzucker et al., 1999).

2.5.5 External concentration

The relationship between the rate of influx (I), or uptake, of an ion and its concentration in solution (C) can usually be described by Michaelis–Menten kinetics in the low concentration range: the flux saturates and transport appears to be selective and closely coupled to metabolism. By contrast, at high external ion concentrations, the uptake rate is often linearly related to solute concentration ($I = kC$) through a proportionality parameter (k), is not very selective, and is not particularly sensitive to temperature or metabolic inhibitors. The typical relationship between external concentration and uptake rate of K^+ is presented schematically in Fig. 2.22, without consideration of C_{min}. Similar uptake isotherms have been obtained for many nutrients, although their kinetic parameters (K_m, I_{max}, C_{min}, k) differ.

The concentrations in soil solutions are usually low for K^+ (<1 mM) and NH_4^+ (<100 μM), and extremely low for P_i (<10 μM), B (<10 μM), Ni, Zn, Cu, Mo, Mn (each <1 μM), and Fe (<0.1 pM in calcareous and alkaline soils). On the other hand, the concentrations of Ca^{2+}, Mg^{2+}, Cl^-, SO_4^{2-}, and NO_3^- are often in the millimolar range, especially in arable soils (Tyler and Olsson, 2001; White and Broadley, 2001, 2009). Therefore, the acquisition of most nutrients requires significant energy input and a selective, high-affinity transport system to enable the plant to satisfy plant demand while avoiding ion toxicities.

The kinetic parameters for a given nutrient are influenced greatly by plant nutritional status and nutrient availability in the rhizosphere. In general, deficiency of a nutrient leads to an increase in the capacity of the root system to take up that nutrient (Britto and Kronzucker, 2013; Britto et al., 2021; Buchner et al., 2004; White and Broadley, 2009; White and Hammond, 2008; White and Karley, 2010). For example, in plants grown in the absence of NO_3^-, noninduced roots possess only a low-capacity (constitutive) influx system for NO_3^- (Table 2.16). This constitutive NO_3^- influx system does not saturate as the external NO_3^- concentration is increased, suggesting that the transport protein has a low affinity for NO_3^-. However, within 20 min of supplying NO_3^-, a high-affinity, high-capacity NO_3^- influx system is induced (Glass et al., 1990). Four days after induction (Table 2.16), the capacity for NO_3^- influx is lower, suggesting negative feedback regulation of the activity of this transport system.

In general, optimal growth can be achieved at concentrations in the range of the high-affinity system for nutrients such as K, P, and N, when supplied continuously to plants in the external solution (Asher and Edwards, 1983; Raddatz

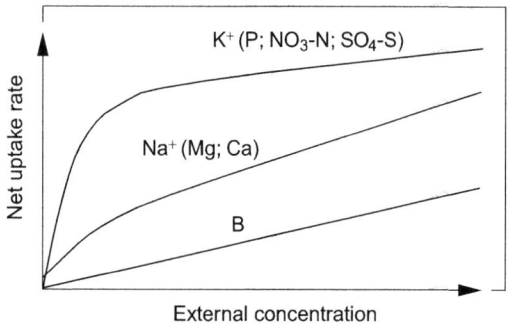

FIGURE 2.22 Schematic representation of the relationships between the rates of uptake of K, Na, and B by barley roots and the concentrations of KCl, NaCl, and B in the external solution. The relationships between the uptake rates of other nutrients are indicated in parentheses.

TABLE 2.16 Influx of nitrate into barley roots without and with induction of a high-capacity nitrate uptake system.

External conc. of NO_3^- (mM)	NO_3^- influx (μmol g^{-1} fresh wt h^{-1})		
	Noninduced	Induced for 1 day[a]	Induced for 4 days[a]
0.02	0.10	2.8	1.5
0.30	0.39	5.3	2.9
20	12	21	8.0

[a]*Pretreatment with nitrate for 1 or 4 days.*
Source: Based on Glass et al. (1990).

et al., 2020). Similarly, for the micronutrients Zn, Cu, Mn, and Fe, the concentration of the free metal species (Zn^{2+}, Cu^{2+}, Mn^{2+}, Fe^{2+}/Fe^{3+}) in the external solution required for optimal growth derived from chelate-buffered nutrient solutions suggests that nutritional requirements can be met by extremely low concentrations of free cations in the external solution ($10^{-9}-10^{-12}$ M for Mn^{2+}, Zn^{2+}, and Cu^{2+}, and perhaps even lower for Fe^{2+}/Fe^{3+}), provided a continuous supply is maintained (Bell et al., 1991; Degryse et al., 2006; Fox et al., 1996; Laurie et al., 1991; Webb et al., 1993).

Thus, under optimal conditions in which a constant nutrient supply is maintained (e.g., in nutrient solutions), only very low concentrations of nutrients are required in the external solution for maximal plant growth. At higher supply, higher uptake rates reflect what is known as "luxury uptake". In soil-grown plants in general, and under field conditions in particular, the conditions in the root environment are far from optimal and the maintenance of a constant nutrient supply to the root surface is unlikely to occur. Higher external concentrations and luxury uptake in preceding periods can be important in providing an internal reserve for periods of high demand or interrupted root supply. This also holds true for natural vegetation subjected to transient nutrient flushes under favorable weather conditions (Millard and Grelet, 2010; Rorison, 1987). Moreover, the pH of the soil solution has a significant impact on exchangeable cation concentrations in the soil and crop yield. In a field experiment over 9 years, liming (a common management strategy to ameliorate acid soils) was found to affect the concentration of most exchangeable cations in the soil (except Cu^{2+} and K^+), and the concentrations of three exchangeable cations (Mn^{2+}, Ca^{2+}, and Al^{3+}) were found to influence barley yields strongly (Holland et al., 2021).

In both soil and nutrient solutions, essential elements can be supplied in concentrations so high that they become toxic to plants [e.g., Mn in acid soils, B in sodic (Na-rich) soils, and Mn and Fe in waterlogged or flooded soils; White and Brown, 2010]. However, there is considerable genotypic variation in the uptake of nutrients, and these differences between and within plant species have important consequences for ecology and agriculture (White and Brown, 2010). The capacity to grow well in soils with high concentrations of elements enables the development of plants for the phytoremediation of contaminated land. For example, in barley, the cultivar Sahara 3771 has a lower B uptake capacity than the cultivar Schooner and therefore requires a higher B concentration in the external solution for optimal growth (Nable et al., 1990). This can be a disadvantage in low-B soils but is an effective mechanism of avoiding B toxicity when plants are grown in soils with high B availability (Sutton et al., 2007).

The capacity to grow well in soils with low availability of nutrients confers an ecological advantage in these soils. Genotypic variation within crop species in the nutrient supply required for optimal growth is well documented, for example, for N, P, and K (Fageria, 2009; Hirel et al., 2007; Wang et al., 2019; White, 2013; White et al., 2021a). Such genotypic variation can be used to improve fertilizer-use efficiency in agriculture through the development of crop varieties that acquire and/or utilize mineral elements more effectively.

2.5.6 Plant nutritional status

The rate of uptake of a nutrient at a given external concentration is often determined by plant growth rate and nutritional status. Nutrient uptake responds rapidly to fluctuations in root nutrient concentrations and more slowly to long-term changes in plant demand. In general, as the tissue concentration of a particular mineral element increases, its influx declines, and vice versa. An example of this feedback regulation is shown for K^+ influx in barley roots in Table 2.17. Similar relationships between internal concentrations and influx are well documented for NO_3^- (Fig. 2.23), P_i

TABLE 2.17 Relationship between root tissue concentration and K influx to barley roots.

Root K concentration (μmol g^{-1} fresh wt)	K$^+$ influx (μmol g^{-1} fresh wt h^{-1})
21	3.1
32	2.7
48	2.2
58	1.6

Source: From Glass and Dunlop (1979).

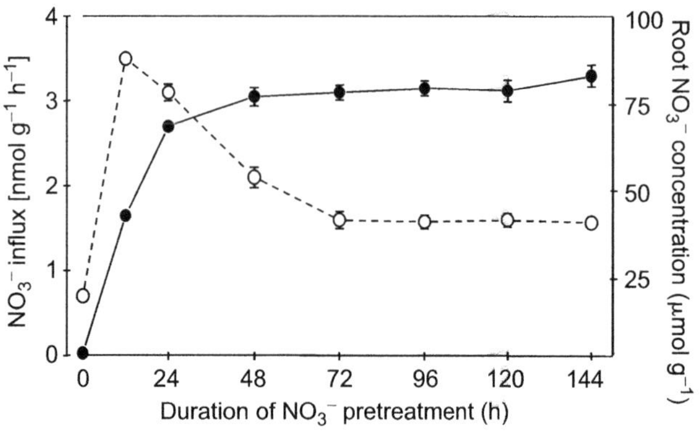

FIGURE 2.23 Root tissue nitrate concentration (●) and nitrate influx (○) to roots of barley plants with different nitrate (NO$_3^-$) pretreatment. Means ± SE. *Adapted from Siddiqi et al. (1989).*

(Table 2.8), SO$_4^{2-}$ (Fig. 2.24), and Fe^{2+}, Zn^{2+}, and Cu^{2+} (Broadley et al., 2007; White and Broadley, 2009). Immediate effects on nutrient influx are often due to posttranslation modifications of regulatory components or transport proteins, whereas longer term effects are mediated by transcriptional responses. The mechanisms that may be involved in this feedback regulation are summarized in Fig. 2.25.

The uptake of NH$_4^+$ and NO$_3^-$ is closely related to the N status of plants. For example, NH$_4^+$ uptake capacity is negatively correlated with the concentrations of NH$_4^+$ and certain amino acids in the roots, such as glutamine and asparagine (Causin and Barneix, 1993; Liu and von Wirén, 2017; Rawat et al., 1999; von Wirén et al., 2000). Accordingly, the NH$_4^+$ uptake capacity increases rapidly within a few days after the withdrawal of N supply (Coskun et al., 2017; Rawat et al., 1999; The et al., 2021). In arabidopsis, the increase in NH$_4^+$ uptake capacity is correlated with a decrease in glutamine and increased expression of the *AtAMT1* gene (Rawat et al., 1999). The regulation of NO$_3^-$ uptake involves the induction of a high-capacity, high-affinity uptake system and the negative feedback regulation of NO$_3^-$ uptake by increasing internal NO$_3^-$ concentrations (Table 2.17). In noninduced plants, NO$_3^-$ supply increases both NO$_3^-$ influx and NO$_3^-$ concentration in root tissues. Later, the increase in NO$_3^-$ concentrations reduces NO$_3^-$ influx. Much recent progress has been made in the transcriptional regulation of this process (Bellegarde et al., 2017; Ueda and Yanagisawa, 2019). In arabidopsis, an intricate "feed-forward loop" network has been described in which two sets of transcription factors belonging to the NLP and NIGT1/HHO family of proteins coordinate the fine-tuned expression of the high-affinity nitrate transporter *NRT2* genes in response to NO$_3^-$ signals and nutritional status (Kiba et al., 2018; Liu et al., 2017; Maeda et al., 2018; Ueda and Yanagisawa, 2019). It is also noteworthy that increased root sucrose concentrations appear to increase the uptake of both NH$_4^+$ and NO$_3^-$ through upregulation of genes encoding high-affinity transport proteins (Girin et al., 2010; Lejay et al., 2003; Louahlia et al., 2008). Other long-distance signals hypothesized to convey information on plant N status and regulate root N acquisition include mRNAs and microRNAs, short peptides (10–20 residues), and phytohormones (including auxin, ABA and cytokinins) (Chapter 3; Bellegarde et al., 2017).

In a similar manner, SO$_4^{2-}$ uptake is regulated by plant S status. The dynamics of this feedback regulation on the uptake of SO$_4^{2-}$ are shown in Fig. 2.24. Without external SO$_4^{2-}$ supply, the capacity of barley roots to take up SO$_4^{2-}$ increases rapidly over 5 days but decreases strongly within a few hours of resupplying SO$_4^{2-}$ and is lost after 1 day. Induction of the transport system for SO$_4^{2-}$ requires transcription and protein synthesis, as does the induction of

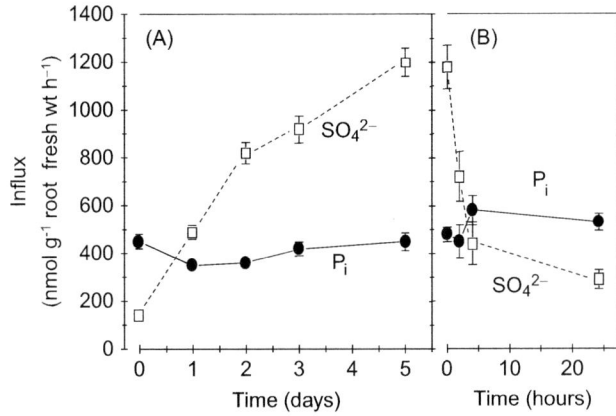

FIGURE 2.24 Time-course of changes in the influx of sulfate (SO_4^{2-}) and phosphate (P_i) in roots of barley plants deprived of external sulfate supply for up to 5 days (A) and then resupplied with sulfate for up to 24 h (B). Means ± SE. *Adapted from Clarkson and Saker (1989).*

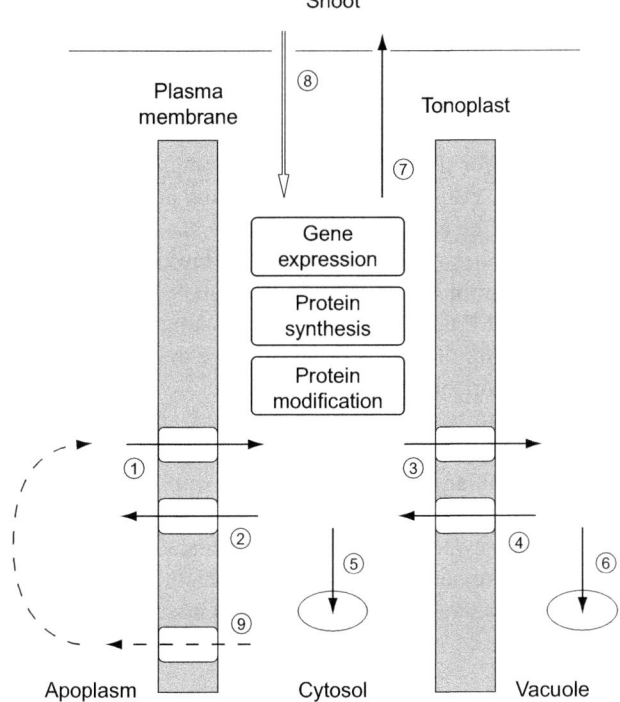

FIGURE 2.25 Model for the regulation of nutrient uptake by roots via plant nutritional status. Changes in gene expression, protein synthesis and protein modification in response to either the cytosolic concentration of the mineral nutrient or its metabolites can regulate: (1) nutrient influx via the number and activity of transporters, and the electrochemical gradient across the plasma membrane; (2) nutrient efflux via nutrient concentration in the cytosol, the number and activity of transporters, and the electrochemical gradient across the plasma membrane; (3) nutrient influx to the vacuole via nutrient concentration in the cytosol, the number and activity of transporters, and the trans-tonoplast electrochemical gradient; (4) nutrient efflux from the vacuole via nutrient concentration in the vacuole, the number and activity of transporters, and the trans-tonoplast electrochemical gradient; (5) complexation in the cytosol, sequestration in organelles, and metabolism of nutrients, which govern the concentrations of the nutrient and its metabolites in the cytosol; (6) complexation or precipitation of the nutrient in the vacuole that influences its capacity to exit the vacuole; (7) transport of a nutrient or metabolite from the root to the shoot in the xylem that influences their cytosolic concentrations; (8) transport of a nutrient or metabolite from the shoot to the root in the phloem, mediating their cytosolic concentrations; and (9) modification of the rhizosphere through the secretion of protons, enzymes, and organic solutes to increase the concentration of the mineral nutrient in the apoplasm, and its availability for influx.

enzymes involved in the S assimilatory pathway (Aarabi et al., 2020; Hawkesford and De Kok, 2006; Hell et al., 2010; Hubberten et al., 2012). The accumulation of reduced S compounds, such as cysteine or reduced glutathione (GSH), is the dominant signal of tissue S status that regulates SO_4^{2-} uptake and assimilation (Hawkesford and De Kok, 2006; Hell et al., 2010). These compounds can be the product of root cell metabolism or, in the case of GSH and its precursor γ-glutamylcysteine, can be translocated from the shoots to the roots as a systemic signal of shoot S status (Hell et al., 2010; Herschbach and Rennenberg, 2001). The substrates for cysteine synthesis are sulfide and *o*-acetylserine (OAS); in response to tissue S depletion, OAS content increases and has been shown to promote the activity of sulfate (SULTR) transporters and the expression of several genes of the sulfur-reducing pathway (Aarabi et al., 2020; Hubberten et al., 2012). The expression of genes involved in S uptake and assimilation is also controlled by the delivery of sucrose (Lejay et al., 2003) and a regulatory microRNA (miRNA395; Kragler, 2010; Kumar et al., 2017; Liang et al., 2010) from the shoot.

Phosphate uptake capacity increases after P_i is withheld from the external solution (White and Hammond, 2008). This is correlated with increased transcription of genes encoding proton-coupled P_i transporters. This transcriptional response is mediated through the interplay of biochemical signals indicating root and shoot P_i status. One major regulatory mechanism involves the PHOSPHATE STARVATION RESPONSE1 (PHR1) transcription factor and its PHR1-LIKE 1 (PHL) homologs (López-Arredondo et al., 2014; Puga et al., 2017). Inositol polyphosphate (InsP) signaling molecules, which act as proxies for cellular P_i content, regulate P_i homeostasis by binding to SPX-domain-containing targets, including PHR1 (Puga et al., 2014, 2017; Wild et al., 2016). When plants lack sufficient P, InsP concentration decreases, releasing PHR1 to promote the transcription of P_i starvation response genes (Wild et al., 2016). Moreover, sucrose and microRNA (miR399) in the phloem act as systemic signals of low shoot P_i status (Buhtz et al., 2010; Hammond and White, 2008; Kragler, 2010; Kumar et al., 2017; Puga et al., 2017; Vance, 2010). Although resupplying P_i to P_i-deficient plants ultimately reduces their capacity for P_i uptake, this response does not occur immediately (Table 2.18). Thus resupplying P_i after a period of deficiency can result in greatly increased tissue P_i concentrations and P_i toxicity (Hammond et al., 2011; Lambers, 2022).

Although the transport systems induced by nutrient deficiencies are generally specific for the nutrient the plant lacks, they can also transport other elements. For example, the uptake of selenate increases in S-deficient plants (White, 2018), and the uptake rate of arsenate increases in P_i-deficient plants (Wang and Duan, 2009). This is likely to be a consequence of increased abundance of the SULTR1 proton-coupled sulfate symporter and of the Pht1;1 and Pht1;2 proton-coupled phosphate symporters, respectively (Buchner et al., 2004; Christophersen et al., 2009; Schunmann et al., 2004; White, 2018). Similarly, Cs^+ uptake increases in K^+-deficient arabidopsis because of the upregulation of the *AtHAK5* gene that encodes a high-affinity plasma membrane proton-coupled K^+ (and Cs^+) symporter (Qi et al., 2008). The capacity for Mn^{2+} uptake can be increased by Fe and Zn deficiencies because of the induction of nonspecific transport proteins (White and Neugebauer, 2021).

In addition to regulating the uptake capacity for nutrients, signals of root and shoot nutritional status can also influence the biomass and morphology of the root system, adaptive responses to nutrient deficiency that result in mobilization of nutrients from recalcitrant compounds, and biochemical pathways for the assimilation of nutrients such as N, S, and P (Giehl et al., 2014; Jia and von Wirén, 2020; Satbhai et al., 2015). For example, plant N status influences plant shoot/root biomass ratio, root morphology, and N assimilation (Garnett et al., 2009; Hermans et al., 2006; Hodge, 2004), plant S status influences plant shoot/root biomass ratio and modulates sulfate assimilation (Hawkesford and De Kok, 2006; Hell et al., 2010), and plant P status influences plant shoot/root biomass ratio, root morphology, the release of protons, phosphatases, and organic acid anions into the rhizosphere, and P metabolism in plants (Wang and Lambers, 2020; White and Hammond, 2008). These responses increase the acquisition of nutrients and improve the physiological efficiency by which plants utilize nutrients for growth when they are in short supply.

Regarding their response to Fe deficiency, plants can be classified into two categories (Strategy I and Strategy II) (Brumbarova et al., 2015; Guerinot, 2010; Miller and Busch, 2021; White and Broadley, 2009) (Chapter 14). In both strategies, the responses are confined to the apical zones of growing roots and are fully repressed within about 1 day after resupply of Fe.

Strategy I is typical of dicotyledonous and nongramineous monocotyledonous plants. It is characterized by an increase in ferric (Fe^{3+}) reduction capacity, acidification of the rhizosphere, and the release of organic acid anions and phenolic compounds into the soil solution (Fig. 2.26). The plasma membrane ferric (chelate) reductases that catalyze

TABLE 2.18 Phosphorus concentrations in tissues of barley plants following the supply of P to plants grown without P.

	P concentration (μmol P g^{-1} dry wt)		
	8 days − P[a]	7 days − P +1 day + P[b]	7 days − P +3 days + P[c]
Shoot	49 (20)[d]	151 (61)	412 (176)
Youngest leaves	26 (5)	684 (141)	1647 (493)
Roots	43 (24)	86 (48)	169 (94)

[a]*Eight days of growth without P.*
[b]*Seven days of growth without P and 1 day of growth upon addition of P (150 μM).*
[c]*Seven days of growth without P and 3 days of growth upon addition of P (10 μM).*
[d]*Numerals in parentheses are relative values; 100 represents control with continuous supply of 150 μM P.*
Source: Based on Clarkson and Scattergood (1982).

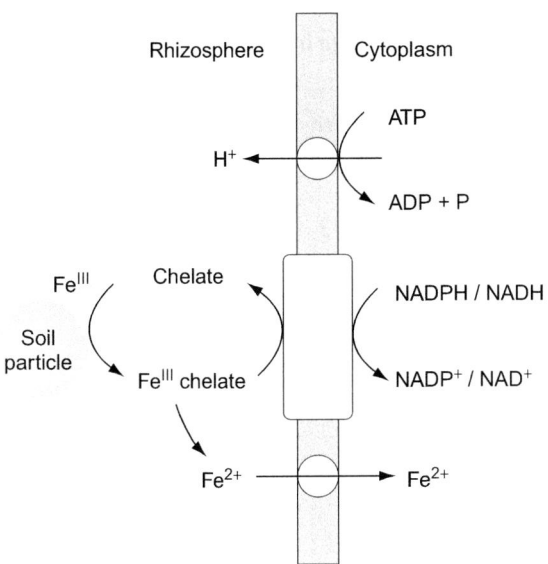

FIGURE 2.26 Model for root responses to iron deficiency in dicots and nongraminaceous monocots (Strategy I): increased acidification of the rhizosphere by H^+-ATPases, induction of ferric reductase activity, reduction of Fe(III)-chelates to Fe^{2+}, and uptake of Fe^{2+} across the plasma membrane by Fe deficiency-inducible, high-affinity Fe^{2+} transporters.

the reduction of Fe^{3+} to Fe^{2+} in roots are encoded by members of the ferric reductase oxidase (*FRO*) gene family (Curie and Mari, 2017; Guerinot, 2010; White and Broadley, 2009). Members of the ZIP transporter family, such as the AtIRT1 transporter of arabidopsis, then mediate Fe^{2+} influx to root cells (Guerinot, 2010; Krishna et al., 2020; Puig et al., 2007; White and Broadley, 2009). The expression of both *AtFRO2* and *AtIRT1* in roots of arabidopsis vary diurnally, being less at night than in the day, and are increased under Fe deficiency (Vert et al., 2003). An example of Strategy I root responses to Fe deficiency, and the corresponding enhanced rates of Fe uptake, of cucumber (*Cucumis sativus*) are shown in Table 2.19.

Strategy II is confined to graminaceous plant species and characterized by an Fe deficiency-induced enhanced release of nonproteinogenic amino acids called phytosiderophores (Chen et al., 2017; Guerinot, 2010; von Wirén et al., 1995; White and Broadley, 2009). Enzymes involved in the synthesis of phytosiderophores from L-methionine include S-adenosyl-methionine synthetase, nicotianamine synthase, nicotianamine aminotransferase, and deoxymugineic acid synthase (Britto and Kronzucker, 2015; Chen et al., 2017; Guerinot, 2010). The expression of genes encoding these enzymes, and also of genes involved in S uptake and methionine synthesis, is often rapidly upregulated in response to Fe deficiency (Britto and Kronzucker, 2015; Guerinot, 2010; White and Broadley, 2009). Phytosiderophores, such as mugineic acid (Fig. 2.27), form highly stable complexes with Fe^{3+}, Zn^{2+}, and Cu^{2+}. The release of phytosiderophores is induced by Fe deficiency (Table 2.20) and rapidly decreases when Fe is resupplied to a Fe-deficient plant (Fig. 2.28A). Both the release of phytosiderophores and the uptake of metal–phytosiderophore complexes follow a distinct diurnal rhythm (Fig. 2.28B) being highest several hours after onset of light. The Fe^{3+}-phytosiderophore complex enters the root cytoplasm via proton-coupled Fe^{3+}-phytosiderophore symporters in the plasma membrane of root cells of grasses (Fig. 2.27; Guerinot, 2010; Ishimaru et al., 2006; Kobayashi et al., 2019; Schaaf et al., 2004; White and Broadley, 2009). Homologs of the maize yellow stripe 1 (ZmYS1) protein belonging to the oligopeptide transporter (OPT) family mediate Fe^{3+}-phytosiderophore uptake by Strategy II plants (Guerinot, 2010; Hanikenne et al., 2021; Ishimaru et al., 2006; Puig et al., 2007; White and Broadley, 2009). In addition to their ability to take up Fe^{3+}-phytosiderophores, graminaceous species can also take up Fe^{2+} (Guerinot, 2010; Kobayashi et al., 2019; White and Broadley, 2009).

2.5.7 Studying nutrition at constant tissue concentration

The common approach of studying plant nutrition in relation to nutrient uptake, nutrient supply, plant nutritional status, and growth rate has been questioned by Ingestad and coworkers (Ingestad, 1997; Ingestad and Ågren, 1992). These authors have argued that, particularly during the exponential phase of plant growth and even in a flowing solution, low external concentrations and plant tissue concentrations are difficult to maintain constant. Thus, the relationships between external supply, uptake rates, and plant nutritional status are difficult to ascertain unequivocally.

To overcome these difficulties, and also to define the effectiveness of a nutrient in terms of biomass production per unit of nutrient at different internal concentrations, Ingestad and colleagues have used a different theoretical and

TABLE 2.19 Proton excretion (pH), reducing capacity of the roots and iron uptake rate in cucumber (Strategy I) with or without Fe preculture.

Fe nutritional status (preculture)	Chlorophyll (mg g^{-1} dry wt)	H$^+$ excretion (solution pH)	Reducing capacity [μmol Fe^{II} g^{-1} root dry wt (4 h)$^{-1}$]	Fe uptake [μmol g^{-1} root dry wt (4 h)$^{-1}$]
+Fe[a]	12	6.2	3.2	0.03
−Fe	7.8	4.8	97	2.6

[a]Supply of 1×10^{-6} M FeEDDHA, pH 6.2.
Source: Data compiled from Römheld and Kramer (1983) and Römheld and Marschner (1990).

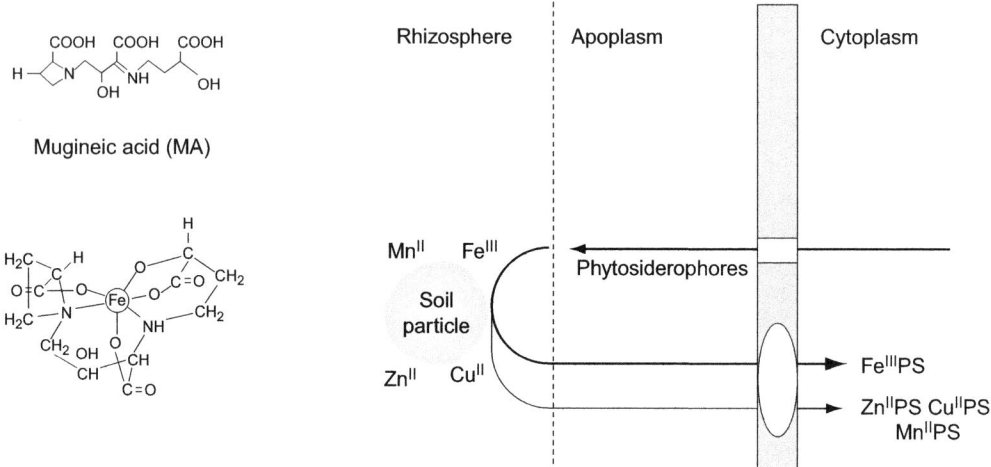

FIGURE 2.27 Model for root responses to Fe deficiency in gramineaceous species (Strategy II): enhanced synthesis and release of phytosiderophores into the rhizosphere, chelation of Fe^{3+}, Zn^{2+}, Cu^{2+}, and Mn^{2+}, and transport of metal−phytosiderophore complexes across the plasma membrane by transport proteins. The structures of the phytosiderophore mucigenic acid and its corresponding Fe(III) complex are also shown.

TABLE 2.20 Release of phytosiderophores (PS; mugineic acid) and uptake of Fe-phytosiderophores in Fe-sufficient and Fe-deficient barley plants.

Fe nutritional status (preculture)	Chlorophyll concentration (mg g^{-1} dry wt)	PS release [μmol g^{-1} root dry wt (4 h)$^{-1}$]	Fe uptake [μmol g^{-1} root dry wt (4 h)$^{-1}$]
+Fe	13	0.4	0.4
−Fe	7.5	8.2	3.4

Source: From Römheld and Marschner (1990).

experimental approach. In principle, this approach is based on relative values. To set a constant relative uptake, the relative addition rates of nutrients (i.e., the supply of nutrients) are divided by the amount of nutrient already in the plant. Accordingly, only the amount of nutrients supplied count, and not the external concentration. Using this approach, a range of different but constant relative growth rates can be achieved at different degrees of nutrient limitations. Interestingly, although the root/shoot dry weight ratio of the nutrient-limited plants is often large, visual deficiency symptoms are absent.

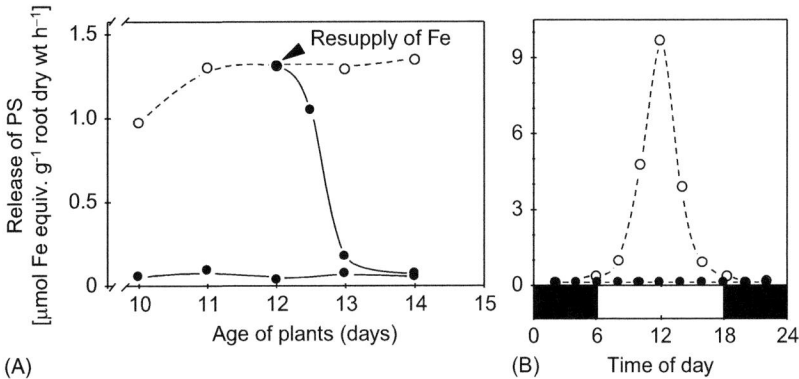

FIGURE 2.28 Release of PS from barley roots as affected by plant Fe nutritional status (A), and diurnal rhythm of release of phytosiderophores (B). Assays were performed on Fe-sufficient (●) and Fe-deficient (○) plants. *PS*, Phytosiderophores. *Data from Römheld (1987a, 1987b)*.

This highly formal concept is an interesting variation to the common approach for studying the nutrition of plants, in which the influence of external concentration and plant nutrient status on nutrient uptake, growth responses, and various physiological and biochemical parameters (e.g., photosynthesis) are studied. This concept allows studying the effects of mineral nutrition under suboptimal but steady-state conditions. However, these steady-state conditions, in which the relative nutrient supply is adjusted to the relative growth rate, are not typical of those experienced by plants growing in the field. For field-grown plants, fluctuations in nutrient supply to the roots are as common as fluctuations in other environmental parameters, such as irradiation, temperature, and water supply. To cope with these fluctuations, plants possess a range of adaptive mechanisms. Fluctuations in nutrient supply are compensated for by modulating uptake capacity, changes in root morphology and physiology, and root/shoot biomass ratio, and the storage and remobilization of mineral nutrients.

2.6 Uptake of ions and water along the root axis

Roots vary both anatomically and physiologically along their longitudinal axes. This should be borne in mind when models for the behavior of root tissue and root cells are based on studies with isolated roots or roots of intact plants. In the apical zone, nonvacuolated cells dominate. These cells differ in many respects from the vacuolated cells in the basal zones. In general, root cells of the apical root zones have higher rates of H^+ efflux (driven primarily by the plasma membrane H^+-ATPase; Siao et al., 2020) and more negative membrane potentials (Hirsch et al., 1998; Lupini et al., 2016) than basal root zones. Thus there is a tendency for the rate of ion uptake per unit root length to decrease with distance from the root apex. However, this tendency depends strongly on the ion, plant nutritional status, and plant species. When K^+ or Ca^{2+} are supplied to different regions of seminal roots of maize (Table 2.21), the uptake rate of K^+ is slightly lower in the apical zone than the subapical zone, despite the high K^+ requirement for growth. The high K^+ concentration in root apical cells of about 200 mM (Hajibagheri et al., 1988; Huang and Van Steveninck, 1989) is maintained not only by uptake from the external solution but also by delivery from more basal root zones (Table 2.21) or from the shoot via the phloem (Gould et al., 2004). Similar observations have been made in other cereals (White et al., 1987; Vallejo et al., 2005) and in nonmycorrhizal roots of perennials, such as Norway spruce (Häussling et al., 1988).

In contrast to K^+, the uptake of Mg^{2+}, and particularly of Ca^{2+}, is higher in apical than in basal root zones (Ferguson and Clarkson, 1976; Häussling et al., 1988; Marschner and Richter, 1973; White, 2001). This is also shown in Table 2.21. Because Ca^{2+} mobility in the phloem is low, apical cells of the root must meet their Ca^{2+} demand for growth by direct uptake from the external solution. Root apical zones also contribute considerably to Ca^{2+} delivery to the shoot (Table 2.21; Clarkson, 1984; White, 2001). At the root tip, Ca^{2+} may reach the xylem through an exclusively apoplasmic pathway or may be transported across the Casparian band through immature, nonsuberized endodermal cells (Moore et al., 2002; White, 2001). Calcium delivery to the xylem is also high in basal root zones, where lateral roots emerge from the pericycle, disrupting the integrity of the Casparian band (Clarkson, 1984; White, 2001). The apoplasmic pathway is also important for the movement of Na^+, Zn^{2+}, Fe^{2+}, and Cd^{2+} to the xylem (Barberon, 2017; Curie and Mari, 2017; Flam-Shepherd et al., 2018; Lux et al., 2011). The delivery of these elements to the xylem is often greatest at the root tip.

TABLE 2.21 Uptake and translocation of K (^{42}K) and Ca (^{45}Ca) supplied to different zones of the seminal roots of maize.

Nutrient (1 mM)	Accumulation and translocation[a]	Root zone supplied (distance from tip, cm)		
		0–3	6–9	12–15
K	Translocation to shoot	3.8	15	16
	Accumulation in zone of supply	12	3.8	1.9
	Translocation to root tip	–	4.3	2.0
	Total	16	23	20
Ca	Translocation to shoot	2.4	2.2	2.4
	Accumulation in zone of supply	4.1	1.6	0.4
	Translocation to root tip	–	0	0
	Total	6.5	3.8	2.8

[a]Data expressed as micromoles per 12 plants in 24 h.
Source: Based on Marschner and Richter (1973).

TABLE 2.22 Rate of P uptake by various root zones of barley plants after different pretreatment.

Pretreatment for 9 days	P uptake rate [nmol mm^{-3} treated root segment (24 h)$^{-1}$]		
	Root zone (distance from root tip, cm)		
	1	2	5
With P	2.0	1.6	0.97
Without P	3.2	4.5	4.6

Source: Based on Clarkson et al. (1978).

The decline in P_i uptake along the root axis is much less striking than that for Ca^{2+} (Rubio et al., 2004; Kanno et al., 2016). In soil-grown maize, this decline is mainly related to a decrease in root-hair viability and, thus, in absorbing root surface area (Ernst et al., 1989). The gradient in P_i uptake along the root axis also depends on the P nutritional status of the plant and may be reversed under deficiency in favor of the basal zones (Table 2.22). The situation is different under Fe deficiency in Strategy I plants where the apical, but not the basal, root zones increase their capacity for Fe uptake by a factor of up to 100 (Römheld and Marschner, 1981). Apical or subapical root zones generally contribute most to NO_3^- and NH_4^+ uptake by intact plants irrespective of their nutritional status, although the magnitude of the decline in uptake with distance from the root apex depends greatly on root anatomy (Colmer and Bloom, 1998; Reidenbach and Horst, 1997; Sorgona et al., 2010). In maize, the root apex has been shown to function primarily in NO_3^- sensing and signal transduction, whereas the subapical region is responsible for NO_3^- acquisition (Trevisan et al., 2015; Lupini et al., 2016).

Formation of cortical gas spaces (*aerenchyma*), particularly in more basal root zones, can often be observed (Fig. 2.29). The formation of aerenchyma is a typical response to oxygen deficiency in the root zone in plant species adapted to wetland conditions, but it can also be induced, for example, in maize roots under fully aerobic conditions by temporary deprivation of N or P_i supply (He et al., 1992; Lynch, 2007; Saengwilai et al., 2014). Despite these anatomical changes, the basal root zones still have a considerable capacity for ion uptake and for radial transport, indicating that the strands of cells bridging the cortex maintain sufficient ion transport capacity from the rhizodermis to the endodermis (Drew and Saker, 1986; Colmer and Greenway, 2011).

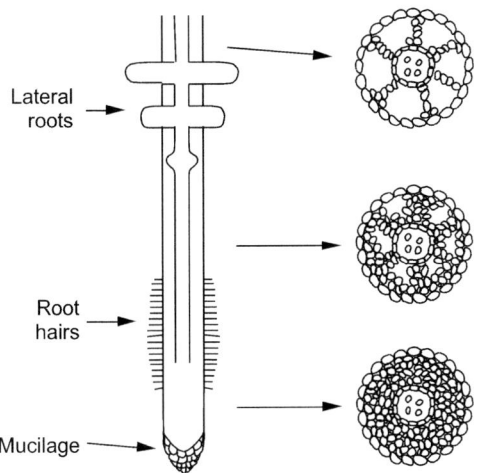

FIGURE 2.29 Schematic representation of anatomical changes along the axis of a maize nodal root. In basal zones there is degeneration of cortical cells and formation of tertiary endodermis.

Water uptake can affect ion uptake both directly (through effects on the rate of radial transport of ions through the apoplasm) and indirectly (by influencing the supply of ions to the plasma membrane of root cells). Water uptake is usually low at the root apex but increases in the elongation zone and reaches a maximum in the root-hair zone, where the endodermis is undergoing suberization (e.g., Häussling et al., 1988; Knipfer et al., 2011; Vetterlein and Doussan, 2016). Water uptake is often reduced strongly following suberization of the endodermis and, particularly, the exodermis. Water can reach the xylem through both the apoplasm and symplasm (Steudle, 2000), the latter pathway being facilitated by aquaporins (Maurel et al., 2015). Aquaporins are found in various membranes of root cells, including the plasma membrane and the tonoplast. Recent data, using mercury to inhibit the activity of aquaporins, suggest that rapid changes in root hydraulic conductivity in response to many stimuli, such as diurnal cycles, nutrient deficiency, salt stress, low temperatures, anoxia, and drought, are the result of changes in cell-membrane permeability achieved by regulation of aquaporin activity (Maurel et al., 2015). The abundance of aquaporins is often greatest in the elongation and mature root zones (Hachez et al., 2006). In these root zones, strong expression of genes encoding aquaporins is observed in the endodermis and exodermis, presumably to allow water to bypass the Casparian bands through a transcellular pathway (Hachez et al., 2006). Similarly, silicon uptake (which is also aquaporin-mediated) is often observed specifically in basal root regions (Coskun et al., 2021).

2.7 Radial transport of ions and water across the root

There are three parallel pathways of movement of solutes and water across the cortex towards the stele (Fig. 2.30): apoplasmic (via cell walls and intercellular spaces), symplasmic (passing from cell to cell via plasmodesmata), and transcellular (passing from cell to cell via membrane transporters expressed in either the distal or proximal membrane surface; i.e., polarly localized). In most of the root, the apoplasmic and transcellular movements to the stele are restricted by the development of the endodermis (Ramakrishna and Barberon, 2019; White, 2001). Initially, the Casparian band joins each endodermal cell (stage I endodermis) and prevents apoplasmic movement of water and solutes. Then, in the basal regions of the root, suberin lamellae cover the entire surface of endodermal cells (stage II endodermis), preventing the uptake of solutes from the apoplasm (Moore et al., 2002). Thick cellulose secondary walls are further deposited over the suberin lamellae, which can also be lignified (stage III endodermis). The nature and extent of these cell-wall modifications are determined by both genetic and environmental factors. The transcellular pathway, mediated by the polar localization of membrane transporters, is becoming increasingly recognized as a major radial transport route for many nutrients (Barberon and Geldner, 2014; Che et al., 2018; Nakamura and Grebe, 2018; Ramakrishna and Barberon, 2019).

In arabidopsis, $B(OH)_3$ influx has been shown to be mediated by AtNIP5;1 channels localized to the soil-facing side of the plasma membrane of epidermal cells and $B(OH)_4^-$ efflux catalyzed by AtBOR1 transporters localized on the stelar side of the plasma membrane of endodermal cells (Takano et al., 2010; Wang et al., 2017; Yoshinari and Takano 2017; Yoshinari et al., 2019). In rice, $Si(OH)_4$ influx is mediated by Lsi1/NIP2;1 channels that are localized to the soil-facing side of the plasma membranes of exodermal and endodermal cells, whereas $Si(OH)_4$ efflux is mediated by Lsi2

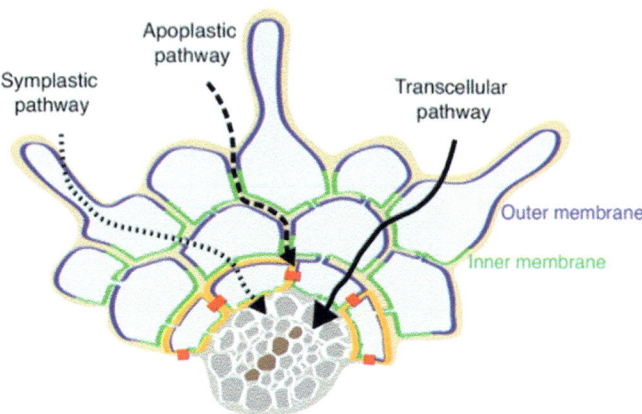

FIGURE 2.30 Segment of a transverse root section showing solute transport via symplastic (through plasmodesmata), apoplastic [between cells, prevented by Casparian bands (in red)], and transcellular [through polarized transporters and prevented by suberin (in yellow)]. *Image taken from Ramakrishna and Barberon (2019).*

transporters localized to the stelar side of the plasma membranes of the same cell types (Coskun et al., 2021; Ma and Yamaji, 2015; Mitani-Ueno and Ma, 2021). Similarly, the Mn^{2+} transporters NRAMP5 (facilitating Mn^{2+} influx) and MTP9 (catalyzing Mn^{2+} efflux) were found to be localized on different sides of the plasma membranes of the same cell types (Sasaki et al., 2012; Ueno et al., 2015). In arabidopsis, Fe^{2+} uptake has been shown to depend on AtIRT1 that is localized to the plasma membrane on the soil-facing side of epidermal cells, but the corresponding efflux transporter has yet to be found (Barberon et al., 2014; Dubeaux et al., 2018; Cointry and Vert, 2019). A similar scenario presents itself for NO_3^- influx through AtNRT2;4 (Kiba et al., 2012). The symplasmic and transcellular pathways allow nutrient uptake and xylem loading to be regulated independently.

In most angiosperms, another apoplasmic barrier, the exodermis, can develop in parallel with the endodermis (Ma and Peterson, 2003). The exodermis develops in the same three stages as the endodermis. Formation of an exodermis is found, for example, in maize, onion (*Allium cepa*), and sunflower (*Helianthus annuus*), but not in broad bean (*Vicia faba*) and pea (*Pisum sativum*) (Enstone and Peterson, 1992). However, there are somewhat different views on the function of the exodermis as an effective barrier for transport of water and solutes in the apoplasm of the root cortex. Termination of the apoplasmic pathway at the exodermis, as suggested by Enstone and Peterson (1992), suggests that the entry of solutes and water into the root symplasm is confined to the rhizodermal cells in basal root zones. Although rhizodermal cells, and in particular root-hair cells, play a key role in the acquisition of mineral nutrients, especially K^+ and P_i (Brown et al., 2013; Gahoonia and Nielsen, 2004; White et al., 2021a; Zhu et al., 2010), the relative importance of the apoplasmic and symplasmic/transcellular pathways for solute transport across the root cortex is unknown. It depends on (1) the external concentration versus the capacity and affinity of the transport system for a particular solute at the plasma membrane of root cells; (2) the root zone considered; and (3) the hydraulic conductivity of the root zone considered and the transpiration rate of the shoot. Depending on environmental conditions and the growth rate of the root, the exodermis can develop within a centimeter of the root apex or remain undeveloped (Ma and Peterson, 2003) and may possess nonsuberized "passage cells" (Holbein et al., 2021; Hose et al., 2001; Peterson and Enstone, 1996). For water, estimates of the contribution of the apoplasmic pathway to radial transport across roots vary between about 10% and 70% (Maurel et al., 2015).

The endodermis is not a perfect barrier to the apoplasmic movement of water and solutes from the cortex to the stele (Fig. 2.30). For instance, passage cells (isolated nonsuberized cells) provide a "bypass route" in some plant species (Andersen et al., 2018b; Holbein et al., 2021; Li et al., 2017c). Moreover, the endodermis may be "leaky" at two other sites along the root axis. At the root apex, where the Casparian band is not fully developed, apoplasmic movement of water and solutes to the stele can occur. However, the movement of some solutes, such as polyvalent cations like Al^{3+}, through the apoplasm of the root apex can be restricted by mucilage formed at the external surface of the rhizodermal cells. An apoplasmic pathway to the stele is also present in basal root zones where the structural continuity of the endodermis is disrupted transiently by the emergence of lateral roots from the pericycle (Faiyue et al., 2010; Krishnamurthy et al., 2011; Ranathunge et al., 2005). This "bypass flow" becomes particularly important for water supply to the shoot at high transpirational demand and in the accumulation of Na^+ and Cl^- in leaves under saline conditions (Flam-Shepherd et al., 2018; Kronzucker and Britto, 2011; Shi et al., 2013; Yeo et al., 1987).

Both genetic and environmental factors influence the movement of water and solutes via the apoplasmic pathway through their effects on the development of the endodermis and exodermis. Accelerated deposition of suberin and lignin

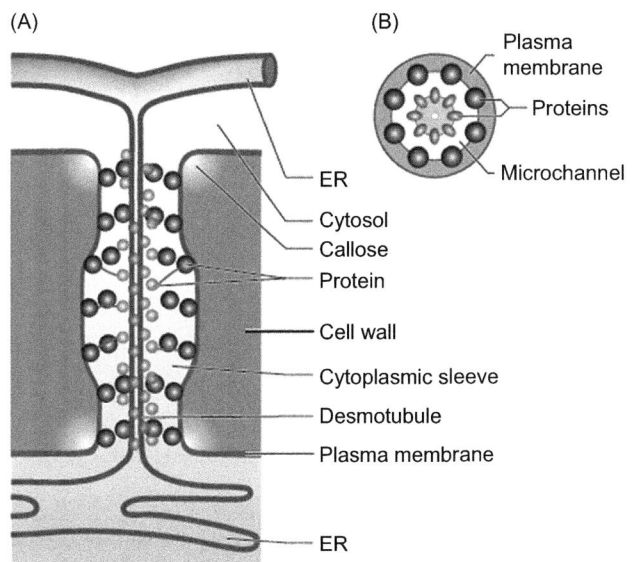

FIGURE 2.31 Schematic representation of (A) longitudinal and (B) transverse sections through plasmodesmata including substructural components. Solute fluxes between adjacent cells occur in the cytoplasmic sleeve, between the plasma membrane and the appressed endoplasmic reticulum (ER) forming the desmotubule. Partial control of solute fluxes occurs by callose deposition in the cell wall. The cytoplasmic sleeve is interrupted by actin and other proteins that create microchannels through which solutes can diffuse. *Modified from Maule (2008).*

restricts the apoplasmic movement of ions and other solutes to the xylem (Andersen, et al., 2015; Barberon, 2017; Krishnamurthy et al., 2009; Lux et al., 2011). Silicon deposition at the endodermis can also block the apoplasmic pathway (Coskun et al., 2019). Moreover, it has recently been shown that changes in apoplasmic barriers can alter the location of transport proteins in root cells and determine the relative extents of apoplasmic and symplasmic/transcellular transport. In a Casparian band-defective arabidopsis mutant (*sgn3*), in which the endodermal bypass is augmented, apoplasmic transport of NH_4^+ decreased (coinciding with a decrease in endodermal AMT1;2 expression) and symplasmic transport increased (coinciding with increased epidermal AMT1;3 expression; Duan et al., 2018). The same mutation (*sgn3*) in *Brassica rapa* resulted in double the Mg^{2+} accumulation and one-third more Ca^{2+} in leaves (Alcock et al., 2021).

The symplasmic pathway plays a key role in delivering most nutrients to the xylem, beginning either at the rhizodermis and the root hairs, at the exodermis, or at the endodermis. Radial transport in the symplasm requires movement through plasmodesmata that connect neighboring root cells (Fig. 2.31). Plasmodesmata have a complex structure (Lucas et al., 2009; Maule, 2008; Peters et al., 2021). The simplest type, which occurs in young tissues, comprises a tube of appressed endoplasmic reticulum (ER) running through the pore, the desmotubule. The transport of solutes and water between cells occurs in the "cytoplasmic sleeve," that is, the cytosol between the desmotubule and the plasma membrane (Fig. 2.31). Protein structures in the cytoplasmic sleeve create microchannels through which solutes can diffuse (Lucas et al., 2009; Maule, 2008; Peters et al., 2021). In more mature tissues, the structure becomes more complex through the addition of branches and the formation of central cavities (Lucas et al., 2009; Maule, 2008; Peters et al., 2021). Plasmodesmata can be closed and opened by the production and degradation of a "collar" of callose (β-1,3-glucan) and they generally have a size-exclusion limit of about 1 kDa, which is regulated physically by the collar and by interactions with cytosolic proteins (Lucas et al., 2009; Maule, 2008; Wu et al., 2018). However, plasmodesmatal microchannels can dilate to allow the passage of solutes larger than 20 kDa. The primary role of plasmodesmata appears to be cell communication, as they regulate the transport of transcription factors and microRNAs that control plant development and responses to biotic and environmental stresses (Lucas et al., 2009; Tilsner et al., 2016). The number of plasmodesmata per cell varies considerably between plant species and cell types (Table 2.23).

The mechanism of symplasmic transport of solutes seems to be chiefly by diffusion, facilitated by radial water flux and cytoplasmic streaming (the intracellular flow of cytoplasm). Understanding the mechanisms for the force generation in cytoplasmic streaming is an active area of research, and recent molecular-genetic approaches have elucidated the role of specific myosin proteins and their movement along cellular actin filaments in this process (Tominaga and Ito, 2015). Genetic modification of myosin proteins not only reduces the velocity of cytoplasmic streaming but also leads to defects in plant development.

During their radial transport through the symplasm, elements can be metabolized and/or sequestered in the vacuoles of root cells. When a nutrient is supplied to roots of a plant that is deficient in that nutrient ("low-salt" roots), it is accumulated in vacuoles of root cells resulting in a delay in its translocation from the roots to the shoots (Fig. 2.32). Thus,

TABLE 2.23 Intracellular K⁺ activity and number of plasmodesmata in tangential walls of hair and hairless cells of the root epidermis.

Plant species	Cell type	K⁺ activity (mM)	Number of plasmodesmata per µm²	per cell junction
Trianea bogotensis	Hair	133	2.1	10,419
	Hairless	74	0.11	693
Raphanus sativus	Hair	129	0.16	273
	Hairless	124	0.07	150

Source: From Vakhmistrov (1981).

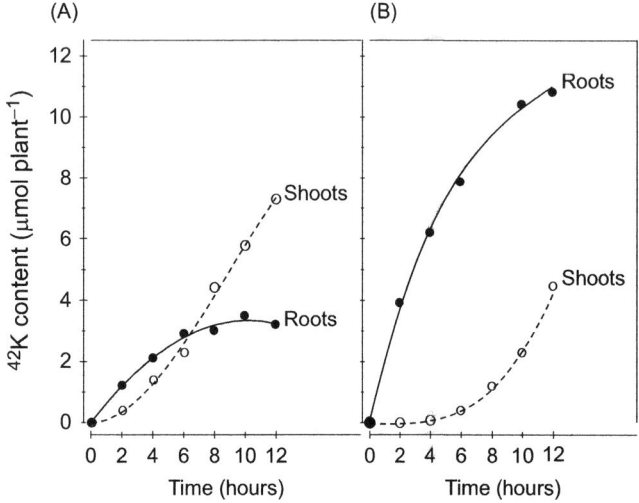

FIGURE 2.32 Accumulation and translocation rates of K⁺ (⁴²K) in barley plants from a solution containing 1 mM KCl (+0.5 mM CaSO₄) after preculture with (A) or without (B) 1 mM KCl.

when the supply of a nutrient is suboptimal, the roots usually have higher tissue concentrations of that particular nutrient than the shoot. In long-term studies, this phenomenon is responsible, in part, for the often-observed shift in the relative growth rates of roots and shoots in favor of the roots under nutrient deficiency.

The vacuoles of root cells also remove potentially toxic elements from the symplasmic pathway. For example, vacuolar sequestration of Na⁺ in the root accounts for the restricted shoot transport of Na⁺ in natrophobic plant species (Chapter 3). Preferential accumulation in roots also restricts the translocation of P_i, Ca, Fe, Cu, Mo, Cd, and Al to the shoot in some circumstances (Conn and Gilliham, 2010; Heuer et al., 2016; Neugebauer et al., 2020). The exchange rate between ions in the vacuoles of cortex cells and those in the symplasm depends on the ion species (e.g., $K^+ > Na^+$; $NO_3^- > SO_4^{2-}$), and the half-time for exchange is generally in the order of at least a few days.

The radial transport of water and solutes is also strongly influenced by maturation of the xylem vessels along the root axis and the presence of an external rhizosheath. Two root zones can be observed in many plant species growing in soil: a "sheathed" zone, which is covered by a layer of strongly adhering soil (the rhizosheath), and a "bare" zone (Fig. 2.33). The development of the rhizosheath requires root hairs (Brown et al., 2017; Burak et al., 2021). In maize, the metaxylem vessels (still having live cells) are nonconducting in the sheathed zones, whereas in the bare zones, the metaxylem is mature (Carminati and Vetterlein, 2013). Accordingly, the hydraulic conductivity of bare roots is greater than that of sheathed roots (Wenzel et al., 1989). This difference in hydraulic conductivity and thus water uptake results in high water contents in the rhizosphere soil of the sheathed zones and low water contents in the rhizosphere soil of the bare zones. The delay in metaxylem maturation not only affects hydraulic conductivity of the roots and plant water relations but also the movement of solutes to the xylem and their translocation to the shoot.

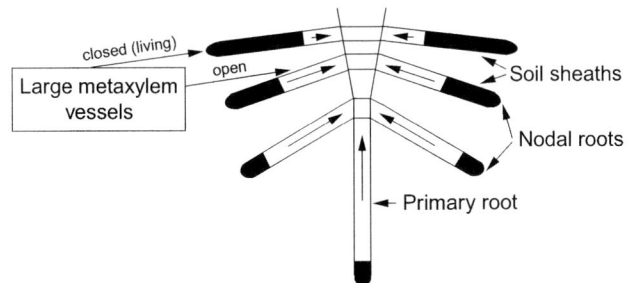

FIGURE 2.33 Model of root hydraulic conductivity and formation of a soil rhizosheath in the root system of maize. *Modified from Wenzel et al. (1989).*

2.8 Release of ions into the xylem

After radial transport through the symplasm to the stele, ions and organic solutes (amino acids, organic acids) are released into the xylem. This release (xylem loading) into fully differentiated, nonliving xylem vessels occurs across the plasma membrane of xylem parenchyma cells. The membrane potential of these cells is slightly negative (Bowling, 1981; Zhu et al., 2007), and the xylem sap has a pH between about 5.2 and 6.0 (Chapter 3). Solutes enter the xylem through ion channels or carriers, if their electrochemical gradients allow this, or their transport is coupled to the proton electrochemical gradient generated by the plasma membrane H^+-ATPase. The xylem parenchyma cells are also responsible for the reabsorption of solutes from the xylem sap by tissues along the pathway to the shoot.

The H^+-ATPases in the plasma membrane of parenchyma cells generate a negative membrane potential and acidify the xylem sap (De Boer and Volkov, 2003). The K^+ electrochemical gradient is often sufficient for K^+ to be loaded into the xylem by voltage-gated, outward-rectifying K^+ channels, such as AtSKOR in arabidopsis (Gaymard et al., 1998) and OsK5.2 in rice (Nguyen et al., 2017). However, K^+/H^+ antiporters for loading K^+ into the xylem have also been discovered, such as AtNRT1.5 (AtNPF7.3) in arabidopsis (Drechsler et al., 2015; Li et al., 2017b) that has also been shown to load NO_3^- into the xylem (Lin et al., 2008). When the NO_3^- concentration in the rhizosphere is low, OsNRT2.3a loads NO_3^- into the xylem of rice (Huang et al., 2019; Tang et al., 2012). In arabidopsis, AtAMT2;1 loads NH_4^+ into the xylem (Giehl et al., 2017). Both NPF2.4 (Li et al., 2016) and SLAH1/SLAH3 heteromeric channels load Cl^- into the xylem (Cubero-Font et al., 2016). Cations present at low concentrations in the root symplasm are loaded into the xylem by active transport mechanisms. Members of the P_{2A} and P_{2B} Ca^{2+}-ATPase families load Ca^{2+} into the xylem, and members of the heavy metal P_{1B}-ATPase family load Zn^{2+} and Cu^{2+} into the xylem (Deng et al., 2013; White and Broadley, 2003, 2009; White and Pongrac, 2017). Members of the magnesium release (MGR) transporters have recently been discovered to load Mg^{2+} into the xylem of arabidopsis (Meng et al., 2022). Boron is loaded into the xylem by orthologues of the arabidopsis AtBOR1 transporter (Miwa and Fujiwara, 2010; Yoshinari and Takano, 2017). In rice, xylem loading of Si is mediated by OsLsi2 and OsLsi3, which are expressed in the endodermis and pericycle, respectively (Huang et al., 2022; Ma et al., 2007).

2.9 Factors governing ion release into the xylem and exudation rate

For technical reasons, it is difficult to measure the release of ions into the xylem directly (Chapter 3). Fluxes of nutrients from the rhizosphere to the xylem are generally estimated by collecting xylem exudate from excised roots.

Because the permeability of plant membranes to water is higher than that to ions, and the endodermis with its Casparian band "seals" the apoplasm of the stele, plant roots behave as osmometers. Ion release into the apoplasm of the stele decreases both the osmotic potential and the water potential (both become more negative) in the stele, and a corresponding net flux of water from the external solution is induced. Water and nutrient flows resulting from such "root pressure" are manifest in intact plants by nighttime guttation from the tips and margins of leaves, which is especially important in Ca nutrition of organs with restricted transpiration, such as enclosed leaves and fruits (Chapter 3). Because of reabsorption along the xylem pathway, and the contribution of solutes from maturing metaxylem vessels, the concentrations of ions at the sites of collection can differ from those at the sites of loading. When interpreting analyses of xylem exudate it should be kept in mind that (1) at least two separately regulated membrane transport processes are involved in symplasmic radial transport of nutrients from the external solution into the xylem (i.e., influx to the symplasm and xylem loading), (2) an apoplasmic pathway can contribute to the delivery of water and solutes to the xylem, and (3) xylem sap volume flow is affected by root hydraulic conductivity and rate of transpiration in intact plants.

As a rule, an increase in the external ion concentration leads to an increase in the concentration of ions in xylem exudate. However, the relative concentration difference decreases as the external concentration is increased (Table 2.24). Thus, the concentration gradient between the external solution and the xylem exudate decreases and can fall even below one (i.e., the concentration of Ca^{2+} in the xylem exudate is lower than that in the external solution). The volume flow of xylem exudation shows a somewhat different pattern and is maximal at an external concentration of 1.0 mM in the experiment reported in Table 2.24. At 0.1 mM, this flow is limited by the ion concentration in the xylem. In contrast, at 10 mM, the flow is limited by water availability (i.e., the low water potential in the external solution) and the small concentration gradient between the external solution and the xylem. The increase in the concentration of nutrients in the xylem exudate with increasing external concentration from 1.0 to 10 mM does not compensate for the decrease in the exudation volume flow. Thus, in contrast to their accumulation in roots, which generally follows a hyperbolic relationship with the external concentration, the rate of root pressure-driven translocation of nutrients to the shoot can decline at high external solute concentrations due to limited water uptake.

An increase in the root-zone temperature often has a greater effect on the exudation volume flow than on the ion concentrations in the exudate (Table 2.25). This is consistent with the expectation that a root behaves as an osmometer. Temperature has a marked effect on the rate of solute uptake, transport through the symplasm and release into the xylem, and water moves along the water potential gradient. Because different transport proteins facilitate the movement of each nutrient across the plasma membrane of root cells, and the relative contributions of symplasmic and apoplasmic transport differ among nutrients, temperature can have differential effects on the delivery of each element to the xylem. For example, an increase in the root temperature results in an increase in K^+ concentration but in a decrease in Ca^{2+} concentration of the exudate (Table 2.25). This shift in the K^+/Ca^{2+} ratio may reflect temperature effects on either membrane transport or the relative contributions of the apoplasmic pathway to radial transport of Ca^{2+} and water (Engels and Marschner, 1992). Shifts in the K^+/Ca^{2+} transport ratio can have important implications for the Ca^{2+} nutrition of plants.

TABLE 2.24 Relationship between external concentration, exudate concentration, and exudate volume flow in decapitated sunflower (*Helianthus annuus*) plants.

External solution KNO₃ and CaCl₂ (mM each)	Exudate (mM)			Exudate concentration/ external concentration			Efflux volume flow [mL (4 h)⁻¹]
	K⁺	Ca²⁺	NO₃⁻	K⁺	Ca²⁺	NO₃⁻	
0.1	7.3	2.8	7.4	73	28	74	4.0
1.0	10	3.2	11	10	3.2	10	4.5
10	17	4.2	10	1.7	0.4	1.0	1.6

TABLE 2.25 Exudate volume flow and K and Ca concentrations in the exudate of decapitated maize (*Zea mays*) plants at different temperatures.

Temperature (°C)	Exudate volume flow (mL h⁻¹)	Exudate concentration (mM)		Ratio K⁺/Ca²⁺
		K⁺	Ca²⁺	
8	1.3	13	1.5	8.9
18	5.5	15	1.0	15
28	7.9	20	0.8	25

Concentration of KNO₃ and CaCl₂ in the external solution: 1 mM each.
Source: From Marschner (1995).

The rate of xylem loading is closely related to root respiration (Table 2.26). A lack of oxygen strongly reduces exudation volume flow but not the concentrations of K^+ and Ca^{2+} in the exudate. Oxygen deficiency seems to affect the release of ions into the xylem and root hydraulic conductivity to a similar degree.

The cation–anion balance in the xylem exudate needs to be maintained. Thus, the accompanying ion can affect xylem exudation flow and sap composition even at low rhizosphere concentrations (Table 2.27). When KNO_3 is supplied, the exudation flow rate is almost twice as high as the flow rate when an equivalent concentration of K_2SO_4 is added. Given that the K^+ concentration in the exudate is the same in both treatments, the translocation rate of K^+ supplied as K_2SO_4 is only about half the rate of K^+ supplied as KNO_3. In contrast to the K^+ concentration, the concentrations of NO_3^- and SO_4^{2-} in the exudate exhibit large differences between treatments (Table 2.27). When plants are supplied SO_4^{2-} rather than NO_3^-, the difference in negative charge in the exudate is compensated for by elevated concentrations of organic acid anions. However, the capacity of the roots to maintain the cation–anion balance by organic acid synthesis in the K_2SO_4 treatment appears to be limited, which leads to a decrease in the rate of K^+ and Ca^{2+} release into the xylem and a corresponding decrease in exudation flow rate when compared to the KNO_3 treatment.

Because of the energy demand for ion transport processes, release of ions into the xylem and the corresponding changes in root pressure are also closely related to the carbohydrate status of the roots (Table 2.28). Variation in the length of the photoperiod 1 day before decapitation affected the carbohydrate status of roots and, correspondingly, the rate and duration of exudation volume flow after decapitation. Both the uptake and transport rate of K^+ in roots with high carbohydrate content are greater than in roots with low carbohydrate content. The higher transport rate is closely related to the exudation volume flow. In roots with low carbohydrate content, reserves are rapidly depleted after

TABLE 2.26 Exudation volume flow and ion concentration in the exudate of decapitated maize plants with (O_2 treatment) or without root respiration (N_2 treatment).

Treatment[a,b]	Exudation volume flow (mL h^{-1})	Exudate concentration (mM)	
		K^+	Ca^{2+}
O_2	8.8	17	1.8
N_2	1.9	15	1.7

[a]Concentration of KNO_3 and $CaCl_2$ in the external solution: 0.5 mM each.
[b]Respiration treatment consisted of bubbling oxygen or nitrogen through the external solution.
Source: From Marschner (1995).

TABLE 2.27 Flow rate and ion concentration in the xylem exudate of wheat seedlings.

Parameter	Treatment	
	KNO_3	K_2SO_4
Exudation flow rate [μL h^{-1} (50 plants)$^{-1}$]	372	180
Ion concentration (mM)		
Potassium	23	25
Calcium	9.1	9.5
Nitrate	18	0.0
Sulfate	0.2	0.8
Organic acids	9.6	26

Seedlings were supplied with either KNO_3 (1 mM) or K_2SO_4 (0.5 mM) in the presence of 0.2 mM $CaSO_4$.
Source: From Triplett et al. (1980).

TABLE 2.28 Relationship between photoperiod, carbohydrate content of roots, and uptake and translocation of K in decapitated maize plants.[a]

Parameter	Photoperiod (h) 12/12[b]	24/0
Carbohydrate in roots (mg)	122 (48)[c]	328 (226)
Total K uptake (mmol)	1.3	5.0
Potassium translocation in exudation volume flow (mmol)	1.0	3.5
Exudation volume flow [mL (8 h)$^{-1}$]	30	89
Relative decline in flow rate within 8 h (%)	60	12

[a]Data per 12 plants.
[b]Hours of light/hours of darkness. The pretreatment with different daylengths was for 1 day (i.e., the day prior to decapitation).
[c]Numbers in parentheses denote carbohydrate content after 8 h (decapitation).
Source: From Marschner (1995).

decapitation, and there is a corresponding decline in the rate of exudation volume flow. This depletion of carbohydrates in the roots of decapitated plants and the consequent decline in xylem exudation flow is one of the factors restricting studies of xylem loading.

It is important to recognize that xylem exudate collected from decapitated plants represents only the root pressure component of xylem volume flow in an intact plant. To evaluate the transpirational component, xylem sap should be collected from intact plants in situ. This can be achieved using insects, such as the meadow spittlebug (*Philaenus spumarius*), that feed on xylem sap (e.g., Malone et al., 2002; Teakle et al., 2007; Watson et al., 2001), but this technique is rarely used. Estimates of fluxes of nutrients across the root to the xylem differ between excised roots and roots of intact plants (Allen et al., 1988; Salim and Pitman, 1984; White, 1997b). In addition, irrespective of the collection method, xylem sap also contains shoot-derived nutrients recycled in the phloem and reloaded into the xylem in the roots, which can account for a significant proportion of the K^+, P_i, N, and S in the xylem sap (Chapter 3). The proportion of recycled nutrients in the xylem sap depends on various factors, including plant species and plant nutritional status and shoot demand for nutrients for growth.

References

Aarabi, F., Naake, T., Fernie, A.R., Hoefgen, R., 2020. Coordinating sulfur pools under sulfate deprivation. Trends Plant Sci. 25, 1227–1239.

Alberts, B., Bray, D., Hopkin, K., Johnson, A., Lewis, J., Raff, M., et al., 1998. Essential cell biology, Garland Science, second ed. Taylor and Francis Group, Boca Raton, FL.

Alcock, T.D., Thomas, C.L., Ó Lochlainn, S., Pongrac, P., Wilson, M., Moore, C., et al., 2021. Magnesium and calcium overaccumulate in the leaves of a *schengen3* mutant of *Brassica rapa*. Plant Physiol. 86, 1616–1631.

Alejandro, S., Höller, S., Meier, B., Peiter, E., 2020. Manganese in plants: from acquisition to subcellular allocation. Front. Plant Sci. 11, 300. Available from: https://doi.org/10.3389/fpls.2020.00300.

Allan, D.L., Jarrell, W.M., 1989. Proton and copper adsorption to maize and soybean root cell walls. Plant Physiol. 89, 823–832.

Allen, S., Raven, J.A., Sprent, J.I., 1988. The role of long-distance transport in intracellular pH regulation in *Phaseolus vulgaris* grown with ammonium or nitrate as nitrogen source, or nodulated. J. Exp. Bot. 39, 513–528.

Amini, S., Arsova, B., Hanikenne, M., 2022. The molecular basis of zinc homeostasis in cereals. Plant Cell Environ. 45, 1339–1361.

Andersen, T.G., Barberon, M., Geldner, N., 2015. Suberization – the second life of an endodermal cell. Curr. Opin. Plant Biol. 28, 9–15.

Andresen, E., Peiter, E., Küpper, H., 2018a. Trace metal metabolism in plants. J. Exp. Bot. 69, 909–954.

Andersen, T.G., Naseer, S., Ursache, R., Wybouw, B., Smet, W., De Rybel, B., et al., 2018b. Diffusible repression of cytokinin signalling produces endodermal symmetry and passage cells. Nature 555, 529–533.

Andersen, T.G., Molina, D., Kilian, J., Franke, R.B., Ragni, L., Geldner, N., 2021. Tissue-autonomous phenylpropanoid production is essential for establishment of root barriers. Curr. Biol. 31, 965–977.

Asher, C.J., Edwards, D.G., 1983. Modern solution culture techniques. In: Läuchli, A., Bieleski, R.L. (Eds.), Encyclopedia of Plant Physiology, New Series, 15A. Springer-Verlag, Berlin and New York, pp. 94–119.

Axelsen, K.B., Palmgren, M.G., 2001. Inventory of the superfamily of P-type ion pumps in Arabidopsis. Plant Physiol. 126, 696–706.

Balkos, K.D., Britto, D.T., Kronzucker, H.J., 2010. Optimization of ammonium acquisition and metabolism by potassium in rice (*Oryza sativa* L. cv. IR-72). Plant Cell Environ. 33, 23−34.

Barber, S.A., 1979. Growth requirement for nutrients in relation to demand at the root surface. In: Harley, J.L., Scott-Russell, R. (Eds.), The Soil-Root Interface. Academic Press, London and Orlando, pp. 5−20.

Barberon, M., 2017. The endodermis as a checkpoint for nutrients. New Phytol. 213, 1604−1610.

Barberon, M., Geldner, N., 2014. Radial transport of nutrients: the plant root as a polarized epithelium. Plant Physiol. 166, 528−537.

Barberon, M., Dubeaux, G., Kolb, C., Isono, E., Zelazny, E., Vert, G., 2014. Polarization of IRON-REGULATED TRANSPORTER 1 (IRT1) to the plant-soil interface plays crucial role in metal homeostasis. Proc. Natl. Acad. Sci. USA 111, 8293−8298.

Barberon, M., Vermeer, J.E.M., De Bellis, D., Wang, P., Naseer, S., Andersen, T.G., et al., 2016. Adaptation of root function by nutrient-induced plasticity of endodermal differentiation. Cell 164, 447−459.

Bell, P.F., Chaney, R.L., Angle, J.S., 1991. Free metal activity and total metal concentrations as indices of micronutrient availability to barley (*Hordeum vulgare* (L.) Klages). Plant Soil 130, 51−62.

Bell, C.I., Clarkson, D.T., Cram, W.J., 1995. Sulfate supply and its regulation of transport in roots of a tropical legume *Macroptilium atropurpureum* cv Siratro. J. Exp. Bot. 46, 65−71.

Bellegarde, F., Gojon, A., Martin, A., 2017. Signals and players in the transcriptional regulation of root responses by local and systemic N signaling in *Arabidopsis thaliana*. J. Exp. Bot. 68, 2553−2565.

Bossi, J.G., Kumar, K., Barberini, M.L., Domínguez, G.D., Rondón Guerrero, Y.D.C., Marino-Buslje, C., et al., 2020. The role of P-type IIA and P-type IIB Ca^{2+}-ATPases in plant development and growth. J. Exp. Bot. 71, 1239−1248.

Bowling, D.J.F., 1981. Release of ions to the xylem in roots. Physiol. Plant. 53, 392−397.

Bravo, F.P., Uribe, E.G., 1981. Temperature dependence of the concentration kinetics of absorption of phosphate and potassium in corn roots. Plant Physiol. 67, 815−819.

Breeze, V.G., Wild, A., Hopper, M.J., Jones, L.H.P., 1984. The uptake of phosphate by plants from flowing nutrient solution. II. Growth of *Lolium perenne* L. at constant phosphate concentrations. J. Exp. Bot 35, 1210−1221.

Breteler, H., Siegerist, M., 1984. Effect of ammonium on nitrate utilization by roots of dwarf bean. Plant Physiol. 75, 1099−1103.

Britto, D.T., Kronzucker, H.J., 2001. Can unidirectional influx be measured in higher plants? A mathematical approach using parameters from efflux analysis. New Phytol. 150, 37−47.

Britto, D.T., Kronzucker, H.J., 2002. NH_4^+ toxicity in higher plants: a critical review. J. Plant Physiol. 159, 567−584.

Britto, D.T., Kronzucker, H.J., 2003. Ion fluxes and cytosolic pool sizes: examining fundamental relationships in transmembrane flux regulation. Planta 217, 490−497.

Britto, D.T., Kronzucker, H.J., 2005. Nitrogen acquisition, PEP carboxylase, and cellular pH homeostasis: new views on old paradigms. Plant Cell Environ. 28, 1396−1409.

Britto, D.T., Kronzucker, H.J., 2006. Futile cycling at the plasma membrane: a hallmark of low-affinity nutrient transport. Trends Plant Sci. 11, 529−534.

Britto, D.T., Kronzucker, H.J., 2009. Ussing's conundrum and the search for transport mechanisms in plants. New Phytol. 183, 243−246.

Britto, D.T., Kronzucker, H.J., 2013. Ecological significance and complexity of N-source preference in plants. Ann. Bot. 112, 957−963.

Britto, D.T., Kronzucker, H.J., 2015. Sodium efflux in plant roots: what do we really know? J. Plant Physiol. 186/187, 1−12.

Britto, D.T., Ruth, T.J., Lapi, S., Kronzucker, H.J., 2004. Cellular and whole-plant chloride dynamics in barley: insights into chloride-nitrogen interactions and salinity responses. Planta 218, 615−622.

Britto, D.T., Ebrahimi-Ardebili, S., Hamam, A.M., Coskun, D., Kronzucker, H.J., 2010. ^{42}K analysis of sodium-induced potassium efflux in barley: mechanism and relevance to salt tolerance. New Phytol. 186, 373−384.

Britto, D.T., Coskun, D., Kronzucker, H.J., 2021. Potassium physiology from Archean to Holocene: a higher-plant perspective. J. Plant Physiol. 262, 153432. Available from: https://doi.org/10.1016/j.jplph.2021.153432.

Broadley, M.R., White, P.J., Hammond, J.P., Zelko, I., Lux, A., 2007. Zinc in plants. New Phytol. 173, 677−702.

Brown, L.K., George, T.S., Dupuy, L., White, P.J., 2013. A conceptual model of root hair ideotypes for future agricultural environments: what combination of traits should be targeted to cope with limited P availability? Ann. Bot. 112, 317−330.

Brown, L.K., George, T.S., Neugebauer, K., White, P.J., 2017. The rhizosheath − a potential trait for future agricultural sustainability occurs in orders throughout the angiosperms. Plant Soil 418, 115−128.

Brumbarova, T., Bauer, P., Ivanov, R., 2015. Molecular mechanisms governing Arabidopsis iron uptake. Trends Plant Sci. 20, 124−133.

Buchner, P., Takahashi, H., Hawkesford, M.J., 2004. Plant sulphate transporters: co-ordination of uptake, intracellular and long-distance transport. J. Exp. Bot. 55, 1765−1773.

Buhtz, A., Pieritz, J., Springer, F., Kehr, J., 2010. Phloem small RNAs, nutrient stress responses, and systemic mobility. BMC Plant Biol. 10, 64. Available from: https://doi.org/10.1186/1471-2229-10-64.

Burak, E., Quinton, J.N., Dodd, I.C., 2021. Root hairs are the most important root trait for rhizosheath formation of barley (*Hordeum vulgare*), maize (*Zea mays*) and *Lotus japonicus* (Gifu). Ann. Bot. 128, 45−57.

Burton, R.A., Gidley, M.J., Fincher, G.B., 2010. Heterogeneity in the chemistry, structure and function of plant cell walls. Nat. Chem. Biol. 6, 724−732.

Cacas, J.L., Buré, C., Grosjean, K., Gerbeau-Pissot, P., Lherminier, J., Rombouts, Y., et al., 2016. Revisiting plant plasma membrane lipids in tobacco: a focus on sphingolipids. Plant Physiol. 170, 367−384.

Cakmak, I., Marschner, H., 1990. Decrease in nitrate uptake and increase in proton release in zinc deficient cotton, sunflower and buckwheat plants. Plant Soil 129, 261–268.

Carminati, A., Vetterlein, D., 2013. Plasticity of rhizosphere hydraulic properties as a key for efficient utilization of scarce resources. Ann. Bot. 112, 277–290.

Cassim, A.M., Gouguet, P., Gronnier, J., Laurent, N., Germain, V., Grison, M., et al., 2019. Plant lipids: key players of plasma membrane organization and function. Prog. Lipid Res. 73, 1–27.

Causin, H.F., Barneix, A.J., 1993. Regulation of NH_4^+ uptake in wheat plants: effect of root ammonium concentration and amino acids. Plant Soil 151, 211–218.

Cesco, S., Nikolic, M., Römheld, V., Varanini, Z., Pinton, R., 2002. Uptake of ^{59}Fe from soluble ^{59}Fe-humate complexes by cucumber and barley plants. Plant Soil 241, 121–128.

Chang, K., Roberts, J.K.M., 1992. Quantitation of rates of transport, metabolic fluxes, and cytoplasmic levels of inorganic carbon in maize root tips during K^+ ion uptake. Plant Physiol. 99, 291–297.

Che, J., Yamaji, N., Ma, J.F., 2018. Efficient and flexible uptake system for mineral elements in plants. New Phytol. 219, 513–517.

Chen, Y.-T., Wang, Y., Yeh, K.-C., 2017. Role of root exudates in metal acquisition and tolerance. Curr. Opin. Plant Biol. 39, 66–72.

Chen, A., Husted, S., Salt, D.E., Schjoerring, J.K., Persson, D.P., 2019. The intensity of manganese deficiency strongly affects root endodermal suberization and ion homeostasis. Plant Physiol. 181, 729–742.

Cheung, A.Y., Wu, H.-M., 2008. Structural and signaling networks for the polar cell growth machinery in pollen tubes. Annu. Rev. Plant Biol. 59, 547–572.

Christophersen, H.M., Smith, F.A., Smith, S.E., 2009. Arbuscular mycorrhizal colonization reduces arsenate uptake in barley via downregulation of transporters in the direct epidermal phosphate uptake pathway. New Phytol. 184, 962–974.

Clark, R.B., 1982. Nutrient solution growth of sorghum and corn in mineral nutrition studies. J. Plant Nutr. 5, 1039–1057.

Clark, G.B., Sessions, A., Eastburn, D.J., Roux, S.J., 2001. Differential expression of members of the annexin multigene family in Arabidopsis. Plant Physiol. 126, 1072–1084.

Clarkson, D.T., 1984. Calcium transport between tissues and its distribution in the plant. Plant Cell Environ. 7, 449–456.

Clarkson, D.T., Scattergood, C.B., 1982. Growth and phosphate transport in barley and tomato plants during the development of, and recovery from, phosphate stress. J. Exp. Bot. 33, 865–875.

Clarkson, D.T., Saker, L.R., 1989. Sulphate influx in wheat and barley roots becomes more sensitive to specific protein-binding reagents when plants are sulphate-deficient. Planta 178, 249–257.

Clarkson, D.T., Sanderson, J., Scattergood, C.B., 1978. Influence of phosphate-stress and phosphate absorption and translocation by various parts of the root system of *Hordeum vulgare* L. (barley). Planta 139, 47–53.

Clarkson, D.T., Robards, A.W., Stephens, J.E., Stark, M., 1987. Suberin lamellae in the hypodermis of maize (*Zea mays*) roots; development and factors affecting the permeability of hypodermal layers. Plant Cell Environ. 10, 83–93.

Clarkson, D.T., Earnshaw, M.J., White, P.J., Cooper, H.D., 1988. Temperature dependent factors influencing nutrient uptake: an analysis of responses at different levels of organization. In: Long, S.P., Woodward, F.I. (Eds.), Symposium of the Society for Experimental Biology 42, Plants and Temperature. Company of Biologists, Cambridge, pp. 281–309.

Clement, C.R., Hopper, M.J., Jones, L.H.P., Leafe, E.L., 1978. The uptake of nitrate by *Lolium perenne* from flowing nutrient solution. II. Effect of light, defoliation, and relationship to CO_2 flux. J. Exp. Bot. 29, 1173–1183.

Cointry, V., Vert, G., 2019. The bifunctional transporter-receptor IRT1 at the heart of metal sensing and signalling. New Phytol. 223, 1173–1178.

Cole, R.A., Fowler, J.E., 2006. Polarized growth: maintaining focus on the tip. Curr. Opin. Plant Biol. 9, 579–588.

Colmenero-Flores, J.M., Martínez, G., Gamba, G., Vázquez, N., Iglesias, D.J., Brumós, J., et al., 2007. Identification and functional characterization of cation–chloride cotransporters in plants. Plant J. 50, 278–292.

Colmer, T.D., Bloom, A.J., 1998. A comparison of NH_4^+ and NO_3^- net fluxes along roots of rice and maize. Plant Cell Environ. 21, 240–246.

Colmer, T., Greenway, H., 2011. Ion transport in seminal and adventitious roots of cereals during O_2 deficiency. J. Exp. Bot. 62, 39–57.

Conn, S., Gilliham, M., 2010. Comparative physiology of elemental distributions in plants. Ann. Bot. 105, 1081–1102.

Cosgrove, D.J., 2018. Diffuse growth of plant cell walls. Plant Physiol. 176, 16–27.

Coskun, D., Kronzucker, H.J., 2013. Complexity of potassium acquisition: how much flows through channels? Plant Signal. Behav. 8, e24799. Available from: https://doi.org/10.4161/psb.24799.

Coskun, D., Britto, D.T., Kronzucker, H.J., 2010. Regulation and mechanism of potassium release from barley roots: an *in planta* $^{42}K^+$ analysis. New Phytol. 188, 1028–1038.

Coskun, D., Britto, D.T., Li, M., Oh, S., Kronzucker, H.J., 2013a. Capacity and plasticity of potassium channels and high-affinity transporters in roots of barley and Arabidopsis. Plant Physiol. 162, 496–511.

Coskun, D., Britto, D.T., Li, M., Becker, A., Kronzucker, H.J., 2013b. Rapid ammonia gas transport accounts for futile transmembrane cycling under NH_3/NH_4^+ toxicity in plant roots. Plant Physiol. 163, 1859–1867.

Coskun, D., Britto, D.T., Jean, Y.K., Kabir, I., Tolay, I., Torun, A.A., et al., 2013c. K^+ efflux and retention in response to NaCl stress do not predict salt tolerance in contrasting genotypes of rice (*Oryza sativa* L.). PLoS One 8, e57767. Available from: https://doi.org/10.1371/journal.pone.0057767.

Coskun, D., Britto, D.T., Hamam, A.M., Kronzucker, H.J., 2014. Measuring fluxes of mineral nutrients and toxicants in plants with radioactive tracers. J. Vis. Exp. 90, e51877. Available from: https://doi.org/10.3791/51877.

Coskun, D., Britto, D.T., Kochian, L.V., Kronzucker, H.J., 2016. How high do ion fluxes go? A re-evaluation of the two-mechanism model of K$^+$ transport in plant roots. Plant Sci. 243, 96–104.

Coskun, D., Britto, D.T., Kronzucker, H.J., 2017. The nitrogen–potassium intersection: membranes, metabolism, and mechanism. Plant Cell Environ. 40, 2029–2041.

Coskun, D., Deshmukh, R., Sonah, H., Menzies, J.G., Reynolds, O., Ma, J.F., et al., 2019. The controversies of silicon's role in plant biology. New Phytol. 221, 67–85.

Coskun, D., Deshmukh, R., Shivaraj, S.M., Isenring, P., Bélanger, R.R., 2021. Lsi2: a black box in plant silicon transport. Plant Soil 466, 1–20.

Cosse, M., Seidel, T., 2021. Plant proton pumps and cytosolic pH-homeostasis. Front. Plant Sci. 12, 672873. Available from: https://doi.org/10.3389/fpls.2021.672873.

Cram, W.J., 1973. Internal factors regulating nitrate and chloride influx in plant cells. J. Exp. Bot. 24, 328–341.

Cramer, G.R., Läuchi, A., Polito, V.S., 1985. Displacement of Ca^{2+} by Na$^+$ from the plasmalemma of root cells. A primary response of salt stress. Plant Physiol. 79, 207–211.

Cramer, M.D., Hawkins, H.J., Verboom, G.A., 2009. The importance of nutritional regulation of plant water flux. Oecologia 161, 15–24.

Cubero-Font, P., Maierhofer, T., Jaslan, J., Rosales, M.A., Espartero, J., Díaz-Rueda, P., et al., 2016. Silent S-type anion channel subunit SLAH1 gates SLAH3 open for chloride root-to-shoot translocation. Curr. Biol. 26, 2213–2220.

Curie, C., Cassin, G., Couch, D., Divol, F., Higuchi, K., Le Jean, M., et al., 2009. Metal movement within the plant: contribution of nicotianamine and yellow stripe 1-like transporters. Ann. Bot. 103, 1–11.

Curie, C., Mari, S., 2017. New routes for plant iron mining. New Phytol. 214, 521–525.

Dannel, F., Pfeffer, H., Römheld, V., 2000. Characterization of root boron pools, boron uptake and boron translocation in sunflower using the stable isotopes ^{10}B and ^{11}B. Aust. J. Plant Physiol. 27, 397–405.

Davies, D.D., 1986. The fine control of cytosolic pH. Physiol. Plant. 67, 702–706.

Davies, J.M., 2014. Annexin-mediated calcium signalling in plants. Plants 3, 128–140.

De Angeli, A., Monachello, D., Ephritikhine, G., Frachisse, J.M., Thomine, S., Gambale, F., et al., 2006. The nitrate/proton antiporter AtCLCa mediates nitrate accumulation in plant vacuoles. Nature 442, 939–942.

De Boer, A.H., Volkov, V., 2003. Logistics of water and salt transport through the plant: structure and functioning of the xylem. Plant Cell Environ. 26, 87–101.

De Michele, R., Loqué, D., Lalonde, S., Frommer, W.B., 2012. Ammonium and urea transporter inventory of the *Selaginella* and *Physcomitrella* genomes. Front. Plant Sci. 3, 62. Available from: https://doi.org/10.3389/fpls.2012.00062.

Deane-Drummond, C.E., 1987. The regulation of sulphate uptake following growth of *Pisum sativum* L. seedlings in S nutrient limiting conditions. Interaction between nitrate and sulphate transport. Plant Sci. 50, 27–35.

Deane-Drummond, C.E., Chaffey, N.J., 1985. Characteristics of nitrate uptake into seedlings of pea (*Pisum sativum* L. cv. Faltham First). Changes in net NO$_3^-$ uptake following inoculation with *Rhizobium* and growth in low nitrate concentrations. Plant Cell Environ. 8, 517–523.

Degryse, F., Smolders, E., Parker, D.R., 2006. Metal complexes increase uptake of Zn and Cu by plants: implications for uptake and deficiency studies in chelator-buffered solutions. Plant Soil 289, 171–185.

Delhaize, E., Gruber, B.D., Ryan, P.R., 2007. The roles of organic anion permeases in aluminium resistance and mineral nutrition. FEBS Lett. 581, 2255–2262.

Demidchik, V., Shabala, S., Isayenkov, S., Cuin, T.A., Pottosin, I., 2018. Calcium transport across plant membranes: mechanisms and functions. New Phytol. 220, 49–69.

Deng, F., Yamaji, N., Xia, J., Ma, J.F., 2013. A member of the heavy metal P-type ATPase OsHMA5 is involved in xylem loading of copper in rice. Plant Physiol. 163, 1353–1362.

Doblas, V.G., Geldner, N., Barberon, M., 2017. The endodermis, a tightly controlled barrier for nutrients. Curr. Opin. Plant Biol. 39, 136–143.

Drechsler, N., Zheng, Y., Bohner, A., Nobmann, B., von Wirén, N., Kunze, R., et al., 2015. Nitrate-dependent control of shoot K homeostasis by the nitrate transporter1/peptide transporter family member NPF7.3/NRT1.5 and the stelar K$^+$ outward rectifier SKOR in Arabidopsis. Plant Physiol. 169, 2832–2847.

Drew, M.C., Saker, L.R., 1986. Ion transport to the xylem in aerenchymatous roots of *Zea mays* L. J. Exp. Bot. 37, 22–33.

Drew, M.C., Saker, L.R., Barber, S.A., Jenkins, W., 1984. Changes in the kinetics of phosphate and potassium absorption in nutrient-deficient barley roots measured by a solution-depletion technique. Planta 160, 490–499.

Dreyer, I., 2017. Plant potassium channels are in general dual affinity uptake systems. AIMS Biophysics 4, 90–106.

Dreyer, I., 2021. Nutrient cycling is an important mechanism for homeostasis in plant cells. Plant Physiol. 187, 2246–2261.

Dreyer, I., Michard, E., 2020. High- and low-affinity transport in plants from a thermodynamic point of view. Front. Plant Sci. 10, 1797. Available from: https://doi.org/10.3389/fpls.2019.01797.

Duan, G., Hakoyama, T., Kamiya, T., Miwa, H., Lombardo, F., Sato, S., et al., 2017. LjMOT1, a high-affinity molybdate transporter from *Lotus japonicus*, is essential for molybdate uptake, but not for the delivery to nodules. Plant J. 90, 1108–1119.

Duan, F., Giehl, R.F.H., Geldner, N., Salt, D.E., von Wirén, N., 2018. Root zone–specific localization of AMTs determines ammonium transport pathways and nitrogen allocation to shoots. PLoS Biol. 16, e2006024. Available from: https://doi.org/10.1371/journal.pbio.2006024.

Dubeaux, G., Neveu, J., Zelazny, E., Vert, G., 2018. Metal sensing by the IRT1 transporter-receptor orchestrates its own degradation and plant metal nutrition. Mol. Cell 69, 953–964.

Dunlop, J., Bowling, D.J.F., 1978. Uptake of phosphate by white clover. II. The effect of pH on the electrogenic phosphate pump. J. Exp. Bot. 29, 1147–1153.

El Mehdawi, A.F., Jiang, Y., Guignardi, Z.S., Esmat, A., Pilon, M., Pilon-Smits, E.A.H., et al., 2017. Influence of sulfate supply on selenium uptake dynamics and expression of sulfate/selenate transporters in selenium hyperaccumulator and nonhyperaccumulator Brassicaceae. New Phytol. 217, 194–205.

Emery, L., Whelan, S., Hirschi, K.D., Pittman, J.K., 2012. Protein phylogenetic analysis of Ca^{2+}/cation antiporters and insights into their evolution in plants. Front. Plant Sci. 3, 1. Available from: https://doi.org/10.3389/fpls.2012.00001.

Engels, C., 1993. Differences between maize and wheat in growth-related nutrient demand and uptake of potassium and phosphorus at suboptimal root-zone temperatures. Plant Soil 150, 129–138.

Engels, C., Marschner, H., 1992. Root to shoot translocation of macronutrients in relation to shoot demand in maize (*Zea mays* L.) grown at different root zone temperatures. Z. Pflanzenernähr. Bodenk. 155, 121–128.

Engels, C., Marschner, H., 1993. Influence of the form of nitrogen supply on root uptake and translocation of cations in the xylem exudate of maize (*Zea mays* L.). J. Exp. Bot. 44, 1695–1701.

Enstone, D.E., Peterson, C.A., 1992. The apoplastic permeability of root apices. Can. J. Bot. 70, 1502–1512.

Epstein, E., 1972. Mineral Nutrition of Plants: Principles and Perspectives. Wiley, New York, NY.

Epstein, E., 1999. Silicon. Annu. Rev. Plant Physiol. Plant Mol. Biol. 50, 641–664.

Epstein, E., Rains, D.W., Elzam, O.E., 1963. Resolution of dual mechanisms of potassium absorption by barley roots. Proc. Natl. Acad. Sci. USA 49, 684–692.

Ernst, M., Römheld, V., Marschner, H., 1989. Estimation of phosphorus uptake capacity by different zones of the primary root of soil-grown maize (*Zea mays* L.). Z. Pflanzenernähr. Bodenk. 152, 21–25.

Fageria, N.K., 2009. The Use of Nutrients in Crop Plants. CRC Press, Boca Raton, FL.

Faiyue, B., Al-Azzawi, M.J., Flowers, T.J., 2012. A new screening technique for salinity resistance in rice (*Oryza sativa* L.) seedlings using bypass flow. Plant Cell Environ. 35, 1099–1108.

Faiyue, B., Vijayalakshmi, C., Nawaz, S, Nagato, Y., Taketa, S., Ichi, M., et al., 2010. Studies on sodium bypass flow in lateral rootless mutants *lrt1* and *lrt2*, and crown rootless mutant *crl1* of rice (*Oryza sativa* L.). Plant Cell Environ. 33, 687–701.

Falhof, J., Pedersen, J.T., Fuglsang, A.T., Palmgren, M., 2016. Plasma membrane H^+-ATPase regulation in the center of plant physiology. Mol. Plant 9, 323–337.

Fan, X., Naz, M., Fan, X., Xuan, W., Miller, A.J., Xu, G., 2017. Plant nitrate transporters: from gene function to application. J. Exp. Bot. 68, 2463–2475.

Felle, H.H., 2001. pH: signal and messenger in plant cells. Plant Biol. 3, 577–591.

Ferguson, I.B., Clarkson, D.T., 1976. Simultaneous uptake and translocation of magnesium and calcium in barley (*Hordeum vulgare* L.) roots. Planta 128, 267–269.

Fitzpatrick, K.L., Tyerman, S.D., Kaiser, B.N., 2008. Molybdate transport through the plant sulfate transporter SHST1. FEBS Lett. 582, 1508–1513.

Flam-Shepherd, R., Huynh, W.Q., Coskun, D., Hamam, A.M., Britto, D.T., Kronzucker, H.J., 2018. Membrane fluxes, bypass flows, and sodium stress in rice: the influence of silicon. J. Exp. Bot. 69, 1679–1692.

Forde, B.G., 2014. Glutamate signalling in roots. J. Exp. Bot. 65, 779–787.

Forgac, M., 2007. Vacuolar ATPases: rotary proton pumps in physiology and pathophysiology. Nat. Rev. Mol. Cell Biol. 8, 917–929.

Fox, T.C., Shaff, J.E., Grusak, M.A., Norvell, W.A., Chen, Y., Chaney, R.L., et al., 1996. Direct measurement of ^{59}Fe-labeled Fe^{2+} influx in roots of pea using a chelator buffer system to control free Fe^{2+} in solution. Plant Physiol. 111, 93–100.

Gahoonia, T.S., Nielsen, N.E., 2004. Barley genotypes with long root hairs sustain high grain yields in low-P field. Plant Soil 262, 55–62.

Garnett, T., Conn, V., Kaiser, B.N., 2009. Root based approaches to improving nitrogen use efficiency in plants. Plant Cell Environ. 32, 1272–1283.

Gaxiola, R.A., Palmgren, M.G., Schumacher, K., 2007. Plant proton pumps. FEBS Lett. 581, 2204–2214.

Gaymard, F., Pilot, G., Lacombe, B., Bouchez, D., Bruneau, D., Boucherez, J., et al., 1998. Identification and disruption of a plant shaker-like outward channel involved in K^+ release into the xylem sap. Cell 94, 647–655.

Geldner, N., 2013. The endodermis. Annu. Rev. Plant Biol. 64, 531–558.

Gerendás, J., Ratcliffe, R.G., Sattelmacher, B., 1990. ^{31}P nuclear magnetic resonance evidence for differences in intracellular pH in the roots of maize seedlings grown with nitrate or ammonium. J. Plant Physiol. 137, 125–128.

Giehl, R.F.H., Gruber, B.D., von Wirén, N., 2014. It's time to make changes: modulation of root system architecture by nutrient signals. J. Exp. Bot. 65, 769–778.

Giehl, R.F.H., Laginha, A.M., Duan, F., Rentsch, D., Yuan, L., von Wirén, N., 2017. A critical role of AMT2;1 in root-to-shoot translocation of ammonium in Arabidopsis. Mol. Plant 10, 1449–1460.

Gigolashvili, T., Kopriva, S., 2014. Transporters in plant sulfur metabolism. Front. Plant Sci. 5, 442. Available from: https://doi.org/10.3389/fpls.2014.00442.

Girin, T., El-Kafafi, E.-S., Widiez, T., Erban, A., Hubberten, H.-M., Kopka, J., et al., 2010. Identification of Arabidopsis mutants impaired in the systemic regulation of root nitrate uptake by the nitrogen status of the plant. Plant Physiol. 153, 1250–1260.

Glass, A.D.M., Dunlop, J., 1979. The regulation of K^+ influx in excised barley roots. Planta 145, 395–397.

Glass, A.D.M., Siddiqi, M.Y., 1985. Nitrate inhibition of chloride influx in barley: implications for a proposed chloride homeostat. J. Exp. Bot. 36, 556–566.

Glass, A.D.M., Siddiqi, M.Y., Ruth, T.J., Rufty, T.W., 1990. Studies on the uptake of nitrate in barley. II. Energetics. Plant Physiol. 93, 1585–1589.

Gojon, A., Nacry, P., Davidian, J.-C., 2009. Root uptake regulation: a central process for NPS homeostasis in plants. Curr. Opin. Plant Biol. 12, 328–338.

Gong, H.J., Randall, D.P., Flowers, T.J., 2006. Silicon deposition in the root reduces sodium uptake in rice (*Oryza sativa* L.) seedlings by reducing bypass flow. Plant Cell Environ. 29, 1970–1979.

Gould, N., Thorpe, M.R., Minchin, P.E.H., Pritchard, J., White, P.J., 2004. Solute is imported to elongating root cells of barley as a pressure driven-flow of solution. Funct. Plant Biol. 31, 391–397.

Gransee, A., Führs, H., 2013. Magnesium mobility in soils as a challenge for soil and plant analysis, magnesium fertilization and root uptake under adverse growth conditions. Plant Soil 368, 5–21.

Gronnier, J., Gerbeau-Pissot, P., Germain, V., Mongrand, S., Simon-Plas, F., 2018. Divide and rule: plant plasma membrane organization. Trends Plant Sci. 23, 899–917.

Guerinot, M.L., 2010. Iron. In: Hell, R., Mendel, R.-R. (Eds.), Cell Biology of Metals and Nutrients, Plant Cell Monographs 17. Springer, Dordrecht, pp. 75–94.

Guo, J., Zeng, W., Chen, Q., Lee, C., Chen, L., Yang, Y., et al., 2016. Structure of the voltage-gated two-pore channel TPC1 from *Arabidopsis thaliana*. Nature 531, 196–201.

Gustin, J.L., Zanis, M.J., Salt, D.E., 2011. Structure and evolution of the plant cation diffusion facilitator family of ion transporters. BMC Evol. Biol. 11, 76. Available from: https://doi.org/10.1186/1471-2148-11-76.

Hachez, C., Moshelion, M., Zelazny, E., Cavez, D., Chaumont, F., 2006. Localization and quantification of plasma membrane aquaporin expression in maize primary root: a clue to understanding their role as cellular plumbers. Plant Mol. Biol. 62, 305–323.

Hajibagheri, M.A., Flowers, T.J., Collins, J.C., Yeo, A.R., 1988. A comparison of the methods of X-ray microanalysis, compartmental analysis and longitudinal ion profiles to estimate cytoplasmic ion concentrations in two maize varieties. J. Exp. Bot. 39, 279–290.

Hamam, A.M., Coskun, D., Britto, D.T., Plett, D., Kronzucker, H.J., 2019. Plasma–membrane electrical responses to salt and osmotic gradients contradict radiotracer kinetics, and reveal Na^+–transport dynamics in rice (*Oryza sativa* L.). Planta 249, 1037–1051.

Hamamoto, S., Horie, T., Hauser, F., Deinlein, U., Schroeder, J.I., Uozumi, N., 2015. HKT transporters mediate salt stress resistance in plants: from structure and function to the field. Curr. Opin. Biotechnol. 32, 113–120.

Hamilton, E.S., Schlegel, A.M., Haswell, E.S., 2015. United in diversity: mechanosensitive ion channels in plants. Annu. Rev. Plant Biol. 66, 113–137.

Hammond, J.P., White, P.J., 2008. Sucrose transport in the phloem: integrating root responses to phosphorus starvation. J. Exp. Bot. 59, 93–109.

Hammond, J.P., Broadley, M.R., Bowen, H.C., Spracklen, W.P., Hayden, R.M., White, P.J., 2011. Gene expression changes in phosphorus deficient potato (*Solanum tuberosum* L.) leaves and the potential for diagnostic gene expression markers. PLoS One 6, e24606. Available from: https://doi.org/10.1371/journal.pone.0024606.

Hanikenne, M., Esteves, S.M., Fanara, S., Rouached, H., 2021. Coordinated homeostasis of essential mineral nutrients: a focus on iron. J. Exp. Bot. 72, 2136–2153.

Haruta, M., Gray, W.M., Sussman, M.R., 2015. Regulation of the plasma membrane proton pump (H^+-ATPase) by phosphorylation. Curr. Opin. Plant Biol. 28, 68–75.

Häussling, M., Jorns, C.A., Lehmbecker, G., Hecht-Buchholz, C., Marschner, H., 1988. Ion and water uptake in relation to root development in Norway spruce (*Picea abies* (L.) Karst.). J. Plant Physiol. 133, 486–491.

Hawkesford, M.J., De Kok, L.J., 2006. Managing sulphur metabolism in plants. Plant Cell Environ. 29, 382–395.

He, C.-J., Morgan, P.W., Drew, M.C., 1992. Enhanced sensitivity to ethylene in nitrogen- or phosphate-starved roots of *Zea mays* L. during aerenchyma formation. Plant Physiol. 98, 137–142.

Hedrich, R., Geiger, D., 2017. Biology of SLAC1-type anion channels – from nutrient uptake to stomatal closure. New Phytol. 216, 46–61.

Heenan, D.P., Campbell, L.C., 1981. Influence of potassium and manganese on growth and uptake of magnesium by soybeans (*Glycine max* (L.) Merr. cv Bragg). Plant Soil 61, 447–456.

Hell, R., Khan, M.S., Wirtz, M., 2010. Cellular biology of sulphur and its functions in plants. In: Hell, R., Mendel, R.-R. (Eds.), Cell Biology of Metals and Nutrients. Plant Cell Monographs 17, Springer, Dordrecht, pp. 243–279.

Henderson, S.W., Wege, S., Gilliham, M., 2018. Plant cation-chloride cotransporters (CCC): evolutionary origins and functional insights. Int. J. Mol. Sci. 19, 492. Available from: https://doi.org/10.3390/ijms19020492.

Hendrix, J.E., 1967. The effect of pH on the uptake and accumulation of phosphate and sulfate ions by bean plants. Am. J. Bot. 54, 560–564.

Hepler, P.K., 2005. Calcium: a central regulator of plant growth and development. Plant Cell 17, 2142–2155.

Hermans, C., Hammond, J.P., White, P.J., Verbruggen, N., 2006. How do plants respond to nutrient shortage by biomass allocation? Trends Plant Sci. 11, 610–617.

Herschbach, C., Rennenberg, H., 2001. Sulfur nutrition of deciduous trees. Naturwissenschaften 88, 25–36.

Heuer, S., Gaxiola, R., Schilling, R., Herrera-Estrella, L., López-Arredondo, D., Wissuwa, M., et al., 2016. Improving phosphorus use efficiency: a complex trait with emerging opportunities. Plant J. 90, 868–885.

Hirel, B., Le Gouis, J., Ney, B., Gallais, A., 2007. The challenge of improving nitrogen use efficiency in crop plants: towards a more central role for genetic variability and quantitative genetics within integrated approaches. J. Exp. Bot. 58, 2369–2387.

Hirsch, R.E., Lewis, B.D., Spalding, E.P., Sussman, M.R., 1998. A role for the AKT1 potassium channel in plant nutrition. Science 280, 918–921.

Hoagland, D.R., 1948. Lectures on the Inorganic Nutrition of Plants. Chronica Botanica, Waltham, MA.

Hodge, A., 2004. The plastic plant: root responses to heterogeneous supplies of nutrients. New Phytol. 162, 9–24.
Holbein, J., Shen, D., Andersen, T.G., 2021. The endodermal passage cell – just another brick in the wall? New Phytol. 230, 1321–1328.
Holland, J.E., White, P.J., Thauvin, J.-N., Jordan-Meille, L., Haefele, S.-M., Thomas, C.L., et al., 2021. Liming impacts barley yield over a wide concentration range of soil exchangeable cations. Nutr. Cycl. Agroecosyst. 120, 131–144.
Hölzl, G., Dörmann, P., 2007. Structure and function of glycoglycerolipids in plants and bacteria. Prog. Lipid Res. 46, 225–243.
Hope, A.B., Stevens, P.G., 1952. Electrical potential differences in bean roots and their relation to salt uptake. Aust. J. Sci. Res. Ser. B 5, 335–343.
Hose, E., Clarkson, D.T., Steudle, E., Schreiber, L., Hartung, W., 2001. The exodermis: a variable apoplastic barrier. J. Exp. Bot. 52, 2245–2264.
Huang, C.X., Van Steveninck, R.F.M., 1989. Longitudinal and transverse profiles of K^+ and Cl^- concentration in "low-" and "high-salt" barley roots. New Phytol. 112, 475–480.
Huang, S., Liang, Z., Chen, S., Sun, H., Fan, X., Wang, C., et al., 2019. A transcription factor, OsMADS57, regulates long-distance nitrate transport and root elongation. Plant Physiol. 180, 882–895.
Huang, S., Yamaji, N., Sakurai, G., Mitani-Ueno, N., Konishi, N., Ma, J.F., 2022. A pericycle-localized silicon transporter for efficient xylem loading in rice. New Phytol. 234, 197–208.
Hubberten, H.-M., Drozd, A., Tran, B.V., Hesse, H., Hoefgen, R., 2012. Local and systemic regulation of sulfur homeostasis in roots of *Arabidopsis thaliana*. Plant J. 72, 625–635.
Huby, E., Napier, J.A., Baillieul, F., Michaelson, L.V., Dhondt-Cordelier, S., 2020. Sphingolipids: towards an integrated view of metabolism during the plant stress response. New Phytol. 225, 659–670.
Ingestad, T., 1997. A shift of paradigm is needed in plant science. Physiol. Plant. 101, 446–450.
Ingestad, T., Ågren, G.I., 1992. Theories and methods on plant nutrition and growth. Physiol. Plant. 84, 177–184.
Isayenkov, S.V., Dabravolski, S.A., Pan, T., Shabala, S., 2020. Phylogenetic diversity and physiological roles of plant monovalent cation/H^+ antiporters. Front. Plant Sci. 11, 573564. Available from: https://doi.org/10.3389/fpls.2020.573564.
Ishida, T., Kurata, T., Okada, K., Wada, T., 2008. A genetic regulatory network in the development of trichomes and root hairs. Annu. Rev. Plant Biol. 59, 365–386.
Ishimaru, Y., Suzuki, M., Tsukamoto, T., Suzuki, K., Nakazono, M., Kobayashi, T., et al., 2006. Rice plants take up iron as an Fe^{3+}-phytosiderophore and as Fe^{2+}. Plant J. 45, 335–346.
Itoh, S., Barber, S.A., 1983. Phosphorus uptake by six plant species as related to root hairs. Agron. J. 75, 457–461.
Jarratt-Barnham, E., Wang, L., Ning, Y., Davies, J.M., 2021. The complex story of plant cyclic nucleotide-gated channels. Int. J. Mol. Sci. 22, 874. Available from: https://doi.org/10.3390/ijms22020874.
Jia, Z., von Wirén, N., 2020. Signaling pathways underlying nitrogen-dependent changes in root system architecture: from model to crop species. J. Exp. Bot. 71, 4393–4404.
Jungk, A., Asher, C.J., Edwards, D.G., Meyer, D., 1990. Influence of phosphate status on phosphate uptake kinetics of maize (*Zea mays*) and soybean (*Glycine max*). Plant Soil 124, 175–182.
Kang, J., Park, J., Choi, H., Burla, B., Kretzschmar, T., Lee, Y., et al., 2011. Plant ABC transporters. The Arabidopsis Book 9. Available from: https://doi.org/10.1199/tab.0153.
Kanno, S., Arrighi, J.-F., Chiarenza, S., Bayle, V., Berthomé, R., Péret, B., et al., 2016. A novel role for the root cap in phosphate uptake and homeostasis. eLife 5, e14577. Available from: https://doi.org/10.7554/eLife.14577.
Kaplan, B., Sherman, T., Fromm, H., 2007. Cyclic nucleotide-gated channels in plants. FEBS Lett. 581, 2237–2246.
Kehelpannala, C., Rupasinghe, T., Pasha, A., Esteban, E., Hennessy, T., Bradley, D., et al., 2021. An Arabidopsis lipid map reveals differences between tissues and dynamic changes throughout development. Plant J. 107, 287–302.
Keller, P., Deuel, H., 1957. Kationenaustauschkapazität und Pektingehalt von Pflanzenwurzeln. Z. Pflanzenernähr. Bodenk. 79, 119–131.
Keltjens, W.G., Nijenstein, J.H., 1987. Diurnal variations in uptake, transport and assimilation of NO_3^- and efflux of OH^- in maize plants. J. Plant Nutr. 10, 887–900.
Kiba, T., Feria-Bourrellier, A.-B., Lafouge, F., Lezhneva, L., Boutet-Mercey, S., Orsel, M., et al., 2012. The Arabidopsis nitrate transporter NRT2.4 plays a double role in roots and shoots of nitrogen-starved plants. Plant Cell 24, 245–258.
Kiba, T., Inaba, J., Kudo, T., Ueda, N., Konishi, M., Mitsuda, N., et al., 2018. Repression of nitrogen starvation responses by members of the Arabidopsis GARP-type transcription factor NIGT1/HRS1 subfamily. Plant Cell 30, 925–945.
Kintzer, A.F., Stroud, R.M., 2016. Structure, inhibition and regulation of two-pore channel TPC1 from *Arabidopsis thaliana*. Nature 531, 258–264.
Kirscht, A., Kaptan, S.S., Bienert, G.P., Chaumont, F., Nissen, P., de Groot, B.L., et al., 2016. Crystal structure of an ammonia-permeable aquaporin. PLoS Biol. 14, e1002411. Available from: https://doi.org/10.1371/journal.pbio.100241.
Knipfer, T., Besse, M., Verdeil, J.-L., Fricke, W., 2011. Aquaporin-facilitated water uptake in barley (*Hordeum vulgare* L.) roots. J. Exp. Bot. 62, 4115–4126.
Kobayashi, T., Nozoye, T., Nishizawa, N.K., 2019. Iron transport and its regulation in plants. Free Rad. Biol. Med. 133, 11–20.
Kochian, L.V., Xin-Zhi, J., Lucas, W.J., 1985. Potassium transport in corn roots. IV. Characterization of the linear component. Plant Physiol. 79, 771–776.
Koevoets, I.T., Venema, J.H., Elzenga, J.T.M., Testerink, C., 2016. Roots withstanding their environment: exploiting root system architecture responses to abiotic stress to improve crop tolerance. Front. Plant Sci. 7, 1335. Available from: https://doi.org/10.3389/fpls.2016.01335.
Kotula, L., Ranathunge, K., Steudle, E., 2009. Apoplastic barriers effectively block oxygen permeability across outer cell layers of rice roots under deoxygenated conditions: roles of apoplastic pores and of respiration. New Phytol. 184, 909–917.

Kragler, F., 2010. RNA in the phloem: a crisis or a return on investment? Plant Sci. 178, 99–104.

Krapp, A., David, L.C., Chardin, C., Girin, T., Marmagne, A., Leprince, A.-S., et al., 2014. Nitrate transport and signalling in Arabidopsis. J. Exp. Bot. 65, 789–798.

Krishna, T.P.A., Maharajan, T., Roch, G.V., Ignacimuthu, S., Ceasar, S.A., 2020. Structure, function, regulation and phylogenetic relationship of ZIP family transporters of plants. Front. Plant Sci. 11, 662. Available from: https://doi.org/10.3389/fpls.2020.00662.

Krishnamurthy, P., Ranathunge, K., Franke, R., Prakash, H.S., Schreiber, L., Mathew, M.K., 2009. The role of root apoplastic transport barriers in salt tolerance of rice (*Oryza sativa* L.). Planta 230, 119–134.

Krishnamurthy, P., Ranathunge, K., Nayak, S., Schreiber, L., Mathew, M.K., 2011. Root apoplastic barriers block Na^+ transport to shoots in rice (*Oryza sativa* L.). J. Exp. Bot. 62, 4215–4228.

Kronzucker, H.J., Britto, D.T., 2011. Sodium transport in plants: a critical review. New Phytol. 189, 54–81.

Kronzucker, H.J., Kirk, G.J.D., Siddiqi, M.Y., Glass, A.D.M., 1998. Effects of hypoxia on $^{13}NH_4^+$ fluxes in rice roots. Kinetics and compartmental analysis. Plant Physiol. 116, 581–587.

Kronzucker, H.J., Siddiqi, M.Y., Glass, A.D.M., Kirk, G.J.D., 1999. Nitrate-ammonium synergism in rice. A subcellular flux analysis. Plant Physiol. 119, 1041–1046.

Kumar, S., Verma, S., Trivedi, P.K., 2017. Involvement of small RNAs in phosphorus and sulfur sensing, signaling and stress: current update. Front. Plant Sci. 8, 285. Available from: https://doi.org/10.3389/fpls.2017.00285.

Kurimoto, K., Day, D.A., Lambers, H., Noguchi, K., 2004. Effect of respiratory homeostasis on plant growth in cultivars of wheat and rice. Plant Cell Environ. 27, 853–862.

Lalonde, S., Wipf, D., Frommer, W.B., 2004. Transport mechanisms for organic forms of carbon and nitrogen between source and sink. Annu. Rev. Plant Biol. 55, 341–372.

Lambers, H., 2022. Phosphorus acquisition and utilization in plants. Annu. Rev. Plant Biol. 73, 17–42.

Lane, T.S., Rempe, C.S., Davitt, J., Staton, M.E., Peng, Y., Soltis, D.E., et al., 2016. Diversity of ABC transporter genes across the plant kingdom and their potential utility in biotechnology. BMC Biotechnol. 16, 47. Available from: https://doi.org/10.1186/s12896-016-0277-6.

Läuchli, A., Grattan, S.R., 2012. Soil pH extremes. In: Shabala, S. (Ed.), Plant Stress Physiology. CABI, Wallingford, pp. 194–209.

Laurie, S.H., Tanock, N.P., McGrath, S.P., Sanders, J.R., 1991. Influence of complexation on the uptake by plants of iron, manganese, copper and zinc. I. Effect of EDTA in a multi-metal and computer simulation study. J. Exp. Bot. 42, 509–513.

Le Bot, J., Kirkby, E.A., 1992. Diurnal uptake of nitrate and potassium during the vegetative growth of tomato plants. J. Plant Nutr. 15, 247–264.

Lee, R.B., Clarkson, D.T., 1986. Nitrogen-13 studies of nitrate fluxes in barley roots. I. Compartmental analysis from measurements of ^{13}N efflux. J. Exp. Bot. 37, 1753–1767.

Lee, R.B., Drew, M.C., 1989. Rapid, reversible inhibition of nitrate influx in barley by ammonium. J. Exp. Bot. 40, 741–752.

Lee, R.B., Ratcliffe, R.G., Southon, T.E., 1990. ^{31}P NMR measurements of the cytoplasmic and vacuolar Pi content of mature maize roots: relationships with phosphorus status and phosphate fluxes. J. Exp. Bot. 41, 1063–1078.

Lejay, L., Gansel, X., Cerezo, M., Tillard, P., Müller, C., Krapp, A., et al., 2003. Regulation of root ion transporters by photosynthesis: functional importance and relation with hexokinase. Plant Cell 15, 2218–2232.

Li, L., Tutone, A.F., Drummond, R.S.M., Gardner, R.C., Luan, S., 2001. A novel family of magnesium transport genes in Arabidopsis. Plant Cell 13, 2761–2775.

Li, W., Wang, Y., Okamoto, M., Crawford, N.M., Siddiqi, M.Y., Glass, A.D.M., 2007. Dissection of the *AtNRT2.1:AtNRT2.2* inducible high-affinity nitrate transporter gene cluster. Plant Physiol. 143, 425–433.

Li, S., Wang, Z., Stewart, B.A., 2013. Responses of crop plants to ammonium and nitrate N. Adv. Agron. 118, 205–397.

Li, B., Byrt, C., Qiu, J., Baumann, U., Hrmova, M., Evrard, A., et al., 2016. Identification of a stelar-localized transport protein that facilitates root-to-shoot transfer of chloride in Arabidopsis. Plant Physiol. 170, 1014–1029.

Li, B., Tester, M., Gilliham, M., 2017a. Chloride on the move. Trends Plant Sci. 22, 236–248.

Li, H., Yu, M., Du, X.Q., Wang, Z.F., Wu, W.H., Quintero, F.J., et al., 2017b. NRT1.5/NPF7.3 Functions as a proton-coupled H^+/K^+ antiporter for K^+ loading into the xylem in Arabidopsis. Plant Cell 29, 2016–2026.

Li, B., Kamiya, T., Kalmbach, L., Yamagami, M., Yamaguchi, K., Shigenobu, S., et al., 2017c. Role of *LOTR1* in nutrient transport through organization of spatial distribution of root endodermal barriers. Curr. Biol. 27, 758–765.

Liang, G., Yang, F., Yu, D., 2010. MicroRNA395 mediates regulation of sulfate accumulation and allocation in *Arabidopsis thaliana*. Plant J. 62, 1046–1057.

Lin, S.-H., Kuo, H.-F., Canivenc, G., Lin, C.-S., Lepetit, M., Hsu, P.-K., et al., 2008. Mutation of the Arabidopsis NRT1.5 nitrate transporter causes defective root-to-shoot nitrate transport. Plant Cell 20, 2514–2528.

Lindberg, S., Banas, A., Stymne, S., 2005. Effects of different cultivation temperatures on plasma membrane ATPase activity and lipid composition of sugar beet roots. Plant Physiol. Biochem. 43, 261–268.

Liu, Y., von Wirén, N., 2017. Ammonium as a signal for physiological and morphological responses in plants. J. Exp. Bot. 68, 2581–2592.

Liu, T.Y., Chang, C.Y., Chiou, T.J., 2009. The long-distance signaling of mineral macronutrients. Curr. Opin. Plant Biol. 12, 312–319.

Liu, K.H., Niu, Y., Konishi, M., Wu, Y., Du, H., Sun Chung, H., et al., 2017. Discovery of nitrate-CPK-NLP signalling in central nutrient-growth networks. Nature 545, 311–316.

Loneragan, J.F., Asher, C.H., 1967. Response of plants to phosphate concentration in solution culture. II. Role of phosphate absorption and its relation to growth. Soil Sci. 103, 311–318.

Longnecker, N., Welch, R.M., 1990. Accumulation of apoplastic iron in plant roots. A factor in the resistance of soybeans to iron-deficiency induced chlorosis? Plant Physiol. 92, 17−22.

López-Arredondo, D.L., Leyva-González, M.A., González-Morales, S.I., López-Bucio, J., Herrera-Estrella, L., 2014. Phosphate nutrition: improving low-phosphate tolerance in crops. Annu. Rev. Plant Biol. 65, 95−123.

López-Pérez, L., del Carmen Martínez-Ballesta, M., Maurel, C., Carvajal, M., 2009. Changes in plasma membrane lipids, aquaporins and proton pump of broccoli roots, as an adaptation mechanism to salinity. Phytochemistry 70, 492−500.

Loqué, D., Ludewig, U., Yuan, L., von Wirén, N., 2005. Tonoplast intrinsic proteins AtTIP2;1 and AtTIP2;3 facilitate NH_3 transport into the vacuole. Plant Physiol. 137, 671−680.

Louahlia, S., Laine, P., MacDuff, J.H., Ourry, A., Humphreys, M., Boucaud, J., 2008. Interactions between reserve mobilization and regulation of nitrate uptake during regrowth of *Lolium perenne* L.: putative roles of amino acids and carbohydrates. Botany 86, 1101−1110.

Lubkowitz, M., 2011. The oligopeptide transporters: a small gene family with a diverse group of substrates and functions? Mol. Plant 4, 407−415.

Lucas, W.J., Ham, B.-K., Kim, J.-Y., 2009. Plasmodesmata – bridging the gap between neighboring plant cells. Trends Cell Biol. 19, 495−503.

Lupini, A., Mercati, F., Araniti, F., Miller, A.J., Sunseri, F., Abenavoli, M.R., 2016. NAR2.1/NRT2.1 functional interaction with NO_3^- and H^+ fluxes in high-affinity nitrate transport in maize root regions. Plant Physiol. Biochem. 102, 107−114.

Lüttge, U., Laties, G.G., 1966. Dual mechanism of ion absorption in relation to long distance transport in plants. Plant Physiol. 41, 1531−1539.

Lux, A., Martinka, M., Vaculík, M., White, P.J., 2011. Root responses to cadmium in the rhizosphere: a review. J. Exp. Bot. 62, 21−37.

Lynch, J.P., 2007. Roots of the second green revolution. Aust. J. Bot. 55, 493−512.

Lynch, J.P., 2011. Root phenes for enhanced soil exploration and phosphorus acquisition: tools for future crops. Plant Physiol. 156, 1041−1049.

Lynch, J.P., 2019. Root phenotypes for improved nutrient capture: an underexploited opportunity for global agriculture. New Phytol. 223, 548−564.

Lynch, D.V., Dunn, T.M., 2004. An introduction to plant sphingolipids and a review of recent advances in understanding their metabolism and function. New Phytol. 161, 677−702.

Ma, F., Peterson, C.A., 2003. Recent insights into the development, structure and chemistry of the endodermis and exodermis. Can. J. Bot. 81, 405−421.

Ma, J.F., Yamaji, N., 2015. A cooperative system of silicon transport in plants. Trends Plant Sci. 20, 435−442.

Ma, J.F., Yamaji, N., Mitani, N., Tamai, K., Konishi, S., Fujiwara, T., et al., 2007. An efflux transporter of silicon in rice. Nature 448, 209−212.

Maathuis, F.J.M., Sanders, D., 1993. Energization of potassium uptake in *Arabidopsis thaliana*. Planta 191, 302−307.

Maathuis, F.J.M., Amtmann, A., 1999. K^+ nutrition and Na^+ toxicity: the basis of cellular K^+/Na^+ ratios. Ann. Bot. 84, 123−133.

Macduff, J.H., Bakken, A.K., Dhanoa, M.S., 1997. An analysis of the physiological basis of commonality between diurnal patterns of NH_4^+, NO_3^- and K^+ uptake by *Phleum pratense* and *Festuca pratensis*. J. Exp. Bot. 48, 1691−1701.

Macklon, A.E.S., Ron, M.M., Sim, A., 1990. Cortical cell fluxes of ammonium and nitrate in excised root segments of *Allium cepa* L.: studies using ^{15}N. J. Exp. Bot. 41, 359−370.

Macklon, A.E.S., Lumsdon, D.G., Sim, A., McHardy, W.J., 1996. Phosphate fluxes, compartmentation and vacuolar speciation in root cortex cells of intact *Agrostis capillaris* seedlings: effect of non-toxic levels of aluminium. J. Exp. Bot. 47, 793−803.

Maeda, Y., Konishi, M., Kiba, T., Sakuraba, Y., Sawaki, N., Kurai, T., et al., 2018. A NIGT1-centred transcriptional cascade regulates nitrate signalling and incorporates phosphorus starvation signals in Arabidopsis. Nat. Commun. 9, 1376. Available from: https://doi.org/10.1038/s41467-018-03832-6.

Maillard, A., Sorin, E., Etienne, P., Diquélou, S., Koprivova, A., Kopriva, S., et al., 2016. Non-specific root transport of nutrient gives access to an early nutritional indicator: the case of sulfate and molybdate. PLoS One 11, e0166910.

Malone, M., Herron, M., Morales, M.-A., 2002. Continuous measurement of macronutrient ions in the transpiration stream of intact plants using the meadow spittlebug coupled with ion chromatography. Plant Physiol. 130, 1436−1442.

Manishankar, P., Wang, N., Köster, P., Alatar, A.A., Kudla, J., 2018. Calcium signaling during salt stress and in the regulation of ion homeostasis. J. Exp. Bot. 69, 4215−4226.

Markham, J.E., Li, J., Cahoon, E.B., Jaworski, J.G., 2006. Separation and identification of major plant sphingolipid classes from leaves. J. Biol. Chem. 281, 22684−22694.

Marschner, H., 1995. Rhizosphere pH effects on phosphorus nutrition. In: Johansen, C., Lee, K.K., Sharma, K.K., Subbarao, G.V., Kueneman, E.A. (Eds.), Proceedings of an FAO/ICRISAT Expert Consultancy Workshop. Genetic Manipulation of Crop Plants to Enhance Integrated Nutrient Management in Cropping Systems. 1. Phosphorus. International Crops Research Institute for the Semi-Arid Tropics, Patancheru, India, pp. 107−115.

Marschner, H., Richter, C., 1973. Akkumulation und Translokation von K^+, Na^+ und Ca^{2+} bei Angebot zu einzelnen Wurzelzonen von Maiskeimpflanzen. Z. Pflanzenernähr. Bodenk. 135, 1−15.

Marschner, H., Häussling, M., George, E., 1991. Ammonium and nitrate uptake rates and rhizosphere-pH in non-mycorrhizal roots of Norway spruce (*Picea abies* (L.) Karst.). Trees 5, 14−21.

Marshansky, V., Futai, M., 2008. The V-type H^+-ATPase in vesicular trafficking: targeting, regulation and function. Curr. Opin. Cell Biol. 20, 415−426.

Martinoia, E., 2018. Vacuolar transporters – companions on a longtime journey. Plant Physiol. 176, 1384−1407.

Martinoia, E., Maeshima, M., Neuhaus, H.E., 2007. Vacuolar transporters and their essential role in plant metabolism. J. Exp. Bot. 58, 83−102.

Mäser, P., Thomine, S., Schroeder, J.I., Ward, J.M., Hirschi, K., Sze, H., et al., 2001. Phylogenetic relationships within cation transporter families of Arabidopsis. Plant Physiol. 126, 1646−1667.

Matheson, L.A., Hanton, S.L., Brandizzi, F., 2006. Traffic between the plant endoplasmic reticulum and Golgi apparatus: to the Golgi and beyond. Curr. Opin. Plant Biol. 9, 601–609.

Maule, A.J., 2008. Plasmodesmata: structure, function and biogenesis. Curr. Opin. Plant Biol. 11, 680–686.

Maurel, C., Boursiac, Y., Luu, D.-T., Santoni, V., Shahzad, Z., Verdoucq, L., 2015. Aquaporins in plants. Physiol. Rev. 95, 1321–1358.

Maurel, C., Tournaire-Roux, C., Verdoucq, L., Santoni, V., 2021. Hormonal and environmental signaling pathways target membrane water transport. Plant Physiol. 187, 2056–2070.

McClure, P., Kochian, L.V., Spanswick, R.M., Shaff, J.E., 1990. Evidence for cotransport of nitrate and protons in maize roots. II. Measurement of NO_3^- and H^+ fluxes with ion-selective microelectrodes. Plant Physiol. 93, 290–294.

McDonald, T.R., Ward, J.M., 2016. Evolution of electrogenic ammonium transporters (AMTs). Front. Plant Sci. 7, 352. Available from: https://doi.org/10.3389/fpls.2016.00352.

Meharg, A.A., Macnair, M.R., 1992. Suppression of the high affinity phosphate uptake system: a mechanism of arsenate tolerance in *Holcus lanatus* L. J. Exp. Bot. 43, 519–524.

Meng, S.-F., Zhang, B., Tang, R.-J., Zheng, X.-J., Chen, R., Liu, C.-G., et al., 2022. Four plasma membrane-localized MGR transporters mediate xylem Mg^{2+} loading for root-to-shoot Mg^{2+} translocation in *Arabidopsis*. Mol. Plant 15, 805–819.

Millard, P., Grelet, G.A., 2010. Nitrogen storage and remobilization by trees: ecophysiological relevance in a changing world. Tree Physiol. 30, 1083–1095.

Miller, A.J., Cramer, M.D., 2004. Root nitrogen acquisition and assimilation. Plant Soil 274, 1–36.

Miller, A.J., Shen, Q., Xu, G., 2009. Freeways in the plant: transporters for N, P and S and their regulation. Curr. Opin. Plant Biol. 12, 284–290.

Miller, C.N., Busch, W., 2021. Using natural variation to understand plant responses to iron availability. J. Exp. Bot. 72, 2154–2164.

Milner, M.J., Seamon, J., Craft, E., Kochian, L.V., 2013. Transport properties of members of the ZIP family in plants and their role in Zn and Mn homeostasis. J. Exp. Bot. 64, 369–381.

Mitani-Ueno, N., Ma, J.F., 2021. Linking transport system of silicon with its accumulation in different plant species. Soil Sci. Plant Nutr. 67, 10–17.

Miwa, K., Fujiwara, T., 2010. Boron transport in plants: coordinated regulation of transporters. Ann. Bot. 105, 1103–1108.

Miwa, K., Kamiya, T., Fujiwara, T., 2009. Homeostasis of the structurally important micronutrients, B and Si. Curr. Opin. Plant Biol. 12, 307–311.

Miyasaka, S.C., Grunes, D.L., 1990. Root temperature and calcium level effects in winter wheat forage: II. Nutrient composition and tetany potential. Agron. J. 82, 242–249.

Moore, C.A., Bowen, H.C., Scrase-Field, S., Knight, M.R., White, P.J., 2002. The deposition of suberin lamellae determines the magnitude of cytosolic Ca^{2+} elevations in root endodermal cells subjected to cooling. Plant J. 30, 457–466.

Munns, R., Tester, M., 2008. Mechanisms of salinity tolerance. Annu. Rev. Plant Biol. 59, 651–681.

Munns, R., Day, D.A., Fricke, W., Watt, M., Arsova, B., Barkla, B.J., et al., 2020. Energy costs of salt tolerance in crop plants. New Phytol. 225, 1072–1090.

Nable, R.O., Lance, R.C.M., Cartwright, B., 1990. Uptake of boron and silicon by barley genotypes with differing susceptibilities to boron toxicity. Ann. Bot. 66, 83–90.

Nakamura, M., Grebe, M., 2018. Outer, inner and planar polarity in the Arabidopsis root. Curr. Opin. Plant Biol. 41, 46–53.

Neugebauer, K., El-Serehy, H.A., George, T.S., McNicol, J.W., Moraes, M.F., Sorreano, M.C.M., et al., 2020. The influence of phylogeny and ecology on root, shoot and plant ionomes of fourteen native Brazilian species. Physiol. Plant. 168, 790–802.

Nguyen, T.H., Huang, S., Meynard, D., Chaine, C., Michel, R., Roelfsema, M.R.G., et al., 2017. A dual role for the OsK5.2 ion channel in stomatal movements and K^+ loading into xylem sap. Plant Physiol. 174, 2409–2418.

Nieves-Cordones, M., Alemán, F., Martínez, V., Rubio, F., 2014. K^+ uptake in plant roots. The systems involved, their regulation and parallels in other organisms. J. Plant Physiol. 171, 688–695.

Nieves-Cordones, M., Lara, A., Silva, M., Amo, J., Rodriguez-Sepulveda, P., Rivero, R.M., et al., 2020. Root high-affinity K^+ and Cs^+ uptake and plant fertility in tomato plants are dependent on the activity of the high-affinity K^+ transporter SlHAK5. Plant Cell Environ. 43, 1707–1721.

Niño-González, M., Novo-Uzal, E., Richardson, D.N., Barros, P.M., Duque, P., 2019. More transporters, more substrates: the Arabidopsis major facilitator superfamily revisited. Mol. Plant 12, 1182–1202.

Nussaume, L., Kanno, S., Javot, H., Marin, E., Pochon, N., Ayadi, A., et al., 2011. Phosphate import in plants: focus on the PHT1 transporters. Front. Plant Sci. 2, 83. Available from: https://doi.org/10.3389/fpls.2011.00083.

Oertli, J.J., Grgurevic, E., 1975. Effect of pH on the absorption of boron by excised barley roots. Agron. J. 67, 278–280.

Oparka, K.J., 2004. Getting the message across: how do plant cells exchange macromolecular complexes? Trends Plant Sci. 9, 33–41.

Opekarová, M., Malinsky, J., Tanner, W., 2010. Plants and fungi in the era of heterogeneous plasma membranes. Plant Biol. 12, 94–98.

Pata, M.O., Hannun, Y.A., Ng, C.K.-Y., 2010. Plant sphingolipids: decoding the enigma of the Sphinx. New Phytol. 185, 611–630.

Penfield, S., 2008. Temperature perception and signal transduction in plants. New Phytol. 179, 615–628.

Peters, W.S., Jensen, K.H., Stone, H.A., Knoblauch, M., 2021. Plasmodesmata and the problems with size: interpreting the confusion. J. Plant Physiol. 257, 153341. Available from: https://doi.org/10.1016/j.jplph.2020.153341.

Peterson, C.A., Enstone, D.E., 1996. Functions of passage cells in the endodermis and exodermis of roots. Physiol. Plant. 97, 592–598.

Peuke, A.D., 2010. Correlations in concentrations, xylem and phloem flows, and partitioning of elements and ions in intact plants. A summary and statistical re-evaluation of modeling experiments in *Ricinus communis*. J. Exp. Bot. 61, 635–655.

Pommerrenig, B., Diehn, T.A., Bernhardt, N., Bienert, M.D., Mitani-Ueno, N., Fuge, J., et al., 2020. Functional evolution of nodulin 26-like intrinsic proteins: from bacterial arsenic detoxification to plant nutrient transport. New Phytol. 225, 1383–1396.

Printz, B., Lutts, S., Hausman, J.-F., Sergeant, K., 2016. Copper trafficking in plants and its implication on cell wall dynamics. Front. Plant Sci. 7, 601. Available from: https://doi.org/10.3389/fpls.2016.00601.

Puga, M.I., Mateos, I., Charukesi, R., Wang, Z., Franco-Zorrilla, J.M., de Lorenzo, L., et al., 2014. SPX1 is a phosphate-dependent inhibitor of Phosphate Starvation Response 1 in Arabidopsis. Proc. Natl. Acad. Sci. USA 111, 14947–14952.

Puga, M.I., Rojas-Triana, M., Lorenzo, L., Leyva, A., Rubio, V., Paz-Ares, J., 2017. Novel signals in the regulation of Pi starvation responses in plants: facts and promises. Curr. Opin. Plant Biol. 39, 40–49.

Puig, S., Andrés-Colás, N., García-Molina, A., Peñarrubia, L., 2007. Copper and iron homeostasis in *Arabidopsis*: responses to metal deficiencies, interactions and biotechnological applications. Plant Cell Environ. 30, 271–290.

Pyo, Y.J., Gierth, M., Schroeder, J.I., Cho, M.H., 2010. High-affinity K^+ transport in Arabidopsis: AtHAK5 and AKT1 are vital for seedling establishment and postgermination growth under low-potassium conditions. Plant Physiol. 153, 863–875.

Qi, Z., Hampton, C.R., Shin, R., Barkla, B.J., White, P.J., Schachtman, D.P., 2008. The high affinity K^+ transporter AtHAK5 plays a physiological role *in planta* at very low K^+ concentrations and provides a caesium uptake pathway in *Arabidopsis*. J. Exp. Bot. 59, 595–607.

Raddatz, N., de los Rios, L.M., Lindahl, M., Quintero, F.J., Pardo, J.M., 2020. Coordinated transport of nitrate, potassium, and sodium. Front. Plant Sci. 11, 247. Available from: https://doi.org/10.3389/fpls.2020.00247.

Ragel, P., Raddatz, N., Leidi, E.O., Quintero, F.J., Pardo, J.M., 2019. Regulation of K^+ nutrition in plants. Front. Plant Sci. 10, 281. Available from: https://doi.org/10.3389/fpls.2019.00281.

Raggi, S., Demes, E., Liu, S., Verger, S., Robert, S., 2020. Polar expedition: mechanisms for protein polar localization. Curr. Opin. Plant Biol. 53, 134–140.

Rai, H., Kawabata, M., 2020. The dynamics of radio-cesium in soils and mechanism of cesium uptake into higher plants: newly elucidated mechanism of cesium uptake into rice plants. Front. Plant Sci. 11, 528. Available from: https://doi.org/10.3389/fpls.2020.00528.

Ramakrishna, P., Barberon, M., 2019. Polarized transport across root epithelia. Curr. Opin. Plant Biol. 52, 23–29.

Ranathunge, K., Steudle, E., Lafitte, R., 2005. Water transport in rice roots. Plant Cell Environ. 28, 121–133.

Raven, J.A., 1986. Biochemical disposal of excess H^+ in growing plants? New Phytol. 104, 175–206.

Rawat, S.R., Silim, S.N., Kronzucker, H.J., Siddiqi, M.Y., Glass, A.D.M., 1999. AtAMT1 gene expression and NH_4^+ uptake in roots of *Arabidopsis thaliana*: evidence for regulation by root glutamine levels. Plant J. 19, 143–152.

Reidenbach, G., Horst, W., 1997. Nitrate-uptake capacity of different root zones of *Zea mays* (L.) in vitro and in situ. Plant Soil 196, 295–300.

Remy, E., Cabrito, T.R., Baster, P., Batista, R.A., Teixeira, M.C., Friml, J., et al., 2013. A major facilitator superfamily transporter plays a dual role in polar auxin transport and drought stress tolerance in Arabidopsis. Plant Cell 25, 901–926.

Remy, E., Cabrito, T.R., Baster, P., Batista, R.A., Teixeira, M.C., Sá-Correia, I., et al., 2015. The major facilitator superfamily transporter ZIFL2 modulates cesium and potassium homeostasis in Arabidopsis. Plant Cell Physiol. 56, 148–162.

Rich, S.M., Watt, M., 2013. Soil conditions and cereal root system architecture: review and considerations for linking Darwin and Weaver. J. Exp. Bot. 64, 1193–1208.

Roberts, S.K., 2006. Plasma membrane anion channels in higher plants and their putative functions in roots. New Phytol. 169, 647–666.

Robinson, D.G., Herranz, M.-C., Bubeck, J., Pepperkok, R., Ritzenthaler, C., 2007. Membrane dynamics in the early secretory pathway. Crit. Rev. Plant Sci. 26, 199–225.

Römheld, V., 1987a. Different strategies for iron acquisition in higher plants. Physiol. Plant. 70, 231–234.

Römheld, V., 1987b. Existence of two different strategies for the acquisition of iron in higher plants. In: Winkelmann, G., van der Helm, D., Neilands, J.B. (Eds.), Iron Transport in Microbes, Plants and Animals. VCH Verlag, Weinheim, Germany, pp. 353–374.

Römheld, V., Marschner, H., 1981. Iron deficiency stress induced morphological and physiological changes in root tips of sunflower. Physiol. Plant. 53, 354–360.

Römheld, V., Kramer, D., 1983. Relationship between proton efflux and rhizodermal transfer cells induced by iron deficiency. Z. Pflanzenphysiol. 113, 73–83.

Römheld, V., Marschner, H., 1990. Genotypical differences among gramineaceous species in release of phytosiderophores and uptake of iron phytosiderophores. Plant Soil 123, 147–153.

Rorison, I.H., 1987. Mineral nutrition in time and space. New Phytol. 106, 79–92.

Rubio, G., Sorgona, A., Lynch, J.P., 2004. Spatial mapping of phosphorus influx in bean root systems using digital autoradiography. J. Exp. Bot. 55, 2269–2280.

Rufty, T.W., Jackson, W.A., Raper, C.D., 1982. Inhibition of nitrate assimilation in roots in the presence of ammonium: the moderating influence of potassium. J. Exp. Bot. 33, 1122–1137.

Ryan, P.R., Tyerman, S.D., Sasaki, T., Furuichi, T., Yamamoto, Y., Zhang, W.H., et al., 2011. The identification of aluminium-resistance genes provides opportunities for enhancing crop production on acid soils. J. Exp. Bot. 62, 9–20.

Sadana, U.S., Sharma, P., Castañeda Ortiz, N., Samal, D., Claassen, N., 2005. Manganese uptake and Mn efficiency of wheat cultivars are related to Mn-uptake kinetics and root growth. J. Plant Nutr. Soil Sci. 168, 581–589.

Saengwilai, P., Nord, E.A., Chimungu, J.G., Brown, K.M., Lynch, J.P., 2014. Root cortical aerenchyma enhances nitrogen acquisition from low-nitrogen soils in maize. Plant Physiol. 166, 726–735.

Salim, M., Pitman, M.G., 1984. Pressure-induced water and solute flow through plant roots. J. Exp. Bot. 35, 869–881.

Santa-María, G.E., Oliferuk, S., Moriconi, J.I., 2018. KT-HAK-KUP transporters in major terrestrial photosynthetic organisms: a twenty years tale. J. Plant Physiol. 226, 77–90.

Santa-María, G.E., Danna, C.H., Czibener, C., 2000. High-affinity potassium transport in barley roots. Ammonium-sensitive and -insensitive pathways. Plant Physiol. 123, 297–306.

Sasaki, A., Yamaji, N., Yokosho, K., Ma, J.F., 2012. Nramp5 is a major transporter responsible for manganese and cadmium uptake in rice. Plant Cell 24, 2155–2167.

Satbhai, S.B., Ristova, D., Busch, W., 2015. Underground tuning: quantitative regulation of root growth. J. Exp. Bot. 66, 1099–1112.

Schaaf, G., Ludewig, U., Erenoglu, B.E., Mori, S., Kitahara, T., von Wirén, N., 2004. ZmYS1 functions as a proton-coupled symporter for phytosiderophore- and nicotianamine-chelated metals. J. Biol. Chem. 279, 9091–9096.

Schapire, A.L., Valpuesta, V., Botella, M.A., 2009. Plasma membrane repair in plants. Trends Plant Sci. 14, 645–652.

Schiavon, M., Pilon-Smits, E.A.H., 2017. The fascinating facets of plant selenium accumulation – biochemistry, physiology, evolution and ecology. New Phytol. 213, 1582–1596.

Schimansky, C., 1981. Der Einfluß einiger Versuchsparameter auf das Fluxverhalten von ^{28}Mg bei Gerstenkeimpflanzen in Hydrokulturversuchen. Landwirtsch. Forsch. 34, 154–165.

Schjørring, J.K., 1986. Nitrate and ammonium absorption by plants growing at a sufficient or insufficient level of phosphorus in nutrient solution. Plant Soil 91, 313–318.

Schumacher, K., 2014. pH in the plant endomembrane system – an import and export business. Curr. Opin. Plant Biol. 22, 71–76.

Schunmann, P.H.D., Richardson, A.E., Smith, F.W., Delhaize, E., 2004. Characterization of promoter expression patterns derived from the *Pht1* phosphate transporter genes of barley (*Hordeum vulgare* L.). J. Exp. Bot. 55, 855–865.

Segami, S., Asaoka, M., Kinoshita, S., Fukuda, M., Nakanishi, Y., Maeshima, M., 2018. Biochemical, structural and physiological characteristics of vacuolar H$^+$-pyrophosphatase. Plant Cell Physiol. 59, 1300–1308.

Sentenac, H., Grignon, C., 1985. Effect of pH on orthophosphate uptake by corn roots. Plant Physiol. 77, 136–141.

Seresinhe, P.S.J.W., Oertli, J.J., 1991. Effects of boron on growth of tomato cell suspensions. Physiol. Plant. 81, 31–36.

Shao, J.F., Yamaji, N., Liu, X.W., Yokosho, K., Shen, R.F., Ma, J.F., 2018. Preferential distribution of boron to developing tissues is mediated by the intrinsic protein OsNIP3. Plant Physiol. 176, 1739–1750.

Sharma, T., Dreyer, I., Riedelsberger, J., 2013. The role of K$^+$ channels in uptake and redistribution of potassium in the model plant *Arabidopsis thaliana*. Front. Plant Sci. 4, 224. Available from: https://doi.org/10.3389/fpls.2013.00224.

Shaul, O., Hilgemann, D.W., de-Almeida-Engler, J., Van Montagu, M., Inze, D., Galili, G., 1999. Cloning and characterization of a novel Mg^{2+}/H$^+$ exchanger. EMBO J. 18, 3973–3980.

Shi, Y., Wang, Y., Flowers, T.J., Gong, H., 2013. Silicon decreases chloride transport in rice (*Oryza sativa* L.) in saline conditions. J. Plant Physiol. 170, 847–853.

Shigaki, T., Hirschi, K.D., 2006. Diverse functions and molecular properties emerging for CAX cation/H$^+$ exchangers in plants. Plant Biol. 8, 419–429.

Shin, H., Shin, H.-S., Dewbre, G.R., Harrison, M.J., 2004. Phosphate transport in *Arabidopsis*: Pht1;1 and Pht1;4 play a major role in phosphate acquisition from both low- and high-phosphate environments. Plant J. 39, 629–642.

Shinmachi, F., Buchner, P., Stroud, J.L., Parmar, S., Zhao, F.J., McGrath, S.P., et al., 2010. Influence of sulfur deficiency on the expression of specific sulfate transporters and the distribution of sulfur, selenium, and molybdenum in wheat. Plant Physiol. 153, 327–336.

Shone, M.G.T., Flood, A.V., 1985. Measurement of free space and sorption of large molecules by cereal roots. Plant Cell Environ. 8, 309–315.

Siao, W., Coskun, D., Baluška, F., Kronzucker, H.J., Xu, W., 2020. Root-apex proton fluxes at the centre of soil-stress acclimation. Trends Plant Sci. 25, 794–804.

Siddiqi, M.Y., Glass, A.D.M., Ruth, T.J., Fernando, M., 1989. Studies of the regulation of nitrate influx by barley seedlings using ^{13}NO$_3^-$. Plant Physiol. 90, 806–813.

Siddiqi, M.Y., Glass, A.D.M., Ruth, T.J., Rufty, T.W., 1990. Studies of the uptake of nitrate in barley. I. Kinetics of ^{13}NO$_3^-$ influx. Plant Physiol. 93, 1426–1432.

Silberbush, M., Barber, S.A., 1984. Phosphorus and potassium uptake of field-grown soybean cultivars predicted by a simulation model. Soil Sci. Soc. Am. J. 48, 592–596.

Socha, A.L., Guerinot, M.L., 2014. Mn-euvering manganese: the role of transporter gene family members in manganese uptake and mobilization in plants. Front. Plant Sci. 5, 106. Available from: https://doi.org/10.3389/fpls.2014.00106.

Sorek, N., Bloch, D., Yalovsky, S., 2009. Protein lipid modifications in signaling and subcellular targeting. Curr. Opin. Plant Biol. 12, 714–720.

Sorgona, A., Cacco, G., Di Dio, L., Schmidt, W., Perry, P.J., Abenavoli, M.R., 2010. Spatial and temporal patterns of net nitrate uptake regulation and kinetics along the tap root of *Citrus aurantium*. Acta Physiol. Plant. 32, 683–693.

Soukup, A., Armstrong, W., Schreiber, L., Franke, R., Votrubova, O., 2007. Apoplastic barriers to radial oxygen loss and solute penetration: a chemical and functional comparison of the exodermis of two wetland species, *Phragmites australis* and *Glyceria maxima*. New Phytol. 173, 264–278.

Staehelin, L.A., 2015. Membrane structure and membranous organelles. In: Buchanan, B.B., Gruissem, W., Jones, R.L. (Eds.), Biochemistry and Molecular Biology of Plants, second ed. Wiley, Chichester, pp. 2–44.

Steudle, E., 2000. Water uptake by plant roots: an integration of views. Plant Soil 226, 45–56.

Stroud, J.L., Li, H.F., Lopez-Bellido, F.J., Broadley, M.R., Foot, I., Fairweather-Tait, S.J., et al., 2010. Impact of sulphur fertilisation on crop response to selenium fertilisation. Plant Soil 332, 31–40.

Sutton, T., Baumann, U., Hayes, J., Collins, N.C., Shi, B.-J., Schnurbusch, T., et al., 2007. Boron-toxicity tolerance in barley arising from efflux transporter amplification. Science 318, 1446–1449.

Szczerba, M.W., Britto, D.T., Kronzucker, H.J., 2006. The face value of ion fluxes: the challenge of determining influx in the low-affinity transport range. J. Exp. Bot. 57, 3293–3300.

Szczerba, M.W., Britto, D.T., Balkos, K.D., Kronzucker, H.J., 2008. Alleviation of rapid, futile ammonium cycling at the plasma membrane by potassium reveals K^+-sensitive and -insensitive components of NH_4^+ transport. J. Exp. Bot. 59, 303–313.

Sze, H., Chanroj, S., 2018. Plant endomembrane dynamics: studies of K^+/H^+ antiporters provide insights on the effects of pH and ion homeostasis. Plant Physiol. 177, 875–895.

Takahashi, H., 2019. Sulfate transport systems in plants: functional diversity and molecular mechanisms underlying regulatory coordination. J. Exp. Bot. 70, 4075–4087.

Takano, J., Tanaka, M., Toyoda, A., Miwa, K., Kasai, K., Fuji, K., et al., 2010. Polar localization and degradation of Arabidopsis boron transporters through distinct trafficking pathways. Proc. Natl. Acad. Sci. USA 107, 5220–5225.

Tang, R.-J., Luan, S., 2017. Regulation of calcium and magnesium homeostasis in plants: from transporters to signaling network. Curr. Opin. Plant Biol. 39, 97–105.

Tang, Z., Fan, X., Li, Q., Feng, H., Miller, A.J., Shen, Q., et al., 2012. Knockdown of a rice stelar nitrate transporter alters long-distance translocation but not root influx. Plant Physiol. 160, 2052–2063.

Tang, R.J., Wang, C., Li, K., Luan, S., 2020a. The CBL–CIPK calcium signaling network: unified paradigm from 20 years of discoveries. Trends Plant Sci. 25, 604–617.

Tang, R.J., Zhao, F.G., Yang, Y., Wang, C., Li, K., Kleist, T.J., et al., 2020b. A calcium signalling network activates vacuolar K^+ remobilization to enable plant adaptation to low-K environments. Nat. Plants 6, 384–393.

Tanoi, K., Saito, T., Iwata, N., Kobayashi, N.I., Nakanishi, T.M., 2011. The analysis of magnesium transport system from external solution to xylem in rice root. Soil Sci. Plant Nutr. 57, 265–271.

Tardieu, F., Simonneau, T., Parent, B., 2015. Modelling the coordination of the controls of stomatal aperture, transpiration, leaf growth, and abscisic acid: update and extension of the Tardieu–Davies model. J. Exp. Bot. 66, 2227–2237.

Teakle, N.L., Tyerman, S.D., 2010. Mechanisms of Cl^- transport contributing to salt tolerance. Plant Cell Environ. 33, 566–589.

Teakle, N., Flowers, T., Real, D., Colmer, T., 2007. *Lotus tenuis* tolerates the interactive effects of salinity and waterlogging by 'excluding' Na^+ and Cl^- from the xylem. J. Exp. Bot. 58, 2169–2180.

Tegeder, M., Masclaux-Daubresse, C., 2018. Source and sink mechanisms of nitrogen transport and use. New Phytol. 217, 35–53.

ten Hoopen, F., Cuin, T.A., Pedas, P., Hegelund, J.N., Shabala, S., Schjoerring, J.K., et al., 2010. Competition between uptake of ammonium and potassium in barley and Arabidopsis roots: molecular mechanisms and physiological consequences. J. Exp. Bot. 61, 2303–2315.

The, S.V., Snyder, R., Tegeder, M., 2021. Targeting nitrogen metabolism and transport processes to improve plant nitrogen use efficiency. Front. Plant Sci. 11, 628366. Available from: https://doi.org/10.3389/fpls.2020.628366.

Tilsner, J., Nicolas, W., Rosado, A., Bayer, E.M., 2016. Staying tight: plasmodesmal membrane contact sites and the control of cell-to-cell connectivity in plants. Annu. Rev. Plant Biol. 67, 337–364.

Tominaga, M., Ito, K., 2015. The molecular mechanism and physiological role of cytoplasmic streaming. Curr. Opin. Plant Biol. 27, 104–110.

Trevisan, S., Manoli, A., Ravazzolo, L., Botton, A., Pivato, M., Masi, A., et al., 2015. Nitrate sensing by the maize root apex transition zone: a merged transcriptomic and proteomic survey. J. Exp. Bot. 66, 3699–3715.

Triplett, E.W., Barnett, N.M., Blevins, D.G., 1980. Organic acids and ionic balance in xylem exudate of wheat during nitrate or sulfate absorption. Plant Physiol. 65, 610–613.

Tsujii, M., Tanudjaja, E., Uozumi, N., 2020. Diverse physiological functions of cation proton antiporters across bacteria and plant cells. Int. J. Mol. Sci. 21, 4566. Available from: https://doi.org/10.3390/ijms21124566.

Tyerman, S.D., McGaughey, S.A., Qiu, J., Yool, A.J., Byrt, C.S., 2021. Adaptable and multifunctional ion-conducting aquaporins. Annu. Rev. Plant Biol. 72, 703–736.

Tyler, G., Olsson, T., 2001. Concentrations of 60 elements in the soil solution as related to the soil acidity. Eur. J. Soil Sci. 52, 151–165.

Tylová, E., Pecková, E., Blascheová, Z., Soukup, A., 2017. Casparian bands and suberin lamellae in exodermis of lateral roots: an important trait of roots system response to abiotic stress factors. Ann. Bot. 120, 71–85.

Ueda, Y., Yanagisawa, S., 2019. Perception, transduction, and integration of nitrogen and phosphorus nutritional signals in the transcriptional regulatory network in plants. J. Exp. Bot. 70, 3709–3717.

Ueno, D., Sasaki, A., Yamaji, N., Miyaji, T., Fujii, Y., Takemoto, Y., et al., 2015. A polarly localized transporter for efficient manganese uptake in rice. Nat. Plants 1, 15170. Available from: https://doi.org/10.1038/nplants.2015.170.

Vakhmistrov, D.B., 1967. On the function of the apparent free space in plant roots. A comparative study of the absorption power of epidermal and cortex cells in barley roots. Sov. Plant Physiol. (English Translation) 14, 123–129.

Vakhmistrov, D.B., 1981. Specialization of root tissues in ion transport. Plant Soil 63, 33–38.

Vallejo, A.J., Peralta, M.L., Santa-Maria, G.E., 2005. Expression of potassium-transporter coding genes, and kinetics of rubidium uptake, along a longitudinal root axis. Plant Cell Environ. 28, 850–862.

van Beusichem, M.L., Kirkby, E.A., Baas, R., 1988. Influence of nitrate and ammonium nutrition and the uptake, assimilation, and distribution of nutrients in *Ricinus communis*. Plant Physiol. 86, 914–921.

Van der Werf, A., Kooijman, A., Welschen, R., Lambers, H., 1988. Respiratory energy costs for the maintenance of biomass, for growth and for iron uptake in roots of *Carex diandra* and *Carex acutiformis*. Physiol. Plant. 72, 483–491.

Vance, C.P., 2010. Quantitative trait loci, epigenetics, sugars, and microRNAs: quaternaries in phosphate acquisition and use. Plant Physiol. 154, 582–588.
Veneklaas, E.J., Lambers, H., Bragg, J., Finnegan, P.M., Lovelock, C.E., Plaxton, W.C., et al., 2012. Opportunities for improving phosphorus-use efficiency in crop plants. New Phytol. 195, 306–320.
Verrier, P.J., Bird, D., Burla, B., Dassa, E., Forestier, C., Geisler, M., et al., 2008. Plant ABC proteins – a unified nomenclature and updated inventory. Trends Plant Sci. 13, 151–159.
Vert, G.A., Briat, J.-F., Curie, C., 2003. Dual regulation of the Arabidopsis high-affinity root iron uptake system by local and long-distance signals. Plant Physiol. 132, 796–804.
Vetterlein, D., Doussan, C., 2016. Root age distribution: how does it matter in plant processes? A focus on water uptake. Plant Soil 407, 145–160.
Vidal, E.A., Alvarez, J.M., Araus, V., Riveras, E., Brooks, M.D., Krouk, G., et al., 2020. Nitrate in 2020: thirty years from transport to signaling networks. Plant Cell 32, 2094–2119.
Viets, F.G., 1944. Calcium and other polyvalent cations as accelerators of ion accumulation by excised barley roots. Plant Physiol. 19, 466–480.
Voesenek, L.A.C.J., Bailey-Serres, J., 2015. Flood adaptive traits and processes: an overview. New Phytol. 206, 57–73.
von der Fecht-Bartenbach, J., Bogner, M., Dynowski, M., Ludewig, U., 2010. CLC-b-mediated NO_3^-/H^+ exchange across the tonoplast of Arabidopsis vacuoles. Plant Cell Physiol. 51, 960–968.
von Wirén, N., Marschner, H., Römheld, V., 1995. Uptake kinetics of iron-phytosiderophores in two maize genotypes differing in iron efficiency. Physiol. Plant. 93, 611–616.
von Wirén, N., Lauter, F.R., Ninnemann, O., Gillissen, B., Walch-Liu, P., Engels, C., et al., 2000. Differential regulation of three functional ammonium transporter genes by nitrogen in root hairs and by light in leaves of tomato. Plant J. 21, 167–175.
Walker, D.J., Leigh, R.A., Miller, A.J., 1996. Potassium homeostasis in vacuolate plant cells. Proc. Natl. Acad. Sci. USA 93, 10510–10514.
Wallis, J.G., Browse, J., 2002. Mutants of *Arabidopsis* reveal many roles for membrane lipids. Prog. Lipid Res. 41, 254–278.
Wang, L., Duan, G., 2009. Effect of external and internal phosphate status on arsenic toxicity and accumulation in rice seedlings. J. Environ. Sci. 21, 346–351.
Wang, Y., Lambers, H., 2020. Root-released organic anions in response to low phosphorus availability: recent progress, challenges and future perspectives. Plant Soil 447, 135–156.
Wang, M.Y., Siddiqi, M.Y., Ruth, T.J., Glass, A.D.M., 1993. Ammonium uptake by rice roots. 2. Kinetics of $^{13}NH_4^+$ influx across the plasmalemma. Plant Physiol. 103, 1259–1267.
Wang, Y.-Y., Hsu, P.-K., Tsay, Y.-F., 2012. Uptake, allocation and signaling of nitrate. Trends Plant Sci. 17, 458–467.
Wang, S., Yoshinari, A., Shimada, T., Hara-Nishimura, I., Mitani-Ueno, N., Ma, J.F., et al., 2017. Polar localization of the NIP5;1 boric acid channel is maintained by endocytosis and facilitates boron transport in Arabidopsis roots. Plant Cell 29, 824–842.
Wang, F., Deng, M., Xu, J., Zhu, X., Mao, C., 2018a. Molecular mechanisms of phosphate transport and signaling in higher plants. Semin. Cell Dev. Biol. 74, 114–122.
Wang, Y.-Y., Cheng, Y.-H., Chen, K.-E., Tsay, Y.-F., 2018b. Nitrate transport, signaling, and use efficiency. Annu. Rev. Plant Biol. 69, 85–122.
Wang, W., Ding, G.-D., White, P.J., Wang, X.-H., Jin, K.-M., Xu, F.-S., et al., 2019. Mapping and cloning of quantitative trait loci for phosphorus efficiency in crops: opportunities and challenges. Plant Soil 439, 91–112.
Wang, C., Xiang, Y., Qian, D., 2021. Current progress in plant V-ATPase: from biochemical properties to physiological functions. J. Plant Physiol. 266, 153525. Available from: https://doi.org/10.1016/j.jplph.2021.153525.
Watson, R., Pritchard, J., Malone, M., 2001. Direct measurement of sodium and potassium in the transpiration stream of salt-excluding and non-excluding varieties of wheat. J. Exp. Bot. 52, 1873–1881.
Webb, M.J., Norvell, W.A., Welch, R.M., Graham, R.D., 1993. Using a chelate-buffered nutrient solution to establish the critical solution activity of Mn^{2+} required by barley (*Hordeum vulgare* L.). Plant Soil 153, 195–205.
Wege, S., Gilliham, M., Henderson, S.W., 2017. Chloride: not simply a 'cheap osmoticum', but a beneficial plant macronutrient. J. Exp. Bot. 68, 3057–3069.
Welti, R., Li, W., Li, M., Sang, Y., Biesiada, H., Zhou, H.-E., et al., 2002. Profiling membrane lipids in plant stress responses. Role of phospholipase Dα in freezing-induced lipid changes in *Arabidopsis*. J. Biol. Chem. 277, 31994–32002.
Wenzel, C.L., McCully, M.E., Canny, M.J., 1989. Development of water conducting capacity in the root system of young plants of corn and some other C_4 species. Plant Physiol. 89, 1094–1101.
White, P.J., 1997a. Cation channels in the plasma membrane of rye roots. J. Exp. Bot. 48, 499–514.
White, P.J., 1997b. The regulation of K^+ influx into roots of rye (*Secale cereale* L.) seedlings by negative feedback via the K^+ flux from shoot to root in the phloem. J. Exp. Bot. 48, 2063–2073.
White, P.J., 1999. The molecular mechanism of sodium influx to root cells. Trends Plant Sci. 4, 245–246.
White, P.J., 2001. The pathways of calcium movement to the xylem. J. Exp. Bot. 52, 891–899.
White, P.J., 2013. Improving potassium acquisition and utilisation by crop plants. J. Plant Nutr. Soil Sci. 176, 305–316.
White, P.J., 2015. Calcium. In: Barker, A.V., Pilbeam, D.J. (Eds.), A Handbook of Plant Nutrition, second ed. CRC Press, Boca Raton, FL, pp. 165–198.
White, P.J., 2016. Selenium accumulation by plants. Ann. Bot. 117, 217–235.
White, P.J., 2017. Ion transport. In: Thomas, B., Murphy, D.J., Murray, B.G. (Eds.), Encyclopedia of Applied Plant Sciences, second ed. Academic Press, Waltham, MA, pp. 238–245.
White, P.J., 2018. Selenium metabolism in plants. Biochim. Biophys. Acta-Gen. Subj. 1862, 2333–2342.
White, P.J., Broadley, M.R., 2000. Mechanisms of caesium uptake by plants. New Phytol. 147, 241–256.

White, P.J., Broadley, M.R., 2001. Chloride in soils and its uptake and movement within the plant: a review. Ann. Bot. 88, 967–988.
White, P.J., Broadley, M.R., 2003. Calcium in plants. Ann. Bot. 92, 487–511.
White, P.J., Hammond, J.P., 2008. Phosphorus nutrition of terrestrial plants. In: White, P.J., Hammond, J.P. (Eds.), The Ecophysiology of Plant-Phosphorus Interactions. Springer, Dordrecht, pp. 51–81.
White, P.J., Broadley, M.R., 2009. Biofortification of crops with seven mineral elements often lacking in human diets — iron, zinc, copper, calcium, magnesium, selenium and iodine. New Phytol. 182, 49–84.
White, P.J., Brown, P.H., 2010. Plant nutrition for sustainable development and global health. Ann. Bot. 105, 1073–1080.
White, P.J., Karley, A.J., 2010. Potassium. In: Hell, R., Mendel, R.-R. (Eds.), Cell Biology of Metals and Nutrients. Plant Cell Monographs 17, Springer, Dordrecht, pp. 199–224.
White, P.J., Pongrac, P., 2017. Heavy-metal toxicity in plants. In: Shabala, S. (Ed.), Plant Stress Physiology, second ed. CABI, Wallingford, pp. 301–331.
White, P.J., Neugebauer, K., 2021. Possible consequences of an inability of plants to control manganese uptake. Plant Soil 461, 63–68.
White, P.J., Clarkson, D.T., Earnshaw, M.J., 1987. Acclimation of potassium influx in rye (*Secale cereale*) to low root temperatures. Planta 171, 377–385.
White, P.J., Cooke, D.T., Earnshaw, M.J., Clarkson, D.T., Burden, R.S., 1990a. Does plant growth temperature modulate the membrane composition and ATPase activities of tonoplast and plasma membrane fractions from rye roots? Phytochemistry 29, 3385–3393.
White, P.J., Marshall, J., Smith, J.A.C., 1990b. Substrate kinetics of the tonoplast H^+-translocating inorganic pyrophosphatase and its activation by free Mg^{2+}. Plant Physiol. 93, 1063–1070.
White, P.J., Earnshaw, M.J., Clarkson, D.T., 1991. Effects of growth and assay temperatures on unidirectional K^+ fluxes in roots of rye (*Secale cereale*). J. Exp. Bot. 42, 1031–1041.
White, P.J., Banfield, J., Diaz, M., 1992. Unidirectional Ca^{2+} fluxes in roots of rye (*Secale cereale* L.). A comparison of excised roots with roots of intact plants. J. Exp. Bot. 43, 1061–1074.
White, P.J., Bowen, H.C., Parmaguru, P., Fritz, M., Spracklen, W.P., Spiby, R.E., et al., 2004. Interactions between selenium and sulphur nutrition in *Arabidopsis thaliana*. J. Exp. Bot. 55, 1927–1937.
White, P.J., Broadley, M.R., Gregory, P.J., 2012. Managing the nutrition of plants and people. Appl. Environ. Soil Sci. 104826. Available from: https://doi.org/10.1155/2012/104826.
White, P.J., George, T.S., Dupuy, L.X., Karley, A.J., Valentine, T.A., Wiesel, L., et al., 2013a. Root traits for infertile soils. Front. Plant Sci. 4, 193. Available from: https://doi.org/10.3389/fpls.2013.00193.
White, P.J., George, T.S., Gregory, P.J., Bengough, A.G., Hallett, P.D., McKenzie, B.M., 2013b. Matching roots to their environment. Ann. Bot. 112, 207–222.
White, P.J., Broadley, M.R., El-Serehy, H.A., George, T.S., Neugebauer, K., 2018. Linear relationships between shoot magnesium and calcium concentrations among angiosperm species are associated with cell wall chemistry. Ann. Bot. 122, 221–226.
White, P.J., Bell, M.J., Djalovic, I., Hinsinger, P., Rengel, Z., 2021a. Potassium use efficiency of plants. In: Murrell, T.S., Mikkelsen, R.L., Sulewski, G., Norton, R., Thompson, M.L. (Eds.), Improving Potassium Recommendations for Agricultural Crops. Springer, Cham, Switzerland, pp. 119–145.
White, P.J., Ding, G., Shi, L., Xu, F., 2021b. Proceedings of the International Fertiliser Society (IFS) 859: Boron in Plant Physiology and Nutrition. IFS, York.
Wild, R., Gerasimaite, R., Jung, J.Y., Truffault, V., Pavlovic, I., Schmidt, A., et al., 2016. Control of eukaryotic phosphate homeostasis by inositol polyphosphate sensor domains. Science 352, 986–990.
Williams, L.E., Mills, R.F., 2005. P_{1B}-ATPases — an ancient family of transition metal pumps with diverse functions in plants. Trends Plant Sci. 10, 491–502.
Wimmer, M.A., Abreu, I., Bell, R.W., Bienert, M.D., Brown, P.H., Dell, B., et al., 2020. Boron: an essential element for vascular plants. New Phytol. 226, 1232–1237.
Wu, J., Seliskar, D.M., Gallagher, J.L., 1998. Stress tolerance in the marsh plant *Spartina patens*: impact of NaCl on growth and root plasma membrane lipid composition. Physiol. Plant. 102, 307–317.
Wu, S.-W., Kumar, R., Iswanto, A.B.B., Kim, J.-Y., 2018. Callose balancing at plasmodesmata. J. Exp. Bot. 69, 5325–5339.
Xu, G., Magen, H., Tarchitzky, J., Kafkafi, U., 2000. Advances in chloride nutrition of plants. Adv. Agron. 68, 97–150.
Xue, S., Yao, X., Luo, W., Jha, D., Tester, M., Horie, T., et al., 2011. AtHKT1;1 mediates Nernstian sodium channel transport properties in Arabidopsis root stelar cells. PLoS One 6, e24725. Available from: https://doi.org/10.1371/journal.pone.0024725.
Yeo, A.R., Yeo, M.E., Flowers, T.J., 1987. The contribution of an apoplastic pathway to sodium uptake by rice roots in saline conditions. J. Exp. Bot. 38, 1141–1153.
Yoshida, S., Uemura, M., 1986. Lipid composition of plasma membranes and tonoplasts isolated from etiolated seedlings of mung bean (*Vigna radiata* L.). Plant Physiol. 82, 807–812.
Yoshinari, A., Takano, J., 2017. Insights into the mechanisms underlying boron homeostasis in plants. Front. Plant Sci. 8, 1951. Available from: https://doi.org/10.3389/fpls.2017.01951.
Yoshinari, A., Hosokawa, T., Amano, T., Beier, M.P., Kunieda, T., Shimada, T., et al., 2019. Polar localization of the borate exporter BOR1 requires AP2-dependent endocytosis. Plant Physiol. 179, 1569–1580.
Yuan, L., Loqué, D., Kojima, S., Rauch, S., Ishiyama, K., Inoue, E., et al., 2007. The organization of high-affinity ammonium uptake in Arabidopsis roots depends on the spatial arrangement and biochemical properties of AMT1-type transporters. Plant Cell 19, 2636–2652.

Zhang, F.S., Römheld, V., Marschner, H., 1991a. Diurnal rhythm of release of phytosiderophores and uptake rate of zinc in iron-deficient wheat. Soil Sci. Plant Nutr. 37, 671–678.

Zhang, F., Römheld, V., Marschner, H., 1991b. Role of the root apoplasm for iron acquisition by wheat plants. Plant Physiol. 97, 1302–1305.

Zhang, L., Hu, B., Li, W., Che, R., Deng, K., Li, H., et al., 2014. OsPT2, a phosphate transporter, is involved in the active uptake of selenite in rice. New Phytol. 201, 1183–1191.

Zhao, F.-J., McGrath, S.P., Meharg, A.A., 2010. Arsenic as a food chain contaminant: mechanisms of plant uptake and metabolism and mitigation strategies. Annu. Rev. Plant Biol. 61, 535–559.

Zhu, J., Raschke, K., Köhler, B., 2007. An electrogenic pump in the xylem parenchyma of barley roots. Physiol. Plant. 129, 397–406.

Zhu, J.M., Zhang, C.C., Lynch, J.P., 2010. The utility of phenotypic plasticity of root hair length for phosphorus acquisition. Funct. Plant Biol. 37, 313–322.

Zifarelli, G., Pusch, M., 2010. CLC transport proteins in plants. FEBS Lett. 584, 2122–2127.

Chapter 3

Long-distance transport in the xylem and phloem

Philip J. White[1] and Guangda Ding[2]

[1]Ecological Sciences, James Hutton Institute, Dundee, Scotland, United Kingdom, [2]College of Resources and Environment, Huazhong Agricultural University, Wuhan, P.R. China

Summary

Long-distance transport of solutes in the xylem and phloem is important for shoot and root nutrition, the redistribution of essential elements between tissues during ontogeny, the maintenance of charge balance in leaves of nitrate-fed plants, the removal of potentially toxic elements from leaf tissues, and the systemic signaling of plant nutritional status. This chapter describes the anatomy of the xylem and phloem, the composition of xylem and phloem saps, and the movement of (1) xylem sap from root to shoot in response to gradients of water potential generated by root pressure and transpiration and (2) phloem sap from source to sink tissues in response to osmotic gradients generated by differences in phloem sucrose concentration. Emphasis is placed on the pathways of solute movement within the plant and recent insight into the transport proteins catalyzing the loading and unloading of elements to and from the xylem and phloem.

3.1 General

The long-distance transport of water and solutes takes place in the vascular system of xylem and phloem (White, 2017).

Long-distance transport from the roots to the shoots occurs predominantly in xylem vessels. Xylem vessels of angiosperms, with their thick, lignified secondary cell walls, are arranged in files and their adjoining end walls are perforated. The diameter of these vessels can vary between 8 and 500 μm. These vessels are associated with narrower tracheids and with living xylem parenchyma cells, which regulate the fluxes of solutes into and out of the vessel elements. Tracheids are the only type of water-conducting cells in most gymnosperms and seedless vascular plants. The juxtaposition of xylem and phloem within the stele allows the transfer of solutes from xylem to phloem.

Xylem transport is driven by the gradient in hydrostatic pressure (root pressure) and by the gradient in water potential (White, 2017). Pure free water is defined as having a water potential of zero. Accordingly, values for water potential are usually negative. The gradient in water potential between roots and shoots is quite steep, particularly during the day when stomata are open. Values become less negative in the following sequence: atmosphere >> leaf cells > xylem sap > root cells > external solution. Solute flow in the xylem from the roots to the shoots is therefore unidirectional (Fig. 3.1). However, under certain conditions, a counterflow of water in the xylem may occur in shoots, for example, from low-transpiring fruits back to the leaves (Lang and Thorpe, 1989).

In contrast to the xylem, long-distance transport in the phloem takes place in the living sieve-tube cells and is bidirectional (White, 2017). The direction of transport is determined primarily by the growth of plant organs or tissues and occurs, therefore, from a source to a sink (Chapter 5). In addition, the phloem is an important pathway for the cycling of nutrients between shoots and roots and for signaling the nutritional status of the shoot to the root. Elements can enter the phloem in either the shoot or the root. The translocation of different elements taken up by a particular zone of the root varies markedly as shown in Table 3.1 for maize seedlings. Long-distance transport from the zone of supply (12–15 cm from the root tip) to the root tip must take place in the phloem. Although ^{45}Ca is rapidly translocated to the shoot, it is not transported to the root tip. By contrast, the translocation

---

[☆]This chapter is a revision of the third edition chapter by P.J. White, pp. 49–70. DOI: https://doi.org/10.1016/B978-0-12-384905-2.00003-0. © Elsevier Ltd.

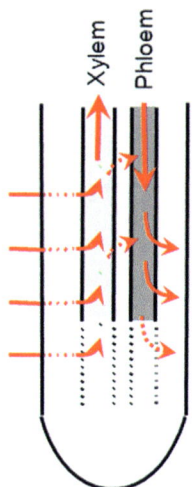

FIGURE 3.1 Direction of long-distance transport of elements in the nonliving xylem vessels and in the phloem of roots.

TABLE 3.1 Accumulation and long-distance transport of ^{45}Ca, ^{22}Na, and ^{42}K in 9-day-old maize (Zea mays) seedlings with 25- to 30-cm-long roots.[a]

Plant part	Rate (nmol plant^{-1} d^{-1})		
	^{45}Ca	^{22}Na	^{42}K
Shoot	183	1	755
Endosperm	15	4	198
Root (cm from tip)			
24–27	1	5	37
21–24	1	7	71
18–21	1	15	108
15–18	1	39	132
12–15 (zone of supply)	34	106	161
9–12	0	3	33
6–9	0	2	31
3–6	0	2	38
0–3	0	1	63
Total	235	184	1772

[a]*Each seedling was supplied with 1 mM of labeled nutrient solution to the root zone 12–15 cm from the root tip. The remainder of the root system was supplied with the same solution in which the nutrients were not labeled.*
Source: Based on Marschner and Richter (1973).

of ^{22}Na toward the shoot is severely restricted. The steep gradient in ^{22}Na content of the root sections in the direction of the shoot reflects retrieval by the surrounding root tissue and is a typical feature of natrophobic plant species (see Chapter 18). Some ^{22}Na is also translocated via the phloem to the root tip. By contrast, ^{42}K is mobile in both the xylem and the phloem, and a high proportion of the ^{42}K taken up in basal root zones is translocated via the xylem to the shoot and via the phloem toward the root tip.

3.2 Xylem transport

3.2.1 Composition of the xylem sap

The composition of xylem sap and concentrations of elements and organic solutes in the xylem sap depend on factors such as plant species, element supply to the roots, assimilation of nutrients in the roots, and nutrient recycling. The concentrations

TABLE 3.2 Xylem volume flow (pressurized exudation at 100 kPa) and nutrient concentrations in the xylem sap of soil-grown nodulated soybean (*Glycine max*) during the reproductive stage.

Parameter	Plant developmental stages			
	Full pod extension	Early-to-mid podfill	Late podfill	Early leaf yellowing
Sap volume (mL plant^{-1} h^{-1})	1.72	1.57	0.64	0.80
Nutrient concentrations				
K (mM)	11.5	8.3	5.7	4.2
Ca (mM)	3.9	3.4	2.3	2.1
Mg (mM)	2.5	2.5	1.6	1.6
P (mM)	3.6	3.6	1.4	1.3
S (mM)	1.8	1.3	1.3	2.3
B (μM)	1.6	2.1	3.6	3.0
Zn (μM)	24.7	22.7	13.0	13.7
Fe (μM)	23.6	31.3	14.5	11.1
Mn (μM)	20.0	22.6	10.3	6.3
Cu (μM)	2.8	4.5	2.3	1.9
Mo (μM)	1.7	1.3	1.1	1.0

Source: Based on Noodén and Mauk (1987).

of solutes are also strongly influenced by dilution by water and are, therefore, dependent on the transpiration rate and the time of the day. The composition and concentration of xylem sap also change during plant development. For example, during the reproductive stage in soybean, the concentrations of most nutrients in xylem sap decrease while that of B increases (Table 3.2). The decline in nutrient concentrations can be prevented by removal of the pods. This eases the competition for photosynthates between pods and roots, leading to greater nutrient uptake and transport to the xylem (Noodén and Mauk, 1987). In perennial species, the composition of xylem sap changes with season; for example, inorganic and organic solutes are mobilized in spring (Glavac et al., 1991; Losso et al., 2018).

The proportions of the various N fractions in xylem sap depend on the form of N supply (NO_3^-, NH_4^+, and N_2 fixation), the predominant site of nitrate reduction (roots or shoots) and the amount of N recycled from the shoot. The nitrate concentration in xylem sap ranges widely, from about 3 μM to 25 mM, depending on environmental conditions and plant species (Cui et al., 2021). Except at very high external NH_4^+ supply, the concentration of NH_4^+ in the xylem sap is about 1 mM, irrespective of whether N is supplied as NH_4^+ or NO_3^- (Finnemann and Schjoerring, 1999). A variety of amino acids and amides are also present in xylem sap, with the most abundant being aspartate, glutamate, asparagine, and glutamine (Tegeder and Masclaux-Daubresse, 2018). In plants that assimilate N in roots or root nodules, the amides glutamine and asparagine, the amino-acid arginine, and the ureides allantoin and allantoic acid can be found in xylem sap, depending on plant species (Tegeder and Masclaux-Daubresse, 2018). The concentration of organic acids in the xylem sap depends primarily on the root cation:anion uptake ratio and the form of N supply. A variety of peptides and proteins are also found in xylem sap, which are thought to act in communicating, for example, N availability and biotic challenges to the shoot (Chapter 5; Notaguchi and Okamoto, 2015; Tegeder and Masclaux-Daubresse, 2018).

In the xylem sap, K and Na are present as free cations, Ca and Mg are found as free cations or as complexes with organic acids, and polyvalent "heavy metal" cations occur mainly in organic forms complexed with organic acids, deoxymugineic acid (DMA), nicotianamine (NA), amino acids, and peptides (Álvarez-Fernández et al., 2014; Clemens, 2019; Flis et al., 2016; Karley and White, 2009). Both the number and abundance of these complexes vary with plant species, nutrition, and plant age. The concentrations of K, Ca, and Mg in the xylem sap are normally in the ranges 5–20, <1–20, and 0.5–1.0 mM, respectively (Karley and White, 2009), and the concentrations of Fe, Zn, Mn, and Cu in the xylem sap are normally in the ranges 1–40, 1–148, 4–80, and 0.2–9 μM, respectively but can reach greater concentrations if the supply of these elements is excessive (Álvarez-Fernández et al., 2014).

In the xylem sap of annual species, high concentrations of sugars may also occur. For example, sugar concentrations of up to 5 mM (mainly as sucrose) can occur in maize (Canny and McCully, 1989), and sugars may account for about 15% w/w of the total organic carbon in the xylem sap of soybean (Cataldo et al., 1988). Phytohormones are normal constituents of xylem sap, particularly cytokinins that regulate shoot growth and delay senescence (Notaguchi and Okamoto, 2015), but also strigolactones that control shoot branching (Notaguchi and Okamoto, 2015), and abscisic acid (ABA) that acts as a signal of root water status and also soil strength (Hussain et al., 1999; Tardieu et al., 2015; Wilkinson and Davies, 2002). The concentrations of phytohormones in the xylem sap can affect the long-distance transport of nutrients, for example, by influencing the flow rate in the xylem, the rate of xylem-to-phloem transfer, and the nutrient distribution within the shoot. In return, the concentrations of phytohormones in the xylem sap are influenced by N, P, and K nutrition (Peuke et al., 1994, 2002; Jeschke et al., 1997b; Jiang and Hartung, 2008), and this has consequences for plant water relations, shoot growth, and responses to nutrient availability (Chapter 5). As the soil dries, stomatal conductance decreases prior to a reduction in leaf turgor, and there is an inverse relationship between xylem sap ABA concentration and stomatal conductance (Wilkinson and Davies, 2002). In addition, there is evidence that high concentrations of ABA in the xylem sap reduce rates of cell extension and cell division and, thereby, reduce the rate of leaf elongation in response to drying or compacted soil (Tardieu et al., 2010). As the soil dries, the ionic composition and pH of the xylem sap increase (Bahrun et al., 2002), which may also alter the distribution of ABA in the shoot and lead to preferential transport of ABA to the guard cells (Jiang and Hartung, 2008).

3.2.2 Xylem loading

Elements that traverse the root via the symplasmic pathway are loaded into the xylem by various transport proteins across the plasma membrane of xylem parenchyma cells. Nitrate can be loaded into the xylem of arabidopsis (*Arabidopsis thaliana*) by (1) members of the nitrate transporter 1/peptide transporter (NPF) family, such as AtNPF7.3 (AtNRT1.5) and AtNPF2.3 (O'Brien et al., 2016; Tegeder and Masclaux-Daubresse, 2018); (2) members of the slow anion channel-associated homologs (SLAC/SLAH) family, such as AtSLAH2 and the dimer AtSLAH1/AtSLAH3 (Hedrich and Geiger, 2017); and (3) members of the amino-acid transporter (UMAMIT) family (Tegeder and Masclaux-Daubresse, 2018; Wang et al., 2018).

Ammonium enters the xylem through members of the ammonium transporter (AMT) family or via nonspecific cation channels (Giehl et al., 2017; Tegeder and Masclaux-Daubresse, 2018). In plants that assimilate N in roots or root nodules, UMAMIT proteins (UMAMIT14, UMAMIT18/SIAR1) and ureide transporters load, respectively, amino acids and ureides into the xylem (Tegeder and Masclaux-Daubresse, 2018).

The phosphate transporter PHO1 is implicated in loading P into the xylem (Liu et al., 2014), and members of the ALMT family of transporters are thought to load sulfate into the xylem (Gigolashvili and Kopriva, 2014). Anion channels facilitate the movement of chloride from the cytosol of xylem parenchyma cells to xylem vessels (Gilliham and Tester, 2005; Hedrich and Geiger, 2017). Boron is loaded into the xylem by orthologs of the arabidopsis AtBOR1 transporter, the activities of which are regulated in response to plant B status to ensure appropriate B concentrations are maintained in the shoot (Miwa and Fujiwara, 2010; Yoshinari and Takano, 2017).

Potassium is loaded into the xylem by voltage-gated, outwardly rectifying K-channels present in the plasma membrane of root pericycle and stelar parenchyma cells, such as orthologs of the AtSKOR protein of arabidopsis (Gaymard et al., 1998). The SOS1 (AtNHX7) Na^+/H^+ antiporter, cation-chloride cotransporters, and nonspecific cation channels have been implicated in loading Na into the xylem (Munns and Tester, 2008; Wu, 2018).

Cations that are present in low concentrations in the cytosol of root cells are loaded into the xylem by active transport mechanisms. Calcium is loaded into the xylem by members of the P_{2A}-Ca^{2+}-ATPase and P_{2B}-Ca^{2+}-ATPase families (White and Broadley, 2003). Members of the "heavy metal" P_{1B}-ATPase family (HMA2, HMA4, HMA5) load Zn and Cu into the xylem (Andresen et al., 2018; Deng et al., 2013; Sinclair et al., 2018; White and Pongrac, 2017). In arabidopsis, Mg might be loaded into the xylem by AtCNGC10 (Hermans et al., 2013). Cation carriers have also been implicated in loading Zn, Mn, Fe, and Co into the xylem (Andresen et al., 2018; White and Pongrac, 2017). Manganese is loaded into the xylem by H^+-coupled antiporters, such as OsMTP9 in rice (Alejandro et al., 2020). A candidate for loading Fe and Co into the xylem of arabidopsis is AtFPN1 (ferroportin1; also known as IREG1, iron-regulated protein1; Morrissey et al., 2009).

Significant amounts of Ca and other potentially cytotoxic elements, such as Zn, Fe, and Na, also reach the xylem through an apoplasmic route when they are present at high concentrations in the soil solution (Broadley et al., 2007; Plett and Møller, 2010; White, 2001). These fluxes are restricted by the development of the Casparian band, which also serves to prevent the loss of nutrients, particularly K, from the stele (Barberon, 2017). The integrity of the Casparian band is sensed from the leakage of small peptides from the stele to the cortex followed by the deactivation of aquaporins together with the subsequent deposition of suberin that rebalance the long-distance transport of water and nutrients (Wang et al., 2019a).

Nitrogen is present in the xylem mostly in its inorganic forms, although amino acids, amides, and ureides have also been recorded (Peuke, 2010; Tegeder and Masclaux-Daubresse, 2018). Similarly, phosphate and sulfate are the dominant forms of P and S in the xylem. Calcium, Mg, Fe, Zn, and Mn are likely to be transported in the xylem as cations or cation complexes with organic acid anions. For example, Fe is transported mainly as Fe^{3+} or Fe^{2+} complexes with citrate or malate (Álvarez-Fernández et al., 2014; Flis et al., 2016; Kutrowska and Szelag, 2014). In arabidopsis, a member of the multidrug and toxin efflux transporter family, AtFRD3, is expressed in the root pericycle and loads citrate into the xylem (Andresen et al., 2018; White and Pongrac, 2017). Zinc and Cu can also be transported as histidine complexes (Álvarez-Fernández et al., 2014; Kutrowska and Szelag, 2014), and Zn, Cu, and Ni can be transported as DMA or NA complexes (Álvarez-Fernández et al., 2014; Clemens, 2019; Kutrowska and Szelag, 2014).

3.2.2.1 Exchange adsorption in xylem vessels

The interactions between cations and the negatively charged groups in the cell walls of the xylem vessels (and tracheides) are similar to those in the apparent free space of the root cortex (Chapter 2). Thus the long-distance transport of cations in the xylem can be compared with ion movement in a cation exchanger. The translocation rates of cations such as Ca^{2+} are less than those of water and anions. This cation-exchange adsorption is not restricted to the xylem vessels; the cell walls of the surrounding tissue also take part in these exchange reactions (Wolterbeek et al., 1984). The degree of hindrance of cation translocation depends on (1) the valency of the cation (trivalent > divalent > monovalent), (2) its concentration and activity, (3) the charge density of the negative groups, (4) the diameter of the xylem vessels, and (5) the pH of the xylem sap. The translocation rate of cations in the xylem is enhanced when they are complexed, as found for Cu (Smeulders and van de Geijn, 1983) and Zn (McGrath and Robson, 1984).

3.2.2.2 Retrieval and release of nutrients by living cells

Solutes can be retrieved from the xylem into the living cells along the pathway of the xylem sap from the roots to the leaves, which is thought to regulate the delivery of elements to the shoot. Nitrate is retrieved from the xylem by members of the NPF family, such as AtNPF7.2/AtNRT1.8 in arabidopsis (O'Brien et al., 2016; Tegeder and Masclaux-Daubresse, 2018; Wang et al., 2018). Sulfate is retrieved from the xylem by members of the SULTR (sulfate transporter) family, including SULTR2;1, SULTR2;2, and SULTR3;5 (Kataoka et al., 2004; Takahashi et al., 2000). Phosphate is retrieved from the xylem by members of the Pht1 (phosphate transporter 1) family, such as Pht1;5 in arabidopsis (Mudge et al., 2002), and by members of the SPDT (SULTR-like P distribution transporter) family, such as AtSPDT (AtSULTR3;4) in arabidopsis (Ding et al., 2020). In arabidopsis, amino acids are retrieved from the xylem by AtAAP6 (Tegeder and Masclaux-Daubresse, 2018). The retrieved elements and molecules can be stored transiently or permanently in the xylem parenchyma and other stem tissue or transferred from xylem to phloem via specialized cells.

In some plant species, the retrieval of certain elements from the xylem sap is very pronounced and can have important consequences for the nutrition of these plants. This is most evident in natrophobic plant species. In these species (e.g., *Phaseolus vulgaris*), Na is retained mainly in the roots and lower stem, whereas in natrophilic species (e.g., sugar beet), Na is readily translocated to the leaves (Fig. 3.2). The restricted upward movement of Na in natrophobic plants is caused by selective Na accumulation in the xylem parenchyma cells (Blom-Zandstra et al., 1998; Drew and Läuchli, 1987) together with re-translocation to the roots (Fig. 3.10). In castor bean, these two components led to a decrease in

FIGURE 3.2 Autoradiogram of distribution of ^{22}Na in bean (*Phaseolus vulgaris*) and sugar beet (*Beta vulgaris*) 24 h after 5 mM ^{22}NaCl was supplied to the roots.

the Na concentration of the xylem stream from 0.8 to 0.2 mM (Jeschke and Pate, 1991). A similar restriction in Cl transport to the shoot has been reported in some species, including maize (Zhang et al., 2020).

Retrieval of Na from the xylem sap is an effective mechanism for restricting translocation to the leaf blades. Homologs of the arabidopsis AtHKT1 transporter retrieve Na from the xylem (Munns and Tester, 2008; Wu et al., 2018). However, this is not the main mechanism enabling salt tolerance of halophytes that accumulate Na as an osmoticum (Flowers et al., 2010; White et al., 2017). Furthermore, for animal nutrition, the Na concentration of forage should be at least 1−2 mg g^{-1} dry mass (Cheeseman, 2015). As shown in Table 3.3, in *Lolium perenne* and *Trifolium repens*, Na is readily translocated to the shoots, whereas in *Phleum pratense* and *Trifolium hybridum*, Na translocation to the shoot is restricted. Thus it is evident that selecting suitable plant species can be just as important as the application of Na fertilizers for increasing Na concentrations in forage.

Retrieval from the xylem sap in roots and stems can also determine shoot concentrations of micronutrients. In certain species, such as common bean (*P. vulgaris*) and sunflower, Mo is preferentially accumulated in the xylem parenchyma of the roots and stems. In these species, a steep gradient occurs in tissue Mo concentrations from the roots to the leaves (Table 3.4). By contrast, in other species, such as tomato, Mo is readily translocated from the roots to the leaves. In accordance with this finding, when the Mo supply in the nutrient medium is high, toxicity occurs much earlier in tomato than in bean or sunflower (Hecht-Buchholz, 1973).

The composition of the xylem sap along the transport pathway can also be changed by the release or secretion of solutes from the surrounding cells. The release of nutrients from the xylem parenchyma (and stem tissue in general) into the xylem is of major importance for the maintenance of a continuous nutrient supply to growing shoot tissues. In periods of ample supply to the roots, nutrients are retrieved from the xylem sap, whereas in periods of insufficient root supply, they are released into the xylem sap. Changes in the K and nitrate concentrations of the stem base reflect this functioning of the tissues along the xylem in response to changes in the nutritional status of a plant. Based on this, a rapid test for nitrate in the stem base has been developed as a means for recommending N fertilizer rates. Magnesium is also often stored in root cells and released to the xylem when shoots become Mg deficient (Hermans et al., 2004). In

TABLE 3.3 Sodium concentration of roots and shoots of pasture plants with and without Na fertilizer.

Plant species	Na concentration (g kg^{-1} dw)			
	Without Na fertilizer		With Na fertilizer	
	Roots	Shoots	Roots	Shoots
Lolium perenne	0.3	2.6	0.6	11.6
Phleum pratense	1.0	0.4	2.8	3.6
Trifolium repens	2.7	2.2	7.7	19.6
Trifolium hybridum	4.5	0.3	7.7	2.2

Source: Based on Saalbach and Aigner (1970).

TABLE 3.4 Distribution of Mo in common bean (*Phaseolus vulgaris*), sunflower (*Helianthus annuus*), and tomato (*Solanum lycopersicum*) plants supplied with Mo in the nutrient solution at 4 mg L^{-1}.

Plant part	Mo concentration (μg g^{-1} dw)		
	Bean	Sunflower	Tomato
Leaves	85	125	325
Stems	210	115	123
Roots	1030	565	470

Source: Based on Hecht-Buchholz (1973).

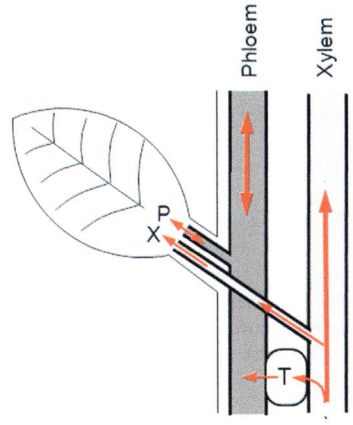

FIGURE 3.3 Schematic diagram of long-distance transport in the xylem and phloem in a stem with a connected leaf, and xylem-to-phloem transfer mediated by a transfer cell (T).

addition, in the nodes of graminaceous species, some nutrients, such as Mn and Cu, can be transferred between the xylem of enlarged vascular bundles (EVBs) and the xylem of diffuse vascular bundles (DVBs) under periods of excess to mitigate against toxicity in growing tissues (Yamaji and Ma, 2014, 2017).

The transfer of nutrients between xylem and phloem also plays an important role in regulating the distribution of nutrients within plants (Section 3.4; Fig. 3.3). The nodes of graminaceous species, such as rice, govern the preferential distribution of nutrients within the shoot (Kakei et al., 2012; Sasaki et al., 2015; Shao et al., 2017, 2018; Yamaji et al., 2013; Yamaji and Ma, 2014, 2017). The regulation of the abundance of transport proteins in the nodes of rice by Fe status controls the distribution of Fe in shoots, such that Fe is delivered preferentially to developing tissues when plants lack sufficient Fe (Ishimaru et al., 2010; Kakei et al., 2012). It has been suggested that OsYSL16 (yellow stripe-like) unloads Fe(III)DMA from the xylem of older leaves (Kakei et al., 2012). Subsequently, OsYSL18 and OsYSL2, the activities of which are increased upon Fe deficiency, load Fe(III)DMA and Fe(II)NA, respectively, from the apoplasm into the phloem (Aoyama et al., 2009; Ishimaru et al., 2010). In arabidopsis, AtOPT3 (oligopeptide transporter 3), a proton-coupled Fe-chelate symporter, the activity of which is upregulated by Fe deficiency, facilitates recirculation of Fe by loading Fe into the phloem, enabling more Fe to be translocated to developing tissues upon Fe deficiency (Zhai et al., 2014).

Similarly, the distribution of Zn in rice shoots is controlled by xylem-to-phloem transport at nodes; the OsZIP3 (zinc iron permease) transporter and the P_{1B}-type ATPase OsHMA2 have been implicated in the preferential transport of Zn to developing tissues by OsZIP3 removing Zn from the xylem in the EVB and OsHMA2 loading Zn into the phloem in both the DVB and EVB (Fig. 3.4) (Sasaki et al., 2015; Yamaji et al., 2013). The activity of natural resistance−associated macrophage protein 3 (OsNramp3), which unloads Mn from the xylem and also loads Mn into the phloem, controls the distribution of Mn in the shoots of rice in response to plant Mn status (Fig. 3.4; Shao et al., 2017). The abundance of OsNramp3 is regulated posttranscriptionally, such that Mn is preferentially delivered to developing tissues when plants lack Mn and to older tissues when plants have excess Mn (Shao et al., 2017).

The abundance of the OsNIP3.1 channel, which is permeable to boric acid and is thought to unload B from the xylem and load B into the phloem of rice, is regulated by plant B status such that B is preferentially delivered to developing tissues when plants lack B and to older tissues when plants have excess B (Fig. 3.4; Shao et al., 2018). The AtNIP6.1 channel performs a similar function in arabidopsis (Tanaka et al., 2008).

In arabidopsis, P transport between xylem and phloem is mediated by AtSPDT (AtSULTR3;4), a proton-coupled phosphate symporter the activity of which increases during P deficiency to deliver more P to developing tissues (Ding et al., 2020). The transport of nitrate between the xylem and phloem is facilitated by AtNFP7.2 (AtNRT1.8), which is present in the xylem parenchyma cells, and AtNPF2.9 (AtNRT1.9), AtNPF1.2 (AtNRT1.11), and AtNPF1.1 (AtNRT1.12) that load nitrate into the phloem to supply developing tissues (Cui et al., 2021; Tegeder and Masclaux-Daubresse, 2018). The transport of amino acids between xylem and phloem is facilitated by AtAAP6 that unloads the xylem, and AtAAP2 and AtAAP3 that load the phloem (Tegeder and Masclaux-Daubresse, 2018).

3.2.2.3 Xylem unloading in leaves

Despite the retrieval of elements along the pathway in the stem, most of the solutes and water are transported in the xylem vessels into the leaves. Here, water is transported preferentially in the major veins to sites of rapid transpiration

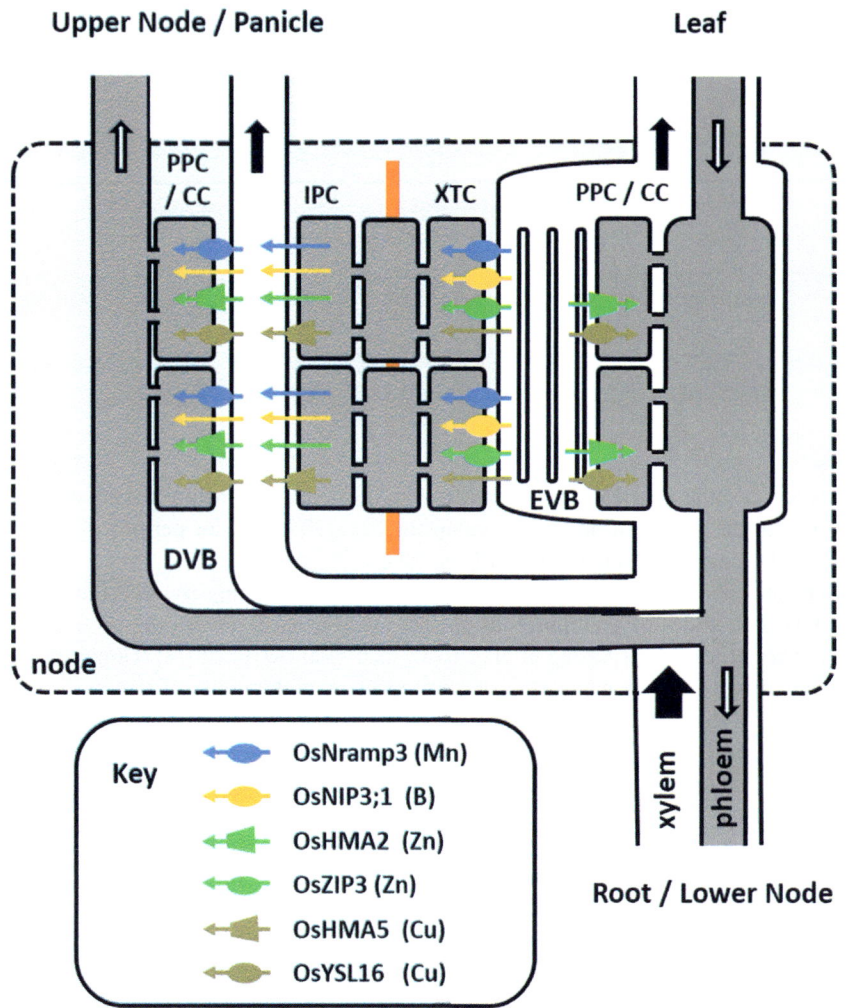

FIGURE 3.4 Schematic representation of a stem node of rice illustrating the mechanisms by which the flows of Mn (blue), B (yellow), Zn (green), and Cu (brown) are decreased to the mature leaf and directed to the upper node/panicle when plants lack these nutrients by removing them from the xylem in the EVB and loading them into the phloem in the DVB. *DVB*, Diffuse vascular bundle; *EVB*, enlarged vascular bundle; *IPC*, intervening parenchyma cell; *Orange strip*, apoplastic barrier; *PPC/CC*, phloem parenchyma/companion cells; *XTC*, xylem transfer cell. *Adapted from Ma and Yamaji (2006).*

such as leaf margins, or from the vein endings in the apoplast or symplast toward the stomata (Karley et al., 2000). Depending on the concentration and composition of solutes in the xylem sap entering the leaf, the rates of solute uptake by leaf cells, and the rate of water loss by transpiration, the solute concentrations may be increased several fold at, for example, the leaf edges. This is particularly true when element concentrations are high in the root medium (e.g., in saline substrates) and for elements such as B and Ca. Unless some of this excessive solute accumulation at the terminal sites of the transpiration stream is removed [e.g., by guttation, as occurs for B (Sutton et al., 2007), or through epidermal glands or bladder cells, as in some halophytes (Flowers et al., 2010)] necrosis of the leaf tips or margins ensues (Fig. 3.5). Some plants accumulate Ca in leaf trichomes (White and Broadley, 2003), or form Ca-oxalate crystals in specific cell types (Franceschi and Nakata, 2005; He et al., 2014).

Elements are distributed heterogeneously among cell types, and this distribution can differ between plant species for an individual element. In many graminaceous species, Ca, Na, and Cl accumulate in the leaf epidermal cells, whereas P, S, and Mg accumulate in the bundle sheath or mesophyll cells, but in many eudicot species, P, Mg, Na, and Cl accumulate in epidermal cells, whereas S and Ca accumulate in the bundle sheath or mesophyll cells (Conn and Gilliham, 2010). However, there can be variation in the distribution of elements within the leaf among both monocots and eudicots, for example, among species of Proteaceae in the location of Ca and P accumulation (Hayes et al., 2019). These differences are determined by the rates of uptake and efflux of an element by individual cell types within the leaf. The sequestration of elements in different cell types has enabled plants to avoid the accumulation of ions that form salts of low solubility in the same cell, such as P and Ca, and to accumulate potentially toxic elements in specific cell types to prevent toxicity (Conn and Gilliham, 2010; Hayes et al., 2019; Schiavon and Pilon-Smits, 2017; White and Pongrac, 2017). The capacity to accumulate large concentrations of potentially toxic elements in epidermal cells can also provide a deterrent to herbivores and pathogens (He et al., 2014; Schiavon and Pilon-Smits, 2017; White and Pongrac, 2017).

FIGURE 3.5 Boron toxicity in lentil (*Lens culinaris*) leaves: control (left); B toxicity (right).

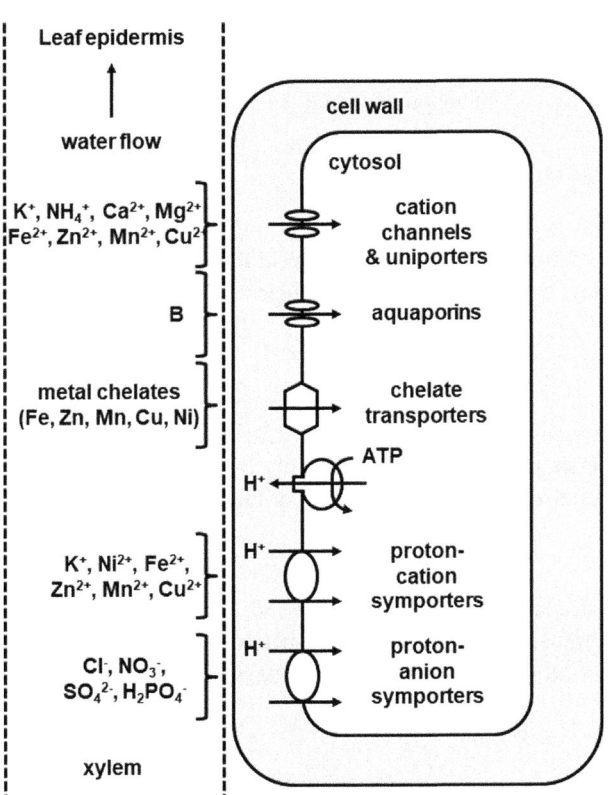

FIGURE 3.6 Mechanisms for the retrieval of major solutes from the xylem sap (xylem unloading) in leaf cells.

Prevention of excessive solute accumulation in the leaf apoplasm by mechanisms other than uptake by the leaf cells can be achieved by the formation of salts of low solubility in the apoplasm. This strategy is utilized for the removal of soluble Ca in gymnosperms (Fink, 1991a). Calcium oxalate crystals are abundant in the needles of various gymnosperms in the cell walls of the epidermis, mesophyll, and phloem. This mechanism of precipitation seems to be a safe way of coping with a continuous xylem import of Ca, which cannot be exported in the phloem. However, the import of solutes to leaves via the xylem and the evaporation of water does not necessarily lead to the accumulation of solutes in the leaf apoplasm. In fast-growing plants at low nutrient supply, the solute concentrations in the xylem sap decline sharply from the roots to the leaves, and within a leaf blade from the base to the tip. For example, in barley, the xylem sap concentration of Mg decreased from 1.1 to 0.1 mM and that of K decreased from 18.0 to 8.0 mM from the leaf base to the tip (Wolf et al., 1990).

The molecular mechanisms responsible for the uptake of solutes from the leaf apoplasm are being identified (Fig. 3.6). The cells of the bundle sheath are sites of intensive net proton excretion that acidifies the apoplasm.

The electrochemical potential and the proton gradient across the plasma membrane of leaf cells act as the driving force for solute uptake. Nitrate is thought to enter leaf cells through members of the nitrate transporter/peptide transporter (NPF) family, such as AtNPF6.2 (AtNRT1.4) in the petiole of arabidopsis, and ammonium is taken up by members of the AMT1 and AMT2 transporter families (Tegeder and Masclaux-Daubresse, 2018). The lysine/histidine-like (LHT1) transporter catalyzes the entry of amino acids into leaf cells, and ureide permease (UPS1) transporters catalyze the entry of ureides into leaf cells (Tegeder and Masclaux-Daubresse, 2018). Phosphate and sulfate are likely to enter leaf cells through proton-coupled transporters encoded by members of the *Pht1* and the *SULTR1* and *SULTR2* gene families, respectively (Gigolashvili and Kopriva, 2014; Hell et al., 2010; Miller et al., 2009). Boric acid channels, encoded by members of the *NIP* (*nodulin-26-like intrinsic protein*) gene family, are likely to facilitate B influx to leaf cells (Miwa and Fujiwara, 2010).

The electrochemical gradient between the shoot apoplasm and the cytosol of shoot cells suggests that the influx of K and Ca to shoot cells can be mediated by cation channels in their plasma membrane (Karley and White, 2009). It is likely that K influx is mediated by voltage-gated, inwardly rectifying K-channels, whereas Ca influx is mediated by nonselective cation channels. The influx of other cations to shoot cells can also be facilitated by nonselective cation channels. Magnesium influx to shoot cells is thought to be catalyzed by members of the mitochondrial RNA splicing 2 (Mrs2) family of transport proteins (Hermans et al., 2013; Karley and White, 2009), whereas members of the ZIP and NRAMP families of transport proteins allow Zn, Fe, and Mn influx to shoot cells (Andresen et al., 2018; White and Pongrac, 2017). Members of the copper transporter family mediate Cu influx to shoot cells (Andresen et al., 2018). In addition, members of the YSL family may catalyze the influx of metal chelates to shoot cells (Andresen et al., 2018).

3.2.3 Effect of transpiration rate on solute transport in the xylem

The rate of water flux across the root (short-distance transport) and in the xylem vessels (long-distance transport) is determined by both root pressure and the rate of transpiration. Whether or not an increase in transpiration affects the flux of an element in the xylem depends on plant age, time of day, concentration of the element in the external medium, and the type of element.

3.2.3.1 Plant age

In seedlings and young plants with a small leaf surface area, increased transpiration rarely affects the accumulation of elements; water uptake and solute transport in the xylem to the shoots are determined mainly by root pressure. As the age and size of plants increase, the relative importance of transpiration for the translocation of elements increases.

3.2.3.2 Time of day

In leaves, up to 90% of the total transpiration occurs via the stomata. During the light period, transpiration rate (and thus the potential for uptake and translocation of elements) is higher than during the dark period. Transient reductions in the translocation rates of elements at the onset of the dark period reflect the change from transpiration-driven to root pressure-driven xylem volume flow.

3.2.3.3 External concentration

An increase in the concentration of elements in the external medium can enhance the effect of transpiration rate on their translocation, especially for elements such as Na, for which there is a significant apoplasmic flux across the root to the xylem (Table 3.5). It is noteworthy that insufficient N, P, or K supply reduces the expression of aquaporin genes in both roots and shoots, thereby decreasing hydraulic conductance and, consequently, transpiration (Armand et al., 2019; Coffey et al., 2018; Ding et al., 2018; Maurel et al., 2016).

3.2.3.4 Type of element

Under otherwise comparable conditions (e.g., plant age and external concentration), the effect of transpiration rate on the uptake and transport of elements follows a defined rank order. It is usually absent, or minor, for K, nitrate, and P, but it may be significant for Na and Ca. As a rule, transpiration enhances the uptake and translocation of uncharged molecules to a greater extent than the uptake and translocation of ions. The uptake and translocation of elements in uncharged forms are of great importance for B (boric acid; Miwa and Fujiwara, 2010; Yoshinari and Takano, 2017) and Si (monosilicic acid; Kaur and Greger, 2019; Ma and Yamaji, 2006).

TABLE 3.5 Uptake and translocation of K and Na from contrasting nutrient solutions at high or low transpiration rates in sugar beet (*Beta vulgaris*) plants.

Nutrient solution concentration (mM)	K Low	K High	Na Low	Na High
Uptake rate (μmol plant^{-1} h^{-1})				
1 K$^+$ + 1 Na$^+$	1.15	1.23	2.10	2.80
10 K$^+$ + 10 Na$^+$	2.58	2.75	3.00	4.78
Translocation rate (μmol plant^{-1} h^{-1})				
1 K$^+$ + 1 Na$^+$	0.73	0.75	0.50	0.98
10 K$^+$ + 10 Na$^+$	1.63	1.75	0.85	2.03

Transpiration in relative values: low transpiration = 100; high transpiration = 650.
Source: Based on Marschner and Schafarczyk (1967).

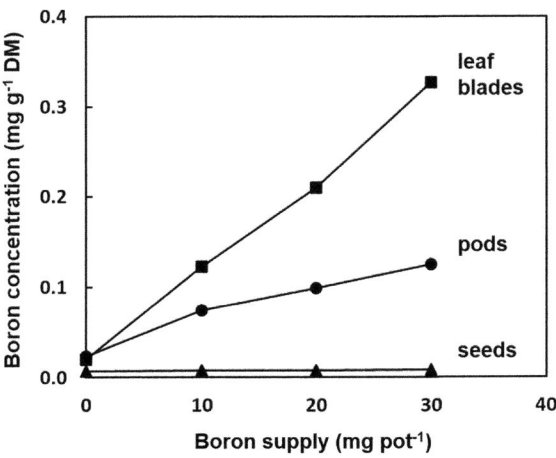

FIGURE 3.7 Distribution of B among leaf blades, pods, and seeds of oilseed rape (*Brassica napus*) with increasing B application to the soil. *DM*, Dry matter. *Recalculated from the data of Gerath et al. (1975).*

3.2.4 Effect of transpiration rate on distribution of elements within the shoot

The distribution of an element that is transported in the xylem but not the phloem should be related solely to transpiration rates (e.g., mL g^{-1} dw day^{-1}) and duration of transpiration (e.g., age of the organ). This is true, for example, for Mn in maple trees in which the "sun leaves" (high transpiration rates) have higher Mn concentrations in their dry matter than "shade leaves" (low transpiration rates) of a similar age (McCain and Markley, 1989). Similarly, the distribution and concentration of Si usually reflect the loss of water from various organs through transpiration. The Si concentration increases with leaf age and is particularly high in spikelets of cereals such as barley. Even within an organ, such as a leaf, the Si distribution often resembles the pathway of transpiration flow in the apoplasm and is deposited in the epidermis (Nawaz et al., 2019).

The distribution of B is also related to the loss of water from the shoot, as shown for oilseed rape in response to an increasing B supply (Fig. 3.7). The typical gradient in transpiration rates in the shoot organs (leaves > pods >> seeds) corresponds to the differences in tissue B concentration. Even for a particular leaf, an excessive supply of B creates a steep gradient in the B concentrations: petioles < middle of the leaf blade < leaf tip (Oertli, 1994). Necrosis on the margins and/or tips of leaves is therefore a typical symptom of B toxicity (Fig. 3.5).

Frequently, there is a close positive correlation between Ca distribution and transpiration rates of shoot organs, as is evident from the low Ca concentrations of low-transpiring fleshy fruits (<1 g Ca kg^{-1} dw) when compared with that of the leaves (30–50 g Ca kg^{-1} dw) of the same plant (Chapter 5). Lower transpiration rates decrease the Ca

TABLE 3.6 Concentrations of K, Mg, and Ca in fruits of red pepper (*Capsicum annuum*) grown at high or low transpiration rates during fruit development.

	Transpiration rate (relative)	
	35	100
Fruit weight (g dw)	0.69	0.62
	Concentration in fruits (mg g^{-1} dw)	
K	88	91
Mg	2.4	3.0
Ca	1.45	2.75

Source: Based on Mix and Marschner (1976).

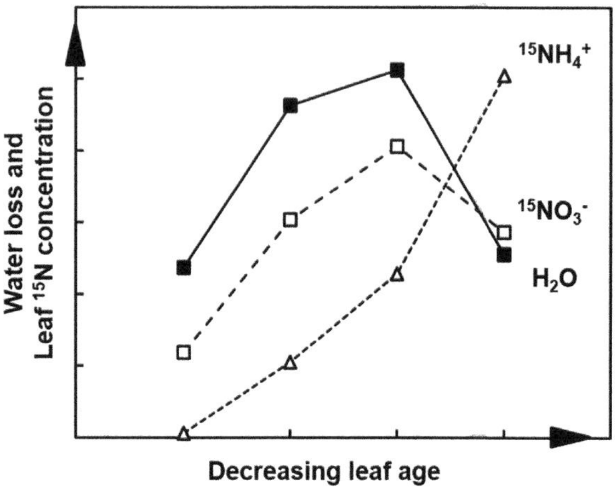

FIGURE 3.8 Water loss and concentrations of labeled nitrogen (^{15}N) in leaves of bean (*Phaseolus vulgaris*) at the four-leaf stage expressed on a dry weight basis following the application of ^{15}NO$_3^-$ and ^{15}NH$_4^+$ to the roots. *Adapted from Martin (1971)*.

concentration of fruits further (Table 3.6). The effect of transpiration on Mg distribution is less than on Ca, and the effect of transpiration on K distribution is negligible.

The influence of transpiration on the distribution of elements differs not only among elements but also between various forms of the same element, as shown in Fig. 3.8 for N. The distribution of ^{15}N within the shoot of ammonium-fed plants is independent of transpiration rates and is directed preferentially to the shoot apex that acts as a sink for reduced N; in contrast, shoot ^{15}N concentration in nitrate-fed plants follows the transpiration rate quite closely (Martin, 1971). The decrease in xylem flux of water and nitrate into older leaves of plant species such as common bean (*P. vulgaris*) is due to a decrease in hydraulic conductivity caused by the plugging of the xylem vessels at the pulvinal junction in older leaves (Neumann, 1987).

3.3 Phloem transport

3.3.1 Principles of phloem transport and phloem anatomy

Long-distance transport in the phloem takes place in living cells, the sieve tubes (Fig. 3.9). The principles of the phloem transport mechanism were proposed as early as 1930 by Münch in a pressure flow hypothesis based on the principle of an osmometer (Chapter 5; White, 2017). Münch suggested that solutes such as sucrose are concentrated in the phloem of leaves (i.e., *phloem loading*) and water is drawn into the phloem by osmosis, creating a positive internal pressure. This pressure induces a mass flow in the phloem to the sites of lower positive pressure caused by removal of solutes from the phloem. Therefore flow rate and direction of flow are closely related to phloem unloading at the sink. Solute transport in the phloem differs from that in the xylem in three important ways: (1) organic compounds are the dominant solutes in the phloem sap, (2) transport takes place in living cells, and (3) the unloading of solutes at the sink has an important role.

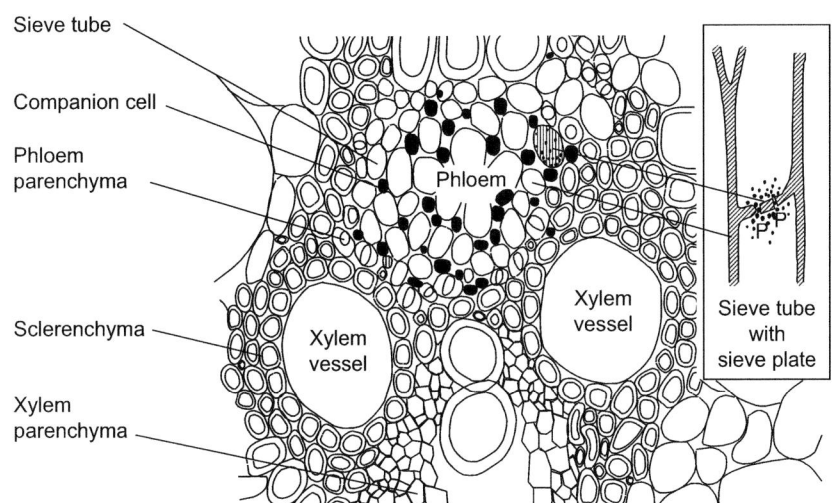

FIGURE 3.9 Cross-section of a vascular bundle from the stem of maize. Inset: Sieve tube with sieve-plate pores and P-protein. *Redrawn from Eschrich (1976).*

FIGURE 3.10 Translocation of labeled P (^{32}P) and Na (^{22}Na) after application to the tip of a cotyledon primary leaf of common bean (*Phaseolus vulgaris*). Autoradiogram, 24 h after application.

For nutrients, the main sites for phloem loading are in the stem and the leaves (sources). These supply nutrients to growth sinks (shoot apices, fruits, roots) and allow nutrient cycling within the plant. An example of sink-regulated transport of a nutrient is shown in Fig. 3.10 for P. After application to one of the two cotyledons, radiolabeled P is transported to the shoot apex and the roots, whereas transport to the other cotyledon is negligible. In contrast, radiolabeled Na is not transported to the shoot apex but moves exclusively downwards (basipetally) to the roots where it is confined to the basal zones (Fig. 3.10) from where it is effluxed to the rhizosphere solution (Lessani and Marschner, 1978). The recirculation of Na to roots via the phloem restricts Na accumulation in shoots and is implicated in salt tolerance of several species, including white lupin (*Lupinus albus*), berseem clover (*Trifolium alexandrinum*), sweet pepper (*Capsicum annuum*), and maize (Wu, 2018).

Within the phloem, the sieve-tube elements are associated with companion cells and parenchyma cells (Fig. 3.9). Some of the individual sieve-tube elements are stacked end-to-end, forming the sieve tubes that are connected by pores (Fig. 3.9 inset) called sieve-plate pores. The sieve tubes are highly specialized vascular systems for the long-distance transport of solutes. The sieve-tube cells contain a thin layer of cytoplasm, which forms transcellular filaments (the so-called P-protein) that pass through the sieve-plate pores. The anatomical features of long-distance transport in the sieve tube across the sieve-plate pores are similar to those of short-distance transport between cells in the symplasm across the plasmodesmata (Chapter 2).

In most plant species, the sieve-plate pores are lined with callose, a highly hydrated polysaccharide (Furch et al., 2010). There is strong evidence that callose can swell rapidly and fill the pores, thus blocking long-distance transport in the sieve tubes. Callose formation is induced by Ca^{2+} even at a low concentration of a few μM. Thus low concentrations of free Ca^{2+} must be maintained in the phloem sap for long-distance transport to occur. Plugging of sieve-tube pores is also induced by factors such as heat treatment or mechanical stress, as well as by mechanical injury of the sieve tubes. This process can be thought of as a "security valve" that prevents "bleeding" when the system is ruptured.

3.3.2 Phloem loading and the composition of phloem sap

Phloem sap has a high pH (7–8) and has large concentrations of solutes, on average 150–250 g kg^{-1} dry matter (Table 3.7). The main component is usually sucrose that may comprise up to 90% of the solids, but the proportion of sucrose to other solutes depends on plant species, the developmental stage of a plant, and the site of phloem sap collection (Gaupels and Ghirardo, 2013; Nalam et al., 2021). In addition to sucrose, amino compounds are usually present in large concentrations in phloem sap (Table 3.7; Peuke, 2010; Tegeder and Masclaux-Daubresse, 2018), with the amides glutamine and asparagine often representing up to 90% of this fraction. On the other hand, the concentrations of nitrate and ammonium are usually low (Cui et al., 2021; Van Beusichem et al., 1988). Organic acid anions such as citrate and malate are also abundant in the phloem sap (Álvarez-Fernández et al., 2014), and, in white lupin, succinate concentrations may reach the same order of magnitude as the concentration of total amino-N (Jeschke et al., 1986). A whole range of other organic compounds are also found in phloem sap, such as secondary metabolites (Turgeon and Wolf, 2009), DMA and NA (chelators of polyvalent cations; Clemens, 2019), phytohormones (including ABA, auxin, cytokinins, gibberellins, jasmonic and salicylic acids plus their derivatives; Koenig and Hoffmann-Benning, 2020; Kudoyarova et al., 2015; Notaguchi and Okamoto, 2015; Turgeon and Wolf, 2009), proteins (such as florigens and antiflorigens that control flowering and tuberization; Koenig and Hoffmann-Benning, 2020; Notaguchi and Okamoto, 2015; Turgeon and Wolf, 2009),

TABLE 3.7 Comparison of concentrations of organic and inorganic solutes in the phloem (stem incision, pH 7.9–8.0) and xylem (tracheal, pH 5.6–5.9) exudates of *Nicotiana glauca*.

	Phloem	Xylem	Ratio phloem/xylem
	(mg L^{-1})		
Dry matter	170–196	1.1–1.2	155–163
Sucrose	155–168	nd	
Amino compounds	10.8	0.28	38.2
	(mM)		
Nitrate	nd	na	
Ammonium	2.51	0.538	4.7
K	93.9	5.23	18.0
P	14.0	2.20	6.4
Cl	13.7	1.80	7.6
S	4.33	1.35	3.2
Ca	2.08	4.72	0.44
Mg	4.29	1.39	3.1
Na	5.06	2.01	2.5
Fe	0.168	0.011	15.7
Zn	0.243	0.023	10.8
Mn	0.016	0.004	3.8
Cu	0.019	0.002	10.9

na, data not available; *nd*, not detectable.
Source: Based on Hocking (1980).

and RNA molecules (regulating nutrient acquisition and mediating systemic acquired resistance; Kehr, 2013; Kehr and Kragler, 2018; Koenig and Hoffmann-Benning, 2020; Morris, 2018; Notaguchi and Okamoto, 2015; Turgeon and Wolf, 2009). The concentrations of these compounds in phloem sap vary during ontogeny and in response to biotic and abiotic stresses.

All plants transport sucrose in the phloem, but some plants also transport raffinose and stachyose and/or sugar alcohols (Chapter 5; Turgeon and Wolf, 2009; Zhang and Turgeon, 2018). These compounds enter the phloem in mature leaves. In plants with an apoplasmic step in phloem loading, sucrose is released to the apoplasm through sucrose uniporters belonging to the SWEET protein family (Jeena et al., 2019; Zhang and Turgeon, 2018) and then loaded into the phloem by H^+/sucrose symporters, encoded by members of the *SUT/SUC* gene family, such as *AtSUC2* from arabidopsis (Kühn and Grof, 2010; Pommerrenig et al., 2020; Sauer, 2007; Zhang and Turgeon, 2018). Sorbitol and mannitol are similarly loaded into the phloem by H^+/sugar alcohol symporters (Juchaux-Cachau et al., 2007; Ramsperger-Gleixner et al., 2004).

In general, nutrients are released into the leaf apoplasm before being loaded into the phloem. Several genes encoding transporters involved in loading nitrate, amino acids, ureides, and short peptides into the phloem have been reported. Nitrate can be loaded into the phloem by transporters of the NPF family, such as AtNPF2.13 (AtNRT1.7), AtNRT2.4, AtNRT2.5, AtNPF1.2 (AtNRT1.11) and AtNPF1.1 (AtNRT1.12) in arabidopsis (Chen et al., 2020; O'Brien et al., 2016; Tegeder and Masclaux-Daubresse, 2018). Amino acids are thought to be released into the apoplasm by uniporters, such as UMAMIT18 (SIAR1) and BAT1 in arabidopsis, and imported to the phloem through H^+/amino-acid symporters of the amino-acid permease (AAP) family, such as AtAAP3 and AtAAP8, and, possibly, also by H^+/amino-acid symporters of the cationic amino acid (CAT) and proline transporter families (Koenig and Hoffmann-Benning, 2020; Tegeder, 2014; Tegeder and Masclaux-Daubresse, 2018). Companion cells also possess transporters for loading ureides [such as UPS1 in French bean (*P. vulgaris*)] and peptides into the phloem (Tegeder and Masclaux-Daubresse, 2018).

Among the mineral nutrients, K is usually present in the phloem sap at the highest concentration, followed by P, Mg, Cl, and S (Table 3.7). In contrast, the concentration of Ca in the phloem sap is always low, regardless of plant species.

Potassium is loaded into the phloem by voltage-gated, inward-rectifying K-channels such as AtAKT2/3 in arabidopsis (Deeken et al., 2002; Hafke et al., 2007). Sulfur occurs in both the reduced form (e.g., glutathione, S-methylmethionine, methionine, cysteine, glucosinolates) and as sulfate (Gigolashvili and Kopriva, 2014; Hell et al., 2010). Sulfate is loaded into the phloem by orthologs of the arabidopsis AtSULTR1;3 transporter, whereas methionine and cysteine are likely to be loaded by amino-acid transporters, and glucosinolates are loaded by proton symporters of the NPF family (Gigolashvili and Kopriva, 2014; Hell et al., 2010; Jørgensen et al., 2017). Sulfate concentrations in the phloem sap can be as high as those of phosphate (Van Beusichem et al., 1988). Boron is loaded into the phloem by homologs of the arabidopsis AtNIP6;1 transporter; an enhanced capacity to transport B in the phloem is associated with (1) the presence of polyols, such as mannitol, sorbitol, and dulcitol in the phloem sap (Brown and Hu, 1998; Wang et al., 2015) or (2) in plants such as wheat and canola, with the presence of bis-sucrose borate and bis-N-acetyl-serine borate complexes (Yoshinari and Takano, 2017).

Chloride and Na may also be present in large concentrations in phloem sap (Table 3.7), but this depends on their external supply and the plant species (Jeschke and Pate, 1991; White and Broadley, 2001). In soybean, GmSALT3 (a proton/cation antiporter in the endoplasmic reticulum) has been implicated in Cl recirculation via the phloem (Qu et al., 2021). The homologs of the arabidopsis H^+/Na^+ symporter AtHKT1 are thought to load Na into the phloem (Wu, 2018).

The concentrations of Fe, Zn, Mn, and Cu in the phloem sap are usually in the range of 40–168, 14–245, 4–76, and 8–43 μM, respectively (Table 3.7; Álvarez-Fernández et al., 2014). Boron concentrations in the range of 200–500 μM have been reported (Huang et al., 2008). Members of the ZIP family are thought to transport Zn into the phloem (Ishimaru et al., 2005), and the YSL proteins are likely to be involved in loading Fe, Mn, Zn, and Cu into the phloem. The concentration of Fe in the phloem is determined, in part, by the activity of OLIGOPEPTIDE TRANSPORTER 3 (OPT3) that loads Fe-chelates into the phloem (Koenig and Hoffmann-Benning, 2020). In general, these micronutrients are transported to sink tissues as complexes with DMA and/or NA or in association with small proteins or amino acids (Álvarez-Fernández et al., 2014; Clemens, 2019; Harris et al., 2012; Kutrowska and Szelag, 2014; Waters and Grusak, 2008; White and Broadley, 2009).

With the exception of Ca, the concentrations of all solutes are usually several times greater in the phloem than in the xylem (Table 3.7). The data in Table 3.7 on phloem composition are similar to those obtained from analyses of phloem saps from a variety of plant species (e.g., Álvarez-Fernández et al., 2014; Harris et al., 2012; Turgeon and Wolf, 2009; van Beusichem et al., 1988).

TABLE 3.8 Mobility of nutrients in the phloem.

Mobility		
High	Intermediate	Low
K	Fe	Ca
Mg	Zn	Mn
P	Cu	
S	B[a]	
N (amino-N)	Mo	
Cl		
(Na)		

[a]*The mobility of B in the phloem varies greatly among species.*

3.3.3 Mobility in the phloem

Most nutrients are found in phloem sap (Table 3.7), which suggests that they have mobility in the phloem. A more direct approach to elucidate the mobility of nutrients in the phloem is to use labeled elements (radioactive or stable isotopes) to follow their long-distance transport after application, for example, to the tip of a leaf blade (Fig. 3.10). Because of the gradient in xylem water potential, translocation from the leaf tips and out of the treated leaf must take place in the phloem. Based on such studies, and with consideration of the data on phloem sap composition, nutrients can be classified into groups with different phloem mobility (Table 3.8). Sodium has been included in this list as it is beneficial for the growth of some plant species and its mobility in the phloem is of particular importance for plants growing in saline substrates.

For the macronutrients, except Ca, phloem mobility is generally high, and for the micronutrients it is at least intermediate, with the exception of Mn (Table 3.8). Some long-distance transport in the phloem can be demonstrated with labeled Mn (El-Baz et al., 1990; Nable and Loneragan, 1984), but the mobility of Mn in the phloem is considered low. The same holds true for Ca. Although Ca can be found in the phloem sap (Table 3.7), Ca has low phloem mobility. Most of the Ca demand of growth sinks has to be covered by import via the xylem.

The mobility of B in the phloem varies among plant species. It is high in species loading polyols into the phloem sap (Brown and Hu, 1998; Hanson, 1991; Wang et al., 2015) and intermediate in species, such as wheat and canola, in which bis-sucrose borate and bis-N-acetyl-serine borate complexes are present in the phloem (Yoshinari and Takano, 2017). In other species, the mobility of B in the phloem can be low.

3.3.4 Transfer between the xylem and phloem

In the vascular bundles, the phloem and xylem are separated by only a few cell layers (Fig. 3.9). Exchange of solutes between the two conducting systems is important for regulation of long-distance transport (Section 3.2.2.2; Figs. 3.3 and 3.4). From the concentration differences shown in Table 3.7, it is evident that transfer from phloem to xylem can occur down a concentration gradient (except for Ca). Conversely, for most organic and inorganic solutes, a transfer from xylem to phloem is usually against a steep concentration gradient. Nevertheless, xylem-to-phloem transfer of nutrients is of particular importance for the mineral nutrition of plants because xylem transport is directed mainly to the sites (organs) of high transpiration that are often not the sites of high demand for nutrients. The transfer of organic and inorganic solutes from the xylem to the phloem can take place along the entire pathway from roots to shoot, and the stem plays an important role in this respect. In particular, stem nodes are sites of intensive xylem-to-phloem transfer of K (Haeder and Beringer, 1984), Fe (Ishimaru et al., 2010; Kakei et al., 2012), Zn (Sasaki et al., 2015; Yamaji et al., 2013), Mn (Shao et al., 2017), and B (Shao et al., 2018) in cereals, and amino acids in soybean (Da Silva and Shelp, 1990).

The proportion of xylem-to-phloem transfer in the stem is influenced by the rate of xylem volume flow, that is, by the transpiration rate, with high rates reducing transfer to the phloem. In tomato, doubling the volume flow rate reduced the transfer of amino acids in the stem resulting in a greater proportion being transported to the mature leaves at the expense of the shoot apex (Van Bel, 1984). Thus a diurnal rhythm in the partitioning of solutes between mature leaves and shoot apex or fruits might be expected from the diurnal variation in transpiration, unless it is compensated for by greater xylem-to-phloem transfer in the leaf blades.

3.3.5 Phloem unloading

In growing organs, such as developing leaves, fruits, seeds, tubers, and root apices, solutes are initially transported symplasmically from the phloem to the neighboring tissue, but apoplasmic steps ultimately follow (Chapter 5). The filial tissues (embryo, endosperm) of seeds are symplasmically isolated from the phloem present in the maternal tissues (seed coat). Members of the SWEET family of transporters are thought to release sucrose to the apoplasm, and sucrose is then retrieved by members of the SUT family of transporters (Jeena et al., 2019; Milne et al., 2018). Members of the UMAMIT family unload amino acids from the phloem (e.g., UMAMIT11, UMAMIT14, UMAMIT18) and also from cells in the seed coat (e.g., UMAMIT28, UMAMIT29; Tegeder and Masclaux-Daubresses, 2018). These are then acquired by cells in the endosperm and embryo by H^+/amino-acid symporters of the AAP and CAT families, such as AtAAP1, AtAAP8, and AtCAT6 in arabidopsis (Tegeder and Masclaux-Daubresses, 2018). In arabidopsis, nitrate is thought to be unloaded from the phloem by AtNPF2.12 (AtNRT1.6) and moved into the seed by AtNPF5.5 (Tegeder and Masclaux-Daubresses, 2018). The transport of other elements into seeds is thought to be mediated by the same mechanisms that facilitate efflux and influx in other cells (Chapter 5). For example, the uptake and release of K are thought to be mediated by cation channels (Zhang et al., 2007), whereas loading seeds with Fe, Mn, and Cu are thought to be facilitated by members of the YSL family (Nikolic and Pavlovic, 2018). Similar relays of transporters are found in developing leaves and root apices, as well as in fruits and tubers.

3.4 Relative importance of phloem and xylem for long-distance transport of nutrients

3.4.1 General

Precise quantitative assessments of the relative importance of solute transport in the phloem and xylem into parts or organs of plants are difficult to make. For such assessments not only are the concentrations of solutes required but also the velocity of transport and the cross-sectional area of the conducting vessels. The specific rate of solute transport (g cm^{-2} h^{-1}) is the product of the velocity of transport (cm h^{-1}) and the solute concentration (g cm^{-3}). The velocity of transport in the xylem and phloem varies greatly. Velocities range from about 10 to 100 cm h^{-1} in the xylem, and average about 0.1 cm h^{-1} in the phloem (Raven, 2017).

3.4.2 Nutrients with high phloem mobility

For nutrients with high phloem mobility, such as N, amino-N, P, or K, the relative importance of phloem and xylem transport into an organ depends on its stage of development. An example of this is the N dynamics of an individual leaf of a nitrate-fed castor bean plant (Table 3.9). Throughout its lifespan, N import by the xylem sap was high and only

TABLE 3.9 Import (+) and export (−) of N during the lifespan of a leaf of nitrate-fed castor bean (*Ricinus communis*) plants.

Days after leaf emergence	N (nmol leaf^{-1})			
	Xylem		Phloem	Net change
	Total N	N as NO$_3^-$	Total N	Total N
1–12 (leaf expansion)	+2.7	+0.23	+1.4	+4.10
13–20	+2.5	+0.43	−1.1	+1.36
21–40	+2.8	+0.63	−3.7	−0.87
41–60 (leaf senescence)	+1.4	+0.48	−4.0	−2.63

Source: Based on Jeschke and Pate (1992).

declined at the onset of senescence. Nitrate represented only a small fraction of the N imported in the xylem. Additional N import by the phloem during rapid leaf expansion was followed by a strong increase in phloem export so that net export became greater than import as the leaf senesced. Similar observations have been made for the import and export of P and K during the lifespan of individual leaves in various plant species, including barley (Greenway and Gunn, 1966; Greenway and Pitman, 1965).

3.4.3 Nutrients with low phloem mobility

Calcium is often used as an example of a nutrient with low phloem mobility. Manganese also falls into this category (Table 3.8). Because of its low mobility in the phloem, the import of Ca into growth sinks such as shoot apices, young leaves or fruits takes place almost exclusively via the xylem, whereas import via the phloem is negligible, as shown for castor bean in Table 3.10. This is in marked contrast to K, for which most (terminal bud) and nearly half (youngest leaves) of the total net import takes place in the phloem. For Mg, phloem import contributes 20% and 28% of the total import into terminal bud and youngest leaves, respectively.

A high xylem volume flow is often required to cover the relatively high Ca demand of growing tissues, particularly in dicotyledonous plant species that have a large apoplasmic cation-exchange capacity. Fruits developing in the soil, such as peanut (Howe et al., 2012), and potato tubers (White, 2018) are exceptions, as they can cover part of their Ca demand by direct uptake from the soil solution. Shoot apices, young leaves (particularly those enclosed by mature leaves, e.g., cabbage) and fleshy fruits are characterized by low rates of transpiration and inherent low rates of xylem volume flow. Calcium deficiency and the so-called Ca deficiency-related disorders, such as tip-burn in lettuce, blossom end-rot in tomato, and bitter pits in apple, are therefore widespread (Ho and White, 2005; Hocking et al., 2016; Marschner, 1983; Shear, 1975).

To increase the Ca concentration in fruits, increasing their transpiration rate is often more effective than increasing the Ca supply in the substrate (Table 3.11). As expected, because of the high mobility of K in the phloem, the K concentration is not affected by varied Ca supply and relative humidity. High transpiration rates of the whole shoot,

TABLE 3.10 Xylem and phloem import of K, Mg, and Ca into the terminal bud and youngest leaves of castor bean (*Ricinus communis*) averaged between 44 and 53 days after sowing.

	Terminal bud			Youngest leaves		
	K	Mg	Ca	K	Mg	Ca
	(μmol plant^{-1} d^{-1})					
Xylem	0.43	0.89	0.47	2.29	0.58	0.27
Phloem	2.27	0.22	0.003	2.14	0.22	0.003

Source: Based on Jeschke and Pate (1991).

TABLE 3.11 Calcium and K concentrations of red pepper (*Capsicum annuum*) fruits as affected by differential Ca supply (with a relative humidity of 60%−80% in the fruit environment) and varied relative humidity in the fruit environment (with a Ca supply of 2 mM).

	Solution Ca concentration (mM)		Relative humidity in fruit environment (%)	
	0.5	5.0	90	40
	Concentration in fruits (μmol g^{-1} dw)			
Ca	26.9	33.2	32.7	55.4
K	1315	1228	1892	1918

Source: Based on Marschner (1983).

TABLE 3.12 Relationships between root pressure (guttation), leaf tip necrosis, and Ca and Mg content of expanding strawberry (*Fragaria* × *ananassa*) leaves.

Solution concentration		Guttation (relative)[a]	Leaf tip necrosis (relative)[b]	Content (μg leaf^{-1})	
Day	Night			Ca	Mg
High	High	0.3	3.8	7	57
High	Low	2.4	0.3	25	77
Low	High	0.8	1.3	14	74
Low	Low	2.3	0.0	62	78

Root pressure was altered by varying the concentration of the nutrient solution. Osmotic potentials of the concentrated and diluted solutions were −65 and −16 kPa, respectively. Guttation was scored in the morning.
[a] 0 = no guttation; 1 = some drops on some leaves; 2 = most leaves guttating; 3 = copious guttation, with drops coalescing on emerging leaves.
[b] 0 = none; 5 = very severe.
Source: Based on Guttridge et al. (1981).

however, often decrease rather than increase the Ca influx to tomato fruit (Ho and White, 2005) and other low-transpiring organs, such as the inner leaves of cabbages (Everaarts and Blom-Zandstra, 2001). In the latter case, the xylem volume flow is directed to the rapidly transpiring outer leaves at the expense of the inner leaves or the rosettes. Inhibition of transpiration (by high relative humidity or during the dark period) usually favors the direction of the xylem sap flow towards low-transpiring organs. For example, in Chinese cabbage, an increase in relative humidity during the night increased the Ca concentrations in the inner leaves by 64% and decreased the proportion of tip-burn in heads by 90% (Van Berkel, 1988). In potato plants subjected to soil drying, Ca deficiency-related tuber necrosis could be reduced significantly by foliar application of antitranspirants, which altered leaf-tuber water potential gradients (Win et al., 1991).

At low transpiration, the rate of xylem sap flow from the roots to the shoots is determined by root pressure (Schenk et al., 2021). The import of water and Ca via the xylem into low-transpiring organs, therefore, depends strongly on root pressure. Consistent with this, low osmotic potential of the soil solution (e.g., soil salinity) decreases both root pressure and Ca influx into young leaves or fruits and induces Ca deficiency symptoms (Ho and White, 2005; Mizrahi and Pasternak, 1985; Van Berkel, 1988). In expanding strawberry leaves, high root pressure at night, but not during the day, was correlated with the intensity of guttation, an increased Ca content of expanding leaves and either the absence of, or only mild symptoms of, Ca deficiency (tip necrosis) (Table 3.12). Leaf content of Mg, which is highly phloem mobile, was only slightly affected by root pressure.

Root pressure also depends strongly on root respiration and oxygen supply to the roots. Interruption of the aeration of the nutrient solution during the night had no effect on Ca accumulation in the roots of tomato but reduced the Ca transport into the stem by 42% and into the leaves by 82% (Tachibana, 1991). The increase in blossom end-rot in tomato by poor aeration of the rooting medium is well documented (Ho and White, 2005).

Increasing Ca import to fruits and leaves is, however, not advantageous under all circumstances. In tomato, for example, environmental factors that enhanced Ca import into fruits increased the incidence of "gold specks," which is a physiological disorder caused by an excess of Ca in the tissue (DeKreij et al., 1992). In the gold-speck tissue, high total Ca concentrations are found together with a high density of Ca-oxalate crystals. In gymnosperms, abundant formation of Ca-oxalate crystals in the apoplast of needles is another example of excessive Ca import into an organ, and is particularly evident in trees growing on calcareous soils (Fink, 1991b).

3.4.4 Re-translocation and cycling of nutrients

With the exception of Ca, and potentially also Mn, the import of nutrients in the xylem and their export in the phloem occurs throughout the life of a leaf (Chapter 5). There is rapid xylem-to-phloem transfer in leaf blades and accelerated phloem loading during senescence. Considerable amounts of nutrients are re-translocated in the phloem from the shoots back to the roots. The amounts of nutrients recycled to roots may be used to convey information about the nutritional status of the shoots and, via feedback regulation, control uptake by the roots. In some plant species, re-translocation in the phloem is an important component in maintaining nontoxic Na concentrations in the leaves (Section 3.2.2.2).

TABLE 3.13 Fluxes of K, Na, Mg, and Ca in white lupin (*Lupinus albus*) and castor bean (*Ricinus communis*) expressed per plant as a proportion of uptake by roots.

	Proportion of total uptake (%)							
	White lupin				Castor bean			
	K	Na	Mg	Ca	K	Na	Mg	Ca
Leaf blade								
Import via xylem	96	45	33	29	138	11	51	39
Export via phloem	72	33	25	12	93	9	13	2
Roots								
Import from shoot via phloem	59	33	20	9	85	9	15	1
Cycling through roots	38	–[a]	10	–	78	–	7	–

The fluxes through roots are the contributions to the xylem flux of the nutrients imported to the root via the phloem.
[a] Could not be quantified.
Source: Based on Jeschke and Pate (1991).

In plant species for which nitrate reduction occurs mostly in shoots, re-translocation of N in reduced form in the phloem from shoot to the roots is required to meet root demand for reduced N. Frequently, however, considerable amounts of re-translocated nutrients are again loaded into the xylem of the roots to be transported back to the shoot, that is, they cycle within the plant. For K, it has been demonstrated that, at least in certain plant species, cycling is an important process for maintaining charge balance in shoots and roots of nitrate-fed plants (see below). In more general terms, cycling of nutrients may smooth out fluctuations in external supply to match a more consistent demand. In addition, cycling of nutrients can compensate for their nonuniform distribution in the rooting zone, as observed for Zn (Loneragan et al., 1987; Webb and Loneragan, 1990) and Mg (Hermans et al., 2004). However, in many instances, cycling is simply the consequence of the mechanism and direction of phloem transport, which are largely governed by sugar transport from leaves as the source to roots as a sink.

Comprehensive studies on nutrient cycling were conducted with white lupin and castor bean by Jeschke and Pate (1991). Some of their data are summarized in Table 3.13. These give the rates of nutrient import and export from leaf blades, re-translocation via the phloem and cycling through the roots. As has already been shown for reduced N (Table 3.9) and is also the case for K, Na, and Mg, export via the phloem can comprise a major fraction of the import via the xylem. Phloem export of Ca is negligible in castor bean but unexpectedly high in white lupin. This high Ca export might relate to the exceptionally high concentrations of organic acid anions (mainly succinate) in the phloem sap of white lupin (Cramer et al., 2005; Jeschke et al., 1986). Organic acid anions chelate Ca and may thereby improve its phloem mobility. In the studies of Jeschke and Pate (1991), between 82% and 91% of the K, 80% and 91% of the Mg, 50% and 75% of the Ca, and all of the Na exported in the phloem by leaves was re-translocated back to the roots, and a high proportion of the K and Mg cycled through the root, that is, they were loaded again into the xylem and transported to the shoots (Table 3.13). For Ca, no precise data can be given, but cycling is of minor importance. In general, these data are consistent with other studies indicating that up to 90% of K (Armstrong and Kirkby, 1979; Cooper and Clarkson, 1989; Karley and White, 2009; Peuke, 2010; White, 1997), 80% of N (Cooper and Clarkson, 1989; Jeschke et al., 1997a; Peuke, 2010), 65% of Mg (Karley and White, 2009; Peuke, 2010; White, 1997), 30% of P (Jeschke et al., 1997a), 30% of S (Cooper and Clarkson, 1989), and 30% of Cl (Peuke, 2010; White and Broadley, 2001) delivered to the shoot via the xylem are exported back to the root via the phloem.

Nutrient cycling is of particular importance for the N nutrition of plants. In nitrate-fed barley plants, up to 79% of the N translocated in the xylem to the shoots was re-translocated in the phloem as reduced N back to the roots; of this, about 21% was incorporated into the root tissue and the remainder cycled back in the xylem to the shoots (Simpson et al., 1982). In young wheat and rye plants, more than 60% of the reduced N in the xylem sap cycles within the plant (Cooper and Clarkson, 1989). In wheat throughout ontogenesis, 10%–17% of N and 12%–33% of S in the xylem sap are derived from the fraction recycled in the phloem from shoots to roots (Larsson et al., 1991).

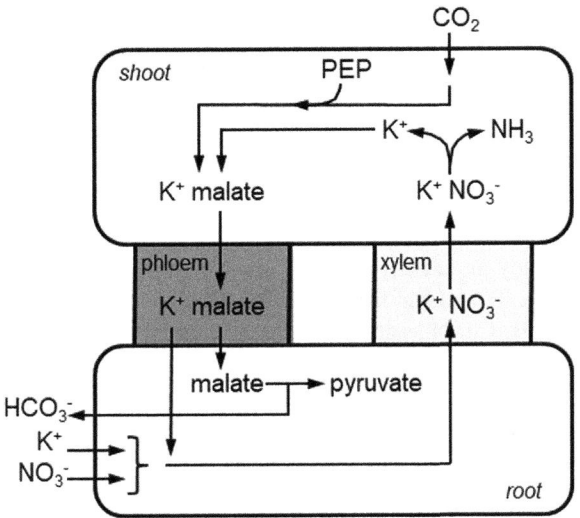

FIGURE 3.11 Model for the circulation of K between root and shoot in relation to nitrate and malate transport. *PEP*, Phosphenolpyruvate. *Based on Ben-Zioni et al. (1971).*

The predominant site of nitrate reduction in plants, whether roots or shoots, can have an important impact on K cycling (Fig. 3.11). Potassium plays an important role as counter-ion for nitrate transport in the xylem (Chapter 5; Van Beusichem et al., 1988). After nitrate reduction in the shoots, charge balance must be maintained by a corresponding net increase in concentration of organic acid anions. As an alternative to their storage in leaf cell vacuoles, organic acid anions (mainly malate) are re-translocated in the phloem to the roots together with K as the accompanying cation. After decarboxylation of the organic acid anions in the roots, K may act again as counter-ion for nitrate transport in the xylem to the shoot. Support for this model was provided by Touraine et al. (1990) in soybean that reduces about 90% of the nitrate in shoots. In soybean, close correlations were found between nitrate reduction in the shoot, re-translocation of K and organic acid anions in the phloem, decarboxylation in the roots, and alkalinization of the medium.

3.5 Remobilization of nutrients

3.5.1 General

Import and export of nutrients occur simultaneously during the lifespan of plant organs such as leaves (Table 3.13). As a rule, aging (senescence) is associated with higher rates of export of nutrients than rates of import and, thus, a decrease in net content (amount per organ) (Table 3.9). In the following discussion, this decrease in net content is denoted by the term "remobilization."

A range of different physiological and biochemical processes contribute to remobilization: (1) release of nutrients stored in vacuoles (K, P, Mg, amino-N, etc.), (2) breakdown of storage proteins (e.g., in vacuoles of paraveinal mesophyll cells of legumes), and (3) breakdown of cell structures (e.g., chloroplasts) and enzymes thereby converting structurally bound nutrients (e.g., Mg in chlorophyll, micronutrients in enzymes) into a mobile form.

Remobilization of nutrients is important during ontogenesis at the following stages: seed germination, periods of insufficient supply to the roots during vegetative growth, reproductive growth, and during the period before leaf drop in perennials.

3.5.2 Seed germination

During the germination of seeds (or storage organs such as tubers), nutrients are remobilized within the tissue and translocated in the phloem and/or xylem to the developing roots and shoots. Consequently, seedlings may grow for at least several days without an external supply of nutrients (White and Veneklaas, 2012). In seeds, many nutrients (e.g., K, Mg, Ca, Mn, Zn, Fe) are usually present as phytate salts, and remobilization of these nutrients, and also of P, is correlated with phytase activity (White and Veneklaas, 2012).

3.5.3 Vegetative stage

During vegetative growth, nutrient supply to roots is often either insufficient (as in the case of low soil availability) or temporarily interrupted (when, e.g., there is a transient lack of soil water). Remobilization of nutrients from mature leaves to areas of new growth is thus of key importance for the completion of the life cycle of plants under such conditions. This behavior is typical for fast-growing crop species, whereas many wild species simply cease to grow under adverse environmental conditions and, therefore, remobilization of nutrients plays a less important role (Chapin, 1983).

The extent to which remobilization takes place differs among nutrients, and this is reflected in the distribution of visible deficiency symptoms in plants. Deficiency symptoms that predominantly occur in young leaves and apical meristems reflect insufficient remobilization. This might be the consequence of either insufficient phloem mobility or that only a relatively small fraction of the nutrients can be converted into a mobile form in older leaves. The extent of remobilization can be used to diagnose the nutritional status of plants (Chapter 11).

Remobilization of several nutrients is accelerated by their deficiency in actively growing plant parts (Etienne et al., 2018; Maillard et al., 2015). In oilseed rape plants, remobilization of K, P, and Mg from older leaves is increased during periods of their deficiency, whereas remobilization of N, S, Fe, Zn, B, Cu, Ni, and Mo is unchanged when these elements are deficient (Maillard et al., 2015). In contrast, Ca and Mn are remobilized from the root system via the xylem during periods of deficiency (Maillard et al., 2015). The capacity to remobilize K from older leaves is positively correlated with greater K utilization efficiency in a number of crops (White et al., 2021); the capacity to remobilize other nutrients is also likely to contribute to their utilization efficiencies (Sylvester-Bradley and Kindred, 2009; Veneklaas et al., 2012).

3.5.4 Reproductive stage

Remobilization of nutrients is particularly important during reproductive growth when seeds, fruits, and storage organs are formed. At this growth stage, root activity and nutrient uptake generally decrease, mainly as a result of decreasing carbohydrate supply to the roots (Chapter 5), although increased nitrogen-use efficiency is linked to postanthesis N uptake in brassica and cereal crops (White et al., 2013). The nutrient concentrations of vegetative parts therefore often decline sharply during the reproductive stage (Fig. 3.12). There is a delicate balance between the timing of senescence, nutrient composition of fruits, seeds and storage organs, nutrient-use efficiency, and yield.

The extent of remobilization of nutrients depends on various factors, including (1) the specific requirement of seeds, fruits, and storage organs for a given nutrient; (2) the nutritional status of the vegetative parts; (3) the ratio between vegetative mass (source size) and number and size of seeds, fruits or tubers (sink size); and (4) the nutrient uptake rate by the roots during the reproductive stage. Remobilization is highly nutrient selective and varies among plant species (Maillard et al., 2015).

A typical example of the differences in the extent of remobilization of nutrients from vegetative shoots is shown in Table 3.14 for pea plants grown under field conditions. The percentage of remobilization of N and P is very high, whereas there is a lack of remobilization of Mg and Ca; instead, a net increase in the concentrations of these nutrients takes place in the vegetative organs, as has also been shown for soil-grown soybean plants (Bender et al., 2015; Wood et al., 1986). Relatively high concentrations of nutrients in the soil solution leading to continuous uptake by the roots and the import of nutrients into leaves after anthesis/flowering are the reasons for the increase in leaf Mg and Ca

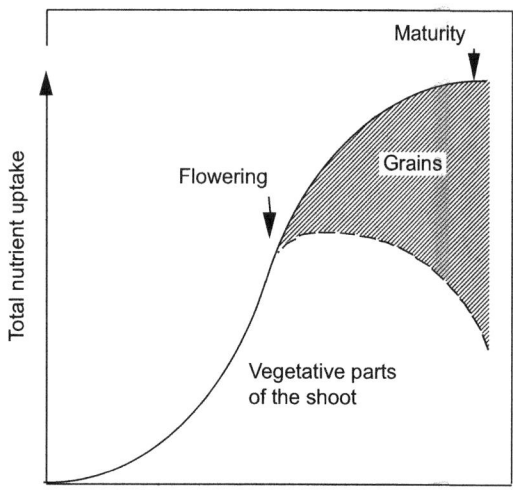

FIGURE 3.12 Schematic representation of the distribution of nutrients in a cereal plant throughout ontogeny.

TABLE 3.14 Remobilization of nutrients in field-grown pea (*Pisum sativum*) between flowering and ripening.

Harvest	N	P	K	Mg	Ca
	Content in stems and leaves (kg ha^{-1})				
June 8 (flowering)	64	7	53	5	31
June 22 (14 d after flowering)	87	10	66	8	60
July 1 (23 d after flowering)	60	7	61	8	69
July 12 (maturity)	32	3	46	9	76
	Increase or decrease 22 June–12 July (%)				
	−63	−71	−30	+10	+21
	In seeds (% of total shoot content on 12 July)				
	76	82	39	26	4

Source: Based on Garz (1966).

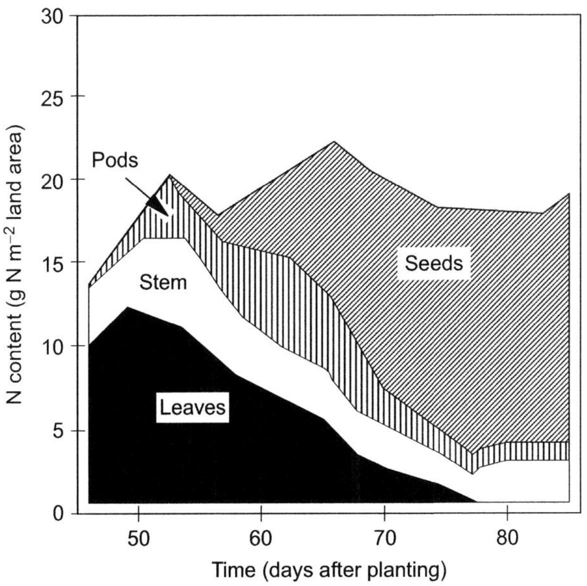

FIGURE 3.13 Nitrogen partitioning in field-grown bean (*Phaseolus vulgaris*) during reproductive growth. *Adapted from Lynch and White (1992).*

concentrations. An inherently low capacity for remobilization of Ca is also a contributing factor. A high remobilization of N from leaves has also been observed in bean (*P. vulgaris*) grown in the field (Fig. 3.13). In potato, N, P, and K are readily mobilized from leaves and stems to tubers in the tuber bulking phase, but Ca, and to a lesser extent, Mg and Mn are not remobilized readily (White, 2018).

In cereals such as wheat, up to about 90% of the total P in grains can be attributed to remobilization from vegetative parts, although lower proportions may be found when the roots are continuously well supplied with P (e.g., in sand culture; Batten et al., 1986). A comparison of N remobilization in different wheat cultivars under field conditions gave an average value of 83%, but values ranged from 51% to 91% depending on the total N uptake by the cultivar (Van Sanford and MacKown, 1987). In maize, N, P, and K are remobilized readily, S, Mg, Zn, and B less readily, and Ca, Cu, Fe, and Mn scarcely (Bender et al., 2013; Maillard et al., 2015).

The capacity to remobilize nutrients from senescing leaves varies among plant species (Etienne et al., 2018; Hocking and Pate, 1977; Maillard et al., 2015; Milla et al., 2005). Nitrogen is readily mobilized from senescing leaves in all plant species, but the efficiency of remobilization can vary substantially. For example, in the study of Maillard et al. (2015), 90% of leaf N was remobilized in wheat, but only 40% in maize. Potassium, P, S, and Mg were remobilized during senescence in most plant species studied by Maillard et al. (2015), whereas Fe, Zn, Cu, Ni, B, and Mo

TABLE 3.15 Changes in fresh and dry weights and nutrient contents of leaf blades of soybean (*Glycine max*) during podfill.

	Early-to-mid podfill (day 64)	Late podfill (day 88)
Fresh weight (g leaflet^{-1})		
	0.653	0.857
Dry weight (g leaflet^{-1})		
	0.183	0.120
Content (mg leaflet^{-1})		
P	0.86	0.21
K	1.60	0.23
Ca	2.97	8.74
Mg	0.84	1.43
S	0.59	0.45
Content (µg leaflet^{-1})		
Fe	31.9	25.9
Zn	29.5	18.5
Mn	23.7	48.1
Cu	0.66	0.75
B	11.4	20.7
Mo	0.29	0.08

Source: Recalculated from Wood et al. (1986).

were remobilized in many, but not all, plant species they studied. Small amounts of Ca and Mn were remobilized in a few plant species, including wheat and barley (Maillard et al., 2015). During the reproductive stage, the degree of remobilization of micronutrients and of Ca is often high compared with that during vegetative growth. In white lupin (*L. albus*), for example, up to 50% of micronutrients and 18% of Ca that originally accumulated in the leaves were remobilized to the fruits (Hocking and Pate, 1978). Substantial remobilization of at least some of the micronutrients during pod filling also occurs in soil-grown plants, for example, Fe, Zn, and Mo in soybean (Table 3.15).

The extent of remobilization of micronutrients is inversely related to their concentrations in fully expanded leaves (Loneragan et al., 1976). This is because there is a greater proportion of firmly bound micronutrients in, for example, enzymes and cell walls in deficient leaves. Thus, when boron (^{10}B) was applied to leaves of apple (*Malus domestica*), pear (*Pyrus communis*), plum (*Prunus domestica*), or cherry (*Prunus cerasus*), which transport B readily in the phloem, it was almost completely exported within 9–33 days, whereas the content of soil-derived leaf B (already present before foliar application of the labeled isotope) remained unchanged (Hanson, 1991). These relationships contrast with those for highly mobile nutrients, such as N, K, P, and Mg, where a greater proportion is remobilized from leaves of deficient plants (Maillard et al., 2015).

The extent of remobilization of the micronutrients Cu, Fe, and Zn, but not Mn, is closely related to leaf senescence (Distelfeld et al., 2014; Nable and Loneragan, 1984; Waters et al., 2009). This is reflected, for example, in the close positive correlation between the remobilization of N and of Cu (Fig. 3.14). The onset of senescence can be accelerated by shading, and this is associated with a more rapid remobilization of both N and Cu; in Cu-deficient plants most of the Cu can then be remobilized. Nitrogen deficiency, like shading, also enhances Cu remobilization (Hill et al., 1978). The same is true for Zn (Hill et al., 1979b).

Remobilization of nutrients requires several steps: (1) mobilization within individual leaf cells, (2) short-distance transport in the apoplasm or symplasm to the phloem, (3) phloem loading, and (4) phloem transport. During vegetative

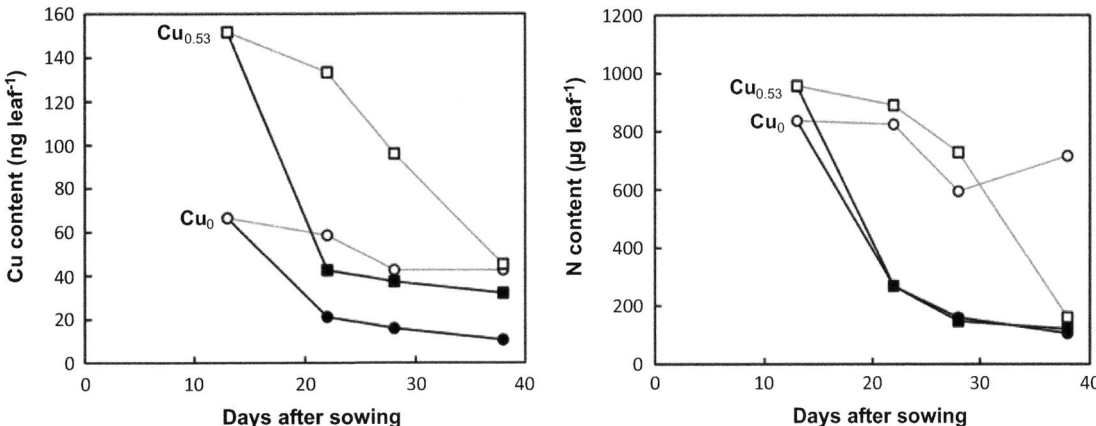

FIGURE 3.14 Copper and N contents of the oldest leaf of wheat plants grown in a sandy soil with no (Cu$_0$; *circles*) or optimal Cu fertilization at 0.53 mg Cu kg^{-1} soil (Cu$_{0.53}$; *squares*) in unshaded (*open symbols*) or shaded (*closed symbols*) conditions. *Adapted from Hill et al. (1979a)*.

growth, low rates of mobilization within individual cells often restrict the remobilization of nutrients from leaves. This is most likely the reason why, despite the intermediate phloem mobility of Fe, Zn, Cu, Mo, and B, deficiency symptoms of these micronutrients first appear in young leaves and the shoot apex during vegetative growth. However, during reproductive growth, leaf senescence promotes the mobilization of nutrients within individual leaf cells and, thereby, enhances their remobilization.

The extent of remobilization of nutrients is attracting increasing attention in connection with the selection and breeding of genotypes with high "nutrient efficiency" (Sylvester-Bradley and Kindred, 2009; Veneklaas et al., 2012; Wang et al., 2019b; White et al., 2021). Better growth on soils with low nutrient availability can be conferred not only by improved acquisition of nutrients but also by increasing the efficiency of their use at the physiological level. Genotypic variation has been observed in tissue utilization of many nutrients, including those commonly used as fertilizers, such as N, P, and K (Sylvester-Bradley and Kindred, 2009; Veneklaas et al., 2012; Wang et al., 2019b; White et al., 2021). In addition, the necessity to increase the concentrations of elements essential for human and animal nutrition in edible produce is driving the selection of genotypes that remobilize a greater proportion of micronutrients from leaves to seeds, fruits, and storage organs (Bouis and Saltzman, 2017; Cakmak, 2008; White and Broadley, 2009, 2011; Zou et al., 2019).

3.5.5 Perennials

Remobilization of nutrients from leaves to woody parts is a typical feature of perennial species before leaf drop in temperate climates and is closely related to the discoloration of leaves in the autumn. As a rule, and similar to annual species, the extent of remobilization is high for N, K, P, and S in most species and for Mg, Cu, Mo, and B in some species (Maillard et al., 2015). The extent of B remobilization depends primarily on the B status and plant capacity to translocate sugar alcohols in the phloem (Wang et al., 2015). The remobilization of Zn, Fe, Ca, and Mn is generally limited (Maillard et al., 2015). During the period preceding leaf drop, typical visible deficiency symptoms are often observed, indicating remobilization of a particular nutrient. In evergreens, the remobilization of N and P is associated with leaf senescence in mid-summer, whereas K is remobilized during winter and over the summer drought period, perhaps to mitigate osmotic stress in growing tissues (Milla et al., 2005).

References

Alejandro, S., Höller, S., Meier, B., Peiter, E., 2020. Manganese in plants: from acquisition to subcellular allocation. Front. Plant Sci. 11, 300. Available from: https://doi.org/10.3389/fpls.2020.00300.

Álvarez-Fernández, A., Diaz-Benito, P., Abadia, A., López-Millán, A.-F., Abadia, J., 2014. Metal species involved in long distance metal transport in plants. Front. Plant Sci. 5, 105. Available from: https://doi.org/10.3389/fpls.2014.00105.

Andresen, E., Peiter, E., Küpper, H., 2018. Trace metal metabolism in plants. J. Exp. Bot. 69, 909–954.

Aoyama, T., Kobayashi, T., Takahashi, M., Nagasaka, S., Usuda, K., Kakei, Y., et al., 2009. OsYSL18 is a rice iron(III)−deoxymugineic acid transporter specifically expressed in reproductive organs and phloem of lamina joints. Plant Mol. Biol. 70, 681−692.

Armand, T., Cullen, M., Boiziot, F., Li, L., Fricke, W., 2019. Cortex cell hydraulic conductivity, endodermal apoplastic barriers and root hydraulics change in barley (*Hordeum vulgare* L.) in response to a low supply of N and P. Ann. Bot. 124, 1091−1107.

Armstrong, M.J., Kirkby, E.A., 1979. Estimation of potassium recirculation in tomato plants by comparison of the rates of potassium and calcium accumulation in the tops with their fluxes in the xylem stream. Plant Physiol. 63, 1143−1148.

Bahrun, A., Jensen, C.R., Asch, F., Mogensen, V.O., 2002. Drought-induced changes in xylem pH, ionic composition, and ABA concentration act as early signals in field-grown maize (*Zea mays* L.). J. Exp. Bot. 53, 251−263.

Barberon, M., 2017. The endodermis as a checkpoint for nutrients. New Phytol. 213, 1604−1610.

Batten, G.D., Wardlaw, I.F., Aston, M.J., 1986. Growth and the distribution of phosphorus in wheat developed under various phosphorus and temperature regimes. Aust. J. Agric. Res. 37, 459−469.

Bender, R.R., Haegele, J.W., Ruffo, M.L., Below, F.E., 2013. Nutrient uptake, partitioning, and remobilization in modern, transgenic insect-protected maize hybrids. Agron. J 105, 161−170.

Bender, R.R., Haegele, J.W., Below, F.E., 2015. Nutrient uptake, partitioning, and remobilization in modern soybean varieties. Agron. J. 107, 563−573.

Ben-Zioni, A., Vaadia, Y., Lips, S.H., 1971. Nitrate uptake by roots as regulated by nitrate reduction products of the shoot. Physiol. Plant. 24, 288−290.

Blom-Zandstra, M., Vogelzang, S.A., Veen, B.W., 1998. Sodium fluxes in sweet pepper exposed to varying sodium concentrations. J. Exp. Bot. 49, 1863−1868.

Bouis, H.E., Saltzman, A., 2017. Improving nutrition through biofortification: A review of evidence from HarvestPlus, 2003 through 2016. Global Food Security 12, 49−58.

Broadley, M.R., White, P.J., Hammond, J.P., Zelko, I., Lux, A., 2007. Zinc in plants. New Phytol. 173, 677−702.

Brown, P.H., Hu, H., 1998. Phloem boron mobility in diverse plant species. Bot. Acta 111, 331−335.

Cakmak, I., 2008. Enrichment of cereal grains with zinc: agronomic or genetic biofortification? Plant Soil 302, 1−17.

Canny, M.J., McCully, M.E., 1989. The xylem sap of maize roots: its collection, composition and formation. Aust. J. Plant Physiol. 15, 557−566.

Cataldo, D.A., McFadden, K.M., Garland, T.R., Wildung, R.E., 1988. Organic constituents and complexation of nickel (II), iron (III), cadmium (II), and plutonium (IV) in soybean xylem exudates. Plant Physiol. 86, 734−739.

Chapin III, F.S., 1983. Adaptation of selected trees and grasses to low availability of phosphorus. Plant Soil 72, 283−297.

Cheeseman, J.M., 2015. The evolution of halophytes, glycophytes and crops, and its implications for food security under saline conditions. New Phytol. 206, 557−570.

Chen, K.E., Chen, H.Y., Tseng, C.S., Tsay, Y.F., 2020. Improving nitrogen use efficiency by manipulating nitrate remobilization in plants. Nat. Plants 6, 1126−1135.

Clemens, S., 2019. Metal ligands in micronutrient acquisition and homeostasis. Plant Cell Environ. 42, 2902−2912.

Coffey, O., Bonfield, R., Corre, F., Sirigiri, J.A., Meng, D., Fricke, W., 2018. Root and cell hydraulic conductivity, apoplastic barriers and aquaporin gene expression in barley (*Hordeum vulgare* L.) grown with low supply of potassium. Ann. Bot. 122, 1131−1141.

Conn, S., Gilliham, M., 2010. Comparative physiology of elemental distributions in plants. Ann. Bot. 105, 1081−1102.

Cooper, H.D., Clarkson, D.T., 1989. Cycling of amino-nitrogen and other nutrients between shoots and roots in cereals - a possible mechanism integrating shoot and root in the regulation of nutrient uptake. J. Exp. Bot. 40, 753−762.

Cramer, M.D., Shane, M.W., Lambers, H., 2005. Physiological changes in white lupin associated with variation in root-zone CO_2 concentration and cluster-root P mobilization. Plant Cell Environ. 28, 1203−1217.

Cui, J., Peuke, A.D., Limami, A.M., Tcherkez, G., 2021. Why is phloem sap nitrate kept low? Plant Cell Environ. 44, 2838−2843. Available from: https://doi.org/10.1111/pce.14116.

Da Silva, M.C., Shelp, B.J., 1990. Xylem-to-phloem transfer of organic nitrogen in young soybean plants. Plant Physiol. 92, 797−801.

Deeken, R., Geiger, D., Fromm, J., Koroleva, O., Ache, P., Langenfeld-Heyser, R., et al., 2002. Loss of the AKT2/3 potassium channel affects sugar loading into the phloem of Arabidopsis. Planta 216, 334−344.

DeKreij, C., Janse, J., Van Goor, B.J., Van Doesburg, J.D.J., 1992. The incidence of calcium oxalate crystals in fruit walls of tomato (*Lycopersicon esculentum* Mill.) as affected by humidity, phosphate and calcium supply. J. Hort. Sci. 67, 45−50.

Deng, F., Yamaji, N., Xia, J., Ma, J.F., 2013. A member of heavy metal P-type ATPase OsHMA5 is involved in xylem loading of copper in rice. Plant Physiol. 163, 1353−1362.

Ding, L., Li, Y., Gao, L., Lu, Z., Wang, M., Ling, N., et al., 2018. Aquaporin expression and water transport pathways inside leaves are affected by nitrogen supply through transpiration in rice plants. Int. J. Mol. Sci. 19, 256. Available from: https://doi.org/10.3390/ijms19010256.

Ding, G., Lei, G.J., Yamaji, N., Yokosho, K., Mitani-Ueno, N., Huang, S., et al., 2020. Vascular cambium-localized AtSPDT mediates xylem-to-phloem transfer of phosphorus for its preferential distribution in Arabidopsis. Mol. Plant. 13, 99−111.

Distelfeld, A., Avni, R., Fischer, A.M., 2014. Senescence, nutrient remobilization, and yield in wheat and barley. J. Exp. Bot. 65, 3783−3798.

Drew, M.C., Läuchli, A., 1987. The role of the mesocotyl in sodium exclusion from the shoot of *Zea mays* L. (cv. Pioneer 3906). J. Exp. Bot. 38, 409−418.

El-Baz, F.K., Maier, P., Wissemeier, A., Horst, W.J., 1990. Uptake and distribution of manganese applied to leaves of *Vicia faba* (cv. Herzfreya) and *Zea mays* (cv. Regent) plants. Z. Pflanzenernähr. Bodenk. 153, 279−282.

Eschrich, W., 1976. Strasburger's Kleines Botanisches Praktikum für Anfänger. Fischer, Stuttgart.

Etienne, P., Diquelou, S., Prudent, M., Salon, C., Maillard, A., Ourry, A., 2018. Macro and micronutrient storage in plants and their remobilization when facing scarcity: the case of drought. Agriculture 8, 14. Available from: https://doi.org/10.3390/agriculture8010014.

Everaarts, A.P., Blom-Zandstra, M., 2001. Internal tipburn of cabbage (*Brassica oleracea* var. *capitata*). J. Hort. Sci. Biotech. 76, 515–521.

Fink, S., 1991a. Comparative microscopical studies on the patterns of calcium oxalate distribution in the needles of various conifer species. Bot. Acta 104, 306–315.

Fink, S., 1991b. The micromorphological distribution of bound calcium in needles of Norway spruce (*Picea abies* (L.) Karst.). New Phytol. 119, 33–40.

Finnemann, J., Schjoerring, J.K., 1999. Translocation of NH_4^+ in oilseed rape plants in relation to glutamine synthetase isogene expression and activity. Physiol. Plant. 105, 469–477.

Flis, P., Ouerdane, L., Grillet, L., Curie, C., Mari, S., Lobinski, R., 2016. Inventory of metal complexes circulating in plant fluids: A reliable method based on HPLC coupled with dual elemental and high-resolution molecular mass spectrometric detection. New Phytol. 211, 1129–1141.

Flowers, T.J., Galal, H.K., Bromham, L., 2010. Evolution of halophytes: multiple origins of salt tolerance in land plants. Funct. Plant Biol. 37, 604–612.

Franceschi, V.R., Nakata, P.A., 2005. Calcium oxalate in plants: formation and function. Ann. Rev. Plant Biol. 56, 41–71.

Furch, A.C.U., Zimmermann, M.R., Will, T., Hafke, J.B., van Bel, A.J.E., 2010. Remote-controlled stop of phloem mass flow by biphasic occlusion in *Cucurbita maxima*. J. Exp. Bot. 61, 3697–3708.

Garz, J., 1966. Menge, Verteilung und Bindungsform der Mineralstoffe (P, K, Mg und Ca) in den Leguminosensamen in Abhängigkeit von der Mineralstoffumlagerung innerhalb der Pflanze und den Ernährungsbedingungen. Kuehn-Arch. 80, 137–194.

Gaupels, F., Ghirardo, A., 2013. The extrafascicular phloem is made for fighting. Front. Plant Sci. 11, 187. Available from: https://doi.org/10.3389/fpls.2013.00187.

Gaymard, F., Pilot, G., Lacombe, B., Bouchez, D., Bruneau, D., Boucherez, J., et al., 1998. Identification and disruption of a plant shaker-like outward channel involved in K^+ release into the xylem sap. Cell 94, 647–655.

Gerath, H., Borchmann, W., Zajonc, I., 1975. Zur Wirkung des Mikronährstoffs Bor auf die Ertragsbildung von Winterraps (*Brassica napus* L. ssp. *oleifera*). Arch. Acker- Pflanzenbau Bodenkd. 19, 781–792.

Giehl, R.F.H., Laginha, A.M., Duan, F.Y., Rentsch, D., Yuan, L.X., von Wirén, N., 2017. A critical role of AMT2;1 in root-to-shoot translocation of ammonium in Arabidopsis. Mol. Plant 10, 1449–1460.

Gigolashvili, T., Kopriva, S., 2014. Transporters in plant sulfur metabolism. Front. Plant Sci. 5, 442. Available from: https://doi.org/10.3389/fpls.2014.00442.

Gilliham, M., Tester, M., 2005. The regulation of anion loading to the maize root xylem. Plant Physiol. 137, 819–828.

Glavac, V., Koenies, H., Ebben, U., Avenhaus, U., 1991. Jahreszeitliche Veränderung der NO3--Konzentrationen im Xylemsaft des unteren Stammteiles von Buchen (*Fagus sylvatica* L.). Z. Pflanzenernähr. Bodenk. 154, 121–125.

Greenway, H., Gunn, A., 1966. Phosphorus retranslocation in *Hordeum vulgare* during early tillering. Planta 71, 43–67.

Greenway, H., Pitman, M.G., 1965. Potassium retranslocation in seedlings of *Hordeum vulgare*. Aust. J. Biol. Sci. 18, 235–247.

Guttridge, C.G., Bradfield, E.G., Holder, R., 1981. Dependence of calcium transport into strawberry leaves on positive pressure in the xylem. Ann. Bot. 48, 473–480.

Haeder, H.-E., Beringer, H., 1984. Long distance transport of potassium in cereals during grain filling in intact plants. Physiol. Plant. 62, 439–444.

Hafke, J.B., Furch, A.C.U., Reitz, M.U., van Bel, A.J.E., 2007. Functional sieve element protoplasts. Plant Physiol. 145, 703–711.

Hanson, E.J., 1991. Movement of boron out of tree fruit leaves. HortScience 26, 271–273.

Harris, W.R., Sammons, R.D., Grabiak, R.C., 2012. A speciation model of essential trace metal ions in phloem. J. Inorg. Biochem. 116, 140–150.

Hayes, P.E., Clode, P.L., Pereira, C.G., Lambers, H., 2019. Calcium modulates leaf cell-specific phosphorus allocation in Proteaceae from southwestern Australia. J. Exp. Bot. 70, 3995–4009.

He, H., Veneklaas, E.J., Kuo, J., Lambers, H., 2014. Physiological and ecological significance of biomineralization in plants. Trends Plant Sci. 19, 166–174.

Hecht-Buchholz, C., 1973. Molybdänverteilung und -verträglichkeit bei Tomate, Sonnenblume und Bohne. Z. Pflanzenernähr. Bodenk. 136, 110–119.

Hedrich, R., Geiger, D., 2017. Biology of SLAC1-type anion channels - from nutrient uptake to stomatal closure. New Phytol. 216, 46–61.

Hell, R., Khan, M.S., Wirtz, M., 2010. Cellular biology of sulfur and its functions in plants. In: Hell, R., Mendel, R.R. (Eds.), Plant Cell Monographs 17, Cell Biology of Metals and Nutrients, Springer, Berlin, Germany, pp. 243–279.

Hermans, C., Johnson, G.N., Strasser, R.J., Verbruggen, N., 2004. Physiological characterisation of magnesium deficiency in sugar beet: acclimation to low magnesium differentially affects photosystems I and II. Planta 220, 344–355.

Hermans, C., Conn, S.J., Chen, J., Xiao, O., Verbruggen, N., 2013. An update on magnesium homeostasis mechanisms in plants. Metallomics 5, 1170. Available from: https://doi.org/10.1039/c3mt20223b.

Hill, J., Robson, A.D., Loneragan, J.F., 1978. The effect of copper and nitrogen supply on the retranslocation of copper in four cultivars of wheat. Aust. J. Agric. Res. 29, 925–939.

Hill, J., Robson, A.D., Loneragan, J.F., 1979a. The effects of Cu supply and shading on Cu retranslocation from old wheat leaves. Ann. Bot. 43, 449–457.

Hill, J., Robson, A.D., Loneragan, J.F., 1979b. The effect of copper supply on the senescence and the retranslocation of nutrients of the oldest leaf of wheat. Ann. Bot. 44, 279–287.

Ho, L., White, P.J., 2005. A cellular hypothesis for the induction of blossom end rot in tomato fruit. Ann. Bot. 95, 571–581.

Hocking, P.J., 1980. The composition of phloem exudate and xylem sap from tree tobacco (*Nicotiana glauca* Groh). Ann. Bot. 45, 633–643.

Hocking, P.J., Pate, J.S., 1977. Mobilization of minerals to developing seeds of legumes. Ann. Bot. 41, 1259–1278.

Hocking, P.J., Pate, J.S., 1978. Accumulation and distribution of mineral elements in annual lupins *Lupinus albus* and *Lupinus angustifolius* L. Aust. J. Agric. Res. 29, 267–280.

Hocking, B., Tyerman, S.D., Burton, R.A., Gilliham, M., 2016. Fruit calcium: transport and physiology. Front. Plant Sci. 7, 569. Available from: https://doi.org/10.3389/fpls.2016.00569.

Howe, J.A., Florence, R.J., Harris, G., van Santen, E., Beasley, J.P., Bostick, J.P., et al., 2012. Effect of cultivar, irrigation, and soil calcium on runner peanut response to gypsum. Agron. J. 104, 1312–1320.

Huang, L.B., Bell, R.W., Dell, B., 2008. Evidence of phloem boron transport in response to interrupted boron supply in white lupin (*Lupinus albus* L. cv. Kiev Mutant) at the reproductive stage. J. Exp. Bot. 59, 575–583.

Hussain, A., Black, C.R., Taylor, I.B., Mulholland, B.J., Roberts, J.A., 1999. Novel approaches for examining the effects of differential soil compaction on xylem sap abscisic acid concentration, stomatal conductance and growth in barley (*Hordeum vulgare* L.). Plant Cell Environ. 22, 1377–1388.

Ishimaru, Y., Suzuki, M., Kobayashi, T., Takahashi, M., Nakanishi, H., Mori, S., et al., 2005. OsZIP4, a novel zinc-regulated zinc transporter in rice. J. Exp. Bot. 56, 3207–3214.

Ishimaru, Y., Masuda, H., Bashir, K., Inoue, H., Tsukamoto, T., Takahashi, M., et al., 2010. Rice metal-nicotianamine transporter, OsYSL2, is required for the long-distance transport of iron and manganese. Plant J. 62, 379–390.

Jeena, G.S., Kumar, S., Shukla, R.K., 2019. Structure, evolution and diverse physiological roles of SWEET sugar transporters in plants. Plant Mol. Biol. 100, 351–365.

Jeschke, W.D., Pate, J.S., 1991. Cation and chloride partitioning through xylem and phloem within the whole plant of *Ricinus communis* L. under conditions of salt stress. J. Exp. Bot. 42, 1105–1116.

Jeschke, W.D., Pate, J.S., 1992. Temporal patterns of uptake, flow and utilization of nitrate, reduced nitrogen and carbon in a leaf of salt-treated castor bean (*Ricinus communis* L.). J. Exp. Bot. 43, 393–402.

Jeschke, W.D., Kirkby, E.A., Peuke, A.D., Pate, J.S., Hartung, W., 1997a. Effects of P deficiency on assimilation and transport of nitrate and phosphate in intact plants of castor bean (*Ricinus communis* L). J. Exp. Bot. 48, 75–91.

Jeschke, W.D., Peuke, A.D., Pate, J.S., Hartung, W., 1997b. Transport, synthesis, and catabolism of abscisic acid (ABA) in intact plants of castor bean (*Ricinus communis* L.) under phosphate deficiency and moderate salinity. J. Exp. Bot. 48, 1737–1747.

Jeschke, W.D., Pate, J.S., Atkins, C.A., 1986. Effects of NaCl salinity on growth, development, ion transport and ion storge in white lupin (*Lupinus albus* L. cv. Ultra). J. Plant Physiol. 124, 257–274.

Jiang, F., Hartung, W., 2008. Long-distance signalling of abscisic acid (ABA): the factors regulating the intensity of the ABA signal. J. Exp. Bot. 59, 37–43.

Jørgensen, M.E., Xu, D., Crocoll, C., Ernst, H.A., Ramírez, D., Motawia, M.S., et al., 2017. Origin and evolution of transporter substrate specificity within the NPF family. eLife 2017, e19466. Available from: https://doi.org/10.7554/eLife.19466.

Juchaux-Cachau, M., Landouar-Arsivaud, L., Pichaut, J.-P., Campion, C., Porcheron, B., Jeauffre, J., et al., 2007. Characterization of AgMaT2, a plasma membrane mannitol transporter from celery, expressed in phloem cells, including phloem parenchyma cells. Plant Physiol. 145, 62–74.

Kakei, Y., Ishimaru, Y., Kobayashi, T., Yamakawa, T., Nakanishi, H., Nishizawa, N., 2012. OsYSL16 plays a role in the allocation of iron. Plant Mol. Biol. 79, 583–594.

Karley, A.J., White, P.J., 2009. Moving cationic minerals to edible tissues: Potassium, magnesium, calcium. Curr. Opin. Plant Biol. 12, 291–298.

Karley, A.J., Leigh, R.A., Sanders, D., 2000. Where do all the ions go? The cellular basis of differential ion accumulation in leaf cells. Trends Plant Sci. 5, 465–470.

Kataoka, T., Hayashi, N., Yamaya, T., Takahashi, H., 2004. Root-to-shoot transport of sulfate in Arabidopsis. Evidence for the role of SULTR3;5 as a component of low-affinity sulfate transport system in the root vasculature. Plant Physiol. 136, 4198–4204.

Kaur, H., Greger, M., 2019. A review on Si uptake and transport system. Plants 8, 81. Available from: https://doi.org/10.3390/plants8040081.

Kehr, J., 2013. Systemic regulation of mineral homeostasis by micro RNAs. Front. Plant Sci. 4, 145. Available from: https://doi.org/10.3389/fpls.2013.00145.

Kehr, J., Kragler, F., 2018. Long distance RNA movement. New Phytol. 218, 29–40.

Kirkby, E.A., Knight, A.H., 1977. Influence of the level of nitrate nutrition on ion uptake and assimilation, organic acid accumulation, and cation-anion balance in whole tomato plants. Plant Physiol. 60, 349–353.

Koenig, A.M., Hoffmann-Benning, S., 2020. The interplay of phloem-mobile signals in plant development and stress response. Biosci. Rep. 40. Available from: https://doi.org/10.1042/BSR20193329.

Kudoyarova, G.R., Dodd, I.C., Veselov, D.S., Rothwell, S.A., Veselov, S.Y., 2015. Common and specific responses to availability of mineral nutrients and water. J. Exp. Bot. 66, 2133–2144.

Kühn, C., Grof, C.P.L., 2010. Sucrose transporters of higher plants. Curr. Opin. Plant Biol. 12, 288–298.

Kutrowska, A., Szelag, M., 2014. Low-molecular weight organic acids and peptides involved in the long-distance transport of trace metals. Acta Physiol. Plant. 36, 1957–1968.

Lang, A., Thorpe, M.R., 1989. Xylem, phloem and transpiration flows in a grape: application of a technique for measuring the volume of attached fruits to high resolution using archimedes' principle. J. Exp. Bot. 40, 1069–1078.

Larsson, C.-M., Larsson, M., Purves, J.V., Clarkson, D.T., 1991. Translocation and cycling through roots of recently absorbed nitrogen and sulphur in wheat (*Triticum aestivum*) during vegetative and generative growth. Physiol. Plant. 82, 345–352.

Lessani, H., Marschner, H., 1978. Relation between salt tolerance and long distance transport of sodium and chloride in various crop species. Aust. J. Plant Physiol. 5, 27–37.

Liu, T.Y., Lin, W.Y., Huang, T.K., Chiou, T.J., 2014. MicroRNA-mediated surveillance of phosphate transporters on the move. Trends Plant Sci. 19, 647–655.

Loneragan, J.F., Snowball, K., Robson, A.D., 1976. Remobilization of nutrients and its significance in plant nutrition. In: Wardlaw, I.F., Passioura, J.B. (Eds.), Transport and Transfer Process in Plants. Academic Press, London, UK, pp. 463–469.

Loneragan, J.F., Kirk, G.J., Webb, M.J., 1987. Translocation and function of zinc in roots. J. Plant Nutr. 10, 1247–1254.

Losso, A., Nardini, A., Dämon, B., Mayr, S., 2018. Xylem sap chemistry: seasonal changes in timberline conifers *Pinus cembra*, *Picea abies*, and *Larix decidua*. Biol. Plant. 62, 157–165.

Lynch, J., White, J.W., 1992. Shoot nitrogen dynamics in tropical common bean. Crop Sci. 32, 392–397.

Ma, J.F., Yamaji, N., 2006. Silicon uptake and accumulation in higher plants. Trends Plant Sci. 11, 392–397.

Maillard, A., Diquélou, S., Billard, V., Laîné, P., Garnica, M., Prudent, M., et al., 2015. Leaf mineral nutrient remobilization during leaf senescence and modulation by nutrient deficiency. Front. Plant Sci. 6, 317. Available from: https://doi.org/10.3389/fpls.2015.00317.

Marschner, H., 1983. General introduction to the mineral nutrition of plants. In: Läuchli, A., Bieleski, R.L. (Eds.), Encyclopedia of Plant Physiology, New Series, Vol. 15A, Springer-Verlag, Berlin, Germany, pp. 5–60.

Marschner, H., Richter, C., 1973. Akkumulation und Translokation von K^+, Na^+ und Ca^{2+} bei Angebot zu einzelnen Wurzelzonen von Maiskeimpflanzen. Z. Pflanzenernähr. Bodenk. 135, 1–15.

Marschner, H., Schafarczyk, W., 1967. Vergleich der Nettoaufnahme von Natrium und Kalium bei Mais- und Zuckerrübenpflanzen. Z. Pflanzenernähr. Bodenk. 118, 172–187.

Martin, P., 1971. Wanderwege des Stickstoffs in Buschbohnenpflanzen beim Aufwärtstransport nach der Aufnahme durch die Wurzel. Z. Pflanzenphysiol. 64, 206–222.

Maurel, C., Verdoucq, L., Rodrigues, O., 2016. Aquaporins and plant transpiration. Plant Cell Environ. 39, 2580–2587.

McCain, D.C., Markley, J.L., 1989. More manganese accumulates in maple sun leaves than in shade leaves. Plant Physiol. 90, 1417–1421.

McGrath, J.F., Robson, A.D., 1984. The movement of zinc through excised stems of seedlings of *Pinus radiata* D. Don. Ann. Bot. 54, 231–242.

Milla, R., Castro-Díez, P., Maestro-Martínez, M., Montserrat-Martí, G., 2005. Relationships between phenology and the remobilization of nitrogen, phosphorus and potassium in branches of eight Mediterranean evergreens. New Phytol. 168, 167–178.

Miller, A.J., Shen, Q., Xu, G., 2009. Freeways in the plant: transporters for N, P and S and their regulation. Curr. Opin. Plant Biol. 12, 284–290.

Milne, R.J., Grof, C.P.L., Patrick, J.W., 2018. Mechanisms of phloem unloading: shaped by cellular pathways, their conductances and sink function. Curr. Opin. Plant Biol. 43, 8–15.

Miwa, K., Fujiwara, T., 2010. Boron transport in plants: coordinated regulation of transporters. Ann. Bot. 105, 1103–1108.

Mix, G.P., Marschner, H., 1976. Einfluß exogener und endogener Faktoren auf den Calciumgehalt von Paprika- und Bohnenfrüchten. Z. Pflanzenernähr. Bodenk. 139, 551–563.

Mizrahi, Y., Pasternak, D., 1985. Effect of salinity on quality of various agricultural crops. Plant Soil 89, 301–307.

Morris, R.J., 2018. On the selectivity, specificity and signalling potential of the long-distance movement of messenger RNA. Curr. Opin. Plant Biol. 43, 1–7.

Morrissey, J., Baxter, I.R., Lee, J., Li, L., Lahner, B., Grotz, N., et al., 2009. The ferroportin metal efflux proteins function in iron and cobalt homeostasis in Arabidopsis. Plant Cell 21, 3326–3338.

Mudge, S.R., Rae, A.L., Diatloff, E., Smith, F.W., 2002. Expression analysis suggests novel roles for members of the Pht1 family of phosphate transporters in *Arabidopsis*. Plant J. 31, 341–353.

Munns, R., Tester, M., 2008. Mechanisms of salinity tolerance. Annu. Rev. Plant Biol. 59, 651–681.

Nable, R.O., Loneragan, J.F., 1984. Translocation of manganese in subterranean clover (*Trifolium subterraneum* L. cv. Seaton Park). II. Effects of leaf senescence and of restricting supply of manganese to part of a split root system. Aust. J. Plant Physiol. 11, 113–118.

Nalam, V.J., Han, J., Pitt, W.J., Acharya, S.R., Nachappa, P., 2021. Location, location, location: Feeding site affects aphid performance by altering access and quality of nutrients. PLoS ONE 16, e0245380. Available from: https://doi.org/10.1371/journal.pone.0245380.

Nawaz, M.A., Zakharenko, A.M., Zemchenko, I.V., Haider, M.S., Ali, M.A., Imtiaz, M., et al., 2019. Phytolith formation in plants: from soil to cell. Plants 8, 249. Available from: https://doi.org/10.3390/plants8080249.

Neumann, P.M., 1987. Sequential leaf senescence and correlatively controlled increase in xylem flow resistance. Plant Physiol. 83, 941–944.

Nikolic, M., Pavlovic, J., 2018. Plant responses to iron deficiency and toxicity and iron use efficiency in plants. In: Hossain, M., Kamiya, T., Burritt, D., Phan Tran, L.-S., Fujiwara, T. (Eds.), Plant Micronutrient Use Efficiency: Molecular and Genomic Perspectives in Crop Plants. Elsevier, Academic Press, London, UK, pp. 55–69.

Noodén, L.D., Mauk, C.S., 1987. Changes in the mineral composition of soybean xylem sap during monocarpic senescence and alterations by depodding. Physiol. Plant. 70, 735–742.

Notaguchi, M., Okamoto, S., 2015. Dynamics of long-distance signaling via plant vascular tissues. Front. Plant Sci. 6, 161. Available from: https://doi.org/10.3389/fpls.2015.00161.

O'Brien, J.A., Vega, A., Bouguyon, E., Krouk, G., Gojon, A., Coruzzi, G., et al., 2016. Nitrate transport, sensing, and responses in plants. Mol. Plant. 6, 837–856.

Oertli, J.J., 1994. Non-homogeneity of boron distribution in plants and consequences for foliar diagnosis. Commun. Soil Sci. Plant Anal. 25, 1133–1147.

Peuke, A.D., 2010. Correlations in concentrations, xylem and phloem flows, and partitioning of elements and ions in intact plants. A summary and statistical re-evaluation of modelling experiments in *Ricinus communis*. J. Exp. Bot. 61, 635–655.

Peuke, A.D., Jeschke, W.D., Hartung, W., 1994. The uptake and flow of C, N and ions between roots and shoots in *Ricinus communis* L. 3. Long-distance transport of abscisic acid depending on nitrogen nutrition and salt stress. J. Exp. Bot. 45, 741–747.

Peuke, A.D., Jeschke, W.D., Hartung, W., 2002. Flows of elements, ions and abscisic acid in *Ricinus communis* and site of nitrate reduction under potassium limitation. J. Exp. Bot. 53, 241–250.

Plett, D.C., Møller, I.S., 2010. Na$^+$ transport in glycophytic plants: what we know and would like to know. Plant Cell Environ. 33, 612–626.

Pommerrenig, B., Müdsam, C., Kischka, D., Neuhaus, H.E., 2020. Treat and trick: common regulation and manipulation of sugar transporters during sink establishment by the plant and the pathogen. J. Exp. Bot. 71, 3930–3940.

Qu, Y., Guan, R., Bose, J., Henderson, S.W., Wege, S., Qiu, L., Gilliham, M., 2021. Soybean CHX-type ion transport protein GmSALT3 confers leaf Na$^+$ exclusion via a root derived mechanism, and Cl$^-$ exclusion via a shoot derived process. Plant Cell Environ. 44, 856–869.

Ramsperger-Gleixner, M., Geiger, D., Hedrich, R., Sauer, N., 2004. Differential expression of sucrose transporter and polyol transporter genes during maturation of common plantain companion cells. Plant Physiol. 134, 147–160.

Raven, J.A., 2017. Evolution and palaeophysiology of the vascular system and other means of long-distance transport. Phil. Trans. R. Soc. B 373, 20160497. Available from: https://doi.org/10.1098/rstb.2016.0497.

Saalbach, E., Aigner, H., 1970. Über die Wirkung einer Natriumdüngung auf Natriumgehalt, Ertrag und Trockensubstanzgehalt einiger Gras- und Kleearten. Landwirtsch. Forsch. 23, 264–274.

Sasaki, A., Yamaji, N., Mitani-Ueno, N., Kashino, M., Ma, J.F., 2015. A node localized transporter OsZIP3 is responsible for the preferential distribution of Zn to developing tissues in rice. Plant J. 84, 374–384.

Sauer, N., 2007. Molecular physiology of higher plant sucrose transporters. FEBS Lett. 581, 2309–2317.

Schenk, H.J., Jansen, S., Holtta, T., 2021. Positive pressure in xylem and its role in hydraulic function. New Phytol. 230, 27–45.

Schiavon, M., Pilon-Smits, E.A.H., 2017. The fascinating facets of plant selenium accumulation - biochemistry, physiology, evolution and ecology. New Phytol. 213, 1582–1596.

Shao, J.F., Yamaji, N., Shen, R.F., Ma, J.F., 2017. The key to Mn homeostasis in plants: regulation of Mn transporters. Trends Plant. Sci. 22, 215–224.

Shao, J.F., Yamaji, N., Liu, X.W., Yokosho, K., Shen, R.F., Ma, J.F., 2018. Preferential distribution of boron to developing tissues is mediated by the intrinsic protein OsNIP3. Plant Physiol. 176, 1739–1750.

Shear, C.B., 1975. Calcium related disorders of fruits and vegetables. HortScience 10, 361–365.

Simpson, R.J., Lambers, H., Dalling, M.J., 1982. Translocation of nitrogen in a vegetative wheat plant (*Triticum aestivum*). Physiol. Plant. 56, 11–17.

Sinclair, S.A., Senger, T., Talke, I.N., Cobbett, C.S., Haydon, M.J., Krämer, U., 2018. Systemic upregulation of MTP2- and HMA2-mediated Zn partitioning to the shoot supplements local Zn deficiency responses. Plant Cell 30, 2463–2479.

Smeulders, F., van de Geijn, S.C., 1983. In situ immobilization of heavy metals with tetraethylenepentamine(tetren) in natural soils and its effect on toxicity and plant growth. III. Uptake and mobility of copper and its tetren-complex in corn plants. Plant Soil 70, 59–68.

Sutton, T., Baumann, U., Hayes, J., Collins, N.C., Shi, B.J., Schnurbusch, T., et al., 2007. Boron-toxicity tolerance in barley arising from efflux transporter amplification. Science 318, 1446–1449.

Sylvester-Bradley, R., Kindred, D.R., 2009. Analysing nitrogen responses of cereals to prioritize routes to the improvement of nitrogen use efficiency. J. Exp. Bot. 60, 1939–1951.

Tachibana, S., 1991. Import of calcium by tomato fruit in relation to the day-night periodicity. Sci. Hort. 45, 235–243.

Takahashi, H., Watanabe-Takahashi, A., Smith, F.W., Blake-Kalff, M., Hawkesford, M.J., Saito, K., 2000. The role of three functional sulphate transporters involved in uptake and translocation of sulphate in *Arabidopsis thaliana*. Plant J. 23, 171–182.

Tanaka, M., Wallace, I.S., Takano, J., Roberts, D.M., Fujiwara, T., 2008. NIP6;1 is a boric acid channel for preferential transport of boron to growing shoot tissues in *Arabidopsis*. Plant Cell 20, 2860–2875.

Tardieu, F., Parent, B., Simonneau, T., 2010. Control of leaf growth by abscisic acid: hydraulic or non-hydraulic processes? Plant Cell Environ. 33, 636–647.

Tardieu, F., Simonneau, T., Parent, B., 2015. Modelling the coordination of the controls of stomatal aperture, transpiration, leaf growth, and abscisic acid: update and extension of the Tardieu-Davies model. J. Exp. Bot. 66, 2227–2237.

Tegeder, M., 2014. Transporters involved in source to sink partitioning of amino acids and ureides: opportunities for crop improvement. J. Exp. Bot. 65, 1865–1878.

Tegeder, M., Masclaux-Daubresse, C., 2018. Source and sink mechanisms of nitrogen transport and use. New Phytol. 217, 35–53.

Touraine, B., Grignon, N., Grignon, C., 1990. Interaction between nitrate assimilation in shoots and nitrate uptake by roots of soybean (*Glycine max*) plants: Role of carboxylate. Plant Soil 124, 169–174.

Turgeon, R., Wolf, S., 2009. Phloem transport: cellular pathways and molecular trafficking. Ann. Rev. Plant Biol. 60, 207–221.

Van Bel, A.J.E., 1984. Quantification of the xylem-to-phloem transfer of amino acids by use of inulin (^{14}C) carboxylic acid as xylem transport marker. Plant Sci. Lett. 35, 81–85.

Van Berkel, N., 1988. Preventing tipburn in chinese cabbage by high relative humidity during the night. Neth. J. Agric. Sci. 36, 301–308.

Van Beusichem, M.L., Kirkby, E.A., Baas, R., 1988. Influence of nitrate and ammonium nutrition and the uptake, assimilation, and distribution of nutrients in *Ricinus communis*. Plant Physiol. 86, 914–921.

Van Sanford, D.A., MacKown, C.T., 1987. Cultivar differences in nitrogen remobilization during grain fill in soft red winter wheat. Crop Sci. 27, 295–300.

Veneklaas, E.J., Lambers, H., Bragg, J., Finnegan, P.M., Lovelock, C.E., Plaxton, W.C., et al., 2012. Opportunities for improving phosphorus-use efficiency in crop plants. New Phytol. 195, 306–320.

Wang, N., Yang, C., Pan, Z., Liu, Y., Peng, S., 2015. Boron deficiency in woody plants: various responses and tolerance mechanisms. Front. Plant Sci. 6, 916. Available from: https://doi.org/10.3389/fpls.2015.00916.

Wang, Y.Y., Cheng, Y.H., Chen, K.E., Tsay, Y.F., 2018. Nitrate transport, signaling, and use efficiency. Annu. Rev. Plant Biol. 69, 85–122.

Wang, P., Calvo-Polanco, M., Reyt, G., Barberon, M., Champeyroux, C., Santoni, V., et al., 2019a. Surveillance of cell wall diffusion barrier integrity modulates water and solute transport in plants. Sci. Rep. 9, 4227. Available from: https://doi.org/10.1038/s41598-019-40588-5.

Wang, W., Ding, G.D., White, P.J., Wang, X.H., Jin, K.M., Xu, F.S., et al., 2019b. Mapping and cloning of quantitative trait loci for phosphorus efficiency in crops: opportunities and challenges. Plant Soil 439, 91–112.

Waters, B.M., Grusak, M.A., 2008. Whole-plant mineral partitioning throughout the life cycle in *Arabidopsis thaliana* ecotypes Columbia, Landsberg *erecta*, Cape Verde Islands, and the mutant line *ysl1ysl3*. New Phytol. 177, 389–405.

Waters, B.M., Uauy, C., Dubcovsky, J., Grusak, M.A., 2009. Wheat (*Triticum aestivum*) NAM proteins regulate the translocation of iron, zinc, and nitrogen compounds from vegetative tissues to grain. J. Exp. Bot. 60, 4263–4274.

Webb, M.J., Loneragan, J.F., 1990. Zinc translocation to wheat roots and its implications for a phosphorus/zinc interaction in wheat plants. J. Plant Nutr. 13, 1499–1512.

White, P.J., 1997. The regulation of K^+ influx into roots of rye (*Secale cereale* L.) seedlings by negative feedback via the K^+ flux from shoot to root in the phloem. J. Exp. Bot. 48, 2063–2073.

White, P.J., 2001. The pathways of calcium movement to the xylem. J. Exp. Bot. 52, 891–899.

White, P.J., 2017. Ion transport. In: Thomas, B., Murphy, D.J., Murray, B.G. (Eds.), Encyclopedia of Applied Plant Sciences, Second Edition, Academic Press, Waltham, MA, USA, pp. 238–245.

White, P.J., 2018. Improving nutrient management in potato cultivation. In: Wale, S. (Ed.), Achieving Sustainable Cultivation of Potatoes, Volume 2: Production, Storage and Crop Protection. Burleigh Dodds Science Publishing, Cambridge, UK, pp. 45–67.

White, P.J., Broadley, M.R., 2001. Chloride in soils and its uptake and movement within the plant: A review. Ann. Bot. 88, 967–988.

White, P.J., Broadley, M.R., 2003. Calcium in plants. Ann. Bot. 92, 487–511.

White, P.J., Broadley, M.R., 2009. Biofortification of crops with seven mineral elements often lacking in human diets – iron, zinc, copper, calcium, magnesium, selenium and iodine. New Phytol. 182, 49–84.

White, P.J., Broadley, M.R., 2011. Physiological limits to zinc biofortification of edible crops. Front. Plant Sci. 2, 80. Available from: https://doi.org/10.3389/fpls.2011.00080.

White, P.J., Pongrac, P., 2017. Heavy-metal toxicity in plants. In: Shabala, S. (Ed.), Plant Stress Physiology, Second Edition, CABI, Wallingford, UK, pp. 301–331.

White, P.J., Veneklaas, E.J., 2012. Nature and nurture: the importance of seed phosphorus. Plant Soil 357, 1–8.

White, P.J., George, T.S., Gregory, P.J., Bengough, A.G., Hallett, P.D., McKenzie, B.M., 2013. Matching roots to their environment. Ann. Bot. 112, 207–222.

White, P.J., Bowen, H.C., Broadley, M.R., El-Serehy, H.A., Neugebauer, K., Taylor, A., et al., 2017. Evolutionary origins of abnormally large shoot sodium accumulation in non-saline environments within the Caryophyllales. New Phytol. 214, 284–293.

White, P.J., Bell, M.J., Djalovic, I., Hinsinger, P., Rengel, Z., 2021. Potassium use efficiency of plants. In: Murrell, T.S., Mikkelsen, R.L., Sulewski, G., Norton, R., Thompson, M.L. (Eds.), Improving Potassium Recommendations for Agricultural Crops. Springer, Cham, Switzerland, pp. 119–145.

Wilkinson, S., Davies, W.J., 2002. ABA-based chemical signalling: the co-ordination of responses to stress in plants. Plant Cell Environ. 25, 195–210.

Win, K., Berkowitz, G.A., Henninger, M., 1991. Antitranspirant-induced increases in leaf water potential increase tuber calcium and decrease tuber necrosis in water-stressed potato plants. Physiol. Plant. 96, 116–120.

Wolf, O., Munns, R., Tonnet, M.L., Jeschke, W.D., 1990. Concentrations and transport of solutes in xylem and phloem along the leaf axis of NaCl-treated *Hordeum vulgare*. J. Exp. Bot. 41, 1133–1141.

Wolterbeek, H.T., van Luipen, J., de Bruin, M., 1984. Non-steady state xylem transport of fifteen elements into the tomato leaf as measured by gamma-ray spectroscopy: A model. Physiol. Plant. 61, 599–606.

Wood, L.J., Murray, B.J., Okatan, Y., Noodén, L.D., 1986. Effect of petiole phloem distribution on starch and mineral distribution in senescing soybean leaves. Am. J. Bot. 73, 1377–1383.

Wu, H.H., 2018. Plant salt tolerance and Na^+ sensing and transport. Crop J. 6, 215–225.

Wu, H.H., Zhang, X.C., Giraldo, J.P., Shabala, S., 2018. It is not all about sodium: revealing tissue specificity and signalling roles of potassium in plant responses to salt stress. Plant Soil 431, 1–17.

Yamaji, N., Ma, J.F., 2014. The node, a hub for mineral nutrient distribution in graminaceous plants. Trends Plant Sci. 19, 556–563.

Yamaji, N., Ma, J.F., 2017. Node-controlled allocation of mineral elements in Poaceae. Curr. Opin. Plant Biol. 39, 18–24.

Yamaji, N., Xia, J.X., Mitani-Ueno, N., Yokosho, K., Ma, J.F., 2013. Preferential delivery of zinc to developing tissues in rice is mediated by P-type heavy metal ATPase OsHMA2. Plant Physiol. 162, 927–939.

Yoshinari, A., Takano, J., 2017. Insights into the mechanisms underlying boron homeostasis in plants. Front. Plant Sci. 8, 1951. Available from: https://doi.org/10.3389/fpls.2017.01951.

Zhai, Z., Gayomba, S.R., Jung, H.-I., Vimalakumari, N.K., Piñeros, M., Craft, E., et al., 2014. OPT3 is a phloem-specific iron transporter that is essential for systemic iron signaling and redistribution of iron and cadmium in Arabidopsis. Plant Cell 26, 2248–2264.

Zhang, C.K., Turgeon, R., 2018. Mechanisms of phloem loading. Curr. Opin. Plant Biol. 43, 71–75.

Zhang, W.H., Zhou, Y., Dibley, K.E., Tyerman, S.D., Furbank, R.T., Patrick, J.W., 2007. Nutrient loading of developing seeds. Funct. Plant Biol. 34, 314–331.

Zhang, X.D., Zorb, C., Geilfus, C.M., 2020. The root as a sink for chloride under chloride-salinity. Plant Physiol. Biochem. 155, 161–168.

Zou, C., Du, Y., Rashid, A., Ram, H., Savasli, E., Pieterse, P.J., et al., 2019. Simultaneous biofortification of wheat with zinc, iodine, selenium, and iron through foliar treatment of a micronutrient cocktail in six countries. J. Agric. Food Chem. 67, 8096–8106.

Chapter 4

Uptake and release of elements by leaves and other aerial plant parts

Thomas Eichert[1] and Victoria Fernández[2]

[1]Department of Horticulture, University of Applied Sciences, Erfurt, Germany, [2]Systems and Natural Resources Department, Technical University of Madrid, Madrid, Spain

Summary

Although leaves and other aerial plant parts are protected by the cuticle and stomata against the uncontrolled exchange of matter with the environment, elements may penetrate the external plant surface either through the cuticle (solutes) or through stomata (gases and solutes). Gases, such as ammonia (NH_3) and sulfur dioxide (SO_2), may be taken up or released through open stomata. Dissolved nutrients may cross the leaf surface in both directions, resulting in foliar uptake of solutes originating from atmospheric deposition or foliar fertilization and leaching of nutrients out of leaves. This chapter gives an overview of the importance of uptake and release of gases through stomata and summarizes the current knowledge about the barrier properties of both the cuticle and stomata against the penetration of solutes. Practical aspects of foliar fertilization and the ecological consequences of nutrient uptake and release are outlined.

4.1 General

To minimize uncontrolled exchange of matter with the environment, leaf surfaces of terrestrial plants are covered by a cuticle. The cuticle is a nonliving, hydrophobic skin with a low permeability for water, gases, and solutes. To enable CO_2 uptake, leaves are equipped with stomata as adjustable apertures in the leaf surface, which optimize the trade-off between CO_2 uptake and water loss by plants (Figs. 4.1 and 4.2). The evolutionary development of the cuticle and stomata as barriers against the uncontrolled exchange of matter was the prerequisite for the colonization of the land surface by higher plants, but these barriers do not fully impede the exchange of gaseous and dissolved nutrients. Gases, such as ammonia (NH_3) and sulfur dioxide (SO_2), may be taken up or released through open stomata. Dissolved nutrients may penetrate the leaf surface in both directions, resulting in uptake as well as leaching of nutrients from leaves.

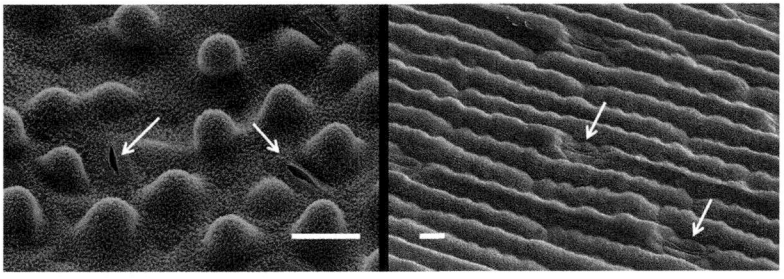

FIGURE 4.1 Scanning electron micrograph of lower (abaxial) leaf surfaces of *Apocynum cannabinum* (left) and *Zea mays* (right). Arrows point to stomata. Scale bars = 20 μm. *Courtesy of W. Barthlott, University of Bonn, Germany.*

☆ This chapter is a revision of the third edition chapter by T. Eichert and V. Fernández, pp. 71–84. DOI: https://doi.org/10.1016/B978-0-12-384905-2.00004-2. © Elsevier Ltd.

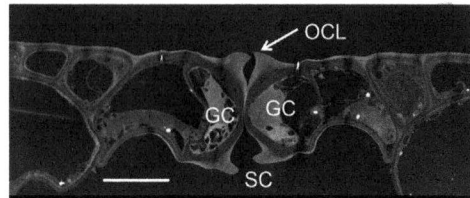

FIGURE 4.2 Cross-section of the epidermis of a *Heliconia choconiana* leaf. The cuticle can be identified as a light-colored layer covering the epidermal cells, including the GCs. Scale bar = 10 μm. *GC*, Guard cell; *OCL*, outer cuticular ledge; *SC*, substomatal cavity. *Courtesy of W. Barthlott, University of Bonn, Germany.*

TABLE 4.1 Plant dry weight (dw), total N content, and N content and percentage derived from NH_3–N in the atmosphere by Italian ryegrass grown at low soil nitrate concentrations and exposed to different atmospheric NH_3 concentrations for 33 days.

NH_3 concentration (μg m^{-3})	Plant dw (g pot^{-1})	Total plant N (mg pot^{-1})	Total plant N derived from NH_3 (mg pot^{-1})	Percentage of total N derived from NH_3
14	16.8	120	5	4.2
123	18.2	157	47	30
297	19.6	221	121	55
498	20.9	304	218	72
709	22.0	340	335	99

Source: Adapted and recalculated from Whitehead and Lockyer (1987).

4.2 Uptake and release of gases and other volatile compounds through stomata

In terrestrial plants, stomata (Figs. 4.1 and 4.2) are the sites of exchange of gases (mainly CO_2, O_2) with the atmosphere. Their number per mm^2 of leaf surface varies between about 20 in succulents (CAM species), 100–200 in most annual species, and more than 800 in certain tree species (e.g., *Acer montanum*). Stomata are usually more abundant in (mostly in annual species), or confined to (in many tree species, e.g., *Fagus sylvatica*), the lower (abaxial) leaf surface. Nutrients in the form of gases, such as SO_2, NH_3, and NO_2, enter leaves predominantly through the stomata and are usually rapidly metabolized in the leaves. Foliar uptake of these gases has attracted much interest as they are major components of air pollution, and their uptake can be substantial. Moreover, depending on their density and the plant species, they may reduce or enhance plant growth. For many gases, plant surfaces can act as both a source and a sink. The compensation point, that is, the external gas concentration at which the net flux is zero, depends mainly on the type of gas, plant species, plant nutritional status, and climatic conditions.

4.2.1 Volatile nitrogen compounds

On a global scale, agriculture is the major source of atmospheric NH_3 emissions (Morán et al., 2016), with an estimated share of 80%–90% of the total (Behera et al., 2013; Chang et al., 2019; Clarisse et al., 2010). Globally, about 18% of the applied mineral nitrogen is lost due to NH_3 volatilization, and annual N losses can be as high as 96 kg ha^{-1} (Pan et al., 2016). In areas away from significant sources, NH_3 concentrations in air can be quite low, <1 μg m^{-3} (Nair and Yu, 2020), whereas in agricultural areas, they may be about three orders of magnitude higher (Krupa, 2003). Concentrations of NH_3 can reach 20–30 μg m^{-3} immediately after the application of mineral N fertilizers (Hensen et al., 2009; Herrmann et al., 2001). Organic N fertilizers may cause even higher NH_3 emissions; after sewage sludge application, NH_3 concentrations of up to 100–2400 μg m^{-3} were measured in the air (Beauchamp et al., 1978). In the past years, research has focused on developing methods to apply organic materials, such as manure and slurry, and minimize the emissions of volatile N compounds (Pan et al., 2016; Pedersen et al., 2020; Webb et al., 2010).

Plants may absorb and utilize NH_3 rapidly at concentrations above the compensation point (Table 4.1), which depends on a range of factors, such as plant species, root N supply, and the form of N applied (Wichink Kruit et al., 2010). Compensation points of 0.1 to >10 μg NH_3 m^{-3} have been reported (David et al., 2009; Herrmann et al., 2001; Massad et al., 2009). With a high root N supply, exposure to NH_3 may reduce biomass production and induce downregulation of root nitrate uptake, whereas in N-deficient plants shoot growth may be increased (Castro et al., 2006; Chen et al., 2020).

Under field conditions leaves may absorb 3%–70% of NH_3 volatilized from soil-applied N fertilizers (Bash et al., 2010; Boaretto et al., 2013; Schoninger et al., 2018). Under greenhouse conditions with elevated NH_3 concentrations, uptake values of 24%–38% have been reported (Huang et al., 2020).

In urban areas, the main source of N oxides [NO_x, i.e., the sum of mainly N oxide (NO) and N dioxide (NO_2)] is fossil fuel combustion, whereas in rural areas the use of N fertilizers is responsible for substantial NO_x emissions (Williams et al., 1992). Soils emit NO_x mostly as NO, which is formed as a by-product of nitrification and denitrification (Vinken et al., 2014). Nitric oxide acts as a gaseous signal in plants (León and Costa-Broseta, 2020). Nitrogen oxides can be both emitted from and deposited on plant surfaces (Teklemariam and Sparks, 2006). The compensation points for NO_x were reported to range from <0.2 to 34 $\mu g\ m^{-3}$ (Raivonen et al., 2009). The uptake of atmospheric NO_2 through stomata is linearly related to the external concentration, and its metabolism is rapid (Hu et al., 2014). Long-term exposure of plants to NO_2 can contribute considerably to their N nutrition (Hu and Sun, 2010). Nitrogen dioxide uptake by maize shoots accounted for more than 25% of the soil-emitted NO_x (Hereid and Monson, 2001).

Peroxyacetyl nitrate (PAN) is a toxic organic nitrate formed by photochemical reactions in the atmosphere (Teklemariam and Sparks, 2004). It was estimated that 3% of global N oxide emissions could be removed by foliar uptake of PAN (Sparks et al., 2003).

4.2.2 Volatile sulfur compounds

Sulfur dioxide (SO_2) is readily taken up through stomata and follows a distinct diurnal pattern related to the stomatal aperture (Rennenberg et al., 1990). In sensitive plants, SO_2 can be phytotoxic at relatively low atmospheric concentrations (0.1 mg m^{-3}), but susceptibility to SO_2 varies greatly among plant species (van der Kooij et al., 1997). In SO_2-fumigated Norway spruce seedlings (Kaiser et al., 1993), *Arabidopsis thaliana* (van der Kooij et al., 1997), Chinese cabbage (Yang et al., 2006), and poplar (Randewig et al., 2014), S accumulation in the leaves or needles increased linearly with increasing atmospheric SO_2 concentrations.

Depending on S availability in the soil, S demand of the crop and S concentration in the atmosphere, foliar S uptake may contribute significantly to crop S supply. Foliar uptake of H_2S can reduce root uptake of sulfate, suggesting the existence of coordinated shoot-to-root signals (Westerman et al., 2001). Under certain conditions, it may even fully replace root uptake (Ausma and De Kok, 2019).

Grassland species growing with a low S supply to the roots acquired 15%–35% of S from the atmosphere (Cliquet and Lemauviel-Lavenant, 2019). In short-term experiments with Chinese cabbage, the exposure to comparatively low SO_2 concentrations (0.3 mg m^{-3}) was estimated to be sufficient to cover the S requirement for growth (Yang et al., 2006).

Plants grown in a nonpolluted atmosphere and supplied only with sulfate in soil may release substantial amounts of volatile S compounds through stomata. Both the amounts and the spectrum of the emitted volatile S compounds vary among plant species, and in the case of oilseed rape may represent up to 1% of the total S in the plant per day. It was estimated that in cotton between a few hundred grams and a few kilograms of S per ha are emitted during the growing season (Grundon and Asher, 1988). Plants may release an average of 2–3 kg H_2S ha^{-1} per year (Schröder, 1993).

It has been suggested that the emission of volatile S may be a mechanism to dispose of excess S taken up by plants. Release of H_2S can also be part of a defense mechanism against pathogen attack (Bloem et al., 2012; Künstler et al., 2020; Vojtovic et al., 2021), the so-called "sulfur-induced resistance" (Bloem et al., 2005) or "sulfur-enhanced defense" (Rausch and Wachter, 2005). It should also be borne in mind that H_2S may act as a signaling molecule (García-Mata and Lamattina, 2010; Zhang et al., 2010) in plant stress responses and defense (Deng et al., 2016; Shi et al., 2015).

Plants are also net sinks for other volatile S compounds, such as carbonyl sulfide (COS) and carbon disulfide (CS_2) (Xu et al., 2002). The uptake of COS is closely correlated with CO_2 uptake because both involve conversion by carbonic anhydrase in plants (Protoschill-Krebs et al., 1996). The measurement of COS fluxes between atmosphere and vegetation can therefore be used for the large-scale estimation of CO_2 uptake, photosynthesis, and gross primary production (Asaf et al., 2013; Campbell et al., 2017; Stimler et al., 2010; Whelan et al., 2018).

4.3 Uptake of solutes

4.3.1 General

Foliar-applied nutrients may penetrate the leaf surface via several pathways such as the cuticle, stomata, cuticular irregularities, trichomes, veins, or other epidermal structures (Fernández et al., 2021). Recently, new imaging techniques, such as synchrotron-based X-ray fluorescence microscopy (Otto et al., 2021; Rodrigues et al., 2018) or high-resolution laser ablation inductively coupled plasma mass spectrometry (Arsic et al., 2020), have been used to evaluate the mechanisms of foliar absorption and transport of nutrients, but the contribution of each penetration route is not easy to determine (Fernández et al., 2021). Most foliar absorption studies have focused on the roles of cuticle and stomata, but the relative importance of the two pathways is still not fully understood. There is evidence that both pathways can be of equal importance (Eichert and Goldbach, 2008), but this also depends on the properties of the compound under consideration (e.g., size and water solubility) and the leaf surface properties (e.g., wettability, composition of the cuticle, abundance of stomata).

The penetration of leaf surfaces by solutes is a passive process driven by the concentration difference between the surface and the leaf interior. Furthermore, uptake into leaves has to be separated from the subsequent uptake of substances into the leaf cells. Both processes may be affected by similar factors, such as light or temperature, but because there is no strict feedback loop between solute uptake rates into the leaf cells and uptake rates through the leaf surface, this chapter will focus exclusively on the initial process of leaf penetration.

In aquatic plants, the leaves (not the roots) are the main sites of nutrient uptake. In terrestrial plants, however, the uptake of solutes by the surface of leaves and other aerial parts is severely restricted by the outer wall of the epidermal cells and the overlaying cuticle.

4.3.2 Structure of the cuticle

The surface of plants is covered by the cuticle, a biopolymer synthesized by epidermal cells (Pollard et al., 2008). Structure and composition of the cuticle vary greatly among plant species, genotypes, organs, and developmental stages (Fernández et al., 2016; Heredia-Guerrero et al., 2008) and are also affected by environmental conditions during development.

The cuticle consists of cell wall material (Guzmán et al., 2014a; Segado et al., 2016), cutin (a polyester of C_{16} and/or C_{18} hydroxy-fatty acids), and embedded waxes (intracuticular waxes) (Pollard et al., 2008). Waxes are a mixture of apolar compounds mainly composed of long-chain aliphatic and aromatic lipids. The cuticle is covered by epicuticular waxes that are often well structured (Barthlott et al., 1998). Variable amounts of polysaccharide fibrils and pectin lamellae may extend from the epidermal cell wall, binding the cuticle to the underlying tissue (Jeffree, 2006). Traditionally, two cuticular layers have been distinguished: the inner layer (cuticular layer) that contains many polysaccharides from the epidermal cell wall, and the outermost layer (cuticle proper) that was thought to be free of cell wall extensions (Fig. 4.3A). Recent findings, however, suggest that this model oversimplifies the complexity and variability of cuticle composition and structure (Fernández et al., 2016, 2017; Guzmán et al., 2014a,b; Yeats and Rose, 2013). In many plant species, cell wall polysaccharides may extend to the outermost leaf surface, and therefore the cuticle should be viewed as a modified epidermal cell wall with embedded lipids (Fig. 4.3B).

The cuticle has diverse functions. A major function is to protect the leaf from excessive water loss by transpiration. The control of water economy in terrestrial higher plants by stomata is dependent on the remaining surface of the plant having low hydraulic conductivity, which is ensured by the presence of the cuticle. The other main function of the cuticle is to protect leaves against excessive leaching of inorganic and organic solutes by rain. The relative importance of these two main functions of the cuticle depends on climatic conditions (arid zones vs humid tropics). In addition, the cuticle is involved in temperature control, optical properties of leaves and plays a role in defense against pests and diseases (Chapter 10).

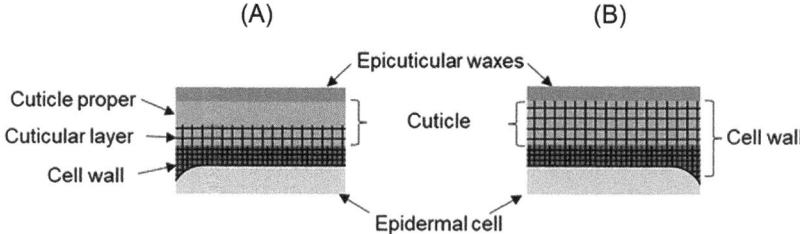

FIGURE 4.3 Models of the structure and composition of plant cuticles. According to the traditional model (A), the cuticle covers the epidermal cell wall and consists of two distinct layers: the lower "cuticular layer" impregnates the cell wall and thus contains significant amounts of cellulose and pectin, whereas the outermost "cuticle proper" is free of cell wall material. A modified model (B) considers the cuticle as the epidermal cell wall impregnated with cuticular waxes. In this model, only the epicuticular waxes are free of cell wall material. Cell wall constituents are indicated by the grid pattern.

4.3.3 Nutrient uptake through the cuticle

The general hydrophobic nature of the cuticle makes it an effective barrier against the penetration by hydrophilic, polar solutes, whereas lipophilic molecules may penetrate cuticles at much higher rates (Schönherr, 2006). The penetration of cuticle by lipophilic molecules is described by the solution-diffusion model (Riederer and Friedmann, 2006). This model predicts penetration rates of a molecule from its solubility and mobility in the cuticle (Zeisler-Diehl et al., 2017) according to Eq. (4.1):

$$P = D * K / \Delta x \tag{4.1}$$

where P is permeability (m s^{-1}), D is the diffusion coefficient in the cuticle (m^2 s^{-1}), K is the partition coefficient of the molecule between the external solution and the cuticle as a measurement of solubility, and Δx is the diffusion path length (m).

On a molecular level, solubilization and diffusion of a molecule in the cuticle can be viewed as moving into and between voids in a three-dimensional cutin network (Schönherr, 2006). The solubility parameter K takes into account the chemical affinity between the permeating molecule and the cutin matrix, whereas the diffusion coefficient D is determined by the size of the molecule as compared to the size of the voids in the cutin matrix. The cuticle is highly size-selective (Buchholz et al., 1998) because the size of voids, which is in the same order of magnitude as the solutes on the leaf surface, hinders diffusion and sets the size limits for penetrating molecules.

With a few exceptions, such as boric acid or urea, foliar fertilizers are applied as ions that have a low solubility in the cuticle (Schönherr, 2006). According to the solution-diffusion model, this should result in low cuticle penetration rates. However, laboratory studies using isolated cuticles and greenhouse or field studies using whole plants have shown that ion uptake can be substantial. To solve this apparent contradiction, a second penetration pathway for hydrophilic solutes, named "polar pores," has been postulated. These "pores" are thought to be created by clusters of water molecules sorbed by the cuticle (Tyree et al., 1990) (Fig. 4.4). Under dry atmospheric conditions, only small amounts of water are absorbed by the outer cuticle, and hence fewer functional "pores" traversing the cuticle exist. With increasing air humidity, more water is absorbed by the cuticle, eventually creating a continuous connection between the leaf surface and the epidermis (Fig. 4.4).

This hypothesis is supported by the observation that penetration rates of ions across isolated cuticles increased greatly with increasing relative humidity (RH) (Schönherr, 2000; Schönherr and Luber, 2001). The structure and composition of the cuticle make it most likely that, contrary to the common perception, the water "pores" are not created by filling previously empty spaces with water. Rather, water molecules get lodged in the cuticle by docking at hydrophilic domains, a process related to cuticle swelling (Arand et al., 2010; Chamel et al., 1991). To account for the dynamics of the humidity-dependent formation of functional aqueous connections across the cuticle, the term "dynamic aqueous continuum" was suggested (Fernández et al., 2017).

Cuticles may also have different size selectivity for lipophilic vs. hydrophilic molecules, which is further evidence for the existence of two spatially separate cuticular penetration pathways for these classes of compounds (Schreiber, 2005). In other studies, however, such differences were less evident (Popp et al., 2005). It is still debated whether polar "pores" actually exist as an independent pathway for hydrophilic solutes. Mathematical models describing and

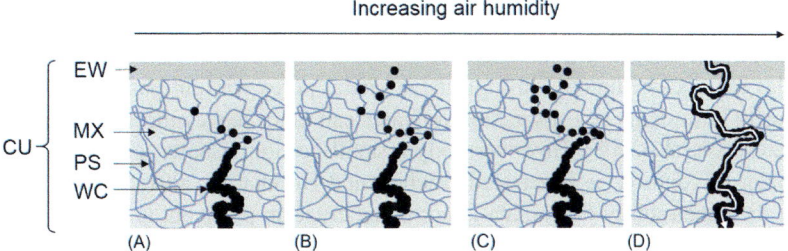

FIGURE 4.4 Model of the formation of a "dynamic aqueous continuum" traversing the cuticle. In this simplified model, the cuticle (CU) consists of a matrix of cutin and embedded waxes (MX) interspersed with hydrophilic domains, mostly provided by polysaccharides (PS) and other hydrophilic constituents. The overlying layer of EW facing the outer side is lacking polysaccharides. WC (*black dots*) are formed by adsorption of water by the hydrophilic domains. If air humidity is low, water clusters originate mainly from the epidermal cells underneath the cuticle (A). With increasing external air humidity, more water is sorbed by the cuticle from the outer surface (A–C). At high humidity, the tortuous connections emerge between the leaf surface and the leaf interior (D). Externally applied solutes may diffuse in these connections through the cuticle (*white arrow* in D). For clarity, other water clusters in the cuticle adjacent to the depicted emerging connection are not shown. *CU*, cuticle; *EW*, Epicuticular waxes; *MX*, matrix of cutin and embedded waxes; *PS*, polysaccharides; *WC*, water clusters.

predicting the uptake of hydrophilic solutes through cuticular "pores" have been developed (Tredenick et al., 2017, 2018). Alternatively, it was suggested that the dependence of ion penetration on cuticle hydration might simply be caused by the resulting increase in overall hydrophilicity of the cuticle, which in turn increases the solubility of polar solutes (Fernández and Eichert, 2009). In this case, the solution-diffusion model may still be valid and can be used to predict the permeability of cuticles for both hydrophilic and lipophilic solutes.

Irrespective of the nature and location of the hydrophilic cuticle penetration pathway, the question arises as to the diameters of the uptake routes in relation to the size of the permeating molecules. The diameter of hydrated ions such as metal cations, NO_3^- or NH_4^+ is well below 1 nm, whereas many organic compounds such as sugars or chelates are larger. Early estimations of "pore" sizes in de-waxed isolated citrus cuticles yielded diameters of about 1 nm (Schönherr, 1976), and a similar value was reported for isolated ivy cuticles (Popp et al., 2005). Such pore diameters would disallow the penetration of larger molecules. It must be considered, however, that these values represent the average pore diameters, which implies that some pores may be larger. In intact poplar or coffee leaves, average diameters of 4−5 nm were found (Eichert and Goldbach, 2008).

A range of studies have shown that the cuticle covering leaf veins and trichomes may represent preferential uptake routes for nutrients applied to leaves. Bahamonde et al. (2018) have shown that calcium (Ca) was preferentially absorbed by veins of beech leaves and provided evidence that the cuticle above veins differed from the surrounding cuticle in structure and composition. Li et al. (2019) found that in sunflower leaves nonglandular trichomes served as important entry sites for foliar-applied zinc (Zn). The reason for, and the ecological significance of, the increased permeability of the cuticle above these leaf structures is poorly understood.

4.3.4 Uptake through stomata

Stomata are adjustable apertures that enable the controlled entry of CO_2 into the leaf mesophyll required to sustain photosynthesis while minimizing water loss via transpiration. Many studies have demonstrated that the presence of stomata promotes foliar solute penetration. In species with hypostomatous leaves, foliar uptake rates were higher through the lower (abaxial) leaf surface that has stomata than through the upper (adaxial) side lacking them (Eichert and Goldbach, 2008; Kannan, 1969). Other studies reported positive correlations between uptake rates and stomatal density (Eichert and Burkhardt, 2001; Schönherr and Bukovac, 1978) or stomatal aperture (Eichert et al., 1998; Fernández et al., 2005; Schlegel and Schönherr, 2002a; Schlegel et al., 2006).

Stomata were initially assumed to be involved in foliar penetration of solutes via mass flow into the leaf mesophyll. However, stomata are protected against capillary infiltration of aqueous solutions due to their specific architecture (Schönherr and Bukovac, 1972). Stomata penetration by mass flow can only be induced by certain surface-active compounds, such as organosilicon surfactants (Field and Bishop, 1988) that lower the surface tension of the foliar-applied solution below a critical threshold. Thus, mass flow of foliar-applied solutions through stomata is negligible in most cases.

To explain the role of stomata in foliar solute uptake under the assumption of restricted mass flow, specific features of the peristomatal cuticle surrounding the guard cells have been postulated.

1. It was suggested that the peristomatal cuticle is more permeable than the rest of the cuticle, or that the cuticular ledges (see Fig. 4.2) could be the preferred entry points for foliar-applied solutes (Schönherr and Bukovac, 1978), but evidence for these hypotheses is still lacking (Fernández et al., 2021).
2. Penetration of solutes may occur directly through the stomatal pores without passing through the cuticle (Fig. 4.5). Hydrophilic particles (43 nm diameter) suspended in water can enter leaves through stomata by diffusion along the walls of the pore (Eichert et al., 2008). However, only a small portion, usually less than 10% of stomata covered by a foliar-applied droplet of solution, is penetrated (Eichert and Burkhardt, 2001; Eichert and Goldbach, 2008). It is therefore likely that the penetrability of stomata is not an inherent property but is acquired possibly by modifications

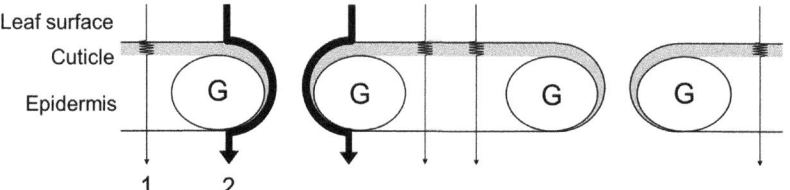

FIGURE 4.5 Schematic diagram of the solute penetration pathways across the leaf surface. (1) penetration of the cuticle, (2) penetration of stomata. Thickness of the arrows indicates the relative permeability of the pathways. Note that not all stomata are penetrable. Not to scale. G, guard cell.

of the pore wall cuticle (otherwise rather hydrophobic) due to the effect of deposited hygroscopic particles, microbes growing in the stomatal chamber, or salts ascending the pore, rendering it more wettable (Eichert et al., 1998; Eichert and Burkhardt, 2001). Burkhardt (2010) suggested that these modifications (increasing the wettability of stomatal pore walls) induced the "hydraulic activation of stomata," that is, the formation of continuous thin water films connecting the leaf apoplast and the leaf surface and enabling the bidirectional transport of substances (Burkhardt et al., 2012). Although not all stomata present on the leaf surface are involved in foliar solute uptake, the contribution of this penetration pathway may be very important and, depending on the type of solute, be as high as, or even higher than, the cuticular route (Eichert and Goldbach, 2008).

As directly demonstrated by Eichert et al. (2008), stomata represent possible penetration pathways for nanoparticles into plants. This pathway therefore may play an important role in a broad range of possible applications such as foliar fertilization (Du et al., 2015; Li et al., 2018a; Servin et al., 2015; Zhang et al., 2018) or plant stimulation (Alidoust and Isoda, 2013; Khan et al., 2019). On the other hand, stomatal uptake of nanoparticles may also serve as a port of entry of potentially toxic substances into plants and the food chain (Chichiricco and Poma, 2015; Larue et al., 2014).

4.3.5 Role of external factors

4.3.5.1 Environmental effects on the barrier properties during ontogenesis

The environmental conditions during plant growth have a direct influence on the leaf surface regarding cuticle thickness or amount and composition of epicuticular waxes (Bird and Gray, 2003; Koch et al., 2006), which, in turn, affect cuticle wettability and permeability. Shading during plant development may induce a decrease in the amount of wax per leaf area (Whitecross and Armstrong, 1972), whereas high temperatures can modify the morphology and composition of epicuticular waxes (Riederer and Schneider, 1990). RH influences the amount of wax per leaf area and wax crystal morphology (Koch et al., 2006).

In general, young, partially expanded leaves may be more penetrable than fully expanded leaves (Sargent and Blackman, 1962). This may also be the case with fruits because, for example, young apples were found to be more permeable to calcium chloride than mature apples (Schlegel and Schönherr, 2002b). The potentially higher permeability of young organs may be linked to the presence of trichomes, increased stomatal densities, and stomatal functionality (Schlegel and Schönherr, 2002b). However, the barrier function of the surfaces of older leaves can be modified by erosion/abrasion and pathogens, as well as by regulated responses to the environment (Jordan and Brodribb, 2007; Munné-Bosch, 2007).

Nutrient deficiencies may affect leaf structure and function, including the leaf surface. For example, in lemon trees grown on calcareous soils in Murcia (Spain), Fe deficiency (mean SPAD = 5 vs mean SPAD = 55 of Fe-sufficient plants) did not alter the upper (adaxial) leaf surface, but decreased the abundance of stomata in the lower (abaxial) surface significantly (Fig. 4.6), potentially altering, for example, wettability by water.

Relatively few studies have analyzed the effect of nutrient deficiency on the leaf epidermis, but general changes in trichome and/or stomatal abundance and morphology have received some attention. For example, compared to Fe-sufficient peach and pear leaves, Fe-chlorotic leaves were found to have decreased stomatal pore size and conductance, in addition to lower leaf cuticle weight per unit surface and concentration of soluble cuticular lipids (Fernández et al., 2008a). Manganese (Mn) deficiency in barley was reported to increase water permeability and alter light reflectance of barley leaves (Hebbern et al., 2009). Boron (B) deficiency was found to affect stomatal morphology and leaf functionality in coffee (Rosolem and Leite, 2007) and soybean (Will et al., 2011). Phosphorus (P) deficiency in wheat led to lower stomatal and trichome abundances (Fernández et al., 2014a; Peirce et al., 2014), similarly to Zn deficiency in sunflower (Li et al., 2018b) and pecan leaves (Ojeda-Barrios et al., 2012). Concerning the absorption of fertilizer solutions by nutrient-deficient compared to healthy leaves, lower rates of uptake have been reported for K-deficient olive (Restrepo-Diaz et al., 2008), B-deficient soybean (Will et al., 2011), P-deficient wheat (Fernández et al., 2014a; Peirce et al., 2014) and Zn-deficient sunflower (Li et al., 2018b) after foliar application of K, B, P, and Zn, respectively.

4.3.5.2 Humidity effects on solute concentration and leaf permeability

Foliar-applied solutions are usually rather dilute and not in equilibrium with the water potential of the atmosphere; therefore, they evaporate quickly from the leaves (or even before reaching the leaf surfaces). This results in decreased droplet volume accompanied by an increase in solute concentrations. During equilibration with the atmosphere, solutions may dry out completely or remain liquid depending on the RH of the air and the type of solute. For each solute, there is a threshold RH called deliquescence RH (DRH) or deliquescence point (Burkhardt and Eiden, 1994) at which

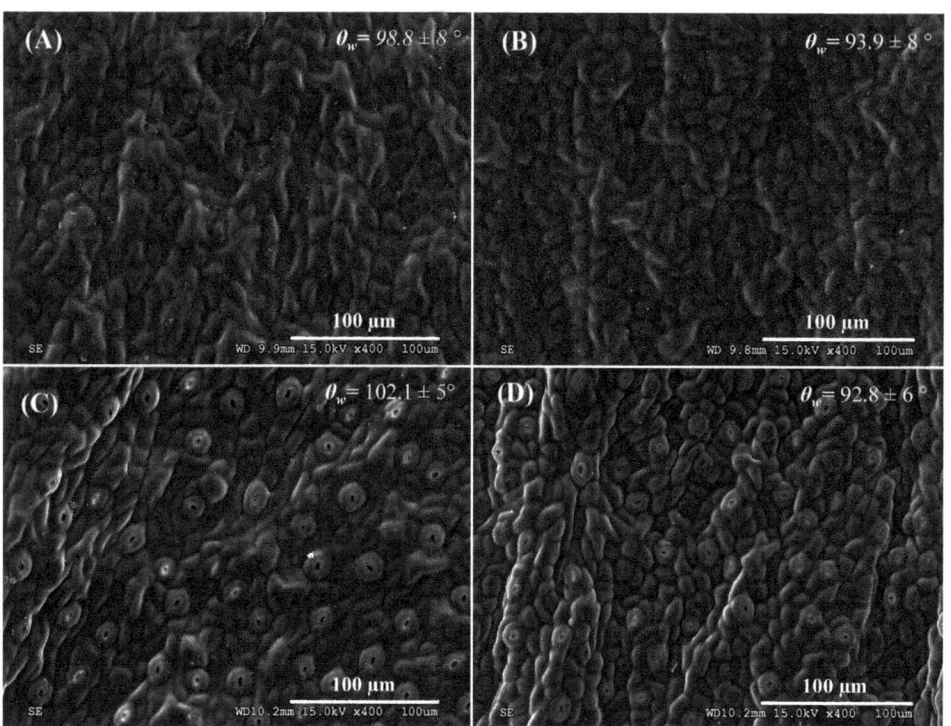

FIGURE 4.6 Scanning electron micrographs of upper (A, B) and lower (C, D) leaf surfaces of iron-sufficient (A, C) and iron-deficient (chlorotic) (B, D) leaves of lemon (*Citrus x limon* "Verna"). The mean contact angles with water (θ_w) are indicated ($n = 30$). *Courtesy of H.A. Bahamonde (Universidad Nacional de La Plata, Argentina).*

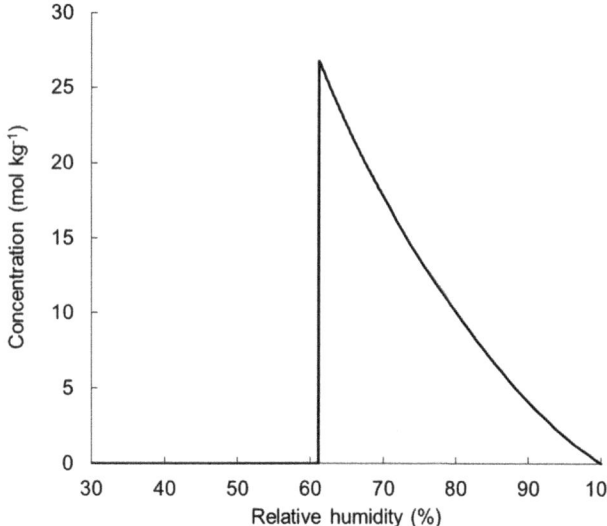

FIGURE 4.7 Effect of RH on the equilibrium concentrations (expressed as molality) of NH_4^+ in a solution of NH_4NO_3. Below 62% RH, corresponding to the DRH of the solution, the solution is dry. At RH = DRH = 62%, the solution has its maximal concentration. Above DRH, equilibrium concentrations decrease with increasing RH, approaching 0 at 100% RH due to (theoretical) infinite dilution by absorbed water. Data were calculated using the Extended AIM Aerosol Thermodynamics Model of Clegg et al. (1998) (http://www.aim.env.uea.ac.uk/aim/aim.php) and a starting value of 1 mmol m^{-3} of air. *DRH*, Deliquescence relative humidity; *RH*, relative humidity.

solutes begin to dissolve in water absorbed from the atmosphere. The value of the DRH is affected by temperature, particle size, and other factors (Fernández et al., 2020).

Due to kinetic inhibition, solutions can become supersaturated. This delays droplet drying that occurs at a RH lower than the DRH. The RH threshold where solutions begin to crystalize is known as the efflorescence RH (ERH) (Cziczo and Abbatt, 2000; Fernández et al., 2020). Both DRH and ERH are important parameters controlling foliar uptake processes: ERH controls the drying of fertilizer spray drops, whereas DRH influences the rehydration of drop deposits due to, for example, dew, fog, or high environmental RH (Fernández et al., 2020). When RH is at, or just above, the DRH or ERH, solute concentrations are maximal. As RH increases above the DRH, concentrations decrease and theoretically approach zero when RH rises to saturation (Fig. 4.7).

Given that the concentration gradient between the solution on the leaf surface and the leaf interior is the driving force for foliar uptake, RH has a strong impact on penetration rates. Moreover, at a RH below the DRH, uptake ceases

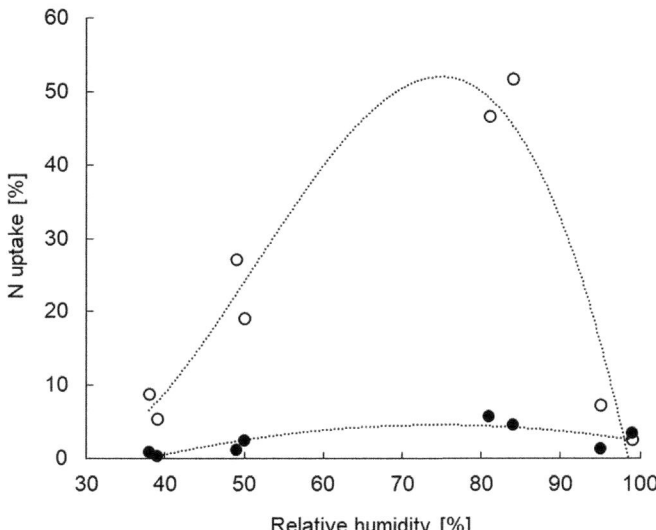

FIGURE 4.8 Cumulative N uptake into upper (adaxial, *filled circles*) and lower (abaxial, *open circles*) leaf surfaces of *Coffea arabica* as affected by RH. Nitrogen was applied as stable isotope-labeled Ca(NO$_3$)$_2$. Uptake was measured 24 h after application and is expressed in the percentage of the applied amount. The deliquescence relative humidity of Ca(NO$_3$)$_2$ is around 50% RH. *RH*, Relative humidity.

due to immobilization on the leaf surface and resumes once RH increases above the DRH, for example, during the night. The weather conditions and the diurnal rhythm of RH and temperature thus result in fluctuating concentrations on the leaf surface, making the prediction of uptake rates difficult under field conditions.

Increasing RH not only decreases the solute concentration on the leaf surface due to dilution by water absorbed from the atmosphere but also increases water sorption by the cuticle and thus the permeability for polar solutes. An increase in RH from 50% to saturation may increase the permeability of isolated cuticles for ions by two orders of magnitude (Fernández and Eichert, 2009). On the other hand, air humidity, more precisely water vapor deficit, may also influence stomatal aperture, which is correlated with foliar penetration rates. Control and adjustment of stomatal aperture, however, is a complex process governed by a range of additional interacting factors, such as soil water availability, plant water status, and irradiation. Therefore, no strict relationship between RH and aperture exists, and thus the stomatal penetration pathway is probably less dependent on RH than the cuticular pathway.

In summary, solute concentrations in solutions present on the leaves (and thus the driving force for uptake) decrease with increasing RH, whereas permeability (at least of the cuticle) increases. This means RH usually has the opposite effects on the solute concentrations on leaves and the permeability of leaf surfaces. The combination of these two effects determines foliar uptake rates depending on RH. In analogy to the Ohm's law, uptake rates can be predicted from the product of driving force and leaf permeability according to Eq. (4.2):

$$F = \Delta c \times P \tag{4.2}$$

where F is flux (mol m^{-2} s^{-1}), Δc is the concentration difference across the leaf surface (mol m^{-3}) and P is permeability (m s^{-1}, see Eq. 4.1).

It can be surmised that at high RH (close to, or at, saturation) when leaf permeability is high, uptake rates are low because of the low solute concentrations on the leaf surface (see Fig. 4.7). Therefore the widespread assumption that high RH results in highest foliar uptake rates is not necessarily true. An example of this is shown in Fig. 4.8, for an experiment in which the uptake of nitrate into coffee leaves was measured at RH ranging from 38% to 99%. Maximal uptake rates were achieved at around 75% RH, and uptake decreased when RH was higher than 80%.

4.3.5.3 Active ingredients and adjuvants

The effect of solubility of nutrient compounds used as fertilizers on foliar absorption is controversial (Alexander and Hunsche, 2016) given the growing use of nano- and micro-particle mineral element suspensions in commercial products. Recently, the absorption of Zn from Zn-oxide (ZnO) nano-particle suspensions was demonstrated by synchrotron-based X-ray fluorescence microscopy and X-ray absorption spectroscopy (Li et al., 2018a,b, 2019). However, higher Zn uptake by sunflower and soybean was recorded for Zn-sulfate compared to ZnO nanoparticles (Figs. 4.9 and 4.10). The absorption of Zn applied foliarly as ZnO nanoparticles varied between the plant species tested (Fig. 4.9).

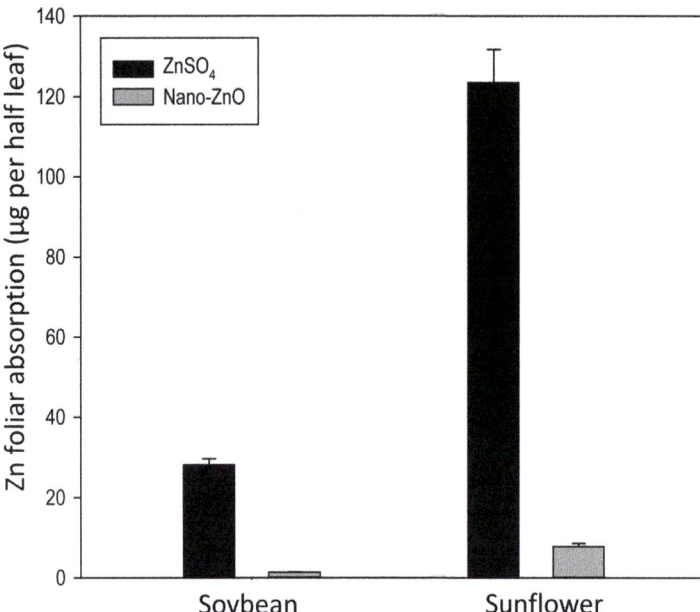

FIGURE 4.9 Comparison of the foliar absorption of Zn after 6-h exposure to 1000 mg Zn L^{-1} supplied as ZnSO$_4$ or nano-ZnO in soybean (cv. Clark) and sunflower (cv. Hyoleic 41). Means + SE. Experimental procedures are described in Li et al. (2018a,b). *Courtesy of C. Li and P. Kopittke, University of Queensland, Brisbane, Australia.*

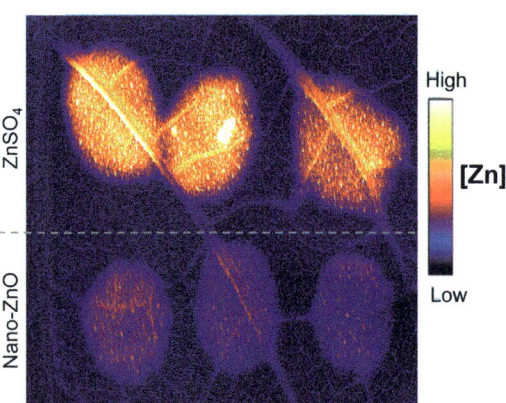

FIGURE 4.10 Micro X-ray fluorescence images showing the distribution of Zn in a hydrated sunflower (cv. Hyoleic 41) leaf that had three droplets of 1000 mg Zn L^{-1} of ZnSO$_4$ (top part) and nano-ZnO (bottom part) applied for 3 h. A color bar on the right corresponds to a range of Zn concentrations. Experimental procedures are described in Li et al. (2018a,b). *Courtesy of C. Li and P. Kopittke, University of Queensland, Brisbane, Australia.*

Adjuvants are chemicals added to foliar-applied solutions to optimize wetting and penetration processes, among other factors. The low rates of uptake of dissolved ions through the cuticle can be increased substantially by such substances, leading to multiple physicochemical changes in the solution and at the leaf surface level. In laboratory studies, surface-active agents increased the penetration of Ca^{2+} by a factor of six (Schönherr, 2000). Some specific adjuvants ("plasticizers") may increase the plasticity of the cuticle, thereby enhancing the penetration rates of solutes (Schönherr and Baur, 1994). Surface-active adjuvants also increase stomatal penetration by improving the contact area between the foliar-applied solution and the leaf surface. Selection of an appropriate adjuvant is often crucial for improving penetration rates, and this depends on a range of factors, such as properties of the target surface, the compound to be applied, and its concentration. However, surfactants may also have an effect on leaf physiology, for example, by increasing leaf transpiration (Räsch et al., 2018).

4.4 Foliar application of nutrients

4.4.1 General

Foliar sprays are widely used in agricultural production as an alternative or complementary strategy to applying fertilizers to soil (Fernández et al., 2013; Wójcik 2004). Theoretically, this method of application is more target-oriented and environment-friendly than soil treatments because nutrients are delivered directly to target organs, and there is a lower risk of environmental contamination. Plant responses to elements supplied via foliar sprays are normally more rapid than to root treatment, but the transport of foliar-applied nutrients through the epidermis and within organs is a complex process of variable efficacy (Fernández and Brown, 2013). Hence, complete reliance on foliar sprays to meet the crop nutrient demand cannot be achieved in commercial plant production, particularly in the case of macronutrients, without compromising crop yield and quality (Johnson et al., 2001).

The efficacy of foliar fertilizers is determined by many environmental, physicochemical, and physiological factors associated with the plant and the properties of the spray formulation, which are currently not fully understood (Fernández et al., 2021). An array of problems limiting the effectiveness of foliar fertilization may occur in response to foliar sprays, such as:

1. Limited leaf wetting and spreading of the applied nutrient solution when treating unwettable and even water-repellent leaves, or if applying unformulated spray solutions (i.e., without surfactants).
2. Spray run-off due to low solution retention by the foliage.
3. Washing off by rain following the application of nutrient sprays.
4. Low penetration rates of the applied nutrient solution due to the nature of the treated leaf surface (e.g., due to ontogeny), or the effect of environmental conditions on the physicochemical properties of fertilizer solution and plant performance (e.g., stomatal opening/closing).
5. Rapid drying of spray solutions, particularly at low RH and high temperature.
6. Following absorption, limited rates of distribution of nutrients (such as Ca or micronutrients) from the site of foliar uptake within the leaf or to other plant organs due to transport constraints.
7. Limitations related to concentration range of foliar-applied macronutrients to be supplied for meeting plant demand without inducing phytotoxicity (leaf burn).
8. Changes in phyllosphere microbial populations or the rate of spore germination and colony growth of pathogens such as *Erysiphe graminis* (powdery mildew), as observed after the application of urea sprays (Gooding and Davies, 1992).
9. Potential transient nutrient imbalances when single nutrient sprays are applied on foliage (e.g., Fernández et al., 2008b).

Leaf burn and even defoliation is a serious risk when applying foliar sprays. Different degrees of damage, ranging from the appearance of necrotic spots to complete defoliation, can occur with, for example, excessive nutrient concentrations, hygroscopic active ingredients (e.g., mineral salts with low DRH), or in the presence of surfactants that significantly increase the rate of penetration of the applied nutrients and are often phytotoxic when applied at high concentrations (Burkhardt, 2010; Fernández and Eichert, 2009). Because plant cuticles are polyelectrolytes (Schönherr and Huber, 1977), the pH of the spray solution and the molecular/ionic charge of active ingredients influence the rate of penetration (Aponte and Baur, 2018), and hence the phytotoxicity risk of foliar sprays.

4.4.2 Practical importance of foliar application of nutrients

Although plant leaves are organs specialized in capturing light and CO_2, their capacity to absorb water and nutrients has long been recognized and exploited in agriculture (Fernández and Eichert, 2009). Despite the constraints described above, foliar fertilization is of increasing importance in agricultural production worldwide.

Used either as a supplement for applying fertilizers to soil or under conditions of limited soil/substrate nutrient availability, the application of foliar sprays can help maintain crop yield and quality, with low environmental impact (Fageria et al., 2009; Fernández-Escobar et al., 2009; Fernández et al., 2013; Zhang and Brown, 1999). Traditionally, foliar fertilization has been used to correct nutrient deficiencies (Kannan, 2010). However, there is an increasing trend of using nutrient sprays in the absence of deficiency symptoms, particularly for elements with limited phloem mobility such as Ca or micronutrients (Fernández and Brown, 2013; Fernández and Eichert, 2009).

4.4.2.1 Low nutrient availability in soils

Many soil properties can limit element solubility and uptake of nutrients by plant roots as well as microbial communities (Trivedi et al., 2016). For example, due to the low P availability in many soils (e.g., high pH, calcareous soils) P is a limiting factor for crop production worldwide (Shen et al., 2011; see also Chapter 6). The limited availability of several micronutrients (e.g., Fe, Mn, Cu, and Zn) in high pH, calcareous soils is also a common problem (Chapter 7). In staple food crops, plant nutrient shortage due to agricultural soil constraints may ultimately lead to deficiency in humans as shown to occur with Zn (Cakmak et al., 2017; Rengel, 2015) (Chapter 9).

The application of micronutrient sprays to control or avoid the occurrence of micronutrient deficiencies under conditions of limited plant availability in the soil is one of the most important practical uses of foliar fertilization (Fageria et al., 2009; Fernández et al., 2013; Kannan, 2010). However, the effectiveness of foliar micronutrient treatments may vary significantly among plant species and in relation to the active ingredients (e.g., salts, complexes, or chelates; Fernández and Ebert, 2005; Haslett et al., 2001; Zhang and Brown, 1999).

In high pH, calcareous soils found in many arid and semiarid areas of the world, Fe deficiency-chlorosis (lime-induced chlorosis) is a common physiological disorder affecting plants (Abadía et al., 2011; Nikolic and Römheld, 2003). Foliar Fe sprays may be effective in regreening chlorotic leaves, particularly when adjuvants are added to the formulation (Fernández and Ebert, 2005; Neumann and Prinz, 1974). Iron-containing sprays have been shown to restore some leaf physiological parameters such as the rate of photosynthesis or transpiration (Eichert et al., 2010). However, there may also be problems due to the limited distribution of Fe within different plant parts and its metabolic functionality (Fernández and Ebert, 2005; Fernández et al., 2008b).

Manganese foliar sprays are also widely used and can be effective in overcoming Mn deficiency (e.g., Dordas, 2009; Papadakis et al., 2005). However, after foliar application, Mn transport to other plant parts is low due to its poor phloem mobility (e.g., Gettier et al., 1985; Papadakis et al., 2007). Similarly, variable rates of Zn transport after foliar application have been reported for different plant species (Cakmak and Kutman, 2018; Haslett et al., 2001).

Zinc foliar sprays are often applied to fruit trees before leaf fall because this can substantially increase the Zn pool (Fernández et al., 2013; Walworth et al., 2006; Zhang and Brown, 1999). A recent study on the seasonal dynamics of Zn storage and mobilization in several fruit species during the growing season (Xie et al., 2020) showed that following root absorption, Zn is transported and allocated to the stem nodes where it is available to supply Zn for budbreak and organ development in the following spring. Hence, autumn Zn sprays can contribute to increased Zn availability in tree tissues before dormancy, which can be used for their growth in the next season (Xie et al., 2021).

4.4.2.2 Dry topsoil

In semiarid regions, there can often be a lack of available water in the topsoil, with a consequent decrease in nutrient availability during the growing season. Even though water may still be available in the subsoil, nutrient shortage becomes a growth-limiting factor, especially when plants are also subjected to drought stress. For example, world areas prone to micronutrient deficiencies often have high pH, calcareous soils, and seasonally dry climates (Ascher-Ellis et al., 2001; Cakmak and Kutman, 2018). Under such conditions, a normal practice to maintain crop yield and quality is to complement soil micronutrient applications with foliar sprays (Ascher-Ellis et al., 2001). Furthermore, soil and foliar fertilization may improve plant drought tolerance (e.g., Ahmad et al., 2014; Ma et al., 2017).

Under dry conditions, soil treatments may be less effective than foliar nutrient sprays (Ahmad et al., 2014; McBeath et al., 2011). For example, foliar K fertilization is recommended in dryland olive orchards of Andalusia (Spain) due to the lower application costs and potentially higher efficacy (Barranco et al., 2012; Restrepo-Diaz et al., 2008). However, water shortage-induced physiological changes (e.g., stomatal closure) may decrease the rate of foliar absorption as shown by Restrepo-Diaz et al. (2008) after applying K sprays to water-stressed olive trees.

4.4.2.3 Decrease in root activity during the reproductive stage

As a result of sink competition for carbohydrates, root activity and thus nutrient uptake by the roots decline with the onset of the reproductive stage. Foliar sprays containing nutrients can compensate for this decline (Fernández et al., 2013). In legumes that rely on symbiotic N_2 fixation, sink competition for carbohydrates between developing seeds and root nodules may cause a marked decrease in the rate of N_2 fixation (Chapter 16). Foliar application of N at key developmental stages can be effective in increasing the yield of nodulated legumes. In soybean, foliar N application during the early reproductive stage decreased the percentage of pod abortion and increased grain yield by between 6% and 68% (Oko et al., 2003). In cereals such as wheat, grain quality (including protein content) and yield can be increased by late-season foliar N applications (e.g., Blandino et al., 2015; Kara and Uysal, 2009; Varga and Svečnjak, 2006). The rate of N absorption after the application of urea sprays can be high (Powlson et al., 1989), but losses due to ammonia volatilization may also occur. Ammonia volatilization has been estimated to range from 4% of the applied N in wheat (Smith et al., 1991) to >30% in Kentucky bluegrass (Wesely et al., 1987).

The effect of foliar P application on wheat yield response may be variable as shown in Table 4.2. Yield increments associated with foliar P supply (sometimes combined with other elements) were reported by, for example,

TABLE 4.2 Summary of the literature on yield responses of wheat to foliar P treatments.

Foliar rates (kg/ha)	Foliar formulation	Grain yield response	Source
N: 0, 55–110[a] P: 0, 6–12	N, P, and NP as urea and phosphoric acid	Glasshouse: nonsignificant for foliar P under low soil water supply, 66% increase for foliar NP with sufficient water	Alston (1979)
P: 1.65	Ammonium polyphosphate, NP blend, phosphoric acid	Glasshouse: 17% decrease with NP blend plus adjuvant in one soil type and up to 25% increase for phosphoric acid plus adjuvant in the other soil type	McBeath et al. (2011)
P: 0, 2.6, 5.2	K-phosphate, β-glycerophosphate, tripolyphosphate, phytic acid or K-phosphate	Field: yield increase in response to all (except phytic acid at low P dose), with up to 0.66 t ha^{-1} (22% for K-phosphate). High dose was less effective than low dose.	Sherchand and Paulsen (1985)
P: 0, 2.2, 4.4, 6.6	K-phosphate	Field: from no response up to 1.04 t ha^{-1} (22%) increase	Benbella and Paulsen (1998)
P: 0, 1, 2, 4	K-phosphate	Field: 0.45 t ha^{-1} decrease (−15%) up to 0.84 kg/ha (45%) increase.	Mosali et al. (2006)
P: 0.4, 0.8, 1.2[b]	Complete foliar fertilizer—N, P, K, Zn, Fe, Mn, Mg, Cu, S, B	Field: from 4% increase at the lowest application rate to 20% increase at the highest application rate	Ahmed et al. (2006)

[a]Rates in Alston (1979) take into account an estimated foliar interception of 15%–30% of spray to give the range of applied rate.
[b]Approximate rates split between two applications based on rates of 0.3, 0.6, and 0.9 w/v P, with 714 L/ha applied each time.
Source: T.M. McBeath and C.A.E. Pierce (University of Adelaide and CSIRO Agriculture and Food, Adelaide, Australia).

Mosali et al. (2006), Ahmed et al. (2006), or McBeath et al. (2011), but some foliar P treatments led to no significant effects (Noack et al., 2010).

Boron plays a crucial role in plant reproduction. Thus, it is important to ensure adequate B status to maintain crop yield and quality (Brown et al., 2002; Chapter 7). The mobility of B in plants is species-dependent (Brown et al., 2002, Chapter 3). In sunflower, yields were improved by the application of B sprays before flowering (Asad et al., 2003). Foliar application of B in cotton led to significant retranslocation to other plant organs within the first 24 h after treatment, with preferential transport to apical tissues (Bogiani et al., 2014). In fruit-tree species with high B phloem mobility, foliar application of B in the autumn is an effective procedure for increasing B concentration in reproductive and vegetative tissues and improving fruit set in the following season (Christensen et al., 2006; Fernández et al., 2013).

4.4.2.4 Avoiding the occurrence of physiological disorders and improving quality of horticultural crops

The occurrence of Ca-related imbalances is a problem of major economic significance in horticultural crops, with more than 30 different Ca-related disorders in fruits and vegetables being recognised, including bitter-pit in apple and blossom-end rot in tomato (Shear, 1975). These physiological disorders that develop during fruit growth have generally been related to localized Ca deficiencies in plant organs (Saure, 2005). The low transpiration rate of fruits and the low mobility of Ca in the phloem (see Chapter 3) make it difficult to enhance the transport of Ca to fruits by applying Ca to the root system (Bangerth, 1979; Montanaro et al., 2014). However, a contribution of phloem transport of Ca from the pedicel into the fruit has recently been reported for several fruit species (Song et al., 2018), but its magnitude is comparatively low and should be clarified in future investigations.

The treatment of aerial plant parts with Ca sprays alone (Lötze et al., 2008; Lurie and Crisosto, 2005) or in combination with other techniques (Torres et al., 2017) is recommended and applied routinely in many fruit-production areas of the world to prevent the occurrence of localized Ca deficiencies and improve quality of fruits and vegetables (Fernández et al., 2009; Liebisch et al., 2009; Peryea et al., 2007; Schmitz-Eiberger et al., 2002). However, in some some studies of foliar Ca sprays, inconsistent results regarding tissue Ca concentrations, fruit quality, and storability have been reported (e.g., Bonomelli and Ruiz, 2010; Sotiropoulos et al., 2010; Val et al., 2008).

Foliar sprays of N and K as well as B and other micronutrients are often applied to fruit crops during the growing season to improve flowering, fruit set and development (Fernández et al., 2013). In addition, better grape and vine quality has been achieved after foliar applications of proline, phenylalanine, and urea, but the timing of foliar fertilization appears to be a key factor affecting plant response to the treatments (Fernández et al., 2013; Garde-Cerdán et al., 2014; Portu et al., 2015).

4.4.2.5 Biofortification

The use of foliar nutrient sprays as an approach to enhance the nutritional value of staple food crops for human consumption has gained importance recently (Cakmak, 2008; Cakmak and Kutman, 2018; Poblaciones and Rengel, 2016; White and Broadley, 2009) (Chapter 9). Foliar sprays applied alone or in combination with soil treatments can be used to increase the concentration of micronutrients in edible crop parts and alleviate micronutrient deficiencies in human populations around the world (Cakmak, 2008; Rengel et al., 1999). For example, in a study performed in seven different countries, foliar Zn application led to a significant increase in Zn concentration in wheat grains, with no negative yield effects (Cakmak et al., 2010; Zou et al., 2012). Application of Fe and B sprays to rice increased the nutritional value and Fe content of rice grains (Zhang et al., 2009). The enrichment of staple food crops or horticultural commodities with foliar sprays of beneficial elements such as selenium (Li et al., 2018c; Puccinelli et al., 2017) or iodine (Lawson et al., 2015) has also been achieved successfully in some studies.

4.4.3 Foliar fertilizers for pest and disease control

The positive effect of mineral element sprays in pest and disease control in sustainable agricultural production systems has been reported (e.g., Dordas, 2008; Singh, 2015). Elements such as N, K, P, Mn, Zn, B, or Cl have been associated with disease resistance and/or tolerance (Dordas, 2008), and their application as foliar sprays may have a synergistic effect with other plant protection mechanisms and/or have an effect on the pathogens. Recently, positive effects of spraying the beneficial element silicon (Si) on plant growth, crop quality, and biotic and abiotic stress tolerance have been shown (e.g., de Souza Alonso et al., 2020; Jang et al., 2020; Laane, 2018; Luyckx et al., 2017; Rodrigues et al., 2015) (Table 4.3).

4.4.4 Foliar uptake and irrigation methods

Shortage of quality water for irrigation purposes is becoming increasingly prevalent, especially in arid and semiarid areas around the world. The competition for water resources leads to the use of low-quality water (Singh et al., 2009) and recycled wastewater for crop irrigation (Devitt et al., 2005; Jordan et al., 2001; Valdez-Aguilar et al., 2009).

Implementation of sprinkler irrigation is increasing because it provides improved irrigation efficiency, automation, and reduced labor costs (Isla and Aragüés, 2010). However, sprinkler irrigation with saline water has been shown to increase the concentration of Na and Cl in crops such as pepper, maize, wheat, barley, alfalfa, and ornamentals (e.g., Devitt et al., 2005; Grattan et al., 1994; Isla and Aragüés, 2009, 2010; Jordan et al., 2001; Singh et al., 2009).

TABLE 4.3 Control of foliar diseases of crops using foliar silicon sprays.

Crops	Disease	Pathogen
Cucumber	Powdery mildew	*Podosphaera xanthii*
Coffee	Coffee leaf rust	*Hemileia vastatrix*
Coffee	Cercospora leaf spot	*Cercospora coffeicola*
Melon	Powdery mildew	*P. xanthii*
Rice	Blast	*Pyricularia grisea*
Rice	Brown spot	*Bipolaris oryzae*
Soybean	Rust	*Phakopsora pachyrhizi*
Tomato	Powdery mildew	*Leveillula taurica*
Wheat	Powdery mildew	*Blumeria graminis* f. sp. *tritici*

Source: Adapted from Rodrigues et al. (2015).

The concentrations of these two ions in the leaves may become toxic quite rapidly, decreasing yield and quality (Francois and Clark, 1979; Isla and Aragüés, 2010; Maas, 1985). Moreover, increasing salt concentrations in the irrigation water may decrease plant K concentration because there is generally a negative relationship between tissue Na and K concentrations (e.g., Isla and Aragüés, 2009, 2010). Salinity problems may be minimized by irrigating due to a smaller contribution of stomata to foliar water absorption (Urrego-Pereira et al., 2013).

In general, the sensitivity to foliar injury by irrigation with saline water depends on the wettability and nature of the leaf surface (e.g., degree of roughness and chemical composition; Urrego-Pereira et al., 2013). For example, deciduous fruit trees (e.g., almond, apricot) are more sensitive to leaf injury by saline sprinkler water than cotton and sunflower (Maas, 1985).

4.5 Leaching of elements from leaves

Compounds can be lost from plants via

1. active excretion to the external surface, such as excretion of lipophilic or hydrophilic compounds by glandular trichomes (Werker, 2000);
2. excretion of inorganic solutes at tips and margins of leaves by guttation (induced by root pressure); and
3. passive leaching from damaged plant parts but also from intact plant tissues.

In this section, the term "leaching" refers to the removal of inorganic and organic substances from aerial plant parts by the action of aqueous solutions such as rain, dew, mist, irrigation, and fog. Substances leached from plants may include inorganic elements and organic compounds such as carbohydrates, amino acids, vitamins, phytohormones, or phenols (Tukey, 1970). However, the quantity and quality of leached materials may vary among plant species and varieties as well as elements; for example, K and Mg are leached more easily than Fe and Zn (Tukey, 1970). However, the processes by which compounds are leached from plant tissues remain unclear (Dawson and Goldsmith, 2018).

Leaching by rain increases with duration and intensity of rainfall as well as with leaf age. It has been hypothesized that the accumulation of substances in the apoplast of mature leaves may result in a steeper concentration gradient across the leaf surface that may favor leaching (Borer et al., 2005; Tukey, 1970; Turner and van Broekhuizen, 1992).

Stress conditions such as prolonged darkness, water shortage, and high temperatures, or air pollutants such as ozone, dry deposition, or high acidity of the rainwater or fog can also increase the leaching of nutrients from leaves (Borer et al., 2005; González-Arias et al., 2000; Runyan et al., 2013; Schaberg et al., 2000; Tukey, 1970). High acidity of rainwater may increase leaching of solutes, possibly due to more rapid leaf senescence and/or structural damage at the cuticular or tissue level. Cations may be particularly affected because they can be replaced on their binding sites by protons present in rainwater. For example, acid mist has been found to deplete the labile and physiologically available pool of membrane-associated Ca in red spruce (Borer et al., 2005; Schaberg et al., 2000). On average, a decrease in the pH of rainwater or fog from about 5.5 to 3.5–3.0 may increase the leaching of K, Ca, Mg, Mn, and Zn by 2- to 10-fold (Leisen and Marschner, 1990; Mengel et al., 1987; Turner and Tingey, 1990).

Although the amounts of cations leached usually do not represent more than 1% of the total cation content of leaves (Pfirrmann et al., 1990), they may reach up to 10% of the annual net incorporation of cations into aerial biomass (Schuepp and Hendershot, 1989). The leaching of specific cations from aerial plant parts may be compensated for by higher rates of uptake of these cations by roots (Kaupenjohann et al., 1988; Mengel et al., 1987). As a consequence of higher cation uptake rate, the rhizosphere pH may decrease (Kaupenjohann et al., 1988; Leonardi and Flückiger, 1989) (see also Chapter 14) and, thus, part of the acid load of the canopy may indirectly be transferred into the rhizosphere.

4.6 Ecological importance of foliar uptake and leaching

4.6.1 Foliar leaching

Uptake of soluted by, and leaching from, leaves and other aerial plant parts are continuous in different ecosystems. Leaching and uptake of solutes can contribute to the cycling of nutrients through plants due to either foliar or root absorption of the leachates (Johnson and Turner, 2019; Tukey, 1970). For example, in forest trees, these processes can be a dominant component in nutrition, influencing nutrient cycling internally as well as in forest ecosystems and their long-term stability (Chabot and Hicks, 1982). Nitrogen deposited as gaseous or dissolved compounds can be taken up readily by the canopy (Bourgeois et al., 2019; Hinko-Najera Umana and Wanek, 2010). In central Europe and northern America, a considerable amount of dry and wet deposition of N compounds (between 3 and 10 kg N ha^{-1} per year;

TABLE 4.4 Element input in precipitation (wet deposition) and dry deposition, throughfall, and leaching in a 40-year-old Scots pine forest in Berlin, Germany.

Parameter	(mm year^{-1})	Element input (kg ha^{-1} year^{-1})				
		Ca	Mg	K	Mn	Na
Wet deposition	550	12.7	1.4	1.9	0.24	4.1
Dry deposition	–	10.2	1.1	1.5	0.19	3.7
Throughfall	397	27.8	4.3	15.1	2.50	7.4
Leaching from canopy	–	4.9	1.8	11.7	2.07	–

Source: Adapted from Marschner et al., (1991).

Brumme et al., 1992) can be taken up by the foliage (Garten and Hanson, 1990; Krupa, 2003). Between 40% and 90% of N deposited in forests is retained by their canopies (Clark et al., 1998; Gaige et al., 2007; Lovett and Lindberg, 1993), which could satisfy a substantial proportion of the annual N demand of temperate and boreal forests (Hinko-Najera Umana and Wanek, 2010; Schulze, 2000). In peatland vegetation, the canopy retained 38%–80% of deposited N (Chiwa et al., 2019).

The leaves of rainforest plants may be more resistant to leaching compared to some agricultural plants adapted to lower rainfall areas (Tukey, 1970). However, due to the frequent and prolonged precipitation in tropical rainforests, the amounts of nutrients leached from the canopy can be high. For example, the annual leaching losses were (kg ha^{-1}): K 100–200, N 12–60, Mn 18–45, Ca 25–29, and P 4–10 (Bernhard-Reversat, 1975; Nye and Greenland, 1960). Under such conditions, the magnitude of nutrient leaching is similar to the annual rate of nutrients supplied to the soil surface from throughfall and is thus an important component of nutrient recycling, particularly in ecosystems with low amounts of available nutrients in the soil, such as in highly weathered tropical soils. Reabsorption of leached elements also enables the direct supply of nutrients to plant parts with high nutrient demand (e.g., new growth) regardless of nutrient mobility in the phloem (e.g., generally low for Ca and Mn) (Tukey, 1970).

In temperate climates where rainfall is lower than in rainforests, losses by leaching from aerial plant parts are lower but still considerable (Table 4.4). Compared with their concentrations in leaves, the amounts of Ca and particularly Mn leached are often high. However, quantification of leaching losses is difficult because "dry deposition" (particulates and gases) may constitute a substantial portion of the nutrients in the throughfall (Table 4.4).

In addition to inorganic elements, substantial amounts of organic solutes can also be leached from forest canopy, reaching amounts between 25 and 60 kg organic C ha^{-1} per year in temperate climates (Bartels, 1990), and several hundreds of kilograms per hectare annually in tropical forests.

4.6.2 Foliar water absorption

Foliar water absorption by epiphytic plants such as members of the Bromeliaceae is well-known (Martin, 1994), but this phenomenon has been a matter of debate in other plant species. It was shown that species of the genus *Crassula* can absorb water, presumably by hydatodes (Martin and von Willert, 2000). Leaves of trees, understorey ferns, and shrubs of the redwood forest in California showed the capacity to rehydrate upon fog exposure (Burgess and Dawson, 2004; Limm et al., 2009). However, foliar water uptake is not restricted to plants growing in fog-dominated climates and may occur in almost all plant families irrespective of their natural habitat (Berry et al., 2019; Boanares et al., 2018; Fernández et al., 2014b, 2021). The ecological significance of foliar water uptake is still not understood fully.

Foliar water uptake is limited by the leaf surface having a greater resistance to water transport than the xylem (Guzmán-Delgado et al., 2018). Nevertheless, additional water delivered by leaf uptake may be decisive under drought conditions and contribute up to 40%–50% of the total water content of leaves (Berry et al., 2014; Eller et al., 2013). During the dry phase, water absorbed by leaves may help sustain photosynthesis, thereby increasing carbon gain (Berry et al., 2014). Leaf-absorbed water may thus be important for plant water economy during periods of water shortage (Binks et al., 2019; Burgess and Dawson, 2004; Oliveira et al., 2005). However, factors such as the contribution of foliar water uptake to plant water balance, the leaf absorption mechanisms of different species in various climates, and the effect of environmental factors on foliar water uptake should be investigated further (Fernández et al., 2021).

References

Abadía, J., Vázquez, S., Rellán-Álvarez, R., El-Jendoubi, H., Abadía, A., Álvarez-Fernández, A., et al., 2011. Towards a knowledge-based correction of iron chlorosis. Plant Physiol. Biochem. 49, 471–482.

Ahmad, R., Waraich, E.A., Ashraf, M.Y., Ahmad, S., Aziz, T., 2014. Does nitrogen fertilization enhance drought tolerance in sunflower? A review. J. Plant Nutr. 37, 942–963.

Ahmed, A.G., Hassanein, M.S., El-Gazzar, M.M., 2006. Growth and yield response of two wheat cultivars to complete foliar fertilizer compound 'Dogoplus'. J. Appl. Sci. Res. 2, 20–26.

Alexander, A., Hunsche, M., 2016. Influence of formulation on the cuticular penetration and on spray deposit properties of manganese and zinc foliar fertilizers. Agronomy 6, 39.

Alidoust, D., Isoda, A., 2013. Effect of γFe_2O_3 nanoparticles on photosynthetic characteristic of soybean (*Glycine max* (L.) Merr.): foliar spray versus soil amendment. Acta Physiol. Plant. 35, 3365–3375.

Alston, A.M., 1979. Effects of soil water content and foliar fertilization with nitrogen and phosphorus in late season on the yield and composition of wheat. Aust. J. Agric. Res. 30, 577–585.

Aponte, J., Baur, P., 2018. The role of pH for ionic solute uptake by the non-aerial hypocotyl of mung bean plants. J. Plant Disease Protec. 125, 433–442.

Arand, K., Stock, D., Burghardt, M., Riederer, M., 2010. pH-dependent permeation of amino acids through isolated ivy cuticles is affected by cuticular water sorption and hydration shell size of the solute. J. Exp. Bot. 61, 3865–3873.

Arsic, M., Le Tougaard, S., Persson, D.P., Martens, H.J., Doolette, C.L., Lombi, E., et al., 2020. Bioimaging techniques reveal foliar phosphate uptake pathways and leaf phosphorus status. Plant Physiol. 183, 1472–1483.

Asad, A., Blamey, F.P.C., Edwards, D.G., 2003. Effects of boron foliar applications on vegetative and reproductive growth of sunflower. Ann. Bot. 92, 565–570.

Asaf, D., Rotenberg, E., Tatarinov, F., Dicken, U., Montzka, S.A., Yakir, D., 2013. Ecosystem photosynthesis inferred from measurements of carbonyl sulphide flux. Nat. Geosci. 6, 186–190.

Ascher-Ellis, J.S., Graham, R.D., Hollamby, G.J., Paull, J., Davies, P., Huang, C., et al., 2001. Micronutrients. In: Reynolds, M.P., Ortiz-Monasterio, J.I., McNab, A. (Eds.), Application of Physiology in Wheat Breeding. International Maize and Wheat Improvement Center. El Batan, Mexico, pp. 219–240.

Ausma, T., De Kok, L.J., 2019. Atmospheric H_2S: Impact on plant functioning. Front. Plant Sci. 10, 743.

Bahamonde, H.A., Gil, L., Fernández, V., 2018. Surface properties and permeability to calcium chloride of *Fagus sylvatica* and *Quercus petraea* leaves of different canopy heights. Front. Plant Sci. 9, 494.

Bangerth, F., 1979. Calcium-related physiological disorders of plants. Ann. Rev. Agric. Food Chem. 11, 204–207.

Barranco, D., Ercan, H., Muñoz-Díez, C., Belaj, A., Arquero, O., 2012. Factors influencing the efficiency of foliar sprays of monopotassium phosphate in the olive. Int. J. Plant Prod. 4, 235–240.

Bartels, U., 1990. Organischer Kohlenstoff im Niederschlag nordrhein-westfälischer Fichten- und Buchenbestände. Z. Pflanzen. Bodenkd. 153, 125–127.

Barthlott, W., Neinhuis, C., Cutler, D., Ditsch, F., Meusel, I., Theisen, I., et al., 1998. Classification and terminology of plant epicuticular waxes. Bot. J. Linn. Soc. 126, 237–260.

Bash, J.O., Walker, J.T., Katul, G.G., Jones, M.R., Nemitz, E., Robarg, W.P., 2010. Estimation of in-canopy ammonia sources and sinks in a fertilized *Zea mays* field. Environ. Sci. Technol. 44, 1683–1689.

Beauchamp, E.G., Kidd, G.E., Thurtell, G., 1978. Ammonia volatilization from sewage sludge applied in field. J. Environ. Qual. 7, 141–146.

Behera, S.N., Sharma, M., Aneja, V.P., Balasubramanian, R., 2013. Ammonia in the atmosphere: a review on emission sources, atmospheric chemistry and deposition on terrestrial bodies. Environ. Sci. Pollut. Res. 20, 8092–8131.

Benbella, M., Paulsen, G.M., 1998. Efficacy of treatments for delaying senescence of wheat leaves: I. Senescence under controlled conditions. Agron. J. 90, 329–332.

Bernhard-Reversat, F., 1975. Nutrients in through fall and their quantitative importance in rain forest mineral cycle. Ecol. Stud. 11, 153–159.

Berry, Z.C., White, J.C., Smith, W.K., 2014. Foliar uptake, carbon fluxes and water status are affected by the timing of daily fog in saplings from the threatened cloud forest. Tree Physiol. 34, 459–470.

Berry, Z.C., Emery, N.C., Gotsch, S.G., Goldsmith, G.R., 2019. Foliar water uptake: Processes, pathways, and integration into plant water budgets. Plant Cell Environ. 42, 410–423.

Binks, O., Mencuccini, M., Rowland, L., da Costa, A.C., de Carvalho, C.J.R., Bittencourt, P., et al., 2019. Foliar water uptake in Amazonian trees: Evidence and consequences. Glob. Change Biol. 25, 2678–2690.

Bird, S.M., Gray, J.E., 2003. Signals from the cuticle affect epidermal cell differentiation. New Phytol. 157, 9–23.

Blandino, M., Vaccino, P., Reyneri, A., 2015. Late-season nitrogen increases improver common and durum wheat quality. Agron. J. 107, 680–690.

Bloem, E., Haneklaus, S., Kesselmeier, J., Schnug, E., 2012. Sulfur fertilization and fungal infections affect the exchange of H_2S and COS from agricultural crops. J. Agric. Food Chem. 60, 7588–7596.

Bloem, E., Haneklaus, S., Salac, I., Wickenhäuser, P., Schnug, E., 2005. Facts and fiction about sulfur metabolism in relation to plant-pathogen interactions. Plant Biol. 9, 596–607.

Boanares, D., Isaias, R.R.M.S., de Sousa, H.C., Kozovits, A.R., 2018. Strategies of leaf water uptake based on anatomical traits. Plant Biol. 20, 848–856.

Boaretto, R.M., Mattos Jr., D., Quaggio, J.A., Cantarella, H., Trivelin, B.C.O., 2013. Absorption of $^{15}NH_3$ volatilized from urea by *Citrus* trees. Plant Soil 365, 283–290.

Bogiani, J.C., Sampaio, T.F., Abreu-Junior, C.H., Rosolem, C.A., 2014. Boron uptake and translocation in some cotton cultivars. Plant Soil 375, 241–253.

Bonomelli, C., Ruiz, R., 2010. Effects of foliar and soil calcium application on yield and quality of table grape cv. 'Thompson seedless'. J. Plant Nutr. 33, 299–314.

Borer, C.H., Schaberg, P.G., DeHayes, D.H., 2005. Acidic mist reduces foliar membrane-associated calcium and impairs stomatal responsiveness in red spruce. Tree Physiol. 25, 673–680.

Bourgeois, I., Clément, J.-C., Caillon, N., Savarino, J., 2019. Foliar uptake of atmospheric nitrate by two dominant subalpine plants: insights from in-situ triple-isotope analysis. New Phytol. 223, 1784–1794.

Brown, P.H., Bellaloui, N., Wimmer, M.A., Bassil, E.S., Ruiz, J., Hu, H., et al., 2002. Boron in plant biology. Plant Biol. 4, 205–223.

Brumme, R., Leimcke, U., Matzner, E., 1992. Interception and uptake of NH_4 and NO_3 from wet deposition by aboveground parts of young beech (*Fagus sylvatica* L.) trees. Plant Soil 142, 273–279.

Buchholz, A., Baur, P., Schönherr, J., 1998. Differences among plant species in cuticular permeabilities and solute mobilities are not caused by differential size selectivities. Planta 206, 322–328.

Burgess, S.S.O., Dawson, T.E., 2004. The contribution of fog to the water relations of *Sequoia sempervirens* (D. Don): foliar uptake and prevention of dehydration. Plant Cell Environ. 27, 1023–1034.

Burkhardt, J., 2010. Hygroscopic particles on leaf surfaces: Nutrients or desiccants? Ecol. Monogr. 80, 369–399.

Burkhardt, J., Eiden, R., 1994. Thin water films on coniferous needles. Atmos. Environ. 28, 2001–2011.

Burkhardt, J., Basi, S., Paryar, S., Hunsche, M., 2012. Stomatal penetration by aqueous solutions - an update involving leaf surface particles. New Phytol. 196, 774–787.

Cakmak, I., 2008. Enrichment of cereal grains with zinc: Agronomic or genetic biofortification? Plant Soil 302, 1–17.

Cakmak, I., Kutman, U.B., 2018. Agronomic biofortification of cereals with zinc: a review. Eur. J. Soil Sci. 69, 172–180.

Cakmak, I., Kalayci, M., Kaya, Y., Torun, A.A., Aydin, N., Wang, Y., et al., 2010. Biofortification and localization of zinc in wheat grain. J. Agric. Food Chem. 58, 9092–9102.

Cakmak, I., McLaughlin, M.J., White, P., 2017. Zinc for better crop production and human health. Plant Soil 411, 1–4.

Campbell, J.E., Whelan, M.E., Berry, J.A., Hilton, T.W., Zumkehr, A., Stinecipher, J., et al., 2017. Plant uptake of atmospheric carbonyl sulphide in coast redwood forests. J. Geophys. Res.-Biogeosci. 122, 3391–3404.

Castro, A., Stulen, I., Posthumus, F.S., De Kok, L.J., 2006. Changes in growth and nutrient uptake in *Brassica oleracea* exposed to atmospheric ammonia. Ann. Bot. 97, 121–131.

Chabot, B.F., Hicks, D.J., 1982. The ecology of leaf life spans. Annu. Rev. Ecol. Syst. 13, 229–259.

Chamel, A., Pineri, M., Escoubes, M., 1991. Quantitative determination of water sorption by plant cuticles. Plant Cell Environ. 14, 87–95.

Chang, Y., Zou, Z., Zhang, Y., Deng, C., Hu, J., Shi, Z., et al., 2019. Assessing contributions of agricultural and nonagricultural emissions to atmospheric ammonia in a Chinese megacity. Environ. Sci. Tech. 53, 1822–1833.

Chen, X., Ren, X., Hussain, S., Hussain, S., Saqib, M., 2020. Effects of elevated ammonia concentration on corn growth and grain yield under different nitrogen application rates. J. Soil Sci. Plant Nutr. 20, 1961–1968.

Chichiricco, G., Poma, A., 2015. Penetration and toxicity of nanomaterials in higher plants. Nanomaterials 5, 851–973.

Chiwa, M., Sheppard, L.J., Leith, I.D., Leeson, S.R., Tang, Y.S., Cape, J.N., 2019. P and K additions enhance canopy N retention and accelerate the associated leaching. Biogeochemistry 142, 413–423.

Christensen, L.P., Beede, R.H., Peacock, W.L., 2006. Fall foliar sprays prevent boron-deficiency symptoms in grapes. California Agric. 60, 100–103.

Clarisse, L., Shephard, M.W., Dentener, F., Hurtmans, D., Cady-Pereira, K., Karagulian, F., et al., 2010. Satellite monitoring of ammonia: A case study of the San Joaquin Valley. J. Geophys. Res. 115, D13302.

Clark, K.L., Nadkarni, N.M., Schaefer, D., Gholz, H.L., 1998. Atmospheric deposition and net retention of ions by the canopy in a tropical montane forest, Monteverde, Costa Rica. J. Trop. Ecol. 14, 27–45.

Clegg, S.L., Brimblecombe, P., Wexler, A.S., 1998. A thermodynamic model of the system H^+ - NH_4^+ - Na^+ - SO_4^{2-} - NO_3^- - Cl^- - H_2O at 298.15 K. J. Phys. Chem. A 102, 2155–2171.

Cliquet, J.B., Lemauviel-Lavenant, S., 2019. Grassland species are more efficient in acquisition of S from the atmosphere when pedospheric S availability decreases. Plant Soil 435, 69–80.

Cziczo, D.J., Abbatt, J.P.D., 2000. Infrared observations of the response of NaCl, $MgCl_2$, NH_4HSO_4, and NH_4NO_3 aerosols to changes in relative humidity from 298 to 238 K. J. Phys. Chem. A 104, 2038–2047.

David, M., Loubet, B., Cellier, P., Mattsson, M., Schjoerring, J.K., Nemitz, E., et al., 2009. Ammonia sources and sinks in an intensively managed grassland canopy. Biogeosciences 6, 1903–1915.

Dawson, T.E., Goldsmith, G.R., 2018. The value of wet leaves. New Phytol. 219, 1156–1169.

Deng, Y.Q., Bao, J., Yuan, F., Liang, X., Feng, Z.T., Wang, B.S., 2016. Exogenous hydrogen sulphide alleviates salt stress in wheat seedlings by decreasing Na^+ content. Plant Growth Regul. 79, 391–399.

de Souza Alonso, T.A., Ferreira Barreto, R., de Mello Prado, R., Pereira de Souza, J., Falleiros Carvalho, R., 2020. Silicon spraying alleviates calcium deficiency in tomato plants, but Ca-EDTA is toxic. J. Plant Nutr. Soil Sci. 183, 659–664.

Devitt, D.A., Morris, R.L., Fenstermaker, L.F., Baghzouz, M., Neuman, D.S., 2005. Foliar damage and flower production of landscape plants sprinkle irrigated with reuse water. HortScience 40, 1871–1878.

Dordas, C., 2008. Role of nutrients in controlling plant diseases in sustainable agriculture. A review. Agron. Sustain. Dev. 28, 33–46.

Dordas, C., 2009. Foliar application of manganese increases seed yield and improves seed quality of cotton grown on calcareous soils. J. Plant Nutr. 32, 160–176.

Du, Y., Li, P., Nguyen, A.V., Xu, Z.P., Mulligan, D., Huang, L., 2015. Zinc uptake and distribution in tomato plants in response to foliar supply of Zn hydroxide-nitrate nanocrystal suspension with controlled Zn solubility. J. Plant Nutr. Soil Sci. 178, 722–731.

Eichert, T., Burkhardt, J., 2001. Quantification of stomatal uptake of ionic solutes using a new model system. J. Exp. Bot. 52, 771–781.

Eichert, T., Goldbach, H.E., 2008. Equivalent pore radii of hydrophilic foliar uptake routes in stomatous and astomatous leaf surfaces – further evidence for a stomatal pathway. Physiol. Plant. 132, 491–502.

Eichert, T., Goldbach, H.E., Burkhardt, J., 1998. Evidence for the uptake of large anions through stomatal pores. Bot. Acta 111, 461–466.

Eichert, T., Kurtz, A., Steiner, U., Goldbach, H.E., 2008. Size exclusion limits and lateral heterogeneity of the stomatal foliar uptake pathway for aqueous solutes and water-suspended nanoparticles. Physiol. Plant. 134, 151–160.

Eichert, T., Peguero-Pina, J.J., Gil-Pelegrín, E., Heredia, A., Fernández, V., 2010. Effects of iron chlorosis and iron resupply on leaf xylem architecture, water relations, gas exchange and stomatal performance of field-grown peach (*Prunus persica*). Physiol. Plant. 138, 48–59.

Eller, C.B., Lima, A.L., Oliveira, R.S., 2013. Foliar uptake of fog water and transport belowground alleviates drought effects in the cloud forest tree species, *Drimys brasiliensis* (Winteraceae). New Phytol. 199, 151–162.

Fageria, N.K., Barbosa Filho, M.P., Moreira, A., Guimarães, C.M., 2009. Foliar fertilization of crop plants. J. Plant Nutr. 32, 1044–1064.

Fernández, V., Brown, P.H., 2013. From plant surface to plant metabolism: the uncertain fate of foliar-applied nutrients. Front. Plant Sci. 4, 289.

Fernández, V., Ebert, G., 2005. Foliar iron fertilisation: a critical review. J. Plant Nutr. 28, 2113–2124.

Fernández, V., Eichert, T., 2009. Uptake of hydrophilic solutes through plant leaves: Current state of knowledge and perspectives of foliar fertilization. Crit. Rev. Plant Sci. 28, 36–68.

Fernández, V., Ebert, G., Winkelmann, G., 2005. The use of microbial siderophores for foliar iron application studies. Plant Soil 272, 245–252.

Fernández, V., Eichert, T., Del Río, V., López-Casado, G., Heredia-Guerrero, J.A., Abadía, A., et al., 2008a. Leaf structural changes associated with iron deficiency chlorosis in field-grown pear and peach: physiological implications. Plant Soil 311, 161–172.

Fernández, V., Del Río, V., Pumariño, L., Igartua, E., Abadía, J., Abadía, A., 2008b. Foliar fertilization of peach (*Prunus persica* (L.) Batsch) with different iron formulations: Effects on re-greening, iron concentration and mineral composition in treated and untreated leaf surfaces. Sci. Hortic. 117, 241–248.

Fernández, V., Diaz, A., Blanco, A., Val, J., 2009. Surface application of calcium-containing gels to improve quality of late maturing peach cultivars. J. Sci. Food Agric. 89, 2323–2330.

Fernández, V., Sotiropoulos, T., Brown, P.H., 2013. Foliar fertilisation: principles and practices, First edition. International Fertiliser Industry Association (IFA).

Fernández, V., Guzmán, P., Peirce, C.A., McBeath, T.M., Khayet, M., McLaughlin, M.J., 2014a. Effect of wheat phosphorus status on leaf surface properties and permeability to foliar-applied phosphorus. Plant Soil 384, 7–20.

Fernández, V., Sancho-Knapik, D., Guzmán, P., Peguero-Pina, J.J., Gil, L., Karabourniotis, G., et al., 2014b. Wettability, polarity, and water absorption of holm oak leaves: Effect of leaf side and age. Plant Physiol. 166, 168–180.

Fernández, V., Guzmán-Delgado, P., Graça, J., Santos, S., Gil, L., 2016. Cuticle structure in relation to chemical composition: Re-assessing the prevailing model. Front. Plant Sci. 7, 427.

Fernández, V., Bahamonde, H.A., Peguero-Pina, J.J., Gil-Pelegrín, E., Sancho-Knapik, D., Gil, L., et al., 2017. Physico-chemical properties of plant cuticles and their functional and ecological significance. J. Exp. Bot. 68, 5293–5306.

Fernández, V., Pimentel, C., Bahamonde, H.A., 2020. Salt hydration and drop drying of two model calcium salts: Implications for foliar nutrient absorption and deposition. J. Plant Nutr. Soil Sci. 183, 592–601.

Fernández, V., Gil-Pelegrín, E., Eichert, T., 2021. Foliar water and solute absorption: an update. Plant J. 105, 870–883.

Fernández-Escobar, R., Marin, L., Sánchez-Zamora, M.A., García-Novelo, J.M., Molina-Soria, C., Parra, M.A., 2009. Long-term effects of N fertilization on cropping and growth of olive trees and on N accumulation in soil profile. Eur. J. Agron. 31, 223–232.

Field, R.J., Bishop, N.G., 1988. Promotion of stomatal infiltration of glyphosate by an organosilicone surfactant reduces the critical rainfall period. Pestic. Sci. 24, 55–62.

Francois, L.E., Clark, R.A., 1979. Accumulation of sodium and chloride in leaves of sprinkler-irrigated grapes. J. Am. Soc. Hortic. Sci. 104, 11–13.

Gaige, E., Dail, D.B., Hollinger, D.Y., Davidson, E.A., Fernandez, I.J., Sievering, H., et al., 2007. Changes in canopy processes following whole-forest canopy nitrogen fertilization of a mature spruce-hemlock forest. Ecosystems 10, 1133–1147.

García-Mata, C., Lamattina, L., 2010. Hydrogen sulphide, a novel gasotransmitter involved in guard cell signalling. New Phytol. 188, 977–984.

Garde-Cerdán, T., López, R., Portu, J., González-Arenzana, L., López-Alfaro, I., Santamaría, P., 2014. Study of the effects of proline, phenylalanine, and urea foliar application to Tempranillo vineyards on grape amino acid content. Comparison with commercial nitrogen fertilisers. Food Chem. 163, 136–141.

Garten Jr., C.T., Hanson, P.J., 1990. Foliar retention of ^{15}N-nitrate and ^{15}N-ammonium by red maple (*Acer rubrum*) and white oak (*Quercus alba*) leaves from simulated rain. Env. Exp. Bot. 30, 333–342.

Gettier, S.W., Martens, D.C., Brumback Jr., T.B., 1985. Timing of foliar manganese application for correction of manganese deficiency in soybeans. Agron. J. 77, 627–629.

González-Arias, A., Amezaga, I., Echeandía, A., Onaindia, M., 2000. Buffering capacity through cation leaching of *Pinus radiata* D. Don canopy. Plant Ecol. 149, 23–42.

Gooding, M.J., Davies, W.P., 1992. Foliar urea fertilization of cereals: a review. Fert. Res. 32, 209–222.

Grattan, S.R., Royo, A., Aragüés, R., 1994. Chloride accumulation and partitioning in barley as affected by differential root and foliar salt absorption under saline sprinkler irrigation. Irrig. Sci. 14, 147–155.

Grundon, N.J., Asher, C.J., 1988. Volatile losses of sulfur from intact plants. J. Plant Nutr. 11, 563–576.

Guzmán, P., Fernández, V., Graça, J., Cabral, V., Kayali, N., Khayet, M., et al., 2014a. Chemical and structural analysis of *Eucalyptus globulus* and *E. camaldulensis* leaf cuticles: a lipidized cell wall region. Front. Plant Sci. 5, 481.

Guzmán, P., Fernández, V., García, M.L., Khayet, M., Fernández, A., Gil, L., 2014b. Localization of polysaccharides in isolated and intact cuticles of eucalypt, poplar and pear leaves by enzyme-gold labeling. Plant Physiol. Biochem. 76, 1–6.

Guzmán-Delgado, P., Earles, J.E., Zwieniecki, M.A., 2018. Insight into the physiological role of water absorption via the leaf surface from a rehydration kinetics perspective. Plant Cell Environ. 41, 1886–1894.

Haslett, B.S., Reid, R.J., Rengel, Z., 2001. Zinc mobility in wheat: Uptake and distribution of zinc applied to leaves or roots. Ann. Bot. 87, 379–386.

Hebbern, C.A., Laursen, K.H., Ladegaard, A.H., Schmidt, S.B., Pedas, P., Bruhn, D., et al., 2009. Latent manganese deficiency increases transpiration in barley (*Hordeum vulgare*). Physiol. Plant 135, 307–316.

Hensen, A., Nemitz, E., Flynn, M.J., Blatter, A., Jones, S.K., Sørensen, L.L., et al., 2009. Inter-comparison of ammonia fluxes obtained using the Relaxed Eddy Accumulation technique. Biogeosciences 6, 2575–2588.

Hereid, D.P., Monson, R.K., 2001. Nitrogen fluxes between corn (*Zea mays* L.) leaves and the atmosphere. Atmos. Environ. 35, 975–983.

Heredia-Guerrero, J.A., Benítez, J.J., Heredia, A., 2008. Self-assembled polyhydroxy fatty acids vesicles: a mechanism for plant cutin synthesis. Bioessays 30, 273–277.

Herrmann, B., Jones, S.K., Fuhrer, J., Feller, U., Neftel, A., 2001. N budget and NH_3 exchange of a grass/clover crop at two levels of N application. Plant Soil 235, 243–252.

Hinko-Najera Umana, N., Wanek, W., 2010. Large canopy exchange fluxes of inorganic and organic nitrogen and preferential retention of nitrogen by epiphytes in a tropical lowland rainforest. Ecosystems 13, 367–381.

Hu, Y., Sun, G., 2010. Leaf nitrogen dioxide uptake coupling apoplastic chemistry, carbon/sulfur assimilation, and plant nitrogen status. Plant Cell Rep. 29, 1069–1077.

Hu, Y., Fernández, V., Ma, L., 2014. Nitrate transporters in leaves and their potential roles in foliar uptake of nitrogen dioxide. Front. Plant Sci. 5, 360.

Huang, H., Li, H., Xiang, D., Liu, Q., Li, F., Liang, B., 2020. Translocation and recovery of ^{15}N-labeled N derived from the foliar uptake of $^{15}NH_3$ by the greenhouse tomato (*Lycopersicon esculentum* Mill.). J. Integr. Agric. 19, 859–865.

Isla, R., Aragüés, R., 2009. Response of alfalfa (*Medicago sativa* L.) to diurnal and nocturnal saline sprinkler irrigations. II: shoot ion content and yield relationships. Irrig. Sci. 27, 507–513.

Isla, R., Aragüés, R., 2010. Yield and plant ion concentrations in maize (*Zea mays* L.) subject to diurnal and nocturnal saline sprinkler irrigations. Field Crops Res. 116, 175–183.

Jang, S.W., Sadiq, N.B., Hamayun, M., Jung, J., Lee, T., Yang, J.S., et al., 2020. Silicon foliage spraying improves growth characteristics, morphological traits, and root quality of *Panax ginseng* CA Mey. Ind. Crops Prod. 156, 112848.

Jeffree, C.E., 2006. The fine structure of the plant cuticle. In: Riederer, M., Müller, C. (Eds.), Biology of the Plant Cuticle, Annual Plant Reviews, Vol. 23. Blackwell Publishing, Oxford, UK, pp. 11–125.

Johnson, D.W., Turner, J., 2019. Nutrient cycling in forests: A historical look and newer developments. Forest Ecol. Manag. 444, 344–373.

Johnson, R.S., Rosecrance, R., Weinbaum, S., Andris, H., Wang, J., 2001. Can we approach complete dependence on foliar-applied urea nitrogen in an early-maturing peach? J. Amer. Soc. Hort. Sci. 126, 364–370.

Jordan, G.J., Brodribb, T.J., 2007. Incontinence in aging leaves: Deteriorating water relations with leaf age in *Agastachys odorata* (Proteaceae), a shrub with very long-lived leaves. Funct. Plant Biol. 34, 918–924.

Jordan, L.A., Devitt, D.A., Morris, R.L., Neuman, D.S., 2001. Foliar damage to ornamental trees sprinkler-irrigated with reuse water. Irrig. Sci. 21, 17–25.

Kaiser, W., Dittrich, A., Heber, U., 1993. Sulfate concentrations in Norway spruce needles in relation to atmospheric SO_2: a comparison of trees from various forests in Germany with trees fumigated with SO_2 in growth chambers. Tree Physiol. 12, 1–13.

Kannan, S., 1969. Penetration of iron and some organic substances through isolated cuticular membranes. Plant Physiol. 44, 517–521.

Kannan, S., 2010. Foliar fertilization for sustainable crop production. In: Lichtfouse, E. (Ed.), Genetic Engineering, Biofertilisation, Soil Quality and Organic Farming. Sustainable Agriculture Reviews, Vol. 4. Springer, Dodrecht, The Netherlands, pp. 371–402.

Kara, B., Uysal, N., 2009. Influence on grain yield and grain protein content of late-season nitrogen application in triticale. J. Anim. Vet. Adv. 8, 579–586.

Kaupenjohann, M., Schneider, B.U., Hantschel, R., Zech, W., Horn, R., 1988. Sulfuric acid rain treatment of *Picea abies* (L.) Karst: effects on nutrient solution, throughfall chemistry, and tree nutrition. Z. Pflanzen. Bodenkd. 151, 123–126.

Khan, M.R., Adam, V., Rizvi, T.F., Zhang, B., Ahamad, F., Josko, I., et al., 2019. Nanoparticle-plant interactions: two-way traffic. Small 15, 1901794.

Koch, K., Hartmann, K.D., Schreiber, L., Barthlott, W., Neinhuis, C., 2006. Influence of air humidity on epicuticular wax chemical composition, morphology and wettability of leaf surfaces. Environ. Exp. Bot. 56, 1–9.

Krupa, S.V., 2003. Effects of atmospheric ammonia (NH$_3$) on terrestrial vegetation: a review. Environ. Pollut. 124, 179–221.
Künstler, A., Gullner, G., Ádám, A.L., Nagy, J.K., Király, L., 2020. The versatile roles of sulfur-containing biomolecules in plant defense - a road to disease resistance. Plants 9, 1705.
Laane, H.M., 2018. The effects of foliar sprays with different silicon compounds. Plants 7, 45.
Larue, C., Castillo-Michel, H., Sobanska, S., Cécillon, L., Bureau, S., Barthès, V., et al., 2014. Foliar exposure of the crop *Lactuca sativa* to silver nanoparticles: Evidence for internalization and changes in Ag speciation. J. Hazard. Mater. 264, 98–106.
Lawson, P.G., Daum, D., Czauderna, R., Meuser, H., Härtling, J.W., 2015. Soil versus foliar iodine fertilization as a biofortification strategy for field-grown vegetables. Front. Plant Sci. 6, 450.
Leisen, E., Marschner, H., 1990. Einfluss von Düngung und saurer Benebelung auf Nadelverluste sowie Auswaschung und Gehalte an Mineralstoffen und Kohlenhydraten in Nadeln von Fichten (*Picea abies* (L.) Karst.). Forstwiss. Cbl. 109, 253–263.
León, J., Costa-Broseta, Á., 2020. Present knowledge and controversies, deficiencies, and misconceptions on nitric oxide synthesis, sensing, and signaling in plants. Plant Cell Environ. 43, 1–15.
Leonardi, S., Flückiger, W., 1989. Effects of cation leaching on mineral cycling and transpiration: Investigations with beech seedlings, *Fagus sylvatica* L. New Phytol. 111, 173–179.
Li, C., Wang, P., Lombi, E., Cheng, M., Tang, C., Howard, D.L., et al., 2018a. Absorption of foliar-applied Zn fertilizers by trichomes in soybean and tomato. J. Exp. Bot. 69, 2717–2729.
Li, C., Wang, P., Lombi, E., Wu, J., Blamey, F.P.C., Fernández, V., et al., 2018b. Absorption of foliar applied Zn is decreased in Zn deficient sunflower (*Helianthus annuus*) due to changes in leaf properties. Plant Soil 433, 309–322.
Li, M., Zhao, Z., Zhou, J., Zhou, D., Chen, B., Huang, L., et al., 2018c. Effects of a foliar spray of selenite or selenate at different growth stages on selenium distribution and quality of blueberries. J. Sci. Food Agric. 98, 4700–4706.
Li, C., Wang, P., van der Ent, A., Cheng, M., Jiang, H., Lund Read, T., et al., 2019. Absorption of foliar-applied Zn in sunflower (*Helianthus annuus*): importance of the cuticle, stomata and trichomes. Ann. Bot. 123, 57–68.
Liebisch, F., Max, J.F.J., Heine, G., Horst, W.J., 2009. Blossom-end rot and fruit cracking of tomato grown in net-covered greenhouses in Central Thailand can partly be corrected by calcium and boron sprays. J. Plant Nutr. Soil Sci. 172, 140–150.
Limm, E.B., Simonin, K.A., Bothman, A.G., Dawson, T.E., 2009. Foliar water uptake: a common water acquisition strategy for plants of the redwood forest. Oecologia 161, 449–459.
Lötze, E., Joubert, J., Theron, K.I., 2008. Evaluating pre-harvest foliar calcium applications to increase fruit calcium and reduce bitter pit in 'Golden Delicious' apples. Sci. Hort. 116, 299–304.
Lovett, G.M., Lindberg, S.E., 1993. Atmospheric deposition and canopy interactions of nitrogen in forests. Can. J. For. Res. 23, 1603–1616.
Lurie, S., Crisosto, C.H., 2005. Chilling injury in peach and nectarine: a review. Postharvest Biol. Technol. 37, 195–208.
Luyckx, M., Hausman, J.-F., Lutts, S., Guerriero, G., 2017. Silicon and plants: Current knowledge and technological perspectives. Front. Plant Sci. 8, 411.
Ma, D., Sun, D., Wang, C., Ding, H., Qin, H., Hou, J., et al., 2017. Physiological responses and yield of wheat plants in zinc-mediated alleviation of drought stress. Front. Plant Sci. 8, 860.
Maas, E.V., 1985. Crop tolerance to saline sprinkling water. Plant Soil 89, 372-284.
Marschner, B., Stahr, K., Renger, M., 1991. Element inputs and canopy interactions in two pine forest ecosystems in Berlin, Germany. Z. Pflanzen. Bodenkd. 145, 147–151.
Martin, C.E., 1994. Physiological ecology of the Bromeliaceae. Bot. Rev. 60, 1–82.
Martin, C.E., von Willert, D.J., 2000. Leaf epidermal hydathodes and the ecophysiological consequences of foliar water uptake in species of *Crassula* from the Namib desert in Southern Africa. Plant Biol 2, 229–242.
Massad, R.S., Loubet, B., Tuzet, A., Autret, H., Cellier, P., 2009. Ammonia stomatal compensation point of young oilseed rape leaves during dark/light cycles under various nitrogen nutritions. Agr. Ecosyst. Environ. 133, 170–182.
McBeath, T.M., McLaughlin, M.J., Noack, S.R., 2011. Wheat grain yield response to and translocation of foliar-applied phosphorus. Crop Pasture Sci. 62, 58–65.
Mengel, K., Lutz, H.J., Breininger, M.T., 1987. Auswaschung von Nährstoffen durch sauren Nebel aus jungen intakten Fichten (*Picea abies*). Z. Pflanzen. Bodenkd. 150, 61–68.
Montanaro, G., Dichio, B., Lang, A., Mininni, A.N., Nuzzo, V., Clearwater, M.J., et al., 2014. Internal versus external control of calcium nutrition in kiwifruit. J. Plant Nutr. Soil Sci. 177, 819–830.
Morán, M., Ferreira, J., Martins, H., Monteiro, A., Borrego, C., González, J.A., 2016. Ammonia agriculture emissions: from EMEP to high resolution inventory. Atmos. Pollut. Res. 7, 786–798.
Mosali, J., Desta, K., Teal, R.K., Freeman, K.W., Martin, K.L., Lawles, J.W., et al., 2006. Effect of foliar application of phosphorus on winter wheat grain yield, phosphorus uptake, and use efficiency. J. Plant Nutr. 29, 2147–2163.
Munné-Bosch, S., 2007. Aging in perennials. Crit. Rev. Plant Sci. 26, 123–138.
Nair, A.A., Yu, F., 2020. Quantification of atmospheric ammonia concentrations: a review of its measurement and modeling. Atmosphere 11, 1092.
Neumann, P.M., Prinz, R., 1974. Evaluation of surfactants for use in the spray treatment of iron chlorosis in citrus trees. J. Sci. Food Agric. 25, 221–226.
Nikolic, M., Römheld, V., 2003. Nitrate does not result in iron inactivation in the apoplast of sunflower leaves. Plant Physiol. 132, 1303–1314.

Noack, S.R., McBeath, T.M., McLaughlin, M.J., 2010. Potential for foliar phosphorus fertilisation of dryland cereal crops: a review. Crop Pasture Sci. 61, 659–669.

Nye, P.H., Greenland, D.J., 1960. The Soil under Shifting Cultivation. Technical Communication No. 51. Commonwealth Agricultural Bureau, Harpenden, UK.

Oko, B.F.D., Eneji, A.E., Binang, W., Irshad, M., Yamamoto, S., Honna, T., et al., 2003. Effect of foliar application of urea on reproductive abscission and grain yield of soybean. J. Plant Nutr. 26, 1223–1234.

Ojeda-Barrios, D., Abadía, J., Lombardini, L., Abadía, A., Vázquez, S., 2012. Zinc deficiency in field-grown pecan trees: changes in leaf nutrient concentrations and structure. J. Sci. Food Agric. 92, 1672–1678.

Oliveira, R.S., Dawson, T.E., Burgess, S.S.O., 2005. Evidence for direct water absorption by the shoot of the desiccation-tolerant plant *Vellozia flavicans* in the savannas of central Brazil. J. Trop. Ecol 21, 585–588.

Otto, R., Marques, J.P.R., Pereira de Carvalho, H.W., 2021. Strategies for probing absorption and translocation of foliar-applied nutrients. J. Exp. Bot. 72, 4600–4603.

Pan, B., Lam, S.K., Mosier, A., Luo, Y., Chen, D., 2016. Ammonia volatilization from synthetic fertilizers and its mitigation strategies: A global synthesis. Agric. Ecosys. Environ. 232, 283–289.

Papadakis, I.E., Protopapadakis, E., Therios, I.N., Tsirakoglou, V., 2005. Foliar treatment of Mn deficient 'Washington navel' orange trees with two Mn sources. Sci. Hortic. 106, 70–75.

Papadakis, I.E., Sotiropoulos, T.E., Therios, I.N., 2007. Mobility of iron and manganese within two citrus genotypes after foliar applications of iron sulfate and manganese. J. Plant Nutr. 30, 1385–1396.

Pedersen, J.M., Feilberg, A., Kamp, J.N., Hafner, S., Nyord, T., 2020. Ammonia emission measurement with an online wind tunnel system for evaluation of manure application techniques. Atmos. Environ. 230, 117562.

Peirce, C.A.E., McBeath, T.M., Fernández, V., McLaughlin, M.J., 2014. Wheat leaf properties affecting the absorption and subsequent translocation of foliar-applied phosphoric acid fertilizer. Plant Soil 384, 37–51.

Peryea, F.J., Neilsen, G.H., Faubion, D., 2007. Start-timing for calcium chloride spray programs influences fruit calcium and bitter pit in 'Braeburn' and 'Honeycrisp' apples. J. Plant Nutr. 30, 1213–1227.

Pfirrmann, T., Runkel, K.H., Schramel, P., Eisenmann, T., 1990. Mineral and nutrient supply, content and leaching in Norway spruce exposed for 14 months to ozone and acid mist. Environ. Pollut. 64, 229–254.

Poblaciones, M.J., Rengel, Z., 2016. Soil and foliar zinc biofortification in field pea (*Pisum sativum* L.): grain accumulation and bioavailability in raw and cooked grains. Food Chem 212, 427–433.

Pollard, M., Beisson, F., Li, Y., Ohlrogge, J.B., 2008. Building lipid barriers: biosynthesis of cutin and suberin. Trends Plant Sci. 13, 236–246.

Popp, C., Burghardt, M., Friedmann, A., Riederer, M., 2005. Characterization of hydrophilic and lipophilic pathways of *Hedera helix* L. cuticular membranes: permeation of water and uncharged organic compounds. J. Exp. Bot. 56, 2797–2806.

Portu, J., González-Arenzana, L., Hermosín-Gutiérrez, I., Santamaría, P., Garde-Cerdán, T., 2015. Phenylalanine and urea foliar applications to grapevine: effect on wine phenolic content. Food Chem. 180, 55–63.

Powlson, D.S., Poulton, P.R., Møller, N.E., Hewitt, M.V., Penny, A., Jenkinson, D.S., 1989. Uptake of foliar-applied urea by winter wheat (*Triticum aestivum*). The influence of application time and the use of a new ^{15}N technique. J. Sci. Food Agric. 48, 429–440.

Protoschill-Krebs, G., Wilhelm, C., Kesselmeier, J., 1996. Consumption of carbonyl sulphide (COS) by higher plant carbonic anhydrase (CA). Atmos. Environ. 30, 3151–3156.

Puccinelli, M., Malorgio, F., Pezzarossa, B., 2017. Selenium enrichment of horticultural crops. Molecules 22, 933.

Räsch, A., Hunsche, M., Mail, M., Burkhardt, J., Noga, G., Pariyar, S., 2018. Agricultural adjuvants may impair leaf transpiration and photosynthetic activity. Plant Physiol. Biochem. 132, 229–237.

Raivonen, M., Vesala, T., Pirjola, L., Altimir, N., Keronen, P., Kumala, M., et al., 2009. Compensation point of NO_x exchange: Net result of NO_x consumption and production. Agric. For. Meteorol. 149, 1073–1081.

Randewig, D., Hamisch, D., Eiblmeier, M., Boedecker, C., Kreuzwieser, J., Mendel, R.R., et al., 2014. Oxidation and reduction of sulfite contribute to susceptibility and detoxification of SO_2 in *Populus x canescens* leaves. Trees 28, 399–411.

Rausch, T., Wachter, A., 2005. Sulfur metabolism: a versatile platform for launching defence operations. Trends Plant Sci. 10, 503–509.

Rengel, Z., 2015. Availability of Mn, Zn and Fe in the rhizosphere. J. Soil Sci. Plant Nutr. 15, 397–409.

Rengel, Z., Batten, G.D., Crowley, D.E., 1999. Agronomic approaches for improving the micronutrient density in edible portions of field crops. Field Crops Res. 60, 27–40.

Rennenberg, H., Huber, B., Schroder, P., Stahl, K., Haunold, W., Georgii, H.W., et al., 1990. Emission of volatile sulfur compounds from spruce trees. Plant Physiol. 92, 560–564.

Restrepo-Diaz, H., Benlloch, M., Fernandez-Escobar, R., 2008. Plant water stress and K^+ starvation reduce absorption of foliar applied K^+ by olive leaves. Sci. Hortic. 116, 409–413.

Riederer, M., Friedmann, A., 2006. Transport of lipophilic non-electrolytes across the cuticle. In: Riederer, M., Müller, C. (Eds.), Biology of the Plant Cuticle, Annual Plant Reviews, Vol. 23. Blackwell Publishing, Oxford, UK, pp. 250–279.

Riederer, M., Schneider, G., 1990. The effect of the environment on the permeability and composition of *Citrus* leaf cuticles. II. Composition of soluble cuticular lipids and correlation with transport properties. Planta 180, 154–165.

Rodrigues, F.A., Dallagnol, L.J., Duarte, H.S.S., Datnoff, L.E., 2015. Silicon control of foliar diseases in monocots and dicots. In: Rodrigues, F.A., Datnoff, L.E. (Eds.), Silicon and Plant Diseases. Springer, Cham, Switzerland, pp. 67–108.

Rodrigues, E.S., Gomes, M.H.F., Duran, N.M., Cassanji, J.G.B., Cruz, T.N.M., Sant'Anna Neto, A., et al., 2018. Laboratory microprobe X-ray fluorescence in plant science: emerging applications and case studies. Front. Plant Sci. 9, 1588.

Rosolem, C.A., Leite, V.M., 2007. Coffee leaf and stem anatomy under boron deficiency. Rev. Bras. Cienc. Solo 31, 477–483.

Runyan, C.W., Lawrence, D., Vandecar, K.L., D'Odorico, P., 2013. Experimental evidence for limited leaching of phosphorus from canopy leaves in a tropical forest. Ecohydrology 6, 806–817.

Sargent, J.A., Blackman, G.E., 1962. Studies on foliar penetration. I. Factors controlling the entry of 2,4-dichlorophenoxyacetic acid. J. Exp. Biol. 13, 348–368.

Saure, M.C., 2005. Calcium translocation to fleshy fruit: its mechanism and endogenous control. Sci. Hortic. 105, 65–89.

Schaberg, P.G., DeHayes, D.H., Hawley, G.J., Strimbeck, G.R., Cumming, J.R., Murakami, P.F., et al., 2000. Acid mist and soil Ca and Al alter the mineral nutrition and physiology of red spruce. Tree Physiol. 20, 73–85.

Schlegel, T.K., Schönherr, J., 2002a. Selective permeability of cuticles over stomata and trichomes to calcium chloride. Acta Hortic. 549, 91–96.

Schlegel, T.K., Schönherr, J., 2002b. Stage of development affects penetration of calcium chloride into apple fruits. J. Plant Nutr. Soil Sci. 165, 738–745.

Schlegel, T., Schönherr, J., Schreiber, L., 2006. Rates of foliar penetration of chelated Fe(III): role of light, stomata, species, and leaf age. J. Agric. Food Chem. 54, 6809–6813.

Schmitz-Eiberger, M., Haefs, R., Noga, G., 2002. Enhancing biological efficacy and rainfastness of foliar applied calcium chloride solutions by addition of rapeseed oil surfactants. J. Plant Nutr. Soil Sci. 165, 634–639.

Schoninger, E.L., González-Villalba, H.A., Bendassolli, J.A., Ocheuze Trivelin, P.C., 2018. Fertilizer nitrogen and corn plants: not all volatilized ammonia is lost. Agron. J. 110, 1111–1118.

Schönherr, J., 1976. Water permeability of isolated cuticular membranes: The effect of pH and cations on diffusion, hydrodynamic permeability and size of polar pores in the cutin matrix. Planta 128, 113–126.

Schönherr, J., 2000. Calcium chloride penetrates plant cuticles via aqueous pores. Planta 212, 112–118.

Schönherr, J., 2006. Characterization of aqueous pores in plant cuticles and permeation of ionic solutes. J. Exp. Bot. 57, 2471–2491.

Schönherr, J., Baur, P., 1994. Modelling penetration of plant cuticles by crop protection agents and effects of adjuvants on their rates of penetration. Pestic. Sci. 42, 185–208.

Schönherr, J., Bukovac, M.J., 1972. Penetration of stomata by liquids. Dependence on surface tension, wettability, and stomatal morphology. Plant Physiol. 49, 813–819.

Schönherr, J., Bukovac, M.J., 1978. Foliar penetration of succinic acid-2,2-dimethylhydrazide: mechanisms and rate limiting step. Physiol. Plant. 42, 243–251.

Schönherr, J., Huber, R., 1977. Plant cuticles are polyelectrolytes with isoelectric points around three. Plant Physiol. 59, 145–150.

Schönherr, J., Luber, M., 2001. Cuticular penetration of potassium salts: effects of humidity, anions, and temperature. Plant Soil 236, 117–122.

Schreiber, L., 2005. Polar paths of diffusion across plant cuticles: new evidence for an old hypothesis. Ann. Bot. 95, 1069–1073.

Schröder, P., 1993. Plants as a source of atmospheric sulfur. In: De Kok, L.J., Stulen, I., Rennenberg, H., Brunold, C., Rauser, W.E. (Eds.), Sulfur Nutrition and Sulfur Assimilation in Higher Plants. SPB Academic Publishing bv, The Hague, The Netherlands, pp. 253–270.

Schuepp, P.H., Hendershot, W.H., 1989. Nutrient leaching from dormant trees at an elevated site. Water Air Soil Pollut. 45, 253–264.

Schulze, E.-D., 2000. Carbon and Nitrogen Cycling in European Forest Ecosystems. Springer, Berlin, Germany.

Segado, P., Domínguez, E., Heredia, A., 2016. Ultrastructure of the epidermal cell wall and cuticle of tomato fruit (*Solanum lycopersicum* L.) during development. Plant Physiol. 170, 935–946.

Servin, A., Elmer, W., Mukherjee, A., De la Torre-Roche, R., Hamdi, H., White, J.C., et al., 2015. A review of the use of engineered nanomaterials to suppress plant disease and enhance crop yield. J. Nanopart. Res. 17, 92.

Shear, C.B., 1975. Calcium-related disorders of fruits and vegetables. HortScience 10, 361–365.

Shen, J., Yuan, L., Zhang, J., Li, H., Bai, Z., Chen, X., et al., 2011. Phosphorus dynamics: from soil to plant. Plant Physiol. 156, 997–1005.

Sherchand, K., Paulsen, G.M., 1985. Response of wheat to foliar phosphorus treatments under field and high temperature regimes. J. Plant Nutr. 8, 1171–1181.

Shi, H., Ye, T., Han, N., Bian, H., Liu, X., Chan, Z., 2015. Hydrogen sulfide regulates abiotic stress tolerance and biotic stress resistance in *Arabidopsis*. J. Integr. Plant Biol. 57, 628–640.

Singh, D.P., 2015. Plant nutrition in the management of plant diseases with particular reference to wheat. In: Awasthi, L.P. (Ed.), Recent Advances in the Diagnosis and Management of Plant Diseases. Springer, New Delhi, India, pp. 273–284.

Singh, R.B., Chauhan, C.P.S., Minhas, P.S., 2009. Water production functions of wheat (*Triticum aestivum* L.) irrigated with saline and alkali waters using double-line source sprinkler system. Agric. Water Manag. 96, 736–744.

Smith, C.J., Freney, J.R., Sherlock, R.R., Galbally, I.E., 1991. The fate of urea nitrogen applied in a foliar spray to wheat at heading. Fert. Res. 28, 129–138.

Song, W., Yi, J., Kurniadinata, O.F., Wang, H., Huang, X., 2018. Linking fruit Ca uptake capacity to fruit growth and pedicel anatomy, a cross-species study. Front. Plant Sci. 9, 575.

Sotiropoulos, T., Therios, I., Voulgarakis, N., 2010. Effect of various foliar sprays on some fruit quality attributes and leaf nutritional status of the peach cultivar 'Andross'. J. Plant Nutr. 33, 471–484.

Sparks, J.P., Roberts, J.M., Monson, R.K., 2003. The uptake of gaseous organic nitrogen by leaves: a significant global nitrogen transfer process. Geophys. Res. Lett. 30, 2189.

Stimler, K., Nelson, D., Yakir, D., 2010. High precision measurements of atmospheric concentrations and plant exchange rates of carbonyl sulfide using mid-IR quantum cascade laser. Glob. Change Biol. 16, 2496–2503.

Teklemariam, T.A., Sparks, J.P., 2004. Gaseous fluxes of peroxyacetyl nitrate (PAN) into plant leaves. Plant Cell Environ. 27, 1149–1158.

Teklemariam, T.A., Sparks, J.P., 2006. Leaf fluxes of NO and NO_2 in four herbaceous plant species: the role of ascorbic acid. Atmos. Environ. 40, 2235–2244.

Torres, E., Recasens, I., Lordan, J., Alegre, S., 2017. Combination of strategies to supply calcium and reduce bitter pit in 'Golden Delicious' apples. Sci. Hortic. 217, 179–188.

Tredenick, E.C., Farrell, T.W., Forster, W.A., Psaltis, S.T.P., 2017. Nonlinear porous diffusion modeling of hydrophilic ionic agrochemicals in astomatous plant cuticle aqueous pores: a mechanistic approach. Front. Plant Sci. 8, 746.

Tredenick, E.C., Farrell, T.W., Forster, W.A., 2018. Mathematical modeling of diffusion of a hydrophilic ionic fertilizer in plant cuticles: surfactant and hygroscopic effects. Front. Plant Sci. 9, 1888.

Trivedi, P., Delgado-Baquerizo, M., Anderson, I.C., Singh, B.K., 2016. Response of soil properties and microbial communities to agriculture: Implications for primary productivity and soil health indicators. Front. Plant Sci. 7, 990.

Tukey Jr., H.B., 1970. The leaching of substances from plants. Ann. Rev. Plant Physiol. 21, 305–324.

Turner, D.P., Tingey, D.T., 1990. Foliar leaching and root uptake of Ca, Mg, and K in relation to acid fog effects on Douglas-fir. Water Air Soil Poll. 49, 205–214.

Turner, D.P., van Broekhuizen, H.J., 1992. Nutrient leaching from conifer needles in relation to foliar apoplast cation exchange capacity. Environ. Pollut. 75, 259–263.

Tyree, M.T., Scherbatskoy, T.D., Tabor, C.A., 1990. Leaf cuticles behave as asymmetric membranes. Evidence from the measurement of diffusion potentials. Plant Physiol. 92, 103–109.

Urrego-Pereira, Y.F., Martinez-Cob, A., Fernández, V., Cavero, J., 2013. Daytime sprinkler irrigation effects on net photosynthesis of maize and alfalfa. Agron. J. 105, 1515–1528.

Val, J., Monge, E., Risco, D., Blanco, A., 2008. Effect of pre-harvest calcium sprays on calcium concentrations in the skin and flesh of apples. J. Plant Nutr. 31, 1889–1905.

Valdez-Aguilar, L.A., Grieve, C.M., Poss, J.A., 2009. Salinity and alkaline pH of irrigation water affect marigold plants. I. Growth and shoot dry weight partitioning. HortScience 44, 1719–1725.

van der Kooij, T.A.W., De Kok, L.J., Haneklaus, S., Schnug, E., 1997. Uptake and metabolism of sulphur dioxide by *Arabidopsis thaliana*. New Phytol. 135, 101–107.

Varga, B., Svečnjak, Z., 2006. The effect of late-season urea spraying on grain yield and quality of winter wheat cultivars under low and high basal nitrogen fertilization. Field Crops Res. 96, 125–132.

Vinken, G.C.M., Boersma, K.F., Maasakkers, J.D., Adon, M., Martin, R.V., 2014. Worldwide biogenic soil NO_x emissions inferred from OMI NO_2 observations. Atmos. Chem. Phys. 14, 10363–10381.

Vojtovic, D., Luhová, L., Petrivalsky, M., 2021. Something smells bad to plant pathogens: Production of hydrogen sulfide in plants and its role in plant defence responses. J. Adv. Res. 27, 199–209.

Walworth, J.L., Pond, A.P., Sower, G.J., Kilby, M.W., 2006. Fall-applied foliar zinc for pecans. HortScience 41, 275–276.

Webb, J., Pain, B., Bittman, S., Morgan, J., 2010. The impacts of manure application methods on emissions of ammonia, nitrous oxide and on crop response - a review. Agric. Ecosyst. Environ. 137, 39–46.

Werker, E., 2000. Trichome diversity and development. Adv. Bot. Res. 31, 1–35.

Wesely, R.W., Shearman, R.C., Kinbacher, E.J., Lowry, S.R., 1987. Ammonia volatilization from foliar-applied urea on field-grown Kentucky bluegrass. HortScience 22, 1278–1280.

Westerman, S., Stulen, I., Suter, M., Brunold, C., De Kok, L.J., 2001. Atmospheric H_2S as sulphur source for *Brassica oleracea*: Consequences for the activity of the enzymes of the assimilatory sulphate reduction pathway. Plant Physiol. Biochem. 39, 425–432.

Whelan, M.E., Lennartz, S.T., Gimeno, T.E., Wehr, R., Wohlfahrt, G., Wang, Y., et al., 2018. Review and synthesis: Carbonyl sulphide as a multi-scale tracer for carbon and water cycles. Biogeosciences 15, 3625–3657.

White, P.J., Broadley, M.R., 2009. Biofortification of crops with seven mineral elements often lacking in human diets – iron, zinc, copper, calcium, magnesium, selenium and iodine. New Phytol. 182, 49–84.

Whitecross, M.I., Armstrong, D.J., 1972. Environmental effects on epicuticular waxes of *Brassica napus* L. Aust. J. Bot 20, 87–95.

Whitehead, D.C., Lockyer, D.R., 1987. The influence of the concentration of gaseous ammonia on its uptake by the leaves of Italian ryegrass with and without an adequate supply of nitrogen to the roots. J. Exp. Bot. 38, 818–827.

Wichink Kruit, R.J., van Pul, W.A.J., Sauter, F.J., van den Broek, M., Nemitz, E., Sutton, M.A., et al., 2010. Modeling the surface-atmosphere exchange of ammonia. Atmos. Environ. 44, 945–957.

Will, S., Eichert, T., Fernández, V., Möhring, J., Müller, T., Römheld, V., 2011. Absorption and mobility of foliar-applied boron in soybean as affected by plant boron status and application as a polyol complex. Plant Soil 344, 283–293.

Williams, E.J., Hutchinson, G.L., Fehsenfeld, F.C., 1992. NO_x and N_2O emissions from soil. Glob. Biogeochem. Cycles 6, 351–388.

Wójcik, P., 2004. Uptake of mineral nutrients from foliar fertilization. J. Fruit Ornam. Plant Res. 12, 201–218.

Xie, R., Zhao, J., Lu, L., Brown, P., Lin, X., Webb, S.M., et al., 2020. Seasonal zinc storage and a strategy for its use in buds of fruit trees. Plant Phys. 183, 1200–1212.

Xie, R., Zhao, J., Lu, L., Jernstedt, J., Guo, J., Brown, P.H., et al., 2021. Spatial imaging reveals the pathways of Zn transport and accumulation during reproductive growth stage in almond plants. Plant Cell Environ. 44, 1858–1868.

Xu, X., Bingemer, H.G., Schmidt, U., 2002. The flux of carbonyl sulfide and carbon disulfide between the atmosphere and a spruce forest. Atmos. Chem. Phys. 2, 171–181.

Yang, L., Stulen, I., De Kok, L.J., 2006. Sulfur dioxide: relevance of toxic and nutritional effects for Chinese cabbage. Environ. Exp. Bot. 57, 236–245.

Yeats, T.H., Rose, J.K.C., 2013. The formation and function of plant cuticles. Plant Physiol. 163, 5–20.

Zeisler-Diehl, V.V., Migdal, B., Schreiber, L., 2017. Quantitative characterization of cuticular barrier properties: methods, requirements, and problems. J. Exp. Bot. 68, 5281–5291.

Zhang, H., Tan, Z.Q., Hu, L.Y., Wang, S.H., Luo, J.P., Jones, R.L., 2010. Hydrogen sulfide alleviates aluminum toxicity in germinating wheat seedlings. J. Integr. Plant Biol. 52, 556–567.

Zhang, J., Wang, M.Y., Wu, L.H., 2009. Can foliar iron-containing solutions be a potential strategy to enrich iron concentration of rice grains (*Oryza sativa* L.)? Acta Agric. Scand. Sect. B-Soil Plant Sci. 59, 389–394.

Zhang, Q., Brown, P.H., 1999. The mechanism of foliar zinc absorption in pistachio and walnut. J. Am. Soc. Hortic. Sci. 124, 312–317.

Zhang, T., Sun, H., Lv, Z., Cui, L., Mao, H., Kopittke, P.M., 2018. Using synchrotron-based approaches to examine the foliar application of $ZnSO_4$ and ZnO nanoparticles for field-grown winter wheat. J. Agric. Food. Chem. 66, 2572–2579.

Zou, C.Q., Zhang, Y.Q., Rashid, A., Ram, H., Savasli, E., Arisoy, R.Z., et al., 2012. Biofortification of wheat with zinc through zinc fertilization in seven countries. Plant Soil 361, 119–130.

Chapter 5

Mineral nutrition, yield, and source−sink relationships

Ernest A. Kirkby[1], Miroslav Nikolic[2], Philip J. White[3], and Guohua Xu[4]

[1]*Faculty of Biological Sciences, University of Leeds, Leeds, United Kingdom,* [2]*Institute for Multidisciplinary Research, University of Belgrade, Belgrade, Serbia,* [3]*Ecological Sciences, James Hutton Institute, Dundee, Scotland, United Kingdom,* [4]*College of Resources and Environmental Sciences, Nanjing Agricultural University, Nanjing, P.R. China*

Summary

This chapter describes the role of nutrients in regulating plant processes underlying yield formation. The yield of crop plants is controlled by biomass production and its partitioning to harvested plant organs. Biomass production is dependent on the capture of light energy, through the photosynthetic activity of leaves (i.e., source activity) and leaf area, to provide carbon and energy for the entire plant. Roots supply plants with water and nutrients from the soil. Nutrients are required for leaf growth and as integral constituents of the photosynthetic apparatus. Nutrient supply also affects photosynthesis and leaf senescence indirectly via photooxidation, hydraulic and phytohormonal signals as well as by sugar signaling. Nutrients impact respiration as constituents of the respiratory electron chain and by their influence on the efficiency of respiratory ATP synthesis. The chapter describes how photosynthate partitioning to plant organs is controlled by the capacity of these organs to utilize assimilates for growth and storage, that is, their sink strength, and how this is influenced by nutrient supply. Nutrients play an important role in regulating sink formation, for example, by their effects on plant architecture, flowering, pollination, and tuber initiation, as well as in controlling storage processes in the sink organs. Nutrient supply also modifies endogenous concentrations of phytohormones that regulate sink−source relationships. The source and sink organs are physically separated. Therefore, long-distance transport of photosynthates and nutrients in the phloem from source to sink is essential for growth and plant yield. The principles of phloem loading of assimilates at source sites, phloem transport, and phloem unloading at the sink sites are described.

5.1 General

More than 90% of plant dry matter consists of organic compounds such as cellulose, starch, lipids, and proteins. The total dry matter production of plants (i.e., the *biological yield*) is directly related to photosynthesis as the primary process of synthesis of organic compounds in green plants. In crops, *economic yield* is defined as the dry matter production of harvested plant organs for which particular crops are cultivated (e.g., grains and tubers). Thus, in many crop plants, it is not only total dry matter production that is of importance but also partitioning of the dry matter. The *harvest index* is the proportion of the total plant dry matter production present in the harvested parts of the crop. The partitioning of biomass among plant organs and its controlling mechanisms are therefore of crucial importance in crop production. For the quality of food plants, in addition to the organic compounds synthesized in primary and secondary metabolism, the contents of mineral nutrients and their distribution among plant organs are also important (Chapter 9; Bouis and Saltzman, 2017; White and Broadley, 2009).

In this chapter some principles of photosynthesis are discussed, as are the related processes of photophosphorylation and photorespiration, and examples of the involvement of mineral nutrients in these processes are given. This discussion includes aspects of photoinhibition and photooxidation and mechanisms protecting the photosynthetic apparatus against this damage. In plants, the main sites of photosynthesis—the source (mature green leaves)—and the sites of consumption and storage—the sink (roots, shoot apices, seeds, and fruits)—are separated. The long-distance transport of photosynthates in the phloem from source to sink is therefore essential for growth and plant yield. It is necessary to have a basic understanding of the processes of phloem loading of photosynthates at the source sites, phloem transport, and phloem unloading at the sink sites and the regulation of these processes, particularly the role of phytohormones. Finally, the relationships between source and sink, and the question of whether yield can be limited by source or sink, are discussed.

5.2 Relationships between nutrient supply and yield

Various factors are required for plant growth such as light, CO_2, water, and nutrients. Increasing the supply of any of these factors from the deficiency range increases growth rate and yield but the response diminishes with the increasing supply of the growth factor. This relationship was formulated mathematically for nutrients by Mitscherlich as the *law of diminishing yield increment* (Mitscherlich, 1954). According to this law, the yield response curves for a particular nutrient are asymptotic: when the supply of one nutrient (or growth factor) is increased, other nutrients (or growth factors) or the genetic potential of crop plants become limiting factors. Typical yield response curves for nutrients are shown in Fig. 5.1. The slopes of the three curves differ. Micronutrients have the steepest and N the gentlest slope when the nutrient supply is expressed in the same mass units. The slopes reflect the different plant demands for particular nutrients.

Some of the assumptions made by Mitscherlich (1954) were incorrect. The slope of the yield response curve for a particular nutrient cannot be described by a constant factor, nor is the curve asymptotic. When there is an abundant supply of nutrients, a point of inversion is obtained, as shown for micronutrients in Fig. 5.1. This inversion can be caused by many factors, including toxicity of a nutrient per se or induced deficiency of another nutrient. In the case of high N supply to cereals, grain yield may be reduced due to lodging or pathogen attack. High N supply may also reduce yield due to its effects on phytohormone concentrations and thus on development processes (Section 5.9).

An example of the interactions between nutrients influencing yield is given in Fig. 5.2. At the lowest K supply, the response to increasing N supply is small, and at high N supply the yield is strongly depressed. Yield response curves differ between grain and straw. In contrast to straw yield, grain yield at medium to high K supply levels off when N supply is high, reflecting sink limitation (e.g., small grain number per ear), sink competition (e.g., enhanced formation of tillers), or source limitation (e.g., mutual shading of leaves). Similar interactions between K and N supply have also been observed in yields of field-grown spring barley (Johnston and Milford, 2012) and other crops (Zhang et al., 2007a).

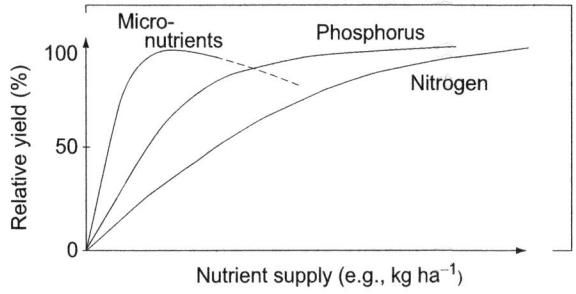

FIGURE 5.1 Yield response curves for N, P, and micronutrients.

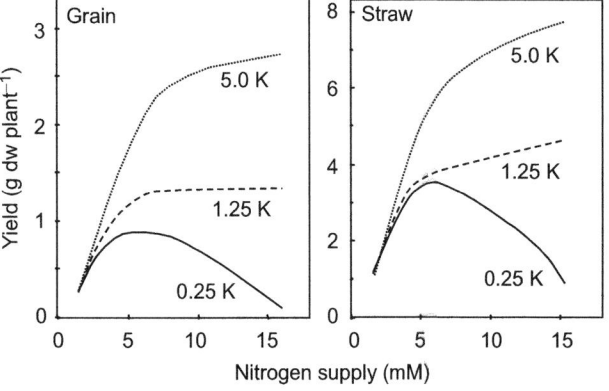

FIGURE 5.2 Grain and straw yields of barley (*Hordeum vulgare*) grown in nutrient solution at three K concentrations (0.25, 1.25, and 5.0 mM) with increasing N concentration. *Based on MacLeod (1969).*

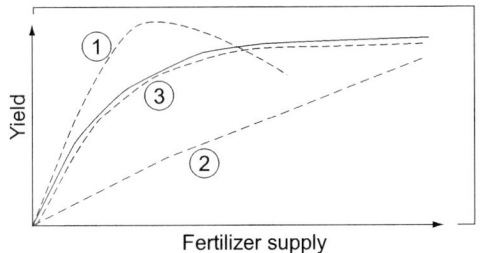

FIGURE 5.3 Schematic representation of yield response curves of harvested products. "———" quantitative yield (e.g., t ha^{-1}); "- - -" qualitative yield (e.g., contents of sugar, protein, and mineral elements). For examples of (1) to (3), see text.

Yield response curves are strongly modulated by interactions between nutrients and other growth factors. Under field conditions, the interaction between N nutrition and water use, in particular, has a global impact on crop productivity (Plett et al., 2020). In maize (*Zea mays*), for example, when water supply was most restricted, optimum yield was obtained at low N, and increasing N supply depressed yield (Shimshi, 1969) (similarly to the response to low K supply in Fig. 5.2). Depression in yield at high N application rates under restricted water supply may be caused by factors such as (1) delay in stomatal response to water deficiency, (2) large water consumption by vegetative biomass, and correspondingly an increased risk of drought stress at critical periods of grain formation, and (3) increase in shoot–root dry weight ratio with increasing N supply, an effect that may be stronger in C_3 than C_4 plant species (Way et al., 2014).

The yield response curves also differ depending on the yield component of harvested products. In most crops, both quantity (e.g., dry matter yield in tons per hectare) and quality (e.g., content of sugars or protein) are important yield components. As shown schematically in Fig. 5.3, maximum quality can be obtained either before [curve (1)] or after [curve (2)] the maximum dry matter yield has been reached, or both yield components can have a synchronous pattern [curve (3)]. Examples for the different curves are (1) accumulation of nitrate in spinach (*Spinacia oleracea*) and sucrose in sugar beet (*Beta vulgaris*) in response to increasing amounts of N fertilizer; (2) increases in protein content of cereal grains or forage plants with increasing N supply; and (3) positive relationship between number of either reproductive sinks (e.g., grains) or vegetative storage sinks (e.g., tubers) with nutrient supply.

5.3 Photosynthetic activity and related processes

To meet the growing demands of the rapidly growing world population for food and renewable primary products including biofuels, bioenergy and biomaterials, yields of agricultural and horticultural crops must be increased significantly. Over the past 50 years the continuously increasing use of N fertilizers in particular has promoted the development of the crop canopy to allow efficient interception of available light and its efficient conversion by photosynthesis to biomass (Long et al., 2006). Increased yields of many crops have been achieved largely by increasing harvest index, that is, by increasing the ratio of biomass in harvestable plant organs (Fischer, 2007; Smith et al., 2018). Modern wheat (*Triticum aestivum*) varieties have harvest indices between 0.45 and 0.50, and those of the best rice and maize varieties exceed 0.50 (Unkovitch et al., 2010). Further increases in yield will also necessitate increased biomass production, that is, net photosynthesis. In principle, photosynthesis can be enhanced by increasing the photosynthetic activity of the leaves or the photosynthetic area.

5.3.1 Photosynthetic energy flow and photophosphorylation

The conversion of light energy to chemical energy in plants is brought about by a flow of electrons through pigment systems embedded in thylakoid membranes of the chloroplasts. Often, the thylakoid membranes are stacked into piles that appear as grains or "grana" under a light microscope. The principles involved in the process of electron flow are illustrated in Fig. 5.4. Light energy is absorbed by two pigment systems: photosystem II (PS II) and PS I. In each of these PSs, between 400 and 500 individual chlorophyll molecules and accessory pigments (e.g., carotenoids) act as "light-harvesting antennae" to funnel protons into chlorophyll molecules with maximum absorbance at 680 nm in PS II and 700 nm in PS I. In both PSs, the absorption of light energy induces the emission and uphill transport of two electrons against the electrical gradients. The electrons required for this process are derived from the photolysis of water mediated by PS II. In plants, PS II and PS I act in a series known as the Z scheme as illustrated in Fig. 5.4 by the thick arrows. The scheme is both an energy diagram and a map of electron flow. For more details, see the reviews of Renger and Renger (2008) and Mohapatra and Singh (2015). At the end of the uphill transport chain, the electrons are taken up by an acceptor chlorophyll *a* molecule and transferred through a chain of redox centers [including phylloquinones and

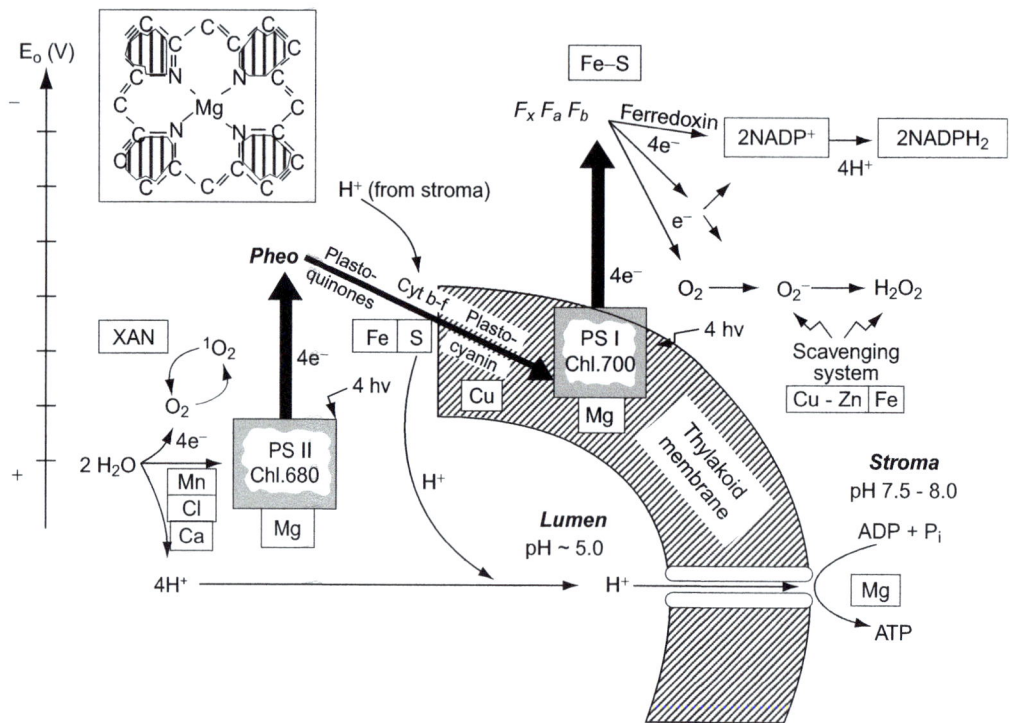

FIGURE 5.4 Photosynthetic electron transport chain with PS II and PS I and photophosphorylation. Inset: section of the porphyrin structure of chlorophyll with the central Mg atom. *Cyt*, Cytochrome; F_x, F_a, F_b, Fe-S clusters that transfer electrons from primary acceptors to ferredoxin; *Pheo*, pheophytins; *PS*, photosystems; *XAN*, xanthophyll cycle.

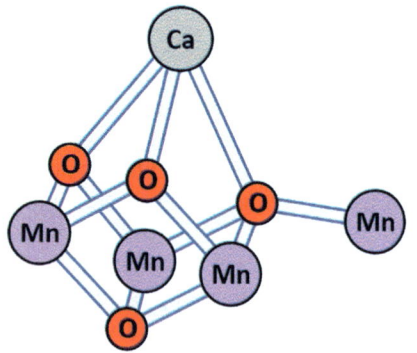

FIGURE 5.5 Mn_4CaO_4 cluster of the photosynthetic oxygen-evolving center within the photosystem II protein. *Adapted from Barber (2016).*

three Fe-S clusters (Fx, Fa, and Fb)] to ferredoxin, an Fe-S protein with a very low redox potential that reduces $NADP^+$ (nicotinamide adenine dinucleotide phosphate) to NADPH, as well as other compounds (see below).

Several nutrients are involved in this photosynthetic electron transport chain directly (Fig. 5.4). The chlorophyll molecules in PS II and PS I with their central Mg atom absorb photons, thereby initiating the electron flow. Photolysis (water splitting) is achieved by PS II. The catalytic water-splitting center consists of four Mn atoms, one Ca atom and four oxygen atoms in the form of Mn_4CaO_4 cubane, with the fourth Mn attached to the cubane structure by one of its oxo bridging bonds (Barber, 2016) (Fig. 5.5). This metalloprotein acts as an energy store and may also be a binding site for the water molecules that are oxidized. Chloride acts as an inorganic cofactor in the water-splitting system. Two molecules of water release one molecule of oxygen and four protons and four electrons that are used in the assimilation of one molecule of CO_2. Cytochromes (Cyt b–f) (with a central Fe atom as well as an Fe-S complex) mediate the electron flow between PS II and PS I (Marder and Barber, 1989). One of the electron acceptors in this chain is plastocyanin, a Cu-containing protein. In PS I, the electrons are transferred via three Fe-S clusters (F_x, F_a, F_b) to ferredoxin. Ferredoxin is a soluble 9-kDa Fe-S protein in the stroma; it acts as transmitter of electrons to $NADP^+$ that is reduced to NADPH by the enzyme ferredoxin-$NADP^+$-reductase anchored to the thylakoid surface.

Reduced ferredoxin in the chloroplasts can also function as an electron donor to other acceptors. The ferredoxin-mediated dependent reduction of nitrite and sulfite is of particular importance for the nutrition of plants.

$$PSI^- \xrightarrow{e^-} \begin{array}{c} | \quad | \\ S \quad S \\ | \diagdown S \diagup | \\ Fe \quad Fe \\ | \diagup S \diagdown | \\ S \quad S \\ | \quad | \end{array} \begin{array}{l} \xrightarrow{e^-} \text{Sulfite reductase} \\ \xrightarrow{e^-} NADP^+ \\ \xrightarrow{e^-} \text{Nitrite reductase} \end{array}$$

In the chloroplasts, both nitrite and sulfite compete with $NADP^+$ for reduction. In leaves, the rates of reduction of nitrite and sulfite are higher during the light period. This coupling of nitrite and sulfite reduction with light is also an example of a more general regulatory mechanism because photosynthesis supplies the structures (carbon skeletons) required for the incorporation of reduced nitrogen ($-NH_2$) and sulfur ($-SH$) into organic compounds such as amino acids.

Water splitting and the passage of electrons through the electron transport chain in the thylakoid membrane are coupled with the pumping of protons into the thylakoid lumen (Fig. 5.4), leading to acidification to about pH 5. The light-induced accumulation of H^+ (positive charge) in the lumen is counterbalanced by efflux of Mg^{2+}. The protons are consumed at the terminal site of the electron transport chain (formation of NADPH), raising the pH of the stroma to 7.5–8.0. The corresponding electrochemical potential gradient across the thylakoid membrane is used for *photophosphorylation*, a proton-driven synthesis of adenosine triphosphate (ATP) by an F-type ATPase. An additional component in the formation of the proton gradient is *cyclic photophosphorylation*, a pumping system for protons between PS II and PS I (Fig. 5.4). The downhill transport of three protons across the thylakoid membrane is thought to result in the production of one ATP molecule. In the stroma, ATP is required at various steps of CO_2 assimilation, carbohydrate synthesis as well as other ferredoxin-mediated processes.

5.3.2 Photoinhibition and photooxidation

Light absorbed by PS II and PS I is not necessarily balanced by a corresponding electron flow and formation of reduced ferredoxin and the consumption of electrons (e.g., in CO_2 assimilation). Imbalances occur under conditions of high light intensity and particularly when high light conditions are combined with other environmental stress factors such as drought, low temperatures, or nutrient deficiencies. Excess excitation energy depresses photosynthesis and quantum yield, which although usually reversible (*dynamic photoinhibition*) and related to impaired D1 protein activity in PS II (Miguez et al., 2015), may also lead in the long term to irreversible damage of the photosynthetic apparatus, thereby inhibiting photosynthesis (*chronic photoinhibition*) and inducing chlorosis and necrosis of the leaves (*photooxidation*). These latter symptoms are associated with the formation of reactive oxygen species (ROS) (Apel and Hirt, 2004; Asada, 1999). ROS in the forms of partially reduced forms of molecular oxygen, for example, $^1O_2$ (singlet), $O_2^{\bullet-}$ (superoxide radical), H_2O_2 (hydrogen peroxide), and $OH^{\bullet}$ (hydroxyl radical), are produced in large amounts during photosynthesis.

Plants possess a range of protective adaptations to reduce damage by ROS. These include, for example, the light-reflecting wax surface of the leaf epidermis, variation in leaf angle, leaf rolling, and chloroplast movement to reduce light absorption. If excess light energy is absorbed, plants can (1) dissipate the energy in the form of heat (Ort, 2001), (2) activate detoxification mechanisms against damage by ROS (Niyogi, 1999), and (3) continuously repair photodamaged PS II by fast and efficient turnover of the D1 protein (Nishiyama et al., 2006).

The primary target for photoinhibition is PS II. This PS produces molecular oxygen to which excessive excitation energy can be transferred to form the highly toxic singlet oxygen (Fig. 5.4). As a self-protecting mechanism, carotenoids (xanthophylls in particular) play an important role in both scavenging singlet oxygen and discharging excess photon flux energy as heat (*thermal dissipation, nonphotochemical quenching*) (Johnson et al., 2007; Ort, 2001). In a process induced by low lumen pH, PS II is transformed from a high-efficiency state, which uses most of the absorbed light energy for photochemical processes, to a photoprotected state, which dissipates excess light energy via the xanthophyll cycle in the form of heat (Eskling et al., 1997). The transition from the photoprotected state back to the high-efficiency state is a relatively slow process, particularly in thermophilic plant species at low temperatures (Zhu et al., 2004). Thus, in leaf canopies in the field, where there are short-term fluctuations in light intensity, ongoing dissipation of thermal energy after transfer from high to low light may cause a decrease in carbon gain by crops (Long et al., 2006).

The capacity of plants to accelerate xanthophyll cycle-dependent energy dissipation is increased by environmental stresses that depress the photosynthetic rate, such as low temperatures and low N supply (Demmig-Adams and Adams, 1996). However, despite this acclimation, the lower CO_2 assimilation capacity of N-deficient plants leads to increased susceptibility of PS II to photoinhibition, for example, in Norway spruce (Grassi et al., 2001) and rice (Kumagai et al., 2010). High ultraviolet-B (UV-B) radiation may also cause inhibition of photosynthesis and photooxidation of pigments (Jordan, 2002). In a field study with maize, N-deficient plants were less sensitive to increased UV-B radiation than N-sufficient plants, suggesting the effects of the two stresses are not additive (Correia et al., 2005).

Another main site of formation of ROS is the stroma of chloroplasts, where reduced ferredoxin can use molecular oxygen (O_2) as an electron acceptor leading to reduction of O_2 to the superoxide anion ($O_2^{\bullet-}$) (Figs. 5.4 and 5.6). This reductive O_2 activation in chloroplasts is greater under conditions that increase the $NADPH/NADP^+$ ratio, for example, low CO_2 supply or impaired CO_2 fixation, caused by a range of environmental stresses including low temperatures in chilling-sensitive plant species (Hodgson and Raison, 1991), salinity, drought, and nutrient deficiency. Reductive O_2 activation is also increased by low or inhibited export of photosynthates from source leaves under nutrient deficiency (Marschner and Cakmak, 1989; Cakmak and Kirkby, 2008). In C_3 species, photorespiration (i.e., the release of CO_2 in the light) may be an important protective mechanism consuming ATP and reducing equivalents, which prevents overreduction of the photosynthetic electron transport chain and photoinhibition (Wingler et al., 2000).

Other systems play a key role in preventing elevated levels of ROS, photoinhibition, and photooxidation by detoxifying $O_2^{\bullet-}$ and related compounds such as H_2O_2. In chloroplasts, where catalase is absent, $O_2^{\bullet-}$ is detoxified by Cu-Zn superoxide dismutase (SOD) producing H_2O_2 that is reduced to H_2O by the ascorbate peroxidase−glutathione reductase cycle (Asada, 1999; Fig. 5.6).

Elevated activity of detoxifying enzymes and increased concentrations of their metabolites (glutathione, ascorbate) are indicators of oxidative stress, particularly under high light intensity, for example, in pine needles during winter (Anderson et al., 1990) and in bean leaves under Mg deficiency (Cakmak and Marschner, 1992). There is substantial evidence that ROS are also involved in senescence of cells and organs such as leaves, and that the appearance of chlorosis and necrosis of leaves as visual symptoms of nutrient deficiency is often causally related to elevated ROS concentrations. An example of this in common bean leaves is shown in Fig. 5.7. Under Mg deficiency and high light intensity, oxidative stress is caused by impaired phloem loading of carbohydrates (Cakmak and Kirkby, 2008; Cakmak and Marschner, 1992). Zinc deficiency increases the concentrations of ROS (Cakmak, 2000) because of both depressed SOD activity and diminished export of carbohydrates as a result of low sink activity (Marschner and Cakmak, 1989). However, the production of photooxidants and thus photooxidation of leaf pigments could be decreased by partial shading of the leaf blades (Fig. 5.7). In agreement with this observation, inhibited phloem loading of sucrose in genetically manipulated tobacco (*Nicotiana tabacum*) and tomato (*Solanum lycopersicum*) plants is associated with severe chlorosis and necrosis of the leaf blades (Von Schaewen et al., 1990). However, chlorosis and necrosis in leaves following sugar accumulation may also be caused by the regulation of photosynthetic and senescence-related genes by sugars (Rolland et al., 2006) and redox signals (Mayta et al., 2019; Pfannschmidt et al., 2009).

Over the past decade, it has become increasingly clear that ROS play a much more intricate role in plant physiology than was formerly thought. For example, ROS may induce signals between chloroplasts and nucleus to convey information about the redox status of the electron transport chain. Foyer (2018) suggested there is specificity in the gene-expression profiles triggered by different ROS signals so that the singlet oxygen triggers programs related to PS II, whereas superoxide and hydrogen peroxide promote the expression of genes related to PS I. Foyer (2018) questioned the veracity of the long-held view that light-induced production of ROS results in damage to PS II and photoinhibition.

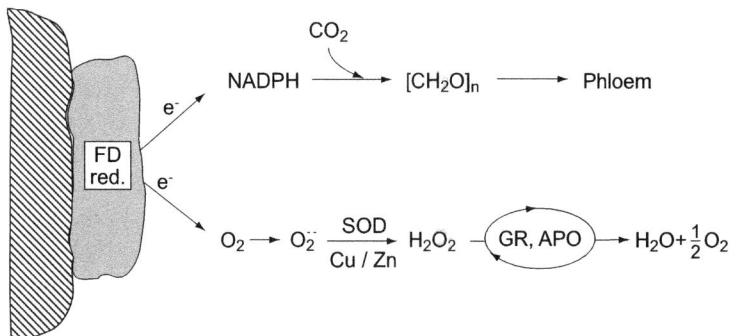

FIGURE 5.6 Alternative utilization of photoreductants for CO_2 assimilation or activation of molecular oxygen and detoxification (scavenger) systems. *APO*, Ascorbate peroxidase; *FD red.*, reduced ferredoxin; *GR*, glutathione reductase; *SOD*, superoxide dismutase.

FIGURE 5.7 Chlorosis and necrosis in partially shaded primary leaves of Mg-deficient (above) and Zn-deficient (below) common bean plants (*Phaseolus vulgaris*) exposed to high light intensity (480 μmol m^{-2} s^{-1}). *Photographs from Cakmak and Kirkby (2008) (top) and Cakmak (2000) (bottom), both with permission from John Wiley and Sons.*

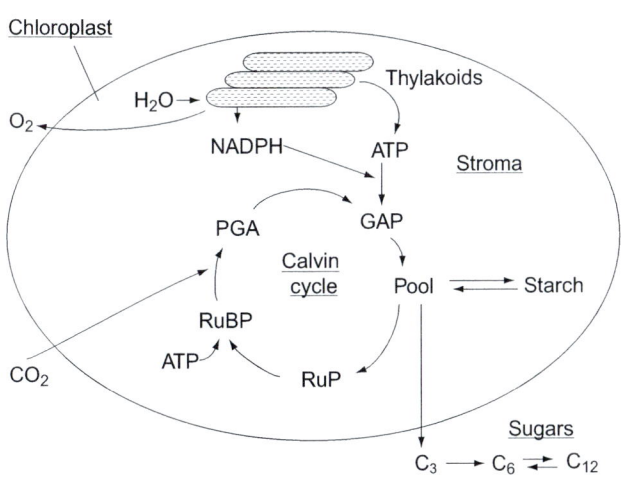

FIGURE 5.8 Simplified scheme of CO$_2$ fixation and carbohydrate synthesis in the Calvin cycle in C$_3$ plants. *Modified from Larcher (2003).*

Mittler (2017) emphasized a role of ROS as signaling molecules in cells. Hence, a basal level of ROS may be necessary for fundamental biological processes including cellular proliferation and differentiation. The literature concerning ROS and photosynthesis was reviewed recently by Khorobrykh et al. (2020).

5.3.3 Carbon dioxide assimilation and photorespiration

The reduction equivalents (NADPH) and ATP produced in the light reactions of photosynthesis are used for CO$_2$ assimilation in the Calvin (also known as the Calvin–Benson–Bassham) cycle (*dark reactions* or *light-independent reactions*). The principles of CO$_2$ fixation in chloroplasts are shown in Fig. 5.8. The enzymes catalyzing the individual steps of the Calvin cycle are located in the stroma of the chloroplasts, whereas NADPH and ATP are supplied by the thylakoids. The first step in the Calvin cycle (carbon fixation phase) is catalyzed by the key enzyme ribulose-1,5-bisphosphate carboxylase/oxygenase (Rubisco), which binds CO_2 to the C_5 compound ribulose-1,5-bisphosphate (RuBP).

$$\begin{array}{c} CH_2-O-\text{\textcircled{P}} \\ | \\ C-OH \\ || \\ C-OH \\ | \\ HC-OH \\ | \\ CH_2-O-\text{\textcircled{P}} \\ \text{Ribulose} \\ \text{bisphosphate (RuBP)} \end{array} \xrightarrow[\boxed{Mg}]{+CO_2\ +H_2O \atop \text{RuBP carboxylase}} \begin{array}{c} CH_2-O-\text{\textcircled{P}} \\ | \\ HC-OH \\ | \\ COOH \\ COOH \\ | \\ HC-OH \\ | \\ CH_2-O-\text{\textcircled{P}} \\ \text{2 x phosphoglycerate} \end{array}$$

An unstable C_6 complex breaks down immediately into two molecules of the C_3 compound phosphoglycerate (PGA). This route of CO_2 assimilation is thus referred to as the C_3 pathway, and those plants for which this is the main form of CO_2 acquisition are known as C_3 species. In the reductive phase of the cycle, in which NADPH and ATP are used, PGA is reduced to glyceraldehyde 3-phosphate and further transformed to triosephosphates. Of the six triosephosphate molecules generated by the carboxylation of three molecules of RuBP, five are used to regenerate three RuBP molecules which can again act as acceptors for CO_2; one is available either for the synthesis of starch in the chloroplast or exported through the chloroplast envelope into the cytosol for further synthesis of mono- and disaccharides. The rate of release of C_3 compounds from the chloroplasts is controlled by the concentration of inorganic phosphate (P_i) in the cytoplasm; therefore, P_i has a strong regulatory effect on the ratio of starch accumulation to sugar released from the chloroplast.

The cycle comprises 11 different enzymes. Molecular approaches to increase the efficiency of photosynthetic CO_2 fixation are focusing on those processes and enzymes that limit photosynthesis. The use of antisense plants with reduced concentrations of Calvin cycle enzymes allows investigation of the contribution of individual enzymes to the control of carbon flux through the cycle ("metabolic control analysis"; see Raines, 2003, 2011). Enzymes exerting considerable control of carbon flux include Rubisco, sedoheptulose-1,7-bisphosphate, plastid aldolase, and transketolase (Raines, 2003, 2011). Rubisco makes up about 50% of leaf soluble protein and 25% of leaf N. The extent to which flux control is exerted by specific enzymes depends greatly on environmental conditions. For example, the impact of decreased Rubisco concentrations on photosynthesis is exacerbated at low N supply or under saturating light conditions (Stitt and Schulze, 1994).

Potential Rubisco activity is determined not only by the concentration of the Rubisco protein but also by activation and inhibition of the enzyme (Fig. 5.9; Parry et al., 2008). The enzyme is activated by carbamylation (i.e., reversible reaction of a molecule of CO_2 with a lysine residue of Rubisco to form a carbamate) and subsequent stabilization by Mg, and is retransformed to its inactivated form by release of CO_2 and Mg. In addition, organic inhibitors can bind to the Rubisco protein and block the active site of the enzyme. Such inhibitors include 2-carboxy-D-arabinitol-1-phosphate formed in the chloroplast during periods of low irradiance or darkness, and pentadiulose-1,5-bisphosphate produced under conditions favoring photorespiration, such as high temperature and drought. Rubisco activity is also regulated by Rubisco activase. This enzyme is required for the removal of inhibitors from the catalytic site of Rubisco (Portis et al., 2008). The activity of Rubisco activase is increased by illumination and is very sensitive to heat stress (Parry et al., 2008). Increasing the efficiency of Rubisco in fixing CO_2 is a challenging target to modulate crop yield because of the complexity of the enzyme (Raines, 2011). However, Lin et al. (2014) have successfully transferred a faster acting Rubisco from cyanobacteria into transplastomic tobacco lines, indicating the potential for increasing photosynthesis in crop plants.

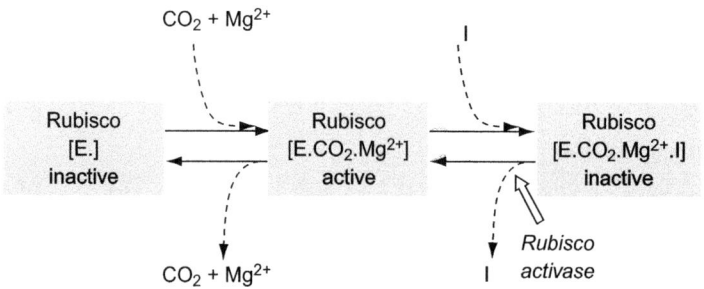

FIGURE 5.9 Principles of regulation of Rubisco activity. The unmodified enzyme [E.] is inactive; reversible reaction with CO_2 and Mg^{2+} leads to the formation of an active state of the enzyme [E.CO_2.Mg^{2+}]. The active state is transformed to an inactive state through binding of inhibitors (I); removal of the inhibitors is mediated by the enzyme Rubisco activase. *Modified from Parry et al. (2008).*

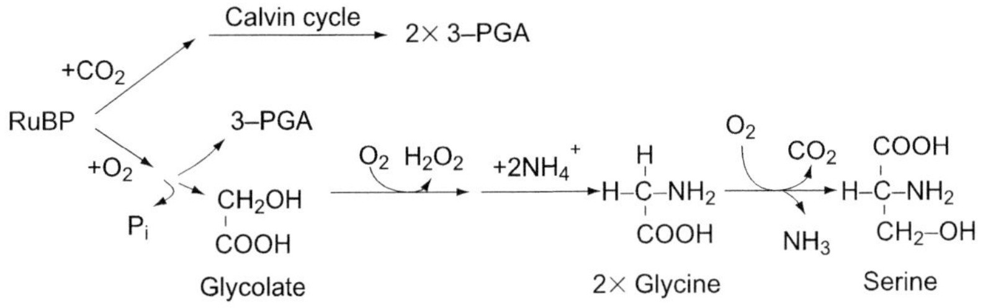

FIGURE 5.10 Photorespiration, glycolate pathway, and synthesis of the amino acids glycine and serine involving chloroplasts, peroxisomes, and mitochondria.

In C_3 species, light increases not only the incorporation of CO_2 but also its evolution, which is stimulated by the presence of O_2. Incorporation and release of CO_2 are dependent on Rubisco. This light-driven evolution of CO_2 (*photorespiration*) has been recognized for over 50 years and occurs simultaneously with the incorporation of CO_2 (Reumann and Weber, 2006). It is a consequence of the oxygenation of RuBP catalyzed by Rubisco. About 25% of the Rubisco activity is directed to the oxygenase reaction (Fig. 5.10), producing the unstable C_5 compound (RuBP) that is cleaved into one molecule of 3-PGA and one molecule of 2-phosphoglycolate (2-PG). The 3-PGA recycles to regenerate RuBP via the Calvin cycle but the 2-PG is immediately removed through the oxidative or photorespiratory C_2 cycle. The 2-PG undergoes a complex and energetically expensive set of biochemical changes that occur across three different interacting organelles: chloroplasts, peroxisomes, and mitochondria. The phosphate group is removed and the glycolate is shuttled from the chloroplasts to the peroxisomes where it is transaminated with glutamate to form the amino acid glycine. Following translocation into the mitochondria, two molecules of glycine are oxidized to produce the amino acid serine, with the simultaneous release of one molecule of CO_2 (photorespiration) and one molecule of NH_3 (Fig. 5.10). For a detailed account of these reactions within various cellular compartments, see Busch (2020). Both ammonia toxicity and losses by volatilization are avoided by the reassimilation of ammonium via the formation of glutamine from glutamate in the chloroplast (Chapter 6). Serine produced in the mitochondria is translocated to the peroxisomes where it is converted to glycerate and transferred into the chloroplasts to be transformed to 3-PGA to enter the Calvin cycle. In total, for every two molecules of glycolate formed during photorespiration, one molecule of 3-PGA is regenerated. Both ATP and NADPH are consumed in the phosphorylation of glycerate and the reassimilation of ammonium in the chloroplast (Busch, 2020; Wingler et al., 2020). The photorespiratory N cycle represents the largest component of ammonium incorporation in leaves of most C_3 plants (Yu and Wo, 1991). Introduction of a glycolate catabolic pathway from *Escherichia coli* into arabidopsis significantly increased biomass formation (Kebeish et al., 2007). This yield increase may be explained by the confinement of the pathway to the chloroplasts, thus avoiding the consumption of ATP and the reducing equivalents for ammonium assimilation, and the release of CO_2 into the stroma in the vicinity of Rubisco.

A key factor regulating the rate of photorespiration is the CO_2/O_2 ratio at the active site of Rubisco. In C_3 species an increase in ambient concentration of CO_2 in the stroma of the chloroplasts decreases the rate of photorespiration (Portis, 1992). On the other hand, high temperatures strongly increase photorespiration of C_3 plants because CO_2 is less soluble and O_2 is more soluble at increased temperatures. Photorespiration is also increased by mild to moderate drought stress that leads to a closure of the stomata and a decrease in internal CO_2 concentration in leaves. Because of carbon and energy losses that restrict photosynthesis and biomass production, decreased photorespiration is regarded as a prime target for crop improvement. Timm and Hagermann (2020) have identified various layers of regulation and suggest glycine decarboxylase as the key enzyme to regulate via genetic engineering to adjust the photorespiratory carbon flow.

Photorespiration takes place in C_4 plants as well (see below), but at lower rates than in C_3 plants. For example, in maize (C_4), under ambient conditions (in $\mu L\ L^{-1}$: 21 O_2 and 0.035 CO_2), photorespiratory CO_2 loss was about 6% of net photosynthesis as compared with 27% in wheat (C_3) (de Veau and Burris, 1989). The lower rates of photorespiration in C_4 plants result from the higher CO_2 concentration in the bundle sheath cells, that is, at the sites of CO_2 assimilation by Rubisco. This higher CO_2 concentration also explains why, under otherwise optimal light and temperature conditions, rates of net photosynthesis and biomass production are considerably higher in C_4 than C_3 plants (Kellogg, 2013). The evolution of C_4 plants from their C_3 ancestors has involved changes in leaf anatomy and physiology accompanied by changes in the expression of thousands of genes favoring photosynthesis at the cost of photorespiration (Bräutigam and Gowik, 2016). These authors see the cost of photorespiration as a major driving force for the evolution of C_4 plants.

Photorespiration is often considered a wasteful process in which CO_2 and ammonium are produced and ATP and reducing equivalents are consumed. However, it may serve as an energy sink, preventing overreduction of the photosynthetic transport chain, especially under stress conditions of drought, salinity, and high light intensity (Wingler et al., 2000). The photorespiratory pathway has many beneficial effects as it is well integrated within other metabolic processes (Busch, 2020). These include nitrogen and sulfur metabolism (Abadie and Tcherkez, 2019) and

CH_2-tetrahydrofolate metabolism (Li et al., 2003). The photorespiratory pathway is the main biosynthetic pathway for the synthesis of the amino acids glycine and serine in leaf cells (Fig. 5.10). In addition to the regeneration of 3-PGA for the photorespiratory cycle, these amino acids can be used in various metabolic processes, including protein synthesis, and the production of other amino acids and phospholipids (Ros et al., 2014) as well as dehydrins (glutathione and glycine betaine) involved in the alleviation of drought stress.

In a diverse range of C_3 species, but not in C_4 species, inhibition of photorespiration (for example, at elevated atmospheric CO_2 concentrations) is associated with a strong depression of nitrate assimilation in shoots, which slows growth (Bloom et al., 2012). During photorespiration, Rubisco, when associated with Mn^{2+}, generates additional reductant that participates in the reduction of nitrate for the synthesis of amino acids and proteins. This phenomenon explains the continued dominance of C_3 plants over the past 23 million years of low CO_2 atmospheres as well as the decline in plant protein concentration linked with the current rise in atmospheric CO_2 (Bloom, 2015). Magnesium and Mn are associated with several key enzymes of the photorespiratory pathway and can influence their function (Bloom and Lancaster, 2018).

5.3.4 C₄ pathway of photosynthesis and Crassulacean acid metabolism

The incorporation of CO_2 into organic compounds is not restricted to the Calvin cycle, that is, the C_3 pathway discussed above. As described earlier (Chapter 2), an imbalance of cation and anion uptake by roots in favor of cations must be compensated for by the incorporation of CO_2 via phosphoenol pyruvate carboxylase (PEPCase) and the formation of organic acids. In principle, the same pathway of CO_2 incorporation occurs in the chloroplasts of C_4 plants.

PEP acts as CO_2 acceptor to form oxaloacetate that is reduced to malate. This fixation of CO_2 in the chloroplasts of mesophyll cells is dependent on the Zn-containing cytosolic enzyme carbonic anhydrase that hydrates CO_2 to produce hydrogen carbonate (Badger and Price, 1994) prior to assimilation in the chloroplasts by PEPCase. The products of this CO_2 incorporation are C_4 compounds, either malate or aspartate. These C_4 compounds are transported from the mesophyll cells via plasmodesmata to the bundle sheath cells where they are decarboxylated, and where Rubisco is located (Fig. 5.11). The suberin layer surrounding bundle sheath cells forms a gas-tight barrier that allows an elevated CO_2 partial pressure around Rubisco, allowing it to function close to its maximum carboxylase activity while simultaneously inhibiting its oxygenase activity. The CO_2 fixed by Rubisco is channeled into the Calvin cycle in the chloroplasts of the bundle sheath cells. The C_4 pathway, in forcing CO_2 into the standard C_3 photosynthetic apparatus, has been likened to the turbocharger of an engine (Kellogg, 2013). Its importance becomes obvious at a global scale where only about 3% of angiosperms use the C_4 pathway, but these few species account for 23% of primary productivity. The remaining C_3 acids in the bundle sheath cells are translocated back to the mesophyll cells where PEP is regenerated via PEPCase to

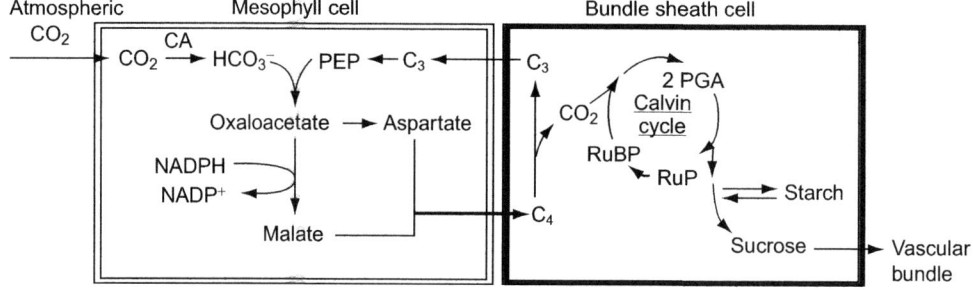

FIGURE 5.11 Simplified scheme of CO_2 fixation and compartmentation in C_4 plants. The thick line surrounding the bundle sheath cell indicates the suberin layer. *CA*, Carbonic anhydrase.

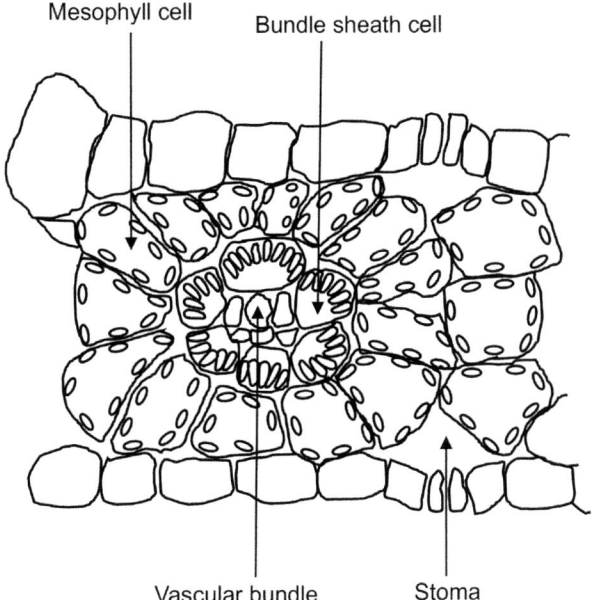

FIGURE 5.12 Diagrammatic representation of a transverse section of a leaf with C_4 Kranz anatomy.

act again as a CO_2 acceptor. The C_4 pathway was first reported in the seminal paper of Hatch and Slack (1966). In most C_4 plants the two forms of CO_2 fixation are usually spatially separated between the mesophyll and bundle sheath cells (Fig. 5.11). However, in some C_4 species, including members of the Chenopodiaceae, the CO_2 concentrating mechanism and the C_4 pathway are separated into different compartments within the photosynthetic cells (King et al., 2012; Voznesenskaya et al., 2001).

In most C_4 species, the two cell types are arranged in a Kranz-type anatomy. The minor veins of the vascular bundles are surrounded by bundle sheath cells forming a Kranz, or wreath, and they are surrounded by a concentric layer of large mesophyll cells (Fig. 5.12; Hibberd et al., 2008). In addition, in C_4 species, the chloroplasts are dimorphic, with those in the bundle sheath cells being larger and having grana that are not as well developed as those of the mesophyll cells. On the other hand, the starch-synthesizing enzymes are confined to the bundle sheath chloroplasts (Spilatro and Preiss, 1987). Both cell types have anatomical features that favor the C_4 pathway, including a high frequency of plasmodesmata linking the mesophyll cell and bundle sheath cell cytosol, and suberin deposition in the cell walls of bundle sheath cells to restrict CO_2 leakage (Kellogg, 2013). The differentiation of chloroplasts between mesophyll and bundle sheath cells to accommodate the C_4 pathway not only influences processes associated with CO_2 assimilation and C and N metabolism but is also important for fatty acid synthesis and isoprenoid and sulfur metabolism (Friso et al., 2010).

Many C_4 species are of tropical and subtropical origin, have high photosynthetic rates, and produce large amounts of dry matter. These species include some of the world's most productive cereal and forage crops. The C_4 species are generally categorized into subtypes based on the enzymes used to release CO_2 into the bundle sheath cells. These enzymes are NADP-malic enzyme (NADP-ME), NAD-malic enzyme (NAD-ME), and PEP carboxykinase (PEPCK) (Buchanan et al., 2000). No pure PEPCK subtype has yet been discovered and it is now considered that NADP-ME and NAD-ME are distinct C_4 biochemical pathways with or without participation of PEPCK. These two enzymes are associated with particular species depending on evolutionary development (Rao and Dixon, 2016). The NADP-ME group includes maize (*Z. mays*), sugarcane (*Saccharum* sp.), sorghum (*Sorghum bicolor*), and miscanthus (*Miscanthus x giganteus*), whereas the NAD-ME subtype includes pearl millet (*Pennisetum glaucum*), amaranths (*Amaranthus* spp.), and switchgrass (*Panicum virgatum*). All these crops are highly efficient in using light, CO_2, water, and mineral nutrients. Radiation-use efficiency in C_4 plants is about 50% greater than in C_3 plants (Long et al., 2006). Large differences, however, occur in the resource use, for example, between C_4 crops miscanthus and switchgrass. Both are bioenergy crops because of their low input requirements, but miscanthus is more efficient than switchgrass (Dohleman et al., 2009; Heaton et al., 2008). The NAD-ME enzyme in bundle sheath cells has a requirement for Mn for activation. To produce maximum biomass the NAD-ME C_4 plants pearl millet and purple amaranth (*Amaranthus blitum*) require an approximately 20-fold greater Mn supply than some other C_3 and C_4 crop plants at similar photosynthetic rates (Kering et al., 2009).

The greater water-use efficiency (WUE, the ratio between the amount of water transpired and dry matter accumulated) of C_4 species can be explained by their lower internal CO_2 partial pressure in leaves and the correspondingly

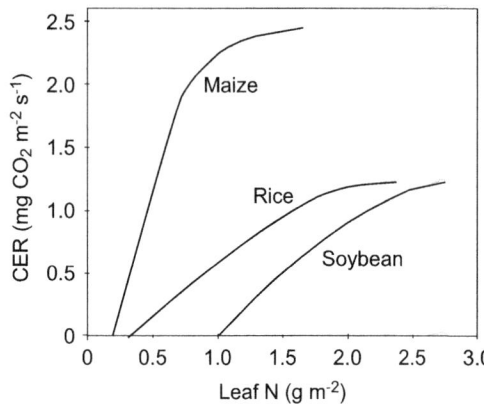

FIGURE 5.13 Leaf CO_2 exchange rate (CER) at light saturation for maize (*Zea mays*), rice (*Oryza sativa*), and soybean (*Glycine max*) as a function of leaf N concentration. *From Sinclair and Horie (1989).*

steeper CO_2 gradient from the ambient atmosphere through the open stomata into the leaf tissue. Lower internal CO_2 concentrations are achieved by efficient conversion of CO_2 to HCO_3^- in the cytosol by carbonic anhydrase and the high affinity of PEPCase for HCO_3^- in the cytosol of mesophyll cells (Badger and Price, 1994). Thus, in C_4 plants, there is a relatively greater inward diffusion of CO_2 through the stomata per unit of water vapor lost, which can be utilized for photosynthesis and dry matter production. In addition, when stomata partially close in response to water deficit, the decrease in CO_2 influx is less in C_4 than in C_3 plants, because the internal recycling of CO_2 maintains a lower CO_2 concentration in the leaf tissue of C_4 plants. Correspondingly, the relative efficiency of water use is around 200–300 g water transpired per g dry matter produced in C_4 species as compared to usually more than 500 g water g^{-1} in C_3 species.

In general, photosynthetic N-use efficiency (PNUE, the CO_2 assimilation rate per unit leaf organic N concentration) is greater in C_4 than C_3 species (Poorter and Evans, 1998). An example of this is shown in Fig. 5.13; maize not only has higher rates of CO_2 fixation than the two C_3 crop species, but these higher rates are achieved at a lower leaf N concentration. In C_3 plants, Rubisco has a slow catalytic rate, a low affinity for atmospheric CO_2 and uses O_2 as an alternative substrate, resulting in photorespiration (Spreitzer and Salvucci, 2002). As a result, C_3 plants require high concentrations of Rubisco to maintain high rates of photosynthesis (Makino et al., 2003). The higher PNUE in C_4 plants is possible because Rubisco in their bundle sheath cells operates at CO_2 saturation with minimal photorespiration. Thus, in C_4 species, a high assimilation rate can be maintained with only one-third to one-quarter of the Rubisco required in C_3 species. In comparison with this saving, the N cost of the C_4 cycle enzymes is low (Friso et al., 2010; Makino et al., 2003). Furthermore, in many C_4 species the catalytic efficiency of Rubisco (mol CO_2 mol^{-1} Rubisco active sites s^{-1}) is higher than in C_3 species (Sage, 2002). PNUE is greater in the NADP-ME subtype than in the NAD-ME subtype C_4 species (Ghannoum et al., 2005) and can vary among cultivars of same species, as observed in maize (Paponov and Engels, 2003).

Phosphorus-use efficiency (PUE) is less well understood in C_4 than C_3 species. The enzymes of the additional steps and membrane transport processes that characterize the C_4 pathway are regulated by phosphate (P_i). These include P_i activation of the enzymes PEPCase and pyruvate:orthophosphate dikinase, as well as PEP/P_i transport in the mesophyll chloroplast envelope (Iglesias et al., 1993). Measurements of CO_2 assimilation rates in leaves of C_3 and C_4 tropical grasses show greater PUE in the C_4 species (higher CO_2 assimilation rates at low leaf P_i concentrations), but the C_4 grasses are more sensitive to P deficiency than C_3 grasses (Ghannoum et al., 2008).

It should be emphasized that the greater nutrient-use efficiency of C_4 species (e.g., maize and sugarcane) in general does not imply a lower fertilizer demand. Indeed, because of their potential to produce high dry matter yields, C_4 crops often have larger fertilizer requirements than C_3 crops. Accordingly, in grasslands, C_4 species may respond more strongly to N and P fertilization than C_3 species (Rubio et al., 2010).

The rise in atmospheric CO_2 concentration stimulates C_3 photosynthesis more than C_4 photosynthesis because in C_4 plants Rubisco is already saturated with CO_2 in the bundle sheath cells. On the other hand, a warmer climate favors C_4 photosynthesis because the greater photorespiration in C_3 plants is expected to constrain net CO_2 fixation at higher temperatures (Sage and Kubien, 2003).

C_4 photosynthesis is attracting great attention in agriculture because of its high energy conversion efficiency and greater PNUE and WUE. There is an opportunity to exploit these attributes both by genetic engineering as well as by conventional plant breeding to create varieties with high efficiencies (Sage and Zhu, 2011). The improvement in photosynthetic efficiency is of worldwide importance to meet the predicted food shortages of the mid-century (Long et al., 2015). Genetic engineering to incorporate C_4 photosynthetic machinery into C_3 crops to increase yields still appears to be a long-term strategy dependent on a more complete understanding of the genomic regulation involved in the Kranz anatomy. Kellogg (2013) suggests that success will probably be achieved by a combination of genetic, genomic, and comparative evolutionary studies (see also Furbank, 2016; Von Caemmerer et al., 2017).

Fixation of CO_2 via the PEPCase pathway is a characteristic feature of plant species in certain families, such as Crassulaceae and Bromeliaceae, that are particularly well adapted to dry and saline habitats and have high WUE. These plants are mostly succulent (i.e., they have a low surface area per unit of fresh weight). The *Crassulacean acid metabolism* (CAM) pathway of CO_2 fixation was first identified by Thomas and his colleagues at Newcastle on Tyne in the United Kingdom more than 60 years ago (Ranson and Thomas, 1960), although at that time, its relevance to photosynthesis was not appreciated. CAM plants differ from C_4 species in a number of features: (1) the stomata of CAM species open at night, (2) CO_2 enters the leaves and is fixed by PEPCase in the cytosol with subsequent reduction to malic acid that is stored in the vacuoles during the night, and (3) during the day malic acid is released from the vacuoles and decarboxylated. The CO_2 released from malate is present at high internal concentrations behind closed stomata and is refixed by Rubisco in the chloroplasts following the C_3 pathway. Accordingly, large fluctuations occur in the vacuolar pH in the leaves of CAM plants, with the lowest values at the onset of the day (Winter and Smith, 2022), and both proton pumps (ATPase and PP_iase) and malate channels are involved in controlling the transport of malate into and out of the vacuole (Lim et al., 2019; Winter and Smith, 2022).

In contrast to the spatial separation of the two steps of CO_2 fixation in C_4 species, the separation of the three steps of CO_2 fixation in CAM species is temporal (*diurnal acid rhythm*). The combination of CAM and succulence is particularly advantageous for adaptation to dry habitats, high salinity, or both. Under these conditions the diurnal patterns of gas exchange and leaf acidification contribute to the capacity of CAM species to maximize biomass, CO_2 uptake, and WUE. On average, CAM species have three- and sixfold greater WUE than C_4 and C_3 species, respectively (Garcia et al., 2014). CAM is expressed in about 6% of vascular species. The most well-known cultivated CAM plants are pineapple (*Ananas comosus*), prickly pear (*Opuntia ficus-indica*), sisal (*Agave sisalana*), and tequila agave (*Agave tequilana*), which are harvested for food, fodder, fiber, and ethanol, respectively. These crops reach near maximum productivity under water-limiting conditions (Borland et al., 2009).

Depending on plant species, developmental stage and environmental conditions, CAM may operate in different modes: (1) obligate CAM; (2) facultative or inducible CAM (C_3-CAM), in which CAM metabolism is induced by factors such as drought, salinity, high photon flux, N and P deficiency; and (3) CAM-cycling, with daytime CO_2 fixation and acid accumulation but closed stomata at night (Cushman, 2001; Herrera, 2009; Winter et al., 2015). In facultative CAM and CAM-cycling species, dark CO_2 fixation not only contributes to water saving but also plays an important role in photo-protection (Herrera, 2009). As global temperatures continue to rise and water-limiting environments become more prevalent, there is much interest in the genetic engineering of drought-resistant crop species by introducing CAM into C_3 plants. This long-term challenge necessitates a comprehensive understanding of the genetic components and regulatory control of CAM, which will require multidisciplinary approaches combining insights from ecophysiology, evolutionary biology, and functional genomics (Borland et al., 2014; Garcia et al., 2014; Winter and Smith, 2022).

5.3.5 Effect of leaf maturation on its sink–source transition

During its life cycle, each leaf changes from a sink to a source for both nutrients and photosynthates. For nutrients, this shift is correlated with a change in the prevailing long-distance transport in the phloem and xylem (Chapter 3). The long-distance transport of sugars such as sucrose is restricted to the phloem, so that the sink–source transition of leaves is associated with a corresponding shift from phloem unloading (import) to phloem loading (export). In leaves of sugar beet, the transition from net import to net export of sugars occurs when the leaf has reached about 40%–50% of its final area and net photosynthetic capacity (Fig. 5.14). Similarly, this transition occurs in other dicotyledonous species when their leaves are 30%–60% expanded.

The sink–source transition of leaves is associated with biochemical, physiological, and anatomical changes. In maturing sugar beet leaves a shift occurs in the incorporation of carbon into sugars, which can be demonstrated by supplying $^{14}CO_2$ to leaves of different ages (Fig. 5.14). This shift in favor of sucrose synthesis is closely correlated with changes in the activity of enzymes associated with carbohydrate metabolism in the leaves: a decrease in activities of acid invertase and sucrose synthase (responsible for the hydrolysis of sucrose) and an increase in sucrose-P-synthase activity (catalyzing sucrose synthesis) (Li et al., 2008; Schurr et al., 2000). The high activity of invertase at the early stages of leaf development reflects the provision of hexoses (e.g., glucose) required for cell wall synthesis rather than any regulatory functions of this enzyme in phloem unloading of sucrose (Haupt et al., 2001). In sink leaves, the activity

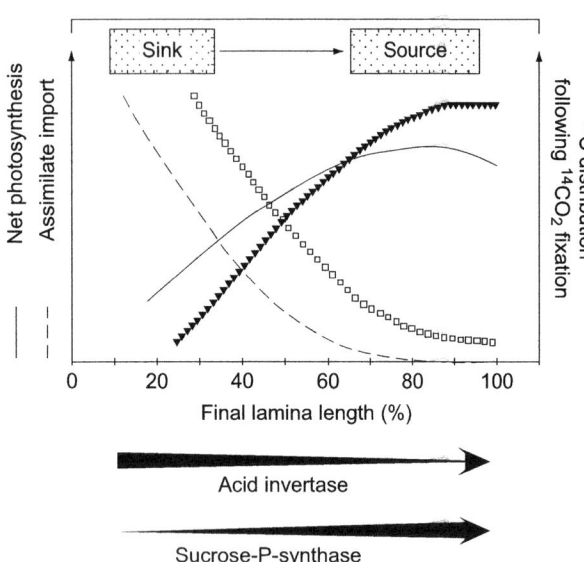

FIGURE 5.14 Relationship between assimilate import, net photosynthesis, rate of sugar synthesis (▼: sucrose; □: glucose), and enzyme activity during maturation of sugar beet (*Beta vulgaris*) leaves. *Based on Giaquinta (1978).*

of the cytosolic enzyme sucrose synthase is also high but declines rapidly during the sink–source transition (Turgeon, 2006). The correlation between a decrease in sucrose synthase and an increase in sucrose-P-synthase is correlated with the transition from sink to source (Fig. 5.14) in plants where sucrose is the dominating sugar in the phloem sap. Results similar to those obtained with sugar beet leaves have also been reported for soybean (Silvius et al., 1978), castor bean (*Ricinus communis*; Schurr et al., 2000), and rice (Li et al., 2008) leaves during maturation. The sink–source transition of leaves is also associated with changes in the frequency and architecture of plasmodesmata in the mesophyll and epidermal cells, influencing symplasmic continuity (Turgeon, 2006).

Much remains to be learned about the mechanisms by which the import and export of nutrients are regulated during leaf maturation. Based on the mechanism of phloem transport (solute volume flow) and the average composition of phloem sap in the stem of plants during vegetative growth (Chapter 3), there should be a positive correlation between the import rate of sugars such as sucrose into a sink leaf and the import rate of nutrients such as K and P, and also amino acids, if phloem unloading of these solutes is regulated by the requirement of growth processes. Phloem transport from source to sink has been investigated using a nonproteinogenic amino acid (α-aminoisobutyric acid) supplied to a source leaf or to the stem (Van Bel, 1984). The finding that this amino acid accumulates in the sink in the soluble N fraction indicates that movement is controlled by the direction and rate of solute volume flow in the phloem rather than by sink demand *per se*. A similar technique using the phloem-mobile fluorophore (5,6-carboxyl fluoresceine) was used to quantify the sink-to-source transition in young pea (*Pisum sativum*) leaves (Ade-Ademilua and Botha, 2007). More recently, Ely et al. (2019) have employed spectroscopic techniques to predict key leaf traits associated with source–sink balance of C and N compounds.

With the onset of leaf maturation and the capacity for synthesis of sucrose and other exported sugars (e.g., mannose), the leaf becomes a new source as loading of sugars into the phloem begins and an increase in the volume flow rate in the phloem from the leaf is induced. This is a highly regulated process in which typically 50%–80% of photoassimilates of a single mature leaf are transported into the phloem (Ainsworth and Bush, 2011). The phloem export of other solutes such as nutrients and amino compounds can also increase. As discussed previously (Chapter 3) for highly phloem-mobile nutrients such as K and P, import via the xylem and export via the phloem can be in equilibrium in mature leaves. The degree to which mature leaves also act as sources of nutrients depends, however, not only on the rate of photosynthate export but also on the nutrient content of the source leaf.

5.3.6 Leaf senescence

Leaf senescence is an important developmental phase critical to plant fitness. Senescence restricts the life span and photosynthetic activity of the leaf and also allows an ordered disassembly of cellular structures and biomolecules and the recycling of C and nutrients within the plant (Mayta et al., 2019; Woo et al., 2019). Leaf senescence involves coordinated action at cellular, tissue, organ, and organism levels under the control of a highly regulated genetic program (Zhang et al., 2021b). The expression of many genes is downregulated, for example, genes encoding

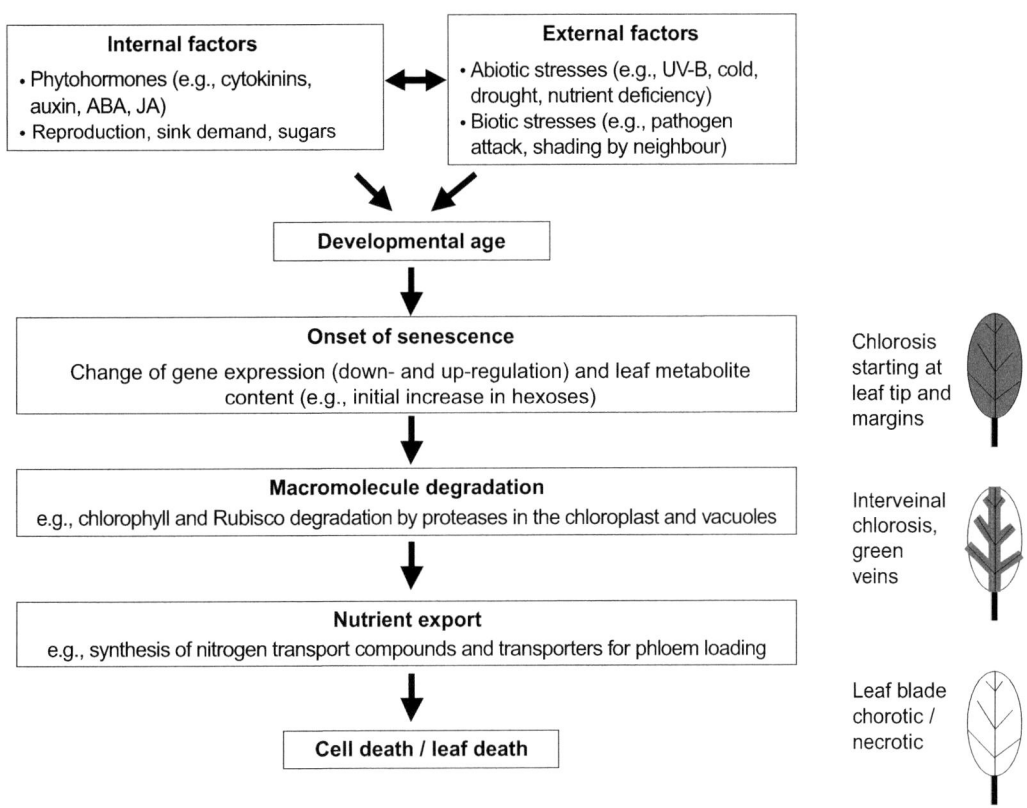

FIGURE 5.15 Overview of the factors involved in the onset of leaf senescence and processes related to senescence-induced nutrient export.

photosynthesis-related proteins. However, the expression of many other genes is upregulated. These include genes for degrading enzymes like proteases and lipases and genes encoding the nutrient transporters needed for remobilization in the phloem (Fig. 5.15). The onset of leaf senescence is not only governed by developmental age but also influenced by various internal and external factors. Internal factors regulating leaf senescence include sugars (Wingler and Roitsch, 2008), phytohormones, and ROS. Phytohormones such as gibberellic acid (GA), auxins, and cytokinins (CYTs) have been shown to delay senescence, whereas ethylene, jasmonic acid (JA), abscisic acid (ABA), and salicylic acid (SA) accelerate it. Phytohormones, ROS, and environmental inputs do not operate through independent pathways but display a significant degree of interaction (Merewitz et al., 2011). Mayta et al. (2019) suggest that chloroplasts could cooperate with other redox sources (e.g., mitochondria) and signaling molecules including phytohormones to induce leaf senescence and regulate nutrient recycling. Both abiotic and biotic external factors may also induce leaf senescence. Abiotic factors include drought, nutrient limitation, extreme temperatures, oxidative stress by UV-B irradiation, and ozone, whereas biotic factors include pathogen infection and shading by neighboring plants. In many cases, environmental cues induce a change in phytohormone synthesis and translocation to the leaves.

Leaf senescence is strongly influenced by nutrition. Limited N availability and a high tissue C/N ratio are important regulators of leaf senescence (Parrott et al., 2010). For example, after 48 h of N starvation, the chlorophyll, protein, and starch concentrations in older leaves of wheat plants were lower than in well-supplied plants (Table 5.1). This decrease was associated with a strong reduction in the concentration of isopentenyl adenosine (an active form of CYT) in leaves. Supplying the CYT 6-benzylaminopurine via the roots increased leaf concentrations of isopentenyl adenosine significantly and also raised chlorophyll, protein, and starch concentrations. This finding suggests that N deficiency-induced acceleration of leaf senescence could be due to reduced supply of root-derived CYT to the leaves. In field-grown cotton, K deficiency also accelerated leaf senescence as indicated by early chlorophyll degradation (Hu et al., 2016). Hosseini et al. (2016) reported that increasing flag leaf K concentration might confer tolerance to drought-induced leaf senescence in barley (*Hordeum vulgare*) by promoting ABA degradation and attenuating starch degradation. The authors suggest that genotypic differences in flag leaf K concentrations may even serve as markers of drought−tolerant lines in plant breeding. In barley, K deficiency is associated with increased expression of genes encoding enzymes involved in JA biosynthesis and increased leaf JA concentrations (Davis et al., 2018).

TABLE 5.1 Leaf composition in wheat (*Triticum aestivum*) with and without N and 6-benzylaminopurine (BAP) addition; +N: continuous N supply; −N: interruption of N supply for 48 h.

	+N	−N	+N/+BAP	−N/+BAP
Protein (mg g^{-1} FW)	18	17	21	19
Chlorophyll (mg g^{-1} FW)	1.5	1.3	1.7	1.6
Starch (μmol glucose eq. g^{-1} FW)	31	23	41	34
Isopentenyl adenosine (pmol g^{-1} FW)	16	1.5	322	200

Source: Based on Criado et al. (2009).

Leaf senescence, in the form of chlorosis of source leaves, can be induced readily by high light combined with Zn, Mg, or K deficiency (Cakmak, 2000). In these conditions, senescence is not induced by a decrease in CYT import but rather by inhibited export of photosynthates and accumulation of large amounts of nonstructural carbohydrates in the source leaves. This type of premature leaf senescence is associated with ROS accumulation and photooxidation of chloroplast pigments. Increased concentrations of ROS, which are the result of both elevated formation of radicals and reduced activity of antioxidative enzymes, play an important role during developmental and stress-induced senescence (Zimmermann and Zentgraf, 2005).

The effect of Mg deficiency in inducing leaf senescence in rice has been reported by Tanoi and Kobayashi (2015). Their findings, supported by observations from the literature, indicate that a lack of Mg induces the following sequence of events: phloem transport dysfunction, accumulation of sugars and starch in leaves, decline in photosynthesis, generation of ROS, oxidative stress, degradation of chlorophyll, chlorosis and, finally, necrosis. Physiological changes in plants occurring in parallel with this sequence were: drought stress, reduced transpiration, inhibited growth and wilting. The sensing of low Mg status appears to be linked to ABA signaling: Many early transcriptomic changes induced by Mg deficiency in arabidopsis are associated with ABA signaling, with 50% of the genes upregulated upon Mg deficiency in leaves being ABA responsive. Remarkably, in contrast to other mineral nutrient deficiencies, lack of Mg does not induce greater expression of genes associated with Mg uptake (Hermans et al., 2010). Further examples of the effect of nutrient supply on leaf senescence are given in Section 5.4.3.

At the whole leaf level, natural senescence usually begins in the tips or margins followed by the base of a leaf (Fig. 5.15). Cell death, which is the final stage of senescence, starts in the mesophyll cells and then proceeds to other cell types. Cells close to the veins are often the last to die, presumably because they are needed for nutrient export (Quirino et al., 2000). At the cellular level, the earliest structural changes occur in chloroplasts, whereas the nucleus and mitochondria, which are essential for gene expression and energy production, respectively, remain intact until the last stages of senescence, when visible disintegration of the plasma and vacuolar membranes appears (Mayta et al., 2019).

The proteolysis of chloroplast proteins and related transamination reactions, including deamination activity of glutamate dehydrogenase, lead to the release of ammonium that is re-assimilated by a cytosolic form of glutamine synthetase (GS1) to form glutamine for export to sink organs via the phloem (Fig. 5.15). Genes encoding a specific form of GS1, aminotransferases, and asparagine synthetase are upregulated during senescence, suggesting a role of these enzymes in N recycling during leaf senescence (Masclaux-Daubresse et al., 2010). Furthermore, the same authors observed that amino acid and peptide transporters are also induced during senescence, and the export of amino acids and other phloem-mobile nutrients from leaf blades is increased (Chapter 3).

Other nutrients are also remobilized during leaf senescence in arabidopsis (Himelblau and Amasino, 2001) and other plant species. In senescing tomato, remobilization of RNA-bound P is associated with the induction of specific ribonucleases (Lers et al., 2006). In arabidopsis, specific members of the Pht1 family of P transporters (Pht1;5) localized in the leaf phloem are induced during senescence, suggesting a role for these transporters in redistributing P from older leaves (Mudge et al., 2002). Expression of genes encoding yellow stripe-like (YSL) transporters also increases during senescence. Because these transporters play an important role in the mobilization of Cu and Zn as metal-nicotianamine complexes, they may be involved in the remobilization of these nutrients from senescing leaves (Curie et al., 2009).

In agriculture and horticulture, leaf senescence may restrict the yield in crop plants by limiting growth duration and may also cause postharvest spoilage such as leaf yellowing and nutrient loss in vegetable crops. In various plant species, including many crops, "stay green" mutants (with delayed senescence) have been identified (Thomas and Howarth, 2000). Possible advantages of "stay green" genotypes include increased biomass production and yield (Spano et al., 2003), increased N acquisition (Martin et al., 2005), and water uptake from deep soil layers (Christopher et al., 2008), as well as increased tolerance to extreme drought (Rivero et al., 2007), and extended shelf-life of vegetables (Barry, 2009). In grain crops such as maize or wheat, however, the "stay green" phenotype may be associated with lower N-use efficiency and N harvest index, if the delay in leaf senescence is not associated with a faster rate of senescence and N remobilization before maturity (Thomas and Howarth, 2000).

The quality of plant products may be improved by accelerated senescence. In durum wheat (*Triticum durum*), a gene that encodes a member of the NAC transcription factor family, belonging to the NAM (No Apical Meristem) subgroup (*NAM-B1*) is involved in the regulation of leaf senescence (Uauy et al., 2006). The gene confers accelerated flag leaf senescence, which is associated with higher grain contents of protein and micronutrients such as Fe and Zn (Diestelfeld et al., 2007).

5.3.7 Feedback regulation of photosynthesis by sink demand for carbohydrates

Photosynthesis in source leaves responds to the demand for carbohydrates in sink organs (Paul and Foyer, 2001; Smith et al., 2018). The rate of photosynthesis of a specific leaf is increased when the photosynthetic capacity of other leaves is reduced, for example, by abscission (Eyles et al., 2013), herbivory (Nabity et al., 2009), or shading (McCormick et al., 2008). Increased carbohydrate drain to rhizobia and mycorrhiza can also increase photosynthesis (Kaschuk et al., 2009). Glanz-Idan et al. (2020) reported a significant and rapid elevation of the photosynthetic rate of a source leaf of tomato when the source-to-sink ratio was decreased by shading or partial defoliation. The leaf turned deep green, its area increased almost threefold within a week, and senescence was inhibited. Leaf carbohydrate concentration decreased together with the expression of genes involved in sucrose export. The findings of Glanz-Idan et al. (2020) provide evidence that on defoliation, a CYT-like substance (trans-zeatin riboside) acts as a signal that moves from the root via the xylem to support the growth of the remaining leaf in which gene expression is altered to elevate photosynthetic activity.

By contrast, reduction of the demand for carbohydrates, for example, by removal of sink organs (Iglesias et al., 2002) can reduce photosynthesis (so-called *feedback inhibition of photosynthesis*). Reduction of photosynthesis induced by low sink demand is often associated with increased leaf carbohydrate concentration. Various mechanisms have been suggested to explain feedback inhibition of photosynthesis by high concentrations of sugars and starch in the leaves (Stitt, 1991): (1) chloroplast damage, (2) negative effects on CO_2 diffusion by excessive starch accumulation, (3) limitation of photosynthesis by P deficiency within the chloroplasts induced by accumulation of sugar phosphates in the cytosol, and (4) sugar-induced repression of photosynthetic genes. High leaf carbohydrate concentrations, particularly hexose, inhibit transcription of genes coding for enzymes involved in photosynthesis (Rolland et al., 2006). Sugar-induced repression of photosynthetic genes is also involved in the reduction of photosynthesis associated with leaf senescence (Rolland et al., 2006).

Sugar-mediated regulation of photosynthetic genes has been found to be dependent on plant nutrition. In tobacco seedlings, chlorophyll content and Rubisco activity were decreased by adding sugar to N-deficient plants, whereas in plants that were either P-deficient or well supplied with N and P, sugar addition had no effect (Nielsen et al., 1998). In contrast to N deficiency, P deficiency is not associated with rapid reduction of chlorophyll content and radiation-use efficiency (Fletcher et al., 2008; Plénet et al., 2000). The rapid loss of chlorophyll and leaf photosynthetic activity in N-deficient plants may also be related to direct effects of N on gene expression. Nitrate deficiency represses the expression of genes involved in photosynthesis, including chlorophyll synthesis, and induces the expression of genes involved in protein degradation and senescence (Peng et al., 2007).

The rate of net photosynthesis often increases after fruit or seed set due to increased sink demand. However, high demand of sink organs for carbohydrates is not necessarily associated with high net photosynthesis and biomass production at the whole plant level because increased biomass allocation to storage sink organs can be at the expense of biomass allocation for the construction of new leaves.

The effect of sink demand on photosynthesis is complicated in legumes fixing atmospheric N (N_2) in root nodules, which represents an additional sink for carbohydrates supplied from the leaves. As shown in Table 5.2, the removal of source leaves leads to a decrease in both nodule growth and N_2 fixation, whereas the removal of flowers and pods (competing sinks) results in an increase in both nodule weight and N_2 fixation to values that are higher than those of untreated control plants. This shows that in legumes the high sink demand of generative organs can decrease the N supply to leaves from the rhizobial symbionts and, thus, can decrease leaf photosynthesis and accelerate leaf senescence.

TABLE 5.2 Nodule weight and N content of pea (*Pisum sativum*) plants after 60 days of growth as influenced by defoliation or removal of flowers and developing pods.

Treatment	Dry weight of root nodules (mg plant^{-1})	Nitrogen (mg plant^{-1})
Control	298	475
Defoliation	176	266
Removal of flowers and pods	430	548

Source: Based on Bethlenfalvay et al. (1978).

The close interaction between shoot and nodule activity was also demonstrated by Kaur et al. (2019) in alfalfa (*Medicago sativa*). The key enzyme involved in the synthesis of sucrose in leaves is sucrose phosphate synthase (SPS), and GS is the key enzyme for the assimilation of ammonium in root nodules. Transgenic plants overexpressing *SPS* or *GS* (or both) showed increased growth and enhanced nodule function. The endogenous genes for the two key enzymes showed increased expression in the leaves and nodules of transformed plants, suggesting a common signaling pathway regulating C/N metabolism in leaves and nodules.

5.3.8 Nutrition and photosynthesis

The rate of net photosynthesis may be influenced by nutrition through various modes of action (Fig. 5.16). The direct involvement of some nutrients in the light and dark reaction of photosynthesis has been discussed in Sections 5.3.1 and 5.3.3. An example of the direct involvement of nutrients in the light reaction is the light-induced efflux of Mg and K from the lumen to the stroma of chloroplasts to maintain charge balance for light-induced influx and generation of protons in the lumen (Fig. 5.4). In the dark reaction, light-induced influx of K into the guard cells leads to the opening of stomata and, thereby, CO_2 flux into the leaf, which is required for CO_2 assimilation. An important example of the direct involvement of nutrients in the dark reaction was the discovery by Heldt et al. (1977) of the Pi-dependent control of triosephosphate transport across the chloroplast envelope into the cytosol.

Over the past decade it has become increasingly clear that Ca plays a significant role in numerous aspects of photosynthesis. It forms part of the photosynthetic apparatus in the Mn_4Ca cluster of the photosynthetic oxygen-evolving center (Barber, 2016). It is also intricately involved in both light and light-independent reactions of photosynthesis (Hochmal et al., 2015). Calcium binding to calmodulin (CaM) in the CaM-binding protein NADK2 regulates the *de novo* synthesis of $NADP^+$, the terminal electron acceptor of photosynthesis. Calcium also plays a pivotal role in cyclic and linear electron flow. Activation of light-induced CO_2 fixation requires an extremely low stromal Ca^{2+} concentration. In the light-independent reactions Ca activates a number of the key enzymes of the C_3 cycle. Perturbations of cytosolic Ca^{2+} concentration also play a central role in intercellular signaling, acting as a *second messenger* linking many environmental and developmental stimuli to an appropriate response (White, 2015; Chapter 6).

Nutrients are also required for biosynthesis of the photosynthetic apparatus, either as cofactors of enzymes involved in biosynthetic pathways (e.g., Fe for chlorophyll synthesis) or as integral components of the photosynthetic apparatus (Fig. 5.4). Magnesium is critical in the activation and chelation of the catalytic cycle of Mg chelatase that brings about the insertion of the Mg^{2+} ion into protoporphyrin 1X in chlorophyll (Grzebisz et al., 2015; Masuda, 2008). Magnesium is also directly involved in the activation and activity of Rubisco. Furthermore, Mg is crucial for protein (and thus Rubisco) synthesis. Lower protein content in Mg-deficient plants is likely related to the involvement of Mg in ribosomal subunit association and activity (Tränkner et al., 2018).

Deficiency of nutrients that are involved in synthesis of protein or chloroplast pigments or electron transfer results in the formation of chloroplasts with altered structure and low photosynthetic efficiency (Chen et al., 2008). In spinach leaves about 24% of the total N is in the thylakoid membranes; N nutrition therefore also affects the amount of thylakoids per unit leaf area. In Mn-deficient leaves, the photosynthetic efficiency per unit chlorophyll is decreased and can be restored within 2 days after foliar application of Mn, indicating a direct effect on PS II (Fig. 5.4) rather than an indirect effect via source–sink relationships (Kriedemann et al., 1985).

In the range between suboptimal and optimal nutrient supply, positive correlations are often observed between nutrient concentrations in leaves and the rate of net photosynthesis (Fletcher et al., 2008; Paponov et al., 2005; Fig. 5.13). In

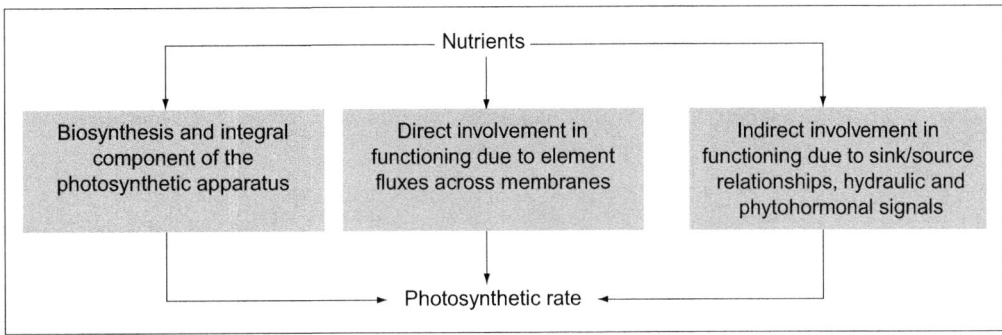

FIGURE 5.16 Modes of action of nutrients in the regulation of photosynthesis.

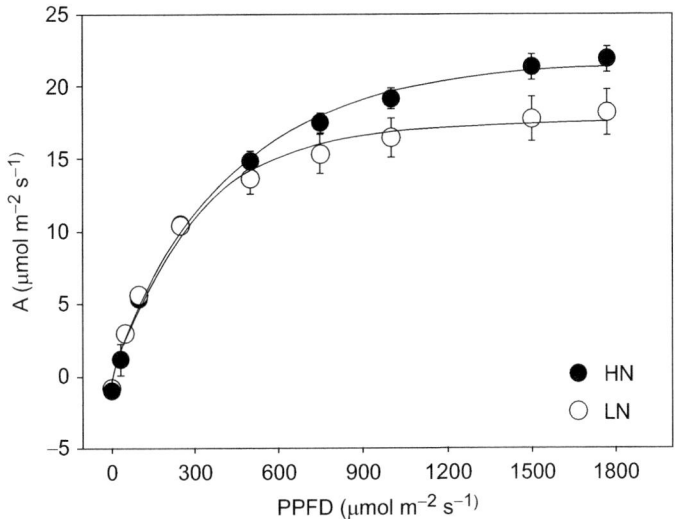

FIGURE 5.17 Light response curves of N-deficient (low N supply: LN) and N-sufficient (high N supply: HN) field-grown wheat (*Triticum turgidum* conv. *durum*) plants. Means ± SE. *A*, Rate of net photosynthesis; *PPFD*, photosynthetically active photon flux density. *Based on Cabrera-Bosquet et al. (2009).*

field-grown wheat, rates of net photosynthesis at low light intensity were similar in N-deficient plants and plants supplied with sufficient N (Fig. 5.17). By contrast, at high light intensity net rates of photosynthesis were lower in N-deficient plants. In low-N plants, with increasing light intensity an increasing proportion of the absorbed light energy is not used in photochemical reactions but dissipated as heat (de Groot et al., 2003). Similar changes in the light response curves are also found under deficiency of P (Lauer et al., 1989), K (Weng et al., 2007), and a range of other nutrients.

At the whole plant level, the rate of net photosynthesis may also be influenced indirectly by nutrient supply via the effects of nutrition on growth and source−sink relationships (Fig. 5.16). Despite poor radiation-use efficiency at high light intensities, carbohydrates may accumulate in both leaves and roots of P-deficient plants (Hermans et al., 2006). Thus, low photosynthetic efficiency of source leaves from nutrient-deficient plants is often due to feedback regulation by a lower demand for photosynthates at sinks (Pieters et al., 2001). An example of this is shown for Zn deficiency in common bean in Table 5.3. With increasing light intensity, plant dry weight increases in the Zn-sufficient but not in the Zn-deficient plants. Although the chlorophyll concentration decreases with increasing light intensity, particularly in the Zn-deficient plants, the carbohydrate concentration increases, indicating that the lack of growth response to increasing light intensity reflects a sink and not a source limitation. Similarly, K deficiency, by restricting water supply to fruiting tomato plants, results in the accumulation of sugars in leaves and decreased expansion of stems and fruit, which is also indicative of a sink rather than source limitation (Kanal et al., 2007).

Accumulation of photosynthates under high light intensity in source leaves of deficient plants not only decreases utilization of light energy but also poses a stress. This high-light stress is indicated, for example, by an increase in the antioxidative defense mechanisms in the nutrient-deficient leaves (Cakmak and Marschner, 1992; Fig. 5.7), photooxidation of chloroplast pigments (Table 5.3) and enhanced leaf senescence. These side effects of nutrient deficiency decrease not only photosynthesis and current *leaf area index* (LAI) but also *leaf area duration* (LAD), that is, the length of time during which the source leaves supply photosynthates to sink sites (discussed in Section 5.4.3).

TABLE 5.3 Shoot growth and concentrations of chlorophyll and carbohydrates in primary leaves of common bean (*Phaseolus vulgaris*) at different light intensities and with or without Zn addition.

Light intensity	Shoot dw (g plant^{-1})		Chlorophyll (mg g^{-1} dw)		Carbohydrates			
					Sucrose		Total[a]	
					(mg glucose equiv g^{-1} dw)			
(μmol m^{-2} s^{-1})	+Zn	−Zn	+Zn	−Zn	+Zn	−Zn	+Zn	−Zn
80	1.24	1.13	19	17	10	11	40	42
230	2.38	1.13	17	7.8	11	54	42	124
490	3.80	1.16	11	4.5	17	82	77	138

[a]Sucrose, reducing sugars, and starch.
Source: Based on Marschner and Cakmak (1989).

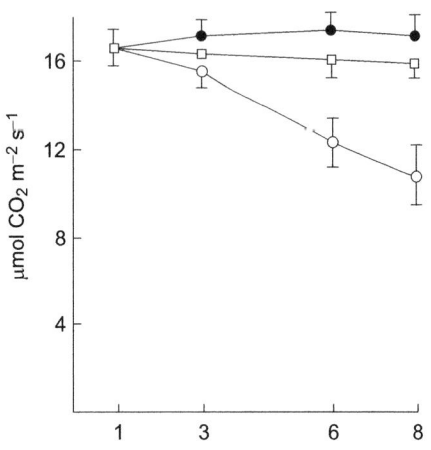

FIGURE 5.18 Rates of photosynthesis in leaves of tobacco (*Nicotiana tabacum*) grown for 8 days without N. ○: unshaded plants; ●: shaded plants; □, unshaded N-sufficient control. Error bars denote SE or ± SE. *From Paul and Driscoll (1997)*.

In N-deficient plants, sugars also accumulate due to low sugar utilization for N assimilation and growth. This N deficiency-induced sugar accumulation may lead to suppression of photosynthesis (Fig. 5.18). In tobacco leaves, photosynthetic rate decreases after withdrawal of N, the decrease being associated with a strong decline in the amount and activity of Rubisco. This decrease can be prevented by shading the leaves (Fig. 5.18). Leaf shading has a negligible effect on amino acids but decreases hexose concentration in the leaves (Paul and Driscoll, 1997). Another example of an indirect effect of nutrients on photosynthesis is the decrease of photosynthetic rates in plants via hydraulic or phytohormonal signals that reduce stomatal conductance, and thus leaf gas exchange (Fig. 5.15; Cramer et al., 2009; Shabala et al., 2016). Transpiration can also be affected by root hydraulic conductance, which can be influenced by N, P, and K supply through the control of aquaporin activities (Armand et al., 2019; Coffey et al., 2018; Ding et al., 2014). Nutrient deficiency can also affect stomatal conductance by altering the supply of root- or leaf-sourced phytohormones, such as ABA and CYT, to guard cells (Wilkinson et al., 2007).

The effects of nutrient deficiencies depend on sensing and signaling between shoot and root and vice versa. Changes in root and shoot architecture resulting from N limitation involve long-distance signaling (Bellegarde et al., 2017; Jia and von Wirén, 2020; Luo et al., 2020; Ruffel et al., 2011; Ruffel, 2018). Downregulation of N uptake in the root requires shoot signals that are linked to alterated CYT signaling (Ruffel et al., 2011). Evidence from arabidopsis mutants with defective CYT synthesis strongly suggests that CYT transport from root to shoot via the xylem is one of the signals mediating root N responses (Miyawaki et al., 2004; Ruffel et al., 2011). Sugars, microRNAs, and auxins are candidates for shoot-to-root signals of nutrient deficiency (Chapter 3; Shabala et al., 2016).

5.4 Photosynthetic area

The plant capacity to produce assimilates is not only related to photosynthetic activity (*source activity*) but also to photosynthetic area (*source size*) including leaves, stems, husks, and other green organs. For example, spikes of grain crops, such as barley, may contribute substantially to plant photosynthesis during grain filling, particularly under drought and in dense crop stands, where mutual shading and senescence limit leaf photosynthesis (Tambussi et al., 2007). In this section only leaves are considered. Control of leaf growth has been associated with many processes, including cell cycle regulation, tissue extensibility, as well as hydraulic, sugar, and phytohormonal signaling (Granier and Tardieu, 2009; Walter et al., 2009), and there is evidence that specific nutrients are involved in these processes (see also Chapters 6 and 7).

5.4.1 Individual leaf area

The area of individual leaves is dependent on leaf position and environmental conditions during leaf development. Environmental stresses, such as low temperatures, drought, salinity, and nutrient deficiency, reduce final leaf area, with this reduction being dependent on genotype (Granier and Tardieu, 2009). An example of two soil-grown maize genotypes differing in N-use efficiency (i.e., grain yield obtained under low N supply in the field) is presented in Fig. 5.19. Regardless of genotype, individual leaf area increased from the basal leaf positions to leaf 8 and decreased from leaf 10 toward the apical leaves. The effect of N fertilization on the leaf area of individual leaves was significant only in leaves 10–15, presumably because, in nonfertilized plants, N supply from seeds and mineralization of soil organic matter was sufficient to meet growth demands in the initial phase of plant development. Nitrogen deficiency-induced reduction of leaf area was less pronounced in the N-efficient genotype, which yielded more than the N-inefficient genotype under low N supply. The lower leaf N concentration of the N-efficient than the N-inefficient genotype indicates a lower sensitivity of leaf growth to suboptimal internal leaf N status (Paponov and Engels, 2003). Similar observations have been made on other crops, such as oilseed rape (*Brassica napus*), in which N-efficient genotypes maintain greater leaf growth under low N supply (Gironde et al., 2015). Deficiency of other nutrients, including K and P, also reduces leaf elongation rates and final plant leaf area (Fletcher et al., 2008; Jordan-Meille and Pellerin, 2004; Sun et al., 2020).

Leaf growth is controlled by cell division and cell expansion. From a biophysical viewpoint, cell expansion is dependent on cell turgor as driving force together with cell wall properties (Cosgrove, 2005; Fricke, 2002). These biophysical cell properties are regulated by an internal circadian clock and external cues such as temperature, light, and nutrient availability via complex signaling cascades involving phytohormones (Section 5.9) and sugars (Poiré et al., 2010; Walter et al., 2009). In the long term, the expansion of cell walls must be matched by synthesis and integration of new wall materials and thus is dependent on assimilate supply. In plants lacking sufficient N or K, leaf elongation rates may decline before there is any reduction in net photosynthesis (Hu et al., 2020; Seufert et al., 2019).

Cell division and expansion in the leaf growth zone are particularly sensitive to the supply of P (Assuero et al., 2004) and N (Luo et al., 2020). For example, in perennial ryegrass (*Lolium perenne*), leaf elongation rate was 43% lower in low-N plants than in plants with an adequate N supply (Kavanová et al., 2008). This reduction was the result of changes in both cell production rate and final cell length. Similar effects on cell production and elongation have been found under P deficiency (Kavanová et al., 2006). An adequate K supply is also required for optimal cell division and cell expansion (Hu et al., 2020). Postmitotic cell expansion is mainly due to an increase in vacuolar volume. Nutrients such as K are required for cell expansion as solutes that decrease the vacuolar osmotic potential and, thereby,

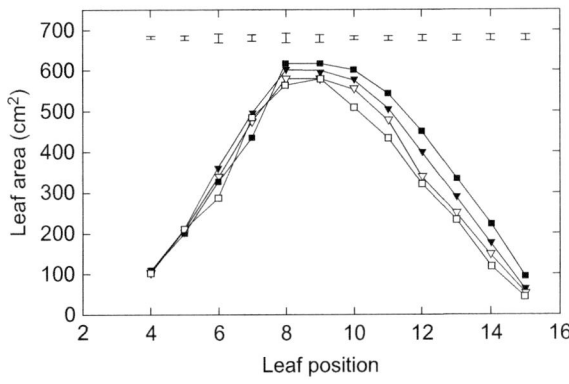

FIGURE 5.19 Individual area of leaves in different positions (leaf position 15 = apical leaf) in an N-efficient maize (*Zea mays*) genotype (*triangles*) and a nonefficient genotype (*squares*) grown at high (*filled symbols*) or low N supply (*open symbols*). Vertical bars denote LSD$_{0.05}$. *From Paponov and Engels (2003), with permission from John Wiley and Sons.*

TABLE 5.4 Inhibition of leaf growth by N deficiency in different plant species.

Plant species	Average growth inhibition (%)	
	Day	Night
Cereals (wheat, barley, maize, sorghum)	16	18
Dicotyledons (sunflower, cotton, soybean, radish)	53	8

Source: Based on Radin (1983).

drive water influx and increase cell turgor (White, 2020). In addition, the supply of N, P, and K can modify the abundance and activity of aquaporins that mediate water influx (Armand et al., 2019; Coffey et al., 2018; Ding et al., 2014).

The effect of N deficiency on leaf expansion differs between monocotyledons and dicotyledons (Table 5.4). In monocotyledons, cell expansion is inhibited to the same extent during the day and night, whereas in dicotyledons inhibition is more severe during the day. In dicotyledons, cell expansion occurs in leaf blades exposed to the atmosphere and having a high rate of transpiration during the day. In cereals, however, cell expansion occurs at the base of the leaf blade, a zone that is protected from the atmosphere by the sheath of the preceding leaf, so that little transpiration occurs during elongation. In contrast to leaf expansion, net photosynthesis per unit leaf area is depressed to a similar extent in both groups of plants by N deficiency. Similar results to those in Table 5.4 have also been obtained for P deficiency in cotton plants (Radin and Eidenbock, 1984) where leaf growth occurs only during the day.

Monocotyledons and dicotyledons differ in the dependence of diel changes in leaf growth on environmental conditions (Poiré et al., 2010). In monocotyledons, diel variation in leaf growth shows no circadian oscillation when plants are grown in continuous light. By contrast, in dicotyledons leaf growth is regulated by an endogenous circadian clock and shows a clear circadian oscillation even under constant day/night conditions (Poiré et al., 2010).

Turgor above a threshold value (*yield threshold*) induces cell expansion which is dependent on cell wall extensibility (Fricke, 2002). Regulators of wall extensibility include cell wall loosening expansins, xyloglucan endotransglucosylase/hydrolase, and hydroxyl radicals that may be produced by wall peroxidases or nonenzymatically by Cu ions bound to the cell wall (Cosgrove, 2005). The activity of expansins is regulated by phytohormones, including auxin, ethylene, gibberellin, and CYT (Downes et al., 2001; Sánchez-Rodríguez et al., 2010). Phytohormone supply to leaves is affected by nutrition and other external cues such as drought and salinity. For example, the stimulation of leaf growth by nitrate and the inhibition by ammonium nutrition are related to xylem transport and leaf contents of active forms of CYTs, which are increased by nitrate and decreased by ammonium in tobacco (Walch-Liu et al., 2001) and tomato (Rahayu et al., 2005).

5.4.2 Leaf area per plant

Leaf area development at the whole plant level also depends on the rate of leaf production, and on tillering and formation of axillary branches. Some nutrient deficiencies can delay plant development. For example, in barley the period to reach the booting stage is about twice as long for Mn-deficient than for Mn-sufficient plants (Longnecker et al., 1991). The rate of leaf production is strongly influenced by temperature and can be described on the basis of thermal time, that is, the product of time and temperature exceeding a minimal threshold below which development is completely arrested (Granier et al., 2002). The thermal time elapsing between the visual appearance of two successive leaf tips (degree days leaf^{-1}) is termed the phyllochron. Water availability and, occasionally, nutrient deficiencies can affect the phyllochron. For example, a lack of water, but not a reduction in N supply, increased the phyllochron of fodder beet (Chakwizira et al., 2016). Similarly, N deficiency and S deficiency generally have little effect on, or slightly increase, the phyllochron of wheat and barley (Alzueta et al., 2012), and insufficient K supply had little effect on the phyllochron of rye (*Secale cereale*; White, 1993) or maize (Usandivaras et al., 2018). However, P deficiency increased the phyllochron of wheat (Table 5.5; Rodríguez et al., 1998), barley, and maize (Usandivaras et al., 2018). Furthermore, the total number of tillers per plant was also significantly reduced in P-deficient wheat. It is well known in agricultural and horticultural plant production that tillering and axillary branching of field-grown plants can be stimulated by N and P fertilization.

TABLE 5.5 Phyllochron and number of tillers per plant in wheat (*Triticum aestivum*) at different rates of P fertilization.

	P fertilization (kg P ha^{-1})			
	7	15	60	300
Phyllochron (degree days leaf^{-1})	124	108	110	94
Tiller number plant^{-1}	1.1	1.6	2.8	3.2

Source: Based on Rodríguez et al. (1998).

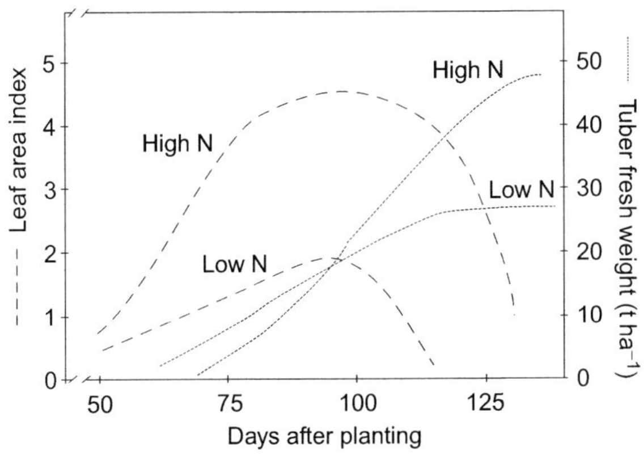

FIGURE 5.20 Time-course of leaf area index and fresh weight of potato (*Solanum tuberosum*) tubers at high or low N supply. *Based on Ivins and Bremner (1964) and Kleinkopf et al. (1981).*

5.4.3 Canopy leaf area (leaf area index and leaf area duration)

At the canopy level, photosynthetic area is often expressed in terms of the *LAI*, which is defined as leaf area of plants per unit area of soil. For example, an LAI of 5 means that there is 5 m^2 leaf area per m^2 soil area. LAI values below 3 are often associated with incomplete interception of incoming solar radiation, whereas LAI values above 6 indicate strong shading, and thus negative net photosynthesis of lower leaves. Light distribution within the canopy is influenced by leaf architecture (Horton, 2000; Long et al., 2006). Compared to horizontal leaves, erect leaves reduce excessive light interception of the top leaf layer in a crop stand, and thus photoinhibition. By contrast, light interception by lower leaves within the canopy is increased. Model simulations have shown that in a canopy with an LAI of 3, the efficiency with which the intercepted light is converted to biomass through photosynthesis is about 40% greater in erect leaves than in horizontal leaves (Long et al., 2006).

Source size is not only determined by leaf area but also *LAD* (the sum of LAI integrated over the period of time in which the source leaves supply photosynthates to sink sites). In crop species, LAD is of crucial importance for the length of the sink filling period and is often positively correlated with yield (Gregersen et al., 2013).

Nutrient supply influences leaf growth and leaf senescence, and thus also LAI and LAD. Nutrient deficiency under high light intensity is often associated with the accumulation of photosynthates in source leaves, which not only decreases utilization of light energy but also poses a stress. This high light stress is indicated by an increase in the antioxidative defense mechanisms in the deficient leaves (Cakmak and Marschner, 1992; Fig. 5.7), photooxidation of chloroplast pigments (Table 5.3), and accelerated leaf senescence. These side effects of nutrient deficiency decrease not only current photosynthesis and LAI but also LAD.

In the initial phase of development, leaves are sink organs that utilize assimilates exported from source organs. In crop species with vegetative storage organs, such as roots and tubers, there is competition for assimilates between leaf area construction and storage processes (Kleinkopf et al., 1981; White, 2018). This has to be considered, for example, in N fertilization of potato. On the one hand, a high N supply is important for rapid leaf expansion and for obtaining a LAI between 4 and 6, a value considered necessary for high tuber yields (Kleinkopf et al., 1981). On the other hand, high N supply delays tuberization and/or the onset of the linear phase of tuber growth. The principles of these interactions are shown in Fig. 5.20. At low N supply, the advantage of earlier tuberization is offset by a smaller LAI and earlier leaf senescence (i.e., shorter LAD) and a correspondingly smaller tuber yield. When the N supply is high, both LAI and LAD, and thus final tuber yield, are higher. However, higher tuber yield induced by a high N supply can be achieved only when the vegetation period is sufficiently long, for example, when there is no early frost or severe drought stress.

The early decline in LAI at low N supply (Fig. 5.20) indicates that the final tuber yield is limited by the source. One of the reasons for this source limitation is that, in potato plants at maturity, between 60% and 80% of the total N is present in the tubers (Kleinkopf et al., 1981; White, 2018). Thus, when the N supply is low, exhaustion of N in the source leaves likely plays a key role in leaf senescence and the termination of tuber growth. However, these simple relationships between N supply, LAI, LAD, and tuber yield (Fig. 5.20) are not only modified by the length of the growing period but also by the mineralization rate of soil N and by soil temperature during tuber growth. At high N supply, a high LAI causes shading of the basal leaves, which not only decreases their net photosynthesis but also LAD by accelerating leaf senescence (Firman and Allen, 1988), a process which is further accelerated at high temperatures (Manrique and Bartholomew, 1991). Thus, a lower, but more continuous, supply of N, which facilitates early tuberization, continuous root growth, and increases LAD rather than LAI, can often lead to greater tuber yields than a higher N supply that promotes rapid establishment of a high LAI during early plant growth.

Both LAI and LAD depend on leaf area per plant and on plant density that is influenced by sowing density. The increase in grain yield of modern wheat and maize varieties has largely been attributed to increases in LAI and LAD (Lee and Tollenaar, 2007; Ning et al., 2013). Interestingly, in modern maize hybrids leaf area per plant has not generally changed in comparison to old hybrids. Improvements in LAI have thus resulted mainly from greater crowding tolerance (tolerance to intraspecific competition among neighboring plants) allowing higher plant densities (Boomsma et al., 2009).

5.5 Respiration and oxidative phosphorylation

In nongreen tissue (e.g., roots, seeds, and tubers) or in green tissue during the dark period, respiratory carbohydrate metabolism provides the main source of energy required for energy-consuming processes such as biosynthesis and transport. Respiration consumes 30%–70% of the carbon assimilated during photosynthesis (Tcherkez and Limami, 2019) and can be partitioned into two functional components: growth and maintenance. Growth respiration is defined as respiratory energy required for biosynthesis of new plant constituents. Maintenance respiration is the respiratory energy required for all processes that maintain cellular structure, for example, turnover of cellular components and maintenance of intracellular ion gradients. For roots, a third functional component is the energy needed to support nutrient acquisition through the biosynthesis and rhizospheric deposition of enzymes and exudates, which can represent 20%–60% of the net fixed carbon. This amounts to 800–4500 kg C per ha per annum (Neumann, 2007; Chapter 14). Respiratory measurements on mature leaves of field-grown maize hybrids suggest there is potential for increasing crop yields through reduction of maintenance respiration (Earl and Tollenaar, 1998). As demonstrated by Lynch (2019), the formation of root cortical aerenchyma (RCA) reduces respiration and the carbon costs of maintaining root tissue and enhances P capture in field-grown maize and beans. Furthermore, maize genotypes with greater RCA formation have greater topsoil foraging, P capture, growth, and yield in low-P soil than genotypes with less RCA formation, notwithstanding the reduction in mycorrhizal colonization caused by the formation of RCA (Galindo-Castañeda et al., 2018). Potassium acquisition also benefits from the formation of RCA (Postma and Lynch, 2011). Potassium moves to roots primarily by diffusion and has the greatest bioavailability in the topsoil. Moreover, K acquisition centers on soil exploration and K solubilization from poorly available K pools (Römheld and Kirkby, 2010; White et al., 2021a).

Respiration can be divided into three major steps (Fig. 5.21A). The first step is glycolysis, whereby sugars are converted into organic acid anions, such as pyruvate, in the cytosol and plastids, which yields a small amount of ATP and reduced nicotinamide dinucleotide (NADH). In the second step, the tricarboxylic acid (TCA) cycle (also called the Krebs cycle or citric acid cycle), pyruvate is completely oxidized to CO_2 in the mitochondrial matrix, and a considerable amount of reducing power (NADH and reduced flavin adenine dinucleotide, $FADH_2$) is produced. In the third step (oxidative phosphorylation), electrons from the donors NADH and $FADH_2$ are transferred along an electron transport chain in the inner mitochondrial membrane to oxygen (Fig. 5.21B). The individual electron transport proteins are organized into four multiprotein complexes (CI–CIV). Electrons are transferred from CI (NADH dehydrogenase) and CII (succinate dehydrogenase) via ubiquinone (UQ) to CIII (cytochrome bc_1). Cytochrome c transfers the electrons to CIV (cytochrome oxidase), the terminal oxidase that transfers the electrons to molecular oxygen.

Several nutrients are directly involved in this mitochondrial electron transport chain (Fig. 5.21B). In CI, CII, and CIII, electrons are transferred via Fe-S proteins. CIV contains two Cu centers. Electron transport via CI, CIII, and CIV is coupled to proton pumping across the inner mitochondrial membrane, and the resultant electrochemical gradient is

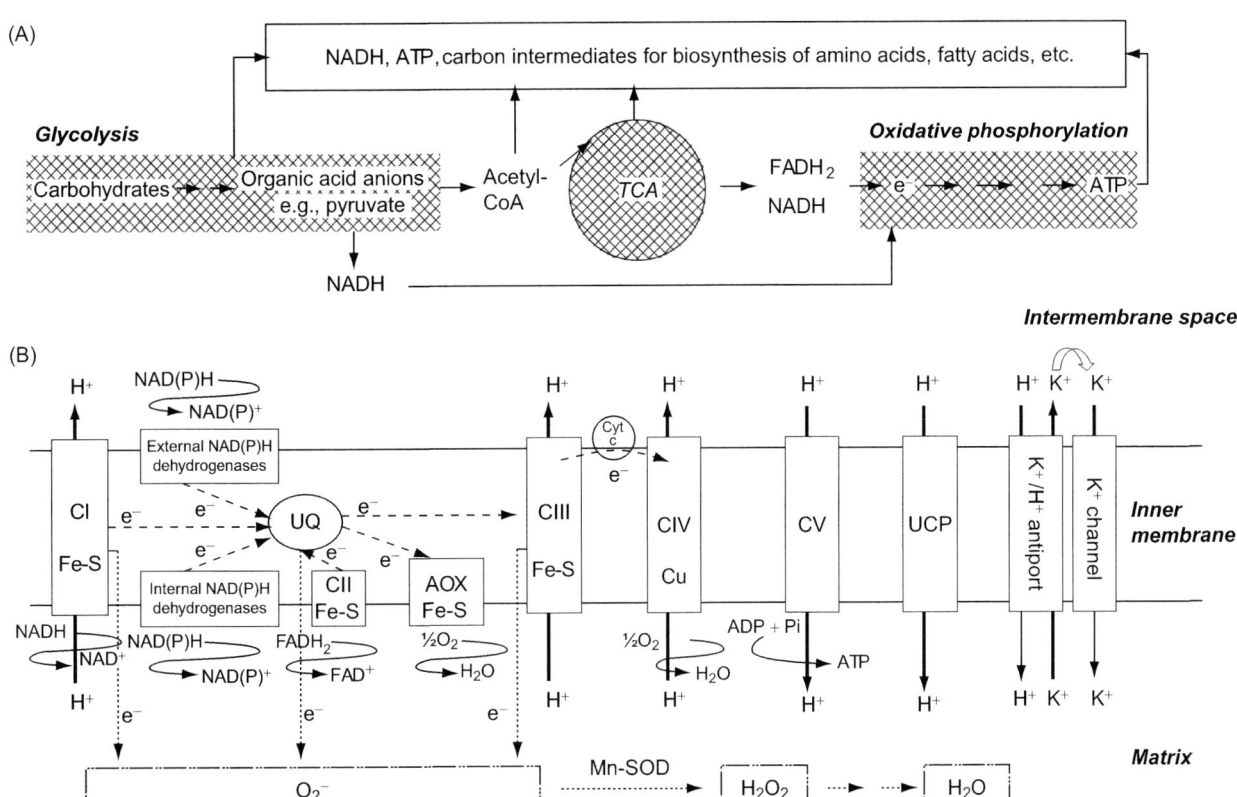

FIGURE 5.21 Scheme of respiration. (A) Main steps of respiration: glycolysis in the cytosol and plastids, TCA, and oxidative phosphorylation in mitochondria. (B) Organization of the electron transport processes on the inner membrane of mitochondria. *CI*, Complex I (NADH dehydrogenase); *CII*, complex II (succinate dehydrogenase); *CIII*, complex III (cytochrome bc₁ complex); *CIV*, complex IV (cytochrome c oxidase); *CV*, complex V (ATP synthase); *cyt c*, cytochrome c; *Mn-SOD*, Mn superoxide dismutase; *TCA*, tricarboxylic acid cycle; *UCP*, uncoupling protein; *UQ*, ubiquinone. Based on Plaxton and Podestá (2006), Navrot et al. (2007), and Atkin and Macherel (2009).

used by an ATP synthase (also called complex V) for ATP production (Fig. 5.21B). Thus, the principles involved in ATP synthesis in the mitochondria are the same as those of ATP synthesis in the chloroplasts: charge separation by a membrane with a corresponding proton (pH) gradient across the membrane constituting the electromotive force for ATP synthesis (Fig. 5.21A). Furthermore, the various intermediates of carbohydrate metabolism are essential structures (carbon skeletons) for, for example, the synthesis of amino acids and fatty acids. The rate of respiration is therefore regulated not only by environmental factors (such as temperature) and energy requirements (e.g., ATP for ion uptake in the roots) but also by the demand for reducing equivalents and intermediates.

Depending on the metabolic process, the demand for ATP (activating agent) relative to that for NADH (reducing agent) and carbon intermediates can vary markedly. For example, for transport processes across membranes, mainly ATP is needed, whereas the biosynthesis of lipids or amino acids requires large amounts of NADH and carbon intermediates. This variable demand for respiratory products is met by metabolic "bypasses" in the three steps of respiration: glycolysis, TCA, and oxidative phosphorylation (Plaxton and Podestá, 2006; Sweetlove et al., 2010). These bypasses yield different amounts of respiratory products. For example, the proton gradient across the inner mitochondrial membrane can be dissipated by an uncoupling protein that allows proton diffusion into the matrix without ATP production (Plaxton and Podestá, 2006; Fig. 5.21B). There is also evidence for a role of K^+ in the dissipation of the transmembrane proton gradient (Pastore et al., 1999) (Fig. 5.21B). It is still unclear, however, exactly how $K^+(Na^+)/H^+$ antiporters function, but they may be crucial players in the pH and cation homeostases (Sze and Chanroj, 2018) that alter plant growth and development and acclimation to stresses. Energy-dissipating pathways may serve to avoid overreduction of the electron transport chain as a major mechanism for ROS production (Fig. 5.21B), for example, under conditions of salt and drought stresses (Pastore et al., 2007).

Another example of plasticity in plant respiration is the engagement of mitochondrial electron transport pathways that allow electron transfer to oxygen, which circumvents proton-pumping sites (complexes I, III, and IV). Electrons from NADH and NADPH can be transferred via external and internal NAD(P)H dehydrogenases to UQ (Fig. 5.21B).

From UQ, electrons can be transferred via the alternative oxidase (AOX) to molecular oxygen. Operation of either of the NAD(P)H dehydrogenases with the AOX thus provides an alternative pathway of electron transport that is not coupled to H^+ transport across the mitochondrial membrane and therefore not associated with ATP synthesis. The lower efficiency of the alternative pathways in energy conversion in the form of ATP results in higher energy dissipation in the form of heat (*thermogenesis*). Besides their role in primary metabolism, plant mitochondria act as signaling organelles, for example, to influence nuclear gene expression. In reducing molecular oxygen to water, AOX lowers the potential for oxygen to be converted to ROS. The alternative pathway thus provides a degree of metabolic homeostasis to carbon and energy metabolism of particular importance during abiotic stresses (such as low temperature, drought, and nutrient deficiency) as well as biotic stresses (such as bacterial infection) (Vanlerberghe, 2013).

The proportion of the alternative pathway can vary between less than 10% and more than 80% of the total respiration in roots and leaves (Florez-Sarasa et al., 2007; Poorter et al., 1991). Factors contributing to this variation include time of day (Tcherkez et al., 2017), P nutritional status (Theodorou and Plaxton, 1993), plant species, developmental stage, and plant organ. The proportion of the alternative pathway is also dependent on the amount (Scheible et al., 2004) and form of N supply (Escobar et al., 2006; Foyer et al., 2011). In N-deficient arabidopsis, the transcription of *AOX* genes was reduced by resupplying nitrate and increased by resupplying ammonium. Similar results were observed at adequate N supply when nitrate and ammonium sources were switched (Escobar et al., 2006). The authors suggest that effects of the form of N supply on AOX capacity are in accordance with a role of the alternative pathway in redox balancing. Nitrate nutrition is associated with a high demand for NADH for nitrate reduction. There is, thus, no excess of NADH for electron transport via the alternative pathway. By contrast, ammonium nutrition is associated with a lower demand for NADH so that NADH is oxidized via the alternative pathway. A decreased transcription of *AOX* in nitrate-fed plants was mediated by the nitrate ion itself, whereas ammonium regulation of *AOX* transcription was dependent upon assimilation and affected by ammonium-induced changes in apoplastic pH (Escobar et al., 2006). Ammonium initiates rapid changes in cytosolic pH and apoplastic acidification, whereas nitrate does not (Liu and von Wirén, 2017). A recent review by Rasmusson et al. (2020) provides further evidence that assimilation of the two N sources, nitrate or ammonium, alters redox homeostasis and antioxidant systems differentially in various cellular compartments, including the mitochondria and the cell wall.

5.6 Transport of assimilates in phloem and its regulation

Long-distance transport of photosynthates, assimilates, and mineral nutrients from source leaves to sink organs occurs in the phloem. The phloem can be subdivided into three sectors (Fig. 5.22). In the collection phloem, photosynthates, assimilates and nutrients are loaded into the minor veins of source leaves. The transport phloem in the main veins, leaf sheaths, petioles, stems, and roots distributes solutes to the sink organs, where they are released into the sink cells via the release phloem (Van Bel, 2003). The processes associated with the transport of photosynthates, amino acids and ureides in the different sectors of the phloem are briefly described below. Ureides, such as allantoin, allantoic acid, or citrulline, are the main N compounds transported in nodulated subtropical and tropical legumes (Atkins and Smith, 2007).

5.6.1 Phloem loading of assimilates

The first step in supplying young leaves and other sinks with assimilates from the source leaves is short-distance transport of the assimilates from individual leaf cells to the phloem parenchyma cells of the vascular bundles, which is followed by loading into the companion cells (CCs) and sieve elements (SEs). The transport of assimilates from mesophyll cells to CCs can be symplasmic or apoplasmic (Knoblauch and Peters, 2013; Zhang and Turgeon, 2018). Symplasmic movement from other leaf cells to CCs is enabled by concentration gradients of assimilates. This can be enhanced by conversion of the assimilate to larger polymers (polymer trapping), as occurs in plants that transport raffinose or stachyose in the phloem (Rennie and Turgeon, 2009). An apoplastic step in the pathway enables large concentrations of assimilates to be accumulated in SEs energized by the activity of a plasma-membrane H^+-ATPase in CCs (Lalonde et al., 2003).

CCs and SEs are symplastically connected by special plasmodesmata (referred to as *plasmodesmata-pore units*) that are highly permeable (Turgeon and Wolf, 2009; Van Bel, 2003). The conductivity of plasmodesmata is dependent on the diameter of their pores. This diameter may be expressed as a size-exclusion limit, that is, the maximum size or mass of a molecule that can pass through the plasmodesmata passively. The branched plasmodesmata connecting SEs and CCs are characterized by a high size exclusion limit that allows transfer of large molecules (20–40 kD; Van Bel, 2003). The SEs

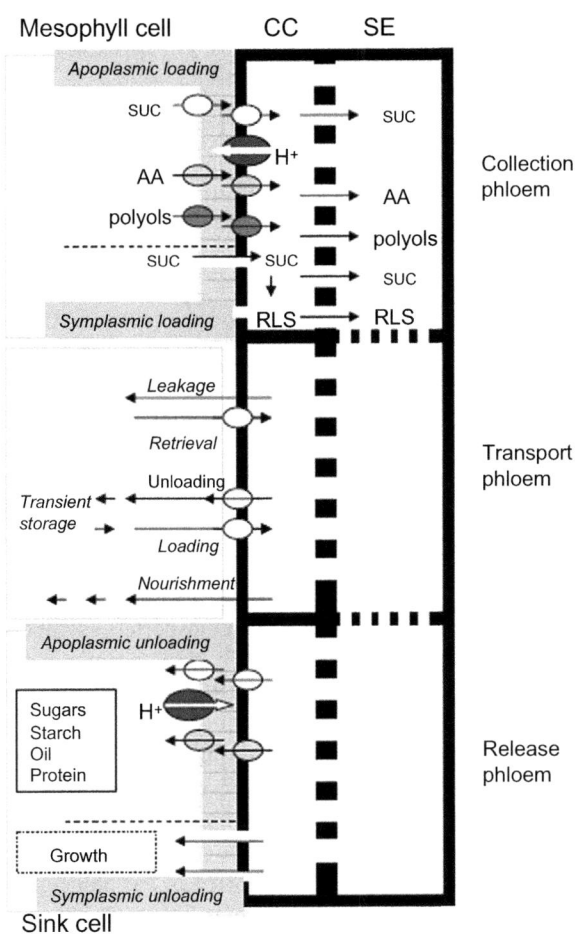

FIGURE 5.22 Long-distance transport of sugars and amino acids via the phloem. *AA*, Amino acids; *CC*, companion cell; *RLS*, raffinose-like sugars; *SE*, sieve element; *suc*, sucrose. Based on Lalonde et al. (2003).

lack many structures normally found in living cells, such as nuclei, ribosomes, and vacuoles. They have characteristic sieve areas with large pores in their cell walls and are highly specialized for the transport of water and solutes by mass flow. The essential metabolic functions lacking in SEs (such as protein synthesis) are taken over by the CCs. Thus, SEs and CCs form a functional unit referred to as the SE−CC complex.

As a rule, sugars represent 80%−90% of the assimilates exported in the SEs from the source leaves (Chapter 3). In most plant species, sucrose is the dominant sugar in the phloem sap, but in some plant species oligosaccharides such as raffinose or stachyose (e.g., in cucurbits) or sugar alcohols like mannitol (e.g., in celery, parsley, carrot, and olive) or sorbitol (e.g., in apple and cherry) are also transported (Turgeon and Wolf, 2009; Zhang and Turgeon, 2018). The preferential sites for phloem loading of sugars are the minor veins of a source leaf.

Depending on the plant species, time of the day, and the site of collection, the sucrose concentrations in the phloem sap range from 200 to 1000 mM. In plant species that additionally transport the sugar alcohols mannitol and sorbitol, concentrations of these polyols of between 300 and 700 mM have been measured (Nadwodnik and Lohaus, 2008). To achieve these high concentrations, a loading step from the mesophyll or phloem parenchyma cells into the SE−CC complex of the minor veins is required in most plant species (Fig. 5.22, collection phloem). In plants with an apoplasmic step in phloem loading, sucrose is released into the apoplasm through sucrose uniporters belonging to the SWEET protein family (Jeena et al., 2019; Zhang and Turgeon, 2018). Estimates of whole-leaf apoplasmic sucrose concentrations are in the range of 1−5 mM, and estimates of apoplasmic sucrose concentrations in the vicinity of minor veins are 27−133 mM (Lalonde et al., 2003). In plant species that additionally transport sugar alcohols, sugar alcohol concentrations in the leaf apoplasm have been found to be substantially lower than in the phloem (Nadwodnik and Lohaus, 2008). Loading of sucrose and sugar alcohols from the apoplasm into the phloem must therefore be energized by the proton motive force generated by a proton-pumping ATPase in the plasma membrane of the SE−CC complex (Lalonde et al., 2003). This ATPase creates a steep transmembrane electrical potential gradient as well as a pH gradient between the lumen ("symplasm") of the SE−CC complex and the apoplasm (Fig. 5.23). This gradient acts as a driving force for

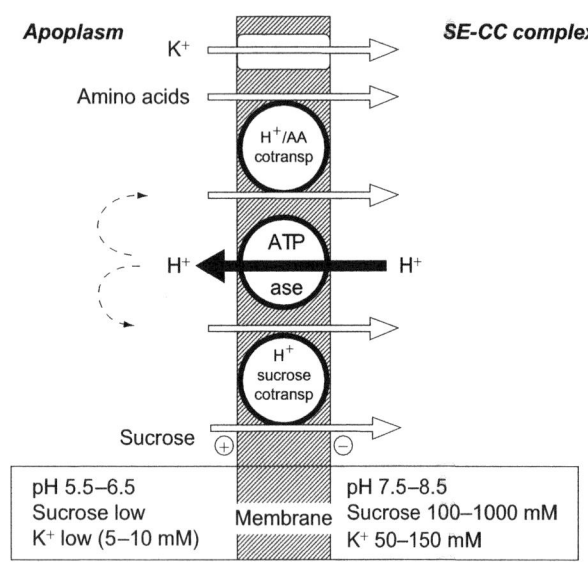

FIGURE 5.23 Model for phloem loading of sucrose and AA mediated by proton-sucrose cotransport, proton-AA cotransport, and uniport of K. *AA*, amino acid; *SE−CC*, sieve element-companion cell.

the transport of sucrose from the apoplasm into the SE−CC complex catalyzed by H$^+$/sucrose cotransporters (symporters) of the SUcrose Transporter (abbreviated as SUT or SUC) protein family (Lalonde et al., 2003; Pommerrenig et al., 2020; Zhang and Turgeon, 2018) and of sorbitol or mannitol by H$^+$/sugar alcohol symporters (Ramsperger-Gleixner et al., 2004) localized in the CCs of the source leaves (Fig. 5.22, collection phloem).

In some plant species, abundant plasmodesmata occur along the possible solute pathway from the mesophyll to the SE−CC complex (Rennie and Turgeon, 2009), suggesting symplasmic phloem loading (Fig. 5.22, collection phloem). In leaves of willow and other woody species, a gradient of sucrose concentration between mesophyll cells and phloem allows diffusion-driven transport of sucrose through plasmodesmata into the phloem (Rennie and Turgeon, 2009). In species (e.g., cucurbits) that load sugars via the symplast, raffinose-like sugars (RLS) are transported in the phloem. In these species, sucrose is transported from the mesophyll cells into specialized CCs in the minor veins known as intermediary cells. In these cells sucrose is converted to raffinose and stachyose (RLS; Fig. 5.22, collection phloem), that is, to molecules that are larger than sucrose. Diffusion back to the mesophyll cells is prevented through the low size-exclusion limit of the plasmodesmata connecting the intermediary and the mesophyll cells (*polymer trapping*), whereas diffusion into the SEs is made possible by the high size-exclusion limit of the plasmodesmata that connect the intermediary cells with the SEs (Rennie and Turgeon, 2009).

Phloem loading of amino acids is selective. In castor bean, for example, glutamine is loaded preferentially compared to glutamate or arginine (Schobert and Komor, 1989). In maize leaves, asparagine is preferentially loaded into the phloem where its concentration is about eightfold greater than in the cytosol of the leaf cells (Weiner et al., 1991). However, as a rule, the concentration gradient of sucrose from mesophyll cells to the phloem is steeper than that of most of the amino acids. For example, in barley leaves sucrose is preferentially loaded into the phloem compared to amino acids, as reflected by the ratio of amino acids/sucrose in the cytosol of about one in contrast to that of about 0.2 in the phloem sap (Table 5.6). In most plants, amino acids are released by parenchyma or bundle sheath cells into the leaf apoplasm by uniporters (such as UmamiT proteins and the amino acid transporter BAT1) and are loaded into CCs by H$^+$/amino acid symporters of the amino acid permease (AAP) protein family and, possibly, also members of the cationic amino acid transporter (CAT) and proline transporter families (Tegeder, 2014; Tegeder and Masclaux-Daubresse, 2018). These transport mechanisms are consistent with the concentrations of amino acids being substantially lower in the leaf apoplasm than in the cytosol of mesophyll cells (Tilsner et al., 2005).

The concentrations of amino acids in the phloem sap are usually lower than those of sugars and are in the range of 50−200 mM. However, in individual SEs of wheat, amino acid concentrations above 1000 mM have been measured (Gattolin et al., 2008), and the phloem sap of oilseed rape can contain amino acid concentrations up to 650 mM, more than four times greater than the cytosol of the mesophyll cells (Tilsner et al., 2005). Thus, in many plant species, loading of amino acids into the SE/CC complex may be as important as sucrose loading for generating osmotic potential. In addition to amino acid transporters, transporters for ureides, such as UPS1 in bean, and peptides have been found in the phloem, suggesting a role for these transporters in loading N compounds into the phloem (Rentsch et al., 2007; Tegeder, 2014).

TABLE 5.6 Concentrations of sucrose and amino acids in different cell compartments of barley (*Hordeum vulgare*) leaves.

	After 8 h of light		After 5 h of dark	
	Cytosol	Phloem	Cytosol	Phloem
Sucrose (mM)	150	1030	43	930
Amino acids (mM)	156	186	58	244
Ratio of amino acids/sucrose	1.04	0.18	1.35	0.26

Source: Based on Winter et al. (1992).

It is well established that phloem loading of sucrose is enhanced by adequate K nutrition (Chapter 3). In arabidopsis, a K channel of the AKT2/3 family is present in the plasma membrane of phloem cells and is required to sustain the loading of sucrose into the phloem (Deeken et al., 2002). It is not clear, however, whether stimulation of N transport in the phloem by K is a direct effect on the loading mechanism (e.g., maintenance of the transmembrane pH gradient) or an indirect one via an increase in osmotic potential of phloem sap and, therefore, the rate of mass flow in the sieve tubes.

5.6.2 Mechanism of phloem transport of assimilates

The principles regulating transport in the sieve tubes, the anatomy of the phloem, and transport direction (from source to sink) have been discussed in Chapter 3 in relation to long-distance transport of nutrients. In brief, according to the Münch *pressure flow hypothesis* (Knoblauch and Peters, 2013) solutes are loaded into the sieve tubes of leaves, and water is moved osmotically into the sieve tubes, creating a positive internal pressure. As sucrose and other sugars, together with amino acids and ureides in some species, are the dominant osmotically active solutes in the sieve tubes of leaves, volume flow rates are determined primarily by phloem loading of photosynthates and assimilates at the source and unloading at the sink. Water availability in the source leaves is also an important factor for volume flow rates in the sieve tubes and phloem loading associated with the lateral transport of water in the leaves toward the phloem (Salmon et al., 2019). In addition, the hydraulic conductivity between the xylem and the phloem affects the sensitivity of phloem transport to drought (Sevanto, 2018).

Of the mineral nutrients, K is usually present at the highest concentrations in the phloem sap (Chapter 3). Thus, K contributes substantially to the volume flow rates in sieve tubes as shown in Table 5.7 for castor bean. In plants well supplied with K, the concentration of K, the osmotic potential of the phloem sap, and, particularly, the volume flow rate (exudation rate), are higher than in plants with a low K supply. Sucrose concentration in the phloem sap remains mostly unaffected, and a high K supply increases the transport rate of sucrose in the phloem by about twofold. There could be several reasons for this enhancement of the volume flow rate by K, including higher rates of sucrose synthesis (White, 2020), enhancement of phloem loading (Deeken et al., 2002), or direct osmotic effects of K within the sieve tubes.

Along the pathway between source and sink, concentrations and composition of the phloem sap may change considerably for various reasons, including leakage, unloading to and reloading from transient storage compartments along the axial pathway (Fig. 5.22, transport phloem) and xylem-phloem transfer (Aubry et al., 2019). Photosynthates may leak from the sieve tubes; hence, retrieval becomes important to drive the pressure flow and to supply the sink. Along the pathway, retrieval of sucrose is mediated by the same mechanism (sucrose-proton cotransport) as phloem loading in the source tissue (Milne et al., 2018). Leakage (or unloading) along the pathway may serve several functions, such as (1) supplying sucrose as an energy source to surrounding tissues; (2) transient storage of nonstructural carbohydrates, for example, in stems of trees (Furze et al., 2018) or in leaf sheaths and stems of cereals and forage grasses (Berthier et al., 2009; Schnyder, 1993); and (3) adjustment of the solute composition in the sieve tubes according to the demand of the sink. In soybean, phloem sucrose concentration was found to decrease from 336 mM in the leaves to 155 mM in the roots as a growth (utilization) sink, with a corresponding decrease in pressure from 600 to 180 kPa (Fisher, 1978). By contrast, in rice the solute concentration in the phloem sap increases from the source leaves toward the panicles as a storage sink (Table 5.8). This increase is due to sucrose because the K concentration decreases. Despite a similar total

TABLE 5.7 Composition of phloem sap and rate of phloem sap exudation of castor bean (*Ricinus communis*) plants at different K supply.

	K supply in the growth medium	
	0.4 mM	1.0 mM
Concentration (mM) in phloem sap		
Potassium	47	66
Sucrose	228	238
Osmotic potential (kPa)	−1250	−1450
Exudation rate [mL (3 h)$^{-1}$]	1.4	2.5

Source: Based on Mengel and Haeder (1977).

TABLE 5.8 Concentration of various solutes in phloem sap of rice (*Oryza sativa*).

Solutes (mM)	Site of collection	
	Leaf sheath (7–8 leaf stage)	Uppermost internode (1 week after anthesis)
Sucrose	206	574
Amino acids	103	125
K	147	40
ATP	1.63	1.76

Source: From Hayashi and Chino (1990).

concentration of amino acids (Table 5.8), the composition differed: toward the sink, the proportion of glutamine and arginine increased at the expense of glutamate and asparagine (Hayashi and Chino, 1990).

The shift in the ratio of sucrose/K in the phloem sap (Table 5.8) reflects the demand at the sink sites. Developing grains of cereals, as starch storing organs with low water content, have a high demand for sucrose but a low K demand, particularly at later stages of grain filling. Thus, the contribution of K to the total osmotic potential of the phloem sap in the peduncles of wheat ears decreases from 8% to 2% within 5 weeks after anthesis (Fisher, 1987). The increase in sucrose/K ratio (Table 5.8) is most likely the result of mobilization of carbohydrates (starch, fructans) in the stem tissue and subsequent sucrose loading into the phloem (Fig. 5.22, transport phloem). Thus, K is replaced by sucrose and presumably transferred into the vacuoles of the stem tissue, suggesting that the turgor of individual sieve tubes may be regulated along the pathway from source to sink and that K plays an important role in this regulation (Lang, 1983). Along the pathway, K may therefore not only fulfill a role in phloem loading of sucrose but can also provide a means for regulating pressure-driven solute flow from source to sink (Martin, 1989).

5.6.3 Phloem unloading

The release of solutes from the phloem into the surrounding tissue at the sink sites is regulated by the sink strength, that is, the capacity of a tissue or organ to accumulate or metabolize photosynthates (Zhou et al., 2009; Fig. 5.22, release phloem). Phloem unloading can be symplasmic or apoplasmic. In this section, only the transport of assimilates from the SE–CC complex to the adjacent sink cells is considered, including the transport processes on the post-SE pathway but excluding uptake into sink cells. For processes related to solute uptake into the sink cells see Section 5.8.

The sinks for solutes (sugars, amino acids, nutrients) delivered in the SEs can be differentiated into (1) utilization sinks, such as root tips, shoot apices, and stem elongation zones in which photosynthates are utilized for growth, or as substrates for root cell exudates, and (2) storage sinks that accumulate photosynthates. Storage sinks include storage roots (e.g., sugar beet, sweet potato, cassava, carrot), stems (e.g., sugarcane) and other vegetative shoot organs (e.g., tubers), as well as generative organs like fleshy fruits and seeds. In utilization sinks there is evidence for symplasmic phloem unloading, and SEs possess many plasmodesmata linking them with adjacent meristem cells (Patrick, 1997; Fig. 5.22, release phloem). For utilization sinks such as vegetative apices, young leaves, and root tips, it has been demonstrated that a 27-kD protein (the jellyfish green fluorescent protein) can be unloaded symplasmically from the phloem into sink tissues (Imlau et al., 1999), indicating the high size-exclusion limit of plasmodesmata on the postphloem pathway of these sinks. It is thought that a variety of specialized plasmodesmatal connections regulate the selective symplasmic movement of solutes from the phloem (Lee and Frank, 2018).

In generative storage organs such as seeds of cereals and legumes, the filial tissues (embryo, endosperm) are symplasmically isolated from the phloem in the maternal seed tissues (seed coat). Therefore, the transport of photoassimilates and nutrients from the phloem to the filial tissues is always associated with unloading from maternal tissues into the apoplasm and subsequent loading into the symplast of filial tissues (Milne et al., 2018; Patrick, 1997; Fig. 5.22, release phloem). Apoplasmic unloading of solutes delivered in the phloem is generally not located at the SE−CC complex, but at sites more distant from the phloem at the interface between maternal and filial tissues (Patrick and Offler, 2001).

In other storage sink organs, the mode of phloem unloading can change during development (Ma et al., 2019). A switch from apoplasmic phloem unloading in early stages of organ development to symplasmic phloem unloading in later stages has been found in vegetative storage sinks such as potato tubers (Viola et al., 2001) and sugar beet storage roots (Godt and Roitsch, 2006). This switch may be caused by modification of the conductivity of plasmodesmata connecting the phloem with the surrounding sink cells (Ruan et al., 2001) and is also associated with changes in the expression of sugar transporters (Lalonde et al., 2003) and metabolism. For example, in sugar beet, the transition from apoplasmic to symplasmic unloading is associated with reduced activity of the cell wall invertase that cleaves sucrose into fructose and glucose in the apoplasm, and thus maintains low apoplasmic sucrose concentrations (Godt and Roitsch, 2006).

Symplasmic unloading from the SE−CC complex through plasmodesmata can take place by diffusion and/or mass flow that is linked to metabolism and compartmentation in sink cells (Lalonde et al., 2003). Sucrose unloading from cells along the post-SE pathway to the apoplasm in some sink organs may also follow a transmembrane concentration gradient established by extracellular invertases that cleave sucrose to hexoses. Members of the SWEET protein family have been implicated in the release of sucrose to the apoplasm, which is retrieved by members of the SUT protein family (Jeena et al., 2019; Milne et al., 2018; Pommerrenig et al., 2020). In the seed coats of bean (*Phaseolus vulgaris*) and faba bean (*Vicia faba*), sucrose release to the apoplasm may account for 50% of the total sucrose flux (Lalonde et al., 2003). The induction of SWEET proteins in the host by pathogens and parasites, which causes a localized increase in the release of sucrose from the phloem, is how these organisms acquire the carbon they require for growth (Jeena et al., 2019; Pommerrenig et al., 2020). Members of the UmamiT protein family have been associated with the release of amino acids from the phloem to the apoplasm in seeds and roots, whereas members of the AAP and CAT families of H^+/amino acid symporters have been linked with the retrieval of amino acids from the apoplasm in these tissues (Tegeder, 2014; Tegeder and Masclaux-Daubresse, 2018).

Negative feedback regulation of phloem unloading is exerted by high sucrose concentrations in a utilization sink. In legume seeds, a turgor-sensitive component is involved in phloem unloading. Enhanced sucrose uptake by filial tissues decreases the osmolality of the seed apoplasmic solution, and, consequently, raises the turgor of seed coat cells (maternal tissue). If seed coat turgor exceeds a set point (about 2 kPa), the activity of the sucrose transporters responsible for release into the apoplasm is enhanced (short-term turgor regulation), and the long-term rates of phloem import are increased (Zhang et al., 2007a). The increase of seed coat turgor also leads to increased efflux of K and accompanying anions (Walker et al., 2000).

5.7 Sink formation

In crop species in which storage organs such as fruits, seeds, and tubers represent yield, the effects of nutrient supply on yield response curves often reflect sink limitations imposed by either a deficiency or an excess supply of nutrients during critical periods of plant development. These periods include flower induction, pollination, and tuber initiation. The effects can be both direct (e.g., deficiency of a nutrient needed for a particular metabolic process) and indirect (e.g., a nutrient deficiency or excess inducing changes in shoot architecture or flowering time through effects on photosynthesis or phytohormonal balance).

5.7.1 Shoot architecture for grain/seed yield formation

For cereal crops, the branch number and panicle structure (inflorescence) are two of the most important traits determining grain yield (Kyozuka et al., 2014). The number of branches (or tillers in rice and wheat) is determined by the initiation of axillary meristems and the elongation of axillary buds (Bennett and Leyser, 2006) that are influenced by the growth environment, particularly light period and intensity, and the supply of nutrients and water (de Wit et al., 2016; Feng et al., 2016; Kudoyarova et al., 2015; Wang et al., 2018). Nitrogen and P supplies influence shoot architecture, particularly tiller number (Luo et al., 2017; Umehara et al., 2010). For example, increasing N to ensure high grain yield of rice occurs through an increase of effective tiller number at low to medium N supply (Fig. 5.24).

Nitrogen limitation can inhibit cell division and the outgrowth of tiller buds, probably via altering expression of cell cycle-related genes (Luo et al., 2017). Increasing N supply accelerates cell division and cell expansion and enables the production of greater biomass. High N also triggers CYT transport from roots to shoots, which regulates cell division (Landrein et al., 2018; Wang et al., 2018). Some transporters of nitrate and amino acids may be involved in regulating shoot branching. For example, nitrate transporters belonging to nitrate and peptide transporter family (NPF) in rice (OsNPF7.1, OsNPF2, OsNPF7.4, and OsNPF7.7) have been reported to participate in modulating plant tiller number and panicle architecture, whereas putative amino acid transporters, such as AAP3 and AAP5, are necessary for maintaining tillering bud outgrowth and effective tiller number in rice (Luo et al., 2020).

Phosphorus deficiency generally decreases plant branching or tiller bud outgrowth. In rice, the influence of P nutrition on shoot architecture appears to be through regulation of strigolactone (SL) biosynthesis (Umehara et al., 2010). High P supply decreases SL biosynthesis and promotes shoot branching or tiller bud outgrowth (Umehara et al., 2010). However, it is noteworthy that leaf concentrations of N and P in rice change during tiller development, with P deficiency more likely during early growth stages and N deficiency suppressing P starvation responses (Takehisa and Sato, 2019).

FIGURE 5.24 Effects of increasing N fertilizer applications on (A) the phenotype, (B) the effective tiller number, and (C) grain yield per panicle of rice (*Oryza sativa*). A japonica cultivar was grown in a paddy field with four different supplies of N fertilizer, and plants were transferred to pots at the late grain-filling stage for photographing. N1, N2, N3, and N4 denote 75, 150, 250, and 350 kg N ha^{-1}, respectively. Data are means ± SD ($n \geq 8$ plants).

5.7.2 Flower initiation and development

Flowering represents the transition from vegetative to reproductive growth. Flowering time is determined by multiple endogenous and exogenous factors to ensure that the transition to reproduction coincides with favorable environmental conditions. In addition to light and temperature, nutritional status also influences flowering time (Cho et al., 2017).

The effect of N on flowering time depends on the form of N supplied, plant N status, and plant species. It is often observed that a low N supply accelerates and a high N supply delays flowering in crops (Vidal et al., 2014). However, the response of flowering time to N supply is a U-shaped relationship, with flowering time being delayed by either deficiency or excess of N in arabidopsis (Lin and Tsay, 2017) and rice (Zhang et al., 2021a; Fig. 5.25). Many studies have shown a lack of response of flowering time to N supply in wheat and barley (Hall et al., 2014). The photoreceptor CRY1 and a flowering activator FLOWERING BHLH 4 (FBH4) in arabidopsis, and a crucial circadian clock component Nhd1 (N-mediated heading date-1) in rice (orthologue of CCA1 in arabidopsis) have been identified as key components in the regulation of flowering time by N (Sanagi et al., 2021; Yuan et al., 2016; Zhang et al., 2021a). In rice, Nhd1 can balance flowering time, N-use efficiency, and photoperiod (Fig. 5.26; Zhang et al., 2021a). Two microRNAs have been implicated in the regulation of flowering time by N: miR156 is upregulated and miR172 is downregulated by N deficiency (Pant et al., 2009). Moreover, miR156 seems to be a potential regulatory hub for integrating the signals of plant age, photosynthesis (sugar), and N supply in the control of flowering (Yu et al., 2013).

In contrast to N, low P supply delays flowering and excess P accelerates flowering (Kant et al., 2011; Ye et al., 2019), indicating an antagonistic interaction between P and N (nitrate) on flowering time (Lin and Tsay, 2017). The effect of P on flowering time is not via tissue P concentration because mutants with decreased P accumulation do not show delayed flowering (Li et al., 2017). However, it has been observed that the ubiquitin-conjugating E2 enzyme PHO2, which functions in P homeostasis, interacts with GIGANTEA (SGI), a key regulator of flowering time, and flowering is delayed in mutants lacking either of these proteins (Kim et al., 2011; Li et al., 2017). It is possible that P nutrition influences flowering time via effects on gibberellin concentration (Lin and Tsay, 2017).

Floret development in wheat and barley is influenced by the availability of photosynthates as well as N and P-containing assimilates during the critical growth period immediately before heading (Abbate et al., 1995; Prystupa et al., 2004). In field experiments with durum wheat (Ferrante et al., 2010) and barley (Arisnabarreta and Miralles, 2010), N fertilizer application increased the number of fertile florets mainly by reducing the degeneration of initiated florets during the late part of stem elongation.

In apple trees, flower formation is influenced to a greater extent by the time and form of N application than by the amount of N supplied. Compared to continuous nitrate supply, a short-term application of ammonium to roots more than doubled both the percentage of buds developing inflorescences and the arginine content in the stem (Table 5.9). Arginine is a precursor of polyamines that accumulate particularly in leaves of plants supplied with high amounts of

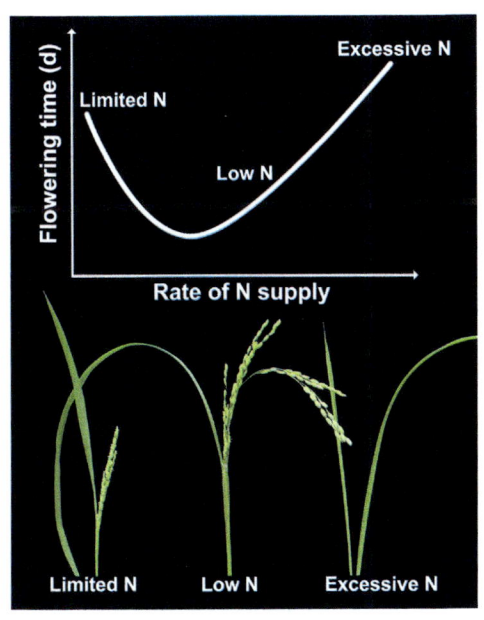

FIGURE 5.25 Effects of nitrogen supply on the flowering time of rice (*Oryza sativa*). *Based on Zhang et al. (2021a).*

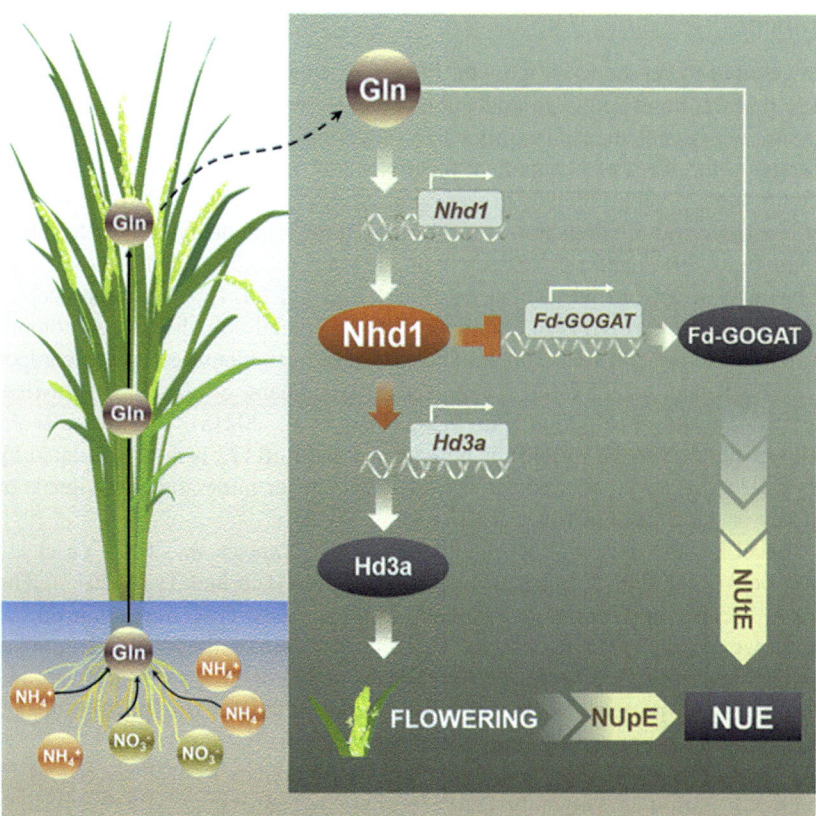

FIGURE 5.26 A rice floral regulator (Nhd1) balances flowering time and NUE by activating the florigen gene *OsHd3a* and downregulating the expression of the N-assimilation gene *OsFd-GOGAT* and the activity of OsFd-GOGAT. *Gln*, Glutamine; *Nhd1*, nitrogen-mediated heading date-1; *NUE*, N-use efficiency; *NupE*, N-uptake efficiency. *From Zhang et al. (2021a), with permission from Elsevier.*

TABLE 5.9 Flower initiation in apple (*Malus domestica*) trees supplied with N and polyamines.

Treatment	Percentage flowering	Stem arginine concentration (mg g^{-1} dry wt)
Control, continuous nitrate supply	15	1.1
Ammonium for 24 h[a]	37	2.6
Ammonium for 1 week	40	2.3
Putrescine[b]	51	
Spermine[b]	47	
Ammonium[b]	50	

[a] 8 mM NH_4^+ in the nutrient solution.
[b] 8 mM petiole infiltration.
Source: Based on Rohozinski et al. (1986).

ammonium (Paschalidis et al., 2019). The involvement of polyamines in ammonium-induced enhancement of inflorescence development in apple trees is indicated by the similar effects obtained by infiltrating polyamines or ammonium into the petioles (Table 5.9). Because the apple trees in this study were well supplied with N throughout the growing season, it is unlikely that these effects on flower initiation (i.e., on developmental processes) are related to a direct nutritional role of N. Instead, it is likely that some N compounds such as polyamines function as secondary messengers in flower initiation. The involvement of polyamines in the biochemical control of the events leading to gametophyte

TABLE 5.10 Shoot growth, flower induction, and cytokinin (CYT) in xylem exudate of apple root stock with different forms of N supply.

Form of N supply	Shoot length (cm)	No. of lateral shoots (spurs)	Flowering bud (% of emerged)	CYT (nmol kg^{-1} shoot fw)
NO_3^-	326	6.4	7.4	0.02
NH_4NO_3	268	6.0	8.2	3.7
NH_4^+	209	8.9	21	8.3

Source: Based on Gao et al. (1992).

formation, to fertilization, and fruit development has been demonstrated, for example, in apricot (Alburquerque et al., 2006), kiwi (Falasca et al., 2010), and maize (Liang and Lur, 2002). In addition to its effects on polyamine metabolism, the application of ammonium also increases CYT concentration in the xylem exudate and the number of flower-bearing lateral branches (Table 5.10). The promotion of flower morphogenesis by CYT is well documented for various plant species (Herzog, 1981; Bonhomme et al., 2000).

Plant nutritional status may also affect flower initiation and seed set by increasing the supply of photosynthates during critical periods of the reproductive phase (Arisnabarreta and Miralles, 2010; Corbesier et al., 1998). Seed number per plant can be increased, for example, by high concentrations of sucrose prior to flower initiation (Waters et al., 1984) and by high light intensity (Reynolds et al., 2005; Stockman et al., 1983).

Flowering initiates the development of new sink organs, that is, flowers and seeds. Therefore, appropriate timing of flowering is critical for ensuring the optimum use of light, heat, water, and nutrient resources. The effects of nutrients on the time of flowering are genetically controlled. For example, the effect of N fertilization on flowering time in rice varies among genotypes, which was associated with a single-nucleotide polymorphism in the promoter of the *Nhd1* gene (Zhang et al., 2021a). Thus, there is potential for breeding for particular responses in flowering time to crop nutrition appropriate to a given environment.

5.7.3 Pollination and seed development

The number of seeds and/or fruits per plant can also be affected by nutrient supply. This is clearly the case with various micronutrients. In cereals, in particular, Cu deficiency affects the reproductive phase (Table 5.11). The critical period in Cu-deficient plants is the early booting stage at the onset of pollen formation (microsporogenesis). When Cu deficiency is severe, no grains are produced even though the straw yield is quite high because of enhanced tiller formation (due to the loss of apical dominance of the main stem). With increasing Cu supply, grain yield increases more strongly than straw yield. These results are a good example of both sink limitation on yield and deviation from the typical response curve (Fig. 5.1) between grain yield and nutrient supply.

The primary causes of failure of grain set in Cu-deficient plants are inhibition of anther formation, the production of a much smaller number of pollen grains per anther, and particularly the loss of pollen viability (Graham, 1975). The percentage of pollen abnormalities was greater in transgenic plants with low expression of the Cu transporters COPT1 or YSL3 than in wild-type plants, but the formation of abnormal pollen was reduced by the addition of Cu (Sancenón et al., 2004; Sheng et al., 2021).

In principle, similar results to those found for Cu deficiency are also obtained with Zn and Mn deficiencies. In maize, Zn deficiency prior to microsporogenesis ($\sim$35 days after germination) decreases pollen viability and cob dry weight by about 75% (Sharma et al., 1990). Zinc-finger proteins are essential for the proper progression of male and possibly also female meiosis (Kapoor and Takatsuji, 2006) and participate in processes that influence shedding of floral organs (Cai and Lashbrook, 2008). Vegetative growth is decreased less than grain yield in Mn-deficient maize plants (Table 5.12). In Mn-deficient plants, anther development is delayed and fewer and smaller pollen grains are produced with low germination rates. By contrast, ovule fertility is not significantly affected by Mn deficiency (Sharma et al., 1991).

There is evidence that Fe is also involved in inflorescence formation and pollen production (Takahashi et al., 2003). In transgenic tobacco with low internal nicotianamine (NA) concentrations resulting from the constitutive expression of the nicotianamine consuming enzyme nicotianamine aminotransferase, flowers were abnormally shaped and sterile.

TABLE 5.11 Number of tillers, straw, and grain yield of wheat (*Triticum aestivum*) (4 plants pot^{-1}) grown in low-Cu soil at varied Cu supply.

	Cu supply (mg pot^{-1})			
	0	0.1	0.4	2
Number of tillers pot^{-1}	22	15	13	10
Straw yield (g pot^{-1})	7.7	9.0	10	11
Grain yield (g pot^{-1})	0.0	0.5	3.5	12

Source: Based on Nambiar (1976).

TABLE 5.12 Growth, fertilization, and grain yield of maize (*Zea mays*) at high or low Mn supply.

Mn supply (µg L^{-1})	Dry weight			Pollen	
	Shoot (g plant^{-1})	Grain (g plant^{-1})	Single grain (mg)	number (no. anther^{-1})	Germination (%)
550	83	69	302	2770	86
5.5	58	12	358	1060	9.4

Source: From Sharma et al. (1991).

Application of a solution containing NA and Fe(III) citrate reversed the morphological abnormalities in flowers and optimized pollen production (Takahashi et al., 2003).

Both production and viability of pollen are also affected by Mo supply (Kaiser et al., 2005). In maize, a decrease in the Mo concentration in pollen was correlated with a decrease in the number of pollen grains per anther as well as a decrease in the size and viability of the pollen grains. It remains unclear whether these changes influence fertilization and grain set. However, it is well documented that preharvest sprouting in maize (Chapter 9) and wheat (Cairns and Kritzinger, 1992) is high in seeds with low Mo concentration and can be decreased by Mo supply to the soil or as foliar spray. In grapevine, Mo deficiency may be the primary cause of the bunch development disorder "millerandage" or "hen and chicken". Millerandage is characterized by unevenly developed grapevine bunches: in the same bunch, fully matured berries are present alongside a large number of fertilized underdeveloped berries as well as unfertilized swollen green ovaries (Kaiser et al., 2005). This grapevine disorder can be prevented by foliar sprays of Mo before flowering (Williams et al., 2005).

Boron is essential for pollen tube growth (White et al., 2021b). Boron deficiency in cereals results in a decrease in grain number and, sometimes, a total absence of fertilization (Matthes et al., 2020). With increasing B supply, vegetative growth is either unaffected or decreased, whereas grain formation is increased (Fig. 5.27). There is a minimum B requirement for fertilization and grain set, which is around 3 mg B in maize. The B requirement of anthers and carpels is higher than that of leaves in cereals such as wheat (Rerkasem and Jamjod, 2004). Fig. 5.27 provides another example of a strict sink limitation induced by nutrient deficiency that would result in a yield response curve quite different from the typical curve. In wheat (Nachiangmai et al., 2004) and barley (Jamjod and Rerkasem, 1999), genetic variation occurs in the degree of pollen sterility or the ability to set grain under B deficiency (White et al., 2021b). In wheat, B concentrations in the ear were greater in an efficient than in an inefficient genotype, and this was associated with a greater ability of the efficient genotype to transport B from the rooting medium to the ear via the xylem (Nachiangmai et al., 2004). Low B supply not only inhibits flowering and seed development but may also result in low B concentration in seeds, even in plants without visual symptoms of B deficiency. Seeds with a low B concentration have a low germination rate and produce a high percentage of abnormal seedlings (Bell et al., 1989).

In some crops, such as grain legumes, drop of flowers and developing pods are major yield-limiting factors (Patrick and Stoddard, 2010). Deficiency of N or P during the flowering period increases flower and pod drop and thus depresses

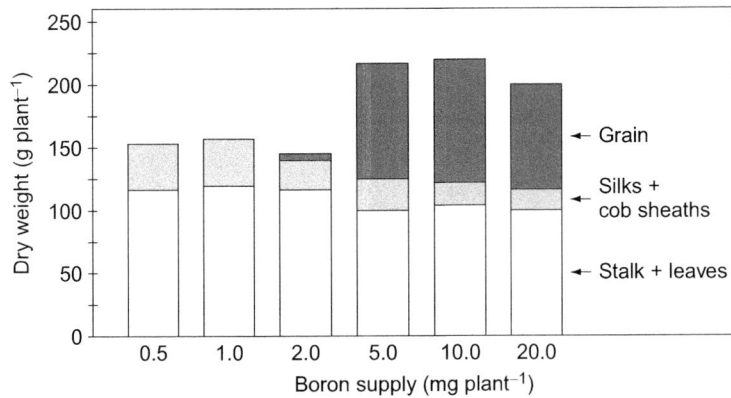

FIGURE 5.27 Effect of B supply on the production and distribution of dry matter in maize (*Zea mays*) plants. *Based on Vaughan (1977).*

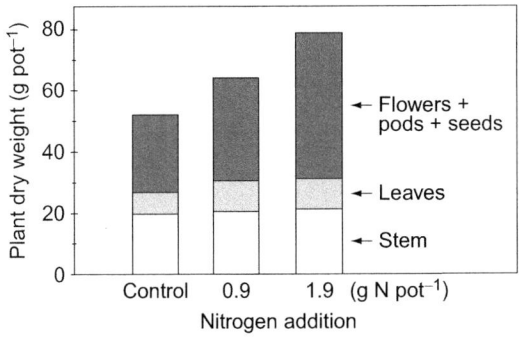

FIGURE 5.28 Total dry weight and dry weight distribution in shoots of white mustard (*Sinapis alba*) plants with addition of various amounts of N at the onset of flowering. *Based on Trobisch and Schilling (1970).*

seed yield (Brevedan et al., 1978; Lauer and Blevins, 1989; Roy et al., 2016). Supplying sufficient amounts of N and P during this critical phase is therefore effective in reducing flower and pod drop and increasing final seed yield.

It is well documented that N deficiency also decreases grain number in cereals such as wheat (Abbate et al., 1995) and maize (Uhart and Andrade, 1995). The reduction of grain number in N-deficient plants is associated with decreased supply of assimilates and N to the generative organs during the critical period determining grain set around flowering, which in turn may result from low photosynthetic activity/area or reduced assimilate partitioning to the generative organs (D'Andrea et al., 2008; Paponov and Engels, 2005). However, competition for N rather than for carbohydrates supplied from the source leaves can be the main limiting factor for seed yield in brassicas, such as mustard and oilseed rape. Developing seeds and leaves compete for N so that seed set, seed growth, and final seed yield are determined primarily by the size of the N pool in the vegetative parts. In brassicas, flower differentiation at the auxiliary stems occurs after the onset of flowering of the main stem and is dependent on the availability of N during this period. Additional N application at the onset of flowering, therefore, leads to an increase in seed number and yield (Fig. 5.28), demonstrating that source limitation can be imposed by N as well as carbohydrates.

5.7.4 Formation of vegetative sink organs

In root and tuber crops such as sugar beet or potato, the induction and growth rate of the storage organs are strongly influenced by environmental factors. In root and tuber crops there is often a strong sink competition between vegetative shoot growth and storage tissue growth for long periods after the onset of storage growth. This competition is particularly evident in so-called indeterminate genotypes of crop species, for example, potato (Kleinkopf et al., 1981). In general, environmental factors (e.g., high N supply) that increase vegetative shoot growth, delay the initiation of the storage process and decrease growth rate and photosynthate accumulation in storage organs, for example, in sugar beet (Forster, 1970) and potato (White, 2018).

A high and continuous N supply to the roots of potato delays or even prevents tuberization (White, 2018). After tuberization, the tuber growth rate is also reduced by high N supply, whereas the growth rate of the vegetative shoot is enhanced. The effect of N supply on tuber growth rate is illustrated in Table 5.13. Resumption of tuber growth rate after interruption of N supply indicates that sink competition between the vegetative shoot and tubers can be manipulated readily by the N supply. The form of N also affects tuber growth, with nitrate producing more and thicker stolons than NH_4-N (Gao et al., 2014).

TABLE 5.13 Growth rate of potato tubers at different rates of nitrate supply to the roots of potato plants.

Nitrate concentration (mM)	Nitrate uptake (mmol day^{-1} plant^{-1})	Tuber growth rate (cm^3 day^{-1} plant^{-1})
1.5	1.18	3.24
3.5	2.10	4.06
7.0	6.04	0.44
Nitrogen supply withheld for 6 days	–	3.89

Source: From Krauss and Marschner (1971).

FIGURE 5.29 Secondary growth and malformation of potato (*Solanum tuberosum*) tubers induced by alternating high and low N supply to the roots. *Courtesy A. Krauss.*

In potato, cessation of tuber growth caused by a sudden increase in N supply to the roots induces "regrowth" of tubers, that is, the formation of stolons on the tuber apex (Krauss and Marschner, 1976, 1982). Interruption and resupply of N, therefore, can result in the production of chain-like tubers or so-called secondary growth (Fig. 5.29). After a temporary cessation of growth, resumption of the normal growth rate is usually restricted to a certain area of the tubers (meristems or "eyes"), leading to typical malformations and knobbly tubers, often observed under field conditions after periods of transient drought. Similar effects on cessation of tuber growth and regrowth occur when growing tubers are exposed to high temperatures, which rapidly inhibit starch synthesis and lead to the accumulation of sugars in the tubers (Krauss and Marschner, 1984; Van den Berg et al., 1991), followed by a decrease in ABA concentrations in the tubers and regrowth.

The effects of N supply on tuber growth rate and regrowth are due, in part, to N-induced changes in the phytohormone balance in the vegetative shoots and tubers. Interruption of N supply results in a decrease in CYT export from roots to shoots as well as in decreased sink strength and growth rate of the vegetative shoot. A corresponding increase in the ABA/GA ratio of the shoots seems to trigger tuberization (Rodríguez-Falcón et al., 2006). In agreement with this, tuberization can also be induced by the application of either ABA or the GA antagonist CCC (chlorocholine chloride) (Krauss and Marschner, 1976) or by the removal of the shoot apices, the main sites of GA synthesis (Hammes and Beyers, 1973). On the other hand, regrowth of tubers induced by a sudden increase in N supply is correlated with a decrease in the ABA/GA ratio not only in the vegetative shoots but also in the tubers, where the GA concentration increases twofold, while the ABA concentration decreases to less than 5% of that in normally growing tubers (Krauss, 1978b).

5.8 Sink activity

Partitioning of phloem-delivered compounds (e.g., sugars, amino acids, nutrients) among competing sink organs is governed by sink activity, that is, the relative capacity of specific sink organs to unload nutrients from the phloem and to use them for growth and storage. Phloem unloading from the SE−CC complex or the post-SE pathway has been discussed above. In this section, uptake of assimilates and nutrients into sink cells of storage organs is considered.

The development of generative sink organs, for example, seeds of cereals and legumes, can be divided into a pre−storage phase that is dominated by cell division and cell extension, and a storage phase in which storage compounds are accumulated (Patrick and Stoddard, 2010; Weber et al., 2005). The pre−storage phase is often characterized by high activity of invertase in the cell wall and vacuoles of sink organs (Weber et al., 2005). Sucrose unloaded from the phloem is cleaved into hexoses, and their uptake into sink cells is mediated by H^+/hexose symporters. A high intracellular glucose/sucrose ratio appears to be a key component of the regulatory network that induces and sustains mitotic activity in this phase and, therefore, determines potential seed size (Ruan, 2014; Ruan et al., 2010; Weber et al., 2005).

In the storage phase, cell wall invertase activity is low. In legumes and temperate cereals, sucrose released from the phloem is imported into sink cells via sucrose/H^+ symporters (SUTs). In plant species such as broad bean, common bean, barley, and wheat, low intracellular glucose concentrations and high ethylene concentrations induce the formation of transfer cells in the filial tissues. These transfer cells are characterized by wall ingrowths to amplify the membrane surface area, and a high density of H^+-ATPases, sucrose transporters, and amino acid transporters in their plasma membrane (Patrick and Offler, 2001). High intracellular sucrose concentrations may induce the expression of key enzymes involved in starch biosynthesis such as sucrose synthase and ADP-glucose-pyrophosphorylase (Weber et al., 2005). For starch synthesis, sucrose is cleaved within the storage cells by sucrose synthase to UDP-glucose and fructose. After further metabolic conversion of sugars, glucose-6 phosphate is imported through the plastid membranes into the amyloplasts and ADP-glucose is synthesized by the plastidic ADP-glucose pyrophosphorylase. From ADP-glucose, the glucose can be transferred to starch by various forms of starch synthase and the starch branching enzyme. In cereals, ADP-glucose is synthesized by cytosolic ADP-glucose pyrophosphorylase and then imported into the amyloplasts for starch synthesis (Smith, 2008).

In sugarcane and sugar beet, sucrose is the main storage compound. The sucrose concentration in the vacuoles of storage cells can exceed 500 mM, which is about 10 times higher than in the cytosol of storage cells (Saftner et al., 1983). Tonoplast-located H^+-ATPases and pyrophosphatases maintain a low pH inside the vacuole and provide the driving force for transport across the membrane against a concentration gradient (Maeshima, 2001). Vacuolar sucrose transporters have been identified in mesophyll cells of various plants, including barley, rice, and *Arabidopsis* (Endler et al., 2006; Neuhaus, 2007). In storage cells of sugar beet roots, the accumulation of sucrose is stimulated by K and, even more so, by Na (Saftner and Wyse, 1980).

The yield of crop plants is influenced by the length of the storage phase that is shortened by, for example, drought stress, high temperature (Barnabás et al., 2008), and nutrient deficiency; these stresses reduce the weight (size) of the sink organs, but not the number of sink organs. There is substantial evidence that phytohormones are involved in premature ripening. An example of this is shown in Table 5.14 for K-deficient wheat. In these plants, and particularly 4−6 weeks after anthesis, the concentration of ABA in the grains is higher than that in grains of plants well supplied with K. Correspondingly, the grain-filling period in K-deficient plants is shorter, and the weight of a single grain at maturity is lower than that in K-sufficient plants. High ABA concentrations in grains coincide with a strong decline in their sink activity. Therefore, the high ABA concentrations in the flag leaves of K-deficient wheat plants (Haeder and Beringer, 1981) and a correspondingly higher ABA import to the developing grains may be responsible for the premature ripening and not the source limitation of a nutrient per se.

In cereals and grain legumes, proteins are important storage compounds in sink organs (Shewry, 2007). Thus, high import of amino acids into sink organs is needed. Amino-N unloaded from the phloem is imported into storage cells by

TABLE 5.14 ABA content and weight of grains of wheat (*Triticum aestivum*) after anthesis at high or low (deficient) K supply.

K supply	ABA content (ng per grain) days after anthesis				Days from anthesis to full ripening	Single grain weight (mg)
	28	35	38	44		
Low	7.7	13	17	2.2	46	16
High	3.7	4.4	nd[a]	9.4	75	34

[a]nd: not determined.
Source: Based on Haeder and Beringer (1981).

facilitated diffusion or amino acid/H$^+$ symport (Lalonde et al., 2003; Zhang and Turgeon, 2018). In the storage phase of seed development, the onset of storage protein accumulation is linked to a sucrose signal from the onset of sucrose/H$^+$ symporter activity (Rosche et al., 2002).

Results from many N fertilization experiments show that in cereal grain, total protein content is strongly associated with N supply to the developing grains (Barneix, 2007). This is also supported by observations in the *opaque 2* mutant of maize, in which downregulation of a specific class of storage proteins is associated with a compensatory increase in N storage in other seed proteins. Thus, the total amount of reduced N stored in mutants and nonmutants does not differ (Tabe et al., 2002). Moreover, the expression of transgenes encoding specific storage proteins does not increase the total amount of amino acids that are stored in seeds (Tabe et al., 2002). In other crop species, however, there is evidence for sink limitation of protein contents in storage organs. In potato, tuber-specific expression of the gene encoding a seed protein from *Amaranth* (*AmA1, Amaranth Albumin 1*) increased tuber protein concentration by up to 60%, indicating that in wild-type potato, protein concentration was limited by sink activity, that is, the capacity of tuber tissue for protein synthesis (Chakraborty et al., 2010).

In addition to organic compounds such as carbohydrates, proteins, and lipids, nutrients are also accumulated in storage organs. The uptake of nutrients across the plasma membrane of cells in storage organs is generally mediated by the same transport mechanisms that mediate nutrient uptake into root and leaf cells (Chapters 2 and 3). For example, in developing seeds of bean (*P. vulgaris*), the uptake of K and other monovalent cations into cells of developing cotyledons is mediated by cation channels (Zhang et al., 2004). For loading Fe and other micronutrients including Mn and Cu into seeds, an important role is played by transporters of the YSL transporter family that mediate transport of metals complexed with nicotianamine (NA), for example, OsYSL2 in rice (Ishimaru et al., 2010) and AtYSL1 and AtTSL3 in arabidopsis (Conte and Walker, 2011; Curie et al., 2009). In rice, OsYSL15 is also involved in phloem Fe unloading and loading of Fe into seeds, whereas OsYSL9 is responsible for Fe(II)-NA translocation from endosperm into embryo in developing seeds at the grain filling stage (Nikolic and Pavlovic, 2018). In arabidopsis, AtOPT3 (a member of the oligopeptide transporter family) is implicated in loading Fe into developing seeds (Zhai et al., 2014). Arabidopsis mutants defective in the expression of AtOPT3 have reduced accumulation of Fe in seeds but not in other tissues (Stacey et al., 2008).

The transport of Fe and other micronutrients in plants is facilitated by chelating compounds, such as citrate, NA, and deoxymugineic acid, that form soluble complexes with micronutrient cations (Chapter 3; Morrissey and Guerinot, 2009), and the availability of such chelators can determine concentrations of micronutrients in the seed. For example, overexpression of the barley *HvNAS1* gene (involved in NA biosynthesis) in rice increased endogenous NA and phytosiderophore concentrations in roots, shoots, and seeds, thereby increasing Fe and Zn concentrations in seeds three- and twofold, respectively (Masuda et al., 2009).

In arabidopsis seeds, Mn, Zn, and Fe are stored in vacuoles in complexes with phytate or other chelators, for example, NA (Morrissey and Guerinot, 2009; Otegui et al., 2002; Roschzttardtz et al., 2009). Iron import into the vacuole may be mediated by VIT1 (vacuolar ion transporter 1) that transports Fe^{2+} (Kim et al., 2006) or YSL4 and YSL6 that transport Fe (and Mn)-nicotianamine complexes (Morrissey and Guerinot, 2009). In rice, OsVIT2 may facilitate vacuolar Fe storage in embryos (Zhang et al., 2012). Iron export from vacuoles during seed germination is mediated by NRAMP3 and NRAMP4 (natural resistance-associated macrophage protein family) (Bastow et al., 2018; Lanquar et al., 2005). Thus, in the provascular strand of the embryo endodermis, NRAMP3/NRAMP4 and VIT1 work together to regulate subcellular partitioning of Fe.

In addition to the activity of transporters mediating nutrient loading of seeds and the availability of chelating compounds enhancing the mobility of micronutrients in plants, the storage capacity of sink tissues also plays an important role in regulating seed micronutrient concentrations. For Fe, ferritin is the principal Fe storage protein in all aerobic organisms and can store up to 4500 Fe(III) atoms in its cavity in a soluble and bioavailable form (Harrison and Arosio, 1996). In seeds of cereals and legumes, ferritins are proposed as a major storage form of Fe (Briat et al., 2010). Transgenic rice plants expressing soybean ferritin under the control of a seed-specific promoter accumulated up to three times more Fe in their seed than wild-type plants (Goto et al., 1999). Transgenic rice plants expressing both *AtNAS1* (an *Arabidopsis* gene encoding a nicotianamine-synthesizing enzyme) and *Pvferritin* (a *Phaseolus* gene encoding ferritin) accumulated six times more Fe in their endosperm than wild-type plants (Wirth et al., 2009).

5.9 Role of phytohormones in the regulation of the sink–source relationships

Phytohormones play an important role in the regulation of the growth and development of plants and, thereby, influence sink–source relationships. The synthesis and action of phytohormones are modulated by environmental factors, such as nutrient supply. At least some of the effects of nutrient deficiencies on plant growth and yield are caused by their influence on phytohormone concentrations. Some examples of these effects are given in the following sections.

Phytohormones are chemical messengers, or "signal" molecules for which sites of synthesis and sites of action are usually physically separated. Transport either from cell to cell or from organ to organ is therefore necessary. With the exception of ethylene and brassinosteroids (BRs), phytohormones can be translocated in the phloem and the xylem (Hirose et al., 2008; Notaguchi and Okamoto, 2015; Robert and Friml, 2009; Wilkinson and Davies, 2002). The prevailing direction of transport depends on their site of synthesis and the developmental stage of the plant. Each phytohormone has a broad spectrum of actions; the same phytohormone can influence various processes depending on its concentration and conditions at the sites of action.

5.9.1 Structure, sites of biosynthesis, and main effects of phytohormones

The importance of the five classical classes of phytohormones in higher plants is well established. These are auxins (indole-3-acetic acid, IAA), CYTs, gibberellins (GA), ABA, and ethylene (ET). More recently, several other molecules have also been recognized as phytohormones. These include JA and its derivates, SA, BRs, polyamines, and SLs. The basic molecular structures of the various phytohormone classes are shown in Fig. 5.30, and some of their major characteristics are summarized in Table 5.15.

Auxins are indole derivates of the amino acid tryptophan, the most prominent being IAA (or "auxin"). They are synthesized in meristems or young expanding tissues (Crozier et al., 2000). They can be transported in the phloem and are redistributed locally from cell to cell, with the direction determined by the polar locations in the plasma membrane of the AUX/LAX auxin influx carriers, the PIN auxin efflux carriers, and auxin efflux transporters of the multidrug-resistant/P-glycoprotein subfamily of ATP-binding cassette proteins (Robert and Friml, 2009). Several pathways for irreversible IAA catabolism have been elucidated, but reversible inactivation by O-glycosylation allows IAA-ester conjugates to be stored (Crozier et al., 2000). Auxin promotes cell expansion and cell division and is implicated in apical dominance, shoot elongation, adventitious root development, xylogenesis, and plant tropisms (Table 5.15).

CYTs are synthesized from purine derivatives (Argueso et al., 2009; Crozier et al., 2000; Hirose et al., 2008). They are readily mobile within plants. Although the major sites of their biosynthesis are in roots, and root-to-shoot xylem transport dominates, CYTs are also mobile in the phloem and are transported from source leaves into inflorescences and developing seeds (Hirose et al., 2008; Sasaki et al., 2019). CYTs are degraded by CYT oxidases that are induced in

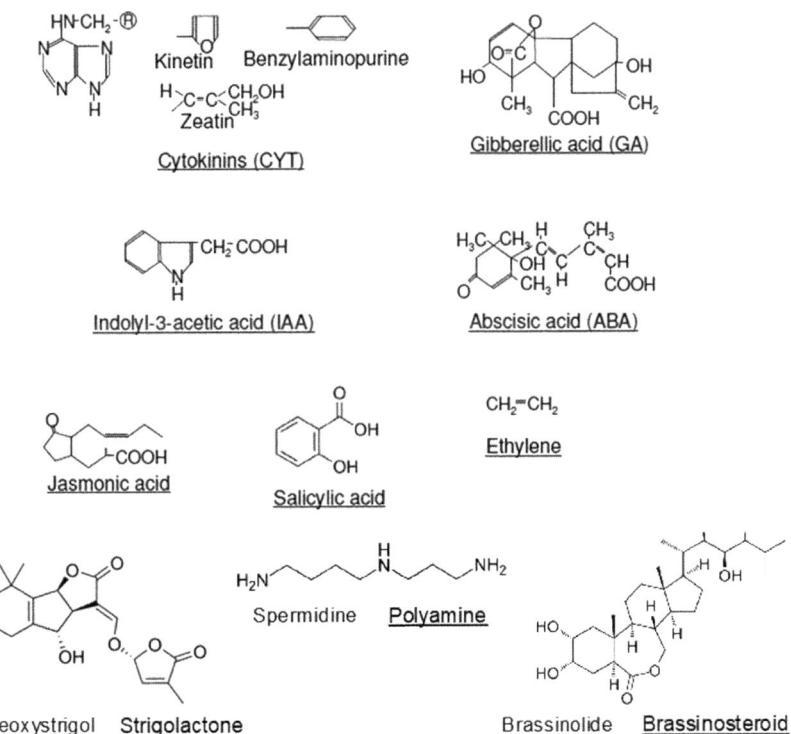

FIGURE 5.30 Molecular structures of phytohormones.

TABLE 5.15 Pathways and main sites of biosynthesis and some major effects of phytohormones.

Auxins (IAA)
Biosynthetic precursors: indole derivates of the amino acid tryptophan.
Main sites of biosynthesis: meristems or young expanding tissues; in dicots mainly the apical meristems and young leaves; prevailing direction of transport basipetally: polar from cell to cell, and some long distance in the vicinity of the phloem.
Effects: promote cell division and expansion, apical dominance, adventitious root development, tropisms.
Antagonists/inhibitors: ABA, coumarins, TIBA, 2,4-D, NAA, and other synthetic auxins.

Cytokinins (CYT)
Biosynthetic precursors: purine derivates (adenine).
Main sites of biosynthesis: primarily root meristems, but also shoot meristems and embryo in seeds. Prevailing long-distance transport via xylem from roots to shoot.
Effects: promote cell division and expansion, stimulate RNA and protein synthesis, suppress auxin-induced apical dominance, delay senescence.

Gibberellins (GA)
Biosynthesis: hemiterpenes to the gibbane carbon skeleton; more than 100 gibberellins with this basic structure have been found.
Main sites of biosynthesis: seeds and developing tissues.
Effects: promote cell expansion, induce enzymic activities (e.g., hydrolases), stimulate shoot elongation, delay leaf and fruit senescence, break dormancy of buds and seeds, induce flowering.
Inhibitors of biosynthesis: CCC, ancymidol, triazoles.

Abscisic acid (ABA)
Biosynthetic precursors: carotenoids violaxanthin and neoxanthin.
Main sites of biosynthesis: fully differentiated tissues of shoots and roots.
Effects: inhibits cell extension, induces stomatal closure, favors abscission of leaves and fruits and enhances or induces dormancy of seeds and buds.
Antagonists/inhibitors: IAA, CYT, GA, fusicoccin.

Ethylene (ET)
Biosynthetic precursor: methionine.
Main sites of biosynthesis: various plant parts and organs.
Effects: promotes seed germination, formation of root hairs, formation of root aerenchyma, epinastic curvature of leaves, flowering, ripening and senescence, defense responses to pests and pathogens.
Antagonists/inhibitors: Co, Ag, polyamines.

Jasmonic acid (JA)
Biosynthetic precursor: linolenic acid.
Main sites of biosynthesis: roots, shoot, fruits.
Effects: inhibits seed germination and root and shoot growth, promotes leaf senescence, fruit ripening and tuber formation, induces tendril coiling, induces defense responses to pests and pathogens.
Antagonist: CYT.

Salicylic acid (SA)
Biosynthetic precursor: phenylalanine.
Main sites of biosynthesis: present in all tissues.
Effects: inhibits leaf senescence and induces flowering, induces thermogenesis in the spadix of voodoo lily (*Dracunculus vulgaris*), induces defense responses to pests and pathogens.

Brassinosteroids (BR)
Biosynthesis: from isopentyl diphosphate, with campesterol as an important intermediate.
Main sites of biosynthesis: pollen, seeds, vegetative tissues.
Effects: influence cell division and cell elongation, promote stem elongation and apical dominance, prevent leaf abscission, enhance stress resistance.
Antagonist: brassinazole.

Polyamines (PA)
Biosynthetic precursors: arginine and ornithine.
Main sites of biosynthesis: present in all tissues.
Effects: stimulate cell division, the synthesis of DNA, RNA and proteins, root initiation, embryogenesis, flower development, fruit ripening and tuber formation, delay leaf senescence.

Strigolactones (SL)
Biosynthetic precursors: carotenoids.
Main sites of biosynthesis: roots.
Effects: inhibit primary root growth, lateral and adventitious root formation and promote root hair formation, stimulate branching and tillering, induce leaf senescence, induce parasitic seed germination, and enhance arbuscular mycorrhizal colonization.
Antagonists/inhibitors: abamine, 2-methoxy-1-naphthaldehyde, N-phenylanthranilic acids, 1,2,3-triazole ureas.

2,4-D, 2,4-Dichlorophenoxyacetic acid; *CCC*, chlorocholine chloride; *IAA*, indole-3-acetic acid; *NAA*, naphthaleneacetic acid; *TIBA*, 2,3,5-tri-iodobenzoic acid.

response to increasing tissue CYT concentrations, and reversibly inactivated by glucosylation (Argueso et al., 2009; Crozier et al., 2000). CYTs promote cell division and differentiation, stimulate transcription and protein synthesis, and delay protein degradation. They suppress auxin-induced apical dominance and delay leaf senescence (Table 5.15).

The terpenoid pathway produces the gibbane carbon skeleton that gives rise to more than 100 giberellin structures in plants (Crozier et al., 2000; Yamaguchi, 2008). Different plant species produce different GAs, and physiological responses are often specific to a subset of GAs, which is likely to be the result of structural specificity in the GA-receptors of the target cells. Gibberellin concentrations are greater in developing seeds than in vegetative tissues. Gibberellins are implicated in stimulating shoot elongation, delaying leaf and fruit senescence, breaking dormancy of buds and seeds, promoting seed germination, and inducing flowering (Table 5.15). In addition to free GAs, plants contain biologically inactive GA-conjugates such as GA-O-β-glucosides and β-glucosyl esters (Crozier et al., 2000; Yamaguchi, 2008).

The synthesis of the "stress phytohormone" ABA occurs rapidly in response to environmental factors, especially a lack of water or N (Wilkinson and Davies, 2002). The precursors for ABA biosynthesis are the carotenoids violaxanthin and neoxanthin (Crozier et al., 2000). Roots and shoots are important sites of ABA biosynthesis, and ABA is highly mobile in both the xylem and phloem and can circulate within the plant (Jiang and Hartung, 2008). In white lupin (*Lupinus albus*), when water supply was adequate 28% of the ABA in xylem sap originated from biosynthesis in the roots, whereas under drought conditions this proportion increased to about 55% (Wolf et al., 1990). In response to water deficit, ABA rapidly induces stomatal closure by opening ion channels in the tonoplast and plasma membrane of guard cells (Amtmann and Blatt, 2009; Wilkinson and Davies, 2002). ABA can be converted to various biologically inactive metabolites, including phaseic acid, dihydrophaseic acid, and glucose conjugates (Crozier et al., 2000; Jiang and Hartung, 2008). In addition to its role in preventing water loss from leaves, ABA promotes desiccation tolerance of seeds and induces dormancy of seeds and buds.

Ethylene is produced in response to abiotic (e.g., hypoxia, chilling, dehydration) and biotic stresses. It is synthesized from methionine by the sequential action of S-adenosyl-L-methionine synthase, 1-aminocyclopropane-1-carboxylic acid (ACC) synthase, and ACC oxidase (Crozier et al., 2000; Dugardeyn and Van Der Straeten, 2008; Lin et al., 2009). Unlike other phytohormones, ET is a gas, and its sites of synthesis and action are in the same tissue. Responses to ET often show concentration optima. Ethylene (1) enhances or represses root, stem, and leaf growth (Dugardeyn and Van Der Straeten, 2008; Pierik et al., 2006), (2) induces aerenchyma formation in roots in response to hypoxia, (3) induces senescence of leaves and flowers (Lin et al., 2009), (4) is required for the ripening of climacteric fruits, such as bananas, apples, avocado and tomatoes (Lin et al., 2009), (5) is important for plant tropisms (Dugardeyn and Van Der Straeten, 2008), (6) accelerates germination, and (7) may play a role in defense responses to pests and pathogens (Bari and Jones, 2009). Enhanced biosynthesis of ET in shoots in response to O_2 deficiency in the rooting medium is thought to be mediated by an increase in xylem transport of the precursor ACC to the shoot (Jackson, 1990). The action of ET as a local signal is demonstrated by its stimulation of root hair development in response to patches of high P availability (White and Hammond, 2008; Zhang et al., 2003).

Jasmonates are synthesized from linolenic acid (Browse, 2009; Crozier et al., 2000). The concentration of jasmonoyl-isoleucine, the active form of JA, increases in response to various abiotic stresses, including drought, UV radiation, and ozone (Browse, 2009; Parthier, 1991). Jasmonate concentrations also increase when plants are challenged by specific pests and pathogens, and JA is thought to be involved in the induction of systemic acquired resistance (Shabala et al., 2016). Jasmonates are highly phloem mobile and are thought to act as systemic signals inducing defense responses (Bari and Jones 2009; Browse, 2009). Jasmonate inhibits seed germination and root and shoot growth, promotes fruit ripening and tuber induction, and accelerates fruit- and seed-induced leaf senescence (Creelman and Mullet, 1997).

SA is synthesized from phenylalanine (Crozier et al., 2000) and is present in all plant tissues. Increasing SA concentration slows leaf senescence and induces flowering, probably by reducing the rate of ET synthesis, and induces thermogenesis in the spadix of voodoo lily (*Dracunculus vulgaris*). The synthesis of SA is also associated with local hypersensitive responses and the induction of systemic resistance to the spread of some fungal, bacterial, and viral diseases (Bari and Jones, 2009; Crozier et al., 2000).

BRs are synthesized from isopentyl diphosphate through campesterol as an important intermediate (Crozier et al., 2000). They have the same basic structure as sterols in plant membranes, such as campesterol, sitosterol, and stigmasterol (Chapter 2). Since their first isolation from the pollen of oilseed rape, more than 60 BRs have been identified in various plant species (Clouse and Sasse, 1998). BRs have strong effects on plant growth and development. They are lipophilic compounds that increase cell elongation and division, acting synergistically to IAA and GA. They promote apical dominance, stem elongation, and the bending of grass leaves even at very low concentrations. For example, at concentrations as low as 10^{-10} M, BRs stimulate elongation growth. Impressive beneficial effects on horticultural crop species have been achieved by application of brassinolids (Rao et al., 2002). Several pathways for BR catabolism have been elucidated, and they can affect BR concentrations in tissues (Crozier et al., 2000). BRs are not transported long distance but can be transported locally between cells (Symons et al., 2008).

Polyamines can be considered another class of phytohormones. The major polyamines are the diamine putrescine (NH_2-CH_2-CH_2-CH_2-NH_2), the triamine spermidine, and the tetramine spermine. In legumes, cadaverine (1,5-diaminopentane) can also be found at high concentrations. They are synthesized from arginine and ornithine (Alcázar et al., 2010), are ubiquitous in plant cells and mobile in both xylem and phloem. Depending upon environmental conditions, their concentrations can range from micromolar to millimolar. In cereals, PA biosynthesis is increased rapidly under a range of environmental stresses, including drought, heat, and salinity (Crozier et al., 2000). Polyamines also accumulate under K deficiency (White and Karley, 2010), or when NH_4^+ is the main N source (Paschalidis et al., 2019). By contrast, PA concentrations are low under N deficiency, even in combination with K deficiency (Altmann et al., 1989). Polyamines stimulate cell division, the synthesis of DNA, RNA and proteins, root initiation, embryogenesis, flower development, fruit ripening, and tuber formation. They also delay senescence, acting synergistically to CYT. They can act as compatible osmotica to protect cells from dehydration when plants are exposed to stresses such as drought, salinity, and chilling (Alcázar et al., 2010) and act as antioxidants to protect cells from oxidative damage, for example when plants are exposed to ozone or metal toxicity stresses (Sharma and Dietz, 2006). Polyamines are effective inhibitors of ethylene biosynthesis; during fruit ripening a decline in PA content is correlated with a steep increase in ethylene production (Gao et al., 2021).

SLs are synthesized from ß-carotene in roots and transported to the shoots via the xylem (Kohlen et al., 2011; Machin et al., 2020). SLs are now recognized as important signaling molecules both in the rhizosphere and in internal phytohormonal functions of higher plants. Their major functions in plants are the regulation of root architecture and development, shoot branching/tillering, secondary growth of plants, and symbiotic and parasitic interactions (Agusti et al., 2011; Brewer et al., 2013; Foo et al., 2013; Rasmussen et al., 2012). SLs interfere with PIN auxin-efflux carriers in roots, thereby altering the auxin optimum needed for lateral root formation (Koltai et al., 2010; Ruyter-Spira et al., 2011). Under low-P conditions, SL biosynthesis is increased in roots (Kapulnik and Koltai, 2012), which increases lateral root formation (Ruyter-Spira et al., 2011), root hair density (Mayzlish-Gati et al., 2012) and promotes AM fungal symbiosis to enhance P uptake (Czarnecki et al., 2013). For a recent review on the biology of SLs and the molecular biology and genomics related to understanding their signaling role, see Machin et al. (2020).

Regardless of the various effects of phytohormones on plant growth and development (Table 5.15) and the effects of environmental factors on their biosynthesis, specific patterns occur in the concentrations of the individual phytohormones in a given organ during its growth and development. One such pattern is shown in Table 5.16 for trifoliate leaves of common bean. The concentrations of IAA, ABA, and CYT are high in very young leaves and decrease rapidly during

TABLE 5.16 Patterns of auxin (indole-3-acetic acid, IAA), abscisic acid (ABA), and cytokinin (CYT; zeatin and zeatin riboside) concentrations during the growth of trifoliate leaves of bean (*Phaseolus vulgaris*).

Area of the trifoliate leaf (cm²)	Phytohormone concentration (ng g⁻¹ dw)		
	IAA	ABA	CYT
1.3	419	568	23
6.8	336	245	19
23	297	146	14
58	217	57	11
110	153	106	10
191[a]	166	158	10

[a]*Fully expanded leaf.*
Source: From Cakmak et al. (1989).

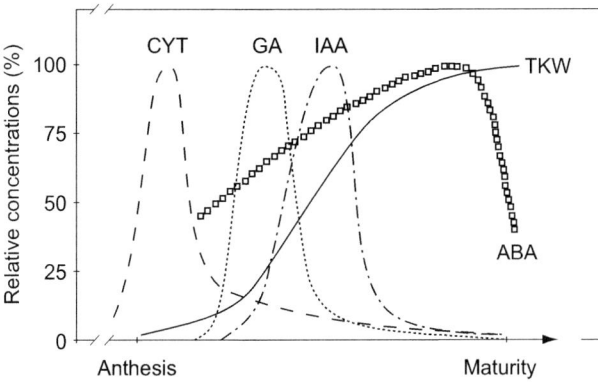

FIGURE 5.31 Generalized patterns of relative phytohormone concentrations (CYT, GA, IAA, ABA) in cereal grains during grain development. *ABA*, abscisic acid; *CYT*, cytokinin; *GA*, gibberellic acid; *IAA*, indole-3-acetic acid; *TKW*, thousand kernel weight. *Data compiled from Rademacher (1978), Radley (1978), Michael and Beringer (1980), Mounla et al. (1980), and Jameson et al. (1982).*

early leaf development, partly due to a dilution effect. Thereafter, IAA and CYT concentrations remain constant, whereas the concentration of ABA increases.

The phytohormones in developing leaves can originate from biosynthesis within the leaves or can be imported from other plant tissues. In view of the main sites of biosynthesis in plants (Table 5.15), IAA most likely originates in leaves, and the gradient in IAA concentrations correlates with the leaf shift from sink to source. On the other hand, ABA is mainly synthesized in mature (source) leaves and exported with the photosynthates in the phloem to young (sink) leaves. The changes in ABA concentration in a leaf (Table 5.16) may reflect the shift in its physiology from being a sink to becoming a source. Leaf CYT concentrations change less than those of IAA and ABA (Table 5.16). The high CYT concentrations in very young leaves may be attributed to a combination of both local biosynthesis and phloem import, and the subsequent decrease may be due to export to the xylem.

Changes in the concentrations of phytohormones during the development of reproductive sinks, such as seeds and fruits, also follow a characteristic sequence (Fig. 5.31) that is different from the one observed in developing leaves (Table 5.16). In cereal grains, maximum CYT concentrations are reached a few days after anthesis, which coincides with the maximum rate of cell division (Jameson et al., 1982). Maxima of GA and IAA concentrations are reached when rates of dry matter accumulation are highest, that is, when both sink activity and rate of phloem unloading are largest. By contrast, ABA concentrations increase later and reach a maximum during the period of rapid decline in the rate of dry matter accumulation. The peak in ABA concentration is correlated with rapid water loss and the corresponding desiccation of the grains. Similar patterns in endogenous phytohormone concentrations also occur in fruits such as tomatoes (Quinet et al., 2019) and grapes (Li et al., 2021).

There is a well-established positive correlation between final grain weight and the number of endosperm cells (Grimberg et al., 2020) as well as the length of the grain-filling period (days between anthesis and maturity). In maize, elevated ABA concentrations during early kernel development decrease the rate of cell division in the endosperm and, therefore, the storage capacity of the kernels (Jiang et al., 2021).

The dependency of developing seeds and fruits on the import of phytohormones from the xylem (e.g., CYT) and phloem (ABA, GA) is unclear. However, at least for cereals such as wheat, it has been demonstrated that there is no such dependency. In cultures of isolated ears, even when isolated prior to anthesis, normal kernel development can be achieved in the absence of phytohormones, with the exogenous supply of only sugars and N (Lee et al., 1989).

Based on knowledge of the effects of phytohormones on plant growth and development, and their typical concentrations during organ development, "bioregulators" that mimic or alter the concentrations or activity of endogenous phytohormones have been developed to improve crop production. For example, synthetic plant hormones, such as kinetin, and growth retardants, such as chlorocholine chloride (CCC) and 2,3,5-tri-iodobenzoic acid (TIBA), can regulate vegetative and reproductive growth, as well as senescence and abscission. Bioregulators are used on a large scale (Yakhin et al., 2017), the most successful being the "antigibberellins" that interfere with the biosynthesis of GAs (Grossmann, 1990) and BRs (Rao et al., 2002).

5.9.2 Phytohormones, signal perception, and signal transduction

There are often poor correlations between the concentrations of endogenous phytohormones, as determined by chemical methods or bioassays, and their actions in plants. These poor correlations are attributed to the capacity of target cells and organs to receive, perceive and transduce the phytohormone signal into a physiological response (Fig. 5.32).

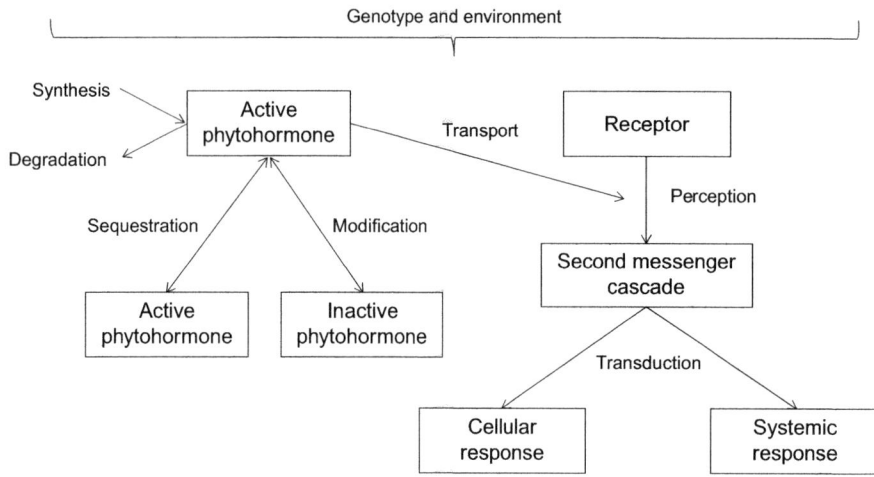

FIGURE 5.32 Relationships controlling the concentration of active phytohormone, its perception by the cell, and its eliciting of a physiological or developmental response.

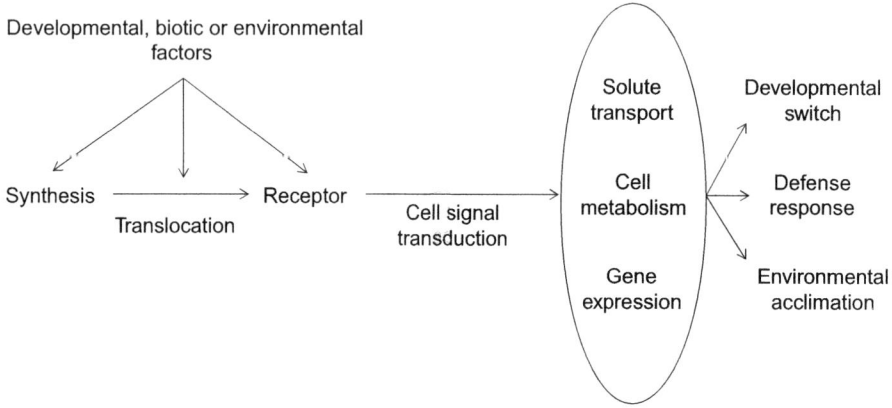

FIGURE 5.33 Responses of phytohormone synthesis, translocation, and perception to developmental, environmental, or biotic factors.

Following their biosynthesis, phytohormones must be transported to their site of action. Usually, only a fraction of the total phytohormone synthesized remains in its biologically active form. The remainder is either degraded or modified, transiently or permanently, to biologically inactive compounds. Phytohormones or their modified products can also be sequestered in the vacuoles of plant cells. Sequestration in cellular organelles is particularly important for ABA, GAs, and IAA (Hartung and Slovik, 1991). In target tissues, cells must be competent to respond to the phytohormone, that is, they must possess a receptor for the phytohormone. The synthesis, degradation, modification, sequestration, and transport of phytohormones are affected by plant genotype and environment, as are the abundance and activity of the phytohormone receptors. Recently, the molecular identity of receptors for IAA, ABA, CYT, GA, ET, JA, BR, and SL have all been revealed (Argueso et al., 2009; Browse, 2009; Chow and McCourt, 2006; Czarnecki et al., 2013; Kline et al., 2010; Lin et al., 2009; Machin et al., 2020; Wolters and Jürgens, 2009). Similarly, the interaction between a phytohormone and its receptor will initiate intracellular signal transduction cascades only in cells competent to respond to the phytohormone. The expression of these biochemical cascades is also determined by genetic and environmental factors. Enzymes, metabolites, ions, and electrical events involved in these signal transduction cascades are called second messengers. These signal transduction cascades alter solute transport across cellular membranes, cell metabolism, and gene expression and thereby the physiological and developmental responses (Fig. 5.33). During cell and tissue differentiation, and organ maturation, both the perception (sensitivity) and the response to a given phytohormone can change.

Several of the signal transduction cascades initiated by phytohormones involve electrical events at the plasma membrane and changes in cytosolic Ca^{2+} activities (Chapter 6). For example, changes in cytosolic Ca^{2+} activities appear to be necessary for cellular responses to alterations in the concentrations of (1) IAA to influence root development and tropisms, (2) ABA to affect stomatal closure and responses to drought and chilling, (3) GA during seed germination, and (4) ET in combination with reactive-oxygen species to influence responses to pests and pathogens, oxidative stress and metals (White and Broadley, 2003). Calcium can enter the cytosol from the apoplast, vacuole, or intracellular organelles through cation channels, and the origin of Ca^{2+} entering into the cytosol can differ between phytohormones (White and Broadley, 2003). In addition, systemic long-distance Ca^{2+} signals enable plants to prepare all their tissues and respond to biotic and abiotic stresses (Shabala et al., 2016).

Other typical components of phytohormone signal transduction are multistep phospho-relay cascades. For example, the CYT signal transduction cascades consist of sensor histidine kinase receptors, histidine phosphotransfer proteins, and cystolic response regulators (Argueso et al., 2009), the ET signal transduction cascades are initiated by sensor histidine kinase receptors (Chow and McCourt, 2006; Lin et al., 2009), and the BR signal transduction cascades are initiated by sensor serine/threonine kinase receptors (Chow and McCourt, 2006; Wolters and Jürgens, 2009). The ABA signal transduction cascades are initiated by protein phosphatases that modulate the activities of diverse protein kinases (Kline et al., 2010). Such signaling cascades are integrated with mitogen-activated protein kinase cascades, as well as other protein kinase and phosphatase enzymes and Ca/calmodulin systems (Alcázar et al., 2010; Browse, 2009). Receptors for IAA, GA, ET, and JA are F-box subunits of E3 ubiquitin ligase complexes that result in the activation of transcriptional regulators (Browse, 2009; Chow and McCourt, 2006; Wolters and Jürgens, 2009). Studies have identified a large number of transcriptional regulators that modulate gene expression in target cells in response to changes in IAA, ABA, GA, CYT, ET, JA, BR, or SL concentrations (Argueso et al., 2009; Brewer et al., 2013; Kline et al., 2010; Wolters and Jürgens, 2009).

Phytohormones and intracellular second messengers are integral parts of an extensive signal transduction chain to induce the appropriate plant responses to environmental challenges. For example, low soil water content in the rhizosphere increases the synthesis and export of ABA from the roots to the shoot (Wilkinson and Davies, 2002). Some of this ABA binds to the ABA-receptors in guard cells and opens K^+ and Cl^- channels in their plasma membrane and tonoplast through an intracellular signal transduction cascade involving changes in cytosolic Ca^{2+}, cytosolic pH, and a variety of kinase cascades (Amtmann and Blatt, 2009; Kashtoh and Baek, 2021). This results in a loss of osmotica from the guard cells, causing them to shrink and close the stomatal pore, thereby preventing further water loss.

In addition to producing cellular or local effects, signaling cascades can also initiate systemic second messengers that induce physiological or developmental responses in other tissues. Examples of this are phloem-mobile signals that influence root biochemistry and morphology that govern root uptake and translocation of elements to the shoot (Chapter 2). Systemic second messengers can include recycling nutrients, sucrose, or specific microRNAs (Chapter 3; Buhtz et al., 2010; Hammond and White, 2008; Liu et al., 2009). The interactions between phytohormones and other systemic signals integrate the physiology of the whole plant. Because phytohormones interact and form parts of an integrated, whole-plant signal transduction system, care is required in interpreting their endogenous concentrations in terms of expected effects on plant growth and development. However, the endogenous concentrations do provide valuable information as to whether, for example, an environmental stress was sufficiently severe to elicit a distinct phytohormonal signal.

5.9.3 Effects of nutrition on the endogenous concentrations of phytohormones

The synthesis, degradation, and action of phytohormones are influenced by environmental factors such as temperature, daylength, and water and nutrient supply. Some of these factors can be altered relatively easily by agronomic and horticultural practices. Thus, growth and development of plants, and ultimately economic yield, can be improved via manipulation of endogenous phytohormone concentrations. The focus of the following discussion is on the effects of nutrition on endogenous phytohormone concentrations.

The main sites of CYT synthesis are root meristems. There is a close relationship between the number of root meristems, root system development, and CYT production in roots (Del Bianco et al., 2013). Local antagonistic interactions between CYT and IAA determine root meristem size and the growth rate of the root system (Perilli et al., 2010). Among the nutrients, N exerts the most obvious influence on root growth as well as the production and export of CYT to the shoots (Argueso et al., 2009). Because CYT is exported mainly in the xylem, collecting xylem exudate is a simple method of obtaining information on this N effect. When the N supply to potato plants is continuous, CYT export increases with plant age, whereas when N supply is interrupted, there is a rapid decrease in CYT export from the roots (Table 5.17). After restoring the N supply, CYT export is rapidly enhanced. When tomato plants precultured in NH_4^+ were supplied NO_3^-, the resultant increase in leaf expansion was associated with an increase in the CYT concentration in the xylem sap (Rahayu et al., 2005). The synthesis and export of CYT from roots are also affected by P, S, and K supply (Hirose et al., 2008), although their effect is not as pronounced as that of N (Table 5.18). Similar results have been obtained in a variety of plants, including both annuals and perennials (Argueso et al., 2009; Sakakibara et al., 2006; Wilkinson et al., 2007). The expression of genes responsible for CYT biosynthesis is downregulated in roots of

TABLE 5.17 Export of cytokinins from potato (*Solanum tuberosum*) roots with continuous or interrupted N supply.

Plant age at zero time[a] (days)	N supply	
	Continuous	Interrupted
	Cytokinins exported [ng plant^{-1} (24 h)$^{-1}$]	
0	196	196
3	420	26
6	561	17
9[b]	–	132

[a] 30 days after sprouting.
[b] Restoration of N supply after 6 days without N.
Source: Based on Sattelmacher and Marschner (1978).

TABLE 5.18 Cytokinin concentration of roots and leaves of sunflower (*Helianthus annuus*) plants grown in nutrient solution with sufficient or insufficient supply of N, P, or K.

Treatment (15 days)	Cytokinins	
	(kinetin equivalents, µg kg^{-1} fresh wt)	
	Roots	Leaves
Control	2.38	3.36
1/10 N[a]	0.94	1.06
1/10 P	1.06	1.28
1/10 K	1.06	2.02

[a] Indicates proportion of nutrients in relation to fully concentrated control solution.
Source: From El-D et al. (1979).

plants with inadequate nitrate, sulfate, or phosphate supply, whereas the expression of genes encoding transporters for these anions is increased (Hirose et al., 2008).

The role of CYT in the reduction of plant growth at low supply of nutrients is shown in Table 5.19. Compared with the high nutrient supply (100%), the growth rate and tissue CYT concentration of broadleaf plantain (*Plantago major*) were lower at continuous low nutrient supply (2%). Two days after transfer from high to low nutrient supply (100→2%), CYT concentrations in shoots and roots and shoot growth rate declined strongly, whereas root growth rate increased slightly. The decline in shoot growth rate could be prevented by adding 10^{-8} M benzyladenine (CYT) to the nutrient solution. During these short-term responses in growth rates, shoot concentrations of nutrients did not change significantly (Kuiper et al., 1988), suggesting that the changes in shoot growth rates in response to altered nutrient supply were mediated indirectly by tissue CYT concentrations (Kuiper et al., 1989). It is likely that the plants in these experiments were responding to an interrupted N supply. In *Urtica dioica*, photosynthates were preferentially allocated to roots at low N supply, whereas at high N supply, or following direct application of BA to the roots, photosynthates were preferentially allocated to the shoot apex (Fetene et al., 1993).

Enhanced synthesis and higher concentrations of ABA in roots and shoots are typical of N-deficient plants (Jiang and Hartung, 2008; Wilkinson and Davies, 2002). When the N supply to sunflower plants was interrupted, ABA concentrations in all parts of the shoot increased strongly within 7 days (Table 5.20; Goldbach et al., 1975). In potato

TABLE 5.19 Relative growth rate of broadleaf plantain (*Plantago major*) and cytokinin [CYT, zeatin (Z) plus zeatin ribose (ZR)] concentration at high or low nutrient supply with and without addition of benzyladenine (BA, 10^{-8} M).

Nutrient supply (%)[a]	BA	Relative growth rate (mg dw g^{-1} day^{-1})		CYT concentration (pmol Z + ZR g^{-1} fw)	
		Shoot	Roots	Shoot	Roots
100	−	208	159	78	105
2	−	49	76	21	39
100→2	−	73	183	34	50
100→2	+	220	163	81	110

[a]Full-strength nutrient solution (100%) or diluted to 2%; treatment 100%→2% was sampled after 2 days.
Source: Compiled from Kuiper (1988) and Kuiper et al. (1988).

TABLE 5.20 ABA (abscisic acid) concentrations in shoots of sunflower (*Helianthus annuus*) plants grown in nutrient solution with or without N.

Plant part	With N	Without N (7 days)
	ABA (μg g^{-1} fresh wt)	
Leaves		
Old	8.1	30
Fully expanded	6.8	21
Young	14	24
Stem	2.5	4.9

Source: Based on Goldbach et al. (1975).

plants, this response can be observed within 3 days, and this effect is even more apparent in roots and xylem exudate than in shoots (Krauss, 1978a). Similar observations have been made on other plant species, but they are not universal (Wilkinson and Davies, 2002).

In many plant species, a reduction in leaf elongation rate is an immediate response to restricted N supply (Chapin et al., 1988; Kavanová et al., 2008; Pantin et al., 2011). Net photosynthesis, however, is not affected immediately, and sugars accumulate (Chapin et al., 1988). This short-term response in leaf elongation rate has been associated with a decrease in CYT translocation from the root to the shoot (Römheld et al., 2008). The expansion of leaf cells is mediated by expansins, whose abundance and activity are influenced by the concentrations of phytohormones, including CYT, IAA, GA, and ET (Downes et al., 2001; Sánchez-Rodríguez et al., 2010). The failure to restore shoot growth in potato plants by foliar application of N, when the root N supply is interrupted, suggests that systemic signals from the roots cause the reduction in shoot growth (Krauss and Marschner, 1982; Poitout et al., 2018; Sattelmacher and Marschner, 1979).

The effects of N supply on ABA production are important for the water balance in plants. Under water-limited conditions (e.g., in dry or saline soils), elevated ABA concentrations in the leaves favor stomata closure and prevent excessive water loss (Fig. 5.34). When plants are N-deficient, or are supplied with suboptimal amounts of N, they respond to a shortage of available water in the substrate (i.e., a decrease in substrate water potential) by a more rapid stomatal closure (indicated by an increase in leaf resistance to water vapor diffusion) than do plants well supplied with N (Fig. 5.34). The faster stomatal response in leaves of N-deficient plants is due not only to higher ABA concentrations in the xylem sap but also to greater responsiveness of the stomata to elevated ABA concentrations (Fig. 5.35). The responsiveness of stomata to ABA concentrations is greater in older leaves than in younger leaves, and (for a given leaf age) greater in leaves from N-deficient than N-sufficient plants.

The higher stomatal responsiveness to ABA in N-limited plants is also related to lower CYT concentration (Wilkinson et al., 2007). It is well documented that CYT and ABA have opposite effects on stomatal aperture (Arnaud et al., 2017). Correspondingly, the greater stomatal responsiveness of the older, N-limited leaves to ABA (Fig. 5.35) could be reversed at least partially by a simultaneous supply of CYT (Radin and Hendrix, 1988). The greater drought resistance of low-N plants (Wang et al., 2016) is therefore the result of not only morphological changes in root growth or leaf anatomy (e.g., smaller leaf blades) but also physiological changes such as an increase in the ABA/CYT ratio.

The well-documented increase in root/shoot biomass ratio that occurs when plants lack N may be explained, at least partly, by an increase in ABA and a decrease in CYT concentrations. This response to N deficiency, and also to drought stress, is, in most instances, advantageous to plants growing in soils with limited availability of N and water.

Somewhat similar relationships to those described for N have been reported for P (Radin, 1984). In response to drought stress, more ABA is accumulated in leaves of P-deficient than P-sufficient cotton plants; in the deficient plants, the stomata close at leaf water potentials of approximately −1.2 MPa compared with −1.6 MPa in the sufficient plants. The high sensitivity of the stomata to ABA under P deficiency can be reversed by CYT.

Tissue concentrations of GA are also influenced by plant nutrition. For example, interrupting the N supply to the roots of potato plants causes a rapid decrease in their shoot GA and ABA concentrations (Krauss and Marschner, 1982). After restoring the N supply, shoot concentrations of GA increase and ABA concentrations fall. Comparable changes in GA and ABA concentrations induced by N supply can also be observed in the tubers of potato plants, where the changes are correlated with differences in tuber growth and development (Sonnewald and Sonnewald, 2014).

The effects of N on GA concentrations are likely indirect. The main sites of GA synthesis during vegetative growth are the shoot apex and the expanding leaves. Thus, environmental factors that favor shoot growth (e.g., high N supply, sufficient water supply), also indirectly favor GA synthesis, which is reflected in changes in plant morphology. For

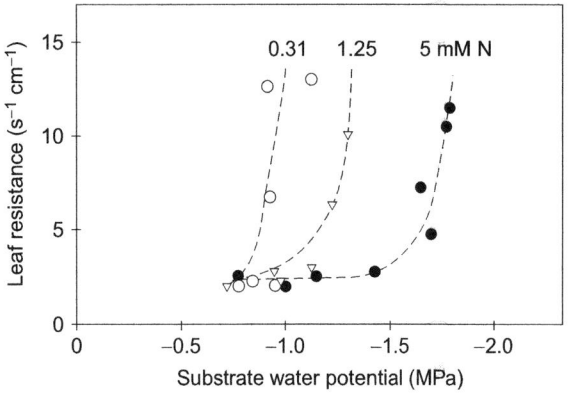

FIGURE 5.34 Relationship between N supply (mM NO$_3$-N), substrate water potential, and leaf resistance to water vapor diffusion in cotton (*Gossypium hirsutum*) plants. *Based on Radin and Ackerson (1981).*

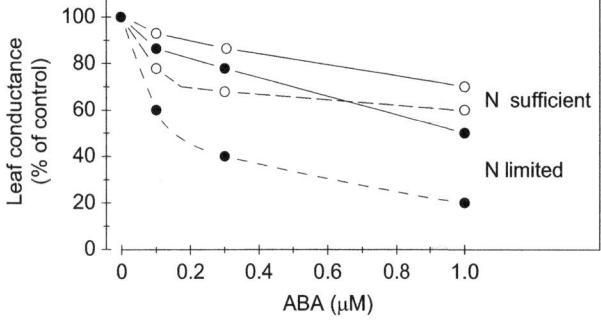

FIGURE 5.35 Stomatal conductance of expanded (○) and old (●) excised cotton (*Gossypium hirsutum*) leaves from N-sufficient (—) and N-deficient (--) plants at different ABA concentrations. *Based on Radin and Hendrix (1988).*

example, at high N supply to cereals, stem elongation is enhanced, and the plants become prone to lodging. To counteract these effects, growth retardants, such as CCC, that depress GA synthesis are often applied to cereals receiving high quantities of N fertilizer.

The phytohormone balance in plants is influenced not only by the quantity but also the form of N fertilizer supplied to the roots (Römheld et al., 2008). Thus, plant growth and development and, ultimately, plant yield can be affected indirectly by plant N nutrition through specific effects on phytohormone synthesis, transport, or perception. For example, deep placement of urea fertilizer close to cereal roots reduces CYT concentration in the xylem sap, which correlates well with decreased tillering (Bauer and von Wirén, 2020).

The B nutritional status in oilseed rape strongly affects the metabolic pathways of phytohormones such as IAA, CK, GA, ABA, and BR (Eggert and von Wirén, 2017). The concentrations of IAA and ABA decreased and the concentrations of CYT and GA increased with increasing B supply, whereas BR appeared to regulate GA biosynthesis and therefore support shoot growth at suboptimal B supply.

An increase in the exudation of SLs from roots occurs when plants are N or P deficient, and it is possible that these compounds are involved in altering root and shoot architecture observed under N and P deficiencies (Dun et al., 2009). In addition, SLs induced by N or P deficiency affect rhizobial and mycorrhizal symbioses (Foo et al., 2013; Yoneyama et al., 2015). Phosphorus deficiency can also repress CYT biosynthesis, and changes in the translocation of SLs and CYT from the root to the shoot are likely to mediate acclimation to P deficiency in plants (Shabala et al., 2016). The shoot-to-root signals involved in this feedback mechanism include, but are not restricted to, phloem-mobile microRNAs, sugars, and possibly P itself (Shabala et al., 2016; Zhang et al., 2014).

5.9.4 Phytohormones and sink action

During the growth and development of a plant organ, the concentrations of different phytohormones vary substantially (Table 5.16; Fig. 5.31) and are usually correlated with the sink strength and, in leaves, the transition from sink to source. Phytohormone concentrations are also important for sink competition, for example, between the reproductive and vegetative sinks (Römheld et al., 2008). In this section examples are given of the involvement of phytohormones in determining the distribution of photosynthates and nutrients within the plant.

Expanding leaves act as a strong sink for photosynthates, and the application of ABA not only reduces leaf expansion (similarly to the effect of drought stress) but also increases the rate of export of photosynthates (Table 5.21). Although ABA reduces dry mass gain, the rate of photosynthesis is not affected. This ABA-induced shift from sink to source may be due to a decrease in IAA concentration diminishing cell elongation and the sink strength of the leaf. As early as 1950, Nitsch demonstrated the role of IAA in the sink activity of developing strawberry fruits (Nitsch, 1950). Removal of the seeds from the developing fruits resulted in the immediate cessation of fruit growth. Application of IAA to the seedless fruits restored the growth rate of the fruits, indicating that solute volume flow via the phloem into developing strawberry fruits depended on IAA produced in the seeds.

Phytohormones in general, and IAA in particular, are also involved in nutrient import into tissues, as shown for P in bean plants in Table 5.22. Removal of the seeds and, especially, the fruit reduced the accumulation of ^{32}P in the peduncles. The movement of ^{32}P to the peduncles could partly be restored by the application of IAA to the cut end and was strongly stimulated by the application of IAA in combination with kinetin. The accumulation of ^{14}C in the peduncle following the exposure of a mature leaf to ^{14}CO$_2$ is also stimulated by the application of IAA.

TABLE 5.21 Rate of photosynthesis and increases in leaf area expansion and dry weight in soybean (*Glycine max*) as influenced by exposure to abscisic acid (ABA) (1 nmol leaflet^{-1}), drought stress, or low light intensity.

Parameter	Treatments			
	Control	ABA	Drought	Low light
Photosynthesis (μmol CO$_2$ m^{-2} s^{-1})	15	14	15	2.4
Area increase (cm^2)	4.5	3.2	2.7	4.2
Dry weight increase (mg leaflet^{-1})	33	26	27	7

Source: Based on Bunce (1990).

TABLE 5.22 Concentrations of leaf-applied ^{32}P and ^{14}C in peduncles of bean (*Phaseolus vulgaris*) with seeds or fruits removed and treated with phytohormones at the cut end after fruit removal.

Treatment	^{32}P	^{14}C from ^{14}CO$_2$-supplied mature leaf
	(cpm)	
Control (intact fruit)	373	
Seeds removed	189	
Fruit removed	34	
Fruit removed and cut end treated with		
Lanoline	6	320
Kinetin	20	
IAA	235	5520
IAA + kinetin	471	

IAA, indole-3-acetic acid.
Source: Based on Seth and Wareing (1967).

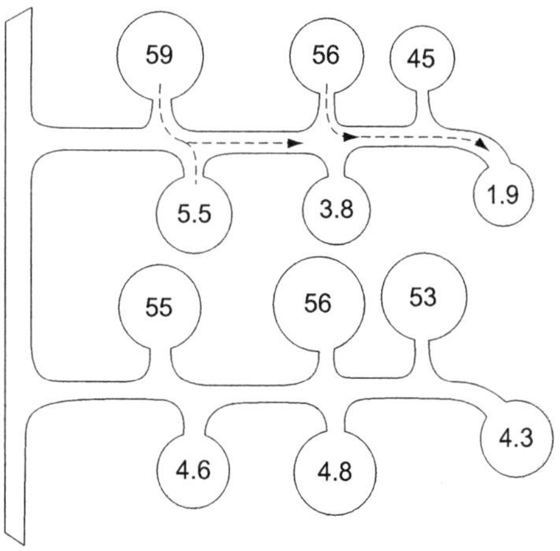

FIGURE 5.36 Tomato (*Solanum lycopersicum*) trusses with natural sequence of fruit development (upper truss) and "synchronized" fruit development by pollination on the same day (lower truss). Numbers in fruits represent final fruit weight in g (upper fruits) and polar IAA export (ng per fruit per day) from 10-day-old fruits (lower fruits). *IAA*, indole-3-acetic acid. Based on Bangerth (1989).

Auxins play an important role in dominance phenomena that are widespread in the plant kingdom and are common in reproductive sinks between fruits (e.g., individual tomato fruits on the same truss), seeds (e.g., grains in proximal, medial, and acropetal positions within an ear) and in vegetative sinks (e.g., terminal vs lateral buds). Competition for photosynthates may be an explanation for these phenomena. However, dominance frequently occurs very early in the ontogeny of reproductive and vegetative sinks, when competition for the limited amount of photosynthates available is less likely and may even be unimportant. A dominance signal may account for this effect (Fig. 5.36). Fruits developing early dominate over those developing later, and dominance is achieved by higher polar, basipetal transport of IAA as a signal of higher sink activity. The higher IAA export from dominating sinks seems to have additional repressing effects ("auto-inhibition") on those fruits that are dominated (Bangerth, 1989). The same trend is observed in the dominance of apical versus lateral buds on a stem (Shimizu-Sato et al., 2009). Auxin from the apical bud inhibits CYT biosynthesis and suppresses axillary bud outgrowth. However, more than one dominance mechanism can be involved. In potato plants, for example, dominance phenomena may have a phytohormonal basis, and competition for photosynthates operates at later stages (Engels and Marschner, 1986).

Although it is relatively easy to modify the activities of endogenous phytohormones by environmental factors such as N supply, these changes are complex, and the plant system is not an easy one to manipulate. Direct application of phytohormones to increase sink activity appears more straight-forward but has been successful in only a few cases. One of the few successful examples is the increased seed yield of faba bean by foliar application of GA at the six-leaf stage (Belucci et al., 1982). The yield increase obtained was mainly the result of an increased number of pods and seeds per plant. In faba beans a high proportion of flowers are aborted, and the application of GA decreases this abortion.

A similar mode of action may be responsible for the increase in grain yield and harvest index in maize after foliar application of CYT (Smiciklas and Below, 1992) that reduced kernel abortion. Interestingly, this effect of CYT was dependent on the form of N supplied. In plants supplied predominantly with nitrate, kernel abortion was higher than in plants supplied with a mixture of ammonium and nitrate. This negative effect of sole nitrate nutrition on kernel abortion was reversed by foliar application of CYT, whereas in plants supplied with a mixture of ammonium and nitrate, CYT application had no effect on kernel number.

In plants with vegetative storage sinks, such as tuber and root crops, manipulations of sink activity by phytohormone application appear somewhat easier, and successful results have been reported on storage roots of winter radish (Starck et al., 1980) and potato tubers (Ahmed and Sagar, 1981). In these experiments, however, phytohormones were not, and could not, be applied to the sink organs directly; therefore, their effects were indirect.

5.10 Source and sink limitations on yield

The growth rate of sink tissues and organs such as roots, shoot apices, and storage organs can be limited either by supply of photosynthates from source organs (*source limitation*) or the capacity of the sink to utilize these photosynthates (*sink limitation*). In storage organs, sink limitation can result from low rates of phloem unloading and conversion of photosynthates to storage compounds (e.g., starch) as well as from a low number of either storage cells per sink organ or sink organs (e.g., grains) per plant or per land area. Sink limitation, in turn, can lead to inhibition of photosynthesis. Sink—source limitations are strongly affected by interactions between genotype and environment. In the following examples, both types of limitation are considered with particular emphasis on different phases of plant development and environmental factors.

The potential sink capacity is determined by the number of (1) sink organs (e.g., grain or tuber number), (2) storage cells per organ (e.g., number of endosperm cells per grain), and (3) storage organelles per cell (e.g., number of amyloplasts per endosperm cell). The transport of assimilates is an important aspect of source—sink interactions and determines yield components, including tiller number, grain/seed number, and 1000-grain weight in cereals (Chang and Zhu, 2017). In rice, loss of a tonoplast sucrose transporter (OsSUT2) responsible for moving sucrose from the vacuole to the cytoplasm of leaf mesophyll cells results in the retention of sucrose in source leaves, less sucrose translocation to sink organs, and a decreased tiller number and 1000-grain weight (Eom et al., 2011). In crop species with vegetative storage organs such as potato, the determination of potential sink capacity is not restricted to a specific phase of plant development. Depending on environmental conditions, additional tubers and/or storage cells in existing tubers can be formed until shortly before maturity. Thus, sink capacity and tuber growth rates can continuously be adjusted to current photosynthate supply, suggesting that tuber growth is mainly source limited (Engels and Marschner, 1987), unless tuber initiation or growth are directly inhibited, for example, by extreme soil temperatures. In sugarcane, by contrast, there is evidence of sink limitation during sugar accumulation (McCormick et al., 2009). Transgenic sugarcane producing the sucrose isomer isomaltulose exhibited a substantial increase in both sugar concentration in the stalk and photosynthetic rates in leaves, suggesting that in wild-type sugarcane, photosynthetic capacity is not completely utilized because of limited sugar storage capacity of the stalks (Wu and Birch, 2007).

In crops with generative storage organs, the potential sink capacity is established in a relatively short period around anthesis (the so-called *critical period*). Processes related to sexual reproduction such as meiosis, pollination, and zygote formation are particularly sensitive to stress conditions such as heat, drought, and nutrient deficiency (Barnabás et al., 2008; Hedhly et al., 2009); these stresses during anthesis reduce seed and fruit set, which leads to sink limitation of yield and decrease of harvest index (Porter and Semenov, 2005). Stress-induced decrease in grain set may be due to elevated concentrations of ABA (Setter and Parra, 2010) or ethylene (Hays et al., 2007) in the reproductive organs. In wheat, for example, drought stress during meiosis of pollen mother cells decreased the proportion of fertile spikelets from 68% in well watered plants to 44%, and simultaneously increased ABA concentrations in the ears threefold (Morgan, 1980). Application of ABA to the ears of well-watered plants also decreased spikelet fertility from 68% to

37%. However, during grain-filling in wheat, grain ABA concentrations were positively correlated to grain filling rate (Yang et al., 2006).

Conclusions about the role of phytohormones as the "signal" of drought stress and also for depressing fertilization and grain set must be drawn with care. In maize, drought is associated with a decrease in photosynthate influx into kernels and depletion of carbohydrate reserves; the quantity of these reserves in the florets provides a sugar signal to genes regulating senescence and kernel abortion (Ruan et al., 2010). Drought at anthesis is associated with upregulation of genes for senescence enzymes in maize ovaries, leading to kernel abortion. This upregulation during water deficit is prevented by feeding sucrose (Ruan et al., 2010).

In the absence of abiotic stresses that inhibit sexual propagation, the establishment of potential sink capacity is often closely related to assimilate supply to generative organs during the critical period (Andrade et al., 1999; Fischer, 2007). Assimilate supply is dependent on photosynthesis (source strength) and partitioning of assimilates to generative organs (strength of alternative sinks relative to that of generative organs). In wheat, considerable genotypic variation exists for dry matter partitioning to sinks at anthesis, including leaves, stems, and roots (Foulkes et al., 2011). Even though lower partitioning to roots and leaves involves the risk of, respectively, reduced acquisition of soil resources (water and mineral nutrients) and lower photosynthesis, there is scope for reducing allocation of assimilates to stems in favor of increased allocation to the spikes (Foulkes et al., 2011).

The actual yield not only depends on the potential sink capacity that is fixed during the critical period but also on assimilate supply after anthesis. Whether yield is limited by source or sink is dependent on environmental conditions during the critical period and the post–anthesis period. Transient stress during the critical period may reduce the potential sink capacity to such an extent that yield is limited mainly by the capacity of sink organs to store assimilates (i.e., sink limitation of yield). By contrast, abiotic and biotic stresses during the post–anthesis period that reduce photosynthesis and accelerate leaf senescence can reduce yield by source limitation. Under these conditions reserve pools accumulated in vegetative organs prior to anthesis become an important source of assimilates for grain filling. In graminaceous species of cool and temperate climate (e.g., *Agrostis* spp., wheat), the main transient storage carbohydrates in stems are fructans accumulated prior to and during the first weeks after anthesis, and utilized thereafter for grain filling (Schnyder, 1993). In wheat, the contribution of reserve pools to individual grain weight and yield varies greatly depending on genotype and environmental conditions (Dreccer et al., 2009). Under drought stress after flowering, remobilization of water-soluble carbohydrates (mainly fructans) may contribute up to 50% of yield (Blum, 1998). In maize, large amounts of N are transiently stored in the stem, and nearly half the N in the grains may derive from this source (Ta, 1991).

In the absence of stresses, grain filling and yield are often sink limited (Borrás et al., 2004; Fischer, 2007). For example, artificially increasing light penetration into the canopy of field-grown wheat for 12 days prior to anthesis increased the kernel number per spike and, thus, sink strength (Table 5.23). The increase in sink strength enhanced light-saturated rates of flag leaf photosynthesis during grain filling and increased grain yield. The extent to which yield of grain crops is limited by sink and source is dependent on species and genotype. During grain filling, wheat yield was mainly sink limited, whereas in soybean, yield was limited by sink as well as source (Borrás et al., 2004). In wheat varieties released from 1940 to 2005, breeding has decreased the degree of sink limitation during postanthesis from nearly complete sink limitation in the oldest varieties to limitation of both sink and source in the most modern varieties

TABLE 5.23 Kernel number per spike, light-saturated net assimilation rate of the flag leaf of wheat (*Triticum aestivum*) during grain filling and grain yield at normal or transiently increased light penetration into the canopy for 12 days during the booting stage.

Treatment	Kernel (no. spike^{-1})	Net assimilation rate (μmol CO_2 m^{-2} s^{-1})	Grain yield (g m^{-2})
Control	40	26	790
Increased light[a]	43	29	950

[a]*Light penetration into the canopy was increased by bending back and holding neighboring plants at an angle of approximately 45 degrees from the vertical from 8 am until 5 pm prior to anthesis.*
Source: Based on Reynolds et al. (2009).

(Acreche and Slafer, 2009). These findings point to the importance of considering plant traits related to the assimilate supply from source organs (source strength) as well as to the assimilate storage in sink organs (i.e., sink strength) in conventional and molecular plant breeding to increase yields of grain crops. Sink strength may become an even more important yield determinant with rising atmospheric CO_2 concentration.

References

Abadie, C., Tcherkez, G., 2019. Plant sulphur metabolism is stimulated by photorespiration. Commun. Biol. 2, 379.
Abbate, P.E., Andrade, F.H., Culot, J.P., 1995. The effects of radiation and nitrogen on number of grains in wheat. J. Agric. Sci. 124, 351–360.
Acreche, M.M., Slafer, G.A., 2009. Grain weight, radiation interception and use efficiency as affected by sink-strength in Mediterranean wheats released from 1940 to 2005. Field Crops Res. 110, 98–105.
Ade-Ademilua, O.E., Botha, C.E.J., 2007. Sink to source transition of *Pisum sativum* leaves in relation to leaf plastochron index. Am. J. Plant Physiol. 2, 27–35.
Agusti, J., Herold, S., Schwarz, M., Sanchez, P., Ljung, K., Dun, E.A., et al., 2011. Strigolactone signaling is required for auxin-dependent stimulation of secondary growth in plants. Proc. Natl. Acad. Sci. USA 108, 20242–20247.
Ahmed, C.M.S., Sagar, G.R., 1981. Effects of a mixture of NAA + BA on numbers and growth rates of tubers of *Solanum tuberosum* L. Potato Res. 24, 267–278.
Ainsworth, E.A., Bush, D.R., 2011. Carbohydrate export from the leaf: a highly regulated process and target to enhance photosynthesis and productivity. Plant Physiol. 155, 64–69.
Alburquerque, N., Egea, J., Burgos, L., Martínez-Romero, D., Valero, D., Serrano, M., 2006. The influence of polyamines on apricot ovary development and fruit set. Ann. Appl. Biol. 149, 27–33.
Alcázar, R., Altabella, T., Marco, F., Bortolotti, C., Reymond, M., Koncz, C., et al., 2010. Polyamines: molecules with regulatory functions in plant abiotic stress tolerance. Planta 231, 1237–1249.
Altman, A., Levin, N., Cohen, P., Schneider, M., Nadel, B., 1989. Polyamines in growth and differentiation of plant cell cultures: the effect of nitrogen nutrition, salt stress and embrogenic media. In: Zappia, V., Pegg, A.E. (Eds.), Progress in Polyamine Research. Springer, Boston, MA, pp. 559–572.
Alzueta, I., Abeledo, L.G., Mignone, C.M., Miralles, D.J., 2012. Differences between wheat and barley in leaf and tillering coordination under contrasting nitrogen and sulfur conditions. Eur. J. Agron. 41, 92–102.
Amtmann, A., Blatt, M.R., 2009. Regulation of macronutrient transport. New Phytol. 181, 35–52.
Anderson, J.V., Hess, J.L., Chevone, B.I., 1990. Purification, characterization, and immunological properties for two isoforms of glutathione reductase from eastern white pine needles. Plant Physiol. 94, 1402–1409.
Andrade, F.H., Vega, C., Uhart, S., Cirilo, A., Cantarero, M., Valentinuz, O., 1999. Kernel number determination in maize. Crop Sci. 39, 453–459.
Apel, K., Hirt, H., 2004. Reactive oxygen species: metabolism, oxidative stress and signal transduction. Annu. Rev. Plant Biol. 55, 373–399.
Argueso, C.T., Ferreira, F.J., Kieber, J.J., 2009. Environmental perception avenues: the interaction of cytokinin and environmental response pathways. Plant Cell Environ. 32, 1147–1160.
Arisnabarreta, S., Miralles, D.J., 2010. Nitrogen and radiation effects during the active spike-growth phase on floret development and biomass partitioning in 2- and 6-rowed barley isolines. Crop Pasture Sci. 61, 578–587.
Armand, T., Cullen, M., Boiziot, F., Li, L., Fricke, W., 2019. Cortex cell hydraulic conductivity, endodermal apoplastic barriers and root hydraulics change in barley (*Hordeum vulgare* L.) in response to a low supply of N and P. Ann. Bot. 124, 1091–1107.
Arnaud, D., Lee, S., Takebayashi, Y., Choi, D., Choi, J., Sakakibara, H., et al., 2017. Cytokinin-mediated regulation of reactive oxygen species homeostasis modulates stomatal immunity in Arabidopsis. Plant Cell 29, 543–559.
Asada, K., 1999. The water–water cycle in chloroplasts: scavenging of active oxygen and dissipation of excess photon. Annu. Rev. Plant Physiol. Plant Mol. Biol. 50, 601–639.
Assuero, S.G., Mollier, A., Pellerin, S., 2004. The decrease in growth of phosphorus-deficient maize leaves is related to a lower cell production. Plant Cell Environ. 27, 887–895.
Atkin, O.K., Macherel, D., 2009. The crucial role of plant mitochondria in orchestrating drought tolerance. Ann. Bot. 103, 581–597.
Atkins, C.A., Smith, P.M., 2007. Translocation in legumes: assimilates, nutrients, and signaling molecules. Plant Physiol. 144, 550–561.
Aubry, E., Dinant, S., Vilaine, F., Bellini, C., Le Hir, R., 2019. Lateral transport of organic and inorganic solutes. Plants 8, 20. Available from: https://doi.org/10.3390/plants8010020.
Badger, M.R., Price, G.D., 1994. The role of carbonic anhydrase in photosynthesis. Annu. Rev. Plant Physiol. Plant Mol. Biol. 45, 369–392.
Bangerth, F., 1989. Dominance among fruits/sinks and the search for a correlative signal. Physiol. Plant. 76, 608–614.
Barber, J., 2016. Mn$_4$Ca cluster of photosynthetic oxygen-evolving center: structure, function and evolution. Biochemistry. 55, 5901–5906.
Bari, R., Jones, J.D.G., 2009. Role of plant hormones in plant defence responses. Plant Mol. Biol. 69, 473–488.
Barnabás, B., Jaeger, K., Fehér, A., 2008. The effect of drought and heat stress on reproductive processes in cereals. Plant Cell Environ. 31, 11–38.
Barneix, A.J., 2007. Physiology and biochemistry of source-regulated protein accumulation in the wheat grain. J. Plant Physiol. 164, 581–590.
Barry, C.S., 2009. The stay-green revolution: recent progress in deciphering the mechanisms of chlorophyll degradation in higher plants. Plant Sci. 176, 325–333.

Bastow, E.L., de la Torre, V.S.G., Maclean, A.E., Green, R.T., Merlot, S., Thomine, S., et al., 2018. Vacuolar iron stores gated by NRAMP3 and NRAMP4 are the primary source of iron in germinating seeds. Plant Physiol. 177, 1267–1276.

Bauer, B., von Wirén, N., 2020. Modulating tiller formation in cereal crops by the signalling function of fertilizer nitrogen forms. Sci. Rep. 10, 20504. Available from: https://doi.org/10.1038/s41598-020-77467-3.

Bell, M.J., Middleton, K.J., Thompson, J.P., 1989. Effects of vesicular-arbuscular mycorrhizae on growth and phosphorus and zinc nutrition of peanut (*Arachis hypogaea* L.) in an oxisol from subtropical Australia. Plant Soil 117, 49–57.

Bellegarde, F., Gojon, A., Martin, A., 2017. Signals and players in the transcriptional regulation of root responses by local and systemic N signaling in *Arabidopsis thaliana*. J. Exp. Bot. 68, 2553–2565.

Belucci, S., Keller, E.R., Schwendimann, F., 1982. Einfluß von Wachstumsregulatoren auf die Entwicklung und den Ertragsaufbau der Ackerbohne (*Vicia faba* L.). I. Wirkung von Gibberellinsäure (GA$_3$) auf die Ertragskomponenten und die Versorgung der jungen Früchte mit ^{14}C. Angew. Bot. 56, 35–53.

Bennett, T., Leyser, O., 2006. Something on the side: Axillary meristems and plant development. Plant Mol. Biol. 60, 843–854.

Berthier, A., Desclos, M., Amiard, V., Morvan–Betrand, A., Demming-Adams, B., Adams III, W.W., et al., 2009. Activation of sucrose transport in defoliated *Lolium perenne* L.: an example of apoplastic phloem loading plasticity. Plant Cell Physiol. 50, 1329–1344.

Bethlenfalvay, G.J., Abu-Shakra, S.S., Fishbeck, K., Phillips, D.A., 1978. The effect of source-sink manipulations on nitrogen fixation in peas. Physiol. Plant. 43, 31–34.

Bloom, A.J., 2015. Photorespiration and nitrate assimilation: a major intersection between plant carbon and nitrogen. Photosynth. Res. 123, 117–128.

Bloom, A.J., Lancaster, K.M., 2018. Manganese binding to Rubisco could drive a photorespiratory pathway that increases the energy efficiency of photosynthesis. Nat. Plants 4, 414–422.

Bloom, A.J., Rubio-Asensio, J.S., Randall, L., Rachmilevitsch, S., Cousins, A.B., Carlisle, E.A., 2012. CO$_2$ enrichment inhibits shoot nitrate assimilation in C$_3$ but not C$_4$ plants and slows down growth under nitrate in C$_3$ plants. Ecology 93, 355–367.

Blum, A., 1998. Improving wheat grain filling under stress by stem reserve mobilisation. Euphytica 100, 77–83.

Bonhomme, F., Kurz, B., Melzer, S., Bernier, G., Jacqmard, A., 2000. Cytokinin and gibberellin activate *SaMADS A*, a gene apparently involved in regulation of the floral transition in *Sinapis alba*. Plant J. 24, 103–111.

Boomsma, C.R., Santini, J.B., Tollenaar, M., Vyn, T.J., 2009. Maize morphophysiological responses to intense crowding and low nitrogen availability: an analysis and review. Agron. J. 101, 1426–1452.

Borland, A.M., Griffiths, H., Hartwell, J., Smith, J.A.C., 2009. Exploiting the potential of plants with Crassulaean acid metabolism for bioenergy production on marginal plants. J. Exp. Bot. 60, 2879–2896.

Borland, A.M., Hartwell, J., Weston, D.J., Schlauch, K.A., Tschaplinski, T.J., Tushan, G.A., et al., 2014. Engineering crassulacean acid metabolism by improved water-use efficiency. Trends Plant Sci. 19, 327–338.

Borrás, L., Slafer, G.A., Otegui, M.E., 2004. Seed dry weight response to source-sink manipulations in wheat, maize and soybean: a quantitative reappraisal. Field Crops Res. 86, 131–146.

Bouis, H.E., Saltzman, A., 2017. Improving nutrition through biofortification: A review of evidence from HarvestPlus, 2003 through 2016. Glob. Food Secur.-Agric. Policy 12, 49–58.

Bräutigam, A., Gowik, K., 2016. Photorespiration connects C$_3$ and C$_4$ photosynthesis. J. Exp. Bot. 67, 2953–2962.

Brevedan, R.E., Egli, D.B., Leggett, J.E., 1978. Influence of N nutrition on flower and pod abortion and yield of soybeans. Agron. J. 70, 81–84.

Brewer, P.B., Koltai, H., Beveridge, C.A., 2013. Diverse roles of strigolactones in plant development. Mol. Plant 6, 18–28.

Briat, J.-F., Duc, C., Ravet, K., Gaymard, F., 2010. Ferritins and iron storage in plants. Biochim. Biophys. Acta 1800, 806–814.

Browse, J., 2009. Jasmonate passes muster: a receptor and targets for the defense hormone. Annu. Rev. Plant Biol. 60, 183–205.

Buchanan, B.B., Gruissem, W., Jones, R.L., 2000. Biochemistry and Molecular Biology of Plants. American Society of Plant Physiologists, Rockville, MD, USA.

Buhtz, A., Pieritz, J., Springer, F., Kehr, J., 2010. Phloem small RNAs, nutrient stress responses, and systemic mobility. BMC Plant Biol. 10, 64. Available from: https://doi.org/10.1186/1471-2229-10-64.

Bunce, J.A., 1990. Abscisic acid mimics effects of dehydration on area expansion and photosynthetic partitioning in young soybean leaves. Plant Cell Environ. 13, 295–298.

Busch, F.A., 2020. Photorespiration in the context of rubisco biochemistry, CO$_2$ diffusion and metabolism. Plant J. 191, 919–939.

Cabrera-Bosquet, L., Albrizio, R., Araus, J.L., Nogués, S., 2009. Photosynthetic capacity of field-grown durum wheat under different N availabilities: A comparative study from leaf to canopy. Environ. Exp. Bot. 67, 145–152.

Cai, S., Lashbrook, C.C., 2008. Stamen abscission zone transcriptome profiling reveals new candidates for abscission control: enhanced retention of floral organs in transgenic plants overexpressing Arabidopsis ZINC FINGER PROTEIN 2. Plant Physiol. 146, 1305–1321.

Cairns, A.L.P., Kritzinger, J.H., 1992. The effect of molybdenum on seed dormancy in wheat. Plant Soil 145, 295–297.

Cakmak, I., 2000. Possible roles of zinc in protecting plant cells from damage by reactive oxygen species. New Phytol. 146, 185–205.

Cakmak, I., Kirkby, E.A., 2008. Role of magnesium in carbon partitioning and alleviating photooxidative damage. Physiol. Plant. 133, 623–806.

Cakmak, I., Marschner, H., 1992. Magnesium deficiency and high light intensity enhance activities of superoxide dismutase, ascorbate peroxidase and glutathione reductase in bean leaves. Plant Physiol. 98, 1222–1227.

Cakmak, I., Marschner, H., Bangerth, F., 1989. Effect of zinc nutritional status on growth, protein metabolism and levels of indole-3-acetic acid and other phytohormones in bean (*Phaseolus vulgaris* L.). J. Exp.Bot. 40, 405–412.

Chakraborty, S., Chakraborty, N., Agrawal, L., Ghosh, S., Narula, K., Shekhar, S., et al., 2010. Next-generation protein-rich potato expressing the seed protein gene *AmA1* is a result of proteome rebalancing in transgenic tuber. Proc. Natl. Acad. Sci. USA 107, 17533–17538.

Chakwizira, E., Dellow, S.J., Teixeira, E.I., 2016. Quantifying canopy formation processes in fodder beet (*Beta vulgaris* subsp *vulgaris* var. *alba* L.) crops. Eur. J. Agron. 74, 144–154.

Chang, T.G., Zhu, X.G., 2017. Source-sink interaction: a century old concept under light of modern molecular systems biology. J. Exp. Bot. 68, 4417–4431.

Chapin, F.S., Walter, C.H.S., Clarkson, D.T., 1988. Growth response of barley and tomato to nitrogen stress and its control by abscisic acid, water relations and photosynthesis. Planta 173, 352–366.

Chen, W., Yang, X., He, Z., Feng, Y., Hu, F., 2008. Differential changes in photosynthetic capacity, 77 K chlorophyll fluorescence and chloroplast ultrastructure between Zn-efficient and Zn-inefficient rice genotypes (*Oryza sativa*) under low zinc stress. Physiol. Plant. 132, 89–101.

Cho, L.H., Yoon, J., An, G., 2017. The control of flowering time by environmental factors. Plant J. 90, 708–719.

Chow, B., McCourt, P., 2006. Plant hormone receptors: perception is everything. Genes Develop. 20, 1998–2008.

Christopher, J.T., Manschadi, A.M., Hammer, G.L., Borrell, A.K., 2008. Developmental and physiological traits associated with high yield and stay-green phenotype in wheat. Aust. J. Agric. Res. 59, 354–364.

Clouse, S.D., Sasse, J.M., 1998. Brassinosteroids: essential regulators of plant growth and development. Annu. Rev. Plant Physiol. Plant Mol. Biol. 49, 427–451.

Coffey, O., Bonfield, R., Corre, F., Sirigiri, J.A., Meng, D., Fricke, W., 2018. Root and cell hydraulic conductivity, apoplastic barriers and aquaporin gene expression in barley (*Hordeum vulgare* L.) grown with low supply of potassium. Ann. Bot. 122, 1131–1141.

Conte, S.S., Walker, E.L., 2011. Transporters contributing to iron trafficking in plants. Mol. Plant 4, 464–476.

Corbesier, L., Lejeune, P., Bernier, G., 1998. The role of carbohydrates in the induction of flowering in *Arabidopsis thaliana*: comparison between the wild type and a starchless mutant. Planta 206, 131–137.

Correia, C.M., Moutinho Pereira, J.M., Coutinho, J.F., Björn, L.O., Torres-Pereira, J.M.G., 2005. Ultraviolet-B radiation and nitrogen affect the photosynthesis of maize: a Mediterranean field study. Eur. J. Agron. 22, 337–347.

Cosgrove, D.J., 2005. Growth of the plant cell wall. Nat. Rev. Mol. Cell Biol. 6, 850–861.

Cramer, M.D., Hawkins, H.-J., Verboom, G.A., 2009. The importance of nutritional regulation of plant water flux. Oecologia 161, 15–24.

Creelman, R.A., Mullet, J.E., 1997. Biosynthesis and action of jasmonates in plants. Annu. Rev. Plant Physiol. Plant Mol. Biol. 48, 355–381.

Criado, M.V., Caputo, C., Roberts, I.N., Castro, M.A., Barneix, A.J., 2009. Cytokinin-induced changes of nitrogen remobilization and chloroplast ultrastructure in wheat (*Triticum aestivum*). J. Plant Physiol. 166, 1775–1785.

Crozier, A., Kamiya, Y., Bishop, G., Yokota, T., 2000. Biosynthesis of hormones and elicitor molecules. In: Buchanan, B.B., Gruissem, W., Jones, R. L. (Eds.), Biochemistry & Molecular Biology of Plants. American Society of Plant Physiologists, Rockville, MD, USA, pp. 850–929.

Curie, C., Cassin, G., Couch, D., Divol, F., Higuchi, K., Le Jean, M., et al., 2009. Metal movement within the plant: contribution of nicotianamine and yellow stripe 1-like transporters. Ann. Bot. 103, 1–11.

Cushman, J.C., 2001. Crassulacean acid metabolism. A plastic photosynthetic adaptation to arid environments. Plant Physiol. 127, 1439–1448.

Czarnecki, O., Yang, J., Weston, D.J., Tuskan, G.A., Chen, J.-G., 2013. A dual role of strigolactones in phosphate acquisition and utilization in plants. Int. J. Mol. Sci. 14, 7681–7701.

Davis, J.L., Armengaud, P., Larson, T.R., Graham, I.A., White, P.J., Newton, A.C., et al., 2018. Contrasting nutrient-disease relationships: Potassium gradients in barley leaves have opposite effects on two fungal pathogens with different sensitivities to jasmonic acid. Plant Cell Environ. 41, 2357–2372.

de Groot, C.C., van den Boogaard, R., Marcelis, L.F.M., Harbinson, J., Lambers, H., 2003. Contrasting effects of N and P deprivation on the regulation of photosynthesis in tomato plants in relation to feedback limitation. J. Exp. Bot. 54, 1957–1967.

de Veau, E.J., Burris, J.E., 1989. Photorespiratory rates in wheat and maize as determined by ^{18}O-labeling. Plant Physiol. 90, 500–511.

de Wit, M., Galvao, V.C., Fankhauser, C., 2016. Light-mediated hormonal regulation of plant growth and development. Annu. Rev. Plant Biol. 67, 513–537.

Deeken, R., Geiger, D., Fromm, J., Koroleva, O., Ache, P., Langenfeld-Heyser, R., et al., 2002. Loss of the AKT2/3 K$^+$ channel affects sugar loading into the phloem of Arabidopsis. Planta 216, 334–344.

Del Bianco, M., Giustini, L., Sabatini, S., 2013. Spatiotemporal changes in the role of cytokinin during root development. New Phytol. 199, 324–338.

Demmig-Adams, B., Adams III, W.A., 1996. The role of xanthophylls cycle carotenoids in the protection of photosynthesis. Trends Plant Sci. 1, 21–26.

Diestelfeld, A., Cakmak, I., Peleg, Z., Ozturk, L., Yazici, A., Budak, H., et al., 2007. Multiple QTL-effects of wheat *Gpc-B1* locus on grain protein and micronutrient concentrations. Physiol. Plant. 129, 459–466.

Ding, C., You, J., Chen, L., Wang, S., Ding, Y., 2014. Nitrogen fertilizer increases spikelet number per panicle by enhancing cytokinin synthesis in rice. Plant Cell Rep. 33, 363–371.

Dohleman, F.G., Heaton, E.A., Leakey, A.D.B., Long, S.P., 2009. Does greater leaf-level photosynthesis explain the larger solar energy conversion efficiency of Miscanthus relative to switchgrass? Plant Cell Environ. 32, 1525–1537.

Downes, B., Steinbaker, C.R., Crowell, D.N., 2001. Expression and processing of a hormonally regulated β-expansin from soybean. Plant Physiol. 126, 244–252.

Dreccer, M.F., van Herwaarden, A.F., Chapman, S.C., 2009. Grain number and grain weight in wheat lines contrasting for stem water soluble carbohydrate concentration. Field Crops Res. 112, 43–54.

Dugardeyn, J., Van Der Straeten, D., 2008. Ethylene: fine-tuning plant growth and development by stimulation and inhibition of elongation. Plant Sci. 175, 59–70.

Dun, E.A., Brewer, P.B., Beveridge, C.A., 2009. Strigolactones: discovery of the elusive shoot branching hormone. Trends Plant Sci. 14, 364–372.

D'Andrea, K.E., Otegui, M.E., Cirilo, A.G., 2008. Kernel number determination differs among maize hybrids in response to nitrogen. Field Crops Res. 105, 228–239.

Earl, H.J., Tollenaar, M., 1998. Differences among commercial maize (*Zea mays* L.) hybrids in respiration rates of mature leaves. Field Crops Res. 59, 9–19.

Eggert, K., von Wirén, N., 2017. Response of the plant hormone network to boron deficiency. New Phytol. 216, 868–881.

El-D, A.M.S.A., Salama, A., Wareing, P.F., 1979. Effects of mineral nutrition on endogenous cytokinins in plants of sunflower (*Helianthus annuus* L.). J. Exp. Bot. 30, 971–981.

Ely, K., Burnett, A.C., Liebermann-Cribbin, W., Serbin, S.P., Rogers, A., 2019. Spectroscopy can predict key leaf traits associated with source sink balance and carbon nitrogen status. J. Exp. Bot. 70, 1789–1799.

Endler, A., Meyer, S., Schelbert, S., Schneider, T., Weschke, W., Peters, S.W., et al., 2006. Identification of a vacuolar sucrose transporter in barley and Arabidopsis mesophyll cells by a tonoplast proteomic approach. Plant Physiol. 141, 196–207.

Engels, C., Marschner, H., 1986. Allocation of photosynthates to individual tubers of *Solanum tuberosum* L. III. Relationship between growth rate of individual tubers, tuber weight and stolon growth prior to tuber initiation. J. Exp. Bot. 37, 1813–1822.

Engels, C., Marschner, H., 1987. Effects of reducing leaf area and tuber number on the growth rates of tubers on individual potato plants. Potato Res. 30, 177–186.

Eom, J.S., Cho, J.I., Reinders, A., Lee, S.W., Yoo, Y.C., Tuan, P.Q., et al., 2011. Impaired function of the tonoplast-localized sucrose transporter in rice, OsSUT2, limits the transport of vacuolar reserve sucrose and affects plant growth. Plant Physiol. 157, 109–119.

Escobar, M., Geisler, D., Rasmusson, A.G., 2006. Reorganization of the alternative pathways of the Arabidopsis respiratory chain by nitrogen supply: opposing effects of ammonium and nitrate. Plant J. 45, 775–788.

Eskling, M., Ávidsson, P.-O., Akerlund, H.-E., 1997. The xanthophyll cycle, its regulation and components. Physiol. Plant. 100, 806–816.

Eyles, A., Pinkard, E.A., Davies, N.W., Corkrey, R., Churchill, K., O'Grady, A.P., et al., 2013. Whole-plant versus leaf-level regulation of photosynthetic responses after partial defoliation in *Eucalyptus globulus* saplings. J. Exp. Bot. 64, 1625–1636.

Falasca, G., Franceschetti, M., Bagni, N., Altamura, M.M., Biasi, R., 2010. Polyamine biosynthesis and control of the development of functional pollen in kiwifruit. Plant Physiol. Biochem. 48, 565–573.

Feng, W., Lindner, H., Robbins, N.E., Dinneny, J.R., 2016. Growing out of stress: The role of cell- and organ-scale growth control in plant water-stress responses. Plant Cell 28, 1769–1782.

Ferrante, A., Savin, R., Slafer, G.A., 2010. Floret development of durum wheat in response to nitrogen availability. J. Exp. Bot. 61, 4351–4359.

Fetene, M., Möller, I., Beck, E., 1993. The effect of nitrogen supply to *Urtica dioica* L. plants on the distribution of assimilate between shoot and roots. Bot. Acta 106, 228–234.

Firman, D.M., Allen, E.J., 1988. Field measurements of the photosynthetic rate of potatoes grown with different amounts of nitrogen fertilizer. J. Agric. Sci. 111, 85–90.

Fisher, D., 1978. An evaluation of the Münch hypothesis for phloem transport in soybean. Planta 139, 25–28.

Fisher, D.B., 1987. Changes in the concentration and composition of peduncle sieve tube sap during grain filling in normal and phosphate-deficient wheat plants. Aust. J. Plant Physiol. 14, 147–156.

Fischer, R.A., 2007. Understanding the physiological basis of yield potential in wheat. J. Agric. Sci. 145, 99–113.

Fletcher, A.L., Moot, D.J., Stone, P.J., 2008. Solar radiation interception and canopy expansion of sweet corn in response to phosphorus. Eur. J. Agron. 29, 80–87.

Florez-Sarasa, I.D., Bouma, T.J., Medrano, H., Azcon-Bieto, J., Ribas-Carbo, M., 2007. Contribution of the cytochrome and alternative pathways to growth respiration and maintenance respiration in *Arabidopsis thaliana*. Physiol. Plant. 129, 143–151.

Foo, E., Yoneyama, K., Hugill, C.J., Quittenden, L.J., Reid, J.B., 2013. Strigolactones and the regulation of pea symbioses in response to nitrate and phosphate deficiency. Mol. Plant 6, 76–87.

Forster, H., 1970. Der Einfluss einiger Ernährungsunterbrechungen auf die Ausbildung von Ertrags- und Qualitätsmerkmalen der Zuckerrübe. Landwirtsch. Forsch. Sonderh. 25, 99–105.

Foulkes, M.J., Slafer, G.A., Davies, W.J., Berry, P.M., Sylvester-Bradley, R., Martre, P., et al., 2011. Raising yield potential of wheat. III. Optimizing partitioning to grain while maintaining lodging resistance. J. Exp. Bot. 62, 469–486.

Foyer, C., 2018. Reactive oxygen species, oxidative signaling and the regulation of photosynthesis. Environ. Exp. Bot. 154, 134–142.

Foyer, C.H., Noctor, G., Hodges, M., 2011. Respiration and nitrogen assimilation: targeting mitochondria-associated metabolism as a means to enhance nitrogen use efficiency. J. Exp. Bot. 62, 1467–1482.

Fricke, W., 2002. Biophysical limitation of cell elongation in cereal leaves. Ann. Bot. 90, 157–167.

Friso, G., Majeran, W., Huang, M.L., Sun, Q., van Wijk, K.J., 2010. Reconstruction of metabolic pathways, protein expression, and homeostasis machineries across maize bundle sheath and mesophyll chloroplasts: large-scale quantitative proteomics using the first maize genome assembly. Plant Physiol. 152, 1219–1250.

Furbank, R.T., 2016. Walking the C_4 pathway: past, present and future. J. Exp. Bot. 68, 4057–4066.

Furze, M.E., Trumbore, S., Hartmann, H., 2018. Detours on the phloem sugar highway: stem carbon storage and remobilization. Curr. Opin. Plant Biol. 43, 89–95.

Galindo-Castañeda, T., Brown, K.M., Lynch, J.P., 2018. Reduced root cortical burden improves growth and grain yield under low phosphorus availability in maize. Plant Cell Environ. 41, 1579–1592.

Gao, Y.-P., Motosugi, H., Sugiura, A., 1992. Rootstock effects on growth and flowering in young apple trees grown with ammonium and nitrate nitrogen. J. Amer. Soc. Hort. Sci. 117, 446–452.

Gao, Y., Jia, L., Hu, B., Alva, A., Fan, M., 2014. Potato stolon and tuber growth induced by nitrogen form. Plant Prod. Sci. 17, 138–143.

Gao, F., Mei, X., Li, Y., Guo, J., Shen, Y., 2021. Update on the roles of polyamines in fleshy fruit ripening, senescence, and quality. Front. Plant Sci. 12, 610313. Available from: https://doi.org/10.3389/fpls.2021.610313.

Garcia, T., Heyduk, K., Kuzmick, E., Mayer, J.A., 2014. Crassulacean acid metabolism biology. New Phytol. 204, 738–740.

Gattolin, S., Newbury, H.J., Bale, J.S., Tseng, H.-M., Barett, D.A., Pritchard, J., 2008. A diurnal component to the variation in sieve tube amino acid content in wheat. Plant Physiol. 147, 912–921.

Ghannoum, O., Evans, J.R., Chow, W.S., Andrews, T.J., Conroy, J.P., von Caemmerer, S., 2005. Faster Rubisco is the key to superior nitrogen-use efficiency in NADP-malic enzyme relative to NAD-malic enzyme C_4 grasses. Plant Physiol. 137, 638–650.

Ghannoum, O., Paul, M.J., Ward, J.L., Beale, M.H., Corol, D.-I., Conroy, J.P., 2008. The sensitivity of photosynthesis to phosphorus deficiency differs between C_3 and C_4 tropical grasses. Funct. Plant Biol. 35, 213–221.

Giaquinta, R.T., 1978. Source and sink leaf metabolism in relation to phloem translocation. Plant Physiol. 61, 380–385.

Gironde, A., Poret, M., Etienne, P., Trouverie, J., Bouchereau, A., Le Caherec, F., et al., 2015. A profiling approach of the natural variability of foliar N remobilization at the rosette stage gives clues to understand the limiting processes involved in the low N use efficiency of winter oilseed rape. J. Exp. Bot. 66, 2461–2473.

Glanz-Idan, N., Tarkowski, P., Turekova, V., Wolf, S., 2020. Root-shoot communication in tomato plants: cytokinins as a signal molecule modulating leaf photosynthetic activity. J. Exp. Bot. 71, 247–257.

Godt, D., Roitsch, T., 2006. The developmental and organ specific expression of sucrose cleaving enzymes in sugar beet suggests a transition between apoplasmic and symplasmic phloem unloading in the tap roots. Plant Physiol. Biochem. 44, 656–665.

Goldbach, E., Goldbach, H., Wagner, H., Michael, G., 1975. Influence of N-deficiency on the abscisic acid content of sunflower plants. Physiol. Plant. 34, 138–140.

Goto, F., Yoshihara, T., Shigemoto, N., Toki, S., Takaiwa, F., 1999. Iron fortification of rice seed by the soybean ferritin gene. Nat. Biotechnol. 17, 282–286.

Graham, R.D., 1975. Male sterility in wheat plants deficient in copper. Nature 254, 514–515.

Granier, C., Tardieu, F., 2009. Multi-scale phenotyping of leaf expansion in response to environmental changes: the whole is more than the sum of parts. Plant Cell Environ. 32, 1175–1184.

Granier, C., Massonnet, C., Turc, O., Muller, B., Chenu, K., Tardieu, F., 2002. Individual leaf development in *Arabidopsis thaliana*: a stable thermal-time-based programme. Ann. Bot. 89, 595–604.

Grassi, G., Colom, M.R., Minotta, G., 2001. Effects of nutrient supply on photosynthetic acclimation and photoinhibition of one-year-old foliage of *Picea abies*. Physiol. Plant. 111, 245–254.

Gregersen, P.L., Culetic, A., Boschian, L., Krupinska, K., 2013. Plant senescence and crop productivity. Plant Mol. Biol. 82, 603–622.

Grimberg, A., Wilkinson, M., Snell, P., De Vos, R.P., González-Thuillier, I., Tawfike, A., et al., 2020. Transitions in wheat endosperm metabolism upon transcriptional induction of oil accumulation by oat endosperm WRINKLED1. BMC Plant Biol. 20, 235. Available from: https://doi.org/10.1186/s12870-020-02438-9.

Grossmann, K., 1990. Plant growth retardants as tools in physiological research. Physiol. Plant. 78, 640–648.

Grzebisz, 2015. Magnesium. In: Barker, A.V., Pilbeam, D.J. (Eds.), Handbook of Plant Nutrition, 2nd ed CRC Press, London, UK, pp. 199–260.

Haeder, H.-E., Beringer, H., 1981. Influence of potassium nutrition and water stress on the abscisic acid content in grains and flag leaves during grain development. J. Sci. Food Agric. 32, 552–556.

Hall, A.J., Savin, R., Slafer, G.A., 2014. Is time to flowering in wheat and barley influenced by nitrogen? A critical appraisal of recent published reports. Eur. J. Agron. 54, 40–46.

Hammes, P.S., Beyers, E.A., 1973. Localization of the photoperiodic perception in potatoes. Potato Res. 16, 68–72.

Hammond, J.P., White, P.J., 2008. Sucrose transport in the phloem: integrating root responses to phosphorus starvation. J. Exp. Bot. 59, 93–109.

Harrison, P.M., Arosio, P., 1996. The ferritins: molecular properties, iron storage function and cellular regulation. Biochim. Biophys. Acta 1275, 161–203.

Hartung, W., Slovik, S., 1991. Physicochemical properties of plant growth regulators and plant tissues determine their distribution and redistribution: stomatal regulation by abscisic acid in leaves. New Phytol. 119, 361–382.

Hatch, M.D., Slack, C.R., 1966. Photosynthesis by sugar-cane leaves: a new carboxylation reaction and pathway of sugar formation. Biochem. J. 101, 103–111.

Haupt, S., Duncan, G.H., Holzberg, S., Oparka, K.J., 2001. Evidence for symplastic phloem unloading in sink leaves of barley. Plant Physiol. 125, 209–218.

Hayashi, H., Chino, M., 1990. Chemical composition of phloem sap from the uppermost internode of the rice plant. Plant Cell Physiol. 31, 247–251.

Hays, D.B., Do, J.H., Mason, R.E., Morgan, G., Finlayson, S.A., 2007. Heat stress induced ethylene production in developing wheat grains induces kernel abortion and increased maturation in a susceptible cultivar. Plant Sci. 172, 1113–1123.

Heaton, E.A., Dohleman, F.G., Long, S.P., 2008. Meeting US biofuel goals with less land: the potential of Miscanthus. Glob. Change Biol. 14, 2000–2014.

Hedhly, A., Hormaza, J.I., Herrero, M., 2009. Global warming and sexual plant reproduction. Trends Plant Sci. 14, 30–36.

Heldt, H.W., Chon, C.J., Maronde, D., Herold, A., Stankovic, Z.S., Walker, D.A., et al., 1977. Role of orthophosphate and other factors in the regulation of starch formation in leaves and isolated chloroplasts. Plant Physiol. 59, 1146–1155.

Hermans, C., Hammond, J.P., White, P.J., Verbruggen, N., 2006. How do plants respond to nutrient shortage by biomass allocation? Trends Plant Sc. 11, 610–617.

Hermans, C., Vuylsteke, M., Coppens, F., Cracium, A., Inzé, D.V., Verbuggen, N., 2010. Early transcriptomic changes induced by magnesium deficiency in *Aribidopsis thaliana* reveal the alteration of circadian clock gene in roots and the triggering of abscisic acid response genes. New Phytol. 187, 119–131.

Herrera, A., 2009. Crassulacean acid metabolism and fitness under water deficit stress: if not for carbon gain, what is facultative CAM good for? Ann. Bot. 103, 645–653.

Herzog, H., 1981. Wirkung von zeitlich begrenzten Stickstoff- und Cytokiningaben auf die Fahnenblatt- und Kornentwickung von Weizen. Z. Pflanzenernähr. Bodenk. 144, 241–253.

Hibberd, J.M., Sheehy, J., Langdale, J.A., 2008. Using C_4 photosynthesis to increase the yield of rice - rationale and feasibility. Curr. Opin. Plant Biol. 11, 228–231.

Himelblau, E., Amasino, R.M., 2001. Nutrients mobilized from leaves of *Arabidopsis thaliana* during leaf senescence. J. Plant Physiol. 158, 1317–1323.

Hirose, N., Takei, K., Kuroha, T., Kamada-Nobusada, T., Hayashi, H., Sakakibara, H., 2008. Regulation of cytokinin biosynthesis, compartmentalization and translocation. J. Exp. Bot. 59, 75–83.

Hochmal, R.K., Schulze, K., Trompelt, K., Hippler, M., 2015. Calcium-dependent regulation of photosynthesis. Biochem. Biophys. Acta-Bioenerg. 1847, 993–1003.

Hodgson, R.A.J., Raison, J.K., 1991. Superoxide production by thylakoids during chilling and its implication in the susceptibility of plants to chilling induced photoinhibition. Planta 183, 222–228.

Horton, P., 2000. Prospects for crop improvement through the genetic manipulation of photosynthesis: morphological and biochemical aspects of light capture. J. Exp. Bot. 51, 475–485.

Hosseini, S.A., Hajirezaei, M.R., Seiler, C., Sreenivasulu, N., von Wirén, N., 2016. A potential role of flag leaf potassium iiinnn conferring tolerance to drought-induced leaf senescence in barley. Front. Plant Sci. 7, 206. Available from: https://doi.org/10.3389/pls.2016.00206.

Hu, W., Lv, X., Yang, J., Chen, B., Zhao, W., Meng, Y., et al., 2016. Effects of potassium deficiency on antioxidant metabolism related to leaf senescence in cotton (*Gossypium hirsutum* L.). Field Crops Res. 191, 139–149.

Hu, W., Lu, Z., Meng, F., Li, X., Cong, R.M., Ren, T., et al., 2020. The reduction in leaf area precedes that in photosynthesis under potassium deficiency: the importance of leaf anatomy. New Phytol. 227, 1749–1763.

Iglesias, A.A., Plaxton, W.C., Podestá, F.E., 1993. The role of inorganic phosphate in the regulation of C_4 photosynthesis. Photosynth. Res. 35, 205–211.

Iglesias, D.J., Lliso, I., Tadeo, F.R., Talon, M., 2002. Regulation of photosynthesis through source:sink imbalance in citrus is mediated by carbohydrate content in leaves. Physiol. Plant. 116, 563–572.

Imlau, A., Truernit, E., Sauer, N., 1999. Cell-to-cell and long-distance trafficking of the green fluorescent protein in the phloem and symplastic unloading of the protein into sink tissues. Plant Cell 11, 309–322.

Ishimaru, Y., Masuda, H., Bashir, K., Inoue, H., Tsukamoto, T., Takahashi, M., et al., 2010. Rice metal-nicotianamine transporter, OsYSL2, is required for the long-distance transport of iron and manganese. Plant J. 62, 379–390.

Ivins, J.D., Bremner, P.M., 1964. Growth, development and yield in the potato. Outlook Agric. 4, 211–217.

Jackson, M.B., 1990. Communication between the root and shoots of flooded plants.. Br. Soc. Plant Growth Regul. Monograph 21, 115–133.

Jameson, P.E., McWha, J.A., Wright, G.J., 1982. Cytokinins and changes in their activity during development of grains of wheat (*Triticum aestivum* L.). Z. Pflanzenphysiol. 106, 27–36.

Jamjod, S., Rerkasem, B., 1999. Genotypic variation in response of barley to boron deficiency. Plant Soil. 215, 65–72.

Jeena, G.S., Kumar, S., Shukla, R.K., 2019. Structure, evolution and diverse physiological roles of SWEET sugar transporters in plants. Plant Mol. Biol. 100, 351–365.

Jia, Z., von Wirén, N., 2020. Signaling pathways underlying nitrogen-dependent changes in root system architecture: from model to crop species. J. Exp. Bot. 71, 4393–4404.

Jiang, F., Hartung, W., 2008. Long-distance signalling of abscisic acid (ABA): the factors regulating the intensity of the ABA signal. J. Exp. Bot. 59, 37–43.

Jiang, Z., Piao, L., Guo, D., Zhu, H., Wang, S., Zhu, H., et al., 2021. Regulation of maize kernel carbohydrate metabolism by abscisic acid applied at the grain-filling stage at low soil water potential. Sustainability 13, 3125. Available from: https://doi.org/10.3390/su13063125.

Johnston, A.E., Milford, G.F., 2012. Potassium and nitrogen interaction in crops. Potash Development Association, Stamford, UK.

Johnson, M.P., Havaux, M., Triantaphylides, C., Ksas, B., Pascal, A.A., Robert, B., et al., 2007. Elevated zeaxanthin bound to oligomeric LHCII enhances the resistance of Arabidopsis to photooxidative stress by a lipid-protective, antioxidant mechanism. J. Biol. Chem. 282, 22605–22618.

Jordan, B.R., 2002. Molecular response of plant cells to UV-B stress. Funct. Plant Biol. 29, 909–916.

Jordan-Meille, L., Pellerin, S., 2004. Leaf area establishment of a maize (*Zea mays* L.) field crop under potassium deficiency. Plant Soil. 265, 75–92.

Kaiser, B.N., Gridley, K.L., Brady, J.N., Phillips, T., Tyerman, S.D., 2005. The role of molybdenum in agricultural plant production. Ann. Bot. 96, 745–754.

Kanal, S., Ohura, K., Adu-Gyamfi, J.J., Mohapatra, P.K., Nguyen, N.T., Saneoko, H., et al., 2007. Depression in sink activity precedes the inhibition of biomass production in tomato plants subjected to potassium deficiency stress. J. Exp. Bot. 58, 2917–2928.

Kant, S., Peng, M., Rothstein, S.J., 2011. Genetic regulation by NLA and microRNA827 for maintaining nitrate-dependent phosphate homeostasis in Arabidopsis. PLoS Genetics. 7, e1002021. Available from: https://doi.org/10.1371/journal.pgen.1002021.

Kapoor, S., Takatsuji, H., 2006. Silencing of an anther-specific zinc-finger gene, MEZ1, causes aberrant meiosis and pollen abortion in petunia. Plant Mol. Biol. 51, 415–430.

Kapulnik, Y., Koltai, H., 2012. Strigolactones are involved in root response to low phosphate conditions in Arabidopsis. Plant Physiol. 160, 1329–1341.

Kaschuk, G., Kuyper, T.W., Leffelaar, P.A., Hungria, M., Giller, K.E., 2009. Are the rates of photosynthesis stimulated by the carbon sink strength of rhizobial and arbuscular mycorrhizal symbioses? Soil Biol. Biochem. 41, 1233–1244.

Kashtoh, H., Baek, K.-H., 2021. Structural and functional insights into the role of guard cell ion channels in abiotic stress-induced stomatal closure. Plants 10, 2774. Available from: https://doi.org/10.3390/plants10122774.

Kaur, H., Peel, A., Acosta, K., Gebril, S., Ortaga, J.L., Sengupta-Gopalan, C., 2019. Comparison of alfalfa plants overexpressing glutamine sythetase with those overexpressing sucrose phosphate synthase demonstrates a signal mechanism linking carbon and nitrogen metabolism between the leaves and nodules. Plant Direct 3, e00115. Available from: https://doi.org/10.1002/pld3.115.

Kavanová, M., Lattanzi, F.A., Grimoldi, A.A., Schnyder, H., 2006. Phosphorus deficiency decreases cell division and elongation in grass leale. Plant Physiol. 141, 766–775.

Kavanová, M., Lattanzi, F.A., Schnyder, H., 2008. Nitrogen deficiency inhibits leaf blade growth in *Lolium perenne* by increasing cell cycle duration and decreasing mitotic and post-mitotic growth rates. Plant Cell Environ. 31, 727–737.

Kebeish, R., Niessen, M., Thiruveedhi, K., Bari, R., Hirsch, H.-J., Rosenkranz, R., et al., 2007. Chloroplastic photorespiratory bypass increases photosynthesis and biomass production in *Arabidopsis thaliana*. Nat. Biotechnol. 25, 593–599.

Kellogg, E.A., 2013. C_4 photosynthesis. Curr. Biol. 23, 594–599.

Kering, M.K., Lukaszewski, K.M., Blevins, D.G., 2009. Manganese requirements for optimum photosynthesis and growth in NAD-malic enzyme C-4 species. Plant Soil 316, 217–226.

Khorobrykh, S., Havurinee, V., Mattila, H., Tyystjärvi, E., 2020. Oxygen and ROS in photosynthesis. Plants 9, 91. Available from: https://doi.org/10.3390/plants9010091.

Kim, S.A., Punshon, T., Lanzirotti, A., Li, L., Alonso, J.M., Ecker, J.R., et al., 2006. Localization of iron in Arabidopsis seed requires the vacuolar membrane transporter VIT1. Science 314, 295–298.

Kim, W., Ahn, H.J., Chiou, T.-J., Ahn, J.H., 2011. The role of the Mir399-Pho2 module in the regulation of flowering time in response to different ambient temperatures in *Arabidopsis thaliana*. Mol. Cells. 32, 83–88.

King, J.L., Edwards, G.E., Cousins, A.B., 2012. The efficiency of the CO_2 concentrating mechanism during single cell C_4 photosynthesis. Plant Cell Environ. 35, 513–523.

Kleinkopf, G.E., Westermann, D.T., Dwelle, R.B., 1981. Dry matter production and nitrogen utilization by six potato cultivars. Agron J. 73, 799–802.

Kline, K.G., Sussman, M.R., Jones, A.M., 2010. Abscisic acid receptors. Plant Physiol. 154, 479–482.

Knoblauch, M., Peters, W.S., 2013. Long-distance translocation of photosynthates: a primer. Photosynth. Res. 117, 189–196.

Kohlen, W., Charnikhova, T., Liu, Q., Bours, R., Domagalska, M.A., Beguerie, S., et al., 2011. Strigolactones are transported through the xylem and play a key role in shoot architectural response to phosphate deficiency in nonarbuscular mycorrhizal host Arabidopsis. Plant Physiol. 155, 974–987.

Koltai, H., Dor, E., Hershenhorn, J., Joel, D.M., Weininger, S., Lekalla, S., et al., 2010. Strigolactones' effect on root growth and root-hair elongation may be mediated by auxin-efflux carriers. J. Plant Growth Regul. 29, 129–136.

Krauss, A., 1978a. Tuberization and abscisic acid content in *Solanum tuberosum* as affected by nitrogen nutrition. Potato Res. 21, 183–193.

Krauss, A., 1978b. Endogenous regulation mechanisms in tuberization of potato plants in relation to environmental factors. European Association of Potato Research. Abstracts of Conference Papers. 7, 47–48.

Krauss, A., Marschner, H., 1971. Einfluß der Stickstoffernährung der Kartoffeln auf Induktion und Wachstumsrate der Knolle. Z. Pflanzenernähr. Bodenk. 128, 153–168.

Krauss, A., Marschner, H., 1976. Einfluss von Stickstoffernährung und Wuchstoffapplikation auf die Knolleninduktion bei Kartoffelpflanzen. Z. Pflanzenernähr. Bodenk. 139, 143–155.

Krauss, A., Marschner, H., 1982. Influence of nitrogen nutrition, day length and temperature on contents of gibberellic and abscisic acid and on tuberization in potato plants. Potato Res. 25, 13–21.

Krauss, A., Marschner, H., 1984. Growth rate and carbohydrate metabolism of potato tubers exposed to high temperatures. Potato Res. 27, 297–303.

Kriedemann, P.E., Graham, R.D., Wiskich, J.T., 1985. Photosynthetic disfunction and in vivo changes in chlorophyll a fluorescence from manganese-deficient wheat leaves. Aust. J. Agric. Res. 36, 157–169.

Kudoyarova, G.R., Dodd, I.C., Veselov, D.S., Rothwell, S.A., Veselov, S.Y., 2015. Common and specific responses to availability of mineral nutrients and water. J. Exp. Bot. 66, 2133–2144.

Kuiper, D., 1988. Growth responses of *Plantago major* L. ssp. *pleiosperma* (Pilger) to changes in mineral supply. Plant Physiol. 87, 555–557.

Kuiper, D., Schuit, J., Kuiper, P.J.C., 1988. Effect of internal and external cytokinin concentrations on root growth and shoot to root ration of *Plantago major* ssp. *pleiosperma* at different nutrient concentrations. Plant Soil 111, 231–236.

Kuiper, D., Kuiper, P.J.C., Lambers, H., Schuit, J., Staal, M., 1989. Cytokinin concentration in relation to mineral nutrition and benzyladenine treatment in *Plantago major* ssp. *pleiosperma*. Physiol. Plant. 75, 511–517.

Kumagai, E., Araki, T., Ueno, O., 2010. Comparison of susceptibility to photoinhibition and energy partitioning of absorbed light in photosystem II in flag leaves of two rice (*Oryza sativa* L.) cultivars that differ in their response to nitrogen-deficiency. Plant Prod. Sci. 13, 11–20.

Kyozuka, J., Tokunaga, H., Yoshida, A., 2014. Control of grass inflorescence form by the fine-tuning of meristem phase change. Curr. Opin. Plant Biol. 17, 110–115.

Lalonde, S., Tegeder, M., Throne-Holst, M., Frommer, W.B., Patrick, J.W., 2003. Phloem loading and unloading of sugars and amino acids. Plant Cell Environ. 26, 37–56.

Landrein, B., Formosa-Jordan, P., Malivert, A., Schuster, C., Melnyk, C.W., Yang, W.B., et al., 2018. Nitrate modulates stem cell dynamics in Arabidopsis shoot meristems through cytokinins. Proc. Natl. Acad. Sci. USA 115, 1382–1387.

Lang, A., 1983. Turgor regulated translocation. Plant Cell Environ. 6, 683–689.

Lanquar, V., Lelièvre, F., Bolte, S., Hamès, C., Alcon, C., Neumann, D., et al., 2005. Mobilization of vacuolar iron by AtNRAMP3 and AtNRAMP4 is essential for seed germination on low iron. EMBO J. 24, 4041–4051.

Larcher, W., 2003. Plant Ecology: Ecophysiology and Stress Physiology of Functional Groups, 4th ed Springer, Berlin, Germany.

Lauer, M.J., Blevins, D.G., 1989. Flowering and podding characteristics on the main stem of soybean grown on varying levels of phosphate nutrition. J. Plant Nutr. 12, 1061–1072.

Lauer, M.J., Pallardy, S.G., Blevins, D.G., Randall, D.D., 1989. Whole leaf carbon exchange characteristics of phosphate deficient soybeans (*Glycine max* L.). Plant Physiol. 91, 848–854.

Lee, E.A., Tollenaar, M., 2007. Physiological basis of successful breeding strategies for maize grain yield. Crop Sci. 47 (S3), S202–S215.

Lee, J.Y., Frank, M., 2018. Plasmodesmata in phloem: different gateways for different cargoes. Curr. Opin. Plant Biol. 43, 119–124.

Lee, B., Martin, P., Bangerth, F., 1989. The effect of sucrose in the levels of abscisic acid, indoleacetic acid and zeatin/zeatin ribose in wheat ears growing in liquid culture. Physiol. Plant. 77, 73–80.

Lers, A., Sonego, L., Green, P.J., Burd, S., 2006. Suppression of LX ribonuclease in tomato results in a delay of leaf senescence and abscission. Plant Physiol. 142, 710–721.

Li, R., Moore, M., King, J., 2003. Investigating the regulation of one carbon metabolism in *Arabidopsis thaliana*. Plant Cell Physiol. 44, 233–241.

Li, J.-Y., Liu, X.-H., Cai, Q.-S., Gu, H., Zhang, S.-S., Wu, Y.-Y., et al., 2008. Effects of elevated CO_2 on growth, carbon assimilation, photosynthate accumulation and related enzymes in rice leaves during sink-source transition. J. Integr. Plant Biol. 50, 723–732.

Li, S., Ying, Y., Secco, D., Wang, C., Narsai, R., Whelan, J., et al., 2017. Molecular interaction between PHO2 and GIGANTEA reveals a new crosstalk between flowering time and phosphate homeostasis in *Oryza sativa*. Plant Cell Environ. 40, 1487–1499.

Li, J., Liu, B., Li, X., Li, D., Han, J., Zhang, Y., et al., 2021. Exogenous abscisic acid mediates berry quality improvement by altered endogenous plant hormones level in "Ruiduhongyu" grapevine. Front. Plant Sci. 12, 739964. Available from: https://doi.org/10.3389/fpls.2021.739964.

Liang, Y.-L., Lur, H.-S., 2002. Conjugated and free polyamine levels in normal and aborting maize kernels. Crop Sci. 42, 1217–1224.

Lim, S.D., Lee, S., Choi, W.-G., Yim, W.C., Cushman, J.C., 2019. Laying the foundation for Crassulacean Acid Metabolism (CAM) biodesign: Expression of the C_4 metabolism cycle genes of CAM in Arabidopsis. Front. Plant Sci. 10, 101. Available from: https://doi.org/10.3389/fpls.2019.00101.

Lin, Y.-L., Tsay, Y.-F., 2017. Influence of differing nitrate and nitrogen availability on flowering control in Arabidopsis. J. Exp. Bot. 68, 2603–2609.

Lin, Z.F., Zhong, S.L., Grierson, D., 2009. Recent advances in ethylene research. J. Exp. Bot. 60, 3311–3336.

Lin, M., Occhialini, A., Andralojic, P., Parry, M.A.J., Hanson, M.R., 2014. A faster Rubisco with potential to increase photosynthesis in crops. Nature 513, 547–550.

Liu, Y., von Wirén, N., 2017. Ammonium as a signal for physiological and morphological responses in plants. J. Exp. Bot. 68, 2581–2592.

Liu, T.Y., Chang, C.Y., Chiou, T.J., 2009. The long-distance signaling of mineral macronutrients. Curr. Opin. Plant Biol. 12, 312–319.

Long, S.P., Zhu, X.-G., Naidu, S.L., Ort, D.R., 2006. Can improvement in photosynthesis increase crop yields? Plant Cell Environ. 29, 315–330.

Long, S.P., Marshall-Colon, A., Zhu, X.-G., 2015. Meeting the global food demand of the future by engineering crop photosynthesis. Cell. 161, 56–61.

Longnecker, N.E., Graham, R.D., Card, G., 1991. Effects of manganese deficiency on the pattern of tillering and development of barley (*Hordeum vulgare* cv. Galleon). Field Crops Res. 28, 85–102.

Luo, L., Pan, S., Liu, X.H., Wang, H.X., Xu, G.H., 2017. Nitrogen deficiency inhibits cell division-determined elongation, but not initiation, of rice tiller buds. Isr. J. Plant Sci. 64, 32–40.

Luo, L., Zhang, Y., Xu, G., 2020. How does nitrogen shape plant architecture? J. Exp. Bot. 71, 4415–4427.

Lynch, J.P., 2019. Root phenotypes for improved nutrient capture: an underexploited opportunity for global agriculture. New Phytol. 223, 548–564.

Ma, S., Li, Y.X., Li, X., Sui, X.L., Zhang, Z.X., 2019. Phloem unloading strategies and mechanisms in crop fruits. J. Plant Growth Regul. 38, 494–500.

Machin, D.C., Hammon-Josse, M., Bennett, T., 2020. Fellowship of the rings: saga of strigolactones and other small signals. New Phytol. 225, 621–636.

MacLeod, L.B., 1969. Effects of N, P and K and their interactions on the yield and kernel weight of barley in hydroponic culture. Agron. J. 61, 26–29.

Maeshima, M., 2001. Tonoplast transporters: organization and function. Annu. Rev. Plant Physiol. Plant Mol. Biol. 52, 469–497.

Makino, A., Sakuma, H., Sudo, E., Mae, T., 2003. Differences between maize and rice in N-use efficiency of photosynthesis and protein allocation. Plant Cell Physiol. 44, 952–956.

Manrique, L.A., Bartholomew, D.P., 1991. Growth and yield performance of potato grown at three elevations in Hawaii: II. Dry matter production and efficiency of partitioning. Crop Sci. 31, 367–371.

Marder, J.B., Barber, J., 1989. The molecular anatomy and function of thylakoid proteins. Plant Cell Environ. 12, 595–614.

Marschner, H., Cakmak, I., 1989. High light intensity enhances chlorosis and necrosis in leaves of zinc, potassium, and magnesium deficient bean (*Phaseolus vulgaris*) plants. J. Plant Physiol. 134, 308–315.

Martin, A., Belastegui-Macadam, X., Quilleré, I., Floriot, M., Valadier, M.-H., Pommel, B., et al., 2005. Nitrogen management and senescence in two maize hybrids differing in the persistence of leaf greenness: agronomic, physiological and molecular aspects. New Phytol. 167, 483–492.

Martin, P., 1989. Long-distance transport and distribution of potassium in crop plants. In: Methods of K-research in Plants. Proc. 21st Colloq. Int. Potash Inst., Bern, Switzerland, pp. 83–100.

Masclaux-Daubresse, C., Daniel-Vedele, F., Dechorgnat, J., Chardon, F., Gaufichon, L., Suzuki, A., 2010. Nitrogen uptake, assimilation and remobilization in plants: challenges for sustainable and productive agriculture. Ann. Bot. 105, 1141–1157.

Masuda, T., 2008. Recent overview of the branch of the tetrapyrrole biosynthesis leading to chlorophylls. Photosynth. Res. 96, 121–143.

Masuda, H., Usuda, K., Kobayashi, T., Ishimaru, Y., Kakei, Y., Takahashi, M., et al., 2009. Overexpression of the barley nicotianamine synthase gene *HvNAS1* increases iron and zinc concentrations in rice grains. Rice 2, 155–166.

Matthes, M.S., Robil, J.M., McSteen, P., 2020. From element to development: the power of the essential micronutrient boron to shape morphological processes in plants. J. Exp. Bot. 71, 1681–1693.

Mayta, M.L., Harjirezaci, M.-H., Carrvillo, N., Lodeyro, R.F., 2019. Leaf senescence: the chloroplast connection comes of age. Plants 8, 495. Available from: https://doi.org/10.3390/plants8110495.

Mayzlish-Gati, E., De Cuyper, C., Goormachtig, G., Beeckman, T., Vuylsteke, M., Brewer, P.B., et al., 2012. Strigolactones are involved in low phosphate conditions in Arabidopsis. Plant Physiol. 160, 1329–1341.

McCormick, A.J., Cramer, M.D., Watt, D.A., 2008. Changes in photosynthetic rates and gene expression of leaves during a source-sink perturbation in sugarcane. Ann. Bot. 101, 89–102.

McCormick, A.J., Watt, D.A., Cramer, M.D., 2009. Supply and demand: sink regulation of sugar accumulation in sugarcane. J. Exp. Bot. 60, 357–364.

Mengel, K., Haeder, H.E., 1977. Effect of potassium supply on the rate of phloem sap exudation and the composition of phloem sap of *Ricinus communis*. Plant Physiol. 59, 282–284.

Merewitz, E.B., Gianfagna, T., Huang, B., 2011. Photosynthesis, water use and root viability under water stress as affected by expression of SAG12-ipt controlling cytokinin synthesis in *Agrostis stolonifera*. J. Exp. Bot. 62, 383–395.

Michael, G., Beringer, H. 1980. The role of hormones in yield formation. In: Physiological Aspects of Crop Productivity. Proc. 15th Colloq. Int. Potash Inst., Bern, Switzerland, pp. 85-115.

Miguez, F., Fernandez-Marin, B., Becerril, J.M., Garcia, H., 2015. Activation of photoprotective winter photoinhibition in plants: a literature compilation and meta-analysis. Physiol. Plant. 153, 414–423.

Milne, R.J., Grof, C.P.L., Patrick, J.W., 2018. Mechanisms of phloem unloading: shaped by cellular pathways, their conductances and sink function. Curr. Opin. Plant Biol. 43, 8–15.

Mitscherlich, E.A., 1954. Bodenkunde für Landwirte, 7th ed. Förster und Gärtner, Parey, Berlin, Germany.

Mittler, R., 2017. ROS are good. Trends Plant Sci. 22, 11–19.

Miyawaki, K., Matsumoto-Kitano, M., Kakimoto, T., 2004. Expression of cytokinin biosynthetic isopentenyltransferase genes in Arabidopsis: tissue specificity and regulation by auxin, cytokinin, and nitrate. Plant J. 37, 128–138.

Mohapatra, P.K., Singh, N.R., 2015. Teaching the Z-scheme of electron transport in photosynthesis: a perspective. Photosynth. Res. 123, 105–114.

Morgan, J.M., 1980. Possible role of abscisic acid in reducing seed set in water stressed plants. Nature 285, 655–657.

Morrissey, J., Guerinot, M.L., 2009. Iron uptake and transport in plants: the good, the bad, and the ionome. Chem. Rev. 109, 4553–4567.

Mounla, M.A.K., Bangerth, F., Stoy, V., 1980. Gibberellin-like substances and indole type auxins in developing grains of normal- and high-lysine genotypes of barley. Physiol. Plant. 48, 568–573.

Mudge, S.R., Rae, A.L., Diatloff, E., Smith, F.W., 2002. Expression analysis suggests novel roles for members of the Pht1 family of phosphate transporters in Arabidopsis. Plant J. 31, 341–353.

Nabity, P.D., Zavala, A., DeLucia, E.H., 2009. Indirect suppression of photosynthesis on individual leaves by arthropod herbivory. Ann. Bot. 103, 655–663.

Nachiangmai, D., Dell, B., Bell, R., Huang, L., Rerkasem, B., 2004. Enhanced boron transport into the ear of wheat as a mechanism for boron efficiency. Plant Soil 264, 141–147.

Nadwodnik, J., Lohaus, G., 2008. Subcellular concentrations of sugar alocohols and sugars in relation to phloem translocation in *Plantago major, Plantago maritima, Prunus persica*, and *Apium graveolens*. Planta 227, 1079–1089.

Nambiar, E.K.S., 1976. Genetic differences in the copper nutrition of cereals. 1. Differential responses of genotypes to copper. Aust. J. Agric. Res. 27, 453–463.

Navrot, N., Rouhier, N., Gelhaye, E., Jacquot, J.-P., 2007. Reactive oxygen species generation and antioxidant systems in plant mitochondria. Physiol. Plant. 129, 185–195.

Neuhaus, H.E., 2007. Transport of primary metabolites across the plant vacuolar membrane. FEBS Lett. 581, 2223–2226.

Neumann, G., 2007. Root exudates and nutrient cycling. In: Marschner, P., Rengel, Z. (Eds.), Nutrient Cycling in Ecosystems. Springer, Berlin, Heidelberg, pp. 123–157.

Nielsen, K.L., Bouma, T.J., Lynch, J.P., Eissenstat, D.M., 1998. Effects of phosphorus availability and vesicular-arbuscular mycorrhizas on the carbon budget of common bean (*Phaseolus vulgaris*). New Phytol. 139, 647–656.

Nikolic, M., Pavlovic, J., 2018. Plant responses to iron deficiency and toxicity and iron use efficiency in plants. In: Hossain, M., Kamiya, T., Burritt, D., Phan Tran, L.-S., Fujiwara, T. (Eds.), Plant Micronutrient Use Efficiency: Molecular and Genomic Perspectives in Crop Plants. Academic Press, London, UK, pp. 55–69.

Ning, P., Li, S., Yu, P., Zhang, Y., Li, C., 2013. Post-silking accumulation and partitioning of dry matter, nitrogen, phosphorus and potassium in maize varieties differing in leaf longevity. Field Crops Res. 144, 19–27.

Nishiyama, Y., Allakhverdiev, S.I., Murata, N., 2006. A new paradigm for the action of reactive oxygen species in the photoinhibition of photosystem II. Biochim. Biophys. Acta. 1757, 742–749.

Nitsch, J.P., 1950. Growth and morphogenesis of strawberry as related to auxin. Am. J. Bot. 37, 211–216.

Niyogi, K.K., 1999. Photoprotection revisited: genetic and molecular approaches. Annu. Rev. Plant Physiol. Plant Mol. Biol. 50, 333–359.

Notaguchi, M., Okamoto, S., 2015. Dynamics of long-distance signaling via plant vascular tissues. Front. Plant Sci. 6, 161. Available from: https://doi.org/10.3389/fpls.2015.00161.

Ort, D.R., 2001. When there is too much light. Plant Physiol. 125, 29–32.

Otegui, M.S., Capp, R., Staehlin, L.A., 2002. Developing seeds of Arabidopsis store different minerals in two types of vacuoles and in the endoplasmic reticulum. Plant Cell 14, 1311–1327.

Pant, B.D., Musialak-Lange, M., Nuc, P., May, P., Buhtz, A., Kehr, J., et al., 2009. Identification of nutrient-responsive Arabidopsis and rapeseed microRNAs by comprehensive real-time polymerase chain reaction profiling and small RNA sequencing. Plant Physiol. 150, 1541–1555.

Pantin, F., Simonneau, T., Rolland, G., Dauzat, M., Muller, B., 2011. Control of leaf expansion: A developmental switch from metabolics to hydraulics. Plant Physiol. 156, 803–815.

Paponov, I.A., Engels, C., 2003. Effect of nitrogen supply on leaf traits related to photosynthesis during grain filling in two maize genotypes with different N efficiency. J. Plant Nutr. Soil Sci. 166, 756–763.

Paponov, I.A., Engels, C., 2005. Effect of nitrogen supply on carbon and nitrogen partitioning after flowering in maize. J. Plant Nutr. Soil Sci. 168, 447–453.

Paponov, I.A., Sambo, P., Schulte auf'm Erley, G., Presterl, T., Geiger, H.H., Engels, C., 2005. Grain yield and kernel weight of two maize genotypes differing in nitrogen use efficiency at various levels of nitrogen and carbohydrate availability during flowering and grain filling. Plant Soil 272, 111–123.

Parrott, D.L., Martin, J.M., Fischer, A.M., 2010. Analysis of barley (*Hordeum vulgare*) leaf senescence and protease gene expression: a family C1A cysteine protease is specifically induced under conditions characterized by high carbohydrate, but low to moderate nitrogen levels. New Phytol. 187, 313–331.

Parry, M.A.J., Keys, A.J., Madgwick, P.J., Carmo-Silva, A.E., Andralojc, P.J., 2008. Rubisco regulation: a role for inhibitors. J. Exp. Bot. 59, 1569–1580.

Parthier, B., 1991. Jasmonates, new regulators of plant growth and development: many facts and few hypotheses on their actions. Bot. Acta. 104, 446–454.

Paschalidis, K., Tsaniklidis, G., Wang, B.-Q., Delis, C., Trantas, E., Loulakakis, K., et al., 2019. The interplay among polyamines and nitrogen in plant stress responses. Plants 8, 315. Available from: https://doi.org/10.3390/plants8090315.

Pastore, D., Stoppelli, M.C., Di Fonzo, N., Passarella, S., 1999. The existence of the K^+ channel in plant mitochondria. J. Biol. Chem. 274, 26683–26690.

Pastore, D., Trono, D., Laus, M.N., Di Fonzo, N., Flagella, Z., 2007. Possible plant mitochondria involvement in cell adaptation to drought stress. A case study: durum wheat mitochondria. J. Exp. Bot. 58, 195–210.

Patrick, J.W., 1997. Phloem unloading: sieve element unloading and post-sieve element transport. Annu. Rev. Plant Physiol. Plant Mol. Biol. 48, 191–222.

Patrick, J.W., Offler, C.E., 2001. Compartmentation of transport and transfer events in developing seeds. J. Exp. Bot. 52, 551–564.

Patrick, J.W., Stoddard, F.L., 2010. Physiology of flowering and grain filling in faba bean. Field Crops Res. 115, 234–242.

Paul, M.J., Driscoll, S.P., 1997. Sugar repression of photosynthesis: the role of carbohydrates in signalling nitrogen deficiency through source:sink imbalance. Plant Cell Environ. 20, 110–116.

Paul, M.J., Foyer, C.H., 2001. Sink regulation of photosynthesis. J. Exp. Bot. 52, 1383–1400.

Peng, M., Bi, Y.M., Zhu, T., Rothstein, S.J., 2007. Genome-wide analysis of *Arabidopsis* responsive transcriptome to nitrogen limitation and its regulation by the ubiquitin ligase gene *NLA*. Plant Mol. Biol. 65, 775–797.

Perilli, P., Mitchell, L.G., Grant, C.A., Pisante, M., 2010. Cadmium concentration in durum wheat grain (*Triticum turgidum*) as influenced by nitrogen rate, seeding date and soil type. J. Sci. Food Agric. 90, 813–822.

Pfannschmidt, T., Bräutigam, K., Wagner, R., Dietzel, L., Schröter, Y., Steiner, S., et al., 2009. Potential regulation of gene expression in photosynthetic cells by redox and energy state: approaches towards better understanding. Ann. Bot. 103, 599–607.

Pierik, R., Tholen, D., Poorter, H., Visser, E.J.W., Voesenek, L.A.C.J., 2006. The Janus face of ethylene: growth inhibition and stimulation. Trends Plant Sci. 11, 176–183.

Pieters, A.J., Paul, M.J., Lawlor, D.W., 2001. Low sink demand limits photosynthesis under Pi deficiency. J. Exp. Bot. 52, 1083–1091.

Plaxton, W.C., Podestá, F.E., 2006. The functional organization and control of plant respiration. Crit. Rev. Plant Sci. 25, 159–198.

Plénet, D., Mollier, A., Pellerin, S., 2000. Growth analysis of maize field crops under phosphorus deficiency. II. Radiation-use efficiency, biomass accumulation and yield components. Plant Soil 224, 259–272.

Plett, D.C., Ranathunga, K., Melino, V.J., Kuya, N., Uga, Y., Kronzucker, H.J., 2020. The intersection of nitrogen nutrition and water use in plants: new paths toward improved crop productivity. J. Exp. Bot. 71, 4452–4468.

Poiré, R., Wiese-Klinkenberg, A., Parent, B., Mielewczik, M., Schurr, U., Tardieu, F., et al., 2010. Diel time-courses of leaf growth in monocot and dicot species: endogenous rhythms and temperature effect. J. Exp. Bot. 61, 1751–1759.

Poitout, A., Crabos, A., Petřík, I., Novák, O., Krouk, G., Lacombe, B., et al., 2018. Responses to systemic nitrogen signaling in Arabidopsis roots involve trans-zeatin in shoots. Plant Cell 30, 1243–1257.

Pommerrenig, B., Müdsam, C., Kischka, D., Neuhaus, H.E., 2020. Treat and trick: common regulation and manipulation of sugar transporters during sink establishment by the plant and the pathogen. J. Exp. Bot. 71, 3930–3940.

Poorter, H., Evans, J.R., 1998. Photosynthetic nitrogen-use efficiency of species that differ inherently in specific leaf area. Oecologia 116, 26–37.

Poorter, H., Van der Werf, A., Atkin, O.K., Lambers, H., 1991. Respiratory energy requirements of root vary with the potential growth rate of a plant species. Physiol. Plant. 83, 469–475.

Porter, J.R., Semenov, M.A., 2005. Crop responses to climatic variation. Phil. Trans. Roy. Soc. London B. 360, 2021–2035.

Portis, A.R., 1992. Regulation of ribulose 1,5-bisphosphate carboxylase/oxygenase activity. Annu. Rev. Plant Physiol. 35, 415–437.

Portis Jr., A.R., Li, C., Wang, D., Salvucci, M.E., 2008. Regulation of Rubisco activase and its interaction with Rubisco. J. Exp. Bot. 59, 1597–1604.

Postma, J.A., Lynch, J.P., 2011. Root cortical aerenchyma enhances the growth of maize on soils with suboptimal availability of nitrogen, phosphorus and potassium. Plant Physiol. 156, 1190–1201.

Prystupa, P., Savin, R., Slafer, G.A., 2004. Grain number and its relationship with dry matter, N and P in the spikes at heading in response to N x P fertilization in barley. Field Crops Res. 90, 245–254.

Quinet, M., Angosto, T., Yuste-Lisbona, F.J., Blanchard-Gros, R., Bigot, S., Martinez, J.-P., et al., 2019. Tomato fruit development and metabolism. Front. Plant Sci. 10, 1554. Available from: https://doi.org/10.3389/fpls.2019.01554.

Quirino, B.F., Noh, Y.-S., Himelblau, E., Amasino, R.M., 2000. Molecular aspects of leaf senescence. Trends Plant Sci. 5, 278–282.

Rademacher, W., 1978. Gaschromatographische Analyse der Veränderungen im Hormongehalt des wachsenden Weizenkorns. Ph.D. Thesis, Universität Göttingen.

Radin, J.W., 1983. Control of plant growth by nitrogen: Differences between cereals and broadleaf species. Plant Cell Environ. 6, 65–68.

Radin, J.W., 1984. Stomatal responses to water stress and to abscisic acid in phosphorus-deficient cotton plants. Plant Physiol. 76, 392–394.

Radin, J.W., Ackerson, R.C., 1981. Water relations of cotton plants under nitrogen deficiency. III. Stomatal conductance. Plant Physiol. 67, 115–119.

Radin, J.W., Eidenbock, M.P., 1984. Hydraulic conductance as a factor limiting leaf expansion of phosphorus-deficient cotton plants. Plant Physiol. 75, 372–377.

Radin, J.W., Hendrix, D.L., 1988. The apoplastic pool of abscisic acid in cotton leaves in relation to stomatal closure. Planta 174, 180–186.

Radley, M., 1978. Factors affecting grain enlargement in wheat. J. Exp. Bot. 29, 919–934.

Rahayu, Y.S., Walch-Liu, P., Neumann, G., Römheld, V., von Wirén, N., Bangerth, F., 2005. Root-derived cytokinins as long-distance signals for NO_3^--induced stimulation of leaf growth. J. Exp. Bot. 56, 1143–1152.

Raines, C.A., 2003. The Calvin cycle revisited. Photosynth. Res. 75, 1–10.

Raines, C.A., 2011. Increasing photosynthetic carbon assimilation in C_3 plants to improve crop yields: current and future strategies. Plant Physiol. 155, 36–42.

Ramsperger-Gleixner, M., Geiger, D., Hedrich, R., Sauer, N., 2004. Differential expression of sucrose transporter and polyol transporter genes during maturation of common plantain companion cells. Plant Physiol. 134, 147–160.

Ranson, S.L., Thomas, M., 1960. Crassulacean acid metabolism. Annu. Rev. Plant Physiol. 11, 81–110.

Rao, X., Dixon, R.A., 2016. The differences between NAD-ME and NADP-ME subtypes of C_4 photosynthesis: more than decarboxylating enzymes. Front. Plant Sci. 7, 1525. Available from: https://doi.org/10.3389/fpls.201601525.

Rao, S.S.R., Vardhini, B.V., Sujatha, E., Anuradha, S., 2002. Brassinosteroids – a new class of phytohormones. Curr. Sci. 82, 1239–1245.

Rasmussen, A., Mason, M.G., de Cuyper, C., Brewer, P.B., Herold, S., Agusti, J., et al., 2012. Strigolactones suppress adventitious rooting in Arabidopsis and pea. Plant Physiol. 158, 1976–1987.

Rasmusson, A.G., Escobar, M.A., Hao, M., Podgórska, A., Szal, B., 2020. Mitochondrial NAD(P)H oxidation pathways and nitrate/ammonium redox balancing in plants. Mitochondrion 53, 158–165.

Renger, G., Renger, T., 2008. Photosystem II: the machinery of photosynthetic water splitting. Photosynth. Res. 98, 53–80.

Rennie, E.A., Turgeon, R., 2009. A comprehensive picture of phloem loading strategies. Proc. Natl. Acad. Sci. USA 106, 14162–14167.

Rentsch, D., Schmidt, S., Tegeder, M., 2007. Transporters for uptake and allocation of organic nitrogen compounds in plants. FEBS Lett. 581, 2281–2289.

Rerkasem, B., Jamjod, S., 2004. Boron deficiency in wheat: a review. Field Crops Res. 89, 173–186.

Reumann, S., Weber, A.P.M., 2006. Plant peroxisomes respire in the light: some gaps of the photorespiratory C_2 cycle have become filled – others remain. Biochim. Biophys. Acta 1763, 1496–1510.

Reynolds, H.L., Hartley, A.E., Vogelsang, K.M., Bever, J.D., Schultz, P.A., 2005. Arbuscular mycorrhizal fungi do not enhance nitrogen acquisition and growth of old-field perennials under low nitrogen supply in glasshouse culture. New Phytol. 167, 869–880.

Reynolds, M.P., Foulkes, J.M., Slafer, G.A., Berry, P., Parry, M.A.J., Snape, J., et al., 2009. Raising yield potential in wheat. J. Exp. Bot. 60, 1899–1918.

Rivero, R.M., Kojima, M., Gepstein, A., Sakakibara, H., Mittler, R., Gepstein, S., et al., 2007. Delayed senescence induces extreme drought tolerance in a flowering plant. Proc. Natl. Acad. Sci. USA 104, 19631–19636.

Robert, H.S., Friml, J., 2009. Auxin and other signals on the move in plants. Nat. Chem. Biol. 5, 325–332.

Rodríguez, D., Keltjens, W.G., Goudriaan, J., 1998. Plant leaf area expansion and assimilate production in wheat (*Triticum aestivum* L.) under low phosphorus conditions. Plant Soil 200, 227–240.

Rodríguez-Falcón, M., Bou, J., Prat, S., 2006. Seasonal control of tuberization in potato: conserved elements with the flowering response. Annu. Rev. Plant Biol. 57, 151–180.

Rohozinski, J., Edwards, G.R., Hoskyns, P., 1986. Effects of brief exposure to nitrogenous compounds on floral initiation in apple trees. Physiol. Veg. 24, 673–677.

Rolland, F., Baena-Gonzalez, E., Sheen, J., 2006. Sugar sensing and signalling in plants: conserved and novel mechanisms. Annu. Rev. Plant Biol. 57, 675–709.

Römheld, V., Kirkby, E.A., 2010. Research on potassium in agriculture: Needs and prospects. Plant Soil 335, 155–180.

Römheld, V., Jiménez-Becker, S., Neumann, G., Gweyi-Onyango, J.P., Puelschen, L., Spreer, W., et al., 2008. Non-nutritional fertigation effects as a challenge for improved production and quality in horticulture. Fertigation: Optimizing the Utilization of Water and Nutrients. International Potash Institute, Horgen, Switzerland, pp. 103–115.

Ros, R., Muñoz-Bertomeu, J., Kreuger, S., 2014. Serine in plants: biosynthesis, metabolism and functions. Trends Plant Sci. 19, 564–569.

Rosche, E., Blackmore, D., Tegeder, M., Richardson, T., Schroeder, H., Higgins, T.J.V., et al., 2002. Seed-specific expression of a potato sucrose transporter increases sucrose uptake and growth rates of developing pea cotyledons. Plant J. 30, 165–175.

Roschzttardtz, H., Conéjéro, G., Curie, C., Mari, S., 2009. Identification of the endodermal vacuole as the iron storage compartment in the Arabidopsis embryo. Plant Physiol. 151, 1329–1338.

Roy, I., Biswas, P.K., Ali, M.H., Haque, M.N., Parvin, K., 2016. Growth and reproductive behaviour of chickpea (*Cicer arietinum* L.) as influenced by supplemental application of nitrogen, irrigation and hormone. Plant Sci. Today 3, 30–40.

Ruan, Y.-L., Jin, Y., Yang, Y.-J., Li, G.-J., Boyer, J.S., 2010. Sugar input, metabolism, and signalling mediated by invertase: roles in development, yield potential, and response to drought and heat. Mol. Plant 3, 942–955.

Ruan, Y.-L., 2014. Sucrose metabolism gateway to diverse carbon use and sugar signaling. Annu. Rev. Plant Biol. 65, 33–37.

Ruan, Y.-L., Llewellyn, D.J., Furbank, R.T., 2001. The control of single-celled cotton fiber elongation by developmentally reversible gating of plasmodesmata and coordinated expression of sucrose and K^+ transporters and expansin. Plant Cell 13, 47–60.

Rubio, G., Gutierrez Boem, F.H., Lavado, R.S., 2010. Responses of C_3 and C_4 grasses to application of nitrogen and phosphorus fertilizer at two dates in the spring. Grass Forage Sci. 65, 102–109.

Ruffel, S., 2018. Nutrient-related long-distance signals: common players and possible cross talk. Plant Cell Physiol. 59, 1723–1732.

Ruffel, S., Krouk, G., Ristova, D., Shasha, D., Birnbaum, K.D., Coruzzi, G.M., 2011. Nitrogen economics of root foraging: Transitive closure of the nitrate-cytokinin relay and distinct systemic signaling for N supply vs. demand. Proc. Natl. Acad. Sci. USA 108, 18524–18529.

Ruyter-Spira, C., Kohlen, W., Charnikhova, T., van Zeijl, A., van Bezouwen, L., de Ruijter, N., et al., 2011. Physiological effects of the synthetic strigolactone analog GR24 on root system architecture in Arabidopsis: another belowground role for strigolactones? Plant Physiol. 155, 721–734.

Saftner, R.A., Wyse, R.E., 1980. Alkali cation/sucrose co-transport in the root sink of sugar beet. Plant Physiol. 66, 884–889.

Saftner, R.A., Daie, J., Wyse, R.E., 1983. Sucrose uptake and compartmentation in sugar beet transport tissue. Plant Physiol. 72, 1–6.

Sage, R.F., 2002. Variation in the K_{cat} of Rubisco in C_3 and C_4 plants and some implications for photosynthetic performance at high and low temperature. J. Exp. Bot. 53, 609–620.

Sage, R.F., Kubien, D.S., 2003. *Quo vadis* C_4? An ecophysiological perspective on global change and the future of C_4 plants. Photosynth. Res. 77, 209–225.

Sage, R.F., Zhu, X.G., 2011. Exploiting the engine of C_4 photosynthesis. J. Exp. Bot. 62, 2989–3000.

Sakakibara, H., Takei, K., Hirose, N., 2006. Interactions between nitrogen and cytokinin in the regulation of metabolism and development. Trends Plant Sci. 11, 440–448.

Salmon, Y., Dietrich, L., Sevanto, S., Holtta, T., Dannoura, M., Epron, D., 2019. Drought impacts on tree phloem: from cell-level responses to ecological significance. Tree Physiol. 39, 173–191.

Sanagi, M., Aoyama, S., Kubo, A., Lu, Y., Sato, Y., Ito, S., et al., 2021. Low nitrogen conditions accelerate flowering by modulating the phosphorylation state of FLOEERING BHLH 4 in *Arabidopsis*. Proc. Natl. Acad. Sci. USA 18, e2022942118. Available from: https://doi.org/10.1073/pnas.2022942118.

Sancenón, V., Puig, S., Mateu-Andrés, I., Dorcey, E., Thiele, D.J., Penarrubia, L., 2004. The Arabidopsis copper transporter COPT1 functions in root elongation and pollen development. J. Biol. Chem. 279, 15348–15355.

Sánchez-Rodríguez, C., Rubio-Somoza, I., Sibout, R., Persson, S., 2010. Phytohormones and the cell wall in Arabidopsis during seedling growth. Trends Plant Sci. 15, 291–301.

Sasaki, T., Suzaki, T., Soyano, T., Kojima, M., Sakakibara, H., Kawaguchi, M., 2019. Shoot-derived cytokinins systemically regulate root nodulation. Nat. Commun. 5, 4983. Available from: https://doi.org/10.1038/ncomms5983.

Sattelmacher, B., Marschner, H., 1978. Nitrogen nutrition and cytokinin activity in *Solanum tuberosum*. Physiol. Plant. 42, 185–189.

Sattelmacher, B., Marschner, H., 1979. Tuberization in potato plants as affected by application of nitrogen to the roots and leaves. Potato Res. 22, 39–47.

Scheible, W.R., Morcuende, R., Czechowski, T., Fritz, C., Osuna, D., Palacios-Rojas, N., et al., 2004. Genome-wide reprogramming of primary and secondary metabolism, protein synthesis, cellular growth processes, and the regulatory infrastructure of Arabidopsis in response to nitrogen. Plant Physiol. 136, 2483–2499.

Schnyder, H., 1993. The role of carbohydrate storage and redistribution in the source-sink relations of wheat and barley during grain filling – a review. New Phytol. 123, 233–245.

Schobert, C., Komor, E., 1989. The differential transport of amino acids into the phloem of *Ricinus communis* L. seedlings as shown by the analysis of sieve-tube sap. Planta 177, 342–349.

Schurr, U., Heckenberger, U., Herdel, K., Walter, A., Feil, R., 2000. Leaf development in *Ricinus communis* during drought stress: dynamics of growth processes, of cellular structure and of sink-source transition. J. Exp. Bot. 51, 1515–1529.

Seth, A.K., Wareing, P.F., 1967. Hormone-directed transport of metabolites and its possible role in plant senescence. J. Exp. Bot. 18, 65–77.

Setter, T.L., Parra, R., 2010. Relationship of carbohydrate and abscisic acid levels to kernel set in maize under postpollination water deficit. Crop Sci. 50, 980–988.

Seufert, V., Granath, G., Muller, C., 2019. A meta-analysis of crop response patterns to nitrogen limitation for improved model representation. PLoS One 14, e0223508. Available from: https://doi.org/10.1371/journal.pone.0223508.

Sevanto, S., 2018. Drought impacts on phloem transport. Curr. Opin. Plant Biol. 43, 76–81.

Shabala, S., White, R.G., Djordjevic, M.R., Ruan, Y.-L., Mathesius, U., 2016. Root to shoot signalling integration of diverse molecules pathways and functions. Funct. Plant Biol. 43, 87–104.

Sharma, S.S., Dietz, K.-J., 2006. The significance of amino acids and amino acid-derived molecules in plant responses and adaptation to heavy metal stress. J. Exp. Bot. 57, 711–726.

Sharma, P.N., Chatterjee, C., Agarwala, S.C., Sharma, C.P., 1990. Zinc deficiency and pollen fertility in maize (*Zea mays*). Plant Soil 124, 221–225.

Sharma, C.P., Sharma, P.N., Chatterjee, C., Agarwala, S.C., 1991. Manganese deficiency in maize affects pollen viability. Plant Soil 138, 139–142.

Sheng, H.J., Jiang, Y.L., Ishka, M.R., Chia, J.C., Dokuchayeva, T., Kavulych, Y., et al., 2021. YSL3-mediated copper distribution is required for fertility, seed size and protein accumulation in *Brachypodium*. Plant Physiol. 186, 655–676.

Shewry, P.R., 2007. Improving the protein content and composition of cereal grain. J. Cereal Sci. 46, 239–250.

Shimizu-Sato, S., Tanaka, M., Mori, H., 2009. Auxin–cytokinin interactions in the control of shoot branching. Plant Mol. Biol. 69, 429–435.

Shimshi, D., 1969. Interaction between irrigation and plant nutrition. In: Transition from Extensive to Intensive Agriculture with Fertilizers. Proc. 7th Colloq. Int. Potash Inst., Bern, Switzerland, pp. 111–120.

Silvius, J.E., Kremer, D.F., Lee, D.R., 1978. Carbon assimilation and translocation in soybean leaves at different stages of development. Plant Physiol. 62, 54–58.

Sinclair, T.R., Horie, T., 1989. Leaf nitrogen, photosynthesis, and crop radiation use efficiency: a review. Crop Sci. 29, 90–98.

Smiciklas, K.D., Below, F.E., 1992. Role of cytokinin in enhanced productivity of maize supplied with NH_4^+ and NO_3^-. Plant Soil 142, 307–313.

Smith, A.M., 2008. Prospects for increasing starch and sucrose yields for bioethanol production. Plant J. 54, 546–558.

Smith, M.R., Idupulapati, M., Merchant, A., 2018. Source-sink relationships in crop plants and their influence on yield. Front. Plant Sci. 9, 1889. Available from: https://doi.org/10.3389/fpls.2018.01889.

Sonnewald, S., Sonnewald, U., 2014. Regulation of potato tuber sprouting. Planta 239, 27–38.

Spano, G., Di Fonzo, N., Perrotta, C., Platani, C., Ronga, G., Lawlor, D.W., et al., 2003. Physiological characterization of 'stay green' mutants in durum wheat. J. Exp. Bot. 54, 1415–1420.

Spilatro, S.R., Preiss, J., 1987. Regulation of starch synthesis in the bundle sheath and mesophyll of *Zea mays* L. Plant Physiol. 83, 621–627.

Spreitzer, R.J., Salvucci, M.E., 2002. RUBISCO: structure, regulatory interactions and possibilities for a better enzyme. Annu. Rev. Plant Biol. 53, 449–475.

Stacey, M.G., Patel, A., McClain, W.E., Mathieu, M., Remley, M., Rogers, E.E., et al., 2008. The Arabidopsis AtOPT3 protein functions in metal homeostasis and movement of iron to developing seeds. Plant Physiol. 146, 589–601.

Starck, Z., Choluj, D., Szczepanska, B., 1980. Photosynthesis and photosynthates distribution in potassium-deficient radish plants treated with indolyl-3-acetic acid or gibberellic acid. Photosynthetica 14, 497–505.

Stitt, M., 1991. Rising CO_2 levels and their potential significance for carbon flow in photosynthetic cells. Plant Cell Enivron. 14, 741–762.

Stitt, M., Schulze, E.-D., 1994. Does Rubisco control the rate of photosynthesis and plant growth? An exercise in molecular ecophysiology. Plant Cell Environ. 17, 465–487.

Stockman, Y.M., Fischer, R.A., Brittain, E.G., 1983. Assimilate supply and floret development within the spike of wheat (*Triticum aestivum* L.). Aust. J. Plant Physiol. 10, 585–594.

Sun, Y., Tong, C., He, S., Wang, K., Chen, L., 2020. Identification of nitrogen, phosphorus, and potassium deficiencies based on temporal dynamics of leaf morphology and color. Sustainability 10, 762. Available from: https://doi.org/10.3390/su10030762.

Sweetlove, L.J., Beard, K.F.M., Nunes-Nesi, A., Fernie, A.R., Ratcliffe, R.G., 2010. Not just a circle: flux modes in the plant TCA cycle. Trends Plant Sci. 15, 462–470.

Symons, G.M., Ross, J.J., Jager, C.E., Reid, J.B., 2008. Brassinosteroid transport. J. Exp. Bot. 59, 17–24.

Sze, H., Chanroj, S., 2018. Plant endomembrane dynamics: Studies of K^+/H^+ antiporters provide insights on the effects of pH and ion homeostasis. Plant Physiol. 177, 875–895.

Ta, C.T., 1991. Nitrogen metabolism in the stalk tissue of maize. Plant Physiol. 97, 1375–1380.
Tabe, L., Hagan, N., Higgins, T.J.V., 2002. Plasticity of seed protein composition in response to nitrogen and sulphur availability. Curr. Opin. Plant Biol. 5, 212–217.
Takahashi, M., Terada, Y., Nakai, I., Nakanashi, H., Yoshimura, E., Mori, S., et al., 2003. Role of nicotianamine in the intracellular delivery of metals and plant reproductive development. Plant Cell 15, 1263–1280.
Takehisa, H., Sato, Y., 2019. Transcriptome monitoring visualizes growth stage-dependent nutrient status dynamics in rice under field conditions. Plant J. 97, 1048–1060.
Tambussi, E.A., Bort, J., Guiamet, J.J., Nogués, S., Araus, J.L., 2007. The photosynthetic role of ears in C_3 cereals: metabolism, water use efficiency and contribution to grain yield. Crit. Rev. Plant Sci. 26, 1–16.
Tanoi, K., Kobayashi, N.I., 2015. Leaf senescence by magnesium deficiency. Plants 4, 756–772.
Tcherkez, G., Limami, A.M., 2019. Net photosynthetic CO_2 assimilation: more than just CO_2 and O_2 reduction cycles. New Phytol. 223, 520–529.
Tcherkez, G., Gauthier, P., Buckley, T.N., Busch, F.A., Barbour, M.M., Bruhn, D., et al., 2017. Leaf day respiration: low CO_2 flux but high significance for metabolism and carbon balance. New Phytol. 216, 986–1001.
Tegeder, M., 2014. Transporters involved in source to sink partitioning of amino acids and ureides: opportunities for crop improvement. J. Exp. Bot. 65, 1865–1878.
Tegeder, M., Masclaux-Daubresse, C., 2018. Source and sink mechanisms of nitrogen transport and use. New Phytol. 217, 35–53.
Theodorou, M.E., Plaxton, W.C., 1993. Metabolic adaptations of plant respiration to nutritional phosphate deprivation. Plant Physiol. 101, 339–344.
Thomas, H., Howarth, C.J., 2000. Five ways to stay green. J. Exp. Bot. 51, 329–337.
Tilsner, J., Kassner, N., Struck, C., Lohaus, G., 2005. Amino acid contents and transport in oilseed rape (*Brassica napus* L.) under different nitrogen conditions. Planta 221, 328–338.
Timm, S., Hagermann, M., 2020. Photorespiration – how is it regulated and how does it regulate overall plant metabolism? J. Exp. Bot. 71, 3955–3965.
Tränkner, M., Tavakol, E., Jákli, B., 2018. Functioning of potassium and magnesium in photosynthesis, photosynthate translocation and photoprotection. Physiol. Plant. 163, 414–431.
Trobisch, S., Schilling, G., 1970. Beitrag zur Klärung der physiologischen Grundlage der Samenbildung bei einjährigen Pflanzen und zur Wirkung später zusätzlicher N-Gaben auf diesen Prozess am Beispiel von *Sinapis alba* L. Albrecht-Thaer-Arch. 14, 253–265.
Turgeon, R., 2006. Phloem loading: how leaves gain their independence. BioScience 56, 15–24.
Turgeon, R., Wolf, S., 2009. Phloem transport: cellular pathways and molecular trafficking. Annu. Rev. Plant Biol. 60, 207–221.
Uauy, C., Distelfeld, A., Fahima, T., Blechl, A., Dubcovsky, J., 2006. A NAC gene regulating senescence improves grain protein, zinc, and iron content in wheat. Science 314, 1298–1301.
Uhart, S.A., Andrade, F.H., 1995. Nitrogen deficiency in maize: I. Effects on crop growth, development, dry matter partitioning, and kernel set. Crop Sci. 35, 1376–1383.
Umehara, M., Hanada, A., Magome, H., Takeda-Kamiya, N., Yamaguchi, S., 2010. Contribution of strigolactones to the inhibition of tiller bud outgrowth under phosphate deficiency in rice. Plant Cell Physiol. 51, 1118–1126.
Unkovitch, M., Baldock, J., Forbes, M., 2010. Variability in harvest index of grain crops and potential significance for carbon accounting: examples from Australian agriculture. Adv. Agron. 105, 173–219.
Usandivaras, L.M.A., Gutierrez-Boom, F.H., Salvagiotti, F., 2018. Contrasting effects of phosphorus and potassium deficiencies on leaf area development in maize. Crop Sci. 58, 2099–2109.
Van Bel, A.J.E., 1984. Quantification of the xylem-to-phloem transfer of amino acids by use of inulin (^{14}C) carboxylic acid as xylem transport marker. Plant Sci. Lett. 35, 81–85.
Van Bel, A.J.E., 2003. The phloem, a miracle of ingenuity. Plant Cell Environ. 26, 125–149.
Van den Berg, H.J., Vreugdenhil, D., Ludford, P.M., Hillman, L.L., Ewing, E.E., 1991. Changes in starch, sugar, and abscisic acid content associated with second growth in tubers of potato (*Solanum tuberosum* L.) one-leaf cuttings. J. Plant Physiol. 139, 86–89.
Vanlerberghe, J.C., 2013. Alternative oxidase: a mitochondrial respiratory pathway to maintain metabolic and signalling homeostasis during abiotic and biotic stress in plants. Int. J. Mol. Sci. 14, 6805–6847.
Vaughan, A.K.F., 1977. The relation between the concentration of boron in the reproductive and vegetative organs of maize plants and their development. Rhod. J. Agric. Res. 15, 163–170.
Vidal, E.A., Moyano, T.C., Canales, J., Gutiérrez, R.A., 2014. Nitrogen control of developmental phase transitions in *Arabidopsis thaliana*. J. Exp. Bot. 65, 5611–5618.
Viola, R., Roberts, A.G., Haupt, S., Gazzani, S., Hancock, R.D., Marmiroli, N., et al., 2001. Tuberization in potato involves a switch from apoplastic to symplastic phloem unloading. Plant Cell 13, 385–398.
Von Caemmerer, S., Ghannoum, O., Furbank, R.T., 2017. C_4 photosynthesis 50 years of discovery and innovation. J. Exp. Bot. 68, 97–102.
Von Schaewen, A., Stitt, M., Schmidt, R., Sonnewald, U., Willmitzer, L., 1990. Expression of yeast-derived invertase in the cell wall of tobacco and Arabidopsis plants leads to accumulation of carbohydrate and inhibition of photosynthesis and strongly influences growth and phenotype of transgenic tobacco plants. EMBO J. 9, 3033–3044.
Voznesenskaya, E.V., Franceschi, V.R., Kilirats, O., Freitag, H., Edwards, G.E., 2001. Kranz anatomy is not essential for terrestrial C_4 plant photosynthesis. Nature 414, 543–546.

Walch-Liu, P., Neumann, G., Engels, C., 2001. Response of shoot and root growth to supply of different nitrogen forms is not related to carbohydrate and nitrogen status of tobacco plants. J. Plant Nutr. Soil Sci. 164, 97–103.

Walker, N.A., Zhang, W.-H., Harrington, G., Holdaway, N., Patrick, J.W., 2000. Effluxes of solutes from developing seed coats of *Phaseolus vulgaris* L. and *Vicia faba* L.: locating the effect of turgor in a coupled chemiosmotic system. J. Exp. Bot. 51, 1047–1055.

Walter, A., Silk, W.K., Schurr, U., 2009. Environmental effects on spatial and temporal patterns of leaf and root growth. Annu. Rev. Plant Biol. 60, 279–304.

Wang, X., Wang, L., Shangguan, Z., 2016. Leaf gas exchange and fluorescence of two winter wheat varieties in response to drought stress and nitrogen supply. PLoS ONE 11, e0165733. Available from: https://doi.org/10.1371/journal.pone.0165733.

Wang, B., Smith, S.M., Li, J.Y., 2018. Genetic regulation of shoot architecture. Annu. Rev. Plant Biol. 69, 437–468.

Waters, S.P., Martin, P., Lee, B.T., 1984. The influence of sucrose and abscisic acid on the determination of grain number in wheat. J. Exp. Bot. 35, 829–840.

Way, D.A., Katul, G.B., Mansoni, S., Vico, G., 2014. Increasing water use efficiency along the C_3 to C_4 evolutionary pathway: a stomatal optimization perspective. J. Exp. Bot. 65, 3683–3692.

Weber, H., Borisjuk, L., Wobus, U., 2005. Molecular physiology of legume seed development. Annu. Rev. Plant Biol. 56, 253–279.

Weiner, H., Blechschmidt-Schneider, S., Mohme, H., Eschrich, W., Heldt, H.W., 1991. Phloem transport of amino acids. Comparison of amino acid content of maize leaves and of the sieve tube exudate. Plant Physiol. Biochem. 29, 19–23.

Weng, X.-Y., Zheng, C.-J., Xu, H.-X., Sun, J.-Y., 2007. Characteristics of photosynthesis and functions of the water-water cycle in rice (*Oryza sativa*) leaves in response to potassium deficiency. Physiol. Plant. 131, 614–621.

White, P.J., 1993. Relationship between the development and growth of rye (*Secale cereale* L.) and the potassium concentration in solution. Ann. Bot. 72, 349–358.

White, P.J., 2015. Calcium. In: Barker, A.V., Pilbeam, D.J. (Eds.), Handbook of Plant Nutrition, 2nd ed CRC Press, London, UK, pp. 165–198.

White, P.J., 2018. Improving nutrient management in potato cultivation. In: Wale, S. (Ed.), Achieving Sustainable Cultivation of Potatoes, Vol. 2. Production, Storage and Crop Protection. Burleigh Dodds Science Publishing, Cambridge, UK, pp. 45–67.

White, P.J., 2020. Potassium in crop physiology. In: Rengel, Z. (Ed.), Achieving Sustainable Crop Nutrition. Burleigh Dodds, Cambridge, UK, pp. 213–236.

White, P.J., Broadley, M.R., 2003. Calcium in plants. Ann. Bot. 92, 487–511.

White, P.J., Broadley, M.R., 2009. Biofortification of crops with seven mineral elements often lacking in human diets – iron, zinc, copper, calcium, magnesium, selenium and iodine. New Phytol. 182, 49–84.

White, P.J., Hammond, J.P., 2008. Phosphorus nutrition of terrestrial plants. In: White, P.J., Hammond, J.P. (Eds.), The Ecophysiology of Plant-Phosphorus Interactions. Springer, Dordrecht, The Netherlands, pp. 51–81.

White, P.J., Karley, A.J., 2010. Potassium. In: Hell, R., Mendel, R.-R. (Eds.), Plant Cell Monographs 17, Cell Biology of Metals and Nutrients. Springer, Dordrecht, The Netherlands, pp. 199–224.

White, P.J., Bell, M.J., Djalovic, I., Hinsinger, P., Rengel, Z., 2021a. Potassium use efficiency of plants. In: Murrell, T.S., Mikkelsen, R.L., Sulewski, G., Norton, R., Thompson, M.L. (Eds.), Improving Potassium Recommendations for Agricultural Crops. Springer, Cham, Switzerland, pp. 119–145.

White, P.J., Ding, G., Shi, L., Xu, F., 2021b. Proceedings of the International Fertiliser Society 859: Boron in Plant Physiology and Nutrition. IFS, York, UK.

Wilkinson, S., Davies, W.J., 2002. ABA-based chemical signalling: the co-ordination of responses to stress in plants. Plant Cell Environ. 25, 195–210.

Wilkinson, S., Bacon, M.A.Z., Davies, W.J., 2007. Nitrate signalling to stomata and growing leaves: interactions with soil drying, ABA, and xylem sap pH in maize. J. Exp. Bot. 58, 1705–1716.

Williams, C.M.J., Maier, N.A., Bartlett, L., 2005. Effect of molybdenum foliar sprays on yield, berry size, seed formation, and petiolar nutrient composition of "Merlot" grapevines. J. Plant Nutr. 27, 1891–1916.

Wingler, A., Roitsch, T., 2008. Metabolic regulation of leaf senescence: interactions of sugar signalling with biotic and abiotic stress responses. Plant Biol. 10, 50–62.

Wingler, A., Lea, P.J., Quick, W.P., Leegood, R.C., 2000. Photorespiration: metabolic pathways and their role in stress protection. Phil. Trans. Roy. Soc. Lond. B. 355, 1517–1529.

Winter, K., Smith, J.A.C., 2022. CAM photosynthesis: the acid test. New Phytol. 233, 599–609.

Winter, H., Lohaus, G., Heldt, H.W., 1992. Phloem transport of amino acids in relation to their cytosolic levels in barley leaves. Plant Physiol. 99, 996–1004.

Wingler, A., Tijero, V., Müller, M., Yuan, B., Munné-Bosch, S., 2020. Interactions between sucrose and jasmonate signalling in the response to cold stress. BMC Plant Biol. 20, 176. Available from: https://doi.org/10.1186/s12870-020-02376-6.

Winter, K., Holtum, J.A.M., Smith, J.A.C., 2015. Crassulacean acid metabolism: a continuous or discrete trait? New Phytol. 208, 73–78.

Wirth, J., Poletti, S., Aeschlimann, B., Yakandawala, N., Drosse, B., Osorio, S., et al., 2009. Rice endosperm iron biofortification by targeted and synergistic action of nicotianamine synthase and ferritin. Plant Biotechnol. J. 7, 631–644.

Wolf, O., Jeschke, W.D., Hartung, W., 1990. Long distance transport of abscisic acid in NaCl-treated intact plants of *Lupinus albus*. J. Exp. Bot. 41, 593–600.

Wolters, H., Jürgens, G., 2009. Survival of the flexible: hormonal growth control and adaptation in plant development. Nat. Rev. Genetics 10, 305–317.

Woo, H.R., Kim, H.J., Lim, P.O., Nam, H.G., 2019. Leaf senescence: Systems and dynamics. Annu. Rev. Plant Biol. 70, 347–376.

Wu, L., Birch, R.G., 2007. Doubled sugar content in sugarcane plants modified to produce a sucrose isomer. Plant Biotechnol. J. 5, 109–117.

Yakhin, O.I., Lubyanov, A.A., Yakhin, I.A., Brown, P.H., 2017. Biostimulants in plant science: A global perspective. Front. Plant Sci. 7, 2049. Available from: https://doi.org/10.3389/fpls.2016.02049.

Yamaguchi, S., 2008. Gibberellin metabolism and its regulation. Annu. Rev. Plant Biol. 59, 225–251.

Yang, J., Zhang, J., Liu, K., Wang, Z., Liu, L., 2006. Abscisic acid and ethylene interact in wheat grains in response to soil drying during grain filling. New Phytol. 171, 293–303.

Ye, T., Li, Y., Zhang, J., Hou, W., Zhou, W., Lu, J., et al., 2019. Nitrogen, phosphorus, and potassium fertilization affects the flowering time of rice (*Oryza sativa* L.). Glob. Ecol. Conserv. 20, e00753. Available from: https://doi.org/10.1016/j.gecco.2019.e00753.

Yoneyama, K., Kisugi, T., Xie, X., Arakawa, R., Ezawa, T., Nomura, T., et al., 2015. Shoot-derived signals other than auxins are involved in systemic regulation of strigolactone production in roots. Planta 241, 687–698.

Yu, J., Wo, K.C., 1991. Correlations between the development of photorespiration and the changes in activities of NH_3 assimilation enzymes in greening oat leaves. Aust. J. Plant Physiol. 18, 583–588.

Yu, S., Cao, L., Zhou, C.M., Zhang, T.-Q., Lian, H., Sun, Y., et al., 2013. Sugar is an endogenous cue for juvenile-to-adult phase transition in plants. eLife 2, e00269. Available from: https://doi.org/10.7554/eLife.00269.

Yuan, S., Zhang, Z.W., Zheng, C., Zhao, Z.Y., Wang, Y., Feng, L.Y., et al., 2016. Arabidopsis cryptochrome 1 functions in nitrogen regulation of flowering. Proc. Natl. Acad. Sci. USA 113, 7661–7666.

Zhai, Z., Gayomba, S.R., Jung, H., Vimalakumari, N.K., Piñeros, M., Craft, E., et al., 2014. OPT3 is a phloem-specific iron transporter that is essential for systemic iron signaling and redistribution of iron and cadmium in Arabidopsis. Plant Cell 26, 2249–2264.

Zhang, C.K., Turgeon, R., 2018. Mechanisms of phloem loading. Curr. Opin. Plant Biol. 43, 71–75.

Zhang, Y.J., Lynch, J.P., Brown, K.M., 2003. Ethylene and phosphorus availability have interacting yet distinct effects on root hair development. J. Exp. Bot. 54, 2351–2361.

Zhang, W.H., Ryan, P.R., Tyerman, S.D., 2004. Citrate-permeable anion channels in the plasma membrane of cluster roots from white lupin. Plant Physiol. 136, 3771–3783.

Zhang, H., Rong, H., Pilbeam, D., 2007a. Signalling mechanisms underlying the morphological responses of the root system to nitrogen in *Arabidopsis thaliana*. J. Exp. Bot. 58, 2329–2338.

Zhang, K., Greenwood, D.J., White, P.J., Burns, I.G., 2007b. A dynamic model for the combined effects of N, P and K fertilizers on yield and mineral composition; description and experimental test. Plant Soil 298, 81–98.

Zhang, Y., Xu, Y.H., Yi, H.Y., Gong, J.M., 2012. Vacuolar membrane transporters OsVIT1 and OsVIT2 modulate iron translocation between flag leaves and seeds in rice. Plant J. 72, 400–410.

Zhang, Z., Liao, H., Lucas, W.J., 2014. Molecular mechanisms underlying phosphate sensing, signaling, and adaptation in plants. J. Integr. Plant Biol. 56, 192–220.

Zhang, S., Zhang, Y., Li, K., Yan, M., Zhang, J., Yu, M., et al., 2021a. Nitrogen mediates flowering time and nitrogen use efficiency via floral regulators in rice. Curr. Biol. 31, 671–683.

Zhang, Y.-M., Guo, P., Xia, X., Guo, H., Li, Z., 2021b. Multi layers of regulation on leaf senescence: new advances and perspectives. Front. Plant Sci. 12, 788996. Available from: https://doi.org/10.3389/fpls.2021/788996.

Zhou, Y., Chan, K., Wang, T.L., Hedley, C.L., Offler, C.E., Patrick, J.W., 2009. Intracellular sucrose communicates metabolic demand to sucrose transporters in developing pea cotyledons. J. Exp. Bot. 60, 71–85.

Zhu, X.-G., Ort, D.R., Whitmarsh, J., Long, S.P., 2004. The slow reversibility of photosystem II thermal energy dissipation on transfer from high to low light intensity may cause large losses in carbon gain by crop canopies: a theoretical analysis. J. Exp. Bot. 55, 1167–1175.

Zimmermann, P., Zentgraf, U., 2005. The correlation between oxidative stress and leaf senescence during plant development. Cell. Mol. Biol. Lett. 10, 515–534.

Chapter 6

Functions of macronutrients

Malcolm J. Hawkesford[1], Ismail Cakmak[2], Devrim Coskun[3], Luit J. De Kok[4], Hans Lambers[5], Jan K. Schjoerring[6], and Philip J. White[7]

[1]*Rothamsted Research, Harpenden, Hertfordshire, United Kingdom*, [2]*Faculty of Engineering and Natural Sciences, Sabanci University, Istanbul, Turkey*, [3]*Department of Phytology, Laval University, Quebec City, Quebec, Canada*, [4]*Laboratory of Plant Physiology, University of Groningen, Groningen, The Netherlands*, [5]*School of Biological Sciences and Institute of Agriculture, University of Western Australia, Perth, WA, Australia*, [6]*Department of Plant and Environmental Sciences, University of Copenhagen, Frederiksberg, Denmark*, [7]*Ecological Sciences, James Hutton Institute, Dundee, Scotland, United Kingdom*

Summary

In this chapter, the roles of the macronutrients nitrogen (N), sulfur (S), phosphorus (P), magnesium (Mg), calcium (Ca), and potassium (K) in plant metabolism and growth are described together with the consequences of deficiency and toxicity. After carbon, N is the element required in the largest quantity by plants and is a major driver for crop yield; it plays a central role in plant metabolism as a constituent of proteins, nucleic acids, chlorophyll, coenzymes, phytohormones, and secondary metabolites. After uptake from the environment mostly as ammonium or nitrate, N is assimilated into amino acids in either roots or shoots. Within the plant, N may be translocated as nitrate or amino acids. Sulfur is taken up as sulfate and assimilated into S-containing amino acids such as cysteine which are used to synthesize S-containing enzymes and coenzymes as well as secondary compounds such as phytochelatins (detoxification of metals) or allicins and glucosinolates (feeding deterrents). Nitrogen and S are both important constituents of seed storage proteins. Phosphorus is a structural element in nucleic acids and plays key roles in energy transfer as a component of adenosine phosphates and in the transfer of carbohydrates between organelles in leaf cells. Phytate is a typical storage pool of P, particularly in seeds. Magnesium is a component of chlorophyll and is required for photosynthesis, transport of photoassimilates, and protein synthesis. Calcium is important for cell wall and membrane stabilization, osmoregulation, and as a second messenger allowing plants to regulate responses to environmental stimuli. The main role of K is osmoregulation, which is important for cell extension and stomata movement. Potassium affects loading of sucrose and the rate of mass flow-driven solute movement within the plant.

6.1 Nitrogen

Nitrogen (N) is among the mineral elements required in the largest quantity by plants: about 1%–5% of total plant dry matter consists of N as an integral constituent of proteins, nucleic acids, chlorophyll, coenzymes, phytohormones, and secondary metabolites. The availability of N to roots is therefore a decisive factor for plant growth. Atmospheric dinitrogen (N_2) is only available to plants that are capable of forming symbiosis with N_2-fixing soil bacteria (Chapter 16). Most plants therefore depend on other N compounds for their growth. The major sources of N taken up by the roots are nitrate (NO_3^-) and ammonium (NH_4^+). To increase crop production, approximately 100 million tons of N fertilizers were applied globally in 2020 (FAO, 2019). Generally, only about 50% of the applied N fertilizer is utilized by the crop in the first year after its application (Jensen et al., 2011). The rest is incorporated into organic compounds in the soil or lost to the environment. The N that is lost from the plant–soil system in the form of nitrate can result in pollution of water bodies, whereas gaseous emission in the form of ammonia, nitric oxide, and nitrous oxide contributes to air pollution and global warming (Fowler et al., 2009; Thompson et al., 2019).

☆ This chapter is a revision of the third edition chapter by M. Hawkesford, W. Horst, T. Kichey, H. Lambers, J. Schjoerring, I. Skrumsager Møller, and P. White, pp. 135–189. DOI: https://doi.org/10.1016/B978-0-12-384905-2.00006-6. © Elsevier Ltd.

This section describes N acquisition by roots, N assimilation, functions of N compounds, and the effect of N on plant growth and composition. Classical physiological observations of N responses are linked with the underlying molecular mechanisms that have recently been unraveled to provide integrated insight into the principles of N nutrition of plants.

Nitrate and ammonium are the major sources of inorganic N taken up by the roots. Nitrate is generally present in higher concentrations (1−5 mM) than ammonium (20−100 μM) in the soil solution of agricultural soils (Kabala et al., 2017). Nitrate is not adsorbed to the soil particles but is confined to the soil solution. This implies that nitrate is more mobile in the soil than ammonium and, thereby, more available to plants. A further consequence is that nitrate concentrations in soil show considerable temporal and spatial variability associated with fluctuations in soil water content.

In unfertilized agricultural soils, ammonium can be present in higher concentrations than nitrate. Amino acids may provide an additional source of N, attaining concentrations between 0.1 and 100 μM in the soil solution and dominating the pool of N bound to soil particles (Jämtgård et al., 2010; Jones et al., 2002). Ammonium and amino acids are the dominant plant-available N forms in acid forest soils (Rennenberg et al., 2009). The rhizosphere of rice paddy soils contains more ammonium than nitrate, reflecting limited nitrification in hypoxic soils (Li et al., 2016).

The availability of N sources in the soil varies substantially in time and space, depending on soil properties such as texture, pH, moisture, and microbial activity (Robinson, 1994). As a consequence, plants have evolved mechanisms to modulate their N acquisition efficiency in response to both availability and form of external N as well as to plant N demand (Forde, 2002; Hawkesford and Riche, 2020; Xu and Takahashi, 2020). This includes having several N transporter systems that mediate uptake at different external concentrations. In addition, plants can change root system architecture in response to nutrient availability to optimize exploration of a particular soil volume or promote exploration of a larger volume.

The uptake of nitrate and ammonium into plant roots is mediated by transporter proteins located in the plasma membrane of the epidermal and cortical root cells. These transporter systems mediate the uptake of nitrate or ammonium with different affinities. The high-affinity transporter systems (HATS) operate at low concentrations (<0.5 mM) of external nitrate or ammonium. At higher concentrations (>0.5 mM), the uptake is primarily via the low-affinity transporter systems (LATS). Both uptake systems have inducible and constitutive components.

6.1.1 Nitrate transport in plants

6.1.1.1 Nitrate uptake by roots

There are two types of transporters involved in root nitrate uptake. These transporters belong to the NPF (NRT1/PTR; nitrate transporter 1/peptide transporter) or the NRT2 (Nitrate Transporter 2) families. It is generally assumed that the NPF (NRT1/PTR) and NRT2 transporters mediate low- and high-affinity transport of nitrate into roots, respectively, with the exception of NPF6.3 (NRT1.1) that can exhibit both high and low affinity (Wang et al., 2018a). Irrespective of the type of nitrate transporter, the inward transport of nitrate across the plasma membrane occurs against a steep electrochemical potential gradient because the negatively charged nitrate ion has to overcome both the negative plasma−membrane potential as well as an uphill concentration gradient. Nitrate influx therefore requires metabolic energy. The transporters of both the NPF (NRT1/PTR) and NRT2 families transport nitrate across the plasma membrane in symport with protons, which in turn requires the expenditure of adenosine triphosphate (ATP) by the H^+-ATPase for proton extrusion to maintain the proton gradient across the plasma membrane (Fig. 6.1).

Plants have a large number of NPF (NRT1/PTR) genes. There are 53 and 93 NPF (NRT1/PTR) genes in arabidopsis and rice, respectively (Wang et al., 2018a), whereas wheat has 331 NPF genes clustered in 113 homologous groups (Wang et al., 2020a). Phylogenetic analysis has indicated that wheat NPF genes are closely clustered with orthologues in arabidopsis, rice and in the grass model plant *Brachypodium* (Wang et al., 2020a).

Seven NRT2 genes have been described in arabidopsis and *Brachypodium*, whereas the rice genome has only four NRT2 genes (Wang et al., 2018a; Zuluaga and Sonnante, 2019). The NRT2 genes in grasses show significant genetic distance and seem to play distinct roles in response to various N conditions, thereby differing from the NRT2 family in arabidopsis. In contrast to the NPF nitrate transporters, NRT2 transporters require an additional protein component for activity (Orsel et al., 2006). This component, named NAR, is a protein with a single transmembrane domain that interacts directly with NRT2. Separately, neither NRT2 nor NAR can mediate nitrate transport. The functional unit for the inducible high-affinity nitrate transport system is a tetramer of two NRT2 and two NAR subunits, required for plasma membrane targeting and for maintaining NRT2 protein stability.

The NPF (NRT1/PTR) and NRT2 proteins both have the same topology of 12 transmembrane domains divided into two sets of six helices connected by a cytosolic loop (Forde, 2000). Although the protein structure is similar, there is no DNA sequence homology between the NPF and NRT2 families of genes (Orsel et al., 2002).

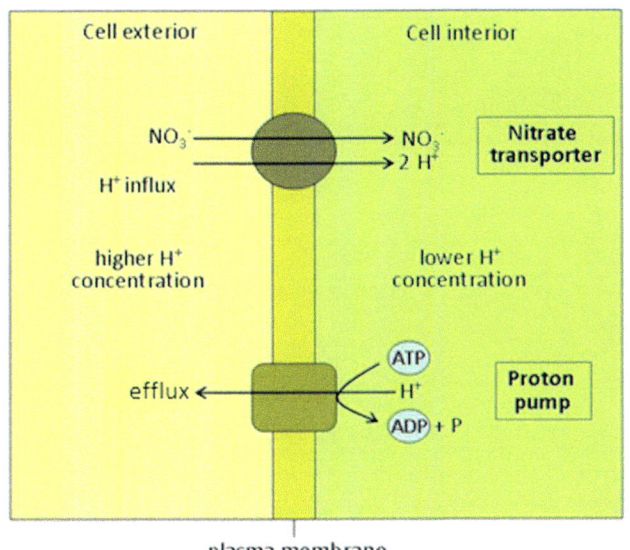

FIGURE 6.1 Schematic diagram of transport of nitrate across the plasma membrane of plant cells.

An overview of the different nitrate transport steps involving NPF (NRT1/PTR) and NRT2 proteins is given in Fig. 6.2. As evident from this Figure, a range of other nitrate transport processes, in addition to high- and low-affinity nitrate uptake from the soil solution, are mediated by NPF (NRT1/PTR) and NRT2 proteins. These include radial transport of nitrate across the root, xylem loading for translocation to shoots, long-distance transport, nitrate storage and remobilization in and out of vacuoles, nitrate reallocation in the phloem, and nitrate efflux from the root.

NPF6.3 (NRT1.1) is expressed in the epidermal cells in young roots and root tips of arabidopsis, which is in accordance with its role in nitrate uptake from the soil. NPF6.3 (NRT1.1) functions in both the high and low-affinity range. The two modes of activity are switched by phosphorylation and dephosphorylation. In the nonphosphorylated state, NPF6.3 (NRT1.1) dimerizes to a homodimer with low nitrate affinity, whereas phosphorylation leads to a monomeric and high-affinity state, conferring the capacity to take up nitrate even when present at low concentrations (Sun et al., 2014). When roots experience an increase in N availability, the high N condition leads to dephosphorylation of the NPF6.3 protein, which then adopts a low affinity to nitrate so that plants can then benefit from the higher capacity of that uptake system (Fig. 6.3). Thus, NPF6.3 senses the external nitrate concentration, thereby controlling its own mode of transport. NPF6.3 is accordingly characterized as a nitrate transceptor (transporter/receptor).

Another NPF (NRT1/PTR) family member, *AtNPF4.6 (NRT1.2)*, is expressed in the root epidermal cells and the root tip and is involved in nitrate uptake from the soil into roots (Fig. 6.2). AtNPF4.6 is solely a low-affinity transporter and is responsible for the constitutive low-affinity nitrate uptake capacity of roots. *AtNPF4.6 (NRT1.2)* is constitutively expressed, even if nitrate is not present, thereby ensuring a basal level of nitrate uptake capacity even in N-starved plants.

Four members of the NRT2 family in arabidopsis are important for nitrate uptake into roots and for nitrate translocation to the shoot. *AtNRT2.1* and *AtNRT2.2* mediate high-affinity nitrate uptake (Fig. 6.2). The main component of nitrate uptake at low nitrate concentrations (<0.5 mM) is AtNRT2.1, which is expressed mainly in the older part of the main root. *AtNRT2.1* displays a regulation pattern in accordance with the primary nitrate response, whereby the expression is induced rapidly upon resupply of nitrate to N-starved roots. This initial induction is followed by feedback repression, and the amino acid glutamine represents the main signal for the shoot to communicate N status to the roots for repression of *AtNRT2.1* (Nazoa et al., 2003). Nitrate uptake in plants is regulated diurnally and gradually increases to a peak toward the end of the light period (Glass et al., 2002). This diurnal regulation of nitrate uptake correlates well with the diurnal regulation of *AtNRT2.1* expression. *AtNRT2.4* and *AtNRT2.5* are involved in high-affinity nitrate uptake in N-starved plants. *AtNRT2.4* is expressed in the younger part of the primary root and the distal region of lateral roots, whereas *AtNRT2.5* is localized in the root hair zone of primary and lateral roots (Kiba et al., 2012; Lezhneva et al., 2014).

6.1.1.2 Nitrate efflux from roots

Even though the influx of nitrogenous compounds into epidermal and cortical root cells is essential for plant growth, efflux back to the soil solution of nitrate, ammonium, and amino acids can also occur. This seemingly energetically wasteful process may happen particularly in conditions of excess N that induces nitrate efflux in barley roots

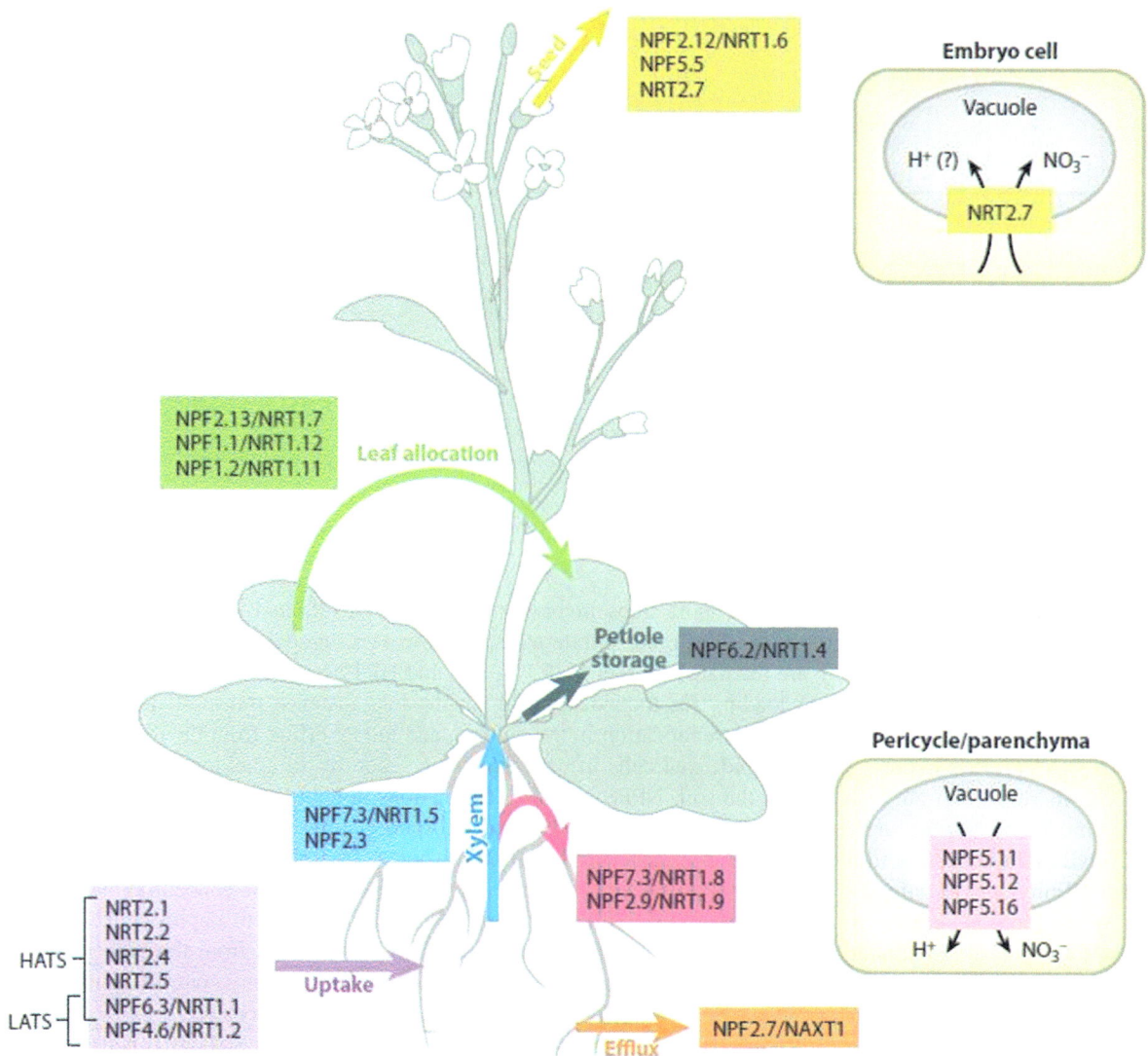

FIGURE 6.2 Physiological functions of arabidopsis NPF and NRT nitrate transporters, showing roles in nitrate uptake by, and efflux from, roots, root-to-shoot transport, nitrate allocation among leaves, and seed development. *HATS*, High-affinity transporter system; *LATS*, low-affinity transporter system; *NAXT*, nitrate excretion transporter; *NPF*, nitrate transporter 1 (NRT1)/peptide transporter (PTR) family; *NRT*, nitrate transporter. *From Wang et al. (2018a), with permission from Annual Reviews, Inc.*

(Kronzucker et al., 1999). The physiological importance of nitrate efflux remains unclear, but the process may serve a role in the sensing of nitrate availability by providing a dynamic and flexible regulation of cytosolic nitrate homeostasis (Miller and Smith, 2008) and nitrate net uptake (influx minus efflux).

In conditions of low external nitrate, both the chemical gradient and the electrical gradient are in favor of passive efflux of nitrate from the cytosol across the plasma membrane (Miller and Smith, 1996). Although this favorable gradient is present for many hours after N deprivation, the efflux process nevertheless decreases and is abolished after a few hours, suggesting that in the absence of external nitrate, nitrate efflux is downregulated (Van der Leij et al., 1998).

A transporter mediating nitrate efflux has been characterized in arabidopsis and belongs to the NPF2 (previously NAXT1) family of transporters. AtNPF2.7 is expressed in the root cortex and is located in the plasma membrane, in accordance with a function in nitrate efflux from roots (Segonzac et al., 2007).

6.1.1.3 Radial transport of nitrate across the root and loading into xylem

Once nitrate has been taken up into the root symplast (the continuum of cell cytoplasm connected via plasmodesmata), nitrate can move radially across the different cell types of the root and pass the endodermal Casparian band. For nitrate

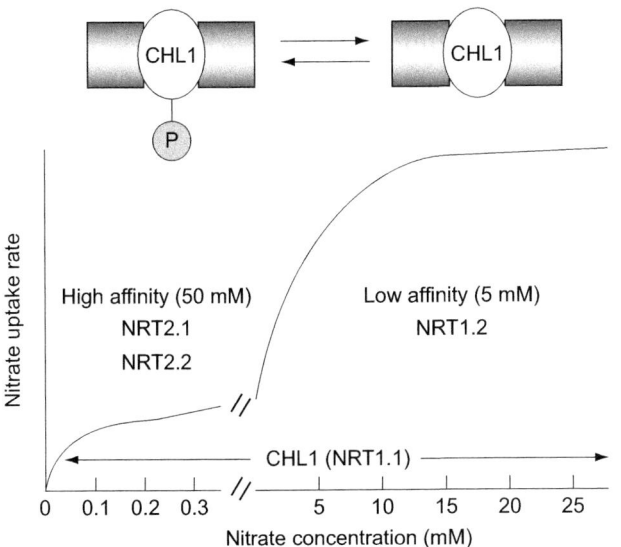

FIGURE 6.3 Nitrate uptake systems in arabidopsis roots in response to increasing nitrate concentration. *From Tsay et al. (2007), with permission from Elsevier.*

to be transported to the shoot, nitrate has to be released from the stelar root cells into the xylem for long-distance transport via the transpiration stream. The negative membrane potentials of the root cells, combined with an outward-directed nitrate concentration gradient, indicate that nitrate entry into the xylem to some extent may be passive. In arabidopsis, the nitrate transporter gene *AtNPF7.3* (*NRT1.5*) is expressed in the pericycle cells adjacent to the protoxylem in the middle of the root, where it mediates low-affinity proton-coupled efflux of nitrate from the root cells into the xylem (Fig. 6.2). Expression of *AtNPF7.3* is induced by nitrate, in accordance with greater transport to the shoot in high-N conditions (Lin et al., 2008). Diurnal regulation of *AtNPF7.3* ensures that more nitrate is loaded in the xylem and transported to the shoot during the day when nitrate reductase (NR) is most active.

Another NPF family member, *AtNPF7.2* (*NRT1.8*), is also expressed in xylem parenchyma cells of the root, but NPF7.2 mediates the opposite transport to NPF7.3 as it is involved in the retrieval of nitrate from the xylem sap (Fig. 6.2). As such, the function of AtNPF7.2 in the root is to allow the roots to remove part of the nitrate loaded in the long-distance transport route before it reaches the shoot (Li et al., 2010). The coexistence in the same tissue of transport systems such as AtNPF7.2 and AtNPF7.3, facilitating nitrate transport in opposite directions, might enable plants to finely regulate nitrate transport to the shoot as well as the distribution of nitrate within the root system.

In addition to the NPF transporters, nitrate release from the xylem parenchyma cells into the xylem vessels may be mediated by transporters belonging to the family of slow anion channel-associated homologs (SLAC/SLAH; Hedrich and Geiger, 2017). Several mechanisms are thus involved in controlling the long-distance transport of nitrate to the shoot.

6.1.1.4 Nitrate transport within the cell

In contrast to ammonium which is predominantly incorporated into organic compounds in the root, nitrate is more readily distributed throughout the plant. Nitrate is accumulated and can be stored in the vacuoles of roots, shoots, and storage organs (Rossato et al., 2001). The storage of nitrate in vacuoles might serve as a reservoir of N to be used when external N supply is low. However, this N reserve is in most cases very small compared to the fraction of organic N in the plant, and furthermore, the store of nitrate in vacuoles is depleted within 12–48 h of nitrate starvation (Richard-Molard et al., 2008). Nitrate is stored in vacuoles transiently, for example, during the night when nitrate is not metabolized by NR. This suggests that storage of nitrate in vacuoles serves as a nitrate buffer for transport processes rather than as N storage.

The nitrate transporters AtNPF5.11, AtNPF5.12, and AtNPF5.16, which are localized in the tonoplast, have been proposed to mediate nitrate efflux from vacuoles and to regulate nitrate partitioning between roots and shoots (He et al., 2017). Some members of the CLC family of voltage-dependent chloride channels may also mediate nitrate transport across the tonoplast (Dechorgnat et al., 2011).

The only NRT2 transporter located in the tonoplast is AtNRT2.7 that is specifically expressed in seeds (Chopin et al., 2007). Overexpression of *AtNRT2.7* in arabidopsis increased seed nitrate concentration and germination, whereas knockout had the opposite effect. In wheat, TaNRT2.5, which is an orthologue of AtNRT2.7, plays a role in nitrate transport to the vacuole and is expressed in the developing grain (Li et al., 2019). TaNRT2.5 is regulated by a nitrate-inducible NAM, ATAF, and CUC (NAC) transcription factor, which binds directly to the promoter of TaNRT2.5 and positively regulates its expression, leading to increased grain nitrate concentration and seed vigor during germination (Li et al., 2019).

6.1.1.5 Nitrate transport within the shoot

Several members of the NPF family of nitrate transporters have been found to control nitrate distribution within the shoot (Fig. 6.2). AtNPF7.2 (NRT1.8) is likely to mediate the unloading of nitrate from the xylem for uptake into leaf cells in the shoot in addition to its role of unloading nitrate from the xylem in the root (Li et al., 2010). AtNPF6.2 (NRT1.4) is involved in controlling the nitrate content of the leaf petiole (Chiu et al., 2004).

The concentration of nitrate in the phloem is generally low (μM), and regulation of nitrate distribution via the phloem has for many years been considered to be of limited importance. However, it was recently reported that *AtNPF2.13* (NRT1.7) encodes a low-affinity nitrate transporter that is expressed in the phloem of the minor veins of older leaves and transports nitrate across the plasma membrane into the phloem (Chen et al., 2020). Such a transport step would potentially allow nitrate remobilization from older (source) leaves to N-demanding (sink) tissues under N starvation. Disruption of NPF2.13 increased the accumulation of nitrate in old leaves, while less nitrate was detected in the phloem exudates of old leaves (Chen et al., 2020). Overexpression of *OsNRT2.3*, coding for a plasma-membrane transporter expressed mainly in the phloem, where it switches nitrate transport activity on or off by a pH-sensing mechanism, significantly improved grain yield and nitrogen-use efficiency (NUE) in rice (Fan et al., 2016).

Nitrate uptake can also occur in the leaves, which is important for epiphytes and for crop plants receiving foliar fertilization (Rossmann et al., 2019).

6.1.2 Ammonium transport into and within plants

Ammonium (NH_4^+) is in equilibrium with ammonia (NH_3), which is a weak base with a pK_a of 9.25. Soil pH is generally considerably lower than this pK_a and therefore very little NH_3 is present in the soil. Thus, NH_4^+ is the main form taken up by roots (Loqué and von Wirén, 2004).

6.1.2.1 Ammonium uptake by roots

The HATS is a saturable ammonium uptake system that operates at ammonium concentrations less than 0.5 mM (Kronzucker et al., 1996). The ammonium LATS dominates at ammonium concentrations above 0.5 mM. All plants express a nonsaturable, low-affinity influx system (Rawat et al., 1999), which is at least partially protein-mediated.

Throughout the plant, ammonium transport is carried out mostly by members of the ammonium transporter family (AMT/MEP/Rh) (von Wirén and Merrick, 2004). This includes the AMT1 subfamily that transports ammonium via NH_4^+ uniport or NH_3/H^+ symport, or the AMT2/MEP subfamily that contains the NH_3 channel, AmtB, from *Escherichia coli* and the Mep1−3 transporters from yeast. The AMT transporters in plants are predicted to have an extracytosolic N-terminus, 11 transmembrane domains, and a cytosolic C-terminus (Loqué and von Wirén, 2004).

Ammonium transporters of the AMT1 family represent the major entry pathway for root uptake of ammonium (Loqué and von Wirén, 2004). In arabidopsis, the AtAMT1;1, AtAMT1;2, and AtAMT1;3 transporters contribute equally to the root ammonium HATS activity (Yuan et al., 2007), whereas AtAMT1;5 plays a minor role in ammonium uptake. An overview of the HATS AMT1s is presented in Fig. 6.4. *AtAMT1;1* and *AtAMT1;3* are expressed in the root cortical and epidermal cells (Loqué et al., 2006) in agreement with a direct role in ammonium uptake from the soil. AtAMT1;2 is expressed in cortical and endodermal root cells, suggesting that AtAMT1.2 is also involved in the uptake of ammonium from the apoplast for radial transport of ammonium (Yuan et al., 2007). AtAMT1;1 ($K_m \approx 50\ \mu M$) and AtAMT1;3 ($K_m \approx 60\ \mu M$) have higher affinity for ammonium than AtAMT1;2 ($K_m \approx 150-230\ \mu M$). As such, the transporters with the highest affinity are located in the outer root cells to acquire ammonium from the soil (Fig. 6.4). The symplastic transport pathway mediated by AtAMT1;3 dominates the radial movement of ammonium across the root tissue when external ammonium is low, whereas apoplastic transport controlled by AMT1;2 at the endodermis takes over at high external ammonium (Duan et al., 2018). Thus, AMT1;2 favors N allocation to the shoot, revealing a major importance of the apoplastic transport pathway for ammonium partitioning to shoots.

The expression levels of the *AMT1* genes are upregulated in N-starved plants but reduced upon resupply of ammonium (Gazzarrini et al., 1999; Rawat et al., 1999). *AtAMT1;1* is upregulated particularly in the portion of the root

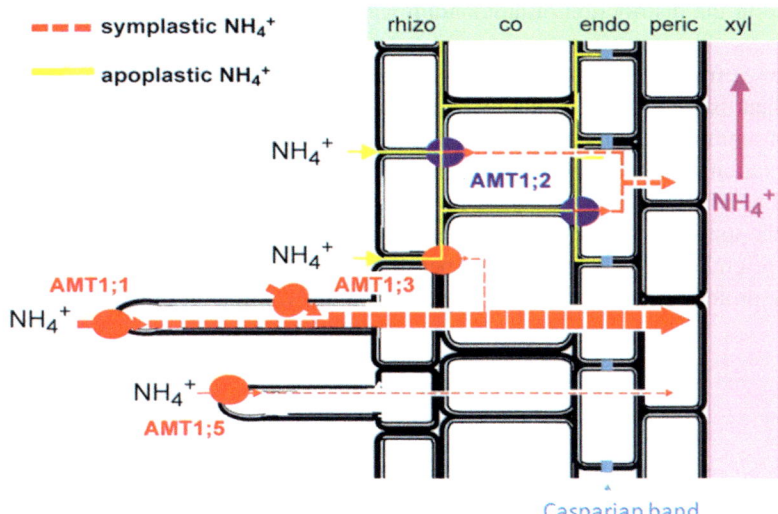

FIGURE 6.4 Schematic representation of the functions of the AMT1 transporters in the ammonium HATS in arabidopsis roots. *Co*, Cortex; *endo*, endodermis; *peric*, pericycle; *rhizo*, rhizodermis; *xyl*, xylem. *From Yuan et al. (2007), with permission from Oxford University Press.*

system that is experiencing N starvation (Gansel et al., 2001). Following the resupply of ammonium to N-starved plants, the expression of the HATS *AMT1* genes is reduced (Rawat et al., 1999). The two rhizodermis-localized transporters ZmAMT1;1a and ZmAMT1;3 are the major components in the root HATS of maize. These two genes are induced by ammonium rather than being upregulated under N deficiency (Gu et al., 2013). In addition to regulation at the transcriptional level, AtAMT1;1 and AtAMT1;2 may be regulated posttranslationally via phosphorylation, inhibiting ammonium uptake at high concentrations (Straub et al., 2017). The kinase involved is CIP23 (Calcineurin B−like Interacting Protein Kinase23), which also regulates the nitrate transporter NPF6;3 and activates channel-mediated uptake of K^+.

The AMT2/MEP subfamily member, AMT2;1, functions mainly in root-to-shoot translocation of ammonium in a concerted action with AMT1 transporters in arabidopsis (Giehl et al., 2017).

Ammonium (NH_4^+) resembles K^+ in terms of ionic radius and size of the hydration shell (Howitt and Udvardi, 2000). This resemblance with K^+ may allow ammonium ions to permeate K^+ channels (ten Hoopen et al., 2010; White, 1996). The low K^+ concentrations often observed in ammonium-fed plants may stimulate upregulation of K^+ channels to improve plant K^+ status, potentially resulting in further ammonium influx in the case of ammonium ions competing with K^+ for transport through the K^+ channels (ten Hoopen et al., 2010). Recently, a mechanism by which NH_3 and H^+ are transported separately across the membrane after ammonium deprotonation has been characterized, providing a two-lane pathway for electrogenic ammonium transport (Williamson et al., 2020). This may be important in achieving transport selectivity against competing ions such as K^+.

Similarly to the diurnal pattern of nitrate uptake, ammonium uptake (and the expression of *AtAMT1;3*) also increases during the day, with a maximum at the end of the light period, after which the uptake decreases (Gazzarrini et al., 1999; Glass et al., 2002). This diurnal pattern shows that N uptake is regulated by C supply (Liu et al., 2009). Indeed, the external supply of photoassimilates leads to ammonium influx in the dark via upregulation of transcription of the *AtAMT1* genes, as well as of the nitrate transporter *AtNRT2.1* (Lejay et al., 2003).

In ammonium-sensitive barley plants, high ammonium influx is counteracted by an active efflux of ammonium back to the soil (Britto et al., 2001; Kronzucker et al., 2001). This results in an apparently futile cycling of ammonium ions across the plasma membrane (Britto et al., 2001), whose effectiveness may be associated with increased sensitivity toward ammonium toxicity in rice seedlings (Chen et al., 2013). Futile NH_3 cycling may also occur, possibly via aquaporins (Coskun et al., 2013a).

6.1.2.2 Ammonium in the shoot

Ammonium taken up by the roots is assimilated or stored in vacuoles in the root or is transported to the aerial parts. Generally, it has been assumed that ammonium is not used for long-distance transport of N within the plant; however, the ammonium concentration in the xylem can be in the millimolar range (Finnemann and Schjoerring, 1999; Yuan et al., 2007), showing that ammonium is transported from roots to shoots.

Ammonium is generated by photorespiration in chloroplasts of illuminated leaf cells, by protein breakdown and amino acid catabolism in senescing tissues, in lignin biosynthesis, and is also supplied from the nodules following N_2

fixation in legumes. AMT transporters involved in the distribution of ammonium within the shoot have been identified in many plant species (Hao et al., 2020).

Generally, cytosolic levels of ammonium range from 1 to 30 mM (Miller et al., 2001). Overaccumulation of ammonium in the cytosol might lead to necrosis of plant tissue. At the cellular level, influx into cells and efflux of ammonium to the apoplast as well as compartmentation of ammonium into vacuoles, or ammonium assimilation in the cytoplasm or plastids all influence the ammonium concentration in the cytosol (Nielsen and Schjoerring, 1998).

In vacuoles, the concentration of ammonium in nonstressed plants ranges from 2 to 45 mM (Miller et al., 2001). Cytosolic NH_3 is passively transported across the tonoplast where the acidic environment traps NH_3 as NH_4^+. NH_3 and water have similar sizes and polarity. This resemblance allows NH_3 to permeate water channels in some cases. Accordingly, members of the tonoplast intrinsic proteins have been shown to play a role in NH_3 import into the vacuole (Jahn et al., 2004; Loqué et al., 2005).

6.1.3 Organic N uptake

In addition to inorganic N acquisition, the uptake of organic N also contributes to plant N nutrition (Näsholm et al., 2009). The organic N fraction of total N in the soil is considerable and is in the form of peptides and proteins, amino acids, and urea.

6.1.3.1 Amino acid uptake

Peptides and proteins are broken down to amino acids in the soil by proteases released by soil microorganisms. The concentration of free amino acids in agricultural soils is in the range of 1–100 μM and is the largest fraction of low-molecular-weight dissolved organic N in the soil (Jones et al., 2005). Several transporters differing in specificity and affinity for amino acids have been shown to play a role in amino acid uptake (Ganeteg et al., 2017). Microorganisms are in strong competition with plants for amino acid uptake, affecting the extent to which plants access organic N from the soil.

6.1.3.2 Urea uptake and metabolism

In agriculture, urea is used as N fertilizer and is also a naturally occurring and readily available N source in soils. Urea is hydrolyzed to ammonium in the soil by urease produced by soil microorganisms, but plants can also take up urea directly (Kojima et al., 2007). Most plants have a single urease gene, with the multiple urease genes in soybean being an exception (Witte, 2011). Urease is activated by incorporation of Ni (Witte, 2011).

Urea transporters involved in uptake from the soil likely include the AtDUR3 transporter in arabidopsis, which appears to mediate high-affinity transport in symport with protons. The expression of *AtDUR3* in arabidopsis roots is upregulated in response to N deficiency (Liu et al., 2003). AtDUR3 has been shown also to play a role in urea retrieval from the leaf apoplast and in urea re-translocation from senescing arabidopsis leaves, where urea is produced in connection with mitochondrial degradation of arginine, the major N storage form (Bohner et al., 2015).

Passive urea transport is mediated by some members of the major intrinsic proteins family of aquaporins. Of those, some are likely to mediate urea transport across the plasma membrane, whereas others mediate urea transport across the tonoplast or mitochondrial membrane (Witte, 2011).

6.1.4 Nitrogen assimilation

Nitrate (NO_3^-) is readily mobile in the xylem and can also be stored in the vacuoles of roots, shoots, and storage organs. For the N in nitrate to be incorporated into organic structures, nitrate has to be reduced to ammonium (NH_4^+). Most of the ammonium, whether originating from nitrate reduction or from direct uptake from the soil solution, is normally incorporated into organic compounds in the roots, although some NH_4^+ may also be translocated to the shoot even in plants receiving nitrate as the sole N form (Schjoerring et al., 2002).

Nitrogen assimilation is intricately regulated. This is necessary to integrate environmental signals with carbon metabolism, so that N assimilation is coupled with the availability of N in the soil, the demand for synthesis of various N-containing compounds as well as with the availability of carbon skeletons, energy, and reductants for the assimilatory pathway (Nunes-Nesi et al., 2010). An overview of N assimilation is given in Fig. 6.5.

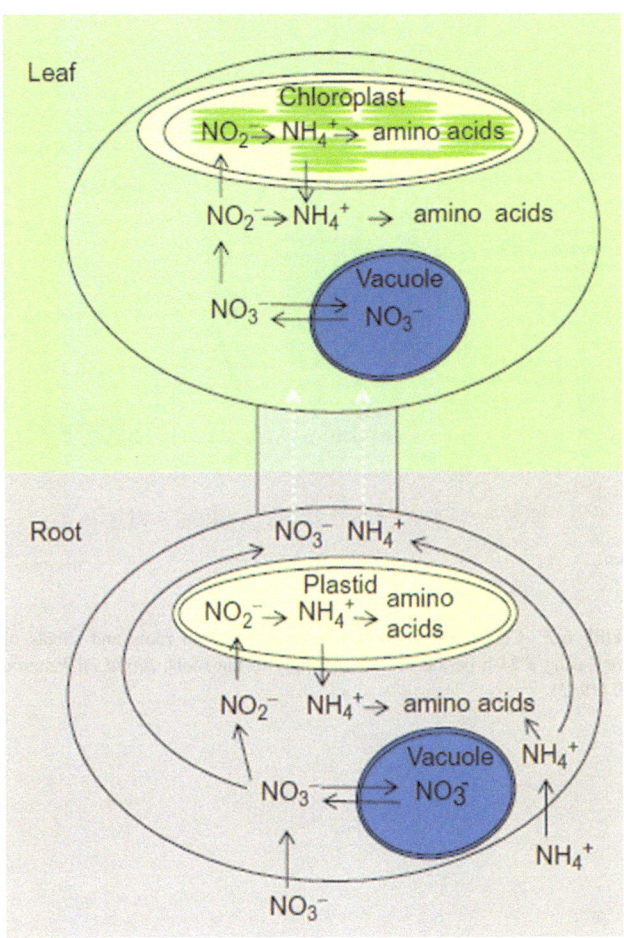

FIGURE 6.5 Overview of N uptake and N assimilation in plants.

6.1.4.1 Nitrate reduction

The reduction of nitrate to ammonium is mediated by two enzymes: *NR*, which catalyzes the two-electron reduction of nitrate to nitrite (NO_2^-), and *nitrite reductase*, which transforms nitrite to ammonium in a six-electron transfer process (Fig. 6.6). The net reaction is

$$NO_3^- + 2\ H^+ + [4NAD(P)H + 4\ H^+] \rightarrow NH_4^+ + [4NAD(P)^+] + 3H_2O \tag{6.1}$$

NR is a cytosolic enzyme consisting of two identical subunits, each with three cofactors covalently bound to specific domains of the enzyme (Fig. 6.6). The three cofactors that participate in the transfer of electrons from NADH/NADPH to nitrate are flavine adenine dinucleotide, a heme (bound to a domain that resembles a family of cytochromes) and molybdopterin (a molybdenum-containing cofactor). Most plant species have two NR (*NIA*) genes (Crawford and Arst, 1993) that are expressed in shoots and roots.

The nitrite generated by NR is transported to the chloroplast for reduction to ammonium by nitrite reductase. Nitrite reductase is encoded by a single gene in higher plants (Kant et al., 2011). It is a plastidial enzyme localized in the chloroplasts of leaves and in the proplastids of roots and other nongreen tissues. In green leaves, the electron donor is reduced ferredoxin, generated by photosystem I during photosynthetic electron transport in the light. Electrons from the reduced ferredoxin are passed to nitrite via a ferredoxin-binding domain, an iron–sulfur cluster, and a siroheme cofactor bound to the nitrite reductase (Fig. 6.6). In the root plastids, reduced ferredoxin is generated via NADPH in the pentose phosphate pathway coupled with ferredoxin-NADP$^+$ reductase (Bowsher et al., 2007).

To prevent accumulation of nitrite that is toxic to plant cells, NR activity is regulated by several mechanisms (Lillo, 2008), including enzyme synthesis, degradation, and reversible inactivation as well as regulation of effectors and the concentration of substrate. The enzyme has a half-life of only a few hours and is absent in plants not receiving nitrate. However, the expression of the NR genes is induced strongly and rapidly by nitrate, leading to active protein within a

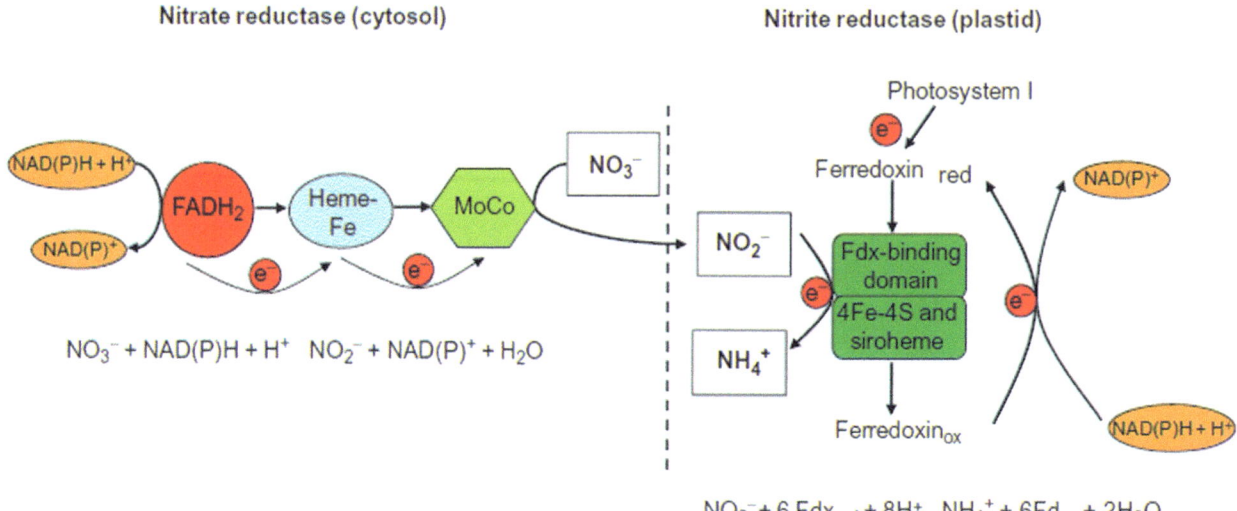

FIGURE 6.6 Schematic representation of the sequence of nitrate assimilation.

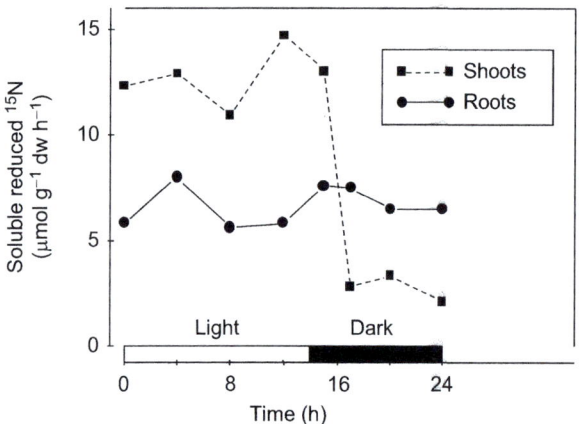

FIGURE 6.7 Concentration of soluble reduced N in roots and shoots of maize during a 24-h period of $^{15}NO_3$ supply to the roots. *Based on Pearson et al. (1981).*

few hours following the addition of nitrate (Patterson et al., 2010). In addition, the concentration of NR protein is increased by light, sucrose, and cytokinin, whereas glutamine, a primary product of N assimilation, represses NR (Krapp et al., 1998). This regulation links the capacity for nitrate assimilation with the availability of sugars to provide carbon skeletons. Elevated atmospheric CO_2 can reduce the assimilation of nitrate by increasing the competition for reductants produced by photosynthesis that are necessary for both carbon and nitrate assimilation (Bloom et al., 2010, 2020).

NR is further regulated by several posttranslational mechanisms. A protein kinase phosphorylates NR and thereby enables the binding of a 14−3-3 protein that inactivates the enzyme. Point mutation of the serine residue prevents phosphorylation, causing constitutive activation of NR and accumulation of nitrite (Lillo et al., 2003). The inactivation of NR by protein binding is inhibited by triose and hexose phosphates. This ensures that NR is maintained in an active state when there is ample supply of carbon skeletons for amino acid synthesis. In addition, enzyme activity can be restored by dephosphorylation by a phosphatase, which prevents protein binding and inhibition. During short-term light−dark transitions, posttranslational inhibition of NR occurs within a few minutes, preventing the accumulation of nitrite (Lea et al., 2006).

The close correlation between light intensity and nitrate reduction in green leaves (Fig. 6.7) may reflect fluctuations in carbohydrate concentrations and in the corresponding supply of reducing equivalents and carbon skeletons (Anjana and Iqbal, 2007). The diurnal fluctuations in NR activity may lead to a decrease in the foliar nitrate concentrations during the light period (Table 6.1; Neely et al., 2010). Plants grown under low-light conditions (e.g., in glasshouses during winter) may contain nitrate concentrations several-fold higher than those of plants grown under high-light conditions

TABLE 6.1 Time course of nitrate content in spinach leaves during the light period from 0900 to 1800 hours.

	Time of day	Nitrate concentration (mg kg^{-1} fresh wt)	
		Leaf blades	Petioles
	0830	228	830
Light	0930	167	725
	1330	101	546
	1730	91	504
	1830	106	578

Source: Based on Steingröver et al. (1982).

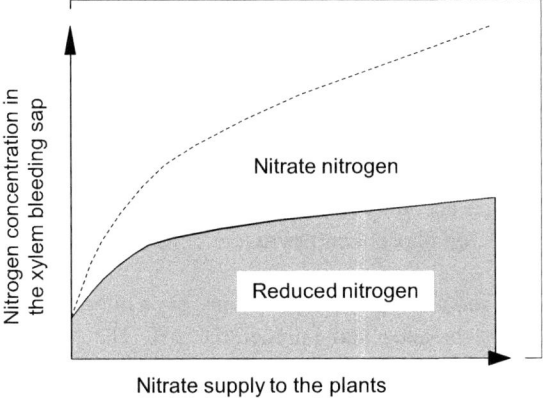

FIGURE 6.8 Schematic representation of the concentration of N in the xylem breeding sap in the form of nitrate and reduced N. *Data recalculated from Wallace and Pate (1965).*

(e.g., in an open field during summer). This is particularly evident in certain vegetables belonging to the Brassicaceae or Chenopodiaceae (Santamaria, 2006); for example, spinach has a high preference for nitrate accumulation in the shoots and uses nitrate accumulation in vacuoles for osmoregulation. Under low-light conditions, nitrate concentrations in spinach leaves can reach around 100 mM, corresponding to 6000 mg kg^{-1} fresh wt (Burns et al., 2011).

Vacuolar nitrate is remobilized rapidly to sustain cytosolic nitrate concentration when plants are deprived of external nitrate or when NR activity is high at high-light intensities (Cookson et al., 2005). Accordingly, the rate of release of nitrate from the vacuoles in leaf cells does not appear to be a rate-limiting step for the utilization of stored nitrate. In roots, NR activity is high in expanding cells of the apical zones and declines rapidly toward the basal root zones (Di Laurenzio et al., 1996).

In most plant species, both roots and shoots are capable of nitrate reduction, and reduction in the roots may be between 5% and 95% of the nitrate taken up. The proportion of reduction carried out in roots and shoots depends on various factors, including the level of nitrate supply, plant species, and plant age (Andrews, 1986). In general, when the external nitrate supply is low, a high proportion of nitrate is reduced in the roots. With an increasing supply of nitrate, the capacity for nitrate reduction in the roots becomes a limiting factor and an increasing proportion of the total N is translocated to the shoots in the form of nitrate (Fig. 6.8).

In temperate perennial species as well as in temperate annual legumes, most of the nitrate is reduced in the roots when the external concentration is relatively low. By contrast, tropical and subtropical annual and perennial species tend to reduce a large proportion of the nitrate in the shoots, even at low external supply, and the proportion between root and shoot reduction remains similar when the external concentration is increased. There are exceptions to this generalization; for example, in Australian open forest plants (Stewart et al., 1990) or woody plants growing in Cerrado and forest communities in Brazil (Stewart et al., 1992), the capacity for nitrate reduction in the leaves is low compared to the roots. With high nitrate availability, shoots appear to be the predominant site of nitrate reduction in both fast- and slow-growing grass species (Scheurwater et al., 2002).

The uptake rate of the accompanying cation also affects the proportion of nitrate reduced in roots. With K$^+$ as accompanying cation, translocation of both K and nitrate to the shoots is rapid; correspondingly, nitrate reduction in the roots is relatively low (Ruiz and Romero, 2002). By contrast, when either Ca^{2+} or Na$^+$ is the accompanying cation, nitrate reduction in the roots is considerably higher (Cramer et al., 1995).

The preferential site of nitrate reduction (roots or shoots) may have an important impact on the carbon economy of plants and probably also has ecological consequences for the adaptation of plants to low-light and high-light conditions. The energy requirements for the reduction and assimilation of nitrate are high (Gavrichkova and Kuzyakov, 2010; Schilling et al., 2006). The energy requirement for the reduction of 1 mol of nitrate is 15 mol ATP, with an additional 5 mol ATP for ammonium assimilation (Salsac et al., 1987). In barley, where a high proportion of nitrate reduction occurs in the roots, up to 23% of the energy from root respiration is required for absorption, reduction, and assimilation of the reduced nitrogen, compared to only 14% for assimilation when ammonium is supplied (Bloom et al., 1992). By contrast, for nitrate reduction in leaves, reducing equivalents can be provided directly by photosystem I and ATP by photophosphorylation. Under low-light conditions or in fruiting plants (Hucklesby and Blanke, 1992), this may lead to competition between CO_2 and nitrate reduction. On the other hand, under high-light conditions and excessive light absorption (photoinhibition, photooxidation), nitrate reduction in leaves may not only use energy reserves but also alleviate high-light stress. Competition with CO_2 reduction under elevated CO_2 may lead to acclimation and reduced CO_2 response when nitrate is the N source, whereas acclimation is less pronounced under ammonium nutrition (Bloom et al., 2010).

In C_4 plants, mesophyll and bundle sheath cells differ in their functions not only in CO_2 assimilation but also in nitrate assimilation. NR and nitrite reductase are localized in the mesophyll cells and are absent in the bundle sheath cells. This division of labor in C_4 plants, whereby mesophyll cells utilize light energy for nitrate reduction and assimilation and bundle sheath cells for CO_2 reduction, contributes to higher photosynthetic NUE in C_4 than C_3 plants (Sage et al., 1987). Due to the CO_2 concentrating mechanism in the bundle sheath cells (see Chapter 5), less RuBP carboxylase (Rubisco) is required in C_4 than C_3 plants. In C_3 plants, N in Rubisco accounts for 20%–30% of the total leaf N compared with less than 10% in C_4 plants [not counting 2%–5% of N for phosphoenolpyruvate (PEP) carboxylase] (Sage et al., 1987).

In plant species where most or all nitrate assimilation occurs in the shoots, organic acid anions are synthesized in the cytoplasm and stored in the vacuole to maintain both cation−anion balance and intracellular pH. The latter is required because nitrate reduction consumes two protons per nitrate reduced. The accumulation of organic acid anions may lead to osmotic problems if nitrate reduction proceeds after the termination of leaf cell expansion (Britto and Kronzucker, 2005; Raven and Smith, 1976). However, several mechanisms exist for the removal of excess osmotic solutes from the shoot tissue: (1) precipitation of excess solutes in an osmotically inactive form, for example, synthesis of oxalate for charge compensation in nitrate reduction and precipitation as Ca oxalate are common in plants, including sugar beet; (2) re-translocation of reduced N (amino acids and amides) together with phloem-mobile cations, such as K$^+$ and Mg^{2+}, to areas of new growth; and (3) re-translocation of organic acid anions, predominately malate, together with K$^+$ into the roots and release of CO_2 after decarboxylation.

6.1.4.2 Ammonium assimilation

Ammonium is a central intermediate in plant N metabolism. As well as uptake from the soil by roots, ammonium is constantly generated in plant tissues by processes such as nitrate reduction, photorespiration, lignin biosynthesis, senescence-induced N remobilization, and N_2 fixation in legumes (Joy, 1988). Irrespective of the source of ammonium, or the organ in which it is assimilated (roots, root nodules, and leaves), the key enzymes involved are glutamine synthetase (GS) and glutamate synthase (GOGAT; glutamine-oxoglutarate aminotransferase). Both enzymes are present in roots, in chloroplasts, and in N_2-fixing microorganisms, and assimilation of most if not all ammonium derived from ammonium uptake, N_2 fixation, nitrate reduction, and photorespiration is mediated by the GS-GOGAT pathway. In this pathway the amino acid glutamate acts as the acceptor for ammonium, and the amide glutamine is formed (Fig. 6.9). The net reaction is:

$$\text{Glutamate} + \text{2-oxoglutarate} + \text{NH}_4^+ \rightarrow \text{glutamine} + \text{H}^+ \tag{6.2}$$

GS exists in multiple enzyme forms located in the cytosol and in plastids (Bernard et al., 2008). Cytosolic GS assimilates ammonium into glutamine for transport and distribution throughout the plant. During leaf senescence, cytosolic GS fulfills a key function in the assimilation and recycling of ammonium generated in various catabolic processes (Masclaux-Daubresse et al., 2010). This role is particularly important after anthesis and during grain development and

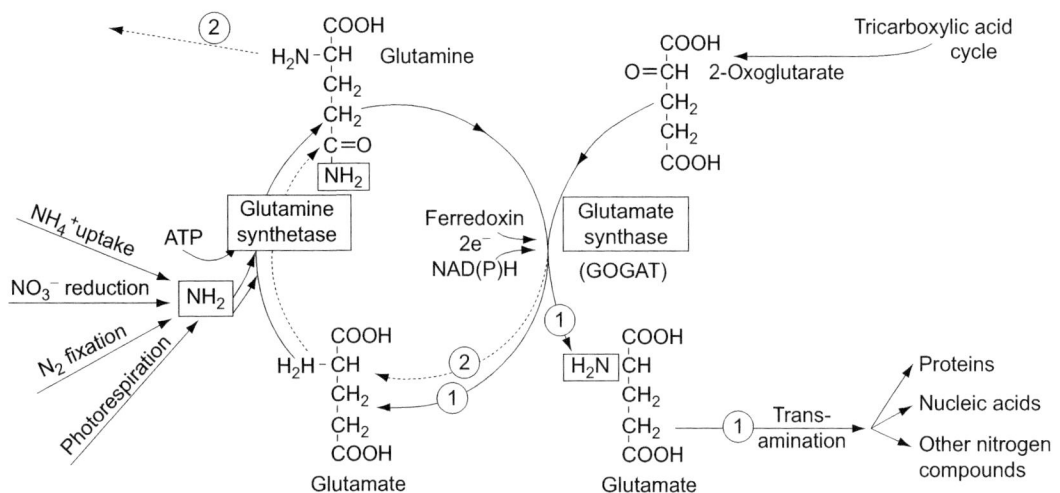

FIGURE 6.9 Model of ammonium assimilation via the glutamine synthetase-glutamate synthase cycle. Pathways at low (1) and high (2) ammonium supply.

filling in cereals, when N is remobilized to the reproductive sinks (Thomsen et al., 2014). Several isoenzymes of the cytosolic GS1 gene family are abundantly expressed in roots and can be classified into high-affinity or low-affinity subtypes differing in V_{max} values (Ishiyama et al., 2004). Some isoenzymes play a redundant role under low ammonium supply and others dominate under high external ammonium supply (Konishi et al., 2018). This dynamic regulation may contribute to the homeostatic control of glutamine synthesis in roots. In arabidopsis shoots, separate isoforms of cytosolic GS play distinct roles for seed yield structure and for coping with ammonium toxicity (Guan et al., 2015; 2016). The same is the case in maize, where the *Gln1−3* and *Gln1−4* genes affect kernel number and kernel size, respectively, with a specific impact on grain production (Martin et al., 2006).

In chloroplasts, light-stimulated nitrate reduction and enhanced ammonium assimilation are coordinated through the import of 2-oxoglutarate from the stroma and export of glutamate from the stroma of chloroplasts into the cytoplasm, thus preventing high ammonium concentrations. Chloroplast GS is activated by high pH and high concentrations of Mg and ATP, and all three factors are increased in the chloroplast stroma upon illumination. In cytosol and chloroplasts, GS is also subject to posttranslational regulation by phosphorylation and subsequent interactions with proteins (Finnemann and Schjoerring, 2000; Lima et al., 2006).

The other enzyme involved in ammonium assimilation, glutamate synthase (GOGAT), catalyzes the transfer of the amide group ($-NH_2$) from glutamine to 2-oxoglutarate, which is a product of the tricarboxylic acid (TCA) cycle. The conversion of glutamine to glutamate takes place in plastids that have two isoforms of GOGAT. One form accepts electrons from reduced ferredoxin (from photosystem I), the other from NADPH from respiration. The ferredoxin-linked GOGAT isoform dominates in leaves, particularly in the chloroplasts of phloem companion cells in leaf veins (Masclaux-Daubresse et al., 2007), whereas the NADPH isoform is prevalent in roots (Tabuchi et al., 2007). Both forms contain an Fe-S cluster that transfers electrons during the reductive synthesis of two glutamate molecules from one 2-oxoglutarate and one glutamine molecule. One of the two molecules of glutamate produced is required for the maintenance of the ammonium assimilation cycle and the other can be transported from the sites of assimilation and utilized elsewhere for the biosynthesis of proteins. When the ammonium supply is high, both glutamate molecules can act as ammonium acceptors, and one molecule of glutamine leaves the cycle (Fig. 6.9). Two Fd-GOGAT genes (*GLU1* and *GLU2*) have been characterized in higher plants, with *GLU1* playing a major role in the assimilation of ammonium derived from photorespiration, whereas the *GLU2* gene may play a major role in the primary N assimilation in roots (Suzuki and Knaff, 2005).

The enzyme glutamate dehydrogenase (GDH) was for many years assumed to be involved in ammonium assimilation. GDH catalyzes the reversible amination of 2-oxoglutarate to form glutamate. However, it is now evident that GDH is involved mainly in the liberation of ammonium during senescence via catalyzing the oxidative deamination of glutamate, providing carbon skeletons for respiration and oxidative phosphorylation. Thereby, the GDH enzyme, in conjunction with NADH-GOGAT, contributes to the control of leaf homeostasis of glutamate, an amino acid that plays a central role in signaling at the interface of the carbon and N assimilatory pathways (Labboun et al., 2009).

6.1.4.3 Low-molecular-weight organic N compounds

The inorganic N assimilated into glutamate and glutamine can be used readily for the synthesis of other amides as well as amino acids, ureides, amines, peptides, proteins, nucleic acids, and other N-containing compounds (Fig. 6.10). In plants, low-molecular-weight organic compounds not only act as intermediates between the assimilation of inorganic N and the synthesis of high-molecular-weight compounds, but they are also important for various other reasons such as transfer of N from source organs to sink tissues and to build up reserves during periods of high N availability.

Plants do not excrete substantial amounts of organically bound N, for example, urea. Although plants can store large amounts of nitrate, they cannot re-oxidize organically bound N to nitrate, which could be a safe storage form in periods of enhanced protein degradation in, for example, senescing leaves. In plants, amino acids and amides act as transient storage, in addition to their function in long-distance transport of reduced N.

Glutamate, glutamine, aspartate, and asparagine occupy a central position in amino acid metabolism and in carbon–N interactions in plants. Indeed, glutamine and glutamate are the major entry points of ammonia into organic compounds, and the amino groups of glutamate and aspartate and the amide group of glutamine are the N source for most plant N compounds, including other amino acids (Morot-Gaudry et al., 2001). Aspartate is a metabolically reactive amino acid that serves as a donor in numerous aminotransferase reactions, whereas asparagine is relatively inert and serves primarily as N storage.

Ammonium assimilation in roots has a large requirement for carbon skeletons for amino acid synthesis. These carbon skeletons are provided by the TCA cycle, and the intermediates removed have to be replenished by increased activity of PEP carboxylase. Compared to nitrate supply, net carbon fixation in roots supplied with ammonium is several-fold higher (Viktor and Cramer, 2005). To minimize the carbon costs for root-to-shoot transport, the bulk of the N assimilated in the roots is transported in the form of N-rich compounds with N/C ratios >0.4. One, rarely two or more, of the following compounds dominate in the xylem exudate of the roots: the amides glutamine (2N/5C) and asparagine (2N/4C), the amino acid arginine (4N/6C), and the ureides allantoin and allantoic acid (4N/4C). In phloem transport to developing fruits, as nonphotosynthetic sinks, amino acids with an N/C ratio >0.4 are the predominant transport forms of nitrogen.

The low-molecular-weight organic N compounds used predominantly for long-distance transport or for storage in individual cells differ among plant families. Glutamine and asparagine are the dominant transport amides in Poaceae. The concentrations of these transport amino acids vary and are modulated by factors such as light and N availability. Glutamine is synthesized mainly in the light, whereas asparagine is synthesized preferentially in the dark. Asparagine is the dominant transport form in legume species such as clover, alfalfa, pea, and lupin, which have indeterminate nodules. In legume species characterized by determinate nodules, such as soybean, cowpea, and common bean, the majority of the fixed N transported in the xylem of nodulated roots is incorporated into the ureides allantoin and allantoic acid (Pelissier et al., 2004).

An important class of low-molecular-weight organic N compounds is amines and polyamines; their biosynthesis is mediated by decarboxylation of amino acids (e.g., serine), and they form the basis for the synthesis of ethanolamine that is included in the lipid fraction of biomembranes. Arginine is the main precursor for polyamines thath are important secondary messengers (Kusano et al., 2008). Putrescine is usually the dominant polyamine in plants and may constitute up to 1.2% of the plant dry matter. The polyamine concentration is particularly high in meristematic tissues of plants supplied with high concentrations of ammonium (Gerendás and Sattelmacher, 1990) and under K deficiency (Watson and Malmberg, 1996).

Another low-molecular-weight organic N compound is betaine (glycine betaine) involved in osmoregulation. Under salt or drought stress, the synthesis of betaine and its accumulation particularly in the cytoplasm are enhanced strongly.

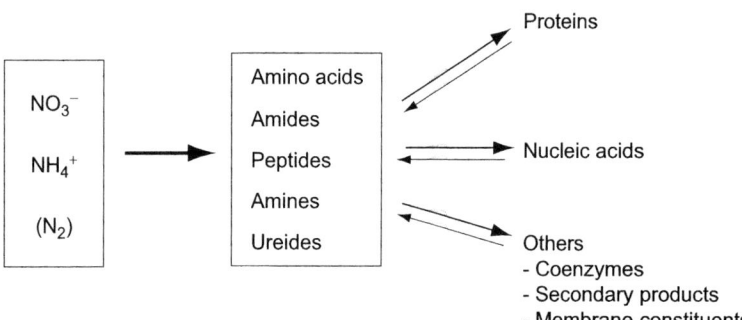

FIGURE 6.10 Major classes of N compounds in plants.

Betaine is important for the adaptation of plants to drought or salinity because it acts as a compatible solute to counteract the osmotic perturbation caused by high vacuolar concentrations of inorganic ions such as Cl$^-$ and Na$^+$ which would inhibit cytoplasmic metabolism (Missihoun et al., 2011).

Although plants may contain up to 200 different amino acids, only about 20 of them are required for protein synthesis. Not much is known about the role of the large number of nonproteinogenic amino acids in plants. However, at least some of them are important for plant nutrition. Nicotianamine is an effective chelator of Fe^{2+} and plays a role in iron homeostasis and in phloem transport of Fe, Zn, and Mn (Suzuki et al., 2008; Ishimaru et al., 2010). In addition, nicotianamine is precursor of a group of other nonproteinogenic amino acids (phytosiderophores) that are of particular importance for the acquisition of iron in graminaceous plant species (Suzuki et al., 2006; see also Chapters 2, 7, and 14).

6.1.5 Nitrogen supply, plant growth, and composition

6.1.5.1 Synergy between ammonium and nitrate nutrition

Whether ammonium or nitrate as a sole source of N supply is better for the growth and yield formation of plants depends on many factors. Generally, plants adapted to soils that are acidic (calcifuge species) or have a low redox potential (e.g., wetlands) prefer ammonium (Lee, 1999). By contrast, plants adapted to calcareous, high pH soils (calcicole species) utilize nitrate preferentially. However, the highest growth rates and plant yields are obtained by combined supply of both ammonium and nitrate.

Ammonium is preferentially taken up by many plants when supplied in equimolar concentrations with nitrate, particularly when the N supply is low (Gazzarrini et al., 1999). The preference for ammonium relative to nitrate increases strongly with decreasing temperatures; below 5°C uptake of ammonium still proceeds, but that of nitrate ceases (MacDuff and Jackson, 1991). This may reflect the greater costs of metabolic energy associated with absorption and assimilation of nitrate compared to ammonium. On the other hand, ammonium is assimilated predominantly in the roots, imposing a direct demand for carbon skeletons, which is reflected in higher activities of PEP carboxylase. Compared with ammonium, nitrate has the advantage of allowing more flexible distribution of assimilation between roots and shoots and can be stored in higher amounts than ammonium in the vacuoles.

As ammonium and nitrate comprise about 80% of the total ions taken up by plants, the form of N has a strong impact on the uptake of other cations and anions, on cellular pH regulation and on rhizosphere pH (see Chapter 14). The assimilation of ammonium in roots produces about one proton per molecule of ammonium (Raven and Smith, 1976). The generated protons are to a large extent excreted into the external medium to maintain cellular pH and electroneutrality, the latter compensating for the excess uptake of cation equivalents over anion equivalents generally associated with ammonium nutrition. Under mixed N nutrition, the proton generated by ammonium assimilation can be used for nitrate reduction; therefore, it is easier for plants to regulate intracellular pH when both forms of nitrogen are supplied.

Rhizosphere chemistry can be affected by the form of N taken up: ammonium supply may reduce rhizosphere pH through a net excretion of protons, whereas nitrate supply may increase rhizosphere pH through a net uptake of protons from the rhizosphere (Hinsinger et al., 2003). The implications of this for the availability of other nutrients such as P and micronutrients are discussed in Chapter 14.

The form in which N is taken up is important for the biosynthesis and function of phytohormones, especially cytokinins (Inoue et al., 2001). For example, the enzymes required for the synthesis of cytokinins are specifically induced by nitrate supply (Miyawaki et al., 2004). In wheat plants supplied with ammonium, the presence of nitrate even at low concentrations (100 μM) can stimulate increases in the concentration of the active cytokinin forms zeatin, transzeatin riboside, and isopentenyl adenosine (Garnica et al., 2010). The higher cytokinin concentrations in nitrate-fed plants may be accompanied by higher shoot concentrations of auxin indole-3-acetic acid (IAA) (Garnica et al., 2010). These results suggest that the beneficial effect of nitrate on the growth of plants predominately supplied with ammonium is mediated by a coordinated effect on the concentrations of cytokinins and IAA in the shoot. Conversely, in nitrate-fed plants, reproductive growth may be delayed due to excessive concentrations of cytokinins. Under such circumstances, the provision of ammonium may induce flowering, probably via increased biosynthesis of polyamines acting as secondary messengers (Rohozinski et al., 1986).

6.1.5.2 Ammonium toxicity

Plant species differ in tolerance to ammonium (Britto and Kronzucker, 2002). Among crop plants, barley is ammonium-sensitive, whereas rice is ammonium-tolerant. The symptoms of ammonium toxicity include leaf chlorosis, stunted growth, and eventually leaf necrosis and plant death.

Various hypotheses have been put forward to explain the physiological processes underlying ammonium toxicity. When whole tissue of ammonium-fed plants is analyzed, several chemical changes are observed. Generally, compared to nitrate-fed plants, there is an accumulation of ammonium ions, inorganic anions such as chloride, sulfate, and phosphate as well as of amino acids. By contrast, there is a reduction in the concentration of the essential cations such as K^+, Ca^{2+}, and Mg^{2+} as well as organic acids such as malate (Britto and Kronzucker, 2002). These and other observations have led to the hypotheses that ammonium toxicity may be the result of (1) decreased uptake of essential cations (Coskun et al., 2013a), (2) ammonium-induced disorders in pH regulation (Walch-Liu et al., 2001), or (3) excessive consumption of sugars for ammonium assimilation causing carbohydrate limitation (Finnemann and Schjoerring, 1999). The presence of low (micromolar) concentrations of nitrate may alleviate ammonium toxicity (Roosta and Schjoerring, 2008). SLAH3, a nitrate efflux channel belonging to SLAC/SLAH gene family of slow anion channels, seems to play a role in the nitrate-dependent alleviation of ammonium toxicity in plants (Zheng et al., 2015b).

6.1.5.3 Nitrogen deficiency

To achieve efficient growth, development, and reproduction, plants require adequate, but not excessive, amounts of N. Therefore, low soil N availability or a decline in root uptake capacity negatively affects plant productivity and ecological competitiveness. Nitrogen-deficient plants are typically stunted, with narrow leaves. Chlorosis caused by N deficiency usually begins in the older leaves as N is remobilized to younger leaves; N-deficient crops appear pale green or yellow. In grasses, tillering as well as the number of seeds per inflorescence are reduced compared to plants growing with adequate N.

With temporary low N supply in the root medium, plants display a two-phase response. In the first phase, the leaf elongation rate is reduced, but photosynthesis is not affected. Root growth is maintained or even stimulated by the transport of assimilated carbon to the roots, which results in a lower shoot/root biomass ratio (Richard-Molard et al., 2008). Concomitantly, N compounds, particularly nitrate, are mobilized to maintain N metabolism, and the capacity to take up nitrate from the soil is increased. In the second phase, upon prolonged N starvation, the breakdown of leaf nucleic acids and proteins is triggered. This is usually associated with leaf senescence (Hortensteiner and Feller, 2002). The breakdown of Rubisco leads to a decrease in the maximum photosynthetic capacity, ultimately inhibiting plant growth.

Plants have evolved multifaceted strategies to respond to variations in N availability in the soil, including metabolic, physiological, and developmental adaptations. As part of the primary nitrate response, plants respond within minutes to changes in external nitrate supply by changing the expression of hundreds to thousands of genes, including NR, nitrite reductase, nitrate transporters, and genes involved in organic acid metabolism. Nitrate and ammonium sensing involve the NPF6.3/NRT1.1 nitrate transporter and the AMT1.1 and AMT1.3 ammonium transporters, acting as transceptors (i.e., combined transporters and receptors) (Liu and von Wirén, 2017; Undurraga et al., 2017). Nitrate sensing by NPF6.3/NRT1.1 triggers Ca^{2+} signaling by calcium-sensor protein kinases (CPKs) that translocate to the nucleus in response to nitrate (Liu et al., 2017). The CPKs then phosphorylate the NLP transcription factor (NLP7) that appears to be a master regulator of the primary N response (Marchive et al., 2013).

Nitrate signaling mediated by AtNPF6.3 leads to upregulation of genes involved in the assimilation of N and regulation of root system architecture. In rice, OsNRT1.1A (OsNPF6.3) functions in the upregulation of the expression of N utilization-related genes not only for nitrate but also for ammonium, as well as flowering-related genes (Wang et al., 2018b). Overexpression of *OsNRT1.1A* in rice greatly improved grain yield and shortened maturation time relative to wild-type plants (Wang et al., 2018b).

6.1.5.4 Changes in root system architecture in response to N supply

One of the most striking examples of plasticity to changing N supply is the modulation of root system architecture. Mild N deficiency leads to elongation of lateral roots and the primary root, whereas severe or prolonged N deficiency inhibits primary root growth and total root length (Gruber et al., 2013). Generally, a uniformly high nutrient supply suppresses root branching. However, when N availability is limited, plants may respond to N-rich patches by enhancing lateral root development into them. An example of such proliferation of lateral roots within a localized nitrate-rich zone is shown in Fig. 6.11 for barley plants. By contrast, elevated local ammonium supply triggers lateral root branching and inhibits primary root elongation (Jia and von Wirén, 2020; Liu and von Wirén, 2017). Nitrogen-deficient plants exposed to a local supply of ammonium respond by developing more second- and third-order lateral roots through a mechanism dependent on the ammonium transporter AMT1;3 (Lima et al., 2010).

When growing under N deficiency, plants develop a more exploratory root system by increasing primary and lateral root length. The stimulation of lateral root elongation in a confined soil volume involves signaling pathways triggered

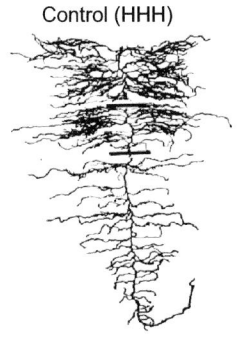 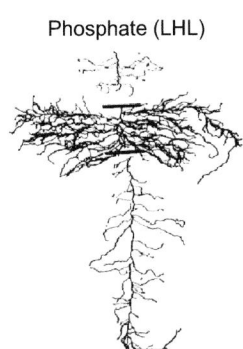 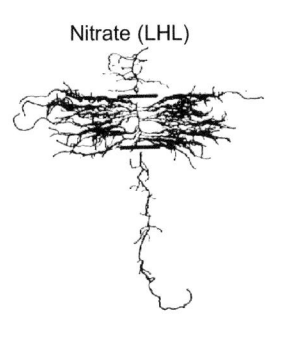

FIGURE 6.11 Root growth of barley plants supplied with complete nutrient solution to all parts of the root system (HHH) or complete nutrient solution in the middle zone only, with the top and bottom parts of the root system supplied with nutrient solution deficient in (LHL) either phosphate (middle) or nitrate (right). *Adapted from Drew (1975).*

by local sensing of nitrate and ammonium. The stimulation of lateral root growth by external nitrate involves the regulation of meristematic activity in the lateral root tip mediated via the nitrate transceptor NRT1.1/NPF6.3 (Krouk et al., 2010; Mounier et al., 2014). In general, lateral root growth is stimulated by an accumulation of auxin at the tip of the lateral root. During prolonged N deficiency, NRT1.1/NPF6.3 acts as an auxin transporter that decreases auxin concentration in the lateral root apex to suppress meristematic activity and lateral root elongation, thus preventing plants from spending energy on root branching in a volume of soil without nitrate. By contrast, at external nitrate concentrations above 1 mM, the auxin transport function of NRT1.1/NPF6.3 is inhibited, which in turn allows auxin to accumulate in the lateral root tip and promote lateral root elongation (Bouguyon et al., 2015; Maghiaoui et al., 2020), thereby promoting further exploitation of the nitrate-rich volume of soil.

The local modulation of lateral root growth during N starvation involves an auxin receptor targeted by a nitrate-inducible microRNA (miR393) (Vidal et al., 2010). In addition, CLE (CLAVATA3/embryo surrounding region−related) peptides are induced to inhibit emergence of lateral roots during N starvation (Araya et al., 2014). The phytohormones abscisic acid (ABA) and brassinosteroids are involved in sensing nitrate transport and linking the derived signals to root branching and root elongation (Jia et al., 2020; Leran et al., 2015; Wang et al., 2020b). Information on local nitrate availability is translocated to the shoot via long-distance mobile signals that include cytokinin and small peptides (Jia and von Wirén, 2020).

The responses of root architecture to changes in nitrogen supply are usually less pronounced for crops growing in soil under field conditions versus controlled environments. The root density of wheat crops increased slightly in response to N fertilizer rate, but root depth was only marginally affected (Rasmussen et al., 2015; Liu et al., 2018). In most cases, shoot growth responds to N fertilization more than root growth, implying a decrease in the root−shoot ratio even when root growth is stimulated.

6.1.5.5 Storage proteins

The amino acids formed by nitrate assimilation can be stored in dedicated storage proteins, which have neither metabolic nor structural roles and often have a relatively high proportion of N-rich amino acids, particularly arginine and amides. Storage proteins accumulate transiently, and upon protein degradation, the amino acids can be used directly for de novo protein synthesis or may be metabolized.

Vegetative storage proteins (VSP) have been identified in many plant species (Staswick, 1994) and can constitute up to 50% of the total soluble proteins in various vegetative storage organs, for example, in the taproot of alfalfa (Ourry et al., 2001).

VSP differ from seed storage proteins in that they accumulate transiently and are degraded within the plant life cycle. The accumulation of VSP can be affected indirectly by changes in source−sink relationships in relation to N within the plant (Staswick, 1994; Ourry et al., 2001), or directly by exogenous stimuli such as methyl-jasmonate (Meuriot et al., 2004; Noquet et al., 2001) or modifications of soil N availability (Meuriot et al., 2003). In alfalfa, the rate of regeneration of new photosynthetic tissues was linearly related to taproot VSP concentration on the day of cutting (Avice et al., 1996). VSP may also be important for fall hardening and overwintering of alfafa (Dhont et al., 2006).

The concentration of protein in seeds varies from 10%−15% of the dry weight in cereals to 40%−50% in some legumes (e.g., soybean). In cereal seeds, protein concentrations on a dry weight basis are in the range of 6%−8% in rice, 8%−15% in barley, 7%−22% in wheat, and 9%−11% in maize (Shewry, 2007). About 50%−85% of these proteins are storage proteins (Shewry, 2007). Seed storage proteins are synthesized during seed development and serve as the principal source of amino acids for germination and seedling growth. They are initially synthesized on the rough endoplasmic reticulum (ER), transported into the lumen, and finally deposited in the protein bodies (Kumamaru et al., 2007).

Seed storage proteins share a number of common properties: (1) high rate of synthesis in specific tissues, (2) presence in mature seeds in discrete deposits called protein bodies, and (3) being mixtures of components that exhibit polymorphism both within single genotypes and among genotypes of the same species.

Due to their abundance and economic importance in agricultural crops, seed storage proteins have been studied for more than 250 years. They were classified initially into albumins (soluble in pure water), globulins (soluble in dilute salt solutions), glutelins (soluble in diluted solutions of alkali and acids), and prolamins (soluble in aqueous ethanol) (Osborne, 1924). The structures of glutelins and prolamins are closely related; therefore, glutelins are now regarded as members of the prolamins (Heldt and Piechulla, 2011).

The predominant storage proteins of cereals are prolamins, except for oats and rice, in which the major storage proteins are globulins (rich in glutamine and asparagine). Globulins are the major storage proteins in legumes. Globulins are the most widely distributed group of storage proteins (Holding and Larkins, 2008). Prolamins are only present in grasses (Holding and Larkins, 2008; Shewry and Halford, 2002). In the major cereals, prolamins usually account for about 50% of the total grain nitrogen, except in oats and rice (5%–10% of the total seed protein). Albumins are a heterogeneous group of proteins for which the only unifying criterion is that they have a sedimentation coefficient of about 2 Svedberg (Holding and Larkins, 2008). They are widely distributed in seeds of dicot species.

Apart from the classical storage proteins, seeds contain additional proteins, such as proteinase inhibitors, lectins, and lectin-like proteins, ribosome-inactivating proteins, lipid transfer proteins, glucanases, and chitinases, that are associated with defense against pests and pathogens. Seed proteins also include hydrolases, such as amylases, proteinases, and lipases, that mobilize several types of associated reserve compounds, the products of which are used during germination for the synthesis of new tissues (Shewry et al., 1995).

In many plant storage proteins, the concentrations of amino acids essential for humans (i.e., cannot be synthesized in the human metabolism) are low. In cereals, the storage proteins are low in threonine, tryptophan, and particularly in lysine, whereas in legumes there is a shortage of methionine. The aleurone and embryo tissues of grains contain higher concentrations of essential amino acids, but these are often removed by milling (Chapter 9).

The increase in grain protein with high N fertilization is due to greater synthesis and accumulation of storage proteins. In wheat, increases in grain N are associated with increased proportions of the monomeric gliadins and a decreased proportion of large glutenin polymers, resulting in increased dough extensibility (Godfrey et al., 2010; Kindred et al., 2008).

6.1.6 Nitrogen-use efficiency

NUE can have several meanings in the context of crop production (Foulkes et al., 2009; Good et al., 2004). In general, NUE is the ratio between the total biomass of output (e.g., grain yield) and the N input (e.g., N supplied in fertilizers and/or residual N present in the soil). NUE is the product of two components: N uptake efficiency (NUpE; the capacity of the plant to remove N from the soil) and the utilization efficiency (NUtE; the capacity to use N to produce biomass or grain yield).

In crops, and particularly in cereals, large amounts of N fertilizer are required to attain maximum yield, and NUE is relatively low. As a consequence, plant breeding aimed at developing new crop genotypes with better NUE has a high priority (Hawkesford, 2014). Plant breeding for better NUE is focused on the different physiological processes that influence N uptake, translocation, assimilation, and redistribution, particularly focusing on the identification of genotypes that grow and yield well under low-N conditions (Hawkesford and Griffiths, 2019; Hirel et al., 2007; Zörb et al., 2018).

With regard to NUpE, the capture of nitrate present in low concentrations in the topsoil requires high rooting density (Dunbabin et al., 2003). The primary root traits affecting nutrient uptake are number of root axes, rooting depth, and rooting density (Thorup-Kristensen et al., 2020). Prolific root systems are more effective at capturing nutrients than sparse systems, but interroot competition sets a natural threshold for optimal root density. Other root traits that could increase N capture include enhanced root longevity for N uptake after flowering (Garnett et al., 2009; Kichey et al., 2007).

Expression of both nitrate and ammonium transporter genes is regulated by supply and demand for N. Higher threshold levels of downregulation of the transporter genes may allow greater influx, which in turn may drive increased N assimilation. Alternatively, decreasing the activity of efflux systems could also improve the efficiency of uptake.

For NUtE, the cytosolic isoforms of the enzyme GS (GS1) appear to play an important role in nitrogen management, plant growth rate, grain yield, and grain filling (Bernard and Habash, 2009). The GS isoforms can be critical for N assimilation and remobilization, and specific manipulation of some isoforms in a developmentally controlled manner may offer prospects for gains in NUtE (Gao et al., 2019; Martin et al., 2006). Increased conversion of N into grain yield may be achieved by improving the efficiency of CO_2 fixation (Long et al., 2006).

To attain maximum yields, modern crop cultivars require large amounts of fertilizers, in particular N, because the genotypes currently cultivated in developed countries have mostly been selected under nonlimiting fertilization. Although plant breeders have consistently targeted improved grain yield under high inputs of fertilizer and crop protection chemicals, NUE per se has only recently become a target. Differences in NUE among varieties show that there is a potential to exploit genotypic differences in N responsiveness of maize, wheat, and rice (Barraclough et al., 2010; Cirilo et al., 2009; Hawkesford and Riche, 2020).

6.2 Sulfur

6.2.1 General

Sulfur is essential for plant functioning and growth and is taken up as sulfate (S^{6+}) by the root and needs to be reduced to sulfide (S^{2-}) in the chloroplasts (plastids in the root) before its assimilation into organic sulfur compounds (Haneklaus et al., 2007; Hawkesford and De Kok, 2006; Fig. 6.12). The uptake and assimilation of S and N are strongly

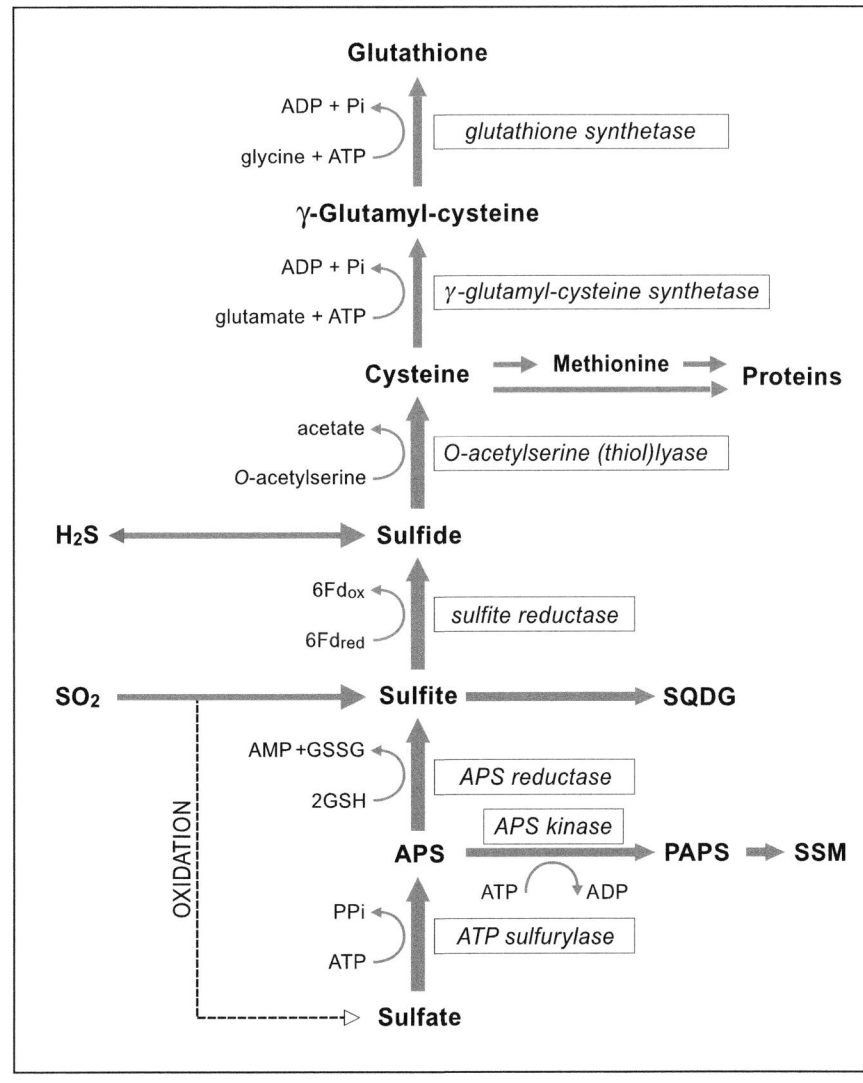

FIGURE 6.12 Sulfur assimilation in plants. *APS*, Adenosine 5′-phosphosulfate; *Fdred, Fdox*, reduced and oxidized ferredoxin; *GSH, GSSG*, reduced and oxidized glutathione; *PAPS*, 3′-phosphoadenosine-5′-phosphosulfate; *SQDG*, sulfoquinovosyldiacylglycerol; *SSM*, secondary sulfur metabolites. *Adapted from De Kok et al. (2007).*

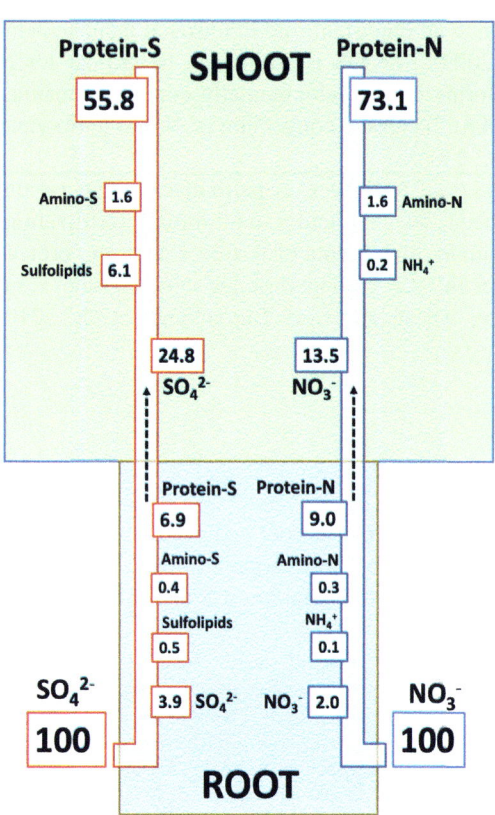

FIGURE 6.13 Distribution of sulfur and nitrogen metabolites in spinach plants as % of the total sulfur and nitrogen content (N/S ratio 20).

interrelated and dependent on each other because the major proportion of the reduced S is utilized for the synthesis of the amino acids cysteine and methionine, which are important for the structure and function of proteins (Stulen and De Kok, 1993; Fig. 6.13). Plants contain a variety of other organic S metabolites, such as thiols (glutathione and phytochelatins), sulfolipids, cofactors (e.g., biotin, thiamine, CoA, and S-adenosyl-methionine) and may contain a variety of secondary S compounds (glucosinolates in Brassicaceae and allyl-cysteine sulfoxides in *Allium* species; Haneklaus et al., 2007). Sulfur deficiency may result in a loss of crop production and quality and a decrease in plant resistance to environmental stresses and pests (Bloem et al., 2015; Haneklaus et al., 2007). Sulfur deficiency also impairs Fe nutritional status of plants (Astolfi et al., 2021). For comprehensive reviews on S in plants, the reader is referred to Kopriva (2006), Haneklaus et al. (2007), Hawkesford and De Kok (2006; 2007), Zhao et al. (2008a,b), Takahashi et al. (2011), and Koprivova and Kopriva (2014).

6.2.2 Sulfate uptake, reduction, and assimilation

The uptake of sulfate across the plasma membrane of the root cells is energy dependent and driven by a proton gradient through a proton/sulfate (presumably $3H^+/SO_4^{2-}$) cotransport and mediated by high-affinity sulfate transporter proteins. The expression and activity of these sulfate transporters are strongly induced by S deficiency (Hawkesford and De Kok, 2006). Low-affinity transporters from the same gene family are involved in cell-to-cell distribution of sulfate across plasma membranes, and in storage and remobilization from vacuoles across the tonoplast (Fig. 6.13). The transporters for delivery of sulfate into the chloroplasts, the site of activation and reduction, are still largely unclear. In vascular plants and in green algae, the first step of S assimilation is the activation of the sulfate ion by ATP (Fig. 6.12). In this reaction the enzyme ATP sulfurylase catalyzes the replacement of two phosphate groups of the ATP by the sulfuryl group, which leads to the formation of adenosine phosphosulfate (APS) and pyrophosphate (Fig. 6.12). This enzyme is regulated by various external (e.g., light) and internal (e.g., reduced S compounds) factors. The activated sulfate (APS) can serve as a substrate for the synthesis of sulfate esters or sulfate reduction. For the synthesis of sulfate esters by sulfotransferases, the enzyme APS kinase catalyzes the formation of phosphoadenosine phosphosulfate (PAPS) in an ATP-dependent reaction (Fig. 6.12). From PAPS, the activated sulfate can be transferred to a hydroxyl group forming a sulfate ester. The PAPS is an S donor for the synthesis of secondary S compounds such as glucosinolates.

For sulfate reduction, the activated sulfate of APS is reduced to sulfite (SO_3^{2-}) by APS reductase requiring two electrons supplied from glutathione (Fig. 6.12). Subsequently, six electrons from ferredoxin are required to produce sulfide (S^{2-}), catalyzed by sulfite reductase, the sole reaction of the pathway that occurs only in the chloroplast (Fig. 6.12). The newly formed sulfide is transferred to O-acetylserine, by the enzyme O-acetylserine(thiol)lyase (OASTL), and cysteine is formed. The synthesis of cysteine is one of the key metabolic links between S and N assimilation in plants. The substrate O-acetylserine is synthesized from serine and acetyl CoA catalyzed by serine actyltransferase (SAT). However, this enzyme is only active when it occurs in a complex with OASTL (by contrast, OASTL is inactive in the complexed state). Excess O-acetylserine (occurring when sulfide is limiting) disrupts the complex, resulting in inactive SAT and limiting further O-acetylserine production and consumption of acetyl CoA (Hell and Wirtz, 2008). In addition, O-acetylserine is thought to be part of a signaling pathway that stimulates the expression of genes for the transporters and APS reductase to enhance sulfate acquisition and S flux to sulfide. Such positive regulation of expression may balance an apparent repression of gene expression of the sulfate transporters and APS reductase caused by reduced S compounds (Hawkesford and De Kok, 2006). Cysteine, the first stable product of the assimilatory SO_4^{2-} reduction, acts as a precursor for the synthesis of most other organic S compounds containing reduced S, including glutathione, and it is the reduced S donor for the synthesis of methionine (Nikiforova et al., 2004), as well as for other biosynthetic pathways, such as the formation of ethylene (Romero et al., 2014).

Sulfate uptake and assimilatory reduction are regulated at various levels (Hawkesford and De Kok, 2006; Stulen and De Kok, 1993; Vauclare et al., 2002) by (1) regulation of expression of the sulfate transporters, (2) modulation of the activity of ATP sulfurylase, (3) the availability of sulfate as a substrate for ATP sulfurylase, (4) change in the level of APS reductase expression and activity, and (5) the state of complexation of SAT and OASTL that may act as both a sensor (of plant S nutritional status) and a regulator (of cysteine biosynthesis).

The isoforms of enzymes of the assimilatory sulfate reduction pathway occur in various subcellular compartments in both leaves and roots (Kopriva, 2006). In many, but not all (Kopriva and Koprivova, 2005), C_4 plants, the bundle sheath chloroplasts are the main sites of sulfate assimilation (Gerwick et al., 1980), whereas the mesophyll chloroplasts are the sites of nitrate assimilation (see Section 6.1), even though they contain at least sulfite reductase and OASTL (Schmidt, 1986). Glutathione biosynthesis occurs in both cell types.

In general, sulfate reduction is several times higher in green leaves than roots, and in leaves the reaction is strongly stimulated by light (Fankhauser and Brunold, 1978). This light enhancement is to be expected because of the requirement for glutathione and ferredoxin as reductants for APS and sulfite, respectively. In addition, the expression of several of the genes for enzymes of the reductive assimilation pathway (e.g., genes encoding ATPS, APR, SiR, and OASTL; Hell et al., 1997) appear to be under light and/or diurnal regulation. The stimulation of sulfate reduction by light may also be related to higher levels of serine (O-acetylserine; Fig. 6.12) synthesized during photorespiration. Reduced S compounds, mainly glutathione, are exported from the leaves via the phloem to sites of demand for protein synthesis (e.g., in the shoot apex, fruits, but also roots) and may also be involved in the regulation of sulfate uptake by roots (Chapter 3). During leaf development, the patterns of sulfate and nitrate reduction are similar, that is, maximal during leaf expansion and declining rapidly after leaf maturation (Schmutz and Brunold, 1982). Similar to the reduction of nitrate, sulfate reduction also seems to be under a strict negative feedback control because high concentrations of reduced S compounds (e.g., cysteine, glutathione) are rare under normal physiological conditions. Secondary S compounds are an exception.

6.2.3 Metabolic functions of S

Sulfur is a constituent of the amino acids cysteine and methionine, and hence of proteins. Both amino acids are also precursors of other S-containing compounds such as coenzymes and secondary plant products. Sulfur is a structural constituent of these compounds (e.g., R_1-C-S-C-R_2) or acts as a functional group (e.g., R-SH) directly involved in metabolic reactions. As a structural component, cysteine has particular effects on structure and function of proteins. The reversible formation of disulfide bonds between two adjacent cysteine residues (cysteinyl moiety) in the polypeptide chain is of fundamental importance for the tertiary structure and thus the function of enzyme proteins. This bond may form a permanent (covalent) cross-link between polypeptide chains or a reversible dipeptide bridge, comparable with the redox functions of glutathione. During dehydration, the number of disulfide bonds in proteins increases at the expense of the sulfhydryl groups, and this shift is associated with protein aggregation and denaturation (Tomati and Galli, 1979).

About 2% of the organically reduced S in plants is present in the water-soluble thiol (sulfhydryl) fraction, and under normal conditions the tripeptide glutathione accounts for more than 90% of this fraction (De Kok and Stulen, 1993; Fig. 6.13). Glutathione has many functions in plants (for a review see Rouhier et al., 2008). The synthesis of glutathione occurs in two steps (Fig. 6.12). In the first step, γ-glutamylcysteine is produced from glutamate and cysteine. In the second step, glycine is coupled to γ-glutamylcysteine, mediated by glutathione synthase, an enzyme that requires Mg for activity (Hell and Bergmann, 1988). In some legume species in the second step, alanine rather than glycine is used by glutathione synthase, forming homo-glutathione that functions similarly to glutathione (Rennenberg et al., 1990).

The glutathione concentration is usually higher in leaves than in roots; in leaves more than 50% of it is localized in the chloroplasts where it may reach millimolar concentrations (Rennenberg et al., 1990). In the chloroplasts, the antioxidants glutathione and ascorbate play a key role in detoxification of oxygen radicals and hydrogen peroxide, for example, in the ascorbate peroxidase-glutathione reductase cycle (see Chapter 5). In the cells, glutathione is maintained in its reduced form by the enzyme glutathione reductase, with NADPH as reductant. The antioxidative role of glutathione is reflected, for example, in the increase in glutathione reductase activity upon exposure of plants to oxidative stresses caused by ozone or sulfur dioxide (Smith et al., 1990). Conjugation of reduced glutathione to a number of xenobiotics, such as herbicide atrazine, is the mechanism of detoxification and resistance (Labrou et al., 2005).

Glutathione may function as a transient storage pool of reduced S (Schütz et al., 1991). Moreover, glutathione is also the precursor of phytochelatins, which may be important in detoxifying certain metals in plants (Cobbett and Goldsbrough, 2002; Grill et al., 1987). Plant cells respond to exposure to high concentrations of metals, such as Cu, Cd, and Zn, by increasing the synthesis of phytochelatins and cysteine-rich polypeptides (metallothioneins; Cobbett and Goldsbrough, 2002).

Chemical structure of phytochelatins, $n = 2 - 11$.

Phytochelatins consist of repetitive glutamylcysteine units (2–11) with a terminal glycine and are synthesized by degradation of glutathione mediated by a carboxypeptidase. Phytochelatins may bind metal cations via thiol coordination and thereby detoxify them (Grill et al., 1987). The synthesis of phytochelatins in roots is most strongly stimulated by Cd, less so by Zn and Cu and negligibly by Ni (Tuckendorf and Rauser, 1990). Synthesis of phytochelatins is strongly increased by exposure of the roots to 3 μM Cd, and this increase is accompanied by a rapid decline in the glutathione concentration (Table 6.2). This inverse relationship is evident after 1–2 h exposure to Cd. Phytochelatin synthesis is induced by exposure to as low as 0.05 μM Cd, and synthesis by far exceeds the amount required for detoxification of the metal (Tuckendorf and Rauser, 1990). Differences in Cd tolerance between ecotypes of *Silene vulgaris* are presumed to be related to differences in phytochelatin synthesis (Verkleij et al., 1990).

Sulfite is the precursor of the sugar sulfonate head of the plant sulfolipid sulfoquinovosyldiacylglycerol (Benning et al., 2008). Sulfolipids are particularly abundant in the thylakoid membranes of chloroplasts where about 5% of the lipids are sulfolipids (Schmidt, 1986). Sulfolipids may also be involved in the regulation of ion transport across

TABLE 6.2 Concentrations of free cysteine, total glutathione, and phytochelatins as well as Cd in the apical 10 cm of maize roots exposed to 0 or 3 μM Cd for 24 h.

Cd (μM)	Thiols (nmol g^{-1} fresh wt)			Cd in roots (nmol g^{-1} fresh wt)
	Cysteine	Glutathione	Phytochelatins	
0	43	421	3	nd
3	44	156	230	13

nd: not determined.
Source: Based on Tuckendorf and Rauser (1990).

biomembranes. Sulfolipid concentrations in roots have been shown to correlate positively with salt tolerance (Erdei et al., 1980).

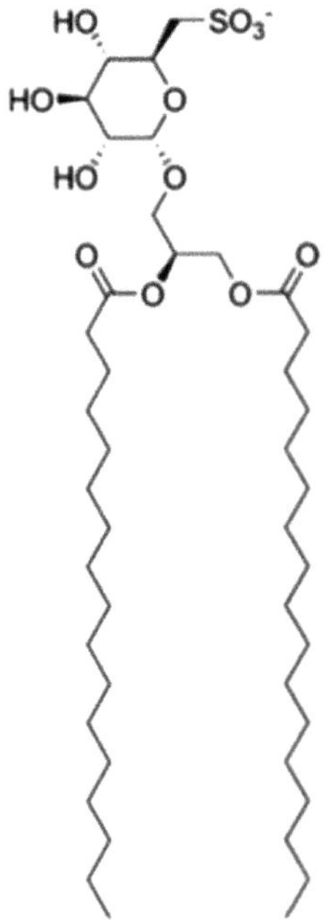

Chemical structure of sulfoquinovosyldiacylglycerol.

Thioredoxins are another important family of thiols. Thioredoxins are low-molecular-weight proteins of about 12 kDa with two well-conserved cysteine residues that form a redox-active, intermolecular disulfide bridge. Plant cells contain two different systems capable of reducing thioredoxins: the ferredoxin/thioredoxin system in chloroplasts and the NADP/thioredoxin system in the cytoplasm (Schürmann, 1993). In chloroplasts, thioredoxins function primarily as regulatory proteins in carbon metabolism. In the reduced form, thioredoxins activate, for example, fructose-1,6-bisphosphatase and several enzymes of the Calvin cycle and thus are a regulatory link between the provision of reducing equivalents (photosystem II) and assimilation of CO_2.

Reduced S is a structural constituent of several coenzymes and prosthetic groups such as ferredoxin, biotin (vitamin H), and thiamine pyrophosphate (TPP) (vitamin B_1). In many enzymes and coenzymes (e.g., urease, sulfotransferases, and coenzyme A), −SH acts as a functional group in the enzyme reaction. In the glycolytic pathway, for example, decarboxylation of pyruvate and the formation of acetyl coenzyme A are catalyzed by a multienzyme complex involving three S-containing coenzymes: TPP, the sulfhydryl-disulfide redox system of lipoic acid, and the sulfhydryl group of coenzyme A:

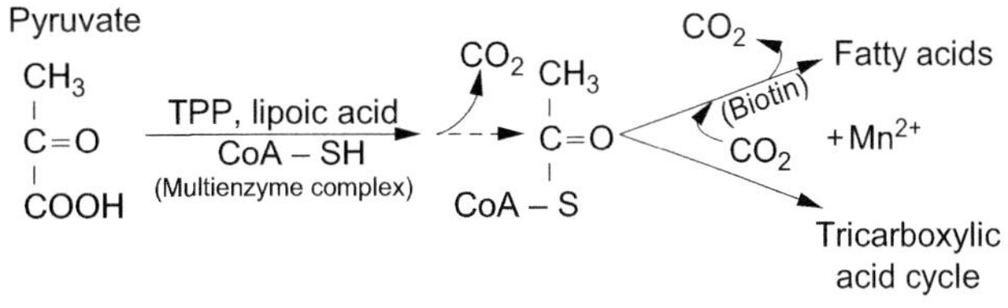

The acetyl group (−CO−CH_3) of coenzyme A is then transferred to the TCA cycle or to the fatty acid synthesis pathway. The coupling of C_2 units in the synthesis of long-chain fatty acids requires transient carboxylation, which is mediated by the S-containing coenzyme biotin and activated by Mn.

The most important S-containing compounds of secondary metabolism are alliins and glucosinolates. They are of particular relevance for horticulture and agriculture (Jones et al., 2004). Glucosinolates are characteristic compounds of the secondary metabolism of at least 15 dicotyledonous taxa, including Brassicaceae (Halkier and Gershenzon, 2006). Glucosinolates contain S both as a sulfhydryl and a sulfo group, and the side chain R varies among plant species:

$$R-C\begin{matrix}S-\text{Glucose}\\\\N-O-S-O^-\\\|\\O\end{matrix}$$

$$R = CH_2=CH-CH_2- \text{Sinigrin (Brassica nigra)}$$

$$R = HO-\langle\;\rangle-CH_2- \text{Glucosinalbin (Sinapis alba)}$$

Glucose ← Myrosinase → R-N=C=S Isothiocyanates (e.g., mustard oil)
Sulfate ←

Glucosinolates are stored in vacuoles and their hydrolysis is catalyzed by the cytosolic enzyme myrosinase that is present in only a small number of cells in a given organ (Höglund et al., 1991; McCully et al., 2008). Hydrolysis leads to the liberation of glucose, sulfate, and volatile compounds such as isothiocyanates in *Brassica napus*. Similarly to alliinase, myrosinase activity in cells is greatly enhanced by mechanical damage of cells.

Alliin is the common name for S-alk(en)ylcysteine sulfoxides that are the characteristic compounds of the genus *Allium*:

$$R-\overset{O}{\underset{\|}{S}}-CH_2-\underset{NH_2}{\overset{}{CH}}-COOH \xrightarrow{\text{Alliinase}} R-\overset{O}{\underset{\|}{S}}-S-R)_1$$

Alliins   Allicins

More than 80% of the total S in *Allium* species may be bound to such compounds, in onion (*Allium cepa*), for example, as S-propylcysteine sulfoxide (R = $-CH_2-CH_2-CH_3$). Enzymatic cleavage of alliins is mediated by alliinase. Loss of cellular compartmentation by mechanical damage of the tissue greatly enhances enzyme activity through increased availability of the substrate and leads to the formation of allicins as precursor of a large number of volatile substances such as mono- and disulfides with a characteristic odor.

The role of many secondary S compounds is not fully understood. They act as defense substances (phytocides, feeding deterrents), although the importance of this defense mechanism may have been overestimated in the past (Ernst, 1993). This is certainly true for glucosinolates, which have important functions as S storage in plants. During periods of low S supply to the roots but high plant demand (e.g., rapid vegetative growth, seed formation), glucosinolates are decomposed by myrosinase, and S is reutilized through the S-assimilatory pathway (Haneklaus et al., 2007). Roles of S-containing compounds in defense against abiotic and biotic stresses have recently been revisited, with sufficient or excess S fertilization having a positive impact on stress resistance (Rausch and Wachter, 2005).

6.2.4 Sulfur supply, plant growth, and plant composition

The S requirement for optimal plant growth varies between 1 and 5 g kg^{-1} dry wt. For the families of crop plants, the requirement increases in the order Poaceae < Fabaceae < Brassicaceae. The protein S concentration varies considerably, both between the protein fractions of individual cells (Table 6.3) and among plant species. On average, proteins from legumes contain less S than proteins from cereals, the N:S ratios being 40:1 and 30:1, respectively (Dijkshoorn and van Wijk, 1967; Haneklaus et al., 2007).

As with N deficiency, under S deficiency shoot growth is decreased more than root growth, leading to a decrease in shoot/root ratio. Interruption of S supply decreases root hydraulic conductivity, stomatal aperture, and net photosynthesis (Karmoker et al., 1991). The reduced leaf area in S-deficient plants is the result of smaller size and particularly the number of leaf cells (Burke et al., 1986).

A decrease in chlorophyll and protein concentrations of leaves is a typical feature of S deficiency (Burke et al., 1986; Dietz, 1989; Gilbert et al., 1997). This is to be expected because a high proportion of the leaf protein is located in the chloroplasts where the chlorophyll molecules comprise prosthetic groups of the chromoprotein complex. Accordingly, under S deficiency, a shortage of the S-containing amino acids cysteine and methionine not only inhibits protein synthesis but also decreases the chlorophyll concentration in leaves (Table 6.3). By contrast, starch may

TABLE 6.3 Leaf composition in tomato with or without S supply.

Sulfur supply	Concentration in leaves (mg kg^{-1} dry wt)			Protein S concentration (μg mg^{-1} protein)	
	Chlorophyll	Protein	Starch	Cytoplasm	Chloroplast
+S	58	480	28	14	7
−S	9	35	270	4	5

Source: Based on Willenbrink (1967).

accumulate as a consequence of either impaired carbohydrate metabolism at the sites of production or low demand at the sink sites (growth inhibition).

In S-deficient plants, inhibition of protein synthesis is correlated with an accumulation of soluble organic N (e.g., amino acids, especially asparagine; Schmidt et al., 2013) and nitrate (Sorin et al., 2015). Sulfur deficiency increases the concentration of amides as well as their proportions in the soluble N fraction (Freney et al., 1978; Karmoker et al., 1991). The sulfate concentration is low in deficient plants and increases markedly when the sulfate supply is sufficient for optimal growth. The sulfate concentration of plants is therefore a more sensitive indicator of S nutritional status than the total S concentration, the best indicators being the proportion of sulfate-S in the total S (Freney et al., 1978), or the ratio of sulfate to malate that accumulates under S deficiency (Blake-Kalff et al., 2000). Sulfur deficiency also leads to the accumulation of the sulfate analogs (selenate and molybdate) in plant tissues due to both decreased competition by sulfate for uptake and enhanced sulfate transporter expression (Shinmachi et al., 2010).

Unlike N deficiency, in S-deficient wheat (but not in all crops) chlorosis is more uniformly distributed between old and new leaves, with S concentration in old and young leaves similarly affected by the low sulfate supply (Freney et al., 1978). Furthermore, the distribution of S in S-deficient plants is also affected by the N supply. Sulfur deficiency symptoms may occur either in young (in combination with sufficient N) or in old (in combination with low N) leaves (Robson and Pitman, 1983), indicating that the extent of remobilization and re-translocation from older leaves depends on the rate of N deficiency−induced leaf senescence, a relationship which is also found for the micronutrients Cu and Zn (see Chapter 3).

In legumes, during the early stages of S deficiency, nitrogenase activity in the root nodules is decreased more strongly than photosynthesis (DeBoer and Duke, 1982). Symptoms of S deficiency in N$_2$-fixing legumes are therefore indistinguishable from N-deficiency symptoms (Anderson and Spencer, 1950). However, in root nodules of S-deficient legumes, the bacteroids may still be well supplied with S (O'Hara et al., 1987). Roots and nodules of pea have a high demand for S, and N$_2$ fixation is very sensitive to S deficiency (Zhao et al., 1999c). The high sensitivity of nitrogenase activity to S deficiency therefore reflects either impaired host plant metabolism or a direct effect on nitrogenase activity (see Chapter 16).

In S-deficient plants, not only the protein concentration but also the S concentration decreases in storage proteins (Table 6.3), indicating that proteins with a lower proportion of methionine and cysteine but higher proportions of other amino acids such as arginine and aspartate are synthesized (Table 6.4). A decrease in S-rich proteins under S deficiency has been shown in wheat (Zhao et al., 1999a,b) and in other cereals and legumes (Randall and Wrigley, 1986). Under S deficiency, the proportion of a low-molecular-weight S-rich polypeptide decreases in wheat (Castle and Randall, 1987), and in maize, the proportion of the major storage protein zein (with low S concentration) increases by about 30%, whereas the proportion of the S-rich glutelin decreases by 36%−71% (Baudet et al., 1986). The lower S concentration of proteins influences nutritional quality: methionine is an essential amino acid in human nutrition and often a limiting factor in diets in which grains are a major source of protein (Arora and Luchra, 1970). Furthermore, a decrease in the cysteine concentration of cereal grains reduces the baking quality of flour because disulfide bridging during dough preparation is responsible for the polymerization of glutelins (Ewart, 1978). There may be prospects to enhance the nutritional quality of grains, that is, methionine concentration, by pathway engineering (Hesse et al., 2004; Tabe and Higgins, 1998).

In Brassicaceae, the concentrations of glucosinolates and their volatile metabolites are closely related to sulfate supply. Their concentrations in plants can be increased beyond the level at which sulfate supply affects growth (Table 6.5). From the qualitative viewpoint, this increase can be favorable (e.g., because it enhances the taste of vegetables, making them spicier) or unfavorable (e.g., because it decreases acceptability as animal feed).

TABLE 6.4 Amino acid composition of endosperm proteins from S-sufficient (2.5 g S kg^{-1} dry wt) and S-deficient (1.0 g S kg^{-1} dry wt) wheat.

Amino acid	Amino acid concentration (nmol g^{-1} protein N)	
	S-sufficient	S-deficient
Methionine	0.9	0.3
Cysteine	1.3	0.4
Arginine	1.7	2.1
Aspartate	2.1	5.8

Source: Based on Wrigley et al. (1980).

TABLE 6.5 Yield and mustard oil concentration of the shoots of *Brassica juncea* at different S supply.

Sulfur supply (mg S pot^{-1})	Shoot fresh weight (g pot^{-1})	Mustard oil concentration (mg kg^{-1} fresh wt)
1.5	80	28
15	208	81
45	285	307
405	261	531
1215	275	521

Source: Based on Marquard et al. (1968).

Sulfur-containing methionine is an important precursor of phytosiderophores in Poaceae, which are produced in response to Fe deficiency (Chapter 7). Therefore, low S supply increases the sensitivity of Poaceae to Fe deficiency by impairing biosynthesis and root release of phytosiderophores (Astolfi et al., 2006; 2021).

In highly industrialized areas the S requirement of plants may be met, fully or to a substantial degree, by atmospheric SO_2 (and locally H_2S) deposition (Ausma and De Kok, 2019; De Kok et al., 2007; Haneklaus et al., 2007). These atmospheric S-containing gases are potentially phytotoxic, but upon foliar absorption, they may be metabolized and be beneficial if the S supply to the root is limited for optimal growth. Throughout the world, however, industrial SO_2 emissions have been significantly reduced during the last two decades, and S deficiency has become a widespread issue in agricultural areas, affecting both yield and quality (De Kok et al., 2007; Haneklaus et al., 2007; Zhao et al., 1999a, 2008a,b). The application of S fertilizers is effective in remediating this problem. Worldwide, S deficiency in crop production is quite common in rural areas, particularly in high rainfall areas, for example, in the humid tropics and temperate climates (Murphy and Boggan, 1988) and in highly leached soils.

6.3 Phosphorus

6.3.1 General

Phosphorus is a macronutrient that frequently limits plant productivity in natural systems in ancient landscapes, whereas N tends to be the most limiting macronutrient in young landscapes (Gallardo et al., 2020; Turner et al., 2018; Walker and Syers, 1976). Species that evolved in N-limited landscapes have received the most attention in the plant science literature because these include most crop plants as well as *Arabidopsis thaliana* (Prodhan et al., 2019). However, vast numbers of species evolved in ancient and megadiverse regions of the world, where P is the major limiting nutrient. Some of these species may harbor traits that may be highly desirable in crops plants and hence deserve further attention (Lambers et al., 2015).

Most of the phosphate that is used for our crops is derived from rock phosphate, which is a nonrenewable resource. We are gradually running out of global phosphate resources in an era when we need more P fertilizers to produce more food and fiber to sustain a growing global population (Fixen and Johnston, 2012; Johnston et al., 2014). In that context, there may be lessons to be learned from species that evolved highly P-efficient traits in ancient landscapes (Prodhan et al., 2019).

Unlike nitrate and sulfate, phosphate is not reduced in plants but remains in its highest oxidized form. Therefore, even though the more reduced oxide of phosphorus (phosphite) is sometimes advertised as a fertilizer, it is distinctly harmful when given to plants that are already short of phosphate because it is an analog of phosphate and inhibits its uptake (Loera-Quezada et al., 2015; Ratjen and Gerendás, 2009). After uptake (at physiological pH mainly as $H_2PO_4^-$), phosphate either remains as inorganic phosphate (Pi) or it is esterified through a hydroxyl group to a carbon chain (C–O–P) as a simple phosphate ester (e.g., sugar phosphate) or attached to another phosphate by the energy-rich pyrophosphate bond P~P (e.g., in ATP). The exchange between Pi and the P in ester and the pyrophosphate bond is fast. For example, Pi taken up by roots is incorporated within minutes into organic P but, thereafter, is released again as Pi into the xylem (see Chapters 2 and 3). Another type of phosphate bond is characterized by the relatively high stability of its diester state (C–P–C). In this association, phosphate forms a bridging group connecting units to more complex or macromolecular structures.

6.3.2 Phosphorus as a structural element

The function of P as a component of macromolecular structures is most prominent in nucleic acids, which, as components of DNA, are the carriers of genetic information, and, as units of RNA, are the structures responsible for the translation of the genetic information. In both DNA and RNA, phosphate forms a bridge between ribonucleoside units to form macromolecules:

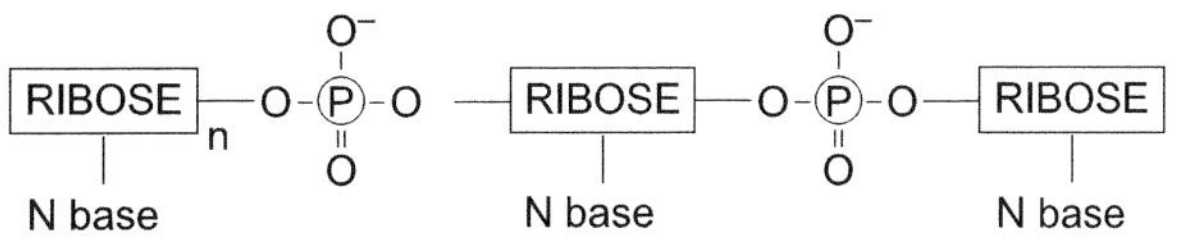

Section of DNA or RNA molecule

Phosphate is responsible for the strongly acidic nature of nucleic acids, and thus for the exceptionally high cation concentrations in DNA and RNA. The proportion of P in ribonucleic acids to total organically bound P differs among organs and cells; it is high in expanding leaves, where a large amount of ribosomal RNA (rRNA) is required to allow rapid protein synthesis, lower in mature leaves, and very low in senescing leaves (Suzuki et al., 2001).

The bridging form of P diester is also abundant in the phospholipids of biomembranes. There, it forms a bridge between a diglyceride and another molecule (amino acid, amine, or alcohol). In biomembranes, amine choline is often the dominant partner, forming phosphatidylcholine (lecithin):

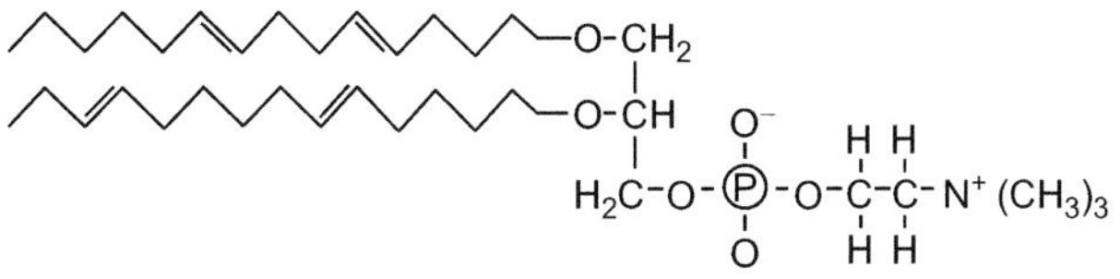

The functions of phospholipids (and sulfolipids; see Section 6.2) are related to their molecular structure. There is a lipophilic region (consisting of two long-chain fatty acid moieties) and a hydrophilic region in one molecule; at a lipid-water interface, the molecules are oriented so that the boundary layer is stabilized. The electrical charges of the hydrophilic region play an important role in the interactions between biomembrane surfaces and ions in the surrounding medium. This is because the charged ions are either attracted or repelled by the charges of the hydrophilic regions, whereas interactions of ions with the hydrophobic regions would be avoided. When P-starved, plants may replace phospholipids by galactolipids (Andersson et al., 2003; Gaude et al., 2008) or sulfolipids (Maathuis, 2009;

Byrne et al., 2011). This is a typical response in species that evolved in young landscapes, where N tends to be most limiting for plant productivity. Species that evolved in ancient landscapes where P is far less available may replace phospholipids during leaf development as part of a developmental process, rather than in response to P shortage (Lambers et al., 2012; Kuppusamy et al., 2014).

6.3.3 Role in energy transfer

Although present in cells in relatively low concentrations, phosphate esters (C-(P)) and energy-rich phosphates ((P)~(P)) represent the metabolic energy of cells. Up to 50 esters formed from phosphate and sugars and alcohols have been identified, about 10 of which, including glucose-6-phosphate and phosphoglyceraldehyde, are present in relatively high concentrations in plant cells. The common structure of phosphate esters is:

$$\text{R}-\overset{\overset{OH}{|}}{\underset{\underset{H}{|}}{C}}-\overset{\overset{H}{|}}{\underset{\underset{OH}{|}}{C}}-O-\overset{\overset{OH}{|}}{\underset{\underset{O}{\|}}{P}}-OH$$

Most phosphate esters are intermediates in metabolic pathways of biosynthesis and catabolism. Their function and formation are directly related to the energy metabolism of the cells and to energy-rich phosphates. The energy required, for example, for the biosynthesis of starch or ion uptake is supplied by an energy-rich intermediate or coenzyme, principally ATP:

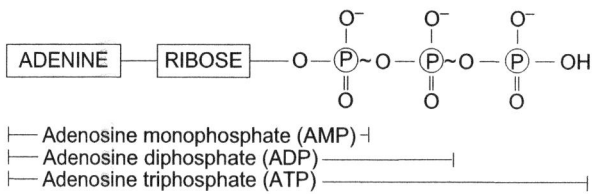

Energy liberated during glycolysis, aerobic respiration, or photosynthesis (Chapter 5) is utilized for the synthesis of the energy-rich pyrophosphate bond, and upon hydrolysis of this bond ~30 kJ mol^{-1} ATP is released. This energy can be transferred with the phosphoryl group in a phosphorylation reaction to another compound, which results in the activation (priming reaction) of this compound:

$$\begin{array}{c}\text{Adenosine}-P\sim P\sim \textcircled{P} \\ [\text{ATP}]\end{array} \begin{array}{c}\text{HO}-\boxed{R}\\ \\ \textcircled{P}-O-\boxed{R} \dashrightarrow\end{array}$$
$$\begin{array}{c}\text{Adenosine}-P\sim P \\ [\text{ADP}]\end{array}$$

ATP is the principal energy-rich phosphate required for starch synthesis. The energy-rich pyrophosphate bonds of ATP can also be transmitted to other coenzymes, which differ from ATP only in the nitrogen base, for example, uridine triphosphate and guanosine triphosphate (GTP), required for the synthesis of sucrose and cellulose, respectively. The activity of ATPases that mediate the hydrolysis of pyrophosphate bonds and thus energy transfer is affected by many factors, including nutrients such as Mg (Section 6.4), Ca (Section 6.5), and K (Section 6.6) (also Chapter 2). In some phosphorylation reactions the energy-rich inorganic pyrophosphate (PPi) is liberated, and the adenosine (or uridine) moiety remains attached to the substrate:

$$\begin{array}{c}\text{Adenosine}-P\sim\boxed{P\sim P} \\ [\text{ATP}]\end{array}\begin{array}{c}\textcircled{P}-O-\boxed{R}\\ \\ \text{Adenosine}-P-\textcircled{P}-O-\boxed{R}\dashrightarrow\end{array}$$
$$\begin{array}{c}\boxed{P\sim P_i}\\ [\text{Pyrophosphate}]\end{array}$$

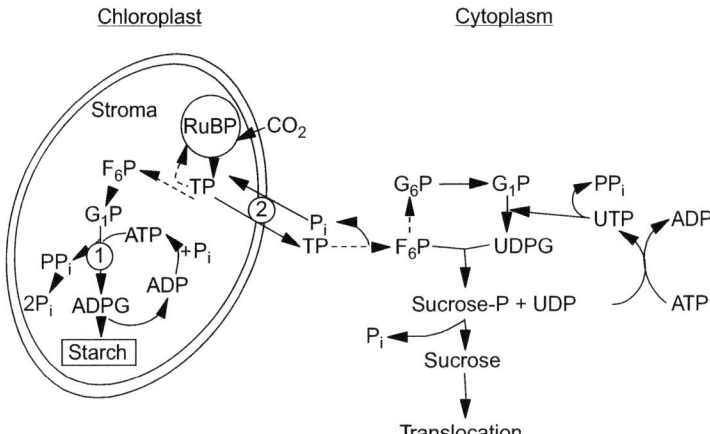

FIGURE 6.14 Involvement and regulatory role of P in starch synthesis and carbohydrate transport in a leaf cell. (1) ADP-glucose pyrophosphatase regulates the rate of starch synthesis and is inhibited by P_i and stimulated by 3-phosphoglycerate. (2) The phosphate translocator regulates the release of photosynthates from chloroplasts and is enhanced by P_i. *ADPG*, ADP-glucose; F_6P, fructose-6-P; G_1P, glucose-1-P; G_6P, glucose-6-P; *RuBP*, ribulose bisphosphate; *TP*, triose phosphate; *UDPG*, uridine disphosphate-glucose; *UTP*, uridine triphosphate. *Based on Walker (1980).*

Liberation of PPi takes place in all of the major biosynthetic pathways, for example, acylation of CoA in fatty acid synthesis, formation of APS in sulfate activation of starch in chloroplasts, and of sucrose in the cytosol (Fig. 6.14). Various enzymes can make use of PPi, including the UDP−glucose phosphorylase and the proton-pumping inorganic pyrophosphatase in the tonoplast (Chapter 2). The cellular concentrations of PPi are in the range of 100−200 nmol g^{-1} fresh wt, a similar range as that of ATP (Duff et al., 1989). In leaves, PPi concentrations are similar in the cytosol and stroma of chloroplasts and are kept stable during the light−dark cycle (Eberl et al., 1992).

In rapidly metabolizing cells, energy-rich phosphates are characterized by high rates of turnover. From pulse-labeling experiments with ^{32}P, the turnover rates of various P compounds can be calculated (Table 6.6). Obviously, a small amount of ATP satisfies the energy requirement of plant cells. For example, 1 g of rapidly metabolizing maize root tips synthesizes about 5 g ATP per day (Pradet and Raymond, 1983). The amounts of phospholipids and RNA are considerably greater, but these are also more stable, with a relatively slow rate of synthesis (Table 6.6).

Phosphorylation of enzyme proteins by ATP, GTP, or ADP is another mechanism by which energy-rich phosphates can modulate enzyme activities:

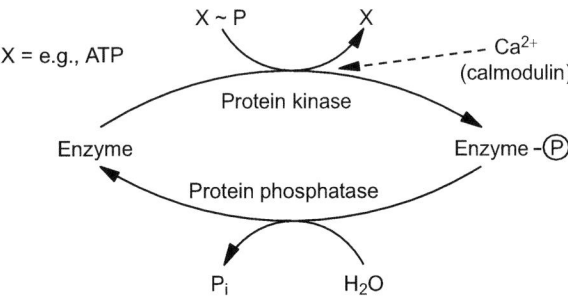

This regulatory phosphorylation is mediated by protein kinases and can result in activation, inactivation, and/or changes in the allosteric properties of the target protein. Phosphorylation is a reversible modification affecting protein activity, stability, interactions, and localization. Phosphorylation is catalyzed by a specific group of enzymes (kinases), and the reverse reaction of dephosphorylation is catalyzed by phosphatases (Xu et al., 2019). Protein phosphorylation is a key factor in signal transduction, for example, in phytochrome-mediated responses of plants (Shen et al., 2009). An example of this is the light-stimulated enhancement of nitrate assimilation in leaves (Fig. 6.6). PEP carboxylase is one of the key enzymes regulated by phosphorylation in both C_3 and C_4 plants. In C_4 and crassulacean acid metabolism (CAM) plants (Chapter 5), phosphorylation increases the activity of PEP carboxylase, and simultaneously the enzyme becomes less sensitive to negative feedback control by high malate concentrations (Budde and Chollet, 1988).

6.3.4 Compartmentation and regulatory role of inorganic phosphate

In many enzyme reactions, Pi is either a substrate or an end-product (e.g., ATP → ADP + Pi). Furthermore, Pi controls some key enzyme reactions. Compartmentation of Pi is therefore essential for the regulation of metabolic pathways in

TABLE 6.6 Turnover times and rates of synthesis of organic P fractions in *Spirodela*.

Phosphorus fraction	Concentration (nmol g^{-1} fresh wt)	Turnover (min)	Synthesis rate (nmol P g^{-1} fresh wt min^{-1})
ATP	170	0.5	340
Glucose-6-P	670	7	95
Phospholipids	2700	130	20
RNA	4900	2800	2
DNA	560	2800	0.2

Source: Based on Bieleski and Ferguson (1983).

the cytosol and chloroplasts. In fruit tissue of tomato, for example, Pi released from the vacuoles into the cytosol can stimulate phosphofructokinase activity (Woodrow and Rowan, 1979) as the key enzyme in the regulation of substrate flux in glycolysis. Thus, the release of Pi from vacuoles can initiate the respiratory burst during fruit ripening.

In vacuolated cells of vascular plants, the vacuole acts as a storage compartment for a "nonmetabolic pool" of P, and at a high P supply ~70%–95% of the total P is located in the vacuoles as Pi (Yang et al., 2017). By contrast, in leaves of P-deficient plants, most Pi is found in the cytosol and chloroplasts, that is, in the "metabolic pool" (Lauer et al., 1989). In leaves, the total P concentration may vary by a factor of 20 without strongly affecting photosynthesis because the Pi concentration in the cytosol is regulated in a narrow range (phosphate homeostasis) whereby the Pi in the vacuole acts as buffer (Mimura et al., 1990). The same is true for roots where the cytosolic Pi concentration is maintained at 6.0 mM (maize) and 4.2 mM (pea), also under P deficiency, unless the vacuolar pool is depleted (Lee et al., 1990). Under severe P deficiency, cytosolic Pi concentrations in leaves may drop from about 5 to less than 0.2 mM, whereas the concentrations of energy-rich phosphates drop to 20%–30% of the original values.

In leaves, photosynthesis and carbon partitioning in the light–dark cycle are strongly affected by the Pi concentrations in the stroma of chloroplasts and the compartmentation between chloroplasts and cytosol. In the light, for maximum photosynthesis, a Pi concentration in chloroplasts in the range of 2.0–2.5 mM is required, and photosynthesis is almost completely inhibited when the Pi concentration falls below 1.4–1.0 mM (Heber et al., 1989). Due to the large demand of Pi for phosphorylated intermediates of photosynthesis (Fig. 6.14), the Pi concentrations in leaves of P-deficient plants (i.e., without vacuolar buffer) may drop to 50% following dark-light transition (Sicher and Kremer, 1988). It should be noted, however, that leaf growth tends to be far more sensitive to a limited P supply than photosynthesis (Assuero et al., 2004; Rodríguez et al., 1998), and slow rates of photosynthesis in P-deficient plants may be due to feedback inhibition, rather than P limitation of photosynthesis (Pieters et al., 2001; Shi et al., 2020).

The role of Pi in carbon partitioning between chloroplasts and cytosol has been demonstrated with isolated chloroplasts (Heldt et al., 1977). An increase in external Pi concentration up to about 1 mM stimulates net photosynthesis but decreases incorporation of the fixed carbon into starch. At the Pi concentration of 5 mM in the stroma, starch synthesis is inhibited. The inhibition of starch synthesis by high concentrations of Pi is caused by two separate mechanisms in the chloroplasts. The key enzyme of starch synthesis in chloroplasts, ADP-glucose pyrophosphorylase [pathway (1), Fig. 6.14], is allosterically inhibited by Pi and stimulated by triose phosphates. The ratio of Pi to triose phosphates, therefore, strongly influences the rate of starch synthesis in chloroplasts (Portis, 1982); at high ratios, the enzyme is "switched off". The other mechanism regulated by Pi is the release of triose phosphates (glyceraldehyde-4-phosphate and dihydroxyacetone phosphate), the main products of CO_2 fixation leaving the chloroplasts. This release is mediated by a phosphate transporter, located in the inner membrane of the chloroplast envelope [pathway (2), Fig. 6.14], facilitating the exchange Pi ↔ triose phosphate (Heldt et al., 1991). In C_4 and CAM plants, this translocator also transports PEP. The net uptake of Pi into the chloroplasts via the phosphate translocator regulates the release of photosynthates from the chloroplast. High Pi concentrations in the cytosol, therefore, deplete the stroma of triose phosphates that serve both as substrates for, and activators of, starch synthesis. Thus, inhibition of starch synthesis by high Pi concentrations is also the result of substrate depletion.

In guard cells of pea, the phosphate transporter in the chloroplast envelope [pathway 2, Fig. 6.14] transports glucose-6-phosphate, similarly as in amyloplasts in storage cells (see below). This mechanism enables guard cells to synthesize starch, although they lack fructose-1,6-bisphosphate synthase, the enzyme required for $C_3 \rightarrow C_6$ biosynthesis (Overlach et al., 1993).

The CO_2 fixation in the Calvin cycle is a process in which five-sixths of the carboxylation products are required in the stroma to regenerate the CO_2 acceptor ribulose bisphosphate (RuBP). Excessive export of triose phosphates induced by high Pi concentrations in the cytosol leads to the depletion of these metabolites required for the regeneration of RuBP (Fig. 6.14). In isolated chloroplasts, high external Pi concentrations, therefore, inhibit CO_2 fixation (Flügge et al., 1980). In intact plants, low Pi concentrations in the cytosol and chloroplasts are common, for example, under severe P deficiency (Lauer et al., 1989). Accumulation of large amounts of starch in the chloroplasts is a typical feature of P deficiency (Table 6.7), and this starch is not completely mobilized at night (Qiu and Israel, 1992) or during reproductive growth (Portis, 1982). Accumulation of starch under P deficiency indicates that leaf initiation and expansion are more sensitive to P shortage than photosynthesis (Kavanová et al., 2006; Pieters et al., 2001).

Accumulation of starch in chloroplasts of P-deficient plants is due to (1) the low Pi concentrations in the cytosol, and thus low export of trioses from the chloroplast, and (2) to the increase in activity of ADPG-pyrophosphorylase (Rao et al., 1990) as a result of low Pi concentrations in the stroma. The shift toward utilizing triose phosphates for starch synthesis may even reduce Calvin cycle activity and CO_2 fixation by limiting the regeneration of RuBP (Fredeen et al., 1990). The accumulation of starch and sugars in leaves of P-deficient plants can also result from lower export due to shortage of ATP for sucrose-proton cotransport in phloem loading (see Chapter 5) and lower demand at the sink sites (Rao et al., 1990). However, under P deficiency shoot growth is suppressed more than photosynthesis (Plénet et al., 2000), especially when plants are grown under low-light conditions (De Groot et al., 2003). The finely tuned Pi homeostasis in the cytosol and chloroplasts is one reason for this, and a higher activity of various enzymes of carbohydrate metabolism (and thus turnover of Pi) may be another (Rao et al., 1990).

In principle, similar regulation of starch synthesis takes place in amyloplasts of storage cells. ADP-glucose pyrophosphorylase is the key enzyme in the regulation of starch synthesis in potato tubers (Zeeman et al., 2010) and wheat grains (Hou et al., 2017). When isolated from these storage tissues, the enzyme is severely inhibited by Pi. By contrast, starch accumulation in the endosperm of wheat grains is not affected by high Pi levels (Rijven and Gifford, 1983), possibly reflecting their large capacity for Pi sequestration (Bilal et al., 2018).

In storage cells, the transport of phosphorylated trioses from the cytosol into the amyloplasts proceeds by strict counter-transport with Pi; however, the P transporter also accepts glucose-6-phosphate and releases Pi in a C_6-Pi shuttle (Heldt et al., 1991).

TABLE 6.7 Growth parameters and concentrations of P and carbohydrates in soybean at high or low P supply.

Parameter		High P	Low P
Leaf area (m^2)		0.121	0.018
Number of primary trifoliates		7	4
Shoot/root ratio		4.2	1.0
Phosphorus concentration (mg g^{-1} dry wt)			
Leaf	Inorganic P	4.4	0.3
	Organic P	2.4	0.6
Total P	Stems and petioles	5.8	1.1
	Roots	10.5	1.3
Total root P/total shoot P		0.5	1.6
Carbohydrates in leaves (g m^{-2} leaf)	Starch	0.4	13
	Sucrose	0.7	0.2
Carbohydrates in roots (mg g^{-1} fresh wt)	Starch	23	160
	Sucrose	16	177

Source: Based on Fredeen et al. (1989).

6.3.5 Phosphorus fractions and the role of phytate

When P supply is increased from the deficiency to the sufficiency range, the major P fractions in vegetative organs also increase (Veneklaas et al., 2012), as shown for leaves in Table 6.8. With further increase in supply, only Pi as the major storage form of P in highly vacuolated tissues increases (Shane et al., 2004b). However, plants may also store P in two other major forms, namely, phytate (Lott et al., 2000) and inorganic polyphosphates (Seufferheld and Curzi, 2010).

The storage of phosphate in cells as inorganic polyphosphates is widespread among bacteria, fungi, and green algae, but it also occurs in vascular plants (Seufferheld and Curzi, 2010). Polyphosphates synthesized by plants are linear polymers of Pi (>500 molecules) with pyrophosphate linkages energetically equivalent to ATP. Polyphosphates may therefore function as energy storage compounds and as compounds controlling the Pi level in the metabolic pool of the cells. Polyphosphate formation in the hyphae of mycorrhizal fungi plays a key role in P nutrition of mycorrhizal plants (Kuga et al., 2008). Hyphae take up Pi from the soil solution and synthesize polyphosphates; these act as a transient storage pool of P in the hyphae, and are subsequently transported as polyphosphates toward the host roots (Nehls and Plassard, 2018; Chapter 15).

Phytate is the typical storage form of P in seeds (Ahn et al., 2010). Phytate is the salt of phytic acid (myo-inositol hexakisphosphate) and is synthesized from the cyclic alcohol myo-inositol by esterification of the hydroxyl groups with phosphoryl groups (Josefsen et al., 2007):

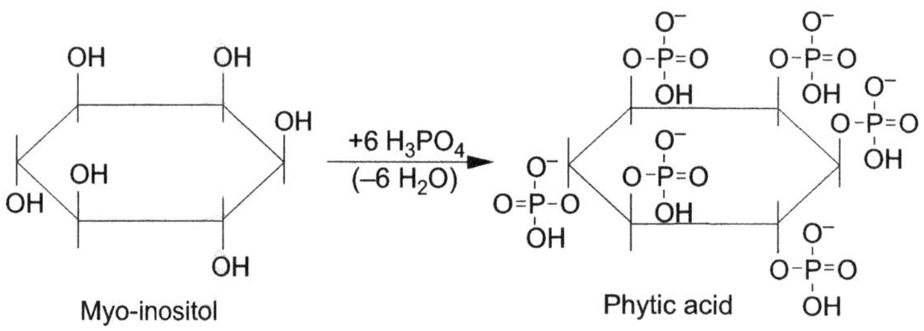

The Ca-Mg salt of phytic acid is sparingly soluble. Phytic acid also has a high affinity for Zn and Fe (Wang et al., 2008). The proportions of K, Mg, and Ca associated with phytic acid vary considerably among plant species and genotypes, and even between different tissues of a seed (Cong et al., 2020). Phytate-P makes up $\sim 50\%$ of the total P in legume and 60%–70% in cereal seeds, and about 86% in wheat bran. In cereals and legumes, phytate is deposited in electron-dense globoid crystals inside membrane-bound intracellular protein bodies, in cereal seeds mainly in the aleurone layer, and in legumes in cotyledons and embryo axes. In seeds, phytates are the main storage form of K and Mg, and in some instances also of Ca and Zn (Lott et al., 1985).

The K–Mg–Ca salt of phytic acid is also the major form of P in pollen (Loewus and Murthy, 2000), where it is deposited as discrete particles, decomposed by phytase during pollen germination (Baldi et al., 1987). Phytate also accumulates in other plant organs such as roots, tubers, stems, and leaves (Alkarawi and Zotz, 2014; Doolette and Smernik, 2016; Takagi et al., 2020).

The high affinity of phytic acid for Zn, Fe, and other metals may be important for metal binding and, therefore, detoxification in roots. In cortical root cells of Zn-tolerant cultivars of a range of crop species, up to 60% of the charges

TABLE 6.8 Phosphorus fractions in tobacco leaves at different P supply.

Phosphorus supply (mg L^{-1})	Leaf dry weight (g plant^{-1})	Phosphorus fraction (g P kg^{-1} leaf dry wt)			
		Lipid	Nucleic acid	Ester	Inorganic
2	0.8	0.32	0.74	0.36	0.33
6	1.1	0.83	1.3	0.9	0.83
8	1.1	0.89	1.3	1.0	1.2
20	1.1	0.91	1.4	1.1	3.4

Source: Based on Kakie (1969).

of phytic acid are occupied by Zn, whereas no Cd is bound by phytate (Van Steveninck et al., 1994). Phytic acid can also be a major component of soil organic P, but this is a different isomer from that stored in roots, tubers, seeds, and pollen; the origin of phytate in soil is unclear (Richardson et al., 2007).

In legumes and cereals, during the early stages of seed development, the concentration of phytate is low (Fig. 6.15), but it rises sharply during the period of rapid starch synthesis (Ogawa et al., 1979; Raboy and Dickinson, 1987). By contrast, the concentration of Pi during the early stages of seed development is low and declines further during rapid phytate formation.

Phytate is presumably involved in the regulation of starch synthesis during seed filling or tuber growth. The synthesis of phytate and a decrease in Pi concentration in the seeds are closely related (Fig. 6.15; Michael et al., 1980). In addition, with the onset of desiccation in seeds in the final stage of the filling period, phytic acid acts as a major cation trap that eliminates excessive cellular concentrations of K and Mg.

Some P is associated with the starch fraction and is incorporated into the starch grains. In cereals this is only a small proportion, but in potato tubers up to 40% of the total P may be incorporated in starch. Starch-bound P may reflect another type of compartmentation of Pi allowing control of its concentration at the sites of starch synthesis. It could also act as a source of P for sugar export from the amyloplasts during sprouting of tubers.

The function of phytate is to provide the germinating seedling with a source of P for synthesis of membrane lipids and nucleic acids. In agreement with this, digestion of the globoid crystals containing phytate is one of the earliest changes in the protein bodies of the cotyledons during germination (Lott and Vollmer, 1973). Decomposition of phytate, catalyzed by phytases, leads to a rapid decline in phytate-bound P. In germinating rice seeds (Table 6.9), within the first 24 h most of the P released from phytate is incorporated into phospholipids, indicating membrane synthesis, which is essential for compartmentation and thus for the regulation of metabolic processes within cells. An increase in Pi and phosphate-ester concentrations reflects the onset of enhanced respiration, phosphorylation, and related processes. The decomposition of phytate continues with time, and finally, the concentrations of DNA and RNA increase, indicating enhanced cell division and net protein synthesis. The rate of phytate decomposition is also controlled by Pi; high concentrations of Pi suppress the synthesis of phytase (Sartirana and Bianchetti, 1967). During decomposition of phytate, various inositol phosphates with a lower P content occur as intermediates, and some of them constitute a

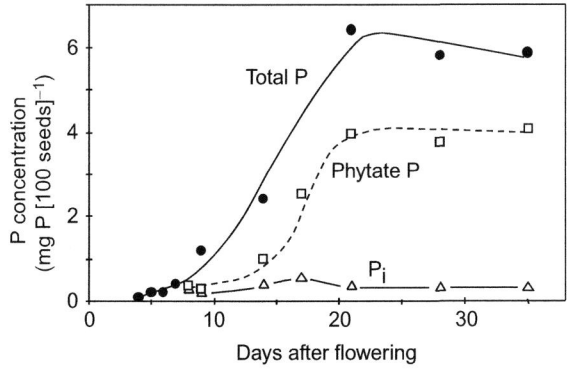

FIGURE 6.15 Total P, phytate-P, and Pi concentration in rice seeds during development. *Based on Ogawa et al. (1979).*

TABLE 6.9 Phosphorus fractions in rice seeds during germination.

Duration of germination (h)	Phosphorus fraction (mg P g^{-1} dry wt)				
	Phytate	Lipid	Inorganic	Ester	RNA + DNA
0	2.7	0.43	0.24	0.08	0.06
24	1.5	1.2	0.64	0.10	0.05
48	1.1	1.5	0.89	0.11	0.08
72	0.80	1.7	0.86	0.12	0.12

Source: From Mukherji et al. (1971).

significant proportion of the phospholipid fraction of membranes. In addition, inositol-1,4,5(tri)phosphate serves as a second messenger regulating Ca channels in membranes of plant cells (Isayenkov et al., 2010).

Phytates interfere with intestinal absorption of mineral elements, especially Zn, Fe, and Ca, thereby causing nutritional deficiencies in monogastric animals and humans (Chapter 9). Dietary phytate/Fe and phytate/Zn molar ratios are used to estimate Fe and Zn bioavailability and to identify dietary Fe and Zn requirements according to diet type (Gibson et al., 2018). In humans on cereal diets, Zn deficiency results from both the low Zn concentration in the grains and the consumption of phytate-rich unleavened whole-meal bread. This problem can be alleviated by Zn supplementation in the diet or by an increase in the Zn/phytate ratio in seeds and grains through the application of Zn fertilizers (Cakmak et al., 2010). Breeding for low phytate concentration is another alternative (Cong et al., 2020; Shi et al., 2007). In rice, low phytate concentration is associated with reduced grain yield and seed viability (Zhao et al., 2008a,b), but it should be possible to overcome this by the treatment of seed before sowing.

6.3.6 Phosphorus supply, plant growth, and plant composition

The P requirement for optimal growth is in the range of 3–5 mg g^{-1} plant dry matter during the vegetative stage of growth, but some plants that have evolved on severely P-impoverished soils contain an order of magnitude less P in their leaves, while exhibiting similar rates of photosynthesis (Guilherme Pereira et al., 2019; Sulpice et al., 2014). The probability of P toxicity increases at concentrations greater than 10 mg g^{-1} dry matter, which is rare because plants downregulate the Pi transporters involved in net P uptake from the root environment when supplied with more P than required for optimum growth (Dong et al., 1999). However, many species from severely nutrient-impoverished soils in Australia and South Africa have a very low capacity to downregulate net P uptake and do show P toxicity symptoms when fertilized with P (Shane et al., 2004c). Some tropical food legumes are also rather sensitive to P toxicity that may occur at P concentrations in the shoot dry matter of 3–4 mg g^{-1} in pigeon pea and 6–7 mg g^{-1} in black gram (Bell et al., 1990). At the other end of the spectrum, *Ptilotus* species (Amaranthaceae), fast-growing nonmycorrhizal Australian native herbs and shrubs, accumulate P to approximately 40 mg g^{-1} shoot dry weight, without signs of P toxicity (Hammer et al., 2020).

In P-deficient plants, reduction in leaf expansion (Fredeen et al., 1989) and number of leaves (Lynch et al., 1991) are the most striking effects (Table 6.7). The average length of the cell division zone is decreased in leaves of P-deficient maize, and both cell production and cell division rates are reduced (Assuero et al., 2004). Leaf expansion is strongly related to the expansion of epidermal cells, and this process might be impaired in P-deficient plants because of diminished root hydraulic conductivity due to a decreased expression of genes encoding aquaporins (Clarkson et al., 2000). The chlorophyll concentration tends to increase under P deficiency, and P-deficient leaves are dark green because leaf expansion is inhibited more strongly than chlorophyll formation (Chiera et al., 2002).

Compared with shoot growth, root growth is less inhibited under P deficiency, leading to a typical decrease in shoot/root ratio (Table 6.7). This decrease in shoot/root ratio is correlated with an increase in partitioning of carbohydrates toward the roots of P-deficient plants (Table 6.7). Phosphorus-starvation responses in plants are mediated via sugar signaling (Karthikeyan et al., 2007). Signaling of the shoot P status also involves specific microRNA molecules (Sunkar et al., 2012). Under P starvation, the elongation rate of individual root cells and of the roots may be enhanced (Anuradha and Narayanan, 1991; Mollier and Pellerin, 1999).

In certain plant species, P deficiency–induced formation of "cluster" or "dauciform" roots is another P-starvation response (Lambers et al., 2018). Cluster roots play an important role when a large fraction of the soil P is poorly available because of a high or low pH. This accounts for the presence of cluster roots in species on young volcanic soils that contain large amounts of P but with low availability due to high concentrations of oxides and hydroxides of Fe and Al and low pH (Lambers et al., 2011). Examples of species on such volcanic soils include *Embothrium coccineum* in Chile (Zúñiga-Feest et al., 2010). Due to the release of carboxylates in an "exudative burst" (Shane et al., 2004a; Watt and Evans, 1999), cluster roots efficiently "mine" P that is not readily available for "scavenging" mycorrhizas (Lambers et al., 2008).

Despite a plethora of adaptive responses in plants (Lambers et al., 2006) triggered by intricate P-starvation signaling pathways (Rolland et al., 2006), shoot growth rate, and formation of reproductive organs are inhibited under P limitation. Flower initiation is delayed (Rossiter, 1978), the number of flowers is decreased (Bould and Parfitt, 1973), and seed formation is restricted (Barry and Miller, 1989). Premature senescence of leaves is another factor limiting seed yield in P-deficient plants.

Challenges for the future, as rock phosphate reserves are being depleted (Cordell and White, 2014; Johnston et al., 2014), include the development of crop and pasture systems that require less P while maintaining productivity. There may well be lessons to be learned from native species that evolved in severely P-impoverished landscapes, but this remains to be explored (Cong et al., 2020).

6.4 Magnesium

6.4.1 General

Magnesium has been suggested as the most versatile cationic nutrient in biological systems because it interacts with a large number of diverse biomolecules, having various critical cellular functions (Weston, 2008). More than 350 enzymes have a specific requirement for Mg (Weston, 2008). The ionic size of Mg is substantially smaller (0.065 nm), and the hydrated radius of Mg is substantially larger (0.476 nm), than those of K^+ and Ca^{2+}. In recent years, various Mg^{2+} transporters in higher plants have been identified, controlling Mg concentrations in different cellular compartments such as the chloroplast and cytosol (Kobayashi and Tanoi, 2015), and affecting critical cellular functions such as activity of enzymes and plant stress tolerance (Chen et al., 2017; Li et al., 2020). The uptake of Mg^{2+} can be depressed strongly by other cations, such as K^+, NH_4^+ (Kurvits and Kirkby, 1980), Ca^{2+}, and Mn^{2+} (Heenan and Campbell, 1981), as well as by H^+ (i.e., by low pH). Potassium-induced Mg deficiency is becoming a common observation in crop production (Senbayram et al., 2015).

The functions of Mg in plants are mainly related to its capacity to interact with strongly nucleophilic ligands. The interactions of Mg^{2+} with proteins fall into two groups: (1) Mg^{2+} binds directly to a protein or enzyme, such as central atom of the chlorophyll molecule or as bridging element for the aggregation of ribosomes; and (2) Mg^{2+} may bind the substrate of an enzyme, thus increasing the efficiency of the catalytic reaction, such as in Mg–ATP phosphorylation and the reaction between Mg-isocitrate and isocitrate lyase (Cowan, 2002). Magnesium forms ternary complexes with enzymes in which bridging cations are required for establishing a precise geometry between enzyme and substrate, for example, in RuBP carboxylase (Pierce, 1986). Magnesium generally binds weakly to proteins and enzymes in the cytosol; hence, their activity depends on the strict control of the cytosolic free Mg^{2+} concentration at around 0.5 mM. Many of the cellular functions of Mg are related to its binding to ATP. The majority of ATP in biological systems exists in the form of Mg–ATP (Maeshima et al., 2018; Weston, 2008). Given that ATP represents the principal source of energy, biological processes depending on ATP show a high dependency on adequate Mg nutrition. A substantial proportion of the total Mg^{2+} in the cell is also involved in the regulation of cellular pH and maintenance of the cation–anion balance.

6.4.2 Binding form, compartmentation, and homeostasis

A major function of Mg in green leaves is as the central atom of the chlorophyll molecule (Chapter 5). In leaves of subterranean clover, the proportion of total Mg bound to chlorophyll ranges from 6% in plants with a high Mg supply to 35% in Mg-deficient plants (Scott and Robson, 1990).

There is a large variation in free Mg^{2+} concentrations among the cell organelles, ranging from 0.2–0.4 mM in cytosol to 15–25 mM in chloroplasts, and most of the metabolically inactive Mg exists in vacuoles at up to 80 mM concentration (Hermans et al., 2013). The large amounts of Mg in the chloroplasts suggest an effective Mg transport system regulating and controlling Mg concentrations in chloroplasts. A specific Mg^{2+} transporter, OsMGT3, has been identified in the chloroplast envelope of rice leaf mesophyll cells, which maintains Mg influx and modulates changes in Mg concentrations of the chloroplasts (Li et al., 2020). As with Pi, the vacuole is the main storage pool required for the maintenance of Mg^{2+} homeostasis in the "metabolic pool". In needles of Mg-sufficient Norway spruce, Mg^{2+} concentrations in the vacuole were 13–17 mM in mesophyll cells and 16–120 mM in endodermis cells. These high concentrations function as a buffer in maintaining Mg^{2+} homeostasis in other cells throughout the season (Stelzer et al., 1990). In addition, vacuolar Mg^{2+} is also important for cation–anion balance and turgor regulation of cells.

6.4.3 Chlorophyll and protein synthesis

Chlorophyll and heme synthesis share a common pathway up to the level of protoporphyrin IX. The first step of chlorophyll biosynthesis (insertion of Mg^{2+} into the porphyrin structure) is catalyzed by Mg chelatase (Walker and Weinstein, 1991). Activation of this enzyme also requires ATP and, thus, additional Mg (Kobayashi et al., 2008). The release of Mg during chlorophyll decomposition requires two steps: chlorophyllase hydrolyzing chlorophyll to

TABLE 6.10 Incorporation of ^{14}C (leucine) into the protein fraction of isolated wheat chloroplasts at different Mg supplies.

Mg concentration (mM)	^{14}C incorporation (cpm mg^{-1} chlorophyll)	Relative value
0	412	12
0.5	688	20
5.0	3550	100

Source: Based on Bamji and Jagendorf (1966).

TABLE 6.11 Magnesium deficiency–induced changes in plastid pigments and leaf dry matter in oilseed rape.

Treatment	Chlorophyll (a + b)	Carotenoids	Leaf dry matter (g kg^{-1})
	(mg g^{-1} fresh wt)		
Control	2.3	0.21	136
Mg-deficient	1.3	0.11	177

Source: Based on Baszynski et al. (1980).

chlorophyllide and phytol (Tsuchiya et al., 1999) and Mg-dechelatase yielding Mg^{2+} and pheophytin (Ougham et al., 2008; Schelbert et al., 2009).

Magnesium has an essential function as a bridging element for the aggregation of ribosome subunits (Cammarano et al., 1972), a process that is necessary for protein synthesis. Under Mg deficiency, or in the presence of high concentrations of K$^+$ (Sperrazza and Spremulli, 1983), the subunits dissociate and protein synthesis ceases. Magnesium is known also for its role in folding and structural stability of RNA molecules, and thus Mg is required for biosynthesis and biological functions of RNA (Sreedhara and Cowan, 2002; Zheng et al., 2015a). The requirement for Mg in protein synthesis was demonstrated in isolated chloroplasts (Bamji and Jagendorf, 1966; Table 6.10). As Mg^{2+} readily permeates the chloroplast envelope, possibly via Mg transporters such as the envelope-localized MGT10 (Sun et al., 2017) or OsMGT3 (Li et al., 2020), a concentration of at least 0.25–0.40 mM Mg^{2+} is required in the cytosol to prevent net efflux of Mg^{2+} from the chloroplast and, thus, to maintain protein synthesis (Deshaies et al., 1984).

In leaf cells at least 25% of the total protein is localized in chloroplasts. This explains why a deficiency of Mg particularly affects the size, structure, and function of chloroplasts, including electron transfer in photosystem II (McSwain et al., 1976). In Mg-deficient plants, Mg transport from mature to young leaves is enhanced and, thus, visual deficiency symptoms typically appear on mature leaves, indicated by enhanced rates of protein degradation, including structural proteins of the thylakoids. The breakdown of the thylakoids also explains why in Mg-deficient plants, the other plastid pigments are often similarly affected as chlorophyll (Baszynski et al., 1980; Table 6.11). Regardless of this decline in chloroplast pigments, starch accumulates in Mg-deficient chloroplasts, which may explain the increase in dry matter of Mg-deficient leaves (Scott and Robson, 1990; Table 6.11). Impaired export of photosynthates is another factor leading to enhanced degradation of chlorophyll in Mg-deficient source leaves.

6.4.4 Enzyme activation, phosphorylation, and photosynthesis

As stated earlier, more than 350 enzymes are Mg-dependent (Weston, 2008), for example, glutathione synthase (Section 6.2) or PEP carboxylase. Regarding PEP carboxylase, the substrate PEP is bound in greater quantities and more tightly in the presence of Mg (Wedding and Black, 1988). Most of the Mg-dependent reactions involve the transfer of phosphate (e.g., phosphatases and ATPases) or carboxyl group (e.g., carboxylase). In these reactions, Mg^{2+} is preferentially bound to N bases and phosphoryl groups, for example, in ATP:

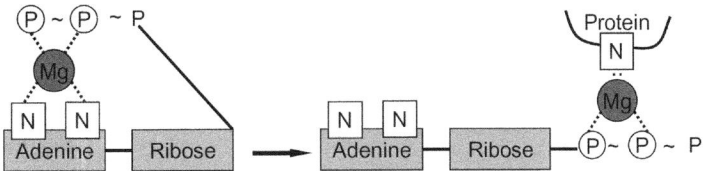

The substrates for ATPases and inorganic PP$_i$ases (Rea and Sanders, 1987) are Mg–ATP, rather than free ATP, and Mg-PPi, rather than free PPi, respectively. The Mg–ATP complex is stable above pH 6, and this complex can be utilized by the active sites of ATPases for the transfer of the energy-rich phosphoryl group (Balke and Hodges, 1975). An example of the Mg^{2+} requirement of membrane-bound ATPases is shown in Fig. 6.16. In meristematic cells of Mg-sufficient roots about 90% of the cytoplasmic ATP is complexed with Mg, and the concentration of free Mg^{2+} is only about 0.4 mM as compared with total Mg concentrations of 3.9 mM in the tissue (Yazaki et al., 1988).

The synthesis of ATP (phosphorylation: ADP + Pi → ATP) has an absolute requirement for Mg^{2+} as a bridging component between ADP and the enzyme. As shown in Table 6.12, ATP synthesis in isolated chloroplasts (photophosphorylation, see Chapter 5) is increased by external supply of Mg^{2+} and inhibited by Ca^{2+}. Hence, a low Ca^{2+} concentration has to be maintained within the chloroplasts at the sites of photophosphorylation (Section 6.5).

Another key reaction of Mg is the modulation of RuBP carboxylase in the stroma of chloroplasts (Pierce, 1986). The activity of this enzyme is highly dependent on both Mg^{2+} and pH (Fig. 6.17A). There is a tight correlation between the daily fluctuations in Mg^{2+} concentration and Rubisco activity. The specific Mg^{2+} transporter, OsMGT3, in the chloroplast envelopes of rice was found to influence the Rubisco activity by controlling daily changes in chloroplast Mg^{2+} concentrations (Li et al., 2020). Binding of Mg to the enzyme increases both its affinity (K_m) for the substrate CO_2 and the turnover rate V_{max} (Sugiyama et al., 1968). Magnesium also shifts the pH optimum of the reaction toward the physiological range (around 7). In chloroplasts, the light-triggered activation of RuBP carboxylase results in increases in pH and Mg^{2+} concentrations in the stroma. As shown in Fig. 6.17B, upon illumination, protons are pumped from the stroma into the interthylakoid space, creating the proton gradient required for ATP synthesis (Kramer et al., 2003). The light-induced transport of protons from the stroma is counterbalanced by transport of Mg^{2+} (and H^+) from the interthylakoid space into the stroma that becomes more alkaline (Oja et al., 1986). In wheat leaf chloroplasts, stroma pH may increase from about 7.6 in the dark to about 8.0 in the light (Heineke and Heldt, 1988). Using an Mg-sensitive fluorescent indicator, free Mg^{2+} concentrations of 0.5 and 2.0 mM have been measured in the stroma of dark and illuminated spinach chloroplasts, respectively (Ishijima, 2003); generally confirming earlier measurements of ~ 2 mM in the dark to ~ 4 mM in the light (Portis and Heldt, 1976). Changes of this magnitude in both pH and Mg^{2+} concentration are sufficient to increase the activity of RuBP carboxylase and other stromal enzymes that depend on high Mg^{2+} concentrations and have a pH optimum above 6.

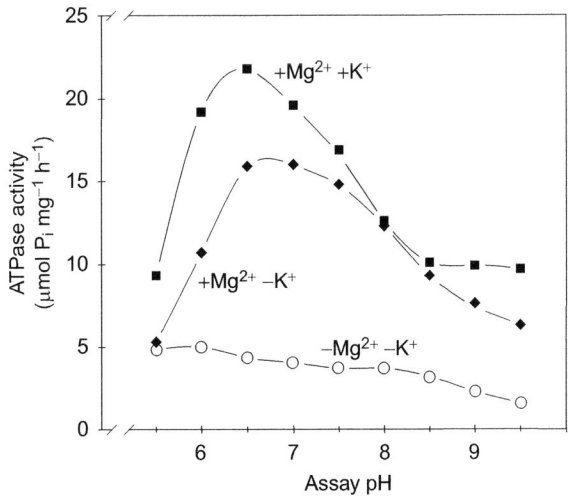

FIGURE 6.16 ATPase activity of the plasma-membrane protein of maize roots at different pH and Mg (3 mM) and K (50 mM) concentrations. *Based on Leonard and Hotchkiss (1976).*

TABLE 6.12 Photophosphorylation of isolated pea chloroplasts with or without Mg or Ca in an incubation medium containing ADP and P_i.

Cation in the incubation medium	Photophosphorylation rate (μmol ATP formed mg^{-1} chlorophyll h^{-1})
None	12
5 mM Mg	34
5 mM Ca	4.3

Source: Based on Lin and Nobel (1971).

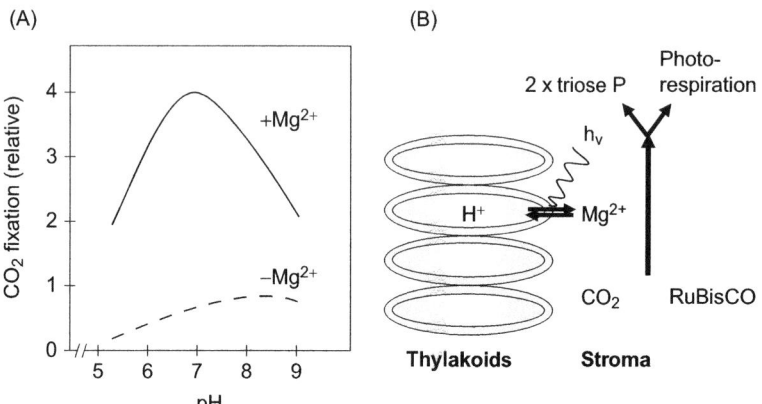

FIGURE 6.17 (A) Activation of ribulose-1,5-bisphosphate (RuBP) carboxylase from spinach leaves by Mg^{2+}. (B) Model for light-induced Mg^{2+} transport from the intra-thylakoid space into the stroma of chloroplasts with subsequent activation of the RuBP carboxylase/oxygenase (RuBisCO). *Modified from Sugiyama et al. (1969).*

One of the key enzymes with a high Mg requirement and high pH optimum is fructose-1,6-bisphosphatase that regulates, for example, assimilate partitioning between starch synthesis and export of triose phosphates in chloroplasts (Gerhardt et al., 1987). Another key enzyme with a high Mg requirement is GS (O'Neal and Joy, 1974). A light-induced increase in nitrite reduction and thus NH_4^+ production requires a simultaneous increase in the activity of enzymes such as GS that regulate ammonium assimilation in the chloroplasts.

6.4.5 Carbohydrate partitioning

The accumulation of nonstructural carbohydrates (starch, sugars) is a typical feature in source leaves of Mg-deficient plants (Fischer and Bussler, 1988; Cakmak et al., 1994a; Table 6.13) and can be detected well before any noticeable change in shoot growth, appearance of Mg deficiency symptoms, and inhibition of photosynthesis (Cakmak et al., 1994a; Hermans et al., 2004; Hermans and Verbruggen, 2005). Thus, inhibition of photosynthesis appears to be a response to increasing sugar concentrations serving as important signals in the regulation of plant metabolism and development (Wingler and Roitsch, 2008). Accumulation of starch is also found in P-deficient leaves, but the latter is associated with high chlorophyll concentrations in the leaves (Table 6.13).

Accumulation of carbohydrates in source leaves of Mg-deficient plants is the result of inhibited export from the leaves via the phloem (Cakmak et al., 1994b; Hermans et al., 2005), leading to lower carbohydrate export to, and thus lower concentrations in, sink organs such as pods and roots in common bean (Cakmak et al., 1994a,b; Fischer and Bussler, 1988; Fig. 6.18), root nodules of soybean plants (Peng et al., 2018) or growing sink leaves in sugar beet (Hermans et al., 2005). In young maize plants, the total amount of soluble sugars was almost equally distributed among the young, middle, and old leaves under sufficient Mg supply, whereas in Mg-deficient plants, there was almost five times more sugars in old (source) leaves when compared to the young (sink) leaves (Fig. 6.19; Mengutay et al., 2013). Accordingly, old leaves of low-Mg plants had higher specific dry weights (i.e., weight per leaf surface area) than those of the Mg-adequate plants, whereas the situation was reverse in the young leaves (Fig. 6.19). These results clearly highlight severe impairment in transport of photoassimilates from the source to the sink organs such as young shoot parts and roots under low Mg supply.

TABLE 6.13 Shoot and root dry weight and carbohydrate content (glucose equivalents) in primary leaves and roots of Mg- and P-deficient common bean.

Treatment	Dry weight (g plant^{-1})			Chlorophyll (mg g^{-1} dry wt)	Carbohydrates (mg g^{-1} dry wt)			
					Leaves		Roots	
	Shoots	Roots	S/R		Starch	Sugars	Starch	Sugars
Control	2.5	0.5	5.0	11	10	27	4	51
−Mg	1.5	0.15	10	4	77	166	4	11
−P	0.9	0.48	1.9	12	43	34	8	35

Source: Compiled from Cakmak et al. (1994a).

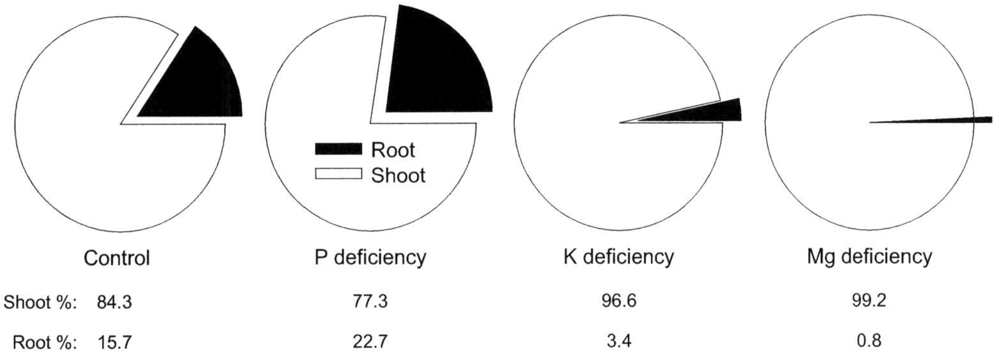

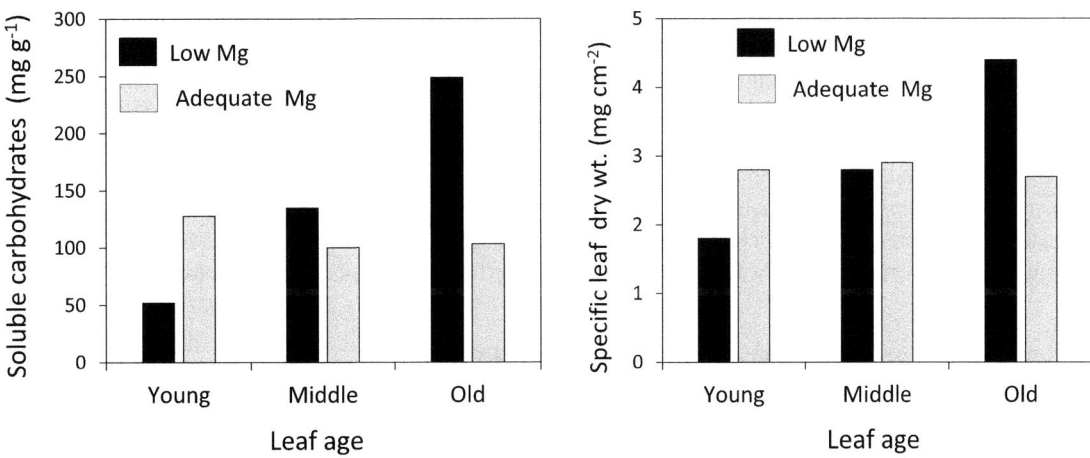

FIGURE 6.18 Relative distribution of carbohydrates (sum of reducing sugars, sucrose, and starch) between shoot and roots of 12-day-old common bean plants grown in nutrient solution with sufficient (control) or deficient supply of P, K, or Mg. *From Cakmak et al. (1994a), with permission from Oxford University Press.*

FIGURE 6.19 Concentration of soluble carbohydrates and specific dry weights of young, middle, and old leaves of 23-day-old maize plants grown in nutrient solutions with low (20 μM) or adequate (450 μM) Mg supply. *Adapted from Mengutay et al. (2013).*

Magnesium nutrition may have a significant effect on N accumulation in plants. For example, Peng et al. (2020) showed that adequate Mg nutrition improves root N uptake by upregulating expression of two nitrate transporter genes in soybean, possibly through enhanced sugar transport from the shoot to the roots.

In soybean plants, when compared to low Mg supply, adequate Mg nutrition significantly promoted nodulation and N$_2$ fixation capacity through enhanced carbohydrate transport from the shoot to the roots (Peng et al., 2018). Similarly, Mg deficiency

TABLE 6.14 Chlorophyll and antioxidant concentrations and activity of oxygen radical and H_2O_2-scavenging enzymes in primary leaves of common bean grown at low or high Mg supply.

Mg supply (μM)	Chlorophyll (mg g^{-1} dry wt)	Ascorbate (μmol g^{-1} fresh wt)	Soluble thiols (nmol g^{-1} fresh wt)	Enzyme activities (relative values)		
				SOD	AsPox	GR
1000	11	0.9	0.6	100	100	100
20	5.3	6.2	2.3	229	752	310

AsPox, Ascorbate peroxidase; GR, glutathione reductase; SOD, superoxide dismutase.
Source: Based on Cakmak and Marschner (1992).

in wheat, impairs transport of carbohydrates to the developing seeds, thereby reducing seed yield by decreasing seed weight, whereas the number of seeds per spike and vegetative growth (i.e., straw yield) were less affected by Mg deficiency (Ceylan et al., 2016). Impairment of carbohydrate supply to the roots by Mg deficiency leads to strongly reduced root growth in young common bean plants (Table 6.13; Fig. 6.18), resulting in a higher shoot/root ratio with a low Mg supply as shown also in wheat, maize (Mengutay et al., 2013) and potato plants (Koch et al., 2020). This effect of Mg deficiency on root growth is similar to the effect of K deficiency (Cakmak et al., 1994a; Ma et al., 2013), but the opposite of what is observed under P deficiency (Fig. 6.18). By contrast, in sugar beet and arabidopsis, Mg deficiency impacted shoot growth more than root growth (Hermans et al., 2004; Hermans and Verbruggen, 2005).

Inhibition of phloem loading of sucrose in Mg-deficient source leaves is most likely the reason for the shift in carbohydrate partitioning. The key role of a proton-pumping ATPase for phloem loading of sucrose (proton-sucrose cotransport) is discussed in Chapter 5. It is known that proton-pumping ATPases require Mg—ATP for their catalytic activity (Apell et al., 2017). The reduced amounts of Mg—ATP because of low concentrations of Mg^{2+} in Mg-deficient leaf tissues likely represent a major reason for the impaired phloem loading of sucrose. This assumption is supported by the upregulation of the gene encoding a phloem companion-cell sucrose/H^+ symporter under Mg deficiency in sugar beet (Hermans et al., 2005; Verbruggen and Hermans, 2013) and potato (Koch et al., 2019).

Accumulation of photosynthates in leaves negatively impacts photosynthetic CO_2 assimilation and limits the use of absorbed light energy in CO_2 fixation. The absorbed light energy then results in enhanced generation of reactive oxygen species (ROS) such as superoxide radicals and hydrogen peroxide, leading to photooxidative damage in chloroplasts (Cakmak and Kirkby, 2008; Peng et al., 2019; Tränkner et al., 2018). In response to ROS formation, antioxidative protective mechanisms are significantly upregulated in Mg-deficient leaves, including increased ascorbate content and the activity of ROS-scavenging enzymes, especially under high-light intensity (Cakmak, 1994; Cakmak and Marschner, 1992; Tränkner et al., 2018; Table 6.14). Magnesium-deficient leaves and needles are, therefore, highly photosensitive, and symptoms of chlorosis and necrosis strongly increase at elevated light intensity (Cakmak and Kirkby, 2008; Marschner and Cakmak, 1989). A transcriptomic study of Mg starvation in arabidopsis not only confirmed the involvement of oxidative stress and the impairment of the photosynthetic apparatus but also revealed a dysfunction of the circadian clock and the triggering of ethylene signaling in response to Mg deficiency (Hermans et al., 2010a, 2010b; Verbruggen and Hermans, 2013). The physiological significance of these findings for the understanding of Mg functions in plants remains to be elucidated.

6.4.6 Magnesium supply, plant growth, and composition

Based on a meta-analysis study, the critical leaf Mg concentration for dry matter production usually ranges between 1 and 2 g kg^{-1} for many plant species, with a tendency to be higher for dicots (i.e., 2—3 g kg^{-1}) (Hauer-Jákli and Tränkner, 2019). Interveinal chlorosis of fully expanded leaves is the most obvious visible symptom of Mg deficiency. The development and severity of leaf Mg deficiency symptoms are induced markedly by high light intensity. This light effect, caused by photooxidative damage (see Section 6.4.5), is not related to a decrease in leaf Mg concentration, indicating that plants may have higher physiological demand for Mg when exposed to the high-light stress (Cakmak and Marschner, 1992; Marschner and Cakmak, 1989). Similarly, plants have a higher demand for Mg (i.e., higher critical leaf Mg concentration) when grown at elevated atmospheric CO_2 (Yilmaz et al., 2017).

With insufficient root supply, Mg is remobilized from mature leaves. For example, in perennials such as Norway spruce, concentrations of Mg and chlorophyll as well as rate of photosynthesis in the older needles decrease in spring when the new needles develop (Lange et al., 1987).

There is increasing evidence that Mg deficiency is widespread in forest ecosystems in Central Europe (Liu and Hüttl, 1991), exacerbated by other stress factors, particularly air pollution (Schulze, 1989) and soil acidification (Marschner, 1992). Impairment of root growth (typical for declining Mg-deficient spruce stands, Roberts et al., 1989) has a considerable impact on the acquisition of not only Mg but also other nutrients and of water and, thus, on drought resistance and adaptation to nutrient-poor sites. Adequate Mg nutrition is also required for better tolerance to acidic soils (i.e., Al toxicity; Bose et al., 2011; Kong et al., 2020) as well as to heat stress conditions (Mengutay et al., 2013).

When Mg is deficient and the export of carbohydrates from source to sink sites is impaired, the starch concentrations in storage tissues, such as potato tubers (Werner, 1959), and the single-grain weight of cereals decrease (Beringer and Forster, 1981; Ceylan et al., 2016). In cereal grains Mg may play an additional role in the regulation of starch synthesis through its effect on the concentration of Pi and phytate. As discussed above, high Pi concentrations inhibit starch synthesis. In Mg-deficient wheat grains, twice as much P remains as Pi, and there is a correspondingly smaller proportion of phytate-P, compared with the grains of plants adequately supplied with Mg (Beringer and Forster, 1981). Low Mg in seeds was found also to be detrimental to seed germination and seedling development, most probably due to decreased concentration of carbohydrates (i.e., starch) in seeds (Ceylan et al., 2016; Zhang et al., 2020).

Increasing the Mg supply beyond growth-limiting levels results in additional Mg being stored mainly in the vacuoles, as a buffer for maintaining Mg^{2+} concentration in the "metabolic pool", and for charge compensation and osmoregulation in the vacuole. However, high Mg concentrations in the leaves (e.g., 15 g kg^{-1}) may be detrimental under drought stress. As the leaf water potential declines, the Mg^{2+} concentration in the "metabolic pool" increases from 3−5 mM up to 8−13 mM in sunflower. Such high concentrations, for example, in the stroma of chloroplasts, inhibit photophosphorylation and photosynthesis (Rao et al., 1987). In pea under drought stress, Mg^{2+} concentrations in the chloroplasts may increase up to 24 mM (Kaiser, 1987).

Generally, high Mg concentrations in edible parts of plants improve the nutritional quality for animals and humans (Chapter 9). For example, hypomagnesemia (grass tetany) is a serious disorder of ruminants associated with low Mg concentrations in feed and reduced efficiency of Mg resorption (Grunes et al., 1970; Schonewille, 2013). An increase in Mg concentrations of forage grasses by Mg fertilization is relatively easy to achieve. Breeding for high leaf Mg concentrations, for example, in Italian ryegrass, could be an alternative (Moseley and Baker, 1991).

In humans, insufficient Mg intake with the diet leading to an Mg-deficiency syndrome has attracted considerable attention (Tong and Rude, 2005; Rosanoff et al., 2015). Magnesium deficiency in human populations is increasing due to reduced dietary intake of Mg as well as increased Ca:Mg intake ratio from foods, especially in western countries (Rosanoff et al., 2015). Boosting Mg concentrations of food crops through fertilization represents a useful approach in improving dietary intake of Mg.

6.5 Calcium

6.5.1 General

Calcium is a relatively large divalent cation with a hydrated ionic radius of 0.412 nm and hydration energy of 1577 J mol^{-1}. In the apoplasm, a portion of Ca is firmly bound in structures, while another portion is exchangeable from the cell walls and the exterior surface of the plasma membrane. A large amount of Ca is often sequestered in vacuoles, whereas its concentration in the cytosol is low. The mobility of Ca in the symplasm and in the phloem is also low. Most of the functions of Ca as a structural or regulatory component of macromolecules are related to its capacity for coordination by providing stable but reversible molecular linkages. Calcium can be supplied at high concentrations and can reach more than 10% of the dry weight, for example, in mature leaves, without symptoms of toxicity or significant inhibition of plant growth. The functions of Ca in plants have been reviewed by White and Broadley (2003) and White (2015). Calcium has attracted much interest in plant physiology and molecular biology because of its role as a second messenger linking environmental and developmental stimuli to their physiological responses. This role is related to perturbations in cytosolic-free Ca^{2+} concentration.

6.5.2 Binding form and compartmentation

In contrast to other macronutrients, a large proportion of the total Ca in plant tissues is often located in cell walls and bound to R−COO^{-} groups of polygalacturonic acids (pectins) in the middle lamella in a readily exchangeable form (White et al., 2018). In eudicotyledons, such as sugar beet, which have a large cation exchange capacity, and particularly when the Ca supply is low, up to 70%−99% of the total Ca in leaves can be present in the cell walls (White et al., 2018).

The Ca requirements of commelinid (grasslike) monocotyledons are less than other plant species because their cell walls contain smaller concentrations of pectate and have lower cation exchange capacity (Fig. 6.20; White and Broadley, 2003; White and Holland, 2018; White et al., 2018). The minimal intracellular Ca concentration required by a plant cell is about 0.82 mg g^{-1} dry wt (White et al., 2018).

When Ca supply is increased, excess Ca is generally accumulated in the vacuole. Three distinct physiological types of plants regarding Ca nutrition exist: calciotrophes, oxalate plants, and potassium plants (Fig. 6.21; Kinzel, 1982; White, 2005). Calciotrophes, such as *Sedum album*, contain high concentrations of water-soluble Ca complexes in their vacuoles, and their accumulation of Ca is stimulated greatly by increasing Ca supply. The oxalate plants are divided into species whose vacuoles contain either soluble oxalate, such as *Oxalis acetosella*, or Ca oxalate crystals, such as *Silene inflata*. Increasing Ca supply increases Ca accumulation in plants that precipitate Ca oxalate, but not in plants containing soluble oxalate. Potassium plants, such as *Carex pendula*, contain little mineralized or water-soluble Ca and maintain a high tissue K:Ca ratio. Calcium can also be precipitated as Ca carbonate or Ca sulfate in vacuoles (Bauer et al., 2011; He et al., 2014; Palacio et al., 2014) and as Ca oxalate or Ca carbonate in the apoplasm (Kinzel, 1989).

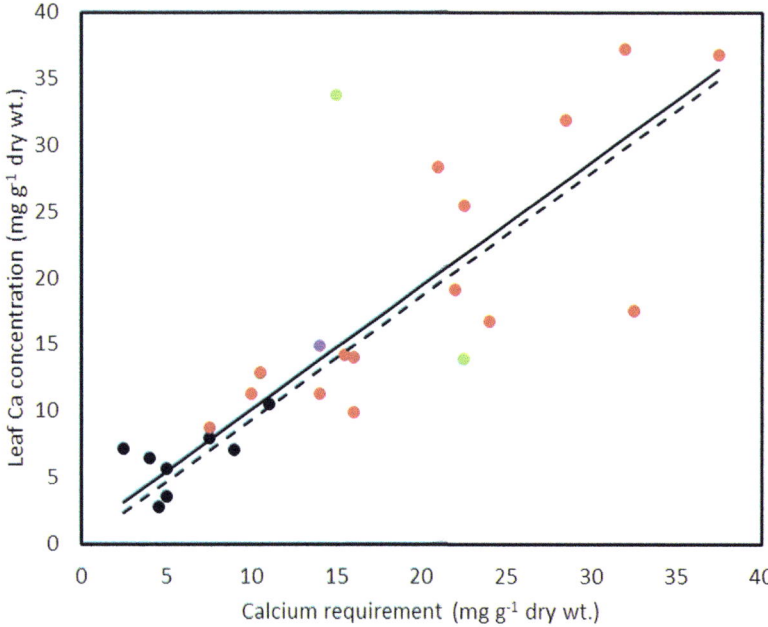

FIGURE 6.20 Relationships between the total (*solid line*) and estimated cell wall (*broken line*) Ca concentrations in leaves of magnoliid (purple), commelinid monocotyledon (black), noncommelinid monocotyledon (green), and eudicotyledon (red) species grown hydroponically and the Ca requirements of these species. *Based on data from White and Holland (2018).*

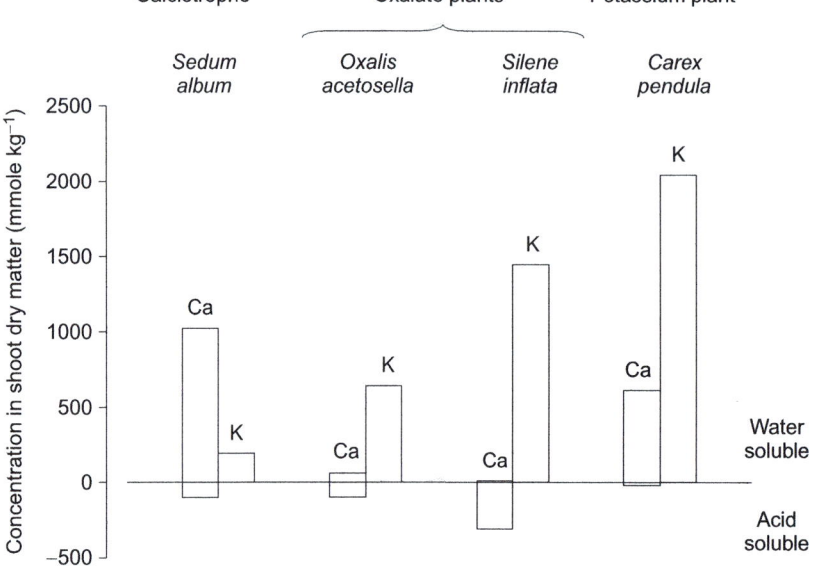

FIGURE 6.21 Physiological plant types regarding Ca nutrition. *Adapted from White (2005).*

The shape and distribution of Ca oxalate crystals differs among plant species and has proven useful as a taxonomic parameter (Prychid and Rudall, 1999; Franceschi and Nakata, 2005).

Calcium is distributed heterogeneously throughout the plant and within plant cells (Conn and Gilliham, 2010; Hayes et al., 2019; Stael et al., 2012; White, 2015; White and Broadley, 2003, 2009; White et al., 2018). Root tissues generally have lower Ca concentrations than stems and leaves; in addition, Ca accumulates at the end of the transpiration stream and in older leaves, and phloem-fed tissues generally have the lowest Ca concentrations. Calcium is partitioned to specific cell types within plant organs to prevent precipitation with P_i (Conn and Gilliham, 2010; Hayes et al., 2019). High Ca concentrations are typically found in the middle lamella of the cell wall, at the exterior surface of the plasma membrane, in the ER, and in the vacuole. Most of the water-soluble Ca in plant tissues is in the vacuoles, accompanied by organic anions (e.g., malate) or inorganic anions (e.g., nitrate, chloride). The vacuolar Ca concentration in cells that do not contain precipitated Ca salts generally lies between 2 and 20 mM. The Ca concentration in the ER, where Ca is associated with Ca^{2+}-binding proteins, is about 2 mM, and the Ca concentrations in mitochondria and nuclei are also about 2 mM (Stael et al., 2012). In contrast to the cell wall and organelles, the concentration of total Ca in the cytosol is low (0.1–1.0 mM), and free Ca^{2+} is buffered at 0.1–0.2 μM by Ca^{2+}-binding proteins and active Ca^{2+} efflux to the apoplasm, vacuole and ER (White and Broadley, 2003; McAinsh and Pittman, 2009; Dodd et al., 2010; White, 2015; Kudla et al., 2018). Such low Ca^{2+} concentrations are essential for various reasons, including (1) preventing P_i precipitation, (2) avoiding competition with Mg^{2+}, and (3) as a prerequisite for the function of Ca^{2+} as a second messenger. The major transporters catalyzing Ca^{2+} efflux from the cytosol to the apoplast and ER are Ca^{2+}-ATPases (Fig. 6.22).

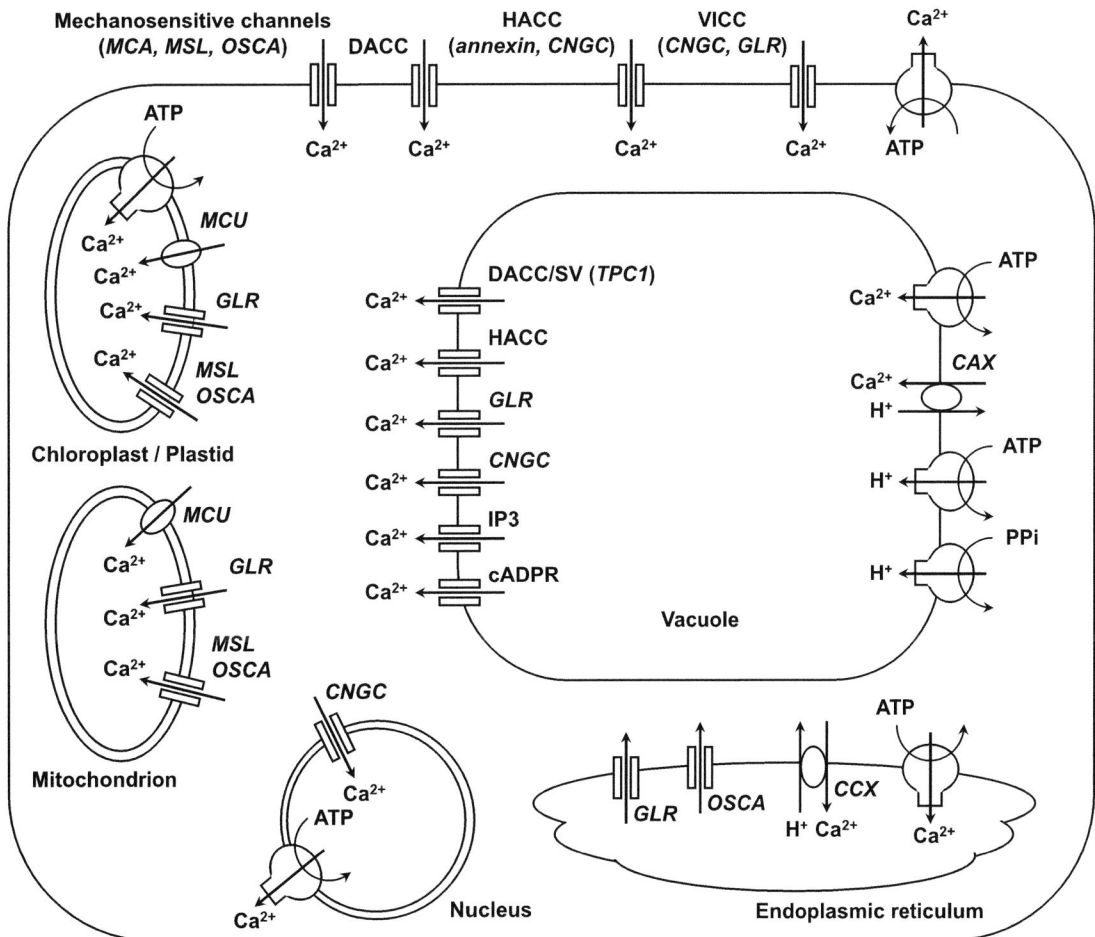

FIGURE 6.22 Calcium transport processes in plant cells. *cADPR*, Cyclic ADP-ribose regulated channel; *CAX*, cation exchanger; *CCX*, cation/Ca^{2+} exchanger; *CNGC*, cyclic nucleotide–gated channel; *DACC*, depolarization-activated Ca^{2+} channel; *GLR*, glutamate receptor channel; *HACC*, hyperpolarization-activated Ca^{2+} channel; *IP3*, inositol-1,4,5-triphosphate-activated channel; *MCA*, *mid1*-complementing activity (Ca^{2+}-permeable mechanosensitive channel); *MCU*, mitochondrial Ca^{2+} uniporter; *MSL*, mechanosensitive channel of small conductance (MscS)-like; *OSCA*, osmolality-sensitive Ca^{2+} channel; *PPi*, pyrophosphate; *SV*, slow vacuolar channel; *TPC1*, two-pore Ca^{2+} channel protein 1; *VICC*, voltage-insensitive Ca^{2+} channel. *Modified from White and Broadley (2003).*

At the tonoplast, both Ca^{2+}-ATPases and Ca^{2+}/H^+ antiporters catalyze Ca^{2+} efflux from the cytosol to the vacuole. The latter is energized by the proton electrochemical gradient generated by tonoplast H^+-ATPase and H^+-PP$_i$ase activities. Chloroplasts can also have high concentrations of Ca (6.5–15 mM total Ca, mostly bound to thylakoid membranes), but in the stroma the resting concentration of free Ca^{2+} is only about 150 nM (Stael et al., 2012; Navazio et al., 2020). Similarly, although mitochondria can contain much bound Ca, the concentration of free Ca^{2+} in the matrix is only about 200 nM (Stael et al., 2012; Costa et al., 2018). A Ca^{2+}-ATPase catalyzes Ca^{2+} uptake by plastids (Navazio et al., 2020), whereas calcium uniporters (mitochondrial calcium uniporter, MCU), glutamate receptor (GLR) channels, and osmolality-sensitive calcium (OSCA) channels are also present in both chloroplasts and mitochondria (Fig. 6.22; Costa et al., 2018; Navazio et al., 2020).

6.5.3 Cell wall stabilization

Calcium bound as Ca pectate in the middle lamella is essential for strengthening cell walls and plant tissues. This function of Ca is reflected clearly in the positive correlation between cation exchange capacity of cell walls and Ca concentration in plant tissues required for optimal growth (White and Holland, 2018; White et al., 2018). The decomposition of pectates is mediated by polygalacturonase that is inhibited strongly by high Ca concentrations (Wehr et al., 2004; Hocking et al., 2016; Yang et al., 2018). Hence, in Ca-deficient tissues, polygalacturonase activity is increased, and a typical symptom of Ca deficiency is the disintegration of cell walls and the collapse of the affected tissues, such as petioles, upper parts of stems, and fruits (Ho and White, 2005; Shear, 1975; White, 2015).

In leaves of plants receiving large amounts of Ca during growth, or when grown under conditions of high light intensity, a large proportion of the pectic material is in the form of Ca pectate. This makes the tissues highly resistant to degradation by polygalacturonase. The proportion of Ca pectate in the cell walls is also of importance for the susceptibility of the tissues to fungal and bacterial infections (Chapter 10) and for the ripening of fruits (Ferguson, 1984; Hocking et al., 2016; Yang et al., 2018). In tomato fruit, the Ca concentration of the cell walls increases to the fully-grown immature stage, but this is followed by a decline in Ca concentration and a change in its bound form just before ripening (Rigney and Wills, 1981). Increasing the Ca concentration in fruits, for example, by spraying several times with Ca salts during fruit development or by postharvest dipping in $CaCl_2$ solution, leads to an increase in the firmness of the fruit and delays fruit ripening (Ferguson, 1984; Oms-Oliu et al., 2010; Dayod et al., 2010).

6.5.4 Cell extension and secretory processes

In the absence of an exogenous Ca supply, root extension ceases within a few hours (Fig. 6.23). This is due to impaired cell elongation, rather than lack of cell division, and is more obvious in a Ca-free nutrient solution than in distilled water, an observation consistent with the role of Ca in counterbalancing the harmful effects of high concentrations of other cations. Cell elongation in roots and shoots requires acidification of the apoplasm and replacement of Ca from the cross-links of the pectic chain, although this is only part of the process (Carpita and McCann, 2000). An increase in cytosolic free Ca^{2+} concentration stimulates the synthesis of cell wall precursors and their secretion into the apoplasm. The latter process is inhibited by removing apoplasmic Ca. The elongation of root hairs and pollen tubes also relies on the availability of apoplasmic Ca. Calcium influx from the apoplasm is restricted to the apex of these cells and increases local cytosolic Ca^{2+} concentration, which acts as focus for the exocytosis of cell wall material and establishes a polarity

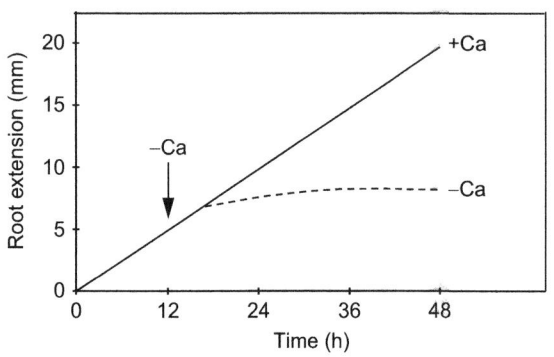

FIGURE 6.23 Extension of primary roots of common bean with or without 2 mM Ca in the nutrient solution. *Based on Marschner and Richter (1974).*

for cell elongation (Michard et al., 2017; Konrad et al., 2018; Pan et al., 2019; Tian et al., 2020). In root caps, the secretion of mucilage also depends on the presence of apoplasmic Ca.

Callose formation is another example of a calcium-induced secretory process. Under normal conditions, cells synthesize cellulose (1,4-β-glucan units). However, in response to injury or the presence of toxic cations such as aluminum, a switch to callose (1,3-β-glucan units) production can occur (Kauss, 1987; Rengel and Zhang, 2003). This switch is triggered by an increase in cytosolic free Ca^{2+} concentration (Kauss, 1987).

Stimulation of α-amylase activity in germinating cereal seeds is one of the few examples of high (millimolar) Ca concentrations increasing enzyme activity. Calcium is a constituent of α-amylase that is synthesized on the rough ER. Transport of Ca^{2+} through the ER membranes is enhanced by gibberellic acid (GA) and inhibited by ABA, leading to the typical stimulation (GA) and inhibition (ABA) of α-amylase activity in the aleurone cells (Lovegrove and Hooley, 2000).

6.5.5 Membrane stabilization

Calcium plays a fundamental role in membrane stability and cell integrity. This is evident in the increased leakage of low-molecular-weight solutes from cells of Ca-deficient tissues and, in severely deficient plants, a general disintegration of membrane structures and loss of cell compartmentation.

Calcium stabilizes cell membranes by bridging phosphate and carboxylate groups of phospholipids and proteins. Calcium can be exchanged for other cations at these binding sites; the exchange of plasma membrane−bound Ca for metals (including Na and Al) can contribute to salinity, metal, and Al toxicity stresses (Cramer, 2002; Horst et al., 2010; White and Pongrac, 2017). To prevent indiscriminate solute leakage and influx of toxic solutes, Ca must always be present in the external solution. The membrane-stabilizing effect of Ca is most prominent under stress conditions such as freezing, low temperature, and anaerobiosis. The loss of low-molecular-weight solutes, such as sugars, in response to chilling or anaerobiosis is reduced by increasing the Ca concentration in the external solution (Table 6.15). In addition to its role in stabilizing membranes, cytosolic Ca^{2+} acts as a second messenger to initiate membrane repair (Schapire et al., 2009) and adaptive responses to freezing, low temperature and anaerobiosis (Ding et al., 2019; Shabala et al., 2014; White and Broadley, 2003).

6.5.6 Cation−anion balance and osmoregulation

In vacuolated cells of leaves in particular, a large proportion of Ca is localized in the vacuoles, where it may contribute to the cation−anion balance by acting as a counter-ion for inorganic and organic anions (White and Broadley, 2003). In plant species that preferentially synthesize oxalate in response to nitrate reduction, the formation of Ca oxalate in vacuoles is important for the maintenance of a low cytosolic free Ca^{2+} concentration (Kinzel, 1989). The same holds true for plant species with preferential formation of Ca oxalate in the apoplasm. The formation of sparingly soluble Ca oxalate is also important for salt accumulation in vacuoles of nitrate-fed plants without increasing the osmotic pressure in the vacuoles (White and Broadley, 2003) and, potentially, for the remobilization of Ca during periods of Ca starvation (Paiva, 2019). In addition, Ca plays a key role in osmoregulation through its involvement as a second messenger in the cell. Stomatal movements and nyctinastic and seismonastic movements are turgor-regulated processes induced by

TABLE 6.15 Carbohydrate loss from cotton roots at different temperatures, aeration, and Ca supply.

Aeration	Treatment temperature (°C)	Solution	Carbohydrate loss (μg seedling^{-1} min^{-1})
Aerobic	31	Distilled water	18
Aerobic	5	Distilled water	57
Aerobic	5	10^{-5} M Ca^{2+}	7
Anoxic	31	Distilled water	89
Anoxic	31	10^{-5} M Ca^{2+}	7

Source: Based on Christiansen et al. (1970).

turgor changes in individual cells (guard cells) or tissues (e.g., motor cells of pulvini). These turgor changes are driven by fluxes of mainly K, Cl, and malate as osmotically active solutes. It is now well established that a transient change of cytosolic free Ca^{2+} concentration is required for transduction of the signals (e.g., light, touch) to the physiological response (Dodd et al., 2010; Jezek and Blatt, 2017; Konrad et al., 2018; Moran, 2007; Saito and Uozumi, 2020; White, 2015).

6.5.7 Calcium as an intracellular second messenger

The capacity of Ca to function as second messenger is based on the very low cytosolic free Ca^{2+} concentrations in plant cells and the chemistry of Ca^{2+} that allows it to alter the conformation of proteins to which it binds (White and Broadley, 2003). Environmental and developmental signals can activate Ca^{2+} channels in cell membranes that catalyze rapid Ca^{2+} influx to the cytosol and increase cytosolic free Ca^{2+} concentrations (Fig. 6.24).

Environmental signals include light intensity and daylength, extreme temperatures, drought, osmotic stress, salinity, Al toxicity, oxidative stress, mechanical stimulation, anoxia, low nutrient availability, symbiotic interactions including nodulation, and attack by pathogens (Dodd et al., 2010; Kudla et al., 2018; McAinsh and Pittman, 2009; Saito and Uozumi, 2020; White, 2015; White and Broadley, 2003). Changes in cytosolic Ca^{2+} concentrations also regulate developmental processes, including cell division, cell elongation, cell polarity after fertilization and in the elongation of pollen tubes and root hairs, germination, circadian rhythms, trophic responses, senescence, and apoptosis.

Calcium influx to the cytosol is mediated by Ca^{2+} channels located in the cell membrane (Fig. 6.22). The specificity of a response to an environmental or developmental signal is encoded by an explicit spatial and temporal perturbation in cytosolic Ca^{2+} concentration (Dodd et al., 2010; Kudla et al., 2018; McAinsh and Pittman, 2009; White, 2015; White and Broadley, 2003). Calcium channels in the plasma membrane have been classified based on their voltage dependence (Fig. 6.22). They include (1) depolarization-activated Ca^{2+} channels (DACCs), whose genetic identity is unknown as yet; (2) hyperpolarization-activated Ca^{2+} channels (HACCs), thought to be formed by members of the cyclic nucleotide–gated channel (CNGC) family or by plant annexins; and (3) voltage-insensitive Ca^{2+} channels (VICCs) that are thought to be encoded by members of the CNGC and GLR gene families (Clark et al., 2012; DeFalco et al., 2016; Demidchik et al., 2018; Dodd et al., 2010; Hedrich et al., 2018; Kudla et al., 2018; McAinsh and Pittman, 2009; Pan et al., 2019; Tian et al., 2020; Weiland et al., 2015; White, 2015; Wudick et al., 2018). In addition, the plasma membrane also contains mechanosensitive Ca^{2+} channels encoded by members of the *mid1*-complementing activity (*MCA*) and, possibly, the mechanosensitive channel of small conductance–like *MSL* gene families and OSCA channels (Demidchik et al., 2018; Kurusu et al., 2013).

Membrane depolarization occurs in response to many environmental cues, and DACCs are thought to initiate general responses to stresses, including adaption to low temperatures (White, 2015; White and Broadley, 2003).

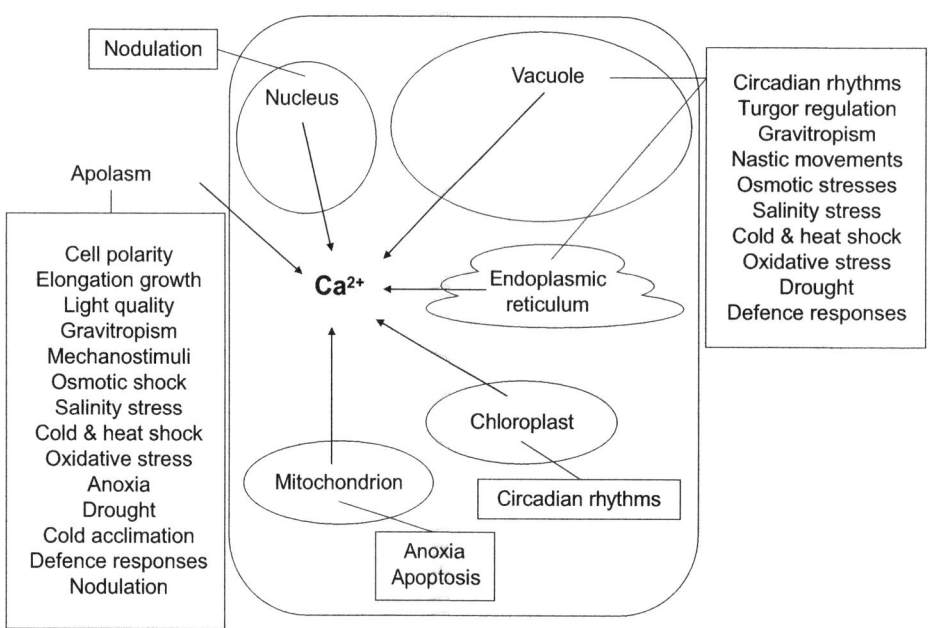

FIGURE 6.24 The origins of Ca influx to the cytosol implicated in plant cell development and responses to environmental signals.

The HACCs are involved in cell elongation, tropisms, symbiotic associations, and responses to biotic and environmental factors causing oxidative stress (Tian et al., 2020; White, 2015; White and Broadley, 2003). The VICCs are thought to be responsible for regulating the basal cytosolic Ca^{2+} concentration of a resting cell (White and Broadley, 2003). Both CNGC and GLR channels have been implicated in cell elongation, tropisms, interactions with microorganisms, responses to nutrient availability in the rhizosphere, and acclimation to extreme temperatures (Demidchik et al., 2018; Pan et al., 2019; Tian et al., 2020; Weiland et al., 2015; White, 2015; Wudick et al., 2018). Mechanosensitive channels have been implicated in generating the growth asymmetries inherent in tropism responses and in regulating cell turgor (Demidchik et al., 2018; Kurusu et al., 2013; White and Broadley, 2003).

Calcium can be released from the vacuole through various cation channels (DeFalco et al., 2016; Hedrich et al., 2018; Kudla et al., 2018; White, 2015; Wudick et al., 2018). These include (1) hyperpolarization-activated channels, which could be formed by annexins; (2) depolarization-activated channels, such as the ubiquitous slow vacuolar channel encoded by homologs of the arabidopsis *AtTPC1* gene; and (3) channels gated by ligands such as inositol-1,4,5-triphosphate (IP3) and cyclic ADP-ribose (cADPR). No genes encoding tonoplast Ca^{2+} channels gated by IP3 or cADPR have yet been identified. However, IP3-gated Ca^{2+} channels have been implicated in regulating stomatal movements, cell elongation, tropisms, responses to salt and hyperosmotic stresses, and defense against pathogens. The cADPR-gated channels have been implicated in regulating stomatal movement, circadian rhythms, cold adaptation, desiccation tolerance, and response to pathogens (Stael et al., 2012; White and Broadley, 2003). A variety of voltage-dependent, ligand-gated, and mechanosensitive channels have also been reported in the ER, plastid, mitochondrial and nuclear membranes, and their roles in signal transduction during, for example, circadian rhythms, photoacclimation, plant development, symbioses, and responses to abiotic stresses and biotic challenges, are being elucidated (Charpentier, 2018; Costa et al., 2018; Kurusu et al., 2013; Leitão et al., 2019; Navazio et al., 2020; Stael et al., 2012). Because Ca^{2+} fluxes through channels are rapid, and an excessive rise in cytoplasmic Ca^{2+} is potentially lethal to a cell, their activity must be regulated precisely.

In the cytosol, the primary targets of Ca signals are Ca^{2+}-binding proteins (Konrad et al., 2018; Kudla et al., 2018; Saito and Uozumi, 2020; Tang et al., 2020; Tian et al., 2020; White, 2015). These include calmodulins (CaMs), CaM-like proteins, calcineurin B–like (CBL) proteins, Ca^{2+}-dependent protein kinases (CDPKs), CDPK-related kinases, and other Ca^{2+}-binding proteins, such as annexins. The binding of Ca^{2+} to these proteins alters their structure or enzymatic properties, which can change solute transport, metabolism, cell morphology, and gene expression. CaMs and CaM-like proteins are involved in the Ca^{2+}-dependent initiation of diverse developmental processes, adaption to numerous adverse environmental conditions and responses to a variety of pathogens. Important targets for CaMs include transport proteins, CaM-dependent protein kinases, and CaM-binding transcription activators (Kudla et al., 2018; White, 2015). Similarly, the CBL proteins, together with their target proteins such as the CIPK protein kinases, play a role in a wide variety of signaling cascades, including those initiated by cold, drought, salinity, anoxia, wounding, or nutrient starvation (Saito and Uozumi, 2020; Tang et al., 2020; White, 2015). The CDPKs, which implement cytosolic Ca^{2+} signals through the phosphorylation of diverse target proteins, are involved in processes such as seed germination, root development, stomatal movements, elongation growth, and responses to salinity, drought, and wounding (Konrad et al., 2018; Saito and Uozumi, 2020; White, 2015). Plant annexins play a role in Ca^{2+}-dependent membrane repair, secretory processes, cell elongation, and responses to drought and salinity (Clark et al., 2012). The spatial and temporal perturbations of cytosolic Ca^{2+} and the types of Ca^{2+}-binding proteins are specific for individual cells and particular stimuli; this is thought to ensure not only an appropriate response to a given stimulus but also phenotypic plasticity of the response.

Calcium-binding proteins in the ER include calreticulin, calsequestrin, calnexin, and binding immunoglobulin protein. These proteins are involved in cellular Ca^{2+} homeostasis, protein folding, and posttranslational modifications.

6.5.8 Calcium as a systemic signal

Calcium also serves as a rapid, systemic signal that primes defense responses in distal tissues to biotic challenges, such as herbivores and pathogens, and initiates acclimatory responses to abiotic stresses, such as drought, salinity, and extreme temperatures (Choi et al., 2016; Hilleary and Gilroy, 2018; Toyota et al., 2018). Systemic Ca^{2+} signals are propagated through the vasculature. Xylem Ca concentration can influence cellular activities and physiological functions of the shoot, such as stomatal closure, by affecting apoplasmic Ca^{2+} concentration (White and Broadley, 2003). In addition, cells in the cortex and endodermis appear to propagate perturbations in cytosolic free Ca^{2+} concentration longitudinally between adjoining cells in response to salinity by triggering Ca^{2+} release from vacuoles (Choi et al., 2014). In the phloem, an electrical signal initiated by Ca^{2+} channels and the generation of ROS transmits rapidly throughout the plant (Choi et al., 2016; Fichman et al., 2019; Hilleary and Gilroy, 2018; Toyota et al., 2018; Wang et al., 2019). It is thought that the release of glutamate from the phloem triggers a systemic Ca^{2+} signal through the stele initiated by GLR channels following wounding damage (Shao et al., 2020; Toyota et al., 2018; Wang et al., 2019).

6.5.9 Calcium supply, plant growth, and plant composition

The Ca concentration of plants varies from about 1 to 50 mg g^{-1} dry wt. depending on the growing conditions, plant species, and plant organ (Mota et al., 2016; Neugebauer et al., 2018; Watanabe et al., 2007; White and Broadley, 2003). The Ca requirement for optimum growth is much lower in commelinid monocotyledons than in other plant species (Table 6.16; Loneragan and Snowball, 1969; Loneragan et al., 1968; White and Holland, 2018). In well-balanced, flowing nutrient solutions with controlled pH, maximal growth rates were obtained at Ca concentrations of 2.5 (perennial ryegrass) and 100 μM (tomato), that is, differing between species by a factor of 40. This difference is mainly a reflection of the Ca demand at the tissue level, which is lower in perennial ryegrass (0.7 mg g^{-1} dry wt.) than in tomato (12.9 mg g^{-1} dry wt.). Differences in Ca requirements between plant species are closely related to Ca^{2+}-binding sites in the cell walls, that is, their cation exchange capacity (Fig. 6.20; White and Broadley, 2003; White and Holland, 2018).

The difference between commelinid monocotyledons and eudicotyledons in Ca demand shown for perennial ryegrass and tomato (Table 6.16) has been confirmed for a large number of plant species (Fig. 6.20; Islam et al., 1987; White and Holland, 2018). However, plants adapted to noncalcareous soils often require less Ca for optimal growth than those adapted to calcareous soils (White, 2015; White and Broadley, 2003); moreover, cultivars of crops, such as rice and common beans, adapted to acid soils have lower Ca requirements than nonadapted cultivars (White, 2015). Accordingly, Ca becomes toxic to calcifuge plants at lower tissue concentrations compared to calcicole plants, which might be related to insufficient capacity for compartmentation and/or physiological inactivation of Ca in calcifuges (White and Broadley, 2003).

Another factor determining the Ca requirement for optimum growth is the concentration of other cations in the external solution. Because Ca is readily replaced by other cations from its binding sites at the exterior surface of the plasma membrane, Ca requirement increases with increasing external concentrations of metals or protons (White, 2015). For example, the Ca^{2+} concentration in the external solution must be several times higher at low pH than at high pH to counteract the adverse effect of high H^+ concentrations on root elongation (Table 6.17). A similar relationship exists between external pH and the Ca requirement for nodulation of legumes (Chapter 16). To protect roots against the adverse effects of high concentrations of other cations in the soil solution, the Ca^{2+} concentrations required for optimal growth have to be substantially higher in soil solutions than in balanced flowing nutrient solutions (Asher and Edwards, 1983).

An increase in the Ca^{2+} concentration in the external solution often leads to an increase in the Ca concentration in transpiring leaves, but not necessarily in low-transpiring organs such as enclosed leaves, fleshy fruits, seeds, or tubers, which are supplied Ca predominantly via the phloem. The mobility of Ca in the phloem is extremely low (Chapter 3), which, coupled with the high growth rates of low-transpiring organs, increases the risk that tissue Ca concentrations in phloem-fed tissues may fall below the critical levels required for cell wall stabilization and membrane integrity, and

TABLE 6.16 Relative growth rates of whole plants of perennial ryegrass and tomato and shoot Ca concentrations at different Ca supplies in the nutrient solution.

	Solution Ca concentration (μM)				
Plant species	0.8	2.5	10	100	1000
	Relative growth rate				
Perennial ryegrass	42	100	94	94	93
Tomato	3	19	52	100	88
	Shoot Ca concentration (mg g^{-1} dry wt)				
Perennial ryegrass	0.6	0.7	1.5	1.7	10.8
Tomato	2.1	1.3	3.0	12.9	24.9

Source: Based on Loneragan and Snowball (1969) and Loneragan et al. (1968).

TABLE 6.17 Growth rate of seminal roots of soybean at different Ca concentrations and solution pH.

Ca^{2+} concentration (mg L^{-1})	Root growth rate (mm h^{-1})	
	pH 5.6	pH 4.5
0.05	2.66	0.04
0.5	2.87	1.36
2.5	2.70	2.38

Source: Based on Lund (1970).

TABLE 6.18 Calcium concentrations and percentage of wastage during storage (3 months at 3.5°C) of "Cox" apples receiving Ca sprays during the growing season or left unsprayed.[a]

	Unsprayed	Sprayed
Calcium concentration (mg kg^{-1} fresh wt)	34	39
Storage disorders (wastage, %)		
Lenticel blotch pit	10	0.0
Senescence breakdown	11	0.0
Internal bitter pit	30	3.4
Gloeosporium rots	9.1	1.7

[a]*Sprays containing 1% w/v Ca nitrate were applied four times during the growing season.*
Source: From Sharpless and Johnson (1977).

perhaps even for intracellular signaling. In rapidly growing tissues, Ca deficiency–related disorders are widespread. These include tipburn in lettuce, blackheart in celery, blossom end rot in tomato or watermelon, and bitter pit in apple (Dayod et al., 2010; Ho and White, 2005; Hocking et al., 2016; Shear, 1975; White, 2015; White and Broadley, 2003). In the case of tomato fruits, water-soluble Ca in the distal part of the fruits plays a key role in the development of the blossom end rot symptoms (when higher than 0.30 μmol Ca g^{-1} fresh wt, the blossom end rot symptoms were not observed; Vinh et al., 2018).

Low Ca concentrations in fleshy fruits and tubers also increase losses caused by accelerated senescence of tissues and by bacterial and fungal infections (Dayod et al., 2010; Hocking et al., 2016; White, 2015, 2018). Even a relatively small increase in the Ca concentration of fruits can be effective in reducing or preventing economic losses caused by storage disorders (Table 6.18). The precipitation of Ca salts in tissues can also deter herbivores (He et al., 2014; Moore and Johnson, 2017; Nakata, 2015).

6.6 Potassium

6.6.1 General

Potassium is critical for all life on Earth (Danchin and Nikel, 2019). In aqueous environments, K loses its lone s-orbital valence electron, producing the univalent cation (K$^+$) with a hydrated ionic radius of 0.331 nm and a hydration energy of 314 J mol^{-1}. Uptake in plants is highly selective (i.e., facilitated by membrane channels and high-affinity transporters) and is closely coupled to metabolic activity (Chapter 2). Potassium is characterized by high mobility at all levels of structural organization—within cells and tissues, as well as in long-distance transport via the xylem and phloem. Potassium forms only weak complexes in which it is readily exchangeable; thus, it exists in plants (as in all biological systems) predominately as a free ion. It is the most abundant cation in the cytosol and is primarily responsible for the maintenance of electrical and osmotic homeostasis in cells and tissues, as well as the chemical coordination and activation of enzymes and nucleic acids. Within vacuoles, it plays critical roles in the maintenance of cell turgor and turgor-related processes (i.e., growth, extension, and movements).

6.6.2 Compartmentation and cellular concentrations

Generally, K^+ concentrations are maintained at 100–200 mM in the cytosol (Britto and Kronzucker, 2008; Leigh and Wyn Jones, 1984; Walker et al., 1996), which, in combination with its electrical- and pH-stabilizing effects, maintains an optimal environment for most enzyme reactions (Clarkson and Hanson, 1980). Similar concentrations can be found in chloroplasts (Schröppel-Meier and Kaiser, 1988). In these compartments, metabolic functions of K cannot be replaced by other inorganic cations, such as Na^+ (Section 8.2). By contrast, vacuolar K^+ concentrations can vary widely, between 20–200 mM (Cuin et al., 2003; Leigh, 2001; Walker et al., 1996), and may even reach up to 500 mM in guard cells (Outlaw, 1983). Cell extension and other turgor-driven processes (e.g., pollen tube development, stomatal aperture, and nastic/tropism movements) are dependent on the K^+ concentration in vacuoles, although it can be replaced to a varying degree by other cations (e.g., Na^+, Mg^{2+}, Ca^{2+}) and organic solutes (e.g., sugars) if K^+ is limiting (Chérel et al., 2014; Rhodes et al., 2006). In contrast to Ca^{2+}, K^+ concentrations in the apoplasm are usually low (e.g., 3–5 mM), with the exception of specialized cells or tissues (e.g., stomata, pulvini), where apoplasmic K^+ concentration may transiently increase up to 100 mM (Grignon and Sentenac, 1991; Mühling and Sattelmacher, 1997).

Under conditions of K deficiency, plant cells prioritize the cytosolic K^+ concentration at the expense of vacuolar K^+ (Amtmann and Armengaud, 2007; Chérel et al., 2014; Walker et al., 1996). When vacuolar K^+ concentrations decline below some critical threshold (c.20 mM), cytosolic concentrations begin to decline, and growth defects become evident (Walker et al., 1996). Abiotic stresses that have been observed to decrease cytosolic and vacuolar K^+ pools include salinity (Cuin et al., 2003; Hajibagheri et al., 1987; Kronzucker et al., 2006; Speer and Kaiser, 1991), ammonium toxicity (Coskun et al., 2013b; Kronzucker et al., 2003), and Al toxicity (Lindberg and Strid, 1997).

6.6.3 Enzyme activation

Potassium is considered essential to the functioning of more than 50 enzymes, serving as either a cofactor or allosteric effector (Leigh and Wyn Jones, 1986; Page and Di Cera, 2006). All macromolecules are highly hydrated and stabilized by firmly bound water molecules forming an electrical double layer. Maximum suppression of this electrical double layer and optimization of protein hydration occur at univalent salt concentrations of about 100–150 mM (Wyn Jones and Pollard, 1983), which corresponds with cellular K^+ concentrations (see above). For example, the activity of starch synthase is highly dependent on univalent cations, of which K^+ is the most effective at typical cytosolic concentrations (Nitsos and Evans, 1969) (Fig. 6.25). The enzyme catalyzes the transfer of glucose to starch molecules:

$$ADP - glucose + starch \leftrightarrow ADP + glycosyl - starch$$

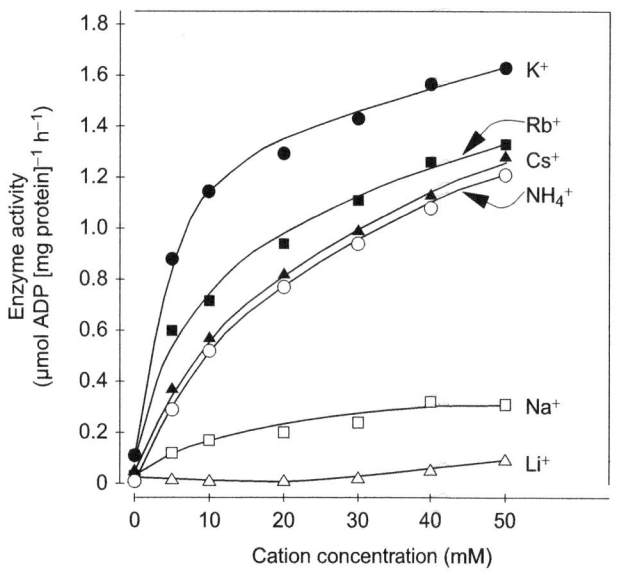

FIGURE 6.25 Activity of ADP-glucose starch synthase from maize with supply of different univalent cations (as chlorides). *Based on Nitsos and Evans (1969).*

The decline in cytosolic K⁺ concentrations with prolonged K deficiency can have severe consequences for the activity of cytosolic enzymes. Many studies from the 1960s and 1970s suggested functional links between K⁺-dependent enzymes (e.g., pyruvate kinase, starch synthase, NR, and Rubisco) and specific metabolic changes under K⁺ deprivation (e.g., Nitsos and Evans, 1966; Peoples and Koch, 1979). With the advent of high-throughput molecular techniques, a clearer understanding of this relationship is beginning to emerge. For example, in low K arabidopsis, the direct inhibition of pyruvate kinase is caused by low cytoplasmic K⁺ in root cells (Armengaud et al., 2009). This corroborates many observations of significant reductions in pyruvate content in K⁺-deprived roots, resulting in reductions in glycolysis and downstream metabolic processes (Amtmann et al., 2008; Armengaud et al., 2009). Other typical changes in response to K⁺ deprivation include an increase in soluble carbohydrates (particularly reducing sugars) and soluble organic N compounds, especially N-rich and positively charged amino acids, whereas the concentrations of nitrate, organic acids, and negatively charged amino acids are typically decreased (Armengaud et al., 2009; Fig. 6.26).

Potassium deficiency alters assimilate partitioning and thus metabolite concentrations in plant organs. For example, the accumulation of sugars in mature leaves is the consequence of inhibited export and a lower demand by sink organs such as growing leaves and fleshy fruits (Gerardeaux et al., 2010; Kanai et al., 2007).

Another major enzyme-related function of K⁺ is the activation of plasma membrane–bound H⁺-ATPase (Gibrat et al., 1990; Morsomme and Boutry, 2000; see Chapter 2) by binding at the cytoplasmic phosphorylation domain and stimulating uncoupled dephosphorylation of an intermediate (E₁P) pump state (Buch-Pedersen et al., 2006). Potassium is also an essential cofactor of tonoplast H⁺-pyrophosphatase (H⁺-PPᵢase) isoforms that pump H⁺ into vacuoles (Darley et al., 1998; Maeshima, 2000).

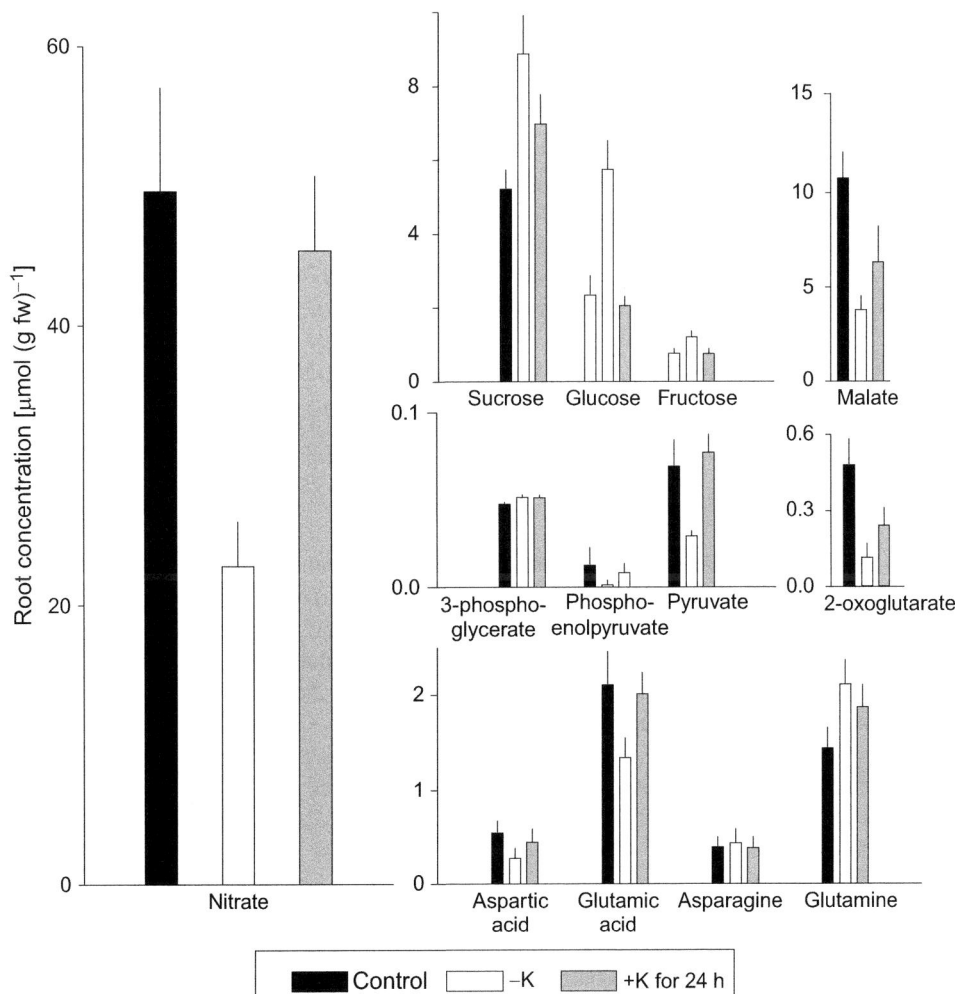

FIGURE 6.26 Concentrations of selected metabolites in roots of *Arabidopsis thaliana* induced by low K supply for 14 days and resupply of K for 24 h. Means + SE. *From Armengaud et al. (2009). An open access article under the terms of Creative Common CC BY licence.*

TABLE 6.19 Incorporation of ^{14}C-leucine into RuBP carboxylase (Rubisco) in the leaves of K-deficient alfalfa plants preincubated at different K concentrations in the light for 20 h.

Preincubation medium (mM KNO$_3$)	^{14}C-leucine incorporation [dpm mg^{-1} RUBP carboxylase (24 h^{-1})]
0.00	99
0.01	167
0.10	220
1.0	274
10	526
Control (K-sufficient plants)	656

Source: From Peoples and Koch (1979).

6.6.4 Protein synthesis

Potassium is critical to ribosome structure and function, and thus protein synthesis. Many forms of functional RNA, including rRNA that catalyzes and regulates protein synthesis, have long been known to depend on metal ions, including K$^+$, for proper functioning (Auffinger et al., 2016; Pyle, 2002). The positions of individual K$^+$ ions within the ribosome functional decoding and peptidyl transferase centers have recently been mapped (Rozov et al., 2019).

Declines in cytosolic K$^+$ concentration due to K deficiency can significantly inhibit protein synthesis (Leigh and Wyn Jones, 1984), but importantly, so too can the cytosolic acidification that accompanies K deficiency (Walker et al., 1998). In green leaves, the chloroplasts account for about half of all leaf RNA and protein, and in C$_3$ species, most of the chloroplast protein is Rubisco. Accordingly, the synthesis of this enzyme is particularly impaired under K deficiency and responds rapidly to resupply of K$^+$ (Peoples and Koch, 1979; Table 6.19).

The role of K$^+$ in protein synthesis is reflected not only in the accumulation of soluble N compounds (e.g., amino acids, amides, and nitrate) in K$^+$-deficient plants (Mengel and Helal, 1968) but can also be demonstrated directly through incorporation of ^{15}N-labeled inorganic N into proteins (Koch and Mengel, 1974). Pflüger and Wiedemann (1977) suggested that K$^+$ both activates NR and is required for the synthesis of this enzyme, a conclusion which is supported by the results of Armengaud et al. (2009).

Protein synthesis, like many enzymatic reactions, is highly sensitive to K$^+$ substitution by Na$^+$, with species such as maize being affected more negatively than sugar beet, which may contribute to the relative salt tolerance of the latter (Faust and Schubert, 2017; see also Sections 8.2.2 and 8.2.5).

6.6.5 Photosynthesis

Potassium influences photosynthesis at various levels, including anatomical, physiological, and biochemical (Tränkner et al., 2018). It plays a prominent role in mesophyll CO$_2$ conductance, chloroplast structure, Rubisco activity, and photoassimilate transport via the phloem. The effect of K$^+$ on CO$_2$ fixation was initially demonstrated with isolated chloroplasts, where an increase in the external K$^+$ concentration up to 100 mM stimulated CO$_2$ fixation more than threefold (Pflüger and Cassier, 1977). Upon illumination, additional influx of K$^+$ from the cytosol is required for the maintenance of a high pH in the stroma necessary for optimal Rubisco activity. Three K$^+$ exchange antiporters in arabidopsis (AtKEA1, -2, and -3) have been identified and characterized in chloroplasts and are thought to mediate the pH, cation, and osmotic homeostasis, as well as light-use (photosystem) efficiency (Galvis et al., 2020; Sze and Chanroj, 2018). AtKEA1 and AtKEA2 are targeted to the inner envelope of chloroplasts, whereas AtKEA3 is targeted to the thylakoid membrane (Kunz et al., 2014). For maximum H$^+$-ATPase activity, an external K$^+$ concentration of about 100 mM is necessary (Wu and Berkowitz, 1992).

As the major inorganic osmolyte of cells, K$^+$ is critical to regulating guard cell turgor and thus stomatal movement and CO$_2$ diffusion (Fischer, 1968; Humble and Raschke, 1971; Tränkner et al., 2018). Potassium deficiency often results in decreased stomatal conductance (Jákli et al., 2017; Zhao et al., 2001), which may have a negative impact on photosynthesis (Fig. 6.27; Battie-Laclau et al., 2014; Huber, 1984). These changes are commonly associated with leaf

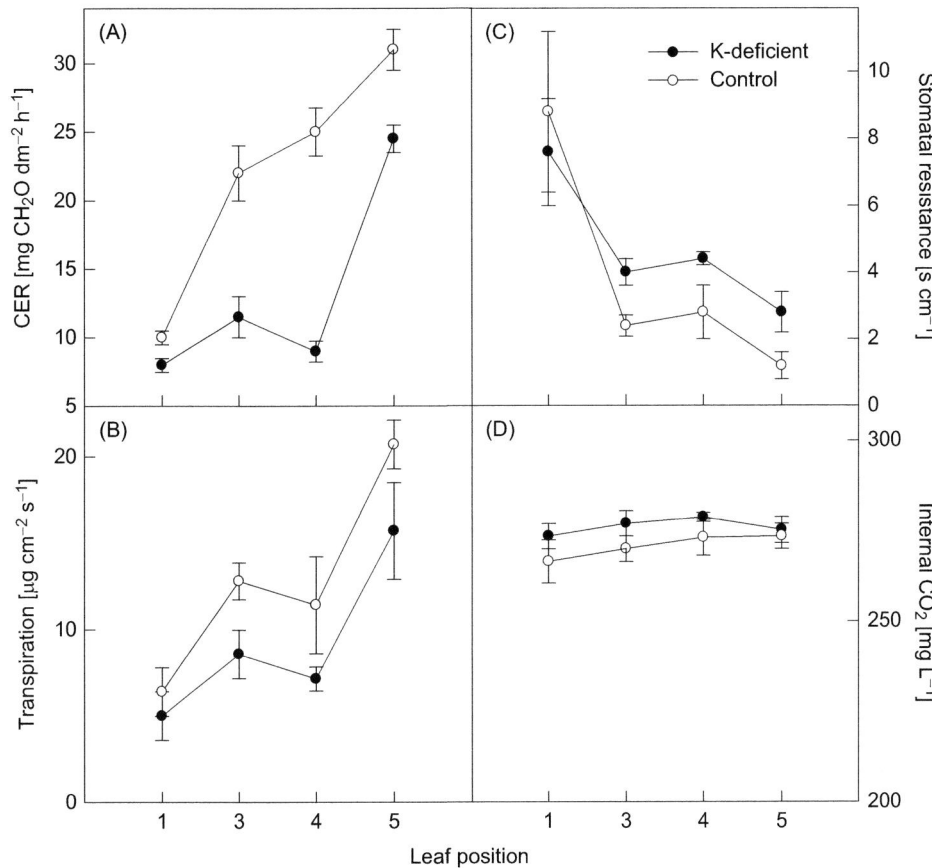

FIGURE 6.27 (A) CER, (B) transpiration, (C) stomatal resistance, and (D) internal CO_2 concentration in soybean plants under K deficiency. *CER*, Carbon exchange rate. Means ± SE. *Adapted from Huber (1984).*

TABLE 6.20 Relationships between K concentration in leaves, CO_2 exchange, RuBP carboxylase activity, photorespiration, and dark respiration in alfalfa.

	Leaf K concentration (mg g^{-1} dry wt)		
	12.8	19.8	38.4
Stomatal resistance (s cm^{-1})	9.3	6.8	5.9
Photosynthesis (mg CO_2 dm^{-2} h^{-1})	11.9	21.7	34.0
RUBP carboxylase activity (μmol CO_2 mg^{-1} protein h^{-1})	1.8	4.5	6.1
Photorespiration (dpm dm^{-2})	4.0	5.9	9.0
Dark respiration (mg CO_2 dm^{-2} h^{-1})	7.6	5.3	3.1

Source: From Peoples and Koch (1979).

anatomical changes, such as decreased leaf thickness, smaller mesophyll cells, and reduced internal leaf airspace (Battie-Laclau et al., 2014; Cakmak et al., 1994a; Gerardeaux et al., 2010; Mengel and Arneke, 1982).

Photorespiration decreases in response to K deficiency (Table 6.20). This may be due to a depletion of CO_2 at the catalytic sites of Rubisco. On the other hand, dark respiration increases under K deficiency and may reflect higher substrate (sugar) availability (Bottrill et al., 1970).

Photosynthetic activity is influenced by the transport and utilization of photosynthates. The marked increases in the accumulation of sucrose in leaves of low-K plants may exert a negative feedback effect on CO_2 assimilation by limiting mesophyll CO_2 uptake and decreasing Rubisco activity (Cakmak, 2005; Tränkner et al., 2018).

Elevated atmospheric CO_2 improves plant photosynthetic activity and productivity, mainly in C_3 plants. However, when plants suffer from K deficiency, the expected positive impact of elevated atmospheric CO_2 is significantly limited, as shown in soybean (Singh and Reddy, 2017) and wheat (Asif et al., 2017). These results highlight the importance of K nutrition under elevated CO_2 conditions.

6.6.6 Osmoregulation

A high osmotic potential in the root stele is a prerequisite for turgor pressure-driven solute transport in the xylem and for the water balance of plants (Chapter 3). The role of K^+ in maintaining xylem-sap flow is evident from the reduced night-time stem expansion and enhanced day-time stem shrinkage in K-deficient tomato plants (Kanai et al., 2007).

6.6.6.1 Cell extension

Cell extension involves the formation of a large central vacuole occupying 80%–90% of the cell volume. There are three major requirements for cell extension: (1) cell extensibility (rearrangement or loosening of the existing cell wall), (2) synthesis and deposition of newly formed wall components, and (3) solute accumulation to create the necessary internal osmotic potential for turgor pressure. In most cases, cell extension is due to K^+ accumulation in the cells, which is required for both stabilizing the pH in the apoplast and the cytoplasm and lowering the osmotic potential in the vacuoles. A decrease in apoplastic pH is necessary to activate enzymes involved in cell-wall loosening (Arsuffi and Braybrook, 2018; Cosgrove, 2016; Hager, 2003). Potassium is required to counterbalance the ATPase-driven H^+ release into the cell wall electrochemically (Stiles and Van Volkenburgh, 2004). In *Avena* coleoptiles, auxin (IAA)-stimulated H^+ efflux was balanced electrochemically by a stoichiometric K^+ influx; in the absence of external K^+, IAA-induced elongation declined and then ceased after a few hours (Haschke and Lüttge, 1975).

Cell extension in leaves and roots is corelated positively with K^+ content (Dolan and Davies, 2004). Potassium deficiency has been shown to reduce turgor, cell size, and leaf area in expanding leaves of bean plants (Mengel and Arneke, 1982). Moreover, reduced leaf extension rate was a highly sensitive indicator of K deficiency in maize grown in the field (Jordan-Meille and Pellerin, 2004) and in hydroponics (Jordan-Meille and Pellerin, 2008). This inverse relationship between K^+ concentration in plants and cell size also holds true for storage tissues such as carrot root (Pfeiffenschneider and Beringer, 1989) and tomato fruits (Kanai et al., 2007).

As shown by De la Guardia and Benlloch (1980) (Table 6.21), the stimulation of stem elongation by GA is also dependent on K^+ supply. Potassium and GA act synergistically, with the highest elongation rate being obtained when both GA and K^+ are applied. Furthermore, the results indicate that K^+ and reducing sugars act in a complementary manner to produce the turgor potential required for cell extension. At low K^+ supply, however, GA-stimulated growth was correlated with a marked increase in K^+ concentration in the elongation zone to a level similar to that of the reducing sugars (De la Guardia and Benlloch, 1980). As K^+ was supplied together with Cl^- (as KCl), a substantial proportion of the effects on plant growth and sugar concentrations may be due to the combined effects of K^+ and Cl^- on osmotic potential.

The extent to which sugars and other low-molecular-weight organic solutes contribute to osmotic potential and turgor-driven cell expansion depends on the plant K nutritional status as well as on plant species and specific organs.

TABLE 6.21 Plant height and concentrations of sugars and K in the shoots of sunflower plants at different K and gibberellic acid (GA) supplies.

Treatment		Plant height (cm)	Concentration (µmol g^{-1} fresh wt)		
KCl (mM)	GA (mg L^{-1})		Reducing sugars	Sucrose	Potassium
0.5	0	7.0	19.1	5.0	10.2
0.5	100	18.5	38.5	5.4	13.2
5.0	0	11.5	4.6	4.1	86.5
5.0	100	26.0	8.4	2.5	77.8

Source: Based on De la Guardia and Benlloch (1980).

For example, in the elongation zone of leaf blades of tall fescue, about half of the imported sugars are used for the accumulation of osmotically active fructans in the vacuoles (Schnyder et al., 1988).

After completion of cell extension, K^+ can be replaced for maintenance of the cell turgor in the vacuoles by other solutes such as Na^+ or reducing sugars. At later stages of leaf extension, sugars even overcompensated leaf-tissue K deficiency in cotton (Gerardeaux et al., 2010). Generally, there is a negative relationship between tissue concentrations of K^+ and sugars, particularly reducing sugars (Pitman et al., 1971), which can also be observed during the growth of storage tissues. As shown by Steingröver (1983), the osmotic potential of the press sap from the storage root of carrot remains constant throughout growth. Before sugar storage begins, K^+ and organic acids are the dominant osmotica. During sugar storage, however, an increase in the concentration of reducing sugars is compensated for by a corresponding decrease in the concentration of K^+ and organic acid anions. The same holds true for the concentrations of sucrose and K^+ in sugar beet storage roots (Beringer et al., 1986).

6.6.6.2 Stomatal movement

In most plant species, K^+ and its accompanying counterions play major roles in turgor changes in guard cells during stomatal movement. Increasing K^+ concentration in guard cells results in decreases in osmotic potential and thus the uptake of water from adjacent cells, resulting in an increase in turgor in the guard cells and thus stomatal opening (Humble and Raschke, 1971; Table 6.22). The accumulation of K^+ in guard cells of open stomata can also be shown by X-ray microprobe analysis (Fig. 6.28).

The metabolic and transport systems involved in stomatal opening are shown schematically in Fig. 6.29 (see also Jezek and Blatt, 2017). In light-induced stomatal opening, K^+ influx into the guard-cell cytoplasm is driven by the electrical potential difference established by the plasma membrane H^+-ATPase (Blatt and Clint, 1989). Approximately 50%–70% of K^+ uptake is thought to be mediated by inward-rectifying (KAT) K^+ channels (Wang et al., 2014), with the remaining uptake mediated by high-affinity K^+ transporters of the HAK/KUP/KT type (Véry et al., 2014). Against the electrical potential, K^+ is then pumped into the vacuole via H^+-driven (NHX-type) antiporters (Barragán et al., 2012; Bassil et al., 2011). The accumulation of K^+ in the vacuoles must be balanced by counterions, mainly malate or Cl^-, depending on the plant species and concentrations of Cl^- in the vicinity of the guard cells. The transport of Cl^- into the guard cell cytoplasm is mediated by Cl^-/H^+ symporters (e.g., NRT1.1 and NRT2.1; Hawkesford and Miller, 2004) at the plasma membrane and down the electrical potential across the tonoplast via anion channels (e.g., AtALMT9; De Angeli et al., 2013).

At low Cl^- availability, or in plant species that do not use Cl^- as an accompanying anion for K^+ in guard cells, the H^+-driven K^+ influx activates PEP carboxylase in the cytoplasm. The malate synthesized in the guard-cell cytosol is transported into the vacuole via anion channels (e.g., ALMT6; Meyer et al., 2011) and serves as an accompanying anion for K^+ in the vacuole. The C_3 compound PEP required for malate synthesis is supplied primarily via starch degradation in the guard cell chloroplasts (Outlaw and Manchester, 1979).

Sugars are considered alternative osmotic solutes for stomatal opening (Tallman and Zeiger, 1988). As guard cells are unable to fix significant amounts of carbon, sugar (sucrose) uptake (via SUC-/STP-type H^+-sugar symporters) in guard cells is dependent largely on apoplastic sources derived from adjacent (mesophyll) cells (Endler et al., 2006; Neuhaus and Trentmann, 2014; Stadler et al., 2003; Wormit et al., 2006; Fig. 6.29). Although sugar uptake and production in guard cells are considered insufficient to meet the high requirement for rapid stomatal opening (Reckmann et al., 1990), sugars

TABLE 6.22 Relationship between stomatal aperture and characteristics of guard cells of faba bean.

		Open stomata	Closed stomata
Stomatal aperture (μm)		12	2
Content per stoma (10^{-14} mol)	K	424	20
	Cl	22	0
Guard cell volume (10^{-12} L per stoma)		4.8	2.6
Guard cell osmotic pressure (MPa)		3.5	1.9

Source: From Humble and Raschke (1971).

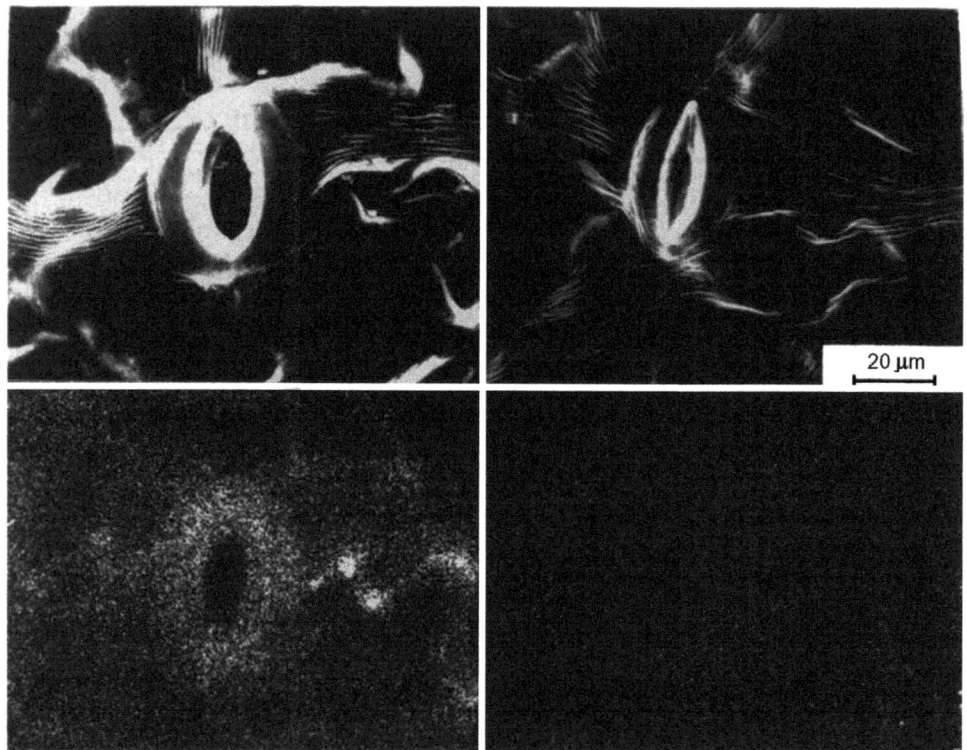

FIGURE 6.28 Electron-probe analyzer image (top) and corresponding X-ray microprobe images of K distribution (bottom) in open (left) and closed (right) stomata of faba bean. *Courtesy B. Wurster.*

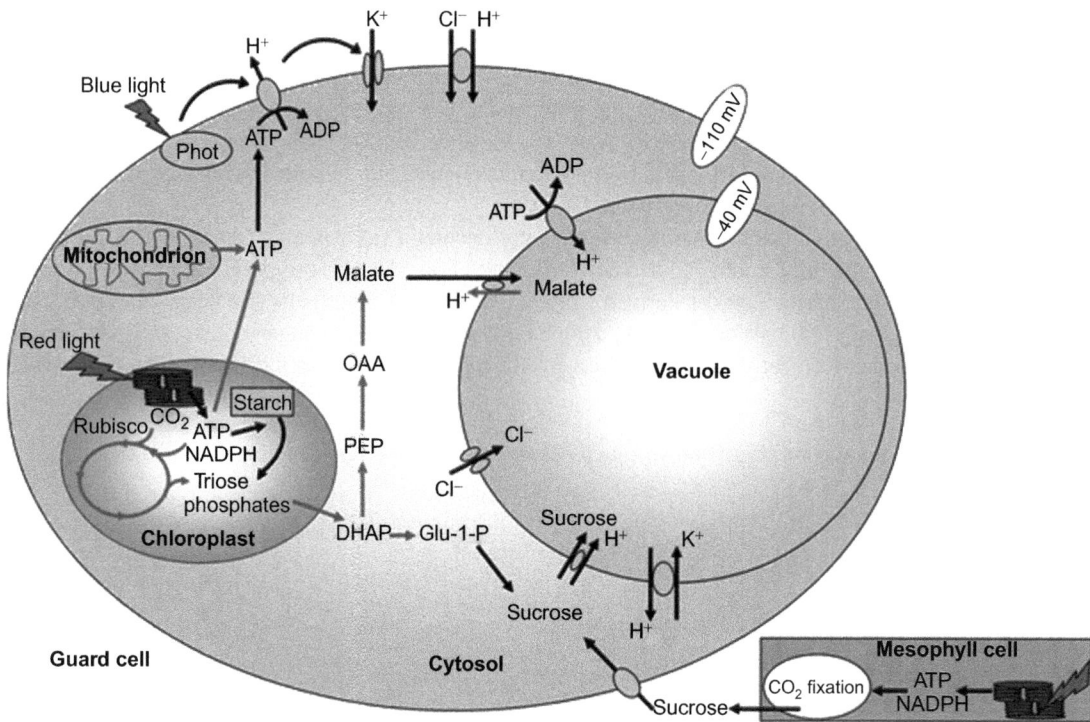

FIGURE 6.29 Schematic diagram of possible osmoregulatory pathways in guard cells governing stomata opening. The diagram is not to scale. *DHAP*, Dihydroxyacetone phosphate; *G-1-P*, glucose-1-phosphate; *OAA*, oxaloacetate; *PEP*, phosphoenolpyruvate. *Redrawn from Roelfsema and Hedrich (2005) and Lawson (2009).*

may nevertheless be important for the sustained opening of stomata (Talbott and Zeiger, 1998), particularly under K deficiency (Poffenroth et al., 1992).

Closure of the stomata is induced by darkness, dehydration, and ABA and is associated with rapid efflux of K^+ from the guard cells (along with its accompanying anions). In contrast to stomatal opening being based on active transport, closure is due to the release of solutes along their concentration gradients (e.g., via outward-rectifying K^+ channels, GORK; Roelfsema and Hedrich, 2005). Stomatal closure is associated with a strong increase in K^+ and Cl^- concentrations in the apoplast of guard cells (Bowling, 1987). In parasitic plants such as *Striga* and *Loranthus*, stomata remain open permanently and do not respond to darkness, ABA, or drought stress. This anomalous behavior is caused by exceptionally high K^+ concentrations in the leaves of these plants that lack a phloem and cannot release K^+ from guard cells (Smith and Stewart, 1990).

Dark-induced stomata closure is initiated by a strong depolarization of the tonoplast and plasma membranes, which activates the outward-rectifying K^+ and anion channels. The membrane depolarization is triggered by (1) the cessation of the blue-light activation of H^+-ATPases and (2) the red light–dependent CO_2 assimilation giving rise to elevated intracellular CO_2 concentrations (Roelfsema and Hedrich, 2005).

The stomatal closure is induced by ABA derived from the roots via the xylem as a "nonhydraulic" signal (Davies and Meinzer, 1990; see Chapter 5). However, endogenous ABA from guard cells may also serve this function, with ABA concentration almost threefold higher in the guard cells compared with other epidermal cells in faba bean (Brinckmann et al., 1990). ABA-induced stomatal closure is triggered by plasma-membrane depolarization via activation of anion channels (Roelfsema et al., 2004), reduced H^+-ATPase activity (Brault et al., 2004), and an increase in cytoplasmic Ca^{2+} concentration through stimulation of Ca^{2+} channels (Roelfsema and Hedrich, 2010).

6.6.6.3 Photonastic and seismonastic movements

In leaves of many plants, particularly in Fabaceae, leaves reorientate their laminae photonastically in response to light signals, either nondirectional (circadian rhythm; e.g., leaf blades fold in the dark and unfold in the light) or directional (e.g., reorientation toward a light source). These photonastic responses either increase light interception or promote avoidance of excess-light damage (Koller, 1990). The movements of leaves and leaflets are brought about by reversible turgor changes in specialized tissues (motor organs or pulvini). Turgor changes cause shrinking and swelling of cells in the opposing regions (extensor and flexor, respectively) of the motor organ. The major solutes involved in osmoregulation are K^+, Cl^-, and malate, inducing water flow through the membrane matrix and particularly aquaporins (Moshelion et al., 2002), followed by volume change and leaflet movement (Satter et al., 1988). The principles of the mechanisms responsible for stomatal movement also apply to leaf and leaflet movement, only the scales are different (individual cells vs specialized tissues).

In leaflet movement the driving force for K^+ influx in the flexor is established by the plasma membrane H^+-ATPase (Satter et al., 1988) and, thus, leaflet movement can be prevented by anaerobiosis or vanadate (Antkowiak et al., 1992). In the primary leaf pulvinus of *Phaseolus vulgaris* during circadian leaf movement, the apoplasmic concentration of H^+ increases and that of K^+ decreases in the extensor at swelling (upward movement of the leaf lamina), and vice versa when the extensor cells shrink (Starrach and Mayer, 1989). The extensor cell walls have a particularly high cation exchange capacity and, thus, are an important reservoir of K^+ and H^+ (Starrach et al., 1985). Similar to stomatal movement, in the leaf movements, environmental signals (e.g., light, mechanical stimulation) activate Ca^{2+} channels in the plasma membrane and/or mobilize Ca^{2+} from internal stores, thereby increasing cytosolic free Ca^{2+} concentrations in the flexor (Roblin et al., 1989).

Although similar mechanisms are responsible for the movement of leaves and other plant parts in response to light and mechanical stimulus, there are differences in the speed of the response to seismonastic signals. In *Mimosa pudica*, the leaflets fold within a few seconds and re-open after about 30 min (Campbell and Thomson, 1977). This turgor-regulated response is correlated with redistribution of K^+ within the motor organ (Allen, 1969) and the sudden release of sucrose from the phloem (Fromm and Eschrich, 1988). In seismonastic reactions, a rapid long-distance transport of the "signal" from the touched leaflet to other leaflets also takes place. This "signal" is an action potential, traveling in the phloem to the motor organs at a speed of $1-10$ cm s^{-1}, inducing phloem unloading of sucrose in the motor organ (Fromm, 1991).

6.6.7 Phloem transport

Potassium has important functions in both the loading of sucrose and the rate of the mass flow-driven solute transport in the sieve tubes of the phloem (Chapter 3). This function of K^+ is related to (1) the necessity of maintaining a high pH in the sieve tubes for sucrose loading, and (2) the contribution of K^+ to the osmotic potential in the sieve tubes and, thus, the transport rates of photosynthates from source to sink. The role of K^+ in phloem loading and assimilate partitioning is evident by comparing the relative distribution of nonstructural carbohydrates between shoots and roots in K^+-sufficient and K^+-deficient plants. Similar to Mg^{2+} deficiency, but unlike P deficiency, assimilate transport to the roots is reduced strongly in K^+-deficient plants (Cakmak et al., 1994a). In K^+-sufficient plants, about half of the ^{14}C-labeled photosynthates are exported from the source leaf to other organs within 90 min, with about 20% transported to the stalk as the main storage organ in sugar cane. By contrast, in K^+-deficient plants, the export rates were much lower, even after 4 h (Hartt, 1969). Hu et al. (2017) found similar impairments in sucrose transport in cotton plants by calculating sucrose export per leaf, per unit leaf fresh weight, or the phloem:leaf ratio.

A lower assimilate transport to sinks is also evident in the reduced root growth in K-deficient plants (Cakmak, 1994; Cakmak et al., 1994b). Compared to K-deficient plants, root nodules in legumes with adequate K supply have a greater supply of sugars, which increases their rates of N_2 fixation and export of fixed N (Collins and Duke, 1981; Mengel et al., 1974).

It has been claimed that phloem transport of amino acids is affected by both sucrose loading processes and osmotic pressure (i.e., mass flow) in phloem sieve tubes (Cakmak et al., 1994b; Winter et al., 1992). Therefore, with any impairment in sucrose loading or mass flow in the phloem due to K deficiency, a corresponding decline in phloem transport of amino acids can be expected. In good agreement with this assumption, K deficiency also reduces phloem transport of amino acids (Cakmak et al., 1994b). One of the positive effects of adequate K^+ nutrition on seed protein concentration is likely related to the enhanced transport of amino acids (Pettigrew, 2008).

6.6.8 Energy transfer

In addition to its role in assimilate transport, K^+ circulating in the phloem may serve as a decentralized energy store that can be used to overcome local energy limitations induced by, for example, shading. This role of K^+ is suggested by a study of the regulation of the K^+ channel AKT2 using an arabidopsis knockout mutant grown under sufficient and limiting K^+ and light (Gajdanowicz et al., 2010). AKT2 mediates K^+ uptake and release from the phloem, which accompanies phloem loading and unloading of assimilates. Simulation of H^+, sucrose, and K^+ transport in the phloem sieve element and companion cells supports the conclusion that posttranslational modification of AKT2 switches on a "K^+ battery" that assists the H^+-ATPase in generating the energy necessary to sustain transmembrane transport processes under energy-limiting conditions (Dreyer et al., 2017).

6.6.9 Cation–anion balance

Potassium plays a fundamental role in charge balance of anions in different organelles such as chloroplasts and vacuoles, as well as in the xylem and phloem (Coskun et al., 2017). Siebrecht and Tischner (1999) showed that reducing K^+ supply to plants reduced NO_3^- concentrations of xylem exudate by 43%. The role of K^+ in cation–anion balance is also reflected in nitrate metabolism, in which K^+ is often the dominant counter-ion for NO_3^- in long-distance transport in the xylem, as well as for storage in vacuoles. As a consequence of NO_3^- reduction in leaves, the remaining K^+ requires the stoichiometric synthesis of organic acid anions for charge balance; thus, K^+ and malate may be transported to the roots for subsequent utilization of K^+ as a counter-ion for NO_3^- within the root cells and for xylem transport (Ben-Zioni et al., 1971; Coskun et al., 2017; Dijkshoorn et al., 1968; Kirkby and Knight, 1977; see also Chapter 3). In nodulated legumes, this recirculation of K^+ may serve a similar function in the xylem transport of amino acids (Jeschke et al., 1985).

6.6.10 Stress resistance

Potassium is more than a nutrient; K^+ has also been recognized as an important signaling factor influencing the adaptive responses of plants to diverse stress conditions (Anschütz et al., 2014). The frequently observed positive effects of K fertilization on crop yields under adverse conditions have been interpreted as evidence that K increases the resistance of plants against biotic (Chapter 10) and abiotic stresses (Cakmak, 2005).

K-deficient plants show exacerbated injury under high-light intensity (Marschner and Cakmak, 1989), drought (Sen Gupta et al., 1989; Fig. 6.30), low temperature (Grewal and Singh, 1980; Table 6.23), Fe toxicity (Li et al., 2001), and pest and disease pressure (Amtmann et al., 2008; see also Chapter 10). Thus, under these stresses, an optimum K

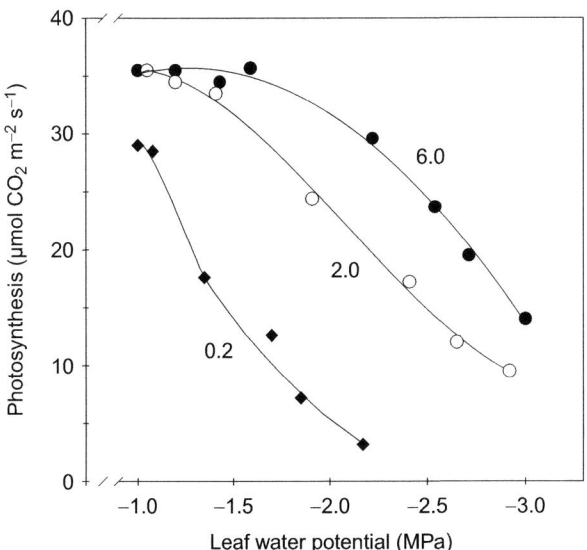

FIGURE 6.30 Photosynthesis in leaves at declining leaf water potentials in wheat grown at different K supplies (0.2, 2.0, and 6.0 mM). *Based on Sen et al. (1989).*

TABLE 6.23 Potato tuber yield, K concentration in leaves, and percentage of leaves damaged by frost at different K supplies.

K supply (kg ha^{-1})	Tuber yield (t ha^{-1})	K concentration in leaves (mg g^{-1} dry wt)	Percentage of foliage damaged by frost
0	2.4	24	30
42	2.7	28	16
84	2.9	30	7

Average values of 14 locations.
Source: Based on Grewal and Singh (1980).

nutritional status is critical for plant stress resistance. One of the reasons for decreased stress resistance under K deficiency is an enhanced production of ROS causing oxidative stress (Cakmak, 2005).

Adequate K nutrition has mitigating effects on multistress conditions. In tomato plants, sufficient K together with Ca supply improved the oxidative stress tolerance of plants in a saline growth medium combined with heat stress, and this positive effect was found to be related to diminished ROS generation and lipid peroxidation, as well as enhanced antioxidative defense potential of leaves (García-Martí et al., 2019). Given that the enhanced generation of ROS is common in plants exposed to drought and/or heat stress (Mittler, 2002; Suzuki et al., 2012), low-K plants exposed to drought or heat stress are more likely to show exacerbated detrimental impacts on cellular structure and functions.

Because K deficiency decreases photosynthetic CO_2 fixation, the photosynthetically produced electrons are transferred increasingly to O_2, boosting ROS production. The ROS production in chloroplasts is highly light-dependent, and K-deficient plants are very sensitive to high light intensity (Marschner and Cakmak, 1989).

Increased drought tolerance of K^+-sufficient plants appears to be associated with (1) oxidative stress avoidance, (2) stomatal regulation, and (3) high osmotic pressure in the vacuoles, maintaining a high tissue water content even under drought conditions (Cakmak, 2005; Lindhauer, 1985). Increased drought tolerance may require leaf K concentrations higher than those for optimal growth. For example, in tea plants, improved drought-stress tolerance was closely related to the maintenance of high concentration of K in mesophyll cells. The tea genotypes with higher drought stress tolerance exhibited lower K efflux from mesophyll cells resulting in higher K concentrations (Zhang et al., 2019). Similarly, genotypes of various plant species exhibiting enhanced salinity-stress tolerance have high genetic capacity to retain K following exposure to NaCl stress. Hence, the genetic capacity to retain intracellular K under salt stress has been recommended as a trait to be used in breeding programs (Anschütz et al., 2014; Chakraborty et al., 2016). However, Coskun et al. (2013c) did not find the same relationship in rice genotypes differing in tolerance to salinity.

The H$^+$/K$^+$ counterflow necessary for pH stabilization in the chloroplast stroma (Section 6.6.5) is impaired under drought stress. A decrease in photosynthesis under drought stress is less severe at high K supply (Sen Gupta et al., 1989). Supply of 2 mM K resulted in maximal photosynthesis in well-watered plants but under drought stress photosynthesis was higher in plants supplied with 6 than 2 mM K (Fig. 6.30). The higher K requirement in leaves of plants exposed to drought or salinity stress (Chow et al., 1990) is primarily caused by the need to maintain high stromal K$^+$ concentrations (Sen Gupta et al., 1989).

The severe frost damage to K-deficient plants is related to water deficiency at the cellular level. Ma et al. (2019) showed that increasing K nutritional status of wheat in the areas with high risk of frost damage was required for better grain set and higher yields, most probably due to the positive impacts of K on cell osmoregulation as well as plant antioxidative defenses.

The greater susceptibility of K-deficient versus K-sufficient plants to pathogens is thought to be due to changes in enzyme activities and metabolite concentrations leading to facilitated entry and development of pathogens in the plant tissue. However, this assumption is difficult to reconcile with the molecular characterization of the response of arabidopsis to K deficiency (Amtmann et al., 2008), whereby K deficiency activates signaling cascades similar to drought, wounding, and biotic stresses involving jasmonic and salicylic acids, both of which are critical for pathogen resistance (Amtmann et al., 2008; Ashley et al., 2006).

6.6.11 Potassium supply, plant growth, and plant composition

Potassium deficiency commonly occurs in plants grown in sandy, saline, and acidic soils, as well as in intensive cropping systems with a high offtake of K (Römheld and Kirkby, 2010; Zörb et al., 2014). After N, K is the nutrient required by plants in the largest amount. The K requirement for optimal plant growth is typically considered to be 20–50 g kg^{-1} in vegetative parts, fleshy fruits, and tubers. The "critical" tissue K concentration, at which growth and development begin to fall below 90% of their maxima, approximates 5–20 µmol g^{-1} fresh wt (Leigh and Wyn Jones, 1984). In natrophilic species, however, the requirement for K can be lower due to replacement by Na. When K is deficient, growth is retarded, and remobilization of K from mature leaves and stems is enhanced. Under severe deficiency, these older leaves become chlorotic, and necrosis develops along the leaf margins. Low K plants are very sensitive to high light, rapidly developing leaf chlorosis (Marschner and Cakmak, 1989). Also, lignification of vascular bundles in such plants is impaired (Pissarek, 1973), a factor that may contribute to the high susceptibility of K-deficient plants to lodging. Potassium is involved in lignin deposition and mechanical strength of stem tissue in wheat (Kong et al., 2014), probably by affecting the deposition of C in the stem tissue.

The changes in plant composition with K deficiency also affect the nutritional and technological (processing) quality of harvested products. This is most obvious in fleshy fruits and tubers that have high K requirement. In tomato fruits, for example, the incidence of so-called ripening disorders ("greenback") increases with inadequate K supply (Lune and van Goor, 1977), and in potato tubers a whole range of quality criteria are affected by low tissue K concentration (see also Chapter 9).

References

Ahn, D.J., Won, J.G., Rico, C.M., Lee, S.C., 2010. Influence of variety, location, growing year, and storage on the total phosphorus, phytate-phosphorus, and phytate-phosphorus to total phosphorus ratio in rice. J. Agric. Sci. Food Chem. 58, 3008–3011.

Alkarawi, H.H., Zotz, G., 2014. Phytic acid in green leaves of herbaceous plants—temporal variation in situ and response to different nitrogen/phosphorus fertilizing regimes. AoB Plants 6, plu048.

Allen, R.D., 1969. Mechanism of the seismonastic reaction in *Mimosa pudica*. Plant Physiol. 44, 1101–1107.

Amtmann, A., Armengaud, P., 2007. The role of calcium sensor-interacting protein kinases in plant adaptation to potassium deficiency: New answers to old questions. Cell Res. 17, 483–485.

Amtmann, A., Troufflard, S., Armengaud, P., 2008. The effect of potassium nutrition on pest and disease resistance in plants. Physiol. Plant 133, 682–691.

Anderson, A.J., Spencer, D., 1950. Sulphur in nitrogen metabolism of legumes and non-legumes. Aust. J. Sci. Res. Ser. B 3, 431–449.

Andersson, M.X., Stridh, M.H., Larsson, K.E., Liljenberg, C., Sandelius, A.S., 2003. Phosphate-deficient oat replaces a major portion of the plasma membrane phospholipids with the galactolipid digalactosyldiacylglycerol. FEBS Lett. 537, 128–132.

Andrews, M., 1986. The partitioning of nitrate assimilation between root and shoot of higher plants. Plant Cell Environ. 9, 511–519.

Anjana, S.U., Iqbal, M., 2007. Nitrate accumulation in plants, factors affecting the process, and human health implications. A review. Agron. Sustain. Dev. 27, 45–57.

Anschütz, U., Becker, D., Shabala, S., 2014. Going beyond nutrition: Regulation of potassium homoeostasis as a common denominator of plant adaptive responses to environment. J. Plant Physiol. 171, 670–687.

Antkowiak, B., Engelmann, W., Herbjornsen, R., Johnsson, A., 1992. Effect of vanadate, N_2, and light on the membrane potential of motor cells and the lateral leaflet movements of *Desmodium motorium*. Physiol. Plant 86, 551–558.

Anuradha, M., Narayanan, A., 1991. Promotion of root elongation by phosphorus deficiency. Plant Soil 136, 273–275.

Apell, H.J., Hitzler, T., Schreiber, G., 2017. Modulation of the Na, K-ATPase by magnesium ions. Biochemistry 56, 1005–1016.

Araya, T., Miyamoto, M., Wibowo, J., Suzuki, A., Kojima, S., Tsuchiya, Y.N., et al., 2014. CLE-CLAVATA1 peptide-receptor signaling module regulates the expansion of plant root systems in a nitrogen-dependent manner. Proc. Natl. Acad. Sci. USA 111, 2029–2034.

Armengaud, P., Sulpice, R., Miller, A.J., Stitt, M., Amtmann, A., Gibon, Y., 2009. Multilevel analysis of primary metabolism provides new insights into the role of potassium nutrition for glycolysis and nitrogen assimilation in Arabidopsis roots. Plant Physiol. 150, 772–785.

Arora, S.K., Luchra, Y.P., 1970. Metabolism of sulphur containing amino acids in *Phaseolus aureus* Linn. Z. Pflanzenernähr. Bodenk. 126, 151–158.

Arsuffi, G., Braybrook, S.A., 2018. Acid growth: An ongoing trip. J. Exp. Bot. 69, 137–146.

Asher, C.J., Edwards, D.G., 1983. Modern solution culture techniques. In: Läuchli, A., Bieleski, R.L. (Eds.), Encyclopedia of Plant Physiology, New Series, 15A. Springer Verlag, Berlin and New York, pp. 94–119.

Ashley, M.K., Grant, M., Grabov, A., 2006. Plant responses to potassium deficiencies: a role for potassium transport proteins. J. Exp. Bot. 57, 425–236.

Asif, M., Yilmaz, O., Ozturk, L., 2017. Potassium deficiency impedes elevated carbon dioxide-induced biomass enhancement in well-watered or drought-stressed bread wheat. J. Plant Nutr. Soil Sci. 180, 474–481.

Assuero, S.G., Mollier, A., Pellerin, S., 2004. The decrease in growth of phosphorus-deficient maize leaves is related to a lower cell production. Plant Cell Environ. 27, 887–895.

Astolfi, S., Celletti, S., Vigani, G., Mimmo, T., Cesco, S., 2021. Interaction between sulfur and iron in plants. Front. Plant Sci. 12, 670308.

Astolfi, S., Cesco, S., Zuchi, S., Neumann, G., Roemheld, V., 2006. Sulfur starvation reduces phytosiderophores release by iron-deficient barley plants. Soil Sci. Plant Nutr. 52, 43–48.

Auffinger, P., D'Ascenzo, L., Ennifar, E., 2016. Sodium and potassium interactions with nucleic acids. In: Sigel, A., Sigel, H., Sigel, R. (Eds.), The Alkali Metal Ions: Their Role for Life. Metal Ions in Life Sciences. Springer, Cham, Switzerland, pp. 167–201.

Ausma, T., De Kok, L.J., 2019. Atmospheric H_2S: Impact on plant functioning. Front. Plant Sci. 10, 743.

Avice, J.C., Ourry, A., Lemaire, G., Boucaud, J., 1996. N and carbon flows estimated by ^{15}N and ^{13}C pulse chase labelling during re-growth of alfalfa. Plant Physiol. 112, 281–290.

Baldi, B.G., Franceschi, V.R., Loewus, F.A., 1987. Localization of phosphorus and cation reserves in *Lilium longiflorum* pollen. Plant Physiol. 83, 1018–1021.

Balke, N.E., Hodges, T.K., 1975. Plasma membrane adenosine triphosphatase of oat roots. Plant Physiol. 55, 83–86.

Bamji, M.S., Jagendorf, A.T., 1966. Amino acid incorporation by wheat chloroplasts. Plant Physiol. 41, 764–770.

Barraclough, P.B., Howarth, J.R., Jones, J., Lopez-Bellido, R., Parmar, S., Shepherd, C.E., et al., 2010. Nitrogen efficiency of wheat: genotypic and environmental variation and prospects for improvement. Eur. J. Agron. 33, 1–11.

Barragán, V., Leidi, E.O., Andrés, Z., Rubio, L., de Luca, A., Fernández, J.A., et al., 2012. Ion exchangers NHX1 and NHX2 mediate active potassium uptake into vacuoles to regulate cell turgor and stomatal function in Arabidopsis. Plant Cell 24, 1127–1142.

Barry, D.A.J., Miller, M.H., 1989. Phosphorus nutritional requirement of maize seedlings for maximum yield. Agron. J. 81, 95–99.

Bassil, E., Tajima, H., Liang, Y.C., Ohto, M., Ushijima, K., Nakano, R., et al., 2011. The arabidopsis Na^+/H^+ antiporters NHX1 and NHX2 control vacuolar pH and K^+ homeostasis to regulate growth, flower development and reproduction. Plant Cell 23, 3482–3497.

Baszynski, T., Warcholowa, M., Krupa, Z., Tukendorf, A., Krol, M., Wolinska, D., 1980. The effect of magnesium deficiency on photochemical activities of rape and buckwheat chloroplasts. Z. Pflanzenphysiol. 99, 295–303.

Battie-Laclau, P., Laclau, J.P., Beri, C., Mietton, L., Muniz, M.R.A., Arenque, B.C., et al., 2014. Photosynthetic and anatomical responses of *Eucalyptus grandis* leaves to potassium and sodium supply in a field experiment. Plant Cell Environ. 37, 70–81.

Baudet, J., Huet, J.-C., Lesaint, C., Mosse, J., Pernollet, J.-C., 1986. Changes in accumulation of seed nitrogen compounds in maize under conditions of sulphur deficiency. Physiol. Plant. 68, 608–614.

Bauer, P., Elbaum, R., Weiss, I.M., 2011. Calcium and silicon mineralization in land plants: Transport, structure and function. Plant Sci. 180, 746–756.

Bell, R.W., Edwards, D.G., Asher, C.J., 1990. Growth and nodulation of tropical food legumes in dilute solution culture. Plant Soil 122, 249–258.

Benning, C., Garavito, R.M., Shimojima, M., 2008. Sulfolipid biosynthesis and functioning in plants. In: Hell, R., Dahl, C., Knaff, D.B., Leustek, T. (Eds.), Sulfur Metabolism in Phototropic Organisms. Springer, Dordrecht, The Netherlands, pp. 185–200.

Ben-Zioni, A., Vaadia, Y., Lips, S.H., 1971. Nitrate uptake by roots as regulated by nitrate reduction products of the shoot. Physiol. Plant 24, 288–290.

Beringer, H., Forster, H., 1981. Einfluss variierter Mg-Ernährung auf Tausendkorngewicht und P-Fraktionen des Gerstenkorns. Z. Pflanzenernähr. Bodenk. 144, 8–15.

Beringer, H., Koch, K., Lindhauer, M.G., 1986. Sucrose accumulation and osmotic potentials in sugar beet at increasing levels of potassium nutrition. J. Sci. Food Agric. 37, 211–218.

Bernard, S.M., Habash, D.Z., 2009. The importance of cytosolic glutamine synthetase in N assimilation and recycling. New Phytol. 182, 608–620.

Bernard, S.M., Møller, A.L., Dionisio, G., Kichey, T., Jahn, T.P., Dubois, F., et al., 2008. Gene expression, cellular localisation and function of glutamine synthetase isozymes in wheat (*Triticum aestivum* L.). Plant Mol. Biol. 67, 89–105.

Bieleski, R.L., Ferguson, I.B., 1983. Physiology and metabolism of phosphate and its compounds. In: Läuchli, A., Bieleski, R.L. (Eds.), Encyclopedia of Plant Physiology, New Series, 15A. Springer Verlag, Berlin and New York, pp. 422–449.

Bilal, H.M., Aziz, T., Maqsood, M.A., Farooq, M., Yan, G., 2018. Categorization of wheat genotypes for phosphorus efficiency. PLoS ONE 13, e0205471.

Blake-Kalff, M.M.A., Hawkesford, M.J., Zhao, F.J., McGrath, S.P., 2000. Diagnosing sulphur deficiency in field-grown oilseed rape (*Brassica napus* L) and wheat (*Triticum aestivum* L.). Plant Soil 225, 95–107.

Blatt, M.R., Clint, G.M., 1989. Mechanisms of fusicoccin action: Kinetic modification and inactivation of K^+ channels in guard cells. Planta 178, 509–523.

Bloem, E., Haneklaus, S., Schnug, E., 2015. Milestones in plant sulfur research on sulfur-induced-resistence (SIR) in Europe. Front. Plant Sci. 5, 779.

Bloom, A.J., Burger, M., Rubio-Asensio, J.S., Cousins, A.B., 2010. Carbon dioxide enrichment inhibits nitrate assimilation in wheat and Arabidopsis. Science 328, 899–903.

Bloom, A.J., Kasemsap, P., Rubio-Asensio, J.S., 2020. Rising atmospheric CO_2 concentration inhibits nitrate assimilation in shoots but enhances it in roots of C_3 plants. Physiol. Plant 168, 963–972.

Bloom, A.J., Sukrapanna, S.S., Warner, R.L., 1992. Root respiration associated with ammonium and nitrate absorption and assimilation by barley. Plant Physiol. 99, 1294–1301.

Bohner, A., Kojima, S., Hajirezaei, M., Melzer, M., von Wirén, N., 2015. Urea retranslocation from senescing Arabidopsis leaves is promoted by DUR3-mediated urea retrieval from leaf apoplast. Plant J. 81, 377–387.

Bose, J., Babourina, O., Rengel, Z., 2011. Role of magnesium in alleviation of aluminium toxicity in plants. J. Exp. Bot. 62, 2251–2264.

Bottrill, D.E., Possingham, J.V., Kriedemann, P.E., 1970. The effect of nutrient deficiencies on photosynthesis and respiration in spinach. Plant Soil 32, 424–438.

Bouguyon, E., Brun, F., Meynard, D., Kubes, M., Pervent, M., Leran, S., et al., 2015. Multiple mechanisms of nitrate sensing by Arabidopsis nitrate transceptor NRT1.1. Nat. Plants 1, 15015.

Bould, C., Parfitt, R.I., 1973. Leaf analysis as a guide to the nutrition of fruit crops. X. Magnesium and phosphorus sand culture experiments with apple. J. Sci. Food Agric. 24, 175–185.

Bowling, D.J.F., 1987. Measurement of the apoplastic activity of K^+ and Cl^- in the leaf epidermis of *Commelina communis* in relation to stomatal activity. J. Exp. Bot. 38, 1351–1355.

Bowsher, C.G., Lacey, A.E., Hanke, G.T., Clarkson, D.T., Saker, L.R., Stulen, I., et al., 2007. The effect of Glc6P uptake and its subsequent oxidation within pea root plastids on nitrite reduction and glutamate synthesis. J. Exp. Bot. 58, 1109–1118.

Brault, M., Amiar, Z., Pennarun, A.M., Monestiez, M., Zhang, Z., Cornel, D., et al., 2004. Plasma membrane depolarization induced by abscisic acid in Arabidopsis suspension cells involves reduction of proton pumping in addition to anion channel activation, which are both Ca^{2+} dependent. Plant Physiol. 135, 231–243.

Brinckmann, E., Hartung, W., Wartinger, M., 1990. Abscisic acid levels of individual leaf cells. Physiol. Plant 80, 51–54.

Britto, D.T., Kronzucker, H.J., 2002. NH_4^+ toxicity in higher plants: a critical review. J. Plant Physiol. 159, 567–584.

Britto, D.T., Kronzucker, H.J., 2005. N acquisition, PEP carboxylase, and cellular pH homeostasis: new views on old paradigms. Plant Cell Environ. 28, 1396–1409.

Britto, D.T., Kronzucker, H.J., 2008. Cellular mechanisms of potassium transport in plants. Physiol. Plant 133, 637–650.

Britto, D.T., Siddiqi, M.Y., Glass, A.D.M., Kronzucker, H.J., 2001. Futile transmembrane NH_4^+ cycling: A cellular hypothesis to explain ammonium toxicity in plants. Proc. Natl. Acad. Sci. USA 98, 4255–4258.

Buch-Pedersen, M.J., Rudashevskaya, E.L., Berner, T.S., Venema, K., Palmgren, M.G., 2006. Potassium as an intrinsic uncoupler of the plasma membrane H^+-ATPase. J. Biol. Chem. 281, 38285–38292.

Budde, R.J.A., Chollet, R., 1988. Regulation of enzyme activity in plants by reversible phosphorylation. Physiol. Plant 72, 435–439.

Burke, J.J., Holloway, P., Dalling, M.J., 1986. The effect of sulfur deficiency on the organization and photosynthetic capability of wheat leaves. J. Plant Physiol. 125, 371–375.

Burns, I.G., Kefeng, Z., Turner, M.K., Edmondson, R., 2011. Iso-osmotic regulation of nitrate accumulation in lettuce. J. Plant Nutr. 34, 283–313.

Byrne, S.L., Foito, A., Hedley, P.E., Morris, J.A., Stewart, D., Barth, S., 2011. Early response mechanisms of perennial ryegrass (*Lolium perenne*) to phosphorus deficiency. Ann. Bot. 107, 243–254.

Cakmak, I., 1994. Activity of ascorbate-dependent H_2O_2-scavenging enzymes and leaf chlorosis are enhanced in magnesium- and potassium-deficient leaves, but not in phosphorus-deficient leaves. J. Exp. Bot. 45, 1259–1266.

Cakmak, I., 2005. The role of potassium in alleviating detrimental effects of abiotic stresses in plants. J. Plant Nutr. Soil Sci. 168, 521–530.

Cakmak, I., Kirkby, E.A., 2008. Role of magnesium in carbon partitioning and alleviating photooxidative damage. Physiol. Plant 133, 692–704.

Cakmak, I., Marschner, H., 1992. Magnesium deficiency and high light intensity enhance activities of superoxide dismutase, ascorbate peroxidase and glutathione reductase in bean leaves. Plant Physiol. 98, 1222–1227.

Cakmak, I., Hengeler, C., Marschner, H., 1994a. Partitioning of shoot and root dry matter and carbohydrates in bean plants suffering from phosphorus, potassium and magnesium deficiency. J. Exp. Bot. 45, 1245–1250.

Cakmak, I., Hengeler, C., Marschner, H., 1994b. Changes in phloem export of sucrose in leaves in response to phosphorus, potassium and magnesium deficiency in bean plants. J. Exp. Bot. 45, 1251–1257.

Cakmak, I., Pfeiffer, W.H., McClafferty, B., 2010. Biofortification of durum wheat with zinc and iron. Cereal Chem. 87, 10–20.

Cammarano, P., Felsani, A., Gentile, M., Gualerzi, C., Romeo, C., Wolf, G., 1972. Formation of active hybrid 80-S particles from subunits of pea seedlings and mammalian liver ribosomes. Biochim. Biophys. Acta 281, 625–642.

Campbell, N.A., Thomson, W.W., 1977. Effects of lanthanum and ethylene-diaminetetraacetate on leaf movements of *Mimosa*. Plant Physiol. 60, 635–639.

Carpita, N., McCann, M., 2000. The cell wall. In: Buchanan, B.B., Gruissem, W., Jones, R.L. (Eds.), Biochemistry and Molecular Biology of Plants. American Society of Plant Physiologists, Rockville, MD, USA, pp. 52–108.

Castle, S.L., Randall, P.J., 1987. Effects of sulfur deficiency on the synthesis and accumulation of proteins in the developing wheat seed. Aust. J. Plant Physiol. 14, 503–516.

Ceylan, Y., Kutman, U.B., Mengutay, M., Cakmak, I., 2016. Magnesium applications to growth medium and foliage affect the starch distribution, increase the grain size and improve the seed germination in wheat. Plant Soil 406, 145–156.

Chakraborty, K., Bose, J., Shabala, L., Shabala, S., 2016. Difference in root K^+ retention ability and reduced sensitivity of K^+-permeable channels to reactive oxygen species confer differential salt tolerance in three *Brassica* species. J. Exp. Bot. 67, 4611–4625.

Charpentier, M., 2018. Calcium signals in the plant nucleus: origin and function. J. Exp. Bot. 69, 4165–4173.

Chen, G., Guo, S., Kronzucker, H.J., Shi, W., 2013. Nitrogen use efficiency (NUE) in rice links to NH_4^+ toxicity and futile NH_4^+ cycling in roots. Plant Soil 369, 351–363.

Chen, K.E., Chen, H.Y., Tseng, C.S., Tsay, Y.F., 2020. Improving nitrogen use efficiency by manipulating nitrate remobilization in plants. Nat. Plants 6, 1126–1135.

Chen, Z.C., Yamaji, N., Horie, T., Che, J., Li, J., An, G., et al., 2017. A magnesium transporter OsMGT1 plays a critical role in salt tolerance in rice. Plant Physiol. 174, 1837–1849.

Chérel, I., Lefoulon, C., Boeglin, M., Sentenac, H., 2014. Molecular mechanisms involved in plant adaptation to low K^+ availability. J. Exp. Bot. 65, 833–848.

Chiera, J., Thomas, J., Rufty, T., 2002. Leaf initiation and development in soybean under phosphorus stress. J. Exp. Bot. 53, 473–481.

Chiu, C.C., Lin, C.S., Hsia, A.P., Su, R.C., Lin, H.L., Tsay, Y.F., 2004. Mutation of a nitrate transporter, AtNRT1.4, results in a reduced petiole nitrate content and altered leaf development. Plant Cell Physiol. 45, 1139–1148.

Choi, W.G., Hilleary, R., Swanson, S.J., Kim, S.H., Gilroy, S., 2016. Rapid, long-distance electrical and calcium signaling in plants. Annu. Rev. Plant Biol. 67, 287–307.

Choi, W.G., Toyota, M., Kim, S.H., Hilleary, R., Gilroy, S., 2014. Salt stress-induced Ca^{2+} waves are associated with rapid, long-distance root-to-shoot signaling in plants. Proc. Natl. Acad. Sci. USA 111, 6497–6502.

Chopin, F., Orsel, M., Dorbe, M.-F., Chardon, F., Truong, H.-N., Miller, A.J., et al., 2007. The Arabidopsis AtNRT2.7 nitrate transporter controls nitrate content in seeds. Plant Cell 19, 1590–1602.

Chow, W.S., Ball, M.C., Naderson, J.M., 1990. Growth and photosynthetic response of spinach to salinity: implications of K^+ nutrition for salt tolerance. Aust. J. Plant Physiol. 17, 563–578.

Christiansen, M.N., Carns, H.R., Slyter, D.J., 1970. Stimulation of solute loss from radicles of *Gossypium hirsutum* L. by chilling, anaerobiosis, and low pH. Plant Physiol. 46, 53–56.

Cirilo, A.G., Dardanelli, J., Balzarini, M., Andrade, F.H., Cantarero, M., Luque, S., et al., 2009. Morpho-physiological traits associated with maize crop adaptations to environments differing in nitrogen availability. Field Crops Res. 113, 116–124.

Clark, G.B., Morgan, R.O., Fernandez, M.P., Roux, S.J., 2012. Evolutionary adaptation of plant annexins has diversified their molecular structures, interactions and functional roles. New Phytol. 196, 695–712.

Clarkson, D.T., Carvajal, M., Henzler, T., Waterhouse, R.N., Smyth, A.J., Cooke, D.T., et al., 2000. Root hydraulic conductance: diurnal aquaporin expression and the effects of nutrient stress. J. Exp. Bot. 51, 61–70.

Clarkson, D.T., Hanson, J.B., 1980. The mineral nutrition of higher plants. Annu. Rev. Plant Physiol. 31, 239–298.

Cobbett, C., Goldsbrough, P., 2002. Phytochelatins and metallothioneins: roles in heavy metal detoxification and homeostasis. Annu. Rev. Plant Biol. 53, 159–182.

Collins, M., Duke, S.H., 1981. Influence of potassium fertilization rate and form on photosynthesis and N_2 fixation of alfalfa. Crop Sci. 21, 481–485.

Cong, W.-F., Suriyagoda, L.D.B., Lambers, H., 2020. Tightening the phosphorus cycle through phosphorus-efficient crop genotypes. Trends Plant Sci. 25, 967–975.

Conn, S., Gilliham, M., 2010. Comparative physiology of elemental distributions in plants. Ann. Bot. 105, 1081–1102.

Cookson, S.J., Williams, L.E., Miller, A.J., 2005. Light-dark changes in cytosolic nitrate pools depend on nitrate reductase activity in Arabidopsis leaf cells. Plant Physiol. 138, 1097–1105.

Cordell, D., White, S., 2014. Life's bottleneck: sustaining the world's phosphorus for a food secure future. Annu. Rev. Env. Res. 39, 161–188.

Cosgrove, D.J., 2016. Plant cell wall extensibility: Connecting plant cell growth with cell wall structure, mechanics and the action of wall-modifying enzymes. J. Exp. Bot. 67, 463–476.

Coskun, D., Britto, D.T., Jean, Y.-K., Kabir, I., Tolay, I., Torun, A.A., et al., 2013c. K^+ efflux and retention in response to NaCl stress do not predict salt tolerance in contrasting genotypes of rice (*Oryza sativa* L.). PLoS ONE 8, e57767.

Coskun, D., Britto, D.T., Kronzucker, H.J., 2017. The nitrogen-potassium intersection: membranes, metabolism and mechanism. Plant Cell Environ. 40, 2029–2041.

Coskun, D., Britto, D.T., Li, M., Becker, A., Kronzucker, H.J., 2013a. Rapid ammonia gas transport accounts for futile transmembrane cycling under NH$_3$/NH$_4^+$ toxicity in plant roots. Plant Physiol. 163, 1859–1867.

Coskun, D., Britto, D.T., Li, M., Oh, S., Kronzucker, H.J., 2013b. Capacity and plasticity of potassium channels and high-affinity transporters in roots of barley and Arabidopsis. Plant Physiol. 162, 496–511.

Costa, A., Navazio, L., Szabo, I., 2018. The contribution of organelles to plant intracellular calcium signaling. J. Exp. Bot. 69, 4175–4193.

Cowan, J.A., 2002. Structural and catalytic chemistry of magnesium-dependent enzymes. Biometals 15, 225–235.

Cramer, G.R., 2002. Sodium-calcium interactions under salinity stress. In: Läuchli, A., Lüttge, U. (Eds.), Salinity: Environment–Plants–Molecules. Kluwer, Dordrecht, The Netherlands, pp. 205–277.

Cramer, M.D., Schierholt, A., Wang, Y.Z., Lips, S.H., 1995. The influence of salinity on the utilization of root anaplerotic carbon and N metabolism in tomato seedlings. J. Exp. Bot. 46, 1569–1577.

Crawford, N.M., Arst Jr., H.N., 1993. The molecular genetics of nitrate assimilation in fungi and plants. Annu. Rev. Genetics 27, 115–146.

Cuin, T.A., Miller, A.J., Laurie, S.A., Leigh, R.A., 2003. Potassium activities in cell compartments of salt-grown barley leaves. J. Exp. Bot. 54, 657–661.

Danchin, A., Nikel, P.I., 2019. Why nature chose potassium. J. Mol. Evol. 87, 271–288.

Darley, C.P., Skiera, L.A., Northrop, F.D., Sanders, D., Davies, J.M., 1998. Tonoplast inorganic pyrophosphatase in *Vicia faba* guard cells. Planta 206, 272–277.

Davies, W.J., Meinzer, F.C., 1990. Stomatal responses of plants in drying soil. Biochem. Physiol. Pflanzen 186, 357–366.

Dayod, M., Tyerman, S.D., Leigh, R.A., Gilliham, M., 2010. Calcium storage in plants and the implications for calcium biofortification. Protoplasma 247, 215–231.

De Angeli, A., Zhang, J., Meyer, S., Martinoia, E., 2013. AtALMT9 is a malate-activated vacuolar chloride channel required for stomatal opening in Arabidopsis. Nat. Commun. 4, 1–10.

De Groot, C.C., Van den Boogaard, R., Marcelis, L.F.M., Harbinson, J., Lambers, H., 2003. Contrasting effects of N and P deprivation on the regulation of photosynthesis in tomato plants in relation to feedback limitation. J. Exp. Bot. 54, 1957–1967.

De Kok, L.J., Durenkamp, M., Yang, L., Stulen, I., 2007. Atmospheric sulfur. In: Hawkesford, M.J., De Kok, L.J. (Eds.), Sulfur in Plants: An Ecological Perspective. Springer, Dordrecht, The Netherlands, pp. 91–106.

De Kok, L.J., Stulen, I., 1993. Role of glutathione in plants under oxidative stress. In: De Kok, L.J., Stulen, I., Rennenberg, H., Brunold, C., Rauser, W.E. (Eds.), Sulfur Nutrition and Assimilation in Higher Plants - Regulatory Agricultural and Environmental Aspects. SPB Academic Publishing bv, The Hague, The Netherlands, pp. 125–138.

De la Guardia, M.D., Benlloch, M., 1980. Effects of potassium and gibberellic acid on stem growth of whole sunflower plants. Physiol. Plant 49, 443–448.

DeBoer, D.L., Duke, S.H., 1982. Effects of sulphur nutrition on nitrogen and carbon metabolism in lucerne (*Medicago sativa* L.). Physiol. Plant 54, 343–350.

Dechorgnat, J., Nguyen, C.T., Armengaud, P., Jossier, M., Diatloff, E., Filleur, S., et al., 2011. From the soil to the seeds, the long journey of nitrate in plants. J. Exp. Bot. 62, 1349–1359.

DeFalco, T.A., Moeder, W., Yoshioka, K., 2016. Opening the gates: insights into cyclic nucleotide-gated channel-mediated signaling. Trends Plant Sci. 21, 903–906.

Demidchik, V., Shabala, S., Isayenkov, S., Cuin, T.A., Pottosin, I., 2018. Calcium transport across plant membranes: mechanisms and functions. New Phytol. 220, 49–69.

Deshaies, R.J., Fish, L.E., Jagendorf, A.T., 1984. Permeability of chloroplast envelopes to Mg^{2+} – Effects on protein synthesis. Plant Physiol. 74, 956–961.

Dhont, C., Castonguay, Y., Avice, J.C., Chalifour, F.P., 2006. VSP accumulation and cold-inducible gene expression during autumn hardening and overwintering of alfalfa, J. Exp. Bot., 57. pp. 2325–2337.

Di Laurenzio, L., Wysocka-Diller, J., Malamy, J.E., Pysh, L., Helariutta, Y., Freshour, G., et al., 1996. The *SCARECROW* gene regulates an asymmetric cell division that is essential for generating the radial organization of the Arabidopsis root. Cell 86, 423–433.

Dietz, K.-J., 1989. Recovery of spinach leaves from sulfate and phosphate deficiency. J. Plant Physiol. 134, 551–557.

Dijkshoorn, W., van Wijk, A.L., 1967. The sulphur requirement of plants as evidenced by the sulphur-nitrogen ratio in the organic matter. A review of published data. Plant Soil 26, 129–157.

Dijkshoorn, W., Lathwell, D.J., De Wit, C.T., 1968. Temporal changes in carboxylate content of ryegrass with stepwise change in nutrition. Plant Soil 29, 369–390.

Ding, Y., Shi, Y., Yang, S., 2019. Advances and challenges in uncovering cold tolerance regulatory mechanisms in plants. New Phytol. 222, 1690–1704.

Dodd, A.N., Kudla, J., Sanders, D., 2010. The language of calcium signaling. Annu. Rev. Plant Biol. 61, 593–620.

Dolan, L., Davies, J., 2004. Cell expansion in roots. Curr. Opin. Plant Biol. 7, 33–39.

Dong, B., Ryan, P.R., Rengel, Z., Delhaize, E., 1999. Phosphate uptake in *Arabidopsis thaliana*: dependence of uptake on the expression of transporter genes and internal phosphate concentrations. Plant Cell Environ. 22, 1455–1461.

Doolette, A.L., Smernik, R.J., 2016. Phosphorus speciation of dormant grapevine (*Vitis vinifera* L.) canes in the Barossa Valley, South Australia. Aust. J. Grape Wine Res. 22, 462–468.

Drew, M.C., 1975. Comparison of the effects of a localized supply of phosphate, nitrate, ammonium, and potassium on the growth of the seminal root system, and the shoot, in barley. New Phytol. 75, 479–490.

Dreyer, I., Gomez-Porras, J.L., Riedelsberger, J., 2017. The potassium battery: A mobile energy source for transport processes in plant vascular tissues. New Phytol. 216, 1049–1053.

Duan, F.Y., Giehl, R.F.H., Geldner, N., Salt, D.E., von Wirén, N., 2018. Root zone-specific localization of AMTs determines ammonium transport pathways and nitrogen allocation to shoots. PLoS Biol. 16, e2006024.

Duff, S.M.G., Moorhead, G.B.G., Lefebvre, D.D., Plaxton, W.C., 1989. Phosphate starvation inducible 'bypasses' of adenylate and phosphate dependent glycolytic enzymes in *Brassica nigra* suspension cells. Plant Physiol. 90, 1275–1278.

Dunbabin, V., Diggle, A., Rengel, Z., 2003. Is there an optimal root architecture for nitrate capture in leaching environments? Plant Cell Environ. 26, 835–844.

Eberl, D., Preissler, M., Steingraber, M., Hampp, R., 1992. Subcellular compartmentation of pyrophosphate and dark/light kinetics in comparison to fructose 2,6-bisphosphate. Physiol. Plant 84, 13–26.

Endler, A., Meyer, S., Schelbert, S., Schneider, T., Weschke, W., Peters, S.W., et al., 2006. Identification of a vacuolar sucrose transporter in barley and Arabidopsis mesophyll cells by a tonoplast proteomic approach. Plant Physiol. 141, 196–207.

Erdei, L., Stuiver, B., Kuiper, P.J.C., 1980. The effect of salinity on lipid composition and on activity of Ca^{2+} and Mg^{2+}-stimulated ATPases in salt-sensitive and salt-tolerant *Plantago* species. Physiol. Plant 49, 315–319.

Ernst, W.H.O., 1993. Ecological aspect of sulfur in higher plants: the impact of SO_2 and the evolution of the biosynthesis of organic sulfur compounds on populations and ecosystems. In: De Kok, J.L., Stulen, I., Rennenberg, H., Brunold, C., Rauser, W.E. (Eds.), Sulfur Nutrition and Assimilation in Higher Plants. SPB Academic Publishing bv, The Hague, The Netherlands, pp. 295–313.

Ewart, J.A.D., 1978. Glutamin and dough tenacity. J. Sci. Food Agric. 29, 551–556.

Fan, X., Tang, Z., Tan, Y., Zhang, Y., Luo, B., Yang, M., et al., 2016. Overexpression of a pH-sensitive nitrate transporter in rice increases crop yields. Proc. Natl. Acad. Sci. USA 113, 7118–7123.

Fankhauser, H., Brunold, C., 1978. Localization of adenosine 5'-phosphosulfate sulfotransferase in spinach leaves. Planta 143, 285–289.

FAO 2019. Current world fertilizer trends and outlook to 2022. In: Food and Agriculture Organization of the United Nations, Rome, Italy.

Faust, F., Schubert, S., 2017. *In vitro* protein synthesis of sugar beet (*Beta vulgaris*) and maize (*Zea mays*) is differentially inhibited when potassium is substituted by sodium. Plant Physiol. Biochem. 118, 228–234.

Ferguson, I.B., 1984. Calcium in plant senescence and fruit ripening. Plant Cell Environ. 7, 477–489.

Fichman, Y., Miller, G., Mittler, R., 2019. Whole-plant live imaging of reactive oxygen species. Mol. Plant 12, 1203–1210.

Finnemann, J., Schjoerring, J.K., 1999. *Translocation of NH_4^+ in oilseed rape plants in relation to glutamine synthetase isogene expression and activity*. Physiol. Plant 105, 469–477.

Finnemann, J., Schjoerring, J.K., 2000. Post-translational regulation of cytosolic glutamine synthetase by reversible phosphorylation and 14-3-3 protein interaction. Plant J. 24, 171–181.

Fischer, E.S., Bussler, W., 1988. Effects of magnesium deficiency on carbohydrates in *Phaseolus vulgaris*. Z. Pflanzenernähr. Bodenk. 151, 295–298.

Fischer, R.A., 1968. Stomatal opening: Role of potassium uptake by guard cells. Science 160, 784–785.

Fixen, P.E., Johnston, A.M., 2012. World fertilizer nutrient reserves: a view to the future. J. Sci. Food Agric. 92, 1001–1005.

Flügge, U.I., Freisl, M., Heldt, H.W., 1980. Balance between metabolite accumulation and transport in relation to photosynthesis by isolated spinach chloroplasts. Plant Physiol. 65, 574–577.

Forde, B.G., 2000. Nitrate transporters in plants, structure, function and regulation. Biochim. Biophys. Acta. 1465, 219–235.

Forde, B.G., 2002. Local and long-range signalling pathways regulating plant responses to nitrate. Annu. Rev. Plant Biol. 53, 203–224.

Foulkes, M.J., Hawkesford, M.J., Barraclough, P.B., Holdsworth, M.J., Kerr, S., Kightley, S., et al., 2009. Identifying traits to improve the N economy of wheat: recent advances and future prospects. Field Crops Res. 114, 329–342.

Fowler, D., Pilegaard, K., Sutton, M.A., Ambus, P., Raivonen, M., Duyzer, J., et al., 2009. Atmospheric composition change: Ecosystems-Atmosphere interactions. Atmos. Environ. 43, 5193–5267.

Franceschi, V.R., Nakata, P.A., 2005. Calcium oxalate in plants: formation and function. Annu. Rev. Plant Biol. 56, 41–71.

Fredeen, A.L., Raab, T.K., Rao, I.M., Terry, N., 1990. Effects of phosphorus nutrition on photosynthesis in *Glycine max* (L.) Merr. Planta 181, 399–405.

Fredeen, A.L., Rao, I.M., Terry, N., 1989. Influence of phosphorus nutrition on growth and carbon partitioning in *Glycine max*. Plant Physiol. 89, 225–230.

Freney, J.R., Spencer, K., Jones, M.B., 1978. The diagnosis of sulphur deficiency in wheat. Aust. J. Agric. Res. 29, 727–738.

Fromm, J., 1991. Control of phloem unloading by action potentials in *Mimosa*. Physiol. Plant 83, 529–533.

Fromm, J., Eschrich, W., 1988. Transport processes in stimulated and non-stimulated leaves of *Mimosa pudica*. III. Displacement of ions during seismonastic leaf movements. Trees 2, 65–72.

Gajdanowicz, P., Michard, E., Sandmann, M., Rocha, M., Guedes Corrêa, L.G., Ramírez-Aguilar, S.J., et al., 2010. Potassium (K^+) gradients serve as a mobile energy source in plant vascular tissues. Proc. Natl. Acad. Sci. USA 108, 864–869.

Gallardo, A., Fernández-Palacios, J.M., Bermúdez, A., de Nascimento, L., Durán, J., García-Velázquez, L., et al., 2020. The pedogenic Walker and Syers model under high atmospheric P deposition rates. Biogeochemistry 148, 237–253.

Galvis, V.C., Strand, D.D., Messer, M., Thiele, W., Bethmann, S., Hübner, D., et al., 2020. H^+ transport by K^+ EXCHANGE ANTIPORTER3 promotes photosynthesis and growth in chloroplast ATP synthase mutants. Plant Physiol. 182, 2126–2142.

Ganeteg, U., Ahmad, I., Jämtgard, S., Aguetoni-Cambui, C., Inselsbacher, E., Svennerstam, H., et al., 2017. Amino acid transporter mutants of Arabidopsis provide evidence that a non-mycorrhizal plant acquires organic nitrogen from agricultural soil. Plant Cell Environ. 40, 413–423.

Gansel, X., Munõs, S., Tillard, P., Gojon, A., 2001. Differential regulation of the NO_3^- and NH_4^+ transporter genes *AtNrt2.1* and *AtAmt1.1* in Arabidopsis: relation with long-distance and local controls by N status of the plant. Plant J. 26, 143–155.

Gao, Y., de Bang, T., Schjoerring, J.K., 2019. Cisgenic overexpression of cytosolic glutamine synthetase improves nitrogen utilization efficiency in barley and prevents grain protein decline under elevated CO_2. Plant Biotech. J. 17, 1209–1221.

García-Martí, M., Piñero, M.C., García-Sanchez, F., Mestre, T.C., López-Delacalle, M., Martínez, V., et al., 2019. Amelioration of the oxidative stress generated by simple or combined abiotic stress through the K^+ and Ca^{2+} supplementation in tomato plants. Antioxidants (Basel) 8, 81.

Garnett, T., Conn, V., Kaiser, B.N., 2009. Root based approaches to improving N use efficiency in plants. Plant Cell Environ. 32, 1272–1283.

Garnica, M., Houdusse, F., Zamarreno, A.M., Garcia-Mina, J.M., 2010. The signal effect of nitrate supply enhances active forms of cytokinins and indole acetic content and reduces abscisic acid in wheat plants grown with ammonium. J. Plant Physiol. 167, 1264–1272.

Gaude, N., Nakamura, Y., Scheible, W.-R., Ohta, H., Dörmann, P., 2008. Phospholipase C5 (NPC5) is involved in galactolipid accumulation during phosphate limitation in leaves of Arabidopsis. Plant J. 56, 28–39.

Gavrichkova, O., Kuzyakov, Y., 2010. Respiration costs associated with nitrate reduction as estimated by $^{14}CO_2$ pulse labeling of corn at various growth stages. Plant Soil 329, 433–445.

Gazzarrini, S., Lejay, L., Gojon, A., Ninnemann, O., Frommer, W.B., von Wirén, N., 1999. Three functional transporters for constitutive, diurnally regulated, and starvation-induced uptake of ammonium into Arabidopsis roots. Plant Cell 11, 937–947.

Gerardeaux, E., Jordan-Meille, L., Constantin, J., Pellerin, S., Dingkuhn, M., 2010. Changes in plant morphology and dry matter partitioning caused by potassium deficiency in *Gossypium hirsutum* (L.). Environ. Exp. Bot. 67, 451–459.

Gerendás, J., Sattelmacher, B., 1990. Influence of nitrogen form and concentration on growth and ionic balance of tomato (*Lycopersicon esculentum*) and potato (*Solanum tuberosum*). In: van Beusichem, M.L. (Ed.), Plant Nutrition – Physiology and Application. Kluwer Acad. Publ., Dordrecht, The Netherlands, pp. 33–37.

Gerhardt, R., Stitt, M., Heldt, H.W., 1987. Subcellular metabolite levels in spinach leaves. Regulation of sucrose synthesis during diurnal alterations in photosynthetic partitioning. Plant Physiol. 83, 399–407.

Gerwick, B.C., Ku, S.B., Black, C.C., 1980. Initiation of sulfate activation: A variation in C_4 photosynthesis in plants. Science 209, 513–515.

Gibrat, R., Grouzis, J.P., Rigaud, J., Grignon, C., 1990. Potassium stimulation of corn root plasmalemma ATPase: II. H-pumping in native and reconstituted vesicles with purified ATPase. Plant Physiol. 93, 1183–1189.

Gibson, R.S., Raboy, V., King, J.C., 2018. Implications of phytate in plant-based foods for iron and zinc bioavailability, setting dietary requirements, and formulating programs and policies. Nutr. Rev. 76, 793–804.

Giehl, R.F.H., Laginha, A.M., Duan, F.Y., Rentsch, D., Yuan, L.X., von Wirén, N., 2017. A critical role of AMT2;1 in root-to-shoot translocation of ammonium in Arabidopsis. Mol. Plant 10, 1449–1460.

Gilbert, S., Clarkson, D.T., Cambridge, M., Lambers, H., Hawkesford, M.J., 1997. Sulfate-deprivation has an early effect on the content of ribulose 1,5-bisphosphate carboxylase/oxygenase and photosynthesis in young leaves of wheat. Plant Physiol. 115, 1231–1239.

Glass, A.D.M., Britto, D.T., Kaiser, M.N., Konghorn, J.R., Kronzucker, H.J., Kumar, A., et al., 2002. The regulation of nitrate and ammonium transport systems in plants. J. Exp. Bot. 53, 855–864.

Godfrey, D., Hawkesford, M.J., Powers, S.J., Millar, S., Shewry, P.R., 2010. Effects of crop nutrition on wheat grain composition and end use quality. J. Agric. Food Chem. 58, 3012–3021.

Good, A.G., Shrawat, A.K., Muench, D.G., 2004. Can less yield more? Is reducing nutrient input into the environment compatible with maintaining crop production? Trends Plant Sci. 9, 597–605.

Grewal, J.S., Singh, S.N., 1980. Effect of potassium nutrition on frost damage and yield of potato plants on alluvial soils of the Punjab (India). Plant Soil 57, 105–110.

Grignon, C., Sentenac, H., 1991. pH and ionic conditions in the apoplast. Annu. Rev. Plant Physiol. 42, 103–128.

Grill, E., Winnacker, E.-L., Zenk, M.H., 1987. Phytochelatins, a class of heavy metal binding peptides from plants are functionally analogous to metallothioneins. Proc. Natl. Acad. Sci. USA 84, 439–443.

Gruber, B.D., Giehl, R.F.H., Friedel, S., von Wirén, N., 2013. Plasticity of the Arabidopsis root system under nutrient deficiencies. Plant Physiol. 163, 161–179.

Grunes, D.L., Stout, P.R., Brownell, J.R., 1970. Grass tetany of ruminants. Adv. Agron. 22, 332–374.

Gu, R., Duan, F., An, X., Zhang, F., von Wirén, N., Yuan, L., 2013. Characterization of AMT-mediated high-affinity ammonium uptake in roots of maize (*Zea mays* L.). Plant Cell Physiol. 54, 1515–1524.

Guan, M., de Bang, T.C., Pedersen, C., Schjoerring, J.K., 2016. Cytosolic glutamine synthetase Gln1;2 is the main isozyme contributing to GS1 activity and can be up-regulated to relieve ammonium toxicity. Plant Physiol. 171, 1921–1933.

Guan, M., Møller, I.S., Schjoerring, J.K., 2015. Two cytosolic glutamine synthetase isoforms play specific roles for seed germination and seed yield structure in Arabidopsis. J. Exp. Bot. 66, 203–212.

Guilherme Pereira, C., Hayes, P.E., O'Sullivan, O., Weerasinghe, L., Clode, P.L., Atkin, O.K., et al., 2019. Trait convergence in photosynthetic nutrient-use efficiency along a 2-million year dune chronosequence in a global biodiversity hotspot. J. Ecol. 107, 2006–2023.

Hager, A., 2003. Role of the plasma membrane H^+-ATPase in auxin-induced elongation growth: historical and new aspects. J. Plant Res. 116, 483–505.

Hajibagheri, M.A., Harvey, D.M.R., Flowers, T.J., 1987. Quantitative ion distribution within root cells of salt-sensitive and salt-tolerant maize varieties. New Phytol. 105, 367–379.

Halkier, B.A., Gershenzon, J., 2006. Biology and biochemistry of glucosinolates. Annu. Rev. Plant Biol. 57, 303–333.

Hammer, T.A., Ye, D., Pang, J., Foster, K., Lambers, H., Ryan, M.H., 2020. Mulling over mulla mullas: revisiting phosphorus hyperaccumulation in the Australian plant genus *Ptilotus (Amaranthaceae)*. Aust. J. Bot. 68, 63–74.

Haneklaus, S., Bloem, E., Schnug, E., De Kok, L.J., Stulen, I., 2007. Sulfur. In: Barker, A.V., Pilbeam, D.J. (Eds.), Handbook of Plant Nutrition. CRC Press, Boca Raton, FL, USA, pp. 183–238.

Hao, D.-L., Zhou, J.-Y., Yang, S.-Y., Qi, W., Yang, K.-J., Su, Y.-H., 2020. Function and regulation of ammonium transporters in plants. Int. J. Mol. Sci. 21, 3557.

Hartt, C.E., 1969. Effect of potassium deficiency upon translocation of ^{14}C in attached blades and entire plants of sugarcane. Plant Physiol. 44, 1461–1469.

Haschke, H.P., Lüttge, K., 1975. Interactions between IAA, potassium, and malate accumulation on growth in *Avena* coleoptile segments. Z. Pflanzenpysiol. 76, 450–455.

Hauer-Jákli, M., Tränkner, M., 2019. Critical leaf magnesium thresholds and the impact of magnesium on plant growth and photo-oxidative defense: A systematic review and meta-analysis from 70 years of research. Front. Plant Sci. 10, 766.

Hawkesford, M.J., 2014. Reducing the reliance on nitrogen fertilizer for wheat production. J. Cereal Sci. 59, 276–283.

Hawkesford, M.J., De Kok, L.J., 2007. Sulfur in Plants: An Ecological Perspective. Springer, Dordrecht, The Netherlands.

Hawkesford, M.J., De Kok, L.J., 2006. Managing sulphur metabolism in plants. Plant Cell Environ. 29, 382–395.

Hawkesford, M.J., Griffiths, S., 2019. Exploiting genetic variation in nitrogen use efficiency for cereal crop improvement. Curr. Opin. Plant Biol. 49, 35–42.

Hawkesford, M.J., Miller, A.J., 2004. Ion-coupled transport of inorganic solutes. In: Blatt, M.R. (Ed.), Membrane Transport in Plants: Annual Plant Reviews. Blackwell Publishing Ltd., Oxford, UK, pp. 105–134.

Hawkesford, M.J., Riche, A.B., 2020. Impacts of G x E x M on nitrogen use efficiency in wheat and future prospects. Front. Plant Sci. 11, 1157.

Hayes, P.E., Clode, P.L., Guilherme Pereira, C., Lambers, H., 2019. Calcium modulates leaf cell-specific phosphorus allocation in Proteaceae from south-western Australia. J. Exp. Bot. 70, 3995–4009.

He, H., Veneklaas, E.J., Kuo, J., Lambers, H., 2014. Physiological and ecological significance of biomineralization in plants. Trends Plant Sci. 19, 166–174.

He, Y.N., Peng, J.S., Cai, Y., Liu, D.F., Guan, Y., Yi, H.Y., et al., 2017. Tonoplast-localized nitrate uptake transporters involved in vacuolar nitrate efflux and reallocation in Arabidopsis. Sci. Rep. 7, 6417.

Heber, U., Viil, J., Neimanis, S., Mimura, T., Dietz, K.J., 1989. Photoinhibitory damage to chloroplasts under phosphate deficiency and alleviation of deficiency and damage by photorespiratory reactions. Z. Naturforschung C 44, 524.

Hedrich, R., Geiger, D., 2017. Biology of SLAC1-type anion channels - from nutrient uptake to stomatal closure. New Phytol. 216, 46–61.

Hedrich, R., Mueller, T.D., Becker, D., Marten, I., 2018. Structure and function of TPC1 vacuole SV channel gains shape. Mol. Plant 11, 764–775.

Heenan, D.P., Campbell, L.C., 1981. Influence of potassium and manganese on growth and uptake of magnesium by soybeans (*Glycine max* (L.) Merr. cv Bragg). Plant Soil 61, 447–456.

Heineke, D., Heldt, H.W., 1988. Measurement of light-dependent changes of the stromal pH in wheat leaf protoplasts. Bot. Acta 101, 45–47.

Heldt, H.W., Piechulla, B., 2011. Products of nitrate assimilation are deposited in plants as storage proteins. In: Heldt, H.W., Piechulla, B. (Eds.), Plant Biochemistry, 4th ed. Academic Press, Oxford, UK, pp. 349–357.

Heldt, H.W., Chon, C.J., Maronde, D., Herold, A., Stankovic, Z.S., Walker, D.A., et al., 1977. Role of orthophosphate and other factors in the regulation of starch formation in leaves and isolated chloroplasts. Plant Physiol. 59, 1146–1155.

Heldt, H.W., Flügge, U.-I., Borchert, S., 1991. Diversity of specificity and function of phosphate translocators in various plastids. Plant Physiol. 95, 341–343.

Hell, R., Bergmann, L., 1988. Glutathione synthetase in tobacco suspension cultures: cotalytic properties and localization. Physiol. Plant 72, 70–76.

Hell, R., Wirtz, M., 2008. Metabolism of cysteine in plants and phototropic bacteria. In: Hell, R., Dahl, C., Knaff, D.B., Leustek, T. (Eds.), Sulfur Metabolism in Phototrophic Organisms. Springer, Dordrecht, The Netherlands, pp. 59–91.

Hell, R., Schwenn, J.D., Bork, C., 1997. Light and sulfur sources modulate mRNA levels of several genes of sulfate assimilation. In: Cram, W.J., De Kok, L.J., Stulen, I., Brunold, C., Rennenberg, H. (Eds.), Sulfur Metabolism in Higher Plants. Backhuys Publishers, Leiden, The Netherlands, pp. 181–185.

Hermans, C., Verbruggen, N., 2005. Physiological characterization of Mg deficiency in *Arabidopsis thaliana*. J. Exp. Bot. 56, 2153–2161.

Hermans, C., Bourgis, F., Faucher, M., Strasser, R.J., Delrot, S., Verbruggen, N., 2005. Magnesium deficiency in sugar beet alters sugar partitioning and phloem loading in young mature leaves. Planta 220, 541–549.

Hermans, C., Conn, S.J., Chen, J., Xiao, Q., Verbruggen, N., 2013. An update on magnesium homeostasis mechanisms in plants. Metallomics 5, 1170–1183.

Hermans, C., Johnson, G.N., Strasser, R.J., Verbruggen, N., 2004. Physiological characterisation of magnesium deficiency in sugar beet: acclimation to low magnesium differentially affects photosystems I and II. Planta 220, 344–355.

Hermans, C., Vuylsteke, M., Coppens, F., Craciun, A., Inzé, D., Verbruggen, N., 2010a. Early transcriptomic changes induced by magnesium deficiency in *Arabidopsis thaliana* reveal the alteration of circadian clock gene expression in roots and the triggering of abscisic acid-responsive genes. New Phytol. 187, 119–131.

Hermans, C., Vuylsteke, M., Coppens, F., Cristescu, S.M., Harren, F.J.M., Inzé, D., et al., 2010b. Systems analysis of the responses to long-term magnesium deficiency and restoration in *Arabidopsis thaliana*. New Phytol. 187, 132–144.

Hesse, H., Kreft, O., Maimann, S., Zeh, M., Hoefgen, R., 2004. Current understanding of the regulation of methionine biosynthesis in plants. J. Exp. Bot. 55, 1799–1808.

Hilleary, R., Gilroy, S., 2018. Systemic signaling in response to wounding and pathogens. Curr. Opin. Plant Biol. 43, 57–62.

Hinsinger, P., Plassard, C., Jaillard, B., Tang, C.X., 2003. Origins of root-mediated pH changes in the rhizosphere and their responses to environmental constraints: a review. Plant Soil 248, 43–59.

Hirel, B., Le Gouis, J., Ney, B., Gallais, A., 2007. The challenge of improving N use efficiency in crop plants: towards a more central role for genetic variability and quantitative genetics within integrated approaches. J. Exp. Bot. 58, 2369–2387.

Ho, L., White, P.J., 2005. A cellular hypothesis for the induction of blossom-end rot in tomato fruit. Ann. Bot. 95, 571–581.

Hocking, B., Tyerman, S.D., Burton, R.A., Gilliham, M., 2016. Fruit calcium: transport and physiology. Front. Plant Sci. 7, 569.

Höglund, A.-S., Lenman, M., Falk, A., Rask, L., 1991. Distribution of myrosinase in rape-seed tissues. Plant Physiol. 95, 213–221.

Holding, D.R., Larkins, B., 2008. Genetic engineering of seed storage proteins. In: Bohnert, H.J., Nguyen, H., Lewis, N.G. (Eds.), Advances in Plant Biochemistry and Molecular Biology, 1. Elsevier, pp. 107–133.

Horst, W.J., Wang, Y.X., Eticha, D., 2010. The role of the root apoplast in aluminium-induced inhibition of root elongation and in aluminium resistance of plants: a review. Ann. Bot. 106, 185–197.

Hortensteiner, S., Feller, U., 2002. N metabolism and remobilization during senescence. J. Exp. Bot. 53, 927–937.

Hou, J., Li, T., Wang, Y., Hao, C., Liu, H., Zhang, X., 2017. ADP-glucose pyrophosphorylase genes, associated with kernel weight, underwent selection during wheat domestication and breeding. Plant Biotechnol. J. 15, 1533–1543.

Howitt, S.M., Udvardi, M.K., 2000. Structure, function and regulation of ammonium transporters in plants. Biochim. Biophys. Acta 1465, 152–170.

Hu, W., Coomer, T.D., Loka, D.A., Oosterhuis, D.M., Zhou, Z., 2017. Potassium deficiency affects the carbon-nitrogen balance in cotton leaves. Plant Physiol. Biochem. 115, 408–417.

Huber, S.C., 1984. Biochemical basis for effects of K-deficiency on assimilate export rate and accumulation of soluble sugars in soybean leaves. Plant Physiol. 76, 424–430.

Hucklesby, D.P., Blanke, M.M., 1992. Limitation of N assimilation in plants. 4. Effect of defruiting on nitrate assimilation, transpiration, and photosynthesis in tomato leaf. Gartenbauwissenschaft 57, 53–56.

Humble, G.D., Raschke, K., 1971. Stomatal opening quantitatively related to potassium transport. Plant Physiol. 48, 447–453.

Inoue, T., Higuchi, M., Hashimoto, Y., Seki, M., Kobayashi, M., Kato, T., 2001. Identification of CRE1 as a cytokinin receptor from Arabidopsis. Nature 409, 1060–1063.

Isayenkov, S., Isner, J.C., Maathuis, F.J.M., 2010. Vacuolar ion channels: roles in plant nutrition and signalling. FEBS Lett. 584, 1982–1988.

Ishijima, S., 2003. Light-induced increase in free Mg^{2+} concentration in spinach chloroplasts: measurement of free Mg^{2+} by using a fluorescent probe and necessity of stromal alkalinization. Arch. Biochem. Biophys. 412, 126–132.

Ishimaru, Y., Masuda, H., Bashir, K., Inoue, H., Tsukamoto, T., Takahashi, M., 2010. Rice metal-nicotianamine transporter, OsYSL2, is required for the long-distance transport of iron and manganese. Plant J. 62, 379–390.

Ishiyama, K., Inoue, E., Watanabe-Takahashi, A., Obara, M., Yamaya, T., Takahashi, H., 2004. Kinetic properties and ammonium-dependent regulation of cytosolic isoenzymes of glutamine synthetase in Arabidopsis. J. Biol. Chem. 279, 16598–16605.

Islam, A.K.M.S., Asher, C.J., Edwards, D.G., 1987. Response of plants to calcium concentration in flowing solution culture with chloride or sulphate as the counter-ion. Plant Soil 98, 377–395.

Jahn, T.P., Moller, A., Zeuthen, T., Holm, L.M., Klaerke, D.A., Mohsin, B., et al., 2004. Aquaporin homologues in plants and mammals transport ammonia. FEBS Lett. 574, 31–36.

Jákli, B., Tavakol, E., Tränkner, M., Senbayram, M., Dittert, K., 2017. Quantitative limitations to photosynthesis in K-deficient sunflower and their implications on water-use efficiency. J. Plant Physiol. 209, 20–30.

Jämtgård, S., Näsholm, T., Huss-Danell, K., 2010. N compounds in soil solutions of agricultural land. Soil Biol. Biochem. 42, 2325–2330.

Jensen, L.S., Schjoerring, J.K., van der Hoek, K., Poulsen, H.D., Zevenbergen, J.F., Pallière, C., et al., 2011. Benefits of nitrogen for food, fibre and industrial production. In: Sutton, M.A., Howard, C.M., Erisman, J.W., Billen, G., Bleeker, A., Grennfelt, P., et al.,The European Nitrogen Assessment. Cambridge University Press, Cambridge, UK, pp. 32–63.

Jeschke, W.D., Atkins, C.A., Pate, J.S., 1985. Ion circulation via phloem and xylem between root and shoot of nodulated white lupin. J. Plant Physiol. 117, 319–330.

Jezek, M., Blatt, M.R., 2017. The membrane transport system of the guard cell and its integration for stomatal dynamics. Plant Physiol. 174, 487–519.

Jia, Z., von Wirén, N., 2020. Signaling pathways underlying nitrogen-dependent changes in root system architecture: from model to crop species. J. Exp. Bot. 71, 4393–4404.

Jia, Z., Giehl, R.F.H., von Wirén, N., 2020. The root foraging response under low nitrogen depends on DWARF1-mediated brassinosteroid biosynthesis. Plant Physiol. 183, 998–1010.

Johnston, A.E., Poulton, P.R., Fixen, P.E., Curtin, D., 2014. Phosphorus: its efficient use in agriculture. Adv. Agron. 123, 177–228.

Jones, D.L., Healey, J.R., Willett, V.B., Farrar, J.F., Hodge, A., 2005. Dissolved organic N uptake by plants: an important N uptake pathway? Soil Biol. Biochem. 37, 413–423.

Jones, D.L., Owen, A.G., Farrar, J.F., 2002. Simple method to enable the high-resolution determination of total free amino acids in soil solutions and soil extracts. Soil Biol. Biochem. 34, 1893–1902.

Jones, M.G., Hughes, J., Tregova, A., Milne, J., Tomsett, A.B., Collin, H.A., 2004. Biosynthesis of the flavour precursors of onion and garlic. J. Exp. Bot. 55, 1903–1918.

Jordan-Meille, L., Pellerin, S., 2004. Leaf area establishment of a maize (*Zea mays* L.) field crop under potassium deficiency. Plant Soil 265, 75–92.

Jordan-Meille, L., Pellerin, S., 2008. Shoot and root growth of hydroponic maize (*Zea mays* L.) as influenced by K deficiency. Plant Soil 304, 157–168.

Josefsen, L., Bohn, L., Sørensen, M.B., Rasmussen, S.K., 2007. Characterization of a multifunctional inositol phosphate kinase from rice and barley belonging to the ATP-grasp superfamily. Gene 397, 114–125.

Joy, K.W., 1988. Ammonium, glutamine, and asparagine – a carbon-N interface. Can. J. Bot. 66, 2103–2109.

Kabala, C., Karczewska, A., Galka, B., Cuske, M., Sowinski, J., 2017. Seasonal dynamics of nitrate and ammonium ion concentrations in soil solutions collected using MacroRhizon suction cups. Environ. Monit. Assess. 189, 304.

Kaiser, W.M., 1987. Effects of water deficit on photosynthetic capacity. Physiol. Plant. 71, 142–149.

Kakie, T., 1969. Phosphorus fractions in tobacco plants as affected by phosphate application. Soil Sci. Plant Nutr. 15, 81–85.

Kanai, S., Ohkura, K., Adu-Gyamfi, J.J., Mohapatra, P.K., Nguyen, N.T., Saneoka, H., et al., 2007. Depression of sink activity precedes the inhibition of biomass production in tomato plants subjected to potassium deficiency stress. J. Exp Bot. 58, 2917–2928.

Kant, S., Bi, Y.-M., Rothstein, S.J., 2011. Understanding plant response to N limitation for the improvement of crop N use efficiency. J. Exp. Bot. 62, 1499–1509.

Karmoker, J.L., Clarkson, D.L., Saker, L.R., Rooney, J.M., Purves, J.V., 1991. Sulphate deprivation depresses the transport of nitrogen to the xylem and the hydraulic conductivity of barley (*Hordeum vulgare* L.) roots. Planta 185, 269–278.

Karthikeyan, A.S., Varadarajan, D.K., Jain, A., Held, M.A., Carpita, N.C., Raghothama, K.G., 2007. Phosphate starvation responses are mediated by sugar signaling in Arabidopsis. Planta 225, 907–918.

Kauss, H., 1987. Some aspects of calcium-dependent regulation in plant metabolism. Annu. Rev. Plant Physiol. 38, 47–72.

Kavanová, M., Grimoldi, A.A., Lattanzi, F.A., Schnyder, H., 2006. Phosphorus nutrition and mycorrhiza effects on grass leaf growth. P status- and size-mediated effects on growth zone kinematics. Plant Cell Environ. 29, 511–520.

Kiba, T., Feria-Bourrellier, A.B., Lafouge, F., Lezhneva, L., Boutet-Mercey, S., Orsel, M., et al., 2012. The Arabidopsis nitrate transporter NRT2.4 plays a double role in roots and shoots of nitrogen-starved plants. Plant Cell 24, 245–258.

Kichey, T., Hirel, B., Heumez, E., Dubois, F., Le Gouis, J., 2007. In winter wheat (*Triticum aestivum* L.), post-anthesis N uptake and remobilisation to the grain correlates with agronomic traits and N physiological markers. Field Crops Res. 102, 22–32.

Kindred, D.R., Verhoeven, T.M.O., Weightman, R.M., Swanston, J.S., Agu, R.C., Brosnan, J.M., et al., 2008. Effects of variety and fertiliser N on alcohol yield, grain yield, starch and protein content, and protein composition of winter wheat. J. Cereal Sci. 48, 46–57.

Kinzel, H., 1982. Pflanzenökologie und Mineralstoffwechsel. Ulmer, Stuttgart, Germany.

Kinzel, H., 1989. Calcium in the vacuoles and cell walls of plant tissue. Forms of deposition and their physiological and ecological significance. Flora 182, 99–125.

Kirkby, E.A., Knight, A.H., 1977. Influence of the level of nitrate nutrition on ion uptake and assimilation, organic acid accumulation and cation-anion balance in whole tomato plants. Plant Physiol. 60, 349–353.

Kobayashi, K., Mochizuki, N., Yoshimura, N., Motohashi, K., Hisabori, T., Masuda, T., 2008. Functional analysis of *Arabidopsis thaliana* isoforms of the Mg-chelatase CHLI subunit. Photochem. Photobiol. Sci. 7, 1188–1195.

Kobayashi, N.I., Tanoi, K., 2015. Critical issues in the study of magnesium transport systems and magnesium deficiency symptoms in plants. Int. J. Mol. Sci. 16, 23076–23093.

Koch, K., Mengel, K., 1974. The influence of the level of potassium supply to young tobacco plants (*Nicotiana tabacum* L.) on short-term uptake and utilisation of nitrate nitrogen. J. Sci. Food Agric. 25, 465–471.

Koch, M., Busse, M., Naumann, M., Jákli, B., Smit, I., Cakmak, I., et al., 2019. Differential effects of varied potassium and magnesium nutrition on production and partitioning of photoassimilates in potato plants. Physiol. Plant. 166, 921–935.

Koch, M., Winkelmann, M.K., Hasler, M., Pawelzik, E., Naumann, M., 2020. Root growth in light of changing magnesium distribution and transport between source and sink tissues in potato (*Solanum tuberosum* L.). Sci. Rep. 10, 8796.

Kojima, S., Bohner, A., Gassert, B., Yuan, L., von Wirén, N., 2007. AtDUR3 represents the major transporter for high-affinity urea transport across the plasma membrane of N-deficient Arabidopsis roots. Plant J. 52, 30–40.

Koller, D., 1990. Light-driven leaf movements. Plant Cell Environ. 13, 615–632.

Kong, L., Sun, M., Wang, F., Liu, J., Feng, B., Si, J., et al., 2014. Effects of high NH_4^+ on K^+ uptake, culm mechanical strength and grain filling in wheat. Front. Plant Sci. 5, 703.

Kong, X., Peng, Z., Li, D., Ma, W., An, R., Khan, D., et al., 2020. Magnesium decreases aluminum accumulation and plays a role in protecting maize from aluminum-induced oxidative stress. Plant Soil 457, 71–81.

Konishi, N., Saito, M., Imagawa, F., Kanno, K., Yamaya, T., Kojima, S., 2018. Cytosolic glutamine synthetase isozymes play redundant roles in ammonium assimilation under low-ammonium conditions in roots of *Arabidopsis thaliana*. Plant Cell Physiol. 59, 601–613.

Konrad, K.R., Maierhofer, T., Hedrich, R., 2018. Spatio-temporal aspects of Ca^{2+} signalling: lessons from guard cells and pollen tubes. J. Exp. Bot. 69, 4195–4214.

Kopriva, S., 2006. Regulation of sulfate assimilation in Arabidopsis and beyond. Ann. Bot. 97, 479–495.

Kopriva, S., Koprivova, A., 2005. Sulfate assimilation and glutathione synthesis in C-4 plants. Photosyn. Res. 86, 363–372.
Koprivova, A., Kopriva, S., 2014. Molecular mechanisms of regulation of sulfate assimilation: first steps on a long road. Front. Plant Sci. 5, 589.
Kramer, D.M., Cruz, J.A., Kanazawa, A., 2003. Balancing the central roles of the thylakoid proton gradient. Trends Plant Sci. 8, 27–32.
Krapp, A., Fraisier, V., Scheible, W.R., Quesada, A., Gojon, A., Stitt, M., et al., 1998. Expression studies of Nrt2:1Np, a putative high-affinity nitrate transporter: evidence for its role in nitrate uptake. Plant J. 14, 723–731.
Kronzucker, H.J., Britto, D.T., Davenport, R.J., Tester, M., 2001. Ammonium toxicity and the real cost of transport. Trends Plant Sci. 6, 335–337.
Kronzucker, H.J., Glass, A.D.M., Siddiqi, M.Y., 1999. Inhibition of nitrate uptake by ammonium in barley. Analysis of component fluxes. Plant Physiol. 120, 283–291.
Kronzucker, H.J., Siddiqi, M.Y., Glass, A.D.M., 1996. Kinetics of NH_4^+ influx in spruce. Plant Physiol. 110, 773–779.
Kronzucker, H.J., Szczerba, M.W., Britto, D.T., 2003. Cytosolic potassium homeostasis revisited: ^{42}K-tracer analysis in *Hordeum vulgare* L. reveals set-point variations in [K^+]. Planta 217, 540–546.
Kronzucker, H.J., Szczerba, M.W., Moazami-Goudarzi, M., Britto, D.T., 2006. The cytosolic Na^+:K^+ ratio does not explain salinity-induced growth impairment in barley: a dual-tracer study using $^{42}K^+$ and $^{24}Na^+$. Plant Cell Environ. 29, 2228–2237.
Krouk, G., Lacombe, B., Bielach, A., Perrine-Walker, F., Malinska, K., Mounier, E., et al., 2010. Nitrate-regulated auxin transport by NRT1.1 defines a mechanism for nutrient sensing in plants. Develop. Cell 18, 927–937.
Kudla, J., Becker, D., Grill, E., Hedrich, R., Hippler, M., Kummer, U., et al., 2018. Advances and current challenges in calcium signalling. New Phytol. 218, 414–431.
Kuga, Y., Saito, K., Nayuki, K., Peterson, R.L., Saito, M., 2008. Ultrastructure of rapidly frozen and freeze-substituted germ tubes of an arbuscular mycorrhizal fungus and localization of polyphosphate. New Phytol. 178, 189–200.
Kumamaru, T., Ogawa, M., Satoh, H., Okita, T.W., 2007. Protein body biogenesis in cereal endosperms. Plant Cell Monogr. 8, 141–158.
Kunz, H.H., Gierth, M., Herdean, A., Satoh-Cruz, M., Kramer, D.M., Spetea, C., et al., 2014. Plastidial transporters KEA1, -2, and -3 are essential for chloroplast osmoregulation, integrity, and pH regulation in Arabidopsis. Proc. Natl. Acad. Sci. USA 111, 7480–7485.
Kuppusamy, T., Giavalisco, P., Arvidsson, S., Sulpice, R., Stitt, M., Finnegan, P.M., et al., 2014. Phospholipids are replaced during leaf development, but protein and mature leaf metabolism respond to phosphate in highly phosphorus-efficient *Hakea prostrata*. Plant Physiol. 166, 1891–1911.
Kurusu, T., Kuchitsu, K., Nakano, M., Nakayama, Y., Iida, H., 2013. Plant mechanosensing and Ca^{2+} transport. Trends Plant Sci. 18, 227–233.
Kurvits, A., Kirkby, E.A., 1980. The uptake of nutrients by sunflower plants (*Helianthus annuus*) growing in a continuous flowing culture system, supplied with nitrate or ammonium as nitrogen source. Z. Pflanzenernähr. Bodenk. 143, 140–149.
Kusano, T., Berberich, T., Tateda, C., Takahashi, Y., 2008. Polyamines: essential factors for growth and survival. Planta 228, 367–381.
Labboun, S., Terce-Laforgue, T., Roscher, A., Bedu, M., Restivo, F.M., Velanis, C.N., et al., 2009. Resolving the role of plant glutamate dehydrogenase. I. *In vivo* real time nuclear magnetic resonance spectroscopy experiments. Plant Cell Physiol. 50, 1761–1773.
Labrou, N.E., Karavangeli, M., Tsaftaris, A., Clonis, Y.D., 2005. Kinetic analysis of maize glutathione S-transferase I catalysing the detoxification from chloroacetanilide herbicides. Planta 222, 91–97.
Lambers, H., Albornoz, F., Kotula, L., Laliberté, E., Ranathunge, K., Teste, F.P., et al., 2018. How belowground interactions contribute to the coexistence of mycorrhizal and non-mycorrhizal species in severely phosphorus-impoverished hyperdiverse ecosystems. Plant Soil 424, 11–34.
Lambers, H., Cawthray, G.R., Giavalisco, P., Kuo, J., Laliberté, E., Pearse, S.J., et al., 2012. Proteaceae from severely phosphorus-impoverished soils extensively replace phospholipids with galactolipids and sulfolipids during leaf development to achieve a high photosynthetic phosphorus-use efficiency. New Phytol. 196, 1098–1108.
Lambers, H., Finnegan, P.M., Jost, R., Plaxton, W.C., Shane, M.W., Stitt, M., 2015. Phosphorus nutrition in Proteaceae and beyond. Nat. Plants 1, 15109.
Lambers, H., Finnegan, P.M., Laliberté, E., Pearse, S.J., Ryan, M.H., Shane, M.W., et al., 2011. Phosphorus nutrition of Proteaceae in severely phosphorus-impoverished soils: are there lessons to be learned for future crops? Plant Physiol. 156, 1058–1066.
Lambers, H., Raven, J.A., Shaver, G.R., Smith, S.E., 2008. Plant nutrient-acquisition strategies change with soil age. Trends Ecol. Evol. 23, 95–103.
Lambers, H., Shane, M.W., Cramer, M.D., Pearse, S.J., Veneklaas, E.J., 2006. Root structure and functioning for efficient acquisition of phosphorus: matching morphological and physiological traits. Ann. Bot. 98, 693–713.
Lange, O.L., Zellner, H., Gebel, J., Schrameli, P., Köstner, B., Czygan, F.-C., 1987. Photosynthetic capacity, chloroplast pigments, and mineral content of the previous year's needles with and without the new flush: analysis of the forest-decline phenomenon of needle bleaching. Oecologia 73, 351–357.
Lauer, M.J., Blevins, D.G., Sierzputowska-Gracz, H., 1989. ^{31}P-Nuclear Magnetic Resonance determination of phosphate compartmentation in leaves of reproductive soybeans (*Glycine max* L.) as affected by phosphate nutrition. Plant Physiol. 89, 1331–1336.
Lawson, T., 2009. Guard cell photosynthesis and stomatal function. New Phytol. 181, 13–34.
Lea, U.S., Leydecker, M.T., Quillere, I., Meyer, C., Lillo, C., 2006. Posttranslational regulation of nitrate reductase strongly affects the levels of free amino acids and nitrate, whereas transcriptional regulation has only minor influence. Plant Physiol. 140, 1085–1094.
Lee, J.A., 1999. The calcicole-calcifuge problem revisited. Adv. Bot. Res. 29, 1–30.
Lee, R.B., Ratcliffe, R.G., Southon, T.E., 1990. ^{31}P NMR measurements of the cytoplasmic and vacuolar Pi content of mature maize roots: relationships with phosphorus status and phosphate fluxes. J. Exp. Bot. 41, 1063–1078.
Leigh, R.A., 2001. Potassium homeostasis and membrane transport. J. Plant Nutr. Soil Sci. 164, 193–198.
Leigh, R.A., Wyn Jones, R.G., 1984. A hypothesis relating critical potassium concentrations for growth to the distribution and functions of this ion in the plant cell. New Phytol. 97, 1–13.

Leigh, R.A., Wyn Jones, R.G., 1986. Cellular compartmentation in plant nutrition: the selective cytoplasm and the promiscuous vacuole. In: Tinker, B., Läuchli, A. (Eds.), Advances in Plant Nutrition 2. Praeger Scientific, New York, NY, USA, pp. 249–279.

Leitão, N., Dangeville, P., Carter, R., Charpentier, M., 2019. Nuclear calcium signatures are associated with root development. Nat. Commun. 10, 4865.

Lejay, L., Gansel, X., Cerezo, M., Tillard, P., Muller, C., Krapp, A., et al., 2003. Regulation of root ion transporters by photosynthesis: functional importance and relation with hexokinase. Plant Cell 15, 2218–2232.

Leonard, R.T., Hotchkiss, C.W., 1976. Cation-stimulated adenosine triphosphatase activity and cation transport in corn roots. Plant Physiol. 58, 331–335.

Leran, S., Edel, K.H., Pervent, M., Hashimoto, K., Corratge-Faillie, C., Offenborn, J.N., et al., 2015. Nitrate sensing and uptake in Arabidopsis are enhanced by ABI2, a phosphatase inactivated by the stress hormone abscisic acid. Sci. Signal. 8 (375), ra43.

Lezhneva, L., Kiba, T., Feria-Bourrellier, A.B., Lafouge, F., Boutet-Mercey, S., Zoufan, P., et al., 2014. The Arabidopsis nitrate transporter NRT2.5 plays a role in nitrate acquisition and remobilization in nitrogen-starved plants. Plant J. 80, 230–241.

Li, H., Yang, X., Luo, A., 2001. Ameliorating effect of potassium on iron toxicity in hybrid rice. J. Plant Nutr. 24, 1849–1860.

Li, J.Y., Fu, Y.L., Pike, S.M., Bao, J., Tian, W., Zhang, Y., et al., 2010. The Arabidopsis nitrate transporter NRT1.8 functions in nitrate removal from the xylem sap and mediates cadmium tolerance. Plant Cell 22, 1633–1646.

Li, J., Yokosho, K., Liu, S., Cao, H.R., Yamaji, N., Zhu, X.G., et al., 2020. Diel magnesium fluctuations in chloroplasts contribute to photosynthesis in rice. Nat. Plants 6, 848–859.

Li, W.J., He, X., Chen, Y., Jing, Y.F., Shen, C.C., Yang, J.B., et al., 2019. A wheat transcription factor positively sets seed vigour by regulating the grain nitrate signal. New Phytol. 225, 1667–1680.

Li, Y.L., Kronzucker, H.J., Shi, W.M., 2016. Microprofiling of nitrogen patches in paddy soil: Analysis of spatiotemporal nutrient heterogeneity at the microscale. Sci. Rep. 6, 27064.

Lillo, C., 2008. Signalling cascades integrating light-enhanced nitrate metabolism. Biochem. J. 415, 11–19.

Lillo, C., Lea, U.S., Leydecker, M.T., Meyer, C., 2003. Mutation of the regulatory phosphorylation site of tobacco nitrate reductase results in constitutive activation of the enzyme *in vivo* and nitrite accumulation. Plant J. 35, 566–573.

Lima, J.E., Kojima, S., Takahashi, H., von Wirén, N., 2010. Ammonium triggers lateral root branching in Arabidopsis in an Ammonium Transporter 1;3-dependent manner. Plant Cell 22, 3621–3633.

Lima, L., Seabra, A., Melo, P., Cullimore, J., Carvalho, H., 2006. Phosphorylation and subsequent interaction with 14-3-3 proteins regulate plastid glutamine synthetase in *Medicago truncatula*. Planta 223, 558–567.

Lin, D.C., Nobel, P.S., 1971. Control of photosynthesis by Mg^{2+}. Arch. Biochem. Biophys. 145, 622–632.

Lin, S.H., Kuo, H.F., Canivenc, G., Lin, C.S., Lepetit, M., Hsu, P.K., et al., 2008. Mutation of the Arabidopsis NRT1.5 nitrate transporter causes defective root-to-shoot nitrate transport. Plant Cell 20, 2514–2528.

Lindberg, S., Strid, H., 1997. Aluminium induces rapid changes in cytosolic pH and free calcium and potassium concentrations in root protoplasts of wheat (*Triticum aestivum*). Physiol. Plant. 99, 405–414.

Lindhauer, M.G., 1985. Influence of K nutrition and drought on water relations and growth of sunflower (*Helianthus annuus* L.). Z. Pflanzenernähr. Bodenk. 148, 654–669.

Liu, J.C., Hüttl, R.F., 1991. Relations between damage symptoms and nutritional status of Norway spruce stands (*Picea abies* Karst.) in southwestern Germany. Fert. Res. 27, 9–22.

Liu, K.H., Niu, Y., Konishi, M., Wu, Y., Du, H., Chung, H.S., et al., 2017. Discovery of nitrate–CPK–NLP signalling in central nutrient–growth networks. Nature 545, 311–316.

Liu, L.H., Ludewig, U., Frommer, W.B., von Wirén, N., 2003. AtDUR3 encodes a new type of high-affinity urea/H^+ symporter in Arabidopsis. Plant Cell 15, 790–800.

Liu, T.Y., Chan, C.Y., Chiou, T.J., 2009. The long-distance signalling of mineral macronutrients. Curr. Opin. Plant Biol. 12, 312–319.

Liu, W.X., Wang, J.R., Wang, C.Y., Ma, G., Wei, Q.R., Lu, H.F., et al., 2018. Root growth, water and nitrogen use efficiencies in winter wheat under different irrigation and nitrogen regimes in North China Plain. Front. Plant Sci. 9, 1798.

Liu, Y., von Wirén, N., 2017. Ammonium as a signal for physiological and morphological responses in plants. J. Exp. Bot. 68, 2581–2592.

Loera-Quezada, M.M., Leyva-González, M.A., López-Arredondo, D., Herrera-Estrella, L., 2015. Phosphite cannot be used as a phosphorus source but is non-toxic for microalgae. Plant Sci. 231, 124–130.

Loewus, F.A., Murthy, P.P.N., 2000. myo-Inositol metabolism in plants. Plant Sci. 150, 1–19.

Loneragan, J.F., Snowball, K., 1969. Calcium requirements of plants. Aust. J. Agric. Res. 20, 465–478.

Loneragan, J.F., Snowball, K., Simmons, W.J., 1968. Response of plants to calcium concentration in solution culture. Aust. J. Agric. Res. 19, 845–857.

Long, S.P., Zhu, X.G., Naidu, S.L., Ort, D.R., 2006. Can improvement in photosynthesis increase crop yields? Plant Cell Environ. 29, 315–330.

Loqué, D., von Wirén, N., 2004. Regulatory levels for the transport of ammonium in plant roots. J. Exp. Bot. 55, 1293–1305.

Loqué, D., Ludewig, U., Yuan, L., von Wirén, N., 2005. Tonoplast aquaporins AtTIP2;1 and AtTIP2;3 facilitate NH_3 transport into the vacuole. Plant Physiol. 137, 671–680.

Loqué, D., Yuan, L., Kojima, S., Gojon, A., Wirth, J., Gazzarrini, S., et al., 2006. Additive contribution of AMT1;1 and AMT1;3 to high-affinity ammonium uptake across the plasma membrane of N-deficient Arabidopsis roots. Plant J. 48, 522–534.

Lott, J.N.A., Vollmer, C.M., 1973. Changes in the cotyledons of *Cucurbita maxima* during germination. Protoplasma 78, 255–271.

Lott, J.N.A., Ockenden, I., Raboy, V., Batten, G.D., 2000. Phytic acid and phosphorus in crop seeds and fruits: a global estimate. Seed Sci. Res. 10, 11–33.

Lott, J., Randall, P., Goodchild, D., Craig, S., 1985. Occurrence of globoid crystals in cotyledonary protein bodies of *Pisum sativum* as influenced by experimentally induced changes in Mg, Ca and K contents of seeds. Funct. Plant Biol. 12, 341–353.

Lovegrove, A., Hooley, R., 2000. Gibberellin and abscisic acid signalling in aleurone. Trends Plant Sci. 5, 102–110.

Lund, Z.F., 1970. The effect of calcium and its relation to several cations in soybean root growth. Soil Sci. Soc. Amer. Proc. 34, 456–459.

Lune, P., van Goor, B.J., 1977. Ripening disorders of tomato as affected by the K/Ca ratio in the culture solution. J. Hortic. Sci. 52, 173–180.

Lynch, J., Läuchli, A., Epstein, E., 1991. Vegetative growth of the common bean in response to phosphorus nutrition. Crop Sci. 31, 380–387.

Ma, Q., Bell, R., Biddulph, B., 2019. Potassium application alleviates grain sterility and increases yield of wheat (*Triticum aestivum*) in frost-prone Mediterranean-type climate. Plant Soil 434, 203–216.

Ma, Q., Scanlan, C., Bell, R., Brennan, R., 2013. The dynamics of potassium uptake and use, leaf gas exchange and root growth throughout plant phenological development and its effects on seed yield in wheat (*Triticum aestivum*) on a low-K sandy soil. Plant Soil 373, 373–384.

Maathuis, F.J.M., 2009. Physiological functions of mineral macronutrients. Curr. Opin. Plant Biol. 12, 250–258.

MacDuff, J.H., Jackson, S.B., 1991. Growth and preferences for ammonium or nitrate uptake by barley in relation to root temperature. J. Exp. Bot. 42, 521–530.

Maeshima, K., Matsuda, T., Shindo, Y., Imamura, H., Tamura, S., Imai, R., et al., 2018. A transient rise in free Mg^{2+} ions released from ATP-Mg hydrolysis contributes to mitotic chromosome condensation. Curr. Biol. 28, 444–451.

Maeshima, M., 2000. Vacuolar H^+-pyrophosphatase. BBA-Biomembranes 1465, 37–51.

Maghiaoui, A., Bouguyon, E., Cuesta, C., Perrine-Walker, F., Alcon, C., Krouk, G., et al., 2020. The Arabidopsis NRT1.1 transceptor coordinately controls auxin biosynthesis and transport to regulate root branching in response to nitrate. J. Exp. Bot. 71, 4480–4494.

Marchive, C., Roudier, F., Castaings, L., Bréhaut, V., Blondet, E., Colot, V., et al., 2013. Nuclear retention of the transcription factor NLP7 orchestrates the early response to nitrate in plants. Nat. Commun. 4, 1713.

Marquard, R., Kühn, H., Linser, H., 1968. Der Einfluß der Schwefelernährung auf die Senfölbildung. Z. Pflanzenernähr. Bodenk. 121, 221–230.

Marschner, H., 1992. Bodenversauerung und Magnesiumernährung der Pflanzen. In: Glatzel, G., Jandl, R., Sieghardt, M., Hager, H. (Eds.), Magnesiummangel in Mitteleuropäischen Waldökosystemen, Forstliche Schriftenreihe, Band 5. Universität für Bodenkultur, Wien, Austria, pp. 1–15.

Marschner, H., Cakmak, I., 1989. High light intensity enhances chlorosis and necrosis in leaves of zinc, potassium, and magnesium deficient bean (*Phaseolus vulgaris*) plants. J. Plant Physiol. 134, 308–315.

Marschner, H., Richter, C., 1974. Calcium-transport in roots of maize and bean seedlings. Plant Soil 40, 193–210.

Martin, A., Lee, J., Kichey, T., Gerentes, D., Zivy, M., Tatout, C., et al., 2006. Two cytosolic glutamine synthetase isoforms of maize are specifically involved in the control of grain production. Plant Cell 18, 3252–3274.

Masclaux-Daubresse, C., Daniel-Vedele, F., Dechorgnat, J., Chardon, F., Gaufichon, L., Suzuki, A., 2010. N uptake, assimilation and remobilization in plants: challenges for sustainable and productive agriculture. Ann. Bot. 105, 1141–1157.

Masclaux-Daubresse, C., Reisdorf-Cren, M., Pageau, K., Lelandais, M., Grandjean, O., Kronenberger, J., et al., 2007. Glutamine synthetase-glutamate synthase pathway and glutamate dehydrogenase play distinct roles in the sink-source N cycle in tobacco. Plant Physiol. 140, 444–456.

McAinsh, M.R., Pittman, J.K., 2009. Shaping the calcium signature. New Phytol. 181, 275–294.

McCully, M.E., Miller, C., Sprague, S.J., Huang, C.X., Kirkegaard, J.A., 2008. Distribution of glucosinolates and sulphur-rich cells in roots of field-grown canola (*Brassica napus*). New Phytol. 180, 193–205.

McSwain, B.D., Tsujimoto, H.Y., Arnon, D.I., 1976. Effects of magnesium and chloride ions on light-induced electron transport in membrane fragments from a blue-green alga. Biochim. Biophys. Acta 423, 313–322.

Mengel, K., Arneke, W.W., 1982. Effect of potassium on the water potential, the osmotic potential, and cell elongation in leaves of *Phaseolus vulgaris*. Physiol. Plant. 54, 402–408.

Mengel, K., Helal, M., 1968. Der Einfluß einer variierten N- und K-Ernährung auf den Gehalt an löslichen Aminoverbindungen in der oberirdischen Pflanzenmasse von Hafer. Z. Pflanzenernähr. Bodenk. 120, 12–20.

Mengel, K., Haghparast, M., Koch, K., 1974. The effect of potassium on the fixation of molecular nitrogen by root nodules of *Vicia faba*. Plant Physiol. 54, 535–538.

Mengutay, M., Ceylan, Y., Kutman, U.B., Cakmak, I., 2013. Adequate magnesium nutrition mitigates adverse effects of heat stress on maize and wheat. Plant Soil 368, 57–72.

Meuriot, F., Avice, J.C., Decau, M.L., Simon, J.C., Ourry, A., 2003. Accumulation of N reserves and vegetative storage protein (VSP) in taproots of non-nodulated alfalfa (*Medicago sativa* L.) are affected by mineral N availability, Plant Sci. 165. pp. 709–718.

Meuriot, F., Noquet, C., Avice, J.C., Volenec, J.J., Cunningham, S.M., Sors, T., et al., 2004. Methyl jasmonate alters N partitioning, N reserves accumulation and induces gene expression of a 32-kDa vegetative storage protein that possess chitinase activity in *Medicago sativa* L. taproots. Physiol. Plant. 119, 1–11.

Meyer, S., Scholz-Starke, J., De Angeli, A., Kovermann, P., Burla, B., Gambale, F., et al., 2011. Malate transport by the vacuolar AtALMT6 channel in guard cells is subject to multiple regulation. Plant J. 67, 247–257.

Michael, B., Zink, F., Lantzsch, H.J., 1980. Effect of phosphate application on phytin-phosphorus and other phosphate fractions in developing wheat grains. Z. Pflanzenernähr. Bodenkd. 143, 369–376.

Michard, E., Simon, A.A., Tavares, B., Wudick, M.M., Feijo, J.A., 2017. Signaling with ions: The keystone for apical cell growth and morphogenesis in pollen tubes. Plant Physiol. 173, 91–111.

Miller, A.J., Smith, S.J., 1996. Nitrate transport and compartmentation in cereal root cells. J. Exp. Bot. 47, 843–854.

Miller, A.J., Cookson, S.J., Smith, S.J., Wells, D.M., 2001. The use of microelectrodes to investigate compartmentation and the transport of metabolized inorganic ions in plants. J. Exp. Bot. 52, 541–549.

Miller, A.J., Smith, S.J., 2008. Cytosolic nitrate ion homeostasis, could it have a role in sensing N status? Ann. Bot. 101, 485–489.

Mimura, T., Dietz, K.J., Kaiser, W., Schramm, M.J., Kaiser, G., Heber, U., 1990. Phosphate transport across biomembranes and cytosolic phosphate homeostasis in barley leaves. Planta 180, 139–146.

Missihoun, T.D., Schmitz, J., Klug, R., Kirch, H.H., Bartels, D., 2011. Betaine aldehyde dehydrogenase genes from Arabidopsis with different sub-cellular localization affect stress responses. Planta 233, 369–382.

Mittler, R., 2002. Oxidative stress, antioxidants and stress tolerance. Trends Plant Sci. 7, 405–410.

Miyawaki, K., Matsumoto-Kitano, M., Kakimoto, T., 2004. Expression of cytokinin biosynthetic isopentenyltransferase genes in Arabidopsis: tissue specificity and regulation by auxin, cytokinin, and nitrate. Plant J. 37, 128–138.

Mollier, A., Pellerin, S., 1999. Maize root system growth and development as influenced by phosphorus deficiency. J. Exp. Bot. 50, 487–497.

Moore, B.D., Johnson, S.N., 2017. Get tough, get toxic, or get a bodyguard: identifying candidate traits conferring belowground resistance to herbivores in grasses. Front. Plant Sci. 7, 1925.

Moran, N., 2007. Osmoregulation of leaf motor cells. FEBS Lett. 581, 2337–2347.

Morot-Gaudry, J.F., Job, D., Lea, P.J., 2001. Amino acid metabolism. In: Lea, P.J., Morot-Gaudry, J.F. (Eds.), Plant N. Springer, Berlin, Heidelberg, New York, pp. 167–211.

Morsomme, P., Boutry, M., 2000. The plant plasma membrane H^+-ATPase: Structure, function and regulation. BBA-Biomembranes 1465, 1–16.

Moseley, G., Baker, D.H., 1991. The efficacy of a high magnesium grass cultivar in controlling hypomagnesaemia in grazing animals. Grass Forage Sci. 46, 375–380.

Moshelion, M., Becker, D., Czempinski, K., Mueller-Roeber, B., Attali, B., Hedrich, R., et al., 2002. Diurnal and circadian regulation of putative potassium channels in a leaf moving organ. Plant Physiol. 128, 634–642.

Mota, J.F., Garrido-Becerra, J.A., Perez-Garcia, F.J., Salmeron-Sanchez, E., Sanchez-Gomez, P., Merlo, E., 2016. Conceptual baseline for a global checklist of gypsophytes. Lazaroa 37, 7–30.

Mounier, E., Pervent, M., Ljung, K., Gojon, A., Nacry, P., 2014. Auxin-mediated nitrate signalling by NRT1.1 participates in the adaptive response of Arabidopsis root architecture to the spatial heterogeneity of nitrate availability. Plant Cell Environ. 37, 162–174.

Mühling, K.H., Sattelmacher, B., 1997. Determination of apoplastic K^+ in intact leaves by ratio imaging of PBFI fluorescence. J. Exp. Bot. 48, 1609–1614.

Mukherji, S., Dey, B., Paul, A.K., Sircar, S.M., 1971. Changes in phosphorus fractions and phytase activity of rice seeds during germination. Physiol. Plant 25, 94–97.

Murphy, M.D., Boggan, J.M., 1988. Sulphur deficiency in herbage in Ireland. 1. Causes and extent. Irish J. Agric. Res. 27, 83–90.

Nakata, P.A., 2015. An assessment of engineered calcium oxalate crystal formation on plant growth and development as a step toward evaluating its use to enhance plant defense. PloS ONE 10, e0141982.

Näsholm, T., Kielland, K., Ganeteg, U., 2009. Uptake of organic N by plants. New Phytol. 182, 31–48.

Navazio, L., Formentin, E., Cendron, L., Szabo, I., 2020. Chloroplast calcium signaling in the spotlight. Front. Plant Sci. 11, 186.

Nazoa, P., Vidmar, J.J., Tranbarger, T.J., Mouline, K., Damiani, I., Tillard, P., et al., 2003. Regulation of the nitrate transporter gene AtNRT2.1 in Arabidopsis thaliana, responses to nitrate, amino acids and developmental stage. Plant Mol. Biol. 52, 689–703.

Neely, H.L., Koenig, R.T., Miles, C.A., Koenig, T.C., Karlsson, M.G., 2010. Diurnal fluctuation in tissue nitrate concentration of field-grown leafy greens at two latitudes. HortScience 45, 1815–1818.

Nehls, U., Plassard, C., 2018. Nitrogen and phosphate metabolism in ectomycorrhizas. New Phytol. 220, 1047–1058.

Neugebauer, K., Broadley, M.R., El-Serehy, H.A., George, T.S., McNicol, J.W., Moraes, M.F., et al., 2018. Variation in the angiosperm ionome. Physiol. Plant 163, 306–322.

Neuhaus, H.E., Trentmann, O., 2014. Regulation of transport processes across the tonoplast. Front. Plant Sci. 5, 460.

Nielsen, K.H., Schjoerring, J.K., 1998. Regulation of apoplastic NH_4^+ concentration in leaves of oilseed rape. Plant Physiol. 118, 1361–1368.

Nikiforova, V.J., Gakière, B., Kempa, S., Adamik, M., Willmitzer, L., Hesse, H., et al., 2004. Towards dissecting nutrient metabolism in plants: a systems biology case study on sulphur metabolism. J. Exp. Bot. 55, 1861–1870.

Nitsos, R.E., Evans, H.J., 1966. Effects of univalent cations on the inductive formation of nitrate reductase. Plant Physiol. 41, 1499–1504.

Nitsos, R.E., Evans, H.J., 1969. Effects of univalent cations on the activity of particulate starch synthetase. Plant Physiol. 44, 1260–1266.

Noquet, C., Avice, J.C., Ourry, A., Volenec, J.J., Cunningham, S.M., et al., 2001. Effects of environmental factors and endogenous signals on N uptake, N partitioning and taproot storage protein accumulation in Medicago sativa L. Aust. J. Plant Physiol. 28, 279–288.

Nunes-Nesi, A., Fernie, A.R., Stitt, M., 2010. Metabolic and signaling aspects underpinning the regulation of plant carbon nitrogen interactions. Mol. Plant 3, 973–996.

O'Hara, G.W., Franklin, M., Dilworth, M.J., 1987. Effect of sulfur supply on sulfate uptake, and alcaline sulfatase activity in free-living and symbiotic bradyrhizobia. Arch. Microbiol. 149, 163–167.

O'Neal, D., Joy, K.W., 1974. Glutamine synthetase of pea leaves. Divalent cation effects, substrate specificity, and other properties. Plant Physiol. 54, 775–779.

Ogawa, M., Tanaka, K., Kasai, Z., 1979. Energy-dispersive x-ray analysis of phytin globoids in aleurone particles of developing rice grains. Soil Sci. Plant Nutr. 25, 437–448.

Oja, V., Laisk, A., Heber, U., 1986. Light-induced alkalization of the chloroplast stroma in vivo as estimated from the CO_2 capacity of intact sunflower leaves. Biochim. Biophys. Acta 849, 355–365.

Oms-Oliu, G., Rojas-Graü, M.A., González, L.A., Alandes, L., Varela, P., Soliva-Fortuny, R., et al., 2010. Recent approaches using chemical treatments to preserve quality of fresh-cut fruit: a review. Postharv. Biol. Technol. 57, 139–148.

Orsel, M., Chopin, F., Leleu, O., Smith, S.J., Krapp, A., Daniel-Vedele, F., et al., 2006. Characterization of a two-component high-affinity nitrate uptake system in Arabidopsis. Physiology and protein-protein interaction. Plant Physiol. 142, 1304–1317.

Orsel, M., Filleur, S., Fraisier, V., Daniel-Vedele, F., 2002. Nitrate transport in plants, which gene and which control? J. Exp. Bot. 53, 825–833.

Osborne, T.B., 1924. The Vegetable Proteins. Longmans, London, UK.

Ougham, H., Hörtensteiner, S., Armstead, I., Donnison, I., King, I., Thomas, H., 2008. The control of chlorophyll catabolism and the status of yellowing as a biomarker of leaf senescence. Plant Biol. 10, 4–14.

Ourry, A., Mac Duff, J., Volenec, J.J., Gaudillère, J.P., 2001. N traffic during plant growth and development. In: Lea, P.J., Morot-Gaudry, J.F. (Eds.), Plant N. Springer, Berlin, Heidelberg, New York, pp. 255–273.

Outlaw Jr., W.H., 1983. Current concepts on the role of potassium in stomatal movements. Physiol. Plant 49, 302–311.

Outlaw Jr., W.H., Manchester, J., 1979. Guard cell starch concentration quantitatively related to stomatal aperture. Plant Physiol. 64, 79–82.

Overlach, S., Diekmann, W., Raschke, K., 1993. Phosphate translocator of isolated guard-cell chloroplasts from *Pisum sativum* L. transports glucose-6-phosphate. Plant Physiol. 101, 1201–1207.

Page, M.J., Di Cera, E., 2006. Role of Na^+ and K^+ in enzyme function. Physiol. Rev. 86, 1049–1092.

Paiva, E.A.S., 2019. Are calcium oxalate crystals a dynamic calcium store in plants? New Phytol. 223, 1707–1711.

Palacio, S., Aitkenhead, M., Escudero, A., Montserrat-Marti, G., Maestro, M., Robertson, A.H.J., 2014. Gypsophile chemistry unveiled: Fourier Transform Infrared (FTIR) spectroscopy provides new insight into plant adaptations to gypsum soils. PloS ONE 9, e107285.

Pan, Y., Chai, X., Gao, Q., Zhou, L., Zhang, S., Li, L., et al., 2019. Dynamic interactions of plant CNGC subunits and calmodulins drive oscillatory Ca^{2+} channel activities. Dev. Cell 48, 710–725.

Patterson, K., Cakmak, T., Cooper, A., Lager, I., Rasmusson, A.G., Escobar, M.A., 2010. Distinct signalling pathways and transcriptome response signatures differentiate ammonium and nitrate-supplied plants. Plant Cell Environ. 33, 1486–1501.

Pearson, C.J., Volk, R.J., Jackson, W.A., 1981. Daily changes in nitrate influx, efflux and metabolism in maize and pearl millet. Planta 152, 319–324.

Pelissier, H.C., Frerich, A., Desimone, M., Schumacher, K., Tegeder, M., 2004. PvUPS1, an allantoin transporter in nodulated roots of french bean. Plant Physiol. 134, 664–675.

Peng, W.T., Qi, W.L., Nie, M.M., Xiao, Y.B., Liao, H., Chen, Z.C., 2020. Magnesium supports nitrogen uptake through regulating NRT2.1/2.2 in soybean. Plant Soil 457, 97–111.

Peng, W.T., Zhang, L.D., Zhou, Z., Fu, C., Chen, Z.C., Liao, H., 2018. Magnesium promotes root nodulation through facilitation of carbohydrate allocation in soybean. Physiol. Plant 163, 372–385.

Peng, Y.Y., Liao, L.L., Liu, S., Nie, M.M., Li, J., Zhang, L.D., et al., 2019. Magnesium deficiency triggers sgr-mediated chlorophyll degradation for magnesium remobilization. Plant Physiol. 181, 262–275.

Peoples, T.R., Koch, D.W., 1979. Role of potassium in carbon dioxide assimilation in *Medicago sativa* L. Plant Physiol. 63, 878–881.

Pettigrew, W.T., 2008. Potassium influences on yield and quality production for maize, wheat, soybean and cotton. Physiol. Plant. 133, 670–681.

Pfeiffenschneider, Y., Beringer, H., 1989. Measurement of turgor potential in carrots of different K-nutrition by using the cell pressure probe. Proc. 21. Colloq. Int. Potash Inst. Bern, 203–217.

Pflüger, R., Cassier, A., 1977. Influence of monovalent cations on photosynthetic CO2 fixation. Proc. 13. Colloq. Int. Potash Inst. Bern, pp. 95–100.

Pflüger, R., Wiedemann, R., 1977. Der Einfluß monovalenter Kationen auf die Nitratreduktion von *Spinacia oleracea* L. Z. Pflanzenphysiol. 85, 125–133.

Pierce, J., 1986. Determinants of substrate specificity and the role of metal in the reaction of ribulosebisphosphate carboxylase/oxygenase. Plant Physiol. 81, 943–945.

Pieters, A.J., Paul, M.J., Lawlor, D.W., 2001. Low sink demand limits photosynthesis under Pi deficiency. J. Exp. Bot. 52, 1083–1091.

Pissarek, H.P., 1973. Zur Entwicklung der Kalium-Mangelsymptome von Sommerraps. Z. Pflanzenernähr. Bodenk. 136, 1–19.

Pitman, M.G., Mowat, J., Nair, H., 1971. Interactions of processes for accumulation of salt and sugar in barley plants. Aust. J. Biol. Sci. 24, 619–631.

Plénet, D., Etchebest, S., Mollier, A., Pellerin, S., 2000. Growth analysis of maize field crops under phosphorus deficiency. I. Leaf growth. Plant Soil 223, 119–132.

Poffenroth, M., Green, D.B., Tallman, G., 1992. Sugar concentrations in guard cells of *Vicia faba* illuminated with red or blue light. Plant Physiol. 98, 1460–1471.

Portis, A.R., 1982. Effects of the relative extrachloroplastic concentrations of inorganic phosphate, 3-phosphoglycerate, and dihydroxyacetone phosphate on the rate of starch synthesis in isolated spinach chloroplasts. Plant Physiol. 70, 393–396.

Portis Jr., A.R., Heldt, H.W., 1976. Light-dependent changes of the Mg^{2+} concentration in the stroma in relation to the Mg^{2+} dependency of CO_2 fixation in intact chloroplasts. Biochim. Biophys. Acta 449, 434–446.

Pradet, A., Raymond, P., 1983. Adenine nucleotide ratios and adenylate energy charge in energy metabolism. Annu. Rev. Plant Physiol. 34, 199–224.

Prodhan, M.A., Finnegan, P.M., Lambers, H., 2019. How does evolution in phosphorus-impoverished landscapes impact plant nitrogen and sulfur assimilation? Trends Plant Sci. 24, 69–82.

Prychid, C.J., Rudall, P.J., 1999. Calcium oxalate crystals in monocotyledons: a review of their structure and systematics. Ann. Bot. 84, 725–739.

Pyle, A.M., 2002. Metal ions in the structure and function of RNA. J. Biol. Inorg. Chem. 7, 679–690.

Qiu, J., Israel, D.W., 1992. Diurnal starch accumulation and utilization in phosphorus-deficient soybean plants. Plant Physiol. 98, 316–323.

Raboy, V., Dickinson, D.B., 1987. The timing and rate of phytic acid accumulation in developing soybean seeds. Plant Physiol. 85, 841–844.

Randall, P.J., Wrigley, C.W., 1986. Effects of sulfur supply on the yield, composition, and quality of grain from cereals, oilseeds and legumes. Adv. Cereal Sci. Technol. 8, 171–206.

Rao, I.M., Fredeen, A.L., Terry, N., 1990. Leaf phosphate status, photosynthesis, and carbon partitioning in sugar beet. III. Diurnal changes in carbon partitioning and carbon export. Plant Physiol. 92, 29–36.

Rao, I.M., Sharp, R.E., Boyer, J.S., 1987. Leaf magnesium alters photosynthetic response to low water potentials in sunflower. Plant Physiol. 84, 1214–1219.

Rasmussen, I.S., Dresbøll, D.B., Thorup-Kristensen, K., 2015. Winter wheat cultivars and nitrogen (N) fertilization—Effects on root growth, N uptake efficiency and N use efficiency. Eur. J. Agron. 68, 38–49.

Ratjen, A.M., Gerendás, J., 2009. A critical assessment of the suitability of phosphite as a source of phosphorus. J. Plant Nutr. Soil Sci. 172, 821–828.

Rausch, T., Wachter, A., 2005. Sulfur metabolism: a versatile platform for launching defence operations. Trends Plant Sci. 10, 503–509.

Raven, J.A., Smith, F.A., 1976. N assimilation and transport in vascular land plants in relation to intracellular pH regulation. New Phytol. 76, 415–431.

Rawat, S.R., Silim, S.N., Kronzucker, H.J., Siddiqi, M.Y., Glass, A.D.M., 1999. *AtAMT1* gene expression and NH_4^+ uptake in roots of *Arabidopsis thaliana*. Evidence for regulation by root glutamine levels. Plant J. 19, 143–152.

Rea, P.A., Sanders, D., 1987. Tonoplast energization: two H^+ pumps, one membrane. Physiol. Plant 71, 131–141.

Reckmann, U., Scheibe, R., Raschke, K., 1990. Rubisco activity in guard cells compared with the solute requirement for stomatal opening. Plant Physiol. 92, 246–253.

Rengel, Z., Zhang, W.-H., 2003. Role of dynamics of intracellular calcium in aluminium toxicity syndrome. New Phytol. 159, 295–314.

Rennenberg, H., Lamoureux, G.L., 1990. Physiological processes that modulate the concentration of glutathione in plant cells. In: Rennenberg, H., Brunold, C., De Kok, L.J., Stulen, I. (Eds.), Sulfur Nutrition and Sulfur Assimilation in Higher Plants. SPB Acad. Publ. bv, The Hague, The Netherlands, pp. 53–65.

Rennenberg, H., Dannenmann, M., Gessler, A., Kreuzwieser, J., Simon, J., Papen, H., 2009. N balance in forest soils: nutritional limitations of plants under climate change stresses. Plant Biol. 14, 4–23.

Rhodes, D., Nadolska-Orczyk, A., Rich, P.J., 2006. Salinity, osmolytes and compatible solutes. In: Läuchlii, A., Lüttge, U. (Eds.), Salinity: Environment – Plants – Molecules. Kluwer Academic Publishers, Dordrecht, The Netherlands, pp. 181–204.

Richard-Molard, C., Krapp, A., Brun, F., Ney, B., Daniel-Vedele, F., Chaillou, S., 2008. Plant response to nitrate starvation is determined by N storage capacity matched by nitrate uptake capacity in two Arabidopsis genotypes. J. Exp. Bot. 59, 779–791.

Richardson, A.E., George, T.S., Jakobsen, I., Simpson, R.J., 2007. Plant utilization of inositol phosphates. In: Turner, B.L., Richardson, A.E., Mullaney, E.J. (Eds.), Inositol Phosphates: Linking Agriculture and the Environment. CABI Publishing, Wallingford, UK, pp. 242–260.

Rigney, C.J., Wills, R.B.H., 1981. Calcium movement, a regulating factor in the initiation of tomato fruit ripening. HortScience 16, 550–551.

Rijven, A.H.G.C., Gifford, R.M., 1983. Accumulation and conversion of sugars by developing wheat grains. 4. Effects of phosphate and potassium ions in endosperm slices. Plant Cell Environ. 6, 625–631.

Roberts, T.M., Skeffington, R.A., Blank, L.W., 1989. Causes of type I spruce decline in Europe. Forestry 62, 179–222.

Robinson, D., 1994. The responses of plants to non-uniform supplies of nutrients. New Phytol. 127, 635–674.

Roblin, G., Fleurat-Lessard, P., Bonmort, J., 1989. Effects of compounds affecting calcium channels on phytochrome- and blue pigment-mediated pulvinar movements of *Cassia fasciculata*. Plant Physiol. 90, 697–701.

Robson, A.D., Pitman, M.G., 1983. Interactions between nutrients in higher plants. In: Läuchli, A., Bieleski, R.L. (Eds.), Encyclopedia of Plant Physiology, New Series, 15A. Springer Verlag, Berlin and New York, pp. 147–180.

Rodríguez, D., Keltjens, W.G., Goudriaan, J., 1998. Plant leaf area expansion and assimilate production in wheat (*Triticum aestivum* L.) growing under low phosphorus conditions. Plant Soil 200, 227–240.

Roelfsema, M.R.G., Hedrich, R., 2005. In the light of stomatal opening: new insights into 'the Watergate'. New Phytol. 167, 665–691.

Roelfsema, M.R.G., Hedrich, R., 2010. Making sense out of Ca^{2+} signals: their role in regulating stomatal movements. Plant Cell Environ. 33, 305–321.

Roelfsema, M.R.G., Levchenko, V., Hedrich, R., 2004. ABA depolarizes guard cells in intact plants through a transient activation of R- and S-type anion channels. Plant J. 37, 578–588.

Rohozinski, J., Edwards, G.R., Hoskyns, P., 1986. Effects of brief exposure to nitrogenous compounds on floral induction in apple trees. Physiol. Veg. 24, 673–677.

Rolland, F., Baena-Gonzalez, E., Sheen, J., 2006. Sugar sensing and signaling in plants: conserved and novel mechanisms. Annu. Rev. Plant Biol. 57, 675–709.

Romero, L.C., Aroca, Á.M., Laureano-Marin, A.M., Moreno, I., Garcia, I., Gotor, C., 2014. Cysteine and cysteine-related signaling pathways in *Arabidopsis thaliana*. Mol. Plant 7, 264–276.

Römheld, V., Kirkby, E.A., 2010. Research on potassium in agriculture: Needs and prospects. Plant Soil 335, 155–180.

Roosta, H.R., Schjoerring, J.K., 2008. Effects of nitrate and potassium on ammonium toxicity in cucumber plants. J. Plant Nutr. 31, 1270–1283.

Rosanoff, A., Capron, E., Barak, P., Mathews, B., Nielsen, F., 2015. Edible plant tissue and soil calcium:magnesium ratios: data too sparse to assess implications for human health. Crop Pasture Sci. 66, 1265–1277.

Rossato, L., Laine, P., Ourry, A., 2001. N storage and remobilization in *Brassica napus* L. during the growth cycle, N fluxes within the plant and changes in soluble protein patterns. J. Exp. Bot. 52, 1655–1663.

Rossiter, R.C., 1978. Phosphorus deficiency and flowering in subterranean clover (*T. subterraneum* L.). Ann. Bot. 42, 325–329.

Rossmann, A., Buchner, P., Savill, G.P., Hawkesford, M.J., Scherf, K.A., Muhling, K.H., 2019. Foliar N application at anthesis alters grain protein composition and enhances baking quality in winter wheat only under a low N fertiliser regimen. Eur. J. Agron. 109, 125909.

Rouhier, N., Lemaire, S.D., Jacquot, J.P., 2008. The role of glutathione in photosynthetic organisms: Emerging functions for glutaredoxins and glutathionylation. Annu. Rev. Plant Biol. 59, 143–166.

Rozov, A., Khusainov, I., El Omari, K., Duman, R., Mykhaylyk, V., Yusupov, M., 2019. Importance of potassium ions for ribosome structure and function revealed by long-wavelength X-ray diffraction. Nat. Commun. 10, 1–12.

Ruiz, J.M., Romero, L., 2002. Relationship between potassium fertilisation and nitrate assimilation in leaves and fruits of cucumber (*Cucumis sativus*) plants. Ann. Appl. Biol. 140, 241–245.

Sage, R.F., Pearcy, R.W., Seemann, J.R., 1987. The N use efficiency of C_3 and C_4 plants. 3. Leaf N effects on the activity of carboxylating enzymes in *Chenopodium album* (L.) and *Amaranthus retroflexus* (L.). Plant Physiol. 85, 355–359.

Saito, S., Uozumi, N., 2020. Calcium regulated phosphorylation systems controlling uptake and balance of plant nutrients. Front. Plant Sci. 11, 44.

Salsac, L., Chaillou, S., Morotgaudry, J.F., Lesaint, C., 1987. Nitrate and ammonium nutrition in plants. Plant Physiol. Biochem. 25, 805–812.

Santamaria, P., 2006. Nitrate in vegetables: toxicity, content, intake and EC regulation. J. Sci. Food Agric. 86, 10–17.

Sartirana, M.L., Bianchetti, R., 1967. The effects of phosphate on the development of phytase in the wheat embryo. Physiol. Plant 20, 1066–1075.

Satter, R., Morse, M.J., Lee, Y., Crain, R.C., Coté, G.G., Moran, N., 1988. Light and clock-controlled leaflet movements in *Samanea saman*: a physiological, biophysical and biochemical analysis. Bot. Acta. 101, 205–213.

Schapire, A.L., Valpuesta, V., Botella, M.A., 2009. Plasma membrane repair in plants. Trends Plant Sci. 14, 645–652.

Schelbert, S., Aubry, S., Burla, B., Agne, B., Kessler, F., Krupinska, K., et al., 2009. Pheophytin pheophorbide hydrolase (pheophytinase) is involved in chlorophyll breakdown during leaf senescence in Arabidopsis. Plant Cell 21, 767–785.

Scheurwater, I., Koren, M., Lambers, H., Atkin, O.K., 2002. The contribution of roots and shoots to whole plant nitrate reduction in fast and slow-growing grass species. J. Exp. Bot. 53, 1635–1642.

Schilling, G., Adgo, E., Schulze, J., 2006. Carbon costs of nitrate reduction in broad bean (*Vicia faba* L.) and pea (*Pisum sativum* L.) plants. J. Plant Nutr. Soil Sci. 169, 691–698.

Schjoerring, J.K., Husted, S., Mäck, G., Mattsson, M., 2002. The regulation of ammonium translocation in plants. J. Exp. Bot. 53, 883–890.

Schmidt, A., 1986. Regulation of sulfur metabolism in plants. In: Behnke, H.-D., Esser, K., Kubitzki, K., Runge, M., Ziegler, H. (Eds.), Progress in Botany, 48. Springer Verlag, Berlin, Heidelberg, pp. 133–150.

Schmidt, F., De Bona, F.D., Monteiro, F.A., 2013. Sulfur limitation increases nitrate and amino acid pools in tropical forages. Crop Pasture Sci. 64, 51–60.

Schmutz, D., Brunold, C., 1982. Regulation of sulfate assimilation in plants. XIII. Assimilatory sulfate reduction during ontogenesis of primary leaves of *Phaseolus vulgaris* L. Plant Physiol. 70, 524–527.

Schnyder, H., Nelson, C.J., Spollen, W.G., 1988. Diurnal growth of tall fescue leaf blades. II. Dry matter partitioning and carbohydrate metabolism in the elongation zone and adjacent expanded tissue. Plant Physiol. 86, 1077–1083.

Schonewille, J.T., 2013. Magnesium in dairy cow nutrition: an overview. Plant Soil 368, 167–178.

Schröppel-Meier, G., Kaiser, W.M., 1988. Ion homeostasis in chloroplasts under salinity and mineral deficiency. I. Solute concentrations in leaves and chloroplasts from spinach plants under NaCl or $NaNO_3$ salinity. Plant Physiol. 87, 822–827.

Schürmann, P., 1993. Plant thioredoxins. In: De Kok, L.J., Stulen, I., Rennenberg, H., Brunold, C., Rauser, W.E. (Eds.), Sulfur Nutrition and Assimilation in Higher Plants. SPB Academic Publ. bv, The Hague, The Netherlands, pp. 153–162.

Schulze, E.-D., 1989. Air pollution and forest decline in a spruce (*Picea abies*) forest. Science 244, 776–783.

Schütz, B., De Kok, L.J., Rennenberg, H., 1991. Thiol accumulation and cysteine desulfurylase activity in H_2S-fumigated leaves and leaf homogenates of cucurbit plants. Plant Cell Physiol. 32, 733–736.

Scott, B.J., Robson, A.D., 1990. Changes in the content and form of magnesium in the first trifoliate leaf of subterranean clover under altered or constant root supply. Aust. J. Agric. Res. 41, 511–519.

Segonzac, C., Boyer, J.-C., Ipotesi, E., Szponarski, W., Tillard, P., Touraine, B., 2007. Nitrate efflux at the root plasma membrane, identification of an Arabidopsis excretion transporter. Plant Cell 19, 3760–3777.

Sen Gupta, A., Berkowitz, G.A., Pier, P.A., 1989. Maintenance of photosynthesis at low leaf water potential in wheat. Plant Physiol. 89, 1358–1365.

Senbayram, M., Gransee, A., Wahle, V., Thiel, H., 2015. Role of magnesium fertilisers in agriculture: plant-soil continuum. Crop Pasture Sci. 66, 1219–1229.

Seufferheld, M., Curzi, M., 2010. Recent discoveries on the roles of polyphosphates in plants. Plant Mol. Biol. Rep. 28, 549–559.

Shabala, S., Shabala, L., Barcelo, J., Poschenrieder, C., 2014. Membrane transporters mediating root signalling and adaptive responses to oxygen deprivation and soil flooding. Plant Cell Environ. 37, 2216–2233.

Shane, M.W., Cramer, M.D., Funayama-Noguchi, S., Cawthray, G.R., Millar, A.H., Day, D.A., et al., 2004a. Developmental physiology of cluster-root carboxylate synthesis and exudation in harsh hakea. Expression of phosphoenolpyruvate carboxylase and the alternative oxidase. Plant Physiol. 135, 549–560.

Shane, M.W., McCully, M.E., Lambers, H., 2004b. Tissue and cellular phosphorus storage during development of phosphorus toxicity in *Hakea prostrata* (Proteaceae). J. Exp. Bot. 55, 1033–1044.

Shane, M.W., Szota, C., Lambers, H., 2004c. A root trait accounting for the extreme phosphorus sensitivity of *Hakea prostrata* (Proteaceae). Plant Cell Environ. 27, 991–1004.

Shao, Q., Gao, Q., Lhamo, D., Zhang, H., Luan, S., 2020. Two glutamate- and pH-regulated Ca^{2+} channels are required for systemic wound signaling in Arabidopsis. Sci. Signal. 13, eaba1453.

Sharpless, R.O., Johnson, D.S., 1977. The influence of calcium on senescence changes in apple. Ann. Appl. Biol. 85, 450–453.

Shear, C.B., 1975. Calcium-related disorders of fruits and vegetables. HortScience 10, 361–365.

Shen, Y., Zhou, Z., Feng, S., Li, J., Tan-Wilson, A., Qu, L.-J., et al., 2009. Phytochrome A mediates rapid red light-induced phosphorylation of Arabidopsis FAR-RED ELONGATED HYPOCOTYL1 in a low fluence response. Plant Cell 21, 494–506.

Shewry, P.R., 2007. Improving the protein content and composition of cereal grain. J. Cereal Sci. 46, 239–250.

Shewry, P.R., Halford, N.G., 2002. Cereal seed storage proteins: structures, properties and role in grain utilization. J. Exp. Bot. 53, 947–958.

Shewry, P.R., Napier, J.A., Tatham, A.S., 1995. Seed storage proteins – structures and biosynthesis. Plant Cell 7, 945–956.

Shi, J., Wang, H., Schellin, K., Li, B., Faller, M., Stoop, J.M., et al., 2007. Embryo-specific silencing of a transporter reduces phytic acid content of maize and soybean seeds. Nat. Biotech. 25, 930–937.

Shi, Q., Pang, J., Yong, J.W.H., Bai, C., Pereira, C.G., Song, Q., et al., 2020. Phosphorus-fertilisation has differential effects on leaf growth and photosynthetic capacity of *Arachis hypogaea* L. Plant Soil 447, 99–116.

Shinmachi, F., Buchner, P., Stroud, J.L., Parmar, S., Zhao, F.J., McGrath, S.P., et al., 2010. Influence of sulfur deficiency on the expression of specific sulfate transporters and the distribution of sulfur, selenium, and molybdenum in wheat. Plant Physiol. 153, 327–336.

Sicher, R.C., Kremer, D.F., 1988. Effects of phosphate deficiency on assimilate partitioning in barley seedlings. Plant Sci. 57, 9–17.

Siebrecht, S., Tischner, R., 1999. Changes in the xylem exudate composition of poplar (*Populus tremula* x *P. alba*)–dependent on the nitrogen and potassium supply. J. Exp. Bot. 50, 1797–1806.

Singh, S.K., Reddy, V.R., 2017. Potassium starvation limits soybean growth more than the photosynthetic processes across CO_2 levels. Front. Plant Sci. 8, 991.

Smith, I.K., Polle, A., Rennenberg, H., 1990. Glutathione. In: Alscher, R.G., Cumming, J.R. (Eds.), Stress Responses in Plants: Adaptation and Acclimation Mechanisms. Wiley-Liss. Inc., New York, NY, USA, pp. 201–215.

Smith, S., Stewart, G.R., 1990. Effect of potassium levels on the stomatal behavior of the hemi-parasite *Striga hermonthica*. Plant Physiol. 94, 1472–1476.

Sorin, E., Etienne, P., Maillard, A., Zamarreno, A.M., Garcia-Mina, J.M., Arkoun, M., et al., 2015. Effect of sulphur deprivation on osmotic potential components and nitrogen metabolism in oilseed rape leaves: Identification of a new early indicator. J. Exp. Bot. 66, 6175–6189.

Speer, M., Kaiser, W.M., 1991. Ion relations of symplastic and apoplastic space in leaves from *Spinacia oleracea* L. and *Pisum sativum* L. under salinity. Plant Physiol. 97, 990–997.

Sperrazza, J.M., Spremulli, L.L., 1983. Quantitation of cation binding to wheat germ ribosomes: influences on subunit association equilibria and ribosome activity. Nucleic Acid Res. 11, 2665–2679.

Sreedhara, A., Cowan, J.A., 2002. Structural and catalytic roles for divalent magnesium in nucleic acid biochemistry. Biometals 15, 211–223.

Stadler, R., Büttner, M., Ache, P., Hedrich, R., Ivashikina, N., Melzer, M., et al., 2003. Diurnal and light-regulated expression of AtSTP1 in guard cells of Arabidopsis. Plant Physiol. 133, 528–537.

Stael, S., Wurzinger, B., Mair, A., Mehlmer, N., Vothknecht, U.C., Teige, M., 2012. Plant organellar calcium signalling: an emerging field. J. Expt. Bot. 63, 1525–1542.

Starrach, N., Mayer, W.-E., 1989. Changes of the apoplastic pH and K^+ concentration in the *Phaseolus* pulvinus *in situ* in relation to rhythmic leaf movements. J. Exp. Bot. 40, 865–873.

Starrach, N., Flach, D., Mayer, W.E., 1985. Activity of fixed negative charges of isolates extensor cell walls of the laminar pulvinus of primary leaves of *Phaseolus*. J. Plant Physiol. 120, 441–455.

Staswick, P.E., 1994. Storage proteins of vegetative plant tissues. Annu. Rev. Plant Physiol. Plant Mol. Biol. 45, 303–322.

Steingröver, E., 1983. Storage of osmotically active compounds in the taproot of *Daucus carota* L. J. Exp. Bot. 34, 425–433.

Steingröver, E., Oosterhuis, R., Wieringa, F., 1982. Effect of light treatment and nutrition on nitrate accumulation in spinach (*Spinacea oleracea* L.). Z. Pflanzenphysiol. 107, 97–102.

Stelzer, R., Lehmann, H., Kramer, D., Lüttge, U., 1990. X-ray microprobe analysis of vacuoles on spruce needle mesophyll, endodermis and transfusion parenchyma cells at different seasons of the year. Bot. Acta. 103, 415–423.

Stewart, G.R., Gracia, C.A., Hegarty, E.E., Specht, R.L., 1990. Nitrate reductase activity and chlorophyll content in sun leaves of subtropical Australian closed-forest (rainforest) and open-forest communities. Oecologia 82, 544–551.

Stewart, G.R., Joly, C.A., Smirnoff, N., 1992. Partitioning of inorganic N assimilation between roots and shoots of cerrado and forest trees of contrasting plant communities of South East Brasil. Oecologia 91, 511–517.

Stiles, K.A., Van Volkenburgh, E., 2004. Role of K^+ in leaf growth: K^+ uptake is required for light-stimulated H^+ efflux but not solute accumulation. Plant Cell Environ. 27, 315–325.

Straub, T., Ludewig, U., Neuhäuser, B., 2017. The kinase CIPK23 inhibits ammonium transport in *Arabidopsis thaliana*. Plant Cell 29, 409–422.

Stulen, I., De Kok, L.J., 1993. Whole plant regulation of sulfur metabolism – a theoretical approach and comparison with current ideas on regulation of nitrogen metabolism. In: De Kok, L.J., Stulen, I., Rennenberg, H., Brunold, C., Rauser, W.E. (Eds.), Sulfur Nutrition and Assimilation in Higher Plants. SPB Academic Publ. bv, The Hague, The Netherlands, pp. 77–91.

Sugiyama, T., Matsumoto, C., Akazawa, T., Miyachi, S., 1969. Structure and function of chloroplast proteins. VII. Ribulose-1,5-diphosphate carboxylase of *Chlorella ellipsoida*. Arch. Biochem. Biophys. 129, 597–602.

Sugiyama, T., Nakyama, N., Akazawa, T., 1968. Structure and function of chloroplast proteins. V. Homotropic effect of bicarbonate in RuBP carboxylase relation and the mechanism of activation by magnesium ions. Arch. Biochem. Biophys. 126, 734–745.

Sulpice, R., Ishihara, H., Schlereth, A., Cawthray, G.R., Encke, B., Giavalisco, P., et al., 2014. Low levels of ribosomal RNA partly account for the very high photosynthetic phosphorus-use efficiency of Proteaceae species. Plant Cell Environ. 37, 1276–1298.

Sun, J., Bankston, J.R., Payandeh, J., Hinds, T.R., Zagotta, W.N., Zheng, N., 2014. Crystal structure of the plant dual-affinity nitrate transporter NRT1.1. Nature 507, 73–77.

Sun, Y., Yang, R., Li, L., Huang, J., 2017. The magnesium transporter MGT10 is essential for chloroplast development and photosynthesis in *Arabidopsis thaliana*. Mol. Plant 10, 1584–1587.

Sunkar, R., Li, Y.-F., Jagadeeswaran, G., 2012. Functions of microRNAs in plant stress responses. Trends Plant Sci. 17, 196–203.

Suzuki, A., Knaff, D.B., 2005. Glutamate synthase: structural, mechanistic and regulatory properties, and role in the amino acid metabolism. Photosynth. Res. 83, 191–217.

Suzuki, M., Takahashi, M., Tsukamoto, T., Watanabe, S., Matsuhashi, S., Yazaki, J., et al., 2006. Biosynthesis and secretion of mugineic acid family phytosiderophores in zinc-deficient barley. Plant J. 48, 85–97.

Suzuki, M., Tsukamoto, T., Inoue, H., Watanabe, S., Matsuhashi, S., Takahashi, M., et al., 2008. Deoxymugineic acid increases Zn translocation in Zn-deficient rice plants. Plant Mol. Biol. 66, 609–617.

Suzuki, N., Koussevitzky, S., Mittler, R., Miller, G., 2012. ROS and redox signalling in the response of plants to abiotic stress. Plant Cell Environ. 35, 259–270.

Suzuki, Y., Makino, A., Mae, T., 2001. An efficient method for extraction of RNA from rice leaves at different ages using benzyl chloride. J. Exp. Bot. 52, 1575–1579.

Sze, H., Chanroj, S., 2018. Plant endomembrane dynamics: studies of K^+/H^+ antiporters provide insights on the effects of pH and ion homeostasis. Plant Physiol. 177, 875–895.

Tabe, L., Higgins, T.J.V., 1998. Engineering plant protein composition for improved nutrition. Trends Plant Sci. 3, 282–286.

Tabuchi, M., Abiko, T., Yamaya, T., 2007. Assimilation of ammonium ions and reutilization of N in rice (*Oryza sativa* L.). J. Exp. Bot. 58, 2319–2327.

Takagi, D., Miyagi, A., Tazoe, Y., Suganami, M., Kawai-Yamada, M., Ueda, A., et al., 2020. Phosphorus toxicity disrupts Rubisco activation and reactive oxygen species defence systems by phytic acid accumulation in leaves. Plant Cell Environ. 43, 2033–2053.

Takahashi, H., Kopriva, S., Giordano, M., Saito, K., Hell, R., 2011. Sulfur assimilation in photosynthetic organisms: molecular functions and regulations of transporters and assimilatory enzymes. Annu. Rev. Plant Biol. 62, 157–184.

Talbott, L.D., Zeiger, E., 1998. The role of sucrose in guard cell osmoregulation. J. Exp. Bot. 49, 329–337.

Tallman, G., Zeiger, E., 1988. Light quality and osmoregulation in *Vicia* guard cells. Evidence for involvement of three metabolic pathways. Plant Physiol. 88, 887–895.

Tang, R.J., Wang, C., Li, K.L., Luan, S., 2020. The CBL-CIPK calcium signaling network: unified paradigm from 20 years of discoveries. Trends Plant Sci. 25, 604–617.

Ten Hoopen, F., Cuin, T.A., Pedas, P., Hegelund, J.N., Shabala, S., Schjoerring, J.K., et al., 2010. Competition between uptake of ammonium and potassium in barley and Arabidopsis roots. Molecular mechanisms and physiological consequences. J. Exp. Bot. 61, 2303–2315.

Thompson, R.L., Lassaletta, L., Patra, P.K., Wilson, C., Wells, K.C., Gressent, A., et al., 2019. Acceleration of global N_2O emissions seen from two decades of atmospheric inversion. Nat. Clim. Change 9, 993–998.

Thomsen, H.C., Eriksson, D., Møller, I.S., Schjoerring, J.K., 2014. Cytosolic glutamine synthetase: A target for improvement of crop nitrogen use efficiency? Trends Plant Sci. 19, 656–663.

Thorup-Kristensen, K., Halberg, N., Nicolaisen, M., Olesen, J.E., Crews, T.E., Hinsinger, P., et al., 2020. Digging deeper for agricultural resources, the value of deep rooting. Trends Plant Sci. 25, 406–417.

Tian, W., Wang, C., Gao, Q.F., Li, L.G., Luan, S., 2020. Calcium spikes, waves and oscillations in plant development and biotic interactions. Nat. Plants 6, 750–759.

Tomati, U., Galli, E., 1979. Water stress and -SH-dependent physiological activities in young maize plants. J. Exp. Bot. 30, 557–563.

Tong, G.M., Rude, R.K., 2005. Magnesium deficiency in critical illness. J. Intensive Care Med. 20, 3–17.

Toyota, M., Spencer, D., Sawai-Toyota, S., Jiaqi, W., Zhang, T., Koo, A.J., et al., 2018. Glutamate triggers long-distance, calcium-based plant defense signaling. Science 361, 1112–1115.

Tränkner, M., Tavakol, E., Jákli, B., 2018. Functioning of potassium and magnesium in photosynthesis, photosynthate translocation and photoprotection. Physiol. Plant. 163, 414–431.

Tsay, Y.F., Chiu, C.C., Tsai, C.B., Ho, C.H., Hsu, P.K., 2007. Nitrate transporters and peptide transporters. FEBS Lett. 581, 2290–2300.

Tsuchiya, T., Ohta, H., Okawa, K., Iwamatsu, A., Shimada, H., Masuda, T., et al., 1999. Cloning of chlorophyllase, the key enzyme in chlorophyll degradation: finding of a lipase motif and the induction by methyl jasmonate. Proc. Natl. Acad. Sci. USA 96, 15362–15367.

Tuckendorf, A., Rauser, W.E., 1990. Changes in glutathione and phytochelatins in roots of maize seedlings exposed to cadmium. Plant Sci. 70, 155–166.

Turner, B.L., Laliberté, E., Hayes, P.E., 2018. A climosequence of chronosequences in southwestern Australia. Eur. J. Soil Sci. 69, 69–85.

Undurraga, S.F., Ibarra-Henriquez, C., Fredes, I., Alvarez, J.M., Gutierrez, R.A., 2017. Nitrate signaling and early responses in Arabidopsis roots. J. Exp. Bot. 68, 2541–2551.

Van der Leij, M., Smith, S.J., Miller, A.J., 1998. Remobilisation of vacuolar stored nitrate in barley root cells. Planta 205, 64–72.

Van Steveninck, R.F.M., Babare, A., Fernando, D.R., Van Steveninck, M.E., 1994. The binding of zinc, but not cadmium, by phytic acid in roots of crop plants. Plant Soil 167, 157–164.

Vauclare, P., Kopriva, S., Fell, D., Suter, M., Sticher, L., von Ballmoos, P., et al., 2002. Flux control of sulphate assimilation in *Arabidopsis thaliana*: adenosine 59-phosphosulphate reductase is more susceptible than ATP sulphurylase to negative control by thiols. Plant J. 31, 729–740.

Veneklaas, E.J., Lambers, H., Bragg, J., Finnegan, P.M., Lovelock, C.E., Plaxton, W.C., et al., 2012. Opportunities for improving phosphorus-use efficiency in crop plants. New Phytol. 195, 306–320.

Verbruggen, N., Hermans, C., 2013. Physiological and molecular responses to magnesium nutritional imbalance in plants. Plant Soil 368, 87–99.

Verkleij, J.A.C., Koevoets, P., van't Riet, J., Bank, R., Nijdam, Y., Ernst, W.H.O., 1990. Poly (γ-glutamylcysteinyl) glycines or phytochelatins and their role in cadmium tolerance of *Silene vulgaris*. Plant Cell Environ. 13, 413–421.

Véry, A.A., Nieves-Cordones, M., Daly, M., Khan, I., Fizames, C., Sentenac, H., 2014. Molecular biology of K^+ transport across the plant cell membrane: What do we learn from comparison between plant species? J. Plant Physiol. 171, 748–769.

Vidal, E.A., Araus, V., Lu, C., Parry, G., Green, P.J., Coruzzi, G.M., et al., 2010. Nitrate-responsive miR393/AFB3 regulatory module controls root system architecture in *Arabidopsis thaliana*. Proc. Natl. Acad. Sci. USA 107, 4477–4482.

Viktor, A., Cramer, M.D., 2005. The influence of root assimilated inorganic carbon on N acquisition/assimilation and carbon partitioning. New Phytol. 165, 157–169.

Vinh, T.D., Yoshida, Y., Ooyama, M., Goto, T., Yasuba, K., Tanaka, Y., 2018. Comparative analysis on blossom-end rot incidence in two tomato cultivars in relation to calcium nutrition and fruit growth. Hort. J. 87, 97–105.

von Wirén, N., Merrick, M., 2004. Regulation and function of ammonium carriers in bacteria, fungi and plants. Topics Curr. Genetics 9, 95–120.

Walch-Liu, P., Neumann, G., Engels, C., 2001. Response of shoot and root growth to supply of different N forms is not related to carbohydrate and N status of tobacco plants. J. Plant Nutr. Soil Sci. 164, 97–103.

Walker, C.J., Weinstein, J.D., 1991. Further characterization of the magnesium chelatase in isolated developing cucumber chloroplasts. Plant Physiol. 95, 1189–1196.

Walker, D.A., 1980. Regulation of starch synthesis in leaves – the role of orthophosphate. Proc. 15th Colloq. Int. Potash Inst. Bern 195–207.

Walker, D.J., Leigh, R.A., Miller, A.J., 1996. Potassium homeostasis in vacuolate plant cells. Proc. Natl. Acad. Sci. USA 93, 10510–10514.

Walker, D.J., Black, C.R., Miller, A.J., 1998. The role of cytosolic potassium and pH in the growth of barley roots. Plant Physiol. 118, 957–964.

Walker, T.W., Syers, J.K., 1976. The fate of phosphorus during pedogenesis. Geoderma 15, 1–9.

Wallace, W., Pate, J.S., 1965. Nitrate reductase in the field pea (*Pisum arvense* L.). Ann. Bot. 29, 655–671.

Wang, G., Hu, C., Zhou, J., Liu, Y., Cai, J., Pan, C., et al., 2019. Systemic root-shoot signaling drives jasmonate-based root defense against nematodes. Curr. Biol. 29, 1–9.

Wang, H.D., Wan, Y.F., Buchner, P., King, R., Ma, H.X., Hawkesford, M.J., 2020a. Phylogeny and gene expression of the complete NITRATE TRANSPORTER 1/PEPTIDE TRANSPORTER FAMILY in *Triticum aestivum*. J. Exp. Bot. 71, 4531–4546.

Wang, M., Zhang, P.L., Liu, Q., Li, G.J., Di, D.W., Xia, G.M., et al., 2020b. TaANR1-TaBG1 and TaWabi5-TaNRT2s/NARs link ABA metabolism and nitrate acquisition in wheat roots. Plant Physiol. 182, 1440–1453.

Wang, W., Hu, B., Yuan, D.Y., Liu, Y.Q., Che, R.H., Hu, Y.C., et al., 2018b. Expression of the nitrate transporter gene OsNRT1.1A/OsNPF6.3 confers high yield and early maturation in rice. Plant Cell 30, 638–651.

Wang, Y.Y., Cheng, Y.H., Chen, K.E., Tsay, Y.F., 2018a. Nitrate transport, signaling, and use efficiency. Annu. Rev. Plant Biol. 69, 85–122.

Wang, Y., Cheng, Y., Ou, K., Lin, L., Liang, J., 2008. *In vitro* solubility of calcium, iron, and zinc in rice bran treated with phytase, cellulase, and protease. J. Agric. Sci. Food Chem. 56, 11868–11874.

Wang, Y., Hills, A., Blatt, M.R., 2014. Systems analysis of guard cell membrane transport for enhanced stomatal dynamics and water use efficiency. Plant Physiol. 164, 1593–1599.

Watanabe, T., Broadley, M.R., Jansen, S., White, P.J., Takada, J., Satake, K., et al., 2007. Evolutionary control of leaf element composition in plants. New Phytol. 174, 516–523.

Watson, M.B., Malmberg, R.L., 1996. Regulation of *Arabidopsis thaliana* (L) Heynh arginine decarboxylase by potassium deficiency stress. Plant Physiol. 111, 1077–1083.

Watt, M., Evans, J.R., 1999. Proteoid roots. Physiology and development. Plant Physiol. 121, 317–323.

Wedding, R.T., Black, M.K., 1988. Role of magnesium in the binding of substrate and effectors to phosphoenolpyruvate carboxylase from a CAM plant. Plant Physiol. 87, 443–446.

Wehr, J.B., Menzies, N.W., Blamey, F.P.C., 2004. Inhibition of cell-wall autolysis and pectin degradation by cations. Plant Physiol. Biochem. 42, 485–492.

Weiland, M., Mancuso, S., Baluska, F., 2015. Signalling via glutamate and GLRs in *Arabidopsis thaliana*. Funct. Plant Biol. 43, 1–25.

Werner, W., 1959. Die Wirkung einer Magnesiumdüngung zu Kartoffeln in Abhängigkeit von Bodenreaktion und Stickstoffform. Kartoffelbau 10, 13–14.

Weston, J., 2008. Biochemistry of magnesium. In: Rappoport, Z., Marek, I. (Eds.), The Chemistry of Organomagnesium Compounds. John Wiley & Sons, Ltd, Hoboken, NJ, USA, pp. 315–367.

White, P.J., 1996. The permeation of ammonium through a voltage-independent K$^+$ channel in the plasma membrane of rye roots. J. Memb. Biol. 152, 89–99.

White, P.J., 2005. Calcium. In: Broadley, M.R., White, P.J. (Eds.), Plant Nutritional Genomics. Blackwell, Oxford, UK, pp. 66–86.

White, P.J., 2015. Calcium. In: Barker, A.V., Pilbeam, D.J. (Eds.), A Handbook of Plant Nutrition, 2nd ed. CRC Press, Boca Raton, FL, USA, pp. 165–198.

White, P.J., 2018. Improving nutrient management in potato cultivation. In: Wale, S. (Ed.), Achieving Sustainable Cultivation of Potatoes. Vol. 2: Production, Storage and Crop Protection. Burleigh Dodds Science Publishing, Cambridge, UK, pp. 45–67.

White, P.J., Broadley, M.R., 2003. Calcium in plants. Ann. Bot. 92, 487–511.

White, P.J., Broadley, M.R., 2009. Biofortification of crops with seven mineral elements often lacking in human diets – iron, zinc, copper, calcium, magnesium, selenium and iodine. New Phytol. 182, 49–84.

White, P.J., Holland, J.E., 2018. Proceedings of The International Fertiliser Society 827: Calcium in Plant Physiology and its Availability from the Soil. International Fertiliser Society, UK.

White, P.J., Pongrac, P., 2017. Heavy-metal toxicity in plants. In: Shabala, S. (Ed.), Plant Stress Physiology, 2nd ed. CABI, Wallingford, UK, pp. 301–331.

White, P.J., Broadley, M.R., El-Serehy, H.A., George, T.S., Neugebauer, K., 2018. Linear relationships between shoot magnesium and calcium concentrations among angiosperm species are associated with cell wall chemistry. Ann. Bot. 122, 221–226.

Willenbrink, J., 1967. Über Beziehungen zwischen Proteinumsatz und Schwefelversorgung der Chloroplasten. Z. Pflanzenphysiol. 56, 427–438.

Williamson, G., Tamburrino, G., Bizior, A., Boeckstaens, M., Mirandela, G.D., Bage, M.G., et al., 2020. A two-lane mechanism for selective biological ammonium transport. Elife 9, e57183.

Wingler, A., Roitsch, T., 2008. Metabolic regulation of leaf senescence: interactions of sugar signalling with biotic and abiotic stress responses. Plant Biol. 10, 50–62.

Winter, H., Lohaus, G., Heldt, H.W., 1992. Phloem transport of amino acids in relation to their cytosolic levels in barley leaves. Plant Physiol. 99, 996–1004.

Witte, C.-P., 2011. Urea metabolism in plants. Plant Sci. 180, 431–438.

Woodrow, I.E., Rowan, K.S., 1979. Change of flux of orthophosphate between cellular compartments in ripening tomato fruits in relation to climacteric rise in respiration. Funct. Plant Biol. 6, 39–46.

Wormit, A., Trentmann, O., Feifer, I., Lohr, C., Tjaden, J., Meyer, S., et al., 2006. Molecular identification and physiological characterization of a novel monosaccharide transporter from Arabidopsis involved in vacuolar sugar transport. Plant Cell 18, 3476–3490.

Wrigley, C.W., du Cros, D.L., Archer, M.J., Downie, P.G., Roxburgh, C.M., 1980. The sulfur content of wheat endosperm and its relevance to grain quality. Aust. J. Plant Physiol. 7, 755–766.

Wu, W., Berkowitz, G.A., 1992. Stromal pH and photosynthesis are affected by electroneutral K$^+$ and H$^+$ exchange through chloroplast envelope ion channels. Plant Physiol. 98, 666–672.

Wudick, M.M., Michard, E., Oliveira Nunes, C., Feijo, J.A., 2018. Comparing plant and animal glutamate receptors: common traits but different fates? J. Exp. Bot. 69, 4151–4163.

Wyn Jones, R.G., Pollard, A., 1983. Proteins, enzymes and inorganic ions. In: Läuchli, A., Bieleski, R.L. (Eds.), Encyclopedia of Plant Physiology, New Series, 15B. Springer Verlag, Berlin and New York, pp. 528–562.

Xu, G.H., Takahashi, H., 2020. Improving nitrogen use efficiency: from cells to plant systems. J. Exp. Bot. 71, 4359–4364.

Xu, X., Zhu, T., Nikonorova, N., De Smet, I., 2019. Phosphorylation-mediated signalling in plants. In: Roberts, J.A. (Ed.), Annual Plant Reviews online. John Wiley & Sons, Chichester, UK, pp. 909–932.

Yang, S.-Y., Huang, T.-K., Kuo, H.-F., Chiou, T.-J., 2017. Role of vacuoles in phosphorus storage and remobilization. J. Exp. Bot. 68, 3045–3055.

Yang, Y., Yu, Y., Liang, Y., Anderson, C.T., Cao, J., 2018. A profusion of molecular scissors for pectins: Classification, expression, and functions of plant polygalacturonases. Front. Plant Sci. 9, 1208.

Yazaki, Y., Asukagawa, N., Ishikawa, Y., Ohta, E., Sakata, M., 1988. Estimation of cytoplasmic free Mg^{2+} levels and phosphorylation potentials in mung bean root tips in vivo ^{31}P NMR spectroscopy. Plant Cell Physiol. 29, 919–924.

Yilmaz, O., Kahraman, K., Ozturk, L., 2017. Elevated carbon dioxide exacerbates adverse effects of Mg deficiency in durum wheat. Plant Soil 410, 41–50.

Yuan, L., Loqué, D., Kojima, S., Rauch, S., Ishiyama, K., Inoue, E., et al., 2007. The organization of high-affinity ammonium uptake in Arabidopsis roots depends on the spatial arrangement and biochemical properties of AMT1-type transporters. Plant Cell 19, 2636–2652.

Zeeman, S.C., Kossmann, J., Smith, A.M., 2010. Starch: its metabolism, evolution, and biotechnological modification in plants. Annu. Rev. Plant Biol. 61, 209–234.

Zhang, B., Cakmak, I., Feng, J., Yu, C., Chen, X., Xie, D., et al., 2020. Magnesium deficiency reduced the yield and seed germination in wax gourd by affecting the carbohydrate translocation. Front. Plant Sci. 11, 797.

Zhang, X., Wu, H., Chen, L., Wang, N., Wei, C., Wan, X., 2019. Mesophyll cells' ability to maintain potassium is correlated with drought tolerance in tea (*Camellia sinensis*). Plant Physiol. Biochem. 136, 196–203.

Zhao, D., Oosterhuis, D.M., Bednarz, C.W., 2001. Influence of potassium deficiency on photosynthesis, chlorophyll content and chloroplast ultrastructure of cotton plants. Photosynthetica 39, 103–109.

Zhao, F.J., De Kok, L.J., Tausz, M., 2008a. Role of sulfur for plant production in agricultural and natural ecosystems. In: Hell, R., Dahl, C., Knaff, D. B., Leustek, T. (Eds.), Sulfur Metabolism in Phototropic Organisms. Springer, Dordrecht, The Netherlands, pp. 91–106.

Zhao, F.-J., Hawkesford, M.J., McGrath, S.P., 1999a. Sulphur assimilation and effects on yield and quality of wheat. J. Cereal Sci. 30, 1–17.

Zhao, F.-J., Salmon, S.E., Withers, P.J.A., Evans, E.J., Monaghan, J.M., Shewry, P.R., et al., 1999b. Responses of breadmaking quality to sulphur in three wheat varieties. J. Sci. Food Agric. 79, 1865–1874.

Zhao, F.J., Wood, A.P., McGrath, S.P., 1999c. Effects of sulphur nutrition on growth and nitrogen fixation of pea (*Pisum sativum* L.). Plant Soil 212, 209–219.

Zhao, H.-J., Liu, Q.-L., Fu, H.-W., Xu, X.-H., Wu, D.-X., Shu, Q.-Y., 2008b. Effect of non-lethal low phytic acid mutations on grain yield and seed viability in rice. Field Crops Res. 108, 206–211.

Zheng, H., Shabalin, I.G., Handing, K.B., Bujnicki, J.M., Minor, W., 2015a. Magnesium-binding architectures in RNA crystal structures: Validation, binding preferences, classification and motif detection. Nucleic Acids Res. 43, 3789–3801.

Zheng, X.J., He, K., Kleist, T., Chen, F., Luan, S., 2015b. Anion channel SLAH3 functions in nitrate-dependent alleviation of ammonium toxicity in Arabidopsis. Plant Cell Environ. 38, 474–486.

Zörb, C., Ludewig, U., Hawkesford, M.J., 2018. Perspective on wheat yield and quality with reduced nitrogen supply. Trends Plant Sci. 11, 1029–1037.

Zörb, C., Senbayram, M., Peiter, E., 2014. Potassium in agriculture – Status and perspectives. J. Plant Physiol. 171, 656–669.

Zuluaga, D.L., Sonnante, G., 2019. The use of nitrogen and its regulation in cereals: structural genes, transcription factors, and the role of miRNAs. Plants-Basel 8, 294.

Zúñiga-Feest, A., Delgado, M., Alberdi, M., 2010. The effect of phosphorus on growth and cluster-root formation in the Chilean Proteaceae: *Embothrium coccineum* (R. et J. Forst.). Plant Soil 334, 113–121.

Chapter 7

Micronutrients

Ismail Cakmak[1], Patrick Brown[2], José M. Colmenero-Flores[3], Søren Husted[4], Bahar Y. Kutman[5], Miroslav Nikolic[6], Zed Rengel[7], Sidsel B. Schmidt[4], and Fang-Jie Zhao[8]

[1]*Faculty of Engineering and Natural Sciences, Sabanci University, Istanbul, Turkey*, [2]*University of California–Davis, Department of Plant Sciences, Davis, California, United States*, [3]*Spanish National Research Council (CSIC), IRNAS, Seville, Spain*, [4]*University of Copenhagen, Department of Plant and Environmental Sciences, Copenhagen, Denmark*, [5]*Gebze Technical University, Institute of Biotechnology, Kocaeli, Turkey*, [6]*Institute for Multidisciplinary Research, University of Belgrade, Belgrade, Serbia*, [7]*UWA School of Agriculture and Environment, University of Western Australia, Perth, WA, Australia*, [8]*College of Resources and Environmental Sciences, Nanjing Agricultural University, Nanjing, P.R. China*

Summary

The functions of iron, manganese, copper, zinc, nickel, molybdenum, boron, and chlorine in plants are discussed. Iron (Fe) plays a crucial role in redox systems in cells and in various enzymes. The strategies of dicotyledonous and gramineous plants to acquire Fe are described. Manganese (Mn) and copper (Cu) are important for redox systems and as activators of various enzymes involved in photosynthesis, detoxification of superoxide radicals, and the synthesis of lignin. Zinc (Zn) plays a key role in the structural and functional integrity of cell membranes, biosynthesis of proteins and detoxification of superoxide radicals. Nickel (Ni) is involved in nitrogen (N) metabolism as metal component of the enzyme urease. Molybdenum (Mo) is important for N metabolism as metal component of nitrogenase (N_2 fixation) and nitrate reductase. Boron (B) is crucial for the stability and function of cell wall and membranes, whereas chlorine (Cl) is essential for the proper functioning of photosystem II (PS II) and cell osmotic regulation. For each micronutrient, the effects of deficiency and toxicity are described.

7.1 Iron

7.1.1 General

Iron is the fourth most abundant element in the Earth's crust after oxygen, silicon, and aluminum. The solubility of Fe is, however, extremely low, especially in aerated alkaline soils. In aerated systems in the physiological pH range, the concentrations of ionic Fe^{3+} and Fe^{2+} are below 10^{-15} M due to formation of Fe hydroxides, oxyhydroxides, and oxides (Lemanceau et al., 2009). Chelates of Fe(III) and occasionally of Fe(II) are therefore the dominant forms of soluble Fe in soil and nutrient solutions. As a rule, Fe(II) is taken up preferentially compared with Fe(III), but this also depends on the plant species (Strategies I and II, see Chapters 2 and 14). In long-distance transport in the xylem, there is a predominance of Fe(III) complexes (Chapter 3).

As a transition element, Fe is characterized by the relative ease by which it may change its oxidation state

$$Fe^{3+} \leftrightarrow Fe^{2+}$$

and by its capacity to form octahedral complexes with various ligands. Depending on the ligand, the redox potential of Fe(II/III) varies widely, explaining the importance of Fe in biological redox systems. Due to the high affinity of Fe for various ligands (e.g., organic acid anions or inorganic phosphate), ionic Fe^{3+} or Fe^{2+} is not transported long distance in

---

☆ This chapter is a revision of the third edition chapter by M. Broadley, P. Brown, I. Cakmak, Z. Rengel, and F.J. Zhao, pp. 191–248. DOI: https://doi.org/10.1016/B978-0-12-384905-2.00007-8. © Elsevier Ltd.

*The individual micronutrient sections were written as follows: Fe (M.N.), Mn (S.H. and S.B.S.), Cu (Z.R.), Zn (I.C.), Ni (P.B. and B.Y.K.), Mo (F-J.Z.), B (P.B. and I.C.), and Cl (J.M.C.F.).

plants. In aerobic systems many low-molecular-weight iron chelates, and free iron in particular (either Fe^{3+} or Fe^{2+}), produce reactive oxygen species (ROS) such as superoxide anion radical ($O_2^{\bullet -}$) and hydroxyl radical ($OH^{\bullet}$) (Halliwell, 2009) and related compounds, for example:

$$O_2 + Fe^{2+} \rightarrow O_2^{\bullet -} + Fe^{3+}$$

or in the Fenton reaction:

$$H_2O_2 + Fe^{2+} \rightarrow Fe^{3+} + OH^- + OH^{\bullet}$$

or in the Haber–Weiss reaction:

$$O_2^{\bullet -} + H_2O_2 \xrightarrow{Fe} O_2 + OH^{\bullet} + OH^-$$

These radicals are highly toxic and responsible for the peroxidation of polyunsaturated fatty acids of membrane lipids and proteins. To prevent oxidative cell damage, Fe has to be either tightly bound (e.g., cell walls) or incorporated into structures (e.g., heme and nonheme proteins) that allow controlled reversible oxidation–reduction reactions

$$Fe(II) \underset{+e^-}{\overset{-e^-}{\rightleftarrows}} Fe(III)$$

including those in antioxidant protection.

7.1.2 Iron-containing constituents of redox systems

7.1.2.1 Heme proteins

The most well-known heme proteins are the cytochromes that contain a heme Fe–porphyrin complex (Fig. 7.1) as a prosthetic group. Cytochromes are constituents of the redox systems in chloroplasts and mitochondria and also a component in the redox chain in nitrate reductase. The role of Fe in leghemoglobin and nitrogenase is discussed in Chapter 16. Small amounts of leghemoglobin may also be present in the roots of plants that do not form root nodules (Appleby et al., 1988). This leghemoglobin may act as a signal molecule indicating O_2 deficiency, initiating a metabolic shift towards fermentation.

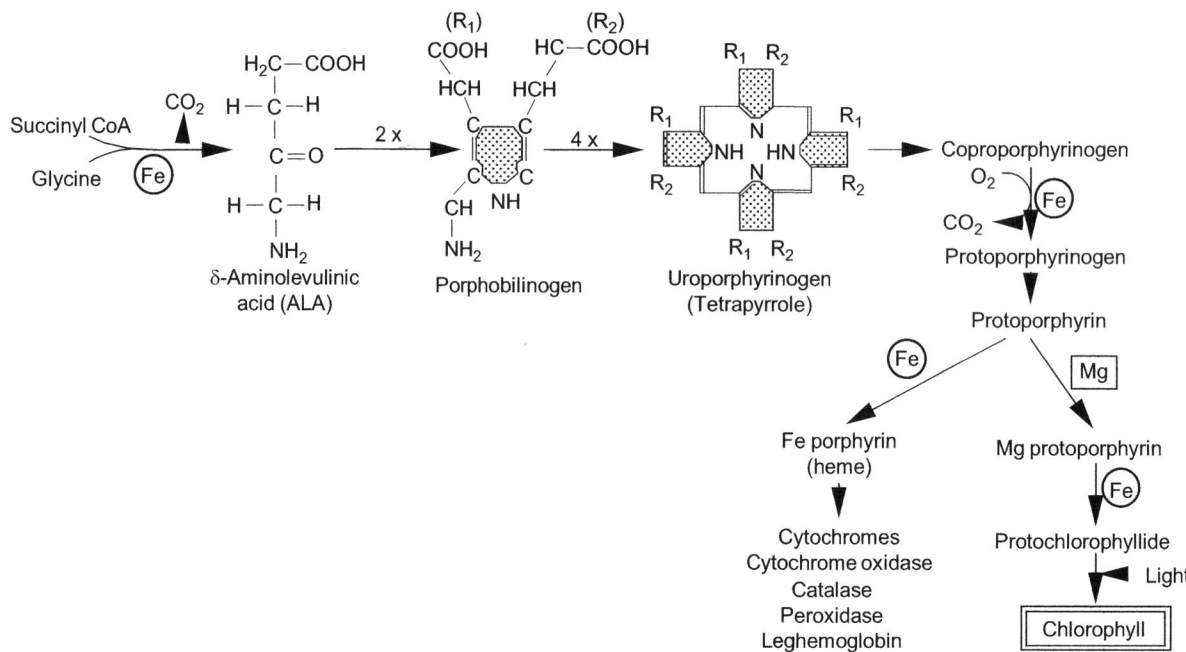

FIGURE 7.1 Role of Fe in the biosynthesis of heme coenzymes and chlorophyll.

TABLE 7.1 Fe concentrations and activities of H_2O_2-scavenging enzymes in leaves of Fe-inefficient (Target) and Fe-efficient (Pakmore) tomato genotypes grown in the nutrient solution with a low or sufficient Fe supply for 50 days.

Parameters	Fe-inefficient		Fe-efficient	
	Sufficient Fe	Low Fe	Sufficient Fe	Low Fe
Leaf Fe concentration (μg g^{-1}dw)	226	21	200	21
Enzyme activity (μmol g^{-1} fw min^{-1})				
Catalase	198	35	244	63
Guaiacol peroxidase	412	136	304	214
Ascorbate peroxidase	613	133	584	192

Source: Based on Dasgan et al. (2003).

Other heme enzymes are catalase and peroxidases. Under Fe deficiency, the activity of these enzymes decreases rapidly in plant tissues, particularly that of catalase in genotypes susceptible to Fe deficiency, for example, in tomato (Table 7.1). The activity of these enzymes is, therefore, an indicator of Fe nutritional status of plants (Chapter 11).

Catalase facilitates detoxification of H_2O_2 to water and O_2 according to the reaction:

$$H_2O_2 \rightarrow H_2O + \tfrac{1}{2}O_2$$

The enzyme plays an important role in association with superoxide dismutase (SOD), as well as in photorespiration and the glycolate pathway.

Various isoenzymes of peroxidases are present in plants. They catalyze the reactions:

$$XH_2 + H_2O_2 \rightarrow X + 2H_2O$$

and

$$XH + XH + H_2O_2 \rightarrow X\text{---}X + 2H_2O$$

An example of the first type of reaction is the detoxification of H_2O_2 in chloroplasts catalyzed by ascorbate peroxidase. In the second type of reaction, cell wall–bound peroxidases catalyze the polymerization of phenols to lignin. The alterations in cell wall formation of rhizodermal cells under Fe deficiency may be related to impaired peroxidase activity. Peroxidases are abundant in cell walls of the epidermis (Hendricks and van Loon, 1990) and rhizodermis (Codignola et al., 1989) and are required for the biosynthesis of lignin and suberin. Both synthetic pathways require phenolic compounds and H_2O_2 as substrates. The formation of H_2O_2 is catalyzed by the oxidation of NADH at the plasma membrane/cell wall interface (Mäder and Füssl, 1982), as follows:

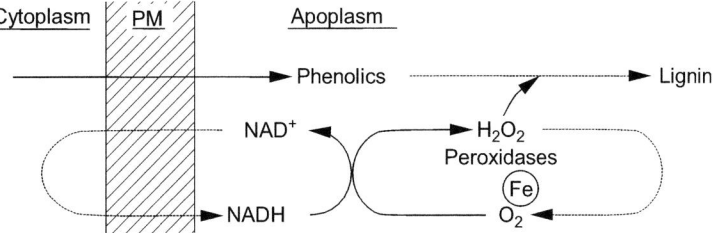

In Fe-deficient roots, peroxidase activity is strongly depressed (Sijmons et al., 1985; Ranieri et al., 2001). Consequently, both substrates (phenolics and H_2O_2) accumulate in root tissues (Ranieri et al., 2001; Römheld and Marschner, 1981a).

7.1.2.2 Fe-S proteins

In the nonheme Fe-S proteins, Fe is coordinated to the thiol group of cysteine or to inorganic S, or to both. The most well-known Fe-S protein is ferredoxin that acts as an electron transmitter in a number of metabolic processes according to the principle:

$$\xrightarrow{e^-} \begin{array}{c} -Cys-S \\ | \\ -Cys-S \end{array} \underset{Fe}{\diagdown} \underset{S}{\diagup} \underset{Fe}{\diagdown} \underset{S}{\diagup} \begin{array}{c} S-Cys- \\ | \\ S-Cys- \end{array} \xrightarrow{e^-} \begin{array}{l} NADP^+ \text{ (photosynthesis)} \\ \text{Nitrite reductase} \\ \text{Sulfite reductase} \\ N_2 \text{ reduction} \\ \text{GOGAT} \end{array}$$

In Fe-deficient leaves, the concentrations of ferredoxin and chlorophyll are decreased to a similar extent (Table 7.2), with the low ferredoxin concentration correlated with decreased nitrate reductase activity (NRA). Both ferredoxin concentration and NRA can be restored by resupplying Fe. The involvement of Fe at various steps in the nitrate reduction is discussed in Section 6.1.

Another example of Fe-S proteins is the isoenzymes of SOD that contain Fe as a metal component of the prosthetic group (Fe-SOD). Superoxide dismutases detoxify $O_2^{\cdot-}$ by the formation of H_2O_2 and may contain Cu, Zn, Mn, or Fe as metal components (Fridovich, 1983). In chloroplasts, Fe-SOD is the main isoenzyme of SOD (Kwiatowsky et al., 1985), but it may also occur in mitochondria and peroxisomes (Droillard and Paulin, 1990). In Fe-deficient plants, Fe-SOD activity is low (Iturbe-Ormaetxe et al., 1995), resulting in the accumulation of H_2O_2 (Tewari et al., 2005). Although Fe-deficient plants have the reduced activity of antioxidative enzymes such as catalase and ascorbate peroxidase and the increased concentrations of H_2O_2, there does not appear to be extensive oxidative cell damage (e.g., lipid peroxidation) (Ranieri et al., 2001), which may be due to the low concentrations of free Fe^{2+} that is involved in generating ROS through the Haber–Weiss and Fenton reactions.

Iron-deficiency stress is associated with enhanced production of organic acids, likely due to decreased aconitase activity. Aconitase is an Fe-S protein (Brouquisse et al., 1986) that catalyzes the isomerization of citrate to isocitrate in the tricarboxylic acid cycle. In Fe-deficient plants, aconitase activity is decreased (De Vos et al., 1986), and reactions in the tricarboxylic acid cycle are disturbed, leading to the accumulation of citric, malic, and other organic acids (Table 7.3). In roots of Fe-deficient tomato plants, the increase in organic acid concentration is correlated with enhanced CO_2 dark fixation and net excretion of H^+ (i.e., acidification of the rhizosphere) (Miller et al., 1990).

The relationship between decreased aconitase activity and organic acid accumulation in roots of Fe-deficient plants is still controversial. In lemon [*Citrus limon* (L.) Burm] fruit, of the two aconitase isoenzymes, only the cytosolic form showed a decreased activity under low Fe conditions, resulting in a slower rate of citrate breakdown and a concomitant increase in citrate concentration (Shlizerman et al., 2007). The loading of citrate into the xylem is mediated by Ferric Reductase Defective 3 (FRD3) in arabidopsis (Green and Rogers, 2004) and its ortholog FRD-Like 1 in rice (Yokosho et al., 2009). Both transporters are present in the plasma membrane of root pericycle cells, and the genes encoding them are upregulated by Fe deficiency. The existence of a tri-Fe(III), tri-citrate complex [Fe(3)Cit(3)] in the xylem sap of tomato plants has been detected (Rellan-Alvarez et al., 2010), confirming previous speculation that Fe is transported in xylem in the form of Fe-citrate complexes.

TABLE 7.2 Concentrations of chlorophyll and ferredoxin, and nitrate reductase activity in leaves of lemon with different Fe concentrations.

Fe concentration (μg g^{-1} dw)	Chlorophyll (mg g^{-1} dw)	Ferredoxin (mg g^{-1} dw)	Nitrate reductase (nmol NO$_2$ g^{-1} fw h^{-1})
96	1.80	0.82	937
62	1.15	0.44	408
47	0.55	0.35	310
47→81[a]	—	0.63	943

[a] 40 h after infiltration of intact Fe-deficient leaves with 0.2% w/v FeSO4.
Source: Based on Alcaraz et al. (1986).

TABLE 7.3 Relationship between Fe supply, chlorophyll concentration in leaves, and organic acid concentration in roots of oats.

Treatment	Chlorophyll concentration (relative)	Organic acid concentration [µg (10 g)$^{-1}$ fw]			
		Malic	Citric	Other	Total
+Fe	100	39	11	23	73
−Fe	12	93	67	78	238

Source: Based on Landsberg (1981).

Iron deficiency–induced CO_2 fixation and high phosphoenolpyruvate carboxylase (PEPC) activity in root cells may be major reasons for the accumulation of organic acids in Fe-deficient plants (Miller et al., 1990; Abadía et al., 2002). Increased PEPC activity may also be linked to acclimatory responses to Fe deficiency in roots, such as proton release and Fe reduction capacity (M'sehli et al., 2009; Rombola et al., 2002). However, Fe deficiency affected rhizosphere acidification and the root PEPC activity differently in two distinct Strategy I species (cucumber and soybean) (Dell'Orto et al., 2013). In Fe-deficient cucumber roots, increased rhizosphere acidification was accompanied by increased PEPC activity, whereas Fe deficiency resulted in reduced acidification and decreased PEPC activity in soybean roots.

Riboflavin also accumulates in most dicotyledonous plant species under Fe deficiency, and its release from roots may be enhanced by a factor of 200 in Fe-deficient plants (Andaluz et al., 2009; Welkie and Miller, 1989). Increased root concentrations of riboflavin are associated with the increased activity of 6,7-dimethyl-8-ribityllumazine synthase that contributes to the final step of riboflavin biosynthesis (Andaluz et al., 2009). The expression of genes involved in the riboflavin biosynthesis was upregulated by Fe deficiency in the roots of cucumber and melon, species known for the high production of flavin compounds (Hsieh and Waters, 2016).

7.1.3 Other Fe-requiring enzymes

There are many enzymes in which Fe acts as either a metal component in redox reactions or a bridging element between enzyme and substrate. In Fe-deficient plants, the activities of some of these enzymes are low, which may result in gross changes in metabolic processes.

In the biosynthetic pathway converting 1-aminocyclopropane-1-carboxylic acid (ACC) to ethylene, a two-step one-electron oxidation takes place, catalyzed by Fe(II) (see Fig. 7.6). Accordingly, ethylene formation is low in Fe-deficient cells and is restored immediately upon resupply of Fe, without the involvement of protein synthesis (Bouzayen et al., 1991). The exogenous addition of ethylene or its precursor ACC to Fe-sufficient plants induces some physiological Fe responses (e.g., enhanced ferric reductase activity) as well as the formation of root hairs (Lucena et al., 2015). Iron-efficient genotypes of *Pisum sativum* and *Medicago truncatula* produce more ethylene than the Fe-inefficient ones (Kabir et al., 2012; Li et al., 2014a,b), suggesting that the higher ethylene production allows these cultivars to activate their root responses to Fe deficiency more efficiently.

Lipoxygenases are enzymes containing one atom of Fe per molecule (Hildebrand, 1989). They catalyze the peroxidation of linoleic and linolenic acids (long-chain polyunsaturated fatty acids) that are important components of cell membranes. Hence, high lipoxygenase activity is typical for fast-growing tissues and organs and may be critical for membrane stability.

Low chlorophyll concentration (chlorosis) of young leaves is the most obvious visible symptom of Fe deficiency. Various factors are responsible for this decrease, the most direct one being the role of Fe in the biosynthesis of chlorophyll (Fig. 7.1). The common precursor of chlorophyll and heme synthesis is aminolevulinic acid (ALA), and the rate of ALA formation is controlled by Fe (Pushnik and Miller, 1989). Iron is also required for the formation of protochlorophyllide from Mg-protoporphyrin (Fig. 7.1; Pushnik et al., 1984).

7.1.4 Chloroplast development and photosynthesis

Iron deficiency has less effect on leaf growth, cell number per unit area, or number of chloroplasts per cell than on the size of the chloroplasts and protein content per chloroplast (Table 7.4). Iron is required for protein synthesis, and the

TABLE 7.4 Properties of leaves of sugar beet (*Beta vulgaris*) with sufficient Fe and mild or severe Fe deficiency.

Parameter	Control	Mild deficiency	Severe deficiency
Chlorophyll concentration (mg cm^{-2})	>40	20–40	<20
Soluble protein (mg cm^{-2} leaf area)	0.57	0.56	0.53
Mean leaf cell volume (10^{-8} cm^3)	2.64	2.78	2.75
Chloroplasts (no. cell^{-1})	72	77	83
Chloroplast volume (μm^3)	42	37	21
Protein-N (pg chloroplast^{-1})	1.88	1.34	1.24

Source: Adapted from Terry (1980).

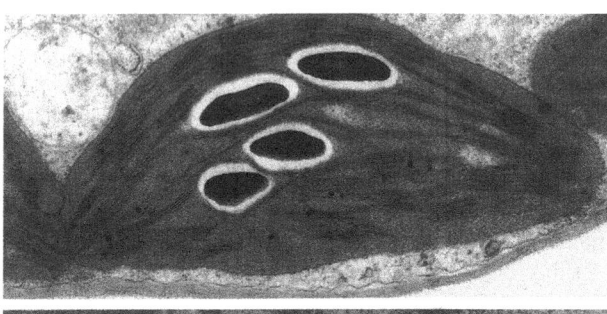
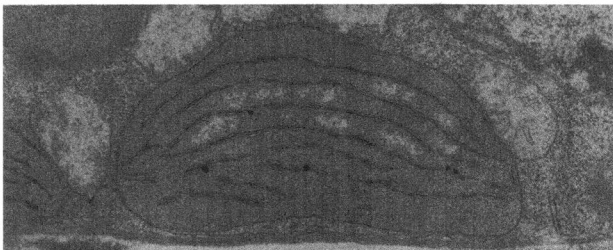

FIGURE 7.2 Fine structure of chloroplasts from Fe-sufficient (top) and Fe-deficient (bottom) soybean (*Glycine max* L.) plants (× 24 000). *Courtesy C. Hecht-Buchholz.*

number of ribosomes—the sites of protein synthesis—decrease in Fe-deficient leaf cells (Lin and Stocking, 1978). Decreases in leaf protein content under Fe deficiency are particularly pronounced for the Rubisco protein that represents nearly 50% of the soluble proteins in the chloroplasts (Ellis, 1979; see also Table 7.6). The concentration of Rubisco protein in the Fe-deficient leaves of sugar beet was regulated by the availability of RNA, and the rate of CO_2 fixation decreased by about 63% in comparison with the Fe-sufficient leaves (Winder and Nishio, 1995).

In thylakoid membranes, Fe is directly involved in the electron transport chain. PS I is a strong sink for Fe due to its higher Fe content (12 atoms of Fe per complex) compared to PSII (three atoms of Fe per complex) and the Cyt *bf* complex (five atoms of Fe per complex) (Raven et al., 1999). The high Fe requirement for the structural and functional integrity of thylakoid membranes, and the additional Fe requirement for ferredoxin and the biosynthesis of chlorophyll explain the particular sensitivity of chloroplasts in general, and the thylakoids in particular, to Fe deficiency (Fig. 7.2; Schmidt et al., 2020).

In Fe-deficient leaves, however, not all photosynthetic pigments and components of the electron transport chain are affected to the same extent (Table 7.5). The activity of PSI is more depressed than that of PSII under Fe deficiency, probably due to the greater Fe requirement of PSI than PSII. Hence, resupplying Fe to chlorotic leaves increases the activity of PSI more strongly than that of PSII. Although PSII efficiency can recover rapidly following Fe resupply to

TABLE 7.5 Concentrations of Fe, chlorophyll, and components of photosystem I (PS I) and photosynthetic electron transport capacity of PSII and PSI in tobacco (*Nicotiana tabacum*) plants grown with a sufficient or deficient Fe supply.

Fe treatment	Fe (μg cm^{-2} leaf)	Chlorophyll	PS I components			Fe transport capacity of isolated chloroplasts (μmol cm^{-2} leaf h^{-1})	
			P700	Cytochromes (pmol cm^{-2})	Protein (μg cm^{-2})	PS II	PS I
+Fe	1.44	89	545	599	108	112	1680
−Fe	0.25	26	220	201	38	60	780
−Fe + Fe[a]	1.16	24	474	474	79	72	1528

[a] 10 days after foliar application of Fe.
Source: Recalculated from Pushnik and Miller (1989).

TABLE 7.6 Concentration of chlorophyll and carotenoids and maximum velocity of Rubisco carboxylation ($V_{c\,max}$) in leaves of hydroponically grown sugar beet (*Beta vulgaris*) and field-grown pear (*Pyrus communis*) and peach (*Prunus persica*) plants as affected by Fe deficiency.

Species	Fe supply	Chlorophyll a + b (μmol m^{-2})		Total carotenoids	$V_{c\,max}$ (μmol m^{-2} s^{-1})
Sugar beet	+Fe	389	3.7	126	22
	Severe Fe deficiency	79	4.6	34	12
	Extreme Fe deficiency	32	5.7	19	7
Pear	+Fe	248	3.5	95	131
	Severe Fe deficiency	83	5.7	60	80
	Extreme Fe deficiency	24	5.9	26	17
Peach	+Fe	199	3.7	103	56
	Severe Fe deficiency	70	4.8	50	15
	Extreme Fe deficiency	37	4.9	29	8

Source: Based on Larbi et al. (2006).

Fe-deficient plants (Hantzis et al., 2018), the recovery becomes more difficult to achieve as Fe deficiency becomes more severe (Table 7.4; Morales et al., 1991). In arabidopsis plants grown under low Fe, the photosynthetic electron transport and the protein concentrations of Fe-dependent enzymes recovered fully after Fe resupply, indicating that Fe deficiency stress did not cause irreversible secondary damage (Hantzis et al., 2018).

Generally, carotenoids are less affected than chlorophylls, and chlorophyll *a* is more sensitive to Fe deficiency than chlorophyll *b*, leading to higher chlorophyll *b*/chlorophyll *a* ratios in Fe-deficient leaf tissues (Table 7.6). Upon Fe deficiency, there is a significant increase in the de-epoxidized xanthophyll pigments zeaxanthin and violaxanthin, whereas the epoxidated form of violaxanthin declines strongly, especially under high light intensity (Jiang et al., 2001; Timperio et al., 2007). Resupplying Fe to Fe-deficient plants increased the concentration of the epoxidated form of violaxanthin rapidly at the expense of zeaxanthin (Larbi et al., 2004). The xanthophyll cycle pigments have photoprotective effects in the chloroplasts by coping with excess light energy through the conversion of zeaxanthin to violaxanthin:

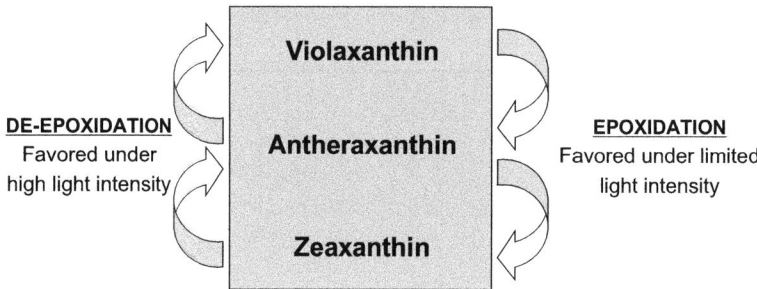

Under Fe deficiency, leaves generally have low photosynthetic activity, but they absorb more light energy per chlorophyll molecule than required for photosynthesis, especially under high radiation (Abadía et al., 1999). This results in a high risk of both photoinhibitory and photooxidative damage in Fe-deficient leaves. Nevertheless, in contrast to Zn-deficient or Mg-deficient plants, there appears to be little photooxidative damage in Fe-deficient plants. The absence of serious photooxidative damage in the Fe-deficient leaves most probably relates to the rapid increases in the concentrations of de-epoxidized xanthophyll pigments and the low concentrations of catalytic Fe required in ROS generation.

Iron-deficient leaves are characterized by low concentrations of starch and sugars (Arulanathan et al., 1990). This is to be expected due to the low concentrations of chlorophyll and ferredoxin, impairment of photosynthetic electron transport, and the decreased regeneration of reduced ferredoxin. Decrease in photosynthesis is a characteristic physiological response of plants to Fe deficiency (Fig. 7.3), attributed to the reduced photosynthetic electron transport and thus impaired carboxylation due to the low availability of adenosine triphosphate (ATP) and nicotinamide adenine dinucleotide phosphate (NADPH) for the Calvin-Benson cycle (Table 7.5). Iron-deficient plants respond to Fe resupply by a rapid increase in photosynthesis (Fig. 7.3).

7.1.5 Localization and binding state of Fe

The apoplasmic space and vacuoles are the two important plant compartments for the storage and sequestration of free Fe. About 80% of total root Fe and about 30% of total leaf Fe is in the apoplast regardless of Fe status (Table 7.7) and is bound to the structural components of cell walls (e.g., pectin and hemicellulose) (Cesco et al., 2002; Nikolic and Römheld, 2003). Root apoplasmic Fe was suggested to be an important Fe storage pool for higher plants, which can be remobilized easily during Fe starvation (Jin et al., 2007; Pavlovic et al., 2013). In response to Fe deficiency, NRAMP3 and NRAMP4 transporters (natural resistance-associated macrophage proteins) are of crucial importance for the mobilization of stored Fe during seed germination (Bastow et al., 2018).

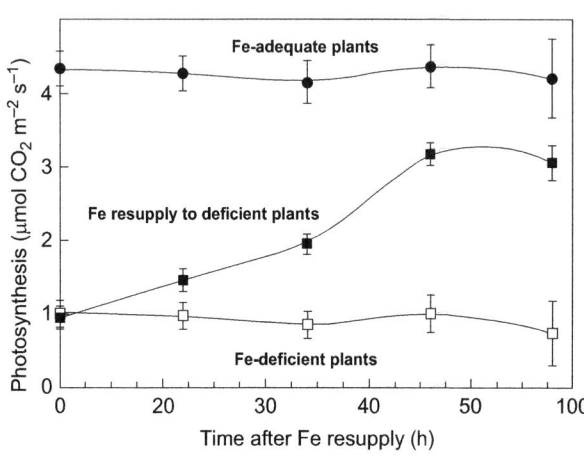

FIGURE 7.3 Net photosynthetic rates in hydroponically grown sugar beet plants: Fe-deficient, Fe-adequate, and Fe-deficient sugar beet plants resupplied with Fe. *Based on Larbi et al. (2004).*

TABLE 7.7 Relative distribution of apoplasmic Fe in roots (grown hydroponically because the precise analysis of root Fe cannot be done on soil-grown roots) and leaves of field-grown plants (showing "chlorosis paradox") of grapevine in relation to Fe status.

Fe status	Chlorophyll (SPAD readings)	Relative apoplasmic Fe (% of total tissue Fe)	
		Root[a]	Leaf[b]
Adequate	24–27	83	30
Low	5–12	85	28

[a]Vitis riparia *(hydroponic conditions); apoplasmic root loading of* ^{59}*Fe during 12 h.*
[b]Vitis vinifera *cv. Pinot Noir (field conditions); young fully developed leaves.*
Source: Based on Nikolic et al. (2000) and Nikolic and Römheld (2002).

Iron can be stored in the stroma of plastids as phytoferritin (plant ferritin). Ferritin consists of a hollow protein shell that can store up to 5000 atoms of Fe(III) (Fe content 12%–23% dw). Ferritin often has a well-defined crystalline form with the proposed formula $(FeO.OH)_8(FeO.OPO_3H_2)$ (Seckbach, 1982). In young leaf tissues, ferritin-bound Fe represents an important Fe source for the biosynthesis of Fe-containing proteins associated with photosynthesis (Briat et al., 2010). Ferritin is a vital compound in the maintenance of Fe homeostasis and the protection against oxidative damage. By sequestration of large amounts of Fe, ferritin exerts a critical protective role against the peroxidative cell damage catalyzed by the Fe-induced formation of ROS (Briat et al., 2010; Ravet et al., 2009). Ferritin is present not only in chloroplasts but also in the xylem and phloem (Smith et al., 1984) and is abundant in seeds. In pea plants, ferritin-bound Fe represents 92% of the total Fe in embryos, indicating that ferritin is a major form of Fe storage in seeds (Marentes and Grusak, 1998). During seed germination, ferritin is rapidly degraded, probably catalyzed by the released Fe^{2+} and generation of $OH^{\bullet}$ radicals that destroy the protein shell (Bienfait, 1989; Lobreaux and Briat, 1991). Ferritin may also serve to store Fe in nodules of legumes for the heme synthesis during nodule development and following the heme degradation during senescence (Ko et al., 1987). The bioavailability of Fe in grains is an important issue for nutritional quality and human nutrition (Chapter 9).

Phytate is also abundant in seeds; it has high affinity to Fe and forms insoluble complexes with Fe (Minihane and Rimbach, 2002). Therefore, phytate-rich diets (e.g., cereal-based foods) may be a key factor in high prevalence of Fe deficiency in humans (Chapter 9).

If plants are grown under controlled conditions (e.g., in nutrient solutions), there is a close positive correlation between total leaf concentrations of Fe and chlorophyll when the Fe supply is suboptimal (Römheld and Marschner, 1981a; Terry and Abadia, 1986). This correlation, however, is often poor or absent in plants grown in calcareous soils (Mengel, 1994; Römheld, 2000) where the Fe concentration in chlorotic leaves may be similar to, or even higher than, that in green leaves. This phenomenon has been termed the "chlorosis paradox" (Römheld, 2000). Initially, physiological inactivation of Fe in the chlorotic leaves of plants grown in calcareous soils has been discussed as a plausible explanation for the same or even higher Fe concentrations in chlorotic compared with green leaves (Mengel, 1994). However, inactivation of Fe in the leaf apoplasm could not be detected in subsequent studies (Nikolic and Römheld, 1999, 2002). Instead, the high Fe concentrations in chlorotic young leaf tissues may be the result of restricted leaf expansion and consequently diminished dilution of Fe concentration by growth (Morales et al., 1998; Römheld, 2000).

7.1.6 Root responses to Fe deficiency

In leaves, the major symptom of Fe deficiency is the inhibition of chloroplast development. In roots, however, Fe deficiency induces morphological and physiological changes that depend on plant species (Strategies I and II, Chapter 2). In dicotyledonous and monocotyledonous plant species, with the exception of the grasses (graminaceous species), Fe deficiency is associated with inhibition of root elongation, increase in the diameter of apical root zones, and abundant root hair formation (Chaney et al., 1992; Römheld and Marschner, 1981a; Schmidt, 2003). These morphological changes are often associated with the formation of cells with a distinct wall labyrinth typical

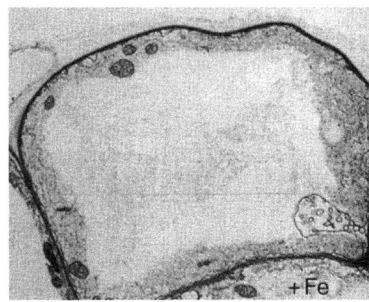 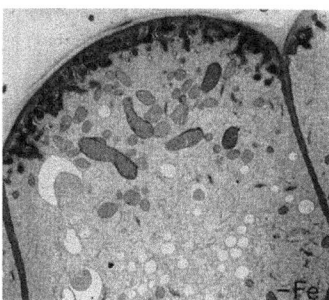

FIGURE 7.4 Sections of rhizodermal cells of sunflower: (left) Fe-sufficient and (right) Fe-deficient. *Courtesy D. Kramer.*

of transfer cells (Kramer et al., 1980; Fig. 7.4). These transfer cells may be induced either in the rhizodermis or in the hypodermis (Landsberg, 1989), serving as the sites of Fe deficiency–induced root responses characteristic of Strategy I.

In recent years, impressive progress has been made not only in the genetic and physiological characterization of root responses to Fe deficiency in Strategy I plants but also in elucidating the molecular mechanisms regulating Fe deficiency responses (Kim et al., 2019; Kobayashi et al., 2014; Li et al., 2016). The components of Strategy I root responses to Fe deficiency (e.g., increased expression of FRO2 and IRT1) are regulated by a FER-like iron deficiency–induced transcription factor, a basic helix–loop–helix protein, that is also involved in the regulation of hormonal and other intracellular signals (Wu and Ling, 2019).

Phenolics are released at greater rates from the roots of Fe-deficient than Fe-sufficient plants (Marschner et al., 1986; Jin et al., 2007; Rodríguez-Celma et al., 2013). Certain phenolics, such as caffeic acid, are very effective in chelation and reduction of inorganic Fe(III), and form a component of Strategy I responses to Fe deficiency; plants release large amounts of phenolic compounds that contribute to the utilization and remobilization of root apoplasmic Fe (Bashir et al., 2011; Jin et al., 2007; Pavlovic et al., 2013). In arabidopsis, feruloyl-coenzyme A 6-hydroxylase is a key enzyme for the synthesis of phenolics in response to Fe deficiency (Schmid et al., 2014) and is released into the rhizosphere by the ABC transporter G family member 37 (ABCG37) (Fourcroy et al., 2014).

In graminaceous species (Strategy II), the Fe deficiency–induced morphological and physiological changes described above for Strategy I plants are less expressed. Instead, roots release phytosiderophores (PSs) as chelators for Fe^{3+}. The pathway of PS biosynthesis is understood relatively well (Fig. 7.5). L-methionine is the dominant precursor (Mori and Nishizawa, 1987), and three molecules of methionine form one molecule of nicotianamine (NA) which, after deamination and hydroxylation, is converted to 2-deoxymugineic acid and further to other PSs (Fig. 7.5) that vary among plant species (Römheld and Marschner, 1990). The release of PSs from roots is facilitated by TOM1, the efflux transporter of the mugineic acid–type PSs (Nozoye et al., 2011). Understanding of the regulation of this transport has been increased further by bioinformatics analysis of methylated DNA fragments (Bocchini et al., 2015) as well as by characterizing the involvement of auxin in the release of PSs (Garnica et al., 2018). The Fe(III)-PS complex is taken up by root cells via YS1 or yellow stripe–like (YSL) transporters characterized in different cereal species.

Atypically, rice plants use both strategies for Fe acquisition, Strategy I (Fe^{2+} uptake by OsIRT1) and Strategy II (uptake of Fe^{3+}-PS by OsYSL); however, rice roots have a weak capacity for Fe(III) reduction and a limited capacity for PS release (Ishimura et al., 2007; Masuda et al., 2019). Transgenic rice plants with the elevated ferric reductase activity have greater tolerance to Fe deficiency and greater grain yield when grown in an Fe-deficient calcareous soil compared with nontransgenic rice plants (Ishimura et al., 2007).

NA is not only a precursor of PS biosynthesis, it is also a strong chelator of Fe^{2+}, but not of Fe^{3+} (Scholz et al., 1988). NA is essential for the proper functioning of Fe(II)-dependent processes (Pich et al., 1991). It plays an important role in Fe homeostasis within cells and cellular compartments as well as in Fe transport in the phloem and to grains (Haydon and Cobbett, 2007). Overexpression of the NA-synthase gene in rice grains resulted in about a threefold increase in grain Fe concentration (Lee et al., 2009).

Root responses to Fe deficiency are additionally regulated by phytohormones and small signaling molecules (see Chapter 5), including auxin (Chen et al., 2010a,b; Römheld and Marschner, 1981b), ethylene (García et al., 2010), cytokinins (CKs) (Seguela et al., 2008), gibberellins (Wang et al., 2017), abscisic acid (ABA) (Lei et al., 2014), jasmonic acid (Kobayashi et al., 2016; Maurer et al., 2011), brassinosteroids (Wang et al., 2012), salicylic acid (Shen et al., 2016) and nitric oxide (NO) (Chen et al., 2010a,b; Graziano and Lamattina, 2007). These phytohormones and small

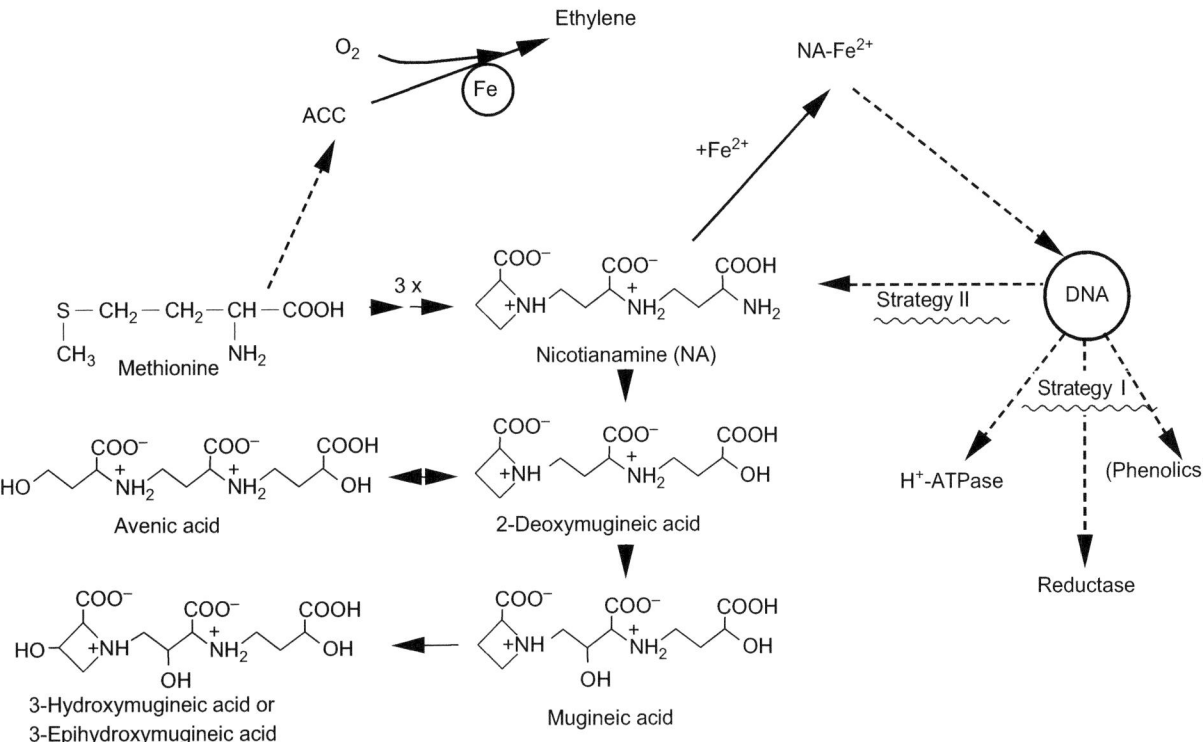

FIGURE 7.5 Model of phytosiderophore biosynthesis and other Fe deficiency-related factors in roots. *NA*, Nicotianamine. *Based on Shojima et al. (1990) and Scholz et al. (1988).*

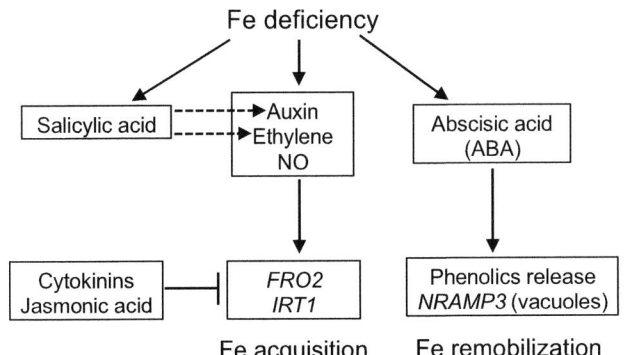

FIGURE 7.6 Model of the effects of phytohormones and small molecules in the regulation of root responses to Fe deficiency in Strategy I plants. Arrow, positive regulation; endline, negative regulation. *Based on Hindt and Guerinot (2012), Lei et al. (2014), and Chen et al. (2010b).*

molecules not only regulate the expression of genes involved in the acquisition and remobilization of Fe in roots (Fig. 7.6) but also modify the root phenotype in response to Fe availability.

7.1.7 Iron deficiency and toxicity

The critical deficiency concentration of Fe in leaves is in the range of 50–150 mg Fe kg^{-1} dw. This refers to total Fe and is, therefore, only of limited value for the characterization of Fe nutritional status of field-grown plants. In general, C$_4$ species require a greater Fe supply than C$_3$ species, but their critical deficiency concentrations are similar (Smith et al., 1984). In fast-growing meristematic and expanding tissues, such as shoot apices, the critical deficiency concentrations are relatively high, in the range of 200 mg total Fe kg^{-1} dw (Häussling et al., 1985). In legumes, the nodule developement has a particularly high Fe demand (Chapter 16).

Iron deficiency is a worldwide problem in crop production on calcareous soils. It is the major factor responsible for so-called lime-induced chlorosis. Iron deficiency also represents an important nutrient deficiency problem in oceans, limiting CO_2 assimilation and N_2 fixation by the phytoplankton (Berman-Frank et al., 2001; Greene et al., 1992).

On the other hand, Fe toxicity ("bronzing") is a serious problem in crop production on waterlogged soils; it is the second-most severe yield-limiting factor in wetland rice. Yield reductions in rice associated with the appearance of bronzing symptoms commonly range from 15% to 30% (Becker and Asch, 2005). The critical toxicity concentrations in leaves are above 500 mg Fe kg^{-1} dw but depend on other factors, such as concentrations of other nutrients (Yamauchi, 1989). Iron toxicity damage is generally associated with the formation of ROS; therefore, the induction of antioxidative enzymes, such as ascorbate peroxidase, and Fe-binding proteins, such as ferritin, are important cellular defense mechanisms mitigating Fe toxicity damage (Fourcroy et al., 2004; Briat et al., 2010). Among the different systems involved in the regulation of Fe homeostasis under Fe excess (reviewed in Aung and Masuda, 2020), NA synthesized by OsNAS3 appears to play an important role in detoxification of excess Fe in the roots and shoot of rice (Aung et al., 2019).

7.2 Manganese

7.2.1 General

Manganese (Mn) is an essential micronutrient with many functional roles in plant metabolism. In biological systems Mn can change oxidation states between +2 and +7. Only the divalent form (Mn^{2+}) is taken up by plants; this is by far the dominant form in biological systems, but it can readily be oxidized to Mn(III) and Mn(IV). As a result, many of the functional roles of Mn in biochemical processes are associated with its unique redox behavior, when utilized as a metallic cofactor in enzyme-catalyzed reactions. Manganese exhibits the rapid ligand exchange kinetics in plants, which means that it readily attaches and detaches from the catalytical center of enzymes. Manganese (II) has a radius of approximately 0.078 nm, which lies between that of magnesium (Mg, 0.066 nm) and calcium (Ca, 0.099 nm), and Mn can therefore be replaced within the coordinating sphere by either of these metal ions in various reactions (Schmidt and Husted, 2019).

7.2.2 Mn-containing enzymes

Manganese has two main functions in enzymes: (1) acting as a Lewis acid (i.e., Mn can accept a pair of electrons from a donor molecule and form a coordinate covalent bond) similarly to Mg, Co, and Zn and (2) being an oxidation catalyst, similarly to other redox-active metal cations, such as Fe and Cu. In plants, at least 101 Mn-activated enzymes have been identified. Among them, 37 contain exclusively Mn as a cofactor, 44 are activated by either Mn or Mg, and 20 enzymes coordinate with Mn, Mg, or another divalent metal cation (Ca, Zn, Fe, Co, Ni, Cu) (Schmidt and Husted, 2019).

The Mn^{2+} cation has a large charge density and a low polarizability. Manganese coordinates with the hard ligands, such as oxygen atoms in water, in the carboxylate groups of the amino acids aspartate (Asp) and glutamate (Glu), and the carbonyl groups of the amino acids asparagine (Asn) and glutamine (Gln). The N atom in the imidazole ring of histidine (His) has a lone electron pair and is another important coordinate ligand for Mn and other metallic cations. By contrast, the S-containing amino acids cysteine (Cys) and methionine (Met) are less likely to coordinate with Mn (Schmidt and Husted, 2019).

7.2.3 The functional role of Mn in photosynthesis

The chloroplasts are rich in transition metals, containing up to 80% of the total leaf Fe and about 30% of the total leaf Cu. By contrast, Mn is stored mostly in the vacuoles and to a relatively smaller extent in the chloroplasts. More than 80% of all Mn contained in the chloroplasts is associated with PS II, reflecting its essential role in photosynthesis (Lysenko et al., 2019; Schmidt et al., 2020). Hence, photosynthesis is sensitive to even mild Mn deficiency (Husted et al., 2009). Manganese deficiency has a strong impact on photosynthetic O_2 evolution [reduced by more than 50% in young leaves of subterranean clover (*Trifolium subterraneum*)], but only a small effect on total chlorophyll concentration (Fig. 7.7) (Nable et al., 1984). This is in agreement with the observation that the abundance of subunits in the major antenna complex, LHCII, is largely unaffected in Mn-deficient plants (Schmidt et al., 2016b), and supports the observation that Mn-deficient plants remain green until the deficiency becomes severe.

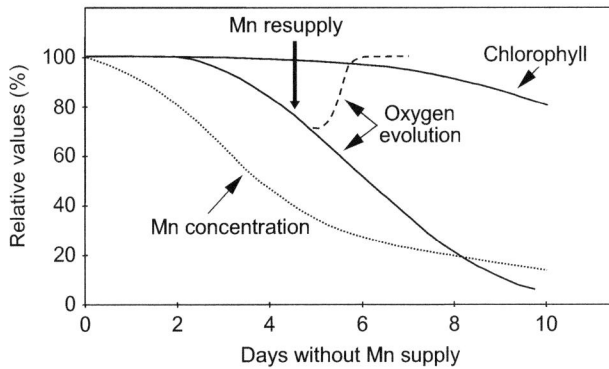

FIGURE 7.7 Concentrations of Mn and chlorophyll, and photosynthetic O$_2$ evolution in young leaves of *Trifolium subterraneum* following withdrawal of Mn supply and resupply of Mn. *Recalculated from Nable et al. (1984).*

Resupplying Mn to deficient leaves of subterranean clover restored the photosynthetic O$_2$ evolution within 1 day to the levels measured in leaves from plants supplied with adequate Mn (Nable et al., 1984). The time for restoration of the PSII functionality depends on the degree of Mn deficiency, but similar results have been obtained in wheat (Kriedemann et al., 1985), maize (Gong et al., 2010), and barley (Schmidt et al., 2016a,b). With increasing severity of Mn deficiency, the capacity of plants to induce nonphotochemical quenching (NPQ) mechanisms in the light decreases, thereby diminishing the controlled dissipation of excess energy. This renders Mn-deficient plants more prone to photoinhibition (Schmidt et al., 2016b).

At the cellular level, the number of chloroplasts and the number of functional PSII complexes within the thylakoid grana stacks are greatly reduced in Mn-deficient plants (Henriques, 2004; Simpson and Robinson, 1984). Upon severe Mn deficiency the ultrastructure of the chloroplasts is drastically changed, and the thylakoid-membrane system becomes disorganized (Papadakis et al., 2007; Simpson and Robinson, 1984). The chloroplast length decreases, and the diameter of the thylakoids increases owing to full relaxation of grana membranes and hyperstacking of thylakoids (Eisenhut et al., 2018), presumably to facilitate PSII repair by providing access for proteases and repair proteins to the damaged PSII complex (Kirchhoff, 2014).

7.2.3.1 Manganese at the active site of water oxidation in photosystem II

The most documented Mn-containing enzyme is the oxygen-evolving complex (Mn-OEC) that catalyzes the oxygen-evolving reaction of photosynthesis, whereby water is split into protons and molecular oxygen.

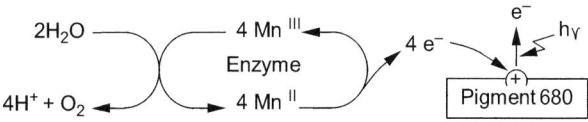

The reaction takes place on the lumenal side of PSII. The configuration and position of the four Mn atoms (Mn 1−4) forming the prosthetic group of OEC has been revealed fully (Umena et al., 2011), and the structure of the Mn$_4$CaO$_5$ cluster resembles a distorted chair (Fig. 7.8). The seat base is formed by three Mn and one Ca atoms in a cubane-like structure. The back of the chair is formed by the fourth Mn, the so-called dangler Mn, which lies outside the cubane, but is linked to two Mn atoms (Mn1 and Mn3) and has been proposed to act as the site of catalysis. The Mn cluster is coordinated by six amino acids from the D1 subunit and one amino acid from the CP43 protein of PSII. The light-induced oxidation of water requires the removal of four electrons and four protons from two water molecules. This necessitates four sequential oxidation events catalyzed by the Mn cluster that cycles through different states, denoted as S_i states ($i = 0-4$). The ability to span such a wide range of redox states makes Mn an ideal element for building the prosthetic group of OEC, with five intermediate states needed to fully oxidize water molecules into molecular oxygen (Kok et al., 1970; McEvoy and Brudvig, 2006).

In vascular plants, the Mn cluster is shielded by the extrinsic proteins PsbO, PsbP, and PsbQ. The extrinsic proteins optimize the efficiency of oxygen evolution in PSII by maintaining the stability of OEC and enhance the binding of Ca^{2+} and Cl$^-$ (Bricker et al., 2012). PsbO and PsbP (23 kDa) are essential for photoautotrophic growth, PSII assembly, and the stabilization of PSII supercomplexes, with PsbP also being involved in the modulation of normal thylakoid architecture (Yi et al., 2005, 2007).

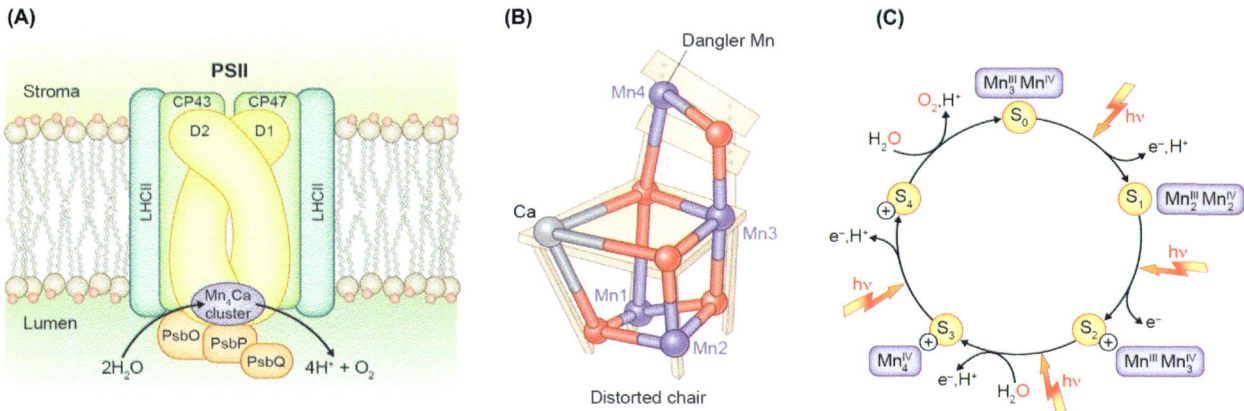

FIGURE 7.8 (A) The structure of the PS II core consisting of the membrane-intrinsic protein subunits D1, D2, CP43, and CP47, including the peripheral antenna system involving a number of light-harvesting pigment−protein complexes (LHCII) that coordinate chlorophylls in different ratios. On the lumenal side of the PSII complex, the Mn prosthetic group catalyzes the oxygen-evolving reaction, whereby water is split into protons and molecular oxygen. In vascular plants, the Mn cluster is shielded by the extrinsic protein subunits PsbO, PsbP, and PsbQ. (B) The chemical structure shows the configuration and position of the metal atoms in the Mn prosthetic group of OEC. Calcium (Ca) is in gray, Mn in purple, and oxygen in red. The structure of the Mn cluster resembles a distorted chair. The distorted seat base is formed by three Mn and one Ca atoms in a cubane-like structure, and the back of the chair is formed by the fourth Mn, the so-called dangler Mn. Seven amino acid ligands supplied by the D1 and CP43 proteins anchor the Mn cluster to PSII. (C) The Kok cycle showing the steps of oxygen evolution in the OEC of PSII with details of the S*i* states, driven by successive absorption of photons in the light-induced steps of the electron (e^-) and proton (H^+) release during water splitting (Schmidt et al., 2016a,b; Schmidt and Husted, 2019).

The abundance of D1 and the extrinsic proteins PsbP and PsbQ (but not PsbO) is significantly decreased under Mn deficiency (de Bang et al., 2015; Husted et al., 2009; Schmidt et al., 2016b), which leads to a destabilization of the PSII supercomplex (Schmidt et al., 2015) and a marked decrease in photosynthetic activity (Schmidt et al., 2016a). Photodamaged PSII undergoes a well-regulated, multistep repair process to replace the damaged D1 protein, involving detachment of Mn from OEC. During PSII repair, the core of PSII is disassembled together with OEC to release the Mn cluster and the extrinsic proteins. The loss of Mn is a prerequisite for the degradation and restoration of the D1 protein (Krieger et al., 1998). It has been proposed that Mn is recycled during PSII repair and that PsbP may act as a Mn storage protein during PSII repair and reassembly (Bondarava et al., 2005). This is supported by the crystal structure of PsbP from spinach, demonstrating a binding site with low affinity to Mn ions that may induce a conformational change of PsbP during the PSII repair cycle (Cao et al., 2015), rendering PsbP a suitable candidate for a Mn chaperone. In addition, the ancestral form of PsbP in cyanobacteria, CyanoP, accumulates as free protein in the thylakoids of the SynPAM71 mutant that is unable to export Mn from the cytoplasm into the periplasmic space (Gandini et al., 2017). The possible recycling and chelation of Mn to ligands therefore appears to be an efficient strategy to buffer the activity of free Mn^{2+} ions to minimize the unfavorable side-reactions causing photooxidative damage to PSII.

7.2.4 Manganese in superoxide dismutase

ROS are inevitable by-products of plant metabolism. To minimize oxidative damage, plant cells have an antioxidative defense system consisting of ROS-scavenging low-molecular-weight compounds and enzymes. Both Mn deficiency and toxicity result in oxidative damage, presented visually as necrotic brown spots in the older leaves, owing to the accumulation of ROS within the chloroplast. Manganese cations are also involved in the antioxidative system in Mn-dependent SOD (Mn-SOD) found in the peroxisomes and mitochondria of plant cells. The Mn-SOD catalyzes the dismutation of superoxide (O^{2-}) radicals into molecular oxygen (O_2) and hydrogen peroxide (H_2O_2). The molecular structure of plant Mn-SOD is homo-tetrameric, with a single Mn atom per subunit. The active site of Mn-SOD is highly conserved, and the plant Mn-SODs share roughly 70% sequence similarity across plant species (Bowler et al., 1994; Miller, 2012). The Mn-SOD dismutation mechanism involves the alternation of Mn-SOD between the oxidized (Mn^{3+}) and less oxidized (Mn^{2+}) state (Abreu and Cabelli, 2010; Miller, 2012). Leaf necrosis induced by prolonged Mn deficiency is, at least partly, due to a decrease in Mn-SOD levels accompanied by an increase in ROS (Allen et al., 2007). However, the regulation of Mn-SOD activity is complex, and Mn deficiency has been reported to increase Mn-SOD activities in some studies, but not in others (Tewari et al., 2013; Yu and Rengel, 1999). In monocot leaves, the Mn-SOD activity increases from the base to the leaf tip, indicating that ROS production increases in the aging mesophyll cells (Leonowicz et al., 2018).

7.2.5 Manganese in oxalate oxidase

Oxalate oxidase (Mn-OxO) is a Mn-dependent enzyme that catalyzes the oxygen-dependent degradation of oxalate, yielding one mole of H_2O_2 and two moles of CO_2 (Whittaker and Whittaker, 2002). The active form of Mn-OxO is a hexamer, and each monomer has a canonical cupin fold. The active site is in the center of the β-barrel and contains a Mn ion (Dunwell et al., 2001). Crystallography of Mn-OxO has shown that the coordinating sphere of the enzyme is quite similar to the one found in Mn-SOD (Woo et al., 2000). The activity of Mn-OxO is localized to the apoplast, and the enzyme has a dual role in the pathogen defense by (1) inhibiting fungal toxins and (2) promoting lignification through the generation of H_2O_2 necessary for cross-linking mono-lignols in lignin biosynthesis (McCay-Buis et al., 1995; Zhang et al., 1995).

7.2.6 Other Mn-dependent enzymes

Manganese activates several enzymes of the shikimic acid and subsequent pathways, leading to the biosynthesis of aromatic amino acids (such as tyrosine), various secondary products (lignin, flavonoids), and the phytohormone indole acetic acid (IAA) (Burnell, 1988; Hughes and Williams, 1988). For example, Mn stabilizes the active conformation of phenylalanine ammonia-lyase (PAL) that catalyzes the deamination of phenylalanine to cinnamic acid (Wall et al., 2008) that is a key metabolite for several secondary products, for example, lignin. Manganese also stimulates peroxidases and works as a diffusible redox shuttle in combination with peroxidases in lignin biosynthesis (Önnerud et al., 2002).

An absolute requirement for Mn exists in the bundle sheath chloroplasts of C_4 plants in which oxaloacetate acts as a carbon shuttle, and where decarboxylation is catalyzed by phosphoenolpyruvate (PEP) carboxykinase. This enzyme has an absolute requirement for Mn that cannot be replaced by Mg (Fig. 7.9). Maximum activity occurs at a Mn:ATP ratio of one, suggesting that the substrate for the enzyme is the Mn−ATP complex (Burnell, 1986) rather than Mg−ATP as in most other reactions (see Section 6.4). In nodulated legumes, such as soybean, that transport N mainly in the form of allantoin and allantoate to the shoot (Chapter 16), the degradation of these ureides in the leaves (Winkler et al., 1985) and in the seed coat (Winkler et al., 1987) is catalyzed by the enzyme allantoate amidohydrolase that has an absolute requirement for Mn (Werner et al., 2008). Arginase is another Mn-dependent enzyme involved in N metabolism (Dabir et al., 2005).

7.2.7 Proteins, carbohydrates, and lipids

Although Mn activates RNA polymerase (Ness and Woolhouse, 1980), protein synthesis is not specifically impaired in Mn-deficient tissues. The protein concentration of deficient plants is either similar to (Table 7.8) or somewhat higher than that of plants adequately supplied with Mn (Lerer and Bar-Akiva, 1976). By contrast, proteomic studies show that the protein synthesis and the degradation of damaged proteins are diminished in roots of tomato plants exposed to Mn

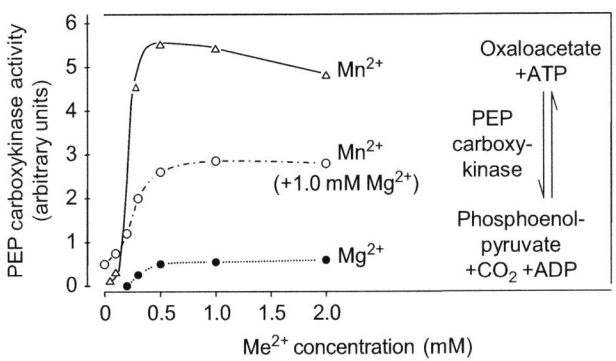

FIGURE 7.9 Activity of PEP carboxykinase from *Urochloa panicoides* with the addition of Mn or Mg or Mn + Mg. ATP concentration was kept constant at 0.25 mM. *Based on Burnell (1986)*.

TABLE 7.8 Growth and composition of bean plants with or without Mn supply.

	Leaves		Stems		Roots	
	+Mn	−Mn	+Mn	−Mn	+Mn	−Mn
Dry weight (g plant^{-1})	0.64	0.46	0.55	0.38	0.21	0.14
Protein-N (mg g^{-1} dw)	53	51	13	14	27	26
Soluble N (mg g^{-1} dw)	6.8	12	10	16	17	22
Soluble carbohydrates (mg g^{-1} dw)	18	4.0	36	15	7.6	0.9

Source: From Vielemeyer et al. (1969).

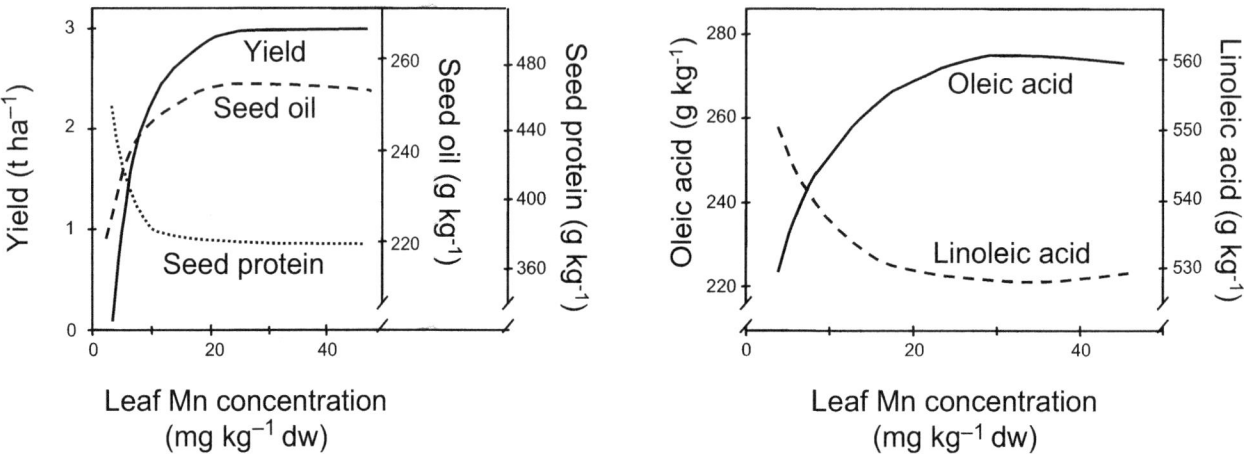

FIGURE 7.10 Relationships between soybean leaf Mn concentration, seed yield, and seed composition. *Adapted from Wilson et al. (1982).*

toxicity (Ceballos-Laita et al., 2018). The accumulation of soluble N in Mn-deficient tissues is due to a shortage of reducing equivalents and carbohydrates for nitrate reduction, as well as a lower demand for reduced N. Manganese deficiency affects the concentration of nonstructural carbohydrates in all plant parts, as shown in Table 7.8 for the soluble carbohydrate fraction. This decrease in carbohydrate concentration is particularly evident in roots and is most likely a key factor responsible for the depression of root growth in Mn-deficient plants (Table 7.8; Marcar and Graham, 1987). In plants without visual leaf symptoms of Mn deficiency, it has been shown that Mn starvation induced a significant depression of root growth, leading to four times lower root dry weight compared with the control plants, whereas shoot dry weight was reduced only by 60% (Hebbern et al., 2009).

The role of Mn in lipid metabolism is complex. In Mn-deficient leaves, the concentration of thylakoid-membrane constituents such as glycolipids and polyunsaturated fatty acids may be decreased by up to 50%. This depression in the lipid concentration in chloroplasts can be attributed to the role of Mn in the biosynthesis of fatty acids, carotenoids, and related compounds. Manganese supply affects the lipid concentration and composition in the seeds (Fig. 7.10). In the deficiency range, the Mn concentration in leaves was positively correlated with the seed yield as well as the oil concentration. The fatty acid composition of the oil was also markedly altered, with the concentrations of linoleic acid (Fig. 7.10) and some other fatty acids increasing (Wilson et al., 1982). This was counteracted by a decrease in the concentration of oleic acid. The lower oil concentration in the seeds of Mn-deficient plants probably resulted mainly from the lower rates of photosynthesis and thus a decreased supply of carbon skeletons for fatty acid synthesis. However, a direct involvement of Mn in the biosynthesis of fatty acids could be a contributing factor.

The lower lignin concentration in Mn-deficient plants (Table 7.9) reflects the requirement for Mn in various steps of lignin biosynthesis. Given that Mn is a cofactor for (1) PAL that mediates the production of cinnamic acid and various other phenolic compounds and (2) the peroxidase involved in the polymerization of cinnamyl alcohols into lignin, deficiency of Mn may decrease the concentrations of phenolics and lignin (Brown et al., 1984; Rengel et al., 1993) that are considered important in defense against fungal infection (Rengel, 2003).

TABLE 7.9 Relationship between Mn and lignin concentration in shoots and roots of young wheat (*Triticum aestivum*) plants.

	Mn concentration (mg kg^{-1} dw)			
	4.2	7.8	12.1	18.9
	Lignin concentration (g kg^{-1} dw)			
Shoots	40	58	60	61
Roots	32	128	150	152

Source: Recalculated from Brown et al. (1984).

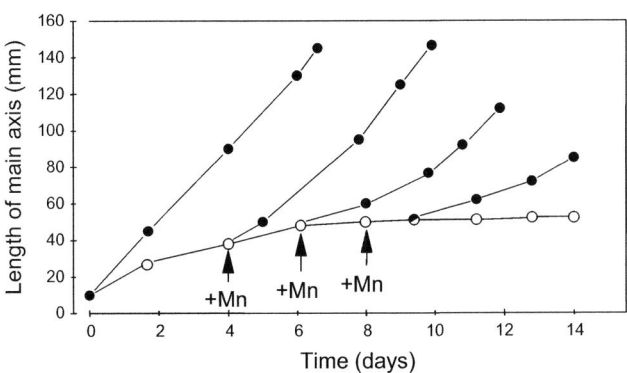

FIGURE 7.11 Growth of main axis of excised tomato roots after transfer from Mn-deficient to complete medium. ○ - Mn; ● + Mn. *Based on Abbott (1967).*

7.2.8 Cell division and extension

Inhibition of root growth in Mn-deficient plants is caused by a shortage of carbohydrates as well as by a direct Mn requirement for growth (Campbell and Nable, 1988; Sadana et al., 2002). The rate of elongation appears to respond more rapidly to Mn deficiency than the rate of cell division. As shown in Fig. 7.11 with isolated tomato roots placed in sterile culture with an ample supply of carbohydrates (but without Mn), a decline in extension of the main axis occurred in less than 2 days. Resupplying Mn rapidly restored the growth rate to normal levels if the deficiency was not too severe. In Mn-deficient plants, the formation of lateral roots ceased completely (Abbott, 1967). Compared to Mn-sufficient plants, there was a greater abundance of small nonvacuolated cells in Mn-deficient roots, indicating that Mn deficiency impairs cell elongation more strongly than cell division, an observation also supported by tissue culture experiments (Neumann and Steward, 1968).

7.2.9 Manganese deficiency

Manganese deficiency is a widespread problem. The amount of plant-available Mn^{2+} in the soil is influenced strongly by changes in pH, soil porosity, soil moisture, and the rhizosphere microbiome. The Mn deficiency symptoms are particularly prevalent in calcareous, alkaline soils, and in sandy soils with high porosity favoring oxidation of Mn^{2+} into plant-unavailable Mn oxides, and in organic matter–rich soil with strong binding of Mn. Examples of particularly problematic regions are the northern parts of Europe and Scandinavia, Northern China, Australia and the Middle East (coarse-textured sandy soils and calcareous soils), and the United Kingdom and northern parts of the United States (peaty soils rich in organic matter) (Schmidt et al., 2016a).

Manganese-deficient plants are particularly sensitive to damage by freezing temperatures that may result in complete crop loss, a phenomenon known as "winterkill" (Schmidt et al., 2013). Because Mn is involved in the biosynthesis of lignin (Barros et al., 2015; Önnerud et al., 2002), Mn-deficient plants are more susceptible to a range of soil-borne root-rotting fungal diseases (e.g., take-all caused by the fungus *Gaeumannomyces graminis*) (Huber and McCay-Buis, 1993;

Rengel et al., 1993). Manganese-deficient plants have reduced biomass production because they require twice as long to reach the booting stage than Mn-sufficient plants (Longnecker et al., 1991), rendering Mn-deficient plants less competitive against weeds. A decrease in grain number and grain yield in Mn-deficient plants is presumably a combination of low pollen fertility (Sharma et al., 1991) and a shortage of carbohydrate supply for grain filling (Longnecker et al., 1991). In barley, it has been demonstrated that Mn deficiency diminishes the deposition of the leaf cuticular wax layer, which leads to increased transpiration rates and eventually poor water-use efficiency (WUE) (Hebbern et al., 2009). Likewise, cuticular wax defects have been observed in arabidopsis (Alejandro et al., 2017). However, plant species may differ in this respect because experiments with sugar beet, wheat, and pecan showed no effect on transpiration under Mn deficiency (Morales and Warren, 2012). In addition, Mn availability affects root suberization, as demonstrated by a greater permeability of the stele in response to Mn deficiency. This is caused by reduction in suberin deposition on the root endodermal cell walls, modulated via the ABA and ethylene signaling pathways (Barberon et al., 2016; Chen et al., 2019).

The typical Mn threshold concentration of young fully expanded leaves is between 10 and 20 mg Mn kg^{-1} dry weight (Reuter et al., 1997), regardless of plant species, cultivar, and environmental conditions. However, *Lupinus angustifolius* appears to have a critical deficiency concentration twice as high as other plant species (Brennan et al., 2001; Hannam and Ohki, 1988). Manganese deficiency symptoms develop as diffuse interveinal chlorosis confined to the youngest leaves because Mn is considered largely immobile in the phloem, and therefore cannot be remobilized from older to younger leaves (Loneragan, 1988; Chapter 3). Under prolonged/severe Mn deficiency, characteristic brown necrotic spots develop on older leaves as a consequence of oxidative stress (Schmidt and Husted, 2020) (Fig. 7.12).

At the field scale, Mn deficiency in cereals is referred to as "light spot disease" because the deficiency often appears patchy due to variability in soil parameters influencing Mn oxidation or reduction. In legumes, the examples of Mn deficiency symptoms include "marsh spot" on the cotyledons in peas or "split seed" disorder in lupins; the visual symptoms of the latter include discoloration, splitting, and deformity of seeds (Campbell and Nable, 1988).

Plant species and genotypes within a species differ considerably in their tolerance towards Mn deficiency when grown on soils with low Mn availability, which is described as Mn efficiency (Hebbern et al., 2005; Rengel, 2001).

FIGURE 7.12 Symptoms of Mn deficiency in spring barley. The characteristic brown necrotic spots are clearly seen.

Oat, barley, wheat, soybean, and peach are susceptible, whereas rye is not (Reuter et al., 1988). Differential Mn efficiency was reported among genotypes of bread wheat (e.g., Sadana et al., 2002), durum wheat (e.g., Khabaz-Saberi et al., 2000), barley (e.g., Hebbern et al., 2005; Leplat et al., 2016), and other crops. A superior capacity to tolerate Mn deficiency has been reported for a subgroup of Scottish barley landraces (Bere barley) that could overcome Mn limitations caused by high-pH soils and produce good grain yields, while modern elite barley varieties completely failed to set seed (George et al., 2014; Schmidt et al., 2019).

7.2.10 Manganese toxicity

Manganese toxicity occurs in plants grown on acid soils because decreasing pH increases soil availability of Mn^{2+}, and in plants grown on soils with a low redox potential (e.g., waterlogged soils), which favors the reduction of Mn oxides to Mn^{2+}. Soils prone to Mn toxicity are abundant in, for example, Australia, Hawaii, Puerto Rico, Brazil, parts of tropical Africa, and the East and South Asia (Fernando and Lynch, 2015). Environmental factors influence the critical toxicity concentrations. Silicon strongly increases the tolerance to high Mn concentration in the shoot. Micro-X-ray-fluorescence imaging has revealed co-location of Mn with Si, supporting the hypothesis that Si facilitates the binding of Mn in cell walls in the nontoxic forms (Rogalla and Römheld, 2002; van der Ent et al., 2020). In a maize genotype tolerant to Mn toxicity, Si substantially increased the thickness of the epidermal leaf layers in which excess Mn was stored (Doncheva et al., 2009).

Manganese toxicity decreases shoot and root fresh weight and affects the root morphological characteristics, with the total root length and the tap root tip length being significantly reduced (Chen et al., 2016a,b). The abundance of peroxidases involved in the cell wall lignification and suberinization is increased, suggesting a Mn-induced increase in lignification that is likely to affect cell wall stiffening and expansion, leading to inhibition of root growth (Ceballos-Laita et al., 2018).

Symptoms of Mn toxicity appear as mild chlorosis and small brown spots on mature leaves. Although these brown speckles contain oxidized Mn, the brown color is derived from oxidized polyphenols, not Mn (Führs et al., 2009; Wissemeier and Horst, 1987). In soybean, the brown spots are prominent near the major veins, and their trifoliate leaves become distorted and show the typical "crinkle leaf" symptoms. Another toxicity symptom is the formation of callose in the brown spot areas (Horst et al., 1999), suggesting toxic effects of Mn on the plasma membrane and enhanced Ca^{2+} influx (Wissemeier and Horst, 1987). The callose production in the tissues on the abaxial side of leaves is significantly correlated with Mn concentration (Wissemeier and Horst, 1992).

In contrast to the narrow range of the critical deficiency concentration of Mn among plant species, the critical toxicity concentration varies widely among plant species. An example of the differences among crop species is given in Table 7.10. Cowpea and soybean are very sensitive to Mn toxicity, whereas sunflower and white lupin tolerate high Mn concentrations. Even within a species, the critical toxicity concentration can vary substantially among cultivars (Edwards and Asher, 1982; Horst, 1988; Khabaz-Saberi et al., 2010; Wang et al., 2002). In leaves of Mn-tolerant plant

TABLE 7.10 Critical toxicity concentrations of Mn (at which dry matter production is reduced by 10%) in the shoots of various plant species.

Species	Critical toxicity concentration (mg Mn kg^{-1} dw)
Maize	200
Pigeon pea	300
Soybean	600
Cotton	750
Sweet potato	1380
Sunflower	5300

Source: Based on Edwards and Asher (1982).

species such as sunflower or stinging nettle (*Urtica dioica*) grown at high Mn concentrations, there is darkening around the base of trichomes (Blamey et al., 1986; Hughes and Williams, 1988), which appears to be a Mn tolerance mechanism associated with decreased soluble Mn concentrations through the accumulation of Mn in the cell walls of the base cells (van der Ent et al., 2020).

Manganese toxicity is accompanied by induced deficiencies of other nutrients such as Ca, Mg, Fe (Horst, 1988), and Zn (de Varennes et al., 2001). Induced deficiency of Fe and Mg is caused by inhibited uptake across the plasma membrane (see Chapter 2) and competition (or imbalance) at the cellular level. Accordingly, Mn toxicity can be counteracted by increasing the supply of Mg (Löhnis, 1960) or Fe, which decreases Mn concentrations in leaves but does not alter leaf Mn distribution or speciation (Blamey et al., 2019). In contrast to Fe and Mg, the induction of Ca deficiency symptoms ("crinkle leaf") by high tissue concentrations of Mn might be an indirect effect on the Ca transport to expanding leaves. The high activities of IAA oxidase and polyphenoloxidase have frequently been reported in tissues with high Mn concentration (Fecht-Christoffers et al., 2007; Horst, 1988). Calcium deficiency symptoms induced by Mn toxicity are therefore probably caused by enhanced degradation of IAA, a process which is aggravated, for example, by high light intensity (Horst, 1988). Another symptom of Mn toxicity is a loss of apical dominance and enhanced formation of auxiliary shoots ("witches' broom") (Bañados et al., 2009; Kang and Fox, 1980), further supporting the hypothesis of a relationship between impaired basipetal IAA transport and Mn toxicity (Gangwar et al., 2010).

7.3 Copper

7.3.1 General

Copper (Cu) is a redox-active transition element with roles in photosynthesis, respiration, C and N metabolism, and protection against oxidative stress (Andresen et al., 2018; Bashir et al., 2019; Schulten et al., 2019; Saeed Ur et al., 2020). Like Fe, it forms highly stable complexes and participates in the electron transfer reactions. Divalent Cu is readily reduced to the unstable monovalent Cu (Zandi et al., 2020). Most of the functions of Cu as a plant nutrient are based on the enzymatically bound Cu that catalyzes redox reactions. In redox reactions of the terminal oxidases, Cu enzymes react directly with molecular oxygen. Terminal oxidation in the living cells is therefore catalyzed by Cu rather than Fe (e.g., Fraudentali et al., 2020a). Nevertheless, some Cu-containing proteins can be substituted by the Fe-containing ones, and the supply of both Cu and Fe governs the expression of a number of genes (e.g., Ramamurthy et al., 2018).

Copper has a high affinity for the peptide and sulfhydryl groups, and thus for the cysteine-rich proteins (e.g., Saeed Ur et al., 2020), as well as for the carboxylic and phenolic groups. Therefore, more than 98% of Cu in plants is present in the complexed forms, and the concentrations of free Cu^{2+} and Cu^+ are extremely low in the cytoplasm.

7.3.2 Copper uptake and transport

Understanding Cu transport into cells and organelles has increased substantially in recent years (Aguirre and Pilon, 2016; Bashir et al., 2019; Burkhead et al., 2009; Yruela, 2013). Under Cu deprivation, Cu^{2+} is reduced to Cu^+ and taken up by members of the high-affinity Cu transporter (COPT) family (Sanz et al., 2019). The expression of genes encoding COPTs is regulated by the Cu-responsive transcription factor (a master regulator of Cu homeostasis) SPL7 (SQUAMOSA promoter–binding protein-like7) (Bernal et al., 2012). There are six COPT members in arabidopsis; among them, COPT1, COPT2, and COPT6 are present in the plasma membrane and therefore may mediate uptake of Cu into cells, whereas other members may mediate intracellular transport (Sanz et al., 2019). Interestingly, COPT3, COPT4, and COPT5 are the high-affinity Cu transporters in arabidopsis and the low-affinity Cu transporters in the grass *Brachypodium distachyon* (Jung et al., 2014), perhaps explaining the poor Cu efficiency of some grass species. The COPT transporters do not transport just Cu, but also other micronutrients (e.g., Fe) and even toxic trace metals (such as Cd and Pb), as demonstrated in *Populus trichocarpa* (Zhang et al., 2015). The maize (*Zea mays*) COPT2 protein appears to transport only Cu, whereas other maize COPTs (such as COPT1 and COPT3) also transport other divalent cations (Wang et al., 2018).

The regulation of *COPT* transcription (and thus Cu uptake, transport, and homeostasis) via SPL7 is complicated by the reciprocal cross-talk between Cu status and the biosynthesis of ABA (Carrio-Segui et al., 2016) and jasmonic acid (Yan et al., 2017); further integrative molecular work is required to elucidate some of the interrelationships. The Zn/Fe permeases (ZIPs) may also be involved in the transport of Cu^{2+} across the plasma membrane. Work on *Vitis vinifera* seedlings showed that ZIP2 was upregulated under Cu deficiency (and downregulated under Cu excess), whereas ZIP4 had the opposite response (Leng et al., 2015), which is consistent with the predominant (but not exclusive) expression of *ZIP2* in roots and *ZIP4* in leaves of *Nicotiana tabacum* (Barabasz et al., 2019).

Divalent Cu^{2+} transport is also facilitated by P_{1B}-type ATPase (HMA) transporters that are selective for both Cu^{2+} and Cu^+ and are localized in the endomembranes and the plasma membrane. In particular, two P1B-type ATPases are responsible for Cu delivery to plastocyanin in the chloroplasts (Sautron et al., 2016; Tapken et al., 2012). YSL proteins are likely to mediate Cu^{2+}-NA transport across the plasma membrane as shown in *Arachis hypogaea* for *AhYSL3.2* expressed in the plasma membrane of leaf cells, but upregulated by the low Cu concentration in roots (Dai et al., 2018). Several Cu chaperones are not only involved in Cu transport but also have a central role in cellular Cu homeostasis (Burkhead et al., 2009; Xu et al., 2020), including insertion of Cu atoms into cytochrome c oxidase (CcO) (Garcia et al., 2019; Llases et al., 2020; Radin et al., 2015) and Cu^+-ATPase in the chloroplast thylakoids (Blaby-Haas et al., 2014).

There are other compounds and transporters involved in Cu homeostasis at various organizational levels (from intracellular to the whole plant). Metallothioneins are involved in Cu remobilization from vegetative organs to developing grains (Benatti et al., 2014). A Cu transporter localized in the Golgi apparatus was reported in bread wheat (*Triticum aestivum*), and it appears to have homologs in other monocotyledonous plants (Li et al., 2014a,b).

7.3.3 Copper proteins

Around 260 different Cu-dependent proteins exist in plants (Schulten et al., 2019). About 50% of plant Cu is present in chloroplasts, bound to plastocyanin that participates in the photosynthetic reactions (Hänsch and Mendel, 2009; Pan et al., 2018). Other proteins that bind Cu include chaperones and numerous enzymes, particularly oxidases containing one of more Cu atoms (Burkhead et al., 2009; Fraudentali et al., 2020a,b; Liao et al., 2006; Llases et al., 2020; Radin et al., 2015). These include the Cu chaperone COX19 that is present in mitochondria and is involved in the biogenesis of CcO (Garcia et al., 2019). Under Cu deficiency, the activity of Cu-dependent enzymes decreases rapidly (e.g., Priyanka and Venkatachalam, 2020), and in most, but not all, cases, these decreases are correlated with the metabolic changes and the inhibition of plant growth.

Copper is also part of the ethylene receptor and is involved in the biosynthesis of the Mo cofactor involved in N_2 fixation (Chapter 16). Hence, in legumes, Cu deficiency may reduce nodulation and N_2 fixation. In *Phaseolus vulgaris*, Cu deficiency resulted in the overexpression of miR398 (one of the small RNAs that govern gene expression posttranscriptionally), which decreased the concentrations of CuZn-SOD and the nodulation factor Nod19 (Naya et al., 2014).

7.3.3.1 Plastocyanin

Plastocyanin is a component of the electron transport chain of PSI (Musiani et al., 2005). This protein has a molecular weight of <10 kDa and contains one Cu atom per molecule. There are three to four molecules of plastocyanin per 1000 molecules of chlorophyll (Sandmann and Böger, 1983).

When plants are Cu deficient, a close relationship exists between the Cu concentration of leaves and the plastocyanin concentration and, thus, the activity of PSI (Shahbaz et al., 2015). Copper allocation to the chloroplasts and the plastocyanin concentration are regulated by miR408, which itself is regulated by a system that includes the Cu-responsive transcription factor SPL7 (Zhang et al., 2014). Overexpression of miR408 (accompanied by increased concentrations of Cu and plastocyanin in the chloroplasts) resulted in the increased grain yield of transgenic rice in the field (Pan et al., 2018). Generally, the activity of PSII is depressed relatively little by Cu deficiency. A potential reason might be that plastocyanin (rather than plastoquinone) is the main electron carrier in the chloroplasts (Hoehner et al., 2020).

7.3.3.2 Superoxide dismutase

The various SOD isoenzymes and their role in the detoxification of superoxide radicals ($O_2^{\bullet-}$) have been discussed in Sections 7.1 (Fe-SOD) and 7.2 (Mn-SOD). The copper–zinc SOD (CuZn-SOD) has a molecular weight of 32.5 kDa; at the active site, one Cu and one Zn atom share a common histidine ligand. The Cu atom in CuZn-SOD is directly involved in the detoxification of $O_2^{\bullet-}$ generated in photosynthesis, with an abundance of polar groups in the copper-binding region of the enzyme essential for channeling ROS to the active center of the enzyme (Shikhi et al., 2020). There are at least three major isoforms of CuZn-SOD in plants (Yruela, 2013), occurring in the cytosol (CSD1), in the chloroplast stroma together with Fe-SOD (CSD2) (Section 7.1), and in the peroxisomes (CSD3).

When plants are Cu deficient, the CuZn-SOD activity declines markedly in the leaves. This decline occurs in the chloroplastic and the cytoplasmic compartments, with a compensatory role played by Fe-SOD in the chloroplasts

(Yamasaki et al., 2008). The Cu status regulates CuZn-SOD by a number of mechanisms, including direct transcriptional control of SOD isoforms by Cu and posttranscriptional activity of microRNAs on CuZn-SOD RNA metabolism; for example, the upregulated expression of miR398 results in downregulation of its *CuZn-SOD* target (Sunkar et al., 2006; Naya et al., 2014; Shahbaz and Pilon, 2019).

7.3.3.3 Cytochrome c oxidase

The CcO is a large integral membrane protein encoded in the mitochondrial genome. It is a terminal oxidase of the mitochondrial electron transport chain and is expressed in the mitochondrial inner membrane (Radin et al., 2015). Assembly of the active oxidase complex is dependent on the insertion of three Cu ions, along with two heme Fe ions and the individual Zn, Mg, and Na ions (Carr and Winge, 2003). The synthesis of CcO (SCO)−like metal chaperone is responsible for inserting Cu ions into the Cu_A site in CcO (Llases et al., 2020). Interestingly, different SCO proteins may have different roles in optimal and stress environments, for example, salinity (Mansilla et al., 2019). Given that respiration is not influenced significantly by Cu deficiency, CcO might be present in excess in the mitochondria (Ayala and Sandmann, 1988).

The activity of CcO can be blocked by cyanide (Sil et al., 2018); the remaining respiratory O_2 consumption of cells is then mediated by the cyanide-insensitive quinol oxidase (Mcdonald and Vanlerberghe, 2014; Young et al., 2016) known as the "alternative oxidase" (in the "alternative pathway," see Chapter 5). This enzyme contains Cu, but no heme Fe; hence, it is unlikely that the alternative respiration can compensate for the low CcO activity in Cu-deficient cells.

7.3.3.4 Ascorbate oxidase

Ascorbate oxidase is a multi-Cu oxidase that catalyzes the oxidation of ascorbic acid to L-dehydroascorbic acid according to the equation (see also Sanmartin et al., 2007).

The enzyme contains at least four Cu atoms per molecule and catalyzes a four-electron reduction of O_2 to water. The enzyme is thought to occur primarily in the cell wall apoplasm and act as a terminal respiratory oxidase; however, it may also act in combination with polyphenol oxidases. Ascorbate oxidase biosynthesis is Cu dependent (Sanmartin et al., 2007). Given that activity of ascorbate oxidase decreases in Cu-deficient plants, it may be a sensitive indicator of the plant's Cu nutritional status. This correlation has been used to develop a rapid and simple colorimetric field test to diagnose Cu deficiency (Delhaize et al., 1982). Resupplying Cu to deficient plants can restore the activity of ascorbate oxidase in young, but not in mature, leaves (Table 7.11), suggesting that the enzyme can only be synthesized in leaf blades during their early development. Similarly, plastocyanin in young (but not mature) leaves is the preferred target for Cu upon its resupply (Shahbaz et al., 2015).

7.3.3.5 Diamine oxidases

Polyamine oxidases are flavoproteins that catalyze the aerobic degradation of polyamines (e.g., spermidine to form putrescine), H_2O_2, and NH_3 (Yu et al., 2019). Polyamine oxidases preferentially degrade tri- and tetraamines. However, the degradation of putrescine (diamine) and, to some extent, spermidine (triamine) is mediated by diamine oxidase, a Cu-containing enzyme. Diamine oxidase is widespread in plants, particularly in the grass and legume families (Tavladoraki et al., 2016). H_2O_2 may be involved in the production of structural defense-related compounds (Xia et al., 2020), as a signal molecule (e.g., Lulai et al., 2020), or as an antimicrobial compound in host resistance (Walters, 2003). The activity of diamine oxidase decreases in Cu-deficient plants and can be restored by resupplying Cu

TABLE 7.11 Cu concentration, ascorbate oxidase activity (AOA), and protein concentration in young and mature leaves of subterranean clover (*Trifolium subterraneum*) with (+Cu) and without Cu supply (−Cu) or Cu resupply to Cu-deficient plants (−Cu→ + Cu).

Leaf age	Parameter	Cu supply		
		−Cu	−Cu→ +Cu	+Cu
Young	Cu concentration (μg Cu g^{-1} dw)	<0.5	18 ± 1.7	13 ± 0.8
	AOA (nmol O$_2$ leaf^{-1} min^{-1})	10 ± 8	245 ± 22	240 ± 5
	Protein (mg g^{-1} dw)	28 ± 2.3	38 ± 1.3	41 ± 2.6
Mature	Cu concentration (μg Cu g^{-1} dw)	1.0 ± 0.2	7.9 ± 2.2	10 ± 2.6
	AOA (nmol O$_2$ leaf^{-1} min^{-1})	5 ± 1	5 ± 5	34 ± 4
	Protein (mg g^{-1} dw)	37 ± 6.2	44 ± 1.6	40 ± 4.5

Means ± SE.
Source: Based on Delhaize et al. (1985).

(Delhaize et al., 1985). Similar to ascorbate oxidase (Table 7.11), this restoration of activity is confined to the young leaves.

Diamine oxidase is located mainly in the apoplasm, including the epidermis and the xylem of mature tissues, where the H$_2$O$_2$ produced can sustain the activity of peroxidases involved in lignification and suberization (Lulai et al., 2020; Prabhjot et al., 2020; Walters, 2003).

7.3.3.6 Polyphenol oxidases

Polyphenol oxidases (also known as catechol oxidases, diphenol oxidases, and tyrosinases) contain two Cu ions. They catalyze the oxygenation reactions of plant phenols in which molecular oxygen is inserted into an aromatic ring, followed by oxidation of dihydroxyphenols to orthoquinones that are powerful oxidants (Parveen et al., 2010).

Tyrosine → Dopa → Quinone → Melanotic substances

Phenol oxidases are abundant in the cell walls but are also located in the thylakoid membranes of chloroplasts (Aguirre and Pilon, 2016). Polyphenol oxidases are involved in the biosynthesis of lignin and alkaloids and in the formation of brown melanotic substances that may be formed when tissues are wounded, for example, in potatoes (Lulai et al., 2020) and sweet potatoes (Liao et al., 2006). The melanotic substances are also active as phytoalexins, inhibiting spore germination and fungal growth (e.g., Ube et al., 2019). Under Cu deficiency, the polyphenol oxidase activity is strongly inhibited, which leads to an accumulation of phenolics and a decrease in the formation of melanotic substances (Davies et al., 1978).

7.3.4 Carbohydrate, lipid, and N metabolism

Due to the role of Cu in PSI, the Cu-deficient plants have low rates of photosynthesis and reduced carbohydrate formation (Shahbaz and Pilon, 2019). In Cu-deficient bread wheat plants, the concentration of soluble carbohydrates during the vegetative stage is lower than in Cu-sufficient plants (Fig. 7.13). However, when grains have developed as a dominant sink after anthesis, the Cu-deficient plants produce few grains, remain green (i.e., still photosynthesizing), and

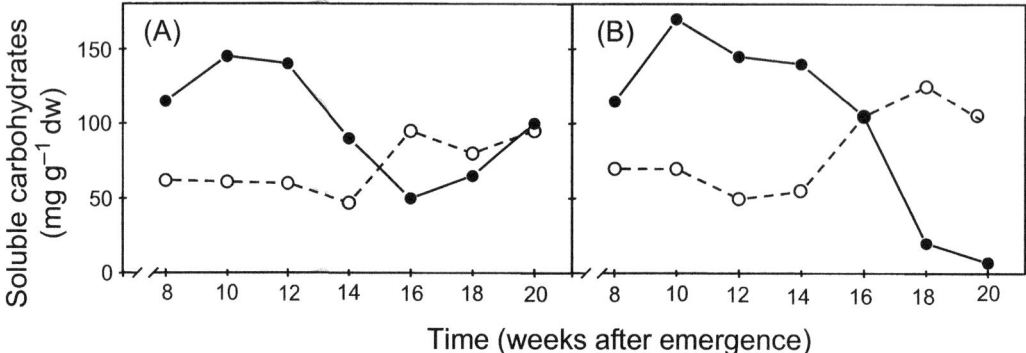

FIGURE 7.13 Concentrations of soluble carbohydrates in flag leaves (A) and roots (B) of wheat (*Triticum aestivum*) plants grown at two Cu levels as a function of plant age. Key: ● + Cu; ○ − Cu. *Modified from Graham (1980)*.

have high concentrations of soluble carbohydrates in the leaves and roots (Fig. 7.13). However, a decrease in the net CO_2 fixation in severely Cu-deficient plants cannot be attributed solely to lower activities of PSI. Under severe Cu deficiency, the polypeptides of PSII are altered (Yruela, 2013), and the lipid composition changes in favor of the less unsaturated fatty acids, for example, 18:3 → 18:2 (Ayala et al., 1992). These changes in fatty acid composition in the thylakoids and in the PSII complex are probably related to the functions of Cu in the desaturation of long-chain fatty acids.

The low carbohydrate concentrations are the main reason for reduced nodulation and N_2 fixation in the Cu-deficient legumes (Cartwright and Hallsworth, 1970). Symptoms of N deficiency in Cu-deficient plants can be overcome by the application of mineral N. However, it has been shown that N application promotes Cu deficiency, and when N supply is high, application of Cu fertilizers may be required for the optimal yield (Robson and Reuter, 1981). In addition to nonspecific growth enhancements, N influences Cu availability and mobility within the plant, including (1) a higher proportion of Cu complexed to amino acids and proteins in the mature tissues and (2) a decrease in the rate of re-translocation of Cu from old leaves to the areas of new growth. Re-translocation of Cu is closely related to leaf senescence (Chapter 3).

7.3.5 Lignification

The impaired lignification of cell walls is a typical anatomical change induced by Cu deficiency in vascular plants. This results in the characteristic distortion of young leaves, bending and twisting of stems and twigs (stem deformation and "pendula" forms in trees; Hopmans, 1990) and an increase in the lodging susceptibility of cereals, particularly in combination with a high N supply. In maize, the monocot-specific microRNA (miR528) targets the Cu-containing laccases (particularly LAC3 and LAC5); miR528 is induced by a luxury N supply, resulting in the downregulation of laccases and the decreased lignin concentration in shoots (Sun et al., 2018).

As shown in Table 7.12, Cu has a strong effect on the formation and chemical composition of cell walls. In Cu-deficient leaves, the ratios of cell wall material to the total dry matter and the lignin concentration decrease, whereas the proportion of α-cellulose and hemicellulose increases compared to leaves adequately supplied with Cu (Robson et al., 1981). In severely Cu-deficient plants, the xylem vessels are insufficiently lignified. A decrease in lignification occurs even with mild Cu deficiency and is, therefore, a suitable indicator of the plant's Cu nutritional status (Rahimi and Bussler, 1974). This poor lignification of the xylem vessels in arabidopsis under Cu deficiency is due to the induction of the SPL7 transcription factor that activates miR857, which downregulates laccase (LAC7), thus decreasing lignin biosynthesis (Zhao et al., 2015).

The inhibition of lignification in the Cu-deficient tissues (Table 7.12) is related to a direct role of at least two Cu enzymes in lignin biosynthesis: polyphenol oxidase catalyzes the oxidation of phenolics as precursors of lignin, and diamine oxidase provides the H_2O_2 required for oxidation by peroxidases. Accordingly, the activity of both enzymes is lower in the Cu-deficient tissues, and phenolics accumulate. In addition, laccase (containing four Cu atoms) is an enzyme involved in the lignification as well as producing phenolics for plant defense against biotic stresses (Liu et al., 2020a,b). Numerous laccase-like multicopper oxidases (such as LAC3, LAC12, and LAC13) are regulated by miR408 that strongly influences lignification (Carrio-Segui et al., 2019).

TABLE 7.12 Cell wall composition of the youngest fully emerged leaves of wheat (*Triticum aestivum*) with or without Cu supply.

	+Cu	−Cu
Cu concentration (μg Cu g^{-1} dw)	7.1	1.0
Cell wall concentration (% of dw)	46	43
Cell wall composition (% of cell walls)		
α-Cellulose	47	55
Hemicellulose	47	41
Lignin	6.5	3.3
Total phenolics (% of dw)	0.73	0.82
Ferulic acid (% of dw)	0.50	0.69

Source: From Robson et al. (1981).

TABLE 7.13 Relationship between Cu supply and growth and dry matter distribution in pepper (*Capsicum annuum*).

Cu supply (μg pot^{-1})	Dry weight (g dw plant^{-1})			
	Roots	Leaves and stems	Buds and flowers	Fruit
0.0	0.8	1.7	0.16	none
0.5	1.6	3.3	0.28	none
1.0	1.5	3.2	0.38	0.87
5.0	1.4	3.0	0.36	1.81
10	1.2	2.0	0.28	1.99

Source: From Rahimi (1970).

Lignification of cell walls is important in disease resistance, for example, against stripe rust in barley (Prabhjot et al., 2020) and against gray mold (*Botrytis cinerea* L.) in tomato (Liu et al., 2020a,b). Increased lignification strengthened (thickened) cell walls, enhancing physical defenses against a pathogen attack (Liu et al., 2020a,b).

7.3.6 Pollen formation and fertilization

Copper deficiency affects grain and fruit formation more strongly than vegetative growth. A typical example is shown in Table 7.13 for pepper (*Capsicum annuum*). Supplying 0.5 μg Cu pot^{-1} produced a maximum dry weight of roots and shoots, but flower formation was impaired, and no fruit was formed. For the fruit formation a higher Cu supply was required (>1.0 μg Cu pot^{-1}), but with 10 μg Cu pot^{-1} toxicity occurred, affecting root and shoot growth.

The main reason for a decrease in the formation of generative organs is the nonviability of pollen from Cu-deficient plants (Graham, 1975). A novel transcription factor CITF1 (Cu deficiency–induced transcription factor1) regulates (together with SPL7) Cu uptake by roots and its transport to the anthers. Copper deficiency upregulates the genes involved in the biosynthesis of jasmonic acid, and jasmonic acid regulates the expression of CITF1 (Yan et al., 2017), but the exact sequence of events and the importance of this observation is yet to be ascertained.

The critical stage of Cu deficiency–induced pollen sterility is microsporogenesis. Reduced grain set in Cu-deficient plants may be the result of the inhibition of pollen release because lignification of the anther cell walls is required to rupture the stamen and release the pollen. In Cu-deficient plants, lignification of the anther cell wall is reduced or absent (Dell, 1981).

7.3.7 Copper deficiency and toxicity

7.3.7.1 Copper deficiency

Copper deficiency is often observed in plants growing on soils either low in total Cu (e.g., ferrallitic and ferruginous coarse–textured soils, or calcareous soils derived from chalk) or high in organic matter where Cu is complexed with organic substances (Alloway and Tills, 1984), even though depletion of organic matter may also be associated with the low Cu availability in soils (Kidron and Zilberman, 2019). Many regions of the world have soils with low Cu availability, including sub-Saharan Africa (Kihara et al., 2020), north-west Africa (Drissi et al., 2018; Fouad et al., 2020), West Africa (Kidron and Zilberman, 2019), and southern Australia (Brennan et al., 2019).

The critical deficiency concentration of Cu in vegetative plant parts is generally in the range of $1-5~\mu g~g^{-1}$ dw, depending on plant species, plant organ, developmental stage, and N supply (Robson and Reuter, 1981). The critical deficiency concentration of the youngest emerged leaf is less affected by the low Cu availability in the root medium than that of older leaves (Billard et al., 2014), partly because of strong remobilization of Cu from the old to young leaves. Stunted growth, distortion of young leaves, chlorosis/necrosis starting at the apical meristem extending down the leaf margins, and bleaching of young leaves are the symptoms of Cu deficiency (Robson and Reuter, 1981).

Plant species as well as genotypes within species differ considerably in their sensitivity to Cu deficiency. Rye (*Secale cereale*) is particularly Cu efficient, and transferring a part of the long arm of chromosome 5 (5RL) to bread wheat enhances Cu efficiency of the wheat-rye translocation lines (Graham et al., 1987a,b; Owuoche et al., 1996). The majority of advanced breeding lines of common bean (*P. vulgaris*) were Cu efficient (53%) or moderately efficient (40%), with only 2 out of 30 tested lines being characterized as having poor Cu efficiency (Fageria et al., 2015).

Copper deficiency impairs the ferric reductase activity, resulting in leaf chlorosis (Waters and Armbrust, 2013). By contrast, the overexpression of Cu transporters COPT1 and COPT3 in arabidopsis inhibited the localized Fe deficiency responses in roots, suppressing Fe translocation to shoots (Perea-Garcia et al., 2020). Similarly in bread wheat, Cu fertilization and accumulation of Cu in the roots decreased Fe uptake and caused chlorosis (Kaur and Manchanda, 2019).

According to Yruela (2013), the molecular responses to Cu deficiency include the increased expression of metal reductases and transporters, and prioritizing Cu to the essential enzymatic pathways, including compensatory increases in Fe-SOD and Mn-SOD in place of CuZn-SOD. Some SPL transcription factors activate transcription of Cu transporters and metabolic reorganization to economize Cu supply under Cu deficiency (e.g., SPL7 in arabidopsis, Schulten et al., 2019). Copper deficiency induced exudation of phytosiderophores (PSs) in wheat, but to a significantly lesser extent than deficiency of the other cationic micronutrients (most notably Fe) (Khobra and Singh, 2018).

The availability of Cu can be low in many soils, and this can be corrected by soil or foliar fertilization. Soil applications of inorganic Cu as sulfate or oxide forms, or slow-release metal compounds, sewage sludges or manures often result in the long-term effects (e.g., up to 30 years after soil application of Cu-amended superphosphate in southern Australia) (Brennan et al., 2019). Soil application of Cu-EDTA or Cu-DTPA resulted in a significant improvement in the Cu uptake by spinach (Obrador et al., 2013). Foliar applications of Cu in the form of inorganic salts, oxides, or chelates can be used to correct Cu deficiency rapidly in the soil-grown plants, including bread wheat (Pakhomova and Daminova, 2020) and barley (Drissi et al., 2018), but care needs to be taken to avoid leaf damage, for example, in bread wheat (Fouad et al., 2020), due to excessive Cu concentration in the foliar spray.

The Cu-containing fertilizers can be used to increase the Cu concentration in the edible portions of crops where there are dietary deficiencies of Cu in humans and livestock (White and Broadley, 2009) (see Chapter 9). However, Cu fertilization must be managed appropriately because high Cu concentrations can be toxic to plants and animals. Selecting genotypes that are highly efficient in Cu uptake, translocation from the roots to the shoots, and retranslocation within the shoot is a promising long-term approach to the prevention of Cu deficiency.

There is an increasing interest in nanofertilizers, including CuO nanoparticles (CuO NPs) that were shown to increase growth and physiological parameters in a number of plant species, including shallot (*Allium ascalonicum*) (Priyanka and Venkatachalam, 2020). In tomato, Cu NPs increased SOD activity and alleviated salinity stress (Hernandez-Hernandez et al., 2018). However, CuO NPs are also used as nanopesticides (e.g., Bonilla-Bird et al., 2018), and increasing the use of nanoparticles may damage plant metabolic functions (e.g., Tighe-Neira et al., 2018) and has led to environmental concerns. In particular, CuO NPs inhibited the root tip growth by inducing physical damage, cytological aberrations, and disrupting biochemistry of the cell wall (Nie et al., 2020; Pramanik et al., 2018). However, decreased uptake of CuO NPs into the root cells of oilseed rape (*Brassica napus*), and thus decreased toxicity, can be achieved by humic acids (Yarmohammadi et al., 2019).

7.3.7.2 Copper toxicity

Toxic soil Cu concentrations can occur under natural conditions or due to anthropogenic inputs. Anthropogenic inputs include those from the long-term use of Cu-containing fungicides (e.g., in vineyards, coffee plantations) (Marastoni et al., 2019; Ruyters et al., 2013), industrial and urban activities (air pollution, urban waste, and sewage sludge), and the application of pig and poultry slurries. There are marked differences in Cu tolerance among plant species, genotypes, and ecotypes (Marastoni et al., 2019; Vijeta et al., 2010; Xv et al., 2020). With respect to the Cu-tolerant metallophytes, particularly among the flora of the Cu-rich soils in the Democratic Republic of Congo (Delhaye et al., 2020), there have been field or herbarium reports that the Cu concentration in leaves can be as high as 1000 µg g^{-1} dw. *Commelina communis* and *Elsholtzia splendens* in China (Li et al., 2018) and *Crassula helmsii* in Europe (Küpper et al., 2009) have been shown to contain "hyperaccumulator concentrations" of Cu in leaves/shoots. However, even though these species may have an elevated requirement for Cu, and are certainly highly tolerant of excess Cu, Cu hyperaccumulation has not been demonstrated under controlled conditions, suggesting that some of the above records may be due to leaf contamination with dust (Chipeng et al., 2010; Macnair, 2003).

A high Cu supply usually inhibits root growth before shoot growth. This does not mean that roots are inherently more sensitive to high Cu concentrations; rather, they are the sites of preferential Cu accumulation when the external Cu supply is high, for example, in *Malus prunifolia* (Wan et al., 2019) and tomato (Table 7.14). This restricted transport to the aerial parts is at least partly due to the downregulation of the transporters YSL3 and HMA5 (Wan et al., 2019).

With high supply, the Cu concentration of the roots increases proportionally to the concentration of Cu in the external medium, whereas transport to the shoot is highly restricted. Critical toxicity concentrations of Cu in the shoots may therefore not necessarily reflect the Cu tolerance of plants. In addition to immobilization of Cu in the roots (Wan et al., 2019; Xv et al., 2020), or reductions in Cu uptake *per se* through binding of extracellular Cu by root exudates (e.g., Yarmohammadi et al., 2019), the cellular mechanisms of Cu tolerance (Burkhead et al., 2009; Hasan et al., 2017; Yruela, 2013) are likely to include (1) enhanced binding to cell walls (e.g., Wan et al., 2019; Xv et al., 2020); (2) restricted influx through the plasma membrane (cf. Marastoni et al., 2019); (3) stimulated efflux from the cytoplasm (Zou et al., 2020); (4) compartmentation of Cu by export to the vacuole; (5) chelation at the cell wall–plasma membrane interface; and (6) intracellular chelation of Cu by organic acid anions, glutathione-derived phytochelatins, and cysteine-rich metallothioneins in the cytoplasm (e.g., Küpper et al., 2009; Saeed Ur et al., 2020; Wan et al., 2019) (Fig. 7.14).

In rice, the P$_{1B}$-type heavy metal ATPases (OsHMA6 and OsHMA9) are present in the plasma membrane, facilitating efflux of excess Cu from cells (Zou et al., 2020). In perennials, root colonization with ectomycorrhiza may play an important role in metal tolerance of the host plant (Chapter 15). An important role in plant tolerance to excess Cu is played by nitric oxide (NO) that decreases accumulation of superoxide radicals and hydrogen peroxide in arabidopsis seedlings exposed to high Cu concentration in the root medium (Kolbert et al., 2015; Peto et al., 2013). Downregulation of miR398 (and thus upregulation of its target CuZn-SOD) contribute to the enhanced capacity to detoxify ROS, as shown in *P. vulgaris* (Naya et al., 2014).

TABLE 7.14 Relationship between Cu supply, dry weight, and Cu concentrations of different plant parts of tomato (*Solanum lycopersicum*) grown in nutrient solution.

Cu supply (µg L^{-1})	Dry weight (g dw plant^{-1})		Cu concentration (mg kg^{-1} dw)		
	Roots	Shoots	Roots	Stems and petioles	Leaves
0	0.3	2.6	4.0	2.8	3.0
2.5	2.5	9.4	3.8	2.1	3.2
5.0	3.2	11	6.4	2.4	4.1
20	3.4	12	64	4.3	15
250	1.6	9.7	360	6.2	20.3

Source: From Rahimi and Bussler (1974).

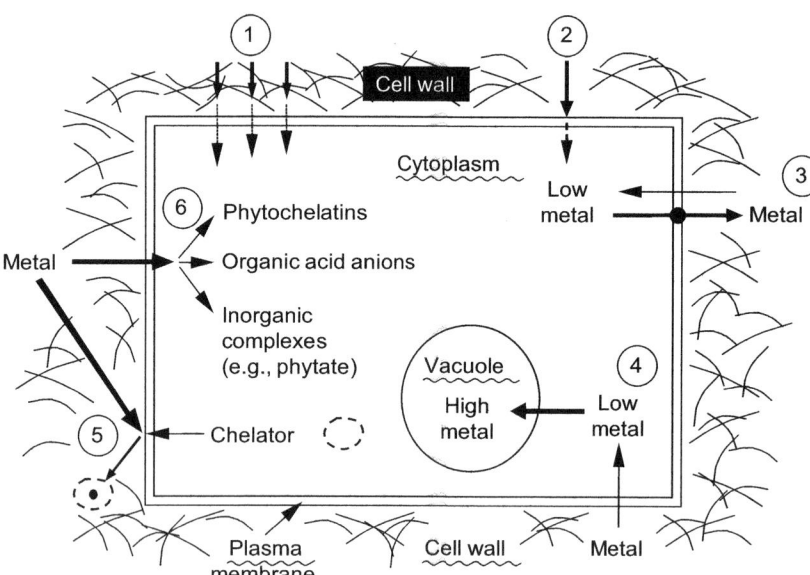

FIGURE 7.14 The mechanisms of metal tolerance in plants. (1) Binding to the cell wall; (2) restricted influx through the plasma membrane; (3) active efflux; (4) compartmentation in the vacuole; (5) chelation at the cell wall–plasma membrane interface; and (6) chelation in the cytoplasm. *Modified from Tomsett and Thurman (1988).*

7.4 Zinc

7.4.1 General

Zinc (Zn) is the second-most abundant transition metal in living organisms after Fe. The average total Zn concentration in cultivated soils is around 65 mg kg^{-1} (Alloway, 2009). Zinc is taken up by plants predominantly as a divalent cation (Zn^{2+}). During long-distance transport in the xylem, Zn is either bound to organic acid anions or occurs as the free divalent cation (Chapter 3). In plants and other biological systems, Zn exists only as ZnII and does not take part in redox reactions.

In recent decades, impressive progress has been made in the identification and characterization of catalytic and structural Zn sites in proteins. Bioinformatic analysis of the human genome showed that Zn can bind to ∼10% of the proteins in the human body, corresponding to about 3000 proteins (Andreini et al., 2006; Kocyla et al., 2021; Maret, 2019). In arabidopsis genome, bioinformatic analysis showed that Zn can bind to ∼13% of the proteins, corresponding to about 2500 proteins (Broadley et al., 2007). Proteins bind to Zn with high affinity, ranging from the picomolar to femtomolar Zn concentrations, to maintain their cellular functions and interactions (Kluska et al., 2018; Kocyla et al., 2021; Krezel and Maret, 2016). As a result, Zn has a high biological significance; importantly, only 2% of all proteins use Fe as a cofactor, whereas 10% of all proteins utilize Zn in the structural and catalytic functions (Maret, 2019).

Zinc has diverse catalytic and structural roles in those proteins and influences several critical biochemical pathways and cellular functions, such as the activities of several hundred enzymes, DNA and protein biosynthesis, cell division, gene expression, and defense processes against oxidative cell damage. Zinc is also critically important for the stabilization of proteins and contributes to the correct folding of protein subunits. Growing evidence has shown that Zn behaves like a signaling molecule in multiple cellular functions (Hara et al., 2017; Maret, 2019). In plant cells, Zn contributes greatly to plant tolerance to environmental stresses by supporting the biosynthesis and function of proteins, as well as the expression and regulation of genes, that are involved in plant defense (Broadley et al., 2007; Cakmak, 2000).

7.4.2 Zn-containing enzymes

In biological systems, Zn is the only metal that is present in all six enzyme classes, including oxidoreductases, transferases, hydrolases, lyases, isomerases, and ligases (Sousa et al., 2009). Zinc is most commonly found in hydrolases and oxidoreductases. In these enzymes, four types of Zn-binding sites have been identified: (1) catalytic, (2) structural, (3) cocatalytic, and (4) protein interface, which determine the biological activity of the enzymes. The most frequent amino acid ligand at the binding sites is histidine followed by glutamine, asparagine, and cysteine (Kochanczyk et al., 2015; Kocyla et al., 2021).

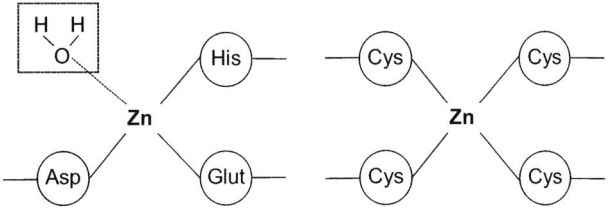

Histidine accounts for 28% of all Zn-binding ligands (Fig. 7.15). Water molecules are also important Zn ligands within the protein structure. Structural Zn sites contribute to the maintenance of the structure of enzymes (e.g., alcohol dehydrogenase, and proteins involved in DNA replication and gene expression). In these proteins, Zn ions are mostly coordinated to four cysteine residues. The co-catalytic Zn sites are present in enzymes containing two or more Zn atoms, with aspartic acid and histidine being the most common ligands in these co-catalytic sites. At the protein interface, Zn bridges proteins or subunits and contributes to protein−protein interactions (Auld and Bergman, 2008; Maret, 2013). As the metal ion associated with the largest number of proteins in biological systems, Zn also has specific roles in the protein−protein communication and interactions, which may have huge impacts on the biological processes in all organisms (Kocyla et al., 2021).

7.4.2.1 Alcohol dehydrogenase

Most enzymes containing Zn have only one Zn atom per molecule, an exception being alcohol dehydrogenase with two Zn atoms per molecule, one with catalytic and the other with structural functions. In plants, alcohol dehydrogenase catalyzes the reduction of acetaldehyde to ethanol during the fermentation of glucose (Auld and Bergman, 2008; Raj et al., 2014). The catalytic Zn sites are bound to two cysteines, one histidine, and one water molecule, while the structural Zn-binding sites are generally complexed by four cysteines.

The enzyme catalyzes the reduction of acetaldehyde to ethanol:

$$\text{Pyruvate} \xrightarrow{\text{CO}_2} \text{Acetaldehyde} \xrightarrow[\text{NAD}^+]{\text{NADH}} \text{Ethanol}$$

Under anaerobic conditions, ethanol formation takes place mainly in the meristematic tissues, such as root apices. Although Zn deficiency affects the activity of alcohol dehydrogenase, the effect is not consistent. In some circumstances Zn deficiency decreases alcohol dehydrogenase activity, whereas in other situations this activity is unaffected or increased (Rose et al., 2012).

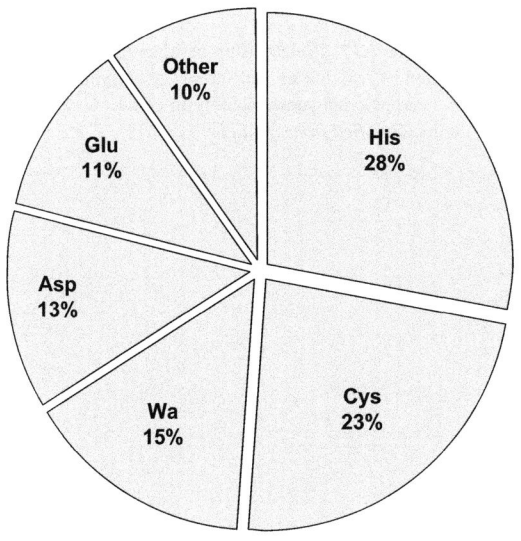

FIGURE 7.15 Percentages of Zn-binding ligands in the Zn proteome found in the Protein Data Bank. *Asp*, Aspartic acid; *Cys*, cysteine; *Glu*, glutamic acid; *His*, histidine; *Wa*, water. *Based on Sousa et al. (2009).*

7.4.2.2 Carbonic anhydrase

Carbonic anhydrase (CA) was the first Zn-containing enzyme discovered (by Keilin and Mann in 1940, cited by Coleman, 1992). The enzyme catalyzes the hydration of CO_2:

$$CO_2 + H_2O \leftrightarrow HCO_3^- + H^+$$

The CA found in dicotyledons comprises six subunits, has a molecular weight of 180 kDa, and six Zn atoms per molecule (Sandmann and Böger, 1983). The enzyme is localized in the chloroplasts and in the cytoplasm (Fig. 7.16).

The role of CA, and particularly its role in chloroplasts, differs between C_3 and C_4 plants. Among C_4 plants, there are further differences between CA roles in the mesophyll and bundle sheath chloroplasts (see Chapter 5). In C_3 plants, CA is required to facilitate the diffusion of CO_2 to the sites of carboxylation by Rubisco. In C_4 plants, CA mediates conversion of CO_2 to HCO_3^- to be used by PEPC (Fig. 7.16; Badger and Price, 1994).

In C_3 plants, there is no direct relationship between CA activity and photosynthetic CO_2 assimilation by plants of different Zn nutritional status (Fig. 7.17). In rice plants, Zn deficiency resulted in a decrease in the expression of mRNAs for CA, suggesting that the decrease in CA activity upon Zn deficiency is due to a reduced amount of the enzyme (Sasaki et al., 1998). With extreme Zn deficiency, CA activity is inhibited completely; however, the maximum net photosynthesis can occur even with low CA activity (Fig. 7.17).

In C_4 plants, however, the situation is different (Burnell and Hatch, 1988; Hatch and Burnell, 1990). A high CA activity is required in the mesophyll chloroplasts to shift the equilibrium in favor of HCO_3^-, the substrate for PEP carboxylase (Fig. 7.16), which forms C4 compounds (e.g., malate) for shuttling into the bundle sheath chloroplasts (Chapter 5) where CO_2 is released and serves as a substrate for RuBP carboxylase. In agreement with this, despite similar total activities in leaves of C_3 and C_4 plants, only 1% of the total CA activity in C4 plants is in the bundle sheath chloroplasts (Burnell and Hatch, 1988), whereas 20%–60% is associated with the plasma membrane (Utsunomiya and Muto, 1993).

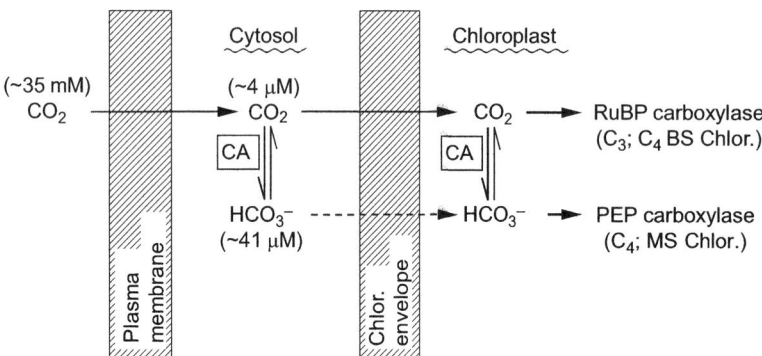

FIGURE 7.16 Functioning of CA in leaf cells of C_3 and C_4 plants. *BS*, Bundle sheath chloroplasts; *CA*, carbonic anhydrase; *MS*, mesophyll chloroplasts. *Based on Edwards and Walker (1983) and Hatch and Burnell (1990).*

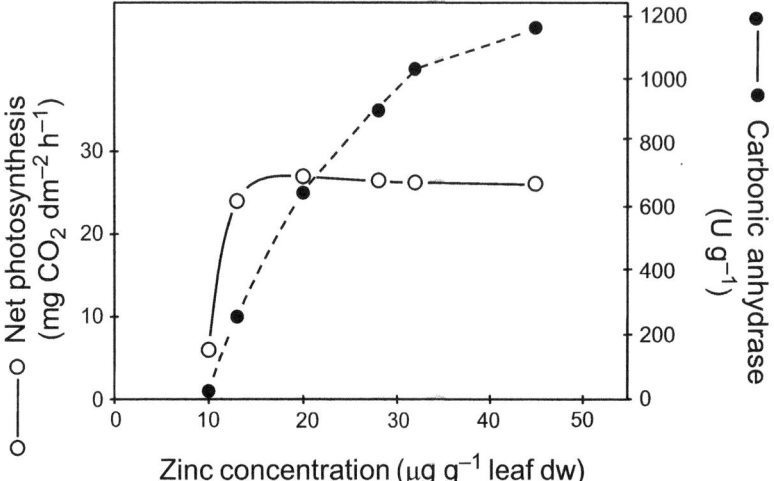

FIGURE 7.17 Relationship between the Zn concentration of leaf blades and the net photosynthesis and the carbonic anhydrase activity in cotton. *U*, Enzyme units. *Modified from Ohki (1976).*

TABLE 7.15 Dry weight of roots and shoots, generation of superoxide radicals ($O_2^{\bullet-}$), and activity of superoxide dismutase (SOD) in roots of cotton (*Gossypium hirsutum*) with or without Zn supply.

Zn supply	Dry weight [g (4 plants)$^{-1}$]		$O_2^{\bullet-}$ generation (nmol mg^{-1} protein min^{-1})	SOD activity (U mg^{-1} protein)
	Shoots	Roots		
+Zn	3.1	0.8	1.3	75
−Zn	1.8	0.5	3.7	35

Source: Based on Cakmak and Marschner (1988a).

At least in C_4 plants, the in vivo activity of CA appears to be just sufficient to ensure that the rate of conversion of CO_2 to HCO_3^- is not limiting photosynthesis (Hatch and Burnell, 1990). Therefore, it can be expected that Zn deficiency may have a greater effect on the rate of photosynthesis in C_4 than C_3 plants (Burnell et al., 1990). Accordingly, in sorghum (C_4), Zn deficiency resulted in strong depression of the CA activity and the gene expression. The activity of CA enzyme and gene expression in sorghum recovered rapidly after Zn resupply, but CuZn-SOD and alcohol dehydrogenase did not show such rapid responses to Zn resupply, indicating that CA was the preferred target protein for the Zn resupply (Li et al., 2013).

7.4.2.3 CuZn superoxide dismutase

The CuZn-SOD is the most abundant SOD in plant cells. Most likely, the Cu atom acts as the catalytic metal and Zn as the structural component. In the enzyme, Zn is bound to two histidines and one aspartate and contributes to structural stability of the enzyme (Abreu and Cabelli, 2010). The localization and role of CuZn-SOD have been discussed in Section 7.3. As reviewed by Cakmak (2000), in many plant species Zn deficiency decreases CuZn-SOD activity, and a resupply of Zn rapidly restores it, indicating that the Zn atom is an essential structural component for the normal functioning of CuZn-SOD. The activity of CuZn-SOD is diminished strongly by a restricted Zn supply and is therefore a better indicator of Zn deficiency tolerance than the total Zn concentration of leaf tissue (Cakmak et al., 1997b; Hacisalihoglu and Kochian, 2003; Yu et al., 1999).

The decrease in CuZn-SOD activity under Zn deficiency is particularly critical because of the simultaneous increase in the rate of $O_2^{\bullet-}$ generation (Table 7.15). High concentrations of the toxic $O_2^{\bullet-}$ radicals and related oxidants lead to peroxidation of membrane lipids and an increase in membrane permeability (Cakmak and Marschner, 1988c). Accordingly, the overexpression of CuZn-SOD in transgenic plants increases their tolerance to various abiotic stresses (Cakmak, 2000; Kim et al., 2010), especially salinity (Wang et al., 2016; Guan et al., 2017).

7.4.2.4 Other Zn-containing enzymes

Zinc is the metal component in many enzymes (Coleman, 1992, 1998; Maret, 2019), including the following:

1. Alkaline phosphatase that is highly sensitive to low amounts of Zn in biological systems.
2. Phospholipase, containing three Zn atoms, of which at least one has catalytic functions.
3. Carboxypeptidase that hydrolyzes peptides and contains one Zn atom with catalytical functions.
4. RNA polymerase, containing two Zn atoms per molecule, becomes inactive when Zn is removed. The effect of Zn nutritional status on RNA polymerase activity has been studied extensively in bacteria, animals, and humans, but there is little information on this relationship in plants, with the exception of green algae.

7.4.3 Zn-activated enzymes

In vascular plants, Zn is either required for, or at least modulates, the activity of a large number of various types of enzymes, including dehydrogenases, aldolases, isomerases, transphosphorylases, and RNA and DNA polymerases (Maret, 2019). Some examples are given below.

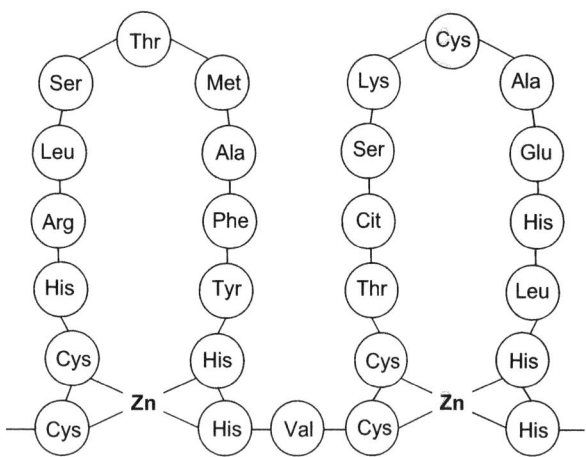

FIGURE 7.18 Schematic presentation of the role of Zn in tertiary structure of the peptide chain in replication proteins ("zinc finger"). *Based on Coleman (1992) and Vallee and Falchuk (1993).*

TABLE 7.16 Shoot dry weight and chemical composition of young leaves and shoot apex of common bean plants with Zn supply (1 μM Zn), without Zn supply, or without Zn supply followed by resupply of 3 μM Zn for 3 days (−Zn → +Zn).

	Zn supply		
	+Zn	−Zn	−Zn → +Zn
Shoot dw (g^{-1} dw plant^{-1})	8.2	3.7	4.5
Zn concentration (μg g^{-1} dw)	52	13	141
Free amino acids (μmol g^{-1} dw)	82	533	118
Protein (mg g^{-1} fw)	28	14	30
Tryptophan (μmol g^{-1} dw)	0.4	1.3	0.3
IAA tryptophan (ng g^{-1} dw)	239	118	198

Source: From Cakmak et al. (1989).

Inorganic pyrophosphatases (PP$_i$ase) are important components of the proton-pumping activity in the tonoplast (Chapter 2). Besides the well-known Mg^{2+}-dependent enzyme (Mg.PP$_i$ase), a PP$_i$ase isoenzyme in leaves is Zn^{2+}-dependent (Zn.PP$_i$ase). In rice leaves, the activity ratios of Mg./Zn.PP$_i$ase vary between three and six (Lin and Kao, 1990).

The role of Zn in DNA and RNA metabolism, in cell division, and protein synthesis has been documented for many years, but only recently has a class of Zn-dependent proteins (Zn metalloproteins) been identified as being involved in DNA replication, transcription, and, thus, regulation of gene expression (Andreini et al., 2009; Krezel and Maret, 2016). For transcription, Zn in these proteins is required for binding to specific genes by forming tetrahedral complexes with amino acid residues of the polypeptide chain (Fig. 7.18). In eukaryotic cells, 44% of the Zn-dependent proteins are used in regulating DNA transcription (Andreini et al., 2009). The Cys(2)His(2) Zn finger is one of the most common DNA-binding motifs in eukaryotic cells, with diverse functions in biological systems (Papworth et al., 2006). In these DNA-binding proteins, Zn is directly involved in the gene expression and the activation or repression of DNA elements.

7.4.4 Protein synthesis

In Zn-deficient plants, the rate of protein synthesis and protein concentration are strongly reduced, whereas free amino acids accumulate (Table 7.16). Upon resupply of Zn to the Zn-deficient plants, protein synthesis resumes quite rapidly.

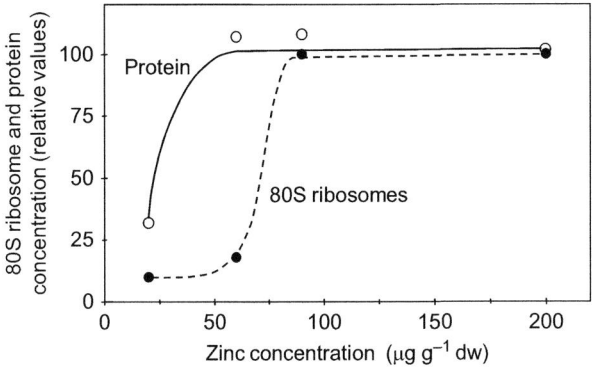

FIGURE 7.19 Relationship between concentration of Zn, 80S ribosomes, and protein in the soluble fraction of rice shoot meristematic tissue. *Based on Kitagishi et al. (1987).*

TABLE 7.17 Element concentration in meristematic tissue of the youngest and mature leaves of Zn-sufficient rice plants.

	Element concentration in dry matter				
	Zn	Mn	Mg	Ca	K
	($\mu g\ g^{-1}$ dw)		($mg\ g^{-1}$ dw)		
Meristem	204	188	4.2	2.3	30
Mature leaves	18	540	8.9	6.0	13

Source: Based on Kitagishi and Obata (1986).

Besides the functions of Zn described above, at least two other functions of Zn in protein metabolism are responsible for these changes. Zinc is a structural component of ribosomes and essential for their structural integrity. In the absence of Zn, ribosomes disintegrate but can be reconstituted after Zn resupply.

In shoot meristems of rice, disintegration of the 80S ribosomes takes place when the Zn concentration is below $100\ \mu g\ g^{-1}$ dw. Considerably lower Zn concentrations are required to decrease protein concentration (Fig. 7.19). In tobacco tissue culture cells, the corresponding concentrations were $70\ \mu g$ Zn for a decrease in the abundance of 80S ribosomes and $50\ \mu g$ Zn for a decrease in protein concentration (Obata and Umebayashi, 1988).

A particularly high Zn requirement for protein synthesis has been shown in pollen tubes, where the Zn concentration at the growing tip was about $150\ \mu g\ g^{-1}$ dw compared with about $50\ \mu g\ g^{-1}$ in the more basal regions (Ender et al., 1983). In the newly emerged root tips of wheat plants, Zn concentrations are about $220\ \mu g\ g^{-1}$ (Ozturk et al., 2006). In the shoot meristems, and presumably also in other meristematic tissues, Zn concentration of at least $100\ \mu g\ g^{-1}$ dw is required for maintenance of protein synthesis. As shown in Table 7.17, this is about 5–10 times more than the adequate Zn concentration in mature leaf blades. For other nutrients, this gradient is usually less steep. To meet the high Zn demand in the shoot meristem, most of the root-supplied Zn is translocated preferentially to the shoot meristem via xylem–phloem transfer in the stem (Kitagishi and Obata, 1986).

Low protein and high amino acid concentrations in Zn-deficient plants are not only the result of reduced transcription and translation but also of enhanced rates of RNA degradation due to the high RNAse activity when plants are Zn deficient (Sharma et al., 1982). There is a negative relationship between Zn supply and RNAse activity, and also between RNAse activity and protein concentration (Table 7.18). Similarly, in rice plants, Zn deficiency enhanced RNA degradation, whereas addition of Zn to Zn-deficient rice plants significantly increased the activity of RNA (Suzuki et al., 2012).

7.4.5 Carbohydrate metabolism

Many Zn-dependent enzymes are involved in carbohydrate metabolism. In Zn-deficient leaves, a rapid decrease in CA activity is the most sensitive and obvious change in activity of enzymes involved in carbohydrate metabolism (Li et al., 2013; Shrotri et al., 1983). Decreases in photosynthetic activity in Zn-deficient plants are attributed to large decreases

TABLE 7.18 Fresh weight, RNAse activity, and protein-N concentration in perennial soybean (*Glycine wightii*) at different rates of Zn supply.

Zn supply (μg L^{-1})	Fresh weight (g fw plant^{-1})	RNAse activity (% hydrolysis)	Protein-N (mg g^{-1} fw)
5	4.0	74	18
10	5.1	58	23
50	6.6	48	28
100	10	40	37

Source: Based on Johnson and Simons (1979).

in CA activity, with no apparent stomatal limitation as shown in sorghum (Li et al., 2013). The activity of fructose-1,6-bisphosphatase also declines rapidly, whereas the activity of other enzymes is affected to a much lesser extent, particularly under mild Zn deficiency (Shrotri et al., 1983). Based on transcriptomic and proteomic analyses, a reduction in photosynthetic activity of Zn-deficient leaves was found due to structural changes and impairments in the chloroplasts in maize (Zhang et al., 2019). Despite a decrease in the enzyme activities and in the rate of photosynthesis (as indicated by the activity of the Hill reaction), sugars and starch often accumulate in Zn-deficient plants (Marschner and Cakmak, 1989; Sharma et al., 1982). Accumulation of sugars in leaves, especially in the source leaves (Marschner and Cakmak, 1989), could be due to reduced sink strength under Zn deficiency.

The accumulation of carbohydrates in the leaves of Zn-deficient plants increases with light intensity (Marschner and Cakmak, 1989) and is associated with impaired new growth (particularly of the shoot apices), that is, lower sink activity probably due to reduced concentrations of the phytohormones that stimulate cell elongation. Increased accumulation of photoassimilates in the Zn-deficient leaves could lead to a reduction in photosynthetic activity (Cakmak, 2000; Lemoine et al., 2013), which in turn may potentiate the generation of ROS in chloroplasts through light-induced activation of molecular oxygen. The well-known susceptibility of Zn-deficient plants to high light intensity is likely related to the light-dependent generation of ROS (i.e., photooxidative damage) as a result of diminished photosynthesis as well as the reduced activity of CuZn-SOD (Cakmak, 2000; Broadley et al., 2007; Marschner and Cakmak, 1989). Zinc deficiency−induced leaf chlorosis and necrosis under high light intensity have been ascribed to the production of hydroxyl radicals (OH$^\bullet$) mediated by a Fenton-like reaction in chloroplasts. Iron is the main player in the Fenton reaction and reacts with H_2O_2 to yield very aggressive OH$^\bullet$ radicals, especially under high light intensity (Shinozaki et al., 2020). Excessive accumulation of Fe in Zn-deficient plants has been reported in several studies (Cakmak, 2000), which further increases the risk of photooxidative damage in plants under low Zn supply.

7.4.6 Tryptophan and indole acetic acid synthesis

The most distinct Zn deficiency symptoms—stunted growth and "little leaf"—are presumably related to disturbance in the metabolism of auxins, IAA in particular. The mode of action of Zn in auxin metabolism is still unclear. In Zn-deficient tomato plants, retarded stem elongation is correlated with a decrease in IAA concentration; upon resupply of Zn, the stem elongation and the IAA concentrations increase. The response to the Zn treatment was more rapid for IAA concentrations than for elongation growth (Tsui, 1948). Low concentrations of IAA in the Zn-deficient plants (Cakmak et al., 1989; Li et al., 2013) may be the result of inhibited synthesis or enhanced degradation of IAA (Cakmak et al., 1989). Biosynthesis of IAA can use tryptophan as the precursor:

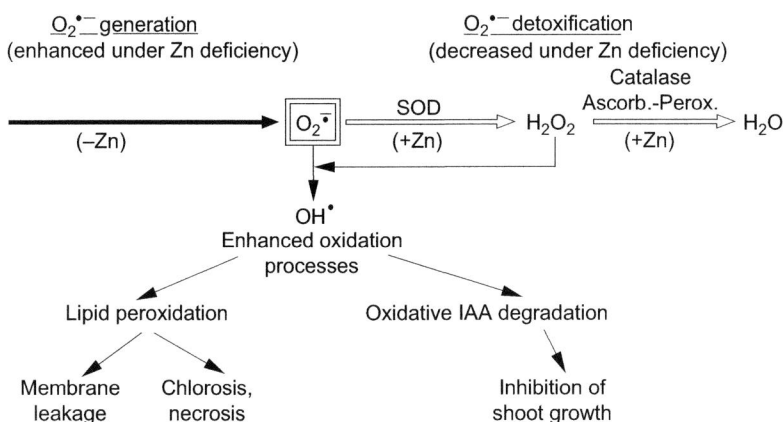

FIGURE 7.20 Involvement of Zn in the generation and detoxification of superoxide radicals, and the effects of oxygen-free radicals on membrane function and IAA metabolism. *IAA*, Indole-3-acetic acid. *Compiled from Cakmak and Marschner (1988a,b) and Cakmak et al. (1989).*

In leaves of Zn-deficient plants, tryptophan concentrations increase similarly to other amino acids (Cakmak et al., 1989; Domingo et al., 1992), most likely due to the impaired protein synthesis (Table 7.16). Although the lower IAA concentration in Zn-deficient leaves may indicate a role for Zn in the biosynthesis of IAA from tryptophan, as postulated by Salami and Kenefick (1970), lower IAA concentrations are more likely the result of enhanced oxidative degradation of IAA (Fig. 7.20). Adequate Zn nutrition also increases the concentrations of endogenous gibberellins (Sekimoto et al., 1997). Low concentrations of IAA and gibberellins may be the cause of the stunted growth and "little leaf" formation under Zn deficiency.

7.4.7 Membrane integrity and lipid peroxidation

Zinc is required for the maintenance of structural and functional integrity of biomembranes. It binds to the phospholipid and sulfhydryl (−SH) groups of membrane constituents or forms the tetrahedral complexes with cysteine residues of polypeptide chains, and thereby protects membrane lipids and proteins against the oxidative damage by ROS. In its function as a metal component in CuZn-SOD, Zn may control the generation of toxic oxygen radicals by interfering with the oxidation of NADPH as well as by scavenging $O_2^{\bullet-}$ (Cakmak and Marschner, 1988a,c). Accordingly, there is a typical increase in the plasma membrane permeability, for example, in roots, under Zn deficiency (Cakmak, 2000; Welch et al., 1982) as indicated by leakage of low-molecular-weight solutes, a decrease in phospholipid concentration and in the degree of unsaturation of fatty acids in the membrane lipids (Table 7.19). As early as 12 h after resupplying Zn to the Zn-deficient plants, some restoration of the membrane integrity can be observed. Plasma membrane vesicles isolated from Zn-deficient roots also have higher passive permeability than the vesicles from Zn-sufficient roots (Pinton et al., 1993).

Genotypic variation in damage to roots of rice plants with a low Zn supply was found to be related to membrane stability. The Zn deficiency−sensitive genotype had higher solute leakage from root cells, which was linked to the Zn deficiency−induced membrane damage by ROS (Lee et al., 2017). Adequate Zn nutrition had positive effects on salt stress tolerance of wheat plants, which could be related to the structural stability of cell membranes and a higher concentration of sulfhydryl groups in the roots of Zn-sufficient plants (Daneshbakhsh et al., 2013).

Increased membrane permeability in Zn-deficient plants is due to higher rates of $O_2^{\bullet-}$ generation (Table 7.15) as a result of the increased activity of an NADPH-dependent $O_2^{\bullet-}$ generating oxidase (Table 7.20). Higher activity of this oxidase is either a reflection of a direct role of Zn in regulation of enzyme activity or an indirect result of the alterations in structure and composition of the membranes (Table 7.19). It is reported that NADPH oxidases are the most commonly investigated source of ROS in biological systems, especially in plants under different stress conditions (Smirnoff and Arnaud, 2019; You and Chan, 2015). Therefore, close attention should be given to the Zn nutritional status of plants when the roles of NADPH oxidases are studied in $O_2^{\bullet-}$ generation.

Many of the most obvious symptoms of Zn deficiency, such as leaf chlorosis and necrosis, inhibited shoot elongation and increased membrane permeability, are the expressions of oxidative stress brought about by higher generation of ROS and an impaired detoxification system in Zn-deficient plants. In Zn-deficient plants, the activity of H_2O_2-scavenging enzymes, such as catalase and ascorbate peroxidase, is also decreased, probably due to the inhibited protein synthesis (Cakmak, 2000; Yu et al., 1999), leading to accumulation of H_2O_2 and stimulation of lipid peroxidation in the Zn-deficient tissues (Chen et al., 2008). These processes are summarized schematically in Fig. 7.20.

TABLE 7.19 Root exudation of low-molecular-weight solutes and lipid composition of roots of cotton plants with or without Zn supply or with the resupply of Zn to deficient plants for 12 h (−Zn → +Zn).

	Zn supply		
	+Zn	−Zn	−Zn → +Zn
Root Zn concentration (µg g^{-1} dw)	258	16	121
Root exudates [g^{-1} dw (6 h)$^{-1}$]			
Amino acids (µg)	48	165	94
Sugars (µg)	375	751	652
Phenolics (µg)	117	161	130
K (mg)	1.7	3.7	2.3
Lipid composition			
Phospholipids (mg g^{-1} fw)	2.2	1.5	nd
Fatty acid ratio (saturated/unsaturated)	0.8	0.9	nd

nd, Not determined.
Source: Based on Cakmak and Marschner (1988c).

TABLE 7.20 Zinc concentration in roots and shoots, chlorophyll concentration and superoxide generation, and NADPH oxidation in root extracts of common bean plants with or without Zn supply or with resupply of Zn to deficient plants for 2 days (−Zn → +Zn).

	Zn supply		
	+Zn	−Zn	−Zn → +Zn
Zn concentration (µg g^{-1} dw)			
Roots	44	11	69
Shoots	37	10	71
Chlorophyll (mg g^{-1} dw)	7.4	3.6	4.1
O$_2$$^{•-}$ generation (nmol mg^{-1} protein min^{-1})	2.2	6.6	4.3
NADPH oxidation (nmol mg^{-1} protein min^{-1})	18	61	40

Source: From Based on Cakmak and Marschner (1988a).

7.4.8 Phosphorus-zinc interactions

High application rates of P fertilizers to soils low in available Zn can induce Zn deficiency (*P-induced Zn deficiency*; Robson and Pitman, 1983). Phosphorus-induced Zn deficiency in plants is often associated with reduced root Zn uptake, imbalance between P and Zn in the leaf tissues, and/or reduced mycorrhizal colonization (Cakmak and Marschner, 1986, 1987; Loneragan et al., 1982; Ova et al., 2015). High P supply is usually responsible for a reduction in root growth and a lesser degree of root colonization by arbuscular mycorrhiza (AM) (Ryan et al., 2008; Chapter 15). Both these factors are important for the acquisition of Zn. In wheat, P fertilization reduced grain Zn concentration by 33%–39%, and root colonization with AM by 33%–75% (Ryan et al., 2008). In greenhouse experiments, Ova et al. (2015) showed that high P fertilization of wheat reduced grain Zn concentrations by about 3-fold under low Zn and 1.6-fold under adequate Zn supply conditions. Mycorrhizal colonization of roots can contribute up to 50% of the plant Zn uptake (Coccina et al., 2019; Li et al., 1991; Ova et al., 2015; Watts-Williams et al., 2015). Accordingly, elimination of root mycorrhizal colonization with high P application in native soil, or by soil sterilization in experimental systems, reduces shoot and grain Zn concentration significantly, which can be restored by reducing P applications or adding mycorrhizal fungi to the sterilized soils (Fig. 7.21; Yazici et al., 2021). These findings indicate a particular role of root AM colonization in P-Zn antagonism (Ova et al., 2015; Yazici et al., 2021).

With increasing P concentration in the shoot, Zn deficiency symptoms become more severe, although the Zn concentration is not decreased in plants grown in nutrient solution (Table 7.21; Cakmak and Marschner, 1987; Loneragan et al., 1982). However, the physiological availability of Zn is decreased as indicated by the lower proportions of water-extractable Zn and the lower SOD activity in leaves (Cakmak and Marschner, 1987). A higher P concentration in the shoot may therefore decrease solubility and mobility of Zn both within cells and in long-distance transport to the shoot apex.

In solution culture at high P but low Zn supply, the P-induced Zn deficiency is often associated with high P concentrations and symptoms of P toxicity in the mature leaves (Cakmak and Marschner, 1986; Loneragan et al., 1982; Parker, 1997), which may be mistaken for evidence of exacerbated Zn deficiency because of the high P/Zn ratio. In the absence of Zn, or with low external concentrations, the P concentration in the shoot is high, leading to P toxicity symptoms. In general, a P concentration greater than 20 mg g^{-1} in leaves can be considered toxic.

The high P concentration in the shoots of Zn-deficient plants is also the result of a specific impairment of re-translocation of P in the phloem (Table 7.22) and, thus, of an important "signal" in the regulation of root P uptake by shoot nutritional status. The mechanism by which Zn deficiency impairs re-translocation of P from the shoots is unclear, as in Zn-deficient plants neither the re-translocation of ^{86}Rb nor ^{36}Cl is impaired (Table 7.22). Zinc deficiency has been also found to enhance the abundance of high-affinity P transporter proteins in barley roots irrespective of the plant P

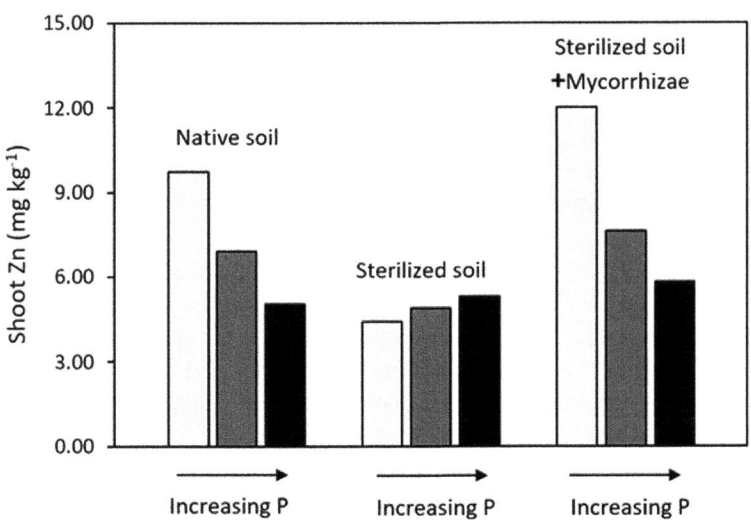

FIGURE 7.21 Effect of increasing P application as Ca-phosphate on shoot Zn concentrations of 61-day-old wheat plants grown with a marginal Zn supply in unsterilized and sterilized soils with and without mycorrhizal inoculum. *Redrawn from Yazici et al. (2021).*

TABLE 7.21 Growth and Zn and P concentrations of the shoots of ochra (*Abelmoschus esculentum* L.) grown in nutrient solution with different Zn and P supplies.

	Dry weight (g plant^{-1})		Zn concentration (μg g^{-1} dw)		P concentration (mg g^{-1} dw)	
P supply (mM)	0.25	2.0	0.25	2.0	0.25	2.0
Zn supply (μM)						
0	8.3	9.5	15	15	11	24
0.25	9.6	9.9	27	27	10	20
1.0	9.8	11.6	54	57	9	12

Source: Based on Loneragan et al. (1982).

TABLE 7.22 Distribution of ^{32}P, ^{86}Rb, and ^{36}Cl between shoots and roots (% of total in plant) of Zn-sufficient and Zn-deficient cotton plants, 19 h after stem application.

Zn supply	^{32}P Shoots	^{32}P Roots	^{86}Rb Shoots	^{86}Rb Roots	^{36}Cl Shoots	^{36}Cl Roots
+Zn	66	34	62	38	29	71
−Zn	92	8	66	34	32	68

Source: From Marschner and Cakmak (1986).

TABLE 7.23 Element concentrations in embryo and protein bodies in the embryo of maize kernels.

	Zn	Fe	Mn	Cu	Ca	K	Mg	P Total	P Phytate
	(μg g^{-1} dw)				(mg g^{-1} dw)				
Embryo	163	186	30	12	449	27	10	30	23
Protein bodies	565	490	170	11	1645	68	44	89	88

Source: From Marschner, Ehret and Haug (unpublished).

nutritional status (Huang et al., 2000). Normally, high plant P downregulates the expression of P transporters, but not under Zn deficiency, leading to high root uptake and shoot accumulation of P (Huang et al., 2000).

7.4.9 Zinc forms and bioavailability in grains

Much is known about the localization and forms of Zn in grains, but less so for vegetative organs. In grains, most Zn and other nutrients are localized in so-called "protein bodies" in the form of discrete particles, the *globoid crystals* (Lott and Buttrose, 1978; Welch, 1986). These globoids mainly consist of phytate, that is, salts of phytic acid (Table 7.23).

In wheat grains, high Zn concentrations (600 μg g^{-1} dw) were also found in the scutellum (Mazzolini et al., 1985). Zinc, Fe, and proteins are generally co-localized within grain tissues (Cakmak et al., 2010), and there is a strong positive correlation between the concentrations of Zn, Fe, and protein in grains of many cereals (Cakmak et al., 2010; Morgounov et al., 2007; Peterson et al., 1986; Zhao et al., 2009), suggesting that proteins are a sink for Zn and Fe. However, a speciation analysis in the barley embryo fraction showed that Fe is bound to phytate, whereas Zn is mainly associated with peptides/proteins (Persson et al., 2009). In sweetcorn kernels, about 79% of Zn in the embryo is present as Zn-phytate, whereas in the endosperm, most Zn is complexed with a N- or S-containing ligand, possibly Zn-histidine or Zn-cysteine (Cheah et al., 2019). Phytate is a negatively charged compound with high affinity for binding divalent cations such as Zn, forming the insoluble or unavailable Zn-phytate complexes in grains (Lönnerdal, 2002; Schlemmer et al., 2009). The strong binding of Zn to phytate is of concern to nutritionists because it reduces the bioavailability of Zn for monogastric animals and humans. Phytate concentrations are low or undetectable in the endosperm, which may indicate that the endosperm Zn is potentially more bioavailable than the Zn present in grain parts rich in phytate such as the embryo and aleurone (Cakmak and Kutman, 2018; Velu et al., 2014).

A negative correlation between phytate concentration and zinc bioavailability was found to occur in soybean products for rats (Lönnerdal, 2002; Zhou et al., 1992). It is possible to reduce the phytate concentration of grains by selection and breeding, or by restricting P supply; however, low phytate concentration in grain is associated with reduced seedling emergence and poor agronomic performance (Oltmans et al., 2005). The formation of phytate is not confined to reproductive organs; therefore, decreased physiological availability of Zn in vegetative plant organs resulting from the binding to phytate may also be important, particularly in the context of P-induced Zn deficiency.

7.4.10 Zinc deficiency and toxicity

7.4.10.1 Zinc deficiency

Zinc deficiency is widespread among plants grown in highly weathered acid soils and in calcareous soils. In the latter case, Zn deficiency is often associated with Fe deficiency ("lime chlorosis"). The low availability of Zn in high-pH soils and diminished root Zn uptake is mainly due to the factors influencing adsorption and immobilization of Zn in soils containing clay minerals, Fe oxides, or $CaCO_3$ (Moreno-Lora and Delgado, 2020; Trehan and Sekhon, 1977). In addition, Zn uptake and translocation to the shoot are inhibited by high concentrations of bicarbonate (HCO_3^-) (Forno et al., 1975). This effect is very similar to the effect of HCO_3^- on Fe uptake and translocation. In contrast to Fe deficiency, however, Zn deficiency in plants grown in calcareous soils can be corrected quite readily by the application of inorganic Zn salts such as $ZnSO_4$ to the soil (Nayyar and Takkar, 1980; Cakmak et al., 1996a).

The most characteristic visible symptoms of Zn deficiency in dicotyledonous plants are stunted growth due to shortening of internodes ("rosetting") and a drastic decrease in leaf size ("little leaf"), as shown in Fig. 7.22 for apple trees. Under severe Zn deficiency, the shoot apices die as shown, for example, in forest plantations in South Australia, and these symptoms are usually more severe at high light intensity than in partial shade (Boardman and McGuire, 1990). Similarly, plants are more susceptible to low Zn supply when exposed to low temperature, heat, or drought stresses (Bagci et al., 2007; Cakmak et al., 1995b; Peck and McDonald, 2010). Enrichment of seeds with Zn using a seed priming approach (Imran et al., 2017) or agronomic biofortification in the field using foliar Zn sprays (Candan et al., 2018) contributes positively to seedling growth and vigor under salt and drought stress, respectively.

In cereals, such as wheat, typical symptoms of Zn deficiency are reduction in shoot elongation and development of whitish-brown necrotic patches on the middle leaves, whereas the young leaves remain yellowish green and show no necrotic lesions (Cakmak et al., 1996a). Symptoms of chlorosis and necrosis in the older leaves of Zn-deficient plants are often secondary effects caused by P or B toxicity, or by photooxidation associated with the impaired export of photosynthates.

When plants are Zn deficient, shoot growth is usually inhibited more than the root growth (Zhang et al., 1991a), and root growth may even be enhanced at the expense of the shoot growth (Cakmak et al., 1996b; Cumbus, 1985). Zinc deficiency increases root exudation of low-molecular-weight solutes. In dicotyledonous plants, the main root exudates are amino acids, sugars, phenolics, and potassium (Table 7.19), whereas in graminaceous species, the main exudates are PSs (Zhang et al., 1991a) released in a distinct diurnal pattern (Cakmak et al., 1994; Zhang et al., 1991b), also typical for Fe deficiency (Chapters 2 and 14). Increased release of PSs under Zn or Fe deficiency is regulated separately and is not related to a Zn deficiency–induced disturbance of Fe metabolism in plants (Suzuki et al., 2006).

In leaves, the critical deficiency concentrations are below $15-20\ \mu g\ Zn\ g^{-1}$ dw (but see Section 7.4.4). Grain yield is depressed to a greater extent by Zn deficiency than the total biomass production, probably due, at least in part, to the impaired pollen fertility in deficient plants. Plant species differ in their sensitivity to Zn deficiency, with maize, rice, and apples being sensitive. Among the cereal species, rye has the highest tolerance to Zn deficiency, followed by triticale, barley, bread wheat, oats, and durum wheat (Cakmak et al., 1997a).

7.4.10.2 Zinc toxicity

Zinc toxicity is observed very rarely in crop plants and occurs mainly in soils contaminated by mining and smelting activities or treated with sewage sludge (Broadley et al., 2007). At a very high Zn supply, Zn toxicity can be induced readily in nontolerant plants, with inhibition of root elongation being a sensitive indicator (Godbold et al., 1983; Ruano et al., 1988). Quite often, Zn toxicity leads to chlorosis in young leaves. This may be an induced deficiency of Fe, because of the similar ionic radii of Zn^{2+} and Fe^{2+} and their competition during root uptake and shoot transport as well

FIGURE 7.22 Symptoms of Zn deficiency in apple with typical inhibition of internode elongation ("rosetting") and reduction in leaf size ("little leaf"). *Photos by V. Römheld and I. Cakmak.*

as in the chelation process (Cakmak, 2000; Kaur and Garg, 2021). Excess Zn also has a significant inhibitory effect on both root uptake and shoot transport of Mn, and also the transport of Mg from roots to shoots (Simon et al., 2021).

In common bean plants, Zn toxicity inhibits photosynthesis through several mechanisms. Decreased RuBP carboxylase activity is likely caused by competition with Mg (Van Assche and Clijsters, 1986a). Excess Zn can also inhibit PSII activity by replacing Mn in the thylakoid membranes (Van Assche and Clijsters, 1986b). In the thylakoid membranes of control plants, six atoms each of Mn and Zn are bound per 400 chlorophyll molecules, but when plants are exposed to excess Zn this proportion shifts to 2 Mn and 30 Zn atoms. A decrease in photosynthetic activity of plants exposed to toxic Zn concentrations can also be ascribed to the effect of excess Zn on root hydraulic conductivity. This is probably due to changes in the xylem structure, leading to detrimental effects on the photosynthetic activity by inducing water deficiency stress and/or reducing stomatal conductance (Fatemi et al., 2020; Kaur and Garg, 2021).

The critical toxicity concentrations in leaves of crop plants range from 100 μg Zn g^{-1} dw (Ruano et al., 1988) to more than 300 μg Zn g^{-1} (Chaney, 1993), the latter values being more typical. Increasing soil pH by liming is the most effective strategy to decrease Zn concentration and Zn toxicity in plants (White et al., 1979).

7.4.10.3 Tolerance to Zn toxicity

The mechanisms responsible for Zn tolerance have been of major interest in ecophysiology for a long time (Verkleij and Schat, 1989). Zinc tolerance is also of interest in agriculture because Zn is the metal occurring in the highest concentrations in a majority of wastes arising in modern, industrialized communities (Boardman and McGuire, 1990; Hall, 2002).

The principal mechanisms of metal tolerance are illustrated in Fig. 7.14 and reviewed comprehensively by Hall (2002). A detailed proteomic, metabolomic, and ionomic analysis of the plant responses to excess metals was published by Singh et al. (2016). A particular mechanism of exclusion may exist in forest tree species and annual crops where certain mycorrhizal fungi strongly increase the Zn tolerance of the host plant through the retention of Zn in the fungal cell walls and the chelation and sequestration of Zn in the fungal vacuoles (Colpaert and van Assche, 1992; Ferrol et al., 2016; Shi et al., 2019; Chapter 15).

Vacuolar sequestration of Zn in plant cells is a well-documented tolerance mechanism to Zn toxicity (Verbruggen et al., 2009; Sharma et al., 2016). For example, when a Zn-tolerant clone of *Deschampsia caespitosa* is exposed to a high rhizosphere Zn concentration, Zn tolerance is achieved mainly by sequestering Zn in the vacuole, whereas in a sensitive clone, Zn is accumulated in the cytoplasm (Table 7.24). In the tolerant clone the Zn concentration in the cytoplasm remains low, protecting cellular metabolism. Vacuolar membranes possess a metal tolerance protein that transports Zn from the cytosol to the vacuole, preventing excessive Zn accumulation in the cytoplasm (Gustin et al., 2009; Kawachi et al., 2009; Sharma et al., 2016). There are positive correlations in tolerant genotypes between the accumulation of organic acid anions, such as malate and citrate, and the accumulation of Zn, suggesting that complexation of Zn with organic acid anions in the vacuoles may be an important mechanism of Zn tolerance (Godbold et al., 1983, 1984; Sinclair and Krämer, 2012; Singh et al., 2016).

There is increasing evidence showing the important roles of phytochelatins in plant tolerance to Zn toxicity. Arabidopsis mutants deficient in phytochelatins were highly susceptible to Zn toxicity (Kühnlenz et al., 2016; Tennstedt et al., 2009). In nonvacuolated meristematic tissues such as root apices, other tolerance mechanisms must be involved, such as sequestration of Zn by binding to phytate as observed in a Zn-tolerant ecotype of *D. caespitosa* (Van Steveninck et al., 1987).

TABLE 7.24 Concentrations of Zn in the cytoplasm and vacuoles of roots of a Zn-tolerant and a Zn-sensitive clone of *Deschampsia caespitosa* grown with a low or high Zn supply.

Zn supply (mM Zn)	Bound Zn in the cytoplasm (mM)		Soluble Zn in the vacuole (mM)	
	Sensitive	Tolerant	Sensitive	Tolerant
0.10	7.1	11	3.7	5.3
0.75	33	6.2	2.1	33

Source: Based on Brookes et al. (1981).

7.5 Nickel

7.5.1 General

Nickel is chemically related to Fe and Co. Its preferred oxidation state in biological systems is Ni^{2+} [Ni (II)], but it can also exist in the redox states Ni(I) and Ni(III) (Cammack et al., 1988). Nickel forms stable complexes, for example, with histidine, cysteine, and citrate (Thauer et al., 1980), and in Ni enzymes, it is coordinated to various ligands (Li and Zamble, 2009). As a metal cofactor of enzymes in biological systems, Ni is classified as a redox-active element like Fe and Cu (Andreini et al., 2008). In the Irving–Williams series, which ranks the biologically relevant divalent metal cations with respect to their affinities to proteins, Ni^{2+}, together with Zn^{2+}, is one of the two most competitive metal ions after Cu^{2+}, and, therefore, the cytoplasmic concentration of free Ni ions must be maintained at extremely low levels to avoid toxicity issues due to mismetallation (Foster et al., 2014). Nickel delivery to Ni-dependent enzymes in biological systems is either assisted by specific metallochaperones or facilitated by the preassembly of complex metal cofactors. Even though Ni is thought to have been a relatively more prevalent cofactor in early anaerobic life, it is one of the least common metal cofactors in extant species, and Ni was the most recent element to be recognized as an essential micronutrient for vascular plants (Brown et al., 1987a,b; Foster et al., 2014; Freitas et al., 2018).

The first clear evidence for the function of Ni in urease in higher plants was provided by Dixon et al. (1975). Later, a requirement of Ni in legumes (Eskew et al., 1984) and subsequently in a number of nonlegumes grown with various N sources was demonstrated (Brown et al., 1987a,b). In barley seeds from plants grown with low-Ni supply, there was a close relationship between Ni concentration, viability, germination rate, and seedling vigor (Brown et al., 1987a). This relationship is shown for germination rate in Fig. 7.23. In this study, viability of the Ni-deficient seeds could not be restored by soaking the seeds in a solution containing Ni, demonstrating that Ni is essential for normal seed development in the maternal plants and, thus, for completing the life cycle of the barley plant. Under field conditions, Ni deficiency in crops was discovered in pecan (*Carya illinoiensis*) trees growing in sandy, poorly draining soils with low cation exchange capacity in south-eastern United States (Wood et al., 2004).

In soils, the availability of Ni is affected by several factors including pH, organic matter content, redox potential, and temperature. Plants can take up Ni from the available Ni pool, which includes soluble and exchangeable Ni, in the form of Ni^{2+} as well as Ni complexes (He et al., 2012). Nickel is taken up by roots via transport processes that require energy directly or indirectly, with the relative importance of each dependent on Ni availability (Dalir and Khoshgoftarmanesh, 2015; He et al., 2012). In wheat, the root plasma membrane H^+-ATPase activity was stimulated by Ni and implicated in root Ni uptake (Dalir et al., 2017). Among biological chelators of Ni, histidine has high affinity for Ni. Both exogenously supplied and endogenously produced histidine are known to enhance root uptake and root-to-shoot translocation of Ni, although direct uptake of a Ni–histidine complex has not been proven unambiguously and may be possible in some species (Dalir and Khoshgoftarmanesh, 2015; Kerkeb and Krämer, 2003).

The specific transporters for root Ni uptake in plants have not been discovered yet. However, studies on both hyperaccumulator and nonhyperaccumulator species indicated that Zn transport systems contribute to Ni uptake by roots

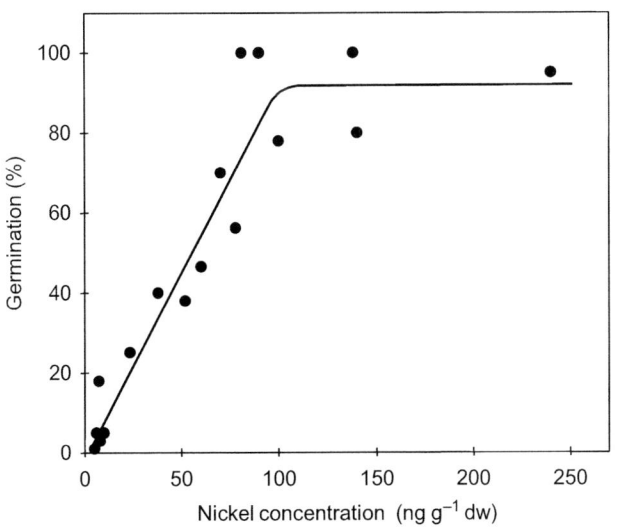

FIGURE 7.23 Relationship between Ni concentration in seeds and germination percentage in barley. *Redrawn from Brown et al. (1987a).*

(Assunção et al., 2008; Nishida et al., 2015). Nickel uptake by roots was shown to be competitively inhibited by other divalent micronutrient cations, including Zn^{2+} and Cu^{2+}, and enhanced by Zn deficiency (Korner et al., 1986; Nishida et al., 2015).

7.5.2 Ni-containing enzymes

Nickel is involved in the function of at least nine enzymes (Li and Zamble, 2009), including methyl-coenzyme M reductase, SOD, Ni-dependent glyoxalase (GLY), aci-reductone dioxygenase, NiFe-hydrogenase, carbon monoxide dehydrogenase, acetyl-CoA decarbonylase synthase, methyleneurease, and urease. Urease has long been considered the only Ni-containing enzyme in vascular plants. In addition, a urease accessory protein called Eu3, possibly a Ni chaperone, was shown to be a Ni-binding protein (Freyermuth et al., 2000). However, symptoms and physiological consequences of Ni deficiency suggest that additional essential roles are likely beyond those associated with urease (Bai et al., 2006; Brown et al., 1990). The suggestion that urease is the only Ni-containing enzyme in plants has been challenged by the demonstration of the strict Ni dependence of a GLY I isozyme in rice (Mustafiz et al., 2014).

Urease isolated from jack bean (*Canavalia ensiformis*) has a molecular weight of 590 kDa and consists of six subunits, each containing two Ni atoms (Dixon et al., 1980). In the subunits, Ni is coordinated to N and O ligands, and it is possible that one of the Ni–O bonds can be displaced by water molecules during hydrolytic reactions.

In the dinuclear active site of urease, the two coordinated Ni ions are bridged via three O ligands provided by a lysine residue and a hydroxy group, and in contrast to the Ni centers of several other Ni-containing enzymes, such as CO dehydrogenase, methyl-coenzyme M reductase, and hydrogenase, the Ni ions of urease do not change their redox states during catalysis (Thauer, 2001).

Nickel is not required for the synthesis of the urease protein (Winkler et al., 1983) but, as the metal component, is essential for the structure and catalytic function of the enzyme (Klucas et al., 1983). In soybean, there is an embryo-specific urease (Eu1, also known as the seed urease), which has relatively high activity, and a ubiquitous urease (Eu4), with relatively low activity but found in all vegetative tissues (Fabiano et al., 2015; Freyermuth et al., 2000). Soybean *Eu3* encodes a plant UreG that is homologous to the bacterial urease accessory protein and is thought to stabilize urease by Ni insertion; it is necessary for the accumulation of functional Eu1 in embryos as well as the activation of Eu4 in nonembryonic tissues (Freyermuth et al., 2000).

The GLY system has been studied extensively in microorganisms and has recently attracted increasing attention in plants. It is responsible for the detoxification of the cytotoxic metabolite methylglyoxal (MG) through the concerted action of GLY I and GLY II (Fabiano et al., 2015; Singla-Pareek et al., 2020). It has been established unambiguously that MG concentrations rise under stress conditions, and that the GLY system is critical for stress tolerance of plants. In agreement with this, the Ni-dependent GLY I in rice (OsGLYI-11.2) was shown to be inducible by MG, resulting in enhanced stress tolerance when heterologously expressed in *Escherichia coli* or tobacco (Mustafiz et al., 2014). It has also been suggested that Ni may play an important role in abiotic stress tolerance of plants due to its involvement in urea turnover (Polacco et al., 2013).

In hydrogenases from sulfate-reducing, photosynthetic and hydrogen-oxidizing bacteria (thus also the hydrogenases of rhizobia) (Chapter 16), Ni is associated with Fe-S clusters (Li and Zamble, 2009). *Rhizobium* and *Bradyrhizobium* produce hydrogen-uptake hydrogenase when free-living and as bacteroids in the root nodules (Maier et al., 1990). In free-living rhizobia, without Ni supply, the hydrogenase activity is low but can be restored within 3 h by resupplying Ni (Maier et al., 1990). In soil-grown soybean, Ni fertilization was shown to increase the number of nodules, nodule dry weight and the nitrogenase activity in nodules and, thereby, improve the N status of host plants (Freitas et al., 2019).

7.5.3 Role of Ni in N metabolism

In plant species such as zucchini (Gerendas and Sattelmacher, 1997), canola (Arkoun et al., 2013), and soybean (Kutman et al., 2014), Ni deficiency was reported to reduce growth, result in N-deficient phenotypes, decrease tissue amino acid concentrations, and lower the rate of urea uptake when plants were grown with urea as the sole N source in the nutrient solution. Despite the essentiality of Ni for all vascular plants, hydroponics experiments on several different crop species documented that an adequate Ni supply was critical for urea-fed plants but had little or no effect on plants grown with ammonium and/or nitrate (Gerendas et al., 1998; Gheibi et al., 2009; Tan et al., 2000).

Foliar application of urea is often associated with urea toxicity. The utilization of foliar urea was impaired in low-Ni plants, which exhibited urea toxicity symptoms (Krogmeier et al., 1991; Kutman et al., 2014). The severity of toxicity symptoms is related to the Ni nutritional status of the plants as shown in Table 7.25 for soybean (Krogmeier et al., 1991).

In plants grown without a Ni supply, urease activity in leaves was low, and foliar application of urea led to the accumulation of urea and severe necrosis of the leaf tips. By contrast, when plants were supplied with Ni, urease activity increased and urea accumulation and necrosis decreased. In addition to Ni applied to the growth medium and the foliage, seed Ni reserves can also make a significant contribution to alleviating the toxicity of foliar urea in soybean (Fig. 7.24).

Urea accumulated in leaves of soybean and cowpea, regardless of the form of N nutrition (urea, ammonium, nitrate, N_2 fixation), and symptoms of leaf tip necrosis were observed when plants were grown without a Ni supply (Eskew et al., 1984). As shown in Table 7.26, there was an accumulation of urea (up to 3% of the dry weight) towards the tip

TABLE 7.25 Leaf tip necrosis, urea concentration, and urease activity in soybean plants with and without Ni supply and with three rates of foliar application of urea.

Ni supply ($\mu g\ L^{-1}$)	Foliar application (mg urea leaf^{-1})	Leaf tip necrosis (%)	Urea concentration ($\mu g\ g^{-1}$ dw)	Urease activity ($\mu mol\ NH_3\ g^{-1}$ dw h^{-1})
0	0	<0.1	64	2.2
	3	5.2	1038	2.7
	6	14	6099	2.4
100	0	0	0	12
	3	2.0	299	11
	6	3.5	1583	10

Source: Based on Krogmeier et al. (1991).

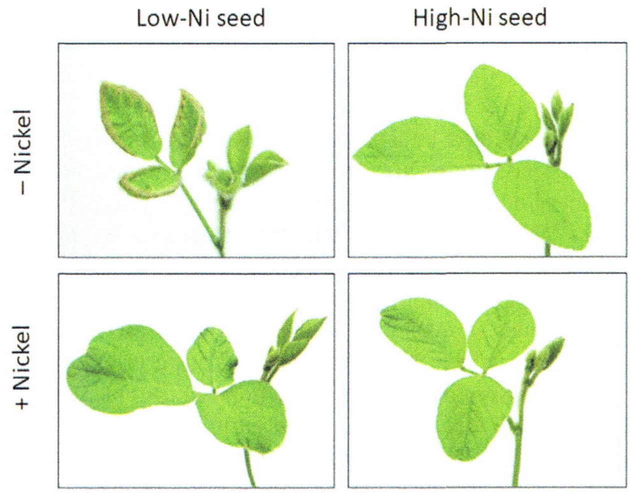

FIGURE 7.24 Toxicity symptoms in growing parts of soybean plants sprayed with urea (2% w/v) raised from low-Ni or high-Ni seeds with or without the addition of Ni to nutrient solution. *Modified from Kutman et al. (2013).*

TABLE 7.26 Concentrations of urea, ureides, and Ni in different parts of mature leaves of cowpea (*Vigna unguiculata*) supplied with NH₄NO₃ with or without Ni supply.

	Urea (μmol g^{-1} dw)		Ureides (μmol g^{-1} dw)		Ni (μg g^{-1} dw)	
	+Ni	−Ni	+Ni	−Ni	+Ni	−Ni
Petiole	0.11	0	nd	nd	nd	nd
Blade base	0.56	18	3.6	4.5	3.7	0.1
Blade tip	2.16	238	nd	nd	nd	nd

nd, Not determined.
Source: Based on Walker et al. (1985).

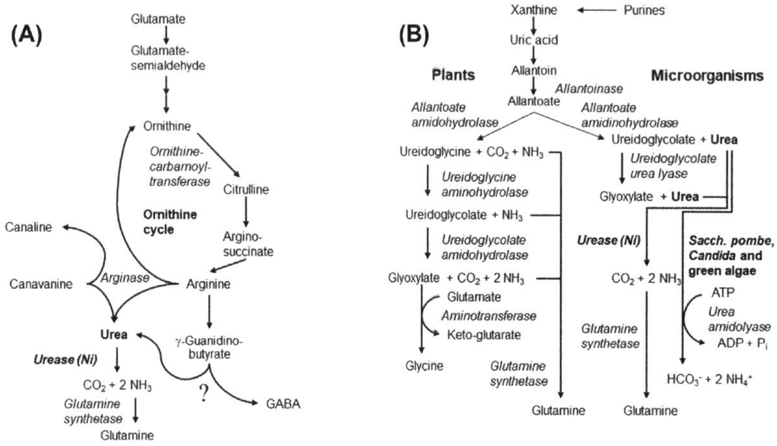

FIGURE 7.25 Urea formation pathways in plants (A) and ureide breakdown routes in plants and microorganisms (B). *Modified from Gerendas et al. (1999).*

of the leaf blade in Ni-deficient plants, suggesting an internal generation of urea. Similarly, wheat, barley, and oat seedlings grown from low-Ni seeds without Ni supply and fed with nitrate were reported to accumulate urea and show severe leaf tip necrosis (Brown et al., 1987b).

Root and shoot growth were significantly lower in the Ni-deficient plants, which were less green, developed interveinal chlorosis and necrosis, and the terminal 2 cm of the leaves failed to unfold. In plants, urea is a normal metabolite that has to be kept at a low concentration to prevent toxicity. Various pathways of urea biosynthesis have been proposed in plants (Fig. 7.25). The ornithine cycle for urea biosynthesis is likely to be of general importance, as well as specifically in urea formation during protein degradation (e.g., in mature leaves) at onset of reproductive growth (Eskew et al., 1984) and in germination of legume seeds (Horak, 1985). Arginine degradation by arginase, which generates ornithine and urea in the mitochondrial matrix, is considered the only confirmed enzymatic source of endogenously produced urea in plants, although some intermediates of ureide catabolism pathway are unstable and can rapidly decay nonenzymatically to generate urea (Werner et al., 2013; Witte, 2011).

In nodulated legumes such as soybean, ureides are the dominant form of N transported to the shoots (Chapters 3 and 16) where they are decomposed to NH₃ and CO₂, most likely independently of urea metabolism. Even though urease-independent ureide catabolism may suggest that the Ni requirement of nodulated soybean is not particularly high simply because soybean is a ureide-transporting species (Winkler et al., 1988), the role of Ni as the cofactor of the uptake hydrogenases of rhizobia implies the opposite. In bacteroids from pea nodules, the hydrogenase activity was shown to be limited by the availability of Ni in the soil (Ureta et al., 2005). There is mounting evidence that an adequate Ni supply is required for efficient biological N₂ fixation in nodulated legumes (Freitas et al., 2019; Lavres et al., 2016). In soybean, ureide concentrations were low and unaffected by Ni supply, which was also true for free purines and uric acid (Walker et al., 1985), whereas in pecan, which is also a ureide-transporting species, Ni deficiency disrupted ureide metabolism, resulting in marked accumulation of xanthine, allantoic acid, ureidoglycolate, and citrulline while reducing the concentrations of total ureide and urea as well as the urease activity (Bai et al., 2006). Whether ureide catabolism in vascular plants involves urea or not is still a controversial topic, but increasing evidence indicates the presence of an enzymatic degradation pathway that can convert allantoate to glyoxylate without generating urea as a by-product in *Arabidopsis thaliana* and other species (Werner et al., 2013; Witte, 2011) (Fig. 7.25). Changes in the concentration of organic acids and other metabolites may result from secondary events of disturbances in N metabolism in Ni-deficient plants (Bai et al., 2006; Brown et al., 1990), although it is not clear if these effects of Ni deficiency are directly related to the function of Ni as the cofactor of urease.

7.5.4 Nickel concentration in plants

The Ni concentration in plants grown on uncontaminated soil ranges from 0.05 to 5.0 μg g^{-1} dw (Brooks, 1980; Welch, 1981). The adequate range for Ni is between 0.01 and >10 μg g^{-1} dw, which is a wide range as compared to other elements (Brown et al., 1987a; Gerendas et al., 1999). This range mainly reflects the differences between plant species in uptake and root-to-shoot transport of Ni (Rebafka et al., 1990). The critical Ni concentration required for seed germination in barley, shoot growth in oats, barley, and wheat, and shoot growth of urea-fed tomato, rice, and zucchini was 100 ng g^{-1} dw (Brown et al., 1987a,b; Gerendas and Sattelmacher, 1997).

7.5.5 Nickel deficiency and toxicity

In pot or container studies, marked responses to foliar and/or soil applications of Ni have been observed particularly, though not exclusively, for various annual crops fed with urea as the sole N source (Gheibi et al., 2009; Kutman et al., 2014; Nicoulaud and Bloom, 1998) as well as ureide-transporting tree crops, including pecan and river birch, irrespective of the N source (Bai et al., 2006; Ruter, 2005). Examples of ureide-transporting woody crop genera are *Annona*, *Carya*, *Diospyros*, *Juglans*, and *Vitis* (Brown, 2008). In addition, crops treated with foliar urea (Eskew et al., 1984; Kutman et al., 2013) and legumes that depend on biological N$_2$ fixation (Freitas et al., 2018) may have an increased Ni requirement and, therefore, exhibit reduced growth, N deficiency−like symptoms and symptoms of urea toxicity under inadequate Ni supply. In pecan, the tissue Fe:Ni ratio is an important determinant of the severity of physiological Ni deficiency, and soil or foliar applications of chelated Fe at budbreak can aggravate Ni deficiency (Wood, 2013).

In graminaceous species, Ni deficiency symptoms include chlorosis similar to that induced by Fe deficiency (Brown et al., 1987a,b), with interveinal chlorosis and patchy necrosis in the youngest leaves. Nickel deficiency also results in a marked acceleration of plant senescence and a decrease in tissue Fe concentrations. In pecan as well as river birch, Ni deficiency results in deformed leaves, a symptom referred to as "mouse-ear" (Wood et al., 2004; Ruter, 2005). The accumulation of urea in leaf tips can be used to detect Ni deficiency in both monocotyledonous and dicotyledonous plants (Eskew et al., 1984).

In crop plants there is generally more concern about Ni toxicity than Ni deficiency. Nickel toxicity may occur as a result of industrial pollution or after applications of high-Ni sewage sludge and excessive applications of manure (Brown et al., 1989; Lešková et al., 2020). Critical toxicity concentrations in crop species are in the range of >10 μg g^{-1} dw in the sensitive to >50 μg g^{-1} dw in the moderately tolerant species (Asher, 1991). In wheat, the critical toxicity levels increased from 63 to 112 μg g^{-1} dw with an increasing supply of urea, suggesting a decreased toxicity of Ni in urea-fed plants (Singh et al., 1990). However, cucumber plants were more severely affected by Ni toxicity when fed with urea rather than NH$_4$NO$_3$ (Khoshgoftarmanesh and Bahmanziari, 2012). In early stages of Ni toxicity, there may be no clear symptoms, although shoot and root growth are decreased. When plants suffer Ni toxicity, leaves tend to accumulate carbohydrates, which may be explained by a decreased sink activity due to inhibition of root and shoot growth (Molas, 2002; Roitto et al., 2005). In Italian catchfly (*Silene italica*), a Ni excluder, root growth was severely inhibited at 7.5 μM Ni, and this effect could be mostly reversed by an adequate Ca supply (Gabbrielli et al., 1990). Nickel was also shown to inhibit root mitotic activity in maize (L'Huillier et al., 1996) and lateral root formation in cowpea (Kopittke et al., 2007). Inhibition of root elongation and gravitropic defects caused by Ni toxicity in arabidopsis were attributed to alterations in cell wall−associated processes and inhibition of polar auxin transport (Lešková et al., 2020). In plants affected by Ni toxicity, leaves may exhibit chlorosis due to excessive Ni accumulation and/or Ni-induced Fe deficiency and eventually turn necrotic (Kopittke et al., 2007; Lešková et al., 2020). Excessive Ni supply can impair root uptake and/or root-to-shoot translocation of different macro- and micronutrients, depending on the species and the mineral concentrations (Yang et al., 1996). Phytotoxic concentrations of Ni in growth medium were reported to decrease the leaf and achene concentrations of Fe, Zn, Mn, and Cu in a dose-dependent manner in sunflower (Ahmad et al., 2011) and the shoot concentrations of the same micronutrients in barley (Rahman et al., 2005).

7.5.6 Tolerance to Ni toxicity

Serpentine (ultramafic) soils are usually high in Fe, Mg, Ni, Cr, and Co but low in Ca. The flora on these soils includes many species exhibiting hyperaccumulation of Ni (e.g., of the genus *Alyssum*), in which the Ni concentration in the leaves may reach 10–30 mg g^{-1} dw (Homer et al., 1991). Nickel hyperaccumulators have attracted much attention not only because their physiology can provide insight into Ni uptake, transport, and tolerance mechanisms but also because they can be used for phytoremediation of Ni-contaminated soils and phytomining of Ni (He et al., 2012). In typical Ni hyperaccumulators, the Ni concentration decreases in the order: leaves > stems > roots, but storage of Ni in senescing root tissues and root bark can also contribute to Ni tolerance (Rabier et al., 2008). Complexation of Ni with organic acids, particularly malic and citric acid, minimizes the free Ni^{2+} concentration in the cytoplasm and is one of the best described mechanisms of Ni detoxification in hyperaccumulators (Homer et al., 1991; Montargès-Pelletier et al., 2008). Other commonly reported Ni-tolerance mechanisms include vacuolar sequestration, apoplastic adsorption, and complexation by phytochelatins and metallothioneins (Eapen, D'souza, 2005; Krämer et al., 2000).

In agricultural crops, the adverse effects of Ni toxicity can be alleviated by different soil amendments and plant growth regulator (PGR) treatments. Organic amendments, including manure, compost, and biochar, improved growth parameters of maize grown in soil containing a toxic Ni concentration, decreasing the shoot Ni concentration, and increasing the concentrations of other metallic micronutrients (Rehman et al., 2016). The oxidative stress caused by Ni toxicity can be ameliorated by applications of PGRs such as salicylic acid, jasmonic acid, nitric oxide, and melatonin, which are reported to increase the activities of antioxidative enzymes, reduce lipid peroxidation and limit the accumulation of ROS in crops, including tomato, canola, and soybean (Jahan et al., 2020; Kazemi et al., 2010; Sirhindi et al., 2016). Exogenous applications of such PGRs may also help plants cope with Ni toxicity by altering gene expression, improving photosynthetic parameters, enhancing the production of secondary metabolites with antioxidant activity, and increasing the accumulation of compatible solutes.

Colonization of roots by arbuscular mycorrhizal fungi can enhance Ni tolerance of plant species, including, but not limited to, serpentine endemic species, by decreasing Ni uptake or Ni translocation to the shoot as well as by improving the uptake of growth-limiting nutrients (Shabani and Sabzalian, 2016; Vivas et al., 2006). In soils underneath the canopy of trees that hyperaccumulate Ni, there is a greater proportion of Ni-resistant bacteria than beyond the canopy, indicating a high rate of Ni cycling in the micro-ecosystem of these trees (Schlegel et al., 1991). Soil Ni availability is the main determinant of the bacterial community diversity in the rhizosphere of Ni hyperaccumulator plants (Lopez et al., 2017). Plant growth–promoting rhizobacteria can not only enhance Ni tolerance and promote growth under Ni-toxic conditions by altering the production of IAA and ethylene biosynthesis (Lucisine et al., 2014; Ma et al., 2009), but also increase Ni uptake by increasing the available Ni concentrations in soils via, for example, siderophore release or P solubilization (Abou-Shanab et al., 2006; Kumar et al., 2008).

7.6 Molybdenum

7.6.1 General

Molybdenum is a transition element and is present in small amounts in the lithosphere (average 2.4 mg kg^{-1}) and in soils (ranging from 0.2 to 36 mg kg^{-1}) (Barber, 1984). In aqueous solution with a pH >4.3, Mo occurs mainly as the molybdate oxyanion, MoO$_4^{2-}$, in its highest oxidized form [Mo(VI)]. At lower pH (<4.3), protonated species [HMoO$_4^-$, MoO$_3$(H$_2$O)$_3$] become the prevailing forms. At high concentrations (>10^{-4} M) and low pH, molybdate can polymerize, but this is unlikely to occur in soil solution because soluble Mo is usually <10^{-6} M (Smith et al., 1997). Due to its electron configuration, Mo(VI) shares many chemical similarities with vanadium and, particularly, tungsten. In fact, many anaerobic archaea and some bacteria require tungsten, but not Mo (Schwarz et al., 2009). Several properties of the molybdate anion MoO$_4^{2-}$ also resemble those of the sulfate anion (SO$_4^{2-}$), which has important implications for Mo availability in soils and uptake by plants. In long-distance transport in plants, Mo is readily mobile in the xylem and phloem (Kaiser et al., 2005; Kannan and Ramani, 1978). The form in which Mo is translocated is unknown, but its chemical properties indicate that it is most likely transported as MoO$_4^{2-}$.

The requirement of plants for Mo is lower than for any of the other nutrients. The biological functions of Mo are related to the valency changes it undergoes as a metal component of enzymes, whereby Mo shuttles between three oxidation states (+4, +5, and +6) and catalyzes the two-electron transfer reactions (Schwarz et al., 2009). In higher plants, only a few enzymes have been found to contain Mo as a cofactor, including nitrate reductase, xanthine dehydrogenase, aldehyde oxidase, and sulfite reductase. In addition, Mo is a cofactor of nitrogenase in N$_2$-fixing bacteria.

The functions of Mo are therefore closely related to N metabolism, and the Mo requirement depends strongly on the mode of N supply.

7.6.2 Molybdenum uptake and transport

A molybdate-specific transporter has been identified in arabidopsis (Baxter et al., 2008; Tomatsu et al., 2007). This transporter, MOT1 (later renamed as MOT1;1), has a high affinity for MoO_4^{2-} ($K_m = 20$ nM) as observed in an uptake assay with yeast expressing the *MOT1;1* gene (Tomatsu et al., 2007). MOT1;1 belongs to the sulfate transporter superfamily but does not appear to transport sulfate. It is expressed in both roots and leaves, and the protein appears to be localized in the mitochondria (Baxter et al., 2008). Mutants lacking MOT1;1 had markedly decreased Mo concentrations in roots and shoots (Tomatsu et al., 2007). Natural variation in Mo accumulation among different accessions of *Arabidopsis thaliana* is, to a large extent, related to the expression level of *MOT1;1* (Baxter et al., 2008). A close paralogue of MOT1;1 in arabidopsis, MOT1;2, is localized at the tonoplast and may play a role in the export of Mo from leaf vacuoles for redistribution to other organs (e.g., grains), especially during leaf senescence (Gasber et al., 2011).

The rice OsMOT1;1 controls the accumulation of Mo in roots, shoots and grains (Huang et al., 2019; Yang et al., 2018). Similar to arabidopsis, natural variation in grain Mo concentration within rice germplasm is attributed to differences in the expression of *OsMOT1;1*, which is likely caused by variation in the promoter sequence (Huang et al., 2019; Yang et al., 2018). Knockout of *OsMOT1;1* leads to hypersensitivity to low pH of the growth medium, in which Mo is poorly available (Huang et al., 2019). In the wild legume *Lotus japonicus*, LjMOT1 is involved in the uptake of Mo by roots (Duan et al., 2017). In barrel medic (*Medicago truncatula*), MtMOT1.2 and MtMOT1.3 appear to be responsible for the transport of Mo into the nodule cells (Gil-Diez et al., 2019; Tejada-Jiménez et al., 2017).

In addition to MOT1, it is likely that some nonspecific transporters also contribute to Mo uptake by plants, particularly sulfate transporters. For example, the high-affinity sulfate transporter from the tropical legume *Stylosanthes hamata*, SHST1, mediates MoO_4^{2-} influx to *SHST1*-expressing yeast cells from the external medium containing nanomollar concentrations of molybdate (Fitzpatrick et al., 2008). There are also numerous reports in the literature showing that molybdate uptake is suppressed by sulfate (reviewed by Macleod et al., 1997). The effect of sulfate can be twofold: a direct competition for the transporters and regulation of the expression of sulfate transporter genes by plant S status. In field-grown wheat, S deficiency resulted in greatly increased transcript abundance of sulfate transporters such as *TaSultr1;1* and *TaSultr4;1* but not *TaSultr5;2* which is the wheat homolog of the *Arabidopsis MOT1*. Molybdenum concentrations in leaves and ears of S-deficient wheat plants were about double those in S-sufficient plants (Shinmachi et al., 2010).

In oilseed rape grown in hydroponic culture, uptake of Mo was increased not only by S deficiency, but also by deficiency of Fe, Mn, Zn, Cu, or B (Maillard et al., 2016). Increased Mo uptake could be explained by increased expression of *BnMOT1* in the Zn- and Cu-deficient plants, and by increased expression of *BnSultr1;1* and *BnSultr1;2* in the S-, Fe-, Mn-, and B-deficient plants.

7.6.3 Nitrogenase

Nitrogenase is the key enzyme complex unique to all N_2-fixing microorganisms. It consists of two Fe proteins, one of which is the FeMo protein containing two unique metal centers, the P-cluster (8Fe-7S) and the FeMo cofactor (Mo-7Fe-9S-X-homocitrate cluster, where X may be C, O, or N) (Schwarz et al., 2009).

Details of the structural arrangement and catalytical functions of Mo in nitrogenase are discussed in Chapter 16. Some free-living diazotrophic bacteria (e.g., *Azotobacter chroococcum*) possess, in addition to the Mo-nitrogenase, a nitrogenase in which Mo is replaced by vanadium (Dilworth et al., 1988).

TABLE 7.27 Leaf N concentration and seed yield of nonnodulated and nodulated soybean (*Glycine max*) at different rates of Mo and N supply.

	N concentration (mg g^{-1} dw)		Grain yield (t ha^{-1})	
Mo supply (g Mo ha^{-1})	0	34	0	34
N supply (kg N ha^{-1})				
Nonnodulated				
0	31	36	1.7	1.6
67	46	47	2.7	2.7
134	53	53	3.0	2.9
201	56	56	3.2	3.2
Nodulated				
0	43	57	2.5	3.1
67	51	55	2.8	3.1
134	54	56	3.1	3.2
201	56	56	3.1	3.1

Source: Based on Parker and Harris (1977).

Legumes and nonlegumes dependent on N$_2$ fixation have a high Mo requirement, particularly in root nodules. When the external supply is low, the Mo concentration of the nodules is usually higher than that of the leaves, whereas when the external supply is high, the concentration in the leaves increases more than that in the nodules (Brodrick and Giller, 1991a). When Mo is limiting, preferential accumulation in root nodules may lead to a lower Mo concentration in the shoot and seeds of nodulated legumes (Ishizuka, 1982). However, the relative allocation of Mo to the various plant organs varies not only among plant species but also among genotypes within a species, for example, in common bean (Brodrick and Giller, 1991b).

As expected, the growth of plants relying on N$_2$ fixation is particularly stimulated by the application of Mo to the low-Mo soils (Becking, 1961). The nodule dry weight can increase 18-fold, which indirectly reflects the increase in the capacity for N$_2$ fixation by improved Mo supply.

In soils low in Mo availability, the effect of Mo application to legumes depends on the form of N supply. As shown in Table 7.27, Mo applied to nodulating and nonnodulating soybean plants increased N concentration and grain yield only in the nodulated plants without or with insufficient supply of N fertilizer. This demonstrates the greater requirement for Mo in N$_2$ fixation than in nitrate reduction. It also indicates that on soils with low Mo availability, it is possible to replace the application of N fertilizer to legumes by the application of Mo fertilizer combined with rhizobium inoculation.

Low availability of Mo in tropical forest soils may limit N$_2$ fixation by free-living heterotrophic bacteria, thus impacting N cycling. Barron et al. (2009) showed that Mo addition to the weathered tropical forest soils from Panama significantly increased N$_2$ fixation.

7.6.4 Nitrate reductase

Nitrate reductase is a homodimeric enzyme with three electron-transferring prosthetic groups per subunit: flavin (FAD), heme, and Mo cofactor (Moco). Moco consists of Mo covalently bound to two S atoms in the tricyclic molecule pterin. Molybdenum in Moco is bound to a third S-ligand either of the cysteine residue (below, top molecule) or of a terminal S (below, bottom molecule). The first type of Moco is used in nitrate reductase and sulfite oxidase. The second type is found in xanthine dehydrogenase and aldehyde oxidase.

During nitrate reduction, electrons are transferred directly from Mo to nitrate (see Section 6.1). Nitrate reductase activity is low in the leaves of Mo-deficient plants, but can be induced within a few hours by infiltrating leaf segments with Mo. As shown in Fig. 7.26, in nitrate-fed plants there are positive relationships between Mo supply, NRA in the leaves, and yield of spinach. Incubation with Mo for 2 h increased NRA only in the leaf segments from Mo-deficient plants. "Inducible NRA" can therefore be used as a test for the Mo nutritional status of plants (Shaked and Bar-Akiva, 1967).

The Mo requirement for plant growth is strongly dependent on whether N is supplied as nitrate or ammonium (Table 7.28). In nitrate-fed plants not supplied with Mo, growth is poor, and the concentrations of chlorophyll and ascorbic acid are low (mainly located in the chloroplasts), but the concentration of nitrate is high. Leaves show typical symptoms of Mo deficiency ("whiptail"). When ammonium is supplied, the response to Mo is less marked, but still measurable regarding plant dry weight and ascorbic acid concentration. Without Mo supply, ammonium-fed plants also develop the whiptail symptoms. However, under the nonsterile culture conditions, nitrification of ammonium may occur in the substrate and nitrate could be taken up. In cauliflower plants grown under sterile conditions, those supplied with ammonium (but not Mo) did not develop deficiency symptoms and seemed to have no Mo requirement (Hewitt and Gundry, 1970).

7.6.5 Other Mo-containing enzymes

Four other Mo-containing enzymes have been identified in plants, including xanthine dehydrogenase, aldehyde oxidase, sulfite oxidase, and the mitochondrial amidoxime reducing component (mARC); they play important roles in the response and resistance to various stresses (Mendel and Kruse, 2012; Schwarz and Mendel, 2006). Xanthine

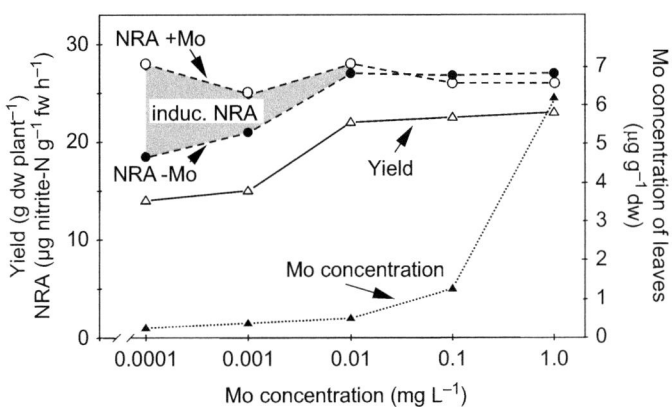

FIGURE 7.26 NRA in spinach (*Spinacia oleracea*) leaves from plants grown with different Mo supply. Leaf segments were incubated with (NRA + Mo) or without (NRA−Mo) Mo for 2 h. Stippled area represents "inducible NRA". *NRA*, Nitrate reductase activity. *Redrawn from Witt and Jungk (1977).*

TABLE 7.28 Growth and the chlorophyll, nitrate, and ascorbic acid concentrations of tomato (*Solanum lycopersicum*) grown with ammonium or nitrate N and with and without Mo supply.

	Nitrate −Mo	Nitrate +Mo	Ammonium −Mo	Ammonium +Mo
Dry weight (g plant^{-1})	9.6	25.0	15.9	19.4
Chlorophyll (mg kg^{-1} fw)	89	158	216	174
Nitrate (mg g^{-1})	73	9	10	9
Ascorbic acid (mg kg^{-1} fw)	990	1950	1260	1840

The pH of the substrate (quartz sand) was buffered with CaCO$_3$.
Source: Based on Hewitt and McCready (1956).

dehydrogenase is a homodimeric metalloflavoprotein, with each subunit containing one Moco together with one molecule of FAD and two [2Fe-2S] centers. Xanthine dehydrogenase catalyzes the oxidation of hypoxanthine to xanthine and of xanthine to uric acid:

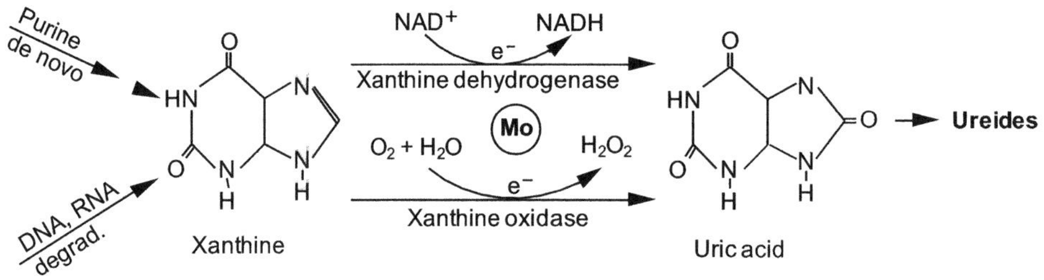

The enzyme is involved in the catabolism of purines and, thus, in the biosynthetic pathway of ureides as oxidation products of purines. In legumes such as soybean and cowpea, in which ureides are the most prevalent N compounds formed in root nodules (Chapter 16), xanthine dehydrogenase plays a key role in N metabolism. In the cytosol of the nodules, purines (e.g., xanthine) are oxidized to uric acid, the precursor of ureides. When nodulated legumes of the ureide type become Mo-deficient, low NRA and/or impaired purine catabolism in the nodules can result in the low N$_2$ fixation rates and the inhibition of growth. Xanthine dehydrogenase may also play a role in plant–pathogen interactions, cell death associated with hypersensitive response, and natural senescence (Schwarz and Mendel, 2006).

Aldehyde oxidase is very similar to xanthine dehydrogenase with respect to amino acid sequence and contains FAD, [2Fe-2S] cluster, and Moco. Aldehyde oxidase catalyzes the conversion of abscisic aldehyde to ABA. ABA is a phytohormone involved in developmental processes and the responses to biotic and abiotic stresses (Chapter 5). An arabidopsis mutant defective in Moco sulfurase that adds the terminal sulfur to Moco had the lower concentrations of ABA and was less tolerant to freezing, salinity, and drought (Xiong et al., 2001). Two rice mutants with the phenotype of preharvest sprouting were found to be caused by mutations in *OsCNX6* and *OsCNX2* that encode enzymes for the biosynthesis of Moco (Liu et al., 2019). These mutants had low activities of Moco-dependent enzymes, including aldehyde oxidase, resulting in the impaired ABA biosynthesis and consequently a lack of seed dormancy. By contrast, the overexpression of *OsCNX6* in rice increased the activity of Moco-dependent enzymes and improved the osmotic and salt stress tolerance, likely by increasing the ABA biosynthesis (Liu et al., 2019). The application of Mo to deficient wheat plants increased ABA concentration and cold tolerance (Sun et al., 2009). Aldehyde oxidase may also be involved in the biosynthesis of the phytohormone indole-3-acetic acid (IAA) by catalyzing the conversion of indole-3-acetaldehyde to IAA (Schwarz and Mendel, 2006).

Compared with other Mo-containing enzymes in plants, sulfite oxidase is smaller and simpler, possessing only Moco as its redox center (Eilers et al., 2001). Sulfite oxidase catalyzes the oxidation of sulfite (SO$_3^{2-}$) to sulfate (SO$_4^{2-}$) inside the peroxisomes, using O$_2$ as the terminal electron acceptor and producing hydrogen peroxide. Sulfite is a toxic metabolite that is produced when plants are exposed to sulfur dioxide (SO_2) or during the decomposition of sulfur-containing amino acids. Therefore, sulfite oxidase plays an important role in protecting plants against the damage caused by SO_2 (Lang et al., 2007).

The mARC is the eukaryotic Mo-enzyme, characterized in humans, pigs, and alga *Chlamydomonas reinhardtii* (Mendel and Kruse, 2012). Vascular plants also contain mARCs, with two isoforms being present in arabidopsis (Tejada-Jiménez et al., 2013). These enzymes catalyze the reduction of a broad range of N-hydroxylated compounds and are likely involved in metabolic detoxification (Llamas et al., 2017). Another possible role of mARCs is the production of nitric oxide (NO) as a signaling molecule in many biological processes. Together with nitrate reductase (also a Moco-enzyme), mARCs from *C. reinhardtii* reduce nitrite to NO (Chamizo-Ampudia et al., 2016). The physiological functions of mARCs in vascular plants require further investigations.

7.6.6 Gross metabolic changes

In arabidopsis, severe Mo deficiency in the *mot1* mutant affects the expression of many genes involved in metabolism, transport, stress responses, and signal transductions and alters the concentrations of amino acids, sugars, organic acids, and purine metabolites, suggesting that Mo deficiency has pleiotropic impacts on metabolism (Ide et al., 2011). Some specific effects are large increases in the expression of the genes encoding nitrate and nitrite reductases, possibly to compensate for the reduced NRA in Mo-deficient plants. In legumes dependent on N_2 fixation, N deficiency and the corresponding metabolic changes are the most prevalent effects of Mo deficiency. This often holds true also for nitrate-fed plants with mild Mo deficiency. With severe Mo deficiency, a range of metabolic changes is different from that of N deficiency. These differences may relate to the role of Mo in xanthine dehydrogenase and aldehyde oxidase. For example, in Mo-deficient plants, organic acids (Höfner and Grieb, 1979) and amino acids (Gruhn, 1961) accumulate, and the activity of ribonuclease is high, whereas that of alanine transferase is low (Agarwala et al., 1978) as are the leaf concentrations of RNA and DNA (Chatterjee et al., 1985). Molybdenum-deficient plants are more sensitive to low-temperature stress and waterlogging (Sun et al., 2009; Vunkova-Radeva et al., 1988) due to the effect on ABA biosynthesis. Molybdenum deficiency also has strong effects on pollen formation in maize (Table 7.29). In deficient plants, tasseling was delayed, a large proportion of flowers failed to open, and the capacity of anthers to produce pollen was diminished. Furthermore, the pollen grains were smaller, lacked starch, had lower invertase activity, and showed poor germination.

The risk of premature sprouting of maize grains in standing crops increases when the Mo concentration is below 0.03 µg g^{-1} in the grains, or below 0.02 µg g^{-1} in the grains and 0.10 µg g^{-1} in the leaves (Fig. 7.27; Farwell et al., 1991). Premature sprouting is also a serious problem in some wheat-growing areas and can be alleviated by foliar sprays of Mo (Cairns and Kritzinger, 1992). In maize, the extent of premature sprouting is related to the time of N application (Tanner, 1978). Little sprouting occurred when top-dressing with ammonium-nitrate took place within 55 days after germination. Later N application increased sprouting of grains low in Mo. Molybdenum deficiency may result in a lack of seed dormancy, thus increased premature sprouting, due to the reduced ABA biosynthesis that stimulates dormancy and reduces germination (Modi and Cairns, 1994). As mentioned in Section 7.6.5, mutations in two genes responsible for Moco biosynthesis led to preharvest sprouting in rice (Liu et al., 2019).

TABLE 7.29 Pollen production and viability in maize plants at different rates of Mo supply.

	Mo suppl (mg kg^{-1})		
	20	0.1	0.01
Mo concentration in pollen grains (µg g^{-1} dw)	92	61	17
Pollen-producing capacity (no. of pollen grains anther^{-1})	2437	1937	1300
Pollen diameter (µm)	94	85	68
Pollen viability (% germination)	86	51	27

Source: From Agarwala et al. (1979).

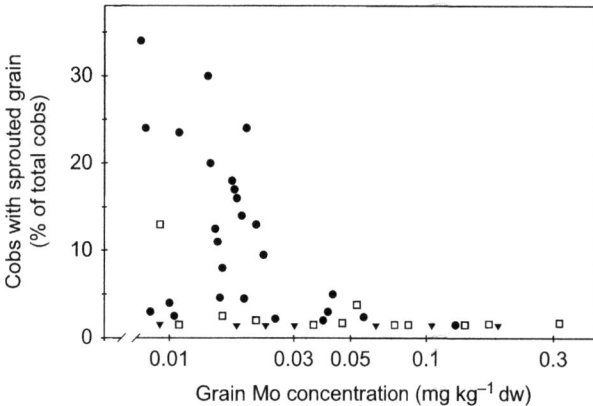

FIGURE 7.27 Relationship between Mo concentration of maize (*Zea mays*) grains, time of N top-dressing, and percentage of sprouted cobs of maize. Top-dressing with N at (▼) 30 days; (□) 40–55 days; (●) 70–85 days after germination. *Based on Tanner (1978).*

TABLE 7.30 Relationship between soil pH, Mo supply, and dry weight and shoot Mo concentration in soybean (*Glycine max*).

	Mo supply (mg pot^{-1})	Soil pH 5.0	Soil pH 6.0	Soil pH 7.0
Dry weight (g pot^{-1})	0	15	19	23
	5	20	20	20
Shoot Mo concentration (μg g^{-1} dw)	0	0.1	0.8	0.9
	5	2.0	6.3	18.5

Source: Based on Mortvedt (1981).

7.6.7 Molybdenum deficiency and toxicity

Depending on plant species and N source, the critical deficiency levels of Mo vary between 0.1 and 1.0 μg g^{-1} leaf dw (Gupta and Lipsett, 1981; Bergmann, 1992). In seeds the Mo concentration is highly variable (see below) but, in general, higher in legumes than in nonlegumes.

In Mo-deficient plants, symptoms of N deficiency and the stunted growth and chlorosis in young leaves are common. In dicotyledonous species, a strong reduction in size and irregularities in leaf blade formation ("whiptail") are the most typical visual symptom caused by local necrosis in the tissue and insufficient differentiation of vascular bundles in the early stages of leaf development (Bussler, 1970).

Local chlorosis and necrosis along the main veins of mature leaves (e.g., "yellow spot" in citrus) and whiptail in young leaves may reflect the same type of local metabolic disturbances but occur at different stages of leaf development (Bussler, 1970). With severe deficiency, the margin chlorosis and necrosis occur on the mature leaves with a high nitrate concentration.

Molybdenum deficiency is widespread in legumes and some other plant species (e.g., cauliflower and maize) grown in acid mineral soils with large concentrations of reactive Fe oxyhydroxides and thus a high capacity for adsorbing MoO_4^{2-}. Furthermore, adsorption of molybdate increases with decreasing soil pH. As shown in Table 7.30, regardless of whether Mo is supplied or not, the Mo concentration of the shoots of soybean increased by a factor of 10 when the soil pH was increased from 5.0 to 7.0 by liming. The effect of the liming treatment alone on plant dry weight is similar to the application of Mo to the unlimed soil. Thus, liming and Mo application might be seen as alternatives for stimulating legume growth on acid mineral soils. Responses of legume growth to liming, therefore, also depend strongly on the Mo availability in the soils (Adams et al., 1990). A combination of both liming and Mo supply often leads to luxury uptake and very high Mo concentrations in the shoots and grain.

A high Mo concentration in seeds ensures proper seedling growth and high grain yields in plants growing in soils low in available Mo (Table 7.31). Hence, the effect of Mo application to a low-Mo soil on plant growth is negatively

TABLE 7.31 Relationship between the Mo concentration of soybean (*Glycine max*) seeds and the subsequent seed yield of plants growing in a Mo-deficient soil.

Mo concentration of seeds (mg kg^{-1} dw)	Grain yield of the subsequent crop (t ha^{-1})
0.05	1.5
19	2.3
48	2.8

Source: Based on Gurley and Giddens (1969).

TABLE 7.32 Dry matter production and N concentration in shoots of the subtropical pasture legume *Desmodium intortum* grown in a soil with pH 4.7 without Mo supply or Mo supplied to soil or seeds.

Mo application (g ha^{-1})	Dry weight (kg ha^{-1})	N concentration (mg g^{-1})
0	70	19
100 (soil application)	1220	32
100 (seed pelleting)	1380	34

Source: From Kerridge et al. (1973).

TABLE 7.33 Dry matter production, N uptake, and Mo concentration in groundnut (*Arachis hypogaea*) grown on a low-Mo, acid sandy soil without Mo supply or Mo supplied to soil or as foliar spray.

Mo application (g ha^{-1})	Dry matter (t ha^{-1})	N uptake (kg ha^{-1})	Mo concentration (µg g^{-1} dw)		
			Shoots	Nodules	Seeds
0	2.7	70	0.02	0.4	0.02
200 (soil)	3.4	90	0.02	1.5	0.20
200 (foliar)	3.7	101	0.05	3.7	0.53

Source: Based on Rebafka (1993).

related to the seed Mo concentration (Tanner, 1982) and the amount of Mo applied to the seed crop (Weir and Hudson, 1966).

Compared with the uptake rates of other micronutrients, the rate of Mo uptake by soybean plants during the first 4 weeks is low; thus, the Mo requirement for growth has to be met mainly by re-translocation from the seed (Ishizuka, 1982). Large-seeded cultivars combined with high Mo availability during the seed-filling period are therefore effective in producing seeds suitable for soils low in available Mo (Franco and Munns, 1981).

Seed pelleting with Mo is another procedure for preventing deficiency during early growth and establishing a vigorous root system for subsequent uptake from soils low in available Mo (Tanner, 1982). As shown in Table 7.32, seed pelleting with the relatively insoluble MoO$_3$ at a rate of 100 g Mo ha^{-1} is as effective as soil application. Seed pelleting with 100 g Mo ha^{-1} in legumes such as groundnut increased dry matter production and the amount of N in the plants more than an application of 60 kg ha^{-1} of mineral fertilizer N (Hafner et al., 1992).

Because Mo is highly phloem-mobile, foliar application is an appropriate procedure for correcting acute Mo deficiency (Gupta and Lipsett, 1981; Kaiser et al., 2005). In legumes, Mo applied as a foliar spray at early growth stages is preferentially translocated into the nodules (Brodrick and Giller, 1991a) and is very effective in increasing final yield, for example, in soybean (Adams et al., 1990) or groundnut (Table 7.33). Compared with soil application, foliar

TABLE 7.34 Dry matter production, N uptake, and Mo concentration in groundnut grown on a low-Mo, acid sandy soil without or with P supply (13 kg ha^{-1}) as single superphosphate (SSP) or triple superphosphate (TSP).

P fertilizer	Dry matter (t ha^{-1})	N uptake (kg ha^{-1})	Mo concentration (µg g^{-1} dw)		
			Shoots[a]	Nodules	Seeds
−P	2.0	52	0.22	4.0	1.0
+SSP	2.6	62	0.09	1.5	0.1
+TSP	3.2	81	0.31	8.2	3.1

[a]At flowering.
Source: Based on Rebafka (1993).

application to groundnut increases not only yield but also the N uptake and the Mo concentration in the shoots, seeds, and nodules. Foliar sprays of Mo applied before flowering are effective in correcting Mo deficiency in grapevine (Williams et al., 2004).

Lower effectiveness of Mo applied to the soil than to leaves may reflect fixation of Mo in the soil; however, it is often also the result of impaired uptake by the roots. Sulfate and molybdate compete strongly during uptake by roots. Therefore, sulfate-containing soil amendments such as gypsum (Pasricha et al., 1977; Stout et al., 1951) as well as single superphosphate (SSP, which contains sulfate), decrease Mo uptake (Table 7.34). Unlike SSP, triple superphosphate (TSP) does not contain sulfate and therefore leads to the greater Mo and N uptake and higher yields than SSP. Moreover, TSP increases seed Mo concentration, thereby increasing seed quality in terms of suitability for use in low-Mo soils.

A unique feature of Mo nutrition is the wide variation between the critical deficiency and toxicity concentrations that may differ by a factor of up to 10^4 (e.g., 0.1–1000 µg Mo g^{-1} dw) as compared with a factor of 10 or less for B or Mn. Plants are generally quite tolerant to Mo toxicity. When present in toxic concentrations, however, excess Mo results in malformation of the leaves and a golden yellow discoloration of the shoot tissue, most likely due to the formation of molybdocatechol complexes in the vacuoles (Hecht-Buchholz, 1973). In oilseed rape and tomato, the most striking symptom of Mo toxicity is a dark blue coloration of stems (McGrath et al., 2010), which is due to the formation of molybdenum–anthocyanin complexes (Hale et al., 2001). Genotypic differences in tolerance to Mo toxicity are closely related to differences in the translocation of Mo from roots to shoots.

High, but nontoxic, concentrations of Mo in plants are advantageous for seed production, but such concentrations in forage plants can be dangerous for animals, particularly ruminants. Molybdenum concentrations above 5–10 mg kg^{-1} dw of forage can induce molybdenosis (or "teart"). This occurs, for example, in western parts of the United States as well as in Australia and New Zealand, often in soils with poor drainage and high in organic matter content (Gupta and Lipsett, 1981). Molybdenosis is caused by an imbalance of Mo and Cu in the ruminant diet, that is, induced Cu deficiency (Miller et al., 1991). The inhibitory effect of sulfate on molybdate uptake (Table 7.34) can be used to reduce the Mo concentrations in plants to nontoxic levels (Chatterjee et al., 1992; Pasricha et al., 1977) either for the plants or for the ruminants.

Molybdenum nutrition of plants growing in mixed pastures of legumes, herbs, and grasses requires special consideration. On the one hand, the relatively large requirement of legumes for N$_2$ fixation and for Mo in the seeds must be met, but at the same time, toxic concentrations in the forage for grazing animals must be avoided.

7.7 Boron

7.7.1 General

Boron (B) is a member of the metalloid group of elements which also includes silicon (Si). The boron atom is small and has a valence of three. Boric acid is a very weak acid, with a pKa of 9.24, and at the pH found in the cytoplasm (pH 7.5), more than 98% of B is in the form of free uncharged boric acid, B(OH)$_3$, and less than 2% as borate,

$B(OH)_4^-$. At the pH of the apoplast (pH 5.5), >99.95% of boron is in the form of $B(OH)_3$ and less than 0.05% in the form of $B(OH)_4^-$.

$$B(OH)_3 + 2H_2O \leftrightarrow B(OH)_4^- + H_3O^+$$

Boric acid forms spontaneous esters with mono-, di-, and polyhydroxy compounds, and both boric acid and borate can react readily with many types of biological molecules. Under normal biological conditions, B-binding molecules available within the plant typically exceed the concentration of free B. Except for B bound in the cell wall, however, all known biological B compounds are relatively unstable.

The molecular characteristics of undissociated boric acid $B(OH)_3$ and the borate anion $B(OH)_4^-$ influence all aspects of boron uptake and transport and are summarized here (for further details on B uptake and transport see Chapters 2 and 3). Boron in solution in soils with pH 3–8 is present largely as undissociated boric acid $B(OH)_3$, with the borate anion $B(OH)_4^-$ becoming more prevalent at extremely high pH. When B is present in adequate or excessive concentrations (>20 μM) in solution, B uptake is linearly related to the external soluble B concentration. This occurs as a result of the high membrane permeability for undissociated boric acid $B(OH)_3$ (Dordas and Brown 2000; Raven, 1980) and because of the abundance of available B-binding molecules within the plant cell that decreases free boric acid/borate and thus provides a favorable uptake gradient. When B is present at low or limiting concentrations (<1 μM), B uptake is regulated through two different classes of B transporters: boric acid channels of the major intrinsic protein family (NIP 5;1) (Dordas et al., 2000; Takano et al., 2006) and borate transporters of the BOR family (BOR1, BOR2) (Noguchi et al., 1997; Takano et al., 2002; reviewed in Yoshinari and Takano, 2017). The characteristics of B transport and related reactions in the cell are illustrated in Fig. 7.28.

The transport and redistribution of B in the xylem and phloem is influenced strongly by the predominant form of carbohydrates produced and transported by a plant species (Brown and Shelp, 1997). In plant species that utilize sucrose as the primary transported carbohydrate, B transport within the plant is primarily xylem-limited and is governed by the transpiration stream, whereas in plant species that transport primary photosynthates as polyols (such as sorbitol or mannitol), B is freely mobile in the xylem and phloem, and distribution patterns of tissue B resemble those of a phloem-mobile element (Brown and Shelp, 1997). Even though this species-dependent differential B transport has been reported widely, evidence of a limited phloem mobility of B is also observed in species that do not produce significant amounts of polyols, particularly during discrete phenological stages such as reproduction and fruit fill, and this may be associated with transient low levels of polyol production (Minchin et al., 2012; Stangoulis et al., 2000) and upregulation of B transporters supplying reproductive and meristematic tissues (Diehn et al., 2019; Durbak et al., 2014).

Boron is a micronutrient for vascular plants, diatoms, yeast, bacteria, and some species of green algae, whereas it is apparently not required by fungi (Loomis and Durst, 1992). Boron is also required by cyanobacteria when dependent on N_2 fixation. The role of B in plant nutrition is still the least understood of all the nutrients, and what is known of B requirement arises mainly from studies in which B was withheld or resupplied after deficiency. This lack of information is surprising because on a molar basis, the requirement for B, at least for dicotyledonous plants, is higher than that for any other micronutrient. Withholding B rapidly induces a range of distinct metabolic changes and visible deficiency symptoms in certain plant species (e.g., sunflower). Boron deficiency in crops is probably more widespread globally than deficiency of any other micronutrient. Visual symptoms of B deficiency generally become evident in dicots, maize, and wheat at tissue concentrations below 30–50, 5–10, and 5 mg kg^{-1} dw, respectively (Gupta, 1993). Nutritional disorders attributed to B deficiency are prevalent among vegetables, fruit, and nut trees. These include brown heart in storage roots of rutabaga (*Brassica napus rapifera*), turnip (*Brassica rapa*) and radish (*Raphanus sativus*), and hollow stem in cauliflower (*Brassica oleracea* var. *botrytis*) and broccoli (*B. oleracea* var. *italica*) (Shelp and Shattuck 1987; Shelp et al., 1987). In fruit and nut trees, B deficiency often results in decreased seed set even when vegetative symptoms are absent (Nyomora et al., 1997 and references therein). The occurrence of these disorders, even when B is in ample supply in the soil, suggests that B deficiency in plants is physiological in nature (e.g., induced by rapid growth, environmental constraints, drought, or other conditions that limit B uptake and transport within the plant to the site of demand).

Boron is neither an enzyme constituent nor is there convincing evidence that it directly affects enzyme activities. Over the years many roles have been proposed for B, including roles in (1) sugar transport, (2) cell wall synthesis, (3) lignification, (4) cell wall structure, (5) carbohydrate metabolism, (6) RNA metabolism, (7) respiration, (8) IAA metabolism, (9) phenol metabolism, and (10) membrane structure. Of these, only a function in cell walls has been demonstrated definitively. Nevertheless, this long list might indicate that (1) B is involved in a number of as yet unproven metabolic pathways, or (2) B deficiency results in a "cascade effect" due to disruption of a critical and central cellular

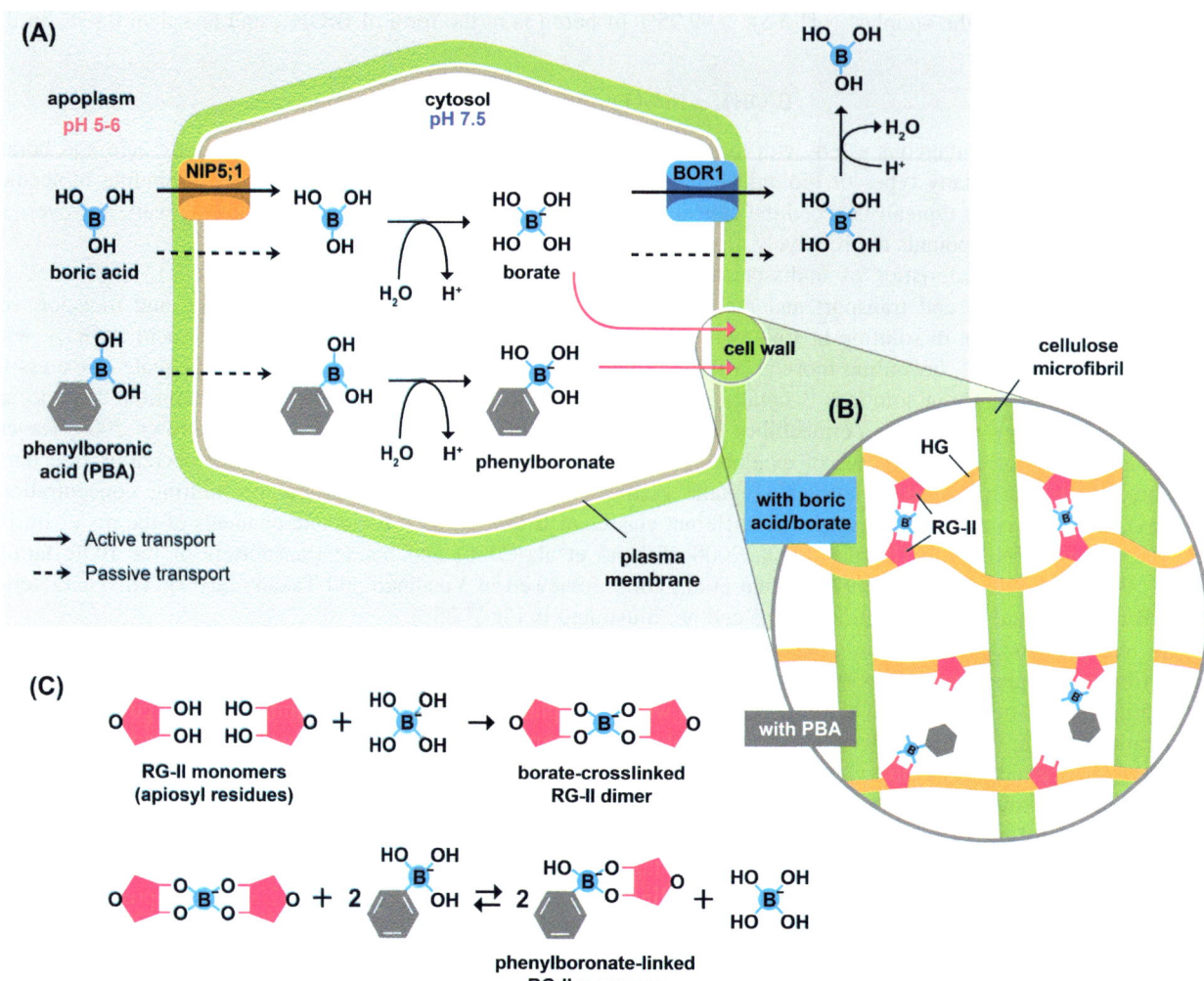

FIGURE 7.28 Boric acid uptake and function in plant cells. (A) Uncharged boric acid is the predominant form of B present in the soil and enters the plant by either direct diffusion through the plasma membrane or facilitated transport through channels (NIP 5;1). In the cytosol at pH 7.5, a small proportion of boron is dissociated to borate anion and exported by borate uniporters driven by the electrochemical gradient. The borate anion may then be incorporated into the cell wall or converted to boric acid and exported into the stele. (B) Close-up of the cell wall pectin matrix with borate crosslinking the apiosyl residues of two rhamnogalacturonan (RGII) monomers to form stable cell wall RGII dimers. (C) The uptake of the synthetic molecule PBA and dissociation of PBA into phenylboronate. Reaction of the phenylboronate with a single RGII monomer in the cell wall prevents the formation of stable RGII dimers and disrupts cellular function. PBA, which cannot form RGII dimers, has been a useful tool for discerning the function of B in plant cells. *PBA*, Phenylboronic acid. *Adapted from Matthes et al. (2020). An open access article under the terms of Creative Commons CC BY license.*

process. There is evidence for the latter; B deficiency disrupts cell wall biosynthesis and structure resulting in a cascade of metabolic disruptions that can explain most, but not all, observed effects of B deficiency. Additional functions of B await discovery, and recent advances have provided evidence of a function of B in phytohormone metabolism and the processes governing the structure and functioning of plant meristems (Matthes et al., 2020). Several reviews on the chemistry and biology of B are available that shed further light on the multitude of effects of B on plant growth (Brown et al., 2002; Cakmak and Römheld, 1997; Dell and Huang, 1997; Eggert and von Wirén, 2017; Loomis and Durst, 1992; Matthes et al., 2020).

7.7.2 Boron complexes with organic structures

Boric acid has an outstanding capacity to form complexes with diols and polyols, particularly with *cis*-diols, either as monoester (Eq. 1) or diester (Eq. 2). Polyhydroxyl compounds with an adjacent *cis*-diol configuration are required for the formation of such complexes; the compounds include various sugars and their derivatives (e.g., sugar alcohols and uronic acids), particularly sorbitol, mannitol, mannan, and polymannuronic acid.

$$(1) \quad \begin{matrix} =C-OH \\ | \\ =C-OH \end{matrix} + \begin{matrix} HO \\ \end{matrix}B-OH \rightleftharpoons \left[\begin{matrix} =C-O \\ | \\ =C-O \end{matrix} B \begin{matrix} OH \\ \\ OH \end{matrix} \right]^{-} + H_3O^+$$

$$(2) \quad \left[\begin{matrix} =C-O \\ | \\ =C-O \end{matrix} B \begin{matrix} OH \\ \\ OH \end{matrix} \right]^{-} + \begin{matrix} OH-C= \\ | \\ OH-C= \end{matrix} \rightleftharpoons \left[\begin{matrix} =C-O \\ | \\ =C-O \end{matrix} B \begin{matrix} O-C= \\ | \\ O-C= \end{matrix} \right]^{-} + 2H_2O$$

These compounds serve, for example, as critical constituents of the hemicellulose fraction of cell walls. By contrast, glucose, fructose and galactose and their derivatives (e.g., sucrose) do not have this *cis*-diol configuration and thus do not form stable borate complexes. The most stable borate diesters are formed with *cis*-diols on a furanoid ring, namely, the pentoses ribose and apiose, the latter being a universal component of the cell walls of vascular plants (Loomis and Durst, 1992). The high B requirement of gum-producing plants is most likely related to the function of B in forming cross-links with the various polyhydroxy polymers such as galactomannan (Loomis and Durst, 1992).

The importance of borate diesters in normal plant function has been demonstrated through the use of phenylboronate [phenylboronic acid (PBA)] that readily binds to *cis*-diols in the cell but unlike borate, cannot form stable diesters. Reaction of the phenylboronate with single RGII monomers in the cell wall prevents the formation of stable RGII dimers and thereby disrupts cellular function (Fig. 7.28; Bassil et al., 2004; Matthes and Torres-Ruiz, 2017). Even though borate has the potential to bind with many biological compounds, including sugar alcohols, RNA, phenols, and NAD^+, the only B-complex currently known to have biological essentiality is the rhamnogalacturonan-II B-complex in the cell wall (see Section 7.7.3).

In plants, a substantial proportion of the total B is complexed as *cis*-diol esters associated with pectins in the cell walls and is often directly correlated with the whole plant B requirement (Hu and Brown, 1994; Hu et al., 1996). The larger B requirement in dicotyledonous plants compared with graminaceous species is related to greater proportions of compounds with the *cis*-diol configuration in the cell walls of the former, namely, pectic substances and polygalacturonans (Loomis and Durst, 1992). The concentration of strongly complexed B in the root cell walls is 3–5 $\mu g\ g^{-1}$ dw in graminaceous species such as wheat, and up to 30 $\mu g\ g^{-1}$ dw in dicotyledonous species such as sunflower (Tanaka, 1967). These differences roughly reflect the differences between the species in B requirement for optimal growth (Hu et al., 1996).

7.7.3 Function of B

Numerous difficulties have hindered progress in understanding the functions of B in plants, including (1) difficulty in measuring low cellular B concentrations, (2) the labile nature of B and its complexes, (3) lack of radioisotopes or fluorescent probes to trace and visualize B, and (4) the capacity of B to bind to diverse molecules rapidly and reversibly. Upon removal of B from growing media, growth (both cell division and cell elongation) is almost immediately inhibited followed by manifestation of numerous secondary effects. Oiwa et al. (2013) showed that deprivation of B for 1 h induced cell death in the root elongation zone, which was associated with incomplete pectin cross-linking and stimulated production of ROS. Previously it was reported that most anatomical symptoms of B deficiency are associated with cell wall abnormalities (Loomis and Durst, 1992; Brown et al., 2002). Cell wall integrity is essential for all cell growth, development, and reproduction; hence, the numerous biochemical and physiological effects often observed upon B deficiency could be the result of secondary effects of cell wall defects and growth disruption (Blevins and Lukaszewski, 1998; Bolanos et al., 2004; Brown et al., 2002; Goldbach, 1997).

Early observable effects of B deficiency on plant processes include altered cell wall function (Goldbach et al., 2001; Hu and Brown, 1994; Yu et al., 2003), changes in cell division and elongation (Poza-viejo et al., 2018), and disruptions of normal meristematic development (Abreu et al., 2014; Matthes et al., 2020). To help define the functions of B in the cell, Bassil et al. (2004) treated plants with PBA and observed that PBA caused a rapid disruption of the cell wall and the cytoskeleton- and plasma membrane–associated processes within 10–15 min, suggesting that the cell wall, plasma membrane, and cytoskeleton are intimately associated with, and collectively disrupted by, B deficiency.

Accumulation of phenolics when plants are B-deficient is typical for many plant species, and it has been proposed that B plays a direct role in phenol detoxification (Lewis, 2019), a hypothesis that has been disputed (Wimmer et al., 2020).

Phenol accumulation in B-deficient tissue, however, may have significant adverse impacts on plants, particularly during the reproductive growth stage and with long-term exposure to high light intensity (Cakmak and Römheld, 1997). Boron deficiency decreases the utilization of absorbed light energy in photosynthesis, inducing oxidation of phenolics and impairing the plant antioxidative defense mechanisms, thereby enhancing the susceptibility of plants to high light intensity and generation of ROS.

7.7.3.1 Cell wall structure

In B-deficient plants, cell walls are altered substantially, which is evident at macroscopic (e.g., "cracked stem," "stem corkiness," "hollow stem disorder") and microscopic levels (Loomis and Durst, 1992; Shorrocks, 1997). Boron deficiency causes a wide range of anatomical, physiological, and biochemical symptoms that are consistent with a function of B in cell walls. Symptoms include inhibition and disruption (Matthes et al., 2020) of apical growth, necrosis of terminal buds, reduction in leaf expansion, breaking of tissues due to brittleness and fragility, abortion of flower initials, and shedding of fruits (Brown et al., 2002; Goldbach, 1997; Rerkasem et al., 2020).

The most prominent symptoms of B deficiency are associated with primary cell walls and include abnormally formed walls that are often thick, brittle, have altered mechanical properties, and do not expand normally. Loomis and Durst (1992) first hypothesized that apiose, a rare sugar specific to the pectic fraction of cell walls, may form esters with borate under physiological conditions and hence influence the cell wall structure. In support of this hypothesis, Hu and Brown (1994) observed that a high proportion of total plant B is associated with the cell wall pectins. Isolation of a B-polysaccharide complex (Matoh et al., 1993), later identified as RGII (Kobayashi et al., 1996; O'Neill et al., 1996), demonstrated that B in the cell wall predominantly cross-links the apiosyl residue in the A side chain of each of two neighboring monomeric RGII molecules to form a dimeric B-dRGII pectin complex (Ishii and Matsunaga, 1996; Ishii et al., 1999; Kobayashi et al., 1996; O'Neill et al., 1996). This role for B in cell walls and its importance to plant growth and development was confirmed by the phenotype of the arabidopsis *mur1* mutant. In *mur1*, shoot RGII, which has a substituted sugar residue, forms B-dRGII less rapidly and, once formed, is less stable than RGII from wild-type plants (O'Neill et al., 2001). The *mur1* plants are dwarfed with brittle stems but show normal growth with added B. A role of B cross-linked RGII (Fig. 7.29) in intercellular attachment of tissues was shown in another RGII biosynthesis mutant (Iwai et al., 2002). The RGII pectic polymer also has a critical role in pollen tube elongation and cell wall stability (Dumont et al., 2014; see Section 7.7.3.4).

Boron does not appear to be involved directly in the synthesis of the cell wall; however, B may influence the incorporation of proteins, pectins, and/or precursors into the existing and extending cell wall (Kobayashi et al., 1996; O'Neill et al., 1996). Fleischer et al. (1999) demonstrated that B deficiency rapidly increased cell wall pore size, which resulted in cell death once cells entered the elongation phase of growth. The inability of B-deficient cells to form a pectic network with appropriate pore size may influence physiologically important processes, including the incorporation of polymers into the wall and the transport of wall-modifying enzymes or proteins to their substrates and the transport of polymers from the protoplast into the cell wall (Fleischer et al., 1999). Brown et al. (2002) hypothesized, and Bassil et al. (2004) demonstrated, that B is involved in the adhesion of cell wall to plasma membrane, cytoskeleton function,

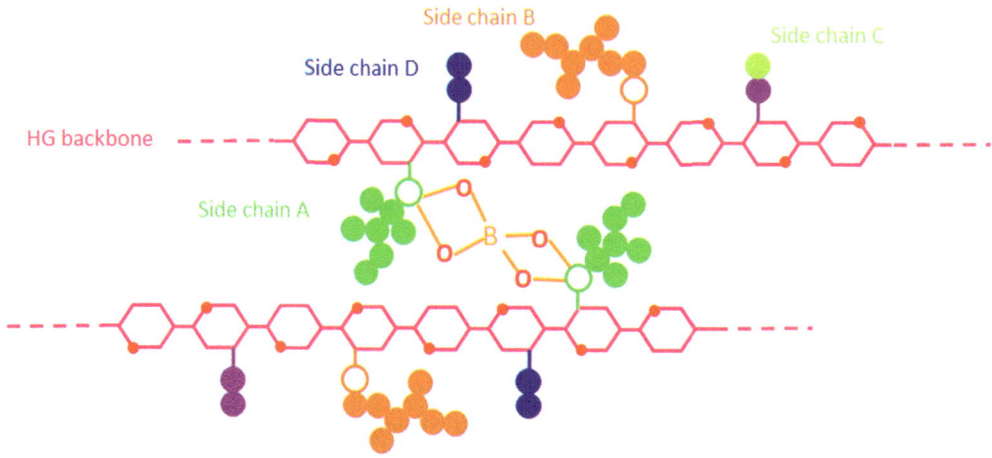

FIGURE 7.29 Structure of borate cross-linked rhamnogalacturonan-II dimer (○○) residues on side chain A. *Courtesy Malcolm O'Neil.*

and the organization of the architectural integrity of the cell, most likely through the binding of B to glycosylinositol phosphorylceramides that are boron-bridged in the plasma membrane and form complexes with RGII in the cell wall (Brown et al., 2002; Voxeur and Fry, 2014; Wimmer et al., 2009).

The changes in cell wall formation and composition under B deficiency result in serious physiological disturbances in plants grown with low B supply. For example, B deficiency enhances the number of aluminum (Al)-binding sites in cell walls, possibly due to increasing the amount of unmethylated pectin in the root tips, resulting in higher Al concentrations and greater Al damage to roots (Stass et al., 2007; Yu et al., 2009). Aberrations in the development and organization of primary cell walls associated with B deficiency have adverse impacts on the form, wood quality, and cold tolerance of trees (Lehto et al., 2010) and impact fruit quality in many species (Rerkasem et al., 2020).

Boron is also required for legume–*Rhizobium* symbiotic interactions (Chapter 16). Boron plays a role in the maintenance of nodule cell wall and membrane structure (Bolanos et al., 1994) and is important for rhizobial infection and the nodule cell invasion processes (Bolanos et al., 1996) as well as for symbiosome development and bacteroid maturation. More recent studies have shown that B participates in nodule organogenesis and in plant–bacteria interactions, suggesting that B has a wide range of functions beyond its role in the cell wall structure (Reguera et al., 2009, 2010).

7.7.3.2 Metabolism

Boron deficiency has a rapid and profound effect on meristematic activity and causes many secondary disruptions to cellular metabolism, which have been interpreted as evidence for a specific function of B in metabolic processes (Brown et al., 2002). Boron deficiency is associated with a range of morphological alterations and changes in differentiation of tissues and causes changes in concentrations of the phytohormones auxin, CK, ethylene, and ABA (Blevins and Lukaszewski, 1998; Matthes and Torres-Ruiz, 2016; Poza-Viejo et al., 2018). The relationships between B nutrition, phytohormone metabolism, and meristematic differentiation and lignification are, however, inconsistent and responses vary depending upon the methodology, plant species, and duration of B deficiency (Matthes et al., 2020). Eggert and von Wirén (2017) provided an elegant illustration of the time-course of B deficiency effects by monitoring active phytohormone levels (CK, IAA, Eth) during the transition of B deficiency–sensitive *Brassica napus* seedlings from B sufficiency to B deficiency. Marked changes in phytohormone activity were evident only after the visual expression of deficiency symptoms (asterisk in Fig. 7.30).

The effect of B nutrition on phytohormones is also dependent upon the plant tissue used for analysis. In a study conducted by Pommerrenig et al. (2019), vascular and nonvascular leaf tissues of common plantain (*Plantago major*) were used to analyze the effect of low and adequate B nutrition on the concentrations of several phytohormones, including ABA and CK. There was a particular decrease in the concentration of ABA in low B plants both in the vascular and nonvascular leaf tissues, while CK accumulated in the vascular tissues of low B plants. The effect of B on leaf CK concentrations was not consistent. Whether the effects of B deficiency on IAA, CYT, ABA, or Eth are direct, or merely a consequence of disruptions in cell wall formation and the subsequent effects on apical dominance, remain to be resolved.

Accumulation of phenols is a typical feature of B-deficient plants. It has been suggested that the formation of borate complexes with certain phenols may be involved in the regulation of the concentration of free phenols and the rate of synthesis of phenol alcohols as precursors of lignin biosynthesis (Lewis, 2019; Pilbeam and Kirkby, 1983).

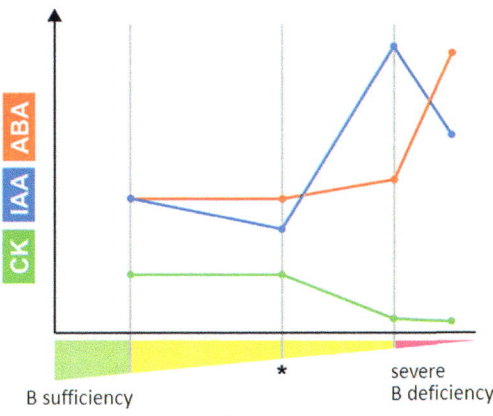

FIGURE 7.30 Trends of IAA, ABA, and CK concentrations with increasing B deficiency in *Brassica napus* plants. The asterisk marks appearance of visual B deficiency symptoms. *ABA*, Abscisic acid; *CK*, cytokinin; *IAA*, indole-3-acetic acid. *Based on Eggert and von Wirén (2017)*.

TABLE 7.35 Phenol concentration, polyphenol oxidase activity, and K efflux from leaf segments of sunflower (*Helianthus annuus*) plants grown at different light intensities and sufficient (10^{-5} M) or deficient (10^{-7} M) B supply.

	B supply (M)	Light intensity (μmol m^{-2} s^{-1})		
		100	250	580
Phenol concentration (μg caffeic acid equiv. per 6 segments)	10^{-5}	30	45	75
	10^{-7}	35	90	265
Polyphenol oxidase activity (relative)	10^{-5}	1.0	0.8	0.6
	10^{-7}	1.4	2.1	4.2
K efflux [μg K (6 segments)$^{-1}$ (2 h)$^{-1}$]	10^{-5}	10	12	25
	10^{-7}	23	63	238

Source: From Cakmak et al. (1995a) and Cakmak and Römheld (1997).

Accordingly, under B deficiency phenols accumulate, and polyphenol oxidase activity is increased (Table 7.35). Even though phenols accumulate in B-deficient plants, simple stoichiometric analysis of cellular B concentrations and the concentrations of potential B-complexing molecules suggest that B concentrations are not sufficient to influence phenol metabolism through complexation (Wimmer et al., 2020). A high proportion of phenols as well as the relevant enzymes are located in the cell walls of the epidermis, thus disruptions of cell wall synthesis upon B deficiency may result in secondary disruptions to the phenol metabolism that may in turn result in tissue necrosis as the toxic products of phenol oxidation accumulate. These processes are probably relevant for the long-term effects of B deficiency, especially for plants grown under field conditions. Accumulation of phenolic compounds and related alterations in lignin concentration may also affect plant defense systems against herbivory and pathogens (Lehto et al., 2010).

Boron deficiency affects several metabolic processes for which the underlying mechanism has not been elucidated adequately. A close relationship between B status and the ascorbate/glutathione cycle has been observed repeatedly (Blevins and Lukaszewski, 1998; Cakmak and Römheld, 1997; Koshiba et al., 2009). Ascorbate and glutathione concentrations are strongly decreased by B deficiency (Cakmak and Römheld, 1997), probably due to the inhibition of ascorbate reductase and glutathione reductase. External application of ascorbate temporarily overcame the root growth reduction caused by B deficiency (Blevins and Lukaszewski, 1998). Boron has been implicated in N metabolism, and both B deficiency and B toxicity can decrease NRA (Bellaloui et al., 2010; Cervilla et al., 2009). The effect of B on NRA may be mediated by a disruption of membrane transport processes (Cervilla et al., 2009). Boron deficiency also influences photosynthesis, resulting in lower quantum yield and a less efficient PSII, probably as a result of lipid oxidation of the thylakoidal membranes.

7.7.3.3 Membrane function

There is considerable evidence in support of a role of B in membrane integrity and functioning. The formation and maintenance of membrane potentials induced by infrared light or by gravity require the presence of B (Tanada, 1978). Boron also influences the turgor-regulated nyctinastic movements of leaflets of *Albizia* (Tanada, 1982) and enhances ^{86}Rb influx and stomata opening in *Commelina communis* (Roth-Bejerano and Itai, 1981).

Uptake rates of P are lower in the root tips of B-deficient compared to sufficient bean and maize plants (Table 7.36). Boron pretreatment of the root tips for only 1 h increases P uptake in both B-sufficient and -deficient roots and restores the uptake rate of the originally B-deficient roots (Pollard et al., 1977). Furthermore, the activity of membrane-bound ATPase that was low in the B-deficient maize roots was restored to the same level as that in the B-sufficient roots within 1 h after resupply of B.

These effects of B on uptake of ions are mediated probably by direct effects of B on the cell wall–plasma membrane structure and hence the function of various membrane transport processes, including the plasma membrane-bound H$^+$-pumping ATPase (Goldbach and Wimmer, 2007) (see Chapter 2). In suspension-cultured tobacco cells the effect of B on the H$^+$-ATPase requires the presence of IAA, and B is required for the enhanced H$^+$ excretion induced by IAA (Goldbach et al., 1990). The role of B in maintaining the plasma membrane integrity and the H$^+$ pumping activity was

TABLE 7.36 Phosphorus uptake by root tip zones (0–2 cm from the apex) of faba bean (*Vicia faba*) and maize (*Zea mays*) grown with or without B supply after pretreatment of the root tips without or with B for 1 h.

	P uptake (nmol g^{-1} h^{-1})			
	Faba bean		Maize	
Growth of plants	+B	−B	+B	−B
Pretreatment of root tips				
−B	112	52	116	66
10^{-5} M B(OH)$_3$	152	108	190	171

Source: From Pollard et al. (1977).

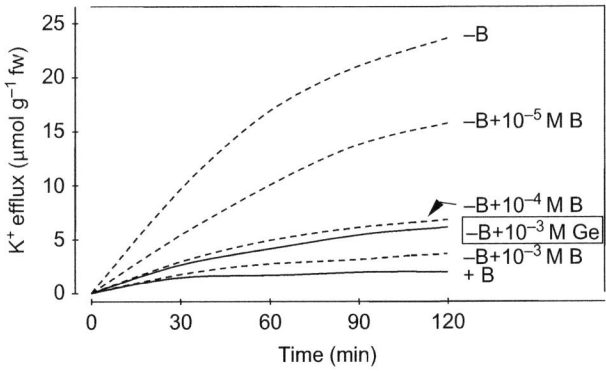

FIGURE 7.31 Potassium efflux from intact B-sufficient (+B) and B-deficient (−B) expanding sunflower leaves and effect of external supply (10^{-5} to 10^{-3} M) of B or germanium (Ge) at time = 0 (−B + B and −B + Ge treatments).

also demonstrated in vitro with membrane vesicles from the B-sufficient and B-deficient roots of several species (Goldbach and Wimmer, 2007).

Although B may have a direct effect on the plasma membrane-bound H$^+$-ATPase, it is more likely that these effects are indirect, for example, mediated by B-complexing the *cis*-diol groups of plasma membrane constituents such as glycoproteins or glycolipids at the cell wall–plasma membrane interface, and thereby acting as a stabilizing and structural factor required for the integrity and functioning of the plasma membrane (Bassil et al., 2004; Brown et al., 2002). This is supported by the high B concentration of isolated plasma membranes and the presence of B in the plasma membrane constituents, including glycosylinositol phosphorylceramides that are B-bridged in the plasma membrane and form complexes with RGII in the cell wall (Brown et al., 2002; Voxeur and Fry, 2014; Wimmer et al., 2009). Adequate B nutrition improves root K uptake by promoting the activity of proton-pumping ATPase of the root cell plasma membrane (Schon et al., 1990). Generally, higher K concentrations in B-sufficient plants, especially in the roots, have been discussed in relation to the positive effects of B on the H$^+$-ATPase activity of root cell plasma membrane (cf. Cakmak and Römheld, 1997; Schon et al., 1990).

Further support for the role of B in the plasma membrane integrity and function is shown in Fig. 7.31 regarding K efflux from the expanding sunflower leaves of B-sufficient and B-deficient plants (Cakmak et al., 1995a). The leaves were isolated and immersed either in distilled water or increasing concentrations of B. Compared with the B-sufficient leaves, K efflux was higher from the B-deficient leaves. Potassium efflux from B-deficient leaves could be decreased by supplying B during the efflux period, with the decrease dependent on the external B concentration and evident after less than 30 min. Similar to K efflux, the efflux of sugars, amino acids, and phenols was also higher from the B-deficient leaves and could be decreased by supplying B externally (Cakmak et al., 1995a). A similar decrease in K efflux could be achieved with external supply of Ge (Fig. 7.31), indicating a substitution of B by Ge not only in achieving the cell wall stability and functions but also the plasma membrane integrity (Cakmak et al., 1995a).

Boron is also essential for organisms that lack a cell wall, such as bacteria and animal embryos, which suggests functions of B beyond those in the cell wall. Additional effects of B deficiency include (1) swelling of liposomes, (2) increased fluidity of microsomes, and (3) disruption of membrane transport processes. In plant cells, B is necessary for the cell wall adhesion and the architectural integrity of the cell wall (Bassil et al., 2004; Fleischer et al., 1998); B may stabilize membrane raft formation through glycolipid binding and hence maintain membrane function (Brown et al., 2002; Voxeur and Fry, 2014).

7.7.3.4 Reproductive growth and development

In many agronomic and horticultural crops, B deficiency results in a decrease in reproductive success as a result of problems with flower formation, pollen production, and pollen viability, as well as infertility and premature flower and fruit drop. These reproductive effects can often be observed in the absence of vegetative symptoms or growth reduction, suggesting that the B requirement is greater for reproductive than vegetative tissues or that delivery of B to reproductive structures is limited. Evidence suggests that both higher demand and restricted delivery contribute to the greater sensitivity of reproduction to B deficiency. The role of B in reproduction and crop productivity is discussed in the excellent review by Rerkasem et al. (2020).

The role of B in cell wall structure and plasma membrane integrity is clearly expressed in pollen tube growth and development. The gene responsible for borate cross-linking of pectin rhamnogalacturonan II is highly active in pollen tubes and is required for the plant reproductive tissue development and fertilization (Dumont et al., 2014; Iwai et al., 2006; Rerkasem et al., 2020). Boron is essential for in vitro pollen cultures of most plant species (Robbertse et al., 1990). In the absence of an adequate B supply, pollen germination is decreased, pollen tubes may burst, and pollen tube extension is slower (Nyomora et al., 2000; Perica et al., 2001). After germination, pollen tubes extend by tip growth through the activity of secretory vesicles that are transported to tube tips by cytoplasmic streaming. Here, they fuse with existing pollen tube plasma membrane, and their contents (polysaccharides and pectins) are discharged to the outside, where they contribute to cell wall formation. Boron may play a critical role in the control of secretory activities in pollen tubes (Jackson, 1989). In growing pollen tubes, abnormal swelling or bursting of the tip region is observed within 2–3 min of removing B from the external solution (Jackson, 1989; Nyomora et al., 2000). Boron deficiency has a greater effect on pollen tube growth than on pollen viability or germination (Fig. 7.32; Nyomora et al., 2000).

In flowers, the B required for pollen tube growth has to be provided by the stigma or the silk. In maize a minimum B concentration of 3 μg g^{-1} dw in the silk is required for pollen germination and fertilization (Lordkaew et al., 2011; Matthes et al., 2018). The critical deficiency concentration in the stigma may, however, vary considerably between species and cultivars (Nyomora et al., 2000). In grapevine (*Vitis vinifera*) known for its high B requirement (Christensen et al., 2006), with sufficient B supply the B concentration of the stigma is 50–60 μg g^{-1} dw, and fertilization is impaired at concentrations lower than 8–20 μg g^{-1} dw (Gärtel, 1974).

Reproductive tissues have a high requirement for B due to their rapid growth rates and pectin-rich cell walls; however, this does not adequately explain why B deficiency occurs during reproductive growth but not during vegetative growth (Rerkasem et al., 2020). Dell and Huang (1997) suggested that the seemingly greater requirement for B by reproductive tissues occurs because reproductive structures are not well supplied by vascular bundles, and low transpiration rates decrease B supply. This is supported by observations that foliar application of B to developing reproductive tissues can increase the reproductive success even in the presence of soil B sufficient for vegetative growth

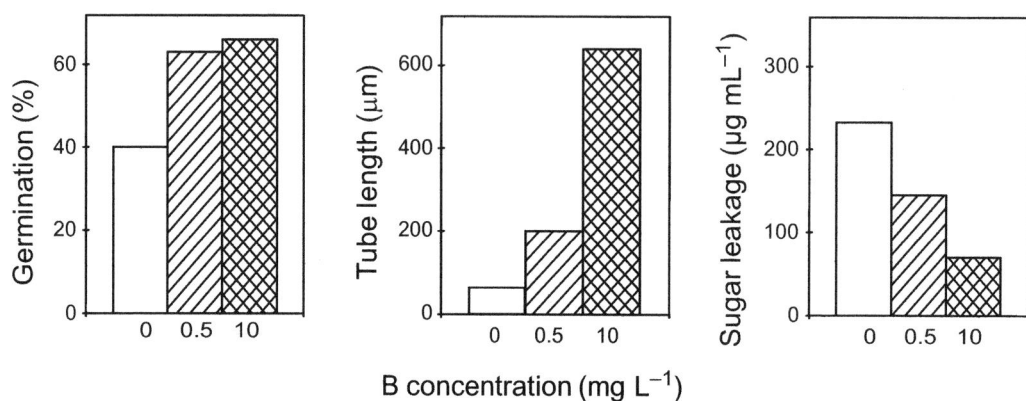

FIGURE 7.32 Germination of lily (*Lilium longiflorum*) pollen, tube length, and leakage of sugars to the medium at different B concentrations. *Based on Dickinson (1978).*

(Dordas, 2006; Nyomora et al., 1999). Factors that influence transpiration, such as temperature, humidity, and water supply, interact to affect the occurrence of reproductive B deficiency (Dell and Huang, 1997).

The role of B in pollen tube growth as well as limitations in the B transport to reproductive structures are the major factors responsible for the greater demand for B supply in the grain production compared to the vegetative growth. This has been shown in, for example, maize (Matthes et al., 2018), white clover (Johnson and Wear, 1967), alfalfa (Dordas, 2006), almond (Nyomora et al., 1999), and olive (Perica et al., 2001). In mango, irregular fruit set caused by suboptimal temperatures during pollination can be, at least in part, compensated for by increasing the B concentrations in the pistil and pollen grains (De Wet et al., 1989). Boron also affects fertilization by increasing pollen production in the anthers and improving pollen grain viability (Dell and Huang, 1997). Indirect effects may also be important, such as an increase in the amount, and changes in the composition, of sugars of the nectar, whereby the flowers of species that rely on pollinating insects become more attractive to insects (Eriksson, 1979).

7.7.3.5 Root elongation and shoot growth

One of the most rapid responses to B deficiency is the cessation of root elongation and shoot meristematic growth (Matthes et al., 2020; Oiwa et al., 2013; Poza-Viejo et al., 2018). As shown in Fig. 7.33 for soybean plants, the growth of both roots and shoots is sensitive to low B supply in growth medium. Boron deficiency results in roots with a stubby appearance, root swelling, and disorganization of the root apical meristem. In some plant species, such as sunflower and cotton, roots darken under B deficiency. In shoots, complete inhibition of meristematic growth occurs in many species, whereas meristem death occurs in others.

Inhibition of root elongation occurs as soon as 3 h after B supply is interrupted, becoming more severe after 6 h, and finally ceasing after 24 h (Fig. 7.34A). Twelve hours after the B supply is restored to roots deprived of B for the same period of time, elongation growth increases. Between 6 and 12 h after the B supply is stopped, there is an increase in the activity of IAA oxidase in the roots (Fig. 7.34B), and a disruption of CK-mediated cell proliferation (Poza-Viejo et al., 2018), which diminishes rapidly when B is resupplied.

The first observed effect of B withdrawal is an inhibition of cell divisions and elongation in the quiescent center (QC) of the root apical meristem (the source of newly dividing cells); this effect precedes the inhibition of primary root growth (Poza-Viejo et al., 2018). Interestingly, maximum root B concentrations are also observed at the root tip, coinciding with the QC and cell division zone (Shimotohno et al., 2015). Further studies suggest that higher B concentrations are required for the growth of newly formed tissues or that the root apical meristem has an abundance of cell wall and plasma membrane binding sites (O'Neill et al., 2001).

The inhibition of shoot growth, specifically shoot apical meristem activity, is a typical early symptom of B deficiency (Blevins and Lukaszewski, 1998; Dell and Huang, 1997). In some species, growth inhibition is followed by tissue death, and B resupply results in a bushy shoot development as lateral shoots emerge. Other responses to B deficiency include (1) growth inhibition and mild chlorosis or (2) growth inhibition with no secondary symptoms. In both cases, B resupply can result in renewed growth of the existing meristem or emergence of lateral meristems. The chlorosis and tissue death seen in some species under B deficiency may be due to the inability to synthesize new cell walls in the absence of adequate B, disruption of cell membrane integrity or changes in patterns of carbon metabolism as a consequence of growth inhibition.

In root tips, B deficiency results in a reduction in elongation growth associated with changes in cell division from a normal longitudinal to a radial direction (Poza-Viejo et al., 2018; Robertson and Loughman, 1974). Given that B is not

FIGURE 7.33 Growth of soybean (*Glycine max*) plants in nutrient solution containing increasing concentration of B (from left to right: 0, 0.05, 0.25, 1, 5, and 10 μM B) for 16 days. *Courtesy Ismail Cakmak.*

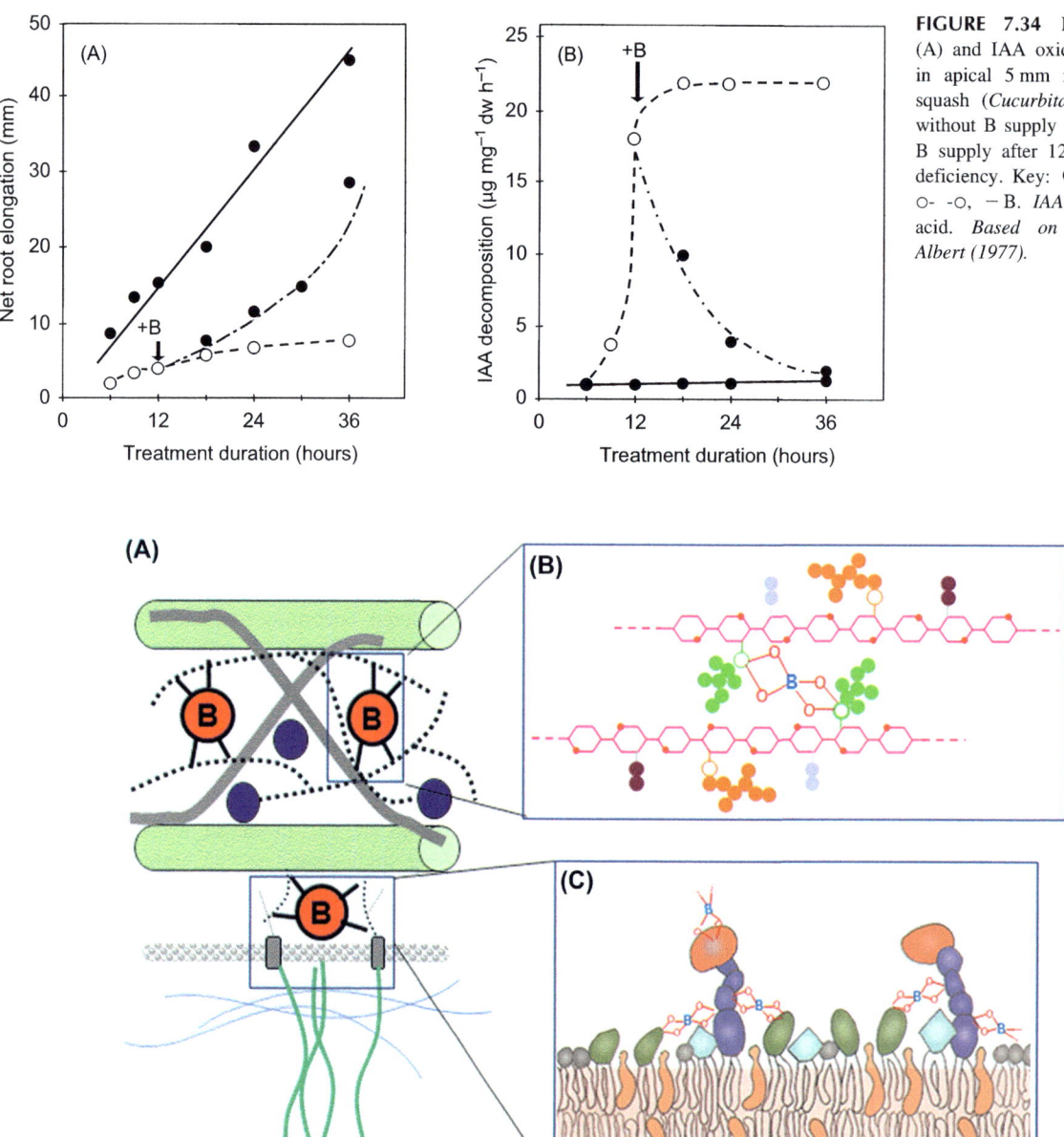

FIGURE 7.34 Root elongation (A) and IAA oxidase activity (B) in apical 5 mm root sections of squash (*Cucurbita pepo*) with or without B supply or resumption of B supply after 12 h (arrow) of B deficiency. Key: ●---●, +B; ○- -○, −B. *IAA*, Indole-3-acetic acid. *Based on Bohnsack and Albert (1977).*

FIGURE 7.35 (A) Network of cellulose fibrils (▭), hemicellulose (◥), pectins (■■•) and cell wall proteins (●). Plasma membrane (▨) with attachment sites (◉◉◉) of actin (—) and tubulin ( ). (B) Galacturonic and backbone (⬡⬡) with various side chains linked by B. (C) Membrane bilayer showing glycosphingolipids (◇), sphingomyelins (◊), glucosylphosphatidyl-inositol anchored proteins ( ) and other membrane components. (B) Modified with permission from Malcolm O'Neill. (C) Modified from Brown et al. (2002).

phloem mobile in most crop species (Section 7.7.1), a continuous supply of B to the elongating roots in soils is required to ensure better root growth.

7.7.3.6 Integrated assessment of the function of B in plants

The specificity of B deficiency effects on root and shoot apical meristems suggests that B exerts its primary influence in the cell wall and at the plasma membrane–cell wall interface, as summarized in the model presented in Fig. 7.35. Changes in the cell wall and at the cell wall–plasma membrane interface are considered as primary effects of B deficiency, leading to a cascade of secondary effects in metabolism, growth, and plant composition. Evidence of a close relationship between B status and the activity and metabolism of phytohormones involved in meristematic activity and patterning (summarized in Matthes et al., 2020) may suggest a role of B in phytohormone sensing mechanisms or may be a consequence of insufficient B in the meristem to cross-link RGII dimers, which inhibits cell wall function, prevents proper cell division and elongation and thus alters phytohormonal activity in meristems.

7.7.4 Boron deficiency and toxicity

7.7.4.1 Boron deficiency

Boron deficiency is a widespread nutritional disorder occurring in many regions of the world (Shorrocks, 1997). In environments with high rainfall, B is readily leached from soils as $B(OH)_3$; hence, B deficiency is common in the tropics. Boron availability to plants also decreases with increasing soil pH, particularly in calcareous soils and soils with a high clay content, presumably due to the formation of $B(OH)_4^-$ and subsequent anion adsorption. Boron deficiency occurs in diverse cropping systems throughout the world and across a wide range of climates and soil types. Boron deficiency is more prevalent on leached sandy, alkaline and heavily limed soils, or in irrigated systems utilizing water with low B concentration (<0.3 μg mL^{-1}). Due to its high mobility in soils and high leaching losses, slow-release B fertilizers are suggested to contribute better to plant B requirement later in the growing season (da Silva et al., 2018). Boron availability is affected significantly by soil water content and becomes limiting in dry soils due to the reduced mass flow (Shorrocks, 1997).

Plant species differ in their capacity to take up B (Table 7.37), which generally reflects differences among species in their requirement for B. For example, the critical deficiency range increases from about 5–10 mg kg^{-1} dw in graminaceous species (e.g., wheat) to 20–70 mg kg^{-1} dw in most dicotyledonous species (e.g., clover), and reaches 80–100 mg kg^{-1} dw in gum-bearing plants such as poppy (Bergmann, 1992). Because B deficiency preferentially disrupts new vegetative and reproductive growth, the critical deficiency concentrations of B can best be evaluated by examining the elongation rate of the youngest leaf, or examining seed set and yield rather than, for example, shoot dry weight (Kirk and Loneragan, 1988).

The differences in B demand, particularly between graminaceous and dicotyledonous species, are most likely related to differences in their cell wall composition. In graminaceous species, the primary cell walls contain little pectic material and also have a lower Ca requirement (Section 6.5) (Brown and Hu, 1998; White et al., 2018). These two plant groups also differ in their capacity for Si uptake, which is usually negatively related to the B and Ca requirement (Hodson et al., 2005; Loomis and Durst, 1992). All three elements are mainly located in the cell walls. Symptoms of B deficiency in shoots are noticeable at the terminal buds or youngest leaves, which become deformed and, depending upon species, may become discolored and die. Internodes are shorter, giving the plants a bushy or rosette appearance. In some species, interveinal chlorosis may occur, and misshaped leaf blades are common. The differences among species in expression of B deficiency are not well understood but may reflect differences in species response to the inhibition of cell wall formation and the changes in cellular metabolism that occur when growth is inhibited. Many plant

TABLE 7.37 Boron concentration of the leaf tissue of plant species from the same location.

Plant species	B concentration (mg kg^{-1} dw)
Wheat	6
Maize	9
Timothy	15
Tobacco	29
Red clover	32
Alfalfa	37
Brussel sprouts	50
Carrots	75
Sugar beet	102

Source: Based on Gupta (1979).

FIGURE 7.36 Stem cracking in oilseed rape plants grown under field conditions without B fertilization (Wuhan region, China). *Photos by I. Cakmak.*

TABLE 7.38 Yield, seed B concentration, seed viability, and germination of black gram (*Vigna mungo* L.) grown with or without B supply.

	Seed yield (g dw plant^{-1})	B concentration (mg kg^{-1} seed)	Percentage of seedlings		
			Normal	Weak/abnormal	Nonviable
−B	5.0	3.4	57	40	3
+B	5.1	7.4	92	6	2

Source: Based on Bell et al. (1989).

species accumulate large amounts of phenols under B deficiency, which can result in increased concentrations of oxidized phenols and other reactive species causing cell death. Some plant species respond to B deficiency with a cessation of shoot growth and deformed leaf blades but do not exhibit chlorosis or necrosis, presumably because a lack of accumulation of toxic metabolites.

An increase in the diameter of petioles and stems is particularly common and may lead to symptoms such as "stem crack" in celery or "hollow stem disorder" in broccoli (Shelp, 1988). Fruit or stem cracking is also common in B-deficient plants (e.g., in oilseed rape; Fig. 7.36). Drop of buds, flowers, and developing fruits is also a typical symptom of B deficiency. In the heads of vegetable crops (e.g., lettuce), water-soaked areas, tipburn, and brown- or blackheart occur (Rerkasem et al., 2020). In stems of celery or storage roots of sugar beet, necrosis of the growing areas leads to heart rot. With severe deficiency, the young leaves also turn brown and die, often followed by rotting and microbial infections of the damaged tissue. In B-deficient fleshy fruits, the growth rate is lower, and the quality may also be affected severely by malformation (e.g., "internal cork" in apple) or, in citrus, by a decrease in the pulp/peel ratio.

Boron deficiency−induced reduction or even failure of seed and fruit set are well known. However, even when seed yield is not depressed in plants grown in a low B soil, the seeds produced may have a lower quality in terms of viability as shown in Table 7.38 for black gram (*Vigna mungo*). Despite having the same seed dry weight, seeds with a lower B concentration had lower viability and produced a higher percentage of abnormal seedlings than seeds with a higher B concentration. A concentration of 6 mg B kg^{-1} seed dw is considered to be critical for normal seedling growth in black gram.

For the application of B either to the soil or as a foliar spray, different sodium borates, including borax or sodium tetraborate, can be used. Boric acid or sodium borate are effective as foliar sprays, for example, to increase flower and fruit set in fruit trees (Hanson, 1991a,b; Nyomora et al., 1997) or in soybean and alfalfa (Dordas, 2006). The amount of B applied varies from 0.3 to 3.0 kg ha^{-1}, depending on the requirement and sensitivity of the crop to B toxicity. The high solubility of many B fertilizers and the possibility of inducing toxicity require special care in the application of B fertilizers.

7.7.4.2 Boron toxicity and tolerance

The mechanism of B toxicity is unknown. Boron toxicity reduces shoot growth, primarily in expanding tissues, followed by chlorosis, beginning at the older leaf tips and margins, before finally causing necrosis (Nable et al., 1997; Reid et al., 2004; Reid and Fitzpatrick, 2009). These symptoms reflect the distribution of B in shoots as related to the transpiration stream. The B concentration in wheat grains can be increased more than 20-fold without negative effects on seed germination and seedlings growth (Paull et al., 1992a).

Boron toxicity is most common in arid and semiarid regions in plants growing on soils formed from parent material of marine origin, or related to the use of irrigation water high in B (Nable et al., 1997). Boron toxicity may also occur when large amounts of municipal compost are applied. Plant species, and also cultivars within a species, differ in their B tolerance. For example, the critical toxicity concentrations (mg kg^{-1} dw) in leaves are in the range of 100 in maize, 400 in cucumber, and 1000 in squash, between 100 and 270 in wheat genotypes (Paull et al., 1992a), about 100 in snap bean, and over 330 in cowpea (Francois and Clark, 1979).

Symptoms of B toxicity differ between species with the restricted versus significant phloem mobility (Brown and Shelp, 1997). In species in which B is immobile in the phloem (e.g. barley and wheat), B moves via the xylem and accumulates at the end of the transpiration stream; hence, foliar symptoms include chlorosis and necrosis spreading from the leaf tips, with brown lesions forming on the margins first, and then covering much of the leaf surface. Leaf tissue analysis, if performed correctly (see below), is an effective measure of B status in species in which B is immobile in the phloem. In species in which B is mobile in the phloem, such as apple, almond (*Prunus amygdala*), and others that produce significant amounts of polyols, B accumulates preferentially at shoot tips and in vascular cambium, causing shoot tip die-back and the formation of gummy exudates. Hence, in these species, leaf sampling is ineffective in diagnosing B status because excess B is preferentially transported to actively growing organs and is not retained in leaves. In almond, fruit tissue analysis has been adopted widely as a sensitive indicator of B status (Nyomora et al., 1999).

Critical toxicity concentrations of B in leaves of species in which B is immobile in the phloem must be interpreted with caution for various reasons. There is a steep gradient in B concentration within a leaf blade (see also Chapter 3). In barley, this gradient from the base to the tip of the leaf blade is from about 80 to 2500 μg B g^{-1} dw, when the average for the leaf is 208 μg g^{-1} (Nable et al., 1990b). Furthermore, the critical toxicity concentrations are often lower in field-grown plants compared with plants grown in a greenhouse. This difference is partially related to leaching of B from leaves by rain (Nable et al., 1990b).

The physiology of B tolerance and B toxicity is not well understood. There is a positive correlation between critical deficiency and toxicity concentrations for a wide range of plant species. In many cases, B concentrations of leaves or whole shoot are, however, not well related to differential tolerance to B toxicity (Choi et al., 2006; Nable et al., 1990a; Torun et al., 2003). Thus, the severity of leaf symptoms of B toxicity and decreases in shoot growth may be better parameters than leaf B concentrations in ranking genotypes for their tolerance to B toxicity (Choi et al., 2006; Torun et al., 2003). Differences in B toxicity tolerance among genotypes that have similarly high leaf B concentration seem to be related to effective redistribution of B by efflux transporters from the sensitive symplastic compartments to the leaf apoplasm (Reid and Fitzpatrick, 2009).

Species with high B demand may also have a higher capacity to sequester B in the cell walls. When B supply is excessive, the inactivation in soluble complexes appears to be less important, with the exception of some halophytes that use compatible solutes (Section 17.6) such as sorbitol for this purpose (Rozema et al., 1992). If these detoxification mechanisms become limiting, the B concentration in the cytosol may increase, causing metabolic disturbances by complexing with, for example, NAD$^+$, or ribose of RNA (Loomis and Durst, 1992) or inhibiting ureide metabolism in the leaves of nodulated soybean (Lukaszewski et al., 1992).

Within species such as barley, wheat, annual medics (*Medicago* spp.), and field peas (*Pisum sativum*), large genotypic differences exist in the capacity to tolerate high B concentrations in soil or nutrient solution (Nable et al., 1997; Paull et al., 1992b). These differences are based on the capacity to restrict B influx to, or increase efflux from, roots (Reid, 2007). In wheat cultivars varying in sensitivity to high soil B, a close correlation between root and shoot B concentrations or root B concentration and shoot yield was observed, suggesting that the main control over B toxicity is exerted at the root level by regulating net B uptake. In wheat and barley, tolerance is due to lower root B concentrations, thereby restricting transfer of B to the shoot, not to high B tolerance of the tissue (Nable et al., 1997). Root B concentration is decreased in tolerant cultivars by B efflux via an efflux transporter from the BOR family (Sutton et al., 2007). This is a different mechanism than in tomato, where root-to-shoot transport of B rather than root B concentration was the most distinct difference between genotypes (Bellaloui and Brown, 1998). In barley, genotypic differences in restriction of uptake by roots and transport of B into the leaves are closely correlated with similar restrictions in uptake and transport of Si (Nable et al., 1990a).

In barley, differences in the capacity to decrease B uptake among genotypes are well defined genetically (Nable et al., 1997; Paull et al., 1988). The mechanisms involved are likely based on both restricted passive movement of B through the plasma membrane of root cells (Huang and Graham, 1990) as well as the function of a B efflux transporter BOR4 (Miwa et al., 2007; Sutton et al., 2007), but not on differences in root anatomy or transpiration rates (Nable et al., 1997). This uptake restriction holds true over a wide range of external B concentrations (Nable et al., 1990a).

7.8 Chlorine

7.8.1 General

Chlorine (Cl) is ubiquitous in nature and occurs in aqueous solution as the monovalent anion chloride. Chloride salts are readily soluble. Chloride mobility in the soil is high, and its concentration in the soil solution varies over a wide range. In plants, the short- and long-distance mobility of Cl is high. Although Cl occurs mainly as a free anion, or is loosely bound to exchange sites, plants contain more than 130 chlorinated organic compounds that may act as antibiotics and fungicides (Engvild, 1986). Given a widespread presence of chloride in seeds, chemicals, water, and air, precautions are required to induce Cl deficiency in most plant species. Using these precautions, Broyer et al. (1954) demonstrated that Cl was an essential micronutrient for plants. As such, Cl plays important roles in PS II stabilization and in regulating the activity of some enzymes. It is also involved in charge balance, stabilizing the electrical potential of cell membranes, and regulating organellar pH. Chloride also regulates osmolarity, turgor, volume changes, and elongation of plant cells.

In most plant species, the minimum Cl requirement for plant growth is in the micronutrient range of $0.2-0.4$ mg g^{-1} dw. However, Cl$^-$ is readily taken up by plants, reaching similar concentrations to macronutrients and becoming, together with NO$_3^-$, the most abundant anion. In the presence of moderate external Cl$^-$ concentrations (e.g., $1-5$ mM Cl), glycophyte plants, including those traditionally considered to be Cl-sensitive species, show positive responses to Cl supply in terms of biomass accumulation, water balance, cell elongation, and leaf expansion. It is now clear that Cl plays quantitatively important roles that cannot be fulfilled by other organic or inorganic anions when supplied in concentrations similar to macronutrients. Chlorine has also been discussed as a beneficial element when present at macronutrient concentrations (Colmenero-Flores et al., 2019; Franco-Navarro et al., 2016, 2019; Maron, 2019; Wege et al., 2017). When present at concentrations similar to a macronutrient, Cl has specific functions that increase plant efficiency in the use of water, N, and CO$_2$ and also improve drought resistance. The Cl concentrations required for these functions may be insufficient in many agricultural soils, particularly in regions far from the marine coast. Therefore, many crops might benefit from Cl fertilization to a greater extent than traditionally believed. In agriculture, there is more concern about Cl$^-$ toxicity in salinized soils than Cl$^-$ deficiency (Section 18.5.3). Chloride toxicity is primarily a consequence of (1) excessive Cl$^-$ accumulation in photosynthetic tissues and (2) Cl$^-$/NO$_3^-$ antagonism that impairs NO$_3^-$ nutrition due to competition for the shared membrane transport mechanisms.

Many studies have focused on the Cl$^-$ exclusion and compartmentalization mechanisms that are important in reducing ion toxicity under salt stress, but little attention has been paid to understanding how Cl$^-$ transport is regulated in plants to ensure adequate nutrition under common environmental conditions. Active transporters involved in Cl$^-$ uptake by roots have been identified only recently. The relative NO$_3^-$/Cl$^-$ selectivity is dependent on the plant species/genotype and can be modified by developmental and environmental factors. When both nutrients are present in the soil, plants prioritize the uptake of NO$_3^-$ over that of Cl$^-$, but when NO$_3^-$ availability is low, plants stimulate Cl$^-$ uptake that results in more efficient utilization of the scarce NO$_3^-$ (see Section 7.8.7.1).

7.8.2 Uptake, transport, and homeostasis

The symplastic pathway dominates Cl$^-$ uptake, which is regulated through changes in the maximum transport capacity and affinity for Cl$^-$ (Brumós et al., 2010; Gong et al., 2011; Wen and Kaiser, 2018; White and Broadley, 2001). Probably as a consequence of a negative feedback mechanism, the uptake capacity is induced by Cl starvation, but inhibited after Cl resupply (Brumós et al., 2010; Cram, 1983).

7.8.2.1 Net Cl$^-$ uptake

Net Cl$^-$ uptake is the consequence of combined activities of influx and efflux transporters. Under most circumstances, the Cl$^-$ electrochemical gradient is strongly favorable to Cl$^-$ efflux from plant cells. On the one hand, cytosolic Cl concentrations of 10−15 mM in glycophyte plants (Lorenzen et al., 2004; Saleh and Plieth, 2013; Teakle and Tyerman, 2010) are normally higher than the external Cl concentration in nonsaline soils (e.g., 0.06-0.25 mM Cl; Geilfus, 2018a). In addition, the plasma membrane H$^+$-ATPase generates a high electrical potential across the plasma membrane (E_m) of −120 to −160 mV (Sze et al., 1999). Therefore, Cl anions must be transported into plant cells against both the chemical and the electrical components of its electrochemical gradient; this transport is driven by the Cl$^-$/H$^+$ symporters with an estimated coupling ratio of 1 Cl$^-$ to 2 H$^+$ (Felle, 1994; Jacoby and Rudich, 1980; Yamashita et al., 1996).

The arabidopsis nitrate transporter 1/peptide transporter 6.3 (AtNPF6.3) is a dual-affinity NO$_3^-$ transporter and receptor (transceptor; Wang et al., 2021). Even though AtNPF6.3 is highly selective for NO$_3^-$, it can transport Cl$^-$ when NO$_3^-$ availability is low (Liu et al., 2020a,b; Xiao et al., 2021). Furthermore, plant species such as maize and barrel medic possess more than one transporter of the NPF6.3 subclade selective for both NO$_3^-$ (ZmNPF6.6 and MtNPF6.7) and Cl$^-$ (ZmNPF6.4 and MtNPF6.5) (Wen et al., 2017; Xiao et al., 2021), suggesting that plant species with different Cl requirements have transport mechanisms that differ in NO$_3^-$ versus Cl$^-$ selectivity (Section 7.8.7.1).

The second component of the net uptake process, Cl$^-$ efflux from cells, is down the electrochemical gradient and, therefore, mediated mostly by anion channels (Britto and Kronzucker, 2006). Chloride release has been proposed to make a considerable contribution to net Cl$^-$ uptake at E_m values more negative than −50 mV (Babourina et al., 1998a), representing a mechanism to fine-tune the cytosolic Cl concentrations. Electrophysiological studies have registered the activity of anion efflux channels that exhibit properties of the rapid (R-type) or slow (S-type) conductances in the plasma membrane of epidermal and/or cortical root cells (Colmenero-Flores et al., 2019; Roberts, 2006). The R- and S-type anion channels are encoded by genes of the ALMT (ALuminum-activated Malate Transporter) and SLAC/SLAH (SLow-type Anion Channel−associated/SLAC1 Homolog) families in plants, respectively. No member of the ALMT or SLAC/SLAH families has been associated yet with the regulation of net Cl$^-$ uptake in plants. Efflux of Cl$^-$ from the roots to the rhizosphere can also be mediated by AtNPF2.5 under salt stress conditions (Li et al., 2016).

7.8.2.2 Root-to-shoot Cl$^-$ translocation

Once in the cytosol of epidermal or cortical root cells, Cl$^-$ follows the symplastic pathway to the xylem-pole pericycle of the root vasculature following its chemical gradient. The loading of Cl$^-$ into the root xylem is electrochemically passive, mainly facilitated by the plasma membrane anion channels (Gilliham and Tester, 2005). Xylem loading is a key step regulating Cl$^-$ accumulation in the aerial part of plants (Brumós et al., 2010; Franzisky et al., 2019; Gong et al., 2011). Several types of anion channels permeable to Cl$^-$ have been identified in the root cells associated with xylem vessels (Gilliham and Tester, 2005; Kohler and Raschke, 2000). The S-type anion channels of arabidopsis, AtSLAH3 and AtSLAH1, regulate the root-to-shoot translocation of NO$_3^-$ and Cl$^-$ (Cubero-Font et al., 2016). Under optimal growing conditions, with plants requiring a synchronized supply of eNO$_3^-$ and Cl$^-$ to the shoot, both anion transporters are coexpressed in the cells of the xylem-pole pericycle. The SLAH3/SLAH1 complex exhibits high Cl$^-$ conductance, allowing significant release of both NO$_3^-$ and Cl$^-$ into xylem vessels. Abiotic stresses such as water deficit or salinity strongly repress the *AtSLAH1* transcription in an ABA-dependent manner (Cubero-Font et al., 2016; Qiu et al., 2016). The subsequent replacement of the SLAH3/SLAH1 heteromer by the SLAH3/SLAH3 homomer significantly reduces SLAH3 Cl$^-$ conductance and Cl$^-$ transport from roots to shoot, but not that of NO$_3^-$. This regulatory mechanism ensures adequate Cl nutrition while adapting Cl homeostasis to the changing environmental conditions, enhancing root osmoregulation under water deficit, or reducing shoot Cl accumulation under salinity stress.

7.8.2.3 Endomembrane Cl$^-$ transporters

Endomembrane Cl$^-$ transporters include members of different anion transporter families in plants. The CLC (chloride channel) family comprises both anion channels and anion/H$^+$ antiporters with variable selectivity for Cl$^-$ and NO$_3^-$ (De Angeli et al., 2009; Zifarelli and Pusch, 2010). Vacuolar CLC channels play important roles in NO$_3^-$ storage (De Angeli et al., 2006), cell osmoregulation and stomatal movement (Jossier et al., 2010), and Cl detoxification under NaCl stress (Nguyen et al., 2016; Wei et al., 2016). The CLC channels are also present in the thylakoid membrane (Section 7.8.4.2) and the Golgi/*trans*-Golgi network, possibly participating in the regulation of lumenal pH (Marmagne et al., 2007; von der Fecht-Bartenbach et al., 2007). The vacuolar DTX/MATE (tonoplast detoxification efflux carrier/multidrug and toxic compound extrusion) anion channels DTX33 and DTX35 are involved in the Cl$^-$ compartmentalization to support cell expansion (Zhang et al., 2017). The vacuolar Cl$^-$ channel AtALMT9 is involved in regulating the stomatal aperture and Cl$^-$ sequestration in roots (Baetz et al., 2016; De Angeli et al., 2013). The Golgi/trans-Golgi network cation-Cl$^-$ cotransporter AtCCC mediates the coordinated symport of K, Na, and Cl and regulates shoot Cl accumulation through a poorly understood mechanism (Colmenero-Flores et al., 2007; Henderson et al., 2015). The endoplasmic reticulum GmNcl/SALT3/CHX1 transporter has been associated with salinity tolerance in soybean, reducing the shoot Na and Cl accumulation (Do et al., 2016; Guan et al., 2014; Liu et al., 2016).

7.8.2.4 Chloride recirculation

Chloride is relatively mobile in the phloem, and its recirculation represents about 20% of the xylem flux (White and Broadley, 2001). In the phloem sap, Cl$^-$ concentrations may be in the order of 120 mM, and Cl$^-$ may play a role in phloem loading and unloading of sugars. The membrane transport proteins involved in Cl$^-$ partitioning through the phloem remain unexplored so far.

7.8.3 Charge balance

As an abundant electrolyte available in soils, highly mobile in the plant and not assimilated in the anabolic processes, Cl$^-$ is the perfect counterion to balance the electric charge of important cations such as K, H, and Ca. Most functions of Cl$^-$ originate from this role, including stabilizing the OEC of PSII (Section 7.8.4), dissipating the electric component of the proton-motive force of endomembranes and the plasma membrane (Section 7.8.2), and as an osmolyte, balancing the charges of cations such as Ca and Na in the vacuole. Therefore, Cl$^-$ plays important roles in the stabilization of the electric potential of cell membranes, in the regulation of pH gradients and electrical excitability, and in the signal perception and transduction because different signals (elicitors, light, pressure) cause membrane depolarization by activating the anion efflux (Hänsch and Mendel, 2009; Spalding, 2000; White and Broadley, 2001).

7.8.4 Photosynthesis and chloroplast performance

Chloride is the most abundant anion in the chloroplast stroma (50–90 mM Cl; Neuhaus and Wagner, 2000), and therefore the chloroplast envelopes and the thylakoid membrane show the high permeability for Cl$^-$ (Bose et al., 2017; Heber and Heldt, 1981). Importantly, the functions of Cl$^-$ in the chloroplast cannot be replaced by other anions such as NO$_3^-$, probably due to the presence of Cl$^-$-selective transporters in the chloroplast membranes.

7.8.4.1 Photosystem II oxygen-evolving complex

In 1946 Warburg and Lüttgens showed that Cl is required in the OEC of PSII; since then, the involvement of Cl in the splitting of water at the oxidizing site of PSII, that is, in O$_2$ evolution (Fig. 7.37), has been confirmed in many studies (see Chapter 5).

Two Cl molecules are required to maintain the coordination structure of the Mn$_4$CaO$_5$ cluster, facilitating the H$^+$ flux from the water oxidation complex to the thylakoid lumen, thereby keeping the OEC fully active (Kawakami et al., 2009; Suzuki et al., 2013). To fulfill this function, the chloroplast Cl concentration needed to meet the Cl requirement in glycophyte plants is 2 mM (Raven, 2020), a demand that is met with micromolar concentrations of external Cl. However, it has been shown recently that photoassembly of the Mn$_4$CaO$_5$ cluster within the PSII-OEC protein requires 40 mM Cl in the chloroplast to saturate the Cl requirement for the PSII-OEC assembly/reassembly (Vinyard et al., 2019).

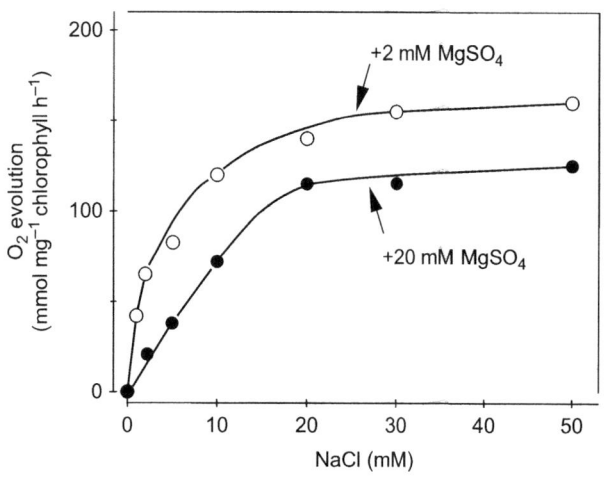

FIGURE 7.37 O$_2$ evolution from Cl-depleted PSII particles of the spinach (*Spinacia oleracea*) chloroplasts at different concentrations of NaCl and 2 or 20 mM MgSO$_4$. *Based on Itoh and Uwano (1986).*

Hence, plants require the tissue Cl concentrations comparable to those of a macronutrient to achieve the maximum rate of PSII-OEC assembly in the ontogeny of thylakoids and for the optimal repair of photodamaged thylakoids (Raven, 2020).

7.8.4.2 Photoprotection and fine-tuning of photosynthesis to changes in light intensity

Chloride is required for adjusting electron transport to changing light conditions and for preventing photoinhibition (Duan et al., 2016; Herdean et al., 2016). Upon illumination and stimulation of photosynthetic electron transport, the electrical component of the resulting proton-motive force (PMF) generated in the thylakoid is dissipated through the activation of AtVCCN1, a voltage-dependent Cl$^-$-selective anion channel that mediates Cl$^-$ influx into the thylakoid lumen. This activity increases the chemical component of the PMF (the thylakoid pH gradient), resulting in (1) downregulation of electron transport at the cytochrome b$_6$f complex and (2) induction of photoprotection through NPQ that results in the dissipation of excess light as heat. In addition, the CLCe Cl$^-$ channel adjusts photosynthesis through a proposed pH-independent mechanism, likely involving Cl homeostasis (Dukic et al., 2019).

7.8.5 Cell osmoregulation and turgor

Plant Cl concentrations normally exceed the critical Cl deficiency concentration (around 0.2 mg g^{-1} dw) by two orders of magnitude, becoming important in osmotic adjustment, plant water balance, and turgor. Tobacco plants supplemented with Cl in the range of macronutrient concentrations (e.g., 1–5 mM Cl in the nutrient media) have higher leaf osmolarity, water content, and turgor than plants supplemented with equivalent concentrations of anionic macronutrients such as NO$_3^-$, phosphate, or sulfate (Franco-Navarro et al., 2016). In the apoplast, a role of Cl in the root pressure and the xylem volume flow is also relevant (Chapter 2).

7.8.5.1 Vacuolar compartmentalization and proton-pumping V-type ATPase

The large hypertonic vacuole regulates cell osmolarity, with Cl$^-$ being the dominant inorganic anion of the bulk tissue (50–150 mM Cl, or higher) at the normal plant Cl concentrations. Although it is generally assumed that Cl$^-$ fulfills a nonspecific osmotic function, recent evidence shows that it is quantitatively and qualitatively a preferred osmolyte in plants and cannot be replaced by other organic or inorganic anions (Franco-Navarro et al., 2016; Wege et al., 2017). The proposed reasons for the superior osmoregulatory capacity of Cl$^-$ are: (1) providing consistency in osmoregulation because Cl$^-$ (contrary to NO$_3^-$ or malate for example) is not metabolically assimilated; (2) the atypically high stability of interactions with water molecules in the solvation shell of halogen anions (Kropman and Bakker, 2001); and (3) the stimulation by Cl of the tonoplast H$^+$-pumping ATPase (H$^+$-V-type ATPase). Membrane-bound H$^+$-pumping ATPases and PP$_i$ases are stimulated by various cations and anions (Table 7.39). The plasma membrane H$^+$-pumping ATPase is stimulated by monovalent cations, K$^+$ in particular, whereas the H$^+$-V-type ATPase in endomembranes is stimulated by Cl$^-$ and other anions (Sze, 1985). In addition, Cl$^-$ dissipates the electrical potential (positive inside) generated by the electrogenic H$^+$-V-type ATPase in vesicles, which is accompanied by an increase in the pH gradient (acid inside). The fluorescently labeled AtClCd Cl$^-$ channel co-localizes with the H$^+$-V-type ATPase in the *trans*-Golgi network,

TABLE 7.39 Proton-pumping ATPase activity of tonoplast vesicles with different salt forms.

Salt (10 mM of monovalent ion)	ATPase stimulation (% of control)
No monovalent ion	10
KCl (control)	100
NaCl	102
NaBr	87
KNO$_3$	21
K$_2$SO$_4$	3

Source: Based on Mettler et al. (1982).

and mutations in either of the two transporters result in phenotypes with impaired root growth and cell elongation, suggesting a functional linkage between Cl$^-$ transport and the H$^+$-V-type ATPase activity (Padmanaban et al., 2004).

7.8.5.2 Cell volume regulation and stomatal function

Turgor is the driving force of cell expansion, requiring a coordinated transport of dominant inorganic ions (e.g., K$^+$ and Cl$^-$) through both the plasma membrane and the tonoplast. Chloride fluxes across the plasma membrane are highly relevant for osmoregulation, not only because of their role in salt partitioning during turgor normalization but also because Cl$^-$ efflux through anion channels strongly depolarizes the plasma membrane. This depolarization activates the outward-rectifying K$^+$ channels, which provides a positive feedback for a massive discharge of salts and water (Sanders and Bethke, 2000). Consistent with this role, the Cl$^-$ transport-based mechanisms are specifically required to regulate cell volume and the activity of specialized motor cells in plants, such as the pulvini of leguminous plants during seismonastic and nyctinastic leaf movements (Fromm and Eschrich, 1988; Iino et al., 2001; Moran, 2007). For example, the asymmetric expression of the rain tree (*Samanea saman*) Cl$^-$ channel SsSLAH1 between the adaxial and abaxial sides of the leaflet plays a key role in nyctinastic opening (Oikawa et al., 2018).

Another important case is the regulation of guard cell volume through the activity of ion transporters (Roux and Leonhardt, 2018). Stomatal opening requires the NO$_3^-$ and/or Cl$^-$ influx through the H$^+$-coupled anion transporters such as AtNPF6.3 (Guo et al., 2003). Intracellular ion accumulation drives osmotic water uptake and a cell volume increase, mainly accounted for by ion compartmentalization and increased vacuolar volume (Section 7.8.2.3). The plasma membrane anion fluxes are of particular relevance for triggering stomatal closure because the guard cell signaling machinery (e.g., ABA perception and signal transduction) directly activates the plasma membrane anion channels SLAC1, SLAH3, and ALMT12 that mediate the efflux of Cl$^-$, NO$_3^-$, and malate, respectively, leading to the plasma membrane depolarization, K$^+$ efflux, and water loss (Saito and Uozumi, 2019).

In plant species such as onion (*Allium cepa*) that do not synthesize malate in their guard cells, Cl is essential for stomatal functioning (Schnabl, 1980). Members of the *Arecaceae* that have chloroplasts containing starch in their guard cells also require Cl for stomatal functioning. Chlorine deficiency in coconut (*Cocos nucifera*) plants delays stomatal opening by about 3 h compared to the Cl-replete plants (Braconnier, d'Auzac, 1990). The guard cells of another palm tree, *Phoenix dactylifera*, release Cl$^-$ rather than NO$_3^-$ during stomatal closure (Müller et al., 2017).

7.8.5.3 Cell elongation and plant growth

When Cl availability is low, tobacco plants prioritize Cl accumulation in the actively growing organs, suggesting a biological role in plant cell growth (Colmenero-Flores et al., 2019). Chloride is required to stimulate cell elongation in the leaf cells of tobacco (Franco-Navarro et al., 2016), epidermal cells from the elongating internodes of pea (Yamagami et al., 2004), the elongating stigma of grasses at the onset of anthesis (Heslop-Harrison and Reger, 1986), the elongating coleoptile of grass seedlings (Burdach et al., 2014), and the elongating pollen tube (Gutermuth et al., 2013). For some of these processes, substituting Cl$^-$ by NO$_3^-$ or other inorganic anions prevents or reduces cell elongation (Burdach et al., 2014; Franco-Navarro et al., 2016), revealing a Cl-specific response. Signaling in some of these processes requires the phytohormone auxin (Burdach et al., 2014) that, in turn, specifically stimulates Cl$^-$ uptake by cells (Babourina et al., 1998b; Iino et al., 2001; Yamagami et al., 2004) as a prerequisite for cell elongation. Chloride may stimulate extension growth in some legume species, such as pea and faba bean, that contain substantial amounts of the chlorinated auxin IAA (Cl-IAA) in their seeds. Cl-IAA enhances hypocotyl elongation 10-fold more than IAA, probably because of its higher resistance to degradation by peroxidases (Hofinger and Böttger, 1979). Different genetic approaches have shown functional linkages between the Cl homeostasis and the plant cell elongation (Chen et al., 2016a,b; Colmenero-Flores et al., 2007; Zhang et al., 2017).

7.8.6 Plant water balance and water relations

7.8.6.1 Water storage capacity

Tissue Cl concentrations equivalent to those of the macronutrients result in larger epidermal and mesophyll leaf cells with higher water content. This implies a greater storage capacity of water in the leaf tissue, which is in agreement with a greater leaf succulence in Cl-treated tobacco plants (Franco-Navarro et al., 2016), confirming the role of Cl in stimulating hydration of plant tissues.

7.8.6.2 Water relations, water-use efficiency, and drought resistance

Another consequence of having larger epidermal cells is the occurrence of fewer and larger guard cells, leading to a significant reduction of stomatal density and water consumption in tobacco plants (Franco-Navarro et al., 2019). However, a decrease in stomatal conductance (g_s) is expected to impair photosynthetic efficiency in C_3 plants. Interestingly, when Cl concentrations are similar to those of the macronutrients, chloroplast performance is enhanced, particularly the mesophyll diffusion conductance to CO_2 (g_m), due to a higher surface area of chloroplasts exposed to the intercellular airspace of mesophyll cells (Franco-Navarro et al., 2019). Therefore, the higher g_m compensates for the lower g_s, resulting in higher water-use efficiency (WUE) (Franco-Navarro et al., 2016, 2019; Maron, 2019). Increasing crop yield while also improving WUE is particularly challenging in C_3 plants, and it is a major focus of research to improve sustainability of agriculture.

Irrigating tobacco and tomato plants with a solution containing 5 mM Cl improved both drought avoidance and dehydration tolerance, making Cl-treated plants more resistant to water deficit than plants treated with equivalent concentrations of NO_3^- or sulfate + phosphate salts (Franco-Navarro et al., 2021). Water deficit avoidance results from lower water consumption, whereas dehydration tolerance is based on the photosynthetic tissues exhibiting greater osmotic adjustment, turgor, and relative water content (Franco-Navarro et al., 2021; Nieves-Cordones et al., 2019).

7.8.7 Interaction with nitrate

7.8.7.1 Regulation of N-use efficiency

Chloride and NO_3^- are the most abundant inorganic anions in plants, having similar physical and osmoregulatory properties, including preferential compartmentalization in the vacuole. They share membrane transport mechanisms (Section 7.8.2) and, therefore, interact competitively (Wege et al., 2017). Thus, many studies have reported Cl^- as being detrimental to agriculture due to the purported negative effect on NO_3^- uptake (Xu et al., 2000). However, contrary to impairing N nutrition, increasing Cl^- availability can improve NO_3^- utilization and N-use efficiency in plants treated with NO_3^- as the sole N source (Rosales et al., 2020). This is largely due to Cl^- improving NO_3^- assimilation efficiency and having moderate effect on NO_3^- uptake. It has been proposed that, when present in tissues at concentrations similar to those of macronutrients, Cl^- is compartmentalized preferentially over NO_3^- that is no longer required to fulfill an osmotic function in the vacuole, making it more available for assimilation into biomass, which ultimately results in higher N-use efficiency.

Chloride inhibits NO_3^- uptake by roots weakly, but NO_3^- inhibits Cl^- uptake by roots strongly. Given the importance of NO_3^- in plant nutrition, plants are equipped with different families of anion transporters that optimize NO_3^- uptake by the roots and its long-distance transport (Section 6.1). Uptake transporters are frequently more selective for NO_3^- than Cl^- (Colmenero-Flores et al., 2019; Wege et al., 2017; Wen and Kaiser, 2018), and Cl^- uptake is inhibited when NO_3^- is available (Glass and Siddiqi, 1985). The NIN-like protein (NLP) transcription factor triggers the transcriptional activation of the NO_3^--responsive genes when NO_3^- is supplied (Castaings et al., 2009; Liu et al., 2017). In barrel medic, the MtNLP1 transcription factor upregulates the NO_3^--selective MtNPF6.7 transporter and downregulates the Cl^--selective MtNPF6.5 transporter in response to the NO_3^- supply (Xiao et al., 2021). This has been interpreted as the mechanism regulating plant preference for NO_3^- over Cl^- when this valuable N source is available in the soil. Conversely, when little NO_3^- is available, Cl^- influx is less inhibited, and/or is induced, increasing the intracellular Cl concentration, which in turn, improves the use efficiency of the scarcely available NO_3^-.

Adequate management of optimal NO_3^-/Cl^- ratios could reduce the NO_3^- input in agriculture without reducing plant performance. In addition, Cl-dependent decrease of plant NO_3^- content is also a strategy to reduce excessive NO_3^- accumulation in vegetables, which may be detrimental to human health (European Food Safety Authority, 2008; Inal et al., 1998; Rosales et al., 2020).

7.8.7.2 Regulation of N metabolism

A role of Cl in N metabolism is indicated by its stimulating effect on asparagine synthetase that uses glutamine as a substrate.

$$\text{Glutamine} \xrightarrow[\text{Asparagine synthetase}]{(NH_3)} \text{Asparagine + Glutamic acid}$$

Chloride enhances this transfer 7-fold, and the affinity of the enzyme for the substrate 50-fold (Rognes, 1980). Therefore, Cl may play a role in N metabolism in plant species in which asparagine is the major compound in the long-distance transport of soluble N (Section 6.1).

7.8.8 Chloride supply, deficiency, plant growth, and crop yield

7.8.8.1 Energy/metabolic saving

When present in similar concentrations as macronutrients, Cl increases the efficiency in the use of water, CO_2, and NO_3-N, therefore increasing the fresh and dry biomass of plants (Franco-Navarro et al., 2016, 2019; Rosales et al., 2020). The cost of energy to generate turgor by the Cl^- influx and compartmentalization is lower than that of synthesizing organic solutes (e.g., malate). It has been proposed that the energy cost of compartmentalizing Cl^-, based on the activity of anion channels, is lower than that of sequestering NO_3^- that is dependent to a greater extent on the secondary active transport mechanisms (Section 7.8.2.3; De Angeli et al., 2006; Wege et al., 2017).

7.8.8.2 Chloride deficiency

In most plants the principal effects of Cl deficiency are wilting and a reduction in leaf surface area and thereby plant biomass. When plants are extremely Cl-deficient (below requirements as a micronutrient), the decrease in leaf area is the result of a reduction in cell division rates (Terry, 1977). At moderate Cl deficiency (i.e., sufficient to meet micronutrient requirements but not present at tissue concentrations similar to a macronutrient), leaf cell elongation, but not cell division, is impaired (Section 7.8.5.3). Wilting of leaves, especially at leaf margins, is a typical symptom of Cl deficiency, even in water culture, when plants are exposed to full sunlight (Broyer et al., 1954), with severe deficiency inducing curling of the youngest leaves followed by shriveling and necrosis (Whitehead, 1985). In roots of Cl-deficient plants, the cell division and especially the cell extension are also impaired, which results in subapical swelling (Smith et al., 1987; Bergmann, 1992) and enhanced formation of short laterals (Johnson et al., 1957). Resupplying Cl to deficient plants restores growth within a few days. The critical level of Cl deficiency is a trait that depends on both plant species and variety (Fig. 7.38).

In plant species (such as red clover) with relatively low Cl requirements (<1 mg Cl g^{-1} leaf dw), the demand can be covered by the concentration of 100 μM Cl in the nutrient solution. At 10 μM Cl supply, shoot dw decreases by 50% (Chisholm and Blair, 1981), indicating that the capacity for Cl^- uptake is not as high as, for example, in the case of phosphate, whereby the greater phosphate requirement in the leaf dw (Section 6.2) can be covered by a supply of less than 10 μM. In plants with high Cl requirements, such as kiwifruit and coconut, there is a greater risk of Cl deficiency. The critical deficiency concentration in kiwifruit leaves is about 2 mg Cl g^{-1} dw and, thus, Cl deficiency can be induced readily in this species. In coconut that has a particularly high Cl requirement of about 6 mg Cl g^{-1} leaf dw, besides wilting and premature senescence of leaves, frond fracture and stem cracking are typical symptoms of Cl deficiency (Table 7.40).

7.8.8.3 Crop yield

In agriculture, there is generally little appreciation of the benefits that adequate Cl nutrition can bring to crop yield. The amounts of Cl^- normally available in nonsaline soils fully satisfy the micronutrient requirements of glycophyte plants but do not afford the benefits that occur at tissue Cl concentrations similar to those of macronutrients, which would

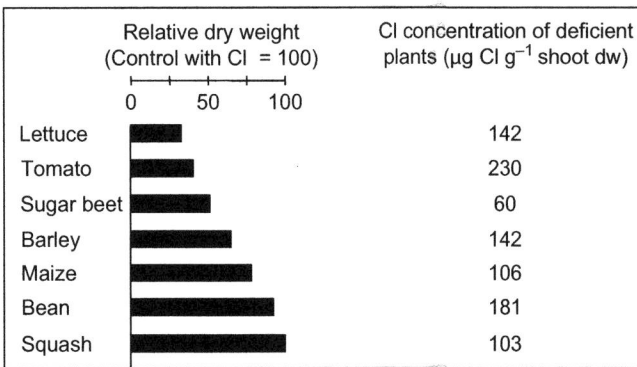

FIGURE 7.38 Relative shoot dry weight and Cl concentrations of Cl-deficient plants. *Redrawn from Johnson et al. (1957).*

TABLE 7.40 The Cl and K concentrations in leaves and the growth disorders in coconut (*Cocos nucifera*) trees at different rates of KCl fertilization.

Fertilization (kg KCl tree^{-1})	Leaf concentration (mg g^{-1} dw) K	Leaf concentration (mg g^{-1} dw) Cl	Growth disorders (%) Frond fracture	Growth disorders (%) Stem cracking
0	1.6	0.07	12	27
2.25	1.6	0.41	1.7	8.1
4.50	1.7	0.51	1.2	4.5

Source: Based on Uexküll (1985).

require 1–5 mM Cl in the soil solution. Assuming a critical micronutrient deficiency concentration of 1 mg Cl g^{-1} shoot dw, a crop requirement would be in the range of 4–8 kg Cl ha^{-1}, which approximates the Cl input from rain in areas distant from oceans and is about 10 times smaller than the input from rain at sites near oceans. In highly leached soils with a low Cl input from rain and other sources, Cl deficiency may occur even in plant species with a low Cl requirement (Ozanne, 1958). In agreement with these observations, the most frequent leaf Cl concentration from 670 species belonging to 138 families of terrestrial seed plants collected from their natural habitats was around 5 mg g^{-1} dw (Watanabe et al., 2007), which is well above the critical requirement as a micronutrient, but below the Cl concentration required for many beneficial functions (e.g., 20–50 mg g^{-1} DW in tobacco plants; Franco-Navarro et al., 2016). Therefore, plants might frequently benefit from Cl fertilization. Indeed, substantial responses to Cl-containing fertilizers have been reported for many different crops around the world (Chen et al., 2010a,b; Xu et al., 2000).

The best documented example of agricultural Cl deficiency is in the Great Plains of the United States. These regions are characterized by low Cl deposition in rain (<0.5 kg ha^{-1}), leached soils with low Cl concentrations, and cropped with species (such as wheat and barley) that have high production/high Cl demand (Fixen, 1993). Yield reductions in the order of 7%–10% in wheat due to Cl deficiency have been reported in North and South America. Yield increases with KCl fertilization may be a combination of various effects, including alleviation of Cl deficiency (Schwenke et al., 2015), suppression of root rot diseases (Timm et al., 1986), or a combination of suppression of diseases and improved plant water relations. Studies have reported that Cl increases the resistance to several diseases when present in tissues at concentrations similar to those of macronutrients (Chen et al., 2010a,b). In a controlled glasshouse experiment, several physiological disorders impairing the growth and yield of durum wheat, including leaf-spotting symptoms, were not caused by plant pathogens but were a direct consequence of low soil Cl (Schwenke et al., 2015).

7.8.9 Chlorine toxicity

Chlorine toxicity occurs worldwide in saline soils, particularly in arid and semiarid regions (Brumós et al., 2009; Colmenero-Flores et al., 2020; Geilfus, 2018a; Li et al., 2017; Teakle and Tyerman, 2010). Excessive Cl$^-$ accumulation impairs NO_3^-, phosphate, and sulfate nutrition; inhibits photosynthetic CO_2 fixation and protein biosynthesis; reduces root and tuber growth; and impairs food quality and safety (Geilfus, 2018b).

On average, Cl concentrations in the external solution of more than 20 mM can lead to Cl toxicity in sensitive plant species, whereas in tolerant species the external concentration can be four to five times greater than this without reducing growth. It is not expected that Cl is more metabolically toxic in Cl-sensitive species. Rather, these species effectively exclude Na from leaf blades, and Cl that is transported to the leaves becomes toxic (Munns and Tester, 2008). More than 3.5 mg Cl g^{-1} leaf dw (10 mM Cl in the leaf water) has been reported as toxic to sensitive species such as most fruit trees, bean, and cotton under salt stress. By contrast, 20–30 mg Cl g^{-1} leaf DW (<60–90 mM Cl in the leaf water) is not harmful to tolerant species such as barley, spinach, lettuce, and sugar beet (White and Broadley, 2001; Xu et al., 2000). However, leaf Cl concentrations in the beneficial range equivalent to macronutrients (20–50 mg g^{-1} DW) clearly exceed these toxicity thresholds. This apparent discrepancy indicates that plant total Cl concentrations cannot be correlated with the resulting impact (beneficial or toxic) when experiments with different Cl treatments are compared.

A high degree of resistance to salinity of halophytes is mostly due to their capacity to accumulate and compartmentalize ions in their aerial tissues, allowing more effective osmotic adjustment and ion detoxification. In glycophyte species, the greater resistance to Cl toxicity under salt stress correlates with a greater capacity to exclude Cl, although compartmentalization also plays a role. Genotypic differences in Cl tolerance are discussed in Sections 18.5.3 and 18.5.4.

7.8.10 Chlorine as micro- and macronutrient - concluding remarks

Relevant functions of Cl as a micronutrient or as a beneficial element when present in tissues at a similar concentration to macronutrients are depicted in Fig. 7.39. In the chloroplast, Cl is an essential cofactor required for stabilization of the OEC of PSII, allowing appropriate organization of the Mn_4CaO_5 cluster, and enabling the oxidation of water. At a concentration similar to that of macronutrients, Cl promotes the assembly of OEC polypeptides for the optimal thylakoid ontogeny and the photodamage repair.

Chloride-dependent photoprotection also relies on Cl^- fluxes that regulate the thylakoid pH gradient, allowing fine-tuning of photosynthetic electron transport to the changes in light intensity. In leaves, Cl concentrations similar to those of macronutrients increase mesophyll diffusion conductance to CO_2 (g_m). These Cl-dependent functions are required for

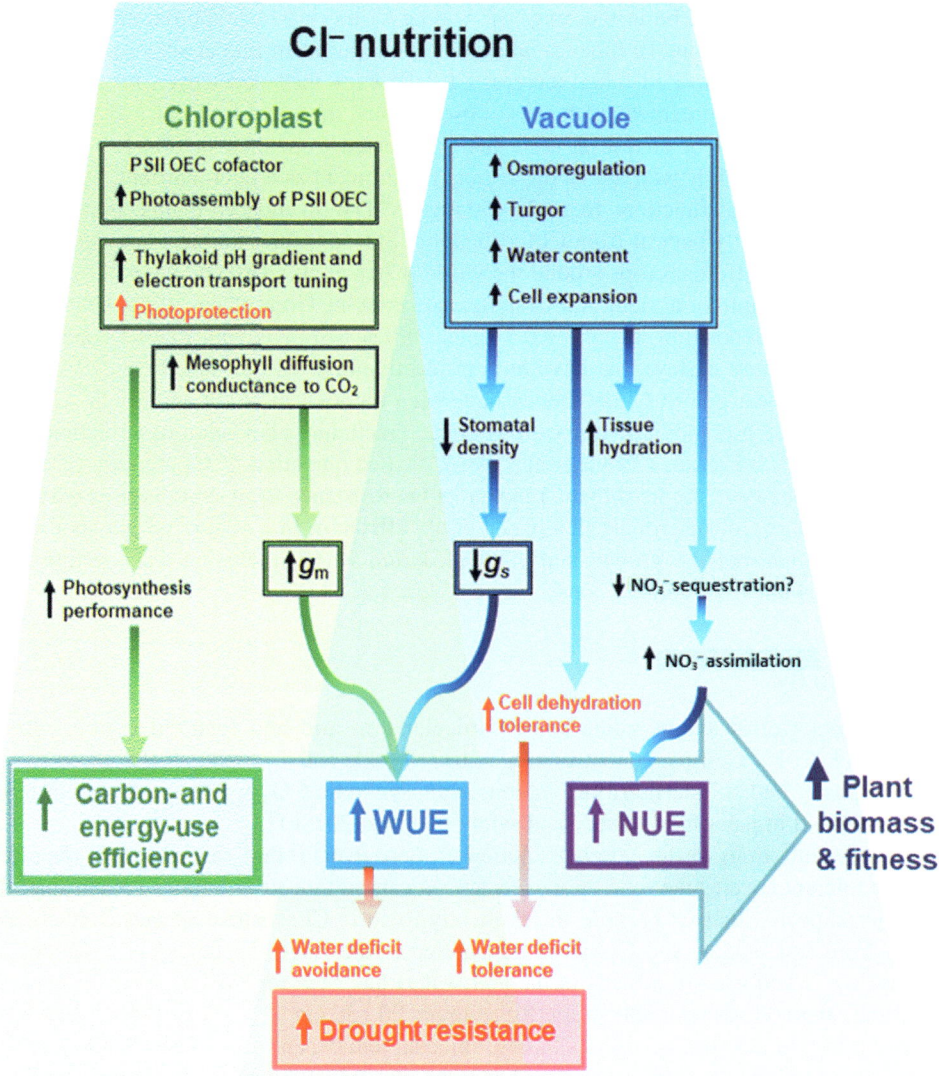

FIGURE 7.39 Relevant roles of Cl in plant growth and performance. Chloroplast and vacuole are the cell organelles where Cl plays important functions when accumulated at concentrations relevant to both micronutrients and macronutrients. Chloride nutrition also improves resistance to stresses associated with drought and high light intensity (red text). g_m, Mesophyll diffusion conductance to CO_2; g_s, stomatal conductance; *NUE*, NO_3^--N use efficiency; *OEC*, oxygen-evolving complex; *PSII*, photosystem II; *WUE*, water-use efficiency.

the optimal chloroplast function and photosynthetic performance. Other functions, such as preventing the diversion of NO_3^- and organic acid anions to the vacuole as osmotica or for charge balance, are expected to improve energy-use efficiency. Millimolar concentrations of Cl in the vacuole also improve osmoregulation, turgor, water relations, and cell expansion, which in turn enables greater water storage capacity, tissue hydration, and succulence. Greater cell expansion, resulting from tissue Cl concentrations similar to those of macronutrients, occurs in the roots and shoot. In the leaf epidermis, a significant increase in cell size results in a reduction of stomatal density, stomatal conductance (g_s) and water consumption. A decrease in photosynthetic capacity caused by lower g_s is compensated for by greater g_m, and probably also by the beneficial effects of Cl on the photosynthetic electron transport function, resulting in greater WUE. Tissue Cl concentrations similar to those of macronutrients improve NO_3^- assimilation, probably as a consequence of decreasing NO_3^- sequestration in the vacuole, resulting in higher NUE. Simultaneous improvement of photosynthetic performance, WUE, and NUE increases plant biomass and improves fitness. Chloride nutrition improves resistance to stresses associated with drought and light intensity. Tissue Cl concentrations similar to those of macronutrients reduce water consumption, avoiding water deficit. In addition, a more efficient osmoregulation increases the relative water content and improves cell dehydration tolerance. Together, water deficit avoidance and tolerance mechanisms improve plant drought resistance. Proper management of Cl fertilization promotes sustainable and resilient agriculture.

References

Abadía, J., Morales, F., Abadía, A., 1999. Photosystem II efficiency in low chlorophyll, iron-deficient leaves. Plant Soil 215, 183–192.

Abadía, J., López-Millán, A.F., Rombola, A., Abadía, A., 2002. Organic acids and Fe deficiency: a review. Plant Soil 241, 75–86.

Abbott, A.J., 1967. Physiological effects of micronutrient deficiencies in isolated roots of *Lycopersicon esculentum*. New Phytol. 66, 419–437.

Abou-Shanab, R.A.I., Angle, J.S., Chaney, R.L., 2006. Bacterial inoculants affecting nickel uptake by *Alyssum murale* from low, moderate and high Ni soils. Soil Biol. Biochem. 38, 2882–2889.

Abreu, I.A., Cabelli, D.E., 2010. Superoxide dismutases-a review of the metal-associated mechanistic variations. Biochim. Biophys. Acta 1804, 263–274.

Abreu, I., Poza, L., Bonilla, I., Bolaños, L., 2014. Boron deficiency results in early repression of a cytokinin receptor gene and abnormal cell differentiation in the apical root meristem of *Arabidopsis thaliana*. Plant Physiol. Bioch. 77, 117–121.

Adams, J.F., Burmester, C.H., Mitchell, C.C., 1990. Long-term fertility treatments and molybdenum availability. Fert. Res. 21, 167–170.

Agarwala, S.C., Sharma, C.P., Farooq, S., Chatterjee, C., 1978. Effect of molybdenum deficiency on the growth and metabolism of corn plants raised in sand culture. Can. J. Bot. 56, 1905–1908.

Agarwala, S.C., Chatterjee, C., Sharma, P.N., Sharma, C.P., Nautiyal, N., 1979. Pollen development in maize plants subjected to molybdenum deficiency. Can. J. Bot. 57, 1946–1950.

Aguirre, G., Pilon, M., 2016. Copper delivery to chloroplast proteins and its regulation. Front. Plant Sci. 6, 1250.

Ahmad, M.S.A., Ashraf, M., Hussain, M., 2011. Phytotoxic effects of nickel on yield and concentration of macro-and micro-nutrients in sunflower (*Helianthus annuus* L.) achenes. J. Hazard. Mater. 185, 1295–1303.

Alcaraz, C.F., Martinez-Sánchez, F., Sevilla, F., Hellin, E., 1986. Influence of ferredoxin levels on nitrate reductase activity in iron deficient lemon leaves. J. Plant Nutr. 9, 1405–1413.

Alejandro, S., Cailliatte, R., Alcon, C., Dirick, L., Domergue, F., Correia, D., et al., 2017. Intracellular distribution of manganese by the *Trans*-Golgi network transporter NRAMP2 is critical for photosynthesis and cellular redox homeostasis. Plant Cell 29, 3068–3084.

Allen, M.D., Kropat, J., Tottey, S., Del Campo, J.A., Merchant, S.S., 2007. Manganese deficiency in *Chlamydomonas* results in loss of photosystem II and MnSOD function, sensitivity to peroxides, and secondary phosphorus and iron deficiency. Plant Physiol. 143, 263–277.

Alloway, B.J., 2009. Soil factors associated with zinc deficiency in crops and humans. Environ. Geochem. Health 31, 537–548.

Alloway, B.J., Tills, A.R., 1984. Copper deficiency in world crops. Outlook Agric. 13, 32–42.

Andaluz, S., Rodriguez-Celma, J., Abadía, A., Abadía, J., López-Millán, A.F., 2009. Time course induction of several key enzymes in *Medicago truncatula* roots in response to Fe deficiency. Plant Physiol. Biochem. 47, 1082–1088.

Andreini, C., Banci, L., Bertini, I., Rosato, A., 2006. Counting the zinc-proteins encoded in the human genome. J. Proteome Res. 5, 196–201.

Andreini, C., Bertini, I., Cavallaro, G., Holliday, G.L., Thorntone, J.M., 2008. Metal ions in biological satalyses: from enzyme databases to general principles. J. Biol. Inorg. Chem. 13, 1205–1218.

Andreini, C., Bertini, I., Rosato, A., 2009. Metalloproteomes: A bioinformatic approach. Acc. Chem. Res. 42, 1471–1479.

Andresen, E., Peiter, E., Kupper, H., 2018. Trace metal metabolism in plants. J. Exp. Bot. 69, 909–954.

Appleby, C.A., Bogusz, D., Dennis, E.S., Peacock, W.J., 1988. A role for haemoglobin in all plant roots? Plant Cell Environ. 11, 359–367.

Arkoun, M., Jannin, L., Laîné, P., Etienne, P., Masclaux-Daubresse, C., Citerne, S., et al., 2013. A physiological and molecular study of the effects of nickel deficiency and phenylphosphorodiamidate (PPD) application on urea metabolism in oilseed rape (*Brassica napus* L.). Plant Soil 362, 79–92.

Arulanathan, A.R., Rao, I.M., Terry, N., 1990. Limiting factors in photosynthesis. VI. Regeneration of ribulose 1,5 bisphophate limits photosynthesis at low photochemical capacity. Plant Physiol. 93, 1466–1475.

Asher, C.J., 1991. Beneficial elements, functional nutrients, and possible new essential elements. In: Mortvedt, J.J., Cox, F.R., Shuman, L.M., Welch, R.M. (Eds.), Micronutrients in Agriculture, 2nd ed. Soil Science Society of America, Madison, WI, USA, pp. 703−723.

Assunção, A.G.L., Bleeker, P., ten Bookum, W.M., Vooijs, R., Schat, H., 2008. Intraspecific variation of metal preference patterns for hyperaccumulation in *Thlaspi caerulescens*: evidence from binary metal exposures. Plant Soil 303, 289−299.

Auld, D.S., Bergman, T., 2008. The role of zinc for alcohol dehydrogenase structure and function. Cell. Mol. Life Sci. 65, 3961−3970.

Aung, M.S., Masuda, H., 2020. How does rice defend against excess iron? Physiological and molecular mechanisms. Front. Plant Sci. 11, 1102.

Aung, M.S., Masuda, H., Nozoye, T., Kobayashi, T., Jeon, J.S., An, G., et al., 2019. Nicotianamine synthesis by OsNAS3 is important for mitigating iron excess stress in rice. Front. Plant Sci. 10, 660.

Ayala, M.B., Sandmann, G., 1988. The role of copper in respiration of pea plants and heterotrophically growing *Scenedesmus* cells. Z. Naturforsch. C. 43, 438−442.

Ayala, M.B., Gorge, J.L., Lachica, M., Sandmann, G., 1992. Changes in carotenoids and fatty acids in photosystem II of copper deficient pea plants. Physiol. Plant. 84, 1−5.

Babourina, O.K., Knowles, A.E., Newman, I.A., 1998a. Chloride uptake by oat coleoptile parenchyma described by combined influx and efflux transport systems. Aust. J. Plant Physiol. 25, 929−936.

Babourina, O., Shabala, S., Newman, I., 1998b. Auxin stimulates Cl⁻ uptake by oat coleoptiles. Ann. Bot. 82, 331−336.

Badger, M.R., Price, G.D., 1994. The role of carbonic anhydrase in photosynthesis. Annu. Rev. Plant Physiol. Plant Mol. Biol. 45, 369−392.

Baetz, U., Eisenach, C., Tohge, T., Martinoia, E., De Angeli, A., 2016. Vacuolar chloride fluxes impact ion content and distribution during early salinity stress. Plant Physiol. 172, 1167−1181. Available from: https://doi.org/10.1104/pp.16.00183.

Bagci, S.A., Ekiz, H., Yilmaz, A., Cakmak, I., 2007. Effects of zinc deficiency and drought on grain yield of field-grown wheat cultivars in Central Anatolia. J. Agron. Crop Sci. 193, 198−206.

Bai, C., Reilly, C.C., Wood, B.W., 2006. Nickel deficiency disrupts metabolism of ureides, amino acids, and organic acids of young pecan foliage. Plant Physiol. 140, 433−443.

Bañados, M.P., Ibáñez, F., Toso, A.M., 2009. Manganese toxicity induces abnormal shoot growth in 'O'Neal' blueberry. Acta Hortic. 810, 509−512.

Barabasz, A., Palusinska, M., Papierniak, A., Kendziorek, M., Kozak, K., Williams, L.E., et al., 2019. Functional analysis of *NtZIP4B* and Zn status-dependent expression pattern of tobacco ZIP genes. Front. Plant Sci. 9, 1984.

Barber, S.A., 1984. Soil Nutrient Bioavailability. John Wiley and Sons, New York, NY, USA.

Barberon, M., Vermeer, J.E.M., De Bellis, D., Wang, P., Naseer, S., Andersen, T.G., et al., 2016. Adaptation of root function by nutrient-induced plasticity of endodermal differentiation. Cell 164, 447−459.

Barron, A.R., Wurzburger, N., Bellenger, J.P., Wright, S.J., Kraepiel, A.L., Hedin, L.O., 2009. Molybdenum limitation of asymbiotic nitrogen fixation in tropical forest soils. Nat. Geosci. 2, 42−45.

Barros, J., Serk, H., Granlund, I., Pesquet, E., 2015. The cell biology of lignification in higher plants. Ann. Bot. 115, 1053−1074.

Bashir, K., Ishimaru, Y., Shimo, H., Kakei, Y., Senoura, T., Takahashi, R., et al., 2011. Rice phenolics efflux transporter 2 (PEZ2) plays an important role in solubilizing apoplasmic iron. Soil Sci. Plant Nutr. 57, 803−812.

Bashir, K., Seki, M., Nishizawa, N.K., 2019. The transport of essential micronutrients in rice. Mol. Breeding 39, 168. Available from: https://doi.org/10.1007/s11032-019-1077-1.

Bassil, E., Hu, H., Brown, P.H., 2004. Use of phenylboronic acids to investigate boron function in plants. Possible role of boron in transvacuolar cytoplasmic strands and cell-to-wall adhesion. Plant Physiol. 136, 3383−3395.

Bastow, E.L., Garcia de la Torre, V.S., Maclean, A.E., Green, R.T., Merlot, S., Thomine, S., et al., 2018. Vacuolar iron stores gated by NRAMP3 and NRAMP4 are the primary source of iron in germinating seeds. Plant Physiol. 177, 1267−1276.

Baxter, I., Muthukumar, B., Park, H.C., Buchner, P., Lahner, B., Danku, J., et al., 2008. Variation in molybdenum content across broadly distributed populations of *Arabidopsis thaliana* is controlled by a mitochondrial molybdenum transporter (MOT1). PLoS Genetics 4, 12. Available from: https://doi.org/10.1371/journal.pgen.1000004.

Becker, M., Asch, F., 2005. Iron toxicity in rice − conditions and management concepts. J. Plant Nutr. Soil Sci. 168, 558−573.

Becking, J.H., 1961. A requirement of molybdenum for the symbiotic nitrogen fixation in alder. Plant Soil 15, 217−227.

Bell, R.W., McLay, L., Plaskett, D., Dell, B., Loneragan, J.F., 1989. Germination and vigour of black gram (*Vigna mungo* (L.) Hepper) seed from plants grown with and without boron. Aust. J. Agric. Res. 40, 273−279.

Bellaloui, N., Brown, P.H., 1998. Cultivar differences in boron uptake and distribution in celery (*Apium graveolens*), tomato (*Lycopersicon esculentum*) and wheat (*Triticum aestivum*). Plant Soil 198, 153−158.

Bellaloui, N., Reddy, K.N., Gillen, A.M., Abel, C.A., 2010. Nitrogen metabolism and seed composition as influenced by foliar boron application in soybean. Plant Soil 336, 143−155.

Benatti, M.R., Yookongkaew, N., Meetam, M., Guo, W.-J., Punyasuk, N., Abuqamar, S., et al., 2014. Metallothionein deficiency impacts copper accumulation and redistribution in leaves and seeds of Arabidopsis. New Phytol. 202, 940−951.

Bergmann, W., 1992. Nutritional Disorders of Plants − Development, Visual and Analytical Diagnosis. Gustav Fischer Verlag, Jena, Germany.

Berman-Frank, I., Cullen, J.T., Shaked, Y., Sherrell, R.M., Falkowski, P.G., 2001. Iron availability, celluar iron quotas, and nitrogen fixation in *Trichodesmium*. Limnol. Oceanogr. 46, 1249−1260.

Bernal, M., Casero, D., Singh, V., Wilson, G., Grande, A., Yang, H., et al., 2012. Transcriptome sequencing identifies SPL7-regulated copper acquisition genes FRO4/FRO5 and the copper dependence of iron homeostasis in Arabidopsis. Plant Cell 24, 738−761.

Bienfait, H.F., 1989. Prevention of stress in iron metabolism of plants. Acta Bot. Nerl. 38, 105–129.

Billard, V., Ourry, A., Maillard, A., Garnica, M., Coquet, L., Jouenne, T., et al., 2014. Copper deficiency in *Brassica napus* induces copper remobilization, molybdenum accumulation and modification of the expression of chloroplastic proteins. PLoS ONE 9, e109889.

Blaby-Haas, C.E., Padilla-Benavides, T., Stube, R., Arguello, J.M., Merchant, S.S., 2014. Evolution of a plant-specific copper chaperone family for chloroplast copper homeostasis. Proc. Nat. Acad. Sci. USA 111, E5480–E5487.

Blamey, F.P.C., Joyce, D.C., Edwards, D.G., Asher, C.J., 1986. Role of trichomes in sunflower tolerance to manganese toxicity. Plant Soil 91, 171–180.

Blamey, F.P.C., Li, C., Howard, D.L., Cheng, M., Tang, C., Scheckel, K.G., et al., 2019. Evaluating effects of iron on manganese toxicity in soybean and sunflower using synchrotron-based X-ray fluorescence microscopy and X-ray absorption spectroscopy. Metallomics 11, 2097–2110.

Blevins, D.G., Lukaszewski, K.M., 1998. Boron in plant structure and function. Annu. Rev. Plant Physiol. Plant Mol. Biol. 49, 481–500.

Boardman, R., McGuire, D.O., 1990. The role of zinc in forestry. II. Zinc deficiency and forest management: Effect on yield and silviculture of *Pinus radiata* plantations in South Australia. Forest Ecol. Manag. 37, 207–218.

Bocchini, M., Bartucca, M.L., Ciancaleoni, S., Mimmo, T., Cesco, S., Pii, Y., et al., 2015. Iron deficiency in barley plants: phytosiderophore release, iron translocation, and DNA methylation. Front. Plant Sci. 6, 514.

Bohnsack, C.W., Albert, L.S., 1977. Early effects of boron deficiency on indole acetic acid oxidase levels of squash root tips. Plant Physiol. 59, 1047–1050.

Bolanos, L., Esteban, E., de Lorenzo, C., Fernández-Pascal, M., De Felipe, M.R., Gárate, A., et al., 1994. Essentiality of boron for symbiotic dinitrogen fixation in pea (*Pisum sativum*) rhizobium nodules. Plant Physiol. 104, 85–90.

Bolanos, L., Brewin, N.J., Bonilla, I., 1996. Effects of boron on rhizobium-legume cell–surface interactions and nodule development. Plant Physiol. 110, 1249–1256.

Bolanos, L., Lukaszewski, K., Bonilla, I., Blevins, D., 2004. Why boron? Plant Physiol. Biochem. 42, 907–912.

Bondarava, N., Beyer, P., Krieger-Liszkay, A., 2005. Function of the 23 kDa extrinsic protein of photosystem II as a manganese binding protein and its role in photoactivation. Biochim. Biophys. Acta 1708, 63–70.

Bonilla-Bird, N.J., Paez, A., Reyes, A., Hernandez-Viezcas, J.A., Li, C., Peralta-Videa, J.R., et al., 2018. Two-photon microscopy and spectroscopy studies to determine the mechanism of copper oxide nanoparticle uptake by sweetpotato roots during postharvest treatment. Environ. Sci. Technol. 52, 9954–9963.

Bose, J., Munns, R., Shabala, S., Gilliham, M., Pogson, B., Tyerman, S.D., 2017. Chloroplast function and ion regulation in plants growing on saline soils: lessons from halophytes. J. Exp. Bot. 68, 3129–3143.

Bouzayen, M., Felix, G., Latché, A., Pech, J.C., Boller, T., 1991. Iron: an essential cofactor for the conversion of 1-aminocyclopropane-1-carboxylic acid to ethylene. Planta 184, 244–247.

Bowler, C., Vancamp, W., Vanmontagu, M., Inze, D., 1994. Superoxide-dismutase in plants. Crit. Rev. Plant Sci. 13, 199–218.

Braconnier, S., d'Auzac, J., 1990. Chloride and stomatal conductance in coconut. Plant Physiol. Biochem. 28, 105–112.

Brennan, R.F., Gartrell, J.W., Adcock, K.G., 2001. Residual value of manganese fertiliser for lupin grain production. Aust. J. Exp. Agric. 41, 1187–1197.

Brennan, R.F., Penrose, B., Bell, R.W., 2019. Micronutrients limiting pasture production in Australia. Crop Pasture Sci. 70, 1053–1064.

Briat, J.F., Duc, C., Ravet, K., Gaymard, F., 2010. Ferritins and iron storage in plants. Biochim. Biophys. Acta 1800, 806–814.

Bricker, T.M., Roose, J.L., Fagerlund, R.D., Frankel, L.K., Eaton-Rye, J.J., 2012. The extrinsic proteins of Photosystem II. Biochim. Biophys. Acta 1817, 121–142.

Britto, D.T., Kronzucker, H.J., 2006. Futile cycling at the plasma membrane: a hallmark of low-affinity nutrient transport. Trends Plant Sci. 11, 529–534.

Broadley, M.R., White, P.J., Hammond, J.P., Zelko, I., Lux, A., 2007. Zinc in plants. New Phytol. 173, 677–702.

Brodrick, S.J., Giller, K.E., 1991a. Root nodules of *Phaseolus*: efficient scavengers of molybdenum for N_2-fixation. J. Exp. Bot. 42, 679–686.

Brodrick, S.J., Giller, K.E., 1991b. Genotypic difference in molybdenum accumulation affects N_2-fixation in tropical *Phaseolus vulgaris*. J. Exp. Bot. 42, 1339–1343.

Brookes, A., Collins, J.C., Thurman, D.A., 1981. The mechanism of zinc tolerance in grasses. J. Plant Nutr. 3, 695–705.

Brooks, R.R., 1980. Accumulation of nickel by terrestrial plants. In: Nriagu, J.O. (Ed.), Nickel in the Environment. Wiley & Sons, New York, NY, USA, pp. 407–430.

Brouquisse, R., Gaillard, J., Douce, R., 1986. Electron paramagnetic resonance characterization of membrane bound iron-sulfur clusters and aconitase in plant mitochondria. Plant Physiol. 81, 247–252.

Brown, P.H., 2008. Micronutrient use in agriculture in the United States of America: current practices, trends and constraints. In: Alloway, B.J. (Ed.), Micronutrient Deficiencies in Global Crop Production. Springer, New York, NY, USA, pp. 267–286.

Brown, P.H., Hu, H., 1998. Phloem boron mobility in diverse plant species. Bot. Acta 111, 331–335.

Brown, P.H., Shelp, B.J., 1997. Boron mobility in plants. Plant Soil 193, 85–101.

Brown, P.H., Graham, R.D., Nicholas, D.J.D., 1984. The effects of manganese and nitrate supply on the level of phenolics and lignin in young wheat plants. Plant Soil 81, 437–440.

Brown, P.H., Welch, R.M., Cary, E.E., 1987a. Nickel: a micronutrient essential for higher plants. Plant Physiol. 85, 801–803.

Brown, P.H., Welch, R.M., Cary, E.E., Checkai, R.T., 1987b. Beneficial effects of nickel on plant growth. J. Plant Nutr. 10, 2125–2135.

Brown, P.H., Dunemann, L., Schulz, R., Marschner, H., 1989. Influence of redox potential and plant species on the uptake of nickel and cadmium from soils. Z. Pflanzenernähr. Bodenk. 152, 85–91.

Brown, P.H., Welch, R.M., Madison, J.T., 1990. Effect of nickel deficiency on soluble anion, amino acid, and nitrogen levels in barley. Plant Soil 125, 19–27.

Brown, P.H., Bellaloui, N., Wimmer, M.A., Bassil, E.S., Ruiz, J., Hu, H., et al., 2002. Boron in plant biology. Plant Biol. 4, 205–223.

Broyer, T.C., Carlton, A.B., Johnson, C.M., Stout, P.R., 1954. Chlorine – a micronutrient element for higher plants. Plant Physiol. 29, 526–532.

Brumós, J., Colmenero-Flores, J.M., Conesa, A., Izquierdo, P., Sánchez, G., Iglesias, D.J., et al., 2009. Membrane transporters and carbon metabolism implicated in chloride homeostasis differentiate salt stress responses in tolerant and sensitive *Citrus* rootstocks. Funct. Integr. Genom. 9, 293–309.

Brumós, J., Talón, M., Bouhlal, R.Y.M., Colmenero-Flores, J.M., 2010. Cl$^-$ homeostasis in includer and excluder citrus rootstocks: transport mechanisms and identification of candidate genes. Plant Cell Environ. 33, 2012–2027.

Burdach, Z., Kurtyka, R., Siemieniuk, A., Karcz, A., 2014. Role of chloride ions in the promotion of auxin-induced growth of maize coleoptile segments. Ann. Bot. 114, 1023–1034.

Burkhead, J.L., Gargolin Reynolds, K.A., Abdel-Ghany, S.E., Cohu, C.M., Pilon, M., 2009. Copper homeostasis. New Phytol. 182, 799–816.

Burnell, J.N., 1986. Purification and properties of phosphoenolpyruvate carboxykinase from C$_4$ plants. Aust. J. Plant Physiol. 13, 577–587.

Burnell, J.N., 1988. The biochemistry of manganese in plants. In: Graham, R.D., Hannam, R.J., Uren, N.C. (Eds.), Manganese in Soils and Plants. Kluwer Academic Publ., Dordrecht, The Netherlands, pp. 125–137.

Burnell, J.N., Hatch, M.D., 1988. Low bundle sheath carbonic anhydrase is apparent by essential for effective C4 pathway operation. Plant Physiol. 86, 1252–1256.

Burnell, J.N., Suzuki, I., Sugiyama, T., 1990. Light induction and the effect of nitrogen status upon the activity of carbonic anhydrase in maize leaves. Plant Physiol. 94, 384–387.

Bussler, W., 1970. Die Entwicklung der Mo-Mangelsymptome an Blumenkohl. Z. Pflanzenernähr. Bodenk. 125, 36–50.

Cairns, A.L.P., Kritzinger, J.H., 1992. The effect of molybdenum on seed dormancy in wheat. Plant Soil 145, 295–297.

Cakmak, I., 2000. Role of zinc in protecting plant cells from reactive oxygen species. New Phytol. 146, 185–205.

Cakmak, I., Kutman, U.B., 2018. Agronomic biofortification of cereals with zinc: a review. Eur. J. Soil Sci. 69, 172–180.

Cakmak, I., Marschner, H., 1986. Mechanism of phosphorus-induced zinc deficiency in cotton. I. Zinc deficiency-enhanced uptake rate of phosphorus. Physiol. Plant. 68, 483–490.

Cakmak, I., Marschner, H., 1987. Mechanism of phosphorus-induced zinc deficiency in cotton. III. Changes in physiological availability of zinc in plants. Physiol. Plant. 70, 13–20.

Cakmak, I., Marschner, H., 1988a. Zinc-dependent changes in ESR signals, NADPH oxidase and plasma membrane permeability in cotton roots. Physiol. Plant. 73, 182–186.

Cakmak, I., Marschner, H., 1988b. Increase in membrane permeability and exudation in roots of zinc deficient plants. J. Plant. Physiol. 132, 356–361.

Cakmak, I., Marschner, H., 1988c. Enhanced superoxide radical production in roots of zinc deficient plants. J. Exp. Bot. 39, 1449–1460.

Cakmak, I., Römheld, V., 1997. Boron-deficiency induced impairments of cellular functions in plants. Plant Soil 193, 71–83.

Cammack, R., Fernandez, V.M., Schneider, K., 1988. Nickel in hydrogenases from sulfate-reducing, photosynthetic, and hydrogenoxidizing bacteria. In: Lancaster Jr., J.R. (Ed.), The Bioorganic Chemistry of Nickel. Verlag Chemie, Weinheim, Germany, pp. 167–190.

Cakmak, I., Marschner, H., Bangerth, F., 1989. Effect of zinc nutritional status on growth, protein metabolism and levels of indole-3-acetic acid and other phytohormones in bean (*Phaseolus vulgaris* L.). J. Exp. Bot. 40, 405–412.

Cakmak, I., Gülüt, K.Y., Marschner, H., Graham, R.D., 1994. Effect of zinc and iron deficiency on phytosiderophore release in wheat genotypes differing in zinc efficiency. J. Plant Nutr. 17, 1–17.

Cakmak, I., Atli, M., Kaya, R., Evliye, H., Marschner, H., 1995a. Association of high light and zinc deficiency in cold induced leaf chlorosis in grapefruit and mandarin trees. J. Plant Physiol. 146, 355–360.

Cakmak, I., Kurz, H., Marschner, H., 1995b. Short-term effects of boron, germanium and high light intensity on membrane permeability in boron-deficient leaves of sunflower. Physiol. Plant. 95, 11–18.

Cakmak, I., Yilmaz, A., Ekiz, H., Torun, B., Erenoglu, B., Braun, H.J., 1996a. Zinc deficiency as a critical nutritional problem in wheat production in Central Anatolia. Plant Soil 180, 165–172.

Cakmak, I., Sari, N., Marschner, H., Kalayci, M., Yılmaz, A., Eker, S., et al., 1996b. Dry matter production and distribution of zinc in bread and durum wheat genotypes differing in zinc efficiency. Plant Soil 180, 173–181.

Cakmak, I., Ekiz, H., Yilmaz, A., Torun, B., Köleli, N., Gültekin, I., et al., 1997a. Differential response of rye, triticale, bread and durum wheats to zinc deficiency in calcareous soils. Plant Soil 188, 1–10.

Cakmak, I., Derici, R., Torun, B., Tolay, I., Braun, H.J., Schlegel, R., 1997b. Role of rye chromosomes in improvement of zinc efficiency in wheat and triticale. Plant Soil 196, 249–253.

Cakmak, I., Pfeiffer, W.H., McClafferty, B., 2010. Biofortification of durum wheat with zinc and iron. Cereal Chem. 87, 10–20.

Campbell, L.C., Nable, R.O., 1988. Physiological functions of manganese in plants. In: Graham, R.D., Hannam, R.J., Uren, N.C. (Eds.), Manganese in Soils and Plants. Kluwer, Dordrecht, The Netherlands, pp. 139–154.

Candan, N., Cakmak, I., Ozturk, L., 2018. Zinc-biofortified seeds improved seedling growth under zinc deficiency and drought stress in durum wheat. J. Plant Nutr. Soil Sci. 181, 388–395.

Cao, P., Xie, Y., Li, M., Pan, X., Zhang, H., Zhao, X., et al., 2015. Crystal structure analysis of extrinsic PsbP protein of photosystem II reveals a manganese-induced conformational change. Mol. Plant 8, 664–666.

Carr, H.S., Winge, D.R., 2003. Assembly of cytochrome c oxidase within the mitochondrion. Acc. Chem. Res. 36, 309–316.

Carrio-Segui, N., Romero, P., Sanz, A., Penarrubia, L., 2016. Interaction between ABA signaling and copper homeostasis in *Arabidopsis thaliana*. Plant Cell Physiol. 57, 1568–1582.

Carrio-Segui, A., Ruiz-Rivero, O., Villamayor-Belinchon, L., Puig, S., Perea-Garcia, A., Penarrubia, L., 2019. The altered expression of microRNA408 influences the Arabidopsis response to iron deficiency. Front. Plant Sci. 10, 324.

Cartwright, B., Hallsworth, E.G., 1970. Effects of copper deficiency on root nodules of subterranean clover. Plant Soil 33, 685–698.

Castaings, L., Camargo, A., Pocholle, D., Gaudon, V., Texier, Y., Boutet-Mercey, S., et al., 2009. The nodule inception-like protein 7 modulates nitrate sensing and metabolism in *Arabidopsis*. Plant J. 57, 426–435.

Ceballos-Laita, L., Gutierrez-Carbonell, E., Imai, H., Abadía, A., Uemura, M., Abadía, J., et al., 2018. Effects of manganese toxicity on the protein profile of tomato (*Solanum lycopersicum*) roots as revealed by two complementary proteomic approaches, two-dimensional electrophoresis and shotgun analysis. J. Proteomics 185, 51–63.

Cervilla, L.M., Blasco, B., Rios, J.J., Rosales, M.A., Rubio-Wilhelmi, M.M., Sanchez-Rodriguez, E., et al., 2009. Response of nitrogen metabolism to boron toxicity in tomato plants. Plant Biol. 11, 671–677.

Cesco, S., Nikolic, M., Römheld, V., Varanini, Z., Pinton, R., 2002. Uptake of ^{59}Fe from soluble ^{59}Fe-humate complexes by cucumber and barley plants. Plant Soil 241, 121–128.

Chamizo-Ampudia, A., Sanz-Luque, E., Llamas, A., Ocana-Calahorro, F., Mariscal, V., Carreras, A., et al., 2016. A dual system formed by the ARC and NR molybdoenzymes mediates nitrite-dependent NO production in *Chlamydomonas*. Plant Cell Environ. 39, 2097–2107.

Chaney, R.L., 1993. Zinc phytotoxicity. In: Robson, A.D. (Ed.), Zinc in Soil and Plants. Kluwer Academic Publishers, Dordrecht, The Netherlands, pp. 135–150.

Chaney, R.L., Chen, Y., Green, C.E., Holden, M.J., Bell, P.F., Luster, D.G., et al., 1992. Root hairs on chlorotic tomatoes are an effect of chlorosis rather than part of the adaptive Fe-stress response. J. Plant Nutr. 15, 1857–1875.

Chatterjee, C., Nautiyal, N., Agarwala, S.C., 1985. Metabolic changes in mustard plant associated with molybdenum deficiency. New Phytol. 100, 511–518.

Chatterjee, C., Nautiyal, N., Agarwala, S.C., 1992. Excess sulphur partially alleviates copper deficiency effects in mustard. Soil Sci. Plant Nutr. 38, 57–64.

Cheah, Z.X., Kopittke, P.M., Harper, S.M., Meyer, G., O'Hare, T.J., Bell, M.J., 2019. Speciation and accumulation of Zn in sweetcorn kernels for genetic and agronomic biofortification programs. Planta 250, 219–227.

Chen, A., Husted, S., Salt, D.E., Schjoerring, J.K., Persson, D.P., 2019. The intensity of manganese deficiency strongly affects root endodermal suberization and ion homeostasis. Plant Physiol. 181, 729–742.

Chen, W., Yang, X., He, Z., Feng, Y., Hu, F., 2008. Differential changes in photosynthetic capacity, 77 K chlorophyll fluorescence and chloroplast ultrastructure between Zn-efficient and Zn-inefficient rice genotypes (*Oryza sativa*) under low zinc stress. Physiol. Plant. 132, 89–101.

Chen, W., He, Z.L., Yang, X.E., Mishra, S., Stoffella, P.J., 2010a. Chlorine nutrition of higher plants: Progress and perspectives. J. Plant Nutr. 33, 943–952.

Chen, W.W., Yang, J.L., Qin, C., Jin, C.W., Mo, J.H., Ye, T., et al., 2010b. Nitric oxide acts downstream of auxin to trigger root ferric-chelate reductase activity in response to iron deficiency in Arabidopsis. Plant Physiol. 154, 810–819.

Chen, Z., Yan, W., Sun, L., Tian, J., Liao, H., 2016a. Proteomic analysis reveals growth inhibition of soybean roots by manganese toxicity is associated with alteration of cell wall structure and lignification. J. Proteomics 143, 151–160.

Chen, Z.C., Yamaji, N., Fujii-Kashino, M., Ma, J.F., 2016b. A cation-chloride cotransporter gene is required for cell elongation and osmoregulation in rice. Plant Physiol. 171, 494–507.

Chipeng, F.K., Hermans, C., Colinet, G., Faucon, M.-P., Ngongo, M., Meerts, P., et al., 2010. Copper tolerance in the cuprophyte *Haumaniastrum katangense* (S. Moore) P. A. Duvign. & Plancke. Plant Soil 328, 235–244.

Chisholm, R.H., Blair, G.J., 1981. Phosphorus uptake and dry weight of stylo and white clover as affected by chlorine. Agron. J. 73, 767–771.

Choi, E.Y., McNeill, A.M., Coventry, D., Stangoulis, J.C.R., 2006. Whole plant response of crop and weed species to high subsoil boron. Aust. J. Agric. Res. 57, 761–770.

Christensen, L.P., Beede, R.H., Peacock, W.L., 2006. Fall foliar sprays prevent boron-deficiency symptoms in grapes. Calif. Agric. 60, 100–104.

Coccina, A., Cavagnaro, T.R., Pellegrino, E., Ercoli, L., McLaughlin, M.J., Watts-Williams, S.J., 2019. The mycorrhizal pathway of zinc uptake contributes to zinc accumulation in barley and wheat grain. BMC Plant Biol. 19, 133.

Codignola, A., Verotta, L., Panu, P., Maffei, M., Scannerini, S., Bonfante-Fasolo, P., 1989. Cell wall bound-phenols in roots of vesicular-arbuscular mycorrhizal plants. New Phytol. 112, 221–228.

Coleman, J.E., 1992. ZINC PROTEINS: Enzymes, storage proteins, transcription factors, and replication proteins. Ann. Rev. Biochem. 61, 897–946.

Coleman, J.E., 1998. Zinc enzymes. Curr. Opin. Chem. Biol. 2, 222–234.

Colmenero-Flores, J.M., Martínez, G., Gamba, G., Vázquez, N., Iglesias, D.J., Brumós, J., et al., 2007. Identification and functional characterization of cation-chloride cotransporters in plants. Plant J. 50, 278–292.

Colmenero-Flores, J.M., Franco-Navarro, J.D., Cubero-Font, P., Peinado-Torrubia, P., Rosales, M.A., 2019. Chloride as a beneficial macronutrient in higher plants: New roles and regulation. Int. J. Mol. Sci. 20, 4686.

Colmenero-Flores, J.M., Arbona, V., Morillon, R., Gómez-Cadenas, A., 2020. Salinity and water deficit. In: Talon, M., Gmitter, F.G., Caruso, M. (Eds.), The Genus Citrus, 1st Ed. Woodhead Publishing Limited, Elsevier, pp. 291–309.

Colpaert, J.V., van Assche, J.A., 1992. Zinc toxicity in ectomycorrhizal *Pinus sylvestris*. Plant Soil 143, 201–211.

Cram, W.J., 1983. Chloride accumulation as a homeostatic system: set points and perturbations: The physiological significance of influx isotherms, temperature effects and the influence of plant growth substances. J. Exp. Bot. 34, 1484–1502.

Cubero-Font, P., Maierhofer, T., Jaslan, J., Rosales, M.A., Espartero, J., Díaz-Rueda, P., et al., 2016. Silent S-type anion channel subunit SLAH1 gates SLAH3 open for chloride root-to-shoot translocation. Cur. Biol. 26, 2213–2220.

Cumbus, I.P., 1985. Development of wheat roots under zinc deficiency. Plant Soil 83, 313–316.

da Silva, R.C., Baird, R., Degryse, F., McLaughlin, M.J., 2018. Slow and fast-release boron sources in potash fertilizers: spatial variability, nutrient dissolution and plant uptake. Soil Sci. Soc. Am. J. 82, 1437–1448.

Dabir, S., Dabir, P., Somvanshi, B., 2005. Purification, properties and alternate substrate specificities of arginase from two different sources: *Vigna catjang* cotyledon and buffalo liver. Int. J. Biol. Sci. 1, 114–122.

Dai, J., Wang, N., Xiong, H., Qiu, W., Nakanishi, H., Kobayashi, T., et al., 2018. The Yellow Stripe-Like (YSL) gene functions in internal copper transport in peanut. Genes 9, 635.

Dalir, N., Khoshgoftarmanesh, A.H., 2015. Root uptake and translocation of nickel in wheat as affected by histidine. Plant Physiol. 184, 8–14.

Dalir, N., Khoshgoftarmanesh, A.H., Massah, A., Shariatmadari, H., 2017. Plasma membrane ATPase and H^+ transport activities of microsomal membranes from wheat roots under Ni deficiency conditions as affected by exogenous histidine. Environ. Exp. Bot. 135, 56–62.

Daneshbakhsh, B., Khoshgoftarmanesh, A.H., Shariatmadari, H., Cakmak, I., 2013. Phytosiderophore release by wheat genotypes differing in zinc deficiency tolerance grown with Zn free nutrient solution as affected by salinity. J. Plant Physiol. 170, 41–46.

Dasgan, H.Y., Ozturk, L., Abak, K., Cakmak, I., 2003. Activities of iron-containing enzymes in leaves of two tomato genotypes differing in their resistance to Fe chlorosis. J. Plant Nutr. 26, 1997–2007.

Davies, J.N., Adams, P., Winsor, G.W., 1978. Bud development and flowering of *Chrysanthemum morifolium* in relation to some enzyme activities and to the copper, iron and manganese status. Commun. Soil Sci. Plant Anal. 9, 249–264.

De Angeli, A., Monachello, D., Ephritikhine, G., Frachisse, J.M., Thomine, S., Gambale, F., et al., 2006. The nitrate/proton antiporter AtCLCa mediates nitrate accumulation in plant vacuoles. Nature 442, 939–942.

De Angeli, A., Monachello, D., Ephritikhine, G., Frachisse, J.-M., Thomine, S., Gambale, F., et al., 2009. CLC-mediated anion transport in plant cells. Phil. Trans. Roy. Soc. London B. 364, 195–201.

De Angeli, A., Zhang, J., Meyer, S., Martinoia, E., 2013. AtALMT9 is a malate-activated vacuolar chloride channel required for stomatal opening in Arabidopsis. Nat. Commun. 4, 1–10.

de Bang, T.C., Petersen, J., Pedas, P.R., Rogowska-Wrzesinska, A., Jensen, O.N., Schjoerring, J.K., et al., 2015. A laser ablation ICP-MS based method for multiplexed immunoblot analysis: applications to manganese-dependent protein dynamics of photosystem II in barley (*Hordeum vulgare* L.). Plant J. 83, 555–565.

de Varennes, A., Carneiro, J.P., Goss, M.J., 2001. Characterization of manganese toxicity in two species of annual medics. J. Plant Nutr. 24, 1947–1955.

De Vos, C.R., Lubberding, H.J., Bienfait, H.F., 1986. Rhizosphere acidification as a response to iron deficiency in bean plants. Plant Physiol. 81, 842–846.

De Wet, E., Robbertse, P.J., Groeneveld, H.T., 1989. The influence of temperature and boron on pollen germination in *Mangifera indica* L. S. Afr. J. Plant Soil 6, 228–234.

Delhaize, E., Loneragan, J.F., Webb, J., 1982. Enzymic diagnosis of copper deficiency in subterranean clover. 2. A simple field test. Aust. J. Agric. Res. 33, 981–988.

Delhaize, E., Loneragan, J.F., Webb, J., 1985. Development of three copper metalloenzymes in clover leaves. Plant Physiol. 78, 4–7.

Delhaye, G., Hardy, O.J., Seleck, M., Ilunga, E.I.W., Mahy, G., Meerts, P., 2020. Plant community assembly along a natural metal gradient in Central Africa: functional and phylogenetic approach. J. Veg. Sci. 31, 151–161.

Dell, B., 1981. Male sterility and outer wall structure in copper-deficient plants. Ann. Bot. 48, 599–608.

Dell, B., Huang, L.B., 1997. Physiological response of plants to low boron. Plant Soil 193, 103–120.

Dell'Orto, M., De Nisi, P., Vigani, G., Zocchi, G., 2013. Fe deficiency differentially affects the vacuolar proton pumps in cucumber and soybean roots. Front. Plant Sci. 4, 326.

Dickinson, D.B., 1978. Influence of borate and pentaerythriol concentrations on germination and tube growth of *Lilium longiflorum* pollen. J. Am. Soc. Hortic. Sci. 103, 413–416.

Diehn, T.A., Bienert, M.D., Pommerrenig, B., Liu, Z., Spitzer, C., Bernhardt, N., et al., 2019. Boron demanding tissues of *Brassica napus* express specific sets of functional Nodulin26-like Intrinsic Proteins and BOR 1 transporters. Plant J. 100, 68–82.

Dilworth, M.J., Eady, R.R., Eldridge, M.E., 1988. The vanadium nitrogenase of *Azotobacter chroococcum*. Reduction of acetylene and ethylene to ethane. Biochem J. 249, 745–751.

Dixon, N.E., Gazola, C., Blakeley, R.L., Zerner, B., 1975. Jack bean urease (EC 3.5.1.5), a metalloenzyme. A simple biological role for nickel? J. Am. Chem. Soc. 97, 4131–4133.

Dixon, N.E., Hinds, J.A., Fihelly, A.K., Gazola, C., Winzor, D.J., Blakeley, R.L., et al., 1980. Jack bean urease (EC 3.5.1.5). IV. The molecular size and the mechanism of inhibition by hydroxamic acids. Spectrophotometric tiration of enzymes with reversible inhibitors. Can. J. Biochem. 58, 1323–1334.

Do, T.D., Chen, H., Hien Thi Thu, V., Hamwieh, A., Yamada, T., Sato, T., et al., 2016. *Ncl* synchronously regulates Na$^+$, K$^+$ and Cl$^-$ in soybean and greatly increases the grain yield in saline field conditions. Sci. Rep. 6, 19147.

Domingo, A.L., Nagatomo, Y., Tamai, M., Takaki, H., 1992. Free-tryptophan and indolacetic acid in zinc-deficient radish shoots. Soil Sci. Plant Nutr. 38, 261–267.

Doncheva, S., Poschenrieder, C., Stoyanova, Z., Georgieva, K., Velichkova, M., Barcelo, J., 2009. Silicon amelioration of manganese toxicity in Mn-sensitive and Mn-tolerant maize varieties. Environ. Exp. Bot. 65, 189–197.

Dordas, C., 2006. Foliar boron application improves seed set, seed yield, and seed quality of alfalfa. Agron. J. 98, 907–913.

Dordas, C., Brown, P.H., 2000. Permeability of boric acid across lipid bilayers and factors affecting it. J. Membr. Biol. 175, 95–105.

Dordas, C., Chrispeels, M.J., Brown, P.H., 2000. Permeability and channel-mediated transport of boric acid across membrane vesicles isolated from squash roots. Plant Physiol. 124, 1349–1362.

Drissi, S., Houssa, A.A., Amlal, F., Dhassi, K., Lamghari, M., Maataoui, A., 2018. Barley responses to copper foliar spray concentrations when grown in a calcareous soil. J. Plant Nut. 41, 2266–2272.

Droillard, M.J., Paulin, A., 1990. Isoenzymes of superoxide dismutase in mitochondria and peroxisomes isolated from petals of carnation (*Dianthus caryophyllus*) during senescence. Plant Physiol. 94, 1187–1192.

Duan, Z., Kong, F., Zhang, L., Li, W., Zhang, J., Peng, L., 2016. A bestrophin-like protein modulates the proton motive force across the thylakoid membrane in Arabidopsis. J. Integr. Plant Biol. 58, 848–858.

Duan, G., Hakoyama, T., Kamiya, T., Miwa, H., Lombardo, F., Sato, S., et al., 2017. LjMOT1, a high-affinity molybdate transporter from *Lotus japonicus*, is essential for molybdate uptake, but not for the delivery to nodules. Plant J. 90, 1108–1119.

Dukic, E., Herdean, A., Cheregi, O., Sharma, A., Nziengui, H., Dmitruk, D., et al., 2019. K$^+$ and Cl$^-$ channels/transporters independently fine-tune photosynthesis in plants. Sci. Rep. 9, 8639. Available from: https://doi.org/10.1038/s41598-019-44972-z.

Dumont, M., Lehner, A., Bouton, S., Kiefer-Meyer, M.C., Voxeur, A., Pelloux, J., et al., 2014. The cell wall pectic polymer rhamnogalacturonan-II is required for proper pollen tube elongation: implications of a putative sialyltransferase-like protein. Ann. Bot. 114, 1177–1188.

Dunwell, J.M., Culham, A., Carter, C.E., Sosa-Aguirre, C.R., Goodenough, P.W., 2001. Evolution of functional diversity in the cupin superfamily. Trends Biochem. Sci. 26, 741–746.

Durbak, A.R., Phillips, K.A., Pike, S., O'Neill, M.A., Mares, J., Gallavoti, A., et al., 2014. Transport of boron by the *tassel-less1* aquaporin is critical for vegetative and reproductive development in maize. Plant Cell 26, 2978–2995.

Eapen, S., D'souza, S.F., 2005. Prospects of genetic engineering of plants for phytoremediation of toxic metals. Biotechnol. Adv. 23, 97–114.

Edwards, D.G., Asher, C.J., 1982. Tolerance of crop and pasture species to manganese toxicity. In: Scaife, A. (Ed.), Proceedings of the Ninth Plant Nutrition Colloquium. Warwick, England Commonwealth Agricultural Bureau, Farnham Royal, UK, pp. 145–150.

Edwards, G., Walker, D., 1983. C3, C4: Mechanisms, and Cellular and Environmental Regulation of Photosynthesis. Blackwell, Oxford, UK.

Eggert, K., von Wirén, N., 2017. Response of the plant hormone network to boron deficiency. New Phytol. 216, 868–881.

Eilers, T., Schwarz, G., Brinkmann, H., Witt, C., Richter, T., Nieder, J., et al., 2001. Identification and biochemical characterization of *Arabidopsis thaliana* sulfite oxidase – a new player in plant sulfur metabolism. J. Biol. Chem. 276, 46989–46994.

Eisenhut, M., Hoecker, N., Schmidt, S.B., Basgaran, R.M., Flachbart, S., Jahns, P., et al., 2018. The plastid envelope CHLOROPLAST MANGANESE TRANSPORTER1 is essential for manganese homeostasis in arabidopsis. Mol. Plant 11, 955–969.

Ellis, R.J., 1979. The most abundant protein in the world. Trends Biochem. Sci. 4, 241–244.

Ender, C., Li, M.Q., Martin, B., Povh, B., Nobiling, R., Reiss, H.-D., et al., 1983. Demonstration of polar zinc distribution in pollen tubes of *Lilium longiflorum* with the Heidelberg proton microprobe. Protoplasma 116, 201–203.

Engvild, K.C., 1986. Chlorine-containing natural compounds in higher plants. Phytochemistry 25, 781–791.

Eriksson, M., 1979. The effect of boron on nectar production and seed setting of red clover (*Trifolium pratense* L.). Swed. J. Agric. Res. 9, 37–41.

Eskew, D.L., Welch, R.M., Norwell, W.A., 1984. Nickel in higher plants. Further evidence for an essential role. Plant Physiol. 76, 691–693.

European Food Safety Authority, 2008. Opinion of the Scientific Panel on Contaminants in the Food Chain on a request from the European Commission to perform a scientific risk assessment on nitrate in vegetables. EFSA J. 689, 1–79.

Fabiano, C., Tezotto, T., Favarin, J.L., Polacco, J.C., Mazzafera, P., 2015. Essentiality of nickel in plants: a role in plant stresses. Front. Plant Sci. 6, 754.

Fageria, N.K., Stone, L.F., Melo, L.C., 2015. Copper-use efficiency in dry bean genotypes. Commun. Soil Sci. Plant Anal. 46, 979–990.

Farwell, A.J., Farina, M.P.W., Channon, P., 1991. Soil acidity effects on premature germination in immature maize grain. In: Baligar, V.C., Murrmann, R.P. (Eds.), Plant–Soil Interactions at Low pH. Kluwer, Dordrecht, The Netherlands, pp. 355–361.

Fatemi, H., Zaghdoud, C., Nortes, P.A., Carvajal, M., Martínez-Ballesta, M.D.C., 2020. Differential aquaporin response to distinct effects of two Zn concentrations after foliar application in pak choi (*Brassica rapa* L.) plants. Agronomy 10, 450.

Fecht-Christoffers, M.M., Maier, P., Iwasaki, K., Braun, H.P., Horst, W.J., 2007. The role of the leaf apoplast in manganese toxicity and tolerance in cowpea (*Vigna unguiculata* L. Walp). In: Sattelmacher, B., Horst, W. (Eds.), Apoplast of Higher Plants: Compartment of Storage, Transport and Reactions. Springer, Dordrecht, The Netherlands, pp. 307–321.

Felle, H.H., 1994. The H$^+$/Cl$^-$ symporter in root-hair cells of *Sinapis alba*. Plant Physiol. 106, 1131–1136.

Fernando, D.R., Lynch, J.P., 2015. Manganese phytotoxicity: new light on an old problem. Ann. Bot. 116, 313–319.

Ferrol, N., Tamayo, E., Vargas, P., 2016. The heavy metal paradox in arbuscular mycorrhizas: from mechanisms to biotechnological applications. J. Exp. Bot. 67, 6253–6265.

Fitzpatrick, K.L., Tyerman, S.D., Kaiser, B.N., 2008. Molybdate transport through the plant sulfate transporter SHST1. FEBS Lett. 582, 1508–1513.
Fixen, P.E., 1993. Crop responses to chloride. Adv. Agron. 50, 107–150.
Fleischer, A., Titel, C., Ehwald, R., 1998. The boron requirement and cell wall properties of growing and stationary suspension-cultured *Chenopodium album* L. cells. Plant Physiol. 117, 1401–1410.
Fleischer, A., O'Neill, M.A., Ehwald, R., 1999. The pore size of non-graminaceaous plant cell walls is rapidly decreased by borate ester cross-linking of the pectic polysaccharide rhamnogalacturonan II. Plant Physiol. 121, 829–838.
Forno, D.A., Yoshida, S., Asher, C.J., 1975. Zinc deficiency in rice. I. Soil factors associated with the deficiency. Plant Soil. 42, 537–550.
Foster, A.W., Osman, D., Robinson, N.J., 2014. Metal preferences and metallation. J. Biol. Chem. 289, 28095–28103.
Fouad, A., Saad, D., Kacem, M., Abdelwahed, M., Khalid, D., Abderrahim, R., et al., 2020. Efficacy of copper foliar spray in preventing copper deficiency of rainfed wheat (*Triticum aestivum* L.) grown in a calcareous soil. J. Plant Nutr. 43, 1617–1626.
Fourcroy, P., Vansuyt, G., Kushnir, S., Inzé, D., Briat, J.F., 2004. Iron-regulated expression of a cytosolic ascorbate peroxidase encoded by the APX1 gene in arabidopsis seedlings. Plant Physiol. 134, 605–613.
Fourcroy, P., Sisó-Terraza, P., Sudre, D., Savirón, M., Reyt, G., Gaymard, F., et al., 2014. Involvement of the ABCG37 transporter in secretion of scopoletin and derivatives by Arabidopsis roots in response to iron deficiency. New Phytol. 201, 155–167.
Franco, A.A., Munns, D.N., 1981. Response of *Phaseolus vulgaris* L. to molybdenum under acid conditions. Soil Sci. Soc. Am. J. 45, 1144–1148.
Francois, L.E., Clark, R.A., 1979. Boron tolerance of 25 ornamental shrub species. J. Am. Soc. Hortic. Sci. 104, 319–322.
Franco-Navarro, J.D., Brumós, J., Rosales, M.A., Cubero-Font, P., Talón, M., Colmenero-Flores, J.M., 2016. Chloride regulates leaf cell size and water relations in tobacco plants. J. Exp. Bot. 67, 873–891.
Franco-Navarro, J.D., Rosales, M.A., Cubero-Font, P., Calvo, P., Álvarez, R., Diaz-Espejo, A., et al., 2019. Chloride as a macronutrient increases water-use efficiency by anatomically driven reduced stomatal conductance and increased mesophyll diffusion to CO_2. Plant J. 99, 815–831.
Franco-Navarro, J.D., Díaz-Rueda, P., Rivero, C., Brumós, J., Rubio-Casal, A., de Cires, A., et al., 2021. Chloride nutrition improves drought resistance by enhancing water deficit avoidance and tolerance mechanisms. J. Exp. Bot. 72, 5246–5261.
Franzisky, B.L., Geilfus, C.-M., Kraenzlein, M., Zhang, X., Zoerb, C., 2019. Shoot chloride translocation as a determinant for NaCl tolerance in *Vicia faba* L. J. Plant Physiol. 236, 23–33.
Fraudentali, I., Ghuge, S.A., Carucci, A., Tavladoraki, P., Angelini, R., Rodrigues-Pousada, R.A., et al., 2020a. Developmental, hormone- and stress-modulated expression profiles of four members of the Arabidopsis copper-amine oxidase gene family. Plant Physiol. Biochem. 147, 141–160.
Fraudentali, I., Rodrigues-Pousada, R.A., Tavladoraki, P., Angelini, R., Cona, A., 2020b. Leaf-wounding long-distance signaling targets AtCuAOβ leading to root phenotypic plasticity. Plants 9, 249. Available from: https://doi.org/10.3390/plants9020249.
Freitas, D., Wurr Rodak, B., Rodrigues dos Reis, A., de Barros Reis, F., Soares de Carvalho, T., Schulze, J., et al., 2018. Hidden nickel deficiency? Nickel fertilization via soil improves nitrogen metabolism and grain yield in soybean genotypes. Front. Plant Sci. 9, 614.
Freitas, D.S., Rodak, B.W., Carneiro, M.A.C., Guilherme, L.R.G., 2019. How does Ni fertilization affect a responsive soybean genotype? A dose study. Plant Soil 441, 567–586.
Freyermuth, S.K., Bacanamwo, M., Polacco, J.C., 2000. The soybean Eu3 gene encodes an Ni-binding protein necessary for urease activity. Plant J. 21, 53–60.
Fridovich, I., 1983. Superoxide radical: an endogenous toxicant. Annu. Rev. Pharmacol. Toxicol. 23, 239–257.
Fromm, J., Eschrich, W., 1988. Transport processes in stimulated and non-stimulated leaves of *Mimosa pudica*. III. Displacement of ions during seismonastic leaf movements. Trees 2, 65–72.
Führs, H., Götze, S., Specht, A., Erban, A., Gallien, S., Heintz, D., et al., 2009. Characterization of leaf apoplastic peroxidases and metabolites in *Vigna unguiculata* in response to toxic manganese supply and silicon. J. Exp. Bot. 60, 1663–1678.
Gabbrielli, R., Pandolfini, T., Vergnano, O., Palandri, M.R., 1990. Comparison of two serpentine species with different nickel tolerance strategies. Plant Soil 122, 271–277.
Gandini, C., Schmidt, S.B., Husted, S., Schneider, A., Leister, D., 2017. The transporter SynPAM71 is located in the plasma membrane and thylakoids, and mediates manganese tolerance in *Synechocystis* PCC6803. New Phytol. 215, 256–268.
Gangwar, S., Singh, V.P., Prasad, S.M., Maurya, J.N., 2010. Differential responses of pea seedlings to indole acetic acid under manganese toxicity. Acta Physiol. Plant. 33, 451–462.
García, M.J., Lucena, C., Romera, F.J., Alcántara, E., Pérez-Vicente, R., 2010. Ethylene and nitric oxide involvement in the up-regulation of key genes related to iron acquisition and homeostasis in Arabidopsis. J. Exp. Bot. 61, 3885–3899.
Garcia, L., Mansilla, N., Ocampos, N., Pagani, M.A., Welchen, E., Gonzalez, D.H., 2019. The mitochondrial copper chaperone COX19 influences copper and iron homeostasis in arabidopsis. Plant Mol. Biol. 99, 621–638.
Garnica, M., Bacaicoa, E., Mora, V., San Francisco, S., Baigorri, R., Zamarreño, A.M., et al., 2018. Shoot iron status and auxin are involved in iron deficiency-induced phytosiderophores release in wheat. BMC Plant Biol. 18, 105. Available from: https://doi.org/10.1186/s12870-018-1324-3.
Gärtel, W., 1974. Die Mikronährstoffe – ihre Bedeutung für die Rebenernährung unter besonderer Berücksichtigung der Mangel- und Überschußerscheinungen. Weinberg Keller 21, 435–507.
Gasber, A., Klaumann, S., Trentmann, O., Trampczynska, A., Clemens, S., Schneider, S., et al., 2011. Identification of an *Arabidopsis* solute carrier critical for intracellular transport and inter-organ allocation of molybdate. Plant Biol. 13, 710–718.

Geilfus, C.M., 2018a. Chloride: from nutrient to toxicant. Plant Cell Physiol. 59, 877–886.

Geilfus, C.M., 2018b. Review on the significance of chlorine for crop yield and quality. Plant Sci. 270, 114–122.

George, T.S., French, A.S., Brown, L.K., Karley, A.J., White, P.J., Ramsay, L., et al., 2014. Genotypic variation in the ability of landraces and commercial cereal varieties to avoid manganese deficiency in soils with limited manganese availability: is there a role for root-exuded phytases? Physiol. Plant. 151, 243–256.

Gerendas, J., Sattelmacher, B., 1997. Significance of Ni supply for growth, urease activity and the concentrations of urea, amino acids and mineral nutrients of urea-grown plants. Plant Soil 190, 153–162.

Gerendas, J., Zhu, Z., Sattelmacher, B., 1998. Influence of N and Ni supply on nitrogen metabolism and urease activity in rice (*Oryza sativa L.*). J. Exp. Bot. 49, 1545–1554.

Gerendas, J., Polacco, J.C., Freyermuth, S.K., Sattelmacher, B., 1999. Significance of nickel for plant growth and metabolism. J. Plant Nutr. Soil Sci. 162, 241–256.

Gheibi, M.N., Malakouti, M.J., Kholdebarin, B., Ghanati, F., Teimouri, S., Sayadi, R., 2009. Significance of nickel supply for growth and chlorophyll content of wheat supplied with urea or ammonium nitrate. J. Plant Nutr. 32, 1440–1450.

Gil-Diez, P., Tejada-Jiménez, M., Leon-Mediavilla, J., Wen, J., Mysore, K.S., Imperial, J., et al., 2019. MtMOT1.2 is responsible for molybdate supply to *Medicago truncatula* nodules. Plant Cell Environ. 42, 310–320.

Gilliham, M., Tester, M., 2005. The regulation of anion loading to the maize root xylem. Plant Physiol. 137, 819–828.

Glass, A.D.M., Siddiqi, M.Y., 1985. Nitrate inhibition of chloride influx in barley - implications for a proposed chloride homeostat. J. Exp. Bot. 36, 556–566.

Godbold, D.L., Horst, W.J., Collins, J.C., Thurman, D.A., Marschner, H., 1984. Accumulation of zinc and organic acids in roots of zinc tolerant and non-tolerant ecotypes of *Deschampsia caespitosa*. J. Plant Physiol. 116, 59–69.

Godbold, D.L., Horst, W.J., Marschner, H., Collins, J.C., Thurman, D.A., 1983. Root growth and Zn uptake by two ecotypes of *Deschampsia caespitosa* as affected by high Zn concentrations. Z. Pflanzenphysiol. 112, 315–324.

Goldbach, H.E., 1997. A critical review on current hypotheses concerning the role of boron in higher plants: suggestions for further research and methodological requirements. J. Trace Microprobe Techniq. 15, 51–91.

Goldbach, H.E., Wimmer, M.A., 2007. Boron in plants and animals: is there a role beyond cell-wall structure? J. Plant Nutr. Soil Sci. 170, 39–48.

Goldbach, H.E., Hartmann, D., Rötzer, T., 1990. Boron is required for the stimulation of the ferric cyanide-induced proton release by auxins in suspension-cultured cells of *Daucus carota* and *Lycopersicon esculentum*. Physiol. Plant. 80, 114–118.

Goldbach, H.E., Yu, Q., Wingender, R., Schulz, M., Wimmer, M., Findeklee, P., et al., 2001. Rapid response reactions of roots to boron deprivation. J. Plant Nutr. Soil Sci. 164, 173–181.

Gong, X., Wang, Y., Liu, C., Wang, S., Zhao, X., Zhou, M., et al., 2010. Effects of manganese deficiency on spectral characteristics and oxygen evolution in maize chloroplasts. Biol. Trace Elem. Res. 136, 372–382.

Gong, H., Blackmore, D., Clingeleffer, P., Sykes, S., Jha, D., Tester, M., et al., 2011. Contrast in chloride exclusion between two grapevine genotypes and its variation in their hybrid progeny. J. Exp. Bot. 62, 989–999.

Graham, R.D., 1975. Male sterility in wheat plants deficient in copper. Nature 254, 514–515.

Graham, R.D., 1980. The distribution of copper and soluble carbohydrates in wheat plants grown at high and low levels of copper supply. Z. Pflanzenernähr. Bodenk. 143, 161–169.

Graham, R.D., Ascher, J., Ellis, P.A.E., Shepherd, K.W., 1987a. Transfer to wheat of the copper efficiency factor carried on rye chromosome arm 5RL. Plant Soil 99, 107–114.

Graham, R.D., Welch, R.M., Grunes, D.L., Cary, E.E., Norvell, W.A., 1987b. Effect of zinc deficiency on the accumulation of boron and other mineral nutrients in barley. Soil Sci. Soc. Am. J. 51, 652–657.

Graziano, M., Lamattina, L., 2007. Nitric oxide accumulation is required for molecular and physiological responses to iron deficiency in tomato roots. Plant J. 52, 949–960.

Green, L.S., Rogers, E.E., 2004. FRD3 controls iron localization in Arabidopsis. Plant Physiol. 136, 2523–2531.

Greene, R.M., Geider, R.J., Kilber, Z., Falkowski, P.G., 1992. Iron-induced changes in light harvesting and photochemical energy conversion processes in eukaryotic marine algae. Plant Physiol. 100, 565–575.

Gruhn, K., 1961. Einfluss einer Molybdän-Düngung auf einige Stickstoff-Fraktionen von Luzerne und Rotklee. Z. Pflanzenernähr. Bodenk. 95, 110–118.

Guan, R., Qu, Y., Guo, Y., Yu, L., Liu, Y., Jiang, J., et al., 2014. Salinity tolerance in soybean is modulated by natural variation in GmSALT3. Plant J. 80, 937–950.

Guan, Q., Liao, X., He, M., Li, X., Wang, Z., Ma, H., et al., 2017. Tolerance analysis of chloroplast *OsCu/Zn-SOD* overexpressing rice under NaCl and NaHCO$_3$ stress. PLoS ONE 12, e0186052.

Guo, F.O., Young, J., Crawford, N.M., 2003. The nitrate transporter AtNRT1.1 (CHL1) functions in stomatal opening and contributes to drought susceptibility in arabidopsis. Plant Cell 15, 107–117.

Gupta, U.C., 1979. Boron nutrition of crops. Adv. Agron. 31, 273–307.

Gupta, U.C., 1993. Boron and Its Role in Crop Production. CRC Press, Boca Raton, FL, USA.

Gupta, U.C., Lipsett, J., 1981. Molybdenum in soils, plants and animals. Adv. Agron. 34, 73–115.

Gurley, W.H., Giddens, J., 1969. Factors affecting uptake, yield response, and carry over of molybdenum on soybean seed. Agron. J. 61, 7–9.

Gustin, J.L., Loureiro, M.E., Kim, D., Na, G., Tikhonova, M., Salt, D.E., 2009. MTP1-dependent Zn sequestration into shoot vacuoles suggests dual roles in Zn tolerance and accumulation in Zn-hyper accumulating plants. Plant J. 57, 1116–1127.

Gutermuth, T., Lassig, R., Portes, M.T., Maierhofer, T., Romeis, T., Borst, J.-W., et al., 2013. Pollen tube growth regulation by free anions depends on the interaction between the anion channel slah3 and calcium-dependent protein kinases CPK2 and CPK20. Plant Cell 25, 4525–4543.

Hacisalihoglu, G., Kochian, L.V., 2003. How do some plants tolerate low levels of soil zinc? Mechanisms of zinc efficiency in crop plants. New Phytol. 159, 341–350.

Hafner, H., Ndunguru, B.J., Bationo, A., Marschner, H., 1992. Effect of nitrogen, phosphorus and molybdenum application on growth and symbiotic N_2-fixation of groundnut in acid sandy soil in Niger. Fert. Res. 31, 69–77.

Hale, K.L., McGrath, S.P., Lombi, E., Stack, S.M., Terry, N., Pickering, I.J., et al., 2001. Molybdenum sequestration in *Brassica* species. A role for anthocyanins? Plant Physiol. 126, 1391–1402.

Hall, J.L., 2002. Cellular mechanisms for heavy metal detoxification and tolerance. J. Exp. Bot. 53, 1–11.

Halliwell, B., 2009. The wanderings of a free radical. Free Radical Biol. Med. 46, 531–542.

Hannam, R.J., Ohki, K., 1988. Detection of manganese deficiency and toxicity in plants. In: Graham, R.D., Hannam, R.J., Uren, N.C. (Eds.), Manganese in Soils and Plants. Kluwer, Dordrecht, The Netherlands, pp. 243–259.

Hänsch, R., Mendel, R.R., 2009. Physiological functions of mineral micronutrients (Cu, Zn, Mn, Fe, Ni, Mo, B, Cl). Curr. Opin. Plant Biol. 12, 259–266.

Hanson, E.J., 1991a. Movement of boron out of tree fruit leaves. HortScience 26, 271–273.

Hanson, E.J., 1991b. Sour cherry trees respond to foliar boron applications. HortScience 26, 1142–1145.

Hantzis, L., Kroh, G.E., Jahn, C.E., Cantrell, M., Peers, G., Pilon, M., et al., 2018. A program for iron economy during deficiency targets specific Fe proteins. Plant Physiol. 176, 596–610.

Hara, T., Takeda, T.A., Takagishi, T., Fukue, K., Kambe, T., Fukada, T., 2017. Physiological roles of zinc transporters: molecular and genetic importance in zinc homeostasis. J. Physiol. Sci. 67, 283–301.

Hasan, M.K., Cheng, Y., Kanwar, M.K., Chu, X.-Y., Ahammed, G.J., Qi, Z.-Y., 2017. Responses of plant proteins to heavy metal stress-a review. Front. Plant Sci. 8, 1492.

Hatch, M.D., Burnell, J.N., 1990. Carbonic anhydrase activity in leaves and its role in the first step of C_4 photosynthesis. Plant Physiol. 93, 825–828.

Häussling, M., Römheld, V., Marschner, H., 1985. Beziehungen zwischen Chlorosegrad, Eisengehalten und Blattwachstum von Weinreben auf verschiedenen Standorten. Vitis 24, 158–168.

Haydon, M.J., Cobbett, C.S., 2007. Transporters of ligands for essential metal ions in plants. New Phytol. 174, 499–506.

He, S., He, Z., Yang, X., Baligar, V.C., 2012. Mechanisms of nickel uptake and hyperaccumulation by plants and implications for soil remediation. Adv. Agron. 117, 117–189.

Hebbern, C.A., Pedas, P., Schjoerring, J.K., Knudsen, L., Husted, S., 2005. Genotypic differences in manganese efficiency: field experiments with winter barley (*Hordeum vulgare* L.). Plant Soil 272, 233–244.

Hebbern, C.A., Laursen, K.H., Ladegaard, A.H., Schmidt, S.B., Pedas, P., Bruhn, D., et al., 2009. Latent manganese deficiency increases transpiration in barley (*Hordeum vulgare*). Physiol. Plant. 135, 307–316.

Heber, U., Heldt, H.W., 1981. The chloroplast envelope: Structure, function, and role in leaf metabolism. Annu. Rev. Plant Physiol. 32, 139–168.

Hecht-Buchholz, C., 1973. Molybdänverteilung und -verträglichkeit bei Tomate, Sonnenblume und Bohne. Z. Pflanzenernähr. Bodenk. 136, 110–119.

Henderson, S.W., Wege, S., Qiu, J., Blackmore, D.H., Walker, A.R., Tyerman, S.D., et al., 2015. Grapevine and arabidopsis cation-chloride cotransporters localize to the golgi and trans-golgi network and indirectly influence long-distance ion transport and plant salt tolerance. Plant Physiol. 169, 2215–2229.

Hendricks, T., van Loon, L.C., 1990. Petunia peroxidase is localized in the epidermis of aerial plant organs. J. Plant Physiol. 136, 519–525.

Henriques, F.S., 2004. Reduction in chloroplast number accounts for the decrease in the photosynthetic capacity of Mn-deficient pecan leaves. Plant Sci. 166, 1051–1055.

Herdean, A., Teardo, E., Nilsson, A.K., Pfeil, B.E., Johansson, O.N., Unnep, R., et al., 2016. A voltage-dependent chloride channel fine-tunes photosynthesis in plants. Nat. Commun. 7, 11654.

Hernandez-Hernandez, H., Juarez-Maldonado, A., Benavides-Mendoza, A., Ortega-Ortiz, H., Cadenas-Pliego, G., Sanchez-Aspeytia, D., et al., 2018. Chitosan-PVA and copper nanoparticles improve growth and overexpress the SOD and JA genes in tomato plants under salt stress. Agronomy 8, 175. Available from: https://doi.org/10.3390/agronomy8090175.

Heslop-Harrison, J.S., Reger, B.J., 1986. Chloride and potassium-ions and turgidity in the grass stigma. J. Plant Physiol. 124, 55–60.

Hewitt, E.J., McCready, C.C., 1956. Molybdenum as a plant nutrient. VII. The effects of different molybdenum and nitrogen supplies on yields and composition of tomato plants grown in sand culture. J. Hortic. Sci. 31, 284–290.

Hewitt, E.J., Gundry, C.S., 1970. The molybdenum requirement of plants in relation to nitrogen supply. J. Hortic. Sci. 45, 351–358.

Hildebrand, D.F., 1989. Lipoxygenases. Physiol. Plant. 76, 249–253.

Hindt, M.N., Guerinot, M.L., 2012. Getting a sense for signals: Regulation of the plant iron deficiency response. Biochim. Biophys. Acta 1823, 1521–1530.

Hodson, M.J., White, P.J., Mead, A., Broadley, M.R., 2005. Phylogenetic variation in the silicon composition of plants. Ann. Bot. 96, 1027–1046.

Hoehner, R., Pribil, M., Herbstova, M., Lopez, L.S., Kunz, H.-H., Li, M., et al., 2020. Plastocyanin is the long-range electron carrier between photosystem II and photosystem I in plants. Proc. Natl. Acad. Sci. USA 117, 15354–15362.

Hofinger, M., Böttger, M., 1979. Identification by GC-MS of 4-chloroindolylacetic acid and its methyl ester in immature *Vicia faba* seeds. Phytochemistry 18, 653–654.

Höfner, W., Grieb, R., 1979. Einfluß von Fe- und Mo-Mangel auf den Ionengehalt mono- und dikotyler Pflanzen unterschiedlicher Chloroseanfälligkeit. Z. Pflanzenernähr. Bodenk. 142, 626–638.

Homer, F.A., Reeves, R.D., Brooks, R.R., Baker, A.J.M., 1991. Characterization of the nickel-rich extract from the nickel hyperaccumulator *Dichapetalum gelonioides*. Phytochemistry 30, 2141–2145.

Hopmans, P., 1990. Stem deformity in *Pinus radiata* plantations in south-eastern Australia: I. Response to copper fertiliser. Plant Soil 122, 97–104.

Horak, O., 1985. Zur Bedeutung des Nickels für *Fabaceae*. II. Nickelaufnahme und Nickelbedarf von *Pisum sativum* L. Phyton (Austria) 25, 301–307.

Horst, W.J., 1988. The physiology of manganese toxicity. In: Graham, R.D., Hannam, R.J., Uren, N.C. (Eds.), Manganese in Soils and Plants. Kluwer, Dordrecht, The Netherlands, pp. 175–188.

Horst, W.J., Fecht, M., Naumann, A., Wissemeier, A.H., Maier, P., 1999. Physiology of manganese toxicity and tolerance in *Vigna unguiculata* (L.) Walp. J. Plant Nutr. Soil Sci. 162, 263–274.

Hsieh, E.J., Waters, B.M., 2016. Alkaline stress and iron deficiency regulate iron uptake and riboflavin synthesis gene expression differently in root and leaf tissue: implications for iron deficiency chlorosis. J. Exp. Bot. 67, 5671–5685.

Hu, H., Brown, P.H., 1994. Localization of boron in cell walls of squash and tobacco and its association with pectin. Plant Physiol. 105, 681–689.

Hu, H., Brown, P.H., Labavitch, J.M., 1996. Species variability in boron requirement is correlated with cell wall pectin. J. Exp. Bot. 47, 227–232.

Huang, C., Graham, R.D., 1990. Resistance of wheat genotypes to boron toxicity is expressed at the cellular level. Plant Soil 126, 295–300.

Huang, C., Barker, S.J., Langridge, P., Smith, F.W., Graham, R.D., 2000. Zinc deficiency up-regulates expression of high-affinity phosphate transporter genes in both phosphate-sufficient and -deficient barley roots. Plant Physiol. 124, 415–422.

Huang, X.Y., Liu, H., Zhu, Y.F., Pinson, S.R.M., Lin, H.X., Guerinot, M.L., et al., 2019. Natural variation in a molybdate transporter controls grain molybdenum concentration in rice. New Phytol. 221, 1983–1997.

Huber, D.M., McCay-Buis, T.S., 1993. A multiple component analysis of the take-all disease of cereals. Plant Dis. 77, 437–447.

Hughes, N.P., Williams, R.J.P., 1988. An introduction to manganese biological chemistry. In: Graham, R.D., Hannam, R.J., Uren, N.C. (Eds.), Manganese in Soils and Plants. Kluwer, Dordrecht, The Netherlands, pp. 7–19.

Husted, S., Laursen, K.H., Hebbern, C.A., Schmidt, S.B., Pedas, P., Haldrup, A., et al., 2009. Manganese deficiency leads to genotype-specific changes in fluorescence induction kinetics and state transitions. Plant Physiol. 150, 825–833.

Ide, Y., Kusano, M., Oikawa, A., Fukushima, A., Tomatsu, H., Saito, K., et al., 2011. Effects of molybdenum deficiency and defects in molybdate transporter MOT1 on transcript accumulation and nitrogen/sulphur metabolism in *Arabidopsis thaliana*. J. Exp. Bot. 62, 1483–1497.

Iino, M., Long, C., Wang, X., 2001. Auxin- and abscisic acid-dependent osmoregulation in protoplasts of *Phaseolus vulgaris* pulvini. Plant Cell Physiol. 42, 1219–1227.

Imran, M., Garbe-Schönberg, D., Neumann, G., Boelt, B., Mühling, K.H., 2017. Zinc distribution and localization in primed maize seeds and its translocation during early seedling development. Environ. Exp. Bot. 143, 91–98.

Inal, A., Gunes, A., Alpaslan, M., Demir, K., 1998. Nitrate versus chloride nutrition effects in a soil-plant system on the growth, nitrate accumulation, and nitrogen, potassium, sodium, calcium, and chloride content of carrot. J. Plant Nutr. 21, 2001–2011.

Ishii, T., Matsunaga, T., 1996. Isolation and characterization of a boron rhamnogalacturonan-II complex from cell walls of sugar beet pulp. Carbohydr. Res. 284, 1–9.

Ishii, T., Matsunaga, T., Pellerin, P., O'Neill, M.A., Darvill, A., Albersheim, P., 1999. The plant cell wall polysaccharide rhamnogalacturonan-II self-assembles into a covalently cross-linked dimer. J. Biol. Chem. 274, 13098–13104.

Ishimura, Y., Kim, S., Tsukamoto, T., Oki, H., Kobayashi, T., Watanabe, S., et al., 2007. Mutational reconstructed ferric chelate reductase confers enhanced tolerance in rice to iron defiency in calcareous soil. Proc. Natl. Acad. Sci. USA 104, 7373–7378.

Ishizuka, J., 1982. Characterization of molybdenum absorption and translocation in soybean plants. Soil Sci. Plant Nutr. 28, 63–78.

Itoh, S., Uwano, S., 1986. Characteristics of the Cl_2 action site in the O_2 evolving reaction in PSII particles: electrostatic interaction with ions. Plant Cell Physiol. 27, 25–36.

Iturbe-Omemaetxe, I., Moran, J.F., Arrese-Igor, C., Gogorcena, Y., Klucas, R.V., Becana, M., 1995. Activated oxygen and antioxidant defences in iron-deficient pea plants. Plant Cell Environ. 18, 421–429.

Iwai, H., Masaoka, N., Ishii, T., Satoh, S., 2002. A pectin glucuronyltransferase gene is essential for intercellular attachment in the plant meristem. Proc. Natl. Acad. Sci. USA 99, 16319–16324.

Iwai, H., Hokura, A., Oishi, M., Chida, H., Ishii, T., Sakai, S., et al., 2006. The gene responsible for borate cross-linking of pectin rhamnogalacturonan-II is required for plant reproductive tissue development and fertilization. Proc. Natl. Acad. Sci. USA 103, 16592–16597.

Jackson, J.F., 1989. Borate control of protein secretion from petunia pollen exhibits critical temperature discontinuities. Sex. Plant Reprod. 2, 11–14.

Jacoby, B., Rudich, B., 1980. Proton-chloride symport in barley roots. Ann. Bot. 46, 493–498.

Jahan, M.S., Guo, S., Baloch, A.R., Sun, J., Shu, S., Wang, Y., et al., 2020. Melatonin alleviates nickel phytotoxicity by improving photosynthesis, secondary metabolism and oxidative stress tolerance in tomato seedlings. Ecotox. Environ. Saf. 197, 110593.

Jiang, C.D., Gao, H.Y., Zou, Q., 2001. Enhanced thermal energy dissipation depending on xanthophyll cycle and D1 protein turnover in iron-deficient maize leaves under high irradiance. Photosynthetica 39, 269–274.

Jin, C.W., You, G.Y., He, Y.F., Tang, C., Wu, P., Zheng, S.J., 2007. Iron deficiency-induced secretion of phenolics facilitates the reutilization of root apoplastic iron in red clover. Plant Physiol. 144, 278–285.

Johnson, A.D., Simons, J.G., 1979. Diagnostic indices of zinc deficiency in tropical legumes. J. Plant Nutr. 1, 123–149.

Johnson, W.C., Wear, J.I., 1967. Effect of boron on white clover (*Trifolium repens* L.) seed production. Agron. J. 59, 205–206.

Johnson, C.M., Stout, P.R., Broyer, T.C., Carlton, A.B., 1957. Comparative chlorine requirements of different plant species. Plant Soil 8, 337–353.

Jossier, M., Kroniewicz, L., Dalmas, F., Le Thiec, D., Ephritikhine, G., Thomine, S., et al., 2010. The Arabidopsis vacuolar anion transporter, AtCLCc, is involved in the regulation of stomatal movements and contributes to salt tolerance. Plant J. 64, 563–576.

Jung, H.-I., Gayomba, S.R., Yan, J., Vatamaniuk, O.K., 2014. *Brachypodium distachyon* as a model system for studies of copper transport in cereal crops. Front. Plant Sci. 5, 236.

Kabir, A.H., Paltridge, N.G., Able, A.J., Paull, J.G., Stangoulis, J.C.R., 2012. Natural variation for Fe-efficiency is associated with up-regulation of Strategy I mechanisms and enhanced citrate and ethylene synthesis in *Pisum sativum* L. Planta 235, 1409–1419.

Kaiser, B.N., Gridley, K.L., Brady, J.N., Phillips, T., Tyerman, S.D., 2005. The role of molybdenum in agricultural plant production. Ann. Bot. 96, 745–754.

Kang, B.T., Fox, R.L., 1980. A methodology for evaluating the manganese tolerance of cowpea (*Vigna unguiculata*) and some preliminary results of field trials. Field Crop Res. 3, 199–210.

Kannan, S., Ramani, S., 1978. Studies on molybdenum absorption and transport in bean and rice. Plant Physiol. 62, 179–181.

Kaur, H., Garg, N., 2021. Zinc toxicity in plants: a review. Planta 253, 129.

Kaur, H., Manchanda, J.S., 2019. Copper induced iron deficiency in wheat (*Triticum aestivum* L). J. Plant Nutr. 42, 2824–2843.

Kawachi, M., Kobae, Y., Mori, H., Tomioka, R., Lee, Y., Maeshima, M., 2009. A mutant strain *Arabidopsis thaliana* that lacks vacuolar membrane zinc transporter MTP1 revealed the latent tolerance to excessive zinc. Plant Cell Physiol. 50, 1156–1170.

Kawakami, K., Umena, Y., Kamiya, N., Shen, J.R., 2009. Location of chloride and its possible functions in oxygen-evolving photosystem II revealed by X-ray crystallography. Proc. Natl. Acad. Sci. USA 106, 8567–8572.

Kazemi, N., Khavari-Nejad, R.A., Fahimi, H., Saadatmand, S., Nejad-Sattari, T., 2010. Effects of exogenous salicylic acid and nitric oxide on lipid peroxidation and antioxidant enzyme activities in leaves of *Brassica napus* L. under nickel stress. Sci. Hortic. 126, 402–407.

Kerkeb, L., Krämer, U., 2003. The role of free histidine in xylem loading of nickel in *Alyssum lesbiacum* and *Brassica juncea*. Plant Physiol. 131, 716–724.

Kerridge, P.C., Cook, B.G., Everett, M.L., 1973. Application of molybdenum trioxide in the seed pellet for sub-tropical pasture legumes. Trop. Grassl. 7, 229–232.

Khabaz-Saberi, H., Graham, R.D., Ascher, J.S., Rathjen, A.J., 2000. Quantification of the confounding effect of seed manganese content in screening for manganese efficiency in durum wheat (*Triticum turgidum* L. var. *durum*). J. Plant Nutr. 23, 855–866.

Khabaz-Saberi, H., Rengel, Z., Wilson, R., Setter, T., 2010. Variation for tolerance to high concentration of ferrous iron (Fe^{2+}) in Australian hexaploid wheat. Euphytica 172, 275–283.

Khobra, R., Singh, B., 2018. Phytosiderophore release in relation to multiple micronutrient metal deficiency in wheat. J. Plant Nutr. 41, 679–688.

Khoshgoftarmanesh, A., Bahmanziari, H., 2012. Stimulating and toxicity effects of nickel on growth, yield, and fruit quality of cucumber supplied with different nitrogen sources. J. Soil Sci. Plant Nutr. 175, 474–481.

Kidron, G.J., Zilberman, A., 2019. Low cotton yield is associated with micronutrient deficiency in West Africa. Agron. J. 111, 1977–1984.

Kihara, J., Bolo, P., Kinyua, M., Rurinda, J., Piikki, K., 2020. Micronutrient deficiencies in African soils and the human nutritional nexus: opportunities with staple crops. Environ. Geochem. Health 42, 3015–3033.

Kim, M.D., Kim, Y.H., Kwon, S.Y., Yun, D.J., Kwak, S.S., Lee, H.S., 2010. Enhanced tolerance to methyl viologen-induced oxidative stress and high temperature in transgenic potato plants overexpressing the Cu Zn SOD, APX and NDPK2 genes. Physiol. Plant. 140, 153–162.

Kim, S.A., LaCroix, I.S., Gerber, S.A., Guerinot, M.L., 2019. The iron deficiency response in *Arabidopsis thaliana* requires the phosphorylated transcription factor URI. Proc. Natl. Acad. Sci. USA 116, 24933–24942.

Kirchhoff, H., 2014. Structural changes of the thylakoid membrane network induced by high light stress in plant chloroplasts. Phil. Trans. Roy. Soc. London B. 369, 1–6.

Kirk, G.J., Loneragan, J.F., 1988. Functional boron requirement for leaf expansion and its use as a critical value for diagnosis of boron deficiency in soybean. Agron. J. 80, 758–762.

Kitagishi, K., Obata, H., 1986. Effects of zinc deficiency on the nitrogen metabolism of meristematic tissues of rice plants with reference to protein synthesis. Soil Sci. Plant Nutr. 32, 397–405.

Kitagishi, K., Obata, H., Kondo, T., 1987. Effect of zinc deficiency on 80S ribosome content of meristematic tissues of rice plant. Soil Sci. Plant Nutr. 33, 423–430.

Klucas, R.V., Hanus, F.J., Russell, S.A., Evans, H.J., 1983. Nickel: a micronutrient element for hydrogen-dependent growth of *Rhizobium japonicum* and expression of urease activity in soybean leaves. Proc. Natl. Acad. Sci. USA 80, 2253–2257.

Kluska, K., Adamczyk, J., Kręzel, A., 2018. Metal binding properties, stability and reactivity of zinc fingers. Coord. Chem. Rev. 367, 18–64.

Ko, M.P., Huang, P.Y., Huang, J.S., Barker, K.R., 1987. The occurrence of phytoferritin and its relationship to effectiveness of soybean nodules. Plant Physiol. 83, 299–305.

Kobayashi, M., Matoh, T., Azuma, J.-I., 1996. Two chains of rhamnogalacturonan II are cross-linked by borate-diol ester bonds in higher plant cell walls. Plant Physiol. 110, 1017–1020.

Kobayashi, T., Naoko, K., Nishizawa, N.K., 2014. Iron sensors and signals in response to iron deficiency. Plant Sci. 224, 36–43.

Kobayashi, T., Itai, R.N., Senoura, T., Oikawa, T., Ishimaru, Y., Ueda, M., et al., 2016. Jasmonate signaling is activated in the very early stages of iron deficiency responses in rice roots. Plant Mol. Biol. 91, 533–547.

Kochanczyk, T., Drozd, A., Krezel, A., 2015. Relationship between the architecture of zinc coordination and zinc binding affinity in proteins--insights into zinc regulation. Metallomics 7, 244–257.

Kocyla, A., Tran, J.B., Krezel, A., 2021. Galvanization of protein–protein interactions in a dynamic zinc interactome. Trends Biochem. Sci. 46, 64–79.

Kohler, B., Raschke, K., 2000. The delivery of salts to the xylem. Three types of anion conductance in the plasmalemma of the xylem parenchyma of roots of barley. Plant Physiol. 122, 243–254.

Kok, B., Forbush, B., McGloin, M., 1970. Cooperation of charges in photosynthetic O_2 evolution. I. A linear four step mechanism. Photochem. Photobiol. 11, 457–475.

Kolbert, Z., Peto, A., Lehotai, N., Feigl, G., Erdei, L., 2015. Copper sensitivity of *nia1nia2noa1-2* mutant is associated with its low nitric oxide (NO) level. Plant Growth Regul. 77, 255–263.

Kopittke, P.M., Asher, C.J., Menzies, N.W., 2007. Toxic effects of Ni^{2+} on growth of cowpea (*Vigna unguiculata*). Plant Soil 292, 283–289.

Korner, L.E., Møller, I.M., Jensén, P., 1986. Free space uptake and influx of Ni^{2+} in excised barley roots. Physiol. Plant. 68, 583–588.

Koshiba, T., Kobayashi, M., Matoh, T., 2009. Boron nutrition of tobacco BY-2 cells. V. Oxidative damage is the major cause of cell death induced by boron deprivation. Plant Cell Physiol. 50, 26–36.

Kramer, D., Römheld, V., Landsberg, E., Marschner, H., 1980. Induction of transfer-cell formation by iron deficiency in the root epidermis of *Helianthus annuus*. Planta 147, 335–339.

Krämer, U., Pickering, I.J., Prince, R.C., Raskin, I., Salt, D.E., 2000. Subcellular localization and speciation of nickel in hyperaccumulator and non-accumulator *Thlaspi* species. Plant Physiol. 122, 1343–1354.

Krezel, A., Maret, W., 2016. The biological inorganic chemistry of zinc ions. Arch. Biochem. Biophys. 611, 3–19.

Kriedemann, P.E., Graham, R.D., Wiskich, J.T., 1985. Photosynthetic dysfunction and *in vivo* changes in chlorophyll *a* fluorescence from manganese-deficient wheat leaves. Aust. J. Agric. Res. 36, 157–169.

Krieger, A., Rutherford, A.W., Vass, I., Hideg, É., 1998. Relationship between activity, D1 loss, and Mn binding in photoinhibition of Photosystem II. Biochemistry (Moscow) 37, 16262–16269.

Krogmeier, M.J., McCarty, G.W., Shogren, D.R., Bremner, J.M., 1991. Effect of nickel deficiency in soybeans on the phytotoxicity of foliar-applied urea. Plant Soil 135, 283–286.

Kropman, M.F., Bakker, H.J., 2001. Dynamics of water molecules in aqueous solvation shells. Science 291, 2118–2120.

Kühnlenz, T., Hofmann, C., Uraguchi, S., Schmidt, H., Schempp, S., Weber, M., et al., 2016. Phytochelatin synthesis promotes leaf Zn accumulation of *Arabidopsis thaliana* plants grown in soil with adequate Zn supply and is essential for survival on Zn-contaminated soil. Plant Cell Physiol. 57, 2342–2352.

Kumar, K.V., Singh, N., Behl, H.M., Srivastava, S., 2008. Influence of plant growth promoting bacteria and its mutant on heavy metal toxicity in *Brassica juncea* grown in fly ash amended soil. Chemosphere 72, 678–683.

Küpper, H.G., Götz, B., Mijovilovich, A., Küpper, F.C., Meyer-Klaucke, W., 2009. Complexation and toxicity of copper in higher plants. I. Characterization of copper accumulation, speciation, and toxicity in *Crassula helmsii* as a new copper accumulator. Plant Physiol. 151, 702–714.

Kutman, B.Y., Kutman, U.B., Cakmak, I., 2013. Nickel-enriched seed and externally supplied nickel improve growth and alleviate foliar urea damage in soybean. Plant Soil 363, 61–75.

Kutman, B.Y., Kutman, U.B., Cakmak, I., 2014. Effects of seed nickel reserves or externally supplied nickel on the growth, nitrogen metabolites and nitrogen use efficiency of urea- or nitrate-fed soybean. Plant Soil 376, 261–276.

Kwiatowsky, J., Safianowska, A., Kaniuga, Z., 1985. Isolation and characterization of an iron-containing superoxide dismutase from tomato leaves, *Lycopersicon exculentum*. Eur. J. Biochem. 146, 459–466.

Landsberg, E.-C., 1981. Organic acid synthesis and release of hydrogen ions in response to Fe deficiency stress of mono- and dicotyledonous plant species. J. Plant Nutr. 3, 579–591.

Landsberg, E.-C., 1989. Proton efflux and transfer cell formation as response to Fe deficiency of soybean in nutrient solution culture. Plant Soil 114, 53–61.

Lang, C., Popko, J., Wirtz, M., Hell, R., Herschbach, C., Kreuzwieser, J., et al., 2007. Sulphite oxidase as key enzyme for protecting plants against sulphur dioxide. Plant Cell Environ. 30, 447–455.

Larbi, A., Abadía, A., Morales, F., Abadía, J., 2004. Fe resupply to Fe-deficient sugar beet plants leads to rapid changes in the violaxanthin cycle and other photosynthetic characteristics without significant de novo chlorophyll synthesis. Photosynth. Res. 79, 59–69.

Larbi, A., Abadía, A., Abadía, J., Morales, F., 2006. Down co-regulation of light absorption, photochemistry, and carboxylation in Fe-deficient plants growing in different environments. Photosynth. Res. 89, 113–126.

Lavres, J., Castro Franco, G., de Sousa Camara, G.M., 2016. Soybean seed treatment with nickel improves biological nitrogen fixation and urease activity. Front. Environ. Sci. 7, 37.

Lee, S., Jeon, U.S., Lee, S.J., Kim, Y.K., Persson, D.P., Husted, S., et al., 2009. Iron fortification of rice seeds through activation of the nicotianamine synthase gene. Proc. Natl. Acad. Sci. USA 106, 22014–22019.

Lee, J.S., Wissuwa, M., Zamora, O.B., Ismail, A.M., 2017. Biochemical indicators of root damage in rice (*Oryza sativa* L.) genotypes under zinc deficiency stress. J. Plant Res. 30, 1071–1077.

Lehto, T., Ruuhola, T., Dell, B., 2010. Boron in forest trees and forest ecosystems. Forest Ecol. Manag. 260, 2053–2069.

Lei, G.J., Zhu, H.F., Wang, Z.W., Dong, F., Dong, N.Y., Zheng, S.J., 2014. Abscisic acid alleviates iron deficiency by promotingroot iron reutilization and transport from root to shoot in Arabidopsis. Plant Cell Environ. 37, 852–863.

Lemanceau, P., Bauer, P., Kraemer, S., Briat, J.F., 2009. Iron dynamics in the rhizosphere as a case study for analyzing interactions between soils, plants and microbes. Plant Soil 321, 513–535.

Lemoine, R., La Camera, S., Atanassova, R., Dédaldéchamp, F., Allario, T., Pourtau, N., et al., 2013. Source to sink transport and regulation by environmental factors. Front. Plant Sci. 4, 272.

Leng, X., Mu, Q., Wang, X., Li, X., Zhu, X., Shangguan, L., et al., 2015. Transporters, chaperones, and P-type ATPases controlling grapevine copper homeostasis. Funct. Integr. Genomics 15, 673–684.

Leonowicz, G., Trzebuniak, K.F., Zimak-Piekarczyk, P., Ślesak, I., Mysliwa-Kurdziel, B., 2018. The activity of superoxide dismutases (SODs) at the early stages of wheat deetiolation. PLoS ONE 13, e0194678.

Leplat, F., Pedas, P.R., Rasmussen, S.K., Husted, S., 2016. Identification of manganese efficiency candidate genes in winter barley (*Hordeum vulgare*) using genome wide association mapping. BMC Genomics 17, 1–15.

Lerer, M., Bar-Akiva, A., 1976. Nitrogen constituents in manganese-deficient lemon leaves. Physiol. Plant. 38, 13–18.

Lešková, A., Zvarík, M., Araya, T., Giehl, R.F., 2020. Nickel toxicity targets cell wall-related processes and PIN2-mediated auxin transport to inhibit root elongation and gravitropic responses in Arabidopsis. Plant Cell Physiol. 61, 519–535.

Lewis, D.H., 2019. Boron: the essential element for vascular plants that never was. New Phytol. 221, 1685–1690.

L'Huillier, L., d'Auzac, J., Durand, M., Michaud-Ferrière, N., 1996. Nickel effects on two maize (*Zea mays*) cultivars: growth, structure, Ni concentration, and localization. Can. J. Bot. 74, 1547–1554.

Li, Y.J., Zamble, D.B., 2009. Nickel homeostasis and nickel regulation: an overview. Chem. Rev. 109, 4617–4643.

Li, X.-L., George, E., Marschner, H., 1991. Extension of the phosphorus depletion zone in VA-mycorrhizal white clover in a calcareous soil. Plant Soil 136, 41–48.

Li, Y., Zhang, Y., Shi, D., Liu, X., Qin, J., Ge, Q., et al., 2013. Spatial-temporal analysis of zinc homeostasis reveals the response mechanisms to acute zinc deficiency in *Sorghum bicolor*. New Phytol. 200, 1102–1115.

Li, G., Wang, B., Tian, Q., Wang, T., Zhang, W.H., 2014a. *Medicago truncatula* ecotypes A17 and R108 differed in their response to iron deficiency. J. Plant Physiol. 171, 639–647.

Li, H., Fan, R., Li, L., Wei, B., Li, G., Gu, L., et al., 2014b. Identification and characterization of a novel copper transporter gene family *TaCT1* in common wheat. Plant Cell Environ. 37, 1561–1573.

Li, B., Qiu, J., Jayakannan, M., Xu, B., Li, Y., Mayo, G.M., et al., 2016. AtNPF2.5 modulates chloride (Cl$^-$) efflux from roots of *Arabidopsis thaliana*. Front. Plant Sci. 7, 2013.

Li, B., Tester, M., Gilliham, M., 2017. Chloride on the move. Trends Plant Sci. 22, 236–248.

Li, J., Gurajala, H.K., Wu, L., van der Ent, A., Qiu, R., Baker, A.J.M., et al., 2018. Hyperaccumulator plants from China: a synthesis of the current state of knowledge. Environ. Sci. Technol. 52, 11980–11994.

Liao, Z., Chen, R., Chen, M., Yang, Y., Fu, Y., Zhang, Q., et al., 2006. Molecular cloning and characterization of the polyphenol oxidase gene from sweetpotato. Mol. Biol. 40, 907–913.

Lin, C.H., Stocking, C.R., 1978. Influence of leaf age, light, dark and iron deficiency on polyribosome levels in maize leaves. Plant Cell Physiol. 19, 461–470.

Lin, M.S., Kao, C.H., 1990. Senescence of rice leaves. XIII. Changes of Zn^{2+}-dependent acid inorganic pyrophosphatase. J. Plant Physiol. 137, 41–45.

Liu, Y., Yu, L., Qu, Y., Chen, J., Liu, X., Hong, H., et al., 2016. GmSALT3, which confers improved soybean salt tolerance in the field, increases leaf Cl$^-$ exclusion prior to Na$^+$ exclusion but does not improve early vigor under salinity. Front. Plant Sci. 7, 1485.

Liu, K., Niu, Y., Konishi, M., Wu, Y., Du, H., Sun Chung, H., et al., 2017. Discovery of nitrate–CPK–NLP signalling in central nutrient–growth networks. Nature 545, 311–316.

Liu, X., Wang, J., Yu, Y., Kong, L., Liu, Y., Liu, Z., et al., 2019. Identification and characterization of the rice pre-harvest sprouting mutants involved in molybdenum cofactor biosynthesis. New Phytol. 222, 275–285.

Liu, J., Zhuang, Y., Huang, X., Zhao, D., Zhao, Y., 2020a. Overexpression of cotton laccase gene LAC1 enhances resistance to *Botrytis cinerea* L in tomato plants. Int. J. Agric. Biol. 24, 221–228.

Liu, X.X., Zhu, Y.X., Fang, X.Z., Ye, J.Y., Du, W.X., Zhu, Q.Y., et al., 2020b. Ammonium aggravates salt stress in plants by entrapping them in a chloride over-accumulation state in an NRT1.1-dependent manner. Sci. Tot. Environ. 746, 141244.

Llamas, A., Chamizo-Ampudia, A., Tejada-Jimenez, M., Galvan, A., Fernandez, E., 2017. The molybdenum cofactor enzyme mARC: Moonlighting or promiscuous enzyme? Biofactors 43, 486–494.

Llases, M.E., Lisa, M.N., Morgada, M.N., Giannini, E., Alzari, P.M., Vila, A.J., 2020. *Arabidopsis thaliana* Hcc1 is a Sco-like metallochaperone for Cu$_A$ assembly in cytochrome c oxidase. FEBS J. 287, 749–762.

Lobreaux, S., Briat, J.F., 1991. Ferritin accumulation and degradation in different organs of pea (*Pisum sativum*) during development. Biochem. J. 274, 601–606.

Löhnis, M.P., 1960. Effect of magnesium on calcium supply on the uptake of manganese by various crop plants. Plant Soil 12, 339–376.

Loneragan, J.F., 1988. Distribution and movement of manganese in plants. In: Graham, R.D., Hannam, R.J., Uren, N.C. (Eds.), Manganese in Soils and Plants. Kluwer Academic Publishers, Dordrecht, The Netherlands, pp. 113–124.

Loneragan, J.F., Grunes, D.L., Welch, R.M., Aduayi, E.A., Tengah, A., Lazar, V.A., et al., 1982. Phosphorus accumulation and toxicity in leaves in relation to zinc supply. Soil Sci. Soc. Am. J. 46, 345–352.

Longnecker, N.E., Graham, R.D., Card, G., 1991. Effects of manganese deficiency on the pattern of tillering and development of barley (*Hordeum vulgare* cv Galleon). Field Crops Res. 28, 85–102.

Lönnerdal, B., 2002. Phytic acid–trace element (Zn, Cu, Mn) interactions. J. Food Sci. Technol. 37, 749–758.

Loomis, W.D., Durst, R.W., 1992. Chemistry and biology of boron. Biofactors 3, 229–239.

Lopez, S., Piutti, S., Vallance, J., Morel, J.L., Echevarria, G., Benizri, E., 2017. Nickel drives bacterial community diversity in the rhizosphere of the hyperaccumulator *Alyssum murale*. Soil Biol. Biochem. 114, 121–130.

Lordkaew, S., Dell, B., Jamjod, S., Rerkasem, B., 2011. Boron deficiency in maize. Plant Soil 342, 207–220.

Lorenzen, I., Aberle, T., Plieth, C., 2004. Salt stress-induced chloride flux: a study using transgenic Arabidopsis expressing a fluorescent anion probe. Plant J. 38, 539–544.

Lott, J.N.A., Buttrose, M.S., 1978. Globoids in protein bodies of legume seed cotyledons. Aust. J. Plant Physiol. 5, 89–111.

Lucena, C., Romera, F.J., García, M.J., Alcántara, E., Pérez-Vicente, R., 2015. Ethylene participates in the regulation of Fe deficiency responses in Strategy I plants and in rice. Front. Plant Sci. 6, 1056.

Lucisine, P., Echevarria, G., Sterckeman, T., Vallance, J., Rey, P., Benizri, E., 2014. Effect of hyperaccumulating plant cover composition and rhizosphere-associated bacteria on the efficiency of nickel extraction from soil. Appl. Soil Ecol. 81, 30–36.

Lukaszewski, K.M., Blevins, D.G., Randall, D.D., 1992. Asparagine and boric acid cause allantoate accumulation in soybean leaves by inhibiting manganese-dependent allantoate amidohydrolase. Plant Physiol. 99, 1670–1676.

Lulai, E.C., Olson, L.L., Fugate, K.K., Neubauer, J.D., Campbell, L.G., 2020. Inhibitors of tri- and tetra- polyamine oxidation, but not diamine oxidation, impair the initial stages of wound-induced suberization. J. Plant Physiol. 246, 153092.

Lysenko, E.A., Klaus, A.A., Kartashov, A.V., Kusnetsov, V.V., 2019. Distribution of Cd and other cations between the stroma and thylakoids: a quantitative approach to the search for Cd targets in chloroplasts. Photosynth. Res. 139, 337–358.

M'sehli, W., Dell'Orto, M., Donnini, S., De Nisi, P., Zocchi, G., Abdelly, C., et al., 2009. Variability of metabolic responses and antioxidant defence in two lines of *Medicago ciliaris* to Fe deficiency. Plant Soil 320, 219–230.

Ma, Y., Rajkumar, M., Freitas, H., 2009. Improvement of plant growth and nickel uptake by nickel resistant-plant-growth promoting bacteria. J. Hazard. Mater. 166, 1154–1161.

Macleod, J.A., Gupta, U.C., Stanfield, B., 1997. Molybdenum and sulfur relationships in plants. In: Gupta, U.C. (Ed.), Molybdenum in Agriculture. Cambridge University Press, Cambridge, UK, pp. 229–244.

Macnair, M.R., 2003. The hyperaccumulation of metals by plants. Adv. Bot. Res. 40, 63–105.

Mäder, M., Füssl, R., 1982. Role of peroxidase in lignification of tobacco cells. II. Regulation by phenolic compounds. Plant Physiol. 70, 1132–1134.

Maier, R.J., Pihl, T.D., Stults, L., Sray, W., 1990. Nickel accumulation and storage in *Bradyrhizobium japonicum*. Appl. Environ. Microbiol. 56, 1905–1911.

Maillard, A., Etienne, P., Diquelou, S., Trouverie, J., Billard, V., Yvin, J.C., et al., 2016. Nutrient deficiencies modify the ionomic composition of plant tissues: a focus on cross-talk between molybdenum and other nutrients in *Brassica napus*. J. Exp. Bot. 67, 5631–5641.

Mansilla, N., Welchen, E., Gonzalez, D.H., 2019. Arabidopsis SCO proteins oppositely influence cytochrome c oxidase levels and gene expression during salinity stress. Plant Cell Physiol. 60, 2769–2784.

Marastoni, L., Sandri, M., Pii, Y., Valentinuzzi, F., Cesco, S., Mimmo, T., 2019. Morphological root responses and molecular regulation of cation transporters are differently affected by copper toxicity and cropping system depending on the grapevine rootstock genotype. Front. Plant Sci. 10, 946.

Marcar, N.E., Graham, R.D., 1987. Genotypic variation for manganese efficiency in wheat. J. Plant Nutr. 10, 2049–2055.

Marentes, E., Grusak, M.A., 1998. Iron transport and storage within the seed coat and embryo of developing seeds of pea (*Pisum sativum* L.). Seed Sci. Res. 8, 367–375.

Maret, W., 2013. Zinc biochemistry: from a single zinc enzyme to a key element of life. Adv Nutr. 4, 82–91.

Maret, W., 2019. The redox biology of redox-inert zinc ions. Free Radic. Biol. Med. 134, 311–326.

Marmagne, A., Vinauger-Douard, M., Monachello, D., de Longevialle, A.F., Charon, C., Allot, M., et al., 2007. Two members of the Arabidopsis CLC (chloride channel) family, AtCLCe and AtCLCf, are associated with thylakoid and Golgi membranes, respectively. J. Exp. Bot. 58, 3385–3393.

Maron, L.G., 2019. From foe to friend: the role of chloride as a beneficial macronutrient. Plant J. 99, 813–814.

Marschner, H., Cakmak, I., 1986. Mechanism of phosphorus-induced zinc deficiency in cotton. II. Evidence for impaired shoot control of phosphorus uptake and translocation under zinc deficiency. Physiol. Plant. 68, 491–496.

Marschner, H., Cakmak, I., 1989. High light intensity enhances chlorosis and necrosis in leaves of zinc, potassium, and magnesium deficient bean (*Phaseolus vulgaris*) plants. J. Plant Physiol. 134, 308–315.

Marschner, H., Römheld, V., Kissel, M., 1986. Different strategies in higher plants in mobilization and uptake of iron. J. Plant Nutr. 9, 695–713.

Masuda, H., Aung, M.S., Kobayashi, T., Hamada, T., Nishizawa, N.K., 2019. Enhancement of iron acquisition in rice by the mugineic acid synthase gene with ferric iron reductase gene and OsIRO2 confers tolerance in submerged and nonsubmerged calcareous soils. Front. Plant Sci. 10, 1179.

Matoh, T., Ishigaki, K., Ohno, K., Azuma, J., 1993. Isolation and characterization of a boron-polysaccharide complex from radish roots. Plant Cell Physiol. 34, 639–642.

Matthes, M., Torres-Ruiz, R.A., 2016. Boronic acid treatment phenocopies monopteros by affecting PIN1 membrane stability and polar auxin transport in *Arabidopsis thaliana* embryos. Development 143, 4053–4062.

Matthes, M., Torres-Ruiz, R.A., 2017. Boronic acids as tools to study (plant) developmental processes? Plant Signal. Behav. 12, e1321190. Available from: https://doi.org/10.1080/15592324.2017.1321190.

Matthes, M.S., Robil, J.M., Tran, T., Kimble, A., Mcsteen, P., 2018. Increased transpiration is correlated with reduced boron deficiency symptoms in the maize tassel-less1 mutant. Physiol. Plant. 163, 344–355.

Matthes, M.S., Robil, J.M., McSteen, P., 2020. From element to development: the power of the essential micronutrient boron to shape morphological processes in plants. J. Exp. Bot. 71, 1681–1693.

Maurer, F., Mueller, S., Bauer, P., 2011. Suppression of Fe deficiency gene expression by jasmonate. Plant Physiol. Biochem. 49, 530–536.

Mazzolini, A.P., Pallaghy, C.K., Legge, G.J.F., 1985. Quantitative microanalysis of Mn, Zn and other elements in mature wheat seed. New Phytol. 100, 483–509.

McCay-Buis, T.S., Huber, D.M., Graham, R.D., Phillips, J.D., Miskin, K.E., 1995. Manganese seed content and take-all of cereals. J. Plant Nutr. 18, 1711–1721.

Mcdonald, A.E., Vanlerberghe, G.C., 2014. Quinol oxidases. In: Hohmannmarriott, M.F. (Ed.), Structural Basis of Biological Energy Generation. Springer, Dordrecht, The Netherlands, pp. 167–185.

McEvoy, J.P., Brudvig, G.W., 2006. Water-splitting chemistry of photosystem II. Chem. Rev. 106, 4455–4483.

McGrath, S.P., Mico, C., Zhao, F.J., Stroud, J.L., Zhang, H., Fozard, S., 2010. Predicting molybdenum toxicity to higher plants: estimation of toxicity threshold values. Environ. Poll. 158, 3085–3094.

Mendel, R.R., Kruse, T., 2012. Cell biology of molybdenum in plants and humans. Biochem. Biophys. Acta-Mol. Cell Res. 1823, 1568–1579.

Mengel, K., 1994. Iron availability in plant tissues – iron chlorosis on calcareous soils. Plant Soil 165, 275–283.

Mettler, I.J., Mandata, S., Taiz, L., 1982. Characterization of in vitro proton pumping by microsomal vesicles isolated from corn coleoptiles. Plant Physiol. 70, 1738–1742.

Miller, A.F., 2012. Superoxide dismutases: ancient enzymes and new insights. FEBS Lett. 586, 585–595.

Miller, G.W., Shigematsu, A., Welkie, G.W., Motoji, N., Szlek, M., 1990. Potassium effect on iron stress in tomato. II. The effects on root CO_2-fixation and organic acid formation. J. Plant Nutr. 13, 1355–1370.

Miller, E.R., Lei, X., Ullrey, D.E., 1991. Trace elements in animal nutrition. In: Mortvedt, J.J., Cox, F.R., Shuman, L.M., Welch, R.M. (Eds.), Micronutrients in Agriculture, 2nd ed. SSSA Book Series No. 4. Soil Science Society of America, Madison, WI, USA, pp. 593–662.

Minchin, P.E.H., Thorp, T.G., Boldingh, H.L., Gould, N., Cooney, J.M., Negm, F.B., et al., 2012. A possible mechanism for phloem transport of boron in 'Hass' avocado (*Persea americana* Mill.) trees. J. Hortic. Sci. Biotech. 87, 23–28.

Minihane, A.M., Rimbach, G., 2002. Iron absorption and the iron binding and anti-oxidant properties of phytic acid. Inter. J. Food Sci. Technol. 37, 741–748.

Miwa, K., Takano, J., Omori, H., Seki, M., Shinozaki, K., Fujiwara, T., 2007. Plants tolerant of high boron levels. Science 318, 1417. 1417.

Modi, A.T., Cairns, A.L.P., 1994. Molybdenum deficiency in wheat results in lower dormancy levels via reduced ABA. Seed Sci. Res. 4, 329–333.

Molas, J., 2002. Changes of chloroplast ultrastructure and total chlorophyll concentration in cabbage leaves caused by excess of organic Ni (II) complexes. Environ. Exp. Bot. 47, 115–126.

Montargès-Pelletier, E., Chardot, V., Echevarria, G., Michot, L.J., Bauer, A., Morel, J.L., 2008. Identification of nickel chelators in three hyperaccumulating plants: an X-ray spectroscopic study. Phytochemistry 69, 1695–1709.

Morales, F., Abadia, A., Abadia, J., 1991. Chlorophyll fluorescence and photon yield of oxygen evolution in iron–deficient sugar beet (*Beta vulgaris* L.) leaves. Plant Physiol. 97, 886–893.

Morales, F., Grasa, R., Abadía, A., Abadía, J., 1998. Iron chlorosis paradox in fruit trees. J. Plant Nutr. 21, 815–825.

Morales, F., Warren, C.R., 2012. Photosynthetic responses to nutrient deprivation and toxicities. In: Flexas, J., Loreto, F., Medrano, H. (Eds.), Terrestrial Photosynthesis in a Changing Environment: A Molecular, Physiological, and Ecological Approach. Cambridge University Press, Cambridge, UK, pp. 312–330.

Moran, N., 2007. Osmoregulation of leaf motor cells. FEBS Lett. 581, 2337–2347.

Moreno-Lora, A., Delgado, A., 2020. Factors determining Zn availability and uptake by plants in soils developed under Mediterranean climate. Geoderma 376, 114509.

Morgounov, A., Gomez-Becerra, H.F., Abugalieva, A., Dzhunusova, M., Yessimbekova, M., Muminjanov, H., et al., 2007. Iron and zinc grain density in common wheat grown in Central Asia. Euphytica 155, 193–203.

Mori, S., Nishizawa, N., 1987. Methionine as a dominant precursor of phytosiderophores in Graminaceae plants. Plant Cell Physiol. 28, 1081–1092.

Mortvedt, J.J., 1981. Nitrogen and molybdenum uptake and dry matter relationship in soybeans and forage legumes in response to applied molybdenum on acid soil. J. Plant Nutr. 3, 245–256.

Müller, H.M., Schäfer, N., Bauer, H., Geiger, D., Lautner, S., Fromm, J., et al., 2017. The desert plant *Phoenix dactylifera* closes stomata via nitrate-regulated SLAC1 anion channel. New Phytol. 216, 150–162.

Munns, R., Tester, M., 2008. Mechanisms of salinity tolerance. Annu. Rev. Plant Biol. 59, 651–681.

Musiani, F., Dikiy, A., Semenov, A.Y., Ciurli, S., 2005. Structure of the intermolecular complex between plastocyanin and cytochrome f from spinach. J. Biol. Chem. 280, 18833–18841.

Mustafiz, A., Ghosh, A., Tripathi, A.K., Kaur, C., Ganguly, A.K., Bhavesh, N.S., et al., 2014. A unique Ni^{2+}-dependent and methylglyoxal-inducible rice glyoxalase I possesses a single active site and functions in abiotic stress response. Plant J. 78, 951–963.

Nable, R.O., Bar-Akiva, A., Loneragan, J.F., 1984. Functional manganese requirement and its use as a critical value for diagnosis of manganese deficiency in subterranean clover (*Trifolium subterraneum* L. cv. Seaton Park). Ann. Bot. 54, 39–49.

Nable, R.O., Cartwright, B., Lance, R.C.M., 1990a. Genotypic differences in boron accumulation in barley: relative susceptibilities to boron deficiency and toxicity. In: El Bassam, N., Dambroth, M., Loughman, B.C. (Eds.), Genetic Aspects of Plant Mineral Nutrition. Kluwer Academic Publishers, Dordrecht, The Netherlands, pp. 243–251.

Nable, R.O., Lance, R.C.M., Cartwright, B., 1990b. Uptake of boron and silicon by barley genotypes with differing susceptibilities to boron toxicity. Ann. Bot. 66, 83–90.

Nable, R.O., Banuelos, G.S., Paull, J.G., 1997. Boron toxicity. Plant Soil 193, 181–197.

Naya, L., Paul, S., Valdes-Lopez, O., Mendoza-Soto, A.B., Nova-Franco, B., Sosa-Valencia, G., et al., 2014. Regulation of copper homeostasis and biotic interactions by microRNA398b in common bean. PLoS ONE 9, e84416.

Nayyar, V.K., Takkar, P.N., 1980. Evaluation of varius zinc sources for rice grown on alkali soil. Z. Pflanzenernähr. Bodenk. 143, 489–493.

Ness, P.J., Woolhouse, H.W., 1980. RNA synthesis in *Phaseolus* chloroplasts. I. Ribonucleic acid synthesis and senescing leaves. J. Exp. Bot. 31, 223–233.

Neuhaus, H.E., Wagner, R., 2000. Solute pores, ion channels, and metabolite transporters in the outer and inner envelope membranes of higher plant plastids. Biochim. Biophys. Acta-Rev. Biomembr. 1465, 307–323.

Neumann, K.H., Steward, F.C., 1968. Investigations on the growth and metabolism of cultured explants of *Daucus carota*. I. Effects of iron, molybdenum and manganese on growth. Planta 81, 333–350.

Nguyen, C.T., Agorio, A., Jossier, M., Depre, S., Thomine, S., Filleur, S., 2016. Characterization of the chloride channel-like, AtCLCg, involved in chloride tolerance in *Arabidopsis thaliana*. Plant Cell Physiol. 57, 764–775.

Nicoulaud, B.A., Bloom, A.J., 1998. Nickel supplements improve growth when foliar urea is the sole nitrogen source for tomato. J. Am. Soc. Hortic. Sci. 123, 556–559.

Nie, G., Zhao, J., He, R., Tang, Y., 2020. CuO nanoparticle exposure impairs the root tip cell walls of *Arabidopsis thaliana* seedlings. Water Air Soil Poll. 231, 1–11.

Nieves-Cordones, M., García-Sánchez, F., Pérez-Pérez, J.G., Colmenero-Flores, J.M., Rubio, F., Rosales, M.A., 2019. Coping with water shortage: An update on the role of K^+, Cl^-, and water membrane transport mechanisms on drought resistance. Front. Plant Sci. 10, 1619.

Nikolic, M., Römheld, V., 1999. Mechanism of Fe uptake by the leaf symplast: is Fe inactivation on leaf a cause of Fe deficiency chlorosis? Plant Soil 215, 229–237.

Nikolic, M., Römheld, V., 2002. Does high bicarbonate supply to roots change availability of iron in the leaf apoplast? Plant Soil 241, 67–74.

Nikolic, M., Römheld, V., 2003. Nitrate does not result in iron inactivation in the apoplast of sunflower leaves. Plant Physiol. 132, 1303–1314.

Nikolic, M., Römheld, V., Merkt, N., 2000. Effect of bicarbonate on uptake and translocation of ^{59}Fe in two grapevine rootstocks differing in their resistance to Fe deficiency chlorosis. Vitis 39, 145–150.

Nishida, S., Kato, A., Tsuzuki, C., Yoshida, J., Mizuno, T., 2015. Induction of nickel accumulation in response to zinc deficiency in *Arabidopsis thaliana*. Int. J. Mol. Sci. 16, 9420–9430.

Noguchi, K., Yasumori, M., Imai, T., Naito, S., Matsunaga, T., Oda, H., et al., 1997. *bor1-1*, an *Arabidopsis thaliana* mutant that requires a high level of boron. Plant Physiol. 115, 901–906.

Nozoye, T., Nagasaka, S., Kobayashi, T., Takahashi, M., Sato, Y., Sato, Y., et al., 2011. Phytosiderophore efflux transporters are crucial for iron acquisition in gramineous plants. J. Biol. Chem. 286, 5446.

Nyomora, A.M., Brown, P.H., Freeman, M., 1997. Fall foliar-applied boron increases tissue boron concentration and nut set of almond. J. Am. Soc. Hortic. Sci. 122, 405–410.

Nyomora, A.M.S., Brown, P.H., Krueger, B., 1999. Rate and time of boron application increase almond productivity and tissue boron concentration. Hortscience 34, 242–245.

Nyomora, A.M.S., Brown, P.H., Pinney, K., Polito, V.S., 2000. Foliar application of boron to almond trees affects pollen quality. J. Am. Soc. Hortic. Sci. 125, 265–270.

O'Neill, M.A., Warrenfeltz, D., Kates, K., Pellerin, P., Doco, T., Darvill, A.G., et al., 1996. Rhamnogalacturonan-II, a pectic polysaccharide in the walls of growing plant cell, forms a dimer that is covalently cross-linked by a borate ester—in vitro conditions for the formation and hydrolysis of the dimer. J. Biol. Chem. 271, 22923–22930.

O'Neill, M., Eberhard, S., Albersheim, P., Darvill, A., 2001. Requirement of borate cross-linking of cell wall rhamnogalacturonan II for Arabidopsis growth. Science 294, 846–849.

Obata, H., Umebayashi, M., 1988. Effect of zinc deficiency on protein synthesis in cultured tobacco plant cells. Soil Sci. Plant Nutr. 34, 351–357.

Obrador, A., Gonzalez, D., Alvarez, J.M., 2013. Effect of inorganic and organic copper fertilizers on copper nutrition in *Spinacia oleracea* and on labile copper in soil. J. Agric. Food Chem. 61, 4692–4701.

Ohki, K., 1976. Effect of zinc nutrition on photosynthesis and carbonic anhydrase activity in cotton. Physiol. Plant. 38, 300–304.

Oikawa, T., Ishimaru, Y., Munemasa, S., Takeuchi, Y., Washiyama, K., Hamamoto, S., et al., 2018. Ion channels regulate nyctinastic leaf opening in *Samanea saman*. Curr. Biol. 28, 2230.

Oiwa, Y., Kitayama, K., Kobayashi, M., Matoh, T., 2013. Boron deprivation immediately causes cell death in growing roots of *Arabidopsis thaliana* (L.) Heynh. Soil Sci. Plant Nutr. 59, 621–627.

Oltmans, S.E., Fehr, W.R., Welke, G.A., Raboy, V., Peterson, K.L., 2005. Agronomic and seed traits of soybean lines with low-phytate phosphorus. Crop Sci. 45, 593–598.

Önnerud, H., Zhang, L., Gellerstedt, G., Henriksson, G., 2002. Polymerization of monolignols by redox shuttle-mediated enzymatic oxidation: a new model in lignin biosynthesis. Plant Cell 14, 1953—1962.

Ova, E.A., Kutman, U.B., Ozturk, L., Cakmak, I., 2015. High phosphorus supply reduced zinc concentration of wheat in native soil but not in autoclaved soil or nutrient solution. Plant Soil 393, 147—162.

Owuoche, J.O., Briggs, K.G., Taylor, G.J., 1996. The efficiency of copper use by 5A/5RL wheat-rye translocation lines and wheat (*Triticum aestivum* L.) cultivars. Plant Soil 180, 113—120.

Ozanne, P.G., 1958. Chlorine deficiency in soils. Nature 182, 1172—1173.

Ozturk, L., Yazici, M.A., Yucel, C., Torun, A., Cekic, C., Bagci, A., et al., 2006. Concentration and localization of zinc during seed development and germination in wheat. Physiol. Plant. 128, 144—152.

Padmanaban, S., Lin, X., Perera, I., Kawamura, Y., Sze, H., 2004. Differential expression of vacuolar H^+-ATPase subunit c genes in tissues active in membrane trafficking and their roles in plant growth as revealed by RNAi. Plant Physiol. 134, 1514—1526.

Pakhomova, V.M., Daminova, A.I., 2020. Chelated micronutrient fertilizers as effective antioxidants applied for foliar plant treatment. BIO Web Conf. 17, 00057. Available from: https://doi.org/10.1051/bioconf/20201700057.

Pan, J., Huang, D., Guo, Z., Kuang, Z., Zhang, H., Xie, X., et al., 2018. Overexpression of microRNA408 enhances photosynthesis, growth, and seed yield in diverse plants. J. Integr. Plant Biol. 60, 323—340.

Papadakis, I.E., Giannakoula, A., Therios, I.N., Bosabalidis, A.M., Moustakas, M., Nastou, A., 2007. Mn-induced changes in leaf structure and chloroplast ultrastructure of *Citrus volkameriana* (L.) plants. J. Plant Physiol. 164, 100—103.

Papworth, M., Kolasinska, P., Minczuk, M., 2006. Designer zinc-finger proteins and their applications. Gene 366, 27—38.

Parker, D.R., 1997. Responses of six crop species to solution Zn^{2+} activities buffered with HEDTA. Soil Sci. Soc. Am. J. 61, 167—176.

Parker, M.B., Harris, H.B., 1977. Yield and leaf nitrogen of nodulating soybeans as affected by nitrogen and molybdenum. Agron. J. 69, 551—554.

Parveen, I., Threadgill, M.D., Moorby, J.M., Winters, A., 2010. Oxidative phenols in forage crops containing polyphenol oxidase enzymes. J. Agric. Food Chem. 58, 1371—1382.

Pasricha, N.S., Nayyar, V.K., Randhawa, N.S., Sinha, M.K., 1977. Influence of sulphur fertilization on suppression of molybdenum uptake by berseem (*Trifolium alexandrinum*) and oats (*Avena sativa*) grown on a molybdenum-toxic soil. Plant Soil 46, 245—250.

Paull, J.G., Cartwright, B., Rathjen, A.J., 1988. Responses of wheat and barley genotypes of toxic concentrations of soil boron. Euphytica 39, 137—144.

Paull, J.G., Nable, R.O., Rathjen, A.J., 1992a. Physiological and genetic control of the tolerance of wheat to high concentrations of boon and implications for plant breeding. Plant Soil 146, 251—260.

Paull, J.G., Nable, R.O., Lake, A.W.H., Materne, M.A., Rathjen, A.J., 1992b. Response of annual medics (*Medicago* spp.) and field peas (*Pisum sativum*) to high concentration of boron: genetic variation and the mechanism of tolerance. Aust. J. Agric. Res. 43, 203—213.

Pavlovic, J., Samardzic, J., Maksimovic, V., Timotijevic, G., Stevic, N., Laursen, K.H., et al., 2013. Silicon alleviates iron deficiency in cucumber by promoting mobilization of iron in the root apoplast. New Phytol. 198, 1096—1107.

Peck, A.W., McDonald, G.K., 2010. Adequate zinc nutrition alleviates the adverse effects of heat stress in bread wheat. Plant Soil 337, 355—374.

Perea-Garcia, A., Andres-Borderia, A., Vera-Sirera, F., Perez-Amador, M.A., Puig, S., Penarrubia, L., 2020. Deregulated high affinity copper transport alters iron homeostasis in Arabidopsis. Front. Plant Sci. 11, 1106.

Perica, S., Brown, P.H., Connell, J.H., Nyomora, A.M.S., Dordas, C., Hu, H.N., et al., 2001. Foliar boron application improves flower fertility and fruit set of olive. Hortscience 36, 714—716.

Persson, D.P., Hansen, T.H., Laursen, K.H., Schjoerring, J.K., Husted, S., 2009. Simultaneous iron, zinc, sulfur and phosphorus speciation analysis of barley grain tissues using SEC-ICP-MS and IP-ICP-MS. J. Metallomics 1, 418—426.

Peterson, C.J., Johnson, V.A., Mattern, P.J., 1986. Influence of cultivar and environment on mineral and protein concentrations of wheat flour, bran and grain. Cereal Chem. 63, 118—186.

Peto, A., Lehotai, N., Feigl, G., Tugyi, N., Oerdoeg, A., Gemes, K., et al., 2013. Nitric oxide contributes to copper tolerance by influencing ROS metabolism in Arabidopsis. Plant Cell Rep. 32, 1913—1923.

Pich, A., Scholz, G., Seifert, K., 1991. Effect of nicotianamine on iron uptake and citrate accumulation in two genotypes of tomato, *Lycopersicon esculentum* Mill. J. Plant Physiol. 137, 323—326.

Pilbeam, D.J., Kirkby, E.A., 1983. The physiological role of boron in plants. J. Plant Nutr. 6, 563—582.

Pinton, R., Cakmak, I., Marschner, H., 1993. Effect of zinc deficiency on proton fluxes in plasma membrane-enriched vesicles isolated from bean roots. J. Exp. Bot. 44, 623—630.

Polacco, J.C., Mazzafera, P., Tezotto, T., 2013. Opinion — nickel and urease in plants: still many knowledge gaps. Plant Sci. 199, 79—90.

Pollard, A.S., Parr, A.J., Loughman, B.C., 1977. Boron in relation to membrane function in higher plants. J. Exp. Bot. 28, 831—841.

Pommerrenig, B., Eggert, K., Bienert, G.P., 2019. Boron deficiency effects on sugar, ionome, and phytohormone profiles of vascular and non-vascular leaf tissues of common plantain (*Plantago major* L.). Int. J. Mol. Sci. 20, 3882.

Poza-Viejo, L., Abreu, I., González-García, M.P., Allauca, P., Bonilla, I., Bolaños, L., et al., 2018. Boron deficiency inhibits root growth by controlling meristem activity under cytokinin regulation. Plant Sci. 270, 176—189.

Prabhjot, S., Bhardwaj, R.D., Simarjit, K., Jaspal, K., Grewal, S.K., 2020. Metabolic adjustments during compatible interaction between barley genotypes and stripe rust pathogen. Plant Physiol. Biochem. 147, 295—302.

Pramanik, A., Datta, A.K., Das, D., Kumbhakar, D.V., Ghosh, B., Mandal, A., et al., 2018. Assessment of nanotoxicity (cadmium sulphide and copper oxide) using cytogenetical parameters in *Coriandrum sativum* L. (Apiaceae). Cytol. Genet. 52, 299—308.

Priyanka, N., Venkatachalam, P., 2020. Accumulation of nutrient contents and biochemical changes in shallot, *Allium ascalonicum* grown in soil amended with CuO nanoparticle. J. Environ. Biol. 41, 563–571.

Pushnik, J.C., Miller, G.W., 1989. Iron regulation of chloroplast photosynthetic function: mediation of PS I development. J. Plant Nutr. 12, 407–421.

Pushnik, J.C., Miller, G.W., Manwaring, J.H., 1984. The role of iron in higher plant chlorophyll biosynthesis, maintenance and chloroplast biogenesis. J. Plant Nutr. 7, 733–758.

Qiu, J., Henderson, S.W., Tester, M., Roy, S.J., Gilliham, M., 2016. SLAH1, a homologue of the slow type anion channel SLAC1, modulates shoot Cl⁻ accumulation and salt tolerance in *Arabidopsis thaliana*. J. Exp. Bot. 67, 4495–4505.

Rabier, J., Laffont-Schwob, I., Notonier, R., Fogliani, B., Bouraïma-Madjèbi, S., 2008. Anatomical element localization by EDXS in *Grevillea exul* var. exul under nickel stress. Environ. Pollut. 156, 1156–1163.

Radin, I., Mansilla, N., Rodel, G., Steinebrunner, I., 2015. The Arabidopsis COX11 homolog is essential for cytochrome c oxidase activity. Front. Plant Sci. 6, 1091. Available from: https://doi.org/10.3389/fpls.2015.01091.

Rahimi, A., 1970. Kupfermangel bei höheren Pflanzen. Landwirtsch. Forsch. 25, 42–47.

Rahimi, A., Bussler, W., 1974. Kupfermangel bei höheren Pflanzen und sein histochemischer Nachweis. Landwirtsch. Forsch. 30, 100–111.

Rahman, H., Sabreen, S., Alam, S., Kawai, S., 2005. Effects of nickel on growth and composition of metal micronutrients in barley plants grown in nutrient solution. J. Plant Nutr. 28, 393–404.

Raj, S.B., Ramaswamy, S., Plapp, B.V., 2014. Yeast alcohol dehydrogenase structure and catalysis. Biochemistry 53, 5791–5803.

Ramamurthy, R.K., Xiang, Q., Hsieh, E.-J., Liu, K., Zhang, C., Waters, B.M., 2018. New aspects of iron-copper crosstalk uncovered by transcriptomic characterization of Col-0 and the copper uptake mutant *spl7* in *Arabidopsis thaliana*. Metallomics 10, 1824–1840.

Ranieri, A., Castagna, A., Baldan, B., Soldatini, G.F., 2001. Iron defiency differently affects peroxidase isoforms in sunflower. J. Exp. Bot. 52, 25–35.

Raven, J.A., 1980. Short- and long-distance transport of boric acid in plants. New Phytol. 84, 231–249.

Raven, J.A., 2020. Chloride involvement in the synthesis, functioning and repair of the photosynthetic apparatus *in vivo*. New Phytol. 227, 334–342.

Raven, J.A., Evans, M.C.W., Korb, R.E., 1999. The role of trace metals in photosynthetic electron transport in O_2-evolving organisms. Photosynth. Res. 60, 111–149.

Ravet, K., Touraine, B., Boucherez, J., Briat, J.F., Gaymard, F., Cellier, F., 2009. Ferritins control interaction between iron homeostasis and oxidative in Arabidopsis. Plant J. 57, 400–412.

Rebafka, F.P., 1993. Deficiency of phosphorus and molybdenum as major growth limiting factors of pearl millet and groundnut on an acid sandy soil in Niger, West Africa. Ph.D. Thesis University Hohenheim. ISSN 0942-0754.

Rebafka, F.P., Schulz, R., Marschner, H., 1990. Erhebungsuntersuchungen zur Pflanzenverfügbarkeit von Nickel auf Böden mit hohen geogenen Nickelgehalten. Angew. Bot. 64, 317–328.

Reguera, M., Espi, A., Bolanos, L., Bonilla, I., Redondo-Nieto, M., 2009. Endoreduplication before cell differentiation fails in boron deficient legume nodules. Is boron involved in signalling during cell cycle regulation? New Phytol. 183, 9–12.

Reguera, M., Bonilla, I., Bolanos, L., 2010. Boron deficiency results in induction of pathogenesis-related proteins from the PR-10 family during the legume-rhizobia interaction. J. Plant Physiol. 167, 625–632.

Rehman, M.Z.U., Rizwan, M., Ali, S., Fatima, N., Yousaf, B., Naeem, A., et al., 2016. Contrasting effects of biochar, compost and farm manure on alleviation of nickel toxicity in maize (*Zea mays* L.) in relation to plant growth, photosynthesis and metal uptake. Ecotox. Environ. Saf. 133, 218–225.

Reid, R., 2007. Identification of boron transporter genes likely to be responsible for tolerance to boron toxicity in wheat and barley. Plant Cell Physiol. 48, 1673–1678.

Reid, R., Fitzpatrick, K., 2009. Influence of leaf tolerance mechanisms and rain on boron toxicity in barley and wheat. Plant Physiol. 151, 413–420.

Reid, R.J., Hayes, J.E., Post, A., Stangoulis, J.C.R., Graham, R.D., 2004. A critical analysis of the causes of boron toxicity in plants. Plant Cell Environ. 27, 1405–1414.

Rellan-Alvarez, R., Giner-Martinez-Sierra, J., Orduna, J., Orera, I., Rodriguez-Castrillon, J.A., Garcia-Alonso, J.I., et al., 2010. Identification of a tri-iron (III), tri-citrate complex in the xylem sap of iron-deficient tomato resupplied with iron: new insights into plant iron long-distance transport. Plant Cell Physiol. 51, 91–102.

Rengel, Z., 2001. Genotypic differences in micronutrient use efficiency in crops. Commun. Soil Sci. Plant Anal. 32, 1163–1186.

Rengel, Z., 2003. Heavy metals as essential nutrients. In: Prasad, M.N.V., Hagemeyer, J. (Eds.), Heavy Metal Stress in Plants: Molecules to Ecosystems, 2nd ed. Springer-Verlag, Berlin, Germany, pp. 271–294.

Rengel, Z., Graham, R.D., Pedler, J.F., 1993. Manganese nutrition and accumulation of phenolics and lignin as related to differential resistance of wheat genotypes to the take-all fungus. Plant Soil 151, 255–263.

Rerkasem, B., Jamjod, S., Pusadee, T., 2020. Productivity limiting impacts of boron deficiency, a review. Plant Soil 455, 23–40.

Reuter, D.J., Edwards, D.G., Wilhelm, N.S., 1997. Temperate and tropical crops. In: Reuter, D.J., Robinson, J.B. (Eds.), Plant Analysis: An Interpretation Manual, 2nd ed. CSIRO Publishing, Melbourne, Australia, pp. 81–279.

Reuter, D.J., Alston, A.M., McFarlane, J.D., 1988. Occurrence and correction of manganese deficiency in plants. In: Graham, R.D., Hannam, R.J., Uren, N.C. (Eds.), Manganese in Soils and Plants. Kluwer, Dordrecht, The Netherlands, pp. 205–224.

Robbertse, P.J., Lock, J.J., Stoffberg, E., Coetzer, L.A., 1990. Effect of boron on directionality of pollen-tube growth in petunia and agapanthus. S. Afr. J. Bot. 56, 487–492.

Roberts, S.K., 2006. Plasma membrane anion channels in higher plants and their putative functions in roots. New Phytol. 169, 647–666.
Robertson, G.A., Loughman, B.C., 1974. Response to boron deficiency: a comparison with responses produced by chemical methods of retarding root elongation. New Phytol. 73, 821–832.
Robson, A.D., Reuter, D.J., 1981. Diagnosis of copper deficiency and toxicity. In: Loneragan, J.F., Robson, A.D., Graham, R.D. (Eds.), Copper in Soils and Plants. Academic Press, London and Orlando, pp. 287–312.
Robson, A.D., Pitman, M.G., 1983. Interactions between nutrients in higher plants. In: Läuchli, A., Bieleski, R.L. (Eds.), Encyclopedia of Plant Physiology, New Series, Vol. 15A. Springer-Verlag, Berlin and New York, pp. 147–180.
Robson, A.D., Hartley, R.D., Jarvis, S.C., 1981. Effect of copper deficiency on phenolic and other constituents of wheat cell walls. New Phytol. 89, 361–373.
Rodríguez-Celma, J., Lin, W.D., Fu, G.M., Abadía, J., López-Millán, A.F., Schmidt, W., 2013. Mutually exclusive alterations in secondary metabolism are critical for the uptake of insoluble iron compounds by Arabidopsis and *Medicago truncatula*. Plant Physiol. 162, 1473–1485.
Rogalla, H., Römheld, V., 2002. Role of leaf apoplast in silicon-mediated manganese tolerance of *Cucumis sativus* L. Plant Cell Environ. 25, 549–555.
Rognes, S.E., 1980. Anion regulation of lupin asparagine synthetase: chloride activation of the glutamine-utilizing reaction. Phytochemistry 19, 2287–2293.
Roitto, M., Rautio, P., Julkunen-Tiitto, R., Kukkola, E., Huttunen, S., 2005. Changes in the concentrations of phenolics and photosynthates in Scots pine (*Pinus sylvestris* L.) seedlings exposed to nickel and copper. Environ. Pollut. 137, 603–609.
Rombola, A.D., Brüggemann, W., López-Millán, A.F., Abadía, J., Tagliavini, M., Marangoni, B., et al., 2002. Biochemical mechanisms of tolerance to Fe-deficiency on kiwifruit. Tree Physiol. 22, 869–875.
Römheld, V., 2000. The chlorosis paradox: Fe inactivation as a secondary event in chlorotic leaves of grapevine. J. Plant Nutr. 23, 1629–1643.
Römheld, V., Marschner, H., 1981a. Rhythmic iron stress reactions in sunflower at suboptimal iron supply. Physiol. Plant. 53, 347–353.
Römheld, V., Marschner, H., 1981b. Iron deficiency stress induced morphological and physiological changes in root tips of sunflower. Physiol. Plant. 53, 354–360.
Römheld, V., Marschner, H., 1990. Genotypical differences among graminaceous species in release of phytosiderophores and uptake of iron phytosiderophores. Plant Soil 123, 147–153.
Rosales, M.A., Franco-Navarro, J.D., Peinado-Torrubia, P., Díaz-Rueda, P., Álvarez, R., Colmenero-Flores, J.M., 2020. Chloride improves nitrate utilization and NUE in plants. Front. Plant Sci. 11, 442.
Rose, M.T., Rose, T.J., Pariasca-Tanaka, J., Yoshihashi, T., Neuweger, H., Goesmann, A., et al., 2012. Root metabolic response of rice (*Oryza sativa* L.) genotypes with contrasting tolerance to zinc deficiency and bicarbonate excess. Planta 236, 959–973.
Roth-Bejerano, N., Itai, C., 1981. Effect of boron on stomatal opening of epidermal strips of *Commelina communis*. Physiol. Plant. 52, 302–304.
Roux, B., Leonhardt, N., 2018. The regulation of ion channels and transporters in the guard cell. In: Maurel, C. (Ed.), Membrane Transport in Plants, Vol. 87. Academic Press-Elsevier Science, New York, NY, USA, pp. 171–214.
Rozema, J., De Bruin, J., Broekman, R.A., 1992. Effect of boron on the growth and mineral ecomony of some halophytes and nonhalophytes. New Phytol. 121, 249–256.
Ruano, A., Poschenrieder, C., Barcelo, J., 1988. Growth and biomass partitioning in zinc-toxic bush beans. J. Plant Nutr. 11, 577–588.
Ruter, J.M., 2005. Effect of nickel applications for the control of mouse ear disorder on river birch. J. Environ. Hortic. 23, 17–20.
Ruyters, S., Salaets, P., Oorts, K., Smolders, E., 2013. Copper toxicity in soils under established vineyards in Europe: a survey. Sci. Total Environ. 443, 470–477.
Ryan, M.H., McInerney, J.K., Record, I.R., Angus, J.F., 2008. Zinc bioavailability in wheat grain in relation to phosphorus fertilizer, crop sequence and mycorrhizal fungi. J. Sci. Food Agric. 88, 1208–1216.
Sadana, U.S., Kusum, L., Claassen, N., 2002. Manganese efficiency of wheat cultivars as related to root growth and internal manganese requirement. J. Plant Nutr. 25, 2677–2688.
Saeed Ur, R., Khalid, M., Hui, N., Kayani, S.I., Tang, K., 2020. Diversity and versatile functions of metallothioneins produced by plants: a review. Pedosphere 30, 577–588.
Saito, S., Uozumi, N., 2019. Guard cell membrane anion transport systems and their regulatory components: an elaborate mechanism controlling stress-induced stomatal closure. Plants 8, 9.
Salami, A.U., Kenefick, D.G., 1970. Stimulation of growth in zinc-deficient corn seedlings by the addition of tryptophan. Crop Sci. 10, 291–294.
Saleh, L., Plieth, C., 2013. A9C sensitive Cl$^-$ accumulation in *A. thaliana* root cells during salt stress is controlled by internal and external calcium. Plant Signal. Behav. 8, e24259. Available from: https://doi.org/10.4161/psb.24259.
Sanders, D., Bethke, P., 2000. Membrane transport. In: Buchanan, B.B., Gruissem, W., Jones, R.L. (Eds.), Biochemistry and Molecular Biology of Plants. American Society of Plant Physiologists, Rockville, MD, USA, pp. 110–158.
Sandmann, G., Böger, P., 1983. The enzymatological function of heavy metals and their role in electron transfer processes of plants. In: Läuchli, A., Bieleski, R.L. (Eds.), Encyclopedia of Plant Physiology, New Series. Springer-Verlag, Berlin and New York, pp. 563–593.
Sanmartin, M., Pateraki, I., Chatzopoulou, F., Kanellis, A.K., 2007. Differential expression of the ascorbate oxidase multigene family during fruit development and in response to stress. Planta 225, 873–885.
Sanz, A., Pike, S., Khan, M.A., Carrio-Segui, A., Mendoza-Cozatl, D.G., Penarrubia, L., et al., 2019. Copper uptake mechanism of *Arabidopsis thaliana* high-affinity COPT transporters. Protoplasma 256, 161–170.

Sasaki, H., Hirose, T., Watanabe, Y., Ohsugi, Y., 1998. Carbonic anhydrase activity and CO_2-transfer resistance in Zn-deficient rice leaves. Plant Physiol. 118, 929–934.

Sautron, E., Giustini, C., Thuyvan, D., Moyet, L., Salvi, D., Crouzy, S., et al., 2016. Identification of two conserved residues involved in copper release from chloroplast PIB-1-ATPases. J. Biol. Chem. 291, 20136–20148.

Schlegel, H.G., Cosson, J.P., Baker, A.J.M., 1991. Nickel-hyperaccumulating plants provide a niche for nickel-resistant bacteria. Bot. Acta 104, 18–25.

Schlemmer, U., Frølich, W., Prieto, R.M., Grases, F., 2009. Phytate in foods and significance for humans: Food sources, intake, processing, bioavailability, protective role and analysis. Mol. Nutr. Food Res. 53, 330–375.

Schmidt, W., 2003. Iron solutions: acquisition strategies and signaling pathways in plants. Trends Plant Sci. 8, 188–193.

Schnabl, H., 1980. Der Anionenmetabolismus in stärkehaltigen und stärkefreien Schließzellenprotoplasten. Ber. Deutsch. Bot. Ges. 93, 595–605.

Schmidt, S.B., Husted, S., 2019. The biochemical properties of manganese in plants. Plants 8, 381.

Schmidt, S.B., Husted, S., 2020. Micronutrients: advances in understanding manganese cycling in soils, acquisition by plants and ways of optimizing manganese efficiency in crops. In: Rengel, Z. (Ed.), Achieving Sustainable Crop Nutrition. Burleigh Dodds Science Publishing, Cambridge, UK, pp. 407–454.

Schmidt, S.B., Pedas, P., Laursen, K.H., Schjoerring, J.K., Husted, S., 2013. Latent manganese deficiency in barley can be diagnosed and remediated on the basis of chlorophyll *a* fluorescence measurements. Plant Soil 372, 417–429.

Schmid, N.B., Giehl, R.F., Döll, S., Mock, H.P., Strehmel, N., Scheel, D., et al., 2014. Feruloyl-CoA 6′-hydroxylase1-dependent coumarins mediate iron acquisition from alkaline substrates in Arabidopsis. Plant Physiol. 164, 160–172.

Schmidt, S.B., Persson, D.P., Powikrowska, M., Frydenvang, J., Schjoerring, J.K., Jensen, P.E., et al., 2015. Metal binding in photosystem II super- and subcomplexes from barley thylakoids. Plant Physiol. 168, 1490–1502.

Schmidt, S.B., Jensen, P.E., Husted, S., 2016a. Manganese deficiency in plants: the impact on photosystem II. Trends Plant Sci. 21, 622–632.

Schmidt, S.B., Powikrowska, M., Krogholm, K.S., Naumann-Busch, B., Schjoerring, J.K., Husted, S., et al., 2016b. Photosystem II functionality in barley responds dynamically to changes in leaf manganese status. Front. Plant Sci. 7, 1–12.

Schmidt, S.B., George, T.S., Brown, L.K., Booth, A., Wishart, J., Hedley, P.E., et al., 2019. Ancient barley landraces adapted to marginal soils demonstrate exceptional tolerance to manganese limitation. Ann. Bot. 123, 831–843.

Schmidt, S.B., Eisenhut, M., Schneider, A., 2020. Chloroplast transition metal regulation for efficient photosynthesis. Trends Plant Sci. 25, 817–828.

Scholz, G., Becker, R., Stephan, U.W., Rudolph, A., Pich, A., 1988. The regulation of iron uptake and possible functions of nicotianamine in higher plants. Biochem. Physiol. Pflanz. 183, 257–269.

Schon, M.K., Novacky, A., Blevins, D.G., 1990. Boron induces hyperpolarization of sunflower root cell membranes and increases membrane permeability to K^+. Plant Physiol. 93, 566–571.

Schulten, A., Bytomski, L., Quintana, J., Bernal, M., Kramer, U., 2019. Do Arabidopsis *Squamosa Promoter Binding Protein-Like* genes act together in plant acclimation to copper or zinc deficiency? Plant Direct 3, e00150.

Schwarz, G., Mendel, R.R., 2006. Molybdenum cofactor biosynthesis and molybdenum enzymes. Annu. Rev. Plant Biol. 57, 623–647.

Schwarz, G., Mendel, R.R., Ribbe, M.W., 2009. Molybdenum cofactors, enzymes and pathways. Nature 460, 839–847.

Schwenke, G.D., Simpfendorfer, S.R., Collard, B.C.Y., 2015. Confirmation of chloride deficiency as the cause of leaf spotting in durum wheat grown in the Australian northern grains region. Crop Pasture Sci. 66, 122–134.

Seckbach, J., 1982. Ferreting out the secrets of plant ferritin – a review. J. Plant Nutr. 5, 369–394.

Seguela, M., Briat, J.F., Vert, G., Curie, C., 2008. Cytokinins negatively regulate the root iron uptake machinery in Arabidopsis through a growth-dependent pathway. Plant J. 55, 289–300.

Sekimoto, H., Hoshi, M., Nomura, T., Yokota, T., 1997. Zinc deficiency affects the levels of endogenous gibberellins in *Zea mays* L. Plant Cell Physiol. 38, 1087–1090.

Shabani, L., Sabzalian, M.R., 2016. Arbuscular mycorrhiza affects nickel translocation and expression of ABC transporter and metallothionein genes in *Festuca arundinacea*. Mycorrhiza 26, 67–76.

Shahbaz, M., Pilon, M., 2019. Conserved Cu-microRNAs in *Arabidopsis thaliana* function in copper economy under deficiency. Plants 8, 141.

Shahbaz, M., Ravet, K., Peers, G., Pilon, M., 2015. Prioritization of copper for the use in photosynthetic electron transport in developing leaves of hybrid poplar. Front. Plant Sci. 6, 407.

Shaked, A., Bar-Akiva, A., 1967. Nitrate reductase activity as an indication of molybdenum level and requirement of citrus plants. Phytochemistry 6, 347–350.

Sharma, C.P., Sharma, P.N., Bisht, S.S., Nautiyal, B.D., 1982. Zinc deficiency induced changes in cabbage. In: Scaife, A. (Ed.), Proceedings of the Ninth Plant Nutrition Colloquium, Warwick, England, Commonwealth Agricicultural Bureau, Farnham Royal, UK, pp. 601–606.

Sharma, C.P., Sharma, P.N., Chatterjee, C., Agarwala, S.C., 1991. Manganese deficiency in maize effects pollen viability. Plant Soil 138, 139–142.

Sharma, S.S., Dietz, K.-J., Mimura, T., 2016. Vacuolar compartmentalization as indispensable component of heavy metal detoxification in plants. Plant Cell Environ. 39, 1112–1126.

Shelp, B.J., 1988. Boron mobility and nutrition in broccoli (*Brassica oleracea* var. *italica*). Ann. Bot. 61, 83–91.

Shelp, B.J., Shattuck, V.I., 1987. Boron nutrition and mobility, and its relation to hollow stem and the elemental composition of greenhouse grown cauliflower. J. Plant Nutr. 10, 143–162.

Shelp, B.J., Shattuck, V.I., Proctor, J.T.A., 1987. Boron nutrition and mobility, and its relation to the elemental composition of greenhouse grown root crops. II. Radish. Commun. Soil Sci. Plant Anal. 18, 203–219.

Shen, C., Yang, Y., Liu, K., Zhang, L., Guo, H., Tao, S., Wang, H., 2016. Involvement of endogenous salicylic acid in iron-deficiency responses in Arabidopsis. J. Exp. Bot. 67, 4179–4193.

Shi, W., Zhang, Y., Chen, S., Polle, A., Rennenberg, H., Luo, Z.-B., 2019. Physiological and molecular mechanisms of heavy metal accumulation in nonmycorrhizal versus mycorrhizal plants. Plant Cell Environ. 42, 1087–1103.

Shikhi, M., Jain, A., Salunke, D.M., 2020. Comparative study of 7S globulin from *Corylus avellana* and *Solanum lycopersicum* revealed importance of salicylic acid and Cu-binding loop in modulating their function. Biochem. Biophys. Res. Commun. 522, 127–132.

Shimotohno, A., Sotta, N., Sato, T., De Ruvo, M., Marée, A.F., Grieneisen, V.A., et al., 2015. Mathematical modeling and experimental validation of the spatial distribution of boron in the root of *Arabidopsis thaliana* identify high boron accumulation in the tip and predict a distinct root tip uptake function. Plant Cell Physiol. 56, 620–630.

Shinmachi, F., Buchner, P., Stroud, J.L., Parmar, S., Zhao, F.J., McGrath, S.P., et al., 2010. Influence of sulfur deficiency on the expression of specific sulfate transporters and the distribution of sulfur, selenium, and molybdenum in wheat. Plant Physiol. 153, 327–336.

Shinozaki, D., Merkulova, E.A., Naya, L., Horie, T., Kanno, Y., Seo, M., et al., 2020. Autophagy increases zinc bioavailability to avoid light-mediated reactive oxygen species production under zinc deficiency. Plant Physiol. 182, 1284–1296.

Shlizerman, L., Marsh, K., Blumwald, E., Sadka, A., 2007. Iron-shortage-induced increase in citric acid content and reduction of cytosolic aconitase activity in *Citrus* fruit vesicles and calli. Physiol. Plant. 131, 72–79.

Shojima, S., Nishizawa, N.-K., Fushiya, S., Nozoe, S., Irifune, T., Mori, S., 1990. Biosynthesis of phytosiderophores. In vitro biosynthesis of 2'-deoxymugineic acid from L-methionine and nicotianamine. Plant Physiol. 93, 1497–1503.

Shorrocks, V., 1997. The occurrence and correction of boron deficiency. Plant Soil 193, 121–148.

Shrotri, C.K., Mohanty, P., Rathore, V.C., Tewari, M.N., 1983. Zinc deficiency limits the photosynthetic enzyme activation in *Zea mays* L. Biochem. Physiol. Pflanz. 178, 213–217.

Sijmons, P.C., Kolattukudy, P.E., Bienfait, H.F., 1985. Iron-deficiency decreases suberization in bean roots through a decrease in suberin-specific peroxidase activity. Plant Physiol. 78, 115–120.

Sil, D., Martinez, Z., Ding, S., Bhuvanesh, N., Darensbourg, D.J., Hall, M.B., et al., 2018. Cyanide docking and linkage isomerism in models for the artificial [FeFe]-hydrogenase maturation process. J. Am. Chem. Soc. 140, 9904–9911.

Simon, M., Shen, Z.J., Ghoto, K., Chen, J., Liu, X., Gao, G.-F., et al., 2021. Proteomic investigation of Zn-challenged rice roots reveals adverse effects and root physiological adaptation. Plant Soil 460, 69–88.

Simpson, D.J., Robinson, S.P., 1984. Freeze-fracture ultrastructure of thylakoid membranes in chloroplasts from manganese-deficient plants. Plant Physiol. 74, 735–741.

Sinclair, S.A., Krämer, U., 2012. The zinc homeostasis network of land plants. Biochim. Biophys. Acta 1823, 1553–1567.

Singh, B., Dang, Y.P., Mehta, S.C., 1990. Influence of nitrogen on the behaviour of nickel in wheat. Plant Soil 127, 213–218.

Singh, S., Parihar, P., Singh, R., Singh, V.P., Prasad, S.M., 2016. Heavy metal tolerance in plants: role of transcriptomics, proteomics, metabolomics, and ionomics. Front. Plant Sci. 6, 1143.

Singla-Pareek, S.L., Kaur, C., Kumar, B., Pareek, A., Sopory, S.K., 2020. Reassessing plant glyoxalases: large family and expanding functions. New Phytol. 227, 714–721.

Sirhindi, G., Mir, M.A., Abd-Allah, E.F., Ahmad, P., Gucel, S., 2016. Jasmonic acid modulates the physio-biochemical attributes, antioxidant enzyme activity, and gene expression in *Glycine max* under nickel toxicity. Front. Plant Sci. 7, 591.

Smirnoff, N., Arnaud, D., 2019. Hydrogen peroxide metabolism and functions in plants. New Phytol. 221, 1197–1214.

Smith, K.S., Balistrieri, L.S., Smith, S.M., Severson, R.C., 1997. Molybdenum in Agriculture. Cambridge University Press.

Smith, G.S., Cornforth, I.S., Henderson, H.V., 1984. Iron requirements of C_3 and C_4 plants. New Phytol. 97, 543–556.

Smith, G.S., Clark, C.J., Holland, P.T., 1987. Chloride requirement of kiwifruit (*Actinidia deliciosa*). New Phytol. 106, 71–80.

Sousa, S.F., Lopes, A.B., Fernandes, P.A., Ramos, M.J., 2009. The zinc proteome: a tale of stability and functionality. Dalton Trans. 7946–7956. Available from: https://doi.org/10.1039/b904404c.

Spalding, E.P., 2000. Ion channels and the transduction of light signals. Plant Cell Environ. 23, 665–674.

Stangoulis, J.C., Webb, M.J., Graham, R.D., 2000. Boron efficiency in oilseed rape: II. Development of a rapid lab-based screening technique. Plant Soil 225, 253–261.

Stass, A., Kotur, Z., Horst, W.J., 2007. Effect of boron on the expression of aluminum toxicity in *Phaseolus vulgaris*. Physiol. Plant. 131, 283–290.

Stout, P.R., Meager, W.R., Pearson, G.A., Johnson, C.M., 1951. Molybdenum nutrition of crop plants. I. The influence of phosphate and sulfate on the absorption of molybdenum from soils and solution cultures. Plant Soil 3, 51–87.

Sun, X.C., Hu, C.X., Tan, Q.L., Liu, J.S., Liu, H.G., 2009. Effects of molybdenum on expression of cold-responsive genes in abscisic acid (ABA)-dependent and ABA-independent pathways in winter wheat under low-temperature stress. Ann. Bot. 104, 345–356.

Sun, Q., Liu, X., Yang, J., Liu, W., Du, Q., Wang, H., et al., 2018. MicroRNA528 affects lodging resistance of maize by regulating lignin biosynthesis under nitrogen-luxury conditions. Mol. Plant 11, 806–814.

Sunkar, R., Kapoor, A., Zhu, Y., 2006. Posttranscriptional induction of two Cu/Zn superoxide dismutase genes in Arabidopsis is mediated by downregulation of miR398 and important for oxidative stress tolerance. Plant Cell 18, 2051–2065.

Sutton, T., Baumann, U., Hayes, J., Collins, N.C., Shi, B.J., Schnurbusch, T., et al., 2007. Boron-toxicity tolerance in barley arising from efflux transporter amplification. Science 318, 1446.

Suzuki, H., Yu, J., Kobayashi, T., Nakanishi, H., Nixon, P.J., Noguchi, T., 2013. Functional roles of D2-Lys317 and the interacting chloride ion in the water oxidation reaction of photosystem II as revealed by fourier transform infrared analysis. Biochemistry 52, 4748–4757.

Suzuki, M., Takahashi, M., Tsukamoto, T., Watanabe, S., Matsuhashi, S., Yazaki, J., et al., 2006. Biosynthesis and secretion of mugineic acid family phytosiderophores in zinc-deficient barley. Plant J. 48, 85–97.

Suzuki, M., Bashir, K., Inoue, H., Takahashi, M., Nakanishi, H., Nishizawa, N.K., 2012. Accumulation of starch in Zn-deficient rice. Rice 5, 9. Available from: https://doi.org/10.1186/1939-8433-5-9.

Sze, H., 1985. H^+-translocating ATPases – advances using membranes vesicles. Annu. Rev. Plant Physiol. Plant Mol. Biol. 36, 175–208.

Sze, H., Li, X.H., Palmgren, M.G., 1999. Energization of plant cell membranes by H^+-pumping ATPases: Regulation and biosynthesis. Plant Cell 11, 677–689.

Takano, J., Noguchi, K., Yasumori, M., Kobayashi, M., Gajdos, Z., Miwa, K., et al., 2002. Arabidopsis boron transporter for xylem loading. Nature 420, 337–340.

Takano, J., Wada, M., Ludewig, U., Schaaf, G., Von Wirén, N., Fujiwara, T., 2006. The Arabidopsis major intrinsic protein NIP5;1 is essential for efficient boron uptake and plant development under boron limitation. Plant Cell 18, 1498–1509.

Tan, X.W., Ikeda, H., Oda, M., 2000. Effects of nickel concentration in the nutrient solution in the nitrogen assimilation and growth of tomato seedlings in hydroponic culture supplied with urea or nitrate as the sole nitrogen source. Sci. Hortic. 84, 265–273.

Tanaka, H., 1967. Boron adsorption by plant roots. Plant Soil 27, 300–302.

Tanada, H., 1978. Boron – key element in actions of phytochrome and gravity. Planta 143, 109–111.

Tanada, T., 1982. Role of boron in the far-red delay of nyctinastic closure of *Albizzia* pinnules. Plant Physiol. 70, 320–321.

Tanner, P.D., 1978. A relationship between premature sprouting on the cob and the molybdenum and nitrogen status of maize grain. Plant Soil 49, 427–432.

Tanner, P.D., 1982. The molybdenum requirements of maize in Zimbabwe. Zimbabwe Agric. J. 79, 61–64.

Tapken, W., Ravet, K., Pilon, M., 2012. Plastocyanin controls the stabilization of the thylakoid Cu-transporting P-type ATPase PAA2/HMA8 in response to low copper in Arabidopsis. J. Biol. Chem. 287, 18544–18550.

Tavladoraki, P., Cona, A., Angelini, R., 2016. Copper-containing amine oxidases and FAD-dependent polyamine oxidases are key players in plant tissue differentiation and organ development. Front. Plant Sci. 7, 824.

Teakle, N.L., Tyerman, S.D., 2010. Mechanisms of Cl^- transport contributing to salt tolerance. Plant Cell Environ. 33, 566–589.

Tejada-Jiménez, M., Chamizo-Ampudia, A., Galvan, A., Fernandez, E., Llamas, A., 2013. Molybdenum metabolism in plants. Metallomics 5, 1191–1203.

Tejada-Jiménez, M., Gil-Diez, P., Leon-Mediavilla, J., Wen, J., Mysore, K.S., Imperial, J., et al., 2017. *Medicago truncatula* Molybdate Transporter type 1 (MtMOT1.3) is a plasma membrane molybdenum transporter required for nitrogenase activity in root nodules under molybdenum deficiency. New Phytol. 216, 1223–1235.

Tennstedt, P., Peisker, D., Böttcher, C., Trampczynska, A., Clemens, S., 2009. Phytochelatin synthesis is essential for the detoxification of excess zinc and contributes significantly to the accumulation of zinc. Plant Physiol. 149, 938–948.

Terry, N., 1977. Photosynthesis, growth, and the role of chloride. Plant Physiol. 60, 69–75.

Terry, N., 1980. Limiting factors in photosynthesis. I. Use of iron stress to control photochemical capacity *in vivo*. Plant Physiol. 65, 114–120.

Terry, N., Abadia, J., 1986. Function of iron in chloroplasts. J. Plant Nutr. 9, 609–646.

Tewari, R.K., Kumar, P., Neetu, Sharma, P.N., 2005. Signs of oxidative stress in the chlorotic leaves of iron starved plants. Plant Sci. 169, 1037–1045.

Tewari, R.K., Kumar, P., Sharma, P.N., 2013. Oxidative stress and antioxidant responses of mulberry (*Morus alba*) plants subjected to deficiency and excess of manganese. Acta Physiol. Plant. 35, 3345–3356.

Thauer, R.K., 2001. Nickel to the fore. Science 293, 1264–1265.

Thauer, R.K., Diekert, G., Schönheit, P., 1980. Biological role of nickel. Trends Biochem. Sci. 5, 304–306.

Tighe-Neira, R., Carmora, E., Recio, G., Nunes-Nesi, A., Reyes-Diaz, M., Alberdi, M., et al., 2018. Metallic nanoparticles influence the structure and function of the photosynthetic apparatus in plants. Plant Physiol. Biochem. 130, 408–417.

Timm, C.A., Goos, R.J., Johnson, B.E., Siobolik, F.J., Stack, R.W., 1986. Effect of potassium fertilizers on malting barley infected with common root rot. Agron. J. 78, 197–200.

Timperio, A.M., D'Amici, G.M., Barta, C., Loreto, F., Zolla, L., 2007. Proteomics pigment composition, and organization of thylakoid membranes in iron-deficiency spinach leaves. J. Exp. Bot. 13, 3695–3710.

Tomatsu, H., Takano, J., Takahashi, H., Watanabe-Takahashi, A., Shibagaki, N., Fujiwara, T., 2007. An *Arabidopsis thaliana* highaffinity molybdate transporter required for efficient uptake of molybdate from soil. Proc. Natl. Acad. Sci. USA 104, 18807–18812.

Tomsett, A.B., Thurman, D.A., 1988. Molecular biology of metal tolerances of plants. Plant Cell Environ. 11, 383–394.

Torun, B., Kalayci, M., Oztürk, L., Torun, A., Aydin, M., Cakmak, I., 2003. Differences in shoot boron concentrations, leaf symptoms and yield of Turkish barley cultivars grown on a boron-toxic soil in field. J. Plant Nutr. 26, 869–881.

Trehan, S.P., Sekhon, G.S., 1977. Effect of clay, organic matter and $CaCO_3$ content on zinc adsorption by soils. Plant Soil 46, 329–336.

Tsui, C., 1948. The role of zinc in auxin synthesis in the tomato plant. Am. J. Bot. 35, 172–179.

Ube, N., Harada, D., Katsuyama, Y., Osaki-Oka, K., Tonooka, T., Ueno, K., et al., 2019. Identification of phenylamide phytoalexins and characterization of inducible phenylamide metabolism in wheat. Phytochemistry 167, 112098.

Uexküll, von H.R., 1985. Chlorine in the nutrition of palm trees. Oleagineux 40, 67–71.

Umena, Y., Kawakami, K., Shen, J.R., Kamiya, N., 2011. Crystal structure of oxygen-evolving photosystem II at a resolution of 1.9 angstrom. Nature 473, 55–65.

Ureta, A.C., Imperial, J., Ruiz-Argüeso, T., Palacios, J.M., 2005. *Rhizobium leguminosarum* biovar viciae symbiotic hydrogenase activity and processing are limited by the level of nickel in agricultural soils. Appl. Environ. Microbiol. 71, 7603–7606.

Utsunomiya, E., Muto, S., 1993. Carbonic anhydrase in the plasma membranes from leaves of C_3 and C_4 plants. Physiol. Plant. 88, 413–419.

Vallee, B.L., Falchuk, K.H., 1993. The biochemical basis of zinc physiology. Physiol. Rev. 73, 79–118.

Van Assche, F., Clijsters, H., 1986a. Inhibition of photosynthesis in *Phaseolus vulgaris* by treatment with toxic concentration of zinc: effect on ribulose-1,5-bisphosphate carboxylase/oxygenase. J. Plant Physiol. 125, 355–360.

Van Assche, F., Clijsters, H., 1986b. Inhibition of photosynthesis in *Phaseolus vulgaris* by treatment with toxic concentrations of zinc: effects on electron transport and photophosphorylation. Physiol. Plant. 66, 717–721.

van der Ent, A., Casey, L.W., Blamey, F.P.C., Kopittke, P.M., 2020. Time-resolved laboratory micro-X-ray fluorescence reveals silicon distribution in relation to manganese toxicity in soybean and sunflower. Ann. Bot. 126, 331–341.

Van Steveninck, R.F.M., Van Steveninck, M.E., Fernando, D.R., Horst, W.J., Marschner, H., 1987. Deposition of zinc phytate in globular bodies in roots of *Deschampsia caespitosa* ecotypes; a detoxification mechanism? J. Plant Physiol. 131, 247–257.

Velu, G., Ortiz-Monasterio, I., Cakmak, I., Hao, Y., Singh, R.P., 2014. Biofortification strategies to increase grain zinc and iron concentrations in wheat. J. Cereal. Sci. 59, 365–372.

Verbruggen, N., Hermans, C., Schat, H., 2009. Mechanisms to cope with arsenic or cadmium excess in plants. Curr. Opin. Plant Biol. 12, 364–372.

Verkleij, J.A.C., Schat, H., 1989. Mechanism of metal tolerance in higher plants. In: Shaw, A.J. (Ed.), Heavy Metal Tolerance in Plants: Evolutionary Aspects. CRC Press Inc., Boca Raton, FL, USA, pp. 179–193.

Vielemeyer, H.P., Fischer, F., Bergmann, W., 1969. Untersuchungen über den Einfluß der Mikronährstoffe Eisen und Mangan auf den Stickstoff-Stoffwechsel landwirtschaftlicher Kulturpflanzen. 2. Mitt.: Untersuchungen über die Wirkung des Mangans auf die Nitratreduktion und den Gehalt an freien Aminosäuren in jungen Buschbohnenpflanzen. Albrecht-Thaer-Arch. 13, 393–404.

Vijeta, S., Indu, B., Anjali, A., Tripathi, B.N., Munjal, A.K., Vinay, S., 2010. Proline improves copper tolerance in chickpea (*Cicer arietinum*). Protoplasma 245, 173–181.

Vinyard, D.J., Badshah, S.L., Riggio, M.R., Kaur, D., Fanguy, A.R., Gunner, M.R., 2019. Photosystem II oxygen-evolving complex photoassembly displays an inverse H/D solvent isotope effect under chloride-limiting conditions. Proc. Natl. Acad. Sci. USA 116, 18917–18922.

Vivas, A., Biro, B., Németh, T., Barea, J.M., Azcón, R., 2006. Nickel-tolerant *Brevibacillus brevis* and arbuscular mycorrhizal fungus can reduce metal acquisition and nickel toxicity effects in plant growing in nickel supplemented soil. Soil Biol. Biochem. 38, 2694–2704.

von der Fecht-Bartenbach, J.V., Bogner, M., Krebs, M., Stierhof, Y.D., Schumacher, K., Ludewig, U., 2007. Function of the anion transporter AtCLC-d in the trans-Golgi network. Plant J. 50, 466–474.

Voxeur, A., Fry, S.C., 2014. Glycosylinositol phosphorylceramides from *Rosa* cell cultures are boron-bridged in the plasma membrane and form complexes with rhamnogalacturonan II. Plant J. 79, 139–149.

Vunkova-Radeva, R., Schiemann, J., Mendel, R.R., Salcheva, G., Georgieva, D., 1988. Stress and activity of molybdenum-containing complex (molybdenum cofactor) in winter wheat seeds. Plant Physiol. 87, 533–535.

Walker, C.D., Graham, R.D., Madison, J.T., Cary, E.E., Welch, R.M., 1985. Effects of Ni deficiency on some nitrogen metabolites in cowpea (*Vigna unguiculata* L. Walp). Plant Physiol. 79, 474–479.

Wall, M.J., Quinn, A.J., D'Cunha, G.B., 2008. Manganese (Mn^{2+})-dependent storage stabilization of *Rhodotorula glutinis* phenylalanine ammonia-lyase activity. J. Agric. Food Chem. 56, 894–902.

Walters, D., 2003. Resistance to plant pathogens: possible roles for free polyamines and polyamine catabolism. New Phytol. 159, 109–115.

Wan, H., Du, J., He, J., Lyu, D., Li, H., 2019. Copper accumulation, subcellular partitioning and physiological and molecular responses in relation to different copper tolerance in apple rootstocks. Tree Physiol. 39, 1215–1234.

Wang, Y.X., Wu, P., Wu, Y.R., Yan, X.L., 2002. Molecular marker analysis of manganese toxicity tolerance in rice under greenhouse conditions. Plant Soil 238, 227–233.

Wang, B., Li, Y., Zhang, W.H., 2012. Brassinosteroids are involved in response of cucumber (*Cucumis sativus*) to iron deficiency. Ann. Bot. 110, 681–688.

Wang, M., Zhao, X., Xiao, Z., Yin, X., Xing, T., Xia, G., 2016. A wheat superoxide dismutase gene TaSOD2 enhances salt resistance through modulating redox homeostasis by promoting NADPH oxidase activity. Plant Mol. Biol. 91, 115–130.

Wang, B., Wei, H., Xue, Z., Zhang, W.H., 2017. Gibberellins regulate iron deficiency-response by influencing iron transport and translocation in rice seedlings (*Oryza sativa*). Ann Bot. 119, 945–956.

Wang, H., Du, H., Li, H., Huang, Y., Ding, J., Liu, C., et al., 2018. Identification and functional characterization of the ZmCOPT copper transporter family in maize. PLoS ONE 13, e0199081.

Wang, X., Feng, C., Tian, L., Hou, C., Tian, W., Hu, B., et al., 2021. A transceptor–channel complex couples nitrate sensing to calcium signaling in *Arabidopsis*. Mol. Plant 14, 774–786.

Warburg, O., Lüttgens, W., 1946. Photochemical reduction of quinone in green cells and granules. Biochimia 11, 303–322.

Watanabe, T., Broadley, M.R., Jansen, S., White, P.J., Takada, J., Satake, K., et al., 2007. Evolutionary control of leaf element composition in plants. New Phytol. 174, 516–523.

Waters, B.M., Armbrust, L.C., 2013. Optimal copper supply is required for normal plant iron deficiency responses. Plant Signal. Behav. 8, e26611. Available from: https://doi.org/10.4161/psb.26611.

Watts-Williams, S.J., Smith, F.A., McLaughlin, M.J., Patti, A.F., Cavagnaro, T.R., 2015. How important is the mycorrhizal pathway for plant Zn uptake? Plant Soil 390, 157–166.

Wege, S., Gilliham, M., Henderson, S.W., 2017. Chloride: not simply a "cheap osmoticum", but a beneficial plant macronutrient. J. Exp. Bot. 68, 3057–3069.

Wei, P.P., Wang, L.C., Liu, A.L., Yu, B.J., Lam, H.M., 2016. GmCLC1 confers enhanced salt tolerance through regulating chloride accumulation in soybean. Front. Plant Sci. 7, 1082.

Weir, R.C., Hudson, A., 1966. Molybdenum deficiency in maize in relation to seed reserves. Aust. J. Exp. Agric. Anim. Husb. 6, 35–41.

Welch, R.M., 1981. The biological significance of nickel. J. Plant Nutr. 3, 345–356.

Welch, R.M., 1986. Effects of nutrient deficiencies on seed production and quality. In: Tinker, B., Läuchli, A. (Eds.), Advances in Plant Nutrition. Praeger Scientific, New York, NY, USA, pp. 205–247.

Welch, R.M., Webb, M.J., Loneragan, J.F., 1982. Zinc in membrane function and its role in phosphorus toxicity. In: Scaife, A. (Ed.), Proceedings of the Ninth Plant Nutrition Colloquium, Warwick, England, Commonwealth Agricultural Bureau, Farnham Royal, UK, pp. 710–715.

Welkie, G.W., Miller, G.W., 1989. Sugar beet responses to iron nutrition and stress. J. Plant Nutr. 12, 1041–1054.

Wen, Z., Kaiser, B.N., 2018. Unraveling the functional role of NPF6 transporters. Front. Plant Sci. 9, 973.

Wen, Z., Tyerman, S.D., Dechorgnat, J., Ovchinnikova, E., Dhugga, K.S., Kaiser, B.N., 2017. Maize NPF6 proteins are homologs of *Arabidopsis* CHL1 that are selective for both nitrate and chloride. Plant Cell 29, 2581–2596.

Werner, A.K., Sparkes, I.A., Romeis, T., Witte, C.P., 2008. Identification, biochemical characterization, and subcellular localization of allantoate amidohydrolases from Arabidopsis and soybean. Plant Physiol. 146, 418–430.

Werner, A.K., Medina-Escobar, N., Zulawski, M., Sparkes, I.A., Cao, F.Q., Witte, C.P., 2013. The ureide-degrading reactions of purine ring catabolism employ three amidohydrolases and one aminohydrolase in Arabidopsis, soybean, and rice. Plant Physiol. 163, 672–681.

White, P.J., Broadley, M.R., 2001. Chloride in soils and its uptake and movement within the plant: A review. Ann. Bot. 88, 967–988.

White, P.J., Broadley, M.R., 2009. Biofortification of crops with seven mineral elements often lacking in human diets - iron, zinc, copper, calcium, magnesium, selenium and iodine. New Phytol. 182, 49–84.

White, M.C., Chaney, R.L., Decker, A.M., 1979. Role of roots and shoots of soybean in tolerance to excess soil zinc. Crop Sci. 19, 126–128.

White, P.J., Broadley, M.R., El-Serehy, H.A., George, T.S., Neugebauer, K., 2018. Linear relationships between shoot magnesium and calcium concentrations among angiosperm species are associated with cell wall chemistry. Ann. Bot. 122, 221–226.

Whitehead, D.C., 1985. Chlorine deficiency in red clover grown in solution culture. J. Plant Nutr. 8, 193–198.

Whittaker, M.M., Whittaker, J.W., 2002. Characterization of recombinant barley oxalate oxidase expressed by *Pichia pastoris*. J. Biol. Inorg. Chem. 7, 136–145.

Williams, C.M.J., Maier, N.A., Bartlett, L., 2004. Effect of molybdenum foliar sprays on yield, berry size, seed formation, and petiolar nutrient composition of 'Merlot' grapevines. J. Plant Nutr. 27, 1891–1916.

Wilson, D.O., Boswell, F.C., Ohki, K., Parker, M.B., Shuman, L.M., Jellum, M.D., 1982. Changes in soybean seed oil and protein as influenced by manganese nutrition. Crop Sci. 22, 948–952.

Wimmer, M.A., Lochnit, G., Bassil, E., Muehling, K.H., Goldbach, H.E., 2009. Membrane-associated, boron-interacting proteins isolated by boronate affinity chromatography. Plant Cell Physiol. 50, 1292–1304.

Wimmer, M.A., Abreu, I., Bell, R.W., Bienert, M.D., Brown, P.H., Dell, B., et al., 2020. Boron: an essential element for vascular plants. New Phytol. 226, 1232–1237.

Winder, T.L., Nishio, J.N., 1995. Early iron deficiency stress response in leaves of sugar beet. Plant Physiol. 108, 1487–1494.

Winkler, R.G., Polacco, J.C., Eskew, D.L., Welch, R.M., 1983. Nickel is not required for apo-urease synthesis in soybean seeds. Plant Physiol. 72, 262–263.

Winkler, R.G., Polacco, J.C., Blevins, D.G., Randall, D.D., 1985. Enzymatic degradation of allantoate in developing soybeans. Plant Physiol. 79, 878. 793.

Winkler, R.G., Blevins, D.G., Polacco, J.C., Randall, D.D., 1987. Ureide catabolism of soybeans. II. Pathway of catabolism in intact leaf tissue. Plant Physiol. 83, 585–591.

Winkler, R.G., Blevins, D.G., Polacco, J.C., Randall, D.D., 1988. Ureide catabolism in nitrogen-fixing legumes. Trends Biochem. Sci. 11, 97–100.

Wissemeier, A.H., Horst, W.J., 1987. Callose deposition in leaves of cowpea (*Vigna unguiculata* L. Walp.) as a sensitive response to high Mn supply. Plant Soil 102, 283–286.

Wissemeier, A.H., Horst, W.J., 1992. Effect of light-intensity on manganese toxicity symptoms and callose formation in cowpea (*Vigna unguiculata* (L) Walp). Plant Soil 143, 299–309.

Witt, H.H., Jungk, A., 1977. Beurteilung der Molybdänversorgung von Pflanzen mit Hilfe der Mo-induzierbaren Nitratreduktase Aktivität. Z. Pflanzenernähr. Bodenk. 140, 209–222.

Witte, C.P., 2011. Urea metabolism in plants. Plant Sci. 180, 431–438.

Woo, E.J., Dunwell, J.M., Goodenough, P.W., Marvier, A.C., Pickersgill, R.W., 2000. Germin is a manganese containing homohexamer with oxalate oxidase and superoxide dismutase activities. Nat. Struct. Biol. 7, 1036–1040.

Wood, B.W., 2013. Iron-induced nickel deficiency in pecan. Hortscience 48, 1145–1153.

Wood, B., Reilly, C., Nyczepir, A., 2004. Mouse-ear of pecan: a nickel deficiency. Hortscience 39, 1238–1242.

Wu, H., Ling, H.Q., 2019. FIT-binding proteins and their functions in the regulation of Fe homeostasis. Front. Plant Sci. 10, 844.

Xia, X., Zhang, H., Offler, C.E., Patrick, J.W., 2020. Enzymes contributing to the hydrogen peroxide signal dynamics that regulate wall labyrinth formation in transfer cells. J. Exp. Bot. 71, 219–233.

Xiao, Q., Chen, Y., Liu, C.-W., Robson, F., Roy, S., Cheng, X., et al., 2021. MtNPF6.5 mediates chloride uptake and nitrate preference in *Medicago* roots. EMBO J. 40, e106847.

Xiong, L.M., Ishitani, M., Lee, H., Zhu, J.K., 2001. The Arabidopsis LOS5/ABA3 locus encodes a molybdenum cofactor sulfurase and modulates cold stress- and osmotic stress-responsive gene expression. Plant Cell 13, 2063–2083.

Xu, G., Magen, H., Tarchitzky, J., Kafkafi, U., 2000. Advances in chloride nutrition of plants. Adv. Agron. 68, 97–150.

Xu, Z., Huang, J., Qu, C., Chang, R., Chen, J., Wang, Q., et al., 2020. Functional characterization and expression patterns of PnATX genes under different abiotic stress treatments in *Populus*. Tree Physiol. 40, 520–537.

Xv, L., Ge, J., Tian, S., Wang, H., Yu, H., Zhao, J., et al., 2020. A Cd/Zn Co-hyperaccumulator and Pb accumulator, *Sedum alfredii*, is of high Cu tolerance. Environ. Pollut. 263, 114401.

Yamagami, M., Haga, K., Napier, R.M., Iino, M., 2004. Two distinct signaling pathways participate in auxin-induced swelling of pea epidermal protoplasts. Plant Physiol. 134, 735–747.

Yamasaki, H., Pilon, M., Shikanai, T., 2008. How do plants respond to copper deficiency. Plant Signal. Behav. 3, 231–232. Available from: https://doi.org/10.4161/psb.3.4.5094.

Yamashita, K., Yamamoto, Y., Matsumoto, H., 1996. Characterization of an anion transporter in the plasma membrane of barley roots. Plant Cell Physiol. 37, 949–956.

Yamauchi, M., 1989. Rice bronzing in Nigeria caused by nutrient imbalances and its control by potassium sulfate application. Plant Soil 117, 275–286.

Yan, J., Chia, J.-C., Sheng, H., Jung, H.-I., Zavodna, T.-O., Zhang, L., et al., 2017. Arabidopsis pollen fertility requires the transcription factors CITF1 and SPL7 that regulate copper delivery to anthers and jasmonic acid synthesis. Plant Cell 29, 3012–3029.

Yang, X., Baligar, V.C., Martens, D.C., Clark, R.B., 1996. Plant tolerance to nickel toxicity: II Nickel effects on influx and transport of mineral nutrients in four plant species. J. Plant Nutr. 19, 265–279.

Yang, M., Lu, K., Zhao, F.J., Xie, W., Ramakrishna, P., Wang, G., et al., 2018. Genome-wide association studies reveal the genetic basis of ionomic variation in rice. Plant Cell 30, 2720–2740.

Yarmohammadi, A., Khoramivafa, M., Honarmand, S.J., 2019. Humic acid reduces the CuO and ZnO nanoparticles cellular toxicity in rapeseed (*Brassica napus*). Cell Mol. Biol. 65, 29–36.

Yazici, M.A., Asif, M., Tutus, Y., Ortas, I., Ozturk, L., Lambers, H., et al., 2021. Reduced root mycorrhizal colonization as affected by phosphorus fertilization is responsible for high cadmium accumulation in wheat. Plant Soil 468, 19–35.

Yi, X.P., McChargue, M., Laborde, S., Frankel, L.K., Bricker, T.M., 2005. The manganese-stabilizing protein is required for photosystem II assembly/stability and photoautotrophy in higher plants. J. Biol. Chem. 280, 16170–16174.

Yi, X.P., Hargett, S.R., Liu, H.J., Frankel, L.K., Bricker, T.M., 2007. The PsbP protein is required for photosystem II complex assembly/stability and photoautotrophy in *Arabidopsis thaliana*. J. Biol. Chem. 282, 24833–24841.

Yokosho, K., Yamaji, N., Ueno, D., Mitani, N., Ma, J.F., 2009. OsFRDL1 is a citrate transporter required for efficient translocation of iron in rice. Plant Physiol. 149, 297–305.

Yoshinari, A., Takano, J., 2017. Insights into the mechanisms underlying boron homeostasis in plants. Front. Plant Sci. 8, 1951.

You, J., Chan, Z., 2015. ROS regulation during abiotic stress responses in crop plants. Front. Plant Sci. 6, 1092.

Young, L., May, B., Shiba, T., Harada, S., Inaoka, D.K., Kita, K., et al., 2016. Structure and mechanism of action of the alternative quinol oxidases. In: Cramer, W.A.K.T. (Ed.), Cytochrome Complexes: Evolution, Structures, Energy Transduction, and Signaling. Springer, Dordrecht, The Netherlands, pp. 375–394.

Yruela, I., 2013. Transition metals in plant photosynthesis. Metallomics 5, 1090–1109.

Yu, Q., Rengel, Z., 1999. Micronutrient deficiency influences plant growth and activities of superoxide dismutases in narrow-leafed lupins. Ann. Bot. 83, 175–182.

Yu, Q., Worth, C., Rengel, Z., 1999. Using capillary electrophoresis to measure Cu/Zn superoxide dismutase concentration in leaves of wheat genotypes differing in tolerance to zinc deficiency. Plant Sci. 143, 231–239.

Yu, Q., Baluska, F., Jasper, F., Menzel, D., Goldbach, H.E., 2003. Short-term boron deprivation enhances levels of cytoskeletal proteins in maize, but not zucchini, root apices. Physiol. Plant. 117, 270–278.

Yu, M., Shen, R.F., Xiao, H.D., Xu, M.M., Wang, H.Z., Wang, H.Y., et al., 2009. Boron alleviates aluminum toxicity in pea (*Pisum sativum*). Plant Soil 314, 87–98.

Yu, Z., Jia, D., Liu, T., 2019. Polyamine oxidases play various roles in plant development and abiotic stress tolerance. Plants 8, 184.

Zandi, P., Yang, J., Mozdzen, K., Barabasz-Krasny, B., 2020. A review of copper speciation and transformation in plant and soil/wetland systems. Adv. Agron. 160, 249–293.

Zhang, F., Römheld, V., Marschner, H., 1991a. Release of zinc mobilizing root exudates in different plant species as affected by zinc nutritional status. J. Plant Nutr. 14, 675–686.

Zhang, F., Römheld, V., Marschner, V., 1991b. Diurnal rhythm of release of phytosiderophores and uptake rate of zinc in iron-deficient wheat. Soil Sci. Plant Nutr. 37, 671–678.

Zhang, Z., Collinge, D.B., Thordal-Christensen, H., 1995. Germin-like oxalate oxidase, a H_2O_2-producing enzyme, accumulates in barley attacked by the powdery mildew fungus. Plant J. 8, 139–145.

Zhang, H., Zhao, X., Li, J.G., Cai, H.Q., Deng, X.W., Li, L., 2014. MicroRNA408 is critical for the HY5-SPL7 gene network that mediates the coordinated response to light and copper. Plant Cell 26, 4933–4953.

Zhang, H., Yang, J., Wang, W., Li, D., Hu, X., Wang, H., et al., 2015. Genome-wide identification and expression profiling of the copper transporter gene family in *Populus trichocarpa*. Plant Physiol. Biochem. 97, 451–460.

Zhang, H., Zhao, F.G., Tang, R.J., Yu, Y., Song, J., Wang, Y., et al., 2017. Two tonoplast MATE proteins function as turgor-regulating chloride channels in Arabidopsis. Proc. Natl. Acad. Sci. USA 114, E2036–E2045.

Zhang, J., Wang, S., Song, S., Xu, F., Pan, Y., Wang, H., 2019. Transcriptomic and proteomic analyses reveal new insight into chlorophyll synthesis and chloroplast structure of maize leaves under zinc deficiency stress. J. Proteom. 199, 123–134.

Zhao, F.J., Su, Y.H., Dunham, S.J., Rakszegi, M., Bedo, Z., McGrath, S.P., et al., 2009. Variation in mineral micronutrient concentrations in grain of wheat lines of diverse origin. J. Cereal Sci. 49, 290–295.

Zhao, Y., Lin, S., Qiu, Z., Cao, D., Wen, J., Deng, X., et al., 2015. MicroRNA857 is involved in the regulation of secondary growth of vascular tissues in Arabidopsis. Plant Physiol. 169, 2539–2552.

Zhou, J.R., Fordyce, E.J., Raboy, V., Dickinson, D.B., Wong, M.-S., Burns, R.A., et al., 1992. Reduction of phytic acid in soybean products improves zinc bioavailability in rats. J. Nutr. 122, 2466–2473.

Zifarelli, G., Pusch, M., 2010. CLC transport proteins in plants. FEBS Lett. 584, 2122–2127.

Zou, W., Li, C., Zhu, Y., Chen, J., He, H., Ye, G., 2020. Rice heavy metal P-type ATPase OsHMA6 is likely a copper efflux protein. Rice Sci. 27, 143–151.

Chapter 8

Beneficial elements[☆]

Jian Feng Ma[1], Fang-Jie Zhao[2], Zed Rengel[3], and Ismail Cakmak[4]
[1]*Institute of Plant Science and Resources, Okayama University, Kurashiki, Japan,* [2]*College of Resources and Environmental Sciences, Nanjing Agricultural University, Nanjing, P.R. China,* [3]*UWA School of Agriculture and Environment, University of Western Australia, Perth, WA, Australia,* [4]*Faculty of Engineering and Natural Sciences, Sabanci University, Istanbul, Turkey*

Summary

The roles of sodium (Na), silicon (Si), cobalt (Co), selenium (Se), and aluminum (Al) are described. These elements are termed beneficial because they stimulate growth only in certain plant species, or under specific conditions, but do not meet the criteria for essential elements. Sodium can stimulate growth of halophytes and some other plants, particularly C_4 species. In C_4 plants, Na aids the movement of substrates between the mesophyll and the bundle sheath. Sodium can also to some extent replace K in its role as osmoticum. Silicon has a number of beneficial effects in many plant species, especially in rice. It can improve leaf erectness and mitigate various biotic and abiotic stresses. Cobalt is essential for N_2-fixing plants because it is part of the coenzyme cobalamin (vitamin B_{12}) that is important in nodule metabolism. Therefore, Co deficiency results in poor nodulation and low N_2 fixation rates. The chemistry of Se is similar to that of sulfur (S), and Se can replace S, to some extent, in proteins, particularly in Se-hyperaccumulating plants. Selenium is essential for animals; therefore, Se fertilization may be beneficial for human and animal health in areas with low Se soils. Aluminum is beneficial to some plants such as tea, but mechanisms of this beneficial effect are poorly understood. It may alleviate proton toxicity and increase the activity of antioxidative enzymes.

8.1 Definition

Elements that stimulate growth only in certain plant species, or under specific conditions, are termed *beneficial elements* (for a definition of essentiality see Chapter 1). The distinction between beneficial and essential is difficult in the case of some trace elements. Developments in analytical chemistry and in methods to minimize contamination during growth experiments may well lead to a lengthening of the list of micronutrient elements and a corresponding shortening in the list of beneficial elements. Nickel is the most recent example of such reclassification.

8.2 Sodium

8.2.1 General

The sodium (Na) concentration of the Earth's crust is ~2.4% (w/w) compared with 2.1%–2.6% (w/w) for K, making Na the seventh most abundant element (cf. Raddatz et al., 2020). In temperate regions, the Na concentration in the soil solution is on average 0.1–1 mM, thus similar to, or higher than, the K concentration. In semiarid and arid regions, particularly under irrigation, concentrations of 50–100 mM Na^+ (mostly as NaCl) in the soil solution are typical and may have a detrimental effect on the growth of most crop plants (Section 18.5). The hydrated Na^+ has a radius of 0.358 nm, whereas that of K^+ is 0.331 nm. Most plants have developed high selectivity in the uptake of K compared to Na, and this is particularly obvious in transport to the shoot (Chapter 3).

Plant species are characterized as *natrophilic* or *natrophobic*, depending on their growth response to Na and their differential capacity to take up Na by roots and transport it to shoots (e.g., Phillips et al., 2000). The differences in capacity for Na uptake

---

[☆] This chapter is a revision of the third edition chapter by M. Broadley, P. Brown, I. Cakmak, J.F. Ma, Z. Rengel, and F.J. Zhao, pp. 249–269. DOI: https://doi.org/10.1016/B978-0-12-384905-2.00008-X. © Elsevier Ltd.

and long-distance transport (for a comprehensive review of relevant transporters, see Raddatz et al., 2020) are large among plant species as well as genotypes within a species (e.g., Hajiboland and Joudmand, 2009; Mari et al., 2018). Genotypic differences in uptake by roots are related to factors such as (1) differential activity/capacity of Na efflux pumps (e.g., Aktas et al., 2006; Flowers and Hajibagheri, 2001; Guo et al., 2009) (Chapter 2), (2) passive Na permeability of the root plasma membranes (Schubert and Läuchli, 1990), and (3) xylem loading of Na (i.e., root-to-shoot transport) (Coban et al., 2020; Davenport et al., 2005). Interestingly, tonoplast pyrophosphatase and V-ATPase were upregulated in salt-resistant, but not in salt-sensitive, genotypes to enhance vacuolar compartmentation of Na in rice (Pons et al., 2011) and sugarcane (Theerawitaya et al., 2020). Genotypic differences in arbuscular mycorrhizal colonization might also be relevant because such colonization decreased Na uptake in both C_3 and C_4 plants as revealed by a meta-analysis of the published literature (Murugesan et al., 2016).

For the role of Na in nutrition of plants, three aspects are important: (1) its essentiality for certain plant species, (2) the extent to which it can replace K in plants, and (3) its growth enhancement effect.

8.2.2 Essentiality: Na as nutrient

Brownell (1965) claimed Na as an essential element (i.e., a nutrient) for the halophyte *Atriplex vesicaria*. The growth response to Na at low concentration (0.02 mM) (Table 8.1) was strong, although the tissue Na concentration (~ 1 g kg^{-1} dw) was in a range more typical for a micronutrient. At higher supply, however, the tissue Na concentration was more typical of a macronutrient, with growth responses presumably related to replacing the functions of K, such as in osmoregulation (Brownell, 1965).

In further studies on various halophytes and nonhalophytes (glycophytes), responses to Na similar to those shown in Table 8.1 were found in species characterized by the C_4 photosynthetic pathway (Brownell and Crossland, 1972) and the crassulacean acid metabolism (CAM) pathway (Brownell and Crossland, 1974), including intermediate species (such as *Sedum kamtschaticum* Fischer), that shift from C_3 to CAM metabolism under exposure to increasing Na concentrations (Moritani et al., 2017). Without Na supply, all C_4 species grow poorly and show visual deficiency symptoms such as chlorosis and necrosis, or even fail to form flowers. The supply of 100 μM Na$^+$ enhanced growth and alleviated the visual symptoms. Photosynthesis was increased 2.5-fold by Na in *Haloxylon aphyllum* (Chenopodiaceae) (Rakhmankulova et al., 2014). Hence, Na may be classified as a nutrient for at least some C_4 species in the families Amaranthaceae, Chenopodiaceae, and Cyperaceae. The amounts of Na required by these plant species are similar to those for a micronutrient rather than a macronutrient. However, Na is essential for many, but not all, C_4 species (e.g., not for natrophobic maize and sugarcane), and it is not essential for C_3 species; however, positive responses to K substitution by Na were recorded in the C_3 perennial tree crop *Theobroma cacao* (Gattward et al., 2012). Nevertheless, the literature on Na as an essential and/or beneficial nutrient is relatively scarce (Kronzucker et al., 2013; Pilon-Smits and Leduc, 2009).

Uptake of Na occurs via the K-uptake pathways, including three families of K channels (Shaker, TPK/KCO, and TPC) and three families of K transporters (HAK, HKT, and CPA), across the plasma membrane as well as endomembranes (Nieves-Cordones et al., 2016), with HKT1 being a particularly important Na transporter in tomato (Almeida et al., 2014). Similarly, in sugar beet exposed to mild salinity (5–50 mM Na), the inward-rectifying channel AKT1 is involved mainly in K transport, whereas HKT1 transports Na (Wu et al., 2015).

TABLE 8.1 Growth and Na and K concentrations in leaves of *Atriplex vesicaria* L. at different Na concentrations in a nutrient solution with 6 mM K.

Na concentration (mM Na)	Dry weight [mg (4 plants)$^{-1}$]	Concentration in leaves (mmol kg^{-1} dw) Na	K
0	86	10	2834
0.02	398	48	4450
0.04	581	78	2504
0.20	771	296	2225
1.20	1101	1129	1688

Source: From Brownell (1965).

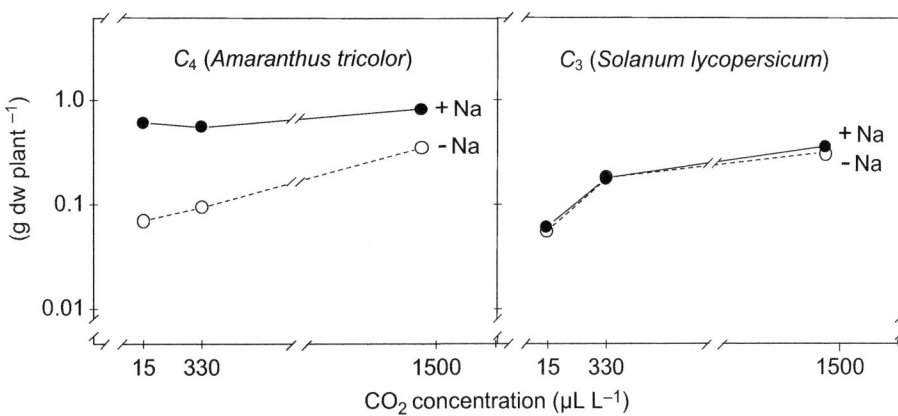

FIGURE 8.1 Growth of C_4 (*Amaranthus tricolor*) and C_3 (*Solanum lycopersicum*) plants grown at increasing ambient CO_2 concentrations with and without Na. *Based on Johnston et al. (1984).*

The growth of many halophytes (C_3 or C_4) is enhanced by high Na concentrations in the substrate (10–100 mM Na, and up to 510 mM Na in extreme cases; Redondo-Gómez et al., 2010). Growth responses of halophytes to Na reflect a high salt requirement for osmotic adjustment, a process in which Na can be more suitable than K in both salt-tolerant (e.g., *Atriplex triangularis*) and salt-sensitive species (e.g., cotton and soybean) (Sun et al., 2020a).

8.2.3 Role in C_4 species

The principle of the C_4 photosynthetic pathway is the shuttle of metabolites between mesophyll and bundle sheath cells (see Chapter 5) and an increase in CO_2 concentration in the bundle sheath cells to optimize the Calvin cycle. This advantage of C_4 over C_3 plants becomes particularly evident at low ambient CO_2 concentrations, provided the C_4 plants are supplied with Na. In the shoots of *Amaranthus tricolor*, Na concentrations as low as 0.2 g kg^{-1} dw were needed for high efficiency in CO_2 utilization at low ambient concentrations (Fig. 8.1) (Johnston et al., 1984). However, in Na-deficient *A. tricolor*, plant growth was poor, and chlorosis was severe at a low ambient CO_2 concentration. Increasing ambient CO_2 concentrations enhanced the growth of *A. tricolor* similarly to the C_3 species tomato, and Na effects on *A. tricolor* growth or CO_2 utilization were absent (Johnston et al., 1984).

The net photosynthetic rate in the C_4 species *Atriplex nummularia* was not stimulated (in contrast to that in the C_3 species *Chenopodium quinoa*) at an elevated CO_2 concentration (540 μL L^{-1}) (Geissler et al., 2015). However, higher CO_2 concentration (700 μL L^{-1}) enhanced photosynthesis and increased chlorophyll fluorescence (indicating suitable growth conditions) in the C_4 species *Spartina maritima* in the presence of Na (Mateos-Naranjo et al., 2010b). In a different species from the same genus (C_4 *Spartina densiflora*), the same elevated CO_2 concentration (700 μL L^{-1}) did not alter the net photosynthetic rate (compared with ambient CO_2) but increased growth under salinity due to enhanced activity of phosphoenolpyruvate carboxylase (Mateos-Naranjo et al., 2010a). Moreover, most of the salt-responsive genes in the C_4 species *Suaeda nudiflora* showed increased expression under elevated CO_2 concentration and Na exposure (Saranya et al., 2020).

The different growth response curves in *A. tricolor* in the presence and absence of Na (Fig. 8.1) suggest that in Na-deficient C_4 plants, the mechanism to concentrate CO_2 in the leaves is impaired or not operating (Johnston et al., 1984). For the mechanism to be operational, the flow of metabolites between mesophyll and bundle sheath cells is mediated through plasmodesmata and driven by the concentration gradient of metabolites in the cytosol, as illustrated below:

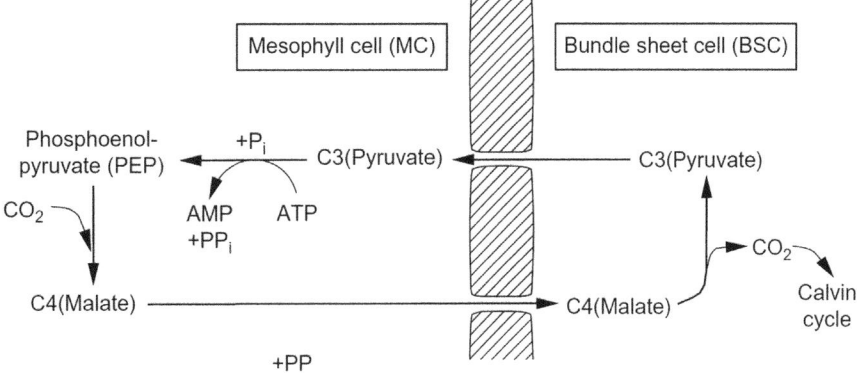

Sodium deficiency particularly impairs the conversion of pyruvate to phosphoenolpyruvate (PEP), which takes place in the mesophyll chloroplasts and has a high-energy requirement. Under Na deficiency in the C$_4$ species *A. tricolor*, the C$_3$ metabolites alanine and pyruvate accumulated, whereas the C$_4$ metabolites PEP, malate, and aspartate decreased (Table 8.2), suggesting that the functioning of the mesophyll chloroplasts is impaired in C$_4$ plants under Na deficiency. Regarding pyruvate, it is known that a small amount of Na is required for Na-dependent pyruvate transport into chloroplasts across the chloroplast envelope in C$_4$ plants (Furumoto et al., 2011). By contrast, in tomato (C$_3$ species), the concentration of these metabolites was not influenced by Na. In Na-deficient *A. tricolor* and *Kochia childsii*, the activity of the PS II in the mesophyll chloroplasts was reduced and the ultrastructure of the chloroplasts altered, whereas these parameters were not affected in the bundle sheath chloroplasts (Grof et al., 1989). Resupplying Na restored the PS II activity and changed metabolite concentrations in less than 3 days. In maize, the activity of mesophyll cell enzymes (pyruvate orthophosphate dikinase, PEP carboxylase, nicotiamide adenine dinucleotide phosphate (NADP)-dependent, and NAD-dependent malate dehydrogenase) increased, and those of bundle sheet cells (NADP-malic enzyme and ribulose-1,5-bisphosphate carboxylase) decreased together with decreased C$_4$ photosynthesis under salt exposure (Omoto et al., 2012). In particular, the activity of PEP carboxylase (a key enzyme in the C$_4$ photosynthetic pathway) increased under Na exposure (Cheng et al., 2016).

The mechanism by which Na affects metabolism and fine structure in the mesophyll chloroplasts of responsive C$_4$ species is unclear. Protection from photo-destruction may be involved (Grof et al., 1989). In C$_4$ species, the CO$_2$ scavenging system as well as nitrate assimilation take place in the mesophyll cells. Thus, in C$_4$ species, such as the grass *Aeluropus littoralis*, exposure to Na increased nitrate reductase activity in leaves (Hajiboland et al., 2015). Importantly, increased nitrate uptake and enhanced nitrate reductase activity were a direct effect of Na (rather than simply associated with Na-induced growth stimulation) in halophytes (e.g., Swiss chard), but not in salt-tolerant nonhalophytes (e.g., barley *Hordeum vulgare*) (Kaburagi et al., 2014). However, a potential role of Cl (accompanying Na) should also be taken into account, given it was specifically shown that Cl rather than Na enhanced nitrate uptake and reduction in the halophyte *Suaeda salsa* (Mori et al., 2008).

Sodium enhances nitrate uptake by the roots and nitrate assimilation in the leaves (Ohta et al., 1989). Nitrate uptake is achieved by an Na/nitrate symporter, for example, in the marine C$_3$ plant *Zostera marina* (García-Sánchez et al., 2000) that is expressed in the plasma membrane and is equivalent to Na-dependent high-affinity nitrate transporter NRT2.5 (Rubio et al., 2019) expressed in roots (epidermis and cortex) and leaves (minor veins) (Lezhneva et al., 2014). Stimulation of nitrate reductase activity and growth enhancement by Na were absent when ammonium was supplied or when nitrate was combined with tungsten, an inhibitor of nitrate reductase (Ohta et al., 1989). Thus, in Na-deficient C$_4$ species, particularly of the aspartate type, N deficiency may be an additional factor associated with impaired functioning of the C$_4$ pathway.

Mesophyll cells of C$_3$ as well as all three types of C$_4$ plants (NADP-malic enzyme, NAD-malic enzyme, and PEP carboxykinase) are similarly damaged by salinity, whereas chloroplasts of bundle sheath cells remained largely unaffected (Omoto et al., 2010). In mesophyll chloroplasts of *Panicum miliaceum*, Na-enhanced pyruvate uptake had a stoichiometry of about 1:1, suggesting Na/pyruvate cotransport through the envelope into the chloroplast, driven by a

TABLE 8.2 Concentration of various metabolites in shoots of *Amaranthus tricolor* (C$_4$) and *Solanum lycopersicum* (C$_3$) with (0.1 mM Na) or without Na supply.

Concentration (μmol g^{-1} fw)	A. tricolor		S. lycopersicum	
	−Na	+Na	−Na	+Na
Alanine	13	6.0	2.5	2.6
Pyruvate	1.7	0.9	0.1	0.1
PEPyruvate	0.9	2.3	0.2	0.2
Malate	2.7	4.8	11	11
Aspartate	1.6	3.7	1.9	1.9

Source: Based on Johnston et al., (1988).

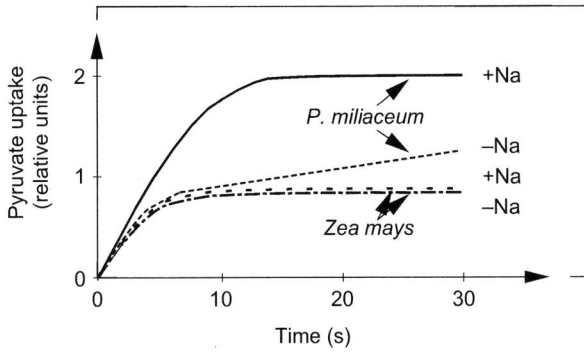 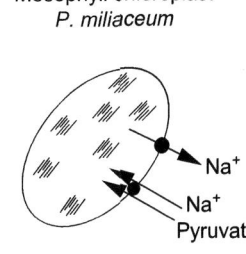

FIGURE 8.2 Pyruvate uptake into mesophyll chloroplasts of *Panicum miliaceum* (NAD$^+$-malic enzyme type) and *Zea mays* (NADP$^+$-malic enzyme type) and proposed Na$^+$/pyruvate cotransport in *P. miliaceum* with and without 1 mM NaCl. Based on Ohnishi et al. (1990).

TABLE 8.3 Variations in the biochemistry of C$_4$ photosynthesis found in some C$_4$ plants.

Major BSC decarboxylases	Energetics of decarboxylation in BSC per CO$_2$	Major substrates moving from MC→BSC	Major substrates moving from BSC→MC	Representative species
NADP$^+$ ME	Production of 1 NADPH	Malate	Pyruvate	*Zea mays*
				Digitaria sanguinalis
NAD$^+$ ME	Production of 1 NADH	Aspartate	Alanine/pyruvate	*Atriplex spongiosa*
				Portulaca oleracea
PEP carboxykinase	Consumption of 1 ATP	Aspartate	PEP	*Panicum maximum*
				Sporobolus poiretti

BSC, Bundle sheath chloroplasts; *MC*, mesophyll chloroplasts; *ME*, malic enzyme.
Source: From Ray and Black (1979).

light-stimulated Na efflux pump (Fig. 8.2) (Ohnishi et al., 1990). By contrast, such a Na effect on pyruvate uptake was absent in mesophyll chloroplasts of *Zea mays*. In C$_4$ species of the NADP-malic enzyme type (Table 8.3), such as *Z. mays* and *Sorghum bicolor*, H$^+$/pyruvate rather than Na$^+$/pyruvate cotransport may operate in the envelope of mesophyll chloroplasts (Ohnishi et al., 1990). However, in a study with 38 monocotyledonous and dicotyledonous C$_4$ species, it was found that majority had Na$^+$/pyruvate cotransport, whereas only species from the tribes Andropogoneae and Arundinelleae (both in family Poaceae) had H$^+$/pyruvate cotransport (Aoki et al., 1992). This result further stresses the necessity of (1) differentiating between the various C$_4$ metabolic types in studying the role of Na and (2) including species for which Na is not essential for metabolic functions in the C$_4$ photosynthetic pathway.

8.2.4 Substitution of K by Na

The beneficial effects of Na on the growth of nonhalophytes (glycophytes) are well known (Pilon-Smits and Leduc, 2009; Wakeel et al., 2011). In particular, the capacity of Na to alleviate K deficiency has important implications in agriculture (Pi et al., 2016; Wakeel et al., 2011), horticulture (Coban et al., 2020; Mcbride et al., 2014), and forestry plantations (particularly regarding *Eucalyptus* species) (Almeida et al., 2010; Mateus et al., 2019; Sette Junior et al., 2013), especially in resource-poor countries where brackish/saline water is more accessible than K fertilizers. Substitution of K by Na is particularly important in plants growing on low-K soils (Huhtanen et al., 2000; Ma and Bell, 2016; Mateus et al., 2019; Milford et al., 2008; Sette Junior et al., 2013; Wakeel et al., 2010). Sodium can partially replace K in barley (Ma et al., 2011; Ma and Bell, 2016), wheat, and canola (Ma and Bell, 2016) and can increase K accumulation in shoots of K-efficient wheat genotypes (Karthika et al., 2014).

Plant species can be classified into four groups according to the differences in their growth response to Na (Fig. 8.3). In group A, a high proportion of K is replaced by Na without a growth decline, and growth stimulation

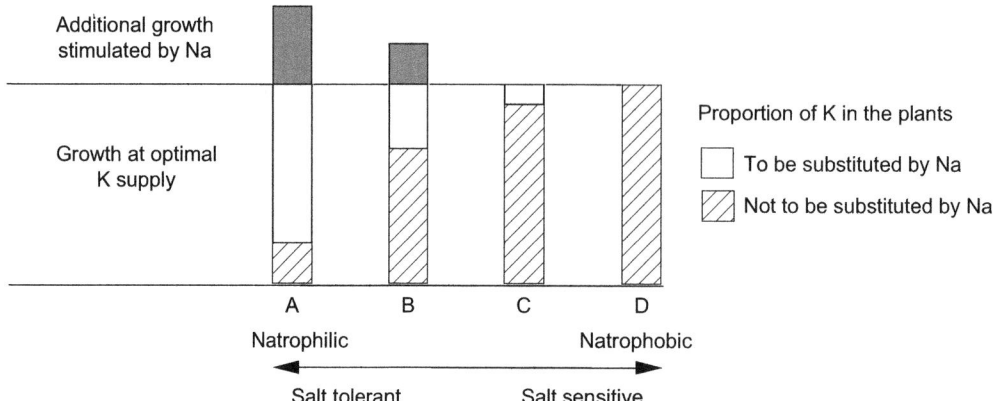

FIGURE 8.3 Tentative schematic diagram for the classification of crop plants according to the extent to which K can be replaced by Na, and additional growth stimulation by Na. Group A: mainly members of Chenopodiaceae (e.g., sugar beet, table beet, turnip, Swiss chard) and many C_4 grasses (e.g., Rhodes grass). Group B: cabbage, radish, cotton, pea, flax, wheat, and spinach. Group C: barley, millet, rice, oat, tomato, potato, and ryegrass. Group D: maize, rye, soybean, *Phaseolus* bean, and timothy.

occurs that cannot be achieved by increasing the K concentration in plant tissues. In group B, specific growth responses to Na are observed, but they are less distinct than in the species of group A. Also, a smaller proportion of K can be replaced without decreasing growth. In group C, the substitution of K can only take place to a limited extent, and Na has no specific effect on growth. In group D, K cannot be replaced by Na. This classification is not absolute because it does not take into account, for example, differences among cultivars within a species in the substitution of K by Na. These genotypic differences can be substantial, as has been shown in tomato (Coban et al., 2020) or cotton (Liaqat et al., 2009).

The differences in the growth responses of natrophilic and natrophobic species to Na are related to differences in uptake, particularly in the translocation of Na to the shoots (Chapter 3). The differential strategies for regulating Na transport to the shoots in pasture plants have important consequences for animal nutrition and in crop plants for salt tolerance (Shabala and Munns, 2017). In sugar beet (a natrophilic species), Na is readily translocated to shoots (Wakeel et al., 2010), where it replaces most of the K (Fig. 8.4) (Hawker et al., 1974). This substitution increased plant dry weight above that of K-deficient plants (grown at 0.05 mM K) and plants receiving a high K supply (5.0 mM K). By contrast, the growth of bean plants (group D species) under K deficiency (grown at 0.5 mM K) was further depressed by Na. A lack of growth response in bean is likely due to an *exclusion mechanism* in roots blocking Na transport to the shoots (Chapter 3). The potential for replacement of K by Na is therefore limited (Valdez-Aguilar and Reed, 2010) or even absent in group D species.

Among forage grasses, ryegrass and cocksfoot are considered natrophilic, and timothy and kikuyu natrophobic (Grieve et al., 2004; Phillips et al., 1999, 2000; Smith et al., 1980). Hence, Na fertilization has positive effects on the growth and nutritional quality of ryegrass but not growth of timothy (Huhtanen et al., 2000). The growth of tropical Guinea grass (*Megathyrsus maximus*, previously classified as *Urochloa maxima*) can be improved by partial substitution of K by Na, particularly when additional Ca is supplied (Carneiro et al., 2017).

Most agriculturally important crops are natrophobic (i.e., *excluders*) (groups C and D; Fig. 8.3) and have low salt tolerance. By contrast, natrophilic species, especially those in group A, have moderate to high salt tolerance and are *includers*. Under saline conditions, they accumulate Na in the shoots, where it is utilized in the vacuoles of leaf cells for osmotic adjustment (Gonzales et al., 2002) (see also Section 18.5.4.3). An interesting exception is *Populus euphratica* that achieves osmotic adjustment by accumulating Na in the leaf cell apoplast rather than the vacuoles (Chen and Polle, 2010; Ottow et al., 2005). Some parasitic plants (e.g., *Cuscuta attenuata*) are also *includers* because they require high internal Na concentrations as osmoticum to aid in water and nutrient extraction from the host plants (Kelly and Horning, 1999), whereas others (e.g., *Cistanche phelypaea*) adjust osmotically using K rather than Na (Fahmy, 2013).

Even in natrophilic species, the substitution of K by Na in the shoots is limited. The extent of substitution differs among individual organs and cell compartments, being large in the vacuoles, but limited in the cytoplasm (Leigh et al., 1986). In sugar beet, the substitution can be high in mature leaves but lower in expanding leaves, leading to an opposite gradient in the K/Na ratios in leaves of different ages (Table 8.4). Hence, average values for substitution in the whole shoot are misleading and underestimate the essentiality of K for growth and metabolism.

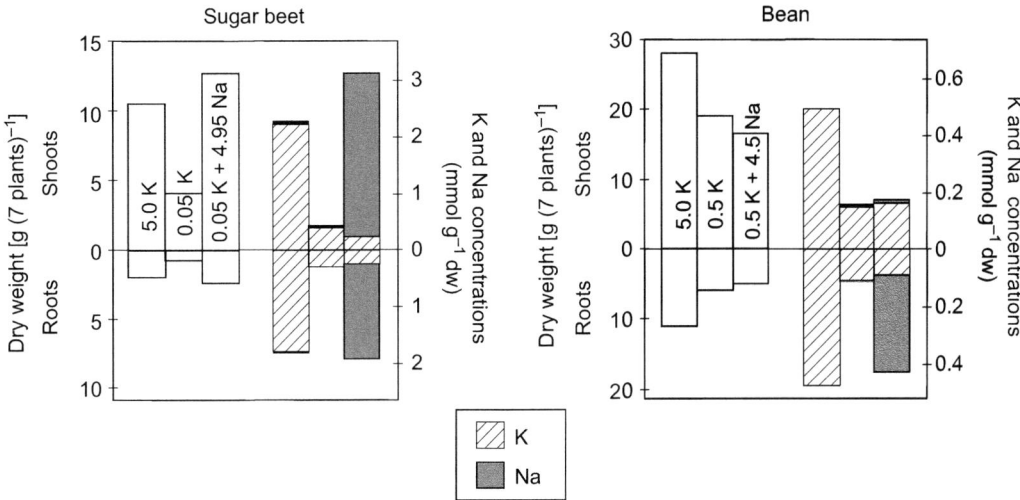

FIGURE 8.4 Dry weight and K and Na concentrations of sugar beet (*Beta vulgaris*) and bean (*Phaseolus vulgaris*) grown in nutrient solutions with different concentrations of K and Na. Concentrations indicated in the columns are in mM. *Based on Hawker et al. (1974).*

TABLE 8.4 Potassium and Na concentrations in sugar beet (cv. Fia) leaves and whole shoots at different K and Na concentrations in the nutrient solution after 9 weeks of growth.

K and Na supply (mM)	mmol g^{-1} dw							
	Whole shoot		Old leaves		Middle leaves		Young leaves	
	K	Na	K	Na	K	Na	K	Na
5.0 K	3.0	<0.03	3.43	<0.03	2.36	<0.03	1.87	<0.03
0.25 K + 4.75 Na	0.24	2.72	0.18	3.05	0.34	2.01	0.52	1.75
0.10 K + 4.90 Na	0.10	3.29	0.05	4.20	0.14	2.97	0.48	1.82

Source: From Marschner et al. (1981a).

In old leaves, nearly all K can be replaced by Na for specific functions in meristematic and expanding tissues. By contrast, in young expanding leaves there is a threshold level of substitution of ~0.5 mmol K g^{-1} dw, which corresponds to a concentration of ~50 mM K kg^{-1} fw and 100–150 mM K in the cytoplasm (Leigh et al., 1986).

In natrophobic species such as maize and bean, there is an absolute requirement for K in most of its metabolic functions (Section 6.6). Replacement of K by Na may occur to some extent in the root vacuoles, whereas such substitution in the cytoplasm causes substantial changes in the fine structure of the cytoplasm and its organelles (Hecht-Buchholz et al., 1971).

Biosynthesis of proteins associated with a wide range of cellular processes was affected by K deficiency and partially restored by Na substitution (Pi et al., 2016). Nevertheless, the most adverse effects of substitution of K by Na occur in protein biosynthesis (Faust and Schubert, 2016), limiting the acceptable extent of substitution. However, there are differences among species, with protein biosynthesis in maize decreased by 20% due to K-by-Na substitution compared with a 40% decrease in sugar beet (Faust and Schubert, 2017).

8.2.5 Growth stimulation by Na

In addition to K substitution, growth stimulation by Na is of practical and scientific interest. It raises the possibility of applying inexpensive, low-grade potash fertilizers with a high proportion of Na (Almeida et al., 2010), and it increases the potential of selecting and breeding crop genotypes adapted to saline soils.

The growth stimulation by salinity is a general phenomenon in halophytic plants, mainly due to improved water relations associated with a role of Na (as well as Cl) in osmotic adjustment (Hussin et al., 2013; Khedr et al., 2011).

However, the roles of Na versus Cl are not easy to distinguish; even though a great majority of studies assume the stimulatory effect is due to Na, there are some reports indicating a positive effect of Cl (rather than Na) on the growth of halophyte plants, for example, *Suaeda salsa* (Mori et al., 2008).

Responses to Na differ not only among plant species but also among genotypes of a species (see also Section 8.2.1). Compared with the effect of K supply only (Table 8.5), the substitution of half the K in the substrate by Na led to a significant increase in the storage root dry weight and sucrose content in two out of three sugar beet genotypes (not in cv. Monohill) (Marschner et al., 1981b). When 95% of the K in the substrate (and ~90% in the plants) was replaced by Na (compared to 100% K in the substrate), storage root dry weight and sucrose content were increased significantly in cv. Fia and were unaffected in cv. Ada (Table 8.5). Salt tolerance differed among the three genotypes, in agreement with the general classification (Fig. 8.3). At 150 mM NaCl in the external medium, growth of cv. Fia was not affected after 9 weeks but was decreased in the other two genotypes (Marschner et al., 1981a).

Growth stimulation by Na is caused mainly by its effect on cell expansion and on plant water balance. Sodium can replace K in its contribution to the solute potential in the vacuoles and consequently in the generation of turgor and cell expansion (Section 6.6) and may even surpass K in this respect because it accumulates preferentially in the vacuoles (Gonzales et al., 2002). The superiority of Na can be demonstrated in intact sugar beet plants, where leaf area, thickness, and succulence all increased when a high proportion of K was replaced by Na (Table 8.6; Hampe and Marschner, 1982). Succulence is a morphological adaptation that is usually observed in salt-tolerant species growing in saline substrates (e.g., Chen and Polle, 2010; Ottow et al., 2005) and is considered an important buffer mechanism against deleterious changes in leaf water potential at moderate drought stress. Better osmotic adjustment by Na compared with K is also a major factor in the growth stimulation of halophytes by high Na supply (Hussin et al., 2013; Khedr et al., 2011).

High Na supply increases leaf area as well as the number of stomata per unit leaf area (Table 8.7), whereas it decreases the chlorophyll concentration in natrophilic species such as sugar beet (Hampe and Marschner, 1982) as well as in natrophobic species such as maize (Turan et al., 2009). Despite decreased chlorophyll content in sugar beet leaves, net photosynthesis rate is not decreased significantly (Table 8.7); in other species, the increased substitution of K by Na (up to a point) may increase the net photosynthetic rate (Gattward et al., 2012; Hajiboland et al., 2015; Mateus et al., 2019). Therefore, the higher growth rates of sugar beet (Hampe and Marschner, 1982) and some other plant species

TABLE 8.5 Genotypic differences in sucrose concentration and content in storage roots of three sugar beet cultivars grown at different K and Na concentrations in the nutrient solution for 9 weeks.

Cultivar	Treatment (mM)		Dry weight of storage root (g root^{-1})	Sucrose in storage roots	
	K	Na		Concentration (g kg^{-1} fw)	Content (g root^{-1})
Monohill	5.0	0	61	92	45
	2.5	2.5	64	119	50
	0.25	4.75	58	76	34
		LSD$_{0.05}$	6.8	29	5.4
Ada	5.0	0	45	49	19
	2.5	2.5	63	71	43
	0.25	4.75	37	77	21
		LSD$_{0.05}$	8.6	20	7.1
Fia	5.0	0	23	100	14
	2.5	2.5	33	104	20
	0.25	4.75	38	112	28
		LSD$_{0.05}$	7.5	13	5.4

Source: From Marschner et al. (1981a).

TABLE 8.6 Leaf properties and K and Na concentrations in sugar beet (cv. Monohill) grown at different K and Na concentrations in the nutrient solution for 50 days.

	Treatment (mM)		LSD$_{0.05}$
	5.0 K	0.25 K + 4.75 Na	
Leaf area (cm² leaf^{-1})	233	302	32
Leaf thickness (μm)	274	319	9
Succulence (g H$_2$O dm^{-1})	3.1	3.7	0.3
Leaf dry weight (g plant^{-1})	7.6	9.7	1.2
	Concentrations in leaves (mmol g^{-1} dw)		
K	2.67	0.43	0.2
Na	<0.03	2.45	–

Source: Based on Hampe and Marschner (1982).

TABLE 8.7 Properties of sugar beet (cv. Monohill) leaves and water consumption at different K and Na concentrations and at different osmotic potential (± mannitol) of the nutrient solution (harvested 50 d after germination).

		Treatment (mM)		LSD$_{0.05}$
		5.0 K	0.25 K + 4.75 Na	
Stomata number on lower leaf surface (no. cm^{-2})		11 805	15 127	1099
Chlorophyll concentration (mg g^{-1} dw)		12.1	9.2	0.6
Net photosynthesis (mg CO$_2$ cm^{-2} h^{-1})		15.2	14.4	2.3
		Water consumption (g H$_2$O g^{-1} fw increment)		
Osmotic potential of the nutrient solution (MPa)	−0.02	17.7	26.5	4.6
	−0.40	28.2	24.6	5.1
	LSD$_{0.05}$	5.6	3.9	

Source: Based on Hampe and Marschner (1982).

(Mateos-Naranjo et al., 2010a) at high Na and low K supply are not due to increased photosynthetic efficiency, but rather to a larger leaf area.

In wheat, low to moderate Na supply enhanced growth (Ma and Bell, 2016) and increased K concentration and content together with stomatal conductance and net photosynthetic rate (which might have been associated with increased root growth through enhanced photosynthate supply), particularly in K-efficient genotypes (Karthika et al., 2014).

When the availability of water in the substrate is high, Na increases the water consumption per unit fresh weight increment in sugar beet (Table 8.7), thus decreasing the water-use efficiency (Hampe and Marschner, 1982). By contrast, partial K substitution by Na increased water-use efficiency in many species, such as *Theobroma cacao* (Gattward et al., 2012), *Atriplex nummularia* (Hussin et al., 2013), eucalyptus (*Eucalyptus urophylla* × *Eucalyptus grandis* hybrid) (Mateus et al., 2019), sugarcane (Theerawitaya et al., 2020), and tomato (Al-Karaki, 2000). However, Na improves the water balance of plants when the water supply is limited via stomatal regulation (Fig. 8.5). With a sudden decrease in the availability of water in the substrate (*drought stress*), the stomata of plants supplied with Na closed more rapidly than those of plants supplied with K only; after stress removal, the opening of stomata of the K-supplied plants was delayed compared to the plants supplied with Na. Thus, in plants supplied with Na, the relative leaf water content

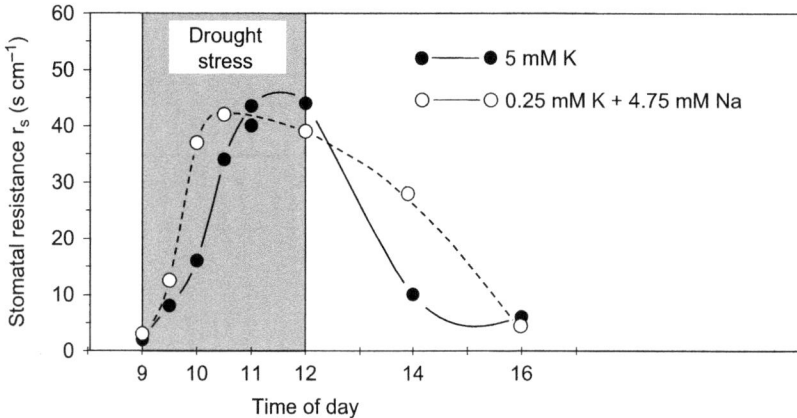

FIGURE 8.5 Stomatal resistance to water vapor exchange in leaves of sugar beet exposed to transient drought stress (decrease in solution water potential to −0.75 MPa by the addition of mannitol). *Based on Hampe and Marschner (1982).*

remained higher, even at low substrate water availability (drought periods, saline soils) (Hampe and Marschner, 1982). A positive effect of Na supply on stomatal conductance has been shown in many species, including sugarcane (Theerawitaya et al., 2020), eucalyptus (*E. urophylla* × *E. grandis* hybrid) (Mateus et al., 2019), and wheat (Karthika et al., 2014), but a decrease was also noted in a number of species, including C$_4$ maize (Omoto et al., 2012) and halophytes (Khedr et al., 2011). Hence, K remains crucial in regulating stomatal opening in some halophytes, such as *Aster tripolium* (Perera et al., 1997).

The replacement at the cellular level of a high proportion of K by Na may also affect the activity of enzymes that respond specifically to K (Section 6.6). For example, K is four times more effective than Na in activating starch synthase that catalyzes the conversion of ADP-glucose to starch (Hawker et al., 1974). Thus, in leaves in which a high proportion of K is replaced by Na, the starch concentration is lower, whereas the concentration of soluble carbohydrates, particularly sucrose (Hawker et al., 1974) or maltose (Kempa et al., 2008), is higher. This shift in carbohydrate metabolism may favor cell expansion in the leaf tissue. Furthermore, Na is more effective than K in stimulating sucrose accumulation in the storage tissue of sugar beet. The import of sucrose across the tonoplast of storage root cells in sugar beet is mediated by proton/sucrose antiporters named BvTST (tonoplast sugar transporters) exchanging protons from vacuole for sucrose (Jung et al., 2015), which is likely to operate together with pyrophosphatases and V-ATPases in the tonoplast (Theerawitaya et al., 2020) to maintain a supply of protons in the vacuole. The expression of V-ATPase was enhanced by Na supply in halophytes (Tran et al., 2020) as well as in rice (Pons et al., 2011).

8.2.6 Application of Na fertilizers

Given the genotypic differences in growth response to Na and the abundance of Na in the biosphere, one can expect the application of Na to have beneficial effects (1) in natrophilic plant species, (2) when the concentrations of available K and/or Na are low, and (3) in areas with irregular rainfall and/or transient drought during the growing season. In addition, Na fertilization and substitution of K may be important in soils that are highly K fixing (Wakeel et al., 2010) or have low K availability (Milford et al., 2008).

The application of Na fertilizers to sugar beet results in an increase in the leaf area index early in the growing season and thus an increase in light interception, improving water-use efficiency under moderate drought stress (Durrant et al., 1978). The potential replacement of K by Na can be considered when applying fertilizers to natrophilic species. When Na concentrations in leaves are high, the leaf K concentrations required for optimal growth decrease from 35 to 8 g kg^{-1} dw in Italian ryegrass (Hylton et al., 1967), from 27 to 5 g kg^{-1} in Rhodes grass (Smith, 1974), and 43 to 10 g kg^{-1} in lettuce (Costigan and Mead, 1987).

Neither sugar beet yield nor white sugar yield decreased with an increased proportion of Na in the K/Na fertilizer mixture (K:Na in fertilizers ranging from 1:0 to 1:2.2) across four field sites (Barlog et al., 2018), suggesting that a large proportion of fertilizer K can be substituted by Na. As expected, no yield responses to NaCl fertilizer occurred on soils well supplied by exchangeable K; such responses were noted only on low K soils (Milford et al., 2008).

The Na concentration of forage and pasture plants is an important factor in animal nutrition. The Na requirement of lactating dairy cows is ∼2.0 g kg^{-1} forage dw (Smith et al., 1978), which is higher than the average Na concentration of natrophobic pasture species (Phillips et al., 2000). By contrast, the K concentration in these natrophobic species is usually at least adequate (in the range of 20 to 25 g kg^{-1} dw) but often in excess of animal needs. The use of Na fertilizer to increase the Na concentration of forage and pasture plants is thus important in many parts of the world (at least on low K soils, Huhtanen et al., 2000). Furthermore, high Na concentration increases the acceptability of forage to animals and enhances daily food intake (Zehler, 1981). However, Na fertilizers are effective only when applied to grassland or mixed pastures with a relatively high proportion of natrophilic species (Phillips et al., 2000).

There is a potential that hazenite [KNaMg$_2$(PO$_4$)$_2$ · 14H$_2$O] (containing 4.2% w/w Na as well as 9.9% w/w Mg) may be a suitable Na fertilizer for natrophilic pasture species such as Italian ryegrass (Watson et al., 2020), particularly because it can also increase Mg content in forage and thus reduce a grass tetany risk.

8.3 Silicon

8.3.1 General

Silicon (Si) is the second most abundant element after oxygen in the Earth's crust. In soil solution at pH below 9.0, the prevailing form is monosilicic acid, Si(OH)$_4$, an uncharged molecule, with a solubility in water (at 25°C) of ~2 mM (equivalent to 56 mg Si L^{-1}) (Fig. 8.6). On average, the concentration in the soil solution is 14–20 mg Si L^{-1} (ranging between 3.5 and 40 mg L^{-1}) with a tendency of lower concentrations at high pH (>7) and when large amounts of sesquioxides dominate anion adsorption (Jones and Handreck, 1965). Such conditions are widespread in highly weathered tropical soils. When the concentrations of silicic acid in aqueous solutions exceed 56 mg Si L^{-1} (2 mM), it polymerizes to silica (SiO$_2$) (Fig. 8.6), which is the major form of Si in plant tissues.

8.3.2 Uptake, concentration, and distribution

All plants rooting in soil contain Si in their tissues. However, the Si concentration in the shoots varies considerably among plant species, ranging from 1 to 100 mg Si g^{-1} dw (Epstein, 1999; Ma and Takahashi, 2002). In general, plants belonging to Bryophyta, Lycopsida, and Equisetopsida in division Pteridophyta have large tissue Si concentrations, whereas those belonging to Filicopsida in Pteridophyta as well as Gymnospermae and Angiospermae have low Si accumulation. Some families and orders have high to moderate concentrations of Si varying between >40 mg kg^{-1} in Poaceae, Cyperaceae, and Balsaminaceae and 20–40 mg kg^{-1} in Cucurbitales, Urticales, and Commelinaceae, whereas most other plant taxa show low Si accumulation (Hodson et al., 2005; Ma and Takahashi, 2002). The differences in Si accumulation between species can be attributed to differential ability of roots to take up Si (Ma and Takahashi, 2002).

Plant roots take up Si in the form of silicic acid [Si(OH)$_4$]. There are three different modes for Si uptake (active, passive, and rejective) depending on plant species. Two different types of Si transporters (Lsi1 and Lsi2) have been identified in higher plants (Fig. 8.7). The Lsi1 was first identified in rice and is a Si-permeable channel-type transporter (Ma et al., 2006) belonging to a Nod26-like major intrinsic protein (NIP) subfamily of aquaporin-like proteins; Lsi1 allows the influx of silicic acid when expressed in *Xenopus* oocytes. The predicted amino acid sequence has six transmembrane domains and two Asn-Pro-Ala (NPA) motifs, which are well conserved in typical aquaporins. On the other hand, Lsi2 is an active efflux transporter of Si in rice (Ma et al., 2007). The Lsi2 is a putative anion transporter without any similarity with the Si influx transporter Lsi1. Both Lsi1 and Lsi2 are localized at the exodermis and endodermis of the roots in rice, where Casparian bands are located (Fig. 8.7). However, Lsi1 and Lsi2 show different polarity; Lsi1 is localized at the distal side, whereas Lsi2 is at the proximal side (Fig. 8.7). Therefore, Lsi1 and Lsi2 form a cooperative uptake system required for efficient Si uptake in rice (Huang et al., 2020; Sakurai et al., 2015). Knocking out either of them results in a significant decrease of root Si uptake. In addition, Casparian bands are required for efficient Si uptake in rice (Wang et al., 2019; Sakurai et al., 2015).

Silicon uptake occurs in the mature regions of the rice roots rather than at the root tips (Yamaji and Ma, 2007). This is associated with high expression of both *Lsi1* and *Lsi2* in the mature regions (Yamaji and Ma, 2007). Furthermore, root hairs do not play any role in Si uptake, but lateral roots do (Ma et al., 2001a). This is attributed to the expression of *Lsi1* and *Lsi2* in the latter, but not the former (Ma et al., 2006, 2007).

Homologs of rice Lsi1 and Lsi2 have also been identified in other plant species such as barley, maize, wheat, pumpkin, cucumber, soybean, grape, and tomato (Chiba et al., 2009; Deshmukh et al., 2013; Mitani et al., 2009a, 2009b, 2011a, 2011b; Montpetit et al., 2012; Noronha et al., 2020; Sun et al., 2017, 2018, 2020b). However, their localization

$$nSiO_2 + nH_2O \underset{>2\ mM}{\overset{<2\ mM}{\rightleftarrows}} nSi(OH)_4 \underset{<pH\ 9}{\overset{>pH\ 9}{\rightleftarrows}} n(OH)_3SiO^- + nH^+$$

FIGURE 8.6 Forms of Si that exist at different concentrations and pH values.

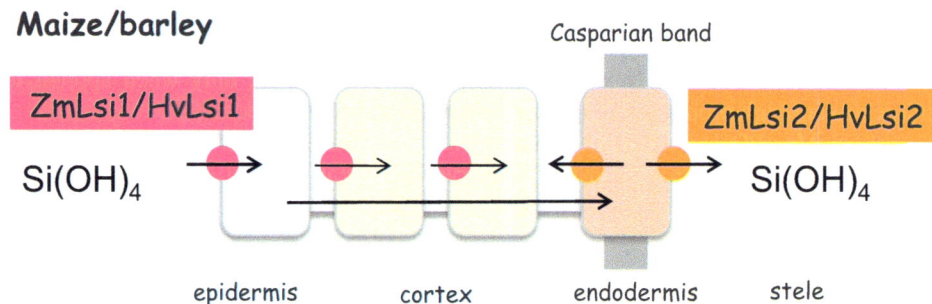

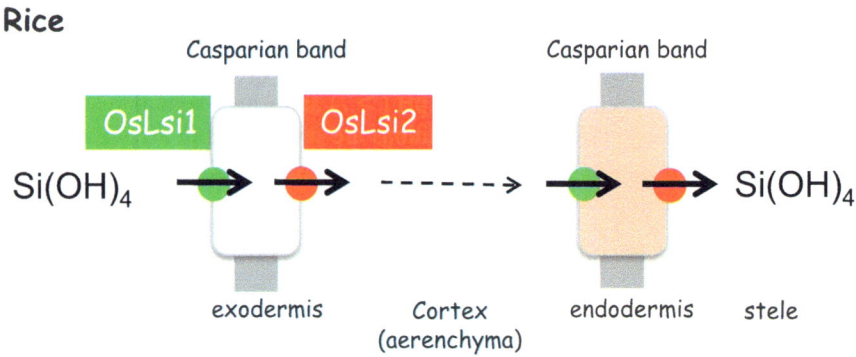

FIGURE 8.7 Different Si uptake systems in the roots of upland crops (e.g., barley and maize) and rice. *Reprinted from Mitani et al. (2009a) with permission from Oxford University Press.*

and expression patterns are different from those in rice (Fig. 8.7). For example, HvLsi1 from barley and ZmLsi1 from maize are localized in the epidermal, hypodermal, and cortical cells (Fig. 8.7), whereas HvLsi2 and ZmLsi2 are localized only at the endodermis and do not show polarity (Mitani et al., 2009a, 2009b). These differences result in differential pathways of Si from the external solution to the xylem between barley/maize and rice. In barley and maize, Si can be taken up from the external solution (soil solution) by HvLsi1/ZmLsi1 at the epidermal, hypodermal, and cortical cells and then transported in the symplasm to the endodermis where it is released to the stele by HvLsi2/ZmLsi2 (Fig. 8.7). By contrast, in rice, Si taken up by Lsi1 at the exodermal cells is released by Lsi2 to the apoplast (aerenchyma) and then transported into the stele by both Lsi1 and Lsi2 again at the endodermal cells (Fig. 8.7).

Silicon accumulation in shoots is determined by the expression of Si transporter genes, cell-specific localization of Si transporters, and co-presence of Lsi1 and Lsi2 (Ma and Yamaji, 2015). The Lsi1 is a passive channel-type Si transporter, whereas Lsi2 is an active transporter driven by the proton gradient (Ma et al., 2011). Therefore, cooperation of Lsi1 and Lsi2 is required for high Si accumulation. In tomato, although Lsi1 is functional, there is no Lsi2-like transporter, resulting in low Si accumulation in its shoots (Sun et al., 2020b).

Following uptake by the roots through Lsi1 and Lsi2, Si is translocated to the shoots in the xylem. More than 90% of Si taken up by the roots is translocated to the shoots in rice, resulting in Si concentrations in the xylem sap as high as 20 mM (present as monosilicic acid; Casey et al., 2003; Mitani et al., 2005). Such high concentrations are probably present only transiently because silicic acid at concentrations above 2 mM in vitro polymerizes into silica gel ($SiO_2 \cdot H_2O$) (Mitani et al., 2005).

Unloading of Si from the xylem is mediated by Lsi6, a homolog of Lsi1 in rice (Yamaji and Ma, 2009). The Lsi6 is mainly localized at the xylem parenchyma cells of the leaf sheaths and blades. Knocking out *Lsi6* does not affect Si uptake by the roots but alters Si deposition in the leaf blades and sheaths and causes increased excretion of Si via guttation (Yamaji et al., 2008).

The highest Si concentration is in the husk of rice grain due to the preferential distribution of Si to the grains, mediated by Si transporters Lsi6, Lsi2, and Lsi3 that are highly expressed in node I (Yamaji et al., 2015) (Fig. 8.8). The Lsi6 is polarly localized at the proximal side of the xylem transfer cells of enlarged vascular bundles (EVBs), whereas Lsi2 is localized at the distal side of bundle sheath cells of EVBs and Lsi3 at parenchyma cells between EVBs and diffuse vascular bundles (DVBs) (Fig. 8.8). Therefore, Si in the xylem of EVBs is selectively unloaded by Lsi6, moved to adjacent cells through plasmodesmata, and then effluxed to the apoplast and reloaded to the xylem of DVBs by Lsi2 and Lsi3. Knocking out *Lsi6*, *Lsi2*, or *Lsi3* decreased Si distribution to the grain but increased Si distribution to the flag leaf in rice.

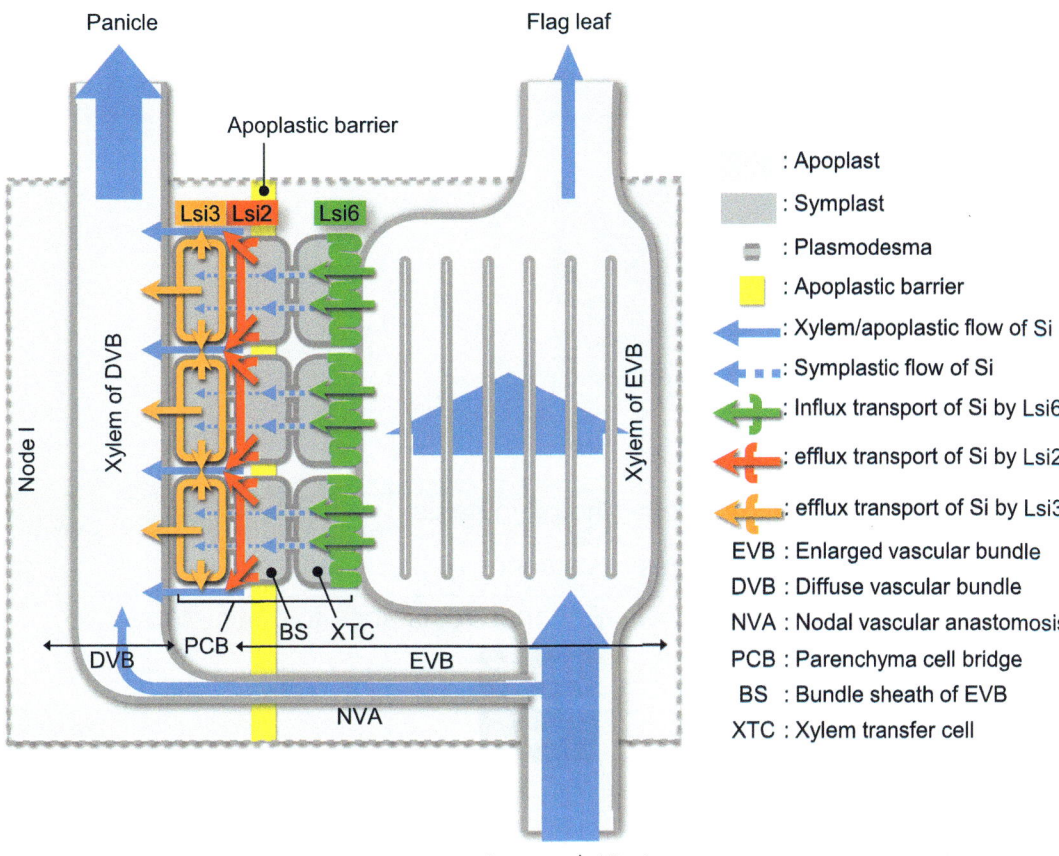

FIGURE 8.8 Transporters involved in the distribution of silicon in node I of rice. *Based on Yamaji et al. (2015).*

Most Si is deposited in the apoplasm of many organs as amorphous silica ($SiO_2 \cdot nH_2O$), including outer walls of the epidermal cells on both surfaces of the leaves as well as in the inflorescence bracts of graminaceous species (Hodson and Sangster, 1989; Ma and Takahashi, 2002) and trichomes (Lanning and Eleuterius, 1989). The epidermal cell walls impregnated with a layer of Si become effective barriers against both water loss by cuticular transpiration and fungal infections (Chapter 10). In grasses, a considerable proportion of Si in the epidermis of both leaf surfaces is also located intracellularly as phytoliths (Fig. 8.9).

The deposition of Si in hairs on leaves, culms, inflorescence bracts, and owns of cereal grains such as wheat pose a potential threat to human health (Hodson and Sangster, 1989). The inflorescence bracts of grasses of the genus *Phalaris* and foxtail millet (*Setaria italica*) contain sharp, elongated siliceous fibers that fall into the critical size range of fibers classified as carcinogenic (Sangster et al., 1983). The occurrence of esophageal cancer is correlated with the consumption of foxtail millet in north China (Parry and Hodson, 1982) or wheat contaminated with *Phalaris* in the Middle East (Sangster et al., 1983).

8.3.3 Beneficial effects

There is still no convincing evidence showing that Si is involved in metabolism, but Si has many beneficial effects on growth and development in various plant species (Ma and Takahashi, 2002). Under field conditions, particularly in dense stands of cereals, Si can stimulate growth and yield directly and indirectly (Ma and Takahashi, 2002). These include decreasing mutual shading by improving leaf erectness and alleviating abiotic and biotic stresses (Fig. 8.10).

Leaf erectness is an important factor affecting light interception in dense plant stands. For a given cultivar, leaf erectness decreases with increasing N supply (Section 6.1). Silicon increases leaf erectness and thus to a large extent counteracts the negative effects of high N supply on light interception. Similarly, Si counteracts the negative effects of an increasing N supply on haulm stability and lodging susceptibility (Idris et al., 1975).

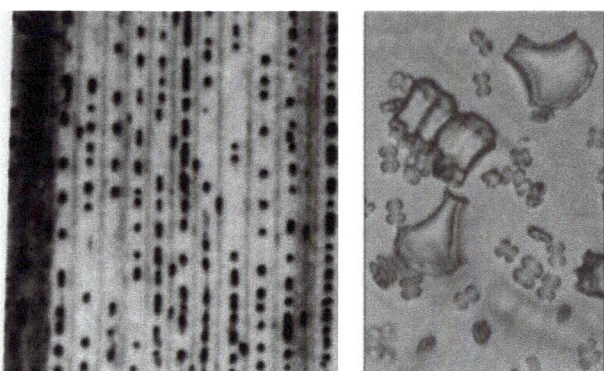

FIGURE 8.9 Deposition of Si in rice leaf (left) and phytoliths (right). *Based on Ma and Yamaji (2006).*

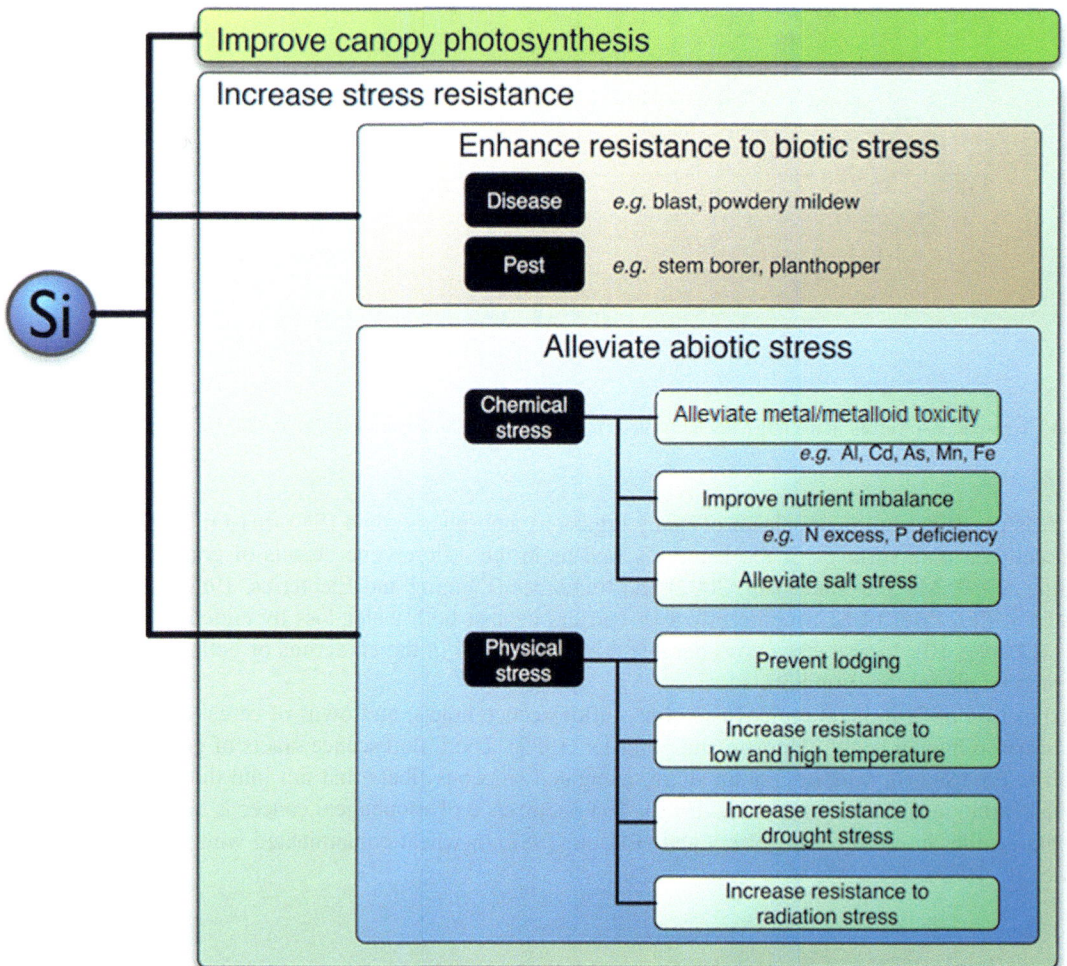

FIGURE 8.10 Beneficial effects of silicon on plant growth. *Reprinted (with some modifications) from Ma and Yamaji (2006).*

Silicon enhances the resistance of plants to diseases caused by fungi and bacteria (see Chapter 10). In rice, Si reduces the severity of both leaf and panicle blast (Fig. 8.11). In low Si soil, the application of silicate fertilizer is as effective as fungicide in controlling rice blast (Datnoff et al., 1997). Silicon also decreases the incidence of powdery mildew in cucumber, barley and wheat, sheath blight in rice, ring spot in sugarcane, rust in cowpea, leaf spot in

FIGURE 8.11 Beneficial effects of silicon on mitigation of various stresses in rice. Wild-type rice (WT) and low Si mutant (*lsi1*) were grown in a field. Effects on rice blast (A), pest damage (B), excess transpiration (C), and fertility (D). *Based on Ma et al. (2011).*

Bermuda grass, and gray leaf spot in St Augustine grass and perennial ryegrass (Fauteux et al., 2005). Silicon suppresses pests such as stem borers, various hoppers, leaf spiders, and mites (Savant et al., 1997).

Two mechanisms for Si-enhanced resistance to diseases and pests have been proposed. One is that Si acts as a physical barrier. Silicon is deposited beneath the cuticle to form a cuticle-Si double layer (Fauteux et al., 2005; Ma and Yamaji, 2006). This layer can mechanically impede penetration by fungi and pests, thereby inhibiting infection. However, according to Heine et al. (2007), inhibition of infection and spread of *Pythium aphanidermatum* in roots of tomato plants by Si is related to symplasmic Si, not apoplasmic Si. Another mechanism is that soluble Si acts as a modulator of host resistance to pathogens. Several studies in monocotyledonous (rice and wheat) and dicotyledonous species (cucumber) have shown that plants supplied with Si produce phenolics, lignin, H_2O_2, and phytoalexins in response to fungal infection (Belanger et al., 2003; Remus-Borel et al., 2005; Rodrigues et al., 2004; Sun et al., 2010). Further studies are required for a better understanding and characterization of the physiological effects of Si in biological systems.

Silicon alleviates various abiotic stresses including physical stresses (lodging, drought, radiation, high and low temperature, freezing, UV irradiation) and chemical stresses (salt, metal toxicity, nutrient imbalance) (Ma, 2004, 2005; Ma and Yamaji, 2006).

The beneficial effect of Si on the alleviation of UV stress in rice may be related to the biosynthesis of phenolic compounds (Goto et al., 2003). Silicon can alleviate water stress by decreasing transpiration in rice (Ma et al., 2001b) (Fig. 8.11). Transpiration from the leaves occurs mainly through the stomata and partly through the cuticle. As Si is deposited beneath the cuticle of the leaves, transpiration through the cuticle may be decreased.

Silicon application in rice is effective in alleviating the damage caused by weather-related stresses such as strong wind, low temperature, and insufficient sunshine during the summer season (Ma et al., 2001b). Strong winds can cause lodging and sterility in rice, resulting in a considerable reduction in grain yield. Deposition of Si in rice enhances the strength of the stem by increasing the thickness of the culm wall and the size of the vascular bundles, thereby preventing lodging.

Silicon may also alleviate Mn toxicity in hydroponically cultured barley (Horiguchi and Morita, 1987; Williams and Vlamis, 1957), bean, and pumpkin (Iwasaki and Matsumura, 1999). Three different mechanisms seem to be involved, depending on the plant species. In rice, Si reduced Mn uptake by downregulation of the Mn transporter gene *OsNramp5* and root-to-shoot translocation of Mn (Che et al., 2016). In bean (Horst and Marschner, 1978) and barley (Williams and Vlamis, 1957), Si did not reduce the Mn uptake but led to a homogeneous distribution of Mn in the leaf blade. The mechanism for this homogeneous distribution is unclear but may be related to a Si-induced increase in the binding capacity of the cell wall, resulting in decreased apoplasmic Mn concentration in cowpea (Horst et al., 1999). Alleviation of Mn toxicity damage in plant cells by Si supply may also be related to stimulation of antioxidative defense systems against oxidative cell damage by reactive oxygen species (Inal et al., 2009).

Silicon is effective in alleviating toxicity of various metals, including Na, Fe, Al, Cd, and Zn, which can be attributed to the interaction between Si and metals in the apoplasm or symplasm. A beneficial effect of Si under salt stress has been observed in rice (Gong et al., 2006; Matoh et al., 1986), wheat (Ahmad et al., 1992), and tomato (Romero-Aranda et al., 2006). This beneficial effect of Si may be due to a Si-induced decrease in transpiration (Matoh et al., 1986) and to the inhibitory effects of Si on Na accumulation in rice plants related to Si deposition in roots. Transport of K in the xylem is not affected by Si application, whereas the Na concentration in the xylem sap is reduced from 6.2 to 2.8 mM in rice plants, which may be explained by inhibited apoplasmic transport of Na across the root (Gong et al., 2006).

Silicon is particularly important for the growth and high production of rice, which can accumulate Si to over 5% w/w in the shoots. When Si concentrations are insufficient, yield is reduced due to decreased fertility (Fig. 8.11). This is well demonstrated by a field test using a rice mutant defective in Si uptake. Insufficient Si accumulation in the rice shoots results in more than 90% loss of grain yield (Tamai and Ma, 2008) (Fig. 8.11). For this reason, Si fertilizers are applied in paddy fields in some countries.

Sugarcane is also a Si accumulator that responds strongly to Si application. Under field conditions, at least 1% w/w Si is required for optimal cane yield, and yield is reduced by 50% at 0.25% w/w Si (Anderson, 1991). Such yield reductions are associated with visible deficiency symptoms ("leaf freckling") on leaf blades directly exposed to full sunlight (Elawad et al., 1982).

Silicon is an essential element for animals (Nielsen, 1984) because it is a constituent of mucopolysaccharides in connective tissues. However, in grazing animals the ingestion of a large amount of phytoliths may lead to excessive abrasion of the rumen wall, and dissolved Si may form secondary deposits in the kidney, thereby causing serious economic losses (Jones and Handreck, 1969).

8.4 Cobalt

Cobalt (Co) is an essential element for prokaryotes (including blue-green algae) and animals, but an essential role in plants has not been demonstrated. There is, however, an increasing number of papers showing that at low levels Co is beneficial for plant growth, especially in legumes. The beneficial effects include delaying leaf senescence through inhibition of ethylene biosynthesis and increased drought resistance of seeds (reviewed by Pilon-Smits and Leduc, 2009).

The requirement of Co for N_2 fixation in legumes and in root nodules of nonlegumes (e.g., alder) was reported by Ahmed and Evans (1960). When *Medicago sativa* was grown under controlled environmental conditions with minimal Co contamination, plants dependent on N_2 fixation grew poorly, and growth was enhanced strongly by Co supply; by contrast, nitrate-fed plants grew equally well without and with the supply of Co (Delwiche et al., 1961). Subsequently, Kliewer and Evans (1963) isolated the cobalamin coenzyme B_{12} from root nodules of legumes and nonlegumes and demonstrated the interdependence of Co supply, the coenzyme cobalamin concentration of *Rhizobium*, the formation of leghemoglobin, and N_2 fixation. Since then, it has been established that *Rhizobium* and other N_2-fixing microorganisms have an absolute Co requirement whether or not they are growing within nodules and regardless of whether they are dependent on an N supply from N_2 fixation or from mineral N.

8.4.1 Role of Co in plants

The main biological role of Co is in the coenzyme cobalamin (vitamin B_{12} and its derivatives). In cobalamin, Co is chelated to four N atoms at the center of a porphyrin-like structure (corrin) and has a similar role to that of Fe in hemoglobin. Cobalamin has a complex biochemistry, and there are a number of cobalamin-dependent enzymes. There are three primary classes of cobalamin enzymes: (1) methylcobalamin-dependent methyltransferase (methionine synthase), (2) adenosylcobalamin-dependent isomerase (methylmalonyl-CoA mutase), and (3) B_{12}-dependent reductive dehalogenase (Zhang and Gladyshev, 2009). Cobalamin-dependent enzymes and Co-induced changes in their activities, nodulation, and N_2 fixation have been identified in *Rhizobium* (and *Bradyrhizobium*) species. One of the well-known cobalamin-dependent enzymes is methionine synthase, which maintains the biosynthesis of methionine in mammalian systems (Koutmos et al., 2009). It is not known whether a similar enzyme is active in plants and contributes to methionine synthesis. The results presented in Table 8.8 indicate that Co deficiency probably reduces methionine synthesis, which may lead to lower protein synthesis and also to the smaller size of the bacteroids (bacteria in the nodules capable of N_2 fixation). Furthermore, methylmalonyl-coA mutase is involved in the synthesis of heme (iron porphyrins) in the bacteroids and thus in the synthesis of leghemoglobin. Therefore, Co deficiency impairs the synthesis of leghemoglobin (see Chapter 16).

Root release of phytosiderophores is a common adaptive response of most monocots to low Fe soils (Chapters 2, 7, and 14), the process that appears to be sensitive to ethylene production. Diminishing biosynthesis of ethylene by Co treatments resulted in increases in root release of phytosiderophores in wheat plants (Divte et al., 2019). It was hypothesized that phytosiderophores and ethylene use the same substrate for their biosynthesis, and therefore any decline in ethylene synthesis may positively influence the synthesis and root release of phytosiderophores.

8.4.2 Cobalt deficiency and toxicity

Soil and plant Co concentrations usually range between 15 and 25 mg kg^{-1} and 0.1 and 10 mg kg^{-1}, respectively. The common leaf concentration of Co is below 0.2 mg kg^{-1} (Pilon-Smits and Leduc, 2009), whereas Co hyperaccumulator plants growing in some Co-contaminated soils may contain up to 5000 mg Co kg^{-1} (Lange et al., 2017). High soil pH represents a major factor in reducing root Co uptake. Soils containing a large amount of Mn oxides and extractable Mn exhibit a high capacity for immobilizing Co and decreasing root Co uptake, respectively (Li et al., 2004; Wendling et al., 2009). In low Co soils, nodule development and function are impaired to different degrees (Table 8.9; Dilworth et al., 1979). When lupins grown in a low Co soil were supplied with Co, the weight and Co concentration of the nodules, the number of bacteroids, and the amounts of cobalamin and leghemoglobin per unit nodule fresh weight all increased.

In legumes grown in low Co soils, the nodule activity is lower in plants without Co addition. This lower activity results in reduced nitrogenase activity or N content of the plants (Fig. 8.12). Furthermore, *Rhizobium* infection is often lower than in plants supplied with Co, and the onset of N$_2$ fixation is delayed for several weeks. In legumes dependent on N$_2$ fixation, Co deficiency is therefore associated with symptoms of N deficiency (Dilworth et al., 1979; Robson and Snowball, 1987). Under Co deficiency, there is a preferential accumulation of Co in the nodules. In deficient plants, the Co concentration in the nodules varies between 20 and 170 μg g^{-1} nodule fw, depending on the plant species (Robson et al., 1979). When grown in low Co soils and dependent on N$_2$ fixation, there is a close relationship between seed Co concentration, plant growth, N concentration, and severity of the visual N deficiency symptoms (Robson and Snowball, 1987). As shown in Fig. 8.13, the shoot growth response to increasing seed Co concentrations is strong up to about 200 ng g^{-1} seed dw.

TABLE 8.8 Characteristics of Co-sufficient and Co-deficient crown nodules of *Lupinus angustifolius*.

Co treatment	Volume of bacteroids (μm^3)	DNA concentration (pg cell^{-1})	Methionine (% of total amino-N)
+Co	3.2	12	1.3
−Co	2.6	8	1.0

Source: Based on Dilworth and Bisseling (1984).

TABLE 8.9 Nodule growth and composition in *Lupinus angustifolius* inoculated with *Rhizobium lupini* grown in a low Co soil with (0.19 mg Co pot^{-1}) or without Co addition.

	Co treatment	
	−Co	+Co
Crown nodule fresh weight (g plant^{-1})	0.1	0.6
Co concentration (ng g^{-1} nodule dw)	45	105
No. bacteroids (× 10^9 g^{-1} nodule fw)	15	27
Cobalamin (ng g^{-1} nodule fw)	5.9	28.3
Leghemoglobin (mg g^{-1} nodule fw)	0.7	1.9

Source: Based on Dilworth et al. (1979).

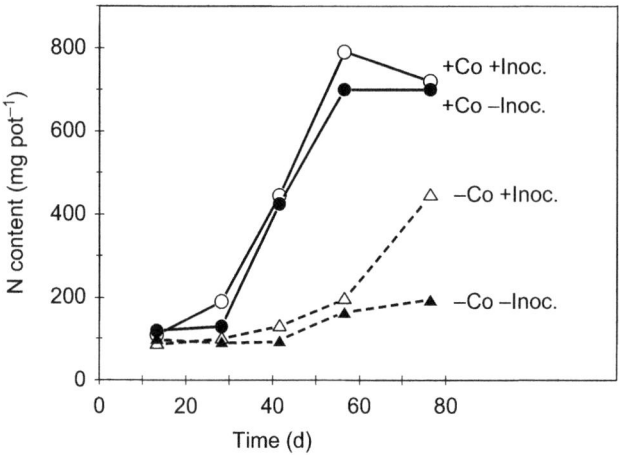

FIGURE 8.12 Time-course of N accumulation in *Lupinus angustifolius* grown in a low Co soil with or without the addition of Co and with or without inoculation with *Rhizobium*. Based on Dilworth et al. (1979).

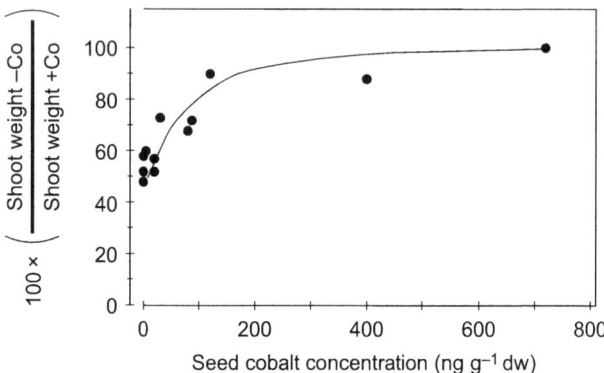

FIGURE 8.13 Relationship between Co concentration of seeds and shoot biomass of *Lupinus angustifolius* grown with or without Co supply. Based on Robson and Snowball (1987).

In large-seeded lupins, a Co concentration of ~100 ng Co g^{-1} seed dw is sufficient to prevent Co deficiency in plants grown in low-Co soils (Gladstones et al., 1977). Field responses to Co fertilization of nodulated legumes are rare but have been demonstrated, for example, on poor siliceous sandy soils (Ozanne et al., 1963).

Treating seeds with Co is an effective procedure for supporting N$_2$ fixation and growth of legumes on low-Co soils (Reddy and Raj, 1975). Foliar sprays can be effective, but less so than combining seed treatments and foliar sprays (Table 8.10). Depending on the concentrations used, seed treatment with Co may have inhibitory effects on nodulation and N$_2$ fixation; therefore, foliar sprays to enrich developing seeds with Co might be a useful agronomic strategy (Cortese et al., 2019). The effectiveness of foliar sprays indicates re-translocation of Co from leaves as has also been shown after the application of labeled Co to clover and alfalfa leaves (Handreck and Riceman, 1969). In the phloem, Co seems to be translocated largely as a negatively charged complex (Wiersma and van Goor, 1979).

Nonruminants, including humans, have a requirement for vitamin B$_{12}$ but not Co. On the other hand, Co is essential for ruminants because they depend on the rumen microflora to synthesize sufficient vitamin B$_{12}$. Cobalt deficiency is widespread in grazing ruminants on soils low in Co, and ruminants respond positively to increased availability of Co in soils or feed by increasing the production of B$_{12}$ (Akins et al., 2013; Huwait et al., 2015). In low-Co soils, Co application may not only enhance the N$_2$ fixation of legumes but also improve the nutritional quality of forage plants. The critical Co concentration for ruminants is about 0.07 mg kg^{-1} dw of forage, which is higher than the critical concentration for N$_2$ fixation in legumes.

A survey of angiosperm species has shown that leaf Co concentrations are typically <0.20 mg kg^{-1}, with a few species having up to 0.50 mg kg^{-1} (Watanabe et al., 2007; Pilon-Smits and Leduc, 2009). However, in a small number of plant species that are highly adapted to metalliferous soils, leaf Co concentration can reach several thousand milligrams per kilogram dw (Brookes and Malaisse, 1989). These species have high Co uptake rates; however, there is debate as to the role of uptake and translocation versus surface contamination in contributing to some of the values reported for Co hyperaccumulators (Faucon et al., 2007).

TABLE 8.10 Peanut yield, total N concentration, and number of nodules with different forms of Co application.

Co treatment	Pod yield (t ha^{-1})	Total N at maturity (g kg^{-1})	Nodulation (no. nodules plant^{-1})
Control (−Co)	1.23	2.4	91
Seed treatment	1.69	2.6	150
Foliar spray (2×)	1.75	3.1	123
Seed treatment + foliar spray (2×)	1.84	3.4	166

Source: Based on Reddy and Raj (1975).

8.5 Selenium

8.5.1 General

The chemistry of selenium (Se) has features in common with sulfur (S). Selenium, like S, can exist in the −2 (selenide Se^{2-}), 0 (elemental Se), +4 (selenite SeO$_3^{2-}$), and +6 (selenate SeO$_4^{2-}$) oxidation states. Selenium is present in soil in small amounts (typically ranging from 0.01 to 2 mg kg^{-1}), but high concentrations (>5 mg kg^{-1}) are found in seleniferous soils (Mayland et al., 1989). Soil pH and Eh influence the proportion of the chemical species of Se present in soil. Thermodynamic calculations show that the predominant form of Se is selenate in alkaline and well-aerated soils (pe + pH > 15), selenite in well-drained mineral soils with pH from acidic to neutral (7.5 < pe + pH < 15), and selenide under reduced soil conditions (pe + pH < 7.5) (Elrashidi et al., 1987).

Selenium is an essential micronutrient for animals, but the essentiality has not been established for higher plants (Sors et al., 2005; Terry et al., 2000). Deficiency of Se in humans is common; it has been estimated that up to 1 billion people worldwide may have an insufficient dietary intake of Se (Combs, 2001; Jones et al., 2017). Because plant-based foods are an important source of Se to humans and domestic animals, it is important to understand how plants take up and metabolize Se.

8.5.2 Uptake and translocation

Selenate is a chemical analog of sulfate; they compete for the same transporters during root uptake and, thus, selenate uptake can be strongly decreased by a high sulfate supply (Mikkelsen and Wan, 1990; Zayed and Terry, 1992). The affinity constants (K$_m$) for sulfate and selenate uptake into barley roots were found to be similar, 19 and 15 μM, respectively (Leggett and Epstein, 1956). Selenate also competitively inhibits sulfate uptake from nutrient solutions, but this inhibition is unlikely to be significant in soil-grown plants because the concentration of selenate in soil solution is much lower than that of sulfate. A number of selenate-resistant mutants of *Arabidopsis thaliana* have been identified; the phenotype of these mutants is caused by a mutation in the high-affinity sulfate transporter Sultr1;2 resulting in decreased uptake of both sulfate and selenate (Shibagaki et al., 2002). Sultr1;2 is localized in the root tip and cortex, and its expression is enhanced by S deficiency.

Sulfur-deficient plants upregulate the expression of sulfate transporter genes, leading to a strong increase in the capacity for selenate uptake (Li et al., 2008; Shinmachi et al., 2010). The nuclear protein MSA1 in *A. thaliana* and its homolog in rice (OsCADT1) act as a negative regulator of sulfate/selenate uptake and assimilation; mutations of the genes encoding these proteins result in a large accumulation of S and Se (Chen et al., 2020; Huang et al., 2016).

Plant species differ markedly in Se uptake and accumulation in the shoots and also in their capacity to tolerate high Se concentrations in the rooting medium and/or in the shoot tissue (Table 8.11). Based on these differences, plants can be classified into *Se accumulators* and *nonaccumulators*, and those between these two types as *Se indicators*. Some species can even accumulate more than 1 mg Se g^{-1} dw in leaves without suffering from Se toxicity; these species are called Se hyperaccumulators (Schiavon and Pilon-Smits, 2017; White, 2016). Some species of the genera *Astragalus*, *Xylorrhiza*, and *Stanleya* are typical Se hyperaccumulators capable of growing on high Se (seleniferous) soils without detrimental effect on growth, with shoot Se concentrations as high as 20–30 mg Se g^{-1} dw (Rosenfeld and Beath,

TABLE 8.11 Selenium concentrations in shoots of accumulator and nonaccumulator species growing on a soil with 2–4 mg Se kg^{-1}.

	Se concentration (mg kg^{-1} dw)
Astragalus pectinalus	4000
Stanleya pinnata	330
Gutierrezia fremontii	70
Zea mays	10
Helianthus annuus	2

Source: Based on Shrift (1969).

1964). However, within the genus *Astragalus* there are large differences among species and ecotypes in their capacity to accumulate Se, with the Se concentration in (hyper)accumulators being 100–200 times higher than in nonaccumulators (Shrift, 1969). Members of the Brassicaceae such as black mustard (*Brassica nigra* L.) and cauliflower (*Brassica oleracea* var. *botrytis* L.) also accumulate relatively large amounts of Se and may contain, and tolerate, several hundred μg Se g^{-1} shoot dw (Zayed and Terry, 1992). On the other hand, most agricultural and horticultural plant species are nonaccumulators (Shrift, 1969), and Se toxicity can occur even at concentrations below 100 μg Se g^{-1} (Mikkelsen et al., 1989).

White et al. (2007) compared selenate and sulfate uptake by 39 plant species grown in hydroponic culture under the same conditions. They found that, among the 37 species of Se nonaccumulators, there was a close positive relationship between leaf S and leaf Se concentrations (Fig. 8.14A), indicating that the accumulation of selenate and sulfate are strongly linked. Two Se hyperaccumulators (*Astragalus racemosus* and *Stanleya pinnata*) included in this study deviated from this relationship by having high Se concentration in the leaves. In general, Brassicaceae species accumulate Se because they have a large capacity to accumulate S.

The capacity to accumulate selenate relative to sulfate can be measured by the selenate/sulfate discrimination index, which is the molar ratio of [leaf Se/leaf S]/[solution selenate/solution sulfate]. Most of the plant species tested by White et al. (2007) have this discrimination index of around 1, indicating no clear discrimination between the two anions (Fig. 8.14B). Fig. 8.14B also includes data from the study of Bell et al. (1992) testing two plant species at different ratios of selenate/sulfate in the nutrient solution.

Some plant species (e.g., *Astragalus glycyphyllos*, *Beta vulgaris*, and *M. sativa*) have a Se/S discrimination index below 1 (Fig. 8.14A), suggesting that the transporter(s) have a higher affinity for sulfate than for selenate. In contrast, Se hyperaccumulators (*Astragalus racemosus*, *S. pinnata*, and *Astragalus bisulcatus*) have a discrimination index between 2 and 10, which is strong evidence that some transporters in these species have a higher selectivity for selenate than sulfate. Several *SULTR* genes are constitutively highly expressed in the roots of Se hyperaccumulators and may be involved in the preferential uptake or translocation of selenate (Schiavon and Pilon-Smits, 2017).

Selenite may be present in soil (Stroud et al., 2010), although its availability to plants is lower than that of selenate because of a stronger adsorption by iron oxides/hydroxides (Barrow and Whelan, 1989). Earlier studies suggested that selenite may enter root cells by diffusion (Terry et al., 2000). However, selenite uptake is, at least partly, energy-dependent (Arvy, 1993; Li et al., 2008). Selenite uptake is inhibited by the presence of phosphate in the medium but is enhanced by plant P deficiency (Hopper and Parker, 1999; Li et al., 2008), suggesting a possible involvement of the phosphate transporters in selenite uptake. Knockdown of *OsPT2* [by RNA interference (RNAi)] encoding a phosphate transporter in rice roots decreases, and the overexpression of *OsPT2* increases, the uptake of selenite (Zhang et al., 2014), indicating that selenite can be taken up by phosphate transporters. At low pH (<4.0), a significant proportion of selenite is undissociated as H_2SeO_3, and this neutral molecule can permeate through the rice NIP2;1 aquaporin channel (Zhao et al., 2010), which is a silicic acid transporter (see Section 8.3).

A marked difference between selenate and selenite is that the former is rapidly translocated from roots to shoots, whereas the latter is readily assimilated into organic forms in roots, with a limited root-to-shoot translocation (Asher et al., 1977; de Souza et al., 1998; Li et al., 2008). This difference, together with a stronger adsorption of selenite in soil, explains why applying selenate is more effective than selenite in increasing the Se concentration in crops.

Selenomethionine can be transported into the cells by OsNRT1.1B, which is a member of the rice peptide transporter family involved in nitrate uptake and transport (Zhang et al., 2019). Although the concentration of

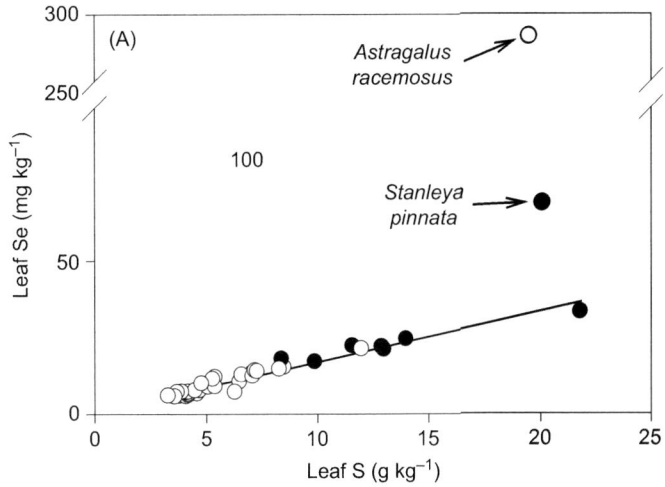
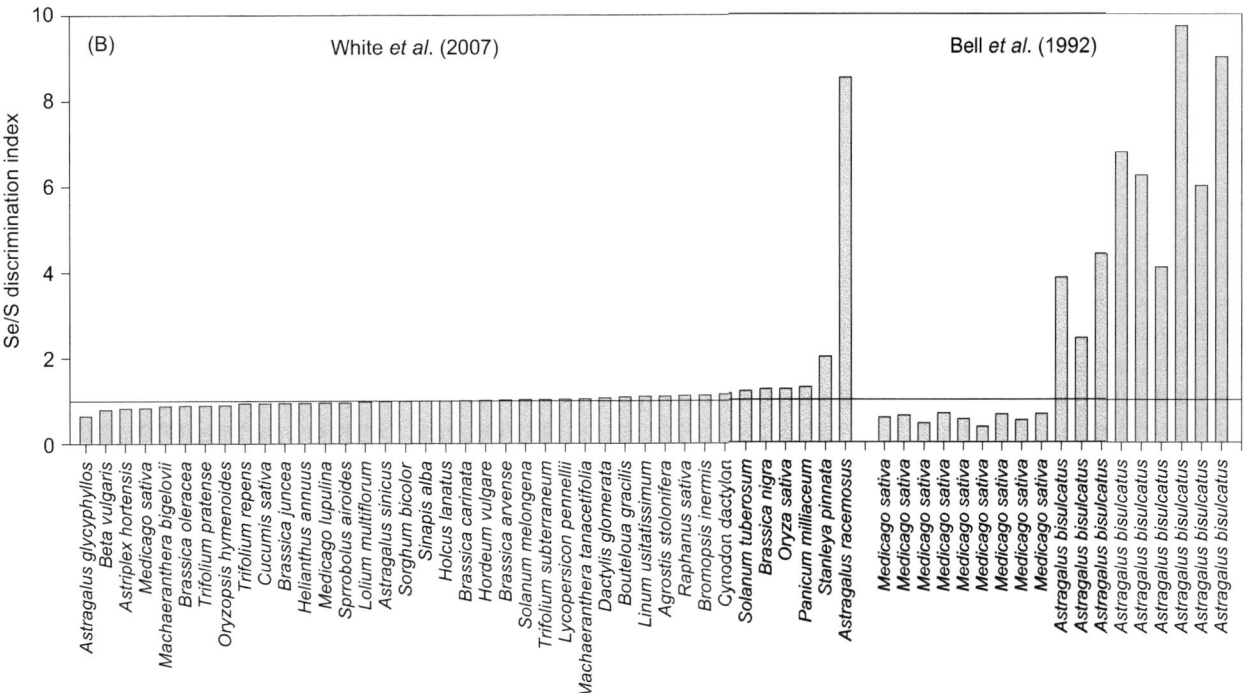

FIGURE 8.14 (A) Relationship between leaf Se and S concentrations in 39 plant species grown hydroponically with 0.91 mM sulfate and 0.63 μM selenate. Closed symbols represent Brassicaceae species. (B) Selenate/sulfate discrimination index calculated from the data of White et al. (2007) and Bell et al. (1992). *(A) Redrawn from White et al. (2007).*

selenomethionine in soil is likely to be low, it is the main form of selenoamino acid synthesized from selenite in the roots (Li et al., 2008; Zhang et al., 2019), and its translocation to the shoots depends on OsNRT1.1B and other amino acid transporters. When grown in a paddy field, rice plants overexpressing *OsNRT1.1B* accumulated significantly more Se in the grain than wild-type plants (Zhang et al., 2019).

8.5.3 Assimilation and metabolism

Selenium is assimilated in plants via the S assimilation pathway (Fig. 8.15) (Sors et al., 2005; Terry et al., 2000). In this pathway, selenate is activated by ATP sulfurylase to adenosine 5′-phosphoselenate (APSe), which is then reduced to selenite by adenosine 5′-phosphosulfate (APS) reductase. Activation of selenate seems to be the rate-limiting step for selenate reduction and can be overcome in transgenic plants overexpressing ATP sulfurylase (Pilon-Smits et al., 1999). This rate-limiting step also explains why selenite is more readily assimilated in plants

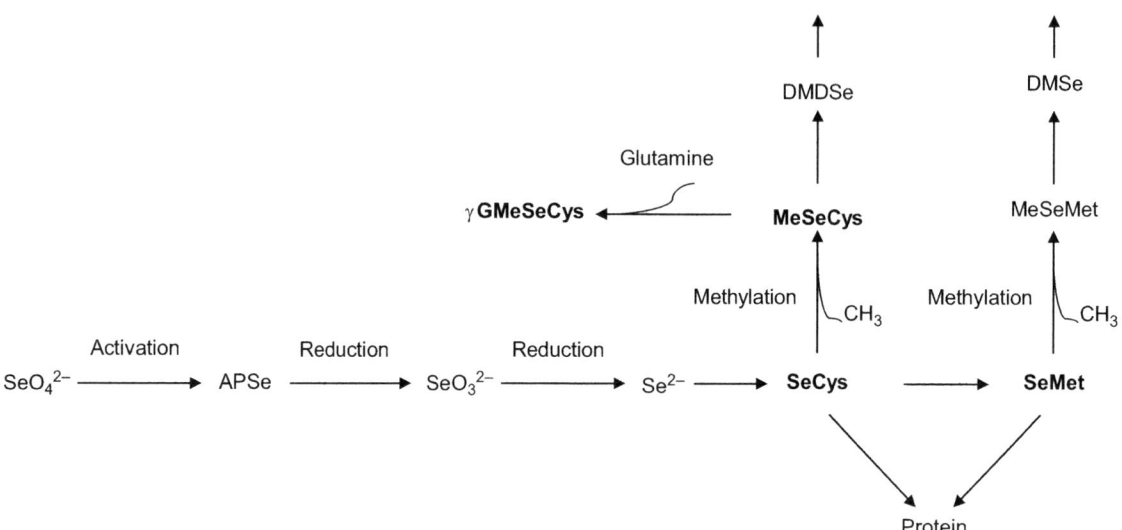

FIGURE 8.15 Selenium assimilation and metabolism in plants. Compounds in bold are common organic Se species in plants. *APSe*, Adenosine 5′-phosphoselenate; *DMDSe*, dimethyldiselenide; *DMSe*, dimethylselenide; *MeSeCys*, Se-methylselenocysteine; *MeSeMet*, methylselenomethionine; *SeCys*, selenocysteine; *SeMet*, selenomethionine; γ-*GMeSeCys*, γ-glutamyl-Se-methylselenocysteine. *Based on Terry et al. (2000) and Sors et al. (2005).*

than selenate (de Souza et al., 1998; Li et al., 2008). Selenite is further reduced to selenide possibly via nonenzymatic reactions using reduced glutathione (GSH) as a reductant (Sors et al., 2005). Selenide is assimilated into the amino acid selenocysteine catalyzed by the cysteine synthase complex and is further assimilated into selenomethionine via the methionine biosynthetic pathway. Both selenocysteine and selenomethionine are readily incorporated into proteins in nonaccumulator plants through the nonspecific substitution of cysteine and methionine, respectively; it is primarily this substitution that causes toxicity to plants because the proteins become nonfunctional or less functional than the corresponding proteins with S-containing amino acids (Brown and Shrift, 1982). Incorporation of selenoamino acids is presumably critical especially in enzymes with a sulfhydryl group (−SH) as a catalytic site.

Both selenocysteine and selenomethionine can be methylated and then are no longer able to substitute for cysteine and methionine in protein synthesis. Selenium hyperaccumulators differ from nonaccumulators in possessing a strong capacity to convert selenocysteine into various nonproteinogenic selenoamino acids, such as Se-methylselenocysteine and γ-glutamyl-Se-methylselenocysteine. Methylation of selenocysteine is an important mechanism of Se detoxification in hyperaccumulator plants. The methylation step is catalyzed by selenocysteine methyltransferase (Neuhierl and Bock, 1996), and the enzyme activity was found to correlate closely with the Se accumulation capacity in eight *Astragalus* species (Sors et al., 2005). In the Se hyperaccumulators *Astragalus bisulcatus* and *Stanleya pinnata*, young leaves contain high concentrations of Se, with Se-methylselenocysteine accounting for more than 70% of the total Se (Freeman et al., 2006; Pickering et al., 2003). Se-methylselenocysteine and its γ-glutamyl-derivatives are found in some edible plants, including garlic, onions, broccoli, and others of the *Allium* and *Brassica* genera, particularly when grown in Se-enriched environments; by contrast, cereal grains contain mainly selenomethionine (Rayman et al., 2008; Whanger, 2002).

Similarities between S and Se metabolism in plants also exist in the production of volatile compounds released by aerial parts (see Chapter 4). The main volatile selenide compound is dimethylselenide, of which selenomethionine is the precursor (Fig. 8.15). Plants can also volatilize dimethyldiselenide that is produced via methylation and subsequent oxidation of selenocysteine. The rates of Se volatilization vary considerably among crop species. With a supply of 20 μM selenate, rice, broccoli, and cabbage volatilized 200–350 μg Se m^{-2} leaf area day^{-1} compared to less than 15 μg Se m^{-2} leaf area day^{-1} in sugar beet, lettuce, and onion (Terry et al., 1992). In broccoli, which accumulates up to several hundred μg Se g^{-1} dw, the release rate of volatile Se compounds is about seven times higher at low S supply compared to high S supply because high S uptake inhibits selenate uptake (Zayed and Terry, 1992). Rhizosphere bacteria also appear to play an important role in Se volatilization (Terry et al., 2000). The capacity of plants and their associated rhizosphere microorganisms to volatilize Se, or to accumulate Se in the plant biomass, may be exploited as a phytoremediation strategy to clean up Se-contaminated soils (Terry et al., 2000).

8.5.4 Beneficial effects on plant growth

Selenium is essential for humans and animals because it is required in a number of enzymes (e.g., glutathione peroxidase) with selenocysteine as the catalytic site. Although there are glutathione peroxidase-like enzymes in vascular plants, they appear to contain cysteine, not selenocysteine, at the active site (Terry et al., 2000). Despite the lack of definitive evidence for Se essentiality in higher plants, there are reports that small doses of Se improve plant growth or reproduction. Small amounts of Se added to soil increased the growth of ryegrass, delayed senescence of lettuce, and enhanced resistance of lettuce and ryegrass to UV irradiation (Hartikainen, 2005). These effects appear to be associated with enhanced activity of glutathione peroxidase and reduced lipid peroxidation. Lyons et al. (2009) showed that an addition of 20–50 nM sodium selenite to the nutrient solution increased seed production of *Brassica rapa* by 43%, while having no effect on the total plant biomass. This experiment was conducted with the + Se treatment and the control plants in two separate growth chambers to preclude transfer of volatile Se from the treatment to the control.

High Se concentrations in (hyper)accumulator plants may offer a protection against herbivory (Galeas et al., 2008; Schiavon and Pilon-Smits, 2017), thus conferring an adaptive advantage. This observation supports the elemental defense hypothesis put forward to explain the evolution of a metal or metalloid hyperaccumulation trait (Boyd, 2007). However, high Se concentrations in plants may cause disorders (Se toxicity) in animals grazing on native vegetation of seleniferous soils (Brown and Shrift, 1982).

8.5.5 Biofortification

A considerable percentage of the population in many countries has inadequate intake of Se (Combs, 2001; Rayman, 2008; Schiavon et al., 2020). Selenium enters the food chain primarily through plant uptake from the soil. Selenium concentration in food crops is highly variable as a result of the variation in the underlying geology and soil conditions; human intake of Se also varies considerably between countries and regions, reflecting the variation in the Se concentrations in foods (Combs, 2001; Rayman, 2008). The minimum Se concentration for animals and humans is about 50–100 μg Se kg^{-1} dw in fodder/food (Gissel-Nielsen et al., 1984). A strategy to increase human intake of Se is to biofortify crops by either using Se fertilizers (agronomic biofortification) or genetic improvement in crop Se accumulation (Prom-u-thai et al., 2020; Schiavon et al., 2020).

Agronomic biofortification has been practiced in Finland since the mid-1980s, with mandatory additions of small amounts of Se as sodium selenate to all multinutrient fertilizers (6–16 mg Se kg^{-1} fertilizer). This practice has raised Se concentrations in cereals, vegetables, silage, and animal products such as meat and milk, more than doubling the Se dietary intake by the Finnish population (Alfthan et al., 2015; Hartikainen, 2005). Compared to direct Se supplementation, agronomic biofortification is considered to be advantageous because inorganic Se is assimilated by plants into organic forms that are more bioavailable to humans. Se-enriched wheat flour contains predominantly selenomethionine (~80%), with selenocysteine, Se-methylselenocysteine, and inorganic Se present in small proportions (Hart et al., 2011). Importantly, plants act as an effective buffer that can prevent accidental excessive Se intake by humans that may occur with direct supplementation (Hartikainen, 2005). Unlike the micronutrients Fe and Zn, it is relatively easy to increase Se concentrations in food crops by fertilization because selenate is highly available to plants and is readily transported from roots to shoots. This is demonstrated in field studies with winter wheat, showing that Se concentration in the grain increased linearly from 30 μg kg^{-1} in the control to 2600 μg kg^{-1} in the treatment receiving 100 g Se ha^{-1} as Na selenate (Broadley et al., 2010) (Fig. 8.16). Total recovery of applied Se by the wheat crop was 20%–35%; the remainder was likely leached out of the rooting depth, leaving little residual Se to the subsequent crop (Broadley et al., 2010; Stroud et al., 2010). In general, applications of selenate are more effective in increasing crop Se content than selenite (Hawkesford and Zhao, 2007; Mikkelsen et al., 1989). Foliar spray of Se as selenate is highly effective in increasing grain Se concentrations of wheat and rice (Prom-u-thai et al., 2020; Zou et al., 2019).

Biofortification of crops with Se through genetic improvement requires genetic variation in the uptake and/or assimilation of Se. There is considerable genetic variation in the concentration of Se in the edible parts of cereals, legumes, and vegetables, although the genetic bases for these variations remain unclear (White, 2016). By contrast, a relatively small genetic variation in grain Se concentration was found among the genotypes of bread wheat tested (Lyons et al., 2005; Zhao et al., 2009).

Many studies have reported several-fold increases in Se uptake by transgenic plants overexpressing various genes involved in S/Se assimilation, but these studies were aimed generally at enhancing Se uptake for phytoremediation of Se-contaminated soils (Pilon-Smits and Leduc, 2009). In rice (Section 8.5.2), the overexpression of *OsNRT 1.1B*

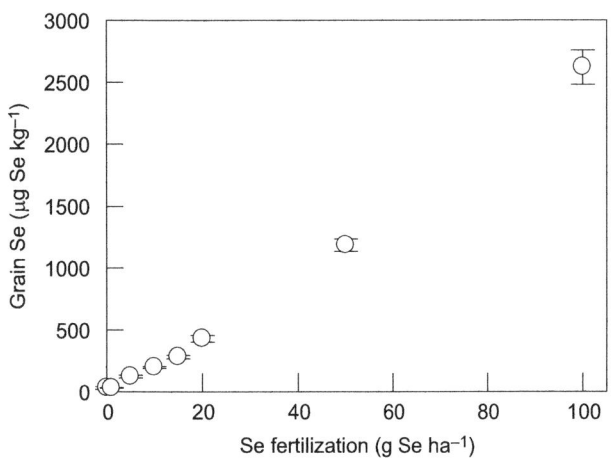

FIGURE 8.16 Relationship between Se fertilization and grain Se concentration of winter wheat. Means ± SE ($n = 4$). *Redrawn from Broadley et al. (2010).*

(Zhang et al., 2019) or mutation of *OsCADT1* (Chen et al., 2020) can lead to substantial increases in Se accumulation in grain without affecting grain yield.

8.6 Aluminum

Aluminum (Al) is the most abundant metal in the Earth's crust at about 8% of total weight. Aluminum concentrations in mineral soil solutions are usually below 1 mg L^{-1} (~37 μM) at pH values higher than 5.5 but rise strongly at lower pH. Even though Al is toxic to most species at relatively low concentrations, causing rapid inhibition of root elongation (Section 18.2.3), some plant species (accumulators) tolerate high Al concentrations in their tissue. Among those, the beneficial effect of Al on plant growth has been observed in tea, Melastoma, hydrangea, and *Quercus serrata* Thunb. These species are usually adapted to acid soils and can accumulate Al in the shoots. In the tea plant, as one of the most Al-tolerant crop species, growth stimulation has been observed at Al concentrations in nutrient solution as high as 6.4 mM (Konishi et al., 1985). Root elongation was enhanced 2.5-fold in tea grown in a nutrient solution containing 0.5 mM Al at pH 4.3 (Ghanati et al., 2005). The root biomass of *Q. serrata* increased as Al concentrations were increased up to 2.5 mM at pH 3.5 (Tomioka et al., 2005). In Melastoma, Al increased the root activity and stimulated the elongation of root cells (Watanabe et al., 2005). The blue color of the petals in hydrangea is caused by Al accumulation (Ma et al., 1997).

The exact mechanisms underlying the beneficial effects of Al are unknown. A study with *Q. serrata* showed that Al-induced growth enhancement is not due to the amelioration of H$^+$ toxicity by Al (Tomioka et al., 2005). One of the positive effects of Al on plant growth has been ascribed to its protective effects against Mn toxicity in plants grown in acidic soils in which Mn toxicity may occur (Fernando and Lynch, 2015). The organic acid anions released from roots in response to Al (see Sections 14.6.3 and 18.3.2) have been suggested to suppress root uptake of Mn by complexing Mn in the growth medium (Yang et al., 2009). The formation of Mn plaques on root surfaces and increase in cell wall Mn binding have been discussed as further potential mechanisms contributing to the alleviation of Mn toxicity by Al (Wang et al., 2015).

Aluminum has been shown to improve root uptake of macronutrients such as N, P, K, and Mg, which might be a further mechanism involved in the stimulation of root growth by Al (Bojórquez-Quintal et al., 2017). Upon exposure of roots to Al, the Mg transporters are activated, improving Mg uptake in Al-resistant plant genotypes (Chen and Ma, 2013). Aluminum increased the activities of superoxide dismutase, catalase, and ascorbate peroxidase in the roots of intact tea plants and cultured cells (Ghanati et al., 2005). Al-induced increase in the activity of these antioxidative enzymes may improve membrane integrity and delay lignification and aging, resulting in growth stimulation. In Melastoma, the primary reason for the Al-induced growth enhancement is proposed to be the alleviation of Fe toxicity (Watanabe et al., 2006). Growth was enhanced by Al more strongly under excess than optimal Fe, and the Fe concentration in both roots and shoots was decreased by Al. However, in plants sensitive to Al, such as soybean, increasing Al supply (up to 40 μM) in the presence of high Fe (40 or 80 μM) strongly depressed root elongation and increased lipid peroxidation in root tips (Cakmak and Horst, 1991). Recently, it has been suggested that Al is required for the maintenance of DNA integrity and root meristematic activity in tea (Sun et al., 2020c). The above overview of the literature indicates that the physiological and molecular mechanisms underpinning the beneficial effects of Al are still poorly understood.

8.7 Other elements

The requirement for elements such as iodine and vanadium (V) is fairly well established for fungi (V) and certain non-vascular plant species, such as marine algae (iodine) freshwater algae (V). Reports of the stimulation of growth of vascular plants by these elements are rare and vague. Recent studies show beneficial effects of iodine on the growth of various vascular plants, probably by improving their stress tolerance and antioxidant capacity (Medrano-Macías et al., 2016). Iodine has been found to result in positive effects on biomass production as well as flowering of arabidopsis at low application rates (i.e., 0.20 and 10 μM) (Kiferle et al., 2019, 2021). These authors suggested that iodine might be a structural component of various plant proteins and involved in the expression of genes affecting plant defense responses.

Low concentrations of V and cerium (Ce) in the growth medium were found to improve the growth and stress tolerance of tomato and pepper plants (Saldaña-Sánchez et al., 2019). A further example is the effect of titanium (Ti) on the growth (Pais, 1983), enzyme activities, and photosynthesis (Dumon and Ernst, 1988) of various crop species.

More recently, interest has increased in the rare earth elements, lanthanum (La) and cerium (Ce), for the enhancement of plant growth. Mixtures of these two elements are used on a large scale in China as foliar sprays or seed treatment of agricultural and horticultural crops. The amounts supplied are in the range typical for micronutrients. There are reports of substantial increases in plant growth and yield under field conditions that require more careful documentation and reproduction under controlled conditions.

References

Ahmad, R., Zaheer, S.H., Ismail, S., 1992. Role of silicon in salt tolerance of wheat (*Triticum aestivum* L.). Plant Sci. 85, 43–50.
Ahmed, S., Evans, H.J., 1960. Cobalt: a micronutrient element for the growth of soybean plants under symbiotic conditions. Soil Sci. 90, 205–210.
Akins, M.S., Bertics, S.J., Socha, M.T., Shaver, R.D., 2013. Effects of cobalt supplementation and vitamin B12 injections on lactation performance and metabolism of Holstein dairy cows. J. Dairy Sci. 96, 1755–1768.
Aktas, H., Abak, K., Cakmak, I., 2006. Genotypic variation in the response of pepper to salinity. Sci. Hortic. 110, 260–266.
Alfthan, G., Eurola, M., Ekholm, P., Venäläinen, E.R., Root, T., Korkalainen, K., et al., 2015. Effects of nationwide addition of selenium to fertilizers on foods, and animal and human health in Finland: From deficiency to optimal selenium status of the population. J. Trace Elem. Med. Biol. 31, 142–147.
Al-Karaki, G.N., 2000. Growth, water use efficiency, and sodium and potassium acquisition by tomato cultivars grown under salt stress. J. Plant Nutr. 23, 1–8.
Almeida, J.C.R., Laclau, J.P., Goncalves, J.L.D.M., Ranger, J., Saint-Andre, L., 2010. A positive growth response to NaCl applications in Eucalyptus plantations established on K-deficient soils. Forest Ecol. Manag. 259, 1786–1795.
Almeida, P., Boer, G.J.D., Boer, A.H.D., 2014. Differences in shoot Na$^+$ accumulation between two tomato species are due to differences in ion affinity of HKT1;2. J. Plant Physiol. 171, 438–447.
Anderson, D.L., 1991. Soil and leaf nutrient interactions following application of calcium silicate slag to sugarcane. Fert. Res. 30, 9–18.
Aoki, N., Ohnishi, J., Kanai, R., 1992. Two different mechanisms for transport of pyruvate into mesophyll chloroplasts of C$_4$ plants - a comparative study. Plant Cell Physiol. 33, 805–809.
Arvy, M.P., 1993. Selenate and selenite uptake and translocation in bean plants (*Phaseolus vulgaris*). J. Exp. Bot. 44, 1083–1087.
Asher, C.J., Butler, G.W., Peterson, P.J., 1977. Selenium transport in root systems of tomato. J. Exp. Bot. 28, 279–291.
Barlog, P., Szczepaniak, W., Grzebisz, W., Poglodzinski, R., 2018. Sugar beet response to different K, Na and Mg ratios in applied fertilizers. Plant Soil Environ. 64, 173–179.
Barrow, N.J., Whelan, B.R., 1989. Testing a mechanistic model. 7. The effects of pH and of electrolyte on the reaction of selenite and selenate with a soil. J. Soil Sci. 40, 17–28.
Belanger, R.R., Benhamou, N., Menzies, J.G., 2003. Cytological evidence of an active role of silicon in wheat resistance to powdery mildew (*Blumeria graminis f. sp tritici*). Phytopathology 93, 402–412.
Bell, P.F., Parker, D.R., Page, A.L., 1992. Contrasting selenate-sulfate interactions in selenium-accumulating and nonaccumulating plant species. Soil Sci. Soc. Am. J. 56, 1818–1824.
Bojórquez-Quintal, E., Escalante-Magaña, C., Echevarría-Machado, I., Martínez-Estévez, M., 2017. Aluminum, a friend or foe of higher plants in acid soils. Front. Plant Sci. 8, 1767.
Boyd, R.S., 2007. The defense hypothesis of elemental hyperaccumulation: status, challenges and new directions. Plant Soil 293, 153–176.
Broadley, M.R., Alcock, J., Alford, J., Cartwright, P., Foot, I., Fairweather-Tait, S.J., et al., 2010. Selenium biofortification of high-yielding winter wheat (*Triticum aestivum* L.) by liquid or granular Se fertilisation. Plant Soil 332, 5–18.

Brookes, R.R., Malaisse, F., 1989. Metal-enriched sites in South Central Africa. In: Shaw, A.J. (Ed.), Heavy Metal Tolerance in Plants: Evolutionary Aspects. CRC Press Inc., Boca Raton, FL, USA, pp. 53–73.

Brown, T.A., Shrift, A., 1982. Selenium: toxicity and tolerance in higher plants. Biol. Rev. Cambridge Philosophic. Soc. 57, 59–84.

Brownell, P.F., 1965. Sodium as an essential micronutrient element for a higher plant (*Atriplex vesicaria*). Plant Physiol. 40, 460–468.

Brownell, P.F., Crossland, C.J., 1972. The requirement for sodium as a micronutrient by species having the C_4 dicarboxylic photosynthetic pathway. Plant Physiol. 49, 794–797.

Brownell, P.F., Crossland, C.J., 1974. Growth responses to sodium by *Bryophyllum tibuflorum* under conditions inducing crassulacean acid metabolism. Plant Physiol. 54, 416–417.

Cakmak, I., Horst, W.J., 1991. Effect of aluminium on lipid peroxidation, superoxide dismutase, catalase, and peroxidase activities in root tips of soybean (*Glycine max*). Physiol. Plant. 83, 463–468.

Carneiro, J.S.D.S., Silva, P.S.S., Santos, A.C.M.D., Freitas, G.A.D., Santos, A.C.D., Silva, R.R.D., 2017. Mombaca grass responds to partial replacement of K^+ by Na^+ with supplemental Ca^{2+} addition in low fertility soil. J. Agr. Sci. 9, 209–219.

Casey, W.H., Kinrade, S.D., Knight, C.T.G., Rains, D.W., Epstein, E., 2003. Aqueous silicate complexes in wheat, *Triticum aestivum* L. Plant Cell Environ. 27, 51–54.

Che, J., Yamaji, N., Shao, J.F., Ma, J.F., Shen, R.F., 2016. Silicon decreases both uptake and root-to-shoot translocation of manganese in rice. J. Exp. Bot. 67, 1535–1544.

Chen, Z.C., Ma, J.F., 2013. Magnesium transporters and their role in Al tolerance in plants. Plant Soil 368, 51–56.

Chen, S., Polle, A., 2010. Salinity tolerance of *Populus*. Plant Biol. 12, 317–333.

Chen, J., Huang, X.-Y., Salt, D.E., Zhao, F.J., 2020. Mutation in OsCADT1 enhances cadmium tolerance and enriches selenium in rice grain. New Phytol. 226, 838–850.

Cheng, G., Wang, L., Lan, H., 2016. Cloning of PEPC-1 from a C_4 halophyte *Suaeda aralocaspica* without Kranz anatomy and its recombinant enzymatic activity in responses to abiotic stresses. Enzyme Microb. Technol. 83, 57–67.

Chiba, Y., Mitani, N., Yamaji, N., Ma, J.F., 2009. HvLsi1 is a silicon influx transporter in barley. Plant J. 57, 810–818.

Coban, A., Akhoundnejad, Y., Dere, S., Dasgan, H.Y., 2020. Impact of salt-tolerant rootstock on the enhancement of sensitive tomato plant responses to salinity. HortScience 55, 35–39.

Combs, G.F., 2001. Selenium in global food systems. Brit. J. Nutr. 85, 517–547.

Cortese, D., Pierozan Junior, C., Walter, J.B., Cortese, D., de Oliveira, S.M., 2019. Enrichment, quality, and productivity of soybean seeds with cobalt and molybdenum applications. J. Seed Sci. 41, 144–150.

Costigan, P.A., Mead, G.P., 1987. The requirements of cabbage and lettuce seedlings for potassium in the presence and absence of sodium. J. Plant Nutr. 10, 385–401.

Datnoff, L.E., Deren, C.W., Snyder, G.H., 1997. Silicon fertilization for disease management of rice in Florida. Crop Protect. 16, 525–531.

Davenport, R., James, R.A., Zakrisson-Plogander, A., Tester, M., Munns, R., 2005. Control of sodium transport in durum wheat. Plant Physiol. 137, 807–818.

de Souza, M.P., Pilon-Smits, E.A.H., Lytle, C.M., Hwang, S., Tai, J., Honma, T.S.U., et al., 1998. Rate-limiting steps in selenium assimilation and volatilization by Indian mustard. Plant Physiol. 117, 1487–1494.

Delwiche, C.C., Johnson, C.M., Reisenauer, H.M., 1961. Influence of cobalt on nitrogen fixation by *Medicago*. Plant Physiol. 36, 73–78.

Deshmukh, R., Vivancos, J., Guerin, V., Sonah, H., Labbe, C., Belzile, F., et al., 2013. Identification and functional characterization of silicon transporters in soybean using comparative genomics of major intrinsic proteins in Arabidopsis and rice. Plant Mol. Biol. 83, 303–315.

Dilworth, M.J., Robson, A.D., Chatel, D.L., 1979. Cobalt and nitrogen fixation in *Lupinus angustifolius* L. II. Nodule formation and functions. New Phytol. 83, 63–79.

Divte, P., Yadav, P., Jain, P.K., Paul, S., Singh, B., 2019. Ethylene regulation of root growth and phytosiderophore biosynthesis determines iron deficiency tolerance in wheat (*Triticum* spp). Environ. Exp. Bot. 162, 1–13.

Dumon, J.C., Ernst, W.H.O., 1988. Titanium in plants. J. Plant Physiol. 133, 203–209.

Durrant, M.J., Draycott, A.P., Milford, G.F.J., 1978. Effect of sodium fertilizer on water status and yield of sugar beet. Ann. Appl. Biol. 88, 321–328.

Elawad, S.H., Stret, J.J., Gascho, G.J., 1982. Response of sugarcane to silicate source and rate. II. Leaf freckling and nutrient content. Agron. J. 74, 484–487.

Elrashidi, M.A., Adriano, D.C., Workman, S.M., Lindsay, W.L., 1987. Chemical equilibria of selenium in soils: a theoretical development. Soil Sci. 144, 141–152.

Epstein, E., 1999. Silicon. Annu. Rev. Plant Physiol. Plant Mol. Biol. 50, 641–664.

Fahmy, G.M., 2013. Ecophysiology of the holoparasitic angiosperm *Cistanche phelypaea* (Orobanchaceae) in a coastal salt marsh. Turk. J. Bot. 37, 908–919.

Faucon, M.-P., Ngoy Shutcha, M., Meerts, P., 2007. Revisiting copper and cobalt concentrations in supposed hyperaccumulators from SC Africa: influence of washing and metal concentrations in soil. Plant Soil 301, 29–36.

Faust, F., Schubert, S., 2016. Protein synthesis is the most sensitive process when potassium is substituted by sodium in the nutrition of sugar beet (*Beta vulgaris*). Plant Physiol. Biochem. 107, 237–247.

Faust, F., Schubert, S., 2017. In vitro protein synthesis of sugar beet (*Beta vulgaris*) and maize (*Zea mays*) is differentially inhibited when potassium is substituted by sodium. Plant Physiol. Biochem. 118, 228–234.

Fauteux, F., Remus-Borel, W., Menzies, J.G., Belanger, R.R., 2005. Silicon and plant disease resistance against pathogenic fungi. FEMS Microbiol. Lett. 249, 1–6.

Fernando, D.R., Lynch, J.P., 2015. Manganese phytotoxicity: new light on an old problem. Ann. Bot. 116, 313–319.
Flowers, T.J., Hajibagheri, M.A., 2001. Salinity tolerance in *Hordeum vulgare*: ion concentrations in root cells of cultivars differing in salt tolerance. Plant Soil 231, 1–9.
Freeman, J.L., Zhang, L.H., Marcus, M.A., Fakra, S., McGrath, S.P., Pilon-Smits, E.A.H., 2006. Spatial imaging, speciation, and quantification of selenium in the hyperaccumulator plants *Astragalus bisulcatus* and *Stanleya pinnata*. Plant Physiol. 142, 124–134.
Furumoto, T., Yamaguchi, T., Ohshima-Ichie, Y., Nakamura, M., Tsuchida-Iwata, Y., Shimamura, M., et al., 2011. A plastidial sodium-dependent pyruvate transporter. Nature 476, 472–475.
Galeas, M.L., Klamper, E.M., Bennett, L.E., Freeman, J.L., Kondratieff, B.C., Quinn, C.F., et al., 2008. Selenium hyperaccumulation reduces plant arthropod loads in the field. New Phytol. 177, 715–724.
García-Sánchez, M.J., Paz Jaime, M., Ramos, A., Sanders, D., Fernández, J.A., 2000. Sodium-dependent nitrate transport at the plasma membrane of leaf cells of the marine higher plant *Zostera marina* L. Plant Physiol. 122, 879–886.
Gattward, J.N., Almeida, A.A.F., Souza-Junior, J.O., Gomes, F.P., Kronzucker, H.J., 2012. Sodium-potassium synergism in *Theobroma cacao*: stimulation of photosynthesis, water-use efficiency and mineral nutrition. Physiol. Plant. 146, 350–362.
Geissler, N., Hussin, S., El-Far, M.M.M., Koyro, H.W., 2015. Elevated atmospheric CO_2 concentration leads to different salt resistance mechanisms in a C_3 (*Chenopodium quinoa*) and a C_4 (*Atriplex nummularia*) halophyte. Environ. Exp. Bot. 118, 67–77.
Ghanati, F., Morita, A., Yokota, H., 2005. Effects of aluminum on the growth of tea plant and activation of antioxidant system. Plant Soil 276, 133–141.
Gissel-Nielsen, G., Gupta, U.C., Lamand, M., Westermarck, T., 1984. Selenium in soils and plants and its importance in livestock and human nutrition. Adv. Agron. 37, 397–460.
Gladstones, J.S., Loneragan, J.F., Goodchild, N.A., 1977. Field responses to cobalt and molybdenum by different legume species, with interferences on the role of cobalt in legume growth. Aust. J. Agric. Res. 28, 619–628.
Gong, H.J., Randall, D.P., Flowers, T.J., 2006. Silicon deposition in the root reduces sodium uptake in rice (*Oryza sativa* L.) seedlings by reducing bypass flow. Plant Cell Environ. 29, 1970–1979.
Gonzales, E.M., Arrese-Igor, C., Aparicio-Tejo, P.M., Royuela, M., Koyro, H.-W., 2002. Osmotic adjustment in different leaf structures of semileafless pea (*Pisum sativum* L.) subjected to water stress. *Develop*. Plant Soil Sci. 92, 374–375.
Goto, M., Ehara, H., Karita, S., Takabe, K., Ogawa, N., Yamada, Y., et al., 2003. Protective effect of silicon on phenolic biosynthesis and ultraviolet spectral stress in rice crop. Plant Sci. 164, 349–356.
Grieve, C.M., Poss, J.A., Grattan, S.R., Suarez, D.L., Benes, S.E., Robinson, P.H., 2004. Evaluation of salt-tolerant forages for sequential water reuse systems. II. Plant-ion relations. Agric. Water Manag. 70, 121–135.
Grof, C.P.L., Johnston, M., Brownell, P.F., 1989. Effect of sodium nutrition on the ultrastructure of chloroplasts of C_4 plants. Plant Physiol. 89, 539–543.
Guo, K.-M., Babourina, O., Rengel, Z., 2009. Na^+/H^+ antiporter activity of the *SOS1* gene: lifetime imaging analysis and electrophysiological studies on Arabidopsis seedlings. Physiol. Plant. 137, 155–165.
Hajiboland, R., Joudmand, A., 2009. The K/Na replacement and function of antioxidant defence system in sugar beet (*Beta vulgaris* L.) cultivars. Acta Agric. Scand. Sect. B-Soil Plant Sci. 59, 246–259.
Hajiboland, R., Dashtebani, F., Aliasgharzad, N., 2015. Physiological responses of halophytic C_4 grass *Aeluropus littoralis* to salinity and arbuscular mycorrhizal fungi colonization. Photosynthetica 53, 572–584.
Hampe, T., Marschner, H., 1982. Effect of sodium on morphology, water relations and net photosynthesis in sugar beet leaves. Z. Pflanzenphysiol. 108, 151–162.
Handreck, K.K., Riceman, D.S., 1969. Cobalt distribution in several pasture species grown in culture solutions. Aust. J. Agric. Res. 20, 213–226.
Hart, D.J., Fairweather-Tait, S.J., Broadley, M.R., Dickinson, S.J., Foot, I., Knott, P., et al., 2011. Selenium concentration and speciation in biofortified flour and bread: retention of selenium during grain biofortification, processing and production of Se-enriched food. Food Chem. 126, 1771–1778.
Hartikainen, H., 2005. Biogeochemistry of selenium and its impact on food chain quality and human health. J. Trace. Elem. Med. Biol. 18, 309–318.
Hawker, J.S., Marschner, H., Downton, W.J.S., 1974. Effect of sodium and potassium on starch synthesis in leaves. Aust. J. Plant Physiol. 1, 491–501.
Hawkesford, M.J., Zhao, F.J., 2007. Strategies for increasing the selenium content of wheat. J. Cereal Sci. 46, 282–292.
Hecht-Buchholz, C., Pflüger, R., Marschner, H., 1971. Einfluss von Natriumchlorid auf Mitochondrienzahl und Atmung von Maiswurzelspitzen. Z. Pflanzenphysiol. 65, 410–417.
Heine, G., Tikum, G., Horst, W.J., 2007. The effect of silicon on the infection by and spread of *Pythium aphanidermatum* in single roots of tomato and bitter gourd. J. Exp. Bot. 8, 569–577.
Hodson, M.J., Sangster, A.G., 1989. Subcellular localization of mineral deposits in the roots of wheat (*Triticum aestivum* L.). Protoplasma 151, 19–32.
Hodson, M.J., White, P.J., Mead, A., Broadley, M.R., 2005. Phylogenetic variation in the silicon composition of plants. Ann. Bot. 96, 1027–1046.
Hopper, J.L., Parker, D.R., 1999. Plant availability of selenite and selenate as influenced by the competing ions phosphate and sulfate. Plant Soil 210, 199–207.
Horiguchi, T., Morita, S., 1987. Mechanism of manganese toxicity and tolerance of plants. VI. Effect of silicon on alleviation of manganese toxicity of barley. J. Plant Nutr. 10, 2299–2310.
Horst, W.J., Marschner, H., 1978. Effect of silicon on manganese tolerance of bean plants (*Phaseolus vulgaris* L.). Plant Soil 50, 287–303.

Horst, W.J., Fecht, M., Naumann, A., Wissemeier, A.H., Maier, P., 1999. Physiology of manganese toxicity and tolerance in *Vigna unguiculata* (L.) Walp. J. Plant Nutr. Soil Sci. 162, 263–274.

Huang, X.Y., Chao, D.Y., Koprivova, A., Danku, J., Wirtz, M., Mueller, S., et al., 2016. Nuclear localised MORE SULPHUR ACCUMULATION1 epigenetically regulates sulphur homeostasis in *Arabidopsis thaliana*. PLoS Genet. 12, e1006298.

Huang, S., Wang, P., Yamaji, N., Ma, J.F., 2020. Plant nutrition for human nutrition: Hints from rice research and future perspectives. Mol. Plant 13, 825–835.

Huhtanen, P., Ahvenjärvi, S., Heikkilä, T., 2000. Effects of sodium sulphate and potassium chloride fertilizers on the nutritive value of timothy grown on different soils. Agric. Food Sci. 9, 105–119.

Hussin, S., Geissler, N., Koyro, H.W., 2013. Effect of NaCl salinity on *Atriplex nummularia* (L.) with special emphasis on carbon and nitrogen metabolism. Acta Physiol. Plant. 35, 1025–1038.

Huwait, E.A., Kumosani, T.A., Moselhy, S.S., Mosaoa, R.M., Yaghmoor, S.S., 2015. Relationship between soil cobalt and vitamin B12 levels in the liver of livestock in Saudi Arabia: Role of competing elements in soils. Afr. Health Sci. 15, 993–1008.

Hylton, L.-O., Ulrich, A., Cornelius, D.R., 1967. Potassium and sodium interrelations in growth and mineral content of Italian ryegrass. Agron. J. 59, 311–314.

Idris, M., Hossain, M.M., Choudhury, F.A., 1975. The effect of silicon on lodging of rice in presence of added nitrogen. Plant Soil 43, 691–695.

Inal, A., Pilbeam, D.J., Gunes, A., 2009. Silicon increases tolerance to boron toxicity and reduces oxidative damage in barley. J. Plant Nutr. 32, 112–128.

Iwasaki, K., Matsumura, A., 1999. Effect of silicon on alleviation of manganese toxicity in pumpkin (*Cucurbita moschata* Duch cv. Shintosa). Soil Sci. Plant Nutr. 45, 909–920.

Johnston, M., Grof, C.P.L., Brownell, P.F., 1984. Responses to ambient CO_2 concentration by sodium-deficient C_4 plants. Aust. J. Plant Physiol. 11, 137–141.

Johnston, M., Grof, C.P.L., Brownell, P.F., 1988. The effect of sodium nutrition on the pool sizes of intermediates of the C_4 photosynthetic pathway. Aust. J. Plant Physiol. 15, 749–760.

Jones, L.H.-P., Handreck, K.A., 1965. Studies of silica in the oat plant. III. Uptake of silica from soils by the plant. Plant Soil 23, 79–96.

Jones, L.H.-P., Handreck, K.A., 1969. Uptake of silica by *Trifolium incarnatum* in relation to the concentration in the external solution and to transpiration. Plant Soil 30, 71–80.

Jones, G.D., Droz, B., Greve, P., Gottschalk, P., Proffet, D., McGrath, S.P., et al., 2017. Selenium deficiency risk predicted to increase under future climate change. Proc. Natl. Acad. Sci. USA 114, 2848–2853.

Jung, B., Ludewig, F., Schulz, A., Meissner, G., Wostefeld, N., Flugge, U.I., et al., 2015. Identification of the transporter responsible for sucrose accumulation in sugar beet taproots. Nat. Plants 1, 14001.

Kaburagi, E., Morikawa, Y., Yamada, M., Fujiyama, H., 2014. Sodium enhances nitrate uptake in Swiss chard (*Beta vulgaris* var. *cicla* L.). Soil Sci. Plant Nutr. 60, 651–658.

Karthika, K., Bell, R., Ma, Q., 2014. Wheat responses to sodium vary with potassium use efficiency of cultivars. Front. Plant Sci. 5, 631.

Kelly, C.K., Horning, K., 1999. Acquisition order and resource value in *Cuscuta attenuata*. Proc. Natl. Acad. Sci. USA 96, 13219–13222.

Kempa, S., Krasensky, J., Dal Santo, S., Kopka, J., Jonak, C., 2008. A central role of abscisic acid in stress-regulated carbohydrate metabolism. PLoS ONE 3, e3935.

Khedr, A.H.A., Serag, M.S., Nemat-Alla, M.M., El-Naga, A.Z.A., Nada, R.M., Quick, W.P., et al., 2011. Growth stimulation and inhibition by salt in relation to Na^+ manipulating genes in xero-halophyte *Atriplex halimus* L. Acta Physiol. Plant. 33, 1769–1784.

Kiferle, C., Ascrizzi, R., Martinelli, M., Gonzali, S., Mariotti, L., Pistelli, L., et al., 2019. Effect of iodine treatments on *Ocimum basilicum* L.: Biofortification, phenolics production and essential oil composition. PLoS One 14, 226559.

Kiferle, C., Martinelli, M., Salzano, A.M., Gonzali, S., Beltrami, S., Salvadori, P.A., et al., 2021. Evidences for a nutritional role of iodine in plants. Front. Plant Sci. 12, 616868.

Kliewer, M., Evans, H.J., 1963. Cobamide coenzyme contents of soybean nodules and nitrogen fixing bacteria in relation to physiological conditions. Plant Physiol. 38, 99–104.

Konishi, S., Miyamoto, S., Taki, T., 1985. Stimulatory effects of aluminum on tea plants grown under low and high phosphorus supply. Soil Sci. Plant Nutr. 31, 361–368.

Koutmos, M., Datta, S., Pattridge, K.A., Smith, J.L., Matthews, R.G., 2009. Insights into the reactivation of cobalamin-dependent methionine synthase. Proc. Natl. Acad. Sci. USA 106, 18527–18532.

Kronzucker, H.J., Coskun, D., Schulze, L.M., Wong, J.R., Britto, D.T., 2013. Sodium as nutrient and toxicant. Plant Soil 369, 1–23.

Lange, B., van der Ent, A., Baker, A.J.M., Echevarria, G., Mahy, G., Malaisse, F., et al., 2017. Copper and cobalt accumulation in plants: A critical assessment of the current state of knowledge. New Phytol. 213, 537–551.

Lanning, F.C., Eleuterius, L.N., 1989. Silica deposition in some C_3 and C_4 species of grasses, sedges and composites in the USA. Ann. Bot. 63, 395–410.

Leggett, J.E., Epstein, E., 1956. Kinetics of sulfate absorption by barley roots. Plant Physiol. 31, 222–226.

Leigh, R.A., Chater, M., Storey, R., Johnston, A.E., 1986. Accumulation and subcellular distribution of cations in relation to growth of potassium-deficient barley. Plant Cell Environ. 9, 595–604.

Lezhneva, L., Kiba, T., Feria-Bourrellier, A.B., Lafouge, F., Boutet-Mercey, S., Zoufan, P., et al., 2014. The Arabidopsis nitrate transporter NRT2.5 plays a role in nitrate acquisition and remobilization in nitrogen-starved plants. Plant J. 80, 230–241.

Li, Z., McLaren, R.G., Metherell, A.K., 2004. The availability of native and applied soil cobalt to ryegrass in relation to soil cobalt and manganese status and other soil properties. New Zeal. J. Agr. Res. 47, 33–43.

Li, H.-F., McGrath, S.P., Zhao, F.-J., 2008. Selenium uptake, translocation and speciation in wheat supplied with selenate or selenite. New Phytol. 178, 92–102.

Liaqat, A., Rahmatullah, Maqsood, M.A., Shamsa, K., Ashraf, M., Hannan, A., 2009. Potassium substitution by sodium in root medium influencing growth behavior and potassium efficiency in cotton genotypes. J. Plant. Nutr. 32, 1657–1673.

Lyons, G., Ortiz-Monasterio, I., Stangoulis, J., Graham, R., 2005. Selenium concentration in wheat grain: is there sufficient genotypic variation to use in breeding? Plant Soil 269, 369–380.

Lyons, G.H., Genc, Y., Soole, K., Stangoulis, J.C.R., Liu, F., Graham, R.D., 2009. Selenium increases seed production in Brassica. Plant Soil 318, 73–80.

Ma, J.F., 2004. Role of silicon in enhancing the resistance of plants to biotic and abiotic stresses. Soil Sci. Plant Nutr. 50, 11–18.

Ma, J.F., 2005. Plant root responses to three abundant soil minerals: silicon, aluminum and iron. Crit. Rev. Plant Sci. 24, 267–281.

Ma, Q., Bell, R., 2016. Canola, narrow-leafed lupin and wheat differ in growth response to low-moderate sodium on a potassium-deficient sandy soil. Crop Pasture Sci. 67, 1168–1178.

Ma, J.F., Takahashi, E., 2002. Soil, Fertilizer, and Plant Silicon Research in Japan. Elsevier, Amsterdam, The Netherlands.

Ma, J.F., Yamaji, N., 2006. Silicon uptake and accumulation in higher plants. Trends Plant Sci. 11, 392–397.

Ma, J.F., Yamaji, N., 2015. A cooperative system of silicon transport in plants. Trends Plant Sci. 20, 435–442.

Ma, J.F., Hiradate, S., Nomoto, K., Iwashita, T., Matsumoto, H., 1997. Internal detoxification mechanism of Al in hydrangea. identification of Al form in the leaves. Plant Physiol. 113, 1033–1039.

Ma, J.F., Goto, S., Tamai, K., Ichii, M., 2001a. Role of root hairs and lateral roots in silicon uptake by rice. Plant Physiol. 127, 1773–1780.

Ma, J.F., Miyake, Y., Takahashi, E., 2001b. Silicon as a beneficial element for crop plants. In: Datnoff, L.E., Snyder, G.H., Korndorfer, G.H. (Eds.), Silicon in Agriculture. Elsevier Science, Amsterdam, The Netherlands, pp. 17–39.

Ma, J.F., Tamai, K., Yamaji, N., Mitani, N., Konishi, S., Katsuhara, M., et al., 2006. A silicon transporter in rice. Nature 440, 688–691.

Ma, J.F., Yamaji, N., Mitani, M., Tamai, K., Konishi, S., Fujiwara, T., et al., 2007. An efflux transporter of silicon in rice. Nature 448, 209–212.

Ma, J.F., Yamaji, N., Mitani-Ueno, N., 2011. Transport of silicon from roots to panicles in plants. Proc. Jpn. Acad. Ser. B-Phys. Biol. Sci. 87, 377–385.

Mari, A.H., Rajpar, I., Zia Ul, H., Tunio, S., Ahmad, S., 2018. Ions accumulation, proline content and juice quality of sugar beet genotypes as affected by water salinity. J. Anim. Plant Sci. 28, 1405–1412.

Marschner, H., Kylin, A., Kuiper, P.J.C., 1981a. Differences in salt tolerance of three sugar beet genotypes. Physiol. Plant. 51, 234–238.

Marschner, H., Kylin, A., Kuiper, P.J.C., 1981b. Genotypic differences in the response of sugar beet plants to replacement of potassium by sodium. Physiol. Plant. 51, 239–244.

Mateos-Naranjo, E., Redondo-Gomez, S., Alvarez, R., Cambrolle, J., Gandullo, J., Figueroa, M.E., 2010a. Synergic effect of salinity and CO_2 enrichment on growth and photosynthetic responses of the invasive cordgrass *Spartina densiflora*. J. Exp. Bot. 61, 1643–1654.

Mateos-Naranjo, E., Redondo-Gomez, S., Andrades-Moreno, L., Davy, A.J., 2010b. Growth and photosynthetic responses of the cordgrass *Spartina maritima* to CO_2 enrichment and salinity. Chemosphere 81, 725–731.

Mateus, N.D.S., Ferreira, E.V.D.O., Arthur Junior, J.C., Domec, J.C., Jordan-Meille, L., Goncalves, J.L.D.M., et al., 2019. The ideal percentage of K substitution by Na in *Eucalyptus* seedlings: evidences from leaf carbon isotopic composition, leaf gas exchanges and plant growth. Plant Physiol. Biochem. 137, 102–112.

Matoh, T., Kairusmee, P., Takahashi, E., 1986. Salt-induced damage to rice plants and alleviation effect of silicate. Soil Sci. Plant Nutr. 32, 295–304.

Mayland, H.F., James, L.F., Panter, K.E., Sonderegger, J.L., 1989. Selenium in seleniferous environments. In: Jacobs, L.W. (Ed.), Selenium in Agriculture and the Environment. Soil Science Society of America, Madison, WI, USA, pp. 15–50.

Mcbride, K.M., Henny, R.J., Mellich, T.A., Chen, J.J., 2014. Mineral nutrition of *Adenium obesum* 'Red'. HortScience 49, 1518–1522.

Medrano-Macías, J., Leija-Martínez, P., González-Morales, S., Juárez-Maldonado, A., Benavides-Mendoza, A., 2016. Use of iodine to biofortify and promote growth and stress tolerance in crops. Front. Plant Sci. 7, 1146.

Mikkelsen, R.L., Wan, H.F., 1990. The effect of selenium on sulfur uptake by barley and rice. Plant Soil 121, 151–153.

Mikkelsen, R.L., Page, A.L., Bingham, F.T., 1989. Factors affecting selenium accumulation by agricultural crops. In: Jacobs, L.W., Chang, A.C., Dowdy, R.H., Severson, R.C., Sommers, L.E., Volk, V.V. (Eds.), Selenium in agriculture and the environment. Soil Science Society of America, Madison, WI, USA, pp. 65–94.

Milford, G.F.J., Jarvis, P.J., Jones, J., Barraclough, P.B., 2008. An agronomic and physiological re-evaluation of the potassium and sodium requirements and fertiliser recommendations for sugar beet. J. Agr. Sci. 146, 1–15.

Mitani, N., Ma, J.F., Iwashita, T., 2005. Identification of the silicon form in xylem sap of rice (*Oryza sativa* L.). Plant Cell Physiol. 46, 279–283.

Mitani, N., Chiba, Y., Yamaji, N., Ma, J.F., 2009a. Identification and characterization of maize and barley Lsi2-like silicon efflux transporters reveals a distinct silicon uptake system from that in rice. Plant Cell 21, 2133–2142.

Mitani, N., Yamaji, N., Ma, J.F., 2009b. Identification of maize silicon influx transporters. Plant Cell Physiol. 50, 5–12.

Mitani, N., Yamaji, N., Ago, Y., Iwasaki, K., Ma, J.F., 2011. Isolation and functional characterization of an influx silicon transporter in two pumpkin cultivars contrasting in silicon accumulation. Plant J. 66, 231–240.

Mitani-Ueno, N., Yamaji, N., Ma, J.F., 2011. Silicon efflux transporters isolated from two pumpkin cultivars contrasting in Si uptake. Plant Signal. Behav. 6, 991–994.

Montpetit, J., Vivancos, J., Mitaniueno, N., Yamaji, N., Remusborel, W., Belzile, F., et al., 2012. Cloning, functional characterization and heterologous expression of *TaLsi1*, a wheat silicon transporter gene. Plant Mol. Biol. 79, 35–46.

Mori, S., Kobayashi, N., Arao, T., Higuchi, K., Maeda, Y., Yoshiba, M., et al., 2008. Enhancement of nitrate reduction by chlorine application in *Suaeda salsa* (L.) Pall. Soil Sci. Plant Nutr. 54, 903–909.

Moritani, S., Yamamoto, T., Andry, H., Saito, H., 2017. Evapotranspiration and mineral content of *Sedum kamtschaticum* Fischer under saline irrigation. Commun. Soil Sci. Plant Anal. 48, 1399–1408.

Murugesan, C., Kim, K., Ramasamy, K., Denver, W., Subbiah, S., Joe, M.M., et al., 2016. Mycorrhizal symbiotic efficiency on C_3 and C_4 plants under salinity stress - a meta-analysis. Frontiers Microbiol. 7, 1246.

Neuhierl, B., Bock, A., 1996. On the mechanism of selenium tolerance in selenium-accumulating plants - purification and characterization of a specific selenocysteine methyltransferase from cultured cells of *Astragalus bisculatus*. Eur. J. Biochem. 239, 235–238.

Nielsen, F.H., 1984. Ultratrace elements in nutrition. Annu. Rev. Nutr. 4, 21–41.

Nieves-Cordones, M., Al Shiblawi, F.R., Sentenac, H., 2016. Roles and transport of sodium and potassium in plants. In: Sigel, A., Sigel, H., Siegel, R.K.O. (Eds.), The Alkali Metal Ions: Their Role for Life. Springer, Dordrecht, The Netherlands, pp. 291–324.

Noronha, H., Silva, A., Mitani-Ueno, N., Conde, C., Sabir, F., Prista, C., et al., 2020. The VvNIP2;1 aquaporin is a grapevine silicon channel. J. Exp. Bot. 71, 6789–6798.

Ohnishi, J., Flugge, U.I., Heldt, H.W., Kanai, R., 1990. Involvement of Na^+ in active uptake of pyruvate in mesophyll chloroplasts of some C_4 plants. Na^+/pyruvate cotransport. Plant Physiol. 94, 950–959.

Ohta, D., Yasuoka, S., Matoh, T., Takahashi, E., 1989. Sodium stimulates growth of *Amaranthus tricolor* L. plants through enhanced nitrate assimilation. Plant Physiol. 89, 1102–1105.

Omoto, E., Taniguchi, M., Miyake, H., 2010. Effects of salinity stress on the structure of bundle sheath and mesophyll chloroplasts in NAD-Malic enzyme and PCK type C_4 plants. Plant Prod. Sci. 13, 169–176.

Omoto, E., Taniguchi, M., Miyake, H., 2012. Adaptation responses in C_4 photosynthesis of maize under salinity. J. Plant Physiol. 169, 469–477.

Ottow, E.A., Brinker, M., Teichmann, T., Fritz, E., Kaiser, W., Brosché, M., et al., 2005. *Populus euphratica* displays apoplastic sodium accumulation, osmotic adjustment by decreases in calcium and soluble carbohydrates, and develops leaf succulence under salt stress. Plant Physiol. 139, 1762–1772.

Ozanne, P.G., Greenwood, E.A.N., Shaw, T.C., 1963. The cobalt requirement of subterranean clover in the field. Aust. J. Agric. Res. 14, 39–50.

Pais, I., 1983. The biological importance of titanium. J. Plant Nutr. 6, 3–131.

Parry, D.W., Hodson, M.J., 1982. Silica distribution in the caryopsis and inflorescence bracts of foxtail millet (*Setaria italica* L. Beauv.) and its possible significance in carcinogenesis. Ann. Bot. 49, 531–540.

Perera, L.K.R.R., De Silva, D.L.R., Mansfield, T.A., 1997. Avoidance of sodium accumulation by the stomatal guard cells of the halophyte *Aster tripolium*. J. Exp. Bot. 48, 707–711.

Phillips, C.J.C., Youssef, M.Y.I., Chiy, P.C., 1999. The effect of introducing timothy, cocksfoot and red fescue into a perennial ryegrass sward and the application of sodium fertilizer on the behaviour of male and female cattle. Appl. Anim. Behav. Sci. 61, 215–226.

Phillips, C.J.C., Chiy, P.C., Arney, D.R., Kart, O., 2000. Effects of sodium fertilizers and supplements on milk production and mammary gland health. J. Dairy Res. 67, 1–12.

Pi, Z., Stevanato, P., Sun, F., Yang, Y., Sun, X., Zhao, H., et al., 2016. Proteomic changes induced by potassium deficiency and potassium substitution by sodium in sugar beet. J. Plant Res. 129, 527–538.

Pickering, I.J., Wright, C., Bubner, B., Ellis, D., Persans, M.W., Yu, E.Y., et al., 2003. Chemical form and distribution of selenium and sulfur in the selenium hyperaccumulator *Astragalus bisulcatus*. Plant Physiol. 131, 1460–1467.

Pilon-Smits, E.A.H., Hwang, S.B., Lytle, C.M., Zhu, Y.L., Tai, J.C., Bravo, R.C., et al., 1999. Overexpression of ATP sulfurylase in Indian mustard leads to increased selenate uptake, reduction, and tolerance. Plant Physiol. 119, 123–132.

Pilon-Smits, E.A.H., Leduc, D.L., 2009. Phytoremediation of selenium using transgenic plants. Curr. Opin. Biotechnol. 20, 207–212.

Pons, R., Cornejo, M.J., Sanz, A., 2011. Differential salinity-induced variations in the activity of H^+-pumps and Na^+/H^+ antiporters that are involved in cytoplasm ion homeostasis as a function of genotype and tolerance level in rice cell lines. Plant Physiol. Biochem. 49, 1399–1409.

Prom-U-Thai, C., Rashid, A., Ram, H., Zou, C., Guilherme, L., Corguinha, A., et al., 2020. Simultaneous biofortification of rice with zinc, iodine, iron and selenium through foliar treatment of a micronutrient cocktail in five countries. Front. Plant Sci. 11, 589835.

Raddatz, N., De Los Rios, L.M., Lindahl, M., Quintero, F.J., Pardo, J.M., 2020. Coordinated transport of nitrate, potassium, and sodium. Front. Plant Sci. 11, 247.

Rakhmankulova, Z.F., Voronin, P.Y., Shuyskaya, E.V., Kuznetsova, N.A., Zhukovskaya, N.V., Toderich, K.N., 2014. Effect of NaCl and isoosmotic polyethylene glycol stress on gas exchange in shoots of the C_4 xerohalophyte *Haloxylon aphyllum* (Chenopodiaceae). Photosynthetica 52, 437–443.

Rayman, M.P., 2008. Food-chain selenium and human health: emphasis on intake. Brit. J. Nutr. 100, 254–268.

Rayman, M.P., Infante, H.G., Sargent, M., 2008. Food-chain selenium and human health: spotlight on speciation. Brit. J. Nutr. 100, 238–253.

Reddy, D.T., Raj, A.S., 1975. Cobalt nutrition of groundnut in relation to growth and yield. Plant Soil 42, 145–152.

Redondo-Gómez, S., Mateos-Naranjo, E., Figueroa, M.E., Davy, A.J., 2010. Salt stimulation of growth and photosynthesis in an extreme halophyte, *Arthrocnemum macrostachyum*. Plant Biol. 12, 79–87.

Remus-Borel, W., Menzies, J.G., Belanger, R.R., 2005. Silicon induces antifungal compounds in powdery mildew-infected wheat. Physiol. Mol. Plant Pathol. 66, 108–115.

Robson, A.D., Snowball, K., 1987. Response of narrow-leafed lupins to cobalt application in relation to cobalt concentration in seed. Aust. J. Exp. Agric. 27, 657–660.

Robson, A.D., Dilworth, M.J., Chatel, D.L., 1979. Cobalt and nitrogen fixation in *Lupinus angustifolius* L. I. Growth nitrogen concentrations and cobalt distribution. New Phytol. 83, 53–62.

Rodrigues, F.A., Mcnally, D.J., Datnoff, L.E., Jones, J.B., Labbe, C., Benhamou, N., et al., 2004. Silicon enhances the accumulation of diterpenoid phytoalexins in rice: a potential mechanism for blast resistance. Phytopathology 94, 177–183.

Romero-Aranda, M.R., Jurado, O., Cuartero, J., 2006. Silicon alleviates the deleterious salt effect on tomato plant growth by improving plant water status. J. Plant Physiol. 163, 847–855.

Rosenfeld, I., Beath, O.A., 1964. Selenium: Geobotany, Biochemistry, Toxicity, and Nutrition. Academic Press, NY, USA.

Rubio, L., Diaz-Garcia, J., Amorim-Silva, V., Macho, A.P., Botella, M.A., Fernandez, J.A., 2019. Molecular characterization of *ZosmaNRT2*, the putative sodium dependent high-affinity nitrate transporter of *Zostera marina* L. Intern. J. Mol. Sci. 20, 3650.

Sakurai, G., Satake, A., Yamaji, N., Mitaniueno, N., Yokozawa, M., Feugier, F.G., et al., 2015. In silico simulation modeling reveals the importance of the Casparian strip for efficient silicon uptake in rice roots. Plant Cell Physiol. 56, 631–639.

Saldaña-Sánchez, W.D., León-Morales, J.M., López-Bibiano, Y., Hernández-Hernández, M., Langarica-Velázquez, E.C., García-Morales, S., 2019. Effect of V, Se, and Ce on growth, photosynthetic pigments, and total phenol content of tomato and pepper seedlings. J. Soil Sci. Plant Nutr. 19, 678–688.

Sangster, A.G., Hodson, M.J., Wynn Parry, D., 1983. Silicon deposition and anatomical studies in the inflorescence with their possible relevance to carcinogenesis. New Phytol. 93, 105–122.

Saranya, J., Benjamin, J.J., Rani, K., Suja, G., Rajalakshmi, S., Ajay, P., 2020. Identification of salt-responsive genes from C_4 halophyte *Suaeda nudiflora* through suppression subtractive hybridization and expression analysis under individual and combined treatment of salt and elevated carbon dioxide conditions. Physiol. Mol. Biol. Plants 26, 163–172.

Savant, N.K., Snyder, G.H., Datnoff, L.E., 1997. Silicon management and sustainable rice production. Adv. Agron. 58, 151–199.

Schiavon, M., Pilon-Smits, E.A.H., 2017. The fascinating facets of plant selenium accumulation - biochemistry, physiology, evolution and ecology. New Phytol. 213, 1582–1596.

Schiavon, M., Nardi, S., dalla Vecchia, F., Ertani, A., 2020. Selenium biofortification in the 21st century: status and challenges for healthy human nutrition. Plant Soil 453, 245–270.

Schubert, S., Läuchli, A., 1990. Sodium exclusion mechanisms at the root surface of two maize cultivars. Plant Soil 123, 205–209.

Sette Junior, C.R., Laclau, J.P., Tomazello Filho, M., Moreira, R.M., Bouillet, J.P., Ranger, J., et al., 2013. Source-driven remobilizations of nutrients within stem wood in *Eucalyptus grandis* plantations. Trees 27, 827–839.

Shabala, S., Munns, R., 2017. Salinity stress: physiological constraints and adaptive mechanisms. In: Shabala, S. (Ed.), Plant Stress Physiology, 2nd ed. CABI, Wallingford, UK, pp. 24–63.

Shibagaki, N., Rose, A., McDermott, J.P., Fujiwara, T., Hayashi, H., Yoneyama, T., et al., 2002. Selenate-resistant mutants of *Arabidopsis thaliana* identify Sultr1;2, a sulfate transporter required for efficient transport of sulfate into roots. Plant J. 29, 475–486.

Shinmachi, F., Buchner, P., Stroud, J.L., Parmar, S., Zhao, F.J., McGrath, S.P., et al., 2010. Influence of sulfur deficiency on the expression of specific sulfate transporters and the distribution of sulfur, selenium, and molybdenum in wheat. Plant Physiol. 153, 327–336.

Shrift, A., 1969. Aspects of selenium metabolism in higher plants. Annu. Rev. Plant Physiol. 20, 475–494.

Smith, F.W., 1974. The effect of sodium on potassium nutrition and ionic relations in Rhodes grass. Aust. J. Agric. Res. 25, 407–414.

Smith, G.S., Middleton, K.R., Edmonds, A.S., 1978. Sodium and potassium contents of top-dressed pastures in New Zealand in relation to plant and animal nutrition. New Zeal. J. Exp. Agric. 6, 217–225.

Smith, G.S., Middleton, K.R., Edmonds, A.S., 1980. Sodium nutrition of pasture plants. II. Effects of sodium chloride on growth, chemical composition and reduction of nitrate nitrogen. New Phytol. 84, 613–622.

Sors, T.G., Ellis, D.R., Salt, D.E., 2005. Selenium uptake, translocation, assimilation and metabolic fate in plants. Photosynt. Res. 86, 373–389.

Stroud, J.L., Broadley, M.R., Foot, I., Fairweather-Tait, S.J., Hart, D.J., Hurst, R., et al., 2010. Soil factors affecting selenium concentration in wheat grain and the fate and speciation of Se fertilisers applied to soil. Plant Soil 332, 19–30.

Sun, W.C., Zhang, J., Fan, Q.H., Xue, G.F., Li, Z.J., Liang, Y.C., 2010. Silicon-enhanced resistance to rice blast is attributed to silicon-mediated defence resistance and its role as physical barrier. Eur. J. Plant Pathol. 128, 39–49.

Sun, H., Guo, J., Duan, Y., Zhang, T., Huo, H., Gong, H., 2017. Isolation and functional characterization of *CsLsi1*, a silicon transporter gene in *Cucumis sativus*. Physiol. Plant. 159, 201–214.

Sun, H., Duan, Y., Qi, X., Zhang, L., Huo, H., Gong, H., 2018. Isolation and functional characterization of *CsLsi2*, a cucumber silicon efflux transporter gene. Ann. Bot. 122, 641–648.

Sun, Y.L., Wang, Y.H., Deng, L.F., Shi, X., Bai, X.F., 2020a. Moderate soil salinity alleviates the impacts of drought on growth and water status of plants. Russ. J. Plant Physiol. 67, 153–161.

Sun, H., Duan, Y., Mitani-Ueno, N., Che, J., Jia, J., Liu, J., et al., 2020b. Tomato roots have a functional silicon influx transporter but not a functional silicon efflux transporter. Plant Cell Environ. 43, 732–744.

Sun, L., Zhang, M., Liu, X., Mao, Q., Shi, C., Kochian, L.V., et al., 2020c. Aluminium is essential for root growth and development of tea plants (*Camellia sinensis*). J. Integr. Plant Biol. 62, 984–997.

Tamai, K., Ma, J.F., 2008. Reexamination of silicon effects on rice growth and production under field conditions using a low silicon mutant. Plant Soil 307, 21–27.

Terry, N., Carlson, C., Raab, T.K., Zayed, A.M., 1992. Rates of selenium volatilization among crop species. J. Environ. Qual. 21, 341–344.

Terry, N., Zayed, A.M., de Souza, M.P., Tarun, A.S., 2000. Selenium in higher plants. Annu. Rev. Plant Physiol. Plant Mol. Biol. 51, 401–432.

Theerawitaya, C., Tisarum, R., Samphumphuang, T., Singh, H.P., Takabe, T., Cha-Um, S., 2020. Expression levels of vacuolar ion homeostasis-related genes, Na$^+$ enrichment, and their physiological responses to salt stress in sugarcane genotypes. Protoplasma 257, 525–536.

Tomioka, R., Oda, A., Takenaka, C., 2005. Root growth enhancement by rhizospheric aluminum treatment in *Quercus serrata* Thunb. seedlings. J. For. Res. 10, 319–324.

Tran, D.Q., Konishi, A., Cushman, J.C., Morokuma, M., Toyota, M., Agarie, S., 2020. Ion accumulation and expression of ion homeostasis-related genes associated with halophilism, NaCl-promoted growth in a halophyte *Mesembryanthemum crystallinum* L. Plant Prod. Sci. 23, 91–102.

Turan, M.A., Hassan, A., Elkarim, A., Taban, N., Taban, S., 2009. Effect of salt stress on growth, stomatal resistance, proline and chlorophyll concentrations on maize plant. Afr. J. Agri. Res. 4, 893–897.

Valdez-Aguilar, L.A., Reed, D.W., 2010. Growth and nutrition of young bean plants under high alkalinity as affected by mixtures of ammonium, potassium, and sodium. J. Plant. Nutr. 33, 1472–1488.

Wakeel, A., Steffens, D., Schubert, S., 2010. Potassium substitution by sodium in sugar beet (*Beta vulgaris*) nutrition on K-fixing soils. J. Plant Nutr. Soil Sci. 173, 127–134.

Wakeel, A., Muhammad, F., Qadir, M., Schubert, S., 2011. Potassium substitution by sodium in plants. Crit. Rev. Plant Sci. 30, 401–413.

Wang, W., Zhao, X.Q., Hu, Z.M., Shao, J.F., Che, J., Chen, R.F., et al., 2015. Aluminium alleviates manganese toxicity to rice by decreasing root symplastic Mn uptake and reducing availability to shoots of Mn stored in roots. Ann. Bot. 116, 237–246.

Wang, Z., Yamaji, N., Huang, S., Zhang, X., Shi, M., Fu, S., et al., 2019. OsCASP1 is required for Casparian strip formation at endodermal cells of rice roots for selective uptake of mineral elements. Plant Cell 31, 2636–2648.

Watanabe, T., Jansen, S., Osaki, M., 2005. The beneficial effect of aluminium and the role of citrate in Al accumulation in *Melastoma malabathricum*. New Phytol. 165, 773–780.

Watanabe, T., Jansen, S., Osaki, M., 2006. Al-Fe interactions and growth enhancement in *Melastoma malabathricum* and *Miscanthus sinensis* dominating acid sulphate soils. Plant Cell Environ. 29, 2124–2132.

Watanabe, T., Broadley, M.R., Jansen, S., White, P.J., Takada, J., Satake, K., et al., 2007. Evolutionary control of leaf element composition in plants. New Phytol. 174, 516–523.

Watson, C., Clemens, J., Wichern, F., 2020. Hazenite: a new secondary phosphorus, potassium and magnesium fertiliser. Plant Soil Environ. 66, 1–6.

Wendling, L.A., Yibing, M., Kirby, J.K., McLaughlin, M.J., 2009. A predictive model of the effects of aging on cobalt fate and behavior in soil. Environ. Sci. Techol. 43, 135–141.

Whanger, P.D., 2002. Selenocompounds in plants and animals and their biological significance. J. Amer. Coll. Nutr. 21, 223–232.

White, P.J., 2016. Selenium accumulation by plants. Ann. Bot. 117, 217–235.

White, P.J., Bowen, H.C., Marshall, B., Broadley, M.R., 2007. Extraordinarily high leaf selenium to sulfur ratios define 'Se-accumulator' plants. Ann. Bot. 100, 111–118.

Wiersma, D., van Goor, B.J., 1979. Chemical forms of nickel and cobalt in phloem of *Ricinus communis*. Physiol. Plant. 45, 440–442.

Williams, D.E., Vlamis, J., 1957. The effect of silicon on yield and manganese-54 uptake and distribution in the leaves of barley plants grown in culture solutions. Plant Physiol. 32, 404–409.

Wu, G., Shui, Q., Wang, C., Zhang, J., Yuan, H., Li, S., et al., 2015. Characteristics of Na$^+$ uptake in sugar beet (*Beta vulgaris* L.) seedlings under mild salt conditions. Acta Physiol. Plant. 37, 70.

Yamaji, N., Ma, J.F., 2007. Spatial distribution and temporal variation of the rice silicon transporter Lsi1. Plant Physiol. 143, 1306–1313.

Yamaji, N., Ma, J.F., 2009. Silicon transporter Lsi6 at the node is responsible for inter-vascular transfer of silicon in rice. Plant Cell 21, 2878–2883.

Yamaji, N., Mitatni, N., Ma, J.F., 2008. A transporter regulating silicon distribution in rice shoots. Plant Cell 20, 1381–1389.

Yamaji, N., Sakurai, G., Mitani-Ueno, N., Ma, J.F., 2015. Orchestration of three transporters and distinct vascular structures in node for intervascular transfer of silicon in rice. Proc. Natl. Acad. Sci. USA 112, 11401–11406.

Yang, Z.B., You, J.F., Xu, M.Y., Yang, Z.M., 2009. Interaction between aluminum toxicity and manganese toxicity in soybean (*Glycine max*). Plant Soil 319, 277–289.

Zayed, A.M., Terry, N., 1992. Selenium volatilization in broccoli as influenced by sulfate supply. J. Plant Physiol. 140, 646–652.

Zehler, E., 1981. Die Natrium-Versorgung von Mensch, Tier und Pflanze. Kali-Briefe 15, 773–792.

Zhang, Y., Gladyshev, V.N., 2009. Comparative genomics of trace elements: emerging dynamic view of trace element utilization and function. Chem. Rev. 109, 4828–4861.

Zhang, L., Hu, B., Li, W., Che, R., Deng, K., Li, H., et al., 2014. OsPT2, a phosphate transporter, is involved in the active uptake of selenite in rice. New Phytol. 201, 1183–1191.

Zhang, L., Hu, B., Deng, K., Gao, X., Sun, G., Zhang, Z., et al., 2019. NRT1.1B improves selenium concentrations in rice grains by facilitating selenomethinone translocation. Plant Biotech. J. 17, 1058–1068.

Zhao, F.J., Su, Y.H., Dunham, S.J., Rakszegi, M., Bedo, Z., McGrath, S.P., et al., 2009. Variation in mineral micronutrient concentrations in grain of wheat lines of diverse origin. J. Cereal Sci. 49, 290–295.

Zhao, X.Q., Mitani, N., Yamaji, N., Shen, R.F., Ma, J.F., 2010. Involvement of silicon influx transporter OsNIP2;1 in selenite uptake in rice. Plant Physiol. 153, 1871–1877.

Zou, C.Q., Du, Y., Rashid, A., Ram, H., Savasli, E., Pieterse, P.J., et al., 2019. Simultaneous biofortification of wheat with zinc, iodine, selenium, and iron through foliar treatment of a micronutrient cocktail in six countries. J. Agric. Food Chem. 67, 8096–8106.

Chapter 9

Mineral nutrition and crop quality

Umit Baris Kutman
Institute of Biotechnology, Gebze Technical University, Kocaeli, Turkey

Summary

Although the first reports linking mineral nutrition of crops to their quality date back to the 1970s, the effects of mineral nutrition on crop quality had not received the attention they deserved until the 2000s. In crop science, the concept of quality has a broad scope, and depending on the crop species as well as the intended use, different sets of physical and chemical parameters may be considered the most important components of overall quality. This chapter summarizes our current understanding of how mineral nutrition influences the technical (i.e., processing), sensory, and nutritional quality of crop plants and discusses the state-of-the-art advances in mineral nutrition aimed at improved sustainability, food safety, and nutrition security through improved crop quality. Considering both the actual and hidden hunger problems globally, the challenges imposed by climate change and other environmental issues, and the increasing awareness of the importance of nutrition for good health, research on mineral nutrition of crops is focusing on developing sustainable methods for simultaneously increasing productivity, profitability, and quality. However, the mineral nutrient requirements to achieve maximum yield may not always coincide with those to attain the best quality; similarly, higher quality does not always automatically translate into higher economic value.

9.1 Introduction

Quality challenges in crop production were for a long time largely overshadowed by yield challenges because

1. fighting hunger and securing the food supply for all were the priorities (and they still are according to the Sustainable Development Goals 2030 Agenda of the UN);
2. economic value of crop production was mostly, if not exclusively, determined by the amount produced and not by the quality of the produce;
3. less was known about the physiology of quality than about the physiology of yield because of the complexity of quality traits; and
4. the awareness of the critical importance of good nutrition and food safety for good health and the social and economic consequences of "hidden hunger" was minimal.

Early research on the effects of mineral nutrition on crop quality focused mostly on the effects of N fertilization on the technical quality traits of a few major crops such as bread wheat (Timms et al., 1981), durum wheat (Dexter et al., 1982), potato (Roe et al., 1990), and sugar beet (Märländer, 1990). Since the 1990s, the link between crop nutrition and human nutrition has attracted an ever-growing attention, and with the increasing awareness of micronutrient malnutrition as a global challenge, and the increasing popularity of biofortification in the last two decades, nutritional quality has become a major focal point in the crop quality research (Bouis and Saltzman, 2017; Cakmak and Kutman, 2018). Today, farmers' earnings depend increasingly on the quality of produce; if it does not meet certain technical and/or sensory quality criteria, it loses economic value, whereas produce with better technical and/or sensory properties may be sold at a premium price. In addition, health-conscious consumers are looking for safer, more nutritious, and functional food, and this changing consumer behavior is shifting research toward quality, not only in soil-based agriculture but also in soilless production systems and vertical indoor farming.

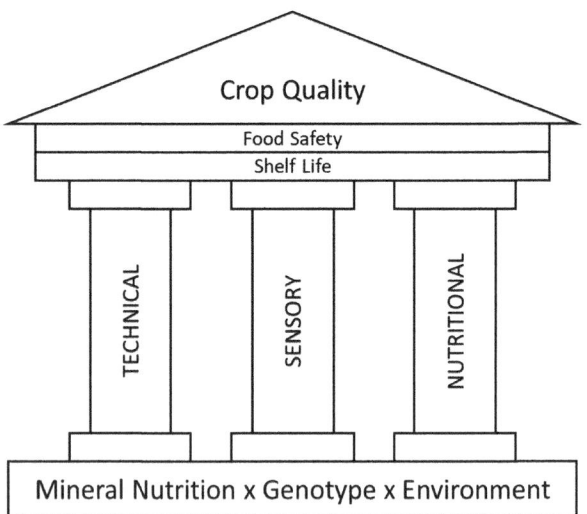

FIGURE 9.1 Aspects of crop quality that are influenced by mineral nutrition, genotype, environment, and their interactions.

The three main pillars of crop quality are technical, sensory, and nutritional quality (Fig. 9.1). Other critical aspects of quality, which do not fall into one of these three categories, include shelf life and food safety. Quality traits are typically quantitative, and like other quantitative traits, they are controlled not only by multiple chromosomal loci but are also influenced strongly by environmental factors and their interactions with the genotype. Even though genotypes are subject to improvement through breeding and/or genetic engineering, mineral nutrient availability and fertilization practices stand out among the exogenous factors that influence the quality of crops. Therefore, genetic improvement of cultivars for better quality often proceeds together with the optimization of fertilization practices.

9.2 Technical quality

The scope of technical quality of crops is very broad because any parameter that influences the suitability of a crop for a specific processing and determines the properties of the processed product can be considered a technical quality parameter. While the other aspects of quality such as nutritional quality, sensory quality, and food safety are mostly relevant to food and feed crops, the concept of technical quality is, by definition, applicable to any crop that is processed to make a product. The link between technical quality and mineral nutrition of crops has been investigated quite extensively in wheat because processed wheat products (including but not limited to bread and pasta) are central to the daily diet of billions of people and thus to global food security, industry, and economy. Technical quality parameters are also of particular importance for industrial crops, including fiber, sugar, oil, and energy crops, and other crops that are processed by the food and beverage industries (e.g., processing tomatoes for tomato paste production, grapes for the wine industry, barley for the malting and brewing industry). Effects of plant nutrition on the technical quality of these crops have attracted considerable attention due to their economic implications.

9.2.1 Bread and pasta

Endosperm storage proteins of wheat, with gluten as the major component, determine the rheological properties of dough. Gluten is composed of prolamins (monomeric proteins) and glutelins (polymeric proteins stabilized with disulfide bridges). Wheat prolamins are called gliadins and determine the viscosity of the dough, whereas wheat glutelins are called glutenins and provide for the elasticity of the dough. When flour is mixed with water and kneaded, a 3-D gluten network is formed, which gives the dough its visco-elastic properties and thus enables it to expand during fermentation by trapping CO_2. Due to their effects on grain protein concentration and composition, mineral nutrients, particularly N and S, are among the most widely studied factors determining the technical quality of bread and durum wheat.

The breadmaking quality of wheat can be improved by optimizing N management practices, which involve optimal N supply, with late N and split N applications (Table 9.1; Rekowski et al., 2019; Wieser and Seilmeier, 1998; Xue et al., 2016). Split and late N applications increase the concentrations and proportions of gluten proteins to a

TABLE 9.1 Protein concentration and gluten composition of two winter wheat (*Triticum aestivum* L.) cultivars Tobak and JB Asano as influenced by N fertilization.

Treatments	Protein concentration (mg g⁻¹ flour)		Gliadin (mg g⁻¹ flour)		Glutenin (mg g⁻¹ flour)		Gliadin/ glutenin (%)		Gluten (mg g⁻¹ flour)	
	Tobak	JB Asano	Tobak	JB Asano	Tobak	JB Asano	Tobak	JB Asano	Tobak	JB Asano
N1 (1 + 0.5 + 0)[a]	82	83	13	13	19	18	65	69	32	31
N2 (1 + 1 + 0)	88	87	16	16	23	20	70	78	39	35
N3 (1 + 0.5 + 0.5N)	92	95	19	20	27	26	71	79	47	46
N4 (1 + 0.5 + 0.5U)	89	90	18	18	25	22	72	82	42	40

[a] Numbers in parentheses show grams of N added per pot before sowing, at EC30 (beginning of stem elongation) and finally at EC45 (late booting) (N = nitrate, U = urea).
Source: Based on Xue et al. (2016).

significantly greater extent than those of other storage proteins, albumins, and globulins. Increases in total glutenin and particularly the proportions of certain hydrophilic high-molecular-weight glutenin subunits (HMW-GS), which can be achieved by N management, may make a greater contribution to the technical quality of wheat flour than the total protein concentration (Wieser and Seilmeier, 1998; Xue et al., 2016).

Sulfur deficiency impairs the technical quality of wheat by causing marked changes in the composition of storage proteins (Zhao et al., 1999). There is a significant positive correlation between grain S concentration and the breadmaking quality of wheat. The percentages of storage proteins rich in S-containing amino acids cysteine and methionine decrease under S-deficient conditions, whereas those of S-poor or S-free proteins increase (Wieser et al., 2004; Zhao et al., 1999). Flour from S-deficient wheat has higher gliadin/glutenin and HMW-GS/LMW-GS ratios, which results in dough with lower extensibility and bread with smaller volume.

For the pasta industry, the grain protein concentration of durum wheat is of critical importance. Durum wheat with a protein concentration higher than 120–130 g kg⁻¹ is preferred because lower-protein semolina decreases the strength and elasticity of dry pasta and the firmness of cooked pasta and increases the cooking losses and stickiness (Sissons et al., 2021). Under both greenhouse and field conditions, improved N nutrition of durum wheat typically leads to a higher grain protein content, which is also influenced by other factors including genotype, water stress, high temperature, and availability of other nutrients (Gagliardi et al., 2020; Kutman et al., 2010).

9.2.2 Sugar and oil crops

In sugar beet, impurities including soluble N, K, and Na decrease sugar recovery by preventing the crystallization of sucrose. Soluble N, known as "harmful N" in the sugar beet industry, is commonly estimated by measuring only the α-amino N, although it also includes betaine, nitrate, and other compounds (Hoffmann and Märländer, 2005). The concentrations of these impurities are important quality parameters. Excessive N applications result in reduced technical quality by increasing the concentrations of these impurities (Abdel-Motagally and Attia, 2009; Kiymaz and Ertek, 2015; Märländer, 1990). Improved Mg nutrition, on the other hand, can enhance juice purity and percent extractable sugar in sugar beet, possibly through positive effects on photosynthesis and source-to-sink translocation of sucrose (Gerendas and Führs, 2013).

For oil crops, N, S, and B stand out among other essential nutrients due to their well-documented effects on oil yield and oil quality. Even though oil quality may also refer to the nutritional quality of vegetable oils, the technical quality parameters include oil content per se, fatty acid composition, oxidative stability, and smoke point. Fatty acid composition, which is critical as both a nutritional and a technical quality parameter depending on the intended use, is to a large extent controlled by the genotype (e.g., high-erucic-acid oilseed rape cultivars grown for industrial-grade oil production vs low-erucic-acid oilseed rape cultivars grown for food-grade oil production) (Knutsen et al., 2016) although environmental factors (such as mineral nutrients) can result in marked changes in fatty acid composition. In sunflower, N

TABLE 9.2 Effect of soil-applied elemental S and foliar-applied thiourea on the seed yield and quality traits in two canola (*Brassica napus* L.) cultivars (Shiralee and Dunkled).

Treatments/cultivars	Seed yield (t ha^{-1})		Oil concentration (g kg^{-1})		Protein concentration (g kg^{-1})		Glucosinolate concentration (µmol g^{-1})	
	Shiralee	Dunkled	Shiralee	Dunkled	Shiralee	Dunkled	Shiralee	Dunkled
Control	1.5	1.5	382	353	228	221	56	51
Soil S (40 kg S ha^{-1})	1.7	1.7	429	418	345	270	69	62
Foliar thiourea (210 mg S L^{-1})	1.6	1.6	405	364	275	224	60	56
Foliar thiourea (420 mg S L^{-1})	1.7	1.6	413	408	321	260	64	59

Source: Based on Rehman et al. (2013).

topdressing applications increased seed protein and content of palmitic and oleic acids but decreased content of crude fat and linoleic acid (Li et al., 2017a). By contrast, in olive, N overfertilization did not significantly alter the fatty acid composition (Fernández-Escobar et al., 2006). However, excess N decreased the technical quality of olive oil by reducing its polyphenol content and thus its oxidative stability. Foliar applications of B to olive trees in the form of boric acid were reported to enhance fruit and oil yield and improve oil quality by decreasing acidity, the peroxide value, and the saponification value (El-Motaium and Hashim, 2020).

Because the biosynthesis of fatty acids and amino acids depends on the same substrate (acetyl-CoA), stimulation of amino acid synthesis by a higher N supply can limit fatty acid biosynthesis, explaining why in oilseed crops seed protein and oil contents are typically negatively correlated (Zhao et al., 1993). For example, in oilseed rape (canola), higher N applications are associated with higher seed protein but lower seed oil content (Ghafoor et al., 2021; Poisson et al., 2019; Zhao et al., 1993). In combination with a sufficiently high N supply, improved S nutritional status of oilseed rape may increase yield while enhancing the seed oil, protein, and glucosinolate (GSL) concentrations (Table 9.2; Poisson et al., 2019; Rehman et al., 2013; Zhao et al., 1993).

9.2.3 Fiber crops

In cotton, technical quality parameters such as fiber strength and whiteness can be improved significantly by K applications, which also maximizes yield and yield components under field conditions (Gormus, 2002). Severe K deficiency in cotton was shown to decrease the micronaire value that is used as a measure of fiber fineness and maturity (Read et al., 2006). By contrast, N deficiency resulted in an increase in fiber micronaire, although the exact effect of N status on quality parameters, including fiber length, strength, and micronaire, depended on the flowering group. In some studies, N stress after flowering was associated with lower fiber quality (Reddy et al., 2004). Moderate N deficiency during reproductive development may accelerate maturity, whereas high levels of N can encourage excess vegetative growth at the expense of yield and quality (Reddy et al., 2004; Tewolde et al., 1994). Optimum N and K supply can enhance cotton yield and quality through positive effects on photosynthesis and source-to-sink translocation of photoassimilates (Pettigrew, 1999; Tewolde and Fernandez, 2003).

9.2.4 Processing tomatoes

The principal technical quality parameters for processing tomatoes that are grown for the tomato paste industry are total solids, total soluble solids (TSS measured as °Brix), titratable acidity, pH, Bostwick viscosity, and color (Renquist and Reid, 1998). In drip-irrigated processing tomatoes, high rates of N fertigation maximized TSS and titratable acidity in addition to yield and fruit size (Kuscu et al., 2014). Fertigation of field-grown processing tomatoes with K improved the blended color and reduced the incidence of yellow shoulder disorder while increasing total and marketable fruit yield (Hartz et al., 2005).

9.2.5 Beer and wine

Depending on the genotype and other environmental factors, N management practices in malting barley determine yield as well as quality parameters (Varvel and Severson, 1987). The upper limit for grain N in malting barley has been set by the malting industry as 21.6 g kg^{-1} and corresponds to 115 g protein kg^{-1}. In barley, the protein concentration of the grain was reported to correlate positively with its phytic acid concentration that needs to be low for high malting and final beer quality (Dai et al., 2007). Therefore, N (and possibly P) availability may affect the technical quality of barley by altering grain size, the percentage of plump kernels, and grain N and phytate concentrations (Dai et al., 2007; Varvel and Severson, 1987).

In white wines, one of the most important quality defects is the protein haze that results from the gradual denaturation, coagulation, and precipitation of proteins during storage (Ferreira et al., 2002; Van Sluyter et al., 2015). Although wine contains varying amounts of different nitrogenous compounds, proteins are present at low concentrations, and most of them are pathogenesis-related proteins. By increasing the soluble protein concentration in wine, higher rates of N fertilization increase the bentonite levels required to remove proteins for enhanced stability and clarity (Spayd et al., 1994).

9.3 Sensory quality

Sensory quality of food crops has a major influence on consumer appeal and market price. Sensory quality can be categorized broadly into two sets of attributes: appearance (visual quality) and flavor. Appearance, which encompasses size, color, shape, uniformity, turgidity, absence/presence of physical or physiological defects, and more, is of particular importance for fresh fruits and vegetables because it determines their marketability and influences consumers' purchasing decisions directly (Kays, 1999). Flavor, on the other hand, is an extremely complex sensory quality that is important for not only fresh produce but also produce that will be cooked or processed in other ways. Although there is no universally agreed definition of flavor, it is not simply a synonym of taste but a complex combination of taste, aroma (smell), texture (mouthfeel), and possibly other sensations (Spence, 2015). Biochemical components that contribute to gustatory or olfactory sensations and physical traits that affect the tactile and even auditory sensations in the mouth determine the flavor of food. Even though flavor perception has a personal component, crops with the superior flavor traits are preferred by consumers, are in high demand, and can be sold at premium prices.

9.3.1 Effects of mineral nutrition on visual quality

Calcium is one of the most critical elements determining fruit and vegetable quality (Sams, 1999). The extreme immobility of Ca in the phloem makes the primarily phloem-fed, rapidly growing sink tissues with low transpiration capacity prone to Ca deficiency (see Chapters 3 and 6). Shoot tips, young and inner leaves of leafy vegetables, and fleshy fruits are among the tissues that are particularly affected by physiological Ca deficiency that leads to quality-impairing defects. Calcium deficiency disorders in susceptible organs are induced by (de Freitas and Mitcham, 2012; Detterbeck et al., 2016; Frantz et al., 2004).

1. high concentrations of competing cations such as NH_4^+, K^+, and Mg^{2+};
2. high growth rates under strong light intensities and high N supply leading to dilution of Ca in target tissues;
3. transpiration-suppressing conditions (cool temperatures, high relative humidity, lack of air movement, etc.) that often prevail in greenhouses; and
4. abiotic stress conditions such as drought and salinity.

Even though a few crops and ornamental species are particularly notorious for their susceptibility to Ca deficiency disorders, there is often also substantial variation with respect to Ca deficiency tolerance within the gene pool of those species. For example, lettuce cultivars with higher tolerance to Ca deficiency (tip-burn) can be bred specifically for greenhouse and soilless vertical farming systems (Birlanga et al., 2021; Frantz et al., 2004).

Blossom end rot (BER) is the most common fruit disorder associated with localized Ca deficiency (Ho and White, 2005; Taylor and Locascio, 2004), although the idea that Ca deficiency is the primary cause of BER was challenged by studies that focused on other stress factors (Saure, 2014). The rot starts with a small, depressed, water-soaked area on the blossom end of the fruit, which then becomes sunken and turns brown or black as it enlarges due to cell wall breakdown. By causing major quality defects, BER can reduce marketable yields of solanaceous crops, including tomato, pepper, and eggplant, as well as cucurbits, such as watermelon, cucumber, and squash.

FIGURE 9.2 Leaf tip-burn caused by physiological Ca deficiency in hydroponically grown lettuce. *Photo by Umit Baris Kutman.*

Other quality-impairing physiological fruit disorders that are induced or aggravated by Ca deficiency and can be alleviated by Ca applications include, but are not limited to, bitter pit of apple (Torres et al., 2017), cork spot of pear (Raese and Drake, 1995) and cracking in pomegranate (Davarpanah et al., 2018). In susceptible leafy vegetables such as lettuce and cabbage, Ca deficiency results in tip-burn (Fig. 9.2), which is characterized by necrosis affecting young leaves, starting from their tips and margins (Aloni et al., 1986; Frantz et al., 2004).

The incidence and severity of quality-impairing defects caused by Ca deficiency can be reduced by direct nutritional strategies such as (1) balanced fertilization of soils, (2) optimization of nutrition solution recipes for soilless systems, and (3) foliar sprays of Ca. In addition, indirect strategies including (1) management of environmental parameters such as humidity, temperature, and light intensity in greenhouses and indoor vertical farms, (2) avoidance or alleviation of abiotic stress conditions such as drought and salinity, and (3) cultivation of relatively more tolerant cultivars (if available) are also important for minimizing losses in visual quality and marketable yield due to Ca deficiency disorders.

Similar to Ca, B is required for the structural integrity of cell walls and is phloem-immobile in most plants, although there are species with polyols in their phloem sap that can translocate B easily via the phloem (Chapter 3). Shape, as a visual quality parameter, is determined predominantly by the genotype but is also affected by environmental factors and cultural practices such as water and nutrient availability. In strawberries, there are several types of shape distortions (multitipping, fasciation, nubbins) that happen due to various environmental factors, such as poor pollination, sub- or supra-optimal temperatures, and pest damage (Carew et al., 2010). By impairing the normal development of stamens, B deficiency may impair pollination and result in small and deformed strawberries (Avigdori-Avidov, 1986). Malformed strawberries tend to have lower B and Ca concentrations than control strawberries (Sharma and Singh, 2008). Similarly, in wine grapes, B deficiency can induce or aggravate millerandage, also known as the "hens and chicken" disorder, which is associated with poor pollination and parthenocarpic fruit production, otherwise caused by cold and wet weather during flowering (Alva et al., 2015). In pome fruits, B deficiency causes internal or external cork formation, which is a different kind of quality-impairing defect than the Ca deficiency-induced bitter pit in apple or cork spot in pear (Nagy et al., 2011). Quality-impairing visual symptoms of B deficiency in other crops include "hollow stem" in broccoli, "hollow heart" in groundnut, "heart rot" in sugar beet, and "cracked stem" in celery (Boersma et al., 2009; Rerkasem et al., 2020; Stockfisch and Koch, 2002).

The rates, forms, and application methods of N fertilizers can enhance or impair crop quality both directly and indirectly by altering the growth rate, vegetative vigor, source-sink relations, and critical nutrient ratios. For most fruits and vegetables, bigger size is often associated with higher quality. Typically, a sufficiently high N supply is required to realize the genetically determined potential size, but excess N can reduce the average size and thus the quality (Carranca et al., 2018; Makino, 2011; Schellenberg et al., 2009). The average size of the harvested unit (fruit, seed, tuber, etc.), however, is not only a quality parameter but also a yield component. Due to source limitations, there is often a negative correlation between unit size and unit number, which is another yield component that responds positively to higher N supply.

In leafy green vegetables, higher N supply and higher nitrate-to-ammonium ratios increase the chlorophyll concentrations and thus enhance the green color, which is highly appealing to consumers (Barickman and Kopsell, 2016; Zhang et al., 2018). Fruit color is one of the most important quality factors determining consumer acceptance in apples, and depending on the color of the apple cultivar, higher N may have a favorable or detrimental effect. In green apple cultivars, urea sprays can enhance the green color (Meheriuk et al., 1996), but in red apple cultivars, higher rates of N

TABLE 9.3 Effects of Ca and K sources and their combinations on the fruit quality parameters of Red Delicious apple.

Treatments	Anthocyanin (mg g^{-1} fresh wt)	Starch (mg g^{-1} dry wt)	Soluble sugars (mg g^{-1} dry wt)	Total acidity (mg g^{-1} fresh wt)	Flavor index[a]
Control	27	39	8.9	4.8	19
CaCl$_2$	29	43	11.4	5.0	23
KNO$_3$	34	36	13.7	5.7	24
K$_2$SO$_4$	36	37	14.4	5.7	26
KCl	32	39	16.0	6.1	27
KNO$_3$ + CaCl$_2$	32	51	14.8	6.1	21
K$_2$SO$_4$ + CaCl$_2$	32	45	13.1	5.4	24
KCl + CaCl$_2$	32	47	15.3	5.1	30

[a]Flavor index = Soluble sugar/total acidity.
Source: Based on Solhjoo et al. (2017).

supply, foliar urea applications, and spring N fertilization as opposed to postharvest N fertilization impair coloration as they result in lower anthocyanin and higher chlorophyll and carotenoid concentrations (De Angelis et al., 2011; Reay et al., 1998). Because the radiation intercepted by the apple fruit is critical for red color formation, increased vegetative vigor and denser foliage due to higher N supply may reduce anthocyanin synthesis by shading the fruit.

Potassium is another mineral nutrient the deficiency of which is associated with poor coloration and other physiological disorders that affect the visual quality of fruits and vegetables. In Red Delicious apple, foliar applications of different K salts alone or in combination with calcium chloride enhanced anthocyanin accumulation and red skin coloration, increased the concentration of soluble sugars more than the acidity of fruit juice, and thus improved the flavor index (Table 9.3). At low K supply, citrus fruits remain small, have a thin rind, and are more prone to splitting and creasing disorders (Alva et al., 2006; Morgan et al., 2005). Uniform ripening and bright red coloration are critical quality issues for most tomato cultivars, and K deficiency can induce or aggravate several quality-impairing physiological disorders in susceptible genotypes. Depending on the cultivar and other environmental factors, improved K nutrition may decrease the incidence of "blotchy ripening" and "yellow shoulder," increase lycopene content, and thus enhance tomato color (Chapagain and Wiesman, 2004; Hartz et al., 2005; Taber et al., 2008). In boiled or fried potato, aftercooking darkening (ACD) is an important visual quality problem caused by the oxidation of ferri-chlorogenic acid (Wang-Pruski and Nowak, 2004). Citric acid can inhibit ACD by competing with the phenolic compound chlorogenic acid for Fe^{3+}. Higher K supply can help reduce ACD by increasing the citric acid concentration and thus the citric acid/chlorogenic acid ratio in potato tubers (Naumann et al., 2020).

9.3.2 Effects of mineral nutrition on flavor

Mineral nutrition is one of the manageable determinants of crop flavor that is also determined by genotype, light intensity, water availability, temperature, and other factors. Nitrogen and S are the two mineral nutrients whose effects on flavor have been studied most extensively (Huchette et al., 2007; McCallum et al., 2005; Yang et al., 2012), but other essential or beneficial minerals, including K, B, and Se, have been reported to influence flavor components (Liu et al., 2020a; Wang et al., 2009; Xu et al., 2021). Because enjoying good flavor is one of the greatest motivations for consuming vegetables, fruits, and beverages, most studies on the effects of mineral nutrition on crop flavor have focused on higher-value and aromatic crops rather than staples.

In carrot, sugars and various terpenes in the essential oil are considered the most important flavor components (Schaller and Schnitzler, 2000). Greater N supply, which increased the shoot biomass, yield, total N content, and tissue nitrate concentrations of carrot plants, were associated with significantly higher concentrations of glucose and fructose, significantly lower concentrations of sucrose, and thus significantly higher monosaccharide-to-disaccharide ratios. The total essential oil concentration in carrots as well as the concentrations of several terpenes in the essential oil decreased markedly with increasing rates of N fertilization, suggesting the aroma of carrots could be enhanced by lower N rates, which is attractive considering the economic, ecological and health benefits of reducing mineral N use in vegetable production. In a study on the effects of low, moderate, and excessive N on flavor components of hydroponically grown strawberries, moderate N maximized the concentrations of the volatile esters butyl and hexyl butanoate and increased the concentrations of soluble carbohydrates. By contrast, excessive N favored vegetative growth, delayed ripening, and significantly increased hexanal concentrations (associated with immaturity), but did not result in improved sensory quality (Ojeda-Real et al., 2009).

Another link between N nutrition and flavor was reported for aromatic rice cultivars, in which the aroma as well as the grain concentrations of total N and L-proline, a precursor of rice aroma compounds, correlated positively with the total soil N content (Yang et al., 2012). In wine grapes (*Vitis vinifera*), depending on the cultivar, moderate or high N supply and mild water deficit were associated with higher levels of aroma compounds and/or their precursors (Mendez-Costabel et al., 2014; Peyrot des Gachons et al., 2005).

Both the N rate and the N form can affect the sensory quality of crops as it has been documented for tea (*Camellia sinensis*) (Huang et al., 2018) and tomato (*Solanum lycopersicum*) (Heeb et al., 2006). The concentrations of flavonoids and amino acids such as theanine, which contribute to the flavor as well as the nutraceutical value of tea, exhibited distinct responses to N deficiency and NO_3^- and NH_4^+ supply (Huang et al., 2018). Amino acid concentrations were maximized by NH_4^+ supply in tea as the NH_4^+-tolerant species adapted to acidic soils. However, in tomato, flavor was enhanced by a NO_3^--dominated nutrient solution (Heeb et al., 2006).

In addition to N, S nutrition has attracted a lot of attention in the context of both sensory and nutritional quality because diverse organosulfur compounds are known to play critical roles in the flavor and nutraceutical value of vegetables and fruits (Dirinck et al., 1981; Huchette et al., 2007). Although cruciferous vegetables (Brassicaceae) and *Allium* crops are particularly rich in aromatic and health-beneficial organosulfur compounds, the importance of S metabolites as aroma components of fruits such as melon and strawberry has also been known for more than four decades (Dirinck et al., 1981; Wyllie and Leach, 1992). The typical flavor of *Allium* crops arises from volatile S compounds formed by the enzymatic hydrolysis of S-alk(en)yl-L-cysteine sulfoxides (ACSO) (Randle et al., 1995). In both glasshouse and field studies on onion (*Allium cepa*), correction of S deficiency and improvement of S nutritional status generally resulted in increased bulb S concentrations, higher ACSO concentrations, altered ACSO composition, and thus enhanced pungency; however, effects were cultivar-dependent and modulated by N supply (McCallum et al., 2005; Randle et al., 1995). An N supply- and cultivar-dependent enhancement of flavor and nutraceutical value by improved S nutrition is also well documented for garlic (*Allium sativum*) wherein the predominant ACSO alliin (S-2-propenyl-L-cysteine sulfoxide) and several other S-containing metabolites tend to show a positive response to increased S supply (Bloem et al., 2010; Huchette et al., 2007; Kacjan Marsic et al., 2019).

The characteristic bitter taste and sulfurous aroma of cruciferous vegetables in the family Brassicaceae are attributed to glucosinolates (GSLs), their breakdown products, particularly isothiocyanates (ITCs), and a diverse set of volatile S metabolites (Bell et al., 2018). Flavors of cruciferous vegetables are influenced by S fertilization that can increase the concentrations of these S-containing metabolites (Kopsell et al., 2007), and by N fertilization that may correlate negatively with their concentrations (Fischer, 1992).

In accordance with the well-documented physiological functions of K in photosynthesis, phloem loading, and source-to-sink translocation of sucrose (Lemoine et al., 2013; Xie et al., 2021), K applications may improve fruit flavor by increasing TSS and sugar content (i.e., sweetness), as documented for muskmelon (Lester et al., 2006), apple (Nava et al., 2008; Solhjoo et al., 2017) and sweet cherry (Yener and Altuntas, 2020). Acidity of fruits is another major flavor component and a key quality factor that may be influenced by K supply. It depends not only on the concentrations of organic acids such as citric, mali and tartaric, but alsoon the ratio of free organic acids to organic acid anions neutralized by K^+ (Villette et al., 2020). In wine grapes, high concentrations of K together with the effects of climate change are associated with high sugar content and low acidity at harvest and result in unbalanced wines with high alcohol content. However, in citrus crops, K supply correlates positively with juice acidity that is conferred mostly by citric acid (Alva et al., 2006). A positive correlation between concentrations of K and citric acid is also well documented in potato tubers, but in this case, it is more a visual quality than a flavor issue because higher levels of citric acid reduce ACD as explained in Section 9.3.1 (Naumann et al., 2020; Wang-Pruski and Nowak, 2004).

9.4 Nutritional quality

The nutritional quality of food crops has a huge impact on public health. While sufficient daily calorie intake prevents hunger, adequate intake of essential nutrients is indispensable for life and good health. In addition to essential nutrients, crop plants contain a wide variety of antioxidants and other health-beneficial phytochemicals. Mineral nutrition of plants can affect the concentrations of essential and beneficial components as well as of antinutrients. Thus, plant nutrition is linked to human nutrition via the nutritional quality of food crops.

9.4.1 Mineral nutrients, hidden hunger, and biofortification

According to recent estimates, more than one-third of people living on our planet today are affected by various micronutrient deficiencies to varying degrees or are at risk of such deficiencies (Bouis and Saltzman, 2017; von Grebmer et al., 2014; White and Broadley, 2009). In this context, the term micronutrient does not necessarily refer to minerals that are essential micronutrients for plants but to any mineral or vitamin essential for human health. If the health and the quality of life of a person are negatively affected by at least one micronutrient deficiency, that person is, by definition, suffering from hidden hunger, irrespective of whether that person has enough to eat purely from a calorie perspective. Globally, hidden hunger is one of the most important nutritional disorders of the 21st century, together with hunger and obesity. In 2016, 5.6 million children under 5 years of age died worldwide according to WHO data, and about half of these deaths were directly or indirectly attributable to malnutrition, including hidden hunger.

There are four main strategies to tackle micronutrient deficiencies in human populations:

1. changing the dietary habits and diversifying the diet,
2. food fortification,
3. supplementation, and
4. biofortification.

Although the first strategy is theoretically the ultimate solution to this problem and may prove very effective at a personal level, its implementation at the population level is extremely difficult, if not impossible, due to social, cultural, logistic, and/or economic constraints. Food fortification, which is the addition of commonly deficient micronutrients to food items during processing, may also contribute significantly to public health. For example, iodization of table salt has proved highly effective for tackling iodine deficiency (Rohner et al., 2014). Supplementation, which depends on the regular use of mineral and/or vitamin supplements in the form of tablets, capsules, syrups, or injections, may be immediately effective to prevent an anticipated deficiency or treat a diagnosed one in an urban setting. It can also contribute to international efforts to manage hidden hunger in developing countries, as exemplified by the vitamin A supplementation campaign by UNICEF, which has been targeting preschool children in high-risk countries for more than two decades (Wirth et al., 2017). However, the consensus is that biofortification is one of the most important strategies in terms of cost-effectiveness, sustainability, and accessibility for target populations in developing countries and rural areas (Bouis and Welch, 2010; de Valenca et al., 2017).

Biofortification is defined as the enrichment of food or feed crops with at least one mineral or vitamin during cultivation. It is considered a biological approach because enrichment occurs prior to harvest and does not depend on industrial food processing. Mostly, the term is applied to crops, but biofortification of livestock and animal-based food is also possible (O'Sullivan et al., 2020). Although the concept of biofortification and the principles behind it date back to the 20th century, the term biofortification entered the literature at the beginning of the 2000s and has attracted an ever-increasing attention since then.

The most common essential micronutrient deficiencies associated with public health issues and targeted by biofortification efforts include vitamin A deficiency and deficiencies of the minerals Fe, Zn, I, and Se (Bouis and Saltzman, 2017; White and Broadley, 2009). In addition, biofortification strategies are discussed, devised, and tested for enriching food crops with other human micronutrients such as Ca, Mg, Cu, B vitamins, and vitamin E (Mène-Saffrané and Pellaud, 2017; Strobbe and van der Straeten, 2018; White and Broadley, 2009). For further information about biofortification of crops with vitamins and the link between mineral nutrition of crops and their vitamin content as a nutritional quality parameter, see Section 9.4.3.

There are two main strategies that are used in biofortification of crops (Cakmak and Kutman, 2018). On the one hand, genetic biofortification aims at developing new conventional or genetically engineered crop cultivars that have increased micronutrient density in their edible parts. On the other hand, agronomic biofortification focuses on agronomic practices, mostly soil and/or foliar applications of micronutrient fertilizers as well as the management of crop macronutrient status to enhance the concentrations of target nutrients in the harvested parts. Recently, microbial-assisted biofortification has emerged as the third biofortification strategy (Singh and Prasanna, 2020; Sun et al., 2021a; Yasin et al., 2015). In this approach, beneficial microorganisms such as plant growth-promoting rhizobacteria, endophytic bacteria, and fungal symbionts are utilized with or without micronutrient applications to enhance micronutrient uptake and accumulation.

At an individual level, there are numerous possible causes of micronutrient deficiencies, including certain infections, genetic disorders, and gastrointestinal problems, but from the population standpoint, the main cause of common micronutrient deficiencies is a micronutrient-poor diet. In developing countries and rural populations, one or a few staple crops typically provide a substantial percentage of daily calories and if these crops are poor in some essential micronutrients, hidden hunger inevitably becomes a widespread public health problem. Three major cereals (wheat, rice, and maize) provide up to 60%−70% of the daily calorie intake of some populations in developing countries that have a high incidence of Zn and Fe deficiencies (Tilman et al., 2002; Timmer, 2014). Despite their critical importance for global food security, these cereal grains are inherently poor in micronutrients, including Zn, Fe, and I, in terms of both concentration and bioavailability. Because soils are the ultimate source of micronutrients in crop production, the situation may be further aggravated when cereals are grown in soils with low micronutrient availability as it was documented for Zn in wheat (Cakmak et al., 1999; Graham et al., 1992). In different target countries, cereals, and other staple crops such as beans, cassava, and sweet potato are often chosen as focus crops in biofortification projects to maximize the potential benefits to public health (Bouis and Saltzman, 2017). However, fruits and vegetables such as strawberries, tomatoes, and leafy greens are also common targets in biofortification projects to explore the potential of these fruits and vegetables in the global fight against hidden hunger and to increasing their economic value by enhancing their nutritional value (Kiferle et al., 2013; Mimmo et al., 2017; Voogt et al., 2010).

In the past, cereal breeders were interested almost exclusively in increasing grain yield. These breeding efforts, together with improved agronomic practices, have resulted in significant increases in the average yields of cereals since the Green Revolution, but the grain concentrations of some essential nutrients, such as Zn, have decreased because of the so-called "dilution effect" (Cakmak and Kutman, 2018; Curtis and Halford, 2016; Davis, 2009; Fan et al., 2008). Cereals that are genetically biofortified through breeding must not only be micronutrient-rich but also have a high yielding potential and other favorable agronomic traits to be adopted by farmers. The inverse relationship between grain yield and Zn concentration, however, makes this a difficult task (see Section 9.7 for the dilution effect and its role in the yield-quality dilemma).

Genetic and agronomic biofortification are complementary strategies (Cakmak and Kutman, 2018). Even though genetic biofortification offers highly sustainable solutions to hidden hunger, reaching the target values through breeding is a long-term goal, provided it is even possible in specific situations given the extent of genetic variation in the gene pool that can be exploited in the breeding efforts (Bouis and Saltzman, 2017; Velu et al., 2014). Genetic studies have led to a substantial improvement in our understanding of Fe and Zn homeostasis and helped with the identification of candidate genes for marker-assisted selection (Stangoulis and Knez, 2022). In contrast to genetic biofortification, agronomic biofortification can result in immediate improvements. Agronomic biofortification can also increase the concentrations of target minerals in genetically biofortified cultivars. Whether agronomic biofortification through fertilizer applications is a viable option depends on a crop-nutrient combination. For example, agronomic biofortification works well for Zn, Se, and I in wheat, but its potential for Fe biofortification is limited (Lyons, 2018; Velu et al., 2014). Foliar applications for simultaneous Zn, Se, and I biofortification were tested successfully on wheat and rice (Prom-u-thai et al., 2020; Zou et al., 2019).

Although soil and/or foliar applications of micronutrients constitute the backbone of agronomic biofortification, an integrated fertilization strategy can offer greater benefits in terms of yield and nutritional quality. In this context, interactions between N management and Zn applications have been investigated extensively. The critical importance of a sufficiently high N supply for maximizing the efficacy of soil and foliar applications of Zn was shown in wheat (Table 9.4). Under low-Zn conditions, increasing the soil N supply did not alter the grain Zn concentration; however, when plants were supplied with high Zn through soil or foliar applications, the grain Zn concentration could be doubled by increasing the N supply. In whole-plant Zn partitioning as well as Zn radioisotope (^{65}Zn) studies, suboptimal N supply was associated with reduced root uptake, root-to-shoot translocation, and remobilization of Zn (Erenoglu et al., 2011; Kutman et al., 2011). Speciation and localization experiments indicated that Zn interacts with proteins in cereal grains and that higher protein concentration in grains may increase the sink strength for Zn (Cakmak et al., 2010; Ozturk et al., 2006; Persson et al., 2009, 2016).

In biofortification, bioavailability of a target micronutrient in the human digestive tract can be as important as its concentration in the edible part. Antinutrients reduce the bioavailability whereas enhancers exert the opposite effect. The concentrations of such antinutrients or enhancers can also be affected by plant mineral nutrition. For example, P deposition in developing seeds is a key determinant of Zn and Fe bioavailability because up to 90% of all P in seeds is stored in the form of phytic acid that forms strong complexes with various divalent cations including Zn and Fe. In

TABLE 9.4 Grain Zn concentration of durum wheat (*Triticum durum* cv. Balcali 2000) as affected by soil and foliar applications of N and Zn in a Zn-deficient soil.

Foliar treatments	Soil N rate	Grain Zn concentration (mg kg^{-1})		
		Low soil Zn	Adequate soil Zn	High soil Zn
None	Low	10	23	37
	Adequate	11	26	60
	High	10	28	61
ZnSO$_4$	Low	45	41	50
	Adequate	94	60	82
	High	92	79	90

Source: Based on Kutman et al. (2010).

monogastric mammals including humans, phytic acid cannot be digested due to the lack of the enzyme phytase. Therefore, phytate-bound micronutrients have low bioavailability. In cereals, the bran including the aleurone layer and the embryo are particularly rich in phytic acid, but the endosperm (which is relatively poor in micronutrients) has negligibly small concentrations of this antinutrient (Lehrfeld and Wu, 1991; Prom-u-thai et al., 2008).

The phytic acid/Zn molar ratio is widely used as a simplified measure of Zn bioavailability in the human diet (Morris and Ellis, 1989). Low-phytic-acid mutants of cereals and legumes accumulate substantially lower amounts of phytic acid in their seeds; however, despite their bioavailability advantages, they have so far found limited use in biofortification applications because the low-phytate trait is typically associated with a yield penalty (Raboy, 2020). Motivations to breed for lower phytic acid in seeds have also been impeded by the health benefits of phytic acid (Silva and Bracarense, 2016), which are unrelated to its role as an antinutrient.

9.4.2 Protein concentration and amino acid composition

Protein-energy malnutrition is a form of malnutrition that involves a deficiency of proteins and/or energy (calories) in the diet. If most of the calories in a person's diet are provided by fats and/or carbohydrates, this person may suffer from protein malnutrition due to insufficient protein intake. A lack of quality protein and essential amino acids in the diet is linked to stunting in children (Semba, 2016). On the other hand, a protein-rich diet and functional foods with high protein levels are recommended to strengthen the immune system, fight metabolic syndrome, decrease insulin resistance, and increase the lean body mass (Iddir et al., 2020; Johnston et al., 2017).

Nitrogen supply is the main mineral nutritional factor affecting crude protein concentration in plants. In general, the N supply levels needed to maximize the crude protein concentration in crop plants are higher than those needed to maximize yield. Protein concentration and composition are important not only for nutritional quality but also for the technical quality of some crops such as bread and durum wheat as explained in Section 9.2. Splitting the N rate and applying late N to the soil or foliage are effective for increasing grain protein concentrations in wheat and rice (Souza et al., 1999; Varga and Svecnjak, 2006). Late N applications increase the efficiency of remobilization of N to the grains (Fuertes-Mendizabal et al., 2012).

Among the 22 proteinogenic amino acids, nine are essential for humans. With arginine (essential for fetuses, neonates, and preterm infants), the number of essential amino acids rises to 10. The quality of dietary protein depends on its digestibility and amino acid composition. An ideal protein source should contain a balanced amino acid profile, providing people with adequate levels of all essential amino acids. However, cereal grains are typically a poor source of lysine (also tryptophan and threonine), whereas legume seeds are typically deficient in methionine (also cysteine, both of which are S-containing amino acids) (Hagan et al., 2003; Leinonen et al., 2019). Breeding and mutagenesis have been used extensively to biofortify cereals and legumes with these essential amino acids (Galili and Amir, 2013). Although high-lysine mutants of cereals often have undesirable agronomic traits such as low grain yield and reduced endosperm hardness, high-lysine genotypes with acceptable yield and quality attributes have been obtained for maize and sorghum (Tesso et al., 2006; Yu and Tian, 2018).

Higher rates of N fertilizer applications may result in a decreased proportion of essential amino acids in the grain protein, thus reducing the quality of the grain protein in cereals. In barley, the proportions of lysine and threonine in the total amino acid pool of the grain protein decreased in response to higher N supply (Bulman et al., 1994). Similarly, in wheat, higher N applications increased the grain yield, the grain protein concentrations and the concentrations of total, nonessential and essential amino acids, but significantly decreased the essential amino acid index and the protein digestibility-corrected amino acid score (Zhang and Ma, 2017a). In legumes, the concentrations of the S-containing amino acids cysteine and methionine are decreased by S deficiency and can be enhanced by S fertilization, as it has been shown for soybean and common bean (Pandurangan et al., 2015; Rushovich and Weil, 2021).

9.4.3 Vitamins and bioactive phytochemicals

Vitamins are essential organic micronutrients that cannot be synthesized by the human body (at all or not in sufficient quantities) and must therefore be taken up in the diet in either the active form or in the form of provitamins (vitamin precursors) to maintain physiological functions. Plants, being autotrophic organisms, can synthesize (except under in vitro conditions) most vitamins (as defined for humans) or their precursors to meet their own physiological needs. Therefore, in a human diet, plant foods serve as important sources of vitamins, except vitamin D and vitamin B12 that are not synthesized by plants in significant quantities, if at all (Jäpelt and Jakobsen, 2013; Watanabe et al., 2014).

Vitamin A deficiency is a global issue that affects more than 100 million children and pregnant women in the developing world and causes hundreds of thousands of children to go blind or die due to an inability of the impaired immune system to fight infections (West and Darnton-Hill, 2008). On the global scale, vitamin A is (together with Fe and Zn) among the top three target micronutrients for biofortification (Bouis and Saltzman, 2017). Plants synthesize provitamin A carotenoids, including α-carotene, β-carotene, and β-cryptoxanthin, with at least one unsubstituted β-ring (von Lintig, 2012). Genetic biofortification through conventional breeding and transgenic approaches has been used successfully to develop high-provitamin A crops. For sweet potato, maize, and cassava, there was sufficient genetic variation in the gene pool, and the high-provitamin A trait was successfully bred into modern biofortified varieties as part of the HarvestPlus Breeding Program (Bouis and Saltzman, 2017); by contrast, in rice and several other crops, genetic engineering was used for provitamin A biofortification (Giuliano, 2017). As of 2022, "golden rice" is the most prominent example among the genetically modified provitamin A-biofortified crops. Although it has a long history dating back to the 1990s, the first approval for cultivation in a target country was granted in the Philippines in 2021 (Wu et al., 2021). While the details and challenges of genetic biofortification of crops with provitamin A are beyond the scope of this chapter, it should be noted that the potential of agronomic biofortification is limited for vitamin A. Nevertheless, higher rates of N and/or S fertilization may result in higher carotenoid concentrations in different crops, including wheat (Fratianni et al., 2013), spinach (Reif et al., 2012), and parsley (Chenard et al., 2005).

L-ascorbic acid (vitamin C) is both a potent free radical scavenger and a cofactor for numerous enzymes involved in plant and human metabolism (Paciolla et al., 2019). Increasing the vitamin C content of plants can provide multiple benefits, including (1) improvement of the nutritional quality of food crops, (2) extension of postharvest shelf life, and (3) enhancement of abiotic stress tolerance. Typically, high light intensities and water stress increase the vitamin C concentration in plant tissues, whereas high N supply decreases it in various fruits and vegetables (Mozafar, 1993; Lee and Kader, 2000). In hydroponically grown spinach, a 3-day N deprivation prior to harvest was effective in increasing the vitamin C concentration and decreasing the nitrate concentration in leaves (Mozafar, 1996).

When the beneficial elements Se and I are applied to crops for biofortification purposes or to enhance stress tolerance and increase growth under specific conditions, they may also bring about positive side effects in terms of increased concentrations of vitamin C, various antioxidants, and other health-beneficial phytochemicals (Medrano-Macías et al., 2018). In soilless lettuce production, the I and Se treatments can boost the vitamin C concentration (Blasco et al., 2008; Rios et al. 2009; Smolen et al., 2019). Similar effects of Se and I on the concentrations of vitamin C and other antioxidants have been reported for various crops, including strawberry, pepper, and tomato (Li et al., 2017b,c; Smolen et al., 2015).

Beta-glucans are bioactive polysaccharides that are readily available from oat and barley grains. They are classified as soluble fibers and are known for their immunomodulatory effects and the beneficial roles in hypercholesterolemia, hypertension, insulin resistance, diabetes, obesity, and metabolic syndrome (El Khoury et al., 2012). High N supply can significantly increase grain β-glucan concentrations in barley (Güler, 2003) and oat (Brunner and Freed, 1994; Yan et al., 2017).

Crop plants contain diverse bioactive phytochemicals (such as phenolic compounds, GSLs/ITCs, carotenoids, and phytic acid) that contribute to their nutritional quality as accessory health factors. In the agronomic approach, mineral nutrient applications, abiotic stresses, and light (only in glasshouses and indoor farming) can be used to manage concentrations of bioactive phytochemicals (Poiroux-Gonord et al., 2010). For example, due to the anticancer and antioxidant effects of certain GSLs and their breakdown products (ITCs), the nutraceutical values of cruciferous vegetables such as broccoli and radish depend on the composition as well as the total concentration of GSLs that were reported to respond positively to N (Omirou et al., 2009), S (Kopsell et al., 2007; Zhou et al., 2013), and Se fertilization (Robbins et al., 2005).

9.5 Shelf life of fresh fruits and vegetables

The shelf life of fresh fruits and vegetables is an important quality parameter that influences the logistics from field to fork and accessibility to consumers, and thus both local and global food and nutrition security. Fresh produce with short shelf life cannot be sustainably transported long distances or stored for a long time before reaching consumers. According to the State of Agriculture and Food Report by Food and Agriculture Organization (2019), around 14% of the world's food is lost after harvest before even reaching the retailers, mostly during storage and transportation. In addition to crop-specific optimization of storage conditions, smart management of fruit maturation, control of postharvest diseases, and advances in packing materials and technologies, plant nutrition can make a substantial contribution to the shelf life of fresh fruits and vegetables and thus reduce food waste and the associated economic losses.

The firmness of fruits and vegetables is one of the key determinants of their shelf life, storability, and suitability for long-distance transportation. It is, however, an important textural property, and as mentioned in Section 9.3.2, texture is also considered a component of flavor. Thus the documented effects of plant nutrition on crop firmness are important not only in the context of shelf life but also in the context of sensory quality.

In apple production, increasing N applications decrease firmness while increasing fruit size (Ferree and Cahoon, 1987; Nava et al., 2008). Fruit size, in general, correlates negatively with firmness, probably because smaller fruits have a similar number of cells as larger fruits but smaller cells and therefore a relatively higher percentage of cell wall material in their volume (Sams, 1999). In green vegetables, including lettuce and spinach, sufficiently high N fertilization ensures a longer shelf life by lowering the weight loss due to respiration and enhancing the membrane integrity during storage (Bonasia et al., 2013; Conversa et al., 2014), whereas in carrot, excessive N fertilization results in greater weight loss during storage and shorter shelf life (Ierna et al., 2020).

Higher K concentration in fruit flesh, accompanied by higher levels of sugars and other osmolytes, can decrease the osmotic potential, attract more water into the fruit and thus increase the turgor. This may explain the positive effects of K on fruit firmness in tomato (Chapagain and Wiesman, 2004), muskmelon (Lester et al., 2006), and apple (Nava et al., 2008).

As Ca is required to cross-link pectins in the middle lamella and maintain the rigidity of the cell wall structure (Chapters 2 and 6), it is a key element in fruit firmness (Sams, 1999; Martins et al., 2020). In the context of shelf life, Ca is likely the most extensively studied mineral nutrient (Martín-Diana et al., 2007). In a variety of crops, both preharvest Ca sprays and postharvest treatments with Ca may increase fruit firmness and delay fruit softening during storage. Foliar applications of $CaCl_2$ to grape vines resulted in a higher pectin deposition in the cell walls of the fruit skin and inhibited pectin degradation and cell wall loosening, which was also documented at the level of gene expression (Martins et al., 2020). Similarly, a postharvest Ca dip treatment, alone or in combination with an early-season B spray, effectively delayed fruit softening in apple, and the associated changes in enzyme activities were evident after a 6-month storage period (Mohebbi et al., 2020). Preharvest sprays of $CaCl_2$ preserved the quality of guava fruit during storage and enhanced its nutritional value (Ribeiro et al., 2020). In cherry tomatoes, foliar applications of Ca and B increased flesh firmness and vitamin C concentration at harvest and after storage (Table 9.5). For greenhouse-grown tomatoes, it was shown that the K/Ca ratio of the fertigation solution can explain variations in quality parameters (including firmness) better than the K or Ca supply rates alone, pointing at the importance of interactions between mineral nutrients (Hernandez-Perez et al., 2020).

Silicon is classified as a beneficial element for plants and is deposited as silica in the apoplast to reinforce cell walls (Chapter 8). Its deposition in the epidermal tissue strengthens the cuticle against biotic and abiotic stress factors, reduces the water loss through the cuticle, and contributes to mechanical rigidity. The use of Si fertilizers to improve the firmness and shelf life of fresh fruits is, therefore, a promising approach. Applying silicate fertilizers to table grapes grown on a calcareous desert soil significantly extended the fruit shelf life by reducing the respiration rate, decay incidence and weight loss after harvest (Zhang et al., 2017b). In hydroponically grown tomato, addition of Si fertilizers to the nutrient solution as well as postharvest Si applications to harvested fruits enhanced the firmness of fruit after storage and increased the shelf life (Costan et al., 2020). Root applications of Si fertilizers to strawberry plants grown in soilless

TABLE 9.5 Effects of foliar B, Ca and their combined application on the firmness, soluble solids, titratable acidity, and vitamin C concentrations of cherry tomatoes (*Solanum lycopersicum* cv. Unicorn) at harvest time, on the 25th day of storage at 5°C, and on the 10th day of storage at 11°C.

Treatment	Firmness (Newton, N)			Vitamin C (mg kg^{-1} fresh wt)		
	Harvest	5°C	11°C	Harvest	5°C	11°C
Control	18	12	13	132	76	119
Boron (B)	22	12	17	149	92	137
Calcium (Ca)	20	14	16	161	90	143
B + Ca	24	15	18	177	108	152

Source: Based on Islam et al. (2016).

culture were reported to improve several quality parameters while extending the shelf life of fruits significantly (Peris-Felipo et al., 2020).

9.6 Food safety

Food safety, which is not to be confused with food security, is about whether eating a food item is safe or not. For plant foods, food safety involves various aspects, such as pesticide residues, microbiological safety, aflatoxin levels, concentrations of toxic elements including metals, and levels of harmful N compounds. Among these, at least the latter two are influenced by the mineral nutrition of crop plants.

9.6.1 Toxic elements

In the context of human and animal nutrition, concentrations of mineral elements in food and feed crops are critical not only because the levels of essential minerals are important determinants of the nutritional value but also because both essential and nonessential elements may be toxic above certain concentrations, thus threatening food safety. As the classic toxicology maxim "The dose makes the poison" by Paracelsus signifies, almost any naturally occurring mineral in the diet may pose a toxicity risk but the "tolerable upper intake levels" are much lower for some elements than for others. Risk is a function of hazard, exposure, and vulnerability. Accordingly, the toxicity risk associated with a potentially toxic mineral in a food or feed item is determined by not only (1) concentration (hazard) but also (2) bioavailability (hazard), (3) chemical speciation (hazard), (4) amounts consumed (exposure), and (5) population parameters such as age and gender (vulnerability).

The most widely discussed "potentially toxic" elements in the diet include (1) metals such as cadmium (Cd), lead (Pb), and mercury (Hg); (2) the metalloids arsenic (As) and boron (B); and (3) the halide fluoride (F$^-$). Drinking water may be a major dietary source of As, B, and F$^-$ (Azara et al., 2018), whereas seafood is typically the principal dietary source of Hg (Moreno-Ortega et al., 2017). On the other hand, plant-based foods are often the principal dietary sources of Cd, As, and Pb (Ma et al., 2021; Wang et al., 2019; Zhao and Wand, 2020), and fertilization practices may affect food safety by increasing or decreasing their concentrations and influencing their bioavailability in the edible parts of crops. In addition to the uptake of metals from contaminated soils and their accumulation in the harvested organs, surface contamination with soil and dust particles may also contribute substantially to the dietary intake of these metals (e.g., Pb) (Finster et al., 2004).

Among the potentially toxic elements for which plant-based foods constitute the main source in the average diet, Cd is of primary concern because of its relatively high soil-to-plant transfer and its comparatively high human but low plant toxicity (Chaney, 1980; Zhao and Wand, 2020). Mining and smelting operations and industrial applications pollute ecosystems with Cd that finds its way into soils mainly through atmospheric deposition and applications of high-Cd biosolids and manure (Haider et al., 2021; Khan et al., 2017). Cadmium also occurs naturally in rock phosphates; therefore, low-quality P fertilizers processed from high-Cd rock phosphates can contain up to 300 mg Cd per kg dry product, and

TABLE 9.6 Effects of increasing applications of analytical-grade Ca(H$_2$PO$_4$)$_2$ and high-Cd (NH$_4$)$_2$HPO$_4$ (containing 28 mg Cd kg^{-1} fertilizer) fertilizers on shoot Cd and Zn concentrations of 61-day-old wheat (*Triticum aestivum* L. cv. Tahirova) plants grown with marginal Zn supply (0.1 mg Zn kg^{-1} soil) in native and sterilized soils with and without arbuscular mycorrhiza inoculation.

Soil treatment	P supply (mg kg^{-1} soil)	Shoot Cd concentration Ca(H$_2$PO$_4$)$_2$	(NH$_4$)$_2$HPO$_4$	Shoot Zn concentration Ca(H$_2$PO$_4$)$_2$	(NH$_4$)$_2$HPO$_4$
		(µg kg^{-1})		(mg kg^{-1})	
Native	20	78	102	9.7	13
	60	112	162	6.9	8.6
	180	163	202	5.0	6.8
Sterilized	20	164	151	4.4	4.8
	60	218	210	4.9	5.3
	180	212	283	5.3	6.9
Sterilized + arbuscular mycorrhiza	20	69	74	12	18
	60	92	114	7.6	11
	180	98	162	5.8	7.7

Source: Based on Yazici et al. (2021).

their long-term use can lead to elevated Cd concentrations in soils and crops (Grant, 2015; Jiao et al., 2004). Because arbuscular mycorrhizae can limit Cd accumulation in crops, presumably by immobilizing Cd in the mycorrhizosphere (Janouskova and Pavlikova, 2010) and by increasing the Zn uptake, the suppression of arbuscular mycorrhizae by P may be another reason behind the elevated Cd concentrations in crops after heavy P fertilization (Yazici et al., 2021). In wheat, both soil sterilization that eliminated native mycorrhizae and higher rates of P fertilizers significantly increased shoot Cd concentrations (Table 9.6). Inoculation of sterilized soil with arbuscular mycorrhizae resulted in the lowest Cd concentration at all P supply rates. In native and inoculated soils, using high-Cd diammonium phosphate instead of analytical-grade calcium dihydrogen phosphate as the P source was associated with increased wheat shoot Cd concentrations. At low and moderate P supplies, shoot Zn concentrations in Zn-deficient plants were reduced by soil sterilization and increased by inoculation with arbuscular mycorrhizae, but this effect disappeared at high P supply that significantly suppressed shoot Zn concentrations in wheat grown on native and inoculated soils.

Heavy N fertilization is also associated with increased Cd concentrations in plant tissues. This effect is usually attributed to soil acidification, increased ionic strength of the soil solution, and enhanced Cd desorption from binding sites, but it could also be explained by higher concentrations of peptides and proteins that are involved in Cd uptake, translocation, sequestration, and storage (Mitchell et al., 2000; Sterckeman and Thomine, 2020).

Because Cd and Zn are chemically similar, plant Zn transporters such as those in the ZIP (zinc-regulated transporter, iron-regulated transporter-like), NRAMP (natural resistance-associated macrophage protein), and heavy metal ATPase families of proteins cannot discriminate between Cd and Zn (Cakmak and Kutman, 2018; Huang et al., 2020). A study of a diverse collection of barley genotypes revealed that higher Zn accumulation in grain correlated positively with higher Cd accumulation (Detterbeck et al., 2016). Competition between Cd and Zn for shared transport mechanisms can, at least partly, explain the antagonistic effects of Zn applications on Cd for root uptake and distribution within the plant (Cakmak et al., 2000; Hart et al., 2002; Jiao et al., 2004). Zinc applications may also reduce Cd accumulation by downregulating the expressions of transporters involved in Cd uptake and translocation (Adil et al., 2020; Zhou et al., 2020). However, the alleviation of Cd toxicity by the correction of Zn deficiency through Zn fertilization is not always associated with decreased tissue Cd concentrations and may be attributed to improved oxidative stress tolerance, dilution of Cd, and physiological competition between Zn and Cd within plant tissues (Köleli et al., 2004; Zhou et al., 2020). When Zn accompanies Cd as a pollutant in the soil, low or moderate Zn contamination can reduce plant Cd uptake, whereas a high level of Zn pollution can aggravate Cd toxicity, presumably by increasing the phytoavailability of Cd (Podar et al., 2004; Shute and Macfie, 2006).

Chloride salinity, heavy use of Cl-containing fertilizers such as KCl, and even the use of Cl-rich water for foliar applications or sprinkler irrigation may aggravate Cd accumulation in crops because Cl^- tends to form relatively stable and highly soluble complexes with Cd^{2+}, thereby mobilizing Cd in soils and plant tissues and enhancing its uptake and translocation (Ozkutlu et al., 2007; Weggler et al., 2004). Liming may decrease Cd uptake by plants by increasing soil pH and exchangeable Ca^{2+} concentration and decreasing the availability of Cd in the soil; however, depending on the conditions, it may also increase Cd uptake by decreasing the availability of Zn that competes with Cd and limits its uptake as explained above (Chaney et al., 2006; Liu et al., 2020b).

Arsenic enters the human diet mainly via drinking water or via food crops among which rice is considered the main food source of As intake, especially in some Asian countries where the population has a rice-based diet, and the irrigation water is contaminated with As (Brackhage et al., 2014). In soils, As exists in the form of arsenite [As(III)] or arsenate [As(V)]. Under flooded conditions, the reducing environment makes As(III) the predominant As species and increases the total soluble As concentration in the soil solution (Zhao and Wand, 2020). Arsenite enters rice mainly through the Si uptake pathway, which is based on the Si influx transporter Lsi1, facilitating root uptake, and the Si efflux transporter Lsi2, facilitating xylem loading (Ma et al., 2008) (Chapter 8). The high efficiency of the Si uptake pathway in rice (a Si accumulator) explains why rice is markedly more efficient in As(III) uptake than wheat or barley (Su et al., 2010). Because Si can compete directly with As(III) for the transporters, application of Si fertilizers or Si-rich amendments is a promising way to mitigate As accumulation in rice grain (Li et al., 2009; Sun et al., 2021b). On the other hand, As(V), which is the predominant As species under aerobic soil conditions, closely resembles phosphate and enters plant roots via phosphate transporters (Brackhage et al., 2014). Although a shared uptake mechanism for As(V) and P suggests competition for the transporters and thus antagonism, higher rates of P fertilizers enhance As uptake of wheat by displacing more As into the soil solution (Brackhage et al., 2014; Tao et al., 2006).

9.6.2 Harmful N compounds

As one of the harmful N compounds threatening food safety, nitrate in plants has received considerable attention, although most of the health problems attributed to nitrate are in fact caused by nitrite, for which nitrate is the precursor. In plant cells, nitrate is stored in the vacuole and used in osmoregulation, especially under low-light conditions that diminish the availability of other osmolytes such as sugars and organic acids and also impair the activity of nitrate reductase due to insufficient supply of electrons from NAD(P)H (Blom-Zandstra and Lampe, 1985; Granstedt and Huffaker, 1982). On the other hand, nitrite (as the product of nitrate reductase) is rapidly converted to ammonium by nitrite reductase and not stored because of its relatively high toxicity. Leafy green vegetables, such as spinach, Swiss chard, and lettuce, and certain root vegetables, such as beetroot, have a particularly high potential to accumulate nitrate, and maximum permissible nitrate levels have been set for these crops (e.g., spinach and lettuce) by national and international food safety authorities (European Food Safety Authority, 2008; Santamaria, 2006). In the human digestive system, dietary nitrate can be reduced to nitrite by oral commensal bacteria and possibly also by the gut microbiome (Lundberg et al., 2004; Tiso and Schechter, 2015).

Nitrate is a competitive inhibitor of iodide (I^-) uptake by the thyroid gland and can therefore impair thyroid hormone production and possibly promote thyroid cancer (Kilfoy et al., 2011). On the other hand, nitrite is associated with two major health concerns: (1) methemoglobinemia in infants and (2) formation of carcinogenic nitrosamines. In nitrite-induced methemoglobinemia, nitrite in the bloodstream oxidizes Fe^{2+} in hemoglobin to Fe^{3+}, forming methemoglobin, which is unable to bind oxygen (Cortazzo and Lichtman, 2014; Knobeloch et al., 2000). Infants younger than 6 months have low methemoglobin reductase (cytochrome-b5 reductase) activity and are therefore particularly susceptible to methemoglobinemia, which results in blue baby syndrome. Nitrate-contaminated drinking water and consumption of foods that are high in nitrate/nitrite are the top two risk factors for methemoglobinemia in infants although methemoglobinemia due to the consumption of plant-based foods is extremely unlikely.

Increased formation of nitrosamines due to high dietary nitrite intake increases the risk of different types of cancer, including gastric cancer (Knobeloch et al., 2000; Song et al., 2015). Metabolic activation of nitrosamines by cytochrome P450 turns them into mutagenic DNA-alkylating agents and thus unleashes their carcinogenic potential (Wong et al., 2005).

Optimization of N management practices in crop and livestock production systems to minimize leaching losses and maximize N-use efficiency can reduce the nitrate concentrations in groundwater that is used as drinking water (Dalgaard et al., 2014). Because plant-based foods, particularly high-nitrate vegetables, are typically the second most important dietary source of nitrate at the population level, lowering the nitrate concentrations of target crops could help reduce the health problems associated with dietary nitrate and nitrite. Avoiding overfertilization with N (Liu et al., 2014) and increasing the light intensity for winter-grown vegetables in greenhouses (Mccall and Willumsen, 1999) can contribute to this goal. Choosing the best harvest time by taking the diurnal rhythm of nitrate uptake and metabolism into account is likely the most practical strategy to harvest vegetables with lower nitrate concentrations (Scaife and Schloemer, 1994).

The health effects of dietary nitrate are, however, complex, and multifaceted. Despite the health risks that are directly or indirectly associated with dietary nitrate, there is also increasing evidence for the safety of nitrate from vegetable sources and even for their health benefits. A meta-analysis by Song et al. (2015) revealed that increased consumption of high-nitrate vegetables was associated with a slightly decreased risk of gastric cancer, possibly owing to the high levels of antioxidants in those vegetables (Kim et al., 2007). Dietary nitrate from high-nitrate vegetables has been documented to reduce cardiovascular disease incidence significantly, protect consumers from metabolic diseases, and improve sports performance by increasing the nitric oxide (NO) levels in the body through the NOS-independent nitrate-nitrite-NO pathway (Bondonno et al., 2021; Gonzales-Soltero et al., 2020; Liu et al., 2020c). Health benefits of high-nitrate vegetables may therefore outweigh their risks, especially in population groups that are not at risk of iodine deficiency, thyroid diseases, or methemoglobinemia.

Allergenic amines, lectins, toxic amino acids found in legumes, alkaloids, cyanogenic glycosides, and GSLs are among other nitrogenous compounds of plant origin potentially affecting human health and are therefore of concern regarding food safety. Their concentrations in foods may be influenced by fertilization practices. In wine, the concentrations of amines such as histamines and tyramine can be elevated by increased N fertilization, causing allergy, migraine, and other symptoms in sensitive individuals (Smit et al., 2014). Changes in plant nutritional status in response to fertilizer applications were demonstrated to influence cyanogenic glucoside production in cassava (Imakumbili et al., 2020).

Acrylamide, yet another harmful N compound, is known for its carcinogenic and neurotoxic properties. It does not occur in plants naturally, but it is formed during processing (heat-treated carbohydrate-rich foods) via the Maillard reaction where the amino acid asparagine reacts with a reducing sugar like glucose. French fries and bakery products including bread have attracted particular attention in this context. Plant nutrition practices may affect acrylamide levels in food products by altering the abundance of asparagine and reducing sugars. The acrylamide-forming potential of potato tubers was shown to increase in response to higher N supply that elevated asparagine concentration, and to decrease in response to K and S fertilization that lowered the concentrations of reducing sugars (Gerendas et al., 2007; Muttucumaru et al., 2013; Sun et al., 2019). In wheat flour, asparagine levels and thus the acrylamide-forming potential increase with increasing N supply and decrease with an adequate S supply (Granvogyl et al., 2007; Martinek et al., 2009; Weber et al., 2008).

9.7 The yield-quality dilemma

The mineral nutrient input required to maximize the yield of a given crop species or cultivar that is grown under a specific set of conditions (soil and climate factors, light, agronomic practices, pest and disease pressure, abiotic stress, etc.) is rarely equal to that required to maximize its quality. There are two different scenarios where the yield versus mineral nutrient curve does not run parallel to the quality versus mineral nutrient curve (Fig. 9.3).

In the first scenario, the highest quality is attained at a mineral nutrient level that is suboptimal for yield (Fig. 9.3A). After this point, further increases in yield can be attained through higher fertilizer rates, albeit at the expense of quality. A common explanation for such an inverse relationship between yield and quality traits is based on the so-called "dilution effect," whereby higher yields cause a dilution of minerals or other critical components in a larger biomass. For example, decreases in grain Zn concentrations of wheat with increasing P applications were attributed primarily to the dilution effect (Hui et al., 2019). Increased P rates increased the grain yield but reduced the nutritional quality of winter wheat by diluting Zn concentration.

In the second scenario, the highest possible yield is obtained at an input level that is insufficient to maximize quality (Fig. 9.3B). Above this level, higher mineral supply continues to increase the quality of the crop up to a certain point but not its yield. Given that in such cases, the additional fertilizer applications are not translated into yield, the economic feasibility of these applications from the growers' perspective depends on whether the premium price for the higher quality produce is worth the extra cost. This scenario is exemplified by preharvest Ca sprays to various crops to improve the visual quality, prevent physiological disorders, and prolong the shelf life (Sections 9.3.1 and 9.5) or S applications to non-S-deficient cruciferous crops that increase the concentrations of GSLs and thus influence both their sensory (flavor) and nutritional quality (Sections 9.3.2 and 9.4.3). Studies on agronomic biofortification also often

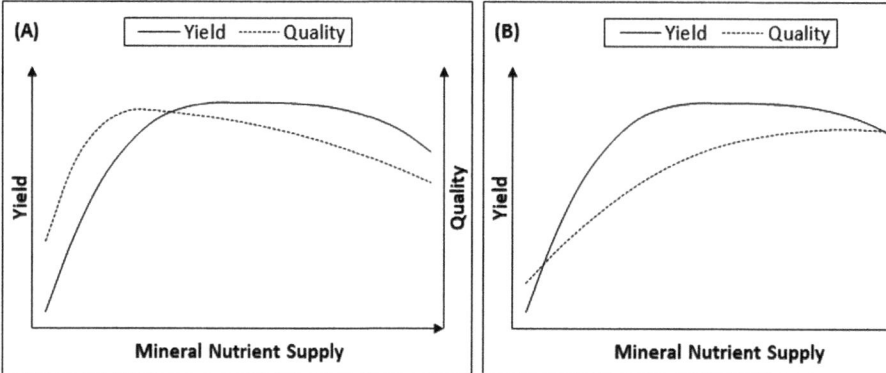

FIGURE 9.3 Differential responses of crop yield and quality parameters to increased mineral nutrient supply in the two common scenarios: (A) and (B).

provide examples for this scenario. In durum wheat grown in a low-Zn calcareous soil, an adequate Zn supply that corrected crop Zn deficiency significantly enhanced both the grain yield and the grain Zn concentration, whereas increasing the Zn fertilization rate above the adequate level boosted the grain Zn concentration but not the yield (Kutman et al., 2010).

It is very difficult, if not impossible, to determine the conditions (including the mineral nutrient input) that maximize all the various aspects of crop quality because management practices that maximize one aspect of quality may not be right for maximizing another aspect. For example, the breadmaking quality of bread wheat and the pasta-making quality of durum wheat can be improved by higher grain protein concentrations, which can be achieved by higher N rates and the split and late N applications (Section 9.2.1). However, from a nutritional perspective, although higher grain protein concentrations may contribute to public health by increasing the proportion of calorie intake from proteins, higher N applications may at the same time decrease the proportions of essential amino acids and thus reduce the protein quality of cereals (Section 9.4.2).

References

Abdel-Motagally, F.M.F., Attia, K.K., 2009. Response of sugar beet plants to nitrogen and potassium fertilization in sandy calcareous soil. Int. J. Agric. Biol. 11, 695–700.

Adil, M.F., Sehar, S., Chen, G., Chen, Z., Jilani, G., Chaundry, A.N., et al., 2020. Cadmium-zinc cross-talk delineates toxicity tolerance in rice via differential gene expression and physiological / ultrastructural adjustments. Ecotox. Environ. Safe. 190, 110076.

Aloni, B., Pashkar, T., Libel, R., 1986. The possible involvement of gibberellins and calcium in tipburn of Chinese cabbage: study of intact plants and detached leaves. Plant Growth Regul. 4, 3–11.

Alva, A.K., Mattos Jr., D., Paramasivam, S., Patil, B., Dou, H., Sajwan, K.S., 2006. Potassium management for optimizing citrus production and quality. Int. J. Fruit Sci. 6, 3–43.

Alva, O., Roa-Roco, R.N., Perez-Diaz, R., Yanez, M., Tapia, J., Moreno, Y., et al., 2015. Pollen morphology and boron concentration in floral tissues as factors triggering natural and GA-induced parthenocarpic fruit development in grapevine. PLoS One 10, e0139503.

Avigdori-Avidov, H., 1986. Strawberry. In: Monselise, S.P. (Ed.), Handbook of Fruit Set and Development. CRC Press Inc., Boca Raton, FL, USA, pp. 419–448.

Azara, A., Castiglia, P., Piana, A., Masia, M.D., Palmieri, A., Arru, B., et al., 2018. Derogation from drinking water quality standards in Italy according to the European Directive 98/83/EC and the Legislative Decree 31/2001 – a look at the recent past. Ann. Ig. Med. Prev. Comunita. 30, 517–526.

Barickman, T.C., Kopsell, D.A., 2016. Nitrogen form and ratio impact Swiss chard (*Beta vulgaris* subsp. cicla) shoot tissue carotenoids and chlorophyll concentrations. Sci. Hortic. 204, 99–105.

Bell, L., Oloyede, O.O., Lignou, S., Wagstaff, C., Methven, L., 2018. Taste and flavour perceptions of glucosinolates, isothiocyanates and related compounds. Mol. Nutr. Food Res. 62, 1700990.

Birlanga, V., Acosta-Motos, J.R., Perez-Perez, J.M., 2021. Genotype-dependent tipburn severity during lettuce hydroponic culture is associated with altered nutrient leaf content. Agronomy 11, 616.

Blasco, B., Rios, J.J., Cervilla, L.M., Sanches-Rodriguez, E., Ruiz, J.M., Romero, L., 2008. Iodine biofortification and antioxidant capacity of lettuce: Potential benefits for cultivation and human health. Ann. Appl. Biol. 152, 289–299.

Bloem, E., Haneklaus, S., Schnug, E., 2010. Influence of fertilizer practices on S-containing metabolites in garlic (*Allium sativum* L.) under field conditions. J. Agric. Food Chem. 58, 10690–10696.
Blom-Zandstra, M., Lampe, J.E.M., 1985. The role of nitrate in the osmoregulation of lettuce (*Lactuca sativa* L.) grown at different light intensities. J. Exp. Bot. 36, 1043–1052.
Boersma, M., Gracie, A.J., Brown, P.H., 2009. Relationship between growth rate and the development of hollow stem in broccoli. Crop Pasture Sci. 60, 995–1001.
Bonasia, A., Conversa, G., Lazzizera, C., Elia, A., 2013. Pre-harvest nitrogen and azoxystrobin application enhances postharvest shelf-life in butterhead lettuce. Postharvest Biol. Technol. 85, 67–76.
Bondonno, C.P., Dalgaard, F., Blekkenhorst, L.C., Murray, K., Lewis, J.R., Croft, K.D., et al., 2021. Vegetable nitrate intake, blood pressure and incident cardiovascular disease: Danish diet, cancer and health study. Eur. J. Epidemiol. 36, 813–825.
Bouis, H.E., Saltzman, A., 2017. Improving nutrition through biofortification: a review of evidence from HarvestPlus, 2003 through 2016. Global Food Secur.-Agric. Policy 12, 49–58.
Bouis, H.E., Welch, R.M., 2010. Biofortification – a sustainable agricultural strategy for reducing micronutrient malnutrition in the global south. Crop Sci. 50, 20–32.
Brackhage, C., Huang, J.-H., Schaller, J., Elzinga, E.J., Dudel, E.G., 2014. Readily available phosphorus and nitrogen counteract for arsenic uptake and distribution in wheat (*Triticum aestivum* L.). Sci. Rep. 4, 4944.
Brunner, B.R., Freed, R.D., 1994. Oat grain beta-glucan content as affected by nitrogen level, location and year. Crop Sci. 34, 473–476.
Bulman, P., Zarkadas, C.G., Smith, D.L., 1994. Nitrogen fertilizer affects amino acid composition and quality of spring barley grain. Crop Sci. 34, 1341–1346.
Cakmak, I., Kutman, U.B., 2018. Agronomic biofortification of cereals with zinc: a review. Eur. J. Soil Sci. 69, 172–180.
Cakmak, I., Kalayci, M., Ekiz, H., Braun, H.J., Yilmaz, A., 1999. Zinc deficiency as an actual problem in plant and human nutrition in Turkey: a NATO-Science for Stability Project. Field Crops Res. 60, 175–188.
Cakmak, I., Welch, R.M., Erenoglu, B., Römheld, V., Norvell, W.A., Kochian, L.V., 2000. Influence of varied zinc supply on re-translocation of cadmium (^{109}Cd) and rubidium (^{86}Rb) applied on mature leaf of durum wheat seedlings. Plant Soil 219, 279–284.
Cakmak, I., Kalayci, M., Kaya, Y., Torun, A.A., Aydin, N., Wang, Y., et al., 2010. Biofortification and localization of zinc in wheat grain. J. Agric. Food Chem. 58, 9092–9102.
Carew, J.G., Morretini, M., Battey, N.H., 2010. Misshapen fruits in strawberry. Small Fruits Rev. 2, 37–50.
Carranca, C., Brunetto, G., Tagliavini, M., 2018. Nitrogen nutrition of fruit trees to reconcile productivity and environmental concerns. Plants 7, 4.
Chaney, R.L., 1980. Health risks associated with toxic metals in municipal sludge. In: Bitton, G. (Ed.), Sludge: Health Risks of Land Applications. Ann Arbor Science Publisher, Ann Arbor, MI, USA, pp. 59–83.
Chaney, R.L., Filcheva, E., Green, C.E., Brown, S.L., 2006. Zn deficiency promotes Cd accumulation by lettuce from biosolids amended soils with high Cd:Zn ratio. J. Residuals Sci. Technol. 3, 79–85.
Chapagain, B.P., Wiesman, Z., 2004. Effect of Nutri-Vant-PeaK foliar spray on plant development, yield and fruit quality in greenhouse tomatoes. Sci. Hortic. 102, 177–188.
Chenard, C.H., Kopsell, D.A., Kopsell, D.E., 2005. Nitrogen concentration affects nutrient and carotenoid accumulation in parsley. J. Plant Nutr. 28, 285–297.
Conversa, G., Bonasia, A., Lazzizera, C., Elia, A., 2014. Pre-harvest nitrogen and azoxystrobin application enhances raw product quality and postharvest shelf-life of baby spinach (*Spinacia oleracea* L.). J. Sci. Food Agric. 94, 3263–3272.
Cortazzo, J.A., Lichtman, A.D., 2014. Methemoglobinemia: a review and recommendations for management. J. Cardiothorac Vasc. Anesthesia 28, 1043–1047.
Costan, A., Stamatakis, A., Chrysargyris, A., Petropoulos, S.A., Tzortzakis, N., 2020. Interactive effects of salinity and silicon application on *Solanum lycopersicum* growth, physiology and shelf-life of fruit produced hydroponically. J. Sci. Food Agric. 100, 732–743.
Curtis, T.Y., Halford, N.G., 2016. Reducing the acrylamide-forming potential of wheat. Food Energy Secur. 5, 153–164.
Dai, F., Wang, J., Zhang, S., Xu, Z., Zhang, G., 2007. Genotypic and environmental variation in phytic acid content and its relation to protein content and malt quality in barley. Food Chem. 105, 606–611.
Dalgaard, T., Hansen, B., Hasler, B., Hertel, O., Hutchings, N.J., Jacobsen, B.H., et al., 2014. Policies for agricultural nitrogen management – trends, challenges and prospects for improved efficiency in Denmark. Environ. Res. Lett. 9, 115002.
Davarpanah, S., Tehranifar, A., Abadia, J., Val, J., Davarynejad, G., Aran, M., et al., 2018. Foliar calcium fertilization reduces fruit cracking in pomegranate (*Punica granatum* cv. Ardestani). Sci. Hortic. 230, 86–91.
Davis, D.R., 2009. Declining fruit and vegetable nutrition composition: what is the evidence? HortScience 44, 15–19.
De Angelis, V., Sanchez, E., Tognetti, J., 2011. Timing of nitrogen fertilization influences color and anthocyanin content of apple (*Malus domestica* Borkh. cv. 'Royal Gala'). Int. J. Fruit Sci. 11, 364–375.
de Freitas, S.T., Mitcham, E.J., 2012. Factors involved in fruit calcium deficiency disorders. Hortic. Rev. 40, 107–143.
de Valenca, A.W., Bake, A., Brouwer, I.D., Giller, K.E., 2017. Agronomic biofortification of crops to fight hidden hunger in sub-Saharan Africa. Global Food Secur. 12, 8–14.
Detterbeck, A., Pongrac, P., Rensch, S., Reuscher, S., Pecovnik, M., Vavpetic, P., et al., 2016. Spatially resolved analysis of variation in barley (*Hordeum vulgare*) grain micronutrient accumulation. New Phytol. 211, 1241–1254.
Dexter, J.E., Matsuo, R.R., Crowle, W.L., Kosmolak, F.G., 1982. Effect of nitrogen fertilization on the quality characteristics of five North American amber durum wheat cultivars. Can. J. Plant Sci. 62, 901–912.

Dirinck, P.J., de Pooter, H.L., Willaert, G.A., Schamp, N.M., 1981. Flavor quality of cultivated strawberries: the role of the sulfur compounds. J. Agric. Food Chem. 29, 316–321.

El Khoury, D., Cuda, C., Luhovyy, B.L., Anderson, G.H., 2012. Beta glucan: health benefits in obesity and metabolic syndrome. J. Nutr. Metab. 2012, 851362.

El-Motaium, R.A., Hashim, M.E., 2020. Boron efficiency in increasing olive (cv. Frantoio) fruit productivity and oil yield and quality. J. Plant Nutr. 43, 2981–2989.

Erenoglu, E.B., Kutman, U.B., Ceylan, Y., Yildiz, B., Cakmak, I., 2011. Improved nitrogen nutrition enhances root uptake, root-to-shoot translocation and remobilization of zinc (^{65}Zn) in wheat. New Phytol. 189, 438–448.

European Food Safety Authority, 2008. Opinion of the Scientific Panel on Contaminants in the Food Chain on a request from the European Commission to perform a scientific risk assessment on nitrate in vegetables. EFSA J. 689, 1–79.

Fan, M.S., Zhao, F.J., Fairweather-Tait, S.J., Poulton, P.R., Sunham, S.J., McGrath, S.P., 2008. Evidence of decreasing mineral density in wheat grain over the last 160 years. J. Trace Elem. Med. Biol. 22, 315–324.

Fernández-Escobar, R., Beltrán, G., Sánchez-Zamora, M.A., García-Novelo, J., Aguilera, M.P., Uceda, M., 2006. Olive oil quality decreases with nitrogen over-fertilization. Hort. Sci. 41, 215–219.

Ferree, D.C., Cahoon, G.A., 1987. Influence of leaf to fruit ratios and nutrient sprays on fruiting, mineral elements, and carbohydrates of apple trees. J. Am. Soc. Hort. Sci. 112, 445–449.

Ferreira, R.B., Picarra-Pereira, M.A., Monteiro, S., Loureiro, V.B., Teixeira, A.R., 2002. The wine proteins. Trends Food Sci. Technol. 12, 230–239.

Finster, M.E., Gray, K.A., Binns, H.J., 2004. Lead levels of edibles grown in contaminated residential soils: a field survey. Sci. Total Environ. 320, 245–257.

Fischer, J., 1992. The influence of different nitrogen and potassium fertilization on the chemical flavour composition of kohlrabi (*Brassica oleracea* var. *gongylodes* L.). J. Sci. Food Agric. 60, 465–470.

Food and Agriculture Organization, 2019. The State of Food and Agriculture 2019. Moving forward on food loss and waste reduction. Rome.

Frantz, J.M., Ritchie, G., Cometti, N.N., Robinson, J., Bugbee, B., 2004. Exploring the limits of crop productivity: beyond the limits of tipburn in lettuce. J. Am. Soc. Hort. Sci. 129, 331–338.

Fratianni, A., Giuzio, L., Criscio, T.D., Zina, F., Panfili, G., 2013. Response of carotenoids and tocols of durum wheat in relation to water stress and sulfur fertilization. J. Agric. Food Chem. 61, 2583–2590.

Fuertes-Mendizabal, T., Gonzales-Murua, C., Gonzalez-Moro, M.B., Estavillo, J.M., 2012. Late nitrogen fertilization affects nitrogen metabolism in wheat. J Plant Nutr. Soil Sci. 175, 115–124.

Gagliardi, A., Carucci, F., Masci, S., Flagella, Z., Gatta, G., Giuliani, M.M., 2020. Effects of genotype, growing season and nitrogen level on gluten protein assembly of durum wheat grown under Mediterranean conditions. Agronomy 10, 755.

Galili, G., Amir, R., 2013. Fortifying plants with the essential amino acids lysine and methionine to improve nutritional quality. Plant Biotechnol. J. 11, 211–222.

Gerendas, J., Führs, H., 2013. The significance of magnesium for crop quality. Plant Soil 368, 101–128.

Gerendas, J., Heuser, F., Sattelmacher, B., 2007. Influence of nitrogen and potassium supply on contents of acrylamide precursors in potato tubers and on acrylamide accumulation in French fries. J. Plant Nutr. 30, 1499–1516.

Ghafoor, A., Karim, H., Asghar, M.A., Raza, A., Hussain, M.I., Javed, H.H., et al., 2021. Carbohydrates accumulation, oil quality and yield of rapeseed genotypes at different nitrogen rates. Plant Prod. Sci. 25, 50–69.

Giuliano, G., 2017. Provitamin A biofortification of crop plants: a gold rush with many miners. Curr. Opin. Biotechnol. 44, 169–180.

Gonzales-Soltero, R., Bailen, M., de Lucas, B., Ramirez-Goercke, M.I., Pareja-Galeano, H., Larrosa, M., 2020. Role of oral and gut microbiota in dietary nitrate metabolism and its impact on sports performance. Nutrients 12, 3611.

Gormus, O., 2002. Effects of rate and time of potassium application on cotton yield and quality in Turkey. J. Agron. Crop Sci. 188, 382–388.

Graham, R.D., Ascher, J.S., Hynes, S.C., 1992. Selecting zinc-efficient cereal genotypes for soils of low zinc status. Plant Soil 146, 241–250.

Granstedt, R.C., Huffaker, R.C., 1982. Identification of the leaf vacuole as a major nitrate storage pool. Plant Physiol. 70, 410–413.

Grant, C.A., 2015. Influence of phosphate fertilizer on cadmium in agricultural soils and crops. In: Selim, H.M. (Ed.), Phosphate in Soils. CRC Press Inc, Boca Raton, FL, USA, pp. 143–155.

Granvogyl, M., Wieser, H., Koehler, P., von Tucher, S., Schieberle, P., 2007. Influence of sulfur fertilization on the amounts of free amino acids in wheat. Correlation with baking properties as well as with 3-aminopropionamide and acrylamide generation during baking. J. Agric. Food. Chem. 55, 4271–4277.

Güler, G., 2003. Barley grain beta-glucan content as affected by nitrogen and irrigation. Field Crops Res. 84, 335–340.

Hagan, N.D., Tabe, L.M., Molvig, L., Higgins, T.J.V., 2003. Modifying the amino acid composition of grains using gene technology. In: Vasil, I.K. (Ed.), Plant Biotechnology 2002 and Beyond. Springer, Dordrecht, The Netherlands, pp. 305–308.

Haider, F.U., Liqun, C., Coulter, J.A., Cheema, S.A., Wu, J., Zhang, R., et al., 2021. Cadmium toxicity in plants: impacts and remediation strategies. Ecotox. Environ. Safe. 211, 111887.

Hart, J.J., Welch, R.M., Norvell, W.A., Kochian, L.V., 2002. Transport interactions between cadmium and zinc in roots of bread and durum wheat seedlings. Physiol. Plant. 116, 73–78.

Hartz, T.K., Johnstone, P.R., Francis, D.M., Miyao, E.M., 2005. Processing tomato yield and fruit quality improved with potassium fertigation. HortScience 40, 1862–1867.

Heeb, A., Lundegardh, B., Savage, G., Ericsson, T., 2006. Impact of organic and inorganic fertilizers on yield, taste, and nutritional quality of tomatoes. J. Plant. Nutr. Soil Sci. 169, 535–541.

Hernandez-Perez, O.I., Valdez-Aguilar, L.A., Alia-Tejacal, I., Cartmill, A.D., Cartmill, D.L., 2020. Tomato fruit yield, quality, and nutrient status in response to potassium: calcium balance and electrical conductivity in the nutrient solution. J. Soil Sci. Plant Nutr. 20, 484−492.

Ho, L.C., White, P.J., 2005. A cellular hypothesis for the induction of blossom-end rot in tomato fruit. Ann. Bot. 95, 571−581.

Hoffmann, C.M., Märländer, B., 2005. Composition of harmful nitrogen in sugar beet (*Beta vulgaris* L.)−amino acids, betaine, nitrate−as affected by genotype and environment. Eur. J. Agron. 22, 255−265.

Huang, H., Yao, Q., Xia, E., Gao, L., 2018. Metabolomics and transcriptomics analyses reveal nitrogen influences on the accumulation of flavonoids and amino acids in young shoots of tea plant (*Camellia sinensis* L.) associated with tea flavor. J. Agric. Food Chem. 66, 9828−9838.

Huang, X., Duan, S., Wu, Q., Yu, M., Shabala, S., 2020. Reducing cadmium accumulation in plants: structure-function relations and tissue-specific operation of transporters in the spotlight. Plants 9, 223.

Huchette, O., Arnault, I., Auger, J., Bellamy, C., Trueman, L., Thomas, B., et al., 2007. Genotype, nitrogen fertility and sulphur availability interact to affect flavour in garlic (*Allium sativum* L.). J. Hortic. Sci. Biotechnol. 82, 79−88.

Hui, X., Luo, L., Wang, S., Cao, H., Huang, M., Shi, M., et al., 2019. Critical concentration of available soil phosphorus for grain yield and zinc nutrition of winter wheat in a zinc-deficient calcareous soil. Plant Soil 444, 315−330.

Iddir, M., Brito, A., Dingeo, G., Del Campo, S.S.F., Samouda, H., La Frano, M.R., et al., 2020. Strengthening the immune system and reducing inflammation and oxidative stress through diet and nutrition: considerations during the Covid-19 crisis. Nutrients 12, 1562.

Ierna, A., Mauro, R.P., Leonardi, C., Giuffrida, F., 2020. Shelf-life of bunched carrots as affected by nitrogen fertilization and leaf presence. Agronomy 10, 1982.

Imakumbili, M.L.E., Semu, E., Semoka, J.M.R., Abass, A., Mkamilo, G., 2020. Plant tissue analysis as a tool for predicting fertilizer needs for low cyanogenic glucoside levels in cassava roots: an assessment of its possible use. PLoS One 15, e0228641.

Islam, M.Z., Mele, M.A., Baek, J.P., Kang, H.-M., 2016. Cherry tomato qualities affected by foliar spraying with boron and calcium. Hortic. Environ. Biotechnol. 57, 46−52.

Janouskova, M., Pavlikova, D., 2010. Cadmium immobilization in the rhizosphere of arbuscular mycorrhizal plants by the fungal extraradical mycelium. Plant Soil 332, 511−520.

Jäpelt, R.B., Jakobsen, J., 2013. Vitamin D in plants: a review of occurrence, analysis, and biosynthesis. Front. Plant Sci. 4, 136.

Jiao, Y., Grant, C.A., Bailey, L.D., 2004. Effects of phosphorus and zinc fertilizer on cadmium uptake and distribution in flax and durum wheat. J. Sci. Food Agric. 84, 777−785.

Johnston, C.S., Sears, B., Perry, M., Knurick, J.R., 2017. Use of novel high-protein functional food products as part of a calorie-restricted diet to reduce insulin resistance and increase lean body mass in adults: a randomized controlled trial. Nutrients 9, 1182.

Kacjan Marsic, N., Necemer, M., Veberic, R., Poklar Ulrih, N., Skrt, M., 2019. Effect of cultivar and fertilization on garlic yield and allicin content in bulbs at harvest and during storage. Turk. J. Agric. For. 43, 414−429.

Kays, S., 1999. Preharvest factors affecting appearance. Postharvest Biol. Technol. 15, 233−247.

Khan, M.A., Khan, S., Khan, A., Alam, M., 2017. Soil contamination with cadmium, consequences and remediation using organic amendments. Sci. Total Environ. 601−602, 1591−1605.

Kiferle, C., Gonzali, S., Holwerda, H.T., Ibaceta, R.R., Perata, P., 2013. Tomato fruits: a good target for iodine biofortification. Front. Plant Sci. 4, 205.

Kilfoy, B.A., Zhang, Y., Park, Y., Holford, T.R., Schatzkin, A., Hollenbeck, A., et al., 2011. Dietary nitrate and nitrite and the risk of thyroid cancer in the NIH-AARP Diet and Health Study. Int. J. Cancer 129, 160−172.

Kim, H.J., Lee, S.S., Choi, B.Y., Kim, M.K., 2007. Nitrate intake relative to antioxidant vitamin intake affects gastric cancer risk: a case-controlled study in Korea. Nutr. Cancer 59, 185−191.

Kiymaz, S., Ertek, A., 2015. Yield and quality of sugar beet (*Beta vulgaris* L.) at different water and nitrogen levels under the climatic conditions of Kirsehir, Turkey. Agric. Water Manage. 158, 156−165.

Knobeloch, L., Salna, B., Hogan, A., Postle, J., Anderson, H., 2000. Blue babies and nitrate-contaminated well water. Environ. Health Perspect. 108, 675−678.

Knutsen, H.K., Alexander, J., Barregård, L., Bignami, M., Brüschweiler, B., Ceccatelli, S., et al., 2016. Erucic acid in feed and food. EFSA J. 14, e04593.

Köleli, N., Eker, S., Cakmak, I., 2004. Effect of zinc fertilization on cadmium toxicity in durum and bread wheat grown in zinc-deficient soil. Environ. Pollut. 131, 453−459.

Kopsell, D.A., Sams, C.E., Charron, C.S., Kopsell, D.E., Randle, W.M., 2007. Kale carotenoids remain stable while glucosinolates and flavor compounds respond to changes in selenium and sulfur fertility. Acta Hortic. 744, 303−310.

Kuscu, H., Turhan, A., Ozmen, N., Aydinol, P., Demir, A.O., 2014. Optimizing levels of water and nitrogen applied through drip irrigation for yield, quality, and water productivity of processing tomato (*Lycopersicon esculentum* Mill.). Hort. Environ. Biotechnol. 55, 103−114.

Kutman, U.B., Yildiz, B., Ozturk, L., Cakmak, I., 2010. Biofortification of durum wheat with zinc through soil and foliar applications of nitrogen. Cereal Chem. 87, 1−9.

Kutman, U.B., Yildiz, B., Cakmak, I., 2011. Effect of nitrogen on uptake, remobilization and partitioning of zinc and iron throughout the development of durum wheat. Plant Soil 342, 149−164.

Lee, S.K., Kader, A.A., 2000. Preharvest and postharvest factors influencing vitamin C content of horticultural crops. Postharvest Biol. Technol. 20, 207−220.

Lehrfeld, J., Wu, Y.V., 1991. Distribution of phytic acid in milled fractions of Scout 66 hard red winter wheat. J. Agric. Food Chem. 39, 1820−1824.

Leinonen, I., Iannetta, P.P.M., Rees, R.M., Russell, W., Watson, C., Barnes, A.P., 2019. Lysine supply is a critical factor in achieving sustainable global protein economy. Front. Sustainable Food Syst. 3, 1–11.

Lemoine, R., Camera, S.L., Atanassova, S., Dédaldéchamp, R., Allario, F., Pourtau, T., et al., 2013. Source-to-sink transport of sugar and regulation by environmental factors. Front. Plant Sci. 4, 272.

Lester, G.E., Jifon, J.L., Makus, D.J., 2006. Supplemental foliar potassium applications with or without a surfactant can enhance netted muskmelon quality. Hort. Sci. 41, 741–744.

Li, R.Y., Stroud, J.L., Ma, J.F., McGrath, S.P., Zhao, F.J., 2009. Mitigation of arsenic accumulation in rice with water management and silicon fertilization. Environ. Sci. Technol. 43, 3778–3783.

Li, W.P., Shi, H.B., Zhu, K., Zheng, Q., Xu, Z., 2017a. The quality of sunflower seed oil changes in response to nitrogen fertilizer. Agron. J. 109, 2499–2507.

Li, R., Liu, H.P., Hong, C.L., Dai, Z.X., Liu, J.W., Zhou, J., et al., 2017b. Iodide and iodate effects on the growth and fruit quality of strawberry. J. Sci. Food Agric. 97, 230–235.

Li, R., Li, D.W., Liu, H.P., Hong, C.L., Song, M.Y., Dai, Z.X., et al., 2017c. Enhancing iodine content and fruit quality of pepper (*Capsicum annuum* L.) through biofortification. Sci. Hortic. 214, 165–173.

Liu, C.W., Sung, Y., Chen, B.C., Lai, H.Y., 2014. Effects of nitrogen fertilizers on the growth and nitrate content of lettuce (*Lactuca sativa* L.). Int. J. Environ. Res. Public Health 11, 4427–4440.

Liu, X., Huang, Z., Li, Y., Xie, W., Li, W., Tang, X., et al., 2020a. Selenium-silicon (Se-Si) induced modulations in physio-biochemical responses, grain yield, quality, aroma formation and lodging in fragrant rice. Ecotox. Environ. Safe. 196, 110525.

Liu, Z., Huang, Y., Ji, X., Xie, Y., Peng, J., Eissa, M.A., et al., 2020b. Effects and mechanism of continuous liming on cadmium immobilization and uptake by rice grown on acid paddy soils. J. Soil Sci. Plant Nutr. 20, 2316–2328.

Liu, Y., Croft, K.D., Hodgson, J.M., Mori, T., Ward, N.C., 2020c. Mechanisms of the protective effects of nitrate and nitrite in cardiovascular and metabolic diseases. Nitric Oxide 96, 35–43.

Lundberg, J.O., Weitzberg, E., Cole, J.A., Benjamin, N., 2004. Nitrate, bacteria and human health. Nat. Rev. Microbiol. 2, 593–602.

Lyons, G., 2018. Biofortification of cereals with foliar selenium and iodine could reduce hypothyroidism. Front. Plant Sci. 9, 730.

Ma, F.J., Yamaji, N., Mitani, N., Xu, X.Y., Su, Y.H., McGrath, S.P., Zhao, F.J., 2008. Transporters of arsenite in rice and their role in arsenic accumulation in rice grain. Proc. Natl. Acad. Sci. USA 105, 9931–9935.

Ma, J.F., Shen, R.F., Shao, J.F., 2021. Transport of cadmium from soil to grain in cereal crops: A review. Pedosphere 31, 3–10.

Makino, A., 2011. Photosynthesis, grain yield, and nitrogen utilization in rice and wheat. Plant Physiol. 155, 125–129.

Märländer, B., 1990. Influence of nitrogen supply on yield and quality of sugar beet. J. Plant Nutr. Soil Sci. 153, 327–332.

Martín-Diana, A.B., Rico, D., Frías, J.M., Barat, J.M., Henehan, G.T.M., Barry-Ryan, C., 2007. Calcium for extending the shelf life of fresh whole and minimally processed fruits and vegetables: a review. Trends Food Sci. Technol. 18, 210–218.

Martinek, P., Klem, K., Váňová, M., Bartáčková, V., Večerková, L., Bucher, P., et al., 2009. Effects of nitrogen nutrition, fungicide treatment and wheat genotype on free asparagine and reducing sugars content as precursors of acrylamide formation in bread. Plant Soil Environ. 55, 187–195.

Martins, V., Garcia, A., Alhinho, A.T., Costa, P., Lanceros-Méndez, S., Costa, M.M.R., et al., 2020. Vineyard calcium sprays induce changes in grape berry skin, firmness, cell wall composition and expression of cell wall-related genes. Plant Physiol. Biochem. 150, 49–55.

Mccall, D., Willumsen, J., 1999. Effects of nitrogen availability and supplementary light on the nitrate content of soil-grown lettuce. J. Hortic. Sci. Biotechnol. 74, 458–463.

McCallum, J., Porter, N., Searle, B., Shaw, M., Bettjeman, B., McManus, M., 2005. Sulfur and nitrogen fertility affects flavour of field-grown onions. Plant Soil 269, 151–158.

Medrano-Macías, J., Mendoza-Villarreal, R., Robledo-Torres, V., Fuentes-Lara, L.O., Ramírez-Godina, F., Pérez-Rodríguez, M.Á., et al., 2018. The use of iodine, selenium, and silicon in plant nutrition for the increase of antioxidants in fruits and vegetables. In: Shalaby, E., Azzam, G.M. (Eds.), Antioxidants in Foods and Its Applications. IntechOpen Limited, London, United Kingdom, pp. 155–168.

Meheriuk, M., McKenzie, D.L., Neilsen, G.H., Hall, J.W., 1996. Fruit pigmentation of four green apple cultivars responds to urea sprays but not to nitrogen fertilization. Hortscience 31, 992–993.

Mendez-Costabel, M.P., Wilkinson, K.L., Bastian, S.E.P., Jordans, C., Mccarthy, M., Ford, C.M., et al., 2014. Effect of increased irrigation and additional nitrogen fertilisation on the concentration of green aroma compounds in *Vitis vinifera* L. Merlot fruit and wine. Aust. J. Grape Wine Res. 20, 80–90.

Mène-Saffrané, L., Pellaud, S., 2017. Current strategies for vitamin E biofortification of crops. Curr. Opin. Biotechnol. 44, 189–197.

Mimmo, T., Tiziani, R., Valentinuzzi, F., Lucini, L., Nicoletto, C., Sambo, P., et al., 2017. Selenium biofortification in *Fragaria × ananassa*: Implications on strawberry fruits quality, content of bioactive health beneficial compounds and metabolomic profile. Front. Plant Sci. 8, 1–12.

Mitchell, L.G., Grant, C.A., Racz, G.J., 2000. Effect of nitrogen application on concentration of cadmium and nutrient ions in soil solution and in durum wheat. Can. J. Soil Sci. 80, 107–115.

Mohebbi, S., Babalar, M., Zamani, Z., Askari, M.A., 2020. Influence of early season boron spraying and postharvest calcium dip treatment on cell-wall degrading enzymes and fruit firmness in 'Starking Delicious' apple during storage. Sci. Hortic. 259, 108822.

Moreno-Ortega, A., Moreno-Rojas, R., Martínez-Álvarez, J.R., González Estecha, M., Castro González, N.P., Amaro López, M.Á., 2017. Probabilistic risk analysis of mercury intake via food consumption in Spain. J. Trace Elem. Med. Biol. 43, 135–141.

Morgan, K.T., Rouse, R.E., Roka, F.M., Futch, S.H., Zekri, M., 2005. Leaf and fruit mineral content and peel thickness of 'Hamlin' orange. Proc. Fla. State Hort. Soc. 118, 19–21.

Morris, E.R., Ellis, R., 1989. Usefulness of the dietary phytic acid/zinc molar ratio as an index of zinc bioavailability to rats and humans. Biol. Trace Elem. Res. 19, 107–117.

Mozafar, A., 1993. Nitrogen fertilizers and the amount of vitamins in plants: A review. J. Plant Nutr. 16, 2479–2506.

Mozafar, A., 1996. Decreasing the NO_3 and increasing the vitamin C contents in spinach by a nitrogen deprivation method. Plant Food Hum. Nutr. 49, 155–162.

Muttucumaru, N., Powers, S.J., Elmore, J.S., Mottram, D.S., Halford, N.G., 2013. Effects of nitrogen and sulfur fertilization on free amino acids, sugars, and acrylamide-forming potential in potato. J. Agric. Food Chem. 61, 6734–6742.

Nagy, P.T., Kincses, I., Nyéki, J., Soltész, M., Szabó, Z., 2011. Importance of boron in fruit nutrition. Int. J. Hort. Sci. 17, 39–44.

Naumann, M., Koch, M., Thiel, H., Gransee, A., Pawelzik, E., 2020. The importance of nutrient management for potato production. Part II: Plant nutrition and tuber quality. Potato Res. 63, 121–137.

Nava, G., Dechen, A.R., Nachtigall, G.R., 2008. Nitrogen and potassium fertilization affect apple fruit quality in southern Brazil. Commun. Soil Sci. Plant Anal. 39, 96–107.

O'Sullivan, S.M., Ball, M.E.E., McDonald, E., Hull, G.L.J., Danaher, M., Cashman, K.D., 2020. Biofortification of chicken eggs with vitamin K - nutritional and quality improvements. Foods 6, 1619.

Ojeda-Real, L.A., Lobit, P., Cárdenas-Navarro, R., Grageda-Cabrera, O., Farías-Rodríguez, R., Valencia-Cantero, E., et al., 2009. Effect of nitrogen fertilization on quality markers of strawberry (*Fragaria* × *ananassa* Duch. cv. Aromas). J. Sci. Food Agric. 89, 935–939.

Omirou, M.D., Papadopoulou, K.K., Papastylianou, I., Constantinou, M., Karpouzas, D.G., Asimakopoulos, I., et al., 2009. Impact of nitrogen and sulfur fertilization on the composition of glucosinolates in relation to sulfur assimilation in different plant organs of broccoli. J. Agric. Food Chem. 57, 9408–9417.

Ozkutlu, F., Ozturk, L., Erdem, H., McLaughlin, M., Cakmak, I., 2007. Leaf-applied sodium chloride promotes cadmium accumulation in durum wheat grain. Plant Soil 290, 323–331.

Ozturk, L., Yazici, M.A., Yucel, C., Torun, A., Cekic, C., Bagci, A., et al., 2006. Concentration and localization of zinc during seed development and germination in wheat. Physiol. Plant. 128, 144–152.

Paciolla, C., Fortunato, S., Dipierro, N., Paradiso, A., De Leonardis, S., Mastropasqua, L., et al., 2019. Vitamin C in plants: From functions to biofortification. Antioxidants 8, 519.

Pandurangan, S., Sandercock, M., Beyaert, R., Conn, K.L., Hou, A., Marsolais, F., 2015. Differential response to sulfur nutrition of two common bean genotypes differing in storage protein composition. Front. Plant Sci. 6, 1–11.

Peris-Felipo, F.J., Benavent-Gil, Y., Hernández-Apaolaza, L., 2020. Silicon beneficial effects on yield, fruit quality and shelf-life of strawberries grown in different culture substrates under different iron status. Plant Physiol. Biochem. 152, 23–31.

Persson, D.P., Hansen, T.H., Laursen, K.H., Schjoerring, J.K., Husted, S., 2009. Simultaneous iron, zinc, sulfur and phosphorus speciation analysis of barley grain tissues using SEC-ICP-MS and IP-ICP-MS. Metallomics 1, 418–426.

Persson, D.P., de Bang, T.C., Pedas, P.R., Kutman, U.B., Cakmak, I., Andersen, B., 2016. Molecular speciation and tissue compartmentation of zinc in durum wheat grains with contrasting nutritional status. New Phytol. 211, 1255–1265.

Pettigrew, W.T., 1999. Potassium deficiency increases specific leaf weights and leaf glucose levels in field-grown cotton. Agron. J. 91, 962–968.

Peyrot des Gachons, C., Van Leeuwen, C., Tominaga, T., Soyer, J.P., Gaudille, J.P., Dubourdieu, D., 2005. Influence of water and nitrogen deficit on fruit ripening and aroma potential of *Vitis vinifera* L cv Sauvignon blanc in field conditions. J. Sci. Food Agric. 85, 73–85.

Podar, D., Ramsey, M.H., Hutchings, M.J., 2004. Effect of cadmium, zinc and substrate heterogeneity on yield, shoot metal concentration and metal uptake by *Brassica juncea*: implications for human health risk assessment and phytoremediation. New Phytol. 163, 313–324.

Poiroux-Gonord, F., Bidel, L.P.R., Fanciullino, A.-L., Gautier, H., Lauri-Lopez, F., Urban, L., 2010. Health benefits of vitamins ad secondary metabolites of fruits and vegetables and prospects to increase their concentrations by agronomic approaches. J. Agric. Food Chem. 58, 12065–12082.

Poisson, E., Trouverie, J., Brunel-Muguet, S., Akmouche, Y., Pontet, C., Pinochet, X., Avice, J.C., 2019. Seed yield components and seed quality of oilseed rape are impacted by sulfur fertilization and its interactions with nitrogen fertilization. Front. Plant Sci. 10, 458.

Prom-u-thai, C., Huang, L., Rerkasem, B., Thomson, G., Kuo, J., Saunders, M., et al., 2008. Distribution of protein bodies and phytate-rich inclusions in grain tissues of low and high iron rice genotypes. Cereal Chem. 85, 257–265.

Prom-u-thai, C., Rashid, A., Ram, H., Zou, C., Guilherme, L.R.G., Corguinha, A.P.B., et al., 2020. Simultaneous biofortification of rice with zinc, iodine, iron and selenium through foliar treatment of a micronutrient cocktail in five countries. Front. Plant Sci. 11, 589835.

Raboy, V., 2020. Low phytic acid crops: Observations based on four decades of research. Plants 9, 140.

Raese, J.T., Drake, S.R., 1995. Calcium sprays and timing affect fruit calcium concentrations, yield, fruit weight, and cork spot of 'Anjou' pears. Hortscience 30, 1037–1039.

Randle, W.M., Lancaster, J.E., Shaw, M.L., Sutton, K.H., Hay, R.L., Bussard, M.L., 1995. Quantifying onion flavor compounds responding to sulfur fertility-sulfur increases levels of alk(en)yl cysteine sulfoxides and biosynthetic intermediates. J. Amer. Soc. Hort. Sci. 120, 1075–1081.

Read, J.J., Reddy, K.R., Jenkins, J.N., 2006. Yield and fiber quality of upland cotton as influenced by nitrogen and potassium nutrition. Eur. J. Agron. 24, 282–290.

Reay, P.F., Fletcher, R.H., Thomas, V.J., 1998. Chlorophylls, carotenoids and anthocyanin concentrations in the skin of 'Gala' apples during maturation and the influence of foliar applications of nitrogen and magnesium. J. Sci. Food Agric. 76, 63–71.

Reddy, K.R., Koti, S., Davidonis, G.H., Reddy, V.R., 2004. Interactive effects of carbon dioxide and nitrogen nutrition on cotton growth, development, yield, and fiber quality. Agron. J. 96, 1148–1157.

Rehman, H., Iqbal, Q., Farooq, M., Wahid, A., Aflaz, I., Basra, S.M.A., 2013. Sulphur application improves the growth, seed yield and oil quality of canola. Acta Physiol. Plant. 35, 2999–3006.

Reif, C., Arrigoni, E., Neuweiler, R., Baumgartner, D., Nyström, L., Hurrell, R.F., 2012. Effect of sulfur and nitrogen fertilization on the content of nutritionally relevant carotenoids in spinach (*Spinacia oleracea*). J. Agric. Food Chem. 60, 5819–5824.

Rekowski, A., Wimmer, M.A., Henkelmann, G., Zörb, C., 2019. Is a change of protein composition after late application of nitrogen sufficient to improve the baking quality of winter wheat? Agriculture 9, 101.

Renquist, A.R., Reid, J.B., 1998. Quality of processing tomato (*Lycopersicon esculentum*) fruit from four bloom dates in relation to optimal harvest timing. N. Z. J. Crop Hortic. Sci. 26, 161–168.

Rerkasem, B., Jamjod, S., Pusadee, T., 2020. Productivity limiting impacts of boron deficiency, a review. Plant Soil 455, 23–40.

Ribeiro, L.R., Leonel, S., Souza, J.M.A., Garcia, E.L., Leonel, M., Monteiro, L.N.H., Silva, M.S., Ferreira, R.B., 2020. Improving the nutritional value and extending shelf life of red guava by adding calcium chloride. LWT Food Sci. Technol. 130, 109655.

Rios, J.J., Blasco, B., Cervilla, L.M., Rosales, M.A., Sanchez-Rodriguez, E., Romero, L., et al., 2009. Production and detoxification of H_2O_2 in lettuce plants exposed to selenium. Ann. Appl. Biol. 154, 107–116.

Robbins, R.J., Keck, A.S., Banuelos, G., Finley, J.W., 2005. Cultivation conditions and selenium fertilization alter the phenolic profile, glucosinolate, and sulforaphane content of broccoli. J. Med. Food 8, 204–214.

Roe, M., Faulks, R.M., Belsten, J.L., 1990. Role of reducing sugars and amino acids in fry colour of chips from potatoes grown under different nitrogen regimes. J. Sci. Food Agric. 52, 207–214.

Rohner, F., Zimmermann, M., Jooste, P., Pandav, C., Caldwell, K., Raghavan, R., et al., 2014. Biomarkers of nutrition for development – iodine review. J. Nutr. 144, 1322S–1342S.

Rushovich, D., Weil, R., 2021. Sulfur fertility management to enhance methionine ad cysteine in soybeans. J. Sci. Food Agric. 101, 6595–6601.

Sams, C.E., 1999. Preharvest factors affecting postharvest texture. Postharvest Biol. Technol. 15, 249–254.

Santamaria, P., 2006. Nitrate in vegetables: toxicity, content, intake and EC regulation. J. Sci. Food Agric. 86, 10–17.

Saure, M.C., 2014. Why calcium deficiency is not the cause of blossom-end rot in tomato and pepper fruit – a reappraisal. Sci. Hortic. 274, 151–154.

Scaife, A., Schloemer, S., 1994. The diurnal pattern of nitrate uptake and reduction by spinach (*Spinacia oleracea* L.). Ann. Bot. 73, 337–343.

Schaller, R.G., Schnitzler, W.H., 2000. Nitrogen nutrition and flavour compounds of carrots (*Daucus carota* L.) cultivated in Mitscherlich pots. J. Sci. Food Agric. 80, 49–56.

Schellenberg, D.L., Bratsch, A.D., Shen, Z., 2009. Large single-head broccoli yield as affected by plant density, nitrogen, and cultivar in a plasticulture system. HortTechnology 19, 792–795.

Semba, R.D., 2016. The rise and fall of protein malnutrition in global health. Ann. Nutr. Metab. 69, 79–88.

Sharma, R.R., Singh, R., 2008. Fruit nutrient content and lipoxygenase activity in relation to the production of malformed and button berries in strawberry (*Fragaria x ananassa* Duch.). Sci. Hortic. 119, 28–31.

Shute, T., Macfie, S.M., 2006. Cadmium and zinc accumulation in soybean: a threat to food safety? Sci. Total Environ. 371, 63–73.

Silva, E.O., Bracarense, A.P.F.R.L., 2016. Phytic acid: from antinutritional to multiple protection factor of organic systems. J. Food Sci. 81, R1357–R1362.

Singh, D., Prasanna, R., 2020. Potential of microbes in the biofortification of Zn and Fe in dietary food grains. A review. Agron. Sustainable Dev. 40, 15.

Sissons, M., Cutillo, S., Marcotuli, I., Gadaleta, A., 2021. Impact of durum wheat protein content on spaghetti in vitro starch digestion and technological properties. J. Cereal Sci. 98, 103156.

Smit, I., Pfliehinger, M., Binner, A., Grossmann, M., Horst, W.J., Lohnertz, O., 2014. Nitrogen fertilisation increases biogenic amines and amino acid concentrations in *Vitis vinifera* var. *Riesling* musts and wines. J. Sci. Food Agric. 94, 2064–2072.

Smolen, S., Wierzbinska, J., Sady, W., Kolton, A., Wiszniewska, A., Liszka-Skoczylas, M., 2015. Iodine biofortification with additional application of salicylic acid affects yield and selected parameters of chemical composition of tomato fruits (*Solanum lycopersicum* L.). Sci. Hortic. 188, 89–96.

Smolen, S., Kowalska, I., Kocacik, P., Sady, W., Grzanka, M., Kutman, U.B., 2019. Changes in the chemical composition of six lettuce cultivars (*Lactuca sativa* L.) in response to biofortification with iodine and selenium combined with salicylic acid application. Agronomy 9, 660.

Solhjoo, S., Gharaghani, A., Fallahi, E., 2017. Calcium and potassium foliar sprays affect fruit skin color, quality attributes, and mineral nutrient concentrations of 'Red Delicious' apples. Int. J. Fruit Sci. 17, 358–373.

Song, P., Wu, L., Guan, W., 2015. Dietary nitrates, nitrites, and nitrosamines intake and the risk of gastric cancer: A meta-analysis. Nutrients 7, 9872–9895.

Souza, S.R., Stark, E.M.L., Fernandes, M.S., 1999. Foliar spraying of rice with nitrogen: effect on protein levels, protein fractions, and grain weight. J. Plant Nutr. 22, 579–588.

Spayd, S.E., Wample, R.L., Evans, R.G., Stevens, R.G., Seymour, B.J., Nagel, C.W., 1994. Nitrogen fertilization of white riesling grapes in Washington. Must and wine composition. Am. J. Enol. Vitic. 45, 34–42.

Spence, C., 2015. Just how much of what we taste derives from the sense of smell? Flavour 4, 30.

Stangoulis, J.C.R., Knez, M., 2022. Biofortification of major crop plants with iron and zinc-achievements and future directions. Plant Soil 474, 57–76.

Sterckeman, T., Thomine, S., 2020. Mechanisms of cadmium accumulation in plants. Crit. Rev. Plant Sci. 39, 322–359.

Stockfisch, N., Koch, H.J., 2002. Reaction of sugar beet to boron fertilizer application in pot experiments. In: Goldbach, H.E., Rerkasem, B., Wimmer, M.A., Brown, P., Thellier, M., Bell, R. (Eds.), Boron in Plant and Animal Nutrition. Kluwer Academic, New York, pp. 381–385.

Strobbe, S., van der Straeten, D., 2018. Toward eradication of B-vitamin deficiencies: considerations for crop biofortification. Front. Plant Sci. 9, 443.

Su, Y.H., McGrath, S.P., Zhao, F.J., 2010. Rice is more efficient in arsenite uptake and translocation than wheat and barley. Plant Soil 328, 27−34.

Sun, N., Wang, Y., Gupta, S.K., Rosen, C.J., 2019. Nitrogen fertility and cultivar effects on potato agronomic properties and acrylamide-forming potential. Soil Fert. Crop Nutr. 111, 408−418.

Sun, Z., Yue, Z., Liu, H., Ma, K., Li, C., 2021a. Microbial-assisted wheat iron biofortification using endophytic *Bacillus altitudinis* WR10. Front. Nutr. 8, 704030.

Sun, G.X., Zhang, L., Chen, P., Yao, B.M., 2021b. Silicon fertilizers mitigate rice cadmium and arsenic uptake in a 4-year field trial. J. Soils Sediments 21, 163−171.

Taber, H., Perkins-Veazie, P., Li, S., White, W., Rodermel, S., Xu, Y., 2008. Enhancement of tomato fruit lycopene by potassium is cultivar dependent. Hort. Sci. 43, 159−165.

Tao, Y., Zhang, S., Jian, W., Yuan, C., Shan, X., 2006. Effects of oxalate and phosphate on the release of arsenic from contaminated soils and arsenic accumulation in wheat. Chemosphere 65, 1281−1287.

Taylor, M.D., Locascio, S.J., 2004. Blossom-end rot: a calcium deficiency. J. Plant. Nutr. 27, 123−139.

Tesso, T., Ejeta, G., Chandrashekar, A., Haung, C.-P., Tandjung, A., Lewamy, M., et al., 2006. A novel modified endosperm texture in a mutant high-protein digestibility/high-lysine grain sorghum (*Sorghum bicolor* (L.) Moench). Cereal Chem. 83, 194−201.

Tewolde, H., Fernandez, C.J., 2003. Fiber quality response of Pima cotton to nitrogen and phosphorus deficiency. J. Plant. Nutr. 26, 223−235.

Tewolde, H., Fernandez, C.J., Foss, D.C., 1994. Maturity and lint yield of nitrogen- and phosphorus-deficient Pima cotton. Agron. J. 86, 303−309.

Tilman, D., Cassman, K.G., Matson, P.A., Naylor, R., Polasky, S., 2002. Review article: agricultural sustainability and intensive production practices. Nature 418, 671−677.

Timmer, C.P., 2014. Food security in Asia and the Pacific: the rapidly changing role of rice. Asia Pac. Policy Stud. 1, 73−90.

Timms, M., Bottomley, R.C., Ellis, J.R.S., Schofield, J.D., 1981. The baking quality and protein characteristics of a winter wheat grown at different levels of nitrogen fertilisation. J. Sci. Food Agric. 32, 684−698.

Tiso, M., Schechter, A.N., 2015. Nitrate reduction to nitrite, nitric oxide and ammonia by gut bacteria under physiological conditions. PLoS One 10, e0119712.

Torres, E., Recasens, I., Lordan, J., Alegre, S., 2017. Combination of strategies to supply calcium and reduce bitter pit in 'Golden Delicious' apples. Sci. Hort. 217, 179−188.

Van Sluyter, S.C., McRae, J.M., Falconer, R.J., Smith, P.A., Bacic, A., Waters, E.J., et al., 2015. Wine protein haze: mechanisms of formation and advances in prevention. J. Agric. Food Chem. 63, 4020−4030.

Varga, B., Svecnjak, Z., 2006. The effect of late-season urea spraying on grain yield and quality of winter wheat cultivars under low and high basal nitrogen fertilization. Field Crops Res. 96, 125−132.

Varvel, G.E., Severson, R.K., 1987. Evaluation of cultivar and nitrogen management options for malting barley. Agron. J. 79, 459−463.

Velu, A., Ortiz-Monasterio, I., Cakmak, I., Hao, Y., Singh, R.P., 2014. Biofortification strategies to increase grain zinc and iron concentrations in wheat. J. Cereal Sci. 59, 365−372.

Villette, J., Cuellar, T., Verdeil, J.L., Delrot, S., Gaillard, I., 2020. Grapevine potassium nutrition and fruit quality in the context of climate change. Front. Plant Sci. 11, 123.

von Grebmer, K., Saltzman, A., Birol, E., Wiesmann, D., Prasai, N., Yin, S., et al., 2014. Global hunger index − the challenge of hidden hunger. International Food Policy Research Institute, Washington D.C., USA.

von Lintig, J., 2012. Provitamin A metabolism and functions in mammalian biology. Am. J. Clin. Nutr. 96, 1234S−1244S.

Voogt, W., Holwerda, H.T., Khodabaks, R., 2010. Biofortification of lettuce (*Lactuca sativa* L.) with iodine: the effect of iodine form and concentration in the nutrient solution on growth, development and iodine uptake of lettuce grown in water culture. J. Sci. Food Agric. 90, 906−913.

Wang, Y.T., Liu, R.L., Huang, S.W., Jin, J.Y., 2009. Effects of potassium application on flavor compounds of cherry tomato fruits. J. Plant. Nutr. 32, 1451−1468.

Wang, M., Liangi, B., Zhang, W., Cheni, K., Zhang, Y., Zhou, H., et al., 2019. Dietary lead exposure and associated health risks in Guangzhou, China. Int. J. Environ. Res. Public Health 16, 1417.

Wang-Pruski, G., Nowak, J., 2004. Potato after-cooking darkening. Am. J. Potato Res. 81, 7−16.

Watanabe, F., Yabuta, Y., Bito, T., Teng, F., 2014. Vitamin B12-containing plant food sources for vegetarians. Nutrients 6, 1861−1873.

Weber, E.A., Koller, W.D., Graeff, S., Hermann, W., Merkt, N., Claupein, W., 2008. Impact of different nitrogen fertilizers and an additional sulfur supply on grain yield, quality, and the potential of acrylamide formation in winter wheat. J. Plant Nutr. Soil Sci. 171, 643−655.

Weggler, K., McLaughlin, M.J., Graham, R.D., 2004. Effect of chloride in soil solution on the plant availability of biosolid-borne cadmium. J. Environ. Qual. 33, 496−504.

West, K.P., Darnton-Hill, I., 2008. Vitamin A deficiency. In: Sembra, R.D., Bloem, M.W., Piot, P. (Eds.), Nutrition and Health in Developing Countries, 2nd Ed. Humana Press, Totowa, NJ, USA.

White, P.J., Broadley, M.R., 2009. Biofortification of crops with seven mineral elements often lacking in human diets − iron, zinc, copper, calcium, magnesium, selenium and iodine. New Phytol. 182, 49−84.

Wieser, Seilmeier, 1998. The influence of nitrogen fertilisation on quantities and proportions of different protein types in wheat flour. J. Sci. Agric. 76, 49−55.

Wieser, H., Gutser, R., von Tucker, S., 2004. Influence of sulphur fertilisation on quantities and proportions of gluten protein types in wheat flour. J. Cereal Sci. 40, 239−244.

Wirth, J.P., Petry, N., Tanumihardjo, S.A., Rogers, L.M., McLean, E., Greig, A., et al., 2017. Vitamin A supplementation programs and country-level evidence of vitamin A deficiency. Nutrients 9, 190.

Wong, H.L., Murphy, S.E., Hecht, S.S., 2005. Cytochrome P450 2A-catalyzed metabolic activation of structurally similar carcinogenic nitrosamines: N'-nitrosonornicotine enantiomers, N'-nitrosopiperidine and N-nitrosopyrrolidine. Chem. Res. Toxicol. 18, 61–69.

Wu, F., Wesseler, J., Zilberman, D., Russell, R.M., Chen, C., Dubock, A.C., 2021. Allow Golden Rice to save lives. Proc. Natl. Acad. Sci. USA 118, e2120901118.

Wyllie, S.G., Leach, D.N., 1992. Sulfur-containing compounds in the aroma volatiles of melons (*Cucumis melo*). J. Agric. Food Chem. 40, 253–256.

Xie, K., Cakmak, I., Wang, S., Zhang, F., Guo, S., 2021. Synergistic and antagonistic interactions between potassium and magnesium in higher plants. Crop J. 9, 249–256.

Xu, W., Wang, P., Yuan, L., Chen, X., Hu, X., 2021. Effects of application methods of boron on tomato growth, fruit quality and flavor. Horticulturae 7, 223.

Xue, C., Erley, G.S., Rossmann, A., Schuster, R., Koehler, P., Mühling, K.H., 2016. Split nitrogen application improves wheat baking quality by influencing protein composition rather than concentration. Front. Plant Sci. 7, 738.

Yan, W., Fregeau-Reid, J., Mai, B.L., Pageau, D., Vera, C., 2017. Nitrogen fertilizer complements breeding in improving yield and quality of milling oat. Crop Sci. 57, 3291–3302.

Yang, S., Zou, Y., Liang, Y., Xia, B., Liu, S., Md, I., et al., 2012. Role of soil total nitrogen in aroma synthesis of traditional regional aromatic rice in China. Field Crops Res. 125, 151–160.

Yasin, M., El-Mehdawi, A.F., Anwar, A., Pilon-Smits, E.A.H., Faisal, M., 2015. Microbial-enhanced selenium and iron biofortification of wheat (*Triticum aestivum* L.) applications in phytoremediation and biofortification. Int. J. Phytorem. 17, 341–347.

Yazici, M.A., Asif, M., Tutus, Y., Ortas, I., Ozturk, L., Lambers, H., et al., 2021. Reduced root mycorrhizal colonization as affected by phosphorus fertilization is responsible for high cadmium accumulation in wheat. Plant Soil 468, 19–35.

Yener, H., Altuntas, O., 2020. Effects of potassium fertilization on leaf nutrient content and quality attributes of sweet cherry fruits (*Prunus avium* L.). J. Plant Nutr. 44, 946–957.

Yu, S., Tian, L., 2018. Breeding major cereal grains through the lens of nutrition sensitivity. Mol. Plant. 11, 23–30.

Zhang, P., Ma, G., 2017a. Effect of irrigation and nitrogen application on grain amino acid composition and protein quality in winter wheat. PLoS One 12, e0178494.

Zhang, M., Liand, Y., Chu, G., 2017b. Applying silicate fertilizer increases both yield and quality of table grape (*Vitis vinifera* L.) grown on calcareous grey desert soil. Sci. Hort. 225, 757–763.

Zhang, J., Liang, Z., Jiao, D., Tian, X., Wang, C., 2018. Different water and nitrogen fertilizer rates effects on growth and development of spinach. Commun. Soil Sci. Plant Anal. 49, 1922–1933.

Zhao, F.J., Wand, P., 2020. Arsenic and cadmium accumulation in rice and mitigation strategies. Plant Soil 446, 1–21.

Zhao, F.J., Evans, E.J., Bilsborrow, P.E., Syers, J.K., 1993. Influence of sulphur and nitrogen on seed yield and quality of low glucosinolate oilseed rape (*Brassica napus* L). J. Sci. Food Agric. 63, 29–37.

Zhao, F.J., Hawkesford, M.J., McGrath, S.P., 1999. Sulphur assimilation and effects on yield and quality of wheat. J. Cereal Sci. 30, 1–17.

Zhou, C., Zhu, Y., Lou, Y., 2013. Effects of sulfur fertilization on the accumulation of health-promoting phytochemicals in radish sprouts. J. Agric. Food Chem. 61, 7552–7559.

Zhou, J., Zhang, C., Du, B., Cui, H., Fan, X., Zhou, D., et al., 2020. Effects of zinc application on cadmium (Cd) accumulation and plant growth through modulation of the antioxidant system and translocation of Cd in low- and high-Cd wheat cultivars. Environ. Pollut. 265, 115045.

Zou, C., Du, Y., Rashid, A., Ram, H., Savasli, E., Pieterse, P.J., et al., 2019. Simultaneous biofortification of wheat with zinc, iodine, selenium, and iron through foliar treatment of a micronutrient cocktail in six countries. J. Agric. Food Chem. 67, 8096–8106.

Chapter 10

Relationship between mineral nutrition, plant diseases, and pests

Markus Weinmann[1], Klára Bradáčová[1], and Miroslav Nikolic[2]
[1]Institute of Crop Science, University of Hohenheim, Stuttgart, Germany, [2]Institute for Multidisciplinary Research, University of Belgrade, Belgrade, Serbia

Summary

Mineral nutrient supply or tissue concentrations affect plant diseases and pests in various ways. The supply of nutrients changes the resistance of plants to pathogens and pests by altering growth and tissue composition (e.g., concentration of soluble compounds or defense compounds), the expression of plant defense and pathogen/pest virulence genes, and the interactions with beneficial soil microorganisms. Depending on the pathogen or pest and nutrient, adequate nutrient supply for optimal plant growth may increase or decrease disease incidence. This chapter discusses the effects of nutrient supply on foliar and soil-borne fungal and bacterial diseases as well as on pests. Calcium and B inhibit pathogen invasion via their stabilizing effect on cell walls and membranes. Silicon and Mn play important roles in defense reactions to pathogen infection or pest attack, whereas N and K exert their effects mainly via modulating the concentrations of soluble compounds in plant tissues. The effects of crop management (e.g., timing and form of fertilizers, liming, and/or biological agents) on disease incidence are outlined.

10.1 General

The effects of nutrients on plant growth and yield are usually explained in terms of the functions of these elements in plant metabolism. However, nutrition may also have secondary, often unpredicted, effects on the growth and yield of crop plants by inducing changes in root and shoot growth patterns, plant morphology, anatomy, and/or chemical composition as well as the plant microbiome, which may alter the resistance of plants to pathogens and pests (Rabelo de Faria et al., 2021). Resistance is mainly determined by the capacity of the host to limit entry, development, and/or reproduction of invading pathogens, or limit the feeding by pests. Tolerance, by contrast, is the capacity of the host plant to maintain its growth despite the infection or pest attack. Depending on the nutrient, the plant nutritional status, plant species, and type of pathogen or pest, nutrition may influence plant resistance or tolerance, virulence of the pathogen or the activity of antagonists and other beneficial organisms in the plant vicinity differently. Diseases can decrease nutrient availability and impair uptake, distribution, and/or utilization by plant; importantly, the symptoms of disease may reflect the altered plant nutritional status (Armijo et al., 2016; Huber, 2021; Mee et al., 2017).

Soil health, as a biological component of soil quality, is vital to both healthy crop growth in agronomic production systems and the ecosystem services provided by soils (Doran, 2002; Harris and Romig, 2005). Healthy soils represent living buffer systems that provide an adequate supply of mineral nutrients to growing plants and limit pathogenesis (Primavesi, 2006; Robertson and Grandy, 2006). Crop rotations are used to manage soil fertility and health to improve the availability and use efficiency of nutrients and avoid the build-up of disease, thereby decreasing the need for fertilizers and pesticides (Shah et al., 2021). The increased incidence of soil-borne pathogens in replant-diseased soils is seen as a sign of biological imbalance in the soil ecosystem (Pankhurst and Lynch, 2005). Agricultural strategies to optimize

---

☆ This chapter is a revision of the third edition chapter by D. Huber, V. Römheld, and M. Weinmann pp. 283–298. DOI: https://doi.org/10.1016/B978-0-12-384905-2.00010-8. © Elsevier Ltd.

the supply of mineral nutrients and/or to promote beneficial soil microorganisms include, for example, (1) adapted rhizosphere management to enhance specific soil indigenous microbial communities, (2) addition of selected microbial strains as the seed or soil inoculants, and (3) application of specific mineral nutrients (Pankhurst and Lynch, 2005; Römheld and Neumann, 2006).

Although the plant susceptibility or resistance depends on the genetics of the host and the pathogen or pest, mineral nutrients play particularly important roles in plant-pathogen and plant-pest interactions (Agrios, 2005; Staskawicz, 2001) by influencing plant growth and development as well as the expression of plant defense and pathogen virulence genes (Huber and Graham, 1999; Walters and Bingham, 2007). For example, increased concentrations of Si and Mn in foliar tissues have been associated with increased resistance against powdery mildew in wheat (Colquhoun, 1940) and grapevine (Fig. 10.1).

As shown in extensive literature surveys (Dordas, 2008; Gupta et al., 2017; Huber and Graham, 1999), all mineral nutrients can influence the severity of plant diseases. Depending on the plant genotype, the environmental conditions, as well as the nutrient amount and chemical form, a particular mineral nutrient may either suppress or exacerbate disease. Importantly, a balanced nutrient supply that is optimal for plant growth should also provide optimal conditions for resistance to a broad spectrum of diseases. By contrast, a deficient or excessive supply of mineral nutrients often predisposes plants to diseases. Such a clear situation is for instance shown in Fig. 10.2 for *Pelargonium* plants. Nevertheless, due to their specific functions in distinct plant-pathogen interactions, the supply of certain mineral elements necessary for an effective expression of resistance mechanisms by plants may exceed their basic needs for an optimal growth (Huber and Graham, 1999). Some pathogens increase plant disease susceptibility by immobilizing mineral nutrients in the rhizosphere or in infected host tissues, whereas others cause disturbances in mineral nutrition by interfering with nutrient uptake, translocation, or use efficiency in plants (Schulze et al., 1995; Thompson et al., 2006). Many of the beneficial microorganisms in soil and the rhizosphere can contribute to the mobilization and plant acquisition of critical mineral nutrients for the successful expression of plant resistance and defense (Kumar and Dubey, 2020; Rengel, 2015).

Considerable progress has been made in selection and breeding of crops for increased resistance or tolerance to diseases and pests. Basically, resistance can be increased by three types of mechanisms, involving complex interactions of multiple components: (1) changes in anatomy (e.g., increased lignification and/or silicification), (2) physiological and biochemical changes leading to higher production of inhibitory or repelling substances, and (3) restriction of nutrient transfer to the pathogen to impede its growth and development (Agrios, 2005; Andersen et al., 2018; Walters and Bingham, 2007). Apparent resistance can be achieved when the most susceptible growth stages of the host plant occur at a different time to the period of highest activity of pathogens and pests (known as "escape from attack" or "outgrowing" the pathogen; Huber, 1980).

Species or strains of pathogens are continuously evolving, enabling them to either evade or suppress the defense mechanisms of their host plants (Anderson et al., 2010). Therefore, virulence (i.e., the capacity of pests to feed successfully on host tissues and of pathogens to induce symptoms of disease) depends on the compatibility of the host and pathogen/pest factors (Jones and Takemoto, 2004).

Although resistance and tolerance are genetically controlled, they are significantly influenced by environmental factors. For mitigation of biotic stresses, nutrition of plants can be considered as an environmental factor that can be manipulated relatively easily by the application of fertilizers. Although frequently unrecognized, fertilization has always been an important component of disease control (Shoaib and Awan, 2021). For example, fertilizers influence microbial activity in the rhizosphere soil, which indirectly induces secondary and cascading effects on plant resistance and tolerance to root and shoot pathogens and pests (Kumar and Dubey, 2020; Raaijmakers et al., 2009).

The impact of nutrition on plant resistance is relatively small in highly susceptible or highly resistant genotypes but can be substantial in moderately susceptible or partially resistant ones. For instance, with increasing N supply, the

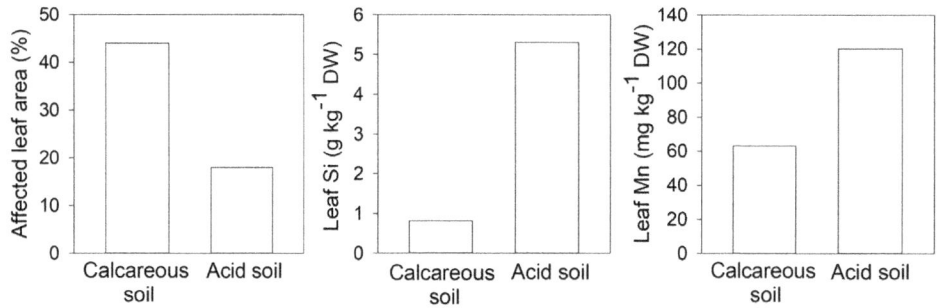

FIGURE 10.1 Percentage of leaf area affected by symptoms of powdery mildew and concentration of Si and Mn in the blades of young fully expanded leaves of *Vitis vinifera* cv. Bacchus grown on calcareous (pH in CaCl$_2$ 7.6) or acid (pH in CaCl$_2$ 4.5) soil. *Based on Weinmann (2019).*

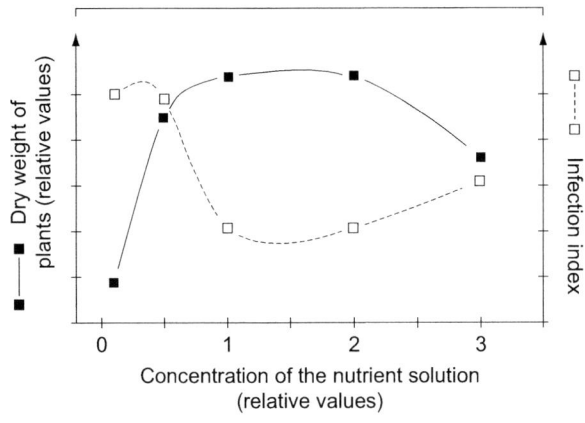

FIGURE 10.2 Growth of noninfected *Pelargonium* plants and degree of infection in plants inoculated with bacterial stem rot (*Xanthomonas pelargonii*) at varied nutrient supplies. Relative values; water only = 0; basal nutrient solution = 1; twofold concentration of nutrient solution = 2; threefold concentration of nutrient solution = 3. *Based on Kivilaan and Scheffer (1958).*

TABLE 10.1 Incidence of leaf blotch (*Rhynchosporium scalis*) in three spring barley cultivars at different rates of N fertilizer supply.

Nitrogen supply (kg ha^{-1})	Flag leaf area infected by leaf blotch (%)		
	Proctor	Cambrinus	Deba Abed
0	0.4	15	3.6
66	1.3	21	21
132	4.5	31	57

Source: Based on Jenkyn (1976).

incidence of leaf blotch (*Rhynchosporium scalis*) increased in all three barley cultivars tested, but the affected leaf area differed among the cultivars (Table 10.1). In the highly resistant cultivar Proctor, the increase was small and unlikely to influence plant growth. In the other two more susceptible cultivars (Cambrinus and Deba Abed), however, the high disease incidence at high N supply would have detrimental effects on photosynthesis and grain yield.

10.2 Relationship between susceptibility and nutritional status of plants

Nutrients play complex roles in the interactions of plants with pathogens and pests (Feldmann et al., 2022). The most intensively studied mineral nutrients with respect to their influence on the plant disease resistance are N, K, Ca, Mn, Cu, Zn, and B (Cabot et al., 2019; Gupta et al., 2017; La Torre et al., 2018; Veresoglou et al., 2013). Furthermore, the beneficial element Si is considered to be particularly important for the healthy growth of plants (Wang et al., 2017). This section provides relevant examples of both the potential and the limitations of disease and pest control by mineral nutrition and fertilizer application.

There are two general relationships between nutritional status of plants (e.g., increasing nutrient supply), growth and disease/pest incidence. The optimal one (Fig. 10.3A), in which optimal nutrient supply stimulates growth and decreases disease/pest incidence, applies to facultative biotrophic pathogens such as *Alternaria* ssp. (leaf spot disease). The other one (Fig. 10.3B) (optimal nutrient supply stimulates both plant growth and disease incidence) is typical for obligate biotrophic pathogens, such as *Rhynchosporium scalis* causing leaf blotch in spring barley (Table 10.1). Further examples are shown in Table 10.2, where optimal N supply increases the severity of diseases caused by obligate biotrophic pathogens but has the opposite effect on diseases caused by facultative ones. By contrast, optimal K supply corresponds with decreased severity of all diseases tested. However, comparing three different pathogens of tomato (*Pseudomonas syringae*, *Fusarium oxysporum*, and *Oidium lycopersicum*) at different tissue N concentrations, Hoffland et al. (2000) suggest that the proposed distinction between obligate and facultative biotrophic pathogens regarding the effect of increasing N supply on disease incidence (Table 10.2) may not always hold true. Generally, plants suffering from nutrient deficiency

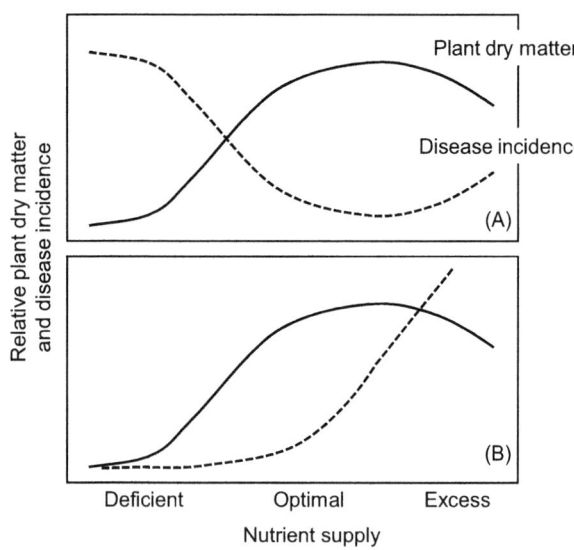

FIGURE 10.3 Schematic presentation of relationships between increasing nutrient supply, plant growth, and disease/pest incidence. (A) Optimal nutrient supply increases plant growth and decreases disease/pest incidence. (B) Optimal nutrient supply increases plant growth and disease incidence.

TABLE 10.2 Incidence of diseases caused by pathogens at optimal or low N and K supplies.

Pathogen and disease	N		K	
	Low	Optimal	Low	Optimal
Obligate biotrophic pathogens				
Puccinia spp. (rust diseases)	+	+++	++++	+
Erysiphe graminis (powdery mildew)	+	+++	++++	+
Facultative biotrophic pathogens				
Alternaria spp. (leaf spot diseases)	+++	+	++++	+
Fusarium oxysporum (wilt and rot disease)	+++	+	++++	+
Xanthomonas spp. (bacterial spots and wilt)	+++	+	++++	+

+, Low disease incidence; ++++, high disease incidence.
Source: Based on Kiraly (1976) and Perrenoud (1977).

have a low tolerance to diseases and pests, and tolerance can be increased by supplying the deficient nutrient because vigorously growing plants usually have a higher capacity to compensate for losses of photosynthates or the leaf and root surface area caused by infection or feeding (Fig. 10.3A).

Both ammonium and nitrate ions are taken up and assimilated by plants, but they frequently have opposite effects on disease severity (Huber and Graham, 1999) due to differential impacts on the rhizosphere soil pH. For example, *Fusarium* wilt of eggplants, *Verticillium* wilt of vegetables, blast (*Pyricularia grisea*) of rice, take-all disease of cereals, and *Streptomyces* scab of potato are less severe in acidic soils and ammonium-supplied plants (Huber and Graham, 1999). By contrast, the severity of *Fusarium* wilt of cucumber, tomato, and pepper, as well as *Rhizoctonia solani*–induced diseases of potato, bean, and pea are decreased by nitrate supply and alkaline soil conditions (Huber and Graham, 1999). Ammonium supply increased plant uptake of P (de Souza Cardoso et al., 2020), Si (Keeping et al., 2015), and micronutrients Fe, Mn, Zn, and Cu (De la Peña et al., 2021; Thomson et al., 1993), particularly on neutral to alkaline soils.

Regarding diseases in the phyllosphere, the effects of N supply generally differ depending on the trophic status of the pathogen. Diseases induced by obligate biotrophs such as powdery mildews (Erysiphales), rusts (Pucciniales), and downy mildews (Peronosporaceae) increase in severity with high N supply, but the opposite response occurs for facultative biotrophs such as *Alternaria* and *Fusarium* species as well as many pathogenic bacteria (Mandal et al., 2008; Riethmüller et al., 2002). These differences in response of the two types of pathogens are related to their specific nutritional requirements. Obligate biotrophic pathogens require assimilates supplied by living cells, whereas facultative biotrophs are semisaprophytes that prefer senescing tissues or produce toxins to kill the host tissue before using it as a growth substrate (Dubery et al., 1994). Therefore, N supply and other factors that support the metabolic activity and delay the senescence of host plants may also enhance their resistance to facultative biotrophic pathogens.

The effects of N on susceptibility to soil-borne pathogens are very complex. Because plant roots and their rhizospheres are colonized by a wide variety of beneficial and pathogenic microorganisms in soils, the effect of N is likely to result not only from its influence on the host plant and its pathogens but also from interactions with other soil organisms and the abiotic soil environment (Huber and Watson, 1974). As an example, the suppressive effect of ammonium supply on take-all disease of wheat in soils with high pH has not only been related to rhizosphere acidification (Brennan, 1992a; Smiley and Cook, 1973) but also to qualitative changes in the rhizosphere populations of fluorescent pseudomonads in favor of pathogen-antagonistic strains (Sarniguet et al., 1992; Smiley, 1978). Furthermore, Mn(IV) reduction and increased uptake of Mn(II) by the host plant seem to be important mechanisms by which such antagonistic rhizobacteria suppress the take-all pathogen (see Section 10.7; Marschner et al., 1991; Ownley et al., 2003).

10.3 Fungal diseases

10.3.1 Principles of infection

The germination of spores on leaf and root surfaces is stimulated by plant exudates (chemotaxis). Potassium deficiency results in high concentrations of sugars and amino acids in leaves, whereas excessive N supply leads to high amino acid concentrations (see also Sections 6.1 and 6.6). The concentrations of amino acids and sugars in the apoplast of leaf and stem tissues are in the range of 1–8 mM (Hancock and Huisman, 1981), but may increase with Ca (de Bang et al., 2021), B (Shireen et al., 2018), or Zn (Tolay, 2021) deficiency (due to increased membrane permeability) or P deficiency (due to impaired phospholipid synthesis) (de Bang et al., 2021).

The concentration of soluble assimilates in the host apoplasm is an important factor for the growth of pathogens during penetration and postinfection because only a few groups of plant pathogens are truly intracellular with direct access to assimilates in the symplasm (Bové and Garnier, 2002; Hancock and Huisman, 1981). Some pathogens, such as powdery mildew of barley, have access only to epidermal cells. Epidermal cells of leaves, stems, and roots (Hutzler et al., 1998) are characterized by increased concentrations of phenolic compounds and flavonoids (i.e., substances with fungistatic properties). The role of nutrients in phenol metabolism is well documented (Bhat et al., 2020), and examples of phenol accumulation have been discussed in relation to Cu and B deficiency (see Sections 7.3 and 7.7). Increasing evidence for the interactions of Si with Mn, other nutrients, and beneficial elements regarding their roles in plant secondary metabolism and plant resistance has accumulated during the last decade (e.g., Dragisic Maksimovic et al., 2012; Hajiboland et al., 2018; Pavlovic et al., 2021).

Many pathogenic fungi and bacteria invade the apoplasm by releasing pectolytic enzymes that dissolve the middle lamella. Calcium, beside its well-documented effectiveness in inhibiting these enzymes and strengthening the plant cell wall, has also been suggested to play an active role in the upregulation of defense-related genes (Debona et al., 2017; Martins et al., 2020). The positive correlation between the Ca concentration of tissues and their resistance to fungal and bacterial diseases, therefore, involves a variety of modes of action.

During fungal infection, inducible resistance mechanisms are associated mainly with the epidermis, and the effectiveness of these mechanisms depends on the type of pathogen and the resistance of the host as well as on the plant nutritional status (Marques et al., 2016; Ziv et al., 2018). Cytosolic Ca^{2+} concentration, as a ubiquitous and versatile second messenger, plays a critical role in many defense-related signaling processes (Chapter 6; Chen et al., 2015). Potassium-deficient plants tend to be more susceptible to pathogen infection than adequately supplied plants, but a surplus K supply does not improve plant resistance further. In the case of leaf spot (*Helminthosporium cynodontis* Marig.), fungal toxins enhance K^+ efflux and thereby deplete cells and infected tissues of K. In general, the severity of disease symptoms (leaf spotting) is negatively correlated with K concentration in leaves (Richardson and Croughan, 1989), but jasmonic acid (JA)-sensitive pathogens such as *Blumeria graminis* show a positive correlation between powdery mildew symptoms and leaf K concentration in barley (Davis et al., 2018).

Several nutrients (particularly Mn, B, and Cu) affect the biosynthesis and binding of phenols (Sections 7.3 and 7.7) and, therefore, defense responses (Pfeffer et al., 1998; Sharma et al., 2019). Low-molecular-weight phenolic compounds, such as flavonoids, may function as signaling molecules in plant defense responses and/or can be directly toxic to pathogens (Sharma et al., 2019; Treutter, 2005). The concentration of phenolics with fungistatic properties is often high in N-deficient plants and may be reduced at optimal N supply (Kiraly, 1964). Phenolic compounds play a key role in the early stages of infection, either as phytoalexins or as precursors of lignin and suberin biosynthesis. For example, glucans in the cell wall of *Phytophthora megasperma* elicit the synthesis of isoflavones that function as phytoalexins and contribute to the rapid accumulation of phenolic polymers at the infection sites (Graham and Graham, 1991). Within a few hours after infection, a signal is transmitted to noninfected leaves, which elicits an increase in their phenol synthesis (Rasmussen et al., 1991).

The production of oxygen radicals (e.g., O$^-$ and HO$^.$) and hydrogen peroxide (H_2O_2) may also increase in response to pathogen infection as a component of the plant defense response (Li et al., 2021). They may contribute to hypersensitive reactions (oxidation of plasma membrane phospholipids, leading to cell death) and initiation of cell wall lignification. The role of Cu, Zn, Fe, and Mn in the generation and detoxification of oxygen radicals and hydrogen peroxide (see Chapter 7) may explain their role in plant resistance to pathogens.

As tissues (particularly leaves) mature, lignification or the accumulation and deposition of Si in epidermal cells may form an effective physical barrier to hyphal penetration. Lignification and Si deposition provide the main structural resistance to diseases and pests, especially in the leaves of grasses (Zeyen et al., 2002) and the endodermis of roots (Yu et al., 2010).

10.3.2 Role of Si

Although Si is not considered essential for most plant species (except members of the Equisetaceae) (Coskun et al., 2019), it mitigates various biotic and abiotic stresses (Liang et al., 2015; Moradtalab et al., 2018) (Section 8.3). It enhances mechanical stability and provides resistance toward insect herbivores and fungal pathogens (Wiese et al., 2007). Generally, grasses (particularly paddy rice) are Si accumulators. In rice, the Si concentration of leaves is negatively correlated with the number of eyespots caused by the rice blast fungus *Magnaporthe grisea* (anamorph, *Pyricularia oryzae* Cav. and *P. grisea*), indicating greater resistance to the disease (Fig. 10.4).

Silicon deposited in the epidermal tissues can provide an effective barrier against water loss by cuticular transpiration (Agarie et al., 1998; Gao et al., 2006), insect herbivores (Massey et al., 2006), and penetration of the fungal pathogens (Bélanger et al., 2003). The Si deposits may hinder infection by the rice blast fungus *Pyricularia oryzae* (Hayasaka et al., 2008) or powdery mildews (Grundhöfer, 1994) that penetrate host tissues directly through the cuticle and epidermal cell wall rather than through stomata (Pryce-Jones et al., 1999). Similar effects can be achieved with foliar Si sprays that form a crust on plant surfaces (Guével et al., 2007; Menzies et al., 1992), which may be a useful approach for crop plants with low Si uptake by roots (Nikolic et al., 2007).

The formation of a physical barrier in epidermal cells against the penetration of hyphae or feeding of insects such as aphids can be localized and relatively rapid. Silicon accumulates at the sites of hyphal penetration of powdery mildew in wheat (Leusch and Buchenauer, 1988a) and barley within 20 h, and this accumulation is 3–4 times higher around unsuccessful infection sites than around successful ones (Carver et al., 1987). The preferential accumulation of Si at the point of pathogen penetration requires a continuous supply of Si, indicating a lack of Si remobilization. Silicon forms a cuticle-Si double layer to prevent pathogen penetration. In cell walls, Si is cross-linked with hemicellulose, which enhances mechanical properties such as elasticity, rigidity, and reinforcement (Guerriero et al., 2016; He et al., 2015). Moreover, Si promotes the formation of cell wall appositions (see also Section 10.3.6) during pathogen infection, which can increase the resistance of various plant species to fungal diseases (Wang et al., 2017). However, inhibition of *Pythium aphanidermatum* spread in roots of bitter gourd (Si-accumulating species) and tomato (Si-excluding species) is dependent on Si accumulation in the symplasm rather than in the root cell walls (Heine et al., 2007).

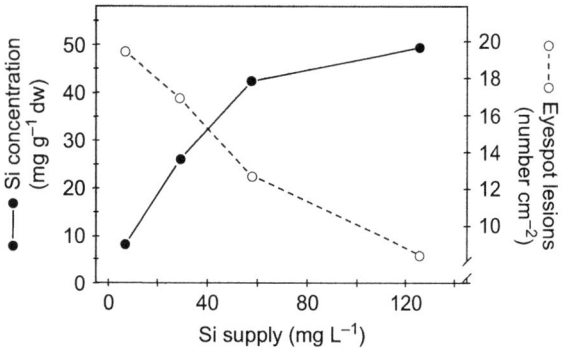

FIGURE 10.4 Silicon concentration and susceptibility to blast fungus (*Pyricularia oryzae* Cav.) of fully expanded rice leaves. *Modified from Volk et al. (1958).*

Despite the positive relationship between Si accumulation at the sites of penetration and inhibition of hyphal invasion, the protective effect is not due to Si alone. Rather, the presence of soluble Si appears to facilitate the rapid deposition of phenolics or phytoalexins at the sites of infection (Fig. 10.5), which is a general defense response to pathogen attack (Menzies et al., 1991).

As further outlined in the section about the plant defense–related roles of Mn in this chapter, Leusch and Buchenauer (1988b) showed that Si accumulates, together with Mn, at the sites of hyphal penetration of powdery mildew, and that high Si concentrations in leaf tissues are necessary for the mobility and short-term allocation of Mn (Fig. 10.6). Hence, the stimulation of phenolic deposition by Si may be indirect via its effect on Mn that plays an important role in the biosynthesis of phenolics and phytoalexins (see Section 7.2).

Silicon appears to modulate the activity of postelicitation intracellular signaling systems that regulate the expression of various defense genes related to structural modifications of cell walls, hypersensitivity responses, and biosynthesis of phytohormones, antimicrobial compounds (e.g., phenolics, flavonoids, phytoalexins), and pathogenesis-related proteins (Wang et al., 2017). A pretreatment with Si can prime plants to respond better to pathogen infections (Chain et al., 2009; Van Bockhaven et al., 2013). Although the indirect effect of Si on regulating the systemic signaling molecules, such as salicylic acid, JA, and ethylene, has been well-established, an open question is whether Si could be involved directly in elicitation of defense responses (e.g., Fortunato et al., 2012; Van Bockhaven et al., 2013). Recently, the importance of effector proteins released by pathogens to inhibit plant defense responses has been characterized (Giraldo and Valent, 2013; Shi et al., 2020). Coskun et al. (2019) hypothesized that the superior protective role of Si is linked with the pathogens' effectors and Si, highlighting that the plant apoplast is a site of intense interactions of Si with many effectors, preventing them from impeding the plant defense response.

Deposition of Si at the root endodermis has been proposed to aid in plant resistance to soil-borne pathogens such as *Gaeumannomyces graminis* (Bennett, 1982) by increasing the lignification of root cells (Fleck et al., 2011). This effect was attributed to the complexation of silicic acid with phenolic compounds that facilitate lignin biosynthesis (Dragisic

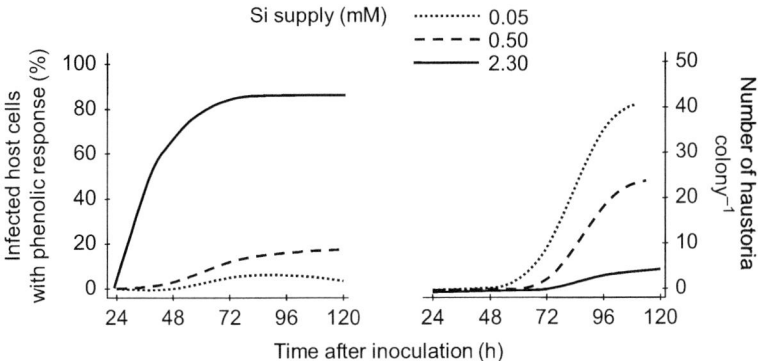

FIGURE 10.5 Percentage of host cells with phenolic response and number of haustoria of *Sphaerotheca fuliginea* in leaf segments of cucumber (*Cucumis sativus*) at different Si supplies. *Based on Menzies et al. (1991).*

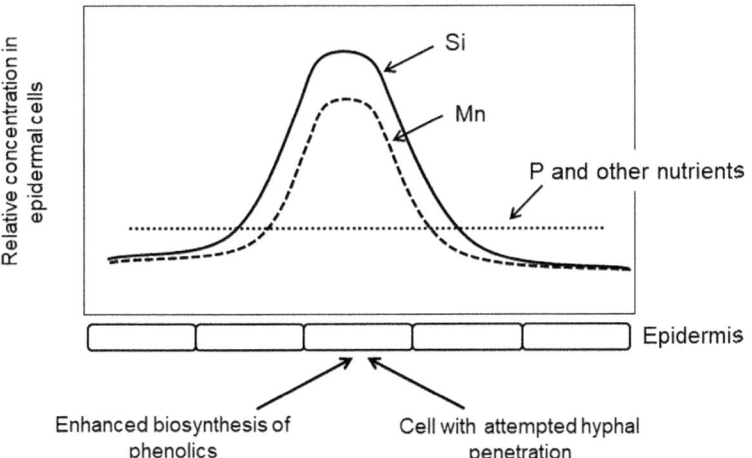

FIGURE 10.6 Schematic presentation of the relative concentrations of Si and Mn at the site of an unsuccessful pathogen penetration (e.g., by powdery mildew) on the leaf epidermis.

Maksimovic et al., 2007), the coprecipitation of Si with lignin (Fang and Ma, 2006), Si promoting Mn accumulation in roots (Greger et al., 2018), and Si-induced alterations in the expression of genes related to the biosynthesis of polyphenols (Fleck et al., 2011; Pavlovic et al., 2013).

10.3.3 Role of N and K

The role of N and K in modulating disease resistance is readily demonstrated and can be of particular importance for fertilizer application. However, the results for N may be inconsistent, and in some cases controversial, because (1) the reports do not state clearly whether the N supply was low, optimal, or excessive (see Figs. 10.2 and 10.3), (2) the effect depended on the form of N supplied (e.g., ammonium or nitrate), or (3) the differences in infection patterns between obligate and facultative biotrophic pathogens were not considered.

The principal differences in the response of obligate and facultative biotrophic pathogens to N are shown in Figs. 10.7 and 10.3B. The susceptibility of wheat plants to stem rust, caused by an obligate biotroph, increased with increasing N supply (Fig. 10.7A). By contrast, the susceptibility of tomato plants to bacterial leaf spot, caused by a facultative biotroph, decreased with increasing N supply (Fig. 10.7B).

The gaseous NO may play a role in the regulation of various processes in plant development and also defense. In systemic resistance, NO acts as a signal molecule inter alia by binding reversibly with cysteine in various regulatory proteins (Ding et al., 2020) and interacting with reactive oxygen species (ROS) (Umbreen et al., 2018).

Potassium deficiency increases the susceptibility of host plants to obligate and facultative biotrophic pathogens. Increasing K supply decreased stem rot incidence and enhanced shoot growth in rice (Fig. 10.8). Similar results have also been reported for coastal Bermuda grass where the severity of leaf spot disease (*Helminthosporium cynodontis*) was decreased with increasing K concentration in the leaves (Matocha and Smith, 1980). Beyond optimal K supply for growth, however, there is no further increase in resistance with increasing K supply or K concentration in the plant tissues.

Potassium deficiency decreases the synthesis of high-molecular-weight compounds (see Section 6.6) and thus leads to the accumulation of low-molecular-weight organic compounds that can serve as easily available nutrient sources for the pathogens (Wang et al., 2013). These effects are alleviated by supplying K to K-deficient plants, whereas an increase in K supply and plant K concentration beyond optimal has no effect on the synthesis of high-molecular-weight

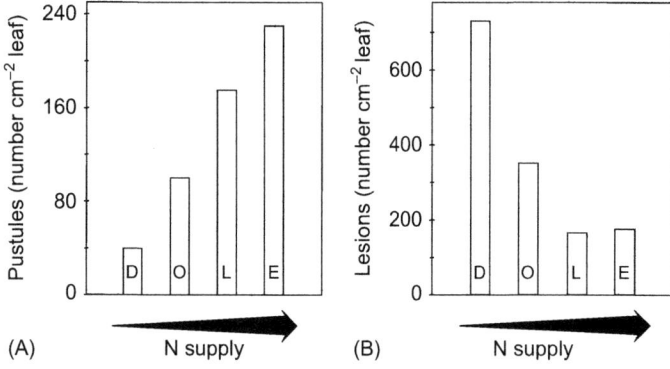

FIGURE 10.7 Number of pustules of stem rust (*Puccinia graminis* spp. *tritici*) in wheat (A) and number of necrotic lesions caused by bacterial leaf spot (*Xanthomonas vesicatoria*) in tomato (B) grown in nutrient solutions with increasing N concentration. *D*, Deficient; *E*, excessive; *L*, luxurious; *O*, optimal. *Based on Kiraly (1976)*.

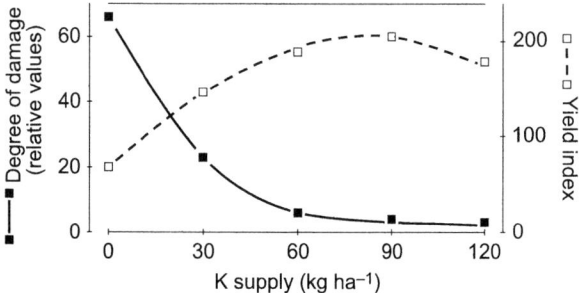

FIGURE 10.8 Relative grain yield of paddy rice and incidence of stem rot (*Helminthosporium sigmoideum*) at varied K supply (all treatments received 120 N kg ha^{-1} and 60 P kg ha^{-1}). *Based on Ismunadji (1976)*.

compounds (Fig. 10.9). However, disease susceptibility and metabolic profiles of K-deficient plants are variable; hence, it is difficult to prove a causal relationship (Amtmann et al., 2008). Recently, Davis et al. (2018) suggested that the sensitivity of pathogens to JA could be an important factor in determining the K-disease relationship. Optimal K supply delayed the appearance of necrotic lesions [due to leaf scald disease (*Rhynchosporium commune*)] on barley leaves, but the symptoms of powdery mildew (caused by *Blumeria graminis*) appeared earlier and were more severe under optimal K supply compared with the − K treatment (Fig. 10.10). These contrasting results could be attributed to the differential sensitivity of the two pathogens to JA as a critical factor in determining the K-disease relationship due to the following observation: (1) the concentrations of JA in barley leaves were generally increased under K deprivation, (2) the powdery mildew fungus was sensitive to methyl-JA, whereas the leaf scald fungus was not. Thus, high concentrations of JA in K-deficient plants might modulate the inherent plant defense responses positively in the case of JA-sensitive pathogens or negatively in JA-insensitive pathogens.

In take-all disease (Trolldenier, 1985) and leaf rust (*Puccinia triticina*) of wheat (Sweeney et al., 2000) and powdery mildew (*Blumeria graminis*) of barley (Brennan and Jayasena, 2007), application of KCl was more effective in disease suppression than K_2SO_4. To explain this effect, it has been suggested that Cl^- can suppress nitrification, and thus affect the pH and availability of other mineral nutrients, particularly Mn, in the rhizosphere (Huber, 1980; Dordas, 2008).

10.3.4 Role of Ca and Mg

The Ca concentration of plant tissues can influence the incidence of diseases by three mechanisms. First, Ca plays a key role in the recognition of pathogenic invaders at the plasma membrane. Within seconds of pathogen invasion, there is a change in membrane potential and an increase in cytosolic Ca concentration, which acts as a secondary messenger. Calcium transporters supporting the rapid influx of Ca^{2+} into the cytosol following pathogen recognition are involved in early plant immune responses (Hilleary et al., 2020). Second, Ca is essential for the stability of biomembranes, and Ca treatment has been reported to enhance the activity of pectin methylesterase, thereby contributing to the maintenance of cell wall structures (Langer et al., 2019). When Ca supply is deficient, the cell membrane becomes more leaky and the efflux of low-molecular-weight compounds (e.g., sugars) into the apoplasm increases (see Section 6.5). Third, Ca-polygalacturonates are required in the middle lamella for cell wall stability and resistance to enzymatic degradation (Laha et al., 2016).

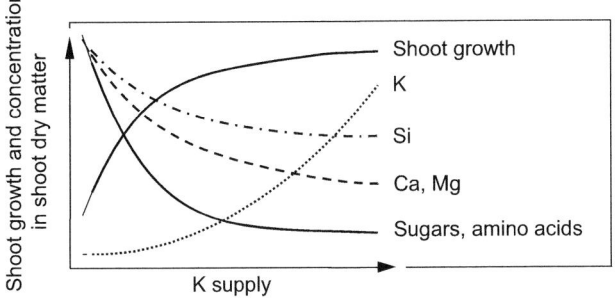

FIGURE 10.9 Schematic diagram of relationship between growth response, changes in plant composition, and K supply.

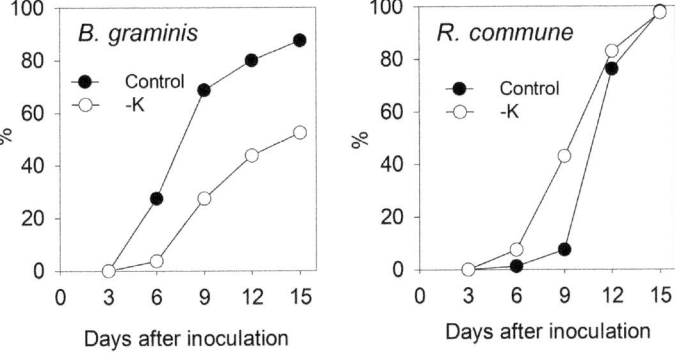

FIGURE 10.10 Percentage of leaf segments showing symptoms of powdery mildew (*Blumeria graminis*) or leaf scald (*Rhynchosporium commune*) in barley plants grown hydroponically with or without a K supply. *Based on Davis et al. (2018).*

Many pathogenic fungi, such as *Fusarium oxysporum* spp. (Blais et al., 1992); *Botrytis cinerea* (Johnston et al., 1994), and *Gaeumannomyces graminis* var. *tritici* (Martyniuk, 1988), and also bacteria (e.g., *Erwinia* spp.), invade the apoplast by releasing pectolytic enzymes, such as polygalacturonases, that cleave structural polymers in the cell wall and middle lamella (Pozza et al., 2015). The activity of some polygalacturonases is inhibited by Ca (Sasanuma and Suzuki, 2016); hence, the susceptibility of plants to infection with pathogens that rely on these enzymes decreases with increasing Ca concentration of the tissue (Table 10.3). Other cell wall–degrading enzymes (e.g., exo-polygalacturonate trans-eliminase) are stimulated by Ca or have even been shown to be Ca-dependent (Bateman and Millar, 1966). Furthermore, some pathogens (e.g., species of *Sclerotium*) produce oxalic acid as a chelating agent that can bind Ca^{2+} and thus enhance the activity of Ca-inhibited pectolytic enzymes (Bateman and Beer, 1965; Stone and Armentrout, 1985).

The induction of defense reactions against pathogenic bacteria (*Pseudomonas syringae*) in tobacco (Atkinson et al., 1990), fungi (*Cladosporium fulvum*) in tomato (Gelli et al., 1997), or oomycetes (*Phytophthora sojae*) in parsley (Blume et al., 2000) has been attributed to an enhanced influx of predominantly extracellular Ca^{2+} mediated by the activation of plasma membrane channels via pathogen-specific elicitors. High concentrations of Ca^{2+} in the cytosol are toxic due to the low solubility of calcium phosphates (White and Broadley, 2003) and are implicated in programmed cell death processes such as the hypersensitive response to pathogen attack (Levine et al., 1996).

Another important role of Ca, based on its low concentrations in the cytosol and high concentrations in organelles, is to function as a secondary messenger (Harper and Harmon, 2005; White and Broadley, 2003). For instance, Ca^{2+} signaling in plants is critically involved in the initiation of mycorrhizal and rhizobial symbioses (Harper and Harmon, 2005; Navazio et al., 2007), but also fungal pathogens have been reported to rely on internal Ca^{2+} signaling pathways during their infection process (Ebbole, 2007).

Various pathogenic fungi preferentially invade the xylem and dissolve the cell walls of conducting vessels. This leads to plugging of the vessels and subsequent wilting symptoms, such as in *Fusarium* wilt, that can effectively be controlled by Ca application as $CaCO_3$ (Datnoff et al., 2007).

Low Ca concentrations in plant tissues increase their disease susceptibility during storage. This is of particular concern for fleshy fruits with typically low Ca concentrations. Calcium treatment of fruits before storage is therefore an effective procedure for preventing losses from fruit rotting. Given that B has similar effects on the stabilization of cell walls and membranes as Ca (see Section 7.7), treatments of fruits before storage with a combination of Ca and B may be more effective than with Ca alone (Liebisch et al., 2009; Xuan et al., 2005).

Regarding Mg, there is limited information about its role in pathogen resistance of plants. Several reports suggest that Mg supply can suppress fungal diseases such as take-all in wheat (Reis et al., 1982), clubroot (*Plasmodiophora brassicae*) in crucifers (Myers and Campbell, 1985), or necrosis (*Phoma* sp.) in potato tubers (Olsson, 1988). The possible mechanisms behind these findings include the degradation of fungal toxins by Mg-activated enzymes (Colrat et al., 1999) or, similarly to Ca, the stabilization of tissues by forming bridges between polymers in the cell wall and middle lamella (Laha et al., 2016; Olsson, 1988). Debona et al. (2016) attributed the increased susceptibility of wheat to leaf blast (*Pyricularia oryzae*) to a decrease in Ca uptake by enhanced Mg supply. Ratios of Mg to other nutrients like Ca and K, therefore, may be more important for the pathogen resistance of plants than Mg status alone (Gerendás and

TABLE 10.3 Relationship between cation concentration in leaves and infection rate of *Botrytis cinerea* Pars. in lettuce.

Cation concentration (mg g^{-1} dw)			Infection with *Botrytis*[a]
K	Ca	Mg	
14	11	3.2	4
24	5.4	4.1	7
34	2.2	4.7	13
49	1.8	4.2	15

[a]*Infection index: 0–5 slight; 6–10 moderate; 11–15 severe.*
Based on Krauss (1971).

Führs, 2013). Depending on the pathogen and other factors, the opposite effects of Mg on pathogenesis may occur in the same host-pathogen system such as *Gaeumannomyces graminis* and cereals (Huber, 1989; Jones and Huber, 2007).

10.3.5 Role of phosphate and phosphite

Optimal P supply enhances disease resistance in plants (Prabhu et al., 2007), and the P status of plants influences not only morphological, metabolic, and physiological changes in roots and shoots but also the complex interactions with fungi and bacteria (Zuccaro, 2020).

Foliar phosphate applications may be effective against some airborne pathogens by conferring local or systemic resistance and controlling fungal diseases such as anthracnose (*Colletotrichum lagenarium*) in cucumber (Gottstein and Kuć, 1989), rust (*Uromyces viciae-fabae*) in faba bean (Walters and Murray, 1992), and powdery mildew in diverse crops (Mitchell and Walters, 2004; Reuveni and Reuveni, 1995). These findings have been attributed to the induction of systemic resistance by the release of elicitor compounds from plant cell walls (Gottstein and Kuć, 1989; Walters and Murray, 1992), or the initiation of localized cell death (Orober et al., 2002), which suggests specific perturbations of plant integrity rather than a nutritional effect of the phosphate salts. However, Reuveni et al. (2000) have shown that root-applied phosphate enhanced resistance to *Sphaerotheca fuliginea* in hydroponically grown cucumber plants, probably by nutrition-based mechanisms.

Phosphorus in the form of phosphite (HPO_3^{2-}) is advertised and marketed as P fertilizer, bio-stimulant, or fungicide for the control of oomycetes (Thao and Yamakawa, 2009). However, there is no scientific evidence that plants can use phosphite as a nutritional P source, unless it is oxidized to monohydrogen phosphate by microorganisms, which is a slow process in soils (Havlin and Schlegel, 2021; Thao and Yamakawa, 2009). Instead, the treatment with phosphite accelerates the severity of P deficiency in plants by disrupting the coordinated expression of plant responses to P starvation, as shown in many studies (Carswell et al., 1996; Lambers et al., 2006, Ratjen and Gerendás, 2009; Varadarajan et al., 2002). Therefore, phosphite is regarded as particularly toxic to P-deficient plants and may not improve the growth of healthy plants (Thao and Yamakawa, 2009). There is growing evidence that foliar application of phosphite, as phosphorous acid (H_3PO_3) or its salts, can inhibit pathogens such as *Phytophthora* and other members of the Peronosporales (Havlin and Schlegel, 2021; Lobato et al., 2008). In south-western Australia, phosphite is considered an effective agent against the dieback of natural forests due to *Phytophthora cinnamomi* (Shearer and Fairman, 2007). It appears that the inhibiting effect of phosphite is due to direct toxic effects on the pathogens and/or inhibition of their metabolism. Of interest is that application of phosphite, for example for suppression of *P. cinnamomi* in Australian forests on soils with low P availability, results in increased P deficiency in plants by disruption of their typical starvation-induced responses for P acquisition (Lambers et al., 2006; Ratjen and Gerendás, 2009).

10.3.6 Role of S

Elemental S has been used for a long time as a fungicide to control powdery mildews (Russell, 2005) and some other pathogenic fungi (Haneklaus et al., 2007; Williams and Cooper, 2004). Improved plant resistance to pathogens results from the synthesis of phytoalexins (Pedras et al., 2000) as secondary metabolites with deterrent or toxic effects. Among the S-containing metabolites that may contribute to the resistance of plants against pathogens are glucosinolates (Tierens et al., 2001), glutathione (Tausz et al., 2004), amino acids and proteins (Bloem et al., 2004; Broekaert et al., 1995), low-molecular-weight antibiotics (Pedras et al., 2000), emitted volatiles such as H_2S (Papenbrock et al., 2007), and elemental S that is formed during active defense responses (Williams and Cooper, 2004).

The mechanisms causing "sulfur-induced resistance (SIR)" have not been elucidated fully and are also host/pathogen specific (Bloem et al., 2015; Haneklaus et al., 2009). The free cysteine pool may be related to SIR as a precursor of all the S-containing metabolites putatively involved in systemic resistance. Furthermore, salicylic acid and H_2O_2 initiate and maintain systemic resistance, and salicylic acid accumulation is linked to S metabolism (Haneklaus et al., 2007).

10.3.7 Role of Mn

Enhanced Mn concentrations in host tissues have been related to enhanced resistance of plants to *Fusarium* spp. (Elmer, 1995), powdery mildews (Brain and Whittington, 1981), *Gaeumannomyces graminis* (Graham and Rovira, 1984), and numerous other fungal and bacterial pathogens (Huber and Wilhelm, 1988). As a cofactor, Mn activates enzymes of primary and secondary metabolism in plants, including key enzymes of the shikimic acid pathway and subsequent metabolic processes leading to the synthesis of important defense-related compounds such as phenolics and alkaloids (Section 7.2). Low-molecular-weight phenolic compounds, such as flavonoids, may function as signaling molecules in plant defense and/or can be directly toxic to pathogens (Nicholson and Hammerschmidt, 1992; Treutter, 2005).

The incorporation of lignin and other phenolic polymers into epidermal cell walls confers mechanical strength and resistance to enzymatic degradation (Nicholson and Hammerschmidt, 1992; Zeyen et al., 2002). Furthermore, epidermal cell walls can actively resist penetration by pathogenic fungi through the rapid deposition of cell wall appositions (lignitubers or papillae), directly subtending the contact area of fungal infection structures (i.e., appressoria and penetration pegs) (Bhuiyan et al., 2009; Yu et al., 2010). Lignitubers, ensheathing the root invading hyphae of *Gaeumannomyces graminis* and other soil-borne pathogens, may slow the colonization process and suppress disease progression (Yu et al., 2010). Papillae that typically develop in leaf cells upon attack by powdery mildews share common characteristics with lignitubers and can effectively block fungal ingress (Yu et al., 2010; Zeyen et al., 2002). Their formation includes the localized deposition of lignin-like phenolics, callose, Si, and proteins with antifungal properties. Furthermore, local concentrations of soluble Ca, Mg, Mn, and Si become elevated at the sites of fungal attack during the formation of papillae (Zeyen et al., 2002). The process of papillae formation also occurs in susceptible plants but at a lower frequency or in a way that is ineffective to stop fungal growth (Basavaraju et al., 2009; Li et al., 2005). In these cases, either the defense system of the plant hosts may not recognize the pathogen early enough or the components of papillae may not be formed at sufficient rates (Conrath and Kauss, 2000).

The deposition of callose (β-1,3 glucan) is a sensitive indicator of high Mn concentrations in plant tissues (Nogueira et al., 2002; Wissemeier and Horst, 1987). Microanalytic investigations by Leusch and Buchenauer (1988b) have proven the accumulation of Mn and Si at the penetration sites of powdery mildew in wheat, indicating that appropriate Si concentrations in leaf tissues are necessary for the rapid translocation of Mn. This agrees with other studies showing a positive influence of Si on the distribution and mobility of Mn in plant tissues (Horst and Marschner, 1978; van der Ent et al., 2020), which may result in an enhanced tolerance toward otherwise toxic Mn concentrations (Rogalla and Römheld, 2002). In addition, the high concentration of Mn^{2+} can be directly toxic to pathogens, as shown for *Streptomyces scabies* of potato (Mortvedt et al., 1963). Manganese may also inhibit the production of pathogenesis-related pectolytic enzymes (especially pectin methylesterase and polygalacturonase) that are released by pathogenic fungi during the infection process to degrade the host cell walls (Kikot et al., 2009). Other, more indirect, effects of Mn on the disease resistance of plants can be through its functions in photosynthesis (Section 7.3).

10.3.8 Role of other micronutrients

Other micronutrients also influence the incidence and severity of diseases, as reviewed by Verbon et al. (2017), Cabot et al. (2019), and Alejandro et al. (2020). Among plant defense mechanisms, those involving phenolics and lignin are the best understood, and the micronutrients B, Cu, and Fe, in addition to Mn, play key roles in phenol metabolism and lignin biosynthesis (see also Chapter 7). As cofactors or components of antioxidative enzymes, Cu, Zn, and Fe contribute to the control of ROS production in plants (Chapter 7), with ROS being important in the establishment of induced defense responses (Bradáčová et al., 2016; Cabot et al., 2019; Lehmann et al., 2015). Iron plays roles as a catalyst of cellular redox processes (Section 7.1), and changes in its homeostasis are critically involved with plant-pathogen interactions. The virulence of plant pathogens depends on sufficient Fe acquisition and can be weakened by Fe-retention strategies of the plant, but also accumulation of Fe to activate a toxic oxidative burst is a defense strategy to inhibit pathogen invasion (Verbon et al., 2017).

In micronutrient-deficient plants, defense mechanisms may be impaired, but the plants may also become more enticing to pathogens. In *Hevea brasiliensis*, Zn deficiency increases the leakage of sugars to the leaf surface, which increases the severity of infection by *Oidium* species (Bolle-Jones and Hilton, 1956). Infection of wheat plants by powdery mildew is several times greater in B-deficient than B-sufficient plants, and the fungus spreads more rapidly over the leaves of deficient plants (Stangoulis and Graham, 2007), which may be due to increased leakage of cell content under B deficiency (Cakmak et al., 1995). In addition to maintaining plasma membrane integrity, B promotes the stability and rigidity of the cell wall (Brown et al., 2002; Camacho-Cristóbal et al., 2015).

Copper has been used extensively as a fungicide, but the amounts required are at least 10–100 times higher than those that are required nutritionally by plants. Increased Cu supply to Cu-deficient plants via either soil or foliar application can decrease leaf infections, for example by powdery mildew and ergot (*Claviceps* sp.) in wheat (Evans et al., 2007), or can control stem pathogens (Table 10.4). For suppression of stem and leaf pathogens, foliar Cu application is often more effective than soil application because of the low phytoavailability of Cu applied to soil (Mackie et al., 2012).

TABLE 10.4 Stem melanosis (caused by *Pseudomonas cichorii*) in wheat grown on a soil with low Cu availability without and with different forms of Cu applied.

Treatment	Cu rate (kg Cu ha^{-1})	Disease (%)	Grain yield (t ha^{-1})
Nil	–	92	0.3
CuSO$_4$, banded	10	76	0.5
CuSO$_4$, incorporated	10	34	2.0
CuSO$_4$, foliar spray	10	6	2.1
Cu-chelate, foliar spray	2	7	2.5

Source: Based on Malhi et al. (1989).

10.4 Bacterial and viral diseases

10.4.1 Bacterial diseases

Root, leaf, and vascular diseases can damage crop plants severely (Maejima et al., 2014; Corrêa et al., 2014). In leaf spot diseases (e.g., bacterial leaf blight, *Xanthomonas oryzae*), pathogens usually enter the host plant through stomata. Having entered the plant, the bacteria spread and multiply in the intercellular spaces (Jiang et al., 2020). The effect of the nutritional status of the host plant on the spread and multiplication of bacterial pathogens is similar to that on facultative fungal pathogens. For instance, multiplication and severity of leaf blight are exacerbated by K and Ca deficiency, and often, but not always, by N deficiency (Huber and Thompson, 2007). Recent findings suggest that the tissue-specificity of different *X. oryzae* strains depends on their ability to break the surface and inner structures of cell walls in mesophyll cells (Cao et al., 2020).

Bacterial vascular wilt diseases spread within plants through the xylem and lead to "slime" formation that plugs the vessels causing plants to wilt. In tomato, alleviation of Ca deficiency suppresses bacterial canker and wilt caused by *Clavibacter michiganensis* subsp. *michiganensis* that spreads through the xylem (Table 10.5). Adequate Ca supply decreases disease severity in both susceptible and resistant cultivars, indicating Ca action may be synergistic to other resistance mechanisms potentially present. By contrast, the capacity to produce cellulase enzyme seems to be an important virulence factor of *C. michiganensis* subsp. *michiganensis* (Gartemann et al., 2003).

Calcium may affect plant resistance to bacterial diseases by stabilization of the middle lamella, through its involvement in hypersensitive responses to bacterial infections and in its function as a signaling ion implicated in defense reactions (Cao et al., 2019). In tobacco, hypersensitive reactions induced by *P. syringae* require a strong influx of Ca from the apoplasm into the cytosol through Ca channels. This leads to enhanced K$^+$/H$^+$ exchange, cytosol acidification and death of the host cells at the infection site (Atkinson et al., 1990) comparable to hypersensitive responses to attacking fungal pathogens (see above). Recent results show similar Ca-mediated defense mechanisms in other pathobacteria-host plant interactions (Cao et al., 2019) as well as the critical roles of Ca signaling and Ca sensors to trigger a series of plant immune responses (Aldon et al., 2018; Zhang et al., 2022).

Infections by endophytic bacteria such as *Xylella* spp. cause little damage, except in plants that are deficient in Mn and Zn (Johal and Huber, 2009). Deficiency of Mn and Zn may be induced by high soil pH or extensive use of the herbicide glyphosate (Kirkby and Römheld, 2004; Martinez et al., 2018).

10.4.2 Viral diseases

Viruses can multiply only in living cells, and their nutritional requirements are restricted to amino acids and nucleotides. Compared with fungal and bacterial diseases and pests, little is known about the effects of plant nutrition on viral diseases. In general, nutritional factors that favor rapid growth and high tissue water content support viral multiplication. This holds true particularly for N and P (Huber and Thompson, 2007; Prabhu et al., 2007), but also for K (Perrenoud, 1990). Alleviation of nutrient deficiency may eliminate symptoms of viral disease because the plants "outgrow" the disease, or the symptoms are hidden.

In many cases, the effect of the nutritional status of the host plant on viral diseases is indirect via their fungal and insect vectors. It is assumed that more than half of plant viruses are spread by aphids (Fingu-Mabola and Francis, 2021),

TABLE 10.5 Relationship between Ca supply, Ca concentration in shoots and bacterial canker disease (*Clavibacter michiganensis* subsp. *michiganensis* Smith) in susceptible (Moneymaker) and resistant (Plovdiv 8/12) tomato cultivars.

Ca supply (mg L^{-1})	Ca concentration (g kg^{-1} dw)		Disease development (% wilted leaves)	
	Moneymaker	Plovdiv 8/12	Moneymaker	Plovdiv 8/12
0	1.2	1.4	84	56
100	3.7	4.2	27	12
200	4.3	5.5	37	6
300	4.4	5.8	27	8

Source: Based on Berry et al. (1988).

and the severity of aphid infestation of plants is strongly influenced by nutritional status. Zellner et al. (2011) reported that elevated Si supply delayed symptoms of ringspot virus in tobacco plants (classified as a low Si accumulator), and the increased Si concentrations in leaves appeared to be associated with specific defense mechanisms. Contrasting results observed with respect to the role of Si in other virus-host plant interactions indicate that the Si effects are virus-specific (Sakr, 2016).

10.5 Soil-borne fungal and bacterial diseases

The population density of microorganisms on the root surface and in the rhizosphere is several times greater than that in the bulk soil (see Chapter 15). The range of root-associated microorganisms that colonize root surface or invade and infect root tissues, includes various pathogens as well as beneficial ones (e.g., plant growth–promoting rhizobacteria, rhizobia, mycorrhizae). High biological diversity and activity have been associated with the long-term capacity of natural and agricultural soils to provide an adequate supply of mineral nutrients to growing plants and to limit the establishment of pathogens (Altieri et al., 2017; Lehmann et al., 2020). The mechanisms by which healthy soils suppress the development of plant diseases include microbial antagonism as well as induced resistance and the role of mineral nutrition in plant defense. However, the huge diversity of soil microorganisms and their multifaceted modes of action have gained increased attention only recently, which to some extent is driven by a growing need to replace agrochemical inputs (Jacoby et al., 2017). By contrast, "soil sickness" has been attributed to complex interactions between biotic and abiotic factors such as the accumulation of toxic materials, unbalanced supply of mineral nutrients, and soil degradation (including the loss of functional biodiversity in the microbial soil community), which are often produced by ecologically unsound crop management practices such as continuous monocropping, use of agrochemicals, and other factors (Cesarano et al., 2017; Donn et al., 2015; Hoestra, 1994). Although competition among microorganisms as well as chemical barriers (e.g., high concentrations of polyphenols in the rhizodermis) and physical barriers (e.g., Si deposition at the endodermis) restrict microbial invasion of roots (and shoots via the roots), endophytes can be found in any plant organ (Ibáñez et al., 2017).

Microbial communities and plant nutrition (especially Mn, N, and Fe) influence soil-borne fungal and bacterial diseases in various ways, whereby synergistic interactions may be more important than individual effects (Legrand et al., 2019; Siddiqui et al., 2015). High Mn and low N supply lead to an increase in the fungistatic activity in the inner bark of *Picea abies* Karst. (Alcubilla et al., 1971). The incidence of common scab infection of potato tubers by *Streptomyces scabies* is suppressed by either lowering the soil pH or applying Mn (Thompson and Huber, 2007). The suppressive effect on Mn is due to (1) increased resistance of the tuber tissue to the pathogen and (2) inhibition of the vegetative growth of *Streptomyces scabies* before the onset of infection (Huber and Wilhelm, 1988; Thompson and Huber, 2007).

Low Ca concentrations in pods of peanut are related to more severe preharvest pod rot caused by infection with *Pythium myriotylum* and *Rhizoctonia solani* and can be suppressed by soil application of Ca (e.g., as gypsum) (Hallock and Garren, 1968). For that purpose, it seems to be necessary that Ca concentrations in the pot shell and seed are increased. This, however, could be dependent on soil factors (e.g., soil pH and availability of other mineral nutrients such as K and Mg), crop rotation, and cultivar-specific characteristics such as pod size, as indicated by contrasting results reported by other authors (Hollowell et al., 1998; Wheeler et al., 2016). Likewise, increasing the Ca concentration in the peel of potato tubers can improve the resistance against bacterial soft rot disease of potato caused by various species of *Erwinia* (Kelman et al., 1989).

The root rot disease take-all of wheat and barley caused by *Gaeumannomyces graminis* limits grain production in many regions of the world, but disease severity can be effectively controlled by the nutrition of the host plant (Huber and McCay-Buis, 1993; Kwak and Weller, 2013; Thompson and Huber, 2007). The fungus has a growth optimum at pH 7 and is sensitive to low pH; hence, liming of acid soils increases the risk of root infections and yield losses due to take-all disease. Fig. 10.11 shows that, in a soil of pH_{CaCl_2} 3.8, inoculation with *G. graminis* was without a significant effect on growth or yield. Liming increased soil pH and enhanced yield in noninfected plants but reduced yield of infected plants.

Manganese availability in the rhizosphere and Mn concentration of root tissues play a key role in root infection and severity of take-all and other soil-borne fungal diseases (Graham and Webb, 1991; Huber and Wilhelm, 1988; Thompson and Huber, 2007). All factors that decrease the availability of Mn increase the severity of take-all (e.g., increase in soil pH by liming, nitrate vs ammonium fertilizer; Table 10.6; see also Section 7.2). Manganese deficiency also increases the severity of rice blast and *Phymatotrichum* root rot of cotton and potato scab (Thompson and Huber, 2007). In Mn-deficient plants, the capacity of the roots to restrict the penetration of fungal hyphae into the root tissue by enhanced lignification at the infection site is impaired because Mn is required for the biosynthesis of phenolics, phytoalexins, and lignin (Graham and Webb, 1991; see also Section 7.2). Furthermore, *Gaeumannmyces graminis* oxidizes Mn, thereby reducing Mn availability to the plant. Differences between isolates in their oxidation power are related to their capacity to decrease Mn availability and cause disease (Thompson and Huber, 2007; Wilhelm et al., 1990), with isolates that cannot oxidize Mn being avirulent and unable to infect wheat roots.

Suppression of take-all by soil application of Mn fertilizers is possible under field conditions (Brennan, 1992a) but has its limitations on calcareous soils because of rapid oxidation and immobilization of Mn. Foliar Mn sprays are not effective in suppression of root pathogens because of the poor phloem mobility of Mn (see Sections 3.3 and 7.2). The herbicide glyphosate used to control weeds between crops can increase take-all disease severity in subsequently grown wheat (van Toor et al., 2017). By contrast, the use of ammonium instead of nitrate N fertilizer is effective in controlling take-all. However, the effect of glyphosate as well as of different forms of nitrogen fertilizer could be associated with alterations in the soil and rhizosphere microbiome (Durán et al., 2018; Smiley, 1978; van Toor et al., 2017). Especially at high soil pH, the decrease of take-all disease severity by applying ammonium fertilizer is probably not only due to rhizosphere acidification but also to quantitative and qualitative changes in *Pseudomonas fluorescens* spp. populations, favoring those that increase Mn availability and are antagonistic to *Gaeumannomyces graminis* (Sarniguet et al., 1992). The suppressive effect of an oat precrop on take-all disease in wheat was associated with an increase in the population density of Mn reducers and a decrease of Mn oxidizers in the rhizosphere as well as corresponding increases in the Mn status of the wheat plants (Table 10.7).

Another approach to control take-all is to apply microbial agents such as *Pseudomonas fluorescens* spp., *Trichoderma* sp., or other Mn-reducing microorganisms that suppress growth of *Gaeumannomycs graminis* in vitro. However, suppression in vivo is related not only to the Mn-reducing capacity of the antagonistic microbial strains, but also to Mn-oxidizing potential of the pathogen and the availability of Mn in the soil (Huber and McCay-Buis, 1993). Furthermore, stimulation of root growth, decrease in rhizosphere pH, and increased Zn availability are important modes of action by which *Trichoderma* spp. could contribute to improved mineral nutrition and resistance of wheat against

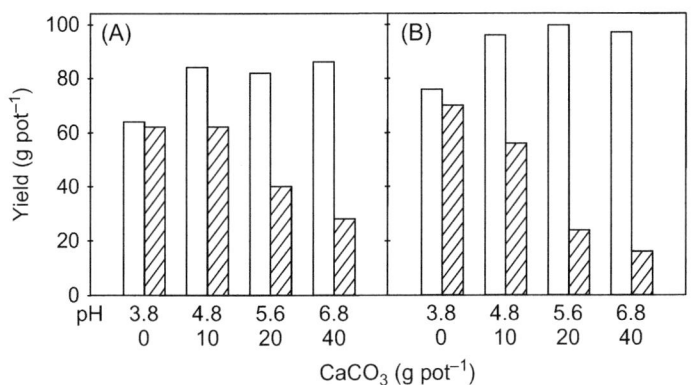

FIGURE 10.11 Straw (A) and grain yield (B) of spring wheat with different lime application rates and without (*open bars*) or with inoculation with *Gaeumannomyces graminis* var. *tritici* (take-all) (*striped bars*). Soil pH measured in a soil:CaCl₂ suspension (1:2.5). *Modified from Trolldenier (1981)*.

TABLE 10.6 Cultural conditions affecting Mn availability, nitrification, and severity of take-all caused by *Gaeumannomyces graminis*.

		Take-all	Nitrification	Mn availability
Soil	Acid	↓	↓	↑
	Alkaline	↑	↑	↓
	Cool, wet	↑		↓
Fertilization	Ammonium N	↓		↑
	Nitrate N	↑		↓
	Cl	↓	↓	↑
	Mn	↓		↑
Inhibition of nitrification		↓	↓	↑
Liming (CaCO$_3$)		↑	↑	↓
Precrop	*Lupinus* sp.	↓	↓	↑
	Glyphosate between crops	↑		
	Paddy rice	↓	↓	↑
	Oat	↓		↑
	Soybean or alfalfa	↑	↑	↓
Seedbed	Firm	↓	↓	↑
	Loose	↑	↑	↓
Dense seeding		↑		↓
Tolerant cultivars		↓		↑
Animal manure		↑	↑	↓

Source: Based on Thompson and Huber (2007) and van Toor et al. (2017).

take-all disease (Akter et al., 2015). The growth suppression of soil-borne pathogens by some *Pseudomonas* and other plant growth–promoting bacterial or fungal strains may also be related to the production of toxic substances such as cyanide and antibiotics (Sehrawat et al., 2022).

The capacity of fluorescent *Pseudomonas* strains to produce siderophores with a high affinity for Fe ions may also be involved in suppression of soil-borne pathogens by restricting Fe availability to the pathogens (Höfte et al., 1991; Kloepper et al., 1980). In addition to their Fe-scavenging function, siderophores act as signaling molecules that have important roles in the regulation of virulence factors that contribute to (1) the capacity of pathogenic pseudomonads to cause diseases, (2) quorum sensing, and (3) the interaction with other organisms including the host plant and mycorrhizal fungi (Deveau et al., 2016; Lamont et al., 2002; Romera et al., 2019). Root infection with arbuscular mycorrhiza is another factor that may suppress soil-borne pathogens and enhance the resistance of plants by improved nutrition and other modes of action (Poveda et al., 2020; Sikes et al., 2010; see Chapter 15).

The severity of take-all in wheat is increased not only by Mn deficiency, but also by deficiency of N, P, or Cu (Brennan, 1989, 1992b). The decrease in severity with the application of N and P fertilizers to deficient plants is most likely due to a greater tolerance by more vigorously growing plants rather than an increase in physiological resistance. By contrast, Cu deficiency results in impaired biosynthesis of lignin, and supplying Cu fertilizer overcomes this impairment and thereby increases resistance, with soil and foliar applications having different effects (Table 10.8). Foliar application increased yield but did not diminish root infection by take-all, indicating that, despite its phloem mobility, Cu concentrations at the infection sites were not high enough to suppress the pathogen. The greatest effect was achieved by a combination of Cu and Ca (gypsum) applied to the soil, probably by enhanced desorption and mobility of Cu in the soil. These examples provide evidence that each nutrient functions as part of a delicately balanced and

TABLE 10.7 Numbers of Mn-reducing and Mn-oxidizing microorganisms isolated from the rhizosphere of wheat (cv. Paragon) grown in a pot experiment on wheat- or oat-precropped soil.

Crop rotation	Ratio of Mn-reducing to Mn-oxidizing microorganisms isolated from the rhizosphere of wheat	Shoot Mn (mg kg^{-1} dw)	Disease index
Wheat–wheat–wheat	0.3	28	9
Wheat–oat–wheat	6.3	55	2
Wheat–oat–wheat + Ggt	6.3	52	2

The inoculum of *Gaeumannomyces graminis* var. *tritici* (+Ggt) was added as indicated.
Source: Based on Schackmann (2005) and Akter (2007).

TABLE 10.8 Growth, yield, and root infection by take-all (*Gaeumannomyces graminis* var. *tritici*) of winter wheat grown in a soil with low Cu availability without and with the application of Cu and gypsum.

Treatment	Dry weight (g pot^{-1})	Grain (g pot^{-1})	Infected plants (%)
Nil	8.5	4.3	100
CuSO$_4$, soil	12.7	6.5	83
CuSO$_4$, foliar	12.8	6.1	100
CaSO$_4$, soil	9.8	5.4	83
CuSO$_4$ + CaSO$_4$, soil	17.0	9.0	0

Source: Based on Gardner and Flynn (1988).

interdependent system comprising the plant and its environment, which can be mismanaged or optimized by cultural practices influencing plant nutrition and health (Huber, 2021).

10.6 Pests

Pests are animals such as insects, mites, and nematodes that are harmful to plants. In contrast to fungal and bacterial pathogens, they have digestive and excretory systems, and their dietary requirements are often less specific. The main types of resistance of host plants to pests are (1) physical (e.g., color, surface properties, hairs), (2) mechanical (e.g., fiber, Si), and (3) chemical/biochemical (e.g., concentration of stimulants, toxins, repellents) (Alhousari and Greger, 2018). Plant nutrition and the balance of nutritional elements in plants can influence these three factors to varying degrees (Altieri and Nicholls, 2003).

Nitrogen in plants is an important factor influencing insect pests, increasing population density and feeding preferences, except in a few examples where the application of N fertilizer decreases herbivory (Lu et al., 2007). Young or rapidly growing plants are more likely to be attacked by pests than old and slow-growing plants. Hence, there is often a positive correlation between N application and pest attack as has been reported for plant hoppers (*Nilaparvata lugens*, *Sogatella furcifera*), leaffolder (*Cnaphalocrocis medinalis*) and stem borers (*Scirpophaga incertulas, Chilo suppressalis, Scirpophaga innotata, Chilo polychrysus, Sesamia inferens*) in rice (Lu et al., 2007; Salim and Saxena, 1991). Field studies, however, suggest that the influence of N fertilization may not follow this pattern due to the influence of several plant factors affecting pest-host interactions. Not only visual cues such as color of leaves but also semiochemical signals, such as the release of volatiles, are important to pests for recognition or orientation (Reddy, 2016; Veromann et al., 2013). For example, many aphid species tend to settle on yellow-reflecting surfaces common in cases of N and other nutrient deficiencies (Döring, 2014). Nitrogen supply, by contrast, can increase the emission of certain volatiles (acetic acid and several products of the lipoxygenase pathway) that are correlated with increased pest abundance (Veromann et al., 2013). Therefore, plants with moderate N status could be less attractive to pests than plants with deficient or excessive N levels.

Potassium deficiency generally increases pest attack, and the application of K fertilizers is recommended to decrease pest incidence. However, in a literature overview comprising more than 500 cases of the effect of K fertilization on pests, it was found that the incidence of insect pests and mites increased in 28% of cases and that of nematodes in 63% of cases (Perrenoud, 1990). Although the increased concentration of sugars in K-deficient plants can act as a feeding stimulant (Beck, 1965), most sucking insects such as the rice brown plant hopper (Sogawa, 1982) depend more strongly on amino acids (Dreyer and Campbell, 1987). This is illustrated in Table 10.9 for squash bugs, whereby the number of squash bugs per plant was positively correlated with the concentration of total soluble N in leaves. By contrast, the protein concentration in leaves did not influence pest density (Benepal and Hall, 1967). The severity of attack by sucking parasites increases with the concentration of amino acids in plants, and high N supply or impaired protein synthesis due to deficiency of K or Zn elevate free amino acid concentrations.

A recent investigation of the interactive influence of N, P, and K fertilizer on the brown planthopper (*Nilaparvata lugens*) and its host plant rice showed that N fertilization resulted in enhanced N status of both rice and planthoppers, whereas P and K supply improved only the plant status of these elements (Table 10.10). These findings could be attributed to increased feeding of the insect on N-fertilized rice plants due to increased concentrations of soluble proteins and decreased concentrations of Si in rice. Phosphorus fertilization did not change the concentrations of N, K, Si, free sugar, and soluble proteins in rice, but K fertilization decreased the concentrations of N, Si, free sugar, and soluble proteins in the plant tissue and thereby improved tolerance to the pest, as measured indirectly by the amount of honeydew excretion by the planthoppers (Rashid et al., 2016).

The determinants of interactions between plants and pests are more complex than just N supply or nutrient imbalances. This is particularly evident in trees where a pest attack often depends more strongly on the presence of repellents or toxic compounds than on N concentration. For example, in *Salix dasyclados* grown at different nutrient concentrations and light intensities, damage to the leaves by the herbivore *Galerucella lineola* was negatively correlated with the phenol concentration (high light >> low light) and the N concentration of the leaves (Larsson et al., 1986). In Scots pine (*Pinus sylvestris*), high N fertilization increased the concentration of N and di-terpenoids in the needles. Di-terpenoids act as deterrents to herbivorous insects and thereby counteract the effect of N supply on amino acid concentrations in the leaves.

Silicon deposited in the epidermal cell walls acts as a mechanical barrier to the penetration by the stylet and mandibles of, respectively, sucking and biting insects. The mandibles of larvae of the rice stem borer are damaged when the Si concentration of rice plants is high (Datnoff et al., 2007; Jeer et al., 2017). The physical properties of leaf surfaces are also of importance in regulating the severity of attack by sucking insects. Labial exploration of the surface takes place before insertion of the stylet into the tissue (Sogawa, 1982). Changes in the surface properties of leaves are likely to be the main reason for a decrease in the attack by aphids on wheat plants when several foliar sprays of sodium silicate were applied (Fig. 10.12). Increasing N supply enhanced the number of *Sitobion avenae* aphids, whereas foliar sprays of Si reduced the number of aphids below that in N-deficient plants. The results of this experiment also illustrate the difficulties of generalizing the relationship between increasing N supply and attack by sucking insects. In contrast

TABLE 10.9 Relationship between nutrient supply, number of squash bugs (*Anasa tristis*) per plant and soluble N concentration in the leaves of squash.

Nutrient supply	Squash bugs (no. plant⁻¹)	Soluble nitrogen (µg g⁻¹ fw)
Complete	1.7	32
−N	0.7	5
−P	2.1	94
−K	2.5	99
−S	3.4	144

Source: From Benepal and Hall (1967).

TABLE 10.10 Effects of N, P, and K supply to rice plants on shoot damage by the brown planthopper (*Nilaparvata lugens*) after 10 days, and the area of honeydew secretion by insects during 48 h, indicating the magnitude of their feeding.

Nutrient supply (kg ha^{-1})	Damage severity index	Honeydew excretion area (mm^2)
N$_0$	2.8	150
N$_{100}$	6	320
P$_0$	6.2	260
P$_{20}$	6.3	260
K$_0$	7.1	270
K$_{60}$	6.3	250

The damage severity ranged from 1 (very slight damage) to 9 (death of all plants).
Source: Based on Rashid et al. (2016).

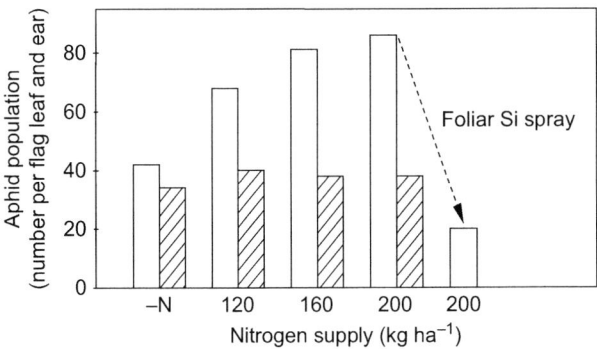

FIGURE 10.12 Population density of two aphid species in winter wheat with different N supply and foliar Si application. Open bars: *Sitobion avenae*; striped bars: *Metopolophium dirhodum*. Based on Hanisch (1980).

to *S. avenae*, which is a typical ear feeder, the density of other aphid species, such as *Metopolophium dirhodum* that prefers leaf blades, did not increase with an increasing N supply. The strong depression of aphid populations on leaves after foliar application of Si (Fig. 10.12) is not only due to changes in surface properties of the leaves but also to an increase in soluble Si within the leaf tissue. Soluble silicic acid, rather than the deposited Si in leaves, is an effective deterrent of the rice brown plant hopper. Soluble Si appears to facilitate the rapid deposition of phenolics (acting as insect repellent and antifeedant; Singh et al., 2021) at the sites of invasion of plant tissue.

Differences in feeding habits and preferences for different plant organs are possible reasons for different responses to N supply. Further investigations regarding the interaction between N and Si and their influence on the resistance of rice against the brown planthopper *Nilaparvata lugens* have shown that high N fertilization rates decrease Si uptake due to a decreased expression of Si transporters in roots (Wu et al., 2017). In plant physiology, Si can induce specific chemical defenses and interacts with the JA signaling pathway, which has a critical role in plant defense against herbivores and mediates the accumulation of Si in rice leaves (Ye et al., 2013). Silicon plays a role in the recruitment of natural enemies of insect pests via JA-triggered release of volatiles by the attacked plant (Kvedaras et al., 2010; Turlings and Wäckers, 2004).

In wetland rice, several species of leaf hoppers pose a more serious threat as vectors of viruses than as sap-sucking pests (Beck, 1965). Thus, another important reason for controlling sucking insects is to reduce the spread of viruses. Yet, the interactions between host plants, plant viruses, and their herbivore vectors are complex. Recent research has shown that plant viruses can alter the host preference and feeding behavior of insect pests. Green rice leaf hoppers carrying the rice dwarf virus preferred rice plants that were not yet infected by the virus and vice versa (Wang et al., 2018).

Nematodes, such as cereal cyst nematodes (*Heterodera avenae*) or root lesion nematodes (*Pratylenchus penetrans*), can strongly depress plant growth (Mokrini et al., 2018). Root exudates might act as signals for recognition or as

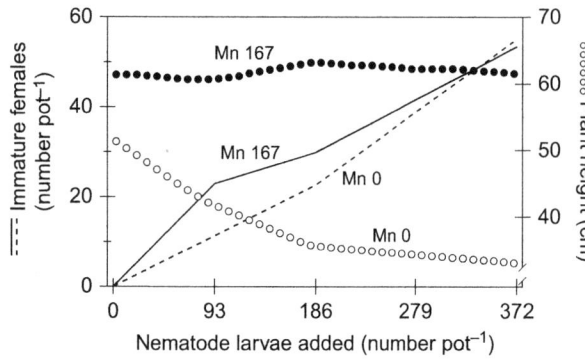

FIGURE 10.13 Number of infections (immature females) and height of barley plants growing in a low Mn soil without (Mn 0) or with 167 mg Mn kg^{-1} soil (Mn 167) and different nematode (*Heterodera avenae*) densities. *Based on Wilhelm et al. (1985).*

repellents, but it is not clear whether nutrition plays an important role in either context. There are, however, many examples showing that nematodes depress root growth and activity, thereby influencing nutrient uptake and the nutritional status of plants. For example, nematodes are mainly responsible for K deficiency in the replant disease of apple trees (Merwin and Stiles, 1989), whereas maize precultivation alleviates soil acidification, improves the availability of K and other nutrients and the soil enzyme activity, and decreases the number of plant-feeding nematodes in replanted apple orchards (Yang and Liu, 2020). Nematode attacks had little or no effect on cotton shoot growth at high K availability, but depressed shoot growth severely in the low-K soils, even though the total number of nematodes was higher on plants at high K supply (Oteifa and Elgindi, 1976). This is a typical example of an increase in tolerance to pests and diseases resulting from a high nutrient supply. This can also be demonstrated for micronutrients (Fig. 10.13). In barley plants grown in a soil with low Mn availability, addition of Mn had no effect on the number of infections (immature females), but growth was severely depressed only in plants not supplied with Mn. In this case, Mn application possibly compensated for the impaired capacity for Mn acquisition caused by nematode infection.

Silicon, mineral nutrients, and other elements appear to be involved directly in induced resistance of plants against nematodes, but also influence indirect responses, such as increased activity of natural antagonists (de Melo Santana-Gomes et al., 2013; Zhan et al., 2018). Selenium (Se), which is not required by higher plants, can enhance the resistance of Se-accumulating plants against Se-sensitive herbivores and interacts with other mineral nutrients to improve plant stress tolerance (El-Ramady et al., 2016; Nawaz et al., 2021; Pilon-Smits, 2019). Due to such complex interdependencies, the distinct influence of certain mineral nutrients on plant diseases caused by nematodes and other pests under variable field conditions is quite challenging to predict reliably.

10.7 Direct and indirect effects of fertilizer application on plants and their pathogens and pests

Under field conditions, fertilizers influence the performance of plants and their pathogens and pests directly via the effects on plant nutrition and indirectly by changing the biotic and abiotic environment that modifies pathogen and pest survival and function (Alabouvette et al., 2004). Dense plant stands and alterations in light interception and humidity within a crop change the microenvironment, thus favoring some foliar pathogens, but the increased plant vigor or hastened maturity may diminish other diseases and pests. In addition, the timing of fertilizer application is important, especially for N. As shown in Table 10.11, the severity of take-all infection of spring wheat is high without N fertilization and is exacerbated by application of ammonium in the autumn, leading to yield depressions because of increased disease severity. By contrast, the same amount of ammonium N supplied in spring suppresses take-all, and high grain yields are obtained. The low yield with split application of N in autumn and spring demonstrates that the effects of N fertilizer application on grain yield were governed more by the effects on take-all than on the N nutritional status of wheat per se. Ammonium N applied in the autumn is rapidly nitrified, and nitrate intensifies take-all in nonsuppressive soils. The use of timed ammonium fertilizer application is therefore a practical approach to suppress take-all, with variations in suppression between years and locations (Christensen et al., 1987; MacNish, 1988) probably related to rate of nitrification prior to N uptake by the crop. An opposite relationship to time of N application is observed with eye spot on winter wheat, with a spring application of N aggravating this disease (Huber, 1980).

The phytoavailability of Si could be improved as the soil, and especially the rhizosphere, pH is decreased (Anggria et al., 2020; Schaller et al., 2021; Sommer et al., 2006). Thus, the Si concentration in plants is dependent on Si fertilization and, at least to some extent, on the form of N fertilizer applied. Compared with Ca nitrate, fertilization with

TABLE 10.11 Take-all (*Gaeumannomyces graminis* var. *tritici*) root infection and grain yield of winter wheat with ammonium N fertilizer application at different times during the growing season.

Time of application	Rate (kg N ha^{-1})	Take-all infection (%)	Grain yield (t ha^{-1})
	0	1.9	2.6
Autumn	83	2.8	1.7
Spring	83	0.1	5.3
Autumn + spring	83 + 28	1.9	2.4

Source: Based on Huber (1989).

TABLE 10.12 Leaf Si concentration and disease incidence of powdery mildew (*Erysiphe graminis*) in spring wheat grown in soil amended with either lime (CaCO$_3$) or blast furnace lime (BFL) and different forms of N fertilizer.

Nitrogen form	Si concentration (g kg^{-1})		Disease incidence (% leaf area affected)	
	CaCO$_3$	BFL	CaCO$_3$	BFL
Ca(NO$_3$)$_2$	5.6	11	28	12
(NH$_4$)$_2$SO$_4$	9.8	34	18	2.0

Source: Recalculated from Leusch and Buchenauer (1988b).

ammonium sulfate increases the Si concentration in spring wheat and diminishes the incidence of powdery mildew (Table 10.12). Ammonium application may result in a decrease in soil pH via nitrification or net proton release by the roots upon ammonium uptake.

The form of nitrogen fertilizer influences the activity and composition of soil microbial populations. As shown by Sarniguet et al. (1992), nitrogen applied as Ca nitrate enhanced the abundance of deleterious fluorescent pseudomonads in the soil, whereas ammonium sulfate favored strains that antagonized the take-all fungus. Moreover, strains of *Pseudomonas* and *Bacillus* species that can alter the Mn redox status have been reported to increase (Mn oxidizers) or decrease (Mn reducers) the severity of take-all under nitrate or ammonium supply, respectively (Huber and McCay-Buis, 1993). Hence, fertilizer supply may influence whether certain rhizobacteria act as deleterious, beneficial, or indifferent root associates, indicating the importance of adapting agricultural management to achieve the intended results (Miransari, 2011).

There are numerous reports that chloride fertilizer application in amounts similar to those of macronutrients may suppress various diseases (Chen et al., 2010): soil-borne diseases such as take-all in wheat (Christensen et al., 1987) or root rot (*Cochliobolus sativus*) in barley (Timm et al., 1986), and leaf diseases such as leaf rust (*Puccinia recondita*) in wheat (Elmer, 2007; Fixen et al., 1986). The mechanism for the disease-suppressive effect of chloride fertilizers is not clear. Chloride may act directly by improving the plant water balance and thus tolerance to disease, or indirectly by inhibiting nitrification or enhancing mobilization of Mn in the soil (Elmer, 2007; Graham and Webb, 1991). It was suggested that Cl$^-$ competes with nitrate for uptake, inhibits nitrification of ammonium, and thereby exacerbates rhizosphere acidification, which can mediate the reduction of Mn oxides and improve the phytoavailability of Mn (Dordas, 2008).

Agronomic measures used to decrease disease severity may exert their effect by modifying the availability or form of nutrients, particularly N and Mn (Table 10.13). Conditions that inhibit nitrification or increase the availability of Mn can reduce the severity of potato scab, rice blast, take-all, maize stalk rot (*Gibberella*), and *Phymatotrichum* root rots (Thompson and Huber, 2007). By contrast, conditions that stimulate nitrification and decrease Mn availability for plant uptake may exacerbate these diseases.

TABLE 10.13 Factors affecting N and Mn availability and severity of some diseases (potato scab, rice blast, take-all of cereals, *Phymatotrichum* root rot, maize stalk rot).

		Nitrification	Mn availability	Disease severity
Soil pH	Low	↓	↑	↓
	High	↑	↓	↑
N fertilizer	Ammonium	↓	↑	↓
	Nitrate	↑	↓	↑
Nitrification inhibitors		↓	↑	↓
Metal sulfides		↓	↑	↓
Liming		↑	↓	↑
Manure	Green (rye)	↓	↑	↓
	Animal	↑	↓	↑
Soil fumigation		↓	↑	↓
Glyphosate herbicide		↑	↓	↑
Seed bed	Loose	↑	↓	↑
	Firm	↓	↑	↓
Irrigation		↓	↑	↓
Low soil water content		↑	↓	↑

Source: Based on Thompson and Huber (2007).

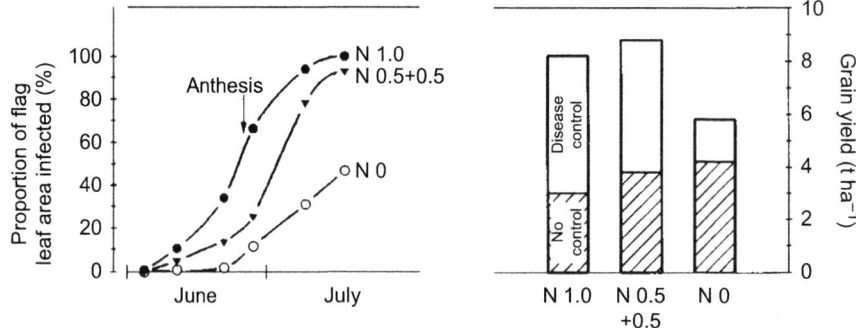

FIGURE 10.14 Yellow rust infections (*Puccinia striiformis* Westend.) and grain yield of winter wheat with and without chemical disease control at different rates and timing of N supply (N 1.0: 160 kg N ha^{-1} as early topdressing, N 0.5 + 0.5: 80 kg N ha^{-1} early and 80 kg N ha^{-1} at anthesis); N0: no N addition. *Based on Darwinkel (1980a)*.

Fertilizer applications may substitute, or at least reduce, the demand for chemical disease control in some cases, but may increase the demand in others. In temperate climates, high N application rates to winter wheat early in the growing season support abundant tillering and dense, tall crop stands that provide conditions favorable for infection. Yellow (=stripe) rust infection was highest with a large early single topdressing (N 1.0) (Fig. 10.14). Split application of N decreased infection in the early growth stages, but fungal growth increased rapidly after the second application at anthesis. In plants not receiving N (N0), infection remained low. Without chemical disease control, stripe rust infection lowered grain yield, especially in the treatments receiving N compared to plants without N addition. Similar results have been reported for wheat infected with powdery mildew (Darwinkel, 1980b).

Agricultural strategies to optimize the supply of mineral nutrients and/or promote beneficial soil microorganisms are regarded as promising approaches to enhance the resistance of crop plants and maintain the microbial balance of

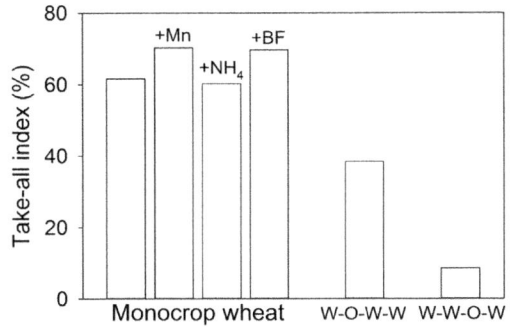

FIGURE 10.15 Effect of foliar manganese (Mn) spray, soil application of ammonium fertilizer (NH$_4^+$) and microbial biofertilizer (BF) in wheat monoculture (monocrop wheat) and wheat (W)/oat (O) rotations on take-all infection index in the field-grown wheat. *Based on Weinmann (2019)*.

pathogen-infested soils, thus mimicking the effect of crop rotations (Buchenauer, 1998; Janvier et al., 2007; Römheld and Neumann, 2006). This may be achieved by the application of specific mineral nutrients, by addition of isolated microbial strains as seed or soil inoculants (biofertilizers), and/or by the augmentation of specific soil indigenous microbial communities through an adapted rhizosphere management (Pankhurst and Lynch, 2005; Römheld and Neumann, 2006). Nevertheless, it is difficult to extrapolate results from the laboratory to the natural plant–soil environment. In a field experiment with Mn, ammonium or selected microbial agents applied to control take-all disease in wheat monoculture, no treatment was effective (Weinmann, 2019). By contrast, crop rotation with oats proved to be a reliable practice to ensure the healthy growth of wheat plants (Fig. 10.15).

References

Agarie, S., Uchida, H., Agata, W., Kubota, F., Kaufman, P.B., 1998. Effects of silicon on transpiration and leaf conductance in rice plants (*Oryza sativa* L.). Plant Prod. Sci. 1, 89–95.

Agrios, G.N., 2005. Plant Pathology, 5th ed. Elsevier Academic Press, Burlington, MN, USA.

Akter, Z., 2007. Investigations on rhizosphere management for a better Mn and Zn acquisition and disease resistance in crop plants in the model system of take-all disease in wheat. Dissertation, Universität Hohenheim, Stuttgart, Germany.

Akter, Z., Neumann, G., Römheld, V., 2015. Effects of biofertilizers on Mn and Zn acquisition and growth of higher plants: a rhizobox experiment. J. Plant Nutr. 38, 596–608.

Alabouvette, C., Backhouse, D., Steinberg, C., Donovan, N.J., Edel-Hermann, V., Burgess, L.W., 2004. Microbial diversity in soil - effects on crop health. In: Schjønning, P., Elmholt, S., Christensen, B.T. (Eds.), Managing Soil Quality: Challenges in Modern Agriculture. CAB International, Wallingford, UK, pp. 121–138.

Alcubilla, M., Diaz-Palacio, M.P., Kreutzer, K., Laatsch, W., Rehfuess, K.E., Wenzel, G., 1971. Beziehungen zwischen dem Ernährungszustand der Fichte (*Picea abies* Karst.), ihrem Kernfäulebefall und der Pilzhemmwirkung ihres Basts. Eur. J. For. Pathol. 1, 100–114.

Aldon, D., Mbengue, M., Mazars, C., Galaud, J.-P., 2018. Calcium signalling in plant biotic interactions. Int. J. Mol. Sci. 19, 1–19.

Alejandro, S., Höller, S., Meier, B., Peiter, E., 2020. Manganese in plants: from acquisition to subcellular allocation. Front. Plant Sci. 11, 300.

Alhousari, F., Greger, M., 2018. Silicon and mechanisms of plant resistance to insect pests. Plants 7, 33.

Altieri, M.A., Nicholls, C.I., 2003. Soil fertility management and insect pests: harmonizing soil and plant health in agroecosystems. Soil Till. Res. 72, 203–211.

Altieri, M.A., Nicholls, C.I., Montalba, R., 2017. Technological approaches to sustainable agriculture at a crossroads: an agroecological perspective. Sustainability 9, 349.

Amtmann, A., Troufflard, S., Armengaud, P., 2008. The effect of potassium nutrition on pest and disease resistance in plants. Physiol Plant. 133, 682–691.

Anderson, J.P., Gleason, C.A., Foley, R.C., Thrall, P.H., Burdon, J.B., Singh, K.B., 2010. Plants versus pathogens: an evolutionary arms race. Funct. Plant Biol. 37, 499–512.

Andersen, E.J., Ali, S., Byamukama, E., Yen, Y., Nepal, M.P., 2018. Disease resistance mechanisms in plants. Genes 9, 339.

Anggria, L., Husnain, H., Masunaga, T., 2020. The controlling factors of silicon solubility in soil solution. Agric 32, 83–94.

Armijo, G., Schlechter, R., Agurto, M., Muñoz, D., Nuñez, C., Arce-Johnson, P., 2016. Grapevine pathogenic microorganisms: understanding infection strategies and host response scenarios. Front. Pant Sci. 7, 382.

Atkinson, M.M., Koppler, L.D., Orlandi, E.W., Baker, C.J., Mischke, C.F., 1990. Involvement of plasma membrane calcium influx in bacterial induction of the K$^+$/H$^+$ and hypersensitive responses in tobacco. Plant Physiol. 92, 215–221.

Basavaraju, P., Shetty, N.P., Shetty, H.S., de Neergaard, E., Jørgensen, H.J.L., 2009. Infection biology and defence responses in sorghum against *Colletotrichum sublineolum*. J. App. Microbiol. 107, 404–415.

Bateman, D.F., Beer, S.V., 1965. Simultaneous production and synergistic action of oxalic acid and polygalacturonase during pathogenesis by *Sclerotium rolfsii*. Phytopathology 55, 204–211.

Bateman, D.F., Millar, R.L., 1966. Pectic enzymes in tissue degradation. Ann. Rev. Phytopath. 4, 119–144.
Beck, S.T., 1965. Resistance of plants to insects. Ann. Rev. Entomol. 10, 207–232.
Bélanger, R.R., Benhamou, N., Menzies, J.G., 2003. Cytological evidence of an active role of silicon in wheat resistance to powdery mildew (*Blumeria graminis* f. sp. *tritici*). Phytopathology 93, 402–412.
Benepal, P.S., Hall, C.V., 1967. The influence of mineral nutrition of varieties of *Cucurbita pepo* L. on the feeding response of squash bug *Anasa tristis* De Geer. Proc. Am. Soc. Hortic. Sci. 90, 304–312.
Bennett, D.M., 1982. Silicon deposition in the roots of *Hordeum sativum* Jess, *Avena sativa* L. and *Triticum aestivum* L. Ann. Bot. 50, 239–245.
Berry, S.Z., Madumadu, G.G., Uddin, M.R., 1988. Effect of calcium and nitrogen nutrition on bacterial cancer disease of tomato. Plant Soil 112, 113–120.
Bhat, B.A., Islam, S.T., Ali, A., Sheikh, B.A., Tariq, L., Islam, S.U., et al., 2020. Role of micronutrients in secondary metabolism of plants. In: Aftab, T., Hakeem, K.R. (Eds.), Plant Micronutrients. Deficiency and Toxicity Management. Springer Nature, Cham, Switzerland, pp. 311–329.
Bhuiyan, N.H., Selvaraj, G., Wei, Y., King, J., 2009. Role of lignification in plant defense. Plant Signal. Behav. 4, 158–159.
Blais, P., Rogers, P.A., Charest, P.M., 1992. Kinetic of the production of polygalacturonase and pectin lyase by two closely related formae speciales of *Fusarium oxysporum*. Exp. Mycol. 16, 1–7.
Bloem, E., Haneklaus, S., Schnug, E., 2015. Milestones in plant sulfur research on sulfur-induced-resistance (SIR) in Europe. Front. Plant Sci. 5, 779.
Bloem, E., Riemenschneider, A., Volker, J., Papenbrock, J., Schmidt, A., Salac, I., et al., 2004. Sulphur supply and infection with *Pyrenopeziza brassicae* influence L-cysteine desulphydrase activity in *Brassica napus* L. J. Exp. Bot. 55, 2305–2312.
Blume, B., Nürnberger, T., Nass, N., Scheel, D., 2000. Receptor-mediated increase in cytoplasmic free calcium required for activation of pathogen defense in parsley. Plant Cell 12, 1425–1440.
Bolle-Jones, E.W., Hilton, R.N., 1956. Zinc-deficiency of *Hevea braziliensis* as a predisposing factor to *Oidium* infection. Nature 177, 619–620.
Bové, J.M., Garnier, M., 2002. Phloem- and xylem-restricted plant pathogenic bacteria. Plant Sci. 163, 1083–1098.
Bradáčová, K., Weber, N.F., Morad-Talab, N., Asim, M., Imran, M., Weinmann, M., et al., 2016. Micronutrients (Zn/Mn), seaweed extracts, and plan growth-promoting bacteria as cold-stress protectants in maize. Chem. Biol. Technol. Agric. 3, 1–10.
Brain, P.J., Whittington, W.J., 1981. The influence of soil pH on the severity of swede powdery mildew infection. Plant Pathol. 30, 105–109.
Brennan, R.F., 1989. Effect of nitrogen and phosphorus deficiency in wheat on the infection of roots by *Gaeumannomyces graminis* var. *tritici*. Aust. J. Agric. Res. 40, 489–495.
Brennan, R.F., 1992a. The role of manganese and nitrogen nutrition in the susceptibility of wheat plants to take-all in Western Australia. Fert. Res. 31, 35–41.
Brennan, R.F., 1992b. Effect of superphosphate and nitrogen on yield and take-all of wheat. Fert. Res. 31, 43–49.
Brennan, R.F., Jayasena, K.W., 2007. Increasing applications of potassium fertiliser to barley crops grown on deficient sandy soils increased grain yields while decreasing some foliar diseases. Aust. J. Agric. Res. 58, 680–689.
Broekaert, W.F., Terras, F.R.G., Cammue, B.P.A., Osborn, R.W., 1995. Plant defensins: novel antimicrobial peptides as components of the host defense system. Plant Physiol. 108, 1353–1358.
Brown, P.H., Bellaloui, N., Wimmer, M.A., Bassil, E.S., Ruiz, J., Hu, H., et al., 2002. Boron in plant biology. Plant Biol. 4, 205–223.
Buchenauer, H., 1998. Biological control of soil-borne diseases by rhizobacteria. J. Plant Dis. Prot. 105, 329–348.
Cabot, C., Martos, S., Llugany, M., Gallego, B., Tolraà, R., Poschenrieder, C., 2019. A role for zinc in plant defense against pathogens and herbivores. Front. Plant Sci. 10, 1171.
Cakmak, I., Kurz, H., Marschner, H., 1995. Short-term effects of boron, germanium and high light intensity on membrane permeability in boron-deficient leaves of sunflower. Physiol. Plant. 95, 11–18.
Camacho-Cristóbal, J.J., Martín-Rejano, E.M., Herrera-Rodríguez, M.B., Navarro-Gichicoa, M.T., Rexach, J., Gonzáles-Fontes, A., 2015. Boron deficiency inhibits root cell elongation via an ethylene/auxin/ROS-dependent pathway in Arabidopsis seedlings. J. Exp. Bot. 66, 3831–3840.
Cao, J., Zhang, M., Zhu, M., He, L., Xiao, J., Li, X., et al., 2019. Autophagy-like cell death regulates hydrogen peroxide and calcium ion distribution in Xa3/Xa26-mediated resistance to *Xanthomonas oryzae* pv. *oryzae*. Int. J. Mol. Sci. 21, 194.
Cao, J., Chu, C., Zhang, M., He, L., Qin, L., Li, X., Yuan, M., 2020. Different cell wall-degradation ability leads to tissue-specificity between *Xanthomonas oryzae* pv. *oryzae* and *Xanthomonas oryzae* pv. *oryzicola*. Pathogens 9, 187.
Carswell, C., Grant, B.R., Theodorou, M.E., Harris, J., Niere, J.O., Plaxton, W.C., 1996. The fungicide phosphonate disrupts the phosphate-starvation response in *Brassica nicra* seedlings. Plant Physiol. 110, 105–110.
Carver, T.L.W., Zeyen, R.J., Ahlstrand, G.G., 1987. The relationship between insoluble silicon and success or failure of attempted primary penetration by powdery mildew (*Erysiphe graminis*) germlings on barley. Physiol. Mol. Plant Pathol. 31, 133–148.
Cesarano, G., Zotti, M., Antignani, V., Marra, R., Scala, F., Bonanomi, G., 2017. Soil sickness and negative plant-soil feedback: a reappraisal of hypotheses. J. Plant Pathol. 99, 545–570.
Chain, F., Côté-Beaulieu, C., Belzile, F., Menzies, J., Beélanger, R., 2009. A comprehensive transcriptomic analysis of the effect of silicon on wheat plants under control and pathogen stress conditions. Mol. Plant Microbe Interact. 22, 1323–1330.
Chen, W., He, Z.L., Yang, X.E., Mishra, S., Stoffella, P.J., 2010. Chlorine nutrition of higher plants: progress and perspectives. J. Plant Nutr. 33, 943–952.
Chen, J., Gutjahr, C., Bleckmann, A., Dresselhaus, T., 2015. Calcium signaling during reproduction and biotrophic fungal interactions in plants. Molec. Plant 8, 595–611.
Christensen, N.W., Powelson, R.L., Brett, M., 1987. Epidemiology of wheat take-all as influenced by soil pH and temporal changes in inorganic soil N. Plant Soil 98, 221–230.

Colquhoun, T.T., 1940. Effect of manganese on powdery mildew of wheat. J. Aust. Inst. Agric. Sci. 6, 54.

Colrat, S., Deswarte, C., Latché, A., Klaébé, A., Bouzayen, M., Fallot, J., Roustan, J.P., 1999. Enzymatic detoxification of eutypine, a toxin from *Eutypa lata*, by *Vitis vinifera* cells: partial purification of an NADPH-dependent aldehyde reductase. Planta 207, 544–550.

Conrath, U., Kauss, H., 2000. Systemisch erworbene Resistenz. Das "Immunsystem" der Pflanze. Biol. Unserer Zeit. 30, 202–208.

Corrêa, B.O., Schafer, J.T., Moura, A.B., 2014. Spectrum of biocontrol bacteria to control leaf, root and vascular diseases of dry bean. Biol. Control 72, 71–75.

Coskun, D., Deshmukh, R., Sonah, H., Menzies, J.G., Reynolds, O., Ma, J.F., et al., 2019. The controversies of silicon's role in plant biology. New Phytol. 221, 67–85.

Darwinkel, A., 1980a. Grain production of winter wheat in relation to nitrogen and diseases. I. Relationship between nitrogen dressing and yellow rust infection. Z. Acker-Pflanzenbau 149, 299–308.

Darwinkel, A., 1980b. Grain production of winter wheat in relation to nitrogen and diseases. II. Relationship between nitrogen dressing and mildew infection. Z. Acker-Pflanzenbau 149, 309–317.

Datnoff, L.E., Elmer, W.E., Huber, D.M., 2007. Mineral Nutrition and Plant Disease. APS Press, Saint Paul, MN.

Davis, J.L., Armengaud, P., Larson, T.R., Graham, I.A., White, P.J., Newton, A.C., et al., 2018. Contrasting nutrient–disease relationships: Potassium gradients in barley leaves have opposite effects on two fungal pathogens with different sensitivities to jasmonic acid. Plant Cell Environ. 41, 2357–2372.

de Bang, T.C., Husted, S., Laursen, K.H., Persson, D.P., Schjoerring, J.K., 2021. The molecular-physiological functions of mineral macronutrients and their consequences for deficiency symptoms in plants. New Phytol. 229, 2446–2469.

De la Peña, M., Marín-Peña, A.J., Urmeneta, L., Coleto, I., Castillo-González, J., van Liempd, S.M., et al., 2021. Ammonium nutrition interacts with iron homeostasis in *Brachypodium distachyon*. J. Exp. Bot. 73, 263–274.

de Souza Cardoso, A.A., Lopes Santos, J.Z., Oka, J.M., da Silva Ferreira, M., Barbosa, T.M.B., Tucci, C.A.F., 2020. Ammonium supply enhances growth and phosphorus uptake of mahogany (*Swietenia macrophylla*) seedlings compared to nitrate. J. Plant Nutr. 44, 1349–1364.

de Melo Santana-Gomes, S., Dias-Arieira, C.R., Roldi, M., Santo Dadazio, T., Marini, M.P., de Oliveira Barizão, D.A., 2013. Mineral nutrition in the control of nematodes. Afr. J. Agric. Res. 8, 2413–2420.

Debona, D., Cruz, M.F.A., Rodrigues, F.A., 2017. Calcium-triggered accumulation of defense-related transcripts enhances wheat resistance to leaf blast. Trop. Plant Pathol. 42, 309–314.

Debona, D., Rios, J.A., Nascimento, K.J.T., Silva, L.C., Rodrigues, F.A., 2016. Influence of magnesium on physiological responses of wheat infected by *Pyricularia oryzae*. Plant Pathol. 65, 114–123.

Deveau, A., Gross, H., Palin, B., Mehnaz, S., Schnepf, M., Leblond, P., et al., 2016. Role of secondary metabolites in the interaction between *Pseudomonas fluorescens* and soil microorganisms under iron-limited conditions. FEMS Microbiol. Ecol. 92, fiw107.

Ding, Y., Gardiner, D.M., Xiao, D., Kazan, K., 2020. Regulators of nitric oxide signaling triggered by host perception in a plant pathogen. Proc. Natl. Acad. Sci. USA 117, 11147–11157.

Donn, S., Kirkegaard, J.A., Perera, G., Richardson, A.E., Watt, M., 2015. Evolution of bacterial communities in the wheat crop rhizosphere. Environ. Microbiol. 17, 610–621.

Doran, J.W., 2002. Soil health and global sustainability: translating science into practice. Agric. Ecosyst. Environ. 88, 119–127.

Dordas, C., 2008. Role of nutrients in controlling plant disease in sustainable agriculture. A review. Agron. Sustain. Dev. 28, 33–46.

Döring, T.F., 2014. How aphids find their host plants, and how they don't. Ann. Appl. Biol. 165, 3–26.

Dragisic Maksimovic, J., Bogdanovic, J., Maksimovic, V., Nikolic, M., 2007. Silicon modulates the metabolism and utilization of phenolic compounds in cucumber (*Cucumis sativus* L.) grown at excess manganese. J. Plant Nutr. Soil Sci. 170, 739–744.

Dragisic Maksimovic, J., Mojovic, M., Maksimovic, V., Römheld, V., Nikolic, M., 2012. Silicon ameliorates manganese toxicity in cucumber by decreasing hydroxyl radical accumulation in the leaf apoplast. J. Exp. Bot. 63, 2411–2420.

Dreyer, D.L., Campbell, B.C., 1987. Chemical basis of host-plant resistance to aphids. Plant Cell Environ. 10, 353–361.

Dubery, I.A., Meyer, D., Bothma, C., 1994. Purification and characterization of cactorein, a phytotoxin secreted by *Phytophthora cactorum*. Phytochemistry 35, 307–312.

Durán, P., Tortella, G., Viscardi, S., Barra, P.J., Carrión, V.J., de la Luz Mora, M., et al., 2018. Microbial community composition in take-all suppressive soils. Front. Microbiol. 9, 2198.

Ebbole, D.J., 2007. Magnaporthe as a model for understanding host-pathogen interactions. Ann. Rev. Phytopathol. 45, 437–456.

Elmer, W.H., 1995. Associations between Mn-reducing root bacteria and NaCl applications in suppression of *Fusarium* crown and root rot of asparagus. Phytopathology 84, 1461–1467.

Elmer, W.H., 2007. Chlorine and plant disease. In: Datnoff, L.E., Elmer, W.H., Huber, D.M. (Eds.), Mineral Nutrition and Plant Disease. APS Press, Saint Paul, MN, USA, pp. 189–202.

El-Ramady, H., Abdalla, N., Taha, H.S., Alshaal, T., El-Henawy, A., Faizy, S.E.-D.A., et al., 2016. Selenium and nano-selenium in plant nutrition. Environ. Chem. Lett. 14, 123–147.

Evans, I., Solberg, E., Huber, D.M., 2007. Copper and plant disease. In: Datnoff, L.E., Elmer, W.H., Huber, D.M. (Eds.), Mineral Nutrition and Plant Disease. APS Press, Saint Paul, MN, USA, pp. 177–188.

Fang, J.-Y., Ma, X.-L., 2006. In vitro simulation studies of silica deposition induced by lignin from rice. J. Zhejiang Univ. Sci. B 7, 267–271.

Feldmann, F., Jehle, J., Bradáčová, K., Weinmann, M., 2022. Biostimulants, soil improvers, bioprotectants: promoters of bio-intensification in plant production. J. Plant Dis. Prot. 129, 707–713. Available from: https://doi.org/10.1007/s41348-022-00567-x.

Fingu-Mabola, J.C., Francis, F., 2021. Aphid-plant-phytovirus pathosystems: Influencing factors from vector behaviour to virus spread. Agriculture 11, 502.

Fixen, P.E., Buchenau, G.W., Gelderman, R.H., Schumacher, T.E., Gerwing, J.R., Cholik, F.A., et al., 1986. Influence of soil and applied chloride on several wheat parameters. Agron. J. 78, 736–740.

Fleck, A.T., Nye, T., Repenning, C., Stahl, F., Zahn, M., Schenk, M.K., 2011. Silicon enhances suberization and lignification in roots of rice (*Oryza sativa*). J. Exp. Bot. 62, 2001–2011.

Fortunato, A.A., Rodrigues, F., Do Nascimento, K.J., 2012. Physiological and biochemical aspects of the resistance of banana plants to *Fusarium* wilt potentiated by silicon. Phytopathology 102, 957–966.

Gao, X., Zou, C., Wang, L., Zhang, F., 2006. Silicon decreases transpiration rate and conductance from stomata of maize plants. J. Plant Nutr. 29, 1637–1647.

Gardner, W.K., Flynn, A., 1988. The effect of gypsum on copper nutrition of wheat grown in marginally deficient soil. J. Plant Nutr. 11, 475–493.

Gartemann, K.-H., Kirchner, O., Engemann, J., Gräfen, I., Eichenlaub, R., Burger, A., 2003. Clavibacter michiganensis subsp. michiganensis: first steps in the understanding of virulence of a Gram-positive phytopathogenic bacterium. J. Biotechnol. 106, 179–191.

Gelli, A., Higgins, V.J., Blumwald, E., 1997. Activation of plant plasma membrane Ca^{2+}-permeable channels by race-specific fungal elicitors. Plant Physiol. 113, 269–279.

Gerendás, J., Führs, H., 2013. The significance of magnesium for crop quality. Plant Soil 368, 101–128.

Giraldo, M.C., Valent, B., 2013. Filamentous plant pathogen effectors in action. Nat. Rev. Microbiol. 11, 800–814.

Gottstein, H.D., Kuć, J., 1989. Induction of systemic resistance to anthracnose in cucumber by phosphates. Phytopathology 79, 176–179.

Graham, R.D., Rovira, A.D., 1984. A role for manganese in the resistance of wheat plants to take-all. Plant Soil 78, 441–444.

Graham, M.Y., Graham, T.L., 1991. Rapid accumulation of anionic peroxidases and phenolic polymers in soybean cotyledon tissues following treatment with *Phytophthora megasperma* f. sp. *Glycinea* wall glucan. Plant Physiol. 97, 1445–1455.

Graham, R.D., Webb, M.J., 1991. Micronutrients and plant disease resistance and tolerance in plants. In: Mortvedt, J.J., Cox, F.R., Shuman, L.M., Welch, R.M. (Eds.), Micronutrients in Agriculture. SSSA Book Series No. 4. Soil Science Society of America, Madison, WI, USA, pp. 329–370.

Greger, M., Landberg, T., Vaculík, M., 2018. Silicon influences soil availability and accumulation of mineral nutrients in various plant species. Plants 7 (41), 1–16.

Grundhöfer, H., 1994. Einfluß von Silikataufnahme und -einlagerung auf den Befall der Rebe mit Echtem Mehltau. Dissertation Universität Hohenheim, Stuttgart, Germany.

Guerriero, G., Hausman, J.F., Legay, S., 2016. Silicon and the plant extracellular matrix. Front. Plant Sci. 7, 463.

Guével, M.-H., Menzies, J.G., Bélanger, R.R., 2007. Effect of root and foliar applications of soluble silicon on powdery mildew control and growth of wheat plants. Eur. J. Plant Pathol. 119, 429–436.

Gupta, N., Debnath, S., Sharma, S., Sharma, P., Purohit, J., 2017. Role of nutrients in controlling the plant diseases in sustainable agriculture. In: Meena, V.S., Mishra, P.K., Bisht, J.K., Pattanayak, A. (Eds.), Agriculturally Important Microbes for Sustainable Agriculture, 2. Springer, Singapore, pp. 217–262.

Hajiboland, R., Moradtalab, N., Eshaghi, Z., Feizy, J., 2018. Effect of silicon supplementation on growth and metabolism of strawberry plants at three developmental stages. N. Z. J. Crop Hortic. Sci. 46, 144–161.

Hallock, D.L., Garren, K.H., 1968. Pod breakdown, yield and grade of Virginia type peanuts as affected by Ca, Mg, and K sulfates. Agron. J. 60, 253–357.

Hancock, J.G., Huisman, O.C., 1981. Nutrient movement in host-pathogen systems. Annu. Rev. Phytopathol. 19, 309–331.

Haneklaus, S., Bloem, E., Schnug, E., 2009. Sulfur induced resistance (SIR): biological and environmentally sound concept for disease control. In: Sirko, A., De Kok, L.J., Haneklaus, S., Hawkesford, M.J., Rennenberg, H., Saito, K., et al., (Eds.), Sulfur Metabolism in Plants. Regulatory Aspects, Significance of Sulfur in the Food Chain, Agriculture and the Environment. Margraf Publishers GmBH, Weikersheim, Germany, pp. 129–133.

Haneklaus, S., Bloem, E., Schnug, E., 2007. Sulfur and plant diseaseIn: Datnoff, L.W., Elmer, W.H., Huber, D.M. (Eds.), Mineral Nutrition and Plant Disease. APS Press, Saint Paul, MN, USA, pp. 101–118.

Hanisch, H.-C., 1980. Zum Einfluss der Stickstoffdüngung und vorbeugender Spritzung von Natronwasserglas zu Weizenpflanzen auf deren Widerstandsfähigkeit gegen Getreideblattläuse. Kali-Briefe 15, 287–296.

Harper, J.F., Harmon, A., 2005. Plants, symbiosis and parasites: a calcium signalling connection. Nature Reviews Molecular Cell Biol. 6, 555–566.

Harris, R.F., Romig, D.E., 2005. Health. In: 2nd ed. Lal, R. (Ed.), Encyclopedia of Soil Science, 1. Taylor and Francis, New York, NY, USA, pp. 807–809.

Havlin, J.L., Schlegel, A.J., 2021. Review of phosphite as a plant nutrient and fungicide. Soil Syst. 5, 52.

Hayasaka, T., Fujii, H., Ishiguro, K., 2008. The role of silicon in preventing appressorial penetration by the rice blast fungus. Phytopathology 98, 1038–1044.

He, C.W., Ma, J., Wang, L.J., 2015. A hemicellulose-bound form of silicon with potential to improve the mechanical properties and regeneration of the cell wall of rice. New Phytol. 206, 1051–1062.

Heine, G., Tikum, G., Horst, W.J., 2007. The effect of silicon on the infection by and spread of *Pythium aphanidermatum* in single roots of tomato and bitter gourd. J. Exp. Bot. 58, 569–577.

Hilleary, R., Paez-Valencia, J., Vens, C., Toyota, M., Palmgren, M., Gilroy, S., 2020. Tonoplast-localized Ca^{2+} pumps regulate Ca^{2+} signals during pattern-triggered immunity in *Arabidopsis thaliana*. Proc. Natl. Acad. Sci. USA 117, 18849–18857.

Hoestra, H., 1994. Ecology and pathology of replant problems. Acta Hort. 363, 1–10.
Hoffland, E., Jeger, M.J., van Beusichem, M.L., 2000. Effect of nitrogen supply rate on disease resistance in tomato depends on the pathogen. Plant Soil 218, 239–247.
Höfte, M., Seong, K.Y., Jurkevitch, E., Verstraete, W., 1991. Pyoverdin production by the plant growth beneficial *Pseudomonas* strain 7NSK$_2$: ecological significance in soil. Plant Soil 130, 249–257.
Hollowell, J.E., Shew, B.B., Beute, M.K., Abad, Z.G., 1998. Occurrence of pod rot pathogens in peanuts grown in North Carolina. Plant Dis. 82, 1345–1349.
Horst, W.J., Marschner, H., 1978. Effect of silicon on manganese tolerance of bean plants (*Phaseolus vulgaris* L.). Plant Soil 50, 287–303.
Huber, D.M., 1980. The role of mineral nutrition in defense. In: Harsfall, J.G., Cowling, E.B. (Eds.), Plant Disease, V. Academic Press, New York, pp. 381–406.
Huber, D.M., 1989. The role of nutrition in the take-all disease of wheat and other small grains. In: Engelhard, A.W. (Ed.), Soilborne Plant Pathogens: Management of Diseases with Macro- and Microelements. APS Press, Saint Paul, MN, USA, pp. 46–75.
Huber, D.M., 2021. Glyphosate's impact on humans, animals, and the environment. In: Wilson, C.L., Huber, D.M. (Eds.), Synthetic Pesticide Use in Africa. CRC Press, Boca Raton, FL, USA. Available from: https://doi.org/10.1201/9781003007036.
Huber, D.M., Watson, R.D., 1974. Nitrogen form and plant disease. Ann. Rev. Phytopathol. 12, 139–165.
Huber, D.M., McCay-Buis, T.S., 1993. A multiple component analysis of the take-all disease of cereals. Plant Dis. 77, 437–447.
Huber, D.M., Wilhelm, N.S., 1988. The role of manganese in resistance to plant diseases. In: Graham, R.D., Hannan, R.J., Uren, N.C. (Eds.), Manganese in Soils and Plants. Kluwer Academic Publ, Dordrecht, The Netherlands, pp. 155–173.
Huber, D.M., Graham, R.D., 1999. The role of nutrition in crop resistance and tolerance to diseases. In: Rengel, Z. (Ed.), Mineral Nutrition of Crops. Fundamental Mechanisms and Implications. Food Products Press, New York, NY, USA, pp. 169–204.
Huber, D.M., Thompson, I.A., 2007. Nitrogen and plant disease. In: Datnoff, L.W., Elmer, W.H., Huber, D.M. (Eds.), Mineral Nutrition and Plant Disease. APS Press, Saint Paul, MN, USA, pp. 31–44.
Hutzler, P., Fischbach, R., Heller, W., Jungblut, T.P., Reuber, S., Schmitz, R., et al., 1998. Tissue localization of phenolic compounds in plants by confocal laser scanning microscopy. J. Exp. Bot. 49, 953–965.
Ibáñez, F., Tonelli, M.L., Muñoz, V., Figueredo, M.S., Fabra, A., 2017. Bacterial endophytes of plants: Diversity, invasion mechanisms and effects on the host. In: Maheshwari, D.K. (Ed.), Endophytes: Biology and Biotechnology. Sustainable Development and Biodiversity, 15. Springer, Cham, Switzerland, pp. 25–40.
Ismunadji, M., 1976. Rice diseases and physiological disorders related to potassium deficiency. In: Fertilizer Use and Plant Health: Proceedings of the 12th Colloquium of the International Potash Institute. Izmir, Turkey, pp. 47–60.
Jacoby, R., Peukert, M., Succurro, A., Koprivova, A., Kopriva, S., 2017. The role of soil microorganisms in plant mineral nutrition – current knowledge and future directions. Front. Plant Sci. 8, 1617.
Janvier, C., Villeneuve, F., Alabouvette, C., Edel-Hermann, V., Mateille, T., Steinberg, C., 2007. Soil health through soil disease suppression: Which strategy from descriptors to indicators? Soil Biol. Biochem. 39, 1–23.
Jeer, M., Telugu, U.M., Voleti, S.R., Padmakumari, A.P., 2017. Soil application of silicon reduces yellow stem borer, *Scirpophaga incertulas* (Walker) damage in rice. J. App. Entomol. 141, 189–201.
Jenkyn, J.F., 1976. Nitrogen and leaf diseases of spring barley. In: Fertilizer Use and Plant Health: Proceedings of the 12th Colloquium of the International Potash Institute. Izmir, Turkey, pp. 119–128.
Jiang, N., Yan, J., Liang, Y., Shi, Y., He, Z., Wu, Y., et al., 2020. Resistance genes and their interactions with bacterial blight/leaf streak pathogens (*Xanthomonas oryzae*) in rice (*Oryza sativa* L.) - an updated review. Rice 13, 3.
Johal, G.S., Huber, D.M., 2009. Glyphosate effects on diseases of plants. Eur. J. Agron. 3, 144–152.
Johnston, D.J., Williamson, B., McMillan, G.P., 1994. The interaction in planta of polygalacturonases from *Botrytis cinerea* with a cell wall-bound polygalacturonase-inhibiting protein (PGIP) in raspberry fruits. J. Exp. Bot. 45 (281), 1837–1843.
Jones, D.A., Takemoto, D., 2004. Plant innate immunity – direct and indirect recognition of general and specific pathogen-associated molecules. Curr. Opin. Immunol. 16, 48–62.
Jones, J.B., Huber, D.M., 2007. Magnesium and plant disease. In: Datnoff, L.E., Elmer, W.H., Huber, D.M. (Eds.), Mineral Nutrition and Plant Disease. APS Press, Saint Paul, MN, USA, pp. 95–100.
Kivilaan, A., Scheffer, R.P., 1958. Factors affecting development of bacterial stem rot of *Pelargonium*. Phytopathology 48, 185–191.
Keeping, M.G., Rutherford, R.S., Sewpersad, C., Miles, N., 2015. Provision of nitrogen as ammonium rather than nitrate increases silicon uptake in sugarcane. AoB Plants 7, plu080.
Kelman, A., McGuire, R.G., Tzeng, K.-C., 1989. Reducing the severity of bacterial soft rot by increasing the concentration of calcium in potato tubers. In: Engelhard, A.W. (Ed.), Soilborne Plant Pathogens: Managements. APS Press, Saint Paul, MN, USA, pp. 102–123.
Kikot, G.E., Hours, R.A., Alconada, T.M., 2009. Contribution of cell wall degrading enzymes to pathogenesis of *Fusarium graminearum*: a review. J. Basic Microbiol. 49, 231–241.
Kiraly, Z., 1964. Effect of nitrogen fertilization on phenol metabolism and stem rust susceptibility of wheat. Phytopathol. Z. 51, 252–261.
Kiraly, Z., 1976. Plant disease resistance as influenced by biochemical effects of nutrients in fertilizers. In: Fertilizer Use and Plant Health: Proceedings of the 12th Colloquium of the International Potash Institute. Izmir, Turkey, pp. 33–46.
Kirkby, E.A., Römheld, V., 2004. Micronutrients in plant physiology: functions, uptake and mobility. Proc. Int. Fert. Soc. York, UK 543, 1–51.

Kloepper, J.W., Leong, J., Teintze, M., Schroth, M.N., 1980. *Pseudomonas* siderophores: a mechanism explaining disease-suppressive soils. Curr. Microbiol. 4, 317−320.

Krauss, A., 1971. Einfluß der Ernährung des Salats mit Massennährstoffen auf den Befall mit *Botrytis cinera* Pers. Z. Pflanzenernähr. Bodenk. 128, 12−23.

Kumar, A., Dubey, A., 2020. Rhizosphere microbiome: Engineering bacterial competitiveness for enhancing crop production. J. Adv. Res. 24, 337−352.

Kvedaras, O.L., An, M., Choi, Y.S., Gurr, G.M., 2010. Silicon enhances natural enemy attraction and biological control through induced plant defences. Bull. Entomol. Res. 100, 367−371.

Kwak, Y.-S., Weller, D.M., 2013. Take-all of wheat and natural disease suppression: a review. Plant Pathol. J. 29, 125−135.

La Tore, A., Iovino, V., Caradonia, F., 2018. Copper in plant protection: current situation and prospects. Phytopathol. Mediterr. 57, 201−236.

Laha, G.S., Prasad, V., Muthuraman, P., Srinivas Prasad, M., Brajendra, Yugander, A., et al., 2016. Mineral nutrition for the management of rice diseases. Intern. J. Plant Prot. 9, 310−313.

Lambers, H., Shane, M.W., Cramer, M.D., Pearse, S.J., Veneklaas, E.J., 2006. Root structure and functioning for efficient acquisition of phosphorus: matching morphological and physiological traits. Ann. Bot. 98, 693−713.

Lamont, I.L., Beare, P.A., Ochsner, U., Vasil, A.I., Vasil, M.L., 2002. Siderophore-mediated signaling regulates virulence factor production in *Pseudomonas aeruginosa*. Proc. Natl. Acad. Sci. USA 99, 7072−7077.

Langer, S.E., Marina, M., Burgos, J.L., Martínez, G.A., Civello, P.M., Villarreal, N.M., 2019. Calcium chloride treatment modifies cell wall metabolism and activates defense responses in strawberry fruit (*Fragaria* × *ananassa*, Duch). J. Sci. Food Agric. 99, 4003−4010.

Larsson, S., Wiren, A., Lundgren, L., Ericsson, T., 1986. Effects of light and nutrient stress on leaf phenolic chemistry in *Salix dasyclados* and susceptibility to *Garlerucella lineola* (COL. Chrysomelidea). Oikos 47, 205−210.

Legrand, F., Chen, W., Francisco Cobo-Díaz, J., Picot, A., Le Floch, G., 2019. Co-occurrence analysis reveal that biotic and abiotic factors influence soil fungistasis against *Fusarium graminearum*. FEMS Microbiol. Ecol. 95, fiz056.

Lehmann, S., Serrano, M., L'Haridon, F., Tjamos, S.E., Metraux, J.-P., 2015. Reactive oxygen species and plant resistance to fungal pathogens. Phytochemistry 112, 54−62.

Lehmann, J., Bossio, D.A., Kögel-Knabner, I., Rillig, M.C., 2020. The concept and future prospects of soil health. Nat. Rev. Earth Environ. 1, 544−553.

Leusch, H.-J., Buchenauer, H., 1988a. Einfluss von Bodenbehandlungen mit siliziumreichen Kalken und Natriumsilikat auf den Mehltaubefall von Weizen. Kali-Briefe (Büntehof) 19, 1−11.

Leusch, H.-J., Buchenauer, H., 1988b. Si-Gehalte und Si-Lokalisation im Weizenblatt und deren Bedeutung für die Abwehr einer Mehltauinfektion. Kali-Briefe (Büntehof) 19, 13−24.

Levine, A., Pennell, R.I., Alvarez, M.E., Palmer, R., Lamb, C., 1996. Calcium-mediated apoptosis in a plant hypersensitive disease resistance response. Curr. Biol. 6 (4), 427−437.

Li, A.L., Wang, M.L., Zhou, R.H., Kong, X.Y., Huo, N.X., Wang, W.S., et al., 2005. Comparative analysis of early H_2O_2 accumulation in compatible and incompatible wheat-powdery mildew interactions. Plant Pathol. 54, 308−316.

Li, H., Zhou, X., Huang, Y., Liao, B., Cheng, L., Ren, B., 2021. Reactive oxygen species in pathogen clearance: The killing mechanisms, the adaption response, and the side effects. Front. Microbiol. 11, 622534.

Liang, Y., Nikolic, M., Bélanger, R., Gong, H., Song, A., 2015. Silicon in Agriculture: From Theory to Practice. Springer, Dordrecht, The Netherlands.

Liebisch, F., Max, J.F.J., Heine, G., Horst, W.J., 2009. Blossom-end rot and fruit cracking of tomato grown in net-covered greenhouses in central Thailand can partly be corrected by calcium and boron sprays. J. Plant Nutr. Soil Sci. 172, 140−150.

Lobato, M.C., Olivieri, F.P., Altamiranda, E.A.G., Wolski, E.A., Daleo, G.R., Caldiz, D.O., et al., 2008. Phosphite compounds reduce disease severity in potato seed tubers and foliage. Eur. J. Plant Pathol. 122, 349−358.

Lu, Z.-X., Yu, X.-P., Heong, K.-L., Hu, C., 2007. Effect of nitrogen fertilizer on herbivores and its stimulation to major insect pests in rice. Rice Sci. 14, 56−66.

Mackie, K.A., Müller, T., Kandeler, E., 2012. Remediation of copper in vineyards - A mini review. Environ. Poll. 167, 16−26.

MacNish, G.C., 1988. Changes in take-all (*Gaeumannomyces graminis* var. *tritici*), rhizoctonia root rot (*Rhicotonia solani*) and soil pH in continuous wheat with annual applications of nitrogenous fertilizer in Western Australia. Austr. J. Exp. Agric. 28, 333−341.

Maejima, K., Oshima, K., Namba, S., 2014. Exploring the phytoplasmas, plant pathogenic bacteria. J. Gen. Plant Pathol. 80, 210−221.

Malhi, S.S., Piening, L.J., Macpherson, D.J., 1989. Effect of copper on stem melanosis and yield of wheat: Sources, rates and methods of application. Plant Soil 119, 199−204.

Mandal, K., Saravanan, R., Maiti, S., 2008. Effect of different levels of N, P and K on downy mildew (*Peronospora plantaginis*) and seed yield of isabgol (*Plantago ovata*). Crop Prot. 27, 988−995.

Marques, J.P.R., Amorim, L., Spósito, M.B., Appezzato da Glória, B., 2016. Ultrastructural changes in the epidermis of petals of the sweet orange infected by *Colletotrichum acutatum*. Protoplasma 253, 1233−1242.

Marschner, P., Asher, J.S., Graham, R.D., 1991. Effect of manganese-reducing rhizosphere bacteria on the growth of *Glaeumanomyces graminis* var. *tritici* and on manganese uptake by wheat (*Triticum aestivum* L.). Biol. Fertil. Soils 12, 33−38.

Martinez, D.A., Loening, U.E., Graham, M.C., 2018. Impacts of glyphosate-based herbicides on disease resistance and health of crops: a review. Environ. Sci. Eur. 30, 2.

Martins, V., Garcia, A., Alhinho, A.T., Costa, P., Lanceros-Méndez, S., Costa, M.M.R., et al., 2020. Vineyard calcium sprays induce changes in grape berry skin, firmness, cell wall composition and expression of cell wall-related genes. Plant Physiol. Biochem. 150, 49–55.

Martyniuk, S., 1988. Pectolytic enzymes activity and pathogenicity of *Gaeumannomyces graminis* var. *tritici* and related *Phialophora*-like fungi. Plant Soil 107, 19–23.

Massey, F.P., Ennos, A.R., Hartley, S.E., 2006. Silica in grasses as a defence against insect herbivores: contrasting effects on folivores and a phloem feeder. J. Anim. Ecol. 75, 595–603.

Matocha, J.E., Smith, L., 1980. Influence of potassium on *Helminthosporium cynodontis* and dry matter yields of coastal Bermuda grass. Agron. J. 72, 565–567.

Mee, C.Y., Balasundram, K.S., Hanif, A.H.M., 2017. Detecting and monitoring plant nutrient stress using remote sensing approaches: A review. Asian J. Plant Sci. 16, 1–8.

Menzies, J.G., Ehret, D.L., Glass, A.D.M., Samuels, A.L., 1991. The influence of silicon on cytological interactions between *Sphaerotheca fuliginea* and *Cucumis sativus*. Physiol. Molec. Plant Pathol. 39, 403–414.

Menzies, J., Bowen, P., Ehret, D., Glass, A.D.M., 1992. Foliar applications of potassium silicate reduce severity of powdery mildew on cucumber, muskmelon, and zucchini squash. J. Am. Soc. Hortic. Sci. 117, 902–905.

Merwin, I.A., Stiles, W.C., 1989. Root-lesion nematodes, potassium deficiency, and prior cover crops as factors in apple replant disease. J. Am. Soc. Hort. Sci. 114, 724–728.

Miransari, M., 2011. Interactions between arbuscular mycorrhizal fungi and soil bacteria. App. Microbiol. Biotechnol. 89, 917–930.

Mitchell, A.F., Walters, D.R., 2004. Potassium phosphate induces systemic protection in barley to powdery mildew infection. Pest Manag. Sci. 60, 126–134.

Mokrini, F., Viaene, N., Waeyenberge, L., Dababat, A.A., Moens, M., 2018. Investigation of resistance to *Pratylenchus penetrans* and *P. thornei* in international wheat lines and its durability when inoculated together with the cereal cyst nematode *Heterodera avenae*, using qPCR for nematode quantification. Eur. J. Plant Pathol. 151, 875–889.

Moradtalab, N., Weinmann, M., Walker, F., Höglinger, B., Ludewig, U., Neumann, G., 2018. Silicon Improves chilling tolerance during early growth of maize by effects on micronutrient homeostasis and hormonal balances. Front. Plant Sci. 9, 420.

Mortvedt, J.J., Berger, K.C., Darling, H.M., 1963. Effect of manganese and copper on the growth of *Streptomyces scabies* and the incidence of potato scab. Am. Potato J. 40, 96–102.

Myers, D.F., Campbell, R.N., 1985. Lime and the control of clubroot of crucifers: effects of pH, calcium, magnesium, and their interactions. Phytopathology 75, 670–673.

Navazio, L., Moscatiello, R., Genre, A., Novero, M., Baldan, B., Bonfante, P., Mariani, P., 2007. A diffusible signal from arbuscular mycorrhizal fungi elicits a transient cytosolic calcium elevation in host plant cells. Plant Physiol. 144, 673–681.

Nawaz, F., Zulfiqar, B., Ahmad, K.S., Majeed, S., Shehzad, M.A., Javeed, H.M.R., et al., 2021. Pretreatment with selenium and zinc modulates physiological indices and antioxidant machinery to improve drought tolerance in maize (*Zea mays* L.). S. Afr. J. Bot. 138, 209–216.

Nicholson, R.L., Hammerschmidt, R., 1992. Phenolic compounds and their role in disease resistance. Ann. Rev. Phytopathol. 30, 369–389.

Nikolic, M., Nikolic, N., Liang, Y., Kirkby, E.A., Römheld, V., 2007. Germanium-68 as an adequate tracer for silicon transport in plants. Characterization of silicon uptake in different crop species. Plant Physiol. 143, 495–503.

Nogueira, M.A., Cardoso, E.J.B.N., Hampp, R., 2002. Manganese toxicity and callose deposition in leaves are attenuated in mycorrhizal soybean. Plant Soil 246, 1–10.

Olsson, K., 1988. Resistance to gangrene (*Phoma exigua* var. *foveata*) and dry rot (*Fusarium solani* var. *coeruleum*) in potato tubers. I. The influence of pectin-bound magnesium and calcium. Potato Res. 31, 413–422.

Orober, M., Siegrist, J., Buchenauer, H., 2002. Mechanisms of phosphate-induced disease resistance in cucumber. Europ. J. Plant Pathol. 108, 345–353.

Oteifa, B.A., Elgindi, A.Y., 1976. Potassium nutrition of cotton, *Gossypium barbadense*, in relation to nematode infection by *Meliodogyne incognita* and *Rotylenchulus reniformis*. In: Fertilizer Use and Plant Health: Proceedings of the 12th Colloquium of the International Potash Institute. Izmir, Turkey, 301–306.

Ownley, B.H., Duffy, B.K., Weller, D.M., 2003. Identification and manipulation of soil properties to improve the biological control performance of phenazine-producing *Pseudomonas fluorescens*. App. Environ. Microbiol. 69, 3333–3343.

Pankhurst, C.E., Lynch, J.M., 2005. Biocontrol of soil-borne plant diseases. In: Hillel, D., Hatfield, J.H., Powlson, D.S., Rosenzweig, C., Scow, K.M., Singer, M.J., et al., (Eds.), Encyclopedia of Soils in the Environment, 1. Elsevier, Oxford, UK, pp. 129–136.

Papenbrock, J., Riemenschneider, A., Kamp, A., Schulz-Vogt, H.N., Schmidt, A., 2007. Characterization of cysteine-degrading and H_2S-releasing enzymes of higher plants - from the field to the test tube and back. Plant Biol. 9, 582–588.

Pavlovic, J., Samardzic, J., Maksimovic, V., Timotijevic, G., Stevic, N., Laursen, K.H., et al., 2013. Silicon alleviates iron deficiency in cucumber by promoting mobilization of iron in the root apoplast. New Phytol. 198, 1096–1107.

Pavlovic, J., Kostic, L., Bosnic, P., Kirkby, E.A., Nikolic, M., 2021. Interactions of silicon with essential and beneficial elements in plants. Front. Plant Sci. 12, 697592.

Pedras, M.S.C., Okanga, F.I., Zaharia, I.L., Khan, A.Q., 2000. Phytoalexins from crucifers: synthesis, biosynthesis, and biotransformation. Phytochemistry 53, 161–176.

Perrenoud, S., 1977. Potassium and plant health. In Research Topics, No. 3, pp. 1–118. Intern. Potash Inst. Bern, Switzerland.

Perrenoud, S., 1990. Potassium and Plant Health, Research Topics, No. 3, 2nd ed. Intern. Potash Inst, Bern, Switzerland.

Pfeffer, H., Dannel, F., Römheld, V., 1998. Are there connections between phenol metabolism, ascorbate metabolism and membrane integrity in leaves of boron-deficient sunflower plants? Physiol. Plant. 104, 479–485.

Pilon-Smits, E.A.H., 2019. On the ecology of selenium accumulation in plants. Plants 8, 197.

Poveda, J., Abril-Urias, P., Escobar, C., 2020. Biological control of plant-parasitic nematodes by filamentous fungi inducers of resistance: trichoderma, mycorrhizal and endophytic fungi. Front. Microbiol. 11, 992.

Pozza, E.A., Pozza, A.A.A., Magna dos Santos Botelho, D., 2015. Silicon in plant disease control. Rev. Ceres Viçosa 62, 323–331.

Prabhu, A.S., Fageria, N.K., Huber, D.M., Rodrigues, F.A., 2007. Potassium nutrition and plant diseases. In: Datnoff, L.E., Elmer, W.H., Huber, D.M. (Eds.), Mineral Nutrition and Plant Disease. APS Press, Saint Paul, MN, USA, pp. 57–78.

Primavesi, A., 2006. Soil system management in the humid and subhumid tropics. In: Uphoff, N., Ball, A.S., Fernandes, E., Herren, H., Husson, O., Laing, M., et al., (Eds.), Biological Approaches to Sustainable Soil Systems. CRC Press, Boca Raton, FL, USA, pp. 15–26.

Pryce-Jones, E., Carver, T., Gurr, S.J., 1999. The roles of cellulase enzymes and mechanical force in host penetration by *Erysiphe graminis* f.sp. *hordei*. Physiol. Mol. Plant Pathol. 55, 175–182.

Raaijmakers, J.M., Paulitz, T.C., Steinberg, C., Alabouvette, C., Moënne-Loccoz, Y., 2009. The rhizosphere: a playground and battlefield for soilborne pathogens and beneficial microorganisms. Plant Soil 321, 341–361.

Rabelo de Faria, M., Costa, L.S.A.S., Chiaramonte, J.B., Bettiol, W., Mendes, R., 2021. The rhizosphere microbiome: functions, dynamics, and role in plant protection. Trop. Plant Pathol. 46, 13–25.

Rashid, M.M., Jahan, M., Islam, K.S., 2016. Impact of nitrogen, phosphorus and potassium on brown planthopper and tolerance of its host rice plants. Rice Sci. 23, 119–131.

Rasmussen, J.B., Hammerschmidt, R., Zook, M.N., 1991. Systemic induction of salicylic acid accumulation in cucumber after inoculation with *Pseudomonas syringae* pv. *syringae*. Plant Physiol. 97, 1342–1347.

Ratjen, A.M., Gerendás, J., 2009. A critical assessment of the suitability of phosphite as a source of phosphorus. J. Plant Nutr. Soil Sci. 172, 821–828.

Reddy, P.P., 2016. Insect pests and their management. In: Meena, V.S., Mishra, P.K., Bisht, J.K., Pattanayak, A. (Eds.), Agriculturally Important Microbes for Sustainable Agriculture, 2. Springer, Singapore, pp. 187–206.

Reis, E.M., Cook, R.J., McNeal, B.L., 1982. Effect of mineral nutrition on take-all of wheat. Phytopathology 72, 224–229.

Rengel, Z., 2015. Availability of Mn, Zn and Fe in the rhizosphere. J. Soil Sci. Plant Nutr. 15, 397–409.

Reuveni, R., Dor, G., Raviv, M., Reuveni, M., Tuzun, S., 2000. Systemic resistance against *Sphaerotheca fuliginea* in cucumber plants exposed to phosphate in hydroponics system, and its control by foliar spray of mono-potassium phosphate. Crop Prot. 19, 355–361.

Reuveni, M., Reuveni, R., 1995. Efficacy of foliar sprays of phosphates in controlling powdery mildews in field-grown nectarine, mango trees and grapevines. Crop Prot. 14, 311–314.

Richardson, M.D., Croughan, S.S., 1989. Potassium influence on susceptibility of Bermuda grass to *Helminthosporium cynodontis* toxin. Crop Sci. 29, 1280–1282.

Riethmüller, A., Voglmayr, H., Göker, M., Weiß, M., Oberwinkler, F., 2002. Phylogenetic relationships of the downy mildews (Peronosporales) and related groups based on nuclear large subunit ribosomal DNA sequences. Mycologia 94, 834–849.

Robertson, G.P., Grandy, A.S., 2006. Soil system management in temperate regions. In: Uphoff, N., Ball, A.S., Fernandes, E., Herren, H., Husson, O., Laing, M., et al., (Eds.), Biological Approaches to Sustainable Soil Systems. CRC Press, Boca Raton, FL, USA, pp. 27–39.

Rogalla, H., Römheld, V., 2002. Role of leaf apoplast in silicon-mediated manganese tolerance of *Cucumis sativus* L. Plant Cell Environ. 25, 549–555.

Romera, F.J., García, M.J., Lucena, C., Martínez-Medina, A., Aparicio, M.A., Ramos, J., et al., 2019. Induced systemic resistance (ISR) and Fe deficiency responses in dicot plants. Front. Plant Sci. 10, 287.

Römheld, V., Neumann, G., 2006. The rhizosphere: contributions of the soil-root interface to sustainable soil systems. In: Uphoff, N., Ball, A.S., Fernandes, E., Herren, H., Husson, O., Laing, M., et al., (Eds.), Biological Approaches to Sustainable Soil Systems. CRC Press, Boca Raton, FL, USA, pp. 91–107.

Russell, P.E., 2005. A century of fungicide evolution. J. Agric. Sci. 143, 11–25.

Sakr, N., 2016. Silicon control of bacterial and viral diseases in plants. J. Plant Prot. Res. 56, 331–336.

Salim, M., Saxena, R.C., 1991. Nutritional stresses and varietal resistance in rice: effects on whitebacked plant-hopper. Crop Sci. 31, 797–805.

Sarniguet, A., Lucas, P., Lucas, M., Samson, R., 1992. Soil conduciveness to take-all of wheat: influence of the nitrogen fertilizers on the structure of populations of fluorescent pseudomonads. Plant Soil 145, 29–36.

Sasanuma, I., Suzuki, T., 2016. Effect of calcium on cell-wall degrading enzymes of *Botrytis cinerea*. Biosci. Biotechnol. Biochem. 80, 1730–1736.

Schackmann S., 2005. Approaches to Control Take-All Disease in Wheat by an optimised Crop and Fertilisation Management Practice. Bachelor Thesis, Universität Hohenheim, Stuttgart, Germany.

Schaller, J., Puppe, D., Kaczorek, D., Ellerbrock, R., Sommer, M., 2021. Silicon cycling in soils revisited. Plants 10, 295.

Schulze, D.G., McCay-Buis, T., Sutton, S.R., Huber, D.M., 1995. Manganese oxidation states in *Gaeumannomyces*-infested wheat rhizospheres probed by micro-XANES spectroscopy. Phytopathology 85, 990–994.

Sehrawat, A., Sindhu, S.S., Glick, B.R., 2022. Hydrogen cyanide production by soil bacteria: Biological control of pests and promotion of plant growth in sustainable agriculture. Pedosphere 32, 15–38.

Shah, K.K., Modi, B., Pandey, H.P., Subedi, A., Aryal, G., Pandey, M., et al., 2021. Diversified crop rotation: An approach for sustainable agriculture production. Adv. Agric. 2021, 8924087.

Sharma, A., Shahzad, B., Rehman, A., Bhardwaj, R., Landi, M., Zheng, B., 2019. Response of phenylpropanoid pathway and the role of polyphenols in plants under abiotic stress. Molecules 24, 2452.

Shearer, B.L., Fairman, R.G., 2007. Application of phosphite in a high-volume foliar spray delays and reduces the rate of mortality of four Banksia species infected with *Phytophthora cinnamomi*. Austral. Plant Pathol. 36, 358–368.

Shi, J., Jia, Y., Fang, D., He, S., Zhang, P., Guo, Y., et al., 2020. Screening and identification of RNA silencing suppressors from secreted effectors of plant pathogens. J. Vis. Exp. 156, e60697.

Shireen, F., Nawaz, M.A., Chen, C., Zhang, Q., Zheng, Z., Sohail, H., et al., 2018. Boron: functions and approaches to enhance its availability in plants for sustainable agriculture. Int. J. Mol. Sci. 19, 1856.

Shoaib, A., Awan, Z.A., 2021. Mineral fertilizers improve defense related responses and reduce early blight disease in tomato (*Solanum lycopersicum* L.). J. Plant Pathol. 103, 217–229.

Siddiqui, S., Almari, S.A., Alrumman, S.A., Meghvansi, M.K., Chaudhary, K.K., Kilany, M., et al., 2015. Role of soil amendment with micronutrients in suppression of certain soilborne plant fungal diseases: A review. In: Meghvansi, M., Varma, A. (Eds.), Organic Amendments and Soil Suppressiveness in Plant Disease Management. Springer, Cham, Switzerland, pp. 363–380.

Sikes, B.A., Powell, J.R., Rillig, M.C., 2010. Deciphering the relative contributions of multiple functions within plant-microbe symbioses. Ecology 91, 1591–1597.

Singh, S., Kaur, I., Kariyat, R., 2021. The multifunctional roles of polyphenols in plant-herbivore interactions. Int. J. Mol. Sci. 22, 1442.

Smiley, R.W., 1978. Colonization of wheat roots by *Gaeumannomyces graminis* inhibited by specific soils, microorganisms and ammonium-nitrogen. Soil Biol. Biochem. 10, 175–179.

Smiley, R.W., Cook, R.J., 1973. Relationship between take-all of wheat and rhizosphere pH in soils fertilized with ammonium vs. nitrate-nitrogen. Phytopathology 63, 882–890.

Sogawa, K., 1982. The rice brown plant hopper: feeding physiology and host plant interactions. Annu. Rev. Entomol. 27, 49–73.

Sommer, M., Kaczorek, D., Kuzyakov, Y., Breuer, J., 2006. Silicon pools and fluxes in soils and landscapes - a review. J. Plant Nutr. Soil Sci. 169, 310–329.

Stangoulis, J.C.R., Graham, R.D., 2007. Boron and plant disease. In: Datnoff, L.E., Elmer, W.H., Huber, D.M. (Eds.), Mineral Nutrition and Plant Disease. APS Press, Saint Paul, MN, USA, pp. 207–214.

Staskawicz, B.J., 2001. Genetics of plant-pathogen interactions specifying plant disease resistance. Plant Physiol. 125, 73–76.

Stone, H.E., Armentrout, V.N., 1985. Production of oxalic acid by *Sclerotium cepivorum* during infection of onion. Mycologia 77, 526–530.

Sweeney, D.W., Granade, G.V., Eversmeyer, M.G., Whitney, D.A., 2000. Phosphorus, potassium, chloride, and fungicide effects on wheat yield and leaf rust severity. J. Plant Nutr. 23, 1267–1281.

Tausz, M., Šircelj, H., Grill, D., 2004. The glutathione system as a stress marker in plant ecophysiology: is a stress-response concept valid? J. Exp. Bot. 55, 1955–1962.

Thao, H.T.B., Yamakawa, T., 2009. Phosphite (phosphorous acid): fungicide, fertilizer or bio-stimulator? Soil Sci. Plant Nutr. 55, 228–234.

Thompson, I.A., Huber, D.M., 2007. Manganese and plant disease. In: Datnoff, L.E., Elmer, W.H., Huber, D.M. (Eds.), Mineral Nutrition and Plant Disease. APS Press, Saint Paul, MN, USA, pp. 139–154.

Thompson, I.A., Huber, D.M., Schulze, D.G., 2006. Evidence of a multicopper oxidase in Mn oxidation by *Gaeumannomyces graminis* var. tritici. Phytopathology 96, 130–136.

Thomson, C.J., Marschner, H., Römheld, V., 1993. Effect of nitrogen fertilizer form on pH of the bulk soil and rhizosphere, and on the growth, phosphorus, and micronutrient uptake of bean. J. Plant Nutr. 16, 493–506.

Tierens, K.F.M.-J., Thomma, B.P.H.J., Brouwer, M., Schmidt, J., Kistner, K., Porzel, A., et al., 2001. Study of the role of antimicrobial glucosinolate-derived isothiocyanates in resistance of *Arabidopsis* to microbial pathogens. Plant Physiol. 125, 1688–1699.

Timm, C.A., Goos, R.J., Johnson, B.E., Siobolik, F.J., Stack, R.W., 1986. Effect of potassium fertilizers on malting barley infected with common root rot. Agron. J. 78, 197–200.

Tolay, I., 2021. The impact of different zinc (Zn) levels on growth and nutrient uptake of basil (*Ocimum basilicum* L.) grown under salinity stress. PLoS ONE 16, e0246493.

Treutter, D., 2005. Significance of flavonoids in plant resistance and enhancement of their biosynthesis. Plant Biol. 7, 581–591.

Trolldenier, G., 1981. Influence of soil moisture, soil acidity and nitrogen source on take-all of wheat. Phytopathol. Z. 102, 163–177.

Trolldenier, G., 1985. Effect of potassium chloride vs. potassium sulphate fertilization at different soil moisture on take-all of wheat. Phytopathol. Z. 112, 56–62.

Turlings, T.C.J., Wäckers, F., 2004. Recruitment of predators and parasitoids by herbivore-injured plants. In: Cardé, R.T., Millar, J.G. (Eds.), Advances in Insect Chemical Ecology. Cambridge University Press, Cambridge, UK, pp. 21–75.

Umbreen, S., Lubega, J., Cui, B., Pan, Q., Jiang, J., Loake, G.J., 2018. Specificity in nitric oxide signaling. J. Exp. Bot. 69, 3439–3448.

Van Bockhaven, J., de Vleesschauwer, D., Höfte, M., 2013. Towards establishing broad-spectrum disease resistance in plants: silicon leads the way. J. Exp. Bot. 64, 1281–1293.

van der Ent, A., Casey, L.W., Blamey, F.P.C., Kopittke, P.M., 2020. Time-resolved laboratory micro-X-ray fluorescence reveals silicon distribution in relation to manganese toxicity in soybean and sunflower. Ann. Bot. 126, 331–341.

van Toor, R.F., Chng, S.F., Warren, R.M., Butler, R.C., 2017. Influence of glyphosate herbicide treatment of couch grass on take-all caused by *Gaeumannomyces graminis* var. *tritici* with the addition of soil-borne microorganisms. N. Z. Plant Prot. 70, 186–195.

Varadarajan, D.K., Karthikeyan, A.S., Matilda, P.D., Raghothama, K.G., 2002. Phosphite, an analog of phosphate, suppresses the coordinated expression of genes under phosphate starvation. Plant Physiol. 129, 1232–1240.

Verbon, E.H., Trapet, P.L., Stringlis, I.A., Kruijs, S., Bakker, P.A.H.M., Pieterse, C.M.J., 2017. Iron and immunity. Annu. Rev. Phytopathol. 55, 355–375.

Veresoglou, S.D., Barto, E.K., Menexes, G., Rillig, M.C., 2013. Fertilization affects severity of disease caused by fungal plant pathogens. Plant Pathol. 62, 961–969.

Veromann, E., Toome, M., Kännaste, A., Kaasik, R., Copolovici, L., Flink, J., et al., 2013. Effects of nitrogen fertilization on insect pests, their parasitoids, plant diseases and volatile organic compounds in *Brassica napus*. Crop Prot. 43, 79–88.

Volk, R.J., Kahn, R.P., Weintraub, R.L., 1958. Silicon content of the rice plant as a factor influencing its resistance to infection by the blast fungus *Pyricularia oryzae*. Phytopathology 48, 179–184.

Walters, D.R., Murray, D.C., 1992. Induction of systemic resistance to rust in *Vicia faba* by phosphate and EDTA: effects of calcium. Plant Pathol. 41, 444–448.

Walters, D.R., Bingham, I.J., 2007. Influence of nutrition on disease development caused by fungal pathogens: implications for plant disease control. Ann. Appl. Biol. 151, 307–324.

Wang, M., Zheng, Q., Shen, Q., Guo, S., 2013. The critical role of potassium in plant stress response. Int. J. Mol. Sci. 14, 7370–7390.

Wang, M., Gao, L., Dong, S., Sun, Y., Shen, Q., Guo, S., 2017. Role of silicon on plant–pathogen interactions. Front. Plant Sci. 8, 701.

Wang, Q., Li, J., Dang, C., Chang, X., Fang, Q., Stanley, D., Ye, G., 2018. Rice dwarf virus infection alters green rice leafhopper host preference and feeding behavior. PLoS ONE 13, 9.

Weinmann, M., 2019. Bio-effectors for Improved Growth, Nutrient Acquisition and Disease Resistance of Crops. In: Raupp, M. (Ed.), Madora GmbH. Lörrach and Lörrach International e.V., Lörrach, Germany.

Wheeler, T.A., Russell, S.A., Anderson, M.G., Serrato-Diaz, L.M., French-Monar, R.D., Woodward, J.E., 2016. Management of peanut pod rot. I: Disease dynamics and sampling. Crop Prot. 79, 135–142.

White, P.J., Broadley, M.R., 2003. Calcium in plants. Ann. Bot. 92, 487–511.

Wiese, H., Nikolic, M., Römheld, V., 2007. Silicon in plant nutrition. In: Sattelmacher, B., Horst, W.J. (Eds.), The Apoplast of Higher Plants: Compartment of Storage, Transport and Reactions. The Significance of the Apoplast for the Mineral Nutrition of Higher Plants. Springer, Dordrecht, The Netherlands, pp. 33–47.

Wilhelm, N.S., Fisher, J.M., Graham, R.D., 1985. The effect of manganese deficiency and cereal cyst nematode infection on the growth of barley. Plant Soil 85, 23–32.

Wilhelm, N.S., Graham, R.D., Rovira, A.D., 1990. Control of Mn status and infection rate by genotype of both host and pathogen in the wheat take-all interaction. Plant Soil 123, 267–275.

Williams, J.S., Cooper, R.M., 2004. The oldest fungicide and newest phytoalexin - a reappraisal of the fungitoxicity of elemental sulphur. Plant Pathol. 53, 263–279.

Wissemeier, A.H., Horst, W.J., 1987. Callose deposition in leaves of cowpea (*Vigna unguiculata* [L.] Walp.) as a sensitive response to high Mn supply. Plant Soil 102, 283–286.

Wu, X., Yu, Y., Baerson, S.R., Song, Y., Liang, G., Ding, C., et al., 2017. Interactions between nitrogen and silicon in rice and their effects on resistance toward the brown planthopper *Nilaparvata lugens*. Front. Plant Sci. 8, 28.

Xuan, H., Streif, J., Saquet, A., Römheld, V., Bangerth, F., 2005. Application of boron with calcium affects respiration and ATP/ADP ratio in 'Conference' pears during controlled atmosphere storage. J. Hort. Sci. Biotech. 80, 633–637.

Yang, Y.-Q., Liu, Q.-Z., 2020. Effects of maize pre-cultivation on soil deterioration indicated by the soil nematode community in replanted apple orchard. Biocontrol Sci. Technol. 30, 878–896.

Ye, M., Song, Y., Long, J., Wang, R., Baerson, S.R., Panc, Z., et al., 2013. Priming of jasmonate-mediated antiherbivore defense responses in rice by silicon. Proc. Natl. Acad. Sci. USA 110, E3631–E3639.

Yu, Y., Kang, Z., Han, Q., Buchenauer, H., Huang, L., 2010. Immunolocalization of 1,3-β-glucanases secreted by *Gaeumannomyces graminis* var. *tritici* in infected wheat roots. J. Phytopathol. 158, 344–350.

Zellner, W., Frantz, J., Leisner, S., 2011. Silicon delays tobacco ringspot virus systemic symptoms in *Nicotiana tabacum*. J. Plant Physiol. 168, 1866–1869.

Zeyen, R.J., Carver, T.L.W., Lyngkjaer, M.F., 2002. Epidermal cell papillae. In: Bélanger, R.R., Bushnell, R.W.R., Dik, A.J., Carver, T.L.W. (Eds.), The Powdery Mildews. A Comprehensive Treatise. APS Press, Saint Paul, MN, USA, pp. 107–125.

Zhan, L.-P., Peng, D.-L., Wang, X.-L., Kong, L.-A., Peng, H., Liu, S.-M., et al., 2018. Priming effect of root-applied silicon on the enhancement of induced resistance to the root-knot nematode *Meloidogyne graminicola* in rice. BMC Plant Biol. 18 (50).

Zhang, J., Zhou, M., Liu, W., Nie, J., Huang, L., 2022. *Pseudomonas syringae* pv. *actinidiae* effector HopAU1 interacts with calcium-sensing receptor to activate plant immunity. Int. J. Mol. Sci. 23 (508), 1–14.

Ziv, C., Zhao, Z., Gao, Y.G., Xia, Y., 2018. Multifunctional roles of plant cuticle during plant-pathogen interactions. Front. Plant Sci. 9, 1088.

Zuccaro, A., 2020. Plant phosphate status drives host microbial preferences: a trade-off between fungi and bacteria. EMBO J. 39, e104144.

Chapter 11

Diagnosis and prediction of deficiency and toxicity of nutrients

Richard Bell
Centre for Sustainable Farming Systems, Food Futures Institute, Murdoch University, Murdoch, WA, Australia

Summary

Plant and soil analyses are complementary tools for the diagnosis and prognosis (prediction) of crop nutrient status and environmental quality. Their use depends on well-established relationships between nutrient concentration (in plants or soils) and plant growth/yield or environmental quality. Based on a calibrated relationship, nutrient concentrations are categorized into deficient, adequate, and toxic ranges. From these relationships, the critical concentration range is defined, corresponding usually to concentrations required to obtain 90% or 95% of maximal growth/yield. In addition, use can be made of visual plant symptoms to recognize nutritional disorders. Fertilizer recommendations are often based on plant or soil analysis results. Critical deficiency ranges are dependent on plant age, plant part, concentrations of other nutrients, as well as on environmental factors. Plant analysis can be used to either diagnose existing disorders in plants or predict future nutrient limitations based on various calibration relationships. Soil tests are complementary to plant tests and are used primarily to predict future nutrient limitations.

11.1 General

The goals for managing crop nutrition are profitable yield, high-quality products for specified markets, low disease impacts, avoidance or alleviation of crop stress, optimal symbiotic N_2 fixation (for legumes), nutritious food, and minimal environmental impact. In many parts of the world, inadequate crop nutrient supply results in pronounced yield losses and low product quality; by contrast, oversupply of nutrients causes unnecessary environmental impacts such as elevated nitrate leaching or gaseous N emissions (Dobermann et al., 2021). This chapter examines tools that can be used to make better decisions about crop nutrition to avoid under- or oversupply of nutrients. The tools considered are crop response to nutrient supply in the field, plant symptoms, and plant and soil analyses.

Optimizing nutrient supply to crops and pastures is critical to (1) satisfying the growing demand in the global market for sufficient amounts of nutritious food, (2) limiting fertilization costs, (3) minimizing potential environmental problems caused by agriculture, and (4) increasing resistance of crops to drought, waterlogging, frost, salinity, and climate variability. For example, in southwest Australia, higher plant concentrations of K were required to maintain grain set in wheat by mitigating the effects of frost events on floret sterility (Ma et al., 2019). Optimization of nutrient supply to crops to meet the above goals, including profit optimization, requires decisions by farmers on fertilizer applications that should be based on the 4Rs concept (Bruulsema et al., 2012). That is, for optimal impact the fertilizer selected should be the right form, applied at the right rate, in the right place, and at the right time.

Better fertilizer decisions by farmers can be made by correct diagnosis of existing nutrient disorders and accurate prognosis of future nutrient limitations. Better tools will support an integrated nutrient management approach to crop production (Bruulsema et al., 2012; Schjoerring et al., 2019).

[☆] This chapter is a revision of the third edition chapter by V. Römheld, pp. 299–312. DOI: https://doi.org/10.1016/B978-0-12-384905-2.00011-X.
© Elsevier Ltd.

11.2 Tools for diagnosis of nutrient disorders

Optimal crop nutrition depends on the adequate supply of nutrients, but supply from soils varies over time, both within a growing season and over successive seasons. Crop nutrient demand also changes over time. The ideal scenario is when the supply of nutrients from the soil matches the crop demand throughout the growing season. However, there is often a mismatch between supply and demand that results in nutritional disorders, restricted crop growth, or economic and/or environmental losses from nutrient under- or oversupply.

Diagnosis is the process of detecting and identifying nutrient disorders. Recognition of visible plant symptoms of nutrient disorders and plant analysis are the main tools for diagnosis. Equally important is the capacity to predict whether a gap between nutrient supply and demand will occur in the future (during the growing season or in the following growing season). Soil analysis of samples taken before sowing a crop or at the start of the growing season can be used for the prediction of nutrient supply, known as prognosis. Similarly, plant samples taken early in the growth cycle can be used to predict the likelihood of a nutrient disorder occurring at a later time. When used for prognosis of nutrient disorders, the prediction is based on a different relationship between plant nutrient concentration and growth (usually as yield) compared with the diagnostic use (Smith and Loneragan, 1997; see also Jongruaysup et al., 1994). Prediction is inherently less certain because it involves unknown future changes in weather, diseases, rainfall, etc. during the growing season. Hence, different critical deficiency ranges (CDRs) apply for prognosis than for diagnosis (Ulrich and Hills, 1990). Over a longer time and including successive crops grown for two or more years, soil and plant analyses can be used to monitor trends in nutrient status so that adjustments in nutrient management programs can be made. Once a gap between nutrient supply and demand is diagnosed or predicted, fertilizer programs can be designed, based on the 4Rs concept, to alleviate shortages or excesses in nutrient supply.

11.2.1 Field responses to nutrient supply

The most powerful diagnostic evidence for nutrient disorders comes from the application of a treatment (usually a fertilizer) in a replicated field experiment that produces a measured crop response (increase in yield, quality, decrease in disease, etc.) (Bell, 1997). The demonstration of crop response in the field is most likely to persuade farmers of both the validity of the diagnosis and the efficacy of the treatments to alleviate the deficiency. Field experiments are also the preferred means for developing calibration relationships between plant nutrient or soil nutrient concentrations and crop yield response. However, field experiments are costly to conduct and that limits the number of sites where a diagnosis can be provided. Other limiting factors for crop growth such as drought, waterlogging, acidity, salinity, disease, and insects can hamper crop growth in field experiments and produce an inconclusive result. Experimental methods applied also need to be considered carefully to avoid confounding influences on crop response. For example, basal NPK fertilizers that are not pure may add sufficient micronutrients to the soil to correct a deficiency and give a false conclusion about the lack of response to the treatment (Bell et al., 1990). Hence, despite the value of well-designed and managed field experiments for the diagnosis of nutrient disorders, they should be supplemented by plant analysis, soil analysis, and symptom recognition to gather information from a large number of soils, seasonal conditions, and crop species for integrated nutrient management.

11.2.2 Diagnosis of nutritional disorders by visible symptoms

Symptoms generally become clearly visible only when a deficiency is acute and the growth rate and yield are substantially reduced. Hence, most cases of nutritional disorders are not characterized by specific visible symptoms. For example, in natural vegetation, many annual and perennial plant species, particularly those adapted to nutrient-poor sites, adjust their growth rate to the most limiting nutrient and, thus, do not develop visual deficiency symptoms (Chapin, 1988). On the other hand, transient visible symptoms of Mg deficiency in cereals, which may be observed under field conditions during intensive growth, may be without detrimental effect on the final grain yield (Pissarek, 1979).

The form of chlorosis or necrosis of leaves, the aberrations in the development of flowers and fruit, and their pattern of expression within that plant part and within the crop canopy, are important criteria for symptom diagnosis (De Bang et al., 2021). Visible deficiency symptoms of individual nutrients are described briefly in Chapters 6 and 7. For descriptions and color images of a broad range of symptoms of nutrient disorders, the reader is referred to Bergmann (1992), Reuter and Robinson (1997), and Zorn et al. (2006) as well as websites such as http://www.tll.de/visuplant. Apps on smartphones now provide easy access to catalogs of visual symptoms for a diverse range of crop species (annual crops, pasture species, vegetable crops, fruit and nut trees), facilitating their recognition in the field (e.g., Yara CheckIT).

Diagnosis based on visible symptoms requires a systematic approach as summarized in Table 11.1. Symptoms appear preferentially on either older or younger leaves, depending on whether the nutrient in question is readily

TABLE 11.1 Principles of visual diagnosis of nutritional disorders in plants.

Plant part	Main symptom		Disorder
			Deficiency
Old and mature leaf blades	Chlorosis → Uniform		N (S)
	Chlorosis → Interveinal or blotched		Mg (Mn)
	Necrosis → Tip and marginal scorch		K
	Necrosis → Interveinal		Mg (Mn)
Young leaf blades and apex	Chlorosis → Uniform		Fe (S)
	Chlorosis → Interveinal or blotched		Zn (Mn)
	Necrosis (chlorosis)		Ca, B, Cu
	Deformations		Mo (Zn, B)
			Toxicity
Old and mature leaf blades	Necrosis → Spots		Mn (B)
	Necrosis → Tip and marginal scorch		B, salt (spray injury)
	Chlorosis, necrosis		Nonspecific toxicity

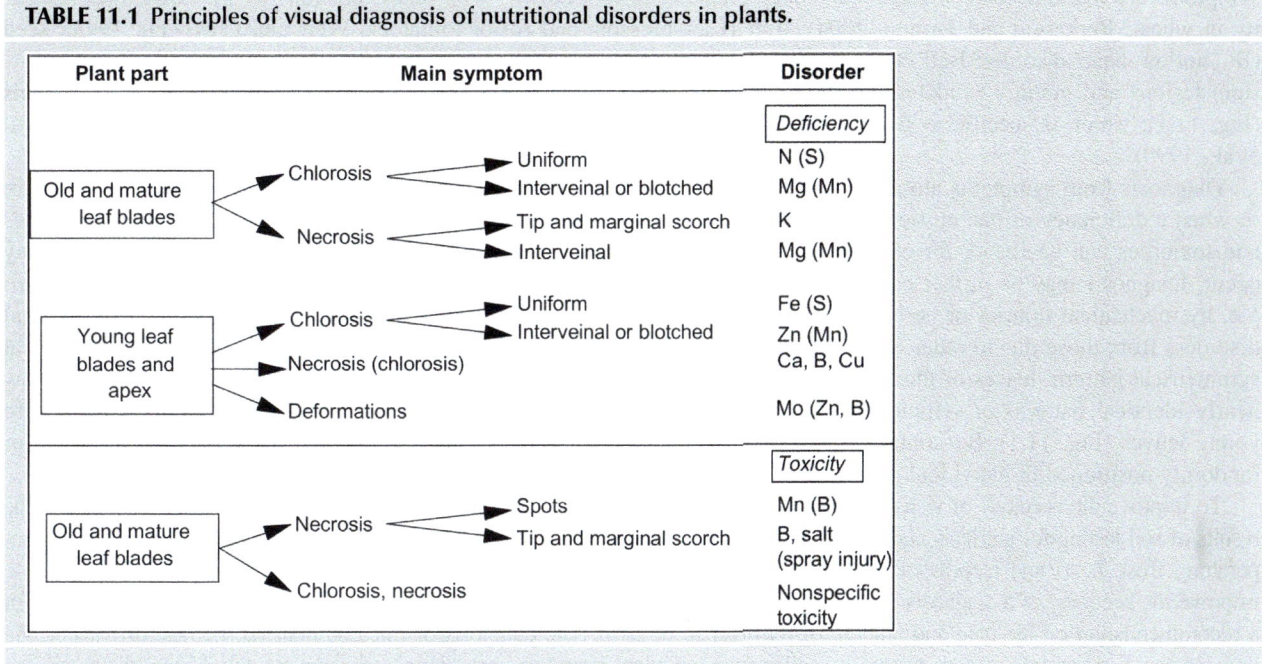

FIGURE 11.1 Deficiency symptoms of (A) zinc in maize; (B) potassium in canola; (C) phosphorus in peanut; and (D) boron in the kernels of peanut seed (referred to as hollow heart).

re-translocated in the phloem (see Chapter 3). Potassium and P deficiency symptoms, for example, are generally expressed in old leaves, whereas Zn deficiency symptoms are more common on young or recently matured leaves (Fig. 11.1). The pattern of symptoms within the canopy may also be modified by how the deficiency is induced: (1) permanent insufficient supply; (2) a sudden decline from adequate supply, or (3) an interruption of adequate nutrient supply by a transient weather event (Bell, 2000). In addition to symptoms expressed in leaves, the deficiency and toxicity symptoms are found in roots or tubers (B—deep cracking of root crown; Weir and Cresswell, 1993), flowers (B—sterility in wheat; Rerkasem and Jamjod, 2004), fruit (Ca—blossom-end rot of tomatoes; Weir and Cresswell, 1993), seed (B—hollow heart disorder; Bell et al., 1990), and stems or wood (Cu—stem twisting; Dell et al., 2001; B toxicity—stem lesions and gummy exudates; El-Motaium et al., 1994). The distinctive hollow heart disorder in peanut kernels (Fig. 11.1), which is specific to B deficiency, was used to develop a map of B deficiency risk across Thailand (Bell et al., 1990).

Diagnosis from symptoms alone may be challenging in field-grown plants when more than one nutrient is deficient or when a deficiency of one nutrient is accompanied by toxicity of another. Such simultaneously occurring deficiencies and toxicities can be found, for example, in waterlogged acid soils, where both Mn toxicity and Mg deficiency may occur. Diagnosis may be further complicated by the presence of diseases, pests and other symptoms caused, for example, by mechanical injuries or herbicide spray damage (Bergmann, 1992). To differentiate the symptoms of nutritional disorders from those due to other causes, it is important to bear in mind that nutritional disorders always have a typical symmetrical pattern: leaves of the same or similar position (physiological age) on a plant for a particular species show nearly identical patterns of symptoms, and there is a marked gradation in the severity of the symptoms from old to young leaves (Fig. 11.1). By contrast, symptoms induced by the presence of diseases and pests are nonsymmetric or randomly positioned in individual plants as well as within a field, particularly at an advanced stage of infections.

To improve the accuracy of visual diagnosis, it is helpful to acquire additional site, crop, weather, and soil information. The results of soil testing for nutrients and pH, observations on soil water status (dry/waterlogged) and weather conditions (low temperature, frost, heat) and records of the application of fertilizers, herbicides, fungicides, and pesticides can all be useful to improve the accuracy of a diagnosis based on plant symptoms. In some cases, visual diagnosis provides enough information for a recommendation on the type and amount of fertilizer to be used. The diagnosis of micronutrient (B, Fe, Zn, or Mn) or Mg deficiencies, for example, can be treated by foliar sprays if plant symptoms are diagnosed early enough. Visual diagnosis can identify a deficiency and the need for a fertilizer treatment but provides insufficient information to determine the appropriate rate, form, or method of fertilizer application. Nevertheless, visual symptoms can be useful to narrow down the range of possible causes of a disorder and to focus further attention on chemical and biochemical analyses of leaves and other plant parts for selected nutrients. This is particularly important for annual crops where the results of the diagnosis are required quickly because timely treatment is essential for it to be effective.

11.2.3 Plant analysis

11.2.3.1 General

The use of chemical analysis of plant material for diagnostic purposes is based on the premise that the growth rate of plants is affected by nutrient concentration in the shoot (Fig. 11.2). Element composition of plant parts can be expressed as *concentration* [e.g., mg g^{-1} dry weight (dw)] or as *content* (e.g., mg leaf^{-1}). Some literature uses the terms concentration and content interchangeably, which can be confusing. Depending on the nutrient, plant species and age, the most

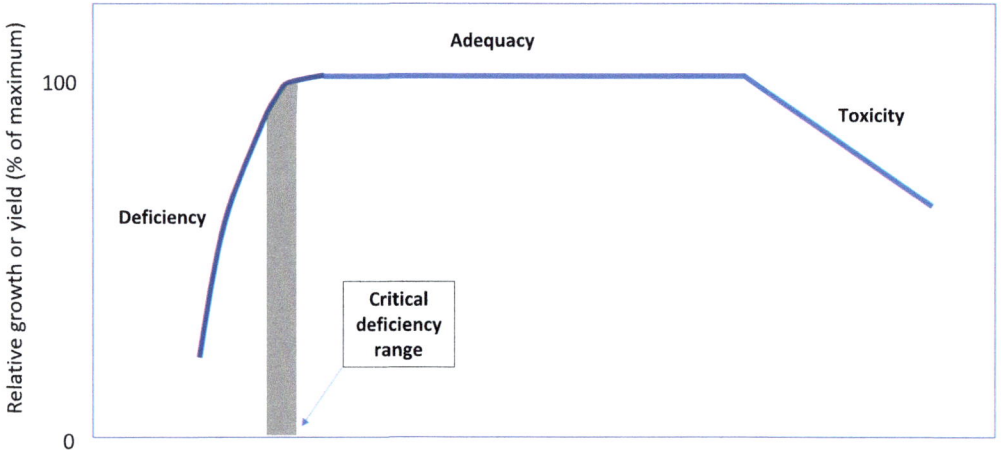

FIGURE 11.2 Relationship between tissue nutrient concentration and relative growth showing the deficiency, adequacy (sufficiency) and toxicity, and the critical deficiency range.

suitable plant part or organ for analysis differs, as well as whether the total concentration or only a certain fraction of the nutrient (e.g., water extractable) should be determined. In general, the nutritional status of a plant is better reflected in the element concentration in leaves than in the other plant organs (e.g., Campbell, 2009; Reuter and Robinson, 1997). Thus, leaves are usually sampled for plant analysis. Nutrient concentrations may differ considerably between leaf blades and petioles (e.g., Zn in canola; Huang et al., 1995), but sometimes the petioles are a more suitable indicator of nutritional status (e.g., P in subterranean clover; Bouma, 1983). In fruit trees, analysis of flowers is thought to be a more sensitive indicator of Fe nutritional status and other disorders than analysis of leaves (Khelil et al., 2010). Analysis of Ca and B in the fruits or even distinct parts of a fruit provides the best indication of quality and storage properties (Liebisch et al., 2009).

Samples from field-grown plants are often contaminated by dust or sprays and require washing. However, washing may result in loss of elements, with the extent of loss differing among nutrients. For B, washing leaves with water for a few minutes can result in high losses due to diffusion of B across plasma membranes (Brown et al., 2002).

The greatest challenge in the use of plant analysis for diagnostic purposes is, however, short-term changes in nutrient concentrations in leaves due to either "dilution effects" by fast growth or transient weather effects such as drought (Bell, 2000). Thus, the nutrient concentrations reflecting deficiency, sufficiency, or toxicity ranges vary with environmental factors as well as with plant genotype, plant part, and developmental stage of plants and leaves. For example, the percentage of dry matter usually increases with the age of plants or organs (Walworth and Sumner, 1988) or at elevated CO_2 concentrations because of starch accumulation (Kuehny et al., 1991), thus resulting in a decline in the critical deficiency concentrations with plant age (Bouma, 1983). By contrast, the K concentration on a dry matter basis declines with plant age, whereas the K concentration in the plant cell sap remains relatively constant during plant growth (Leigh and Johnston, 1983; Römheld and Kirkby, 2010). Strict standardization of sampling procedure and availability of critical deficiency or toxicity ranges that match the circumstances of the plant when sampled (plant growth stage, plant part) are therefore crucial for accurate diagnosis by plant analysis. For additional reviews on plant analysis for diagnostic purposes, the reader is referred to Smith and Loneragan (1997) and De Mello Prado and Rozanne (2020).

The critical deficiency range (CDR) is estimated from a calibration relationship fitted to the plot of growth or yield vs plant or leaf nutrient concentration. The data may be drawn from a single experiment or multiple experiments, in either a glasshouse or the field. The general pattern of relationships between plant growth and nutrient concentrations in plant parts is shown schematically in Fig. 11.2. There is an ascending part of the curve where growth either increases without much change in nutrient concentration or where increases in growth and nutrient concentration are closely related. This is followed by a transition, sometimes called the marginal zone (Smith and Loneragan, 1997) or the CDR (Dow and Roberts, 1982). At plant concentrations above the CDR, further increases in nutrient concentration have no effect on growth (not nutrient limited); finally, where the excessive nutrient concentration causes toxicity or detrimental effects, there is a corresponding decline in growth or quality.

Originally, the calibration lines were drawn freeform, as the best fit to the data according to the author, paying particular attention to the fit in the transition from deficient to adequate supply (e.g., Fig. 11.2). Subsequently, regression equations fitted to the data provided an estimate of the goodness of the fit (Ware et al., 1982). Moreover, rather than deriving a single critical concentration that implies a level of uniformity in crops that is not experimentally justifiable, especially in a field-grown crop, the emphasis has shifted to the estimate of a CDR that is a more realistic representation of the variability that exists within field data sets (Dow and Roberts, 1982). The CDR is defined as "that range of nutrient concentration above which we are reasonably confident the crop is amply supplied, and below which we are reasonably confident the crop is deficient" (Dow and Roberts, 1982). If a single value is used, it should be borne in mind that the probability of deficiency or sufficiency increases with the extent of deviation from this single value.

Given the spread of the data can vary among studies, species and elements, there is no single regression model that is universally superior to others. Jongruaysup et al. (1994) compared several approaches (replicated hand-fitted curves, Cate-Nelson graphical and statistical models, two-phase linear model, and Mitscherlich model) to estimate the CDR for diagnosis and prognosis of Mo deficiency in black gram. Based on a number of criteria, it was concluded that the Mitscherlich model was most accurate for the study. By contrast, Huang et al. (1995) found both the two-phase linear and the Mitscherlich models gave similar estimates of the concentration associated with near maximum growth, and they recommended a CDR based on a combination of the two approaches. Hence, it seems advisable to test a number of models to fit a calibration relationship rather than rely on a single model from which to derive the CDR. Models that provide an estimate of the variability associated with 90% or 95% of maximum growth can be used to directly estimate the CDR.

The actual shape of the relationship can vary from the generalized form shown in Fig. 11.2 (Ulrich and Hills, 1990; Smith and Loneragan, 1997). For example, with an extreme deficiency of Cu (Reuter et al., 1981) or Zn (Howeler et al., 1982), a C-shaped response curve is obtained in which a nutrient-induced increase in growth rate is accompanied by a decrease in its concentration, which is often referred to as the "Piper-Steenbjerg" effect (Bates, 1971). A possible explanation for this type of response is a lack of remobilization of the nutrient from old leaves and stem of the deficient plants (Reuter et al., 1981).

For fruit trees, it is common to focus on the sufficiency range of nutrients in leaves, rather than the CDR (e.g., Campbell, 2009). The sufficiency range draws attention to both the safe lower and upper limits of leaf nutrient concentration that is important in fruit trees because above-optimal supply of N, P, or K can impair fruit quality and/or storability (Righetti et al., 1990). The use of sufficiency ranges for fruit trees is generally a pragmatic choice to deal with the large number of species and nutrient elements, the fact that fruit trees may take a few years before their yield responds to fertilizer treatments, and because the quality of fruit is as important as yield. The sufficiency ranges selected for fruit trees are generally based on the typical concentrations found in leaves of trees that produce a high yield in a particular environment. However, there is a degree of subjectivity in the ranges chosen by different researchers or agencies, and the values relate to the varieties grown in a particular location or region (De Mello Prado and Rozanne, 2020).

11.2.3.2 Plant analysis for diagnosis of nutrient disorders

Diagnosis is the process of using plant analysis to identify a disorder that currently impairs plant growth (Smith and Loneragan, 1997). A representative example of the relationship between plant growth and tissue nutrient concentration is shown in Fig. 11.3 for Mn. The CDR for Mn in the youngest emerged leaf blade of barley plants is $10-15$ mg kg^{-1} dw. However, the CDR varies with plant part, for example, it may differ between the youngest emerged leaf blade and the sites of new growth (shoot meristem) in which the values may be substantially higher.

Usually, 90% or 95% of the maximum shoot dry matter is taken as a reference point for definition of the CDR of a nutrient (Smith and Loneragan, 1997). In low-input systems where farmers have limited capacity to purchase fertilizers, the reference point may be 80% of the maximum dry matter yield. Hence, because of the lower yield potential, the CDRs for those production systems are considerably lower (Smyth and Cravo, 1990).

Growth is maximal between the critical deficiency and toxicity concentration ranges, but for statistical and practical reasons the nutrient concentration resulting in 90%–95% of maximal growth rather than maximal growth is used to define the adequate range. Statistically, there is often no significant difference in growth between 90% and 100% of maximum growth. Practically in agriculture, farmers are reluctant to oversupply fertilizer because it is not profitable. With high-value crops, where fertilizer costs are low relative to the loss of income from decreased yield, the CDR may be set at 99% of maximum yield or the upper end of the CDR range associated with 95% of maximum yield. If nutrient concentrations are in the adequate range, there is a high probability these nutrients are not growth-limiting factors. Concentrations in the luxury range further decrease the risk that these nutrients will become deficient even under conditions unfavorable for root uptake (e.g., dry topsoil) or when the demand is very high (e.g., re-translocation to fruits). However, in the luxury concentration range there is a greater risk of growth reduction by direct toxicity of these nutrients or by inducing a deficiency of other nutrients, that is, nutrient imbalance.

The CDR for plants provided with a continuous low supply of nutrients may be quite different to those in which a high supply is suddenly interrupted (Scott and Robson, 1990). Such a sudden interruption may occur, for example, at the onset of transient drought. This leads not only to a changed pattern of symptoms but also to relatively high CDRs for various nutrients (Burns, 1992) because plants become more dependent on remobilization and re-translocation of nutrients stored within the plant.

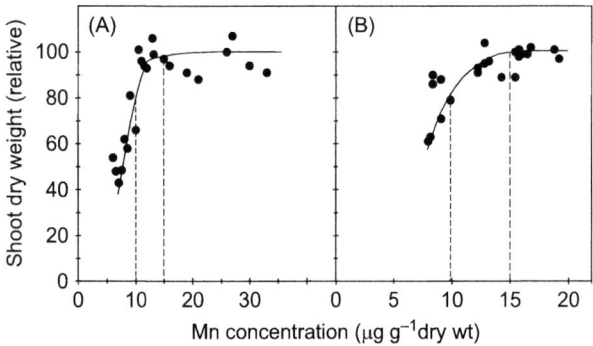

FIGURE 11.3 Relationship between Mn concentrations in youngest emerged leaf blades and shoot dry weight in barley grown in a growth chamber (A) and under field conditions (B). Dashed lines define the CDR. *CDR*, Critical deficiency range. *Based on Hannam et al., (1987).*

11.2.3.3 Developmental stage of plant and age of leaves

To make a recommendation for fertilizer application based on plant analysis, leaves or needles are generally collected during the period of the most intensive growth (i.e., the highest nutrient demand) (Bergmann, 1992, 1993). However, after nutrient supply, the physiological age of a plant or plant part is the most important factor affecting tissue nutrient concentration. With the exception of Ca and B, there is usually a decrease in nutrient concentration as plants and organs age. This decline is caused mainly by a relative increase in the proportion of structural material (cell walls and lignin) and of storage compounds (e.g., starch) in leaves over time, as shown, for example, for Fe (Venkat-Raju and Marschner, 1981) and other micronutrients (Drossopoulos et al., 1994) in fast-growing seedlings.

Due to the increasing concentration of structural material in tissues with plant age, the CDR is lower in old than in young plants. For example, in rice the optimum shoot N concentration decreased from 43 g kg^{-1} at the initiation of tillering to 11 g kg^{-1} at flowering (Fageria, 2003). In field-grown barley, the shoot K concentration decreased from 50–60 g kg^{-1} in young plants to about 10 g kg^{-1} at maturation, although the plants were well supplied with K throughout (Leigh et al., 1982). In this case, the decline in concentration was exclusively a "dilution effect," given that the K concentration in the tissue water (i.e., the vacuolar solution) remained fairly constant at $\sim$100 mM throughout the season.

Complications arising from changes in the CDR with age can be minimized by sampling specific plant parts and at specific physiological ages. As shown in Table 11.2, the CDR of Cu in the whole shoots of subterranean clover decreases with age but remains relatively constant at $\sim$2.5 mg kg^{-1} in the youngest open leaf blades throughout the season.

The use of the youngest leaves, however, is suitable only for nutrients that are re-translocated either not at all or only to a limited extent from the mature leaves to areas of new growth, that is, when deficiency occurs first in young leaves and at the shoot apex (Table 11.1). The situation is different for K, N, P, and Mg because the concentrations of these nutrients are maintained relatively constant in the youngest leaves by translocation from the mature leaves. Thus, the recently matured leaves are a better indicator of the nutritional status of a plant, as shown for K in Fig. 11.4. The K concentration in the youngest leaf is not a suitable indicator because the range of K concentrations between deficiency and toxicity was only between 30 and 35 g kg^{-1} compared with 15 and 55 g kg^{-1} in recently matured leaves. This

TABLE 11.2 Critical deficiency concentration of Cu (mg kg^{-1} dw) for 90% of maximum yield in subterranean clover whole shoots or youngest fully open (most recently matured) leaf blades at different plant ages.

	Age of plants (days after sowing)				Early flowering
	26	40	55	98	
Whole plant shoots	2.8	2.5	2.1	1.4	1.0
Youngest open leaf blades	2.5	2.5	2.7	2.6	2.4

Source: Based on Reuter et al. (1981).

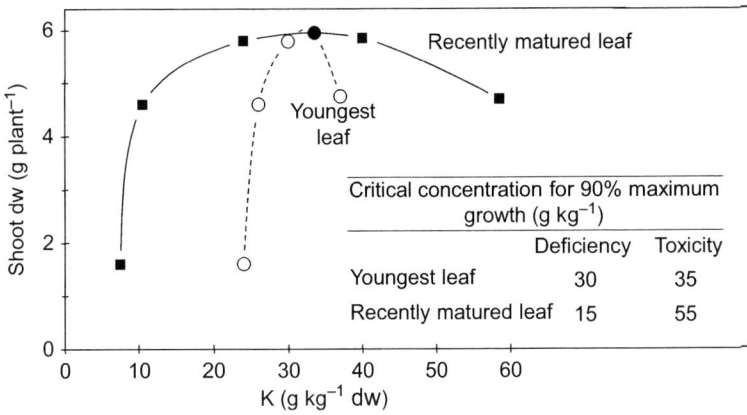

FIGURE 11.4 Relationship between shoot dry weight and K concentrations of recently matured and youngest leaves of tomato plants grown in nutrient solution with different K concentrations showing critical concentrations for 90% of maximum growth.

illustrates the necessity of using recently matured leaves to assess the nutritional status for nutrients that are readily retranslocated in plants.

If young and old leaves of the same plant are analyzed separately, additional information can be obtained on the nutritional status. For example, a higher K concentration in the mature leaves than in young leaves indicates luxury consumption or even toxicity. The reverse gradient (a higher concentration in the young leaves) is an indicator of the transition stage between the adequate and deficient ranges; if this gradient is steep, a latent or even acute deficiency may exist. Such comparisons between young and old leaves are particularly useful when relevant CDR values are lacking (e.g., for a species or cultivar). If toxicity is suspected, the old leaves are the most suitable organs for plant analysis because excess elements often accumulate in the older leaves.

When choosing a given plant organ such as the most recently developed, or fully expanded, leaf for analysis, it should be recognized that the CDR value may decline throughout plant development, even when expressed as a concentration in the plant sap. In soybean, for example, the critical deficiency concentration of K decreased from 65 to 29 mM between podset and podfilling (Bell et al., 1987). This decline during plant development is particularly evident for nitrate in leaf or petiole sap that is often used as a rapid indicator of the N nutritional status of field-grown vegetables and potatoes (Sparrow and Chapman, 2003). For changes in nutrient concentrations during development in various crops, the reader is referred to Bergmann (1993) and Reuter and Robinson (1997).

The importance of shoot nutrient storage for buffering nutrient supply to growing parts is particularly significant for woody perennial plants such as fruit trees (Muhammad et al., 2020). Compared with the changes in the nutrient concentration in annual species, the fluctuations throughout the growing season in the nutrient concentration of leaves and needles of trees are relatively small because of the nutrient buffering capacity of twigs and trunk. In evergreen trees, the simultaneous analysis of leaves or needles differing in age provides reliable data less affected by short-term fluctuations (Table 11.3). With the increasing age of the needles, the concentration of macronutrients (except Ca) decreased. This decrease may in part indicate remobilization but is mainly an expression of a dilution effect resulting from increased lignification of the old needles. By contrast, dilution of Ca is more than offset by the continued Ca influx into the old needles via the transpiration stream. With the exception of Mg, the data in Table 11.3 indicate that the trees are well supplied with macronutrients. In Norway spruce, the silicon (Si) concentration in the needles also increases with needle age (Wyttenbach et al., 1991) because it is transported via the transpiration stream. Leaf Ca concentrations in *Leucaena leucocephala* were useful to define the physiological age of the youngest fully expanded leaves because only leaves <21 days old provided meaningful values for assessing plant nutrient status (Radrizzani et al., 2011).

For deciduous tree crops, the concentrations of most nutrients (apart from the phloem-immobile Ca, B, and Mn) decline as leaves increase in size and weight until mid-to-late summer when they stabilize (Robinson, 1993). Hence, the sampling of mature leaves for such species should be in mid-summer. In citrus, nutrient concentrations are relatively stable when mature leaves are 6–8 months old, so this has become the recommended sampling time (Mattos et al., 2020). The nutrient analysis of mature leaves obtained late in the season may be useful to adjust the fertilizer program in the following year (Righetti et al., 1990) but would be too late to guide fertilizer decisions within that season (Saa et al., 2014). In the case of almond production in California, growers need protocols to adjust N fertilizer rate earlier in the season to account for fruit load and minimize the risk of nitrate leaching. Saa et al. (2014) developed a spring

TABLE 11.3 Nutrient concentration (g kg^{-1}) of Norway spruce (*Picea abies* Karst) needles of different ages.

Nutrient	Age of needles (years)			
	1	2	3	4
N	17.9	17.6	14.6	12.2
P	2.0	1.7	1.4	1.3
K	6.3	5.6	4.7	4.4
Mg	0.4	0.4	0.3	0.3
Ca	2.8	4.0	5.0	5.9

Source: Based on Bosch, (1983).

(April) leaf sampling protocol that predicts the mid-season (July) almond leaf N concentration based on leaf N and B concentrations and the sum of K + Mg + Ca equivalents.

The rapid decline in nutrient concentrations in shoots as they accumulate dry matter and age is a challenge for plant analysis, especially in annual crops with short growth duration. An alternative to standardizing the time of sampling and the plant part sampled is the use of nutrient dilution curves for whole shoots so that nutrient deficiency can be assessed at any time during growth using a nutrient index. Greenwood et al. (1990) reported that the decline in shoot N concentration with plant age in a number of C_3 and C_4 species can be described as a function of shoot dry matter by the relationship:

$$[N] = a\, DM^{-b}$$

where [N] is the shoot nitrogen concentration, DM is shoot dry matter, and a represents the shoot [N] at DM = 1 t ha^{-1} (equivalent to the crop N requirement during early growth; Lemaire et al., 2007). Parameter b describes the decline in shoot [N] as the crop grows and is determined by the relative shoot accumulation rates of N and biomass (Gastal and Lemaire, 2002). Using this approach, N dilution curves have been developed for a range of crops such as spring wheat (Justes et al., 1994), maize (Ziadi et al., 2008), sunflower (Debaeke et al., 2012), sugarcane (De Oliveira et al., 2013), and potato (Gómez et al., 2019). The nutrient dilution curve approach has also been applied to P and K (Gómez et al., 2019; Soratto et al., 2020). The critical nutrient dilution curve describes the shoot nutrient concentration that is just adequate for maximum growth, usually corresponding to 95% of maximum shoot growth. Hence, for any plant dry weight value, the shoot nutrient concentration relative to the value on the critical nutrient dilution curve, known as a nutrition index, is a measure of deficiency or adequacy of that nutrient at that time. Nutrient dilution curves need to be calibrated for specific cultivars and growing environments; they provide the flexibility of sampling at any time during crop growth, rather than at strictly defined stages. Nutrient dilution curves may also be calibrated against a nondestructive leaf area index measurement to avoid the need for cumbersome plant sampling (Wang et al., 2017).

11.2.3.4 Plant species and genotypes

Adequate concentration ranges and CDRs differ among plant species even when comparing the same organs at the same physiological age. These variations are mainly based on differences in plant metabolism and plant composition, for example, differences in the requirement for Ca and B in cell walls. When grown under the same conditions, the critical deficiency concentration of B in a fully expanded youngest leaf is (in mg kg^{-1} dw) 3 in wheat, 5 in rice, but as high as 25 in soybean and 34 in sunflower (Rerkasem et al., 1988). Native plant species from nutrient-rich habitats may have higher critical deficiency concentrations of K in the shoots ($\sim$100 mM) than species from nutrient-poor habitats ($\sim$50 mM; Hommels et al., 1989). Representative data for adequate nutrient ranges of selected species are given in Table 11.4. More extensive and detailed data, including deficiency and toxicity concentrations, can be found in Righetti et al. (1990), Bergmann (1992), Mills and Benton Jones Jr. (1996), Reuter and Robinson (1997), Campbell (2009), and Hochmuth et al. (2018).

Generally, the concentrations of macronutrients in the adequate range are of similar order of magnitude in all plant species (Table 11.4). In all species, the adequate range is relatively narrow for N because luxury concentrations of N have negative effects on growth and plant composition (see Section 6.1). In apple, for example, leaf N concentration is negatively related to fruit color, bitter pits, and scald (Raese and Drake, 1997). On the other hand, the adequate range for Mg is usually broader, mainly due to competing effects of K; at high K concentrations, high Mg concentrations are also required to ensure an adequate Mg nutritional status.

The concentrations of micronutrients in the adequate range vary by a factor of two or more (Table 11.4). Manganese and Mo show the greatest variation, indicating that leaf tissue may buffer fluctuations in the root uptake of these elements. For Mn, this is probably an evolutionary adaptation because in soil-grown plants fluctuations in the uptake of Mn may be greater than those of other nutrients due to variations in soil redox potential that strongly influences the concentrations of plant-available Mn^{2+} (Section 7.2). The data in Table 11.4 should be regarded as a guide as to whether a nutrient is in the deficient, adequate, or toxic range.

Genotypic differences in the CDR of a nutrient can be due to differences in the utilization of that nutrient (see Chapter 17), which may be expressed in terms of unit dry matter produced per unit nutrient in the dry matter (e.g., kg dw g^{-1} P). As an example, the difference in N efficiency between C_3 and C_4 grasses is shown in Table 11.5. Per unit leaf N, C_4 grasses produce more dry matter than C_3 species; this is also observed in other comparisons of C_3 and C_4 grasses (Brown, 1985). The higher N efficiency of C_4 species may be related to the lower investment of N in enzymes used in chloroplasts for CO_2 fixation. In C_4 species, only 5%–10% of the soluble leaf protein is found in RuBP

TABLE 11.4 Nutrient concentrations in the adequate range of some annual and perennial species.

	N	P	K	Ca	Mg	B	Mo	Mn	Zn	Cu
	(g kg^{-1} dw)					(mg kg^{-1} dw)				
Spring wheat (whole shoot, booting stage)	30–45	3.0–5	29–38	4–10	1.5–3	5–10	0.1–0.3	30–100	20–70	5–10
Ryegrass (whole shoot)	30–42	3.5–5	25–35	6–12	2–5	6–12	0.15–0.5	40–100	20–50	6–12
Sugar beet (mature leaf)	40–60	3.5–6	35–60	7–20	3–7	40–100	0.25–1.0	35–100	20–80	7–15
Cotton (mature leaf)	36–47	3–5	17–35	6–15	3.5–8	20–80	0.6–2.0	35–100	25–80	8–20
Tomato (mature leaf)	40–55	4–6.5	30–60	3–4	3.5–8	40–80	0.3–1.0	40–100	30–80	6–12
Alfalfa (upper shoot)	35–50	3–6	25–38	1–2.5	3–8	35–80	0.5–2.0	30–100	25–70	6–15
Apple (mature leaf)	22–28	1.8–3	11–15	13–22	2–3.5	30–50	0.1–0.3	35–100	20–50	5–12
Orange (*Citrus* spp.) (mature leaf)	24–35	1.5–3	12–20	30–70	2.5–7	30–70	0.2–0.5	25–125	25–60	6–15
Norway spruce (1- to 2-year-old needles)	14–17	1.3–2.5	5–12	3.5–8	1–2.5	15–50	0.04–0.2	50–500	15–60	4–10
Oak, beech (mature leaves)	19–30	1.5–3	10–15	3–5	1.5–3	15–40	0.05–0.2	35–100	15–50	6–12

Source: Based on Bergmann (1992).

carboxylase, compared with 30%–60% in C$_3$ species (see also Chapter 5). The lower internal requirement of N in C$_4$ plants is of advantage for biomass production on N-poor sites but not necessarily of advantage for the nutritional quality of forage (Brown, 1985).

11.2.3.5 Nutrient interactions and ratios

There are many nonspecific as well as specific interactions between nutrients in plants (Robson and Pitman, 1983) that affect the interpretation of plant analysis. For example, in maize, at low P concentration an increase in N concentration

TABLE 11.5 Relationship between dry matter production and N concentration of C₃ (*Lolium perenne* and *Phalaris tuberosa*) and C₄ grasses (*Digitaria macroglossa* and *Paspalum dilatatum*).

N supply (kg ha⁻¹)	Dry matter (g pot⁻¹)		N concentration (g kg⁻¹)	
	C₃	C₄	C₃	C₄
0	11	22	18.2	9.1
67	20	35	26.3	11.8
134	27	35	27.7	16.1
269	35	48	27.8	20.0

Source: Based on Colman and Lazemby (1970).

of the ear leaf from 21 to 29 g kg⁻¹ had little effect on yield, but at high P concentration yield continued to increase as ear leaf N concentrations rose well above 30 g kg⁻¹ (Sumner and Farina, 1986). This type of positive interaction may occur for the supply of any pair of growth-limiting nutrients.

Optimal ratios between nutrients in plants may be useful to consider in addition to absolute concentrations in some circumstances. However, optimal ratios considered alone are insufficient because they can be obtained when both nutrients are in the deficiency, sufficiency, or toxicity ranges (Jarrell and Beverly, 1981). Specific interactions that affect CDR include the replacement of K by Na in natrophilic species, which has to be considered in the evaluation of the critical leaf K concentration (Ivahupa et al., 2006) and an optimal N-to-S ratio that reflects the need for both N and S in essential amino acids. For example, an optimal N-to-S ratio of ∼17 is considered to be adequate for the S nutrition of wheat (Rasmussen et al., 1977) and soybean (Bansal et al., 1983).

Due to the decline in CDRs during plant development, and the significance of nutrient ratios in plant analysis for diagnostic purposes, the Beaufils' Diagnosis and Recommendation Integrated System (DRIS) was developed. For certain crops and under certain conditions (high yielding sites, large-scale farming), the higher analytical input may pay off by permitting a refinement in the interpretation of the data in terms of fertilizer recommendations, as has been reported for sugarcane (Elwali and Gascho, 1984), maize and fruit trees (Walworth and Sumner, 1988). However, recommendations based on DRIS have not proven to be substantially better than the conventional, less demanding approach (Righetti et al., 1990; Smith and Loneragan, 1997).

11.2.3.6 Environmental factors

Fluctuations in environmental factors such as temperature and soil water content can affect the nutrient concentration of leaves considerably. These factors can influence leaf nutrient concentrations directly by altering both the availability and uptake of nutrients by the roots and indirectly by changing the shoot growth rate. Published CDRs, especially if developed under nonlimiting conditions, may not be calibrated to account for fluctuations in environmental conditions such as a period of drought during crop growth (Bell, 2000). Environmental effects are more distinct in shallow-rooted annual species than in deep-rooted perennial species because of the differential amounts of stored nutrients. This aspect must be considered in interpreting critical deficiency and toxicity ranges in leaf analysis. If fluctuations in soil water content are high, the CDR of nutrients such as K and P are higher, ensuring a higher capacity for re-translocation during periods of limited root supply.

The effects of irradiation and temperature on the nutrient concentration of leaves are described in detail by Bates (1971). Increasing light intensity accentuates the symptoms of K, Mg, and Zn deficiency in plants (Cakmak and Marschner, 1987). Shading that reduced light intensity from 70% to 30% of full sunlight decreased the critical deficiency concentrations for leaf blade elongation in youngest open leaves of black gram from 15 to 10 mg B kg⁻¹ (Noppakoonwong et al., 1993).

11.2.3.7 Total analysis versus fractionated extraction

In plant analysis, the total concentration of a nutrient is usually determined (e.g., after wet or dry ashing). However, determination of only a fraction of the total amount—for example, the fraction that is soluble in water or in dilute acids or chelators—sometimes provides a better indication of the nutritional status. In terms of plant analysis as a basis for fertilizer recommendations, this is particularly true for nitrate as an important storage form of N in many plant species (see Section 6.1). In those species, the nitrate concentration is usually a better indicator of the N nutritional status than the total N concentration, and a better basis for recommendations of topdressing of N fertilizers. This method is a satisfactory predictor of crop responsiveness to N fertilizer in winter cereals (Wollring and Wehrmann, 1990), irrigated wheat (Knowles et al., 1991), potato (Sparrow and Chapman, 2003), cabbage (Gardner and Roth, 1989), and other vegetable crops (Scaife, 1988; Hochmuth et al., 2018). This method is not suitable for plant species that preferentially reduce nitrate in the roots (e.g., members of the Rosaceae), or when ammonium is supplied and taken up prior to nitrification in soils. The latter may occur in soils high in organic N and with high mineralization rates during the stages of high N demand of the crop.

In some cases, the ratio of SO_4-S to total S may be a preferred indicator of the S status, for example in wheat (Freney et al., 1978) or rice (Islam and Ponnamperuma, 1982). For P, there are conflicting reports on the suitability of using the inorganic (or readily extractable) fraction of P, instead of total P, as diagnostic criteria of the P nutritional status of plants. This approach seems to be suitable in grapevine (Skinner et al., 1987), but not in subterranean clover (Lewis, 1992).

Determination of only a fraction of a nutrient may allow better characterization of the physiological availability of a nutrient in the plant tissue. For example, extraction of leaves with diluted acids or chelators of Fe^{2+} for determination of the so-called "active iron" may improve the correlations between Fe and chlorophyll concentrations in leaves in field-grown plants, but not necessarily in plants grown under controlled conditions in nutrient solutions (Lucena et al., 1990). Determination of water-extractable Zn in leaves may better reflect the Zn nutritional status of plants than total Zn (Rahimi and Schropp, 1984), particularly in plants suffering from P-induced Zn deficiency (Cakmak and Marschner, 1987).

Another example of the advantage of determining a fraction of a nutrient for the characterization of physiological availability is shown in Table 11.6. Differences in the susceptibility of tobacco cultivars to Ca deficiency were not related to the total Ca concentration but to the soluble fraction in the buds. These differences were caused by variations in oxalic acid synthesis and thus in the precipitation of sparingly soluble Ca oxalate. Accordingly, the critical deficiency concentration of total Ca was higher in genotype B 21 than Ky 10.

11.2.3.8 Histochemical, biochemical, and spectral methods

Nutritional disorders are generally related to changes in the fine structure of cells and their organelles (Vesk et al., 1966; Niegengerd and Hecht-Buchholz, 1983). Light microscopy studies of changes in anatomy and morphology of leaf and stem tissue can be helpful in the diagnosis of deficiencies of cobalt (Co), B, and Mo (Pissarek, 1980). A combination of chlorophyll fluorescence plus confocal and electron microscopy was useful in the diagnosis of the early stages of Mg deficiency in radish (Samborska et al., 2018).

Methods involving marker enzymes offer another approach to assessing the nutritional status of plants. These methods are based on the fact that the activity of certain enzymes is lower or higher (depending on the nutrient) in deficient than in normal tissue. Examples are given in Chapter 7 for Cu and ascorbate oxidase, Zn and aldolase or carbonic anhydrase, and Mo and nitrate reductase. Typically, the actual enzyme activity is determined in the tissue after extraction, or the leaves are incubated with the nutrient in question to determine the inducible enzyme activities of, for example,

TABLE 11.6 Concentrations of total and soluble Ca (calculated as total Ca minus oxalic acid) and oxalic acid in two cultivars of burley tobacco differing in susceptibility to Ca deficiency.

Cultivar	Plants with Ca deficiency symptoms (% of total)	Concentration in buds (mmol kg^{-1} dw)			Concentration in upper leaves (mmol kg^{-1} dw)		
		Ca	Oxalic acid	Soluble Ca	Ca	Oxalic acid	Soluble Ca
Ky 10	0	125	40	85	140	55	85
B 21	50	115	80	35	150	75	75

Source: Based on Brumagen and Hiatt, (1966).

TABLE 11.7 Growth, P concentration, and phosphatase activity of young wheat plants.

P supply	Shoot dw (mg plant^{-1})	P concentration (g kg^{-1})	Phosphatase activity (μmol NPP g^{-1} fw min^{-1})[a]		
			Total	Fraction A	Fraction B
High	223	8.0	5.6	4.4	0.5
Low	135	3.0	11.1	6.7	2.9

[a]NPP, p-nitrophenylphosphate.
Source: From Barrett-Lennard and Greenway, (1982).

peroxidase activity by Fe (Bar-Akiva et al., 1970) and nitrate reductase by Mo. For assessing the Mn nutritional status, the activity of Mn superoxide dismutase (Section 7.3) in leaves may be used as biochemical marker (Leidi et al., 1987).

Biochemical methods can also be used for assessing nutritional status of plants in relation to macronutrients. The accumulation of putrescine in K-deficient plants (Section 6.6) is a biochemical indicator of the K requirement in alfalfa (Smith et al., 2006). Inducible nitrate reductase activity can be used as an indicator of N nutritional status (Witt and Jungk, 1974). Pyruvate kinase activity in leaf extracts is related to the K and Mg concentration of the leaf tissue (Besford, 1978). In P-deficient tissue, phosphatase activity is increased (Table 11.7), indicating enhanced P turnover rates or enhanced remobilization of P (Smyth and Chevalier, 1984). In eucalyptus, acid phosphatase activity was a more sensitive parameter for diagnosis of growth limitations by P than total P in leaves and stems (O'Connell and Grove, 1985), whereas in maize, acid phosphatase activity increased significantly only under severe P deficiency and therefore may be suitable as a means of confirming visual diagnosis but is not sensitive enough to indicate latent P deficiency (Elliott and Läuchli, 1986).

Enzymatic, biochemical, and biophysical methods can be very valuable if the total concentration or the soluble fraction of a nutrient is poorly correlated with its physiological availability. Whether these enzymatic, biochemical, and biophysical methods can realistically be used as alternatives to chemical analysis as a basis for making fertilizer recommendations depends on their selectivity, accuracy, and particularly whether they are sufficiently simple and robust to provide a rapid and cheap test. These requirements may be met in the case of Fe and peroxidase (Bar-Akiva et al., 1978; Bar-Akiva, 1984; Dasgan et al., 2003), Cu and ascorbate oxidase (Delhaize et al., 1982), or, as a nondestructive method, chlorophyll *a* fluorescence measured using a handheld portable fluorimeter for Mn deficiency (Schmidt et al., 2013). Nevertheless, calibration of the methods remains a problem when a suitable standard (nondeficient plants) is not available and there are no visible deficiency symptoms. To date, these methods of foliar analysis for diagnostic purposes have been useful for solving particular problems of nutritional disorders and to supplement total and fractionated foliar analysis but have not gained widespread use to replace the conventional method.

Methods that are nondestructive, field-based, and cheap could revolutionize plant analysis and shift the focus away from soil testing that is currently used more widely by farmers than laboratory-based plant analysis (Van Maarschalkerweerd and Husted, 2015). A range of new methods for plant analysis are being developed involving fast spectroscopy of leaves. Some (such as X-ray fluorescence) measure the element directly, whereas others (such as visible-near infrared spectroscopy) rely on secondary correlations between spectral signatures and the element concentration in plants. Much work still needs to be done to develop specific calibrations for a range of elements and to validate them for numerous crop species. Near infrared spectra were demonstrated to predict Cu deficiency (Van Maarschalkerweerd et al., 2013). The handheld portable fluorimeter measures the change in chlorophyll *a* fluorescence as a direct response to leaf Mn concentration (Schmidt et al., 2013). Laser-induced breakdown spectroscopy directly measures a wide range of elements in plant material but is still restricted to laboratory use because sample preparation protocols need to be carefully followed.

11.3 Plant analysis for prognosis of nutrient deficiency

The distinction between plant analysis for diagnosis and for prognosis is often not well understood or explicit in the literature (Munson and Nelson, 1990). Generally, the CDRs differ depending on whether they are calibrated for diagnosis or prognosis. Calibrations of CDRs for diagnosis may be accomplished in glasshouse studies, but the calibration for

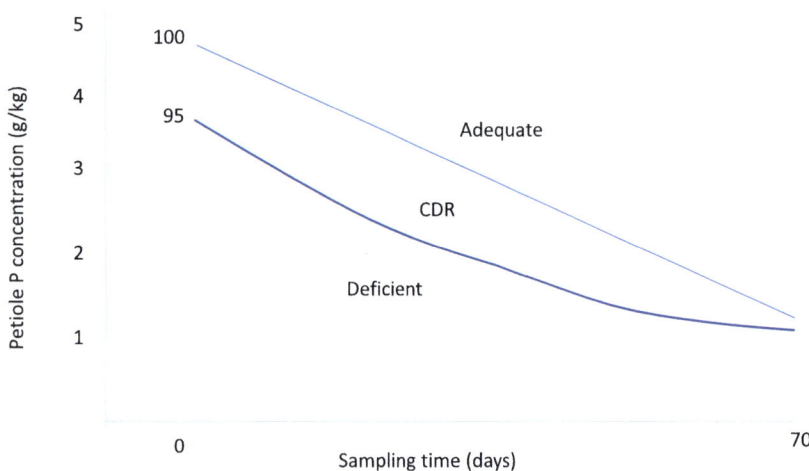

FIGURE 11.5 Change in adequate, deficient, and the CDR of phosphorus concentrations in potato petiole during the growing season. *CDR*, Critical deficiency range. *Based on Roberts and Dow (1982).*

prognosis should be developed under field conditions relevant to the production system being managed. Similarly to the prognosis of deficiency based on soil testing, for plant analysis different CDRs may be needed for the same crop species grown in different soils and environments that affect the crop growth rate, nutrient demand, and soil nutrient supply.

The CDR for prognosis is generally higher than for diagnosis, and values decline with plant age (Fig. 11.5). In the case of potato, Roberts and Dow (1982) demonstrated a well-defined decline in the petiole P concentrations over time was related to final crop yield. Petiole P concentrations that remained above the 100% line ensured adequate P supply to the potato crop throughout, whereas concentrations below the 95% line resulted in P limitations for potato yield. The CDR of P concentrations throughout the growing season fell between those two lines. Similar examples of declines in CDRs for prognosis of deficiency as the crop aged have been reported in peanut for P (Bell, 1985), in potato for N (Sparrow and Chapman, 2003), and in black gram for B (Bell et al., 1990) and Mo (Jongruaysup et al., 1994). Hence, the CDRs for prognosis can be interpreted correctly only if applied strictly to the specified growth stage and plant part. The alternative approach, as outlined above, is to develop critical nutrient dilution curves.

11.4 Plant analysis versus soil analysis

Plant and soil analyses should be viewed as complementary tools for the diagnosis and prognosis of nutrient disorders and for monitoring of nutrient management practices. Both methods have advantages and limitations, and they also give qualitatively different results (Schlichting, 1976). Currently, soil testing is in greater demand than plant testing (Van Maarschalkerweerd and Husted, 2015). Both methods rely in a similar manner on a well-defined calibration, that is, the determination of the relationship between concentrations in soils or plants and the corresponding growth and yield response curves, usually obtained in field experiments using different rates of fertilizers. Chemical soil analysis indicates the potential availability of nutrients that roots may take up under conditions favorable for root growth and root activity. That is, soil testing is used to make a prognosis, but its use has to recognize the uncertainty at the time of sampling about the supply of nutrients to crops during the growing season and about the crop demand. By contrast, plant analysis in the diagnostic mode reflects the actual nutritional status of plants at the time of sampling. Plant analysis used for prognosis is the prediction of whether nutrient supply will be adequate through to the end of the period of crop nutrient demand. Therefore, a combination of both methods provides a more robust basis for recommending fertilizer applications than either one alone. In pastures, plant analysis has some advantages over soil analysis because of the direct relevance of the nutrient composition of pasture and forage plants for animal nutrition. However, the relative importance of each method for making recommendations differs depending on plant species, soil properties, and the nutrient in question.

In fruit or forest trees, soil analysis alone is not a satisfactory guide for fertilizer recommendations, mainly because of the difficulty of determining with sufficient accuracy the depths from which deep-rooting plants take up most of their nutrients. On the other hand, in these perennial plants, seasonal fluctuations in the nutrient concentration of leaves and needles are buffered compared with those in annual species because of the potential for redistribution of nutrients stored in the trunks and branches (Muhammad et al., 2020). The nutrient concentration of mature leaves and needles can therefore be an accurate reflection of the long-term nutritional status of a perennial plant. Furthermore, calibrations of CDR and the definition of the adequate range can be refined for a special location, plant species, and even cultivar. Therefore, in perennial species, foliar and needle analysis is the method of choice in most cases. However, chemical soil analysis is necessary for characterizing the level of potentially available nutrients and the existence of other soil constraints (including those in subsoil) that might hamper nutrient uptake by roots.

In annual crops, the short duration of the cropping season and the fluctuations of nutrient concentrations over time place a severe limitation on plant analysis as a basis for making in-season fertilizer recommendations. Delays of even 1−2 weeks between leaf sampling and laboratory analysis of the sample may render the results useless for tactical in-season nutrient management. This is particularly the case for short-term vegetable crops that are harvested after 8−12 weeks (Geraldson and Tyler, 1990). Field-based plant tests of stem or leaf sap can be useful for the rapid application of treatments once a diagnosis is made (e.g., for nitrate in potato; Sparrow and Chapman, 2003). Spectral analysis of leaves in the field using portable instruments may also become more useful for rapid tactical decision-making, provided the measured values can be interpreted from a well-calibrated relationship with crop growth (Van Maarschalkerweerd and Husted, 2015).

Soil analysis from samples taken before sowing can be used for predicting the likely availability of nutrients to plants throughout the growing season. In annual crops, a large proportion of the nutrients are taken up from the topsoil, which makes soil sampling easier and increases the importance of soil testing as a tool for making fertilizer recommendations. However, depths of soil sampling vary from 10 to 30 cm, so that the latter depth includes part of the subsoil. Indeed, crops can acquire up to 75% of N, 85% of P, and 70% of K from the subsoil (Krautz et al., 2013); hence, its contribution to crop nutrition should not be overlooked. For cationic micronutrients, root modification of the rhizosphere can increase the plant availability of micronutrients substantially, which renders soil analysis less reliable (Bell and Dell, 2008).

Nutrient imbalances in plants, especially latent micronutrient deficiencies, continue to be a worldwide problem particularly in intensive agriculture (Schjoerring et al., 2019; Dobermann et al., 2021), with consequences for plant yield and also for plant tolerance to diseases and pests (Chapter 10), as well as for animal and human nutrition (Bell and Dell, 2008; Chapter 9).

References

Bansal, K.N., Motiramani, D.P., Pal, A.R., 1983. Studies on sulphur in vertisols. I. Soil and plant tests for diagnosing sulphur deficiency in soybean (*Glycine max* L. Merr.). Plant Soil 70, 133−140.

Bar-Akiva, A., 1984. Substitutes for benzidine as H-donors in the peroxidase assay, for rapid diagnosis of iron deficiency in plants. Commun. Soil Sci. Plant Anal. 15, 929−934.

Bar-Akiva, A., Sagiv, J., Leshem, J., 1970. Nitrate reductase activity as an indicator for assessing the nitrogen requirement of grass crops. J. Sci. Food. Agric. 21, 405−407.

Bar-Akiva, A., Maynard, D.N., English, J.E., 1978. A rapid tissue test for diagnosing iron deficiencies in vegetable crops. Hortic. Sci. 13, 284−285.

Barrett-Lennard, E.G., Greenway, H., 1982. Partial separation and characterization of soluble phosphatases from leaves of wheat grown under phosphorus deficiency and water deficit. J. Exp. Bot. 33, 694−704.

Bates, T.E., 1971. Factors affecting critical nutrient concentrations in plants and their evaluation: a review. Soil Sci. 112, 116−126.

Bell, M.J., 1985. Phosphorus nutrition of peanut (*Arachis hypogaea* L.) on Cockatoo sands of the Ord River irrigation area. Aust. J. Exp. Agric. 25, 649−653.

Bell, R.W., 1997. Diagnosis and prediction of boron deficiency for plant production. Plant Soil 193, 149−168.

Bell, R.W., 2000. Temporary nutrient deficiency - A difficult case for diagnosis and prognosis by plant analysis. Commun. Soil Sci. Plant Anal. 31, 1847−1861.

Bell, R.W., Dell, B., 2008. Micronutrients for Sustainable Food, Feed, Fibre and Bioenergy Production. IFA, Paris, France.

Bell, R.W., Brady, D., Plaskett, D., Loneragan, J.F., 1987. Diagnosis of potassium deficiency in soybean. J. Plant Nutr. 10, 1947−1953.

Bell, R.W., Rerkasem, B., Keerati-Kasikorn, P., Phetchawee, S., Hiranburana, N., Ratanarat, S., et al., 1990. Mineral Nutrition of Food Legumes in Thailand with Particular Reference to Micronutrients. In: ACIAR Technical Report 19, ACIAR, Canberra.

Bergmann, W., 1992. Nutritional Disorders of Plants — Development, Visual and Analytical Diagnosis. Gustav Fischer, Verlag, Jena, Germany.

Bergmann, W., 1993. Ernährungsstörungen bei Kulturpflanzen, 3rd ed. Gustav Fischer, Verlag, Jena, Germany.

Besford, R.T., 1978. Use of pyruvate kinase activity of leaf extracts for the quantitative assessment of potassium and magnesium status of tomato plants. Ann. Bot. 42, 317−324.

Bosch, C., 1983. Ernährungskundliche Untersuchung über die Erkrankung der Fichte (*Picea abies* Karst.) den Hochlagen des Bayrischen Waldes. Diplomarbeit, Universität München, München, Germany.

Bouma, D., 1983. Diagnosis of mineral deficiencies using plant tests. In: Läuchli, A., Bieleski, R.L. (Eds.), Encyclopedia of Plant Physiology, New Series. Springer-Verlag, Berlin, Heideberg and New York, pp. 120−146.

Brown, R.H., 1985. Growth of C_3 and C_4 grasses under low N levels. Crop Sci. 25, 954−957.

Brown, P.H., Bellaloui, N., Wimmer, M.A., Bassil, E.S., Ruiz, J., Hu, H., et al., 2002. Boron in plant biology. Plant Biol. 4, 205–223.

Brumagen, D.M., Hiatt, A.J., 1966. The relationship of oxalic acid to the translocation and utilization of calcium in *Nicotiana tabacum*. Plant Soil 24, 239–249.

Bruulsema, T.W., Fixen, P.E., Sulewski, G.D. (eds.), 2012. 4R Plant Nutrition Manual: a Manual for Improving the Management of Plant Nutrition. International Plant Nutrition Institute (IPNI), Norcross, GA, USA.

Burns, I.G., 1992. Influence of plant nutrient concentration on growth rate: Use of a nutrient interruption technique to determine critical concentrations of N, P and K in young plants. Plant Soil 142, 221–233.

Cakmak, I., Marschner, H., 1987. Mechanism of phosphorus-induced zinc deficiency in cotton. III. Changes in physiological availability of zinc in plants. Physiol. Plant. 70, 13–20.

Campbell, C.R., 2009. Reference sufficiency ranges for plant analysis in the southern region of the United States. In: Campbell, C.R. (Ed.), Southern Cooperative Series Bulletin #394. Southern Region Agricultural Experiment Station, Raleigh, NC. Available from: https://aesl.ces.uga.edu/sera6/PUB/scsb394.pdf.

Chapin III, F.S., 1988. Ecological aspects of plant mineral nutrition. In: Tinker, B., Läuchli, A. (Eds.), Advances in Plant Nutrition. Praeger Publications, New York, USA, pp. 161–191.

Colman, R.L., Lazemby, A., 1970. Factors affecting the response of tropical and temperate grasses to fertilizer nitrogen. 11th International Grassland Conference, 1970 Surfers Paradise, Queensland, Australia. pp. 393–397.

Dasgan, H.M., Ozturk, L., Abak, K., Cakmak, I., 2003. Activities of iron-containing enzymes in leaves of two tomato genotypes differing in their resistance to Fe chlorosis. J. Plant Nutr. 26, 1997–2007.

De Bang, T.C., Husted, S., Laursen, K.H., Persson, D.P., Schjoerring, J.K., 2021. The molecular–physiological functions of mineral macronutrients and their consequences for deficiency symptoms in plants. New Phytol. 229, 2446–2469.

De Mello Prado, R., Rozanne, D., 2020. Leaf analysis as diagnostic tool for balanced fertilization in tropical fruits. In: Srivastava, A.K., Hu, C. (Eds.), Fruit Crops: Diagnosis and Management of Nutrient Constraints. Elsevier, Amsterdam, The Netherlands, pp. 131–143.

De Oliveira, E.C.A., De Castro Gava, G.J., Trivelin, P.C.O., Otto, R., Franco, H.C.J., 2013. Determining a critical nitrogen dilution curve for sugarcane. J. Plant Nutr. Soil Sci. 176, 712–723.

Debaeke, P., Oosterom, E.J.V., Justes, E., Champolivier, L., Merrien, A., Aguirrezabal, L.A.N., et al., 2012. A species-specific critical nitrogen dilution curve for sunflower (*Helianthus annuus* L.). Field Crops Res. 136, 76–84.

Delhaize, E., Loneragan, J., Webb, J., 1982. Enzymic diagnosis of copper deficiency in subterranean clover. II. A simple field test. Aust. J. Agric. Res. 33, 981–987.

Dell, B., Malajczuk, N., Xu, D., Grove, T.S., 2001. Nutrient disorders in plantation eucalypts, 2nd ed. Australian Centre for International Agricultural Research, Canberra, Australia.

Dobermann, A., Bruulsema, T., Cakmak, I., Gerard, B., Majumdar, K., McLaughlin, M., et al., 2021. A new paradigm for plant nutrition. Food systems summit brief prepared by research partners of the scientific group for the food systems summit February 10, 2021. Scientific Panel on Responsible Plant Nutrition, Paris, France.

Dow, A.I., Roberts, S., 1982. Proposal: critical nutrient ranges for crop diagnosis. Agron. J. 74, 401–403.

Drossopoulos, J.B., Bouranis, D.L., Bairaktari, B.D., 1994. Patterns of mineral nutrient fluctuations in soybean leaves in relation to their position. J. Plant Nutr. 17, 1017–1035.

Elliott, G.C., Läuchli, A., 1986. Evaluation of an acid phosphatase assay for detection of phosphorus deficiency in leaves of maize (*Zea mays* L.). J. Plant Nutr. 9, 1469–1477.

El-Motaium, R., Hu, H., Brown, P.H., 1994. The relative tolerance of six *Prunus* rootstocks to boron and salinity. J. Am. Soc. Hortic. Sci. 119, 1169–1175.

Elwali, A.M.O., Gascho, G.J., 1984. Soil testing, foliar analysis, and DRIS as guides for sugarcane fertilization. Agron. J. 76, 466–470.

Fageria, N.K., 2003. Plant tissue test for determination of optimum concentration and uptake of nitrogen at different growth stages in lowland rice. Commun. Soil Sci. Plant Anal. 34, 259–270.

Freney, J., Spencer, K., Jones, M., 1978. The diagnosis of sulphur deficiency in wheat. Aust. J. Agric. Res. 29, 727–738.

Gardner, B.R., Roth, R.L., 1989. Midrib nitrate concentration as a means for determining nitrogen needs of cabbage. J. Plant Nutr. 12, 1073–1088.

Gastal, F., Lemaire, G., 2002. N uptake and distribution in crops: an agronomical and ecophysiological perspective. J. Exp. Bot. 53, 789–799.

Geraldson, C.M., Tyler, K.B., 1990. Plant analysis as an aid in fertilizing vegetable crops. In: Westermann, R.L. (Ed.), Soil Testing and Plant Analysis. Soil Science Society of America, Madison, WI, USA, pp. 549–562.

Gómez, M.I., Magnitskiy, S., Rodríguez, L.E., 2019. Critical dilution curves for nitrogen, phosphorus, and potassium in potato group Andigenum. Agron. J. 111, 419–427.

Greenwood, D.J., Lemaire, E.G., Gosse, G., Cruz, P., Draycott, A., Neeteson, J.J., 1990. Decline in percentage N of C3 and C4 crops with increasing plant mass. Ann. Bot. 66, 425–436.

Hannam, R.J., Riggs, J.L., Graham, R.D., 1987. The critical concentration of manganese in barley. J. Plant Nutr. 10, 2039–2048.

Hochmuth, G., Maynard, D., Vavrina, C., Hanlon, E., Simonne, E., 2018. Plant tissue analysis and interpretation for vegetable crops in Florida. Horticultural Sciences Department, University of Florida/IFAS Extension, Florida, USA.

Hommels, C.H., Kuiper, P.J.C., Haan, A.D., 1989. Responses to internal potassium ion concentrations of two *Taraxacum* microspecies of contrasting mineral ecology: the role of inorganic ions in growth. Physiol. Plant. 77, 562–568.

Howeler, R., Edwards, D.G., Asher, C.J., 1982. Micronutrient deficiencies and toxicities of cassava plants grown in nutrient solutions. I. Critical tissue concentrations. J. Plant Nutr. 5, 1059–1076.

Huang, L., Hu, D., Bell, R.W., 1995. Diagnosis of zinc deficiency in canola by plant analysis. Commun. Soil Sci. Plant Anal. 26, 3005–3022.

Islam, M.M., Ponnamperuma, F.N., 1982. Soil and plant tests for available sulfur in wetland rice soils. Plant Soil 68, 97–113.

Ivahupa, S.R., Asher, C.J., Blamey, F.P.C., O'Sullivan, J.N., 2006. Effects of sodium on potassium nutrition in three tropical root crop species. J. Plant Nutr. 29, 1095–1108.

Jarrell, W.M., Beverly, R.B., 1981. The dilution effect in plant nutrition studies. In: Brady, N.C. (Ed.), Advances in Agronomy. Academic Press, pp. 197–224.

Jongruaysup, S., Bell, R.W., Dell, B., 1994. Diagnosis and prognosis of molybdenum deficiency in black gram (*Vigna mungo* L. Hepper) by plant analysis. Aust. J. Agric. Res. 45, 195–201.

Justes, E., Mary, B., Meynard, J.-M., Machet, J.-M., Thelier-Huche, L., 1994. Determination of a critical nitrogen dilution curve for winter wheat crops. Ann. Bot. 74, 397–407.

Khelil, B.M., Sanaa, M., Msallem, M., Larbi, A., 2010. Floral analysis as a new approach to evaluate the nutritional status of olive trees. J. Plant Nutr. 33, 627–639.

Knowles, T.C., Doerge, T.A., Ottman, M.J., 1991. Improved nitrogen management in irrigated durum wheat using stem nitrate analysis. II. Interpretation of nitrate-nitrogen concentrations. Agron. J. 83, 353–356.

Krautz, T., Amelung, W., Ewert, F., Gaiser, T., Horn, R., Jahn, R., et al., 2013. Nutrient acquisition from arable subsoils in temperate climates: A review. Soil Biol. Biochem. 57, 1003–1022.

Kuehny, J.S., Peet, M.M., Nelson, P.V., Willits, D.H., 1991. Nutrient dilution by starch in CO_2-enriched chrysanthemum. J. Exp. Bot. 42, 711–716.

Leidi, E.O., Gómez, M., Del Río, L.A., 1987. Evaluation of biochemical indicators of Fe and Mn nutrition for soybean plants. II. Superoxide dismutases, chlorophyll contents and photosystem II activity. J. Plant Nutr. 10, 261–271.

Leigh, R.A., Johnston, A.E., 1983. Concentrations of potassium in the dry matter and tissue water of field-grown spring barley and their relationships to grain yield. J. Agric. Sci. 101, 675–685.

Leigh, R.A., Stribley, D.P., Jonston, A.E., 1982. How should tissue nutrient concentrations be expressed? In: Scaife, A. (Ed.), Proceedings of Ninth International Plant Nutrition Colloquium, Warwick, England. Commonwealth Agricultutural Bureaux Royal, Bucks, Farnham, pp. 39–44.

Lemaire, G., Van Oosterom, E., Sheehy, J., Jeuffroy, M.H., Massignam, A., Rossato, L., 2007. Is crop N demand more closely related to dry matter accumulation or leaf area expansion during vegetative growth? Field Crops Res. 100, 91–106.

Lewis, D.C., 1992. Effect of plant age on the critical inorganic and total phosphorus concentrations in selected tissues of subterranean clover (cv. Trikkala). Aust. J. Agric. Res. 43, 215–223.

Liebisch, F., Max, J.F.J., Heine, G., Horst, W.J., 2009. Blossom end rot and fruit cracking of tomato grown in net-covered greenhouses in central Thailand can partly be corrected by calcium and boron sprays. J. Plant Nutr. Soil Sci. 172, 140–150.

Lucena, J.J., Garate, A., Ramon, A.M., Manzanares, M., 1990. Iron nutrition of a hydroponic strawberry culture (*Fragaria vesca* L.) supplied with different Fe chelates. Plant Soil 123, 9–15.

Ma, Q., Bell, R.W., Biddulph, B., 2019. Potassium application alleviates grain sterility and increases yield of wheat (*Triticum aestivum*) in frost-prone Mediterranean-type climate. Plant Soil 434, 203–216.

Mattos, J., Padyampakeni, D., Oliver, D.M., Boaretto, A.Q., Morgan, R.M., Quaggio, J.A., K.T., 2020. Soil and nutrient interactions. In: Talon, M., Caruso, M., Gmitter Jr, F.G. (Eds.), The Citrus Genus. Elsevier, Duxford, UK, pp. 317–331.

Mills, H.A., Benton Jones Jr., J., 1996. Plant Analysis Handbook II. MicroMacro Publishing, Inc., Athens, Georgia, USA.

Muhammad, S., Sanden, B.L., Lampinen, B.D., Smart, D.R., Saa, S., Shackel, K.A., et al., 2020. Nutrient storage in the perennial organs of deciduous trees and remobilization in spring – a study in almond (*Prunus dulcis*) (Mill.) D. A. Webb. Front. Plant Sci. 11, 658.

Munson, R.D., Nelson, W.L., 1990. Principles and practices in plant analysis. In: Westermann, R.L. (Ed.), Soil Testing and Plant Analysis. Soil Science Society of America, Madison, WI, USA, pp. 359–388.

Niegengerd, E., Hecht-Buchholz, C., 1983. Elektronenmikroskopische Untersuchungen einer Virusinfektion (BYMV) von *Vicia faba* beigleichzeitigem Mineralstoffmangel. Z. Pflanzenernähr. Bodenk. 146, 589–603.

Noppakoonwong, R.N., Bell, R.W., Dell, B., Loneragan, J.F., 1993. An effect of light on the B requirement for leaf blade elongation in black gram (*Vigna mungo*). Plant Soil 155/156, 413–416.

O'Connell, A.M., Grove, T.S., 1985. Acid phosphatase activity in karri (*Eucalyptus diversicolor* F. Muell.) in relation to soil phosphate and nitrogen supply. J. Exp. Bot. 36, 1359–1372.

Pissarek, H.P., 1979. Der Einfluß von Grad und Dauer des Mg-Mangels auf den Kornertrag von Hafer. Z. Acker- Pflanzenbau 148, 62–71.

Pissarek, H.P., 1980. Makro- und Mikrosymptome des Bormangels bei Sonnenblumen, Chinakohl und Mais. Z. Pflanzenernähr. Bodenk. 143, 150–160.

Radrizzani, A., Dalzell, S.A., Shelton, H.M., 2011. Effect of environment and plant phenology on prediction of plant nutrient deficiency using leaf analysis in *Leucaena leucocephala*. Crop Pasture Sci. 62, 248–260.

Raese, J.T., Drake, S.R., 1997. Nitrogen fertilization and elemental composition affects fruit quality of 'Fuji' apples. J. Plant Nutr. 20, 1797–1809.

Rahimi, A., Schropp, A., 1984. Carboanhydraseaktivität und extrahierbares Zink als Maßstab für die Zink-Versorgung von Pflanzen. Z. Pflanzenernähr. Bodenk. 147, 572–583.

Rasmussen, P.E., Ramig, R.E., Ekin, L.G., Rhode, C.R., 1977. Tissue analyses guidelines for diagnosing sulfur deficiency in white wheat. Plant Soil 46, 153–163.

Rerkasem, B., Jamjod, S., 2004. Boron deficiency in wheat: a review. Field Crops Res. 89, 173–186.
Rerkasem, B., Netsangtip, R., Bell, R.W., Loneragan, J.F., Hiranburana, N., 1988. Comparative species responses to boron on a Typic Tropaqualf in Northern Thailand. Plant Soil 106, 15–21.
Reuter, D.J., Robinson, J.B. (Eds.), 1997. Plant Analysis. An Interpretation Manual. 2nd ed. CSIRO Publishing, Melbourne, Australia..
Reuter, D.J., Robson, A.D., Loneragan, J.F., Tranthim-Fryer, D.J., 1981. Copper nutrition of subterranean clover (*Trifolium subterraneum* L. cv. Seaton Park). II. Effects of copper supply on distribution of copper and the diagnosis of copper deficiency by plant analysis. Aust. J. Agric. Res. 32, 267–282.
Righetti, T., Wilder, K.L., Cummings, G.A., 1990. Plant analysis as an aid in fertilizing orchards. In: Westermann, R.L. (Ed.), Soil Testing and Plant Analysis. Soil Science Society of America, Madison, WI, USA, pp. 563–601.
Roberts, S., Dow, A.I., 1982. Critical nutrient ranges for petiole phosphorus levels of sprinkler-irrigated Russet Burbank potatoes. Agron. J. 74, 583–585.
Robinson, J.B., 1993. Plant sampling: a review. Aust. J. Exp. Agric. 33, 1007–1014.
Robson, A.D., Pitman, M.G., 1983. Interactions between nutrients in higher plants. In: Läuchli, A., Bieleski, R.L. (Eds.), Encyclopedia of Plant Physiology, New Series. Springer, Verlag, Berlin and New York, pp. 147–180.
Römheld, V., Kirkby, E.A., 2010. Research on potassium in agriculture: needs and prospects. Plant Soil 335, 155–180.
Saa, S., Brown, P., Muhammad, S., Andres, M., Olivos-Del Rio, A., Sanden, B., et al., 2014. Prediction of leaf nitrogen from early season samples and development of field sampling protocols for nitrogen management in almond (*Prunus dulcis* [Mill.] DA Webb). Plant Soil 380, 153–163.
Samborska, I.A., Kalaji, H.M., Sieczko, L., Goltsev, V., Borucki, W., Jajoo, A., 2018. Structural and functional disorder in the photosynthetic apparatus of radish plants under magnesium deficiency. Funct. Plant Biol. 45, 668–679.
Scaife, A., 1988. Derivation of critical nutrient concentrations for growth rate from data from field experiments. Plant Soil 109, 159–169.
Schjoerring, J.K., Cakmak, I., White, P.J., 2019. Plant nutrition and soil fertility: synergies for acquiring global green growth and sustainable development. Plant Soil 434, 1–6.
Schlichting, E., 1976. Pflanzen- und Bodenanalysen zur Charakterisierung des Nährstoffzustandes von Standorten. Landwirtsch. Forsch. 29, 317–321.
Schmidt, S., Pedas, P., Laursen, K., Schjoerring, J., Husted, S., 2013. Latent manganese deficiency in barley can be diagnosed and remediated on the basis of chlorophyll *a* fluorescence measurements. Plant Soil 372, 417–429.
Scott, B.J., Robson, A.D., 1990. Distribution of magnesium in subterranean clover (*Trifolium subterraneum* L.) in relation to supply. Aust. J. Agric. Res. 41, 499–510.
Skinner, P.W., Matthews, M.A., Carlson, R.M., 1987. Phosphorus requirements of wine grapes: extractable phosphate of leaves indicates phosphorus status. J. Am. Soc. Hortic. Sci. 112, 449–454.
Smith, F.W., Loneragan, J.F., 1997. Interpretation of plant analysis: concepts and principles. In: Reuter, D.J., Robinson, J.B. (Eds.), Plant Analysis. An Interpretation Manual, 2nd ed. CSIRO Publishing, Melbourne, Australia, pp. 1–33.
Smith, G., Lauren, D., Cornforth, I., Agnew, M., 2006. Evaluation of putrescine as a biochemical indicator of potassium requirements of lucerne. New Phytol. 91, 419–428.
Smyth, D.A., Chevalier, P., 1984. Increases in phosphatase and β-glucosidase activities in wheat seedlings in response to phosphorus-deficient growth. J. Plant Nutr. 7, 1221–1231.
Smyth, T.J., Cravo, M.S., 1990. Critical phosphorus levels for corn and cowpea in a Brazilian Amazon Oxisol. Agron. J. 82, 309–312.
Soratto, R.P., Sandana, P., Fernandes, A.M., Martins, J.D.L., Job, A.L.G., 2020. Testing critical phosphorus dilution curves for potato cropped in tropical Oxisols of southeastern Brazil. Eur. J. Agron. 115, 126020.
Sparrow, L.A., Chapman, K.S.R., 2003. Effects of nitrogen fertiliser on potato (*Solanum tuberosum* L., cv. Russet Burbank) in Tasmania. 2. Petiole and soil analysis. Aust. J. Exp. Agric. 43, 643–650.
Sumner, M.E., Farina, P.M.W., 1986. Phosphorus interactions with other nutrients and lime in field cropping systems. Adv. Soil Sci. 5, 210–236.
Ulrich, A., Hills, F.J., 1990. Plant analysis as an aid in fertilizing sugarbeet. In: Westermann, R.L. (Ed.), Soil Testing and Plant Analysis. Soil Science Society of America, Madison, WI, USA, pp. 429–448.
Van Maarschalkerweerd, M., Husted, S., 2015. Recent developments in fast spectroscopy for plant mineral analysis. Front. Plant Sci. 6, 169.
Van Maarschalkerweerd, M., Bro, R., Egebo, M., Husted, S., 2013. Diagnosing latent copper deficiency in intact barley leaves (*Hordeum vulgare* L.) using near infrared spectroscopy. J. Agric. Food Chem. 61, 10901–10910.
Venkat-Raju, K., Marschner, H., 1981. Inhibition of iron-stress reactions in sunflower by bicarbonate. Z. Pflanzenernähr. Bodenk. 144, 339–355.
Vesk, M., Possingham, V., Mercer, F.V., 1966. The effect of mineral nutrient deficiency on the structure of the leaf cells of tomato, spinach and maize. Aust. J. Bot. 14, 1–18.
Walworth, J.L., Sumner, M.E., 1988. Foliar diagnosis: a review. Adv. Plant Nutr. 3, 193–241.
Wang, X., Ye, T., Ata-Ul-Karim, S.T., Zhu, Y., Liu, L., Cao, W., et al., 2017. Development of a critical nitrogen dilution curve based on leaf area duration in wheat. Front. Plant Sci. 8, 1517.
Ware, G.O., Ohki, K., Moon, L.C., 1982. The Mitscherlich plant growth model for determining critical nutrient deficiency levels. Agron. J. 74, 88–91.
Weir, R.G., Cresswell, G.C., 1993. Plant Nutrient Disorders. 3. Vegetable Crops. InKata Press, Sydney, Australia.
Witt, H.H., Jungk, A., 1974. The nitrate inducible nitrate reductase activity in relation to nitrogen nutritional status of plants. 7th Int. Colloq. Plant Anal. Fert. Probl., Hannover, Germany, pp. 519–527.

Wollring, J., Wehrmann, J., 1990. Der Nitratgehalt in der Halmbasisals Maßstab für den Stickstoffdüngerbedarf bei Wintergetreide. Z. Pflanzenernähr. Bodenk. 153, 47–53.

Wyttenbach, A., Tobler, L., Bajo, S., 1991. Silicon concentration in spruce needles. Z. Pflanzenernähr. Bodenk. 154, 253–258.

Ziadi, N., Brassard, M., Bélanger, G., Cambouris, A., Tremblay, N., Nolin, M., et al., 2008. Critical nitrogen curve and nitrogen nutrition index for corn in Eastern Canada. Agron. J. 100, 271–276.

Zorn, W., Marks, G., Heß, H., Bergmann, W., 2006. Handbuch zur Visuellen Diagnose von Ernährungsstörungen bei Kulturpflanzen. Elsevier Spektrum Akademischer, Verlag, München, Germany.

Part II

Plant–soil relationships

Chapter 12

Nutrient availability in soils

Petra Marschner[1], and Zed Rengel[2]

[1]School of Agriculture, Food & Wine, The University of Adelaide, Adelaide, SA, Australia, [2]UWA School of Agriculture and Environment, University of Western Australia, Perth, WA, Australia

Summary

Only a proportion of the total nutrient amount in soil can be taken up and utilized by plants. This proportion varies with nutrient and is influenced by a range of soil, plant, and environmental factors. In this chapter, methods for estimating nutrient availability and their limitations are discussed. In the soil, nutrients move to the soil surface by mass flow and diffusion, and the relative importance of these two processes varies with nutrient. Nutrient movement by diffusion is slow; therefore depletion zones develop around roots. The factors affecting the extent of this depletion zone, namely, root hair length and release of nutrient-mobilizing exudates, are described. The roles of root density, soil structure, and soil water content on nutrient availability are discussed. The chapter concludes with a critical assessment of the usefulness of soil tests and modeling approaches to improve the understanding of nutrient availability in soils.

12.1 General

Availability of soil nutrients to plants is governed by many soil, plant, and environmental factors. For practical purposes, estimating the plant-available nutrient fraction in soil involves chemical extractions that have been calibrated against field responses of crops to increasing fertilization rates. However, such extractions cannot simulate the temporal and spatial dynamics of the soil nutrient supply or plant nutrient demand. Such dynamics are particularly important in the rhizosphere, where root exudation and microbial activity can substantially alter nutrient availability (see also Chapter 14). Nutrient availability is also dependent on soil water content that influences nutrient movement in soil. This chapter discusses various factors influencing the proportion of the total amount of soil nutrients that is plant available.

12.2 Chemical soil analysis

The most direct way of determining nutrient availability in soils is to measure the growth responses of plants by means of field fertilizer trials. However, this is time- and labor-intensive, and the results are not easily extrapolated from one location to another. In contrast, chemical soil analysis—soil testing—is a comparatively rapid and inexpensive procedure for obtaining an estimate of nutrient availability in soils as a basis for recommending fertilizer application. Soil testing has been practiced in agriculture and horticulture for many years. The effectiveness of the procedure depends on (1) the extent to which the data can be calibrated with field fertilizer trials and (2) the interpretation of the analysis. It is important to keep in mind that all methods used in characterizing the availability of a given nutrient to plants, and thus to predict fertilizer response, must be evaluated by multi-rate fertilization trials in the field. Quite often, far more is expected from soil testing than the methods allow. The limitations of soil testing are discussed in detail in this chapter, with special reference to P and K.

Most soil testing uses batch extraction methods involving different forms of dilute acids, salts, or complexing agents, as well as water. Depending on the method used, quite different amounts of plant nutrients may be extracted from a given soil, as shown for P in Table 12.1. As a guide, 10 mg P kg^{-1} soil is equivalent to ~30 kg P ha^{-1} in the top 20 cm of the profile (at soil bulk density of 1.5 t m^{-3}, the weight of the top 20 cm of soil is 3 million kg soil ha^{-1}).

This chapter is a revision of the third edition chapter by P. Marschner and Z. Rengel, pp. 315–330. DOI: https://doi.org/10.1016/B978-0-12-384905-2.00012-1. © Elsevier Ltd.

TABLE 12.1 Mean concentrations of readily extractable (estimated plant-available) P in 214 soils (0–15 cm) using various extractants.

Soil test	Extractant	pH	Readily extractable P (mg kg^{-1} soil)
Olsen	0.5 M NaHCO$_3$	8.5	33 c
Mehlich 3	0.2 M CH$_3$COOH + 0.25 M NH$_4$NO$_3$ + 0.015 M NH$_4$F + 0.013 M HNO$_3$ + 0.001 M EDTA	2.5	59 a
Kelowna 1	0.25 M CH$_3$COOH + 0.015 M NH$_4$F	Not reported	57 a
Kelowna 2	0.25 M CH$_3$COOH + 0.25 M CH$_3$COONH$_4$ + 0.015 M NH$_4$F	Not reported	43 b
Kelowna 3	0.25 M CH$_3$COOH + 1 M CH$_3$COONH$_4$ + 0.015 M NH$_4$F	Not reported	42 b
Bray 1	0.03 M NH$_4$F + 0.025 M HCl		39 b
Miller and Axley	0.03 M NH$_4$F + 0.015 M H$_2$SO$_4$		35 bc
Water	Deionized water		6.9 d
CaCl$_2$	0.01 M CaCl$_2$		2.0 d
Filter paper impregnated with FeO	Soaked in FeCl$_2$.6H$_2$O, dried, then quickly passed through NH$_4$OH		37 bc

Means followed by the same letter are not different at $P \leq 0.05$.
Source: Based on Kumaragamage et al. (2007) and references therein.

Weak extractants such as water or sodium bicarbonate (Table 12.1) reflect mainly the intensity of supply (concentration in soil solution), whereas strong extractants (e.g., various acids) primarily indicate the capacity of the soil to replenish nutrients in the soil solution (buffering capacity) because they release also less available nutrient fractions.

Particularly for poorly mobile nutrients such as P, conventional soil tests may overestimate plant availability. Novel methods that mimic the root as a nutrient sink and are therefore better indicators of nutrient availability to plants include anion-exchange resins (Kouno et al., 1995) and diffusive gradients in thin films (DGT) (Mason et al., 2010, 2013). Mason et al. (2013) found that about 25% of P extracted by the Colwell method (0.5 M bicarbonate) was nonlabile P. By contrast, only 9% and 2% of nonlabile P were extracted by anion-exchange resin and DGT, respectively. This indicates that Colwell P may overestimate plant-available P.

Among the nutrients, P availability is particularly dependent on adsorption and transport processes rather than the total amount of P present, making batch extraction methods for assessing P availability not very useful. Nevertheless, the speed and low costs of such methods will continue warranting their use in commercial applications. Microdialysis, based on diffusional exchange through a semipermeable membrane, has shown some promise in realistically estimating P availability and complementing the existing analytical procedures (Demand et al., 2017). It has been proved useful in estimating the role of citrate in influencing P availability and root uptake (e.g., McKay Fletcher et al., 2019). Instead of "flooding" the soil with citrate solution as used in batch experiments, controlled release of citrate from the microdialysis probe into soil mimicked the expected root exudation. Microdialysis has also been used to estimate nitrate and ammonium availability in soil (Brackin et al., 2017).

Typically, as is the case for P, soil testing methods provide a good indication of nutrient status of the soil, and a likelihood of fertilizer response, when the soil is either low or high in P (Bolland and Gilkes, 1992). However, in the middle part of the response curve relating nutrient supply to plant growth, soil chemical analyses alone are inadequate for predicting the effects of fertilizer application. There are numerous practical ways to deal with such situations depending on the prevalent soil types. For example, in a range of Australian soils, combining soil test *P* values with soil P buffering capacity (which represents the soil capacity to sorb P and thus influence P availability after fertilizer addition) correlated better with fertilizer responses than the soil P tests alone (Bell et al., 2013). However, crop rotation should also be taken into account (Neuhaus et al., 2015). There are a number of methods to measure P buffering capacity (cf. Burkitt et al., 2002), most of which are quite time-consuming. Forrester et al. (2015) found that mid-infra-red spectroscopy, a method used to estimate various soil properties, can also estimate P buffering capacity.

Soil analysis mainly provides an indication of the capacity of a soil to supply nutrients to plants but does not adequately (and in some cases not at all) characterize the mobility of the nutrients in the soil. Additionally, it fails to provide information about the effects of soil structure, microbial activity, and plant aspects (such as root growth, root-induced changes in the rhizosphere, and plant genotypic differences), which are all critical factors influencing nutrient uptake under field conditions. These factors are discussed next, beginning with nutrient availability in relation to mobility in soils and root growth. For comprehensive overview of nutrient movement in soil, the reader is referred to Jungk (2002).

12.3 Movement of nutrients to the root surface

12.3.1 Principles of calculations

The importance of the mobility of nutrients in soils in relation to plant availability was emphasized by Barber (1962); these ideas were refined and further developed and summarized in a concept of nutrient availability (Barber, 1995). Although this concept is focused on aerated soils, its principles can also be applied to waterlogged/submerged soils. The three components in the concept are root interception, mass flow, and diffusion (Fig. 12.1). As roots grow through the soil, they come into areas containing available nutrients, for example, adsorbed to clay surfaces. This is referred to as interception of nutrients (Barber, 1995). Calculations showing the importance of root interception for plant nutrient uptake are based on (1) the amounts of available nutrients in the soil volume occupied by roots; (2) root volume as a percentage of the total soil volume—on average 1% for the topsoil; and (3) the proportion of the total soil volume occupied by pores, on average 50%, but strongly dependent on soil bulk density (Section 12.6). In general, only a small proportion of the total nutrient requirement can be met by root interception (Barber, 1995).

The second component of nutrient supply to roots is mass flow, representing movement of water and dissolved nutrients to the root surface driven by transpiration. Estimates of the nutrient amount supplied to plants by mass flow are based on the nutrient concentration in the soil solution and the amount of water transpired either per unit weight of shoot tissue (transpiration coefficient, e.g., 300–600 L H_2O kg^{-1} shoot dw) or per unit area.

The contribution of diffusion, the third component influencing nutrient supply to the root surface, can be calculated on the basis of the effective diffusion coefficients. Such data for various nutrients are more difficult to obtain than those on mass flow. Hence, the estimates of the contribution of diffusion can also be based on the difference between total uptake by plants and the sum of the amounts supplied by root interception + mass flow.

Relative contributions of the various modes of transport are given in Table 12.2. Typical for many soils, N and Mg were supplied mainly by mass flow, whereas the supply of K and P depended mainly on diffusion. The supply of Ca by mass flow was greater than uptake; therefore this nutrient would be expected to accumulate at the root surface, particularly in calcareous soils (see also Chapter 14). Similarly, Mg may also accumulate at the root surface in some soils (Barber, 1995).

Frequently, solute movement in the soil–root system considers only mass flow and diffusion, and root interception is included in the diffusion component (Barber, 1995; Tinker and Nye, 2000). However, in a number of aspects, conditions at the soil–root interface are often very different from those at a distance from the roots (Chapters 14 and 15). These conditions are insufficiently described by mechanistic models that treat roots primarily as a sink for nutrients supplied by mass flow or diffusion or both. In addition, nutrient availability at the soil–root interface and nutrient uptake are influenced by root and microbial exudates. An additional complication is that the soil structure at the root–soil interface plays an important role in determining soil water content and thus nutrient movement.

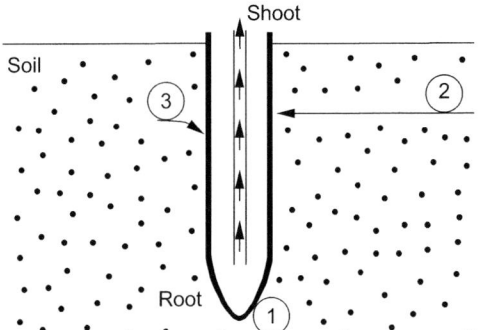

FIGURE 12.1 Schematic presentation of the movement of elements to the root surface of soil-grown plants. (1) Root interception: direct contact of roots and soil particles. (2) Mass flow: transport of soil solution along the water potential gradient (driven by transpiration). (3) Diffusion: element transport along a concentration gradient. ● = available nutrients (as determined by soil testing).

TABLE 12.2 Nutrient demand of a maize crop and estimates of nutrient supply from the soil by mass flow and diffusion [soil water potential −10 kPa, equivalent to field capacity in sandy soil, plants grown in soil (2- to 15-cm layer) collected from the area with native vegetation].

Nutrient	Plant uptake (mg pot^{-1})	Estimated supply by (% of the amount taken up)	
		Mass flow	Diffusion
N	340	59	41
P	22	5	95
K	160	11	89
Ca	150	340	0
Mg	131	73	27
S	23	11	89

Interception was not taken into account.
Source: Based on Oliveira et al. (2010).

12.3.2 Concentration of nutrients in the soil solution

To meet the nutrient demand of soil-grown plants, nutrients must reach the root surface, and this is mainly mediated by movement in (diffusion), as well as of (mass flow), the soil solution. The concentration of nutrients in the soil solution is therefore critical for nutrient supply to roots. However, root growth and extension of root hairs or hyphae of mycorrhizal fungi are also important because they reduce the distance nutrients have to travel from the soil to the root. The concentrations of nutrients in the soil solution vary widely, depending on factors such as soil water content, soil depth, pH, cation-exchange capacity, redox potential, soil organic matter content, microbial activity, season, and fertilizer application (Asher, 1978). The concentrations of nutrients in the soil solution are usually very low in many natural ecosystems, for example, in tundra (Chapin, 1988) compared to arable soils.

The nutrient concentration in the soil solution is an indicator of the mobility of nutrients toward the root surface and down the soil profile (i.e., leaching). The phosphate concentration in the soil solution is very low because, in contrast to other anions such as nitrate and sulfate, phosphate binds to surface-active sesquioxides and hydroxides of clay minerals. The P concentration in the soil solution and P mobility are enhanced by complexation of sesquioxides with organic ligands because that increases their solubility and thereby reduces the number of potential binding sites. Organic ligands may further increase P availability by anion exchange, that is, replacing P from the binding sites (Gerke, 1993, 1994). Hence, diffusion is the main form of phosphate transport in soils, whereas leaching or transport by mass flow to root surfaces is generally of minor importance for P in mineral soils (Table 12.2).

Shortly after fertilizer application, soil solution concentrations of nutrients, particularly N, can be high (Table 12.3). In such conditions, N transport to the root surface does not limit uptake by the crop (Barraclough, 1989).

The concentration of Mn and Fe in the soil solution mainly depends on soil pH and redox potential and, in temperate climate, may fluctuate over time, with a maximum in early summer (Sinclair et al., 1990). A decrease in pH or redox potential can increase the concentration of Mn and Fe in the soil solution (Miao et al., 2006; Sanders, 1983). For example, a decrease in redox potential due to waterlogging can result in soil extractable concentrations of Fe doubling and those of Mn tripling (Khabaz-Saberi et al., 2012).

Chelation by low-molecular-weight organic substances strongly affects the concentration of micronutrient cations in the soil solution and their transport to the root surface by mass flow and diffusion. For example, citric acid decreased and humic acid increased Zn sorption in soils (Piri et al., 2019). In the soil solution of calcareous soils, between 40% and 75% of Zn and 98% and 99% of Cu are in organic complexes (Hodgson et al., 1966; Sanders, 1983), with higher percentages in soils rich in organic matter (McGrath et al., 1988).

In nutrient solution experiments, the rate of uptake of metal cations from metal−organic complexes is lower than that of free cations (Table 12.4) and decreases with increasing size of the organic ligand (Jarvis, 1987). In soil, however, chelation of micronutrient cations such as Cu and Ni increases plant uptake (Alvarez and Rico, 2003), even though stability of ligands in soils (particularly in case of chelated Mn in oxidized conditions) is inversely related to plant uptake (López-Rayo et al., 2015).

TABLE 12.3 Time-course of nutrient concentrations in the soil solution of the topsoil (0–20 cm) of a high-yielding winter oilseed rape (*Brassica napus*) crop.

Nutrient	Concentration (μM)		
	22 February	28 March[a]	15 May
$NO_3 - N$	620	12,300	1843
$NH_4 - N$	29	1100	<1
$H_2PO_4^-$	14	14	10
K	91	202	133
Ca	1106	5258	1558
Mg	34	84	52

[a]Split application of N on 25 February (90 kg N ha^{-1}) and 25 March (185 kg N ha^{-1}).
Source: Based on Barraclough (1989).

TABLE 12.4 Trace element concentration in leaves of bean plants grown in nutrient solution or soil without or with the metal chelator diethylenetriamine pentaacetate (DPTA).

	Concentration in leaves (mg kg^{-1} dw)				
	Zn	Cu	Fe	Mn	Ni
Nutrient solution					
Control	34	37	125	132	33
+10^{-4} M DTPA	19	4	149	118	0
Soil					
Control	23	8	124	108	2
+10^{-3} M DTPA	27	19	230	136	13

Source: Based on Wallace (1980a,b).

Recognizing the importance of concentration of nutrients in the soil solution for transport of nutrients to the roots as well as for leaching from the rooting zone, various techniques have been developed to sample the soil solution. In older methods, soils were dried and ground prior to rewetting and collection of soil solution by displacement or centrifugation. However, drying and grinding may increase nutrient availability by releasing nutrients from inside of aggregates or decreasing it by adsorption and precipitation. Hence, to characterize soil solutions of relevance to field-grown plants, collection by suction cups or from undisturbed soil cores by centrifugation or percolation is preferable. For example, microlysimeters can be placed close to the root surface to extract soil solution (Braun et al., 2001). Short-term fluctuations in nutrient concentration can also be assessed using microdialysis tubes (Inselsbacher and Nasholm, 2012).

12.3.3 Role of mass flow

Mass flow is the convective transport of nutrients dissolved in the soil solution from the surrounding soil to the root surface. The average contribution of mass flow to total supply differs not only among nutrients but also among plant species. In Table 12.5 (based on a 4-year field study with spring wheat and sugar beet), mass flow was more than sufficient to supply Ca to both plant species as well as to supply Mg to spring wheat, but not to sugar beet. By contrast, due to low K concentration in the soil solution, mass flow is negligible for K supply; K is supplied mainly via diffusion.

TABLE 12.5 Plant uptake of K, Mg, and Ca by spring wheat and sugar beet grown in silty loam soil and the estimates of supply to the roots by mass flow.

	Spring wheat			Sugar beet		
	K	Mg	Ca	K	Mg	Ca
Plant uptake (kg ha^{-1})	215	13	35	326	44	104
Mass flow (kg ha^{-1})	5	17	272	3	10	236
Mass flow (% of total uptake)	2	131	777	1	23	227

From Strebel and Duynisveld (1989).

Therefore, the soil around the roots is depleted of K, whereas Ca and Mg accumulate at the root surface (Barber, 1995; Chapter 14). In the abovementioned 4-year study, the average contribution of mass flow to the total N supply was between 15% and 33% (Strebel and Duynisveld, 1989).

The contribution of mass flow to transporting nutrients to roots depends on the plant species (Table 12.5) because they differ in water uptake per unit root length; for example, it is higher in onion than maize (Baligar and Barber, 1978). The relative contribution of mass flow also varies with plant age (Brewster and Tinker, 1970) and time of the day because both influence transpiration and thus water uptake rate.

When the soil water content is relatively high (e.g., at field capacity), mass flow is unrestricted, and the water content (potential) at the root surface is similar to that of the bulk soil. As the soil water content decreases, the rate of water uptake by the roots can exceed the supply by mass flow, and the soil at the soil–root interface may become dry. This is observed around roots, particularly when the transpiration rate is high (Doussan et al., 2003, 2006; Nye and Tinker, 1977), and often occurs in the topsoil during the growing season. The dry soil surrounding the roots will limit or even eliminate transport of nutrients via mass flow. Under field conditions, the rainfall pattern (or irrigation cycle) therefore strongly affects the contribution of mass flow to the total nutrient supply.

Mass flow and diffusion to the root surface usually occur simultaneously; therefore it is not possible to strictly separate these processes. The term "apparent mass flow" has been recommended to define the amount of solutes transported to the root by mass flow (Nye and Tinker, 1977). A principal limitation of these calculations by mechanistic models is the assumption that uptake rates of nutrients and water are uniform along the length of individual roots, which is not the case (Chapter 2).

12.3.4 Role of diffusion

Diffusion is the main mechanism for the movement to the root surface of P and K and other nutrients with low concentration in the soil solution. The driving force for diffusion is a concentration gradient. In soil-grown plants a concentration gradient between the adjacent soil and the root surface is formed when the uptake rate of ions exceeds the supply by mass flow (Roose and Kirk, 2009). Depletion profiles develop over time and their shape depends mainly on the balance between uptake by roots, replenishment from soil, and mobility of ions by diffusion (Fig. 12.2). The shape of the depletion profile is also influenced by root hair length and mycorrhizal colonization.

The mobility of ions is defined by the diffusion coefficient. Diffusion coefficients in homogeneous media such as water (D_l) are similar for different ions and orders of magnitude higher than in nonuniform porous media such as aerated soils (Table 12.6). This is true particularly for P. In aerated soils, ions diffuse only in pore spaces that are filled with water or in the water film surrounding soil particles, interacting with the solid phase of the soil. For describing the diffusion of ions in soils, the term "effective diffusion coefficient" D_e was introduced by Nye and Tinker (1977), which is much smaller than the diffusion coefficient in water, because of the many physical and chemical interactions that ions encounter within the soil solid phase (Tinker and Nye, 2000):

$$D_e = D_l \theta \frac{1}{f} \frac{dC_l}{dC_s}$$

where D_e is the effective diffusion coefficient in soil (m^2 s^{-1}), D_l is the diffusion coefficient in water (m^2 s^{-1}), θ is the volumetric water content in soil (m^3 m^{-3}), f is the impedance factor that takes into account a tortuous pathway of ions

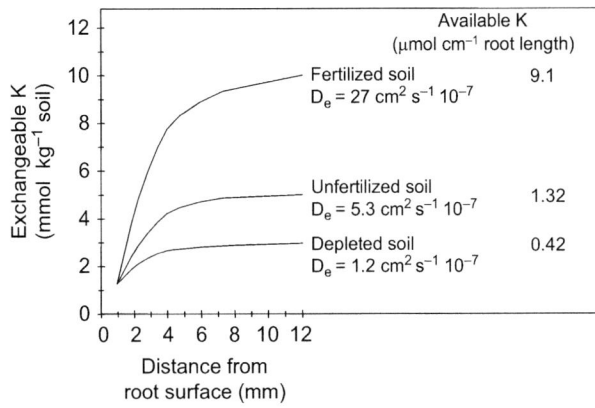

FIGURE 12.2 Exchangeable K concentration gradients in the rhizosphere of roots of 7-day-old oilseed rape (*Brassica napus*) seedlings grown in a soil with different concentrations of exchangeable K. *Modified from Kuchenbuch and Jungk (1984).*

TABLE 12.6 Estimates of diffusion coefficients (m² s⁻¹) of ions in water (D_l) and soil (D_e) and of movement per day at average values of D_e.

Ion	Diffusion coefficient			Movement in soil (mm d⁻¹)
	Water (D_l)	Soil (D_e)	Average D_e in soil	
NO_3^-	1.9×10^{-9}	10^{-10}–10^{-11}	5×10^{-11}	3.0
K^+	2.0×10^{-9}	10^{-11}–10^{-12}	5×10^{-12}	0.9
$H_2PO_4^-$	0.9×10^{-9}	10^{-12}–10^{-15}	1×10^{-13}	0.13

Source: From Jungk (1991).

and other solutes through water-filled soil pores, increasing the path length. The tortuosity (and thus impedance factor) becomes larger as the soil water content decreases.

$$\frac{dC_l}{dC_s} = \text{reciprocal of the soil buffering capacity for an ion}$$

where C_l is the concentration of an ion in the soil solution, and C_s is the sum of ions in the soil solution and those that can be released from the solid phase (e.g., exchangeable K).

12.3.4.1 Soil factors

Generally, the K and P concentrations are substantially lower at the root surface than in the bulk soil, creating a depletion zone around roots (Gahoonia and Nielsen, 1991; Wang et al., 2005). For P, depletion zones can exist not only for inorganic, but also for organic P (Gahoonia and Nielsen, 1992). Depletion of organic P is due to the release of phosphatases by roots and rhizosphere microorganisms that mineralize organic into inorganic P (Tarafdar and Jungk, 1987; Chapter 14), followed by uptake of inorganic P.

As shown in Fig. 12.2, D_e increases with increasing K concentration in the bulk soil. The extent of the K-depletion zone surrounding the roots increased from ~4 mm in soil that was depleted by previous intensive cropping to 5.3 mm in unfertilized and 6.3 mm in fertilized soils. Hence, raising the concentration of exchangeable K by fertilizer application enhanced the gradient between the bulk soil and the root surface and, therefore, increased the amount of K supplied via diffusion by a factor of more than 20. Application of $NaCl$ or $MgCl_2$ also increased the extent of the depletion zone and thus transport of K to the root surface (Kuchenbuch and Jungk, 1984). This is probably because Na^+ and Mg^{2+} diminished K binding by occupying the potential K^+ sorption sites.

For K, the shape and width of the depletion zone in different soils depends on their clay content, which is an important parameter of the buffering capacity for K (Fig. 12.3) because most clays have high cation-exchange capacity. In the soil with 21% w/w clay and an expected high cation-exchange capacity, the equilibrium concentration of K in the

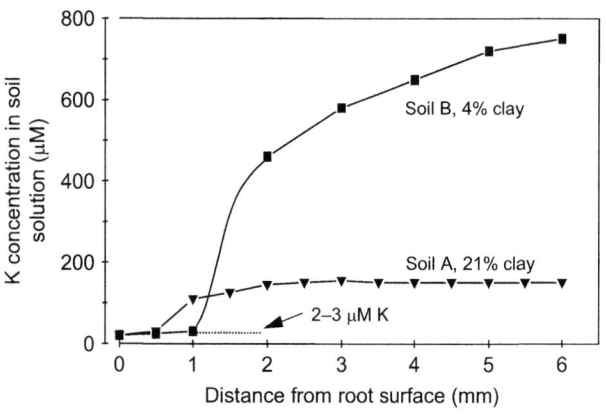

FIGURE 12.3 Concentration gradients of K in the soil solution in the rhizosphere of maize roots growing in soils with different clay contents. *Modified from Claassen and Jungk (1982).*

TABLE 12.7 Reciprocal of the impedance factor, buffering capacity and effective diffusion coefficient of K at different volumetric water content (refer to the equation in Section 12.3.4).

Water content [$\varnothing$ (cm^3 cm^{-3})]	Reciprocal of the impedance factor ($\frac{1}{f}$)	Buffering capacity (b)	Effective diffusion coefficient [D_e (cm^2 s^{-1})]
0.19	0.20	2.7	2.6×10^{-7}
0.26	0.30	3.1	4.9×10^{-7}
0.34	0.45	4.4	6.4×10^{-7}

Source: Based on Kuchenbuch et al. (1986).

soil solution away from the root surface was lower than in the soil with only 4% w/w clay. In both soils the K concentration in the soil solution at the root surface was about 2–3 μM K$^+$. However, the depletion zone was wider in the soil with low clay content, reflecting the lower capacity of that soil to replenish K in the soil solution compared to the soil with high clay content.

In soils low in exchangeable K, plant demand may exceed K supply, and some plant species have the capacity to derive a large proportion of the K taken up from the nonexchangeable fraction (see Rengel and Damon, 2008). Hence, plants may derive a large proportion of K from a fraction that is characterized by the standard soil test method as either unavailable or only available to a minor extent. For example, a high proportion of K from the nonexchangeable fraction was taken up by ryegrass, and part of this K originated from the interlayers of clay minerals (Hinsinger et al., 1992; Kong and Steffens, 1989). The release of K is influenced by K uptake by roots that decreases solution K concentration, resulting in selective dissolution of clay minerals (Hinsinger and Jaillard, 1993).

Soil water content is another important factor influencing D_e. Increasing the volumetric soil water content increases the volume available for ion diffusion, which results in a decrease in the impedance factor and an increase in the soil buffering capacity (Table 12.7). As a consequence, D_e increases more than twofold. For P, this effect of soil water content on D_e is even more pronounced.

This pronounced effect of water content on D_e is important in comparing soils of different texture (e.g., sandy vs clay soils) because soils differing in texture differ in water content at the same soil water potential (pF or $-$kPa). At the same pF, water content increases with an increase in clay content. At -33 kPa the water content of four different soils varied between 0.13 and 0.4 g cm^{-3} (Cox and Barber, 1992). To achieve the same P uptake by maize plants, the soil solution of only 10 μM P was necessary in the soil with water content of 0.4 g cm^{-3} as compared with 200 μM P in the soil with water content of 0.13 g cm^{-3} (Cox and Barber, 1992).

At low soil water content, the soil mechanical impedance increases and root growth is inhibited, which further limits nutrient supply to the root surface by diffusion. However, root hair growth is enhanced at low soil water content (Mackay and Barber, 1985, 1987; Watt et al., 1994) and may partly compensate for a decrease in surface area from impeded root growth.

At high transpiration rates (Garrigues et al., 2006), the water content at the root surface can be lower than in the bulk soil. Hence, the contact between root surface and soil via the soil solution can be lost. Root hairs contribute substantially to water uptake from drying soil (Carminati et al., 2017). Moreover, root hairs and exopolysaccharides produced by roots and microorganisms in the rhizosphere (mucilage) can improve soil−root contact and modify water content and water movement and thus impact on mass flow and diffusion. Indeed, the water content can be higher in the rhizosphere compared to the bulk soil (Carminati et al., 2010; Young, 1995) because exopolysaccharides in mucilage can increase water retention and aggregate stability (Benard et al., 2019; Czarnes et al., 2000; Morel et al., 1991), thereby maintaining continuity of the water film around soil particles in drying soil and thus P diffusion rate (Liebersbach et al., 2004).

Roots may also increase the water content in the rhizosphere by redistributing water from wetter soil layers deeper in the soil profile to drier topsoil by hydraulic lift. Water taken up by deep roots is released from roots in the dry topsoil when transpiration ceases (e.g., at night) and soil water potential in the dry top soil is more negative than plant water potential (Horton and Hart, 1998; Liste and White, 2008). The release of water in the topsoil improves nutrient uptake (e.g., Rose et al., 2008; Valizadeh et al., 2003) and thereby plant drought tolerance.

12.3.4.2 Plant factors

The volume of the root hair cylinder (as an indicator of root hair length) per unit of root length is positively correlated with K uptake (Fig. 12.4, Jungk et al., 1982). In nonmycorrhizal plants the extension of the P depletion zone is also positively related to root hair length (Gahoonia and Nielsen, 1997; Itoh and Barber, 1983). For example, in maize and oilseed rape, the extent of maximum P depletion in the rhizosphere was closely related to average root hair length, which was 0.7 mm in maize and 1.3 mm in oilseed rape (Hendriks et al., 1981). In agreement with this, the extent of P depletion in the rhizosphere and plant P uptake was greater in barley genotype Satka (long root hairs) than in Zita (short root hairs) (Fig. 12.5; Gahoonia and Nielsen, 1997). Moreover, P uptake in the field was greater by Satka than Zita (Gahoonia and Nielsen, 1997). Similar differences among genotypes in root hair length and P uptake were also found in white clover (Caradus, 1982). In addition to being dependent on genotype, root hair length increases under low P availability (Zhu et al., 2010).

The importance of root hairs for P uptake from soils was confirmed in simulation models for P uptake by different plant species (Föhse et al., 1991; Ma et al., 2001; Leitner et al., 2010). The inclusion of root hairs led to a better agreement between modeled and measured values. In soils low in extractable P, the contribution of root hairs can account for 50%−90% of the total P uptake (Föhse et al., 1991; Nigussie et al., 2003). Moreover, influx per unit area is greater in root hairs than the root cylinder because of the smaller diameter and the specific geometric arrangement of the root hairs that maintains high P diffusion rates ((Jungk and Claassen, 1986).

A relationship between root hair length and the width of the depletion zone of P and K may not always be found. For example, in the same soil, the width of the P depletion zone was confined mainly to the root hair cylinder in oilseed rape that has long root hairs (>0.5 mm), but exceeded the root hair cylinder in species with short root hairs (~0.2 mm), for example, cotton (Misra et al., 1988). The extension of the depletion zone beyond the root hair cylinder

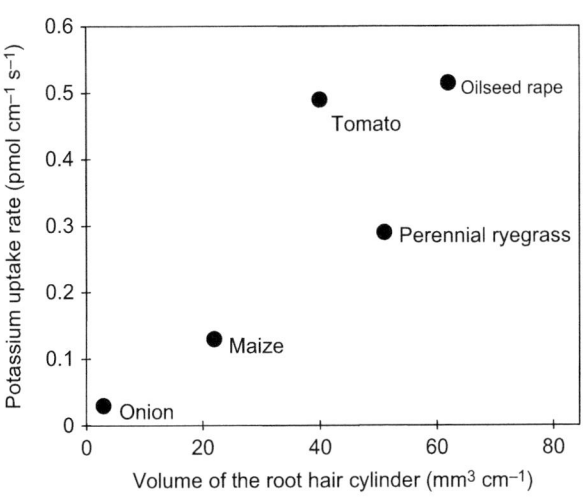

FIGURE 12.4 Rate of K uptake per unit root length in relation to the volume of the root hair cylinder. *Modified from Jungk et al. (1982).*

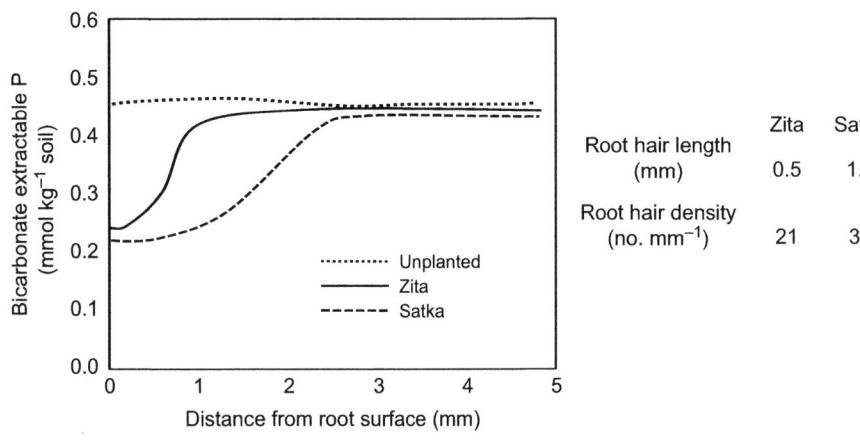

FIGURE 12.5 Depletion of bicarbonate-extractable P in the rhizosphere of two barley cultivars differing in root hair length and density. *From Gahoonia and Nielsen (1997) with permission from Springer Science & Business Media.*

TABLE 12.8 Estimates of proportions of soil supplying P and K to field-grown maize at two root length densities.

Root length density (cm cm^{-3})	Proportion of soil supplying P or K (%)	
	P	K
>2	20	50
<2	5	12

Source: From Fusseder and Kraus (1986).

in nonmycorrhizal plants can be explained by root-induced changes in the rhizosphere (e.g., release of root exudates, pH changes; Chapter 14) or higher efficiency in uptake parameters (K_m, I_{max}; Chapter 2). In mycorrhizal plants the P depletion zone extends beyond the root hair cylinder (Jungk and Claassen, 1986; Schnepf et al., 2008) because mycorrhizal hyphae grow beyond the root depletion zone and take up P. For example, in mycorrhizal white clover, the P depletion zone can be up to 11 cm from the root surface (Li et al., 1991).

In addition to mycorrhizal hyphae contributing to root P uptake, in most ecosystems (other than managed monocultures), roots of different species may intermingle, with a potential for positive or negative interactions. For example, faba bean roots exuded citrate and alkaline phosphatase that increased P availability, and roots of maize intercropped with faba bean were observed growing along faba bean roots (Zhang et al., 2016, 2019), apparently taking advantage of increased P availability induced by faba bean.

Both the degree of depletion within the root hair cylinder and the width of the depletion zone are influenced by the minimum concentration of nutrients (C_{min}) to which the roots can deplete the soil. C_{min} differs among plant species and even genotypes within a given species (Akhtar et al., 2007). In general, compared with solution culture, C_{min} in soil-grown plants is usually higher due to the soil buffering capacity that counteracts a decrease in nutrient concentration in the rhizosphere soil solution caused by plant uptake. Average C_{min} values for soil-grown plants are 2–3 μM for K (Claassen and Jungk, 1982) and 1 μM for P (Hendriks et al., 1981).

In 10 calcareous soils, Heidari et al. (2017) found the longest response time for replenishment of soil solution P from the solid phase to be 39 min (in a soil with high clay content). For K the rate of replenishment in the root hair cylinder also has to be high. Within 2.5 days, more than half of the K taken up by maize was derived from the nonexchangeable fraction of the soil in the root hair cylinder (Claassen and Jungk, 1982). From these data, it can be concluded that field-grown plants do not uniformly deplete even the densely rooted topsoil. In the rhizosphere a high proportion of the nonexchangeable K contributes to the total uptake, whereas in the bulk soil, even the readily exchangeable K is not utilized. An example giving estimates of the proportion of soil delivering P and K to maize roots is shown in Table 12.8. Because of the lower D_e of P than K, the proportion of soil supplying P is lower than that supplying K. Table 12.8 also demonstrates the importance of root length density (the length of roots per unit volume of soil) for nutrient uptake, with greater root density increasing the proportion of soil supplying P and K to plants.

12.4 Role of root density

The relationship between root density and uptake rate may be linear (e.g., Kristensen and Thorup-Kristensen, 2004), but as shown in Fig. 12.6, this is not always the case. At high root density, the uptake rate levels off. This is due to overlap of the depletion zones of individual roots and reflects root competition for nutrients (Fig. 12.7). For a given interroot distance, the degree of competition depends mainly on the diffusion coefficient D_e; in case of maize, competition is usually higher for nitrate than K and is of minor importance for P (Fusseder et al., 1988). However, modeling by Ge et al. (2000) suggested that root competition was important for P uptake in common bean. Moreover, in poorly structured soils, root density may be high in some areas; in those zones, root competition for nutrients can be important even for P (Fusseder and Kraus, 1986). Such competition for P may occur in the zones of high root density induced by localized fertilizer placement (Li et al., 2019), but not in all crops (Li et al., 2014), and may not have consequences for shoot growth (Li et al., 2016). Genotypes of the same species may react differently to fertilizer placement because root density is genetically controlled, with a number of relevant root morphology quantitative trait loci identified, for example, in *Arabidopsis* (Loudet et al., 2005), wheat (Soriano and Alvaro, 2019), and barley (Jia et al., 2019).

In some plant species (e.g., Proteaceae, white lupin), P deficiency induces the formation of cluster roots that are characterized by dense clusters of short laterals covered by root hairs (Lambers et al., 2006; Watt and Evans, 1999). Competition among lateral roots and also among root hairs is likely to be intense, effectively depleting available P in the soil volume occupied by the cluster roots (Neumann et al., 2000). Cluster roots release P-mobilizing exudates that further increase P availability (Chapter 14). However, cluster roots are active for only a few days (Watt and Evans, 1999), indicating that continuous P uptake is only possible by the formation of new cluster roots in as yet nondepleted

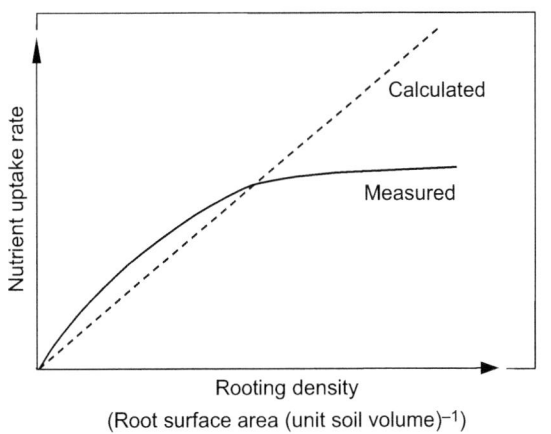

FIGURE 12.6 Relationship between rooting density and uptake rate of nutrients supplied by diffusion.

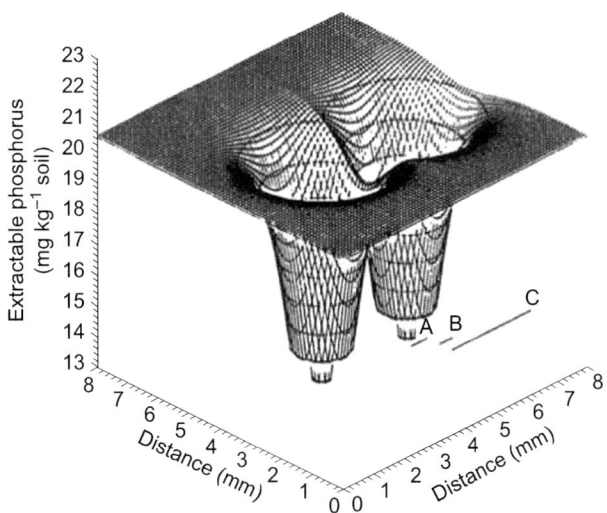

FIGURE 12.7 Profile of extractable P around two maize roots with overlapping depletion zones. Root cylinder (A), root hair cylinder (B), and maximal depletion zone (C). *From Fusseder and Kraus (1986) with permission from Elsevier.*

soil. Similarly for plant species not forming cluster roots, modeling by Steingrobe et al. (2001) suggested that root turnover, that is, death of old roots and formation of new ones in nondepleted soil, was important for P and K uptake.

In field-grown plants, root length density decreases from topsoil to subsoil (Table 12.9). The high root length density in the topsoil (e.g., Liedgens et al., 2000) is associated with the usually more favorable physical, chemical, and biological conditions in the topsoil compared with the subsoil (e.g., Adcock et al., 2007; Lofkvist et al., 2005). On average, the logarithm of root length density of crops declines linearly with increasing depth (Greenwood et al., 1982). However, in cereals, this gradient becomes less steep during the growing season, and root length density in the subsoil increases (Barber and Mackay, 1986; Vincent and Gregory, 1989) as the topsoil is depleted of nutrients and, in drier climates, of water (e.g., Tang et al., 2002). An example of the average root length density of cereals at heading is shown in Fig. 12.8.

Despite the relatively low root density, nutrient uptake from subsoil can be considerable. The importance of subsoil nitrate for N uptake by crop plants is widely established (see Dunbabin et al., 2003), especially in dry regions (Zhang et al., 2007). For cereal crops such as winter wheat growing in deep loess soils, on average 30% of total N uptake can be derived from the subsoil (Kuhlmann et al., 1989).

Uptake from the subsoil is also important for other nutrients such as Mg, K, and P. The relative importance of subsoil supply depends not only on root length density in the subsoil (Fig. 12.9), but also on root length density and nutrient availability in the topsoil (Barber and Mackay, 1986). Roots in the subsoil can be important in preventing nutrient leaching and maximizing nutrient capture. While root proliferation in the topsoil early in the season is important in reducing total nitrate leached, deep roots are important for capturing leached nitrate (Dunbabin et al., 2003).

TABLE 12.9 Root length distribution of maize at flowering in a Luvisol.

Soil depth (cm)	Root length density (cm cm^{-3})
0–15	6.2
15–30	3.1
30–45	1.2
45–60	0.5
60–75	0.4
75–90	0.3
90–135	0.2

Source: From Horlacher (1991).

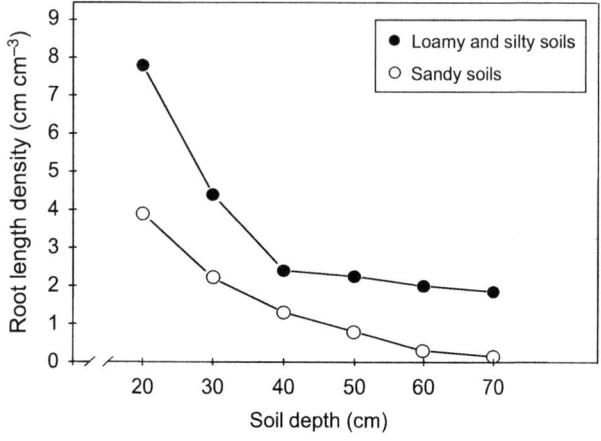

FIGURE 12.8 Modeled root length densities of various cereals (rye, oats, wheat, and barley) at heading in different soils as a function of soil depth. *From Gäth et al. (1989) with permission from Wiley.*

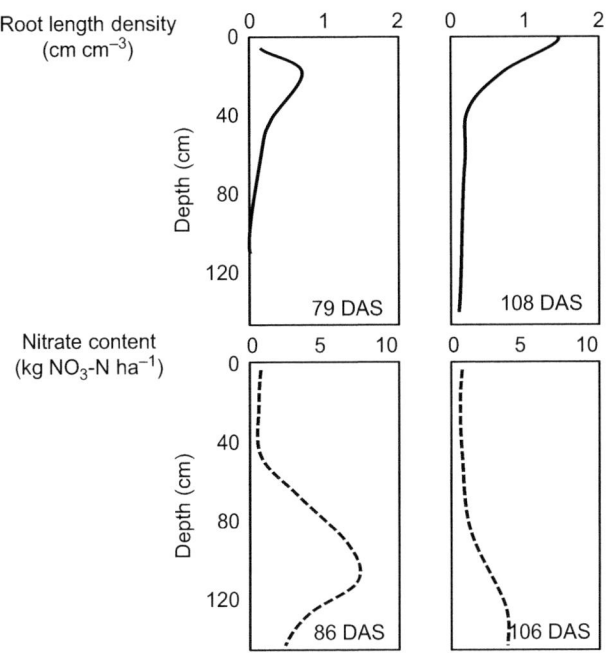

FIGURE 12.9 Root length density of *Lupinus angustifolius* and nitrate contents in the soil profile at different times after sowing (days after sowing, DAS). *Based on Dunbabin et al. (2003).*

Accessibility of nutrients in the subsoil can also depend on the activity of the soil fauna, particularly earthworms. For example, in barley and sugar beet, 20%–40% of roots in the subsoil (>0.65 m depth) were found to follow earthworm channels (Meuser, 1991). Root channels left by previous crops can also aid root growth at depth, particularly in compacted soils (Williams and Weil, 2004).

12.5 Nutrient availability and distribution of water in soils

In dry climates, nutrient availability in the topsoil can be diminished because low soil water content becomes a limiting factor for nutrient transport to the root surface. Nutrient uptake is further decreased by impaired root growth in dry soil.

In spring barley exposed to high rainfall (82 mm) in the first month after planting, more than 70% of the total root mass was in the topsoil (0–0.13 m) 2 months after planting, and only ~10% of the roots were below 0.23 m (Scott-Russell, 1977). By contrast, in the following year, with only 24 mm in the first month after planting, 40% of the root mass was in the topsoil and 30% in the subsoil. This differential root distribution has important consequences for nutrient uptake from different soil horizons.

Depending on the rainfall during the growing season, the percentage of plant-accumulated K in spring wheat taken up from the subsoil may vary between 60% in a dry year and 30% in a wet year (Fleige et al., 1983). Hydraulic lift of water from moist subsoil to dry topsoil can facilitate uptake of K from dry topsoil, at least in oilseed rape (Rose et al., 2008). However, in 10 duplex soils (sand over clay) in Western Australia, subsoil K was not important for wheat growth (Wong et al., 2000), possibly due to the high density of the subsoil clay limiting root growth at depth.

Using a combination of field experiments, measuring root length distribution in various cereal crops (Fig. 12.8) and exchangeable K, and considering water balance at a given location, models have been established that predict K transport to roots of cereal crops (Gäth et al., 1989). However, such models fail to adequately account for mobilization of K in the rhizosphere (e.g., of sugar beet) (El-Dessougi et al., 2002) because of inadequate knowledge of the underlying mechanisms (Rengel and Damon, 2008).

The uptake of P from different soil horizons varies during plant growth due to altered water availability in topsoil versus deeper soil layers (Table 12.10). Despite a higher concentration of available P in the topsoil, between 30% and 40% of the total P taken up by spring wheat in the later stages of growth came from the subsoil due to low water content in the topsoil. Nevertheless, plants can compensate for reduced diffusion caused by low water content in the topsoil by exuding low- and high-molecular-weight exudates to enhance P uptake (Liebersbach et al., 2004). In addition, hydraulic lift of water to the P-containing topsoil may increase P uptake by wheat (Valizadeh et al., 2003).

The prediction of the N supply as nitrate from the topsoil and subsoil is different from that for K and P. Transport of nitrate by mass flow contributes substantially to total nitrate delivery to the root surface (Strebel et al., 1983; Strebel

TABLE 12.10 Phosphorus uptake by spring wheat and the proportion of total P taken up that was supplied from different soil depths during plant development.

Available P[a] (mg kg^{-1})	Soil depth (cm)	Developmental stage		
		Booting	Anthesis	Milk development
		P uptake (kg ha^{-1} d^{-1})		
		0.35	0.27	0.15
		Proportion of P supplied from different soil depths (%)		
115	0–30	83	59	67
45	30–50	8	18	16
25	50–75	6	16	12
20	75–90	3	7	5

[a]Extraction with calcium ammonium lactate.
Source: Based on Fleige et al. (1981).

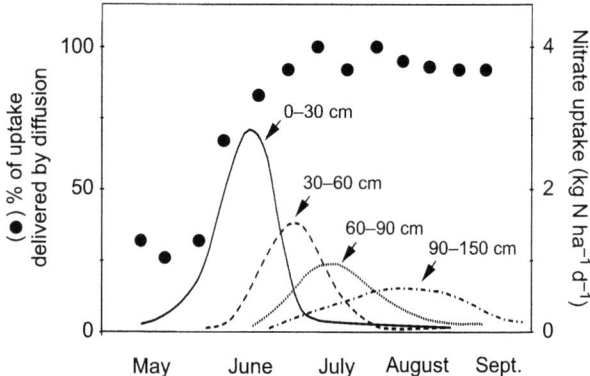

FIGURE 12.10 Nitrate uptake rate and transport by diffusion to sugar beet plants as a function of soil depth and time. *Based on Strebel et al. (1983).*

and Duynisveld, 1989). However, unless a high nitrate concentration is maintained in the soil solution by fertilizer applications during the growing season, the relative proportions of nitrate supplied by mass flow and diffusion may shift considerably because the soil buffering capacity for nitrate is generally low. A decrease in nitrate concentration in the soil solution decreases the proportion of nitrate supplied to spring wheat by mass flow and increases nitrate supply via diffusion to more than 50% of the total (Strebel et al., 1980). The soil depth from which nitrate is taken up also changes during the growing season (Dunbabin et al., 2003). In the early growth stages, nitrate is taken up mainly from the topsoil. However, after depletion of nitrate there, roots grow deeper and deplete subsoil nitrate (Fig. 12.9).

During the growing season of sugar beet, the supply of nitrate by mass flow may be low (32 kg nitrate-N ha^{-1}) compared to that by diffusion (181 kg nitrate-N ha^{-1}) (Strebel et al., 1983). A time-course study (Fig. 12.10) demonstrated that the supply by mass flow was restricted to the early growth period and to the topsoil that had relatively high nitrate concentration in the soil solution. Upon depletion of nitrate in the topsoil and root proliferation into the subsoil, nitrate was supplied exclusively by diffusion. Hence, the average (during the season) contribution of mass flow and diffusion (as well as of different soil horizons) to total nitrate supply may be misleading. Further, there can be considerable genotypic differences in the extent of root growth and nitrate depletion in the subsoil among genotypes of maize (Wiesler and Horst, 1993) and wheat (Liao et al., 2004). Therefore simulation models for predicting nutrient uptake in field-grown plants have to take into account soil heterogeneity, water and nutrient dynamics in space and time, root plasticity in terms of structure and function as well as differences among plant species and genotypes within a species (Dunbabin et al., 2004; Nakhforoosh et al., 2021).

12.6 Role of soil structure

Soil structure plays an important role in determining the amounts of nutrients that are available for uptake by roots as well as growth of roots laterally and down the profile (Bodner et al., 2021). Location of roots and the properties of the surrounding microenvironment influence nutrient uptake. Various techniques have been used for characterizing root properties and soil structure (Blaser et al., 2021), for example, (1) X-ray-computed micro-tomography for determining properties of soil aggregates and mesopores in the 27–67 μm range (Gryze et al., 2006), for characterizing macro- and mesoporosity down to 19 μm pore resolution at interfaces of texture-contrast soils (Jassogne et al., 2007; Helliwell et al., 2013) and for imaging roots in 3D in undisturbed soil columns (Tracy et al., 2010; Flavel et al., 2012); (2) a combination of X-ray absorption and phase-contrast imaging (Moran et al., 2000); (3) high-resolution 2D X-ray imaging (Pierret et al., 2003) to characterize root properties and the soil structure in intact soil cores, which provided a good fit with simulation models (cf. Doussan et al., 2006); and (4) root electrical capacitance measurements in field-grown plants (Streda et al., 2020). X-ray microscopy is becoming the method of choice because of its capacity to image particles in the nanometer size range with submicrometer spatial resolution and the option of combining with high spectral resolution for spectromicroscopy studies (Thieme et al., 2010).

In structured soils, not all roots have complete contact with the soil matrix, and in nonmycorrhizal plants the degree of root–soil contact at various positions along the root axis may vary from 0% to 100% (Van Noordwijk et al., 1992). Soil–root contact can be improved by mucilage exuded by roots (Rabbi et al., 2018; Read et al., 1999).

High bulk density increases root–soil contact but reduces root elongation (Table 12.11) (see also Shah et al., 2017). This reduction in root elongation can be, in part, compensated for by higher uptake rates per unit root length, for example, of nitrate and water (Table 12.11), as well as of P, particularly in soils high in available P (Cornish et al., 1984).

In soils with high bulk density, restricted root penetration through the soil matrix is accompanied by increased proportion of roots in macropores (diameter >30 μm) (Pankhurst et al., 2002). When soil bulk density was increased, resulting in the average soil–root contact increasing from 25% to 75%, root aggregation in certain zones was associated with localized high O_2 demand, which could lead to O_2 depletion (Asady and Smucker, 1989). The degree of soil–root contact and soil bulk density for optimal nutrient uptake and plant growth thus depend not only on soil fertility, but also on aeration (Van Noordwijk et al., 1992).

The conventional methods for determining available nutrients in soil are based on well-mixed and sieved soil samples. Hence, they not only ignore the importance of spatial heterogeneity in nutrient availability (as discussed earlier) but also destroy the soil structure and thereby eliminate gradients that occur in cation-exchange capacity and base saturation between external and internal surfaces of soil aggregates (Horn, 1987, 1989; Kaupenjohann and Hantschel, 1989).

More realistic data on nutrient availability in the soil can be obtained by collecting soil solution by lysimeters or suction cups in the field (e.g., Liedgens et al., 2000), or from undisturbed soil cores. Soil solution can be obtained from such cores either by circulation of a percolating solution or by centrifuging after adjustment to field capacity.

TABLE 12.11 Soil porosity, root length, estimated root–soil contact, and uptake rate of nitrate and water per unit root length of maize at different soil bulk densities.

	Bulk density (g cm^{-3})		
	1.1	1.3	1.5
Soil porosity (%)	60	51	44
Root length (m pot^{-1})	114	83	50
Proportion of root surface in contact with soil (%)	60	72	87
	Uptake (mmol m^{-1} root length)		
Nitrate	14	15	19
Water	18	21	24

Source: Compiled from Kooistra et al. (1992), Van Noordwijk et al. (1992), and Veen et al. (1992).

TABLE 12.12 Cation concentrations in soil equilibrium solutions from homogenized soil and in percolation solution from an undisturbed soil column [Brown earth, pH (CaCl$_2$) 3.1].

	Concentration in solution (μM)				
	K	Ca	Mg	Al	Fe
Equilibrium solution	55	41	39	104	39
Percolation solution	13	15	17	52	17

Source: Based on Hantschel et al. (1988) and Kaupenjohann and Hantschel (1989).

Cation concentrations differ between equilibrium soil solution from homogenized soil and percolation solution of the same but undisturbed soil (Table 12.12). The concentration of cations (except H$^+$) is usually higher in the homogenized samples because of the destruction of aggregates and exposure of internal surfaces from which cations are released into the extractant. Accordingly, concentrations of K and Mg in the needles of Norway spruce correlated poorly with the concentrations of the two nutrients in the soil extraction solution, but correlated strongly with concentrations of the two nutrients in undisturbed soil (Kaupenjohann and Hantschel, 1989).

12.7 Intensity/quantity ratio, plant factors, and consequences for soil testing

Routine soil testing methods determine a fraction of "chemically available" nutrients. Depending on the extraction method, routine soil testing mainly characterizes the intensity of nutrient supply (e.g., water extraction), representing the labile pool (Fig. 12.11). Measuring P in water extracts [at 1:60 (v/v) soil:water ratio and shaking for 22 h] is a reasonable compromise between measuring intensity and capacity of P supply in soil (Van Noordwijk et al., 1992). More detailed information concerning binding strengths, rate of replenishment and the intensity/quantity ratios for different nutrients can be obtained with the electro-ultrafiltration method (EUF), which involves the use of different electrical field strengths and temperatures in an aqueous soil suspension (Nemeth, 1982). However, for routine soil testing, the EUF method is not necessarily superior to the conventional extraction methods (e.g., with CaCl$_2$) in estimating fertilizer requirements (Houba et al., 1986; Rao et al., 2000) or in characterizing organic N mineralization (Mengel et al., 2006).

There is a large number of extraction methods used in routine soil testing for micronutrients that, as a rule, mainly measure nutrient quantity (Fig. 12.11). These methods successfully predict fertilizer requirement only when the extracted amounts are much higher or lower than those considered adequate (Sims and Johnson, 1991). Predictions can sometimes be improved by considering other soil properties such as pH, redox potential, and clay and organic matter content (Brennan, 1992; Moraghan and Mascagni, 1991).

Ion-exchange resins can be used to determine not only ion concentrations in the soil solution, but also the rates of replenishment of these ions, for example, P (Wang et al., 2007), K (Shenker and Huang, 2001), or various cations and anions simultaneously (e.g., Castle and Neff, 2009). In experiments with bean and maize under field conditions, prediction of Zn uptake was more accurate using ion-exchange resins than the conventional DTPA extraction (Hamilton and Westermann, 1991).

There is a voluminous literature on sequential extraction of various nutrients, for example, P (Hedley et al., 1982, with various modifications, e.g., Wang et al., 2007), K (Moody and Bell, 2006) and Zn (Alvarez and Gonzalez, 2006). In principle, sequential extraction determines the distribution of a nutrient among fractions of different chemical or binding characteristics, based on the properties of selected extractants. However, relating different fractions to plant availability remains a difficult and unresolved task (e.g., Herencia et al., 2008; Li et al., 2010; Moody and Bell, 2006; Wang et al., 2007; Yang and Post, 2011) that may be dependent on soil type and organic versus inorganic fractions considered (Braos et al., 2020).

Recommendations for N fertilization of various agricultural and horticultural field crops have been improved by the N_{min} method that measures the amount of mineralized N, mainly nitrate, in the soil profile at the beginning of the growing season. The method takes into account various components of availability, such as the high mobility of nitrate in the soil profile (mass flow) and N uptake from the subsoil (root growth). The N_{min} method can improve fertilizer

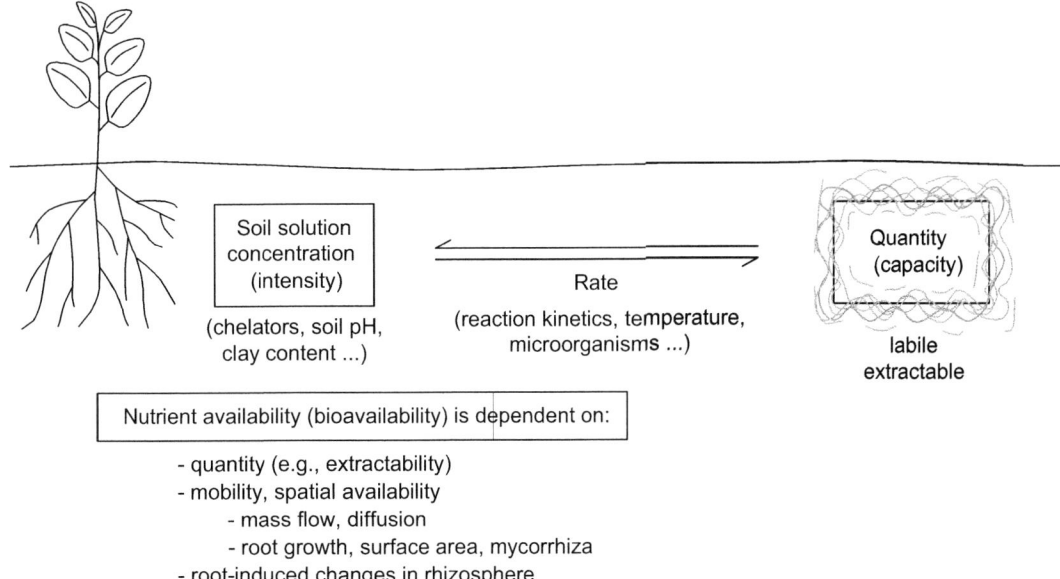

FIGURE 12.11 Intensity/quantity ratio of nutrient availability and factors determining nutrient availability. *From Marschner (1993).*

recommendations in rain-fed agriculture (Soltanpour et al., 1989) as well as in a variety of other plant systems (e.g., Cui et al., 2008; Khayyo et al., 2004; Liu et al., 2005), and may be as good as the EUF method in cereal cropping systems (Mengel et al., 2006).

In humid and semihumid climates, most nitrates in soil originate from mineralization of organic nitrogen (N_{org}). Various attempts have been made to characterize this mineralizable N_{org} fraction in the topsoil, for example, by the EUF method or $CaCl_2$ extraction (Mengel et al., 2006). For cereals, both EUF (N_{org}) and $CaCl_2$ extraction appear to be suitable alternatives to the N_{min} method (Appel and Mengel, 1992). Benbi and Richter (2002) cautioned that the experimental data on N mineralization are influenced by temperature, soil water content as well as site-specific parameters.

12.7.1 Modeling of nutrient availability and crop nutrient uptake

The principal limitation of soil testing methods is that they characterize only some of the factors that govern nutrient supply to the roots of field-grown plants. Improving the reliability of fertilizer recommendations based on chemical soil analysis does not depend on the extraction method used, but rather on consideration of the root factors (species, genotype) as well as environmental factors such as soil water content and soil structure. Current models for predicting nutrient availability and nutrient uptake under field conditions are therefore based on both soil and plant factors (Fig. 12.12) in which root parameters are the key element (e.g., Doussan et al., 2006; Dunbabin et al., 2002, 2013; Leitner et al., 2010). These models have been refined in recent years (e.g., De Moraes et al., 2019; Postma et al., 2017; Postma and Black, 2021), and predictions of nutrient uptake are often, but not always, in good agreement with actual uptake by crops, for example, in case of P (Fig. 12.12) or K (Ali Roshani et al., 2009). As shown in Fig. 12.12 for P, predicted and measured uptake were closely related in the soil with high P concentration. However, in the soil with relatively low P concentration, predicted uptake was lower than the measured uptake. Similar results were obtained for P uptake by maize (Mollier et al., 2008), indicating that the plants in the low-P soil had access to soil P sources that were not considered in the model because of mycorrhizal colonization and/or root-induced changes in the rhizosphere (discussed in detail in Chapters 14 and 15).

For K, predictions were in close agreement with measured uptake by wheat only in low-K soils, whereas in the K-sufficient soil the models overpredicted K uptake by as much as fourfold (Seward et al., 1990). This overprediction may be the result of poor characterization of plant demand and thus an underestimation of the role of negative feedback regulation of K uptake by the roots with high internal content (Chapter 2) or due to overestimation of K release from nonexchangeable pools (Ali Roshani et al., 2009). Using a different model, the actual K uptake by a range of crops was below the predicted values under poor K supply (Samal et al., 2010). However, prediction was improved by increasing soil K concentration or soil K buffering power, suggesting that release of K from nonexchangeable pools may have

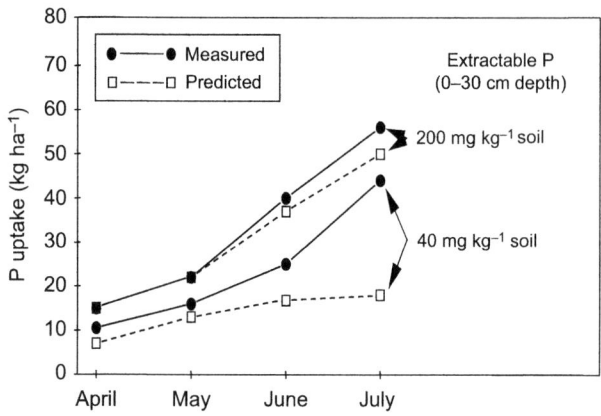

FIGURE 12.12 Phosphorus uptake: predicted by the Claassen–Barber model and measured in winter wheat in a field experiment after long-term application of either 100 kg P ha^{-1} per year (extractable P at 200 mg kg^{-1} soil) or without P fertilizer application (extractable P at 40 mg kg^{-1} soil). *From Jungk and Claassen (1989) with permission from Wiley.*

been underestimated in that particular model. The K adsorption isotherm was the most influential parameter in the K simulation model developed to work in the APSIM structure (Scanlan et al., 2015); the model allowed good prediction of crop responses to low K supply rates.

In conclusion, mechanistic simulation models are instrumental in increasing the understanding of the nutrient dynamics at the soil–root interface, but continuous improvements to account better for the relevant processes are needed. Validation of models against independently produced experimental data is crucial to maintain confidence in the simulation and modeling studies.

References

Adcock, D., McNeill, A.M., McDonald, G.K., Armstrong, R.D., 2007. Subsoil constraints to crop production on neutral and alkaline soils in southeastern Australia: a review of current knowledge and management strategies. Aust. J. Exp. Agric. 47, 1245–1261.

Akhtar, M.S., Oki, Y., Adachi, T., Murata, Y., Khan, M.H.R., 2007. Relative phosphorus utilization efficiency, growth response, and phosphorus uptake kinetics of Brassica cultivars under a phosphorus stress environment. Comm. Soil Sci. Plant Anal. 38, 1061–1085.

Ali Roshani, G., Narayanasamy, G., Datta, S.C., 2009. Modelling potassium uptake by wheat. Intern. J. Plant Prod. 3, 55–68.

Alvarez, J.M., Gonzalez, D., 2006. Zinc transformations in neutral soil and zinc efficiency in maize fertilization. J. Agric. Food Chem. 54, 9488–9495.

Alvarez, J.M., Rico, M.I., 2003. Effects of zinc complexes on the distribution of zinc in calcareous soil and zinc uptake by maize. J. Agric. Food Chem. 51, 5760–5767.

Appel, T., Mengel, K., 1992. Nitrogen uptake of cereals grown on sandy soils as related to nitrogen fertilizer application and soil nitrogen fractions obtained by electro-ultrafiltration (EUF) and CaCl$_2$ extraction. Eur. J. Agron. 1, 1–9.

Asady, G.H., Smucker, A.J.M., 1989. Compaction and root modifications of soil aeration. Soil Sci. Soc. Am. J. 53, 251–254.

Asher, C.J., 1978. Natural and synthetic culture media for Spermatophytes. CRC Handb. Ser. Nutr. Food, Sect. G 3, 575–609.

Baligar, V.C., Barber, S.A., 1978. Potassium uptake by onion roots characterized by potassium/rubidium ratio. Soil Sci. Soc. Am. J. 42, 618–622.

Barber, S.A., 1962. A diffusion and massflow concept of soil nutrient availability. Soil Sci. 93, 39–49.

Barber, S.A., 1995. Soil Nutrient Bioavailability: A Mechanistic Approach, 2nd ed. John Wiley, New York.

Barber, S.A., Mackay, A.D., 1986. Root growth and phosphorus and potassium uptake by two corn genotypes in the field. Fert. Res. 10, 217–230.

Barraclough, P.B., 1989. Root growth, macronutrient uptake dynamics and soil fertility requirements of a high-yielding winter oilseed rape crop. Plant Soil 119, 59–70.

Bell, R., Reuter, D., Scott, B., Sparrow, L., Strong, W., Chen, T.L.W., 2013. Soil phosphorus–crop response calibration relationships and criteria for winter cereal crops grown in Australia. Crop Pasture Sci. 64, 480–498.

Benard, P., Zarebanadkouki, M., Brax, M., Kaltenbach, R., Jerjen, I., Marone, F., et al., 2019. Microhydrological niches in soils: how mucilage and EPS alter the biophysical properties of the rhizosphere and other biological hotspots. Vadose Zone J. 18, 50–56.

Benbi, D.K., Richter, J., 2002. A critical review of some approaches to modelling nitrogen mineralization. Biol. Fertil. Soils 35, 168–183.

Blaser, S.R.G.A., Koebernick, N., Schlüter, S., Vetterlein, D., 2021. The 3-D imaging of roots growing in soil. In: Rengel, Z., Djalovic, I. (Eds.), The Root Systems in Sustainable Agricultural Intensification. Wiley, Hoboken, New Jersey, pp. 329–353.

Bodner, G., Mentler, A., Keiblinger, K., 2021. Plant roots for sustainable soil structure management in cropping systems. In: Rengel, Z., Djalovic, I. (Eds.), The Root Systems in Sustainable Agricultural Intensification. Wiley, Hoboken, New Jersey, pp. 45–90.

Bolland, M.D.A., Gilkes, R.J., 1992. Evaluation of the Bray 1, calcium acetate lactate (CAL), Truog and Colwell soil test as predictors of triticale grain production on soil fertilized with superphosphate and rock phosphate. Fert. Res. 31, 363–372.

Brackin, R., Atkinson, B.S., Sturrock, C.J., Rasmussen, A., 2017. Roots-eye view: Using microdialysis and microCT to non-destructively map root nutrient depletion and accumulation zones. Plant Cell Environ. 40, 3135–3142.

Braos, L.B., Bettiol, A.C.T., Di Santo, L.G., Ferreira, M.E., Cruz, M.C.P., 2020. Dynamics of phosphorus fractions in soils treated with dairy manure. Soil Res. 58, 289–298.

Braun, M., Dieffenbach, A., Matzner, E., 2001. Soil solution chemistry in the rhizosphere of beech (*Fagus silvatica* L.) roots as influenced by ammonium supply. J. Plant Nutr. Soil Sci. 164, 271–277.

Brennan, R.F., 1992. The relationship between critical concentrations of DTPA-extractable zinc from the soil for wheat production and properties of south-western Australian soils responsive to applied zinc. Commun. Soil Sci. Plant Anal. 23, 747–759.

Brewster, J.L., Tinker, P.B., 1970. Nutrient cation flows in soil around plant roots. Soil Sci. Soc. Amer. Proc. 34, 421–426.

Burkitt, L.L., Moody, P.W., Gourley, C.J.P., Hannah, M.C., 2002. A simple phosphorus buffering index for Australian soils. Soil Res. 40, 497–513.

Caradus, J.R., 1982. Genetic differences in the length of root hairs in white clover and their effect on phosphorus uptake. In: Scaife, A. (Ed.), Proceedings of the Ninth International Plant Nutrition Colloquium. Farnham Royal: Commonwealth Agricultural Bureau, Warwick, England, pp. 84–88.

Carminati, A., Moradi, A.B., Vetterlein, D., Vontobel, P., Lehmann, E., Weller, U., et al., 2010. Dynamics of soil water content in the rhizosphere. Plant Soil 332, 163–176.

Carminati, A., Passioura, J.B., Zarebanadkouki, M., Ahmed, M.A., Ryan, P.R., Watt, M., et al., 2017. Root hairs enable high transpiration rates in drying soils. New Phytol. 216, 771–781.

Castle, S.C., Neff, J.C., 2009. Plant response to nutrient availability across variable bedrock geologies. Ecosystems 12, 101–113.

Chapin III, F.S., 1988. Ecological aspects of plant mineral nutrition. In: Tinker, B., Läuchli, A. (Eds.), Advances in Plant Nutrition, Vol. 3. Praeger Publ, New York, pp. 161–191.

Claassen, N., Jungk, A., 1982. Kaliumdynamik im wurzelnahen Boden in Beziehung zur Kaliumaufnahme von Maispflanzen. Z. Pflanzenernähr. Bodenk. 145, 513–525.

Cornish, P.S., So, H.B., McWilliam, J.R., 1984. Effects of soil bulk density and water regimen on root growth and uptake of phosphorus by ryegrass. Aust. J. Agric. Res. 35, 631–644.

Cox, M.S., Barber, S.A., 1992. Soil phosphorus levels needed for equal P uptake from four soils with different water contents at the same water potential. Plant Soil 143, 93–98.

Cui, Z., Zhang, F.S., Chen, X.P., Miao, Y., Li, J., Shi, L., et al., 2008. On-farm evaluation of an in-season nitrogen management strategy based on soil N min test. Field Crops Res. 105, 48–55.

Czarnes, S., Hallett, P.D., Bengough, A.G., Young, I.M., 2000. Root- and microbial derived mucilages affect soil structure and water transport. Eur. J. Soil Sci. 51, 435–443.

De Moraes, M.T., Debiasi, H., Franchini, J.C., Bonetti, J.D.A., Levien, R., Schnepf, A., et al., 2019. Mechanical and hydric stress effects on maize root system development at different soil compaction levels. Front. Plant Sci. 10. Available from: https://doi.org/10.3389/fpls.2019.01358.

Demand, D., Schack-Kirchner, H., Lang, F., 2017. Assessment of diffusive phosphate supply in soils by microdialysis. J. Plant Nutr. Soil Sci. 180, 220–230.

Doussan, C., Pages, L., Pierret, A., 2003. Soil exploration and resource acquisition by plant roots: an architectural and modelling point of view. Agronomie 23, 419–431.

Doussan, C., Pierret, A., Garrigues, E., Pages, L., 2006. Water uptake by plant roots: II – Modelling of water transfer in the soil root-system with explicit account of flow within the root system – comparison with experiments. Plant Soil 283, 99–117.

Dunbabin, V.M., Diggle, A.J., Rengel, Z., van Hugten, R., 2002. Modelling the interactions between water and nutrient uptake and root growth. Plant Soil 239, 19–38.

Dunbabin, V., Diggle, A., Rengel, Z., 2003. Is there an optimal root architecture for nitrate capture in leaching environments? Plant Cell Environ. 26, 835–844.

Dunbabin, V., Rengel, Z., Diggle, A.J., 2004. Simulating form and function of root systems: efficiency of nitrate uptake is dependent on root system architecture and the spatial and temporal variability of nitrate supply. Funct. Ecol. 18, 204–211.

Dunbabin, V.M., Postma, J.A., Schnepf, A., Pages, L., Javaux, M., Wu, L., et al., 2013. Modelling root-soil interactions using three-dimensional models of root growth, architecture and function. Plant Soil 372, 93–124.

El-Dessougi, H., Claassen, N., Steingrobe, B., 2002. Potassium efficiency mechanisms of wheat, barley, and sugar beet grown on a K fixing soil under controlled conditions. J. Plant. Nutr. Soil Sci. 165, 732–737.

Flavel, R.J., Guppy, C.N., Tighe, M., Watt, M., McNeill, A., Young, I.M., 2012. Non-destructive quantification of cereal roots in soil using high-resolution X-ray tomography. J. Exp. Bot. 63, 2503–2511.

Fleige, H., Strebel, O., Renger, M., Grimme, H., 1981. Die potentielle P-Anlieferung durch Diffusion als Funktion von Tiefe, Zeit und Durchwurzelung bei einer Parabraunerde aus Löss. Mitt. Dtsch. Bodenkd. Ges. 32, 305–310.

Fleige, H., Grimme, H., Renger, M., Strebel, O., 1983. Zur Erfassung der Nährstoffanlieferung durch Diffusion im effektiven Wurzelraum. Mitt. Dtsch. Bodenkd. Ges. 38, 381–386.

Föhse, D., Claassen, N., Jungk, A., 1991. Phosphorus efficiency of plants. II. Significance of root hairs and cation-anion balance for phosphorus influx in seven plant species. Plant Soil 132, 261–272.

Forrester, S.T., Janik, L.J., Soriano-Disla, J.M., Mason, S., Burkitt, L., Moody, P., et al., 2015. Use of handheld mid-infrared spectroscopy and partial least-squares regression for the prediction of the phosphorus buffering index in Australian soils. Soil Res. 53, 67–80.

Fussder, A., Kraus, M., 1986. Individuelle Wurzelkonkurrenz und Ausnutzung der immobilen Makronährstoffe im Wurzelraum von Mais. Flora 178, 11–18.

Fusseder, A., Kraus, M., Beck, E., 1988. Reassessment of root competition for P of field-grown maize in pure and mixed cropping. Plant Soil 106, 299–301.

Gahoonia, T.S., Nielsen, N.E., 1991. A method to study rhizosphere processes in thin soil layers of different proximity to roots. Plant Soil 135, 143–146.

Gahoonia, T.S., Nielsen, N.E., 1992. The effects of root-induced pH changes on the depletion of inorganic and organic phosphorus in the rhizosphere. Plant Soil 143, 185–191.

Gahoonia, T.S., Nielsen, N.E., 1997. Variation in root hairs of barley cultivars doubled soil phosphorus uptake. Euphytica 98, 177–182.

Garrigues, E., Doussan, C., Pierret, A., 2006. Water uptake by plant roots: I – Formation and propagation of a water extraction front in mature root systems as evidenced by 2D light transmission imaging. Plant Soil 283, 83–96.

Gäth, S., Meuser, H., Abitz, C.-A., Wessolek, G., Renger, M., 1989. Determination of potassium delivery to the roots of cereal plants. Z. Pflanzenernähr. Bodenk. 152, 143–149.

Ge, Z.Y., Rubio, G., Lynch, J.P., 2000. The importance of root gravitropism for inter-root competition and phosphorus acquisition efficiency: results from a geometric simulation model. Plant Soil 218, 159–171.

Gerke, J., 1993. Phosphate absorption by humic/Fe-oxide mixtures aged at pH 4 and 7 and by poorly ordered Fe-oxide. Geoderma 59, 279–288.

Gerke, J., 1994. Kinetics of soil phosphate desorption as affected by citric acid. Z. Pflanzenernähr. Bodenk. 157, 17–22.

Greenwood, D.J., Gerwitz, A., Stone, D.A., Barnes, A., 1982. Root development of vegetable crops. Plant Soil 68, 75–96.

Gryze, S., Jassogne, L., Six, J., Bossuyt, H., Wevers, M., Merckx, R., 2006. Pore structure changes during decomposition of fresh residue: X-ray tomography analyses. Geoderma 134, 82–96.

Hamilton, M.A., Westermann, D.T., 1991. Comparison of DTPA and resin extractable soil Zn to plant zinc uptake. Commun. Soil Sci. Plant Anal. 22, 517–528.

Hantschel, R., Kaupenjohann, M., Horn, R., Gradl, J., Zech, W., 1988. Ecologically important differences between equilibrium and percolation soil extracts. Geoderma 43, 213–227.

Hedley, M.J., Nye, P.H., White, R.E., 1982. Plant-induced changes in the rhizosphere of rape (*Brassica napus* var. Emerald) seedlings. II. Origin of the pH change. New Phytol. 91, 31–44.

Heidari, S., Reyhanitabar, A., Oustan, S., 2017. Kinetics of phosphorus desorption from calcareous soils using DGT technique. Geoderma 305, 275–280.

Helliwell, J.R., Sturrock, C.J., Grayling, K.M., Tracy, S.R., Flavel, R.J., Young, I.M., et al., 2013. Applications of X-ray computed tomography for examining biophysical interactions and structural development in soil systems: a review. Eur. J. Soil Sci. 64, 279–297.

Hendriks, L., Claassen, N., Jungk, A., 1981. Phosphatverarmung des wurzelnahen Bodens und Phosphataufnahme von Mais und Raps. Z. Pflanzenernähr. Bodenk. 144, 486–499.

Herencia, J.F., Ruiz, J.C., Morillo, E., Melero, S., Villaverde, J., Maqueda, C., 2008. The effect of organic and mineral fertilization on micronutrient availability in soil. Soil Sci. 173, 69–80.

Hinsinger, P., Jaillard, B., 1993. Root-induced release of interlayer potassium and vermiculitization of phlogopite as related to potassium depletion in the rhizosphere of ryegrass. J. Soil Sci 44, 525–534.

Hinsinger, P., Jaillard, B., Dufey, J.E., 1992. Rapid weathering of a trioctahedral mica by the roots of ryegrass. Soil Sci. Soc. Am. J. 56, 977–982.

Hodgson, J.F., Lindsay, W.L., Trierweiler, J.T., 1966. Micronutrient cation complexing in soil solution. II. Complexing of zinc and copper in displaced solution from calcareous soils. Soil Sci. Soc. Amer. Proc. 30, 723–726.

Horlacher, D., 1991. Einfluß organischer und mineralischer N-Dünger auf Sproßwachstum und Nitratauswaschung bei Silomais sowie Quantifizierung der Ammoniakverluste nach Ausbringung von Flüssigmist. Ph.D. Thesis, University of Hohenheim, Hohenheim, Germany.

Horn, R., 1987. Die Bedeutung der Aggregierung für die Nährstoffsorption in Böden. Z. Pflanzenernähr. Bodenk. 150, 13–16.

Horn, R., 1989. Die Bedeutung der Bodenstruktur für die Nährstoffverfügbarkeit. Kali-Briefe 19, 505–515.

Horton, J.L., Hart, S.C., 1998. Hydraulic lift: a potentially important ecosystem process. Trees 13, 232–235.

Houba, V.J.G., Novozamsky, I., Huybregts, A.W.M., van der Lee, J.J., 1986. Comparison of soil extractions by 0.01 $CaCl_2$, by EUF and by some conventional extraction procedures. Plant Soil 96, 433–437.

Inselsbacher, E., Nasholm, T., 2012. A novel method to measure the effect of temperature on diffusion of plant-available nitrogen in soil. Plant Soil 354, 251–257.

Itoh, S., Barber, S.A., 1983. Phosphorus uptake by six plant species as related to root hairs. Agron. J. 75, 457–461.

Jarvis, S.C., 1987. The uptake and transport of silicon by perennial ryegrass and wheat. Plant Soil 97, 429–437.

Jassogne, L., McNeill, A., Chittleborough, D., 2007. 3D-visualization and analysis of macro- and meso-porosity of the upper horizons of a sodic, texture-contrast soil. Eur. J. Soil Sci. 58, 589–598.

Jia, Z., Liu, Y., Gruber, B.D., Neumann, K., Kilian, B., Graner, A., et al., 2019. Genetic dissection of root system architectural traits in spring barley. Front. Plant Sci. 10. Available from: https://doi.org/10.3389/fpls.2019.00400.

Jungk, A., 1991. Dynamics of nutrient movement at the soil-root interface. In: Waisel, J., Eshel, A., Kafkafi, U. (Eds.), Plant Roots: The Hidden Half. Marcel Dekker Inc, New York, pp. 455–481.

Jungk, A., Claassen, N., 1986. Availability of phosphate and potassium as the result of interactions between root and soil in the rhizosphere. Z. Pflanzenernähr. Bodenk. 149, 411–427.

Jungk, A., 1984. Phosphatdynamik in der Rhizosphäre und Phosphatverfügbarkeit für Pflanzen. Die Bodenkultur (Wien) 35, 99–107.

Jungk, A.O., 2002. Dynamics of nutrient movement at the soil-root interface. In: Waisel, Y., Eshel, A., Beeckman, T., Kafkaki, U. (Eds.), Plant Roots: The Hidden Half. CRC Press, Boca Raton, Florida, pp. 587–616.

Jungk, A., Claassen, N., 1989. Availability in soil and acquisition by plants as the basis for phosphorus and potassium supply to plants of phosphate and potassium as the result of interactions between root and soil in the rhizosphere. Z. Pflanzenernähr. Bodenk. 152, 151–157.

Jungk A., Claassen, N., Kuchenbuch, R., 1982. Potassium depletion of the soil-root interface in relation to soil parameters and root properties. In: Scaife, A. (Ed.), Proceedings of the Ninth International Plant Nutrition Colloquium. Farnham Royal: Commonwealth Agricultural Bureau, Warwick, England, pp. 250–255..

Khabaz-Saberi, H., Barker, S.J., Rengel, Z., 2012. Tolerance to ion toxicities enhances wheat (*Triticum aestivum* L.) grain yield in waterlogged acidic soils. Plant Soil 354, 371–381.

Kaupenjohann, M., Hantschel, R., 1989. Nährstofffreisetzung aus homogenen und in situ Bodenproben: Bedeutung für die Waldernährung und Gewässerversauerung. Kali-Briefe 19, 557–572.

Khayyo, S., Pérez-Lotz, J., Ramos, C., 2004. Application of the Nmin nitrogen fertilizer recommendation system in artichoke in the Valencian community. Acta Hortic. 660, 261–266.

Kong, T., Steffens, D., 1989. Bedeutung der Kalium-Verarmung in der Rhizosphäre und der Tonminerale für die Freisetzung von nichtaustauschbarem Kalium und dessen Bestimmung mit CHl. Z. Pflanzenernähr. Bodenk. 152, 337–343.

Kooistra, M.J., Schoonderbeek, D., Boone, F.R., Veen, B.W., Van Noordwijk, M., 1992. Root-soil contact of maize, as measured by a thin-section technique. II. Effects of soil compaction. Plant Soil 139, 119–129.

Kristensen, H.L., Thorup-Kristensen, K., 2004. Root growth and nitrate uptake of three different catch crops in deep soil layers. Soil Sci. Soc. Am. J. 68, 529–537.

Kuchenbuch, R., Jungk, A., 1984. Wirkung der Kaliumdüngung auf die Kaliumverfügbarkeit in der Rhizosphäre von Raps. Z. Pflanzenernähr. Bodenk. 147, 435–448.

Kuchenbuch, R., Claassen, N., Jungk, A., 1986. Potassium availability in relation to soil moisture. I. Effect of soil moisture on potassium diffusion, root growth and potassium uptake of onion plants. Plant Soil 95, 221–231.

Kuhlmann, H., Barraclough, P.B., Weir, A.H., 1989. Utilization of mineral nitrogen in the subsoil by winter wheat. Z. Pflanzenernähr. Bodenk. 152, 291–295.

Kumaragamage, D., Akinremi, O.O., Flaten, D., Heard, J., 2007. Agronomic and environmental soil test phosphorus in manured and non-manured Manitoba soils. Can. J. Soil Sci. 87, 73–83.

Kouno, K., Tuchiya, Y., Ando, T., 1995. Measurement of soil microbial biomass phosphorus by an anion exchange membrane method. Soil Biol. Biochem. 27, 1353–1357.

Lambers, H., Shane, M.W., Cramer, M.D., Pearse, S.J., Veneklaas, E.J., 2006. Root structure and functioning for efficient acquisition of phosphorus: matching morphological and physiological traits. Ann. Bot. 98, 693–713.

Leitner, D., Klepsch, S., Ptashnyk, M., Marchant, A., Kirk, G.J.D., Schnepf, A., et al., 2010. A dynamic model of nutrient uptake by root hairs. New Phytol. 185, 792–802.

Li, X.-L., George, E., Marschner, H., 1991. Extension of the phosphorus depletion zone in VA-mycorrhizal white clover in a calcareous soil. Plant Soil 136, 41–48.

Li, J., Lu, Y., Shim, H., Deng, X., Lian, J., Jia, Z., et al., 2010. Use of the BCR sequential extraction procedure for the study of metal availability to plants. J. Environ. Monit. 12, 466–471.

Li, H., Ma, Q., Li, H., Zhang, F., Rengel, Z., Shen, J., 2014. Root morphological responses to localized nutrient supply differ among crop species with contrasting root traits. Plant Soil 376, 151–163.

Li, H., Wang, X., Rengel, Z., Ma, Q., Zhang, F., Shen, J., 2016. Root over-production in heterogeneous nutrient environment has no negative effects on *Zea mays* shoot growth in the field. Plant Soil 409, 405–417.

Li, H., Wang, X., Brooker, R.W., Rengel, Z., Zhang, F., Davies, W.J., Shen, J., 2019. Root competition resulting from spatial variation in nutrient distribution elicits decreasing maize yield at high planting density. Plant Soil 439, 219–232.

Liao, M., Fillery, I.R.P., Palta, J.A., 2004. Early vigorous growth is a major factor influencing nitrogen uptake in wheat. Funct. Plant Biol. 31, 121–129.

Lieberbach, H., Steingrobe, B., Claassen, N., 2004. Roots regulate ion transport in the rhizosphere to counteract reduced mobility in dry soil. Plant Soil 260, 79–88.

Liedgens, M., Richner, W., Stamp, P., Soldati, A., 2000. A rhizolysimeter facility for studying the dynamics of crop and soil processes: description and evaluation. Plant Soil 223, 87–97.

Liste, H.H., White, J.C., 2008. Plant hydraulic lift of soil water - implications for crop production and land restoration. Plant Soil 313, 1–17.

Liu, X.J., Ju, X.T., Chen, X.P., Zhang, F.S., Römheld, V., 2005. Nitrogen recommendations for summer maize in northern China using the Nmin test and rapid plant tests. Pedosphere 15, 246–254.

Lofkvist, J., Whalley, W.R., Clark, L.J., 2005. A rapid screening method for good root-penetration ability: comparison of species with very different root morphology. Acta Agric. Scand. Sect. B, Soil Plant Sci. 55, 120–124.

López-Rayo, S., Nadal, P., Lucena, J., 2015. Reactivity and effectiveness of traditional and novel ligands for multi-micronutrient fertilization in a calcareous soil. Front. Plant Sci. 6, 752.

Loudet, O., Gaudon, V., Trubuil, A., Daniel-Vedele, F., 2005. Quantitative trait loci controlling root growth and architecture in *Arabidopsis thaliana* confirmed by heterogeneous inbred family. Theor. Appl. Genet. 110, 742–753.

Ma, Z., Walk, T.C., Marcus, A., Lynch, J.P., 2001. Morphological synergism in root hair length, density, initiation and geometry for phosphorus acquisition in *Arabidopsis thaliana*: a modeling approach. Plant Soil 236, 221–235.

Mackay, A.D., Barber, S.A., 1985. Effect of soil moisture and phosphate level on root hair growth of corn roots. Plant Soil 86, 321–331.

Mackay, A.D., Barber, S.A., 1987. Effect of cyclic wetting and drying of a soil on root hair growth of maize roots. Plant Soil 104, 291–293.

Marschner, H., 1993. Zinc uptake from soils. In: Robson, A.D. (Ed.), Zinc in Soils and Plants. Kluwer Academic Publishers, Dordrecht, The Netherlands, pp. 59–77.

Mason, S., McNeill, A., McLaughlin, M.J., Zhang, H., 2010. Prediction of wheat response to an application of phosphorus under field conditions using diffusive gradients in thin-films (DGT) and extraction methods. Plant Soil 337, 243–258.

Mason, S.D., McLaughlin, M.J., Johnston, C., McNeill, A., 2013. Soil test measures of available P (Colwell, resin and DGT) compared with plant P uptake using isotope dilution. Plant Soil 373, 711–722.

McGrath, S.P., Sanders, J.R., Shalaby, M.H., 1988. The effect of soil organic matter levels on soil solution concentrations and extractabilities of manganese, zinc and copper. Geoderma 42, 177–188.

McKay Fletcher, D.M., Shaw, R., Sánchez-Rodríguez, A.R., Daly, K.R., Van Veelen, A., Jones, D.L., et al., 2019. Quantifying citrate-enhanced phosphate root uptake using microdialysis. Plant Soil 461, 68–89.

Mengel, K., Hütsch, B., Kane, Y., 2006. Nitrogen fertilizer application rates on cereal crops according to available mineral and organic soil nitrogen. Eur. J. Agron. 24, 343–348.

Meuser, H., 1991. Bodenkundliche Aspekte bei Wurzeluntersuchungen an Kulturpflanzen. Die Geowissenschaf. 9, 247–250.

Miao, S.Y., DeLaune, R.D., Jugsujinda, A., 2006. Influence of sediment redox conditions on release/solubility of metals and nutrients in a Louisiana Mississippi River deltaic plain freshwater lake. Sci. Total Environ. 371, 334–343.

Misra, R.K., Alston, A.M., Dexter, A.R., 1988. Role of root hairs in phosphorus depletion from a macro-structured soil. Plant Soil 107, 11–18.

Mollier, A., De Willigen, P., Heinen, M., Morel, C., Schneider, A., Pellerin, S., 2008. A two-dimensional simulation model of phosphorus uptake including crop growth and P-response. Ecol. Model. 210, 453–464.

Moody, P.W., Bell, M.J., 2006. Availability of soil potassium and diagnostic soil tests. Aust. J. Soil Res. 44, 265–275.

Moraghan, J.T., Mascagni Jr, H.J., 1991. Environmental and soil factors affecting micronutrient deficiencies and toxicities. In: Mortvedt, J.J., Cox, F.R., Shuman, L.M., Welch, R.M. (Eds.), Micronutrients in Agriculture. SSSA Book Series, No. 4. Soil Science Society of America, Madison, WI, pp. 371–425.

Moran, C.J., Pierret, A., Stevenson, A.W., 2000. X-ray absorption and phase contrast imaging to study the interplay between plant roots and soil structure. Plant Soil 223, 99–115.

Morel, J.L., Habib, L., Plantureux, S., Guckert, A., 1991. Influence of maize root mucilage on soil aggregate stability. Plant Soil 136, 111–119.

Nemeth, K., 1982. Electro-ultrafiltration of aqueous soil suspension with simultaneously varying temperature and voltage. Plant Soil 64, 7–23.

Neuhaus, A., Easton, J., Walker, C., 2015. Phosphorus requirements for cereals: what role does crop rotation play? Better Crops 99, 20–22.

Neumann, G., Massonneau, A., Langlade, N., Dinkelaker, B., Hengeler, C., Römheld, V., et al., 2000. Physiological aspects of cluster root function and development in phosphorus-deficient white lupin (*Lupinus albus* L.). Ann. Bot. 85, 909–919.

Nigussie, D., Schenk, M.K., Claassen, N., Steingrobe, B., 2003. Phosphorus efficiency of cabbage (*Brassica oleracea* L. var. *capitata*), carrot (*Daucus carota* L.), and potato (*Solanum tuberosum* L.). Plant Soil 250, 215–224.

Nakhforoosh, A., Nagel, K.A., Fiorani, F., Bodner, G., 2021. Deep soil exploration vs. topsoil exploitation: distinctive rooting strategies between wheat landraces and wild relatives. Plant Soil 459, 397–421.

Nye, P.H., Tinker, P.B., 1977. Solute Movements in the Root-Soil System. Blackwell, Oxford, UK.

Oliveira, E.M.M., Ruiz, H.A., Alvarez, V.V.H., Ferreira, P.A., Costa, F.O., Almeida, I.C.C., 2010. Nutrient supply by mass flow and diffusion to maize plants in response to soil aggregate size and water potential. Rev. Brasil. Ciên. Solo 34, 317–328.

Pankhurst, C.E., Pierret, A., Hawke, B.G., Kirby, J.M., 2002. Microbiological and chemical properties of soil associated with macropores at different depths in a red-duplex soil in NSW Australia. Plant Soil 238, 11–20.

Pierret, A., Doussan, C., Garrigues, E., McKirby, J., 2003. Observing plant roots in their environment: current imaging options and specific contribution of two-dimensional approaches. Agronomie 23, 471–479.

Piri, M., Sepehr, E., Rengel, Z., 2019. Citric acid decreased and humic acid increased Zn sorption in soils. Geoderma 341, 39–45.

Postma, J.A., Black, C.K., 2021. Advances in root architectural modeling. In: Gregory, P. (Ed.), Understanding and Improving Crop Root Function. Burleigh Dodds Science Publishing Limited, Cambridge, UK, pp. 3–32.

Postma, J.A., Kuppe, C., Owen, M.R., Mellor, N., Griffiths, M., Bennett, M.J., et al., 2017. OpenSimRoot: widening the scope and application of root architectural models. New Phytol. 215, 1274–1286.

Rabbi, S.M.F., Tighe, M.K., Flavel, R.J., Kaiser, B.N., Guppy, C.N., Zhang, X.X., et al., 2018. Plant roots redesign the rhizosphere to alter the three-dimensional physical architecture and water dynamics. New Phytol. 219, 542–550.

Rao, C.S., Rao, A.S., Swarup, A., Bansal, S.K., Rajagopal, V., 2000. Monitoring the changes in soil potassium by extraction procedures and electroultrafiltration (EUF) in a Tropaquept under twenty years of rice-rice cropping. Nutr. Cycl. Agroecosys. 56, 277–282.

Read, D.B., Gregory, P.J., Bell, A.E., 1999. Physical properties of axenic maize root mucilage. Plant Soil 211, 87–91.

Rengel, Z., Damon, P.M., 2008. Crops and genotypes differ in efficiency of potassium uptake and use. Physiol. Plant. 133, 624–636.

Roose, T., Kirk, G.J.D., 2009. The solution of convection–diffusion equations for solute transport to plant roots. Plant Soil 316, 257–264.

Rose, T.J., Rengel, Z., Ma, Q., Bowden, J.W., 2008. Hydraulic lift by canola plants aids P and K uptake from dry topsoil. Aust. J. Agric. Res. 59, 38–45.

Samal, D., Kovar, J.L., Steingrobe, B., Sadana, U.S., Bhadoria, P.S., Claassen, N., 2010. Potassium uptake efficiency and dynamics in the rhizosphere of maize, wheat, and sugar beet evaluated with a mechanistic model. Plant Soil 332, 105–121.

Sanders, J.R., 1983. The effect of pH on the total and free ionic concentrations of manganese, zinc and cobalt in soil solutions. J. Soil Sci. 34, 315–323.

Scanlan, C.A., Huth, N.I., Bell, R.W., 2015. Simulating wheat growth response to potassium availability under field conditions with sandy soils. I. Model development. Field Crops Res. 178, 109–124.

Schnepf, A., Roose, T., Schweiger, P., 2008. Impact of growth and uptake patterns of arbuscular mycorrhizal fungi on plant phosphorus uptake — a modelling study. Plant Soil 312, 85–99.

Scott-Russell, R., 1977. Plant Root Systems: Their Function and Interaction with the Soil. McGraw-Hill, New York.

Seward, P., Barraclough, P.B., Gregory, P.J., 1990. Modelling potassium uptake by wheat (*Triticum aestivum*) crops. Plant Soil 124, 303–307.

Shah, A.N., Tanveer, M., Shahzad, B., Yang, G.Z., Fahad, S., Ali, S., et al., 2017. Soil compaction effects on soil health and crop productivity: an overview. Environ. Sci. Poll. Res. 24, 10056–10067.

Shenker, M., Huang, X., 2001. Potassium availability indices and plant response. In: Horst, W., Schenk, M.K., Bürkert, A., Claassen, N., Frommer, W.B., Goldbach, H., et al., (Eds.), Plant Nutrition — Food Security and Sustainability of Agro-ecosystems Through Basic and Applied Research. Kluwer Academic Publishers, Dordrecht, The Netherlands, pp. 742–743.

Sims, J.T., Johnson, G.V., 1991. Micronutrient soil tests. In: Mortvedt, J.J., Cox, F.R., Shuman, L.M., Welch, R.M. (Eds.), Micronutrients in Agriculture. 2nd ed. Soil Science Society of America, Madison, WI, pp. 427–476.

Sinclair, A.H., Mackie-Dawson, L.A., Linehan, D.L., 1990. Micronutrient inflow rates and mobilization into soil solution in the root zone of winter wheat (*Triticum aestivum* L.). Plant Soil 122, 143–146.

Soltanpour, P.N., El Gharous, M., Azzaouri, A., Abdelmonum, M., 1989. A soil test based N recommendation model for dryland wheat. Commun. Soil Sci. Plant Anal. 20, 1053–1068.

Soriano, J.J., Alvaro, F., 2019. Discovering consensus genomic regions in wheat for root-related traits by QTL meta-analysis. Sci. Reports 9, 10537.

Steingrobe, B., Schmid, H., Gutser, R., Claasen, N., 2001. Root production and root mortality of winter wheat grown on sandy and loamy soils in different farming systems. Biol. Fertil. Soils 33, 331–339.

Strebel, O., Duynisveld, W.H.M., 1989. Nitrogen supply to cereals and sugar beet by mass flow and diffusion on a silty loam soil. Z. Pflanzenernähr. Bodenk. 152, 135–141.

Strebel, O., Grimme, H., Renger, M., Fleige, H., 1980. A field study with nitrogen-15 of soil and fertilizer nitrate uptake and of water withdrawal by spring wheat. Soil Sci. 130, 205–210.

Strebel, O., Duynisveld, W.H.M., Grimme, H., Renger, M., Fleige, H., 1983. Wasserentzug durch Wurzeln und Nitratanlieferung (Massenfluss, Diffusion) als Funktion von Bodentiefe und Zeit bei einem Zuckerrübenbestand. Mitt. Dtsch. Bodenkd. Ges. 38, 153–158.

Streda, T., Haberle, J., Klimesova, J., Klimek-Kopyra, A., Stoedova, H., Bodner, G., et al., 2020. Field phenotyping of plant roots by electrical capacitance - a standardized methodological protocol for application in plant breeding: a review. Int. Agrophys. 34, 173–184.

Tang, C., Rengel, Z., Abrecht, D., Tennant, D., 2002. Aluminium-tolerant wheat uses more water and yields higher than aluminium-sensitive one on a sandy soil with subsurface acidity. Field Crops Res. 78, 93–103.

Tarafdar, J.C., Jungk, A., 1987. Phosphatase activity in the rhizosphere and its relation to the depletion of soil organic phosphorus. Biol. Fertil. Soils 3, 199–204.

Thieme, J., Sedlmair, J., Gleber, S.C., Prietzel, J., Coates, J., Eusterhues, K., et al., 2010. X-ray spectromicroscopy in soil and environmental sciences. J. Synchrotron Radiat. 17, 149–157.

Tinker, B.T., Nye, P.H., 2000. Solute Movement in the Rhizosphere. Oxford University Press, Oxford, UK.

Tracy, S.R., Roberts, J.A., Black, C.R., McNeill, A., Davidson, R., Mooney, S.J., 2010. The X-factor: visualizing undisturbed root architecture in soils using X-ray computed tomography. J. Exp. Bot. 61, 311–313.

Valizadeh, G.R., Rengel, Z., Rate, A.W., 2003. Response of wheat genotypes efficient in P utilisation and genotypes responsive to P fertilisation to different P banding depths and watering regimes. Aust. J. Agric. Res. 54, 59–65.

Van Noordwijk, M., Kooistra, J.J., Boone, F.R., Veen, B.W., Schoonderbeek, D., 1992. Root-soil contact of maize, as measured by a thin-section technique. I. Validity of the method. Plant Soil 139, 108–118.

Veen, B.W., Van Noordwijk, M., De Willigen, P., Boone, F.R., Kooistra, M.J., 1992. Root-soil contact maize, as measured by a thin-section technique. III. Effects on shoot growth, nitrate and water uptake efficiency. Plant Soil 139, 131–138.

Vincent, C.D., Gregory, P.J., 1989. Effects of temperature on the development and growth of winter wheat roots. II. Field studies of temperature, nitrogen and irradiance. Plant Soil 199, 99–110.

Wallace, A., 1980a. Effect of excess chelating agent on micronutrient concentrations in bush beans grown in solution culture. J. Plant Nutr. 2, 163–170.

Wallace, A., 1980b. Effect of chelating agents on uptake of trace metals when chelating agents are supplied to soil in contrast to when they are applied to solution culture. J. Plant Nutr. 2, 171–175.

Wang, Z.Y., Kelly, J.M., Kovar, J.L., 2005. Depletion of macro-nutrients from rhizosphere soil solution by juvenile corn, cottonwood, and switchgrass plants. Plant Soil 270, 213–221.

Wang, X., Lester, D.W., Guppy, C.N., Lockwood, P.V., Tang, C., 2007. Changes in phosphorus fractions at various soil depths following long-term P fertiliser application on a Black Vertosol from south-eastern Queensland. Aust. J. Soil Res. 45, 524–532.

Watt, M., Evans, J., 1999. Linking development and determinacy with organic acid efflux from proteoid roots of white lupin grown with low phosphorus and ambient or elevated atmospheric CO_2 concentration. Plant Physiol. 120, 705–716.

Watt, M., McCully, M.E., Canny, M.J., 1994. Formation and stabilisation of rhizosheaths of *Zea mays* L. Plant Physiol. 106, 179–186.

Wiesler, F., Horst, W.J., 1993. Differences among maize cultivars in the utilization of soil nitrate and the related losses of nitrate through leaching. Plant Soil 151, 193–203.

Williams, S.M., Weil, R.R., 2004. Crop cover root channels may alleviate soil compaction effects on soybean crop. Soil Sci. Soc. Am. J. 68, 1403–1409.

Wong, M.T.F., Edwards, N.K., Barrow, N.J., 2000. Accessibility of subsoil potassium to wheat grown on duplex soils in the south-west of Western Australia. Aust. J. Soil Res. 38, 745–751.

Yang, X., Post, W.M., 2011. Phosphorus transformations as a function of pedogenesis: A synthesis of soil phosphorus data using Hedley fractionation method. Biogeosciences 8, 2907–2916.

Young, I.M., 1995. Variation in moisture contents between bulk soil and rhizosheath of wheat (*Triticum aestivum* L. cv Wembley). New Phytol. 130, 135–139.

Zhang, L., Ju, X., Gao, Q., Zhang, F., 2007. Recovery of ^{15}N-labeled nitrate injected into deep subsoil by maize in a calcareous alluvial soil on the North China Plain. Commum. Soil Sci. Plant Anal. 38, 1563–1577.

Zhang, D., Zhang, C., Tang, X., Li, H., Zhang, F., Rengel, Z., et al., 2016. Increased soil phosphorus availability induced by faba bean root exudation stimulates root growth and phosphorus uptake in neighbouring maize. New Phytol. 209, 823–831.

Zhang, D., Lyu, Y., Li, H., Tang, X., Hu, R., Rengel, Z., et al., 2019. Neighbouring plants modify maize root foraging for phosphorus: coupling nutrients and neighbours for improved nutrient-use efficiency. New Phytol. 226, 244–253.

Zhu, J., Zhang, C., Lynch, J.P., 2010. The utility of phenotypic plasticity of root hair length for phosphorus acquisition. Funct. Plant Biol. 37, 313–322.

Chapter 13

Genetic and environmental regulation of root growth and development

Peng Yu[1] and Frank Hochholdinger[2]

[1]*Emmy Noether Root Functional Biology Group, Faculty of Agriculture, Institute for Crop Science and Resource Conservation (INRES), University of Bonn, Bonn, Germany,* [2]*Crop Functional Genomics, Faculty of Agriculture, Institute for Crop Science and Resource Conservation (INRES), University of Bonn, Bonn, Germany*

Summary

Roots of vascular plants serve important functions in anchorage and the acquisition of nutrients and water from soil. Root systems continuously adjust their architectural, physiological, and metabolic performance to adapt to alterations of their biotic and abiotic environment. In this chapter, we highlight the current knowledge of the genetic control and molecular regulation of the formation and function of individual root types and whole root systems. We summarize the role of phytohormone homeostasis and the effects of environmental signals, such as nutrient and water availability as well as soil physical and chemical factors, on root development. In addition, we discuss alterations of root morphology and architecture during the interaction with beneficial microbes, such as plant growth–promoting rhizobacteria and arbuscular mycorrhizal fungi, and illustrate the importance of phytohormone production and signaling crosstalk between microorganisms and plant roots. The chapter concludes with a special emphasis on the phytohormone auxin as the key biological molecule mediating growth and development of the root system and root responses to environmental variables.

13.1 General

Plant roots make up between a third and a half of the global plant biomass (Robinson, 2007). The major functions of root systems are to extract water and nutrients from soil, thus linking the soil environment to the energy-delivering green parts of the plant. Root systems are complex three-dimensional structures composed of different root types that form at different stages of plant development. Plant root system architecture varies widely within and among plant species and is regulated by an endogenous genetic program and temporary and spatially fluctuating external triggers. To exploit limited and heterogeneously distributed soil resources and to respond to alterations of the biotic and abiotic environment, root systems continuously adjust their architecture and their physiological, biochemical, and metabolic performance. This enormous developmental plasticity is possible because plant root systems are not fully predetermined during embryogenesis and are able to form new root types or components such as root hairs throughout their life cycle (Leyser and Day, 2003).

13.2 Genetic control of root growth and development

13.2.1 Root system architecture

The evolution and adaptation of roots was an important prerequisite for the successful colonization of land by early plants (Lynch, 1995; Pires and Dolan, 2012). The establishment of an elaborate root system plays a major role in plant fitness, crop performance, and grain yield (Lynch, 1995; Rogers and Benfey, 2015). In general, there are two categories of roots: embryonic and postembryonic. The root system of the dicot model plant *Arabidopsis thaliana* forms only one major root type, the primary root, which is established during embryogenesis (Schiefelbein et al., 1997; Fig. 13.1).

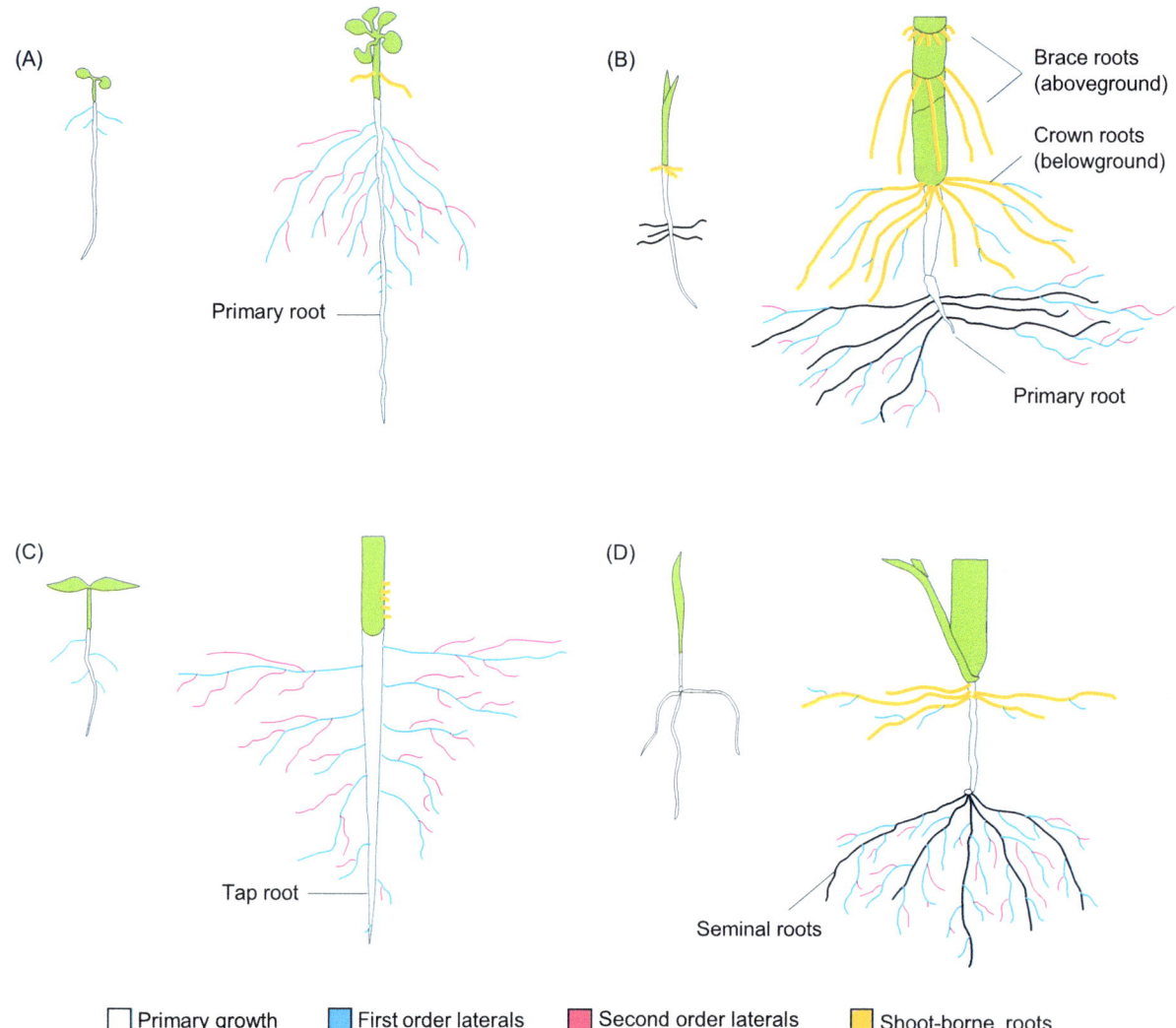

FIGURE 13.1 Diversity in root system architecture in the seedling (left) and mature (right) stages in eudicots (A, arabidopsis; C, tomato) and monocots (B, maize; D, wheat). *Modified from Atkinson et al. (2014), with permission from Oxford University Press.*

By contrast, monocots, such as cereals, display a more complex root structure comprising additional root types (Hochholdinger et al., 2004; Smith and De Smet, 2012; Fig. 13.1). A common characteristic of all cereals is the formation of shoot-borne roots from consecutive underground and aboveground stem nodes. In addition, some (but not all) cereals form a variable number of seminal roots, which are already preformed during embryogenesis (Yu et al., 2018a; Fig. 13.1). All root types form postembryonic lateral roots and root hairs (Hochholdinger et al., 2018). The overall morphology of the root system is consistent within a species, whereas the spatial configuration of root system architecture such as number, position and growth rate of individual roots is highly variable, even among genetically identical plants.

The major difference between the root systems of cereals and arabidopsis is that the embryonic root system in cereals is functionally important only during the early stages of plant development (Hochholdinger et al., 2018). In cereals, an extensive postembryonic shoot-borne root system makes up a major backbone of the root system within a few weeks after germination, whereas in arabidopsis the embryonic primary root remains dominant throughout the whole life cycle (Hochholdinger et al., 2004). In addition to the root types that are part of the basic blueprint of a plant root system, adventitious roots are formed under stress conditions (e.g., flooding, nutrient deprivation, or wounding) at unusual positions such as the mesocotyl (Steffens and Rasmussen, 2016). Variation among species reflects the evolution of root systems from holdfasts in algae, to simple rhizoids in primitive land plants, to the increasingly complex root systems of ferns, gymnosperms, and angiosperms (Lynch, 2005). Variation in root traits among genotypes of a given species likely reflects adaptation to diverse soil environments because root traits have a strong influence on water and nutrient acquisition.

13.2.2 Root anatomy and structure: from arabidopsis to crops

Root formation is a multistep process, including the initiation, elongation, and development of new root axes. The anatomical organization of root axes varies to a certain extent among different plant species but follows a general pattern (Hochholdinger et al., 2004). The root cap protects the root apical meristem and facilitates root penetration through soil by the production of mucilage. The root apical meristem produces cells that differentiate to form the primary tissues of the root axis: epidermis, cortex, endodermis, and various tissues in the stele, including xylem, phloem, parenchyma, sclerenchyma, and collenchyma, with their associated specialized cell types (Evert and Eichorn, 2007; Fig. 13.2). In plants that produce secondary growth, such as dicotyledonous angiosperms and gymnosperms, root development continues by production of new cells from the vascular cambium, leading to radial thickening and woody roots (Spicer and Groover, 2010). The production of lateral branches and root hairs contributes significantly to the absorbing surface for water and nutrient acquisition.

Cereal and arabidopsis roots show distinct anatomies (Hochholdinger et al., 2004; Fig. 13.2). First, different tissue organization is observed in radial patterning. The tissue of maize and rice roots consists of 8–15 layers of cortical cells and one endodermal cell layer (Coudert et al., 2010), whereas young arabidopsis roots form only one endodermal and one cortical cell layer. The radial number of cortical cells in maize and rice is variable, whereas the single arabidopsis cortex cell layer contains a fixed number of eight cells. Furthermore, cereal roots show some differences from arabidopsis roots longitudinally beyond the common organization into root cap, meristematic, elongation, and differentiation zones (Ishikawa and Evans, 1995). In maize and rice the quiescent center (QC), a central region of the root tip with reduced mitotic activity, consists of 800–1200 cells (Jiang et al., 2003) and is surrounded by the proximal and distal meristems that have several hundred cells (Feldman, 1994). These meristematic cells initiate the emerging root and the root cap. Arabidopsis forms a small QC that always contains four cells surrounded by a limited number of initial cells

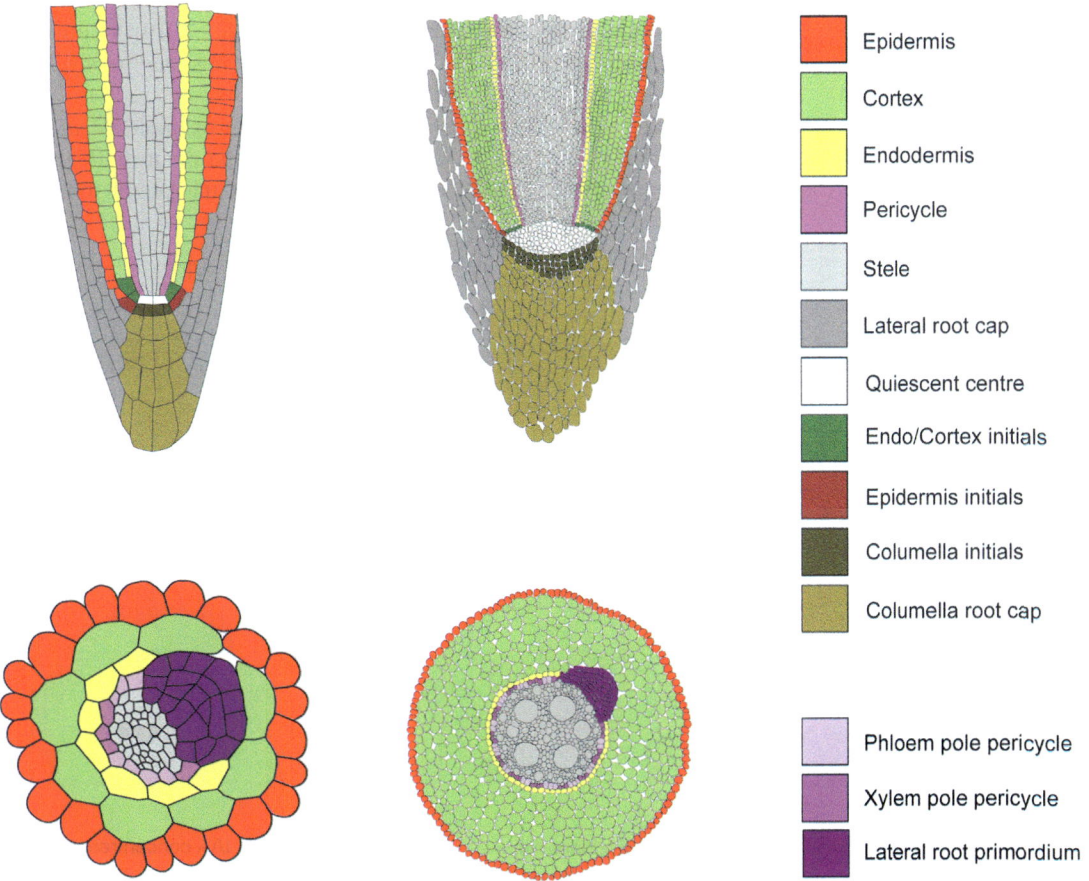

FIGURE 13.2 Longitudinal and transverse root sections in the two model species arabidopsis and maize. For the longitudinal sections, see Yu et al. (2016). *Adapted from Péret et al. (2009a), with permission from Oxford University Press.*

that give rise to the surrounding root tissues (Schiefelbein et al., 1997). The functional analysis of the arabidopsis QC using laser ablation of individual QC cells has demonstrated that the major function of the QC is the maintenance of stem cells by suppressing the differentiation of initial cells (Van den Berg et al., 1997). The functional dissection of the maize and rice QCs via laser ablation experiments is difficult due to their large size. However, it has been demonstrated that the cells in and around the QC contain patterning information that enables maize root apices to be regenerated after the complete removal of the root cap and the QC (Kerk and Feldman, 1994). In maize, the inactivity of these QC cells, and therefore the establishment of the QC, is mainly due to the maintenance of a highly oxidized environment and the root cap containing the maximum auxin concentration, which prevents cell division (Jiang et al., 2003).

13.2.3 Embryonic and postembryonic root branching

Root branching is a major determinant of root system architecture, and its plasticity underpins a plant potential to adapt to the heterogeneity of the environment. Increased root branching in crops enables them to exploit soil efficiently for nutrients and to increase stress tolerance, improving yield while decreasing the need for heavy fertilizer application (Atkinson et al., 2014; Smith and De Smet, 2012). Monocots, such as maize, form various types of roots that can all proliferate lateral roots. However, maize mutants that do not form lateral roots on the embryonic but only on the postembryonic roots indicate that at least part of the genetic program necessary for lateral root formation is root-type-specific (Hochholdinger and Feix, 1998; Woll et al., 2005). Moreover, in maize, several genes are known that control the formation of various root types, which illustrates the complex regulation of root system development in cereals (Hetz et al., 1996; Woll et al., 2005).

In dicotyledonous plants, such as arabidopsis, the root system develops mainly postembryonically by the formation of numerous lateral roots from the primary root established during embryogenesis. These new lateral roots are comparable to the primary root in their anatomical structure and can initiate additional lateral roots. In arabidopsis, lateral roots are formed exclusively from pericycle cells that are unique in their capacity to be reprogrammed into stem cells (Dubrovsky et al., 2000; Fig. 13.3). This has evolved only in the Brassicaceae (Xiao et al., 2019). In other seed plants, the formation of lateral root primordia occurs by mitotic activation of the cortex, endodermis, and pericycle cells, which is a common and ancestral trait.

The endodermis can be reprogrammed into lateral root stem cells in some species, for example, *Medicago truncatula* and pea (Xiao et al., 2019). In maize and probably also in rice, the endodermis of the main roots gives rise to the epidermis and columella, whereas the pericycle contributes to the remaining tissues of the lateral root (Fahn, 1990).

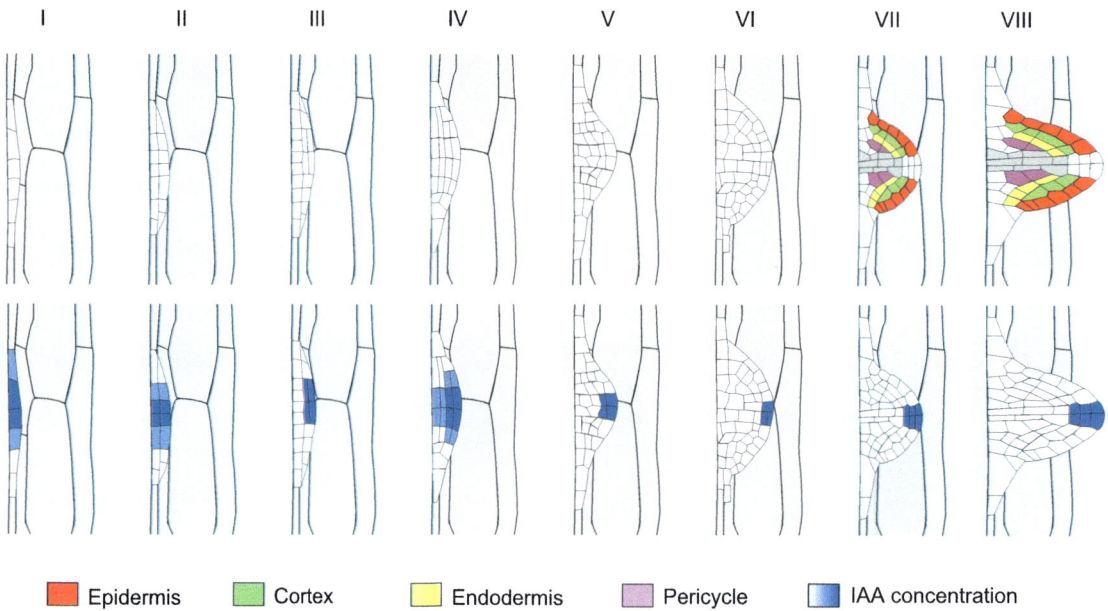

FIGURE 13.3 Morphological changes during lateral root development in *Arabidopsis*. Lateral root primordia development (top row) is divided into eight stages (I–VIII), and the increased auxin concentrations are highlighted in blue in the bottom row. *Adapted from Péret et al. (2009b), with permission from Springer Nature.*

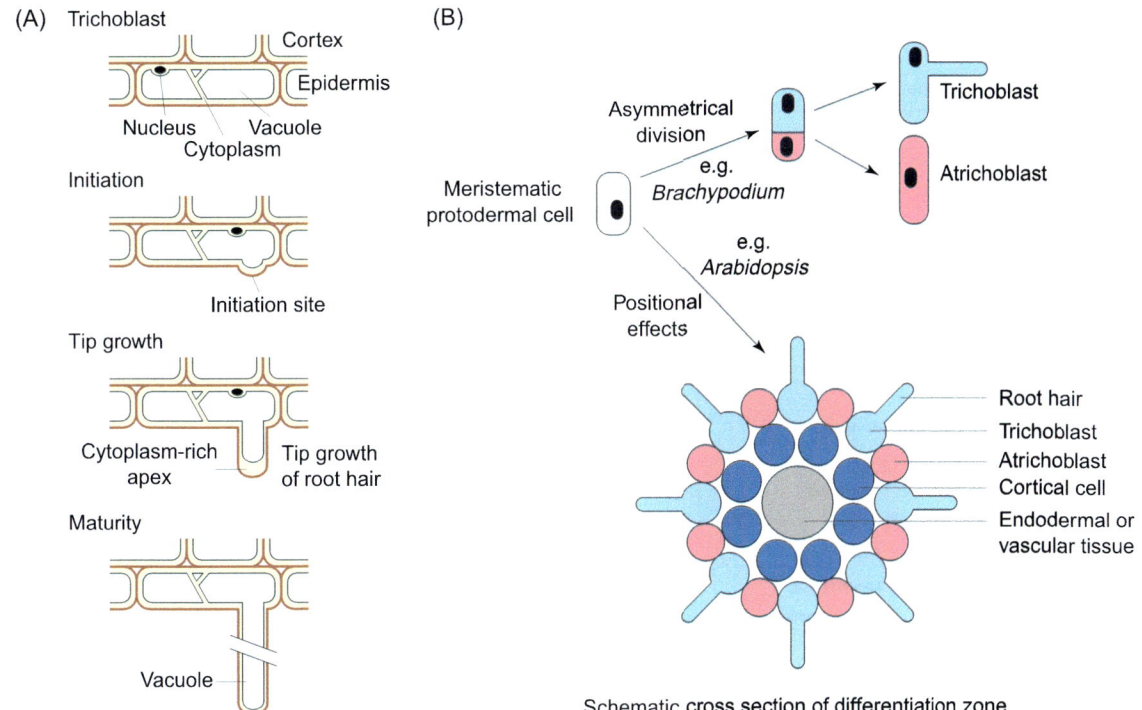

FIGURE 13.4 Patterns of root hair development. (A) Stages of root hair development in trichoblasts. Note the nuclear movement accompanying root hair emergence and changes in the organization of the cytoplasm at each new developmental phase. (B) Divergence of a root epidermal cell into trichoblast or atrichoblast might be determined by asymmetrical division and differences in subsequent differentiation of the daughter cells or by positional effects relative to underlying cell layers. *Reproduced from Gilroy and Jones (2000), with permission from Elsevier.*

Root hairs are tubular outgrowths of epidermal cells that substantially extend the absorbing surface of the root system. They play an essential function in nutrient acquisition and are particularly important for the uptake of the poorly mobile nutrient ions (e.g., mono- and dihydrogen phosphates) (Gilroy and Jones, 2000). Moreover, root hairs and root exudation of adhesive molecules are involved in rhizosheath formation, which contributes to plant adaptation to abiotic and biotic stresses such as drought tolerance, nutrient deficiency, mechanical impedance, and pathogens (Pang et al., 2017).

Root hairs arise from epidermal cells. The epidermis comprises two types of cells: trichoblasts and atrichoblasts (Fig. 13.4). Only trichoblasts give rise to tubular extensions that develop into root hairs. Root hair formation is a three-step process that is initiated by cell swelling to form a bulge, followed by initiation of tip growth and elongation through polarized exocytosis (Dolan, 2017; Fig. 13.4A).

Different ways of epidermal cells differentiating into trichoblasts and atrichoblasts exist in plants. In arabidopsis, only epidermal cells that are situated in a cleft between two underlying cortical cells become trichoblasts and form root hairs (Fig. 13.4B). By contrast, maize epidermal cells can differentiate into trichoblasts irrespective of their position (Marzec et al., 2015).

13.2.4 Phytohormonal control of root growth and development

Among the intrinsic regulators, phytohormones play a central role in root growth and development. Different phytohormones target distinct root tissues and regions in arabidopsis (Fig. 13.5). Gibberellic acid targets the endodermis as the primary response tissue that controls cell elongation and division (Ubeda-Tomás et al., 2008, 2009). Auxin targets elongating epidermal cells during the gravitropic response (Swarup et al., 2005) and regulates cell division in the meristem and the stem cell niche (Blilou et al., 2005). Cytokinin promotes vascular differentiation in the transition zone (Ioio et al., 2007) and antagonizes the effect of auxin on cell division in the transition zone to control root meristem size (Ioio et al., 2008). Brassinosteroid targets the epidermis (Hacham et al., 2011) in the meristematic zone, whereas abscisic acid (ABA) acts on QC and stem cells to regulate root meristem size (Zhang et al., 2010). Ethylene regulates cell division in QC as well as auxin biosynthesis in columella cells (Swarup et al., 2007).

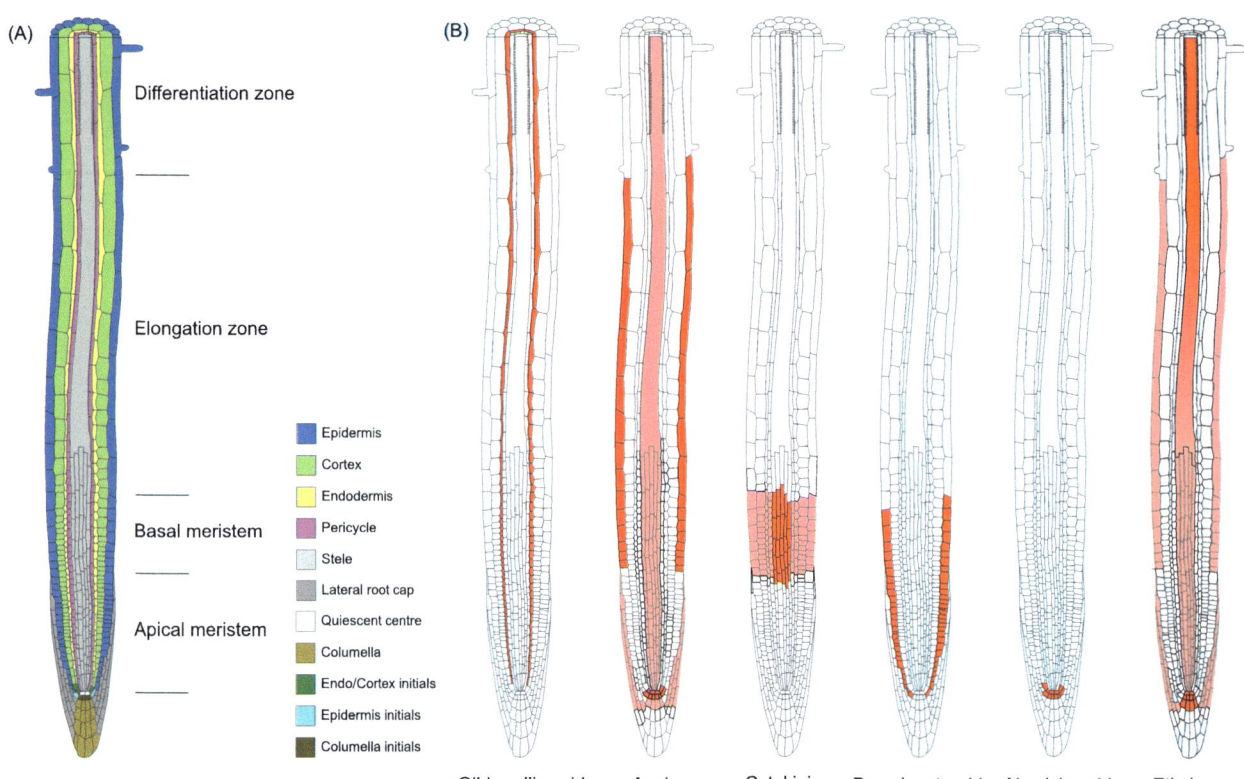

FIGURE 13.5 The major phytohormonal responses in tissues/zones along the arabidopsis primary root. (A) Cell types of the primary arabidopsis root. (B) Direct targets of the specific phytohormones. Red denotes tissues that are direct targets of the designated phytohormones, whereas pink denotes indirect target tissues. *Adapted from Ubeda-Tomás et al. (2012), with permission from Elsevier.*

Auxin is a key phytohormone in the regulation of cell division, cell differentiation, organogenesis, and morphogenesis. For example, numerous root developmental mutants with defects in root system architecture, root angle, or lateral root formation are impaired in auxin-related processes. Moreover, other phytohormones, notably cytokinin and ethylene, often alter root system morphogenesis similarly, that is, by modulating auxin concentration, transport, or signaling via crosstalk between phytohormonal pathways.

13.2.4.1 Auxin and its role in root growth and development

Roots are organized in patterns of functionally different cell types and different developmental zones along the longitudinal axis. Auxin orchestrates almost every aspect of root growth and development (Vanneste and Friml, 2009). A broad range of processes in root system development, ranging from initiation and organization of root apical meristems to initiation of lateral roots, depends on auxin signaling and homeostasis (Benková and Hejátko, 2009). The most typical auxin-associated root phenotypes are dose-dependent changes in root hair length and bimodal effects of auxin concentration on primary root length and formation of lateral roots (Péret et al., 2009a,b). Auxin effects on root growth and development depend on the interplay of its biosynthesis, transport, and signal transduction (McSteen, 2010). The pivotal role of auxin is conserved across the root systems of monocotyledons and dicotyledons (de Dorlodot et al., 2007).

Shoot-derived auxin contributes to root development and organization by polar auxin transport. Auxin distribution is directed based on coordinated activities of several membrane transport protein families that facilitate auxin influx (Paponov et al., 2005). Formation of root hairs is driven by auxin gradients, with auxin import into epidermal cells inhibiting root hair development (Jones et al., 2009). Cellular concentration gradients of auxin in turn contribute to the regulation of gene expression that defines cell fate for root patterning and primordium initiation (Benjamins and Scheres, 2008). The polar auxin transport-triggered auxin maximum in the QC creates a highly organized center and guides asymmetrical division of stem cells and production of new cells for establishment of root meristem (van den Berg et al., 1997). Therefore, pharmacological or genetic disruption of polar auxin transport or auxin movement affects root growth and development, such as patterning of the root tip (e.g., Reed et al., 1998).

The PLETHORA (PLT) genes that encode AP2 transcriptional factors play an important role in defining the root stem cell niche (Aida et al., 2004) by directing auxin distribution and gradients (Blilou et al., 2005). In particular, many genes involved in the synthesis of auxin are expressed in roots, and the root-derived auxin pools contribute to the maintenance of the gradients and the maxima required for normal root development (Ljung et al., 2005). In addition to *de novo* biosynthesis of auxin, the formation and hydrolysis of auxin conjugates provide an additional mechanism to control cellular concentration of free auxin (Ljung et al., 2005). Disturbance of auxin homeostasis can have strong effects on root growth and development, for instance, by impairing initiation of lateral roots and root hairs and root hair elongation (Quint et al., 2009).

Among plant hormones, auxin is the dominant regulator of lateral root development (Benková and Hejátko, 2009; Fig. 13.3). Lateral root initiation in arabidopsis starts with divisions of pericycle cells situated adjacent to the protoxylem pole in response to elevated auxin levels (Himanen et al., 2002). Auxin promotes lateral root initiation by activating the cell cycle-related genes such as cyclins and cyclin-dependent kinases (Himanen et al., 2002). Spacing and positioning of lateral root primordia depend on the priming events of pericycle founder cells in response to auxin maxima in the basal meristem (De Smet et al., 2007). Application of the indole-3-acetic acid (IAA) transport inhibitor N-1-naphthylphthalamic acid (NPA) demonstrated that auxin movement through the root tip is necessary for lateral root initiation (Casimiro et al., 2001). Functional analysis of AUX1 suggests a role of this protein in basipetal auxin transport. Auxin transport in roots is directed from the tip via the outer cell layers toward the basal region (Swarup et al., 2001). Shoot-derived auxin is essential for lateral root emergence, but not for the initiation of lateral roots. Genetic disruption of this auxin flow inhibits lateral root formation (Reed et al., 1998). Lateral root emergence through the cortex and epidermis cell layers that overlie the developing primordium is facilitated by auxin influx (Swarup et al., 2008). In general, distinct auxin transport systems are active at different stages of lateral root development (Fukaki and Tasaka, 2009). Members of the Aux/IAA and ARF families regulate auxin signal transduction transcriptionally (Fukaki et al., 2002) and have distinct and overlapping functions in the regulation of root growth and development (Overvoorde et al., 2005; Tatematsu et al., 2004 Fig. 13.3).

13.2.4.2 Crosstalk between auxin and other phytohormones

Tissue- or cell-specific regulation of auxin biosynthesis, transport, and response is important for understanding the hormonal crosstalk with cytokinin and ethylene in root development. The best-known example of hormonal crosstalk during root growth and development is the balance of the antagonistic activities between auxin and cytokinin signaling during early growth of the primary root and organogenesis of lateral roots. For example, in the arabidopsis root meristem, auxin promotes cell division (Ioio et al., 2008), whereas cytokinin promotes cell differentiation (Perilli et al., 2010). A role of cytokinin in antagonizing auxin action has also been indicated from lateral root phenotypes. Decreased cytokinin levels produce a higher density of lateral roots (Werner et al., 2003), whereas exogenous cytokinin application inhibits lateral root formation (Li et al., 2006). A general theme regarding auxin and cytokinin crosstalk is that they reciprocally regulate their biosynthetic pathways (Ioio et al., 2008). In addition, cytokinin also directs auxin transport. During signal transduction, auxin inhibits the activity of negative regulators of cytokinin signaling, whereas cytokinin signaling promotes the expression of inhibitors of auxin signaling.

Another important phytohormone interacting with auxin is ethylene, a volatile compound that has been identified as a general modulator of root development (Swarup et al., 2002). The most characteristic effect of ethylene in root development is the inhibition of primary root elongation (Swarup et al., 2002) by reducing the expansion rate of cells in the elongation zone of the primary root (Swarup et al., 2007). Transcriptomic analyses of ethylene-treated roots demonstrate that ethylene inhibition of cell elongation is synchronized with the induction of auxin-dependent genes (Markakis et al., 2012). In agreement with this observation, auxin–ethylene crosstalk appears to act mainly by ethylene regulating auxin biosynthesis in arabidopsis roots (Swarup et al., 2007). In general, ethylene increases auxin transport to roots by increasing transcription and translation of auxin efflux transporters in the central cylinder, which may then deplete auxin concentration in the lateral root-forming zone while increasing auxin accumulation in the root apex or the meristematic zone. At the root tip, increases in transcription and translation of AUX1 and PIN2 enhance proximal transport of auxin into the elongation zone, thereby reducing primary root elongation (Muday et al., 2012). Auxin and ethylene stimulate root hair elongation synergistically through their canonical signaling pathways (Rahman et al., 2001). Although both phytohormones stimulate root hair initiation, genetic and pharmacological experiments suggest that the action of ethylene on this process is through enhancing auxin synthesis and signaling (Muday et al., 2012).

13.3 Regulation of root growth and development by environmental cues

Root systems can perceive extrinsic environmental signals and translate them into changes in architecture and physiology throughout their life cycle. Plasticity of lateral root growth and development can readjust root system architecture in response to limited availability of macronutrients, including nitrate and phosphate (Rogers and Benfey, 2015), or water (Robbins and Dinneny, 2015), and in beneficial interactions with plant growth−promoting bacteria (Lugtenberg and Kamilova, 2009) and arbuscular mycorrhizal (AM) fungi (Gutjahr and Paszkowski, 2013). Manipulating genes that influence root system architecture might be pivotal for future global crop improvement.

13.3.1 Nutritional control of root development

As soil mineral nutrients are essential for plant survival and reproduction, root systems have evolved the capacity to adapt to nutrient stresses by eliciting a complex array of (1) morphological changes to capture soil resources and (2) physiological adaptions to enhance uptake of nutrients heterogeneously distributed in soil (Ruiz Herrera et al., 2015). Nutrient supply and distribution can influence the morphology of root systems and the distribution of roots in soil greatly, either directly via changes in the external concentration of the nutrient or indirectly via changing the internal nutrient status of the plant (Forde and Lorenzo, 2001). The direct pathway results in developmental responses in the part of the root directly exposed to a given nutrient supply. The indirect pathway produces systemic responses dependent on long-distance signals derived from the shoot (Forde and Lorenzo, 2001). Root growth and development are sensitive to variations in the supply and distribution of inorganic nutrients as illustrated for arabidopsis plants grown on agar (Gruber et al., 2013). Systematic comparison of root system architectural responses to nutrient deficiencies provides a comprehensive view of the overall changes in root plasticity induced by the deficiency of single nutrients (Gruber et al., 2013; Fig. 13.6). Root development in response to nutrient deficiency is mediated by phytohormones, specific transporter proteins, transcription factors regulating nutrient sensing, and microRNAs and peptides as nutrient signals. In the following section, we discuss the genetic and molecular control of root growth and development by the mineral nutrients nitrogen, phosphorus, potassium, and iron that are critical for agriculture.

13.3.1.1 Nitrogen

Nitrate and ammonium are the major forms of nitrogen acquired by plant roots and can influence root and shoot growth by regulating a series of physiological and metabolic processes (Alvarez et al., 2012). The following paragraphs will start with root system architectural changes in response to homogeneous nitrogen supply and will subsequently discuss

FIGURE 13.6 Arabidopsis root system architecture responses to nitrogen availability. *Reproduced from Giehl and von Wirén (2014), with permission from Oxford University Press.*

localized nutrient supply. Root growth responses to homogenous nitrate supply underlie dose-dependent regulation. During a short-term homogeneous nitrate deficiency, young plants tend to increase carbon partitioning to roots, which accelerates root growth and results in fewer, but longer main roots with longer lateral roots (Gaudin et al., 2011; Zhang et al., 1999; Fig. 13.6). Another general response of roots to systemic nitrate deficiency is the increase in rooting depth (Gastal and Lemaire, 2002) and steeper root angles (Gaudin et al., 2011). Conversely, excessive nitrogen supply inhibits root growth and produces a shallow root system (Durieux et al., 1994; Fig. 13.6). In field studies, genotypes of maize with sparsely spaced and long lateral roots were optimal for nitrate acquisition, whereas those with densely spaced and short lateral roots were optimal for phosphate acquisition (Lynch, 2013).

In arabidopsis a nitrate-responsive gene network, including nitrate transporters, modulates lateral root initiation by controlling cell cycle progression in the pericycle via auxin signaling (Gifford et al., 2008; Vidal et al., 2013). Under relatively mild nitrogen deficiency, auxin biosynthesis is increased at later stages of lateral root primordia development, thus promoting lateral root emergence (Stepanova et al., 2008). In severe nitrogen-deficient plants, CLE (CLAVATA3/ESR-related) peptides move from the pericycle to phloem companion cells, where they interact with CLV1 (CLAVATA1) to inhibit the outgrowth and emergence of lateral roots (Araya et al., 2014).

The stimulation of lateral root development by nitrate-rich patches is a classic example of a nutrient-induced alteration of root system architecture in plants (Drew et al., 1973; Drew and Saker, 1975). In arabidopsis, genetic and molecular analyses using split-root growth systems reveal that adaptive responses of lateral roots to local nitrate supply involve both local and systemic signals (Zhang and Forde, 1998; Zhang et al., 1999). Local nitrate supply modulates lateral root elongation by regulating meristematic activity in the lateral root tip via nitrate transport and signaling (Krouk et al., 2010; Mounier et al., 2014). Lateral root development depends on regulatory networks that integrate both local and systemic signals such as cytokinin and small peptides to coordinate the root responses to nitrate availability, linking nitrate assimilation and signaling with cell cycle progression (Malamy, 2005; Ruffel et al., 2011). Local ammonium and nitrate supplies act synergistically on lateral root proliferation. It has been demonstrated in arabidopsis that ammonium transporter—dependent ammonium uptake is required for higher order lateral root branching (Lima et al., 2010; Meier et al., 2020). These findings suggest that local ammonium supply influences pH-dependent radial auxin mobility and its regulatory function in arabidopsis lateral root development (Meier et al., 2020).

13.3.1.2 Phosphorus

Phosphorus is taken up by plants roots as soluble phosphate that is a key component of many biomolecules, such as nucleic acids and phospholipids, and participates in various enzymatic reactions and metabolic pathways (López-Arredondo et al., 2014). Studies in arabidopsis and crops have shown that typical morphological changes of root system architecture in response to phosphate deficiency are a decrease in primary root length and an increase in the lateral root and root hair length and density (Williamson et al., 2001; Fig. 13.7). On the anatomical level, phosphorus availability typically regulates the size and number of cortical cell files and the formation of aerenchyma in the cortex (Lynch, 2011; Lynch and Brown, 2008). Poorly mobile phosphate is mainly distributed in the topsoil layer and is therefore preferentially accessible by shallow roots (Postma et al., 2014).

Several studies suggest that auxin signaling plays an important role in mediating the phosphate starvation effects on root system architecture (López-Bucio et al., 2002). Genetic analyses suggest that the root cap is involved in sensing phosphate deficiency (Svistoonoff et al., 2007) and that root hairs and lateral roots can increase the root absorptive capacity upon phosphate deficiency (Sánchez-Calderón et al., 2005, 2006; Fig. 13.7). Changes in root hair formation under low-phosphate supply are linked to auxin signaling. Under low-phosphate conditions, auxin is synthesized in the root tip and transported into root hair cells, promoting their elongation (Bhosale et al., 2018). In summary, phosphate-dependent regulation of root growth and development involves monitoring of the environmental phosphate status, maintenance and fine tuning of meristematic activity, and adjustment of root system architecture to maximize phosphate acquisition.

13.3.1.3 Potassium

In general, plant roots respond to low potassium availability with the inhibition of root growth (Hodge, 2004). Several studies highlighted that the concentration and distribution of auxin in plant roots are influenced by external potassium availability (Vicente-Agullo et al., 2004). Moreover, potassium deficiency inhibits auxin transport in the root, which leads to auxin misdistribution and accumulation in central cylinder cells in the upper part of the root tip (Vicente-Agullo et al., 2004). Auxin signaling interacts with potassium transporter proteins that might function as auxin efflux facilitators and regulator of root hair development and root gravitropism (Rigas et al., 2001). Under potassium

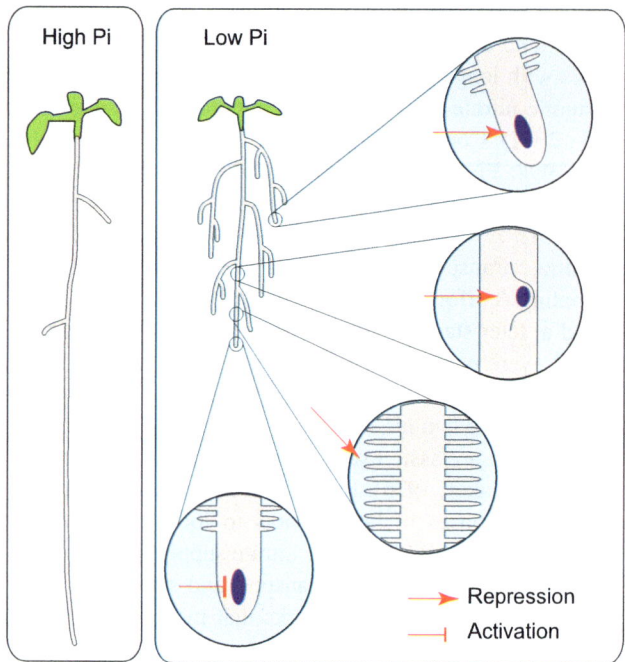

FIGURE 13.7 Phosphorus starvation signaling pathways regulate the primary root, lateral root, and root hair development. *Reproduced from Péret et al. (2011), with permission from Elsevier.*

deprivation, ethylene production and ethylene-related genes are induced (Shin and Schachtman, 2004). These genes play important roles in root hair elongation and primary root growth in arabidopsis in response to potassium deficiency (Jung et al., 2009).

13.3.1.4 Iron

Iron is an essential micronutrient for all organisms as a key component of many enzymes and proteins. Iron deficiency can block the synthesis of chlorophyll precursors, induce leaf yellowing, and reduce photosynthesis (Samira et al., 2013). Typically, root architectural changes in response to low availability of iron include ectopic formation of root hairs due to modulation of their position and abundance (Schmidt and Schikora, 2001). Lateral root elongation is highly responsive to local iron supply, and the high symplastic iron pool in lateral roots favors local auxin accumulation (Giehl et al., 2012). Auxin transport is a major iron-sensitive component in auxin signaling, directing the rootward auxin stream mainly into lateral roots that have access to iron.

13.3.2 Soil physical and chemical factors

13.3.2.1 Mechanical impedance

Roots in soil grow in pores or channels or penetrate the soil matrix. Mechanical impedance describes the resistance of the soil matrix to alterations, and it influences root growth substantially (Bengough et al., 2011). Increased soil impedance is associated with high bulk density, soil drying and soil compaction. Root elongation is related inversely to soil impedance. To cope with soil impedance, plants have evolved several morphological adaptations such as shorter and thicker roots (Rich and Watt, 2013) and stiffened cell walls in the direction of root growth. Moreover, roots display typical physiological responses to soil impedance (such as secretion of mucilage from root tips, increased release of border cells from the root cap, and root exudation) to improve root penetration in soil (Bengough et al., 2011). An increased diameter of roots has the potential to strengthen the mechanical properties via greater axial root growth pressure, radial expansion, and elongation rate (Pagès et al., 2010). Specifically, soil compaction significantly inhibits lateral root formation in tomato (Tracy et al., 2012), wheat (Colombi et al., 2017), triticale and soybean (Colombi and Walter, 2016). The heterogeneous distribution of soil impedance is associated with poorly effective root system architecture, for example, a shallower and narrower root system (Tardieu, 1994), resulting in potential growth defects by hampering water uptake from deeper soil layers. In particular, genetic studies using root hair mutants have revealed positive effects of root hair formation on root penetration rates in crop species such as maize and barley (Haling et al., 2013). An adaptive

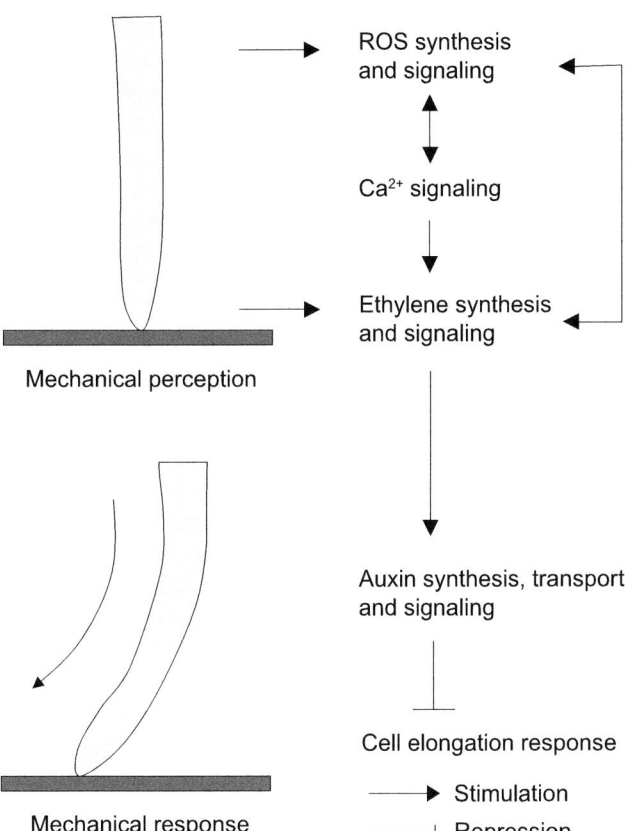

FIGURE 13.8 Pathways involved in the response of the primary root to a barrier.

phenotype to maintain root growth in compacted soil with limited oxygen supply is the formation of root cortical aerenchyma, as shown in triticale and soybean (Colombi and Walter, 2016).

A strategy called "obstacle avoidance" describes the phenomenon of roots bending away from obstacles and finding routes around, thus adapting to the spatially heterogeneous mechanical stress (Massa and Gilroy, 2003). Mechanistic links between asymmetrical auxin responses, calcium signaling, and root morphogenesis have been found in arabidopsis (Fig. 13.8; Lee et al., 2020; Nakagawa et al., 2007). In addition to calcium and auxin, several other signals, such as pH and reactive oxygen species (ROS) have been implicated in the regulation of mechanical sensing (Nakagawa et al., 2007). During root bending, the cytosolic calcium ion concentration increases rapidly in the epidermal cells on the convex side of the root, and the increased calcium ion concentrations are linked with the production of apoplastic ROS and cytosolic acidification that promote cell elongation (Monshausen et al., 2009).

13.3.2.2 Soil temperature

Temperature is one of the most important environmental variables that influences plant growth and development. In natural soils, temperature varies significantly over the seasons and with depth. In the topsoil layer, temperature is highly variable depending on air temperature, time of day, and thermal radiation. By contrast, in deeper soil layers, the temperature is more stable. Alterations of root system architecture in response to changes in soil temperature are species-specific and genotype-dependent, as different species or genotypes often have different optimum temperatures for root growth and development. Morphological traits affected by gradual temperature changes are primary root elongation, radial expansion of roots, and the length of root zones (Pahlavanian and Silk, 1988). In wheat, the diameter of metaxylem vessels was reduced at suboptimal temperatures, thus decreasing the capacity to conduct water (Huang et al., 1991). In arabidopsis, it was demonstrated that a sudden increase of ambient growth temperatures was linked to the increased synthesis of auxin, resulting in an auxin-dependent acceleration of root growth (Hanzawa et al., 2013). Indeed, at high temperature, roots counterbalance the elevated level of intracellular auxin by promoting shootward auxin efflux (Hanzawa et al., 2013).

Various heat stress response factors and heat-shock proteins play central roles in heat sensing and signaling (Kotak et al., 2007) by integrating temperature and auxin signaling to regulate root growth at increased temperatures

(Wang et al., 2016). Interestingly, prolonged elevation of ambient temperature enhances root growth by a brassinosteroid-dependent pathway independently of auxin (Martins et al., 2017). Therefore, phytohormonal pathways involved in the temperature-dependent increase of root growth are regulated by the extent and duration of high-temperature conditions.

Plants exhibit reduced root growth when exposed to low temperature as manifested by the reduction of root elongation (Frey et al., 2020) and branching angles (Nagel et al., 2009). Moreover, because low temperatures inhibit enzyme activity, root respiration is slowed (Covey-Crump et al., 2002). In maize the reduction of root elongation rate at low temperature is genotype-dependent and can decrease to less than 20% of the control growth rate (Frey et al., 2020). At the anatomical level, cold stress affects the lignification of late metaxylem elements and increases the hydraulic conductivity of roots (Huang et al., 1991). Moreover, low temperature reduces root meristem size and cell number (thus repressing the division potential of meristematic cells) by reducing auxin accumulation (Zhu et al., 2015).

Cold stress inhibits root growth partly by modulation of auxin biosynthesis, transport, and signaling. In arabidopsis, reduced root growth is associated with inhibition of root basipetal auxin transport (Shibasaki et al., 2009). Cytokinin is also involved in cold stress signaling and response (Jeon et al., 2016).

13.3.2.3 Soil water

Water deficit results in impaired plant development and inhibition of growth. The development of roots is less sensitive to water deficit than the development of the aboveground vegetative parts (Sharp et al., 1988; Westgate and Boyer, 1985). Maintenance of root growth is an important adaptive trait ensuring that plants can access deep water and nutrient resources to survive (Rodrigues et al., 1995). Deep rooting has been proposed as a key trait for drought tolerance because it permits access to unexploited water resources when the soil surface dries out (Lynch, 2013; Uga et al., 2013). For example, in rice, the gene DEEPER ROOTING 1 (DRO1) determines the root growth angle by favoring deep rooting, thereby enhancing plant performance under drought conditions (Uga et al., 2013).

Across grass species, a water deficit-triggered crown root arrest is a mechanism to conserve water under drought conditions (Sebastian et al., 2016). Substantial phenotypic variation exists in maize for this trait, which might be a useful breeding target to improve drought tolerance (Trachsel et al., 2011). Root system architecture can be reshaped by reduced crown root number, leading to water acquisition by increased rooting depth under drought conditions in maize (Gao and Lynch, 2016).

Water uptake is determined by the intrinsic water transport capacity. Water is first transported from the soil to the stele through concentric layers of root cells, loaded into xylem vessels and then transported axially to the shoot. Therefore several root anatomical traits (such as root cortical aerenchyma, cortical cell file number and cortical senescence) that reduce the metabolic cost of the root cortex can improve soil exploration and therefore water acquisition from drying soil (Lynch et al., 2014). For instance, greater water uptake by the maize crown roots compared with other root types is explained by their higher axial conductivity of xylem tissue in the proximal parts (Ahmed et al., 2018). Quantitative trait loci (QTL) and genome-wide association studies (GWAS) using natural populations have identified major checkpoints of root hydraulic conductivity in arabidopsis (Shahzad et al., 2016; Tang et al., 2018). Root hydraulics and lateral root development are interconnected via auxin-controlled and aquaporin-dependent water flows in lateral primordia, thereby reducing the mechanical resistance of overlaying cells in arabidopsis (Péret et al., 2012).

Water availability exerts multiple short- and long-term effects on root growth and hydraulics in a dose-dependent manner. For instance, moderate water deprivation enhances primary root growth and lateral root formation, whereas strong water deficit exerts inhibitory effects (Jung and McCouch, 2013). Root sensitivity to drought also varies among root types and among different segments of the same root (Rosales et al., 2019). The growth zone of the root tip perceives the spatial distribution of water and influences the developmental competence for lateral root formation (Robbins and Dinneny, 2018).

Abscisic acid has been identified as the main phytohormone that mediates dose-dependent developmental and hydraulic responses of roots to water deficit (Yamaguchi and Sharp, 2010). Moreover, water deficit-triggered continuous root xylem formation and vascular acclimation are associated with endodermal ABA signaling (Ramachandran et al., 2018). Root growth and lateral root patterning along the main root can be oriented toward regions of higher water availability (Bao et al., 2014). Local water deficit is regulated via auxin signaling (Orosa-Puente et al., 2018). Therefore, auxin and ABA-dependent adaptive mechanisms allow lateral roots to respond rapidly to changes in water availability in their local microenvironment (Bao et al., 2014).

13.3.2.4 Soil aeration

Root morphological and anatomical traits determine root growth and functioning in hypoxic waterlogged soils (Voesenek and Bailey-Serres, 2015). Plants combat waterlogging stress by growing adventitious roots, forming aerenchyma and having roots with a greater cortex-to-stele ratio and a smaller surface area-to-volume to enhance the internal

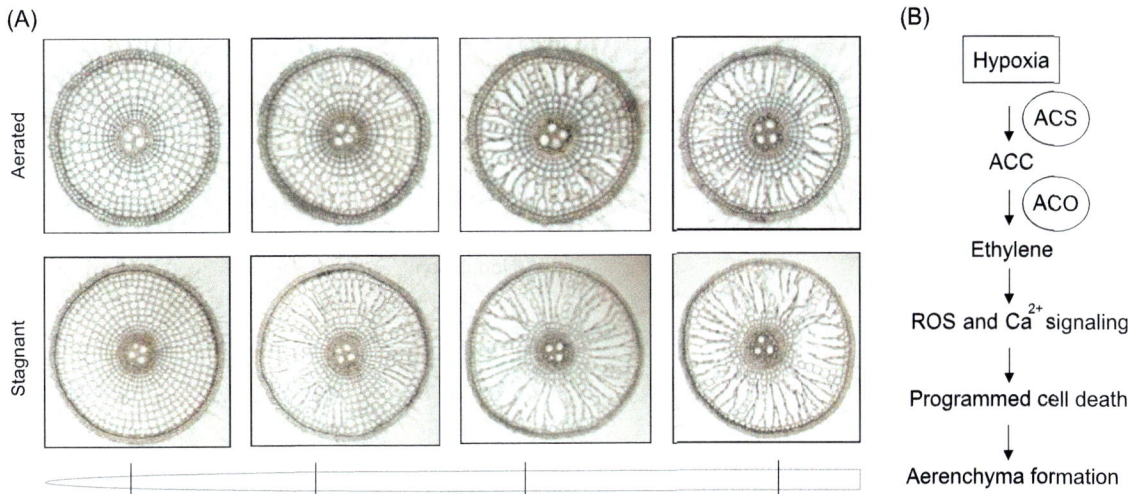

FIGURE 13.9 Rice root morphological and anatomical adaptations to hypoxic soils. (A) Constitutive and inducible aerenchyma formation along rice crown roots grown in aerated and stagnant (hypoxic) conditions. (B) Ethylene signaling pathway is involved in aerenchyma formation in rice. *Part (A) courtesy Dr. Takaki Yamauchi (Nagoya University).*

supply and diffusion of oxygen along the roots (Sauter, 2013) (see also Section 18.4). Adventitious roots improve nutrient uptake and plant fitness, particularly during hypoxic stress caused by long-term waterlogging (Sauter, 2013). Therefore, the formation of adventitious roots and the capacity to form aerenchyma are major quantitative traits of crop tolerance to waterlogging (Mustroph, 2018).

In maize and arabidopsis, survival after long-term waterlogging is regulated via ethylene and auxin signaling (Yu et al., 2019) via adjustment of the root structure under unfavorable oxygen conditions (Eysholdt-Derzsó and Sauter, 2017).

Constitutive aerenchyma formation in roots is of adaptive significance to plants, such as rice, that are faced with transient soil waterlogging (Fig. 13.9A). Constitutive and inducible formation of aerenchyma promotes internal oxygen movement to the root tips (Yamauchi et al., 2014). Cortical cell death underpinning inducible aerenchyma formation is regulated by the production of ethylene and ROS (Wany et al., 2017; Fig. 13.9B). Auxin signaling is required for the formation of both constitutive aerenchyma and lateral roots in rice (Yamauchi et al., 2019).

Induction of a barrier within the outer cell layers preventing radial oxygen loss from the cortex to the rhizosphere further enhances oxygen movement to the growing apex of roots in oxygen-stressed soils (Pedersen et al., 2020). Suberin deposited in the walls of hypodermis/exodermis cells likely strengthens the barrier to the radial oxygen loss (e.g., Armstrong et al., 2000). The barrier to root radial oxygen loss enhances the role of aerenchyma in promoting longitudinal diffusion of oxygen within roots and thus root growth in hypoxic waterlogged soils (Shukla et al., 2019).

13.3.3 Root–soil biotic interactions

It is estimated that 1 g of soil contains up to 10^{10}–10^{11} bacteria (Horner-Devine et al., 2003) and up to 200 m of fungal hyphae (Leake et al., 2004). Most of soil microbes are decomposing the organic matter in the soil, and many soil microorganisms interact with plants and colonize root niches during the plant life cycle. Some of these microorganisms play important roles in plant performance by improving mineral nutrition (Jacoby et al., 2017). Root morphogenesis can be reshaped considerably by microorganisms, including beneficial plant growth–promoting rhizobacteria (PGPR) and AM fungi.

Many microorganisms influence root growth and development by the production of phytohormones, improving nutrient availability and suppressing soil pathogens (Bloemberg and Lugtenberg, 2001). The plant growth–promoting effects of microorganisms depend on how they associate with each other. They can be categorized into rhizospheric and endophytic associations.

13.3.3.1 Beneficial rhizosphere bacteria

One of the best characterized beneficial interactions between bacteria and the roots is the symbiosis between rhizobia and legumes, such as soybean, chickpea, pea, common bean, alfalfa, and clover. Rhizobia are nitrogen-fixing bacteria that elicit the formation of root nodules in which they differentiate into bacteroids (Poole et al., 2018). In this symbiosis, plants provide carbon and energy in the form of dicarboxylic acids, and in return, the bacteroids secrete ammonium that

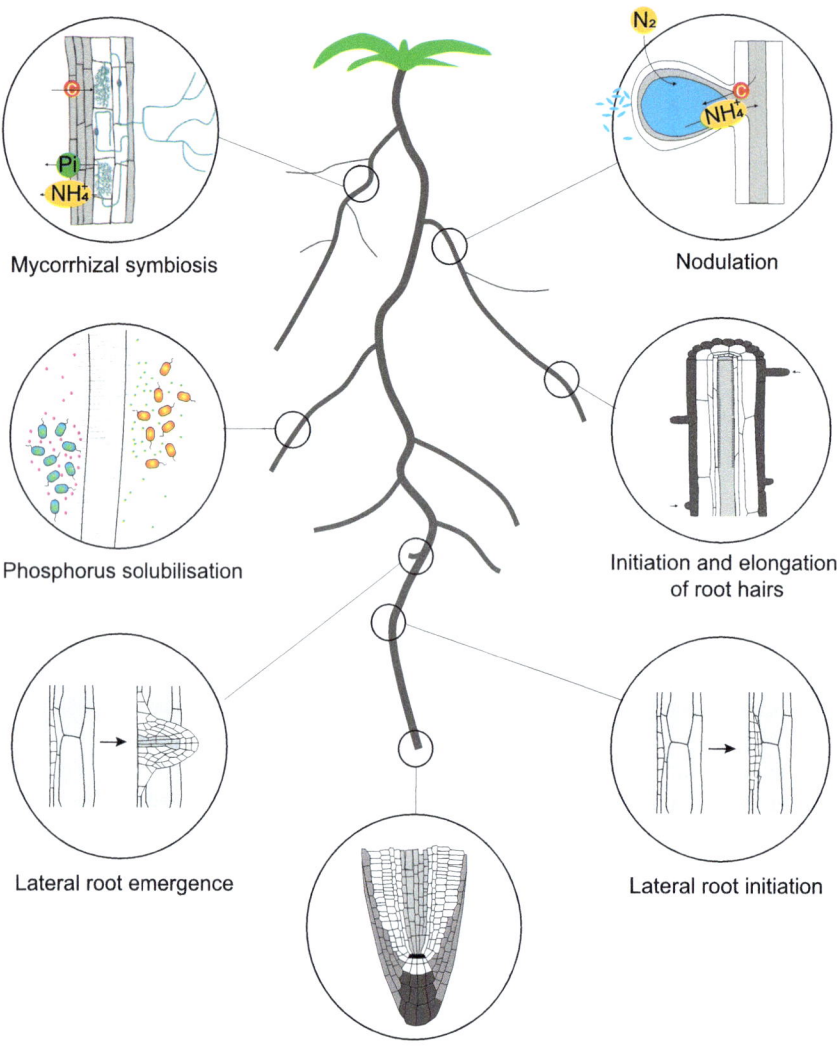

FIGURE 13.10 Impact of plant growth–promoting rhizobacteria on root system architecture, nutrient acquisition, and root functioning. *Reproduced from Den Herder et al. (2010), with permission from Elsevier.*

the plant uses to synthesize amino acids (see also Chapter 16). In addition, nodulation triggered by the rhizobia influences the development of the host root system, indicating that the microorganisms can alter host developmental pathways (Desbrosses and Stougaard, 2011).

Examples of the effects of inoculation with PGPR on root biomass production and architecture have been reported in different plant species (for a review, see Carvalho et al., 2013). In arabidopsis, PGPR influence root growth and development by modulating cell division and differentiation of primary and lateral roots by production of phytohormones (Verbon and Liberman, 2016; Fig. 13.10). The most common effect of PGPR on the root phenotype is a suppression of primary root growth coupled with proliferation of lateral roots and root hairs (Persello-Cartieaux et al., 2001; Ryu et al., 2005). In particular, specific PGPR species can increase the number of lateral root primordia and lateral roots by influencing the transition from proliferation to differentiation, but without inhibition of primary root growth (López-Bucio et al., 2007; Fig. 13.10). The auxin IAA is the most studied phytohormone mediating plant-microbe interactions (Sukumar et al., 2013), with auxin production being a major root growth–promoting trait (Spaepen et al., 2007). The auxin production and signaling pathways (López-Bucio et al., 2007) mediated by various plant-associated bacteria are well-defined mechanisms by which bacteria modulate root development in Arabidopsis.

13.3.3.2 Beneficial plant–fungus interactions in mycorrhizal symbiosis

Mycorrhizal fungi, a heterogeneous group of species spread over diverse fungal taxa, associate with over 80% of plant species. Mycorrhizal fungi improve the nutrient status of their host plants, influencing mineral nutrition, water absorption, and growth and disease resistance; in exchange, the host plant is a source of carbon for fungal growth and reproduction (Bonfante and Genre, 2010) (see Section 15.6). Numerous studies report root system changes in response to AM fungi, leading to an increased root branching and root system volume (reviewed by Hodge et al., 2009). In both dicotyledonous and monocotyledonous root systems, AM fungi show selective colonization of lateral roots and negligible effects in primary roots of dicots or crown roots of monocots (Gutjahr et al., 2009; Yu et al., 2018b). Lateral root induction by AM fungi is regulated at multiple levels, including presymbiotic signaling and intraroot colonization (Gutjahr and Parniske, 2013).

Establishment of AM interactions involves plant recognition of diffusible signals from the fungus, including lipochitooligosaccharides (LCOs) (Maillet et al., 2011) and chitooligosaccharides (COs) (Genre et al., 2013). In the legume *Medicago truncatula*, diffusible signals [e.g., LCO Myc factors (Myc-LCOs)] released by germinating AM fungal spores activate lateral root development (Oláh et al., 2005). The AM symbiosis-induced lateral root formation involves presymbiotic and/or intraradical microbial chitin-based signaling cues (Gutjahr and Paszkowski, 2013). Stimulation of lateral root emergence in *M. truncatula* occurred following treatment with Myc-LCOs, but not CO4, whereas both Myc-LCOs and CO4 were active in rice (Sun et al., 2015a).

Both AM and ectomycorrhizal fungi trigger lateral root production through the stimulation of auxin signaling and the production and release of auxin and ethylene or other volatile compounds (reviewed by Gutjahr and Paszkowski, 2013). AM colonization of roots has been reported to induce changes in the amount of phytohormones such as cytokinins, jasmonic acid, certain auxins, ABA, ethylene, salicylic acid, and strigolactones (reviewed by Foo et al., 2013). These phytohormones are all involved in the regulation of root system architecture (Nibau et al., 2008).

Sesquiterpenes derived from ectomycorrhizal fungi have been reported as volatile signals modulating lateral root formation in poplar and arabidopsis (Ditengou et al., 2015). In fact, volatile organic compounds released by the germinating AM fungal spores may stimulate lateral root formation independently of the symbiotic signaling pathway and the host, whereas exudates from the germinated spore stimulate lateral root formation in a symbiosis- and host-dependent manner (Sun et al., 2015b).

In conclusion, this chapter summarizes the current state of the genetic and molecular analysis of root growth and development, including interactions with external abiotic and biotic factors. Understanding the regulation of root system architecture and function will facilitate the improvement of resource-use efficiency in crops.

References

Ahmed, M.A., Zarebanadkouki, M., Meunier, F., Javaux, M., Kaestner, A., Carminati, A., 2018. Root type matters: measurement of water uptake by seminal, crown, and lateral roots in maize. J. Exp. Bot. 69, 1199–1206.

Aida, M., Beis, D., Heidstra, R., Willemsen, V., Blilou, I., Galinha, C., et al., 2004. The PLETHORA genes mediate patterning of the *Arabidopsis* root stem cell niche. Cell 119, 109–120.

Alvarez, J.M., Vidal, E.A., Gutiérrez, R.A., 2012. Integration of local and systemic signaling pathways for plant N responses. Curr. Opin. Plant Biol. 15, 185–191.

Araya, T., Miyamoto, M., Wibowo, J., Suzuki, A., Kojima, S., Tsuchiya, Y.N., et al., 2014. CLE-CLAVATA1 peptide-receptor signaling module regulates the expansion of plant root systems in a nitrogen-dependent manner. Proc. Natl. Acad. Sci. U.S.A. 111, 2029–2034.

Armstrong, W., Cousins, D., Armstrong, J., Turner, D.W., Beckett, P.M., 2000. Oxygen distribution in wetland plant roots and permeability barriers to gas-exchange with the rhizosphere: a microelectrode and modelling study with *Phragmites australis*. Ann. Bot. 86, 687–703.

Atkinson, J.A., Rasmussen, A., Traini, R., Voß, U., Sturrock, C., Mooney, S.J., et al., 2014. Branching out in roots: uncovering form, function, and regulation. Plant Physiol. 166, 538–550.

Bao, Y., Aggarwal, P., Robbins, N.E., Sturrock, C.J., Thompson, M.C., Tan, H.Q., et al., 2014. Plant roots use a patterning mechanism to position lateral root branches toward available water. Proc. Natl. Acad. Sci. U.S.A. 111, 9319–9324.

Bengough, A.G., McKenzie, B.M., Hallett, P.D., Valentine, T.A., 2011. Root elongation, water stress, and mechanical impedance: a review of limiting stresses and beneficial root tip traits. J. Exp. Bot. 62, 59–68.

Benjamins, R., Scheres, B., 2008. Auxin: the looping star in plant development. Ann. Rev. Plant Biol. 59, 443–465.

Benková, E., Hejátko, J., 2009. Hormone interactions at the root apical meristem. Plant Mol. Biol. 69, 383.

Bhosale, R., Giri, J., Pandey, B.K., Giehl, R.F., Hartmann, A., Traini, R., et al., 2018. A mechanistic framework for auxin dependent *Arabidopsis* root hair elongation to low external phosphate. Nat. Commun. 9, 1–9.

Blilou, I., Xu, J., Wildwater, M., Willemsen, V., Paponov, I., Friml, J., et al., 2005. The PIN auxin efflux facilitator network controls growth and patterning in *Arabidopsis* roots. Nature 433, 39–44.

Bloemberg, G.V., Lugtenberg, B.J., 2001. Molecular basis of plant growth promotion and biocontrol by rhizobacteria. Curr. Opin. Plant Biol. 4, 343–350.

Bonfante, P., Genre, A., 2010. Mechanisms underlying beneficial plant–fungus interactions in mycorrhizal symbiosis. Nat. Commun. 1, 1–11.

Carvalho, T.L.G., Ferreira, P.C.G., Hemerly, A.S., 2013. Plant growth promoting rhizobacteria and root architecture. In: Crespi, M. (Ed.), Root Genomics and Soil Interactions. Wiley-Blackwell, New Delhi, India, pp. 227–242.

Casimiro, I., Marchant, A., Bhalerao, R.P., Beeckman, T., Dhooge, S., Swarup, R., et al., 2001. Auxin transport promotes *Arabidopsis* lateral root initiation. Plant Cell 13, 843–852.

Colombi, T., Walter, A., 2016. Root responses of triticale and soybean to soil compaction in the field are reproducible under controlled conditions. Funct. Plant Biol. 43, 114–128.

Colombi, T., Kirchgessner, N., Walter, A., Keller, T., 2017. Root tip shape governs root elongation rate under increased soil strength. Plant Physiol. 174, 2289–2301.

Coudert, Y., Perin, C., Courtois, B., Khong, N.G., Gantet, P., 2010. Genetic control of root development in rice, the model cereal. Trends Plant Sci. 15, 219–226.

Covey-Crump, E.M., Attwood, R.G., Atkin, O.K., 2002. Regulation of root respiration in two species of Plantago that differ in relative growth rate: the effect of short-and long-term changes in temperature. Plant Cell Environ. 25, 1501–1513.

de Dorlodot, S., Forster, B., Pagès, L., Price, A., Tuberosa, R., Draye, X., 2007. Root system architecture: opportunities and constraints for genetic improvement of crops. Trends Plant Sci. 12, 474–481.

De Smet, I., Tetsumura, T., De Rybel, B., Frei dit Frey, N., Laplaze, L., Casimiro, I., et al., 2007. Auxin-dependent regulation of lateral root positioning in the basal meristem of Arabidopsis. Development 134, 681–690.

Den Herder, G., Van Isterdael, G., Beeckman, T., De Smet, I., 2010. The roots of a new green revolution. Trends Plant Sci. 15, 600–607.

Desbrosses, G.J., Stougaard, J., 2011. Root nodulation: a paradigm for how plant-microbe symbiosis influences host developmental pathways. Cell Host Microbe 10, 348–358.

Ditengou, F.A., Müller, A., Rosenkranz, M., Felten, J., Lasok, H., Van Doorn, M.M., et al., 2015. Volatile signalling by sesquiterpenes from ectomycorrhizal fungi reprogrammes root architecture. Nat. Commun. 6, 1–9.

Dolan, L., 2017. Root hair development in grasses and cereals (Poaceae). Curr. Opin. Genet. Dev. 45, 76–81.

Drew, M.C., Saker, L.R., 1975. Nutrient supply and the growth of the seminal root system in barley: II. Localized, compensatory increases in lateral root growth and rates op nitrate uptake when nitrate supply is restricted to only part of the root system. J. Exp. Bot. 26, 79–90.

Drew, M.C., Saker, L.R., Ashley, T.W., 1973. Nutrient supply and the growth of the seminal root system in barley: I. The effect of nitrate concentration on the growth of axes and laterals. J. Exp. Bot. 24, 1189–1202.

Dubrovsky, J.G., Doerner, P.W., Colón-Carmona, A., Rost, T.L., 2000. Pericycle cell proliferation and lateral root initiation in *Arabidopsis*. Plant Physiol. 124, 1648–1657.

Durieux, R.P., Kamprath, E.J., Jackson, W.A., Moll, R.H., 1994. Root distribution of corn: the effect of nitrogen fertilization. Agron. J. 86, 958–962.

Evert, R.F., Eichorn, S.E., 2007. Esau's Plant Anatomy: Meristems, Cells, and Tissues of the Plant Body: Their Structure, Function, and Development, third ed. John Wiley Sons Inc, Hoboken, New Jersey, USA.

Eysholdt-Derzsó, E., Sauter, M., 2017. Root bending is antagonistically affected by hypoxia and ERF-mediated transcription via auxin signaling. Plant Physiol. 175, 412–423.

Fahn, A., 1990. Plant Anatomy. Pergamon Press, Oxford, UK.

Feldman, L., 1994. The maize root. In: Freeling, M., Walbot, V. (Eds.), The Maize Handbook. Springer, New York, USA, pp. 29–37.

Foo, E., Ross, J.J., Jones, W.T., Reid, J.B., 2013. Plant hormones in arbuscular mycorrhizal symbioses: an emerging role for gibberellins. Ann. Bot. 111, 769–779.

Forde, B., Lorenzo, H., 2001. The nutritional control of root development. Plant Soil 232, 51–68.

Frey, F.P., Pitz, M., Schön, C.C., Hochholdinger, F., 2020. Transcriptomic diversity in seedling roots of European flint maize in response to cold. BMC Genomics 21, 1–15.

Fukaki, H., Tasaka, M., 2009. Hormone interactions during lateral root formation. Plant Mol. Biol. 69, 437.

Fukaki, H., Tameda, S., Masuda, H., Tasaka, M., 2002. Lateral root formation is blocked by a gain-of-function mutation in the SOLITARY-ROOT/IAA14 gene of *Arabidopsis*. Plant J. 29, 153–168.

Gao, Y., Lynch, J.P., 2016. Reduced crown root number improves water acquisition under water deficit stress in maize (*Zea mays* L.). J. Exp. Bot. 67, 4545–4557.

Gastal, F., Lemaire, G., 2002. N uptake and distribution in crops: an agronomical and ecophysiological perspective. J. Exp. Bot. 53, 789–799.

Gaudin, A.C., McClymont, S.A., Holmes, B.M., Lyons, E., Raizada, M.N., 2011. Novel temporal, fine-scale and growth variation phenotypes in roots of adult-stage maize (*Zea mays* L.) in response to low nitrogen stress. Plant Cell Environ. 34, 2122–2137.

Genre, A., Chabaud, M., Balzergue, C., Puech-Pagès, V., Novero, M., Rey, T., et al., 2013. Short-chain chitin oligomers from arbuscular mycorrhizal fungi trigger nuclear Ca^{2+} spiking in *Medicago truncatula* roots and their production is enhanced by strigolactone. New Phytol. 198, 190–202.

Giehl, R.F., von Wirén, N., 2014. Root nutrient foraging. Plant Physiol. 166, 509–517.

Giehl, R.F., Lima, J.E., von Wirén, N., 2012. Localized iron supply triggers lateral root elongation in *Arabidopsis* by altering the AUX1-mediated auxin distribution. Plant Cell 24, 33–49.

Gifford, M.L., Dean, A., Gutierrez, R.A., Coruzzi, G.M., Birnbaum, K.D., 2008. Cell-specific nitrogen responses mediate developmental plasticity. Proc. Natl. Acad. Sci. U.S.A. 105, 803–808.

Gilroy, S., Jones, D.L., 2000. Through form to function: root hair development and nutrient uptake. Trends Plant Sci. 5, 56–60.

Gruber, B.D., Giehl, R.F., Friedel, S., von Wirén, N., 2013. Plasticity of the *Arabidopsis* root system under nutrient deficiencies. Plant Physiol. 163, 161–179.

Gutjahr, C., Parniske, M., 2013. Cell and developmental biology of arbuscular mycorrhiza symbiosis. Ann. Rev. Cell Dev. Biol. 29, 593–617.

Gutjahr, C., Paszkowski, U., 2013. Multiple control levels of root system remodeling in arbuscular mycorrhizal symbiosis. Front. Plant Sci. 4, 204.

Gutjahr, C., Casieri, L., Paszkowski, U., 2009. *Glomus intraradices* induces changes in root system architecture of rice independently of common symbiosis signaling. New Phytol. 182, 829–837.

Hacham, Y., Holland, N., Butterfield, C., Ubeda-Tomas, S., Bennett, M.J., Chory, J., et al., 2011. Brassinosteroid perception in the epidermis controls root meristem size. Development 138, 839–848.

Haling, R.E., Brown, L.K., Bengough, A.G., Young, I.M., Hallett, P.D., White, P.J., et al., 2013. Root hairs improve root penetration, root–soil contact, and phosphorus acquisition in soils of different strength. J. Exp. Bot. 64, 3711–3721.

Hanzawa, T., Shibasaki, K., Numata, T., Kawamura, Y., Gaude, T., Rahman, A., 2013. Cellular auxin homeostasis under high temperature is regulated through a SORTING NEXIN1–dependent endosomal trafficking pathway. Plant Cell 25, 3424–3433.

Hetz, W., Hochholdinger, F., Schwall, M., Feix, G., 1996. Isolation and characterisation of *rtcs*, a mutant deficient in the formation of nodal roots. Plant J. 10, 845–857.

Himanen, K., Boucheron, E., Vanneste, S., de Almeida Engler, J., Inzé, D., Beeckman, T., 2002. Auxin-mediated cell cycle activation during early lateral root initiation. Plant Cell 14, 2339–2351.

Hochholdinger, F., Feix, G., 1998. Early post-embryonic root formation is specifically affected in the maize mutant *lrt1*. Plant J. 16, 247–255.

Hochholdinger, F., Park, W.J., Sauer, M., Woll, K., 2004. From weeds to crops: genetic analysis of root development in cereals. Trends Plant Sci. 9, 42–48.

Hochholdinger, F., Yu, P., Marcon, C., 2018. Genetic control of root system development in maize. Trends Plant Sci. 23, 79–88.

Hodge, A., 2004. The plastic plant: root responses to heterogeneous supplies of nutrients. New Phytol. 162, 9–24.

Hodge, A., Berta, G., Doussan, C., Merchan, F., Crespi, M., 2009. Plant root growth, architecture and function. Plant Soil 321, 153–187.

Horner-Devine, C.M., Leibold, M.A., Smith, V.H., Bohannan, B.J., 2003. Bacterial diversity patterns along a gradient of primary productivity. Ecol. Lett. 6, 613–622.

Huang, B.R., Taylor, H.M., McMichael, B.L., 1991. Effects of temperature on the development of metaxylem in primary wheat roots and its hydraulic consequence. Ann. Bot. 67, 163–166.

Ioio, R.D., Linhares, F.S., Scacchi, E., Casamitjana-Martinez, E., Heidstra, R., Costantino, P., et al., 2007. Cytokinins determine *Arabidopsis* root-meristem size by controlling cell differentiation. Curr. Biol. 17, 678–682.

Ioio, R.D., Nakamura, K., Moubayidin, L., Perilli, S., Taniguchi, M., Morita, M.T., et al., 2008. A genetic framework for the control of cell division and differentiation in the root meristem. Science 322, 1380–1384.

Ishikawa, H., Evans, M.L., 1995. Specialized zones of development in roots. Plant Physiol. 109, 725–727.

Jacoby, R., Peukert, M., Succurro, A., Koprivova, A., Kopriva, S., 2017. The role of soil microorganisms in plant mineral nutrition—current knowledge and future directions. Front. Plant Sci. 8, 1617.

Jeon, J., Cho, C., Lee, M.R., Van Binh, N., Kim, J., 2016. CYTOKININ RESPONSE FACTOR2 (CRF2) and CRF3 regulate lateral root development in response to cold stress in *Arabidopsis*. Plant Cell 28, 1828–1843.

Jiang, K., Meng, Y.L., Feldman, L.J., 2003. Quiescent center formation in maize roots is associated with an auxin-regulated oxidizing environment. Development 130, 1429–1438.

Jones, A.R., Kramer, E.M., Knox, K., Swarup, R., Bennett, M.J., Lazarus, C.M., et al., 2009. Auxin transport through non-hair cells sustains root-hair development. Nat. Cell Biol. 11, 78–84.

Jung, J.K.H.M., McCouch, S.R.M., 2013. Getting to the roots of it: genetic and hormonal control of root architecture. Front. Plant Sci. 4, 186.

Jung, J.Y., Shin, R., Schachtman, D.P., 2009. Ethylene mediates response and tolerance to potassium deprivation in *Arabidopsis*. Plant Cell 21, 607–621.

Kerk, N., Feldman, L., 1994. The quiescent center in roots of maize: initiation, maintenance and role in organization of the root apical meristem. Protoplasma 183, 100–106.

Kotak, S., Vierling, E., Bäumlein, H., von Koskull-Döring, P., 2007. A novel transcriptional cascade regulating expression of heat stress proteins during seed development of *Arabidopsis*. Plant Cell 19, 182–195.

Krouk, G., Lacombe, B., Bielach, A., Perrine-Walker, F., Malinska, K., Mounier, E., et al., 2010. Nitrate-regulated auxin transport by NRT1.1 defines a mechanism for nutrient sensing in plants. Dev. Cell 18, 927–937.

Leake, J., Johnson, D., Donnelly, D., Muckle, G., Boddy, L., Read, D., 2004. Networks of power and influence: the role of mycorrhizal mycelium in controlling plant communities and agroecosystem functioning. Can. J. Bot. 82, 1016–1045.

Lee, H.J., Kim, H.S., Park, J.M., Cho, H.S., Jeon, J.H., 2020. PIN-mediated polar auxin transport facilitates root-obstacle avoidance. New Phytol. 225, 1285–1296.

Leyser, O., Day, S., 2003. Mechanisms in Plant Development. Blackwell Science, Ltd, Oxford, UK.

Li, X., Mo, X., Shou, H., Wu, P., 2006. Cytokinin-mediated cell cycling arrest of pericycle founder cells in lateral root initiation of *Arabidopsis*. Plant Cell Physiol. 47, 1112–1123.

Lima, J.E., Kojima, S., Takahashi, H., von Wirén, N., 2010. Ammonium triggers lateral root branching in Arabidopsis in an AMMONIUM TRANSPORTER1;3-dependent manner. Plant Cell 22, 3621–3633.

Ljung, K., Hull, A.K., Celenza, J., Yamada, M., Estelle, M., Normanly, J., et al., 2005. Sites and regulation of auxin biosynthesis in *Arabidopsis* roots. Plant Cell 17, 1090–1104.

López-Arredondo, D.L., Leyva-González, M.A., González-Morales, S.I., López-Bucio, J., Herrera-Estrella, L., 2014. Phosphate nutrition: improving low-phosphate tolerance in crops. Ann. Rev. Plant Biol. 65, 95–123.

López-Bucio, J., Hernández-Abreu, E., Sánchez-Calderón, L., Nieto-Jacobo, M.F., Simpson, J., Herrera-Estrella, L., 2002. Phosphate availability alters architecture and causes changes in hormone sensitivity in the *Arabidopsis* root system. Plant Physiol. 129, 244–256.

López-Bucio, J., Campos-Cuevas, J.C., Hernández-Calderón, E., Velásquez-Becerra, C., Farías-Rodríguez, R., Macías-Rodríguez, L.I., et al., 2007. *Bacillus megaterium* rhizobacteria promote growth and alter root-system architecture through an auxin-and ethylene-independent signaling mechanism in *Arabidopsis thaliana*. Mol. Plant Microbe Interact. 20, 207–217.

Lugtenberg, B., Kamilova, F., 2009. Plant-growth-promoting rhizobacteria. Ann. Rev. Microbiol. 63, 541–556.

Lynch, J.P., 1995. Root architecture and plant productivity. Plant Physiol. 109, 7–13.

Lynch, J.P., 2005. Root architecture and nutrient acquisition. In: BassiriRad, H. (Ed.), Nutrient Acquisition by Plants. Springer, Berlin, Heidelberg, Germany, pp. 147–183.

Lynch, J.P., 2011. Root phenes for enhanced soil exploration and phosphorus acquisition: tools for future crops. Plant Physiol. 156, 1041–1049.

Lynch, J.P., 2013. Steep, cheap and deep: an ideotype to optimize water and N acquisition by maize root systems. Ann. Bot. 112, 347–357.

Lynch, J.P., Brown, K.M., 2008. Root strategies for phosphorus acquisition. In: White, P.J., Hammond, J.P. (Eds.), The Ecophysiology of Plant-Phosphorus Interactions. Springer, Dordrecht, The Netherlands, pp. 83–116.

Lynch, J.P., Chimungu, J.G., Brown, K.M., 2014. Root anatomical phenes associated with water acquisition from drying soil: targets for crop improvement. J. Exp. Bot. 65, 6155–6166.

Maillet, F., Poinsot, V., André, O., Puech-Pagès, V., Haouy, A., Gueunier, M., et al., 2011. Fungal lipochitooligosaccharide symbiotic signals in arbuscular mycorrhiza. Nature 469, 58–63.

Malamy, J.E., 2005. Intrinsic and environmental response pathways that regulate root system architecture. Plant Cell Environ. 28, 67–77.

Markakis, M.N., De Cnodder, T., Lewandowski, M., Simon, D., Boron, A., Balcerowicz, D., et al., 2012. Identification of genes involved in the ACC-mediated control of root cell elongation in *Arabidopsis thaliana*. BMC Plant Biol. 12, 1–11.

Martins, S., Montiel-Jorda, A., Cayrel, A., Huguet, S., Paysant-Le Roux, C., Ljung, K., et al., 2017. Brassinosteroid signaling-dependent root responses to prolonged elevated ambient temperature. Nat. Commun. 8, 1–11.

Marzec, M., Melzer, M., Szarejko, I., 2015. Root hair development in the grasses: what we already know and what we still need to know. Plant Physiol. 168, 407–414.

Massa, G.D., Gilroy, S., 2003. Touch modulates gravity sensing to regulate the growth of primary roots of *Arabidopsis thaliana*. Plant J. 33, 435–445.

McSteen, P., 2010. Auxin and monocot development. Cold Spring Harb. Perspect. Biol. 2, a001479.

Meier, M., Liu, Y., Lay-Pruitt, K.S., Takahashi, H., von Wirén, N., 2020. Auxin-mediated root branching is determined by the form of available nitrogen. Nat. Plants 6, 1136–1145.

Monshausen, G.B., Bibikova, T.N., Weisenseel, M.H., Gilroy, S., 2009. Ca^{2+} regulates reactive oxygen species production and pH during mechanosensing in *Arabidopsis* roots. Plant Cell 21, 2341–2356.

Mounier, E., Pervent, M., Ljung, K., Gojon, A., Nacry, P., 2014. Auxin-mediated nitrate signalling by NRT 1.1 participates in the adaptive response of Arabidopsis root architecture to the spatial heterogeneity of nitrate availability. Plant Cell Environ. 37, 162–174.

Muday, G.K., Rahman, A., Binder, B.M., 2012. Auxin and ethylene: collaborators or competitors? Trends Plant Sci. 17, 181–195.

Mustroph, A., 2018. Improving flooding tolerance of crop plants. Agronomy 8, 160.

Nagel, K.A., Kastenholz, B., Jahnke, S., Van Dusschoten, D., Aach, T., Mühlich, M., et al., 2009. Temperature responses of roots: impact on growth, root system architecture and implications for phenotyping. Funct. Plant Biol. 36, 947–959.

Nakagawa, Y., Katagiri, T., Shinozaki, K., Qi, Z., Tatsumi, H., Furuichi, T., et al., 2007. *Arabidopsis* plasma membrane protein crucial for Ca^{2+} influx and touch sensing in roots. Proc. Natl. Acad. Sci. U.S.A. 104, 3639–3644.

Nibau, C., Gibbs, D.J., Coates, J.C., 2008. Branching out in new directions: the control of root architecture by lateral root formation. New Phytol. 179, 595–614.

Oláh, B., Brière, C., Bécard, G., Dénarié, J., Gough, C., 2005. Nod factors and a diffusible factor from arbuscular mycorrhizal fungi stimulate lateral root formation in *Medicago truncatula* via the DMI1/DMI2 signalling pathway. Plant J. 44, 195–207.

Orosa-Puente, B., Leftley, N., von Wangenheim, D., Banda, J., Srivastava, A.K., Hill, K., et al., 2018. Root branching toward water involves posttranslational modification of transcription factor ARF7. Science 362, 1407–1410.

Overvoorde, P.J., Okushima, Y., Alonso, J.M., Chan, A., Chang, C., Ecker, J.R., et al., 2005. Functional genomic analysis of the AUXIN/INDOLE-3-ACETIC ACID gene family members in *Arabidopsis thaliana*. Plant Cell 17, 3282–3300.

Pagès, L., Serra, V., Draye, X., Doussan, C., Pierret, A., 2010. Estimating root elongation rates from morphological measurements of the root tip. Plant Soil 328, 35–44.

Pahlavanian, A.M., Silk, W.K., 1988. Effect of temperature on spatial and temporal aspects of growth in the primary maize root. Plant Physiol. 87, 529–532.

Pang, J., Ryan, M.H., Siddique, K.H., Simpson, R.J., 2017. Unwrapping the rhizosheath. Plant Soil 418, 129–139.

Paponov, I.A., Teale, W.D., Trebar, M., Blilou, I., Palme, K., 2005. The PIN auxin efflux facilitators: evolutionary and functional perspectives. Trends Plant Sci. 10, 170–177.

Pedersen, O., Sauter, M., Colmer, T.D., Nakazono, M., 2020. Regulation of root adaptive anatomical and morphological traits during low soil oxygen. New Phytol. 229, 42–49.

Péret, B., Larrieu, A., Bennett, M.J., 2009a. Lateral root emergence: a difficult birth. J. Exp. Bot. 60, 3637–3643.

Péret, B., De Rybel, B., Casimiro, I., Benková, E., Swarup, R., Laplaze, L., et al., 2009b. Arabidopsis lateral root development: an emerging story. Trends Plant Sci. 14, 399–408.

Péret, B., Clément, M., Nussaume, L., Desnos, T., 2011. Root developmental adaptation to phosphate starvation: better safe than sorry. Trends Plant Sci. 16, 442–450.

Péret, B., Li, G., Zhao, J., Band, L.R., Voß, U., Postaire, O., et al., 2012. Auxin regulates aquaporin function to facilitate lateral root emergence. Nat. Cell Biol. 14, 991–998.

Perilli, S., Moubayidin, L., Sabatini, S., 2010. The molecular basis of cytokinin function. Curr. Opin. Plant Biol. 13, 21–26.

Persello-Cartieaux, F., David, P., Sarrobert, C., Thibaud, M.C., Achouak, W., Robaglia, C., et al., 2001. Utilization of mutants to analyze the interaction between *Arabidopsis thaliana* and its naturally root-associated *Pseudomonas*. Planta 212, 190–198.

Pires, N.D., Dolan, L., 2012. Morphological evolution in land plants: new designs with old genes. Philos. Trans. R. Soc. Lond. B Biol. Sci. 367, 508–518.

Poole, P., Ramachandran, V., Terpolilli, J., 2018. Rhizobia: from saprophytes to endosymbionts. Nat. Rev. Microbiol. 16, 291–303.

Postma, J.A., Dathe, A., Lynch, J.P., 2014. The optimal lateral root branching density for maize depends on nitrogen and phosphorus availability. Plant Physiol. 166, 590–602.

Quint, M., Barkawi, L.S., Fan, K.T., Cohen, J.D., Gray, W.M., 2009. *Arabidopsis* IAR4 modulates auxin response by regulating auxin homeostasis. Plant Physiol. 150, 748–758.

Rahman, A., Amakawa, T., Goto, N., Tsurumi, S., 2001. Auxin is a positive regulator for ethylene-mediated response in the growth of *Arabidopsis* roots. Plant Cell Physiol. 42, 301–307.

Ramachandran, P., Wang, G., Augstein, F., de Vries, J., Carlsbecker, A., 2018. Continuous root xylem formation and vascular acclimation to water deficit involves endodermal ABA signalling via miR165. Development 145, dev159202.

Reed, R.C., Brady, S.R., Muday, G.K., 1998. Inhibition of auxin movement from the shoot into the root inhibits lateral root development in *Arabidopsis*. Plant Physiol. 118, 1369–1378.

Rich, S.M., Watt, M., 2013. Soil conditions and cereal root system architecture: review and considerations for linking Darwin and Weaver. J. Exp. Bot. 64, 1193–1208.

Rigas, S., Debrosses, G., Haralampidis, K., Vicente-Agullo, F., Feldmann, K.A., Grabov, A., et al., 2001. TRH1 encodes a potassium transporter required for tip growth in *Arabidopsis* root hairs. Plant Cell 13, 139–151.

Robbins, N.E., Dinneny, J.R., 2015. The divining root: moisture-driven responses of roots at the micro-and macro-scale. J. Exp. Bot. 66, 2145–2154.

Robbins, N.E., Dinneny, J.R., 2018. Growth is required for perception of water availability to pattern root branches in plants. Proc. Natl. Acad. Sci. U.S.A. 115, E822–E831.

Robinson, D., 2007. Implications of a large global root biomass for carbon sink estimates and for soil carbon dynamics. Philos. Trans. R. Soc. Lond. B Biol. Sci. 274, 2753–2759.

Rodrigues, M.L., Pacheco, C.M.A., Chaves, M.M., 1995. Soil-plant water relations, root distribution and biomass partitioning in *Lupinus albus* L. under drought conditions. J. Exp. Bot. 46, 947–956.

Rogers, E.D., Benfey, P.N., 2015. Regulation of plant root system architecture: implications for crop advancement. Curr. Opin. Biotechnol. 32, 93–98.

Rosales, M.A., Maurel, C., Nacry, P., 2019. Abscisic acid coordinates dose-dependent developmental and hydraulic responses of roots to water deficit. Plant Physiol. 180, 2198–2211.

Ruffel, S., Krouk, G., Ristova, D., Shasha, D., Birnbaum, K.D., Coruzzi, G.M., 2011. Nitrogen economics of root foraging: transitive closure of the nitrate–cytokinin relay and distinct systemic signaling for N supply vs. demand. Proc. Natl. Acad. Sci. U.S.A. 108, 18524–18529.

Ruiz Herrera, L.F., Shane, M.W., López-Bucio, J., 2015. Nutritional regulation of root development. Wiley Interdiscip. Rev. Dev. Biol. 4, 431–443.

Ryu, C.M., Hu, C.H., Locy, R.D., Kloepper, J.W., 2005. Study of mechanisms for plant growth promotion elicited by rhizobacteria in *Arabidopsis thaliana*. Plant Soil 268, 285–292.

Samira, R., Stallmann, A., Massenburg, L.N., Long, T.A., 2013. Ironing out the issues: integrated approaches to understanding iron homeostasis in plants. Plant Sci. 210, 250–259.

Sánchez-Calderón, L., López-Bucio, J., Chacón-López, A., Cruz-Ramírez, A., Nieto-Jacobo, F., Dubrovsky, J.G., et al., 2005. Phosphate starvation induces a determinate developmental program in the roots of *Arabidopsis thaliana*. Plant Cell Physiol. 46, 174–184.

Sánchez-Calderón, L., López-Bucio, J., Chacón-López, A., Gutiérrez-Ortega, A., Hernández-Abreu, E., Herrera-Estrella, L., 2006. Characterization of low phosphorus insensitive mutants reveals a crosstalk between low phosphorus-induced determinate root development and the activation of genes involved in the adaptation of *Arabidopsis* to phosphorus deficiency. Plant Physiol. 140, 879–889.

Sauter, M., 2013. Root responses to flooding. Curr. Opin. Plant Biol. 16, 282–286.

Schiefelbein, J.W., Masucci, J.D., Wang, H., 1997. Building a root: the control of patterning and morphogenesis during root development. Plant Cell 9, 1089–1098.

Schmidt, W., Schikora, A., 2001. Different pathways are involved in phosphate and iron stress-induced alterations of root epidermal cell development. Plant Physiol. 125, 2078–2084.

Sebastian, J., Yee, M.C., Viana, W.G., Rellán-Álvarez, R., Feldman, M., Priest, H.D., et al., 2016. Grasses suppress shoot-borne roots to conserve water during drought. Proc. Natl. Acad. Sci. U.S.A. 113, 8861–8866.

Shahzad, Z., Canut, M., Tournaire-Roux, C., Martiniere, A., Boursiac, Y., Loudet, O., et al., 2016. A potassium-dependent oxygen sensing pathway regulates plant root hydraulics. Cell 167, 87–98.

Sharp, R.E., Silk, W.K., Hsiao, T.C., 1988. Growth of the maize primary root at low water potentials: I. Spatial distribution of expansive growth. Plant Physiol. 87, 50–57.

Shibasaki, K., Uemura, M., Tsurumi, S., Rahman, A., 2009. Auxin response in *Arabidopsis* under cold stress: underlying molecular mechanisms. Plant Cell 21, 3823–3838.

Shin, R., Schachtman, D.P., 2004. Hydrogen peroxide mediates plant root cell response to nutrient deprivation. Proc. Natl. Acad. Sci. U.S.A. 101, 8827–8832.

Shukla, V., Lombardi, L., Iacopino, S., Pencik, A., Novak, O., Perata, P., et al., 2019. Endogenous hypoxia in lateral root primordia controls root architecture by antagonizing auxin signaling in *Arabidopsis*. Mol. Plant 12, 538–551.

Smith, S., De Smet, I., 2012. Root system architecture: insights from *Arabidopsis* and cereal crops. Philos. Trans. R. Soc. Lond. B Biol. Sci. 367, 1441–1452.

Spaepen, S., Versées, W., Gocke, D., Pohl, M., Steyaert, J., Vanderleyden, J., 2007. Characterization of phenylpyruvate decarboxylase, involved in auxin production of *Azospirillum brasilense*. J. Bacteriol. 189, 7626.

Spicer, R., Groover, A., 2010. Evolution of development of vascular cambia and secondary growth. New Phytol. 186, 577–592.

Steffens, B., Rasmussen, A., 2016. The physiology of adventitious roots. Plant Physiol. 170, 603–617.

Stepanova, A.N., Robertson-Hoyt, J., Yun, J., Benavente, L.M., Xie, D.Y., Doležal, K., et al., 2008. TAA1-mediated auxin biosynthesis is essential for hormone crosstalk and plant development. Cell 133, 177–191.

Sukumar, P., Legue, V., Vayssieres, A., Martin, F., Tuskan, G.A., Kalluri, U.C., 2013. Involvement of auxin pathways in modulating root architecture during beneficial plant-microorganism interactions. Plant Cell Environ. 36, 909–919.

Sun, J., Miller, J.B., Granqvist, E., Wiley-Kalil, A., Gobbato, E., Maillet, F., et al., 2015a. Activation of symbiosis signaling by arbuscular mycorrhizal fungi in legumes and rice. Plant Cell 27, 823–838.

Sun, X.G., Bonfante, P., Tang, M., 2015b. Effect of volatiles versus exudates released by germinating spores of *Gigaspora margarita* on lateral root formation. Plant Physiol. Biochem. 97, 1–10.

Svistoonoff, S., Creff, A., Reymond, M., Sigoillot-Claude, C., Ricaud, L., Blanchet, A., et al., 2007. Root tip contact with low-phosphate media reprograms plant root architecture. Nat. Genet. 39, 792–796.

Swarup, R., Friml, J., Marchant, A., Ljung, K., Sandberg, G., Palme, K., et al., 2001. Localization of the auxin permease AUX1 suggests two functionally distinct hormone transport pathways operate in the *Arabidopsis* root apex. Genes Dev. 15, 2648–2653.

Swarup, R., Parry, G., Graham, N., Allen, T., Bennett, M., 2002. Auxin cross-talk: integration of signalling pathways to control plant development. In: Perrot-Rechenmann, C., Hagen, G. (Eds.), Auxin Molecular Biology. Springer, Dordrecht, The Netherlands, pp. 411–426.

Swarup, R., Kramer, E.M., Perry, P., Knox, K., Leyser, H.O., Haseloff, J., et al., 2005. Root gravitropism requires lateral root cap and epidermal cells for transport and response to a mobile auxin signal. Nat. Cell Biol. 7, 1057–1065.

Swarup, R., Perry, P., Hagenbeek, D., Van Der Straeten, D., Beemster, G.T., Sandberg, G., et al., 2007. Ethylene upregulates auxin biosynthesis in *Arabidopsis* seedlings to enhance inhibition of root cell elongation. Plant Cell 19, 2186–2196.

Swarup, K., Benková, E., Swarup, R., Casimiro, I., Péret, B., Yang, Y., et al., 2008. The auxin influx carrier LAX3 promotes lateral root emergence. Nat. Cell Biol. 10, 946–954.

Tang, N., Shahzad, Z., Lonjon, F., Loudet, O., Vailleau, F., Maurel, C., 2018. Natural variation at XND1 impacts root hydraulics and trade-off for stress responses in *Arabidopsis*. Nat. Commun. 9, 1–12.

Tardieu, F., 1994. Growth and functioning of roots and of root systems subjected to soil compaction. Towards a system with multiple signalling? Soil Till. Res. 30, 217–243.

Tatematsu, K., Kumagai, S., Muto, H., Sato, A., Watahiki, M.K., Harper, R.M., et al., 2004. MASSUGU2 encodes Aux/IAA19, an auxin-regulated protein that functions together with the transcriptional activator NPH4/ARF7 to regulate differential growth responses of hypocotyl and formation of lateral roots in *Arabidopsis thaliana*. Plant Cell 16, 379–393.

Trachsel, S., Kaeppler, S.M., Brown, K.M., Lynch, J.P., 2011. Shovelomics: high throughput phenotyping of maize (*Zea mays* L.) root architecture in the field. Plant Soil 341, 75–87.

Tracy, S.R., Black, C.R., Roberts, J.A., Sturrock, C., Mairhofer, S., Craigon, J., et al., 2012. Quantifying the impact of soil compaction on root system architecture in tomato (*Solanum lycopersicum*) by X-ray micro-computed tomography. Ann. Bot. 110, 511–519.

Ubeda-Tomás, S., Swarup, R., Coates, J., Swarup, K., Laplaze, L., Beemster, G.T., et al., 2008. Root growth in *Arabidopsis* requires gibberellin/DELLA signalling in the endodermis. Nat. Cell Biol. 10, 625–628.

Ubeda-Tomás, S., Federici, F., Casimiro, I., Beemster, G.T., Bhalerao, R., Swarup, R., et al., 2009. Gibberellin signaling in the endodermis controls *Arabidopsis* root meristem size. Curr. Biol. 19, 1194–1199.

Ubeda-Tomás, S., Beemster, G.T., Bennett, M.J., 2012. Hormonal regulation of root growth: integrating local activities into global behaviour. Trends Plant Sci. 17, 326–331.

Uga, Y., Sugimoto, K., Ogawa, S., Rane, J., Ishitani, M., Hara, N., et al., 2013. Control of root system architecture by DEEPER ROOTING 1 increases rice yield under drought conditions. Nat. Genet. 45, 1097–1102.

van den Berg, C., Willemsen, V., Hendriks, G., Weisbeek, P., Scheres, B., 1997. Short-range control of cell differentiation in the *Arabidopsis* root meristem. Nature 390, 287–289.

Vanneste, S., Friml, J., 2009. Auxin: a trigger for change in plant development. Cell 136, 1005–1016.

Verbon, E.H., Liberman, L.M., 2016. Beneficial microbes affect endogenous mechanisms controlling root development. Trends Plant Sci. 21, 218–229.

Vicente-Agullo, F., Rigas, S., Desbrosses, G., Dolan, L., Hatzopoulos, P., Grabov, A., 2004. Potassium carrier TRH1 is required for auxin transport in *Arabidopsis* roots. Plant J. 40, 523–535.

Vidal, E.A., Moyano, T.C., Riveras, E., Contreras-López, O., Gutiérrez, R.A., 2013. Systems approaches map regulatory networks downstream of the auxin receptor AFB3 in the nitrate response of *Arabidopsis thaliana* roots. Proc. Natl. Acad. Sci. U.S.A. 110, 12840–12845.

Voesenek, L.A., Bailey-Serres, J., 2015. Flood adaptive traits and processes: an overview. New Phytol. 206, 57–73.

Wang, R., Zhang, Y., Kieffer, M., Yu, H., Kepinski, S., Estelle, M., 2016. HSP90 regulates temperature-dependent seedling growth in *Arabidopsis* by stabilizing the auxin co-receptor F-box protein TIR1. Nat. Commun. 7, 1–11.

Wany, A., Kumari, A., Gupta, K.J., 2017. Nitric oxide is essential for the development of aerenchyma in wheat roots under hypoxic stress. Plant Cell Environ. 40, 3002–3017.

Werner, T., Motyka, V., Laucou, V., Smets, R., Van Onckelen, H., Schmülling, T., 2003. Cytokinin-deficient transgenic *Arabidopsis* plants show multiple developmental alterations indicating opposite functions of cytokinins in the regulation of shoot and root meristem activity. Plant Cell 15, 2532–2550.

Westgate, M.E., Boyer, J.S., 1985. Osmotic adjustment and the inhibition of leaf, root, stem and silk growth at low water potentials in maize. Planta 164, 540–549.

Williamson, L.C., Ribrioux, S.P., Fitter, A.H., Leyser, H.O., 2001. Phosphate availability regulates root system architecture in *Arabidopsis*. Plant Physiol. 126, 875–882.

Woll, K., Borsuk, L.A., Stransky, H., Nettleton, D., Schnable, P.S., Hochholdinger, F., 2005. Isolation, characterization, and pericycle-specific transcriptome analyses of the novel maize lateral and seminal root initiation mutant *rum1*. Plant Physiol. 139, 1255–1267.

Xiao, T., van Velzen, R., Kulikova, O., Franken, C., Bisseling, T., 2019. Lateral root formation involving cell division in both pericycle, cortex and endodermis is a common and ancestral trait in seed plants. Development 146, dev182592.

Yamaguchi, M., Sharp, R.E., 2010. Complexity and coordination of root growth at low water potentials: recent advances from transcriptomic and proteomic analyses. Plant Cell Environ. 33, 590–603.

Yamauchi, T., Watanabe, K., Fukazawa, A., Mori, H., Abe, F., Kawaguchi, K., et al., 2014. Ethylene and reactive oxygen species are involved in root aerenchyma formation and adaptation of wheat seedlings to oxygen-deficient conditions. J. Exp. Bot. 65, 261–273.

Yamauchi, T., Tanaka, A., Inahashi, H., Nishizawa, N.K., Tsutsumi, N., Inukai, Y., et al., 2019. Fine control of aerenchyma and lateral root development through AUX/IAA-and ARF-dependent auxin signaling. Proc. Natl. Acad. Sci. U.S.A. 116, 20770–20775.

Yu, P., Gutjahr, C., Li, C., Hochholdinger, F., 2016. Genetic control of lateral root formation in cereals. Trends Plant Sci. 21, 951–961.

Yu, P., Marcon, C., Baldauf, J.A., Frey, F., Baer, M., Hochholdinger, F., 2018a. Transcriptomic dissection of maize root system development. In: Bennetzen, J., Flint-Garcia, S., Hirsch, C., Tuberosa, R. (Eds.), The Maize Genome. Springer, Cham, Switzerland, pp. 247–257.

Yu, P., Wang, C., Baldauf, J.A., Tai, H., Gutjahr, C., Hochholdinger, F., et al., 2018b. Root type and soil phosphate determine the taxonomic landscape of colonizing fungi and the transcriptome of field-grown maize roots. New Phytol. 217, 1240–1253.

Yu, F., Liang, K., Fang, T., Zhao, H., Han, X., Cai, M., et al., 2019. A group VII ethylene response factor gene, ZmEREB180, coordinates waterlogging tolerance in maize seedlings. Plant Biotechnol. J. 17, 2286–2298.

Zhang, H., Forde, B.G., 1998. An *Arabidopsis* MADS box gene that controls nutrient-induced changes in root architecture. Science 279, 407–409.

Zhang, H., Jennings, A., Barlow, P.W., Forde, B.G., 1999. Dual pathways for regulation of root branching by nitrate. Proc. Natl. Acad. Sci. U.S.A. 96, 6529–6534.

Zhang, H., Han, W., De Smet, I., Talboys, P., Loya, R., Hassan, A., et al., 2010. ABA promotes quiescence of the quiescent centre and suppresses stem cell differentiation in the *Arabidopsis* primary root meristem. Plant J. 64, 764–774.

Zhu, J., Zhang, K.X., Wang, W.S., Gong, W., Liu, W.C., Chen, H.G., et al., 2015. Low temperature inhibits root growth by reducing auxin accumulation via ARR1/12. Plant Cell Physiol. 56, 727–736.

Chapter 14

Rhizosphere chemistry influencing plant nutrition[☆]

Günter Neumann and Uwe Ludewig
Department of Nutritional Crop Physiology, Institute of Crop Science, University of Hohenheim, Stuttgart, Germany

Summary

This chapter describes the physicochemical processes determining the rhizosphere as a soil compartment influenced by the activity of roots and the consequences for plant nutrition. An introductory section discusses properties and the spatial extent of the rhizosphere as well as the temporal variability of rhizosphere processes. The mechanisms determining root-induced modifications of the pH and the redox conditions in the rhizosphere are discussed with respect to effects on the solubility and plant availability of nutrients and toxic elements and the impact on plant–microbe interactions. A final section describes the composition, quantity, and the release mechanisms of organic compounds by plant roots (rhizodeposition) and their role in rhizosphere processes, covering turnover of fine root structures and mycorrhizal hyphae, diffusion-mediated losses of organic substances from root cells but also the controlled release of root border cells, mucilage, ectoenzymes, and various low-molecular-weight compounds with specific adaptive functions regarding nutrient acquisition and stress tolerance.

14.1 General

In 1904 the German phytopathologist Lorenz Hiltner defined the soil–root interface termed the rhizosphere as the volume of soil surrounding the roots that is influenced by root activity. Hiltner described the so-called "rhizosphere effect," with selective stimulation of microbial growth in the soil near the root surface, as well as plant interactions with beneficial or pathogenic microorganisms and potential implications for nutrient cycling in soils and health and nutrition of plants (Hiltner, 1904). The stimulation of microbial activity and density as well as alterations in microbial diversity in the rhizosphere are mainly due to the release of easily decomposable organic rhizodeposits with selective effects depending on plant species, nutritional status, soil type, agricultural management, and exposure to biotic and abiotic stresses. This also involves feedback loops on root growth and activity with systemic effects on whole plant performance induced by the respective rhizosphere microbiota (Kuppardt et al., 2018; Windisch et al., 2021). The effects of rhizodeposition on rhizosphere microorganisms are discussed in detail in Chapter 15.

Plant roots can modify the rhizosphere chemistry in a number of ways: (1) by release and uptake of organic compounds, (2) by gas exchange (CO_2/O_2) and production of organic volatiles [volatile organic compounds (VOC)] released from roots and rhizosphere microorganisms, and (3) by uptake and release of water and mineral nutrients. These processes can be associated with uptake or extrusion of protons and modifications of the redox potential. Roots and associated rhizosphere microorganisms also modify the physical properties of the rhizosphere soil, such as aggregate stability, hydrophobicity, and number and size of micropores by their growth through the soil as well as the presence of polymeric substances. These root-induced changes determine the availability of water and nutrients with impact on plant growth, health, stress resilience, and yield formation. Therefore, it is appropriate to state that life on earth is sustained by the small volume of soil surrounding roots and influenced by them, called the rhizosphere (Hinsinger et al., 2009), with major implications for climate and environment change, greenhouse gas emissions, carbon sequestration, soil fertility management, and food security (York et al., 2016).

[☆] This chapter is a revision of the third edition chapter by G. Neumann and V. Römheld, pp. 347–368. DOI: https://doi.org/10.1016/B978-0-12-384905-2.00014-5. © Elsevier Ltd.

14.1.1 Rhizosphere sampling

Determining the properties of the rhizosphere is challenging. Using destructive methods, loosely adhering soil is first detached from roots by shaking, and soil still attached to the root surface (rhizosphere soil) is then removed by root washings. However, this simple technique has a range of shortcomings: the amount of rhizosphere soil obtained depends on root morphology (particularly length and density of root hairs) and physiology (release of binding agents, such as mucilage) as well as on soil properties (texture, water content, organic matter content). Importantly, contaminations of the samples by rupturing of root structures (i.e., root hairs or mycorrhizal hyphae) cannot be avoided. This technique works well for subsequent DNA extraction and sequencing for the characterization of rhizosphere microbial communities, but the recovery of exudate compounds in the rhizosphere soil samples or even in washings of excavated root samples bears a high risk of contamination from damaged root tissues and is also limited by adsorption of certain exudate compounds to the soil matrix (Neumann, 2006).

In another approach, root and bulk soil compartments can be separated by mesh (Fig. 14.1A) in vertical or horizontal orientation (Engels et al., 2000). After the formation of a root mat along the mesh of the rhizosphere compartment, soil analysis in the bulk soil compartment is possible at defined distances from the root surface, for example, after microtome slicing of frozen soil columns. Although this method can provide accurate data on gradients in the rhizosphere with a high spatial resolution (mm range), the root-induced changes may be overestimated due to the high root density formed along the mesh separating the root and the bulk soil compartments.

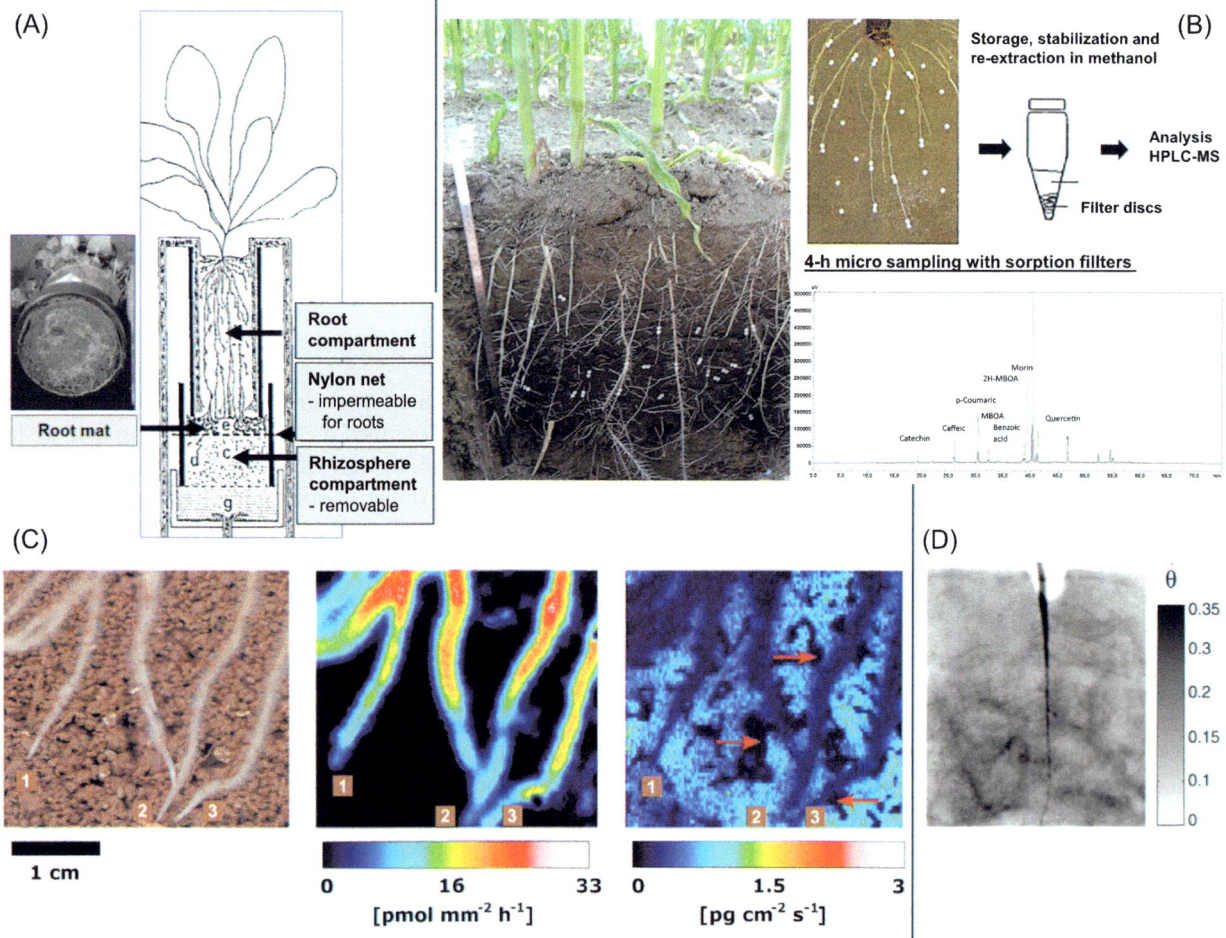

FIGURE 14.1 Techniques for rhizosphere sampling: (A) compartmented growth system; (B) root observation windows installed in the field (left) and in rhizoboxes (right) for micro-sampling of root exudates and rhizosphere soil solution using sorption filters; (C) root growth (left), zymography of acid phosphatase (middle) and DGT imaging of rhizosphere P depletion (right) along a root observation window with *Lupinus angustifolius* plants; (D) neutron autoradiography of water content in the *Lupinus albus* rhizosphere. *DGT*, Diffusive gradients in thin films. *Adapted from: part (A) Engels et al. (2000) and Kuchenbuch and Jungk (1982); (B) Neumann (2006), photos courtesy Narges Moradtalab; (C) Hummel et al. (2021); and (D) Carminati et al. (2010).*

Less invasive methods include isotope techniques with radioactive nutrient tracers mixed into the soil substrate in combination with autoradiography, using X-ray films placed onto soil-grown plant roots at the surface of planar root observation windows in specialized microcosms, so-called rhizoboxes (Fig. 14.1A and B). $^{14}CO_2$ isotope pulse labeling of plant shoots has been used to trace and quantify photoassimilates released into the rhizosphere and incorporated into microbial biomass (Haase et al., 2007). Similarly, the diffusive gradient in thin films (DGT) technique has been used in the rhizosphere by deploying gels, containing ion-selective adsorbent phases, on root sections using thin (10 μm) membrane filters for separating the soil and the gel. The element composition of the gel is subsequently analyzed by laser ablation inductively coupled plasma mass spectrometry (LA–ICP–MS), allowing the assessment of two-dimensional DGT-labile element fluxes in the rhizosphere (Fig. 14.1C) with resolutions down to the submillimeter range (Smolders et al., 2020). Recently, a positron-emitting tracer imaging system has been described for visualizing temporal changes in the release of ^{11}C-labeled photoassimilates of soil-grown plants (white lupin, soybean) in a specific rhizobox system (Yin et al., 2020).

Root observation windows have been employed for monitoring rhizosphere pH, nutrient uptake, and gas exchange with ion-selective microelectrodes, gel matrices (agar, agarose) with pH or redox indicators (Neumann, 2006), or with fluorescence indicators immobilized in a planar membrane matrix combined with fiber optics detection (termed imaging optodes) (Elberling et al., 2011; Faget et al., 2013). Similarly, nondestructive rhizosphere activity-imaging of enzymes originating from plant roots and soil microorganisms along root observation windows, termed zymography (Fig. 14.1C), uses fluorogenic indicator substrates or color reagents applied in a filter or membrane matrix to observe spatiotemporal rhizosphere gradients (Razavi et al., 2019).

The root-window approaches are employed to collect soluble rhizodeposits and rhizosphere soil solution of soil-grown roots by the use of ceramic microsuction cups, or by the application of sorption materials (filter papers, membrane filters, ion-exchange resin foils, Fig. 14.1B) or even microcapillaries (Neumann, 2006; Oburger and Jones, 2018). Similarly, microdialysis tubes inserted into the rhizosphere have been used successfully to collect low-molecular-weight (LMW) organic and inorganic compounds in the soil solution (Inselsbacher et al., 2011), but also in simulating root exudation of selected compounds applied via microdialysis probes (McKay Fletcher et al., 2021; Schack-Kirchner et al., 2020). The spatial resolution of the described methods reaches the μm-to-mm range, with the subsequent detection and quantitative determination based mainly on mass-spectrometry-coupled high performance liquid chromatography (HPLC–MS), gas chromatography (GC–MS), or capillary electrophoresis (CE).

Although installation of root windows is possible in microcosms as well as under field conditions (Engels et al., 2000; Vetterlein et al., 2020) allowing repeated nondestructive samplings, they do not represent a completely undisturbed system and provide only a two-dimensional observation plane (Fig. 14.1B). Therefore, in the recent years, nondestructive imaging techniques have increasingly been adopted from applications in medical and material sciences (York et al., 2016). Using X-ray computed tomography (Mooney et al., 2011; Mairhofer et al., 2013), magnetic resonance imaging (Schulz et al., 2013), and neutron radiography (Carminati et al., 2010), nondestructive 3D imaging of root development in microcosms (and in excavated soil cores) allows characterization of changes in soil porosity and water relationships in the rhizosphere (Fig. 14.1D) (Holz et al., 2019).

At a submicrometer scale, stable isotope (^{15}N, ^{13}C) labeling and use of nanoscale secondary ion MS (nanoSIMS) and transmission electron microscopy have been used to investigate nutrient partitioning between root cells and native soil microorganisms (Clode et al., 2009). In situ analysis of root exudate compounds along tomato roots grown in a soil-free system on transparent agar slides by matrix-assisted laser desorption/ionization mass spectrometry imaging has been described recently by Korenblum et al. (2020). However, these techniques require highly sophisticated and expensive instrumentation. In most cases, they are still applicable only in model systems with artificial or even soil-free culture conditions and illustrate the enormous methodological challenges related to sampling and data acquisition for investigating rhizosphere processes. Major knowledge gaps are related to the difficulty in linking mechanistically the physical, chemical, and biological processes taking place at different spatial and temporal scales in the rhizosphere and then upscaling them to the root system, soil profile, and the field scale, requiring the integration of modeling approaches (Vetterlein et al., 2020).

14.2 Spatial extent of the rhizosphere

The extent of the rhizosphere in space and time is highly variable. Gradients exist in both the radial direction toward the bulk soil and longitudinally along the roots; these gradients are also influenced by temporal changes in root activity (Fig. 14.2).

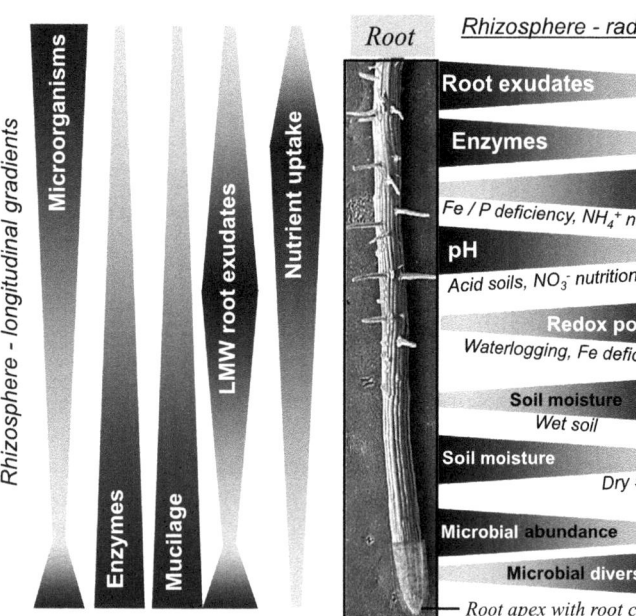

FIGURE 14.2 Physicochemical and biological gradients in the rhizosphere. *Based on Kuzyakov and Razavi (2019) and Neumann and Römheld (2002).*

14.2.1 Radial gradients

The space around the roots can be subdivided into three regions: (1) the apoplast within the cell walls of the rhizodermis and outer root cortex cells represents the so-called *endosphere*; (2) the outer surface of the rhizodermis, the *rhizoplane*, and (3) the *rhizosphere*. Depending on the rhizosphere processes considered (exudation of organic and inorganic compounds, respiration, uptake of nutrients and water) but also on the sampling techniques (Kuzyakov and Razavi, 2019), the radial extent of the rhizosphere has been reported from less than 1 mm to several centimeters (Gregory, 2006; Hinsinger et al., 2005). In this context, nondestructive imaging techniques usually show smaller radial rhizosphere extensions (frequently in the range of 2–4 mm) than destructive samplings (Kuzyakov and Razavi, 2019). However, nutrient extraction zones can extend up to several meters from the root surface when mycorrhizal hyphae (hyphosphere) are considered.

The concentration of organic compounds released from plant roots declines with increasing distance from the root surface. The distance of diffusion largely depends on soil properties and adsorption characteristics of the respective compounds. Polar compounds such as uncharged LMW sugars or simple amino acids can diffuse several millimeters from the root surface, whereas di- and tricarboxylates (such as malate, citrate, or oxalate) may be adsorbed preferentially to positively charged sorption sites on the soil matrix (Jones et al., 2003) and are only detectable close to the rhizoplane. This holds true also for root-secreted proteins (such as phosphatases), polygalacturonic acids, and phenolic compounds (Jones et al., 2003).

The radial gradients of nutrients in the rhizosphere are determined by their solubility and mobility, and the uptake capacity of the roots. Poorly mobile nutrients, such as P, K, ammonium, and micronutrients with low concentrations in the soil solution, are frequently depleted in the rhizosphere by rapid root uptake, whereas soluble nutrients (e.g., Ca and Mg) and toxic compounds (e.g., Na and Cl) may accumulate close to roots (see also Chapter 12).

Nutrient uptake is closely coupled to uptake or release of protons and therefore is frequently associated with root-induced changes in the rhizosphere pH (see Section 14.4). In some plant species, increased rhizosphere acidification is also an adaptive response to improve acquisition of Fe and P (Neumann and Römheld, 2002). However, the spatial extent of these pH changes in the rhizosphere strongly depends on the buffering capacity of the soil. Fig. 14.3 shows that the extent of the acidification by chickpea roots decreases with increasing concentrations of $CaCO_3$ with a high pH buffering potential in the soil. Under field conditions, localized patches of organic matter and the distribution of $CaCO_3$ particles may contribute to the variability in soil pH that is not easily distinguishable from root-induced pH changes (Schöttelndreier and Falkengren-Grerup, 1999).

Gas exchange due to root and microbial respiration also leads to the formation of gradients in the rhizosphere. CO_2 dissolved in the soil solution results in the formation of H^+ and HCO_3^-, thereby contributing to rhizosphere

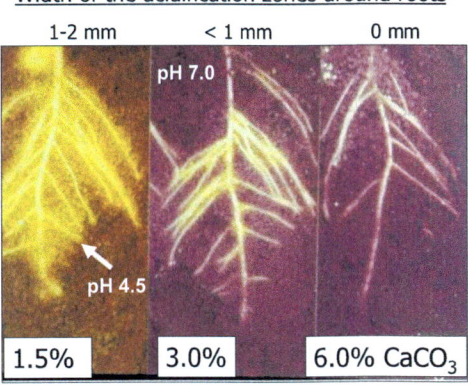

FIGURE 14.3 Extension of ammonium-induced rhizosphere acidification in chickpea, depending on the soil buffering capacity modified by addition of CaCO₃ (detected by embedding the roots of soil-grown plants in agar containing the pH-indicator bromocresol purple). *Adapted from Römheld (1986).*

acidification; by contrast, respiratory O₂ consumption leads to depletion of O₂ close to the root surface, particularly in young root tissues characterized by high respiration rates, strongly influencing the redox potential (Bidel et al., 2000). However, at low O₂ concentrations in soil (e.g., in waterlogged soils), adapted plant species can also release O₂ (transported from aerial plant parts), with gradients extending 0.4–4.0 mm from the root surface (Revsbech et al., 1999; Kuzyakov and Razavi, 2019; see also Section 14.5).

14.2.2 Longitudinal gradients

Along single roots, gradients are formed between apical root zones (root meristem, elongation zone, root hair formation zone) and the older, more basal parts of the root. Water and nutrient uptake is usually highest in the apical root zones (Häussling et al., 1988; Reidenbach and Horst, 1997). This is in part attributed to incomplete development of the endodermis and exodermis in these root zones, facilitating uptake of water and dissolved nutrients via the apoplastic pathway (Chapter 2) and also the release of organic and inorganic compounds (Canarini et al., 2019). The presence of root hairs in the subapical root zones increases the surface area available for nutrient absorption; root-hairless mutants have a decreased release of protons and organic compounds in this zone (Holz et al., 2018). Additionally, subapical root zones are characterized by lower rhizosphere microbial abundance as compared with older root parts (von Wirén et al., 1993), which may reduce the rate of microbial degradation of organic compounds in the apical root zones (see Chapter 15). On the other hand, mucilage (mainly galacturonic acid polymers with high viscosity), covering the surface of apical root zones, may limit diffusion of LMW organic compounds into the surrounding soil, which in turn can stimulate microbial colonization of the mucilage layer. The gradients and their orientation in the rhizosphere are summarized in Fig. 14.2.

14.2.3 Temporal variability

Apart from spatial variation, rhizosphere processes also exhibit a temporal variability (Hinsinger et al., 2009), including seasonal variation dependent on root activity (Turpault et al., 2006). Characteristic patterns of rhizodeposits have been recorded during different developmental stages of *Arabidopsis thaliana* grown on sterile agar media, with predominance of sugars during early growth and amino acids and phenolics in the later stages of plant development. Despite the artificial nature of these growth conditions, the expression analysis of rhizosphere microbial genes involved in metabolism of the respective compounds showed corresponding changes during growth stages of soil-grown plants (Chaparro et al., 2013). Apart from variation over longer time periods, short-term changes, such as diurnal variations, have also been demonstrated, for example, regarding the release of root exudates involved in Fe (Fig. 14.4) and P mobilization (Nishizawa and Mori, 1987; Watt and Evans, 1999) and rhizosphere acidification (Blossfeld and Gansert, 2007). However, these diurnal variations cannot be generalized for all compounds released by plant roots (Badri et al., 2010). In *Lupinus albus*, root exudation of organic acid anions induced by P limitation is strongly influenced by the root development (Neumann and Martinoia, 2002) (Fig. 14.4) and is associated with changes in the rhizosphere microbial communities (Marschner et al., 2002; Weisskopf et al., 2006).

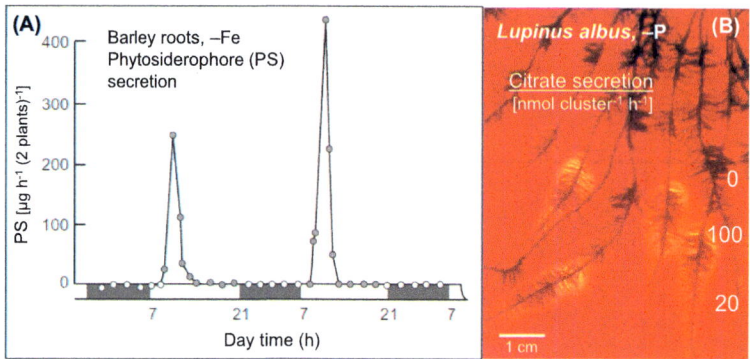

FIGURE 14.4 Temporal variability of rhizosphere processes: (A) diurnal pulses of phytosiderophore secretion in Fe-deficient barley and (B) secretion of citrate and protons in different developmental stages of cluster roots in P-deficient white lupin. *Adapted from Tagaki (1984) and Neumann et al. (2000).*

14.3 Inorganic elements in the rhizosphere

Although the total soil content of nutrients frequently exceeds the plant requirements by several orders of magnitude, plant availability is often limited by low solubility of nutrients such as P, K, ammonium, Fe, Zn, Mn, Cu, and Mo. These nutrients reach the root surface mainly by diffusion. As described in Chapter 12, the rate of diffusion may be too low to meet plant demand for these nutrients. However, plants have developed several strategies for acquisition of nutrients with limited solubility, including:

1. adaptations to exploit a larger soil volume to improve acquisition of the available nutrient fractions by root and root hair growth (see Chapter 12) or in association with mycorrhizal fungi (see Chapter 15), and
2. modifications of the rhizosphere chemistry to increase the solubility of poorly available nutrients (see below).

The concentration of a particular ion in the rhizosphere can be lower or higher than, or similar to, that in the bulk soil, depending on the concentration in the bulk soil solution, the rate of delivery of the ion to the root surface, and its rate of uptake by the root (see Chapter 12). When the flow of nutrients to the root surface is lower than root uptake, their concentrations decrease in the rhizosphere, generating a diffusion gradient toward the root surface (Barber, 1995; Hinsinger, 2004; Jungk, 2002; Lorenz et al., 1994). This typically occurs for nutrients with low concentrations in the soil solution, which are delivered slowly by diffusion to the soil solution, especially P and K (Fig. 14.5). In soils low in available K, this can lead to the disaggregation of shale particles and the accumulation of amorphous Fe and Al oxyhydrates, indicative of enhanced weathering of minerals at the soil–root interface (Kong and Steffens, 1989).

In Italian ryegrass (*Lolium multiflorum*), the K concentration in the rhizosphere soil solution can decrease below 80 μM, which enhances the release of interlayer K within a few days (Fig. 14.6) and concomitant transformation of trioctahedral mica into vermiculite in the rhizosphere (Hinsinger and Jaillard, 1993). In the rhizosphere of oilseed rape (*Brassica napus*), depletion of both K and Mg, together with a decrease in pH to about 4, increases not only the release of interlayer K, but also of octahedral Mg and, thus, induces irreversible transformations of the mica (Hinsinger et al., 1993).

Greater uptake of water than ions leads to ion accumulation in the rhizosphere. This can be predicted from calculations based on models of solute transport to the root surface by diffusion and mass flow for ions that are present in the soil solution in high concentrations. After 2-month growth in a sandy loam soil, the concentration of Ca in the rhizosphere was increased two- to threefold compared to the bulk soil (Fig. 14.7), and similar observations were made for Mg.

Mass flow to the roots and rate of nutrient uptake by roots differ among plant species. For example, in ryegrass and lupin grown in the same soil type, Ca supply by mass flow was 2.8 and 8 mg Ca, respectively, but Ca uptake was 0.8 mg in ryegrass (*Lolium rigidum*) and 9.0 mg in lupin (*Lupinus digitatus*). Thus, despite the higher supply, Ca was depleted in the rhizosphere of lupin but accumulated in that of ryegrass (Barber and Ozanne, 1970). At high Ca^{2+} and SO_4^{2-} concentrations in the soil solution, $CaSO_4$ may precipitate at the root surface (Jungk, 1991) and, over a long period, form a solid mantle around the roots (*pedotubules*) with diameters of a few millimeters to more than 1 cm (Barber, 1995).

In calcareous soils, calcification of fine roots may occur (Fig. 14.7), in which the calcite elements retain the structure of the original cortex cells (Jaillard et al., 1991). These cytomorphic calcite elements (<60–80 μm) are initially

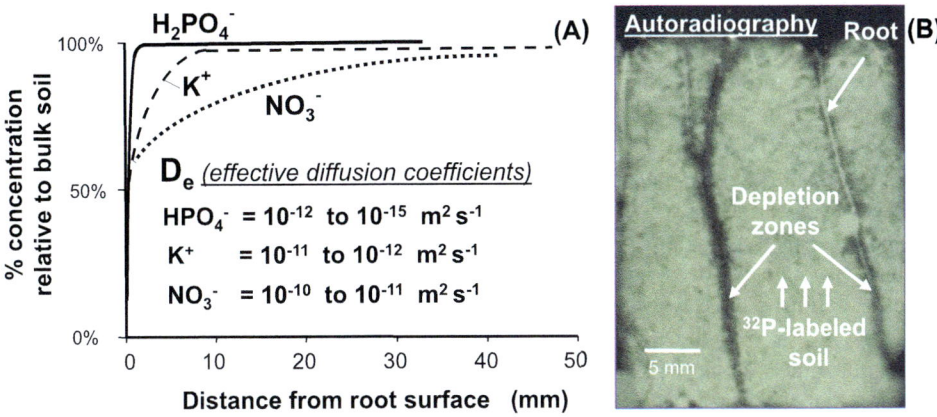

FIGURE 14.5 Root-induced depletion of nutrients in the rhizosphere: (A) dependence on the effective soil diffusion coefficients (D_e) and (B) autoradiographic visualization of ^{32}P depletion in the rhizosphere of soil-grown barley roots. *Adapted from Hinsinger (2004) and Jungk (2002).*

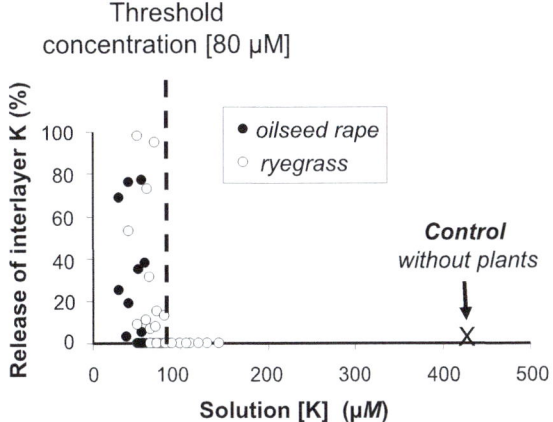

FIGURE 14.6 Release of interlayer K^+ from clay minerals and vermiculitization by root-induced depletion of K^+ in the rhizosphere of oilseed rape and Italian ryegrass. *Adapted from Hinsinger and Jaillard (1993).*

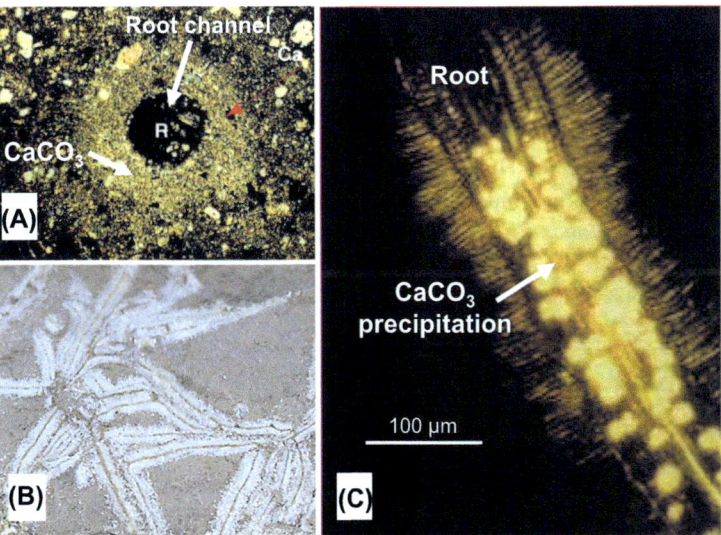

FIGURE 14.7 Nutrient accumulation in the rhizosphere. (A) Accumulation of $CaCO_3$ in the rhizosphere around a peach tree root channel; (B) $CaCO_3$ precipitation at the root surface of oilseed rape; (C) Root calcification by calcite precipitation in root cortex cells. *Adapted from Jaillard et al. (1991) and Callot et al. (1992).*

formed by high respiratory CO_2 concentrations in the rhizosphere and Ca solubilization induced by rhizosphere acidification, leading to $CaCO_3$ precipitation at the root surface. After excess Ca uptake, precipitation can occur to some extent even within cortex cells. This may inhibit root function, leading to death of fine roots. However, the calcification process can proceed during decay of dead roots, associated with increased respiratory CO_2 formation and a decline in pH due to microbial activity involved in degradation of root tissues (Huguet et al., 2021). In agreement with this, calcified roots are surrounded by a decalcified rhizocylinder with a silicon—aluminum matrix (Jaillard, 1985; Jaillard et al., 1991). This is an example of the role of root-induced changes in the rhizosphere that can be important in pedogenesis because the cytomorphic calcite fraction in certain locations may represent up to a quarter of the soil mass (Jaillard et al., 1991).

Accumulation of some salts of low solubility in the rhizosphere (e.g., $CaCO_3$, $CaSO_4$) may not be harmful to plants. This is different, however, in saline soils with high concentrations of water-soluble salts such as NaCl where Cl and Na can accumulate in the rhizosphere creating a concentration gradient to the bulk soil (Yi et al., 2007; Zhang et al., 2005), and this gradient becomes steeper as the transpiration rate increases. Increasing the salt concentration and osmotic potential of the soil solution decreases water availability to plants and can impair plant water relations (see also Section 18.6).

Gradients in ion uptake rates along the root axis are also important for ion competition and selectivity in uptake (see Chapter 2). The strong reduction of Mg uptake by K, which can be demonstrated readily in nutrient solution culture, occurs in soil-grown plants only if the rhizosphere K concentration is high. Depletion of K in the rhizosphere soil solution below 20 μM increases the uptake rate of Mg by ryegrass twofold. The increasing extent of K depletion from the apical to the basal zones allows greater uptake of Mg in the basal zone. However, in saline soil with high Na concentrations, the preferential K uptake in apical zones may also increase Na uptake rates in basal zones and, thus, decrease the K/Na selectivity (Hinsinger and Jaillard, 1993; Hinsinger et al., 1993).

14.4 Rhizosphere pH

The rhizosphere pH may differ from the bulk soil pH by up to two units, depending on plant and soil factors, with important consequences for the pH-dependent solubility of nutrients and toxic elements in the soil solution (see Chapter 2). The most important factor for root-induced changes in rhizosphere pH is the uptake of nutrients that is coupled to, or driven by, proton (H^+) transport. The driving force for nutrient uptake by root cells is H^+ extrusion, mediated by the activity of a plasma membrane (PM)-bound H^+-pumping ATPase (PM-H^+-ATPase) that creates an outward positive gradient in electrochemical potential and pH between the cytosol (pH 7—7.5) and the apoplasm (pH 5—6). The electrochemical potential gradient provides the energy (proton-motive force) for anion uptake by proton—anion cotransport and for cation uptake via (1) uniport (with proton extrusion via PM-H^+-ATPase to balance charges) and (2) cation and H^+ cotransport. Solute accumulation inside cells promotes a steep gradient of the water potential that drives water uptake to provide the turgor pressure to govern cell expansion (Siao et al., 2020).

Due to differences in plant requirements and in the availability of nutrients, uptake of cations and anions is often not balanced. Excess uptake of anions over cations leads to net removal of protons from the rhizosphere and consequently to an increase in rhizosphere pH. By contrast, excessive uptake of cations is balanced by a net release of protons, leading to rhizosphere acidification (Fig. 14.8).

Rhizosphere pH may be influenced by the release and uptake of HCO_3^-, respiratory CO_2 production by roots and rhizosphere microorganisms, and the release of LMW organic compounds that can also be coupled with proton transport (Canarini et al., 2019). In aerated soils, CO_2 is of minor importance for rhizosphere pH because it rapidly diffuses away

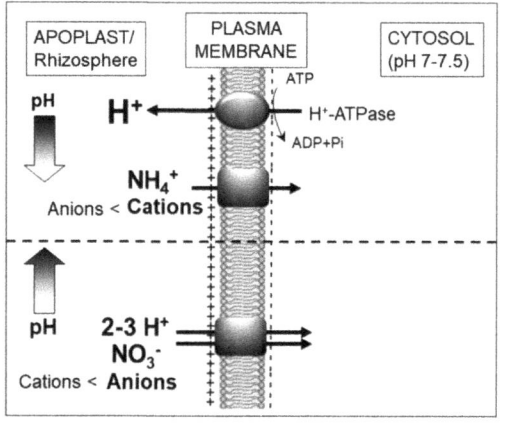

FIGURE 14.8 Impact of nutrient uptake on root-induced alterations in the rhizosphere pH.

from the roots through air-filled pores (Nye, 1986). At high soil water content, CO_2 is dissolved in the soil solution (forming H^+ and HCO_3^-), which affects rhizosphere pH because the mobility of H^+ and HCO_3^- is relatively low in the soil solution. This can result in the establishment of efficient bicarbonate-based pH buffering systems reducing the solubility of Fe, Zn, and Mn and even inducing problems with bicarbonate toxicity, particularly in alkaline soils (Lee, 1999; Poschenrieder et al., 2018).

The pH buffering capacity of soil and the initial soil pH are the main factors determining the extent to which plant roots can change the rhizosphere pH (Fig. 14.3). The pH buffering capacity of soils depends primarily on initial pH and organic matter content, but also on clay and $CaCO_3$ content; the pH buffering capacity is lowest at about pH 6 and increases at both lower and higher pH values (Nye, 1986; Schaller and Fischer, 1985). However, a lack of significant pH change in soils with high pH buffering capacity does not necessarily mean the absence of proton flux in the rhizosphere. Indeed, the protons may replace other cations on the cation-exchange sites of the soil and thereby affect the mobilization/immobilization of nutrients (Hinsinger et al., 2009).

14.4.1 Source of nitrogen supply and rhizosphere pH

Mineral N is plant available both in cationic (ammonium, NH_4^+) and anionic (nitrate, NO_3^-) forms and can comprise up to 80% of the total ion uptake due to the demand for N as major macronutrient. Therefore, the form of N supply determines the cation/anion uptake ratio and thus the rhizosphere pH (Fig. 14.9) in both annual (Marschner and Römheld, 1983) and perennial plant species (Rollwagen and Zasoski, 1988). Nitrate is the major form of inorganic N available to plants in many well-aerated agricultural soils. Nitrate uptake results in excess uptake of anions over cations, coupled with a net uptake of protons and thus an increase in rhizosphere pH. Furthermore, nitrate assimilation in the root tissue is associated with consumption of H^+ and may contribute to some extent to rhizosphere alkalinization.

In acid soils, the pH increase induced by nitrate supply can enhance to some extent P uptake by exchange of phosphate adsorbed to Fe and Al by HCO_3^- (Gahoonia et al., 1992) or by stimulation of microbial P mineralization (Bagayoko et al., 2000). For pasture grasses grown in acid soils with low P availability, the increase in rhizosphere pH results in P depletion in the rhizosphere (Armstrong and Helyar, 1992). Rhizosphere alkalinization may also alleviate the negative effects of soil acidity on plant growth (see Section 18.3) by increasing the availability of Ca and Mg, but reducing the concentration of toxic Al species in the rhizosphere soil solution (Fig. 14.10) (Bagayoko et al., 2000; Degenhardt et al., 1998; Pineros et al., 2005).

Ammonium uptake and H^+ production during ammonium assimilation in root tissues ($3NH_4^+ \rightarrow 4H^+$; see Chapter 2, Section 6.1) result in enhanced net extrusion of H^+ and rhizosphere acidification (Fig. 14.9). Preferential uptake of

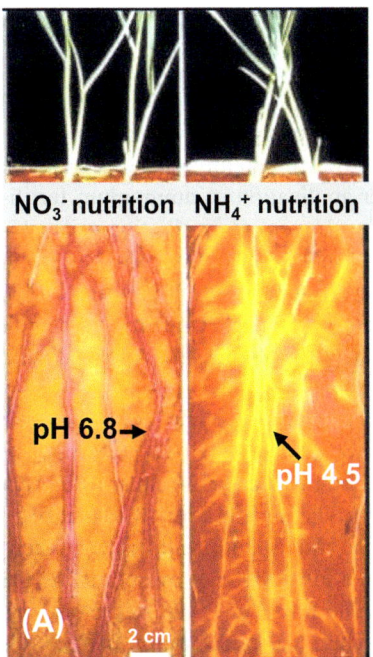

FIGURE 14.9 (A) Root-induced alterations in rhizosphere pH as affected by the form of nitrogen supply, detected by overlying of soil-grown roots by an agar slab containing a pH indicator. (B) Root imprints in limestone as a consequence of root-induced rhizosphere acidification. *Adapted from Römheld (1986) and Jaillard and Hinsinger (1993).*

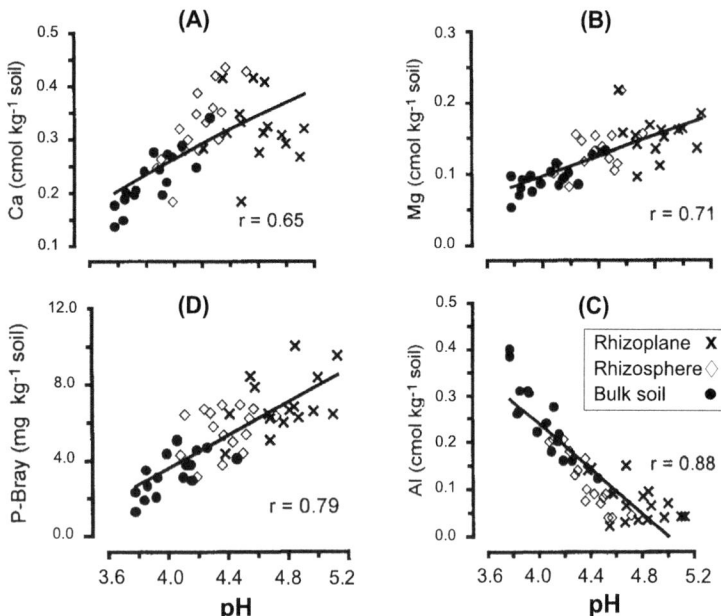

FIGURE 14.10 Effects of a root-induced increase in rhizosphere pH on availability of nutrients (Ca, Mg, P) and toxic elements (Al) to field-grown pearl millet (*Pennisetum glaucum*) in an acidic sandy soil in West Africa. (A) Exchangeable Ca, (B) exchangeable Mg, (C) exchangeable Al, and (D) available P (Bray). *Adapted from Bagayoko et al. (2000).*

TABLE 14.1 Nitrogen supply, rhizosphere pH, and concentration of mineral nutrients in shoots of bean (*Phaseolus vulgaris* L.) plants grown in a Luvisol (pH$_{water}$ 6.8).

Nitrogen supply	Rhizosphere pH	Concentration in shoot dry matter				
		K	P	Fe	Mn	Zn
		mg g^{-1}			µg g^{-1}	
NO$_3$-N	7.3	13.6	1.5	130	60	34
NH$_4$-N	5.4	14.0	2.9	200	70	49

Source: Adapted from Thomson et al. (1993).

ammonium occurs when nitrification is inhibited or delayed by oxygen limitation (particularly in wetland soils), by inhibition of microbial nitrifiers in acid soils, and by inhibited microbial activity at low temperatures in soils of arctic tundras (Chapin et al., 1993). High ammonium availability also occurs shortly after application of ammonium fertilizers (especially if combined with nitrification inhibitors) and many organic fertilizers. In neutral or alkaline soils, rhizosphere acidification due to ammonium supply can enhance solubilization of sparingly soluble Ca phosphates and thereby increase uptake of P (Gahoonia et al., 1992) as well as of micronutrients such as B (Reynolds et al., 1987), Fe, Mn, and Zn (Table 14.1). However, these benefits are largely dependent on the pH buffering capacity of the soil and the capacity of the root system to extrude protons (Jing et al., 2010), including genotypic differences in that capacity.

Due to the lower soil mobility of ammonium in comparison with nitrate and the capacity of plant roots to exploit patches of certain nutrients (i.e., N, P) via localized lateral root proliferation (Drew, 1975), fertilizer placement of ammonium phosphates has been identified as a suitable approach to enhance the ammonium-induced acidification potential, for example, of field-grown maize roots even in alkaline soils ($\sim$ pH$_{water}$ 8). The proliferation of lateral roots around patches of ammonium and phosphate fertilizers elevates the ammonium-induced rhizosphere acidification potential. Accordingly, the proton concentrations at the fertilizer placement zone can reach a level that can overcome the pH buffering capacity of alkaline soil, thus contributing to mobilization of sparingly soluble forms of P and micronutrients and increasing the agronomic N-use efficiency (Jing et al., 2010; Ma et al., 2015). The local proliferation of root tips around fertilizer patches also creates rhizosphere hot spots with increased concentrations of rhizodeposits, which can further stimulate root growth by supporting rhizosphere bacteria with root growth-promoting properties applied as inoculants (Nkebiwe et al., 2016).

The root-induced pH changes, mediated by supply of different N forms, can also impact on the composition and activity of native microbial communities in the rhizosphere, depending on the genotype of the host plant (Gallart et al., 2018; Wang and Tang, 2018). Rhizosphere acidification due to ammonium fertilization favors accumulation of soil organic carbon (SOC) in the rhizosphere by decreasing microbial activity and growth (a negative rhizosphere priming effect). By contrast, alkalization in response to the supply of nitrate-N stimulates SOC mineralization (a positive rhizosphere priming effect), associated with temporary N sequestration in the microbial biomass, with potential consequences for N availability to the host plant (Wang and Tang, 2018).

Rhizosphere acidification induced by preferential ammonium uptake may also contribute to enhanced resistance to plant diseases, such as take-all (*Gaeumannomyces graminis*) in wheat (Brennan, 1992), *Rhizoctonia fragariae* in strawberry (Elmer and LaMondia, 1999), and *Pseudomonas syringae* in tomato (González-Hernández et al., 2019). This has been related to improved micronutrient availability due to ammonium-induced rhizosphere acidification. Some micronutrients (Mn, Cu) are co-factors in the biosynthesis of lignin that acts as a mechanical barrier against pathogen attack, in the biosynthesis of phenolics with antibiotic properties, and in the biosynthesis of enzymes involved in defense reactions, such as superoxide dismutase (Zn, Cu, Mn, Fe), ascorbate peroxidase (Fe), diaminoxidase (Cu), polyphenol oxidase (Cu), ascorbate oxidase (Cu), peroxidase (Mn), and lipoxygenase (Fe) (see Chapter 7). Moreover, ammonium fertilization improved the root colonizing potential of various plant growth-promoting microorganisms with pathogen-antagonistic properties and Mn-mobilizing potential, belonging to the genera *Trichoderma*, *Pseudomonas*, and *Bacillus* (Mpanga et al., 2019; Sarniguet et al., 1992). Rhizosphere acidification due to ammonium uptake may also enhance the mechanical resistance of the cell walls due to increased incorporation of SiO_2 (Graham and Webb, 1991) (see Section 8.3). On the other hand, soil acidification also promotes some diseases, such as clubroot in cabbage and *Fusarium* wilt in cotton (Huber and Wilhelm, 1988). The N-form-dependent effects on pathogen sensitivity of host plants have been attributed to differential contents of stress-related metabolites, such as polyamines, salicylic acid, γ-aminobutyric acid, NO, and apoplasm sugars and amino acids (Mur et al., 2017).

Ammonium-induced rhizosphere acidification may also increase the availability of toxic elements such as Cd (Table 14.2) (Wu et al., 1989); hence, it has been proposed as a bioremediation strategy to improve the solubility and uptake of metals in neutral and alkaline soils by accumulator plants (phytoextraction), with the pH remaining high in the bulk soil, which prevents metal leaching (Zaccheo et al., 2006). Similar effects have been reported for uptake of Cd by, and storage in, woody plant species (phytostabilization) such as poplar (Qasim et al., 2015) or As in hyperaccumulating fern (*Pteris vittata*) (Liao et al., 2007), but the demonstrations that these strategies have a practical impact under real field conditions are still rare.

On acid soils, a pH decrease induced by ammonium fertilization would not necessarily enhance mobilization of nutrients and may even cause adverse effects on plant growth as a consequence of enhanced P adsorption to Fe and Al oxides, solubilization of toxic Al species, competition with Ca and Mg uptake or acid-induced root injury (see Section 18.3). However, under these conditions, it is possible to improve P acquisition by simultaneous application of acid-soluble Ca−P compounds (i.e., rock phosphates), mobilized by ammonium-induced rhizosphere acidification (Kanabo and Gilkes, 1987; Mpanga et al., 2018).

In waterlogged soils the inhibition of nitrification results in ammonium uptake and thus low rhizosphere pH. Moreover, in flooded soils, rhizosphere oxidation by release of O_2 from plant roots is an essential adaptation to prevent

TABLE 14.2 Effect of N form on rhizosphere soil pH and Cd concentration in *Lolium perenne* shoots.

Nitrogen form	Rhizosphere pH	Cadmium concentration (mg kg^{-1} shoot dw)		
		1st cut	2nd cut	3rd cut
NO$_3$	6.8	6.5	5.5	4.2
NH$_4$NO$_3$	6.8	9.2	8.2	7.6
NH$_4$	5.5	12.4	12.8	12.2

Source: Adapted from Wu et al. (1989).

the accumulation of Fe^{2+}, Mn^{2+}, H_2S, and monocarboxylic acids to phytotoxic levels (see Section 14.5.1). Oxidation of Fe^{2+} further promotes rhizosphere acidification according to the reaction:

$$4Fe^{2+} + O_2 + 10H_2O \rightarrow 4Fe(OH)_3 + 8H^+,$$

which can enhance (1) mobilization of Zn adsorbed to Fe^{III} hydroxides (Kirk and Bajita, 1995), (2) solubilization of acid-soluble soil P fractions (Saleque and Kirk, 1995), and (3) release of fixed NH_4^+ (Schneider and Scherer, 1998).

For rhizosphere pH measurements, average values integrated over the whole root system can be misleading and may result in erroneous conclusions about nutrient relationships in the rhizosphere. For example, within the root system of an individual plant, pH differences of more than two pH units may occur between primary and lateral roots or along the root axis (Marschner and Römheld, 1983; Marschner et al., 1986a). In Norway spruce growing in acid soil, the pH is high at the root apex and decreases in the subapical (extension) zone, irrespective of the form of N in the soil solution. By contrast, in the more basal root zones, a pH increase with nitrate supply and a pH decrease with ammonium supply occur (Häussling et al., 1988).

Rhizosphere acidification can take place despite the presence of high nitrate concentrations, particularly at high soil water content that facilitates diffusion of ammonium (Gijsman, 1991) and inhibits nitrification. A relatively high pH at the root apex is a common feature of plants grown in acid soils and may be related to the release of root exudates (see below) or to high nitrate reductase activity in root apical zones of nitrate-fed plants (Klotz and Horst, 1988). Studies with *Arabidopsis* mutants demonstrated that a high activity of the NRT1.1 transporter coupled with proton uptake was responsible for rhizosphere alkalinization in the apical root zones of plants exposed to a low-pH growth medium (Fang et al., 2016). Moreover, in Al-stressed *Pisum sativum*, sufficient B availability promoted polar auxin transport, leading to downregulation of the PM H^+-ATPase and elevated root surface pH in the Al-sensitive subapical transition zone of lateral roots (Li et al., 2018a).

Legumes and actinorhizal plants that meet their N requirement by symbiotic N_2 fixation take up more cations than anions because uncharged N_2 enters the roots and most other macronutrients required in large amounts (K, Ca, Mg) are cations (see Chapter 16). The high cation/anion uptake ratio of N_2-fixing plants results in net release of H^+, although per unit assimilated N, it is less than in ammonium-fed plants (Raven et al., 1991). The N_2-fixing plants have a higher capacity to utilize P from acid-soluble rock phosphate than nitrate-fed plants (Table 14.3). In soybean, Fe and Mn concentrations in N_2-fixing plants were higher than in nitrate-fed plants, indicating mobilization of these micronutrients by rhizosphere acidification (Wallace, 1982).

On soils with extremely low P availability, utilization of rock phosphate as a P source for legumes can be low when nodulation is limited by P deficiency. Thus, a starter supply of soluble P can enhance nodulation, N_2 fixation, and rhizosphere acidification and then utilization of rock phosphate (Swart and van Diest, 1987). When N_2-fixing legumes are grown together with nonlegumes (intercropping), rhizosphere acidification by legumes can increase P uptake from rock phosphate by nonlegumes (Xue et al., 2016).

In the long term, symbiotic N_2 fixation also affects the acidification of the bulk soil and thus the lime requirement. An alfalfa crop, fixing N_2 with an annual shoot dry matter production of 10 t ha^{-1}, produces soil acidity equivalent to

TABLE 14.3 Effects of N sources on acidity and alkalinity generated by roots of alfalfa (*Medicago sativa*), on soil pH and the utilization of rock phosphate (Aguilar and Van Diest, 1981).

Treatment						
Nitrogen source	Rock phosphate	Acidity	Alkalinity	Soil pH (H_2O)	Phosphorus uptake (mg pot^{-1})	Yield (g dw pot^{-1})
		mmol g^{-1} dw				
Nitrate	−	0	1.1	6.3	1	2.5
	+	0	0.8	7.3	23	19
N_2 fixation	−	0.5	0	6.2	4	4.7
	+	1.4	0	5.3	49	27

Source: Adapted from Adnan et al. (2017).

600 kg $CaCO_3$ ha^{-1} annually (Nyatsanaga and Pierre, 1973). In humid climate the loss of symbiotically fixed N through leaching of nitrate (and of an equivalent amount of countercations such as Ca and Mg) contributes to soil acidification under leguminous pastures (Bolan et al., 1991). A similar impact on long-term soil acidification by N_2 fixation can be observed in forest ecosystems where the pH under the actinorhizal red alder is lower than under Douglas fir (van Miegroet and Cole, 1984), and in crop rotations with a high proportion of legumes (Coventry and Slattery, 1991).

14.4.2 Nutritional status of plants and rhizosphere pH

Root-induced changes in rhizosphere pH are related to the nutritional status of plants (Cakmak and Marschner, 1990). Examples are rhizosphere acidification in cotton and other dicotyledonous plant species under Zn and P deficiency (Cakmak and Marschner, 1990; Hoffland et al., 1989; Neumann and Römheld, 1999), and in nongraminaceous species under Fe deficiency (Römheld, 1987). In P- and Zn-deficient plants, uptake of nitrate is inhibited due to nutrient stress, resulting in increased cation/anion uptake ratio and thus net release of H^+ (Cakmak and Marschner, 1990; Gniazdowska et al., 1999) (Table 14.4).

A substantial fraction of organic soil nitrogen is adsorbed to mineral surfaces of clay minerals in so-called mineral-associated organic matter (MAOM). Limited solubility delays N mineralization in this fraction. However, destabilization of metal-oxygen bonds in metal oxides and silicate clays by root-induced rhizosphere acidification can contribute to solubilization of MAOM, which stimulates microbial mineralization as a priming effect and thereby increases plant N availability (Jilling et al., 2018).

Direct stimulation of rhizosphere acidification in response to nutrient limitation may occur in nongraminaceous plant species (Strategy I plants, see Chapter 2) under Fe deficiency, whereby a strong local acidification occurs in the subapical root zones. This reaction is part of a coordinated response to Fe deficiency, including upregulation of (1) PM H^+-ATPase-mediated proton extrusion, (2) Fe^{2+} transporters (encoded by *IRT1* genes), (3) PM Fe^{3+}-chelate reductase encoded by *FRO2* genes, (4) the biosynthesis and release of organic iron chelators (coumarins, flavonoids, flavins, carboxylates), and (5) increased root surface area by the formation of root epidermal transfer cells and proliferation of root hairs in the relevant root zones (Fig. 14.11, see Chapters 2 and 7). This response facilitates Fe^{3+} solubilization by rhizosphere acidification and chelation, subsequent transport of chelated Fe^{3+} to the root surface, reduction and splitting of

TABLE 14.4 Effect of the P-nutritional status on the cation/anion uptake ratio, uptake of NO_3^- and net extrusion of protons, phosphoenolpyruvate carboxylase (PEPC) activity, and carboxylate accumulation in roots of various plant species.

Plant species and P treatment	Cation/anion uptake ratio	NO_3^- uptake (% change)	ΔpH (growth medium)	PEPC activity (nmol NADH min^{-1} mg^{-1} protein)	Carboxylates (mmol g^{-1} root fresh weight)
Tomato					
+P	0.78			90	8.5
−P	1.33	−83	−1.4	375	12
Chickpea					
+P	1.17			144	8.5
−P	1.26	−48	−0.6	302	17
White lupin					
+P	n.d.			120	8.7
−P	1.38	−56	−1.1	270	22
Wheat					
+P	0.39			426	1.2
−P	0.29	n.d.	+1.6	703	5.4

Source: Data compiled from Dinkelaker et al. (1989), Le Bot et al. (1990), Heuwinkel et al. (1992), Pilbeam et al. (1993), and Neumann et al. (1999).

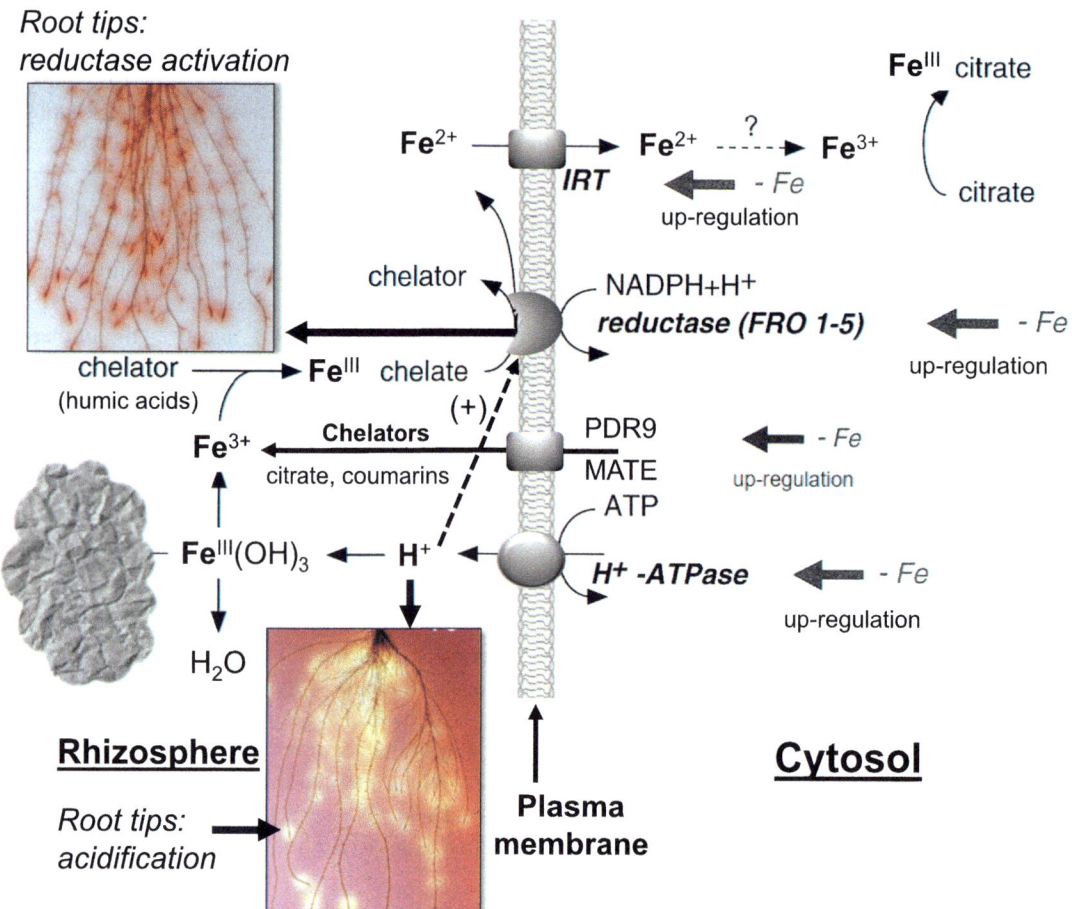

FIGURE 14.11 Model for iron (Fe) deficiency-induced changes in root physiology and rhizosphere chemistry associated with Fe acquisition by Strategy I plants. *Adapted from Marschner et al. (1986b)*.

this complex by the PM Fe^{3+}-chelate reductase that has a low pH optimum in the root apoplast, and finally uptake as Fe^{2+} via the IRT1 transporters.

For the root system as a whole, the rates of Fe deficiency-induced net H^+ release per unit root weight are of a similar order of magnitude as in the Fe-sufficient, ammonium-fed plants. However, average values are misleading because, under Fe deficiency, enhanced net release of H^+ is confined to the apical root zones where the actual rates are nearly eight times higher than in the ammonium-fed plants. This highly localized acidification may enable roots to decrease the rhizosphere pH in apical zones to enhance Fe mobilization. Nevertheless, in highly calcareous soils or in the presence of high bicarbonate concentrations in the soil solution, the soil pH buffering capacity can exceed the root-induced acidification potential, leading to Fe deficiency in Strategy I plants (Grillet and Schmidt, 2017; Poschenrieder et al., 2018).

Protons released into the rhizosphere are replaced intracellularly via upregulation of phosphoenolpyruvate carboxylase (PEPC) (Table 14.4) and glycolysis, which provides the protons for pH maintenance in the cytosol of the root cells (Sakano, 1998). The PEPC is involved in biosynthesis of carboxylates (oxaloacetate, malate, citrate) by catalyzing the carboxylation of PEP via nonphotosynthetic CO_2 fixation. Accordingly, PEPC activity usually increases in response to Fe and P limitation (Hoffland et al., 1992; Johnson et al., 1996; Neumann and Römheld, 1999). The carboxylate anions may be (1) further metabolized (in ammonium assimilation), (2) stored in the vacuoles of the root cells (Table 14.4), (3) translocated to the shoot, or (4) released into the rhizosphere (see Chapters 2 and 3). In some plant species, such as *L. albus* and the members of Proteaceae and Cyperaceae, a strong release of carboxylates and the concomitant release of H^+ or K^+ as counter ions is part of a strategy for mobilization of sparingly soluble P forms (Neumann and Römheld, 2007).

The enhanced extrusion of protons is mediated by upregulation of the PM H^+-ATPase in the root cells at the transcriptional, translational, and posttranslational levels (Fig. 14.12) (Rabotti and Zocchi, 1994; Tomasi et al., 2009; Yan et al., 2002). Moreover, external factors may modulate the activity of PM H^+-ATPase: for example, humic

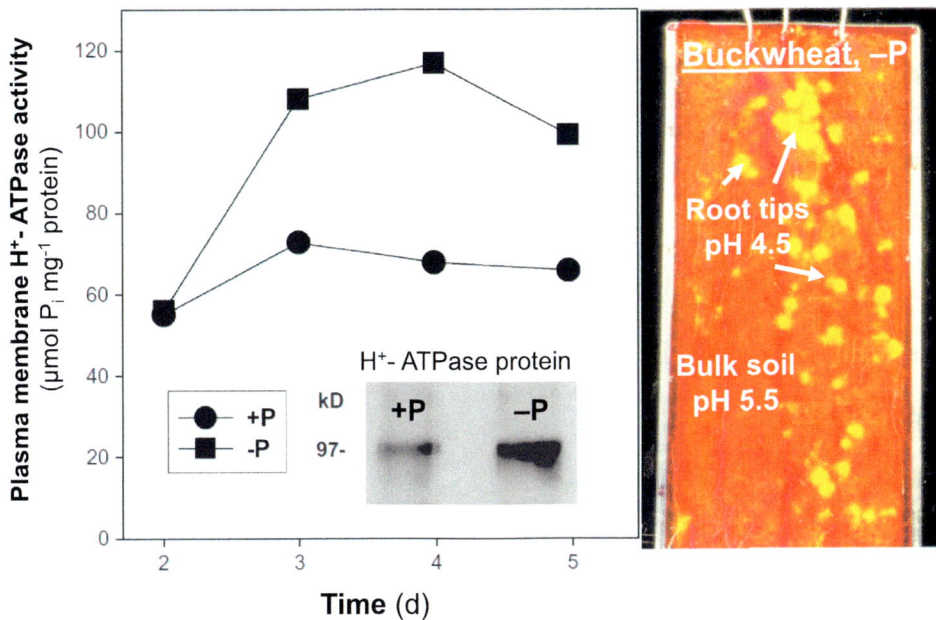

FIGURE 14.12 Modifications of rhizosphere pH induced by P limitation. Phosphorus deficiency-induced upregulation of the plasma membrane H$^+$-ATPase in white lupin (left) and localized rhizosphere acidification in apical root zones of P-deficient buckwheat (*Fagopyrum esculentum*) (right) detected by overlying the roots of soil-grown plants with an agar slab containing a pH indicator. *Adapted from Römheld (1986).*

substances (HSs) can stimulate root extrusion of H$^+$ (Pinton et al., 1997). The HS effects on ion uptake may be explained, at least partly, by interactions of phenolics such as HS and the root PM H$^+$-ATPase (Varanini et al., 1993; Pinton et al., 1999). The HS-mediated activation of the PM H$^+$-ATPase induces acidification of the rhizosphere, promotion of root growth, activation of nutrient transporters, and exudation of carboxylates (Canellas et al., 2015).

14.5 Redox potential and reducing processes

14.5.1 Effect of waterlogging

As soil water content increases, redox potential decreases. A decrease in redox potential is causally linked to the depletion of oxygen and correlated with a range of changes in the solubility of nutrients (e.g., N, Mn, Fe, and, occasionally, P). LMW organic acids as products of microbial fermentation processes and Fe^{2+}, Mn^{2+}, and H$_2$S can accumulate at phytotoxic concentrations after oxygen limitation lasting for several days or weeks. Moreover, in poorly aerated soils with low pH, a high microbial activity in response to a high supply of stress-induced root-exuded carbohydrates can further promote a decline in rhizosphere redox potential, which may result in increased Mn and Fe solubility and Mn and Fe toxicity to plants. However, at higher soil pH, trapped carbon dioxide originating from the fermentation process may form bicarbonate ions that have a high pH buffering capacity and may lead to Fe unavailability and leaf chlorosis.

As redox potential and O$_2$ concentration decrease, nitrate is rapidly used by microorganisms as an alternative electron acceptor, followed by reduction of Mn and Fe oxides. Due to the greater O$_2$ consumption in the rhizosphere than in the bulk soil, the risk of N losses by denitrification or incomplete nitrification (Klemedtsson et al., 1988) is higher in the vicinity of plant roots than in bulk soil. Rhizosphere denitrification is promoted by input of organic carbon from the roots into the rhizosphere (Bakken, 1988), particularly in K-deficient plants (von Rheinbaben and Trolldenier, 1984).

Plants adapted to waterlogging and submerged soils (e.g., lowland rice) maintain high redox potentials in the rhizosphere by the transport of O$_2$ from the shoot through aerenchyma in the roots and release of O$_2$ into the rhizosphere (Fig. 14.13). Aerenchymas are formed by programmed cell death of the cortex cells (lysigenic aerenchyma). This process involves local ethylene-induced reactive oxygen species (ROS) production mediated by NADPH oxidase (RBOH) and Ca signaling. This is characteristic not only for conditions of O$_2$ shortage but also occurs in response to drought or deficiency of N or P to reduce the energy and nutrient demand of the root tissue (Lynch, 2015). By contrast, so-called schizogenous aerenchyma is formed by separation of adjacent files of cells, increasing the volume of intercellular spaces (Takahashi et al., 2014). For many wetland plants, such as lowland rice, the formation of lysigenic aerenchyma is a constitutive trait, further stimulated in waterlogged soils, whereas in nonadapted crops (e.g., maize, wheat), the formation of aerenchyma is induced by external stimuli (Yamauchi et al., 2018). In waterlogged soils the existing root system can partially be replaced by aerenchymatous adventitious roots close to the surface or even submerged below the

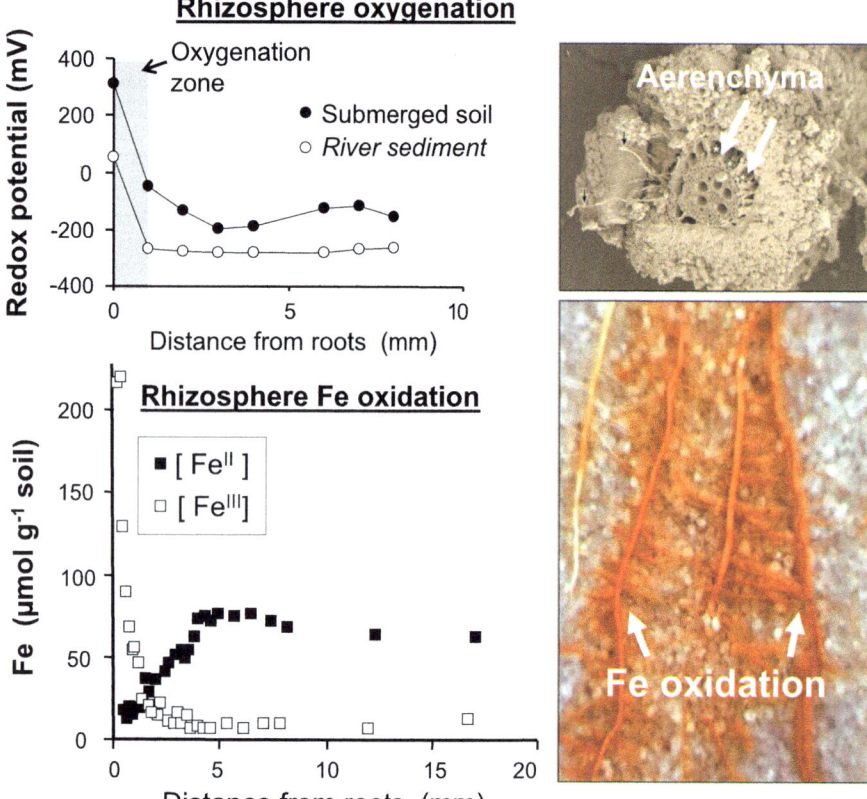

FIGURE 14.13 Rhizosphere oxygenation via aerenchyma in paddy rice associated with oxidation of Fe(II) in the rhizosphere. *Adapted from Flessa and Fischer (1992), Begg et al. (1994), and Watt et al. (2006).*

water table (aquatic adventitious roots), induced by an interplay of ethylene, abscisic acid, and auxin signaling (Fukao et al., 2019). In some plant species, even chloroplasts can develop in aquatic roots, and photosynthesis can further contribute to oxygen production (Rich et al., 2011).

The oxygenation of the rhizosphere is essential for avoiding phytotoxic concentrations of organic solutes and excess Fe^{2+} and Mn^{2+} uptake in poorly aerated or waterlogged soils. Oxygen transport within the roots and the rate of O_2 consumption in the roots (and particularly in the rhizosphere) are strongly influenced by nutrition.

The oxidation zone extends between 0.4 and 4 mm from the rhizoplane into the bulk soil, depending on O_2 supply and consumption, and on the redox buffer capacity of the soil (Fig. 14.13). The distance also varies along the axis of individual roots (Flessa and Fischer, 1992; Revsbech et al., 1999). In flooded rice the redox potential strongly increases behind the root apex (e.g., from -250 to about $+100$ mV), is low in more basal zones, and is high again at sites where lateral roots penetrate the cortex. This pattern of redox potential along the root axis may be related to differential density of rhizosphere microorganisms (as main O_2 consumers), which is low at the apex and increases in the basal zones prior to the emergence of lateral roots (Murakami et al., 1990). However, many wetland plants also form subepidermal barriers of suberinized cell layers in the basal root zones to restrict radial O_2 losses from aerenchyma to enhance longitudinal diffusion of O_2 toward the growing root tip that has a very high O_2 demand (Yanauchi et al., 2018). In rice this process is further stimulated by Si (Fleck et al., 2011). Oxygen released from the roots of wetland plants such as *Typha latifolia* L. can even be used for respiration by neighboring plants that otherwise would not withstand the low O_2 in the root environment (Callaway and King, 1996).

In aerated soils, average redox potentials are in the range of 500–700 mV. However, aerated soils are nonuniform, and hypoxic microsites may occur. Such microsites are more abundant in the rhizosphere than in the bulk soil (Fischer et al., 1989) and are particularly important for the acquisition of Mn and Fe, and for gaseous N losses (e.g., N_2, N_2O).

14.5.2 Manganese mobilization

Given that Mn is plant-available only in the reduced form (Mn^{2+}), in aerated soils root-induced reduction of Mn oxides may be a mechanism for increasing Mn availability. Reduction is mediated by the combined effects of (1) enzymatic reduction at the root surface, (2) chemical reduction by release of reductants, such as phenolics and malate, and (3) Mn reduction by Mn-reducing rhizosphere microorganisms (Godo and Reisenauer, 1980). Co-mobilization of Mn has frequently been reported for plant species efficient in P mobilization via release of carboxylates (Dinkelaker et al., 1989; Neumann and Martinoia, 2002), and leaf Mn concentrations may serve as a proxy for the capacity of carboxylate-mediated P mobilization in the rhizosphere (Wang and Lambers, 2019). The activity of rhizosphere microorganisms is of particular significance in Mn nutrition of plants because microorganisms can mediate Mn immobilization by oxidation as well as Mn solubilization by Mn reduction. Thus the balance of Mn-oxidizing bacteria (e.g., *Arthrobacter* spp.) to Mn reducers (e.g., fluorescent pseudomonads) strongly influences Mn availability in the rhizosphere (Posta et al., 1994; Rengel, 1997) (see Chapter 15).

14.5.3 Iron mobilization

Enhanced reducing activity at the root surface of subapical root zones is a typical feature of roots of Fe-deficient dicotyledons and nongraminaceous monocots. The reductive capacity is facilitated by PM-bound Fe^{3+}-chelate reductase (ferric reductase/oxidase; FRO) with a low pH optimum (Brüggemann et al., 1991; Holden et al., 1991), encoded by the *FRO2* gene (Robinson et al., 1997). The PM reductase–oxidase is activated by acidification of the apoplast (Römheld and Kramer, 1983) associated with increased activity of PM H^+-ATPase in the subapical root zones (see Section 14.4.2; Fig. 14.11). *FRO2* overexpression studies and a strict relationship between Fe^{3+} reduction and Fe^{2+} uptake indicated that Fe^{3+} reduction is the rate-limiting step for Fe uptake (Connolly et al., 2003; Grusak et al., 1990). However, activities of FRO and PM H^+-ATPase can be compromised by high soil pH and high buffering capacity and may be insufficient to match the plant demand for Fe (Grillet and Schmidt, 2017).

In Strategy I plants, increased reduction capacity and rhizosphere acidification are also associated with enhanced release of Fe-chelating and Fe-reducing compounds (Fig. 14.11), such as phenolics, flavins, and carboxylates (Römheld, 1987; Sisó-Terraza et al., 2016a,b). Certain phenolics, such as coumarins, flavonoids, and also flavins in root exudates of Fe-limited Strategy I plants exhibit both Fe^{3+}-chelating and Fe^{3+}-reducing properties even at high soil pH, and are preferentially exuded under these conditions (Rajniak et al., 2018; Sisó-Terraza et al., 2016a,b). These compounds can therefore supplement the enzymatic systems of Fe mobilization and reduction (PM H^+-ATPase, FRO) to extend the pH range of Fe acquisition in Strategy I plants (Grillet and Schmidt, 2017). For coumarins, apart from upregulation of the general phenylpropanoid metabolism, the oxoglutarate-dependent oxidase F6′H and the ABCG-type transporter PDR9 are involved in biosynthesis of coumarin glycosides and export as aglycones after cleavage of sugar moieties by β-glucosidase BGLU42. The related processes are under coordinated control of the FIT-regulated MYB47 transcription factor (Fourcroy et al., 2014; Tsai and Schmidt, 2017; Verbon et al., 2017). LMW phenolic acids, such as protocatechuic and caffeic acids, released from Fe-deficient plant roots via phenolic efflux transporters (e.g., the MATE transporter PEZ1 in rice) have also been implicated in the remobilization of Fe, and precipitation on the root surface and in the root apoplast (Ishimaru et al., 2011; Jin et al., 2007). However, considerable genotypic variation exists in the expression of these Strategy I responses, which is positively correlated with the resistance of plant species and cultivars to Fe deficiency under field conditions (Römheld, 1987).

Complexation of Fe^{3+} with soluble humic acids in soils and subsequent splitting of the complex by Fe^{3+} reduction may contribute to Fe acquisition by Strategy I plants (Pinton et al., 1998). HSs may also enhance the root-induced adaptations to Fe deficiency (Pinton et al., 1999), including the central components of the adaptive Strategy I responses, such as H^+ extrusion, release of organic Fe^{3+} chelators, and activation of the PM Fe^{3+}-chelate reductase, and Fe^{2+} transporters (Canellas et al., 2015; Tomasi et al., 2013). Moreover, volatile compounds released from beneficial rhizobacteria and fungi (e.g., strains of *Bacillus*, *Pseudomonas*, and *Trichoderma*) can activate the whole array of Strategy I responses, including the release of protons, FRO-mediated Fe^{3+} reduction, IRT1-mediated Fe^{2+} uptake, and synthesis and exudation of coumarins with Fe-mobilizing and antimicrobial properties and the capacity to induce systemic defense responses of the host plant (induced systemic resistance) against pathogens directly or indirectly (Verbon et al., 2017). Because Fe is an important nutrient limiting growth of both plants and microbes in soils, it is not surprising that Fe availability also influences the outcome of plant–pathogen interactions.

14.6 Rhizodeposition and root exudates

A substantial proportion (20%–60%) of photosynthetic C is allocated belowground (Grayston et al., 1996; Kuzyakov and Domanski, 2002). Depending on root activity, 15%–60% of this carbon allocation is used for root respiration and is released as CO_2 (Lambers et al., 2002). However, a substantial proportion of the assimilates reaches the rhizosphere as organic *rhizodeposition* (Fig. 14.14). Amount and composition of the released compounds are highly variable and influenced by multiple factors, including plant nutritional status, environmental conditions, biotic and abiotic stresses, and also sampling techniques. Estimates of rhizodeposition range from 800 to 4500 kg C ha^{-1} y^{-1} (Kuzyakov and Domanski, 2002; Lynch and Whipps, 1990) and can comprise up to 70% of the C partitioned belowground in perennials and up to 40% in annual plants. This is associated with a rhizodeposition of N ranging between 15 and 60 kg ha^{-1} y^{-1} (Hooker et al., 2000). Free amino acids and proteins usually make up only a minor fraction of organic compounds released from intact plant roots (typically 1%−2% of released C) (Jones and Darrah, 1993). Therefore it is assumed that N rhizodeposition may be related mainly to root turnover or efflux of inorganic N forms such as ammonium or nitrate (Feng et al., 1994; Jones et al., 2009; Scheurwater et al., 1999).

In contrast to the bulk soil where microbial growth is C-limited (Wardle, 1992), microbial growth in the rhizosphere is N-limited. Therefore, rhizodeposition by growing roots enhances the microbial turnover of soil organic C in the rhizosphere (*priming effect*; Helal and Sauerbeck, 1989; Kuzyakov, 2002), particularly in plants well supplied with N (Liljeroth et al., 1990). This induces temporary N sequestration in microbial biomass via NH_4^+ uptake and can also decrease nitrification (Leptin et al., 2021). Rhizodeposition is also influenced by N availability in soils. However, the effects reported in the literature are highly variable due to the influence of root morphology and physiology, the composition of plant tissues in response to different N forms, and the use of various approaches to characterize rhizodeposits (Bowsher et al., 2018).

A portion of N temporarily immobilized in the rhizosphere microbial biomass can be released and mineralized via excretions of protozoa and nematodes grazing on rhizosphere microbial populations, the so-called *microbial loop* increasing plant N availability (Bonkowski et al., 2009) (see Chapter 15). A similar relationship has been described in pine trees for utilization of phytate P as a major, albeit highly insoluble, organic P form in soils, preferentially mobilized by phytase-secreting bacteria and fungi. In this case, sequestered microbial P was converted into plant-available forms by grazing of nematodes on the respective microbiota (Irshad et al., 2012). The importance of rhizosphere in cycling of C and nutrients in soils is further illustrated by the fact that organic rhizodeposition, which can account for 30%−40% of the total soil organic matter input, is released into the rhizosphere soil that comprises only 2%−3% of the total soil volume (Grayston et al., 1996).

Many internal and external factors determine the amount and composition of rhizodeposits. Qualitative and quantitative variability arises from plant genotypic differences, diurnal variations in release patterns at different root zones, and

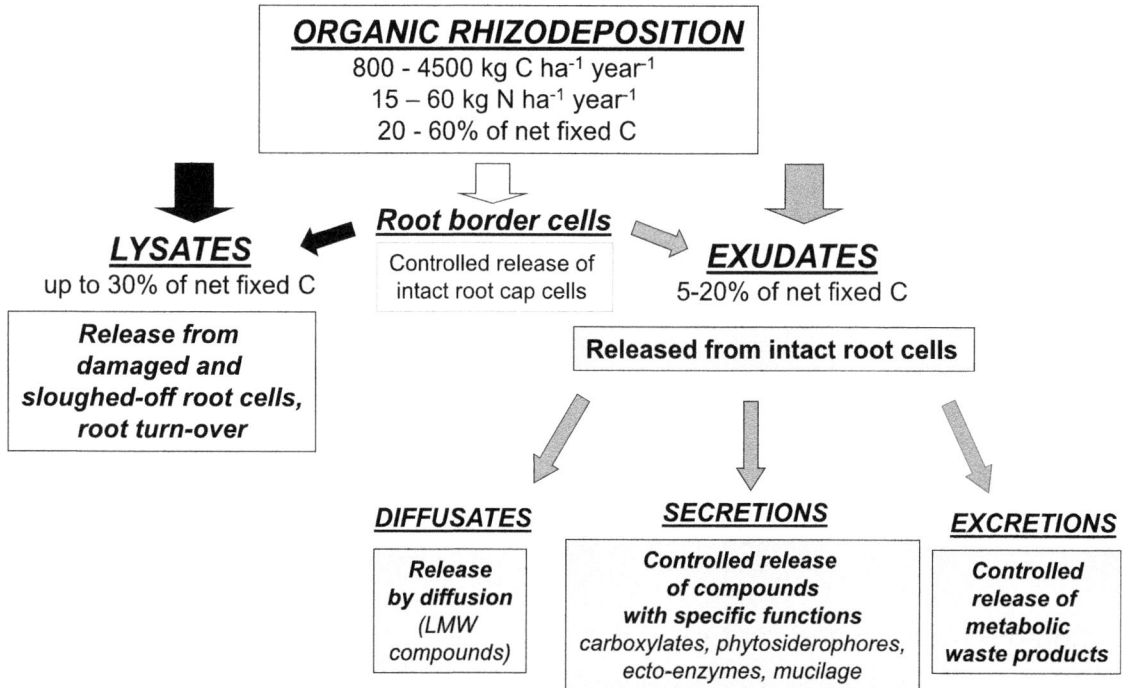

FIGURE 14.14 Classification and quantities of organic rhizodeposition. *Adapted from Neumann (2007).*

various environmental factors (Neumann and Römheld, 2007). Rhizodeposition can be stimulated by increased mechanical impedance of the growth substrate (Boeuf-Tremblay et al., 1995), toxic elements and low pH in the soil solution (Costa et al., 1997; Kochian, 1995; Römheld and Marschner, 1983), nutrient limitation (Marschner, 1998; Neumann and Römheld, 2007), high light intensity (Cakmak et al., 1998; Rovira, 1959), elevated atmospheric CO_2 concentrations (Haase et al., 2007), temperature extremes (Rovira, 1959), and the presence of microorganisms (Meharg and Kilham, 1995). On the other hand, the quantity, composition, and variation of rhizodeposits are the major factors shaping rhizosphere microbial communities (rhizosphere effect), frequently leading to a higher population density, but lower species diversity, of rhizosphere microorganisms (Berg and Smalla, 2009; Mendes et al., 2013). Because nearly all soil processes are influenced by microbial activities (Mäder et al., 2002), plant—microbe interactions in the rhizosphere driven by rhizodeposition are of outstanding importance for growth and development, nutrient acquisition, health status, and stress resilience of plants (see Chapter 15).

Depending on the origin and release mechanisms, rhizodeposition may include two main fractions: (1) lysates of sloughed-off cells and tissues originating from root turnover (up to 50% of the belowground C translocation) (Grayston et al., 1996) and (2) organic compounds released from intact root cells as *root exudates*. The root exudate fraction may be subdivided into (1) LMW organic compounds permanently lost from root cells by diffusion (*diffusates*); (2) *root secretions* with special functions in nutrient mobilization, compound detoxification, defense reactions, or root signaling, released by controlled mechanisms via membrane channels or transport proteins; and (3) metabolic waste products released as *root excretions* (Fig. 14.14). However, due to overlapping release mechanisms, it is not always easy to differentiate among these fractions.

Carbon input into soils via root exudation may comprise 5%—10% of the net fixed carbon in soil-grown plants (Jones et al., 2004a). A significant proportion of C also reaches the soil via mycorrhizal hyphae (Johnson et al., 2002). In the case of ectomycorrhizal fungi, C input via large extended hyphal systems may even dominate C transfer into the soil (Godbold et al., 2006).

Organic rhizodeposition also includes nutrients previously taken up by the plant and bound in organic molecules. In young wheat plants, for example, this contributes 1%—5% of plant P (McLaughlin et al., 1987) and over the whole growth period, it can comprise up to 18% of the total N in low-N plants and 33% in high-N plants (Janzen, 1990). Significant inputs (up to 5%) of soil N may also originate from glomalin proteins associated with arbuscular mycorrhizal hyphae (Rillig et al., 2003).

14.6.1 Sloughed-off cells and tissues

In soil-grown plants, parts of the rhizodermis, including root hairs and cortical cells, may degenerate and release their content into the rhizosphere (Fusseder, 1984; McCully, 1999). During the growing season, approximately 25% of the roots turn over each month (Jones et al., 2009), with the diameter and life span of fine roots being positively correlated, as shown, for example, in forest ecosystems (Gill and Jackson, 2000). The life span of arbuscular mycorrhizal hyphae has been estimated at 5—6 days (Staddon et al., 2003).

The release of sloughed-off root cells may be a genetically controlled process. The *root border cells* are produced by the peripheral cells of the root cap (Fig. 14.15). The number of released border cells varies among genotypes and ranges between approximately 100 cells d^{-1} in tobacco (*Nicotiana tabaccum*) to up to 10,000 cells d^{-1} in cotton (*Gossypium hirsutum*). Morphotypic variation of border cells has been observed in different root zones of the same plant species, ranging from single cells of different shapes to multilayered stacks of cells. The production and release are stimulated by various environmental factors, including water availability, elevated CO_2 concentrations, and mechanical impedance, but are also regulated by phytohormonal signals, such as auxins and ethylene (Driouich et al., 2007; Ropitaux et al., 2020) or the presence of humic acids (Canellas and Olivares, 2017).

The release of root border cells starts with increased activity of pectolytic enzymes, such as polygalacturonidase and pectin methyl esterase, responsible for the hydrolysis of the pectin matrix of the cell wall. This is associated with the extrusion of protons; the resulting acidification may activate other cell wall-degrading enzymes (Driouich et al., 2007). After detachment a yet unknown feedback signal released by the root border cells leads to downregulation of the hydrolytic processes in the cell walls of the root cap. Embedded into a layer of mucilage polysaccharides, the cells are viable after detachment from the root cap for up to 1 week or even longer, and with the progression of root growth can be found around more basal root parts (Hawes et al., 2016; McCully, 1999). After release of border cells, mitosis in the root meristem is initiated within minutes to replace the displaced cells (Hawes et al., 2016).

For border cell populations, specific attraction by root pathogens, such as parasitic nematodes, fungal zoospores, and pathogenic bacteria has been reported, which may serve as pathogen distraction and reduce infection of the apical

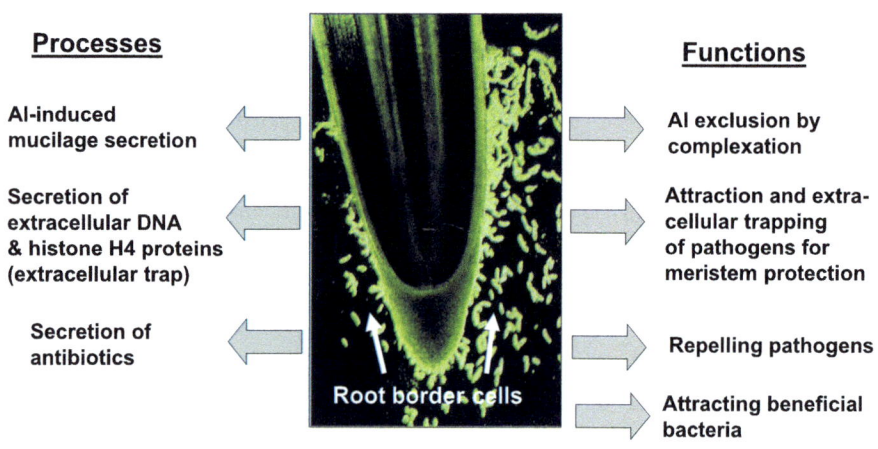

FIGURE 14.15 Liberation and potential functions of root border cells. *Adapted from Hawes et al. (2016).*

root meristem. However, despite selective attraction by pathogens, border cells were either not infected or, in cases of cell penetration, intracellular development of the pathogen was inhibited (Hawes et al., 2016). Secretion of a certain set of proteins during detachment of root border cells appears to be involved in this process (Wen et al., 2007). Accordingly, border cells exhibit characteristic transcriptome and proteome with more than 100 secretory proteins, mainly with stress- and defense-related functions, including chitinases, glucanases, and peptidases.

Border cells may produce and release antibiotics. Moreover, the secretion of extracellular DNA (exDNA) linked to histone H4 proteins was detected in the border cell secretome, known to play a critical role in defense against microbial pathogens in mammalian blood cells (Hawes et al., 2016). The whole plethora of antimicrobial phytochemicals accumulate in a matrix of secretory polysaccharides (mucilage) with characteristic polymers (e.g., xyloglucan, heteromannan, homogalacturonan, arabinogalactan) forming a biofilm-like protective structure termed the *root extracellular trap* that provides first defense against root pathogens (Baetz and Martinoia, 2013; Driouich et al., 2010; Hawes et al., 2016; Ropitaux et al., 2020). Apart from pathogen defense, root border cells may also be involved in the establishment of beneficial plant–microbe interactions with plant growth-promoting bacteria (Canellas and Olivares, 2017).

In response to toxic Al concentrations, root border cells of Al-resistant plants secrete large amounts of mucilage, which may be involved in Al detoxification due to complexation with galacturonates (Miyasaka and Hawes, 2001) as an example of protection against abiotic stress factors. In response to increased mechanical impedance of the growth substrate, stimulation of border cell production may indicate their functions as lubricant producers to support root penetration in compacted or drying soils (Ijima et al., 2004).

14.6.2 High-molecular-weight compounds in root exudates

High-molecular-weight (HMW) secretions, such as mucilage polysaccharides and ectoenzymes, are released from roots by vesicle transport (exocytosis) (Battey et al., 1999).

14.6.2.1 Mucilage and mucigel

Mucilage is secreted mainly via the Golgi apparatus of hypersecretory root cap cells as a gelatinous polygalacturonic acid polysaccharide (Fig. 14.16A); in addition, epidermal cells and root hairs can also secrete mucilage (Vermeer and McCully, 1981). In nonsterile media, mucilage includes substances produced by microbial degradation of the cell walls (Rovira et al., 1983). In soil-grown plants, mucilage is usually invaded by microorganisms, and both organic and inorganic soil particles are embedded in it. Entrapped root exudates and root debris are processed by the invading microorganisms forming biofilms of exopolysaccharides. Mucilage, microbial exopolysaccharides as well as glomalin proteins linked with hyphae of arbuscular mycorrhizal fungi can act as binding agents in the formation of soil microaggregates (Chenu and Cosentino, 2011). This mixture of gelatinous material, microorganisms, and soil particles (Fig. 14.16b) is termed *mucigel* (Bowen and Rovira, 1991). After cell death and lysis of microbial cells, the microaggregates (including mucigel polymers) remain stable (Totsche et al., 2018).

Mucilage and mucigel may not only protect the root meristem but also improve the root–soil contact by inclusion and aggregation of soil particles (Carminati et al., 2017; McCully, 1999). A putative direct function as lubricant, suggested in

FIGURE 14.16 Release and functions of mucilage from plant roots: (A) swelling of the mucilage layer covering the root cap by rapid water uptake; (B) formation of mucigel by mucilage with enclosed soil particles; (C) mucilage-mediated binding of soil particles in the rhizosphere of *Noccaea* (formerly *Thlaspi*) *caerulescens*; (D) mucilage-mediated formation of soil rhizosheaths around roots of field-grown maize proliferating around a band of ammonium phosphate fertilizer. *Adapted parts (A, B, and C) from Ingwersen et al. (2006) and Neumann (2007) and (D) courtesy Prof. Jianbo Shen (China Agricultural University, Beijing).*

earlier studies, appears to be unlikely because mucilage does not retain water or swell at water potentials lower than zero (McCully, 1999). However, mucilage moved during root elongation to more basal root parts and also mucilage released from root hairs can form *rhizosheaths* by inclusion of soil particles (McCully, 1999) (Fig. 14.16C and D). Apart from the intensity of mucilage secretion, the length of the root hairs determines the extension of the rhizosheaths as demonstrated with root-hairless mutants and plant genotypes differing in root hair length (Haling et al., 2014; Holz et al., 2018). Shrinking of mucilage with declining water potentials leads to a tighter association of soil particles within the rhizosheaths.

Mucilage increases the viscosity and decreases the surface tension of the soil solution. This contributes to improved water-holding capacity of the rhizosphere soil and increases the proportion of water-stable soil aggregates from about 2% to nearly 40% (Carminati et al., 2017; Morel et al., 1991). However, mucilage turns hydrophobic upon drying under severe drought stress, causing water repellency in the rhizosphere, which may counteract further water losses from root cells. This process depends on the nature and amount of root exudates and microbial metabolites accumulating in the mucigel matrix and can vary in different plant species and root developmental stages (Carminati, 2012; Carminati et al., 2017).

In desert plants (*Opuntia*), the formation of rhizosheaths or so-called sand-binding roots under drought stress can reduce water loss from roots by approximately 30% (Huang et al., 1993) and is a widespread feature in plants adapted to semiarid and arid conditions, including members of the families Poaceae, Cactaceae, Restionaceae, and Haemodoraceae (Smith et al., 2011). Recently, higher mucilage exudation and differences in composition have been reported in maize genotypes from semiarid regions in comparison with those originating from temperate climates (Nazari et al., 2020).

The close contact between soil particles and the root surface via mucilage can be of considerable importance in uptake of nutrients, particularly micronutrients and P. At the soil/root interface, processes may be different from those occurring in the free solution ("two-phase-effect"; Matar et al., 1967). In low-P soil, plants take up P that is not in equilibrium with the soil solution but is mobilized at the root/soil interface presumably via P desorption from clay surfaces by polygalacturonates in mucilage (Nagarajah et al., 1970). Possible functions of phosphatidylcholine surfactants in root mucilage in P mobilization, inhibition of nitrification, and modification of soil water retention have been discussed by Read et al. (2003). The two-phase effects may supply only a minor fraction of the total demand for macronutrients such as P but can have greater importance in uptake of micronutrients. In dry soils, stimulation of mucilage secretion in response to increased soil mechanical impedance can contribute to the maintenance of Zn^{2+} uptake by facilitating Zn^{2+} transport from the embedded soil particles to the root surface (Nambiar, 1976).

Mucilage may contribute to exclusion of toxic elements such as Al (Table 14.5) (Horst et al., 1982) and metals (Cd, Pb; Morel et al., 1986) by complexation with galacturonates, mainly in exchange for Ca^{2+}. In roots of cowpea (*Vigna unguiculata*) exposed to Al toxicity, a high proportion of Al is bound to the mucilage (see Section 18.3). On a dry

TABLE 14.5 Root growth and Al concentration and content in roots and mucilage of cowpea grown in nutrient solution with or without Al in the presence or absence of mucilage (removed three times per day).

Aluminum treatment (mg L^{-1})	Mucilage	Root growth (cm d^{-1})	Aluminum in root tips			
			Content [μg (25 tips)$^{-1}$]		Concentration (mg g^{-1} dw)	
			Roots	Mucilage	Roots	Mucilage
0	Present	6.3	–	–	–	–
	Removed	5.9	–	–	–	–
5	Present	4.8	12	17	2.1	17
	Removed	2.1	21	3.6	3.2	15

Source: Adapted from Horst et al. (1982).

weight basis, the mucilage contains about eight times more Al than the root tissue, and removal of the mucilage leads to an increase in the Al concentration of the root tissue and inhibition of root extension. The enhancement of mucilage production by mechanical impedance is therefore a major contributing factor to the higher Al tolerance of roots growing in solid substrates compared with nutrient solution culture.

14.6.2.2 Secretory proteins

Plant roots release a wide range of proteins, including various enzymes. Secretory proteins are synthesized by polysomes attached to the endoplasmic reticulum (ER) and are segregated into the ER lumen during the translation process. During the passage through the Golgi apparatus, transfer vesicles containing the secretory proteins are separated from vesicles with vacuolar destination. After reaching the PM, the proteins in the vesicles are released into the apoplast via exocytosis (Chrispeels, 1991). All processes linked with exocytosis strongly depend on Ca^{2+} supply (Battey et al., 1999).

A wide range of enzymes involved in the hydrolysis of organic P esters, such as phytase, nuclease, pyrophosphatase, apyrase, and acid and alkaline phosphatases, have been detected in the rhizosphere (Neumann and Römheld, 2007). These enzymes may originate from plant roots as well as from rhizosphere microorganisms. However, acid phosphatase activity in the rhizosphere of *L. albus* and other Fabaceae as well as species of Poaceae and Brassicaceae was shown to be predominantly of plant origin (Wasaki et al., 2005). Acid phosphatases with a pH optimum ranging between 4 and 7 are thought to be released mainly from plant roots and fungal hyphae, whereas alkaline phosphatases with a pH optimum between 8 and 10 are thought to be of bacterial origin (Nannipieri et al., 2011).

In most agricultural soils, between 30% and 70% of the total soil P is present in organic form. In forest soils, this proportion may rise to 80%–90% (Häussling and Marschner, 1989). Organic P may comprise phytate (myo-inositolhexaphosphate), the least soluble and frequently the dominant organic P fraction in many soils, whereas sugar, lipid, or nucleotide phosphates exhibit higher solubility and thus higher rates of mineralization, mediated by soil phosphatases (Richardson et al., 2005). Phytate mineralization via secretion of phytases has been attributed mainly to soil microorganisms (Richardson and Simpson, 2011), but considerable phytase activities have been reported recently also for purple acid phosphatases released from roots of soybean and *Stylosanthes guianensis* (Liu et al., 2018).

Phosphorus limitation frequently leads to a stimulation of phosphatase secretion from plant roots (Fig. 14.17), but considerable variation exists among plant species and cultivars (Li et al., 1997; Römer et al., 1995). Phosphorus deficiency-induced root secretion of acid phosphatases is regulated at the transcriptional and posttranscriptional level (Tran et al., 2010; Wasaki et al., 1997) and may involve sensing of external P concentrations in the growth medium (Ma et al., 2021; Wasaki et al., 1999) and differential induction of various iso-enzymes (Gilbert et al., 1999; Tran et al., 2010).

Plants grown in nutrient solution or sand culture can use organic P forms to a similar extent as inorganic P. However, in many soils, enzymatic hydrolysis by root-secreted phosphatases is limited by the low solubility of organic P forms in soils (Table 14.6) (Adams and Pate, 1992; Hübel and Beck, 1993) due to adsorption and precipitation by forming sparingly soluble salts and complexes with Ca, Fe, and Al. Furthermore, the hydrolysis of organic P esters in

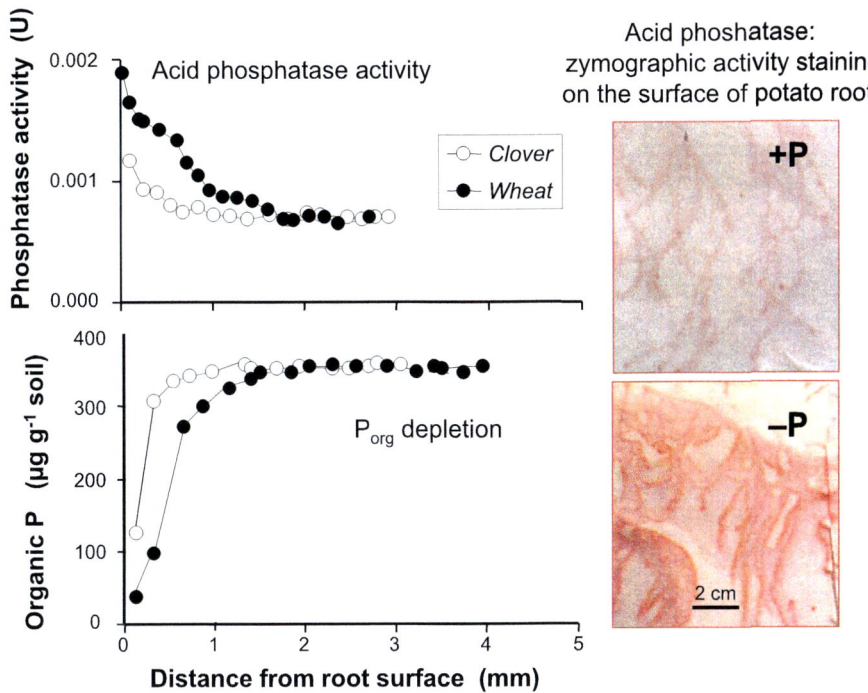

FIGURE 14.17 Phosphatase activity and depletion of organic soil P in the rhizosphere of red clover (*Trifolium pratense*) and bread wheat (*Triticum aestivum*) (left) and activity staining of root-secreted acid phosphatase (right) in P-sufficient and P-deficient potato (*Solanum tuberosum*). Adapted from Tarafdar and Jungk (1987) and Dinkelaker and Marschner (1992).

TABLE 14.6 Utilization of organic P (phytate) and KH_2PO_4 in potato grown in hydroponics and in soil culture (mean ± SE).

Parameter	No P	Organic P (phytate)	Soluble P (KH_2PO_4)
\multicolumn{4}{c}{Dry matter (g plant^{-1})}
Nutrient solution	6.5 ± 1.4	14 ± 2.7	23 ± 2.6
Soil culture	–	8.5 ± 1.9	13 ± 1.8
\multicolumn{4}{c}{Shoot P concentration (mg g^{-1} dw)}
Nutrient solution	1.6 ± 0.1	4.3 ± 0.7	6.9 ± 0.4
Soil culture	–	1.8 ± 0.1	4.5 ± 0.2

Source: Adapted from Neumann et al. (2001).

the rhizosphere may be limited by (1) immobilization of the secretory phosphatases in the root cell wall or the mucilage layer (Dinkelaker et al., 1997) and (2) adsorption and inactivation on clay minerals and organo-mineral complexes (Rao et al., 1996). Accordingly, attempts to increase the acquisition of phytate P by transgenic expression of secretory phytase genes from *Aspergillus niger* or *Bacillus subtilis* in *Arabidopsis* were successful in artificial growth media (agar, sand, perlite) with low adsorption potential (Belgaroui et al., 2016; Richardson et al., 2005; Valeeva et al., 2018) but largely failed in soil culture (Richardson et al., 2005) or worked only after application of soluble Na-phytate (Wang et al., 2013).

At least in some plant species, root secretion of carboxylates, such as oxalate and citrate, may enhance the solubility of organic P forms, making them available for hydrolysis by phosphohydrolases in the rhizosphere (Fig. 14.18) (Beissner, 1997; Otani and Ae, 1993) and the same holds true for microbial P mineralization in soils (Alori et al., 2017). In dry soils, enhanced water retention in the rhizosphere due to mucigel production may support organic P mineralization by increasing diffusion of phosphatases and organic P forms near roots (Holz et al., 2019). However, in many cases, P acquisition from sparingly soluble organic soil P forms via root-secreted acid phosphatases appears to be insufficient to satisfy the plant P demand and requires the presence of higher concentrations of soluble organic P forms, for example, after application of manure-based organic fertilizers (Hummel et al., 2021; Mehra et al., 2017). Organic

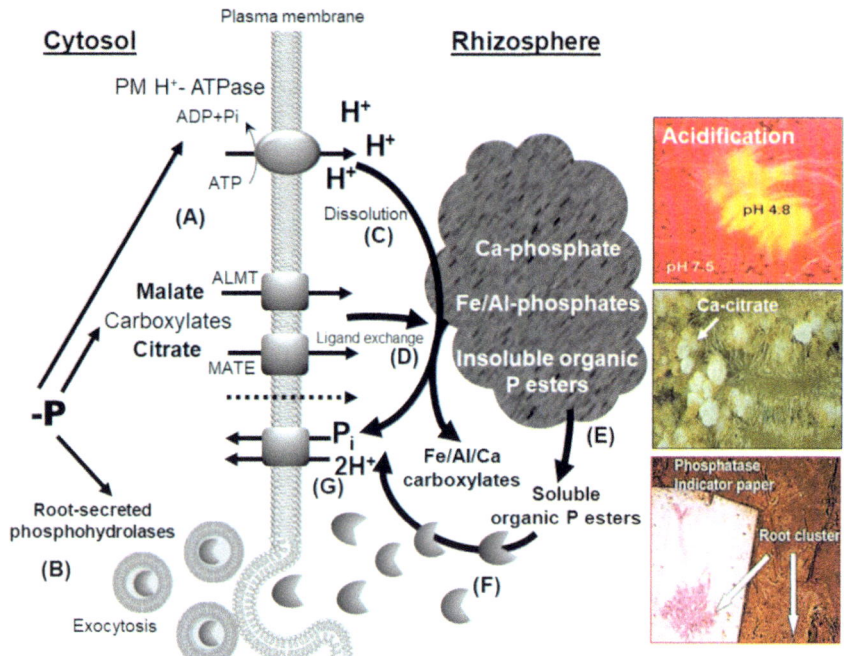

FIGURE 14.18 Model for the role of root exudates in chemical P mobilization in the rhizosphere. Phosphorus deficiency-induced secretion of carboxylates with concomitant H$^+$ extrusion (A) and of root-secreted acid phosphatases (B). Dissolution of acid-soluble Ca–P by H$^+$ release from roots (C) and displacement of phosphate anions from the anion sorption sites (Fe/Al/Ca) on the soil matrix by carboxylates (D). Displacement of organic P esters from the anion sorption sites on the soil matrix by carboxylates (E). Enzymatic hydrolysis of organic P esters in the soil solution by the activity of phosphatases released by roots and microorganisms (F). Root uptake of mobilized inorganic P by P transporters via H$^+$ cotransport (G). *Adapted from Dinkelaker et al. (1989), Neumann and Martinoia (2002), and Neumann (2007).*

soil amendments providing soluble organic P sources can increase plant P availability by stimulating P mineralization via microbial phosphatases and can contribute even to direct solubilization of sparingly soluble mineral forms of soil P (Adnan et al., 2017; Gichangi, 2019).

The P mobilization in the rhizosphere is a common trait of plants and microorganisms, but they also compete for soluble P. Therefore transformation into soluble organic P forms, available for hydrolysis by root-secreted phosphatases, is rather a consequence of turnover of the microbial biomass over longer periods than a direct contribution of microbial P mobilization to plant P uptake (Raymond et al., 2020). Liberation of P sequestered in the microbial biomass has been identified as a major factor determining gross P fluxes in soils with low inorganic P availability (Bünemann et al., 2012). This process can be further stimulated by abiotic perturbations, such as freezing-thawing or drying-rewetting cycles that can kill up to 70% of the total microbial biomass (Blackwell et al., 2010), or by predators grazing on bacterial populations (Irshad et al., 2012). Additionally, root-secreted acid phosphatases may also contribute to P retrieval by hydrolysis of organic P lost into the rhizosphere from sloughed-off and damaged root cells (Yadav and Tarafdar, 2001).

Many enzymes are located in the root apoplasm, particularly in the epidermal cells of apical root zones. These include enzymes potentially involved in defense reactions, such as chitinases, glucanases, peroxidases, and phenoloxidases, as well as those needed for cell wall biosynthesis and C supply to mycorrhizal fungi (e.g., invertase). Their contributions to nutrient dynamics in the rhizosphere and nutrient acquisition may be mainly indirect. A role of proteases released from plant roots in N cycling in the rhizosphere is unclear, but proteases secreted by some ericoid and ectomycorrhizal fungi have been shown to increase access of the host plant to HMW organic sources of N such as proteins (see Chapter 15). Because the host plants themselves have little or no access to these resources, the associated fungi may play a crucial role in the host plant growth on substrates containing complex organic N (Hutchison, 1990). Protease secretion involved in soil N mineralization is a feature of many other soil microorganisms (Zaman et al., 1999) and has been demonstrated unequivocally also for many plant species in axenic culture (Adamczyk et al., 2010; Greenfield et al., 2020).

Plant roots can take up amino acids and small peptides from the external medium (Section 14.6.3.2), and endocytosis of large proteins or bacterial cells has been reported (Paungfoo-Lonhienne et al., 2010). However, it remains unclear to what extent these processes can contribute to plant N nutritional status. In maize, wheat, and tomato, secretory proteases were almost completely attached to the root surface, and secretion was not upregulated under N deficiency, suggesting no close relationship with N acquisition (Holzgreve et al., 2019; Greenfield et al., 2020). However, this does not exclude potential functions in retrieval of protein N from, for example, inactive enzymes and proteins in the root apoplasm during root turnover, in pathogen defense, or in plant species adapted to ecosystems dominated by organic N forms in soils. In addition, a contribution to N acquisition after severe drought stress is possible, leading to liberation of large amounts of protein N from decaying microorganisms (Blackwell et al., 2010; Homyak et al., 2017; Kohli et al., 2012).

14.6.3 Low-molecular-weight root exudates

14.6.3.1 Diffusion-mediated release of low-molecular-weight compounds

Even in intact root cells, any soluble LMW compound present in the cytosol may be lost into the rhizosphere (Oburger and Jones, 2018). A large concentration gradient of LMW solutes usually exists between the cytosol (millimolar concentrations) and the rhizosphere (micromolar concentrations because of microbial degradation), promoting outward diffusion of LMW compounds. Moreover, the outward positive electrochemical potential gradient created by proton extrusion via PM H^+-ATPases further promotes outward diffusion of those LMW compounds that are negatively charged at the cytosolic pH of 7.0–7.5, such as organic acids and amino acids (Neumann, 2007). Diffusion-mediated passage through model biomembranes has been reported mainly for gases or small molecules (e.g., urea or glycerol), but not for larger, polar compounds such as most sugars, amino acids, carboxylates, and nucleosides (Canarini et al., 2019; Yang and Hinner, 2015). However, under natural growth conditions, PM of root epidermal cells is permanently exposed to a wide range of biotic and abiotic stress factors of varying intensity (such as extremes in temperature and water availability, hypoxia, nutrient limitations, ion toxicities, oxidative stress, microbial metabolites) and with variable effects on membrane integrity and permeability and thereby also on diffusion-mediated loss of cell contents. Diffusion-mediated release of LMW compounds is particularly intense in the apical root zones where phloem unloading of solutes occurs through plasmodesmata by a combination of mass flow and diffusion. Given that diffusion barriers, such as the suberinized Casparian band or lignification, are absent in this early stage of root development, solutes can move out of the root via the apoplastic pathway (Canarini et al., 2019; Ross-Elliott et al., 2017). In the rhizosphere, these compounds may contribute to nutrient cycling as C and N sources, serve as signals for rhizosphere microorganisms, and influence rhizosphere microbial communities.

Major LMW compounds detected in root exudates include sugars, organic acid anions, amino acids, and various secondary metabolites such as phenolics, terpenoids, glucosinolates, and alkaloids. Due to rapid microbial decomposition, the half-life of many LMW compounds in the rhizosphere is only 1–5 h (Jones et al., 2005). However, immobilization of exudates on the soil matrix by adsorption and complexation can mitigate against biodegradation, substantially increasing their residence time in soils (Boudot, 1992).

Root VOCs are another important group of LMW exudates released from plant roots via diffusion. High mobility in air-filled and also water-filled soil pores determines their roles in interplant and even interkingdom signaling between plants and other soil organisms. Apart from ubiquitous volatile signaling molecules, such as ethylene and NO, various species-specific volatiles have been identified, including terpenoids, fatty acid derivatives, or sulfur-containing compounds (Delory et al., 2016). Although recent studies suggest that soil volatilomes may be dominated by microbial VOCs (Schenkel et al., 2019) with well-documented functions as rhizosphere signals (see Chapter 15), important ecological functions in multitrophic interactions have also been attributed to root-emitted volatile compounds (antimicrobials, allelochemicals, chemo-attractants, or infochemicals) (Delory et al., 2016; van Dam et al., 2016). Examples include the release of sesquiterpene β-caryophyllene from maize roots in response to attack by root-feeding beetle larvae, which in turn attracts entomopathogenic nematodes countering the insect pest. Similar interactions have been reported for *Brassica* roots infected by cabbage root fly larvae (*Delia radicum* L.) and simultaneous attraction of entomoparasitic insects, related to the release of volatile methanethiol, methyl sulfides, and breakdown products of glucosinolates (Delory et al., 2016). Another example is the attraction of bacterial pathogen antagonists by VOCs released from roots of *Carex arenaria* upon infection by the fungal pathogen *Fusarium culmorum* (Schulz-Bohm et al., 2018).

Root volatiles can interact with interbacterial quorum sensing (Ahmad et al., 2014) and influence plant community structures by triggering positive and negative interactions with neighboring plants (Chen et al., 2020; Gfeller et al., 2019). However, surprisingly limited knowledge exists on the effects of root VOCs on plant nutrient availability and acquisition. As an example, in tropical agroecosystems, release of the allelopathic volatile compound hydrogen cyanide from roots of cassava triggered the release of ethylene from intercropped peanut roots, which resulted in modifications of the bacterial microbiome (e.g., boosted abundance of *Actinomycetes*) and increased N and P availability in the peanut rhizosphere (Chen et al., 2020).

14.6.3.2 Retrieval mechanisms

For sugars and particularly for N-containing compounds (such as amino acids, polyamines, and small peptides), efficient uptake systems have been identified in roots of many plant species (Fischer et al., 1998; Warren, 2015; Xia and Saglio, 1988). There is even some evidence for root uptake of HMW compounds and bacterial cells via endocytosis (Samaj et al., 2005; Paungfoo-Lonhienne et al., 2010). Moreover, internalization of bacteria (*Pseudomonas*) and fungi (*Trichoderma*) has been reported in root hairs (Harman, 2000; Prieto et al., 2011).

Retrieval of up to 90% of the amino acids and sugars lost by plant roots via diffusion has been demonstrated in hydroponic culture (Jones and Darrah, 1993). Uptake is mediated by transporters and involves H$^+$ cotransport (Chapter 2). Amino acid and peptide transporters frequently show enhanced expression under limited N supply (Hirner et al., 2006; Nazoa et al., 2003; Svennerstam et al., 2011).

In natural ecosystems, N is a major limiting nutrient, and rhizosphere microorganisms not only mineralize organic N but also compete for N uptake with plant roots and immobilize N in their biomass (Bonkowski et al., 2009, see also Chapter 15). Secondary metabolites released by various microorganisms (e.g., phenazine, 2,4-diacetylphloroglucinol, zearalenone) can substantially increase exudation of amino acids from plant roots (Moe, 2013; Phillips et al., 2004). Therefore, efficient reuptake of amino acids may be a successful strategy in N competition with rhizosphere microorganisms (Fig. 14.19). These transporters are located in the PM and may retrieve compounds from the root apoplasm and the rhizosphere densely colonized by competing microorganisms. This view is in agreement with K_m values for amino acid uptake by crop roots, which are in the millimolar range, according to the concentrations of LMW compounds in the apoplasmic fluid. By contrast, only micromolar concentrations of amino acids usually occur in the rhizosphere soil solution, suggesting a rather low efficiency of crop roots in competing for amino acid uptake with microorganisms in the rhizosphere (Jones et al., 2005). However, in arabidopsis roots, amino acid transporters with K_m values in the micromolar range have been reported (Svennerstam et al., 2011). Root uptake of organic N forms may significantly contribute to the N supply in forests, arctic tundra, and waterlogged and acidic soils characterized by high concentrations of dissolved organic N in the soil solution (Chapin et al., 1993). The same holds true for patches of nutrient-rich organic matter in soils and after application of organic fertilizers. In decomposing organic matter, glutamate is one of the dominant amino acids and can even exert signal functions in stimulating lateral root growth to promote targeted exploration of nutrient-rich organic patches in soils (Forde and Lea, 2007).

Under certain stress conditions, such as nutrient deficiency, drought, or oxidative damage, sugar supply may be a limiting factor for plant growth due to reduced photosynthesis. Thus, reuptake of sugars lost by diffusion may be a strategy to minimize C losses. The expression of retrieval mechanisms for LMW sugars in plant roots (Xia and Saglio, 1988) may also enable plants to control microbial colonization at the rhizoplane and in the rhizosphere by modifying the supply of easily available carbohydrates to beneficial and pathogenic rhizosphere microorganisms (Fig. 14.19) (Hennion et al., 2019; Jones et al., 2004b). In addition, a tonoplast sugar transporter of the SWEET family (AtSWEET2) has been identified in roots of *A. thaliana* that limits sugar exudation (as a carbon supply to the root pathogen *Phytium*) by sequestering sugar into the vacuole (Chen et al., 2015). Similarly, the analysis of mutants revealed that amino acid exudation is indirectly affected by several transporters expressed in the root vascular tissues, such as Umami transporters or Glutamine dumpers localized in the phloem and xylem, respectively (Sasse et al., 2018).

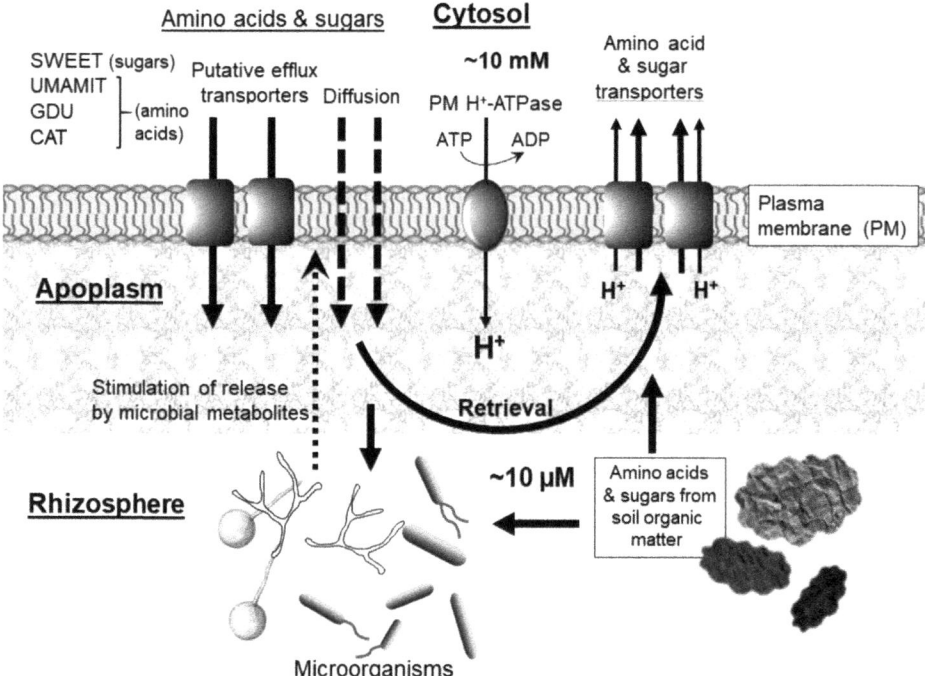

FIGURE 14.19 Model for uptake and retrieval of low-molecular-weight sugars and amino acids by plant roots. *Adapted from Jones et al. (2005).*

Importantly, no retrieval mechanism has been reported for organic acid anions (carboxylates) from the rhizosphere, likely due to a high carboxylate adsorption potential in soils and the outward positive electrochemical potential gradient at the PM energetically limiting efficient reuptake (Oburger and Jones, 2018).

14.6.3.3 Controlled release of low-molecular-weight compounds

Apart from continuous outward diffusion of LMW compounds and the release from damaged root cells, there is also evidence for a controlled excretion of metabolic waste products and secretion of specific compounds into the rhizosphere. Release patterns regarding composition, timing and spatial localization along the root as influenced by external stimuli suggest the involvement of transport proteins at least in some cases (Canarini et al., 2019; Sasse et al., 2018). As an example of excretion of metabolic waste products, the release of lactate from root tips of maize seedlings adapted to low oxygen environments can prevent excessive intracellular accumulation of lactic acid as a product of fermentation induced by hypoxia (Xia and Roberts, 1994).

Secretion of organic acid anions is a mechanism to mobilize poorly available inorganic as well as organic P forms (adsorbed to Fe and Al oxides/hydroxides or Fe-, Al-, and Ca-phosphates) via solubilization and chelation of metal cations (Fig. 14.18) (Jones et al., 2003; Wang and Lambers, 2019).

Citrate, oxalate, malonate, and malate are the most efficient organic acid anions with respect to P mobilization in soils (Neumann and Römheld, 2007). However, the effectiveness of this mechanism has been questioned because the rates of release as well as concentration of organic acid anions in the rhizosphere of most plants are frequently low. MacKay-Fletcher et al. (2021) used a modeling approach combined with microdialysis tubes to simulate root efflux of citrate and P mobilization in an Eutric Cambisol with limited P availability. Characteristic efflux rates for single roots (approx. 4 nmol cm^{-1} h^{-1}), as previously reported for the apical root zones of many crops (e.g., oilseed rape, wheat, maize, chickpea, potato) under P limitation (McKay Fletcher et al., 2021; Neumann and Römheld, 2002), were far below the critical level of 730 nmol cm^{-1} h^{-1} needed for significant P solubilization (McKay Fletcher et al., 2021).

To mobilize significant amounts of P, organic acid anion concentrations in the soil solution in the millimolar range are required (Gerke et al., 2000; Jones, 1998), equivalent to carboxylate accumulation of >5–10 μmol g^{-1} soil (Neumann and Römheld, 2007; Wang and Lambers, 2019). This fits well with the model calculations of McKay Fletcher et al. (2021), whereby critical citrate efflux rates of 730 nmol h^{-1} cm^{-1} root length would translate into rhizosphere soil concentration of 5.5–11 μmol g^{-1} soil during exudation period, assuming 150 mg rhizosphere soil (1-mm-thick layer around roots) with a volumetric water content of 20%, half-life of 3–6 h for carboxylates released into the soil solution, and the residence time of an apical root zone in a given soil compartment of about 5 h. Such carboxylate concentrations have been measured only in a limited range of plant species and only in the rhizosphere of plants forming cluster roots or similar cluster-like root structures, such as dauciform roots, sand-binding roots or capillaroid roots in the members of Proteaceae, Fabaceae, Casuarinaceae, Myricaceae, Cyperaceae, Cactaceae, and Anarthriaceae (Lambers et al., 2015; Neumann and Martinoia, 2002) (see Section 6.3 and Chapter 13).

Compared with regular lateral roots, the formation of the bottlebrush-like cluster roots with densely spaced, second-order lateral rootlets (reaching a length of about 5 mm) covered with root hairs strongly increase the root surface area available for secretion of organic acid anions into a small soil volume. Organic acid anion secretion is increased further by upregulating biosynthetic pathways. In cluster roots of white lupin, the inhibition of citrate turnover contributes to preferential accumulation of citrate (Neumann and Martinoia, 2002). All these adaptations together can explain the reported citrate concentrations of >50 μmol g^{-1} soil in the rhizosphere of cluster roots (Neumann and Römheld, 2007).

Transcriptomic profiling and physiological analysis revealed a highly coordinated developmental setup determining cluster root morphology and the metabolic changes required for the specific functions of cluster roots. This setup comprises increased expression of auxin-, cytokinin-, and brassinosteroid-related genes required for the induction and outgrowth of lateral rootlet primordia that also involves sucrose signaling. Subsequently, increased production of ethylene and induction of cytokinin oxidase involved in cytokinin degradation probably mediates growth inhibition and meristem inactivation of the lateral rootlets as well as senescence-related P retranslocation to the young growing root tissues. The resulting P limitation in the cluster root tissue induces intense expression of P deficiency-related metabolic changes, including increased accumulation of carboxylates, release of secretory acid phosphatases, and induction of P uptake transporters (O'Rourke et al., 2013; Wang et al., 2014, 2015). In the white lupin cluster roots, citrate secretion is high over a period of 2–3 days, charge-balanced by a concomitant extrusion of protons via activation of the PM H$^+$-ATPase (Yan et al., 2002) and also other cations such as K$^+$, Na$^+$, and Mg^{2+} (Zhu et al., 2005). The resulting acidification of the rhizosphere can inhibit microbial growth, thereby minimizing microbial degradation of carboxylates. Chitinases, glucanases, and flavonoids, which are simultaneously secreted from cluster roots, may have a similar antimicrobial effect (Weisskopf et al., 2006).

Although transcriptome analysis of cluster roots revealed increased expression of various transporter genes (such as ALMT-anion channels, MATE, and ABCG-type transporters, known to be involved in the release of carboxylates and secondary metabolites), it is still not entirely clear which transporters are mediating the simultaneous release of various exudates involved in P solubilization (Sasse et al., 2018; Wang et al., 2014).

Apart from mobilization of sparingly soluble organic P forms, organic P with adsorption characteristics similar to those of organic acid anions is mobilized for hydrolysis by simultaneously secreted acid phosphatase. Additionally, co-solubilization of various metal cations (such as Fe, Zn, Mn, and Al) occurs, frequently requiring the expression of adaptive traits to avoid respective toxicities (Dinkelaker et al., 1989; Neumann and Martinoia, 2002). Interestingly, many cluster-root plants are nonmycorrhizal; they are considered P-mining plant species and are characteristic for plant communities adapted to extremely P-impoverished soils. By contrast, the proportion of mycorrhizal plant species with improved capacity for spatial P acquisition increases in soils with moderate P availability (Zemunik et al., 2015).

Under P limitation, malate exudation can mediate Fe^{3+} solubilization, which is reduced by apoplastic ascorbate. However, subsequent precipitation of the extracellular Fe, triggered by P deficiency-induced ferroxidases (LPR1, LPR2), coincides with the generation of ROS and intracellular deposition of callose, blocking the carbon supply to the root meristem. This results in inhibition of primary root elongation and leads to a higher density of root hairs and stimulation of lateral root growth for improved spatial P acquisition (Grillet and Schmidt, 2017; Müller et al., 2015).

Root secretion of carboxylates is an important component determining Al resistance in plants (Fig. 14.20) (see also Section 18.3). At soil pH_{water} below 5, solubilization of mononuclear Al species limits root growth. In many Al-resistant plant species and cultivars, Al-induced root secretion of carboxylates (particularly malate, citrate, or oxalate) is important in Al^{3+} detoxification by external complexation in the root apoplast and the rhizosphere soil (Kochian et al., 2004; Zhang et al., 2019). The carboxylates are released in response to Al toxicity, particularly in the root transition

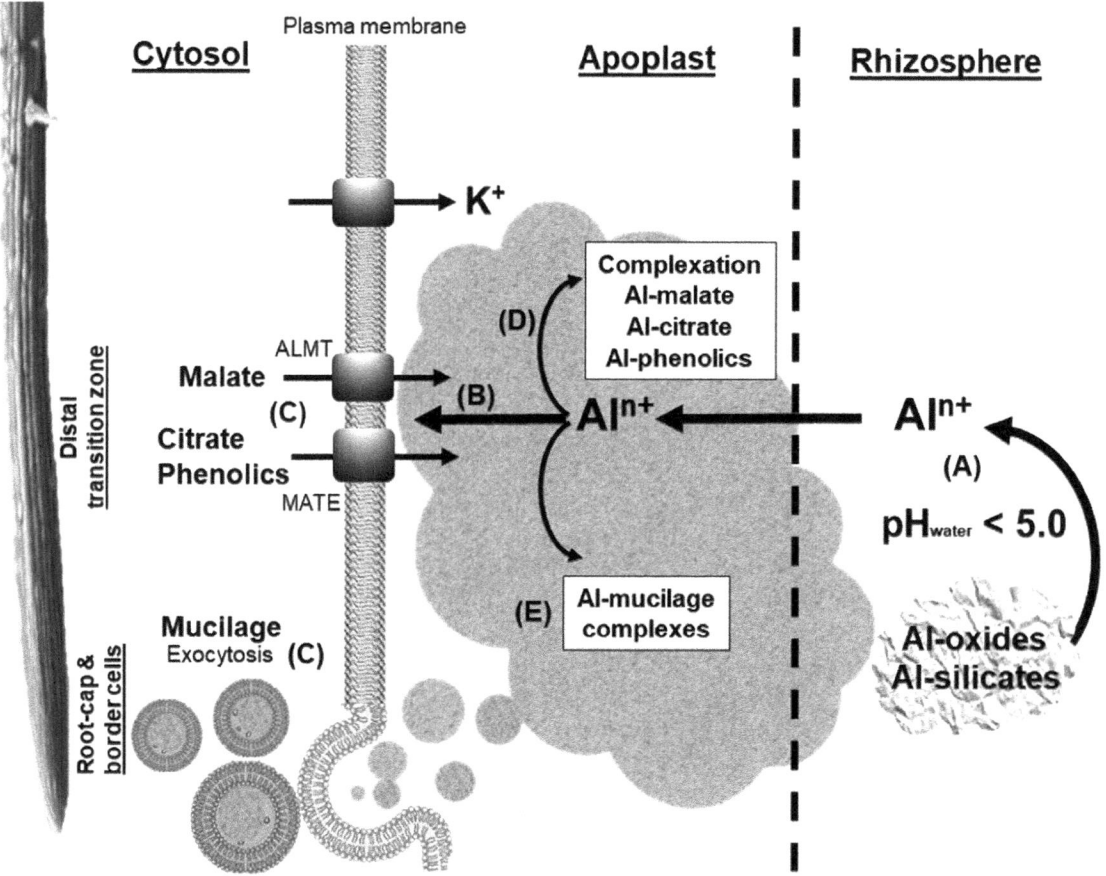

FIGURE 14.20 Model of root-induced aluminum (Al) detoxification by secretion of organic Al chelators. Solubilization of toxic Al^{n+} species in acid mineral soils at pH_{water} <5.0 (A). Aluminum-induced activation of organic anion transporters in the Al-sensitive apical root zones (B). Release of organic Al chelators (C). Detoxification of Al^{n+} by complexation with low-molecular-weight chelators (D) and mucilage (E). *Adapted from Neumann (2007).*

zone (where cells start elongating) as the most Al-sensitive part of the root (Kollmeier et al., 2000). In Al-resistant wheat, malate is released immediately after exposure to high Al concentrations by facilitated diffusion via an anion channel (ALMT1) expressed in the root tips (Zhang et al., 2001; Sasaki et al., 2004). Transgenic expression of wheat ALMT1 confers Al resistance to Al-sensitive barley (Delhaize et al., 2004); hence, ALMT1 may be a tool to increase Al resistance in transgenic plants. By contrast, Al-mediated citrate release in Al-resistant genotypes of sorghum and barley is mediated by MATE transporters (Magalhaes et al., 2007; Wang et al., 2007). Some members of the ALMT and MATE families mediate carboxylate transport independently of Al stress and may therefore be candidate genes for a general manipulation of carboxylate exudation (Ryan et al., 2009). Complexation of Al can be achieved not only by exuded organic acid anions but also by exuded phenolics (catechol, catechin, quercetin) and benzoxazinoids. However, genes involved in Al-induced secretion of phenolics and benzoxazinoids have not been identified in any plant species yet (Zhang et al., 2019).

In graminaceous plant species, secretion of phytosiderophores (PS) (Fig. 14.21) is induced by the limited supply of Fe and Zn (Neumann and Römheld, 2007) (see Chapter 2 and Section 7.1). Derived from nicotianamine [an ubiquitous intracellular metal chelator in higher plants, synthesized from methionine under control of the FIT and MYB (MYB10, MYB72) transcription factors (Palmer et al., 2013)], PS are synthesized at high rates in the roots of Fe-deficient graminaceous plants (Ma and Nomoto, 1996). In barley, diurnal pulses of high PS secretion are restricted to the young tissues in the apical root zones with limited microbial colonization, which minimizes microbial degradation of PS (Nishizawa and Mori, 1987; von Wirén et al., 1995). PS are released into the rhizosphere by the TOM1 transporter. The *TOM1* gene is a member of the MFS gene family that also contains genes encoding transporters involved in the release of bacterial siderophores (Nozoye et al., 2011). In the rhizosphere, PS mobilize Fe^{3+}, but also other micronutrients, such as Zn, Mn, and Cu, by the formation of stable complexes even at high soil pH (Treeby et al., 1989). The soluble Fe^{3+}−PS complex is subsequently taken up by H^+ cotransport via the YS1 transporter (Curie et al., 2001). Upregulation of *ZmYS1* in maize is induced by Fe deficiency, but not by deficiency of Zn, Mn, or Cu (Roberts et al., 2004), suggesting that Fe mobilization is the primary

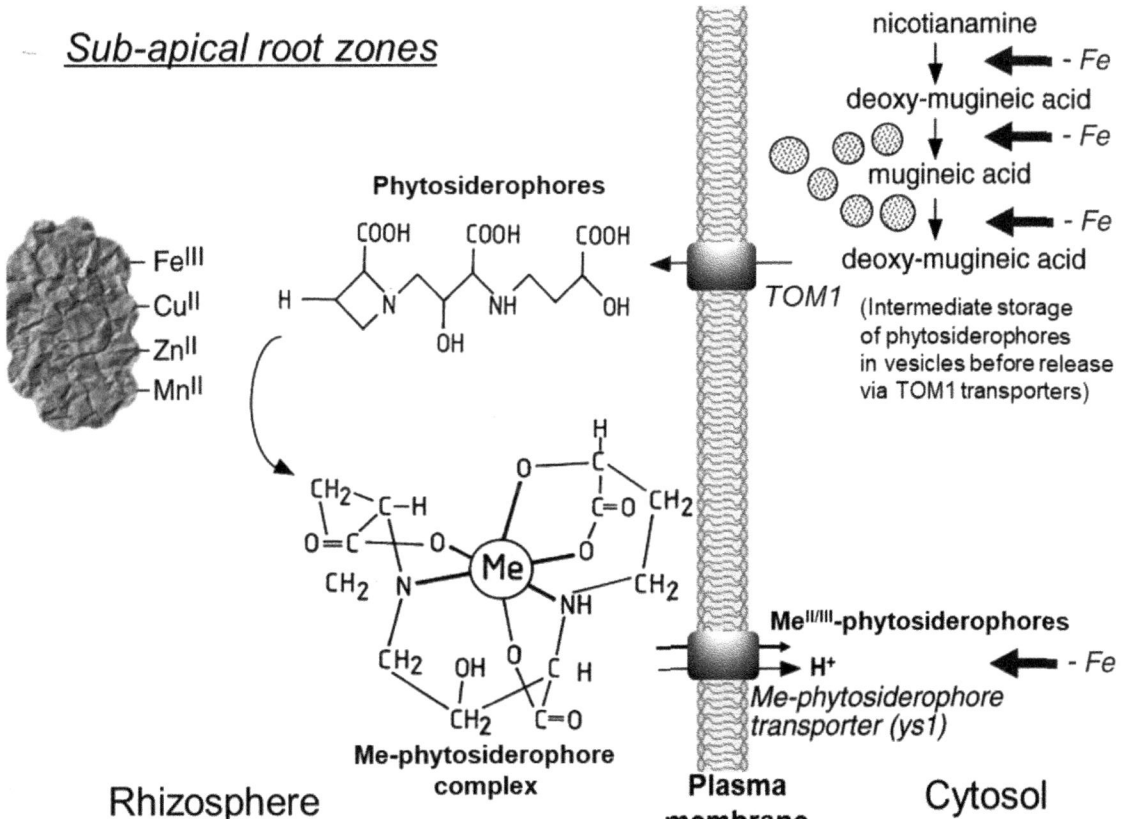

FIGURE 14.21 Model for root-induced mobilization of Fe and other metallic (Me) micronutrients (Zn, Mn, Cu) in the rhizosphere of graminaceous (Strategy II) plants, mediated by release of PS and reuptake of PS−metal complexes. *PS*, Phytosiderophores. *Adapted from Neumann (2007).*

function of PS release. Homologous YS-like PS transporter genes (YSL) induced under Fe limitation have been identified also in other grasses, such as barley (Ueno et al., 2009), rice (Inoue et al., 2009), and *Brachypodium distachyon* (Yordem et al., 2009). These YSL homologs have important functions in internal Fe homeostasis in both monocotyledonous and dicotyledonous plant species (Curie et al., 2009; Grillet and Schmidt, 2019).

Large genotypic differences in the capacity for PS secretion occur among plant species and cultivars. Graminaceous species originating from the humid tropics with abundance of acid soils, in which Fe availability is high, are usually less efficient in PS secretion and more susceptible to Fe deficiency chlorosis. Phytosiderophore secretion and tolerance to Fe limitation in upland rice was improved in pot experiments and under field conditions by transgenic expression of the barley nicotianamine aminotransferase gene as the key enzyme in PS biosynthesis (Takahashi et al., 2001).

Iron acquisition via PS secretion in graminaceous plant species (Strategy II) is less dependent on rhizosphere pH than the Strategy I mechanism expressed in dicotyledonous plants and nongraminaceous monocots (see Section 14.5.3). However, some Strategy I components are also found in certain graminaceous species. This includes the expression of IRT transporters mediating Fe^{2+} uptake and the exudation of P-mobilizing phenolics via phenolic efflux transporters (PEZ1) in rice (Ishimaru et al., 2006, 2011; Jin et al., 2007). Similarly, ZmIRT1 and ZmZIP3 were identified as functional zinc and iron transporters in the maize genome as were the homolog genes encoding the FER-like iron deficiency induced transcription factor (FIT), PM H^+-ATPase and ferric reductases (Li et al., 2018b), suggesting that components of the Fe acquisition system in the phylogenetically older Strategy I plants are at least partially conserved in the graminaceous species.

LMW phenolics, flavonoids, and strigolactones as well as VOCs and potentially other compounds released by plant roots are important signals in plant–microbe interactions, such as symbiosis with N_2-fixing microorganisms and mycorrhiza, but also in the interactions with bacterial communication systems (quorum sensing) and with weeds (Akiyama and Hayashi, 2006; Bauer et al., 2005; Martin et al., 2001; Werner, 2007) (see Chapter 15). Particularly in the rhizobium symbiosis with leguminous plants, the molecular events involved in the infection by rhizobial microsymbiont are well characterized (Werner, 2007) (see Chapter 16). However, much less is known concerning the mechanisms controlling release of these signals, which are expected to require a highly coordinated regulation in space and time. At least in some cases, transporters have been characterized: ATP-dependent ABC transporters have been implicated in the release of strigolactones, flavonoids, and coumarins involved in the establishment of arbuscular mycorrhizal symbiosis, *Rhizobium* symbiosis, and Fe acquisition (Sasse et al., 2018; Tsai and Schmidt, 2017).

References

Adamczyk, B., Smolander, A., Kitunen, V., Godlewski, M., 2010. Proteins as nitrogen source for plants: a short story about exudation of proteases by plant roots. Plant Signal. Behav. 5, 817–819.

Adams, M.A., Pate, J.S., 1992. Availability of organic and inorganic forms of phosphorus to lupins (*Lupinus* spp.). Plant Soil 145, 107–113.

Adnan, M., Shah, Z., Fahad, S., Arif, M., Alam, M., Imtiaz, A.K., et al., 2017. Phosphate-solubilizing bacteria nullify the antagonistic effect of soil calcification on bioavailability of phosphorus in alkaline soils. Sci. Rep. 8, 4339.

Aguilar, S.A., Van Diest, A., 1981. Rock-phosphate mobilization induced by the alkaline uptake pattern of legumes utilizing symbiotically fixed nitrogen. Plant Soil 61, 27–42.

Ahmad, A., Viljoen, A.M., Chenia, H.Y., 2014. The impact of plant volatiles on bacterial quorum sensing. Lett. Microbiol. 60, 8–19.

Akiyama, K., Hayashi, H., 2006. Strigolactones: chemical signals for fungal symbionts and parasitic weeds in plant roots. Ann. Bot. 97, 925–931.

Alori, E.T., Glick, B.R., Babalola, O.O., 2017. Microbial phosphorus solubilization and Its potential for use in sustainable agriculture. Front. Microbiol. 8, 971.

Armstrong, R.D., Helyar, K.R., 1992. Changes in soil phosphate fractions in the rhizosphere of semi-arid pasture grasses. Aust. J. Soil Res. 30, 131–143.

Badri, D.V., Loyola-Vargas, V.M., Broeckling, C.D., Vivanco, J.M., 2010. Root secretion of phytochemicals in Arabidopsis Is predominantly not Influenced by diurnal rhythms. Molecular Plant 3, 491–498.

Baetz, U., Martinoia, E., 2013. Root exudates: the hidden part of plant defense. Trends Plant Sci. 19, 90–98.

Bagayoko, M., Alvey, S., Neumann, G., Buerkert, A., 2000. Root induced increases in soil pH and nutrient availability to field-grown cereals and legumes on acid sandy soils of Sudano-Sahelian West Africa. Plant Soil 225, 117–127.

Barber, S.A., 1995. Soil Nutrient Bioavailability: A Mechanistic Approach, 2nd ed. John Wiley, New York, USA.

Barber, S.A., Ozanne, P.G., 1970. Autoradiographic evidence for the differential effect of four plant species in altering the calcium content of the rhizosphere soil. Soil Sci. Soc. Amer. Proc. 34, 635–637.

Battey, N.H., James, N.C., Greenland, A.J., Brownlee, C., 1999. Exocytosis and endocytosis. Plant Cell 11, 643–659.

Bauer, W.D., Mathesius, U., Teplitski, M., 2005. Eukaryotes deal with bacterial quorum sensing. ASM News 71, 129–135.

Begg, C.B.M., Kirk, G.J.D., Mackenzie, A.F., Neue, H.U., 1994. Root-induced iron oxidation and pH changes in the lowland rice rhizosphere. New Phytol. 128, 469–477.

Beissner, L., 1997. Mobilisierung von Phosphor aus organischen und anorganischen P-Verbindungen durch Zuckerrübenwurzeln. PhD thesis. Georg-August University, Göttingen, Cuvillier Verlag, Germany.

Belgaroui, N., Berthomieu, P., Rouached, H., Hanin, M., 2016. The secretion of the bacterial phytase PHY-US417 by Arabidopsis roots reveals its potential for increasing phosphate acquisition and biomass production during co-growth. Plant Biotechnol. J. 14, 1914–1924.

Berg, G., Smalla, K., 2009. Plant species and soil type cooperatively shape the structure and function of microbial communities in the rhizosphere. FEMS Microbiol. Ecol. 68, 1–13.

Bidel, L.P.R., Renault, P., Pages, L., Riviere, L.M., 2000. Mapping meristem respiration of *Prunus persica* (L.) Batsch seedlings: potential respiration of the meristems, O_2 diffusional constraints and combined effects on root growth. J. Exp. Bot. 51, 755–768.

Blackwell, M.S.A., Brookes, P.C., de la Fuente-Martinez, N., Gordon, H., Murray, P.J., Snars, K.E., et al., 2010. Phosphorus solubilization and potential transfer to surface waters from the soil microbial biomass following drying–rewetting and freezing–thawing. Adv. Agron. 106, 1–35.

Blossfeld, S., Gansert, D., 2007. A novel non-invasive optical method for quantitative visualization of pH dynamics in the rhizosphere of plants. Plant Cell Environ. 30, 176–186.

Boeuf-Tremblay, V., Plantureux, S., Guckert, A., 1995. Influence of mechanical impedance on root exudation of maize seedlings at two developmental stages. Plant Soil 172, 279–287.

Bolan, N.S., Hedley, M.J., White, R.E., 1991. Processes of soil acidification during nitrogen cycling with emphasis on legume based pastures. Plant Soil 134, 53–63.

Bonkowski, M., Villenave, C., Griffiths, B., 2009. Rhizosphere fauna: the functional and structural diversity of intimate interactions of soil fauna with plant roots. Plant Soil 321, 213–233.

Boudot, J.P., 1992. Relative efficiency of complexed aluminium, noncrystalline Al hydroxide, allophane and imogolite in retarding the biodegradation of citric acid. Geoderma 52, 29–39.

Bowen, G.D., Rovira, A.D., 1991. The rhizosphere, the hidden half of the hidden half. In: Waisel, Y., Eshel, A., Kafkafi, U. (Eds.), Plant Roots, the Hidden Half. Marcel Dekker, Inc., New York, USA, pp. 641–669.

Bowsher, A.W., Evans, S., Tiemann, L.K., Friesen, M.L., 2018. Effects of soil nitrogen availability on rhizodeposition in plants: a review. Plant Soil 423, 59–85.

Brennan, R.F., 1992. The role of manganese and nitrogen nutrition in the susceptibility of wheat plants to take-all in Western Australia. Fert. Res. 31, 35–41.

Brüggemann, W., Moog, P.R., Nakagawa, H., Janiesch, P., Kuiper, P.J.C., 1991. Plasma membrane-bound NADH:Fe^{3+}-EDTA reductase and iron deficiency in tomato (*Lycopersicon esculentum*). Is there a turbo reductase? Physiol. Plant. 79, 339–346.

Bünemann, E.K., Oberson, A., Liebisch, F., Keller, F., Annaheim, K.E., Huguenin-Elie, O., et al., 2012. Rapid microbial phosphorus immobilization dominates gross phosphorus fluxes in a grassland soil with low inorganic phosphorus availability. Soil Biol. Biochem. 51, 84–95.

Cakmak, I., Marschner, H., 1990. Decrease in nitrate uptake and increase in proton release in zinc deficient cotton, sunflower and buckwheat plants. Plant Soil 129, 261–268.

Cakmak, I., Erenoglu, B., Gülüt, K.Y., Derici, R., Römheld, V., 1998. Light-mediated release of phytosiderophores in wheat and barley under iron or zinc deficiency. Plant Soil 202, 309–315.

Callaway, R.M., King, L., 1996. Temperature-driven variation in substrate oxygenation and the balance of competition and facilitation. Ecology 77, 1189–1195.

Callot, G., Chauvel, A., Arvieu, J.C., Chamayou, H., 1992. Mise en evidence de kaolinite et de silice dans les structures cellulaires de l'epiderme et du cortex de racines de palmier, en for et d'Amazonie. Bull. Soc. Bot. Fr. 139, 7–14.

Canarini, A., Kaiser, C., Merchant, A., Richter, A., Wanek, W., 2019. Root exudation of primary metabolites: mechanisms and their roles in plant responses to environmental stimuli. Front. Plant Sci. 10, 157.

Canellas, L.P., Olivares, F.L., 2017. Production of border cells and colonization of maize root tips by *Herbaspirillum seropedicae* are modulated by humic acid. Plant Soil 417, 403–413.

Canellas, L.P., Olivares, F.L., Aguiar, N.O., Jones, D.L., Nebioso, A., Mazzei, P., et al., 2015. Humic and fulvic acids as biostimulants in horticulture. Sci. Hortic. 30, 15–27.

Carminati, A., 2012. A model of root water uptake coupled with rhizosphere dynamics. Vadose Zone J. 11, vzj2011.0106.

Carminati, A., Moradi, A.B., Vetterlein, D., Vontobel, P., Lehmann, E., Weller, U., et al., 2010. Dynamics of soil water content in the rhizosphere. Plant Soil 332, 163–176.

Carminati, A., Benard, P., Ahmed, M., Zarebanadkouki, M., 2017. Liquid bridges at the root-soil interface. Plant Soil 417, 1–15.

Chaparro, J.M., Badri, D.V., Bakker, M.G., Sugiyama, A., Manter, D.K., et al., 2013. Root exudation of phytochemicals in Arabidopsis follows specific patterns that are developmentally programmed and correlate with soil microbial functions. PLoS ONE 8, e55731.

Chapin III, F.S., Moilanen, L., Kielland, K., 1993. Preferential use of organic nitrogen for growth by a non-mycorrhizal artic sedge. Nature 361, 150–153.

Chen, H.Y., Huh, J.H., Yu, Y.C., Ho, L.H., Chen, L.Q., Tholl, D., et al., 2015. The Arabidopsis vacuolar sugar transporter SWEET2 limits carbon sequestration from roots and restricts *Pythium* infection. Plant J. 83, 1046–1058.

Chen, Y., Bonkowski, M., Shen, Y., Griffiths, B.S., Jiang, Y., Wang, X., et al., 2020. Root ethylene mediates rhizosphere microbial community reconstruction when chemically detecting cyanide produced by neighbouring plants. Microbiome 8, 4.

Chenu, C., Cosentino, D., 2011. Microbial regulation of soil structural dynamics. In: Ritz, K., Young, I. (Eds.), The Architecture and Biology of Soils: Life in Inner Space. CABI, Wallingford, UK, pp. 37–70.

Chrispeels, M., 1991. Sorting of proteins in the secretory system. Annu. Rev. Plant Physiol. 42, 21–55.
Clode, P.L., Kilburn, M.R., Jones, D.L., Stockdale, E.A., Cliff III, J.B., Herrmann, A.M., et al., 2009. In situ mapping of nutrient uptake in the rhizosphere using nanoscale secondary ion mass spectrometry. Plant Physiol. 151, 1751–1757.
Connolly, E.L., Campbell, N.H., Grotz, N., Prichard, C.L., Guerinot, M.L., 2003. Overexpression of the FRO2 ferric chelate reductase confers tolerance to growth on low iron and uncovers posttranscriptional control. Plant Physiol. 133, 1102–1110.
Costa, G., Michaut, J.C., Guckert, A., 1997. Amino acids exuded from axenic roots of lettuce and white lupin seedlings exposed to different cadmium concentrations. J. Plant Nutr. 20, 883–900.
Coventry, D.R., Slattery, W.J., 1991. Acidification of soil associated with lupins grown in a crop rotation in north-eastern Victoria. Aust. J. Agric. Res. 42, 391–397.
Curie, C., Panaviene, Z., Loulergue, C., Dellaporta, S.L., Briat, J.F., Walker, E.L., 2001. Maize yellow stripe1 encodes a membrane protein directly involved in Fe(III) uptake. Nature 409, 346–349.
Curie, C., Cassin, G., Couch, D., Divol, F., Higuchi, K., Le Jean, M., et al., 2009. Metal movement within the plant: contribution of nicotianamine and yellow stripe 1-like transporters. Ann. Bot. 103, 1–11.
Degenhardt, J., Larsen, P.B., Howell, S.H., Kochian, L.V., 1998. Aluminum resistance in the Arabidopsis mutant *alr-104* is caused by an aluminum-induced increase in rhizosphere pH. Plant Physiol. 117, 19–27.
Delhaize, E., Ryan, P.R., Hebb, D.M., Yamamoto, Y., Sasaki, T., Matsumoto, H., 2004. Engineering high-level aluminum tolerance in barley with the ALMT1 gene. Proc. Natl. Acad. Sci. USA 101, 15249–15254.
Delory, B.M., Delaplace, P., Fauconnier, M.L., du Jardin, P., 2016. Root-emitted volatile organic compounds: can they mediate belowground plant-plant interactions? Plant Soil 402, 1–26.
Dinkelaker, B., Marschner, H., 1992. In vivo demonstration of acid phosphatase activity in the rhizosphere of soil-grown plants. Plant Soil 144, 199–205.
Dinkelaker, B., Römheld, V., Marschner, H., 1989. Citric acid excretion and precipitation of calcium citrate in the rhizosphere of white lupin (*Lupinus albus* L.). Plant Cell Environ. 12, 285–292.
Dinkelaker, B., Hengeler, C., Neumann, G., Eltrop, L., Marschner, H., 1997. Root exudates and mobilization of nutrients. In: Rennenberg, H., Eschrich, W., Ziegler, H. (Eds.), Trees - Contributions to Modern Tree Physiology. Backhuys, Leiden, The Netherlands, pp. 441–452.
Drew, M.C., 1975. Comparison of the effects of a localized supply of phosphate, nitrate, ammonium, and potassium on the growth of the seminal root system, and the shoot, in barley. New Phytol. 75, 479–490.
Driouich, A., Durand, C., Cannesan, M.A., Percoco, G., Gibouin, M.V., 2010. Border cells versus border-like cells: are they alike? J. Exp. Bot. 61, 3827–3831.
Driouich, A., Durand, C., Vicre-Gibouin, M., 2007. Formation and separation of root border cells. Trends Plant Sci. 12, 14–19.
Elmer, W.H., LaMondia, J.A., 1999. Influence of ammonium sulfate and rotation crops on strawberry black root rot. Plant Dis. 83, 119–123.
Engels, C., Neumann, G., Gahoonia, T., George, E., Schenk, M., 2000. Assessment of the ability of roots for nutrient acquisition. In: Smit, A.L., Bengough, A.G., Engels, C., Van Noordwijk, M., Pellerin, S., van de Geijn, S.C. (Eds.), Root Methods. A Handbook. Springer, Heidelberg, Germany, pp. 403–459.
Faget, M., Blossfeld, S., von Gillhaussen, P., Schurr, U., Temperton, V.M., 2013. Disentangling who is who during rhizosphere acidification in root interactions: combining fluorescence with optode techniques. Front. Plant Sci. 4, 392.
Feng, J.N., Volk, R.J., Jackson, W.A., 1994. Inward and outward transport of ammonium in roots of maize and sorghum contrasting effects of methionine sulfoximine. J. Exp. Bot. 45, 429–439.
Fang, X.Z., Tian, W.H., Liu, X.X., Lin, X.Y., Jin, C.W., Zheng, S.J., 2016. Alleviation of proton toxicity by nitrate uptake specifically depends on nitrate transporter 1.1 in *Arabidopsis*. New Phytol. 211, 149–158.
Fischer, W., Felssa, H., Schaller, G., 1989. pH values and redox potentials in microsites of the rhizosphere. Z. Pflanzenernähr. Bodenk. 152, 191–195.
Fischer, W.N., André, B., Rentsch, D., Krolkiewicz, S., Tegeder, M., Breitkreuz, K., et al., 1998. Amino acid transport in plants. Trends Plant Sci. 3, 188–195.
Flessa, H., Fischer, W.R., 1992. Plant-induced changes in the redox potential of rice rhizospheres. Plant Soil 143, 55–60.
Forde, B.G., Lea, P.J., 2007. Glutamate in plants: metabolism, regulation, and signalling. J. Exp. Bot. 58, 2339–2358.
Fourcroy, P., Sisó-Terraza, P., Sudre, D., Savirón, M., Reyt, G., Gaymard, F., et al., 2014. Involvement of the ABCG37 transporter in secretion of scopoletin and derivatives by Arabidopsis roots in response to iron deficiency. New Phytol. 201, 155–167.
Fukao, T., Barrera-Figueroa, B.E., Juntawong, P., Peña-Castro, J.M., 2019. Submergence and waterlogging stress in plants: a review highlighting research opportunities and understudied aspects. Front. Plant Sci. 10, 340.
Fusseder, A., 1984. Der Einfluß von Bodenart, Durchlüftung des Bodens, N-Ernährung und Rhizosphärenflora auf die Morphologie des seminalen Wurzelsystems von Mais. Z. Pflanzenernähr. Bodenk. 147, 553–565.
Gahoonia, T.S., Claassen, N., Jungk, A., 1992. Mobilization of phosphate in different soils by ryegrass supplied with ammonium or nitrate. Plant Soil 140, 241–248.
Gallart, M., Adair, K.L., Love, J., Meason, D., Clinton, P.W., Xue, J.M., et al., 2018. Host genotype and nitrogen form shape the root microbiome of *Pinus radiata*. Microb. Ecol. 75, 419–433.
Gerke, J., Römer, W., Beißner, L., 2000. The quantitative effect of chemical phosphate mobilization by carboxylate anions on P uptake by a single root. II. The importance of soil and plant parameters for uptake of mobilized P. J. Plant Nutr. Soil Sci. 163, 213–219.
Gfeller, V., Huber, M., Förster, C., Huang, W., Köllner, T.G., Erb, M., 2019. Root volatiles in plant-plant interactions. I. High root sesquiterpene release is associated with increased germination and growth of plant neighbours. Plant Cell Environ. 42, 1950–1963.

Gichangi, E.M., 2019. Effects of organic amendments on the transformations and bioavailability of phosphorus in soils: a review. Discovery Agric. 5, 41–50.

Gijsman, A.J., 1991. Soil water content as a key factor determining the source of nitrogen (NH_4^+ or NO_3^-) absorbed by Douglas fir (*Pseudotsuga menziesii*) along its roots. Can. J. For. Res. 21, 616–625.

Gilbert, G.A., Knight, J.D., Vance, C.P., Allan, D.L., 1999. Acid phosphatase in phosphorus-deficient white lupin roots. Plant Cell Environ. 21, 801–810.

Gill, R.A., Jackson, R.B., 2000. Global patterns of root turnover for terrestrial ecosystems. New Phytol. 147, 13–31.

Gniazdowska, A., Krawczak, A., Mikukska, M., Rychter, A.M., 1999. Low phosphate nutrition alters bean plants' ability to assimilate and translocate nitrate. J. Plant Nutr. 21, 551–563.

Godbold, D.L., Hoosbeek, M.R., Lukac, M., Cotrufo, M.F., Janssens, I.A., Ceulemans, R., et al., 2006. Mycorrhizal hyphal turnover as a dominant process for carbon input into soil organic matter. Plant Soil 281, 15–24.

Godo, G.H., Reisenauer, H.M., 1980. Plant effects on soil manganese availability. Soil Sci. Soc. Am. J. 44, 993–995.

González-Hernández, A.I., Fernández-Crespo, E., Scalschi, L., Jairezaei, M.R., von Wiren, N., Garcia-Agustin, P., et al., 2019. Ammonium-mediated changes in carbon and nitrogen metabolisms induce resistance against *Pseudomonas syringae* in tomato plants. J. Plant Physiol. 239, 28–37.

Graham, R.D., Webb, M.J., 1991. Micronutrients and plant disease resistance and tolerance in plants. In: Mortvedt, J.J., Cox, F.R., Shuman, L.M., Welch, R.M. (Eds.), Micronutrients in Agriculture. Soil Science Society of America, Madison, WI, USA, pp. 329–370.

Grayston, S.J., Vaughan, D., Jones, D., 1996. Rhizosphere carbon flow in trees, in comparison with annual plants: the importance of root exudation and its impact on microbial activity and nutrient availability. Appl. Soil Ecol. 5, 29–56.

Greenfield, L.M., Hill, P.W., Paterson, E., Baggs, E.M., Jones, D.L., 2020. Do plants use root-derived proteases to promote the uptake of soil organic nitrogen? Plant Soil 456, 355–367.

Gregory, P.J., 2006. Roots, rhizosphere, and soil: the route to a better understanding of soil science? Eur. J. Soil Sci. 57, 2–12.

Grillet, L., Schmidt, W., 2017. The multiple facets of root iron reduction. J. Exp. Bot. 68, 5021–5027.

Grillet, L., Schmidt, W., 2019. Iron acquisition strategies in land plants: not so different after all. New Phytol. 224, 11–18.

Grusak, M.A., Welch, R.M., Kochian, L.V., 1990. Does iron deficiency in *Pisum sativum* enhance the activity of the root plasmalemma iron transport protein? Plant Physiol. 94, 1353–1357.

Haase, S., Neumann, G., Kania, A., Kuzyakov, Y., Römheld, V., Kandeler, E., 2007. Atmospheric CO_2 and the N-nutritional status modify nodulation, nodule-carbon supply and root exudation of *Phaseolus vulgaris* L. Soil Biol. Biochem. 39, 2208–2221.

Haling, R.E., Brown, L.K., Bengough, A.G., Valentine, T.A., White, P.J., Young, I.M., et al., 2014. Root hair length and rhizosheath mass depend on soil porosity, strength and water content in barley genotypes. Planta 239, 643–651.

Harman, G.E., 2000. Changes in perceptions derived from research on *Trichoderma harzianum* T-22. Plant Dis. 84, 377–292.

Häussling, M., Marschner, H., 1989. Organic and inorganic soil phosphates and acid phosphatase activity in the rhizosphere of 80-year-old Norway spruce (*Picea abies* (L.) Karst.) trees. Biol. Fertil. Soils 8, 128–133.

Häussling, M., Jorns, C.A., Lehmbecker, G., Hecht-Buchholz, C., Marschner, H., 1988. Ion and water uptake in relation to root development in Norway spruce (*Picea abies* (L.) Karst.). J. Plant Physiol. 133, 486–491.

Hawes, M., Allen, C., Turgeon, B.G., Curlango-Rivera, G., Minh Tran, T., Huskey, D.A., et al., 2016. Root border cells and their role in plant defense. Annu. Rev. Phytopathol. 54, 143–161.

Helal, H.M., Sauerbeck, D., 1989. Input and turnover of plant carbon in the rhizosphere. Z. Pflanzenernähr. Bodenk. 152, 211–216.

Hennion, N., Durand, M., Vriet, C., Doidy, J., Maurousset, L., Lemoine, R., et al., 2019. Sugars en route to the roots. Transport, metabolism and storage within plant roots and towards microorganisms of the rhizosphere. Physiol. Plant. 165, 44–57.

Heuwinkel, H., Kirkby, E.A., Le Bot, J., Marschner, H., 1992. Phosphorus deficiency enhances molybdenum uptake by tomato plants. J. Plant Nutr. 15, 549–568.

Hiltner, L., 1904. Über neuere Erfahrungen und Probleme auf dem Gebiete der Bodenbakteriologie unter besonderer Berücksichtigung der Gründüngung und Brache. Arbeiten der Deutschen Landwirtschaftlichen Gesellschaft 98, 59–78.

Hinsinger, P., 2004. Nutrient availability and transport in the rhizosphere. In: Goodman, R.M. (Ed.), Encyclopedia of Plant and Crop Science. Marcel Dekker, New York, USA, pp. 1094–1097.

Hinsinger, P., Jaillard, B., 1993. Root-induced release of interlayer potassium and vermiculitization of phlogopite as related to potassium depletion in the rhizosphere of ryegrass. J. Soil Sci. 44, 525–534.

Hinsinger, P., Elsass, F., Jaillard, B., Robert, M., 1993. Root-induced irreversible transformation of a trioctahedral mica in the rhizosphere of rape. Eur. J. Soil Sci. 44, 535–545.

Hinsinger, P., Gobran, G.R., Gregory, P.J., Wenzel, W.W., 2005. Rhizosphere geometry and heterogeneity arising from root-mediated physical and chemical processes. New Phytol. 168, 293–303.

Hinsinger, P., Bengough, A.G., Vetterlein, D., Young, I.M., 2009. Rhizosphere: biophysics, biogeochemistry and ecological relevance. Plant Soil 32, 117–152.

Hirner, A., Ladwig, F., Stransky, H., Okumoto, S., Keinath, M., Harms, A., et al., 2006. Arabidopsis LHT1 is a high-affinity transporter for cellular amino acid uptake in both root epidermis and leaf mesophyll. Plant Cell 18, 1931–1946.

Hoffland, E., Findenegg, G.R., Nelemans, J.A., 1989. Solubilization of rock phosphate by rape. II. Local root exudation of organic acids as a response to P-starvation. Plant Soil 113, 161–165.

Holden, M.J., Luster, D.G., Chaney, R.L., Buckhout, T.J., Robinson, C., 1991. Fe^{3+}-chelate reductase activity of plasma membranes isolated from tomato (*Lycopersicon esculentum* Mill.) roots. Comparison of enzymes from Fe-deficient and Fe-sufficient roots. Plant Physiol. 97, 537–544.

Hoffland, E., van den Boogaard, R., Nelemans, J., Findenegg, G., 1992. Biosynthesis and root exudation of citric and malic acid in phosphate-starved rape plants. New Phytol. 122, 675–680.

Holz, M., Zarebanadkouki, M., Kuzyakov, Y., Pausch, J., Carminati, A., 2018. Root hairs increase rhizosphere extension and carbon input to soil. Ann. Bot. 121, 61–69.

Holz, M., Zarebanadkouki, M., Carminati, A., Hovind, J., Kaestner, A., Spohn, M., 2019. Increased water retention in the rhizosphere allows for high phosphatase activity in drying soil. Plant Soil 443, 259–271.

Holzgreve, H., Eick, M., Stoehr, C., 2019. Protease activity in the rhizosphere of tomato plants is independent from nitrogen status. In: Ohyama, T. (Ed.), Root Biology – Growth, Physiology, and Functions. IntechOpen, London, UK.

Homyak, P.M., Allison, S.D., Huxman, T.E., Goulden, M.L., Treseder, K.K., 2017. Effects of drought manipulation on soil nitrogen cycling: a meta-analysis. J. Geophys. Res.-Biogeosci. 122, 3260–3272.

Hooker, J.E., Hendrick, R., Atkinson, D., 2000. The measurement and analysis of fine root longevity. In: Smit, A.L., Bengough, A.G., Engels, C., van Noordwijk, M., Pellerin, S., van de Geijn, S.C. (Eds.), Root Methods. A Handbook. Springer-Verlag, Heidelberg, Germany, pp. 403–459.

Horst, W.J., Wagner, A., Marschner, H., 1982. Mucilage protects root meristems from aluminium injury. Z. Pflanzenphysiol. 105, 435–444.

Huang, B., North, G., Nobel, P.S., 1993. Soil sheaths, photosynthate distribution to roots and rhizosphere water relations for *Opuntia ficus-indica*. Int. J. Plant Sci. 154, 425–431.

Hübel, F., Beck, E., 1993. *In-situ* determination of the P-relations around the primary root of maize with respect to inorganic and phytate-P. Plant Soil 157, 1–9.

Huber, D.M., Wilhelm, N.S., 1988. The role of manganese in resistance to plant diseases. In: Graham, R.D., Hannan, R.J., Uren, N.C. (Eds.), Manganese in Soils and Plants. Kluwer Academic Publishers, Dordrecht, The Netherlands, pp. 155–173.

Huguet, A., Bernard, S., El Khatib, R., Gocke, M.I., Wiesenberg, G.L.B., Derenne, S., 2021. Multiple stages of plant root calcification deciphered by chemical and micromorphological analyses. Geobiology 19, 75–86.

Hummel, C., Boitt, G., Santner, J., Lehto, N.J., Condron, L., Wenzel, W.W., 2021. Co-occurring increased phosphatase activity and labile P depletion in the rhizosphere of *Lupinus angustifolius* assessed with a novel, combined 2D-imaging approach. Soil Biol. Biochem. 153, 107963.

Hutchison, L.J., 1990. Studies on the systematics of ectomycorrhizal fungi in axenic culture. II. The enzymatic degradation of selected carbon and nitrogen compounds. Can. J. Bot. 68, 1522–1530.

Ijima, M., Higuchi, T., Barlow, P.W., 2004. Contribution of root cap mucilage and presence of an intact root cap in maize (*Zea mays*) to the reduction of soil mechanical impedance. Ann. Bot. 94, 473–477.

Ingwersen, J., Bücherl, B., Neumann, G., Streck, T., 2006. Experimental modelling of kinetic desorption in Cd hyperaccumulation by *Thlaspi caerulescens*. J. Env. Qual. 35, 2055–2065.

Inoue, H., Kobayashi, T., Nozoye, T., Takahashi, M., Kakei, Y., Suzuki, K., et al., 2009. Rice OsYSL15 is an iron-regulated iron(III)-deoxymugineic acid transporter expressed in the roots and is essential for iron uptake in early growth of the seedlings. J. Biol. Chem. 284, 3470–3479.

Inselsbacher, E., Öhlund, J., Sämtgård, S., Huss-Danell, K., Näsholm, T., 2011. The potential of microdialysis to monitor organic and inorganic nitrogen compounds in soil. Soil Biol. Biochem. 43, 1321–1332.

Irshad, U., Brauman, A., Villenave, C., Plassard, C., 2012. Phosphorus acquisition from phytate depends on efficient bacterial grazing, irrespective of the mycorrhizal status of *Pinus pinaster*. Plant Soil 358, 155–168.

Ishimaru, Y., Kakei, Y., Shimo, H., Bashir, K., Sato, Y., Sato, Y., et al., 2011. A rice phenolic efflux transporter is essential for solubilizing precipitated apoplasmic iron in the plant stele. J. Biol. Chem. 286, 24649–24655.

Ishimaru, Y., Suzuki, M., Tsukamoto, T., Suzuki, K., Nakazono, M., Kobayashi, T., et al., 2006. Rice plants take up iron as an Fe^{3+}-phytosiderophore and as Fe^{2+}. Plant J. 45, 335–446.

Jaillard, B., 1985. Activité racinaire et rhizostructures en milieu carbonate. Pedologie 35, 297–313.

Jaillard, B., Hinsinger, P., 1993. Alimentation minérale des végétaux dans le sol. Encyclopédie des Techniques Agricoles 87 (6-93), 1–13. fasc. 1210.

Jaillard, B., Guyon, A., Maurin, A.F., 1991. Structure and composition of calcified roots, and their identification in calcareous soils. Geoderma 50, 197–210.

Janzen, H.H., 1990. Deposition of nitrogen into the rhizosphere by wheat roots. Soil Biol. Biochem. 22, 1155–1160.

Jilling, A., Keiluweit, M., Contosta, A.R., Frey, S., Schimel, J., Schnecker, J., et al., 2018. Minerals in the rhizosphere: overlooked mediators of soil nitrogen availability to plants and microbes. Biogeochemistry 139, 103–122.

Jin, C.W., You, G.Y., He, Y.F., Tang, C., Wu, P., Zheng, S.J., 2007. Iron deficiency-induced secretion of phenolics facilitates the reutilization of root apoplastic iron in red clover. Plant Physiol. 144, 278–285.

Jing, J., Rui, Y., Zhang, F., Rengel, Z., Shen, J., 2010. Localized application of phosphorus and ammonium improves growth of maize seedlings by stimulating root proliferation and rhizosphere acidification. Field Crops Res. 119, 355–364.

Johnson, J.F., Vance, C.P., Allan, D.L., 1996. Phosphorus deficiency in *Lupinus albus*: altered lateral root development and enhanced expression of phosphoenolpyruvate carboxylase. Plant Physiol. 111, 31–41.

Johnson, D., Leake, J.R., Ostle, N., Ineson, P., Read, D.J., 2002. *In situ* $^{13}CO_2$ pulse-labelling of upland grassland demonstrates that a rapid pathway of carbon flux from arbuscular mycorrhizal mycelia to the soil. New Phytol. 153, 327–334.

Jones, D.L., 1998. Organic acids in the rhizosphere — a critical review. Plant Soil 205, 25—44.
Jones, D.L., Darrah, P.R., 1993. Re-absorption of organic compounds by roots of Zea mays L. and its consequences in the rhizosphere. II. Experimental and model evidence for simultaneous exudation and re-absorption of soluble C compounds. Plant Soil 153, 47—59.
Jones, D.L., Shannon, D., Murphy, D.V., Farrar, J., 2004a. Role of dissolved organic nitrogen (DON) in soil N cycling in grassland soils. Soil Biol. Biochem. 36, 749—756.
Jones, D.L., Hodge, A., Kuzyakov, Y., 2004b. Plant and mycorrhizal regulation of rhizodeposition. New Phytol. 163, 459—480.
Jones, D.L., Dennis, P.G., Owen, A.G., van Hees, P.A.W., 2003. Organic acid behaviour in soils: misconceptions and knowledge gaps. Plant Soil 248, 31—41.
Jones, D.L., Healey, J.R., Willet, V.B., Farrar, J.F., Hodge, A., 2005. Dissolved organic nitrogen uptake by plants — an important N uptake pathway? Soil Biol. Biochem. 37, 413—423.
Jones, D.L., Nguyen, C., Finlay, R.D., 2009. Carbon flow in the rhizosphere: carbon trading at the soil—root interface. Plant Soil 321, 5—33.
Jungk, A., 1991. Dynamics of nutrient movement at the soil—root interface. In: Waisel, Y., Eshel, A., Kafkafi, U. (Eds.), Plant Roots: The Hidden Half. Marcel Dekker, New York, USA, pp. 455—481.
Jungk, A., 2002. Dynamics of nutrient movement at the soil-root interface. In: Waisel, Y., Eshel, A., Kafkafi, U. (Eds.), Plant Roots: The Hidden Half, 3rd Ed. Marcel Dekker, New York, USA, pp. 587—616.
Kanabo, A.I., Gilkes, R., 1987. The role of soil pH in the dissolution of phosphate rock fertilizers. Fert. Res. 12, 165—173.
Kirk, G.J.D., Bajita, J.B., 1995. Root-induced iron oxidation, pH changes and zinc solubilization in the rhizosphere of lowland rice. New Phytol. 131, 129—137.
Klemedtsson, L., Svensson, B.H., Rosswall, T., 1988. Relationships between soil moisture content and nitrous oxide production during nitrification and denitrification. Biol. Fert. Soils 6, 106—111.
Klotz, F., Horst, W.J., 1988. Effect of ammonium- and nitrate-nitrogen nutrition on aluminium tolerance of soybean (*Glycine max* L.). Plant Soil 111, 59—65.
Kochian, L.V., 1995. Cellular mechanisms of aluminum toxicity and resistance of plants. Annu. Rev. Plant Physiol. Plant Mol. Biol. 46, 237—260.
Kochian, L.V., Hoekenga, A.O., Piñeros, M.A., 2004. How do plants tolerate acid soils? Mechanisms of aluminum tolerance and phosphorus efficiency. Ann. Rev. Plant Biol. 55, 459—493.
Kohli, A., Narciso, J.O., Miro, B., Raorane, M., 2012. Root proteases: reinforced links between nitrogen uptake and mobilization and drought tolerance. Physiol. Plant. 145, 165—179.
Kollmeier, M., Felle, H.H., Horst, W.J., 2000. Genotypical differences in aluminum resistance in maize are expressed in the distal part of the transition zone. Is reduced basipetal auxin flow involved in inhibition of root elongation by aluminium? Plant Physiol. 122, 945—956.
Kong, T., Steffens, D., 1989. Bedeutung der Kalium-Verarmung in der Rhizosphäre und der Tonminerale für die Freisetzung von nichtaustauschbarem Kalium und dessen Bestimmung mit CHl. Z. Pflanzenernähr. Bodenk. 152, 337—343.
Korenblum, E., Donga, Y., Szymanski, J., Pandaa, S., Jozwiak, A., Massalha, H., et al., 2020. Rhizosphere microbiome mediates systemic root metabolite exudation by root-to-root signaling. Proc. Natl. Acad. Sci. USA 117, 3874—3883.
Kuchenbuch, R., Jungk, A., 1982. A method for determining concentration profiles at soil-root interface by thin slicing rhizosphere soil. Plant Soil 68, 391—394.
Kuppardt, A., Fester, T., Härtig, C., Chatzinotas, A., 2018. Rhizosphere protists change metabolite profiles in *Zea mays*. Front. Microbiol. 9, 857.
Kuzyakov, Y., 2002. Separating microbial respiration of exudates from root respiration in non-sterile soils: a comparison of four methods. Soil Biol. Biochem. 34, 1621—1631.
Kuzyakov, Y., Domanski, G., 2002. Model for rhizodeposition and CO_2 efflux from planted soil and its validation by ^{14}C pulse labelling of ryegrass. Plant Soil 239, 87—102.
Kuzyakov, Y., Razavi, B.S., 2019. Rhizosphere size and shape: temporal dynamics and spatial stationarity. Soil Biol. Biochem. 135, 343—360.
Lambers, H., Atkin, O.K., Millenaar, F.F., 2002. Respiratory patterns in roots in relation to their functioning. In: Waisel, Y., Eshel, A., Kafkafi, U. (Eds.), Plant Roots: The Hidden Half, 3rd ed. Marcel Dekker, New York, USA, pp. 521—552.
Lambers, H., Martinoia, E., Renton, M., 2015. Plant adaptations to severely phosphorus-impoverished soils. Curr. Opin. Plant Biol. 25, 23—31.
Le Bot, J., Kirkby, E.A., van Beusichem, M.L., 1990. Manganese toxicity in tomato plants: effects on cation uptake and distribution. J. Plant Nutr. 13, 513—525.
Lee, J.A., 1999. The calcicole-calcifuge problem revisited. Adv. Bot. Res. 29, 1—30.
Leptin, A., Whitehead, D., Anderson, C.R., Cameron, K.C., Lehto, N.J., 2021. Increased soil nitrogen supply enhances root-derived available soil carbon leading to reduced potential nitrification activity. Appl. Soil Ecol. 159, 103842.
Li, X., Li, Y., Mai, J., Tao, L., Qu, M., Liu, J., et al., 2018a. Boron alleviates aluminum toxicity by promoting root alkalization in transition zone via polar auxin transport. Plant Physiol. 177, 1254—1266.
Li, S., Zhou, X., Chen, J., Chen, R., 2018b. Is there a strategy I iron uptake mechanism in maize? Plant Signal. Behav. 13, 4.
Li, M., Osaki, M., Rao, I.M., Tadano, T., 1997. Secretion of phytase from the roots of several plant species under phosphorus-deficient conditions. Plant Soil 195, 161—169.
Liao, X.Y., Chen, T.B., Xiao, X.Y., Xie, H., Yan, X.L., Zhai, L.M., et al., 2007. Selecting appropriate forms of nitrogen fertilizer to enhance soil arsenic removal by *Pteris vittata*: a new approach in phytoremediation. Int. J. Phytoremediation 9, 269—280.
Liljeroth, E., van Veen, J.A., Miller, H.J., 1990. Assimilate translocation to the rhizosphere of two wheat lines and subsequent utilization by rhizosphere microorganisms at two soil nitrogen concentrations. Soil Biol. Biochem. 22, 1015—1021.

Liu, P., Cai, Z., Chen, Z., Mo, X., Ding, X., Liang, C., et al., 2018. A root-associated purple acid phosphatase, SgPAP23, mediates extracellular phytate-P utilization in *Stylosanthes guianensis*. Plant Cell Environ. 41, 2821–2834.

Lorenz, S.E., Hamon, R.E., McGrath, S.P., 1994. Differences between soil solutions obtained from rhizosphere and non-rhizosphere soils by water displacement and soil centrifugation. Eur. J. Soil Sci. 45, 431–438.

Lynch, J.P., 2015. Root phenes that reduce the metabolic costs of soil exploration: opportunities for 21st century agriculture. Plant Cell Environ. 38, 1775–1784.

Lynch, J.M., Whipps, J.M., 1990. Substrate flow in the rhizosphere. Plant Soil 129, 1–10.

Ma, J.F., Nomoto, K., 1996. Effective regulation of iron acquisition in graminaceous plants. The role of mugineic acids as phytosiderophores. Physiol. Plant. 97, 609–617.

Ma, Q., Wang, X., Li, H.B., Li, H.G., Zhang, F., Rengel, Z., et al., 2015. Comparing localized application of different N fertilizer species on maize grain yield and agronomic N-use efficiency on a calcareous soil. Field Crops Res. 180, 72–79.

Ma, B., Zhang, L., Gao, Q., Wang, J., Li, X., Wang, H., et al., 2021. A plasma membrane transporter coordinates phosphate reallocation and grain filling in cereals. Nat. Genet. 53, 906–915.

Mäder, P., Fließbach, A., Dubois, D., Gunst, L., Fried, P., Niggli, U., 2002. Soil fertility and biodiversity in organic farming. Science 296, 1694–1697.

Magalhaes, J.V., Liu, J., Guimaraes, C.T., Lana, U.G.P., Alves, V.M.C., Wang, Y.H., et al., 2007. A gene in the multidrug and toxic compound extrusion (MATE) family confers aluminium tolerance in sorghum. Nat. Genet. 39, 1156–1161.

Mairhofer, S., Zappala, S., Tracy, S., Sturrock, C., Bennett, M.J., Mooney, S.J., et al., 2013. Recovering complete plant root system architectures from soil via X-ray μ-Computed Tomography. Plant Methods 9, 8.

Marschner, H., 1998. Soil-root interface: biological and biochemical processes. In: Huang, P.M. (Ed.), Soil Chemistry and Ecosystem Health. Soil Science Society of America, Madison, WI, USA, pp. 191–231.

Marschner, H., Römheld, V., 1983. *In vivo* measurement of root induced pH changes at the soil-root interface: effect of plant species and nitrogen source. Z. Pflanzenphysiol. 111, 241–251.

Marschner, H., Römheld, V., Kissel, M., 1986a. Different strategies in higher plants in mobilization and uptake of iron. J. Plant Nutr. 9, 695–713.

Marschner, H., Römheld, V., Horst, W.J., Martin, P., 1986b. Root-induced changes in the rhizosphere: importance for the mineral nutrition of plants. Z. Pflanzenernähr. Bodenk. 149, 441–456.

Marschner, P., Neumann, G., Kania, A., Weiskopf, L., Lieberei, R., 2002. Spatial and temporal dynamics of the microbial community structure in the rhizosphere of cluster roots of white lupin (*Lupinus albus* L.). Plant Soil 246, 167–174.

Martin, F., Duplessis, S., Ditengou, F., Lagrange, H., Voiblet, C., Lapeyrie, F., 2001. Developmental cross talking in the ectomycorrhizal symbiosis: signals and communication genes. New Phytol. 151, 145–154.

Matar, A.E., Paul, J.L., Jenny, H., 1967. Two phase experiments with plants growing in phosphate-treated soil. Soil Sci. Soc. Amer. Proc. 31, 235–237.

McCully, M.E., 1999. Roots in soil: unearthing the complexities of roots and their rhizospheres. Ann. Rev. Plant Physiol. Plant Mol. Biol. 50, 695–718.

McKay Fletcher, D.M., Shaw, R., Sánchez-Rodríguez, A.R., Daly, K.R., van Veelen, A., Jones, D.L., et al., 2021. Quantifying citrate-enhanced phosphate root uptake using microdialysis. Plant Soil 461, 69–89.

McLaughlin, M.J., Alston, A.M., Martin, J.K., 1987. Transformations and movement of P in the rhizosphere. Plant Soil 97, 391–399.

Meharg, A.A., Kilham, K., 1995. Loss of exudates from the roots of perennial ryegrass inoculated with a range of microorganisms. Plant Soil 170, 345–349.

Mehra, P., Pandey, B.K., Giri, J., 2017. Improvement in phosphate acquisition and utilization by a secretory purple acid phosphatase (OsPAP21b) in rice. Plant Biotech. J. 15, 1054–1067.

Mendes, R., Garbeva, P., Raaijmakers, J.M., 2013. The rhizosphere microbiome: significance of plant beneficial, plant pathogenic, and human pathogenic microorganisms. FEMS Microbiol. Rev. 37, 634–663.

Miyasaka, S.C., Hawes, M.C., 2001. Possible role of root border cells in detection and avoidance of aluminum toxicity. Plant Physiol. 125, 1978–1987.

Moe, L.A., 2013. Amino acids in the rhizosphere: from plants to microbes. Am. J. Bot. 100, 1692–1705.

Mooney, S.J., Pridmore, T.P., Helliwell, J., Bennett, M.J., 2011. Developing X-ray computed tomography to non-invasively image 3-D root systems architecture in soil. Plant Soil 352, 1–22.

Morel, J.L., Habib, L., Plantureux, S., Guckert, A., 1991. Influence of maize root mucilage on soil aggregate stability. Plant Soil 136, 111–119.

Morel, J.L., Mench, M., Guckert, A., 1986. Measurement of Pb^{2+}, Cu^{2+} and Cd^{2+} binding with mucilage exudates from maize (*Zea mays* L.) roots. Biol. Fert. Soils 2, 29–34.

Mpanga, I.K., Dapaah, H.K., Geistlinger, J., Ludewig, U., Neumann, G., 2018. Soil type-dependent interactions of P-solubilizing microorganisms with organic and inorganic fertilizers mediate plant growth promotion in tomato. Agronomy 8, 213.

Mpanga, I.K., Nkebiwe, P.M., Kuhlmann, K., Cozzolino, V., Piccolo, A., Geistlinger, J., et al., 2019. The form of N supply determines plant growth promotion by P-solubilizing microorganisms in maize. Microorganisms 7, 38.

Müller, J., Toev, T., Heisters, M., Teller, J., Moore, K.L., Hause, G., et al., 2015. Iron-dependent callose deposition adjusts root meristem maintenance to phosphate availability. Dev. Cell 33, 216–230.

Mur, L.A., Simpson, C., Kumari, A., Gupta, A.K., Gupta, K.J., 2017. Moving nitrogen to the centre of plant defence against pathogens. Ann. Bot. 119, 703–709.

Murakami, H., Kimura, M., Wada, H., 1990. Microbial colonization and decomposition processes in rice rhizoplane. II. Decomposition of young and old roots. Soil Sci. Plant Nutr. 36, 441–450.

Nagarajah, S., Posner, A.M., Quirk, J.P., 1970. Competitive adsorptions of phosphate with polygalacturonate and other organic anions on kaolinite and oxide surfaces. Nature 228, 83–84.

Nambiar, E.K.S., 1976. Uptake of Zn-65 from dry soil by plants. Plant Soil 44, 267–271.

Nannipieri, P., Giagnoni, L., Landi, L., Renella, G., 2011. Role of phosphatase enzymes in soil. In: Bünemann, E., Oberson, A., Frossard, E. (Eds.), Phosphorus in Action. Springer, Berlin, Heidelberg, Germany, pp. 2165–2243.

Nazari, M., Riebeling, S., Banfield, C.C., Akale, A., Crosta, M., Mason-Jones, K., et al., 2020. Mucilage composition and exudation in maize from contrasting climatic regions. Front. Plant Sci. 11, 587610.

Nazoa, P., Vidmar, J.J., Tranbarger, T.J., Mouline, K., Damiani, I., Tillard, P., et al., 2003. Regulation of the nitrate transporter gene AtNRT2.1 in *Arabidopsis thaliana*: responses to nitrate, amino acids and developmental stage. Plant Mol. Biol. 52, 689–703.

Neumann, G., 2006. Root exudates and organic composition of plant roots. In: Luster, J., Finlay, R., Brunner, I., Fitz, W.J., Frey, B., Goettlein, A., et al.,Handbook of Methods used in Rhizosphere Research. Swiss Federal Institute for Forest, Snow, and Landscape Research, Birmensdorf, Switzerland, pp. 52–63.

Neumann, G., 2007. Root exudates and nutrient cycling. In: Marschner, P., Rengel, Z. (Eds.), Nutrient Cycling in Ecosystems. Springer, Berlin, Heidelberg, Germany, pp. 123–157.

Neumann, G., Römheld, V., 1999. Root excretion of carboxylic acids and protons in phosphorus-deficient plants. Plant Soil 211, 121–130.

Neumann, G., Martinoia, E., 2002. Cluster roots - an underground adaptation for survival in extreme environments. Trends Plant Sci. 7, 162–167.

Neumann, G., Römheld, V., 2002. Root-induced changes in the availability of nutrients in the rhizosphere. In: Waisel, Y., Eshel, A., Kafkafi, U. (Eds.), Plant Roots. The Hidden Half, 3rd ed. Marcel Dekker, New York, USA, pp. 617–649.

Neumann, G., Römheld, V., 2007. The release of root exudates as affected by the plant physiological status. In: Pinton, R., Varanini, Z., Nannipieri, Z. (Eds.), The Rhizosphere: Biochemistry and Organic Substances at the Soil-Plant Interface, 2nd ed CRC Press, Boca Raton, FL, USA, pp. 23–72.

Neumann, G., Massonneau, A., Martinoia, E., Römheld, V., 1999. Physiological adaptations to phosphorus deficiency during proteoid root development in white lupin. Planta 208, 373–382.

Neumann, G., Massonneau, A., Langlade, N., Dinkelaker, B., Hengeler, C., Römheld, V., et al., 2000. Physiological aspects of cluster root function and development in phosphorus-deficient white lupin (*Lupinus albus* L.). Ann. Bot. 85, 909–919.

Neumann, G., Schulze, C., George, E., Römheld, V., et al., 2001. Acquisition of phosphorus in potato (*Solanum tuberosum* L. cv. Désirée) with altered carbohydrate partitioning between shoot and roots. Plant Nutrition: Food Security and Sustainability of Agro-ecosystems through Basic and Applied Research. XIV International Plant Nutrition Colloquium. Kluwer Academic Publishers, Dordrecht, The Netherlands, pp. 134–135.

Nishizawa, N., Mori, S., 1987. The particular vesicle appearing in barley root cells and its relation to mugineic acid secretion. J. Plant Nutr. 10, 1013–1020.

Nkebiwe, P.M., Weinmann, M., Müller, T., 2016. Improving fertilizer depot exploitation and maize growth by inoculation with plant growth-promoting bacteria: from lab to field. Chem. Biol. Technol. Agric. 3, 15.

Nozoye, T., Nagasaka, S., Kobayashi, T., Takahashi, M., Sato, Y., Sato, Y., et al., 2011. Phytosiderophore efflux transporters are crucial for iron acquisition in graminaceous plants. J. Biol. Chem. 286, 5446–5454.

Nyatsanaga, T., Pierre, W.H., 1973. Effect of nitrogen fixation by legumes on soil acidity. Agron. J. 65, 936–940.

Nye, P.H., 1986. Acid–base changes in the rhizosphere. In: Tinker, B., Läuchli, A. (Eds.), Advances in Plant Nutrition 2. Praeger Scientific, New York, USA, pp. 129–153.

Oburger, E., Jones, D.L., 2018. Sampling root exudates — mission impossible? Rhizosphere 6, 116–133.

O'Rourke, J.A., Yang, S.S., Miller, S.S., Bucciarelli, B., Liu, J., Rydeen, A., et al., 2013. An RNA-Seq transcriptome analysis of orthophosphate-deficient white lupin reveals novel insights into phosphorus acclimation in plants. Plant Physiol. 161, 705–724.

Otani, T., Ae, N., 1993. Ethylene and carbon dioxide concentrations of soils as influenced by rhizosphere of crops under field and pot conditions. Plant Soil 150, 255–262.

Palmer, C.M., Hindt, M.N., Schmidt, H., Clemens, S., Guerinot, M.L., 2013. MYB10 and MYB72 are required for growth under iron-limiting conditions. PLoS Genet. 9, e1003953.

Paungfoo-Lonhienne, C., Rentsch, D., Robatzek, S., Webb, R.I., Sagulenko, E., Näsholm, T., et al., 2010. Turning the table: plants consume microbes as a source of nutrients. PloS One 5, e11915.

Phillips, D.A., Fox, T.C., King, M.D., Bhuvaneswari, T.V., Teuber, L.R., 2004. Microbial products trigger amino acid exudation from plant roots. Plant Physiol. 136, 2887–2894.

Pilbeam, D.J., Cakmak, I., Marschner, H., Kirkby, E.A., 1993. Effect of withdrawal of phosphorus on nitrate assimilation and PEP carboxylase activity in tomato. Plant Soil 154, 111–117.

Pineros, M.A., Shaff, J.E., Holly, S., Manslank, V.M., Carvalho, A., Kochian, L.V., 2005. Aluminum resistance in maize cannot be solely explained by root organic acid exudation. A comparative physiological study. Plant Physiol. 137, 231–241.

Pinton, R., Cesco, S., De Nobili, M., Santi, S., Varanini, Z., 1998. Water and pyrophosphate-extractable humic substances fractions as a source of iron for Fe-deficient cucumber plants. Biol. Fert. Soil 26, 23–27.

Pinton, R., Cesco, S., Iacolettig, G., Astolfi, S., Varanini, Z., 1999. Modulation of NO_3^- uptake by water-extractable humic substances: involvement of root plasmalemma H^+ ATPase. Plant Soil 215, 155–161.

Pinton, R., Cesco, S., Santi, S., Varanini, Z., 1997. Soil humic substances stimulate proton release by intact oat seedling roots. J. Plant Nutr. 20, 857–869.

Poschenrieder, C., Fernández, J., Rubio, A., Pérez, L., Terés, L., Barceló, J., J., 2018. Transport and use of bicarbonate in plants: current knowledge and challenges ahead. Int. J. Mol. Sci. 19, 1352.

Posta, K., Marschner, H., Römheld, V., 1994. Manganese reduction in the rhizosphere of mycorrhizal and non-mycorrhizal maize. Mycorrhiza 5, 119–124.

Prieto, P., Schiliro, E., Maldonado-Gonzalez, M.M., Valderrama, R., Barroso-Albarracın, J.B., Mercado-Blanco, J., 2011. Root hairs play a key role in the endophytic colonization of olive roots by *Pseudomonas* spp. with biocontrol activity. Microb. Ecol. 62, 435–445.

Qasim, B., Motelica-Heino, M., Bourgerie, S., Gauthier, A., Morabito, D., 2015. Effect of nitrate and ammonium fertilization on Zn, Pb, and Cd phytostabilization by *Populus euramericana* Dorskamp in contaminated technosol. Environ. Sci. Pollut. Res. Int. 22, 18759–18771.

Rabotti, G., Zocchi, G., 1994. Plasma membrane-bound H^+-ATPase and reductase activities in Fe-deficient cucumber roots. Physiol. Plant. 90, 779–785.

Rajniak, J., Giehl, R.F.H., Chang, E., Murgia, I., von Wirén, N., Sattely, E.S., 2018. Biosynthesis of redox-active metabolites in response to iron deficiency in plants. Nat. Chem Biol. 14, 442–450.

Rao, A.-M., Gianfreda, L., Palmiero, F., Violante, A., 1996. Interactions of acid phosphatase with clays, organic molecules and organo-mineral complexes. Soil Sci. 161, 751–760.

Raven, J.A., Rothemund, C., Wollenweber, B., 1991. Acid-base regulation by *Azolla* spp. with N_2 as sole N source and with supplementation by NH_4 or NO_3^-. Acta Bot. 104, 132–138.

Raymond, N.S., Gómez-Muñoz, B., van der Bom, F.J.T., Nybroe, O., Jensen, L.S., Müller-Stöver, D.S., et al., 2020. Phosphate-solubilising microorganisms for improved crop productivity: a critical assessment. New Phytol. 229, 1268–1277.

Razavi, B.S., Zhang, X., Bilyera, N., Guber, A., Zarebanadkouki, M., 2019. Soil zymography: simple and reliable? Review of current knowledge and optimization of the method. Rhizosphere 11, 100161.

Read, D.B., Bengough, A.G., Gregory, P.J., Crawford, J.W., Robinson, D., Scrimgeour, C.M., et al., 2003. Plant roots release phospholipid surfactants that modify the physical and chemical properties of soil. New Phytol. 157, 315–321.

Reidenbach, G., Horst, W., 1997. Nitrate-uptake capacity of different root zones of *Zea mays* (L.) *in vitro* and *in situ*. Plant Soil 196, 295–300.

Rengel, Z., 1997. Root exudation and microflora populations in rhizosphere of crop genotypes differing in tolerance to micronutrient deficiency. Plant Soil 196, 255–260.

Revsbech, N.P., Pedersen, O., Reichardt, W., Briones, A., 1999. Microsensor analysis of oxygen and pH in the rice rhizosphere under field and laboratory conditions. Biol. Fert. Soils 29, 379–385.

Reynolds, S.B., Scaife, A., Turner, M.K., 1987. Effect of nitrogen form on boron uptake by cauliflower. Commun. Soil Sci. Plant Anal. 18, 1143–1153.

Rich, S.M., Ludwig, M., Pedersen, O., Colmer, T.D., 2011. Aquatic adventitious roots of the wetland plant *Meionectes brownii* can photosynthesize: implications for root function during flooding. New Phytol. 190, 311–319.

Richardson, A.E., Simpson, R.J., 2011. Soil microorganisms mediating phosphorus availability update on microbial phosphorus. Plant Physiol. 156, 989–996.

Richardson, A.E., George, T.S., Hens, M., Simpson, R.J., 2005. Utilization of soil organic phosphorus by higher plants. In: Turner, B.L., Frossard, E., Baldwin, D.S. (Eds.), Organic phosphorus in the environment. CAB International, Cambridge, MA, USA, pp. 165–184.

Rillig, M.C., Ramsey, P.W., Morris, S., Pau, E.A., 2003. Glomalin, an arbuscular-mycorrhizal fungal soil protein, responds to land-use change. Plant Soil 253, 293–299.

Roberts, L.A., Pierson, A.J., Panaviene, Z., Walker, E.L., 2004. Yellow stripe 1. Expanded roles for the maize iron-phytosiderophore transporter. Plant Physiol. 135, 112–120.

Robinson, N.J., Sadjuga, M.R., Groom, Q.J., 1997. The *froh* gene family from *Arabidopsis thaliana*: putative iron-chelate reductases. Plant Soil 196, 245–248.

Rollwagen, B.A., Zasoski, R.J., 1988. Nitrogen source effects on rhizosphere pH and nutrient accumulation by Pacific Northwest conifers. Plant Soil 105, 79–86.

Römer, W., Beißner, L., Schenk, H., Jungk, A., 1995. Einfluß von Sorte und Phosphordüngung auf den Phosphorgehalt und die Aktivität der sauren Phosphatasen von Weizen und Gerste – Ein Beitrag zur Diagnose der P-Versorgung von Pflanzen. Z. Pflanzenernähr. Bodenk. 158, 3–8.

Römheld, V., 1986. pH changes in the rhizosphere of various crop plants, in relation to the supply of plant nutrients. Potash Review 12, 1–16.

Römheld, V., 1987. Different strategies for iron acquisition in higher plants. Physiol. Plant. 70, 231–234.

Römheld, V., Kramer, D., 1983. Relationship between proton efflux and rhizodermal transfer cells induced by iron deficiency. Z. Pflanzenphysiol. 113, 73–83.

Römheld, V., Marschner, H., 1983. Mechanisms of iron uptake by peanut plants. 1. Reduction, chelate splitting, and release of phenolics. Plant Physiol. 71, 949–954.

Ropitaux, M., Bernard, S., Schapman, D., Follet-Gueye, M.-L., Vicré, M., Boulogne, I., et al., 2020. Root border cells and mucilage secretions of soybean, *Glycine max* (Merr) L.: Characterization and role in interactions with the oomycete *Phytophthora parasitica*. Cells 9, 2215.

Ross-Elliott, T.J., Jensen, K.H., Haaning, K.S., Wager, B.M., Knoblauch, J., Howell, A.H., et al., 2017. Phloem unloading in Arabidopsis roots is convective and regulated by the phloem-pole pericycle. elife 6, e24125.

Rovira, A.D., 1959. Root excretions in relation to the rhizosphere effect. IV. Influence of plant species, age of plant, light, temperature and calcium nutrition on exudation. Plant Soil 11, 53–64.

Rovira, A.D., Bowen, G.D., Foster, R.C., 1983. The significance of rhizosphere microflora and mycorrhizas in plant nutrition. In: Läuchli, A., Bieleski, R.L. (Eds.), Encyclopedia of Plant Physiology, New Series, 15A. Springer-Verlag, Berlin and New York, pp. 61–89.

Ryan, P.R., Dessaux, Y., Thomashow, L.S., Weller, D.M., 2009. Rhizosphere engineering and management for sustainable agriculture. Plant Soil 321, 363–383.

Sakano, K., 1998. Revision of biochemical pH-stat: involvement of alternative pathway metabolisms. Plant Cell Physiol. 39, 467–473.
Saleque, M.A., Kirk, G.J.D., 1995. Root-induced solubilisation of phosphate in the rhizosphere of lowland rice. New Phytol. 129, 325–336.
Samaj, J., Read, N.D., Volkmann, D., Menzel, D., Baluška, F., 2005. The endocytic network in plants. Trends Cell Biol. 15, 425–433.
Sarniguet, A., Lucas, P., Lucas, M., Samson, R., 1992. Soil conduciveness to take-all of wheat: influence of the nitrogen fertilizers on the structure of populations of fluorescent pseudomonads. Plant Soil 145, 29-26.
Sasaki, T., Yamamoto, Y., Ezaki, B., Katsuhara, M., Ahn, S.J., Ryan, P., et al., 2004. A wheat gene encoding an aluminum-activated malate transporter. Plant J. 37, 645–653.
Sasse, J., Martinoia, E., Northen, T., 2018. Feed your friends: do plant exudates shape the root microbiome? Trends Plant Sci. 23, 25–41.
Schack-Kirchner, H., Loew, C.A., Lang, F., 2020. The cumulative amount of exuded citrate controls its efficiency to mobilize soil phosphorus. Front. For. Glob. Change 3, 550884.
Schaller, G., Fischer, W.R., 1985. Kurzfristige pH-Pufferung von Böden. Z. Pflanzenernähr. Bodenk. 148, 471–480.
Schenkel, D., Deveau, A., Niimi, J., Mariotte, P., Vitra, A., Meisser, M., et al., 2019. Linking soil's volatilome to microbes and plant roots highlights the importance of microbes as emitters of belowground volatile signals. Environ. Microbiol. 21, 3313–3327.
Scheurwater, I., Clarkson, D.T., Purves, J.V., van Rijt, G., Saker, L.R., Welschen, R., et al., 1999. Relatively large nitrate efflux can account for the high specific respiratory costs for nitrate transport in slow-growing grass species. Plant Soil 215, 123–134.
Schneider, M., Scherer, H.W., 1998. Fixation and release of ammonium in flooded rice soils as affected by redox potential. Eur. J. Agron. 8, 181–189.
Schöttelndreier, M., Falkengren-Grerup, U., 1999. Plant induced alteration in the rhizosphere and the utilisation of soil heterogeneity. Plant Soil 209, 297–309.
Schulz, H., Postma, J.A., van Dusschoten, D., Scharr, H., Behnke, S., 2013. Plant root system analysis from MRI images. Comm. Comp. Inf. Sci. 359, 411–425.
Schulz-Bohm, K., Gerards, S., Hundscheid, M.P.J., Melenhorst, J., de Boer, W., Garbeva, P.V., 2018. Calling from distance: attraction of soil bacteria by plant root volatiles. ISME J. 12, 1252–1262.
Siao, W., Coskun, D., Baluška, F., Kronzucker, H.J., Xu, W., 2020. Root-apex proton fluxes at the centre of soil-stress acclimation. Trends Plant Sci. 25, 794–804.
Sisó-Terraza, P., Luis-Villarroya, A., Fourcroy, P., Briat, J.F., Abadía, A., Gaymard, F., et al., 2016a. Accumulation and secretion of coumarinolignans and other coumarins in Arabidopsis thaliana roots in response to iron deficiency at high pH. Front. Plant Sci. 7, 1711.
Sisó-Terraza, P., Rios, J.J., Abadía, J., Abadía, A., Álvarez-Fernández, A., 2016b. Flavins secreted by roots of iron-deficient *Beta vulgaris* enable mining of ferric oxide via reductive mechanisms. New Phytol. 209, 733–745.
Smith, R.J., Hopper, S.D., Shane, M.W., 2011. Sand-binding roots in *Haemodoraceae*: global survey and morphology in a phylogenetic context. Plant Soil 348, 453–470.
Smolders, E., Wagner, S., Prohaska, T., Irrgeher, J., Santner, J., 2020. Sub-millimeter distribution of labile trace element fluxes in the rhizosphere explains differential effects of soil liming on cadmium and zinc uptake in maize. Sci. Total Environ. 738, 140311.
Staddon, P.L., Bronk Ramsey, C., Ostle, N., Ineson, P., Fitter, A.H., 2003. Rapid turnover of hyphae of mycorrhizal fungi determined by AMS microanalysis of ^{14}C. Science 300, 1138–1140.
Svennerstam, H., Jämtgård, S., Ahmad, I., Huss-Danell, K., Näsholm, T., Ganeteg, U., 2011. Transporters in Arabidopsis roots mediating uptake of amino acids at naturally occurring concentrations. New Phytol. 191, 459–467.
Swart, P.H., van Diest, A., 1987. The rock-phosphate solubilizing capacity of *Pueraria javanica* as affected by soil pH, superphosphate priming effect and symbiotic N_2 fixation. Plant Soil 100, 135–147.
Tagaki, S., 1984. Mechanism of iron uptake regulation in roots and genetic differences. Agriculture, Soil Science and Plant Nutrition in the Northern Part of Japan. Japanese Society of Soil Science and Plant Nutrition, Tokyo, Japan, pp. 190–195.
Takahashi, M., Nakanishi, H., Kawasaki, S., Nishizawa, N.K., Mori, S., 2001. Enhanced tolerance of rice to low iron availability in alkaline soils using barley nicotianamine aminotransferase genes. Nat. Biotechnol. 19, 466–469.
Takahashi, H., Yamauchi, T., Colmer, T.D., Nakazono, M., 2014. Aerenchyma formation in plants. Plant Cell Monogr. 21, 247–265.
Tarafdar, J.C., Jungk, A., 1987. Phosphatase activity in the rhizosphere and its relation to the depletion of soil organic phosphorus. Biol. Fert. Soils 3, 199–204.
Thomson, C.J., Marschner, H., Römheld, V., 1993. Effect of nitrogen fertilizer form on pH of the bulk soil and rhizosphere, and on the growth, phosphorus, and micronutrient uptake of bean. J. Plant Nutr. 16, 493–506.
Tomasi, N., Kretzschmar, T., Espen, L., Weisskopf, L., Fuglsang, A.T., Palmgren, M.G., et al., 2009. Plasma membrane H^+-ATPase-dependent citrate exudation from cluster roots of phosphate-deficient white lupin. Plant Cell Environ. 32, 465–475.
Tomasi, N., Nobili, M., Gottardi, S., Zanin, L., Mimmo, T., Varanini, Z., et al., 2013. Physiological and molecular aspects of Fe acquisition by tomato plants treated with natural Fe complexes. Biol. Fert. Soils 49, 187–200.
Totsche, K.U., Amelung, W., Gerzabek, M.H., Guggenberger, G., Klumpp, E., Knief, C., et al., 2018. Microaggregates in soils. J. Plant Nutr. Soil Sci. 181, 104–136.
Tran, H.T., Qian, W., Hurley, B.A., She, Y.M., Wang, D., Plaxton, W.C., 2010. Biochemical and molecular characterization of AtPAP12 and AtPAP26: the predominant purple acid phosphatase isozymes secreted by phosphate-starved *Arabidopsis thaliana*. Plant Cell Environ. 33, 1789–1803.
Treeby, M., Marschner, H., Römheld, V., 1989. Mobilization of iron and other micronutrient cations from a calcareous soil by plant borne, microbial, and synthetic metal chelators. Plant Soil 114, 217–226.

Tsai, H.H., Schmidt, W., 2017. Mobilization of iron by plant-borne coumarins. Trends Plant Sci. 22, 538–548.

Turpault, M.P., Gobran, G., Bonnaud, P., 2006. Temporal variations of rhizosphere and bulk soil chemistry in a Douglas fir stand. Geoderma 137, 490–496.

Ueno, D., Yamaji, N., Ma, J.F., 2009. Further characterization of ferric-phytosiderophore transporters ZmYS1 and HvYS1 in maize and barley. J. Exp. Bot. 60, 3513–3520.

Valeeva, L.R., Nyamsuren, C., Sharipova, M.R., Shakirov, E.V., 2018. Heterologous expression of secreted bacterial BPP and HAP phytases in plants stimulates *Arabidopsis thaliana* growth on phytate. Front. Plant Sci. 9, 186.

van Dam, N.M., Weinhold, A., Garbeva, P., 2016. Calling in the dark: the role of volatiles for communication in the rhizosphere. In: Blande, J.D., Glinwood, R. (Eds.), Deciphering Chemical Language of Plant Communication. Springer International Publishing, Cham, Switzerland, pp. 175–210.

van Miegroet, H., Cole, D.W., 1984. The impact of nitrification on soil acidification and cation leaching in a red alder ecosystem. J. Environ. Qual. 13, 586–590.

Varanini, Z., Pinton, R., DeBiasi, M.G., Astolfi, S., Maggioni, A., 1993. Low molecular weight humic substances stimulate H^+-ATPase activity of plasma membrane vesicles isolated from oat *(Avena sativa* L.) roots. Plant Soil 153, 61–69.

Verbon, E.H., Trapet, P.L., Stringlis, I.A., Kruijs, S., Bakker, P.A.H.M., Pieterse, M.J., 2017. Iron and immunity. Annu. Rev. Phytopathol. 55, 255–275.

Vermeer, J., McCully, M.E., 1981. Fucose in the surface deposits of axenic and field grown roots of *Zea mays* L. Protoplasma 109, 233–248.

Vetterlein, D., Carminati, A., Kögel-Knabner, I., Bienert, G.P., Smalla, K., Oburger, E., et al., 2020. Rhizosphere spatiotemporal organization – a key to rhizosphere functions. Front. Agron. 2, 8.

von Rheinbaben, W., Trolldenier, G., 1984. Influence of plant growth on denitrification in relation to soil moisture and potassium nutrition. Z. Pflanzenernähr. Bodenk. 147, 730–738.

von Wirén, N., Marschner, H., Römheld, V., 1995. Uptake kinetics of iron-phytosiderophores in two maize genotypes differing in iron efficiency. Physiol. Plant. 93, 611–616.

von Wirén, N., Morel, J.L., Guckert, A., Römheld, V., Marschner, H., 1993. Influence of soil microorganisms on iron acquisition in maize. Soil Biol. Biochem. 25, 371–376.

Wallace, A., 1982. Effect of nitrogen fertilizer and nodulation on lime induced chlorosis in soybean. J. Plant Nutr. 5, 363–368.

Wang, X., Tang, C., 2018. The role of rhizosphere pH in regulating the rhizosphere priming effect and implications for the availability of soil-derived nitrogen to plants. Ann. Bot. 121, 143–151.

Wang, Y., Lambers, H., 2019. Root-released organic anions in response to low phosphorus availability: recent progress, challenges and future perspectives. Plant Soil 447, 135–156.

Wang, J., Raman, H., Zhou, M., Ryan, P.R., Delhaize, E., Hebb, D.M., et al., 2007. High-resolution mapping of Alp, the aluminium tolerance locus in barley (*Hordeum vulgare* L.), identifies a candidate gene controlling tolerance. Theor. Appl. Genet. 115, 265–276.

Wang, Y., Ye, X., Ding, G., Xu, F., 2013. Overexpression of *phyA* and *appA* genes improves soil organic phosphorus utilisation and seed phytase activity *in Brassica napus*. PloS One 8, e60801.

Wang, Z., Straub, D., Yang, H., Kania, A., Shen, J., Ludewig, U., et al., 2014. The regulatory network of cluster-root function and development in phosphate-deficient white lupin (*Lupinus albus*) identified by transcriptome sequencing. Physiol. Plant. 151, 323–338.

Wang, Z., Shen, J., Ludewig, U., Neumann, G., 2015. A re-assessment of sucrose signalling involved in cluster-root formation and function in phosphate-deficient white lupin (*Lupinus albus* L.). Physiol. Plant. 151, 323–338.

Wardle, D.A., 1992. A comparative assessment of factors which influence microbial growth carbon and nitrogen levels in soil. Biol. Rev. Camb. Philos. Soc. 67, 321–358.

Warren, C.R., 2015. Wheat roots efflux a diverse array of organic N compounds and are highly proficient at their recapture. Plant Soil 397, 147–162.

Wasaki, J., Michiko, A., Ozawa, K., Omura, M., Osaki, M., Ito, H., et al., 1997. Properties of secretory acid phosphatase from lupin roots under phosphorus-deficient conditions. Soil Sci. Plant Nutr. 43, 981–986.

Wasaki, J., Omura, M., Ando, M., Shinano, T., Osaki, M., Tadano, T., 1999. Secreting portion of acid phosphatase in roots of lupin (*Lupinus albus* L.) and a key signal for the secretion from the roots. Soil Sci. Plant Nutr. 45, 937–945.

Wasaki, J., Rothe, A., Kania, A., Neumann, G., Römheld, V., Shinano, T., et al., 2005. Root exudation, phosphorus acquisition, and microbial diversity in the rhizosphere of white lupin as affected by phosphorus supply and atmospheric carbon dioxide concentration. J. Environ. Qual. 34, 2157–2166.

Watt, M., Evans, J., 1999. Linking development and determinacy with organic acid efflux from proteoid roots of white lupin grown with low phosphorus and ambient or elevated atmospheric CO_2 concentration. Plant Physiol. 120, 705–716.

Watt, M., Kirkegaard, J.A., Passioura, J.B., 2006. Rhizosphere biology and crop productivity – a review. Aust. J. Soil Res. 44, 299–317.

Weisskopf, L., Abou-Mansour, E., Fromin, N., Tomasi, N., Santelia, D., Edelkott, I., et al., 2006. White lupin has developed a complex strategy to limit microbial degradation of secreted citrate required for phosphate nutrition. Plant Cell Environ. 29, 919–927.

Wen, F., Van Etten, H.D., Tsaprailis, G., Hawes, M.C., 2007. Extracellular proteins in pea root tip and border cell exudates. Plant Physiol. 143, 773–783.

Werner, D., 2007. Molecular biology and ecology of the Rhizobia legume symbiosis. In: Pinton, R., Varanini, Z., Nannipieri, Z. (Eds.), The Rhizosphere: Biochemistry and Organic Substances at the Soil-plant Interface. 2nd ed. CRC Press, Boca Raton, FL, USA, pp. 237–266.

Windisch, S., Sommermann, L., Babin, D., Chowdhury, S.P., Grosch, R., Moradtalab, N., et al., 2021. Impact of long-term organic and mineral fertilization on rhizosphere metabolites, root-microbial interactions and plant health of lettuce. Front. Microbiol. 11, 597745.

Wu, Q.T., Morel, J.L., Guckert, A., 1989. Effect of nitrogen source on cadmium uptake by plants. Compt. Rend. Acad. Sci. 309, 215–220.
Xia, J., Saglio, P.H., 1988. Characterization of the hexose transport system in maize root tips. Plant Physiol. 88, 1015–1020.
Xia, J.H., Roberts, J.K.M., 1994. Improved cytoplasmic pH regulation, increased lactate efflux, and reduced cytoplasmic lactate levels are biochemical traits expressed in root tips of whole maize seedlings acclimated to a low-oxygen environment. Plant Physiol. 105, 651-637.
Xue, Y., Xia, H., Christie, P., Zhang, Z., Li, L., Tang, C., 2016. Crop acquisition of phosphorus, iron and zinc from soil in cereal/legume intercropping systems: a critical review. Ann. Bot. 117, 363–377.
Yadav, R., Tarafdar, J., 2001. Influence of organic and inorganic phosphorus supply on the maximum secretion of acid phosphatase by plants. Biol. Fert. Soils 34, 140–143.
Yamauchi, T., Colmer, T.D., Pedersen, O., Nakazonoa, M., 2018. Regulation of root traits for internal aeration and tolerance to soil waterlogging-flooding stress. Plant Physiol. 176, 1118–1130.
Yan, F., Zhu, Y., Müller, C., Zorb, C., Schubert, S., 2002. Adaptation of H^+-pumping and plasma membrane H^+ ATPase activity in proteoid roots of white lupin under phosphate deficiency. Plant Physiol. 129, 50–63.
Yang, N., Hinner, M., 2015. Getting across the cell membrane: an overview for small molecules, peptides, and proteins. Methods Mol. Biol. 1266, 29–53.
Yi, L., Ma, J., Li, Y., 2007. Soil salt and nutrient concentration in the rhizosphere of desert halophytes. Acta Ecol. Sinica 27, 3565–3571.
Yin, Y.G., Suzui, N., Kurita, K., Miyoshi, Y., Unno, Y., Fujimaki, S., et al., 2020. Visualising spatio-temporal distributions of assimilated carbon translocation and release in root systems of leguminous plants. Sci. Rep. 10, 8446.
Yordem, B.K., Conte, S.S., Ma, J.F., Yokosho, K., Vasques, K.A., Gopalsamy, S.N., et al., 2009. *Brachypodium distachyon* as a new model system for understanding iron homeostasis in grasses: phylogenetic and expression analysis of Yellow Stripe-Like (YSL) transporters. Ann. Bot. 108, 821–833.
York, L.M., Carminati, A., Mooney, S.J., Ritz, K., Bennett, M.J., 2016. The holistic rhizosphere: integrating zones, processes, and semantics in the soil influenced by roots. J. Exp. Bot. 67, 3629–3643.
Zaccheo, P., Crippa, L., Di Muzio Pasta, V., 2006. Ammonium nutrition as a strategy for cadmium mobilisation in the rhizosphere of sunflower. Plant Soil 283, 43–56.
Zaman, M., Di, H.J., Cameron, K.C., Frampton, C.M., 1999. Gross nitrogen mineralization and nitrification rates and their relationships to enzyme activities and the soil microbial ammonium fertilizer at different water potentials. Biol. Fert. Soils 29, 178–186.
Zemunik, G., Turner, B., Lambers, H., Laliberté, E., 2015. Diversity of plant nutrient-acquisition strategies increases during long-term ecosystem development. Nat. Plants 1, 15050.
Zhang, W.H., Ryan, P.R., Tyerman, S.D., 2001. Malate-permeable channels and cation channels activated by aluminum in the apical cells of wheat roots. Plant Physiol. 125, 1459–1472.
Zhang, Z., Tang, C., Rengel, Z., 2005. Salt dynamics in rhizosphere of *Puccinellia ciliata* Bor. in a loamy soil. Pedosphere 15, 784–791.
Zhang, X., Long, Y., Huang, J., Xia, J., 2019. Molecular mechanisms for coping with Al toxicity in plants. Int. J. Mol. Sci. 20, 1551.
Zhu, Y., Yan, F., Zörb, C., Schubert, S., 2005. A link between citrate and proton release by proteoid roots of white lupin (*Lupinus albus* L.) grown under phosphorus-deficient conditions? Plant Cell Physiol. 46, 892–901.

Chapter 15

Rhizosphere biology

Petra Marschner
School of Agriculture, Food & Wine, The University of Adelaide, Adelaide, SA, Australia

Summary

The release of easily decomposable root exudates leads to higher microbial abundance and activity in the rhizosphere compared to the bulk soil. In this chapter the colonization of roots by microorganisms is outlined followed by discussion about how these microorganisms either enhance or reduce nutrient availability to plants. Important microorganisms with respect to plant nutrition are mycorrhizal fungi, which improve plant uptake of poorly mobile nutrients such as P and Zn via the network of hyphae extending into the surrounding soil. This chapter also discusses other beneficial effects of mycorrhiza such as increased tolerance to metals and drought and resistance to diseases. The reasons for differential responsiveness of plants to mycorrhizal colonization are outlined.

15.1 General

The rhizosphere is a dynamic system, changing over time scales ranging from diurnal to annual or longer (York et al., 2016). It is the site of signal exchange between plants and soil microbes that affect plant growth and microbial community structure and activity (Venturi and Keel, 2013; Reinhold-Hurek et al., 2015). The release of easily decomposable low-molecular-weight exudates by roots attracts soil microorganisms into the rhizosphere where they proliferate to densities that can be several orders of magnitude higher than in the bulk soil. The high density of microorganisms, in turn, attracts predators such as nematodes and protozoa. Soil microbes play a pivotal role in nutrient turnover and thus nutrition of plants by decomposing and mineralizing organic material and releasing as well as transforming inorganic nutrients by solubilization, chelation, and oxidation/reduction. Grazing by predators releases nutrients from the microbial biomass and enhances microbial growth rates. Rhizosphere microorganisms may also affect plant nutrient uptake indirectly by enhancing root growth (see also Chapter 13).

Almost all plant species form an association with mycorrhizal fungi which can increase plant uptake of poorly mobile nutrients such as P and Zn. The improved uptake of P and Zn is due to the extensive network of external hyphae accessing nutrients beyond the rhizosphere. Other benefits of mycorrhizal colonization include increased access to organic N, increased heavy metal and Al tolerance, and increased resistance to diseases. However, mycorrhizal colonization comes at a cost: plants have to supply the fungus with organic carbon. Under conditions where the fungus provides little or no benefit (e.g., high soil nutrient availability), the cost of the symbiosis may outweigh its benefit, and mycorrhizal colonization can result in growth depression.

Another symbiosis that plays a key role in plant nutrition is the one between certain plant species and N_2-fixing microorganisms, which will be discussed in Chapter 16.

15.2 The rhizosphere as dynamic system

Not only root exudates but also other plant-induced changes in the rhizosphere chemical and physical properties can influence microorganisms. The rhizosphere often has a distinct gradient in water availability due to water uptake by roots (York et al., 2016). Despite water uptake, the soil water content in the rhizosphere may be higher than expected because mucilage can hold water (Carminati et al., 2010). Further, water uptake by roots usually decreases in the afternoon, which may be an adaptive response to prevent excessive drying of rhizosphere soil (Caldeira et al., 2014). This circadian water uptake may be regulated by aquaporins (Caldeira et al., 2014). Similarly, nutrient uptake follows a diurnal rhythm (LeBot and Kirkby, 1992; Chapter 2). It has been suggested that internal nutrient concentrations drive transcript abundance and thereby number of nutrient transporters (Ono et al., 2000). Changes in rhizosphere pH can also follow a diurnal pattern (Rudolph et al., 2013; Rudolph-Mohr et al., 2014).

A number of methods have been developed to study short-term changes of rhizosphere properties. These include microtensiometers for water (Segal et al., 2008) and zymography for enzymes (Spohn and Kuzyakov, 2013). Optodes are fiber optic devices that can measure color changes in indicator dyes to measure gases (Elberling et al., 2011), pH (Faget et al., 2013), or nutrients (Strömberg, 2008; Warwick et al., 2013). Microdialysis is used to extract soil solution from the rhizosphere (Inselsbacher et al., 2011).

15.3 Rhizosphere microorganisms

15.3.1 Root colonization

In soil, growth and activity of microorganisms is mainly limited by organic carbon availability (Demoling et al., 2007; De Nobili et al., 2001) because of the poor decomposability of soil organic matter due to spatial inaccessibility (in small pores and aggregates) and the complex nature of organic matter (Von Luetzow et al., 2006). By contrast, root exudates generally have low molecular weight and are easily decomposable; therefore microbial density is considerably higher in the rhizosphere than in the bulk soil.

As the root grows through the soil, the new root surface just behind the meristematic tissue is colonized by microorganisms that are attracted to the root surface. Root exudates released in the zone immediately behind the root tip and in the distal elongation zone stimulate microbial growth and attract more soil microorganisms to the root surface (Fig. 15.1). Further from the root tip, in the root hair and adjacent zones, root exudation is lower, leading to lower microbial growth rates (Dennis et al., 2010; Nguyen and Guckert, 2001). Along the more mature root parts exudation is even lower, and the primary substrates for microorganisms are cellulose and other recalcitrant cell wall materials; thus microbial growth rates and activity are low (Nguyen and Guckert, 2001).

In fast-growing roots, there is usually a steep gradient of rhizoplane and rhizosphere microorganisms along the root axis from apical to basal zones (Bowen and Rovira, 1991). The differences in type and quantity of organic carbon available in different root zones also lead to distinct rhizosphere communities (Baudoin et al., 2003; Chiarini et al., 2000; Marschner et al., 2001; Yang and Crowley, 2000). Changes in microbial density along the root axis are important for nutrient turnover within the microbial biomass, which influences nutrient availability to plants (Marschner et al., 2011). An increase in microbial biomass may result in net immobilization of nutrients, whereas microbial decay is associated with net release of nutrients.

Equally important as microbial density for root growth and physiology as well as nutrient dynamics in the rhizosphere is the microbial community composition [i.e., which genotypes (species, strains) are present at which abundance] because each genotype has certain physiological characteristics. They may be, for example, ammonifiers, nitrifiers, producers of phytohormones, N_2 fixers, pathogens, or antagonists. Different plant species have a distinct rhizosphere microflora with respect to abundance and physiological characteristics (Marschner et al., 2001; Marschner and

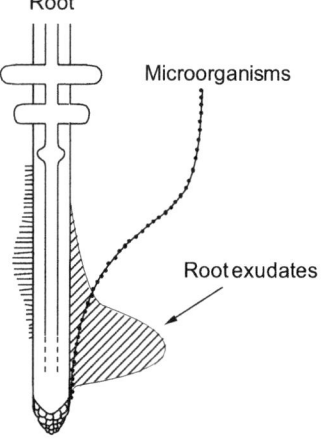

FIGURE 15.1 Schematic diagram of spatial differences in release of low-molecular-weight root exudates and microbial density in the rhizosphere of soil-grown plants.

Timonen, 2004; Miethling et al., 2003) that can be modified further by soil properties (Marschner et al., 2001; Miethling et al., 2000), plant age (Gomes et al., 2001; Smalla et al., 2001), and plant nutritional status (Marschner et al., 2004; Solaiman et al., 2007; Yang and Crowley, 2000). For example, Matthews et al. (2019) found that the composition of the root-associated bacterial community was distinct in the four crop species: onion (*Allium fistulosum*), pea (*Pisum sativum*), tomato (*Solanum lycopersicum*), and maize (*Zea mays*), but that it was also influenced by the soil in which the plants grew (Fig. 15.2).

Root exudates influence microbes not only by providing organic carbon, but also indirectly. Microbes can sense the size of their population by quorum sensing and adjust their behavior accordingly (Waters and Bassler, 2005). Quorum sensing is considered to be important for rhizosphere colonization by beneficial and pathogenic microbes. Plants can interfere with quorum sensing, for example, by releasing compounds that either quench or mimic compounds used for quorum sensing (Gao et al., 2003; Teplitski et al., 2000). Roots have been shown to release denitrification inhibitors and thereby regulate N loss from soil (Bardon et al., 2014). In response to root exudates, microbes can express genes involved in chemotaxis, mobility, and biofilm formation, all of which are important for root colonization (Zhang et al., 2015).

Nutrient availability also influences microbial community composition in the rhizosphere. In wheat, depending on whether N is supplied as ammonium or nitrate, there is a shift in the proportion of pathogen (*Gaeumannomyces graminis*) and antagonists (*Pseudomonas* spp.) in the rhizosphere (Sarniguet et al., 1992a, b). Bacterial community composition is also altered by the amount and form of P fertilization (Marschner et al., 2004).

Microorganisms decompose root exudates and may therefore reduce the efficiency of the exudates in nutrient mobilization. The gradient in microbial populations along the root axis and the rapid decomposition of exudates by rhizosphere microorganisms have important implications for the efficiency of root exudates released in response to nutrient deficiency. The half-life of organic acid anions (such as citrate and malate) or amino acids in soils is 6–12 h (Jones et al., 1994; Jones, 1998). Consequently, Jones (1998) questioned the role of organic acid anions in mobilization of P and other nutrients (see Chapter 14). However, high exudation rates immediately behind the root tip where microbial density is low maximize effectiveness of the exudates. For example, in grasses under Fe deficiency, phytosiderophore release is confined to the zone immediately behind the root tips (see also Chapters 2 and 14). In addition, phytosiderophores are released during a short period between 2 and 8 h after onset of light. This pulse of phytosiderophore release minimizes microbial decomposition and thereby maximizes their effectiveness (Crowley and Gries, 1994; Römheld, 1991). Similarly, under P deficiency, release of organic acid anions is highest immediately behind the root tip or around specialized root structures such as cluster roots (Chapter 14). Additionally, phenolic compounds released together with

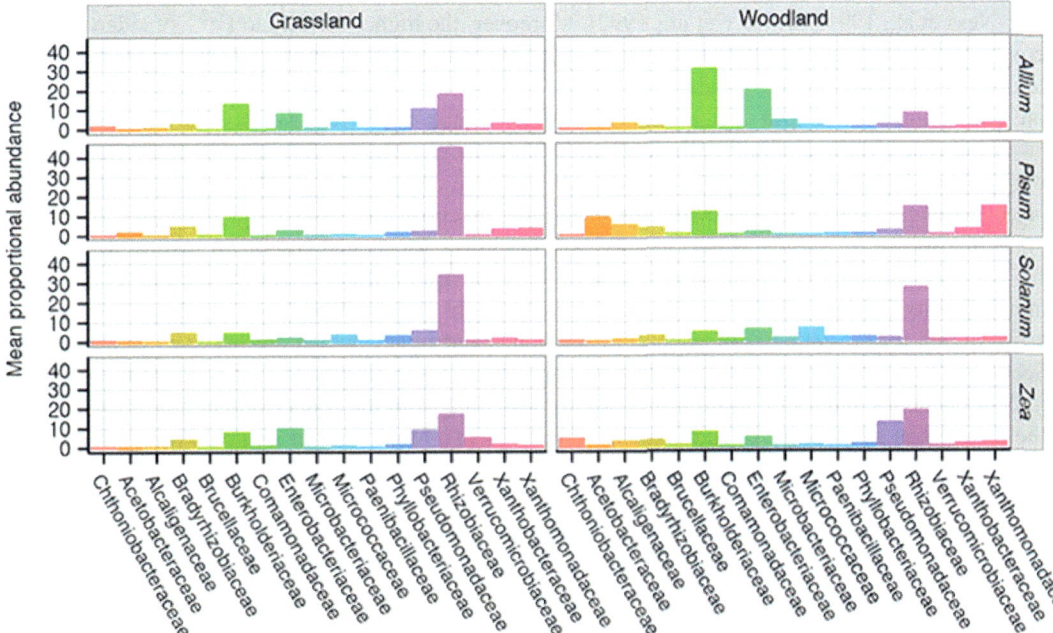

FIGURE 15.2 Taxonomic composition of bacterial communities associated with roots of onion (*Allium fistulosum*), pea (*Pisum sativum*), tomato (*Solanum lycopersicum*), and maize (*Zea mays*) grown in grassland or woodland soil. *From Matthews et al. (2019).*

organic acid anions can inhibit their decomposition by microorganisms (Weisskopf et al., 2006). Therefore model calculations of effectiveness of root exudates in nutrient acquisition have to consider this spatial separation of root exudation and microbial activity (Darrah, 1991, 1993). Moreover, decomposition of root exudates can be reduced by sorption to soil particles while still remaining effective in nutrient mobilization (Jones and Edwards, 1998). On the other hand, factors that favor a more uniform distribution of rhizosphere microorganisms along the roots, such as reduced root growth (Watt et al., 2003), reduce the effectiveness of root-released phytosiderophores (Crowley and Gries, 1994).

15.3.2   Role in nutrition of plants

Rhizosphere microorganisms may affect nutrient uptake by plants through their influence on (1) growth, morphology, and physiology of roots (Chapter 13); (2) physiology and development of plants; (3) availability of nutrients; and (4) nutrient uptake processes.

Rhizosphere microorganisms are the main drivers of turnover of organic C, N, and P and thus recycling of organically bound nutrients, for example, by ammonification and nitrification, but may also increase N loss via denitrification (Butterbach-Bahl et al., 2013; Kuzyakov and Xu, 2013). Mineralization of organic N (more than 90% of total N in soils) to ammonium and nitrate by soil microorganisms is critical for plant N uptake because most plants have a limited capacity to take up organic N (Dunn et al., 2006; Xu et al., 2008). Another important source of N for plants is atmospheric N_2; the role of N_2-fixing microorganisms in plant nutrition is discussed in Chapter 16.

Microorganisms can mobilize P by solubilization of poorly soluble inorganic P such as apatite (Banik and Dey, 1983; Jorquera et al., 2008), mineralization of organic P (George et al., 2007; Jorquera et al., 2008; Richardson and Hadobas, 1998; Tarafdar and Jungk, 1987) and, in alkaline soils, by decreasing the pH. There has been considerable speculation as to whether inoculation with such microorganisms may allow increased utilization of soil and fertilizer P. Although these bacteria are capable of mobilizing P in vitro, it is unlikely that this mechanism operates to any great extent in the rhizosphere. First, because microorganisms capable of mobilizing P are already present in most soils (Gyaneshwar et al., 2002). Second, when introduced into the soil as "biofertilizer," microbial density declined rapidly (1) because introduced microorganisms were not adapted to the soil conditions and (2) because of competition with other rhizosphere microorganisms for organic carbon as an energy source (Gyaneshwar et al., 2002; Postma et al., 1990). Hence, effectiveness of "biofertilizers" in the field is highly variable, although there are some examples where inoculation increased plant growth and yield (e.g., (Bajpai and Sundara-Rao, 1971; Dhillion, 1992; Zahir et al., 2009).

Microorganisms release organic acid anions or siderophores that chelate and thus mobilize Fe^{3+} (Neilands, 1984). However, Fe bound to bacterial siderophores is usually a poor Fe source for both monocotyledonous and dicotyledonous plants (Bar-Ness et al., 1991; Crowley et al., 1992). Moreover, the higher affinity to Fe^{3+} of siderophores compared to phytosiderophores (Yehuda et al., 1996) may cause ligand exchange with Fe moving from phytosiderophores to siderophores, thereby decreasing plant Fe availability.

Reduction and oxidation of Mn by microorganisms are important for Mn availability in soil. Reduction ($Mn^{3+} \rightarrow Mn^{2+}$) increases Mn availability, whereas oxidation ($Mn^{2+} \rightarrow Mn^{3+}$) decreases it. Interestingly, Mn reducers appear to be more abundant in the rhizosphere of some Mn-efficient compared with Mn-inefficient wheat (*Triticum aestivum*) genotypes (Rengel, 1997). Colonization by arbuscular mycorrhizal (AM) fungi decreased the density of Mn reducers in the rhizosphere of maize, which may explain the lower Mn concentration in mycorrhizal plants (Kothari et al., 1990b). The root pathogen *G. graminis* var. *tritici* (Ggt) is a strong Mn oxidizer (Marschner et al., 1991), thereby reducing Mn availability to plants. The lower Mn availability could also facilitate the infection of roots by the fungus because Mn is required for biosynthesis of phenolic compounds and lignin, which are involved in the defense reaction by plants (Section 7.2). Consequently, Mn fertilization can reduce susceptibility to Ggt because the plants are able to produce sufficient phenolics and lignin to limit the spread of the fungus in the roots (Rengel et al., 1993).

Due to the differential availability of root exudates (Chapter 14), Marschner et al. (2011) proposed that the role of rhizosphere microorganisms in nutrient uptake by plants changes along the root (Fig. 15.3). Just behind the root tip in the distal elongation zone, exudation rates are high, and microbial density in the rhizosphere is relatively low. Therefore, root exudates can mobilize nutrients without strong competition from microorganisms or substantial decomposition of exudates by microorganisms. The high rate of exudation just behind the root tip stimulates the growth of rhizosphere microorganisms, resulting in high microbial density in the proximal elongation zone, accompanied by strong nutrient mobilization. However, most of the mobilized nutrients are taken up by the rapidly multiplying microorganisms, leading to net immobilization in the microbial biomass. Compared to the root zone immediately behind the root tip, root exudation is lower in the root hair zone and the adjacent root zone. Therefore microbial activity and growth are reduced, nutrient mobilization may be equal to or less than immobilization, and some of the nutrients mobilized by root

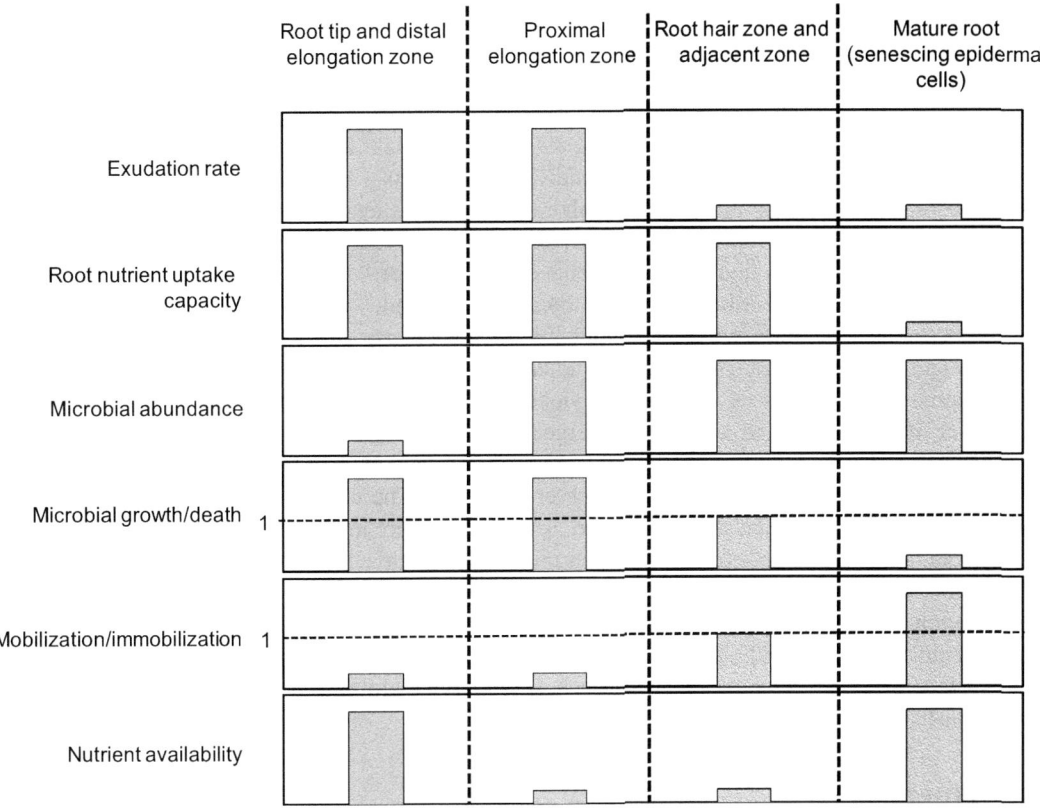

FIGURE 15.3 Model of plant–microbe interactions in the rhizosphere in relation to nutrient availability in different root zones. It should be noted that the boundaries between the different root zones are gradual in nature. *From Marschner et al. (2011), with permission from Elsevier.*

exudates can be taken up by the plant. In the mature root zones where epidermal cells start to senesce, the concentration of easily decomposable organic C sources is low, resulting in low microbial growth rates and thus low nutrient demand. Consequently, part of the microbial biomass may die, releasing previously immobilized nutrients. However, plant uptake of these nutrients may be relatively low because of the low capacity of root tissues in these root zones to take up nutrients (Colmer and Bloom, 1998; Fang et al., 2007; Häussling et al., 1988; Chapter 2).

The scenario outlined earlier does not take into account the effect of other soil biota on microbial turnover. For example, protozoa can increase microbial turnover and thus nutrient release by grazing the microbial biomass (Gao et al., 2019). Little is known about the role of viruses in soil, although it has been shown that a large proportion of soil bacteria are infected by phages (Kuzyakov and Mason-Jones, 2018). Viral infection of soil bacteria would induce cell lysis and thus nutrient release, thereby playing an important role in biogeochemical cycling (Kuzyakov and Mason-Jones, 2018).

15.3.3 Root exudates as signals and phytohormone precursors

Root exudates may act as signals for microbial recognition. For example, some flavonoids (released by legume roots) attract rhizobia (Chapter 16), whereas other flavonoids may be suppressors of certain pathogenic fungi (Hartwig et al., 1991). Root exudates can also indirectly suppress pathogens by attracting antagonists of plant pathogens (Jousset et al., 2011).

Strigolactones are phytohormones that regulate plant physiology but can also be released from roots (De Cuyper and Goormachtig, 2017; Chapter 2). They are known to stimulate germination of parasitic plants. Strigolactone receptors have been identified in AM fungi and can induce AM spore germination and hyphal branching (Akiyama et al., 2005; De Cuyper and Goormachtig, 2017; Xie et al., 2010). They may also affect the growth of other fungi, for example, by inhibiting some pathogens. Strigolactones may be involved in recognition of host plants by rhizobia, but their role and recognition by rhizobia is unclear. Certain root exudates act as signal not only for establishment of symbiotic interactions, but also for parasitic flowering plants. A hydroquinone (sorgolactone) in root exudates of *Sorghum bicolor* stimulates germination of *Striga asiatica*, and thus the formation of the parasitic interaction (Fate et al., 1990; Hauck et al., 1992). Root cap cells, also called "border cells" when detached from the roots, carry host-specific traits into the rhizosphere and contribute to establishment of a characteristic rhizosphere bacterial community or suppress certain soil-borne root pathogens (Gochnauer et al., 1990; Hawes, 1990; Hawes et al., 2002). In *Eucalyptus*, border cells attract the ectomycorrhizal fungus *Pisolithus tinctorius* (Horan and Chilvers, 1990).

Some rhizosphere bacteria are producers of phytohormones such as cytokinin (CYT) and indole acetic acid (IAA). However, in the absence of the appropriate precursors, synthesis of phytohormones by microorganisms is low. Several precursors for phytohormone production are components of root exudates or of lysates from decaying root tissues, thus playing an important role in phytohormone synthesis by rhizosphere microorganisms. Production of CYT and IAA by *Azotobacter chroococcum* was enhanced when supplied with either maize root exudates (González-Lopez et al., 1991) or adenine (Nieto and Frankenberger, 1990). Root exudates also contain amino acids such as tryptophan and L-methionine that are required as precursors for IAA and ethylene (C_2H_4) production by rhizosphere microorganisms (Ahmed et al., 2010; Ali et al., 2009; Arshad and Frankenberger, 1991; Arshad et al., 1993). Under stress, roots produce 1-aminocyclopropane-1-carboxylate (ACC) that is converted into ethylene that, in turn, reduces root growth; it also plays a key role in plant defense mechanisms (Ravanbakhsh et al., 2018). The concentration of ethylene in plants has to be regulated in a specific way to improve stress tolerance. Some rhizosphere microbes and endophytes increase plant stress resistance by modulating ethylene concentration through either ethylene production or decomposition of ACC (Ravanbakhsh et al., 2018).

Release of phytohormones is most likely the reason for changes in root and shoot morphology and growth that are observed after inoculating plants with so-called plant growth−promoting rhizobacteria (Ahmed et al., 2010; Ali et al., 2009; Martin, 1989), although other mechanisms such as nutrient mobilization or N_2 fixation can also be involved (Bashan and de Bashan, 2010).

15.4 Endophytes

Root colonization by microorganisms is not confined to the rhizoplane but may also occur to a varying degree within the plant. These endophytes can be bacteria, actinobacteria, or fungi that may affect plant growth and nutrient uptake (Rahman and Saiga, 2007; Rodriguez et al., 2009). They often enter plants from the soil (Liu et al., 2017). The endophyte community is less diverse than the rhizosphere community (Bulgarelli et al., 2013), indicating that plants can, to some extent, control which microbes enter the roots. To colonize the endosphere, bacteria have to have the ability to adhere to root cells. Usually, they colonize the root in the zone of lateral root emergence where both epidermis and endodermis are broken (Bulgarelli et al., 2013). Endophytes then move within the plant to other plant parts via phloem and xylem where they colonize extra- or intracellularly (Compant et al., 2010).

The composition of the endophyte community depends on plant genotype, plant organ, soil type as well as environmental conditions such as nutrient availability, temperature, and atmospheric CO_2 concentration (Liu et al., 2017). Traits of endophytes include motility, chemotaxis, production of cell wall−degrading enzymes, and lipopolysaccharide formation. They can also suppress the plant's defense reaction (Piromyou et al., 2015). Some endophytes produce phytohormones or improve plant stress tolerance, for example, by decomposition of ACC, the precursor of ethylene, and by ROS detoxification (Sessitsch et al., 2012). Endophytes can also increase disease resistance by accelerating the plant's defense reaction to pathogen attack (Liu et al., 2017). Regarding plant nutrition, siderophore-producing endophytes can improve Fe uptake. N_2-fixing endophytes have been shown to increase N accumulation of nonlegumes. However, most of these studies have been carried out under controlled conditions. The importance of endophytes for plant stress tolerance and nutrition in the field is still unclear (Liu et al., 2017).

15.5 Methods to study rhizosphere microorganisms

Traditionally, soil microbes were enumerated and identified by cultivation on specific media. However, these methods detect <5% of soil microbes, that is, only those that grow quickly and suppress slower growing genotypes. Nowadays, there are a range of methods based on DNA or RNA that can detect a large proportion of the soil microbial community. Methods based on polymerase chain reaction (PCR) amplification are used to study selected groups of microorganisms by using primers specific to the groups, such as microorganisms involved in ammonification, nitrification or N_2 fixation. After amplification, DNA fragments can be separated and visualized in gels or cloned and sequenced. To determine abundance of target groups, quantitative PCR can be used, whereas RNA amplification is required to assess expression of target genes (Biswas and Sarkar, 2018). An important limitation is that PCR-based methods rely on knowledge of the target DNA sequence for primer design and may miss genotypes that have slightly different sequence.

More recently, a number of "omics" tools have been used to study soil microorganisms. Unlike PCR-based methods, metagenomics does not require knowledge of microbial community composition. Without prior amplification, DNA isolated from soil can be sequenced. Sequences can then be compared with data bases of known microbes to determine microbial community composition. Metatranscriptomics uses a similar approach, but based on RNA, to assess gene expression. Protein expression of soil microborganisms can be measured by metaproteomics that provides insights into metabolic functions (Biswas and Sarkar, 2018). These methods have revealed that soil microbial communities are much more diverse than previously thought and highly variable, both in space and time.

15.6 Mycorrhiza

15.6.1 General

The roots of most soil-grown plants are mycorrhizal. Hence, mycorrhiza is the most widespread association between microorganisms and plants (Smith and Read, 2008). Arbuscular mycorrhizal fungi (AMF) are widespread, colonizing 72% of plant species; by contrast, ectomycorrhiza (ECM), ericoid, and orchid mycorrhiza are found in 2%, 1.5%, and 10% of plants, respectively (Brundrett and Tedersoo, 2018). Only about 8% of plants are nonmycorrhizal (NM).

NM plants occur in habitats where the soils are either very dry, saline, waterlogged, severely disturbed (e.g., mining activities), or where soil fertility is extremely high or extremely low (Brundrett and Abbott, 1991). Mycorrhiza are absent in Brassicaceae and Chenopodiaceae (Brundrett and Tedersoo, 2018; Smith and Read, 2008) and are also quite rare or absent in many members of the Proteaceae or other cluster root-forming plant species (Brundrett and Abbott, 1991).

Generally, the mycorrhizal fungi are strongly or wholly dependent on the plant, whereas the plant may or may not benefit. Only in some plants (orchids), mycorrhizal colonization is essential. For plants, mycorrhizal associations are therefore either mutualistic, neutral or parasitic (Johnson et al., 1997), depending on a range of factors, for example, fungal and plant genotype, P availability, and light intensity. This suggests there is a delicate balance between the benefits to the host in terms of nutrient acquisition and the cost associated with supporting the fungus. For a comprehensive review on mycorrhiza in natural ecosystems, the reader is referred to Brundrett and Abbott (1991), Fitter (1991), Smith and Read (2008), and van der Heijden and Horton (2009).

15.6.2 Mycorrhizal groups, morphology, and structure

There are two major mycorrhizal groups based on differential morphology and physiology: endomycorrhiza and ECM (Fig. 15.4).

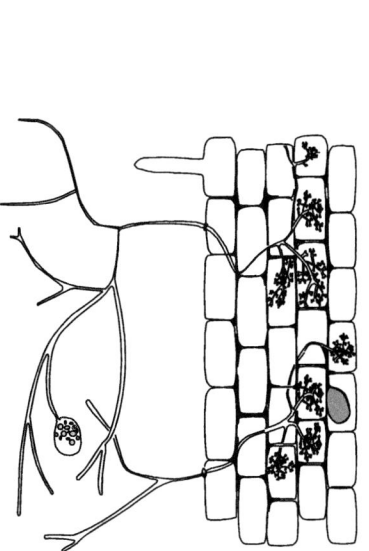
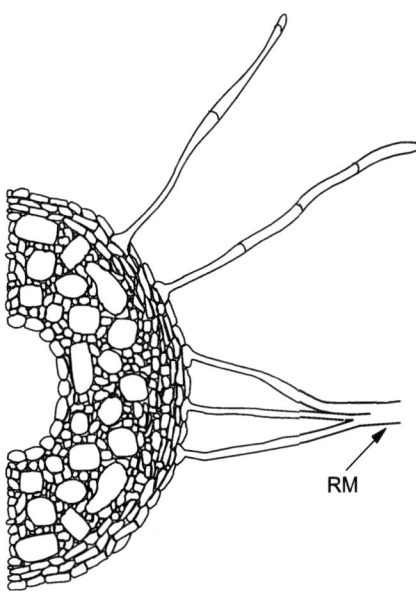

FIGURE 15.4 Schematic diagram of the main structural features of AM mycorrhiza (*left*) and ectomycorrhiza (*right*). Arrow indicates rhizomorphs.

15.6.2.1 Endomycorrhiza

The fungi form structures within the cortical cells and also grow in the intercellular space. Hence, at the fungus–plant interface, the membranes of the fungus and the plant are in direct contact with each other. There are several types of endomycorrhiza, the best known being *arbuscular mycorrhiza* (AM, formerly called vesicular-arbuscular mycorrhiza), *ericoid*, and *orchid mycorrhiza*.

Arbuscular mycorrhiza is by far the most abundant of the endomycorrhiza (Smith and Read, 2008). The fungi forming AM were classified as *Glomeromycotina*. Increasing information about the AM genome showed that they share many genes with other *Mucoromycotina*. Therefore, AM may not be separate phylum *Glomeromycotina* but belong to the *Mucoromycota* (Bruns et al., 2018). However, species identification in AM remains difficult, despite sequencing of ribosomal RNA genes because of a limited number of AMF reference cultures and lack of specific primers for identification (Wipf et al., 2019). Furthermore, genetic variation within individual fungi is high. AM fungi were thought to be asexual, but recent molecular evidence suggests they have sexual reproduction and homokaryotic spores (Wipf et al., 2019).

Arbuscular mycorrhiza is characterized by the formation of (1) intracellular structures (arbuscules or hyphal coils) within the root cortex cells, (2) intercellular hyphae in the cortex, and (3) a mycelium that extends well into the surrounding soil (external mycelium; Fig. 15.5). There are two types of AM with respect to the structures formed in the cortex cells: *Arum* type mycorrhiza characterized by arbuscules and *Paris* type that form hyphal coils (Smith and Read, 2008). A given AM fungus can form either arbuscules or hyphal coils depending on the host plant (Dickson, 2004). Arbuscules and coils are the main sites of solute exchange with the host; they are short lived, being active for about 7 days. Many, but not all, AM fungi form vesicles as lipid-rich storage organs (Fig. 15.5).

Mycoplasma-related endobacteria living in AM cytoplasm are widespread (Wipf et al., 2019), but their role in AM is unknown so far.

Ericoid mycorrhizae occur in Ericales; they are characterized by coils of hyphae within rhizodermal (epidermal) cells and individual hyphae extending into the soil as in the case of AM (Smith and Read, 2008).

FIGURE 15.5 Mycorrhizal root systems. Root of soil-grown potato with external hyphae of *Glomus mosseae* (*top*), ectomycorrhizal short roots of soil-grown Norway spruce (*bottom*). *Courtesy G. Hahn, University of Hohenheim.*

Orchid mycorrhizae are formed between plants of the family Orchidaceae and a variety of fungi. All orchids are myco-heterotrophic at some point in their life cycle. The colonization by mycorrhizal fungi is critically important during seedling development because orchid seeds have virtually no energy reserve, and seedlings obtain their carbon from the fungal symbiont. Hence, during this stage of the symbiosis, the flow of C is from the fungus to the host, which is distinctly different from the other mycorrhiza where C is supplied by the host plant to the fungus. Many adult orchids retain their fungal symbionts, although the benefits to the adult photosynthetic orchid and the fungus remain largely unexplored (Smith and Read, 2008).

15.6.2.2 Ectomycorrhiza

ECM fungi are Basidiomycetes or Ascomycetes. ECMs occur mainly on roots of woody plants (Brundrett and Tedersoo, 2018; Smith and Read, 2008). ECMs are most common in the northern hemisphere, especially in Pinaceae, Betulaceae, Fagaceae, and Salicaceae. However, ECM may also occur in some tropical and subtropical forest trees (Högberg, 1986). They are characterized by (1) an interwoven mantle of hyphae (fungal sheath) around the lateral roots; (2) hyphae that penetrate the intercellular space in the root cortex to form a network of fungal mycelium, the *Hartig net*, that surrounds the cortex cells and increases the surface area at the fungus–root interface; and (3) an extensive network of external hyphae (Figs. 15.4 and 15.5). In contrast to endomycorrhiza, the hyphae of the Hartig net remain intercellular; they do not penetrate the host cells. Because the fungus remains in intercellular space, nutrient transfer across the fungus–plant interface has to occur through the cell walls and membranes of both partners, which is different from the interface in AM, where nutrients only have to be transferred across the plasma membranes of the fungus and the plant.

Some ECMs produce hyphal strands or *rhizomorphs* that extend far into the surrounding soil (Fig. 15.5). Rhizomorphs are differentiated multihyphal organs with a diameter of up to 200 μm and are important for solute transport over large distances.

In some tree species, both AM and ECM occur simultaneously, for example, in *Salix*, *Populus* (Lodge, 1989), or *Eucalyptus* (Gardner and Malajczuk, 1988; Teste et al., 2020), and the proportion of the two types appears to depend on external factors such as soil water content and aeration, and internal factors such as tree age.

On a global scale, ECMs are abundant in boreal and temperate forests with a distinct surface humus horizon, and in N-limited ecosystems. On the other hand, AMs are more abundant in warmer climates with drier soils, in pastures and deciduous forests with high turnover of organic material, and where P supply is limited (Read, 1991). AMs are usually the only form of mycorrhiza in crop plants, pastures, and fruit trees.

Besides the differences in distribution, morphology, and structure (Fig. 15.4), there is another principal difference between ECM and AM. Most ECM fungi can be grown in pure culture (in vitro), but this is not possible for AM fungi. Therefore, knowledge of physiology of AM fungi is based on studies of fungal structures and fungal functions associated with the host roots (Smith and Gianinazzi-Pearson, 1988).

15.6.3 Root colonization, photosynthate demand, and host plant growth

15.6.3.1 Root colonization

Root colonization by mycorrhizal fungi is initiated from either soil-borne propagules (spores, colonized root fragments) or roots of neighboring plants. Colonization is enhanced by a preexisting hyphal network in the soil. Therefore, severe soil disturbance, for example, clear-cut logging or vigorous soil mixing (Jasper et al., 1989), as well as tillage compared to no-till (Garcia et al., 2007; Miller and McGonigle, 1992), may severely reduce and delay mycorrhizal colonization. Tillage can also alter the community composition of AM fungi, suggesting that AM fungal species differ in sensitivity to soil disturbance (Jansa et al., 2003).

Root exudates play an important role in colonization by AM fungi. Roots release not only strigolactones to attract AM fungi, but also N-acetylglucosamine derivatives (also released by AM fungi) (Lanfranco et al., 2018) that have been shown to activate symbiotic responses in plants and induce transcription of AM-specific genes, for example, Pi transporters or Ca channels (MacLean et al., 2017). Volatile compounds released by roots may also be involved in signaling (e.g., Dreher et al., 2019). AM fungi have been shown to release effectors, for example, proteins, to reduce host defense reactions upon colonization (Lanfranco et al., 2018).

Rhizosphere bacteria may enhance or suppress mycorrhizal colonization. The enhancers are referred to as helper bacteria and have been shown to stimulate colonization by AM (Duponnois and Plenchette, 2003; Pacovsky et al., 1985) and ECM (Duponnois and Garbaye, 1991).

In plants that are not hosts to AM mycorrhiza, for example, members of the Chenopodiaceae and Brassicaceae, incompatibility may be caused by the composition of root exudates, toxins, or enhanced defense reactions of the host against colonization, similar to the response to pathogens (Akiyama et al., 2010; Anderson, 1988; Parra-Garcia et al., 1992).

Soil nutrient supply may enhance or suppress mycorrhizal colonization. At extremely low-soil P availability, root colonization by AM is low (Bolan et al., 1984) because low P may limit the growth of the fungi. With increasing P supply, root growth and the proportion of colonized root length increase until an optimum supply of P is attained; beyond this level, the colonization is depressed to a varying degree, depending on AM (Bolan et al., 1984) or ECM species (Jones, 1990), as well as the host species (Davis et al., 1984) and cultivar (Baon et al., 1993). For example, increasing P supply reduced the percentage colonization of soybean by *Glomus custos* and *Rhizophagus irregularis* but had no effect on colonization by *Glomus aggregatum* that had a generally low colonization percentage (Wang et al., 2016; Table 15.1). Plant Pi concentration appears to regulate not only colonization, but also arbuscule formation and turnover (Lanfranco et al., 2018).

High N supply also depresses AM and ECM colonization, particularly in combination with high P availability and when N is supplied as NH_4^+ (Baath and Spokes, 1988). In ECM, in particular, the mass of the mycelium decreases at high N supply (Wallander and Nylund, 1991). A decrease in AM-colonized root length or the proportion of ECM root tips at high supply of P or N is, however, not necessarily an expression of a specific regulation mechanism; it may be the result of enhanced root growth outpacing the fungus. The negative effect of high P (or N) availability can be explained by a number of factors: (1) reduced colonization due to suppression of hyphal growth in the soil, (2) reduced carbohydrate supply to the fungus, and (3) increased root growth (Bruce et al., 1994; Jasper et al., 1979; Smith and Read, 2008).

Total AM-colonized root length or total number of ECM root tips is often an appropriate parameter, but for evaluation of effectiveness in nutrient acquisition, quantification of the external mycelium is likely to be the most important parameter (see below). However, the growth rate of extraradical hyphae ranges from a few mm to hundreds of mm per day, depending on fungal species, host species, and environmental conditions (Lanfranco et al., 2018). It should be noted that most studies on mycorrhiza are carried out with only one fungus, whereas in the field, a plant may be

TABLE 15.1 Shoot biomass, shoot P content, and percentage root length colonized in nonmycorrhizal soybean or soybean colonized by the AM fungi *Glomus custos, Rhizophagus irregularis*, or *Glomus aggregatum* at low, medium, or high P supply (25, 100, or 300 μM P).

		Shoot biomass (g plant^{-1})	Shoot P content (mg plant^{-1})	AM colonization (% root length)
Low P	NM	2.5	7.5	–
	G. custos	3.0	13	25
	R. irregularis	4.5	23	70
	G. aggregatum	2.3	7.5	20
Medium P	NM	4.0	13	–
	G. custos	3.8	13	15
	R. irregularis	5.0	25	60
	G. aggregatum	2.8	8.0	25
High P	NM	6.5	23	–
	G. custos	6.8	23	10
	R. irregularis	6.8	27	55
	G. aggregatum	6.2	23	20

NM, Nonmycorrhizal.
Source: Based on Wang et al. (2016).

colonized by several fungal species. Therefore, the response of a plant in the field is likely to be a mixture of those observed in pot studies.

15.6.3.2 Photosynthate demand

In mycorrhizal roots a substantial proportion of the photosynthates allocated to the roots are used for fungal growth and maintenance. In AM plants, root + fungal respiration may be 20%−30% higher than in roots of NM plants, and most (87%) of that increase can be attributed to the fungus (Baas et al., 1989). This agrees well with the estimates by Lambers et al. (2002) that, generally, the C cost of the AM symbiosis ranges between 4% and 20% of the C fixed by a plant, the majority of which is respired by the roots and fungi. For example, in cucumber, 20% of the net photosynthates were allocated belowground in NM plants, whereas it was 43% in AM plants (Jakobsen and Rosendahl, 1990). The influence of AM as a C sink on plant photosynthesis was shown by Gavito et al. (2019). Excision of part of the extraradical mycelium from the roots caused a sustained decrease (by 10%−40%) in photosynthetic rates. The high C requirement of mycorrhiza can explain why unfavorable environmental conditions such as shading and defoliation depress mycorrhizal development (Same et al., 1983; Son and Smith, 1988).

Recent studies found that AM receive from plants not only carbohydrates (mainly glucose), but also lipids (Roth and Paszkowski, 2017). Gene sequencing of AM fungi indicates that they lack the capacity for lipid synthesis, suggesting they are dependent on plants for lipid supply (MacLean et al., 2017; Wipf et al., 2019). Fatty acids play an important role in AM because they are involved in arbuscule development and are important for the completion of the fungal life cycle, that is, production of spores and vesicles (Roth and Paszkowski, 2017).

The costs in terms of C supply to mycorrhizal fungi are not relevant when the mycorrhizal plants can compensate for the higher demand by an increase in rate of photosynthesis per unit leaf area (Mortimer et al., 2008; Wright et al., 1998). Enhanced rates of photosynthesis in mycorrhizal plants are therefore often an expression of a higher sink activity (Dosskey et al., 1990) rather than a specific stimulatory effect of the mycorrhizal association. The costs of photosynthates must be compared with the benefits such as enhanced uptake of nutrients like P when they limit photosynthesis and growth in NM plants. However, despite the beneficial effect of mycorrhiza on plant growth, root dry mass is enhanced only slightly or is even depressed (Dosskey et al., 1991) compared with shoot growth, resulting in a decreased root/shoot ratio.

In trees, ECMs are also an important C sink. In ECM Douglas fir stands, about 60%−70% of the net photosynthates are allocated belowground for growth of roots and mycorrhiza and for respiration (Fogel, 1988). Other estimates on the proportion of carbon flow to the ECM in forest stands vary between 5% and 30% of the net photosynthates (Hobbie, 2006; Söderström, 1992). Thus, ECM fungi play an important role in carbon import into the soil via the external mycelium, particularly considering the high turnover rate of the fungal C that is about five times greater than that of litter (Fogel and Hunt, 1979).

In ECM and AM plant communities, neighboring plants can be connected via hyphal networks (Griffiths et al., 1991; Wipf et al., 2019). In forest stands, ECM hyphae can act as a conduit for photosynthate transfer from overstorey plants to seedlings shaded by these plants (Griffiths et al., 1991). The amount of C transported to the seedling is small (<10% of C assimilated by the donor), with more being transferred to receiver plants in shade than those in full light. Only 13% of received C was transported into the shoots of the receiver plant (Simard et al., 1997; Table 15.2). Nevertheless, this C transfer (the so-called "nurse plant effect") could be important for the establishment of seedlings in forests.

TABLE 15.2 Net carbon transfer from the donor plant (*Betula papyrifera*) to the receiver plant (*Pseudotsuga menziesii*) grown in full light, partial shade, or full shade.

Light treatment of receiver plant	Net C transfer (% of total isotope in the donor plant)
Full light	2.7
Partial shade	4.3
Full shade	9.5

Net carbon transfer is expressed as percentage of total isotope in the donor.
Source: Based on Simard et al. (1997).

Common AM hyphal networks can improve seedling establishment and facilitate nutrient exchange (Wipf et al., 2019). They may also be involved in plant communication via phytohormones and as conduit for signaling compounds for pathogen defense reactions. Common hyphal networks amplify intraspecific competition by influencing the distribution of plant size classes (Wipf et al., 2019). For example, AM may provide more P to larger plants at the expense of smaller plants, which is likely regulated by C supply to the fungus (Walder and van der Heijden, 2015; Weremijewicz and Janos, 2013). Mycorrhizal nutrient supply may also be influenced by the plant because the largest plants have the highest nutrient demand and take up most nutrients from AM networks (Merrild et al., 2013). In ECM, plants appear to be able to favor fungi that provide more nutrients by allocating more C to them compared with less nutrient-efficient fungi colonizing the same plant (Hortal et al., 2017). Among plants connected by common hyphal networks there can be cheaters, that is, plants that benefit from mycorrhizal nutrient supply while the network is maintained by C allocation by another plant (Grime et al., 1987).

The soil/root interface in mycorrhizal plants is termed *mycorrhizosphere* (Fogel, 1988; Linderman, 1988; Timonen and Marschner, 2005). As most soil-grown plants are mycorrhizal, the mycorrhizosphere may be the rule, rather than the exception. Mycorrhizal colonization affects rhizosphere microbial community composition (Kothari et al., 1991; Marschner et al., 2001). This effect appears to be mediated by the plant because the rhizosphere microbial community composition of AM plants is affected not only in roots colonized by the fungus, but also in NM roots of the same plant (Marschner and Baumann, 2003). Moreover, the community composition of mycorrhizosphere is not only influenced by mycorrhizal colonization per se, but also by the mycorrhizal species (Marschner et al., 2001). Given that rhizosphere microorganisms can influence root morphology and nutrient availability, this alteration may affect nutrient acquisition as well as root and shoot growth.

Mycorrhiza influences not only the microorganisms in the vicinity of the roots, but also the density and activity of microorganisms at greater distance from the root surface may be enhanced because organic carbon is provided by the external mycelium, especially in ECM plants. This interface between external hyphae and the soil has similarities to the rhizosphere and has been termed the "hyphosphere" (Linderman, 1988). In forest stands, ECM can form hyphal mats; soil underneath these mats has been shown to have higher microbial biomass and nutrient availability, suggesting increased microbial activity compared to nonmat soil (Entry et al., 1992; Guevara and Romero, 2004).

15.6.3.3 Host plant growth

Mycorrhizal colonization affects plant root and shoot growth differently. In a nutrient-poor substrate, the external mycelium increases the surface area. Hence, compared with NM plants, mycorrhizal plants have greater access to growth-limiting nutrients, for example P. As a typical plant response to increased nutrient supply, shoot growth of mycorrhizal plants is enhanced more than root growth, leading to a decrease in root/shoot ratio (Oliver et al., 1983). At a given nutritional status of the host plant, this shift is more pronounced in mycorrhizal plants (Bell et al., 1989) as the fungus competes with the roots for photosynthates.

If mycorrhiza is either ineffective in delivering nutrients or nutrients are not limiting the growth of NM plants, mycorrhizal colonization depresses root growth, primarily by sink competition. As mentioned earlier, mycorrhizal structures are strong sinks for photosynthates, irrespective of their contribution to host plant growth (Douds et al., 1988; Lambers et al., 2002). In principle, growth depression can be predicted when root colonization remains high at high P supply and when there is limited photosynthetic source capacity to compensate for the extra costs of mycorrhizal colonization (Gerdemann, 1975; Sanders, 1993).

In addition to root growth depression caused by competition for photosynthates, phytohormones may also be involved in root growth reduction in mycorrhizal plants. In ECM plants, elongation of the short lateral roots is inhibited by IAA production by the fungi. In AM plants, total root length is decreased, but branching and number of lateral roots per unit root length or per plant can be increased (Barker et al., 1998; Berta et al., 1990). A decrease in root surface area and root activity as well as in root/shoot dry weight ratios is, however, not necessarily harmful for shoot growth and plant yield as long as the external mycelium of the mycorrhizal fungi can fully compensate for root functioning in uptake of nutrients and water.

15.6.4 Mycorrhizal responsiveness

A major beneficial effect of mycorrhizal colonization on host plant growth is due to the increase in belowground surface area (roots and mycorrhizal hyphae) for acquisition of nutrients. The mycorrhiza colonization is therefore particularly important for plants with a coarse and poorly branched root system (Hetrick, 1991; Van der Heijden et al., 1998), and those that lack the capacity to mobilize P by root exudates (Chapter 14). The beneficial effect of mycorrhiza on host plant growth is referred to as *mycorrhizal responsiveness* (Alexander, 1989; Smith and Read, 2008). In most soils, roots are colonized by indigenous AM; thus, studies on mycorrhizal responsiveness require soil sterilization (fumigation, steaming) and reinoculation either with the indigenous soil microflora but not AM or with both indigenous soil microflora and AM.

Responsiveness can differ among plant species (Van der Heijden et al., 1998). Therefore, elimination of AM by soil fumigation may reduce or increase plant growth or have no effect. Klironomos (2003) showed a high variability in mycorrhizal responsiveness (from a strongly positive to a strongly negative effect) among a large number of plant species. This can affect plant community composition in natural ecosystems. In a semiarid herbland, suppression of AM fungi by a fungicide strongly reduced shoot biomass and the relative abundance in the community of *Medicago minima*, had no effect on *Carrichtera annua*, but increased biomass and proportion of plants in the community of *Salvia verbenaca* as an NM plant species (O'Connor et al., 2001, Table 15.3).

Root morphology is important for mycorrhizal responsiveness. Grasses with relatively large root surface area respond little to AM inoculation, whereas legumes with short roots and short root hairs have high mycorrhizal responsiveness. Schweiger et al. (2002) showed an inverse relationship between root hair length and mycorrhizal dependency.

An example of mycorrhizal responsiveness is shown in Table 15.4. In Queensland (Australia), farmers used black fallow (soil was maintained plant-free) for up to 3 years to store water and nutrients (Thompson, 1987). However, they found that most crops grew less well after long fallow than after continuous cropping. This long-fallow disorder occurred in various crops, making diseases an unlikely cause. The experiment conducted by Thompson (1987) showed that mycorrhizal colonization was lower after long than after short fallow, particularly in chickpea, sorghum, and sunflower. The low colonization after long plant-free fallow can be explained by a lack of infective mycorrhizal propagules. During long fallow, the soil remained moist and warm, which triggered germination of AM spores. However, due to the absence of host plants, the spores died. When weeds were allowed to grow during the fallow, percentage colonization in the following crop was higher than after black fallow (Thompson, 1987).

Seed size, and thus seed reserves of P, as well as other nutrients, is another important factor influencing AM responsiveness. In a comparison of 15 wild species grown in a low P soil, there was a negative correlation between AM responsiveness and seed size (Allsopp and Stock, 1992).

Responsiveness is influenced by genotype of plants (Sawers et al., 2017) as well as mycorrhizal fungi (Lanfranco et al., 2018). For example, in maize, the genotypes in which *Funneliformis mosseae* produced extensive extraradical hyphae benefitted from AM more than the maize genotypes with short extraradical hyphae (Sawers et al., 2017). This suggests that plant genotype influences fungal performance, possibly by regulating carbon transfer to the fungus. On the other hand, the extent of root colonization by AM is poorly related to plant responsiveness (Lanfranco et al., 2018). This may also be because AM fungal species differ in the ratio of C received from the host to P delivered to the host (Smith et al., 2011).

TABLE 15.3 Aboveground biomass of major plant species and their relative abundance in the community as influenced by treatments with active AM (control) or AM suppressed by a fungicide in a semiarid herbland.

	Aboveground biomass (g m^{-2})		Species abundance (% of total number of plants)	
	Control	Fungicide	Control	Fungicide
Medicago minima	100	38	43	25
Carrichtera annua	40	58	35	31
Salvia verbenaca	4	36	4.2	33
Velleia arguta	10	10	4.7	2.8
Vittadinia gracilis	3	2	4.4	2.2
Erodium crinitum	10	6	0.8	1.5
Others			7.9	4.5

Source: Based on O'Connor et al. (2001).

TABLE 15.4 Shoot dry weight (dw) and percentage AM colonization of various crops in soil from long or short fallow.

Crop	Length of fallow	Shoot dw (g plant^{-1})	AM colonization (% of root length)
Chickpea	Long	0.5	17
	Short	2.9	73
Sorghum	Long	12	3
	Short	58	32
Sunflower	Long	5.2	64
	Short	14.5	84
Wheat	Long	25	39
	Short	28	44

Source: Based on Thompson (1987).

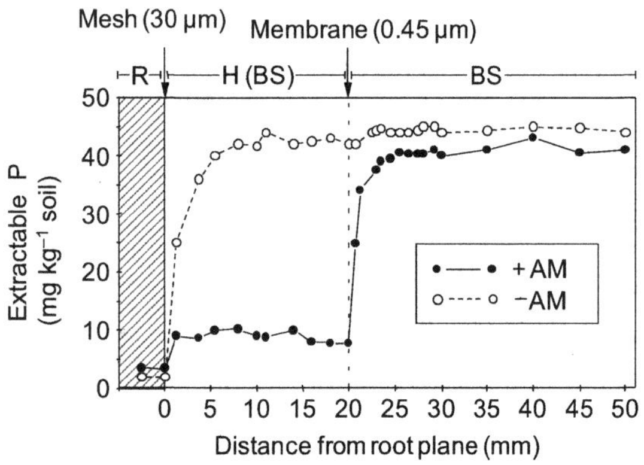

FIGURE 15.6 Concentration of water-extractable P in the root (R), hyphal (H), and bulk soil (BS) compartments of nonmycorrhizal (−AM) and mycorrhizal (AM, *Glomus mosseae*) white clover. *From Li et al. (1991a).*

15.6.5 Role of AM in nutrition of their host plant

The most distinct effect of AM on plant growth is the improved supply of nutrients with low mobility in soil, particularly P. External hyphae can take up and transfer P to the host from soil beyond the rhizosphere depletion zone (Ezawa and Saito, 2018; Smith and Read, 2008; Tinker et al., 1992). Given the key importance of the root hair length to the extent of the P depletion zone and P acquisition (see also Chapter 12), such an enhancing effect of AM on P uptake is to be expected. In mycorrhizal plants, P uptake rate per unit root length can be 2-3 times higher than in NM plants (Tinker et al., 1992). Moreover, the small diameter of the hyphae (1−12 μm) allows them to enter soil pores not accessible to roots.

An example of the differential extension of the P depletion zones in mycorrhizal and NM roots is shown in Fig. 15.6 (Li et al., 1991a). By restricting root extension by a mesh, and hyphal extension by a membrane, P depletion could be measured at the root/soil interface, in the hyphal compartment, and at the hyphae/soil interface. In NM plants the depletion zone extended about 1 cm from the rhizoplane, whereas in the mycorrhizal plants P was uniformly depleted in the hyphal compartment (2 cm from the rhizoplane). At the hyphae/soil interface, a new depletion zone was formed, extending several millimeters into the bulk soil. In the mycorrhizal plants the hyphae contributed between 70% and 80% of total P uptake (Li et al., 1991a). In another study with larger hyphal compartments, P was uniformly depleted up to 12 cm from the rhizoplane of mycorrhizal white clover (Li et al., 1991b).

Similarly to plant roots, the external hyphae of AM fungi release acid phosphatase and therefore can also mineralize organic P (Ezawa and Saito, 2018; Tarafdar and Marschner, 1994). This suggests that AM plants have access to similar soil P pools as non-AM plants, but AM plants can acquire P from a greater soil volume. Extraradical hyphae of AM may also stimulate microbial solubilizers of native soil P (Ezawa and Saito, 2018).

The Pi taken up by hyphae is polymerized to poly-P and transported within hyphae via water flow (Ezawa and Saito 2018; Uetake et al., 2002). A fraction of this transported poly-P is used by the fungus for ATP synthesis. At the fungus—plant interface, poly-P is depolymerized and released into the apoplast (Ezawa and Saito, 2018; Smith and Gianinazzi-Pearson, 1988). Poly-P may also aid the transfer of cations to the host/fungus interface because poly-P is a strongly negatively charged polyanion that can bind cations such as Mg and K, and basic amino acids such as arginine and glutamine (Jennings, 1987).

The effectiveness of AM fungi in providing P to the host plants depends on the AM species. Percentage colonization may be positively correlated with plant P uptake (Raju et al., 1990), but this is not always the case (Lanfranco et al., 2018; Smith et al., 2003), possibly because AM fungal species differ in the exchange ratio of C-to-P (Smith et al., 2011). The extent of the external mycelium may be a better indicator for the capacity of AM fungi to improve P uptake by plants than percentage colonization (Sawers et al., 2017). In subterranean clover growing in soil with low P availability, shoot dry weight was more strongly enhanced by *Acaulospora* than by *Scutellospora* (Schweiger et al., 2007). The stronger growth increase by *Acaulospora* was not related to colonization rate that was similar to that of *Scutellospora*. However, length of the external hyphae was about threefold greater in *Acaulospora* than in *Scutellospora* (Jacobsen et al., 1992).

Transfer of P from the fungus to the plant is mediated by mycorrhiza-induced P transporters at the fungus/host interface (Ezawa and Saito, 2018; Glassop et al., 2005; Smith and Read, 2008). These transporters are not expressed in NM plants, and they are distinct from the P transporters in the root epidermis. Moreover, AM colonization leads to a down-regulation of the P transporters in the root epidermis, suggesting that in mycorrhizal plants, most P is taken up via the fungal pathway (Glassop et al., 2005), even if shoot growth is not increased by mycorrhizal colonization (Smith et al., 2003).

In AM plants the uptake and concentrations of Zn and Cu are also usually higher than in NM plants (Kothari et al., 1990a; Lambert and Weidensaul, 1991). In contrast to Zn and Cu, shoot Mn concentrations are often lower in AM than in NM plants. In red clover, there is a negative relationship between percentage of root colonization with AM and Mn concentration in roots and shoots (Arines et al., 1989). The decrease in Mn uptake in mycorrhizal plants could be due to low uptake and transport of Mn in the external hyphae, but may be explained also by decreased Mn acquisition by AM roots. In maize, AM plants had lower shoot and root Mn concentrations, lower density of Mn-reducing bacteria, and smaller amount of exchangeable manganese (Mn^{2+}) in the rhizosphere soil (Kothari et al., 1991). In red clover, lower Mn concentrations in roots and shoots of AM plants were associated with higher abundance of Mn-oxidizing bacteria in the rhizosphere (Arines et al., 1992), which would also decrease Mn availability.

Little is known about the role of AM in uptake of K, Mg, and S. In *Agropyron repens*, about 10% of the total K in mycorrhizal plants was attributed to hyphal uptake and delivery (George et al., 2007). Although hyphal transport has been demonstrated for Ca by using radioisotopes, the amounts transported are small, as indicated by the lower total Ca concentrations in shoots of mycorrhizal compared with NM plants (Azcon and Barea, 1992; Kothari et al., 1990b).

Transfer of N from hyphae to the plant (Jacobsen et al., 1992) occurs as ammonium or nitrate (Wipf et al., 2019). The effect of AM colonization on N uptake from organic N sources is variable, ranging from no effect in a number of grassland perennials (Reynolds et al., 2005) to twofold higher N uptake by a model grass *Brachypodium distachyon* (Hestrin et al., 2019). Due to the high P requirement for nodulation, a high AM dependency in legumes is to be expected, but the interactions between N_2-fixing Rhizobium and AM are complex (Bethlenfalvay, 1992). In bean (*Phaseolus vulgaris*), AM increased respiration by about 10% and N_2 fixation by up to 40%, but also increased root respiration (root + symbionts) by 30% (Mortimer et al., 2008) and delayed nodulation.

The external AM hyphae bridge individual plants of the same species, or different plant species in mixed stands, as a potential pathway of nutrient transfer between plants. For example, He et al. (2019) demonstrated that connection of root systems of donor plants fertilized with ^{15}N by AM hyphae resulted in about fourfold higher ^{15}N concentration in leaves of receiver plants compared to treatments where roots were separated by a wall.

Native soil microbes can have positive or negative effects on AM. In a study with 21 soils, Svenningsen et al. (2018) found that P uptake by the extraradical mycelium of AM decreased in four soils. The suppressive effect was, at least partly, due to the presence of antagonistic bacteria, particularly *Acidobacteria*, but other bacterial taxa also were involved. On the other hand, acquisition of organic N by AM grass was increased several-fold when inoculated with native soil microbes compared to AM grass in sterile soil or NM grass inoculated with microorganisms (Hestrin et al., 2019). This suggests there can be a synergism between AM and soil microorganisms to enhance N mineralization and uptake.

Soil animals such as earthworms and collembola can increase nutrient uptake by AM plants via several mechanisms, including greater mineralization rates, suppression of pathogenic fungi, and aiding AM spore dispersal (Ngosong et al., 2014; Vasutova et al., 2019).

15.6.6 Role of AM in agriculture

There are contrasting opinions about the role of AM in intensive agriculture. Ryan and Graham (2018) suggest that AM colonization and diversity may have little or inconsistent effects on crop growth in the field even under P limitation. They argue that the variable effect of AM colonization is due to interactions between the environment (climate, soil nutrient availability) and genotype of plants and fungi. One reason for the inconsistent effect of AM on crop growth is that the relationship between percentage root colonization and nutrient delivery to the plant varies among AM fungi. However, Rillig et al. (2019) argue that AM are important for crop performance and sustainability, for example, by improving soil structural stability. AM colonization can also enhance crop quality, for example, by increasing nutrient uptake (biofortification), or storage organ properties. They warn that the combination of management practices such as fertilization, pesticides, tillage, and crop rotation can have negative impacts on AM abundance and diversity.

Comparisons between wild progenitors and modern cultivars (e.g., Bryla and Koide, 1990a, 1990b; Manske, 1989; Zhu et al., 2001) suggest that breeding has selected plant genotypes that only respond to AM with increased growth at low P availability. In the wild types, on the other hand, AM increased growth at both low and high P availability (Lanfranco et al., 2018). This is partly due to differences in root morphology and root/shoot dry weight ratio, but also in growth rate, growth potential, and P-use efficiency. In the field a lack of effect of AM may be due to management (tillage, fertilization, pesticides) (Sawers et al., 2018). However, there are indications that changes in the plant genome during breeding, such as dwarf or high-yielding varieties that can affect phytohormone metabolism, may also influence plant responsiveness to AM, for example, via reduction in signaling compounds released by roots or restriction of fungal development within the roots (Sawers et al., 2018).

15.6.7 Role of ectomycorrhiza in nutrition of plants

With respect to their role in nutrition of their host plant, ECM fungi have many common features with AM. However, there are some principal differences in terms of structural arrangements with the roots and mechanisms of nutrient acquisition. Depending on the tree species as well as on root growth rate and season of the year, a varying proportion of plant nutrients may be taken up via the fungal hyphae of the external mycelium and the sheath. However, ECM fungi differ substantially in thickness of the sheath (Agerer, 1992; Smith and Read, 2008) and its hydraulic resistance to solute flow. The fungal sheath may be more or less sealed and prevent an apoplastic route of solute and water flux into the root cortex, for example, in *Pinus banksiana* with *Hebeloma cylindrosporum* (Taylor and Peterson, 2005), whereas it provides a relatively unrestricted apoplastic route in others, for example, *Pinus sylvestris* with *Suillus bovinus* (Behrmann and Heyser, 1991).

The extent of the external mycelium varies substantially among ECM species, with 300 m m^{-1} colonized root length in *Salix* seedlings (Jones, 1990) and 500 m m^{-1} in *Pinus taeda* (Rousseau et al., 1994). In contrast to AM, many ECM fungal species form rhizomorphs (Agerer, 1992) that can be the main routes for bidirectional solute transport. As in individual hyphae, solute transport in rhizomorphs is driven by cytoplasmic streaming and concentration gradients. However, their large diameter ($\sim$100 μm) and hollow center may also allow rapid apoplastic solute transport (Cairney, 1992; Jennings, 1987). Similarly to AM, ECM hyphae contain poly-P (Bucking and Heyser, 1999; Orlovich et al., 1989) and increase P uptake, with nearly threefold greater P influx in mycorrhizal *Salix* compared to NM plants (Jones et al., 1991). ECM may also increase plant growth when supplied with the poorly soluble apatite; however, there is considerable variation among ECM fungi in utilizing apatite, which could be related to their capacity to release oxalate (Wallander, 2000). This variation among ECM fungi may explain why the contribution of ECM to plant P uptake can vary between 50% and 70% (Nehls and Plassard, 2018). The ECM diversity has been shown to be important for nutrient uptake. In European birch, Koehler et al. (2018) found that P uptake efficiency increased with ECM richness and diversity, particularly when water availability was low.

ECM hyphae can transport K and have been shown to increase plant tissue concentrations of K (Jentschke et al., 2001) and Mg (Van Scholl et al., 2006), possibly due to the weathering of minerals by organic acid anions released by EMC hyphae (see below).

The release of acid phosphatase by ECM fungi is well established, both along the external mycelium (Dinkelaker and Marschner, 1992) and at the surface of mycorrhizal roots (Gourp and Pargney, 1991). In contrast to AM fungi, ECM fungi have a considerable capacity to produce and release organic acid anions. These organic acid anions, and perhaps also siderophores, are involved in the enhanced weathering of micas by ECM as compared with NM pine (Leyval and Berthelin, 1991). This so-called "ectomycorrhizal weathering" or "rock eating" has been confirmed in several studies. Bonneville et al. (2011) showed that hyphae of *Paxillus involutus* colonizing *P. sylvestris* released K, Mg,

Fe, and Al from biotite, likely by acidification. Oxalic acid anion release from hyphae is important for weathering of Ca-bearing minerals by *P. involutus* (Schmalenberger et al., 2015).

Similarly to many plants, ECM fungi prefer ammonium compared to nitrate as N source (Plassard et al., 1991). Accordingly, when both ammonium and nitrate are supplied (e.g., as ammonium nitrate), ECM fungi take up ammonium preferentially and therefore acidify their substrate similarly to the host roots (see also Section 6.1 and Chapters 2 and 14). After uptake of ammonium or reduction of nitrate in the cells of the external mycelium and the fungal sheath, ammonium is incorporated into glutamate and glutamine by glutamate dehydrogenase (GDH) and glutamine synthase (GS), respectively. Nehls and Plassard (2018) report that both GS and GDH genes are expressed at low external ammonium concentrations. However, at high ammonium concentrations, GDH expression was low, indicating that GS was the main enzyme for ammonium assimilation.

The extent to which inorganic N is either assimilated in the fungal cells or passes the sheath to be assimilated in the host root cells may depend on the relative enzyme activities, carbohydrate supply, and thickness of the sheath. Due to this variation, the contribution of ECM to plant N uptake can vary between 1% and 40% (Nehls and Plassard, 2018).

Colonization by some ECM fungi can improve N uptake from organic N compared to NM pine seedlings. Wallander (2002) labeled fungal mycelium with ^{15}N and then used it as organic N source. Compared to NM *P. sylvestris*, colonization by *P. involutus* and one *Suillus variegatus* strain increased ^{15}N uptake about fivefold, whereas another *S. variegatus* strain increased ^{15}N uptake only twofold (Table 15.5).

The capacity to utilize organic N directly may increase the competitiveness of mycorrhizal plants compared to NM ones in ecosystems with high organic matter and thus organic N content (Northup et al., 1995). Using ^{15}N labeling in alpine forests, Zhang et al. (2019) showed that about 80% of N supplied in inorganic form was taken up directly by the roots, whereas when organic N was added, 44% of N uptake was attributed to ECM hyphae.

ECM can store N in the vacuoles of hyphae as basic amino acids, such as arginine and histidine, which can counterbalance the charge of poly P (Nehls and Plassard, 2018). Transport of N and P within hyphae may be via diffusion, vesicles, cytoplasmic streaming, or mass flow, the latter mainly in rhizomorphs, possibly driven by plant transpiration. However, the relative importance of different transport mechanisms is still unclear and may depend on plant and fungal genotypes as well as on environmental conditions (Nehls and Plassard, 2018).

Nutrient transfer from fungus to host requires several steps, including poly-P/amino acid breakdown in the vacuole, export of N and P into the cytoplasm, and N/P metabolization for export from hyphae to plant cells. The mechanisms involved are still poorly understood in ECM. The export of N from the fungus was thought to be in the form of amino acids, but some studies suggested that fungi release N as ammonium (Dietz et al., 2011). Identification of transporter genes in the fungal membrane suggests that P export occurs as Pi.

15.6.8 Role of mycorrhiza in tolerance to high metal concentrations

Colonization by ECM fungi can increase tolerance of host plants to high metal concentrations (Colpaert and van Assche, 1993; Wilkins, 1991). For example, tolerance to high Zn concentrations was enhanced by inoculation with *P. involutus* (Denny and Wilkins, 1987) or *S. bovinus* (Adriaensen et al., 2006). ECM colonization can also alleviate Mn toxicity. Compared to NM *Eucalyptus grandis*, colonization by *P. tinctorius* increased shoot dry weight irrespective of Mn concentration, but the increase by ECM was twofold at 0 Mn, threefold at 200 μM Mn, and sixfold at 1000 μM Mn (Canton et al., 2016; Table 15.6). Colonization by ECM also increased net C assimilation and diminished plant Mn uptake at 200 and 1000 μM Mn.

TABLE 15.5 Total uptake of ^{15}N from labeled lyophilized *Suillus variegatus* mycelium by nonmycorrhizal (NM) *Pinus sylvestris* seedlings or seedlings colonized by ectomycorrhiza fungi *Paxillus involutus* or two strains of *S. variegatus*.

	Total ^{15}N taken up (μg plant^{-1})
NM	4
Paxillus involutus	19
S. variegatus strain 1	18
S. variegatus strain 2	7.5

Source: Based on Wallander (2002).

TABLE 15.6 Shoot dry weight (dw), N and Mn shoot content, and net C assimilation by *Eucalyptus grandis* grown in a sand:vermiculite (3:1) mixture supplemented with 0.28 mg Mn kg^{-1} and watered with nutrient solutions containing 0, 200, or 1000 μM Mn for 90 days.

	Concentration of Mn in the watering solution (μM)					
	0		200		1000	
	NM	Myc	NM	Myc	NM	Myc
Shoot dw (g plant^{-1})	0.4	0.8	0.3	0.9	0.2	1.2
N content (mg plant^{-1})	8.1	9.5	6.4	11.5	1.7	10.8
Mn content (μg plant^{-1})	874	381	1897	1106	18,632	1269
Net C assimilation (μmol m^{-1}s^{-1})	3.8	4.1	4.8	6.2	2.2	4.2

Plants were either colonized by ectomycorrhizal fungus *Pisolithus tinctorius* (Myc) or were nonmycorrhizal (NM).
Source: Based on Canton et al. (2016).

Fungal tolerance to high metal concentrations and a decrease in plant metal toxicity are due to several processes (Bellion et al., 2006; Shi et al., 2019): (1) extracellular binding on the external mycelium or the fungal mantle by excreted ligands, (2) surface sequestration by binding to the fungal cell walls, (3) enhanced metal efflux from the fungal cells, (4) binding to metallothionein or glutathione in the fungal cytoplasm, and (5) sequestration of the glutathione−metal complex in the vacuoles. These processes decrease the soil solution concentration of metals in the mycorrhizosphere as well as in root and particularly shoot tissues. Most heavy metals and Al exert their toxic influence by damaging root apical zones. Therefore, preventing heavy metals and Al from reaching the root tips increases host tolerance.

The heavy metal binding capacities of the external mycelium, and its mass, are important for the effectiveness of heavy metal retention in ECM (Colpaert and van Assche, 1992). However, tolerance of fungi to high metal concentrations also plays a role. At high Zn supply, only the Zn-tolerant *S. bovinus* strain increased shoot dry weight and chlorophyll content of *P. sylvestris* compared to the NM plants, whereas the Zn-sensitive *S. bovinus* strain had little effect (Adriaensen et al., 2006; Table 15.7).

The retention capacity of the fungi can be exceeded over time. In Norway spruce grown at 800 μM Al, *P. involutus* decreased Al toxicity (as indicated by increased chlorophyll content compared to the NM plants) in 5-week-old plants, but had no ameliorating effect when plants were grown for 10 weeks (Hentschel et al., 1993).

In contrast to ECM, there are only a few reports on the effect of AM on metal tolerance of the host plant. This is not surprising given the importance of the ECM sheath in retaining metals. Indirect effects may occur, for example, by improving P nutritional status and growth of the host plant on a low-P soil high in heavy metals or Al, that is, by a dilution effect. Colonization by AM can be beneficial for plants at both low and high Zn supply. In *Medicago truncatula*, colonization by *R. irregularis* increased shoot dry weight both under Zn deficiency and at very high Zn concentrations of 20 and 40 mg Zn kg^{-1} (Watts-Williams et al., 2017). At high Zn supply the mycorrhizal plants had lower shoot Zn concentrations than NM plants, but there was no difference in shoot Zn concentrations at low Zn supply. Watts-Williams et al. (2017) measured plant gene expression and found that at low Zn supply, AM colonization increased expression of a Zn transporter. By contrast, at high Zn supply, AMF-induced expression of P transporter gene, indicating that AM colonization improved plant growth by increasing P uptake. The defense mechanisms induced by mycorrhizal colonization may also improve plant tolerance to metals (Shi et al., 2019).

15.6.9 Other mycorrhizal effects

15.6.9.1 Phytohormonal effects and plant water relations

Mycorrhiza may alter host plant growth and development through direct and indirect effects on secondary metabolism that, in turn, can influence mycorrhizal colonization. Adolfsson et al. (2017) reported that mycorrhizal colonization of *Medicago* led to upregulation of genes involved in synthesis of flavonoids, terpenoids, jasmonic acid, and abscisic acid. This resulted in higher concentrations of anthocyanins, flavonoids, and ABA compared to NM plants at the same

TABLE 15.7 Shoot dry weight (dw) and N, P, Zn, Fe, and chlorophyll a + b concentrations in needles of nonmycorrhizal Scots pine (*Pinus sylvestris*) grown at high Zn supply (150 μM Zn) without ectomycorrhiza (ECM) (non-myc) or inoculated with a Zn-tolerant or a Zn-sensitive ECM strain of *Suillus bovinus*.

	Shoot dw (g plant^{-1})	N (mg g^{-1} dw)	P (mg g^{-1} dw)	Zn (μg g^{-1} dw)	Fe (μg g^{-1} dw)	Chl a + b (μg g^{-1} fresh weight)
Non-myc	0.4	5.1	0.6	716	10	201
Zn-sensitive ECM strain	0.5	7.4	0.7	508	62	218
Zn-tolerant ECM strain	1.0	15.4	1.1	400	223	687

Source: Based on Adriaensen et al. (2006).

P concentration. In relation to the effect of secondary metabolites on mycorrhizae, AM colonization is decreased by gibberellins but enhanced by brassinosteroids (McGuiness et al., 2019). Using tomato mutants defective in ABA and ethylene synthesis, Fracetto et al. (2017) found that both ethylene overproduction and ABA deficiency reduced root colonization.

In poplar colonized by *Laccaria bicolor*, Plett et al. (2014) found that ethylene and jasmonic acid reduced colonization. They suggested that ethylene and jasmonic acid synthesis was enhanced in the late colonization stages to limit fungal growth in roots, thereby decreasing C allocation to the fungus.

Mycorrhizal colonization can also affect water uptake by plants. For example, root hydraulic conductivity may be enhanced by ECM (Lehto and Zwiazek, 2011). Indirect mycorrhizal effects on water relations include improved nutrient status of the plant, altered carbohydrate assimilation, increased sink strength of mycorrhizal roots, and enhanced antioxidant metabolism (Lehto and Zwiazek, 2011).

An increase in drought stress tolerance has also been observed in AM plants compared with NM plants. AM colonization can increase transpiration and stomatal conductance during water stress and recovery (Santander et al., 2017) and enhance water-use efficiency that may be due to upregulation of ABA synthesis (Xu et al., 2018). Further, under drought stress, mycorrhizal plants can maintain higher rates of photosynthesis and higher photosynthetic efficiency than NM plants (Mo et al., 2016).

Glomalin, a glycoprotein associated with AM hyphae (Wright and Jawson, 2001) that includes polysaccharides, mucilage, and hydrophobins, can improve soil structural stability and water retention and thus water uptake by AM-colonized plants (Mardhiah et al., 2016; Santander et al., 2017). Using high-resolution monitoring of soil water content, Ruth et al. (2011) determined that hyphae contribute about 20% of total plant water uptake.

15.6.9.2 Suppression of root pathogens and nematodes

Mycorrhizal colonization can suppress soil-borne root pathogens and nematodes. For example, inoculation with AM fungi can increase resistance of barley against the take-all fungus *G. graminis* var. *tritici* (Khaosaad et al., 2007). Using a split-root system where one side was inoculated by *G. mosseae* and the other by the pathogen, Khaosaad et al. (2007) showed that high AM colonization suppressed pathogen infection, whereas low AM colonization had no effect.

Colonization by AM fungi can also reduce root infection by nematodes. In tomato, colonization by *G. mosseae* reduced nematode infection as indicated by the strong reduction of number of juveniles, females, and eggs in the roots and the lower gall index (Vos et al., 2012). The AM fungi were equally effective in reducing nematode infection when AM fungi and nematodes were present in the same root system (local AM) and in split-root plants where roots colonized by AM fungi were separated from those challenged by nematodes (systemic AM) (Vos et al., 2012; Table 15.8).

Schouteden et al. (2015) discussed possible mechanisms by which AM colonization may enhance plant resistance to parasitic nematodes: (1) competition for space and nutrients, (2) enhanced tolerance due to improved nutrient uptake by plants, (3) altered rhizosphere interactions due to changes in root exudation and rhizosphere microbial community composition, and (4) induced systemic resistance by priming the plant's response to root invasion (via modulating expression of genes involved in primary and secondary metabolism, as well as in defense responses).

TABLE 15.8 Root nematode infection (number of juveniles, females, and eggs and gall index of roots) after addition of the nematode *Pratylenchus penetrans* to nonmycorrhizal (non-myc) tomato (*Solanum lycopersicum*) or tomato colonized by AM fungus *Glomus mosseae*.

Treatment	Nematode infection number (root system)$^{-1}$			Gall index
	Juveniles	Females	Eggs	
Non-myc	2165	334	7265	5
Local AM	626	133	3853	3
Systemic AM	682	179	4120	3

Local AM: AM fungi and nematodes in the same root system, systemic AM: AM fungi on one side of the split root system, nematodes on the other. Gall index is proportional to the severity of nematode-induced root knotting.
Source: Based on Vos et al. (2012).

15.6.9.3 Suppression of leaf pathogens

An example of systemic resistance conferred by AM colonization is reduced leaf pathogen infection (Mustafa et al., 2017), whereby expression of several defense genes such as peroxidase, phenylalanine ammonia lyase, and chitinase was increased in mycorrhizal wheat that had 80% lower leaf infection by powdery mildew (*Blumeria graminis*) compared to NM wheat. Nair et al. (2015) showed that mycorrhizal colonization in tomato increased lipid peroxidase activity (involved in jasmonic acid synthesis); tomato leaf infection by *Alternaria alternata* was reduced by mycorrhizal colonization, likely due to the fourfold higher concentration of jasmonic acid in the leaves. Recently, Rivero et al. (2021) showed that AM colonization increased leaf concentrations of alkaloid and fatty acid derivatives, which was associated with increased mortality of insects feeding on tomato leaves.

Cameron et al. (2013) suggest that systemic resistance by AM colonization is due to enhanced defense reaction induced by AM (e.g., increased biosynthesis of jasmonic acid, salicylic acid, phenolics, ABA) and improved nutrient uptake, but also suggest that changes in rhizosphere microbial community composition are involved. They pointed out the mechanisms behind systemic resistance vary among combinations of plant and AM fungus species.

Little is known about the effect of AM on plant virus infections. However, some studies showed that AM colonization exacerbated susceptibility of plants to viruses and increased virus replication, possibly due to the improved nutrient uptake (Miozzi et al., 2019).

References

Adolfsson, L., Nziengui, H., Abreu, I.N., Simura, J., Beebo, A., Herdean, A., et al., 2017. Enhanced secondary- and hormone metabolism in leaves of arbuscular mycorrhizal *Medicago truncatula*. Plant Physiol. 175, 392–411.

Adriaensen, K., Vangronsveld, J., Colpaert, J.V., 2006. Zn-tolerant *Suillus bovinus* improves growth of Zn-exposed *Pinus sylvestris* seedlings. Mycorrhiza 16, 553–558.

Agerer, R., 1992. Ectomycorrhizal rhizomorphs: organs of contact. In: Read, D.J., Lewis, D.H., Fitter, A.H., Alexander, I.J. (Eds.), Mycorrhizas in Ecosystems. CAB International, Wallingford, UK, pp. 84–90.

Ahmed, M., Stal, L.J., Hasnain, S., 2010. Production of indole-3-acetic acid by the cyanobacterium *Arthrospira platensis* strain MMG-9. J. Microbiol. Biotechnol. 20, 1259–1265.

Akiyama, K., Matsuzaki, K., Hayashi, H., 2005. Plant sesquiterpenes induce hyphal branching in arbuscular mycorrhizal fungi. Nature 435, 824–827.

Akiyama, K., Tanigawa, F., Kashihara, T., Hayashi, H., 2010. Lupin pyranoisoflavones inhibiting hyphal development in arbuscular mycorrhizal fungi. Phytochemistry 71, 1865–1871.

Alexander, I., 1989. Mycorrhizas in tropical forests. In: Protector, J. (Ed.), Mineral Nutrients in Tropical Forests and Savannah Ecosystems. Blackwell, Oxford, UK, pp. 169–188.

Ali, B., Sabri, A.N., Ljung, K., Hasnain, S., 2009. Auxin production by plant associated bacteria: impact on endogenous IAA content and growth of *Triticum aestivum* L. Lett. Appl. Microbiol. 48, 542–547.

Allsopp, N., Stock, W.D., 1992. Mycorrhizas, seed size and seedling establishment in a low nutrient environment. In: Read, D.J., Lewis, D.H., Fitter, A.H., Alexander, I.J. (Eds.), Mycorrhizas in Ecosystems. CAB International, Wallingford, UK, pp. 59–64.

Anderson, A.J., 1988. Mycorrhizae-host specificity and recognition. Phytopathology 78, 375–378.

Arines, J., Vilarino, A., Sainz, M., 1989. Effect of different inocula of vesicular-arbuscular mycorrhizal fungi on manganese content and concentration in red clover (*Trifolium pratense* L.) plants. New Phytol. 112, 215–219.

Arines, J., Porto, M.E., Vilarino, A., 1992. Effect of manganese on vesicular-arbuscular mycorrhizal development in red clover plants and on soil Mn-oxidizing bacteria. Mycorrhiza 1, 127–131.

Arshad, M., Frankenberger Jr., W.T., 1991. Microbial production of plant hormones. Plant Soil 133, 1–8.

Arshad, M., Hussain, A., Javed, M., Frankenberger Jr., W.T., 1993. Effect of soil applied L-methionine on growth, nodulation and chemical composition of *Albizia lebbeck* L. Plant Soil 148, 129–135.

Azcon, R., Barea, J.M., 1992. The effect of vesicular-arbuscular mycorrhizae in decreasing Ca acquisition by alfalfa plants in calcareous soils. Biol. Fertil. Soils 13, 155–159.

Baas, R., van der Werf, A., Lambers, H., 1989. Root respiration and growth in *Plantago major* as affected by vesicular-arbuscular mycorrhizal infection. Plant Physiol. 91, 227–232.

Baath, E., Spokes, J., 1988. The effect of added nitrogen and phosphorus on mycorrhizal growth response and infection in *Allium schoenoprasum*. Can. J. Bot. 67, 3227–3232.

Bajpai, P.D., Sundara-Rao, W.V.B., 1971. Phosphate solubilizing bacteria. III. Soil inoculation with phosphorus solubilizing bacteria. Soil Sci. Plant Nutr. 17, 46–53.

Banik, S., Dey, B.K., 1983. Phosphate-solubilizing potentiality of the microorganisms capable of utilizing al phosphate as a sole phosphorus source. Zentralbl. Mikrobiol. 138, 17–23.

Baon, J.B., Smith, S.E., Alston, A.M., 1993. Mycorrhizal response of barley cultivars differing in P efficiency. Plant Soil 157, 97–105.

Bardon, C., Piola, F., Bellvert, F., Haichar, F.Z., Comte, G., Meiffren, G., et al., 2014. Poly F: Evidence for biological denitrification inhibition (BDI) by plant secondary metabolites. New Phytol 204, 620–630.

Barker, S.J., Tagu, D., Delp, G., 1998. Regulation of root and fungal morphogenesis in mycorrhizal symbiosis. Plant Physiol. 116, 1201–1207.

Bar-Ness, E., Chen, Y., Hadar, Y., Marschner, H., Römheld, V., 1991. Siderophores of *Pseudomonas putida* as an iron source for dicot and monocot plants. Plant Soil 130, 231–241.

Bashan, Y., de Bashan, L.E., 2010. How the plant growth-promoting bacterium *Azospirillum* promotes plant growth – a critical assessment. Adv. Agron. 108, 77–136.

Baudoin, E., Benizri, E., Guckert, A., 2003. Impact of artificial root exudates on the bacterial community structure in bulk soil and maize rhizosphere. Soil Biol. Biochem. 35, 1183–1192.

Behrmann, P., Heyser, W., 1991. Apoplastic transport through the fungal sheath of *Pinus sylvestris/Suillus bovinus* ectomycorrhizae. Bot. Acta 105, 427–434.

Bell, M.J., Middleton, K.J., Thompson, J.P., 1989. Effects of vesicular-arbuscular mycorrhizae on growth and phosphorus and zinc nutrition of peanut (*Arachis hypogaea* L.) in an oxisol from subtropical Australia. Plant Soil 117, 49–57.

Bellion, M., Courbot, M., Jacob, C., Blaudez, D., Chalot, M., 2006. Extracellular and cellular mechanisms sustaining metal tolerance in ectomycorrhizal fungi. FEMS Microbiol. Lett. 254, 173-18.

Berta, G., Fusconi, A., Trotta, A., Scannerini, S., 1990. Morphogenetic modifications induced by the mycorrhizal fungus Glomus strain E_3 in the root system of *Allium porrum* L. New Phytol. 114, 207–215.

Bethlenfalvay, G.J., 1992. Vesicular-arbuscular mycorrhizal fungi in nitrogen fixing legumes – problems and prospects. Meth. Microbiol. 24, 375–389.

Biswas, R., Sarkar, A., 2018. "Omics" tools in soil microbiology: the state of the art. In: Adhya, T.K., Lal, B., Mohapatra, B., Paul, D., Das, S. (Eds.), Advances in Soil Microbiology: Recent Trends and Future Prospects. Springer Nature, Singapore, pp. 35–63.

Bolan, N.S., Robson, A.D., Barrow, N.J., 1984. Increasing phosphorus supply can increase the infection of plant roots by vesicular-arbuscular mycorrhizal fungi. Soil Biol. Biochem. 16, 419–420.

Bonneville, S., Morgan, D.J., Schmalenberger, A., Bray, A., Brown, A., Banwart, S.A., et al., 2011. Tree-mycorrhiza symbiosis accelerate weathering: Evidences from nanometer scale elemental fluxes at the hypha-mineral interface. Geochim. Cosmochim. Acta 75, 6988–7005.

Bowen, G.D., Rovira, A.D., 1991. The rhizosphere, the hidden half of the hidden half. In: Waisel, Y., Eshel, A., Kafkafi, U. (Eds.), The Plant Roots, the Hidden Half. Marcel Dekker Inc, New York, pp. 641–669.

Bruce, A., Smith, S.E., Tester, M., 1994. The development of mycorrhizal infection in cucumber – effects of P supply on root growth, formation of entry points and growth of infection units. N. Phytol. 127, 507–514.

Brundrett, M.C., Abbott, L.K., 1991. Roots of jarrah forest plants. I. Mycorrhizal associations of shrubs and herbaceous plants. Aust. J. Bot. 39, 445–457.

Brundrett, M.C., Tedersoo, L., 2018. Evolutionary history of mycorrhizal symbioses and global host plant diversity. N. Phytol. 220, 1108–1115.

Bruns, T.D., Corradi, N., Redecker, D., Taylor, J.W., Opik, M., 2018. *Glomeromycotina*: what is a species and why should we care? New Phytol. 220, 963–967.

Bryla, D.R., Koide, R.T., 1990a. Regulation of reproduction in wild and cultivated *Lycopersicon esculentum* Mill. by vesicular-arbuscular mycorrhizal infection. Oecologia 84, 74–81.

Bryla, D.R., Koide, R.T., 1990b. Role of mycorrhizal infection in the growth and reproduction of wild vs. cultivated plants. II. Eight wild accessions and two cultivars of *Lycopersicon esculentum* Mill. Oecologia 84, 82–92.

Bucking, H., Heyser, W., 1999. Elemental composition and function of polyphosphates in ectomycorrhizal fungi – an X-ray microanalytical study. Mycol. Res. 103, 31–39.

Bulgarelli, D., Schlaeppi, K., Spaepen, S., Ver Loren van Themaat, E., Schulze-Lefert, P., 2013. Structure and functions of the bacterial microbiota of plants. Annu. Rev. Plant Biol. 64, 807–838.

Butterbach-Bahl, K., Baggs, E.M., Dannenmann, M., Kiese, R., Zechmeister-Boltenstern, S., 2013. Nitrous oxide emissions from soils: how well do we understand the processes and their controls? Phil. Trans. Royal Soc. B 368.

Cairney, J.W.G., 1992. Translocation of solutes in ectomycorrhizal and saprophytic rhizomorphs. Mycol. Res. 96, 135–141.

Caldeira, C.F., Jeanguenin, L., Chaumont, F., Tardieu, F., 2014. Circadian rhythms of hydraulic conductance and growth are enhanced by drought and improve plant performance. Nat. Comm. 5, 5365.

Cameron, D.D., Neal, A.L., Van Wees, S.C.M., Ton, J., 2013. Mycorrhiza-induced resistance: more than the sum of its parts? Trends Plant Sci. 18, 539–545.

Canton, G.C., Bertolazi, A.A., Cogo, A.J.D., Eutropio, F.J., Melo, J., de Souza, S.B., et al., 2016. Biochemical and ecophysiological responses to manganese stress by ectomycorrhizal fungus *Pisolithus tinctorius* and in association with *Eucalyptus grandis*. Mycorrhiza 26, 475–487.

Carminati, A., Moradi, A.B., Vetterlein, D., Vontobel, P., Lehmann, E., Weller, U., et al., 2010. Dynamics of soil water content in the rhizosphere. Plant Soil 332, 163–176.

Chiarini, L., Giovanelli, V., Bevivino, A., Dalmastri, C., Tabacchioni, S., 2000. Different proportions of the maize root system host *Burkholderia cepacia* populations with different degrees of genetic polymorphism. Environ. Microbiol. 2, 111–118.

Colmer, T.D., Bloom, A.J., 1998. A comparison of NH_4 and NO_3 net fluxes along roots of rice and maize. Plant Cell Env. 21, 240–246.

Colpaert, J.V., van Assche, J.A., 1992. Zinc toxicity in ectomycorrhizal *Pinus sylvestris*. Plant Soil 143, 201–211.

Colpaert, J.V., van Assche, J.A., 1993. The effect of cadmium on ectomycorrhizal *Pinus sylvestris* L. N. Phytol. 123, 325–333.

Compant, S., Clément, C., Sessitsch, A., 2010. Plant growth-promoting bacteria in the rhizo-and endosphere of plants: their role, colonization, mechanisms involved and prospects for utilization. Soil Biol. Biochem. 42, 669–678.

Crowley, D.E., Gries, D., 1994. Modeling of iron availability in the plant rhizosphere. In: Manthey, J.A., Crowley, D.E., Luster, D.G. (Eds.), Biochemistry of Metal Micronutrients in the Rhizosphere. Lewis Publishers, Boca Raton, FL, pp. 199–224.

Crowley, D.E., Römheld, V., Marschner, H., Szaniszlo, P.J., 1992. Root-microbial effects on plant iron uptake from siderophores and phytosiderophores. Plant Soil 142, 1–7.

Darrah, P.R., 1991. Models of the rhizosphere. I. Microbial population dynamics around a root releasing soluble and insoluble carbon. Plant Soil 133, 187–199.

Darrah, P.R., 1993. The rhizosphere and plant nutrition: a quantitative approach. In: Barrow, N.J. (Ed.), Plant Nutrition – From Genetic Engineering to Field Practice. Kluwer Academic Publishers, Dordrecht, The Netherlands, pp. 3–22.

Davis, E.A., Young, J.L., Rose, S.L., 1984. Detection of high-phosphorus tolerant VAM-fungi colonizing hops and peppermint. Plant Soil 81, 29–36.

De Cuyper, C., Goormachtig, S., 2017. Strigolactones in the rhizosphere: friend or foe? Mol. Plant-Microbe Inter. 30, 683–690.

De Nobili, M., Contin, M., Mondini, C., Brookes, P.C., 2001. Soil microbial biomass is triggered into activity by trace amounts of substrate. Soil Biol. Biochem. 33, 1163–1170.

Demoling, F., Figueroa, D., Baath, E., 2007. Comparison of factors limiting bacterial growth in different soils. Soil Biol. Biochem. 39, 2485–2495.

Dennis, P.G., Miller, A.J., Hirsch, P.R., 2010. Are root exudates more important than other sources of rhizodeposits in structuring rhizosphere bacterial communities? FEMS Microbiol. Ecol. 72, 313–327.

Denny, H.J., Wilkins, D.A., 1987. Zinc tolerance in *Betula* spp. IV. The mechanism of ectomycorrhizal amelioration of zinc toxicity. New Phytol. 106, 545–553.

Dhillion, S.S., 1992. Dual inoculation of pretransplant stage *Oryza sativa* L. plants with indigenous vesicular-arbucular mycorrhizal fungi and flourescent *Pseudomonas* spp. Biol. Fertil. Soils 13, 147–151.

Dickson, S., 2004. The Arum-Paris continuum of mycorrhizal symbioses. New Phytol. 163, 187–200.

Dietz, S., von Bulow, J., Beitz, E., Nehls, U., 2011. The aquaporin gene family of the ectomycorrhizal fungus *Laccaria bicolor*: lessons for symbiotic functions. New Phytol. 190, 927–940.

Dinkelaker, B., Marschner, H., 1992. In vivo demonstration of acid phosphatase activity in the rhizosphere of soil-grown plants. Plant Soil 144, 199–205.

Dosskey, M.G., Linderman, R.G., Boersma, L., 1990. Carbon-sink stimulation of photosynthesis in Douglas-fir seedlings by some ectomycorrhizas. New Phytol. 115, 269–274.

Dosskey, M.G., Boersma, L., Linderman, R.G., 1991. Role for the photosynthate demand of ectomycorrhizas in the response of Douglas fir seedlings to drying soil. New Phytol. 117, 327–334.

Douds Jr., D.D., Johnson, C.R., Koch, K.E., 1988. Carbon cost of the fungal symbiont relative to net leaf P accumulation in a split-root VA mycorrhizal symbiosis. Plant Physiol. 86, 491–496.

Dreher, D., Baldermann, S., Schreiner, M., Hause, B., 2019. An arbuscular mycorrhizal fungus and a root pathogen induce different volatiles emitted by *Medicago truncatula* roots. J. Adv. Res. 19, 85–90.

Dunn, R.M., Mikola, J., Bol, R., Bardgett, R.D., 2006. Influence of microbial activity on plant-microbial competition for organic and inorganic nitrogen. Plant Soil 289, 321–334.

Duponnois, R., Garbaye, J., 1991. Effect of dual inoculation of Douglas fir with the ectomycorrhizal fungus *Laccaria laccata* and mycorrhization helper bacteria (MHB) in two bare-root forest nurseries. Plant Soil 138, 169–176.

Duponnois, R., Plenchette, C., 2003. A mycorrhiza helper bacterium enhances ectomycorrhizal and endomycorrhizal symbiosis of Australian Acacia species. Mycorrhiza 13, 85–91.

Elberling, B., Askaer, L., Jørgensen, C.J., Joensen, H.P., Kühl, M.C., Glud, R.N.E.F., et al., 2011. Linking soil O_2, CO_2, and CH_4 concentrations in a wetland soil: implications for CO_2 and CH_4 fluxes. Environ. Sci. Technol. 45, 3393–3399.

Entry, J.A., Rose, C.L., Cromack, K., 1992. Microbial biomass and nutrient concentrations in hyphal mats of the ectomycorrhizal fungus *Hysterangium setchelii* in a coniferous forest soil. Soil Biol. Biochem. 24, 447–453.
Ezawa, T., Saito, K., 2018. How do arbuscular mycorrhizal fungi handle phosphate? New insight into fine-tuning of phosphate metabolism. New Phytol. 220, 1116–1121.
Faget, M., Blossfeld, S., von Gillhaussen, P., Schurr, U., Temperton, V.M., 2013. Disentangling who is who during rhizosphere acidification in root interactions: combining fluorescence with optode techniques. Front. Plant Sci. 4, 392.
Fang, Y.Y., Babourina, O., Rengel, Z., Yang, X.E., Pu, P.M., 2007. Spatial distribution of ammonium and nitrate fluxes along roots of wetland plants. Plant Sci 173, 240–246.
Fate, G., Chang, M., Lynn, D.G., 1990. Control of germination in *Striga asiatica*: chemistry of spatial definition. Plant Physiol. 93, 201–207.
Fitter, A.H., 1991. Costs and benefits of mycorrhizas: implications for functioning under natural conditions. Experientia 47, 350–355.
Fogel, R., 1988. Interactions among soil biota in coniferous ecosystems. Agric. Ecosyst. Environ. 24, 69–85.
Fogel, R., Hunt, G., 1979. Fungal and arboreal biomass in a western Oregon Douglas-fir ecosystem: distribution patterns and turnover. Can. J. For. Res. 9, 245–256.
Fracetto, G.G.M., Peres, L.E.P., Lambais, M.R., 2017. Gene expression analyses in tomato near isogenic lines provide evidence for ethylene and abscisic acid biosynthesis fine-tuning during arbuscular mycorrhiza development. Arch. Microbiol. 199, 787–798.
Gao, M.S., Teplitski, M., Robinson, J.B., Bauer, W.D., 2003. Production of substances by *Medicago truncatula* that affect bacterial quorum sensing. Mol. Plant-Microbe Inter. 16, 827–834.
Gao, Z.L., Karlsson, I., Geisen, S., Kowalchuk, G., Jousset, J., 2019. Protists: puppet masters of the rhizosphere microbiome. Trends Plant Sci. 24, 165–176.
Garcia, J.P., Wortmann, C.S., Mamo, M., Drijber, R., Tarkalson, D., 2007. One-time tillage of no-till: effects on nutrients, mycorrhizae, and phosphorus uptake. Agron. J. 99, 1093–1103.
Gardner, J.H., Malajczuk, N., 1988. Recolonization of rehabilitated bauxite mine sites in Western Australia by mycorrhizal fungi. For. Ecol. Manag. 24, 27–42.
Gavito, M.E., Jakobsen, I., Mikkelsen, T.N., Mora, F., 2019. Direct evidence for modulation of photosynthesis by an arbuscular mycorrhiza-induced carbon sink strength. New Phytol. 223, 896–907.
George, T.S., Simpson, R.J., Gregory, P.J., Richardson, A.E., 2007. Differential interaction of *Aspergillus niger* and *Peniophora lycii* phytases with soil particles affects the hydrolysis of inositol phosphates. Soil Biol. Biochem. 39, 793–803.
Gerdemann, J.W., 1975. Vesicular-arbuscular mycorrhizae. In: Torrey, J.C., Clarkson, D.T. (Eds.), The Development and Function of Roots. Academic Press, London, UK, pp. 575–591.
Glassop, D., Smith, S.E., Smith, F.W., 2005. Cereal phosphate transporters associated with the mycorrhizal pathway of phosphate uptake into roots. Planta 222, 688–698.
Gochnauer, M.B., Sealey, L.J., McCully, M.E., 1990. Do detached root-cap cells influence bacteria associated with maize roots? Plant Cell Env. 13, 793–801.
Gomes, N.C.M., Heuer, H., Schoenfeld, J., Costa, R., Mendoca-Hagler, L., Smalla, K., 2001. Bacterial diversity of the rhizosphere of maize (*Zea mays*) grown in tropical soil studied by temperature gradient gel electrophoresis. Plant Soil 232, 167–180.
González-Lopez, J., Martinez-Toledo, M.V., Reina, S., Salmeron, V., 1991. Root exudates of maize and production of auxins, gibberellins, cytokinins, amino acids and vitamins by *Azotobacter chroococcum* in chemically-defined media and dialysed-soil media. Technol. Environ. Chem. 33, 69–78.
Gourp, V., Pargney, J.-C., 1991. Immuno-cyto localisation des phosphatases acides de *Pisolithus tinctorius* L. lors de sa confrontation avec le système racinaire de *Pinus sylvestris* (Pers.) Desv. Cryptogam. Mycol. 12, 293–304.
Griffiths, R.P., Castellano, M.A., Caldwell, B.A., 1991. Hyphal mats formed by two ectomycorrhizal fungi and their association with Douglas-fir seedlings: a case study. Plant Soil 134, 255–259.
Guevara, R., Romero, I., 2004. Spatial and temporal abundance of mycelial mats in the soil of a tropical rain forest in Mexico and their effects on the concentration of mineral nutrients in soils and fine roots. New Phytol. 163, 361–370.
Gyaneshwar, P., Naresh Kumar, G., Parekh, L.J., Poole, P.S., 2002. Role of soil microorganisms in improving P nutrition of plants. Plant Soil 245, 83–93.
Hartwig, U.A., Joseph, C.M., Phillips, D.A., 1991. Flavonoids released naturally from alfalfa seeds enhance growth rate of *Rhizobium meliloti*. Plant Physiol. 95, 797–803.
Hauck, C., Müller, S., Schildknecht, H., 1992. A germination stimulant from parasitic flowering plants from Sorghum bicolor, a genuine host plant. J. Plant Physiol. 139, 474–478.
Häussling, M., Jorns, C.A., Lehmbecker, G., Hecht-Buchholz, C., Marschner, H., 1988. Ion and water uptake in relation to root development in Norway spruce (*Picea abies* (L.) Karst.). J. Plant Physiol. 133, 486–491.
Hawes, M.C., 1990. Living plant cells released from the root cap: a regulator of microbial populations in the rhizosphere. Plant Soil 129, 19–27.
Hawes, M.C., Bengough, G., Cassab, G., Ponce, G., 2002. Root caps and rhizosphere. J. Plant Growth Regul. 21, 352–367.
He, Y., Cornelissen, J.H.C., Wang, P., Dong, M., Ou, J., 2019. Nitrogen transfer from one plant to another depends on plant biomass production between conspecific and heterospecific species via a common arbuscular mycorrhizal network. Environ. Sci. Pollut. Res. 26, 8828–8837.
Hentschel, E., Godbold, D.L., Marschner, P., Schlegel, H., Jentschke, G., 1993. The effect of *Paxillus involutus* Fr. on aluminum sensitivity of Norway spruce seedlings. Tree Physiol. 12, 379–390.
Hestrin, R., Hammer, E.C., Mueller, C.W., Lehmann, J., 2019. Synergies between mycorrhizal fungi and soil microbial communities increase plant nitrogen acquisition. Commun. Biol. 2, 233.
Hetrick, B.A.D., 1991. Mycorrhizas and root architecture. Experientia 47, 355–362.
Hobbie, E.A., 2006. Carbon allocation to ectomycorrhizal fungi correlates with belowground allocation in culture studies. Ecology 87, 563–569.

Högberg, P., 1986. Soil nutrient availability, root symbioses and tree species composition in tropical Africa: a review. J. Trop. Ecol. 2, 359–372.
Horan, D.P., Chilvers, G.A., 1990. Chemotropism – the key to ectomycorrhizal formation? New Phytol. 116, 297–302.
Hortal, S., Plett, K.L., Plett, J.M., Cresswell, T., Johansen, M., Pendall, E., et al., 2017. Role of plant-fungal nutrient trading and host control in determining the competitive success of ectomycorrhizal fungi. ISME J. 11, 2666–2676.
Inselsbacher, E., Öhlund, J., Jämtgård, S., Huss-Danell, K., Näsholm, T., 2011. The potential of microdialysis to monitor organic and inorganic nitrogen compounds in soil. Soil Biol. Biochem. 43, 1321–1332.
Jacobsen, K.R., Fisher, D.G., Maretzki, A., Moore, P.H., 1992. Developmental changes in the anatomy of the sugarcane stem in relation to phloem unloading and sucrose storage. Bot. Acta 105, 70–80.
Jakobsen, I., Rosendahl, L., 1990. Carbon flow into soil and external hyphae from roots of mycorrhizal cucumber plants. New Phytol. 115, 77–83.
Jansa, J., Mozafar, A., Kuhn, G., Anken, T., Ruh, R., Sanders, I.R., et al., 2003. Soil tillage affects the community structure of mycorrhizal fungi in maize roots. Ecol. Appl. 13, 1164–1176.
Jasper, D.A., Robson, A.D., Abbott, L.K., 1979. Phosphorus and the formation of vesicular-arbuscular mycorrhiza. Soil Biol. Biochem. 11, 501–505.
Jasper, D.A., Abbott, L.K., Robson, A.D., 1989. Hyphae of a vesicular-arbuscular mycorrhizal fungus maintain infectivity in dry soil, except when the soil is disturbed. N. Phytol. 112, 101–107.
Jennings, D.H., 1987. Translocation of solutes in fungi. Biol. Rev. 62, 215–243.
Jentschke, G., Drexhage, M., Fritz, H.W., Fritz, E., Schella, B., Lee, D.H., et al., 2001. Does soil acidity reduce subsoil rooting in Norway spruce (*Picea abies*)? Plant Soil 237, 91–108.
Johnson, M.G., Tingey, D.T., Storm, M.J., Ganio, L.M., Phillips, D.L., 1997. Effects of elevated carbon dioxide and nitrogen fertilization on the life span of *Pinus ponderosa* fine roots. In: Flores, H.E., Lunch, J.P., Eissenstat, D. (Eds.), Radical Biology: Advances and Perspectives on the Function of Plant Roots. American Society of Plant Physiology, Rockville, MD, pp. 370–373.
Jones, A.M., 1990. Do we have the auxin receptor yet? Physiol. Plant. 80, 154–158.
Jones, D.L., 1998. Organic acids in the rhizosphere – a critical review. Plant Soil 205, 25–44.
Jones, D.L., Edwards, A.C., 1998. Influence of sorption on the biological utilization of two simple carbon substrates. Soil Biol. Biochem. 30, 1895–1902.
Jones, D.L., Edwards, A.C., Donachie, K., Darrah, P.R., 1994. Role of proteinaceous amino acids released in root exudates in nutrient acquisition from the rhizosphere. Plant Soil 158, 183–192.
Jones, M.D., Durall, D.M., Tinker, P.B., 1991. Fluxes of carbon and phosphorus between symbionts in willow ectomycorrhizas and their changes with time. New Phytol. 119, 99–106.
Jorquera, M., Martinez, O., Maruyama, F., Marschner, P., Mora, M.L., 2008. Current and future biotechnological applications of bacterial phytases and phytase-producing bacteria. Microbes Env. 23, 182–191.
Jousset, A., Rochat, L., Lanoue, A., Bonkowski, M., Keel, C., Scheu, S., 2011. Plants respond to pathogen infection by enhancing the antifungal gene expression of root-associated bacteria. Mol. Plant-Microbe Inter. 24, 352–358.
Khaosaad, T., Garcia-Garrido, J.M., Steinkeller, S., Vierheilig, H., 2007. Take-all disease is systematically reduced in roots of mycorrhizal barley plants. Soil Biol. Biochem. 39, 727–734.
Klironomos, J.N., 2003. Variation in plant response to native and exotic arbuscular mycorrhizal fungi. Ecology 84, 2292–2301.
Koehler, J., Yang, N., Pena, R., Raghavan, V., Polle, A., Meier, I.C., 2018. Ectomycorrhizal fungal diversity increases phosphorus uptake efficiency of European beech. New Phytol. 220, 1200–1210.
Kothari, S.K., Marschner, H., Römheld, V., 1990a. Contribution of the VA mycorrhizal hyphae in acquisition of phosphorus and zinc by maize grown in a calcareous soil. Plant Soil 131, 177–185.
Kothari, S.K., Marschner, H., Römheld, V., 1990b. Direct and indirect effects of VA mycorrhiza and rhizosphere microorganisms on mineral nutrient acquisition by maize (*Zea mays* L.) in a calcareous soil. New Phytol. 116, 637–645.
Kothari, S.K., Marschner, H., Römheld, V., 1991. Effect of a vesicular-arbuscular mycorrhizal fungus and rhizosphere microorganisms on manganese reduction in the rhizosphere and manganese concentrations in maize (*Zea mays* L.). New Phytol. 117, 649–655.
Kuzyakov, Y., Mason-Jones, K., 2018. Viruses in soil: Nano-scale undead drivers of microbial life, biogeochemical turnover and ecosystem functions. Soil Biol. Biochem. 127, 305–317.
Kuzyakov, Y., Xu, X.L., 2013. Competition between roots and microorganisms for nitrogen: mechanisms and ecological relevance. New Phytol. 198, 656–669.
Lambert, D.H., Weidensaul, T.C., 1991. Element uptake by mycorrhizal soybean from sewage-sludge-treated soil. Soil Sci. Soc. Am. J. 55, 393–397.
Lambers, H., Atkin, O.K., Millenaar, F.F., 2002. Respiratory patterns in roots in relation to their functioning. In: Waisel, Y., Eshel, A., Kafkafi, U. (Eds.), Plant Roots: The Hidden Half, 3rd ed. Marcel Dekker Inc, New York, pp. 521–552.
Lanfranco, L., Fiorilli, V., Gutjahr, C., 2018. Partner communication and role of nutrients in the arbuscular mycorrhizal symbiosis. New Phytol. 220, 1031–1046.
Le Bot, J., Kirkby, E.A., 1992. Diurnal uptake of nitrate and potassium during the vegetative growth of tomato plants. J. Plant Nutr. 15, 247–264.
Lehto, T., Zwiazek, J.J., 2011. Ectomycorrhizas and water relations of trees: a review. Mycorrhiza 21, 71–90.
Leyval, C., Berthelin, J., 1991. Weathering of a mica by roots and rhizosphere microorganisms of pine. Soil Sci. Soc. Am. J. 55, 1009–1016.
Li, X.L., George, E., Marschner, H., 1991a. Phosphorus depletion and pH decrease at the root-soil and hyphae-soil interfaces of VA mycorrhizal white clover fertilized with ammonium. N. Phytol. 119, 397–404.
Li, X.L., George, E., Marschner, H., 1991b. Extension of the phosphorus depletion zone in VA-mycorrhizal white clover in a calcareous soil. Plant Soil 136, 41–48.

Linderman, R.G., 1988. Mycorrhizal interactions with the rhizosphere microflora: the mycorrhizosphere effect. Phytopathology 78, 366–371.

Liu, H.W., Carvalhais, L.C., Crawford, M., Singh, E., Dennis, P.G., Pieterse, C.M.J., et al., 2017. Inner plant values: diversity, colonization and benefits from endophytic bacteria. Front. Microbiol. 8, 2552.

Lodge, D.J., 1989. The influence of soil moisture and flooding on formation of VA-*endo*- and ectomycorrhizae in *Populus* and *Salix*. Plant Soil 117, 243–253.

MacLean, A.M., Bravo, A., Harrison, M.J., 2017. Plant signaling and metabolic pathways enabling arbuscular mycorrhizal symbiosis. Plant Cell 29, 2319–2335.

Manske, C.G.B., 1989. Genetical analysis of the efficiency of VA mycorrhiza with spring wheat. Agric. Ecosyst. Environ. 29, 273–280.

Mardhiah, U., Caruso, T., Gurnell, A., Rillig, M., 2016. Arbuscular mycorrhizal fungal hyphae reduce soil erosion by surface water flow in a greenhouse experiment. Appl. Soil Ecol. 99, 137–140.

Marschner, P., Baumann, K., 2003. Changes in bacterial community structure induced by mycorrhizal colonisation in split-root maize. Plant Soil 251, 279–289.

Marschner, P., Timonen, S., 2004. Interactions between plant species and mycorrhizal colonization on the bacterial community composition in the rhizosphere. Appl. Soil Ecol. 28, 23–36.

Marschner, P., Asher, J.S., Graham, R.D., 1991. Effect of manganese-reducing rhizosphere bacteria on the growth of *Gaeumannomyces graminis* var. *tritici* and on manganese uptake by wheat (*Triticum aestivum* L.). Biol. Fertil. Soils 12, 33–38.

Marschner, P., Crowley, D.E., Lieberei, R., 2001. Arbuscular mycorrhizal infection changes the bacterial 16S rDNA community composition in the rhizosphere of maize. Mycorrhiza 11, 297–302.

Marschner, P., Crowley, D.E., Yang, C.H., 2004. Development of specific rhizosphere bacterial communities in relation to plant species, nutrition and soil type. Plant Soil 261, 199–208.

Marschner, P., Crowley, D., Rengel, Z., 2011. Rhizosphere interactions between microorganisms and plants govern iron and phosphorus acquisition along the root axis — model and research methods. Soil Biol. Biochem. 43, 883–894.

Martin, P., 1989. Long-distance transport and distribution of potassium in crop plants. Proc. 21st Coll. Int. Potash Inst. Bern. pp. 83–100.

Matthews, A., Pierce, S., Hipperson, H., Raymond, B., 2019. Rhizobacterial community assembly patterns vary between crop species. Front. Microbiol. 10, 581.

McGuiness, P.N., Reid, J.B., Foo, E., 2019. The role of gibberellins and brassinosteroids in nodulation and arbuscular mycorrhizal associations. Front. Plant Sci. 10, 269.

Merrild, M.P., Ambus, P., Rosendahl, S., Jacobsen, I., 2013. Common arbuscular mycorrhizal networks amplify competition for phosphorus between seedlings and established plants. N. Phytol. 200, 229–240.

Miethling, R., Wieland, G., Backhaus, H., Tebbe, C.C., 2000. Variation of microbial rhizosphere communities in response to crop species, soil origin, and inoculation with *Sinorhizobim meliloti* L33. Microb. Ecol. 40, 43–56.

Miethling, R., Ahrends, K., Tebbe, C.C., 2003. Structural differences in the rhizosphere communities of legumes are not equally reflected in community-level physiological profiles. Soil Biol. Biochem. 35, 1405–1410.

Miller, M.H., McGonigle, T.P., 1992. Soil disturbance and the effectiveness of arbuscular mycorrhizas in an agricultural ecosystem. In: Read, D.J., Lewis, D.H., Fitter, A.H., Alexander, I.J. (Eds.), Mycorrhizas in Ecosystems. CAB International, Wallingford, UK, pp. 156–163.

Miozzi, L., Vaira, A.M., Catoni, M., Fiorilli, V., Accotto, G.P., Lanfranco, L., 2019. Arbuscular mycorrhizal symbiosis: plant friend or foe in the fight against viruses. Front. Microbiol. 10, 1238.

Mo, Y., Wang, Y., Yang, R., Zheng, J., Liu, C., Li, H., et al., 2016. Regulation of plant growth, photosynthesis, antioxidation and osmosis by an arbuscular mycorrhizal fungus in watermelon seedlings under well-watered and drought conditions. Front. Plant Sci. 7, 644.

Mortimer, P.E., Pérez-Fernández, M.A., Valentine, A.J., 2008. The role of arbuscular mycorrhizal colonization in the carbon and nutrient economy of the tripartite symbiosis with nodulated *Phaseolus vulgaris*. Soil Biol. Biochem. 40, 1019–1027.

Mustafa, G., Khong, N.G., Tisserant, B., Randoux, B., Fontaine, J., Magnin-Robert, M., et al., 2017. Defence mechanisms associated with mycorrhiza-induced resistance in wheat against powdery mildew. Funct. Plant Biol. 44, 443–454.

Nair, A., Kolet, S.P., Thulasiram, H.V., Bhargava, S., 2015. Systemic jasmonic acid modulation in mycorrhizal tomato plants and its role in induced resistance against *Alternaria alternata*. Plant Biol. 17, 625–631.

Nehls, U., Plassard, C., 2018. Nitrogen and phosphate metabolism in ectomycorrhizas. New Phytol. 220, 1047–1058.

Neilands, J.B., 1984. Siderophores of bacteria and fungi. Microbiol. Sci. 1, 9–14.

Ngosong, C., Gabriel, E., Ruess, L., 2014. Collembola grazing on arbuscular mycorrhiza fungi modulates nutrient allocation in plants. Pedobiologia 57, 171–179.

Nguyen, C., Guckert, A., 2001. Short-term utilisation of ^{14}C(U)glucose by soil microorganisms in relation to carbon availability. Soil Biol. Biochem. 33, 53–60.

Nieto, K.F., Frankenberger Jr., W.T., 1990. Influence of adenine, isopentyl alcohol and *Azotobacter chroococcum* on the growth of *Raphanus sativus*. Plant Soil 127, 147–157.

Northup, R.R., Yu, Z., Dahlgren, R.A., Vogt, K.A., 1995. Polyphenol control of nitrogen release from pine litter. Nature 377, 227–229.

O'Connor, P.J., Smith, S.E., Smith, F.A., 2001. Arbuscular mycorrhizas influence plant diversity and community structure in a semi-arid herbland. New Phytol. 154, 209–218.

Oliver, A.J., Smith, S.E., Nicholas, D.J.D., Wallace, W., Smith, F.A., 1983. Activity of nitrate reductase in *Trifolium subterraneum* — effects of mycorrhizal infection and phosphate nutrition. N. Phytol. 94, 63–79.

Ono, F., Frommer, W.B., Wirén, N., 2000. Coordinated diurnal regulation of low- and high-affinity nitrate transporters in tomato. Plant Biol. 2, 17–23.

Orlovich, D.A., Ashford, A.E., Cox, G.C., 1989. A reassessment of polyphosphate granule composition in the ecto-mycorrhizal fungus *Pisolithus tinctorius*. Aust. J. Plant Physiol. 16, 107–115.

Pacovsky, R.S., Fuller, G., Paul, E.A., 1985. Influence of soil on the interactions between endomycorrhizae and *Azospirillum* in sorghum. Soil Biol. Biochem. 17, 525–531.

Parra-Garcia, M.D., Lo Giudice, V., Ocampo, J.A., 1992. Absence of VA colonization in *Oxalis pes-caprae* inoculated with *Glomus mosseae*. Plant Soil 145, 298–300.

Piromyou, P., Songwattana, P., Greetatorn, T., Okubo, T., Kakizaki, K.C., Prakamhang, J., et al., 2015. The type III secretion system (T3SS) is a determinant for rice-endophyte colonization by non-photosynthetic *Bradyrhizobium*. Microbes Env. 30, 291–300.

Plassard, C., Scheromm, P., Mousain, D., Salsac, L., 1991. Assimilation of mineral nitrogen and ion balance in the two partners of ectomycorrhizal symbiosis: data and hypothesis. Experientia 47, 340–349.

Plett, J.M., Khachane, A., Ouassou, M., Sundberg, B., Kohler, A., Martin, F., 2014. Ethylene and jasmonic acid act as negative modulators during mutualistic symbiosis between *Laccaria bicolor* and *Populus* roots. New Phytol. 202, 270–286.

Postma, J., Hok, A.H.C.H., Oude Voshaar, J.H., 1990. Influence of the inoculum density on the growth and survival of *Rhizobium leguminosarum* biovar *trifolii* introduced into sterile and non-sterile loamy sand and silt loam. FEMS Microb. Ecol. 73, 49–58.

Rahman, M.H., Saiga, S., 2007. Endophyte effects on nutrient acquisition in tall fescue grown in andisols. J. Plant Nutr. 30, 2141–2158.

Raju, P.S., Clark, R.B., Ellis, J.R., Maranville, J.W., 1990. Effects of species of VA-mycorrhizal fungi on growth and mineral uptake of sorghum at different temperatures. Plant Soil 121, 165–170.

Ravanbakhsh, M., Sasidharan, R., Voesenek, L., Kowalchuk, G.A., Jousset, A., 2018. Microbial modulation of plant ethylene signaling: ecological and evolutionary consequences. Microbiome 6, 52.

Read, D.J., 1991. Mycorrhizas in ecosystems. Experientia 47, 376–391.

Reinhold-Hurek, B., Buenger, W., Burbano, C.S., Sabale, M., Hurek, T., 2015. Roots shaping their microbiome: global hotspots for microbial activity. Annu. Rev. Phytopathol. 53, 403–424.

Rengel, Z., 1997. Root exudation and microflora populations in rhizosphere of crop genotypes differing in tolerance to micronutrient deficiency. Plant Soil 196, 255–260.

Rengel, Z., Graham, R.D., Pedler, J.F., 1993. Manganese nutrition and accumulation of phenolics and lignin as related to differential resistance of wheat genotypes to the take-all fungus. Plant Soil 151, 255–263.

Reynolds, H.L., Hartley, A.E., Vogelsang, K.M., Bever, J.D., Schultz, P.A., 2005. Arbuscular mycorrhizal fungi do not enhance nitrogen acquisition and growth of old-field perennials under low nitrogen supply in glasshouse culture. New Phytol. 167, 869–880.

Richardson, A.E., Hadobas, P.A., 1998. Soil isolates of *Pseudomonas* spp. that utilize inositol phosphates. Can. J. Microbiol. 43, 509–516.

Rillig, M.C., Aguilar-Trigueros, C.A., Camenzind, T., Cavagnaro, T.R., Degrune, F., Hohmann, P., et al., 2019. Why farmers should manage the arbuscular mycorrhizal symbiosis. New Phytol. 222, 1171–1175.

Rivero, J., Lidoy, J., Llopis-Giminez, A., Herrero, S., Flors, V., Pozo, M.J., 2021. Mycorrhizal symbiosis primes the accumulation of antiherbivore compounds and enhances herbivore mortality in tomato. J. Exp. Bot. 72, 5038–5050.

Rodriguez, R.J., White, J.F., Arnold, A.E., Redman, R.S., 2009. Fungal endophytes: diversity and functional roles. N. Phytol. 182, 314–330.

Römheld, V., 1991. The role of phytosiderophores in acquisition of iron and other micronutrients in graminaceous species: an ecological approach. Plant Soil 130, 127–134.

Roth, R., Paszkowski, U., 2017. Plant carbon nourishment of arbuscular mycorrhizal fungi. Curr. Opin. Plant Biol. 39, 50–56.

Rousseau, J.V.D., Sylvia, D.M., Fox, A.J., 1994. Contribution of ectomycorrhiza to the potential nutrient-absorbing surface of pine. N. Phytol. 128, 639–644.

Rudolph, N., Voss, S., Moradi, A.B., Nagl, S., Oswald, S.E., 2013. Spatiotemporal mapping of local soil pH changes induced by roots of lupin and soft-rush. Plant Soil 369, 669–680.

Rudolph-Mohr, N., Vontobel, P., Oswald, S.E., 2014. A multi-imaging approach to study the root-soil interface. Ann. Bot. 114, 1779–1787.

Ruth, B., Khalvati, M., Schmidhalter, U., 2011. Quantification of mycorrhizal water uptake via high-resolution on-line water content sensors. Plant Soil 342, 459–468.

Ryan, M.H., Graham, J.H., 2018. Little evidence that farmers should consider abundance or diversity of arbuscular mycorrhizal fungi when managing crops. N. Phytol. 220, 1092–1107.

Same, B.I., Robson, A.D., Abbott, L.K., 1983. Phosphorus, soluble carbohydrates and endomycorrhizal infection. Soil Biol. Biochem. 15, 593–597.

Sanders, F.E., 1993. Modelling plant growth responses to vesicular arbuscular mycorrhizal infection. Adv. Plant Pathol. 9, 135–166.

Santander, C., Aroca, R., Ruiz-Lozano, J.M., Olave, J., Cartes, P., Borie, F., et al., 2017. Arbuscular mycorrhiza effects on plant performance under osmotic stress. Mycorrhiza 27, 639–657.

Sarniguet, A., Lucas, P., Lucas, M., 1992a. Relationships between take-all, soil conduciveness to the disease, populations of fluorescent pseudomonads and nitrogen fertilizer. Plant Soil 145, 17–27.

Sarniguet, A., Lucas, P., Lucas, M., Samson, R., 1992b. Soil conduciveness to take-all of wheat: influence of the nitrogen fertilizers on the structure of populations of fluorescent pseudomonads. Plant Soil 145, 29-26.

Sawers, R.J.H., Svane, S.F., Quan, C., Gronlund, M., Wozniak, B., Gebreselassie, M.N., et al., 2017. Phosphorus acquisition efficiency in arbuscular mycorrhizal maize is correlated with the abundance of root-external hyphae and the accumulation of transcripts encoding PHT1 phosphate transporters. New Phytol. 214, 632–643.

Sawers, R.J.H., Ramirez-Flores, M.R., Olaide-Portugal, V., Paszkowski, U., 2018. The impact of domestication and crop improvement on arbuscular mycorrhizal symbiosis in cereals: insights from genetics and genomics. New Phytol. 220, 1135–1140.

Schmalenberger, A., Duran, A.L., Bray, A.W., Bridge, J., Bonneville, S., Benning, L.G., et al., 2015. Oxalate secretion by ectomycorrhizal *Paxillus involutus* is mineral-specific and controls calcium weathering from minerals. Sci. Rep. 5, 12187.

Schouteden, N., De Waele, D., Panis, B., Vos, C.M., 2015. Arbuscular mycorrhizal fungi for the biocontrol of plant-parasitic nematodes: a review of the mechanisms involved. Front. Microbiol. 6, 1280.

Schweiger, P., Robson, A., Barrow, J., 2002. Root hair length determines beneficial effect of a *Glomus* sp. on shoot growth of some pasture species. New Phytol. 131, 247–254.

Schweiger, P.F., Robson, A.D., Barrow, N.J., Abbott, L.K., 2007. Arbuscular mycorrhizal fungi from three genera induce two-phase plant growth responses on a high P-fixing soil. Plant Soil 292, 181–192.

Segal, E., Kushnir, T., Mualem, Y., Shani, U., 2008. Microsensing of water dynamics and root distributions in sandy soils. Vadose Zone J. 7, 1018–1026.

Sessitsch, A., Hardoim, P., Döring, J., Weilharter, A., Krause, A., Woyke, T.B., et al., 2012. Functional characteristics of an endophyte community colonizing rice roots as revealed by metagenomic analysis. Mol. Plant-Microbe Inter. 25, 28–36.

Shi, W.G., Zhang, Y.H., Chen, S.L., Polle, A., Rennenberg, H., Luo, Z.B., 2019. Physiological and molecular mechanisms of heavy metal accumulation in nonmycorrhizal vs mycorrhizal plants. Plant Cell Env. 42, 1087–1103.

Simard, S.W., Perry, D.A., Jones, M.D., Myrold, D.D., Durall, D.M., Molina, R., 1997. Net transfer of carbon between ectomycorrhizal tree species in the field. Nature 388, 579–582.

Smalla, K., Wieland, G., Buchner, A., Zock, A., Parzy, J., Kaiser, S., et al., 2001. Bulk and rhizosphere soil bacterial communities studied by denaturing gradient gel electrophoresis: plant-dependent enrichment and seasonal shifts revealed. Appl. Environ. Microbiol. 67, 4742–4751.

Smith, S.E., Read, D.J., 2008. Mycorrhizal symbiosis. Academic Press, Amsterdam.

Smith, S.E., Smith, F.A., Jacobsen, I., 2003. Mycorrhizal fungi can dominate phosphate supply to plants irrespective of growth responses. Plant Physiol. 133, 16–20.

Smith, S.E., Gianinazzi-Pearson, V., 1988. Physiological interactions between symbionts in vesicular-arbuscular mycorrhizal plants. Ann. Rev. Plant Physiol. Plant Biol. 39, 221–244.

Smith, S.E., Jacobsen, I., Gronlund, M., Smith, F.A., 2011. Roles of arbuscular mycorrhizas in plant phosphorus nutrition: Interactions between pathways of phosphorus uptake in arbuscular mycorrhizal roots have important implications for understanding and manipulating plant phosphorus acquisition. Plant Physiol. 156, 1050–1057.

Söderström, B., 1992. The ecological potential of the ectomycorrhizal mycelium. In: Read, D.J., Lewis, D.H., Fitter, A.H., Alexander, I.J. (Eds.), Mycorrhizas in Ecosystems. C.A.B. International, Wallingford, UK, pp. 77–83.

Solaiman, M.Z., Marschner, P., Wang, D., Rengel, Z., 2007. Growth, P uptake and rhizosphere properties of wheat and canola genotypes in an alkaline soil with low P availability. Biol. Fertil. Soils 44, 143–153.

Son, C., Smith, S.E., 1988. Mycorrhizal growth responses: interactions between photon irradiance and phosphorus nutrition. N. Phytol. 108, 305–314.

Spohn, M., Kuzyakov, Y., 2013. Distribution of microbial- and root-derived phosphatase activities in the rhizosphere depending on P availability and C allocation - Coupling soil zymography with ^{14}C imaging. Soil Biol. Biochem. 67, 106–113.

Strömberg, N., 2008. Determination of ammonium turnover and flow patterns close to roots using imaging optodes. Environ. Sci. Technol. 42, 1630–1637.

Svenningsen, N.B., Watts-Williams, S.J., Joner, E.J., Battini, F., Efthymiou, A., Cruz-Paredes, C., et al., 2018. Suppression of the activity of arbuscular mycorrhizal fungi by the soil microbiota. ISME J. 12, 1296–1307.

Tarafdar, J.C., Jungk, A., 1987. Phosphatase activity in the rhizosphere and its relation to the depletion of soil organic phosphorus. Biol. Fertil. Soils 3, 199–204.

Tarafdar, J.C., Marschner, H., 1994. Phosphatase activity in the rhizosphere of VA-mycorrhizal wheat supplied with inorganic and organic phosphorus. Soil Biol. Biochem. 26, 387–395.

Taylor, J.H., Peterson, C.A., 2005. Ectomycorrhizal impacts on nutrient uptake pathways in woody roots. New For. 30, 203–214.

Teplitski, M., Robinson, J.B., Bauer, W.D., 2000. Plants secrete compounds that mimic bacterial N-acyl homoserine lactone signal activities and affect population density-dependent behaviours in associated bacteria. Mol. Plant-Microbe Inter. 13, 637–648.

Teste, F.P., Jones, M.D., Dickie, I.A., 2020. Dual-mycorrhizal plants: their ecology and relevance. N. Phytol. 225, 1835–1851.

Thompson, J.P., 1987. Decline of vesicular-arbuscular mycorrhizae in long fallow disorder of field crops and its exrpression in phosphorus deficiency of sunflower. Aust. J. Agric. Res. 38, 847–867.

Timonen, S., Marschner, P., 2005. Mycorrhizosphere concept. In: Mukerji, K.G., Manoharachary, C., Singh, J. (Eds.), Microbial Activity in the Rhizosphere. Springer, Heidelberg, Germany, pp. 155–172.

Tinker, P.B., Jones, M.D., Durall, D.M., 1992. A functional comparison of ecto- and endomycorrhizas. In: Read, D.J., Lewis, D.H., Fitter, A.H., Alexander, I.J. (Eds.), Mycorrhizas in Ecosystems. CAB International, Wallingford, UK, pp. 303–310.

Uetake, Y., Kojima, T., Ezawa, T., Saito, M., 2002. Extensive tubular vacuole system in an arbuscular mycorrhizal fungus, *Gigaspora margarita*. New Phytol. 154, 761–768.

van der Heijden, M.G.A., Horton, T.R., 2009. Socialism in soil? The importance of mycorrhizal fungal networks for facilitation in natural ecosystems. J. Ecol. 97, 1139–1150.

Van der Heijden, M.G.A., Klironomos, J.N., Ursic, M., Moutoglis, P., Streitwolf-Engel, R., Boller, T., et al., 1998. Mycorrhizal fungal diversity determines plant biodiversity, ecosystem variability and productivity. Nature 396, 69–72.

Van Scholl, L., Smits, M.M., Hoffland, E., 2006. Ectomycorrhizal weathering of the soil minerals muscovite and hornblende. New Phytol. 171, 805–814.

Vasutova, M., Mleczko, P., Lopez-Garcia, A., Macek, I., Boros, G., Sevicik, J., et al., 2019. Taxi drivers: the role of animals in transporting mycorrhizal fungi. Mycorrhiza 29, 413–434.

Venturi, V., Keel, C., 2013. Signaling in the rhizosphere. Trends Plant Sci. 2016 21, 187–198.

Von Luetzow, M., Koegel-Knabner, I., Ekschmitt, K., Matzner, E., Guggenberger, G., Marschner, B., et al., 2006. Stabilisation of organic matter in temperate soils: mechanisms and their relevance under different soil conditions - a review. Eur. J. Soil Sci. 57, 426–445.

Vos, C.M., Tesfahun, A.N., Panis, B., De Waele, D., Elsen, A., 2012. Arbuscular mycorrhizal fungi induce systemic resistance in tomato against the sedentary nematode *Meloidogyne incognita* and the migratory nematode *Pratylenchus penetrans*. Appl. Soil Ecol. 61, 1–6.

Walder, F., van der Heijden, M.G.A., 2015. Regulation of resource exchange in the arbuscular mycorrhizal symbiosis. Nat. Plants 3, 159.

Wallander, H., 2000. Uptake of P from apatite by *Pinus sylvestris* seedlings colonised by different ectomycorrhizal fungi. Plant Soil 218, 249–256.

Wallander, H., 2002. Utilization of organic nitrogen at two different substrate pH by different ectomycorrhizal fungi growing in symbiosis with *Pinus sylvestris* seedlings. Plant Soil 243, 23–30.

Wallander, H., Nylund, J.E., 1991. Effects of excess nitrogen on carbohydrate concentration and mycorrhizal development of *Pinus sylvestris* L. seedlings. New Phytol. 119, 405–411.

Wang, X., Zhao, S., Buecking, H., 2016. Arbuscular mycorrhizal growth responses are fungal specific but do not differ between soybean genotypes with different phosphate efficiency. Ann. Bot. 118, 11–21.

Warwick, C., Guerreiro, A., Soares, A., 2013. Sensing and analysis of soluble phosphates in environmental samples: a review. Biosens. Bioelectron. 41, 1–11.

Waters, C.M., Bassler, B.L., 2005. Quorum sensing: communication in bacteria. Annu. Rev. Cell Dev. Biol. 21, 319–346.

Watt, M., McCully, M.E., Kirkegaard, J.A., 2003. Soil strength and rate of root elongation alter the accumulation of *Pseudomonas* spp. and other bacteria in the rhizosphere of wheat. Funct. Plant Biol. 30, 483–491.

Watts-Williams, S.J., Tyerman, S.D., Cavagnaro, T.R., 2017. The dual benefit of arbuscular mycorrhizal fungi under soil zinc deficiency and toxicity: linking plant physiology and gene expression. Plant Soil 420, 375–388.

Weisskopf, L., Abou-Mansour, E., Fromin, N., Tomasi, N., Santelia, D., Edelkott, I., et al., 2006. White lupin has developed a complex strategy to limit microbial degradation of secreted citrate required for phosphate acquisition. Plant Cell Env. 29, 919–927.

Weremijewicz, J., Janos, D.P., 2013. Common mycorrhizal networks amplify size inequality in *Andropogon gerardii* populations. New Phytol. 198, 203–213.

Wilkins, D.A., 1991. The influence of sheathing (ecto-) mycorrhizas of trees on the uptake and toxicity of metals. Agric. Ecosyst. Environ. 35, 245–260.

Wipf, D., Krajinski, F., van Tuinen, D., Recorbet, G., Courty, P.E., 2019. Trading on the arbuscular mycorrhiza market: from arbuscules to common mycorrhizal networks. New Phytol. 223, 1127–1142.

Wright, S.F., Jawson, L., 2001. A pressure cooker method to extract glomalin from soils. Soil Sci. Soc. Am. J. 65, 1734–1735.

Wright, D.P., Scholes, J.D., Read, D.J., 1998. Effects of VA mycorrhizal colonization on photosynthesis and biomass production of *Trifolium repens* L. Plant Cell Env. 21, 209–216.

Xie, X.N., Yoneyama, K., Yoneyama, K., 2010. The strigolactone story. Annu. Rev. Phytopath. 48, 93–117.

Xu, X., Stange, C.F., Richter, A., Wanek, W., Kuzyakov, Y., 2008. Light affects competition for inorganic and organic nitrogen between maize and rhizosphere microorganisms. Plant Soil 304, 59–72.

Xu, L.J., Li, T., Wu, Z.X., Feng, H.Y., Yu, M., Zhang, X., et al., 2018. Arbuscular mycorrhiza enhances drought tolerance of tomato plants by regulating the 14-3-3 genes in the ABA signalling pathway. Appl. Soil Ecol. 125, 213–221.

Yang, C.H., Crowley, D.E., 2000. Rhizosphere microbial community structure in relation to root location and plant iron nutritional status. Appl. Environ. Microbiol. 66, 345–351.

Yehuda, Z., Shenker, M., Römheld, V., Marschner, H., Hadar, Y., Chen, Y., 1996. The role of ligand exchange in the uptake of iron from microbial siderophores by gramineous plants. Plant Physiol. 112, 1273–1280.

York, L.M., Carminati, A., Mooney, S.J., Ritz, K., Bennett, M.J., 2016. The holistic rhizosphere: integrating zones, processes, and semantics in the soil influenced by roots. J. Exp. Bot. 67, 3629–3643.

Zahir, Z.A., Ghani, U., Naveed, M., Nadeem, S.M., Asghar, H.N., 2009. Comparative effectiveness of *Pseudomonas* and *Serratia* sp. containing ACC-deaminase for improving growth and yield of wheat (*Triticum aestivum* L.) under salt-stressed conditions. Arch. Microbiol. 191, 415–424.

Zhang, N., Yang, D., Wang, D., Miao, Y., Shao, J., Zhou, X., et al., 2015. Whole transcriptomic analysis of the plant beneficial rhizobacterium *Bacillus amyloliquefaciens* SQR9 during enhanced biofilm formation regulated by maize root exudates. BMC Genomics 16, 685.

Zhang, Z.L., Yuan, Y.S., Liu, Q., Yin, H.J., 2019. Plant nitrogen acquisition from inorganic and organic sources via roots and mycelia pathways in ectomycorrhizal alpine forests. Soil Biol. Biochem. 136, 107517.

Zhu, Y.G., Smith, S.E., Baritt, A.R., Smith, F.A., 2001. Phosphorus (P) efficiencies and mycorrhizal responsiveness of old and modern wheat cultivars. Plant Soil 237, 249–255.

Chapter 16

Nitrogen fixation

Mariangela Hungria[1,2] and Marco Antonio Nogueira[1,2]
[1]Embrapa Soja, Soil Biotechnology Laboratory, Londrina, Paraná, Brazil, [2]INCT-Plant-Growth Promoting Microorganisms for Agricultural Sustainability and Environmental Responsibility, Brazil

Summary

This chapter begins with an assessment of the contribution of biological N_2 fixation (BNF) to the N economy of terrestrial ecosystems, and the diversity of N_2-fixing systems. The following sections address the biochemistry of nitrogenase and the significance of symbiotic microorganisms as suppliers of N to higher plants. Legume–rhizobia symbioses are treated in detail: range of partnerships, interactive signaling, root infection, nodule formation and metabolism. The influence of mineral N and of macro and micronutrients on BNF is presented together with the main environmental limitations (soil acidity and salinity, temperature, water stress) to a successful symbiosis. The contribution of nonsymbiotic diazotrophs to global agriculture is discussed. Methods for quantification of the amounts of fixed N in legumes in single crop, intercropped or in crop rotations are presented. The chapter finishes by discussing the increasing global perception of the importance of BNF, the growing inoculant market, and some research challenges for the next decades.

16.1 General

Nitrogen (N) is required in larger amounts for plant production than any other nutrient and is almost always the most limiting in agroecosystems. Nitrogen is structural component of DNA, RNA, chlorophyll, amino acids, and proteins. In the biogeochemical cycle, four N sources participate in N balance, including abiotic nitrogen fixation (e.g., by lightning and volcanism) and soil N (mineralization of soil organic matter, a limited resource in most soils), both of which are far behind the third source, the synthetic fertilizers. The fourth and environmentally most important one is biological nitrogen fixation (BNF), which represents a key supply route in both natural and managed ecosystems, being considered the second most important biological process on Earth after photosynthesis (de Bruijn, 2015; Stewart and Lal, 2017).

Modern agriculture with high-yielding crops that have high nutrient demands is not practicable without N fertilizers. They are produced by the catalytic reduction of nonreactive atmospheric nitrogen (N_2) to ammonia (NH_3) under high temperature (350°C–550°C) and pressure (15–35 MPa) in the industrial process named Haber-Bosch (H-B) ($N_2 + 3H_2 \rightarrow 2NH_3$); this process impacted agriculture so much that both inventors received the Nobel Prizes (Fritz Haber in 1918 for inventing it and Carl Bosch in 1931 for improving it and putting it into production). The importance of large-scale application of synthetic N fertilizers to increase rice (*Oryza sativa* L.) and wheat (*Triticum aestivum* L.) yields was highlighted and recognized worldwide for reducing world hunger and malnutrition, saving millions of lives since the 1960s in the Green Revolution that also led to the Nobel Prize awarded to Norman Borlaug in 1970 (Smil, 2001; Stewart and Lal, 2017).

Nitrogen fertilizers have high economic and environmental costs. The first limitation is the high-energy requirement of the industrial N_2 reduction, representing more than 50% of the total energy used in commercial agriculture (Woods et al., 2010). Alternative sources of energy and improvement of synthesis efficiency are needed and have been searched for, but there is still no prospect of significant innovation.

The second limitation is the emission of greenhouse gases (GHG), comprising both carbon dioxide (CO_2) released from fossil fuel combustion and nitrous oxide (N_2O) (estimated 1% of applied fertilizer N is lost to the atmosphere as N_2O) that

---

☆ This chapter is a revision of the third edition chapter by J.E. Cooper and H.W. Scherer, pp. 389–408. DOI: https://doi.org/10.1016/B978-0-12-384905-2.00016-9. © Elsevier Ltd.

has around 300 times more impact than CO_2 (Stewart and Lal, 2017; Woods et al., 2010). Based on the methodology proposed by Van Amstel (2006) and considering the direct and indirect GHG emissions in the process of synthesis, transport, and application of N fertilizers, we estimate approximately 10 kg CO_2-eq kg^{-1} of N fertilizer.

Third, the N-use efficiency of N fertilizers by plants is low (30%–70%); losses occur not only by emissions to the atmosphere, but mainly by leaching, resulting in a variety of environmental problems, such as eutrophication of water bodies (Bindraban et al., 2015; Stewart and Lal, 2017). Concerns about N fertilizers are so great that Keeler et al. (2016) proposed an estimate of the social cost of nitrogen, defined as the monetary damage caused by the use of N, that would be larger than the social cost of carbon due to the potential damage to water quality in addition to climate change. Therefore, research aiming to minimize N fertilizer use is needed by improving plant use efficiency, not only by determining the right time, place, amount, and composition of the applied N fertilizer for each crop, but also by acquiring better physiological knowledge on the N translocation and metabolism in plants (Bindraban et al., 2015).

In contrast to the industrial production of N fertilizers, the biological reduction of N_2 to NH_3 takes place at ambient temperature and subambient pressure and can be performed by a few prokaryotes that possess a key enzyme, nitrogenase. The BNF occurs in all terrestrial ecosystems as well as in oceans (de Bruijn, 2015). Despite the difficulties in quantifying BNF on a global scale, mainly due to inaccurate statistics on areas under legume cultivation and the sparseness of data on BNF for nonlegume crops and natural ecosystems (Cleveland et al., 1999; Herridge et al., 2008), estimates suggest inputs from 107 Mt N $year^{-1}$ (Galloway et al., 2004) to 195 Mt N $year^{-1}$ (Cleveland et al., 1999) to natural terrestrial ecosystems, and a range from 40 (Galloway et al., 2004) to 50–70 Mt N $year^{-1}$ (Herridge et al., 2008) into agricultural systems. For comparison, in 2014, the world fertilizer N usage was estimated at 109 Mt (Stewart and Lal, 2017).

16.2 Biological nitrogen-fixing systems

The ability to fix atmospheric N_2 to NH_3 is restricted to a small subset of taxonomically diverse prokaryotes. They are called diazotrophs (*di* = two; *azote* = nitrogen; and *trophicos* = food) and inhabit not only external (soils, rivers, lakes, and oceans) but also internal environments (e.g., insects, cattle rumen, and human intestine). However, the presence of diazotrophs is not necessarily related to BNF, as very specific conditions are necessary for the process to take place. Some are symbionts of plants, animals, and protists. When associated with higher plants, they have been classified as symbiotic or associative (rhizospheric or endophytic) depending on the relationship with host (Ormeño-Orrillo et al., 2013). Symbionts are the most significant N_2 fixers for plants; they are represented primarily by rhizobia (α- and β-Proteobacteria), *Frankia* (Actinobacteria), and *Nostoc/Anabaena* (Cyanobacteria). These organisms live in specialized structures where they have access to energy from plant photosynthates, and an environment that is conducive to both the nitrogenase activity [e.g., via a protective mechanism against excessive oxygen (O_2)] and the translocation of fixed N to the host plant.

The symbioses between rhizobia and plants of the Fabaceae (old name: Leguminosae) family are widespread in managed and natural ecosystems. Traditionally, the Fabaceae family was divided into three subfamilies: Papilionoideae, Mimosoideae, and Caesalpinioideae; more recently, the last subfamily was subdivided further (Sprent et al., 2017). The Papilionoideae subfamily comprises the majority of the host symbionts, including agronomically important grain legumes such as soybean [*Glycine max* (L.) Merr], chickpea (*Cicer arietinum* L.), cowpea [*Vigna unguiculata* (L.) Walp], pigeon pea [*Cajanus cajan* (L.) Millsp.], etc., green manures such as jack bean [*Canavalia ensiformis* (L.) DC.], and forages such as alfalfa (*Medicago sativa* L.) (Table 16.1). *Parasponia* is the only nonlegume species that can establish N_2-fixing endosymbiosis with rhizobium, and it is the only nodulating plant in the Cannabaceae family.

In some forest and woodland ecosystems, most of the N input from BNF comes from nonrhizobial symbionts in root nodules of nonleguminous tree species, that is, Actinobacteria of the genus *Frankia* in symbiosis with woody perennials such as *Alnus*, *Casuarina*, and *Caenothus*. Approximately 200 species of woody shrubs and trees, mostly from temperate regions, form symbioses with *Frankia* (Vessey et al., 2005).

Members of the gymnosperm order Cycadales form cyanobacterial N_2-fixing root symbioses, mainly with filamentous *Nostoc* species (Rasmussen and Nilsson, 2002). *Nostoc* can also form a stem/petiole symbiosis with a tropical angiosperm *Gunnera*. In aquatic ecosystems, such as rice paddies, the symbiosis between the heterocyst-forming cyanobacterium *Anabaena azollae* and the fern *Azolla* represents an important source of BNF.

The roots of higher plants are colonized by other diazotrophs that do not reside in specialized organs but might, in some cases, invade the root cortex as endophytes; however, the majority of these relationships involve free-living bacteria in soil, or those growing on the rhizoplane or in the rhizosphere, with relatively low rates of N_2 fixation. There are numerous examples of associative bacteria harboring N_2-fixing capacity isolated from the rhizosphere; however, in many cases increased plant growth cannot solely be attributed to fixed N_2 (see Section 16.8) (Bashan and de-Bashan, 2010; Fukami et al., 2018).

TABLE 16.1 Estimated rates of biological N$_2$ fixation in some legumes.

Species	Common name	N$_2$ fixation rates (kg N ha^{-1} year^{-1})
Acacia spp.	Acacia	5–50
Arachis hypogaea	Peanut, groundnut	32–206
Cajanus cajan	Pigeon pea	68–88
Calliandra calothyrsus	Calliandra	24
Calopogonium mucunoides	Calopogonium	64–182
Centrosema spp.	Centrosema	41–280
Cicer arietinum	Chickpea	0–141
Crotalaria grahamiana	Rattlepod	142
Desmodium spp.	Desmodium	25–380
Gliricidia sepium	Gliricidia	26–75
Glycine max	Soybean	0–450
Lathyrus sativus	Lathyrus	172–227
Lens culinaris	Lentil	5–191
Leucaena leucocephala	Leucaena	98–274
Lupinus albus	White lupin	40–160
Lupinus angustifolius	Narrowleaf lupin	19–327
Lupinus mutabilis	Bitter lupin	95–527
Macroptilium atropurpureum	Siratro	46–167
Medicago sativa	Alfalfa	45–470
Melilotus officinalis	Yellow sweet clover	84
Neonotonia wightii	Perennial soybean	126
Phaseolus vulgaris	Common bean	0–165
Pisum sativum	Field pea	4–244
Pueraria phaseoloides	Tropical kudzu	115
Sesbania spp.	Sesbania	7–109
Stylosanthes spp.	Stylosanthes	4–263
Trifolium spp.	Clover	67–260
Vicia benghalensis	Vetch	125–147
Vicia faba	Faba bean	12–330
Vigna mango	Black gram	119–140
Vigna radiata	Green gram	58–107
Vigna unguiculata	Cowpea	9–201
Zornia glabra	Zornia	61

Source: Based on Giller (2001), Werner (2005), and Ormeño-Orrillo et al. (2013).

The BNF contributed by associative bacteria, such as *Azospirillum* sp., has been estimated at about 10 kg N ha^{-1} (Okon et al., 2015), representing 5%–18% of the total N accumulated by plants (Bashan and de-Bashan, 2010). Some diazotrophic bacteria have been classified as endophytes, including *Gluconacetobacter*, *Herbaspirillum*, and *Azoarcus*, having a closer relationship with plants and higher rates of BNF than non-endophytes (Section 16.8).

Free-living N₂ fixers are extensively distributed in soils, but in the case of heterotrophic bacteria, they are usually restricted in their BNF capacity by a lack of organic substrates for energy generation. Examples exist among aerobes (*Azotobacter*), anaerobes (*Clostridium*), and facultative anaerobes (*Klebsiella*). A few N₂ fixers occur among chemolithotrophic (e.g., *Acidithiobacillus ferrooxidans*) and photolithotrophic bacteria (e.g., *Chlorobium, Chromatium,* and *Rhodospirillum*). Some heterocyst-forming cyanobacteria (e.g., *Anabaena, Nostoc, Calothrix,* and *Cylindrospermum*) can also fix N₂ without symbiosis with an eukaryotic host.

16.3 Biochemistry of nitrogen fixation

Biological reduction of N_2 to NH_3 is a highly energy-demanding process with a minimum requirement of about 960 kJ mol^{-1} of fixed N (Sprent and Raven, 1985). The main reaction steps are the same in all N_2-fixing microorganisms (Fig. 16.1). The key enzyme, nitrogenase, is unique to N_2-fixing microorganisms and is classified into two groups. In the first group, there are three distinct O_2-sensitive nitrogenase systems: molybdenum nitrogenase (Mo-nitrogenase; encoded by the *nifHDK* genes, Table 16.2), vanadium nitrogenase (V-nitrogenase, *vnfHDGK* genes), and iron-only nitrogenase (Fe-nitrogenase, *anfHDGK* genes). They have high similarity, suggesting a common ancestor, and can cross-complement each other, although there are exceptions. All diazotrophs have Mo-nitrogenase, but the presence of either V- or Fe-nitrogenase seems to be incidental, and the presence of all the three nitrogenases has been reported only in *Azotobacter vinelandii, Azotobacter paspali, Methanosarcina acetivorans,* and *Rhodopseudomonas palustris*. The expression of each type of nitrogenase depends on the availability of the metal in the growth environment; when Mo is available, only the Mo-nitrogenase is detected. One hypothesis is that the MoFe protein is the ancestor, and that the V- and Fe-proteins are derivatives. The second group includes a completely different enzyme [insensitive to O_2, CO_2, and hydrogen (H_2)] that occurs only in *Streptomyces thermoautotrophicus* (Newton, 2015).

The Mo-nitrogenase consists of two proteins: dinitrogenase (Component I) and dinitrogenase reductase (Component II). Component I, which contains the active site for N_2 reduction, has a molecular mass of approximately 230 kDa; it is a MoFe-protein comprising α2β2-tetramer associated with two copies each of two metalloclusters: the FeMo-cofactor and the P-cluster. The former contains the site of the substrate (N_2) reduction and the latter is thought to be the initial acceptor of electrons from Component II. Component II is a homodimer Fe-protein that contains one metallocluster per dimer. It has a molecular mass of approximately 64 kDa and couples ATP hydrolysis to interprotein electron transfer (Newton, 2015).

The nitrogenase reaction requires energy (ATP) and reducing equivalents (electrons) or flavodoxin. A basic four-stage mechanism for the Mo-nitrogenase has been proposed:

1. formation of a complex among the reduced Fe-protein, two ATP molecules and the MoFe-protein;
2. electron transfer between the two proteins coupled with the hydrolysis of ATP;
3. dissociation of the Fe-protein accompanied by reduction and conversion of ATP to ADP; and
4. repetition of this cycle until sufficient electrons and protons have accumulated to reduce the substrate (Rees and Howard, 2000; Newton, 2015).

The overall stoichiometry of the reaction catalyzed by the Mo-nitrogenase is:

$$N_2 + 8e^- + 8H^+ + 16MgATP \rightarrow 2NH_3 + H_2 + 16MgADP + 16Pi$$

The Mo-nitrogenase also catalyzes the reduction of other substrates, such as conversion of acetylene (C_2H_2) to ethylene (C_2H_4) (Fig. 16.1). All substrates (including N_2) have the same requirements, that is, the supply of

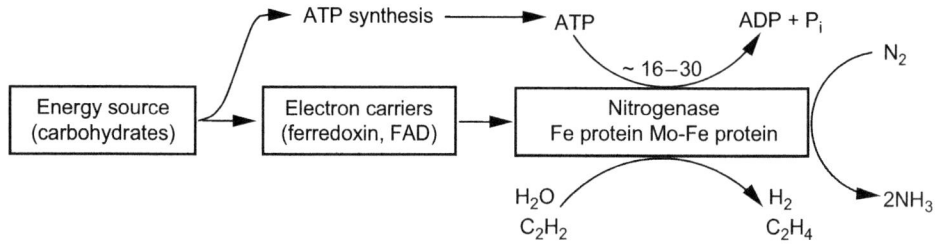

FIGURE 16.1 Main reactions of the nitrogenase system in the process of biological nitrogen fixation. *Modified based on Evans and Barber, (1977).*

TABLE 16.2 Main gene products required for the synthesis of the nodulation (Nod) factors, and for the synthesis and activity of the nitrogenase.

Protein	Function
	Regulation
NodD	Transcriptional regulator of common *nod* genes
	Biosynthesis of glucosamine (chitin) oligosaccharide backbone
NodC	N-acetylglucosamine transferase
NodB	Deacetylase, acting at the nonreducing end of glucosamine oligosaccharide
NodM	Glucosamine synthase
	Biosynthesis and transfer of fatty acid moiety at nonreducing terminus
NodA	Acyl transferase involved in N-acylation of deacetylated nonreducing terminus of glucosamine oligosaccharide
NodE	β-ketoacyl synthase
NodF	Acyl carrier protein
	Modification of nonreducing terminus
NodS	S-adenosyl methionine methyltransferase
NodU	Carbamoyl transferase
NolO	Carbamoyl transferase
NodL	O-acetyl transferase acetylating at 6-C position
NodPQ	ATP sulfurylase and adenosine 5′-phosphosulfate (APS) kinase, providing activated sulfur for sulfated Nod factors
NodH	Sulfotransferase
NoeE	Sulfotransferase involved in sulfation of fucose
NolK	Guanosine diphosphate (GDP) fucose synthesis
NodZ	Fucosyl transferase
NolL	O-acetyltransferase, involved in acetyl-fucose formation
NodX	O-acetyltransferase, specifically acetylating the 6-C of the terminal reducing sugar of the penta-N-acetylglucosamine of *Rhizobium leguminosarum* biovar *viciae* strain TOM from Afghanistan pea
NoeI	2-O-methyltransferase involved in 2-O-methylation of fucose
	Nitrogen fixation
NifH	Dinitrogenase reductase (Fe-protein)
NifD	α subunits of dinitrogenase (MoFe-protein)
NifK	β subunits of dinitrogenase (MoFe-protein)
NifA	Transcriptional regulator of other *nif* genes
NifBEN	Biosynthesis of FeMo-cofactor
FixABCX	Electron transport chain to nitrogenase
FixNOPQ	Cytochrome oxidase
FixIJ	Transcriptional regulators
FixK	Transcriptional regulator
FixGHIS	Copper uptake and metabolism
FdxN	Ferredoxin

MgATP, a low-potential reductant, and an anaerobic environment. The FeMo-cofactor is considered the main binding site for the substrates.

Because nitrogenase is the key enzyme for BNF, studies on the comparison of the key *nifH* gene may clarify evolutionary questions. Zehr et al. (2003) grouped the *nifH* sequences as follows:
- Cluster I includes bacterial and some other *nifH* sequences;
- Cluster II comprises bacterial and methanogenic archaeal *anfH* sequences;
- Cluster III includes anaerobic bacteria and archaea; and
- Cluster IV has nonfunctional paralogous *nifH* genes.

Because Cluster I nitrogenases are very sensitive to O_2, various strategies have evolved to protect the enzyme from irreversible inactivation by O_2 in vivo, including:

1. living and fixing N_2 exclusively under anaerobic conditions (e.g., *Clostridium*);
2. living under aerobic or anaerobic conditions, but fixing N_2 only in the latter (e.g., *Klebsiella*);
3. respiratory protection, providing a microaerophilic environment at the enzyme site (e.g., *Azotobacter vinelandii*);
4. living in colonies covered by slime layers that restrict O_2 diffusion (e.g., *Derxia gummosa* and *Gluconacetobacter diazotrophicus*);
5. spatial separation of nitrogenase and sites of photosynthesis/O_2 evolution (e.g., heterocysts of cyanobacteria such as *Anabaena*);
6. controlling O_2 diffusion with physical barriers and by binding to leghemoglobin (e.g., root nodules of legumes; Fig. 16.2); and
7. synthesis of enzymes scavenging reactive O_2 species (ROS) and H_2O_2 (e.g., ascorbate peroxidase in root nodules of legumes) (Becana and Rodriguez-Barrueco, 1989).

The high demand for ATP, which can be provided in large amounts only by aerobic catabolism of carbohydrates, coupled with the need to protect nitrogenase from O_2, requires that BNF is highly regulated at the transcriptional level by the networks that respond to changes in various environmental parameters.

The cost involved in the reduction of N_2 includes a loss of energy resulting from the reduction of H^+ to H_2 (Fig. 16.1), consuming at least 25% of the energy supplied to the nitrogenase, even at unrestricted N_2 supply (Simpson, 1987). Some diazotrophs possess the H_2-uptake NiFe-hydrogenase (Hup) that recycles part of the energy (ATP) spent in the BNF process via oxidation of the released H_2, improving the energy efficiency (Neves and Hungria, 1987). In addition to improving the efficiency of the BNF process, two other roles have been suggested for the Hup enzyme: (1) auxiliary mechanism for respiratory protection of nitrogenase, and (2) protection of nitrogenase from inhibition by the

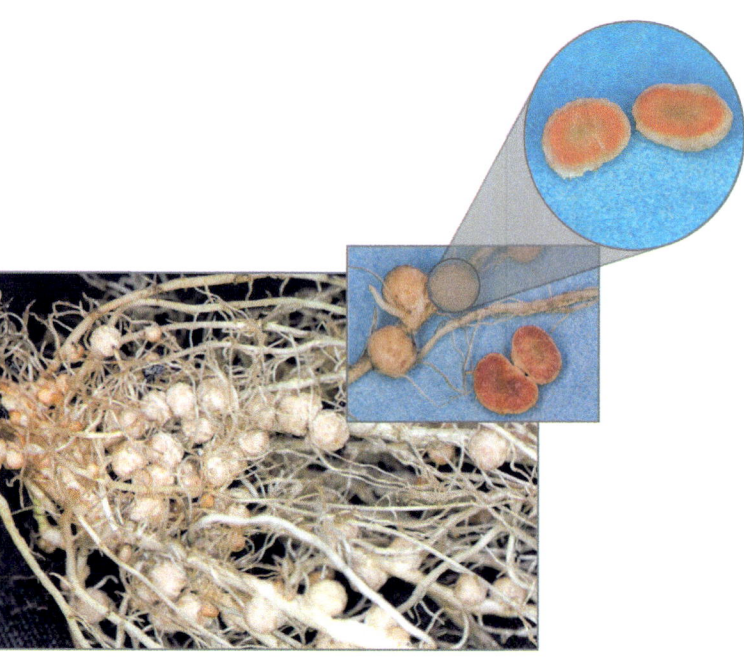

FIGURE 16.2 Nodulated roots of soybean, with typical internal pink color (due to leghemoglobin) of nodules active in fixing N_2.

produced H$_2$. The importance of the hydrogenases has been highlighted mainly in rhizobia–legume symbioses, making the symbiosis more energy efficient, for example, in soybean, lupins (*Lupinus* spp.), and cowpea, in contrast to rhizobia lacking the Hup enzyme, for example, in common bean (*Phaseolus vulgaris* L.) (Neves and Hungria, 1987). Genetic manipulation to introduce and express exotic Hup genes in rhizobia lacking the enzyme has proven to be feasible, improving growth performance of the host plant (Torres et al., 2020).

16.4 Symbiotic systems: how do they work?

16.4.1 General

Two broad categories of symbiotic systems can be identified, based mainly on the type and location of the structure that houses the microsymbiont: (1) nodulated legumes and nonlegumes, and (2) symbioses with cyanobacteria.

In Category I the N$_2$-fixing microorganisms are either rhizobia (in legumes) or actinobacteria of the genus *Frankia* (in nonlegumes). Nodules are located on the roots of the host plant, as shown in Fig. 16.2 for soybean. There are a few exceptions, however, found in legumes growing in regions subjected to waterlogging (e.g., *Sesbania rostrata* and *Aeschynomene*, *Discolobium*, and *Neptunia* species), whereby rhizobia develop stem nodules. In Category I symbiotic systems, photoassimilates supplied by the host plant are the source of substrates used by the microsymbiont to produce ATP via aerobic respiration.

In Category II, N$_2$-fixing cyanobacteria have a small (but taxonomically diverse) host range relative to that of rhizobia (Usher et al., 2007). They occur extracellularly on fungal hyphae in lichens, in cavities of *Azolla* leaves, in the thalli of bryophytes, and in the coralloid roots of cycads. Moreover, they are found intracellularly in stem glands of *Gunnera*. In some cyanobacterial symbioses (e.g., in cycad roots in the absence of light), the cyanobiont must switch from photoautotrophy to chemoautotrophy to be symbiotically competent (Vessey et al., 2005). In other hosts that receive light and remain photosynthetically active, the photosynthetic host (e.g., *Azolla*, *Gunnera*) is the major provider of carbohydrates for generating the ATP required in the N$_2$ fixation.

In agricultural production systems, *A. azollae* in symbiosis with the freshwater fern *Azolla* has long been recognized for its contribution to the N balance of paddy soils. Average inputs of up to 103 kg N ha^{-1} year^{-1} have been reported (App et al., 1984).

16.4.2 Range of legume–rhizobia symbioses

The capacity of legumes to capture (fix) atmospheric N$_2$ via rhizobia in root nodules makes their growth theoretically independent of both soil N and the addition of N fertilizer. Among them are important grains, forages, and cover crops that grow in tropical and temperate climatic zones, with variable BNF contributions (Table 16.1). Nitrogen fixation occurs when these plants are in the symbiotic state with the N$_2$-fixing rhizobia. Molecular biology tools have expanded the range of "rhizobia," including bacteria classified as α-Proteobacteria [genera *Allorhizobium*, *Aminobacter*, *Azorhizobium*, *Blastobacter*, *Bradyrhizobium*, *Devosia*, *Ensifer* (=*Sinorhizobium*), *Microvirga*, *Mesorhizobium*, *Methylobacterium*, *Neorhizobium*, *Ochrobactrum*, *Pararhizobium*, *Phyllobacterium*, *Rhizobium*, and *Shinella*, in addition to some *Agrobacterium* spp.] and also β-Proteobacteria [genera *Paraburkholderia* (former *Burkholderia*), *Cupriavidus* (former *Raltsonia*), and *Trinickia*] (de Lajudie et al., 2019; Peix et al., 2015; Velázquez et al., 2017) (Table 16.3).

In 1984 (after more than 50 years of research), there were only five rhizobial species recognized by the official committee of bacteriology: *Rhizobium leguminosarum*, *Rhizobium loti*, *Rhizobium meliloti*, *Rhizobium fredii*, and *Bradyrhizobium japonicum* (Jordan, 1984). In 2020, thanks to new molecular tools used in microbial taxonomy and systematics, we have more than 250 rhizobial species, some of which have been reclassified (e.g., *R. meliloti* and *R. fredii* into the genus *Sinorhizobium* and then into *Ensifer*). However, not all species in the genera *Rhizobium*, *Bradyrhizobium*, *Mesorhizobium*, and others can fix N$_2$. Only the main symbionts of the economically and environmentally important legumes are listed in Table 16.3, and specialized websites are available for the up-to-date list, for example, http://bacterio.net or https://www.ncbi.nlm.nih.gov/guide/taxonomy.

Rhizobia exhibit varying degrees of specificity to their hosts. Some rhizobial species or biovars (currently named symbiovars) have been classified based on the range of hosts and nucleotide sequences of nodulation genes, particularly *nodC* (Rogel et al., 2011), and are specific for a single genus or small groups of legume genera. For example, *Rhizobium leguminosarum* has three symbiovars, sv. *phaseoli*, sv. *trifolii*, and sv. *viciae*, that form root nodules on *Phaseolus*, *Trifolium*, and *Vicia* species, respectively (Table 16.3), whereas symbiovars of *Bradyrhizobium* are related

TABLE 16.3 Examples of rhizobial species as symbionts of selected legumes of economic and environmental importance.

Host legume	Rhizobial species
Acacia spp.	Ensifer americanus, E. arboris, E. kostiensis, E. mexicanus, E. saheli, E. terangae, Mesorhizobium abyssinicae, M. plurifarium, M. shonense, and Bradyrhizobium ganzhouense
Albizia spp.	Mesorhizobium albiziae, Rhizobium mesosinicum
Amorpha spp.	Mesorhizobium amorphae
Anthyllis vulneraria	Aminobacter anthyllidis
Arachis spp.	Bradyrhizobium arachidis, B. lablabi
Astragalus spp.	Mesorhizobium gobiense, M. huakuii, M. qingshengii, M. sangaii, M. septentrionale, M. silamurrunense, M. temperatum, Rhizobium herbae, R. loessense, and R. multihospitium
Calliandra spp.	Rhizobium calliandrae, R. jaguaris, R. mayense
Caragana spp.	Rhizobium alkalisoli, Mesorhizobium caraganae, M. shangrilense
Centrolobium	Bradyrhizobium centrolobii, B. macuxiense
Centrosema spp.	Bradyrhizobium viridifuturi
Chamaecrista	Bradyrhizobium canariense, B. frederickii, B. niftali, Mesorhizobium plurifarium
Cicer spp.	Mesorhizobium ciceri, M. mediterraneum, M. muleiense
Crotalaria spp.	Methylobacterium nodulans
Dalea spp.	Rhizobium grahamii
Deguelia spp.	Bradyrhizobium mercantei
Desmodium spp.	Rhizobium hainanense, Bradyrhizobium embrapense
Galega officinalis	Neorhizobium galegae
Glycine max	Bradyrhizobium daquingense, B. diazoefficiens, B. huanghuaihaiense, B. elkanii, B. japonicum, B. liaoningense, Ensifer fredii, E. sojae, E. xinjiangensis
Glycyrrhiza spp.	Mesorhizobium tianshanense
Indigofera spp.	Rhizobium indigoferae, R. vallis
Kummerowia stipulacea	Shinella kummerowiae
Lens spp.	Rhizobium leguminosarum sv. viciae
Lespedeza spp.	Rhizobium miluonense, Bradyrhizobium yuanmingense
Leucaena spp.	Rhizobium leucaenae, R. tropici, Ensifer morelense
Listia spp.	Microvirga lotononidis, M. zambiensis
Lotus spp.	Mesorhizobium loti, M. tarimense, Rhizobium multihospitium
Lupinus spp.	Mesorhizobium loti, Rhizobium lupini, Ochrobactrum lupine, Microvirga lupine
Medicago spp.	Ensifer medicae, E. meliloti, Rhizobium alamii, R. cellulosilyticum, R. daejeonense, R. mongolense
Mimosa spp. and Piptadeniae group	Paraburkholderia atlantica, P. caribensis, P. diazotrophica, P. franconis, P. gonoacantha, P. guartelaensis, P. mimosarum, P. nodosa, P. phenoliruptrix, P. piptadeniae, P. phymatum, P. ribeironis, P. rhynchosiae, P. sabiae, P. sediminicola, P. sprentiae, P. strydomiana, P. symbiotica, P. tuberum, Cupriavidus taiwanensis, C. necator, Rhizobium altiplani, R. vallis
Neonotonia spp.	Bradyrhizobium tropiciagri
Neptunia natans	Devosia neptuniae, Rhizobium undicola
Pachyrhizus erosus	Bradyrhizobium jicamae, B. pachyrhizi

(Continued)

TABLE 16.3 (Continued)

Host legume	Rhizobial species
Phaseolus vulgaris	*Rhizobium azibense, R. ecuadorense, R. esperanzae, R. etli, R. freirei, R. gallicum, R. giardinii, R. leguminosarum* sv. *phaseoli, R. leucaenae, R. lusitanum, R. mesoamericanum, R. miluonense, R. paranaense, R. phaseoli, R. pisi, R. tropici, R. vallis, Mesorhizobium atlanticum, Paraburkholderia nodosa, P. phymatum, Cupriavidus necator*
Phaseolus lunatus	*Bradyrhizobium icense, B. paxllaeri*
Pisum spp.	*Rhizobium leguminosarum* sv. *viciae, R. pisi*
Prosopis spp.	*Mesorhizobium chacoense, M. plurifarium*
Sesbania spp.	*Azorhizobium caulinodans, A. doebereinerae, Rhizobium huautlense, Ensifer saheli, E. sesbaniae, Mesorhizobium hawassense*
Sophora spp.	*Rhizobium multihospitium, Mesorhizobium tianshanense*
Stylosanthes spp.	*Bradyrhizobium stylosanthis*
Trifolium spp.	*Rhizobium leguminosarum* sv. *trifolii*
Vicia spp.	*Rhizobium fabae, R. laguerrae, R. leguminosarum* sv. *viciae, R. multihospitium*
Vigna spp.	*Bradyrhizobium manausense, Rhizobium vignae, Microvirga vignae*

Source: Based on Rivas et al. (2009), Peix et al. (2015), and List of Prokaryotic names with Standing Nomenclature (LPSN) (http://bacterio.net).

to several host legume species (Delamuta et al., 2017). Symbiotic promiscuity can be found in some other rhizobia and is perhaps more widespread than previously thought, for example, an impressive diversity of species can nodulate common bean (Table 16.3). Another example is one strain from New Guinea (*Ensifer fredii* NGR234) with a broad legume host range (at least 112 genera) (Pueppke and Broughton, 1999). From the host plant perspective, some legumes (e.g., *Vigna, Phaseolus*) are considered to be nonselective for rhizobia, whereas others (e.g., *Pisum, Trifolium*) are nodulated by a single rhizobial species or symbiovar.

16.4.3 Legume root infection by rhizobia

At the biochemical and molecular genetic level, much is known about the interactions between rhizobia and their host plants that lead to the formation of root nodules, and the main steps are summarized in Fig. 16.3. The first step to nodule formation involves root exudate–driven chemotaxis of rhizobia toward roots. Rhizobia are positively chemotactic to a range of molecules released by the host plant after seed germination, including sugars, amino acids, various dicarboxylic acid anions (such as succinate, malate, and fumarate), and aromatic compounds, including shikimate, quinate, protocatechuate, vanillate, acetosyringone, gallate, catechol, luteolin, and others. Bacterial genes, such as *chvE*, are responsible for the perception of these molecules; the ensuing chemotaxis is critical for bacterial adhesion to the root surface (Brencic and Winans, 2005).

By the early 1990s a series of studies revealed an intricate pattern involving the generation, transmission, recognition, and processing of signals between the rhizobia and legumes, establishing a "molecular dialogue" (Hungria and Stacey, 1997; Oldroyd, 2013). First, secondary metabolites (mostly flavonoids) are released from legume seed coats or rootlets (Fig. 16.3) in micromolar or even nanomolar concentrations and are sensed by free-living rhizobia in the rhizosphere and on root hair surfaces. Their main function is to act, along with NodD proteins (that belong to the LysR-type transcriptional-regulator family) of rhizobia (Table 16.2), as coinducers of the rhizobial nodulation genes. In the rhizobia a flavonoid–protein complex is formed with the constitutively expressed regulatory *nodD* gene product (NodD) that is bound to conserved DNA sequences (*nod*-boxes) in the promoter regions of structural nodulation genes (*nod*, *noe*, and *nol*; collectively known as *nod* genes) (Table 16.2). The presence of a specific flavonoid at the NodD binding site in the *nod*-box activates the transcription of these genes.

Traditionally, nodulation genes (Table 16.2) have been classified as:

1. regulatory genes, responsible for the transcription of the remaining nodulation genes, mainly *nodD*, but also few other genes, such as *nolR* in *Ensifer meliloti*, *nolA* and *nodVW* in *Bradyrhizobium japonicum/B. diazoefficiens*, and *nrcR* in *Rhizobium tropici*;

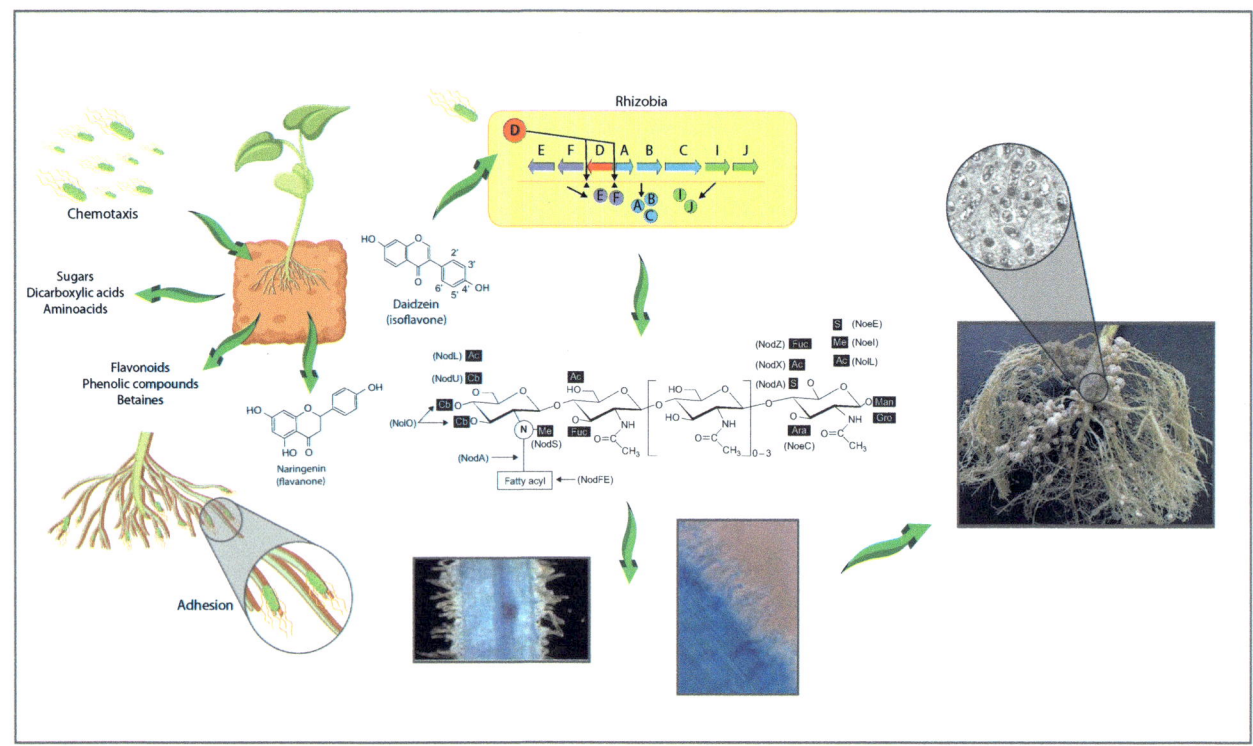

FIGURE 16.3 Schematic representation of the main events that lead to nodule formation, including chemotaxis of rhizobia toward compounds released by the plant, release of *nod*-gene inducers by the host plant, *nod*-gene expression in rhizobia and synthesis of Nod factors, effects on root hairs and nodule formation, and transformation of rhizobia into bacteroids.

2. "common genes" (*nodA*, *nodB*, and *nodC*) that are host-inducible and essential for nodulation (showing high degree of nucleotide sequence similarity and functional homology among rhizobial species) and are involved in the biosynthesis of the backbone structure of the reciprocal bacterial signal molecules—the lipochitooligosaccharide Nod factors;
3. "host-specific" genes (encoding enzymes that modify the basic Nod factor molecules), with some (e.g., *nodFE*, *nodL*, and *nodM*), but not others (e.g., *nodO*, *nodH*, and *nodPQ*), showing sequence homology among rhizobial species; and
4. genotype-specific nodulation genes that appear to govern the capacity to nodulate only selected genotypes within a legume species (e.g., *nodX* of *Rhizobium leguminosarum* sv. *viceae* strain TOM and *nolA* of *Bradyrhizobium japonicum*).

The nodulation and core nitrogen fixation genes are located on transmissible genetic elements, in plasmids (symbiotic plasmids, pSym) or chromids (considered to be a "second chromosome" or "megaplasmid") in fast-growing rhizobia. In "slow-growers" (classification based on growth in culture medium with mannitol as carbon source), such as *Bradyrhizobium* and *Azorhizobium*, these genes are located in a symbiotic island on the chromosome, whereas in *Mesorhizobium* they are located either on a megaplasmid or in the symbiotic island; however, there are some exceptions of nodulation genes outside these regions. As the genes are in transmissible elements, horizontal gene transfer is very common. The Nod factors elicit several changes in root metabolism and morphogenesis and are the essential signals for rhizobia entering into legume roots (Lindström and Mousavi, 2020; Oldroyd, 2013). Although it was initially thought that all rhizobia required common *nod* genes and Nod factors for nodulation, with the genome sequencing of the stem-nodulating *Bradyrhizobium* strains BTAi1 and ORS 278 it has been shown that there are exceptions to this rule (Giraud et al., 2007).

The flavones luteolin from alfalfa seed coats and 7,4′-dihydroxyflavone from roots of *Trifolium repens* were the first *nod* gene–inducing flavonoids to be revealed; subsequently, dozens of compounds have been identified in several legumes that are either glycones or aglycones of flavonoid subgroups, such as chalcones, flavones, flavanones, isoflavones, and coumestans (Hungria and Stacey, 1997). Most legumes release several compounds, and in the case of *Phaseolus vulgaris*, 11 *nod*-gene inducers have been isolated from seeds and four from roots (Hungria and Stacey,

1997). The mechanism of action of flavonoids as coinducers of *nod*-gene transcription has not been resolved completely yet. There is evidence that the presence of an appropriate flavonoid changes the angle of bend in the DNA at the points where a NodD protein is bound to the *nod*-box site in a *nod*-gene promoter region, thereby allowing the RNA polymerase to initiate gene transcription (Chen et al., 2005). Different rhizobia respond to different sets of flavonoid inducers; in some cases the interaction is consistent with host specificity, whereas in others there is no specificity. For example, the *nod* genes of *Ensifer fredii* NGR234 are induced by many flavonoids, and this bacterium has a very broad range of hosts; by contrast, *Rhizobium leguminosarum* sv. *viciae*, which also responds to many inducers, has a narrow host range (Table 16.3).

A few nonflavonoid compounds from plants can also induce nod-gene expression, usually at higher (millimolar) concentrations, and none has been shown to replace flavonoids in this role. Examples are betaines, stachydrine, and trigonelline from alfalfa and the aldonic, tetronic, and erytronic acids from lupins (Brencic and Winans, 2005). Plant flavonoids not only induce *nod* genes, but also several other rhizobial genes (e.g., Batista and Hungria, 2012).

The Nod factors are lipochitooligosaccharides composed of $\beta - 1,4$-linked *N*-acetyl-D-glucosamine residues with a fatty acyl chain at the nonreducing end. The first characterized Nod factor was isolated from *Ensifer meliloti* by Lerouge et al. (1990). Variations on the basic structure arise from several sources: (1) the number of acetylglucosamine residues in the oligosaccharide (chitin) backbone (between two and six); (2) the type of fatty acid at the nonreducing end (common saturated/monounsaturated or specific highly unsaturated); and (3) the number and types of substituent groups (acetyl, arabinosyl, fucosyl, mannosyl, sulfate, etc.). A basic Nod factor structure is shown in Fig. 16.3, and the main *nod*-gene products needed for the synthesis of a variety of Nod factors are listed in Table 16.2. Extra gene products, such as NodI and NodJ, are required for the secretion of Nod factors from bacteria (Hungria and Stacey, 1997).

Although great progress has been achieved in the knowledge of molecular signaling between rhizobia and legumes, we are still far from understanding the complexity of this process. One remarkable example has been documented in *Rhizobium tropici* strain CIAT 899 that is able to nodulate common bean, leucaena (*Leucaena* spp.), and other legume species. The strain carries five copies of the regulatory *nodD* gene and three copies of *nodA* and synthesizes more than 50 different Nod factors; interestingly, the synthesis of some Nod factors is induced by salinity stress in the absence of flavonoids, but the biological implications of such variety of signals are still poorly understood (del Cerro et al., 2015; Morón et al., 2005).

When applied to legume roots in nanomolar or femtomolar concentrations, the appropriate Nod factors elicit a number of responses, including:

1. deformation and plasma membrane depolarization in root hairs;
2. rapid increases in cytosolic Ca^{2+} activity (so-called calcium spiking) in root hairs;
3. preinfection thread formation in deformed root hairs (Fig. 16.3); and
4. cytokinin-stimulated cortical cell divisions at nodule primordia.

The Nod factors, even in the absence of the bacteria that produce them, can induce some of the many plant nodulin genes that are expressed in several phases of the symbiosis (preinfection, infection, nodule development, and nodule functioning) (Oldroyd, 2013).

The interaction between Nod factors and legumes influences the host specificity. For some Nod factors, certain features (e.g., sulfation in those produced by *Ensifer meliloti* and an arabinosyl substitution in the Nod factors of *Azorhizobium caulinodans*) are needed for nodule formation on the hosts (*Medicago* and *Sesbania*, respectively) (D'Haeze et al., 2000; Lerouge et al., 1990). Substitutions and other aspects of structure, such as the length of the oligosaccharide backbone and the size and degree of saturation of its acyl chains, may also be host-range determinants (del Cerro et al., 2015; Oldroyd and Downie, 2008) (Fig. 16.3). Nevertheless, for many Nod factors there appears to be no correlation between their structure and the nodulation of a particular host or a group of hosts, and, as mentioned earlier, rhizobia may synthesize dozens of Nod factor variants, without having a large range of hosts (Morón et al., 2005). The signaling pathways associated with nodulation by bradyrhizobia that do not synthesize Nod factors (e.g., in the case of nodulation of the aquatic legume *Aeschynomene*) have not yet been elucidated, with a suggestion that perhaps it would occur through a cytokinin-type signal (Masson-Boivin et al., 2009).

In partnerships where nodulation is dependent on Nod factors (i.e., crop legumes such as soybean), the nature of the plant receptors for these molecules and the signal-transduction pathways leading to symbiosis-related plant gene activation have been investigated intensively (Oldroyd, 2013; Oldroyd and Downie, 2008), as have the genes encoding a symbiosis receptor–like kinase (SYMRK) in *Lotus* (Stracke et al., 2002) and a nodulation receptor kinase in *Medicago* (Endre et al., 2002). SYMRK is a common element in legumes and nonlegumes, regardless of whether they form root endosymbioses with rhizobia, *Frankia*, or arbuscular mycorrhizal fungi (AMF) (Gherbi et al., 2008). Genes encoding

LysM receptor–like kinases that function upstream of SYMRK and could be direct receptors for Nod factors are found in *Lotus japonicus* (Desbrosses and Stougaard, 2011; Jones et al., 2007). Likewise, *Medicago truncatula* has receptor-like kinase genes that encode potential Nod factor receptors (Arrighi et al., 2006), as well as other genes required for the transduction of rhizobial Nod factor signals, but not for mycorrhizal colonization (Oldroyd, 2013). However, there are several receptors of both rhizobial and mycorrhizal molecules, as the machinery regulating interactions with N_2-fixing bacteria greatly overlaps with the pathways mediating the more ancient symbiosis with mycorrhizal fungi (Desbrosses and Stougaard, 2011; Oldroyd, 2013).

Nod factors are not the only rhizobial compounds participating in the molecular dialog with legumes, as other signal molecules influence the successful progression to a functioning nodule (Jones et al., 2007). Particularly important are the cell surface polysaccharides found in all rhizobia, such as extracellular, lipo, and capsular polysaccharides and cyclic glucans. They are involved in various phases of the symbiotic development, including root colonization, host recognition, infection thread formation, and nodule invasion (Brencic and Winans, 2005; Jones et al., 2007). Additionally, numerous proteins are released by rhizobia that also influence the course of symbiotic infection (Brencic and Winans, 2005; Jones et al., 2007). For example, proteins released through a type III secretion system influence the legume host range (Bartsev et al., 2004). Concerning plant factors other than the flavonoid and nonflavonoid *nod*-gene inducers, the carbohydrate-binding lectin proteins, found on legume root hair surfaces, have been proposed as determinants of host specificity (Hirsch, 1999). Another category of plant proteins (the flotillins) has been shown to play a critical role in legume infection by rhizobia (Haney and Long, 2010). Nodulation also requires a subset of plant phytohormones, including auxins, cytokinin, and ethylene, that are required for root development (Desbrosses and Stougaard, 2011).

16.4.4 Nodule formation and functioning in legumes

The mode of infection by rhizobia may be inter- or intracellular or a combination of both (Sprent and James, 2007; Vessey et al., 2005). In many cases, including crop plants such as common bean, pea (*Pisum sativum* L.), soybean, alfalfa, and clover (*Trifolium* spp.), intracellular invasion involves rhizobia entering root hairs via infection threads. Nodule morphogenesis is coordinated with bacterial infection (Oldroyd and Downie, 2008), beginning with root cortical cell divisions at the sites of nodule primordia and meristem initiation. Following attachment around the tip of a root hair, rhizobia become entrapped in a pocket when the root tip curls backward. Hollow, cylindrical infection threads (built by the plant in response to the Nod factors) develop along the length of the root hair and terminate at nodule primordia in the root cortex (Gage, 2004) and are colonized by the invading rhizobia. Branches of infection threads penetrate cells of the nodule primordium and release rhizobia into them (with some plant cells in the nodule remaining free of rhizobia throughout the life of the nodule). Once inside a host cell, rhizobia differentiate from Gram-negative motile bacteria into nonmotile bacteroids within an organelle-like structure—the symbiosome. A symbiosome is surrounded by a membrane and may contain one or several bacteroids, and each infected nodule cell can contain several symbiosomes (Werner, 2007). The region between the peribacteroid membrane(s) and the symbiosome membrane is termed symbiosome or peribacteroid space (Fig. 16.4).

Depending on the host plant, nodules may be initiated in the inner or outer root cortex. In pasture and crop legumes such as *Medicago*, *Lens*, *Trifolium*, *Pisum*, and *Vicia*, nodules are initiated in the inner cortex and are of the indeterminate type (maintaining an active apical meristem and distinct developmental zones) with mature bacteroids at the growing tip, where N_2 fixation takes place (Vessey et al., 2005). Nodules originating in the outer root cortex, as in *Glycine*, *Lotus*, *Phaseolus*, and *Vigna*, are of determinate type; they do not maintain an active meristem and have a limited lifespan. Another difference between the two types is that bacteroids from determinate nodules can regenerate the free-living form of the bacterium, whereas those from indeterminate nodules cannot (Zhou et al., 1985).

The root system of a single plant, and perhaps also individual nodules, can be infected by more than one strain of a rhizobial genus, species, or symbiovar (Hagen and Hamrick, 1996). Various specificity of interactions between legumes and rhizobia has been highlighted by Andrews and Andrews (2017). For example, *Bradyrhizobium* spp. seems to be exclusive rhizobial symbionts of Caesalpinioideae species, whereas two Mimosoideae tribes (Ingeae and Mimoseae) are nodulated by different rhizobial genera. In general, Papilionoideae species with indeterminate nodules are promiscuous in relation to rhizobial symbionts, but high specificity is found at the tribe level [e.g., for the Fabeae (*Rhizobium* spp.)] and at the species level [for *Galega* spp. (*Neorhizobium galegae*) and *Hedysarum coronarium* (*Rhizobium sullae*)]. On the other hand, those tribes with determinate nodules, such as Dalbergieae, are nodulated primarily by *Bradyrhizobium* spp., whereas Desmodieae, Phaseoleae, Psoraleae, and Loteae are promiscuous across different rhizobial genera. Concerning β-rhizobia, for *Mimosa* (the largest genus of the Mimoseae tribe) either *Paraburkholderia* or *Cupriavidus* are the prevalent symbionts (Dall'Agnol et al., 2017).

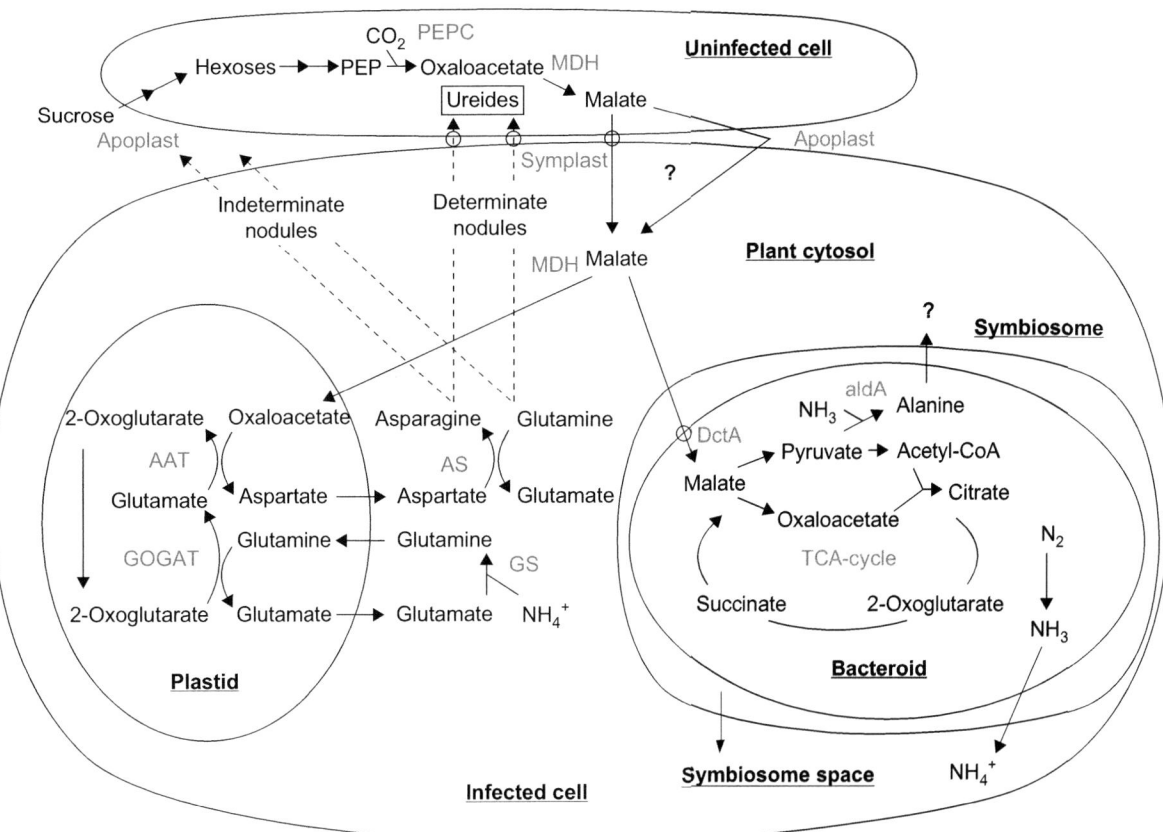

FIGURE 16.4 Schematic representation of the main reactions and processes in carbohydrate metabolism in legume root nodules and export of fixed nitrogen. *Reproduced from White et al. (2007), with permission of the American Society of Plant Biologists.*

The great potential of nodulating legumes for N_2 fixation is based on three main factors: (1) direct supply of photosynthetically fixed C to the bacteroids in the nodules; (2) effective maintenance of very low O_2 concentrations within the nodule to protect nitrogenase; and (3) rapid export of fixed N via the xylem. Nutrient exchanges in root nodules are shown in Fig. 16.4. In all legume–rhizobia symbioses, the energy substrate for N_2 fixation is derived from photosynthates transported to the nodule cytosol as sucrose via the phloem (Fig. 16.4). After entering uninfected nodule cells, sucrose synthase cleaves it to monosaccharides; some of these undergo glycolysis to produce phosphoenolpyruvate (PEP) that is carboxylated to oxaloacetate by PEP carboxylase followed by reduction to the C4-dicarboxylic acid malate by malate dehydrogenase. Monosaccharides that do not enter the glycolytic pathway are channeled into cellulose or starch synthesis. Malate, along with other C4-dicarboxylic acids such as succinate and fumarate, is transported from uninfected nodule cells into the cytosol of infected cells and then across the symbiosome and peribacteroid membranes into the bacteroids by a dicarboxylic acid transport system. In the bacteroids, oxidation in the tricarboxylic acid (TCA) cycle generates the reducing equivalents and ATP needed for nitrogenase activity. Via metabolism outside the bacteroid, malate also provides carbon skeletons required for the assimilation of fixed N in the nodule cytosol (Fig. 16.4) (Lindström and Mousavi, 2020; Lodwig and Poole, 2003; White et al., 2007).

Nodules have to be equipped to deal with two seemingly incompatible physiological requisites for N_2 fixation in bacteroids—ensuring a plentiful supply of O_2 for oxidative phosphorylation to provide energy for nitrogenase activity, while at the same time, and in more or less the same location, protecting nitrogenase from the damaging effect of O_2. Low O_2 concentration in nodules is achieved in two ways: (1) an O_2 diffusion barrier in densely packed cells in the inner nodule cortex; and (2) high respiration rates of the bacteroids. In the barrier the permeability to O_2 adjusts rapidly to changes in external O_2 concentration or internal O_2 demand (Vessey et al., 2005). The barrier is one to five cell layers thick, and the intercellular spaces can be filled with air or water. Because the diffusion coefficient of O_2 in air is about 10^4 times higher than in water, a water barrier is an effective means of limiting O_2 diffusion to the interior of nodules (Blevins, 1989).

In this O₂-limited environment, leghemoglobin plays an important role in ensuring sufficient O₂ supply to the bacteroids. Leghemoglobin is encoded by at least four *lb* genes and constitutes about 5% of total nodule protein (e.g., in mature soybean nodules) (Cvitanich et al., 2000), giving the typical pink color that characterizes an active nodule (Fig. 16.2). This protein, with a central Fe atom in a porphyrin ring (identical to cytochromes), acts by binding O₂ from intercellular spaces in the infected zone of the nodule and delivering it by diffusion along an oxyleghemoglobin concentration gradient to a high-affinity *cbb3*-type cytochrome oxidase in the bacteroids (Preisig et al., 1996).

The high O₂ consumption in nodules, necessary for the provision of energy, can also lead to the production of ROS as regular by-products in metabolic processes and deleterious at high concentration. For protection against ROS toxicity, rhizobia employ a variety of antioxidant defense mechanisms, including ROS scavengers and reductants, that appear to be essential for normal nodule development (Tavares et al., 2007). In the bacteroids, NH₃ produced in N₂ fixation diffuses into the acidic symbiosome space where it is protonated to ammonium and prevented from being recycled back into the bacteroid by the suppression of an ammonium-transporting system that otherwise operates in free-living rhizobia (Tate et al., 1998). A monovalent cation channel transports ammonium across the symbiosome membrane into the host cytosol where it is assimilated into glutamine via the glutamine synthetase (GS)/glutamate synthase (GOGAT) pathway (Fig. 16.4) (Vance and Gantt, 1992). Another amino acid, alanine, may be synthesized in bacteroids and then transported to the host cytosol. There is also evidence of exchange (cycling) of some amino acids between the bacteroid and the host plant (Prell and Poole, 2006).

Fixed N is delivered via the xylem to the shoot mainly as asparagine and glutamine in legumes with indeterminate nodules such as Vicieae and Trifolieae (usually including temperate legumes) or as ureides (allantoin and allantoic acid) from determinate nodules of Phaseoleae (most tropical and subtropical legumes) through complex pathways involving at least 20 different enzymes. Ureides are synthesized in uninfected cells from asparagine and glutamine received from the cytosol of adjacent infected cells (Fig. 16.4). Ureides have a lower C:N ratio than amino acids; hence, tropical and subtropical legumes transport more N per unit of C than temperate legumes, but a disadvantage of ureides is poor solubility at low temperatures. For temperate legumes grown at temperatures of 15°C–20°C, the transport of nitrogen as allantoin would not be possible because it would precipitate. Ureides are exported to aerial organs by the transpiration water flow via the xylem; in leaves, ureides are broken down enzymatically to yield NH_4^+ that is used in metabolism, whereas glutamine and asparagine can be used directly for protein synthesis and other metabolic processes (Baral et al., 2016; Neves and Hungria, 1987; Sprent, 1980).

Many attempts have been made to calculate the carbon costs of N₂ fixation to plants versus the uptake of mineral N from soil. The energy costs of N₂ fixation may exceed those of mineral N uptake by more than 10% (Kaschuk et al., 2009). Minchin and Witty (2005) estimated the costs of nitrate assimilation at up to 2.5 g C g^{-1} of assimilated N, whereas N₂ fixation would cost 5.2–18.8 g C g^{-1} of N. However, such comparisons may be misleading because they do not consider the costs of synthesizing N fertilizers, their low use efficiency in agriculture, and the environmental costs arising from N losses (Section 16.1).

In efficient combinations of rhizobia and legumes, there is a source:sink synchronicity, and the higher C sink strength and efficient N₂ fixation result in higher photosynthetic rates and delayed leaf senescence (Kaschuk et al., 2010a; Neves and Hungria, 1987). A meta-analysis, including 348 data points from 12 legumes, provided evidence that legumes are not C-limited in the efficient symbiotic combinations based on increased yield and higher protein and lipid mass fractions in grains (Kaschuk et al., 2010b).

Not all rhizobia fix large amounts of N₂ once they have nodulated their host legume. The N₂-fixing capacity varies substantially among rhizobial species and symbiovars, ranging from zero to high (Table 16.1). For example, all *Rhizobium giardinii* strains colonizing nodules of *Phaseolus vulgaris* are ineffective (Depret and Laguerre, 2008), whereas other species and strains, such as *Rhizobium tropici* CIAT 899, are very effective (Gomes et al., 2015). Therefore, high N₂ fixation and yield may not be achieved even under ideal conditions for plant growth (e.g., regarding soil pH and availability of water and nutrients other than N). As discussed in Section 16.9, seed inoculation with elite rhizobial strains has increasingly been used to ensure effective legume nodulation in soils containing inefficient N₂-fixing indigenous strains. Unfortunately, knowledge of the mechanisms governing interstrain competition for root infection and nodulation is still poor, and inoculant strains chosen for their high N₂-fixing properties may not outcompete the indigenous rhizobia population in the soil.

The benefit to a legume from an effective symbiosis with rhizobia is obvious: an adequate supply of N. However, an effective rhizobial strain can survive and multiply in soil as a free-living heterotroph for decades, even in the absence of a suitable host legume, without the need to fix N₂ to survive. In symbiosis, the rhizobia sacrifice a significant percentage of their respiratory potential to supply its host with N, whereas an ineffective strain can nodulate a host plant without having such demand placed on its metabolism. Furthermore, the bacteroids in indeterminate nodules lose their capacity to replicate and be released from senescent nodules as free-living bacteria. Indeed, questions have been posed—"why do rhizobia fix nitrogen?" (West et al., 2002) and "why are most rhizobia beneficial to their host plants, rather than parasitic?" (Denison and Kiers, 2004). A possible answer is that the host plant imposes metabolic sanctions on nodules containing weakly effective or ineffective strains of rhizobia. One strategy (particularly in determinate nodules containing replicable bacteroids) could involve restricting O_2 supply to nonfixing nodules, thereby limiting the multiplication of rhizobia and reducing the number of viable cells eventually released from senescing nodules (Kiers et al., 2003). Such restrictions could inhibit parasitism and stabilize the N_2-fixing symbiosis by favoring the effective rhizobial strains (Denison, 2000; Oono et al., 2009).

At present, the bulk of the literature on restrictions is usually of a theoretical nature [studies on pea nodulation by Depret and Laguerre (2008) and partner choice in *Medicago truncatula–Ensifer* symbiosis by Gubry-Rangin et al. (2010) are exceptions], and some of the complexities of nodulation have not yet been taken into account, such as:

1. Rhizobial strains may be ineffective in one host but effective in another.
2. There are individual root nodules that harbor both ineffective and effective strains.
3. Ineffective strains sometimes produce far more nodules on a root system than effective ones.
4. A host legume may influence rhizobium strain selection without employing restrictions (Gubry-Rangin et al., 2010).

Understanding the nodulation process may be key to achieving an old dream of transfering the nodulation and BNF capacity of legumes to cereals. In this context, it is important to determine the evolutionary history of the symbioses. A detailed study with an N_2-fixation database of 9156 angiosperm species suggested that about 200 million years ago (MYA) a single evolutionary event drove the evolution of BNF in a specific group of angiosperms (the nitrogen-fixing clade), comprising Fagales, Fabales, Rosales, and Cucurbitales (Werner et al., 2014). This was followed by multiple losses and gains of the symbiotic gene, resulting in the diversification of nodulation types, and the emergence of "stable fixers" about 100 MYA (in the Papilionoideae). Although it has often been proposed that the evolution of BNF was triggered by an environmental event, such as climatic change, we are still far from understanding what the precursor was. Unfortunately, none of the major cereals was identified among the precursor group of species, suggesting that modifying them for symbiosis with N_2-fixing bacteria will be difficult (Werner et al., 2014).

Rhizobia comprise a group of distantly related bacteria that apparently acquired capacity to infect legumes by horizontal and vertical transfer of genes located in plasmids, chromids, or symbiotic islands in the chromosome, with *Bradyrhizobium* probably representing the common ancestor (Broughton and Perret, 1999; Hungria et al., 2015).

16.5 Effects of nutrients on the biological nitrogen fixation

16.5.1 Nutrients other than nitrogen

Nutrients are key for maintenance and multiplication of diazotrophs, host plant growth, and the symbiosis. The three main roles or effects of each nutrient relevant to BNF are depicted in Fig. 16.5.

Nutrient deficiencies can adversely affect legume root nodule symbioses at a very early stage of development, including multiplication of the microsymbiont in the host plant rhizosphere, its capacity to detect nodulation signals, and its production and excretion of Nod factors (McKay and Djordjevic, 1993). Essential nutrients required by rhizobia are directly involved in the structure and metabolic functioning of microbial cells (O'Hara, 2001). It should be kept in mind that all plant nutritional requirements must be fulfilled independently of the symbiotic association; therefore, factors influencing the host plant also influence the symbiosis (Santachiara et al., 2019). Specific nutrients are required for N_2 fixation at various stages of the symbiotic interaction, infection, and nodule development and functioning (Bonilla et al., 2011; O'Hara, 2001).

16.5.1.1 Phosphorus

Phosphorus is a key nutrient for bacteria, plants, and the symbiosis, taking part at the early signaling events (Fig. 16.5), in nonkinase and kinase receptors, and in other metabolic processes requiring phosphorylation steps, such as regulation of intracellular enzymes and binding of proteins to DNA for gene regulation (O'Hara, 2001).

Plants take up P mainly as $H_2PO_4^-$, but also as HPO_4^{2-}, and its concentration in plants is usually around 60 μmol g^{-1} or 2 g kg^{-1} of shoot or leaf dry weight (Bonilla et al., 2011). Because of the role of P in several symbiotic steps, nodulated legumes have higher P requirements than nonsymbiotic plants (Vadez et al., 1996). Under P limitation, rhizobia respond in the same way as the host plants (by increased expression of genes involved in P acquisition) (Sadowsky, 2005). Low P availability is often the most limiting constraint to the symbiosis, especially in acidic and calcareous soils,

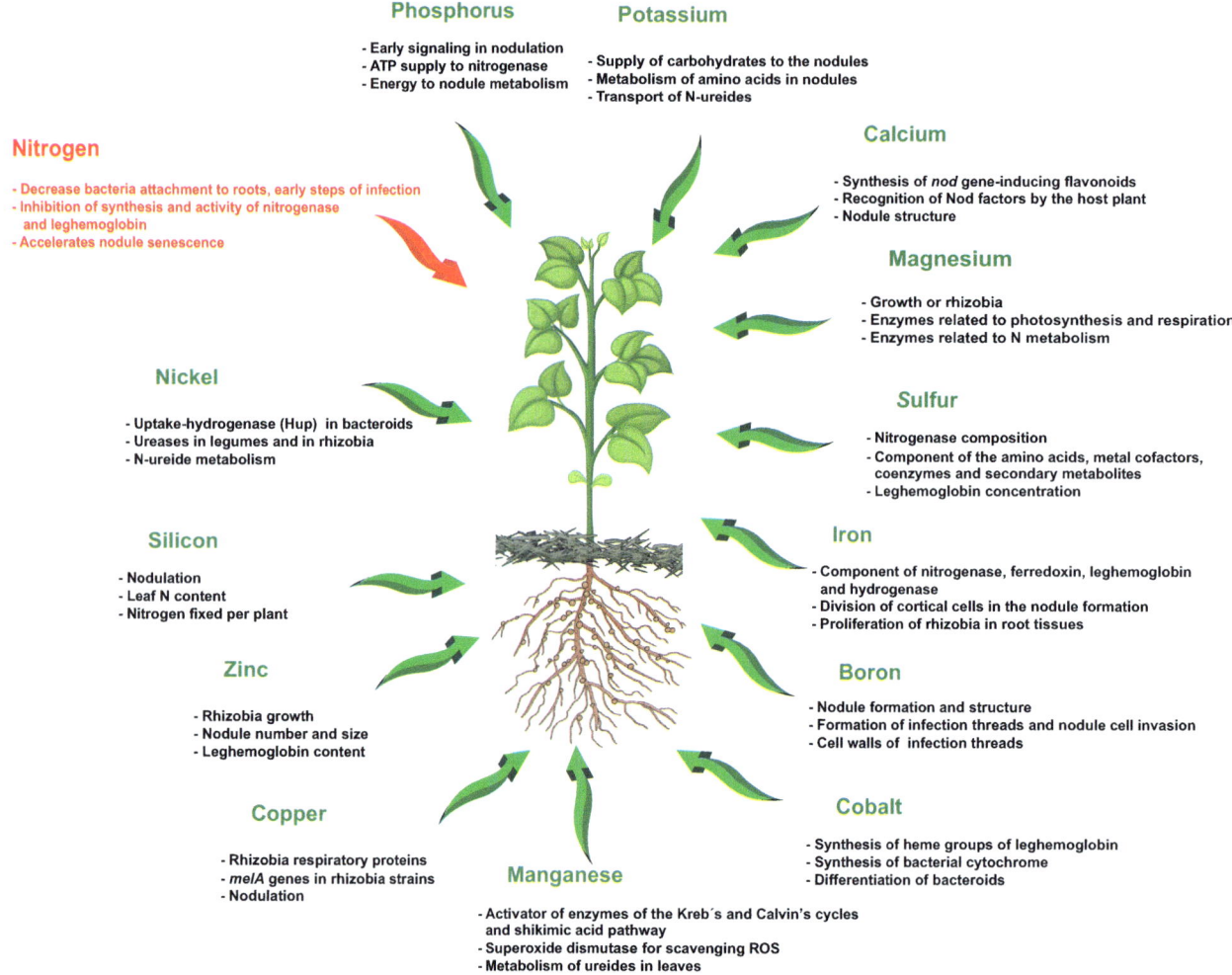

FIGURE 16.5 The main roles of each nutrient that influences (directly or indirectly) the symbiosis and the biological nitrogen fixation.

and in low-input agriculture. Phosphorus fertilizers tend to be four- to fivefold more expensive than the rhizobial inoculum; however, the additional yield obtained when rhizobial inoculants and P fertilizers are applied together, comprising an integrated soil fertility management package, usually results in high return on the investment (Ronner et al., 2016; Vanlauwe et al., 2019; Wolde-Meskel et al., 2018).

Phosphorus has essential roles in plant energy metabolism and maintaining the energy status of legume nodules (Fig. 16.5) (Olivera et al., 2004). Assimilation of ammonium in the GS/GOGAT cycle and further conversion into amino acids or ureides are energy demanding (Cullimore and Bennett, 1988). The syntheses of nucleic acids and phospholipids are dependent on P (O'Hara, 2001). Hence, nodules are a strong P sink; their P concentration is up to threefold higher compared with roots and shoots, particularly when external P supply is low (Adu-Gyamfi et al., 1989; Hart, 1989). The competitiveness of developing nodules compared with other plant sinks (root and shoot meristems) for P under limited external supply differs among legume species. This may be partly responsible for disagreements regarding the amounts of P required for nodulation (Jakobsen, 1985), suggesting the need for research to establish critical P concentrations in plant and nodule tissues (Divito and Sadras, 2014).

Rhizobia have the high-affinity active uptake systems for acquiring phosphate from the external environment (O'Hara, 2001). High P consumption during nodulation can, to some extent, be satisfied by mycorrhizal colonization; tripartite symbioses among legumes, rhizobia, and AMF are common (and perhaps the rule) (Bonilla et al., 2011).

Nodulation and plant growth can be improved by the application of P fertilizers (Abbasi et al., 2010; Divito and Sadras, 2014; Ronner et al., 2016; Wolde-Meskel et al., 2018). However, the magnitude of the response varies with the experimental conditions. For example, P addition increased BNF by 5% under field conditions, whereas responses in greenhouse trials averaged 132%, suggesting that the BNF response to P addition is saturated at low rates in the field

(Santachiara et al., 2019). Genotypic variation in P-use efficiency (PUE) has also been reported in several legume species, indicating the importance of plant breeding programs aimed at producing genotypes adapted to low-P soils (Drevon et al., 2015; Tajini and Drevon, 2011). The regulation of nodule permeability to O_2 is among the mechanisms associated with genotypes that have high PUE (Drevon et al., 2015), suggesting osmoregulation. Also, a correlation between O_2 permeability and phytase activity in common bean nodules indicates that the internal use of P may be increased by a mechanism involving the regulation of nodule permeability to O_2 related to the BNF process (Lazali and Drevon, 2014).

Shoot N and P concentrations are positively correlated in legumes (Kuang et al., 2005). Therefore, legumes dependent on BNF may exhibit N deficiency under limiting P supply. At low P availability, P addition increases the root nodule number (Abbasi et al., 2010; Gunawardena et al., 1993), nodule mass (Gunawardena et al., 1993), nodule size (Kuang et al., 2005), and yield (Ronner et al., 2016; Wolde-Meskel et al., 2018). Increases in soil P availability improved nodulation and P and N contents in alfalfa, and these effects were magnified by coinoculation with AMF (Azcon and Barea, 1992) (Table 16.4). Despite the improvement in nodulation, a major effect of P application and AMF coinoculation may rely on increases in total N uptake from soil rather than on the percentage of N derived from the atmosphere (%Ndfa) (Azcon and Barea, 1992; Somado et al., 2006).

16.5.1.2 Potassium

This nutrient is relevant in several plant physiological processes, but there are few studies on its direct effect on BNF. A pioneering study (Mengel et al., 1974) pointed out that K increased the soluble fractions of amino acids and proteins in various plant organs, especially in nodules; in addition, K influenced the supply of carbohydrates to the nodules that could impact the provision of ATP and the reducing power required by the nitrogenase (Fig. 16.5). In both pasture and grain legumes, there are reports of K increasing root growth, nodulation, nitrogenase activity, and the activity of N-assimilation enzymes GS and GOGAT (Divito and Sadras, 2014; Thomas and Hungria, 1988). In common bean, K increased concentration of ureides in pod walls, suggesting stimulation of the transport of ureides to developing fruits (Fig. 16.5). A meta-analysis based on 20 trials showed a 180% increase in BNF due to K supply compared with the nonfertilized control, which was attributed to improved plant growth (Santachiara et al., 2019).

16.5.1.3 Calcium

In the symbiosis, Ca is important from the early infection events (Fig. 16.5) (Bonilla et al., 2011; Shailes and Oldroyd, 2015) by playing a key role in molecular signaling. First, it influences the synthesis of the *nod* gene–inducing flavonoids by the host plants (Bonilla et al., 2011), such as white (*T. repens* L.) and subterranean (*Trifolium subterraneum* L.) clovers, in which increasing Ca supply increased the release of *nod* gene–inducing compounds in root exudates (Richardson et al., 1988). Second, Ca has a key role in the plant perception of the Nod factors released by rhizobia

TABLE 16.4 Number of nodules and N and P contents in alfalfa (*Medicago sativa*) shoots grown in various soils with different P supply or inoculated with arbuscular mycorrhizal fungus (AMF).

Parameter	Zaidin soil, available P (mg kg^{-1})				
	6.1	12.6	17.4	25.8	6.1 + AMF
Nodulation (number plant^{-1})	75	91	110	123	100
N content (mg plant^{-1})	25	35	46	57	66
P content (mg plant^{-1})	1.0	1.5	1.8	2.2	2.6
	Arenales soil, available P (mg kg^{-1})				
	4.5	9.1	19.2	28.2	4.5 + AMF
Nodulation (number plant^{-1})	41	82	130	135	178
N content (mg plant^{-1})	9	26	31	53	57
P content (mg plant^{-1})	0.5	1.0	1.9	2.2	2.5

Source: Based on Azcon and Barea (1992).

(Shailes and Oldroyd, 2015). Calcium signaling can be represented by dynamics of Ca^{2+} ionic activity in a cellular compartment via either a single peak or a specific pattern of oscillations. In the symbiosis the perception of the Nod factors has been associated with Ca nuclear oscillation (spiking) and Ca influx in the root tips, both activating early events in bacterial infection (Shailes and Oldroyd, 2015). Calcium also plays an essential role as a secondary messenger via a unique Ca-activated kinase (Oldroyd and Downie, 2008). The deficiency of Ca impairs nodule structure and nodulation and decreases the N_2 fixation rate (Banath et al., 1996; Lowther and Loneragan, 1968).

16.5.1.4 Magnesium

Bergersen (1961) was one of the first researchers to highlight the importance of Mg for rhizobia (Fig. 16.5). Among its many roles, Mg is an activator of enzymes involved in the energy metabolism and plays important roles in metabolism of N. There are few studies on the direct role of Mg in the symbiosis, with reports of positive effects on root architecture, nodule number, dry weight, and N_2 fixation (Egamberdieva et al., 2018; Kiss et al., 2004).

16.5.1.5 Sulfur

Sulfur is a constituent of amino acids cysteine and methionine, metal cofactors, coenzymes, and secondary metabolites (Becana et al., 2018). Deficiency of S-containing amino acids cysteine and methionine may restrict the nutritional value of seeds and vegetative parts (Sexton et al., 1998). Sulfur deficiency may also decrease N_2 fixation by affecting nodule development and function (Pacyna et al., 2006; Scherer, 2008), reducing leghemoglobin concentration in nodules (Scherer et al., 2008; Singh and Raj, 1988), lowering ATP concentration in bacteroids as well as in root nodule mitochondria, and decreasing the concentration of ferredoxin in nodules (Fig. 16.5); these effects may be caused by low carbohydrate supply to nodules of S-starved legumes (Becana et al., 2018; Scherer et al., 2006). Sulfur is important for nitrogenase activity (Fig. 16.5) because the smaller of the two O_2-sensitive proteins contains a single Fe_4S_4 unit (Jeong and Jang, 2006) and also because of the role of S in the activity of several other important enzymes directly or indirectly involved in BNF; in addition, S plays a major role in photosynthesis, influencing the supply of carbon substrates to BNF (Becana et al., 2018). It has been shown that S application to low-S soils may not only increase the quality of grain legumes but also enhance root growth, nodule number, nodule dry weight, nitrogenase activity, leghemoglobin synthesis, N accumulation in plants, proportion of N derived from N_2 fixation, biomass production, and grain yield (Becana et al., 2018; Divito and Sadras, 2014; Scherer, 2008; Scherer et al., 2006).

16.5.1.6 Molybdenum

Because Mo is a metal component of nitrogenase, the legume–rhizobia symbiosis has high Mo requirement (Bambara and Ndakidemi, 2010), and a deficiency of this micronutrient impairs BNF (Fig. 16.5). Although Mo is not specifically needed for nodulation, Mo deficiency affects nodule development by reducing bacteroid multiplication (O'Hara, 2001); at low supply, Mo is preferentially transported into the nodules (Brodrick and Giller, 1991). Because measuring Mo concentration in soils is challenging, the fact that Mo accumulates in nodules represents an alternative strategy to detect its deficiency, with the establishment of critical levels in nodules (Jacob-Neto and Franco, 1989). The Mo deficiency–induced N deficiency in legumes relying on N_2 fixation is particularly widespread in acid mineral soils in the humid and subhumid tropics but may also be associated with continuous cropping, soil erosion, and decrease in soil organic matter. Under Mo deficiency, there are decreases in nodulation, BNF rates, plant growth, yield, and protein content in legumes (Campo et al., 2009; Hegazy et al., 1990).

Seed pelleting with Mo was used to meet the needs of BNF; however, toxic effects of Mo on rhizobia applied to the seeds have been reported (Burton and Curley, 1966). Therefore, soil Mo application and foliar sprays are commonly used (Campo et al., 2009; Hegazy et al., 1990), avoiding the direct contact with the seed-applied rhizobia. It has also been shown that the response to Mo fertilization depends on the Mo concentration in seeds (Brodrick et al., 1992). In this context, Campo et al. (2009) verified that the production of Mo-enriched seeds (by Mo application via foliar spray at the pod-filling stage) is a useful approach, increasing BNF, grain yield (Table 16.5), and protein concentration in grains; in addition, considerable increases in seed Mo concentration were obtained (as much as 3000%) in comparison with seeds obtained from plants not sprayed with Mo.

For soybean in Brazil, where soils are low in Mo, the regular recommendation for each crop season consists of 12–25 g ha^{-1} of Mo via foliar sprays at the early vegetative growth stage (V3–V5), to avoid toxicity to seed-applied rhizobia (Hungria and Nogueira, 2019). To enrich soybean seeds with Mo, the recommendation is for 200 g ha^{-1} of Mo in two spray applications around the R3 and R5.4 reproductive stages (Hungria and Nogueira, 2019).

TABLE 16.5 Grain yield of soybean grown from seeds with low, medium, high, and very high Mo concentrations in response to the application of 0, 10, or 20 g ha^{-1} of Mo to seeds at sowing.

Seed Mo concentration (µg g^{-1})	Mo application (g ha^{-1})		
	0	10	20
	Grain yield (t ha^{-1}), year 1		
Low (0.0)	2.8 cB[a]	3.1 bA	3.0 bA
Medium (0.3)	3.0 bA	3.2 bA	3.0 bA
High (7.6)	3.4 aB	3.5 aAB	3.6 aA
	Grain yield (t ha^{-1}), year 2		
Low (0.73)[b]	2.3 bB	2.6 bA	2.8 bA
Medium (7.5)	3.2 aB	3.8 aA	3.8 aA
High (13.3)	3.6 aA	3.9 aA	3.8 aA
	Grain yield (t ha^{-1}), year 3		
Low (2.4)	2.4 bB	2.7 aA	2.7 aA
Medium (9.8)	2.6 abA	2.6 aA	2.6 aA
High (15.6)	2.6 abA	2.7 aA	2.6 aA
Very high (31.6)	2.8 aA	2.7 aA	2.8 aA

[a]Means (n = 6) followed by different letters in the same column (lowercase) or row (uppercase) are significantly different at P ≤ 0.05.
[b]Note progressive enrichment of seed in Mo from one cropping season to the next.
Source: Based on Campo et al. (2009).

16.5.1.7 Iron

In N$_2$ fixation, Fe is a component of several enzymes including nitrogenase, several hydrogenases, electron carrier ferredoxin, and leghemoglobin (Fig. 16.5) (Abdelmajid et al., 2008). The heme component of the leghemoglobin has a particularly high Fe concentration, resulting in a higher requirement for nodule formation than for host plant growth (Abdelmajid et al., 2008). Iron deficiency does not impair the infection process but hampers division of cortical cells in the early stages of nodule development and the proliferation of invading rhizobia in root tissue (Tang et al., 1992) and affects the synthesis of nitrogenase (O'Hara et al., 1988). Based on the data from eight trials, increasing iron supply raised the average nitrogenase activity by 23% in soybean nodules (Santachiara et al., 2019).

16.5.1.8 Boron

Boron plays an important role in nodule structure in all legume and actinorhizal N$_2$-fixing symbioses (Fig. 16.5) (Bonilla and Bolaños, 2009). Legumes require a four- to fivefold higher B concentration for nodule development than for root growth (Carpena et al., 2000). Hence, deficiency of B in the rooting medium reduces nodulation and impairs nodule development (Bolaños et al., 1994).

Boron is required for the development of infection threads and nodule cell invasion (Fig. 16.5). Under B deficiency the binding of rhizobial cells to the infection thread wall is inhibited; bacteria cannot progress through the infection thread and are unable to establish inside the plant. Boron deficiency also causes abortion of infection threads and degeneration of cell walls and membranes surrounding the intracellular bacteroids (Bolaños et al., 1996). These impairments to nodule development result in decreased N$_2$ fixation. A meta-analysis of 28 trials showed that nitrogenase activity increased by 30% in soybean nodules in plants fertilized with B compared with the no-B control (Santachiara et al., 2019).

The physiological B−Ca interactions influence transport processes across the plasma membrane and impacts on the structure and function of the cell wall. This interaction was also observed in cyanobacteria, influencing the maintenance of the envelope of specialized N$_2$-fixing heterocysts (Bonilla and Bolaños, 2009).

16.5.1.9 Cobalt

Cobalt is closely related to BNF as a component of the cobalamin coenzyme B12 that has Co(III) as the metal component and is required for enzymes such as methionine synthase and ribonucleotide reductase involved in bacteroid differentiation, and the methylmalonil coenzyme A mutase involved in the synthesis of heme groups in leghemoglobin and bacterial cytochromes (Fig. 16.5) (Bonilla and Bolaños, 2009). Severe Co deficiency reduces root infection and delays nodule formation, whereas nodule growth rate is usually not influenced by Co supply (O'Hara, 2001). Cobalt deficiency may affect nodule development and function in several ways; for example, in lupins relying on symbiotic N_2 fixation, Co deficiency depresses host plant growth, leghemoglobin concentration, and nitrogenase activity (Riley and Dilworth, 1985) (Table 16.6). The most sensitive indicator of Co deficiency is the bacteroid content in nodules. The synthesis of leghemoglobin is enhanced by Co supply, but an increase in nitrogenase activity per unit of leghemoglobin is relatively small.

In Brazil the recommended dose of Co to maximize BNF in soybean is 2–3 g ha^{-1}. However, symptoms of Co toxicity have frequently been reported in both bacteria and plantlets when applied directly to the seeds; by contrast, foliar sprays with Mo at early soybean growth stages (V3–V5) can avoid toxicity (Hungria and Nogueira, 2019).

16.5.1.10 Manganese

Because Mn is an activator of several enzymes in the Krebs and Calvin cycles and the shikimic acid pathway, Mn deficiency affects plant growth and consequently the BNF process; however, Mn can be replaced by Mg in some of these functions (Bonilla et al., 2011). Superoxide dismutases (SOD) are important in protecting cells against ROS, such as the superoxide radical (O_2^-). Two phylogenetically unrelated SOD families, CuZnSOD (Cu and Zn SOD) and FeMnCamSOD (iron, manganese, or cambialistic SOD), can be found in different locations within legume cells and rhizobial bacteroids, with FeMnCamSOD being of bacterial origin (Bannister et al., 1987; Rubio et al., 2004). In indeterminate nodules of pea, FeMnCamSOD was the main SOD present in the bacteroids and in the cytosol, whereas FeSOD was found only in determinate nodules, for example, in cowpea and soybean; both SODs might also have specific functions related to signaling in the symbiosis (Asensio et al., 2011). Manganese is also required for ureide metabolism in leaves; soybean genotypes tolerant to drought showed increased leaf Mn^{2+} concentrations, promoting ureide breakdown that avoids the feedback effect in nodules and prolongs N_2 fixation under water deficit (Purcell et al., 2000).

In tropical acid soils, Mn toxicity is a frequent limitation to BNF, which can be alleviated by liming. Differences in tolerance to Mn toxicity, reflected in differential decreases in nodulation and BNF, were reported for tropical pasture legume species (Souto and Döbereiner, 1968) and common bean cultivars (Döbereiner, 1966), indicating the feasibility of selecting more Mn-tolerant hosts. Furthermore, mycorrhizal symbiosis alleviates Mn toxicity (Nogueira et al., 2007) and increases nodulation in soybean (Sakamoto et al., 2019) and alfalfa (Azcon and Barea, 1992).

16.5.1.11 Copper, zinc, and silicon

Even though deficiencies of Cu and Zn have been reported in many plants, there are very few such studies on BNF. Copper has a role in rhizobial respiratory proteins (Delgado et al., 1998) and the expression of *melA*-genes (melanin) in

TABLE 16.6 Nodule development and function in *Lupinus angustifolius* at different Co supply.

Parameter	Co supply (mg CoSO$_4$.7H$_2$O per 6 kg of soil)				
	0	0.01	0.05	0.10	0.50
Foliage mass (g^{-1} plant fw)	5.0	6.1	7.5	9.6	14.0
Nodule mass (g^{-1} plant fw)	2.9	2.8	2.5	2.3	1.1
Bacteroid number ($\times 10^9$ per nodule)	6.0	12.0	12.5	20.5	22.5
Leghemoglobin concentration (nmol g^{-1} lateral root fw)	–	1	11	20	120
Nitrogenase activity (nmol C$_2$H$_2$ reduced g^{-1} nodule fw min^{-1})	10	21	58	104	172
Nitrogenase activity (nmol C$_2$H$_2$ reduced nmol^{-1} leghemoglobin min^{-1})	1.1	2.5	3.7	3.8	3.2

Source: Recalculated from Riley and Dilworth (1985).

some rhizobial strains (Hawkins and Johnston, 1988) (Fig. 16.5). One of the few reports on the effects of Cu supply on BNF indicated no improvement in nodulation of soybean grown in an Inceptisol of Indonesia (Johannes and Gunarto, 1987). On the other hand, excess Cu impairs BNF (Tindwa et al., 2014).

Zinc availability affects not only rhizobial growth, but also nodule number and size, leghemoglobin content (Fig. 16.5), plant dry weight, and the amount of N fixed in soybean, cowpea, and chickpea (Kryvoruchko, 2017). In alfalfa, Zn deficiency affects nodule number and dry weight (Grewal, 2001).

Even though a direct role of Si in BNF has not been shown yet, Si has been increasingly applied to legume crops. In cowpea, Si improved nodule number, dry weight, and the amount of nitrogen fixed per plant (Nelwamondo and Dakora, 1999). In soybean the application of 200 mg kg^{-1} of Si to soil increased nodule size, chlorophyll content, and leaf N concentration (Steiner et al., 2018) (Fig. 16.5), whereas in chickpea Si enhanced the symbiosis by improving water-use efficiency (Kurdali et al., 2013).

16.5.1.12 Nickel

Nickel is a constituent of hydrogenase. In pea, Brito et al. (2000) reported that Ni addition was necessary to achieve maximum hydrogenase activity. By contrast, the addition of NiCl$_2$ to the nutrient solution did not increase hydrogenase activity in common bean, indicating that the plant provided enough Ni for the synthesis of hydrogenase in the bacteroids (Torres et al., 2020). Nickel is also a component of ureases in both legumes and bacteria. For ureide-transporting legumes, urea is an intermediate of ureide metabolism and has to be broken down by urease to prevent leaf necrosis (Bonilla and Bolaños, 2009).

Under field conditions, increases in soybean grain yield with the application of Ni have been reported along with improvements in N metabolism, concentrations of leaf N, chlorophyll, ammonia, ureides and urea, urease activity, nodulation, and nitrogenase activity (Freitas et al., 2018, 2019; Lavres et al., 2016). Freitas et al. (2018) concluded that the application of 0.5 mg Ni kg^{-1} to soil (equivalent to the application of 1 kg ha^{-1} of Ni) increased yield without exceeding safe levels of this element in grains for human consumption. In relation to application to seeds, potential Ni toxicity to the rhizobia must be considered. A likelihood of soybean response to Ni is relatively high in sandy soils with low concentration of Ni, or in medium- to coarse-textured soils that received lime (Lavres et al., 2016).

16.5.1.13 Nutritional and other disorders related to soil acidity

Acidity is a major problem especially in tropical, ancient, highly weathered, and leached soils, frequently containing toxic concentrations of available Al or Mn. Acidity affects not only the free-living rhizobia but also the host plants and the symbiosis (Slattery et al., 2001). In acid soils, factors such as high concentrations of H$^+$ and monomeric Al (Alva et al., 1990) contribute to poor nodulation and inhibition of plant growth (Hungria and Vargas, 2000). Moreover, Ca and P availability are adversely affected by low soil pH, affecting rhizobial growth and survival (Sadowsky, 2005).

The microsymbiont is usually the most pH-sensitive partner (with the optimum pH for growth between 6.0 and 7.0), although there are some rhizobial species and strains that tolerate lower or higher pH (Hungria and Vargas, 2000). Rhizobial survival and abundance, in addition to their competitiveness in nodulation, are affected strongly by soil acidity (Ferguson et al., 2013; Schubert et al., 1990). In general, nodule formation is very sensitive to soil acidity (Ferguson et al., 2013), affecting all steps of the infection process, including the exchange of molecular signals between the symbiotic partners (Hungria and Vargas, 2000) and the inhibition of root hair formation due to low concentrations of Ca (Ewens and Leigh, 1985).

In acid soils the net release of H$^+$ into the rhizosphere (inherently coupled with N$_2$ fixation by nodulated legumes) may exacerbate the effects of acidity. Depending on the legume species, between 37 and 49 mg of H$^+$ is released per gram of fixed N, amounting to an annual production of 4.6 kg H$^+$ ha^{-1} in sweet clover (*Melilotus albus* Medik.) and 15.2 kg H$^+$ ha^{-1} in alfalfa (Lui et al., 1989). Soil pH was lowered from 6.0 to 5.6 during 45 days of faba bean (*Vicia faba* L.) growth (Yan et al., 1996). In soils containing high concentration of CaCO$_3$, the H$^+$ released is neutralized, and no acidification occurs (Mengel, 1994). Otherwise, an equivalent of 80−96 kg of lime (CaCO$_3$) may be required to neutralize the acidity formed in the production of 1 t of legume dry matter (Jarvis and Hatch, 1985).

Increasing soil pH by liming is an effective way of mitigating the effects of acidity. There are reports that liming increases nodulation and the BNF contribution in common bean (Buerkert et al., 1990), alfalfa (Pijnenberg and Lie, 1990), and peanut (Angelini et al., 2005). Another (relatively short-term) option to improve nodulation in low-pH soils is to use acid-tolerant legume cultivars (Ferguson et al., 2013).

16.5.2 Effect of mineral nitrogen

The effect of mineral N (from either mineralization of soil organic matter or fertilizers) on BNF is well documented (Peoples and Baldock, 2001). In legumes (and other symbiotic N_2-fixing systems), mineral N may enhance or depress N_2 fixation, depending on a range of factors, including the rate of N supply. However, the most common effect of mineral N is to decrease BNF, and the magnitude of this effect depends on the timing, rates, and conditions of application. Mineral N may decrease nodulation (by reducing the bacterial attachment to the infection sites) and root infection, inhibit leghemoglobin synthesis and nitrogenase activity, and accelerate nodule senescence (Fig. 16.5) (Bonilla and Bolaños, 2009). An enhancing effect of low supply of mineral N on BNF has been related to the lag phase between root infection and the onset of N_2 fixation. Nitrogen deficiency during this phase may be detrimental to the formation of a leaf area sufficiently large to supply the photosynthates needed for nodule growth and activity; however, the N pool in the cotyledons is usually sufficient to supply the plant needs in that early phase (Hungria and Thomas, 1987).

In Brazilian soils, the majority of which are very poor in available N, starter N doses as low as 20–40 kg ha^{-1} decreased both nodulation and N_2 fixation in soybean (Hungria et al., 2006; Hungria and Mendes, 2015; Hungria and Nogueira, 2019; Kaschuk et al., 2016; Saturno et al., 2017). Interestingly, in common bean, the early period of N deficiency before the full establishment of the symbiosis was not limited by low N_2 fixation activity, but rather by the low expression of enzymes involved in N assimilation and transport of fixed N to the xylem vessels; however, some rhizobial strains promoted an earlier establishment of the BNF process, showing that the selection of elite rhizobial strains is key for early and effective BNF (Hungria et al., 1991).

Nitrogenase activity and nodule numbers decrease with increasing supply of N (Scherer and Danzeisen, 1980). Shoot growth, on the other hand, continues to increase, indicating a shift from symbiotic to inorganic N nutrition (Hungria and Thomas, 1987). The highest N content in the shoot coincides with the highest nitrogenase activity, but not with the highest plant dry weight. Although this suggests that at maximum N_2 fixation dry matter production was source-limited (photosynthate supply) and that the amount of photosynthates used for N_2 fixation may restrict plant growth, there is also evidence that N_2 fixation stimulates photosynthesis, which, in turn, supports higher rates of BNF (Kaschuk et al., 2009). In addition, although nitrogenase activity drops earlier than that of nitrate reductase in common bean, suggesting the need for N supplementation after flowering (Franco et al., 1979), this has not been confirmed in studies with soybean in Brazil, wherein late applications of N fertilizers did not result in any increase in N accumulation in shoots and grains, or in grain yield (Hungria and Mendes, 2015; Hungria and Nogueira, 2019; Kaschuk et al., 2016; Saturno et al., 2017).

Dozens of experiments performed in Brazil confirmed no benefits of supplemental N fertilizer application in soybean in different soil management systems (no-till or conventional tillage), using cultivars of different maturity groups, genetically modified or conventional, of indeterminate or determinate growth types, with different sources and ways of N fertilizer applications at sowing, early flowering, full flowering, and pod-filling stages (Hungria et al., 2006; Hungria and Mendes, 2015; Hungria and Nogueira, 2019; Saturno et al., 2017). A meta-analysis indicated that the application of mineral N led to a relative decrease in N derived from N_2 fixation in both field (by 44%) and greenhouse experiments (by about 70%) (Santachiara et al., 2019).

The extent to which nodulation and nodule activity are decreased by mineral N may depend on the plant genotype and the source of supplied N. High rates of nitrate have a more inhibitory effect on nodulation in soybean than chickpea or lupins, whereas nitrogenase activity is severely inhibited in chickpea and lupins but is only slightly affected in subterranean clover (Harper and Gibson, 1984).

The regulation of root metabolism to ensure an adequate supply of N to plants involves feedback systems, whereby the N status of the whole plant influences root growth, transport activity, and, in the case of legumes, nodule growth and activity (Parsons and Sunley, 2001). Although the precise signals that communicate plant N status to the root nodules are not clear yet, N-rich amino acids transported from shoots play a role in regulating the nodule formation and function; hence, sensing may occur in the shoot cells and signals may be communicated to roots and nodules via the phloem (Parsons and Sunley, 2001). These signals appear to be operating in a quantitative way, allowing N uptake and N_2 fixation to be matched to the plant demand.

The inhibitory effect of nitrate on nitrogenase activity in nodules may operate at several levels: (1) nitrate reduction in nodules could lead to competition for reducing power with bacteroid or mitochondrial respiration inside the nodule (Heckmann et al., 1989); (2) nitrite toxicity (Becana et al., 1985); and (3) O_2 deficiency (Vessey et al., 1988). The last two factors are considered particularly relevant. Nitrate supply rapidly induces nitrate reductase activity in bacteroids, but in some strains of rhizobia nitrite reductase is induced only after a considerable delay, leading to accumulation of nitrite (Arrese-Igor et al., 1990), which can inactivate leghemoglobin by forming nitrosyl-leghemoglobin (Kanayama and Yamamoto, 1991). Inhibition of nitrogenase by nitrate may also be due to O_2 deficiency because the effect can be alleviated by increasing the O_2 partial pressure in the rhizosphere (Vessey et al., 1988). Similarly to stem girdling and defoliation, nitrate also significantly increases the resistance of nodules to O_2 diffusion (Vessey et al., 1988).

16.6 Soil and environmental limitations

The rhizosphere environment strongly influences the symbiotic interaction between rhizobia and their host legumes. Favorable conditions for plant growth and establishment of bacterial populations enhance the inoculum success and promote the development of infection sites on root hairs. Therefore, in addition to soil fertility as discussed in Section 16.5, other soil factors that influence plant and rhizobial growth, such as salinity, temperature, moisture, and physical structure, also influence root infection, nodulation, and nitrogen fixation efficiency (Hungria and Vargas, 2000; Slattery et al., 2001).

16.6.1 Salinity

Most legumes are salt-sensitive (El-Hamdaoui et al., 2003), and effects of salt stress on N_2 fixation have been reported widely (Zaran, 1999). Salinity may influence the symbiosis directly by reducing the growth of the host plant, or indirectly by impairing interactions between rhizobia and the host plant, leading to inhibition of nodule formation (Anthraper and DuBois, 2003). Bacterial attachment to the roots is impaired under salinity, and root hairs do not show the characteristic response (deformation and curling) to Nod factors (El-Hamdaoui et al., 2003). However, one intriguing exception occurs with *Rhizobium tropici* strain CIAT 899, a symbiont of common bean and other legumes, which synthesizes a variety of Nod factors under saline conditions, even in the absence of *nod* gene−inducing signals released by the host legume (del Cerro et al., 2015; Guasch-Vidal et al., 2013). Salt stress also inhibits bacterial invasion and proliferation inside the host cells but can be alleviated by supplementing legumes with B and Ca (Bolaños et al., 2006). However, these effects vary among legume species, for example, common bean is more salt-sensitive than soybean and alfalfa (Serraj et al., 1998). Salt-tolerant rhizobial strains may be valuable for symbiosis in saline environments (Ogutcu et al., 2010), but the tolerance of the host legume might be even more important for the establishment of successful symbiosis. The search for tolerant rhizobia and plant breeding for tolerance to salinity represent valuable tools for achieving efficient BNF in saline environments (Dwivedi et al., 2015).

16.6.2 Soil water content

Drought limits agricultural production worldwide. Due to rising temperatures in recent and future years, evaporative demands will increase and dry spells will become more frequent (5−10 times) (Naumann et al., 2018). This scenario represents a serious limitation to BNF, which is affected by drought even before transpiration and photosynthesis (Arrese-Igor et al., 2011; King and Purcell, 2005; Sinclair et al., 2007). Low soil water content limits rhizobial survival in the free-living stage as well as in symbiosis by impairing plant growth, root architecture, and root exudation (Sadowsky, 2005). Tolerance to and recovery after drought also depend on the type of nodule structure; because indeterminate nodules have prolonged meristematic activity, they show higher drought tolerance than determinate ones (Kantar et al., 2010). The timing of drought stress relative to plant growth stage also has important effects on nodulation and N_2 fixation; an extended period of drought stress during the vegetative stage delays both processes. Once nodules are established, drought reduces N_2 fixation (Peña-Cabriales and Castellanos, 1993). The negative effect of drought on nodule activity is related to limitation of O_2 diffusion to bacteroids (Durand et al., 1987); however, other factors are involved, as raising the external O_2 concentration cannot restore nodule activity (Kantar et al., 2010).

Recovery of nodule activity involves both the rehydration of the existing N_2-fixing nodule tissues and a resumption of nodule meristem growth (del Castillo et al., 1994). In soybean, decreased N_2 fixation at low water availability is associated with increased concentrations of ureides and free amino acids in the plant tissues, indicating a potential feedback inhibition of BNF (King and Purcell, 2005). Restricting O_2 diffusion to the bacteroids in the nodules has also been proposed as one of the mechanisms limiting BNF by constraining the source of energy required for N_2 reduction by the nitrogenase (King and Purcell, 2001).

Additional microbial strategies to mitigate the impacts of drought on BNF have been considered. Coinoculation with plant growth−promoting bacteria (PGPB) (Cerezini et al., 2016) or AMF (Dwivedi et al., 2015) under field conditions has led to increased root growth and more effective water uptake. There is variation in the performance of soybean genotypes under drought that can be explored in breeding for more tolerant varieties (Sinclair et al., 2007). In soybean, drought-tolerant genotypes accumulate more N in shoots and have higher BNF rates under drought than sensitive genotypes (Chen et al., 2007), which can be explored in breeding (Cerezini et al., 2016, 2019; Sinclair and Nogueira, 2018). Tolerance to drought stress has also been reported in common bean and groundnut genotypes (Dwivedi et al., 2015).

Waterlogging in poorly drained fields and those with shallow water tables is limiting the growth of many crops. Flooding decreased the incorporation of $^{15}N_2$ into amino acids, activity of nitrogenase, and the expression of *nifH* and *nifD* genes in soybean nodules; it also decreased the concentration of asparagine (the most abundant amino acid) and increased that of γ-aminobutyric acid (GABA), suggesting that GABA has a storage role during flooding stress (Souza et al., 2016).

The presence of the *nosZ* gene confers a competitive advantage to *Bradyrhizobium* microsymbionts of soybean in flooded soils, potentially favoring saprophytic survival or enhancing competitiveness during waterlogging or flooding, such as in paddy−soybean rotations (Saeki et al., 2017).

In natural ecosystems, legumes can adapt to different water conditions, as in the periodically flooded forest soils of the central Amazon floodplain, where no differences in N_2 fixation rates among various legumes were found between dry and wet seasons (Kreibich et al., 2006).

16.6.3 Temperature

The BNF process is more sensitive to high temperatures than leaf gas exchange or plant growth (Montañez et al., 1995). High temperatures impair the survival and persistence of rhizobial strains in soils, as well as all stages of nodule formation and functioning, starting with the release of *nod*-gene inducers (Hungria and Vargas, 2000). High temperature also decreases root hair formation, bacteria adherence to root hairs, root hair infection and formation of the infection thread, rhizobial release from the infection thread into symbiosomes, bacteroid development, leghemoglobin synthesis (Hungria and Vargas, 2000; Sadowsky, 2005), *nif* gene expression (Michiels et al., 1994), nitrogenase activity (Hungria and Kaschuk, 2014; Hungria and Vargas, 2000; Montañez et al., 1995), N metabolism (including the nodule enzymes GS and GOGAT), ureide transport in the xylem sap (Hungria and Kaschuk, 2014), and nodule senescence (Hungria and Vargas, 2000). However, it is possible to identify and select rhizobia (Hungria et al., 1993; Hungria and Vargas, 2000) and plant genotypes (Dwivedi et al., 2015) tolerant to high temperature; hence, breeding programs for improving yields under high temperatures must consider the best combinations of rhizobia and host plants (Hungria and Kaschuk, 2014).

Crop and soil management influence the temperature in the root zone. In the tropics, soil temperature in the no-till system (sowing directly into the residues of the previous crop) may be 5°C−10°C lower (with minimal soil temperature fluctuations) than in conventional tillage systems. Lower temperatures in Brazilian tropical soils under no-till management increase the rhizobia survival, promoting nodulation, nitrogen fixation rates, and grain yields (Hungria and Vargas, 2000).

The BNF is also affected by low temperatures. In *Trifolium repens*, compared with 12°C, N_2 fixation decreased two- and fourfold at 9°C and 6°C, respectively (Bouchart et al., 1998).

16.7 Methods to quantify the contribution of BNF, amounts of N fixed by legumes, and N transfer to other plants in intercropping and crop rotations

Several methods have been used to quantify the contribution of BNF to plant N nutrition. The N balance and N difference methods are simple and based on comparing total N in two treatments or sets of conditions, considering outputs and inputs (Unkovich et al., 2008).

The acetylene reduction assay (ARA) has been used for decades to detect and quantify BNF. As mentioned in Section 16.3, acetylene is an alternative substrate for the nitrogenase, which converts it to ethylene that is detected by gas chromatography at a low cost. However, as also mentioned in Section 16.3, at least 25% of the total electrons is used by nitrogenase to produce H_2, and this gas must also be quantified to guarantee the accuracy of the method. Studies with legumes have used mainly detached roots or nodules (Boddey et al., 1987; Hardy et al., 1968) but Minchin et al. (1983, 1986) demonstrated that physical disturbance affects the ARA, mainly by affecting gas exchanges in the nodules. Flow-through gas systems were then used to overcome the effects of disturbance, but physiological limitations caused by acetylene were also detected (Minchin et al., 1986); hence, the use of this method has been discouraged since the 1980s (Unkovich et al., 2008). As an alternative, if N_2 is replaced by an inert gas such as argon (Ar), all electrons directed to the nitrogenase will lead to H_2 production, without inhibiting the nitrogenase, but the method has the limitation of Ar-induced inhibition lasting only about 2 hours (Unkovich et al., 2008).

For legumes transporting mainly N-ureides, quantification of the concentrations of N-ureides in the xylem sap, or in plant tissues such as petioles, has been used to quantify the contribution of BNF (Boddey et al., 1987; Herridge et al., 2008; Unkovich et al., 2008). However, the methods of ^{15}N dilution and natural isotopic abundance are preferred, particularly if quantification is the main objective (Unkovich et al., 2008).

Legumes are commonly used as a source of N in intercropping or in rotation with nonlegume crops (Glasener et al., 2002). In monocropping, the total amount of fixed N_2 and the proportion of plant N coming from fixation vary greatly. In some cases, nearly all plant N is derived from atmospheric N_2 (Table 16.1). On average, around 20–25 kg of N are fixed per ton of shoot dry matter in legumes across a broad range of environments (Peoples and Baldock, 2001).

In low-fertility soils, intercropped legumes may accumulate relatively small amounts of fixed N due to low yield. However, the proportion of fixed N is usually increased because uptake of soil N by the intercropped nonlegume can reduce the mineral N concentration in the soil, stimulating BNF in the legume (Hauggaard-Nielsen et al., 2009; Vanlauwe et al., 2019). For example, in field trials with sole crops or intercropping in Ghana, the proportion of N derived from fixation was greater in intercropping on poorly fertile (55%–94%) than fertile soils (23%–85%), although the total amounts of fixed N were lower in the low- (15–123 kg ha^{-1}) than high-fertility soils (16–145 kg ha^{-1}) (Kermah et al., 2018).

In addition to soil fertility, other environmental factors, and the rhizobia population, the amount of fixed N depends on the host species and the architecture and density of the legume component in species mixtures. For example, in mixed stands of *Trifolium* spp. and *Lolium* spp., annual N_2 fixation has been estimated to be 232–308 kg ha^{-1} (75%–86% of the total plant N) and may reach 390 kg N ha^{-1} in mixed stands of *Trifolium* spp. and *Festuca arundinacea* (Mallarino et al., 1990).

In polyculture, direct transfer of fixed N from legumes to nonlegumes may occur during the growing season. With no application of N fertilizer the grain yield of maize intercropped with cowpea increased by 72% over the sole maize crop, likely due to the transfer of N from cowpea (Remison, 1978). A hyphal network of AMF can link the roots of neighboring plants and may be involved in the transfer of N from legumes to nonlegumes (Frey and Schüepp, 1992). However, the amounts transferred may be relatively small; for example, in soybean-maize intercropping, maize received 15% of its N content from soybean through N transfer via the mycorrhizal hyphal network (Wang et al., 2016).

Several studies have reported increased yields when cereals were grown after legumes due to the N left in the soil as shoot and root residues and nodules. According to Peoples et al. (2009) and Herridge (2011), about one-third of the total N fixed by legumes is assumed to remain in the roots and may be either mineralized and used by the following crop or converted into soil organic matter. In field-grown narrow-leafed lupin, more than 80% of the N requirement can be derived from BNF (Herridge and Doyle, 1988), with amounts exceeding the N exported in grains. In a meta-analysis of 44 studies in Africa, maize yield after a legume crop was on average 500 kg ha^{-1} higher than after maize (Franke et al., 2018). In another study in Africa, maize grain yields were related to the amount of N input to the soil by legumes in rotation across three agroecological zones in Western Kenya (Ojiem et al., 2014).

In general, incorporation of legume residues into soil causes faster N mineralization in comparison with surface placement. Ladd et al. (1981) found that wheat recovered, on average, only 14% of the N from residues of *Medicago littoralis* that were incorporated into three different soil layers under field conditions, but Yaacob and Blair (1980) reported an increase from 13.4% to 55.5% in the N recovered by pasture grass grown for 1–6 years after legumes. In addition to the fixed N, the benefits of rotation with legumes can be attributed to other yield-promoting effects, including improvements in soil organic matter, structure, water holding capacity, P mobilization, and lower incidence of pests and diseases (Vanlauwe et al., 2019).

16.8 Significance of free-living and associative nitrogen fixation

Free-living N_2-fixing microorganisms are ubiquitous in soils. However, because of carbon limitation, especially in nonrhizosphere soil, amounts of fixed N are usually small, estimated at <1 kg ha^{-1} year^{-1}. Transfer of fixed N to plants occurs mainly after mineralization, and adding plant residues with high C:N ratios may temporarily stimulate higher rates of BNF by free-living diazotrophs. Based on a mathematical model, the carbon supply near roots due to rhizodeposition has been estimated to support BNF rates between 0.2 and 4 kg N ha^{-1} year^{-1} in natural ecosystems, reaching up to 20 kg ha^{-1} year^{-1} under optimal conditions (Jones et al., 2003). These estimates agree with experimentally derived values (Bremer et al., 1995).

In the surface soils of temperate zones, BNF rates between 13 and 38 kg N ha^{-1} year^{-1} have been recorded for photosynthetic cyanobacterial diazotrophs (Witty et al., 1979), slightly higher than in cyanobacteria–rhizosphere interactions in rice (10–30 kg ha^{-1} year^{-1}), but lower than estimates for the *Azolla–Anabaena* symbiosis (20–100 kg N ha^{-1} year^{-1})

(Roger and Ladha, 1992). Cyanobacteria are also found on leaf surfaces (phylloplane), with estimates of their BNF rates being in the range of 10–20 kg N ha^{-1} year^{-1} in temperate zone forests (Favilli and Messini, 1990) and up to 90 kg N ha^{-1} year^{-1} in tropical rain forests.

Other diazotrophs form closer relationships with plants, with preferential colonization of the rhizosphere and/or root surface, or the occupation of intercellular or intracellular spaces of various plant tissues. They are usually placed into the broader category of PGPB or plant growth–promoting rhizobacteria, in recognition of the likelihood that their beneficial effects can be due to factors other than, or in addition to, BNF. The benefits encompass BNF, synthesis of phytohormones [such as auxins (mainly indole-acetic acid), cytokinins, abscisic acid, and salicylic acid]; inhibition of ethylene, synthesis of vitamins, improvement of nutrient uptake by increased root growth, P solubilization, synthesis of siderophores, and mechanisms that increase tolerance to abiotic and biotic stresses by the induction of systemic tolerance and systemic resistance (Bashan and de-Bashan, 2010; Fukami et al., 2018).

The frequently reported detection of diazotrophs in the rhizosphere of various nonlegume crops from the 1970s to the 1980s (e.g., Day et al., 1975; Döbereiner, 1983), in addition to evidence of yield increases when these organisms were applied as inoculants, created enthusiasm among researchers about the possibilities of associative N$_2$ fixation in nonlegumes. Subsequently, the same high level of interest was dedicated to the endophytic diazotrophs. Although it is still not clear that BNF is the main cause of improved plant growth in most rhizospheric and endophytic bacteria–plant associations (e.g., Bashan and de-Bashan, 2010; Fukami et al., 2018), there is strong evidence of important contributions to the plant N supply. For the endophyte *Gluconacetobacter diazotrophicus*, initially reported to contribute up to 80% of the N requirement in some varieties of sugarcane (*Saccharum* spp.) (Boddey et al., 1991), further research brought that contribution down to a more realistic, but still important, estimate of about 30% (Boddey et al., 2003). *Azoarcus* is an endophyte of kallar grass [*Leptochloa fusca* (L.) Kunth] (Reinhold-Hurek and Hurek, 1998) that expresses *nif* genes and nitrogenase proteins *in planta* and may supply up to 26% of the plant N requirement via BNF (Malik et al., 1997). *Herbaspirillum* is another endophyte that has been reported to supply significant amounts of fixed N to several hosts, particularly rice (Kennedy et al., 2004).

Apart from rhizobia, *Azospirillum* is the most studied and reported PGPB, with hundreds of studies reporting that applications improve the growth of cereals and other nonlegume plants, as well as legumes (e.g., Bashan and de-Bashan, 2010; Fukami et al., 2018; Okon et al., 2015; Pereg et al., 2016). Indeed, in a survey reported by Pereg et al. (2016) *Azospirillum* was identified as a general PGPB for every plant species tested so far. Plant-growth promotion by *Azospirillum* has been attributed to practically all mechanisms cited at the beginning of this section, with the predominance of phytohormone synthesis over BNF (Fukami et al., 2018; Kennedy et al., 2004).

16.9 Microbial inoculation to promote BNF and improve plant nutrition

The use of rhizobial inoculants to increase the contribution of BNF to plant N nutrition was practiced even before the isolation of the first N$_2$-fixing bacteria by transferring soil from a field previously cropped with legumes to new areas to be planted with the same crop, for example, in alfalfa pastures in England. Beijerinck isolated the first rhizobia from nodules in 1888, and soon after in 1890 two other German scientists, Nobbe and Hiltner, demonstrated the advantages of adding pure bacteria to the seeds at sowing. The first rhizobia-based commercial inoculant was produced in 1896 in the United States by the Nitragin Company (Fred et al., 1932; Voelcker, 1896).

Currently, there is a variety of other well-characterized PGPB, mainly *Azospirillum*, *Bacillus*, and *Pseudomonas*, that have also been commercialized as inoculants, with reports of improvements in plant nutrient uptake and grain yield; for example, there are many inoculants containing *Azospirillum* in South America in a market that has grown impressively (Hungria and Nogueira, 2019; Okon et al., 2015; Santos et al., 2019).

Several studies have investigated the mainly positive responses of legumes to inoculation with elite rhizobial strains. However, there are no doubts about the limitations of inoculation in soils with high populations of indigenous or naturalized rhizobia. A classical study with seven legumes [soybean, lima bean (*Phaseolus lunatus*), cowpea, common bean, peanut, *Leucaena leucocephala*, tinga pea (*Lathyrus tingeatus*), alfalfa, and white clover] indicated that rhizobia populations in the soil as low as about 50 cells g^{-1} would limit the establishment of new strains (Thies et al., 1991). In addition, it has been suggested that a successful inoculant expected to form at least 50% of the total number of nodules in soybean should contain at least 1000 times more cells than the rhizobial population in the soil (Weaver and Frederick, 1974). However, there are successful examples of soybean in South America showing impressive increases in nodulation and grain yield even in soils harboring populations of more than 1 million cells of naturalized soybean bradyrhizobia g^{-1} (Hungria and Mendes, 2015; Hungria and Nogueira, 2019). In Brazil, consistent average increases of 8% in soybean grain yields have been achieved by annual inoculation with elite *Bradyrhizobium* strains (Hungria and Mendes, 2015; Hungria

TABLE 16.7 Nodule dry weight (NDW, mg plant^{-1}) at the V4 stage of soybean and grain yield (Y, t ha^{-1}) at two sites in Brazil in soils with an established population of *Bradyrhizobium* spp. in response to seed inoculation with *Bradyrhizobium* and in-furrow inoculation with *Azospirillum brasilense* for two consecutive summer crop seasons.

Treatment	*Bradyrhizobium*[a] (cells g^{-1} soil)	Year 1, site 1 NDW[c]	Y	Year 1, site 2 NDW	Y	Year 2, site 1 NDW	Y	Year 2, site 2 NDW	Y
Not inoculated (NI)	3.57 × 10^3	101 a	2.7 c	106 a	2.0 c	112 a	3.4 c	56 b	2.6 c
NI + 200 kg N ha^{-1}	4.27 × 10^3	33 b	2.9 b	53 b	2.3 ab	24 b	3.8 a	44 b	3.6 a
Inoculated (I) with *Bradyrhizobium* spp.[b]	1.79 × 10^4	102 a	2.9 b	125 a	2.2 b	123 a	3.5 b	195 a	2.9 bc
I + *A. brasilense*[b]	2.14 × 10^4	108 a	3.0 a	121 a	2.5 a	136 a	3.8 a	212 a	3.0 a

[a]*Bradyrhizobium* population in soil before each sowing.
[b]Inoculation at sowing: Bradyrhizobium applied to the seeds at the concentration of 1.2 × 10^6 cells seed^{-1} and A. brasilense applied in-furrow at the concentration of 2.5 × 10^5 cells seed^{-1}.
[c]Means (n = 6) in a column followed by different letters are significantly different (P ≤ 0.05).
Source: Based on Hungria et al. (2013).

and Nogueira, 2019). In Africa, several reports have also indicated improvements in soybean and other legumes by inoculation with elite rhizobial strains, especially when combined with P fertilization (N2Africa, 2019; Vanlauwe et al., 2019). Beneficial results have also been attributed to inoculation with PGPB, especially *Azospirillum*, in a variety of crops (Okon et al., 2015). Moreover, coinoculation of legumes with PGPB can result in further benefits; for example, in soybean, grain yield increases from 8% by a single inoculation with *Bradyrhizobium* to 16% by coinoculation with *Azospirillum brasilense* have been reported (Hungria et al., 2013; Hungria and Nogueira, 2019) (Table 16.7).

Reports of unsuccessful inoculation should be investigated, as failures could be attributed to low-quality inoculants, or to the excess of N in soils due to heavy applications of N fertilizers (Hungria et al., 2006; Hungria and Mendes, 2015; van Kessel and Hartley, 2000). Nevertheless, the reality is that the inoculant market has increased impressively, reaching more than 130 million doses sold in South America in 2020, indicating the farmers' willingness to pay for the benefits of using microbe-based inoculants in agriculture.

16.10 Final remarks

There are no doubts about the importance of the contribution of rhizobia–legume symbioses to global N cycling and N nutrition in agriculture (de Bruijn, 2015). Although on a smaller scale than rhizobia, nonsymbiotic diazotrophic bacteria and other PGPB also contribute to global N input and to the improvement in plant nutrition (Bashan and de-Bashan, 2010; de Bruijn, 2015; Pereg et al., 2016). The protocols for isolation, characterization, and maintenance of rhizobia and other diazotrophs, quantification of their N contribution, as well as for the analysis of microbial inoculants have been developed over many decades (Bergersen, 1980; Howieson and Dilworth, 2016; Somasegaran and Hoben, 1994; Vincent, 1970). These protocols have contributed to the improvement in the quality of microbial inoculants commercially available around the world.

The contribution of BNF to plant N nutrition is dependent on good performance under ever-changing field conditions, with an emphasis on adaptation to changing environmental conditions. However, it is feasible to select and improve both rhizobial and plant genotypes for tolerance to biotic and abiotic stresses (Dwivedi et al., 2015), a strategy that should be emphasized in the following years. Further work could improve BNF, for example, by enhancing mechanisms of initial recognition and infection events that are controlled by the plant genome, and by selecting elite strains able to compete against inefficient indigenous rhizobia. Improvement of rhizobia strains through genetic manipulation such as the introduction of hydrogenase genes (Torres et al., 2020) might result in higher rates of BNF and can now be accelerated with the CRISPR–Cas and other genome editing technologies (Wang et al., 2017).

Great progress in the understanding of plant–microbe signaling has been achieved, which may help in realising the long-standing dream of transferring the nodulation capacity to nonlegume crops. However, there are simple problems to be solved and more effective strategies to consider that would certainly bring about higher and faster impacts of BNF on world agriculture. Among them, the search for elite strains within the local soil microbiome, and improvement in the quality of inoculants reaching many farmers as well as in the strategies for effective inoculation, can improve the benefits of BNF. A practice as simple as the use of high-quality inoculants containing elite rhizobia strains together with P application can result in benefits continent-wide, as described in Africa (Vanlauwe et al., 2019).

References

Abbasi, M.K., Manzoor, M., Tahir, M.M., 2010. Efficiency of rhizobium inoculation and P fertilization in enhancing nodulation, seed yield, and phosphorus use efficiency by field grown soybean under hilly region of Ramalakot Azad Jammu and Kashmir, Pakistan. J. Plant Nutr. 33, 1080–1102.

Abdelmajid, K., Karim, B.H., Chedley, A., 2008. Symbiotic response of common bean (*Phaseolus vulgaris* L.) to iron deficiency. Acta Physiol. Plant. 30, 27–34.

Adu-Gyamfi, J.J., Fujita, K., Ogata, S., 1989. Phosphorus absorption and utilization efficiency of pigeon pea (*Cajanus cajan* (L) Millsp.) in relation to dry matter production and dinitrogen fixation. Plant Soil 119, 315–324.

Alva, A.K., Asher, C.J., Edwards, D.G., 1990. Effect of solution pH, external calcium concentration, and aluminium activity on nodulation and early growth of cowpea. Aust. J. Agric. Res. 41, 359–365.

Andrews, M., Andrews, M.E., 2017. Specificity in legume-rhizobia aymbioses. Int. J. Mol. Sci. 18, 705.

Angelini, J., Taurian, T., Morgante, C., Ibanez, F., Castro, S., Fabra, A., 2005. Peanut nodulation kinetics in response to low pH. Plant Physiol. Biochem. 43, 754–759.

Anthraper, A., DuBois, J.D., 2003. The effect of NaCl on growth, N_2 fixation (acetylene reduction), and percentage total nitrogen in *Leucaena leucocephala* (Leguminosae) var. K-8. Am. J. Bot. 90, 683–692.

App, A., Santiago, T., Daez, C., Menguito, C., Ventura, W., Tirol, A., et al., 1984. Estimation of the nitrogen balance for irrigated rice and the contribution of phototropic nitrogen fixation. Field Crops Res. 9, 17–27.

Arrese-Igor, C., Garcia-Plazaola, J.I., Hernandez, A., Aparicio-Tejo, P.M., 1990. Effect of low nitrate supply to nodulated lucerne on time course of activities of enzymes involved in inorganic nitrogen metabolism. Physiol. Plant. 80, 185–190.

Arrese-Igor, C., González, E.M., Marino, D., Ladrera, R., Larrainzar, E., Gil-Quintana, E., 2011. Physiological responses of legume nodules to drought. Plant Stress 5, 24–31.

Arrighi, J.-F., Barre, A., Ben Amor, B., Bersoult, A., Soriano, L.C., Mirabella, R., et al., 2006. The *Medicago truncatula* lysine motif-receptor-like kinase gene family includes *NFP* and new nodule-expressed genes. Plant Physiol. 142, 265–279.

Asensio, A.C., Marino, D., James, E.K., Ariz, I., Arrese-Igor, C., Aparício-Tejo, C., et al., 2011. Expression and localization of a *Rhizobium*-derived cambialistic superoxide dismutase in pea (*Pisum sativum*) nodules subjected to oxidative stress. Mol. Plant Microb. Interact. 24, 1247–1257.

Azcon, R., Barea, J.M., 1992. Nodulation, N_2 fixation (^{15}N) and N nutrition relationships in mycorrhical or phosphate-amended alfalfa plants. Symbiosis 12, 33–41.

Bambara, S., Ndakidemi, P.A., 2010. The potential roles of lime and molybdenum on the growth, nitrogen fixation and assimilation of metabolites in nodulated legumes: a special reference to *Phaseolus vulgaris* L. A review. Afr. J. Biotechnol. 8, 2482–2489.

Banath, C.L., Greenwood, E.A.N., Loneragan, J.F., 1996. Effects of calcium deficiency on symbiotic nitrogen fixation. Plant Physiol. 41, 760–763.

Bannister, J.V., Bannister, W.H., Rotilio, G., 1987. Aspects of the structure, function and applications of superoxide dismutase. CRC Crit. Rev. Biochem. 22, 111–180.

Baral, B., Silva, J.A.T., Izaguirre-Mayoral, M., 2016. Early signaling, synthesis, transport and metabolism of ureides. J. Plant Physiol. 193, 97–109.

Bartsev, A., Kobayashi, H., Broughton, W.J., 2004. Rhizobial signals convert pathogens to symbionts at the legume interface. In: Gillings, M., Holmes, A. (Eds.), Plant Microbiology. Garland Science/BIOS Scientific, Abingdon, UK, pp. 19–31.

Bashan, Y., de-Bashan, L.E., 2010. How the plant growth-promoting bacterium *Azospirillum* promotes plant growth — a critical assessment. Adv. Agron. 108, 77–136.

Batista, J.S.S., Hungria, M., 2012. Proteomics reveals differential expression of proteins related to a variety of metabolic pathways by genistein-induced *Bradyrhizobium japonicum* strains. J. Proteomics 75, 1211–1219.

Becana, M., Rodriguez-Barrueco, C., 1989. Protective mechanisms of nitrogenase against oxygen excess and partially-reduced oxygen intermediates. Physiol. Plant. 75, 429–438.

Becana, M., Aparicio-Tejo, P.M., Sánchez-Diaz, M., 1985. Nitrate and nitrite reduction by alfalfa root nodules: accumulation of nitrite in *Rhizobium meliloti* bacteroids and senescence of nodules. Physiol. Plant. 64, 353–358.

Becana, M., Wienkoop, S., Matamoros, M.A., 2018. Sulfur transport and metabolism in legume root nodules. Front. Plant Sci. 9, 1434.

Bergersen, F.J., 1961. The growth of *Rhizobium* in synthetic media. Aust. J. Biol. Sci 14, 349–360.

Bergersen, F.J., 1980. Methods for Evaluating Biological Nitrogen Fixation. John Wiley & Sons, Chichester, 702p.

Bindraban, P.S., Dimkpa, C., Nagarajan, L., Roy, M., Rabbinge, R., 2015. Revisiting fertilisers and fertilisation strategies for improved nutrient uptake by plants. Biol. Fertil. Soils 51, 897–911.

Blevins, D.G., 1989. An overview of nitrogen metabolism in higher plants. In: Poulton, J.E., Romeo, J.T., Conn, E.E. (Eds.), Plant Nitrogen Metabolism. Plenum, New York, pp. 1–41.

Boddey, R.M., Pereira, J.A.R., Hungria, M., Thomas, R.J., Neves, M.C.P., 1987. Methods for the study of nitrogen assimilation and transport in grain legumes. MIRCEN J. 3, 3–32.

Boddey, R.M., Urquiaga, S., Reis, V., Döbereiner, J., 1991. Biological nitrogen fixation associated with sugar cane. Plant Soil 137, 111–117.

Boddey, R.M., Urquiaga, S., Alves, B.J., Reis, V., 2003. Endophytic nitrogen fixation in sugarcane: present knowledge and future applications. Plant Soil 252, 139–149.

Bolaños, L., Brewin, N.J., Bonilla, I., 1996. Effects of boron on rhizobium-legume cell–surface interactions and nodule development. Plant Physiol. 110, 1249–1256.

Bolaños, L., Esteban, E., de Lorenzo, C., Fernández-Pascal, M., De Felipe, M.R., Gárate, A., et al., 1994. Essentiality of boron for symbiotic dinitrogen fixation in pea (*Pisum sativum*) rhizobium nodules. Plant Physiol. 104, 85–90.

Bolaños, L., Martin, M., El-Hamdaoui, A., Rivilla, R., Bonilla, I., 2006. Nitrogenase inhibition in nodules from pea plants grown under salt stress occurs at the physiological level and can be alleviated by B and Ca. Plant Soil 280, 135–142.

Bonilla, I., Bolaños, L., 2009. Mineral nutrition for legume-rhizobia symbiosis: B, Ca, N, P, S, K, Fe, Mo, Co, and Ni: a review. In: Lichtfouse, E. (Ed.), Organic Farming, Pest Control and Remediation of Soil Pollutants. Springer, Dordrecht, The Netherlands, pp. 253–274.

Bonilla, A., Reguera, M., Abreu, I., Bolaños, L., Bonilla, I., 2011. Importancia de la nutrición mineral em las relaciones planta-microorganismo, con especial interés em la fijación simbiótica del nitrógeno, y su proyección sobre la agricultura sostenible. In: Guijo, M.M., Palma, R.R., Misffut, M.J.S., Igeño, M.J.D., García, E.G., González, P.F.M., et al., (Eds.), Fundamentos y Aplicaciones Agroambientales de las interacciones beneficiosas plantas-microrganismos. Sociedad Espanola de Fijación de Nitrógeno (SEFN), Sevilla, Spain, pp. 93–110.

Bouchart, V., Macduff, J.H., Ourry, A., Svenning, M.M., Gay, A.P., Simon, J.C., et al., 1998. Seasonal pattern of accumulation and effects of low temperatures on storage compounds in *Trifolium repens*. Physiol. Plant. 104, 65–74.

Bremer, E., Janzen, H.H., Gilbertson, C., 1995. Evidence against associative N_2 fixation as a significant N source in long-term wheat plots. Plant Soil 175, 13–19.

Brencic, A., Winans, S.C., 2005. Detection of and response to signals involved in host-microbe interactions by plant-associated bacteria. Microbiol. Mol. Biol. Rev. 69, 155–194.

Brito, B., Monza, J., Imperial, J., Ruiz-Argüeso, T., Palacios, J.M., 2000. Nickel availability and *hupSL* activation by heterologous regulators limit symbiotic expression of the *Rhizobium leguminosarum* bv *viciae* hydrogenase system in Hup(-) rhizobia. Appl. Environ. Microbiol. 66, 937–942.

Brodrick, S.J., Giller, K.E., 1991. Root nodules of *Phaseolus*: efficient scavengers of molybdenum for N_2-fixation. J. Exp. Bot. 42, 679–686.

Brodrick, S.J., Sakala, M.K., Giller, K.E., 1992. Molybdenum reserves of seed, and growth and N_2 fixation by *Phaseolus vulgaris* L. Biol. Fertil. Soils 13, 39–44.

Broughton, W.J., Perret, X., 1999. Genealogy of legume-*Rhizobium* symbioses. Curr. Opin. Plant Biol. 2, 305–311.

Buerkert, A., Cassmann, K.G., de la Piedra, R., Munns, D.N., 1990. Soil acidity and liming affects on stand, nodulation, and yield of common bean. Agron. J. 82, 749–754.

Burton, J.C., Curley, R.L., 1966. Compatibility of *Rhizobium japonicum* and sodium molybdate when combined in a peat carrier medium. Agron. J. 58, 327–330.

Campo, R.J., Araujo, R.S., Hungria, M., 2009. Molybdenum-enriched soybean seeds enhance N accumulation, seed yield, and seed protein content in Brazil. Field Crops Res. 110, 219–224.

Carpena, R., Esteban, E., Sarro, M., Penalosa, J., Gárate, A., Lucena, J., et al., 2000. Boron and calcium distribution in nitrogen-fixing pea plants. Plant Sci. 151, 163–170.

Cerezini, P., Kuwano, B., Santos, M., Terassi, F., Hungria, M., Nogueira, M.A., 2016. Strategies to promote early nodulation in soybean under drought. Field Crops Res. 196, 160–167.

Cerezini, P., Kuwano, B.H., Neiverth, W., Grunvald, A.K., Pípolo, A.E., Hungria, M., et al., 2019. Physiological and N_2-fixation-related traits for tolerance to drought in soybean progenies. Pesq. Agropec. Bras. 54, e00839.

Chen, X.-C., Feng, J., Hou, B.-H., Li, F.-Q., Li, Q., Hong, G.-F., 2005. Modulating DNA bending affects NodD-mediated transcriptional control in *Rhizobium leguminosarum*. Nucleic Acids Res. 33, 2540–2548.

Chen, P., Sneller, C.H., Purcell, L.C., Sinclair, T.R., King, C.A., Ishibashi, T., 2007. Registration of soybean germplasm lines R01-416F and R01-581F for improved yield and nitrogen fixation under drought stress. J. Plant Regist. 1, 166–167.

Cleveland, C.C., Townsend, A.R., Fisher, H., Howarth, R.W., Hedin, L.O., Perakis, S.S., et al., 1999. Global patterns of terrestrial biological nitrogen (N_2) fixation in natural ecosystems. Glob. Biogeochem. Cycle 13, 623–645.

Cullimore, J., Bennett, M.J., 1988. The molecular biology and biochemistry of plant glutamine synthetase from root nodules of *Phaseolus vulgaris* L. and other legumes. J. Plant Physiol. 132, 387–393.

Cvitanich, C., Pallisgaard, N., Nielsen, K.A., Chemnitz Hansen, A., Larsen, K., Pihakaski-Maunsbach, K., et al., 2000. CPP1, a DNA-binding protein involved in the expression of a soybean *leghemoglobin c3* gene. Proc. Natl. Acad. Sci. U.S.A. 97, 8163–8168.

Dall'Agnol, R.F., Bournaud, C., De Faria, S.M., Béna, G., Moulin, L., Hungria, M., 2017. Genetic diversity of symbiotic *Paraburkholderia* species isolated from nodules of *Mimosa pudica* (L.) and *Phaseolus vulgaris* (L.) grown in soils of the Brazilian Atlantic Forest (Mata Atlântica). FEMS Microbiol. Ecol. 93, fix027.

Day, J.M., Neves, M.C.P., Döbereiner, J., 1975. Nitrogenase activity on the roots of tropical forage grasses. Soil Biol. Biochem. 7, 107–112.

de Bruijn, F.J. (Ed.), 2015. Biological Nitrogen Fixation, Vol. 1 and Vol. 2. John Wiley & Sons Inc., Hoboken, New Jersey.

de Lajudie, P.M., Andrews, M., Ardley, J., Eardly, B., Jumas-Bilak, E., Kuzmanović, N., et al., 2019. Minimal standards for the description of new genera and species of rhizobia and agrobacteria. Int. J. Syst. Evol. Microbiol. 69, 1852–1863.

del Castillo, L.D., Hunt, S., Layzell, D.B., 1994. The role of oxygen in the regulation of nitrogenase activity in drought-stressed soybean nodules. Plant Physiol. 106, 949–955.

del Cerro, P., Rolla-Santos, A.A.P., Gomes, D.F., Marks, B.B., Espuny, M.R., Rodriguez-Carvajal, M., et al., 2015. Opening the "black box" of *nodD3*, *nodD4* and *nodD5* genes of *Rhizobium tropici* strain CIAT 899. BMC Genomics 16, 864.

Delamuta, J.R.M., Menna, P., Ribeiro, R.A., Hungria, M., 2017. Phylogenies of symbiotic genes of *Bradyrhizobium* symbionts of legumes of economic and environmental importance in Brazil support the definition of the new symbiovars pachyrhizi and sojae. Syst. Appl. Microbiol. 40, 254–265.

Delgado, M.J., Bedmar, E.J., Downie, J.A., 1998. Genes involved in the formation and assembly of rhizobial cytochromes and their role in symbiotic nitrogen fixation. Adv. Microb. Physiol 40, 191–231.

Denison, R.F., 2000. Legume sanctions and the evolution of symbiotic cooperation by rhizobia. Am. Nat. 156, 567–576.

Denison, R.F., Kiers, E.T., 2004. Why are most rhizobia beneficial to their host plants, rather than parasitic? Microbes Infect. 6, 1235–1239.

Depret, G., Laguerre, G., 2008. Plant phenology and genetic variability in root and nodule development strongly influence genetic structuring of *Rhizobium leguminosarum* biovar *viciae* populations nodulating pea. New Phytol. 179, 224–235.

Desbrosses, G.J., Stougaard, J., 2011. Root nodulation: A paradigma for how plant-microbe symbiosis influences host development pathways. Cell Host Microbe 10, 348–358.

Divito, G.A., Sadras, V.O., 2014. How do phosphorus, potassium and sulphur affect plant growth and biological nitrogen fixation in crop and pasture legumes? A meta-analysis. Field Crops Res. 156, 161–171.

Döbereiner, J., 1966. Manganese toxicity effects on nodulation and nitrogen fixation of beans (*Phaseolus vulgaris* L.) in acid soils. Plant Soil 24, 153–166.

Döbereiner, J., 1983. Dinitrogen fixation in rhizosphere and phyllosphere associations. In: Läuchli, A., Bieleski, R.L. (Eds.), Encyclopedia of Plant Physiology, New Series, Vol. 15A. Springer-Verlag, Berlin and New York, pp. 330–350.

Drevon, J.-J., Abadie, J., Alkama, N., Andriamananjara, A., Amenc, L., Bargaz, A., et al., 2015. Phosphorus use efficiency for N_2 fixation in the rhizobial symbiosis with legumes. In: de Bruijn, F.J. (Ed.), Biological Nitrogen Fixation, Vol. 1. John Wiley & Sons Inc., Hoboken, New Jersey, pp. 455–464.

Durand, J.L., Sheehy, J.E., Michin, F.R., 1987. Nitrogenase activity, photosynthesis and nodule water potential in soya beans experiencing water deprivation. J. Exp. Bot. 38, 311–321.

Dwivedi, S.L., Sahrawat, K.L., Upadhyaya, H.D., Mengoni, A., Galardini, M., Bazzicalupo, M., et al., 2015. Advances in host plant and *Rhizobium* genomics to enhance symbiotic nitrogen fixation in grain legumes. Adv. Agron. 129, 1–116.

D'Haeze, W., Mergaert, P., Promé, J.-C., Holsters, M., 2000. Nod factor requirements for efficient stem and root nodulation of the tropical legume *Sesbania rostrata*. J. Biol. Chem. 275, 15676–15684.

Egamberdieva, D., Jabborova, D., Wirth, S.J., Alam, P., Alyemeni, M.N., Ahmad, P., 2018. Interactive effects of nutrients and *Bradyrhizobium japonicum* on the growth and root architecture of soybean (*Glycine max* L.). Front. Microbiol. 9, 1000.

El-Hamdaoui, A., Redondo-Nieto, M., Rivilla, R., Bonilla, I., Bolanos, L., 2003. Effects of boron and calcium nutrition on the establishment of the *Rhizobium leguminosarum*-pea (*Pisum sativum*) symbiosis and nodule development under salt stress. Plant Cell Environ. 26, 1003–1011.

Endre, G., Kereszt, A., Kevei, Z., Mihacea, S., Kalo, P., Kiss, G.B., 2002. A receptor kinase gene regulating symbiotic nodule development. Nature 417, 962–966.

Evans, H.J., Barber, L.E., 1977. Biological nitrogen fixation for food and fiber production. Science 197, 332–339.

Ewens, M., Leigh, R.A., 1985. The effect of nutrient solution composition on the length of root hairs of wheat (*Triticum aestivum* L.). J. Exp. Bot. 36, 713–724.

Favilli, F., Messini, A., 1990. Nitrogen fixation at phyllospheric level in coniferous plants in Italy. Plant Soil 128, 91–95.

Ferguson, B.J., Lin, M.H., Gresshoff, P.M., 2013. Regulation of legume nodulation by acidic growth conditions. Plant Signal. Behav. 8, e23426.

Franco, A.A., Pereira, J.C., Neyra, C.A., 1979. Seasonal patterns of nitrate reductase and nitrogenase activity in *Phaseolus vulgaris* L. Plant Physiol. 63, 421–424.

Franke, A.C., van den Brand, G.J., Vanlauwe, B., Giller, K.E., 2018. Sustainable intensification through rotations with grain legumes in Sub-Saharan Africa: A review. Agric. Ecosyst. Environ. 261, 172–185.

Fred, E.B., Baldwin, I.L., McCoy, E., 1932. Root Nodule Bacteria and Leguminous Plants. University of Wisconsin Press, Madison, Wisconsin.

Freitas, D.S., Rodak, B.W., Reis, A.R., Reis, F.B., Carvalho, T.S., Schulze, J., et al., 2018. Hidden nickel deficiency? Nickel fertilization via soil improves nitrogen metabolism and grain yield in soybean genotypes. Front. Plant Sci. 9, 614.

Freitas, D.S., Rodak, B.W., Carneiro, M.A.C., Guilherme, L.R.G., 2019. How does Ni fertilization affect a responsive soybean genotype? A dose study. Plant Soil 441, 567–586.

Frey, B., Schüepp, H., 1992. Transfer of symbiotically fixed nitrogen from berseem (*Trifolium alexandrinum* L.) to maize via vesicular-arbuscular mycorrhizal hyphae. New Phytol. 122, 447–454.

Fukami, J., Cerezini, P., Hungria, M., 2018. *Azospirillum*: benefits that go far beyond biological nitrogen fixation. AMB Express 8, 73.

Gage, D.J., 2004. Infection and invasion of roots by symbiotic, nitrogen-fixing rhizobia during nodulation of temperate legumes. Microbiol. Mol. Biol. Rev. 68, 280–300.

Galloway, J.N., Dentener, F.J., Capone, D.G., Boyer, E.W., Howarth, R.W., Seitzinger, S.P., et al., 2004. Nitrogen cycles: past, present and future. Biogeochemistry 70, 153–226.

Gherbi, H., Markmann, K., Svistoonoff, S., Estevan, J., Autran, D., Giczey, G., et al., 2008. SymRK defines a common genetic basis for plant root endosymbioses with arbuscular mycorrhiza fungi, rhizobia and *Frankia* bacteria. Proc. Natl. Acad. Sci. U.S.A. 105, 4928–4932.

Giller, K.E., 2001. Nitrogen Fixation in Tropical Cropping Systems, 2nd ed. CABI Publishing, Wallingford, UK, p. 448.

Giraud, E., Moulin, L., Vallenet, D., Barbe, V., Cytryn, E., Avarre, J.-C., et al., 2007. Legume symbioses: absence of *nod* genes in photosynthetic bradyrhizobia. Science 316, 1307–1312.

Glasener, K.M., Wagger, M.G., MacKown, C.T., 2002. Contributions of shoot and root nitrogen-15 labeled legume nitrogen sources to a sequence of three cereal crops. Soil Sci. Soc. Am. J. 66, 523–530.

Gomes, D.F., Ormeño-Orrillo, E., Hungria, M., 2015. Biodiversity, symbiotic efficiency and genomics of *Rhizobium tropici* and related species. In: de Bruijn, F.J. (Ed.), Biological Nitrogen Fixation, Vol. 2. John Wiley & Sons Inc., Hoboken, New Jersey, pp. 747–756.

Grewal, H.S., 2001. Zinc influences nodulation, disease severity, leaf drop and herbage yield of alfalfa cultivars. Plant Soil 234, 47–59.

Guasch-Vidal, B., Estévez, J., Dardanelli, M.S., Soria-Díaz, M.E., de Córdoba, F.F., Balog, C.I., et al., 2013. High NaCl concentrations induce the *nod* genes of *Rhizobium tropici* CIAT899 in the absence of flavonoid inducers. Mol. Plant Microbe Interact. 26, 451–460.

Gubry-Rangin, C., Garcia, M., Béna, G., 2010. Partner choice in *Medicago truncatula–Sinorhizobium* symbiosis. Proc. Roy. Soc. Lond. B 217, 1947–1951.

Gunawardena, S.F.B.N., Danso, S.K.A., Zapata, F., 1993. Phosphorus requirement and sources of nitrogen in three soybean (*Glycine max*) genotypes, Brag, nts 382 and Chippewa. Plant Soil 151, 1–9.

Hagen, M.J., Hamrick, J.L., 1996. Population level processes in *Rhizobium leguminosarum* bv. *trifolii*: the role of founder effects. Mol. Ecol. 5, 707–714.

Haney, C.H., Long, S.R., 2010. Plant flotillins are required for infection by nitrogen-fixing bacteria. Proc. Natl. Acad. Sci. U.S.A. 107, 478–483.

Hardy, R.W.F., Holsten, R.D., Jackson, E.K., Burns, R.C., 1968. The acetylene-ethylene assay for measurement of nitrogen fixation. Soil Biol. Biochem. 5, 47–81.

Harper, J.E., Gibson, A.H., 1984. Differential nodulation tolerance to nitrate among legume species. Crop Sci. 24, 797–801.

Hart, A., 1989. Distribution of phosphorus in nodulated white clover plants. J. Plant Nutr. 12, 159–171.

Hauggaard-Nielsen, H., Gooding, M., Ambus, P., Corre-Hellou, G., Crozat, Y., Dahlmann, C., et al., 2009. Pea-barley intercropping for efficient symbiotic N_2 fixation, soil N acquisition and use of other nutrients in European organic cropping systems. Field Crop Res. 113, 64–71.

Hawkins, F.K.L., Johnston, A.W.B., 1988. Transcription of a *Rhizobium leguminosarum* biovar phaseoli gene needed for melanin synthesis is activated by *nifA* of *Rhizobium* and *Klebsiella pneumoniae*. Mol. Microbiol. 2, 331–337.

Heckmann, M.-O., Drevon, J.-J., Saglio, P., Salsac, L., 1989. Effect of oxygen and malate on NO_3^- inhibition of nitrogenase in soybean nodules. Plant Physiol. 90, 224–229.

Hegazy, M.H., El Hawary, F.I., Ghobrial, W.N. 1990. Effect of micronutrients application on *Bradyrhizobium japonicum* inoculation on soybean. Annals of the Third Conference of Agriculture, Development and Evaluation, Ain Sham University. Published in Annals Agric. Sci., Special Issue, 381-398.

Herridge, D.F., 2011. Managing Legumes and Fertiliser N for Northern Grains Cropping Grains Research and Development Corporation, Project UNE00014. Grains Research and Development Corporation, Kingston, Australia.

Herridge, D.F., Doyle, A.D., 1988. The narrow-leafed lupin (*Lupinus angustifolius* L.) as a nitrogen-fixing rotation crop for cereal production. II. Estimates of fixation by fieldgrown crops. Aust. J. Agric. Res. 39, 1017–1028.

Herridge, D.F., Peoples, M.B., Boddey, R.M., 2008. Global inputs of biological nitrogen fixation in agricultural systems. Plant Soil 311, 1–18.

Hirsch, A.M., 1999. Role of lectins and rhizobial exopolysaccharides in legume nodulation. Curr. Opin. Plant Biol. 2, 320–326.

Howieson, J.G., Dilworth, M.J., 2016. Working with Rhizobia, ACIAR Monograph 173. Australian Center for International Agricultural Research (ACIAR), Canberra, Australia, p. 312.

Hungria, M., Kaschuk, G., 2014. Regulation of N_2 fixation and NO_3^-/NH_4^+ assimilation in nodulated and N-fertilized *Phaseolus vulgaris* L. exposed to high-temperature stress. Env. Exp. Bot. 98, 32–39.

Hungria, M., Mendes, I.C., 2015. Nitrogen fixation with soybean: the perfect symbiosis. In: de Bruijn, F.J. (Ed.), Biological Nitrogen Fixation, Vol. 2. John Wiley & Sons Inc., Hoboken, New Jersey, pp. 1009–1023.

Hungria, M., Nogueira, M.A., 2019. Tecnologias de inoculação da cultura da soja: Mitos, verdades e desafios. Boletim de Pesquisa 2019/2020. Fundação MT, Rondonópolis, MT, Brazil, pp. 50–62.

Hungria, M., Stacey, G., 1997. Molecular signals exchanged between host plants and rhizobia, basic aspects and potential application in agriculture. Soil Biol. Biochem. 29, 519–530.

Hungria, M., Thomas, R.J., 1987. Effects of cotyledons and nitrate on the nitrogen assimilation of *Phaseolus vulgaris*. MIRCEN J. 3, 411–419.

Hungria, M., Vargas, M.A.T., 2000. Environmental factors affecting N_2 fixation in grain legumes in the tropics, with emphasis on Brazil. Field Crops Res. 65, 151–164.

Hungria, M., Barradas, C.A.A., Wallsgrove, R.M., 1991. Nitrogen fixation, nitrogen assimilation and transport during the initial growth stage of *Phaseolus vulgaris* L. J. Exp. Bot. 42, 839–844.

Hungria, M., Franchini, J.C., Campo, R.J., Crispino, C.C., Moraes, J.Z., Sibaldelli, R.N.R., et al., 2006. Nitrogen nutrition of soybean in Brazil: contributions of biological N_2 fixation and of N fertilizer to grain yield. Can. J. Plant Sci. 86, 927–939.

Hungria, M., Franco, A.A., Sprent, J.I., 1993. New sources of high temperature tolerant rhizobia for *Phaseolus vulgaris* (L.). Plant Soil 149, 103–109.

Hungria, M., Menna, P., Delamuta, J.R.M., 2015. *Bradyrhizobium*, the ancestor of all rhizobia: phylogeny of housekeeping and nitrogen-fixation genes. In: de Bruijn, F.J. (Ed.), Biological Nitrogen Fixation, Vol. 1. John Wiley & Sons Inc., Hoboken, New Jersey, pp. 191–202.

Hungria, M., Nogueira, M.A., Araujo, R.S., 2013. Co-inoculation of soybeans and common beans with rhizobia and azospirilla: strategies to improve sustainability. Biol. Fertil. Soils 49, 791–801.

Jacob-Neto, J., Franco, A.A., 1989. Determinação do nível crítico de Mo nos nódulos de feijoeiro (*Phaseolus vulgaris* L.). Turrialba 39, 215–223.

Jakobsen, I., 1985. The role of phosphorus in nitrogen fixation by young pea plants (*Pisum sativum*). Physiol. Plant. 64, 190–196.

Jarvis, S.C., Hatch, D.J., 1985. Rates of hydrogen ion efflux by nodulated legumes grown in flowing solution culture with continuous pH monitoring and adjustment. Ann. Bot. 55, 41–51.

Jeong, M.S., Jang, S.B., 2006. Electron transfer and nano-scale motions in nitrogenase Fe-protein. Curr. Nanosci. 2, 33–41.

Johannes, E., Gunarto, L., 1987. Nodulation and uptake of nitrogen and phosphorus by soybean inoculated with four strains of *Bradyrhizobium japonicum* and applied with phorphorus, molybdenum and copper. Philipp. Agric. 70, 193–201.

Jones, D.L., Farrar, J., Giller, K.E., 2003. Associative nitrogen fixation and root exudation – what is theoretically possible in the rhizosphere? Symbiosis 35, 19–38.

Jones, K.M., Kobayashi, H., Davies, B.W., Taga, M.E., Graham, C.W., 2007. How rhizobial symbionts invade plants: The *Sinorhizobium–Medicago* model. Nat. Rev. Microbiol. 5, 619–633.

Jordan, D.C., 1984. Rhizobiaceae Conn 1938. In: Krieg, N.R., Holt, J.G. (Eds.), Bergey's Manual of Systematic Bacteriology. Williams & Wilkins Co., Baltimore-London, pp. 235–244.

Kanayama, Y., Yamamoto, Y., 1991. Formation of nitrosylleghemoglobin in nodules of nitrate-treated cowpea and pea plants. Plant Cell Physiol. 32, 19–23.

Kantar, F., Shivakumar, B.G., Arrese-Igor, C., Hafeez, F.Y., González, E.M., Imran, A., et al., 2010. Efficient biological nitrogen fixation under warming climates. In: Yadav, S.S., McNeil, D.L., Redden, R., Patil, S.A. (Eds.), Climate Change and Management of Cool Season Grain Legume Crops. Springer Science + Business Media B.V, New York, pp. 283–306.

Kaschuk, G., Kuyper, T.W., Leffelaar, P.A., Hungria, M., Giller, K.E., 2009. Are the rates of photosynthesis stimulated by the carbon sink strength of rhizobial and arbuscular mycorrhizal symbioses? Soil Biol. Biochem. 41, 1233–1244.

Kaschuk, G., Hungria, M., Leffelaar, P.A., Giller, K.E., Kuyper, T.W., 2010a. Differences in photosynthetic behavior and leaf senescence of soybean *Glycine max* [L.] Merrill relying on N_2 fixation or supplied with nitrate. Plant Biol. 12, 60–69.

Kaschuk, G., Leffelaar, P.A., Giller, K.E., Alberton, O., Hungria, M., Kuyper, T.W., 2010b. Responses of grain legumes to rhizobia and arbuscular mycorrhizal fungi: a meta-analysis of potential photosynthate limitation of symbioses. Soil Biol. Biochem. 42, 125–127.

Kaschuk, G., Nogueira, M.A., de Luca, M.J., Hungria, M., 2016. Response of determinate and indeterminate soybean cultivars to basal and topdressing N fertilization compared to sole inoculation with *Bradyrhizobium*. Field Crops Res. 195, 21–27.

Keeler, B.L., Gourevitch, J.D., Polasky, S., Isbell, F., Tessum, C.W., Hill, J.D., et al., 2016. The social costs of nitrogen. Sci. Adv. 2, e1600219.

Kennedy, I.R., Choudhury, A.T.M.A., Kecskés, M.L., 2004. Non-symbiotic bacterial diazotrophs in crop-farming systems: can their potential for plant growth promotion be better exploited. Soil Biol. Biochem. 36, 1229–1244.

Kermah, M., Franke, A.C., Adjei-Nsiah, S., Ahiabor, B.D.K., Abaidoo, R.C., Giller, K.E., 2018. N_2-fixation and N contribution by grain legumes under different soil fertility status and cropping systems in the Guinea savanna of northern Ghana. Agric. Ecosyst. Environ. 261, 201–210.

Kiers, E.T., Rousseau, R.A., West, S.A., Denison, R.F., 2003. Host sanctions and the legume-rhizobium symbiosis. Nature 425, 78–81.

King, C.A., Purcell, L.C., 2001. Soybean nodule size and relationship to nitrogen fixation response to water deficit. Crop Sci. 41, 1099–1107.

King, C.A., Purcell, L.C., 2005. Inhibition of N_2 fixation in soybean is associated with elevated ureides and amino acids. Plant Physiol. 137, 1389–1396.

Kiss, A., Stefanovits-Bányai, E., Takács-Hájos, M., 2004. Magnesium-content of *Rhizobium* nodules in different plants: the importance of magnesium in nitrogen-fixation of nodules. J. Am. Coll. Nutr. 23, 751S–753S.

Kreibich, H., Kern, J., de Camargo, P.B., Moreira, M.Z., Victoria, R.L., Werner, D., 2006. Estimation of symbiotic N_2 fixation in an Amazon floodplain forest. Oecologia 147, 359–368.

Kryvoruchko, I.S., 2017. Zn-use efficiency for optimization of symbiotic nitrogen fixation in chickpea (*Cicer arietinum* L.). Turk. J. Bot. 41, 423–441.

Kuang, R.B., Liao, H., Yan, X.L., Dong, Y.S., 2005. Phosphorus and nitrogen interactions in field-grown soybean as related to genetic attributes of root morphological and nodular traits. J. Int. Plant Biol. 47, 549–559.

Kurdali, F., Al-Chammaa, M., Mouasess, A., 2013. Growth and nitrogen fixation in silicon and/or potassium fed chickpeas grown under drought and well watered conditions. J. Stress Physiol. Biochem. 9, 385–406.

Ladd, J.N., Oades, J.M., Amato, M., 1981. Distribution and recovery of nitrogen from legume residues decomposing in soils sown to wheat in the field. Soil Biol. Biochem. 13, 251–256.

Lavres, J., Franco, G.C., Câmara, G.M.S., 2016. Soybean seed treatment with nickel improves biological nitrogen fixation and urease activity. Front. Environ. Sci. 4, 37.

Lazali, M., Drevon, J.-J., 2014. The nodule conductance to O_2 diffusion increases with phytase activity in N_2-fixing *Phaseolus vulgaris* L. Plant Physiol. Biochem. 80, 53–59.

Lerouge, P., Roche, P., Faucher, C., Maillet, F., Truchet, G., Promé, J.-C., et al., 1990. Symbiotic host-specificity of *Rhizobium meliloti* is determined by a sulphated and acylated glucosamine oligosaccharide signal. Nature 344, 781–784.

Lindström, K., Mousavi, S.A., 2020. Effectiveness of nitrogen fixation in rhizobia. Microbial Biotechnol. 13, 1314–1335.

Lodwig, E., Poole, P., 2003. Metabolism of *Rhizobium* bacteroids. Crit. Rev. Plant Sci. 22, 37–78.

Lowther, W.L., Loneragan, J.F., 1968. Calcium and nodulation in subterranean clover *Trifolium subterraneum* L. Plant Physiol. 43, 1362–1366.

Lui, W.C., Lund, L.J., Page, A.L., 1989. Acidity produced by leguminous plants through symbiotic dinitrogen fixation. J. Environ. Qual. 18, 529–534.
Malik, K.A., Rakhshanda, B., Mehnaz, S., Rasul, G., Mirza, M.S., Ali, S., 1997. Association of nitrogen-fixing plant-growth-promoting rhizobacteria PGPR with kallar grass and rice. Plant Soil 194, 37–44.
Mallarino, A.P., Wedin, W.F., Perdomo, C.H., Goyenola, R.S., West, C.P., 1990. Nitrogen transfer from white clover, red clover, and birdsfoot trefoil to associated grass. Agron. J. 82, 790–795.
Masson-Boivin, C., Giraud, E., Perret, X., Batut, J., 2009. Establishing nitrogen-fixing symbiosis with legumes: how many recipes? Trends Microbiol. 17, 458–466.
McKay, I.A., Djordjevic, M.A., 1993. Production and excretion of Nod metabolites by *Rhizobium leguminosarum* biovar *trifolii* disrupted by the same environmental factors that reduce nodulation in the field. Appl. Environ. Microbiol. 59, 3385–3392.
Mengel, K., 1994. Symbiotic dinitrogen fixation – its dependence on plant nutrition and its ecophysiological impact. Z. Pflanzenernähr. Bodenk. 157, 233–241.
Mengel, K., Haghparast, M., Koch, K., 1974. The effect of potassium on the fixation of molecular nitrogen by root nodules of *Vicia faba*. Plant Physiol. 54, 535–538.
Michiels, J., Verreth, C., Vanderleyden, J., 1994. Effects of temperature stress on bean-nodulating *Rhizobium* strains. Appl. Environ. Microbiol. 60, 1206–1212.
Minchin, F.R., Witty, J.F., 2005. Respiratory/carbon costs of symbiotic nitrogen fixation in legumes. In: Lambers, H., Ribas-Carbo, M. (Eds.), Plant Respiration. Springer, Dordrecht, The Netherlands, pp. 195–205.
Minchin, F.R., Witty, J.F., Sheehy, J.E., 1983. A major error in the acetylene reduction assay: decreases in nodular nitrogenase activity under assay conditions. J. Exp. Bot. 34, 641–649.
Minchin, F.R., Sheehy, J.E., Witty, J.F., 1986. Further errors in the acetylene reduction assay: effects of plant disturbance. J. Exp. Bot. 37, 1581–1591.
Montañez, A., Danso, S.L.A., Hardarson, G., 1995. The effect of temperature on nodulation and nitrogen fixation by five *Bradyrhizobium japonicum* strains. Appl. Soil Ecol. 2, 165–174.
Morón, B., Soria-Diaz, M.E., Ault, J., Verroios, G., Sadaf, N., Rodríguez-Navarro, D.N., et al., 2005. Low pH changes the profile of nodulation factors produced by *Rhizobium tropici* CIAT899. Chem. Biol. 12, 1029–1040.
N2Africa, 2019. N2Africa: Putting Nitrogen Fixation to Work for Smallholder Farmers in Africa. <https://www.n2africa.org/>.
Naumann, G., Alfieri, L., Wyser, K., Mentaschi, L., Betts, R.A., Carrao, H., et al., 2018. Global changes in drought conditions under different levels of warming. Geophys. Res. Lett. 45, 3285–3296.
Nelwamondo, A., Dakora, F.D., 1999. Silicon promotes nodule formation and nodule function in symbiotic cowpea (*Vigna unguiculata*). New Phytol. 142, 463–467.
Neves, M.C.P., Hungria, M., 1987. The physiology of nitrogen fixation in tropical grain legumes. CRC Crit. Rev. Plant Sci. 6, 267–321.
Newton, W.E., 2015. Recent advances in understanding nitrogenases and how they work. In: de Bruijn, F.J. (Ed.), Biological Nitrogen Fixation, Vol. 1. John Wiley & Sons Inc., Hoboken, New Jersey, pp. 7–20.
Nogueira, M.A., Nehls, U., Hampp, R., Poralla, K., Cardoso, E.J.B.N., 2007. Mycorrhiza and soil bacteria influence extractable iron and manganese in soil and uptake by soybean. Plant Soil 298, 273–284.
O'Hara, G.W., 2001. Nutritional constraints on root nodule bacteria affecting symbiotic nitrogen fixation: a review. Aust. J. Exp. Agric. 41, 417–433.
O'Hara, G.W., Dilworth, M.J., Parkpian, P., 1988. Iron deficiency specifically limits nodule development in peanut inoculated with *Bradyrhizobium* sp. New Phytol. 108, 51–57.
Ogutcu, H., Kasimoglu, C., Elkoca, E., 2010. Effect of rhizobium strains isolated from wild chickpeas on the growth and symbiotic performance of chickpea (*Cicer arietinum* L.) under salt stress. Turk. J. Agric. 34, 361–371.
Ojiem, J.O., Franke, A.C., Vanlauwe, B., de Ridder, N., Giller, K.E., 2014. Benefits of legume–maize rotations: assessing the impact of diversity on the productivity of smallholders in Western Kenya. Field Crops Res. 168, 75–85.
Okon, Y., Labandera-Gonzales, C., Lage, M., Lage, P., 2015. Agronomic applications of *Azospirillum* and other PGPR. In: de Bruijn, F.J. (Ed.), Biological Nitrogen Fixation, Vol. 3. John Wiley & Sons Inc., Hoboken, New Jersey, pp. 925–937.
Oldroyd, G.E.D., 2013. Speak, friend, and enter: signaling systems that promote beneficial symbiotic associations in plants. Nat. Rev. Microbiol. 11, 252–263.
Oldroyd, G.E.D., Downie, J.A., 2008. Coordinating nodule morphogenesis with rhizobial infection in legumes. Annu. Rev. Plant Biol. 59, 519–546.
Olivera, M., Tejera, N., Iridarne, C., Ocana, A., Lluch, C., 2004. Growth, nitrogen fixation and ammonium assimilation in common bean *Phaseolus vulgaris*: effect of phosphorus. Physiol. Plant. 121, 498–505.
Oono, R., Denison, R.F., Kiers, E.T., 2009. Controlling the reproductive fate of rhizobia: how universal are legume sanctions? New Phytol. 183, 967–979.
Ormeño-Orrillo, E., Hungria, M., Martinez-Romero, E., 2013. Dinitrogen-fixing prokaryotes. In: Rosenberg, E., DeLong, E.F., Lory, S., Stackebrandt, E., Thompson, F. (Eds.), The Prokaryotes: Prokaryotic Physiology and Biochemistry. Springer Berlin Heidelberg, pp. 427–451.
Pacyna, S., Schulz, M., Scherer, H.W., 2006. Influence of sulphur supply on glucose and ATP concentrations of inoculated broad beans (*V. faba minor* L.). Biol. Fertil. Soils 42, 324–329.
Parsons, R., Sunley, R.J., 2001. Nitrogen nutrition and the role of root-shoot nitrogen signalling particularly in symbiotic systems. J. Exp. Bot. 52, 435–443.

Peix, A., Ramírez-Bahena, M.H., Velázquez, E., Bedmar, E.J., 2015. Bacterial associations with legumes. CRC Crit. Rev. Plant Sci. 34, 17–42.
Peña-Cabriales, J.J., Castellanos, J.Z., 1993. Effects of water stress on N_2 fixation and grain yield of *Phaseolus vulgaris* L. Plant Soil 152, 151–155.
Peoples, M.B., Baldock, J.A., 2001. Nitrogen dynamics of pastures: nitrogen fixation inputs, the impact of legumes on soil nitrogen fertility, and the contributions of fixed nitrogen to Australian farming systems. Aust. J. Exp. Agric. 41, 27–346.
Peoples, M.B., Brockwell, J., Herridge, D., Rochester, I.J., Alves, B.J.R., Urquiaga, S., et al., 2009. The contributions of nitrogen fixing crop legumes to the productivity of agricultural systems. Symbiosis 48, 1–17.
Pereg, L., Luz, E., Bashan, Y., 2016. Assessment of affinity and specificity of *Azospirillum* for plants. Plant Soil 399, 389–414.
Pijnenberg, J.W.M., Lie, T.A., 1990. Effect of lime-pelleting on the nodulation of lucerne *Medicago sativa* L. in acid soil: a comparative study carried out in the field, in pots and in rhizotrons. Plant Soil 121, 225–234.
Preisig, O., Zufferey, R., Thony-Meyer, L., Appleby, C.A., Hennecke, H., 1996. A high-affinity cbb_3-type cytochrome oxidase terminates the symbiosis-specific respiratory chain of *Bradyrhizobium japonicum*. J. Bacteriol. 178, 1532–1538.
Prell, J., Poole, P., 2006. Metabolic changes of rhizobia in legume nodules. Trends Microbiol. 14, 161–168.
Pueppke, S.G., Broughton, W.J., 1999. *Rhizobium* sp. strain NGR234 and *R. fredii* USDA257 share exceptionally broad, nested host ranges. Mol. Plant Microbe Interact. 12, 293–318.
Purcell, L.C., King, C.A., Ball, R.A., 2000. Soybean cultivar differences in ureides and the relationship to drought tolerant nitrogen fixation and manganese nutrition. Crop Sci. 40, 1062–1070.
Rasmussen, U., Nilsson, M., 2002. Cyanobacterial diversity and specificity in plant symbioses. In: Rai, A.N., Bergman, B., Rasmussen, U. (Eds.), Cyanobacteria in Symbiosis. Springer, Dordrecht, The Netherlands, pp. 312–328.
Rees, D.C., Howard, J.B., 2000. Nitrogenase: standing at the crossroads. Curr. Opin. Chem. Biol. 4, 559–566.
Reinhold-Hurek, B., Hurek, T., 1998. Interactions of gramineous plants with *Azoarcus* spp. and other diazotrophs: identification, localization and perspectives to study their function. Crit. Rev. Plant Sci. 17, 29–54.
Remison, S.U., 1978. Neighbour effects between maize and cowpea at various levels of N and P. Expl. Agric. 14, 205–212.
Richardson, A.E., Djordjevic, M.A., Rolfe, B.G., Simpson, R.J., 1988. Effects of pH, Ca and Al on the exudation from clover seedlings of compounds that induce the expression of nodulation genes in *Rhizobium trifolii*. Plant Soil 109, 37–47.
Riley, I.T., Dilworth, M.J., 1985. Cobalt requirement for nodule development and function in *Lupinus angustifolius* L. New Phytol. 100, 347–359.
Rivas, R., García-Fraile, P., Velázquez, E., 2009. Taxonomy of bacteria nodulating legumes. Microbiol. Insights 2, 51–59.
Rogel, M.A., Ormeño-Orrillo, E., Martinez Romero, E., 2011. Symbiovars in rhizobia reflect bacterial adaptation to legumes. Syst. Appl. Microbiol. 34, 96–104.
Roger, P.A., Ladha, J.K., 1992. Biological N_2 fixation in wetland rice fields: estimation and contribution to nitrogen balance. Plant Soil 141, 41–55.
Ronner, E., Franke, A.C., Vanlauwe, B., Dianda, M., Edeh, E., Ukem, B., et al., 2016. Understanding variability in soybean yield and response to P-fertilizer and rhizobium inoculants on farmers' fields in northern Nigeria. Field Crops Res. 186, 133–145.
Rubio, M.C., James, E.K., Clemente, M.R., Bucciarelli, B., Fedorova, M., Vance, C.P., et al., 2004. Localization of superoxide dismutasesand hydrogen peroxide in legume root nodules. Mol. Plant Microbe Interact. 17, 1294–1305.
Sadowsky, M.J., 2005. Soil stress factors influencing symbiotic nitrogen fixation. In: Werner, D., Newton, W.E. (Eds.), Nitrogen Fixation in Agriculture, Forestry, Ecology, and the Environment. Springer, Dordrecht, The Netherlands, pp. 89–112.
Saeki, Y., Nakamura, M., Mason, M.L.T., Yano, T., Shiro, S., Sameshima-Saito, R., et al., 2017. Effect of flooding and the *nosZ* gene in bradyrhizobia on bradyrhizobial community structure in the soil. Microbes Environ. 32, 154–163.
Santachiara, G., Salvagiotti, F., Rotundo, J.L., 2019. Nutritional and environmental effects on biological nitrogen fixation in soybean: A meta-analysis. Field Crops Res. 240, 106–115.
Santos, M.S., Nogueira, M.A., Hungria, M., 2019. Microbial inoculants: Reviewing the past, discussing the present and previewing an outstanding future for the use of beneficial bacteria in agriculture. AMB Express 9, 205.
Sakamoto, K., Ogiwara, N., Kaji, T., Sugimoto, Y., Ueno, M., Sonoda, M., et al., 2019. Transcriptome analysis of soybean *Glycine max* root genes differentially expressed in rhizobial, arbuscular mycorrhizal, and dual symbiosis. J. Plant Res. 132, 541–568.
Saturno, D.F., Cerezini, P., Silva, P.M., Oliveira, A.B., Oliveira, M.C.N., Hungria, M., et al., 2017. Mineral nitrogen impairs the biological nitrogen fixation in soybean of determinate and indeterminate growth types. J. Plant Nutr. 40, 1690–1701.
Scherer, H.W., 2008. Impact of sulfur on N_2 Fixation of legumes. In: Khan, N.A., Singh, S., Umar, S. (Eds.), Sulfur Assimilation and Abiotic Stress in Plants. Springer, Berlin, Heidelberg, Germany, pp. 43–54.
Scherer, H.W., Danzeisen, L., 1980. Der einfluß gesteigerter stickstoffgaben auf die entwicklung der wurzelknöllchen, auf die symbiontische stickstoffassimilation sowie auf das wachstum und den ertrag von ackerbohnen (*Vicia faba* L.). Z. Pflanzenernaehr. Bodenk. 143, 464–470.
Scherer, H.W., Pacyna, S., Manthey, N., Schulz, M., 2006. Sulphur supply to peas *Pisum sativum* L. influences symbiotic N_2 fixation. Plant Soil Environ. 52, 72–77.
Scherer, H.W., Pacyna, S., Spoth, K.R., Schulz, M., 2008. Low levels of ferredoxin, ATP and leghemoglobin contribute to limited N_2 fixation of peas *Pisum sativum* L. and alfalfa *Medicago sativa* L. under S deficiency conditions. Biol. Fertil. Soils 44, 909–916.
Schubert, E., Mengel, K., Schubert, S., 1990. Soil pH and calcium effect on nitrogen fixation and growth of broad bean. Agron. J. 82, 969–972.
Serraj, R., Vasquez-Diaz, H., Drevon, J.J., 1998. Effects of salt stress on nitrogen fixation and ion distribution in soybean, common bean, and alfalfa. J. Plant Nutr. 21, 475–488.
Sexton, P.J., Batchelor, W.D., Shibles, R.M., 1998. Effects of nitrogen source and timing of sulfur deficiency on seed yield and expression of 11S and 7S seed storage proteins of soybean. Field Crops Res. 59, 1–8.

Shailes, S., Oldroyd, G.E.D., 2015. Nod factor-induced calcium signalling in legumes. In: de Bruijn, F.J. (Ed.), Biological Nitrogen Fixation, Vol. 2. John Wiley & Sons Inc., Hoboken, New Jersey, pp. 533–546.

Simpson, F.B., 1987. The hydrogen reactions of nitrogenase. Physiol. Plant. 69, 187–190.

Sinclair, T.R., Nogueira, M.A., 2018. Selection of host-plant genotype: the next step to increase grain legume N_2 fixation activity. J. Exp. Bot. 69, 3523–3530.

Sinclair, T.R., Purcell, L.C., King, A., Sneller, C.H., Chen, P., Vadez, V., 2007. Drought tolerance and yield increase of soybean resulting from improved symbiotic N_2 fixation. Field Crops Res. 101, 68–71.

Singh, P., Raj, B., 1988. Sulphur fertilization in relation to yield and trend of production of leghemoglobin in the nodules of pea *Pisum sativum* var. Arvense. Ann. Agric. Res. 9, 13–19.

Slattery, J.F., Coventry, D.R., Slattery, W.J., 2001. Rhizobial ecology as affected by the soil environment. Aust. J. Exp. Agric. 41, 289–298.

Smil, V., 2001. Enriching the Earth: Fritz Haber, Carl Bosch and the Transformation of World Food Production. MIT Press, Cambridge, Massachusetts, p. 360.

Somado, E.A., Sahrawat, K.L., Kuehne, R.F., 2006. Rock phosphate-P enhances biomass and nitrogen accumulation by legumes in upland crop production systems in humid West Africa. Biol. Fertil. Soils 43, 124–130.

Somasegaran, P., Hoben, H.J., 1994. Handbook for Rhizobia. Methods in Legume-Rhizobium Technology. Springer-Verlag, New York, p. 472.

Souto, S.M., Döbereiner, J., 1968. Toxidez de manganês em leguminosas forrageiras tropicais. Pesq. Agropec. Bras. 4, 129-128.

Souza, S.C., Mazzafera, P., Sodek, L., 2016. Flooding of the root system in soybean: biochemical and molecular aspects of N metabolism in the nodule during stress and recovery. Amino Acids 48, 1285–1295.

Sprent, J.I., 1980. Root nodule anatomy, type of export product and evolutionary origin in some Leguminosae. Plant Cell Environ. 3, 35–43.

Sprent, J.I., James, E.K., 2007. Legume evolution: where do nodules and mycorrhizas fit in? Plant Physiol. 144, 575–581.

Sprent, J.I., Raven, J.A., 1985. Evolution of nitrogen-fixing symbioses. Proc. R. Soc. Edinb. 85B, 215–237.

Sprent, J.I., Ardley, J., James, E.K., 2017. Biogeography of nodulated legumes and their nitrogen-fixing symbionts. New Phytol. 215, 40–56.

Steiner, F., Zuffo, A.M., Bush, A., Santos, D.M.S., 2018. Silicate fertilization potentiates the nodule formation and symbiotic nitrogen fixation in soybean. Pesq. Agrop. Trop. 48, 212–221.

Stewart, R.A., Lal, R., 2017. The nitrogen dilemma: Food or the environment. J. Soil Water Conserv. 72, 124A–128A.

Stracke, S., Kistner, C., Yoshida, S., Mulder, L., Sato, S., Kaneko, T., et al., 2002. A plant receptor-like kinase required for both bacterial and fungal symbiosis. Nature 417, 959–962.

Tajini, F., Drevon, J.–J., 2011. Phosphorus use efficiency for symbiotic nitrogen fixation varies among common bean recombinant inbred lines under P deficiency. J. Plant Nutr. 37, 532–545.

Tang, C., Robson, A.D., Dilworth, M.J., 1992. The role of iron in the (brady)rhizobium legume symbiosis. J. Plant Nutr. 15, 2235–2252.

Tate, R., Riccio, A., Merrick, M., Patriarca, E.J., 1998. The *Rhizobium etli amtB* gene coding for an NH_4^+ transporter is down-regulated early during bacteroid differentiation. Mol. Plant Microbe Interact. 11, 188–198.

Tavares, F., Santos, C.L., Sellstedt, A., 2007. Reactive oxygen species in legume and actinorhizal nitrogen-fixing symbioses: the microsymbiont's responses to an unfriendly reception. Physiol. Plant. 130, 344–356.

Thies, J.E., Singleton, P.W., Bohlool, B.B., 1991. Influence of size of indigenous rhizobial populations on establishment and symbiotic performance of introduced rhizobia on field-grown legumes. Appl. Environ. Microbiol. 57, 19–28.

Thomas, R.J., Hungria, M., 1988. Effect of potassium on nitrogen fixation, nitrogen transport, and nitrogen harvest index of bean. J. Plant Nutr. 11, 175–188.

Tindwa, H., Semu, E., Msumali, G.P., 2014. Effects of elevated copper levels on biological nitrogen fixation and occurrence of rhizobia in a Tanzanian coffee-cropped soil. J. Agric. Sci. Appl. 3, 13–19.

Torres, A.R., Brito, B., Imperial, J., Palacions, J.M., Ciampitti, I.A., Ruíz-Argüeso, T., Hungria, M., 2020. Hydrogen-uptake genes improve symbiotic efficiency in common beans (*Phaseolus vulgaris* L.). Antonie van Leeuwenhoek 13, 687–696.

Unkovich, M., Herridge, D., Peoples, M., Cadish, G., Boddey, B., Giller, K., et al., 2008. Measuring Plant-Associated Nitrogen Fixation in Agricultural Systems, ACIAR Monograph 136. Australian Center for International Agricultural Research (ACIAR), Canberra, Australia, p. 258.

Usher, K.M., Bergman, B., Raven, J.A., 2007. Exploring cyanobacterial mutualisms. Annu. Rev. Ecol. Evol. Syst. 38, 255–273.

Vadez, V., Rodier, F., Payre, H., Devron, J.J., 1996. Nodule permeability to O_2 and nitrogenase-linked respiration in bean genotypes varying in the tolerance of N_2 fixation to P deficiency. Plant Physiol. Biochem. 34, 872–878.

Van Amstel, A., 2006. In: Eggleston, S., Buendia, L., Miwa, K., Ngara, T., Tanabe, K. (Eds.), IPCC Guidelines for National Greenhouse Gas Inventories. Institute for Global Environmental Strategies (IGES), Hayama, Japan.

van Kessel, C., Hartley, C., 2000. Agricultural management of grain legumes: has it led to an increase in nitrogen fixation? Field Crops Res. 65, 165–181.

Vance, C.P., Gantt, J.S., 1992. Control of nitrogen and carbon metabolism in root nodules. Physiol. Plant. 85, 266–274.

Vanlauwe, B., Hungria, M., Kanampiu, F., Giller, K.E., 2019. The role of legumes in the sustainable intensification of African smallholder agriculture: Lessons learnt and challenges for the future. Agric. Ecosyst. Environ. 294, 106583.

Velázquez, E., García-Fraile, P., Ramírez-Bahena, M.H., Rivas, R., Martínez-Molina, E., 2017. Current status of the taxonomy of bacteria able to establish nitrogen-fixing legume symbiosis. In: Zaidi, A., Khan, M.S., Musarrat, J. (Eds.), Microbes for Legume Improvement. Springer, Cham, Switzerland, pp. 1–43.

Vessey, J.K., Walsh, K.B., Layzell, D.B., 1988. Oxygen limitation of N$_2$ fixation in stem-girdled and nitrate-treated soybean. Physiol. Plant. 73, 113–121.

Vessey, J.K., Pawlowski, K., Bergman, B., 2005. Root-based N$_2$-fixing symbioses: legumes, actinorhizal plants, *Parasponia* sp. and cycads. Plant Soil 274, 51–78.

Vincent, J.M., 1970. A Manual for the Practical Study of Root-Nodule Bacteria. IBP Handbook 15. Blackwell Scientifica Publication, Oxford and Ediburgh, England, 164p.

Voelcker, J.A., 1896. "Nitragin" or the use of "pure cultivation" bacteria for leguminous crops. J. Roy. Agr. Soc. 3rd Ser. 7, 253–264.

Wang, G., Sheng, L., Zhao, D., Sheng, J., Wang, X., Liao, H., 2016. Allocation of nitrogen and carbon is regulated by nodulation and mycorrhizal networks in soybean/maize intercropping system. Front. Plant Sci. 7, 1901.

Wang, L., Wang, L., Zhu, Y., Duanmu, D., 2017. Use of CRISPR/Cas9 for symbiotic nitrogen fixation research in legumes. Prog. Mol. Biol. Transl. Sci. 149, 187–213.

Weaver, R.W., Frederick, L.R., 1974. Effect of inoculum rate on competitive nodulation of *Glycine max* (L.) Merrill. II. Field studies. Agron. J. 58, 233–236.

Werner, D., 2005. Production and biological nitrogen fixation of tropical legumes. In: Werner, D., Newton, W.E. (Eds.), Nitrogen Fixation in Agriculture, Forestry, Ecology, and the Environment. Springer, Dordrecht, The Netherlands, pp. 1–13.

Werner, D., 2007. Molecular biology and ecology of the *Rhizobia*-legume symbiosis. In: Pinton, R., Varanini, Z., Nannipieri, P. (Eds.), The Rhizosphere: Biochemistry and Organic Substances at the Soil-Plant Interface, 2nd ed. CRC Press, Boca Raton, Florida, pp. 237–266.

Werner, G.D.A., Cornwell, W.K., Sprent, J.I., Kattge, J., Kiers, E.T., 2014. A single evolutionary innovation drives the deep evolution of symbiotic N$_2$-fixation in angiosperms. Nature Comm. 5, 4087.

West, S.A., Kiers, E.T., Simms, E.L., Denison, R.F., 2002. Sanctions and mutualism stability: why do rhizobia fix nitrogen. Proc. R. Soc. Lond. B 269, 685–694.

White, J.W., Prell, J., James, E.K., Poole, P., 2007. Nutrient sharing between symbionts. Plant Physiol. 144, 604–614.

Witty, J.F., Keay, P.J., Frogatt, P.J., Dart, P.J., 1979. Algal nitrogen fixation on temperate arable fields. The Broadbalk experiment. Plant Soil 52, 151–164.

Wolde-Meskel, E., van Heerwaarden, J., Abdulkadir, B., Kassa, S., Aliyi, I., Degefu, T., et al., 2018. Additive yield response of chickpea (*Cicer arietinum* L.) to rhizobium inoculation and phosphorus fertilizer acrosssmallholder farms in Ethiopia. Agric. Ecosyst. Environ. 261, 144–152.

Woods, J., Williams, A., Hughes, J.K., Black, M., Murphy, R., 2010. Energy and the food system. Philos. Trans. R. Soc. Lond. B Biol. Sci. 365, 2991–3006.

Yaacob, O., Blair, G.J., 1980. Mineralization of ^{15}N-labelled legume residues in soils with different nitrogen contents and its uptake by Rhodes grass. Plant Soil 57, 237–248.

Yan, F., Schubert, S., Mengel, K., 1996. Soil pH changes during legume growth and application of plant material. Biol. Fertil. Soils 23, 236–242.

Zaran, H.H., 1999. *Rhizobium*-legume symbiosis and nitrogen fixation under severe conditions and in an arid climate. Microbiol. Mol. Biol. Rev. 63, 968–989.

Zehr, J.P., Jenkins, B.D., Short, S.M., Steward, G.E., 2003. Nitrogenase gene diversity and microbial community structure: a cross-system comparison. Environ. Microbiol. 5, 539–554.

Zhou, J.C., Tchan, Y.T., Vincent, J.M., 1985. Reproductive capacity of bacteroids in nodules of *Trifolium repens* L. and *Glycine max* (L.) Merr. Planta 163, 473–482.

Chapter 17

Nutrient-use efficiency

Hans Lambers
School of Biological Sciences and Institute of Agriculture, University of Western Australia, Perth, WA, Australia

Summary

All plants require the same macro- and micronutrients, but the concentrations of all these elements vary greatly among plant species. In addition, some plants in specific taxa contain significant concentrations of elements that are not essential or even beneficial at those higher concentrations. In this chapter, I explore the variation among taxa and what the significance of that variation is.

Monocots and dicots differ in their requirement for calcium (Ca) and boron (B), which reflects the composition of their cell walls. Monocots have smaller amounts of cell wall components that bind Ca and B.

A particularly low leaf phosphorus (P) concentration is common in species that evolved in ancient landscapes, where P is a major limiting nutrient for primary productivity. Despite these low leaf P concentrations, these plants exhibit relatively rapid rates of photosynthesis because they invest little in major leaf P fractions such as nucleic acids and phospholipids and preferentially allocate P to photosynthetically active cells.

The high rate of photosynthesis per unit leaf N (PNUE) of C_4 plants compared with C_3 species can be explained by a low investment in Rubisco, whose oxygenation reaction is suppressed by a high CO_2 concentration in its vicinity. In addition, C_4 plants have a Rubisco with high catalytic activity.

Plant species also vary in their concentrations of elements that are not essential nutrients, for example leaf fluoride (which they accumulate as highly toxic fluoroacetate), selenium and silicon (Si). These elements play a role in defense against herbivores. Silicon plays an additional role in defense against diseases and abiotic stresses. Leaf Si concentrations are particularly high in plants on severely P-impoverished soils because Si is mobilized by carboxylates released by plants that depend on a carboxylate-based P-acquisition strategy.

17.1 General

All plants require the same essential nutrients and some also acquire beneficial elements such as silicon (Si). However, species differ in their nutrient demand. For example, many pasture legumes have a higher demand for phosphorus (P) than co-occurring grasses (Bolan et al., 1987; Ozanne et al., 1969), which reflects their lower capacity to acquire soil P, rather than a high leaf P requirement (Ozanne et al., 1969). In this chapter the focus is not on nutrient acquisition, which has been covered in Chapters 2, 12, and 14, but on nutrient utilization. I will explore why some species can function at a much lower concentration of specific nutrients, for example, P, in their leaves than in the leaves of other species, and why some require relatively high concentrations of specific nutrients, for example, calcium (Ca) or other elements, for example, fluoride (F).

When Justus von Liebig did his groundbreaking work toward what we now know as the Law of the Minimum (Von Liebig, 1855), little was known about the chemical nature of the nutrients he showed were needed by plants. Whilst many who are familiar with this Law of the Minimum tend to refer to it as Liebig's law, in all fairness, we should refer to it as "Sprengel-Liebig's Law of the Minimum," so as to give credit to Carl Sprengel [(Sprengel, 1828), as cited in Jungk (2009)]. Sprengel (1787–1859) elucidated the mineral theory and the Law of the Minimum, and thus laid the foundation of modern plant nutrition (Van Der Ploeg et al., 1999). Much later, it was discovered that, while all species require the same nutrients, the quantity they require for their leaf functioning varies substantially. In this chapter, I explore why that is the case.

TABLE 17.1 Effect of the calcium concentration in the nutrient solution on growth and calcium concentration in the shoots of a monocotyledonous [*Lolium perenne* (perennial ryegrass)] and a dicotyledonous [*Solanum lycopersicum* (tomato)] species.

Species	Calcium supply (μM)				
	0.8	2.5	10	100	1000
	Growth rate (% of maximum value)				
L. perenne	42	100	94	94	93
S. lycopersicum	3	19	52	100	80
	Calcium concentration (μmol g^{-1} dry mass)				
L. perenne	15	18	37	92	270
S. lycopersicum	50	32	75	322	621

Adapted from Loneragan (1968) and Loneragan and Snowball (1969).

17.2 Calcium and boron requirements of monocots and dicots

The leaf Ca concentration at which 90% of the maximum yield is achieved is about twice as high for dicots as for monocots (Table 17.1) (Loneragan, 1968; Loneragan and Snowball, 1969). When comparing graminoids and forbs at similar sites, forbs invariably have higher concentrations of both Ca and magnesium (Mg) (Meerts, 1997). Shoot Ca and Mg concentrations tend to be linearly correlated among angiosperm species growing in the same environment, except for species of the Caryophyllales (White et al., 2018). Commelinid monocots have lower Ca concentrations than other monocots or eudicots (Broadley et al., 2003). The root cation-exchange capacity in monocots is significantly smaller than that in dicots (White et al., 2018; Woodward et al., 1984). The Ca requirement of monocots is lower because of their lower concentration of cell wall pectate (Gigli-Bisceglia et al., 2020; White and Broadley, 2003).

In terrestrial plants a substantial proportion of the total amount of boron (B) is associated with cell wall pectins, and this tends to be correlated with the whole plant B requirement (Hu and Brown, 1994; Hu et al., 1996). The higher B requirement in dicots than in monocots (Asad et al., 2001) is associated with a higher proportion of cell wall pectic substances and polygalacturonase (Loomis and Durst, 1992; Gigli-Bisceglia et al., 2020). The concentration of strongly complexed B in root cell walls is 3–5 μg g^{-1} dry weight in graminaceous species, and up to 30 μg g^{-1} dry weight in dicotyledonous species (Tanaka, 1967). These differences reflect the differences in B requirement for optimal growth (Hu et al., 1996).

Legumes require B in relatively high concentrations for nodule development, and the B concentration in nodules is about four to five times greater than that in roots (Carpena et al., 2000). Boron deficiency reduces nodulation and impacts nodule development (Bolanos et al., 1994) because it is required for the development of infection threads (Bolanos et al., 1996). In *Pisum sativum* (pea) the development of infection threads is arrested at very early stages, and cell invasion by endocytosis is precluded, leading to nodules that contain almost no bacteria in B-deficient plants (Redondo-Nieto et al., 2001). Impaired nodule development decreases symbiotic N$_2$ fixation.

17.3 Phosphorus and nitrogen requirements of plant species that evolved in severely phosphorus-impoverished landscapes

Plant species from severely P-impoverished environments tend to have very low leaf P and nitrogen (N) concentrations (Guilherme Pereira et al., 2019; Lambers et al., 2010). The very low P concentrations [~0.2 mg g^{-1} dry weight, about a tenth of what is considered adequate for crop growth; Chapter 6; Epstein and Bloom (2005)] are partly caused by large investments in sclerenchymatous tissues in leaves (Fig. 17.1) that contain no P and N; this effectively dilutes the leaf nutrient concentrations, but that is only a part of the story (Sulpice et al., 2014). The main story is that these plants use their P and N extremely efficiently, as evidenced by a very high photosynthetic P- and N-use efficiency [i.e., the rate of photosynthesis per unit leaf P photosynthetic P-use efficiency (PPUE) and N photosynthetic N-use efficiency

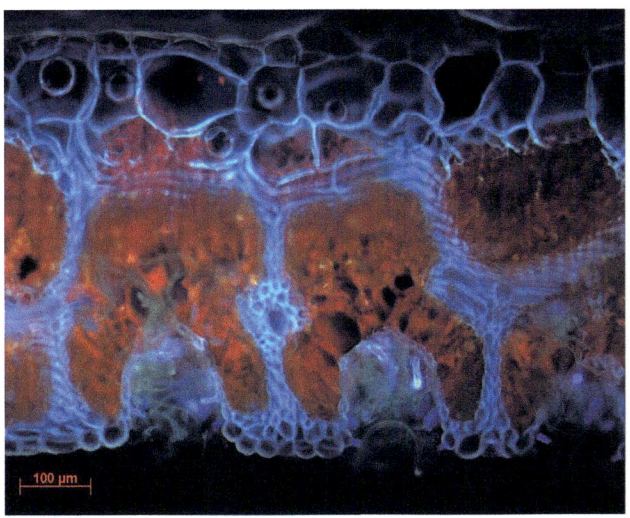

FIGURE 17.1 Fluorescent image of a cross section of a scleromorphic leaf of *Banksia ilicifolia* (holly-leaved banksia). Chloroplast-containing cells appear red; epidermal and sclerenchymatous cells are blue. Note that lignified and thick-walled fibrous bundles are dividing mesophyll tissue into segments. *Dr. Foteini Kakulas (formerly Hassiotou), University of Western Australia.*

(PNUE)] (Guilherme Pereira et al., 2019; Sulpice et al., 2014). To understand the physiological basis of a high PPUE, it is important to measure four major fractions of P in mature leaves: inorganic P, phospholipids, nucleic acids [most of which is ribosomal RNA (rRNA)], and small P-containing metabolites (Veneklaas et al., 2012). Next, I discuss how species invest P in some of these fractions to underpin a high PPUE (Lambers, 2022).

Conn and Gilliham (2010) suggested that monocots typically exhibit preferential allocation of leaf P to their mesophyll cells, and of Ca to epidermal cells, whereas dicots exhibit the opposite pattern. In this way precipitation of Ca phosphates is prevented. However, the pattern in eudicots is more complex, with species that evolved in severely P-impoverished landscapes preferentially allocating P to mesophyll cells, whereas species in the same family that evolved in landscapes where P is more abundant show no such preferential allocation (Guilherme Pereira et al., 2018; Hayes et al., 2018). Preferential allocation of P to photosynthetically active mesophyll cells, rather than metabolically inactive epidermal cells, enhances PPUE (Guilherme Pereira et al., 2019), whereas a lack of preferential P allocation allows accumulation of large amounts of P in leaves without exhibiting P-toxicity symptoms (Ye et al., 2021).

Proteaceae that evolved in severely P-impoverished environments invest a significant amount of P in phospholipids during early stages of leaf development, but then replace most of these by galactolipids and sulfolipids during later developmental stages, without exhibiting slow photosynthetic rates (Kuppusamy et al., 2014; Lambers et al., 2012). Mature leaves of *Melaleuca systena*, a co-occurring Myrtaceae, show a similar low investment in phospholipids (Yan et al., 2019). Replacement of phospholipids by membrane lipids that do not contain P contributes to a high PPUE in species from severely P-impoverished habitats.

The main leaf P fraction with very low concentration in Proteaceae from P-impoverished habitats is that associated with rRNA (Sulpice et al., 2014). Because this is the largest organic P fraction in leaves, a low investment in rRNA allows a greatly enhanced PPUE. The consequence of low rRNA abundance is a low capacity to synthesize proteins, resulting in low leaf concentrations of protein and N, and thus high PNUE. Remarkably, photosynthetic enzymes including Rubisco, do not exhibit low abundance, which explains why these plants exhibit relatively fast rates of photosynthesis and high PNUE. That prompts the question: which proteins exist in very low abundance in these P-efficient plants? The most likely candidates are ribosomal proteins, given the low abundance of ribosomes in these plants; ribosomal proteins represent a large fraction of total leaf proteins because eukaryotic ribosomes comprise 79–80 ribosomal proteins (Wilson and Doudna Cate, 2012).

The fourth major leaf fraction (small P-containing metabolites that are intermediates of the Calvin cycle and glycolysis) does not exhibit a low concentration in P-efficient Proteaceae (Sulpice et al., 2014). Metabolite P is therefore the only one of the four major fractions not present at low concentrations in P-efficient Proteaceae, relative to other tested species. A decrease in the concentrations of substrates in carbon metabolism might limit the activity of these pathways, unless compensated for by greater abundance of enzymes (Lambers et al., 2015b) in these pathways, but this would incur P costs associated with rRNA and thus not lead to a more efficient use of P. This suggests a trade-off between metabolite (enzyme substrate) and protein (enzyme) concentrations in achieving rapid metabolic fluxes (Lambers et al., 2015b).

A common trait in species from P-impoverished environments is delayed leaf greening (Lambers et al., 2015a; Sulpice et al., 2014). Although it has been suggested that this is a strategy to reduce herbivory because the young expanding leaves contain little protein (Kursar and Coley, 1992), it is also a photoprotective and nutrient-saving strategy (Kuppusamy et al., 2020). First, P is used for RNA in cytosolic ribosomes to build leaf cells, and then most of these ribosomes are recycled, and P is used for RNA in plastidic ribosomes (Sulpice et al., 2014). This allows the leaves to function at much lower ribosomal RNA abundance than would be required if leaf cells and chloroplasts were produced at the same time, as in plants that do not exhibit delayed greening (Gibon, 2014). Delayed greening is common in Proteaceae (Lambers et al., 2015a) and in co-occurring species from other families that evolved in severely P-impoverished landscapes (Fig. 17.2).

What does variation in PPUE and PNUE as underpinned by variation in leaf P fractions really mean for desirable crop genotypes and ecosystem functioning, bearing in mind that we are dealing with mature leaves that have finished expanding? These nutrient-efficient leaves do not directly impact the growth rate of the plants, but perhaps do so indirectly; this deserves to be explored further (Lambers, 2022). There is insufficient information about the significance of differences in P fractions for photosynthetic nutrient-use efficiency in contrasting species (Lambers, 2022). Some Fabaceae function at higher leaf P concentrations than global averages (Moir et al., 2016; Wright et al., 2004), but this is by no means a general trend among crop and pasture species (Ozanne et al., 1969). Moreover, Fabaceae from severely P-impoverished habitats typically exhibit low leaf P concentration (Guilherme Pereira et al., 2019; Sprent, 1999). It would be interesting to know more about the allocation of P to the leaf P fractions in a range of Fabaceae that function at low leaf P concentrations (Lambers, 2022), especially in the *Daviesia* group, because cluster roots appear to be pervasive in *Daviesia* and close relatives (Nge et al., 2020). It would also be of great interest to find out if a high PPUE, similar to that in Proteaceae, is a desirable trait in crop plants, or if it is traded-off against another desirable trait such as acclimation capacity (Lambers, 2022). If there is a trade-off, a compromise might be desirable if crops can be managed in a way to minimize stress.

FIGURE 17.2 Examples of delayed greening in species from severely phosphorus-impoverished environments. Delayed greening is a strategy that saves both phosphorus and nitrogen. Proteaceae: *Hakea amplexicaulis* (A), *Synaphea spinulosa* (E); Myrtaceae: *Corymbia calophylla* (B), *Calothamnus hirsutus* (F); Fabaceae: *Acacia glaucoptera* (C), *Gastrolobium spinosum* (G); Ericaceae: *Leucopogon verticillatus* (D); Podocarpaceae: *Podocarpus drouynianus* (H). *Courtesy Part (C) Marion Cambridge, University of Western Australia; (F) Hongtao Zhong, University of Western Australia; (G) Patrick Hayes, University of Western Australia; (H) Dean Beaver; other images, Hans Lambers.*

17.4 Micronutrient requirements of plant species that evolved in severely phosphorus-impoverished landscapes

It has long been known that the concentrations of iron (Fe), zinc (Zn), copper (Cu), and nickel (Ni) are very low in leaves of Proteaceae and co-occurring plants (Denton et al., 2007; Fisher et al., 2006; Kuo et al., 1982), but the background for this was unclear. Now we know that protein concentrations are also very low in these species (Prodhan et al., 2019). Because many micronutrients are associated with enzymes (Rasheed and Seeley, 1966), the low concentrations of protein-associated micronutrients reflect low leaf protein concentrations, which are the result of low abundance of ribosomes (Sulpice et al., 2014). Given that the concentrations of these micronutrients in leaves remain low when plants are grown with a higher supply than they encounter in the field, micronutrient uptake must be tightly controlled in these highly efficient species (Prodhan et al., 2019), as is their uptake of nitrate (Prodhan et al., 2016) and sulfate (Prodhan et al., 2017).

In contrast to the micronutrients referred to previously, leaf manganese (Mn) concentrations of plants on P-impoverished soils are often relatively high, but with a large variation among species (Denton et al., 2007; Fisher et al., 2006; Kuo et al., 1982). High leaf Mn concentrations are typically associated with a carboxylate-releasing P-acquisition strategy (Lambers et al., 2015c; Pang et al., 2018). Conversely, low Mn concentrations are the result of interception of Mn by mycorrhizal hyphae (Bethlenfalvay and Franson, 1989; Lambers et al., 2015c). In contrast to the tightly controlled uptake of the micronutrients discussed earlier (Prodhan et al., 2019), Mn uptake predominantly involves nonspecific transporters and is poorly controlled (Baxter et al., 2008; Korshunova et al., 1999). Leaf Mn concentration can therefore be used as a tool to screen for a belowground P-acquisition trait in plants growing in the field (Fig. 17.3) (Lambers et al., 2021; Zhong et al., 2021) or for carboxylate-releasing genotypes in glasshouse studies (Pang et al., 2018). In plants growing on severely nutrient-impoverished soils, leaf Mn tends not to accumulate to extreme concentrations, as it does in metallophytes with cluster roots in New Caledonia (Fernando et al., 2010). It is therefore the leaf

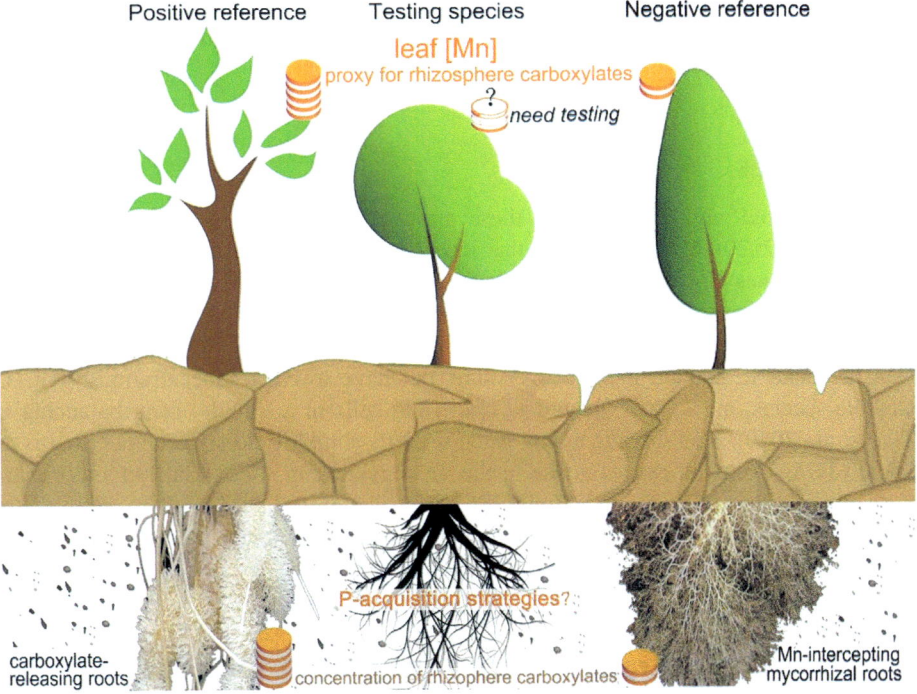

FIGURE 17.3 Diagram illustrating an experimental design to assess if a species whose phosphorus (P)-acquisition strategy is unknown (middle) uses a P-mobilizing strategy that is based on the release of large amounts of carboxylates. What is needed is a "positive reference" (left), which is known to release large amounts of carboxylates, and a "negative reference" (right), which depends on mycorrhizas and does not release large amounts of carboxylates. The positive reference has a high leaf manganese concentration [Mn], whereas the negative reference has a low leaf [Mn]. The leaf [Mn] of the species to be assessed will give a strong indication whether it depends on carboxylate release. If it grows in proximity to a species that releases large amounts of carboxylates, then it may depend on facilitation by this species, so it is important to sample it at some distance from a positive reference. *Modified after Zhou et al. (2020).*

Mn concentration relative to that of reference species that is informative, rather than the absolute concentration (Zhou et al., 2020). The references species to be selected must include a positive reference that is known to release large amounts of carboxylates, for example, most Proteaceae and Cyperaceae (Fig. 17.3). The other is a negative reference that is known to be mycorrhizal and not depend on carboxylates for P acquisition (Fig. 17.3) (Zhou et al., 2020). Indications provided by leaf Mn concentration of plants growing in the field can then be followed up by glasshouse trials to assess the release of P-mobilizing exudates (Lambers et al., 2021).

Metallophytes may accumulate a range of micronutrients other than Mn such as Cu, Ni, and Zn (Alford et al., 2010; Deng et al., 2018; Séleck et al., 2013) that used to be referred to as "heavy metals." Because heavy metals are ill-defined, this term has been replaced simply by "metals" (Pourret and Bollinger, 2018). Metal accumulation in metallophytes is not restricted to micronutrients, but also includes metals and metalloids that are not required by plants, such as cadmium (Cd), lead (Pb), and arsenic (As) (Baker and Proctor, 1990; Ditusa et al., 2015; Li et al., 2018). Metallophytes are usually restricted to habitats characterized by soils with high concentrations of one or more metals, for instance, serpentine and ultramafic soils (Hulshof and Spasojevic, 2020; Van Der Ent et al., 2015) and mine sites, some dating back to Roman times (Baker et al., 2010).

17.5 Nitrogen requirements of C$_3$ and C$_4$ plants

Compared with C$_3$ plants, C$_4$ plants function at much higher CO$_2$ concentrations at the site of Rubisco, 1000−2000 μmol mol^{-1} in C$_4$ plants as opposed to about 200 μmol mol^{-1} in C$_3$ plants (Christin and Osborne, 2014; Lambers and Oliveira, 2019). Rubisco catalyzes both a carboxylating and an oxygenating reaction, depending on CO$_2$ and O$_2$ concentrations (Bathellier et al., 2018). As a result of the much higher CO$_2$ concentrations at the site of Rubisco, the oxygenation reaction is virtually completely suppressed in C$_4$ plants, which contributes to their higher PNUE, even though some of the effect of a high CO$_2$ concentration is offset by the requirement of additional enzymes needed in C$_4$ photosynthesis (Lambers and Oliveira, 2019). The higher CO$_2$ concentration has allowed C$_4$ plants to evolve a Rubisco enzyme with lower affinity (higher K_m) for CO$_2$ but a higher catalytic activity (k_{cat}) (Tcherkez et al., 2006). These kinetic properties allow a smaller investment in Rubisco for the same catalytic activity and, thus, further contribute to the higher PNUE of C$_4$ plants than that of C$_3$ plants (Sage and Pearcy, 1987).

17.6 Calcicole species

Calcicole species are associated with Ca-rich soils; these may be calcareous, but that is not invariably the case (De Silva, 1934; De Souza et al., 2020). Many calcicoles that have high leaf Ca concentrations accumulate Ca in nonstomatal epidermal cells and trichomes (Charisios et al., 2003; De Silva et al., 1996). Excess Ca occurs in crystals (Charisios et al., 2003; He et al., 2012) or is excreted as Ca salts via stomata (Wu et al., 2011).

Grevillea thelemanniana (Proteaceae) is a rare calcicole species in which very little Ca occurs in crystals, in contrast to other Proteaceae that tend to contain numerous crystals but have low Ca concentrations (Fig. 17.4) (Gao et al., 2020). In this calcicole species, most Ca is located in epidermal cells in a soluble form, balancing an organic anion, *trans*-aconitate (Gao et al., 2020). That organic anion, because of its similarity to *cis*-aconitate that is an intermediate of the tricarboxylic acid (TCA) cycle, blocks aconitase in the TCA cycle (Fig. 17.5) (Saffran and Prado, 1949), and thus acts as a putative feeding deterrent (Burau and Stout, 1965; Katsuhara et al., 1993). It is likely that this specific trait underpins the rarity of *G. thelemanniana* because in the ancient landscapes where this species evolved, high Ca availability and neutral to acidic pH are rather uncommon (Gao et al., 2020).

17.7 Variation in leaf sulfur requirement among plant species

Plants vary considerably in their leaf sulfur (S) concentrations (Chen et al., 2021). While this partly reflects S availability in the soil, there are also distinct species effects. For example, Brassicaceae accumulate S-containing antiherbivore compounds (glucosinolates; Falk et al., 2007; Halkier and Gershenzon, 2006) and hence tend to exhibit relatively high leaf S concentrations (Dijkshoorn and Van Wijk, 1967). In *Brassica* species, glucosinolates comprise 1.7%−8.0% of total plant S, but the proportion is even greater in other species (Falk et al., 2007). The buds of *Schouwia purpurea* contain 12% glucosinolates per dry weight (300 μmol glucosinolates g^{-1} dry weight), and in some plants, glucosinolates may make up 10%−30% of the total sulfur content (Falk et al., 2007).

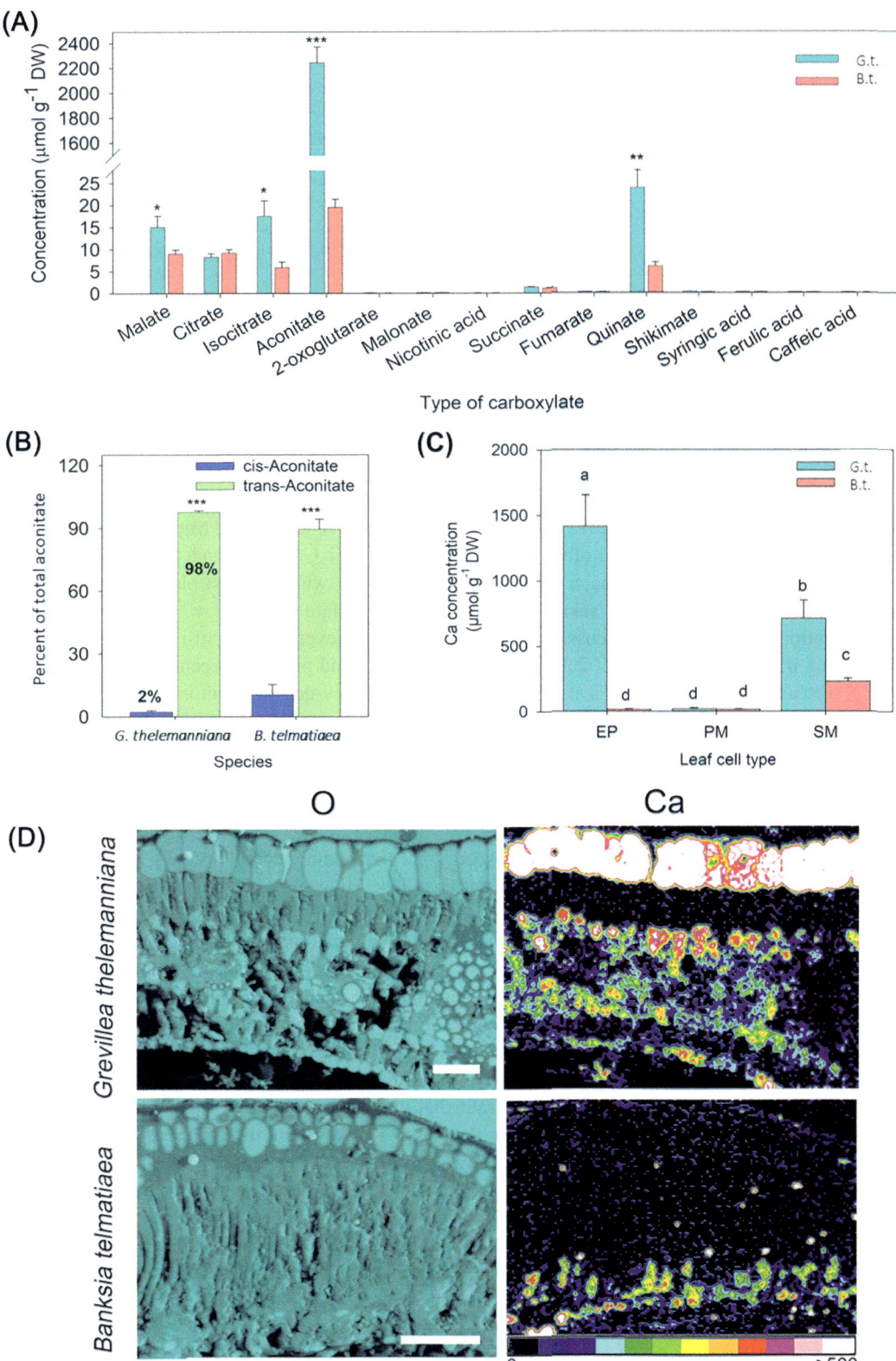

FIGURE 17.4 (A) Concentrations of a range of metabolites, including several carboxylates, in leaves of *Grevillea thelemanniana* (G.t.; spider-net grevillea) and *Banksia telmatiaea* (B.t.; swamp fox banksia) (both Proteaceae) collected in their natural habitat. (B) The *cis*- and *trans*-aconitate as percentages of the total amount of aconitate in leaves of *G. thelemanniana* and *B. telmatiaea*. (C) Mean leaf cell-specific calcium (Ca) concentrations quantified in different cell types. (D) Element maps showing oxygen distribution (O; to show leaf anatomy) and cellular Ca concentrations of *G. thelemanniana* and *B. telmatiaea* leaves. Bars, 50 μm. Calcium concentration scale is shown for both species at the bottom right (black, not detectable; white >500 mmol kg^{-1}). All leaves are isobilateral. *EP*, Epidermis; *PM*, palisade mesophyll; *SM*, spongy mesophyll. *Reproduced from Gao et al. (2020), with permission from John Wiley and Sons.*

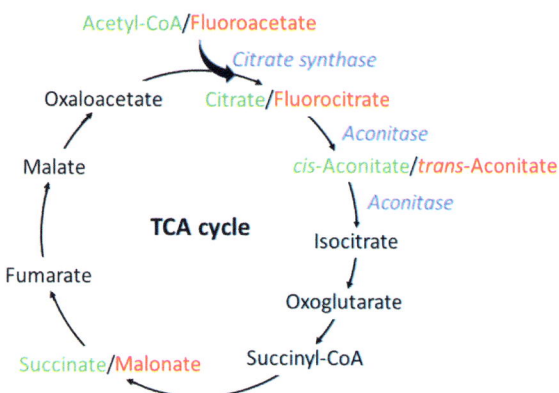

FIGURE 17.5 Simplified scheme of the TCA cycle, showing key metabolites (green) and antimetabolites (red) produced in some plant species. Fluoroacetate is metabolized via citrate synthase, producing fluorocitrate, which blocks the TCA cycle. Some plants accumulate *trans*-aconitate (Burau and Stout, 1965; Gao et al., 2020; MacLennan and Beevers, 1964), a stereoisomer of *cis*-aconitate; it also blocks the TCA cycle. Several plant species produce malonate (Bentley, 1952) that has a structure similar to succinate and competitively inhibits the conversion into fumarate (Dervartanian and Veeger, 1964).

Other plants that exhibit relatively high S concentrations are those that accumulate Ca sulfate crystals, as found in needles of *Pinus palustris* (Pinaceae) (Pritchard et al., 2000) and phyllodes (modified petioles that function as leaves) of *Acacia robeorum* (Fabaceae) (He et al., 2012). The crystals may remove excess Ca, Mg, and S, protect plants against herbivory, and detoxify aluminum (Al) and other metals. Based on their Ca and S concentrations, phyllodes of other *Acacia* species likely also produce Ca sulfate crystals (Reid et al., 2016), whereas most plants tend to accumulate Ca in crystals containing oxalate (Franceschi and Nakata, 2005) rather than sulfate.

Leaf S concentrations also deserve a discussion for Proteaceae from severely nutrient-impoverished soils, where S is not a growth-limiting nutrient (Denton et al., 2007). Given their low N and protein concentrations (Sulpice et al., 2014), one might expect Proteaceae to exhibit low leaf S concentrations, but they do not (Denton et al., 2007). This is because a large proportion of their phospholipids is replaced by sulfolipids during leaf development (Lambers et al., 2012). This is a strategy to save P at the expense of S that is in relative abundance.

17.8 Fluoride in leaves of plants occurring on soils containing little fluoride

Some plants that naturally occur on nutrient-impoverished soils that contain very little F, for example, in Australia (McEwan, 1964; Twigg et al., 1996), southern Africa (Marais, 1944; Minnaar et al., 2000), and tropical South America (De Oliveira, 1963; Krebs et al., 1994), accumulate F as fluoroacetate. This compound is extremely toxic, and fluoroacetate-accumulating plants can cause the death of farm animals that ate only small amounts of them (McEwan, 1964; Steyn, 1928; Ubiali et al., 2020). Yet, fluoride is not an essential or beneficial plant nutrient. Fluoroacetate only occurs in a small number of genera in Bignoniaceae, Dichapetalaceae, Fabaceae, Malpighiaceae, and Rubiaceae (Lee et al., 2014, with major variation in fluoroacetate concentration in leaves of species within a genus (Chandler et al., 2002; Lee et al., 2012).

Fluoroacetate is metabolized via citrate synthase that normally uses acetyl-CoA as a substrate and produces citrate. With fluoroacetate as a substrate, it produces fluorocitrate that cannot be converted by aconitase and thus blocks the TCA cycle (Morrison and Peters, 1954); *trans*-aconitate blocks the TCA cycle at the same enzyme (Fig. 17.5).

17.9 Selenium in leaves of some plants

Selenium (Se) is essential for animal nutrition (Lima and Schiavon, 2021; Reis et al., 2019), but it is not an essential plant nutrient; yet some plants (hyper)accumulate Se (Broyer et al., 1972; Lima and Schiavon, 2021; Pinto Irish et al., 2021). Selenium may act as an antioxidant in plants at low concentrations (1–10 µg Se g^{-1} dry weight) because of its capacity to enhance the activity of radical-scavenging enzymes and the synthesis of nonenzymatic antioxidant compounds (Chauhan et al., 2019). It is therefore sometimes considered a beneficial element for plants (Silva et al., 2020). In *Brassica juncea*, Se accumulation offers no protection against snails but does protect from invertebrate herbivory and fungal infection (Hanson et al., 2003).

17.10 Silicon as a beneficial element in leaves of some plants

The Si concentrations in aboveground plant parts vary significantly among species (De Tombeur et al., 2020a; Hodson et al., 2005) and with environmental conditions (Carey, 2020; Minden et al., 2021) (Section 8.3). In some plants, Si is

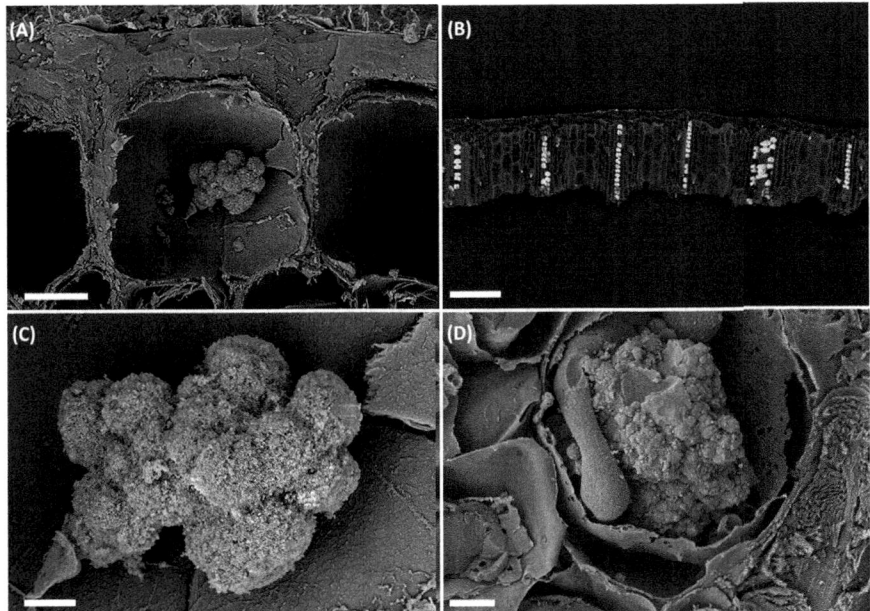

FIGURE 17.6 Phytoliths in leaves of a range of plant species. (A, C) Scanning electron micrographs of a cross section through a *Banksia attenuata* (slender banksia; Proteaceae) leaf showing a phytolith in an upper epidermal cell; scale bar in (A) = 10 μm and in (C) = 2 μm. (B) Backscattered scanning electron micrograph of a cross section through a *Microlaena stipoides* (weeping grass; Poaceae) leaf showing abundant phytoliths throughout the leaf structure; scale bar = 100 μm. (D) Scanning electron micrograph of a cross section through a *Petrophile macrostachya* (Proteaceae) leaf showing a phytolith adjacent to vascular tissue; scale bar = 5 μm. *Courtesy Nicolas Honvault and Peta Clode (A, C, and D) and Kosala Ranathunge and Peta Clode (B); all from University of Western Australia.*

barely detectable in leaves, whereas in others it may comprise 10% of the leaf dry weight; it is predominantly taken up as $Si(OH)_4$ via specific influx channels (Coskun et al., 2019). Soluble Si is transported in the transpiration stream until it reaches a site where it precipitates to form a solid silica body or phytolith, mainly in cell walls but also in the cell lumen (Hodson, 2016). Phytoliths may have various shapes (Fig. 17.6). Silicon is involved in mitigation of several plant biotic and abiotic stresses (De Tombeur et al., 2021b); it can be used as an inexpensive structural component (Raven, 2003), and in some plant species, it may increase primary productivity and crop yield (Liang et al., 2015; Tamai and Ma, 2008).

In *Brachypodium distachyon* (false brome; Poaceae), Si accumulation and deposition are enhanced at growth-limiting N availability (Johnson et al., 2021). In addition to its role as a long-term defense against herbivores, on a short-term scale, Si accumulation responds to herbivore signals and impacts plant defense (Waterman et al., 2020). The physical defense is presumably based on SiO_2 deposits (phytoliths) interfering with herbivore digestion, wearing down herbivore mouthparts, and increasing leaf toughness (Johnson et al., 2019; Vandegeer et al., 2021a). In *Oryza sativa* (rice), Si is essential for defense against chewing herbivores; the leaves of cultivar Nipponbare are covered with sharp nonglandular Si-impregnated trichomes, whereas these defense structures are virtually absent in the herbivory-prone NERICA cultivar (Andama et al., 2020).

Silicon also enhances plant resistance to water stress, partly by Si deposition on stomatal guard cells, which is associated with increased stomatal sensitivity to ABA mediated by guard cell K^+ efflux (Vandegeer et al., 2021b). In *O. sativa*, Si mitigates Al-induced inhibition of root elongation by reducing the deposition of Al in cell walls (Xiao et al., 2021) and toxicity in shoots by decreasing Al transport to the shoots, without reducing Al uptake by roots (De Freitas et al., 2017). Aluminum can also coprecipitate in phytoliths (Hodson and Evans, 2020).

Silicon can ameliorate the phytotoxicity induced by excess metal(loid)s, in part by restricting root uptake and immobilization in the rhizosphere, and by retention of elements in the root apoplast (Bokor et al., 2019). At the cellular level, the formation of insoluble complexes of Si and metal(loid)s and their storage in cell walls decreases available element concentrations and restricts symplastic uptake (Vaculík et al., 2020).

Coskun et al. (2019) argue that the various roles of Si in plants in abiotic stress tolerance are at odds with the rather limited biochemical properties of this element. They suggest that most of the roles are associated with a prevention of the deregulation caused by stress. The authors propose a unifying model for the multitude of beneficial effects of Si which they call the "apoplastic obstruction hypothesis," to stimulate advances toward a better understanding of the roles of Si.

Silicon accumulation in plants increases sharply with increasing soil age, despite a very low availability of Si in ancient soils (De Tombeur et al., 2020b). This conundrum may be explained by increased abundance of plants species with a carboxylate-releasing P-mobilization strategy on such soils (Zemunik et al., 2015). These carboxylates not only mobilize P and Mn, but also Si (Bennett et al., 1988; De Tombeur et al., 2021a). The wide variation in leaf Si concentrations among species on ancient soils is likely explained by the extent to which the species express Si transporters (Ma et al., 2006).

Given the role of Si in plant stress tolerance, high-Si crop genotypes are desirable. In rice, there is substantial variation among cultivars, indicating a potential to breed cultivars that accumulate Si and have greater disease resistance (Bryant et al., 2011).

17.11 Leaf longevity and nutrient remobilization

In slow-growing native plants, increased leaf longevity with a high leaf mass per area (LMA) is a strategy to enhance long-term use of leaf nutrients (Lambers and Poorter, 1992; Wright et al., 2004). Conversely, in fast-growing native plants and crops, leaves do not live long, but turn over quickly and have a low LMA. The long-term effect of leaf longevity on nutrient economy is far greater than that of functioning at a low nutrient concentration, which impacts short-term nutrient-use efficiency (e.g., PNUE, PPUE).

Efficient remobilization of nutrients from senescing organs is another way to use acquired nutrients efficiently. Nutrient-remobilization efficiency (the proportion of nutrients present in mature leaves that is resorbed during senescence) varies substantially among nutrients. It is lowest for nutrients that are poorly mobile in the phloem because of the high pH of phloem sap (Lambers and Oliveira, 2019), for example, B, Ca, Fe, and Mn (Chen et al., 2021; Vergutz et al., 2012) (see Chapter 3). It is high for phloem-mobile nutrients in plants growing on nutrient-impoverished soils (Reed et al., 2012; Urbina et al., 2021). Especially for leaf P, remobilization strongly increases with decreasing nutrient availability in the habitat (Vergutz et al., 2012). Efficient remobilization of nutrients from senescing leaves and roots would be a desirable trait in crops, but very little is known about the genetic variation for this trait as related to crop yield (Chen and Liao, 2017; Cong et al., 2020; Zhao et al., 2014).

References

Alford, É., Pilon-Smits, E.H., Paschke, M., 2010. Metallophytes—a view from the rhizosphere. Plant Soil 337, 33–50.

Andama, J.B., Mujiono, K., Hojo, Y., Shinya, T., Galis, I., 2020. Nonglandular silicified trichomes are essential for rice defense against chewing herbivores. Plant Cell Environ. 43, 2019–2032.

Asad, A., Bell, R.W., Dell, B., 2001. A critical comparison of the external and internal boron requirements for contrasting species in boron-buffered solution culture. Plant Soil 233, 31–45.

Baker, A.J.M., Proctor, J., 1990. The influence of cadmium, copper, lead, and zinc on the distribution and evolution of metallophytes in the British Isles. Plant Syst. Evol. 173, 91–108.

Baker, A.J., Ernst, W.H., Van Der Ent, A., Malaisse, F., Ginocchio, R., 2010. Metallophytes: the unique biological resource, its ecology and conservational status in Europe, central Africa and Latin America. In: Batty, L.C., Hallberg, K.B. (Eds.), Ecology of Iindustrial Pollution. Cambridge University Press, Cambridge, UK, pp. 7–39.

Bathellier, C., Tcherkez, G., Lorimer, G.H., Farquhar, G.D., 2018. Rubisco is not really so bad. Plant Cell Environ. 41, 705–716.

Baxter, I.R., Vitek, O., Lahner, B., Muthukumar, B., Borghi, M., Morrissey, J., et al., 2008. The leaf ionome as a multivariable system to detect a plant's physiological status. Proc. Natl. Acad. Sci. U.S.A. 105, 12081–12086.

Bennett, P.C., Melcer, M.E., Siegel, D.I., Hassett, J.P., 1988. The dissolution of quartz in dilute aqueous solutions of organic acids at 25°C. Geochim. Cosmochim. Acta 52, 1521–1530.

Bentley, L.E., 1952. Occurrence of malonic acid in plants. Nature 170, 847–848.

Bethlenfalvay, G.J., Franson, R.L., 1989. Manganese toxicity alleviated by mycorrhizae in soybean. J. Plant Nutr. 12, 953–970.

Bokor, B., Soukup, M., Vaculík, M., Vd'ačný, P., Weidinger, M., Lichtscheidl, I., et al., 2019. Silicon uptake and localisation in date palm (*Phoenix dactylifera*) – A unique association with sclerenchyma. Front. Plant Sci. 10, 988.

Bolan, N.S., Robson, A.D., Barrow, N.J., 1987. Effects of vesicular-arbuscular mycorrhiza on the availability of iron phosphates to plants. Plant Soil 99, 401–410.

Bolanos, L., Esteban, E., De Lorenzo, C., Fernandez-Pascual, M., De Felipe, M.R., Garate, A., et al., 1994. Essentiality of boron for symbiotic dinitrogen fixation in pea (*Pisum sativum*) rhizobium nodules. Plant Physiol. 104, 85–90.

Bolanos, L., Brewin, N.J., Bonilla, I., 1996. Effects of boron on rhizobium-legume cell-surface interactions and nodule development. Plant Physiol. 110, 1249–1256.

Broadley, M.R., Bowen, H.C., Cotterill, H.L., Hammond, J.P., Meacham, M.C., Mead, A., et al., 2003. Variation in the shoot calcium content of angiosperms. J. Exp. Bot. 54, 1431–1446.

Broyer, T.C., Johnson, C.M., Huston, R.P., 1972. Selenium and nutrition of *Astragalus*. I. Effects of selenite or selenate supply on growth and selenium content. Plant Soil 36, 635–649.

Bryant, R., Proctor, A., Hawkridge, M., Jackson, A., Yeater, K., Counce, P., et al., 2011. Genetic variation and association mapping of silica concentration in rice hulls using a germplasm collection. Genetica 139, 1383–1398.

Burau, R., Stout, P.R., 1965. Trans-aconitic acid in range grasses in early spring. Science 150, 766–767.

Carey, J., 2020. Soil age alters the global silicon cycle. Science 369, 1161–1162.

Carpena, R.O., Esteban, E., Sarro, M.J., Peñalosa, J., Gárate, A.N., Lucena, J.J., et al., 2000. Boron and calcium distribution in nitrogen-fixing pea plants. Plant Sci. 151, 163–170.

Chandler, G.T., Crisp, M.D., Cayzer, L.W., Bayer, R.J., 2002. Monograph of *Gastrolobium* (Fabaceae: Mirbelieae). Aust. Syst. Bot. 15, 619–739.

Charisios, T.D., Cotsopoulos, B., Psaras, G.K., 2003. Distribution of calcium in epidermal cell types of Mediterranean xeromorphic calcicoles. Flora 198, 341–348.

Chauhan, R., Awasthi, S., Srivastava, S., Dwivedi, S., Pilon-Smits, E.A.H., Dhankher, O.P., et al., 2019. Understanding selenium metabolism in plants and its role as a beneficial element. Crit. Rev. Environ. Sci. Technol. 49, 1937–1958.

Chen, L., Liao, H., 2017. Engineering crop nutrient efficiency for sustainable agriculture. J. Integr. Plant Biol. 59, 710–735.

Chen, H., Reed, S.C., Lü, X., Xiao, K., Wang, K., Li, D., 2021. Global resorption efficiencies of trace elements in leaves of terrestrial plants. Funct. Ecol. 35, 1596–1602.

Christin, P.-A., Osborne, C.P., 2014. The evolutionary ecology of C_4 plants. New Phytol. 204, 765–781.

Cong, W.-F., Suriyagoda, L.D.B., Lambers, H., 2020. Tightening the phosphorus cycle through phosphorus-efficient crop genotypes. Trends Plant Sci. 25, 967–975.

Conn, S., Gilliham, M., 2010. Comparative physiology of elemental distributions in plants. Ann. Bot. 105, 1081–1102.

Coskun, D., Deshmukh, R., Sonah, H., Menzies, J.G., Reynolds, O., Ma, J.F., et al., 2019. The controversies of silicon's role in plant biology. New Phytol. 221, 67–85.

De Freitas, L.B., Fernandes, D.M., Maia, S.C.M., Fernandes, A.M., 2017. Effects of silicon on aluminum toxicity in upland rice plants. Plant Soil 420, 263–275.

De Oliveira, M., 1963. Chromatographic isolation of monofluoroacetic acid from *Palicourea marcgravii* St. Hil. Experientia 19, 586–587.

De Silva, B.L.T., 1934. The distribution of "calcicole" and "calcifuge" species in relation to the content of the soil in calcium carbonate and exchangeable calcium, and to soil reaction. J. Ecol. 22, 532–553.

De Silva, D.L.R., Hetherington, A.M., Mansfield, T.A., 1996. Where does all the calcium go? Evidence of an important regulatory role for trichomes in two calcicoles. Plant Cell Environ. 19, 880–886.

De Souza, M.C., Williams, T.C.R., Poschenrieder, C., Jansen, S., Pinheiro, M.H.O., Soares, I.P., et al., 2020. Calcicole behaviour of *Callisthene fasciculata* Mart., an Al-accumulating species from the Brazilian Cerrado. Plant Biol 22, 30–37.

De Tombeur, F., Cornelis, J.T., Lambers, H., 2021a. Silicon mobilisation by root-released carboxylates. Trends Plant Sci. 26, 1116–1125.

De Tombeur, F., Roux, P., Cornelis, J.T., 2021b. Silicon dynamics through the lens of soil-plant-animal interactions: perspectives for agricultural practices. Plant Soil 467, 1–28.

De Tombeur, F., Turner, B.L., Laliberté, E., Lambers, H., Cornélis, J.-T., 2020a. Silicon dynamics during 2 million years of soil development in a coastal dune chronosequence under a Mediterranean climate. Ecosystems 23, 1614–1630.

De Tombeur, F., Turner, B.L., Laliberté, E., Lambers, H., Mahy, G., Faucon, M.-P., et al., 2020b. Plants sustain the terrestrial silicon cycle during ecosystem retrogression. Science 369, 1245–1248.

Deng, T.-H.-B., Van Der Ent, A., Tang, Y.-T., Sterckeman, T., Echevarria, G., Morel, J.-L., et al., 2018. Nickel hyperaccumulation mechanisms: a review on the current state of knowledge. Plant Soil 423, 1–11.

Denton, M.D., Veneklaas, E.J., Freimoser, F.M., Lambers, H., 2007. *Banksia* species (Proteaceae) from severely phosphorus-impoverished soils exhibit extreme efficiency in the use and re-mobilization of phosphorus. Plant Cell Environ. 30, 1557–1565.

Dervartanian, D.V., Veeger, C., 1964. Studies on succinate dehydrogenase: I. Spectral properties of the purified enzyme and formation of enzyme-competitive inhibitor complexes. Biochim. Biophys. Acta. 92, 233–247.

Dijkshoorn, W., Van Wijk, A.L., 1967. The sulphur requirements of plants as evidenced by the sulphur-nitrogen ratio in the organic matter a review of published data. Plant Soil 26, 129–157.

Ditusa, S.F., Fontenot, E.B., Wallace, R.W., Silvers, M.A., Steele, T.N., Elnagar, A.H., et al., 2015. A member of the Phosphate transporter 1 (Pht1) family from the arsenic-hyperaccumulating fern *Pteris vittata* is a high-affinity arsenate transporter. New Phytol. 209, 762–772.

Epstein, E., Bloom, A.J., 2005. Mineral Nutrition of Plants: Principles and Perspectives. Sinauer, Sunderland, MA, USA.

Falk, K.L., Tokuhisa, J.G., Gershenzon, J., 2007. The effect of sulfur nutrition on plant glucosinolate content: physiology and molecular mechanisms. Plant Biol. 9, 573–581.

Fernando, D.R., Mizuno, T., Woodrow, I.E., Baker, A.J.M., Collins, R.N., 2010. Characterization of foliar manganese (Mn) in Mn (hyper)accumulators using X-ray absorption spectroscopy. New Phytol. 188, 1014–1027.

Fisher, J.L., Veneklaas, E.J., Lambers, H., Loneragan, W.A., 2006. Enhanced soil and leaf nutrient status of a Western Australian *Banksia* woodland community invaded by *Ehrharta calycina* and *Pelargonium capitatum*. Plant Soil 284, 253–264.

Franceschi, V.R., Nakata, P.A., 2005. Calcium oxalate in plants: formation and function. Annu. Rev. Plant Biol. 56, 41–71.

Gao, J., Wang, F., Ranathunge, K., Arruda, A.J., Cawthray, G.R., Clode, P.L., et al., 2020. Edaphic niche characterization of four Proteaceae reveals unique calcicole physiology linked to hyper-endemism of *Grevillea thelemanniana*. New Phytol. 228, 869–883.

Gibon, Y., 2014. Why bring post genomics into the phosphorus-impoverished bush? Plant Cell Environ. 37, 1273–1275.

Gigli-Bisceglia, N., Engelsdorf, T., Hamann, T., 2020. Plant cell wall integrity maintenance in model plants and crop species-relevant cell wall components and underlying guiding principles. Cell. Mol. Life Sci. 77, 2049–2077.

Guilherme Pereira, C., Clode, P.L., Oliveira, R.S., Lambers, H., 2018. Eudicots from severely phosphorus-impoverished environments preferentially allocate phosphorus to their mesophyll. New Phytol. 218, 959–973.

Guilherme Pereira, C., Hayes, P.E., O'Sullivan, O., Weerasinghe, L., Clode, P.L., Atkin, O.K., et al., 2019. Trait convergence in photosynthetic nutrient-use efficiency along a 2-million year dune chronosequence in a global biodiversity hotspot. J. Ecol. 107, 2006–2023.

Halkier, B.A., Gershenzon, J., 2006. Biology and biochemistry of glucosinolates. Annu. Rev. Plant Biol. 57, 303–333.

Hanson, B., Garifullina, G.F., Lindblom, S.D., Wangeline, A., Ackley, A., Kramer, K., et al., 2003. Selenium accumulation protects *Brassica juncea* from invertebrate herbivory and fungal infection. New Phytol. 159, 461–469.

Hayes, P.E., Clode, P.L., Oliveira, R.S., Lambers, H., 2018. Proteaceae from phosphorus-impoverished habitats preferentially allocate phosphorus to photosynthetic cells: an adaptation improving phosphorus-use efficiency. Plant Cell Environ. 41, 605–619.

He, H., Bleby, T.M., Veneklaas, E.J., Lambers, H., Kuo, J., 2012. Morphologies and elemental compositions of calcium crystals in phyllodes and branchlets of *Acacia robeorum* (Leguminosae: Mimosoideae). Ann. Bot. 109, 887–896.

Hodson, M.J., 2016. The development of phytoliths in plants and its influence on their chemistry and isotopic composition. Implications for palaeoecology and archaeology. J. Archaeol. Sci. 68, 62–69.

Hodson, M.J., Evans, D.E., 2020. Aluminium–silicon interactions in higher plants: an update. J. Exp. Bot. 71, 6719–6729.

Hodson, M.J., White, P.J., Mead, A., Broadley, M.R., 2005. Phylogenetic variation in the silicon composition of plants. Ann. Bot. 96, 1027–1046.

Hu, H., Brown, P.H., 1994. Localization of boron in cell walls of squash and tobacco and its association with pectin (evidence for a structural role of boron in the cell wall). Plant Physiol. 105, 681–689.

Hu, H., Brown, P.H., Labavitch, J.M., 1996. Species variability in boron requirement is correlated with cell wall pectin. J. Exp. Bot. 47, 227–232.

Hulshof, C.M., Spasojevic, M.J., 2020. The edaphic control of plant diversity. Glob. Ecol. Biogeogr. 29, 1634–1650.

Johnson, S.N., Ryalls, J.M.W., Barton, C.V.M., Tjoelker, M.G., Wright, I.J., Moore, B.D., 2019. Climate warming and plant biomechanical defences: silicon addition contributes to herbivore suppression in a pasture grass. Funct. Ecol. 33, 587–596.

Johnson, S.N., Waterman, J.M., Wuhrer, R., Rowe, R.C., Hall, C.R., Cibils-Stewart, X., 2021. Siliceous and non-nutritious: nitrogen limitation increases anti-herbivore silicon defences in a model grass. J. Ecol. 109, 3767–3778.

Jungk, A., 2009. Carl Sprengel—The founder of agricultural chemistry: A re-appraisal commemorating the 150th anniversary of his death. J. Plant Nutr. Soil Sci. 172, 633–636.

Katsuhara, M., Sakano, K., Sato, M., Kawakita, H., Kawabe, S., 1993. Distribution and production of trans-aconitic acid in barnyard grass (*Echinochloa crus-galli* var. *oryzicola*) as putative antifeedant against brown planthoppers. Plant Cell Physiol. 34, 251–254.

Korshunova, Y.O., Eide, D., Clark, W.G., Guerinot, M.L., Pakrasi, H.B., 1999. The IRT1 protein from *Arabidopsis thaliana* is a metal transporter with a broad substrate range. Plant Mol. Biol. 40, 37–44.

Krebs, H.C., Kemmerling, W., Habermehl, G., 1994. Qualitative and quantitative determination of fluoroacetic acid in *Arrabidea bilabiata* and *Palicourea marcgravii* by ^{19}F-NMR spectroscopy. Toxicon 32, 909–913.

Kuo, J., Hocking, P.J., Pate, J.S., 1982. Nutrient reserves in seeds of selected Proteaceous species from south-western Australia. Aust. J. Bot. 30, 231–249.

Kuppusamy, T., Giavalisco, P., Arvidsson, S., Sulpice, R., Stitt, M., Finnegan, P.M., et al., 2014. Lipid biosynthesis and protein concentration respond uniquely to phosphate supply during leaf development in highly phosphorus-efficient *Hakea prostrata*. Plant Physiol. 166, 1891–1911.

Kuppusamy, T., Hahne, D., Ranathunge, K., Lambers, H., Finnegan, P.M., 2020. Delayed greening in phosphorus-efficient *Hakea prostrata* (Proteaceae) is a photoprotective and nutrient-saving strategy. Funct. Plant Biol. 48, 218–230.

Kursar, T.A., Coley, P.D., 1992. Delayed greening in tropical leaves: an antiherbivore defense? Biotropica 24, 256–262.

Lambers, H., 2022. Phosphorus acquisition and utilization in plants. Annu. Rev. Plant Biol. 73, 17–42.

Lambers, H., Poorter, H., 1992. Inherent variation in growth rate between higher plants: a search for physiological causes and ecological consequences. Adv. Ecol. Res. 22, 187–261.

Lambers, H., Oliveira, R.S., 2019. Plant Physiological Ecology, 3rd edition Springer, Cham, Switzerland.

Lambers, H., Brundrett, M.C., Raven, J.A., Hopper, S.D., 2010. Plant mineral nutrition in ancient landscapes: high plant species diversity on infertile soils is linked to functional diversity for nutritional strategies. Plant Soil 334, 11–31.

Lambers, H., Cawthray, G.R., Giavalisco, P., Kuo, J., Laliberté, E., Pearse, S.J., et al., 2012. Proteaceae from severely phosphorus-impoverished soils extensively replace phospholipids with galactolipids and sulfolipids during leaf development to achieve a high photosynthetic phosphorus-use-efficiency. New Phytol. 196, 1098–1108.

Lambers, H., Clode, P.L., Hawkins, H.-J., Laliberté, E., Oliveira, R.S., Reddell, P., et al., 2015a. Metabolic adaptations of the non-mycotrophic Proteaceae to soil with a low phosphorus availability. In: Plaxton, W.C., Lambers, H. (Eds.), Annual Plant Reviews, Volume 48, Phosphorus Metabolism in Plants. John Wiley & Sons, Chicester, NY, USA, pp. 289–336.

Lambers, H., Finnegan, P.M., Jost, R., Plaxton, W.C., Shane, M.W., Stitt, M., 2015b. Phosphorus nutrition in Proteaceae and beyond. Nat. Plants 1, 15109.

Lambers, H., Hayes, P.E., Laliberté, E., Oliveira, R.S., Turner, B.L., 2015c. Leaf manganese accumulation and phosphorus-acquisition efficiency. Trends Plant Sci. 20, 83–90.

Lambers, H., Guilherme Pereira, C., Wright, I.J., Bellingham, P.J., Bentley, L.P., Boonman, A., et al., 2021. Leaf manganese concentrations as a tool to assess belowground plant functioning in phosphorus-impoverished environments. Plant Soil 461, 43–61.

Lee, S.T., Cook, D., Riet-Correa, F., Pfister, J.A., Anderson, W.R., Lima, F.G., et al., 2012. Detection of monofluoroacetate in *Palicourea* and *Amorimia* species. Toxicon 60, 791–796.

Lee, S.T., Cook, D., Pfister, J.A., Allen, J.G., Colegate, S.M., Riet-Correa, F., et al., 2014. Monofluoroacetate-containing plants that are potentially toxic to livestock. J. Agric. Sci. Food Chem. 62, 7345–7354.

Li, J.-T., Gurajala, H.K., Wu, L.-H., Van Der Ent, A., Qiu, R.-L., Baker, A.J.M., et al., 2018. Hyperaccumulator plants from China: a synthesis of the current state of knowledge. Environ. Sci. Technol. 52, 11980–11994.

Liang, Y., Nikolic, M., Bélanger, R., Gong, H., Song, A., 2015. Silicon in Agriculture. From Theory to Practice. Springer, Dordrecht, The Netherlands.

Lima, L.W., Schiavon, M., 2021. Selenium hyperaccumulation in plants. In: Lens, P.N.L., Pakshirajan, K. (Eds.), Environmental Technologies to Treat Selenium Pollution: Principles and Engineering. IWA Publishing, London, UK, pp. 245–264.

Loneragan, J.F., 1968. Nutrient requirements of plants. Nature 220, 1307–1308.

Loneragan, J.F., Snowball, K., 1969. Rate of calcium absorption by plant roots and its relation to growth. Aust. J. Agric. Res. 20, 479–490.

Loomis, W.D., Durst, R.W., 1992. Chemistry and biology of boron. Biofactors 3, 229–239.

Ma, J.F., Tamai, K., Yamaji, N., Mitani, N., Konishi, S., Katsuhara, M., et al., 2006. A silicon transporter in rice. Nature 440, 688–691.

MacLennan, D.H., Beevers, H., 1964. *Trans*-aconitate in plant tissues. Phytochemistry 3, 109–113.

Marais, J.S.C., 1944. Monofluoroacetic acid, the toxic principle of "gifblaar", *Dichapetalum cymosum* (Hook) Engl. Onderstepoort J. Vet. Sci. Anim. Ind. 20, 67–73.

McEwan, T., 1964. Isolation and identification of the toxic principle of *Gastrolobium grandiflorum*. Nature 201, 827-827.

Meerts, P., 1997. Foliar macronutrient concentrations of forest understorey species in relation to Ellenberg's indices and potential relative growth rate. Plant Soil 189, 257–265.

Minden, V., Schaller, J., Olde Venterink, H., 2021. Plants increase silicon content as a response to nitrogen or phosphorus limitation: a case study with *Holcus lanatus*. Plant Soil 462, 95–108.

Minnaar, P.P., Swan, G.E., Mccrindle, R.I., De Beer, W.H.J., Naudé, T.W., 2000. A High-Performance Liquid Chromatographic method for the determination of monofluoroacetate. J. Chromatogr. Sci. 38, 16–20.

Moir, J., Jordan, P., Moot, D., Lucas, R., 2016. Phosphorus response and optimum pH ranges of twelve pasture legumes grown in an acid upland New Zealand soil under glasshouse conditions. J. Soil Sci. Plant Nutr. 16, 438–460.

Morrison, J.F., Peters, R.A., 1954. Biochemistry of fluoroacetate poisoning: the effect of fluorocitrate on purified aconitase. Biochem. J. 58, 473–479.

Nge, F.J., Cambridge, M.L., Ellsworth, D.S., Zhong, H., Lambers, H., 2020. Cluster roots are common in *Daviesia* and allies (Mirbelioids; Fabaceae). J. R. Soc. W. Austr. 103, 111–118.

Ozanne, P.G., Keay, J., Biddiscombe, E.F., 1969. The comparative applied phosphate requirements of eight annual pasture species. Aust. J. Agric. Res. 20, 809–818.

Pang, J., Ruchi, B., Zhao, H., Bansal, R., Bohuon, E., Lambers, H., et al., 2018. The carboxylate-releasing phosphorus-mobilising strategy could be proxied by foliar manganese concentration in a large set of chickpea germplasm under low phosphorus supply. New Phytol. 219, 518–529.

Pinto Irish, K., Harvey, M.-A., Erskine, P.D., Van Der Ent, A., 2021. Root foraging and selenium uptake in the Australian hyperaccumulator *Neptunia amplexicaulis* and non-accumulator *Neptunia gracilis*. Plant Soil 462, 219–233.

Pourret, O., Bollinger, J.-C., 2018. "Heavy metal" - What to do now: to use or not to use? Sci. Total Environ. 610–611, 419–420.

Pritchard, S.G., Prior, S.A., Rogers, H.H., Peterson, C.M., 2000. Calcium sulfate deposits associated with needle substomatal cavities of container-grown longleaf pine (*Pinus palustris*) seedlings. Int. J. Plant Sci. 161, 917–923.

Prodhan, M.A., Jost, R., Watanabe, M., Hoefgen, R., Lambers, H., Finnegan, P.M., 2016. Tight control of nitrate acquisition in a plant species that evolved in an extremely phosphorus-impoverished environment. Plant Cell Environ. 39, 2754–2761.

Prodhan, M.A., Jost, R., Watanabe, M., Hoefgen, R., Lambers, H., Finnegan, P.M., 2017. Tight control of sulfur assimilation: an adaptive mechanism for a plant from a severely phosphorus-impoverished habitat. New Phytol. 215, 1068–1079.

Prodhan, M.A., Finnegan, P.M., Lambers, H., 2019. How does evolution in phosphorus-impoverished landscapes impact plant nitrogen and sulfur assimilation? Trends Plant Sci. 24, 69–82.

Rasheed, M.A., Seeley, R.C., 1966. Relationship between the protein and copper contents of some plants. Nature 212, 644–645.

Raven, J.A., 2003. Cycling silicon – the role of accumulation in plants. New Phytol. 158, 419–421.

Redondo-Nieto, M., Rivilla, R., El-Hamdaoui, A., Bonilla, I., Bolaños, L., 2001. Boron deficiency affects early infection events in the pea-*Rhizobium* symbiotic interaction. Funct. Plant Biol. 28, 819–823.

Reed, S.C., Townsend, A.R., Davidson, E.A., Cleveland, C.C., 2012. Stoichiometric patterns in foliar nutrient resorption across multiple scales. New Phytol. 196, 173–180.

Reid, N., Robson, T.C., Radcliffe, B., Verrall, M., 2016. Excessive sulphur accumulation and ionic storage behaviour identified in species of *Acacia* (Leguminosae: Mimosoideae). Ann. Bot. 117, 653–666.

Reis, H.P.G., Barcelos, J.P.D.Q., Silva, V.M., Santos, E.F., Tavanti, R.F.R., Putti, F.F., et al., 2019. Agronomic biofortification with selenium impacts storage proteins in grains of upland rice. J. Sci. Food Agric. 100, 1990–1997.

Saffran, M., Prado, J.L., 1949. Inhibition of aconitase by *trans*-aconitate. J. Biol. Chem. 180, 1301–1309.

Sage, R.F., Pearcy, R.W., 1987. The nitrogen use efficiency of C_3 and C_4 pants. II. Leaf nitrogen effects on the gas exchange characteristics of *Chenopodium album* (L.) and *Amaranthus retroflexus* (L.). Plant Physiol. 84, 959–963.

Séleck, M., Bizoux, J.-P., Colinet, G., Faucon, M.-P., Guillaume, A., Meerts, P., et al., 2013. Chemical soil factors influencing plant assemblages along copper-cobalt gradients: implications for conservation and restoration. Plant Soil 373, 455–469.

Silva, V.M., Rimoldi Tavanti, R.F., Gratão, P.L., Alcock, T.D., Reis, A.R.D., 2020. Selenate and selenite affect photosynthetic pigments and ROS scavenging through distinct mechanisms in cowpea (*Vigna unguiculata* (L.) Walp) plants. Ecotox. Environ. Safe 201, 110777.

Sprengel, C., 1828. Von den Substanzen der Ackerkrume und des Untergrundes (About the substances in the plough layer and the subsoil). J. Techn. Ökon. Chem. 2, 397–421.

Sprent, J.I., 1999. Nitrogen fixation and growth of non-crop legume species in diverse environments. Perspect. Plant Ecol. Evol. Syst. 2, 149–162.

Steyn, D.G., 1928. Gifblaar poisoning: a summary of our present knowledge in respect of poisoning by *Dichapetalum cymosum*. Government Printer and Stationery Office, Pretoria, South Africa.

Sulpice, R., Ishihara, H., Schlereth, A., Cawthray, G.R., Encke, B., Giavalisco, P., et al., 2014. Low levels of ribosomal RNA partly account for the very high photosynthetic phosphorus-use efficiency of Proteaceae species. Plant Cell Environ. 37, 1276–1298.

Tamai, K., Ma, J.F., 2008. Reexamination of silicon effects on rice growth and production under field conditions using a low silicon mutant. Plant Soil 307, 21–27.

Tanaka, H., 1967. Boron adsorption by plant roots. Plant Soil 27, 300–302.

Tcherkez, G.G.B., Farquhar, G.D., Andrews, T.J., 2006. Despite slow catalysis and confused substrate specificity, all ribulose bisphosphate carboxylases may be nearly perfectly optimized. Proc. Natl. Acad. Sci. U.S.A. 103, 7246–7251.

Twigg, L.E., King, D.E., Bowen, L.H., Wright, G.R., Eason, C.T., 1996. Fluoroacetate content of some species of the toxic Australian plant genus, *Gastrolobium*, and its environmental persistence. Nat. Toxins 4, 122–127.

Ubiali, D.G., Cardoso, L.F.C.G., Pires, C.A., Riet-Correa, F., 2020. *Palicourea marcgravii* (Rubiaceae) poisoning in cattle grazing in Brazil. Trop. Anim. Health Prod. 52, 3527–3535.

Urbina, I., Grau, O., Sardans, J., Margalef, O., Peguero, G., Asensio, D., et al., 2021. High foliar K and P resorption efficiencies in old-growth tropical forests growing on nutrient-poor soils. Ecol. Evol. 11, 8969–8982.

Vaculík, M., Lukačová, Z., Bokor, B., Martinka, M., Tripathi, D.K., Lux, A., 2020. Alleviation mechanisms of metal(loid) stress in plants by silicon: a review. J. Exp. Bot. 71, 6744–6757.

Van Der Ent, A., Jaffré, T., L'huillier, L., Gibson, N., Reeves, R.D., 2015. The flora of ultramafic soils in the Australia–Pacific Region: state of knowledge and research priorities. Aust. J. Bot. 63, 173–190.

Van Der Ploeg, R.R., Böhm, W., Kirkham, M.B., 1999. On the origin of the theory of mineral nutrition of plants and the Law of the minimum. Soil Sci. Soc. Am. J. 63, 1055–1062.

Vandegeer, R.K., Cibils-Stewart, X., Wuhrer, R., Hartley, S.E., Tissue, D.T., Johnson, S.N., 2021a. Leaf silicification provides herbivore defence regardless of the extensive impacts of water stress. Funct. Ecol. 35, 1200–1211.

Vandegeer, R.K., Zhao, C., Cibils-Stewart, X., Wuhrer, R., Hall, C.R., Hartley, S.E., et al., 2021b. Silicon deposition on guard cells increases stomatal sensitivity as mediated by K^+ efflux and consequently reduces stomatal conductance. Physiol. Plant. 171, 358–370.

Veneklaas, E.J., Lambers, H., Bragg, J., Finnegan, P.M., Lovelock, C.E., Plaxton, W.C., et al., 2012. Opportunities for improving phosphorus-use efficiency in crop plants. New Phytol. 195, 306–320.

Vergutz, L., Manzoni, S., Porporato, A., Novais, R.F., Jackson, R.B., 2012. Global resorption efficiencies and concentrations of carbon and nutrients in leaves of terrestrial plants. Ecol. Monogr. 82, 205–220.

Von Liebig, J., 1855. Principles of Agricultural Chemistry with Special Reference to the Late Researches made in England. Walton & Maberly, London, UK.

Waterman, J.M., Hall, C.R., Mikhael, M., Cazzonelli, C.I., Hartley, S.E., Johnson, S.N., 2020. Short-term resistance that persists: rapidly induced silicon anti-herbivore defence affects carbon-based plant defences. Funct. Ecol. 35, 82–92.

White, P.J., Broadley, M.R., 2003. Calcium in plants. Ann. Bot. 92, 487–511.

White, P.J., Broadley, M.R., El-Serehy, H.A., George, T.S., Neugebauer, K., 2018. Linear relationships between shoot magnesium and calcium concentrations among angiosperm species are associated with cell wall chemistry. Ann. Bot. 122, 221–226.

Wilson, D.N., Doudna Cate, J.H., 2012. The structure and function of the eukaryotic ribosome. Cold Spring Harb. Perspect. Biol. 4, a011536.

Woodward, R.A., Harper, K.T., Tiedemann, A.R., 1984. An ecological consideration of the significance of cation-exchange capacity of roots of some Utah range plants. Plant Soil 79, 169–180.

Wright, I.J., Reich, P.B., Westoby, M., Ackerly, D.D., Baruch, Z., Bongers, F., et al., 2004. The worldwide leaf economics spectrum. Nature 428, 821–827.

Wu, G., Li, M., Zhong, F., Fu, C., Sun, J., Yu, L., 2011. *Lonicera confusa* has an anatomical mechanism to respond to calcium-rich environment. Plant Soil 338, 343–353.

Xiao, Z., Yan, G., Ye, M., Liang, Y., 2021. Silicon relieves aluminum-induced inhibition of cell elongation in rice root apex by reducing the deposition of aluminum in the cell wall. Plant Soil 462, 189–205.

Yan, L., Zhang, X., Han, Z., Lambers, H., Finnegan, P.M., 2019. Responses of foliar phosphorus fractions to soil age are diverse along a 2 Myr dune chronosequence. New Phytol. 223, 1621–1633.

Ye, D., Clode, P.L., Hammer, T.A., Pang, J., Lambers, H., Ryan, M.H., 2021. Accumulation of phosphorus and calcium in different cells protects the phosphorus-hyperaccumulator *Ptilotus exaltatus* from phosphorus toxicity in high-phosphorus soils. Chemosphere 264, 128438.

Zemunik, G., Turner, B.L., Lambers, H., Laliberté, E., 2015. Diversity of plant nutrient-acquisition strategies increases during long-term ecosystem development. Nat. Plants 1, 15050.

Zhao, X., Zheng, S.H., Fatichin, S.A., Arima, S., 2014. Varietal difference in nitrogen redistribution from leaves and its contribution to seed yield in soybean. Plant Prod. Sci. 17, 103–108.

Zhong, H., Zhou, J., Azmi, A., Arruda, A.J., Doolette, A.L., Smernik, R.J., et al., 2021. *Xylomelum occidentale* (Proteaceae) accesses relatively mobile soil organic phosphorus without releasing carboxylates. J. Ecol. 109, 246–259.

Zhou, J., Zúñiga-Feest, A., Lambers, H., 2020. In the beginning, there was only bare regolith - then some plants arrived and changed the regolith. J. Plant Ecol. 13, 511–516.

Chapter 18

Plant responses to soil-borne ion toxicities

Zed Rengel

UWA School of Agriculture and Environment, University of Western Australia, Perth, WA, Australia

Summary

This chapter focuses on the constraints imposed on plants by soil-associated abiotic stresses, including toxicity of aluminum (Al), protons (H), and manganese (Mn) in acid soils; Mn and iron (Fe) in waterlogged and flooded soils; and sodium (Na) and chloride (or sulfate) in saline/sodic soils, emphasizing the mechanisms of adaptation and resistance. The two components of resistance (avoidance and internal tolerance) are highlighted. Resistance to Al is based mainly on exudation of malate, citrate, or other organic acid anions that complex Al into nontoxic forms in the rhizosphere. In waterlogged or flooded soils, anaerobic conditions are conducive to the reduction of Fe and Mn into their most mobile forms (Fe^{2+} and Mn^{2+}), potentially exceeding the nutritional or even tolerable concentrations in the soil solution. The main constraints to plant growth in saline soils (in addition to ion toxicity) are low osmotic potential and ion imbalances (low K/Na ratio). Salt tolerance mechanisms can be divided into salt exclusion (reduced uptake, increased efflux) and salt inclusion (compartmentation into the vacuole, release of salts via salt glands, salt-tolerant enzymes). To avoid water loss, plants accumulate osmotically active compounds to retain water in the cells.

18.1 Introduction

The spatial distribution of species and ecotypes, for example, acidophobes and acidophiles, halophytes and glycophytes, is dependent on soil properties. Soils are not just a source of water and nutrients essential for growth but may also cause a range of abiotic stresses to plants, including ion toxicities. Numerous recent reviews have been published on ecophysiological and molecular aspects of plant responses to soil-borne ion toxicities associated with acidity (Jaiswal et al., 2018; Kopittke, 2016; Rao et al., 2016; Rasheed et al., 2020; Ryan, 2018), hypoxia (waterlogging/flooding) (Yusuff et al., 2020) and salinity (Tarun et al., 2020; Wani et al., 2020; Yolcu et al., 2021).

Even though toxicities of various elements (ions) are associated with specific soil properties (Al and Mn in acid mineral soils, Mn and Fe in waterlogged or flooded soils, and Na and Cl (or sulfate) in saline soils), *multiple stress resistance* is often necessary for adaptation to specific environmental conditions (e.g., in acidic soils, Zhao et al., 2014), even though different rhizotoxic ions can induce expression of different combinations of genes (e.g., Sawaki et al., 2016), including Al toxicity and low pH stress (Kobayashi et al., 2007).

Resistance to soil-borne toxicity is defined as a set of plant traits that allows a genotype/ecotype to grow and form reproductive organs (i.e., produce a yield in terms of human-grown species) better than the population mean when grown in a soil with an excessive supply of a specific ion. The plant traits contributing to soil-borne toxicity resistance are shown in Fig. 18.1. Such resistance is an important part of the adaptation of plants to acid mineral soils (Section 18.2), waterlogged and flooded soils (Section 18.4), and saline soil (Section 18.5).

Resistance has external and internal components (Fig. 18.1). The external component includes exclusion from uptake (or minimized uptake), whereas the internal component comprises avoidance (generally by minimizing the accumulation of toxic elements in shoots) and tolerance (capacity to live with the element present in shoots).

Exclusion of the toxic ion from uptake may be achieved by modifying root anatomy/morphology or root physiology/biochemistry. The formation of a suberized exodermis may restrict the flow of ions into the root cortex (Peckova

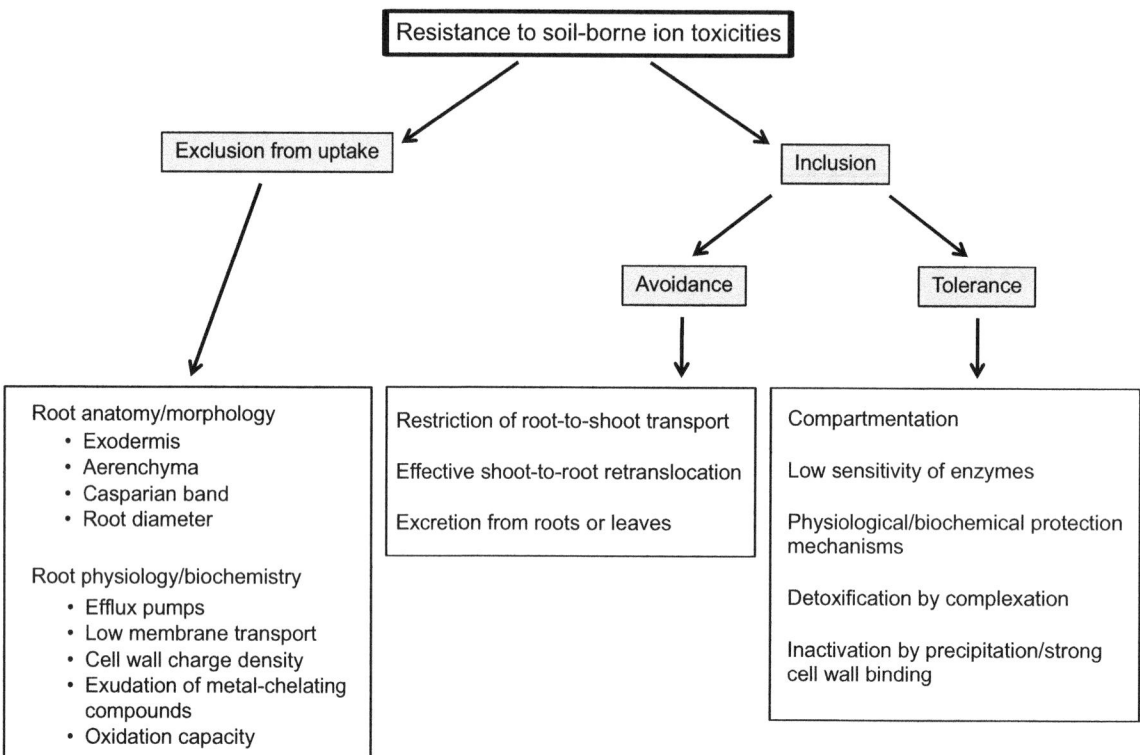

FIGURE 18.1 Mechanisms of adaptation of plants to excessive supply of ions.

et al., 2016), and further movement of ions into the stele can be prevented by the Casparian band in the endodermis (Pan et al., 2020).

Low membrane transport and/or efficient efflux pumps can restrict the accumulation of ions (particularly Na^+ and Cl^-) intracellularly. The cation exchange capacity (CEC) of the root apoplast can be decreased by (1) lowering pectin concentration (Szatanik-Kloc et al., 2017) and/or (2) a high degree of methylation of pectin, thus decreasing the binding of di- and trivalent cations such as Cd^{2+} and Al^{3+} in the root apoplast (e.g., Li et al., 2017c), which may be associated with decreased uptake across the plasma membrane. An effective way of metal detoxification in the rhizosphere soil and the root apoplast as well as decreasing their uptake is the release of metal-chelating ligands from roots, particularly organic acid anions that play a major role in Al resistance. A high oxidation capacity of roots contributes to the inactivation of Fe^{2+} in waterlogged/flooded soils.

When potentially toxic ionic elements are taken up (inclusion), plants may avoid toxicity in the shoots by restricting the translocation from roots, effectively retranslocating the element from the shoot back to the roots, and excreting it from the roots, as has been shown for Na^+ in natrophobic glycophytes (see Section 5.4.1). In halophytes, excessive concentrations of Na^+ in the shoots can be avoided by its excretion through specialized leaf cells (salt glands).

Tolerance of the shoot tissue to specific ions requires sequestration of the toxic ions into the plant compartments where ions no longer affect the metabolism, such as the vacuole that plays a major role in tolerance to Na^+ and other metals (e.g., Mn). Natrophilic glycophytes and the halophytes have evolved several strategies to deal with high shoot Na concentrations (see also Chapter 8): (1) Na can replace K in certain functions, (2) Na has metabolic functions, and (3) enzymes are tolerant to Na. Element toxicity is often related to oxidative stress; therefore, more efficient detoxification/scavenging of reactive oxygen species (ROS) may contribute to tolerance, particularly to excess Fe and Mn. For metal tolerance, internal detoxification by complexation with organic ligands, particularly organic acid anions, may even allow safe hyperaccumulation in the shoots. Alternatively, strong binding in the cell walls and precipitation (particularly in vacuoles) may also confer enhanced metal tolerance.

18.2 Acid mineral soils

18.2.1 Major constraints

Acid soils, defined by a pH_{water} lower than 5.5 in the surface layer, comprise about 30% of the total ice-free land on the planet (Brunner and Sperisen, 2013), primarily in humid climates but also in areas and production systems with excessive applications of ammonium-based fertilizers (e.g., see *meta*-analysis on the data from tea plantations, Qiao et al., 2018). Plant growth inhibition and yield reduction on acid soils result from a variety of specific chemical factors and their interactions (Marschner, 1991). The major constraints to plant growth are toxicities of protons, Al, and/or Mn as well as deficiencies of Mg, Ca, P, and Mo. The relative importance of these constraints depends on plant species and genotype, soil type and horizon, parent material, soil pH, concentration and ionic species of Al, soil structure and aeration, and climate. Aluminum toxicity as well as Ca and Mg deficiencies occur in more than 70% of the acid soils of tropical America, and nearly all these soils are low in P or have a high P-fixing capacity (Sanchez and Salinas, 1981). In the surface layer, where the organic matter content is higher than in the subsoil, H^+ toxicity may dominate, but in the subsoil root growth may be depressed by Al toxicity. Subsoil acidity is a potential growth-limiting factor throughout many areas of the United States (Blumenschein et al., 2018), the tropics (Nair et al., 2019), south-west Australia (Anderson et al., 2020), etc.

Manganese toxicity may become a major stress in slightly acidic and/or waterlogged/flooded soils high in Mn-containing minerals and exchangeable Mn^{2+}. Depending on the soil pH and the redox level (pe = activity of electrons), the expected Mn^{2+} concentration in the soil solution can be calculated as (Schmidt and Husted, 2020):

$$\log[Mn^{2+}] = 41.39 - 2(pe + pH) - 2pH.$$

Given different ways in which soil acidity can restrict plant growth, plants adapted to acid mineral soils require a variety of mechanisms to cope with the adverse soil chemical factors. On a worldwide scale, high concentrations of Al, H^+ for some plant species, and Mn in some locations are key factors of soil acidity stress; therefore, high resistance to these factors is required for plant adaptation to acid soils.

18.2.2 Proton toxicity

Proton toxicity is primarily expressed as inhibition of root elongation (Ikka et al., 2007) and root death (Koyama et al., 1995), but the pH at which H^+ toxicity occurs differs among plant species. The physiological and molecular mechanisms of H^+ toxicity are not yet understood fully (Shavrukov and Hirai, 2016), but there are principally three mechanisms: (1) disruption of cell wall integrity, (2) interference with the maintenance of the cytosolic pH, and (3) inhibition of cation uptake.

High H^+ concentrations in the root apoplasm disturb the stability of the pectic polysaccharide network by displacing Ca^{2+} that plays a key role in the maintenance of the network (Koyama et al., 2001). Numerous genes related to the cell wall are upregulated upon exposure to H^+ toxicity in the seedlings of acidophile *Rhododendron protistum* var. *giganteum* (Zhou et al., 2020b). At high H^+ concentrations in the apoplasm, the plasma membrane H^+-ATPase cannot maintain the cytosolic pH (Yan et al., 1992). Increased tolerance to high H^+ concentrations has been explained by a higher ATPase H^+ pumping capacity (Rossini Oliva et al., 2018; Yan et al., 1998), which also commonly occurs under combined low pH/Al toxicity stresses (Bose et al., 2010; Zhang et al., 2017b) due to enhanced activation (i.e., phosphorylation) of the H^+-ATPase (Wang et al., 2016).

Studies with an H^+-hypersensitive arabidopsis mutant suggested that the Zn-finger protein sensitive to proton toxicity (STOP1) is involved in metabolic pathways controlling the cytosolic pH (Sawaki et al., 2009) in addition to its numerous other roles, the most relevant of which for discussion here is the regulation of the Al-inducible malate transporter (ALMT1) (e.g., Sawaki et al., 2016). In addition to STOP1, other transcription factors related to the cytosolic pH-stat regulation, such as STOP2 (e.g., in soybean, *Glycine max*) (Chen et al., 2019b) and WRKY (e.g., in wheat, *Triticum aestivum*) (Hu et al., 2018), are upregulated under acidic stress to shore-up the capacity to maintain cytosolic pH. The F-box RAE1 (regulation of Atalmt1 expression) protein is involved in the inactivation and degradation of STOP1, providing another layer of regulation of plant responses to the proton and Al toxicities (Zhang et al., 2019b).

High proton concentrations inhibit the uptake of cations by depolarization of the plasma membrane (Bose et al., 2010; Shabala et al., 1997) and decrease loading of polyvalent cations (Mg^{2+}, Ca^{2+}, Zn^{2+}, Mn^{2+}) in the apoplast of root cortical cells, which then reduces their uptake into the symplast (see also Chapter 2). In particular, increased Mg^{2+} uptake enhances plant resistance to low pH stress (Bose et al., 2013).

The uptake of nitrate is accompanied by an increase in the rhizosphere pH because of the involvement of nitrate/H^+ cotransporters. In arabidopsis, NRT1.1-mediated nitrate/H^+ cotransport is particularly important for tolerance to proton toxicity (Fang et al., 2016).

18.2.3 Aluminum toxicity

18.2.3.1 Aluminum solution chemistry

In acid mineral soils below pH$_{water}$ 5.5, an increasing proportion of the cation exchange sites is occupied by Al^{3+} that replaces divalent cations Mg^{2+} and Ca^{2+}. Thus, with decreasing soil pH, the percentage of exchangeable Al (=Al saturation) increases (e.g., Nair et al., 2019). Therefore, Al saturation is often used to predict Al excess, and decreased Al saturation is a target in the process of soil liming (Abdulaha-Al Baquy et al., 2018; Somavilla et al., 2021). The total Al concentration in the soil extract or soil solution may (Carr and Ritchie, 1993) or may not (Richter et al., 2011) correlate well with the inhibition of root growth (the most sensitive parameter of Al toxicity) depending on ionic strength, activity of sulfate ions, and plant species/genotypes.

The phytotoxicity of Al depends primarily on the Al speciation in solution. Aluminum released from soil minerals into the soil solution under acid conditions, or the Al in nutrient solutions of pH ≤ 4.0, is mainly Al(H$_2$O)$_6^{3+}$ (referred to as Al^{3+}). As the pH increases, the total Al concentration of the solution decreases, but the mononuclear hydrolysis products such as Al(OH)$^{2+}$ and Al(OH)$_2^+$ are formed as intermediates, preceding the precipitation of solid Al(OH)$_3^0$ (Kinraide, 1991). Above pH 7, the solution Al concentration increases again due to the formation of the aluminate ion Al(OH)$_4^-$ that predominates at pH above 9.2, exerting a significant toxic effect on field pea (*Pisum sativum*) grown in potted soil over and above the toxicity associated with high pH (Brautigan et al., 2012). However, in hydroponic culture with pH adjusted to 9.5, no rhizotoxicity of aluminate was found (Kopittke et al., 2004); instead, the formation of polynuclear hydroxyl Al species such as AlO$_4$Al$_{12}$(OH)$_{24}$(H$_2$O)$_{12}^{7+}$ (referred to as Al$_{13}$) (Parker et al., 1989) in the root apoplasm due to root acidification might be a cause of Al rhizotoxicity (Kopittke et al., 2004). At elevated OH$^-$/Al ratios in solution, Al$_{13}$ may form (Parker et al., 1988). However, a role (if any) of Al$_{13}$ in Al toxicity in acid soils remains unclear.

There have been contradicting results regarding the relative phytotoxicity of the various mononuclear cationic Al species (Kinraide and Parker, 1990) at low pH, but Al^{3+} is considered the most phytotoxic Al mononuclear species. Stass et al. (2006) provided evidence that at pH 4.3, Al^{3+} inhibits root growth of maize (*Zea mays*) through binding to sensitive sites in the apoplast of the epidermis and the outer cortex, whereas at pH 8.0, with Al(OH)$_4^-$ as the dominant Al species, a strong decrease of the apoplasm pH leads to Al(OH)$_3$ precipitation in the epidermis forming a mechanical barrier that impairs root functioning.

Some mononuclear Al species associated with inorganic ligands such as AlF^{2+}, AlF$_2^+$, or AlSO$_4^+$ are less, or not at all, phytotoxic compared to Al^{3+} (Kinraide, 1997). The low phytotoxicity of AlSO$_4^+$ is particularly important because it explains the amelioration of Al phytotoxicity by application of gypsum (CaSO$_4$) (Wright et al., 1989) (Fig. 18.2). Because of its sulfate component and higher water solubility compared with lime (CaCO$_3$), CaSO$_4$ is suitable for amelioration of subsoil Al toxicity, the claim strengthened by the *meta*-analysis of grain yield of five cereals and soybean grown in no-till systems (Pias et al., 2020); however, leaching of sulfate may result in relatively short duration of the beneficial effects (Anderson et al., 2020).

Aluminum readily forms complexes with organic ligands, particularly organic acid anions, which reduces phytotoxicity of Al. In many soils, the prevalent forms of Al are the complexes with organic compounds (e.g., Martins et al.,

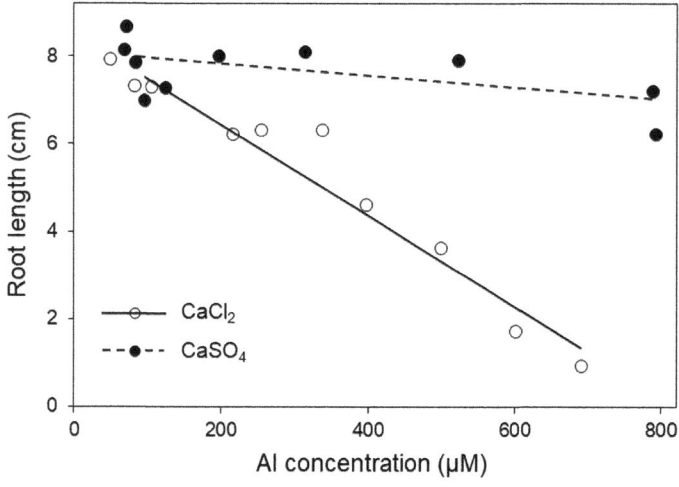

FIGURE 18.2 Root length of wheat as a function of Al concentration in the soil solution from soils treated with CaSO$_4$ or CaCl$_2$. *Modified from Wright et al. (1989).*

2020; Shi et al., 2020). Their detoxifying capacity decreases in the order citrate > oxalate > malate > succinate. Because of high stability of Al complex with citrate, 1:1 Al:ligand ratio eliminates Al rhizotoxicity almost completely, whereas for malate, 1:8 ratio is necessary (Li et al., 2002) (Fig. 18.3). Fulvic acid (Ciarkowska and Miechowka, 2019) and phenols may also detoxify Al (Barcelo and Poschenrieder, 2002; Zhang et al., 2016).

Detoxification of plant-available Al in the soil is one of the reasons for the amelioration of Al toxicity by soil organic matter. The application of various crop residue materials as well as organic fertilizers contributed to a decrease in the concentration of phytotoxic Al species and an increase in crop yield (Lauricella et al., 2020; Zhao et al., 2020b). There is increasing popularity of using biochar (e.g., Mehmood et al., 2017; Shi et al., 2019; Shi et al., 2020) or biomass ash (e.g., Ondrasek et al., 2021a, 2021b) in amelioration of soil acidity and Al toxicity because these amendments may show at least as good, if not better, elimination of Al toxicity than liming. An increase in dissolved organic carbon in soil and binding of Al to organic compounds underpin the ameliorating effect of biochar (e.g., Shi et al., 2020).

The phytotoxicity of Al depends not only on its speciation and solution concentration but also on the ionic strength of the solution and, particularly, the $Al^{3+}:Ca^{2+}$ and $Al^{3+}:H^+$ ratios. The ameliorating effect of high H^+ concentrations (i.e., low pH) on Al toxicity has been explained by the higher competitiveness of H^+ than Al^{3+} for apoplastic binding sites and/or the reduction of the cell surface negativity (Kinraide et al., 1992) of solution-grown plants. Proton amelioration of Al toxicity is expected to be less relevant for plants grown in acid soils because high concentrations of protons enhance the release of Al^{3+} from the solid phase and can reduce Ca^{2+} and Mg^{2+} uptake.

Nitrification contributes to soil acidification (Han et al., 2015; Xiao et al., 2014), and is therefore progressively inhibited with decreasing soil pH (cf. Lan et al., 2018). However, the relative abundance of bacterial and archaeal genes associated with nitrification decreased in the field soil amended by lime or lime + gypsum (Bossolani et al., 2020).

Given relatively low rates of nitrification (and thus general prevalence of ammonium as the N form) in acidic soils, plant tolerance to Al is generally associated with plant preference for ammonium rather than nitrate uptake (uptake of both N forms is inhibited by Al, but that of nitrate to a lesser extent) (Zhao and Shen, 2018). Ammonium uptake generates a net increase in the concentration of protons in the apoplasm, which may contribute to a partial alleviation of Al toxicity (Zhao and Shen, 2018) through decreased binding of highly charged Al ionic species to the sensitive sites in the cell wall and on the plasma membrane. By contrast, a higher microbial capacity for nitrification was found in the rhizosphere soil of the Al-tolerant than the Al-sensitive soybean genotype (Li et al., 2020c), suggesting the reliance on nitrate rather than ammonium nutrition by the Al-tolerant genotype. Whether this phenomenon is restricted to soybean (or even just the single pair of Al-tolerant and Al-sensitive genotypes tested) remains to be elucidated.

18.2.3.2 Inhibition of root growth

Inhibition of root elongation is the primary response of plants to Al toxicity (Fig. 18.4). It can be measured within the first hour or less after exposing roots to Al (see Rengel and Zhang, 2003), with 5 min being the shortest reported time for measurable Al inhibition of root growth (Kopittke et al., 2015). Ryan et al. (1993) were the first to demonstrate

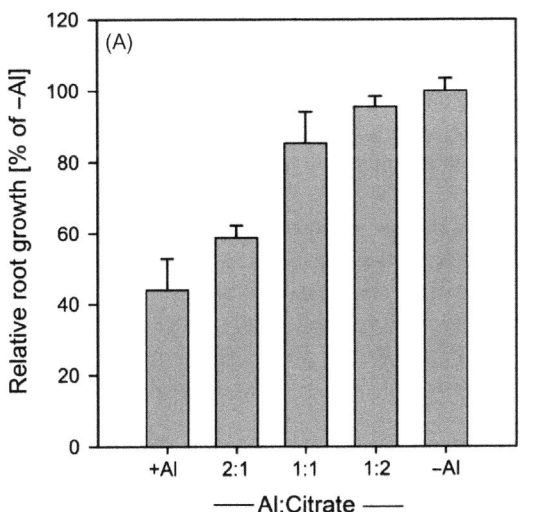
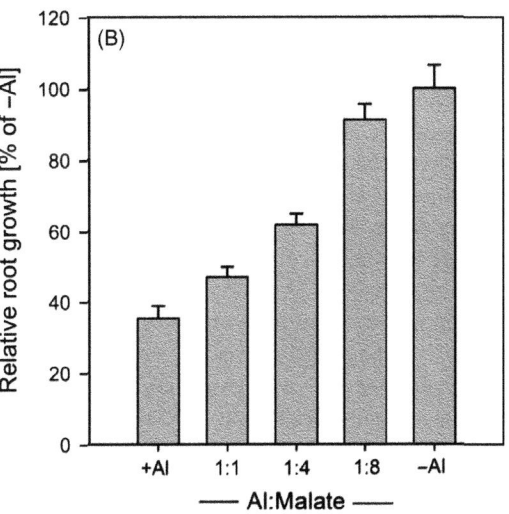

FIGURE 18.3 Relative root growth of wheat (cv Scout-66) exposed to 50 μM Al for 9 h without or with citrate or malate supplied at different Al: organic anion ratios. Means + standard deviation (n = 12). *From Li et al. (2002), with permission from Springer Nature.*

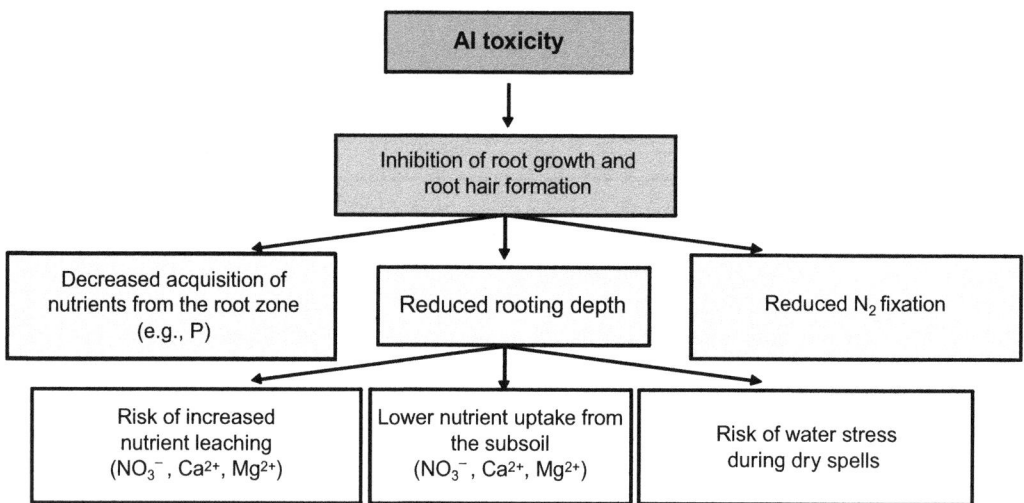

FIGURE 18.4 Consequences of root growth inhibition by Al on nutrient and water uptake.

unequivocally the role of the root apex in the perception of Al toxicity in maize. Sivaguru and Horst (1998) showed that the distal part of the transition zone (DTZ, 1−2 mm) is the most Al-sensitive apical root zone in maize. Application of Al only to the DTZ reduced cell elongation in the elongation zone (EZ) to the same extent as application to the entire 10 mm of the root apex (Kollmeier et al., 2000) (Fig. 18.5). However, application of Al only to the EZ did not inhibit root elongation. This indicates that signal transduction between the DTZ and the EZ is involved in the inhibition of root growth by Al, possibly via Al interference with the basipetal auxin transport to the EZ through the rhizodermis and the outer cortex (Kollmeier et al., 2000; Kopittke, 2016) increasing auxin concentration in the root tip and decreasing it in the EZ, thus inhibiting root elongation. The increased auxin concentration enhanced transcription of the gene encoding the multidrug and toxic compound extrusion *(MATE)* transporter in soybean (thus increasing citrate exudation) and boosted the activation of H^+-ATPase by phosphorylation (Wang et al., 2016), both contributing to alleviation of the Al-related root growth inhibition in soybean. In rice (*Oryza sativa*), OsAUX3 (localized in the plasma membrane and facilitating acropetal auxin transport) was upregulated by Al, thus starving the EZ of auxin and inhibiting cell elongation and root growth (Wang et al., 2019c). Aluminum-induced inhibition of root elongation may also involve the inhibition of polar auxin transport by rapid Al-enhanced ethylene production (Sun et al., 2010).

Treatment with Al results in the development of transverse ruptures in subapical regions of the root (see Section 3.3) through the breaking and separation of the rhizodermis and the outer cortical from inner cortical cell layers (Blamey et al., 2004; Kopittke et al., 2008). It was proposed that these ruptures were due to the binding of Al to the cell wall that increases cell wall rigidity. However, the relationship between these ruptures and inhibition of root elongation is poorly understood.

Another sensitive indicator of Al injury to roots is the induction of callose synthesis (e.g., Lin and Chen, 2019; Yu et al., 2019). Aluminum-induced callose formation is an indicator of Al sensitivity and a reliable parameter for the classification of genotypes of different plant species for Al resistance (Wissemeier et al., 1992; Zhang et al., 2015). The overexpression of *SbGlu1* (β-1,3-glucanase from *Sorghum bicolor*) in arabidopsis results in lower callose production and greater Al resistance, suggesting callose may be associated with the pathway leading to the Al-related inhibition of root growth (Zhang et al., 2015).

The primary target site of Al phytotoxicity causing inhibition of root elongation appears to be apoplastic (Horst et al., 2010) due to binding of Al to the cell walls in the root EZ (Kopittke et al., 2015), thus interfering with the wall loosening required for cell elongation. Cell elongation entails (1) cell turgor pressure driving expansion, (2) the release of cell wall components from the symplasm to the apoplasm for cell wall synthesis, and (3) the formation and cleavage of Ca^{2+} bonds with the pectic matrix that control cell wall extensibility (Boyer, 2009). Strong binding of Al to the pectic matrix may prevent cell wall extension physically and/or physiologically by decreasing the effectiveness of cell wall−loosening enzymes (Wehr et al., 2004).

18.2.3.3 Interference with root-cell plasma membrane properties

Aluminum not only rapidly affects properties of the cell wall but also those of the plasma membrane. Interaction of Al with membrane lipids and proteins induces modifications of the plasma membrane structural properties such as fluidity

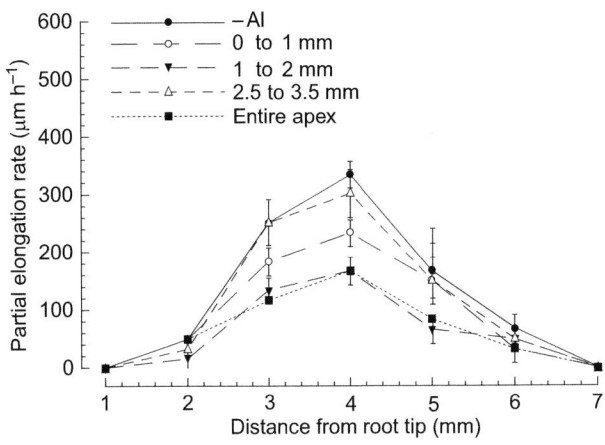

FIGURE 18.5 Partial elongation rates of 1-mm root segments of the primary roots of maize (cv Lixis) with Al supply to the entire root apex or to specific 1-mm root zones. Means ± standard deviation (n = 5). *Redrawn from Kollmeier et al. (2000).*

and permeability (Khan et al., 2009; Wagatsuma et al., 2005; Yamamoto, 2019). Given that monogalactosyldiacylglycerol is important in maintaining plasma membrane integrity, the arabidopsis monogalactosyldiacylglycerol synthase mutants were hypersensitive to Al toxicity, showing impaired stability and permeability of root-cell plasma membrane (Liu et al., 2020a). Binding of Al to the plasma membrane alters its surface negativity (Kinraide, 2006) as shown by Ahn et al. (2001, 2002, 2004) in squash (*Cucurbita pepo*) and wheat. Hence, it was proposed that decreasing plasma membrane negativity might be a practical way to increase Al resistance by minimizing Al uptake (Maejima et al., 2017; Wagatsuma, 2017). In particular, decreased proportion of phospholipids and increased proportion of galactolipids and sterols would decrease plasma membrane negativity. Indeed, increased sterol biosynthesis (together with decreased formation of stigmasterol) facilitated increased resistance to Al in rice (Wagatsuma et al., 2018).

Aluminum rapidly induced membrane depolarization (Bose et al., 2010; Shabala et al., 1997), especially in the most Al-sensitive distal elongation root zone (Sivaguru et al., 1999). This may be related to inhibition of the H^+-ATPase activity (Ahn et al., 2001). These Al-related changes in the plasma membrane properties affect ion transport.

An Al-induced impairment of plasma membrane functions may be related to Al-enhanced oxidative stress through the formation of ROS leading to lipid peroxidation (Jones et al., 2006; Yamamoto et al., 1997) and protein oxidation (Boscolo et al., 2003). Aluminum-induced lipid peroxidation is exacerbated in the presence of Fe (Cakmak and Horst, 1991); it should be borne in mind that acid soils have high concentrations of soluble Al as well as Fe. Oxidative stress genes are strongly expressed after Al exposure (Ezaki et al., 2005). Transformation of *Arabidopsis thaliana* with these genes conferred Al resistance (Ezaki et al., 2001); increased activity of antioxidants and lower lipid peroxidation were associated with the Al-resistant compared with Al-sensitive varieties of rice (Awasthi et al., 2019) and wheat (Sun et al., 2017). However, oxidative stress in roots may not be the primary cause of Al-induced inhibition of root elongation because in most cases oxidative stress occurs only after prolonged Al treatment (Cakmak and Horst, 1991; Liu et al., 2008), or may be more involved in the plant responses to low pH than Al toxicity (Borgo et al., 2020). Nevertheless, sustained Al resistance may require protection against oxidative stress.

Aluminum triggers the signal transduction pathways leading to the physiological disorders in the symplast. In this regard, the effect of Al on cytosolic Ca concentrations appears to be particularly important (Jones et al., 2006; Rengel and Zhang, 2003). An increase in cytosolic Ca activity is an immediate response to Al exposure in a range of plant species (Jones et al., 1998; Rengel, 1992; Zhang and Rengel, 1999). Increasing cytosolic Ca activity can explain callose formation and the disorganization of the cytoskeleton (Rengel and Zhang, 2003). Enhanced callose deposition in the cell wall may be responsible for an Al-induced blockage of the cell-to-cell transport via plasmodesmata (Sivaguru et al., 2000). Although a direct effect of cytosolic Al on the cytoskeleton cannot be ruled out, an interaction of apoplasmic Al with the cell wall–plasma membrane–cytoskeleton continuum appears more likely (Horst et al., 1999), particularly in the case of boron (B) alleviation of Al toxicity, whereby formation of rhamnogalacturonan II-B complexes contributes to the stabilization of the cytoskeleton (Li et al., 2017c).

18.2.3.4 Inhibited nutrient and water uptake

The primary and specific toxic effect of Al is inhibition of root growth, but longer-term Al toxicity may also affect plant capacity to take up nutrients and water, potentially inducing nutrient deficiency and water stress. The influence of Al on the uptake of nutrients and water may be direct (through competing with divalent cations for binding sites on the transporters) or indirect through the inhibition of root growth, thus impairing the capacity to explore soil volume across various soil layers (see George et al., 2012).

18.2.3.5 Importance of Mg nutrition in alleviating Al toxicity

The role of Mg in enhancing the plant capacity to withstand Al toxicity stress has been suggested many decades ago, mainly because of competition between Al and Mg ions that have similar hydrated radii (for references see Rengel et al., 2015). Adding Mg to soil and spraying Mg foliarly significantly improved root and shoot growth of wheat (similarly to liming) (Kibria et al., 2021b) and other species. Legumes show the responses that differ from those of cereals (Rengel et al., 2015) (see also Section 3.2). In legumes, micromolar Mg concentrations enhance biosynthesis of organic acid anions, increase expression of citrate transporters (e.g., GmMATE) and increase the activity of H^+-ATPase by posttranslational activation/phosphorylation (Zhang et al., 2017b). Interestingly, foliar Mg application to wheat growing in acidic soil enhanced root exudation of organic acid anions, more so in the Al-tolerant than Al-sensitive near-isogenic genotype (Kibria et al., 2021a). Adequate Mg nutrition plays also a key role in the export of photosynthates to the roots (Chapter 6), influencing the production and exudation of organic acid anions under Al toxicity. Increased Mg supply also maintained cytosolic Mg^{2+} activity in arabidopsis, promoting the H^+-ATPase activity (Bose et al., 2013). In poplar (*Populus tomentosa*), adequate Mg supply promoted polar auxin transport and distribution, which might have contributed to root surface alkalinization in the transition zone, thus ameliorating Al toxicity (Zhang et al., 2020c).

18.2.3.6 Nitric oxide

Nitric oxide (NO) is an important signaling molecule, and its endogenous concentration in various plant species can either increase or decrease upon exposure to Al toxicity. The measurable changes may occur relatively quickly after exposure, such as the suppressed production of NO in arabidopsis roots after 60 min (Illes et al., 2006) or the increased rate of NO biosynthesis in bean (*Phaseolus vulgaris*) after 6 h of exposure to Al (Wang et al., 2010). There is a discrepancy in the reported effects of NO on Al resistance. There are reports of Al-enhanced NO production decreasing a degree of methylation of cell wall pectins in the root tips, resulting in the increased Al binding capacity and thus Al accumulation in the roots, exacerbating toxicity (Sun et al., 2016). By contrast, there is a substantial body of literature showing that increased endogenous concentrations of NO enhance the antioxidative capacity under Al stress, thus minimizing Al toxicity (e.g., He et al., 2012; Wang et al., 2019b). However, Al exposure results in the increased NO production and the exacerbated oxidative stress (due to production of ROS) in the Al-sensitive wheat genotype, whereas decreased production of NO and ROS occurred in the Al-resistant genotype (Sun et al., 2017). In soybean roots (similarly to the Al-sensitive wheat genotype mentioned above), Al exposure increased the NO production via nitrate reductase that resulted in the increased activity of cytosolic glucose-6-phosphate dehydrogenase producing NADPH, and the increased activity of NADPH oxidase resulting in ROS accumulation and the exacerbated oxidative stress (Wang et al., 2017). Further work in soybean suggested that NO mediated the *GmMATE* expression, H^+-ATPase activity, and thus citrate exudation, as well as that H_2S is a likely downstream effector (Wang et al., 2019a). Consistent with these observations, NO also activated citrate synthase in soybean exposed to Al stress, contributing to the enhanced citrate exudation and alleviation of Al toxicity (Wang et al., 2019b).

The two important signaling molecules (NO and phytomelatonin) interact in modulating the severity of arabidopsis root growth inhibition caused by Al. Exposure to Al downregulated *A. thaliana* SNAT (serotonin N-acetyltransferase), thus decreasing phytomelatonin concentration, which was associated with enhanced NO production and exacerbated Al sensitivity. Exogenous addition of phytomelatonin decreased NO synthesis via nitrate reductase and nitric oxide synthase, thus alleviating root growth inhibition (Zhang et al., 2019a). Similarly, Mg decreased the activities of the NO-producing enzymes in arabidopsis under Al stress, thus decreasing NO production and alleviating Al-caused root inhibition (Li et al., 2020a).

18.2.4 Manganese toxicity

With decreasing pH, the concentration of exchangeable Mn^{2+} increases in many soils. This increase is also a function of the redox potential (Rengel, 2015; Schmidt and Husted, 2020):

$$MnO_2 + 4H^+ + 2e^- \leftrightarrow Mn^{2+} + 2H_2O.$$

High concentrations of Mn^{2+} at the exchange sites and in the soil solution are, therefore, to be expected only in acid soils with relatively large amounts of readily reducible Mn in combination with a high concentration of organic matter, elevated microbial activity (thus, high oxygen consumption during decomposition of organic matter by soil microorganisms), and anaerobiosis due to waterlogging/flooding (see also Section 18.4). However, many acid soils in the tropics are highly weathered, and their total Mn concentration is often low because of prolonged mobilization and leaching.

Thus, in these soils, there is a lower risk of Mn toxicity than of Al toxicity, and even Mn deficiency may be observed when these soils are limed to pH$_{water}$ > 5.0.

In contrast to Al, Mn is readily transported from roots to shoots, even though no increase in the Mn concentration in stele compared with the surrounding root tissues was noted in cowpea (*Vigna unguiculata*) after 24-h exposure to toxic Mn concentration (Kopittke et al., 2013). Root growth is severely inhibited upon exposure to Mn toxicity (e.g., Chen et al., 2016; Inostroza-Blancheteau et al., 2017; Liu et al., 2020e; Zhao et al., 2017). Decreased root growth under Mn toxicity is due to preferential accumulation of Mn in the root tips (Wang et al., 2013a), and is associated with the altered cell wall structure and lignification (Ceballos-Laita et al., 2018; Chen et al., 2016), strong oxidative stress (Santos et al., 2017) and decreased energy production and protein turnover in roots (Ceballos-Laita et al., 2018). The inhibition of primary root growth in arabidopsis was associated mainly with decreased auxin biosynthesis and downregulation of the auxin efflux transporters (Zhao et al., 2017). In leaves and shoots, Mn toxicity leads to decreased chlorophyll and carotenoid biosynthesis, impaired photosystem I stability and structure (Li et al., 2015) and decreased net photosynthesis and stomatal conductance (Santos et al., 2017; Tavanti et al., 2020). Manganese accumulation in leaves is uneven, with dark spots representing areas of Mn accumulation (e.g., potentially exceeding 10 g Mn kg^{-1} leaf fw in cowpea) that are limited to around 4% of the total surface area after 48-h exposure (Blamey et al., 2018b).

The effects of excessive Mn supply on nutrient uptake, metabolism, and phytohormonal balance have been summarized by Li et al. (2019) (see also Section 7.2). Of particular importance for plant growth in acid mineral soils is the inhibition of Ca and Mg uptake by high Mn concentrations (see Inostroza-Blancheteau et al., 2017). Crinkle leaf and chlorosis in young leaves may be related to induced deficiency of, respectively, Ca and Fe, whereas chlorosis or brown speckling in mature leaves are the symptoms of Mn toxicity in dicotyledonous species growing in acid soils. Under these conditions, visible symptoms of Mn toxicity are observed even at concentrations that may decrease growth only slightly. This contrasts with Al toxicity that severely inhibits growth without producing visible shoot symptoms.

Accumulation of Mn in the root parts is quick (e.g., after 5 min, substantial accumulation was recorded in the root cap using synchrotron-based in situ techniques) (Kopittke et al., 2013). Inhibition of root growth by Mn toxicity was related to the Mn^{2+} activity at the outer surface of the plasma membrane rather than to Mn^{2+} activity in the bulk solution (Kopittke et al., 2011), clearly indicating the importance of charge effects in the apoplast to Mn toxicity.

In legumes, Mn toxicity depends on the form of N nutrition, but there are conflicting reports in the literature, whereby Mn toxicity was reported to be more severe in N$_2$-fixing plants or those fed inorganic N. Similarly, there are conflicting reports regarding whether growth or N$_2$ fixation is more affected by Mn toxicity. Aluminum toxicity decreased nodulation and growth of both cowpea and siratro (*Macroptilium atropurpureum*), whereas Mn toxicity had a negative effect only on cowpea (Manet et al., 2016). Hence, further work is required with legumes regarding the effects of Mn and/or Al toxicity as well as accompanying nutrient deficiencies (e.g., Ca and P) on nodulation and N$_2$ fixation in acid mineral soils.

18.3 Mechanisms of adaptation to acid mineral soils

18.3.1 General

Plants adapted to acid mineral soils utilize a variety of mechanisms to cope with the adverse chemical soil factors. These mechanisms are regulated separately (e.g., those of Al and Mn resistance) or are interrelated (e.g., those of Al resistance and P acquisition efficiency). From the agronomic viewpoint, the sum of the individual mechanisms is important because it determines the required inputs for amelioration of acid soils.

Large differences occur among crop species in their adaptation to acid soils. For example, regarding the annual root crop species, cassava (*Manihot esculenta* Crantz) is known for its high tolerance to acid soils compared to, for example, sweet potato, taniers and yam (Abruña-Rodríguez et al., 1982) (Fig. 18.6). Other acid soil-tolerant crop species are rye, yellow lupin, rice, cowpea, peanut, and potato, whereas barley, faba bean, maize, common bean, and wheat are nontolerant species (George et al., 2012).

Differences in acid soil tolerance among the genotypes of a given species can be quite large. For example, in a nonlimed soil of pH 4.5 and 80% Al saturation, a traditional, adapted dryland rice cultivar produced ~2.3 tons of grain ha^{-1}, compared with an introduced nonadapted cultivar, which produced only 1 ton; the latter required ~6 tons of lime ha^{-1} and a corresponding decrease in Al saturation to 15% to achieve the grain yield of the traditional, adapted cultivar in the nonlimed soil (George et al., 2012).

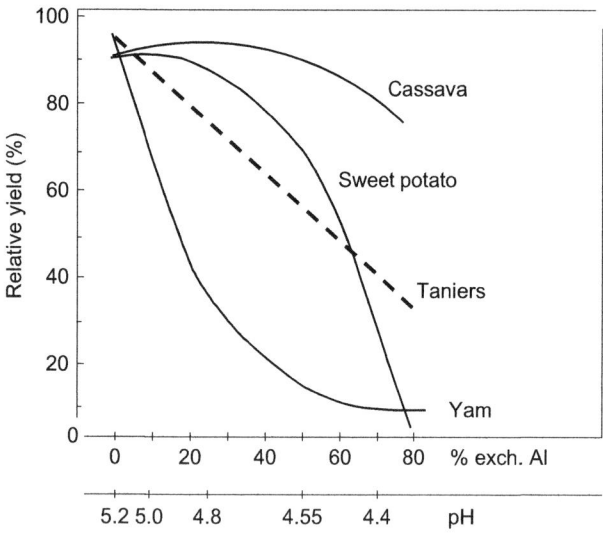

FIGURE 18.6 Relationship between exchangeable Al (% of Al saturation), soil pH, and yield of four tropical root crops: cassava (*Manihot esculenta* Crantz), sweet potato (*Ipomoea batatas* L. Lam.), taniers (*Xanthosoma* sp.), and yam (*Dioscorea alata* L.). Redrawn from Abruña-Rodríguez et al. (1982).

18.3.2 Aluminum resistance by avoidance

18.3.2.1 Aluminum detoxification by root exudates

The Al-activated release of Al-complexing solutes, particularly organic acid anions, in the Al-sensitive apical root zone is the most effective way to reduce Al uptake into the root apoplasm, the impact of Al on apoplasm functions, and thus inhibition of root elongation (Delhaize et al., 2007b). Ma et al. (2001) described two patterns of organic acid anion secretion: pattern I plants release organic anions immediately after the onset of Al exposure, whereas in pattern II plants, organic acid anion release starts after a lag phase of several hours. This suggests that the organic acid anion exudation mechanism is expressed constitutively in pattern I plants and is inducible in pattern II plants (involving gene expression and new protein synthesis).

The Al-induced release of organic acid anions is mediated by plasma-membrane anion channels. The genes encoding these channel proteins belong to two families: *ALMT* and *MATE*. The ALMT (Al-activated malate transporter) facilitates malate efflux (Hoekenga et al., 2006; Sasaki et al., 2004). The MATE (multidrug and toxin extrusion) proteins are citrate transporters (Furukawa et al., 2007; Magalhaes et al., 2007). The decisive role of organic acid anion transporters in enhancing Al resistance was shown by transforming barley with *ALMT1* from wheat (Delhaize et al., 2004) (Fig. 18.7).

The role of the metabolism of organic acids in Al resistance remains unclear. Older studies that generally concentrated on only one organic acid anion and the corresponding biosynthetic machinery found either no clear relationship between the root concentration and release of organic acid anions and the activities of enzymes involved in their biosynthesis (Ryan et al., 2001) or showed the importance of the constitutively high activity of citrate synthase fueled by the high phosphoenolpyruvate (PEP) carboxylase activity (Rangel et al., 2010) underpinning citrate exudation in the Al-resistant common bean genotype (Eticha et al., 2010) (Fig. 18.8). By contrast, newer studies using transcriptomic or proteomic approaches clearly show the complexity of the metabolic changes (e.g., Chen et al., 2019a; Li et al., 2017d; Ma and Lin, 2019; Zhao et al., 2020a) that occur after exposure to Al (even after 1 h only) (Xu et al., 2018). Among differentially expressed genes/proteins, there are those involved in the biosynthesis of organic acid anions but also many others associated with cell wall components, glucan biosynthesis, ATPases, cellular transport, signal transduction, transcription factors, phytohormones, lipid metabolism, antioxidant activity, etc.

Two common bean genotypes distinct in Al resistance differed in the upregulation of *MATE* genes coding for citrate permeases, but the Al-sensitive genotype showed 2−3 orders of magnitude greater expression than the Al-resistant one (Eticha et al., 2010) (Fig. 18.8). However, the Al-resistant genotype had significantly higher citrate content at any time (including before Al exposure) and higher citrate exudation after 24 h compared with the Al-sensitive genotype.

Many studies using transgenic plants with modified organic acid metabolism confirmed the importance of organic acid anion exudation in conferring Al resistance (e.g., Chen and Liao, 2016). Aluminum-activated citrate exudation driven by the Al-inducible expression of mitochondrial citrate synthase has been demonstrated in leguminous tree *Paraserianthes falcataria* (Osawa and Kojima, 2006) and two species of tobacco: *Nicotiana benthamiana* (Deng et al.,

Plant responses to soil-borne ion toxicities **Chapter | 18** 675

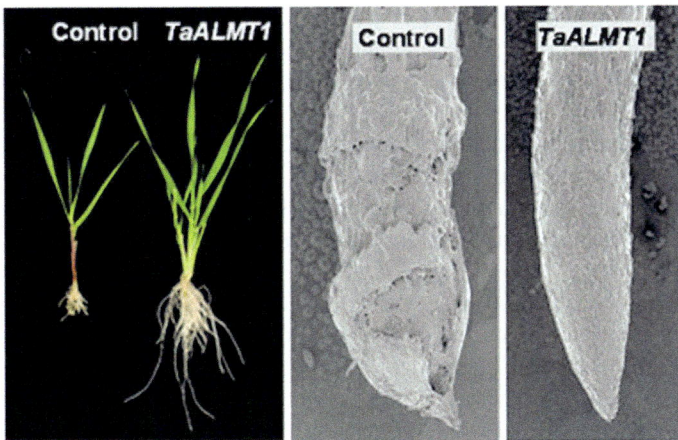

FIGURE 18.7 Growth (left) and morphology of the root apex (right) of the control (empty vector) barley line and a transformant containing wheat malate transporter gene *TaALMT1* grown in nutrient solution with 3 μM Al for 10 days. *Courtesy E. Delhaize (CSIRO, Canberra, Australia).*

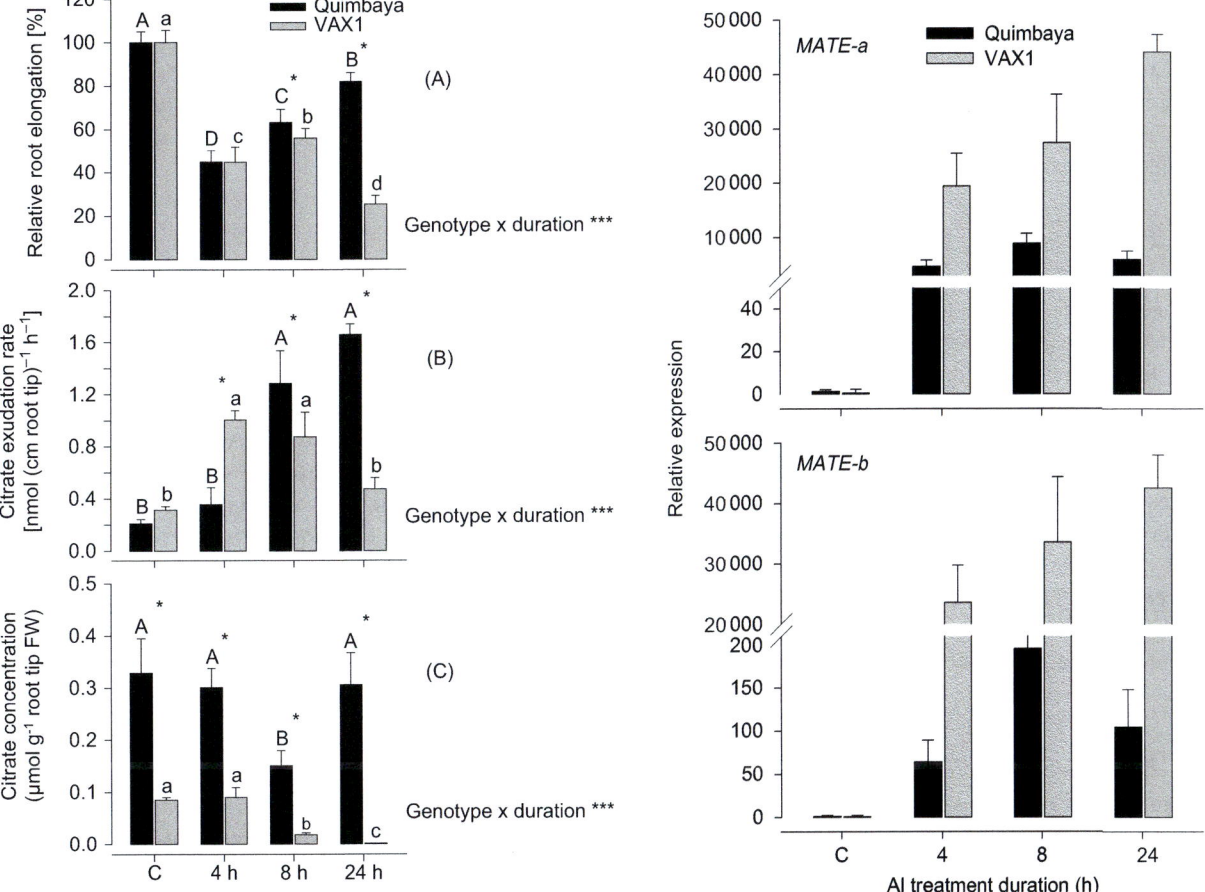

FIGURE 18.8 Root growth (A), citrate exudation rate (B), citrate concentration (C), and expression of two *MATE* genes (right) in 1-cm root tips of two common bean genotypes [Quimbaya (Al-resistant) and VAX 1 (Al-sensitive)] grown without (control, C) or with 20 μM Al at pH 4.5 for up to 24 h. In A-C, different capital (Quimbaya) and lowercase letters (VAX1) indicate significant ($P \leq 0.05$) differences among treatment durations. The asterisk indicates a significant difference ($P \leq 0.05$) between the two genotypes at a given Al treatment duration. Means + standard deviation (n = 4-6). *From Eticha et al. (2010), with permission from Oxford University Press.*

2006) and *Nicotiana tabacum* (Han et al., 2009). Not only the overexpression of citrate synthase but also that of malate dehydrogenase (Tesfaye et al., 2001), PEP carboxylase (Begum et al., 2009; Ermolayev et al., 2003), or pyruvate phosphate dikinase (Trejo-Tellez et al., 2010) enhanced plant Al resistance. Similarly, overexpressing the organic acid anion transporters has resulted in increased organic acid exudation and enhanced Al resistance (e.g., Liu et al., 2018; Zhou et al., 2013b). Hence, the maintenance of cytosolic concentrations of organic acid anions and their release into the root tip apoplasm via activation of anion permeases are the key factors (albeit not the only ones) conferring Al resistance in many plant species.

In addition to organic acid anions, the release into the apoplasm of phenolics by maize (Kidd et al., 2001) and barley (Vega et al., 2020) may be involved in Al resistance, given strong complexing affinity of phenolic substances (especially polyphenols, Zhang et al., 2016) for Al. For Al exclusion, mucilage may also play a key role (Section 14.6.2.1). Mucilage is mainly secreted at the root cap and root apical zones and has a high capacity for Al binding and complexation (Nagayama et al., 2019).

Enhanced release of organic acid anions under P deficiency occurs in many dicotyledonous plants (see Chapter 17) and may be an important component in the plant strategies of adaptation to acid mineral soils for both increased efficiency in nutrient acquisition and avoidance of Al toxicity (Chen and Liao, 2016; Magalhaes et al., 2018). There are examples of the pleiotropic effects in the field, such as Al-resistant wheat showing greater efficiency of P acquisition from both soil and fertilizer P compared with Al-sensitive wheat (Seguel et al., 2017) as well as transgenic canola overexpressing the *Pseudomonas aeruginosa* citrate synthase gene and showing enhanced citrate exudation under Al toxicity as well as low P in the soil (Wang et al., 2013b).

The complexity of the Al resistance mechanisms is illustrated by comparing wheat cultivars that carry the *ALMT1* gene with noncarriers that may or may not show adaptation to acid soils. The grain yield of *ALMT1* carriers was greater than that of nonadapted noncarriers but was similar to that of adapted noncarriers (Fordyce et al., 2020). To complicate matters even more, resistances to toxicities of Al and protons are quite different (even though these two stresses coincide in nature), with only a relatively small number of shared loci, as determined by genome-wide association analysis in arabidopsis (Nakano et al., 2020).

18.3.2.2 Rhizosphere pH

At pH$_{water}$ 4–4.5, even a 0.1–0.2 increase in rhizosphere pH strongly decreases the concentration of Al^{3+} but may simultaneously decrease competition with H$^+$ for binding sites and increase the negativity of the apoplast, enhancing Al phytotoxicity. An increase in the rhizosphere pH was proposed as an Al exclusion and detoxification mechanism, but the literature contains conflicting experimental results; for example, depending on the wheat genotype studied, Al-resistant wheat either did not have (Taylor, 1988) or had increased rhizosphere pH around the root tips (Yang et al., 2011). A higher rhizosphere pH at the root tip may be a consequence rather than a cause of Al resistance (Kollmeier et al., 2000). Moreover, the findings from nutrient solution experiments have to be interpreted with caution in relation to Al resistance of soil-grown plants, and no reports of the rhizosphere pH changes between soil-grown genotypes differing in Al resistance using the modern imaging and spatially sensitive analytical techniques could be found.

The pot-grown Al-resistant soybean genotypes had higher relative abundance of Al-resistant bacterial genera (such as *Tumebacillus*, *Granulicella*, and *Burkholderia*) in the rhizosphere soil compared with the Al-sensitive genotype (Lian et al., 2019). In addition, the Al-tolerant soybean genotype had a higher microbial capacity for denitrification in the rhizosphere than the Al-sensitive genotype, which was linked to increased exudation of succinate in the former (Li et al., 2020c).

18.3.2.3 Reduced aluminum binding in the root apoplast

Aluminum binds readily to negative sites of the cell wall and the plasma membrane in the Al-sensitive zones of the root apex (Kinraide et al., 1992); decreasing that binding may contribute significantly to Al resistance (e.g., Kopittke et al., 2016; Li et al., 2020b), even though that may be applicable only to dicots with relatively high root CEC, and not to monocots such as maize with low root CEC (Barbosa et al., 2018). However, the role of Mg present in relatively high (millimolar) concentrations in reducing Al saturation and activity at the cell wall and plasma membrane binding sites is more prominent in monocots than dicots (Rengel et al., 2015). Binding of Al in the root apoplast of *Sorghum bicolor* can also be reduced by additional Si supply, whereby the formation of the nontoxic Si–Al complexes prevents Al binding to the cell wall (Kopittke et al., 2017a).

The negativity of the cell wall depends mainly on the pectin concentration and its degree of methylation. Increased methyl-esterification excludes Al from binding in the apoplast (Mimmo et al., 2009) and thus contributes to Al resistance in a range of plant species (El-Moneim et al., 2014; Li et al., 2017b). Accumulation of Al in alkali-soluble pectin may increase Al resistance of root border cells in pea (Li et al., 2017c), but boron alleviated Al toxicity despite decreasing the alkali-soluble pectin content in barley roots (Yan et al., 2018). In common bean, the initially high Al sensitivity and Al accumulation by roots of the Al-resistant cultivar Quimbaya was related to a higher concentration of nonmethylated pectin in the 5-mm root tips (Rangel et al., 2009).

18.3.3 Aluminum tolerance

18.3.3.1 Aluminum accumulation

An influx transporter facilitating the uptake of Al^{3+} (but not Al-citrate complex) was described first in the plasma membrane of the root tip cells (except in the epidermis) in rice (Xia et al., 2010). It was designated Nrat1; it belongs to the Nramp (natural resistance-associated macrophage protein) family and is inducible by Al. The Nrat1 locus was mapped to QTL associated with Al tolerance on chromosome 2 in rice (Tao et al., 2018; Xia et al., 2014). Strong Al induction of Nrat1 activity (and thus uptake of Al across the root-cell plasma membrane) is associated with Al tolerance in rice genotypes (Li et al., 2014; Xia et al., 2014), followed by Al sequestration in the vacuoles (Negishi et al., 2012). In Al-accumulator hydrangea (*Hydrangea macrophylla*), there are Al transporters from the aquaporin (HmPALT1) (Negishi et al., 2012) and anion permease families (HmPALT2) localized in the plasma membrane and facilitating the Al uptake into the symplast (Negishi et al., 2013). It should be borne in mind that the transport of Al into root symplast decreases binding of Al in the cell walls, thus contributing to alleviation of Al toxicity, emphasizing the coordination between the exclusion and internal detoxification mechanisms (Kopittke et al., 2016; Zhu et al., 2013).

In contrast to the Al-excluder species, many plant species (particularly woody ones in tropical areas) are not just resistant to Al but are also Al accumulators (Brunner and Sperisen, 2013) and even Al hyperaccumulators, such as some Proteaceae species from southern South America accumulating more than 1 g Al kg^{-1} leaf dw (e.g., *Gevuina avellana* with up to 6.3 g Al kg^{-1} of mature leaves) (Delgado et al., 2019). In tropical rainforests, Al includers and excluders coexist at the same sites, varying in Al concentrations in the leaf press sap from <10 to 4780 mg L^{-1} (Cuenca et al., 1990). Only a few agricultural and horticultural species are Al includers/accumulators, such as tea (*Camellia sinensis* (L.) Kuntze), buckwheat (*Fagopyrum esculentum*), and hortensia (*Hortensia macrophylla*). Tea plants not just tolerate high Al concentrations but also their growth is enhanced by Al supply within a particular range (similarly to nutrients, too little or too much Al results in growth depression) (Li et al., 2017b). There are also reports on stimulatory effects of Al on the growth of other Al-accumulator species (e.g., Fan et al., 2020), but the mechanism of this stimulation is unclear. However, in tea plants (Li et al., 2017b) as well as in another Al-accumulator *Qualea grandiflora* (Cury et al., 2020), the supply of Al was critical for the cell wall and lignin biosynthesis (that includes pectin methylesterase) and the organic acid metabolism (featuring malate dehydrogenase and citrate synthase).

Sequestration of Al in vacuoles lessens Al toxicity and allows plants to tolerate relatively high internal Al concentrations. Aluminum tolerance is attributed to intracellular complexation of Al by organic ligands, particularly organic acid anions (Ma et al., 1998). When exposed to Al, Al accumulators such as buckwheat, tea, and hortensia not only complex Al in the symplasm but also release organic acid anions from the Al-sensitive root tips and complex Al in the root apoplasm (George et al., 2012). In buckwheat, there was a close relationship between Al and oxalate concentrations in the symplasm and the apoplasm of root tips (Klug and Horst, 2010), with the formation of the 1:1 oxalate:Al complex in the root apoplasm protecting the sensitive root sites from binding Al. Similarly in wheat, sequestration of Al in the malate:Al complex occurred in the apoplasm of roots (particularly in the Al-tolerant in comparison with the Al-sensitive near-isogenic genotype) (Kopittke et al., 2017b). By contrast, different tissues of tea roots have differential patterns of exclusion/accumulation of Al (Hajiboland and Poschenrieder, 2015).

18.3.4 Screening for aluminum resistance

Field screening of genotypes in acid soils is a long-term and labor-intensive process that is often influenced by the secondary factors such as genotypic differences in resistance to diseases and pests. Given that Al toxicity is the main factor limiting plant growth in most acid mineral soils, many rapid screening methods for Al resistance in nutrient solution were developed (see George et al., 2012). However, the main problem in screening for Al resistance is the potential confounding effect of H^+ toxicity in the H^+-sensitive plant species such as arabidopsis (Bose et al., 2010) and common bean (Rangel et al., 2005).

Eticha et al. (2005a) showed that Al-induced callose concentration in the Al-treated root apices was negatively correlated with the relative grain yield (limed soil = 100%) of maize genotypes evaluated across five tropical environments. In addition, the diallel analysis revealed a strong genetic correlation between callose formation in nutrient solution and yield on acid soils (Fig. 18.9). These findings suggest that Al-induced callose formation is a powerful tool to enhance the breeding of maize cultivars adapted to acid soils.

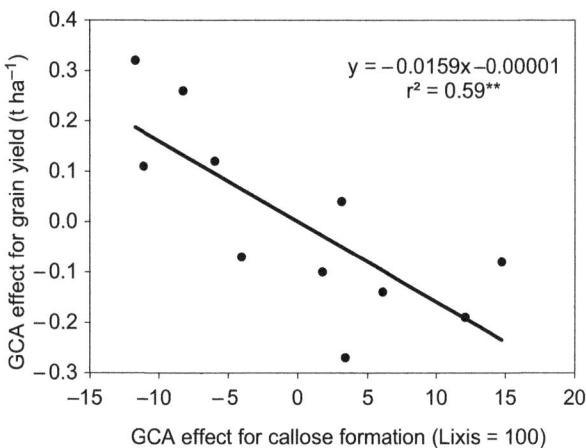

FIGURE 18.9 Relationship between GCA effects for grain yield on acid soils (field experiments) and for relative Al-induced callose formation (nutrient solution experiments, cv. Lixis used as the reference) of 11 maize cultivars (Eticha et al., 2005b). *GCA*, General combining ability. *From Eticha et al. (2005b), with permission from Elsevier.*

Given that the specific genome parts (QTL or genes) associated with Al resistance have been identified in many plant species, such as rice (Rasheed et al., 2020), maize (Coelho et al., 2019), wheat (Navakode et al., 2009; Raman et al., 2005), barley (Ma et al., 2016), soybean (Cai et al., 2019), sorghum (Melo et al., 2019), lentil (Singh et al., 2018), etc., molecular marker−assisted breeding can be used to increase the speed, efficiency, and accuracy of the selection and breeding process. Using such an approach in combination with backcrossing to transfer *TaALMT1* from the Al-resistant donor to an Al-sensitive (but otherwise well-adapted) local cultivar, Soto-Cerda et al. (2015) shortened the breeding cycle to 2 years.

18.3.5 Manganese tolerance

The occurrence of Mn toxicity is not only a function of soil pH, concentrations of Mn^{2+} and other polyvalent cations in the soil solution, plant species and genotype, and microbial activity in the rhizosphere (Section 18.2.4) but also of the availability of Si. Silicon strongly increases tolerance of the shoot tissue to high Mn concentrations (see also Section 8.3) by apoplasmic binding of Mn, thus decreasing Mn accumulation in the cytoplasm (Blamey et al., 2018a) as well as the apoplasm, diminishing the ROS generation (Maksimovic et al., 2012). In rice, additional Si supply decreased Mn uptake or transport (depending on the cultivar) and diminished oxidative stress (Li et al., 2012). Similarly, increased Fe supply ameliorated Mn toxicity in soybean by decreasing Mn uptake and transport (Blamey et al., 2019). Calcium and Mg supply may alleviate Mn toxicity stress (e.g., in rice; Lu et al., 2021), probably because Mn toxicity results in large decreases in Ca and Mg uptake in the Mn-sensitive perennial ryegrass genotypes (Inostroza-Blancheteau et al., 2017). Thus, on acid mineral soils, the harmful effects of excessive Mn concentrations on plant growth may depend on the availability and concentration of Si, Fe, Ca, and/or Mg at the cell wall and the plasma membrane due to interference with Mn uptake and accumulation. In addition to the expected electrical effects imposed by the cations charged similarly to Mn, ammonium also alleviated Mn toxicity by inducing acidification of the root environment (and H^+ has a strong capacity to bind to the negative charges on the cell wall and the plasma membrane), but lowered pH also inhibited the OsNram5 Mn influx transporter and decreased Mn accumulation in rice (Hu et al., 2019).

Even though considerably less work was done on a role of nitric oxide (NO) in Mn tolerance compared to that of Al, the limited literature suggests that the increased NO supply enhanced the antioxidative capacity of rice leaves, thus alleviating Mn toxicity (Sarita and Dubey, 2012). In addition, increased NO supply (by using an exogenous NO donor) not just ameliorated oxidative stress but also decreased Mn accumulation; however, that ameliorative effect was dissipated at high NO concentrations (Kovacik et al., 2014).

Plant species, and genotypes within plant species, may differ considerably in tolerance to excess Mn (e.g., Inostroza-Blancheteau et al., 2017; Khabaz-Saberi et al., 2010b, 2012; Lu et al., 2021; Zemunik et al., 2020; Zhou et al., 2013a, 2017). Differences in Mn tolerance among species and genotypes may be associated with Mn-tolerant ones retaining more Mn in the roots and transporting proportionally less to the shoots (e.g., Ducic and Polle, 2007; Zhou et al., 2013a). However, there are numerous examples of a lack of difference in Mn uptake among the genotypes differing in Mn tolerance (e.g., Ribera et al., 2013).

In contrast to Al toxicity, whereby exudation of organic acid anions into the rhizosphere soil via the ALMT1 malate or the MATE citrate transporters represents a major and well-characterized resistance mechanism (Section 18.3.2.1), tolerance to excess Mn involves a number of processes and pathways that are not understood as well as the Al-related ones. The recent review (Li et al., 2019) concluded that Mn tolerance is underpinned by activation of the antioxidant system and the regulation of Mn uptake and subcellular compartmentation, with numerous genes being involved in these processes and pathways. Hence, the work toward identifying specific markers that can be used in breeding for enhanced Mn tolerance is nowhere near as advanced as it is regarding Al resistance (Section 18.3.4).

Presently, there are two major lines of evidence for the regulation of Mn tolerance. Based on the assumption that cytosolic Mn^{2+} activity has to be kept low to avoid Mn interfering with essential metabolic functions, pumping Mn^{2+} from the cytosol into the other cell compartments may confer Mn tolerance (Blamey et al., 2018a; Zhang et al., 2020a). Some cation diffusion facilitators (CDFs), such as *Camelia sinensis* CsMTP8.2, are the Mn-specific transporters that facilitate efflux of Mn from the cytosol; when expressed heterologously in onion and tobacco, CsMTP8.2 was localized in the plasma membrane of the epidermal cells (Zhang et al., 2020a).

Manganese tolerance is mainly due to the sequestration of Mn in the vacuoles and other organelles (Shao et al., 2017). Molecular studies on cation/Mn transporters confirmed that sequestration of Mn in the vacuoles of rice root cells via two CDFs (MTP8.1 and MTP8.2) and the cation exchanger (OsCAX) plays a crucial role in Mn tolerance in rice (Shao et al., 2017; Tsunemitsu et al., 2018) and a number of other plant species (Hirschi et al., 2000). In addition, transporters sequestering Mn into the Golgi-like compartments in arabidopsis (AtMTP11) (Delhaize et al., 2007a; Peiter et al., 2007) or into organelles of *Stylosanthes hamata* (ShMTP1) (Delhaize et al., 2003) have been characterized.

The second line of evidence for the mechanism of Mn tolerance in plants is based mainly on experimental work with cowpea. The Mn toxicity symptoms are cell wall localized; with excess Mn supply, the apoplasm activity of H_2O_2-producing and H_2O_2-consuming peroxidases is strongly enhanced in the leaf tissues of the Mn-sensitive but not in the Mn-tolerant genotypes (Fecht-Christoffers et al., 2006). Therefore, it has been proposed that the leaf apoplasm is the crucial compartment for the avoidance of Mn toxicity in cowpea (Fecht-Christoffers et al., 2007).

Rice nodes regulate the Mn fluxes and distribution, depending on the external Mn supply. Under excess Mn supply, OsNramp3 (which otherwise transports Mn from the xylem to the phloem to ensure Mn supply to the young and developing tissues) is degraded, with a consequence of diminished capture of Mn from the xylem sap, thus protecting young tissues from excessive Mn accumulation (Shao et al., 2017).

Manganese tolerance may or may not be correlated with Al resistance (e.g., Khabaz-Saberi et al., 2010b). Separate germplasm screening for Mn tolerance and Al resistance is, therefore, necessary to identify the genotypes suitable for soils with toxic Mn and Al concentrations.

18.3.5.1 Breeding for Mn tolerance

There is relatively little work reported on screening germplasm for tolerance to Mn toxicity. Tolerant genotypes were identified in wheat (Khabaz-Saberi et al., 2010b) and barley (Huang et al., 2015). The QTL and molecular markers associated with tolerance to Mn toxicity were detected in barley (Huang et al., 2018) and *Brassica napus* (Raman et al., 2017). Interestingly, relatively close to the major *B. napus* QTL, there is an orthologue of arabidopsis cation efflux facilitator, and this type of transporters plays a major role in Mn tolerance (Raman et al., 2017).

18.3.5.2 Hyperaccumulation of Mn

Manganese hyperaccumulators accumulate shoot/leaf tissue Mn to more than 10 g kg^{-1}. Many plant species with vastly different phylogenetic positions fall into the category of Mn hyperaccumulators, including *Ilex paraguariensis* A. St. Hill. (family Aquifoliaceae) with up to 13.5 g kg^{-1} in leaves (Magri et al., 2020) and *Socratea exorrhiza* (Mart.) H. Wendl. (family Arecaceae) with $27.8 \text{ g Mn kg}^{-1}$ leaf dw (albeit growing on the mine tailings rich in Mn rather than on natural soils) (Silva et al., 2019). *Polygonum hydropiper* was recommended for phytoremediation of Mn-polluted paddy soils (Yang et al., 2018). *Phytolacca americana*, a common weed not specifically associated with high Mn soils, hyperaccumulates Mn due to the strong rhizosphere soil acidification associated with P acquisition (Degroote et al., 2018). However, the molecular and biochemical mechanisms governing the Mn hyperaccumulation capacity are still poorly understood.

18.4 Waterlogged and flooded (hypoxic) soils

18.4.1 Soil chemical factors

Waterlogging and flooding often occur following heavy rainfall or excessive irrigation on slowly draining soils. In the current global climate change scenarios, precipitation is projected to occur as more intense events in the future, causing more frequent and/or more severe waterlogging or flooding (Ashu and Lee, 2020; Dougherty and Rasmussen, 2020).

TABLE 18.1 Sequence of redox reactions in soil in relation to declines in soil redox potential.

Redox reaction	Redox potential Eh (mV)[a]
Reduction of O_2 $O_2 + 4H^+ + 4e^- \rightarrow 2H_2O$	812
Nitrate reduction (denitrification) $NO_3^- + 2H^+ + 2e^- \rightarrow NO_2^- + H_2O$	747
Reduction of Mn^{4+} to Mn^{2+} $MnO_2 + 4H^+ + 2e^- \rightarrow Mn_2^+ + 2H_2O$	526
Reduction of Fe^{3+} to Fe^{2+} $Fe(OH)_3 + 3H^+ + e^- \rightarrow Fe_2^+ + 3H_2O$	−47
Sulfate reduction to H_2S $SO_4^{2-} + 10H^+ + 8e^- \rightarrow H_2S + 4H_2O$	−221
Reduction of CO_2 to CH_4 $CO_2 + 8H^+ + 8e^- \rightarrow CH_4 + 2H_2O$	−244

[a]Assumptions: the coupling to the oxidation reaction:
$CH_2O + H_2O \rightarrow CO_2 + 4H^+ + 4e^-$ is complete, and energy released = RT ln (K), whereby R is universal gas constant, T is temperature, and K is equilibrium constant.
Source: From Chapin III et al. (2011).

Waterlogging of soils constitutes a major abiotic stress to plants, affecting plant growth, productivity, and species distribution in many areas of the world (Kaur et al., 2020; Zhou et al., 2020a). Flooding stress (when plant parts normally above-ground are covered with water) (George et al., 2012) is even more severe abiotic stress than waterlogging (Mullan and Barrett-Lennard, 2010; Striker et al., 2019). However, excess water generally occurs along a gradient (spatial/temporal), blurring a distinction between waterlogging and flooding; therefore the term "hypoxic" is adopted henceforth in this chapter to cover various types of environments with limited oxygen availability.

Oxygen diffuses in air about 10^4 times faster than in water. Hence, oxygen is depleted rapidly by the respiration of soil microorganisms and plant roots in waterlogged soils; then, other terminal electron acceptors are used by various microorganisms for respiration. A sequence of reduction takes place at specific redox potentials (Table 18.1). As soils are nonuniform and characterized by microsites differing in pore size, water content and microbial activity, the redox potentials often vary widely over short distances. A change from oxygen sufficiency to deficiency can occur within a few millimeters, and even in aerobic soils, the interior of soil aggregates may be hypoxic (Renault and Stengel, 1994).

After nitrate, Mn oxides (mainly Mn^{4+}) are the next electron acceptors (Table 18.1). In acid soils high in Mn oxides and organic matter but low in nitrate, high concentrations of water-soluble and exchangeable Mn^{2+} can build up within a few days. After prolonged waterlogging, Fe^{3+} is reduced to Fe^{2+} that may accumulate to toxic concentrations.

Various products of microbial carbon metabolism, such as ethylene, accumulate in hypoxic soils (e.g., Najeeb et al., 2015). During prolonged hypoxia, volatile fatty acids and phenolics accumulate in soils high in readily decomposable organic matter (e.g., after application of green manure or straw), which may have a detrimental effect on the root metabolism and growth.

In hypoxic soils at low redox potentials, large amounts of methane (CH_4) may be formed. Indeed, paddy rice fields are a major source of CH_4 emission (15%−20% of total anthropogenic methane emissions are attributed to rice production, Tian et al., 2021). Methane emissions are higher in planted than unplanted paddy fields because root exudates may increase methane production in the rhizosphere (Philippot et al., 2009). Another potent greenhouse gas (N_2O) is also produced in paddy fields because nitrate (in the absence of oxygen) is used as an alternative electron acceptor, leading to nitrate reduction (denitrification) (Kaur et al., 2020) (Fig. 18.10).

Changes in the agricultural management offer possibilities for substantial mitigation of CH_4 from paddy soils (Binh Thanh et al., 2020; Sander et al., 2020; Tian et al., 2021). In addition, molecular approaches have improved our

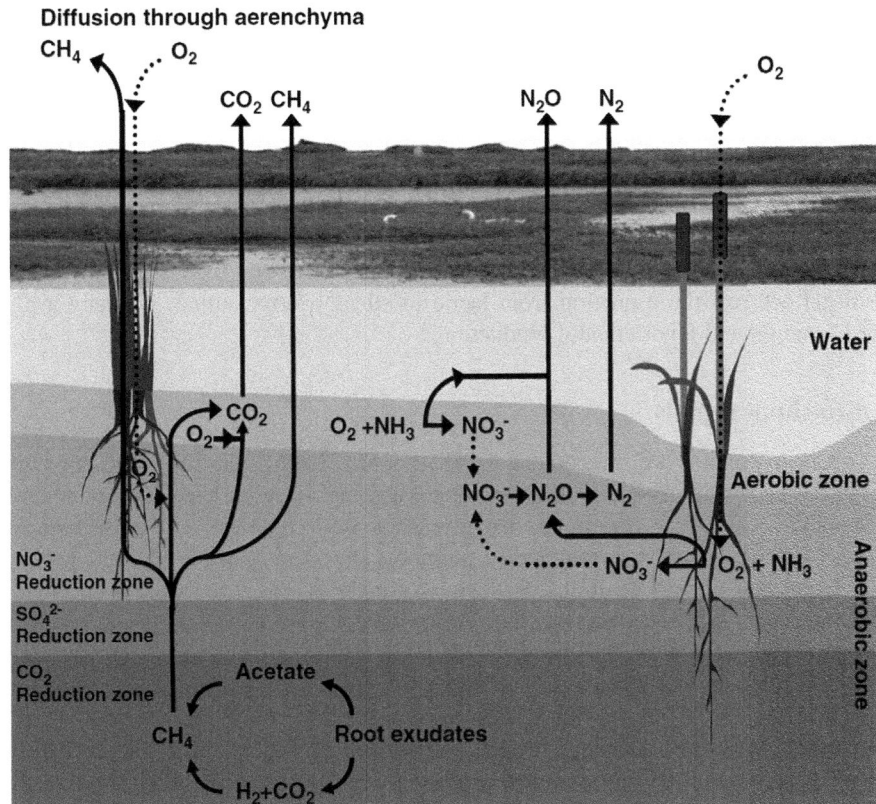

FIGURE 18.10 Production and consumption of N₂O and CH₄ in the rhizosphere of wetland plants. *From Philippot et al. (2009), with permission from Springer Nature.*

understanding of the microbial communities involved in greenhouse gas emissions from hypoxic soils (Yoneyama et al., 2019), with a reduction in the abundance of methanogenic and an increase in the abundance of methanotrophic genes resulting in decreased methane emissions (Binh Thanh et al., 2020). The abundance and the diversity of nitrifiers, denitrifiers, methanogens, and methanotrophs are influenced by the presence of plant roots (Philippot et al., 2009).

18.4.2 Hypoxia stress

The severity of the effects of hypoxia on growth and yield depends on the plant species, developmental stage of the plants, soil properties (e.g., pH, organic matter content), and soil temperature (Kaur et al., 2020). Wilting, leaf senescence, and, in herbaceous species, epinasty (downward bending of leaves) are often the first symptoms of hypoxia stress due to a decrease in hydraulic conductivity of the roots (wilting) and accumulation of ethylene in the shoots (epinasty) (Drew, 1990). A decrease in hydraulic conductivity is, at least partly, related to a rapid reduction in aquaporin synthesis in roots under hypoxia, which is preceded by a decrease in cytosolic pH (Tornroth-Horsefield et al., 2006).

Hypoxia stress is related to oxygen deficiency in the root environment, but the actual cause of stress may differ depending on the circumstances (for the recent reviews see Kaur et al., 2020; Khan et al., 2020b; Loreti and Perata, 2020; Yusuff et al., 2020; Zhou et al., 2020a). The hypoxia impairments vary greatly among species, tissues, and experimental conditions; the time taken for tissues or plants to die can range from a few hours to months. When death is rapid, the cellular malfunctions may be due to a decline in the ATP concentrations leading to the impairment of the H^+ efflux pump and acidification of the cytosol. In tissues and plants in which damage develops slowly and ATP concentrations are maintained, carbohydrate shortage may limit survival under hypoxia.

18.4.3 Phytotoxic metabolites under hypoxia

Anaerobic metabolism is enhanced in the roots of most plant species under hypoxia, regardless of their tolerance. Fermentation to ethanol is inefficient carbon utilization, with two moles of ATP per mole of hexose produced compared with 36 moles of ATP per mole of hexose in the aerobic tricarboxylic acid cycle. The enhanced rate of glycolysis under hypoxia (Keska et al., 2021) may be considered a compensatory action.

Under oxygen deficiency, cytochrome oxidase activity becomes oxygen limited, and ATP has to be generated by fermentation. Pyruvate decarboxylase converts pyruvate to acetaldehyde that is metabolized by alcohol dehydrogenase to ethanol. NAD^+ is regenerated to sustain glycolysis. Ethanol is not detrimental because of rapid diffusion out of cells, but acetaldehyde is toxic. Acetaldehyde dehydrogenase catalyzes the conversion of acetaldehyde to acetate, together with the concomitant reduction of NAD^+ to NADH (Bailey-Serres and Voesenek, 2008).

In addition to ethanol, lactate is produced in plant cells under oxygen deficiency by lactate dehydrogenase (Da-Silva and Amarante, 2020). Under hypoxia, the pH of the cytosol in rice and wheat declines by about 1 pH unit on average (Yemelyanov et al., 2020). The cytosol pH controls the transition from lactic to ethanol fermentation, whereby a pH decline in the cytosol may limit lactate formation and favor ethanol production.

18.4.4 Phytohormones and root-to-shoot signals

The accumulation of ethylene in roots under hypoxia is well documented (Khan et al., 2020b). The increased ethylene concentration in the root tissue has numerous effects on root growth and morphology, including the formation of adventitious roots (e.g., in *Cucumis melo*, Zhang et al., 2021), as a common adaptive response to hypoxia. The expression of the genes encoding the ethylene biosynthesis enzymes, for example, 1-aminocyclopropane carboxylic acid (ACC) synthase, increased after a few hours of hypoxia (Geisler-Lee et al., 2010). Interestingly, wheat inoculation with fungal endophyte *Trichoderma asperellum* (that produces ACC-cleaving and thus ethylene-depleting ACC deaminase) resulted in downregulation of the expression of the ethylene biosynthesis genes (Rauf et al., 2021), increasing wheat tolerance to hypoxia. The flooding-induced ethylene responsive factors have been identified in rice and characterized in arabidopsis (Loreti and Perata, 2020) (Section 18.4.6).

Increased formation of adventitious roots (that is governed by inherently complex interactions, Mhimdi and Perez-Perez, 2020) in hypoxic plants is linked to decreased concentrations of cytokinin (e.g., in wheat, Tran Nguyen et al., 2018) and gibberellins (e.g., in soybean, Kim et al., 2015). Correspondingly, foliar application of cytokinin may counteract the hypoxia stress and increase grain yield (e.g., in maize, Ren et al., 2019). In soybean, the genotype tolerant to hypoxia stress had higher gibberellin concentration and better developed adventitious roots and aerenchyma than the sensitive genotype (Kim et al., 2015).

The phytohormonal crosstalks that govern plant responses, adaptation, and tolerance to hypoxia are still poorly understood. However, it appears that ethylene, gibberellins, and abscisic acid are the main phytohormones in the hypoxia stress response (Bashar et al., 2019), whereas in the posthypoxic stress (reoxygenation) the main regulators are ethylene and ABA (Bashar, 2018). Numerous other factors (as well as other phytohormones, such as jasmonic acid, Shukla et al., 2020) may also influence the phytohormonal effects, including specific metabolites (e.g., ATP, sugars, and pyruvate). Unsurprisingly, there are differences among the species and plant growth stages. For example, ethylene, gibberellic acid, and indole-3-acetic acid (IAA) influenced shoot morphology of flooded *Rumex palustris* (Cox et al., 2004), whereas the concentration of ethylene-regulated ABA in petioles, and the responsiveness to gibberellic acid of these petioles, explained the differences in shoot elongation among submerged *R. palustris* accessions (Chen et al., 2010b).

The formation of aerenchyma (the large air spaces in the root cortex formed in anoxic/hypoxic conditions) (Fig. 18.11) is associated with ethylene accumulation (see Section 14.5.1). This response to elevated ethylene concentrations is not restricted to the roots but is also observed at the shoot base and basal parts of the stem (Shimamura et al., 2010); the shoot aerenchyma is essential for oxygen supply to the roots growing in hypoxic environments.

18.4.5 Element toxicity as a component of hypoxia stress

During hypoxia, toxic concentrations of Mn, Fe, Na, Al, and B may occur (Setter et al., 2009; Shabala, 2011). Arsenic (As) solubilization is increased in hypoxic/anoxic conditions, especially if soils contain a relatively high content of organic matter serving as electron donor (Lewinska et al., 2019; Verbeeck et al., 2020). In addition, As reduction from arsenate [As(V)] to arsenite [As(III)] occurs under hypoxia, and the plant populations that preferentially take up arsenite would accumulate it almost an order of magnitude more than in the nonhypoxic conditions (Wan et al., 2015).

Hypoxia injury caused primarily by Mn toxicity occurs in the wheat genotypes sensitive to Mn toxicity but not in the Mn-resistant genotypes grown in slightly acidic soil with a high concentration of DTPA-extractable Mn (Table 18.2) (see Section 18.2.4). The uptake of Mn under hypoxia can be decreased by high soil Fe concentrations (Khabaz-Saberi and Rengel, 2010).

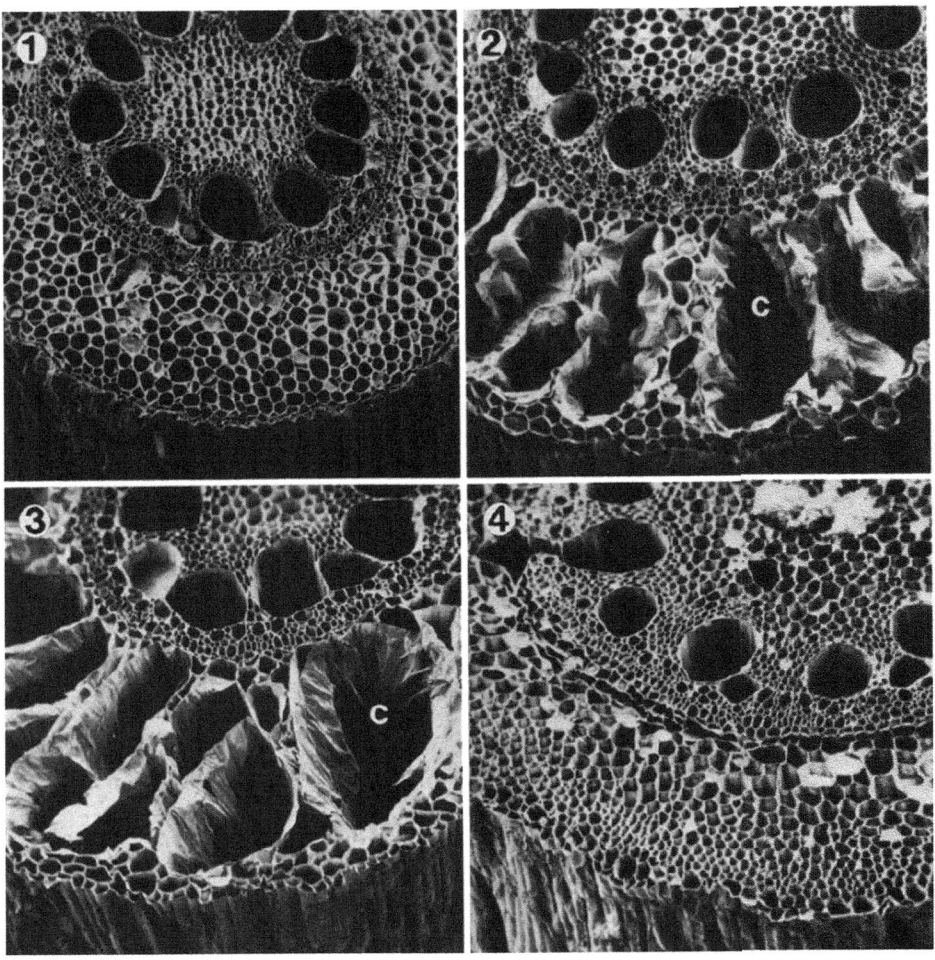

FIGURE 18.11 Transverse sections of maize roots under a scanning electron microscope. (1) Control grown in well-aerated solution; (2) root receiving 5 μL ethylene in the air (l:1); (3) root from nonaerated solution; (4) root receiving nitrogen gas (anoxic treatment). C, Cortical air space. *From Drew et al. (1979), with permission from Springer Nature.*

For adaptation to waterlogging, high Mn tolerance of the shoot tissue is important, for example, in wetland rice compared to the waterlogging-sensitive barley (Table 18.3). Less than 200 mg Mn kg^{-1} leaf dw is toxic to barley, whereas a 10-fold higher concentration is tolerated by rice without growth depression.

In some wetland species such as rice, tissue tolerance to Fe is considerably higher than in nonwetland species. In wheat, there is a considerable genetic variation in tolerance to Fe and Mn toxicity, and high tolerance improves the performance of wheat in waterlogged acid soils (Khabaz-Saberi and Rengel, 2010; Khabaz-Saberi et al., 2012, 2014).

In wetland species (or upland ones upon soil waterlogging), excessive Fe uptake may cause Fe toxicity. "Bronzing" of leaves is a typical nutritional disorder in wetland rice (Dufey et al., 2009) but also in wheat (Khabaz-Saberi et al., 2010a). "Bronzing" is due to Fe toxicity and occurs at leaf concentrations of ≥ 700 mg Fe kg^{-1}, increasing the activity of peroxidases and inducing high concentrations of oxidized polyphenols similarly to "brown speckles" caused by Mn toxicity (George et al., 2012) (see also Section 7.2). The high peroxidase activity can be explained by the formation of oxygen radicals:

$$Fe^{2+} + O_2 \rightarrow Fe^{3+} + O_2^{\cdot -}$$
$$O_2^{\cdot -} + O_2^{\cdot -} \xrightarrow{SOD} H_2O_2$$
$$Fe^{2+} + H_2O_2 \rightarrow Fe^{3+} + HO^{\cdot} + OH^-$$

ROS such as hydrogen peroxide (H_2O_2), singlet oxygen ($^1O_2$), and free radicals (superoxide radical, $O_2^{\cdot -}$; hydroxyl radical, $HO^{\cdot}$) are produced in many cellular reactions, including the Fe-catalyzed Fenton reaction (Farooq et al., 2019). Hydroxyl radicals are highly phytotoxic, causing peroxidation of membrane lipids and protein degradation. Superoxide dismutase (SOD) is responsible for the dismutation of superoxide anions to H_2O_2 that is then detoxified by peroxidases or catalase.

In general, waterlogging under saline conditions, for example after irrigation with saline water, causes increased Na and Cl shoot concentrations, and rapid leaf senescence (Barrett-Lennard, 2003; Barrett-Lennard and Shabala, 2013). At

TABLE 18.2 Shoot dry weight and Mn concentration in the shoots of bread wheat grown in loamy sand soil from Bromehill (Western Australia) in pots.

Genotype	Waterlogging (WL)	Soil pH$_{CaCl_2}$	DTPA-extractable Mn (mg/kg soil)	After 42 days of WL (63 DAS) Relative shoot dw (%)	After 42 days of WL (63 DAS) Shoot Mn concentration (mg/kg)	Days to maturity relative to non-WL control (%)	Relative grain yield (% of non-WL control)
Sensitive to Mn toxicity	No	4.3	5.9		168		
	Yes	5.1[a]	18	78	202[a]	107	53
Resistant to Mn toxicity	No	4.3	5.9		168		
	Yes	5.1[a]	18	91[a]	167	110	80[a]

WL was imposed 21 days after germination and lasted for 42 days. Plants were harvested 63 DAS and at maturity. *DAS*, days after sowing; *WL*, waterlogging.
[a]*Different at P ≤ 0.05.*
Source: Based on Khabaz-Saberi et al. (2012).

TABLE 18.3 Growth and Mn concentration of mature leaves of barley and wetland rice at different Mn supply.

Mn supply (mg L^{-1})	Shoot dry weight (g plant^{-1}) Barley	Shoot dry weight (g plant^{-1}) Rice	Mn concentration (mg kg^{-1} dw) Barley	Mn concentration (mg kg^{-1} dw) Rice
0.2	14	15	70	100
0.5	12	16	190	400
2.0	7	15	310	2200
5.0	6	12	960	5300

Source: Data recalculated from Vlamis and Williams (1964).

low O_2 concentration in the rooting medium, the selectivity of K^+/Na^+ uptake by roots decreases in favor of Na, diminishing the transport of K to the shoots. In most crop species, salt tolerance is based on mechanisms that prevent, or at least restrict, salt accumulation in the shoots (exclusion mechanisms, Section 18.5.4). These mechanisms rely on a high metabolic activity of roots and thus, in nonwetland species, on soil aeration. Crop species or cultivars with better root aeration under hypoxia may also have a greater tolerance to combined salt and waterlogging stress, as shown with different *Lotus* species (Teakle et al., 2010) or *Melilotus siculus* accessions (Striker et al., 2015).

18.4.6 Mechanisms underpinning tolerance to, and avoidance of, hypoxic stress

Plant species differ widely in their capacity to adapt to hypoxia. The differences in adaptation also exist among cultivars of cereals (e.g., Farkas et al., 2020; Prasanna et al., 2021), legumes (e.g., soybean, Kim et al., 2019), vegetables (e.g., carrot, Schmid et al., 2021), fruits (e.g., sweet cherry, Perez-Jimenez et al., 2018), etc. Differential tolerance to hypoxia among wheat (Khabaz-Saberi and Rengel, 2010; Khabaz-Saberi et al., 2012) and barley genotypes (Huang et al., 2015) is based on differential tolerance to hypoxia-induced ion toxicities (primarily of Mn and Fe) (see Sections 18.3.4 and 18.3.4.1).

Plant adaptations to hypoxia have been classified into two main strategies (Colmer and Voesenek, 2009; Loreti and Perata, 2020; Yusuff et al., 2020): the Low Oxygen Quiescence Syndrome (LOQS) and the Low Oxygen Escape Syndrome (LOES). Plants with the LOQS are characterized by traits that (1) allow economical usage of ATP, (2) increase the abundance of enzymes required to produce ATP without oxygen, and (3) enhance cell components that act against harmful cellular changes associated with hypoxia. In plants with the LOQS, shoots do not elongate when flooded, and shoot growth is arrested. Examples are rice genotypes used in rain-fed rice production. By contrast, LOES plants can increase the rate of shoot growth and adjust the growth direction to emerge above the water surface. In addition, they invest more resources into the formation of aerenchyma or other structures that improve internal gas transport, or gas exchange between plants and the hypoxic soil (Colmer and Voesenek, 2009). Extensively studied examples of plants with the LOES syndrome are *Rumex palustris* and deep-water rice genotypes.

The relative benefit of the two strategies (tolerance vs avoidance) depends on various factors, especially the duration of hypoxia (Table 18.4). For short-term oxygen deficit (e.g., after heavy rains), the stress tolerance is required, whereas for long-term oxygen deficit the stress avoidance is also needed. The stress avoidance may start with direct or indirect sensing of low oxygen supply in some plant tissues, followed by low-oxygen signaling within the plant, and a specific pattern of gene expression (Fig. 18.12) (Loreti and Perata, 2020).

Plants have a range of sensors of oxygen concentrations (Bailey-Serres et al., 2012; Licausi and Perata, 2009), for example, various ethylene response transcription factors (Loreti and Perata, 2020) (Fig. 18.12), with the oxygen sensing machinery being influenced by a complex crosstalk of phytohormones (e.g., including jasmonic acid, Shukla et al., 2020).

Hypoxic soils are not characterized only by oxygen deficit but also by many other properties that may be detrimental to plants (see above). For example, to cope with the high concentrations of Fe^{2+} in submerged soils, wetland species may require the mechanisms to minimize Fe uptake (exclusion in oxygenated rhizosphere) or taking it up (includers) and detoxifying the resulting ROS produced in the cells (Fig. 18.13).

Increased SOD activity under hypoxia may be an important protection mechanism in preventing oxidative damage during recovery from hypoxia stress in normoxic conditions (Lamers et al., 2020; Salah et al., 2019). However, some regulatory and adaptive mechanisms are triggered by ROS (Malathi and Pandey, 2019; Waszczak et al., 2018), including hypoxia stress (Fukao et al., 2019). Hence, exhaustive scavenging of ROS during hypoxia could be detrimental because of interfering with plant adaptation to the stress (Licausi and Perata, 2009).

TABLE 18.4 Overview of the hypothesized importance of various biophysical traits associated with plant tolerance to hypoxic soils.

Traits	Waterlogged		Submerged		
	Short	Long	Short	Long/shallow	Long/deep
Adventitious roots (in sediment)	*	***	*	***	*
Adventitious roots (in water)	na	na	na	**	***
Aerenchyma	**	***	**	***	***
Radial O_2 loss barrier	*	***	*	***	*
Anaerobic energy production	***	*	**	*	***
Energy conservation	**	*	***	*	***
Prevention of ROS formation/ROS defense system	***	***	***	***	**
Tolerance to toxic soil constituents	na	***	*	***	***
Nastic movements	*	**	*	***	**
Shoot elongation	na	na	−ve	***	*
Aquatic leaf traits	na	na	na	***	***
Leaf gas films	na	na	***	***	**
Convective gas movement	*	***	na	***	na

*, little importance; **, moderate importance; ***, high importance; −ve, costs outweigh benefits; this trait can decrease fitness in the specific environments; na, not applicable; short duration, <2 weeks; prolonged duration, >2 weeks; shallow, <1 m (i.e., water levels that plants may "outgrow"); deep, >1 m.
Five contrasting types of hypoxic environments inhabited by some terrestrial plant species are shown. *ROS*, Reactive oxygen species.
Source: Based on Colmer and Voesenek (2009).

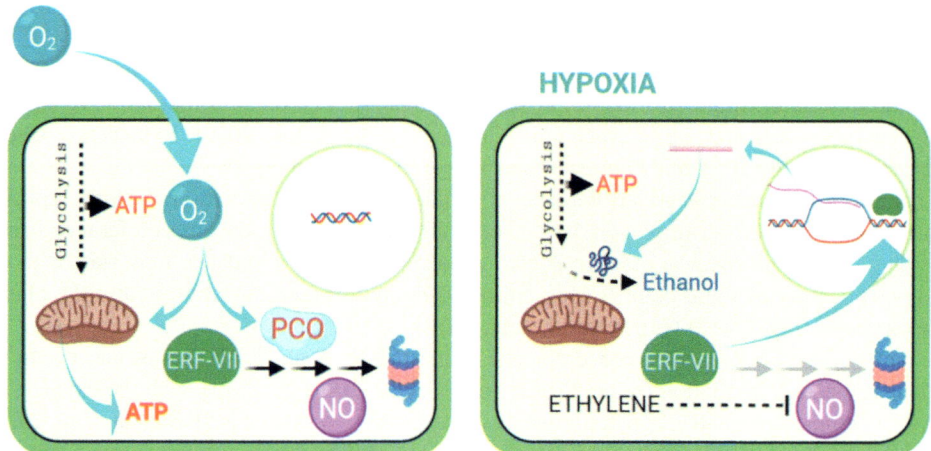

FIGURE 18.12 The mechanisms for sensing oxygen concentration based on NO and the ethylene response factor ERF-VII. Under normoxia, the stability of ERF-VII transcription factor is compromised by PCOs that oxidize the N-terminus Cys residue, channeling the ERF-VII proteins to the proteasome in a process that requires NO. Under hypoxia, the ERF-VII proteins are stabilized by the oxygen absence and the ethylene production that diminishes NO in the cell. The stable ERF-VII proteins migrate to the nucleus where they activate the transcription of hypoxia-responsive genes, including those associated with alcoholic fermentation. *NO*, Nitric oxide; *PCOs*, plant cysteine oxidases. *Reproduced from Loreti and Perata (2020). An open access article under the terms of Creative Commons CC BY license.*

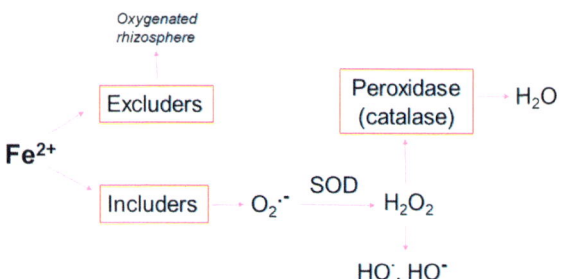

FIGURE 18.13 The two strategies of dealing with excess Fe^{2+} in hypoxic root environments. The uptake of Fe^{2+} results in the generation of reactive oxygen species that can be detoxified via antioxidative enzymes.

In the absence of oxygen, nitrite may become an electron acceptor, yielding nitric oxide (NO). The concentration of NO in oilseed rape roots increases upon exposure to hypoxia because the genes for nitrate reductase and nitrite reductase are upregulated but, to keep the balance, the nonsymbiotic hemoglobin synthesis gene (hemoglobin can decrease NO concentration by converting it back to nitrate) is also overexpressed (Lee et al., 2014). The overexpression of the hemoglobin gene from *Vitreoscilla* (a Gram-negative aerobic bacterium) increases tolerance to hypoxia in many plant species, for example, arabidopsis, maize (Du et al., 2016), and *Populus alba* x *glandulosa* (Li et al., 2020d).

18.4.6.1 Phenotypic adaptation

The possibility to transport oxygen from the shoots to the roots and into the rhizosphere is the basis of most avoidance strategies in response to hypoxia. Oxygen transport takes place to a limited extent in air-filled intercellular spaces; the main pathway, however, is the aerenchyma in the root cortex (Fig. 18.11). Although in many wetland species aerenchyma formation is constitutive, hypoxia enhances aerenchyma formation, with ethylene involved in this effect (Tran Nguyen et al., 2018). Aerenchyma formation in the basal part of the stem connects the root aerenchyma with hypertrophic lenticels on the stem just above the water surface, serving as oxygen entry points in crops (e.g., soybean, Shimamura et al., 2010) as well as various tree species, with hypoxia tolerance positively associated with the area of hypertrophied lenticels (Almeida et al., 2016).

Root porosity (air-filled intercellular spaces as a proportion of total root volume) increases under hypoxia but differs among plant species and genotypes (Malik et al., 2015; Xiao and Jespersen, 2019). Greater formation of aerenchyma and increased root porosity are generally associated with increased tolerance to hypoxia stress (Striker and Colmer, 2017).

Long-distance transport of oxygen in the aerenchyma to the apical zones of roots growing in hypoxic soils requires restriction of oxygen loss by diffusion into the rhizosphere along the transport pathway toward root tips, where released oxygen creates the aerobic rhizosphere conducive to root growth. Hence, in many wetland species, the basal zones of roots have a barrier to radial oxygen loss, either hypoxia-induced, such as in flooded rice (Colmer and Voesenek, 2009), or constitutive (Ejiri et al., 2020).

The barrier to radial oxygen loss is mainly due to suberization of cell walls in the outer layers of the basal part (Ejiri et al., 2021) of both adventitious and lateral roots in many species (Noorrohmah et al., 2020; Pedersen et al., 2020). The root barrier to oxygen loss can be measured by a range of well-established techniques (Jiménez et al., 2021), which can also include measuring restricted radial diffusion of other gases, including water vapor (Ogorek et al., 2021).

Upon exposure to hypoxia, adventitious roots with well-developed aerenchyma emerge from the base of the stem and grow to a limited extent into the hypoxic soil. Tolerance to hypoxia is generally correlated with the adventitious root formation (Herzog et al., 2016; Kim et al., 2019) that is regulated by phytohormones (Tran Nguyen et al., 2018).

18.5 Saline soils

18.5.1 General

Saline soils are abundant in semiarid and arid regions where the amount of rainfall is insufficient for substantial leaching. Salt enters soils mainly via rainfall, irrigation water, and rising groundwater. The water is lost by evaporation or transpiration, and salts may accumulate on the soil surface or within the solum. About 1/3 of the land in arid areas is affected by salinity (Ondrasek and Rengel, 2021). Anthropogenic soil salinization is the result of inappropriate irrigation and drainage practices. Out of the 230 million ha of irrigated agricultural land worldwide, around 45 million ha are salt-affected (Athar and Ashraf, 2009). The use of poor-quality irrigation water is one reason for the increasing salinization of agricultural land. Even good-quality water may contain from 100 to 1000 g salt m^{-3}. With an annual application of 10 000 m^3 ha^{-1}, between 1 and 10 tons of salt are added to the soil. To prevent salinization, the accumulated salts must be removed periodically by leaching and drainage.

Salt tolerance of most crop species is relatively low. Given a growing world population, the strategies to maintain or increase crop production on saline soils are required. Progress in the utilization of genetic variability among plant genotypes for breeding salt-tolerant cultivars has been relatively slow (Jha et al., 2019; Mujeeb-Kazi et al., 2019; Sun et al., 2019; Tarun et al., 2020). Molecular studies have recently shed light on some mechanisms involved in plant salt tolerance, and this may translate into more rapid selection of salt-tolerant genotypes and development of suitable transgenic cultivars (Arzani and Ashraf, 2016; Schroeder et al., 2013; Shah et al., 2018; Sun et al., 2019). Some halophytes have been explored for their potential to be used as crop plants, for example in the production of oil, animal fodder, and even grains (Centofanti and Banuelos, 2019; Esfahan et al., 2010; Glenn et al., 2013). Halophytes as well as salt-tolerant species may also be used in phytoremediation of saline soils (Hayat et al., 2020; Shao et al., 2019; Yue et al., 2020).

18.5.2 Soil characteristics and classification

Salt-affected soils are characterized by high concentrations of soluble salts in the solution phase (saline soils) and/or a considerable fraction of the cation exchange sites being occupied by Na$^+$ (sodic soils). The electrical conductivity of the soil saturation extract (EC$_e$) is commonly used as a measure of soil salinity, that is, the concentration of soluble salts (Fig. 18.14). The saturation extract comprises the soil solution extracted from a soil at its saturation water content. According to the *Glossary of Soil Science Terms* published by the Soil Science Glossary Terms Committee (2008), a soil is considered saline when the EC$_e$ is above 4 dS m^{-1}, which is equivalent to approximately 40 mM NaCl (George et al., 2012); at that EC$_e$ growth of most crops is negatively affected. However, the effects of soil salinity also depend on soil texture, water content, and the type of salts.

The sodium adsorption ratio (SAR), which provides information on the concentration of Na$^+$ in relation to Ca^{2+} and Mg^{2+} in the soil solution in equilibrium with the adsorbed fraction of these ions, can be used to further describe ion relations in salt-affected soils. Plant growth can be affected by high SAR even if the EC$_e$ is below 4 dS m^{-1}. Nonsaline soils with SAR above 13 are termed "sodic soils" (Soil Science Glossary Terms Committee, 2008).

The extent to which the cation exchange complex in a soil is occupied by Na$^+$ is reflected in the exchangeable sodium percentage (ESP = exchangeable Na × 100/CEC). Saline soils with ESP greater than 15% are termed "saline−sodic" soils (Soil Science Glossary Terms Committee, 2008). A high ESP or SAR causes clay minerals and organic matter to disperse; therefore, saline−sodic and sodic soils are often characterized by poor soil structure that can result in high soil density, clogging of pores, and the formation of surface crusts, making soils impermeable to air and water.

In the context of constraints to plant growth, it is necessary to make a clear distinction between salinity and sodicity. Saline soils are not necessarily alkaline, and plant growth on saline soils is affected mainly by high concentrations of

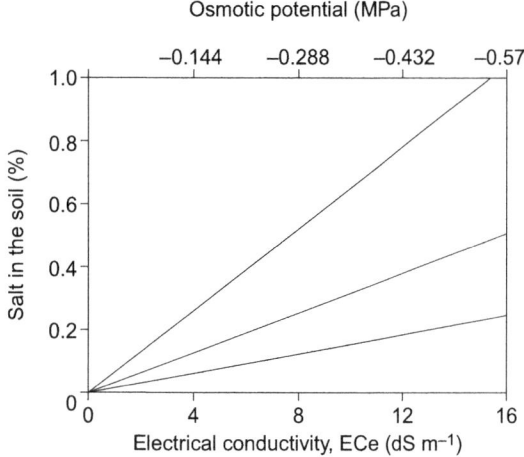

FIGURE 18.14 Relationship between the salt concentration in soil and the electrical conductivity of the extract at 25°C for different amounts of water in the paste. *Based on US Salinity Laboratory Staff (1954).*

NaCl (ion toxicity, ion imbalance) and impairment of water balance (Section 18.5.3.1). By contrast, sodic soils are alkaline, and plant growth is impaired by high pH, high bicarbonate, poor aeration, and frequently Na and B toxicity.

In most saline soils, Na^+, Ca^{2+}, and Mg^{2+} are the main cations. The most abundant anions are Cl^-, SO_4^{2-}, HCO_3^-/CO_3^{2-}, and NO_3^- (George et al., 2012). All salts with solubility greater than that of gypsum (15 mmol L^{-1}) contribute to the osmotic potential of the soil solution. The osmotic potential of the saturation extract can be calculated from the EC_e:

$$\text{Osmotic potential (MPa)} = -0.036 \times EC_e$$

Measurement of EC_e is an easy and commonly used tool to characterize the salt concentration in soil, nutrient solution, or irrigation water. The salt concentration in the soil solution at field capacity is about twice that of the saturation extract and correspondingly higher when the soil water content declines below field capacity. It should also be borne in mind that salts accumulate in the rhizosphere of salt-tolerant plants when their transport towards the root surface via mass flow exceeds plant uptake (Zhang et al., 2005). Thus, measurement of the EC_e alone is not sufficient to assess the effect of a saline soil on plant growth. This is not only because it may underestimate the actual salinity level the plant is exposed to but also because it does not provide any information on the identity of the ions present in the soil solution.

Plants growing on saline soils are often exposed to additional environmental stresses, such as shallow groundwater tables or excessive concentrations of available B. The sensitivity of many plant species to soil salinity increases significantly upon exposure of their roots to hypoxia (Barrett-Lennard, 2003; Barrett-Lennard and Shabala, 2013). Soil salinity and excess B may (Karimi and Tavallali, 2017) or may not (Tripler et al., 2007; Yermiyahu et al., 2008) have an additive effect in hampering the growth depending on the plant species.

18.5.3 Salinity and plant growth

18.5.3.1 Major constraints

There are three major constraints for plant growth on saline substrates (Fig. 18.15): (1) water deficit ("osmotic stress") arising from the low (strongly negative) water potential of the rooting medium; (2) ion toxicity resulting from excessive uptake mainly of Cl^- and/or Na^+; and (3) nutrient imbalance by decreased uptake and/or shoot transport and impaired internal distribution of nutrients.

It is often not possible to assess the relative contribution of these three major salinity-related constraints to growth inhibition because of differences among individual plant organs, and dependence on the duration of the exposure to salinity, the plant developmental stage, genotype, and various environmental parameters.

18.5.3.2 Water deficit

Plant water availability is determined by the water potential of the soil, which is the sum of matric and osmotic potentials, in relation to water potential of the root tissues. As soil salinity increases, plants must overcome an increasing gradient in water potential between the soil (more negative) and their roots. For most plants, the threshold value for sufficient water extraction lies at a soil water potential of around −1500 kPa. The water content at which this critical value is reached is lower for saline soils because of the greater contribution of osmotic potential to water potential

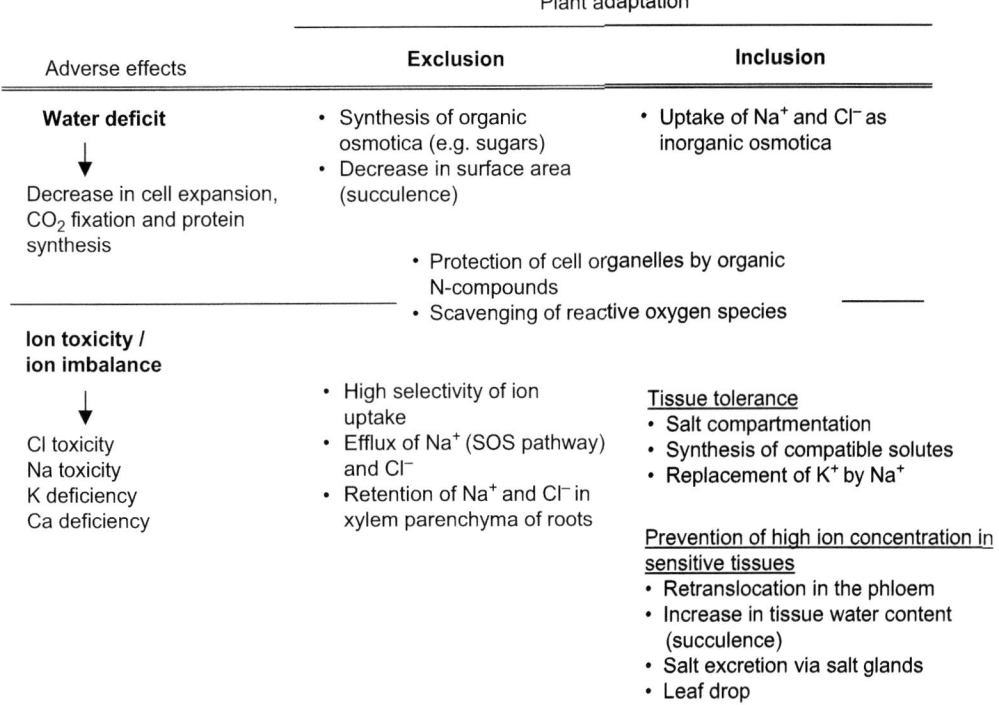

FIGURE 18.15 Adverse effects of salinity and mechanisms of adaptation. *Modified after Greenway and Munns (1980).*

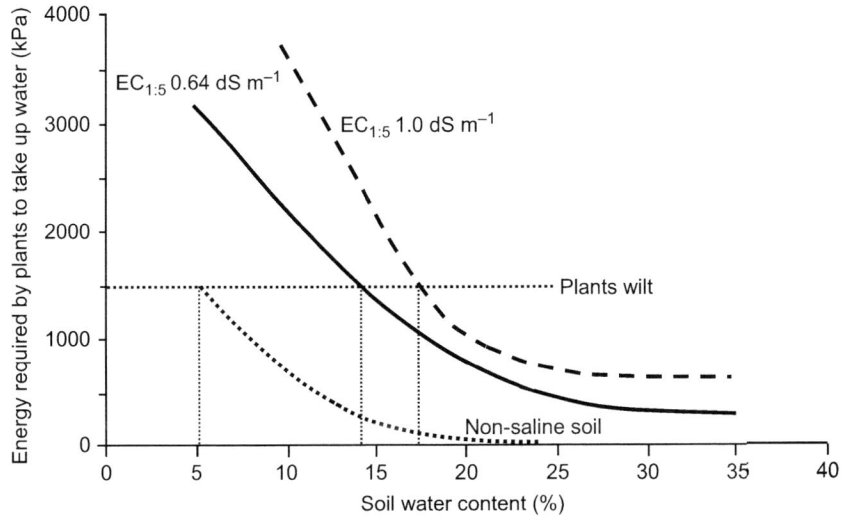

FIGURE 18.16 Energy required by plants to take up water (equivalent to soil matric + osmotic potentials) from a loamy soil at different $EC_{1:5}$ (EC measured in a 1:5 water extract) and soil water content. *Redrawn from Rengasamy (2006).*

(Fig. 18.16). Moreover, at a given EC_e, the water potential decreases with decreasing soil water content (i.e., matric potential becoming more negative).

A rapid decrease in leaf and root expansion is observed in many plant species upon sudden exposure of roots to salinity (Neumann, 1993), and development of high-throughput 3D modeling to assess early responses of leaf elongation to salinity is an important methodological development (Ward et al., 2019). A decrease in leaf expansion upon exposure to salinity is dependent on the genotypic tolerance to salinity; for example, salt-tolerant genotypes sustain leaf growth under salinity due to maintaining apoplasmic pH suitable for acid growth and a relatively high content of expansin proteins (e.g., in maize, Zorb et al., 2015) as well as higher expression of the genes coding for expansin proteins and xyloglucan endotransglucosylase (e.g., in tall fescue, Xu et al., 2016).

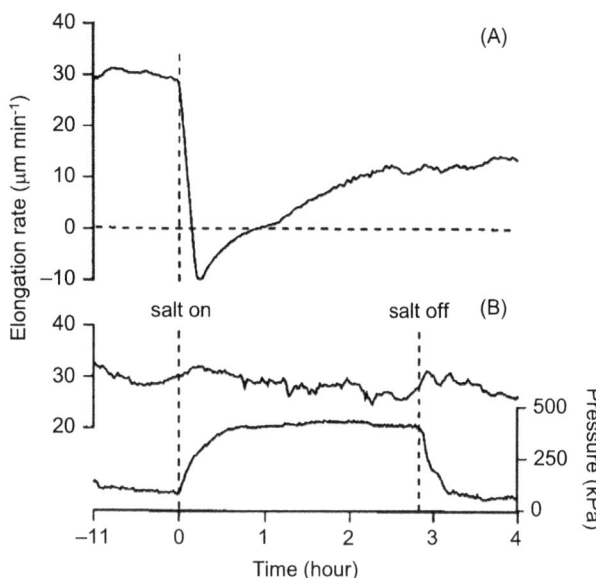

FIGURE 18.17 Effect of salinity (80 mM NaCl) on the leaf elongation rate of maize seedlings (A) without balancing pressure and (B) with balancing pressure to maintain leaf water status. The vertical dashed lines indicate times at which the roots were flushed with saline or nonsaline solution. *From Munns et al. (2000), with permission from Oxford University Press.*

The rapid responses in leaf elongation rate to substrate salinity are mainly due to changes in the leaf water status. Upon removal of root-zone salinity within a few hours after the onset of salinity, leaf extension rate promptly reverts to the presalinity level, suggesting that water deficit (rather than salt toxicity) is the main reason for growth reduction due to root-zone salinity. Following an initial strong decrease upon sudden exposure to salinity, leaf expansion rates recover gradually, and reach a new steady state (Fig. 18.17).

When the leaf water status of plants exposed to salinity is maintained by a pressurization technique, only the early growth reduction in response to salinity is prevented (Fig. 18.17). This suggests that, after a certain duration of exposure to salinity, factors other than the cell turgor govern tissue expansion. Phytohormonal signals similar to those in roots (e.g., ABA) in response to a decreased influx of water at low soil water content may be an explanation (Munns, 2002).

18.5.3.3 Sodium and chloride uptake and toxicity in plants

In saline substrates, Na^+ and Cl^- are usually the dominant ions. Despite the essentiality of Cl as a micronutrient for all higher plants and of Na as nutrient for many halophytes and some C_4 species (see also Sections 7.8 and 8.2), the concentrations of both ions in saline substrates are far in excess. Sodium and Cl^- are toxic to plants when accumulated in the cytoplasm at high concentrations. They may, for example, displace other ions from the enzyme binding sites, and thus impair cellular functions.

In salt-sensitive plants, growth inhibition and injury to the foliage (chlorosis and necrosis on the margins of mature leaves) occur even at low concentrations of NaCl, with water deficit not limiting growth; instead, high Cl sensitivity and thus Cl toxicity impair plant growth (Geilfus, 2018; Hanh Duy and Hirai, 2018). Many plant species (arabidopsis, rice, *Nicotiana benthamiana*) are hypersensitive to salinity and accumulate more Cl when fed ammonium compared with nitrate; this Cl accumulation is facilitated by upregulation of the NRT1.1-mediated Cl uptake in ammonium-fed plants due to the absence of competition between nitrate and Cl for the active sites on the NRT1.1 nitrate transporter (Liu et al., 2020d). In many plants, particularly legumes such as tree medics (Sibole et al., 2003) and soybean (Cox et al., 2018), and many fruit trees such as citrange (Ruiz et al., 2016), there is a positive correlation between the capacity to exclude Cl from the shoot and root tissues, and growth under salinity. This suggests that Cl toxicity is the major limitation for these plants grown in saline substrates.

The chloride channel (CLC) family comprises both channels and exchangers (Cl^-/H^+) (Nedelyaeva et al., 2019) and is present in all domains of life; some members of the family facilitate nitrate/H^+ exchange as well (Subba et al., 2021). In plants, the CLC family members have highly differential expression in various tissues and are localized in diverse cellular organelles, such as Golgi vesicles, thylakoids, mitochondria, tonoplast, and other membranes, and function in nutrient uptake, maintenance of cell turgor, cellular action potential, cytosol pH control, stomata movements, signal transduction, salt and drought stress resistance, etc. (Angeli et al., 2013; Guo et al., 2014; Herdean et al., 2016; Um et al., 2018; Wei et al., 2019; Zhang et al., 2017a). A common way of the CLC family members to alleviate salt stress is by sequestration of Cl⁻ in the root-cell vacuoles, thus minimizing Cl transport into sensitive shoot tissues (e.g., Chi Tam et al., 2016; Wei et al., 2016, 2019), even though enhanced activation in leaves of halophyte *Suaeda altissima* (L.) Pall was recorded upon exposure to salinity (Nedelyaeva et al., 2019). Chloride can also be transported by other anion transporters such as DTX/MATE, functioning in Cl influx across the tonoplast into the vacuoles of various cells (from root hairs to stomatal guard cells) (Zhang et al., 2017a).

Salinity may decrease the uptake of anions due to competition with Cl⁻ for the uptake sites because chloride channels also transport nitrate. However, different members of the chloride channel family may be specific for either chloride or nitrate, for example, in halophyte *Suaeda altissima* (Nedelyaeva et al., 2019). In maize, the uptake of nitrate was decreased by Cl⁻ (Hessini et al., 2019). However, in tobacco, Cl⁻ had a small or moderate effect on nitrate uptake and a significant positive effect on nitrate utilization efficiency (Rosales et al., 2020). In summary, interference of Cl⁻ with nitrate uptake and metabolism is unlikely to be an important factor in growth depression caused by salinity.

Because K⁺ and Na⁺ have similar ionic properties, they may enter the cytoplasm through the same ion channels (Zhang et al., 2009). Moreover, the high-affinity Na uptake may occur via the high-affinity K transporters (HKT) in most land plants (Haro et al., 2009). Expressed in the endomembranes, HKT may facilitate sequestration of Na and thus enhance salt tolerance (Wu et al., 2020).

The net uptake of Cl and Na into the cells depends not only on the respective influxes but also on the efflux rates. Under salinity, influx and efflux of Na and Cl occur simultaneously. The energy requirements for these "futile cycles" are most likely similar, with energy spent mainly for efflux of Na and influx of Cl (Teakle and Tyerman, 2010).

The efflux of Na⁺ from the cytoplasm into the apoplasm is mediated by the plasma membrane Na⁺/H⁺ antiporter SOS1 (salt overly sensitive 1) (e.g., Wang et al., 2019e) and from the cytosol into the vacuole by the tonoplast Na⁺/H⁺ antiporter NHX1 (e.g., Fukuda et al., 1999; Wang et al., 2020b) (Fig. 18.18). The Na⁺ efflux ensures the maintenance of relatively low cytosolic concentration (e.g., El-Mahi et al., 2019).

Long-distance distribution of Na and Cl within the shoot is mainly a function of the transpiration stream. Thus, toxicity symptoms specific to Na and Cl usually become visible as necrosis in older leaves, starting from the leaf tip and the margins. When salinity-induced leaf loss is considerable, the decreased photosynthetic capacity of the plant may contribute to growth depression.

18.5.3.4 Ion imbalances

Sodium toxicity is based mainly on its competition with K. Therefore, the low cytosolic K/Na ratio, rather than the Na concentration alone, causes the deleterious effects associated with elevated plant Na uptake.

Potassium homeostasis can be achieved by high selectivity of the K uptake systems or efflux of Na from the cytoplasm into vacuoles or apoplasm. After only 1 h of exposure to salinity, considerable efflux of K from roots may occur

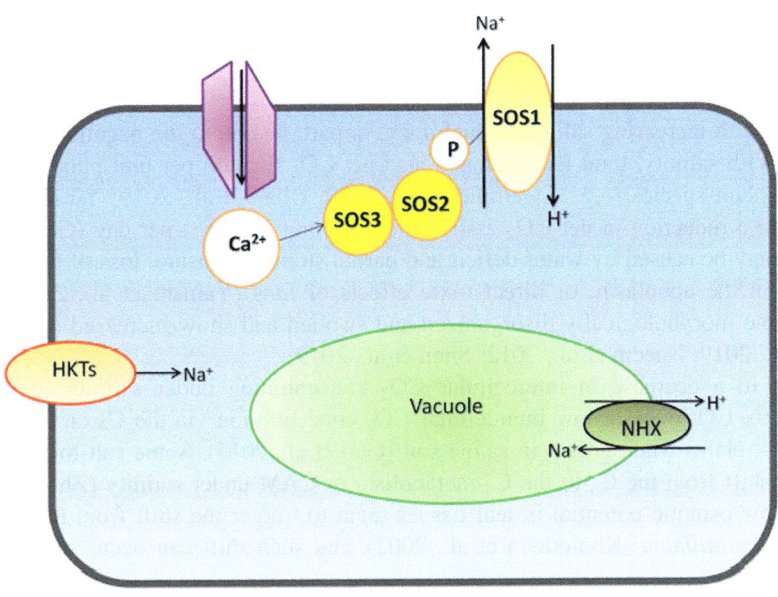

FIGURE 18.18 Na⁺ fluxes across the PM and tonoplast. For simplicity, some related transporters were omitted (e.g., H⁺-ATPase, nonspecific cation channel in the PM and ATPases/PPases in the tonoplast, etc.) (shown in Fig. 18.24). *HKT*, High-affinity potassium transporter; *NHX1*, Na⁺/H⁺ exchanger; *PM*, plasma membrane; *SOS1*, Na⁺/H⁺ antiporter; *SOS2*, serine/threonine protein kinase; *SOS3*, calcium-binding protein. *From Tarun et al. (2020), with permission from Elsevier.*

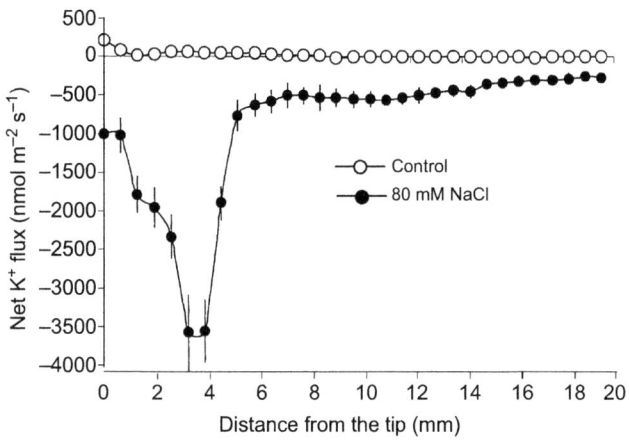

FIGURE 18.19 K$^+$ flux profiles along the root axis of barley seedlings without or with 80 mM NaCl. Positive and negative values indicate influx and efflux, respectively. *From Chen et al. (2005), with permission from John Wiley and Sons.*

(Fig. 18.19), likely due to decreased membrane integrity under salt stress. A strong negative correlation between K efflux upon salt exposure and salinity resistance was found in wheat (Chen et al., 2005) and barley (Wu et al., 2015).

Calcium appears to be involved in the maintenance of the K/Na homeostasis. The conductance of nonselective cation channels that may be involved in Na entry into the cells is inhibited strongly by Ca (Demidchik et al., 2002). Cyclic nucleotide-gated channels (CNGCs), as nonselective cation channels transporting not just Na$^+$ and K$^+$ (Guo et al., 2008; Mori et al., 2018) but also Ca^{2+} (Guo et al., 2010), are involved in the crosstalk between Ca and Na signaling (Koster et al., 2019), with compartmentation of Na in vacuoles being dependent on Ca accumulation and the expression of the Ca-signaling genes in root cells (Liu et al., 2019).

In its function as a secondary messenger (see also Section 6.5), Ca is involved in the salinity perception and induction of physiological responses (Kader and Lindberg, 2010). Plant species, cell types, and even cell organelles appear to differ in their specific Ca signals upon perception of salinity. Different Ca^{2+} signals are induced by osmotic stress and elevated concentrations of Na or Cl within and outside the cells (Tracy et al., 2008). The salinity-induced Ca signal is perceived by the Ca-sensor SOS3 (Fig. 18.18), which then interacts with the protein kinase SOS2 to activate the Na$^+$/H$^+$ antiporter SOS1 (Koster et al., 2019; Mahajan et al., 2008). The perception of the initial salinity signal is based on the Na-bound glycosyl inositol phosphorylceramide sphingolipids at the external surface of the plant cell plasma membrane activating a Ca^{2+} influx channel to increase cytosolic Ca^{2+} activity (Jiang et al., 2019) that is then sensed by the SOS pathway, resulting in the activation of SOS1.

At high Na concentrations, increasing Ca supply can strongly enhance growth (e.g., Al-Khateeb, 2006; Tan et al., 2019) and prevent a Na-induced decrease in Ca tissue accumulation (Bolat et al., 2006). Increasing the ratio of plant-available Ca/Na in the soil can promote plant growth (Bolat et al., 2006) and tissue Na/K homeostasis (Table 18.5).

18.5.3.5 Photosynthesis and respiration

Water loss per plant by transpiration decreases with increasing salinity, which may, in part, be due to the negative relationship between salinity level and leaf area. With salinity, total leaf area and also net CO$_2$ fixation per unit photosynthetic tissue may decline in the salt-sensitive plant species (e.g., Essemine et al., 2020; Ullah et al., 2019; Yu et al., 2020), whereas respiration increases, leading to a reduction in net CO$_2$ assimilation per unit leaf area per day (George et al., 2012). Lower rates of net CO$_2$ fixation may be caused by water deficit and partial stomatal closure, loss of turgor of mesophyll cells due to salt accumulation in the apoplasm, or direct toxic effects of ions (Yamshi et al., 2020). Chloroplasts of salt-stressed plants often become morphologically disorganized and swollen and show increased starch accumulation (Goussi et al., 2018; Kotula et al., 2019; Naeem et al., 2012; Shen et al., 2019).

Low rates of photosynthesis may be due to a decrease in intercellular CO$_2$ concentration under salinity (e.g., Benedict et al., 2020). Hence, plants that can fix CO$_2$ even at low intercellular CO$_2$ concentration via the C$_4$ or CAM pathway often have higher growth rates than C$_3$ plants when grown in saline soil (Guo et al., 2015). Some salt-tolerant plants, for example, *Atriplex lentiformis*, may shift from the C$_3$ to the C$_4$ metabolism or CAM under salinity (Zhu and Meinzer, 1999). Low CO$_2$ concentration and low osmotic potential in leaf tissues seem to trigger the shift from the C$_3$ to CAM in salt-exposed *Mesembryanthemum crystallinum* (Kholodova et al., 2002), and such shift can occur specifically in the guard cells (Kong et al., 2020).

TABLE 18.5 Cation concentrations and electrolyte leakage in leaves of plum grown on three rootstocks (Marianna GF 8-1, Myrobolan B, and Pixi) differing in salinity tolerance and grown in a sandy substrate supplied with nutrient solution containing NaCl without or with CaSO$_4$.

Treatment	Leaf element concentration (mmol kg^{-1} dw)				Electrolyte leakage (%)
	Na	K	Na:K	Ca	
Marianna GF 8–1					
Control	34	605	0.6	242	20
40 mM NaCl	1969	433	4.6	154	52
40 mM NaCl + 5 mM CaSO$_4$	1243	530	2.4	187	42
Myrobolan B					
Control	46	623	0.08	330	17
40 mM NaCl	1559	492	3.2	250	50
40 mM NaCl + 5 mM CaSO$_4$	1176	564	2.1	295	27
Pixi					
Control	24	564	0.04	380	21
40 mM NaCl	407	545	0.75	345	25
40 mM NaCl + 5 mM CaSO$_4$	200	516	0.39	397	16

Source: From Bolat et al. (2006).

Moderate salinity increases the carbohydrate requirement for maintenance of respiration, which is most likely due to the energy costs of the compartmentation of ions, ion secretion (e.g., Na efflux transporters), or the repair of cellular damage (Kotula et al., 2015; Malagoli et al., 2008; Munns and Gilliham, 2015; Otgonsuren et al., 2016). However, when salinity levels exceed a certain threshold, root respiration may also decrease because ion toxicity impairs cell metabolism (Epron et al., 1999).

18.5.3.6 Protein biosynthesis

Salinity impairs protein biosynthesis (e.g., Hajiboland et al., 2014), and the effects may be measurable within 3 h of exposure to salinity (Glenn et al., 2017). The effects of NaCl salinity on protein synthesis may be due to Cl$^-$ toxicity or Na/K imbalance in leaves. The synthesis of amino acids does not appear to be a major limitation to protein synthesis under salinity. Instead, protein synthesis may be reduced in favor of the accumulation of many amino acids and other N-containing organic compounds (e.g., glycine betaine or proline) that are involved in osmotic adjustment (Annunziata et al., 2019; Nie et al., 2020) (see Section 18.5.4.3) and protection of enzymes or detoxification of oxygen radicals (Hildebrandt, 2018). It has also been suggested that Na$^+$ in the cytoplasm impairs ribosomal attachment to rRNA by competing with K$^+$ for the binding sites (Tester and Davenport, 2003).

18.5.3.7 Phytohormones

In response to salinity, the cytokinin and auxin concentrations decrease, whereas those of ABA and ethylene increase (Fig. 18.20) in a similar way as under drought stress. The high ABA concentrations are important for rapid osmotic adjustment to salinity (Kanchan et al., 2017), regulating Na retrieval from the xylem to decrease Na transport to shoots (Zhu et al., 2017). Salt exposure increases the biosynthesis of proteins associated with ABA signaling, with jasmonic acid acting as a positive regulator of ABA biosynthesis (Ahmad et al., 2019). Application of ABA may therefore increase salt tolerance by enhancing the mechanisms of rapid adaptation to salinity (Fig. 18.21), for example, by increasing the expression of various genes involved in the PEP carboxylate biosynthesis (Muhammad and Fiaz, 2019), which may increase CO$_2$ fixation rate, or by hampering the catabolism of PEP carboxylase (Monreal et al., 2007). Pretreatment with ABA prior to exposure to elevated NaCl may improve tolerance of plants to salt stress, probably indirectly via upregulating various resistance-related pathways (e.g., Palma et al., 2014; Ren et al., 2018).

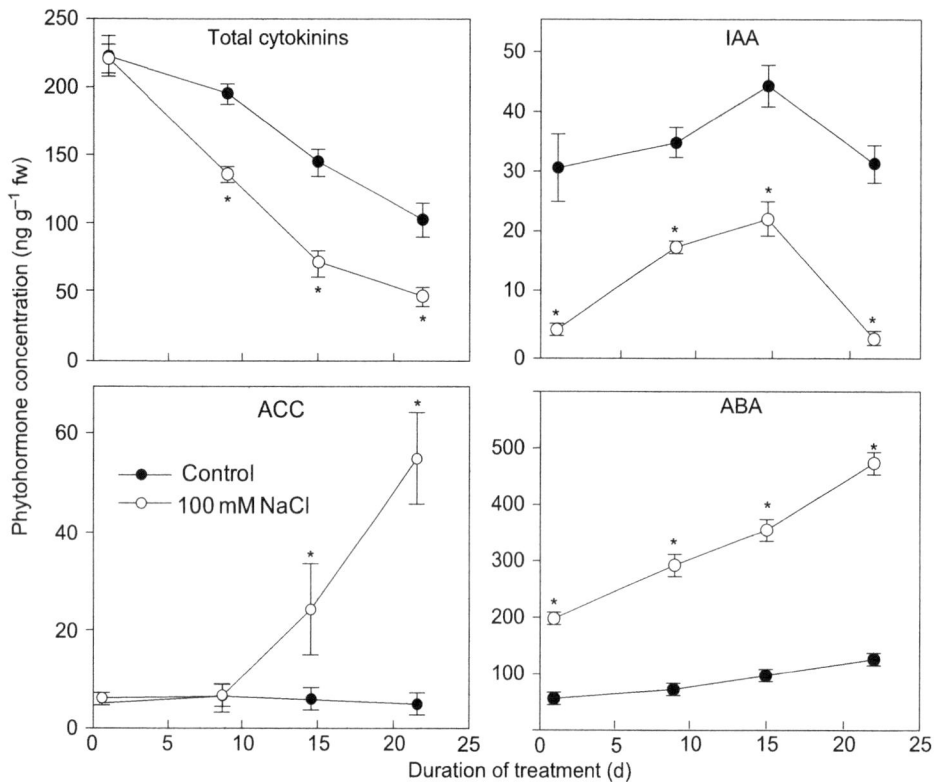

FIGURE 18.20 Phytohormone concentrations in young expanding leaves of tomato plants grown in a nutrient solution with or without 100 mM NaCl. Total cytokinins: zeatin + zeatin riboside; IAA: indole-3-acetic acid (auxin); ACC: 1-aminocyclopropane-1-carboxylic acid (ethylene precursor); ABA: abscisic acid. The asterisks indicate significant differences between the control and saline treatment. Means ± SE (n = 3). *Based on Ghanem et al. (2008).*

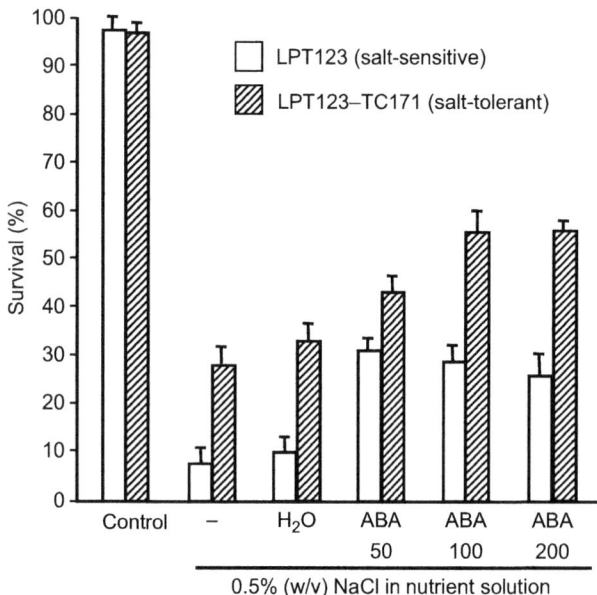

FIGURE 18.21 Survival rate of plants of a salt-sensitive and a salt-tolerant rice genotype after 4 weeks in either nonsaline (control) or saline growth substrate. The salt-stressed plants remained either nontreated (−) or were sprayed daily with water (H$_2$O) or 50, 100, or 200 μM ABA solution. Means + standard deviation (n = 4). *Redrawn from Sripinyowanich et al. (2013).*

The polyamine growth regulators putrescine or spermidine accumulate in plants under salinity stress. They may stabilize plant cell membranes and enhance protein synthesis. They may also play an important role in ion homeostasis under salinity by blocking the nonselective cation channels (Shabala et al., 2007). The overexpression of transglutaminases that protect polyamine metabolism resulted in increased polyamine concentration and enhanced tolerance of transgenic tobacco to salt stress (Zhong et al., 2020). Similarly, the application of polyamines can increase plant growth under salinity (Todorova et al., 2015). Exogenously added ABA improves salt resistance by enhancing the biosynthesis of spermine and spermidine in alfalfa (*Medicago sativa*) (Palma et al., 2014) and apple (*Malus domestica*) (Sales et al., 2018).

The precise action of endogenous phytohormone concentrations on salt tolerance mechanisms is highly complex and not yet fully understood. Moreover, the observed effects may depend not only on the temporal dynamics of tissue-specific activity of exact combinations of phytohormones but also on the presence (abundance) of corresponding receptors.

18.5.4 Mechanisms of adaptation to saline substrates

Plant species differ in salt tolerance (Fig. 18.22, Tables 18.6 and 18.7), even though the putative classification is dependent not just on the conditions imposed but also on the choice of genotypes tested, given large genotypic differences in salt tolerance.

18.5.4.1 Salt exclusion versus salt inclusion

Salt tolerance can be achieved by salt exclusion or salt inclusion (Fig. 18.15). Adaptation by salt exclusion requires mechanisms for avoidance of an internal water deficit. Adaptation by salt inclusion requires high tissue tolerance to Na^+ and Cl^-. A clear distinction is often made between salt-excluding and salt-including plant species; however, there is a continuum of different degrees of exclusion and inclusion, differing between Na^+ and Cl^-, and among parts and organs of plants.

Terrestrial halophytes (around 1% of all land plants) belong mainly to the Chenopodiaceae and Poaceae families. In Chenopodiaceae halophytes, high salt tolerance is based mainly on the inclusion of salts and their utilization for turgor maintenance or for replacement of K by Na in various metabolic functions. Across 32 species of Chenopodiaceae, Na and Cl accounted for around 70% of the solute concentration in the shoot water (Albert et al., 2000). Nevertheless, even within one species (*Atriplex halimus*; Chenopodiaceae), different ecotypes may be either includers or excluders (Belkheiri and Mulas, 2013). Among Poaceae, the highly salt-tolerant kallar grass (*Leptochloa fusca*) is a salt includer, although it also shows some properties of excluders (Gorham, 1987). Even within a single (homogeneous) population of sorghum, both includers and excluders were identified during plant development due to variable sensitivity to gibberellic acid and cytokinin (Amzallag, 2001). Among various *Vigna* species, there are both includers and excluders (Yoshida et al., 2020). The capacity to regulate influx as well as efflux of Na and Cl is crucial for all halophytes (Hasegawa et al., 2000).

In glycophytes, which include most crop species, there is generally a negative relationship between salt uptake and salt tolerance, meaning exclusion is the predominant strategy (Hasegawa et al., 2000; Kotula et al., 2019; Shabala and Munns, 2017; Yamshi et al., 2020).

Differences among crop species in response to NaCl salinity in terms of growth and the element concentration in shoots are shown in Table 18.7. Sugar beet shows the typical features of a salt-tolerant halophytic includer. Growth is enhanced by NaCl, and the concentrations of Cl and especially Na in the shoot increase with increasing external supply. Maize is less salt-tolerant than sugar beet; its growth is inhibited despite relatively low shoot concentrations of Cl and

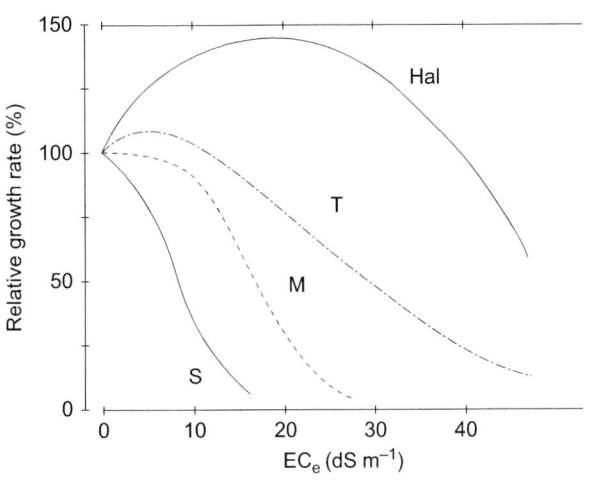

FIGURE 18.22 Typical growth responses of plant species with differential salt tolerance to increasing soil salinity. *Hal*, Halophytes; *M*, moderately salt-tolerant crop species; *S*, salt-sensitive crop species; *T*, salt-tolerant crop species. *Modified after Greenway and Munns (1980).*

TABLE 18.6 Tolerance of crop species to soil salinity.

Crop species	EC$_e$ Threshold (dS m^{-1})	EC$_e$ Slope (% per dS m^{-1})	EC$_e$ Range (dS m^{-1})	EC (dS m^{-1})	Rating
Bean (*Phaseolus vulgaris*)	1.0	19.0			S
Carrot (*Daucus carota*)	1.0	14.0			
Apricot (*Prunus armeniaca*)	1.6	24.0			
Grapevine (*Vitis* sp.)	1.5	9.6			MS
Maize (*Zea mays*)	1.7	12.0			
Tomato (*Solanum lycopersicum*)	2.5	9.9			
Oats (*Avena sativa*)				2–4	
Narrow-leaf lupin (*Lupinus angustifolius*)				2–4	
Canola (*Brassica napus*)				2–4	
Couch grass (*Cynodon dactylon*)				2–4	
Kikuyu grass (*Pennisetum clandestinum*)				2–4	
Alfalfa (*Medicago sativa*)				2–4	
Phalaris (*Phalaris aquatica*)				2–4	
Berseem clover (*Trifolium alexandrinum*)				2–4	
White clover (*Trifolium repens*)				2–4	
Soybean (*Glycine max*)	5.0	20			MT
Perennial ryegrass (*Lolium perenne*)	5.6	7.8		4–8	
Wheat (*Triticum aestivum*)	6.0	7.1		4	
Sorgum (*Sorghum bicolor*)				4–8	
Balansa clover (*Trifolium michelinum*)				4–8	
Persian cover (*Trifolium resupinatum*)				4–8	
Tall fescue (*Festuca arundinacea*)				4–8	
Date palm (*Phoenix dactylifera*)	4.0	3.6	8–16	Up to 12.8	T
Sugar beet (*Beta vulgaris*)	7.0	5.9			
Barley (*Hordeum vulgare*)	8.0	5.0		4–8	
Rhodes grass (*Chloris gayana*)				5–14	

Threshold EC$_e$ (saturation extract) = maximum soil salinity that does not reduce yield; slope = yield reduction per unit increase in EC$_e$ beyond the threshold; rating = classification of the plant as either sensitive (S), moderately sensitive (MS), moderately tolerant (MT) or tolerant (T) to soil salinity. Tolerance of various genotypes within a species may vary significantly.
Source: Based on Maas and Hoffman (1977), Barrett-Lennard et al. (2013), Yaish and Kumar (2015), and Bicknell and Simons (2020).

especially Na. Of the three species in Table 18.7, bean (*Phaseolus vulgaris*) has the lowest salt tolerance, with Cl$^-$ toxicity the main reason for growth depression at the low salinity, given the restricted shoot transport of Na. Many other salt-sensitive crop species are also effective excluders of Na but not Cl$^-$; however, there are notable examples to the contrary, with cucumber (*Cucumis sativus*) excluding Cl$^-$ but not Na (Chen et al., 2020b), and salt tolerance of grapevine rootstocks being dependent on effective Cl$^-$ exclusion (Henderson et al., 2014). The major processes that contribute (either positively or negatively) to exclusion of Cl$^-$ from the cytosol of leaf cells and thus to Cl$^-$ tolerance are detailed in Fig. 18.23 (Li et al., 2017a).

TABLE 18.7 Dry weight and shoot concentrations of Na, Cl, K, and Ca in sugar beet, maize, and bean grown at different NaCl concentrations.

Species	Treatment (mM NaCl)	Relative shoot dry weight (%)	Shoot concentration (mmol g^{-1} dw)			
			Na	Cl	K	Ca
Sugar beet (*Beta vulgaris*)	0	100	0.07	0.04	3.3	0.8
	25	108	1.7	1.0	2.2	0.24
	50	115	2.1	1.2	2.0	0.20
	100	101	2.6	1.5	1.9	0.16
Maize (*Zea mays*)	0	100	0.0	0.0	1.6	0.25
	25	90	0.2	0.5	1.8	0.14
	50	70	0.2	0.6	2.0	0.13
	100	62	0.3	0.8	2.0	0.12
Bean (*Phaseolus vulgaris*)	0	100	0.0	0.0	1.7	1.4
	25	64	0.0	1.0	2.2	1.8
	50	47	0.2	1.4	1.9	1.7
	100	37	0.4	1.5	2.2	1.8

Source: From Lessani and Marschner (1978).

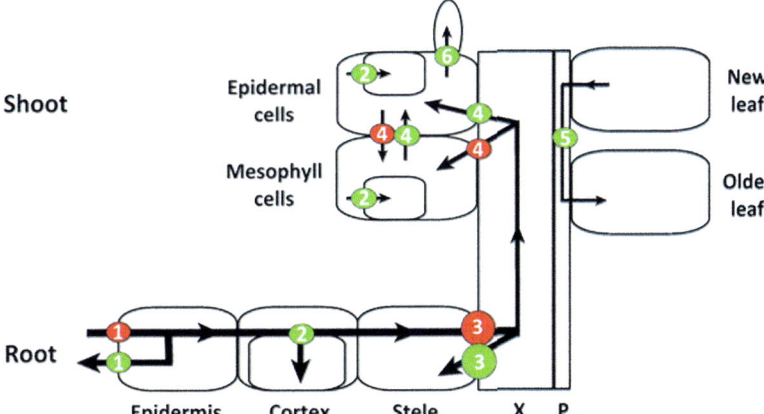

FIGURE 18.23 Mechanisms contributing to Cl$^-$ exclusion from the leaf cytosol leading to Cl$^-$ tolerance. Green circles: processes that improve Cl$^-$ tolerance. Red circles: processes that require inhibition to reduce leaf cytosolic Cl$^-$ load. In the root: (1) net uptake across the root epidermis and cortex is minimized by increasing Cl$^-$ efflux and decreasing its influx. (2) Maximizing intracellular compartmentation in vacuoles to reduce cytoplasmic Cl$^-$. (3) Minimizing net xylem loading. In the shoot: (4) compartmentalizing Cl$^-$ within leaf epidermis to protect the mesophyll cells where photosynthesis occurs. (5) Maximizing phloem translocation from the newly expanded leaves to older leaves. (6) Salt glands and bladders that store or excrete Cl$^-$ out in halophytes. *P*, Phloem; *X*, xylem. *From Li et al. (2017a), with permission from Elsevier.*

Differences in the capacity to exclude Na and Cl$^-$ also exist among cultivars within species (Table 18.8). For example, the higher salt tolerance of specific cultivars of wheat (Munns and James, 2003; Saddiq et al., 2018), barley (Shavrukov et al., 2013), and citrus (Ruiz et al., 2016) is related to increased restriction of shoot transport of Na and/or Cl$^-$. Interestingly, among the pistachio rootstock studied, there were Na or Cl$^-$ excluders, and both types of excluders were more salt-tolerant than the nonexcluder rootstocks (Akbari et al., 2018). In wheat, two gene loci (*Nax1* and *Nax2*) confer salinity tolerance by encoding Na$^+$ transporters of the HKT gene family (Huang et al., 2006). They are expressed in the xylem parenchyma and retrieve Na from the xylem sap of the root (Nax1 and Nax2) and the leaf sheath

TABLE 18.8 Concentrations of Na$^+$ and Cl$^-$ in vacuoles of the epidermis and the first mesophyll layer in the leaves of two barley cultivars after 1-day exposure to 100 mM NaCl.

Cultivar	Organ	Tissue	Concentration in vacuoles (mM)	
			Na$^+$	Cl$^-$
California Mariout (salt-tolerant)	Blade	Epidermis	35	110
		First mesophyll cell layer	42	4
	Sheath	Epidermis	134	204
		First mesophyll cell layer	72	223
Clipper (salt-sensitive)	Blade	Epidermis	41	170
		First mesophyll cell layer	58	44
	Sheath	Epidermis	171	238
		First mesophyll cell layer	157	191

Source: From Huang and Van Stevenick (1989).

(Nax2), thus reducing the amount of Na entering the shoot and the leaf blades (James et al., 2011). Nevertheless, wheat genetic material has been developed that shows all three types of tolerance (tissue ion tolerance, osmotic stress tolerance, and Na$^+$ exclusion) (Mujeeb-Kazi et al., 2019). Similarly to wheat, the barley HKT1 transporter is associated with salt tolerance by regulating K and Na homeostasis and decreasing shoot Na concentration (Han et al., 2018). The difference in shoot accumulation of Na was due to differential affinity of HKT1;2 for Na$^+$ (lower affinity in salt-tolerant (Na includer) tomato *Solanum pennellii* compared with salt-sensitive *Solanum lycopersicum*, resulting in higher Na concentration in xylem sap, shoots, and leaves in the former species) (Almeida et al., 2014).

Among grapevine cultivars differing in the capacity to exclude Cl$^-$ from shoots, root uptake of Cl$^-$ was similar but transfer of Cl$^-$ into the xylem was lower in the efficient excluders (Tregeagle et al., 2010). Similarly, in citrange seedlings, differences in morphology and histology of roots influencing hydraulic conductance (e.g., higher suberin deposition in cell walls) were associated with lower Cl concentration in leaves (Ruiz et al., 2016). Chloride recirculation (from leaves to roots) in phloem was associated with increased salt tolerance in pistachio (Godfrey et al., 2019). The cation-Cl$^-$ cotransporters have been proposed to function in Cl$^-$ retrieval from the xylem, for example, in arabidopsis and rice (Colmenero-Flores et al., 2007; Kong et al., 2011).

Retranslocation of Na from shoots to roots via phloem can contribute to low Na concentrations in the shoots of salt-sensitive species (Wu, 2018). In particular, the HKT1 transporter mediates Na loading into phloem in shoots and unloading in roots (Berthomieu et al., 2003). However, large differences exist among plant species and genotypes in the shoot-to-root recirculation of Na (Davenport et al., 2005; Wu, 2018).

Many halophytes have wider Casparian band than glycophytes. In salt-treated glycophytes (such as *Juncus articulatus*), Casparian band thickens and is strongly lignified (Al-Hassan et al., 2015). Thicker Casparian band and increased suberin deposition in root cells in rice resulted in increased salt tolerance (Vishal et al., 2019), with salts accumulating at the cortical side of the Casparian band (Alassimone et al., 2012). Modeling ion transport in roots suggested that the Casparian band is more important than suberin lamellae in minimizing radial Na transport and enhancing salt tolerance (Foster and Miklavcic, 2017). Increased lignification of the protoxylem in the root tips of halophyte *Aeluropus littoralis* upon exposure to salinity represented an effective barrier to Na transport (Barzegargolchini et al., 2017). In addition to the physical aspect of the root apoplast structure (salt-induced increased deposition of lignin and suberin), the chemical aspects (binding of Na$^+$ to the cell walls) are also important in restricting Na transport from roots to shoots (Byrt et al., 2018; Hajiboland et al., 2014).

18.5.4.2 Salt distribution in shoot tissues

In includers, Na and Cl$^-$ must be partitioned effectively between old and young leaves, leaf sheaths and leaf blades, cell types within leaf blades, and vegetative and reproductive organs. Generally, restricted import of Na and Cl$^-$ into young leaves is characteristic of salt-tolerant species (e.g., *Vigna luteola*) (Yoshida et al., 2020), even though there are many exceptions (e.g., salt-tolerant *Vigna marina*, being an includer, had high Na concentration not just in roots but also in stems and leaves) (Yoshida et al., 2020). In *Kosteletzkya virginica*, a dicotyledonous halophyte, Na concentration in the leaf water decreased from 230 to 25 mM from the oldest to the youngest leaf, whereas the K concentration increased from 100 to 320 mM (Blits and Gallagher, 1990). However, with vastly different critical concentrations of Na in the cytosol reported in various plant species and cells (ranging from around 30 to 200 mM) (see references in Wu, 2018), it remains unclear which Na concentrations may be harmless in a particular cell type under a given set of conditions.

For salt tolerance of crop species, the total salt concentration in the shoot is less important than the capacity to restrict the import into young leaves (e.g., Liu et al., 2020c; Munoz-Mayor et al., 2012). In rice, OsHKT1;5 plays a crucial role in excluding Na$^+$ from the phloem sap, thus restricting Na transport into young leaves and contributing to Na tolerance (Kobayashi et al., 2017), whereas OsHKT1;4 decreases xylem sap Na concentration (even during exposure to nontoxic Na concentrations in the growth medium), thus minimizing the accumulation of Na in leaves (Khan et al., 2020a). In contrast to rice and many other species, barley HvHKT1;5 (localized in xylem parenchyma) increases Na loading into the xylem and thus transport of Na from the roots to the shoots, negatively affecting salt tolerance (Huang et al., 2020), indicating complexities of the salt toxicity syndrome. In sugar beet as a salt-tolerant crop, and also in halophytes, the steep inverse Na$^+$/K$^+$ gradients between old and young leaves are maintained (e.g., Hariadi et al., 2011).

Chloride partitioning within individual leaves is important for salt tolerance (Fig. 18.23). In barley (Table 18.8), Cl is accumulated particularly in the leaf sheath and in the epidermal cells of leaf blades, whereas concentrations are low in the mesophyll (Huang and Van Steveninck, 1989). Hence, the average values of salt concentrations in the shoots are meaningless in terms of interpreting mechanisms of salt tolerance.

In roots, accumulation of Na$^+$ and Cl$^-$ in the root-cell apoplasm can occur as a result of exclusion of these ions from uptake. Moreover, accumulation may occur also in the leaf apoplasm, when leaf cells pump out excessive Na or Cl$^-$. It has been suggested that increasing accumulation of Na and/or Cl$^-$ in the apoplasm may cause dehydration of the cytoplasm and eventually death of leaf tissues commonly observed under salinity (Oertli, 1968). The Oertli's hypothesis was either not supported [e.g., in studies with different maize cultivars (Lohaus et al., 2000) and maize and cotton (*Gossypium hirsutum*) (Muhling and Lauchli, 2002)] or it was in faba bean (*Vicia faba*) (e.g., Shahzad et al., 2013), depending on whether Na or Cl$^-$ accumulated in the apoplasm sufficiently to affect cell turgor. The increases in Na$^+$ concentration in leaf cell apoplasm can be alleviated by Si in faba bean (Shahzad et al., 2013) and okra (Abdul et al., 2017).

18.5.4.3 Osmotic adjustment

With a sudden increase in salinity, osmotic adjustment is achieved initially by a decrease in tissue water content (partial dehydration). In Fig. 18.17, the negative growth rate of maize seedlings during the first 30 min after the onset of the salt treatment was due to dehydration and shrinking of cells. Salt tolerance and further growth in a saline substrate require a net increase in the concentration of osmotically active solutes in the tissue. In genotypes in which salt exclusion is the predominant mechanism of salt tolerance, either the synthesis of organic solutes such as sugars, alcohols, and amino acids (Table 18.9) or the uptake of K, Ca, or nitrate, is increased. Accumulation of these solutes decreases the osmotic potential in the cells and therefore allows uptake of water from the surrounding solution with low osmotic potential. This is an energy-demanding mechanism, and growth rates of such genotypes under salinity are therefore relatively low.

In genotypes in which salt inclusion is the predominant strategy, osmotic adjustment is achieved by the accumulation of salts (mainly NaCl) in the leaf tissue (Chen and Jiang, 2010; Yolcu et al., 2021, 2022). In natrophilic species, Na can replace K not only in its function as an osmotically active solute in the vacuoles but to some extent also in specific functions in the cell metabolism (see also Section 8.2).

When halophytes from the family Chenopodiaceae are exposed to salinity (40–500 mM), Na concentration in the cytosol is commonly in the range of 150–220 mM (contrasting most glycophytes in which cytosolic Na concentrations above 10 mM are detrimental), and the cytosolic Cl concentration in *Salicornia maritima* was 86 mM (see Flowers and Colmer, 2008). In contrast, glycophytes have cytosolic Cl concentration in the range of 5–20 mM (Teakle and Tyerman, 2010).

18.5.4.4 Vacuolar compartmentation and compatible solutes

In saline substrates, osmotic adjustment in includers requires Na and Cl concentrations in the symplasm of 300–500 mM (Gorham et al., 1985). This implies the transfer of considerable amounts of these ions into the vacuole to avoid the toxic concentration in the cytoplasm.

TABLE 18.9 Examples of taxonomic distribution of compatible organic solutes associated with osmotic adjustment under salinity.

Solute	
D-sorbitol	Rosaceae, Plantaginaceae, Stylonematophyceae
D-pinitol	Caryophyllaceae, Fabaceae
Mannitol	Rhodellophyceae
Glycerol	Chlorophyta
Glycine betaine	Amaranthaceae, Chenopodiaceae, Plantaginaceae, Poaceae, Solanaceae
Proline	Amaranthaceae, Asteraceae, Chenopodiaceae, Juncaceae, Plantaginaceae, Poaceae
Trehalose	Fabaceae
3-Dimethylsulfoniopropionate	Asteraceae, Poaceae
Total soluble sugars (glucose, fructose, and sucrose)	Poaceae

Source: Based on Gorham et al. (1985), Tipirdamaz et al. (2006), and Chen and Jiang (2010).

Sodium enters the vacuole via Na/H$^+$ antiporters using the proton motive force generated by the vacuolar H$^+$-ATPase (V-ATPase) and the H$^+$ pyrophosphatase (V-PPase). Exposure to salinity increases V-ATPase/V-PPase activity in glycophytes such as apple and tomato (Hu et al., 2015) as well as in halophytes (Chen et al., 2015; Yi et al., 2014) (Fig. 18.24). However, overexpression of V-PPase, while conferring increased salinity tolerance under salinity, was detrimental under nonsaline conditions (Graus et al., 2018), emphasizing the need for understanding the strict control mechanisms based on the specific environmental cues.

In arabidopsis, the vacuolar Na/H$^+$ transporter is encoded by the *AtNHX1* gene (Gaxiola et al., 1999). Homologous genes have been identified in numerous glycophytes (e.g., Cui et al., 2020) and halophytes (e.g., Hsouna et al., 2020) (Fig. 18.24). The expression of the *NHX1* gene increased in response to salinity in all plants investigated so far. Overexpression of the rice *OsNHX1* gene in maize (Chen et al., 2007) or poplar (Yang et al., 2017) resulted in increased Na accumulation and improved growth of the transformants under salinity.

For osmotic adjustment of the cytoplasm and its organelles, organic solutes have to be synthesized upon accumulation of ions in the vacuole. These "compatible solutes" do not interfere with plant metabolic processes in the cytoplasm. Proline, glycine betaine, and several polyols are the most common compatible solutes found in glycophytes and halophytes (e.g., Annunziata et al., 2019; Khalid et al., 2020; Tipirdamaz et al., 2006; Walid et al., 2020; Yolcu et al., 2021), with the chemical nature of compatible solutes varying among taxonomic groups (Tipirdamaz et al., 2006) (Table 18.9).

Glycine betaine (amphoteric quaternary amine) is an effective compatible solute because it is highly water-soluble and without a net charge. Glycine betaine protects pyruvate kinase and decreases the K requirement for enzyme activation under salinity (Annunziata et al., 2019). Glycine betaine and other compatible solutes may decrease the leakage of K from tissues exposed to salinity by either improving membrane integrity or increasing ion efflux selectivity (e.g., Chen et al., 2020a).

The importance of compatible solute synthesis for plant salt tolerance has been demonstrated in transgenic plants overproducing a range of different compatible solutes (Ke et al., 2016; Wani et al., 2020; Zhifang and Loescher, 2003). Arabidopsis plants transformed with the mannose-6-phosphate reductase gene from celery with constitutive expression under the CaMV35S promoter accumulated mannitol (Zhifang and Loescher, 2003). The transformants did not differ from the wild type regarding growth in the absence of salinity but had greater biomass under saline conditions, with the difference increasing with increasing salt concentrations (Fig. 18.25).

Osmotic adjustment in plants via salt inclusion or exclusion has important implications for the energy balance. Given that NaCl and other soluble salts are abundant in saline substrates, they can be regarded as "cheap," although potentially dangerous, osmotica. According to Wyn Jones (1981), the approximate energy costs of accumulating 1

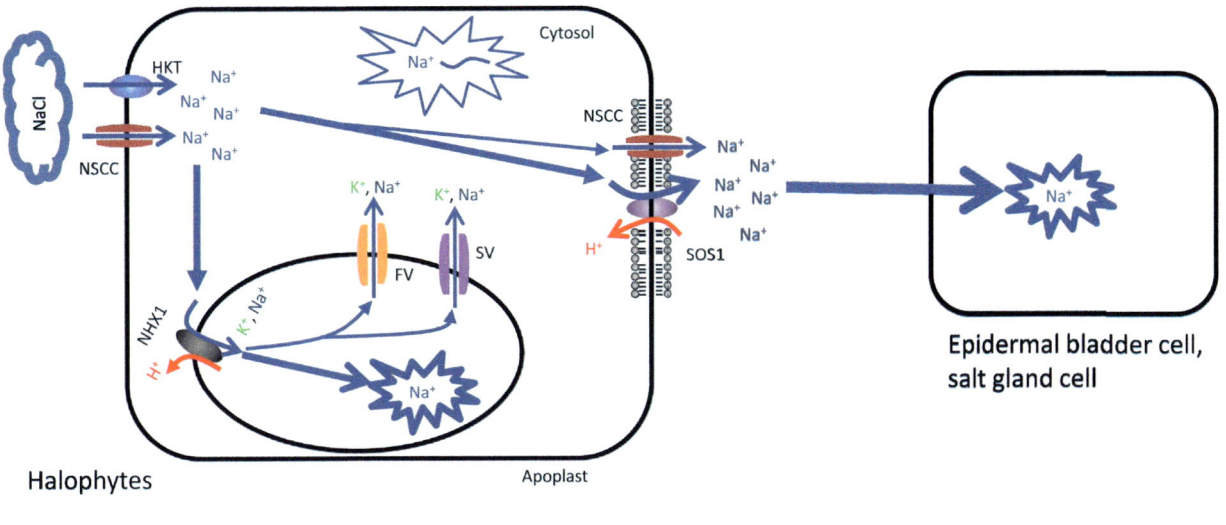
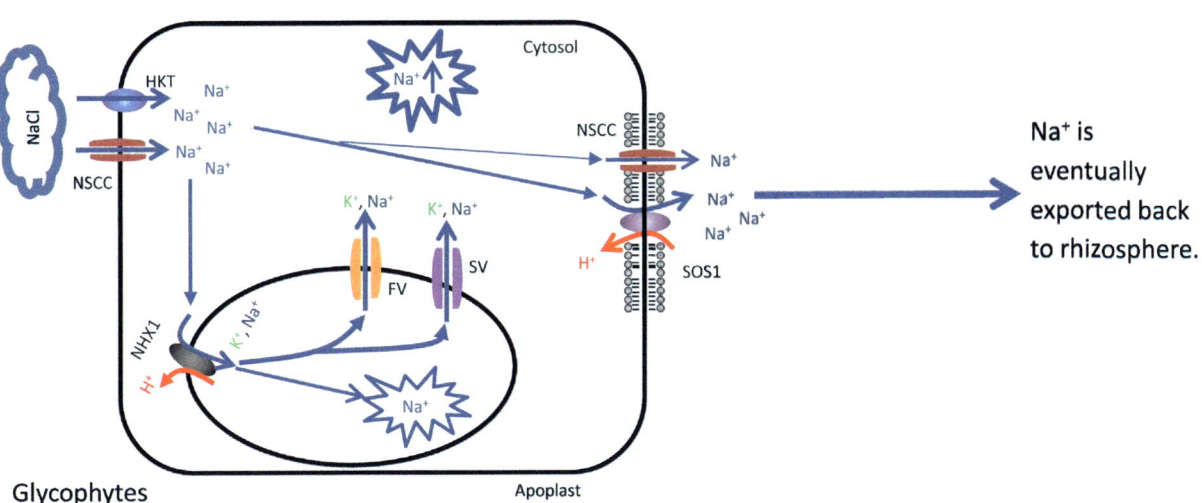

FIGURE 18.24 Different strategies of halophytes and glycophytes in response to salt stress. The line thickness is proportional to the proposed contribution. NSCC = nonselective cation channel; FV and SV = fast and slow vacuolar channels, respectively. For other acronyms, see Fig. 18.18. *From Wu (2018). An open access article under the terms of Creative Commons CC BY license.*

osmol of solute for osmoregulation are 0.54, 13, and 54 mole ATP for NaCl uptake, synthesis of K-malate, and accumulation of C_6 sugars, respectively.

Due to osmotic adjustment via synthesis of organic osmotica being energetically expensive, thus reducing the energy available for growth (Bandehagh and Taylor, 2020), the form of osmotic adjustment may vary between organs of the same plant species, balancing the energy requirements with the importance of protecting particular organs (e.g., young leaves) from the salinity damage. For example, in barley grown on saline substrates, osmotic adjustment was achieved by the accumulation of Na^+ and Cl^- in mature leaves and by the accumulation of sugars in expanding leaves (Delane et al., 1982).

18.5.4.5 Detoxification of reactive oxygen species

In plant tissues, ROS such as hydrogen peroxide (H_2O_2), superoxide (O_2^-), or hydroxyl radicals (OH•) are continuously generated in the cytosol, chloroplasts, and mitochondria by various metabolic processes (Farooq et al., 2019). ROS play a key role in signal transduction but may also damage cells, for example, by membrane peroxidation, protein degradation, and DNA mutation. Plants, therefore, scavenge excessive amounts of ROS by enzymes (e.g., SOD, catalase, or

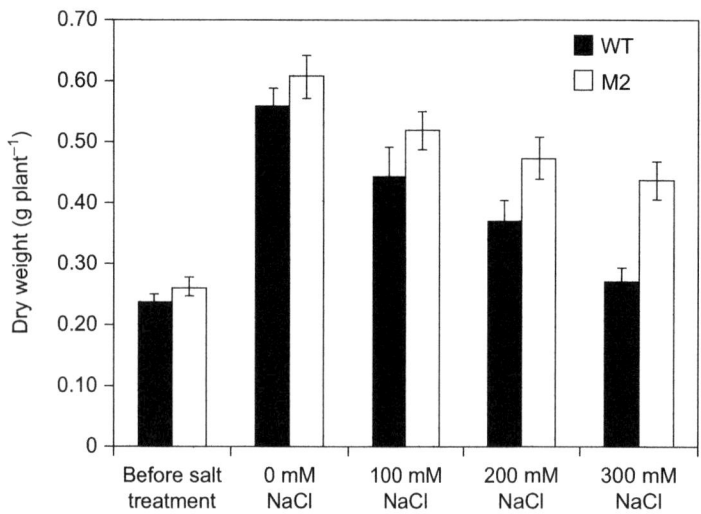

FIGURE 18.25 Dry weights of WT and transgenic (M2, transformed to synthesize mannitol by insertion of the *M6PR* gene from celery, *Apium graveolens* arabidopsis plants grown at different salinity concentrations. Plants were grown for 10 days in the absence of salinity before treatments were established and maintained for 15 days. Means ± SE (n = 3). *WT*, Wild type. *From Zhifang and Loescher (2003), with permission from Wiley and Sons.*

glutathione peroxidase). Antioxidants such as ascorbic acid, tocopherol, or glutathione also contribute to detoxification of ROS.

Salinity and drought cause a rapid increase in cellular ROS concentrations in plants (e.g., Annunziata et al., 2019; Farooq et al., 2019; Parvin et al., 2019), most likely due to a limited CO_2 reduction in the Calvin cycle during osmotic stress. This causes a decrease in the amount of reduced $NADP^+$ (an electron acceptor in the light reaction), resulting in electrons from PS1/ferredoxin being transferred to O_2 instead of $NADP^+$, which leads to the formation of O_2^- (Farooq et al., 2019). Chloroplasts are thus the primary sites of ROS formation under salinity.

The transcription of genes coding for antioxidative enzymes and their activity in plant tissues are usually increased in response to salinity stress (e.g., Ozfidan-Konakci et al., 2020; Wang et al., 2019d). In many plant species, salinity tolerance is correlated positively with the capacity to upregulate the cellular ROS scavenging system (e.g., Nabanita and Soumen, 2020; Parvin et al., 2019). When exposed to salinity, the antioxidant scavenging system in C_4 plants appears to be more effective than that in C_3 plants (Stepien and Klobus, 2005).

The important role of ROS scavenging in plant salinity tolerance was confirmed in transgenic plants (Wani et al., 2020). For example, salinity tolerance was improved in a number of plant species by overexpressing SOD (e.g., Bogoutdinova et al., 2020; Jing et al., 2015; Wang et al., 2004) and/or other antioxidative enzymes (Jin et al., 2019; Negi et al., 2016).

Salinity stress decreases the concentration of antioxidant ascorbic acid (Akbari et al., 2018; Muhammad et al., 2019); hence, external supply of ascorbic acid improves the plant survival under salinity by diminishing oxidative stress (Krupa-Malkiewicz and Smolik, 2019). Similarly, transgenic arabidopsis lines with increased ascorbate concentration have increased tolerance to salinity stress (Acosta-Gamboa et al., 2020).

The precise mechanisms leading to the activation of antioxidant responses in plants upon salinity stress are still not completely clear, but a number of phytohormones may be included (Yamshi et al., 2020), more specifically ethylene (Riyazuddin et al., 2020) and ABA (Szymanska et al., 2019; Zheng et al., 2019) in addition to Ca^{2+} signals, salicylic acid and hydrogen peroxide (Sarika et al., 2005). Pretreatment of plants with a low concentration of H_2O_2 can also improve their salinity tolerance (Ding et al., 2019; Latef et al., 2019; Liu et al., 2020b).

18.5.4.6 Salt excretion

Halophytes may reduce the salt concentration of the photosynthetically active tissues by various mechanisms: accumulation in bladder cells, excretion via salt glands, shedding of salt-saturated leaves, and retranslocation to other organs (Flowers and Colmer, 2008; Flowers et al., 2015). Salt glands vary greatly in anatomy and efficiency (Yuan et al., 2016). They may be simple glands comprising only two cells, for example, in kallar grass (*Leptochloa fusca*) (Wieneke et al., 1987) or multicellular organs of highly specialized cells, for example, in *Tamarix usneoides* (Wilson et al., 2017). Salt glands excrete large amounts of salt to the leaf surface, where they can be washed off by rain or dew. Excretion of Na^+ is mediated by a homolog of the Na/H^+ antiporter SOS1 driven by the plasma membrane H^+-ATPase (Chen et al., 2010a). The salt gland distribution and the excretion rates correlate positively with salt tolerance of various species (Leng et al., 2018; Marcum and Pessarakli, 2006).

TABLE 18.10 Mean cover-abundance percentages of C_3 and C_4 + CAM species in the halophyte and nonhalophyte plant communities of three phytogeographical provinces of Santa Fe (Argentina) differing in soil salinity.

Communities		Phytogeographical provinces of Santa Fe		
		Pampean (low salinity)	Espinal (medium salinity)	Chaquenian (high salinity)
C_3	Nonhalophytes	78	36	8
	Halophytes	28	8	6
C_4 + CAM	Nonhalophytes	22	64	92
	Halophytes	72	92	94

For either nonhalophytes or halophytes, the percentages of C_3 and C_4 + CAM equal 100%.
Source: After Feldman et al. (2008).

18.5.5 Exploiting salt-affected soils

Combinations of traits underpinning tolerance to salinity are highly diverse among plant species and genotypes, their developmental stage, and environmental factors. Understanding the mechanisms of salinity tolerance is important in selecting adapted crop genotypes and interpreting how soil salinity affects natural plant communities and ecosystem functions. For example, Feldman et al. (2008) found that the relative cover abundance of C_4 and CAM species in the halophyte and nonhalophyte communities increased strongly with increasing soil salinity (Table 18.10).

Anthropogenic activities may not affect salinity of only agricultural soils but also that of natural ecosystems, for example, due to rising saline groundwater (Rengasamy, 2006), whereby increasing soil salinity may lead to the replacement of native plant species by salt-tolerant invasive species.

Given the projected increase in salt-affected areas in the future, it is important to develop management strategies for these soils. Amelioration (e.g., gypsum application or drainage) may be one option; the other is using salt-tolerant plants. Recently, plant species with high salinity tolerance and a potential to be utilized in the production of food, feed, biofuel, or pharmaceuticals have been used on saline soils. One such success story is Jerusalem artichoke (*Helianthus tuberosus*; Asteraceae), a perennial with edible inulin-rich tubers that can be used in bioethanol production as well as in the pharmaceutical and other industries. Soils planted to Jerusalem artichoke had improved chemical and biological properties, including decreased salinity and increased abundance and diversity of soil microbiome (Li et al., 2018; Long et al., 2014; Shao et al., 2019; Yue et al., 2020). An additional benefits of planting Jerusalem artichoke in saline soils is increased soil organic matter (especially with moderate amounts of N fertilizers applied), contributing to carbon sequestration (Li et al., 2016). The traditional crops (such as rice, Xu et al., 2020) can also be used on saline soils to decrease soil salinity, increase soil fertility, and enhance the abundance and diversity of soil microbiome.

18.5.6 Genotypic differences in growth response to salinity

Plant species differ greatly in their growth response to salinity, as shown schematically in Fig. 18.22. The growth of halophytes is optimal at relatively high concentrations of NaCl, a response that can be explained in part by the role of Na as a nutrient in these species (see also Section 8.2). Among crops, very few species are stimulated (and even then, only slightly) by low salinity levels. These belong to the relatively small group of agricultural plants classified as salt-tolerant (Table 18.6). Most crops are nonhalophytes (i.e., glycophytes), and their salt tolerance is either moderate, or their growth is severely inhibited even at low salinity (sensitive).

Generally, the classification of the salt tolerance of crops, forage species and fruit trees is based on two parameters: the threshold EC_e and the slope, that is, percentage of yield decrease beyond the threshold. In addition, salt tolerance is often measured as a relative growth decrease in response to salinity compared with a nonsaline control (Table 18.11). This may not necessarily lead to the selection of the plant genotype with the highest yield under salinity. For example, among the rootstocks shown in Table 18.5, the one with the highest relative decrease in growth (Marianna GF 8−1) still produced the highest biomass under saline conditions (Bolat et al., 2006).

Plant salt tolerance is a quantitative trait determined by a relatively large number of genes. Chromosome sections (QTL) with genes relevant for different aspects of salt tolerance were first identified in tomato (Breto et al., 1994). This

TABLE 18.11 Range of relative decrease (% of nonsaline control) of cultivars within crop species in response to salinity.

Crops	Treatment (mM NaCl)	Parameter	Range of relative decrease in the parameter under salinity (%)	References
Winter and spring wheat (*Triticum aestivum*)	100	Shoot length	100–12	Zulfiqar et al. (2007)
Bread wheat (*Triticum eastivum*)	250 mM in hydroponics	Leaf dry weight	100–23	Saddiq et al. (2018)
Barley (*Hordeum vulgare*)	150 mM NaCl	Seed germination	100–54	Mwando et al. (2020)
Sugar beet (*Beta vulgaris*)	150	Total dry weight	92–49	Marschner et al. (1981)
Pepper (*Capcium annuum*)	150	Shoot dry weight	86–42	Aktas et al. (2006)
Olive (*Olea europea*)	100	Shoot length	70–16	Marin et al. (1995)
Cotton (*Gossypium hirsutum*)	5–19 g NaCl/kg soil in the field	Boll number per plants	83–28	Zhu et al. (2020)

has been an active research area recently, with the QTL and single nucleotide polymorphism (SNP) markers for salt tolerance identified in all major crops, including rice (Jahan et al., 2020; Zhang et al., 2020b), wheat (Gupta et al., 2020), maize (Cao et al., 2019), barley (Mwando et al., 2020), sorghum (Wang et al., 2020a), lentils (*Lens culinaris*) (Dharmendra et al., 2020), and cotton (Zhu et al., 2020). However, their use in marker-assisted plant breeding has been limited, generating at best a slow progress in breeding for enhanced salt tolerance (e.g., Mujeeb-Kazi et al., 2019; Tarun et al., 2020) due to the complexities of the salt toxicity syndrome that is dependent on many environmental parameters, plant developmental stage, etc. Salt tolerance may be controlled by numerous mechanisms (explained in Section 18.5.4) that may interact to a variable extent over the course of plant growth and development depending on the environmental conditions; hence, a large number of loci harbor the relevant genes. Pyramiding various tolerance-related loci is a practical way to build up tolerance to salt stress in specific crops grown in the target environments.

References

Abdul, Q., Khan, S.A., Rafiq, A., Sajid, M., Muhammad, I., Fawad, K., et al., 2017. Exogenous Ca$_2$SiO$_4$ enrichment reduces the leaf apoplastic Na$^+$ and increases the growth of okra (*Abelmoschus esculentus* L.) under salt stress. Sci. Hortic. 214, 1–8.

Abdulaha-Al Baquy, M., Li, J., Shi, R., Kamran, M.A., Xu, R., 2018. Higher cation exchange capacity determined lower critical soil pH and higher Al concentration for soybean. Environ. Sci. Pollut. Res. 25, 6980–6989.

Abruña-Rodríguez, F., Vicente-Chandler, J., Rivera, E., Rodríguez, J., 1982. Effect of soil acidity factors on yields and foliar composition of tropical root crops. Soil Sci. Soc. Am. J. 46, 1004–1007.

Acosta-Gamboa, L.M., Liu, S., Creameans, J.W., Campbell, Z.C., Torres, R., Yactayo-Chang, J.P., et al., 2020. Characterization of the response to abiotic stresses of high ascorbate Arabidopsis lines using phenomic approaches. Plant Physiol. Biochem. 151, 500–515.

Ahmad, R.M., Cheng, C., Sheng, J., Wang, W., Ren, H., Aslam, M., et al., 2019. Interruption of jasmonic acid biosynthesis causes differential responses in the roots and shoots of maize seedlings against salt stress. Intern. J. Mol. Sci. 20, 6202.

Ahn, S.J., Sivaguru, M., Osawa, H., Chung, G.C., Matsumoto, H., 2001. Aluminum inhibits the H$^+$-ATPase activity by permanently altering the plasma membrane surface potentials in squash roots. Plant Physiol. 126, 1381–1390.

Ahn, S.J., Sivaguru, M., Chung, G.C., Rengel, Z., Matsumoto, H., 2002. Aluminium-induced growth inhibition is associated with impaired efflux and influx of H$^+$ across the plasma membrane in root apices of squash (*Cucurbita pepo*). J. Exp. Bot. 53, 1959–1966.

Ahn, S.J., Rengel, Z., Matsumoto, H., 2004. Aluminum-induced plasma membrane surface potential and H$^+$-ATPase activity in near-isogenic wheat lines differing in tolerance to aluminum. New Phytol. 162, 71–79.

Akbari, M., Mahna, N., Ramesh, K., Bandehagh, A., Mazzuca, S., 2018. Ion homeostasis, osmoregulation, and physiological changes in the roots and leaves of pistachio rootstocks in response to salinity. Protoplasma 255, 1349–1362.

Aktas, H., Abak, K., Cakmak, I., 2006. Genotypic variation in the response of pepper to salinity. Sci. Hortic. 110, 260–266.

Alassimone, J., Roppolo, D., Geldner, N., Vermeer, J.E.M., 2012. The endodermis-development and differentiation of the plant's inner skin. Protoplasma 249, 433–443.

Albert, R., Pfundner, G., Hertenberger, G., Kastenbauer, T., Watzka, M., 2000. The physiotype approach to understanding halophytes and xerophytes. In: Breckle, S.-W., Schweizer, B., Arndt, U. (Eds.), Ergebnisse Weltweiter Ökologischer Forschung. Verlag G. Heinbach, Stuttgart, Germany, pp. 69–87.

Al-Hassan, M., Gohari, G., Boscaiu, M., Vicente, O., Grigore, M.N., 2015. Anatomical modifications in two Juncus species under salt stress conditions. Not. Bot. Horti Agrobot. Cluj-Na 43, 501–506.

Al-Khateeb, S.A., 2006. Effect of calcium/sodium ratio on growth and ion relations of alfalfa (*Medicago sativa* L.) seedling grown under saline condition. Journal of Agronomy 5, 175–181.

Almeida, P., Boer, G.J.D., Boer, A.H.D., 2014. Differences in shoot Na^+ accumulation between two tomato species are due to differences in ion affinity of HKT1;2. J. Plant Physiol. 171, 438–447.

Almeida, J.D., Tezara, W., Herrera, A., 2016. Physiological responses to drought and experimental water deficit and waterlogging of four clones of cacao (*Theobroma cacao* L.) selected for cultivation in Venezuela. Agric. Water Manag. 171, 80–88.

Amzallag, G.N., 2001. Developmental changes in effect of cytokinin and gibberellin on shoot K^+ and Na^+ accumulation in salt-treated *Sorghum* plants. Plant Biol. 3, 319–325.

Anderson, G.C., Shahab, P., Easton, J., Hall, D.J.M., Rajesh, S., 2020. Short- and long-term effects of lime and gypsum applications on acid soils in a water-limited environment: 1. Grain yield response and nutrient concentration. Agronomy 10, 1213.

Angeli, A.D., Zhang, J., Meyer, S., Martinoia, E., 2013. AtALMT9 is a malate-activated vacuolar chloride channel required for stomatal opening in Arabidopsis. Nat. Commun. 4, 1804.

Annunziata, M.G., Ciarmiello, L.F., Woodrow, P., Dell'aversana, E., Carillo, P., 2019. Spatial and temporal profile of glycine betaine accumulation in plants under abiotic stresses. Front. Plant Sci. 10, 230.

Arzani, A., Ashraf, M., 2016. Smart engineering of genetic resources for enhanced salinity tolerance in crop plants. Crit. Rev. Plant Sci. 35, 146–189.

Ashu, A.B., Lee, S., 2020. Assessing climate change effects on water balance in a monsoon watershed. Water 12, 2564.

Athar, H.R., Ashraf, M., 2009. Strategies for crop improvement against salinity and drought stress. An overview. In: Ashraf, M., Ozturk, M., Athar, H.R. (Eds.), Salinity and Water Stress. Springer, Heidelberg, Germany, pp. 1–18.

Awasthi, J.P., Bedabrata, S., Jogeswar, P., Yanase, E., Koyama, H., Panda, S.K., 2019. Redox balance, metabolic fingerprint and physiological characterization in contrasting North East Indian rice for aluminum stress tolerance. Sci. Rep. 9, 8681.

Bailey-Serres, J., Voesenek, La.C.J., 2008. Flooding stress: acclimations and genetic diversity. Annu. Rev. Plant Biol. 59, 313–339.

Bailey-Serres, J., Fukao, T., Gibbs, D.J., Holdsworth, M.J., Lee, S.C., Licausi, F., et al., 2012. Making sense of low oxygen sensing. Trends Plant Sci. 17, 129–138.

Bandehagh, A., Taylor, N.L., 2020. Can alternative metabolic pathways and shunts overcome salinity induced inhibition of central carbon metabolism in crops? Front. Plant Sci. 11, 1072.

Barbosa, J.Z., Motta, A.C.V., Consalter, R., Pauletti, V., 2018. Boron alters cation exchange properties of corn roots but does not decrease aluminum toxicity. Bioscience J. 34, 917–926.

Barcelo, J., Poschenrieder, C., 2002. Fast root growth responses, root exudates, and internal detoxification as clues to the mechanisms of aluminium toxicity and resistance: a review. Environ. Exp. Bot. 48, 75–92.

Barrett-Lennard, E.G., 2003. The interaction between waterlogging and salinity in higher plants: causes, consequences and implications. Plant Soil 253, 35–54.

Barrett-Lennard, E.G., Shabala, S.N., 2013. The waterlogging/salinity interaction in higher plants revisited - focusing on the hypoxia-induced disturbance to K^+ homeostasis. Funct. Plant Biol. 40, 872–882.

Barrett-Lennard, E.G., Bennett, S.J., Altman, M., 2013. Survival and growth of perennial halophytes on saltland in a Mediterranean environment is affected by depth to watertable in summer as well as subsoil salinity. Crop Pasture Sci. 64, 123–136.

Barzegargolchini, B., Movafeghi, A., Dehestani, A., Mehrabanjoubani, P., 2017. Increased cell wall thickness of endodermis and protoxylem in *Aeluropus littoralis* roots under salinity: the role of *LAC4* and *PER64* genes. J. Plant Physiol. 218, 127–134.

Bashar, K.K., 2018. Hormone dependent survival mechanisms of plants during post-waterlogging stress. Plant Signal. Behav. 13, e1529522.

Bashar, K.K., Tareq, M.Z., Amin, M.R., Ummay, H., Tahjib-Ul-Arif, M., Sadat, M.A., et al., 2019. Phytohormone-mediated stomatal response, escape and quiescence strategies in plants under flooding stress. Agronomy 9, 43.

Begum, H.H., Osaki, M., Watanabe, T., Shinano, T., 2009. Mechanisms of aluminum tolerance in phosphoenolpyruvate carboxylase transgenic rice. J. Plant Nutr. Soil Sci. 32, 84–96.

Belkheiri, O., Mulas, M., 2013. The effects of salt stress on growth, water relations and ion accumulation in two halophyte Atriplex species. Environ. Exp. Bot. 86, 17–28.

Benedict, A., Akhil, M., Kavya, B., Dinakar, C., 2020. Cytochrome oxidase and alternative oxidase pathways of mitochondrial electron transport chain are important for the photosynthetic performance of pea plants under salinity stress conditions. Plant Physiol. Biochem. 154, 248–259.

Berthomieu, P., Conejero, G., Nublat, A., Brackenbury, W.J., Lambert, C., Savio, C., et al., 2003. Functional analysis of AtHKT1 in Arabidopsis shows that Na^+ recirculation by the phloem is crucial for salt tolerance. EMBO J. 22, 2004–2014.

Bicknell, D., Simons, J., 2020. Salinity tolerance of plants for agriculture and revegetation. Department of Primary Industries and Regional Development, Perth, Western Australia.

Binh Thanh, N., Nam Ngoc, T., Quang Vu, B., 2020. Methane emissions and associated microbial activities from paddy salt-affected soil as influenced by biochar and cow manure addition. Appl. Soil Ecol. 152, 152.

Blamey, F.P.C., Nishizawa, N.K., Yoshimura, E., 2004. Timing, magnitude, and location of initial soluble aluminum injuries to mung bean roots. Soil Sci. Plant Nutr. 50, 67–76.

Blamey, F.P.C., Mckenna, B.A., Li, C., Cheng, M., Tang, C., Jiang, H., et al., 2018a. Manganese distribution and speciation help to explain the effects of silicate and phosphate on manganese toxicity in four crop species. New Phytol. 217, 1146–1160.

Blamey, F.P.C., Paterson, D.J., Walsh, A., Afshar, N., Mckenna, B.A., Cheng, M., et al., 2018b. Time-resolved X-ray fluorescence analysis of element distribution and concentration in living plants: an example using manganese toxicity in cowpea leaves. Environ. Exp. Bot. 156, 151–160.

Blamey, F.P.C., Li, C., Howard, D.L., Cheng, M., Tang, C., Scheckel, K.G., et al., 2019. Evaluating effects of iron on manganese toxicity in soybean and sunflower using synchrotron-based X-ray fluorescence microscopy and X-ray absorption spectroscopy. Metallomics 11, 2097–2110.

Blits, K.C., Gallagher, J.L., 1990. Salinity tolerance of *Kosteletzkya virginica*. I. Shoot growth, ion and water relations. Plant Cell Environ. 13, 409–418.

Blumenschein, T.G., Nelson, K.A., Motavalli, P.P., 2018. Impact of a new deep vertical lime placement practice on corn and soybean production in conservation tillage systems. Agronomy 8, 104.

Bogoutdinova, L.R., Lazareva, E.M., Chaban, I.A., Kononenko, N.V., Dilovarova, T., Khaliluev, M.R., et al., 2020. Salt stress-induced structural changes are mitigated in transgenic tomato plants over-expressing superoxide dismutase. Biology 9, 297.

Bolat, I., Kaya, C., Almaca, A., Timucin, S., 2006. Calcium sulfate improves salinity tolerance in rootstocks of plum. J. Plant Nutr. 29, 553–564.

Borgo, L., Rabelo, F.H., Carvalho, G., Ramires, T., Righetto, A.J., Piotto, F.A., et al., 2020. Antioxidant performance and aluminum accumulation in two genotypes of *Solanum lycopersicum* in response to low pH and aluminum availability and under their combined stress. Sci. Hortic. 259, 108813.

Boscolo, P.R., Menossi, M., Jorge, R.A., 2003. Aluminum-induced oxidative stress in maize. Phytochemistry 62, 181–189.

Bose, J., Babourina, O., Shabala, S., Rengel, Z., 2010. Aluminum-dependent dynamics of ion transport in Arabidopsis: specificity of low pH and aluminum responses. Physiol. Plant. 139, 401–412.

Bose, J., Babourina, O., Shabala, S., Rengel, Z., 2013. Low-pH and aluminum resistance in Arabidopsis correlates with high cytosolic magnesium content and increased magnesium uptake by plant roots. Plant Cell Physiol. 54, 1093–1104.

Bossolani, J.W., Costa Crusciol, C.A., Merloti, L.F., Moretti, L.G., Costa, N.R., Tsai, S.M., et al., 2020. Long-term lime and gypsum amendment increase nitrogen fixation and decrease nitrification and denitrification gene abundances in the rhizosphere and soil in a tropical no-till intercropping system. Geoderma 375, 114476.

Boyer, J.S., 2009. Cell wall biosynthesis and the molecular mechanism of plant enlargement. Funct. Plant Biol. 36, 383–394.

Brautigan, D.J., Rengasamy, P., Chittleborough, D.J., 2012. Aluminium speciation and phytotoxicity in alkaline soils. Plant Soil 360, 187–196.

Breto, M.P., Asins, M.J., Carbonell, E.A., 1994. Salt tolerance in Lycopersicon species. III. Detection of quantitative trait loci by means of molecular markers. Theor. Appl. Genet. 88, 395–401.

Brunner, I., Sperisen, C., 2013. Aluminum exclusion and aluminum tolerance in woody plants. Front. Plant Sci. 4, 172.

Byrt, C.S., Munns, R., Burton, R.A., Gilliham, M., Wege, S., 2018. Root cell wall solutions for crop plants in saline soils. Plant Sci. 269, 47–55.

Cai, Z., Cheng, Y., Xian, P., Lin, R., Xia, Q., He, X., et al., 2019. Fine-mapping QTLs and the validation of candidate genes for aluminum tolerance using a high-density genetic map. Plant Soil 444, 119–137.

Cakmak, I., Horst, W.J., 1991. Effect of aluminium on lipid peroxidation, superoxide dismutase, catalase, and peroxidase activities in root tips of soybean (*Glycine max*). Physiol. Plant. 83, 463–468.

Cao, Y., Liang, X., Yin, P., Zhang, M., Jiang, C., 2019. A domestication-associated reduction in K^+-preferring HKT transporter activity underlies maize shoot K^+ accumulation and salt tolerance. New Phytol. 222, 301–317.

Carr, S.J., Ritchie, G.S.P., 1993. Al toxicity of wheat grown in acidic subsoils in relation to soil solution properties and exchangeable cations. Aust. J. Soil Res. 31, 583–596.

Ceballos-Laita, L., Gutierrez-Carbonell, E., Imai, H., Abadia, A., Uemura, M., Abadia, J., et al., 2018. Effects of manganese toxicity on the protein profile of tomato (*Solanum lycopersicum*) roots as revealed by two complementary proteomic approaches, two-dimensional electrophoresis and shotgun analysis. J. Proteomics 185, 51–63.

Centofanti, T., Banuelos, G., 2019. Practical uses of halophytic plants as sources of food and fodder. In: Hasanuzzaman, M., Shabala, S., Fujita, M. (Eds.), Halophytes and Climate Change. CABI, Walford, UK, pp. 324–342.

Chapin Iii, F.S., Matson, P.A., Vitousek, P.M., 2011. Principles of Terrestrial Ecosystem Ecology. Springer, New York, NY, USA.

Chen, H., Jiang, J., 2010. Osmotic adjustment and plant adaptation to environmental changes related to drought and salinity. Environ. Rev. 18, 309–319.

Chen, Z., Liao, H., 2016. Organic acid anions: an effective defensive weapon for plants against aluminum toxicity and phosphorus deficiency in acidic soils. J. Genet. Genomics 43, 631–638.

Chen, Z., Newman, I., Zhou, M., Mendham, N., Zhang, G., Shabala, S., 2005. Screening plants for salt tolerance by measuring K^+ flux: a case study for barley. Plant Cell Environ. 28, 1230–1246.

Chen, M., Chen, Q.J., Niu, X.G., Zhang, R., Lin, H.Q., Xu, C.Y., et al., 2007. Expression of *OsNHX1* gene in maize confers salt tolerance and promotes plant growth in the field. Plant Soil Environ. 53, 490–498.

Chen, J., Xiao, Q., Wu, F., Dong, X.J., He, J., Pei, Z.M., et al., 2010a. Nitric oxide enhances salt secretion and Na^+ sequestration in a mangrove plant, *Avicennia marina*, through increasing the expression of H^+-ATPase and Na^+/H^+ antiporter under high salinity. Tree Physiol. 30, 1570–1585.

Chen, X., Pierik, R., Peeters, A.J.M., Poorter, H., Visser, E.J.W., Huber, H., et al., 2010b. Endogenous abscisic acid as a key switch for natural variation in flooding-induced shoot elongation. Plant Physiol. 154, 969–977.

Chen, Y., Li, L., Zong, J., Chen, J., Guo, H., Guo, A., et al., 2015. Heterologous expression of the halophyte *Zoysia matrella* H^+-pyrophosphatase gene improved salt tolerance in *Arabidopsis thaliana*. Plant Physiol. Biochem. 91, 49–55.

Chen, Z., Yan, W., Sun, L., Tian, J., Liao, H., 2016. Proteomic analysis reveals growth inhibition of soybean roots by manganese toxicity is associated with alteration of cell wall structure and lignification. J. Proteomics 143, 151–160.

Chen, J., Duan, R., Hu, W., Zhang, N., Lin, X., Zhang, J., et al., 2019a. Unravelling calcium-alleviated aluminium toxicity in *Arabidopsis thaliana*: insights into regulatory mechanisms using proteomics. J. Proteomics 199, 15–30.

Chen, Q., Wu, W., Zhao, T., Tan, W., Tian, J., Liang, C., 2019b. Complex gene regulation underlying mineral nutrient homeostasis in soybean root response to acidity stress. Genes 10, 402.

Chen, L., Liu, L., Lu, B., Ma, T., Jiang, D., Li, J., et al., 2020a. Exogenous melatonin promotes seed germination and osmotic regulation under salt stress in cotton (*Gossypium hirsutum* L.). PLoS ONE 15, 228241.

Chen, T., Pineda, I.M.G., Brand, A.M., Stutzel, H., 2020b. Determining ion toxicity in cucumber under salinity stress. Agronomy 10, 677.

Chi Tam, N., Agorio, A., Jossier, M., Depre, S., Thomine, S., Filleur, S., 2016. Characterization of the chloride channel-like, AtCLCg, involved in chloride tolerance in *Arabidopsis thaliana*. Plant Cell Physiol. 57, 764–775.

Ciarkowska, K., Miechowka, A., 2019. The effect of understory on cation binding reactions and aluminium behaviour in acidic soils under spruce forest stands (Southern Poland). Biogeochemistry 143, 55–66.

Coelho, C.D.J., Gardingo, J.R., Almeida, M.C.D., Matiello, R.R., 2019. Mapping of QTL for aluminum tolerance in tropical maize. Crop Breed. Appl. Biotechnol. 19, 86–94.

Colmenero-Flores, J.M., Martinez, G., Gamba, G., Vazquez, N., Iglesias, D.J., Brumos, J., et al., 2007. Identification and functional characterization of cation-chloride cotransporters in plants. Plant J. 50, 278–292.

Colmer, T.D., Voesenek, La.C.J., 2009. Flooding tolerance: suites of plant traits in variable environments. Funct. Plant Biol. 36, 665–681.

Cox, M.C.H., Benschop, J.J., Vreeburg, Ra.M., Wagemaker, Ca.M., Moritz, T., Peeters, A.J.M., et al., 2004. The roles of ethylene, auxin, abscisic acid, and gibberellin in the hyponastic growth of submerged *Rumex palustris* petioles. Plant Physiol. 136, 2948–2960.

Cox, D.D., Slaton, N.A., Ross, W.J., Roberts, T.L., 2018. Trifoliolate leaflet chloride concentrations for characterizing soybean yield loss from chloride toxicity. Agron. J. 110, 1589–1599.

Cuenca, G., Herrera, R., Medina, E., 1990. Aluminium tolerance in trees of a tropical cloud forest. Plant Soil 125, 169–175.

Cui, J., Hua, Y., Zhou, T., Liu, Y., Huang, J., Yue, C., 2020. Global landscapes of the Na$^+$/H$^+$ antiporter (NHX) family members uncover their potential roles in regulating the rapeseed resistance to salt stress. Intern. J. Mol. Sci. 21, 3429.

Cury, N.F., Silva, R.C.C., Andre, M.S.F., Fontes, W., Ricart, Ca.O., Castro, M.S., et al., 2020. Root proteome and metabolome reveal a high nutritional dependency of aluminium in *Qualea grandiflora* Mart. (Vochysiaceae). Plant Soil 446, 125–143.

Da-Silva, C.J., Amarante, L.D., 2020. Time-course biochemical analyses of soybean plants during waterlogging and reoxygenation. Environ. Exp. Bot. 180, 104242.

Davenport, R., James, R.A., Zakrisson-Plogander, A., Tester, M., Munns, R., 2005. Control of sodium transport in durum wheat. Plant Physiol. 137, 807–818.

Degroote, K.V., Mccartha, G.L., Pollard, A.J., 2018. Interactions of the manganese hyperaccumulator *Phytolacca americana* L. with soil pH and phosphate. Ecol. Res. 33, 749–755.

Delane, R., Greenway, H., Munns, R., Gibbs, J., 1982. Ion concentration and carbohydrate status of the elongating leaf tissue of *Hordeum vulgare* growing at high external NaCl. I. Relationship between solute concentration and growth. J. Exp. Bot. 33, 557–573.

Delgado, M., Valle, S., Barra, P.J., Reyes-Diaz, M., Zuniga-Feest, A., 2019. New aluminum hyperaccumulator species of the Proteaceae family from southern South America. Plant Soil 444, 475–487.

Delhaize, E., Kataoka, T., Hebb, D.M., White, R.G., Ryan, P.R., 2003. Genes encoding proteins of the cation diffusion facilitator family that confer manganese tolerance. Plant Cell 15, 1131–1142.

Delhaize, E., Ryan, P.R., Hebb, D.M., Yamamoto, Y., Sasaki, T., Matsumoto, H., 2004. Engineering high-level aluminum tolerance in barley with the *ALMT1* gene. Proc. Nat. Acad. Sci. USA 101, 15249–15254.

Delhaize, E., Gruber, B.D., Pittman, J.K., White, R.G., Leung, H., Miao, Y., et al., 2007a. A role for the *AtMTP11* gene of Arabidopsis in manganese transport and tolerance. Plant J. 51, 198–210.

Delhaize, E., Gruber, B.D., Ryan, P.R., 2007b. The roles of organic anion permeases in aluminium resistance and mineral nutrition. FEBS Lett. 581, 2255–2262.

Demidchik, V., Davenport, R.J., Tester, M., 2002. Nonselective cation channels in plants. Annu. Rev. Plant Biol. 53, 67–107.

Deng, W., Luo, K., Li, D., Zheng, X., Wei, X., Smith, W., et al., 2006. Overexpression of an Arabidopsis magnesium transport gene, *AtMGT1*, in *Nicotiana benthamiana* confers Al tolerance. J. Exp. Bot. 57, 4235–4243.

Dharmendra, S., Singh, C.K., Tomar, R.S.S., Shristi, S., Sourabh, K., Madan, P., et al., 2020. Genetics and molecular mapping for salinity stress tolerance at seedling stage in lentil (*Lens culinaris* Medik). Crop Sci. 60, 1254–1266.

Ding, H., Ma, D., Huang, X., Hou, J., Wang, C., Xie, Y., et al., 2019. Exogenous hydrogen sulfide alleviates salt stress by improving antioxidant defenses and the salt overly sensitive pathway in wheat seedlings. Acta Physiol. Plant. 41, 123.

Dougherty, E., Rasmussen, K.L., 2020. Changes in future flash flood-producing storms in the United States. J. Hydrometeorol. 21, 2221–2236.

Drew, M.C., 1990. Sensing soil oxygen. Plant Cell Environ. 13, 681–693.

Drew, M.C., Jackson, M.B., Giffard, S., 1979. Ethylene-promoted adventitious rooting and development of cortical air spaces (aerenchyma) in roots may be adaptive responses to flooding in *Zea mays* L. Planta 147, 83–88.

Du, H., Shen, X., Huang, Y., Huang, M., Zhang, Z., 2016. Overexpression of *Vitreoscilla* hemoglobin increases waterlogging tolerance in Arabidopsis and maize. BMC Plant Biol. 16, 35.

Ducic, T., Polle, A., 2007. Manganese toxicity in two varieties of Douglas fir (*Pseudotsuga menziesii* var. *viridis* and *glauca*) seedlings as affected by phosphorus supply. Funct. Plant Biol. 34, 31–40.

Dufey, I., Hakizimana, P., Draye, X., Lutts, S., Bertin, P., 2009. QTL mapping for biomass and physiological parameters linked to resistance mechanisms to ferrous iron toxicity in rice. Euphytica 167, 143–160.

Ejiri, M., Sawazaki, Y., Shiono, K., 2020. Some accessions of amazonian wild rice (*Oryza glumaepatula*) constitutively form a barrier to radial oxygen loss along adventitious roots under aerated conditions. Plants 9, 880.

Ejiri, M., Fukao, T., Miyashita, T., Shiono, K., 2021. A barrier to radial oxygen loss helps the root system cope with waterlogging-induced hypoxia. Breed. Sci. 71, 40–50.

El-Mahi, H., Perez-Hormaeche, J., Luca, A.D., Villalta, I., Espartero, J., Gamez-Arjona, F., et al., 2019. A critical role of sodium flux via the plasma membrane Na^+/H^+ exchanger SOS1 in the salt tolerance of rice. Plant Physiol. 180, 1046–1065.

El-Moneim, D.A., Contreras, R., Silva-Navas, J., Gallego, F.J., Figueiras, A.M., Benito, C., 2014. Pectin methylesterase gene and aluminum tolerance in *Secale cereale*. Environ. Exp. Bot. 107, 125–133.

Epron, D., Toussaint, M.L., Badot, P.M., 1999. Effects of sodium chloride salinity on root growth and respiration in oak seedlings. Ann. For. Sci. 56, 41–47.

Ermolayev, V., Weschke, W., Manteuffel, R., 2003. Comparison of Al-induced gene expression in sensitive and tolerant soybean cultivars. J. Exp. Bot. 54, 2745–2756.

Esfahan, E.Z., Assareh, M.H., Jafari, M., Jafari, A.A., Javadi, S.A., Karimi, G., 2010. Phenological effects on forage quality of two halophyte species *Atriplex leucoclada* and *Suaeda vermiculata* in four saline rangelands of Iran. J. Food Agric. Env. 8, 999–1003.

Essemine, J., Qu, M., Lyu, M., Song, Q., Khan, N., Chen, G., et al., 2020. Photosynthetic and transcriptomic responses of two C_4 grass species with different NaCl tolerance. J. Plant Physiol. 253, 153244.

Eticha, D., Stass, A., Horst, W.J., 2005a. Localization of aluminium in the maize root apex: can morin detect cell wall-bound aluminium? J. Exp. Bot. 56, 1351–1357.

Eticha, D., The, C., Welcker, C., Narro, L., Stass, A., Horst, W.J., 2005b. Aluminium-induced callose formation in root apices: inheritance and selection trait for adaptation of tropical maize to acid soils. Field Crops Res. 93, 252–263.

Eticha, D., Zahn, M., Bremer, M., Yang, Z., Rangel, A.F., Rao, I.M., et al., 2010. Transcriptomic analysis reveals differential gene expression in response to aluminium in common bean (*Phaseolus vulgaris*) genotypes. Ann. Bot. 105, 1119–1128.

Ezaki, B., Katsuhara, M., Kawamura, M., Matsumoto, H., 2001. Different mechanisms of four aluminum (Al)-resistant transgenes for Al toxicity in Arabidopsis. Plant Physiol. 127, 918–927.

Ezaki, B., Sasaki, K., Matsumoto, H., Nakashima, S., 2005. Functions of two genes in aluminium (Al) stress resistance: repression of oxidative damage by the AtBCB gene and promotion of efflux of Al ions by the NtGDI1 gene. J. Exp. Bot. 56, 2661–2671.

Fan, Y., Ouyang, Y., Pan, Y., Hong, T., Wu, C., Lin, H., 2020. Effect of aluminum stress on the absorption and transportation of aluminum and macronutrients in roots and leaves of *Aleurites montana*. Forest Ecol. Manag. 458, 117813.

Fang, X., Tian, W., Liu, X., Lin, X., Jin, C., Zheng, S., 2016. Alleviation of proton toxicity by nitrate uptake specifically depends on nitrate transporter 1.1 in Arabidopsis. New Phytol. 211, 149–158.

Farkas, Z., Varga-Laszlo, E., Anda, A., Veisz, O., Varga, B., 2020. Effects of waterlogging, drought and their combination on yield and water-use efficiency of five Hungarian winter wheat varieties. Water 12, 1318.

Farooq, M.A., Niazi, A.K., Javaid, A., Saifullah, Farooq, M., Souri, Z., et al., 2019. Acquiring control: the evolution of ROS-induced oxidative stress and redox signaling pathways in plant stress responses. Plant Physiol. Biochem. 141, 353–369.

Fecht-Christoffers, M.M., Fuehrs, H., Braun, H.-P., Horst, W.J., 2006. The role of hydrogen peroxide-producing and hydrogen peroxide-consuming peroxidases in the leaf apoplast of cowpea in manganese tolerance. Plant Physiol. 140, 1451–1463.

Fecht-Christoffers, M.M., Maier, P., Iwasaki, K., Braun, H.P., Horst, W.J., 2007. The role of the leaf apoplast in manganese toxicity and tolerance in cowpea (*Vigna unguiculata* L. Walp). In: Sattelmacher, B., Horst, W. (Eds.), Apoplast of Higher Plants: Compartment of Storage, Transport and Reactions. Springer, Dordrecht, Netherlands, pp. 307–321.

Feldman, S.R., Bisaro, V., Biani, N.B., Prado, D.E., 2008. Soil salinity determines the relative abundance of C_3/C_4 species in Argentinean grasslands. Global Ecol. Biogeog. 17, 708–714.

Flowers, T.J., Colmer, T.D., 2008. Salinity tolerance in halophytes. New Phytol. 179, 945–963.

Flowers, T.J., Munns, R., Colmer, T.D., 2015. Sodium chloride toxicity and the cellular basis of salt tolerance in halophytes. Ann. Bot. 115, 419–431.

Fordyce, S., Jones, C.A., Dahlhausen, S.J., Lachowiec, J., Eberly, J.O., Sherman, J.D., et al., 2020. A simple cultivar suitability index for low-pH agricultural soils. Agric. Environ. Lett. 5, e20036.

Foster, K.J., Miklavcic, S.J., 2017. A comprehensive biophysical model of ion and water transport in plant roots. I. Clarifying the roles of endodermal barriers in the salt stress response. Front. Plant Sci. 8, 1326.

Fukao, T., Barrera-Figueroa, B.E., Juntawong, P., Pena-Castro, J.M., 2019. Submergence and waterlogging stress in plants: a review highlighting research opportunities and understudied aspects. Front. Plant Sci. 10, 340.

Fukuda, A., Nakamura, A., Tanaka, Y., 1999. Molecular cloning and expression of the Na^+/H^+ exchanger gene in *Oryza sativa*. Biochim. Biophys. Acta-Gene Struct. Expression 1, 149–155.

Furukawa, J., Yamaji, N., Wang, H., Mitani, N., Murata, Y., Sato, K., et al., 2007. An aluminum-activated citrate transporter in barley. Plant Cell Physiol. 48, 1081–1091.

Gaxiola, R.A., Rao, R., Sherman, A., Grisafi, P., Alper, S.L., Fink, G.R., 1999. The *Arabidopsis thaliana* proton transporters, AtNhx1 and Avp1, can function in cation detoxification in yeast. Proc. Nat. Acad. Sci. USA 96, 1480–1485.

Geilfus, C.M., 2018. Chloride: from nutrient to toxicant. Plant Cell Physiol. 59, 877–886.
Geisler-Lee, J., Caldwell, C., Gallie, D.R., 2010. Expression of the ethylene biosynthetic machinery in maize roots is regulated in response to hypoxia. J. Exp. Bot. 61, 857–871.
George, E., Horst, W.J., Neumann, E., 2012. Adaptation of plants to adverse chemical soil conditions. In: Marschner, P. (Ed.), Marschner's Mineral Nutrition of Higher Plants, 3rd ed Elsevier, pp. 409–472.
Ghanem, M.E., Albacete, A., Martinez-Andujar, C., Acosta, M., Romero-Aranda, R., Dodd, I.C., et al., 2008. Hormonal changes during salinity-induced leaf senescence in tomato (*Solanum lycopersicum* L.). J. Exp. Bot. 59, 3039–3050.
Glenn, E.P., Anday, T., Chaturvedi, R., Martinez-Garcia, R., Pearlstein, S., Soliz, D., et al., 2013. Three halophytes for saline-water agriculture: an oilseed, a forage and a grain crop. Environ. Exp. Bot. 92, 110–121.
Glenn, W.S., Stone, S.E., Ho, S.H., Sweredoski, M.J., Moradian, A., Hess, S., et al., 2017. Bioorthogonal noncanonical amino acid tagging (BONCAT) enables time-resolved analysis of protein synthesis in native plant tissue. Plant Physiol. 173, 1543–1553.
Godfrey, J.M., Ferguson, L., Sanden, B.L., Tixier, A., Sperling, O., Grattan, S.R., et al., 2019. Sodium interception by xylem parenchyma and chloride recirculation in phloem may augment exclusion in the salt tolerant *Pistacia* genus: context for salinity studies on tree crops. Tree Physiol. 39, 1484–1498.
Gorham, J., 1987. Photosynthesis, transpiration and salt fluxes through leaves of *Leptochloa fusca* L. Kunth. Plant Cell Environ. 10, 191–196.
Gorham, J., Wyn Jones, R.G., Mcdonnell, E., 1985. Some mechanisms of salt tolerance in crop plants. Plant Soil 89, 15–40.
Goussi, R., Manaa, A., Derbali, W., Cantamessa, S., Abdelly, C., Barbato, R., 2018. Comparative analysis of salt stress, duration and intensity, on the chloroplast ultrastructure and photosynthetic apparatus in *Thellungiella salsuginea*. J. Photochem. Photobiol. B-Biol. 183, 275–287.
Graus, D., Konrad, K.R., Bemm, F., Nebioglu, M.G.P., Lorey, C., Duscha, K., et al., 2018. High V-PPase activity is beneficial under high salt loads, but detrimental without salinity. New Phytol. 219, 1421–1432.
Greenway, H., Munns, R., 1980. Mechanisms of salt tolerance in nonhalophytes. Annu. Rev. Plant Physiol. 31, 149–190.
Guo, K.-M., Babourina, O., Christopher, D.A., Borsics, T., Rengel, Z., 2008. The cyclic nucleotide-gated channel, AtCNGC10, influences salt tolerance in Arabidopsis. Physiol. Plant. 134, 499–507.
Guo, K., Babourina, O., Christopher, D.A., Borsic, T., Rengel, Z., 2010. The cyclic nucleotide-gated channel AtCNGC10 transports Ca^{2+} and Mg^{2+} in *Arabidopsis*. Physiol. Plant. 139, 303–312.
Guo, W., Wang, C., Zuo, Z., Qiu, J., 2014. The roles of anion channels in Arabidopsis immunity. Plant Signal. Behav. 9, e29230.
Guo, C.Y., Wang, X.Z., Chen, L., Ma, L.N., Wang, R.Z., 2015. Physiological and biochemical responses to saline-alkaline stress in two halophytic grass species with different photosynthetic pathways. Photosynthetica 53, 128–135.
Gupta, P.K., Balyan, H.S., Shailendra, S., Rahul, K., 2020. Genetics of yield, abiotic stress tolerance and biofortification in wheat (*Triticum aestivum* L.). Theor. Appl. Genet. 133, 1569–1602.
Hajiboland, R., Poschenrieder, C., 2015. Localization and compartmentation of Al in the leaves and roots of tea plants. Phyton 84, 86–100.
Hajiboland, R., Norouzi, F., Poschenrieder, C., 2014. Growth, physiological, biochemical and ionic responses of pistachio seedlings to mild and high salinity. Trees-Struct. Funct. 28, 1065–1078.
Han, Y., Zhang, W., Zhang, B., Zhang, S., Wang, W., Ming, F., 2009. One novel mitochondrial citrate synthase from *Oryza sativa* L. can enhance aluminum tolerance in transgenic tobacco. Mol. Biotechnol. 42, 299–305.
Han, J., Shi, J., Zeng, L., Xu, J., Wu, L., 2015. Effects of nitrogen fertilization on the acidity and salinity of greenhouse soils. Environ. Sci. Pollut. Res. 22, 2976–2986.
Han, Y., Yin, S., Huang, L., Wu, X., Zeng, J., Liu, X., et al., 2018. A sodium transporter HvHKT1;1 confers salt tolerance in barley via regulating tissue and cell ion homeostasis. Plant Cell Physiol. 59, 1976–1989.
Hanh Duy, D., Hirai, Y., 2018. Cl^- more detrimental than Na^+ in rice under long-term saline conditions. J. Agric. Sci. 10, 66–75.
Hariadi, Y., Marandon, K., Tian, Y., Jacobsen, S.E., Shabala, S., 2011. Ionic and osmotic relations in quinoa (*Chenopodium quinoa* Willd.) plants grown at various salinity levels. J. Exp. Bot. 62, 185–193.
Haro, R., Bañuelos, M.A., Rodríguez-Navarro, A., 2009. High-affinity sodium uptake in land plants. Plant Cell Physiol. 51, 68–79.
Hasegawa, P.M., Bressan, R.A., Zhu, J., Bohnert, H.J., Zhu, J.K., 2000. Plant cellular and molecular responses to high salinity. Annu. Rev. Plant Physiol. Plant Mol. Biol. 51, 463–499.
Hayat, K., Bundschuh, J., Farooq, J., Menhas, S., Hayat, S., Haq, F., et al., 2020. Combating soil salinity with combining saline agriculture and phytomanagement with salt-accumulating plants. Crit. Rev. Environ. Sci. Technol. 50, 1085–1115.
He, H., Zhan, J., He, L., Gu, M., 2012. Nitric oxide signaling in aluminum stress in plants. Protoplasma 249, 483–492.
Henderson, S.W., Baumann, U., Blackmore, D.H., Walker, A.R., Walker, R.R., Gilliham, M., 2014. Shoot chloride exclusion and salt tolerance in grapevine is associated with differential ion transporter expression in roots. BMC Plant Biol. 14, 273.
Herdean, A., Nziengui, H., Zsiros, O., Solymosi, K., Garab, G., Lundin, B., et al., 2016. The Arabidopsis thylakoid chloride channel AtCLCe functions in chloride homeostasis and regulation of photosynthetic electron transport. Front. Plant Sci. 7, 115.
Herzog, M., Striker, G.G., Colmer, T.D., Pedersen, O., 2016. Mechanisms of waterlogging tolerance in wheat - a review of root and shoot physiology. Plant Cell Environ. 39, 1068–1086.
Hessini, K., Issaoui, K., Ferchichi, S., Saif, T., Abdelly, C., Siddique, K.H.M., et al., 2019. Interactive effects of salinity and nitrogen forms on plant growth, photosynthesis and osmotic adjustment in maize. Plant Physiol. Biochem. 139, 171–178.
Hildebrandt, T.M., 2018. Synthesis versus degradation: directions of amino acid metabolism during Arabidopsis abiotic stress response. Plant Mol. Biol. 98, 121–135.

Hirschi, K.D., Korenkov, V.D., Wilganowski, N.L., Wagner, G.J., 2000. Expression of Arabidopsis CAX2 in tobacco. Altered metal accumulation and increased manganese tolerance. Plant Physiol. 124, 125–133.

Hoekenga, O.A., Maron, L.G., Pineros, M.A., Cancado, G.M.A., Shaff, J., Kobayashi, Y., et al., 2006. AtALMT1, which encodes a malate transporter, is identified as one of several genes critical for aluminum tolerance in Arabidopsis. Proc. Nat. Acad. Sci. USA 103, 9738–9743.

Horst, W.J., Schmohl, N., Kollmeier, M., Baluska, F., Sivaguru, M., 1999. Does aluminium affect root growth of maize through interaction with the cell wall - plasma membrane - cytoskeleton continuum? Plant Soil 215, 163–174.

Horst, W.J., Wang, Y., Eticha, D., 2010. The role of the root apoplast in aluminium-induced inhibition of root elongation and in aluminium resistance of plants: a review. Ann. Bot. 106, 185–197.

Hsouna, A.B., Ghneim-Herrera, T., Romdhane, W.B., Amira, D., Rania Ben, S., Faical, B., et al., 2020. Early effects of salt stress on the physiological and oxidative status of the halophyte *Lobularia maritima*. Funct. Plant Biol. 47, 912–924.

Hu, D., Sun, M., Sun, C., Liu, X., Zhang, Q., Zhao, J., et al., 2015. Conserved vacuolar H^+-ATPase subunit B1 improves salt stress tolerance in apple calli and tomato plants. Sci. Hortic. 197, 107–116.

Hu, H., He, J., Zhao, J., Ou, X., Li, H., Ru, Z., 2018. Low pH stress responsive transcriptome of seedling roots in wheat (*Triticum aestivum* L.). Genes Genom. 40, 1199–1211.

Hu, A.Y., Zheng, M., Sun, L., Zhao, X., Shen, R., 2019. Ammonium alleviates manganese toxicity and accumulation in rice by down-regulating the transporter gene *OsNramp5* through rhizosphere acidification. Front. Plant Sci. 10, 1194.

Huang, C.X., Steveninck, R.F.M., 1989. Maintenance of low Cl^- concentrations in mesophyll cells of leaf blades of barley seedlings exposed to salt stress. Plant Physiol. 90, 1440–1443.

Huang, S., Spielmeyer, W., Lagudah, E.S., James, R.A., Platten, J.D., Dennis, E.S., et al., 2006. A sodium transporter (HKT7) is a candidate for *Nax1*, a gene for salt tolerance in durum wheat. Plant Physiol. 142, 1718–1727.

Huang, X., Shabala, S., Shabala, L., Rengel, Z., Wu, X., Zhang, G., et al., 2015. Linking waterlogging tolerance with Mn^{2+} toxicity: a case study for barley. Plant Biol. 17, 26–33.

Huang, X., Fan, Y., Shabala, L., Rengel, Z., Shabala, S., Zhou, M.X., 2018. A major QTL controlling the tolerance to manganese toxicity in barley (*Hordeum vulgare* L.). Mol. Breed. 38, 16.

Huang, L., Kuang, L., Wu, L., Shen, Q., Han, Y., Jiang, L., et al., 2020. The HKT transporter HvHKT1;5 negatively regulates salt tolerance. Plant Physiol. 182, 584–596.

Ikka, T., Kobayashi, Y., Iuchi, S., Sakurai, N., Shibata, D., Kobayashi, M., et al., 2007. Natural variation of *Arabidopsis thaliana* reveals that aluminum resistance and proton resistance are controlled by different genetic factors. Theor. Appl. Genet. 115, 709–719.

Illes, P., Schlicht, M., Pavlovkin, J., Lichtscheidl, I., Baluska, F., Ovecka, M., 2006. Aluminium toxicity in plants: internalization of aluminium into cells of the transition zone in Arabidopsis root apices related to changes in plasma membrane potential, endosomal behaviour, and nitric oxide production. J. Exp. Bot. 57, 4201–4213.

Inostroza-Blancheteau, C., Reyes-Diaz, M., Berrios, G., Rodrigues-Salvador, A., Nunes-Nesi, A., Deppe, M., et al., 2017. Physiological and biochemical responses to manganese toxicity in ryegrass (*Lolium perenne* L.) genotypes. Plant Physiol. Biochem. 113, 89–97.

Jahan, N., Zhang, Y., Lv, Y., Song, M., Zhao, C., Hu, H., et al., 2020. QTL analysis for rice salinity tolerance and fine mapping of a candidate locus qSL7 for shoot length under salt stress. Plant Growth Regul. 90, 307–319.

Jaiswal, S.K., Naamala, J., Dakora, F.D., 2018. Nature and mechanisms of aluminium toxicity, tolerance and amelioration in symbiotic legumes and rhizobia. Biol. Fertil. Soils 54, 309–318.

James, R.A., Blake, C., Byrt, C.S., Munns, R., 2011. Major genes for Na^+ exclusion, *Nax1* and *Nax2* (wheat HKT1;4 and HKT1;5), decrease Na^+ accumulation in bread wheat leaves under saline and waterlogged conditions. J. Exp. Bot. 62, 2939–2947.

Jha, U.C., Abhishek, B., Rintu, J., Parida, S.K., 2019. Salinity stress response and 'omics' approaches for improving salinity stress tolerance in major grain legumes. Plant Cell Rep. 38, 255–277.

Jiang, Z., Zhou, X., Tao, M., Yuan, F., Liu, L., Wu, F., et al., 2019. Plant cell-surface GIPC sphingolipids sense salt to trigger Ca^{2+} influx. Nature 572, 341–346.

Jiménez, J.D.L.C., Pellegrini, E., Pedersen, O., Nakazono, M., 2021. Radial oxygen loss from plant roots—methods. Plants 10, 2322.

Jin, T., Sun, Y., Zhao, R., Shan, Z., Gai, J., Li, Y., 2019. Overexpression of peroxidase gene GsPRX9 confers salt tolerance in soybean. Intern. J. Mol. Sci. 20, 3745.

Jing, X., Hou, P., Lu, Y., Deng, S., Li, N., Zhao, R., et al., 2015. Overexpression of copper/zinc superoxide dismutase from mangrove *Kandelia candel* in tobacco enhances salinity tolerance by the reduction of reactive oxygen species in chloroplast. Front. Plant Sci. 6, 23.

Jones, D.L., Gilroy, S., Larsen, P.B., Howell, S.H., Kochian, L.V., 1998. Effect of aluminum on cytoplasmic Ca^{2+} homeostasis in root hairs of *Arabidopsis thaliana* (L.). Planta 206, 378–387.

Jones, D.L., Blancaflor, E.B., Kochian, L.V., Gilroy, S., 2006. Spatial coordination of aluminium uptake, production of reactive oxygen species, callose production and wall rigidification in maize roots. Plant Cell Environ. 29, 1309–1318.

Kader, M.A., Lindberg, S., 2010. Cytosolic calcium and pH signaling in plants under salinity stress. Plant Signal. Behav. 5, 233–238.

Kanchan, V., Neha, U., Nitin, K., Gaurav, Y., Jaspreet, S., Mishra, R.K., et al., 2017. Abscisic acid signaling and abiotic stress tolerance in plants: a review on current knowledge and future prospects. Front. Plant Sci. 8, 161.

Karimi, S., Tavallali, V., 2017. Interactive effects of soil salinity and boron on growth, mineral composition and CO_2 assimilation of pistachio seedlings. Acta Physiol. Plant. 39, 242.

Kaur, G., Singh, G., Motavalli, P.P., Nelson, K.A., Orlowski, J.M., Golden, B.R., 2020. Impacts and management strategies for crop production in waterlogged or flooded soils: a review. Agron. J. 112, 1475–1501.

Ke, Q., Wang, Z., Ji, C., Jeong, J., Lee, H., Li, H., et al., 2016. Transgenic poplar expressing codA exhibits enhanced growth and abiotic stress tolerance. Plant Physiol. Biochem. 100, 75–84.

Keska, K., Szczesniak, M.W., Makalowska, I., Czernicka, M., 2021. Long-term waterlogging as factor contributing to hypoxia stress tolerance enhancement in cucumber: comparative transcriptome analysis of waterlogging sensitive and tolerant accessions. Genes 12, 189.

Khabaz-Saberi, H., Rengel, Z., 2010. Aluminum, manganese, and iron tolerance improves performance of wheat genotypes in waterlogged acidic soils. J. Plant Nutr. Soil Sci. 173, 461–468.

Khabaz-Saberi, H., Rengel, Z., Wilson, R., Setter, T.L., 2010a. Variation for tolerance to high concentration of ferrous iron (Fe^{2+}) in Australian hexaploid wheat. Euphytica 172, 275–283.

Khabaz-Saberi, H., Rengel, Z., Wilson, R., Setter, T.L., 2010b. Variation of tolerance to manganese toxicity in Australian hexaploid wheat. J. Plant Nutr. Soil Sci. 173, 103–112.

Khabaz-Saberi, H., Barker, S.J., Rengel, Z., 2012. Tolerance to ion toxicities enhances wheat (*Triticum aestivum* L.) grain yield in waterlogged acidic soils. Plant Soil 354, 371–381.

Khabaz-Saberi, H., Barker, S.J., Rengel, Z., 2014. Tolerance to ion toxicities enhances wheat grain yield in acid soils prone to drought and transient waterlogging. Crop Pasture Sci. 65, 862–867.

Khalid, M.F., Sajjad, H., Anjum, M.A., Shakeel, A., Ali, M.A., Shaghef, E., et al., 2020. Better salinity tolerance in tetraploid vs diploid volkamer lemon seedlings is associated with robust antioxidant and osmotic adjustment mechanisms. J. Plant Physiol. 244, 153071.

Khan, M.S.H., Tawaraya, K., Sekimoto, H., Koyama, H., Kobayashi, Y., Murayama, T., et al., 2009. Relative abundance of Delta^5-sterols in plasma membrane lipids of root-tip cells correlates with aluminum tolerance of rice. Physiol. Plant. 135, 73–83.

Khan, I., Mohamed, S., Regnault, T., Mieulet, D., Guiderdoni, E., Sentenac, H., et al., 2020a. Constitutive contribution by the rice OsHKT1;4 Na^+ transporter to xylem sap desalinization and low Na^+ accumulation in young leaves under low as high external Na^+ conditions. Front. Plant Sci. 11, 1130.

Khan, R.M.I., Trivellini, A., Himanshu, C., Priyanka, C., Ferrante, A., Khan, N.A., et al., 2020b. The significance and functions of ethylene in flooding stress tolerance in plants. Environ. Exp. Bot. 179, 179.

Kholodova, V.P., Neto, D.S., Meshcheryakov, A.B., Borisova, N.N., Aleksandrova, S.N., Kuznetsov, V.V., 2002. Can stress-induced CAM provide for performing the developmental program in *Mesembryanthemum crystallinum* plants under long-term salinity? Russian J. Plant Physiol. 49, 336–343.

Kibria, M.G., Barton, L., Rengel, Z., 2021a. Applying foliar magnesium enhances wheat growth in acidic soil by stimulating exudation of malate and citrate. Plant Soil 446, 621–634.

Kibria, M.G., Barton, L., Rengel, Z., 2021b. Foliar application of magnesium mitigates soil acidity stress in wheat. J. Agron. Crop Sci. 207, 378–389.

Kidd, P.S., Llugany, M., Poschenrieder, C., Gunse, B., Barcelo, J., 2001. The role of root exudates in aluminium resistance and silicon-induced amelioration of aluminium toxicity in three varieties of maize (*Zea mays* L.). J. Exp. Bot. 52, 1339–1352.

Kim, Y., Hwang, S., Muhammad, W., Khan, A.L., Lee, J., Lee, J., et al., 2015. Comparative analysis of endogenous hormones level in two soybean (*Glycine max* L.) lines differing in waterlogging tolerance. Front. Plant Sci. 6, 714.

Kim, K., Cho, M., Kim, J., Lee, T., Heo, J., Jeong, J., et al., 2019. Growth response and developing simple test method for waterlogging stress tolerance in soybean. Journal of Crop Science and Biotechnology 22, 371–378.

Kinraide, T.B., 1991. Identity of the rhizotoxic aluminum species. Plant Soil 134, 167–178.

Kinraide, T.B., 1997. Reconsidering the rhizotoxicity of hydroxyl, sulphate, and fluoride complexes of aluminium. J. Exp. Bot. 48, 1115–1124.

Kinraide, T.B., 2006. Plasma membrane surface potential Ψ(PM) as a determinant of ion bioavailability: A critical analysis of new and published toxicological studies and a simplified method for the computation of plant Ψ(PM). Environol. Toxic. Chem. 25, 3188–3198.

Kinraide, T.B., Parker, D.R., 1990. Apparent phytotoxicity of mononuclear hydroxy-aluminum to four dicotyledonous species. Physiol. Plant. 79, 283–288.

Kinraide, T.B., Ryan, P.R., Kochian, L.V., 1992. Interactive effects of Al^{3+}, H^+, and other cations on root elongation considered in terms of cell-surface electrical potential. Plant Physiol. 99, 1461–1468.

Klug, B., Horst, W.J., 2010. Oxalate exudation into the root-tip water free space confers protection from aluminum toxicity and allows aluminum accumulation in the symplast in buckwheat (*Fagopyrum esculentum*). New Phytol. 187, 380–391.

Kobayashi, Y., Hoekenga, O.A., Itoh, H., Nakashima, M., Saito, S., Shaff, J.E., et al., 2007. Characterization of AtALMT1 expression in aluminum-inducible malate release and its role for rhizotoxic stress tolerance in Arabidopsis. Plant Physiol. 145, 843–852.

Kobayashi, N.I., Yamaji, N., Yamamoto, H., Okubo, K., Ueno, H., Costa, A., et al., 2017. OsHKT1;5 mediates Na^+ exclusion in the vasculature to protect leaf blades and reproductive tissues from salt toxicity in rice. Plant J. 91, 657–670.

Kollmeier, M., Felle, H.H., Horst, W.J., 2000. Genotypical differences in aluminum resistance of maize are expressed in the distal part of the transition zone. Is reduced basipetal auxin flow involved in inhibition of root elongation by aluminum? Plant Physiol. 122, 945–956.

Kong, X., Gao, X., Sun, W., An, J., Zhao, Y., Zhang, H., 2011. Cloning and functional characterization of a cation-chloride cotransporter gene *OsCCC1*. Plant Mol. Biol. 75, 567–578.

Kong, W., Yoo, M., Zhu, D., Noble, J.D., Kelley, T.M., Li, J., et al., 2020. Molecular changes in *Mesembryanthemum crystallinum* guard cells underlying the C_3 to CAM transition. Plant Mol. Biol. 103, 653–667.

Kopittke, P.M., 2016. Role of phytohormones in aluminium rhizotoxicity. Plant Cell Environ. 39, 2319–2328.

Kopittke, P.M., Menzies, N.W., Blamey, F.P.C., 2004. Rhizotoxicity of aluminate and polycationic aluminium at high pH. Plant Soil 266, 177–186.

Kopittke, P.M., Blamey, F.P.C., Menzies, N.W., 2008. Toxicities of soluble Al, Cu, and La include ruptures to rhizodermal and root cortical cells of cowpea. Plant Soil 303, 217–227.

Kopittke, P.M., Blamey, F.P.C., Wang, P., Menzies, N.W., 2011. Calculated activity of Mn^{2+} at the outer surface of the root cell plasma membrane governs Mn nutrition of cowpea seedlings. J. Exp. Bot. 62, 3993–4001.

Kopittke, P.M., Lombi, E., Mckenna, B.A., Wang, P., Donner, E., Webb, R.I., et al., 2013. Distribution and speciation of Mn in hydrated roots of cowpea at levels inhibiting root growth. Physiol. Plant. 147, 453–464.

Kopittke, P.M., Moore, K.L., Lombi, E., Gianoncelli, A., Ferguson, B.J., Blamey, F.P.C., et al., 2015. Identification of the primary lesion of toxic aluminum in plant roots. Plant Physiol. 167, 1402–1411.

Kopittke, P.M., Menzies, N.W., Wang, P., Blamey, F.P.C., 2016. Kinetics and nature of aluminium rhizotoxic effects: a review. J. Exp. Bot. 67, 4451–4467.

Kopittke, P.M., Gianoncelli, A., Kourousias, G., Green, K., Mckenna, B.A., 2017a. Alleviation of Al toxicity by Si is associated with the formation of Al-Si complexes in root tissues of sorghum. Front. Plant Sci. 8, 2189.

Kopittke, P.M., Mckenna, B.A., Karunakaran, C., Dynes, J.J., Arthur, Z., Gianoncelli, A., et al., 2017b. Aluminum complexation with malate within the root apoplast differs between aluminum resistant and sensitive wheat lines. Front. Plant Sci. 8, 1377.

Koster, P., Wallrad, L., Edel, K.H., Faisal, M., Alatar, A.A., Kudla, J., 2019. The battle of two ions: Ca^{2+} signalling against Na$^+$ stress. Plant Biol. 21, 39–48.

Kotula, L., Clode, P.L., Striker, G.G., Pedersen, O., Lauchli, A., Shabala, S., et al., 2015. Oxygen deficiency and salinity affect cell-specific ion concentrations in adventitious roots of barley (*Hordeum vulgare*). New Phytol. 208, 1114–1125.

Kotula, L., Clode, P.L., Jimenez, J.D.L.C., Colmer, T.D., 2019. Salinity tolerance in chickpea is associated with the ability to 'exclude' Na from leaf mesophyll cells. J. Exp. Bot. 70, 4991–5002.

Kovacik, J., Babula, P., Hedbavny, J., Svec, P., 2014. Manganese-induced oxidative stress in two ontogenetic stages of chamomile and amelioration by nitric oxide. Plant Sci. 215, 1–10.

Koyama, H., Toda, T., Yokota, S., Dawair, Z., Hara, T., 1995. Effects of aluminium and pH on root growth and cell viability in *Arabidopsis thaliana* strain Landsberg in hydroponic culture. Plant Cell Physiol. 36, 201–205.

Koyama, H., Toda, T., Hara, T., 2001. Brief exposure to low-pH stress causes irreversible damage to the growing root in *Arabidopsis thaliana*: pectin-Ca interaction may play an important role in proton rhizotoxicity. J. Exp. Bot. 52, 361–368.

Krupa-Malkiewicz, M., Smolik, B., 2019. Alleviative effects of chitosan and ascorbic acid on *Petunia x atkinsiana* D. Don under salinity. Eur. J. Hortic. Sci. 84, 359–365.

Lamers, J., Meer, T.V.D., Testerink, C., 2020. How plants sense and respond to stressful environments. Plant Physiol. 182, 1624–1635.

Lan, Z., Chen, C., Rashti, M.R., Yang, H., Zhang, D., 2018. High pyrolysis temperature biochars reduce nitrogen availability and nitrous oxide emissions from an acid soil. Global Change Biol. Bioener. 10, 930–945.

Latef, Aa.H.A., Kordrostami, M., Zakir, A., Zaki, H., Saleh, O.M., 2019. Eustress with H$_2$O$_2$ facilitates plant growth by improving tolerance to salt stress in two wheat cultivars. Plants 8, 303.

Lauricella, D., Butterly, C.R., Clark, G.J., Sale, P.W.G., Li, G., Tang, C., 2020. Effectiveness of innovative organic amendments in acid soils depends on their ability to supply P and alleviate Al and Mn toxicity in plants. J. Soils Sediments 20, 3951–3962.

Lee, Y., Kim, K., Jang, Y., Choi, I., 2014. Nitric oxide production and scavenging in waterlogged roots of rape seedlings. Genes Genom. 36, 691–699.

Leng, B.Y., Yuan, F., Dong, X.X., Wang, J., Wang, B.S., 2018. Distribution pattern and salt excretion rate of salt glands in two recretohalophyte species of *Limonium* (Plumbaginaceae). S. Afr. J. Bot. 115, 74–80.

Lessani, H., Marschner, H., 1978. Relation between salt tolerance and long-distance transport of sodium and chloride in various crop species. Aust. J. Plant Physiol. 5, 27–37.

Lewinska, K., Karczewska, A., Siepak, M., Szopka, K., Galka, B., Iqbal, M., 2019. Effects of waterlogging on the solubility of antimony and arsenic in variously treated shooting range soils. Appl. Geochem. 105, 7–16.

Li, X.F., Ma, J.F., Matsumoto, H., 2002. Aluminum-induced secretion of both citrate and malate in rye. Plant Soil 242, 235–243.

Li, P., Song, A., Li, Z., Fan, F., Liang, Y., 2012. Silicon ameliorates manganese toxicity by regulating manganese transport and antioxidant reactions in rice (*Oryza sativa* L.). Plant Soil 354, 407–419.

Li, J.Y., Liu, J.P., Dong, D.K., Jia, X.M., Mccouch, S.R., Kochian, L.V., 2014. Natural variation underlies alterations in Nramp aluminum transporter (NRAT1) expression and function that play a key role in rice aluminum tolerance. Proc. Nat. Acad. Sci. USA 111, 6503–6508.

Li, P., Song, A., Li, Z., Fan, F., Liang, Y., 2015. Silicon ameliorates manganese toxicity by regulating both physiological processes and expression of genes associated with photosynthesis in rice (*Oryza sativa* L.). Plant Soil 397, 289–301.

Li, N., Chen, M., Gao, X., Long, X., Shao, H., Liu, Z., et al., 2016. Carbon sequestration and Jerusalem artichoke biomass under nitrogen applications in coastal saline zone in the northern region of Jiangsu. China. Sci. Total Env. 568, 885–890.

Li, B., Tester, M., Gilliham, M., 2017a. Chloride on the move. Trends Plant Sci. 22, 236–248.

Li, D., Shu, Z., Ye, X., Zhu, J., Pan, J., Wang, W., et al., 2017b. Cell wall pectin methyl-esterification and organic acids of root tips involve in aluminum tolerance in *Camellia sinensis*. Plant Physiol. Biochem. 119, 265–274.

Li, X., Liu, J., Fang, J., Tao, L., Shen, R., Li, Y., et al., 2017c. Boron supply enhances aluminum tolerance in root border cells of pea (*Pisum sativum*) by interacting with cell wall pectins. Front. Plant Sci. 8, 742.

Li, Y., Huang, J., Song, X., Zhang, Z., Jiang, Y., Zhu, Y., et al., 2017d. An RNA-Seq transcriptome analysis revealing novel insights into aluminum tolerance and accumulation in tea plant. Planta 246, 91–103.

Li, N., Shao, T., Zhu, T., Long, X., Gao, X., Liu, Z., et al., 2018. Vegetation succession influences soil carbon sequestration in coastal alkali-saline soils in southeast China. Sci. Rep. 8, 9728.

Li, J., Jia, Y., Dong, R., Huang, R., Liu, P., Li, X., et al., 2019. Advances in the mechanisms of plant tolerance to manganese toxicity. Intern. J. Mol. Sci. 20, 5096.

Li, D.X., Ma, W., Wei, J., Mao, Y.W., Peng, Z.P., Zhang, J.R., et al., 2020a. Magnesium promotes root growth and increases aluminium tolerance via modulation of nitric oxide production in Arabidopsis. Plant Soil 457, 83–95.

Li, J., Su, L., Lv, A., Li, Y., Zhou, P., An, Y., 2020b. MsPG1 alleviated aluminum-induced inhibition of root growth by decreasing aluminum accumulation and increasing porosity and extensibility of cell walls in alfalfa (*Medicago sativa*). Environ. Exp. Bot. 175, 175.

Li, Y., Li, Y., Yang, M., Chang, S.X., Qi, J., Tang, C., et al., 2020c. Changes of microbial functional capacities in the rhizosphere contribute to aluminum tolerance by genotype-specific soybeans in acid soils. Biol. Fertil. Soil 56, 771–783.

Li, Y., Zhang, W., Zhu, W., Zhang, B., Huang, Q., Su, X., 2020d. Waterlogging tolerance and wood properties of transgenic *Populus alba* x *glandulosa* expressing *Vitreoscilla* hemoglobin gene (Vgb). J. For. Res. 32, 831–839.

Lian, T., Ma, Q., Shi, Q., Cai, Z., Zhang, Y., Cheng, Y., et al., 2019. High aluminum stress drives different rhizosphere soil enzyme activities and bacterial community structure between aluminum-tolerant and aluminum-sensitive soybean genotypes. Plant Soil 440, 409–425.

Licausi, F., Perata, P., 2009. Low oxygen signaling and tolerance in plants. Adv. Bot. Res. 50, 139–198.

Lin, Y., Chen, J., 2019. Effects of aluminum on the cell morphology in the root apices of two pineapples with different Al-resistance characteristics. Soil Sci. Plant Nutr. 65, 353–357.

Liu, Q., Yang, J.L., He, L.S., Li, Y.Y., Zheng, S.J., 2008. Effect of aluminum on cell wall, plasma membrane, antioxidants and root elongation in triticale. Biol. Plant. 52, 87–92.

Liu, M., Lou, H., Chen, W., Pineros, M.A., Xu, J., Fan, W., et al., 2018. Two citrate transporters coordinately regulate citrate secretion from rice bean root tip under aluminum stress. Plant Cell Environ. 41, 809–822.

Liu, T., Zhuang, L., Huang, B., 2019. Metabolic adjustment and gene expression for root sodium transport and calcium signaling contribute to salt tolerance in Agrostis grass species. Plant Soil 443, 219–232.

Liu, C., Liu, Y., Wang, S., Ke, Q., Yin, L., Deng, X., et al., 2020a. Arabidopsis mgd mutants with reduced monogalactosyldiacylglycerol contents are hypersensitive to aluminium stress. Ecotox. Environ. Safe. 203, 203.

Liu, L., Huang, L., Lin, X., Sun, C., 2020b. Hydrogen peroxide alleviates salinity-induced damage through enhancing proline accumulation in wheat seedlings. Plant Cell Rep. 39, 567–575.

Liu, S., Constable, G., Stiller, W., 2020c. Using leaf sodium concentration for screening sodicity tolerance in cotton (*Gossypium hirsutum* L.). Field Crops Res. 246, 107678.

Liu, X., Zhu, Y., Fang, X., Ye, J., Du, W., Zhu, Q., et al., 2020d. Ammonium aggravates salt stress in plants by entrapping them in a chloride over-accumulation state in an NRT1.1-dependent manner. Sci. Total Env. 746, 746.

Liu, Y., Xue, Y., Xie, B., Zhu, S., Lu, X., Liang, C., et al., 2020e. Complex gene regulation between young and old soybean leaves in responses to manganese toxicity. Plant Physiol. Biochem. 155, 231–242.

Lohaus, G., Hussmann, M., Pennewiss, K., Schneider, H., Zhu, J., Sattelmacher, B., 2000. Solute balance of a maize (*Zea mays* L.) source leaf as affected by salt treatment with special emphasis on phloem retranslocation and ion leaching. J. Exp. Bot. 51, 1721–1732.

Long, X.-H., Zhao, J., Liu, Z.-P., Rengel, Z., Liu, L., Shao, H.-B., et al., 2014. Applying geostatistics to determine the soil quality improvement by Jerusalem artichoke in coastal saline zone. Ecol. Eng. 70, 319–326.

Loreti, E., Perata, P., 2020. The many facets of hypoxia in plants. Plants 9, 745.

Lu, H., Nkoh, J.N., Biswash, M.R., Hua, H., Dong, G., Li, J., et al., 2021. Effects of surface charge and chemical forms of manganese(II) on rice roots on manganese absorption by different rice varieties. Ecotox. Environ. Safe. 207, 111224.

Ma, Z., Lin, S., 2019. Transcriptomic revelation of phenolic compounds involved in aluminum toxicity responses in roots of *Cunninghamia lanceolata* (Lamb.). Hook. Genes 10, 835.

Ma, J.F., Hiradate, S., Matsumoto, H., 1998. High aluminum resistance in buckwheat. II. Oxalic acid detoxifies aluminum internally. Plant Physiol. 117, 753–759.

Ma, J.F., Ryan, P.R., Delhaize, E., 2001. Aluminium tolerance in plants and the complexing role of organic acids. Trends Plant Sci. 6, 273–278.

Ma, Y., Li, C., Ryan, P.R., Shabala, S., You, J., Liu, J., et al., 2016. A new allele for aluminium tolerance gene in barley (*Hordeum vulgare* L.). BMC Genomics 17, 186.

Maas, E.V., Hoffman, G.J., 1977. Crop salt tolerance - current assessment. J. Irrig. Drainage Eng-ASCE 103, 115–134.

Maejima, E., Osaki, M., Wagatsuma, T., Watanabe, T., 2017. Contribution of constitutive characteristics of lipids and phenolics in roots of tree species in Myrtales to aluminum tolerance. Physiol. Plant. 160, 11–20.

Magalhaes, J.V., Liu, J., Guimaraes, C.T., Lana, U.G.P., Alves, V.M.C., Wang, Y.-H., et al., 2007. A gene in the multidrug and toxic compound extrusion (MATE) family confers aluminum tolerance in sorghum. Nature Genetics 39, 1156–1161.

Magalhaes, J.V., Pineros, M.A., Maciel, L.S., Kochian, L.V., 2018. Emerging pleiotropic mechanisms underlying aluminum resistance and phosphorus acquisition on acidic soils. Front. Plant Sci. 9, 1420.

Magri, E., Gugelmin, E.K., Grabarski, Fa.P., Barbosa, J.Z., Auler, A.C., Wendling, I., et al., 2020. Manganese hyperaccumulation capacity of *Ilex paraguariensis* A. St. Hil. and occurrence of interveinal chlorosis induced by transient toxicity. Ecotox. Environ. Safe. 203, 203.

Mahajan, S., Pandey, G.K., Tuteja, N., 2008. Calcium- and salt-stress signaling in plants: Shedding light on SOS pathway. Arch. Biochem. Biophys. 471, 146–158.

Maksimovic, J.D., Mojovic, M., Maksimovic, V., Romheld, V., Nikolic, M., 2012. Silicon ameliorates manganese toxicity in cucumber by decreasing hydroxyl radical accumulation in the leaf apoplast. J. Exp. Bot. 63, 2411–2420.

Malagoli, P., Britto, D.T., Schulze, L.M., Kronzucker, H.J., 2008. Futile Na$^+$ cycling at the root plasma membrane in rice (*Oryza sativa* L.): kinetics, energetics, and relationship to salinity tolerance. J. Exp. Bot. 59, 4109–4117.

Malathi, B., Pandey, G.K., 2019. Protein phosphatases meet reactive oxygen species in plant signaling networks. Environ. Exp. Bot. 161, 26–40.

Malik, A.I., Ailewe, T.I., Erskine, W., 2015. Tolerance of three grain legume species to transient waterlogging. AoB Plants 7, plv040.

Manet, L., Boyomo, O., Ngonkeu, E.L.M., Begoude, A.D.B., Sarr, P.S., 2016. Diversity and dynamics of rhizobial populations in acidic soils with aluminum and manganese toxicities in forest zones. International Journal of Agricultural Research, Innovation and Technology 6, 12–23.

Marcum, K.B., Pessarakli, M., 2006. Salinity tolerance and salt gland excretion efficiency of bermudagrass turf cultivars. Crop Sci. 46, 2571–2574.

Marin, L., Benlloch, M., Fernández-Escobar, R., 1995. Screening of olive cultivars for salt tolerance. Sci. Hortic. 64, 113–116.

Marschner, H., 1991. Mechanisms of adaptation of plants to acid soils. Plant Soil 134, 1–20.

Marschner, H., Kylin, A., Kuiper, P.J.C., 1981. Differences in salt tolerance of three sugar beet genotypes. Physiol. Plant. 51, 234–238.

Martins, A.P., Denardin, L.G.D.O., Tiecher, T., Borin, J.B.M., Schaidhauer, W., Anghinoni, I., et al., 2020. Nine-year impact of grazing management on soil acidity and aluminum speciation and fractionation in a long-term no-till integrated crop-livestock system in the subtropics. Geoderma 359, 113986.

Mehmood, K., Li, J.-Y., Jiang, J., Shi, R.-Y., Liu, Z.-D., Xu, R.-K., 2017. Amelioration of an acidic Ultisol by straw-derived biochars combined with dicyandiamide under application of urea. Environ. Sci. Pollut. Res. 24, 6698–6709.

Melo, J.O., Martins, L.G.C., Barros, B.A., Pimenta, M.R., Lana, U.G.P., Duarte, C.E.M., et al., 2019. Repeat variants for the SbMATE transporter protect sorghum roots from aluminum toxicity by transcriptional interplay in *cis* and *trans*. Proc. Nat. Acad. Sci. USA 116, 313–318.

Mhimdi, M., Perez-Perez, J.M., 2020. Understanding of adventitious root formation: what can we learn from comparative genetics? Front. Plant Sci. 11, 582020.

Mimmo, T., Marzadori, C., Gessa, C.E., 2009. Does the degree of pectin esterification influence aluminium sorption by the root apoplast? Plant Soil 314, 159–168.

Monreal, J.A., Feria, A.B., Vinardell, J.M., Vidal, J., Echevarria, C., Garcia-Maurino, S., 2007. ABA modulates the degradation of phosphoenolpyruvate carboxylase kinase in sorghum leaves. FEBS Lett. 581, 3468–3472.

Mori, I.C., Nobukiyo, Y., Nakahara, Y., Shibasaka, M., Furuichi, T., Katsuhara, M., 2018. A cyclic nucleotide-gated channel, HvCNGC2-3, is activated by the co-presence of Na$^+$ and K$^+$ and permeable to Na$^+$ and K$^+$ non-selectively. Plants 7, 61.

Muhammad, W., Fiaz, A., 2019. The phosphoenolpyruvate carboxylase gene family identification and expression analysis under abiotic and phytohormone stresses in *Solanum lycopersicum* L. Gene 690, 11–20.

Muhammad, I., Nabeela, Muhammad, I., Rahman, K.U., 2019. Effects of ascorbic acid against salt stress on the morphological and physiological parameters of *Solanum melongena* (L.). Pure and Applied Biology 8, 1425–1443.

Muhling, K.H., Lauchli, A., 2002. Effect of salt stress on growth and cation compartmentation in leaves of two plant species differing in salt tolerance. J. Plant Physiol. 159, 137–146.

Mujeeb-Kazi, A., Munns, R., Rasheed, A., Ogbonnaya, F.C., Niaz, A., Hollington, P., et al., 2019. Breeding strategies for structuring salinity tolerance in wheat. Adv. Agron. 155, 121–187.

Mullan, D.J., Barrett-Lennard, E.G., 2010. Breeding crops for tolerance to salinity, waterlogging and inundation. In: Reynolds, M.P. (Ed.), Climate Change and Crop Production. CABI, Wallingford, UK, pp. 92–114.

Munns, R., 2002. Comparative physiology of salt and water stress. Plant Cell Environ. 25, 239–250.

Munns, R., Gilliham, M., 2015. Salinity tolerance of crops - what is the cost? New Phytol. 208, 668–673.

Munns, R., James, R.A., 2003. Screening methods for salinity tolerance: a case study with tetraploid wheat. Plant Soil 253, 201–218.

Munns, R., Passioura, J.B., Guo, J., Chazen, O., Cramer, G.R., 2000. Water relations and leaf expansion: importance of time scale. J. Exp. Bot. 51, 1495–1504.

Munoz-Mayor, A., Pineda, B., Garcia-Abellan, J.O., Anton, T., Garcia-Sogo, B., Sanchez-Bel, P., et al., 2012. Overexpression of dehydrin tas14 gene improves the osmotic stress imposed by drought and salinity in tomato. J. Plant Physiol. 169, 459–468.

Mwando, E., Han, Y., Angessa, T.T., Zhou, G., Hill, C.B., Zhang, X., et al., 2020. Genome-wide association study of salinity tolerance during germination in barley (*Hordeum vulgare* L.). Front. Plant Sci. 11, 118.

Nabanita, B., Soumen, B., 2020. Complementation of ROS scavenging secondary metabolites with enzymatic antioxidant defense system augments redox-regulation property under salinity stress in rice. Physiol. Mol. Biol. Plants 26, 1623–1633.

Naeem, M.S., Hasitha, W., Liu, H., Liu, D., Rashid, A., Waraich, E.A., et al., 2012. 5-Aminolevulinic acid alleviates the salinity-induced changes in *Brassica napus* as revealed by the ultrastructural study of chloroplast. Plant Physiol. Biochem. 57, 84–92.

Nagayama, T., Nakamura, A., Yamaji, N., Satoh, S., Furukawa, J., Iwai, H., 2019. Changes in the distribution of pectin in root border cells under aluminum stress. Front. Plant Sci. 10, 1216.

Nair, K.M., Kumar, K.S.A., Lalitha, M., Shivanand, Kumar, S.C.R., Srinivas, S., et al., 2019. Surface soil and subsoil acidity in natural and managed land-use systems in the humid tropics of Peninsular India. Curr. Sci. 116, 1201–1211.

Najeeb, U., Bange, M.P., Tan, D.K.Y., Atwell, B.J., 2015. Consequences of waterlogging in cotton and opportunities for mitigation of yield losses. AoB Plants 7, plv080.

Nakano, Y., Kusunoki, K., Hoekenga, O.A., Tanaka, K., Iuchi, S., Sakata, Y., et al., 2020. Genome-wide association study and genomic prediction elucidate the distinct genetic architecture of aluminum and proton tolerance in *Arabidopsis thaliana*. Front. Plant Sci. 11, 405.

Navakode, S., Weidner, A., Lohwasser, U., Roder, M.S., Borner, A., 2009. Molecular mapping of quantitative trait loci (QTLs) controlling aluminium tolerance in bread wheat. Euphytica 166, 283–290.

Nedelyaeva, O.I., Shuvalov, A.V., Karpichev, I.V., Beliaev, D.V., Myasoedov, N.A., Khalilova, L.A., et al., 2019. Molecular cloning and characterisation of SaCLCa1, a novel protein of the chloride channel (CLC) family from the halophyte *Suaeda altissima* (L.) Pall. J. Plant Physiol. 240.

Negi, N.P., Divya, S., Shashi, S., Vinay, S., Sarin, N.B., 2016. Simultaneous overexpression of CuZnSOD and cAPX from *Arachis hypogaea* leads to salinity stress tolerance in tobacco. In Vitro Cell. Dev. Biol.-Plant 52, 484–491.

Negishi, T., Oshima, K., Hattori, M., Kanai, M., Mano, S., Nishimura, M., et al., 2012. Tonoplast- and plasma membrane-localized aquaporin-family transporters in blue hydrangea sepals of aluminum hyperaccumulating plant. PLoS ONE 7, e43189.

Negishi, T., Oshima, K., Hattori, M., Yoshida, K., 2013. Plasma membrane-localized Al-transporter from blue hydrangea sepals is a member of the anion permease family. Genes Cells 18, 341–352.

Neumann, P.M., 1993. Rapid and reversible modifications of extension capacity of cell walls in elongating maize leaf tissues responding to root addition and removal of NaCl. Plant Cell Environ. 16, 1107–1114.

Nie, H., Xu, L., Zhao, Q., Wang, N., You, C., Zhang, F., et al., 2020. Mechanisms of plant response to salinity stress: current understanding and recent progress. Pak. J. Bot. 52, 1879–1883.

Noorrohmah, S., Takahashi, H., Nakazono, M., 2020. Formation of a barrier to radial oxygen loss in L-type lateral roots of rice. Plant Root 14, 33–41.

Oertli, J.J., 1968. Extracellular salt accumulation, a possible mechanism of salt injury in plants. Agrochimica 12, 461–469.

Ogorek, L.L.P., Pellegrini, E., Pedersen, O., 2021. Novel functions of the root barrier to radial oxygen loss - radial diffusion resistance to H_2 and water vapour. New Phytol. 231, 1365–1376.

Ondrasek, G., Rengel, Z., 2021. Environmental salinization processes: detection, implications & solutions. Sci. Total Env. 754, 142432.

Ondrasek, G., Kranjcec, F., Filipovic, L., Filipovic, V., Bubalo Kovacic, M., Badovinac, I.J., et al., 2021a. Biomass bottom ash & dolomite similarly ameliorate an acidic low-nutrient soil, improve phytonutrition and growth, but increase Cd accumulation in radish. Sci. Total Env. 753, 141902.

Ondrasek, G., Zovko, M., Kranjcec, F., Savic, R., Romic, D., Rengel, Z., 2021b. Wood biomass fly ash ameliorates acidic, low-nutrient hydromorphic soil & reduces metal accumulation in maize. J. Cleaner Produc. 283, 124650.

Osawa, H., Kojima, K., 2006. Citrate-release-mediated aluminum resistance is coupled to the inducible expression of mitochondrial citrate synthase gene in *Paraserianthes falcataria*. Tree Physiol. 26, 565–574.

Otgonsuren, B., Rewald, B., Godbold, D.L., Goransson, H., 2016. Ectomycorrhizal inoculation of *Populus nigra* modifies the response of absorptive root respiration and root surface enzyme activity to salinity stress. Flora 224, 123–129.

Ozfidan-Konakci, C., Yildiztugay, E., Alp, F.N., Kucukoduk, M., Turkan, I., 2020. Naringenin induces tolerance to salt/osmotic stress through the regulation of nitrogen metabolism, cellular redox and ROS scavenging capacity in bean plants. Plant Physiol. Biochem. 157, 264–275.

Palma, F., Lopez-Gomez, M., Tejera, N.A., Lluch, C., 2014. Involvement of abscisic acid in the response of *Medicago sativa* plants in symbiosis with *Sinorhizobium meliloti* to salinity. Plant Sci. 223, 16–24.

Pan, C., Lu, H., Liu, J., Yu, J., Wang, Q., Li, J., et al., 2020. SODs involved in the hormone mediated regulation of H_2O_2 content in *Kandelia obovata* root tissues under cadmium stress. Environ. Pollut. 256.

Parker, D.R., Zelazny, L.W., Kinraide, T.B., 1988. Comparison of three spectrophotometric methods for differentiating mono- and polynuclear hydroxy-aluminum complexes. Soil Sci. Soc. Am. J. 52, 67–75.

Parker, D.R., Kinraide, T.B., Zelazny, L.W., 1989. On the phytotoxicity of polynuclear hydroxy-aluminum complexes. Soil Sci. Soc. Am. J. 53, 789–796.

Parvin, K., Hasanuzzaman, M., Bhuyan, M.H.M.B., Mohsin, S.M., Fujita, M., 2019. Quercetin mediated salt tolerance in tomato through the enhancement of plant antioxidant defense and glyoxalase systems. Plants 8, 247.

Peckova, E., Tylova, E., Soukup, A., 2016. Tracing root permeability: comparison of tracer methods. Biol. Plant. 60, 695–705.

Pedersen, O., Nakayama, Y., Yasue, H., Kurokawa, Y., Takahashi, H., Floytrup, A.H., et al., 2020. Lateral roots, in addition to adventitious roots, form a barrier to radial oxygen loss in *Zea nicaraguensis* and a chromosome segment introgression line in maize. New Phytol. 229, 94–105.

Peiter, E., Montanini, B., Gobert, A., Pedas, P., Husted, S., Maathuis, F.J.M., et al., 2007. A secretory pathway-localized cation diffusion facilitator confers plant manganese tolerance. Proc. Nat. Acad. Sci. USA 104, 8532–8537.

Perez-Jimenez, M., Hernandez-Munuera, M., Pinero, M.C., Lopez-Ortega, G., Amor, F.M.D., 2018. Are commercial sweet cherry rootstocks adapted to climate change? Short-term waterlogging and CO_2 effects on sweet cherry cv. 'Burlat'. Plant Cell Environ. 41, 908–918.

Philippot, L., Hallin, S., Borjesson, G., Baggs, E.M., 2009. Biochemical cycling in the rhizosphere having an impact on global change. Plant Soil 321, 61–81.

Pias, O.H.D.C., Tiecher, T., Cherubin, M.R., Silva, A.G.B., Bayer, C., 2020. Does gypsum increase crop grain yield on no-tilled acid soils? A meta-analysis. Agron. J. 112, 675–692.

Prasanna, B.M., Cairns, J.E., Zaidi, P.H., Beyene, Y., Makumbi, D., Gowda, M., et al., 2021. Beat the stress: breeding for climate resilience in maize for the tropical rainfed environments. Theor. Appl. Genet. 134, 1729–1752.

Qiao, C., Xu, B., Han, Y., Wang, J., Wang, X., Liu, L., et al., 2018. Synthetic nitrogen fertilizers alter the soil chemistry, production and quality of tea. A meta-analysis. Agron. Sustain. Dev. 38, 10.

Raman, H., Zhang, K., Cakir, M., Appels, R., Garvin, D.F., Maron, L.G., et al., 2005. Molecular characterization and mapping of ALMT1, the aluminium-tolerance gene of bread wheat (*Triticum aestivum* L.). Genome 48, 781–791.

Raman, H., Raman, R., Mcvittie, B., Orchard, B., Qiu, Y., Delourme, R., 2017. A major locus for manganese tolerance maps on chromosome A09 in a doubled haploid population of *Brassica napus* L. Front. Plant Sci. 8, 1952.

Rangel, A.F., Mohammad, M., Rao, I.M., Horst, W.J., 2005. Proton toxicity interferes with the screening of common bean (*Phaseolus vulgaris* L.) genotypes for aluminium resistance in nutrient solution. J. Plant Nutr. Soil Sci. 168, 607–616.

Rangel, A.F., Rao, I.M., Horst, W.J., 2009. Intracellular distribution and binding state of aluminum in root apices of two common bean (*Phaseolus vulgaris*) genotypes in relation to Al toxicity. Physiol. Plant. 135, 162–173.

Rangel, A.F., Rao, I.M., Braun, H.P., Horst, W.J., 2010. Aluminum resistance in common bean (*Phaseolus vulgaris*) involves induction and maintenance of citrate exudation from root apices. Physiol. Plant. 138, 176–190.

Rao, I.M., Miles, J.W., Beebe, S.E., Horst, W.J., 2016. Root adaptations to soils with low fertility and aluminium toxicity. Ann. Bot. 118, 593–605.

Rasheed, A., Fahad, S., Hassan, M.U., Tahir, M.M., Aamer, M., Wu, Z.M., 2020. A review on aluminum toxicity and Quantitative Trait Loci mapping in rice (*Oryza sativa* L). Appl. Ecol. Environ. Res. 18, 3951–3964.

Rauf, M., Awais, M., Ud-Din, A., Ali, K., Gul, H., Rahman, M.M., et al., 2021. Molecular mechanisms of the 1-aminocyclopropane-1-carboxylic acid (acc) deaminase producing *Trichoderma asperellum* MAP1 in enhancing wheat tolerance to waterlogging stress. Front. Plant Sci. 12, 614971.

Ren, C., Kong, C., Xie, Z., 2018. Role of abscisic acid in strigolactone-induced salt stress tolerance in arbuscular mycorrhizal *Sesbania cannabina* seedlings. BMC Plant Biol. 18, 74.

Ren, B., Hu, J., Zhang, J., Dong, S., Liu, P., Zhao, B., 2019. Spraying exogenous synthetic cytokinin 6-benzyladenine following the waterlogging improves grain growth of waterlogged maize in the field. J. Agron. Crop Sci. 205, 616–624.

Renault, P., Stengel, P., 1994. Modeling oxygen diffusion in aggregated soils: I. Anaerobiosis inside the aggregates. Soil Sci. Soc. Am. J. 58, 1017–1023.

Rengasamy, P., 2006. World salinization with emphasis on Australia. J. Exp. Bot. 57, 1017–1023.

Rengel, Z., 1992. Role of calcium in aluminium toxicity. New Phytol. 121, 499–513.

Rengel, Z., 2015. Availability of Mn, Zn and Fe in the rhizosphere. J. Soil Sci. Plant Nutr. 15, 397–409.

Rengel, Z., Zhang, W.H., 2003. Role of dynamics of intracellular calcium in aluminium toxicity syndrome. New Phytol. 159, 295–314.

Rengel, Z., Bose, J., Chen, Q., Tripathi, B.N., 2015. Magnesium alleviates plant toxicity of aluminium and heavy metals. Crop Pasture Sci. 66, 1298–1307.

Ribera, A.E., Reyes-Diaz, M.M., Alberdi, M.R., Alvarez-Cortez, D.A., Rengel, Z., Mora, M.D.L.L., 2013. Photosynthetic impairment caused by manganese toxicity and associated antioxidative responses in perennial ryegrass. Crop Pasture Sci. 64, 696–707.

Richter, A.K., Hirano, Y., Luster, J., Frossard, E., Brunner, I., 2011. Soil base saturation affects root growth of European beech seedlings. J. Plant Nutr. Soil Sci. 174, 408–419.

Riyazuddin, R., Radhika, V., Kalpita, S., Nisha, N., Monika, K., Bhati, K.K., et al., 2020. Ethylene: a master regulator of salinity stress tolerance in plants. Biomolecules 10, 959.

Rosales, M.A., Franco-Navarro, J.D., Peinado-Torrubia, P., Diaz-Rueda, P., Alvarez, R., Colmenero-Flores, J.M., 2020. Chloride improves nitrate utilization and NUE in plant. Front. Plant Sci. 11, 442.

Rossini Oliva, S., Mingorance, M.D., Sanhueza, D., Fry, S.C., Leidi, E.O., 2018. Active proton efflux, nutrient retention and boron-bridging of pectin are related to greater tolerance of proton toxicity in the roots of two Erica species. Plant Physiol. Biochem. 126, 142–151.

Ruiz, M., Quinones, A., Martinez-Cuenca, M.R., Aleza, P., Morillon, R., Navarro, L., et al., 2016. Tetraploidy enhances the ability to exclude chloride from leaves in Carrizo citrange seedlings. J. Plant Physiol. 205, 1–10.

Ryan, P.R., 2018. Assessing the role of genetics for improving the yield of Australia's major grain crops on acid soils. Crop Pasture Sci. 69, 242–264.

Ryan, P.R., Ditomaso, J.M., Kochian, L.V., 1993. Aluminium toxicity in roots: an investigation of spatial sensitivity and the role of the root cap. J. Exp. Bot. 44, 437–446.

Ryan, P.R., Delhaize, E., Jones, D.L., 2001. Function and mechanism of organic anion exudation from plant roots. Annu. Rev. Plant Physiol. Plant Mol. Biol. 52, 527–560.

Saddiq, M.S., Irfan, A., Basra, S.M.A., Zulfiqar, A., Ibrahim, A.M.H., 2018. Sodium exclusion is a reliable trait for the improvement of salinity tolerance in bread wheat. Arch. Agron. Soil Sci. 64, 272–284.

Salah, A., Zhan, M., Cao, C., Han, Y., Ling, L., Liu, Z., et al., 2019. γ-aminobutyric acid promotes chloroplast ultrastructure, antioxidant capacity, and growth of waterlogged maize seedlings. Sci. Rep. 9, 484.

Sales, L., Ohara, H., Ohkawa, K., Saito, T., Todoroki, Y., Kondo, S., 2018. Salt tolerance in apple seedlings is affected by exogenous ABA application. Acta Hortic. Available from: https://doi.org/10.17660/ActaHortic.2018.1206.17.

Sanchez, P.A., Salinas, J.G., 1981. Low-input technology for managing Oxisol and Ultisol in tropical America. Adv. Agron. 34, 279–406.

Sander, B.O., Schneider, P., Romasanta, R., Samoy-Pascual, K., Sibayan, E.B., Asis, C.A., et al., 2020. Potential of alternate wetting and drying irrigation practices for the mitigation of GHG emissions from rice fields: two cases in Central Luzon (Philippines). Agriculture 10, 350.

Santos, E.F., Santini, J.M.K., Paixao, A.P., Junior, E.F., Lavres, J., Campos, M., et al., 2017. Physiological highlights of manganese toxicity symptoms in soybean plants: Mn toxicity responses. Plant Physiol. Biochem. 113, 6–19.

Sarika, A., Sairam, R.K., Srivastava, G.C., Aruna, T., Meena, R.C., 2005. Role of ABA, salicylic acid, calcium and hydrogen peroxide on antioxidant enzymes induction in wheat seedlings. Plant Sci. 169, 559–570.

Sarita, S., Dubey, R.S., 2012. Nitric oxide alleviates manganese toxicity by preventing oxidative stress in excised rice leaves. Acta Physiol. Plant. 34, 819–825.

Sasaki, T., Yamamoto, Y., Ezaki, B., Katsuhara, M., Ahn, S., Ryan, P.R., et al., 2004. A wheat gene encoding an aluminum-activated malate transporter. Plant J. 37, 645–653.

Sawaki, Y., Iuchi, S., Kobayashi, Y., Kobayashi, Y., Ikka, T., Sakurai, N., et al., 2009. STOP1 regulates multiple genes that protect Arabidopsis from proton and aluminum toxicities. Plant Physiol. 150, 281–294.

Sawaki, K., Sawaki, Y., Zhao, C., Kobayashi, Y., Koyama, H., 2016. Specific transcriptomic response in the shoots of *Arabidopsis thaliana* after exposure to Al rhizotoxicity: - potential gene expression biomarkers for evaluating Al toxicity in soils. Plant Soil 409, 131–142.

Schmid, C., Sharma, S., Stark, T.D., Gunzkofer, D., Hofmann, T.F., Ulrich, D., et al., 2021. Influence of the abiotic stress conditions, waterlogging and drought, on the bitter sensometabolome as well as agronomical traits of six genotypes of *Daucus carota*. Foods 10, 1607.

Schmidt, S.B., Husted, S., 2020. Micronutrients: advances in understanding manganese cycling in soils, acquisition by plants and ways of optimizing manganese efficiency in crops. In: Rengel, Z. (Ed.), Achieving Sustainable Crop Nutrition. Burleigh Dodds Science Publishing Limited, Cambridge, UK, pp. 407–454.

Schroeder, J.I., Delhaize, E., Frommer, W.B., Guerinot, M.L., Harrison, M.J., Herrera-Estrella, L., et al., 2013. Using membrane transporters to improve crops for sustainable food production. Nature 497, 60–66.

Seguel, A., Cornejo, P., Ramos, A., Baer, E.V., Cumming, J., Borie, F., 2017. Phosphorus acquisition by three wheat cultivars contrasting in aluminium tolerance growing in an aluminium-rich volcanic soil. Crop Pasture Sci. 68, 305–316.

Setter, T.L., Waters, I., Sharma, S.K., Singh, K.N., Kulshreshtha, N., Yaduvanshi, N.P.S., et al., 2009. Review of wheat improvement for waterlogging tolerance in Australia and India: the importance of anaerobiosis and element toxicities associated with different soils. Ann. Bot. 103, 221–235.

Shabala, S., 2011. Physiological and cellular aspects of phytotoxicity tolerance in plants: the role of membrane transporters and implications for crop breeding for waterlogging tolerance. New Phytol. 190, 289–298.

Shabala, S., Munns, R., 2017. Salinity stress: physiological constraints and adaptive mechanisms. In: Shabala, S. (Ed.), Plant Stress Physiology, 2nd ed. CABI, Wallingford, UK, pp. 24–63.

Shabala, S.N., Newman, I.A., Morris, J., 1997. Oscillations in H^+ and Ca^{2+} ion fluxes around the elongation region of corn roots and effects of external pH. Plant Physiol. 113, 111–118.

Shabala, S., Cuin, T.A., Pottosin, I., 2007. Polyamines prevent NaCl-induced K^+ efflux from pea mesophyll by blocking non-selective cation channels. FEBS Lett. 581, 1993–1999.

Shah, N., Anwar, S., Xu, J., Hou, Z., Salah, A., Khan, S., et al., 2018. The response of transgenic *Brassica* species to salt stress: a review. Biotech. Lett. 40, 1159–1165.

Shahzad, M., Zorb, C., Geilfus, C.M., Muhling, K.H., 2013. Apoplastic Na^+ in *Vicia faba* leaves rises after short-term salt stress and is remedied by silicon. J. Agron. Crop Sci. 199, 161–170.

Shao, J., Yamaji, N., Shen, R., Ma, J., 2017. The key to Mn homeostasis in plants: regulation of Mn transporters. Trends Plant Sci. 22, 215–224.

Shao, T., Gu, X., Zhu, T., Pan, X., Zhu, Y., Long, X., et al., 2019. Industrial crop Jerusalem artichoke restored coastal saline soil quality by reducing salt and increasing diversity of bacterial community. Appl. Soil Ecol. 138, 195–206.

Shavrukov, Y., Hirai, Y., 2016. Good and bad protons: genetic aspects of acidity stress responses in plants. J. Exp. Bot. 67, 15–30.

Shavrukov, Y., Bovill, J., Afzal, I., Hayes, J.E., Roy, S.J., Tester, M., et al., 2013. HVP10 encoding V-PPase is a prime candidate for the barley *HvNax3* sodium exclusion gene: evidence from fine mapping and expression analysis. Planta 237, 1111–1122.

Shen, J., Wang, Y., Shu, S., Jahan, M.S., Zhong, M., Wu, J., et al., 2019. Exogenous putrescine regulates leaf starch overaccumulation in cucumber under salt stress. Sci. Hortic. 253, 99–110.

Shi, R.-Y., Ni, N., Nkoh, J.N., Li, J.-Y., Xu, R.-K., Qian, W., 2019. Beneficial dual role of biochars in inhibiting soil acidification resulting from nitrification. Chemosphere 234, 43–51.

Shi, R., Ni, N., Nkoh, J.N., Dong, Y., Zhao, W., Pan, X., et al., 2020. Biochar retards Al toxicity to maize (*Zea mays* L.) during soil acidification: the effects and mechanisms. Sci. Total Env. 719, 719.

Shimamura, S., Yamamoto, R., Nakamura, T., Shimada, S., Komatsu, S., 2010. Stem hypertrophic lenticels and secondary aerenchyma enable oxygen transport to roots of soybean in flooded soil. Ann. Bot. 106, 277–284.

Shukla, V., Lombardi, L., Pencik, A., Novak, O., Weits, D.A., Loreti, E., et al., 2020. Jasmonate signalling contributes to primary root inhibition upon oxygen deficiency in *Arabidopsis thaliana*. Plants 9, 1046.

Sibole, J.V., Cabot, C., Poschenrieder, C., Barcelo, J., 2003. Efficient leaf ion partitioning, an overriding condition for abscisic acid-controlled stomatal and leaf growth responses to NaCl salinization in two legumes. J. Exp. Bot. 54, 2111–2119.

Silva, E.F.L., Moreira, F.M.D.S., Siqueira, J.O., 2019. Mn concentration and mycorrhizal colonization in understory native species grown at areas of manganese mine tailings disposal. Intern. J. Phytoremed. 21, 564–576.

Singh, C.K., Dharmendra, S., Tomar, R.S.S., Sourabh, K., Upadhyaya, K.C., Madan, P., 2018. Molecular mapping of aluminium resistance loci based on root re-growth and Al-induced fluorescent signals (callose accumulation) in lentil (*Lens culinaris* Medikus). Mol. Biol. Rep. 45, 2103–2113.

Sivaguru, M., Horst, W.J., 1998. The distal part of the transition zone is the most aluminum-sensitive apical root zone of maize. Plant Physiol. 116, 155–163.

Sivaguru, M., Baluska, F., Volkmann, D., Felle, H.H., Horst, W.J., 1999. Impacts of aluminum on the cytoskeleton of the maize root apex. Short-term effects on the distal part of the transition zone. Plant Physiol. 119, 1073–1082.

Sivaguru, M., Fujiwara, T., Samaj, J., Baluska, F., Yang, Z.M., Osawa, H., et al., 2000. Aluminum-induced 1,3-β-D-glucan inhibits cell-to-cell trafficking of molecules through plasmodesmata. A new mechanism of aluminum toxicity in plants. Plant Physiol. 124, 991–1005.

Soil Science Glossary Terms Committee, 2008. Glossary of Soil Science Terms. Soil Science Society of America, Madison, WI, USA.

Somavilla, A., Caner, L., Silva, I.C.B.D., Bastos, M.C., Moro, L., Schaefer, G.L., et al., 2021. Chemical pattern of vegetation and topsoil of rangeland fertilized over 21 years with phosphorus sources and limestone. Soil Till. Res. 20547.

Soto-Cerda, B.J., Inostroza-Blancheteau, C., Mathias, M., Penaloza, E., Zuniga, J., Munoz, G., et al., 2015. Marker-assisted breeding for TaALMT1, a major gene conferring aluminium tolerance to wheat. Biol. Plant. 59, 83–91.

Sripinyowanich, S., Klomsakul, P., Boonburapong, B., Bangyeekhun, T., Asami, T., Gu, H., et al., 2013. Exogenous ABA induces salt tolerance in indica rice (*Oryza sativa L.*): the role of *OsP5CS1* and *OsP5CR* gene expression during salt stress. Environ. Exp. Bot. 86, 94–105.

Stass, A., Wang, Y., Eticha, D., Horst, W.J., 2006. Aluminium rhizotoxicity in maize grown in solutions with Al^{3+} or $Al(OH)_4^-$ as predominant solution Al species. J. Exp. Bot. 57, 4033–4042.

Stepien, P., Klobus, G., 2005. Antioxidant defense in the leaves of C3 and C4 plants under salinity stress. Physiol. Plant. 125, 31–40.

Striker, G.G., Colmer, T.D., 2017. Flooding tolerance of forage legumes. J. Exp. Bot. 68, 1851–1872.

Striker, G.G., Teakle, N.L., Colmer, T.D., Barrett-Lennard, E.G., 2015. Growth responses of *Melilotus siculus* accessions to combined salinity and root-zone hypoxia are correlated with differences in tissue ion concentrations and not differences in root aeration. Environ. Exp. Bot. 109, 89–98.

Striker, G.G., Kotula, L., Colmer, T.D., 2019. Tolerance to partial and complete submergence in the forage legume *Melilotus siculus*: an evaluation of 15 accessions for petiole hyponastic response and gas-filled spaces, leaf hydrophobicity and gas films, and root phellem. Ann. Bot. 123, 169–180.

Subba, A., Tomar, S., Pareek, A., Singla-Pareek, S.L., 2021. The chloride channels: Silently serving the plants. Physiol. Plant. 171, 688–702.

Sun, P., Tian, Q.-Y., Chen, J., Zhang, W.-H., 2010. Aluminium-induced inhibition of root elongation in Arabidopsis is mediated by ethylene and auxin. J. Exp. Bot. 61, 347–356.

Sun, C., Lu, L., Yu, Y., Liu, L., Hu, Y., Ye, Y., et al., 2016. Decreasing methylation of pectin caused by nitric oxide leads to higher aluminium binding in cell walls and greater aluminium sensitivity of wheat roots. J. Exp. Bot. 67, 979–989.

Sun, C., Liu, L., Zhou, W., Lu, L., Jin, C., Lin, X., 2017. Aluminum induces distinct changes in the metabolism of reactive oxygen and nitrogen species in the roots of two wheat genotypes with different aluminum resistance. J. Agric. Food Chem. 65, 9419–9427.

Sun, X., Lin, L., Sui, N., 2019. Regulation mechanism of microRNA in plant response to abiotic stress and breeding. Mol. Biol. Rep. 46, 1447–1457.

Szatanik-Kloc, A., Szerement, J., Cybulska, J., Jozefaciuk, G., 2017. Input of different kinds of soluble pectin to cation binding properties of roots cell walls. Plant Physiol. Biochem. 120, 194–201.

Szymanska, K.P., Polkowska-Kowalczyk, L., Lichocka, M., Maszkowska, J., Dobrowolska, G., 2019. SNF1-related protein kinases SnRK2.4 and SnRK2.10 modulate ROS homeostasis in plant response to salt stress. Intern. J. Mol. Sci. 20, 143.

Tan, W., Li, Q., Li, X., Zhao, Q., 2019. Effects of exogenous $CaCl_2$ and calcium inhibitors on the reactive oxygen species metabolism and Ca^{2+} transport of tamina (*Vitis vinifera*) grapevines under NaCl stress. Int. J. Agric. Biol. 21, 1263–1270.

Tao, Y., Niu, Y., Wang, Y., Chen, T., Naveed, S.A., Zhang, J., et al., 2018. Genome-wide association mapping of aluminum toxicity tolerance and fine mapping of a candidate gene for Nrat1 in rice. PLoS ONE 13, e0198589.

Tarun, B., Aditi, S., Sanjeev, P., Minhas, A.P., 2020. Salt tolerance mechanisms and approaches: future scope of halotolerant genes and rice landraces. Rice Sci. 27, 368–383.

Tavanti, R.F.R., Queiroz, G.D., Rocha Silva, A.C.D., Peres, W.M., Paixao, A.P., Galindo, F.S., et al., 2020. Changes in photosynthesis and antioxidant metabolism of cotton (*Gossypium hirsutum* L.) plants in response to manganese stress. Arch. Agron. Soil Sci. 66, 743–762.

Taylor, G.J., 1988. Mechanisms of aluminum tolerance in *Triticum aestivum* (wheat). V. Nitrogen nutrition, plant-induced pH, and tolerance to aluminum; correlation without causality? Can. J. Bot. 66, 694–699.

Teakle, N.L., Tyerman, S.D., 2010. Mechanisms of Cl^- transport contributing to salt tolerance. Plant Cell Environ. 33, 566–589.

Teakle, N.L., Amtmann, A., Real, D., Colmer, T.D., 2010. *Lotus tenuis* tolerates combined salinity and waterlogging: maintaining O_2 transport to roots and expression of an NHX1-like gene contribute to regulation of Na^+ transport. Physiol. Plant. 139, 358–374.

Tesfaye, M., Temple, S.J., Allan, D.L., Vance, C.P., Samac, D.A., 2001. Overexpression of malate dehydrogenase in transgenic alfalfa enhances organic acid synthesis and confers tolerance to aluminum. Plant Physiol. 127, 1836–1844.

Tester, M., Davenport, R., 2003. Na^+ tolerance and Na^+ transport in higher plants. Ann. Bot. 91, 503–527.

Tian, Z., Fan, Y., Wang, K., Zhong, H., Sun, L., Fan, D., et al., 2021. Searching for "win-win" solutions for food-water-GHG emissions tradeoffs across irrigation regimes of paddy rice in China. Resources, Conservation and Recycling 166, 105360.

Tipirdamaz, R., Gagneul, D., Duhaze, C., Ainouche, A., Monnier, C., Ozkum, D., et al., 2006. Clustering of halophytes from an inland salt marsh in Turkey according to their ability to accumulate sodium and nitrogenous osmolytes. Environ. Exp. Bot. 57, 139–153.

Todorova, D., Katerova, Z., Alexieva, V., Sergiev, I., 2015. Polyamines - possibilities for application to increase plant tolerance and adaptation capacity to stress. Genetics and Plant Physiology 5, 123–144.

Tornroth-Horsefield, S., Wang, Y., Hedfalk, K., Johanson, U., Karlsson, M., Tajkhorshid, E., et al., 2006. Structural mechanism of plant aquaporin gating. Nature 439, 688–694.

Tracy, F.E., Gilliham, M., Dodd, A.N., Webb, Aa.R., Tester, M., 2008. NaCl-induced changes in cytosolic free Ca^{2+} in *Arabidopsis thaliana* are heterogeneous and modified by external ionic composition. Plant Cell Environ. 31, 1063–1073.

Tran Nguyen, N., Pham Anh, T., Shalini, M., Son, S., Ayele, B.T., 2018. Hormonal regulation in adventitious roots and during their emergence under waterlogged conditions in wheat. J. Exp. Bot. 69, 4065–4082.

Tregeagle, J.M., Tisdall, J.M., Tester, M., Walker, R.R., 2010. Cl^- uptake, transport and accumulation in grapevine rootstocks of differing capacity for Cl^--exclusion. Funct. Plant Biol. 37, 665–673.

Trejo-Tellez, L.I., Stenzel, R., Gomez-Merino, F.C., Schmitt, J.M., 2010. Transgenic tobacco plants overexpressing pyruvate phosphate dikinase increase exudation of organic acids and decrease accumulation of aluminum in the roots. Plant Soil 326, 187–198.

Tripler, E., Ben-Gal, A., Shani, U., 2007. Consequence of salinity and excess boron on growth, evapotranspiration and ion uptake in date palm (*Phoenix dactylifera* L., cv. Medjool). Plant Soil 297, 147–155.

Tsunemitsu, Y., Yamaji, N., Ma, J., Kato, S., Iwasaki, K., Ueno, D., 2018. Rice reduces Mn uptake in response to Mn stress. Plant Signal. Behav. 13, e1422466.

Ullah, A., Li, M., Noor, J., Tariq, A., Liu, Y., Shi, L., 2019. Effects of salinity on photosynthetic traits, ion homeostasis and nitrogen metabolism in wild and cultivated soybean. PeerJ 7, e8191.

Um, T., Lee, S., Kim, J., Jang, G., Choi, Y., 2018. CHLORIDE CHANNEL 1 promotes drought tolerance in rice, leading to increased grain yield. Plant Biotechnol. Rep. 12, 283–293.

Us Salinity Laboratory Staff, 1954. Diagnosis and improvement of saline and alkali soils. U.S. Department of Agriculture.

Vega, I., Rumpel, C., Ruiz, A., Luz Mora, M.D.L., Calderini, D.F., Cartes, P., 2020. Silicon modulates the production and composition of phenols in barley under aluminum stress. Agronomy 10, 1138. Available from: https://doi.org/10.3390/agronomy10081138.

Verbeeck, M., Thiry, Y., Smolders, E., 2020. Soil organic matter affects arsenic and antimony sorption in anaerobic soils. Environ. Pollut. 257, 113566.

Vishal, B., Krishnamurthy, P., Ramamoorthy, R., Kumar, P.P., 2019. OsTPS 8 controls yield-related traits and confers salt stress tolerance in rice by enhancing suberin deposition. New Phytol. 221, 1369–1386.

Vlamis, J., Williams, D.E., 1964. Iron and manganese relations in rice and barley. Plant Soil 20, 221–231.

Wagatsuma, T., 2017. The membrane lipid bilayer as a regulated barrier to cope with detrimental ionic conditions: making new tolerant plant lines with altered membrane lipid bilayer. Soil Sci. Plant Nutr. 63, 507–516.

Wagatsuma, T., Uemura, M., Mitsuhashi, W., Maeshima, M., Ishikawa, S., Kawamura, T., et al., 2005. A new and simple technique for the isolation of plasma membrane lipids from root-tips. Soil Sci. Plant Nutr. 51, 135–139.

Wagatsuma, T., Maejima, E., Watanabe, T., Toyomasu, T., Kuroda, M., Muranaka, T., et al., 2018. Dark conditions enhance aluminum tolerance in several rice cultivars via multiple modulations of membrane sterols. J. Exp. Bot. 69, 567–577.

Walid, D., Rahma, G., Hans-Werner, K., Chedly, A., Arafet, M., 2020. Physiological and biochemical markers for screening salt tolerant quinoa genotypes at early seedling stage. J. Plant Interact. 15, 27–38.

Wan, X., Lei, M., Chen, T., Yang, J., Zhou, X., Zhou, G., 2015. Impact of waterlogging on the uptake of arsenic by hyperaccumulator and tolerant plant. Chem. Ecol. 31, 53–63.

Wang, Y., Ying, Y., Chen, J., Wang, X., 2004. Transgenic Arabidopsis overexpressing Mn-SOD enhanced salt-tolerance. Plant Sci. 167, 671–677.

Wang, H., Huang, J., Bi, Y., 2010. Nitrate reductase-dependent nitric oxide production is involved in aluminum tolerance in red kidney bean roots. Plant Sci. 179, 281–288.

Wang, P., Menzies, N.W., Lombi, E., Mckenna, B.A., De Jonge, M.D., Donner, E., et al., 2013a. Quantitative determination of metal and metalloid spatial distribution in hydrated and fresh roots of cowpea using synchrotron-based X-ray fluorescence microscopy. Sci. Total Env. 463, 131–139.

Wang, Y., Xu, H., Kou, J., Shi, L., Zhang, C., Xu, F., 2013b. Dual effects of transgenic *Brassica napus* overexpressing CS gene on tolerances to aluminum toxicity and phosphorus deficiency. Plant Soil 362, 231–246.

Wang, P., Yu, W., Zhang, J., Rengel, Z., Xu, J., Han, Q., et al., 2016. Auxin enhances aluminium-induced citrate exudation through upregulation of GmMATE and activation of the plasma membrane H$^+$-ATPase in soybean roots. Ann. Bot. 118, 933–940.

Wang, H., Hou, J., Li, Y., Zhang, Y., Huang, J., Liang, W., 2017. Nitric oxide-mediated cytosolic glucose-6-phosphate dehydrogenase is involved in aluminum toxicity of soybean under high aluminum concentration. Plant Soil 416, 39–52.

Wang, H., Ji, F., Zhang, Y., Hou, J., Liu, W., Huang, J., et al., 2019a. Interactions between hydrogen sulphide and nitric oxide regulate two soybean citrate transporters during the alleviation of aluminium toxicity. Plant Cell Environ. 42, 2340–2356.

Wang, H., Zhang, Y., Hou, J., Liu, W., Huang, J., Liang, W., 2019b. Nitric oxide mediates aluminum-induced citrate secretion through regulating the metabolism and transport of citrate in soybean roots. Plant Soil 435, 127–142.

Wang, M., Qiao, J., Yu, C., Chen, H., Sun, C., Huang, L., et al., 2019c. The auxin influx carrier, OsAUX3, regulates rice root development and responses to aluminium stress. Plant Cell Environ. 42, 1125–1138.

Wang, M., Wang, Y., Zhang, Y., Li, C., Gong, S., Yan, S., et al., 2019d. Comparative transcriptome analysis of salt-sensitive and salt-tolerant maize reveals potential mechanisms to enhance salt resistance. Genes Genom. 41, 781–801.

Wang, Q., Guan, C., Wang, P., Ma, Q., Bao, A., Zhang, J., et al., 2019e. The effect of AtHKT1;1 or AtSOS1 mutation on the expressions of Na$^+$ or K$^+$ transporter genes and ion homeostasis in *Arabidopsis thaliana* under salt stress. Intern. J. Mol. Sci. 20, 1085.

Wang, H., Wang, R., Liu, B., Yang, Y., Qin, L., Chen, E., et al., 2020a. QTL analysis of salt tolerance in *Sorghum bicolor* during whole-plant growth stages. Plant Breed. 139, 455–465.

Wang, W., Liu, Y., Duan, H., Yin, X., Cui, Y., Chai, W., et al., 2020b. SsHKT1;1 is coordinated with SsSOS1 and SsNHX1 to regulate NA$^+$ homeostasis in *Suaeda salsa* under saline conditions. Plant Soil 449, 117–131.

Wani, S.H., Vinay, K., Tushar, K., Rajasheker, G., Maheshwari, P., Solymosi, K., et al., 2020. Engineering salinity tolerance in plants: progress and prospects. Planta 251, 76.

Ward, B., Brien, C., Oakey, H., Pearson, A., Negrao, S., Schilling, R.K., et al., 2019. High-throughput 3D modelling to dissect the genetic control of leaf elongation in barley (*Hordeum vulgare*). Plant J. 98, 555–570.

Waszczak, C., Carmody, M., Kangasjarvi, J., 2018. Reactive oxygen species in plant signaling. Annu. Rev. Plant Biol. 69, 209–236.

Wehr, J.B., Menzies, N.W., Pax, F., Blamey, C., 2004. Inhibition of cell-wall autolysis and pectin degradation by cations. Plant Physiol. Biochem. 42, 485–492.

Wei, P., Wang, L., Liu, A., Yu, B., Lam, H., 2016. *GmCLC1* confers enhanced salt tolerance through regulating chloride accumulation in soybean. Front. Plant Sci. 7, 1082.

Wei, P., Che, B., Shen, L., Cui, Y., Wu, S., Cheng, C., et al., 2019. Identification and functional characterization of the chloride channel gene, *GsCLC-c2* from wild soybean. BMC Plant Biol. 19, 121.

Wieneke, J., Sarwar, G., Roeb, M., 1987. Existence of salt glands on leaves of kallar grass (*Leptochloa fusca* L. Kunth.). J. Plant. Nutr. 10, 805–819.

Wilson, H., Mycock, D., Weiersbye, I.M., 2017. The salt glands of *Tamarix usneoides* E. Mey. ex Bunge (South African salt cedar). Intern. J. Phytoremed. 19, 587–595.

Wissemeier, A.H., Diening, A., Hergenroder, A., Horst, W.J., Mix-Wagner, G., 1992. Callose formation as parameter for assessing genotypical plant tolerance of aluminium and manganese. Plant Soil 146, 67–75.

Wright, R.J., Baligar, V.C., Ritchey, K.D., Wright, S.F., 1989. Influence of soil solution aluminium on root elongation of wheat seedlings. Plant Soil 113, 294–298.

Wu, H., 2018. Plant salt tolerance and Na$^+$ sensing and transport. Crop J. 6, 215–225.

Wu, H., Zhu, M., Shabala, L., Zhou, M., Shabala, S., 2015. K$^+$ retention in leaf mesophyll, an overlooked component of salinity tolerance mechanism: a case study for barley. J. Integr. Plant Biol. 57, 171–185.

Wu, Y., Henderson, S.W., Wege, S., Zheng, F., Walker, A.R., Walker, R.R., et al., 2020. The grapevine NaE sodium exclusion locus encodes sodium transporters with diverse transport properties and localisation. J. Plant Physiol. 246, 247.

Wyn Jones, R.G., 1981. Salt tolerance. In: Johnson, C.B. (Ed.), Physiological Processes Limiting Plant Productivity. Butterworth, London, UK, pp. 271–292.

Xia, J., Yamaji, N., Kasai, T., Ma, J., 2010. Plasma membrane-localized transporter for aluminum in rice. Proc. Nat. Acad. Sci. USA 107, 18381–18385.

Xia, J., Yamaji, N., Che, J., Shen, R., Ma, J., 2014. Differential expression of Nrat1 is responsible for Al-tolerance QTL on chromosome 2 in rice. J. Exp. Bot. 65, 4297–4304.

Xiao, B., Jespersen, D., 2019. Morphological and physiological responses of seashore paspalum and bermudagrass to waterlogging stress. J. Am. Soc. Hortic. Sci. 144, 305–313.

Xiao, K., Yu, L., Xu, J., Brookes, P.C., 2014. pH, nitrogen mineralization, and KCl-extractable aluminum as affected by initial soil pH and rate of vetch residue application: results from a laboratory study. J Soils Sediments 14, 1513–1525.

Xu, Q., Burgess, P., Xu, J., Meyer, W., Huang, B., 2016. Osmotic stress- and salt stress-inhibition and gibberellin-mitigation of leaf elongation associated with up-regulation of genes controlling cell expansion. Environ. Exp. Bot. 131, 101–109.

Xu, L., Liu, C., Cui, B., Wang, N., Zhao, Z., Zhou, L., et al., 2018. Transcriptomic responses to aluminum (Al) stress in maize. J. Integr. Agric. 17, 1946–1958.

Xu, Z., Shao, T., Lv, Z., Yue, Y., Liu, A., Long, X., et al., 2020. The mechanisms of improving coastal saline soils by planting rice. Sci. Total Env. 703, 135529.

Yaish, M.W., Kumar, P.P., 2015. Salt tolerance research in date palm tree (*Phoenix dactylifera* L.), past, present, and future perspectives. Front. Plant Sci. 6, 348.

Yamamoto, Y., 2019. Aluminum toxicity in plant cells: mechanisms of cell death and inhibition of cell elongation. Soil Sci. Plant Nutr. 65, 41–55.

Yamamoto, Y., Hachiya, A., Matsumoto, H., 1997. Oxidative damage to membranes by a combination of aluminum and iron in suspension-cultured tobacco cells. Plant Cell Physiol. 38, 1333–1339.

Yamshi, A., Priyanka, S., Husna, S., Bajguz, A., Shamsul, H., 2020. Salinity induced physiological and biochemical changes in plants: an omic approach towards salt stress tolerance. Plant Physiol. Biochem. 156, 64–77.

Yan, F., Schubert, S., Mengel, K., 1992. Effect of low root medium pH on net proton release, root respiration, and root growth of corn (*Zea mays* L.) and broad bean (*Vicia faba* L.). Plant Physiol. 99, 415–421.

Yan, F., Feuerle, R., Schaffer, S., Fortmeier, H., Schubert, S., 1998. Adaptation of active proton pumping and plasmalemma ATPase activity of corn roots to low root medium pH. Plant Physiol. 117, 311–319.

Yan, L., Riaz, M., Wu, X., Du, C., Liu, Y., Jiang, C., 2018. Ameliorative effects of boron on aluminum induced variations of cell wall cellulose and pectin components in trifoliate orange (*Poncirus trifoliate* (L.) Raf.) rootstock. Environ. Pollut. 240, 764–774.

Yang, Y., Wang, Q.L., Geng, M.J., Guo, Z.H., Zhao, Z., 2011. Rhizosphere pH difference regulated by plasma membrane H$^+$-ATPase is related to differential Al-tolerance of two wheat cultivars. Plant Soil Environ. 57, 201–206.

Yang, L., Liu, H., Fu, S.M., Ge, H.M., Tang, R.J., Yang, Y., et al., 2017. Na$^+$/H$^+$ and K$^+$/H$^+$ antiporters AtNHX1 and AtNHX3 from Arabidopsis improve salt and drought tolerance in transgenic poplar. Biol. Plant. 61, 641–650.

Yang, Q., Ke, H., Liu, S., Zeng, Q., 2018. Phytoremediation of Mn-contaminated paddy soil by two hyperaccumulators (*Phytolacca americana* and *Polygonum hydropiper*) aided with citric acid. Environ. Sci. Pollut. Res. 25, 25933–25941.

Yemelyanov, V.V., Chirkova, T.V., Shishova, M.F., Lindberg, S.M., 2020. Potassium efflux and cytosol acidification as primary anoxia-induced events in wheat and rice seedlings. Plants 9, 1216.

Yermiyahu, U., Ben-Gal, A., Keren, R., Reid, R.J., 2008. Combined effect of salinity and excess boron on plant growth and yield. Plant Soil 304, 73–87.

Yi, X., Sun, Y., Yang, Q., Guo, A., Chang, L., Wang, D., et al., 2014. Quantitative proteomics of *Sesuvium portulacastrum* leaves revealed that ion transportation by V-ATPase and sugar accumulation in chloroplast played crucial roles in halophyte salt tolerance. J. Proteomics 99, 84–100.

Yolcu, S., Alavilli, H., Ganesh, P., Panigrahy, M., Song, K., 2021. Salt and drought stress responses in cultivated beets (*Beta vulgaris* L.) and wild beet (*Beta maritima* L.). Plants 10, 1843.

Yolcu, S., Alavilli, H., Ganesh, P., Asif, M., Kumar, M., Song, K., 2022. An insight into the abiotic stress responses of cultivated beets (*Beta vulgaris* L.). Plants 11, 12.

Yoneyama, T., Terakado-Tonooka, J., Bao, Z., Minamisawa, K., 2019. Molecular analyses of the distribution and function of diazotrophic rhizobia and methanotrophs in the tissues and rhizosphere of non-leguminous plants. Plants 8, 408.

Yoshida, J., Tomooka, N., Khaing, T.Y., Sunil Shantha, P.G., Naito, H., Matsuda, Y., et al., 2020. Unique responses of three highly salt-tolerant wild *Vigna* species against salt stress. Plant Production Sci. 23, 114–128.

Yu, Y., Zhou, W., Liang, X., Zhou, K., Lin, X., 2019. Increased bound putrescine accumulation contributes to the maintenance of antioxidant enzymes and higher aluminum tolerance in wheat. Environ. Pollut. 252, 941–949.

Yu, L., Dong, H., Li, Z., Han, Z., Korpelainen, H., Li, C., 2020. Species-specific responses to drought, salinity and their interactions in *Populus euphratica* and *P. pruinosa* seedlings. J. Plant Ecol. 13, 563–573.

Yuan, F., Leng, B., Wang, B., 2016. Progress in studying salt secretion from the salt glands in recretohalophytes: how do plants secrete salt? Front. Plant Sci. 7, 977.

Yue, Y., Shao, T., Long, X., He, T., Gao, X., Zhou, Z., et al., 2020. Microbiome structure and function in rhizosphere of Jerusalem artichoke grown in saline land. Sci. Total Env. 724, 138259.

Yusuff, O., Rafii, M.Y., Fatai, A., Samuel Chibuike, C., Ismaila, M., Kareem, I., et al., 2020. Submergence tolerance in rice: review of mechanism, breeding and, future prospects. Sustainability 12, 1632.

Zemunik, G., Winter, K., Turner, B.L., 2020. Toxic effects of soil manganese on tropical trees. Plant Soil 453, 343–354.

Zhang, W.H., Rengel, Z., 1999. Aluminium induces an increase in cytoplasmic calcium in intact wheat root apical cells. Aust. J. Plant Physiol. 26, 401–409.

Zhang, Z., Tang, C., Rengel, Z., 2005. Salt dynamics in rhizosphere of *Puccinellia ciliata* Bor. in a loamy soil. Pedosphere 15, 784–791.

Zhang, J.-L., Flowers, T.J., Wang, S.-M., 2009. Mechanisms of sodium uptake by roots of higher plants. Plant Soil 326, 45.

Zhang, H., Shi, W., You, J., Bian, M., Qin, X., Yu, H., et al., 2015. Transgenic *Arabidopsis thaliana* plants expressing a beta-1,3-glucanase from sweet sorghum (*Sorghum bicolor* L.) show reduced callose deposition and increased tolerance to aluminium toxicity. Plant Cell Environ. 38, 1178–1188.

Zhang, L., Liu, R., Gung, B.W., Tindall, S., Gonzalez, J.M., Halvorson, J.J., et al., 2016. Polyphenol-aluminum complex formation: implications for aluminum tolerance in plants. J. Agric. Food Chem. 64, 3025–3033.

Zhang, H., Zhao, F., Tang, R., Yu, Y., Song, J., Wang, Y., et al., 2017a. Two tonoplast MATE proteins function as turgor-regulating chloride channels in Arabidopsis. Proc. Nat. Acad. Sci. USA 114, E2036–E2045.

Zhang, J., Wei, J., Li, D., Kong, X., Rengel, Z., Chen, L., et al., 2017b. The role of the plasma membrane H^+-ATPase in plant responses to aluminum toxicity. Front. Plant Sci. 8, 1757.

Zhang, J., Li, D., Wei, J., Ma, W., Kong, X., Rengel, Z., et al., 2019a. Melatonin alleviates aluminum-induced root growth inhibition by interfering with nitric oxide production in Arabidopsis. Environ. Exp. Bot. 161, 157–165.

Zhang, Y., Zhang, J., Guo, J., Zhou, F., Somesh, S., Xu, X., et al., 2019b. F-box protein RAE1 regulates the stability of the aluminum-resistance transcription factor STOP1 in Arabidopsis. Proc. Nat. Acad. Sci. USA 116, 319–327.

Zhang, X., Li, Q., Xu, W., Zhao, H., Guo, F., Wang, P., et al., 2020a. Identification of MTP gene family in tea plant (*Camellia sinensis* L.) and characterization of CsMTP8.2 in manganese toxicity. Ecotox. Environ. Safe. 202, 110904.

Zhang, Y., Ponce, K.S., Meng, L., Chakraborty, P., Zhao, Q., Guo, L., et al., 2020b. QTL identification for salt tolerance related traits at the seedling stage in indica rice using a multi-parent advanced generation intercross (MAGIC) population. Plant Growth Regul. 92, 365–373.

Zhang, Z., Liu, D., Meng, H., Li, S., Wang, S., Xiao, Z., et al., 2020c. Magnesium alleviates aluminum toxicity by promoting polar auxin transport and distribution and root alkalization in the root apex in populus. Plant Soil 448, 565–585.

Zhang, H., Li, G., Yan, C., Cao, N., Yang, H., Le, M., et al., 2021. Depicting the molecular responses of adventitious rooting to waterlogging in melon hypocotyls by transcriptome profiling. 3 Biotech 11, 351.

Zhao, J., Wang, W., Zhou, H., Wang, R., Zhang, P., Wang, H., et al., 2017. Manganese toxicity inhibited root growth by disrupting auxin biosynthesis and transport in Arabidopsis. Front. Plant Sci. 8, 272.

Zhao, X., Chen, R., Shen, R., 2014. Coadaptation of plants to multiple stresses in acidic soils. Soil Science 179, 503–513.

Zhao, X.Q., Shen, R.F., 2018. Aluminum-nitrogen interactions in the soil-plant system. Front. Plant Sci. 9, 807.

Zhao, L., Cui, J., Cai, Y., Yang, S., Liu, J., Wang, W., et al., 2020a. Comparative transcriptome analysis of two contrasting soybean varieties in response to aluminum toxicity. Int. J. Mol. Sci. 21, 4316.

Zhao, W.-R., Li, J.-Y., Jiang, J., Lu, H.-L., Hong, Z.-N., Qian, W., et al., 2020b. The mechanisms underlying the reduction in aluminum toxicity and improvements in the yield of sweet potato (*Ipomoea batatas* L.) after organic and inorganic amendment of an acidic ultisol. Agric. Ecosys. Env. 288, 106716.

Zheng, W., Li, D., Chen, F., Liu, X., Li, H., 2019. Abiotic stress tolerance and ABA responses of transgenic *Glycine max* plants with modulated RACK1 expression. Can. J. Plant Sci. 99, 250–267.

Zhifang, G., Loescher, W.H., 2003. Expression of a celery mannose 6-phosphate reductase in Arabidopsis thaliana enhances salt tolerance and induces biosynthesis of both mannitol and a glucosyl-mannitol dimer. Plant Cell Environ. 26, 275–283.

Zhong, M., Wang, Y., Shu, S., Sun, J., Guo, S., 2020. Ectopic expression of *CsTGase* enhances salt tolerance by regulating polyamine biosynthesis, antioxidant activities and Na^+/K^+ homeostasis in transgenic tobacco. Plant Sci. 296, 110492.

Zhou, C., Qi, Y., You, X., Yang, L., Guo, P., Ye, X., et al., 2013a. Leaf cDNA-AFLP analysis of two citrus species differing in manganese tolerance in response to long-term manganese-toxicity. BMC Genomics 14, 621.

Zhou, G., Delhaize, E., Zhou, M., Ryan, P.R., 2013b. The barley MATE gene, HvAACT1, increases citrate efflux and Al^{3+} tolerance when expressed in wheat and barley. Ann. Bot. 112, 603–612.

Zhou, C., Li, C., Liang, W., Guo, P., Yang, L., Chen, L., 2017. Identification of manganese-toxicity-responsive genes in roots of two citrus species differing in manganese tolerance using cDNA-AFLP. Trees-Struct. Funct. 31, 813–831.

Zhou, W., Chen, F., Meng, Y., Umashankar, C., Luo, X., Yang, W., et al., 2020a. Plant waterlogging/flooding stress responses: from seed germination to maturation. Plant Physiol. Biochem. 148, 228–236.

Zhou, X., Wang, Y., Shen, S., 2020b. Transcriptomic comparison reveals modifications in gene expression, photosynthesis, and cell wall in woody plant as responses to external pH changes. Ecotox. Environ. Safe. 203, 111007.

Zhu, J., Meinzer, F.C., 1999. Efficiency of C_4 photosynthesis in *Atriplex lentiformis* under salinity stress. Aust. J. Plant Physiol. 26, 79–86.

Zhu, X., Lei, G., Wang, Z., Shi, Y., Braam, J., Li, G., et al., 2013. Coordination between apoplastic and symplastic detoxification confers plant aluminum resistance. Plant Physiol. 162, 1947–1955.

Zhu, M., Zhou, M., Shabala, L., Shabala, S., 2017. Physiological and molecular mechanisms mediating xylem Na^+ loading in barley in the context of salinity stress tolerance. Plant Cell Environ. 40, 1009–1020.

Zhu, G., Gao, W., Song, X., Sun, F., Hou, S., Liu, N., et al., 2020. Genome-wide association reveals genetic variation of lint yield components under salty field conditions in cotton (*Gossypium hirsutum* L.). BMC Plant Biol. 20, 23.

Zorb, C., Muhling, K.H., Kutschera, U., Geilfus, C.M., 2015. Salinity stiffens the epidermal cell walls of salt-stressed maize leaves: is the epidermis growth-restricting? PLoS ONE 10, e0118406.

Zulfiqar, A., Abdus, S., Azhar, F.M., Khan, I.A., 2007. Genotypic variation in salinity tolerance among spring and winter wheat (*Triticum aestivum* L.) accessions. S. Afr. J. Bot. 73, 70–75.

Chapter 19

Nutrition of plants in a changing climate

Sylvie M. Brouder and Jeffrey J. Volenec
Department of Agronomy, Purdue University, West Lafayette, IN, United States

Summary

Increasing levels of greenhouse gasses (e.g., CO_2, N_2O, CH_4) in Earth's atmosphere are elevating the planet's temperature and modifying precipitation patterns. When coupled with additional, human-induced changes in nutrient cycles, including global increases in reactive nitrogen (N), these environmental changes can change biogeochemical cycling of nutrients and plant growth and modify plant nutrient uptake. Plant responses to elevated CO_2 (eCO_2) vary with functional group, with C_3 plants impacted more than those possessing C_4 photosynthesis. In general, uptake of most nutrients scales with changes in plant growth. Thus, CO_2-enhanced growth of C_3 plants generally requires a proportional increase in nutrient uptake. The exception to this is N; protein and N concentrations in leaves of C_3 plants are often reduced at eCO_2. This can translate to greater N-use efficiency. Soil nutrient cycles are also changing with global climate change (GCC). Decreased nutrient availability in root zones induced by GCC may restrict plant growth response to eCO_2, and nutrient imbalances may lead to positive GCC feedbacks. Mineral composition of edible plant tissues (grains, fruits, etc.) can be altered in plants grown in eCO_2, with commonly observed low concentrations of N, Fe, and Zn (mainly in C_3 plants) leading to potential reductions of the nutritional value of food and feed.

19.1 General

Global climate change (GCC) is a present condition and a concern for the future. Energy redistribution from warmer equatorial regions to extreme latitudes is driven by large-scale oceanic and atmospheric circulation that forms Earth's weather and climate (Jia et al., 2019). These circulation and energy patterns have been relatively stable for the recorded history, creating an array of eco-zones (tropics, subtropics, temperate, subpolar, polar) with well-described and predictable local environments, and biota and cropping systems adapted to the prevailing conditions. Burning fossil fuels release gasses that trap radiation in the atmosphere, warming the planet (i.e., the greenhouse effect), changing the Earth's energy balance, and driving unprecedented and rapid environmental change at a global scale. These changes are manifested in numerous ways, including loss of polar ice, increased frequency and intensity of hurricanes, increased average temperatures, greater frequency/duration of heat waves, longer and more widespread drought, larger and more frequent wildfires, and changes in intensity, duration, and seasonality of precipitation.

At regional scales, GCC is expected to increasingly exacerbate food and nutrition insecurity, especially where instability in agricultural productivity already drives hunger and undernutrition (Brevik, 2013; Lipper et al., 2014). Plant nutrition can play a role in either mitigation or exacerbation of GCC when plants drive increased C sequestration in plant–soil systems or cause a net release of greenhouse gases (GHGs: carbon dioxide, CO_2; nitrous oxide, N_2O; and methane, CH_4). Current non-CO_2 contributions of agriculture to global GHG emissions are estimated at 6.2 Gt CO_2 eq yr^{-1} (10%–14%) with 2–3 Gt CO_2 eq yr^{-1} sourced to croplands (Mbow et al., 2019). Carlson et al. (2017) estimate 20% of cropland emissions can be sourced directly to N fertilizer application and losses of N_2O, which has a global warming potential 300 times that of CO_2. However, maintaining crop productivity in future climates may require increased fertilizer inputs but the use efficiency of these inputs may decline, and associated nutrient losses may have positive and/or negative feedbacks to natural plant community responses to GCC and to global stocks of soil carbon (C) and plant mineral nutrients. Thus, nutrient management is a critical consideration in sustainability and GCC mitigation efforts focused on achieving C-neutral crop production. Empirical studies of the interactive impacts of GCC on plant nutrition are increasingly available, adding important evidence to a wealth of model projections. This chapter explores the impact of GCC on nutrient dynamics and availability in soil, demand and uptake of nutrients by plants, system-level nutrient losses with positive feedbacks to GCC, and composition of plant material used for human food and other purposes. This chapter also discusses the role of adequate plant nutrition in mitigation of climate change-linked stresses such as heat, drought, and high irradiation. Given that many potential, nutrient-related effects are not thoroughly understood at present, this chapter provides an illustrative nonexhaustive examination of key considerations.

19.2 The changing climate

19.2.1 Historical climate trends

Climate scientists agree that increasing CO_2 and other GHGs like N_2O are altering Earth's temperature and hydrologic cycle (Jia et al., 2019). From 1960 to 2019 the CO_2 concentrations in Earth's atmosphere have increased from 320 to 415 µL L^{-1} (Fig. 19.1). Current CO_2 levels are higher than any time in the last 800,000 years at least. These GHGs absorb infra-red radiation and warm the earth by amplifying its natural greenhouse effect. As a result, the temperature of Earth has increased approximately 1°C in the last 60 years. This increased temperature is not the result of changes in solar radiation output that has remained unchanged during this time. The warmer temperatures are impacting plants and plant nutrition by changing precipitation patterns, increasing the frequency and duration of drought, and changing the length of the growing season. For example, since 1970, there has been a steady decline in snow cover in the Northern Hemisphere. Snow is melting earlier when compared to the long-term average, and plant growth is resuming sooner in spring. This climate-driven loss of snow and earlier start to the growing season have added from 5 (Europe and North America) to as many as 9.5 (Asia) days to the growing season between 1950 and 2004 (Cornes et al., 2019; Linderholm, 2006).

19.2.2 Soil temperature

Not surprisingly, increases in Earth's air temperature are also increasing the land surface temperature (Yan et al., 2020). In a grassland ecosystem in the Netherlands, mean soil temperatures have increased approximately 1°C from 1987 to 2009 (Jacobs et al., 2011). Soil temperatures in the Tibetan Plateau have increased on average 0.35°C per decade

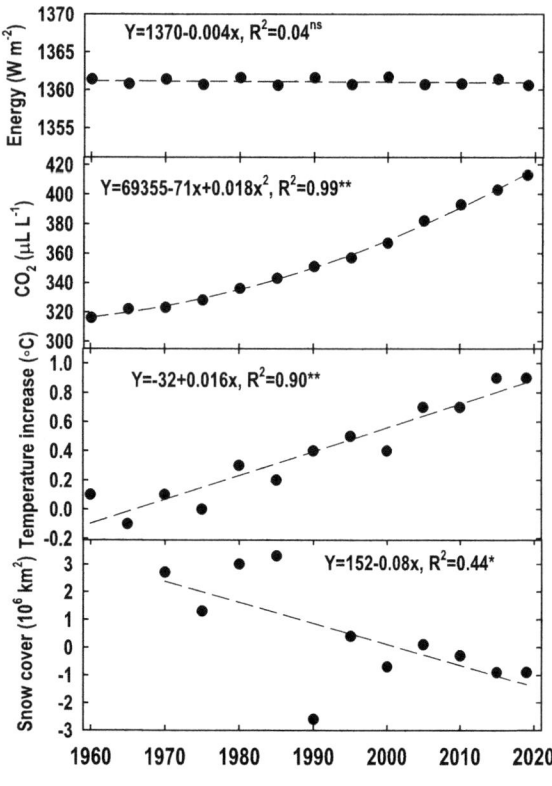

FIGURE 19.1 Trends in solar radiation, atmospheric carbon dioxide (CO_2), air temperature, and spring snow cover in the Northern Hemisphere from 1960 to 2019. Data were plotted at 5-year intervals. Snow cover data represent changes over mid-latitudes between 1970 and 2019 relative to the average for this time interval (set = 0). Trend lines are provided along with the coefficient of determination (R^2) and level of statistical significance of the regression (* and ** represent P levels of 5 and 1% probability, respectively). *Adapted from the National Oceanic and Atmospheric Administration (NOAA) climate website (https://www.climate.gov/maps-data).*

between 1960 and 2014 (Fang et al., 2019). However, the temporal pattern of these increases varies markedly with soil depth and season (Fig. 19.2). For example, surface soil temperatures have largely paralleled those of air temperature, with greater changes in late autumn and winter when compared to summer. By comparison, the temperature of soils 40−320 cm below the surface has increased more in summer (nearly 4°C), whereas winter soil temperatures at this depth have increased less than 1°C. Soil at the 5−20 cm depth also exhibited greater changes in temperature in summer than in autumn and winter months. As plant growth and nutrient acquisition are generally higher in summer months, these changes in root zone temperature impact both root growth (Chapter 13) and soil processes like organic matter mineralization and ion diffusion (Chapter 12) that influence nutrient release and movement.

In the future, soil temperatures are expected to rise in concert with increases in air temperatures. By the end of the 21st century, climate models predict increases of between 2.3 and 4.5°C irrespective of soil depth (Soong et al., 2020). The greatest warming is expected to occur at high altitudes and latitudes, with relatively less warming of tropical soils. In addition, models predict that soil moisture at the 10-cm depth will decline by 5 kg m^{-3} in temperate soils, and up to 10 kg m^{-3} in boreal and some tropical soils.

19.2.3 Precipitation and soil moisture

Precipitation patterns are also expected to be altered by GCC, and like soil temperatures, these patterns will exhibit significant spatiotemporal variation (Jia et al., 2019). In general, wet regions are expected to get wetter and dry regions drier. The increased temperature and reduced precipitation in already dry environments will limit plant evapotranspiration (ET) and further strain water resources for irrigation. Reduced plant productivity in dry environments is expected because low soil moisture decreases stomatal conductance and photosynthesis and increases tissue temperatures (Green et al., 2019). Plant tolerance of high temperatures can interact with plant nutritional status. Mengutay et al. (2013) showed that magnesium (Mg)-deficient plants were more susceptible to high temperature stress than Mg-sufficient plants, associated with the poorer growth and exacerbated oxidative stress in low-Mg plants. Similarly, Zn- and K-deficient plants show increased susceptibility to high irradiation and heat stress (Cakmak, 2000; García-Martí et al., 2019; Peck and McDonald, 2010).

Extra precipitation at high latitudes and in the humid tropics is expected to increase water availability and this may benefit terrestrial ecosystems, including agriculture, but elevated temperatures that can enhance ET may moderate or eliminate these potential benefits. For example, soil moisture at 80 drainage basins was monitored for three decades, and spatiotemporal variation was found to be associated closely with variation in the ratio of precipitation to potential ET (Chen et al., 2016). These authors also reported the vast majority of eastern China had experienced significant reductions in soil moisture between 1979 and 2010, with GCC being the underlying cause at many of the locations. Recent modeling results illustrate the precipitation-ET linkage, and how this can impact future soil moisture levels. By 2090 precipitation in the Great Lakes region of the United States is predicted to be reduced greatly in summer and increased markedly in autumn and winter (Fig. 19.3) (Wang et al., 2018). The reduction in summer precipitation could reduce runoff and erosion, but runoff is expected to increase during autumn and winter when precipitation increases. This temporal shift is expected to increase nutrient losses to lakes and streams; by 2050 P losses to surface waters are expected to increase by as much as 30% in winter (Ockenden et al., 2017). Increased frequency and intensity of extreme

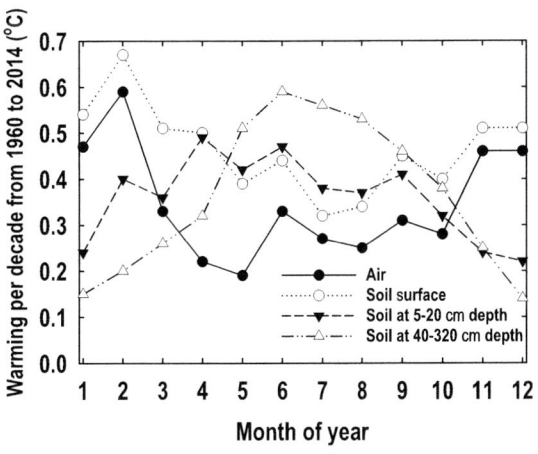

FIGURE 19.2 Monthly warming trends from 1960 to 2014 for air, the soil surface and at 5−20 and 40−320 cm soil depths. Data were compiled from 50 meteorological stations located within the Tibetan Plateau. *Adapted from Fang et al. (2019).*

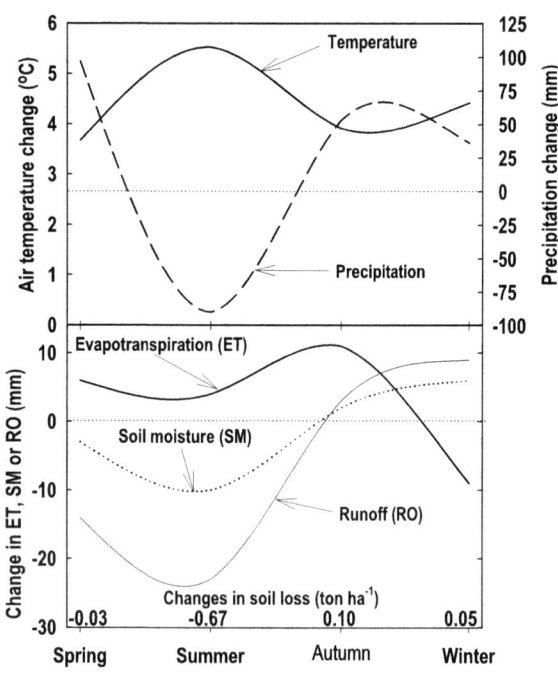

FIGURE 19.3 Predicted changes in air temperature, precipitation, ET, SM, RO, and soil erosion for croplands by season in the Great Lakes region of the United States. Changes are relative to conditions in 2000 and are predictions for the 2090s using the A2 climate scenario and three general circulation models. *ET*, Evapotranspiration; *RO*, runoff; *SM*, soil moisture. *Adapted from Wang et al. (2018).*

TABLE 19.1 Influence of climate change on frequency, duration, and spatial extent of severe and extreme drought events predicted for Europe.

Climate scenario	Period	All events			Events >10^6 km²		
		Events, no.	Event duration, months	Average event area, 10^3 km²	Events, no.	Event duration, months	Average event area, 10^6 km²
Historic	1966–2006	40	10.8	287	0.3	11.0	1.11
RCP 6.0	2020–59	48	12.8	351	2.0	20.4	1.26
RCP 6.0	2060–99	88	16.0	363	7.3	38.0	1.28

A moderate Representative Concentration Pathways (RCP 6.0) scenario was used to predict the frequency, duration, and spatial impact of droughts in Europe in the near (2020–59) and the far future (2060–99). Results for large events (>10^6 km²) are identified separately from all events.
Source: Adapted from Grillakis (2019).

precipitation events are also expected to increase flooding, surface runoff, soil erosion, and nutrient losses to surface waters and reduce groundwater recharge. Significant changes in ET and soil moisture are likely to impact processes critical for root-nutrient contact like diffusion and mass flow (Chapter 12).

Extended periods of low precipitation, high ET, and reduced soil moisture are expected to increase the frequency and intensity of droughts. Using historical data from 1948 to 2005, Huang et al. (2016) show a climate-induced increase in dryland expansion rate and predicted that drought-susceptible lands will cover half of the global land surface by 2100. Both the intensity and duration of these droughts are expected to increase (Grillakis, 2019). Reductions in soil moisture are expected to increase the frequency, duration and spatial expansion of drought in Europe (Table 19.1). Relative to the historic data (1966–2006), the number of droughts in Europe from 2020 to 2059 is predicted to increase from 40 to 48, with modest increases in both duration and event area. However, by 2060–2099 the number of drought events is predicted to more than double to 88, with each event lasting 16 months. Though far less frequent, large-scale events (>10^6 km²) are predicted to increase sevenfold from 2020 to 2059 and 20-fold from 2060 to 2099, and double or nearly quadruple, respectively, in duration. Plant survival under these conditions will be difficult without extensive irrigation systems that will impact local and regional hydrology.

19.3 Plant responses to global climate change

19.3.1 C₃ and C₄ plants

To understand fully how plant nutrition is impacted by climate, it is critical to understand how climate impacts plant physiology. Large differences in plant responses to eCO₂ exist between plants with the Calvin–Benson (C₃) photosynthetic pathway and plants possessing the Hatch–Slack (C₄) photosynthetic pathway. The latter group evolved from C₃ plants beginning about 25 million years ago as atmospheric CO₂ concentrations declined from approximately 1500 to 300 $\mu L\ L^{-1}$ (Sage et al., 2012). Rapid expansion of C₄ plants has occurred under the relatively low CO₂ conditions that have persisted in the last 10 million years. Evolutionary pressures for the appearance of C₄ plants include higher photorespiratory C losses of C₃ plants under low CO₂ conditions, especially when accompanied by high temperatures. Plants with the C₄ pathway are also well-suited to high solar radiation and have high water-use efficiency but have reduced photosynthesis at temperatures lower than 10°C; hence, they are poorly adapted to extreme latitudes. Although C₄ plants account for only 7500 of Earth's 250,000 plant species, approximately 23% of gross primary productivity is attributed to this group. Representative C₄ species include maize (*Zea mays* L.), sorghum (*Sorghum bicolor* L.), sugarcane (*Saccharum officinarum* L.), millet (*Panicum miliaceum* L.), red root amaranth (*Amaranthus retroflexus* L.), switchgrass (*Panicum virgatum* L.), bahiagrass (*Paspalum notatum* L.), big bluestem (*Andropogon gerardii* Vitman), *Miscanthus* spp., papyrus (*Cyperus papyrus* L.), chinchweed (*Pectis papposa* L.), and several tree species in the genus *Euphorbia*. The C₃ group includes numerous agronomic plants, such as rice (*Oryza sativa* L.), wheat (*Triticum aestivum* L.), barley (*Hordeum vulgare* L.), clovers (*Trifolium* spp.), soybean (*Glycine max* L. Merrill), cool-season forage and turf grasses, many vegetable and food crops including tomato (*Solanum lycopersicum* L.), pepper (*Capsicum annuum* L.), beans (*Phaseolus vulgaris* L.), cucumbers (*Cucumis sativus* L.), and most tree species.

Biochemical mechanisms in leaves of C₄ plants that concentrate CO₂ have effectively eliminated photorespiratory C losses. At current CO₂ concentrations, however, C₃ plants can lose 40% or more of their photosynthetic output via photorespiration, especially at elevated temperatures (Sage et al., 2012). At eCO₂ ($>1000\ \mu L\ L^{-1}$), photorespiration is low irrespective of temperature. As expected, increasing CO₂ levels (Fig. 19.1) reduced photorespiration of C₃ plants, effectively enhancing net photosynthesis, dry weight accumulation of both roots and shoots, and economic yield (Table 19.2). With their biochemical CO₂ concentrating mechanism that essentially eliminates photorespiration, increasing atmospheric CO₂ results in a small increase in photosynthesis and no increase in growth or yield of C₄ plants. The greater growth of C₃ plants in a high CO₂ environment will require additional nutrients as discussed later (Section 19.4.1). Elevated CO₂ reduces stomatal conductance and ET of C₃ and C₄ plants equally. Coupled with the equal (C₄) or greater (C₃) growth rates in eCO₂, the reduced water use increases water-use efficiency of both plant groups. However, lower stomatal conductance and ET have the potential to diminish mass flow and reduce root-nutrient contact (Lynch and Clair, 2004). Finally, tissue nutrient concentrations, especially of C₃ plants, can be reduced in eCO₂. This effect is attributed to several causes discussed in detail next. Decreased mineral concentrations in plants under eCO₂ have important implications for animal and human nutrition (Section 19.7).

TABLE 19.2 Relative changes in net photosynthesis, biomass, economic yield, stomatal conductance, evapotranspiration (ET)/water use, and tissue N concentrations in C₃ and C₄ plants grown at elevated CO₂ compared to ambient CO₂.

Photosynthetic process	Leaf net photosynthesis	Biomass Shoots	Biomass Roots	Economic yield	Stomatal conductance	ET/water use	Tissue N concentration
C₃	34	25	45	24	−30	−10	−8
C₄	9	3	nd	2	−34	−10	−2

nd, Not determined.
Source: Compiled from Kimball et al. (2002), Ainsworth and Long (2005), and Hatfield et al. (2011).

19.3.2 Adaptation of C_3 and C_4 plants to future climates

Modeling future climates and their impact on C_3 and C_4 plants reflects experimentally determined differences in growth and physiology obtained from free-air CO_2 exchange (FACE) experiments (Makowski et al., 2020). These experiments release CO_2 into the atmosphere of field-grown plants, often doubling the CO_2 level during the day. Liu et al. (2020) synthesized results from 667 model simulations and found that enhanced CO_2 increased (+9.23%) predicted yield of rice and wheat (both C_3), but yield of maize (a C_4) was unchanged. In a systematic review of future climate simulations of crop yield in Europe, Knox et al. (2016) reported an average increase in crop yield of 8%. This included increases ranging from 8% to 15% for wheat, potato (*Solanum tuberosum* L.), and sugar beet (*Beta vulgaris* spp) (all C_3), and a projected decline of 6% for maize, the only C_4 in the study. However, the maize response varied with latitude, exhibiting large declines in yield in Central and Southern Europe (average of 10%), and an increase in yield (14%) in Northern Europe as future environments became more conducive to growth of this C_4 plant. This underscores the importance of downscaling regional climate predictions to better understand the local impacts of GCC on agriculture, forestry, and other plant-based industries. A large reduction in maize yield in future climates at lower latitudes was predicted also by Aggarwal et al. (2019), especially without technology adoption. In this study, deploying climate-adapted cultivars and adjusting planting dates had modest impact on lessening the impact of GCC. By comparison, improved nutrient management, especially with irrigation, effectively negated climate-related yield reductions in equatorial regions and increased yield by approximately 20% at high latitudes. Thus, improved fertilizer management may have a significant impact on food security in future climates.

19.4 Nutrient accumulation

19.4.1 C_3 versus C_4 plants

Given growth and development of C_3 and C_4, plants differ markedly in their responses to CO_2 and temperature, nutrient accumulation in a changing climate is also likely to differ between these groups. Several experimental approaches have been used to study the impact of GCC on plant mineral nutrition, including growth chambers, greenhouses, FACE experiments, and open-top chambers (OTC) that, like FACE, elevate CO_2 levels by injecting it into the air under field conditions. For C_3 plants, the positive growth responses to eCO_2 in growth chambers can be as much as fivefold greater than those observed for the same species grown in FACE and OTC systems (Wand et al., 1999). Plant responses in FACE systems also compare favorably to responses of plants grown in the natural environment adjacent to natural CO_2 springs (Saban et al., 2019), suggesting these systems may represent real-world conditions better than greenhouses and growth chambers. Based on a synthesis of FACE studies, Leakey et al. (2009) suggested that eCO_2 enhances the growth of C_4 plants primarily during drought because of improved water-use efficiency, and not by enhancing photosynthesis per se; Kimball (2016) reached the same conclusion.

19.4.2 Plant response to fertilization

Recent analyses using results from 138 studies (primarily FACE) concluded that soil N and P levels have an important role globally in modulating plant responses to eCO_2 (Terrer et al., 2019). In general, nutrient accumulation under eCO_2 scales with plant growth responses; as growth increases, so do nutrient needs. For example, irrespective of grassland, cropland, and forest ecosystems, N accumulation in biomass increases in a linear manner with CO_2-enhanced dry weight accumulation (Fig. 19.4) (Feng et al., 2015), with the slope of the line near one (1.02). For grassland and cropland combined, the line is displaced to the right of the 1:1 line because of the lower N concentration/content observed in plants grown at eCO_2. These findings were independent of N application rate (low, medium, high). This indicates that even high N fertilizer applications were not effective in altering the consistently lower N accumulation in plants grown at eCO_2. These authors concluded that N limitation of eCO_2-enhanced growth is associated with negative effects of eCO_2 on plant N acquisition rather than with growth dilution of plant N or processes leading to progressive N limitation (Section 19.6.1.1).

The interaction of eCO_2 and N fertility on plant N uptake and tissue N concentrations is illustrated with several species grown with adequate (high) and suboptimal (low) N fertility in a FACE experiment in Germany (Fig. 19.5). This experiment included winter cereals (barley and wheat), a cover crop (ryegrass, *Lolium multiflorum* Lam.) and sugar beet (all C_3). Nitrogen accumulation in biomass of all species increased in a linear fashion as aboveground biomass responded to changes in CO_2. The slope of the line was near one (1.04) and displaced to the right of the 1:1 line,

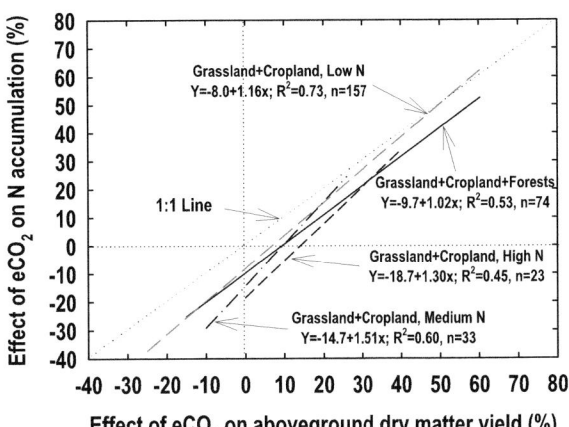

FIGURE 19.4 Relationship between effects of elevated CO_2 (eCO_2) on annual aboveground net primary production and tissue N accumulation for forest, grassland and cropland ecosystems. For the perennial systems (forests and grasslands), data are means over all sampling years of the experiment. The impact of N fertilizer on this relationship is plotted for grassland and cropland ecosystems. Data were grouped into three categories: 0–12.5 g N m^{-2} yr^{-1} (low N), 12.6–25 g N m^{-2} yr^{-1} (medium N), and >25 g N m^{-2} yr^{-1} (high N). *Adapted from Feng et al. (2015).*

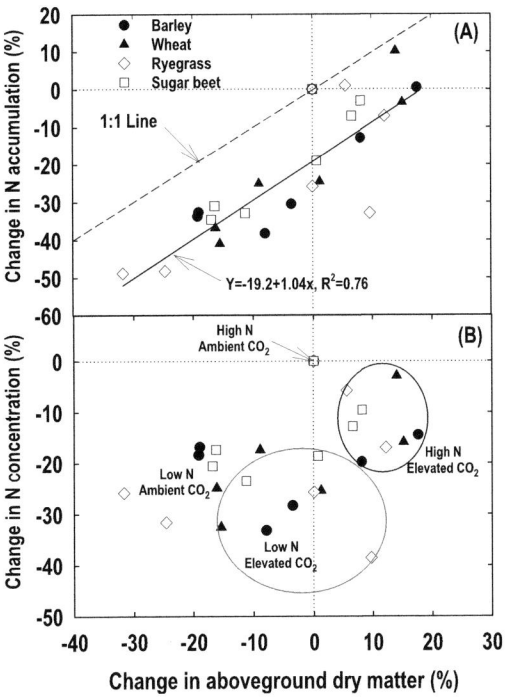

FIGURE 19.5 Effect of elevated CO_2 (575 μL L^{-1}) and N fertility on N accumulation (A) and N concentration (B) in aboveground biomass of barley, wheat, ryegrass, and sugar beet grown in a free-air CO_2 experiment. Data are expressed relative to plants grown at high N (adequate for high yield of each specie) and aCO_2 (375 μL L^{-1}). Species were grown in a rotation system beginning in 1999 and concluding in 2005 that permitted two production years (data points) for each species at both high and low N (approximately ½ of high N) and each CO_2 level. Data for the control treatment (ambient CO_2 at high N) are co-located for each species at the intersection of the dotted lines and were not included in the regression. *Adapted from Weigel and Manderscheid (2012).*

indicating a consistent decrease in N accumulation of these species across the range in biomass (Fig. 19.5A), a result similar to that observed with diverse functional groups in Fig. 19.4. No combination of N fertilization and CO_2 treatments equaled or exceeded the tissue N concentrations of the high N-ambient CO_2 (aCO_2) treatment (Fig. 19.5B). As expected, the low N-aCO_2 data (not circled) clustered at approximately the −20/−20 coordinates, exhibiting proportional reductions in both yield and tissue N. Increasing CO_2 under low N increases yield to near that observed in control plants, but with a further depletion of tissue N concentrations. Adding high N to plants growing in eCO_2 increases yield above that observed in control plants, but tissue N concentrations are lower than the control plants, indicating higher N-use efficiency for this treatment combination.

19.4.3 Nitrogen assimilation in future climates

Nitrogen assimilation under eCO_2 has received more intensive analysis than other nutrients, in part, because of its importance to plant growth in many agro-ecosystems. Early work in greenhouses by Wong (1979) reported rapid growth and high leaf photosynthetic rates of cotton (*Gossypium hirsutum* L.) (C_3) in eCO_2 despite low N and protein concentrations in leaves, responses not noted in maize (C_4). These low leaf N/protein concentrations were not overcome with high rates of N fertilizer, suggesting that N dilution associated with rapid growth is not the main cause of low tissue N. Instead, downregulation of ribulose bis-phosphate carboxylase (RuBisCO), the enzyme that fixes CO_2 via photosynthesis in C_3 plants and can comprise more than 50% of leaf protein (Ku et al., 1979), accompanied the decline in leaf N and protein. The stimulation of photosynthesis in eCO_2 despite a reduction in RuBisCO activity has been termed photosynthetic acclimation (Leakey et al., 2009).

Other mechanisms also may contribute to the low N concentrations often reported in tissues of C_3 plants grown at eCO_2. Reductions in transpiration rate and stomatal conductance associated with eCO_2 (Table 19.2) would be expected to reduce mass flow and plant N uptake. Another factor that may contribute to lower tissue N concentrations is suppression of photorespiration of C_3 plants grown at eCO_2 (Bloom et al., 2014) that decreases the supply of reductants necessary to reduce nitrate and may impede nitrate assimilation and decrease tissue N concentrations. As a result, plants receiving ammonium N have been shown to respond more positively to eCO_2 than those provided nitrate N (Rubio-Asensio and Bloom, 2017), although recent reports suggest this result may not be generalized (Andrews et al., 2019) and may be of limited practical importance because of the rapid conversion of fertilizer ammonium to nitrate in soils (Norton and Ouyang, 2019). Bloom et al. (2014) have proposed shifting nitrate reduction from photosynthetic tissue to roots as a strategy to lessen the negative consequences of eCO_2 on N nutrition of nitrate-fed C_3 plants. Regardless of the underlying cause, the enhanced dry matter production of C_3 plants grown at eCO_2, despite having low tissue N concentration (Fig. 19.5b), results in improved N-use efficiency as described in Section 19.5.

19.4.4 Leguminous plants and N_2 fixation

In general, N_2 fixation (Chapter 16) functions in-concert with photosynthesis to buffer changes in plant (Fig. 19.6) and ecosystem C/N ratios (Lüscher et al., 2000). Leguminous plants possess the C_3 photosynthetic mechanism and exhibit the expected positive response of photosynthesis to eCO_2 (Pastore et al., 2019). Generally, eCO_2 results in higher sugar production in leaves, greater nodule biomass, higher rates of N_2 fixation, and often greater yields (Rogers et al., 2006, 2009), although important species-specific exceptions have been reported (West et al., 2005). This general response also extends to associative N_2 fixing systems, including endophytic diazotrophs resident in roots of nonleguminous plants. Dakora and Drake (2000) reported that eCO_2 increased photosynthesis of a C_3 sedge resulting in greater carbohydrate production, higher N_2 fixation, and more biomass, enhancements not observed in a C_4 grass in which photosynthesis did not respond to eCO_2. This demonstrates the close linkage between photosynthesis and both symbiotic and associative N_2 fixation systems. As a result, N accumulation in leguminous plants generally scales with CO_2-driven increases in plant dry matter accumulation without an accompanying decrease in tissue N concentrations. Hao et al. (2016) reported a linear relationship between soybean growth and N, P, and K accumulation in plants grown at two CO_2 levels and sampled from early reproductive growth through grain harvest (Fig. 19.6). As expected, dry matter accumulation was generally greatest in 550 $\mu L\ L^{-1}$ eCO_2 environment than the aCO_2 environment. Accumulations of N, P, and K were all described by linear relationships irrespective of CO_2 level. These authors suggested that the greater growth of leguminous plants like soybean in the eCO_2 environments of the future may require additional P and K fertilizer applications.

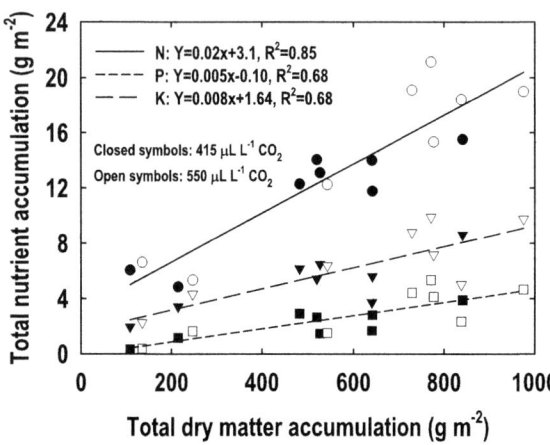

FIGURE 19.6 Effect of CO_2 enrichment on accumulation of plant dry matter and N, P, and K in soybean sampled from flowering to maturity. Data were averaged over two years of sampling beginning at the R1 growth stage and continuing until grain maturity. Plants were grown in a free-air CO_2 enrichment system that provided 415–550 $\mu L\ L^{-1}$ CO_2. *Adapted from Hao et al. (2016).*

19.5 Nutrient-use efficiency

19.5.1 General nutrient-use efficiency concepts

Nutrient-use efficiency (NUE) can be characterized using several indices (Dobermann, 2005). Agronomic efficiency (AE) is the efficiency of applied nutrients in increasing grain or biomass yield. It is calculated as the increase in yield per unit of nutrient applied. For N application to cereals, typical AE values range from 10 to 30 kg grain kg^{-1} N fertilizer applied. The AE is the mathematical product of two other NUE concepts: recovery efficiency (RE) and physiological efficiency (PE). The RE is the proportion of the nutrient applied as fertilizer that is taken up by the plant and is influenced by fertilizer management and crop nutrient needs. For N fertilizer, typical values for RE are in the range of 0.30–0.50 kg kg^{-1}. The PE describes the ability of plants to transform acquired nutrients into economic yield and is impacted by partitioning, environment, and management. For N fertilizer applications to cereals, typical PE values range from 30 to 60 kg yield kg^{-1} N taken up by the plant.

19.5.2 Nutrient-use efficiency of cereals

Given that environment can alter these NUE indices, it is not surprising that GCC has been shown to impact plant NUE. Early work with cotton and wheat grown at varying N and P supplies in glasshouses maintained at 350, 550, and 900 μL L^{-1} CO$_2$ suggested nutrient-specific differences in NUE (Rogers et al., 1993). As expected, eCO$_2$ increased dry weight of these C$_3$ species, but only when adequate N and P were available. For both species, estimates of the "critical level" (tissue nutrient concentration required for 90%–95% of maximum growth) for N decreased as CO$_2$ increased, whereas the critical P level increased with increasing CO$_2$. This indicates that physiological N-use efficiency increased (more dry matter per unit of plant N taken up), but that physiological P-use efficiency declined as CO$_2$ increased. Reddy and Zhao (2005) reported that cotton was more sensitive to K deficiency and also had higher K critical level at eCO$_2$, indicating lower K-use efficiency would be expected in this species in future climates.

In FACE-grown rice, GCC improved N-use efficiency, in part, because the higher NUE associated with eCO$_2$ more than offset the lower NUE caused by elevated temperatures, especially in late-maturing cultivars (Wang et al., 2020). Species and cultivar differences in growth, tissue mineral concentrations, and NUE in response to eCO$_2$ have been reported also for wheat and barley (Weigel et al., 1994; Manderscheid et al., 1995). Both wheat cultivars studied exhibited the often-reported decrease in tissue N in plants grown with eCO$_2$, whereas one of two barley cultivars had nearly twofold greater tissue N accumulation under eCO$_2$. As expected, the N-use efficiency of wheat was higher when grown in eCO$_2$ but was unchanged for the barley cultivar that accumulated high N in eCO$_2$. NUE of Mg, K, S, and P was generally higher in both wheat cultivars in eCO$_2$ than aCO$_2$ but was largely unaffected in the barley cultivars in eCO$_2$. This variation may be useful in plant breeding programs aimed at improving NUE of wheat in future climates.

Given neither photosynthetic rate nor yield is likely to increase in response to eCO$_2$ in C$_4$ plants, integrated soil and nutrient management will be necessary to sustain yields, NUE, and water-use efficiency. Although NUE and GCC studies of C$_4$ crops are sparse, Amouzou et al. (2019) modeled maize and sorghum productivity for future climate scenarios in Benin and found reductions in biomass/grain yield due to higher temperatures and water deficit stress that were not offset by eCO$_2$ fertilization.

19.5.3 Nutrient-use efficiency of forage and pasture species

As expected, NUE of perennial forage and pasture species varies by functional group. Early work with alpine bluegrass (*Poa alpina* L.) and colonial bentgrass (*Agrostis capillaris* L.) (both C$_3$) in outdoor OTC showed the expected increase in dry matter production when grown at 680 versus 340 μL L^{-1} CO$_2$ (Baxter et al., 1994). As noted for other C$_3$ plants, nutrient accumulation scaled with increases in plant dry weight, and there were no differences in PE (g dry matter g^{-1} nutrient accumulated) for N, P, K, Mg, and Ca. By contrast, sheep's fescue (*Festuca vivipara* L.) had the growth reduced in eCO$_2$ and, consistent with the other C$_3$ species, nutrient accumulation decreased as well. PE of N and Ca was lower in leaf tissues of this species exposed to eCO$_2$ compared with aCO$_2$.

In FACE-grown *Stylosanthes capitata* Vogel (C$_3$ legume), nutrient concentrations were not altered markedly by elevated temperatures, but Mg and Ca concentrations were lower in eCO$_2$ (Carvalho et al., 2020a). Elevated temperatures increased both plant growth and content of most nutrients, and the combination of eCO$_2$ and warm temperatures increased NUE of N, S, and K. These authors concluded that new fertilization recommendations may not be warranted in future climates as this species simultaneously increased both productivity and NUE. By contrast, shoot mass of Guinea grass (*Panicum maximum* Jacq.) (C$_4$ perennial) increased when grown at elevated temperatures and CO$_2$ for 30 days (Carvalho et al., 2020b). Accumulation and use efficiency of N, P, and K mirrored shoot mass accumulation. The tight coupling of yield and nutrient accumulation in this species led the authors to conclude that additional fertilizer application may be warranted in warmer climates of the future.

19.5.4 Nutrient-use efficiency of forest species

The NUE of forest systems also has been explored in FACE experiments. Initial experiments with maple (*Acer saccharum* Marsh), paper birch (*Betula papyrifera* Marshall), aspen (*Populus tremuloides* Michx), loblolly pine (*Pinus taeda* L.), sweetgum (*Liquidambar styraciflua* L.), and poplar (*Populus alba* L.; *P. nigra* L.; *P. x euramericana* Dode Guinier) indicated a positive growth response of these species (all C$_3$) to eCO$_2$ (Finzi et al., 2007). Even in presumed N-limited systems, N uptake by all species (except poplar) increased in eCO$_2$. This response was attributed to greater fine root development, enhanced mineralization of N from organic matter, and/or increased N acquisition via associations with mycorrhizal fungi (Sections 19.6.1 and 19.6.2). The greater growth of poplar in eCO$_2$ occurred without additional N uptake and resulted from higher N-use efficiency.

The strategy of greater N uptake to augment fast growth under high CO$_2$ conditions may not be sustainable. Norby et al. (2010) reported that CO$_2$-stimulated growth responses of sweetgum decline markedly after 6 years at eCO$_2$ as rapid plant growth depleted soil N concentrations. Addition of N fertilizer caused an immediate and sustained return to rapid tree growth. Under eCO$_2$, this species maximized growth at a lower leaf N concentration than plants grown under aCO$_2$, indicating that CO$_2$ fertilization increased N-use efficiency. A similar response was noted for poplar species in European FACE studies (Calfapietra et al., 2007). Elevated CO$_2$ enhanced growth and decreased N levels in some soil layers. Total N uptake was unchanged by eCO$_2$, but N concentration of most tissues decreased, and this resulted in greater N-use efficiency at high CO$_2$ levels. This study was relatively short, and depletion of soil N to levels that restricted plant growth was not observed. Even though most climate models accurately represent tree growth and N uptake in future climates, predictions of NUE are inconsistent regarding both direction and magnitude of response (Zaehle et al., 2014).

19.6 Global climate change and root zone nutrient availability

To predict GCC impacts on plant nutrition, changes in ecophysiology must be considered in the context of interlinked changes in soil biogeochemistry. The GCC-induced plant–soil feedbacks are anticipated to be diverse, site-specific, and an important determinant of whether plant–soil systems can sequester new C with further increases in atmospheric CO$_2$. The potential feedback pathways are complex and include (1) the responses of nutrient cycles and various biotic components that drive net primary productivity (NPP) in natural and managed systems and (2) the plant community composition in natural systems (van der Putten et al., 2016). Initial studies suggested that warming would increase heterotrophic respiration, CO$_2$ fertilization would disproportionately stimulate NPP (De Graaff et al., 2006), and accelerated microbial turnover with higher temperatures would lead to increases in soil C stocks (Hagerty et al., 2014). However, more recent reports suggest estimates of the magnitude of the CO$_2$ fertilization effect at global scales are substantially modulated by limited availability of nutrients in the root zone (Bader et al., 2013; Kuzyakov et al., 2019; Wieder et al., 2015). Hungate et al. (2003) reported that simulations of the terrestrial C sink by 2100 required between 2.3 and 37.5 Gt of N depending on the model and assumptions of C:N ratios but estimated that only 1.2–6.1 Gt of new N from anthropogenic deposition and N$_2$ fixation would be available. In a retrospective analysis of 40 years of gross forest production accompanied by a 50 μL L^{-1} increase in atmospheric CO$_2$ concentrations, Peñuelas et al. (2011) found that growth was not increased as expected, likely due to drought, nutrient availability, and/or acclimation limiting the CO$_2$-fertilization effect across tropical, arid, Mediterranean, wet temperate, and boreal ecosystems. Wieder et al. (2015) modeled future N and N + P limitations of NPP, with NPP decreasing by 19% and 25%, respectively (Table 19.3).

Increases in CO$_2$, temperature, and the prevalence of extreme drought and precipitation will alter one or more of the major factors governing nutrient availability in the root zone, including (1) the quantities and qualities of soil organic matter, residue, and rhizosphere C addition and their influence on microbial activity and nutrient availability; (2) the role of mycorrhizae; (3) nutrient-specific chemistry and biogeochemistry that control relationships between soil solution concentration, lability, and mobility by mass flow and diffusion; and (4) net inputs and losses from managed or native ecosystems. As N and P are commonly considered most limiting to primary productivity in terrestrial ecosystems (Vitousek et al., 2010), evaluation of potential GCC impacts on N and P cycling has been extensive. However,

TABLE 19.3 Modeled estimates of global net primary productivity (NPP) and terrestrial C storage for historical (1995–2004) and future (2090–99) conditions as influenced by N and P limitations.

	Period	Nutrient limitations		
		None	N limited	N and P limited
Global NPP (Gt C yr^{-1})	1995–2004	63 (15.5)	61 (14.2)	58 (13.0)
	2090–99	88 (27.7)	69 (16.4)	64 (15.1)
Global terrestrial storage (Gt C)	1995–04	1873 (727)	1868 (724)	1868 (724)
	2090–99	1998 (784)	1757 (661)	1717 (653)

Values are means (standard deviation in parentheses) of 11 models from the Coupled Model Intercomparison Project (CNIP5).
Source: Extracted from Wieder et al. (2015).

representation of most macro- and micronutrients in GCC models remains poor and a critical research need (Brouder and Volenec, 2017).

19.6.1 Impact on coupled carbon-nutrient cycling

Soil C cycling and sequestration responses to GCC are indirect and mediated by plant NPP. Increasing soil C stocks is considered by many as essential to mitigation of both GCC and nutrient pollution of air and water (Keesstra et al., 2016), and initial assumptions were that eCO$_2$ would increase soil C stocks, especially when N and P are not limiting. Most of the total C stocks are organic, and turnover rates vary from decades to centuries (Conant et al., 2011). Lavallee et al. (2019) advocate partitioning particulate from mineral-associated organic matter (POM vs MAOM, respectively) in the context of GCC forecasting because these fractions differ greatly in chemical composition, mean residence time, and their nutritional role for microorganisms and plants. The MAOM is more nutrient rich and stable than POM. When compared to microbial and POM composition, the ratio of C to N, P, and S in MOAM is consistent across a wide range of global soils. Sequestering C requires adequate availability of these nutrients, leading to the suggestion that soil C sequestration can be managed with nutrient additions (Kirkby et al., 2011, 2016). However, recent syntheses demonstrate that when N, P, and other limitations to an NPP response to eCO$_2$ are alleviated, C sequestered in living plants can increase (Section 19.3), but soil C stocks may not. For example, in a meta-analysis of studies dealing with managed grasslands, Sillen and Dieleman (2012) found eCO$_2$ did increase C storage in plant biomass but contingent on concomitant increases in N inputs. In this study, moderate N inputs also promoted decomposition processes that were hypothesized to have limited the soil C storage. As reviewed by Kuzyakov et al. (2019), acceleration of coupled C and N cycling is a common observation, and consensus is emerging that eCO$_2$ accelerates cycles and increases fluxes, but the C input by roots fosters an intense competition between microorganisms and plants for nutrients that may result in no net addition to the sum of soil C pools.

19.6.1.1 Rhizosphere processes

Rhizosphere processes mediated by roots and their associated microorganisms are seminal to understanding impacts of GCC on soil C stocks and vegetation feedbacks to climate (Finzi et al., 2015); to what extent an input of C alters soil C stocks is hypothesized to depend on whether microbial activity is N or P limited (Dijkstra et al., 2013). Numerous studies have shown that eCO$_2$ shifts plant C allocation from shoots to roots (Phillips et al., 2012) as recently demonstrated in Danish heaths where warmer temperatures and eCO$_2$ increased fine root length by 44%, but drought had no effect (Arndal et al., 2018). This eCO$_2$-induced increase in fine root production is accompanied by increases in rhizodeposition, including root exudation (Phillips et al., 2011). Plant allocations of C belowground are not routinely measured, but existing reports suggest total root rhizodeposits (fine root turnover, exudation of organic compounds, sloughing off of root cells, and root cap mucigel) (Section 14.6) are large and a significant, but highly variable, proportion of total plant allocation of C belowground. For example, Amos and Walters (2006) reviewed reports of C partitioning belowground in maize and found approximately 29 ± 13% of shoot C was deposited belowground. Of this, 5%–65% of the belowground C allocation was identified as a net rhizodeposit; the wide variation in net rhizodeposition was associated with environmental drivers of photosynthesis and soil respiration as well as plant nutritional status. In a recent meta-analysis, Dong et al. (2021) reported a proportionate increase in both root growth and organic exudates for plants grown in eCO$_2$; these changes were associated with greater nutrient uptake. Detailed analysis of exudates revealed that citrate (and other carboxylates) and soluble sugars increased, whereas amino acids and malate were not altered in eCO$_2$. However, the eCO$_2$-induced increases in rhizodeposits increase microbial activity and competition with plants for nutrients, which, in turn, may shift the limitation to plant growth from atmospheric CO$_2$ to availability of N, P, or other nutrients in the soil.

Provided nutrients are not limiting, it is hypothesized that microorganisms preferentially use root exudates to meet their energy requirements for growth. However, when eCO$_2$ increases NPP, the system-level N demand is expected to increase due to coupled increases in plant–microbe N demand and to the conserved stoichiometry of MAOM, potentially leading to progressive N limitations (Luo et al., 2004). When nutrients are limiting, microorganisms need to access immobilized nutrients. Soil POM enters the rhizosphere primarily via fragmentation, is only partially processed by organisms, and is less N-rich, with few or no compounds that can be directly assimilated by microorganisms (Lavallee et al., 2020). In N-limited soils, microorganisms can use root exudates to produce extracellular enzymes to oxidize MAOM and meet their N requirements. This rhizosphere priming effect (stimulation of net decomposition) results in a net release of soil C from MAOM to the atmosphere causing a positive feedback to GCC (Fig. 19.7) (Dijkstra et al., 2013; Luo et al., 2004). When P is low, positive feedback may not occur because P can be mobilized from both organic and inorganic sources without net decomposition of MAOM (Dijkstra et al., 2013). Microorganisms can use root exudates to produce phosphatase enzymes to release organic P by hydrolysis (Dakora and Phillips, 2002;

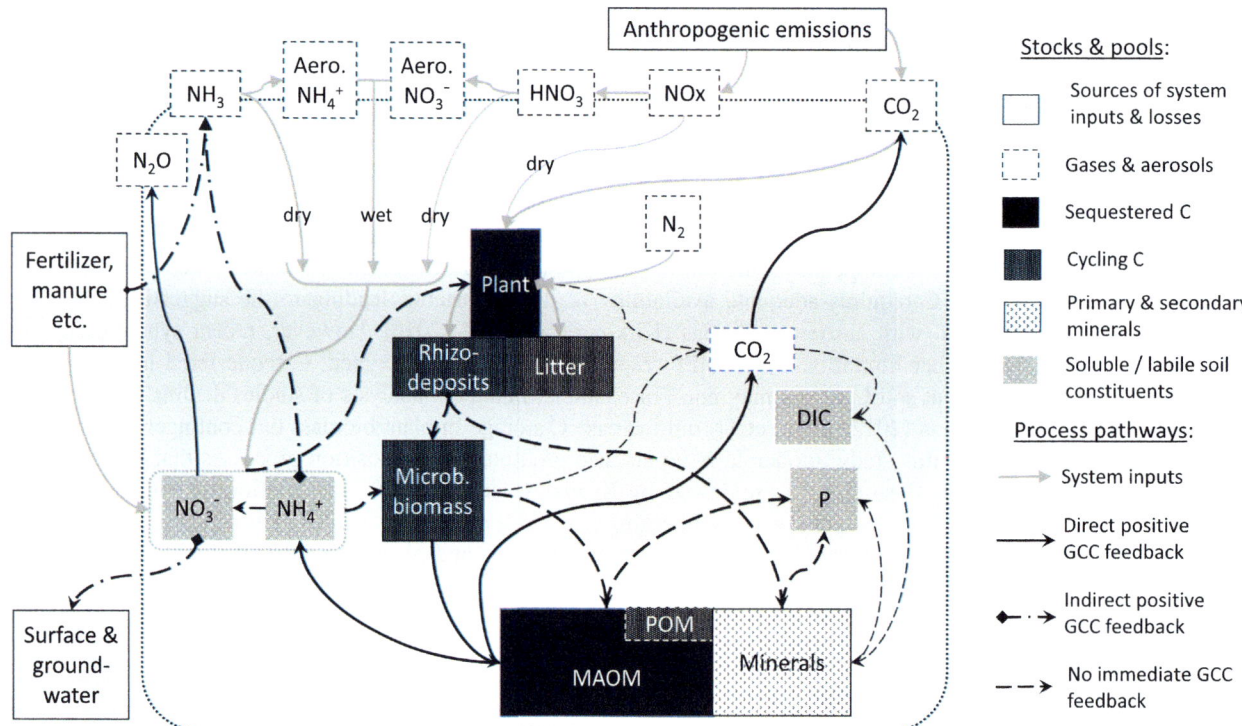

FIGURE 19.7 Enriched atmospheric CO$_2$ can increase rhizodeposition as well as the mass of new plant litter. In N-limited soils, microorganisms use exudates to oxidize a portion of the large pool of older, stable, MAOM that is more nutrient-rich than the much smaller pool of young POM created from fresh litter. MAOM oxidation releases sequestered N and C, creating a positive feedback to GCC (*solid black arrows*). In P-limited soils the exudates released by microorganisms and plants release P (and other nutrients) from minerals and MAOM without oxidation, resulting in nutrient mining but no immediate net depletion of soil C stocks (*dashed black arrows*). Excess N can be lost, resulting in direct (NO$_3^-$ denitrification) and/or indirect positive feedbacks (NH$_3$ volatilization and NO$_3^-$ leaching) (*dash-dot black arrow*). New additions of N that may drive progressive P limitation include N$_2$ fixation, fertilizers, and other anthropogenic (e.g., industry) activities (*gray arrow*). Major atmospheric depositions of N are NH$_3$ (dry deposition) associated with agriculture and NH$_4^+$ aerosols (Aero.) formed from multiple sources in the troposphere (wet deposition). Dry depositions of NO$_x$ and HNO$_3$ are minor. Excessive deposition onto vegetation can be toxic, but all N introduced to soil can enter pathways of positive feedbacks if systems are N saturated. Elevated CO$_2$ enhances root and microbial respiration, which increases concentration of DIC, but the effect DIC has on mineral weathering and direct nutrient release appears limited. Arrow weight indicates relative importance of a process. *DIC*, Dissolved inorganic carbon; *GCC*, global climate change; *MAOM*, mineral-associated organic matter; *POM*, particulate organic matter.

Nannipieri et al., 2011). Root exudates may also directly increase labile P by desorption and solubilization from mineral surfaces (George et al., 2011), potentially benefiting both plants and microorganisms.

The results from empirical research have been mixed regarding the extent to which live roots and eCO$_2$ drive the rhizosphere priming effect (suppression or stimulation of decomposition) to alter root zone availability of nutrients. Meta-analysis of studies under aCO$_2$ revealed rhizosphere priming is influenced by plant species and soil texture and is associated with shoot and microbial biomass (Huo et al., 2017). Mining of N and P from MAOM and associated net losses of CO$_2$ could be offset by new inputs of nutrients in managed systems. In natural systems, new sources are less certain and unpredictable in magnitude. Increases of atmospheric CO$_2$ elevate the partial pressure of CO$_2$ in soil, which may be linked to increased fine root production, respiration and turnover, and the associated increases in microbial respiration. This, in turn, increases the dissolved inorganic carbon (DIC) in the soil solution (Bader et al., 2013). Karberg et al. (2005) hypothesized DIC increases could drive faster weathering of silicates than previously assumed, and this effect could be meaningful for nutrient availability with GCC, although evidence of this effect being quantitatively important is lacking. Other sources of new inputs include atmospheric deposition (Section 19.6.4.1). Additional factors that need to be considered include microbial acclimation to increased temperatures and complexity of introduced C substrates, which may counter the positive GCC feedbacks associated with microbial respiration (Frey et al., 2013). Microbial community structure and metabolic activity are also sensitive to soil moisture, but limited evidence suggests changes in GCC-induced enzymatic activity may not alter soil C stocks (Guenet et al., 2012). In general, Kuzyakov et al. (2019) suggest that, even though quantities of C in older C stocks will be reduced, soil total C stocks will be compensated for by increases in younger stocks and a rapidly growing and cycling microbial pool. Thus, at the global scale, any stimulated terrestrial C sequestration with new inputs of nutrients would not be sufficient to constitute a viable pathway for mitigation of current levels of increases in atmospheric CO$_2$.

19.6.2 Mycorrhizae and nutrient uptake

Mycorrhizal fungi form symbiotic associations with plants, especially in native ecosystems, and can have a significant impact on plant nutrition, particularly P, Zn, and N uptake (see Chapters 7 and 15). The three most common are arbuscular mycorrhizal (AM), ectomycorrhizal (EM), and ericoid mycorrhizal fungi. Because carbohydrates from photosynthesis mediate, in part, the extent of the association between host plants and these fungi, it is logical to anticipate that GCC could impact this symbiotic relationship. In a recent synthesis of how mycorrhizae are impacted by GCC, Bennett and Classen (2020) found that variation in rainfall patterns and soil temperature associated with GCC are more likely to alter mycorrhizal–plant interactions than is eCO$_2$. They also indicated that EM fungi might be more susceptible to these climate perturbations than AM fungi. However, meta-analyses revealed that both mycorrhizal fungi and their host plants can respond separately to eCO$_2$ (Alberton et al., 2005; Treseder, 2004). Even though host-plant responses to CO$_2$ were similar for both EM- and AM-colonized plants, EM fungi exhibited a more positive response to eCO$_2$ compared to AM fungi (Alberton et al., 2005). This observation was confirmed in a recent meta-analysis (Dong et al., 2018). These authors also reported that N and P concentrations of mycorrhizal plants in eCO$_2$ decreased approximately 12%. This decrease occurred because the 26% increase in plant growth of the mycorrhizal plants was greater than uptake of N (+2.5%) and P (+10.7%) in eCO$_2$. Association with EM fungi leads to a positive host-plant response to eCO$_2$ irrespective of soil N concentrations, whereas plants associated with AM fungi were unlikely to respond to eCO$_2$ under low soil N conditions (Terrer et al., 2016). Finally, host responses to GCC also can alter colonization patterns by mycorrhizae. For example, Fernandez et al. (2017) reported that when warming reduced photosynthesis of paper birch and balsam fir (*Abies balsamea* L.) saplings, mycorrhizal populations on roots shifted to less carbon-demanding mycorrhizal species.

19.6.3 Diffusivity and mass flow

In the soil solution, movement of nutrients is directly proportional to their concentrations in the soil solution that are in turn, directly proportional to solid-phase buffering of solution concentrations through organic matter oxidation and/or desorption or dissolution of nutrients from minerals (Chapter 12). Once in the soil solution, nutrients move by mass flow (with the convective flow of water) and diffusion (from regions of high to low concentrations). Convective flow of the soil solution can be created by plant transpiration but also by gravity and/or capillary action in soil. Thus, GCC factors that influence plant transpiration also influence water movement in the soil profile, which can increase or decrease the availability of nutrients at the root surface for plant uptake. Diffusivity is directly proportional to temperature and volumetric water content but indirectly proportional to tortuosity (a function of soil moisture) (Chapter 12). Provided soils remain above a volumetric soil moisture content of approximately 0.1– 0.13, tortuosity is considered an inverse linear function of soil moisture and the primary reason for changes in effective diffusivity (De) with soil moisture (Warncke and Barber, 1972). Below this critical soil moisture content, the liquid phase begins to break down, and the increased viscosity of the liquid next to the soil particles and the interaction factor become more important in determining De in soils. In terms of soil solution buffering, temperature increases the rate of diffusivity to roots and into adsorbing particles in the soil where nutrients may be immobilized in forms or locations that would render them unavailable to plants. Barrow (1992) suggested the impact of temperature is on the slower reaction process that typically follows initial adsorption and generalized that higher temperatures increase adsorption of cations and decrease that of anions, but exceptions occur. In the case of P, immobilization of labile forms into forms that are not plant available is anticipated to accelerate with higher temperatures from GCC (Pilbeam, 2015). Extreme wetting and drying cycles can exacerbate fixation of newly added K and NH_4 at interlayer locations in clay minerals, thereby reducing their ability to buffer the soil solution. However, in low-fertility soils without nutrient amendments, extreme wetting and drying may result in a net increase in K and NH_4 released from interlayer locations (Brouder, 2011). Either effect may be exacerbated with GCC.

In the movement of nutrients to root surfaces, drying of soils has a disproportionate impact on nutrients that have lower De values and relatively high plant requirements (e.g., P and K) as compared to N for which uptake is more sensitive to root morphology (Barber, 1995). Older studies examining the relative importance of soil moisture content, nutrient level, and temperature are informative in predicting GCC impacts on De in regions where the frequency and intensity of soil drying is expected to increase along with soil temperatures (Brouder and Volenec, 2008). Schaff and Skogley (1982) demonstrated that a 17% reduction in soil moisture content (e.g., volumetric water content (θ) of 35% vs 29%) reduced the daily quantity of K arriving at a proxy root by 19%, a reduction that was mitigated to 15% if soil temperatures simultaneously increased by 4°C, an increase consistent with some GCC projections (Section 19.2.2).

For managed lands, large increases in the quantities of a diffusion-limited nutrient such as K in the soil can offset some of the GCC-driven impact of drier soils. Adding 50 μg K g^{-1} (roughly equivalent to a fertilizer dose of 100 kg ha^{-1}) can double the daily linear diffusivity of K in well-moistened soils (Ching and Barber, 1979); further simulations demonstrated this level of addition could maintain diffusivity greater than, or similar to, that in a nonamended soil over a reduction in volumetric soil moisture content by approximately ¼ (i.e., θ of 33 vs 25%; Fig. 19.8A). However, smaller increases in K or other diffusion-limited nutrients such as those associated with enhanced mineral weathering by root exudates will likely do little to overcome increases in drought, especially for generally cooler soils (i.e., temperate and subarctic; Fig. 19.8B).

The reduced transpiration (and higher water productivity) under eCO_2 that results in soil moisture conservation (Bader et al., 2013; Van Vuuren et al., 1997) is anticipated to have a greater impact on nutrient movement to root surfaces compared to the effects of soil temperature and mineral weathering, especially where the prevalence of drought is only moderately increased. As illustrated by Van Vuuren et al. (1997), wheat produced more biomass per unit of water used when grown at eCO_2 than aCO_2 (33% and 41% under dry and well-watered conditions, respectively). These

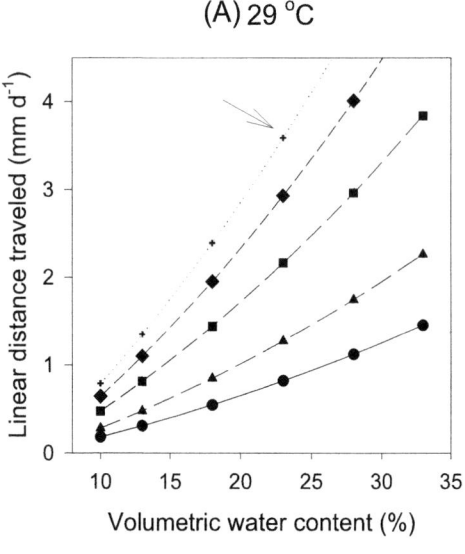
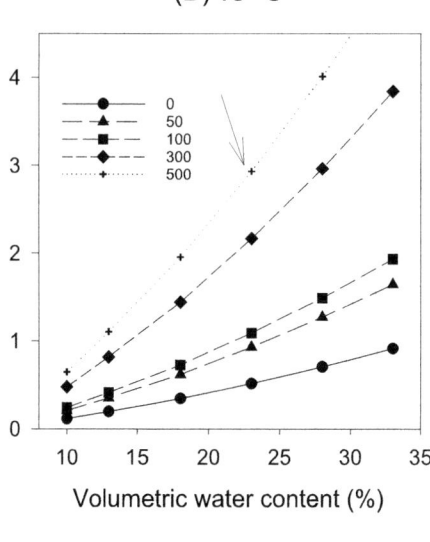

FIGURE 19.8 The benefits of K fertilizer additions as soils dry are illustrated by the linear diffusivity (mm d^{-1}) of K as a function of volumetric soil moisture content (%). Soils were incubated at 29 (A) and 15°C (B) after addition and equilibration of 0, 50, 100, 300, and 500 μg K g^{-1} soil. Values shown for 23% soil moisture (arrows) are as reported in Ching and Barber (1979). Values at all other moisture contents are estimated according to Barber (1995), with tortuosity estimated from soil moisture content after Warncke and Barber (1972).

differences corresponded to observations of consistently higher soil moisture contents immediately prior to rewetting and on average during wet–dry cycles. With water stress, root zone depletion prior to irrigation was reduced by 7%–12% under eCO$_2$ as compared to aCO$_2$; on average, soil moisture was 14% higher (Fig. 19.9A). Water conserved at eCO$_2$ will facilitate diffusivity even as mass flow is reduced, thereby sustaining relatively greater root zone mobility. Under well-watered conditions, the water conservation effect of eCO$_2$ is minimal (Fig. 19.9B).

Whether increased nutrient diffusivity can compensate fully for reductions in nutrient movement to roots via mass flow when eCO$_2$ reduces plant transpiration remains uncertain. Early reports on GCC and plant mineral nutrition theorized that nutrients delivered primarily by mass flow would be affected negatively, resulting in nutrient deficiency (Lynch and Clair, 2004). McGrath and Lobell (2013) concluded that reduction in mass flow under eCO$_2$ could account for 10% of the decreases in grain concentrations of nutrients delivered primarily by mass flow (e.g., N, Ca, and Mg). However, this analysis did not appear to consider the water conservation effect of eCO$_2$ (e.g., reduced transpiration/stomatal conductance at eCO$_2$; Table 19.2).

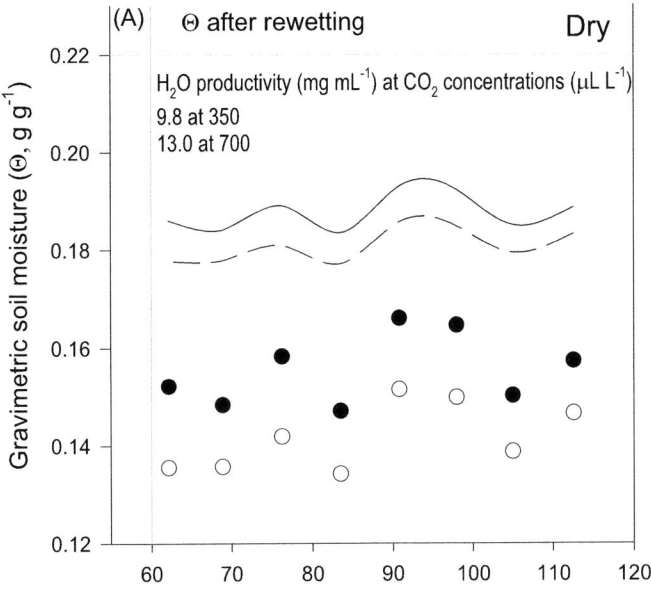

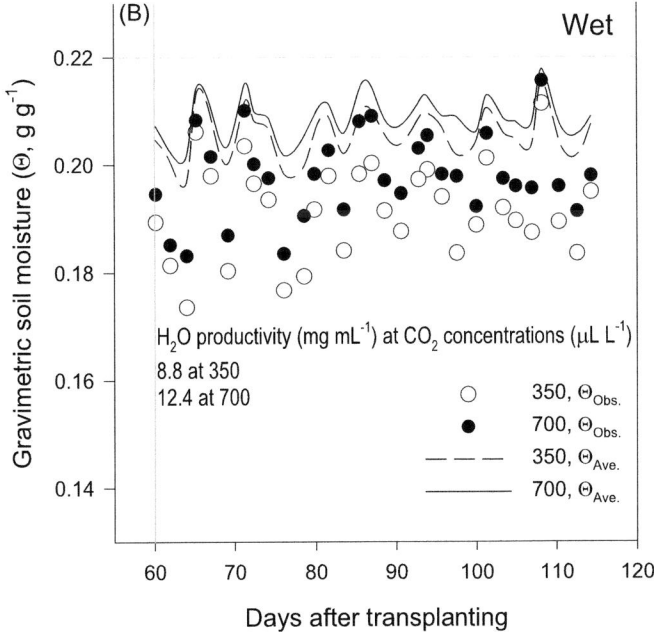

FIGURE 19.9 Impact of elevated CO$_2$ (700 μL L^{-1}) as compared to ambient (350 μL L^{-1}) on soil moisture conservation under dry (A) and wet (B) conditions. Observed gravimetric soil moisture just prior to rewetting is shown as symbols, and the average moisture (a line) is estimated from the difference between the symbols and 0.22 g H$_2$O g^{-1} soil (the moisture immediately after rewetting). Wheat plants were grown with CO$_2$ treatments for 116 d after seedling transplantation, but water treatments were initiated at day 25. The data shown are for the established treatment phase (60–116 d post-transplanting). Water use and productivity and nutrient uptake for this period are provided in Table 19.4. *Adapted from Van Vuuren et al. (1997).*

Mass flow and diffusion are not mutually exclusive delivery mechanisms (Chapter 12), and faster diffusion alone or coupled with greater fine root development because of maintained soil moisture may offset smaller quantities delivered by mass flow. In the Van Vuuren et al. (1997) experiment, moisture conservation under eCO_2 in the drought treatment (Fig. 19.9) was associated with higher nutrient uptake (Table 19.4). However, others have observed a positive association between low transpiration rate and reduced uptake of several nutrients, but nutrient uptake per unit water transpired was nonetheless higher under eCO_2 than under aCO_2 (Houshmandfar et al., 2018). This suggests that, even though reduction in transpiration-driven mass flow of nutrients can contribute to decreased nutrient concentrations under eCO_2, it cannot solely explain the declines discussed earlier (Section 19.4.3). Further analyses are needed to clarify this effect, including additional plant species and/or nutrient-specific variation.

19.6.4 System-level nutrient inputs and losses

Atmospheric emission, transport, and deposition can redistribute significant quantities of mineral nutrients among terrestrial, freshwater and marine ecosystems, and confound predictions of plant, crop, and ecosystem responses to GCC. In terms of the global scale of GCC effects, CO_2 fertilization and deposition of biologically available or reactive N are widely recognized as interconnected (Cusack et al., 2016). Atmospheric redistribution can shift the nutrient constraints on local NPP and potentially confound system responses to GCC and eCO_2. Enrichment via deposition of a limiting nutrient could underpin increased NPP and associated C sequestration, but deposition of a nonlimiting nutrient could result in positive GCC feedbacks through enhanced edge-of-field loss mechanisms (leaching, volatilization, and erosion). For example, in an analysis of potential biodiversity shifts by 2021, Sala et al. (2000) predicted N deposition would have a large impact on biodiversity by advantaging fast-growing species in the N-limited Arctic and Alpine and Boreal biomes but would have a limited impact on species composition and NPP in deserts and tropical forests where CO_2 growth stimulation is restricted by moisture and P, respectively. In a 15-year study of tree responses to simulated N deposition, Braun et al. (2010) observed a minimal response to N but low foliar P concentrations, suggesting a progressive P limitation in Swiss forests. A similar conclusion for humid tropical forests was reached by Cusack et al. (2016) who postulated that eCO_2 would be likely to decrease availability of soil P and base cations, thereby restricting the NPP response.

19.6.4.1 Atmospheric deposition

Of the mineral nutrients which may enter an ecosystem via deposition, N enrichment via wet and dry deposition has been the most extensively studied for its impacts on both natural and anthropogenic ecosystems. The maximum load of introduced N that an ecosystem can tolerate and still keep the same status (e.g., maintain biodiversity) is termed the critical load (Hertel et al., 2006). Empirical estimates of critical loads of N for terrestrial ecosystems are $10-20$ kg N ha^{-1} yr^{-1} with the system-specific value dependent on both the type of ecosystem and the extent of P limitations (Achermann and Bobbink, 2003). Global modeling of total N deposition identifies Western Europe, South Asia, and Eastern China as regions with high deposition ($20-40$ kg ha^{-1} yr^{-1}), while Central Europe, East Asia, and the United States have regions receiving $10-20$ kg ha^{-1} yr^{-1} (Vet et al., 2014).

TABLE 19.4 Wheat biomass increase in the last 56 d of the 116-d treatment of eCO_2 (700 μL L^{-1}) or aCO_2 (350 μL L^{-1}) applied to plants grown without (dry) or with sufficient water (wet).

Treatment	Initial biomass (g plant^{-1}) (SE)	Water use (mL)	Water productivity (mg mL^{-1})	Nutrient uptake (mg plant^{-1})		
				N	P	K
700/dry	5.3 (0.3)	700	13.0	87	15	88
350/dry	5.1 (0.5)	775	9.8	61	11	60
700/wet	6.7 (0.1)	900	12.4	95	20	57
350/wet	6.3 (0.3)	1200	8.8	79	22	69

The CO_2 treatment was applied from seedling transplantation, but water treatments were initiated 25 d after transplanting. Total biomass (standard error in parentheses) measured after 60 d of CO_2 treatment is reported in Van Vuuren et al. (1997); all other values for the established treatment phase (60–116 d) are estimated as the difference between the cumulative reported values for the entire experimental period and values reported at 60 d.

Primary deposition sources of reactive atmospheric N are NO_x (NO and NO_2) and NH_y (NH_3 and aerosols of particulate NH_4) that can be of natural origin but are anthropogenically enriched by fossil fuel combustion (NO_x) and agricultural management of manure and synthetic fertilizers (NH_y; Fig. 19.7) (Hertel et al., 2006). Agricultural ammonia can be returned to soil via dry deposition or reach the troposphere where it can form particulate aerosols with industry-emitted NO_x and SO_x, and their aerosols reflect incoming solar radiation causing a cooling effect (Henze et al., 2012). Eventually, both aerosols and NO_x are deposited on downwind ecosystems (Robertson et al., 2013). Excess N in soils transformed to NO_3^- can be leached or denitrified and released as N_2O. Nitrous oxide is a potent GHG (300 times more potent than CO_2 on a molar basis) and a significant driver of warming (positive feedback).

Current analyses suggest atmospheric NH_4 will remain an important driver of GCC for decades to come. The predominant source for dry deposition of NH_3 is proximate (<50 km) agricultural activity (Bouwman et al., 1997) and, where dry NH_3 depositions occur, they are the major source of introduced N (Ferm, 1998). In the troposphere, particulate NH_4 is a significant contributor to atmospheric particle mass concentrations that are collectively a known health hazard (Pope et al., 2002). Thus, reductions in point-source emissions of the anthropogenic precursors of particulate aerosols (e.g., NO_x and SO_x) have been a common public policy objective that has decreased some environmental N loading. However, decreases in atmospheric NO_x and SO_x can cause NH_3 increases that are not associated with increased soil emissions. In a recent analysis of trends in atmospheric NH_3, Warner et al. (2017) document increases over the world's major agricultural areas of 2.61%, 1.83%, and 2.27% per year for the United States, the European Union (EU), and China, respectively (Table 19.5). These authors attributed their NH_3 observations to concomitant reductions in point-source emissions of NO_x (United States) and SO_x (United States, EU, and China), increased fertilizer use (China) and hot dry conditions (United States), and commented that higher air temperatures with decreased aerosol loads are expected to reduce the stability of ammonium and nitrate aerosols, thereby increasing atmospheric NH_3 concentrations. Additionally, NH_3 emissions can double with an increase in soil temperatures between 27°C and 33°C (Riddick et al., 2016). Thus, even in regions where NO_x and SO_x emissions have diminished, enhanced wet and dry deposition of N will likely continue, albeit shifted to wet and dry deposition of NH_y, a trade-off already observed in deposition patterns across the United States from 1990 to 2010 (Zhang et al., 2018). In a multimodel evaluation, Dentener et al. (2006) predict that, in a future (2030) scenario where all technologically feasible reductions in NO_x emission have been achieved, large reductions in total N deposition may occur but reductions in NO_x deposition will be offset by increases in NH_y (NH_3 and NH_4 aerosols) deposition associated with warmer temperatures and on-going and/or increased agricultural emissions.

Atmospheric deposition can also be a source of ecosystem enrichment with other macro- and micronutrients. Sources of atmospheric P deposits include dusts from soils and deserts, biomass burning, combustion of oil and coal, and emissions from phosphate manufacturing. A key uncertainty in forecasting the extent to which atmospheric P deposition may stimulate NPP in P-limited systems is the scale of transport, with discrepancies associated with transport of small (<10 μm) particles capable of long-distance transport versus larger particles that primarily undergo short-distance transport (Tipping et al., 2014). Nevertheless, these authors suggest that atmospheric transport can redistribute P over timescales that are relatively short compared to ecosystem development. Okin et al. (2004) predict eolian deposition will be critical to maintaining NPP in the Amazon and Congo even if dust sources in Africa are diminished by a warmer

TABLE 19.5 Increases (+) and decreases (−) in atmospheric concentrations of gases, surface skin temperature, fertilizer use, and SO_2 and NO_2 emissions for major agricultural regions from 2002 to 2016.

	% change in concentration			Surface skin temperature (°C yr^{-1})	% change in quantity		
	NH_3	SO_2	NO_2		Fertilizer use	SO_2 emission	NO_2 emission
USA	+2.6	− (N.A.)	−1.5	+0.056	+1.3	−5.3	−9.0
China	+2.3	−8.5		+0.097	+3.5	− (N.A.)	− (N.A.)
Western Europe	+1.8	−14.1	−2.4	N.S.	−0.3	− (N.A.)	− (N.A.)

Missing values are either not significant (N.S.) or not available (N.A.) for the study period, although the general trend is reported.
Source: Extracted from Warner et al. (2017).

climate and with increased precipitation. However, from a global perspective, P deposition measurements are sparse (Vet et al., 2014) and therefore the potential of P deposition to drive NPP in P-limited ecosystems enriched with CO_2 remains entirely uncertain. An additional uncertainty reflects the frequency and extent of large-scale fire events, which are being exacerbated in regions already experiencing GCC-induced droughts and temperature increases (Abatzoglou and Williams, 2016). By contrast, S deposition has been extensively studied, especially in industrialized regions, and is widely known to enhance agricultural productivity in industrial airsheds (De Kok et al., 2007). Dentener et al. (2006) estimated that approximately 12% of global sulfate deposition was to agricultural and urban land, whereas 32% was on natural vegetation. For natural ecosystems, excessive deposition at eCO_2 is considered deleterious because of soil acidification and negative impacts on the soil microbiome and therefore likely to reduce the terrestrial C sequestration potential (Fernández-Martínez et al., 2017). In agricultural systems, reduced atmospheric deposition of S has led to crop deficiency that now must be met with fertilizer applications (Hinckley et al., 2020); uncertainty remains regarding the implications of fertilizer S applications on soil health.

Vet et al. (2014) reviewed wet and dry deposition of basic cations, which can be significant, but a synergistic role of these depositions with CO_2 fertilization is largely unstudied. As with P, local enrichment of micronutrients such as iron can occur with wet and dry deposition of dusts. Mahowald et al. (2006) modeled global dust loading under preindustrial and future climatic conditions and forecast a 60% reduction in dust loading when compared to present conditions, but implications for plant micronutrient nutrition remain unexplored in GCC research.

19.6.4.2 Biological N_2 fixation

Increases in biological N_2 fixation could also increase terrestrial reactive N to either drive NPP or contribute to positive GCC feedbacks. At the ecosystem level, the ability of soil microorganisms to maintain or increase N_2 fixation in N-limited environments with eCO_2 remains highly uncertain both because of the sparseness of data for nonlegume systems and for reasons similar to the uncertainties regarding the CO_2 fertilization effect on the NPP of plant communities. In early simulation work, Gifford et al. (1995) postulated that, for N_2 fixation to meet the N demands for continued plant and/or soil C sequestration with GCC, more available P would be required, which might be facilitated by plant root exudation of organic acid anions and phosphatases (Section 19.6.1). As noted in Chapter 16, N_2 fixation in plants requires relatively high quantities of nutrients other than P such as Fe and Mo. Hungate et al. (2004) suggested Mo may be the limiting factor governing the response of a scrub oak community to eCO_2. De Graaff et al. (2006) performed a meta-analysis of 117 CO_2 enrichment studies on legumes and found eCO_2 did not increase total soil N, N mineralization, or N_2 fixation without supplementation of non-N nutrients, a result supported by a separate synthesis examining all forms of N_2 fixation in both natural and managed communities (van Groenigen et al., 2006). These studies identified P and Mo as potentially limiting N_2 fixation, although K may also have a role, and plant community-microsymbiont feedbacks add complexity. The plant Mg nutritional status may also influence N_2 fixation by affecting the pool of assimilates in roots and nodules (Peng et al., 2018).

Lett and Michelsen (2014) found warming temperatures could increase N_2 fixation in subarctic regions, provided community shifts produced more litter to supply other nutrients, but also hypothesized that the projected increases in snowfall and a later snow melt may ultimately inhibit N_2 fixation in the spring. In a review of soil stress factors influencing symbiotic N_2 fixation, Sadowsky (2005) observed that, in general, rhizobia are more drought resistant than their host plants, their temperature sensitivity is interactive with that of their host, and stress conditions (including drought) increase the nutrient requirements (e.g., Ca, P, and N) in both plants and microsymbionts. Additionally, many (but not all) species downregulate N_2 fixation in response to increases in exogenous N, leading some to speculate that such obligate N_2 fixation species will exacerbate positive GCC feedbacks of the N cycle in conjunction with atmospheric deposition (Kou-Giesbrecht and Menge, 2019). Although eCO_2 may initially stimulate N_2 fixation, nutrient availability as well as temperature, moisture, plant community composition, and host feedbacks are expected to govern whether newly fixed N will drive an NPP response and/or exert positive GCC feedbacks.

19.6.4.3 Surface erosion and leaching

For P, understanding the nonpoint surface erosion and its interaction with GCC factors will be critical to predicting losses of NPP in ecosystems, especially in arid regions expected to become more drought prone. Katra et al. (2016) examined eolian losses associated with cultivation of loess soils and concluded that, at regional scales, current P losses from agricultural soils can be on the order of hundreds of kg km^{-2}, requiring increased P fertilization to maintain yields; in grazing areas with no P supplementation, soil degradation can lead to desertification. Increased drought will amplify and accelerate these phenomena.

In regions expected to become more humid, increased intensity of precipitation can be expected to increase the magnitude and frequency of large runoff events, thereby increasing losses of P and other nutrients that are transport- but not source-limited (Brevik, 2013; Robertson et al., 2013). In the Midwest United States, Michalak et al. (2013) associated record-breaking P loads to Lake Erie with a series of large spring runoff events, predicting such loads will increase because the GCC models show the probability of high intensity rainfall events of >30 mm d^{-1} doubling by 2050. Both particulate and dissolved P may contribute to surface runoff loads (Hansen et al., 2002); dissolved P can originate from soil desorption, fertilizer, organic amendments, and decomposing plant residues on the soil surface (Reid et al., 2018). Rapid thawing of frozen residues followed by enhanced rainfall in temperate regions may increase runoff losses of P and other soluble nutrients (Lozier et al., 2017; Van Esbroeck et al., 2016). Although system losses of N and P via erosion and runoff are most studied in the context of GCC, other solubilized nutrients and nutrients chemically adsorbed, held electrostatically by exchange surfaces, and/or complexed by soil organic matter to soil particles may also be affected.

Where increases in precipitation are accompanied by increases in infiltration that leads to soil moisture contents in excess of field capacity, the risks of nutrient losses via leaching can be greatly increased on moderate- to well-drained soils. The effect may be exacerbated by indirect effects of eCO$_2$ on soil moisture status via increases in plant water productivity. Higher initial soil moisture combined with greater amounts of infiltrating precipitation may result in higher leaching losses even in unfertilized systems, especially of N when it is in surplus with respect to plant demand for other essential nutrients. For example, Sardans et al. (2016) demonstrated that N deposition can initially drive tree growth in European forests but found evidence of progressive P and K limitations to growth and excess N in the soil solution, suggesting an increased potential for surface runoff. However, most soils have a net negative charge and retain cationic nutrients (NH$_4^+$, K$^+$, Mg^{2+}, and Ca^{2+}), whereas the major mineral nutrient anions are in the soil solution, making NO$_3^-$ leaching a likelier loss mechanism than runoff. In a review, Stuart et al. (2011) found projections for climate-induced changes in NO$_3^-$ leaching to groundwater in the United Kingdom ranged from limited impact to possible doubling. In central European hardwood ecosystems receiving 20–25 kg N ha^{-1} yr^{-1} in wet N deposition, eCO$_2$-induced increases in soil moisture were associated with an additional 1.2 kg NO$_3$–N ha^{-1} yr^{-1} lost via leaching, a 50% increase over losses at aCO$_2$ (Bader et al., 2013). Nitrate leaching losses greater than 4–5 kg ha^{-1} yr^{-1} have been observed across a network of Swiss observation plots ($n = 135$) in beech (*Fagus sylvatica*) and Norway spruce (*Picea abies*). These observation plots receive atmospheric depositions of 12–46 kg N ha^{-1} yr^{-1}, and the observed leaching losses suggest deposition has saturated the systems (Braun et al., 2010).

Although cations are less likely to leach, N deposition acidifies soil and may reduce cation-exchange capacity. In their study of European pine forests, Sardans et al. (2016) observed positive correlations between N deposition and soil solution concentrations of K$^+$ and Ca^{2+} as well as NO$_3^-$ and SO$_4^{2-}$ and suggested nutrient loss risks were proportional to N deposition. Liu et al. (2008) modeled subtropical forest ecosystems in OTC and observed that eCO$_2$ increased soil moisture and leachate volumes that were associated with a trend toward increased mass loss of NO$_3^-$ but also significant increases in losses of soluble K, Mg, Na, and Ca (Fig. 19.10). The soils used in this experiment were lateritic and

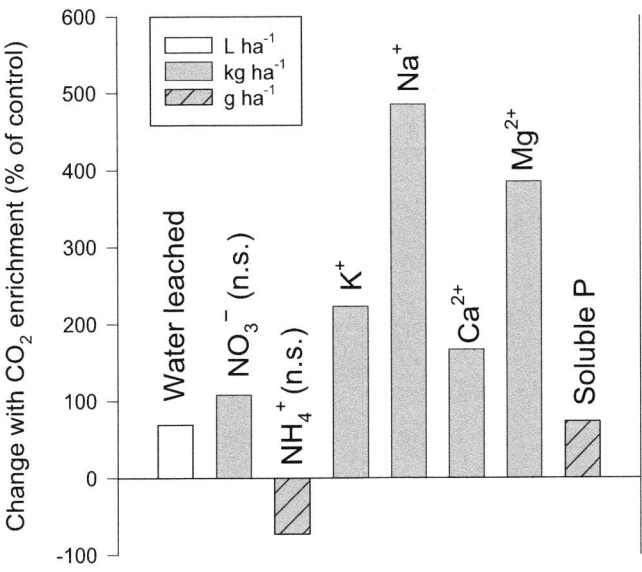

FIGURE 19.10 Illustration of the potential for increased leaching of anions and cations with CO$_2$ enrichment. Data are from an open-top chamber CO$_2$ enrichment experiment in a Chinese subtropical forest ecosystem on highly acidic soil. Data for leachate volume and nutrient leaching losses with CO$_2$ enrichment are shown as the percent change from the ambient CO$_2$ control. Losses of NH$_4^+$ and NO$_3^-$ were not impacted significantly. Percent changes estimated from Liu et al. (2008); bar fill pattern denotes the original units reported by the authors.

highly acidic, illustrating the potential for GCC to drive leaching losses of cations from already highly weathered, tropical soils that typically have lower cation-exchange capacity.

Quantitatively, GCC-driven leaching losses of nutrients are most pronounced in anthropogenic systems, particularly in poorly drained soils with installed drainage enhancements. Nitrate leaching losses from arable lands vary with soil type and cropping system and have been observed to range from <10 to >150 kg N ha^{-1} yr^{-1} (Cameron et al., 2013; Stuart et al., 2011). Installing drains in poorly drained soils shifts the predominant N loss mechanisms from denitrification (and a direct positive GCC feedback) to leaching and an indirect positive feedback. Subsurface drains are prevalent in agricultural soils of North America and Western Europe and create direct conduits for export of surplus nutrients in naturally high-fertility, heavier textured soils (Blann et al., 2009). Even though there is some evidence that N fertilizer in excess of crop need may result in accumulation of soil organic N (Van Meter et al., 2016), leaching loss of NO_3^- is common (Robertson et al., 2013), and load losses typically increase proportionally to drainage volumes in drained systems. For example, Christianson and Harmel (2015) reviewed approximately 400 North American studies and found the increases in drainage discharge were associated with wet years and corresponded to increased NO_3^- losses to surface water because NO_3^- was transport- and not source-limited in these systems.

Initially, P leaching losses from agricultural fields were thought to be negligible (Logan et al., 1980), but installed drains can facilitate export of environmentally significant quantities of dissolved reactive P and are an ongoing concern for freshwater eutrophication. Welikhe et al. (2020) demonstrated that when soil P exceeds the amounts needed for crop nutrition, the soil capacity to sorb P is saturated, and concentrations in drain flows can exceed environmental guidelines (Fig. 19.11). Further, routine application of fertilizer and manure in accordance with agronomic guidelines can result in soils that have lost their capacity to sorb P additions and leaching losses can be equivalent in magnitude to P losses in surface runoff (Ruark et al., 2012). Such soils are hypothesized to function as sources of labile P similar to a new fertilizer addition (Williams et al., 2016) and may continue to be a source of leached P even when new applications cease, with magnitudes of loss proportional to volume of infiltration.

In summary, leaching losses of legacy N and P from anthropogenic landscapes and N and P in excess of plant demand in agricultural and natural systems can be expected to increase with GCC-related increases in precipitation volumes. Nitrate leached to freshwater drives additional denitrification, and N_2O emissions can occur (Beaulieu et al., 2019). When excess P enters freshwater, it drives eutrophication because P is typically the most limiting nutrient in freshwater NPP. Eutrophication enhances methane emission. Methane is a far more potent GHG than CO_2, and CH_4 emission is the major driver of positive GCC feedbacks from lakes (Beaulieu et al., 2019).

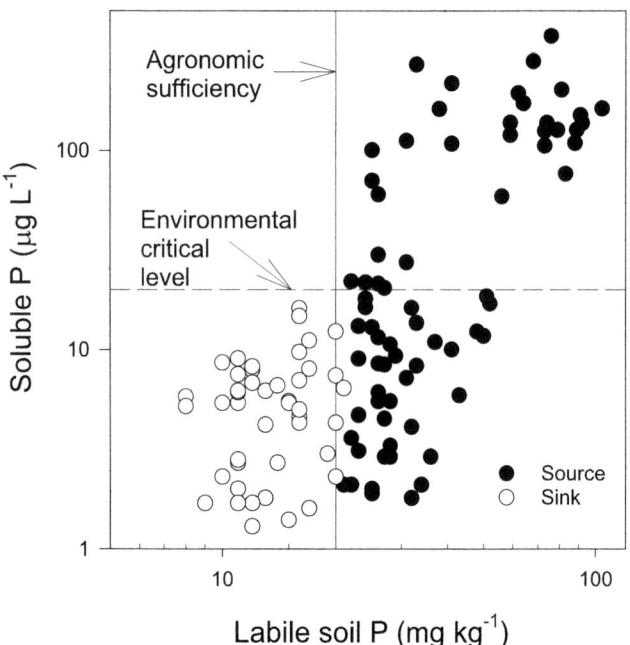

FIGURE 19.11 Dissolved, reactive P in drainage water shown as a function of labile soil P in cropped soils. Source soils are P-saturated, whereas sink soils retain the capacity to sorb new P amendments. Reference lines show the environmental critical level for lake eutrophication and the target soil test level to optimize crop production. *Data are shown on a log-log plot. Extracted from Welikhe et al. (2020).*

19.7 Mineral composition of food/feed

19.7.1 Mineral composition of grains and fruits

Plant functional group differences in nutrient accumulation in biomass described earlier (Section 19.4) are also evident in grains and fruits. Because nutrient composition of grains and fruits tends to be more highly conserved than that of vegetative tissues (Lee and Fenner, 1989; Riedell et al., 2009; Bellaloui et al., 2015), the magnitude of differences is less. Nevertheless, the same general patterns emerge from data compiled from numerous FACE experiments (Dietterich et al., 2015; Al-Hadeethi et al., 2019). Wheat and rice (C_3 cereals) exhibit a consistent reduction in seed N concentration when grown at eCO_2 (Table 19.6). By comparison, grain/fruit N concentrations of soybean/pea (*Pisum sativum* L.) (C_3 legumes) and maize/sorghum (C_4 cereals) exhibit little or no change. This agrees with Jin et al. (2019) who reported that eCO_2 consistently reduced grain N concentrations of wheat but had no impact on pea grain N. Canola (*Brassica napus* L.) (C_3) grain N also was decreased by eCO_2 in one of two years, responses that were consistent across three soil groups. Concentrations of other macronutrients in grains were largely unaffected by eCO_2 within species and across functional groups. However, slightly decreased grain/fruit Zn and Fe concentrations were observed in most C_3, but not C_4 species, at eCO_2. Jin et al. (2019) reported consistently low Zn (but unchanged Fe) concentrations in grains of C_3 plants grown at eCO_2 in FACE systems.

Soares et al. (2019) reviewed how eCO_2 impacts concentration of minerals in edible plant parts (grains, fruits) used as food. Mineral composition of C_4 plant tissues was largely unchanged by eCO_2, whereas Fe and Zn concentrations of edible portions of C_3 cereals, legumes, and vegetables were consistently lower. This agrees with Myers et al. (2014) who reported that these low levels of Zn and Fe could impair human nutrition in future climates. Nonlegumes had lower N, and most C_3 species also had decreased concentrations of S, Mg, and Cu. These authors suggested breeding for greater nutrient concentrations under eCO_2 may improve nutritional quality of C_3 food and feed crops. This strategy is plausible because varieties differ in their relative decrease in mineral nutrients in response to eCO_2. In FACE-grown wheat, reduction in grain N concentrations ranged from 4.6% to 11.6% depending on the variety. Similar varietal ranges were evident for Ca, Mg, Cu, and Zn, but varietal differences in Fe and S were limited (Dietterich et al., 2015). Varietal differences in protein, minerals and vitamins were reported also for 18 rice cultivars (Zhu et al., 2018). These

TABLE 19.6 Concentrations of mineral nutrients in grain of plants possessing the C_3 or C_4 photosynthetic (Ps) system and grown in free-air CO_2 enhancement experiments.

Species	Ps	CO_2	Obs.	N	P	K	Ca	Mg	S	Zn	Fe	B	Mn	Cu
				\multicolumn{7}{c}{g kg^{-1}}	\multicolumn{5}{c}{mg kg^{-1}}
Wheat	C_3	amb	256	30	3.9	5.0	0.5	1.4	1.9	31	41	1.8	49	5.6
		elev	256	28	3.8	5.0	0.5	1.3	1.8	28	39	1.8	48	5.4
Rice	C_3	amb	131	13	3.4	3.0	0.1	1.3	1.0	28	11	2.1	35	3.1
		elev	132	12	3.4	3.0	0.1	1.3	0.9	27	11	2.2	33	2.7
Soybean	C_3	amb	104	63	5.8	19	2.9	2.3	3.1	40	76	46	28	13
		elev	113	63	5.7	19	2.8	2.2	3.0	38	73	43	28	13
Pea	C_3	amb	45	34	3.4	10	0.7	1.1	1.8	26	42	7.1	9.4	5.9
		elev	44	33	3.2	10	0.7	1.1	1.8	24	40	7.0	9.0	5.7
Maize	C_4	amb	18	10	2.5	3.3	0.05	0.9	0.9	19	14	4.3	3.8	2.1
		elev	18	10	2.4	3.2	0.05	0.9	0.9	18	14	4.5	3.7	1.9
Sorghum	C_4	amb	20	19	3.6	4.6	0.2	1.6	1.2	27	32	5.9	25	3.9
		elev	15	20	3.8	4.8	0.2	1.6	1.2	28	34	6.2	26	4.0

Ambient (amb) CO_2 concentrations ranged between 364 and 386 µL L^{-1}, and elevated (elev) CO_2 concentrations were 546–586 µL L^{-1} across all study sites. The number of observations (Obs.) for each species x CO_2 combination is shown. The numbers of cultivars evaluated by species were rice (18), wheat (8), maize (2), soybeans (7), field peas (5), and sorghum (1). All nutrient concentrations are reported on a dry weight basis.
Source: Adapted from Dietterich et al. (2015).

differences suggest the possibility of breeding for improved nutrient concentrations in a high CO_2 world, but trait heritability and other genetic studies are necessary to validate this approach.

19.7.2 Forage and pasture composition and mineral nutrition

Climate change is also expected to alter composition of perennial grasses and legumes used for forage and pasture, impacting nutrition of livestock. As expected, plant growth, tissue N (protein) concentrations, fiber composition, and nutritive value of C_4 forage plants are similar between plants grown in ambient and eCO_2 (Abdalla Filho et al., 2019). In FACE experiments, eCO_2 and warm temperatures increased yield of C_3 forage grasses in a rangeland ecosystem (Wyoming, United States) nearly 40% and decreased forage protein (N) concentration by 13% (Augustine et al., 2018). Tissues of plants grown in eCO_2 also contained higher concentrations of acid detergent fiber (cellulose, lignin) and had lower digestibility, which, along with the lower N concentrations, is predicted to reduce animal growth rates. In a recent meta-analysis, Dellar et al. (2018) found that, averaged across agro-ecosystems in Europe, eCO_2 and warming are expected to increase plant growth and decrease tissue N concentrations of forbs, shrubs and grasses, but not legumes. These authors indicated that the lower tissue N concentrations may require supplementation with high protein concentrates to maintain livestock performance. Forage yield, tissue N (protein) concentrations, and nutritive value in future climates have been estimated for several scenarios in contrasting ecosystems in Canada (Thivierge et al., 2016). Only minor changes in tissue N concentrations of timothy (*Phleum pratense* L.) and alfalfa (*Medicago sativa* L.) (both C_3) were observed, but the longer growing season permitted additional annual harvests, increasing yield by up to 35%. Nutrient needs of these systems would increase in proportion with this greater annual dry matter production.

19.7.3 Composition of trees and timber

Elevated CO_2 and N fertilizer applications can interact to alter the mineral composition and structural integrity of wood of some tree species. Watanabe et al. (2008) grew oak (*Quercus mongolica* Fisch. ex Ledeb.) and alder (*Alnus hirsuta* L.) in phytotrons at aCO_2 (360 μL L^{-1}) and eCO_2 (720 μL L^{-1}) and tested a 10-fold range in N supply. Elevated CO_2 increased total biomass of both species at the high-, but not low-N, treatment. The combination of high N and eCO_2 increased the proportion of large vessels in wood and reduced wood structural quality. Similar to oak and alder, high N and eCO_2 were both required to maximize biomass production of poplar species in a FACE experiment (Luo et al., 2005). Cell wall area of wood was reduced, and cell wall lumen area increased with eCO_2 and additional N. These authors predicted these anatomical effects will have negative consequences on wood quality in future climates. In addition to anatomical changes, eCO_2 also has been shown to increase the concentrations of mineral nutrients in wood of sweetgum (Kim et al., 2015). Increases in ash, Ca and Mg with eCO_2 could impact chemical and thermal utilization of wood of this species as a feedstock for biofuels and other chemicals.

References

Abatzoglou, J.T., Williams, A.P., 2016. Impact of anthropogenic climate change on wildfire across western US forests. Proc. Natl. Acad. Sci. USA 113, 11770–11775.

Abdalla Filho, A.L., Lima, P.D.M.T., Sakita, G.Z., Silva, T.P.D.E., da Costa, W.D.S., Ghini, R., et al., 2019. CO_2 fertilization does not affect biomass production and nutritive value of a C_4 tropical grass in short timeframe. Grass Forage Sci. 74, 670–677.

Achermann, B., Bobbink, R., 2003. Empirical critical loads for nitrogen. Environ. Doc. 164, 327.

Aggarwal, P., Vyas, S., Thornton, P., Campbell, B.M., 2019. How much does climate change add to the challenge of feeding the planet this century? Environ. Res. Lett. 14, 043001.

Ainsworth, E.A., Long, S.P., 2005. What have we learned from 15 years of free-air CO_2 enrichment (FACE)? A meta-analytic review of the responses of photosynthesis, canopy properties and plant production to rising CO_2. New Phytol. 165, 351–372.

Alberton, O., Kuyper, T.W., Gorissen, A., 2005. Taking mycocentrism seriously: mycorrhizal fungal and plant responses to elevated CO_2. New Phytol. 167, 859–868.

Al-Hadeethi, I., Li, Y., Odhafa, A.K.H., Al-Hadeethi, H., Seneweera, S., Lam, S.K., 2019. Assessment of grain quality in terms of functional group response to elevated [CO_2], water, and nitrogen using a meta-analysis: Grain protein, zinc, and iron under future climate. Ecol. Evol. 9, 7425–7437.

Amos, B., Walters, D.T., 2006. Maize root biomass and net rhizodeposited carbon. Soil Sci. Soc. Am. J. 70, 1489–1503.

Amouzou, K.A., Lamers, J.P.A., Naab, J.B., Borgemeister, C., Vlek, P.L.G., Becker, M., 2019. Climate change impact on water- and nitrogen-use efficiencies and yields of maize and sorghum in the northern Benin dry savanna, West Africa. Field Crops Res. 235, 104–117.

Andrews, M., Condron, L.M., Kemp, P.D., Topping, J.F., Lindsey, K., Hodge, S., et al., 2019. Elevated CO_2 effects on nitrogen assimilation and growth of C_3 vascular plants are similar regardless of N-form assimilated. J. Exp. Bot. 70, 683–690.

Arndal, M.F., Tolver, A., Larsen, K.S., Beier, C., Schmidt, I.K., 2018. Fine root growth and vertical distribution in response to elevated CO_2, warming and drought in a mixed heathland–grassland. Ecosystems 21, 15–30.

Augustine, D.J., Blumenthal, D.M., Springer, T.L., LeCain, D.R., Gunter, S.A., Derner, J.D., 2018. Elevated CO_2 induces substantial and persistent declines in forage quality irrespective of warming in mixed grass prairie. Ecol. Appl. 28, 721–735.

Bader, M.K.F., Leuzinger, S., Keel, S.G., Siegwolf, R.T., Hagedorn, F., Schleppi, P., et al., 2013. Central European hardwood trees in a high-CO_2 future: synthesis of an 8-year forest canopy CO_2 enrichment project. J. Ecol. 101, 1509–1519.

Barber, S.A., 1995. Soil Nutrient Bioavailability: A Mechanistic Approach. John Wiley & Sons, New York, USA.

Barrow, N.J., 1992. A brief discussion on the effect of temperature on the reaction of inorganic ions with soil. J. Soil Sci. 43, 37–45.

Baxter, R., Ashenden, T.W., Sparks, T.H., Farrar, J.F., 1994. Effects of elevated carbon dioxide on three montane grass species: I. Growth and dry matter partitioning. J. Exp. Bot. 45, 305–315.

Beaulieu, J.J., DelSontro, T., Downing, J.A., 2019. Eutrophication will increase methane emissions from lakes and impoundments during the 21st century. Nature Comm. 10, 1–5.

Bellaloui, N., Bruns, H.A., Abbas, H.K., Mengistu, A., Fisher, D.K., Reddy, K.N., 2015. Agricultural practices altered soybean seed protein, oil, fatty acids, sugars, and minerals in the Midsouth USA. Front. Plant Sci. 6, 31.

Bennett, A.E., Classen, A.T., 2020. Climate change influences mycorrhizal fungal-plant interactions, but conclusions are limited by geographical study bias. Ecology 101, e02978.

Blann, K.L., Anderson, J.L., Sands, G.R., Vondracek, B., 2009. Effects of agricultural drainage on aquatic ecosystems: a review. Crit. Rev. Environ. Sci. Technol. 39, 909–1001.

Bloom, A.J., Burger, M., Kimball, B.A., Pinter Jr, P.J., 2014. Nitrate assimilation is inhibited by elevated CO_2 in field-grown wheat. Nature Clim. Change 4, 477–480.

Bouwman, A.F., Lee, D.S., Asman, W.A.H., Dentener, F.J., Van Der Hoek, K.W., Olivier, J.G.J., 1997. A global high-resolution emission inventory for ammonia. Glob. Biog. Cycl. 11, 561–587.

Braun, S., Thomas, V.F., Quiring, R., Flückiger, W., 2010. Does nitrogen deposition increase forest production? The role of phosphorus. Environ. Poll. 158, 2043–2052.

Brevik, E.C., 2013. The potential impact of climate change on soil properties and processes and corresponding influence on food security. Agriculture-Basel 3, 398–417.

Brouder, S., 2011. Potassium cycling. In: Hatfield, J.L., Sauer, T.J. (Eds.), Soil Management: Building a Stable Base for Agriculture. American Society of Agronomy and Soil Science Society of America, Madison, WI, USA, pp. 79–102.

Brouder, S.M., Volenec, J.J., 2008. Impact of climate change on crop nutrient and water use efficiencies. Physiol. Plant. 133, 705–724.

Brouder, S.M., Volenec, J.J., 2017. Future climate change and plant macronutrient use efficiency. In: Hossain, M.A., Kamiya, T., Burrit, D., Tran, L.-S.P., Fuyiwara, T. (Eds.), Plant Macronutrient Use Efficiency: Molecular and Genomic Perspectives. Elsevier, New York, USA, pp. 357–379.

Cakmak, I., 2000. Role of zinc in protecting plant cells from reactive oxygen species. New Phytol. 146, 185–205.

Calfapietra, C., Angelis, P.D., Gielen, B., Lukac, M., Moscatelli, M.C., Avino, G., et al., 2007. Increased nitrogen-use efficiency of a short-rotation poplar plantation in elevated CO_2 concentration. Tree Physiol. 27, 1153–1163.

Cameron, K.C., Di, H.J., Moir, J.L., 2013. Nitrogen losses from the soil/plant system: a review. Ann. Appl. Biol. 162, 145–173.

Carlson, K.M., Gerber, J.S., Mueller, N.D., Herrero, M., MacDonald, G.K., Brauman, K.A., et al., 2017. Greenhouse gas emissions intensity of global croplands. Nature Clim. Change 7, 63–68.

Carvalho, J.M., Barreto, R.F., Prado, R.D.M., Habermann, E., Martinez, C.A., Branco, R.B.F., 2020a. Elevated [CO_2] and warming increase the macronutrient use efficiency and biomass of *Stylosanthes capitata* Vogel under field conditions. J. Agron. Crop Sci. 206, 597–606.

Carvalho, J.M., Barreto, R.F., Prado, R.D.M., Habermann, E., Branco, R.B.F., Martinez, C.A., 2020b. Elevated CO_2 and warming change the nutrient status and use efficiency of *Panicum maximum* Jacq. PLoS one 15, e0223937.

Chen, X., Su, Y., Liao, J., Shang, J., Dong, T., Wang, C., et al., 2016. Detecting significant decreasing trends of land surface soil moisture in eastern China during the past three decades (1979–2010). J. Geophys. Res.-Atmos. 121, 5177–5192.

Ching, P.C., Barber, S.A., 1979. Evaluation of temperature effects on K uptake by corn. Agron. J. 71, 1040–1044.

Christianson, L.E., Harmel, R.D., 2015. The MANAGE Drain Load database: Review and compilation of more than fifty years of North American drainage nutrient studies. Agric. Water Manage. 159, 277–289.

Conant, R.T., Ryan, M.G., Ågren, G.I., Birge, H.E., Davidson, E.A., Eliasson, P.E., et al., 2011. Temperature and soil organic matter decomposition rates—synthesis of current knowledge and a way forward. Glob. Change Biol. 17, 3392–3404.

Cornes, R.C., van der Schrier, G., Squintu, A.A., 2019. A reappraisal of the thermal growing season length across Europe. Inter. J. Climatol. 39, 1787–1795.

Cusack, D.F., Karpman, J., Ashdown, D., Cao, Q., Ciochina, M., Halterman, S., et al., 2016. Global change effects on humid tropical forests: Evidence for biogeochemical and biodiversity shifts at an ecosystem scale. Rev. Geophys. 54, 523–610.

Dakora, F.D., Drake, B.G., 2000. Elevated CO_2 stimulates associative N_2 fixation in a C_3 plant of the Chesapeake Bay wetland. Plant Cell Environ. 23, 943–953.

Dakora, F.D., Phillips, D.A., 2002. Root exudates as mediators of mineral acquisition in low-nutrient environments. In: Adu-Gyamfi, J.J. (Ed.), Food Security in Nutrient-stressed Environments: Exploiting Plants' Genetic Capabilities. Kluwer Academic Publishers, Dordrecht, The Netherlands, pp. 201–213.

De Graaff, M.A., Van Groenigen, K.J., Six, J., Hungate, B., van Kessel, C., 2006. Interactions between plant growth and soil nutrient cycling under elevated CO_2: a meta-analysis. Glob. Change Biol. 12, 2077–2091.

De Kok, L.J., Durenkamp, M., Yang, L., Stulen, I., 2007. Atmospheric sulfur. In: Hawkesford, M.J., De Kok, L.J. (Eds.), Sulfur in Plants. An Ecological Perspective. Springer, Dordrecht, The Netherlands, pp. 91–106.

Dellar, M., Topp, C.F.E., Banos, G., Wall, E., 2018. A meta-analysis on the effects of climate change on the yield and quality of European pastures. Agric. Ecosys. Environ. 265, 413–420.

Dentener, F., Drevet, J., Lamarque, J.F., Bey, I., Eickhout, B., Fiore, A.M., et al., 2006. Nitrogen and sulfur deposition on regional and global scales: A multimodel evaluation. Global Biogeochem. Cycl. 20, GB4003.

Dietterich, L.H., Zanobetti, A., Kloog, I., Huybers, P., Leakey, A.D., Bloom, A.J., et al., 2015. Impacts of elevated atmospheric CO_2 on nutrient content of important food crops. Sci. Data 2, 1–8.

Dijkstra, F.A., Carrillo, Y., Pendall, E., Morgan, J.A., 2013. Rhizosphere priming: a nutrient perspective. Front. Microbiol. 4, 216.

Dobermann, A.R., 2005. Nitrogen Use Efficiency – State of the Art. Agronomy & Horticulture -- Faculty Publications. 316. http://digitalcommons.unl.edu/agronomyfacpub/316. (accessed 16 Oct 2020).

Dong, Y., Wang, Z., Sun, H., Yang, W., Xu, H., 2018. The response patterns of arbuscular mycorrhizal and ectomycorrhizal symbionts under elevated CO_2: A meta-analysis. Front. Microbiol. 9, 1248.

Dong, J.L., Hunt, J., Delhaize, E., Zheng, S.J., Jin, C.W., Tang, C.X., 2021. Impacts of elevated CO_2 on plant resistance to nutrient deficiency and toxic ions via root exudates: a review. Sci. Total Environ. 754, 142434.

Fang, X., Luo, S., Lyu, S., 2019. Observed soil temperature trends associated with climate change in the Tibetan Plateau, 1960–2014. Theor. Appl. Climatol. 135, 169–181.

Feng, Z.Z., Rutting, T., Pleijel, H., Wallin, G., Reich, P.B., Kammann, C.I., et al., 2015. Constraints to nitrogen acquisition of terrestrial plants under elevated CO_2. Glob. Change Biol. 21, 3152–3168.

Ferm, M., 1998. Atmospheric ammonia and ammonium transport in Europe and critical loads: a review. Nutr. Cycl. Agroecosys. 51, 5–17.

Fernandez, C.W., Nguyen, N.H., Stefanski, A., Han, Y., Hobbie, S.E., Montgomery, R.A., et al., 2017. Ectomycorrhizal fungal response to warming is linked to poor host performance at the boreal-temperate ecotone. Glob. Change Biol. 23, 1598–1609.

Fernández-Martínez, M., Vicca, S., Janssens, I.A., Ciais, P., Obersteiner, M., Bartrons, M., et al., 2017. Atmospheric deposition, CO_2, and change in the land carbon sink. Sci. Rep. 7, 1–13.

Finzi, A.C., Norby, R.J., Calfapietra, C., Gallet-Budynek, A., Gielen, B., Holmes, W.E., et al., 2007. Increases in nitrogen uptake rather than nitrogen-use efficiency support higher rates of temperate forest productivity under elevated CO_2. Proc. Natl. Acad. Sci. USA 104, 14014–14019.

Finzi, A.C., Abramoff, R.Z., Spiller, K.S., Brzostek, E.R., Darby, B.A., Kramer, M.A., et al., 2015. Rhizosphere processes are quantitatively important components of terrestrial carbon and nutrient cycles. Glob. Change Biol. 21, 2082–2094.

Frey, S.D., Lee, J., Melillo, J.M., Six, J., 2013. The temperature response of soil microbial efficiency and its feedback to climate. Nature Clim. Change 3, 395–398.

García-Martí, M., Piñero, M.C., García-Sanchez, F., Mestre, T.C., López-Delacalle, M., Martínez, V., et al., 2019. Amelioration of the oxidative stress generated by simple or combined abiotic stress through the K^+ and Ca^{2+} supplementation in tomato plants. Antioxidants 8 (8), 81.

George, T.S., Fransson, A.M., Hammond, J.P., White, P.J., 2011. Phosphorus nutrition: rhizosphere processes, plant response and adaptations. In: Bünemann, E., Oberson, A., Frossard, E. (Eds.), Phosphorus in Action. Springer, Heidelberg, Germany, pp. 245–271.

Gifford, R.M., Lutze, J.L., Barrett, D., 1995. Global atmospheric change effects on terrestrial carbon sequestration: Exploration with a global C-and N-cycle model (CQUESTN). Plant Soil 187, 369–387.

Green, J.K., Seneviratne, S.I., Berg, A.M., Findell, K.L., Hagemann, S., Lawrence, D.M., et al., 2019. Large influence of soil moisture on long-term terrestrial carbon uptake. Nature 565, 476–479.

Grillakis, M.G., 2019. Increase in severe and extreme soil moisture droughts for Europe under climate change. Sci. Total Environ. 660, 1245–1255.

Guenet, B., Lenhart, K., Leloup, J., Giusti-Miller, S., Pouteau, V., Mora, P., et al., 2012. The impact of long-term CO_2 enrichment and moisture levels on soil microbial community structure and enzyme activities. Geoderma 170, 331–336.

Hagerty, S.B., Van Groenigen, K.J., Allison, S.D., Hungate, B.A., Schwartz, E., Koch, G.W., et al., 2014. Accelerated microbial turnover but constant growth efficiency with warming in soil. Nature Clim. Change 4, 903–906.

Hansen, N.C., Daniel, T.C., Sharpley, A.N., Lemunyon, J.L., 2002. The fate and transport of phosphorus in agricultural systems. J. Soil Water Cons. 57, 408–417.

Hao, X., Li, P., Han, X., Norton, R.M., Lam, S.K., Zong, Y., et al., 2016. Effects of free-air CO_2 enrichment (FACE) on N, P and K uptake of soybean in northern China. Agric. For. Meteor. 218, 261–266.

Hatfield, J.L., Boote, K.J., Kimball, B.A., Ziska, L.H., Izaurralde, R.C., Ort, D., et al., 2011. Climate impacts on agriculture: implications for crop production. Agron. J. 103, 351–370.

Henze, D.K., Shindell, D.T., Akhtar, F., Spurr, R.J., Pinder, R.W., Loughlin, D., et al., 2012. Spatially refined aerosol direct radiative forcing efficiencies. Environ. Sci. Technol. 46, 9511–9518.

Hertel, O., Skjøth, C.A., Løfstrøm, P., Geels, C., Frohn, L.M., Ellermann, T., et al., 2006. Modelling nitrogen deposition on a local scale—a review of the current state of the art. Environ. Chem. 3, 317–337.

Hinckley, E.L.S., Crawford, J.T., Fakhraei, H., Driscoll, C.T., 2020. A shift in sulfur-cycle manipulation from atmospheric emissions to agricultural additions. Nature Geosci. 13, 597–604.

Houshmandfar, A., Fitzgerald, G.J., O'Leary, G., Tausz-Posch, S., Fletcher, A., Tausz, M., 2018. The relationship between transpiration and nutrient uptake in wheat changes under elevated atmospheric CO_2. Physiol. Plant. 163, 516–529.

Huang, J., Yu, H., Guan, X., Wang, G., Guo, R., 2016. Accelerated dryland expansion under climate change. Nature Clim. Change 6, 166–171.

Hungate, B.A., Dukes, J.S., Shaw, M.R., Luo, Y., Field, C.B., 2003. Nitrogen and climate change. Science 302, 1512–1513.

Hungate, B.A., Stiling, P.D., Dijkstra, P., Johnson, D.W., Ketterer, M.E., Hymus, G.J., et al., 2004. CO_2 elicits long-term decline in nitrogen fixation. Science 304, 1291.

Huo, C., Luo, Y., Cheng, W., 2017. Rhizosphere priming effect: a meta-analysis. Soil Biol. Biochem. 111, 78–84.

Jacobs, A.F., Heusinkveld, B.G., Holtslag, A.A., 2011. Long-term record and analysis of soil temperatures and soil heat fluxes in a grassland area. The Netherlands. Agric. For. Meteor. 151, 774–780.

Jia, G., Shevliakova, E., Artaxo, P., De Noblet-Ducoudré, N., Houghton, R., House, J., et al., 2019. Land–climate interactions. In: Shukla, P.R., Skea, J., Calvo Buendia, E., Masson-Delmotte, V., Pörtner, H.-O., Roberts, D.C., et al.,Climate Change and Land: an IPCC special report on climate change, desertification, land degradation, sustainable land management, food security, and greenhouse gas fluxes in terrestrial ecosystems. Intergovernmental Panel on Climate Change, Geneva, Switzerland, pp. 1–118.

Jin, J., Armstrong, R., Tang, C., 2019. Impact of elevated CO_2 on grain nutrient concentration varies with crops and soils—a long-term FACE study. Sci. Total Environ. 651, 2641–2647.

Karberg, N.J., Pregitzer, K.S., King, J.S., Friend, A.L., Wood, J.R., 2005. Soil carbon dioxide partial pressure and dissolved inorganic carbonate chemistry under elevated carbon dioxide and ozone. Oecologia 142, 296–306.

Katra, I., Gross, A., Swet, N., Tanner, S., Krasnov, H., Angert, A., 2016. Substantial dust loss of bioavailable phosphorus from agricultural soils. Sci. Rep. 6, 1–7.

Keesstra, S.D., Bouma, J., Wallinga, J., Tittonell, P., Smith, P., Cerdà, A., Montanarella, L., et al., 2016. The significance of soils and soil science towards realization of the United Nations Sustainable Development Goals. Soil 2, 111–128.

Kim, K., Labbé, N., Warren, J.M., Elder, T., Rials, T.G., 2015. Chemical and anatomical changes in *Liquidambar styraciflua* L. xylem after long term exposure to elevated CO_2. Environ. Poll. 198, 179–185.

Kimball, B.A., 2016. Crop responses to elevated CO_2 and interactions with H_2O, N, and temperature. Curr. Opin. Plant Biol. 31, 36–43.

Kimball, B.A., Kobayashi, K., Bindi, M., 2002. Responses of agricultural crops to free-air CO_2 enrichment. Adv. Agron. 77, 293–368.

Kirkby, C.A., Kirkegaard, J.A., Richardson, A.E., Wade, L.J., Blanchard, C., Batten, G., 2011. Stable soil organic matter: a comparison of C: N: P: S ratios in Australian and other world soils. Geoderma 163, 197–208.

Kirkby, C.A., Richardson, A.E., Wade, L.J., Conyers, M., Kirkegaard, J.A., 2016. Inorganic nutrients increase humification efficiency and C-sequestration in an annually cropped soil. PLoS One 11, e0153698.

Knox, J., Daccache, A., Hess, T., Haro, D., 2016. Meta-analysis of climate impacts and uncertainty on crop yields in Europe. Environ. Res. Lett. 11, 113004.

Kou-Giesbrecht, S., Menge, D., 2019. Nitrogen-fixing trees could exacerbate climate change under elevated nitrogen deposition. Nature Comm. 10, 1–8.

Ku, M.S., Schmitt, M.R., Edwards, G.E., 1979. Quantitative determination of RuBP carboxylase–oxygenase protein in leaves of several C_3 and C_4 plants. J. Exp. Bot. 30, 89–98.

Kuzyakov, Y., Horwath, W.R., Dorodnikov, M., Blagodatskaya, E., 2019. Review and synthesis of the effects of elevated atmospheric CO_2 on soil processes: No changes in pools, but increased fluxes and accelerated cycles. Soil Biol. Biochem. 128, 66–78.

Lavallee, J.M., Conant, R.T., Haddix, M.J., Follett, R.F., Bird, M.I., Paul, E.A., 2019. Selective preservation of pyrogenic carbon across soil organic matter fractions and its influence on calculations of carbon mean residence times. Geoderma 354, 113866.

Lavallee, J.M., Soong, J.L., Cotrufo, M.F., 2020. Conceptualizing soil organic matter into particulate and mineral-associated forms to address global change in the 21st century. Glob. Change Biol. 26, 261–273.

Leakey, A.D., Ainsworth, E.A., Bernacchi, C.J., Rogers, A., Long, S.P., Ort, D.R., 2009. Elevated CO_2 effects on plant carbon, nitrogen, and water relations: six important lessons from FACE. J. Exp. Bot. 60, 2859–2876.

Lee, W.G., Fenner, M., 1989. Mineral nutrient allocation in seeds and shoots of twelve *Chionochloa* species in relation to soil fertility. J. Ecol. 77, 704–716.

Lett, S., Michelsen, A., 2014. Seasonal variation in nitrogen fixation and effects of climate change in a subarctic heath. Plant Soil 379, 193–204.

Linderholm, H.W., 2006. Growing season changes in the last century. Agric. For. Meteorol. 137, 1–14.

Lipper, L., Thornton, P., Campbell, B.M., Baedeker, T., Braimoh, A., Bwalya, M., et al., 2014. Climate-smart agriculture for food security. Nature Clim. Change 4, 1068–1072.

Liu, J.X., Zhang, D.Q., Zhou, G.Y., Faivre-Vuillin, B., Deng, Q., Wang, C.L., 2008. CO_2 enrichment increases nutrient leaching from model forest ecosystems in subtropical China. Biogeoscience 5, 1783–1795.

Liu, Y., Li, N., Zhang, Z., Huang, C., Chen, X., Wang, F., 2020. The central trend in crop yields under climate change in China: A systematic review. Sci. Total Environ. 704, 135355.

Logan, T.J., Randall, G.W., Timmons, D.R., 1980. Nutrient content of tile drainage from cropland in the North Central Region. Research Bulletin 1119. Ohio Agricultural Research and Development Center, Wooster, OH. Available from: https://kb.osu.edu/bitstream/handle/1811/62940/1/OARDC_research_bulletin_n1119.pdf, Verified 26 Jan. 2022.

Lozier, T.M., Macrae, M.L., Brunke, R., Van Eerd, L.L., 2017. Release of phosphorus from crop residue and cover crops over the non-growing season in a cool temperate region. Agric. Water Manage. 189, 39–51.

Luo, Y., Su, B.O., Currie, W.S., Dukes, J.S., Finzi, A., Hartwig, U., et al., 2004. Progressive nitrogen limitation of ecosystem responses to rising atmospheric carbon dioxide. Bioscience 54, 731–739.

Luo, Z.B., Langenfeld-Heyser, R., Calfapietra, C., Polle, A., 2005. Influence of free air CO_2 enrichment (EUROFACE) and nitrogen fertilisation on the anatomy of juvenile wood of three poplar species after coppicing. Trees 19, 109–118.

Lüscher, A., Hartwig, U.A., Suter, D., Nösberger, J., 2000. Direct evidence that symbiotic N_2 fixation in fertile grassland is an important trait for a strong response of plants to elevated atmospheric CO_2. Glob. Change Biol. 6, 655–662.

Lynch, J.P., Clair, S.B., 2004. Mineral stress: the missing link in understanding how global climate change will affect plants in real world soils. Field Crops Res. 90, 101–115.

Mahowald, N.M., Muhs, D.R., Levis, S., Rasch, P.J., Yoshioka, M., Zender, C.S., et al., 2006. Change in atmospheric mineral aerosols in response to climate: Last glacial period, preindustrial, modern, and doubled carbon dioxide climates. J. Geophys. Res.-Atmos. 111, D10202. Available from: https://doi.org/10.1029/2005JD006653.

Makowski, D., Marajo-Petitzon, E., Durand, J.L., Ben-Ari, T., 2020. Quantitative synthesis of temperature, CO_2, rainfall, and adaptation effects on global crop yields. Eur. J. Agron. 115, 126041.

Manderscheid, R., Bender, J., Jäger, H.J., Weigel, H.J., 1995. Effects of season long CO_2 enrichment on cereals. II. Nutrient concentrations and grain quality. Agric. Ecosyst. Environ. 54, 175–185.

Mbow C., Rosenzweig C., Barioni L.G., Benton T.G., Herrero M., Krishnapillai M., et al., 2019. Climate Change and Land: an IPCC Special Report on Climate Change, Desertification, Land Degradation, Sustainable Land Management, Food Security and Greenhouse Gas Fluxes in Terrestrial Ecosystems. IPCC; Food security; pp. 437–550. https://www.ipcc.ch/site/assets/uploads/2019/11/08_Chapter-5.pdf.

McGrath, J.M., Lobell, D.B., 2013. Reduction of transpiration and altered nutrient allocation contribute to nutrient decline of crops grown in elevated CO_2 concentrations. Plant Cell Environ. 36, 697–705.

Mengutay, M., Ceylan, Y., Kutman, U.B., Cakmak, I., 2013. Adequate magnesium nutrition mitigates adverse effects of heat stress on maize and wheat. Plant Soil 368, 57–72.

Michalak, A.M., Anderson, E.J., Beletsky, D., Boland, S., Bosch, N.S., Bridgeman, T.B., et al., 2013. Record-setting algal bloom in Lake Erie caused by agricultural and meteorological trends consistent with expected future conditions. Proc. Natl. Acad. Sci. USA 110, 6448–6452.

Myers, S.S., Zanobetti, A., Kloog, I., Huybers, P., Leakey, A.D., Bloom, A.J., et al., 2014. Increasing CO_2 threatens human nutrition. Nature 510, 139–142.

Nannipieri, P., Giagnoni, L., Landi, L., Renella, G., 2011. Role of phosphatase enzymes in soil. In: Bünemann, E., Oberson, A., Frossard, E. (Eds.), Phosphorus in Action. Springer, Heidelberg, Germany, pp. 215–243.

Norby, R.J., Warren, J.M., Iversen, C.M., Medlyn, B.E., McMurtrie, R.E., 2010. CO_2 enhancement of forest productivity constrained by limited nitrogen availability. Proc. Natl. Acad. Sci. USA 107, 19368–19373.

Norton, J., Ouyang, Y., 2019. Controls and adaptive management of nitrification in agricultural soils. Front. Microbiol. 10, 1931.

Ockenden, M.C., Hollaway, M.J., Beven, K.J., Collins, A.L., Evans, R., Falloon, P.D., et al., 2017. Major agricultural changes required to mitigate phosphorus losses under climate change. Nature Comm. 8, 1–9.

Okin, G.S., Mahowald, N., Chadwick, O.A., Artaxo, P., 2004. Impact of desert dust on the biogeochemistry of phosphorus in terrestrial ecosystems. Glob. Biogeochem. Cycles 18, GB2005. Available from: https://doi.org/10.1029/2003GB002145.

Pastore, M.A., Lee, T.D., Hobbie, S.E., Reich, P.B., 2019. Strong photosynthetic acclimation and enhanced water-use efficiency in grassland functional groups persist over 21 years of CO_2 enrichment, independent of nitrogen supply. Glob. Change Biol. 25, 3031–3044.

Peck, A.W., McDonald, G.K., 2010. Adequate zinc nutrition alleviates the adverse effects of heat stress in bread wheat. Plant Soil 337, 355–374.

Peng, W.T., Zhang, L.D., Zhou, Z., Fu, C., Chen, Z.C., Liao, H., 2018. Magnesium promotes root nodulation through facilitation of carbohydrate allocation in soybean. Physiol. Plant. 163, 372–385.

Peñuelas, J., Canadell, J.G., Ogaya, R., 2011. Increased water-use efficiency during the 20th century did not translate into enhanced tree growth. Glob. Ecol. Biogeogr. 20, 597–608.

Phillips, R.P., Finzi, A.C., Bernhardt, E.S., 2011. Enhanced root exudation induces microbial feedbacks to N cycling in a pine forest under long-term CO_2 fumigation. Ecol. Lett. 14, 187–194.

Phillips, R.P., Meier, I.C., Bernhardt, E.S., Grandy, A.S., Wickings, K., Finzi, A.C., 2012. Roots and fungi accelerate carbon and nitrogen cycling in forests exposed to elevated CO_2. Ecol. Lett. 15, 1042–1049.

Pilbeam, D.J., 2015. Breeding crops for improved mineral nutrition under climate change conditions. J. Exp. Bot. 66, 3511–3521.

Pope III, C.A., Burnett, R.T., Thun, M.J., Calle, E.E., Krewski, D., Ito, K., et al., 2002. Lung cancer, cardiopulmonary mortality, and long-term exposure to fine particulate air pollution. J. Amer. Med. Assoc. 287, 1132–1141.

Reddy, K.R., Zhao, D., 2005. Interactive effects of elevated CO_2 and potassium deficiency on photosynthesis, growth, and biomass partitioning of cotton. Field Crops Res. 94, 201–213.

Reid, K., Schneider, K., McConkey, B., 2018. Components of phosphorus loss from agricultural landscapes, and how to incorporate them into risk assessment tools. Front. Earth Sci. 6, 135.

Riddick, S., Ward, D., Hess, P., Mahowald, N., Massad, R., Holland, E., 2016. Estimate of changes in agricultural terrestrial nitrogen pathways and ammonia emissions from 1850 to present in the Community Earth System Model. Biogeoscience 13, 3397–3426.

Riedell, W.E., Pikul, J.L., Jaradat, A.A., Schumacher, T.E., 2009. Crop rotation and nitrogen input effects on soil fertility, maize mineral nutrition, yield, and seed composition. Agron. J. 101, 870–879.

Robertson, G.P., Bruulsema, T.W., Gehl, R.J., Kanter, D., Mauzerall, D.L., Rotz, C.A., et al., 2013. Nitrogen–climate interactions in US agriculture. Biogeochemistry 114, 41–70.

Rogers, A., Ainsworth, E.A., Leakey, A.D., 2009. Will elevated carbon dioxide concentration amplify the benefits of nitrogen fixation in legumes? Plant Physiol. 151, 1009–1016.

Rogers, A., Gibon, Y., Stitt, M., Morgan, P.B., Bernacchi, C.J., Ort, D.R., et al., 2006. Increased C availability at elevated carbon dioxide concentration improves N assimilation in a legume. Plant Cell Environ. 29, 1651–1658.

Rogers, G.S., Payne, L., Milham, P., Conroy, J., 1993. Nitrogen and phosphorus requirements of cotton and wheat under changing atmospheric CO_2 concentrations. Plant Soil 155, 231–234.

Ruark, M., Madison, A., Madison, F., Cooley, E., Frame, D., Stuntebeck, T., et al., 2012. Phosphorus loss from tile drains: Should we be concerned. Univ. of Wisconsin, Madison, WI. https://fyi.uwex.edu/drainage/files/2015/09/P-Loss-from-Tile-Drains-ppt.pdf. (Verified 26 Jan. 2022).

Rubio-Asensio, J.S., Bloom, A.J., 2017. Inorganic nitrogen form: a major player in wheat and Arabidopsis responses to elevated CO_2. J. Exp. Bot. 68, 2611–2625.

Saban, J.M., Chapman, M.A., Taylor, G., 2019. FACE facts hold for multiple generations. Evidence from natural CO_2 springs. Glob. Change Biol. 25, 1–11.

Sadowsky, M.J., 2005. Soil stress factors influencing symbiotic nitrogen fixation. In: Werner, D., Newton, W.E. (Eds.), Nitrogen Fixation in Agriculture, Forestry, Ecology, and the Environment. Springer, Dordrecht, The Netherlands, pp. 89–112.

Sage, R.F., Sage, T.L., Kocacinar, F., 2012. Photorespiration and the evolution of C_4 photosynthesis. Annu. Rev. Plant Biol. 63, 19–47.

Sala, O.E., Chapin, F.S., Armesto, J.J., Berlow, E., Bloomfield, J., Dirzo, R., et al., 2000. Global biodiversity scenarios for the year 2100. Science 287, 1770–1774.

Sardans, J., Alonso, R., Janssens, I.A., Carnicer, J., Vereseglou, S., Rillig, M.C., et al., 2016. Foliar and soil concentrations and stoichiometry of nitrogen and phosphorous across European *Pinus sylvestris* forests: relationships with climate, N deposition and tree growth. Func. Ecol. 30, 676–689.

Schaff, B.E., Skogley, E.O., 1982. Diffusion of potassium, calcium, and magnesium in Bozeman silt loam as influenced by temperature and moisture. Soil Sci. Soc. Am. J. 46, 521–524.

Sillen, W.M.A., Dieleman, W.I.J., 2012. Effects of elevated CO_2 and N fertilization on plant and soil carbon pools of managed grasslands: a meta-analysis. Biogeoscience 9, 2247–2258.

Soares, J.C., Santos, C.S., Carvalho, S.M., Pintado, M.M., Vasconcelos, M.W., 2019. Preserving the nutritional quality of crop plants under a changing climate: importance and strategies. Plant Soil 443, 1–26.

Soong, J.L., Phillips, C.L., Ledna, C., Koven, C.D., Torn, M.S., 2020. CMIP5 models predict rapid and deep soil warming over the 21st century. J. Geophys. Res.-Biogeosci. 125, e2019JG005266.

Stuart, M.E., Gooddy, D.C., Bloomfield, J.P., Williams, A.T., 2011. A review of the impact of climate change on future nitrate concentrations in groundwater of the UK. Sci. Total Environ. 409, 2859–2873.

Terrer, C., Vicca, S., Hungate, B.A., Phillips, R.P., Prentice, I.C., 2016. Mycorrhizal association as a primary control of the CO_2 fertilization effect. Science 353, 72–74.

Terrer, C., Jackson, R.B., Prentice, I.C., Keenan, T.F., Kaiser, C., Vicca, S., et al., 2019. Nitrogen and phosphorus constrain the CO_2 fertilization of global plant biomass. Nature Clim. Change 9, 684–689.

Thivierge, M.N., Jégo, G., Bélanger, G., Bertrand, A., Tremblay, G.F., Rotz, C.A., et al., 2016. Predicted yield and nutritive value of an alfalfa–timothy mixture under climate change and elevated atmospheric carbon dioxide. Agron. J. 108, 585–603.

Tipping, E., Benham, S., Boyle, J.F., Crow, P., Davies, J., Fischer, U., et al., 2014. Atmospheric deposition of phosphorus to land and freshwater. Environ. Sci.-Processes Impacts 16, 1608–1617.

Treseder, K.K., 2004. A meta-analysis of mycorrhizal responses to nitrogen, phosphorus, and atmospheric CO_2 in field studies. New Phytol. 164, 347–355.

van der Putten, W.H., Bradford, M.A., Pernilla Brinkman, E., Van de Voorde, T.F., Veen, G.F., 2016. Where, when and how plant–soil feedback matters in a changing world. Func. Ecol. 30, 1109–1121.

Van Esbroeck, C.J., Macrae, M.L., Brunke, R.I., McKague, K., 2016. Annual and seasonal phosphorus export in surface runoff and tile drainage from agricultural fields with cold temperate climates. J. Great Lakes Res. 42, 1271–1280.

van Groenigen, K.J., Six, J., Hungate, B.A., de Graaff, M.A., Van Breemen, N., Van Kessel, C., 2006. Element interactions limit soil carbon storage. Proc. Natl. Acad. Sci. USA 103, 6571–6574.

Van Meter, K.J., Basu, N.B., Veenstra, J.J., Burras, C.L., 2016. The nitrogen legacy: emerging evidence of nitrogen accumulation in anthropogenic landscapes. Environ. Res. Lett. 11, 035014.

Van Vuuren, M.M., Robinson, D., Fitter, A.H., Chasalow, S.D., Williamson, L., Raven, J.A., 1997. Effects of elevated atmospheric CO_2 and soil water availability on root biomass, root length, and N, P and K uptake by wheat. New Phytol. 135, 455–465.

Vet, R., Artz, R.S., Carou, S., Shaw, M., Ro, C.U., Aas, W., et al., 2014. A global assessment of precipitation chemistry and deposition of sulfur, nitrogen, sea salt, base cations, organic acids, acidity and pH, and phosphorus. Atmos. Environ. 93, 3–100.

Vitousek, P.M., Porder, S., Houlton, B.Z., Chadwick, O.A., 2010. Terrestrial phosphorus limitation: mechanisms, implications, and nitrogen–phosphorus interactions. Ecol. Appl. 20, 5–15.

Wand, S.J., Midgley, G.F., Jones, M.H., Curtis, P.S., 1999. Responses of wild C_4 and C_3 grass (*Poaceae*) species to elevated atmospheric CO_2 concentration: a meta-analytic test of current theories and perceptions. Glob. Change Biol. 5, 723–741.

Wang, B., Guo, C., Wan, Y., Li, J., Ju, X., Cai, W., et al., 2020. Air warming and CO_2 enrichment increase N use efficiency and decrease N surplus in a Chinese double rice cropping system. Sci. Total Environ. 706, 136063.

Wang, L., Flanagan, D.C., Wang, Z., Cherkauer, K.A., 2018. Climate change impacts on nutrient losses of two watersheds in the Great Lakes region. Water 10, 442.

Warncke, D.D., Barber, S.A., 1972. Diffusion of zinc in soil: I. The influence of soil moisture. Soil Sci. Soc. Am. J. 36, 39–42.

Warner, J.X., Dickerson, R.R., Wei, Z., Strow, L.L., Wang, Y., Liang, Q., 2017. Increased atmospheric ammonia over the world's major agricultural areas detected from space. Geophys. Res. Lett. 44, 2875–2884.

Watanabe, Y., Tobita, H., Kitao, M., Maruyama, Y., Choi, D., Sasa, K., et al., 2008. Effects of elevated CO_2 and nitrogen on wood structure related to water transport in seedlings of two deciduous broad-leaved tree species. Trees 22, 403–411.

Weigel, H.J., Manderscheid, R., 2012. Crop growth responses to free air CO_2 enrichment and nitrogen fertilization: Rotating barley, ryegrass, sugar beet and wheat. Eur. J. Agron. 43, 97–107.

Weigel, H.J., Manderscheid, R., Jäger, H.J., Mejer, G.J., 1994. Effects of season-long CO_2 enrichment on cereals. I. Growth performance and yield. Agric. Ecosyst. Environ. 48, 231–240.

Welikhe, P., Brouder, S.M., Volenec, J.J., Gitau, M., Turco, R.F., 2020. Development of phosphorus sorption capacity-based environmental indices for tile-drained systems. J. Environ. Qual. 49, 378–391.

West, J.B., HilleRisLambers, J., Lee, T.D., Hobbie, S.E., Reich, P.B., 2005. Legume species identity and soil nitrogen supply determine symbiotic nitrogen-fixation responses to elevated atmospheric [CO_2]. New Phytol. 167, 523–530.

Wieder, W.R., Cleveland, C.C., Smith, W.K., Todd-Brown, K., 2015. Future productivity and carbon storage limited by terrestrial nutrient availability. Nat. Geosci. 8, 441–444.

Williams, M.R., King, K.W., Ford, W., Buda, A.R., Kennedy, C.D., 2016. Effect of tillage on macropore flow and phosphorus transport to tile drains. Water Resour. Res. 52, 2868–2882.

Wong, S.C., 1979. Elevated atmospheric partial pressure of CO_2 and plant growth. I. Interactions of nitrogen nutrition and photosynthetic capacity in C_3 and C_4 plants. Oecologia 44, 68–74.

Yan, Y., Mao, K., Shi, J., Piao, S., Shen, X., Dozier, J., et al., 2020. Driving forces of land surface temperature anomalous changes in North America in 2002–2018. Sci. Rep. 10, 1–13.

Zaehle, S., Medlyn, B.E., De Kauwe, M.G., Walker, A.P., Dietze, M.C., Hickler, T., et al., 2014. Evaluation of 11 terrestrial carbon–nitrogen cycle models against observations from two temperate Free-Air CO_2 Enrichment studies. New Phytol. 202, 803–822.

Zhang, Y., Mathur, R., Bash, J.O., Hogrefe, C., Xing, J., Roselle, S.J., 2018. Long-term trends in total inorganic nitrogen and sulfur deposition in the US from 1990 to 2010. Atmos. Chem. Physics 18, 9091–9106.

Zhu, C., Kobayashi, K., Loladze, I., Zhu, J., Jiang, Q., Xu, X., et al., 2018. Carbon dioxide (CO_2) levels this century will alter the protein, micronutrients, and vitamin content of rice grains with potential health consequences for the poorest rice-dependent countries. Sci. Adv. 4, eaaq1012. Available from: https://doi.org/10.1126/sciadv.aaq1012.

Chapter 20

Nutrient and carbon fluxes in terrestrial agroecosystems[☆]

Andreas Buerkert[1], Rainer Georg Joergensen[2], and Eva Schlecht[3]

[1]*Organic Plant Production in the Tropics and Subtropics (OPATS), University of Kassel, Witzenhausen, Germany,* [2]*Soil Biology and Plant Nutrition (SBPN), University of Kassel, Witzenhausen, Germany,* [3]*Animal Husbandry in Tropics and Subtropics (AHTS), University of Kassel and University of Göttingen, Witzenhausen, Germany*

Summary

This chapter summarizes the soil biological, biochemical, and physical factors that determine the turnover of N, P, K, and C in soils and decomposing soil amendments (such as plant residues, manure, and composts) in arctic, temperate, subtropical, and tropical agroecosystems. It emphasizes the role of bacterial and fungal decomposer communities in soils, the importance of pH and other soil properties, the possible effects of global warming, the quality of organic substrates, and the role that earthworms, termites, and animal husbandry systems play in nutrient cycling at different scales. A particular emphasis is placed on carbon and nutrient fluxes as influenced by the source-sink relationships in rapidly growing rural–urban transformation zones where agriculture is often intensive until urbanization leads to new land uses.

20.1 Microbiological factors determining carbon and nitrogen emissions

Carbon is emitted from soil mainly as CO_2 and CH_4 (IPCC, 2007), although a variety of volatile organic compounds, such as terpenes, are also emitted from soil to the atmosphere (Ludley et al., 2009; Scheller, 2001). CO_2 is a trace gas with a concentration of 407 μL L^{-1} in the atmosphere (in 2018; UBA, 2020), which amounts to approximately 800 Gt of C in the atmospheric pool (IPCC, 2013). Lately, the CO_2 concentration is increasing by 3 μL L^{-1} per year, contributing considerably to global warming and subsequent climate change (IPCC, 2013). Plants, algae, and cyanobacteria absorb CO_2 and water and use sunlight for energy to produce carbohydrates as assimilates and O_2 as a by-product. These photosynthetic organisms convert roughly 120 Gt of C into their biomass per year (Beer et al., 2010; Field et al., 1998), that is, 15% of the atmospheric CO_2. This means that CO_2 is not only a trace gas in the atmosphere but also an essential component for plant growth. An increase in CO_2 concentration enhances plant biomass production (Manderscheid et al., 2010), decreases the water demand of plants, and increases their water-use efficiency (Chun et al., 2011; Durand et al., 2018; Qiao et al., 2010). At present concentrations, some plant species (especially N_2-fixing legumes) have repeatedly been found to be limited by CO_2 (Bourgault et al., 2016; Hao et al., 2016).

Gaseous N emissions from agricultural ecosystems are economically relevant. Nitrogen is emitted mainly as N_2 via microbial denitrification but also as N oxides, especially N_2O, or as NH_3 at high soil pH (Adviento-Borbe et al., 2010; Niraula et al., 2019). N_2O is the third most important anthropogenic greenhouse gas (GHG) (IPCC, 2013) and also is the most important single ozone-depleting molecule (Ravishankara et al., 2009). It has a global warming potential approximately 300 times that of CO_2 (IPCC, 2013). Denitrification in the vadose zone and in wastewater treatment facilities contributes to removing excessive (and easily leachable) NO_3^-, thereby protecting groundwater and aquatic surface ecosystems from eutrophication (Deurer et al., 2008; Leu et al., 2010).

[☆] This chapter is a revision of the third edition chapter by A. Buerkert, R.G. Joergensen, B. Ludwig, and E. Schlecht, pp. 473–482. DOI: https://doi.org/10.1016/B978-0-12-384905-2.00002-9. © Elsevier Ltd.

20.1.1 CO$_2$ emission

Gaseous emissions of carbon from soil consist mainly of CO$_2$ derived from plant roots and soil microorganisms. The contribution of autotrophic plants to the CO$_2$ emission from soils shows diurnal and seasonal variations, depending on photosynthetic activity, stage of development, and type of species. Actively N$_2$-fixing legumes respire especially high amounts of CO$_2$ to fulfill the energy demand of rhizobia (Merbach et al., 1999; Wichern et al., 2004a). The C input via plant residues (Poll et al., 2010), and especially by rhizodeposition, fuels microbial processes in the soil (De Graaff et al., 2013; Hupe et al., 2019; Wichern et al., 2008). The microbial contribution to CO$_2$ emissions depends on soil temperature, soil moisture, and C availability (Wichern et al., 2004a). The C availability depends primarily on the quality of the organic C input but also on soil properties such as pH (Anderson and Domsch, 1993), clay content (Müller and Höper, 2004), soil structure (Farquharson and Baldock, 2008), and other factors controlling microbial activity (Joergensen and Emmerling, 2006) and gas diffusion (Müller et al., 2011; Šimůnek and Suarez, 1993). The emission of CO$_2$ from soils is a sensitive indicator of soil organic C available to the microbial decomposer community (Al-Kaisi and Yin, 2005; Faust et al., 2019a, 2019b). Microbial CO$_2$ production from the decomposition of soil organic matter and plant residues is an important contributor to GHG emissions (Kump, 2002). The catabolic use of soil organic matter by soil microorganisms drives the release of organically bound nutrients, that is, N, S, and P, but also of Ca and Mg (Joergensen et al., 2009; Rottmann et al., 2011).

In laboratory incubation studies, CO$_2$ evolution is usually measured in sieved and preincubated soils using constant temperature and moisture conditions. In these experiments, CO$_2$ derived from the microbial mineralization of soil organic matter is termed basal respiration (Anderson and Domsch, 1990, 2010). In most soils, a close positive relationship exists between basal respiration and microbial biomass (Jörgensen, 1995). The reasons for this are not fully understood as a decrease in the microbial biomass in neutral and aerobic soils due to CHCl$_3$ fumigation does not necessarily lead to a strong shift in basal respiration (Kemmitt et al., 2008). These authors suggested that the rate-limiting step is governed by abiotic processes that convert soil organic matter into bioavailable compounds. However, the abiotic gate hypothesis of Kemmitt et al. (2008) cannot explain the positive correlation between basal respiration and microbial biomass obtained in most agricultural soils.

20.1.2 Fungal and bacterial contributions to CO$_2$ emissions

Soil microorganisms encompass mainly fungi and bacteria, but also archaea and many eukaryotic protists. They are associated with most enzymatic processes in soils and preserve energy and nutrients in their biomass (Jenkinson and Ladd, 1981). The diversity of soil microorganisms is enormous (Torsvik et al., 1990) and most species are still unknown, although microbial diversity has received much research interest and activity in the 21st century (Hartmann et al., 2015; Torsvik and Øvreås, 2007). Soil ecology, in describing the interactions of soil animals and soil microorganisms in food webs, generally separates the microbial community into fungi and bacteria (Coleman, 2008; Hedlund et al., 2004; Holtkamp et al., 2008), which are the two largest functional microbial subgroups in the soil. Archaea and protozoa contribute only approximately 1% and 2%, respectively, to the soil microbial biomass (Bardgett and Griffiths, 1997; Fierer, 2017; Gattinger et al., 2002). The reason for separating the soil microbial community into fungi and bacteria is their different functions in decomposition (Joergensen and Wichern, 2008). Aerobic saprotrophic fungi are the main primary and secondary decomposers of cell wall residues, especially lignin-cellulose complexes, but also hemicelluloses and pectin (Eastwood et al., 2011; Voříšková et al., 2014) by excretion of enzymes (Burns et al., 2013; Schneider et al., 2012). For this reason, fungi, in particular, dominate the microbial community in acidic soils with organic matter that has high C/N ratios (e.g., recalcitrant plant residue) and is low in available nutrients (Högberg et al., 2007; Khan et al., 2016). By contrast, bacterial abundance increases with an enhanced supply of more easily decomposable substrates, such as proteins, lipids, and sugars (Koranda et al., 2014), and is high in specific N and S cycling processes, especially under anaerobic soil conditions (Paul, 2007; Van Elsas et al., 2007).

In the past, fungal energy channels were considered slow cycles (Bardgett et al., 1996) characterized by a higher carbon-use efficiency (CUE) of the organic matter input and a lower maintenance requirement in comparison with bacteria (Holland and Coleman, 1987; Sakamoto and Oba, 1994). Fungi form more biomass than bacteria from the same amount of substrate; hence, the enhancement of fungal populations may be an important tool to foster C sequestration in soils (Bailey et al., 2002; Jastrow et al., 2007). However, this general view has been questioned (Heinze et al., 2010; Khan et al., 2016; Scheller and Joergensen, 2008). The repeated application of straw to soil causes a shift of the soil fungal community toward saprotrophic fungi that decompose organic material and soil organic matter more rapidly than in soils supplemented with farmyard manure (Heinze et al., 2010; Scheller and Joergensen, 2008), leading to significant soil organic matter losses and increased CO_2 emissions.

The view that fungal activity in soils is low has been based mainly on decreasing metabolic quotients qCO_2 (the ratio of basal respiration to microbial biomass) (Anderson and Domsch, 1990, 2010) with an increasing proportion of fungi in the microbial biomass of incubation experiments (Sakamoto and Oba, 1994). However, arbuscular mycorrhizal fungi contribute a large (but difficult to measure) percentage to the total microbial biomass (Faust et al., 2017). These fungi are obligate biotrophs rather than saprotrophs (and are thus unable to decompose soil organic matter actively) (Smith and Read, 2008), exhibiting low metabolic activity (or none) without host plants (Faust et al., 2017; Joergensen and Wichern, 2008). No difference is found in the microbial turnover of fungal and bacterial necromass (Derrien and Amelung, 2011; Joergensen and Wichern, 2008), although fungal hyphae are still considered to be more resistant to decomposition than bacterial biomass as suggested decades ago by Webley and Jones (1971) and Guggenberger et al. (1999). Metal contamination, especially by Zn, Pb, and Ni (Chander et al., 2001; Frostegård et al., 1996), promotes fungal populations over bacterial ones (Tang et al., 2019). The same is true for reduction in tillage intensity (Faust et al., 2019a, 2019b; Frey et al., 1999; Murugan et al., 2014; Thiet et al., 2006).

20.1.3 CH₄ emissions

Methane (CH_4) is one of the main hydrocarbons in the atmosphere and is responsible for approximately 20% of global warming (IPCC, 2013). It has a global warming potential 25 times that of CO_2 on a 100-year timescale. Nowadays, the atmospheric CH_4 concentration is at 1.92 μL L^{-1} (in 2017; UBA, 2020). CH_4 emissions indicate strong reductive soil conditions with a respective consequence for soil pH as well as Fe and P mobility in soils (Kögel-Knabner et al., 2010). Under anaerobic conditions, CH_4 is produced solely by prokaryotic archaea (Chaban et al., 2006; Mondav et al., 2014). Although most biogenic methane is the result of autotrophic CO_2 reduction (Noll et al., 2010), some CH_4 might be derived from acetate by acetotrophic archaea (Noll et al., 2010) or methylated components (methylamines, methanol, and methanethiol) by methylotrophic archaea (Zhang et al., 2008). Methanogenic archaea play a vital ecological role by removing excess hydrogen and fermentation products from soil and typically grow in strictly anaerobic environments in which all electron acceptors other than CO_2 (such as O_2, NO_3^-, SO_4^{2-}, and Fe-III) have been depleted (Dubey, 2005). Their activity strongly increases with soil temperature (Bergman et al., 2000; Chin et al., 1999; Mondav et al., 2014) and is especially high in water-saturated soils, such as fens, bogs, swamps, marshland, and paddy rice fields (Liu et al., 2010; Mondav et al., 2014; Zhang et al., 2018a). CH_4 is produced in a variety of soils due to anaerobic microsites created after heavy rainfall events (Kammann et al., 2009; Sey et al., 2008) or by the incorporation of easily decomposable substrates (Brewer et al., 2018; Gregorich et al., 2006).

20.1.4 N₂ and N₂O emissions

Denitrification is the microbial process of nitrate reduction that may ultimately produce N_2 through a series of intermediate gaseous nitrogen oxide products. The preferred N electron acceptors in order of decreasing energy yield are $NO_3^- > NO_2^- > NO > N_2O$. Denitrification completes the N cycle by returning N_2 to the atmosphere. It occurs mainly in environments where O_2 consumption exceeds the rate of O_2 supply, such as in wetlands or around plant residues in the soil. This creates anoxic microsites induced by the enhanced respiratory O_2 consumption by microorganisms, which favors N_2O production (Flessa and Beese, 1995; Ruser et al., 2017; Velthof et al., 2003). N_2O has become the third most important anthropogenic GHG, and soils are its major source to the atmosphere (IPCC, 2013). N_2O is also the single most important ozone-depleting gas molecule (Ravishankara et al., 2009). A wide range of processes have the potential to produce N_2O, but in soil its production is primarily by denitrification and, to a lesser extent, by nitrification (Baggs, 2008; Kool et al., 2011; Wrage-Mönnig et al., 2018).

Denitrification is usually attributed to heterotrophic bacteria, such as *Paracoccus denitrificans* (α-Proteobacteria) and various *Pseudomonas* species (γ-Proteobacteria), although autotrophic denitrifiers such as *Thiobacillus denitrificans* (β-Proteobacteria) have also been identified. However, nitrifying microorganisms such as NH_4^+-oxidizing bacteria (McLain and Martens, 2005; Venterea, 2007; Wrage et al., 2005; Wrage-Mönnig et al., 2018) and archaea (Jung et al., 2014) also contribute significantly to N_2O emission from soils. Although N_2O production is commonly attributed to bacterial activity, eukaryotes are also capable of producing this important GHG (Maeda et al., 2015; Wu et al., 2019), as demonstrated in grassland and desert soils (Crenshaw et al., 2008; Laughlin and Stevens, 2002). Shoun et al. (1992) reported that the capacity to produce N_2O during the reduction of NO_2^- and NO_3^- was relatively widespread among filamentous fungi. Most of these fungi lack the capacity to reduce N_2O to N_2; thus, in contrast to bacterial denitrification (that may produce N_2), fungal NO_3^- reduction only yields N_2O. Consequently, if a population of bacteria or fungi exhibits equivalent rates of denitrification, fungi have the potential to release a greater quantity of N_2O to the atmosphere. Important N_2O-producing fungal species are, for example, *Cylindrocarpon tonkinense*, *Fusarium oxysporum* (Sutka et al., 2008), and many other *Fusarium* species (Xu et al., 2019).

It is often assumed that increased C sequestration results in increased N_2O emissions. Indeed, Kilian et al. (1998) reported a promoting effect of C and N enrichment in arable soils on N_2O release, which is in line with a recent meta-study (Xia et al., 2018) and modeling results (Li et al., 2005; Qiu et al., 2009). However, in some cases, the inclusion of a 1-year grass ley increased soil organic carbon (SOC) contents and decreased N_2O emissions from cereal-dominated rotations (Prade et al., 2017). The long-term effects of C sequestration on N_2O emissions may be partly due to site-dependent differences in environmental conditions.

20.2 Effects of organic soil amendments on gaseous fluxes

The quality and composition of organic amendments has a strong impact on emission fluxes as well as on organic matter sequestration in soils. High nutrient concentrations in plant residues usually lead to higher decomposition rates, especially during the initial phases of degradation (Swift et al., 1979). The most important quality index for organic amendments is the N content, often expressed as the C/N ratio, although P (Cleveland et al., 2002; Hartman and Richardson, 2013; Salamanca et al., 2006), S, Ca, Mg, and K (Cleveland et al., 2006; Jannoura et al., 2014; Tyler, 2005) concentrations may also influence decomposition rates. Organic amendments with low C/N ratios are usually considered more easily decomposable than those with high C/N ratios (Henriksen and Breland, 1999a/b; Potthoff et al., 2005; Swift et al., 1979). They thus supply not only more inorganic N to plants but also more CO_2 and N_2O to the atmosphere. High initial N concentration in plant residues increases the amount and activity of microbial exo-cellulases, endo-cellulases, and xylanases (Henriksen and Breland, 1999a), thus enhancing cellulose degradation (Berg, 2000; Henriksen and Breland, 1999b; Koranda et al., 2014; Potthoff et al., 2008). Nitrogen immobilization in the microbial biomass usually occurs after soil amendment with freshly decaying plant residues with C/N > 25, whereas N mineralization is fast in the case of C/N < 15 (Powlson et al., 2001). By contrast, the addition to soil of highly processed and digested organic substrates, such as sugar cane filter cake (Rasul et al., 2009) and cow feces (Jost et al., 2013), led to strong initial N immobilization, although the total C/N ratio of the substrate was < 13. Excessive supply of inorganic N to forest stands in humid temperate climates has been shown to depress litter decomposition due to a partial inhibition of lignolytic white-rot fungi (Berg, 2000; Whalen et al., 2018; Zhou et al., 2018). It is unknown whether this process occurs also in tropical forest ecosystems or in arable and grassland soils. High Zn and Cu concentrations (e.g., in sewage sludge, compost, farmyard manure, and animal dung) may also hamper decomposition (Khan and Joergensen, 2006).

Important litter quality indices are the chemical composition of organic matter, especially its concentrations of lipids, carbohydrates, protein, and lignin. Carbohydrates are usually subdivided into a soluble fraction, starch, and structural components, that is, hemicellulose and cellulose. Many plants contain not only lipids but also polyphenols (Hättenschwiler and Vitousek, 2000; Quarmby and Allen, 1989). Polyphenolic components that inhibit N mineralization (Fig. 20.1) are found in legumes (Seneviratne, 2000) as well as in foliage of many tropical and subtropical shrubs and trees such as *Eucalyptus* sp. (Salamanca et al., 2006) or *Acacia* spp. (Tian et al., 2001).

In subtropical and tropical agroecosystems, termites often play a major role in decomposing recalcitrant plant material, thereby contributing to the recycling and redistribution of plant nutrients and loosening of surface soil crusts

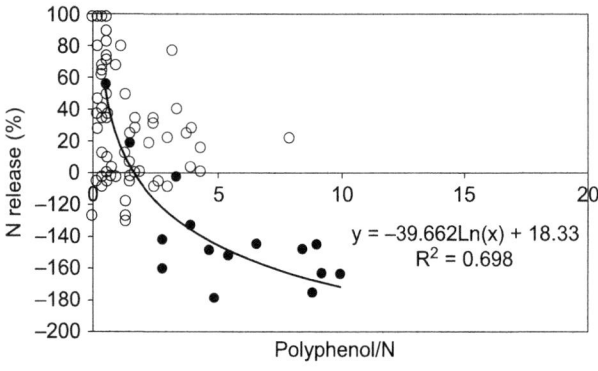

FIGURE 20.1 Effect of polyphenol/nitrogen (N) ratio on N release from plant residues (data from 11 studies). ● plant residues with N concentrations <10 g kg^{-1}; ○ plant residues with N concentrations >10 g kg^{-1}. R^2 was significant at $P < 0.001$. *From Seneviratne (2000), with permission from Springer Nature.*

(Buerkert et al., 2000; Cheik et al., 2019; Esse et al., 2001; Mando and Brussaard, 1999). Termites are crucial in the first step of decomposing organic materials with a high concentration of lignin that protects cellulose against microbial degradation (Robinson et al., 2011; Todaka et al., 2007). However, termites may also contribute to global warming by the decomposition-related emissions of CO_2 and CH_4 on the vast savannah and humid rainforest areas of the tropics where they predominate (Martius et al., 1996; Nauer et al., 2018).

In organic amendments, N can be stabilized by association with lignin- or phenol-rich compounds (Powlson et al., 2001), creating a pool of soluble high-molecular-weight organic N rich in humic substances (Jones et al., 2004). The mineralization of litter depends not only on the quality of organic components but also on the amount and composition of the litter-colonizing microbial community (Dilly and Munch, 2004; Flessa et al., 2002; Potthoff et al., 2008; Santonja et al., 2018; Zhang et al., 2018b). Intensive N fertilization of wheat and other cereals leads to increasing grain yield and N concentration in the grain as well as to elevated N concentrations in leaves and stems (Kernan et al., 1984) and subsequently in straw (Campbell et al., 1993; Raun and Johnson, 1995; Scheller and Joergensen, 2008). The higher protein concentration in the leaf and stem tissues (Wolf and Opitz von Boberfeld, 2003) slows down their maturation (Kichey et al., 2006). Both factors are responsible for a stronger degree of microbial, especially fungal, infection of plant tissues as estimated by the concentrations of ergosterol (Lukas et al., 2018; Scheller and Joergensen, 2008; Wolf and Opitz von Boberfeld, 2003) and fungal glucosamine (Lukas et al., 2018; Scheller and Joergensen, 2008). A higher rate of fungal infection may cause problems with the *Fusarium* toxins deoxynivalenol and zearalenone at harvest (Beukes et al., 2018; Heier et al., 2005; Lemmens et al., 2004). On the other hand, higher fungal colonization of straw leads to increased decomposition rates after incorporation into the soil (Henriksen and Breland, 1999a, 1999b; Jacobs et al., 2011; Rottmann et al., 2011; Scheller and Joergensen, 2008). During maturation, plant species often preferentially carry decomposer communities as surface colonizers that break down their own residues (Keiser et al., 2011; Pouyat and Carreiro, 2003), which has been referred to as the "home field advantage" (Ayres et al., 2009; Di Lonardo et al., 2018). Biogenic composts typically contain highly diverse microbial communities (Franke-Whittle et al., 2005; Meng et al., 2019), and the processing conditions of the waste materials contained therein have a strong impact on C and N mineralization after incorporation into the soil (Cambardella et al., 2003).

Temperature effects on decomposition differ between plant residues and soil organic matter (Kätterer et al., 1998). Only the readily available fractions are decomposed at low temperatures, whereas the more recalcitrant fractions are increasingly decomposed at higher temperatures after the readily available fractions have been depleted (Azmal et al., 1996; Gu et al., 2004; Nicolardot et al., 1994). For this reason, the decomposition of plant residues may be more sensitive to temperature as the recalcitrance of the organic fractions increases. In some experiments, the recalcitrant fractions show a stronger increase in decomposition rates with increasing temperature than labile fractions (Bol et al., 2003; Fierer et al., 2005). By contrast, Kirschbaum (2006) reported no systematic temperature effect on the decomposition of organic fractions differing in recalcitrance, which is in accordance with Bååth and Wallander (2003); Fang et al. (2005, 2006) and Tang et al. (2017). Strong interactions between residue quality, microbial colonization, and temperature have been detected (Hoffmann et al., 2010; Moorhead et al., 2014), warranting further investigation.

20.3 Effects of pH, soil water content, and temperature on organic matter turnover

The adjustment of soil pH by liming improves Ca supply to plants, optimizes the C-use efficiency of soil microorganisms as well as living conditions for earthworms (Holland et al., 2018). High Ca concentration in the soil solution leads to flocculation of clay particles and formation of Ca bridges between the negative binding sites on soil organic matter. Both processes stabilize aggregates and thus soil structure (Holland et al., 2018; Kaiser et al., 2012). Shortly after the application of organic substrates, an increase in soil pH often enhances mineralization and release of N and C (Clay et al., 1993; Wachendorf, 2015), especially in acidic soils with a high proportion of less decomposed plant residues (Curtin et al., 1998). In the long term, adequate Ca supply stabilizes soil organic matter and improves C sequestration (Holland et al., 2018; Wachendorf, 2015) by the increased formation of microbial biomass and microbial residues (Khan et al., 2016).

Soil water content and temperature strongly influence organic matter decomposition and CO_2 release (Faust et al., 2019a, 2019b). Decomposition rates are low in dry and wet soil and strongly increase after rewetting of dry soil as has been demonstrated for rain-fed, seasonally cropped dry savannah soils (Formowitz et al., 2007) or dry-wet cycles in irrigated agriculture (Wichern et al., 2004a, 2004b). At high water content, soils become increasingly anaerobic. Soil respiration strongly declines with decreasing redox potential (Eh) of soils, and microbial activity turns from oxidation of C materials to facultative and subsequently anaerobic fermentation (Ben-Noah and Friedman, 2018; Bhanja and Wang, 2020). With decreasing O_2 availability, the following processes predominate: nitrification and denitrification facilitated by *Nitrosomonas/Nitrobacter* (at Eh ≥ 300 mV, $NH_3 \rightarrow NO_2^- \rightarrow NO_3^- \rightarrow N_2$), Mn^{4+} reduction (at Eh = 300 to 100 mV), Fe^{3+} reduction (at Eh = 100 to −100 mV), SO_4^{2-} reduction (Eh = −100 to −200 mV) and finally methanogenesis (at Eh < −200 mV). These anaerobic processes yield less energy than aerobic decomposition. Hence, decomposition rates are slow in anaerobic soils such as paddy rice fields, protecting organic matter from mineralization (Kögel-Knabner et al., 2010).

Microbial activity and thus C and N mineralization are strongly temperature dependent (De Neve et al., 1996). The strong long-term accumulation of SOC in arctic ecosystems (Rodionow et al., 2006) is combined with the adaptation of microbial decomposer communities to below-freezing conditions (Lukas et al., 2013; Mikan et al., 2002; Steven et al., 2006). Mineralization is more sensitive to low temperatures than photosynthesis, thereby leading to a positive net ecosystem productivity (Schulze et al., 2000), e.g., in boreal zones (see below). Consequently, SOC storage is a function of factors such as temperature, soil water content, and soil chemistry. These factors influence microbial activity not only directly, but also indirectly via effects on plant growth and nutrient uptake. The soil C content is a function of the interactions of plant input and decomposition rate (Table 20.1).

20.4 Global warming effects

The 0−100 cm soil layers of temperate, boreal, and arctic ecosystems, which together occupy 43% of the world surface, store an estimated 64% of SOC and 53% of soil N globally (Batjes, 1996; Nieder and Benbi, 2008). Cool and temperate zone peatlands contain about 450,000 Mt of organic C (out of 1,462,000 Mt of C globally stored in soils) and sequester 200−400 kg C ha^{-1} year^{-1} (Paul, 2016; Tolonen and Turunen, 1996). In (sub-)polar areas above 60 degrees latitude, human-made increases in the concentration of atmospheric greenhouse gases have led to a particularly pronounced effect on the GHG balance (Jahn et al., 2010), with heavily debated consequences for the global climate (AMAP, 2009; IPCC, 2019). On the one hand, warming effects increase the C-sink strength by enhanced net primary production of the forest, tundra, and taiga vegetation following an increase (at least temporary) in photosynthesis, temperature, CO_2, and nutrient availability as well as longer vegetation cycles (Hobbie et al., 2000; Robinson, 2002; Schimel, 1995). Thawing of permafrost soil layers increases microbial turnover processes (Jahn et al., 2010; Steven et al., 2007; Sjögersten et al., 2016). This enhances the release of CO_2 (and CH_4 under anaerobic conditions) from the stored organic matter in northern Histosols that have been protected in the past by cooler conditions (Davidson and Janssens, 2006; Field et al., 2007; McGuire et al., 2006; Uhlířová, et al., 2007) (Table 20.2). These additional gaseous emissions further exacerbate global warming processes, which may be associated with a significant northward movement of permafrost soils (Anisimov et al., 2002; Yeung et al., 2019).

TABLE 20.1 Carbon (C) storage of grasslands compared with forests and agroecosystems. The ranges represent minimum and maximum estimates.

Ecosystem	Vegetation	Soils	Total
	(10^9 t ha^{-1})		
Grasslands			
High-latitude	14−48	281	295−329
Mid-latitude	17−56	140	158−197
Low-latitude	40−126	158	197−284
Total grasslands	71−231	579	650−810
Forests	132−457	481	613−938
Agroecosystems	49−142	264	313−405
Others[a]	16−72	160	177−232
Global total	268−901	1484	1752−2385

[a]*Includes wetlands, barren areas, and human settlements.*
Source: Modified from White et al. (2000).

TABLE 20.2 Estimated stocks of soil organic carbon (SOC) in the northern high latitudes and on the global scale.

Ecosystem	Range of estimated SOC stock (10^9 t C)
Wetlands	
Global[a]	120–460
Northern[b]	202–535
Tundra ecosystems[c]	43–200
Boreal ecosystems[d]	200–750
Northern high latitudes >45 degree N[e]	1400–1850

[a]Gorham (1991), Hobbie et al. (2000), and Turunen et al. (2002).
[b]Several authors in Batjes (1996) and Mitra et al. (2005).
[c]Hobbie et al. (2000) and Stieglitz et al. (2006).
[d]Hobbie et al. (2000), Rodionow et al. (2006), and Stieglitz et al. (2006).
[e]AMAP (2009).
Source: From Jahn et al. (2010).

20.5 Plant–animal interactions affecting nutrient fluxes at different scales

20.5.1 Species-specific relationship between feed intake and excreta quality

20.5.1.1 Quantitative aspects of intake and excretion

The amount and quality of livestock excreta (feces and urine) are primarily determined by the amount and quality of feed ingested. When feed is abundant, the voluntary feed intake by an animal depends on its requirements for energy and nutrients and the so-called "palatability" of the feed, which is a function of its qualitative properties as well as other physicochemical characteristics such as odor and taste (Provenza, 1995; Van Soest, 1994). Energy and nutrient requirements depend on animal species, or even breed [a wool sheep has a higher protein requirement (because of wool growth) than a hair sheep], physiological stage of an animal, its production level, health status, and environmental variables (CSIRO, 2007).

General estimates of the quantity of excreta originating from different livestock species in different regions of the world have been published by FAO (2006). However, given the interdependency of the numerous variables that are modulating feed intake and location-specific aspects of livestock farming, such standard values are often of limited use in predicting the quantity of excreta per animal unit as well as spatiotemporal variation in deposition.

20.5.1.2 Quality of ruminant excreta

In weaned ruminants, ingested feed is at first fermented in the rumen by its microflora (Krause et al., 2013; Van Soest, 1994). They break down crude fiber (cellulose, hemicellulose, and lignin) and other nonstarch polysaccharides (oligosaccharides, pectin, beta-glucan, etc.) that cannot be broken down by the ruminant animal's own digestive enzymes. Sugars and starch as well as lipids and protein and nonprotein N compounds are also fermented in the rumen (Van Soest, 1994). The microbial fermentation processes yield varying proportions of the short-chain fatty acids (SCFA) (acetic, propionic, and butyric acid) and NH_3 that are absorbed through the rumen wall and used in the animal metabolism. The multiplication and decay of the rumen microbial population yield microbial protein that can be utilized in the host's metabolism after postruminal digestion in the abomasum and small intestine (Van Soest, 1994). Feed constituents, microbial debris as well as epithelial cells and mucus shed from the gastrointestinal tract and not digested in the small intestine may undergo fermentative microbial breakdown in the colon ("hindgut fermentation"). The principles of these processes are similar to those in the rumen (Van Soest, 1994). However, in the colon, only the released SCFA, amides, NH_3, and water can be absorbed, whereas microbial protein synthesized in the hindgut is excreted along with any unfermented feed residues and endogenous N (Breves et al., 2009).

The degradability and actual microbial degradation of N-containing feed constituents in the rumen, digestion of feed and microbial protein in the lower gastrointestinal tract, and the extent of hindgut fermentation determine the form and proportion of N excretion in feces and urine, which may then be available for plant uptake. Urea accounts for >70% of urine-N (Bristow et al., 1992; FAO, 2006), and depending on ambient temperature, urease contained in urine quickly breaks down urea to NH_3 and CO_2, leading to large volatilization losses of N. Low ruminal and postruminal degradation of feed N and high microbial activity in the hindgut increase the proportion of N excreted in feces. If feces dry quickly, the N they contain is largely preserved; for example, goat feces dried for 48 h (at 60°C and <30% humidity) contained only 2% less N than the fresh material (Schlecht et al., unpublished data).

Due to the greater stability of N in feces compared to urine, diverting N excretion from urine to feces seems advantageous to plant nutrition. The typical degradability of nitrogenous feed compounds is low if animals are consuming mature and thus strongly lignified grasses, or if their diet contains tannins that are prevalent in many tropical ligneous and legume feeds (Makkar, 2003). Polyphenols have a strong binding affinity to proteins and can inhibit protein degradation in the rumen as well as in the postruminal gastrointestinal tract. The affinity of tannins to proteins depends on the specific tannin-protein combination. Tannins in ruminant diets can also slow down N release from feces after application to the soil (Ingold et al., 2018; Powell et al., 1999; Somda et al., 1995).

The major proportion of organic C contained in ruminant (and nonruminant) feces originates from undigested cell wall constituents. However, coarsely milled grain particles, especially of maize, may be small enough to escape rumen fermentation and also withstand enzymatic breakdown in the small intestine and colon. In such cases, feces may contain a considerable proportion of starch (Kirchgeßner et al., 2008). Strongly lignified cell walls, especially of mature C_4 grasses, but also of legume stalks, can withstand digestion by the enzymatic systems of mammals and their intestinal microflora (Van Soest, 1994). Even though fecal C contained in microbial and epithelial debris along with C from undigested sugars, starch and nonlignified cellulose and hemicellulose can be degraded easily by soil microbes after fecal excretion (leading to CO_2 and CH_4 emissions), the lignified cell wall constituents may contribute (temporarily) to soil organic matter build-up.

In ruminants, large quantities of P are secreted with the saliva, especially when diets rich in fibrous material are fed. With increasing concentration of fibrous material in the diet, the partitioning of gastrointestinally absorbed P between salivary secretion and urinary P excretion shifts toward the salivary route; as a consequence, the primary excretion route for P is via feces (Table 20.3). Urinary P excretion is negligible unless energy-rich pelleted diets are fed, or P is oversupplied (Boeser et al., 2002). Fecal P concentration mainly depends on total dry matter intake, feed-P concentration, P availability, and the animal's P requirements (Underwood and Suttle, 2001).

Phytate, the major form of P stored in cereal grains, is poorly available in monogastrics such as humans, poultry, and pigs; it is therefore largely excreted, causing large losses in nutrient cycles. As an antinutrient to Zn, phytate also decreases the uptake of Zn in humans and other monogastrics (Brugger and Windisch, 2017; Buerkert et al., 1998; Gibson, 2012). For ruminant diets, however, the presence of high phytate concentrations does not pose problems because rumen microbes secrete phytases and are thus able to mineralize phytate-bound P to their host's advantage (Underwood and Suttle, 2001).

In ruminants, 85% of total K is excreted in the urine and 15% in feces (Lhoste et al., 1993). Given that K availability from fibrous materials, cereals, and legume grains typically exceeds 80%, K excretion is mainly determined by the animal's K requirements (Underwood and Suttle, 2001).

20.5.1.3 Quality of pig and poultry excreta

Through their cloaca, birds excrete feces and urine together; the two types of excrements are therefore often treated as one. In both pigs and poultry, microbial fermentation of undigested feed components only takes place in the hindgut (colon in pigs and cecum in poultry) from where resulting SCFA, NH_3, and potentially nonprotein N can be absorbed,

TABLE 20.3 Typical live weight, dry matter excretion, and nitrogen (N) and phosphorus (P) concentrations in urine and feces (slurry) of different livestock species.

Species	Animal weight (kg)	Dry matter (kg year^{-1})	Water content (%)	N (g kg^{-1})	P (g kg^{-1})
Cattle	230	860	87	38	7
Pigs	90	249	88	60	20
Sheep, goats	45	165	75	42	6
Chicken	1.8	9	75	53	20

Source: From Eshenaur (1984).

whereas other (end)products of fermentation remain unavailable to the animal (Mead, 1989; Van Soest, 1994). Because birds and mammals do not secrete phytases, undigested phytate-P is excreted in the feces unless the nonruminant animal's diet is supplemented with phytase (Kirchgeßner et al., 2008). An increasing concentration of grain-bound phytate-P in the diet thus increases fecal P excretion. Depending on feeding practices, fresh matter P concentration may range between 3.6 and 3.9 g kg^{-1} in the slurry of sows (containing 55 g dry matter kg^{-1}) and growing pigs (75 g dry matter kg^{-1}) (De Wit et al., 1997). In poultry manure, fresh matter concentration of P varies from 5 to 10 g kg^{-1} (De Wit et al., 1997). Similarly, typical concentrations of K in slurry fresh matter are 4 g kg^{-1} for sows and 7 g kg^{-1} for growing pigs, whereas poultry manure may contain 5–11 g K kg^{-1} in fresh matter.

Fresh chicken excreta contain on average 16 g N kg^{-1} fresh weight, which consists of 60% uric acid, 2% urea, 6% total ammonium-N, and 32% protein decomposition products. Although the dominance of uric acid over urea in poultry excreta does not affect the release of NH_3 from poultry manure (Rothrock et al., 2010), the acidity of poultry litter may decrease soil pH if litter is applied regularly.

In pigs, the partitioning of N excretion between feces and urine depends on the amino acid pattern of the diet, the structure of feed proteins, the presence of secondary plant metabolites such as protease inhibitors or tannins, the pretreatment of protein-rich feeds, and on the concentration of structural carbohydrates in the diet (Velthof et al., 2005; Kirchgeßner et al., 2008). In feeds with high digestibility, high protein availability, and high biological value of the protein, about 78% of N excreted by a growing pig may be in the urine, whereas at low digestibility, low protein availability, and/or low biological value of the feed protein, the urinary N excretion may decrease to 59% (De Wit et al., 1997).

As in ruminants, organic C in pig and poultry feces originates from undigested cell wall constituents plus nonstarch polysaccharides that cannot be digested by the animal's own enzymes and have escaped hindgut fermentation. This may lead to high C concentration in feces of pigs and poultry fed diets rich in nonstarch polysaccharides (Hadorn, 1994), which are frequently used for sows in organic pig farming (Abel and Breves, 2005) and in many smallholder production systems in the tropics.

20.5.2 Nutrient and carbon losses from livestock excreta

Although C and N are easily lost from stored livestock excreta through gaseous emissions (Predotova et al., 2010a, 2010b; Sommer, 2001) and from the upper soil horizons of sandy soils via leaching, losses of K and P are less frequently reported given their rapid adsorption on clay particles or other ligands. Seepage of urine K into the ground and leaching of P bound to (dissolved) organic matter to deeper soil layers may, however, occur if excreta are stored unprotected or applied to fields in large quantities. For sandy subtropical soils, a few authors (Brouwer and Powell, 1998; Passaglia Azevedo et al., 2018; Siegfried et al., 2011) reported losses of likely C-bound P to deeper soil layers as a consequence of large application rates. Nitrogen leaching and NH_3 emission often lead to a complete loss of urine-N. In feces, only a small fraction of fecal N is lost with initial gaseous N emissions (as NH_3 or N_2O), whereas more substantial volatilization may occur after microbial degradation of N compounds, which depends on environmental conditions and management (FAO, 2006; Ingold et al., 2018).

20.5.3 Spatial aspects of livestock-mediated nutrient fluxes and modeling

More than 55% of the world's pigs and around 70% of the world's poultry are kept in large-scale systems (Robinson et al., 2011). These typically store excreta in slurry tanks, lagoons, and pits near the confinement area; dung heaps may also be found where excreta and bedding materials are mixed. The degree of coupling between such collection systems and manure application to agricultural land largely determines nutrient losses and potential negative environmental impacts (Ju et al., 2005; Mendoza Huaitalla et al., 2010).

More than 70% of global beef cattle, >85% of dairy cattle, and >65% of sheep and goats are kept in mixed cropping-livestock systems (Steinfeld et al., 2006), where animals are stall-fed (zero-grazing systems) or graze grassland, rangeland or harvested fields on a daily or seasonal basis. In zero-grazing systems, slurry or dung are often stored near the animal stables as described for the large-scale systems. In grazing systems, with excreta voided in the grazed areas, excretion frequencies differ among species, with small ruminants urinating and defecating about twice as often as cattle (Schlecht et al., 2006).

Excretions of feces and urine show considerable diurnal variation, but both are particularly frequent at the start of a meal, during drinking, and at getting up after resting (Schlecht et al., 2006). This implies that in grazing systems, excreted C and nutrients are usually concentrated around resting places and watering points, contributing to human-made variations in soil fertility (Fig. 20.2). Apart from these particular events, however, excretions are more or less equally distributed across the day and are therefore proportional to the time spent per land unit.

FIGURE 20.2 Aerial photograph showing residual effects of changes in soil fertility on the growth of pearl millet (*Pennisetum glaucum* L.) due to longevity of farmers' settlements and tethering of animals (see insert, lower right). Numbers indicate the years during which the farmers' settlement remained at a particular site. The photo was taken 75 days after sowing from an altitude of about 300 m in south-west Niger, West Africa. Hardpans (lacking plant growth) within the boundaries of former settlement areas are the result of clay applications to the foundations of the five houses (see the area 1993–94) belonging to one extended family. The increases in millet growth in former settlement areas lasted 4–5 years. The insert at lower left shows traditional strategies of "precision agriculture" using indigenous organic resources (Buerkert et al., 1996). *Modified from Buerkert et al. (1996), with permission from Springer Nature.*

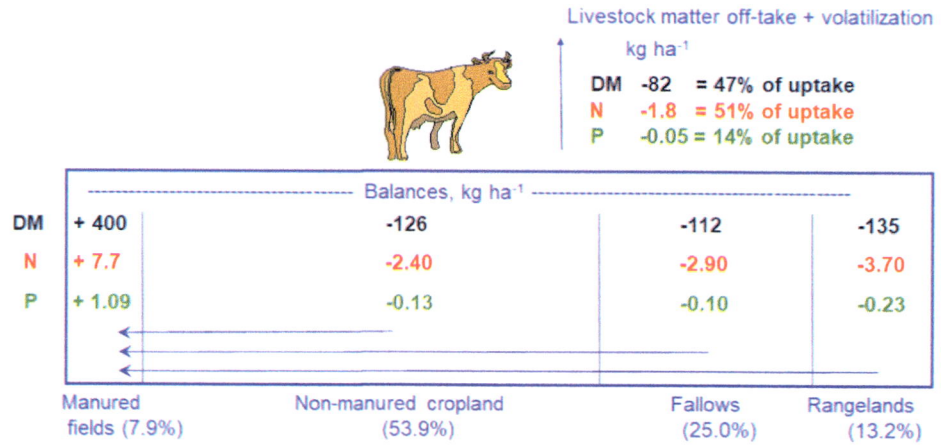

FIGURE 20.3 Livestock-related organic matter and nutrient transfers within village lands of Kodey, Niger. Shown are DM, N, and P stored in animal tissue and voided by the animal (top right), or transferred via intake and excretion from grazed rangelands, fallows, and nonmanured croplands to the manured ones. Values at the bottom indicate the proportion of the land-use types. Assessed area = 75 km^2; observation period = 12 months. For the calculation of animal-related transfers, the respective relative surface areas indicated in the figure must be taken into account (for DM: $400 \times 0.079 = [(126 \times 0.539) + (112 \times 0.25) + (135 \times 0.132)] - 81$). Mean annual stocking rate is 12 tropical livestock units km^{-2}, and average annual rainfall is about 510 mm (Buerkert and Hiernaux, 1998). *DM*, Dry matter; *N*, nitrogen; *P*, phosphorus. *Modified from Buerkert and Hiernaux (1998), with permission from Wiley-VCH Verlag GmbH & Co KGaA.*

The grazing itinerary and behavior of animals (Moreau et al., 2009) are modulated by the location of salt licks and shading trees, grazing and watering regimes, and the supplement feeding and herding practices (Ganskopp, 2001; Turner et al., 2005). Allocation of grazing and resting time to individual land units leads to distinct spatial patterns of nutrient off-take and deposit, and eventually to the build-up of nutrient gradients along livestock routes (Cech et al., 2008, 2010; Turner, 1998; Turner and Hiernaux, 2015; Fig. 20.3). To concentrate manure on a field scheduled for cultivation, livestock can also be corralled or tethered overnight. Across five village territories in Western Niger, herds of approximately 25–60 animals spend between 15 and 46 nights on one field, thereby depositing 3.4–15.5 t ha^{-1} of fecal dry matter in the case of cattle and 1.3–7.2 t ha^{-1} in the case of small ruminants (Schlecht et al., 2004) (Table 20.4).

Since Stoorvogel and Smaling (1994) presented their frequently cited (but rarely verified at the local level) modeling results on substantial nutrient losses in selected African countries, a large number of studies using higher resolutions have been conducted to quantify C and nutrient flows at the field level (Fig. 20.4).

From detailed horizontal balances of crop rotations (Bationo et al., 1998; Buerkert et al., 2005) (Table 20.5) to farm balances (De Jager et al., 1998; Haas et al., 2007; Hiernaux et al., 1997) and the measurement of fluxes of matter in agroecosystems (Hoffmann et al., 2008; Titlyanova, 2007), modeling of the matter turnover processes and losses at different scales has become increasingly important (Fig. 20.5). In addition to the scaling problems and uncertainties about the use of transfer functions for non-measured flux components in agro-ecosystem models, there are also limitations

TABLE 20.4 Average nitrogen (N) and phosphorus (P) availability after crop harvest and the aggregated yearly rates of N and P intake and fecal excretion by two village herds of grazing cattle, sheep, and goats on different land-use types at two locations in south-west Niger (weighted annual averages).

Parameter	Land-use type	Banizoumbou	Kodey
Nitrogen availability[a] (kg ha^{-1})	Rangeland	9.6	7.4
	Fallow	10.5	11.5
	Cropland	19.7	19.2
	Weighted average per site	*17.9*	*17.1*
Nitrogen intake (kg ha^{-1} year^{-1})	Rangeland	3.7	5.3
	Fallow	3.2	4.0
	Cropland	2.6	3.9
	Weighted average per site	*3.0*	*3.7*
Nitrogen excretion (kg ha^{-1} year^{-1})	Rangeland	1.1	1.5
	Fallow	0.8	0.8
	Cropland	0.7	0.8
	Weighted average per site	*0.8*	*0.8*
Phosphorus availability[a] (kg ha^{-1})	Rangeland	0.70	0.54
	Fallow	0.77	0.85
	Cropland	1.13	1.10
	Weighted average per site	*1.26*	*1.15*
Phosphorus intake (kg ha^{-1} year^{-1})	Rangeland	0.26	0.37
	Fallow	0.23	0.28
	Cropland	0.11	0.16
	Weighted average per site	*0.19*	*0.21*
Phosphorus excretion (kg ha^{-1} year^{-1})	Rangeland	0.13	0.17
	Fallow	0.09	0.10
	Cropland	0.08	0.10
	Weighted average per site	*0.09*	*0.11*

[a]N and P availability determined at the end of the rainy season (September for fallow) and after crop harvest (October).
Source: Modified from Schlecht et al. (2004).

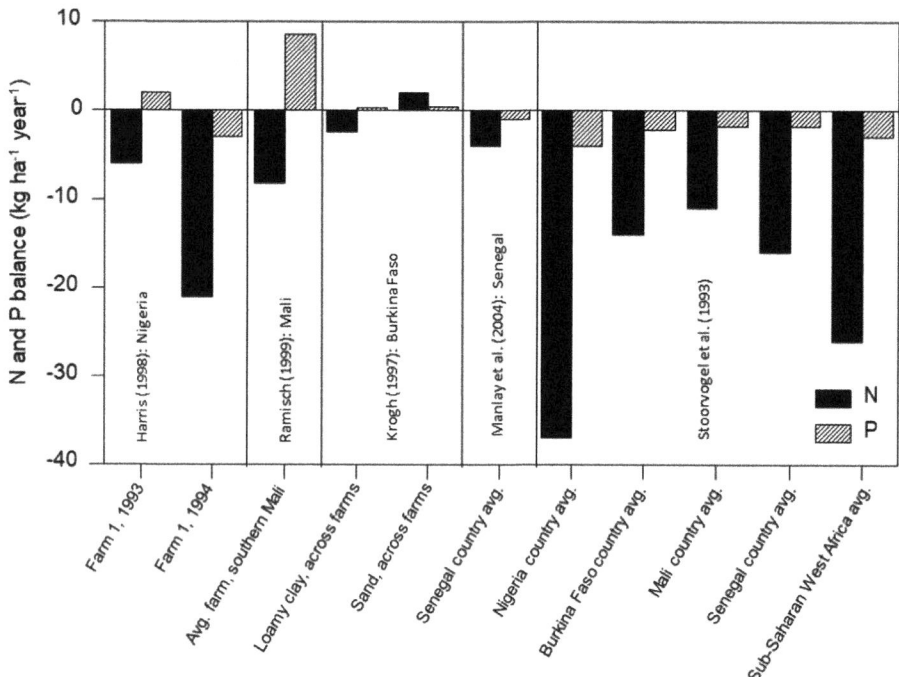

FIGURE 20.4 Nitrogen (N) and phosphorus (P) balances determined at different scales in various locations of sub-Saharan West Africa. *Modified after Schlecht and Hiernaux (2004), with permission from Springer Nature.*

TABLE 20.5 Annual inputs and outputs (kg ha^{-1} year^{-1}) and partial (horizontal) balances (kg year^{-1}) of nitrogen (N), phosphorus (P), and potassium (K) in cropland and palm groves at Balad Seet (Oman).

Land use	Source/process	Input or output[a]			Partial balance[a]		
		N	P	K	N	P	K
Cropland (4.6 ha)	Synthetic fertilizer	143	24	45	658	120	207
	Animal manure	180	40	267	828	184.9	1228
	Irrigation water[b]	10	5.2	17	46	24	78
	Symbiotic N$_2$-fixation	63	n.a.	n.a.	290	n.a.	n.a.
	Crop harvest	−265	−33	−245	−1219	−151	−1127
	Cumulative partial balance	*131*	*36.5*	*84*	*603*	*178*	*386*
Palm groves (8.8 ha)	Synthetic fertilizer	59	1.8	4	519	16	35
	Animal manure + ash	141	8.0	289	1241	70	2543
	Irrigation water	10	5.2	17	88	46	150
	Human feces	170	37	50	1496	326	440
	Date harvest[c]	−63	−13	−176	−554	−112	−1549
	Harvested understory fodder	−14	−1.5	−11	−123	−16	−97
	Cumulative partial balance	*303*	*38*	*173*	*2632*	*324*	*1469*
Oasis (13.4 ha)	Oasis partial balance	244	37	142	3235	502	1855

n.a., Not applicable. The data represent annual averages of a 24-month measurement period from October 2000 to October 2002.
[a]Positive values indicate gains and negative ones losses.
[b]Total annual water flow of 228,587 m^3 was multiplied by nutrient concentrations of 0.57 mg N L^{-1}, 0.30 mg P L^{-1}, and 1.0 mg K L^{-1} and was adjusted to the respective irrigated surface area. The average irrigation intensity was assumed to be similar in both types of land use.
[c]Dates + stems + leaves.

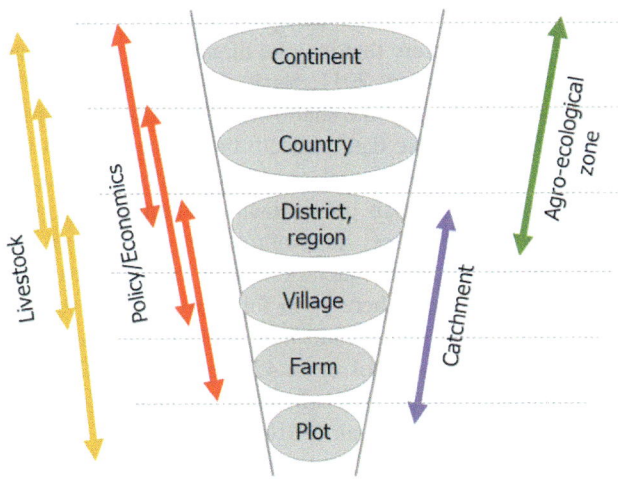

FIGURE 20.5 Diagram indicating various scales of nutrient cycling research. *Adapted from Schlecht and Hiernaux (2004), with permission from Springer Nature.*

imposed by the use of (standard) values for feed digestibility, leaching, volatilization, or nutrient deposition that simple tool-boxes such as NUTMON or FARMSIM (Rufino et al., 2007; Van den Bosch et al., 1998; Van Wijk et al., 2007, 2009) rely on (Bateki and Dickhoefer, 2019). In recent times, process-oriented tools such as the DeNitrification DeComposition (DNDC) model capable of simulating the carbon and nitrogen biogeochemistry in agroecosystems have been tested widely. Initially developed to simulate C and N turnover in US agricultural soils (Li et al., 1992a, 1992b), the updated versions of DNDC have been used successfully across a range of agroecosystems, climates, crops, and on different soil types (Budiman et al., 2020; Gilhespy et al., 2014; Li, 2009; Ludwig et al., 2007, 2011; Zhang et al., 2015) to understand key bottlenecks for improved crop growth and nutrient-use efficiency, and to estimate GHG emissions.

20.6 Scale issues in modeling

To upscale the fluxes of matter from the plot, field, farm, or village level to the regional scale, and from the weekly or seasonal to yearly scale, the restrictions and interdependencies used have to be translated into mathematical relationships. Empirical models are direct reflections of available data, which are set at a defined level of organization and described mathematically (Dijkstra and France, 1995). Mechanistic models instead depict processes occurring within a system and account for organizational hierarchy (France and Thornley, 1984). Numerous models exist to evaluate nutrient balances; the soil, crop, and livestock management strategies; farm economics, and regional land use (De Jager et al., 1998; Fust and Schlecht, 2018; Rufino et al., 2007; Tittonell et al., 2005; Van den Bosch et al., 1998). Which model to choose depends on the aim of the work and the level of detail required for the modeling approach or the level of detail of available spatiotemporal data. The extrapolation or interpolation of processes between scales must consider the degree of heterogeneity specific to each scale, which is generally increasing from the micro- to the macro-scale. Flows and budgets calculated at more aggregated scales such as administrative districts or agro-ecological zones may be useful for high-level policy decisions (Schlecht and Hiernaux, 2004) or may constitute a component in global environmental assessments such as nitrogen and carbon cycling in crop-livestock systems (Herrero et al., 2013).

20.7 Nutrient fluxes in rural−urban systems

Since 2010, considerable research has been undertaken to address special constraints to, and consequences of, urban and peri-urban agriculture (UPA) thriving in many countries within and around cities where the classical agricultural resources (space, labor, and water) are particularly scarce, but market demands for agricultural products and nutrient availability are high (Buerkert and Schlecht, 2019; Drechsel and Keraita, 2014; Lydecker and Drechsel, 2010; Thebo et al., 2014). The production-oriented UPA systems are particularly widespread in Africa and SE-Asia and around European and US cities but are likely less frequent in South America and Australia for cultural and legal reasons. Most UPA systems are operating on land only temporarily and often unofficially allocated to agricultural use (Arnold and Rogé, 2018); therefore, farmers' investments must have a particularly high rate of return. Combined with the proximity of consumer markets, this leads to a predominance of short-duration crops such as vegetables, flowers, and fruits that are usually managed very intensively with respect to nutrient application and irrigation, partly with untreated wastewater (Cofie and Drechsel, 2007). The latter has raised concerns about produce safety that may be threatened by metal and microbial contamination (Abdu et al., 2011a; Graefe et al., 2019; Safi and Buerkert, 2011; Vazhacharickal et al., 2019). Equally debated have been losses of N and C by gaseous emissions and leaching to the shallow aquifers that are of particular interest in densely populated areas (Predotova et al., 2011). The available data indicate that, depending on the location, metal contamination of irrigation water can be high; however, given high soil organic C contents and near-neutral soil pH, their transfer into the food chain is limited, and a risk of human contamination often is negligible (Abdu et al., 2011b). Of more concern is bacterial contamination on fresh produce, for example, by *Escherichia coli* and *Salmonella* spp., which require either effective treatment of (gray) irrigation water or careful product cleansing along the marketing chain (Dao et al., 2018). In African (peri-)urban horticulture where often solid city waste, livestock manure, and mineral N fertilizers are applied to multiple crops per year, detailed studies of nutrient fluxes indicate total annual losses (application surpluses) of >2000 kg N ha^{-1}, with gaseous emissions of up to 420 kg N ha^{-1} and 36 t C ha^{-1} (Lompo et al., 2012; Sangaré et al., 2012). Similar inefficiencies in urban and peri-urban systems, where livestock and crop production are often poorly coupled, have also been reported by Diogo et al. (2010a, 2010b). They indicate the need for regulation or certification of these systems that often network across large distances (Akoto-Danso et al., 2019; Karg et al., 2016) and are increasingly important for vegetable production, income generation, and maintenance of biodiversity in rapidly developing urban environments, particularly in poor income countries of the Global South.

References

Abdu, N., Agbenin, J., Buerkert, A., 2011a. Geochemical assessment, distribution and dynamics of trace elements in contaminated urban soil under long-term wastewater irrigation in Kano, northern Nigeria. J. Plant Nutr. Soil Sci. 174, 447–458.

Abdu, N., Agbenin, J., Buerkert, A., 2011b. Phytoavailability, human risk assessment and transfer characteristics of cadmium and zinc contamination from urban gardens in Kano, Nigeria. J. Sci. Food Agric. 91, 2722–2730.

Abel, H., Breves, G., 2005. Ernährungsphysiologische Bewertung von Öko-Futtermitteln für Schweine. [Physiological feed evaluation for pigs in organic farming.]. Organic eprints; available at: http://orgprints.org/8905/1/8905-02OE209-F-uni-goettingen-abel-2005-schweine.pdf.

Adviento-Borbe, M.A.A., Kaye, J.P., Bruns, M.A., McDaniel, M.D., McCoy, M., Harkcom, S., 2010. Soil greenhouse gas and ammonia emissions in long-term maize-based cropping systems. Soil Sci. Soc. Am. J. 74, 1623–1634.

Akoto-Danso, E.K., Manka'abusi, D., Steiner, C., Werner, S., Häring, V., Lompo, D.J.-P., et al., 2019. Nutrient flows and balances in intensively-managed urban vegetable production of two West African cities. J. Plant Nutr. Soil Sci. 182, 229–243.

Al-Kaisi, M.M., Yin, X., 2005. Tillage and crop residue effects on soil carbon and carbon dioxide emission in corn-soybean rotation. J. Environ. Qual. 34, 437–445.

AMAP, 2009. Update on selected climate issues of concern. Arctic Monitoring and Assessment Programme, Oslo, Norway.

Anderson, T.-H., Domsch, K.H., 1990. Application of eco-physiological quotients (qCO$_2$ and qD) on microbial biomasses from soils of different cropping histories. Soil Biol. Biochem. 22, 251–255.

Anderson, T.-H., Domsch, K.H., 1993. The metabolic quotient for CO$_2$ (qCO$_2$) as a specific activity parameter to assess the effects of environmental conditions, such as pH, on the microbial biomass of forest soils. Soil Biol. Biochem. 25, 393–395.

Anderson, T.-H., Domsch, K.H., 2010. Soil microbial biomass: the eco-physiological approach. Soil Biol. Biochem. 42, 2039–2043.

Anisimov, O.A., Velichko, A.A., Demchenko, P.F., Eliseev, A.V., Mokhov, I.I., Nechaev, V.P., 2002. Effect of climate change on permafrost in the past, present, and future. Izvestiya Atmosph. Ocean. Physics. 38 (Suppl. 1), 25–39.

Arnold, J., Rogé, P., 2018. Indicators of land insecurity for urban farms: institutional affiliation, investment, and location. Sustainability 10, 1963.

Ayres, E., Steltzer, H., Simmons, B.L., Simpson, R.T., Steinweg, J.M., Wallenstein, M.D., et al., 2009. Home-field advantage accelerates leaf litter decomposition in forests. Soil Biol. Biochem. 41, 606–610.

Azmal, A.K.M., Marumoto, T., Shindo, H., Nishiyama, M., 1996. Mineralization and microbial biomass formation in upland soil amended with some tropical plant residues at different temperatures. Soil Sci. Plant Nutr. 42, 463–473.

Bååth, E., Wallander, H., 2003. Soil and rhizosphere microorganisms have the same Q$_{10}$ for respiration in a model system. Global Change Biol. 9, 1788–1791.

Baggs, E.M., 2008. A review of stable isotope techniques for N$_2$O source partitioning in soils: recent progress, remaining challenges and future considerations. Rapid Commun. Mass Spectr. 22, 1664–1672.

Bailey, V.L., Smith, J.L., Bolton, H., 2002. Fungal-to-bacterial ratios in soils investigated for enhanced C sequestration. Soil Biol. Biochem. 34, 997–1007.

Bardgett, R.D., Griffiths, B.S., 1997. Ecology and biology of soil protozoa, nematodes, and microarthropods. In: van Elsas, J.D., Trevors, J.T., Wellington, E.M.H. (Eds.), Modern Soil Microbiology. Marcel Dekker, New York, pp. 129–163.

Bardgett, R.D., Hobbs, P.J., Frostegård, Å., 1996. Changes in fungal:bacterial biomass ratios following reductions in the intensity of management on an upland grassland. Biol. Fertil. Soils 22, 261–264.

Bateki, C.A., Dickhoefer, U., 2019. Predicting dry matter intake using conceptual models for cattle kept under tropical and subtropical conditions. J. Anim. Sci. 97, 3727–3740.

Bationo, A., Lompo, F., Koala, S., 1998. Research on nutrient flows and balances in West Africa: state-of-the-art. Agric. Ecosyst. Environ. 71, 19–36.

Batjes, N.H., 1996. Total carbon and nitrogen in the soils of the world. Eur. J. Soil Sci. 47, 151–163.

Ben-Noah, I., Friedman, S.P., 2018. Review and evaluation of root respiration and of natural and agricultural processes of soil aeration. Vadose Zone J. 17, 170119.

Beer, C., Reichstein, M., Tomelleri, E., Ciais, P., Jung, M., Carvalhais, N., et al., 2010. Terrestrial gross carbon dioxide uptake: Global distribution and covariation with climate. Science 329, 834–838.

Berg, B., 2000. Litter decomposition and organic matter turnover in northern forest soils. Forest Ecol. Managem. 133, 13–22.

Bergman, I., Klarqvist, M., Nilsson, M., 2000. Seasonal variation in rates of methane production from peat of various botanical origins: effects of temperature and substrate quality. FEMS Microbiol. Ecol. 33, 181–189.

Beukes, I., Rose, L.J., van Coller, G.J., Viljoen, A., 2018. Disease development and mycotoxin production by the *Fusarium graminearum* species complex associated with South African maize and wheat. Eur. J. Plant Path. 150, 893–910.

Bhanja, S.N., Wang, J., 2020. Estimating influences of environmental drivers on soil heterotrophic respiration in the Athabasca River Basin, Canada. Environ. Poll. 257, 113630.

Boeser, U., Hovenjürgen, M., Huber, K., Breves, G., Pfeffer, E., 2002. Pathway of regulative excretion of excessive Ca or P by newborn Saanen goat kids. Proc. Soc. Nutr. Phys. 11, 53.

Bol, R., Bolger, T., Cully, R., Little, D., 2003. Recalcitrant soil organic materials mineralize more efficiently at higher temperatures. J. Plant Nutr. Soil Sci. 166, 300–307.

Bourgault, M., Brand, J., Tausz, M., Fitzgerald, G.J., 2016. Yield, growth and grain nitrogen response to elevated CO_2 of five field pea (*Pisum sativum* L.) cultivars in a low rainfall environment. Field Crops Res. 196, 1–9.

Breves, G., Abel, H.-J., Seip, K., Isselstein, J., 2009. Rumen microbial protein synthesis in response to species rich forages in organic farming. Organic eprints; available at: http://orgprints.org/16542/1/16542-06OE139-tiho_hannover-breves-2009-proteinsynthese.pdf.

Brewer, P.E., Calderon, F., Vigil, M., von Fischer, J.C., 2018. Impacts of moisture, soil respiration, and agricultural practices on methanogenesis in upland soils as measured with stable isotope pool dilution. Soil Biol. Biochem. 127, 239–251.

Bristow, A.W., Whitehead, D.C., Cockburn, J.E., 1992. Nitrogenous constituents in the urine of cattle, sheep and goats. J. Sci. Food Agric. 59, 387–394.

Brouwer, J., Powell, J.M., 1998. Increasing nutrient use efficiency in West-African agriculture: the impact of micro-topography on nutrient leaching from cattle and sheep manure. Agric. Ecosyst. Environ. 71, 229–239.

Brugger, D., Windisch, W.M., 2017. Strategies and challenges to increase the precision in feeding zinc to monogastric livestock. Anim. Nutr. 3, 103–108.

Buerkert, A., Schlecht, E., 2019. Rural-urban transformation: a key challenge of the 21st century. Nutr. Cyc. Agroecosyst. 115, 137–142.

Budiman, Steiner, C., Topp, C.F.E., Buerkert, A., 2020. Soil-climate contribution to DNDC model uncertainty in simulating biomass accumulation under urban vegetable production on a Petroplinthic Cambisol in Tamale, Ghana. J. Plant Nutr. Soil Sci. 183, 306–315.

Buerkert, A., Hiernaux, P., 1998. Nutrients in the West African Sudano-Sahelian zone: losses, transfers and role of external inputs. J. Plant Nutr. Soil Sci. 161, 365–383.

Buerkert, A., Mahler, F., Marschner, H., 1996. Soil productivity management and plant growth in the Sahel: Potential of an aerial monitoring technique. Plant Soil 180, 29–38.

Buerkert, A., Haake, C., Ruckwied, M., Marschner, H., 1998. Phosphorus application affects the nutritional quality of millet grain in the Sahel. Field Crops Res. 57, 223–235.

Buerkert, A., Bationo, A., Dossa, K., 2000. Mechanisms of residue mulch-induced cereal growth increases in West Africa. Soil Sci. Soc. Am. J. 64, 346–358.

Buerkert, A., Nagieb, M., Siebert, S., Khan, I., Al-Maskri, A., 2005. Nutrient cycling and field-based partial nutrient balances in two mountain oases of Oman. Field Crops Res. 94, 149–164.

Burns, R.G., DeForest, J.L., Marxsen, J., Sinsabaugh, R.L., Stromberger, M.E., Wallenstein, M.D., et al., 2013. Soil enzymes in a changing environment: current knowledge and future directions. Soil Biol. Biochem. 58, 216–234.

Cambardella, C.A., Richard, T.L., Russell, A., 2003. Compost mineralization in soil as a function of composting process conditions. Eur. J. Soil Biol. 39, 117–127.

Campbell, C.A., Selles, F., Zentner, R.P., McConkey, B.G., 1993. Nitrogen management for zero-till spring wheat: disposition in plant and utilization efficiency. Commun. Soil Sci. Plant Anal. 24, 2223–2239.

Cech, P.G., Kuster, T., Edwards, P.J., Venterink, H.O., 2008. Effects of herbivory, fire and N_2-fixation on nutrient limitation in a humid African savanna. Ecosystems 11, 991–1004.

Cech, P.G., Venterink, H.O., Edwards, P.J., 2010. N and P cycling in Tanzanian humid savanna: influence of herbivores, fire, and N_2-fixation. Ecosystems 13, 1079–1096.

Chaban, B., Ng, S.Y.M., Jarrell, K.F., 2006. Archaeal habitats - from the extreme to the ordinary. Can. J. Microbiol. 52, 73–116.

Chander, K., Dyckmans, J., Joergensen, R.G., Meyer, B., Raubuch, M., 2001. Different sources of heavy metals and their long-term effects on soil microbial properties. Biol. Fertil. Soils 34, 241–247.

Cheik, S., Bottinelli, N., Soudan, B., Harit, A., Chaudhary, E., Raman Sukumar, R., et al., 2019. Effects of termite foraging activity on topsoil physical properties and water infiltration in Vertisol. Appl. Soil Ecol. 133, 132–137.

Chin, K.-J., Lukow, T., Conrad, R., 1999. Effect of temperature on structure and function of the methanogenic archaeal community in an anoxic rice field soil. Appl. Environ. Microbiol. 65, 2341–2349.

Chun, J.A., Wang, Q., Timlin, D., Fleisher, D., Reddy, V.R., 2011. Effect of elevated carbon dioxide and water stress on gas exchange and water use efficiency in corn. Agric. Forest Meteorol. 151, 378–384.

CISRO, 2007. Nutrient Requirements of Domesticated Ruminants. CSIRO Publishing, Commonwealth Scientific and Industrial Research Organization, Melbourne, p. 296.

Clay, D.E., Clapp, C.E., Dowdy, R.H., Molina, J.A.E., 1993. Mineralization of nitrogen in fertilizer-acidified lime-amended soils. Biol. Fertil. Soils 15, 249–252.

Cleveland, C.C., Reed, S.C., Townsend, A.R., 2006. Nutrient regulation of organic matter decomposition in a tropical rain forest. Ecology 87, 492–503.

Cleveland, C.C., Townsend, A.R., Schmidt, S.K., 2002. Phosphorus limitation of microbial processes in moist tropical forests: evidence from short-term laboratory incubations and field studies. Ecosystems 5, 680–691.

Cofie, O.O., Drechsel, P., 2007. Water for food in the cities: The growing paradigm of irrigated (peri)-urban agriculture and its struggle in sub-Saharan Africa. Afr. Water J. 1, 23–32.

Coleman, D.C., 2008. From peds to paradoxes: linkages between soil biota and their influences on ecological processes. Soil Biol. Biochem. 40, 271–289.

Crenshaw, C.L., Lauber, C., Sinsabaugh, R.L., Stavely, L.K., 2008. Fungal control of nitrous oxide production in semiarid grassland. Biogeochemistry 87, 17–27.

Curtin, D., Campbell, C.A., Jalil, A., 1998. Effects of acidity on mineralization: pH-dependence of organic matter mineralization in weakly acidic soils. Soil Biol. Biochem. 30, 57–64.

Dao, J., Stenchly, K., Traore, O., Amoah, P., Buerkert, A., 2018. Effects of water quality and post-harvest handling on microbiological contamination of lettuce at urban and peri-urban locations of Ouagadougou, Burkina Faso. Foods 7, 206.

Davidson, E.A., Janssens, I.A., 2006. Temperature sensitivity of soil carbon decomposition and feedbacks to climate change. Nature 440, 165–173.

De Graaff, M.A., Six, J., Jastrow, J.D., Schadt, C.W., Wullschleger, S.D., 2013. Variation in root architecture among switchgrass cultivars impacts root decomposition rates. Soil Biol. Biochem. 58, 198–206.

De Jager, A., Nandwa, S.M., Okoth, P.F., 1998. Monitoring nutrient flows and economic performance in African farming systems (NUTMON). I. Concepts and methodologies. Agric. Ecosyst. Environ. 71, 37–48.

De Neve, S., Pannier, J., Hofman, G., 1996. Temperature effects on C- and N-mineralization from vegetable crop residues. Plant Soil 181, 25–30.

Derrien, D., Amelung, W., 2011. Computing the mean residence time of soil carbon fractions using stable isotopes: impacts of the model framework. Eur. J. Soil Sci. 62, 237–252.

Deurer, M., von der Heide, C., Böttcher, J., Duijnisveld, W.H.M., Weymann, D., Well, R., 2008. The dynamics of N_2O near the groundwater table and the transfer of N_2O into the unsaturated zone: a case study from a sandy aquifer in Germany. Catena 72, 362–373.

De Wit, J., van Keulen, H., van der Meer, H.G., Nell, A.J., 1997. Animal manure: asset or liability? World Animal Review 88, 1997/1 [available at. Available from: http://www.fao.org/docrep/w5256t/W5256t05.html.

Dijkstra, J., France, J., 1995. Modelling and methodology in animal science. In: Danfær, A., Lescoat, P. (Eds.), Proceedings of the 4th International Workshop on Modelling Nutrient Utilization in Farm Animals. National Institute of Animal Science, Foulum, Denmark, pp. 9–18.

Di Lonardo, D.P., Manrubia, M., De Boer, W., Zweers, H., Veen, G.F., van der Wal, A., 2018. Relationship between home-field advantage of litter decomposition and priming of soil organic matter. Soil Biol. Biochem. 126, 49–56.

Dilly, O., Munch, J.C., 2004. Litter decomposition and microbial characteristics in agricultural soils in northern, central and southern Germany. Soil Sci. Plant Nutr. 50, 843–853.

Diogo, R.V.A., Buerkert, A., Schlecht, E., 2010a. Horizontal nutrient fluxes and food safety in urban and peri-urban vegetable and millet cultivation of Niamey. Niger. Nutr. Cyc. Agroecosyst. 87, 81–102.

Diogo, R.V.A., Buerkert, A., Schlecht, E., 2010b. Resource use efficiency in urban and peri-urban livestock enterprises of Niamey. Niger. Animal 4, 1725–1738.

Drechsel, P., Keraita, B., 2014. Irrigated urban vegetable production in Ghana: characteristics, benefits and risk mitigation, 2nd ed. International Water Management Institute (IWMI), Colombo, Sri Lanka. Available from: https://doi.org/10.5337/2014.219.

Dubey, S.K., 2005. Microbial ecology of methane emission in rice agroecosystem: a review. Appl. Ecol. Environ. Res. 3, 1–27.

Durand, J.-L., Delusca, K., Boote, K., Lizaso, J., Manderscheid, R., Weigel, H.J., Ruane, A.C., et al., 2018. How accurately do maize crop models simulate the interactions of atmospheric CO_2 concentration levels with limited water supply on water use and yield? Eur. J. Agron. 100, 67–75.

Eastwood, D.C., Floudas, D., Binder, M., Majcherczyk, A., Schneider, P., Aerts, A., et al., 2011. The plant cell wall–decomposing machinery underlies the functional diversity of forest fungi. Science 333, 762–765.

Eshenaur, W., 1984. Understanding agricultural waste recycling. Technical paper. Volunteers in Technical Assistance Inc., Arlington, VA, USA, p. 21.

Esse, P.C., Buerkert, A., Hiernaux, P., Assa, A., 2001. Decomposition of and nutrient release from ruminant manure on acid sandy soils in the Sahelian zone of Niger, West Africa. Agric. Ecosyst. Environ. 83, 55–63.

Fang, C., Smith, P., Moncrieff, J.B., Smith, J.U., 2005. Similar response of labile and resistant soil organic matter pools to changes in temperature. Nature 433, 57–59.

Fang, C., Smith, P., Smith, J.U., 2006. Is resistant soil organic matter more sensitive to temperature than the labile organic matter? Biogeoscience 3, 65–68.

FAO, 2006. Livestock's long shadow - environmental issues and options. Food and Agriculture Organization of the United Nations, Rome, Italy, p. 390.

Farquharson, R., Baldock, J., 2008. Concepts in modelling N_2O emissions from land use. Plant Soil 309, 147–167.

Faust, S., Heinze, S., Ngosong, C., Sradnick, A., Oltmanns, M., Raupp, J., et al., 2017. Effect of biodynamic soil amendments on microbial communities in comparison with inorganic fertilization. Appl. Soil Ecol. 114, 82–89.

Faust, S., Koch, H.-J., Dyckmans, J., Joergensen, R.G., 2019a. Response of maize leaf decomposition in litterbags and soil bags to different tillage intensities in a long-term field trial. Appl. Soil Ecol. 141, 38–44.

Faust, S., Koch, H.-J., Joergensen, R.G., 2019b. Respiration response to different tillage intensities in transplanted soil columns. Geoderma 352, 289–297.

Field, C.B., Behrenfeld, M.J., Randerson, J.T., Falkowski, P., 1998. Primary production of the biosphere: integrating terrestrial and oceanic components. Science 281, 237–240.

Field, C.B., Lobel, D.B., Peters, H.A., Chiariello, N.R., 2007. Feedbacks of terrestrial ecosystems to climate change. Ann. Rev. Environ. Resour. 32, 1–29.

Fierer, N., 2017. Embracing the unknown: disentangling the complexities of the soil microbiome. Nat. Rev. Microbiol. 15, 579–590.

Fierer, N., Craine, J.M., McLauchlan, K., Schimel, J.P., 2005. Litter quality and the temperature sensitivity of decomposition. Ecology 86, 320–326.

Flessa, H., Beese, F., 1995. Effects of sugar beet residues on soil redox potential and nitrous oxide emission. Soil Sci. Soc. Am. J. 59, 1044–1051.

Flessa, H., Potthoff, M., Loftfield, N., 2002. Greenhouse estimates of CO_2 and N_2O emissions following surface application of grass mulch: importance of indigenous microflora of mulch. Soil Biol. Biochem. 34, 875–879.

Formowitz, B., Schulz, M.-C., Buerkert, A., Joergensen, R.G., 2007. Reaction of microorganisms to rewetting in dry continuous cereal and legume rotation soils of semi-arid Sub-Saharan Africa. Soil Biol. Biochem. 39, 1512–1517.

France, J., Thornley, J.H.M., 1984. Mathematical Models in Agriculture. Butterworths, London.

Franke-Whittle, I.H., Klammer, S.H., Insam, H., 2005. Design and application of an oligonucleotide microarray for the investigation of compost microbial communities. J. Microbiol. Meth. 62, 37–56.

Frey, S.D., Elliott, E.T., Paustian, K., 1999. Bacterial and fungal abundance and biomass in conventional and no-tillage agroecosystems along two climatic gradients. Soil Biol. Biochem. 31, 573–585.

Frostegård, Å., Tunlid, A., Bååth, E., 1996. Changes in microbial community structure during long-term incubation in two soils experimentally contaminated with metals. Soil Biol. Biochem. 28, 55–63.

Fust, P., Schlecht, E., 2018. Integrating spatio-temporal variation in resource availability and herbivore movements into rangeland management: RaMDry - an agent-based model on livestock feeding ecology in a dynamic, heterogeneous, semi-arid environment. Ecol. Model. 369, 13–41.

Ganskopp, D., 2001. Manipulating cattle distribution with salt and water in large arid-land pastures: a GPS/GIS assessment. Appl. Anim. Behav. Sci. 73, 251–262.

Gattinger, A., Ruser, R., Schloter, M., Munch, J.C., 2002. Microbial community structure varies in different soil zones of a potato field. J. Plant Nutr. Soil Sci. 165, 421–428.

Gibson, R.S., 2012. A historical review of progress in the assessment of dietary zinc intake as an indicator of population zinc status. Adv Nutr. 3, 772–782.

Gilhespy, S.L., Anthony, S., Cardenas, L., Chadwick, D., del Prado, A., Li, C., et al., 2014. First 20 years of DNDC (DeNitrification DeComposition): model evolution. Ecol. Model. 292, 51–62.

Gorham, E., 1991. Northern peatlands: Role in the carbon cycle and probable responses to climatic warming. Ecol. Appl. 1, 182–195.

Graefe, S., Buerkert, A., Schlecht, E., 2019. Trends and gaps in scholarly literature on urban and peri-urban agriculture. Nutr. Cycl. Agroecosyst. 115, 143–158.

Gregorich, E.G., Rochette, P., Hopkins, D.W., McKim, U.F., St-Georges, P., 2006. Tillage-induced environmental conditions in soil and substrate limitation determine biogenic gas production. Soil Biol. Biochem. 38, 2614–2628.

Gu, L., Post, W.M., King, A.W., 2004. Fast labile carbon turnover obscures sensitivity of heterotrophic respiration from soil to temperature: a model analysis. Glob. Biogeochem. Cycle 18, 1022–1032.

Guggenberger, G., Frey, S.D., Six, J., Paustian, K., Elliott, E.T., 1999. Bacterial and fungal cell wall residues in conventional and no-tillage agroecosystems. Soil Sci. Soc. Am. J. 63, 1188–1198.

Haas, G., Deittert, C., Köpke, U., 2007. Farm gate nutrient balances of organic dairy farms at different intensity levels in Germany. Renew. Agric. Food Syst. 22, 223–232.

Hadorn, R., 1994. Einfluss unterschiedlicher Nahrungsfaserträger (Soja und Hirseschalen) im Vergleich zu Weizenquellstärke auf die Nährstoff- und Energieverwertung von wachsenden Schweinen und Broilern. Diss. ETH Nr. 10946. Eidgenössisch Technische Hochschule Zurich, Switzerland, p. 184.

Hättenschwiler, S., Vitousek, P.M., 2000. The role of polyphenols in terrestrial ecosystem nutrient cycling. Tree 15, 238–243.

Hao, X., Li, P., Han, X., Norton, R.M., Lam, S.K., Zong, Y., et al., 2016. Effects of free-air CO_2 enrichment (FACE) on N, P and K uptake of soybean in northern China. Agric. Forest Meteorol. 218/219, 261–266.

Hartman, W.H., Richardson, C.J., 2013. Differential nutrient limitation of soil microbial biomass and metabolic quotients (qCO$_2$): Is there a biological stoichiometry of soil microbes? PLOS ONE 8, e57127.

Hartmann, M., Frey, B., Mayer, J., Mäder, P., Widmer, F., 2015. Distinct soil microbial diversity under long-term organic and conventional farming. ISME J. 9, 1177–1194.

Hedlund, K., Griffiths, B., Christensen, S., Scheu, S., Setälä, H., Tscharntke, T., et al., 2004. Trophic interactions in changing landscapes: responses of soil food webs. Basic Appl. Ecol. 5, 495–503.

Heier, T., Jain, S.K., Kogel, K.-H., Pons-Kühnemann, J., 2005. Influence of N-fertilization and fungicide strategies on *Fusarium* head blight severity and mycotoxin content in winter wheat. J. Phytopath. 153, 551–557.

Heinze, S., Raupp, J., Joergensen, R.G., 2010. Effects of fertilizer and spatial heterogeneity in soil pH on microbial biomass indices in a long-term field trial of organic agriculture. Plant Soil 328, 203–215.

Henriksen, T.M., Breland, T.A., 1999a. Nitrogen availability effects on carbon mineralization, fungal and bacterial growth, and enzyme activities during decomposition of wheat straw in soil. Soil Biol. Biochem. 31, 1121–1134.

Henriksen, T.M., Breland, T.A., 1999b. Decomposition of crop residues in the field: evaluation of a simulation model developed from microcosm studies. Soil Biol. Biochem. 31, 1423–1434.

Herrero, M., Havlík, P., Valin, H., Notenbaert, A., Rufino, M.C., Thornton, P.K., et al., 2013. Biomass use, production, feed efficiencies, and greenhouse gas emissions from global livestock systems. Proc. Natl. Acad. Sci. USA 110, 20888–20893.

Hiernaux, P., Fernández-Rivera, S., Schlecht, E., Turner, M.D., Williams, T.O., 1997. Livestock-mediated nutrient transfers in Sahelian agroecosystems. In: Renard, G., Neef, A., Becker, K., von Oppen, M. (Eds.), Soil Fertility Management in West African Land Use Systems. Proceedings of a Regional Workshop, University of Hohenheim, ICRISAT, INRAN, Niamey, Niger, 4-8 March 1997. Margraf Verlag, Weikersheim, Germany, pp. 339–347.

Hobbie, S.E., Schimel, J.P., Trumbore, S.E., 2000. Controls over carbon storage and turnover in high-latitude soils. Glob. Change Biol. 6, 196–210.

Hoffmann, C., Funk, R., Wieland, R., Li, Y., Sommer, M., 2008. Effects of grazing and topography on dust flux and deposition in the Xilingele grassland, Inner Mongolia. J. Arid Environ. 72, 792–807.

Hoffmann, B., Müller, T., Joergensen, R.G., 2010. Carbon dioxide production and oxygen consumption during the early decomposition of different litter types over a range of temperatures in soil-inoculated quartz sand. J. Plant Nutr. Soil Sci. 173, 217–223.

Högberg, M.N., Högberg, P., Myrold, D.D., 2007. Is microbial community composition in boreal forest soils determined by pH, C-to-N ratio, the trees, or all three? Oecologia 150, 590–601.

Holland, E.A., Coleman, D.C., 1987. Litter placement effects on microbial and organic matter dynamics in an agroecosystem. Ecology 68, 425–433.

Holland, J.E., Bennett, A.E., Newton, A.C., White, P.J., McKenzie, B.M., George, T.S., et al., 2018. Liming impacts on soils, crops and biodiversity in the UK: A review. Sci. Total Environ. 610/611, 316–332.

Holtkamp, R., Kardol, P., van der Wal, A., Dekker, S.C., van der Putten, W.H., de Ruiter, P.C., 2008. Soil food web structure during ecosystem development after land abandonment. Appl. Soil Ecol. 39, 23–34.

Hupe, A., Schulz, H., Bruns, C., Haase, T., Heß, J., Joergensen, R.G., et al., 2019. Get on your boots: Estimating root biomass and rhizodeposition of peas under field conditions reveals the necessity of field experiments. Plant Soil 443, 449–462.

Ingold, M., Schmidt, S., Dietz, H., Joergensen, R., Schlecht, E., Buerkert, A., 2018. Tannins in goat diets modify manure turnover in a subtropical soil. Exp. Agric. 54, 655–669.

IPCC, 2007. Summary for policymakers. In: Solomon, S., Qin, D., Manning, M., Chen, Z., Marquis, M., Averyt, K.B., et al., (Eds.), Climate Change 2007: The Physical Science Basis. Contribution of Working Group I to the Fourth Assessment Report of the Intergovernmental Panel on Climate Change. Cambridge University Press, Cambridge, UK and New York, NY.

IPCC, 2013. Summary for policymakers. In: Stocker, T.F., Qin, D., Plattner, G.-K., Tignor, M., Allen, S.K., Boschung, J., et al., (Eds.), Climate Change 2013: The Physical Science Basis. Contribution of Working Group I to the Fifth Assessment Report of the Intergovernmental Panel on Climate Change. Cambridge University Press, UK and New York, NY.

IPCC, 2019. Summary for Policymakers. In: Shukla, P.R., Skea, J., Calvo Buendia, E., Masson-Delmotte, V., Pörtner, H.-O., Roberts, D.C., et al., (Eds.), Climate Change and Land: an IPCC special report on climate change, desertification, land degradation, sustainable land management, food security, and greenhouse gas fluxes in terrestrial ecosystems. Intergovernmental Panel on Climate Change IPCC, WMO/UNEP, Geneva, Switzerland. Available from: https://www.ipcc.ch/srccl.

Jacobs, A., Kaiser, K., Ludwig, B., Rauber, R., Joergensen, R.G., 2011. Application of biochemical degradation indices to the microbial decomposition of maize leaves and wheat straw in soils under different tillage systems. Geoderma 162, 207–214.

Jahn, M., Sachs, T., Mansfeldt, T., Overesch, M.J., 2010. Global climate change and its impacts on the terrestrial Arctic carbon cycle with special regards to ecosystem components and the greenhouse-gas balance. J. Plant Nutr. Soil Sci. 173, 627–643.

Jannoura, R., Joergensen, R.G., Bruns, C., 2014. Organic fertilizer effects on growth, crop yield, and soil microbial biomass indices in sole and intercropped peas and oats under organic farming conditions. Eur. J. Agron. 52, 259–270.

Jastrow, J.D., Amonette, J.E., Bailey, V.L., 2007. Mechanisms controlling soil carbon turnover and their potential application for enhancing carbon sequestration. Climatic Change 80, 5–23.

Jenkinson, D.S., Ladd, J.N., 1981. Microbial biomass in soil: measurement and turnover. In: Paul, E.A., Ladd, J.N. (Eds.), Soil Biochemistry, Vol. 5. Marcel Dekker, New York, pp. 415–471.

Jörgensen, R.G., 1995. Die quantitative Bestimmung der mikrobiellen Biomasse in Böden mit der Chloroform-Fumigations-Extraktions-Methode. Institut für Bodenwissenschaft, University of Göttingen.

Joergensen, R.G., Emmerling, C., 2006. Methods for evaluating human impact on soil microorganisms based on their activity, biomass, and diversity in agricultural soils. J. Plant Nutr. Soil Sci. 169, 295–309.

Joergensen, R.G., Wichern, F., 2008. Quantitative assessment of the fungal contribution to microbial tissue in soil. Soil Biol. Biochem. 40, 2977–2991.

Joergensen, R.G., Scholle, G.A., Wolters, V., 2009. Dynamics of mineral components in the forest floor of an acidic beech (*Fagus sylvatica* L.) forest. Eur. J. Soil Biol. 45, 285–289.

Jones, D.L., Shannon, D., Murphy, D.V., Farrar, J., 2004. Role of dissolved organic nitrogen (DON) in soil N cycling in grassland soils. Soil Biol. Biochem. 36, 749–756.

Jost, D.I., Joergensen, R.G., Sundrum, A., 2013. Effect of cattle faeces with different microbial biomass content on soil properties, gaseous emissions and plant growth. Biol. Fertil. Soils 49, 61–70.

Ju, X., Zhang, F., Bao, X., Römheld, V., Roelcke, M., 2005. Utilization and management of organic wastes in Chinese agriculture: past, present and perspectives. Sci. China C Life Sci. 48, 965–979.

Jung, M.-Y., Well, R., Min, D., Giesemann, A., Park, S.-J., Kim, J.-G., et al., 2014. Isotopic signatures of N_2O produced by ammonia-oxidizing archaea from soils. ISME J. 8, 1115–1125.

Kaiser, M., Ellerbrock, R.H., Wulf, M., Dultz, S., Hierath, C., Sommer, M., 2012. The influence of mineral characteristics on organic matter content, composition, and stability of topsoils under long-term arable and forest land use. J. Geophys. Res. 117, G02018.

Kammann, C., Hepp, S., Lenhart, K., Müller, C., 2009. Stimulation of methane consumption by endogenous CH_4 production in aerobic grassland soil. Soil Biol. Biochem. 41, 622–629.

Karg, H., Drechsel, P., Akoto-Danso, E.K., Glaser, R., Nyarko, G., Buerkert, A., 2016. Foodsheds and city region food systems in two West-African cities. Sustainability 8, 1175.

Kätterer, T., Reichstein, M., Andren, O., Lomander, A., 1998. Temperature dependence of organic matter decomposition: A critical review using literature data analysed with different models. Biol. Fertil. Soils 27, 258–262.

Kemmitt, S.J., Lanyon, C.V., Waite, I.S., Wen, Q., Addiscott, T.M., Bird, N.R.A., et al., 2008. Mineralization of native soil organic matter is not regulated by the size, activity or composition of the soil microbial biomass-a new perspective. Soil Biol. Biochem. 40, 61–73.

Kernan, J.A., Coxworth, E.C., Crowle, W.L., Spurr, D.T., 1984. The nutritional value of crop residue components from several wheat cultivars grown at different fertilizer levels. Anim Feed Sci. Technol. 11, 301–311.

Khan, S., Joergensen, R.G., 2006. Decomposition of heavy metal-contaminated nettles (*Urtica dioica* L.) in soils differently subjected to heavy metal pollution by river sediments. Chemosphere 65, 981–987.

Khan, K.S., Mack, R., Castillo, X., Kaiser, M., Joergensen, R.G., 2016. Microbial biomass, fungal and bacterial residues, and their relationships to the soil organic matter C/N/P/S ratios. Geoderma 271, 115–123.

Keiser, A.D., Strickland, M.S., Fierer, N., Bradford, M.A., 2011. The effect of resource history on the functioning of soil microbial communities is maintained across time. Biogeoscience 8, 1477–1486.

Kichey, T., Heumez, E., Pocholle, D., Pageau, K., Vanacker, H., Dubois, F., et al., 2006. Combined agronomic and physiological aspects of nitrogen management in wheat highlight a central role for glutamine synthetase. New Phytol. 169, 265–278.

Kilian, A., Gutser, R., Claasen, N., 1998. N_2O emissions following long-term organic fertilization at different levels. Agribiol. Res. 51, 27–36.

Kirchgeßner, M., Roth, F.X., Schwarz, F.J., Stangl, G.I., 2008. Tierernährung, 12th edition, DLG Verlag, Frankfurt am Main, Germany.

Kirschbaum, M.U.F., 2006. The temperature dependence of organic matter decomposition – still a topic of debate. Soil Biol. Biochem. 38, 2510–2518.

Kögel-Knabner, I., Amelung, W., Cao, Z., Fiedler, S., Frenzel, P., Jahn, R., et al., 2010. Biogeochemistry of paddy soils. Geoderma 157, 1–14.

Kool, D.M., Dolfing, J., Wrage, N., Van Groenigen, J.W., 2011. Nitrifier denitrification as a distinct and significant source of nitrous oxide from soil. Soil Biol. Biochem. 43, 174–178.

Koranda, M., Kaiser, C., Fuchslueger, L., Kitzler, B., Sessitsch, A., Zechmeister-Boltenstern, S., et al., 2014. Fungal and bacterial utilization of organic substrates depends on substrate complexity and N availability. FEMS Microbiol. Ecol. 87, 142–152.

Krause, D.O., Nagaraja, T.G., Wright, A.D.G., Callaway, T.R., 2013. Board-invited review: Rumen microbiology: Leading the way in microbial ecology. J. Anim. Sci. 91, 331–341.

Kump, L.R., 2002. Reducing uncertainty about carbon dioxide as a climate driver. Nature 419, 188–190.

Laughlin, R.J., Stevens, R.J., 2002. Evidence for fungal dominance of denitrification and codenitrification in a grassland soil. Soil Sci. Soc. Am. J. 66, 1540–1548.

Lemmens, M., Haim, K., Lew, H., Rückenbauer, P., 2004. The effect of nitrogen fertilization on *Fusarium* head blight development and deoxynivalenol contamination in wheat. J. Phytopath. 152, 1–8.

Leu, S.-Y., Libra, J.A., Stenström, M.K., 2010. Monitoring off-gas O_2/CO_2 to predict nitrification performance in activated sludge processes. Water Res. 44, 3434–3444.

Lhoste, P., Dollé, V., Rousseau, J., Soltner, D., 1993. Manuel de zootechnie des régions chaudes. Les systèmes d'élevage. Ministère de la coopération, Paris, France.

Li, C.S., 2009. User's Guide for the DNDC Model (Version 9.3). Report of the Institute for the Study of Earth, Oceans and Space, Durham, NH, USA.

Li, C., Frolking, S., Frolking, T.A., 1992a. A model of nitrous oxide evolution from soil driven by rainfall events: 1. Model structure and sensitivity. J. Geophys. Res. 97, 9759–9776.

Li, C., Frolking, S., Frolking, T.A., 1992b. A model of nitrous oxide evolution from soil driven by rainfall events: 2. Model applications. J. Geophys. Res. 97, 9777–9783.

Li, C., Frolking, S., Butterbach-Bahl, K., 2005. Carbon sequestration in arable soils is likely to increase nitrous oxide emissions, offsetting reductions in climate radiative forcing. Climatic Change 72, 321–338.

Liu, D., Ding, W., Jia, Z., Cai, Z., 2010. Influence of niche differentiation on the abundance of methanogenic archaea and methane production potential in natural wetland ecosystems across China. Biogeosci. Disc. 7, 7629–7655.

Lompo, D.J.-P., Sangaré, S.A.K., Compaoré, E., Sedogo, M.P., Predotova, M., Schlecht, E., et al., 2012. Gaseous emissions of nitrogen and carbon from urban vegetable gardens of Bobo Dioulasso, Burkina Faso. J. Plant Nutr. Soil Sci. 175, 846–853.

Ludley, K.E., Jickells, S.M., Chamberlain, P.M., Whitaker, J., Robinson, C.H., 2009. Distribution of monoterpenes between organic resources in upper soil horizons under monocultures of *Picea abies*, *Picea sitchensis* and *Pinus sylvestris*. Soil Biol. Biochem. 41, 1050–1059.

Ludwig, B., Schulz, E., Merbach, I., Rethemeyer, J., Flessa, H., 2007. Predictive modelling of the C dynamics for eight variants of the long-term static fertilization experiment in Bad Lauchstädt using the Rothamsted Carbon Model. Eur. J. Soil Sci. 58, 1155–1163.

Ludwig, B., Bergstermann, A., Priesack, E., Flessa, H., 2011. Modelling of crop yields and N_2O emissions from silty arable soils with differing tillage in two long-term experiments. Soil Till. Res. 112, 114–121.

Lukas, S., Potthoff, M., Dyckmans, J., Joergensen, R.G., 2013. Microbial use of ^{15}N-labelled maize residues affected by different winter temperature scenarios. Soil Biol. Biochem. 65, 22–32.

Lukas, S., Abbas, S.J., Kössler, P., Karlovsky, P., Potthoff, M., Joergensen, R.G., 2018. Fungal plant pathogens on inoculated maize residues in a simulated soil warming experiment. Appl. Soil Ecol. 124, 75–82.

Lydecker, M., Drechsel, P., 2010. Urban agriculture and sanitation services in Accra, Ghana: the overlooked contribution. Int. J. Agric. Sustain. 8, 94–103.

Maeda, K., Spor, A., Edel-Hermann, V., Heraud, C., Breuil, M.-C., Bizouard, F., et al., 2015. N_2O production, a widespread trait in fungi. Sci. Rep. 5, 09697.

Makkar, H.P.S., 2003. Effects and fate of tannins in ruminant animals, adaptation to tannins, and strategies to overcome detrimental effects of feeding tannin-rich feeds. Small Ruminant Res. 49, 241–256.

Manderscheid, R., Pacholski, A., Weigel, H.-J., 2010. Effect of free air carbon dioxide enrichment combined with two nitrogen levels on growth, yield and yield quality of sugar beet: evidence for a sink limitation of beet growth under elevated CO_2. Eur. J. Agron. 32, 228–239.

Mando, A., Brussaard, L., 1999. Contribution of termites to the breakdown of straw under Sahelian conditions. Biol. Fertil. Soils 29, 332–334.

Martius, C., Fearnside, P.M., Bandeira, A.G., Wassmann, R., 1996. Deforestation and methane release from termites in Amazonia. Chemosphere 33, 517–536.

McGuire, A.D., Chapin III, F.S., Walsh, J.E., Wirth, C., 2006. Integrated regional changes in Arctic climate feedbacks: Implications for the global climate system. Annu. Rev. Environ. Resour. 31, 61–91.

McLain, J.E.T., Martens, D.A., 2005. Nitrous oxide flux from soil amino acid mineralization. Soil Biol. Biochem. 37, 289–299.

Mead, G.C., 1989. Microbes of the avian cecum: Types present and substrates utilized. J. Exp. Zool. 252, 48–54.

Mendoza Huaitalla, R., Gallmann, E., Zheng, K., Liu, X., Hartung, E., 2010. Pig husbandry and solid manures in a commercial pig farm in Beijing, China. World Acad. Sci. Engin. Techn. 65, 18–27.

Meng, Q., Yang, W., Men, M., Bello, A., Xu, X., Xu, B., et al., 2019. Microbial community succession and response to environmental variables during cow manure and corn straw composting. Front. Microbiol. 10, 529.

Merbach, W.M., Mirus, E., Knof, G., Remus, R., Ruppel, R., Russow, R., et al., 1999. Release of carbon and nitrogen compounds by plant roots and their possible ecological importance. J. Plant Nutr. Soil Sci. 162, 373–383.

Mikan, C.J., Schimel, J.P., Doyle, A.P., 2002. Temperature controls of microbial respiration in Arctic tundra soils above and below freezing. Soil Biol. Biochem. 34, 1785–1795.

Mitra, S., Wassmann, R., Vlek, P.L.G., 2005. An appraisal of global wetland area and its organic carbon stock. Curr. Sci. 88, 25–35.

Mondav, R., Woodcroft, B.J., Kim, E.-H., McCalley, C.K., Hodgkins, S.B., Crill, P.M., et al., 2014. Discovery of a novel methanogen prevalent in thawing permafrost. Nature Comm. 5, 3212.

Moorhead, D., Lashermes, G., Recous, S., Bertrand, I., 2014. Interacting microbe and litter quality controls on litter decomposition: a modeling analysis. PLOS ONE 9, e0108769.

Moreau, M., Siebert, S., Buerkert, A., Schlecht, E., 2009. Use of a tri-axial accelerometer for automated recording and classification of goats' grazing behaviour. J. Appl. Anim. Behav. Sci. 119, 158–170.

Müller, T., Höper, H., 2004. Soil organic matter turnover as a function of the soil clay content: consequences for model applications. Soil Biol. Biochem. 36, 877–888.

Müller, E., Rottmann, N., Bergstermann, A., Wildhagen, H., Joergensen, R.G., 2011. Soil CO_2 evolution rates in the field – a comparison of three methods. Arch. Agron. Soil Sci. 57, 597–608.

Murugan, R., Koch, H.-J., Joergensen, R.G., 2014. Long-term influence of different tillage intensities on soil microbial biomass, residues and community structure at different depths. Biol. Fertil. Soils 50, 487–498.

Nauer, P.A., Hutley, L.B., Arndt, S.K., 2018. Termite mounds mitigate half of termite methane emissions. Proc. Natl. Acad. Sci. USA 115, 52.

Nicolardot, B., Fauvet, G., Cheneby, D., 1994. Carbon and nitrogen cycling through soil microbial biomass at various temperatures. Soil Biol. Biochem. 26, 253–261.

Nieder, R., Benbi, D.K., 2008. Carbon and Nitrogen in the Terrestrial Environment. Springer Science + Business Media B.V., Dordrecht, The Netherlands.

Niraula, S., Rahman, S., Chatterjee, A., Cortus, E.L., Mehata, M., Spiehs, M.J., 2019. Beef manure and urea applied to corn show variable effects on nitrous oxide, methane, carbon dioxide, and ammonia. Agron. J. 11, 1448–1467.

Noll, M., Klose, M., Conrad, R., 2010. Effect of temperature change on the composition of the bacterial and archaeal community potentially involved in the turnover of acetate and propionate in methanogenic rice field soil. FEMS Microbiol. Ecol. 73, 215–225.

Passaglia Azevedo, R., Hernán Salcedo, I., Alves Lima, P., da Silva Fraga, V., Quintão Lana, R.M., 2018. Mobility of phosphorus from organic and inorganic source materials in a sandy soil. Int. J. Recycl. Org. Waste Agric. 7, 153–163.

Paul, E.A. (Ed.), 2007. Soil Microbiology Ecology and Biochemistry. 3rd edition Academic Press, San Diego, California.

Paul, E.A., 2016. The nature and dynamics of soil organic matter: plant inputs, microbial transformations, and organic matter stabilization. Soil Biol. Biochem. 98, 109–126.

Poll, C., Brune, T., Begerow, D., Kandeler, E., 2010. Small-scale diversity and succession of fungi in the detritusphere of rye residues. Microb. Ecol. 59, 130–140.

Potthoff, M., Dyckmans, J., Flessa, H., Muhs, A., Beese, F., Joergensen, R.G., 2005. Dynamics of maize (*Zea mays* L.) leaf straw mineralization as affected by the presence of soil and the availability of nitrogen. Soil Biol. Biochem. 37, 1259–1266.

Potthoff, M., Dyckmans, J., Flessa, H., Beese, F., Joergensen, R.G., 2008. Decomposition of maize residues after manipulation of colonization and its contribution to the soil microbial biomass. Biol. Fertil. Soils 44, 891–895.

Pouyat, R.V., Carreiro, M.M., 2003. Controls on mass loss and nitrogen dynamics of oak leaf litter along an urban-rural land-use gradient. Oecology 135, 288–298.

Powell, J.M., Ikpe, F.N., Somda, Z.C., 1999. Crop yield and the fate of nitrogen and phosphorus following application of plant material and feces to soil. Nutr. Cycl. Agroecosyst. 54, 215–226.

Powlson, D.S., Hirsch, P.R., Brookes, P.C., 2001. The role of soil microorganisms in soil organic matter conservation in the tropics. Nutr. Cycl. Agroecosyst. 61, 41–51.

Prade, T., Kätterer, T., Björnsson, L., 2017. Including a one-year grass ley increases soil organic carbon and decreases greenhouse gas emissions from cereal-dominated rotations — A Swedish farm case study. Biosyst. Engin. 164, 200–212.

Predotova, M., Gebauer, J., Schlecht, E., Buerkert, A., 2010a. Gaseous nitrogen and carbon emissions from urban gardens in Niamey, Niger. Field Crops Res. 115, 1–8.

Predotova, M., Schlecht, E., Buerkert, A., 2010b. Nitrogen and carbon losses from dung storage in urban gardens of Niamey, Niger. Nutr. Cycl. Agroecosyst. 87, 103–114.

Predotova, M., Bischoff, W.-A., Buerkert, A., 2011. Mineral nitrogen and phosphorus leaching in vegetable gardens of Niamey, Niger. J. Plant Nutr. Soil Sci. 174, 47–55.

Provenza, F.D., 1995. Postingestive feedback as an elementary determinant of food preference and intake in ruminants. J. Range Manag. 48, 2–17.

Qiao, Y., Zhang, H., Dong, B., Shi, C., Li, Y., Zhai, H., et al., 2010. Effects of elevated CO_2 concentration on growth and water use efficiency of winter wheat under two soil water regimes. Agric. Water Manag. 97, 1742–1748.

Qiu, J., Li, C., Wang, L., Tang, H., Li, H., Van Rast, E., 2009. Modeling impacts of carbon sequestration on net greenhouse gas emissions from agricultural soils in China. Glob. Biogeochem. Cycle 23, GB1007.

Quarmby, C., Allen, S.E., 1989. Organic constituents. In: Allen, S.E. (Ed.), Chemical Analysis of Ecological Materials, 2nd ed. Blackwell Scientific Publisher, London, England, pp. 189–191.

Rasul, G., Khan, A.A., Khan, K.S., Joergensen, R.G., 2009. Immobilization and mineralization of nitrogen in a saline and alkaline soil during microbial use of sugarcane filter cake amended with glucose. Biol. Fertil. Soils 45, 289–296.

Raun, W.R., Johnson, G.V., 1995. Soil-plant buffering of inorganic nitrogen in continuous winter wheat. Agronomy J. 87, 827–834.

Ravishankara, A.R., Daniel, J.S., Portmann, R.W., 2009. Nitrous oxide (N_2O): the dominant ozone-depleting substance emitted in the 21st century. Science 326, 56–57.

Robinson, C.H., 2002. Controls on decomposition and soil nitrogen availability at high latitudes. Plant Soil 242, 65–81.

Robinson, T.P., Thornton, P.K., Franceschini, G., Kruska, R.L., Chiozza, F., Notenbaert, A., et al., 2011. Global Livestock Production Systems. FAO and ILRI, Rome, Italy, p. 152.

Rodionow, A., Flessa, H., Kazansky, O.A., Guggenberger, G., 2006. Organic matter composition and potential trace gas production of permafrost soils in the forest tundra in Northern Siberia. Geoderma 135, 49–62.

Rothrock Jr., M.J., Cook, K.L., Warren, J.G., Eiteman, M.A., Sistani, K., 2010. Microbial mineralization of organic nitrogen forms in poultry litters. J. Environ. Qual. 39, 1848–1857.

Rottmann, N., Siegfried, K., Buerkert, A., Joergensen, R.G., 2011. Litter decomposition in fertilizer treatments of vegetable crops under irrigated subtropical conditions. Biol. Fertil. Soils 47, 71–80.

Rufino, M.C., Tittonell, P., van Wijk, M.T., Castellanos-Navarrete, A., Delve, R.J., de Ridder, N., et al., 2007. Manure as a key resource within smallholder farming systems: analysing farm-scale nutrient cycling efficiencies with the NUANCES framework. Livestock Sci. 112, 273–287.

Ruser, R., Fuß, R., Andres, M., Hegewald, H., Kesenheimer, K., Köbke, S., et al., 2017. Nitrous oxide emissions from winter oilseed rape cultivation. Agric. Ecosyst. Environ. 249, 57–69.

Safi, Z., Buerkert, A., 2011. Heavy metal and microbial loads in sewage irrigated vegetables of Kabul, Afghanistan. J. Agric. Res. Trop. Subtrop. 112, 29–36.

Sakamoto, K., Oba, Y., 1994. Effect of fungal to bacterial biomass ratio on the relationship between CO_2 evolution and total soil microbial biomass. Biol. Fertil. Soils 17, 39–44.

Salamanca, E.F., Raubuch, M., Joergensen, R.G., 2006. Microbial reaction of secondary tropical forest soils to the addition of leaf litter. Appl. Soil Ecol. 31, 53–61.

Sangaré, S.K., Compaore, E., Buerkert, A., Vanclooster, M., Sedogo, M.P., Bielders, C.L., 2012. Field-scale analysis of water and nutrient use efficiency for vegetable production in a West African urban agricultural system. Nutr. Cycl. Agroecosyst. 92, 207–224.

Santonja, M., Foucault, Q., Rancon, A., Gauquelin, T., Fernandez, C., Baldy, V., et al., 2018. Contrasting responses of bacterial and fungal communities to plant litter diversity in a Mediterranean oak forest. Soil Biol. Biochem. 125, 27–36.

Scheller, E., 2001. Amino acids in dew - Origin and seasonal variation. Atmos. Environ. 35, 2179–2192.

Scheller, E., Joergensen, R.G., 2008. Decomposition of wheat straw differing in N content in soils under conventional and organic farming management. J. Plant Nutr. Soil Sci. 171, 886–892.

Schimel, D.S., 1995. Terrestrial ecosystems and the carbon cycle. Glob. Change Biol. 1, 77–91.

Schlecht, E., Hiernaux, P., 2004. Beyond adding up inputs and outputs: process assessment and upscaling in modelling nutrient flows. Nutr. Cycl. Agroecosyst. 70, 303–319.

Schlecht, E., Hiernaux, P., Achard, F., Turner, M.D., 2004. Livestock related nutrient budgets within village territories in western Niger. Nutr. Cyc. Agroecosyst. 68, 199–211.

Schlecht, E., Hiernaux, P., Kadaouré, I., Hülsebusch, C., Mahler, F., 2006. A spatio-temporal analysis of forage availability and grazing and excretion behaviour of herded and free grazing cattle, sheep and goats in Western Niger. Agric. Ecosyst. Environ. 113, 226–242.

Schneider, T., Keiblinger, K.M., Schmid, E., Sterflinger-Gleixner, K., Ellersdorfer, G., Roschitzki, B., et al., 2012. Who is who in litter decomposition? Metaproteomics reveals major microbial players and their biogeochemical functions. ISME J. 6, 1749–1762.

Schulze, E.-D., Wirth, C., Heiman, M., 2000. Managing forests after Kyoto. Science 289, 2058–2059.

Seneviratne, G., 2000. Litter quality and nitrogen release in tropical agriculture: a synthesis. Biol. Fertil. Soils 31, 60–64.

Sey, B.K., Manceur, A.M., Whalen, J.K., Gregorich, E.G., Rochette, P., 2008. Small-scale heterogeneity in carbon dioxide, nitrous oxide and methane production from aggregates of a cultivated sandy-loam soil. Soil Biol. Biochem. 40, 2468–2473.

Shoun, H., Kim, D.-H., Uchiyama, H., Sugiyama, J., 1992. Denitrification by fungi. FEMS Microbiol. Letters 94, 277–281.

Siegfried, K., Dietz, H., Amthauer Gallardo, D., Schlecht, E., Buerkert, A., 2011. Effects of manure with different C/N ratios on yields, yield components and matter balances of organically grown vegetables on a sandy soil in northern Oman. Org. Agric. 3, 9–22.

Šimůnek, J., Suarez, D.L., 1993. Modeling of carbon dioxide transport and production in soil. 1. Model development. Water Resour. Res. 29, 487–497.

Sjögersten, S., Caul, S., Daniell, T.J., Jurd, A.P.S., O'Sullivan, O.S., Stapleton, C.S., et al., 2016. Organic matter chemistry controls greenhouse gas emissions from permafrost peatlands. Soil Biol. Biochem. 98, 42–53.

Smith, S.E., Read, D.J., 2008. Mycorrhizal Symbiosis, 3rd edition Elsevier, Amsterdam, The Netherlands.

Somda, Z.C., Powell, J.M., Férnandez-Rivera, S., Reed, J.D., 1995. Feed factors affecting nutrient excretion by ruminants and the fate of nutrients when applied to soil. In: Powell, J.M., Fernández-Rivera, S., Williams, T.O., Renard, C. (Eds.), Livestock and Sustainable Nutrient Cycling in Mixed Farming Systems of sub-Saharan Africa, Vol. 2. International Livestock Center for Africa, Addis Ababa, Ethiopia, pp. 227–243.

Sommer, S.G., 2001. Effect of composting on nutrient loss and nitrogen availability of cattle deep litter. Eur. J. Agron. 14, 123–133.

Steinfeld, H., Wassenaar, T., Jutzi, S., 2006. Livestock production systems in developing countries: status, drivers, trends. Rev. Scientif. Techn. (Intern. Office Epiz.) 25, 505–516.

Steven, B., Briggs, G., McKay, C.P., Pollard, W.H., Greer, C.W., Whyte, L.G., 2007. Characterization of the microbial diversity in a permafrost sample from the Canadian high Arctic using culture-dependent and culture-independent methods. FEMS Microbiol. Ecol. 59, 513–523.

Steven, B., Léveillé, R., Pollard, W.H., Whyte, L.G., 2006. Microbial ecology and biodiversity in permafrost. Extremophiles 10, 259–267.

Stieglitz, M., McKane, R.B., Klausmeier, C.A., 2006. A simple model for analyzing climatic effects on terrestrial carbon and nitrogen dynamics: An Arctic case study. Glob. Biogeochem. Cycle 20, GB3016.

Stoorvogel, J.J., Smaling, E.M.A., 1994. Assessment of soil nutrient depletion in sub-Saharan Africa: 1983–2000. Vol. 1. Main Report. The Winand Staring Centre, Wageningen.

Sutka, R.L., Adams, G.C., Ostrom, N.E., Ostrom, P.H., 2008. Isotopologue fractionation during N_2O production by fungal denitrification. Rapid Commun. Mass Spectrom. 22, 3989–3996.

Swift, M.J., Heal, O.W., Anderson, J.M., 1979. Decomposition in Terrestrial Ecosystems. Blackwell, Oxford.

Tang, J., Cheng, H., Fang, C., 2017. The temperature sensitivity of soil organic carbon decomposition is not related to labile and recalcitrant carbon. PLOS ONE 12, e0186675.

Tang, J., Zhang, J., Ren, L., Zhou, Y., Gao, J., Luo, L., et al., 2019. Diagnosis of soil contamination using microbiological indices: a review on heavy metal pollution. J. Environ. Manag. 242, 121–130.

Thebo, A.L., Drechsel, P., Lambin, E.F., 2014. Global assessment of urban and peri-urban agriculture: irrigated and rainfed croplands. Environ. Res. Lett. 9, 114002.

Thiet, R.K., Frey, S.D., Six, J., 2006. Do growth yield efficiencies differ between soil microbial communities differing in fungal:bacterial ratios? Reality check and methodological issues. Soil Biol. Biochem. 38, 837–844.

Tian, G., Salako, F., Ishida, F., 2001. Replenishment of C, N, and P in a degraded Alfisol under humid tropical conditions: Effect of fallow species and litter polyphenols. Soil Sci. 166, 614–621.

Titlyanova, A.A., 2007. Nutrient budget in ecosystems. Euras. Soil Sci. 40, 1270–1278.

Tittonell, P., Vanlauwe, B., Leffelaar, P.A., Shepherd, K.D., Giller, K.G., 2005. Exploring diversity of smallholder farms in western Kenya. Agric. Ecosyst. Environ. 110, 166–184.

Todaka, N., Moriya, S., Saita, K., Hondo, T., Kiuchi, I., Takasu, H., et al., 2007. Environmental cDNA analysis of the genes involved in lignocellulose digestion in the symbiotic protist community of *Reticulitermes speratus*. FEMS Microbiol. Ecol. 59, 592–599.

Tolonen, K., Turunen, J., 1996. Accumulation rates of carbon in mires in Finland and implications for climate change. Holocene 6, 171–178.

Torsvik, V., Øvreås, L., 2007. Microbial phylogeny and diversity in soil. In: van Elsas, J.D., Jansson, J.K., Trevors, J.T. (Eds.), Modern Soil Microbiology, 2nd Edition CRC Press, Boca Raton, Florida, pp. 23–54.

Torsvik, V., Goksøyr, J., Daae, F.L., 1990. High diversity in DNA of soil bacteria. Appl. Environ. Microbiol. 56, 782–787.

Turner, M.D., 1998. Long term effects of daily grazing orbits on nutrient availability in Sahelian West Africa: 1. Gradients in the chemical composition of rangeland soils and vegetation. J. Biogeog. 25, 669–682.

Turner, M.D., Hiernaux, P., 2015. The effects of management history and landscape position on inter-field variation in soil fertility and millet yields in southwestern Niger. Agric. Ecosyst. Environ. 211, 73–83.

Turner, M.D., Hiernaux, P., Schlecht, E., 2005. The distribution of grazing pressure in relation to vegetation resources in semi-arid West Africa: The role of herding. Ecosystems 8, 668. 281.

Turunen, J., Tomppo, E., Tolonen, K., 2002. Estimating carbon accumulation rates of undrained mires in Finland – application to boreal and subArctic regions. Holocene 12, 69–80.

Tyler, G., 2005. Changes in the concentrations of major, minor and rare-earth elements during leaf senescence and decomposition in a *Fagus sylvatica* forest. Forest Ecol. Managem. 206, 167–177.

UBA, 2020. Atmosphärische Treibhausgas-Konzentrationen. https://www.umweltbundesamt.de/daten/klima/atmosphaerische-treibhausgas-konzentrationen (31.01.2020).

Uhlířová, E., Šantrůčková, H., Davidov, S.P., 2007. Quality and potential biodegradability of soil organic matter preserved in permafrost of Siberian tussock tundra. Soil Biol. Biochem. 39, 1978–1989.

Underwood, E.J., Suttle, N., 2001. Mineral Nutrition of Livestock, 3rd. edition CABI Publishing, Wallingford, UK.

Van den Bosch, H., De Jager, A., Vlaming, J., 1998. Monitoring nutrient flows and economic performance in African farming systems (NUTMON). II. Tool development. Agric. Ecosyst. Environ. 71, 49–62.

Van Elsas, J.D., Jansson, J.K., Trevors, J.T., 2007. Modern Soil Microbiology, 2nd edition CRC Press, Boca Raton, Florida.

Van Soest, P.J., 1994. Nutritional Ecology of the Ruminant, 2nd edition Cornell University Press, Ithaca, New York.

Van Wijk, M.T., Rufino, M.C., Tittonell, P.A., Herrero, M., Pacini, C., de Ridder, N., et al., 2007. NUANCES-FARMSIM: a tool to analyse entry points for improved management of smallholder farming systems in sub-Saharan Africa. In: Farming Systems Design 2007: an International Symposium on Methodologies for Integrated Analysis of Farm Production Systems, Book 1 - Farm-regional Scale Design and Improvement, 10-12 September 2007, Catania, Sicily, Italy.

Van Wijk, M.T., Tittonell, P.A., Rufino, M.C., Herrero, M., Pacini, C., de Ridder, N., et al., 2009. Identifying key entry-points for strategic management of smallholder farming systems in sub-Saharan Africa using the dynamic farm-scale simulation model NUANCES-FARMSIM. Agric. Syst. 102, 89–101.

Velthof, G., Kuikman, P.J., Oenema, O., 2003. Nitrous oxide emission from animal manures applied to soil under controlled conditions. Biol. Fertil. Soils 37, 221–230.

Venterea, R.T., 2007. Nitrite-driven nitrous oxide production under aerobic soil conditions: kinetics and biochemical controls. Glob. Change Biol. 13, 1798–1809.

Vazhacharickal, P.J., Gurav, T., Chandrasekharam, D., 2019. Heavy metal signatures in urban and peri-urban agricultural soils across the Mumbai Metropolitan Region, India. Nutr. Cycl. Agroecosyst. 115, 295–312.

Velthof, G.L., Nelemans, J.A., Oenema, O., Kuikman, P.J., 2005. Gaseous nitrogen and carbon losses from pig manure derived from different diets. J. Environ. Qual. 34, 698–706.

Voříšková, J., Brabcová, V., Cajthaml, T., Baldrian, P., 2014. Seasonal dynamics of fungal communities in a temperate oak forest soil. New Phytol. 201, 269–278.

Wachendorf, C., 2015. Effects of liming and mineral N on initial decomposition of soil organic matter and post harvest root residues of poplar. Geoderma 259/260, 243–250.

Webley, D.M., Jones, D., 1971. Biological transformation of microbial residues. In: McLaren, A.D., Skujins, J.J. (Eds.), Soil Biochemistry, Volume 2. Marcel Dekker, New York, pp. 446–485.

Whalen, E.D., Smith, R.G., Grandy, A.S., Frey, S.D., 2018. Manganese limitation as a mechanism for reduced decomposition in soils under atmospheric nitrogen deposition. Soil Biol. Biochem. 127, 252–263.

White, R.P., Murray, S., Rohweder, M., 2000. Pilot analysis of global ecosystems. Grassland ecosystems. World Resource Institute, Washington, USA.

Wichern, F., Luedeling, E., Müller, T., Joergensen, R.G., Buerkert, A., 2004a. Field measurements of the CO_2 evolution rate under different crops during an irrigation cycle in a mountain oasis of Oman. Appl. Soil Ecol. 25, 85–91.

Wichern, F., Müller, T., Joergensen, R.G., Buerkert, A., 2004b. Effects of manure quality and application forms on soil C and N turnover of a subtropical oasis soil under laboratory conditions. Biol. Fertil. Soils 39, 165–171.

Wichern, F., Eberhardt, E., Mayer, J., Joergensen, R.G., Müller, T., 2008. Nitrogen rhizodeposition in agricultural crops: methods, estimates and future prospects. Soil Biol. Biochem. 40, 30–48.

Wolf, D., Opitz von Boberfeld, W., 2003. Effects of nitrogen fertilization and date of utilization on the quality and yield of tall fescue in winter. J. Agron. Crop Sci. 189, 47–53.

Wrage, N., van Groenigen, J.W., Oenema, O., Baggs, E.M., 2005. A novel dual-isotope labelling method for distinguishing between soil sources of N_2O. Rapid Commun. Mass Spectrom. 19, 3298–3306.

Wrage-Mönnig, N., Horn, M.A., Well, R., Müller, C., Velthoff, G., Oenema, O., 2018. The role of nitrifier denitrification in the production of nitrous oxide revisited. Soil Biol. Biochem. 123, A3–A16.

Wu, D., Well, R., Cárdenas, L.M., Fuß, R., Lewicka-Szczebak, D., Köster, J.R., et al., 2019. Quantifying N_2O reduction to N_2 during denitrification in soils via isotopic mapping approach: model evaluation and uncertainty analysis. Environ. Res. 179, 108806.

Xia, L., Lam, S.K., Wolf, B., Kiese, R., Chen, D., Butterbach-Bahl, K., 2018. Trade-offs between soil carbon sequestration and reactive nitrogen losses under straw return in global agroecosystems. Glob. Change Biol. 24, 5919–5932.

Xu, H., Sheng, R., Xing, X., Zhang, W., Hou, H., Liu, Y., et al., 2019. Characterization of fungal nirK-containing communities and N_2O emission from fungal denitrification in arable soils. Front. Microbiol. 10, 117.

Yeung, A.C.Y., Paltsev, A., Daigle, A., Duinker, P.N., Creed, I.F., 2019. Atmospheric change as a driver of change in the Canadian boreal zone. Environ. Rev. 27, 346–376.

Zhang, G., Jiang, N., Liu, X., Dong, X., 2008. Methanogenesis from methanol at low temperatures by a novel psychrophilic methanogen, *Methanolobus psychrophilus* sp. nov., prevalent in Zoige wetland of the Tibetan plateau. Appl. Environ. Microbiol. 74, 6114–6120.

Zhang, Y., Wang, H., Liu, S., Lei, Q., Liu, J., He, J., et al., 2015. Identifying critical nitrogen application rate for maize yield and nitrate leaching in a Haplic Luvisol soil using the DNDC model. Sci. Total Environ. 514, 388–398.

Zhou, G., Zhang, J., Qiu, X., Wei, F., Xu, X., 2018. Decomposing litter and associated microbial activity responses to nitrogen deposition in two subtropical forests containing nitrogen-fixing or non-nitrogen-fixing tree species. Sci. Rep. 8, 12934.

Zhang, W., Sheng, R., Zhang, M., Xiong, G., Hou, H., Li, S., et al., 2018a. Effects of continuous manure application on methanogenic and methanotrophic communities and methane production potentials in rice paddy soil. Agric. Ecosyst. Environ. 258, 121–128.

Zhang, N., Li, Y., Wubet, T., Bruelheide, H., Liang, Y., Purahong, W., et al., 2018b. Tree species richness and fungi in freshly fallen leaf litter: unique patterns of fungal species composition and their implications for enzymatic decomposition. Soil Biol. Biochem. 127, 120–126.

Index

Note: Page numbers followed by "*t*" refer to tables.

A

AAP. *See* Amino acid permease (AAP)
ABA. *See* Abscisic acid (ABA)
ABC. *See* ATP-binding cassette (ABC)
Abiotic stress, 155–156, 247–248, 250, 423, 680
Abscisic acid (ABA), 76, 144–145, 150, 171, 292–293, 534
  ABA-mediated signaling, 29
  grain ABA concentrations, 183–184
  signal transduction cascades, 177
  stomatal closure, 257
Accumulation, 12
  of carbohydrates, 238
  of photosynthates, 149, 240
  of starch, 231
ACD. *See* After-cooking darkening (ACD)
Acetyl group (−CO−CH$_3$), 223
Acetyl-CoA decarbonylase synthase, 324
Acetylene reduction assay (ARA), 638
Aci-reductone dioxygenase, 324
Acid mineral soils, 666–673
  aluminum toxicity, 666–672
    aluminum detoxification by root exudates, 674–676
    aluminum solution chemistry, 668–669
    importance of Mg nutrition in alleviating Al toxicity, 672
    inhibited nutrient and water uptake, 671–672
    inhibition of root growth, 669–670
    interference with root-cell plasma membrane properties, 670–671
    nitric oxide, 672
    reduced aluminum binding in root apoplast, 676–677
    rhizosphere pH, 676
  major constraints, 666–667
  manganese
    tolerance, 678–679
    toxicity, 672–673
  mechanisms of adaptation to, 673–679
    aluminum resistance by avoidance of, 674–677
    aluminum tolerance, 677
  proton toxicity, 667
Acid soils, plants grown on, 301
Acidity, 426
Acidobacteria, 601
Aconitase, 286

Actinobacteria, 616
Actinorhizal plants, 556
Active ingredients, 113–114
Acute deficiency, 484
Adaptation
  of C3 and C4 plants to future climates, 728
  mechanisms of adaptation to acid mineral soils, 673–679
  mechanisms of adaptation to saline substrates, 695–702
Adenosine 5′-phosphoselenate (APSe), 407–408
Adenosine phosphosulfate (APS), 220–221
Adenosine triphosphate (ATP), 18, 228–229, 618–620
Adequate K nutrition, 259
Adequate Mg nutrition, 241, 672
Adjuvants, 113–114
ADP-glucose, 396
  pyrophosphorylase, 169, 231
Adsorption processes, 500
AE. *See* Agronomic efficiency (AE)
Aerated soils, 504–505
Aerenchyma formation, 686
Aerenchymatous adventitious roots, 559–560
Aerobic saprotrophic fungi, 752
AFS. *See* Apparent free space (AFS)
After-cooking darkening (ACD), 425
Agave sisalana. *See* Sisal (*Agave sisalana*)
Agave tequilana. *See* Tequila agave (*Agave tequilana*)
Age of leaves, 483–485
Agricultural ammonia, 739
Agricultural management, 545, 680–681
Agricultural production systems, 621
Agricultural strategies, 466–467
Agriculture, 106, 146–147, 208, 356–357, 391, 499, 755–756
  role of AM in, 602
Agro-ecosystems, 729–730, 761–763
Agronomic biofortification, 409, 428
Agronomic efficiency (AE), 731
Agronomic production systems, 445–446
Airborne pathogens, 455
AKT2, 258
Al-activated malate transporter (ALMT), 20–23, 674
Albumins, 218
Alcohol dehydrogenase, 311
Aldehyde oxidase, 332

Alfalfa (*Medicago sativa*), 214, 402
Alkaline phosphatase, 313
Allantoic acid, 156, 326–327
Allantoin, 156
Alliin, 224
Allium ascalonicum. *See* Shallot (*Allium ascalonicum*)
Allium cepa. *See* Onion (*Allium cepa*)
Allium species, 219–220, 224
Almond (*Prunus amygdala*), 349
Alternative oxidase (AOX), 155–156
Aluminum (Al), 5, 245, 341, 410–411, 659
  accumulation, 677
  Al-resistant bacterial genera, 676
  detoxification by root exudates, 674–676
  importance of Mg nutrition in alleviating Al toxicity, 672
  resistance, 677–678
    aluminum detoxification by root exudates, 674–676
    by avoidance, 674–677
    reduced aluminum binding in root apoplast, 676–677
    rhizosphere pH, 676
    solution chemistry, 668–669
  tolerance, 677
    aluminum accumulation, 677
    screening for aluminum resistance, 677–678
  toxicity, 666–672
    aluminum solution chemistry, 668–669
    importance of Mg nutrition in alleviating Al toxicity, 672
    inhibited nutrient and water uptake, 671–672
    inhibition of root growth, 669–670
    interference with root-cell plasma membrane properties, 670–671
    nitric oxide, 672
AM. *See* Arbuscular mycorrhiza (AM)
Amaranth (*Amaranthus* spp.), 141, 170
AMF. *See* Arbuscular mycorrhizal fungi (AMF)
Amine choline, 227
Amino acid permease (AAP), 87, 89, 158
Amino acid(s), 75, 202, 217, 225, 310–311, 315, 421, 570
  composition, 429–430
  uptake, 208
Amino-K, 89–90

Amino-N, 89–90
Amino-P, 89–90
1-aminocyclopropane-1-carboxylic acid (ACC), 173, 287, 592, 682
Ammonia (NH$_3$), 105, 206, 208
Ammonium (NH$_4^+$), 34, 76, 202, 206–207
  assimilation, 213–214
    in roots, 214
  ions, 216
  NH$_4^+$-based fertilizers, 666–667
  NH$_4^+$-sensitive barley plants, 207
  synergy between NH$_4^+$ and nitrate nutrition, 215
  toxicity, 215–216
  transport into and within plants, 206–208
    in shoot, 207–208
    uptake by roots, 206–207
  uptake, 553–554
Ammonium transporters, 206
AMT1 genes, 206–207
AMT2/MEP subfamily member, 207
Amylases, 218
Anaerobic metabolism, 681–682
Anaerobic processes, 755–756
Anaerobiosis, 672–673
Ananas comosus. *See* Pineapple (*Ananas comosus*)
Anatomical organization, 525
Angiosperm species, 404, 652
  variation in angiosperm ionome, 7–8
Angiospermae, 397
Animal husbandry systems, 751
Annexins, 247
Annona, 327
Annual medics (*Medicago* spp.), 349–350
Anthocyanin, 424–425
Antiflorigens, 86–87
Antigibberellins, 175
Antinutrients, 428–429
Antioxidative defense systems, 401
Antioxidative enzymes, 286, 294
AOX. *See* Alternative oxidase (AOX)
Apical root zones, 549
Apoplasm, 13, 399
  functions, 674
  influx to, 12–14
Apoplasmic AFS, 14
Apoplasmic solute movement, 29–30
Apoplasmic space, 290
Apparent free space (AFS), 13
Apple (*Malus domestica*), 96, 163–165, 173
APS. *See* Adenosine phosphosulfate (APS)
Aquaporins, 24, 49
Aquatic ecosystems, 616
Aquatic plants, nutrient uptake in, 108
Aqueous solutions, 119
ARA. *See* Acetylene reduction assay (ARA)
Arabidopsis, 6, 49–50, 76–77, 79, 169, 206, 329, 524, 670
  A. thaliana, 15, 107, 326–327, 405, 549, 671
    seeds, 170
    shoots, 212–213

Arabidopsis nitrate transporter 1/peptide transporter 6.3 (AtNPF6.3), 351
Arabidopsis thaliana. *See* Dicot model plant (*Arabidopsis thaliana*)
Arbuscular mycorrhiza (AM), 318–319, 530, 594, 735
  in agriculture, 602
  in nutrition of host plant, 600–601
Arbuscular mycorrhizal fungi (AMF), 590, 593, 625–626
Archaea, 752
Arginase, 297
Arginine, 163–165, 214, 225, 326
Arsenic (As), 434
  solubilization, 682
Ascorbate oxidase, 304
Ascorbate peroxidase, 286, 294
Ascorbic acid, 304
Asparagine (Asn), 42, 214, 294
Asparagine synthetase, 146, 355–356
Aspartate (Asp), 214, 225, 294
Assimilatory reduction, 221
Associative nitrogen fixation, 639–640
AtAAP1, 89
AtAAP3, 87
AtAAP8, 87, 89
AtAKT2/3, 87
AtAMT1, 34
AtCAT6, 89
AtHAK5, 34
AtIRT1 transporter of arabidopsis, 44–45
Atmospheric circulation, 723
Atmospheric deposition, 738–740
Atmospheric dinitrogen (N$_2$), 201
AtNAS1, 170
AtNFP7.2, 79
AtNPF1.1 (transporters of NPF family), 79, 87
AtNPF1.2 (transporters of NPF family), 87
AtNPF2.13 (transporters of NPF family), 87
AtNPF4.6, 203
AtNPF6.2, 81–82
AtNPF7.2, 205
AtNPF7.3, 204–205
AtNRT1.8, 77
AtNRT2.1, 203
AtNRT2.4 (transporters of NPF family), 87, 203
AtNRT2.5 (transporters of NPF family), 87, 203
AtOPT3 (member of oligopeptide transporter family), 170
ATP. *See* Adenosine triphosphate (ATP)
ATP-binding cassette (ABC), 24
ATPase
  vacuolar compartmentalization and proton-pumping V-type ATPase, 353–354
Atriplex spp. *See* Saltbush (*Atriplex* spp.)
AtSKOR in arabidopsis, 53
AtSLAH2, 76
AtSUC2, 87
Autoradiography, 545
Autotrophic organisms, 430
Auxin(s), 144–145, 150, 171, 182, 292–293, 528–529

biosynthesis, 673
and phytohormones, crosstalk between, 529
targets, 527
Avena coleoptiles, 254
Avocado, 173
Azolla leaves, 621

B

B-deficient plants, 340
Bacterial diseases, 457
  soil-borne fungal and, 458–461
  and viral diseases, 457–458
Bacterial infections, 244, 457
Banana, 173
Barley (*Hordeum vulgare*), 83, 145, 151, 169, 349–350
Barrel medic, 329
Bean (*Phaseolus vulgaris*), 12, 78, 84, 94–95, 161, 170
Beer, 423
Beet (*Beta vulgaris*), 5
Beneficial element(s), 6, 118, 411
  aluminum, 410–411
  cobalt, 402–404
  for plant growth, 6
  selenium, 405–410
  silicon, 397–402
  sodium, 387–397
BER. *See* Blossom end rot (BER)
Berseem clover (*Trifolium alexandrinum*), 85
β-glucosylesters, 173
Beta vulgaris. *See* Beet (*Beta vulgaris*); "Natrophilic" sugar beet (*Beta vulgaris*); Sugar beet (*Beta vulgaris*)
Beta-glucan, 430
Betaine, 214–215
Bicarbonate toxicity, 552–553
Binding form
  of calcium, 241–244
  of magnesium, 235
Bioactive phytochemicals, 430–431
Bioactive polysaccharides, 430
Biochemical methods, 488–489
Biodiversity, 763–764
Biofertilizer, 590
Biofortification, 118, 409–410, 427–429
  agronomic biofortification, 409, 428
  genetic biofortification, 428
Biogeochemical cycle, 615
Biological membranes, composition of, 15–18
  lipid and fatty-acid composition of plasma membrane and tonoplast of mung bean, 16t
Biological N$_2$ fixation, 740
Biological nitrogen fixation (BNF), 615, 620–621
Biological yield, 131
Biologically inactive GA-conjugates, 173
Biomass, 523
Biomembranes, 227
Bioregulators, 175
Biosynthetic machinery, 674
Biotic stresses, 155–156, 446

Biotin, 223
Biotrophic pathogens, 447–448
Bitter pits in apple, 90
Black gram (*Vigna mungo*), 348
Blossom end rot (BER), 423
    in tomato, 90, 117
BNF. *See* Biological nitrogen fixation (BNF)
Borate anion B(OH)$_4^-$, 337
Border cells. *See* Root, cap cells
Boric acid, 109, 336–338
    channels, 81–82
Boron (B), 4, 76, 111, 117, 166, 336–350, 633
    atom, 336–337
    availability, 347
    complexes with organic structures, 338–339
    deficiency, 341, 347–348, 652
    function of, 339–347
        cell wall structure, 340–341
        integrated assessment of function of B in plants, 346–347
        membrane function, 342–344
        metabolism, 341–342
        reproductive growth and development, 344–345
        root elongation and shoot growth, 345–346
    requirements of monocots and dicots, 652
    tolerance, 349–350
    toxicity, 349–350
Botrytis cinerea L. *See* Gray mold (*Botrytis cinerea* L.)
Brachypodium, 202
    B. distachyon, 302
Brachypodium distachyon. *See* Grass (*Brachypodium distachyon*)
Brackish water, 391
Bradyrhizobium, 324
Branched root system, 598–599
Brassica napus. *See* Oilseed rape (*Brassica napus*)
Brassica oleracea var. *botrytis*. *See* Cauliflower (*Brassica oleracea* var. *botrytis*)
Brassica rapa. *See* Turnip (*Brassica rapa*)
Brassicaceae. *See* Cruciferous vegetables (Brassicaceae)
Brassinosteroids (BRs), 171, 173–174, 217, 292–293
Bread, 420–421
    making, 420–421
Bread wheat (*Triticum aestivum*), 303
Breeding, 429
Broad bean (*Vicia faba*), 4–5, 50, 169
Broadleaf plantain (*Plantago major*), 178
Broccoli, 337
Bromeliaceae, 143
    members of, 120
BRs. *See* Brassinosteroids (BRs)
Bundle sheath
    cells, 212
    chloroplasts, 390

C

C sequestration, 733
C$_3$ plants, 142, 727–728
C$_3$ species, 136, 139, 141, 389–391
C$_4$ plants, 212, 221, 727–728
    adaptation of, 728
    photosynthesis, 140–143, 388
C$_5$ compound ribulose-1,5-bisphosphate (RuBP), 137–138
CA. *See* Carbonic anhydrase (CA)
Ca-polygalacturonates, 453
Ca-related imbalances, 117
Ca^{2+} mobility, 47
Ca^{2+}-dependent protein kinases (CDPKs), 247
    CDPK-related kinases, 247
Cadmium, 432–433
cADPR. *See* Cyclic ADP-ribose (cADPR)
Calcareous soils, 115–116, 501, 550–552
Calcicole species, 656
Calcification process, 550–552
Calcineurin B-like protein Ca^{2+} sensors (CBL Ca^{2+} sensors), 24
Calcineurin B–like proteins (CBL proteins), 247
Calciotrophes, 242–243
Calcium (Ca), 3–4, 6–7, 36, 76, 90, 110, 115, 241–249, 453–455, 631–632
    binding form, 241–244
    bound, 244
    calcium-binding proteins, 247
    cation–anion balance and osmoregulation, 245–246
    cell extension and secretory processes, 244–245
    cell wall stabilization, 244
    compartmentation, 241–244
    content, 91
    deficiency disorders, 90, 423
    influx, 246
    as intracellular second messenger, 246–247
    membrane stabilization, 245
    oxalate crystals, 81, 242–243
    requirements of monocots and dicots, 652
    supply, plant growth, and plant composition, 248–249
    as systemic signal, 247–248
    uniporters, 243–244
Calcium-sensor protein kinases (CPKs), 216
Callose, 301
    synthesis, 670
Callose formation, 85–86, 245
Calmodulin (CaM), 148, 247
    CaM-dependent protein kinases, 247
Calvin cycle, 137–138, 230–231
Calvin–Benson cycle, 653
Calvin–Benson–Bassham (CBB), 137–138
CAM. *See* Crassulacean acid metabolism (CAM)
Campesterol, 15, 173–174
Canavalia ensiformis. *See* Jack bean (*Canavalia ensiformis*)
Canola, 88
Canopy leaf area, 153–154

Capsicum annuum. *See* Pepper (*Capsicum annuum*); Sweet pepper (*Capsicum annuum*)
Carbohydrates, 32, 170, 297–299, 305–306, 316, 327, 754
    feedback regulation of photosynthesis by sink demand for, 147–148
    metabolism, 315–316
    partitioning, 238–240
Carbohydrates, accumulation of, 238
Carbon (C), 6
    fluxes in terrestrial agroecosystems
        effects of organic soil amendments on gaseous fluxes, 754–755
        effects of pH, soil water content, and temperature on organic matter turnover, 755–756
        global warming effects, 756
        microbiological factors determining carbon and nitrogen emissions, 751–754
        nutrient fluxes in rural–urban systems, 763–764
        plant–animal interactions affecting nutrient fluxes at different scales, 757–763
        scale issues in modeling, 763
    losses from livestock excreta, 759
    metabolism, 680
    microbiological factors determining carbon emissions, 751–754
        CH$_4$ emissions, 753
        CO$_2$ emission, 752
        fungal and bacterial contributions to CO$_2$ emissions, 752–753
        N$_2$ and N$_2$O emissions, 753–754
    monoxide dehydrogenase, 324
Carbon dioxide (CO$_2$), 615–616
    assimilation, 137–140
    concentrations, 481
    emission, 752
        fungal and bacterial contributions to, 752–753
Carbon disulfide (CS$_2$), 107
Carbon-use efficiency (CUE), 752–753
Carbonic anhydrase (CA), 312–313
Carbonyl sulfide (COS), 107
Carboxylic group (COO$^-$), 13
Carboxypeptidase, 313
Carex pendula, 242–243
Carotenoids, 135, 289–290
Carya, 327
Carya illinoiensis. *See* Pecan (*Carya illinoiensis*)
Casparian band-defective arabidopsis mutant (*sgn3*), 50–51
Casparian bands, 13–14, 29, 49, 397
Castor bean (*Ricinus communis*), 77–78, 143–144, 158
CAT. *See* Cationic aminoacid transporter (CAT)
Catalase, 285–286
Catalytic water-splitting center, 134
Cate-Nelson graphical model, 481

Catechol oxidases. *See* Polyphenol oxidases
Cation diffusion facilitators (CDFs), 679
Cation exchange capacity (CEC), 13, 513
Cation−anion balance, 245−246, 258
Cation−anion relationships, 37−40
  accompanying anions influence rates of K^+ and Cl^- uptake by maize plants, 37t
  ionic balance in shoots of castor oil plants grown with different forms of N supply, 39t
Cationic aminoacid transporter (CAT), 158
  families, 89
Cations, 76
Cauliflower (*Brassica oleracea* var. *botrytis*), 331, 337
CBB. *See* Calvin−Benson−Bassham (CBB)
CBL Ca^{2+} sensors. *See* Calcineurin B-like protein Ca^{2+} sensors (CBL Ca^{2+} sensors)
CBL proteins. *See* Calcineurin B−like proteins (CBL proteins)
CC. *See* Companion cells (CC)
CCC. *See* Chlorocholine chloride (CCC)
CcO. *See* Cytochrome c oxidase (CcO)
CDFs. *See* Cation diffusion facilitators (CDFs)
CDPKs. *See* Ca^{2+}-dependent protein kinases (CDPKs)
CDRs. *See* Critical deficiency ranges (CDRs)
CEC. *See* Cation exchange capacity (CEC)
Celery stem disorder, 348
Cell walls, 13, 306−307, 454, 667, 758
  integrity, 339
  loosening, 431
  pectins, 652
  stabilization, 244
  structure, 340−341
Cells, 302
  death, 146
  division and extension, 299
  elongation, 529
  extension processes, 244−245, 254−255
  nitrate transport within, 205−206
  osmoregulation and turgor, 353−354
    cell elongation and plant growth, 354
    cell volume regulation and stomatal function, 354
    vacuolar compartmentalization and proton-pumping V-type ATPase, 353−354
  secretory processes, 244−245
Cellular compartments, 235
  of potassium, 250
Cellular membranes, 18
Cellular processes, 392
Cellular redox processes, 456
Cellulase enzyme, 457
Cellulose, 131
Ceramic microsuction cups, 547
Cereals, 95, 169−170, 233, 428, 478
  crops, 162
  NUE of, 731
  roots, 525−526
  seeds, 217−218
Cesium (Cs^+), 34

Chain-like tubers, 168
Chelates, 110
Chelating compounds, 170
Chelation, 6−7, 502
Chemical soil analysis, 480, 499−501
Chenopodiaceae (*Haloxylon aphyllum*), 140−141, 388, 593
Cherry (*Prunus cerasus*), 96
Chilling, 174
Chinese cabbage, 107
Chitinases, 218
Chlamydomonas reinhardtii, 333
Chloride, 87, 134, 216
  concentrations, 35
  fertilizer, 464
  salinity, 434
  supply, deficiency, plant growth, and crop yield, 356−357
    chloride deficiency, 356
    crop yield, 356−357
    energy/metabolic saving, 356
  uptake, 690−691
Chloride channel (CLC), 351−352, 690−691
Chlorinated auxin IAA (Cl-IAA), 354
Chlorine (Cl), 4, 350−359
  cell osmoregulation and turgor, 353−354
  charge balance, 352
  chloride supply, deficiency, plant growth, and crop yield, 356−357
  chlorine toxicity, 357−358
  interaction with nitrate, 355−356
  photosynthesis and chloroplast performance, 352−353
  plant water balance and water relations, 354−355
  uptake, transport, and homeostasis, 350−352
    chloride recirculation, 352
    endomembrane Cl^- transporters, 351−352
    net Cl^- uptake, 350−351
    root-to-shoot Cl^- translocation, 351
Chlorocholine chloride (CCC), 168, 175
Chlorophyll, 235−236
  concentrations, 287, 424−425
  fluorescence, 389
Chloroplasts, 135, 213, 222, 235, 258, 294
  development, 287−290
Chlorosis, 334, 673
  of leaves, 478
  paradox, 291
Chronic photoinhibition, 135
Cinnamic acid, 298
Cinnamyl alcohols, 298
CIPK protein kinases, 247
Circadian water, 587−588
Citrate, 86−87, 170
Citric acid, 425
Citric acid cycle. *See* Tricarboxylic acid cycle (TCA cycle)
Citrulline, 156, 326−327
Cl-IAA. *See* Chlorinated auxin IAA (Cl-IAA)
CLAVATA3/embryo surrounding region−related peptides (CLE peptides), 217

CLC. *See* Chloride channel (CLC)
CLE peptides. *See* CLAVATA3/embryo surrounding region−related peptides (CLE peptides)
Climacteric fruits, 173
Climate change, 724−726, 751
  historical climate trends, 724
  precipitation and soil moisture, 724−725
Clover, 214
  roots, 234, 509−510
C_{min} concentration, 26
CNGCs. *See* Cyclic nucleotide-gated channels (CNGCs)
Cobalamin, 402
Cobalt (Co), 5−7, 402−404, 634
  biological role in plants, 402−403
  deficiency, 403−404
  toxicity, 403−404
Coconut (*Cocos nucifera*), 354
Cocos nucifera. *See* Coconut (*Cocos nucifera*)
Coenzymes, 221
Cold stress, 534
Colonization process, 456, 735
Commelina communis, 309
Commelinid eudicotyledons, 248
Commelinid monocotyledons, 248
Common bean, 169, 214, 248, 308
Companion cells (CC), 156
Compartmentation
  of calcium, 241−244
  of magnesium, 235
  of potassium, 250
  and regulatory role of inorganic phosphate, 229−231
Competition, 34
Concentrations of amino acids, 158
Constant tissue concentration, studying nutrition at, 45−47
Content of soil-derived leaf B, 96
Conundrum, 659
Copper (Cu), 4, 78−79, 302−309, 634−635
  carbohydrate, lipid, and N metabolism, 305−306
  copper uptake and transport, 302−303
  Cu-containing fertilizers, 308
  Cu-deficient plants, 165, 306
  deficiency, 307−308
  lignification, 306−307
  pollen formation and fertilization, 307
  proteins, 303−305
    ascorbate oxidase, 304
    Cu concentration, 305t
    cytochrome c oxidase, 304
    diamine oxidases, 304−305
    plastocyanin, 303
    polyphenol oxidases, 305
    superoxide dismutase, 303−304
  toxicity, 309
Copper−zinc SOD (CuZn-SOD), 303
Corn salad (*Valerianella locusta*), 5
Cortical gas spaces (*aerenchyma*), 48
COS. *See* Carbonyl sulfide (COS)
Cotton (*Gossypium hirsutum* L.), 729−730

Coupled carbon-nutrient cycling, impact on, 733–735
  rhizosphere processes, 733–735
Coupled transporters, 18
Cowpea, 214, 301–302, 325–326
CPKs. See Calcium-sensor protein kinases (CPKs)
Crassula, 120
Crassula helmsii, 309
Crassulaceae, 143
Crassulacean acid metabolism (CAM), 24, 143
  C4 pathway of photosynthesis and, 140–143
Crinkle leaf, 673
Critical deficiency concentration, 485
Critical deficiency ranges (CDRs), 478
Cropping system, 742
Crops, 218
  canopy, 478
  modeling of crop nutrient uptake, 515–516
  nutrition, 419, 477
  plants, 322, 327, 430–431
  production, 419
  rotations, 445–446
  species, 161, 301–302
  yield, 138, 356–357
Cruciferous vegetables (Brassicaceae), 225, 406, 426, 526, 593
Crystallization, 421
Cu transporter (COPT), 302
CUE. See Carbon-use efficiency (CUE)
CuO nanoparticles (CuO NPs), 308
CuO NPs. See CuO nanoparticles (CuO NPs)
Cuticle, 105
  nutrient uptake through, 109–110
  structure of, 108
Cuticular "pores", 109–110
CuZn superoxide dismutase, 313, 313t
CuZn-SOD. See Copper–zinc SOD (CuZn-SOD)
Cyanobacteria, 639–640
Cyanobiont, 621
Cyclic ADP-ribose (cADPR), 247
Cyclic nucleotide-gated channels (CNGCs), 246, 692
Cyclic photophosphorylation, 135
Cycling of nutrients, 91–93
Cylindrocarpon tonkinense, 753–754
Cyperaceae, 655–656
Cys. See Cysteine (Cys)
Cysteine (Cys), 42–43, 150, 221, 294, 452
  cysteine-rich proteins, 302
  signal transduction cascades, 177
Cytochrome c oxidase (CcO), 303–304
Cytochromes, 284
Cytochromes (Cyt b–f), 134
Cytokinins (CYTs), 144–145, 171–173, 292–293, 527, 682
Cytomorphic calcite elements, 550–552
Cytoplasm, 38–39
  passage into, 14–15
Cytoplasmic sleeve, 51
Cytoplasmic streaming, 51
Cytosol, 235, 303
  acidification, 457
Cytosolic GS, 212–213

Cytosolic K$^+$ concentrations, 251
Cytosolic PPi concentrations, 24
Cytotoxic elements, 76
CYTs. See Cytokinins (CYTs)

D

DACCs. See Depolarization-activated Ca^{2+} channels (DACCs)
Dark reactions, 137–138
Dark-induced stomata closure, 257
"Dauciform" roots, 234
Denitrification decomposition model (DNDC model), 761–763
Decarboxylation, 297
Deciduous fruit trees, 119
Defense reactions, 568
Deliquescence point, 111–112
Deliquescence RH (DRH), 111–112
Denitrification, 590, 753
Deoxymugineic acid (DMA), 75, 86–87, 170
  synthase, 45
Depletion zone, 505
Depolarization-activated Ca^{2+} channels (DACCs), 246
Depolarization-activated R-type anion channels, 20–23
Depolarization-activated S-type anion channels, 20–23
Deschampsia caespitosa, 322
Detoxification, 669
  of reactive oxygen species, 701–702
Developmental stage of plant, 483–485
DFS. See Donnan-free space (DFS)
DGDG. See Digalactosyldiacylglycerol (DGDG)
Diagnosis and Recommendation Integrated System (DRIS), 487
Diagnosis of nutrient disorders
  plant analysis for, 482
  tools for, 478–489
    field responses to nutrient supply, 478
    plant analysis, 480–489
    by visible symptoms, 478–480
Diamine oxidases, 304–305
1,5-diaminopentane (cadaverine), 174
Diazotrophs, 616
DIC. See Dissolved inorganic carbon (DIC)
Dicot model plant (*Arabidopsis thaliana*), 523–524
Dicotyledonous species, 347–348, 401, 652
  angiosperms, 525
  gymnosperms, 525
  plant species, 291–292
  plants, 590
Dicotyledons, 152
Dietary nitrate, 434
Dietary phytate/Fe and phytate/Zn molar ratios, 234
Diffuse vascular bundles (DVBs), 78–79, 398
Diffusion, 504–508
  diffusion-mediated release of low-molecular-weight compounds, 569
  plant factors, 507–508
  soil factors, 505–507

Digalactosyldiacylglycerol (DGDG), 15
"Dilution effect", 428, 481
Diospyros, 327
Diphenol oxidases. See Polyphenol oxidases
Diseases, 445–446
  bacterial and viral diseases, 457–458
  direct and indirect effects of fertilizer application on plants and pathogens and pests, 464–467
  foliar fertilizers for disease control, 118
  fungal diseases, 449–456
  pests, 461–464
  relationship between susceptibility and nutritional status of plants, 447–449
  soil-borne fungal and bacterial diseases, 458–461
Disorders related to soil acidity, 635
Dissolved inorganic carbon (DIC), 735
Dissolved nutrients, 105
Distribution of water in soils, 511–512
Disulfides, 224
DMA. See Deoxymugineic acid (DMA)
DNA
  DNA-binding proteins, 314
  extraction, 546
  fragments, 592–593
DNDC model. See De Nitrification De Composition model (DNDC model)
Donnan-free space (DFS), 13
Dracunculus vulgaris. See Voodoo lily (*Dracunculus vulgaris*)
DRH. See Deliquescence RH (DRH)
DRIS. See Diagnosis and Recommendation Integrated System (DRIS)
Drought, 174, 247–248, 726
  resistance, 355
Dry deposition, 120
Dry topsoil, 116
Dry weight (dw), 480–481
  ratios, 598
DTX/MATE (Tonoplast detoxification efflux carrier/multidrug and toxic compound extrusion transporters), 351–352
Dulcitol, 87
Durum wheat (*Triticum durum*), 147
DVBs. See Diffuse vascular bundles (DVBs)
dw. See Dry weight (dw)
Dynamic aqueous continuum, 109
Dynamic photoinhibition, 135
Dynamic system, 587
  rhizosphere as, 587–588

E

Earthworm, 511
ECM. See Ectomycorrhiza (ECM)
Economic yield, 131
Ecosystem services, 445–446
Ectomycorrhiza (ECM), 593, 595
  nutrition of plants, role of, 602–603
Ectomycorrhizal, 735
  fungi, 563
  weathering, 602–603
Efflorescence RH (ERH), 112
Electro-ultrafiltration method (EUF), 514

Electrochemical potential gradient, 552
Electron microscopy, 488
Elements, 6, 76, 80, 82, 118
　leaching of elements from leaves, 119
　toxicity as component of hypoxia stress, 682–684
　effect of transpiration rate on distribution of elements within shoot, 83–84
Elevated CO_2, 727
Elongation zone (EZ), 588, 669–670
Elsholtzia splendens, 309
Embothrium coccineum, 234
Embryo-specific urease (Eu1), 324
Embryonic root
　branching, 526–527
　system, 524
Enclosed leaves, 248–249
Endodermis, 526–527
Endogenous concentrations of phytohormones, effects of nutrition on, 177–181
Endogenous genetic program, 523
Endomembrane Cl⁻ transporters, 351–352
Endomycorrhiza, 594–595
Endophytes, 592
Endoplasmic reticulum (ER), 51, 217–218, 566
Endosperm storage proteins, 420
Energy, 228
　demand for solute transport, 25
　energy-demanding process, 618
　energy-rich phosphates, 229
　metabolism, 630
　substrate, 18
　transfer, 258
　role in, 228–229
Energy-rich inorganic pyrophosphate (Energy-rich PPi), 228–229
Energy-rich PPi. *See* Energy-rich inorganic pyrophosphate (Energy-rich PPi)
Enhanced ammonium assimilation, 213
Enlarged vascular bundles (EVBs), 78–79, 398
Environmental factor, 419–422, 446, 481, 487
Environmental limitations, 637–638
　salinity, 637
　soil water content, 637–638
Environmental problems, 616
Environmental signals, 246
Environmental stresses, 151
Enzymatic degradation, 456
Enzymes, 285, 304, 332, 547
　activation of magnesium, 236–238
　antioxidative, 286, 294
　extracellular, 734–735
　isoforms of, 221
Epidermal cell, 449
　walls, 399, 456
Epidermal tissues, 450
Epiphytic plants, 120
ER. *See* Endoplasmic reticulum (ER)
ERH. *See* Efflorescence RH (ERH)
Ericoid mycorrhizae, 594
Escherichia
　E. coli, 139, 206, 324, 763–764
Esophageal cancer, 399

ESP. *See* Exchangeable sodium percentage (ESP)
Essential mineral elements, 14
Essential nutrient for plant growth, 6
Ethylene (ET), 144–145, 171, 173, 292–293
EU. *See* European Union (EU)
Eu1. *See* Embryo-specific urease (Eu1)
Eu3 protein, 324
Eudicotyledons, 241–242
EUF. *See* Electro-ultrafiltration method (EUF)
Euhalophytes, 5
Eukaryotic cells, 314
European Union (EU), 739
Evapotranspiration (ET), 723
EVBs. *See* Enlarged vascular bundles (EVBs)
Exchange adsorption in xylem vessels, 77
Exchangeable sodium percentage (ESP), 687
Excreta quality, 757–759
　species-specific relationship between feed intake and, 757–759
　quality of pig and poultry excreta, 758–759
　quality of ruminant excreta, 757–758
　quantitative aspects of intake and excretion, 757
Excretion, quantitative aspects of intake and, 757
exDNA. *See* Extracellular DNA (exDNA)
Exodermis, 665–666
Exogenous addition, 672
Exogenous stimuli, 217
Exopolysaccharides, 507
Experimental methods, 478
External concentration, 40–41
　influx of nitrate into barley roots without and with induction of high-capacity nitrate uptake system, 41*t*
External solution into root cells, pathway of solutes from, 12–15
　effects of extracellular calcium, 36–37
　cation–anion relationships, 37–40
　K^+/Na^+ selectivity of roots from external solution, 36*t*
　rates of net K^+ and Cl^- uptake by barley roots with or without Ca^{2+} supply, 36*t*
Extracellular DNA (exDNA), 564
Extracellular enzymes, 734–735
Extractants, 499–500
Extraction methods, 499–500, 514
Extreme temperatures, 247–248
Exudation rate, factors governing ion release into, 53–56
EZ. *See* Elongation zone (EZ)

F

Faba bean (*Vicia faba*), 161
Fabaceae, 653
FACE. *See* Free-air CO_2 exchange (FACE)
Farmyard manure, 4
Fatty acid, 298, 421–422, 597
FBH4. *See* FLOWERING BHLH 4 (FBH4)
Fd-GOGAT genes, 213
Fe deficiency, 289–290

　root responses to, 291–293
Fe-binding proteins, 294
Fe-deficient leaves, 286, 288–289
Fe-deficient roots, 285
"Feed-forward loop" network, 42
Feedback inhibition of photosynthesis, 147
Feedback regulation of photosynthesis by sink demand for carbohydrates, 147–148
Fenton reaction, 283–284
Fermentation, 681–682, 758
Ferredoxin, 134, 223
　thioredoxin system, 223
Ferric Reductase Defective 3 (FRD3), 286
Ferric reductase oxidase (FRO), 44–45, 561
Ferritin, 170, 291, 294
Fertigation, 422
Fertilization, 307
　plant response to, 728–729
Fertilizers, 97, 723–724
　management, 728
　programs, 478
Festuca vivipara L. *See* Sheep's fescue (*Festuca vivipara* L.)
Fiber crops, 422
Field peas, 349–350
Field-grown plants, 481
Field-grown processing, 422
Flavin (FAD), 330–331
Flavor, effects of mineral nutrition on, 425–426
Fleshy fruits, 90, 248–249
Flooded soils, 679–687
　element toxicity as component of hypoxia stress, 682–684
　hypoxia stress, 681
　mechanisms underpinning tolerance to, and avoidance of, hypoxic stress, 684–687
　phenotypic adaptation, 686–687
　phytohormones and root-to-shoot signals, 682
　phytotoxic metabolites under hypoxia, 681–682
　soil chemical factors, 679–681
Floret development, 163
Florigens, 86–87
Flower initiation and development, 163–165
FLOWERING BHLH 4 (FBH4), 163
Fluoride in leaves of plants occurring on soils containing little fluoride, 658
Fluoroacetate, 658
Foliar fertilization, 111
Foliar fertilizers for pest and disease control, 118
　control of foliar diseases of crops using foliar silicon sprays, 118*t*
Foliar irrigation methods, 118–119
Foliar leaching, 119–120
　ecological importance of, 119–120
　foliar water absorption, 119–120
Foliar nutrient sprays, 118
Foliar phosphate, 455
Foliar sprays, 114–115, 117–118, 404
Foliar uptake, 118–119
　ecological importance of, 119–120

Foliar water
  absorption, 119–120
  uptake, 120
Foliar-applied nutrients, 107–108
Foliar-applied solutions, 111–112
Food fortification, 427
Food safety, 420, 432–435
  harmful N compounds, 434–435
  toxic elements, 432–434
Forage
  composition, 744
  NUE of, 731–732
  plants, 396–397
Fossil fuels, 723
Fractionated extraction, 487–488
FRD3. *See* Ferric Reductase Defective 3 (FRD3)
Free-air CO_2 exchange (FACE), 728
Free-living nitrogen fixation, 639–640
Fresh chicken excreta, 759
Fresh fruits, 431–432
FRO. *See* Ferric reductase oxidase (FRO)
Fructose-1,6-bisphosphatase, 315–316
Fruits, 90, 161, 175, 182
  coloration, 424–425
  disorder, 423
Fulvic acid, 668–669
Fungal diseases, 399, 449–456
  principles of infection, 449–450
  role of Ca and Mg, 453–455
  role of Mn, 455–456
  role of N and K, 452–453
  role of other micronutrients, 456
  role of phosphate and phosphite, 455
  role of S, 455
  role of Si, 450–452
Fungal pathogens, 454
Fungal tolerance, 604
Fusarium oxysporum, 753–754
"Futile cycling", 25, 691

G

GA. *See* Gibberellic acid (GA)
GA-O-β-glucosides, 173
GABA. *See* γ-aminobutyric acid (GABA)
Gaeumannomyces graminis, 299–300, 451–452
γ-aminobutyric acid (GABA), 638
γ-glutamylcysteine, 222
Gas exchange, 548–549
Gaseous emissions, 752
Gaseous fluxes, effects of organic soil amendments on, 754–755
Gases, 105
  uptake and release of, 106–107
Gastrointestinal tract, 757
GCC. *See* Global climate change (GCC)
GDH. *See* Glutamate dehydrogenase (GDH)
Generative sink organs, 169
Generative storage organs, 161
Genes encoding transporters, 87
Genetic biofortification, 428
Genetic engineering, 419–420

Genotypes, 12, 485–486, 508, 670
Genotypic differences, 391–392
  in growth response to salinity, 703–704
Genotypic variation, 41, 97, 630–631
GHGs. *See* Greenhouse gases (GHGs)
Gibberellic acid (GA), 144–145, 171, 245, 527
Global climate change (GCC), 723
  plant responses to, 727–728
  and root zone nutrient availability, 732–742
Global environmental assessments, 763
Global food security, 420
Global market, 477
Global warming, 723–724, 751, 756
Globoid crystals, 320
Globulins, 218
GLR channels. *See* Glutamate receptor channels (GLR channels)
GLU1 gene, 213
GLU2 gene, 213
Glucanases, 218
Glucosinolate (GSL), 220–221, 223, 421–422
Glutamate (Glu), 294
Glutamate dehydrogenase (GDH), 213, 603
Glutamate receptor channels (GLR channels), 243–244
Glutamate synthase (GOGAT), 212–213
Glutamine (Gln), 42, 214, 294
Glutamine synthetase (GS), 146, 212, 603, 628
Glutathione, 222
  concentration, 222
  synthase, 236–237
Gluten, 420
Glycerolipids, 15
Glycerophospholipids, 15
Glycine betaine, 700
Glycolipids, 15, 343
Glycolysis, 154
Glycophytes, 388
Glycoprotein, 343
Glyoxalase (GLY), 324
GmSALT3 (proton/cation antiporter), 87
GOGAT. *See* Glutamate synthase (GOGAT)
Golgi apparatus, 564
Gossypium hirsutum L. *See* Cotton (*Gossypium hirsutum* L.)
Gradients, 552
Grain crops, 146–147
Grain legumes, 166–167, 169–170
Grain/seed yield formation, shoot architecture for, 162
Graminaceous plant species, 573–574
Graminaceous species, 79, 292, 327, 347–348, 399
Grapevine (*V. vinifera*), 344
Grass (*Brachypodium distachyon*), 601
Grass-like monocotyledonous plants, 5
Grassland ecosystem, 724
Grassland species, 107
Gray mold (*Botrytis cinerea* L.), 307
Grazing systems, 759
Green revolution, 428, 615
Green rice leaf hoppers, 463

Greenhouse gases (GHGs), 615–616, 723–724, 728, 751
  emissions, 680–681
Gross metabolic changes, 333
Growth respiration, 154
Growth response to salinity, genotypic differences in, 703–704
Growth retardants, 175
Growth stimulation by Na, 393–396
GS. *See* Glutamine synthetase (GS)
GSL. *See* Glucosinolate (GSL)
GTP. *See* Guanosine triphosphate (GTP)
Guanosine triphosphate (GTP), 228–229
Guard cells of pea, 230–231
Gymnospermae, 397
Gypsum ($CaSO_4$), 460–461, 668

H

H-B process. *See* Haber-Bosch process (H-B process)
H^+-coupled antiporters, 76
H^+/K^+ counterflow, 260
Haber-Bosch process (H-B process), 615
Haber–Weiss reaction, 283–284
HACCs. *See* Hyperpolarization-activated Ca^{2+} channels (HACCs)
Halophytes, 388
Haloxylon aphyllum. *See* Chenopodiaceae (*Haloxylon aphyllum*)
Harmful N compounds, 434–435
Harvest index, 131
HATS. *See* High-affinity transporter systems (HATS)
Heart disorder, 478–480
Heat treatment, 85–86
Heavy metal, 604, 656, 763–764
  P1B-ATPase family, 76
  tolerance, 322
Helianthus annuus. *See* Sunflower (*Helianthus annuus*)
Heme enzymes, 285
Heme proteins, 284–285
Hemicellulose, 758
Herbicide atrazine, 222
Herbicide glyphosate, 459
Herbivores, 247–248
Herbivorous insects, 462
Heterotrophic respiration, 732
Hidden hunger, 419, 427–429
High ethylene concentrations, 169
High-affinity K transporters (HKT), 691
High-affinity transporter systems (HATS), 202, 206
High-molecular-weight (HMW), 564
  compounds in, 564–568
    low-molecular-weight root exudates, 569–574
    mucilage and mucigel, 564–566
    secretory proteins, 566–568
    organic solutes, 15
High-molecular-weight glutenin subunits (HMW-GS), 420–421

High-resolution laser ablation inductively coupled plasma mass spectrometry, 107–108
Histidine (His), 294
Histochemical methods, 488–489
Historical climate trends, 724
HKT. *See* High-affinity K transporters (HKT)
HMW. *See* High-molecular-weight (HMW)
HMW-GS. *See* High-molecular-weight glutenin subunits (HMW-GS)
Holcus lanatus. *See* Yorkshire fog (*Holcus lanatus*)
Hollow stem disorder, 348
Homeostasis of magnesium, 235
Homogeneous media, 504–505
Hordeum vulgare. *See* Barley (*Hordeum vulgare*)
Horticultural crops, 344
  avoiding occurrence of physiological disorders and improving quality of, 117–118
Horticultural plant species, 405–406
Horticulture, 146–147, 499
Host apoplasm, 449
Host plant growth, 598
  root colonization, photosynthate demand, and, 595–598
HSs. *See* Humic substances (HSs)
Human digestive tract, 428–429, 434
Human health, 410–411
Human nutrition, 743–744
Humic substances (HSs), 558–559
Humidity effects on solute concentration and leaf permeability, 111–113
Hydrated ions, 12–13
Hydraulic conductivity of bare roots, 52
Hydrogen (H), 6
Hydrogen peroxide (H_2O_2), 135, 240
Hydrogenases, 324
Hydrolases, 218
Hydrologic cycle, 724
Hydrolytic processes, 563
Hydroxyl radicals (OH•), 135, 283–284, 316
Hyperpolarization-activated Ca^{2+} channels (HACCs), 246
Hypersensitive reactions, 450
Hypomagnesemia, 241
Hypoxia stress, 681
  element toxicity as component of, 682–684
  mechanisms underpinning tolerance to, and avoidance of, 682–684

I

IAA. *See* Indole-3-acetic acid (IAA)
Imaging system, 107–108, 547
Immune system, 429
Immunomodulatory effects, 430
Indeterminate genotypes of crop species, 167
Indole-3-acetic acid (IAA), 171, 297, 332, 529
  synthesis, 316–317
Inducible NRA, 331
Industrial food processing, 427
Influx to apoplasm, 12–14
cation exchange capacity of root dry matter of different plant species, 14t
cell-wall thickness and diameter of cell-wall pores, sucrose, and hydrated cations, 13t
Inhibited nutrient, 671–672
Inorganic anions, 216
Inorganic cations, 250
Inorganic elements, 119
  in rhizosphere, 550–552
Inorganic ions, 214–215
Inorganic phosphate (Pi), 138, 227
  compartmentation and regulatory role of, 229–231
Inorganic pyrophosphate (PPi), 24, 232
Inositol polyphosphate (InsP), 43–44
Inositol-1,4,5-triphosphate (IP3), 233–234, 247
InsP. *See* Inositol polyphosphate (InsP)
Intensity ratio, 514–516
Intermediary cells, 158
Interveinal chlorosis, 240
Intracellular second messengers, 177
  calcium as, 246–247
Invasive methods, 545
Inward-rectifying K-channels, 87
Iodine (I), 5, 118
  deficiency, 427
Ion-uptake mechanisms of individual cells and roots, 11–12
  changes in ion concentration of external solution and in root sap of maize and bean, 12t
  composition of biological membranes, 15–18
  factors governing ion release into xylem and exudation rate, 53–56
  factors influencing ion uptake by roots, 29–47
  pathway of solutes from external solution into root cells, 12–15
  radial transport of ions and water across root, 49–52
  relationship between ion concentrations in substrate, 11t
  release of ions into xylem, 53
  root influx
    apoplasm, 12–14
    cytoplasm, 14–15
  solute transport across membranes, 18–29
  uptake of ions and water along root axis, 15–18
Iron, 283–294, 316, 532, 633
  chloroplast development and photosynthesis, 287–290
  deficiency, 287–288
  deficiency and toxicity, 293–294
  Fe-requiring enzymes, 287
  imbalances, 691–692
  interactions among ions in rhizosphere, 33–40
    competition, 34–36
    effects of extracellular calcium, 36–37
    interactions between uptake of NH_4^+ and K^+ by maize roots, 34t
ion-exchange resins, 514
iron-containing constituents of redox systems, 284–287
  Fe-S proteins, 286–287
  heme proteins, 284–285
iron-containing sprays, 116
iron-deficiency stress, 286
iron-deficient leaves, 290
iron-efficient genotypes, 287
localization and binding state of Fe, 290–291
mobilization, 561
radial transport of ions across root, 49–52
root responses to Fe deficiency, 291–293
uptake of ions along root axis, 15–18
into xylem, release of, 53
Irrigation systems, 726
Isothiocyanates (ITCs), 224, 426
Italian catchfly (*Silene italica*), 327
Italian ryegrass (*Lolium multiflorum*), 550

J

JA. *See* Jasmonic acid (JA)
Jack bean (*Canavalia ensiformis*), 324
Jasmonates, 173
Jasmonic acid (JA), 144–145, 292–293, 328
  JA-sensitive pathogens, 449
Juglans, 327

K

Kinetic inhibition, 112
Kinetics of solute transport in plant roots, 25–29
  short-term P uptake parameters of soybean plants with different P nutritional status, 28t
Kinetin, 175
K–Mg–Ca salt, 232
Krebs cycle. *See* Tricarboxylic acid cycle (TCA cycle)

L

L-dehydroascorbic acid, 304
L-methionine, 292
^{14}C-labeled photosynthates, 257–258
Labor-intensive process, 677
Lactate dehydrogenase, 682
LAD. *See* Leaf area duration (LAD)
LAI. *See* Leaf area index (LAI)
LA–ICP–MS. *See* Laser ablation inductively coupled plasma mass spectrometry (LA–ICP–MS)
Large-seeded cultivars, 335
Laser ablation inductively coupled plasma mass spectrometry (LA–ICP–MS), 547
LATS. *See* Low-affinity transporter systems (LATS)
Leaching
  ecological importance of foliar uptake and leaching, 119–120
  of elements from leaves, 119

erosion, 740–742
  foliar leaching, 119–120
  surface erosion and, 740–742
Leaf area duration (LAD), 149, 153–154
Leaf area index (LAI), 149, 153–154
Leaf area per plant, 152
Leaf blotch (*Rhynchosporium scalis*), 446–447
Leaf burn, 115
Leaf content of Mg, 91
Leaf erectness, 399
"Leaf freckling", 402
Leaf growth, 151
Leaf longevity, 660
Leaf margins, 79–80
Leaf mass per area (LMA), 660
Leaf maturation on sink–source transition, effect of, 143–144
Leaf necrosis, 296–297
Leaf pathogens, 606
Leaf permeability, humidity effects on, 111–113
Leaf scald disease (*Rhynchosporium commune*), 452–453
Leaf senescence, 96, 144–147
Leaf sulfur requirement among plant species, variation in, 656–658
Leaf tissue, 488
Leaf-absorbed water, 120
Leaflet movement, 257
Leafy greens, 424–425
Leaves, 93, 136, 230, 232, 291–292
  and aerial plant parts, uptake and release of elements by
    ecological importance of foliar uptake and leaching, 119–120
    foliar application of nutrients, 114–119
    leaching of elements from leaves, 119
    uptake and release of gases and volatile compounds through stomata, 106–107
    uptake of solutes, 107–114
  xylem unloading in, 79–82
Lectin-like proteins, 218
Lectins, 218
Leghemoglobin, 284, 403
Legumes, 174, 225, 233, 326–327, 330, 332, 402, 556, 670–671
  root infection by rhizobia, 623–626
  seeds, 161
  species, 214
Leguminosae, 616
Leguminous plants, 730
Lemon [*Citrus limon* (L.)], 286
Leucaena leucocephala, 484
LHT1 transporter. *See* Lysine/histidine-like transporter (LHT1 transporter)
Liebig's law, 651
Light microscopy, 488
Light spot disease, 300
Light-harvesting antennae, 133–134
Light-stimulated nitrate reduction, 213
Lignification, 446
Lignin biosynthesis, 212, 306
Lime-induced chlorosis, 294
Linoleic acid, 298

Lipases, 218
Lipid(s), 131, 170, 297–298, 305–306
  composition, 17
  metabolism, 298
  peroxidation, 317, 671
  transfer proteins, 218
Lipochitooligosaccharides, 625
Lipoxygenases, 287
Livestock excreta, nutrient and carbon losses from, 759
Livestock-mediated nutrient fluxes and modelling, spatial aspects of, 759–763
Living cells, retrieval and release of nutrients by, 77–79
  distribution of Mo, 78t
  sodium concentration of roots and shoots of pasture plants with and without Na fertilizer, 78t
LMA. *See* Leaf mass per area (LMA)
LMW. *See* Low-molecular-weight (LMW)
LOES. *See* Low Oxygen Escape Syndrome (LOES)
Lolium multiflorum. *See* Italian ryegrass (*Lolium multiflorum*)
Lolium perenne. *See* Perennial ryegrass (*Lolium perenne*)
Lolium rigidum. *See* Ryegrass (*Lolium rigidum*)
Long-distance transport, 73, 84
  of nutrients, relative importance of phloem and xylem for, 89–93
    nutrients with high phloem mobility, 89–90
    nutrients with low phloem mobility, 90–91
    re-translocation and cycling of nutrients, 91–93
  of photosynthates, 156
  of sugars, 143
  in xylem, 283
Longitudinal gradients, 549
LOQS. *See* Low Oxygen Quiescence Syndrome (LOQS)
Loranthus, 257
Lotus japonicus, 329
Low Oxygen Escape Syndrome (LOES), 684–685
Low Oxygen Quiescence Syndrome (LOQS), 684–685
Low-affinity transporter systems (LATS), 202, 206
Low-molecular-weight (LMW), 547
  controlled release of, 571–574
  diffusion-mediated release of, 569
  organic N compounds, 214–215
  root exudates, 569–574
    controlled release of low-molecular-weight compounds, 571–574
    diffusion-mediated release of low-molecular-weight compounds, 569
    retrieval mechanisms, 569–571
Low-transpiring organs, 90–91, 248–249
Lupin, 214
Lupinus albus. *See* White lupin (*Lupinus albus*)
Lupinus angustifolius, 300

Lysigenic aerenchyma, 559
Lysine/histidine-like transporter (LHT1 transporter), 81–82

M

Macronutrients, 4, 88, 484
  calcium, 241–249
  magnesium, 235–241
  nitrogen, 201–219
  phosphorus, 226–235
  potassium, 249–260
  sulfur, 219–226
Magnesium (Mg), 3–4, 6–7, 24, 78, 148, 235–241, 453–455, 632, 725
  binding form, 235
  carbohydrate partitioning, 238–240
  chlorophyll synthesis, 235–236
  compartmentation, 235
  enzyme activation, phosphorylation, and photosynthesis, 236–238
  homeostasis, 235
  importance of Mg nutrition in alleviating Al toxicity, 672
  magnesium-deficient leaves, 240
  Mg-deficient plants, 236
  nutrition, 239
  protein synthesis, 235–236
  supply, plant growth, and composition, 240–241
  transporters, 236
Magnesium release (MGR), 53
Maize (*Zea mays*), 12, 85, 133, 141, 146–147, 252, 302
  genotypes, 677
Maize yellow stripe 1 (ZmYS1), 45
Malate, 86–87, 216
Malus domestica. *See* Apple (*Malus domestica*)
Malus prunifolia, 309
Mammalian systems, 402
Manganese (Mn), 4, 78–79, 90, 111, 294–302, 455–456, 634, 655–656
  cell division and extension, 299
  cluster, 295
  coordinates, 294
  deficiency, 297–301
  foliar sprays, 116
  functional role of Mn in photosynthesis, 294–296
    manganese at active site of water oxidation in photosystem II, 295–296
    manganese-deficient plants, 299–300
  Mn-containing enzymes, 294–295
  Mn-deficient roots, 299
  Mn-dependent enzymes, 297
  Mn-oxygen-evolving complex (Mn-OEC), 295
  Mn-SOD dismutation mechanism, 296–297
  Mn-sufficient plants, 299
  Mn^{2+} cation, 294
  mobilization, 560–561
  in oxalate oxidase, 297
  proteins, carbohydrates, and lipids, 297–298
  in superoxide dismutase, 296–297

Manganese (Mn) (*Continued*)
  tolerance, 678–679
    breeding for, 679
      hyperaccumulation of Mn, 679
  toxicity, 301–302, 667, 672–673
Manganese-oxalate oxidase (Mn-OxO), 297
Mannitol, 87
Manure, 4, 106
MAOM. *See* Mineral-associated organic matter (MAOM)
mARC. *See* Mitochondrial amidoxime reducing component (mARC)
Marginal zone, 481
"Marsh spot", 300
Mass flow, 503–504, 735–738
MATE. *See* Multidrug and toxin extrusion (MATE)
MCA. *See* *mid1*-complementing activity (*MCA*)
Meadow spittlebug (*Philaenus spumarius*), 56
Mechanical stress, 85–86
Medicago sativa. *See* Alfalfa (*Medicago sativa*)
Medicago spp. *See* Annual medics (*Medicago* spp.)
Medicago truncatula, 287
Meiosis, 183–184
Melaleuca systena, 653
Melastoma, 410–411
Melatonin, 328
Membranes, 15
  depolarization, 246–247
  domains, 16
  integrity, 317
  lipids, 15
  matrix, 547
  potential, 453
  proteins function, 16
  solute transport across, 18–29
  stabilization of calcium, 245
Mesembryanthemum crystallinum, 692
Mesophyll, 212
  cells, 158, 653
  chloroplasts, 390
Metabolic functions, 390–391
Metabolic pool, 230, 241
Metabolic process, 155, 286, 530–531, 628
Metabolic systems, 255
Metabolism, 407–408, 666
Metal cations, 110
Metal tolerance protein, 322
Metallophytes, 656
Metallothioneins, 303
Methane (CH$_4$), 680
Methanotrophic genes, 680–681
Methionine (Met), 225, 294
  synthesis, 402
Methyl-jasmonate, 217
Methylene urease, 324
Methylglyoxal (MG), 324
MG. *See* Methylglyoxal (MG)
MGR. *See* Magnesium release (MGR)
Michaelis–Menten equation, 25–27
Micro-X-ray-fluorescence imaging, 301
Microbial density, 588
Microbial fermentation processes, 559, 757

Microbial inoculation to promote BNF, 640–641
Microbial mineralization, 752
Microdialysis, 500, 588
Microlysimeters, 503
Micronutrients, 4, 115, 132, 335, 456
  boron, 336–350
  chlorine, 350–359
  copper, 302–309
  deficiency, 427
  elements, 387
  iron, 283–294
  manganese, 294–302
  molybdenum, 328–336
  nickel, 323–328
  requirements of plant species evolved in severely phosphorus-impoverished landscapes, 655–656
  zinc, 310–322
Microorganisms, 734–735
  decompose root, 589–590
MicroRNAs on CuZn-SOD RNA metabolism, 303
Microtensiometers, 588
mid1-complementing activity (*MCA*), 246
Millerandage, 166
Million years ago (MYA), 629
Mimosa pudica, 257
Mineral composition of food/feed, 743–744
  composition of trees and timber, 744
  forage and pasture composition and mineral nutrition, 744
  of grains and fruits, 743–744
"Mineral element theory", 4, 651
Mineral nitrogen
  effect of, 636–637
Mineral nutrients, 87, 159, 228–229, 419, 427–429, 446, 744
  on crop quality
    food safety, 432–435
    nutritional quality, 426–431
    sensory quality, 423–426
    shelf life of fresh fruits and vegetables, 431–432
    technical quality, 420–423
    yield-quality dilemma, 435–436
  on flavor, effects of, 425–426
  on visual quality, effects of, 423–425
Mineral plant nutrient, 6
Mineral-associated organic matter (MAOM), 557
Mineralization, 515, 590, 755
miRNAs, 150
Miscanthus (*Miscanthus × giganteus*), 141
Mitigation, 446
Mitochondria, 15
Mitochondrial amidoxime reducing component (mARC), 331–332
Mitochondrial citrate synthase, 674–676
Mitochondrial RNA splicing 2 (Mrs2), 82
Mitscherlich model, 481
Mn-OEC. *See* Mn-oxygen-evolving complex (Mn-OEC)

Mn-OxO. *See* Manganese-oxalate oxidase (Mn-OxO)
Mo cofactor (Moco), 330–331
Mobility in phloem, 88
Modeling of nutrient availability and crop nutrient uptake, 515–516
Molecular biology, 621
Molecular mechanisms, 81–82
Molybdate, 35
Molybdate-specific transporter, 329
Molybdenum (Mo), 4, 7, 328–336, 632
  deficiency and toxicity, 334–336
  gross metabolic changes, 333
  Mo Fe protein, 618
  Mo-containing enzymes, 331–333
  Mo-deficient plants, 334
  molybdenum-deficient plants, 333
  nitrate reductase, 330–331
  nitrogenase, 329–330
  uptake and transport, 329
Monocalcium phosphate [Ca(H$_2$PO$_4$)$_2$], 4
Monocarboxylic acids, 555–556
Monocot-specific microRNA (miR528), 306
Monocots, 652
Monocotyledonous plants, 590
  species, 291–292, 401
Monocotyledons, 152
Monogalactosyldiacylglycerol, 670–671
Monogastrics, 758
Mononuclear hydrolysis, 668
Monosulfides, 224
Morphological traits, 533
Morphology groups, 593–595
mot1 mutant, 333
MOT1;1 gene, 329
Mrs2. *See* Mitochondrial RNA splicing 2 (Mrs2)
Mucigel, 564–566
Mucilage, 507, 564–566, 676
Mucopolysaccharides, 402
Mugineic acid, 45
Multidrug and toxin extrusion (MATE), 20–23, 674
Münch *pressure flow hypothesis*, 159
Mustard, 167
MYA. *See* Million years ago (MYA)
Mycorrhiza, 593–606
  general, 593
  mycorrhizal effects, 604–606
  mycorrhizal groups, morphology, and structure, 593–595
  mycorrhizal responsiveness, 598–599
  role of AM in agriculture, 602
  role of AM in nutrition of host plant, 600–601
  role of ectomycorrhiza in nutrition of plants, 602–603
  role of mycorrhiza in tolerance to high metal concentrations, 603–604
  root colonization, photosynthate demand, and host plant growth, 595–598
  in tolerance to high metal concentrations, role of, 603–604
Mycorrhizal colonization, 515, 587

Mycorrhizal effects, 604–606
  phytohormonal effects and water relations, 604–605
  suppression of leaf pathogens, 606
  suppression of root pathogens and nematodes, 605
Mycorrhizal fungi, 550, 587, 732–735
Mycorrhizal groups, 593–595
  ectomycorrhiza, 595
  endomycorrhiza, 594–595
Mycorrhizal hyphae, 507–508, 548
Mycorrhizal plants, 590
Mycorrhizal responsiveness, 598–599
Mycorrhizal symbiosis, beneficial plant–fungus interactions in, 536–537

N

N-mediated heading date-1 (Nhd1), 163
NA. *See* Nicotianamine (NA)
^{22}Na content, 73–74
NAD-malic enzyme (NAD-ME), 141
NAD-ME. *See* NAD-malic enzyme (NAD-ME)
NADH. *See* Nicotianamide dinucleotide (NADH)
NADP-malic enzyme (NADP-ME), 141, 390–391
NADP-ME. *See* NADP-malic enzyme (NADP-ME)
NAM. *See* No Apical Meristem (NAM)
Nanofertilizers, 308
Nanoscale secondary ion MS (nanoSIMS), 547
Native soil microbes, 601
Natrophilic species, 392
"Natrophilic" sugar beet (*Beta vulgaris*), 36–37
Natrophobic glycophytes, 666
Natrophobic species, 392–393
"Natrophobic" maize, 36–37
Natural ecosystems, 502, 570, 616, 638
Natural resistance-associated macrophage protein (NRAMP), 433–434
  Nramp3, 79
Natural systems, 735
Natural vegetation, 478
Necrosis, 334
  of leaves, 478
Negative feedback regulation of phloem unloading, 161
Nematodes, 463–464
  suppression of, 605
Nernst equation, 18–19
Net Cl$^-$ uptake, 350–351
Net content, 93
Net photosynthesis, 179
Net primary productivity (NPP), 732
Net reaction, 209
Neutron radiography, 547
Nhd1. *See* N-mediated heading date-1 (Nhd1)
Nickel (Ni), 4, 7, 323–328, 387, 635
  deficiency and toxicity, 327
  general, 323–324
  hyperaccumulators, 328
  Ni-containing enzymes, 324

nickel concentration in plants, 327
role in N metabolism, 325–327
tolerance to Ni toxicity, 328
Nicotiana tabacum. *See* Tobacco (*Nicotiana tabacum*)
Nicotianamide dinucleotide (NADH), 154
Nicotianamine (NA), 75, 165–166, 170, 215, 292
  aminotransferase, 45
  synthase, 45
NiFe-hydrogenase, 324
NIN-like protein (NLP), 355
NIP gene. *See* Nodulin-26-like intrinsic protein gene (*NIP* gene)
Nitella, 12
Nitrate (NO$_3^-$), 76, 202, 208
  interaction with, 355–356
    nitrate transport within shoot, 206
    regulation of N metabolism, 355–356
    regulation of N-use efficiency, 355
  nitrate-fed barley plants, 92
  nitrate-fed plants, 331
  nutrition, 156
  primary uptake, 206
    synergy between ammonium and, 215
  reductase, 330–331
  reduction, 209–212, 297–298
  signaling, 216
  transport in plants, 202–206
    nitrate efflux from roots, 203–204
    nitrate transport within cell, 205–206
    nitrate uptake by roots, 202–203
    radial transport of nitrate across root and loading into xylem, 204–205
    transporters, 205
  uptake, 390, 553
Nitrate and peptide transporter family (NPF), 162, 202
  transporters, 205
Nitrate of soda (NaNO$_3$), 4
Nitrate reductase activity (NRA), 286
Nitrate Transporter 2 (NRT2), 202
Nitrate transporter/peptide transporter (NPF), 81–82
Nitrate-inducible microRNA (miR393), 217
Nitric oxide (NO), 107, 292–293, 309, 328, 333, 435, 672
Nitrification, 669
Nitrite (NO$_2^-$), 209
  reductase, 209
Nitrogen (N), 3–4, 77, 89–90, 95–96, 139–140, 201–219, 452–453, 530–531
  accumulation, 728–729
  ammonium transport into and within plants, 206–208
  assimilation, 208–215
    ammonium assimilation, 212–213
    in future climates, 729–730
    low-molecular-weight organic N compounds, 214–215
    nitrate reduction, 209–212
    compounds, 163–165
  deficiency, 216

emissions, microbiological factors determining, 751–754
  CH$_4$ emissions, 753
  CO$_2$ emission, 752
  fungal and bacterial contributions to CO$_2$ emissions, 752–753
  N$_2$ and N$_2$O emissions, 753–754
fertilizers, 615
fixation, 402, 477, 730
  biochemistry of, 618–621
  biological nitrogen-fixing systems, 616–618
  effects of nutrients on biological nitrogen fixation, 629–637
  in legumes, 212
  methods to quantify contribution of BNF, amounts of N fixed by legumes, and N transfer to plants in intercropping and crop rotations, 638–639
  microbial inoculation to promote BNF and improve plant nutrition, 640–641
  significance of free-living and associative nitrogen fixation, 639–640
  soil and environmental limitations, 637–638
  symbiotic systems, 621–629
immobilization, 754
metabolism, 305–306
N-deficient plants, 150
nitrate transport in plants, 202–206
nitrogen-use efficiency, 218–219
nutrition, 39
organic N uptake, 208
requirements
  of C$_3$ and C$_4$ plants, 656
  of plant species evolved in severely phosphorus impoverished landscapes, 652–654
supply, 429
source of, 553–557
supply, plant growth, and composition, 215–218
  ammonium toxicity, 215–216
  changes in root system architecture in response to N supply, 216–217
  nitrogen deficiency, 216
  storage proteins, 217–218
  synergy between ammonium and nitrate nutrition, 215
uptake efficiency (NUpE), 218
Nitrogen dioxide (NO$_2$), 107
N$_2$O emissions, 753–754
Nitrogen oxide (NO), 107, 615–616
Nitrogen-use efficiency (NUE), 206
Nitrogenase, 329–330
  reaction, 618–620
  systems, 618
Nitrosomonas, 755–756
NLP. *See* NIN-like protein (NLP)
NM. *See* Nonmycorrhizal (NM)
NO. *See* Nitric oxide (NO)
No Apical Meristem (NAM), 147
Nod26-like major intrinsic protein (NIP), 397
NodD. *See nodD* gene product (NodD)

nodD gene product (NodD), 621
Nodulation process, 402, 629
Nodule formation in legumes, 626–629
Nodulin-26-like intrinsic protein gene (*NIP* gene), 35, 81–82
"Nonessential" elements, 4
Nongreen tissue, 154
Nonheme Fe-S proteins, 286
Nonlegumes, 323, 330
"Nonmetabolic pool" of P, 230
Nonmycorrhizal (NM), 593
Nonphotochemical quenching (NPQ), 295
Nonproteinogenic amino acids, 45
Nonselective cation channels, 34
Norway spruce, 47, 240
Nostoc, 616
Novel methods, 500
NPF. *See* Nitrate and peptide transporter family (NPF); Nitrate transporter/peptide transporter (NPF)
NPF6.3, 203
NPP. *See* Net primary productivity (NPP)
NPQ. *See* Nonphotochemical quenching (NPQ)
NRA. *See* Nitrate reductase activity (NRA)
NRAMP. *See* Natural resistance-associated macrophage protein (NRAMP)
NRT2. *See* Nitrate Transporter 2 (NRT2)
NRT2 family in arabidopsis, 203
NRT2 genes, 202
NRT2 proteins, 202
Nucleic acids, 652–653
NUE. *See* Nitrogen-use efficiency (NUE); Nutrient-use efficiency (NUE)
Nutrient-use efficiency (NUE), 731–732
  calcicole species, 656
  calcium and boron requirements of monocots and dicots, 652
  of cereals, 731
  fluoride in leaves of plants occurring on soils containing little fluoride, 658
  of forage and pasture species, 731–732
  of forest species, 732
  general nutrient-use efficiency concepts, 731
  leaf longevity and nutrient remobilization, 660
  micronutrient requirements of plant species evolved in severely phosphorus-impoverished landscapes, 655–656
  nitrogen requirements of C_3 and C_4 plants, 656
  phosphorus and nitrogen requirements of plant species evolved in severely P-impoverished landscapes, 652–654
  selenium in leaves of some plants, 658
  silicon as beneficial element in leaves of plants, 658–660
  variation in leaf sulfur requirement among plant species, 656–658
Nutrient(s), 78–79, 85, 144, 148, 170, 302, 315
  accumulation, 728–730
    C_3 *vs.* C_4 plants, 728
    leguminous plants and N_2 fixation, 730
    nitrogen assimilation in future climates, 729–730
    plant response to fertilization, 728–729
  acquisition, 651
  availability, 589
  beneficial elements for plant growth, 5–6
  biochemical properties and physiological functions of nutrient elements in plants, 6–7
  on biological nitrogen fixation, effects of, 629–637
  concentrations, 480–481
  cycling, 92
  deficiency, 111, 485, 629, 671–672
  diagnosis and prediction of deficiency and toxicity of
    plant analysis for prognosis of nutrient deficiency, 489–490
    plant analysis *vs.* soil analysis, 490–491
    tools for diagnosis of nutrient disorders, 478–489
  dilution curve, 485
  disorders, 478
    plant analysis for diagnosis of, 482
    tools for diagnosis of, 478–489
  elements in plants, biochemical properties and physiological functions of, 6–7
  essential elements for plant growth, 4–5
  field responses to nutrient supply, 478
  fluxes in rural, 763–764
  foliar application of, 114–119
    general, 114–115
  foliar fertilizers for pest and disease control, 118
  foliar uptake and irrigation methods, 118–119
  general, 3–4
  with high phloem mobility, 89–90
  interactions, 486–487
  by living cells, retrieval and release of, 77–79
  losses from livestock excreta, 759
  with low phloem mobility, 90–91
  management, 478, 723–724, 728
  effect of mineral nitrogen, 636–637
  mineral plant nutrient, 6
  mobilization, 589–590
  modeling of, 515–516
  movement of nutrients to root surface, 501–508
  Na as, 388–389
  nutrient-remobilization efficiency, 660
  plant analysis for prognosis of, 489–490
  practical importance of foliar application of nutrients, 115–118
    avoiding occurrence of physiological disorders and improving quality of horticultural crops, 117–118
    biofortification, 118
    decrease in root activity during reproductive stage, 116–117
    dry topsoil, 116
    low nutrient availability in soils, 115–116
  ratios, 486–487
  re-translocation and cycling of, 91–93
  relationships between yield and nutrient supply, 132–133
  relative importance of phloem and xylem for long-distance transport of, 89–93
  remobilization, 93–97, 660
  in soil solution, concentration of, 502–503
  in soils, 511–512
    chemical soil analysis, 499–501
    intensity/quantity ratio, plant factors, and consequences for soil testing, 514–516
    movement of nutrients to root surface, 501–508
    nutrient availability and distribution of water in soils, 511–512
    role of root density, 509–511
    role of soil structure, 513–514
  solutions, 5, 41
  temperature, 638
  in terrestrial agroecosystems
  uptake, 548, 733–735
    through cuticle, 109–110
  variation in angiosperm ionome, 7–8
Nutrients to root surface, movement of, 501–508
  concentration of nutrients in soil solution, 502–503
  principles of calculations, 501
  role of diffusion, 504–508
  role of mass flow, 503–504
Nutrition, 148–150
  at constant tissue concentration, studying, 45–47
  effects of nutrition on endogenous concentrations of phytohormones, 177–181
  of host plant, role of AM in, 600–601
  index, 485
  of plants, role in, 590–591
  of plants in changing climate
    global climate change and root zone nutrient availability, 732–742
    mineral composition of food/feed, 743–744
    nutrient accumulation, 728–730
    nutrient-use efficiency, 731–732
    plant responses to global climate change, 727–728
Nutritional control of root development, 530–532
  iron, 532
  nitrogen, 530–531
  phosphorus, 531
  potassium, 531–532
Nutritional quality, 421–422, 426–431
  mineral nutrients, hidden hunger, and biofortification, 427–429
  protein concentration and amino acid composition, 429–430
  related to soil acidity, 635
  status of plants, 447–449, 557–559
  vitamins and bioactive phytochemicals, 430–431

O

O-acetylserine (OAS), 42–43
O-acetylserine(thiol)lyase (OASTL), 221
OAS. See O-acetylserine (OAS)
OASTL. See O-acetylserine(thiol)lyase (OASTL)
Obligate biotrophic pathogens, 447–448
Obstacle avoidance strategy, 533
Oil crops, 421–422
Oilseed rape (Brassica napus), 94, 151, 167, 224
Oligopeptide transporter (OPT), 45
Oligosaccharides, 157
Onion (Allium cepa), 50, 224, 354
Ontogenesis, environmental effects on barrier properties during, 111
Ontogeny, 86–87
Open-top chambers (OTC), 728
OPT. See Oligopeptide transporter (OPT)
Opuntia ficus-indica. See Prickly pear (Opuntia ficus-indica)
Orchid mycorrhizae, 595
Organelles, 258
Organic acid anions, 86–87, 154, 410, 589–590, 666, 668–669
Organic acids, 426
Organic compounds, 110, 119, 131, 170
Organic fertilizers, 669
Organic ligands, 502
Organic matter content, 323
Organic N fertilizers, 106
Organic N uptake, 208
    amino acid uptake, 208
    urea uptake and metabolism, 208
Organic rhizodeposition, 563
Organic S metabolites, 219–220
Organic soil amendments on gaseous fluxes, effects of, 754–755
Organic structures, B complexes with, 338–339
Organosilicon surfactants, 110
Organosulfur compounds, 426
Ornithine cycle, 326
Orthologs, 76
Oryza sativa L. See Rice (Oryza sativa L.)
OSCA. See Osmolality-sensitive calcium (OSCA)
OsCNX2 mutation, 332
OsCNX6 mutation, 332
Osmolality-sensitive calcium (OSCA), 243–244
Osmoregulation, 245–246, 254–257, 388
Osmotic adjustment, 392, 699–701
OsMTP9, 76
OsYSL2, 79
OsYSL15, 170
OsYSL18, 79
OsZIP3, 79
OTC. See Open-top chambers (OTC)
Oxalate oxidase, manganese in, 297
Oxalic acid, 488
Oxalis acetosella, 242–243
Oxidative phosphorylation, 154–156
Oxidative stress, 300, 328, 673
Oxygen (O), 6, 32
Oxygenation, 560
Ozone-depleting gas emission, 753

P

P_{1B}-type ATPase transporters (HMA transporters), 303
PAL. See Phenylalanine ammonia-lyase (PAL)
Panicum virgatum. See Switchgrass (Panicum virgatum)
Papilionoideae, 616
PAPS. See Phosphoadenosine phosphosulfate (PAPS)
Paracoccus denitrificans, 753–754
Parasitic plants, 257, 392
Passage cells, 50
Passive urea transport, 208
Pasta, 420–421
    industry, 421
Pasture
    composition, 744
    NUE of pasture species, 731–732
    plants, 396–397
Pathogenesis, 445–446
Pathogenic bacteria, 563–564
Pathogenic microorganisms, 545
Pathogens, 247–248
    direct and indirect effects of fertilizer application on, 464–467
PBA. See Phenylboronic acid (PBA)
PE. See Physiological efficiency (PE)
Pea (Pisum sativum), 50, 214, 652
Peanut, 90
Pear (Pyrus communis), 96
Pearl millet (Pennisetum glaucum), 141
Pecan (Carya illinoiensis), 323
Pectic polysaccharide network, 667
Pectin methylesterase, 453, 456
Pectolytic enzymes, 449, 454
Pelargonium plants, 446
Pennisetum glaucum. See Pearl millet (Pennisetum glaucum)
PEP. See Phosphoenolpyruvate (PEP)
PEP carboxykinase (PEPCK), 141
PEP carboxylase, 236–237
PEPC. See Phosphoenolpyruvate carboxylase (PEPC)
PEPCase. See Phosphoenol pyruvate carboxylase (PEPCase)
PEPCK. See PEP carboxykinase (PEPCK)
Pepper (Capsicum annuum), 307
Peptides, 75, 208, 302
Perennial ryegrass (Lolium perenne), 78, 151–152
Perennials, 97, 240
    species, 487
Peri-urban agriculture (UPA), 763–764
Peroxidases, 301
Peroxyacetyl nitrate (PAN), 107
Pests, 401, 461–464
    direct and indirect effects of fertilizer application on plants and pathogens and, 464–467
    foliar fertilizers for pest control, 118
Petioles, 244
PG. See Phosphatidylglycerol (PG)
PGA. See Phosphoglycerate (PGA)
PGPB. See Plant growth–promoting bacteria (PGPB)
PGPR. See Plant growth–promoting rhizobacteria (PGPR)
PGR. See Plant growth regulator (PGR)
pH
    buffering capacity, 553
    on organic matter turnover, 755–756
Phaseolus vulgaris, 257, 303, 672
Phaseolus vulgaris. See Bean (Phaseolus vulgaris)
Phenol metabolism, 449
Phenol oxidases, 305
Phenolics, 339–340
Phenols, 341–342
Phenotypic adaptation, 686–687
Phenylalanine ammonia-lyase (PAL), 297
Phenylboronic acid (PBA), 339
Philaenus spumarius. See Meadow spittlebug (Philaenus spumarius)
PHL. See PHR1-LIKE 1 (PHL)
Phleum pretense, 78
Phloem, 73–74, 337
    loading of amino acids, 158
    mobility, 115
    phloem-delivered compounds, 168–169
    phloem-mobile cations, 212
    phloem-mobile fluorophore, 144
    phloem-mobile nutrients, 144
    principles of phloem transport and phloem anatomy, 84–86
    transport of assimilates in phloem and regulation, 156–161
        mechanism of phloem transport of assimilates, 159–160
        phloem loading of assimilates, 156–159
        phloem unloading, 160–161
Phloem transport, 84–89, 257–258. See also Xylem transport
    mechanism of assimilates, 159–160
    mobility in phloem, 88
    phloem loading and composition of phloem sap, 86–87
    phloem unloading, 89
    principles of phloem transport and phloem anatomy, 84–86
    relative importance of phloem and xylem for long-distance transport of nutrients, 89–93
    remobilization of nutrients, 93–97
    transfer between xylem and phloem, 88–89
Phosphate, 43–44, 81–82, 216, 227, 455
Phosphate concentration, 502
PHOSPHATE STARVATION RESPONSE1 (PHR1), 43–44
Phosphate transporter 1 (Pht1), 77
Phosphatidic acid, 15
Phosphatidylglycerol (PG), 15
Phosphatidylinositol (PI), 15
Phosphatidylserine, 15

Phosphite, 455
Phosphoadenosine phosphosulfate (PAPS), 220–221
Phosphoenol pyruvate carboxylase (PEPCase), 140
Phosphoenolpyruvate (PEP), 38–39, 212, 627
Phosphoenolpyruvate carboxylase (PEPC), 287, 389
Phosphoglycerate (PGA), 138
2-phosphoglycolate (2-PG), 139
Phospholipase, 313
Phospholipids, 227–228, 317
Phosphorus (P), 3–4, 6, 111, 226–235, 531, 629–631
  compartmentation and regulatory role of inorganic phosphate, 229–231
  deficiency, 162
  fractions and role of phytate, 232–234
  requirements of plant species evolved in severely P-impoverished landscapes, 652–654
  role in energy transfer, 228–229
  as structural element, 227–228
  supply, plant growth, and plant composition, 234–235
Phosphorus-use efficiency (PUE), 142
Phosphorus-zinc interactions, 318–320
Phosphorylation, 229
  of enzyme proteins, 229
  of magnesium, 236–238
Photoinhibition, 135–137
Photolysis, 134
Photonastic movements, 257
Photooxidation, 135–137
Photophosphorylation, 133–135
Photorespiration, 137–140, 212, 253, 727
Photorespiratory pathway, 139–140
Photosynthate demand, 597–598
  and host plant growth, root colonization, 595–598
Photosynthates, accumulation of, 149, 240
Photosynthesis, 148–150, 252–254, 287–290, 388, 426, 636, 653, 692–693, 727, 756
  C_4 pathway of photosynthesis and Crassulacean acid metabolism, 140–143
  and chloroplast performance, 352–353
  photoprotection and fine-tuning of photosynthesis to changes in light intensity, 353
  photosystem II oxygen-evolving complex, 352–353
  functional role of Mn in, 294–296
  of magnesium, 236–238
  by sink demand for carbohydrates, feedback regulation of, 147–148
Photosynthesis per unit leaf P (PPUE), 652–653
Photosynthetic activity and related processes, 133–150
Photosynthetic area, 151–154
  canopy leaf area, 153–154
  individual leaf area, 151–152
  leaf area per plant, 152

Photosynthetic energy flow, 133–135
Photosynthetic N-use efficiency (PNUE), 142
Photosynthetic organisms, 751
Photosystem II (PS II), 133–134
  manganese at active site of water oxidation in, 295–296
  oxygen-evolving complex, 352–353
PHR1-LIKE 1 (PHL), 43–44
PHR1. *See* PHOSPHATE STARVATION RESPONSE1 (PHR1)
Pht1 gene, 81–82
Pht1. *See* Phosphate transporter 1 (Pht1)
Phylogenetic analysis, 202
Physiological disorders, 425
Physiological efficiency (PE), 731
Phytate, 291
  phosphorus fractions and role of, 232–234
  phytate-rich diets, 291
Phytochelatins, 222
Phytohormonal effects, 604–605
Phytohormones, 76, 86–87, 144–145, 177, 292–293, 682, 693–695
  abscisic acid, 217
  crosstalk between auxin and, 529
  precursors, 591–592
  role in regulation of sink–source relationships, 170–183
  effects of nutrition on endogenous concentrations of phytohormones, 177–181
  phytohormones, signal perception, and signal transduction, 175–177
  phytohormones and sink action, 181–183
  structure, sites of biosynthesis, and main effects of phytohormones, 171–175
Phytoliths, 658–659
Phytosiderophores (PSs), 14, 45, 292, 573–574
Phytostabilization, 555
Phytotoxic metabolites under hypoxia, 681–682
PI. *See* Phosphatidylinositol (PI)
Pig and poultry excreta, 758–759
Pigment systems, 133–134
Pineapple (*Ananas comosus*), 143
Piper-Steenbjerg effect, 481–482
Pisum sativum, 287
Pisum sativum. *See* Pea (*Pisum sativum*)
Plant analysis, 480–491
  developmental stage of plant and age of leaves, 483–485
  for diagnosis of nutrient disorders, 482
  environmental factors, 487
  general, 480–482
  histochemical, biochemical, and spectral methods, 488–489
  nutrient interactions and ratios, 486–487
  plant species and genotypes, 485–486
  for prognosis of nutrient deficiency, 489–490
  total analysis *vs.* fractionated extraction, 487–488
Plant breeding programs, 731
Plant cells, 15, 322
  sap, 481

Plant composition
  calcium, 248–249
  nitrogen, 215–218
  phosphorus, 234–235
  potassium, 260
  sulfur, 224–226
Plant defense reactions, 451
Plant dry matter, 131
Plant factors, 507–508, 514–516
Plant growth, 688–695
  beneficial effects on, 409
  beneficial elements for, 5–6
  calcium, 248–249
  essential elements for, 4–5
    discovery of essentiality of micronutrients for plants, 5t
  inhibition, 666–667
  ion imbalances, 691–692
  magnesium, 240–241
  major constraints, 688
  nitrogen, 215–218
  phosphorus, 234–235
  photosynthesis and respiration, 692–693
  phytohormones, 693–695
  potassium, 260
  protein biosynthesis, 693
  sodium and chloride uptake and toxicity in plants, 690–691
  sulfur, 224–226
  water deficit, 688–690
Plant growth regulator (PGR), 328
Plant growth–promoting bacteria (PGPB), 637–638
Plant growth–promoting rhizobacteria (PGPR), 328, 535
Plant membranes, 15, 173–174
Plant nutrition, improving, 640–641
Plant nutritional status, 41–45, 165
  phosphorus concentrations in tissues of barley plants following supply of P to plants grown without P, 44t
  relationship between root tissue concentration and K influx to barley roots, 42t
Plant responses
  to fertilization, 728–729
  to global climate change, 727–728
    adaptation of C_3 and C_4 plants to future climates, 728
    C_3 and C_4 plants, 727
  to soil-borne ion toxicities
    acid mineral soils, 666–673
    mechanisms of adaptation to acid mineral soils, 673–679
    saline soils, 687–704
    waterlogged and flooded soils, 679–687
Plant roots, 523, 545
  kinetics of solute transport in, 25–29
Plant species, 74–75, 77–78, 91, 158, 169, 179, 212, 215, 299–301, 308, 325, 345, 387–388, 485–486, 504
  evolved in severely phosphorus impoverished landscapes
    micronutrient requirements of, 655–656

nitrogen requirements of, 652–654
phosphorus requirements of, 652–654
Plant sterols, 15
Plant stimulation, 111
Plant water
　balance and water relations, 354–355
　　water relations, water-use efficiency, and drought resistance, 355
　　water storage capacity, 354
　relations, 604–605
Plant-based foods, 405
Plantain (*Plantago major*), 341
Plant–animal interactions affecting nutrient fluxes at different scales, 757–763
　nutrient and carbon losses from livestock excreta, 759
　spatial aspects of livestock-mediated nutrient fluxes and modeling, 759–763
　species-specific relationship between feed intake and excreta quality, 757–759
Plant–fungus interactions in mycorrhizal symbiosis, 536–537
Plasma membrane (PM), 11, 15, 454, 481, 552, 666
Plasmodesmata, 51, 156–157
Plastocyanin, 303
Pleiotropic effects, 676
PLETHORA genes (PLT genes), 529
PLT genes. *See* PLETHORA genes (PLT genes)
Plugging of sieve-tube pores, 85–86
Plum (*Prunus domestica*), 96
PM. *See* Plasma membrane (PM)
PMF. *See* Proton-motive force (PMF)
PNUE. *See* Photosynthetic N-use efficiency (PNUE)
Pods, 238
Polar "pores", 109–110
Polar compounds, 548
Pollen
　formation, 307
　tubes, 315
Pollination, 165–167, 183–184
Polyamine oxidases, 304–305
Polyamines, 174
　function, 163–165
Polyethylene glycol, 15
Polygalacturonases, 454, 652
Polygonum hydropiper, 679
Polyhydroxyl compounds, 337–338
Polyols, 87
Polyphenol oxidases, 305
Polyphenolic components, 753–754
Polyvalent metal cations, 75
Polyvalent cations, 50, 667
Poplar, 107
Populus trichocarpa, 302
Postembryonic root branching, 526–527
Postharvest spoilage, 146–147
Potash (KCl), 4
Potassium (K$^+$), 3–4, 34, 76, 87, 93, 95–96, 249–260, 452–453, 478–480, 531–532, 631
　acquisition, 154

cellular concentrations, 250
compartmentation, 250
deficiency, 149, 251, 260, 449
enzyme activation, 250–251
homeostasis, 691–692
K-deficient plants, 258–259
K$^+$ battery, 258
K$^+$ channels, 34
K$^+$-sufficient plants, 257–258
osmoregulation, 254–257
　cation–anion balance, 258
　cell extension, 254–255
　energy transfer, 258
　phloem transport, 257–258
　photonastic and seismonastic movements, 257
　stomatal movement, 255–257
　stress resistance, 258–260
　supply, plant growth, and plant composition, 260
photosynthesis, 252–254
plants, 242–243
potassium-induced Mg deficiency, 235
protein synthesis, 252
substitution of K by Na, 391–393
Potato (*Solanum tuberosum* L.), 167–168, 183, 728
　plants, 178–179, 182
　tubers, 161, 241, 458–459
"Potentially toxic" elements, 432
PPi. *See* Pyrophosphate (PPi)
PPiase. *See* Pyrophosphatases (PPiase)
PPUE. *See* Photosynthesis per unit leaf P (PPUE)
Precipitation, 724–725
Preferential allocation, 653
Premature sprouting, 333
Pressure flow hypothesis, 84
Prickly pear (*Opuntia ficus-indica*), 143
Primary cell wall, 12–13
Processing tomatoes, 422
Production systems, 482
Prognosis, 478
　plant analysis for prognosis of nutrient deficiency, 489–490
Programmed cell death processes, 454
Prokaryotes, 402, 616
Proliferation, 554–555
Prosthetic groups, 223
Proteaceae, 653, 656
Proteinases, 218
　inhibitors, 218
Proteins, 86–87, 131, 170, 208, 297–298, 310, 315
　biosynthesis, 693
　bodies, 320
　concentration, 429–430
　kinase phosphorylates NR, 210
　phosphorylation, 229
　protein-energy malnutrition, 429
　protein–protein communication and interactions, 311
　synthesis, 235–236, 252, 314–315, 402
Proton pumps, 23–24

Proton toxicity, 667
Proton-coupled K$^+$ symporters, 34
Proton-motive force (PMF), 353
Proton-pumping V-type ATPase, 353–354
Protons, 558
Protozoa, 591, 752
Prunus amygdala. *See* Almond (*Prunus amygdala*)
Prunus cerasus. *See* Cherry (*Prunus cerasus*)
Prunus domestica. *See* Plum (*Prunus domestica*)
PS II. *See* Photosystem II (PS II)
PSs. *See* Phytosiderophores (PSs)
Pteridophyta, 397
PUE. *See* Phosphorus-use efficiency (PUE)
Pure free water, 73
Putrescine, 214
Pvferritin, 170
Pyrophosphatases (PPiase), 314
Pyrophosphate (PPi), 18, 220–221
Pyrus communis. *See* Pear (*Pyrus communis*)
Pyruvate, 154
Pyruvate decarboxylase, 682
Pyruvate kinase, 489
Pythium aphanidermatum, 401, 450

Q

QC. *See* Quiescent center (QC)
Quality traits, 419–420
Quantity ratio, 514–516
Quiescent center (QC), 345, 525–526
Quorum sensing, 460, 589

R

Radial gradients, 548–549
Radial transport
　of ions and water across root, 49–52
　　intracellular K$^+$ activity and number of plasmodesmata in tangential walls, 52t
　of nitrate across root and loading into xylem, 204–205
　solutes, 52
　of water, 52
Radioactive nutrient, 547
Radioactive rubidium (^{86}Rb), 34
Radish (*Raphanus sativus*), 337
Raffinose, 157
Raffinose-like sugars (RLS), 158
Raphanus sativus. *See* Radish (*Raphanus sativus*)
RCA. *See* Root cortical aerenchyma (RCA)
RE. *See* Recovery efficiency (RE)
Re-translocation
　of nutrients, 91–93
　in phloem, 91
Reactive oxygen species (ROS), 135, 240, 283–284, 452, 533, 559–560, 666
Recovery efficiency (RE), 731
Redox potential, 502, 559–561
　iron mobilization, 561
　manganese mobilization, 560–561
　effect of waterlogging, 559–560

Redox systems, iron-containing constituents of, 284–287
Redox-active metal cations, 294
Redox-active transition element, 302
Reduced glutathione, 42–43
Reducing processes, 559–561
 iron mobilization, 561
 manganese mobilization, 560–561
 effect of waterlogging, 559–560
Reducing sugars, 255
Regression equations, 481
Relationship between susceptibility and nutritional status of plants, 447–449
Relative humidity (RH), 109
Remobilization of nutrients, 93–97
 general, 93
 perennials, 97
 reproductive stage, 94–97
 seed germination, 93
 vegetative stage, 93–94
Reproductive organs, 665
Reproductive stage
 decrease in root activity during, 116–117
 remobilization of nutrients, 94–97
Resistance, 446
Resistant genotypes, 446–447
Respiration, 154–156, 692–693
Retrieval mechanisms, 569–571
RH. See Relative humidity (RH)
Rhizobia, 616
 legume root infection by, 623–626
 symbioses, 621–623
Rhizobial species, 621
Rhizobium, 324, 402
Rhizodeposition, 561–574
Rhizoplane, 548
Rhizosphere
 acidification, 548, 555
 alkalinisation, 553
 bacteria, 535–536
 biology
  endophytes, 592
  methods to study rhizosphere microorganisms, 592–593
  mycorrhiza, 593–606
  rhizosphere as dynamic system, 587–588
  rhizosphere microorganisms, 588–592
 Ca^{2+} concentration, 36–37
 chemistry, 215
 effect, 545
 general, 545–547
  rhizosphere sampling, 546–547
 inorganic elements in rhizosphere, 550–552
 interactions among ions in, 33–40
 microorganisms, 505, 587–592
  methods to study, 592–593
  role in nutrition of plants, 590–591
  root colonization, 588–590
  root exudates as signals and phytohormone precursors, 591–592
 pH, 552–559, 676
  nutritional status of plants and, 557–559
  source of nitrogen supply and, 553–557
  processes, 733–735

redox potential and reducing processes, 559–561
rhizodeposition and root exudates, 561–574
rhizosphere pH, 552–559
sampling, 546–547
spatial extent of, 547–549
 longitudinal gradients, 549
 radial gradients, 548–549
 temporal variability, 549
spatial extent of rhizosphere, 547–549
Rhododendron protistum, 667
Rhynchosporium commune. See Leaf scald disease (*Rhynchosporium commune*)
Rhynchosporium scalis. See Leaf blotch (*Rhynchosporium scalis*)
Riboflavin, 287
Ribosomal proteins, 653
Ribosomal RNA (rRNA), 227, 652–653
Ribosomes, 156–157, 314–315
 ribosome-inactivating proteins, 218
Ribulose bis-phosphate carboxylase, 729–730
Ribulose bisphosphate (RuBP), 231
Rice (*Oryza sativa* L.), 5, 79, 248, 615, 659
 mutant, 402
 nodes, 679
 plants, 292
Ricinus communis. See Castor bean (*Ricinus communis*)
Ripening disorders, 260
RLS. See Raffinose-like sugars (RLS)
RNA interference (RNAi), 406
RNA polymerase, 313
RNAi. See RNA interference (RNAi)
Root cortical aerenchyma (RCA), 154
Root-to-shoot Cl⁻ translocation, 351
Root-to-shoot signals, 682
Roots, 183, 232, 238. See also Rhizosphere
 ammonium uptake by, 206–207
 anatomy, 525–526
 apoplasmic Fe, 290
 architecture, 150
 border cells, 563
 cap cells, 591–592
 cation exchange, 652
 cells, 317
 colonization, 588–590, 595–597
  photosynthate demand, and host plant growth, 595–598
 crops, 167, 183
 decrease in root activity during reproductive stage, 116–117
 density, 509–511
 effects of pH, 30–32
 by environmental cues, regulation of, 530–537
  nutritional control of root development, 530–532
  root–soil biotic interactions, 535–537
  soil physical and chemical factors, 532–535
 exudates, 561–574
  high-molecular-weight compounds in, 564–568
  as signals, 591–592

sloughed-off cells and tissues, 563–564
factors influencing ion uptake by, 29–47
formation, 525
genetic control of, 523–529
 embryonic and postembryonic root branching, 526–527
 root anatomy and structure, 525–526
 root system architecture, 523–524
hairs, 527
infection, 628
influx to apoplasm, 29–30
inhibition of root growth, 669–670
interference with root-cell plasma membrane properties, 670–671
ion influx
 apoplasm, 12–14
 cytoplasm, 14–15
metabolic activity, 32–33
 carbohydrates, 32
 external concentration, 40–41
 interactions among ions in rhizosphere, 33–40
 oxygen, 32
 plant nutritional status, 41–45
 studying nutrition at constant tissue concentration, 45–47
 temperature, 32–33
morphological characteristics, 301
morphology, 599
movement of nutrients to root surface, 501–508
nitrate efflux from, 203–204
nitrate uptake by, 202–203
pathway of solutes from external solution into root cells, 12–15
phytohormonal control of, 527–529
 auxin and role in, 528–529
 crosstalk between auxin and phytohormones, 529
plasma membranes, 387–388
pressure, 91
proliferation, 512
radial oxygen, 535
radial transport of ions and water across, 49–52
radial transport of nitrate across root and loading into xylem, 204–205
reduced aluminum binding in root apoplast, 676–677
responses to Fe deficiency, 291–293
root-hairless mutants, 549
suppression of root pathogens, 605
systems, 523, 556, 626
 architecture, 523–524
 architecture in response to N supply, changes in, 216–217
tips, 160–161
uptake of ions and water along root axis, 15–18
vascular tissues, 570
zone nutrient availability, 732–742
 diffusivity and mass flow, 735–738
 impact on coupled carbon-nutrient cycling, 733–735

mycorrhizae and nutrient uptake, 733–735
system-level nutrient inputs and losses, 738–742
Root–soil biotic interactions, 535–537
beneficial plant–fungus interactions in mycorrhizal symbiosis, 536–537
beneficial rhizosphere bacteria, 535–536
ROS. *See* Reactive oxygen species (ROS)
rRNA. *See* Ribosomal RNA (rRNA)
Rubidium (Rb$^+$), 34
Rubisco, 656
RuBP. *See* Ribulose bisphosphate (RuBP)
Rumen microflora, 404
Ruminant excreta, quality of, 757–758
Rutabaga, 337
Rye (*Secale cereale*), 308
Ryegrass (*Lolium rigidum*), 550

S

S-adenosyl-methionine synthetase, 45
S-alk(en)yl-L-cysteine sulfoxides (ACSO), 426
S-alk(en)ylcysteine sulfoxides, 224
SA. *See* Salicylic acid (SA)
Saccharum officinarum. *See* Sugarcane (*Saccharum officinarum*)
Salicylic acid (SA), 144–145, 173, 328, 455
Saline soils, 665, 687–704
exploiting salt-affected soils, 703
genotypic differences in growth response to salinity, 703–704
mechanisms of adaptation to saline substrates, 695–702
salinity and plant growth, 688–695
soil characteristics and classification, 687–688
Saline substrates
detoxification of reactive oxygen species, 701–702
mechanisms of adaptation to, 695–702
osmotic adjustment, 699
salt distribution in shoot tissues, 698–699
salt exclusion *vs.* salt inclusion, 695–698
salt excretion, 702
vacuolar compartmentation and compatible solutes, 699–701
Saline water, 391
Salinity, 174, 247–248, 393–394, 637
growth, 688–695
Salt distribution in shoot tissues, 698–699
Salt exclusion, 695–698
Salt excretion, 702
Salt inclusion, 695–698
Salt tolerance, 695
Salt-affected soils, exploiting, 703
Saltbush (*Atriplex* spp.), 5
Sand-binding roots, 565
SAR. *See* Sodium adsorption ratio (SAR)
SAT. *See* Serine actyltransferase (SAT)
SCFA. *See* Short-chain fatty acids (SCFA)
Schizogenous aerenchyma, 559–560
SCO. *See* Synthesis of of cytochrome c oxidase (SCO)

SE. *See* Sieve elements (SE)
Secale cereale. *See* Rye (*Secale cereale*)
Second messengers, 176
Secondary active transporters, 18
Secondary growth, 168
Secondary metabolites, 86–87
Secondary plant products, 221
Secondary S compounds, 220–221
Secretory proteins, 566–568
Sedum album, 242–243
Sedum kamtschaticum, 388
Seeds, 161, 175, 248–249
development, 165–167
germination, 93
pelleting with Mo, 335
proteins, 218
storage proteins, 217–218
urease, 324
Seismonastic movements, 257
Selective ion uptake, 12
Selectivity, 12
Selenate uptake, 405
Seleniferous soils, 405
Selenium (Se), 5, 118, 405–410, 658
assimilation and metabolism, 407–408
beneficial effects on plant growth, 409
biofortification, 409–410
in leaves of some plants, 658
uptake and translocation, 405–407
Selenoamino acid, 406–407
Selenocysteine, 408
Selenomethionine, 406–408
Semiarid regions, 116
Senescence, 91, 94
senescence-induced N remobilization, 212
Sensitive plants, 107
Sensory quality, 423–426
effects of mineral nutrition on flavor, 425–426
effects of mineral nutrition on visual quality, 423–425
Serine, 139
Serine actyltransferase (SAT), 221
Serotonin N-acetyltransferase (SNAT), 672
Serpentine soils, 328
sgn3. *See* Casparian band-defective arabidopsis mutant (*sgn3*)
Shade leaves, 83
Shallot (*Allium ascalonicum*), 308
"Sheathed" zone, 52
Sheep's fescue (*Festuca vivipara* L.), 731
Shelf life of fresh fruits and vegetables, 431–432
Shikimic acid, 297
Shoot, 652
ammonium in, 207–208
apices, 90, 160–161, 183
architecture, 150
for grain/seed yield formation, 162
nitrate transport within, 206
Short-chain fatty acids (SCFA), 757
Sieve elements (SE), 156
Sieve-plate pores, 85
Signal perception, 175–177

Signal transduction, 175–177
Silene inflata, 242–243
Silene italica. *See* Italian catchfly (*Silene italica*)
Silene vulgaris, 222
Silicic acid (Si(OH)$_4$), 397
Silicon (Si), 5, 118, 301, 336–337, 397–402, 449–450, 634–635, 651, 659
beneficial effects, 399–402
as beneficial element in leaves of plants, 658–660
general, 397
uptake, concentration, and distribution, 397–399
Single protein-mediated mechanism of ion transport, 26
Single superphosphate (SSP), 336
Singlet oxygen ($^1O_2$), 135
Sink
activity, 168–170
feedback regulation of photosynthesis by sink demand for carbohydrates, 147–148
formation, 161–168
flower initiation and development, 163–165
formation of vegetative sink organs, 167–168
pollination and seed development, 165–167
shoot architecture for grain/seed yield formation, 162
limitations on yield, 183–185
organs, 238
phytohormones and SINK action, 181–183
for solutes, 160–161
Sink–source relationships, role of phytohormones in regulation of, 170–183
Sink–source transition
effect of leaf maturation on, 143–144
of leaves, 143–144
SIR. *See* Sulfur-induced resistance (SIR)
Sisal (*Agave sisalana*), 143
Sitosterol, 15, 173–174
Six-electron transfer process, 209
SL. *See* Strigolactone (SL)
SLAC/SLAH. *See* Slow anion channel-associated homologs (SLAC/SLAH)
SLAH3, 216
Sloughed-off cells and tissues, 563–564
Slow anion channel-associated homologs (SLAC/SLAH), 76
Slow vacuolar (SV), 24
Slow-type Anion Channel–associated/ SLAC1Homolog (SLAC/SLAH), 351
Slurry, 106
Small signaling molecules, 292–293
SNAT. *See* Serotonin N-acetyltransferase (SNAT)
SOC. *See* Soil organic carbon (SOC)
SOD. *See* Superoxide dismutase (SOD)
Sodium (Na), 5, 34, 88, 387–397
application of Na fertilizers, 396–397

Sodium (Na) (*Continued*)
　deficiency, 390
　essentiality, 388–389
　growth stimulation by, 393–396
　role in C$_4$ species, 389–391
　substitution of K by, 391–393
　toxicity, 691
　uptake, 690–691
Sodium adsorption ratio (SAR), 687
Sodium bicarbonate, 499–500
Soil organic carbon (SOC), 555
Soil-borne fungal diseases, 458–461
Soil-borne pathogens, 445, 451–452
Soil-borne toxicity, 665
Soil-grown soybean, 324
Soil
　aeration, 534–535
　analysis, 478, 490–491
　buffering capacity, 511–512
　bulk density, 501
　carbon, 723–724
　chemical factors, 532–535, 679–681
　　mechanical impedance, 532–533
　　soil aeration, 534–535
　　soil temperature, 533–534
　　soil water, 534
　ecosystem, 445–446
　factors, 505–507
　health, 445–446
　limitations, 637–638
　low nutrient availability in, 115–116
　microorganisms, 445–446, 587, 752
　moisture, 724–725
　physical factors, 532–535
　　mechanical impedance, 532–533
　　soil aeration, 534–535
　　soil temperature, 533–534
　　soil water, 534
　potential, 506
　respiration, 733–734
　sickness, 458
　solutions, 41, 513–514
　　concentration of nutrients in, 502–503
　structure, 513–514
　temperature, 533–534
　testing, 480, 490, 499
　　consequences for, 514–516
　　modeling of nutrient availability and crop nutrient uptake, 515–516
　treatments, 116
　water, 534
　　content, 499, 506, 637–638
　　content on organic matter turnover, 755–756
Solanum lycopersicum. *See* Tomato (*Solanum lycopersicum*)
Solanum tuberosum L. *See* Potato (*Solanum tuberosum* L.)
Solar radiation, 724
Soluble oxalate, 242–243
Solute concentrations, 29–30
　humidity effects on, 111–113
Solute transport, energy demand for, 25

Solute transport across membranes, 18–29
　energy demand for solute transport, 25
　kinetics of solute transport in plant roots, 25–29
　thermodynamics of solute transport, 18–24
Solute transport in xylem, effect of transpiration rate on, 82
Solutes, 77, 144, 255
Solutes, uptake of, 107–114
　general, 107–108
　nutrient uptake through cuticle, 109–110
　role of external factors, 111–114
　　active ingredients and adjuvants, 113–114
　　environmental effects on barrier properties during ontogenesis, 111
　　humidity effects on solute concentration and leaf permeability, 111–113
　　structure of cuticle, 108
　　uptake through stomata, 110–111
Solutes from external solution into root cells, pathway of, 12–15
Solution-diffusion model, 109
Sorbitol, 87
Sorghum (*Sorghum bicolor*), 39–40, 141, 591–592
Source on yield, 183–185
Source–sink relationship, 131–132, 149
Soybean, 214, 297, 301–302, 325–327
　Eu3 protein, 324
　plants, 239–240
Spatial distribution, 665
SPDT. *See* SULTR-like P distribution transporter (SPDT)
Spectral methods, 488–489
Sphingolipids, 15
Spinach (*Spinacia oleracea*), 133
Spinacia oleracea. *See* Spinach (*Spinacia oleracea*)
"Split seed" disorder, 300
Sprengel-Liebig's Law, 651
SPS. *See* Sucrose phosphate synthase (SPS)
SQDG. *See* Sulfoquinovosyldiacylglycerol (SQDG)
SSP. *See* Single superphosphate (SSP)
Stachyose, 157
Standard functional ionome, 8
Starch, 131
　accumulation of, 231
　biosynthesis, 169
Statistical model, 481
"Stay green" genotypes, 146–147
"Stay green" mutants, 146–147
Stems, 232
　elongation zones, 160–161
　tissue, 488
Sterols, 15
Stigmasterol, 15, 173–174
Stinging nettle (*Urtica dioica*), 301–302
Stomata
　uptake and release of gases and volatile compounds through, 106–107
　volatile nitrogen compounds, 106–107

　volatile sulfur compounds, 107
　uptake through, 110–111
Stomatal aperture, 113
Stomatal closure, 257
Stomatal movement, 255–257
Storage organs, 183
Storage phase, 169
Storage proteins, 217–218
Storage sinks, 160–161
Stress
　conditions, 119
　phytohormone, 173
　resistance, 258–260
Striga, 257
Strigolactone (SL), 162
　receptors, 591–592
Stylosanthes capitata, 731–732
Stylosanthes hamata, 329
Suberization, 29
Substrate O-acetylserine, 221
Subterranean clover (*Trifolium subterraneum*), 294
Succulence, 394
Sucrose, 84, 143, 157, 169
Sucrose phosphate synthase (SPS), 147–148
Sucrose synthase, 169
Sugar beet (*Beta vulgaris*), 133, 167, 169, 241–242, 728
Sugarcane (*Saccharum officinarum*), 5, 141, 169
Sugars, 110, 150, 157, 255–257, 421–422
　alcohols, 157
　sugar-induced repression of photosynthetic genes, 147
　sugar-mediated regulation of photosynthetic genes, 147
Sulfate, 35, 77, 81–82, 87, 216
　reduction, 221
　uptake, 221
　reduction, and assimilation, 220–221
Sulfate of ammonia ((NH$_4$)$_2$SO$_4$), 4
Sulfate of magnesia (MgSO$_4$), 4
Sulfate of soda (Na$_2$SO$_4$), 4
Sulfate transporter (SULTR), 77
SULTR expression, 35
SULTR1 gene, 81–82
SULTR2 gene, 81–82
Sulfhydryl groups, 302, 317
Sulfide (S^{2-}), 221
Sulfite (SO$_3^{2-}$), 221–223
Sulfite oxidase, 332–333
Sulfolipids, 219–220, 222–223
Sulfoquinovosyldiacylglycerol (SQDG), 15, 223
Sulfur (S), 3–4, 87, 219–226, 455, 632
　deficiency, 219–220, 421
　metabolic functions of S, 221–224
　metabolism, 139–140
　sulfate uptake, reduction, and assimilation, 220–221
　sulfur-containing methionine, 226
　sulfur-deficient plants, 405
　sulfur-enhanced defense, 107

supply, plant growth, and plant composition, 224–226
Sulfur dioxide (SO$_2$), 105, 107
Sulfur-induced resistance (SIR), 107, 455
SULTR. *See* Sulfate transporter (SULTR)
SULTR-like P distribution transporter (SPDT), 77
Sun leaves, 83
Sunflower (*Helianthus annuus*), 50, 78, 301–302
Superoxide dismutase (SOD), 136, 285–286, 303–304, 634, 683
 manganese in, 296–297
Superoxide radical (O$_2^{\bullet-}$), 135, 240
Superphosphate, 4
Surface erosion, 743–744
Surface-active adjuvants, 114
Surface-active compounds, 110
Susceptibility of plants, 447–449
Susceptible leafy vegetables, 424
SUT/SUC gene family, 87
SV. *See* Slow vacuolar (SV)
Sweet pepper (*Capsicum annuum*), 85
Switchgrass (*Panicum virgatum*), 141
Symbiosis receptor–like kinase (SYMRK), 625–626
Symbiosome, 626
Symbiotic plasmids (pSym), 624
Symbiotic promiscuity, 621–623
Symbiotic systems, 621–629
 general, 621
 legume root infection by rhizobia, 623–626
 nodule formation and functioning in legumes, 626–629
 range of legume–rhizobia symbioses, 621–623
Symplasmic movement, 156
Symplasmic pathway, 51
Symplasmic unloading, 161
SYMRK. *See* Symbiosis receptor–like kinase (SYMRK)
Synchrotron-based X-ray fluorescence microscopy, 107–108
Synergy between ammonium and nitrate nutrition, 215
Synthesis of cytochrome c oxidase (SCO), 304
Synthetic plant hormones, 175
System-level nutrient inputs and losses, 738–742
 atmospheric deposition, 738–740
 biological N$_2$ fixation, 740
 surface erosion and leaching, 740–742
Systemic second messengers, 177
Systemic signal, calcium as, 247–248

T

Tannins, 758
TaSultr1;1 transporter, 329
TaSultr4;1 transporter, 329
TaSultr5;2 transporter, 329
TCA cycle. *See* Tricarboxylic acid cycle (TCA cycle)

Technical quality, 420–423
 beer and wine, 423
 bread and pasta, 420–421
 fiber crops, 422
 parameters, 420
 processing tomatoes, 422
 sugar and oil crops, 421–422
Temperature, 32–33
 on organic matter turnover, 755–756
 temperate perennial species, 211
Temporal variability, 549
Tequila agave (*Agave tequilana*), 143
Terpenoid pathway, 173
Terrestrial ecosystems, 725–726
Terrestrial plants, 106, 652
Thermodynamics of solute transport, 18–24
 families of selected plasma membrane and tonoplast ion transporters in *Arabidopsis thaliana*, 21*t*
 respiratory energy costs for ion uptake in roots of *Carex diandra*, 25*t*
Thiamine pyrophosphate (TPP), 223
Thick cellulose secondary walls, 49
Thiobacillus denitrificans, 753–754
Thiols, 219–220
Thioredoxins, 223
Threshold RH, 111–112
Thylakoid membranes, 288
Timber, composition of, 744
Tip-burn in lettuce, 90
Tobacco (*Nicotiana tabacum*), 136, 302–303, 324
 leaves, 150
 tissue culture cells, 315
Tolerance, 447–448, 666
Tomato (*Solanum lycopersicum*), 78, 136, 173, 175
 paste industry, 422
Tonoplast, 11, 23–24
 H$^+$-ATPases, 23–24
 of vacuole, 23
Tonoplast pyrophosphatase, 387–388
Tortuosity, 504–505
Total analysis, 487–488
Total soluble solids (TSS), 422
Toxic cations, 245
Toxic elements, 432–434, 555
Toxic mineral, 432
Toxic soil Cu concentrations, 309
Toxic substance, 459–460
Toxicity
 in plants, 690–691
 symptom, 301
TPP. *See* Thiamine pyrophosphate (TPP)
Transgenic rice plants, 170
Transgenic tobacco, 165–166
Transition element, 283–284
Transpiration, 401
 effect of
  transpiration rate on distribution of elements within shoot, 83–84
  transpiration rate on solute transport in xylem, 82

Transport proteins, 34
Transport systems, 44, 205, 255
Transporters, 202
Trees, composition of, 744
2,3,5-tri-iodobenzoic acid (TIBA), 175
Tricarboxylates, 548
Tricarboxylic acid cycle (TCA cycle), 154, 213
Trifolium alexandrinum. *See* Berseem clover (*Trifolium alexandrinum*)
Trifolium hybridum, 78
Trifolium repens, 78, 624–625
Trifolium subterraneum. *See* Subterranean clover (*Trifolium subterraneum*)
Triple superphosphate (TSP), 336
Triticum aestivum. *See* Bread wheat (*Triticum aestivum*)
Triticum durum. *See* Durum wheat (*Triticum durum*)
Tryptophan synthesis, 316–317
TSP. *See* Triple superphosphate (TSP)
TSS. *See* Total soluble solids (TSS)
Tuberization, 167
Tubers, 161, 232, 248–249
 crops, 167, 183
Turgor, 152
Turnip (*Brassica rapa*), 4, 337
Type-I cell walls, 12–13
Type III secretion system, 626
Tyramine, 435
Tyrosinases. *See* Polyphenol oxidases

U

Ubiquinone (UQ), 154
Ultraviolet-B radiation (UV-B radiation), 136
Undissociated boric acid (B(OH)$_3$), 337
Unfertilized agricultural soils, 202
UPA. *See* Peri-urban agriculture (UPA)
UPS1. *See* Ureide permease (UPS1)
UQ. *See* Ubiquinone (UQ)
Urban systems, 763–764
Urbanization, 751
Urea, 109, 325
 transporters, 208
 uptake and metabolism, 208
Urease, nickel dependence, 208
Ureide permease (UPS1), 81–82
Ureides, 156
Ureidoglycolate, 326–327
Urtica dioica. *See* Stinging nettle (*Urtica dioica*)
Utilization sinks, 160–161
UV-B radiation. *See* Ultraviolet-B radiation (UV-B radiation)

V

V-type ATPase, 353–354
Vacuolar compartmentalization V-type ATPase, 353–354
Vacuolar compartmentation and compatible solutes, 699–701
Vacuolar nitrate, 211

Vacuolar sucrose transporters, 169
Vacuolated cells, 230
Vacuoles, 156–157, 208, 258
　of root cells, 52
Valerianella locusta. *See* Corn salad (*Valerianella locusta*)
Valonia, 12
Vanadate, 257
Variation in angiosperm ionome, 7–8
Vascular bundles, 88
Vascular diseases, 457
Vegetables, 431–432
Vegetative apices, 160–161
Vegetative growth, 93–94, 165
Vegetative sink organs, formation of, 167–168
Vegetative stage, remobilization of nutrients, 93–94
Vegetative storage organs, 183
Vegetative storage proteins (VSP), 217
Vegetative storage sinks, 161, 183
Vesicular-arbuscular mycorrhiza, 594
VICCs. *See* Voltage-insensitive Ca^{2+} channels (VICCs)
Vicia faba. *See* Broad bean (*Vicia faba*); Faba bean (*Vicia faba*)
Vigna mungo. *See* Black gram (*Vigna mungo*)
Viral diseases, 457–458
　bacterial and, 457–458
Viral infection of soil bacteria, 591
Virulence factor, 457
Virus-host plant, 457–458
Visual deficiency symptoms, 388
Visual quality, effects of mineral nutrition on, 423–425
Visual symptoms, 487
Vitamin A deficiency, 427, 430
Vitamins, 430–431
Vitis, 327
　V. vinifera, 302–303
Vitis vinifera. *See* Wine grapes (*Vitis vinifera*)
VOC. *See* Volatile organic compounds (VOC)
Volatile compounds, 224
　through stomata, 106–107
Volatile nitrogen compounds, 106–107
Volatile organic compounds (VOC), 545
Volatile substances, 224
Volatile sulfur compounds, 107
Voltage-insensitive Ca^{2+} channels (VICCs), 246
Volumetric water content, 504–505
Voodoo lily (*Dracunculus vulgaris*), 173
VSP. *See* Vegetative storage proteins (VSP)

W

Water
　conservation effect, 737
　deficit, 688–690
　manganese at active site of water oxidation in photosystem II, 295–296
　molecules, 311
　radial transport of water across root, 49–52
　relations, 355
　along root axis, 15–18
　splitting, 135
　storage capacity, 354
　uptake, 49, 534, 671–672
Water-free space (WFS), 13
Water-use efficiency (WUE), 141–142, 299–300, 355
Waterlogged soils, 555–556, 679–687
Waterlogging, effect of, 559–560
WFS. *See* Water-free space (WFS)
Wheat, 88, 95, 146–147, 166–167, 169, 175, 349–350
　grains, 320
White lupin (*Lupinus albus*), 85, 95–96
Wine, 423
Wine grapes (*Vitis vinifera*), 426
Winterkill, 299–300
Woody perennial plants, 484
Woody plant species, 555
WUE. *See* Water-use efficiency (WUE)

X

X-ray computed tomography, 547
X-ray-computed micro-tomography, 513
Xanthine, 326–327
Xanthine dehydrogenase, 331–332
Xanthophyll cycle pigments, 289–290
Xenobiotics, 222
Xylem, 337, 387–388
　factors governing ion release into, 53–56
　flow rate and ion concentration in xylem exudate of wheat seedlings, 55t
　parenchyma cells, 79, 398
　radial transport of nitrate across root and loading into, 204–205
　release of ions into, 53
　sap, 75
　　composition of, 74–76
　　xylem volume flow and nutrient concentrations in xylem sap of soil-grown nodulated soybean, 75t
　structure, 322
　transfer between phloem and, 88–89
　unloading in leaves, 79–82
　vessels, 73, 306, 534
　　exchange adsorption in, 77
　xylem-to-phloem transfer of nutrients, 88
Xylem transport, 73–84. *See also* Phloem transport
　composition of xylem sap, 74–76
　relative importance of phloem and xylem for long-distance transport of nutrients, 89–93
　remobilization of nutrients, 93–97
　effect of transpiration rate on distribution of elements within shoot, 83–84
　effect of transpiration rate on solute transport in xylem, 82
　　external concentration, 82
　　plant age, 82
　　time of day, 82
　　type of element, 82
　xylem loading, 76–82
　　exchange adsorption in xylem vessels, 77
　　retrieval and release of nutrients by living cells, 77–79
　　xylem unloading in leaves, 79–82
Xylem volume flow, 75t, 90

Y

Yellow stripe–like transporters (YSL transporters), 146, 292
Yield, 131–132
　biological, 131
　of crop plants, 169
　economic, 131
　nutrient supply relationships and, 132–133
　photosynthetic activity and related processes, 133–150
　　C_4 pathway of photosynthesis and Crassulacean acid metabolism, 140–143
　　carbon dioxide assimilation and photorespiration, 137–140
　　feedback regulation of photosynthesis by sink demand for carbohydrates, 147–148
　　effect of leaf maturation on sink–source transition, 143–144
　　leaf senescence, 144–147
　　nutrition and photosynthesis, 148–150
　　photoinhibition and photooxidation, 135–137
　　photosynthetic energy flow and photophosphorylation, 133–135
　　photosynthetic area, 151–154
　　respiration and oxidative phosphorylation, 154–156
　　response curves, 133
　　role of phytohormones in regulation of sink–source relationships, 170–183
　　sink activity, 168–170
　　sink formation, 161–168
　　source and sink limitations on yield, 183–185
　　transport of assimilates in phloem and regulation, 156–161
　yield-quality dilemma, 435–436
Yorkshire fog (*Holcus lanatus*), 35
Young leaf tissues, 160–161, 291
YSL proteins, 303
YSL transporters. *See* Yellow stripe–like transporters (YSL transporters)

Z

Z scheme, 133–134
Zea mays. *See* Maize (*Zea mays*)
Zero-grazing systems, 759
Zinc (Zn), 4, 111, 310–322, 634–635
　carbohydrate metabolism, 315–316
　deficiency, 136, 317, 319–321
　foliar sprays, 116
　membrane integrity and lipid peroxidation, 317
　phosphorus-zinc interactions, 318–320
　protein synthesis, 314–315

tolerance to Zn toxicity, 322
toxicity, 321–322
tryptophan and indole acetic acid synthesis, 316–317
zinc forms and bioavailability in grains, 320
zinc-deficient plants, 314–315, 319–320

zinc-finger proteins, 165
Zn-activated enzymes, 313–314
Zn-containing enzymes, 310–313
    alcohol dehydrogenase, 311
    CA, 312–313
    CuZn superoxide dismutase, 313

Zn-dependent enzymes, 315–316
Zinc-oxide (ZnO), 113
Zinc/Fe permeases (ZIPs), 302–303
    transporter family, 44–45
Zygote, 183–184

9780128197738